W9-BMN-855

Our world is constantly changing. Recently in math, teachers are being asked to respond to a variety of new learning styles and to deal with multiple classroom environments. Prentice Hall's all-new Math Series—*Algebra, Geometry, and Advanced Algebra: Tools for a Changing World* ©1998—has been specifically developed to help teachers manage these key changes, and many others. We do it through more worked-out examples, better ties to the real world, more small group work, and accessible technology. Finally, a math series that can motivate today's student, cover the basic skills, support the NCTM guidelines, and ultimately lead to success.

Teaching Tools
Skills Practice
Assessment Options
Students' Experiences
Group Work
Interactive Questioning
Real World Contexts
Technology Options

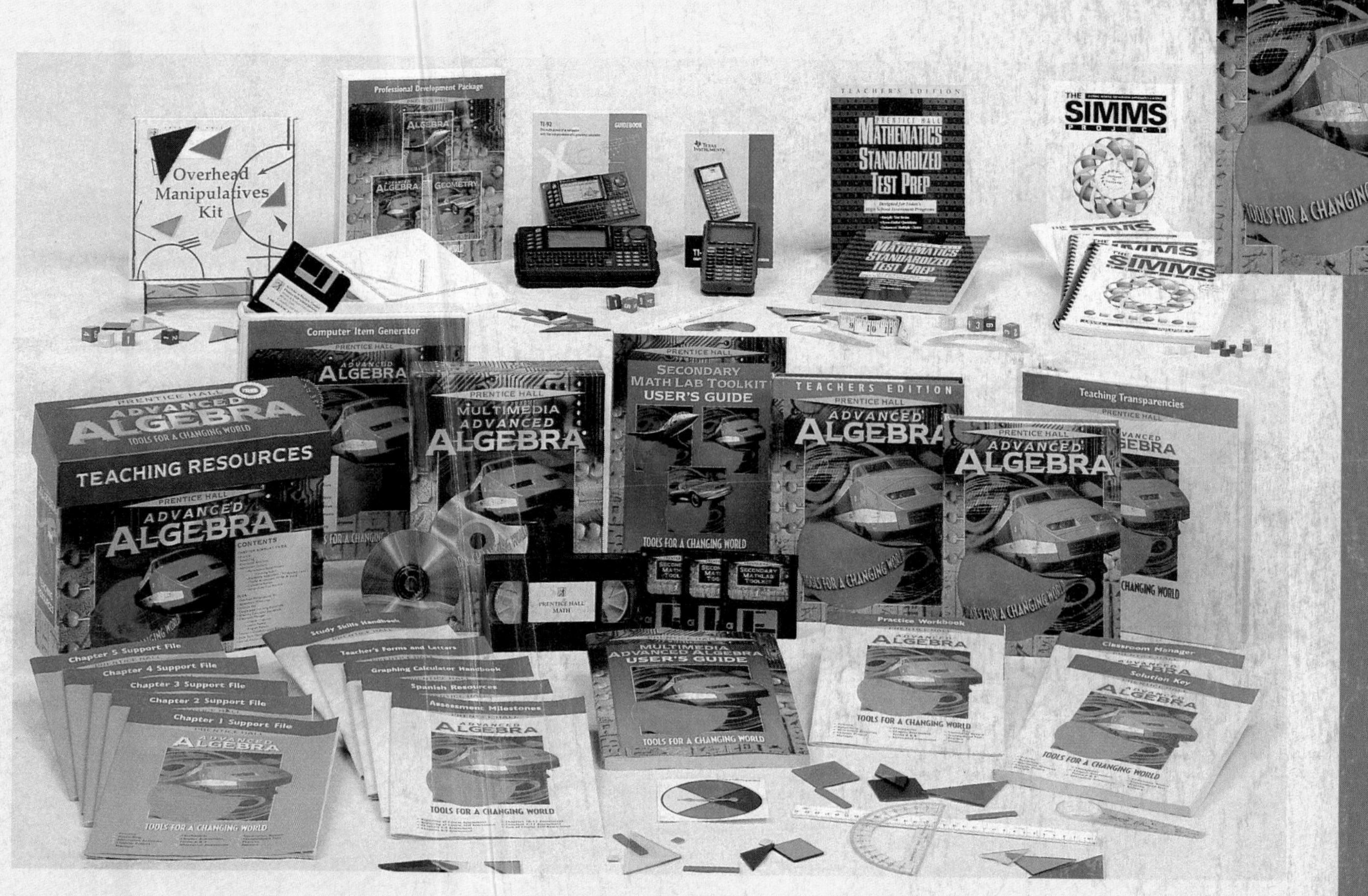

PRENTICE HALL
ADVANCED
ALGEBRA
TOOLS FOR A CHANGING WORLD

Technology Components:
Video Field Trips
Graphing Calculator Handbook
Calculator-Based Lab (CBL) Activities
Graphing Calculator TI-83 10-Pack
Graphing Calculator TI-92 10-Pack
Computer Item Generator with Dial-A-Test™
Secondary Math Lab Toolkit™
Multimedia Algebra CD-ROM
Multimedia Advanced Algebra CD-ROM
Resource Pro™ CD-ROM

TABLE OF CONTENTS ON NEXT PAGE

ALL NEW!

PRENTICE HALL MATH

ALGEBRA

Chapter 1: Tools of Algebra
Chapter 2: Functions and Their Graphs
Chapter 3: Algebraic Concepts and Simple Equations
Chapter 4: Equations and Inequalities
Chapter 5: Graphing and Writing Linear Equations
Chapter 6: Systems of Equations and Inequalities
Chapter 7: Quadratic Equations and Functions
Chapter 8: Exponents and Exponential Functions
Chapter 9: Right Triangles and Radical Expressions
Chapter 10: Polynomials
Chapter 11: Rational Expressions and Functions

ADVANCED ALGEBRA

Chapter 1: Models, Functions, and Permutations
Chapter 2: Linear Relationships and Functions
Chapter 3: Matrices
Chapter 4: Linear Systems
Chapter 5: Quadratic Equations and Functions
Chapter 6: Polynomials and Polynomial Functions
Chapter 7: Exponential and Logarithmic Functions
Chapter 8: Rational Functions
Chapter 9: Periodic Functions and Trigonometry
Chapter 10: Quadratic Relations
Chapter 11: More Probability and Statistics
Chapter 12: Sequences and Series
Appendix: Additional Topics on Trigonometry

© 1998

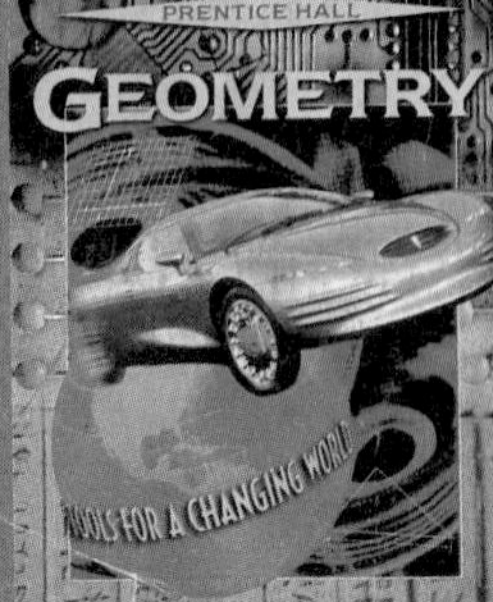

GEOMETRY

Ch. 1: Tools of Geometry
Ch. 2: Investigating Geometric Figures
Ch. 3: Transformations: Shapes in Motion
Ch. 4: Triangle Relationships
Ch. 5: Measuring in the Plane
Ch. 6: Measuring in Space
Ch. 7: Reasoning and Parallel Lines
Ch. 8: Proving Triangles Congruent
Ch. 9: Quadrilaterals
Ch. 10: Similarity
Ch. 11: Right Triangle Trigonometry
Ch. 12: Chords, Secants, and Tangents

PRENTICE HALL

ALGEBRA

AUTHORS

Allan Bellman

Sadie Chavis Bragg

Suzanne H. Chapin

Theodore J. Gardella

Bettye C. Hall

William G. Handlin, Sr.

Edward Manfre

Geometry Authors

Laurie E. Bass
Basia Rinesmith Hall

Art Johnson
Dorothy F. Wood

Contributing Author

Simone W. Bess

TOOLS FOR A CHANGING WORLD

PRENTICE HALL
Needham, Massachusetts
Upper Saddle River, New Jersey

REVIEWERS

Series Reviewers

James Gates, Ed.D.
Executive Director Emeritus, National Council of Teachers of Mathematics
Reston, Virginia

Vinetta Jones, Ph.D.
National Director, Equity 2000, The College Board, New York, New York

Algebra

John J. Brady III
Hume-Fogg High School
Nashville, Tennessee

Elias P. Rodriguez
Leander Junior High School
Leander, Texas

Dorothy S. Strong, Ed.D.
Chicago Public Schools
Chicago, Illinois

Art W. Wilson, Ed.D.
Abraham Lincoln High School
Denver, Colorado

Advanced Algebra

Eleanor Boehner
Methacton High School
Norristown, Pennsylvania

Laura Price Cobb
Dallas Public Schools
Dallas, Texas

William Earl, Ed.D.
Formerly Mathematics
Education Specialist
Utah State Office of Education
Salt Lake City, Utah

Robin Levine Rubinstein
Shorewood High School
Shoreline, Washington

Geometry

Sandra Argüelles Daire
Miami Senior High School
Miami, Florida

Priscilla P. Donkle
South Central High School
Union Mills, Indiana

Tom Muchlinski, Ph.D.
Wayzata High School
Plymouth, Minnesota

Bonnie Walker
Texas ASCD
Houston, Texas

Karen Doyle Walton, Ed.D.
Allentown College of
Saint Francis de Sales
Center Valley, Pennsylvania

Staff Credits

The editors, designers, marketers, managers, production and manufacturing buyers, page production manager, and advertising and promotion manager who made up the Algebra *team are listed below.*

Alison Anholt-White, Jackie Zidek Bedoya, Barbara A. Bertell, Ellen Brown, Judith D. Buice, Kathy Carter, Christine Deliee, Gabriella Della Corte, Robert G. Dunn, Barbara Flockhart, Audra Floyd, David Graham, Bridget A. Hadley, Elizabeth A. Jordan, Martha G. Smith, Stuart Wallace, Cynthia A. Weedel

PRENTICE HALL
Simon & Schuster Education Group
A VIACOM COMPANY

ISBN: 0-13-838673-0

1 2 3 4 5 6 7 8 9 10 03 02 01 00 99 98 97 96

Algebra & Advanced Algebra Authors

Allan Bellman is a classroom teacher at Watkins Mill High School in Gaithersburg, Maryland, and is also an instructor in the Woodrow Wilson National Fellowship Foundation Outreach program for mathematics teachers. Mr. Bellman has particular expertise in the use of the graphing calculator in the classroom, and also serves as a Texas Instruments workshop leader. Advanced Algebra chapters 1, 2, 5, 9

Sadie Chavis Bragg, Ed.D., is Vice President of Academic Affairs at the Borough of Manhattan Community College of The City University of New York. Dr. Bragg is a member of the Mathematical Sciences Education Board (MSEB), is President Elect of the Academic Assembly Council, and is an active member of the Benjamin Bannecker Association. Algebra chapters 3, 6, 11 and Advanced Algebra chapter 3

Suzanne H. Chapin, Ed.D., is Professor of Mathematics Education at Boston University. Dr. Chapin also directs all mathematics professional development in a landmark Boston University/Chelsea Public Schools Partnership program, working closely with teachers, administrators, and parents to provide opportunities for all students to learn mathematics. Algebra chapters 2, 8, and the probability strand

Theodore J. Gardella, formerly of the Bloomfield Hills Public Schools, Bloomfield Hills, Michigan, was awarded the Michigan Presidential Award for Excellence in Science and Mathematics Teaching. Mr. Gardella is also an originator of the "Tune in Mathematics and Science" project, delivering satellite video courses for students and staff development sessions for teachers. Teacher's Editions, and continuity of strands across the books

Bettye C. Hall is the former Director of Mathematics in the Houston Unified School District in Houston, Texas. Ms. Hall, admired as a "teacher's teacher" with her very practical perspective on students' and teachers' needs, is active as a mathematics consultant, speaker, and workshop leader throughout the United States. Algebra chapters 4, 7, 10 and Advanced Algebra chapters 6, 11

William G. Handlin, Sr., is a classroom teacher at Spring Woods High School in Houston, Texas. Awarded Life Membership in the Texas Congress of Parent and Teachers Association for his contributions to the well-being of children, Mr. Handlin is also a frequent workshop and seminar leader in professional meetings throughout the state. Advanced Algebra chapters 4, 8, 10, 12

Edward Manfre is a mathematics consultant from Albuquerque, New Mexico. Mr. Manfre has an extensive teaching background in the elementary, middle, and secondary schools, and for over twenty years has been developing educational materials that encourage thinking and self-reliance. Algebra chapters 1, 5, 9

Geometry Authors

Laurie E. Bass
The Fieldston School
Riverdale, New York.

Basia Rinesmith Hall
Alief Hastings High School
Alief Independent School District
Alief, Texas.

Art Johnson
Nashua High School
Nashua, New Hampshire.

Dorothy F. Wood
Formerly, Kern High School District
Bakersfield, California.

Contributing Author

Simone W. Bess
University of Cincinnati
College of Education
Cincinnati, Ohio

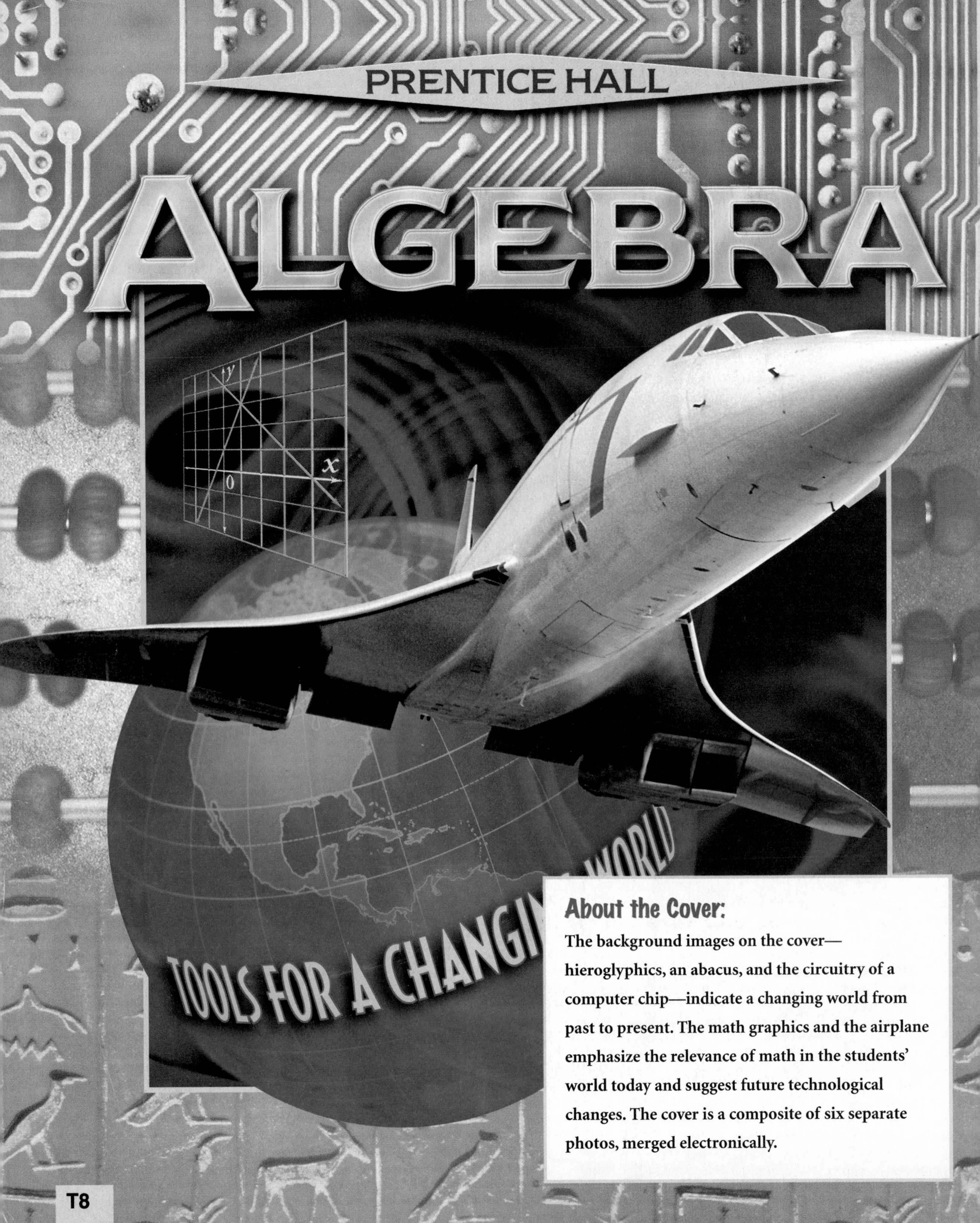

About the Cover:

The background images on the cover—hieroglyphics, an abacus, and the circuitry of a computer chip—indicate a changing world from past to present. The math graphics and the airplane emphasize the relevance of math in the students' world today and suggest future technological changes. The cover is a composite of six separate photos, merged electronically.

PRENTICE HALL MATH

ALGEBRA

TOOLS FOR A CHANGING WORLD

CONTENTS

Algebra Contents

The "To the Student" section, appearing before Chapter 1, helps students understand how they learn and how the textbook relates to their daily lives. At the end of the book there are a number of useful resources. The Skills Handbook provides an in-text tutorial of math skills that students may need to review. Extra Practice pages for each chapter contain additional exercises correlated to each lesson. The Glossary/Study Guide provides not only a definition and a page reference but also an example of each vocabulary term.

CHAPTER

1 Tools of Algebra

Connections and Applications

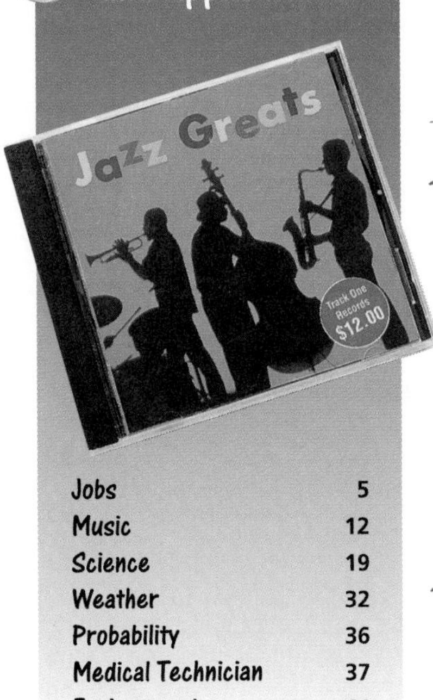

Jobs	5
Music	12
Science	19
Weather	32
Probability	36
Medical Technician	37
Environment	41

. . . and More!

While reviewing skills from middle grades math, students build on their basic understandings of math to explore relationships shown in a variety of ways. Students use integers and fractions to represent relationships in graphs (bar, line, etc.) and with variables (as expressions or equations). New concepts of matrices and spreadsheets are introduced as additional ways to represent relationships among sets of data. Students also use probability experiments to explore data.

CHAPTER

2 Functions and Their Graphs

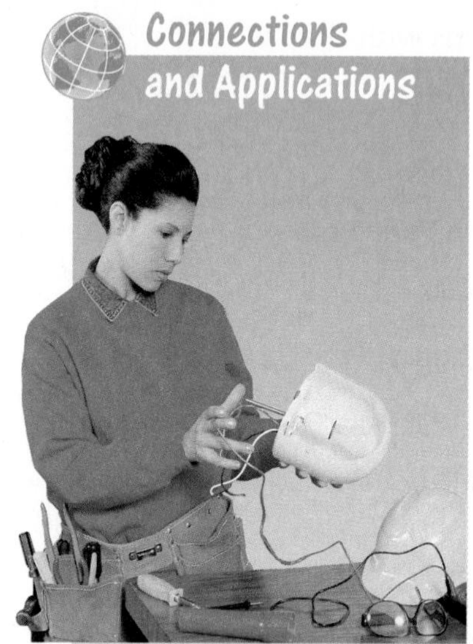

Connections and Applications

After seeing in Chapter 1 how variables can represent relationships, students continue to look at visual representations of relationships by interpreting scatter plots, graphs without scales, and functions. Functions are introduced so that students can model real-world situations with graphs, tables, and equations. Families of functions are briefly covered in order to give students an intuitive grasp of how various types of functions differ visually.

CHAPTER 3

Algebraic Concepts and Simple Equations

Connections and Applications

Checks and Balances
Making a Personal Budget

Using their basic understandings of variables, integers, and fractions, students solve and graph equations with one variable. Building on the functional relationship work of Chapters 1 and 2, students model and solve many real-world word problems by first writing equations.

CHAPTER 4

Equations and Inequalities

Connections and Applications

Chapter 4 continues with solving equations with one variable. Inequalities are also included in this chapter to show students that the procedure for solving inequalities is similar to and an extension of the procedure for solving equations.

CHAPTER 5

Graphing and Writing Linear Equations

Connections and Applications

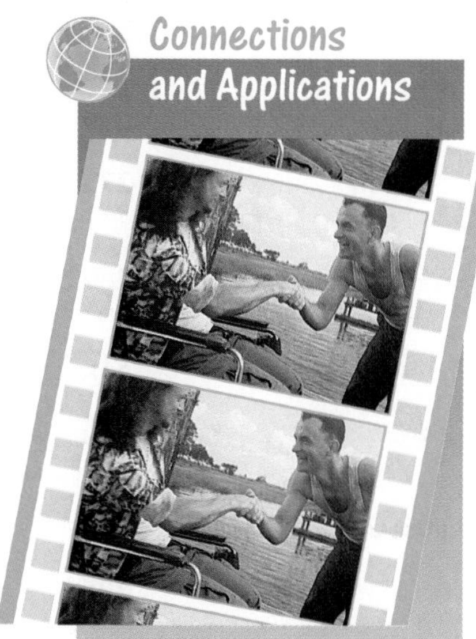

Moving from the specific case of equations with one variable to the general case of functions, Chapter 5 introduces students to graphing linear equations with two variables. By doing work on rate of change, students understand how the slope of a line can be interpreted in real-world situations. Students also learn about the forms of a linear equation and connect them to both statistics and geometry.

CHAPTER 6

Systems of Equations and Inequalities

Connections and Applications

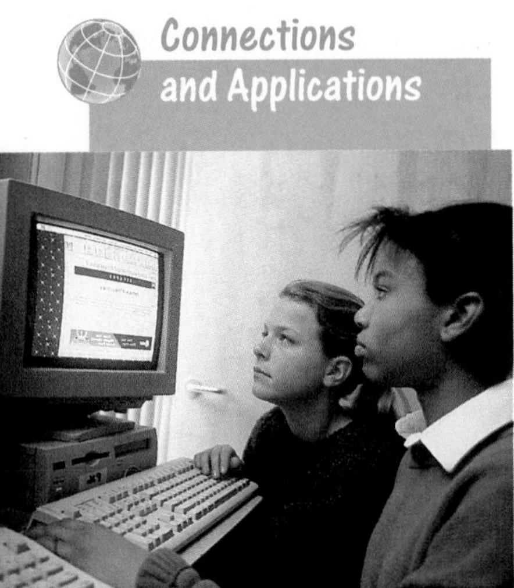

ASSESSMENT

Chapter Project

Let's Dance!
Planning a Dance

In Chapter 6 students build upon their work in Chapter 5 by extending linear equations to linear systems. Students use several methods to solve systems and see how systems are used in many situations. Using their knowledge of inequalities from Chapter 4, students now work with linear inequalities. The basic concepts of linear programming are also introduced.

CHAPTER 7

Quadratic Equations and Functions

Connections and Applications

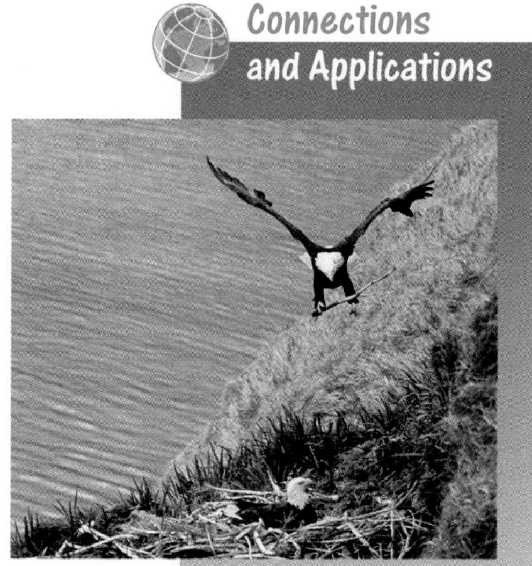

Chapter 7 builds on the quadratic family of functions that was introduced in Chapter 2. Students explore quadratic functions to see how the equation is related to the graph. After reviewing how to find square roots, students learn how to solve quadratic equations by two methods: using square roots and using the quadratic formula. The quadratic formula is introduced here to provide students with a visual representation of the solutions to a quadratic equation.

CHAPTER 8

Exponents and Exponential Functions

Connections and Applications

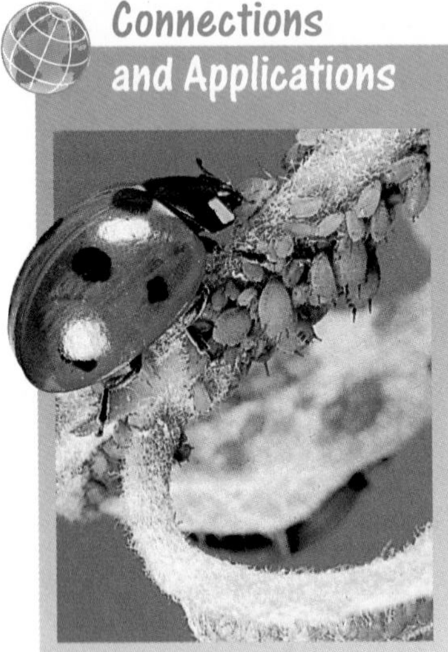

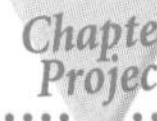

Chapter Project *Moldy Oldies*
Measuring the Growth of Mold

Continuing with the function theme, Chapter 8 covers exponential functions. Starting the chapter with real-world applications, such as growth and decay, gives students a foundation for more abstract concepts of exponents. Using this knowledge and building on their work with exponents in Chapter 1, students simplify expressions using the multiplication, power, and division properties of exponents.

CHAPTER 9

Right Triangles and Radical Expressions

Connections and Applications

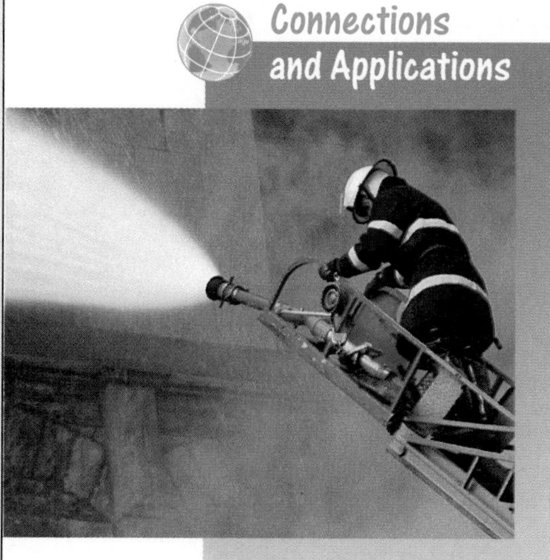

Building on their work in Chapters 7 and 8, students apply square roots and exponents to solving problems involving the Pythagorean theorem, the distance formula, and standard deviation. Using properties similar to ones learned before, students perform operations on radical expressions. Students also learn to solve radical equations and then keep the function theme alive by graphing simple square root functions.

CHAPTER

10 Polynomials

Connections and Applications

Chapter 10 combines work on multiplying and factoring polynomials to show students how these two operations are closely related as "inverse" operations. Students then learn to use factoring to solve quadratic equations. Having learned how the solutions are visually represented on a graph in Chapter 7, students now have an opportunity to choose the appropriate method for solving each quadratic equation that they encounter.

CHAPTER 11

Rational Expressions and Functions

Connections and Applications

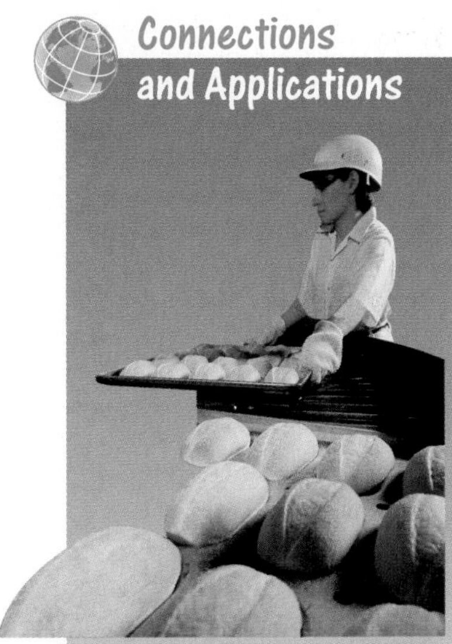

Building on their knowledge of reciprocals and functions, students explore the graphs of simple rational functions. Using what they know about fractions, students simplify rational expressions and then solve rational equations. Fraction work also helps students as they explore more probability topics, such as permutations and combinations.

What people are saying about...

PRENTICE HALL MATH

"After reviewing all of the chapters, I believe I can say that you have done an excellent job interpreting the NCTM Standards, both the process standards as well as the content standards."

James Gates, Ed.D.
Executive Director Emeritus
National Council of Teachers of Mathematics

"I am glad to see that the Prentice Hall textbook will contain information that will answer the most popular questions asked by students: <u>When and where are we going to use this?</u> and <u>Do I really have to know this?</u>"

Elias P. Rodriguez
Mathematics Teacher
Leander, Texas

"The Prentice Hall book was fun, and I learned a lot! I had trouble reading our old book's examples. The examples in the Prentice Hall book were much easier to understand. I learn more by using the examples, and this helped a lot! My grades went up this quarter."

Heidi S.
Field-test student

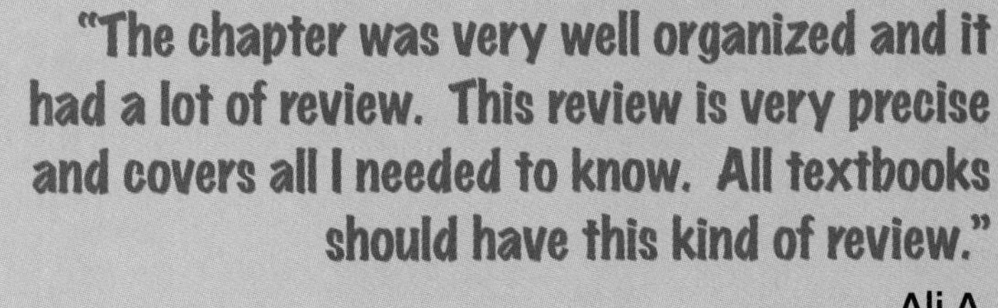

"The chapter was very well organized and it had a lot of review. This review is very precise and covers all I needed to know. All textbooks should have this kind of review."

Ali A.
Field-test student

Teachers and students across the country helped Prentice Hall develop this program. They identified the teaching tools that you can use to motivate students and manage instruction. We would like to hear from you. Write us at Prentice Hall Math, 160 Gould Street, Needham, MA 02194, or visit us on the Internet at http://www.phschool.com.

Tools to Help You Succeed

Why did we name our program "Tools for a Changing World"?

Because you've said

...not all **STUDENTS** have the same learning style;

...not all **TEACHERS** have the same teaching style;

...and no two **CLASSROOM** environments are the same.

It is for these reasons that we offer you a wide variety of teaching tools to help you MOTIVATE your students and MANAGE instruction.

PRENTICE HALL MATH

TOOLS FOR A CHANGING WORLD

brings together the widest variety of Teaching Tools.

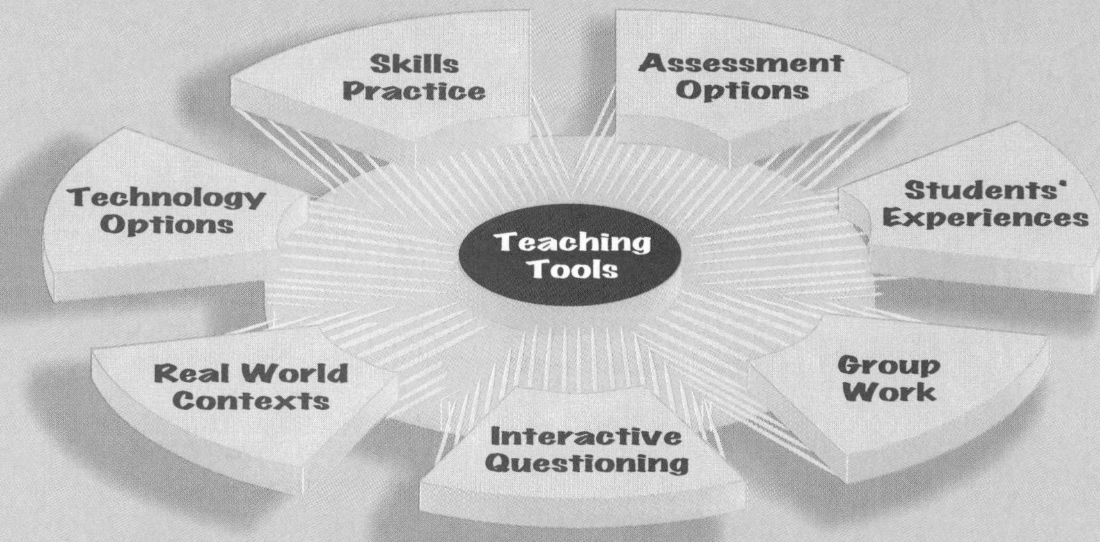

Using these tools, Prentice Hall Math helps you address the *motivation* and *management* issues you've told us are foremost on your mind.

How can I motivate students to *want* to learn math?

BY USING ANY OR ALL OF THESE RESOURCES.	*Where do I look?*
The Relating to the Real World completely worked-out Examples and practice exercises show students just how often math is used in their lives today and in the future.	Student Edition
Chapter Projects are long-term projects that provide a real-world connection to the math content of the chapter.	Student Edition
Video Field Trips support the chapter projects or other real-world connections for each chapter in the Student Edition.	Video Field Trips
Suggestions for Connecting to the Students' World provide opportunities for discussing how students apply math in their lives.	Teacher's Edition
Connecting to Prior Knowledge sections show students how new math concepts are related to material previously learned.	Teacher's Edition
The Student Manipulatives Kit engages students in a hands-on experience when discovering and connecting math concepts.	Student Manipulatives Kit
Multimedia Math Labs combine animation, graphics, video, and Math Tools in chapter-long Math Labs to spark the students' interest in the mathematics of each chapter.	Multimedia Math Labs
Secondary Math Lab Toolkit™ gives you and students software-based Integrated Math Labs with Math Tools for exploration.	Secondary Math Lab Toolkit™
Prentice Hall's Internet Home Page offers a special section on Math, including projects of the month, exciting links, and a calendar of events.	Internet http://www .phschool.com

How can I help *all* students become lifelong learners?

BY USING ANY OR ALL OF THESE RESOURCES.

	Where do I look?
Work Together activities encourage students to collaborate to solve problems, just as they will in the workplace.	Student Edition
Think and Discuss interactive questions offer students opportunities to communicate with their peers by practicing their listening, thinking, talking, and writing skills.	Student Edition
ESL and Diversity suggestions help reach students of various backgrounds to make their math experience fulfilling.	Teacher's Edition
Learning Styles sections provide suggestions for teaching concepts to students of different learning styles (tactile, auditory, kinesthetic, visual).	Teacher's Edition
Alternative Activities offer students of varying learning styles another way of understanding the lesson concepts.	Teaching Resources
Spanish Resources help students whose first language is Spanish to understand and apply concepts.	Teaching Resources
Study Skills Handbook offers students tips for successful studying that can be applied to other subject areas.	Teaching Resources
Secondary Math Lab Toolkit™ gives students the opportunity to integrate software into their learning, just as they will be using technology in most jobs.	Secondary Math Lab Toolkit™

Skills Practice
Assessment Options
Students' Experiences
Technology Options
Teaching Tools
Real World Contexts
Interactive Questioning
Group Work

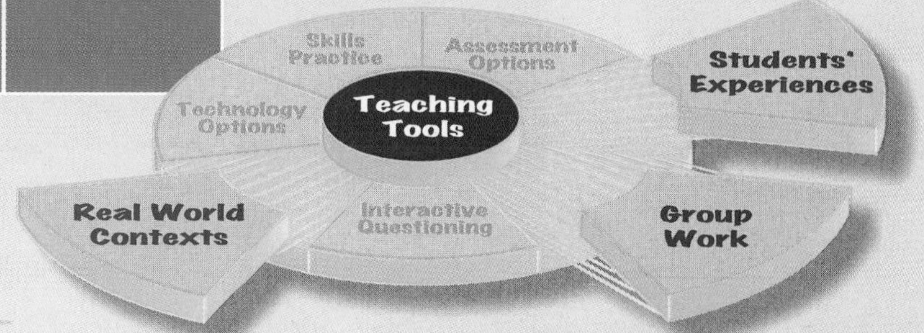

PRENTICE HALL MATH

Helping you manage the widest variety of motivational tools to

Reach All Students

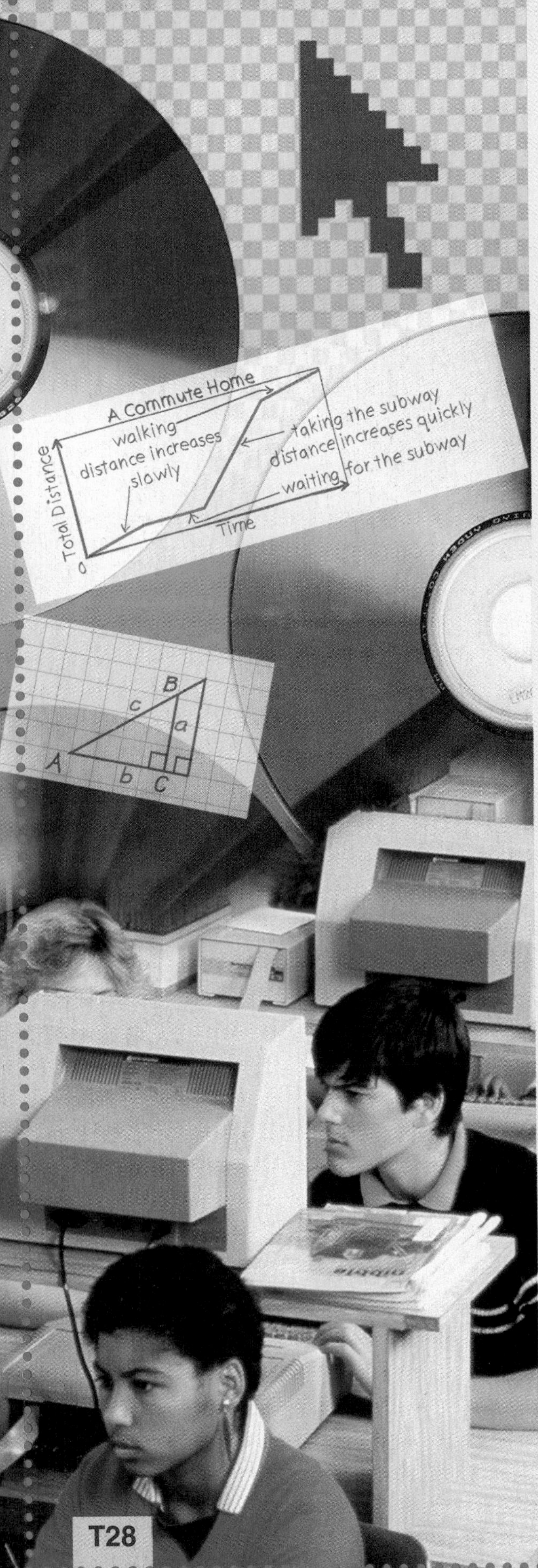

What technology tools are available for me to use?

USE ANY OR ALL OF THESE RESOURCES.	*Where do I look?*
The Student Edition incorporates the use of scientific or graphing calculators during concept development to empower students of all learning styles to succeed.	Student Edition
A Technology Options side column correlates all the technology components available in the program to each lesson.	Teacher's Edition
Graphing Calculator Handbook provides step-by-step instructions to guide students using a variety of calculators as they work on topics in some Student Edition lessons.	Graphing Calculator Handbook
Calculator-Based Lab Activities takes advantage of powerful CBL equipment and provides activities that relate to the math concepts in the Student Edition.	CBL Activities
Multimedia Math Labs feature CD-ROM–based Labs with QuickTime™ movies, animations, and the Secondary Math Lab Toolkit™ — all keyed to the program objectives.	Multimedia Math Labs
Secondary Math Lab Toolkit™ gives you and students software-based Integrated Math Labs with Math Tools for exploration.	Secondary Math Lab Toolkit™
Prentice Hall's Internet Home Page offers a special section on Math, including projects of the month, exciting links, and a calendar of events.	Internet http://www.phschool.com
The Resource Pro™ CD-ROM package helps you to plan by including many of the supplemental materials in an easy-to-access format.	Resource Pro™

Technology Options

Skills Practice

Assessment Options

Teaching Tools

Students' Experiences

Group Work

Interactive Questioning

Real World Contexts

PRENTICE HALL MATH

Helping you manage the many and varied options for

Using Technology

Where do I look for options to evaluate students' success?

USE ANY OR ALL OF THESE RESOURCES.

	Where do I look?
Think and Discuss sections with their interactive questions and Try This practice exercises offer ongoing assessment as lessons are developed.	Student Edition
Preparing for Standardized Tests provides practice with open-ended, multiple-choice, and free-response questions and allows teachers to assess students' learning of the math concepts.	Student Edition and Teaching Resources
Alternative Assessment includes Self-Assessment sections, Portfolio activities, Journal questions, and Chapter Assessments in a variety of testing formats.	Student Edition and Teaching Resources
Lesson Quizzes offer you yet another opportunity to assess students' understanding of the lesson concepts.	Teacher's Edition and Transparencies
Ongoing Assessment activities provide daily alternatives for assessing the students' learning.	Teacher's Edition
Computer Item Generator enables you to customize assessment for individual needs.	Computer Item Generator with Dial-A-Test™
Chapter Project and Scoring Rubric helps students understand what is expected of them as they work on the chapter projects in the Student Edition.	Teaching Resources
Cumulative Assessment offers the option of long-term assessments (mid-book, second-half-of-book, end-of-book).	Teaching Resources

PRENTICE HALL MATH

Helping you manage
the many and varied options for

Assessing Learning

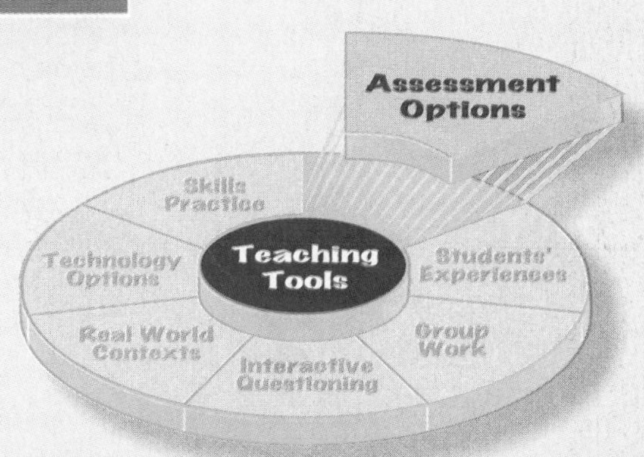

Managing Time Constraints

How can I free up my time for teaching?

BY USING ANY OR ALL OF THESE RESOURCES.

	Where do I look?
Work Together activities in the lesson itself provide cooperative group work without having to supplement the textbook.	Student Edition
Lessons divided into Parts, with titles announcing each new math concept, break the lesson into manageable sections.	Student Edition
Assignment Options that correlate to the lesson parts in the Student Edition offer you a quick and handy assignment, no matter how much of the lesson you cover during a class period.	Teacher's Edition
Chapter Support Files are organized chapter-by-chapter to keep the shuffling of booklets to a minimum and to help you easily find the resources you need.	Teacher's Edition
Classroom Manager pulls together Lesson Planners and Chapter Organizers that streamline your planning time.	Classroom Manager
Transparencies provide clear visual support to facilitate instruction and learning.	Transparencies
Computer Item Generator is a data bank of test items that allows you to quickly customize practice and assessment.	Computer Item Generator with Dial-A-Test™
Resource Pro™ is a CD-ROM–based package, containing supplemental materials and planning tools, that enables you to customize each lesson quickly and easily.	Resource Pro™

PRENTICE HALL MATH

Helping you manage your time!

How can I cover my objectives while accommodating different schedules?

BY USING THE FOLLOWING COURSE PACING GUIDE.

	Chapter 1 (9 lessons)	Chapter 2 (8 lessons)	Chapter 3 (8 lessons)	Chapter 4 (9 lessons)	Chapter 5 (9 lessons)	Chapter 6 (8 lessons)
Traditional Class Periods (40–45 minutes)	13	15	15	16	15	13
Two-Year Algebra Class Periods (40–45 minutes)	28	30	30	32	30	26
Block Scheduling Class Periods (90 minutes)	8	6	7	8	8	7

	Chapter 7 (7 lessons)	Chapter 8 (8 lessons)	Chapter 9 (8 lessons)	Chapter 10 (7 lessons)	Chapter 11 (7 lessons)	TOTALS
Traditional Class Periods (40–45 minutes)	14	16	15	14	14	**160**
Two-Year Algebra Class Periods (40–45 minutes)	26	32	30	28	28	**320**
Block Scheduling Class Periods (90 minutes)	7	8	7	7	7	**80**

AND BY USING THESE PACING RESOURCES.

	Where do I look?
Detailed **Chapter Pacing Options** precede each chapter and give you lesson-by-lesson pacing suggestions for that specific chapter.	Teacher's Edition
Two-Year Algebra Handbook contains worksheets for each lesson part, enabling you to teach day-by-day, one part of a lesson at a time.	Two-Year Algebra Handbook

How can I find ideas and materials that will help me grow professionally?

BY TAKING ADVANTAGE OF SOME OR ALL OF THESE ACTIVITIES.

	Where do I look?
Lesson-specific or chapter-specific Teaching Notes at point of use offer you professional support.	Teacher's Edition
Professional Development Package offers you all the training tools necessary for customized professional development.	Professional Development Package
The Prentice Hall Internet Home Page provides, among other things, a calendar of professional events and a variety of links to other sites on the Web that offer content ideas as well as growth opportunities.	Internet http://www .phschool.com
Prentice Hall Summer Seminars, focusing on math issues and technology, bring together math educators and experts for a variety of interactive and hands-on sessions.	Watch your mail for special flyers.
Prentice Hall Consultant Workshops offer you experiences during in-service training to learn about specific math issues.	Contact your local Prentice Hall sales representative.

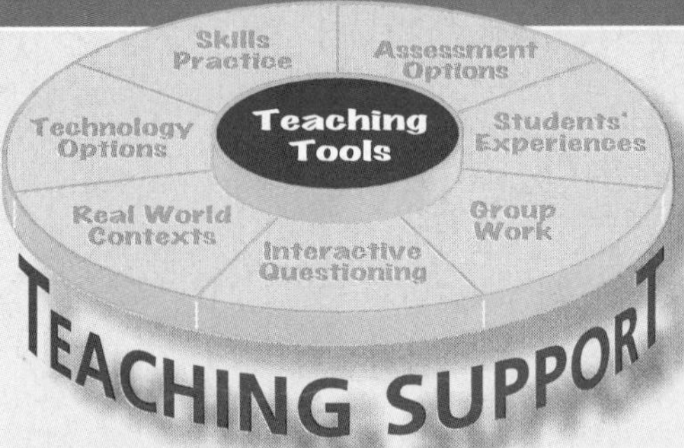

PRENTICE HALL MATH

Helping you manage the many and varied options for effective

Professional Development

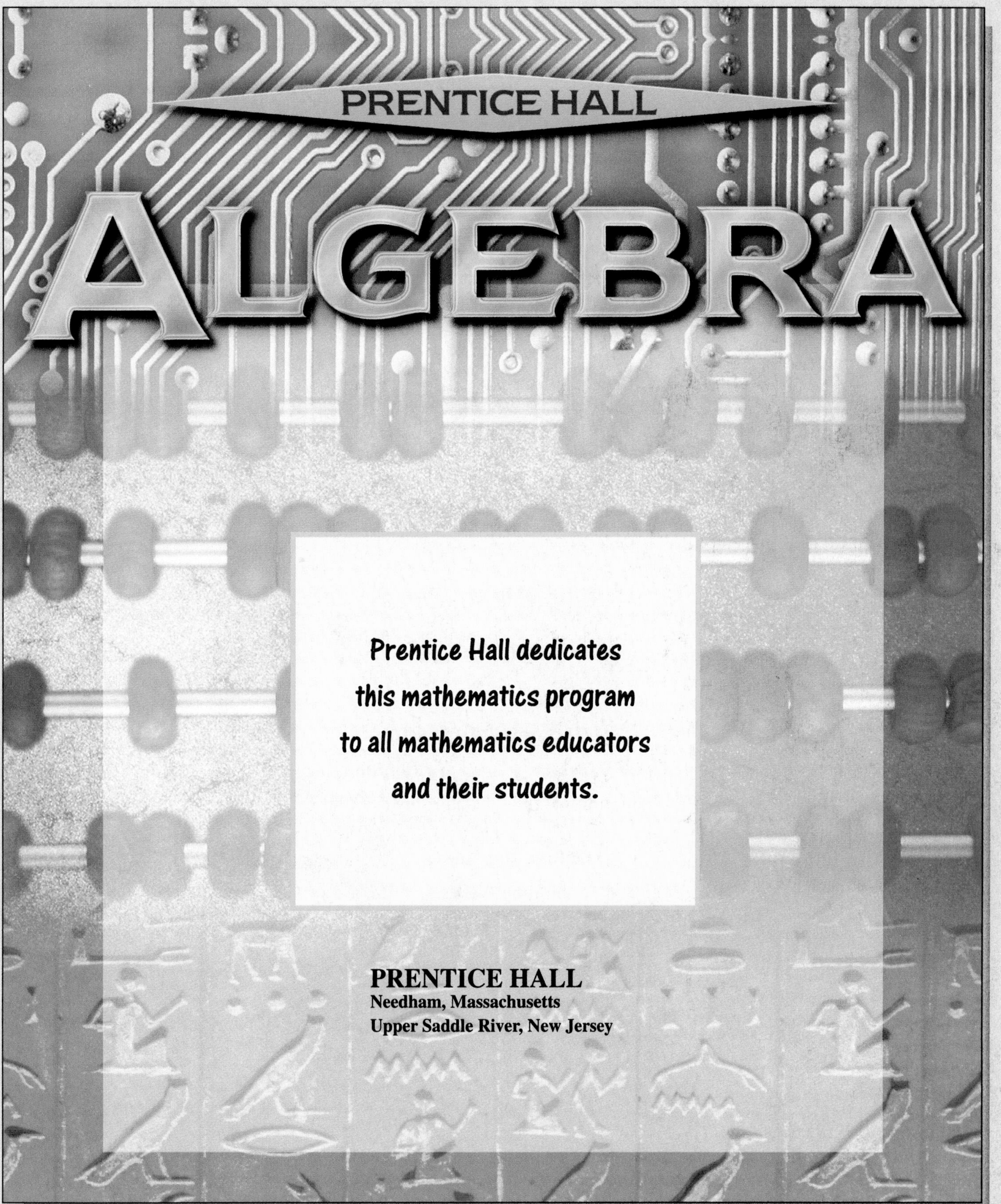
PRENTICE HALL
ALGEBRA
Prentice Hall dedicates
this mathematics program
to all mathematics educators
and their students.
PRENTICE HALL
Needham, Massachusetts
Upper Saddle River, New Jersey

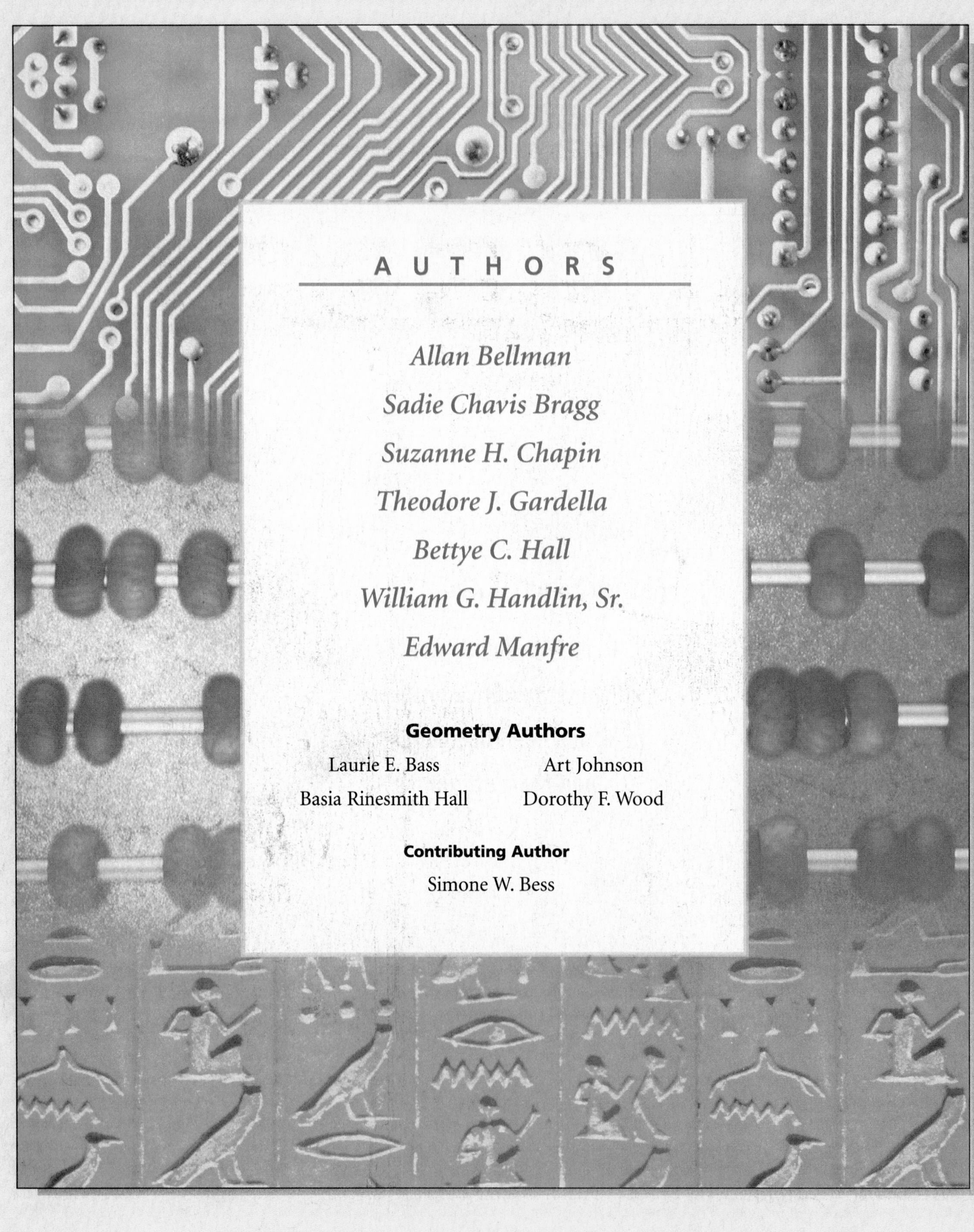

AUTHORS

Allan Bellman

Sadie Chavis Bragg

Suzanne H. Chapin

Theodore J. Gardella

Bettye C. Hall

William G. Handlin, Sr.

Edward Manfre

Geometry Authors

Laurie E. Bass
Art Johnson
Basia Rinesmith Hall
Dorothy F. Wood

Contributing Author

Simone W. Bess

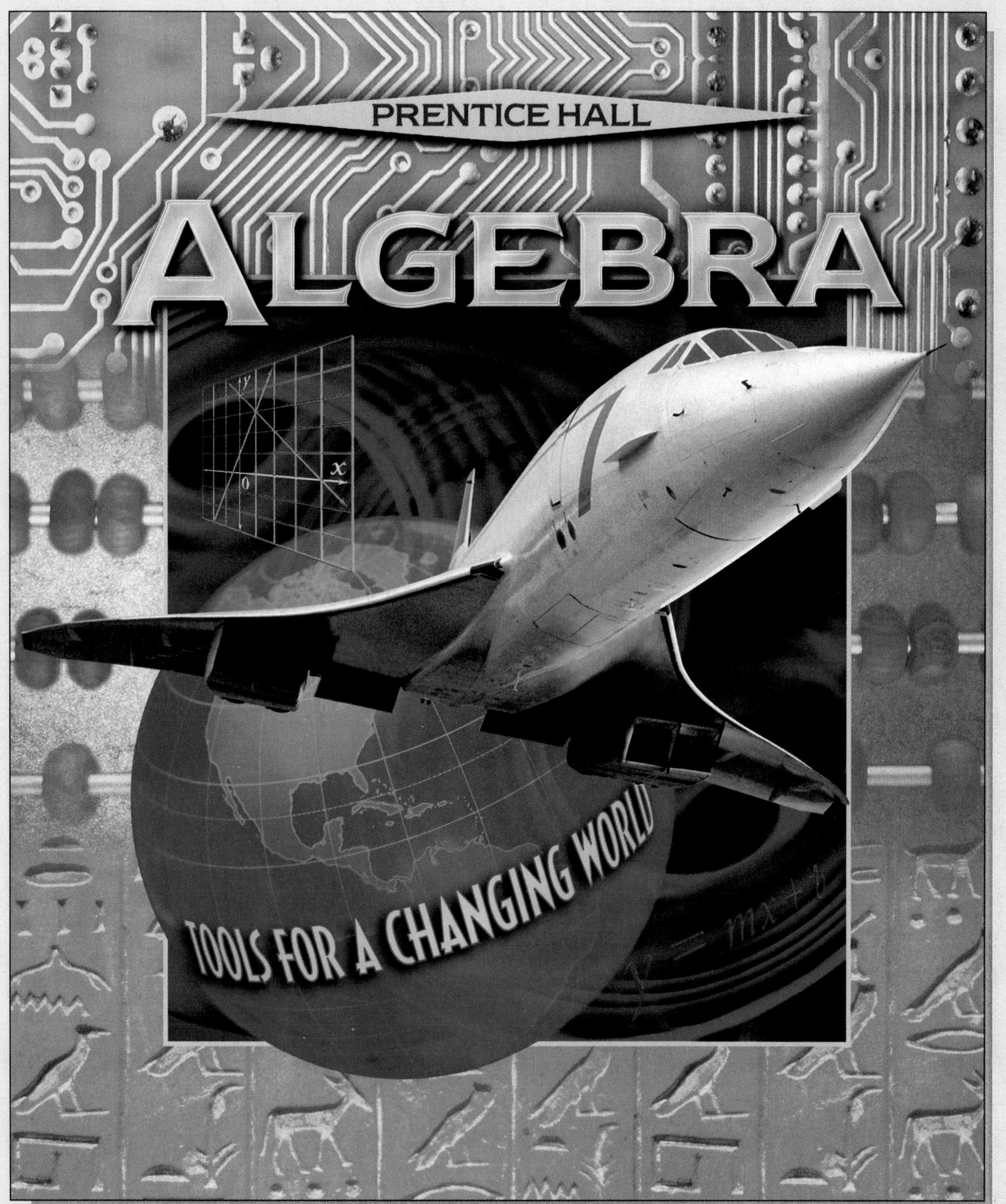
PRENTICE HALL
ALGEBRA
TOOLS FOR A CHANGING WORLD

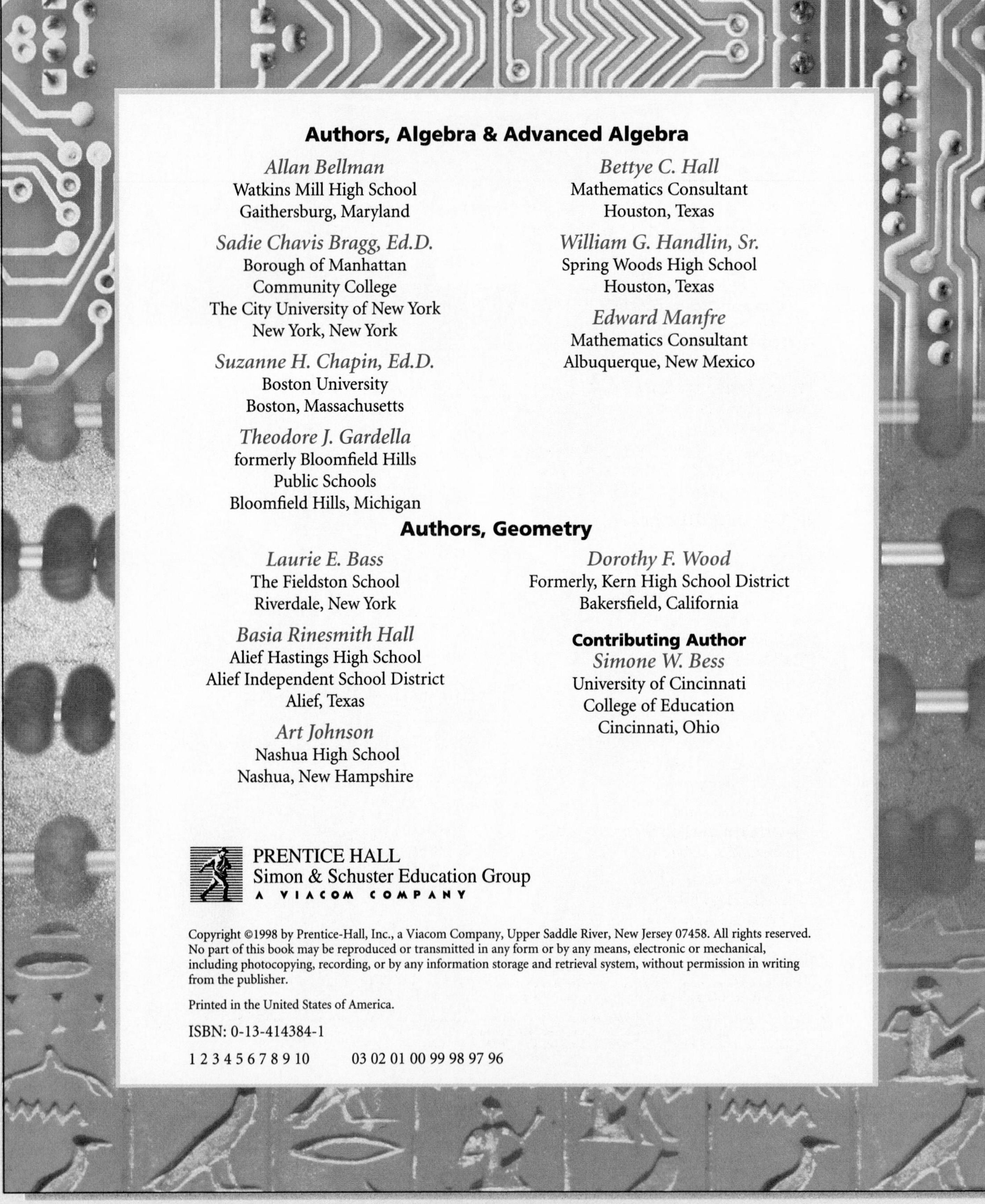

Authors, Algebra & Advanced Algebra

Allan Bellman
Watkins Mill High School
Gaithersburg, Maryland

Sadie Chavis Bragg, Ed.D.
Borough of Manhattan
Community College
The City University of New York
New York, New York

Suzanne H. Chapin, Ed.D.
Boston University
Boston, Massachusetts

Theodore J. Gardella
formerly Bloomfield Hills
Public Schools
Bloomfield Hills, Michigan

Bettye C. Hall
Mathematics Consultant
Houston, Texas

William G. Handlin, Sr.
Spring Woods High School
Houston, Texas

Edward Manfre
Mathematics Consultant
Albuquerque, New Mexico

Authors, Geometry

Laurie E. Bass
The Fieldston School
Riverdale, New York

Basia Rinesmith Hall
Alief Hastings High School
Alief Independent School District
Alief, Texas

Art Johnson
Nashua High School
Nashua, New Hampshire

Dorothy F. Wood
Formerly, Kern High School District
Bakersfield, California

Contributing Author
Simone W. Bess
University of Cincinnati
College of Education
Cincinnati, Ohio

PRENTICE HALL
Simon & Schuster Education Group
A VIACOM COMPANY

Printed in the United States of America.

ISBN: 0-13-414384-1

1 2 3 4 5 6 7 8 9 10 03 02 01 00 99 98 97 96

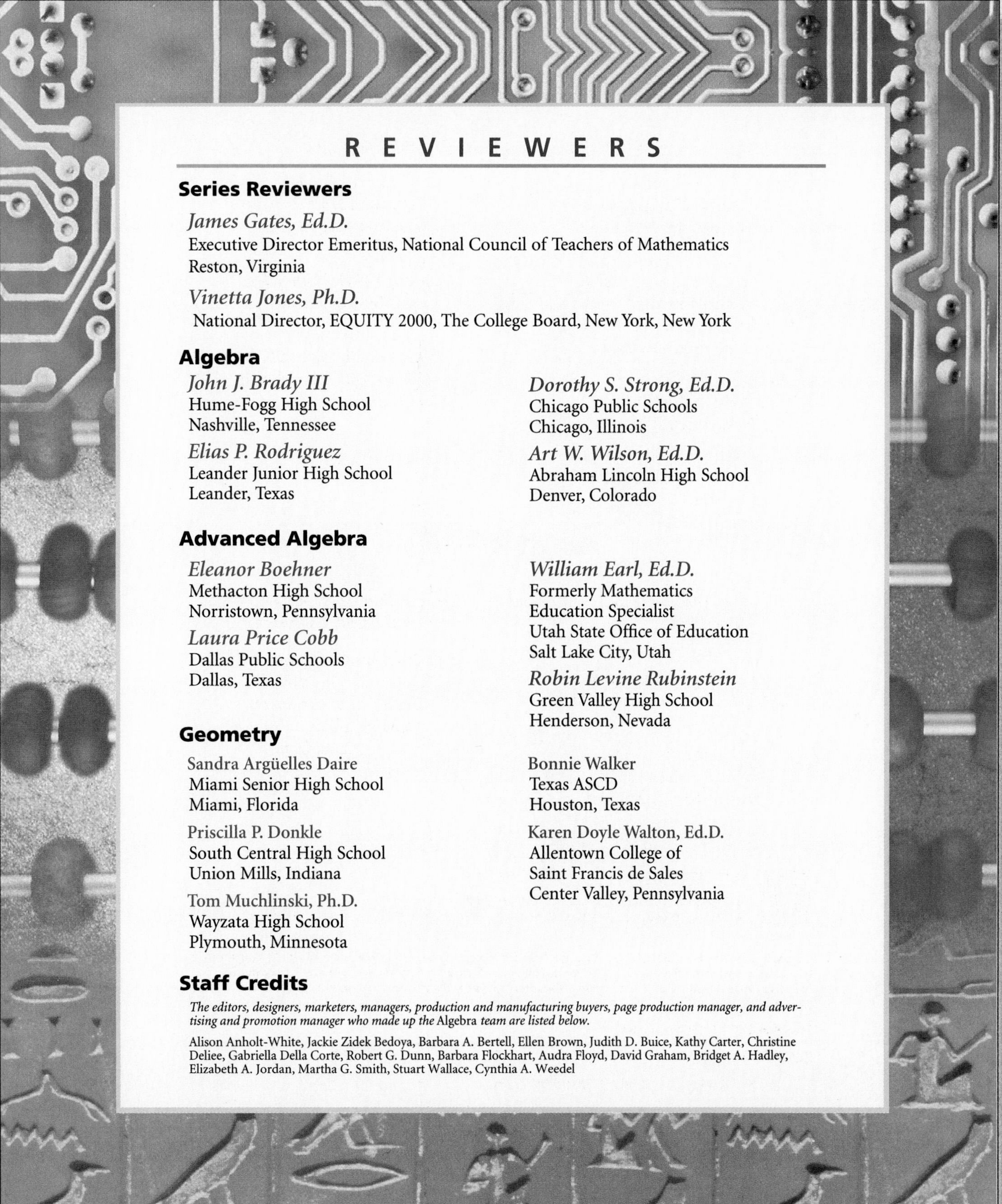

REVIEWERS

Series Reviewers

James Gates, Ed.D.
Executive Director Emeritus, National Council of Teachers of Mathematics
Reston, Virginia

Vinetta Jones, Ph.D.
National Director, EQUITY 2000, The College Board, New York, New York

Algebra

John J. Brady III
Hume-Fogg High School
Nashville, Tennessee

Elias P. Rodriguez
Leander Junior High School
Leander, Texas

Dorothy S. Strong, Ed.D.
Chicago Public Schools
Chicago, Illinois

Art W. Wilson, Ed.D.
Abraham Lincoln High School
Denver, Colorado

Advanced Algebra

Eleanor Boehner
Methacton High School
Norristown, Pennsylvania

Laura Price Cobb
Dallas Public Schools
Dallas, Texas

William Earl, Ed.D.
Formerly Mathematics
Education Specialist
Utah State Office of Education
Salt Lake City, Utah

Robin Levine Rubinstein
Green Valley High School
Henderson, Nevada

Geometry

Sandra Argüelles Daire
Miami Senior High School
Miami, Florida

Priscilla P. Donkle
South Central High School
Union Mills, Indiana

Tom Muchlinski, Ph.D.
Wayzata High School
Plymouth, Minnesota

Bonnie Walker
Texas ASCD
Houston, Texas

Karen Doyle Walton, Ed.D.
Allentown College of
Saint Francis de Sales
Center Valley, Pennsylvania

Staff Credits

The editors, designers, marketers, managers, production and manufacturing buyers, page production manager, and advertising and promotion manager who made up the Algebra *team are listed below.*

Alison Anholt-White, Jackie Zidek Bedoya, Barbara A. Bertell, Ellen Brown, Judith D. Buice, Kathy Carter, Christine Deliee, Gabriella Della Corte, Robert G. Dunn, Barbara Flockhart, Audra Floyd, David Graham, Bridget A. Hadley, Elizabeth A. Jordan, Martha G. Smith, Stuart Wallace, Cynthia A. Weedel

To the Student

Students like you helped Prentice Hall develop this program. They identified tools you can use to help you learn now in this course and beyond. In this special "To the Student" section and throughout this program, you will find the **tools you need to help you succeed.** We'd like to hear how these tools work for you. Write us at Prentice Hall Mathematics, 160 Gould Street, Needham, MA 02194 or visit us at http://www.phschool.com.

"... Instead of problem after problem of pointless numbers, we should have a chance to think and to truly understand what we are doing. I personally think that we all should be taught this way."
Chris, Grade 9
Carson City, NV

"... I learn mathematics best when I draw a diagram or make a graph that helps show what the problem is that I will solve."
Amy, Grade 11
Columbia, SC

"... I like to review what I learn as I go, rather than cramming the night before a test."
Ali, Grade 10
St. Paul, MN

Use this "To the Student" section to help students understand *how* they learn. It also shows how this textbook relates what they experience daily to the math they're learning and will need in the workplace of the future.

LEARN About Learning!

What comes to your mind when you hear the word **style**? Maybe it's hair style, or style of dress, or walking style. Have you ever thought about your learning style? Just like your hair or your clothes or your walk, everybody has a learning style that they like best because it works best for them. Look around you now. What do you see? Different styles … some like yours, some different from yours. That's the way it is with learning styles, too.

What's Your Best Learning Style?

I understand math concepts best when I...

- ❑ **A. Read about them.**
- ❑ **B. Look at and make illustrations, graphs, and charts that show them.**
- ❑ **C. Draw sketches or handle manipulatives to explore them.**
- ❑ **D. Listen to someone explain them.**

When I study, I learn more when I...

- ❑ **A. Review my notes and the textbook.**
- ❑ **B. Study any graphs, charts, diagrams, or other illustrations.**
- ❑ **C. Write ideas on note cards; then study the ideas.**
- ❑ **D. Explain what I know to another person.**

When I collaborate with a group, I am most comfortable when I...

- ❑ **A. Take notes.**
- ❑ **B. Make visuals for display.**
- ❑ **C. Demonstrate what I know to others.**
- ❑ **D. Give presentations to other groups or the whole class.**

Look for a pattern in your responses.
"A" responses suggest that you learn best by reading;
"B" responses indicate a visual learning style;
"C" responses suggest a tactile, or hands-on, learning style;
"D" responses signal that you probably learn best by listening and talking about what you are learning.

Students determine their dominant learning style by taking this quick and easy survey.

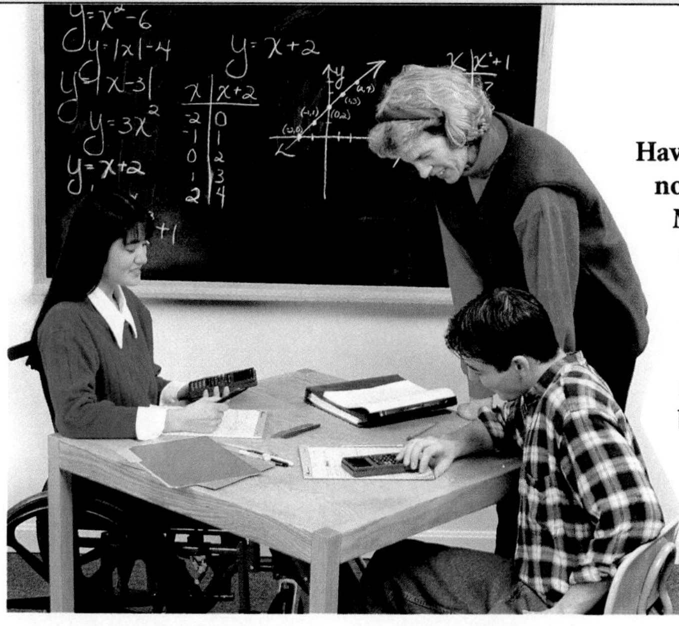

Having a preferred learning style does not limit you to using just that one. Most people learn by using a combination of learning styles. You'll be amazed by the ways that knowing more about yourself and how you learn will help you be successful — not only successful in mathematics, but successful in all your subject areas. When you know how you learn best, you will be well equipped to enter the work place.

Use this chart to help you strengthen your different learning styles.

Learning Style	Learning Tips
Learning by ***reading***	* Schedule time to read each day. * Carry a book or magazine to read during wait time. * Read what you like to read—it's OK not to finish a book.
Learning by using ***visual*** cues	* Visualize a problem situation. * Graph solutions to problems. * Let technology, such as computers and calculators, help you.
Learning by using ***hands-on*** exploration	* Make sketches when solving a problem. * Use objects to help you solve problems. * Rely on technology as a tool for exploration and discovery.
Learning by ***listening and talking***	* Volunteer to give presentations. * Explain your ideas to a friend. * Listen intently to what others are saying.

Most important, believe in yourself and your ability to learn!

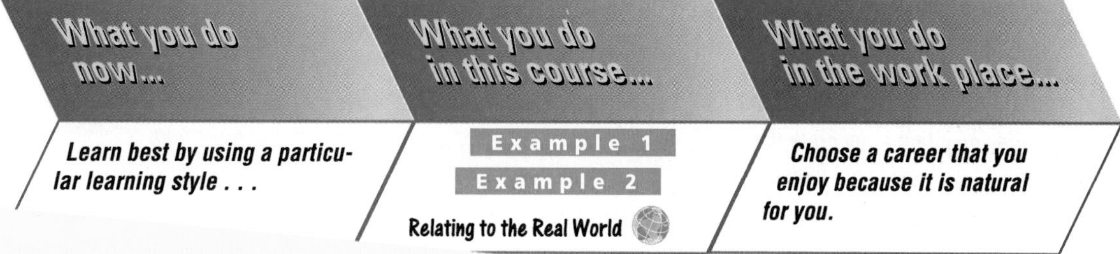

What you do now...
What you do in this course...
What you do in the work place...
Learn best by using a particular learning style . . .
Example 1
Example 2
Relating to the Real World
Choose a career that you enjoy because it is natural for you.

Note the diagram at the bottom of the page that connects instructional methods utilized throughout the program with what students experience today and in their future.

Help Teamwork Work for YOU!

Each of us works with other people on teams throughout our lives. What's your job? Your job on a team, that is. Maybe you play center on your basketball team, maybe you count votes for your school elections, perhaps you help decorate the gym for a school function, or maybe you help make scenery for a community play. From relay races to doing your part of the job in the work place, teamwork is required for success.

TEAMWORK CHECKLIST

- **Break apart the large task into smaller tasks, which become the responsibility of individual group members.**
- **Treat the differences in group members as a benefit.**
- **Try to listen attentively when others speak.**
- **Stay focused on the task at hand and the goal to be accomplished.**
- **Vary the tasks you do in each group and participate.**
- **Recognize your own and others' learning styles.**
- **Offer your ideas and suggestions.**
- **Be socially responsible and act in a respectful way.**

What you do now...	What you do in this course...	What you do in the work place...
Play on a team, decorate the gym, or perform in the band...	WORK TOGETHER	*Collaborate with coworkers on projects.*

Both teachers and students can reference this handy Teamwork Checklist while students are working in pairs or in small groups.

Reading + Writing + Talking + Listening:

It's All COMMUNICATION

We communicate in songs. We communicate in letters. We communicate with our body movements. We communicate on the phone. We communicate in cyberspace. It's all talking about ideas and sharing what you know. It's the same in mathematics — we communicate by reading, writing, talking, and listening. Whether we are working together on a project or studying with a friend for a test, we are communicating.

Ways to Communicate What You Know and Are Able to Do

- ✔ Explain to others how you solve a problem.
- ✔ Listen carefully to others.
- ✔ Use mathematical language in your writing in other subjects.
- ✔ Pay attention to the headings in textbooks — they are signposts that help you.
- ✔ Think about videos and audiotapes as ways to communicate mathematical ideas.
- ✔ Be on the lookout for mathematics when you read, watch television, or see a movie.
- ✔ Communicate with others by using bulletin boards and chat rooms found on the Internet.

What you do now...	What you do in this course...	What you do in the work place...
Teach a young relative a sport...	THINK AND DISCUSS	*Written and verbal communication at work.*

Build communication skills with these hints on how students can communicate more effectively with their teachers and their classmates.

Solving PROBLEMS — *a* SKILL You USE Every DAY

Problem solving is a skill — a skill that you probably use without even knowing it. When you think critically in social studies to draw conclusions about pollution and its stress on the environment, or when a mechanic listens to symptoms of trouble and logically determines the cause, you are both using a mathematical problem-solving skill. Problem solving also involves logical reasoning, wise decision making, and reflecting on our solutions.

Tips for Problem Solving

Recognize that there is more than one way to solve most problems.
When solving a word problem, read it, decide what to do, make a plan, look back at the problem, and revise your answer.
Experiment with various solution methods.
Understand that it is just as important to know how to solve a problem as it is to actually solve it.
Be aware of times you are using mathematics to solve problems that do not involve computation, such as when you reason to make a wise decision.

What you do now...	What you do in this course...	What you do in the work place...
Make decisions based on changing conditions, such as weather...	***PROBLEM SOLVING***	***Synchronize the timing of traffic lights to enhance traffic flow.***

Establish problem solving success by helping students understand that there isn't always one answer to or one way of solving a problem.

Studying for the TEST Whatever It May Be

SATs, ACTs, chapter tests, and weekly quizzes — they all test what you know and are able to do. Have you ever thought about **how** you can take these tests to your advantage? You are evaluated now in your classes and you will be evaluated when you hold a job.

Pointers for Gaining Points

- Study as you progress through a chapter, instead of cramming for a test.
- Recognize when you are lost and seek help before a test.
- Review important graphs and other visuals when studying for a test, then picture them in your mind.
- Study for a test with a friend or study group.
- Take a practice test.
- Think of mnemonic devices to help you, such as Please Excuse My Dear Aunt Sally, which is one way to remember order of operations (parentheses, exponents, multiply, divide, add, subtract).
- Reread test questions before answering them.
- Check to see if your answer is reasonable.
- Think positively and visualize yourself doing well on the test.
- Relax during the test… there is nothing there that you have not seen before.

What you do now...	What you do in this course...	What you do in the work place...
Study notes in preparation for tests and quizzes...	SELF ASSESSMENT How am I doing? Exercises ON YOUR OWN Exercises CHECKPOINT	*Prepare for and participate in a job interview.*

These test-taking strategies highlight techniques that students can use to prepare for various assessments leading to success.

CHAPTER 1 Tools of Algebra

To accommodate flexible scheduling, some lessons are divided into parts. Assignment Options are given in the Lesson Planning Options for each lesson.

PACING OPTIONS

This chart suggests pacing only for the core lessons and their parts, and it is provided merely as a possible guide. It will help you determine how much time you have in your schedule to cover other features, such as the Chapter Project, Math Toolboxes, Wrap Up, and Assessment.

	1 Class Period	1 Class Period	1 Class Period	1 Class Period	1 Class Period	1 Class Period	1 Class Period	1 Class Period	1 Class Period
Traditional (40–45 min class periods)	1-1 Part 1, 1-1 Part 2	1-2	1-3 Part 1	1-3 Part 2	1-4 Part 1, 1-4 Part 2	1-5 Part 1, 1-5 Part 2	1-6 Part 1, 1-6 Part 2	1-7 Part 1, 1-7 Part 2	1-8 Part 1, 1-8 Part 2
Two Year Algebra (40–45 min class periods)	1-1 Part 1	1-1 Part 2	1-2	1-3 Part 1	1-3 Part 1	1-3 Part 2	1-3 Part 2	1-4 Part 1	1-4 Part 2
Block Scheduling (90 min class periods)	1-1 Part 1, 1-1 Part 2, 1-2	1-3 Part 1, 1-3 Part 2	1-4 Part 1, 1-4 Part 2	1-5 Part 1, 1-5 Part 2	1-6 Part 1, 1-6 Part 2	1-7 Part 1, 1-7 Part 2	1-8, 1-9		

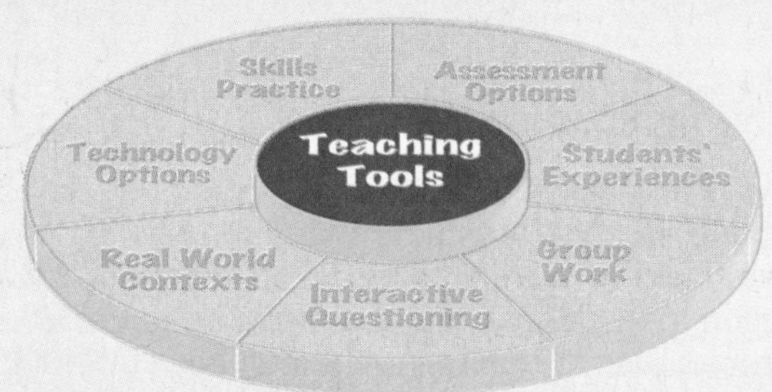

What Students Will Learn and Why

In this chapter, students will build on their knowledge of interpreting data by learning to analyze real-world data to find the mean, median, and mode.

Discussing the Chapter/Building on Experience

The concept map below relates chapter topics to real-world applications. You and your class may wish to add to the map or develop maps of your own. The center oval describes the topic of the chapter. The next level displays topics within the lessons. The outer ovals reflect applications of the content. As you and your class build a concept map, invite students to discuss applications with which they are familiar.

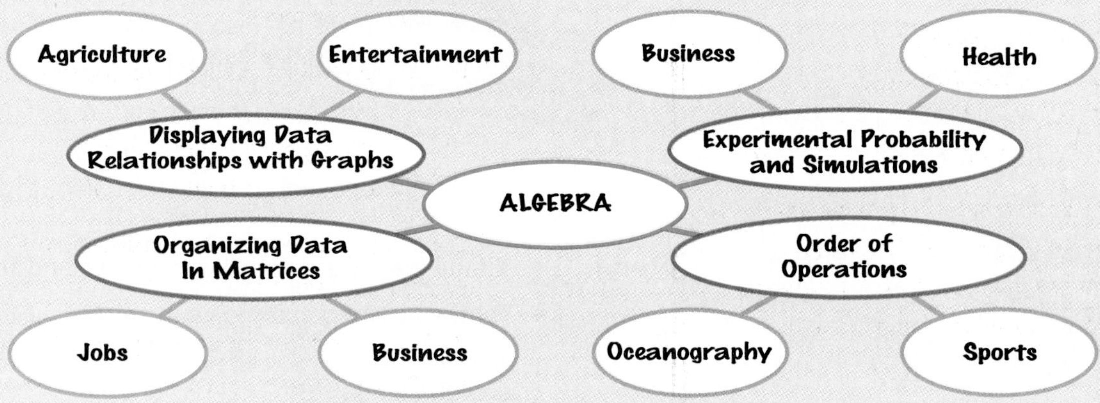

1 Class Period	1 Class Period	1 Class Period	1 Class Period	1 Class Period	1 Class Period	1 Class Period	1 Class Period	1 Class Period	1 Class Period	1 Class Period
1-9										
1-4 2	1-5 1	1-5 2	1-6 1	1-6 2	1-7 1	1-7 2	1-8 1	1-8 2	1-9	1-9

Interactive Questioning Tips

A question is interactive when there is "give and take" between the questioner (teacher or student) and the respondent. Key questions should be planned and asked to provide structure and direction to the lesson. For example, on page 36, Example 1 uses probabilities to determine the number of working light bulbs in stock. After checking a randomly selected group of bulbs, students are asked to apply the information and predict how many of the bulbs in stock are good. It should be noted that the key questions should be used as a platform from which spontaneous questions can be asked based on student responses.

Skills Practice

Every lesson provides skill practice with Exercises On Your Own and Exercises Mixed Review. The Student Edition includes Checkpoints (pp. 29 and 43) and Preparing for Standardized Tests (p. 55). In the Teacher's Edition, the Lesson Planning Options section for each lesson lists Prerequisite Skills students should know for that lesson. At the back of the Student Edition is the Skills Handbook—mini-lessons on math the students may need to review. The Chapter Support File for Chapter 1 in the Teaching Resources box includes two Practice worksheets per lesson, a worksheet for two Checkpoints, and worksheets for Cumulative Review and Standardized Test Preparation.

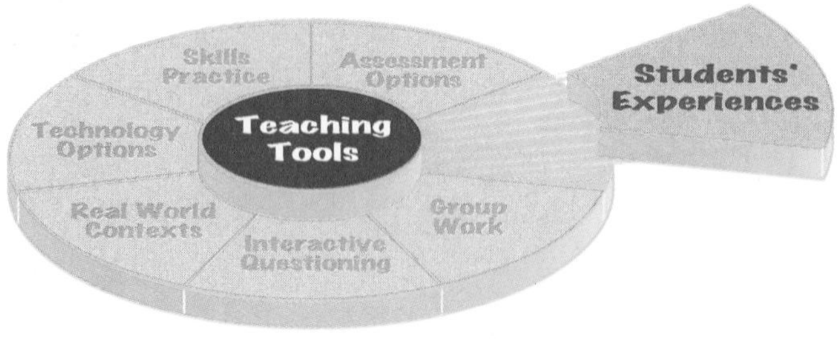

Diverse Learning and Teaching Styles

In your Teacher's Edition, you will find suggestions such as:

- **Visual learning** using tiles to understand zero (p. 19)
- **Tactile learning** making matrix puzzles with clues (p. 41)
- **Auditory learning** reading aloud the rules of multiplication and division (p. 26)
- **Kinesthetic learning** demonstrating sequential actions (p. 16)

Alternative Activity for Lesson 1-1

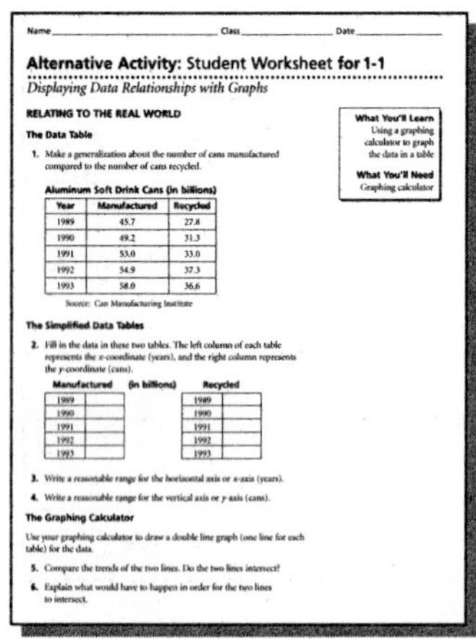

Name________ Class________ Date________

Alternative Activity: Student Worksheet for 1-1

Displaying Data Relationships with Graphs

RELATING TO THE REAL WORLD

What You'll Learn Using a graphing calculator to graph the data in a table

What You'll Need Graphing calculator

The Data Table

1. Make a generalization about the number of cans manufactured compared to the number of cans recycled.

Aluminum Soft Drink Cans (in billions)

Year	Manufactured	Recycled
1989	45.7	27.8
1990	49.2	31.3
1991	53.0	33.0
1992	54.9	37.3
1993	58.0	36.6

Source: Can Manufacturing Institute

The Simplified Data Tables

2. Fill in the data in these two tables. The left column of each table represents the x-coordinate (years), and the right column represents the y-coordinate (cans).

Manufactured (in billions)

1989	
1990	
1991	
1992	
1993	

Recycled

1989	
1990	
1991	
1992	
1993	

3. Write a reasonable range for the horizontal axis or x-axis (years).
4. Write a reasonable range for the vertical axis or y-axis (cans).

The Graphing Calculator

Use your graphing calculator to draw a double line graph (one line for each table) for the data.

5. Compare the trends of the two lines. Do the two lines intersect?
6. Explain what would have to happen in order for the two lines to intersect.

for use with Example 3, addresses visual and tactile learning by using a graphing calculator to make a double-line graph.

Alternative Activity for Lesson 1-4

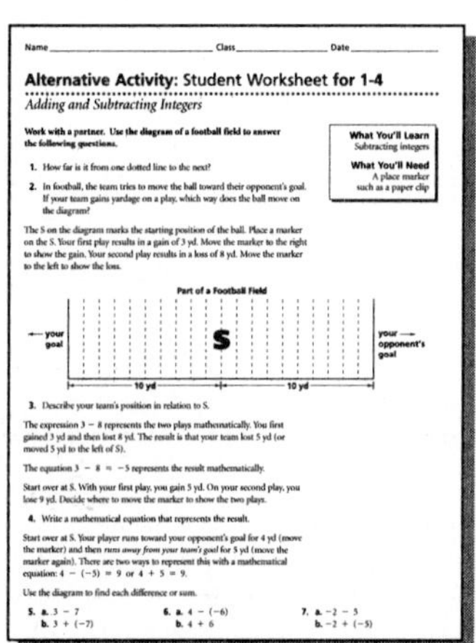

Name________ Class________ Date________

Alternative Activity: Student Worksheet for 1-4

Adding and Subtracting Integers

Work with a partner. Use the diagram of a football field to answer the following questions.

What You'll Learn Subtracting integers

What You'll Need A place marker such as a paper clip

1. How far is it from one dotted line to the next?
2. In football, the team tries to move the ball toward their opponent's goal. If your team gains yardage on a play, which way does the ball move on the diagram?

The S on the diagram marks the starting position of the ball. Place a marker on the S. Your first play results in a gain of 3 yd. Move the marker to the right to show the gain. Your second play results in a loss of 8 yd. Move the marker to the left to show the loss.

3. Describe your team's position in relation to S.

The expression $3 - 8$ represents the two plays mathematically. You first gained 3 yd and then lost 8 yd. The result is that your team lost 5 yd (or moved 5 yd to the left of S).

The equation $3 - 8 = -5$ represents the result mathematically.

Start over at S. With your first play, you gain 5 yd. On your second play, you lose 9 yd. Decide where to move the marker to show the two plays.

4. Write a mathematical equation that represents the result.

Start over at S. Your player *runs* toward your opponent's goal for 4 yd (move the marker) and then *runs away from your team's goal* for 5 yd (move the marker again). There are two ways to represent this with a mathematical equation: $4 - (-5) = 9$ or $4 + 5 = 9$.

Use the diagram to find each difference or sum.

5. a. $3 - 7$ b. $3 + (-7)$
6. a. $4 - (-6)$ b. $4 + 6$
7. a. $-2 - 5$ b. $-2 + (-5)$

for use with Example 2, addresses tactile learning by moving place markers on a number line to help students discover the Subtraction Rule.

Alternative Activity for Lesson 1-7

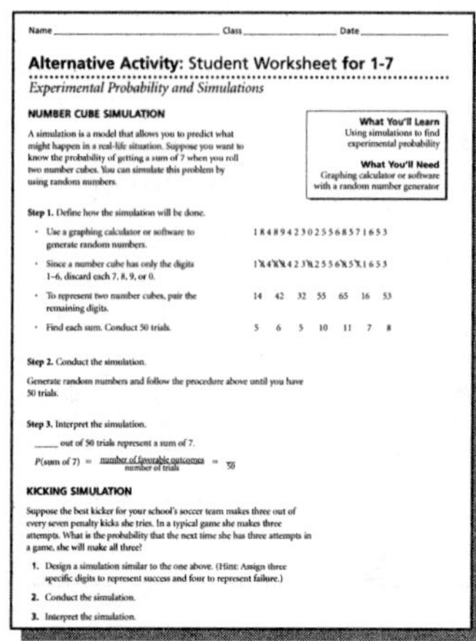

Name________ Class________ Date________

Alternative Activity: Student Worksheet for 1-7

Experimental Probability and Simulations

NUMBER CUBE SIMULATION

A simulation is a model that allows you to predict what might happen in a real-life situation. Suppose you want to know the probability of getting a sum of 7 when you roll two number cubes. You can simulate this problem by using random numbers.

What You'll Learn Using simulations to find experimental probability

What You'll Need Graphing calculator or software with a random number generator

Step 1. Define how the simulation will be done.

- Use a graphing calculator or software to generate random numbers. 1 8 4 8 9 4 2 3 0 2 5 5 6 8 5 7 1 6 5 3
- Since a number cube has only the digits 1–6, discard each 7, 8, 9, or 0.
- To represent two number cubes, pair the remaining digits. 14 42 32 55 65 16 53
- Find each sum. Conduct 50 trials. 5 6 5 10 11 7 8

Step 2. Conduct the simulation.

Generate random numbers and follow the procedure above until you have 50 trials.

Step 3. Interpret the simulation.

____ out of 50 trials represent a sum of 7.

$P(\text{sum of } 7) = \frac{\text{number of favorable outcomes}}{\text{number of trials}} = \frac{}{50}$

KICKING SIMULATION

Suppose the best kicker for your school's soccer team makes three out of every seven penalty kicks she tries. In a typical game she makes three attempts. What is the probability that the next time she has three attempts in a game, she will make all three?

1. Design a simulation similar to the one above. (Hint: Assign three specific digits to represent success and four to represent failure.)
2. Conduct the simulation.
3. Interpret the simulation.

for use with the section Conducting a Simulation (including Example 2), addresses tactile learning by having students calculate probabilities using random numbers.

Skills Practice
Assessment Options
Technology Options
Teaching Tools
Students' Experiences
Real World Contexts
Interactive Questioning
Group Work

Cooperative Learning Tips

When used effectively, cooperative learning can help students develop interpersonal skills, learn to perform specific roles in a group, and learn to carry out specific responsibilities. The components of Chapter 1 provide a range of cooperative learning opportunities.

- In the Student Edition, the ***Work Together*** parts of lessons are specifically designed for cooperative learning activities.
- In the Teacher's Edition, you will find helpful hints for addressing diverse learning styles (see page C for Chapter 1). For every lesson, you will find a ***Reteaching Activity***, which may involve cooperative learning.

Materials and Manipulatives

The authors expect all students to use scientific calculators. Calculator use is integrated throughout the course.

- **calculator (1-3, 1-6)**
- **two number cubes (1-7)**

Technology Options
Skills Practice
Assessment Options
Teaching Tools
Students' Experiences
Real World Contexts
Interactive Questioning
Group Work

TECHNOLOGY OPTIONS

Technology Tools		Chapter Project	1-1	1-2	1-3	1-4	1-5	1-6	1-7	1-8	1-9
Calculator		Assumed that students have scientific calculators at any time.									
Graphing Calculator	Handbook		✔			✔	✔		✔	✔	
	Student Edition				✔	✔		✔	✔	✔	
Software	Secondary Math Lab Toolkit		✔	✔	✔	✔	✔	✔	✔	✔	✔
	Integrated Math Lab					✔					✔
	Computer Item Generator		✔	✔	✔	✔	✔	✔	✔	✔	✔
	Student Edition								✔		✔
Video	Video Field Trip	✔									
CD-ROM	Multimedia Algebra Lab		✔	✔					✔	✔	
Internet		See the Prentice Hall site. (http://www.phschool.com)									

The Prentice Hall Algebra program offers you a rich variety of technology options. Be assured that all these options are provided as a means of enriching the program and are not essential for the successful completion of the course.

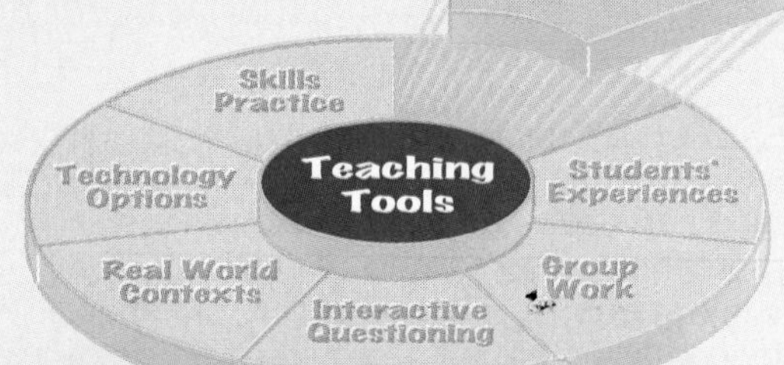

Assessment Options

The Prentice Hall Algebra Program provides you with many options. From these options, you may choose instructional materials and techniques appropriate for your students, or those necessary to meet your district's curriculum requirements. As the chart indicates, the program also supports your teaching efforts by offering you many choices for assessment.

ASSESSMENT OPTIONS

Assessment Support Materials	Chapter Opener	1-1	1-2	1-3	1-4	1-5	1-6	1-7	1-8	1-9	Chapter End
Chapter Project	● ▲ ■	▲ ■				▲ ■		▲ ■		▲ ■	▲ ■
Checkpoints						▲ ■			▲ ■		
Self-Assessment			▲ ■	▲ ■				▲ ■		▲ ■	▲ ■
Writing Assignment		▲	▲	▲	▲	▲	▲	▲	▲	▲	▲ ●
Chapter Assessment											▲ ●
Alternative Assessment		■	■	■	■	■	■	■	■	■	● ■
Cumulative Review											●
Standardized Test Prep		▲ ■	▲ ■		▲ ■						▲ ●
Computer Item Generator	Can be used to create custom-made practice or assessment at any time.										

▲ = Student Edition ■ = Teacher's Edition ● = Teaching Resources

Checkpoints

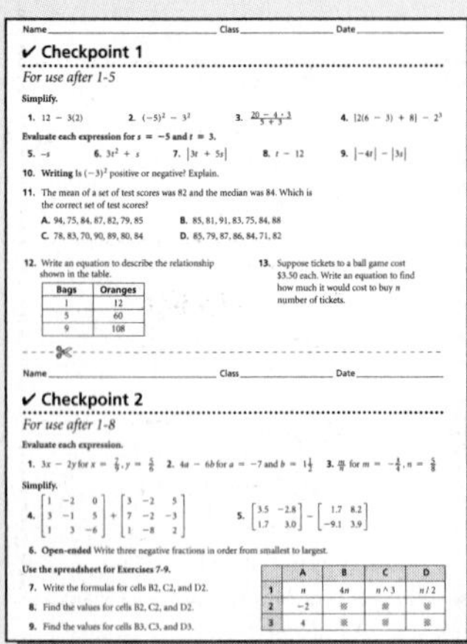
✔ Checkpoint 1

For use after 1-5

✔ Checkpoint 2

For use after 1-8

Alternative Assessment

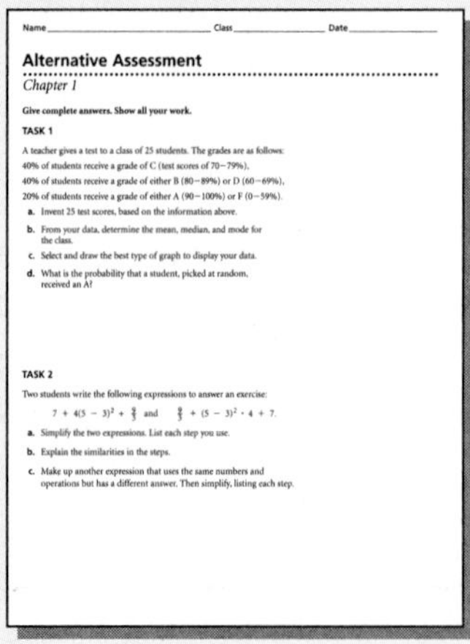
Alternative Assessment

Chapter 1

Give complete answers. Show all your work.

TASK 1

TASK 2

Chapter Assessment

Chapter Assessment Form A

Chapter 1

Available in both Form A and Form B

Making the Right Connections

Mathematics is imbedded in nearly every walk of life. The National Council of Teachers of Mathematics (NCTM) encourages educators to recognize these connections and to emphasize them for the purpose of better educating students for success in life and in a global economy. The ***Connections*** chart below highlights these connections for Chapter 1.

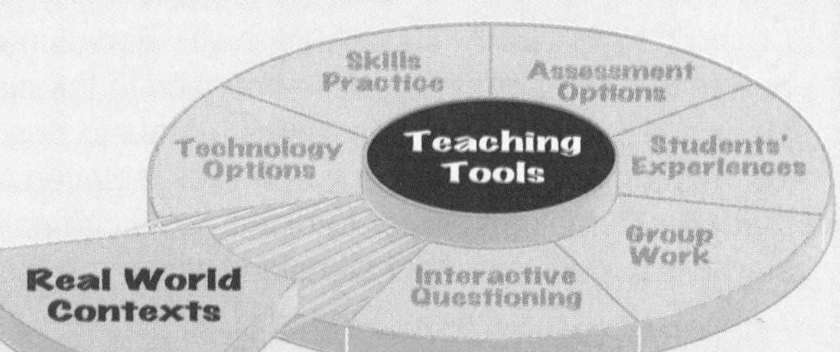

CONNECTIONS

Lesson	Interdiciplinary Connections	Career Prep	Other Real World Connections	Math Integration	NCTM Standards
Chapter Project			Archaeology		Problem Solving
1-1	Languages	Jobs Agriculture	Entertainment Record Industry Sports		Algebra Communication Problem Solving
1-2		Sales	Music	Geometry	Algebra Communication Problem Solving Geometry
1-3		Sales	Community Entertainment	Geometry	Algebra Communication Problem Solving Geometry
1-4	Physics Geography Chemistry		Sports	Statistics	Algebra Communication Problem Solving
1-5	Oceanography		Space Flight Entertainment	Statistics	Algebra Communication Problem Solving
1-6	Science	Real Estate	Weather		Algebra Communication Problem Solving
1-7	Health		Business		Algebra Communication Problem Solving Statistics
1-8			Environment Sports Jobs		Algebra Communication Problem Solving
1-9			Package Delivery Business Computers	Geometry Statistics	Algebra Communication Problem Solving Geometry

CHAPTER 1 Tools of Algebra

CONNECTING TO PRIOR LEARNING Have volunteers bring newspapers, magazines, and almanacs to the classroom. Ask students to choose a graph. Ask students to describe how information and data in the graph are displayed. Elicit the fact that graphs make it easier to display and compare data. Ask students to describe other situations where graphs are valuable.

CULTURAL CONNECTIONS Archaeologists provide important information about the cultures of past civilizations. Begin a discussion with students about the significance of these findings. Discuss how archaeologists in the future might view present-day cultures based on their findings.

INTERDISCIPLINARY CONNECTIONS The Chapter Opener demonstrates the importance of algebra in other disciplines, such as forensic science and archaeology. Help students make the connection that vast quantities of important information would remain unknown without the ability to use algebraic equations.

ABOUT THE PROJECT The Chapter Project will give students an opportunity to explore the connection between math and the bones of the human body. The Find Out questions throughout the chapter will help students understand how to measure, display, and evaluate data.

Technology Options

Prentice Hall Technology

Video Video Field Trip 1, "The Memory of Bones," a look at how archaeologists go about their work.

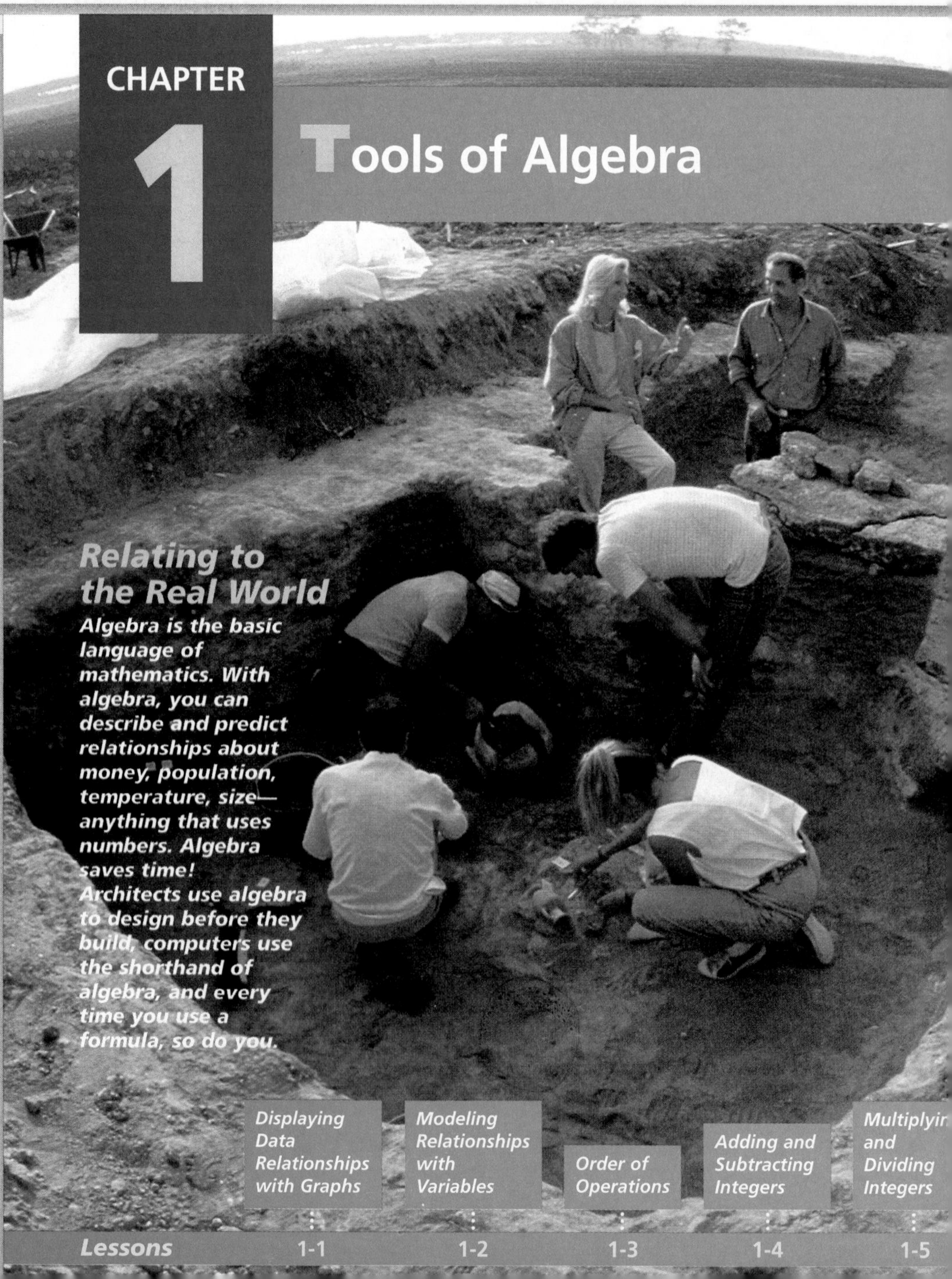

CHAPTER 1 Tools of Algebra

Relating to the Real World

Algebra is the basic language of mathematics. With algebra, you can describe and predict relationships about money, population, temperature, size—anything that uses numbers. Algebra saves time! Architects use algebra to design before they build, computers use the shorthand of algebra, and every time you use a formula, so do you.

Lessons	1-1	1-2	1-3	1-4	1-5
	Displaying Data Relationships with Graphs	*Modeling Relationships with Variables*	*Order of Operations*	*Adding and Subtracting Integers*	*Multiplyin[g] and Dividing Integers*

Launching the Project

PROJECT NOTEBOOK When discussing the project, encourage students to keep all project-related materials in a separate folder or notebook. See Chapter Project Manager and Scoring Rubric in the Chapter Support File.

- Assign students to work with partners or in small groups. Students need to use the chapter visual to locate the tibia, humerus, and radius bones. tibia—the inner and thicker of the two bones from the knee to the ankle; humerus—from the shoulder to the elbow; radius—from the wrist to the elbow
- Discuss what instruments are available for measuring. Determine which would be the best for this project and why.
- What do you think would be the best way to organize and display the data? Answers may vary.
- Suggest students create a spreadsheet for calculating and displaying their information.

TRACKING THE PROJECT Have students read Finishing the Chapter Project on page 50 for an overview of the project. Set benchmark deadlines for students so they can show you their work in progress.

CHAPTER PROJECT

The BIG DIG!

Your bones tell a lot about your body. Archaeologists and forensic scientists study bones to estimate a person's height, build, and age. These data are helpful in learning about ancient people and in solving crimes. The lengths of major bones such as the humerus, radius, or tibia can be substituted into formulas to find a person's height.

As you work through the chapter, you will collect data about bones from your classmates and from adults. You will use formulas to analyze the data and predict heights. Then you will decide how to organize and display your results in graphs and spreadsheets.

To help you complete the project:

- p. 9 *Find Out by Graphing*
- p. 29 *Find Out by Calculating*
- p. 39 *Find Out by Analyzing*
- p. 49 *Find Out by Creating*
- p. 50 *Finishing the Project*

Real Numbers and Rational Numbers 1-6

Experimental Probability and Simulations 1-7

Organizing Data in Matrices 1-8

Variables and Formulas in Spreadsheets 1-9

Project Resources

Teaching Resources

Chapter Support File, Ch. 1
- Chapter Project Manager and Scoring Rubric

Transparencies
31

Using the Rubric

Sharing the scoring rubric for the project with your students will alert them to your expectations before they begin work on the project.

As students complete each Find Out question in the chapter, you may wish to have them evaluate their own work or a partner's work, based on the scoring rubric. Students should have the opportunity to revise their work after it has been reviewed.

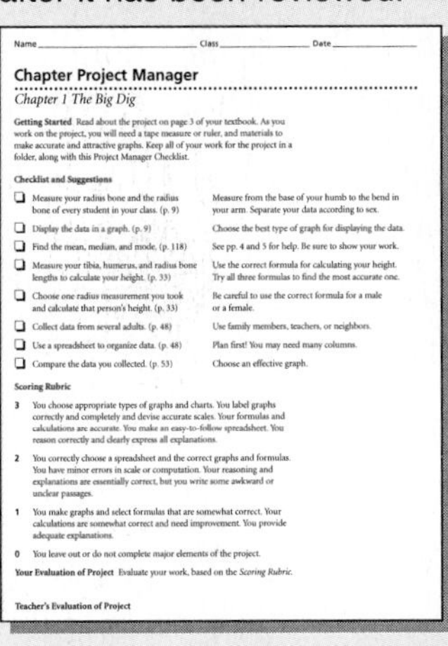

Name ______ Class ______ Date ______

Chapter Project Manager

Chapter 1 The Big Dig

Getting Started Read about the project on page 3 of your textbook. As you work on the project, you will need a tape measure or ruler, and materials to make accurate and attractive graphs. Keep all of your work for the project in a folder, along with this Project Manager Checklist.

Checklist and Suggestions

Checklist	Suggestions
❑ Measure your radius bone and the radius bone of every student in your class. (p. 9)	Measure from the base of your humb to the bend in your arm. Separate your data according to sex.
❑ Display the data in a graph. (p. 9)	Choose the best type of graph for displaying the data.
❑ Find the mean, median, and mode. (p. 118)	See pp. 4 and 5 for help. Be sure to show your work.
❑ Measure your tibia, humerus, and radius bone lengths to calculate your height. (p. 33)	Use the correct formula for calculating your height. Try all three formulas to find the most accurate one.
❑ Choose one radius measurement you took and calculate that person's height. (p. 33)	Be careful to use the correct formula for a male or a female.
❑ Collect data from several adults. (p. 48)	Use family members, teachers, or neighbors.
❑ Use a spreadsheet to organize data. (p. 48)	Plan first! You may need many columns.
❑ Compare the data you collected. (p. 53)	Choose an effective graph.

Scoring Rubric

3 You choose appropriate types of graphs and charts. You label graphs correctly and completely and devise accurate scales. Your formulas and calculations are accurate. You make an easy-to-follow spreadsheet. You reason correctly and clearly express all explanations.

2 You correctly choose a spreadsheet and the correct graphs and formulas. You have minor errors in scale or computation. Your reasoning and explanations are essentially correct, but you write some awkward or unclear passages.

1 You make graphs and select formulas that are somewhat correct. Your calculations are somewhat correct and need improvement. You provide adequate explanations.

0 You leave out or do not complete major elements of the project.

Your Evaluation of Project Evaluate your work, based on the *Scoring Rubric.*

Teacher's Evaluation of Project

1-1 Teaching Notes

PROBLEM OF THE DAY

Separate the digits 1 through 9 into three groups which have the same sum. 1, 5, 9; 2, 6, 7; 3, 4, 8

Problem of the Day is also available in Transparencies.

CONNECTING TO PRIOR KNOWLEDGE Have students write the name of a food anywhere they choose on the board. Ask if the data makes sense in this unorganized form. Have students mention ways they have seen data displayed. Brainstorm ways of organizing the data on the board in a meaningful way.

WORK TOGETHER

Before you organize students in groups of three to five, discuss with the class the sample line plot. Make sure students understand the labels and markings. After the groups collect data from their members, ask a student to draw a master line plot on the board. Ask each group in turn to add their data to the board. Then have each group complete their group's plot from this master.

TACTILE LEARNING Drawing line plots will appeal to students with a tactile learning style. Hands-on activities and activities with manipulatives are particularly effective for these students.

Lesson Planning Options

Prerequisite Skills

- Interpreting data
- Performing operations with integers

Assignment Options

To provide flexible scheduling, this lesson can be subdivided into parts.

 **Core** 4–9
Extension 14–17

 **Core** 1–3, 10, 11
Extension 12, 13

Use Mixed Review to maintain skills.

Resources

 **Student Edition**

Skills Handbook, pp. 567, 579
Extra Practice, p. 556
Glossary/Study Guide

 **Teaching Resources**

Chapter Support File, Ch. 1
- Practice 1-1 (two worksheets)
- Reteaching 1-1
- Alternative Activity 1-1

Classroom Manager 1-1
Glossary, Spanish Resources
Two-Year Algebra Handbook 1-1

 **Transparencies**
2, 32, 37

1-1 Displaying Data Relationships with Graphs

What You'll Learn
- Finding the mean, median, and mode of sets of data
- Drawing and interpreting graphs

...And Why
To analyze real-world data, such as age and employment statistics

WORK TOGETHER

Work in groups.

Line plots are simple graphs that help you see the relationship between data items. At the right is the start of a line plot for data collected on birth months of a group of students. Each × represents one student.

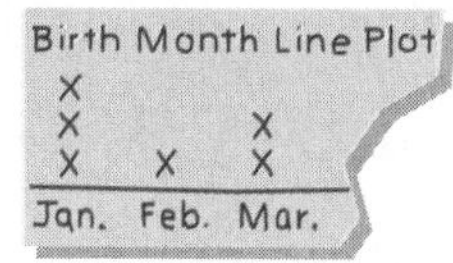

1. What do you think a line plot for the birth months of the students in your class would look like?
Check students' work.
2. a. Data Collection Find the birth month of each student in your group. Share this information with other groups.
b. Draw a line plot of your class data.
c. Does the line plot of your class data look similar to what you expected? Explain.
2a–c. Check students' work.

THINK AND DISCUSS

Part 1 **Finding Mean, Median, and Mode**

You can use a line plot or *histogram* to show the frequency, or number of times, a data item occurs. The data item with the greatest frequency is the **mode.**

Birth Months of Presidents of the United States

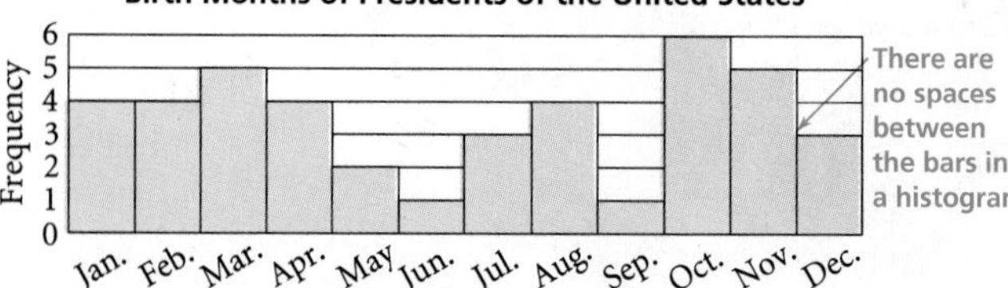

Mount Rushmore in South Dakota has carvings of four Presidents: Washington, Jefferson, T. Roosevelt, and Lincoln. Washington's head is as tall as a five-story building.

3. What is the mode of the data in the histogram above? October
4. In which months were the fewest Presidents born? June and September
5. a. Draw a histogram of the data you collected in the Work Together. Check students' work.
b. Can you predict birth-month patterns for all the people in the United States based on your class data? Why or why not? See below.
6. What is the mode (or modes) of your class data? Check students' work.

5b. Answers may vary. Sample: No; there are too few people in class to make a prediction.

ERROR ALERT! Students may omit data from the chart or repeat the same data. **Remediation:** Encourage students to double check their input to be certain they haven't repeated or omitted data.

THINK AND DISCUSS

Example 1 **Relating to the Real World** 

Question 9 Have students guess how much the median and mean will change before they calculate. Then compare estimates. This is a good time to discuss the importance of estimation skills. As students calculate the mean, median, and mode after the change, ask these questions:

- *Will the mode be affected by the change?* No, the amount that occurs most often has not changed.
- *Why do you think the mean is more affected than the median by the change?* The change in salary does not affect the middle two numbers used to find the mean.

CONNECTING TO THE STUDENTS' WORLD Ask students what similar data they could collect in their neighborhood. Design a plan together. Have them collect and share their data with the class. Have students calculate the mean, median, and mode of the data. This activity may be useful if you have an extended class period or block scheduling.

CRITICAL THINKING **Question 10** As you discuss possible answers with the class, encourage higher-order thinking by not focusing on a *correct* answer. Focus on the quality of students' logic and justifications.

Age When Elected

First Presidents	Recent Presidents
Washington—57	Ford—61
J. Adams—61	Carter—52
Jefferson—57	Reagan—69
Madison—57	Bush—64
Monroe—58	Clinton—46

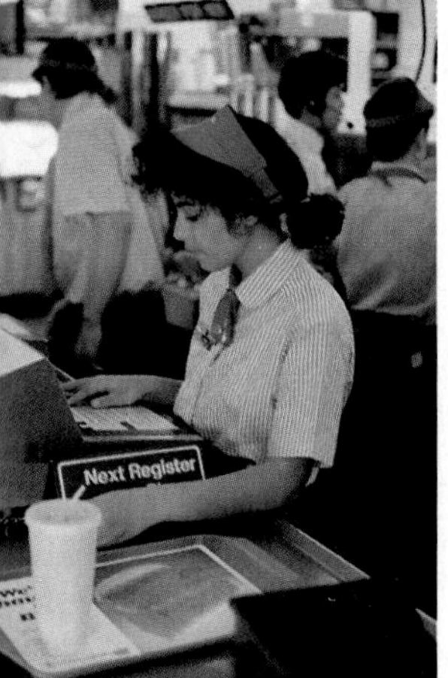

QUICK REVIEW

You can indicate multiplication of two numbers in any of these ways:

6(5.25) (6)(5.25)
6 · 5.25 6 × 5.25

Do you think that the ages of the Presidents have changed a lot since the beginning of the United States in 1776? One way to find out is to compare the mean ages of the first Presidents with those of recent Presidents. Use this ratio to find the mean.

$$\textbf{Mean} = \frac{\text{sum of the data items}}{\text{total number of data items}}$$

7. **a.** Find the mean of each group of ages. 58, 58.4
   b. Has there been much change in the mean age of the Presidents? Explain. Answers may vary. Sample: No; the change in the mean age is less than half a year.
8. Would the mode be as useful as the mean in comparing the ages of the first Presidents to recent Presidents? Explain. No; "First Presidents" has one mode and "Recent Presidents" has no mode. In any case, mode is not helpful in comparing ages.

A third type of measure is the median. The **median** is the middle value in an ordered set of numbers. The mean, median, and mode may give a different picture of data—sometimes slightly different, sometimes very different.

Example 1 **Relating to the Real World** 

Jobs Find the mean, median, and mode of the data below.

What Employees Earn at a Local Fast Food Restaurant

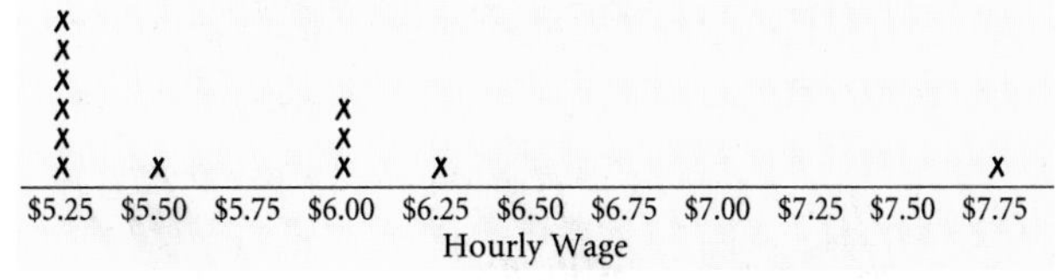

You can multiply 6 times 5.25 as a shortcut for adding 5.25 + 5.25 + 5.25 + 5.25 + 5.25 + 5.25.

$$\text{Mean: } \frac{6(5.25) + 5.50 + 3(6.00) + 6.25 + 7.75}{12} = 5.75$$

12 ← total number of employees

Median: 5.25 5.25 5.25 5.25 5.25 5.25 5.50 6.00 6.00 6.00 6.25 7.75

For an even number of data items, find the mean of the middle terms. → $\frac{5.25 + 5.50}{2} \approx 5.38$

Mode: 5.25 ← the data item that occurs most often

The mean is $5.75/h. The median is about $5.38/h.

The mode is $5.25/h.

Additional Examples

FOR EXAMPLE 1

Find the mean of these odd-job payments: mowing $15, baby-sitting $10, delivering circulars $12.
About $12.33

Discussion: *Is the calculated amount realistic?*

FOR EXAMPLE 3

Rafaela owns two shoe stores. The profits from Rafaela's I during the years 1991–1995 were $12,000; $16,000; $16,500; $18,000; and $20,500. The profits from Rafaela's II over the same period were $16,000; $17,000; $16,000; $14,500; and $14,000. Graph the data on a double line graph.

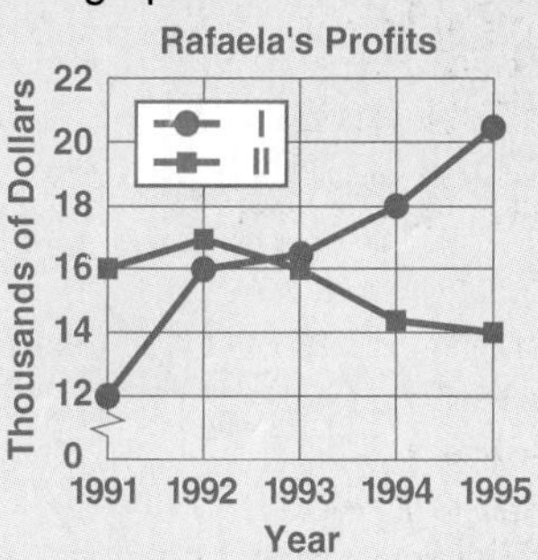

Discussion: *Which store is more profitable and how is this revealed by the graph?*

Example 2 **Relating to the Real World** 

CONNECTING TO THE STUDENTS' WORLD Help students find, in local newspapers and almanacs, similar social studies data about their community and state.

ERROR ALERT! Students may overlook the labels on the graph and try to answer questions before they understand the data. **Remediation:** Help students relate the numbers in the boxes to the graph by asking interactive questions like these.

- *Use the table. What is the median household income for California in 1987?* $37,231
- *Which set of columns in the graph refers to California?* the columns on the far left

ESL **CRITICAL THINKING Question 14** Discuss with the class the meaning of *economic recession* and ask students for examples of possible effects of a recession. Have students add this term to their vocabulary list.

Example 3 **Relating to the Real World** 

CONNECTING TO THE STUDENTS' WORLD Ask students what recycling efforts are currently practiced in their community. Discuss ways to gather and evaluate data about recycling.

CRITICAL THINKING Question 17

- *How does the amount of waste relate to the lines on the graph?* Closer lines represent less waste.

Technology Options

For Exercises 1, 2, and 10, students may use spreadsheet software to analyze and graph data.

Prentice Hall Technology 

Calculator • Graphing Calculator Handbook: Procedure 21, 22

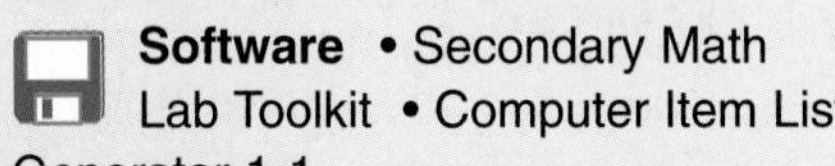
Software • Secondary Math Lab Toolkit • Computer Item List Generator 1-1

CD-ROM • Multimedia Algebra Lab 1

PRENTICE HALL
MULTIMEDIA
ALGEBRA

Internet • See the Prentice Hall site. (http://www.phschool.com)

6

9. **Try This** Suppose the worker earning $7.75/h resigns. Her replacement earns $5.75/h. How does this change affect the mode? the median? the mean? Mode and median do not change; mean decreases to $5.58.

10. *Critical Thinking* Does the mean, median, or mode best describe the set of wages in the line plot? **Justify** your answer. Median; median shows the middle wage. It is not affected by a wage that is much higher than the other wages.

Part 2

Drawing and Interpreting Graphs

Bar graphs are useful when you wish to compare amounts. Whenever you make a graph to display data, you must choose an appropriate scale so that the graph is neither too big for your paper nor too small to read.

Example 2 **Relating to the Real World** 

Social Studies Draw a multiple bar graph for the median income data.

Median Household Income

	1987	1989	1991	1993
Calif.	$37,231	$37,348	$34,677	$33,083
Conn.	$40,586	$47,884	$43,423	$38,369
Ind.	$27,812	$29,302	$27,904	$28,618
Tex.	$30,531	$29,289	$28,568	$27,892
Utah	$32,764	$34,755	$28,859	$34,746

Source: *U. S. Bureau of the Census*

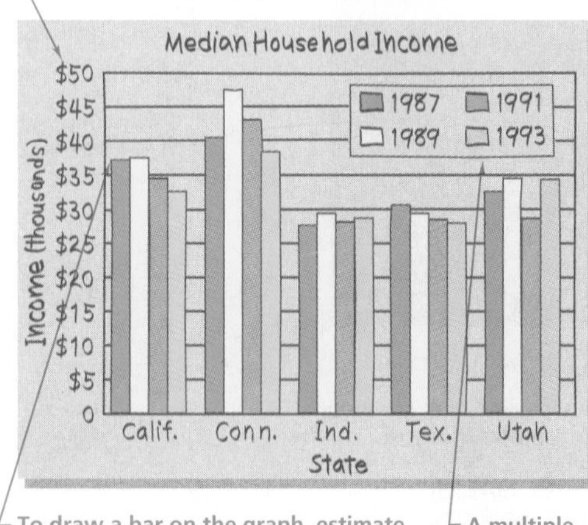
The highest projected income is $47,884. So a reasonable range for the vertical scale is from 0 to $50,000 with every $5000 labeled on the axis.
Median Household Income
1987
1991
1989
1993
Income (thousands)
$50
$45
$40
$35
$30
$25
$20
$15
$10
$5
0
Calif.
Conn.
Ind.
Tex.
Utah
State
To draw a bar on the graph, estimate its placement based on the vertical scale. The value $37,231 is a little less than $\frac{1}{2}$ the distance from $35,000 to $40,000.
A multiple bar graph must include a key.

11. **Try This** Suppose you were drawing a graph of the data. How would you estimate where to put the top of the bar showing the median income for Texas in 1991? The value $28,568 is about $\frac{2}{3}$ of the distance from $25,000 to $30,000.

12. Why is the key necessary for the graph above? because each state has four bars for the four years in the table

13. a. Which states decreased in median income for each of the last three years shown on the graph? Calif., Conn., Tex.
 b. Did you use the table or the graph to answer part (a)? Why? See left.

13b. Answers may vary. Sample: The graph is better because you can see the decrease without comparing the numbers.

- *For which year are the lines closest together?* 1992 Have students calculate the amount of waste for each of the five years in order to confirm this answer.

To help students develop team skills needed in the workplace, have them collaborate on a definition for "amount of waste" and possible ways to calculate this.

Exercises ON YOUR OWN

Exercise 1 Students may confuse a response with how many times that response occurred. Have students say to themselves "two children, one child, three children, no children" and so on as they enter the data onto their line plots.

CONNECTING TO THE STUDENTS' WORLD Exercise 2 Have students individually record how many hours they spend watching television in one week. At the end of the week, compare results. Have students find the mean, median, and mode of their own data. Ask students why they think the times varied. Answers may vary. Sample: Some students have very little time for watching television because they are in extracurricular activities or perhaps they work.

Exercise 5 Stress the importance of including the zero when calculating mean, median, and mode. Have students carry out a similar exercise, once with the zero and once without. Compare the results.

EXTENSION Exercise 7

- *What can you say about the mean, median, and mode when all the numbers in the data set are the same?* The mean, median, and mode will always be identical.

---

14. Answers may vary. Sample: Indiana seemed to be the least affected because the median income changed very little. Also, Utah seems to have recovered in 1993.

14. *Critical Thinking* The United States had an economic recession in the late 1980s and early 1990s. Which states seemed to be least affected by the recession? Explain. See left.

Line graphs allow you to see how a set of data changes over time. You can use line graphs to look for trends in data.

Example 3 Relating to the Real World 

Aluminum Soft Drink Cans (in billions)		
Year	**Manufactured**	**Recycled**
1989	45.7	27.8
1990	49.2	31.3
1991	53.0	33.0
1992	54.9	37.3
1993	58.0	36.6

Source: *Can Manufacturing Institute*

Recycling Make a double line graph of the table at the left.

The least and greatest numbers in the chart are 27.8 and 58.0. A reasonable vertical scale is from 20 to 60 with every 10 units labeled on the axis.

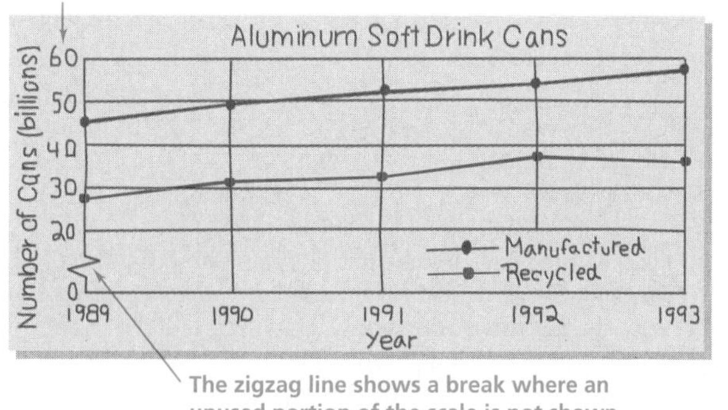

The zigzag line shows a break where an unused portion of the scale is not shown.

15a. The number of manufactured cans increased steadily.
15b. The number of recycled cans increased and then decreased from 1992 to 1993.
16a. The graph would take more space vertically, but it would not affect the data lines.
16b. Answers may vary. Sample: No; both graphs would represent the same data, but the graph without a break would have a larger amount of unused space.

15. a. Describe the trend for the number of cans manufactured.
b. Describe the trend for the number of cans recycled. See left.

16. a. *Critical Thinking* Suppose you draw the graph without a break in the vertical scale. How would this affect the way the graph looks?
b. Would drawing the graph without a break in the scale improve the graph? Why or why not? See left.

17. *Critical Thinking* From 1989–1993, did the amount of waste from soft drink cans increase, decrease, or stay the same? Explain.

17. Answers may vary. Sample: The numbers of wasted cans in 1989–1993 were 17.9, 17.9, 20.0, 17.6, and 21.4 billion. Overall, there was an increase in waste.

Exercises ON YOUR OWN

1. An interviewer asked 25 people, "How many children live in your home?" Here are their responses.
2, 1, 3, 0, 4, 2, 2, 1, 0, 3, 2, 6, 4,
3, 3, 2, 1, 1, 3, 0, 0, 2, 1, 1, 2 See margin.
a. Draw a line plot for the data. Include a title for the line plot.
b. Find the mean, median, and mode of the data. 1.96; 2; 2

page 7–9 On Your Own

1a.

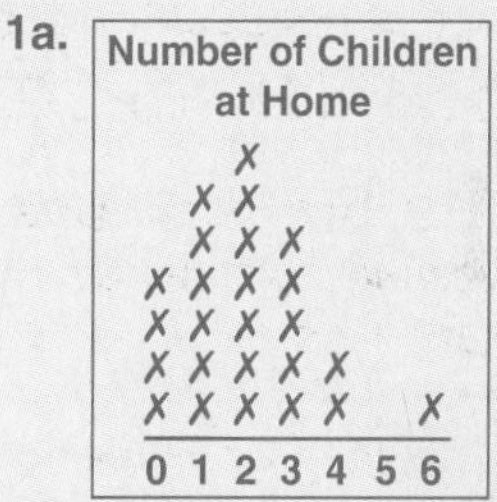

2a.

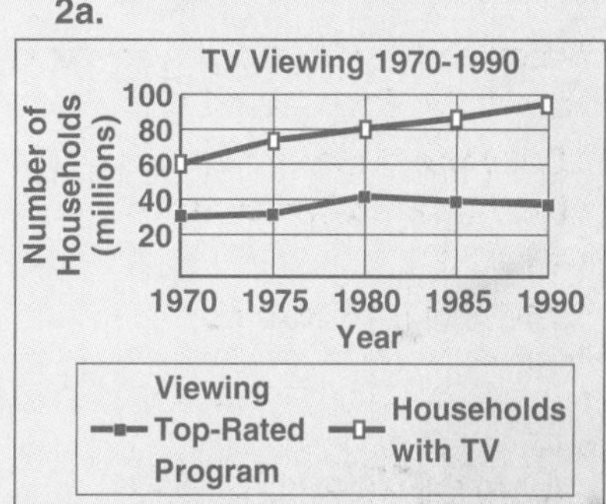

10a. Answers may vary. Samples:

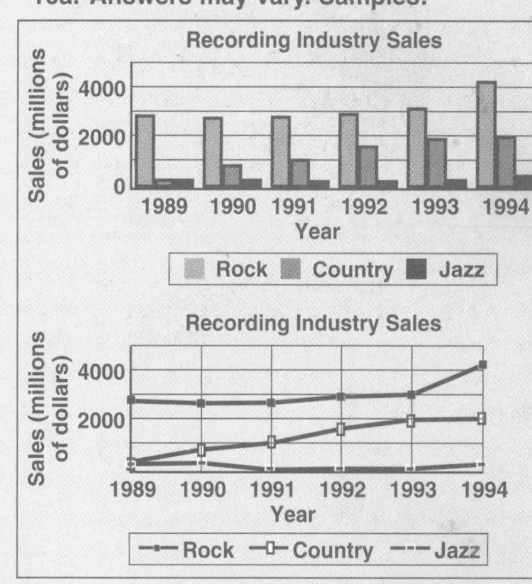

- *Write a set of numbers having the same mean and median but a different mode.* Answers may vary. Sample: 3, 3, 3, 7, 9, 12, 12

STANDARDIZED TEST TIP **Exercise 3** Encourage students to use mental math to estimate the correct answer. They need to validate their choice by solving the problem.

DIVERSITY Exercise 13 Ask teachers and students who are fluent in languages other than Spanish and English to translate the passage into another language. Have students compare the word lengths to those of Spanish and English and share the results with the class.

WRITING Exercise 13c Have students decide whether the excerpt is a good example of typical language usage. Ask them to justify their answer.

ALTERNATIVE ASSESSMENT **Exercises 14–17** Have students work in pairs. Each student makes up a set of data. Partners exchange the data and find the mean, median, and mode.

Chapter Project FIND OUT BY GRAPHING Students are asked to measure the radius bone of each student in the class, graph the data, and calculate the mean, median, and mode of the data. This information is essential to their work on the Chapter Project introduced in the Chapter Opener. Have students add this task to work they may have already completed for the project. You may wish to check students' progress on the project by having each group member write a sentence or two describing his or her duties within the group.

10b. Answers may vary depending on the type of graph chosen. Samples: The bar graph compares yearly amounts directly. The line graph shows the changes over time more clearly.

c. Answers may vary. Sample: Country; sales have increased over 4 times, while rock sales have not quite doubled and jazz sales have decreased.

11b. The 10th, 11th, and 12th girls are 5 ft 7 in. So there are 9 girls who are less then 5 ft 7 in. tall.

12a. The number of farms has decreased.

b. The average size of farms has increased

c. Answers may vary. Sample: Yes; as the number of farms decreased, the size of farms increased.

13a. See back of book.

2. a. *Entertainment* Use the data about television viewing to draw a double line graph. See margin.

b. What trend do you see in the total number of households viewing the top-rated program? increased from 1970 to 1980, then decreased slightly

c. What trend do you see in the number of households with TV?

d. *Critical Thinking* What could explain the differing trends you see in your graph? See below right. 2c. increased steadily

TV Viewing 1970–1990 (in millions)

Year	Total Viewing Top-Rated Program	Households with TV
1970	28.0	60.1
1975	30.3	71.5
1980	41.5	77.8
1985	39.4	84.9
1990	39.0	92.1

Source: *Nielsen Media Research*

3. *Standardized Test Prep* Suppose your mean on four literature tests is 78. Which score would raise the mean to 80? D

A. 95 **B.** 80 **C.** 100 **D.** 88 **E.** not possible

Find the mean, median, and mode of each set of data. Round to the nearest tenth.

2d. Answers may vary. Samples: There are more programs/channels to choose from; more people watch cable/videotapes; people have things to do besides watching TV.

4. 2, 6, 3, 2, 4
3.4; 3; 2

5. 10, 10, 20, 0
10; 10; 10

6. 80, 90, 85, 80, 90, 90, 40, 85
80; 85; 90

7. 25, 25, 25, 25
25; 25; 25

8. 22, 23, 28, 33, 24, 27, 24, 26, 23, 26, 25, 29, 21, 30
25.8; 25.5; 23, 24, and 26

9. 35.2, 42.6, 41.0, 37.2, 34.5, 35.0, 36.8, 41.0, 37.9, 42.1, 41.5
38.6; 37.9; 41.0

10. a. *Recording Industry* Use a triple bar graph or a triple line graph to display the data at the right. 10a–c. See margin.

b. Why did you choose the type of graph you did?

c. Which type of music showed the most growth in sales over the six-year period? **Justify** your answer.

Recording Industry Sales (in millions of dollars)

Year	Rock	Country	Jazz
1989	$2823	$447	$375
1990	$2722	$724	$362
1991	$2726	$1003	$298
1992	$2852	$1570	$313
1993	$3034	$1879	$311
1994	$4236	$1967	$362

Source: *Recording Industry Association of America*

11. a. *Sports* The median height of the 21 players on a girls' soccer team is 5 ft 7 in. What is the greatest possible number of girls who are less than 5 ft 7 in. tall? 10

b. *Critical Thinking* Suppose three girls are 5 ft 7 in. tall. How would this change your answer to part (a)? Explain. See margin.

12. a. *Agriculture* What trend do you see in the graph showing the number of farms in the United States? number decreased

b. What trend do you see in the graph showing the average size of farms in the United States? average size increased

c. *Writing* Do you think the two trends are related? Explain.
Answers may vary. Sample: Yes; as the number of farms decreased, their sizes increased.

United States Farms

Number of Farms (millions) — Year: 1940 1950 1960 1970 1980 1990

Average Size (acres) — Year: 1940 1950 1960 1970 1980 1990

Source: United States Department of Agriculture

Exercises MIXED REVIEW

Exercises 18–23 Review moving the decimal point when writing decimals as percents.

OPEN-ENDED Exercise 24 Remind students that there are many correct answers and that the total purchase can be \$100 or less.

GETTING READY FOR LESSON 1-2 These problems help students think about the relationships between numbers in patterns.

Wrap Up

THE BIG IDEA Ask students to list reasons why displaying data relationships in a graph is better than randomly listing data.

RETEACHING ACTIVITY Students draw and interpret graphs based on given sets of data. (Reteaching worksheet 1-1)

13. *Languages* Do Spanish words have more letters than English words? Compare the word lengths in the following excerpt from the Preamble to the United Nations Charter.

We the peoples of the United Nations determined to save succeeding generations from the scourge of war, which twice in our lifetime has brought untold sorrow to mankind, and
To reaffirm faith in fundamental human rights, in the dignity and worth of the human person, in the equal rights of men and women and of nations large and small . . . have resolved to combine our efforts to accomplish these aims.

Nosotros los pueblos de las Naciones Unidas resueltos a preservar a las generaciones venideras del flagelo de la guerra que dos veces durante nuestra vida ha infligido a la Humanidad sufrimientos indecibles,
a reafirmar la fe en los derechos fundamentales del hombre, en la dignidad y el valor de la persona humana, en la igualdad de derechos de hombres y mujeres y de las naciones grandes y pequeñas . . . hemos decidido aunar nuestros esfuerzos para realizar estos designios.

a. For each version, draw a line plot of word lengths. See margin.
b. Find the mean, median, and mode of the data in each line plot. English: 4.68, 4, 3; Spanish: 4.93, 5, 2
c. *Writing* Which language do you think has the longer word length? **Justify** your answer. Answers may vary. Sample: Spanish; both the mean and the median are greater for Spanish.

Chapter Project ***Find Out by Graphing***

- Measure the length of your radius bone to the nearest half inch.
- Collect the data for the class, and display the data in a graph.
- Find the mean, median, and mode for the data you collected. Write a description of the data.

Open-ended **Create a set of data for the given mean, median, and mode.**

14–17. Samples:

14. mean = 10, median = 10, mode = 10 1, 2, 5, 10, 10, 10, 16, 18, 18
15. mean = 100, median = 80, mode = 70 160, 120, 80, 70, 70
16. mean = 3, median = 2, no mode 8, 0, 4, 1, 2
17. mean = 5, median = 5, mode = 8 and 3 0, 3, 3, 5, 8, 7, 8, 5, 3, 8

Exercises MIXED REVIEW

Write as a decimal, a fraction or mixed number, and a percent.

18. $\frac{5}{8}$ 0.625; 62.5%
19. 40% 0.4; $\frac{2}{5}$
20. 0.75 $\frac{3}{4}$; 75%
21. 25% $\frac{1}{4}$; 0.25
22. 4.6 $4\frac{3}{5}$; 460%
23. $1\frac{4}{5}$ 1.8; 180%
24. *Open-ended* Suppose a pair of shoes costs \$25, a shirt costs \$15, and a pair of pants costs \$30. You have a budget of \$100 and the sales tax is 6%. How many of each item would you buy? Explain. Answers may vary. Sample: 1 pair of shoes, 2 shirts, and 1 pair of pants cost \$90.10, which is less than \$100.

Getting Ready for Lesson 1-2

Write the next three numbers in each pattern.

25. 1, 3, 5, ■, ■, ■ 7, 9, 11
26. 28, 25, 22, ■, ■, ■ 19, 16, 13
27. 4, 12, 36, ■, ■, ■ 108, 324, 972

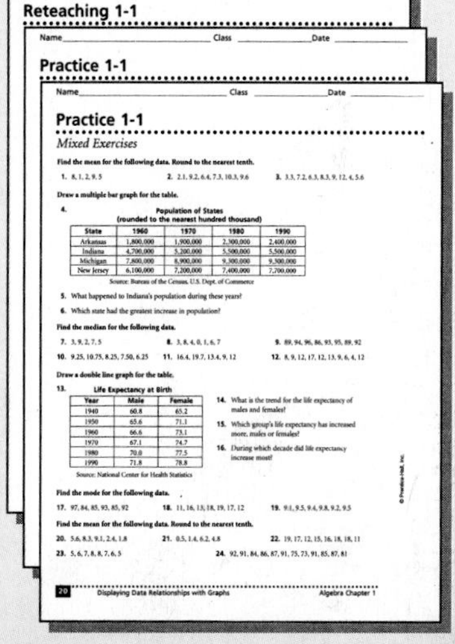

Name______ Class ______ Date ______

Reteaching 1-1

Name______ Class ______ Date ______

Practice 1-1

Name______ Class ______ Date ______

Practice 1-1

Mixed Exercises

Find the mean for the following data. Round to the nearest tenth.

1. 8, 1, 2, 9, 5
2. 2.1, 9.2, 6.4, 7.3, 10.3, 9.6
3. 3.3, 7.2, 6.3, 8.3, 9, 12, 4, 5.6

Draw a multiple bar graph for the table.

4. **Population of States (rounded to the nearest hundred thousand)**

State	1960	1970	1980	1990
Arkansas	1,800,000	1,900,000	2,300,000	2,400,000
Indiana	4,700,000	5,200,000	5,500,000	5,500,000
Michigan	7,800,000	8,900,000	9,300,000	9,300,000
New Jersey	6,100,000	7,200,000	7,400,000	7,700,000

Source: Bureau of the Census, U.S. Dept. of Commerce

5. What happened to Indiana's population during these years?
6. Which state had the greatest increase in population?

Find the median for the following data.

7. 3, 9, 2, 7, 5
8. 3, 8, 4, 0, 1, 6, 7
9. 89, 94, 96, 86, 93, 95, 89, 92
10. 9.25, 10.75, 8.25, 7.50, 6.25
11. 16.4, 19.7, 13.4, 9, 12
12. 8, 9, 12, 17, 12, 13, 9, 6, 4, 12

Draw a double line graph for the table.

13. **Life Expectancy at Birth**

Year	Male	Female
1940	60.8	65.2
1950	65.6	71.1
1960	66.6	73.1
1970	67.1	74.7
1980	70.0	77.5
1990	71.8	78.8

Source: National Center for Health Statistics

14. What is the trend for the life expectancy of males and females?
15. Which group's life expectancy has increased more, males or females?
16. During which decade did life expectancy increase most?

Find the mode for the following data.

17. 97, 84, 85, 93, 85, 92
18. 11, 16, 13, 18, 19, 17, 12
19. 9.1, 9.5, 9.4, 9.8, 9.2, 9.5

Find the mean for the following data. Round to the nearest tenth.

20. 5.6, 8.3, 9.1, 2.4, 1.8
21. 0.5, 1.4, 6.2, 4.8
22. 19, 17, 12, 15, 16, 18, 18, 11
23. 5, 6, 7, 8, 8, 7, 6, 5
24. 92, 91, 84, 86, 87, 91, 75, 73, 91, 85, 87, 81

20 Displaying Data Relationships with Graphs Algebra Chapter 1

Lesson Quiz

Lesson Quiz is also available in Transparencies.

1. Find the mean of these basketball scores: 88, 75, 78, 91, 89. 84.2
2. The hourly wages of eight students are three at \$5.25, two at \$5.50, two at \$5.75, and one at \$6.00. What is the median wage? \$5.50
3. On a test, six students scored 80, three scored 85, one scored 92, one scored 75, and one scored 62. What is the mode score for the test? 80

Math ToolboX

In Lesson 1-1, students organized data from least to greatest to find the median. Stem-and-leaf plots allow students to order a set of numbers quickly without the tediousness of a long list. The plots preserve the individual data while displaying the overall shape. A stem-and-leaf plot turned on its side closely resembles the corresponding histogram. By locating the median and mean in a stem-and-leaf plot, the student can begin to get an intuitive sense of normal distributions versus skewed distributions.

ERROR ALERT! Students often neglect to arrange the leaves from least to greatest. **Remediation:** Students should first sort the data into the correct stems and then arrange each set of numbers in ascending order.

WRITING Exercise 6 Before students can describe the advantages and disadvantages, they may need help comparing the circle graph to the stem-and-leaf plot. Point out the relationship between the heights of the rectangles in the histogram and the corresponding sections in the circle graph.

ADDITIONAL PROBLEM The ages of 25 randomly selected runners in a local 10K run are recorded below. Make a stem-and-leaf plot from the data. What are the median and the mode? **median = 33; mode = 27**

33 9 17 20 41 42 43 48 27 33 39 53 62 9 57 29 15 37 21 66 27 27 31 50 38

page 10 Math Toolbox

4a. To find the median, divide the number of data items, 27 by 2. The result is 13.5. This means that the 14th item will be the median. Start with the first leaf and count to the 14th leaf, which is 3 on the stem 4. So the median is 43. To find the mode, find the digit that appears most often on a single stem.

5a. 1.9, 2.0, 2.1, 2.2, 2.3, 2.4, 2.5

b. Answers may vary. Sample: 2.2 | 5 means $2.25

c.

1.9	5 7 9
2.0	0 5 9
2.1	6
2.2	1 5 9
2.3	6
2.4	0 9
2.5	0 0 7

6. The stem-and-leaf plot shows all of the data and the frequency for each time period, like 20–29 min. The circle graph shows the percent of responses for each time period, which the stem-and-leaf plot does not show. Like the stem-and-leaf plot, the histogram shows the frequency for each time period but does not show each data item.

Math ToolboX — Skills Review

Stem-and-Leaf Plots

After Lesson 1-1

For a class project, a student gathered data from her classmates. She made this request: "Tomorrow morning, find out how many minutes it takes you to get ready—from the time you get up until the time you leave your house for school." Here are the responses:

47 28 78 47 58 93 34 76 35 72 45 53 23
43 75 27 23 87 33 43 25 35 49 35 48 37 28

You can use a *stem-and-leaf plot* to give you a better picture of this information. A stem-and-leaf plot displays the data items in order. The digit (or digits) to the left is the *stem*. The digit farthest to the right is the *leaf*.

58 — stem, leaf

For Exercises 1–4, use the stem-and-leaf plot at the right.

Time to Get Ready

2	3 3 5 7 8 8
3	3 4 5 5 5 7
4	3 3 5 7 7 8 9
5	3 8
6	
7	2 5 6 8
8	7
9	3

key: 2|3 means 23 min

1. How many students responded? **27 students**
2. What does the stem 2 and leaf 8 represent? **28 min**
3. How many students took more than 50 min to get ready in the morning? **8 students**
4. a. Explain how you would use the stem-and-leaf plot to find the median and the mode of the data. **See margin.**
   b. What is the median? the mode? **43; 35**
5. Consider the data on notebook prices. **5a–c. See margin.**
   a. What would be the stems for the notebook data?
   b. What would be the key?
   c. Make a stem-and-leaf plot for this set of data.
   d. Find the median and mode of this set of data. **2.23; 2.50**

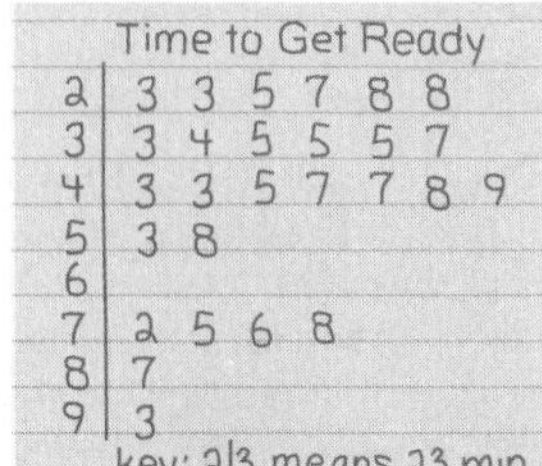

Notebook Prices

$2.00	$2.50	$2.25	$2.50
$2.29	$1.97	$2.16	$2.49
$2.21	$2.36	$2.09	$1.95
$2.05	$2.57	$2.40	$1.99

6. *Writing* The two graphs below display the information on the time it takes to get ready in the morning. Compare the graphs with the stem-and-leaf plot. Describe the advantages and disadvantages of using each to display the data. **See margin.**

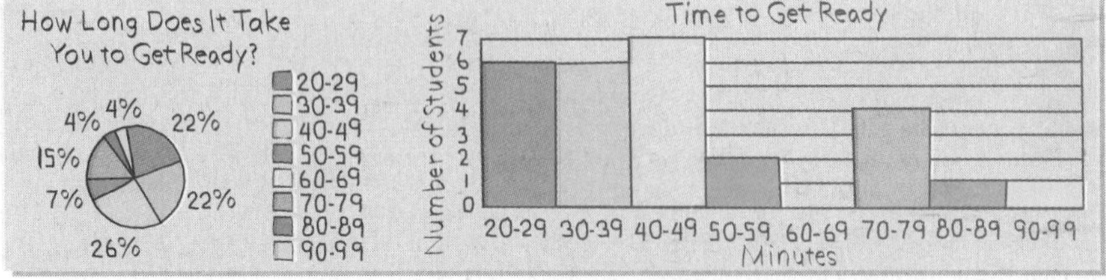

1-2 Teaching Notes

PROBLEM OF THE DAY

A piece of rope 68 in. long is to be cut into two pieces. How long will each piece be if one piece is cut three times longer than the other piece? 17 in. and 51 in.

Problem of the Day is also available in Transparencies.

CONNECTING TO PRIOR KNOWLEDGE Ask students: *A neighbor offers to pay you $5 per hour to babysit. What relationship exists between the number of hours worked and the amount of money earned?* the more hours worked, the more money earned Ask students to express the relationship mathematically in as many different ways as they can. Answers may vary. Sample: $5h = m$

THINK AND DISCUSS

Question 3 Students will not need to draw a 12-sided figure to answer this question. If they doubt that the rule will hold, encourage them to sketch 7-sided and 8-sided figures to check whether the rule correctly predicts the number of triangles in those figures.

ERROR ALERT! Students may think that $\frac{b}{2}$ is two terms. **Remediation:** Remind students that $\frac{b}{2}$ is the quotient of a variable and a number, so it is one term.

Connections Music . . . and more

1-2 Modeling Relationships with Variables

What You'll Learn

- Describing relationships between sets of data
- Using variables as a shorthand way to express relationships

...And Why

To model relationships in areas such as geometry, music, and sales

THINK AND DISCUSS

Geometry The band Numerica is designing a stage to use in its new music video. Their set designer made the sketches shown below.

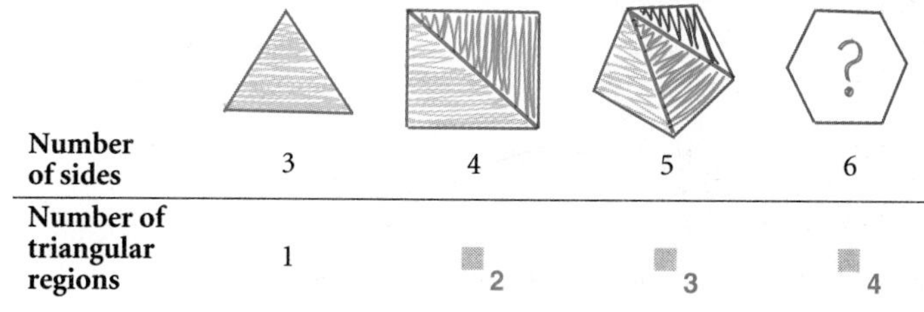

Number of sides	3	4	5	6
Number of triangular regions	1	2	3	4

1. Use the sketches to help you complete the table. See table.
2. **Analyze** the data in the table. What relationship do you see between the number of sides a stage has and the number of triangular regions created by the line segments? The number of triangular regions is two less than the number of sides.
3. Numerica plans to use a 12-sided stage. How many triangular regions can the designer form with the segments starting at one corner? 10 regions

You could use the letter s to represent the number of sides of the stage. Since s can change in value, it is a **variable.** A **variable expression** is a mathematical phrase that uses numbers, variables, and operation symbols. The variable expression $s - 2$ is a short way to represent the number of triangular regions created by the segments.

4. Could you use a letter other than s to represent the number of sides of the stage? Explain. Yes; any letter may be used to represent a variable.

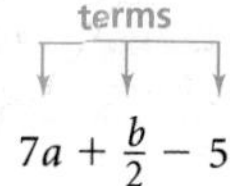

QUICK REVIEW

$7a$ is a short way to write $7 \times a$ or $7 \cdot a$.

Variable expressions are made up of one or more terms. A **term** is a number, a variable, or the product or quotient of a number and a variable.

5. How many terms does the expression $7a + \frac{b}{2} - 5$ have? 3 terms
6. How many terms does the expression $s - 2$ have? 2 terms

You can use variable expressions to write an equation. The **equation** $t = s - 2$ indicates that the two expressions t and $s - 2$ are equal.

Lesson Planning Options

Prerequisite Skills

- Finding number patterns

Assignment Options

Core 1–4, 7–18, 20–22
Extension 5, 6, 19, 23–25
Use Mixed Review to maintain skills.

Resources

Student Edition

Skills Handbook, pp. 569, 573, 582
Extra Practice, p. 556
Glossary/Study Guide

Teaching Resources

Chapter Support File, Ch. 1
- Practice 1-2 (two worksheets)
- Reteaching 1-2

Classroom Manager 1-2
Glossary, Spanish Resources
Two-Year Algebra Handbook 1-2

Transparencies

10, 32

Example 1 Relating to the Real World 

AUDITORY LEARNING Saying and hearing information is appealing to students with an auditory learning style. Write $c = 12n$. Ask students: *Name the variables. Tell what each one represents.* c represents the cost and n represents the number of CDs. Tell students to repeat the answer to themselves numerous times. Have students say out loud what $c = 12n$ stands for. Cost equals 12 times the number of CDs. Encourage students to form the habit of saying to themselves what the variables represent.

ESL Emphasize that any letter may be chosen to represent the unknown amount. However, it is best to avoid using *O* as a variable because *O* can be easily confused with 0 (zero).

Example 2 Relating to the Real World 

Write $a = 20 - c$ on the board. Have students look at the data in the table and answer these questions.

- *How are the two columns related?* Answers may vary but students should recognize that the amount of change is \$20 minus the cost or that the cost added to the change equals \$20.
- *What does a represent?* amount of change
- *What does c represent?* cost of items purchased
- *How does the equation on the board relate to the data in the table?* It describes the relationship between the two columns.

Additional Examples

FOR EXAMPLE 1

Write an equation to show the revenue for selling tickets at \$3.50 each.
$r = 3.5n$

Discussion: *For what other school subjects would equations be useful?*

FOR EXAMPLE 2

Write an equation to show the relationship of the balance *b* in a checking account, starting with \$47.82 and purchasing four items for a total cost of *c*. $b = 47.82 - c$

Discussion: *Is there another equation that will model this relationship?*

Example 1 Relating to the Real World 

Music Each CD at Track One Records costs \$12. Write an equation you can use to find the total cost when you know the number of CDs bought.

The plan below shows how you can go from a word problem to a short statement of the problem using variables.

Relate The total cost is 12 times the number of CDs bought.

cost is 12 times a number

Write $c = 12 \cdot n$

$c = 12n$

7. What do the variables *c* and *n* represent in Example 1? c = total cost; n = number of CDs
8. A worker at Track One Records says that 5 CDs cost \$125 because $c = 12n$ and $n = 5$. Explain the worker's error. $12n$ means $12 \times n$; n is not an additional digit.
9. **Try This** Suppose the manager at Track One Records raises the price of CDs to \$15. Write an equation to find the cost of *n* CDs. $c = 15n$
10. Suppose the manager at Track One Records uses the equation $c = 10.99n$. What does this mean? The price of CDs is \$10.99 each.

Before writing an equation for data in a table, it is helpful to write a short sentence describing the relationship between the data. Then translate the sentence to an equation. Be sure to tell what each variable represents.

Example 2 Relating to the Real World 

Sales Write an equation for the data in the table.

Change from a \$20 Bill

Cost of Items Purchased	Amount of Change
\$20.00	\$0
\$19.00	\$1.00
\$17.50	\$2.50
\$11.59	\$8.41

Define c = cost of items purchased
a = amount of change

Relate Amount of change equals \$20.00 minus cost of items purchased.

Write $a = 20 - c$

$a = 20 - c$

11. Could you define the variables after writing the sentence describing the relationship between the data? Explain. See left.
12. In Example 2, could the letter *c* represent the cost and also represent the change? Why or why not? No; each letter must represent only one quantity in an expression or equation.
13. Suppose you wrote "The [illegible] plus the amount of change equals \$20" for the relationship between [illegible] data in the table. [illegible] this statement with an equation. $c + a = 20$

11. Yes; the variables may be defined before or after describing the relationship.

Lead students to understand that an equation containing a variable describes a relationship between data. If there is no clear relationship between data, it may not be possible to write an equation to describe the data.

Exercises ON YOUR OWN

ALTERNATIVE ASSESSMENT **Exercise 7–10** Instruct students to describe another situation that can be modeled by their equations. This exercise will help you assess students' understanding of modeling relationships with variables.

OPEN-ENDED Exercise 5 Encourage students to include terms that involve products and quotients as well as variables.

Exercise 17 Students' equations should also work when the number of sales is not a multiple of 5. Remind students to check the validity of their equations by working backward. Have them substitute each of the values 5, 10, 15, and 20 for the sales variable to check that the total earnings calculated by the equation matches the earnings in the table.

Exercise 22 Suggest to students that the variable h would not be a good variable to use for this exercise because students could easily become confused about whether h represents the height of the first bounce or the height of the drop.

STANDARDIZED TEST TIP **Exercise 24** Remind students to consider all answer options for each choice before deciding on the appropriate one.

Exercises ON YOUR OWN

State the number of terms in each expression.

1. $3x + 5y - 12$ 3
2. $9x^3 + 2x^2 + 5x + 7$ 4
3. $\frac{2y}{5} - 4$ 2
4. 15 1
5. *Open-ended* Write an expression that has four terms. Answers may vary. Sample: $-2a + 3b - 4 + 5a^2$
6. *Writing* Use an example to explain the meaning of *variable.* Answers may vary. Sample: The letter h may represent the hours left in a school day, which is a value that changes.

Use an equation to model each situation.

7. Total cost equals number of cans times $.70. c = cost, n = number of cans; $c = 0.7n$
8. The perimeter of a square equals four times a side. p = perimeter, s = side; $p = 4s$
9. The total amount of rope, in feet, used to put up scout tents is 60 times the number of tents. r = amount of rope, t = number of tents; $r = 60t$
10. What is the number of slices of pizza left from an 8-slice pizza after you have eaten some slices? n = number of slices left, e = number of slices eaten; $n = 8 - e$

Use an equation to model the relationship in each table.

11.

Hours	Distance
1	50 mi
2	100 mi
3	150 mi
4	200 mi

$d = 50h$

12.

Hours	Pay
4	$24
6	$36
8	$48
10	$60

$p = 6h$

13.

Days	Growth
1	0.165 in.
2	0.330 in.
3	0.495 in.
4	0.660 in.

$g = 0.165d$

14.

Tapes	Cost
1	$8.50
2	$17.00
3	$25.50
4	$34.00

$c = 8.50t$

15.

Workers	Radios
1	13
2	26
3	39
4	52

$r = 13w$

16.

Earned	Saved
$15	$7.50
$20	$10.00
$25	$12.50
$30	$15.00

$s = \frac{1}{2}e$

17.

Number of Sales	Total Earnings
5	$2.00
10	$4.00
15	$6.00
20	$8.00

$e = 0.4s$

18.

Time (months)	Length (inches)
1	4.1
2	8.2
3	12.3
4	16.4

$\ell = 4.1t$

19. *Open-ended* Choose one table from Exercises 11–18. Describe a situation for which the data is reasonable. Answers may vary. Sample for Ex. 12: Wage is $6/h.
20. Does each statement fit the data at the right? Explain.
 a. hours worked = lawns mowed · 2 yes; $3 \cdot 2 = 6$
 b. hours worked = lawns mowed + 3 yes; $3 + 3 = 6$

Lawns Mowed	Hours
1	
2	
3	6

21. a. Which statement in Exercise 20 better describes the relationship between hours worked and lawns mowed? Explain. See below.
 b. Use your answer to part (a). Copy and complete the table. How many hours would it take to mow 5 lawns? b. 1 lawn: 2 hours, 2 lawns: 4 hours; 5 lawns would take 10 hours

21a. Answers may vary. Sample: statement (a) is better because it indicates that each lawn takes 2 h to mow.

Technology Options

For Exercises 11–18, students may use spreadsheet software to validate their answers.

Prentice Hall Technology

Software • Secondary Math Lab Toolkit • Computer Item List Generator 1-2

CD-ROM Multimedia Algebra Lab 1

Internet • See the Prentice Hall site. (http://www.phschool.com)

Exercises MIXED REVIEW

JOURNAL Have students write a sample word problem and then describe in their journals how the problem could be written in a simpler way using variables.

GETTING READY FOR LESSON 1-3 These problems give students practice in simplifying expressions. They also prepare students for simplifying expressions involving variables.

Wrap Up

THE BIG IDEA Ask students to list ways in which using variables is more efficient than using numbers.

RETEACHING ACTIVITY Students translate word expressions into mathematical variable expressions. (Reteaching worksheet 1-2)

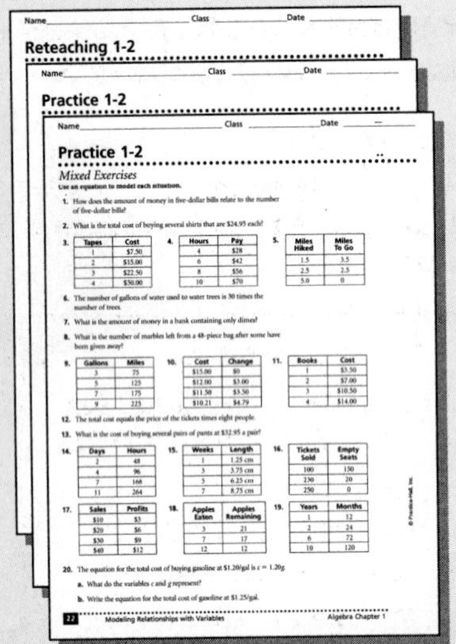
Reteaching 1-2

Practice 1-2

Practice 1-2

Mixed Exercises

Lesson Quiz

Lesson Quiz is also available in Transparencies.

1. Express the total cost c of four tuna sandwiches t. $c = 4t$
2. How many terms does $2 \cdot 3 + 5 \cdot 2 - 3x + 1$ contain? 4
3. Use an equation to model the perimeter p of a hexagon. $p = 6s$
4. Each student needs five pieces of paper for a report. Write an equation to show how many pages will be needed for all the reports. $r = 5s$

page 14 Mixed Review

26a. See back of book.

Use the table at the right for Exercises 22 and 23. The table shows the result when a ball is dropped from different heights.

Drop Height (ft)	Height of First Bounce (ft)
1	$\frac{1}{2}$
2	1
3	$1\frac{1}{2}$
4	2
5	$2\frac{1}{2}$

d = drop height, f = height of first bounce;

22. a. Write an equation to describe the relationship between the height of the first bounce and the drop height. $f = \frac{1}{2}d$
    b. Suppose you drop the ball from a window 20 ft above the ground. **Predict** how high the ball will bounce. 10 ft
23. Suppose the second bounce is $\frac{1}{4}$ of the original drop height. Write an equation to represent the height of the second bounce. s = height of second bounce; $s = \frac{1}{4}d$
24. *Standardized Test Prep* Compare the quantities in Column A and Column B. D

Column A	Column B
the total cost of items bought by a customer	the change the customer receives when paying with a $10 bill

A. The quantity in Column A is greater.
B. The quantity in Column B is greater.
C. The quantities are equal.
D. The relationship cannot be determined from the information given.

25. a. Write an equation to show how the amount of money in a bag of quarters relates to the number of quarters in the bag. n = number of quarters, m = money in the bag (in dollars); $m = 0.25n$
    b. The bag contains 13 quarters. How much money is this? $3.25

Exercises MIXED REVIEW

26. For a project, a student asked 25 classmates the number of television news programs they watched in a week. She recorded these results:
    3 5 1 4 2 0 4 5 3 6 5 5
    2 0 0 1 2 5 3 1 1 4 6 3 1
    a. Make a histogram of the data. See margin.
    b. Find the mean, median, and mode(s) of the data. 2.88; 3; 1 and 5
27. List four prime numbers between 20 and 50. Answers may vary. Complete list: 23, 29, 31, 37, 41, 43, 47
28. Write the prime factorization of 105. $3 \cdot 5 \cdot 7$

FOR YOUR JOURNAL

Write a paragraph explaining what variables are and how you use them to model the relationship between sets of data.

Getting Ready for Lesson 1-3

Find the value of each expression.

29. $3 + 12 - 7$ 8
30. $4 \cdot 3 - 5$ 7
31. $15 \div 3 + 2$ 7
32. $6 \cdot 1 \div 2$ 3
33. $4 - 2 + 9$ 11

1-3 Teaching Notes

PROBLEM OF THE DAY

Shaun told two students a secret. They each told three other students who told two other students. How many students, including Shaun, know Shaun's secret? **21**

Problem of the Day is also available in Transparencies.

CONNECTING TO PRIOR KNOWLEDGE Bring in sales receipts for multiple purchases of items that are taxable. Be sure the sales tax is shown. Ask students: *When you buy several items at a store, why is the tax calculated on the total of the items instead of each item individually?* **Answers may vary. Sample: Figuring sales tax on the total involves less calculating.**

WORK TOGETHER

Question 4 After students have completed Question 4, write on the board $l = \frac{5}{8}$ and $w = \frac{3}{8}$. Ask students which formula for the perimeter they would prefer to use for these values of l and w. Lead students to notice that solving the problem will be easier using the formula $P = 2(l + w)$.

THINK AND DISCUSS

- *In the United States, do people drive on the left or the right side of the road?* **right**
- *What would happen if there was no rule about whether to drive on the left or on the right?* **Answers may vary.**

Connections Community Service . . . and more

1-3 Order of Operations

What You'll Learn

- Using the order of operations
- Evaluating variable expressions

...And Why

To find the total cost of items with tax

What You'll Need

- calculator

WORK TOGETHER

Work with a partner.

Geometry Two formulas for the perimeter of a rectangle are $P = 2l + 2w$ and $P = 2(l + w)$.

1. Let $l = 12$ and $w = 8$. Find the perimeter of the rectangle using each formula. **40**
2. When you used the formula $P = 2l + 2w$, did you add first or did you multiply first? **multiply**
3. When you used the formula $P = 2(l + w)$, did you add first or did you multiply first? **add**
4. Which formula do you prefer to use? Why? **Check students' work.**

Who? Astronaut Ellen Ochoa served on space flights in 1993 and 1994. She studied the solar corona and the effect of solar changes on Earth's environment.

THINK AND DISCUSS

Part 1

Evaluating Expressions

In the Work Together activity, you used your past experience to simplify the expressions to find the perimeter. Look at this new expression and the two ways that it has been simplified.

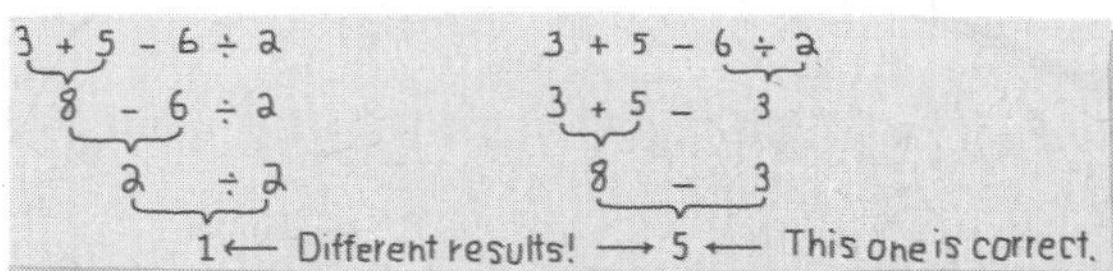

To avoid having two results for the same problem, mathematicians have agreed on an order for doing the operations when simplifying.

QUICK REVIEW

An *exponent* indicates repeated multiplication.

exponent

base → $3^4 = 3 \cdot 3 \cdot 3 \cdot 3$

The base 3 is used as a factor four times.

You read 3^4 as "three to the fourth power."

Order of Operations

1. Perform any operation(s) inside grouping symbols.
2. Simplify any term with exponents.
3. Multiply and divide in order from left to right.
4. Add and subtract in order from left to right.

5. Which operation should you do first to simplify each expression?
   a. $3 + 6 \cdot 4 \div 2$ **multiply**
   b. $3 \cdot 6 - 4^2$ **simplify term with exponent**
   c. $3 \cdot (6 - 4) \div 2$ **subtract**

Lesson Planning Options

Prerequisite Skills

- Simplifying expressions
- Substituting numbers for variables in equations

Assignment Options

To provide flexible scheduling, this lesson can be subdivided into parts.

1 **Core** 1–5, 7, 9
Extension 13, 14, 29–32

2 **Core** 6, 8, 10–12, 15–26
Extension 27, 28

Use Mixed Review to maintain skills.

Resources

Student Edition

Skills Handbook, pp. 580, 582
Extra Practice, p. 556
Glossary/Study Guide

Teaching Resources

Chapter Support File, Ch. 1
- Practice 1-3 (two worksheets)
- Reteaching 1-3

Classroom Manager 1-3
Glossary, Spanish Resources
Two-Year Algebra Handbook 1-3

Transparencies
23, 33

- *If everybody agreed to drive on the left side instead of the right, would that agreement work?* yes

Explain to students that whether people drive on the left or right side is not important so long as everybody follows the same agreement. Similarly, the rules for the order of operations are neither right nor wrong but simply a convention that everybody follows to avoid confusion.

DIVERSITY Have students share examples of times when they met someone who used a familiar word differently than they do. Have them relate the confusion they felt to the class. For example, people in England use the word *biscuit* to mean *cookie* and could be confused by seeing *biscuits and gravy* on a menu in the United States.

Example 1 Relating to the Real World 

CONNECTING TO THE STUDENTS' WORLD Have each student calculate the total cost they would pay for the tennis shoes using the percent of your city's sales tax.

Example 2

KINESTHETIC LEARNING Demonstrations and movement will appeal to students with a kinesthetic learning style. Have students suggest and demonstrate a series of three actions that are effective only if performed in the correct order. For example, the series (1) put on socks, (2) put shoes on feet, and (3) tie shoelaces, will be effective only if completed in that order. Students may act out their series in the wrong order to demonstrate why the order of operations assures that everyone can achieve the same result.

Additional Examples

FOR EXAMPLE 1

A CD costs $14.95 with a sales tax rate of 7%. Use the expression $p + r \cdot p$ to calculate the total cost of the CD.

$$p + r \cdot p = 14.95 + .07(14.95)$$
$$= 14.95 + 1.0465$$
$$\approx 14.95 + 1.05$$
$$\approx \$16.00$$

Discussion: *Would $C = P(1 + r)$ be a simpler expression to use?*

FOR EXAMPLE 2

Evaluate $5x + 12 \div p$ for $x = 3$ and $p = 6$

$5(3) + 12 \div 6 =$

$15 + 2 = 17$

Discussion: *How is the order of operations important in this calculation?*

FOR EXAMPLE 3

Evaluate (a) $\frac{a^2}{x}$, and (b) $\frac{a^2}{x^2}$ for $a = 6, x = 12$

a. $\frac{a^2}{x} = \frac{6^2}{12} = \frac{36}{12} = 3$

b. $\frac{a^2}{x^2} = \frac{6^2}{12^2} = \frac{36}{144} = \frac{1}{4}$

Discussion: *How does the exponent in the denominator affect the answer?*

SALE

$59.00

State sales tax 6%

You **evaluate** an expression with variables by substituting a number for each variable. Then simplify the expression using the order of operations.

Example 1 Relating to the Real World 

Sales Find the total cost of the sneakers shown in the ad. Use the expression at the right.

$p + r \cdot p$ (sales tax rate: r; original price: p; sales tax: $r \cdot p$)

$p + r \cdot p = 59 + (0.06)59$ ← Substitute 59 for p. Change 6% to 0.06 and substitute 0.06 for r.

$= 59 + 3.54$ ← Multiply first.

$= 62.54$ ← Then add.

The total cost of the sneakers is $62.54.

6. **Calculator** Some calculators have the order of operations programmed into them. To check your calculator, use the key sequence below. Does your calculator use the order of operations? Explain. **Yes, if the answer is 62.54. If the answer is 3484.54, it completes the operations left to right.**

59 [+] .06 [×] 59 [=]

Keep in mind that the base for an exponent is the number, variable, or expression directly to the left of the exponent.

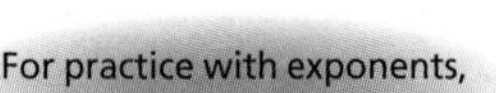
For practice with exponents, see Skills Handbook page 580.

Example 2

Evaluate $3a^2 - 12 \div b$ for $a = 7$ and $b = 4$.

$3a^2 - 12 \div b = 3 \cdot 7^2 - 12 \div 4$ ← Substitute 7 for a and 4 for b.

$= 3 \cdot 49 - 12 \div 4$ ← Simplify 7^2.

$= 147 - 3$ ← Multiply and divide from left to right.

$= 144$ ← Subtract.

7. A student evaluated the expression above for $a = 8$ and $b = 6$. Her result is 141. Is her answer correct? If not, what error did she make? **No; she calculated from left to right.**

8. **Try This** Evaluate $25 \div p + 2q^2$ for $p = 5$ and $q = 7$. **103**

Example 3

Evaluate each expression for $c = 15$ and $d = 12$.

a. cd^2

$cd^2 = 15(12)^2$
$= 15(144)$
$= 2160$

b. $(cd)^2$

$(cd)^2 = (15 \cdot 12)^2$
$= (180)^2$
$= 32{,}400$

CALCULATOR HINT

In part (b), you can use two different key sequences to find 180^2:

180 [x^2] [ENTER] or

180 [^] 2 [ENTER].

9. Write an expression for each phrase. Then simplify your expression.

a. four times three, squared **$(4 \cdot 3)^2$; 144**

b. the square of three, times four **$3^2 \cdot 4$; 36**

Example 3

Question 9a Students may be unsure whether four times three squared means $4 \cdot 3^2$ or $(4 \cdot 3)^2$. Use this ambiguity to help them appreciate that in mathematics, expressions are often more effective than words at communicating meaning.

Example 4 Relating to the Real World

ERROR ALERT! Students may be confused by the two b variables. **Remediation:** Help them understand that b_1 and b_2 are different variables and that the subscripts 1 and 2 are part of the name of each variable.

Exercises ON YOUR OWN

ALTERNATIVE ASSESSMENT **Exercises 1–12** Working in pairs, assign each group an exercise from Exercises 1–12. One student calculates the answer using the order of operations while the other student calculates in left-to-right order. Students should compare their results to see the differences. This activity will reinforce the importance of using the order of operations.

CRITICAL THINKING Question 27 Students may be unsure that 0 is a poss

Exercise 28 F

in the year 200

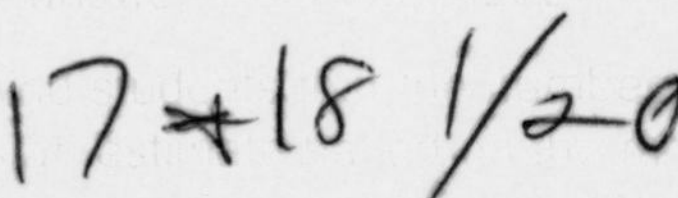

#14 PI = 3.14

Software • Secondary Math Lab Toolkit • Computer Item Generator 1-3

Internet • See the Prentice Hall site. (http://www.phschool.com)

Part 2

Evaluating Expressions With Grouping Symbols

When you evaluate expressions work within the parentheses first. A fra bar is also a grouping symbol. Do any calculations above or below a fraction bar before simplifying the fraction.

Example 4 Relating to the Real World

Community Students participating in a neighborhood clean-up project are cleaning a vacant lot, which has the shape of a trapezoid. They plan to turn the vacant lot into a park. Use the expression $h\left(\frac{b_1 + b_2}{2}\right)$ to find the lot's area.

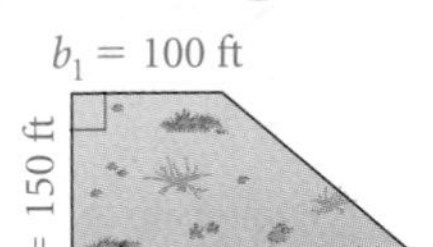

$h\left(\frac{b_1 + b_2}{2}\right) = 150\left(\frac{100 + 290}{2}\right)$ ← Substitute 150 for h, 100 for b_1, and 290 for b_2.

$= 150\left(\frac{390}{2}\right)$ ← Simplify the numerator.

$= 150(195)$ ← Simplify the fraction.

$= 29{,}250$

The area of the lot is 29,250 ft^2.

PROBLEM SOLVING

Look Back You can also use the expression $\frac{1}{2}h(b_1 + b_2)$ to find the area of a trapezoid. Explain why $h\left(\frac{b_1 + b_2}{2}\right)$ is equivalent to $\frac{1}{2}h(b_1 + b_2)$. **Dividing by 2 is the same as multiplying by $\frac{1}{2}$.**

10. Would the result be the same if you substituted 290 for b_1 and 100 for b_2? Why or why not? **yes; 100 + 290 = 290 + 100**
11. Describe the steps you would use to simplify $\frac{12 + 18}{3 + 9}$. **Add 12 and 18. Add 3 and 9. Divide 30 by 12.**
12. Explain the difference in the meaning of the expressions $5\frac{1}{6}$ and $5\left(\frac{1}{6}\right)$. **$5\frac{1}{6}$ means $5 + \frac{1}{6}$. $5\left(\frac{1}{6}\right)$ means $5 \times \frac{1}{6}$.**

You can also use brackets [ ] as grouping symbols. When an expression has several grouping symbols, simplify the innermost expression first.

13. *Try This* Simplify each expression.
    a. $5[4^2 + 3(2 + 1)]$ **125**
    b. $12 + 3[18 + 5(16 - 3^2)]$ **171**

Exercises ON YOUR OWN

Simplify each numerical expression.

1. $18 + 20 \div 4$ **23**
2. $(2.4 - 1.6) \div 0.4$ **2**
3. $\frac{6 \cdot 2 - 1}{9 + 2}$ **1**
4. $(5^2 - 3)6$ **132**
5. $(10 - 2)^2$ **64**
6. $6\left(\frac{4 + 10}{2}\right)$ **42**
7. $24.6 \div 2 \cdot 4.1$ **50.43**
8. $25 - [2(3 + 7)]$ **5**
9. $3 \cdot 5^2$ **75**
10. $(3 \cdot 5)^2$ **225**
11. $(5^2 + 3) \div 2$ **14**
12. $\frac{(2 + 3)^2}{2}$ **12.5**
13. *Open-ended* Write an expression that includes addition, subtraction, multiplication, and parentheses. Simplify your expression. **Answers may vary. Sample: $3 + (12 - 4) \cdot 5 = 43$**

Exercises MIXED REVIEW

JOURNAL Students could write a complex expression and then show how they can use the order of operations to simplify the expression.

GETTING READY FOR LESSON 1-4 These questions prepare students for adding and subtracting integers.

Wrap Up

THE BIG IDEA Ask students to list the advantages of using the rules for the order of operations.

RETEACHING ACTIVITY Students will practice using the order of operations to simplify expressions by randomly choosing operations signs. (Reteaching worksheet 1-3)

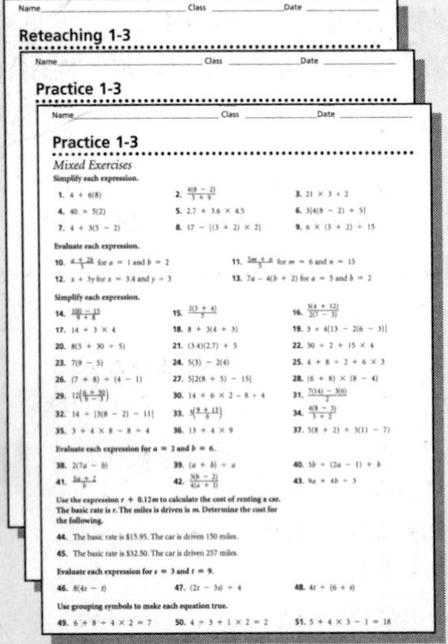
Reteaching 1-3
Practice 1-3
Practice 1-3

Lesson Quiz

Lesson Quiz is also available in Transparencies.

In Exercises 1–3, evaluate each expression.

1. $4x + 3$ for $x = 8$ 35
2. Evaluate $(2)(3^2) + 4x$ for $x = 3$. 30
3. $(d + 2) \div (d - 2)$ for $d = 6$. 2
4. Evaluate $(4 + p)^2$ for $p = 5$. 81

pages 17–18 On Your Own

28b. See back of book.

14. a. *Geometry* What is the volume of the juice can at the right? The formula for the volume of a cylinder is $V = \pi r^2 h$.
   b. About how many cubic inches does an ounce of juice fill?
   a. about 30.8 in.^{3} b. about 2.6 in.^{3}

Evaluate each expression.

15. $a - 7 \cdot 2$ for $a = 15$ 1
16. $\frac{q}{q + 8}$ for $q = 4$ $\frac{1}{3}$
17. $r + 2s$ for $r = 5.2$ and $s = 3.8$ 12.8
18. $5a^2 - 4$ for $a = 3$ 41
19. $2b^2 + 4b$ for $b = 6.3$ 104.58
20. $(5x)^2$ for $x = 3$ 225
21. $2\left(\frac{5d - 6}{3}\right)$ for $d = 9$ 26
22. $[(5.2 + a) + 4]10$ for $a = 3.5$ 127
23. $\frac{m^2}{m + 9}$ for $m = 6$ 2.4

Use grouping symbols to make each equation true.

24. $(10 + 6) \div 2 - 3 = 5$
25. $14 - (2 + 5) - 3 = 4$
26. $8 + 4 \div (3 - 1) = 10$

27. *Critical Thinking* Use the problem solving strategy *Guess and Test* to find two values of n that make the equation $2n = n^2$ true.
   0 and 2

28. a. *Entertainment* You can use the expression $2.4t + 0.779$ to model the number of subscribers, in millions, to cable television. Copy and complete the table. 0.779; 24.779; 48.779; 72.779
   b. *Statistics* Use your table to draw a line graph. See margin.

Cable TV Subscribers (in millions)

Year	Subscribers
1970 ($t = 0$)	■
1980 ($t = 10$)	■
1990 ($t = 20$)	■
2000 ($t = 30$)	■

Writing **Tell if each equation is *true* or *false*. If false, use the order of operations to explain why.** 31. F: simplify within parentheses first

29. $3(2^3) = 6^3$
30. $2^4 = (1 + 1)^4$ T
31. $(4 + 5)^2 = 4^2 + 5^2$

29. F; simplify the power before multiplying

32. *Calculator* Which key sequence could you use to simplify $\frac{3 + 8^2}{5}$? Explain why your choice works and the other choices do not. C

   A. (3 + 8) ^ 2 ÷ 5 ENTER
   B. 3 + 8 ^ 2 ÷ 5 ENTER
   C. (3 + 8 ^ 2) ÷ 5 ENTER
   D. (3 + 8 ÷ 5 ^ 2) ENTER

Exercises MIXED REVIEW

Write each fraction as a fraction or mixed number in simplest form.

33. $\frac{4}{6}$ $\frac{2}{3}$
34. $\frac{15}{10}$ $1\frac{1}{2}$
35. $\frac{27}{6}$ $4\frac{1}{2}$
36. $\frac{9}{12}$ $\frac{3}{4}$
37. $\frac{12}{15}$ $\frac{4}{5}$

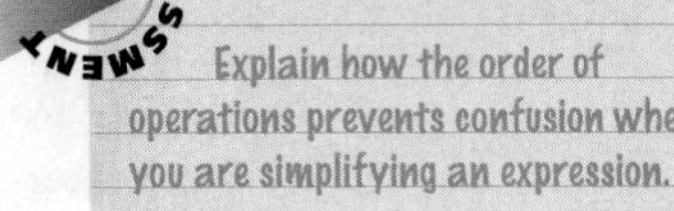

FOR YOUR JOURNAL
Explain how the order of operations prevents confusion when you are simplifying an expression.

Getting Ready for Lesson 1-4

Use the number line at the right.

38. What number corresponds to point A on the number line? −4
39. How far is B from zero? 2 units
40. How far is D from zero? 4 units
41. How far is C from zero? 3 units
42. How far is A from D? 8 units

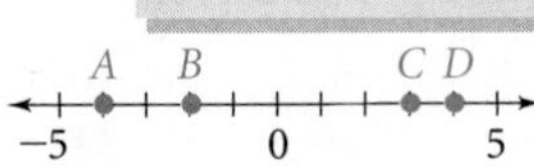

1-4 Teaching Notes

PROBLEM OF THE DAY

A video store offers two membership plans. Plan A costs \$15 to join and charges \$2.50 per video rental. Plan B costs \$20 to join and charges \$1.75 per video rental. How many videos must a customer rent before plan B is the better deal? 7 videos

Problem of the Day is also available in Transparencies.

CONNECTING TO PRIOR KNOWLEDGE Ask students to brainstorm different meanings of the word *negative* while you list them on the board. If any student has experience developing photographs, have the student explain what photographic negatives are and how they are developed.

WORK TOGETHER

Point out to students that 1994 is at the top of the diagram and that the years go back in time as they go down the diagram.

THINK AND DISCUSS

VISUAL LEARNING Encourage students to use algebra tiles until they are comfortable with the concept that the sum of a number and its opposite equals zero.

MAKING CONNECTIONS Some physicists study the particles of which atoms are made. They theorize that, for every type of particle, there exists a corresponding antiparticle with an opposite electric charge. The first antiparticle was observed in 1932 and was a positron, the opposite of an electron.

Connections Science . . . and more

1-4 Adding and Subtracting Integers

What You'll Learn
- Adding and subtracting with integers and decimals
- Finding absolute value
- Evaluating expressions

...And Why

To calculate with data that include negative numbers

2 Samples: temperature, football, accounting, reservoir water levels

WORK TOGETHER

Science Work with a partner. Falling snow has chemical traces of the environmental events that occur each year. In 1994, scientists in Antarctica drilled a core of ice 7000 ft deep. The diagram shows some of what scientists found.

1. Draw a diagram. Place the events listed below where you think they would appear in the core of ice. See right.
   a. 1906: San Francisco earthquake
   b. 1923: Tokyo earthquake and fires
   c. 1935–36: "Dust Bowl" of the U.S.
   d. 1986: nuclear accident at Chernobyl, in the former U.S.S.R.
2. *Open-ended* Name two other situations for which you would use negative numbers. See left.

History in Ice

0 ft — 1994: ice surface
d.
−26.6 ft — 1964: peak of radioactive fallout from atomic testing
c.
b.
a.
−67.3 ft — 1883: eruption of Krakatoa, a volcanic island in Indonesia

Source: *Boston Globe*

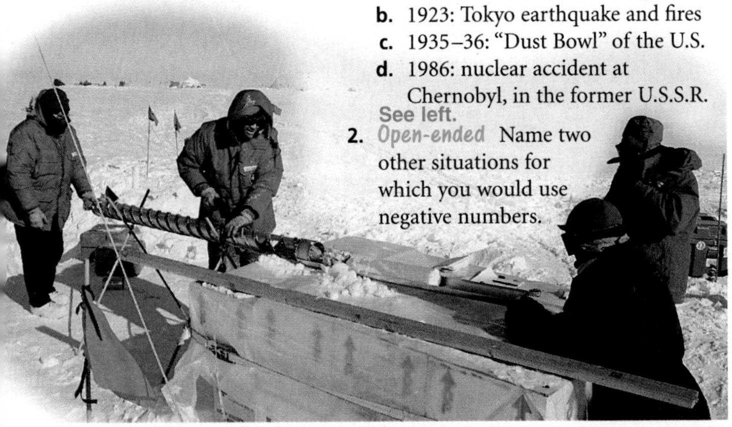

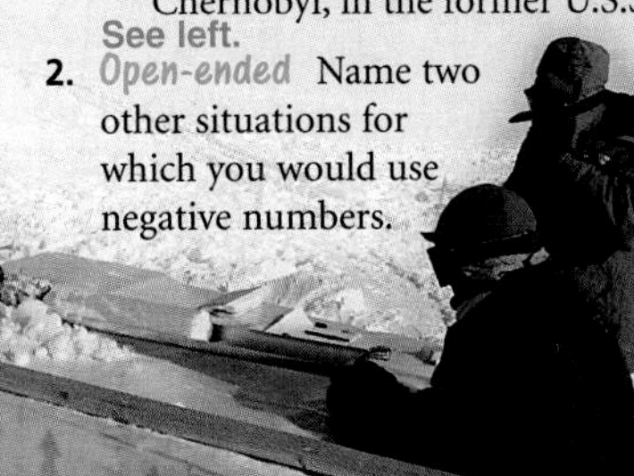

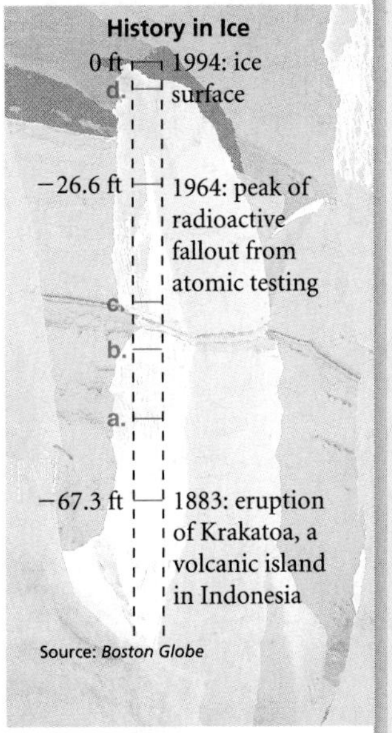

THINK AND DISCUSS

Part 1 **Adding Integers and Decimals**

A number line can help you see the relationship between numbers.

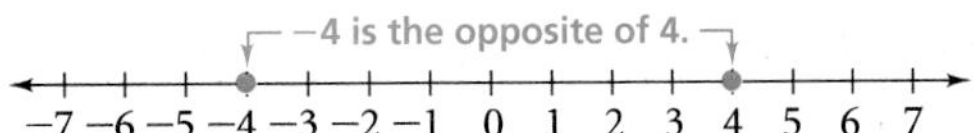

Two numbers are **opposites** if they are the same distance from zero on the number line. Since −4 and 4 are both a distance of 4 units from 0, they are opposites. **Integers** are whole numbers and their opposites.

QUICK REVIEW

Whole numbers are 0, 1, 2, 3 . . . .

3. **Try This** What is the opposite of 3? −5? −100? 0? −3; 5; 100; 0
4. **Try This** Name the two integers that are a distance of 7 units from 0 on the number line. −7 and 7

Lesson Planning Options

Prerequisite Skills
- Using algebra tiles
- Finding distance on a number line

Assignment Options

To provide flexible scheduling, this lesson can be subdivided into parts.

1 **Core** 3–12, 29–31, 38, 42, 44–45
Extension 1, 2, 32, 33, 46, 51–54

2 **Core** 13–28, 36, 37, 39–41, 47–50, 55
Extension 34, 35, 43, 56

Use Mixed Review to maintain skills.

Resources

Student Edition

Skills Handbook, pp. 569, 575
Extra Practice, p. 556
Glossary/Study Guide

Teaching Resources

Chapter Support File, Ch. 1
- Practice 1-4 (two worksheets)
- Reteaching 1-4

Classroom Manager 1-4
Glossary, Spanish Resources
Two-Year Algebra Handbook 1-4

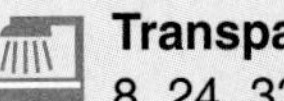

Transparencies
8, 24, 33

Antiparticles for most of the known particles have since been observed. However, scientists do not yet understand why there is far more matter than antimatter in our universe.

Example 1 **Relating to the Real World** 

KINESTHETIC LEARNING Students may need help understanding that distance traveled is always a positive number, although the direction traveled can be either positive or negative. Use masking tape to make a large number line on the floor. Write large numerals on paper and mark 2, 1, 0, –1, and –2 on the number line. Have a student stand on 0 and take two paces to get to 2. Ask students: *In paces, how far did you go from 0?* two paces Now stand on 0 and take two paces to get to –2. Ask students: *In paces, how far did you go from 0?* two paces Stress that the distance traveled is always positive.

TACTILE LEARNING Have students work in pairs to draw a number line from –10 through +10. Give each pair one place marker, which they place on 0, and a number cube. One student represents addition and the other represents subtraction. Students take turns rolling the number cube. The addition student rolls the number cube, adds the number rolled to the number where the marker is sitting. The student then moves the marker to the position determined by the sum. The subtraction student rolls the number cube, subtracts the number rolled from the number where the marker is sitting. Now this student moves the marker to the position determined by the difference. After each roll, students should record the arithmetic (for example, 4 – 5 = –1). The first student to reach +10 or –10 is the winner.

DIVERSITY Write on the board *not, un-, anti-*. Ask students to name other ways we communicate opposites in the English

Additional Examples

FOR EXAMPLE 1

A frog hops forward five times, then turns around and hops backward twice. If each hop is ten inches, how far does he end from his original position?

$(5 \cdot 10) - (2 \cdot 10) =$
$50 - 20 = 30$ in.

Discussion: *If the frog does not return along the same line, will he be closer to or farther from his original position? Explain.*

FOR EXAMPLE 2

Use tiles to find each difference.

a. $2 - 3$ –1

b. $4 - (-2)$ 6

FOR EXAMPLE 3

Evaluate $2p - x$ for $p = 8$ and $x = -2$.

$2(8) - (-2) =$
$16 + 2 = 18$

Tiles to Represent Integers

■ 1 ■ –1

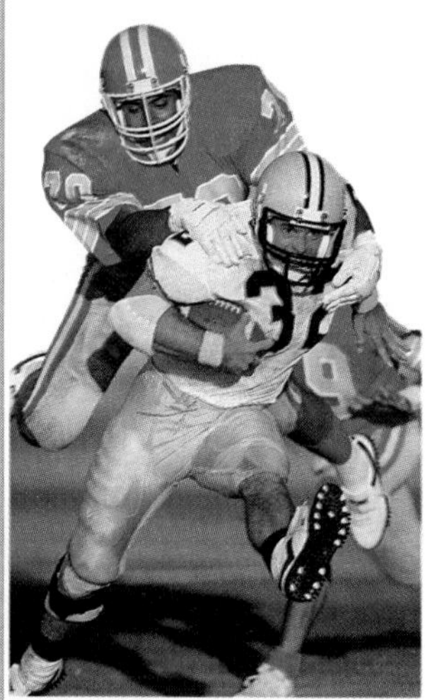

PROBLEM SOLVING

Look Back Which do you find more helpful when adding: using tiles or using a number line? Explain.

Answers may vary.

The sum of a number and its opposite, or *additive inverse*, is zero.

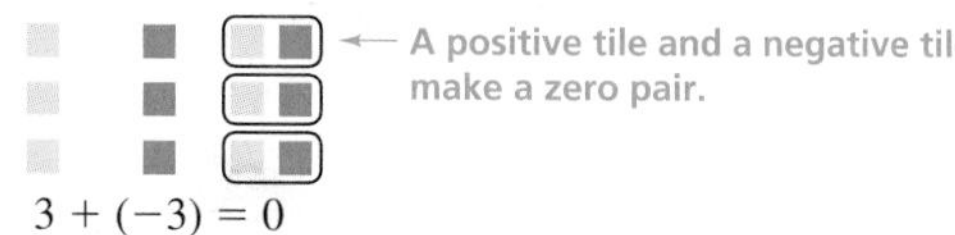
A positive tile and a negative tile make a zero pair.
3 + (−3) = 0

Example 1 **Relating to the Real World** 

Sports On two plays a football team gains 2 yd and then loses 7 yd. What is the result of the two plays? You need to find $2 + (-7)$.

Method 1: Use a number line.

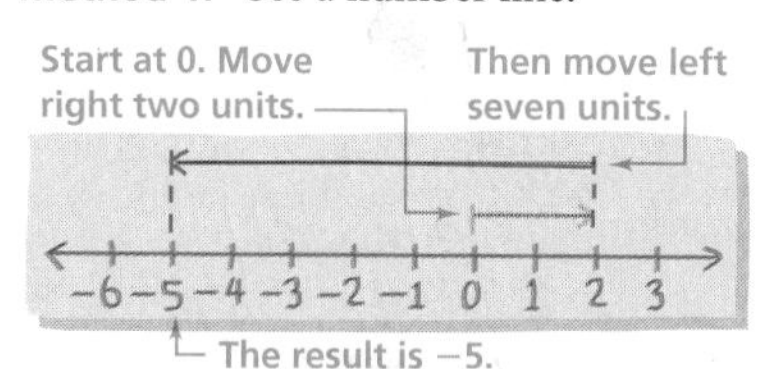
Start at 0. Move right two units.
Then move left seven units.
−6 −5 −4 −3 −2 −1 0 1 2 3
The result is −5.

$2 + (-7) = -5$

Method 2: Use tiles.

Start with 2 positive tiles and 7 negative tiles.

Make zero pairs.

There are 5 negative tiles left.

$2 + (-7) = -5$

The result of the two plays is a loss of 5 yd.

5. **Try This** Use a model to find each sum.
   a. $-6 + 4$ −2 **b.** $4 + (-6)$ −2 **c.** $-3 + (-8)$ −11 **d.** $9 + (-3)$ 6

To understand the rules for addition, you need to be familiar with absolute value. The **absolute value** of a number is its distance from zero on a number line. Using symbols, you write "the absolute value of −6" as $|-6|$.

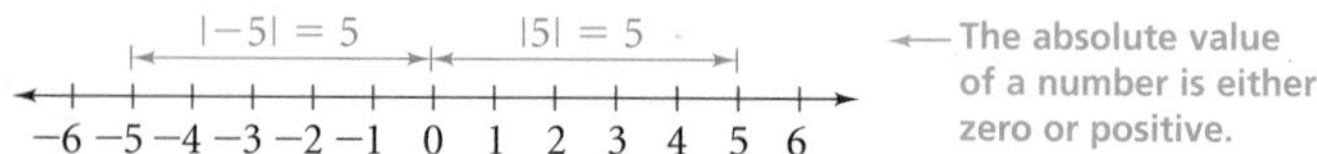
|−5| = 5
|5| = 5
−6 −5 −4 −3 −2 −1 0 1 2 3 4 5 6
The absolute value of a number is either zero or positive.

Absolute value symbols are also grouping symbols. Simplify expressions within absolute value symbols before finding the absolute value.

6. **Try This** Find the value of each expression.
   a. $|-12|$ 12 **b.** $|2.5|$ 2.5 **c.** $|-5 + 3|$ 2 **d.** $|5 + (-5)|$ 0
7. What is the sign of each sum?
   a. $6 + 4$ positive **b.** $4 + 6$ positive **c.** $-6 + (-4)$ negative **d.** $-4 + (-6)$ negative
8. Use your answers to Question 7. When adding two numbers that have the same sign, how can you **predict** the sign of their sum? The sign is the same as the numbers being added.
9. For each expression below, which number has the greater absolute value? What is the sign of each sum?
   a. $6 + (-4)$ 6; positive **b.** $-4 + 6$ 6; positive **c.** $-6 + 4$ −6; negative **d.** $4 + (-6)$ −6; negative

language. Have students who speak languages other than English explain to the class how to express opposites in their languages.

Question 6c Remind students to simplify the expression within the absolute value symbols before finding the absolute value.

Question 6d Students may not realize that the absolute value of zero is zero.

Example 2

Students may think that when they subtract, their answer should be smaller than the number they started with. Have students practice subtraction with algebra tiles or number lines until they are comfortable with subtraction producing greater numbers.

CONNECTING TO THE STUDENTS' WORLD Have each student think of an item they would like to buy that costs more than \$6 but less than \$9. Tell students they each have \$10. Have them calculate how much money they will have left after buying the item they want. Then have them calculate how much money they will need to borrow in order to buy two of these items.

ERROR ALERT! **Question 16a** Students may expect $-n$ to be a negative number. **Remediation:** Tell students that the negative sign in front of the n does not tell you in itself whether $-n$ is going to be positive or negative. As an analogy, point out that the word *not* does not necessarily mean something negative or bad. You could use *not* in a sentence and say something good, for example, "Today we will not have homework."

10. **The sign of the sum is the same as the sign of the number with the greater absolute value.**

10. Use your answers to Question 9. When adding two numbers that have different signs, how can you **predict** the sign of their sum?

Addition Rules

To add two numbers with the same sign, *add* their absolute values. The sum has the same sign as the numbers.

Example: $3 + 5 = 8$ $\quad -3 + (-5) = -8$

To add two numbers with different signs, find the *difference* between their absolute values. The sum has the same sign as the number with the greater absolute value.

Example: $-3 + 5 = 2$ $\quad 3 + (-5) = -2$

CALCULATOR HINT

To find $-12.5 + 4.8$, use this sequence:

11. Try This Find each sum.

a. $-12 + 4$ **−8** b. $26 + (-8)$ **18** c. $-12.5 + 4.8$ **−7.7** d. $14.7 + (-8.3)$ **6.4**

12. Try This Evaluate each expression for $n = 3.5$.

a. $5.2 + n$ **8.7** b. $-5.2 + n$ **−1.7** c. $-n + 5.2$ **1.7** d. $-n + (-5.2)$ **−8.7**

Part 2 Subtracting Integers and Decimals

Example 2

Find each difference.

a. $4 - 7$

← Start with 4 positive tiles.

← Add zero pairs until there are 7 positive tiles.

← Remove 7 positive tiles.

There are 3 negative tiles left.

$4 - 7 = -3$

b. $3 - (-5)$

← Start with 3 positive tiles.

← Add zero pairs until there are 5 negative tiles.

← Remove 5 negative tiles.

There are 8 positive tiles left.

$3 - (-5) = 8$

Technology Options

For Exercise 33, students may use spreadsheet software to create examples to illustrate their answer. Also for Exercises 39–54, students may use spreadsheet software to evaluate each expression.

Prentice Hall Technology

Software • Secondary Math Lab Toolkit • Integrated Math Lab 1 • Computer Item Generator 1-4

Internet • See the Prentice Hall site. (http://www.phschool.com)

Question 16a Students may be confused by the two negative signs in –(–4). On the board, write *I am not unhealthy.* Ask: *How could I simplify this?* **I am healthy.** Cross out the *not* and the *un*. Lead students to understand that the *not* and the *un* cancel each other and can be removed without changing the meaning of the sentence. Now write –(–4) on the board and cross out the two negative signs. Explain that, similarly, the two negative signs cancel each other and can be removed without changing the meaning of the expression.

Example 3

OPEN-ENDED Question 17 Encourage students to write their expressions clearly with plenty of space between characters so they do not confuse a negative sign with a minus sign, or a parenthesis with the numerals 1 or 7.

Exercises ON YOUR OWN

ALTERNATIVE ASSESSMENT Exercises 3–7 These exercises will help you assess students' understanding of adding and subtracting integers.

Exercises 13–28 If students choose to use a calculator to add or subtract these exercises, remind them to use [–] for minus and [(-)] for negative.

CRITICAL THINKING Exercise 29 Have students place one finger on *P* and another on *T*. Move them toward each other simultaneously, one unit at a time.

pages 22–24 On Your Own

38. Find the difference of absolute values of the numbers. Use the sign of the number with the greater absolute value.

page 24 Mixed Review

57a. Answers may vary. Sample:

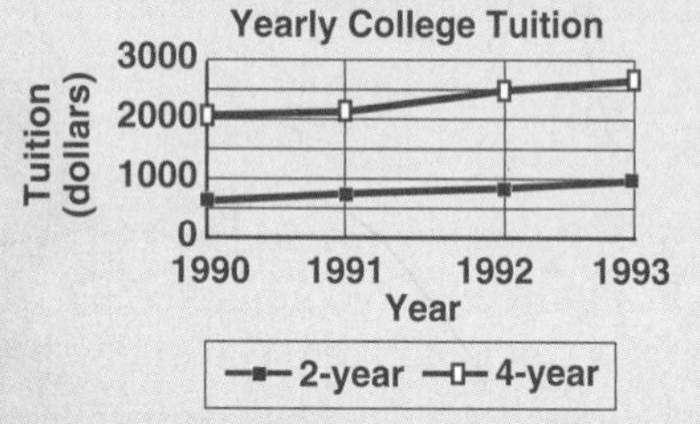

c. Answers may vary. Sample:

Yearly College Tuition Projections

Year	2-Year	4-Year
1998	$1455	$3568
1999	$1542	$3760
2000	$1639	$3952
2001	$1717	$4143
2002	$1804	$4335
2003	$1891	$4527
2004	$1979	$4718

Assume proportionate growth.

13. Use tiles to find each difference and sum.

a. $2 - 6$ **−4**
$2 + (-6)$ **−4**

b. $5 - (-9)$ **14**
$5 + 9$ **14**

c. $-3 - 8$ **−11**
$-3 + (-8)$ **−11**

d. $7 - 2$ **5**
$7 + (-2)$ **5**

14. Use your answers to Question 13. What do you notice about each pair of sums and differences? **The results are the same.**

The relationship you found between subtraction and addition is summarized in a rule for subtraction.

> **Subtraction Rule**
>
> To subtract a number, add its opposite.
>
> Example: $3 - 5 = 3 + (-5)$
> $= -2$

15. Try This Find each difference.

a. $-6 - 2$ **−8** **b.** $8 - (-4)$ **12** **c.** $7.1 - (-5.4)$ **12.5** **d.** $11.5 - 15.6$ **−4.1**

The expression $-n$ means the opposite of n. The expression $-n$ can represent a negative number, zero, or a positive number.

16. a. What is the value of $-n$ when $n = -4$? **4**

b. What is the value of $-n$ when $n = 4$? **−4**

c. For what values of n will $-n$ be positive? negative? **$-n$ will be positive when n is negative; $-n$ will be negative when n is positive.**

Example 3

Evaluate $-a - b$ for $a = -3$ and $b = -5$.

$-a - b = -(-3) - (-5)$ ← Substitute −3 for *a* and −5 for *b*.
$= 3 - (-5)$ ← The opposite of −3 is 3.
$= 3 + 5$ ← To subtract −5, add its opposite, 5.
$= 8$

17. *Open-ended* Choose negative values for a and b other than −3 and −5. Evaluate the expression in Example 3 using the values you chose. **Answers may vary. Sample: $a = -8$, $b = -1$; $-a - b = 9$**

Exercises ON YOUR OWN

1. *Critical Thinking* Which is greater, the sum of −227 and 319 or the sum of 227 and −319? Explain. **−227 + 319 is greater because the sum is positive and 227 + (−319) is negative.**

2. *Open-ended* Use a number line or tiles to show why the sum of a number and its opposite is zero. **Answers may vary. Sample:**

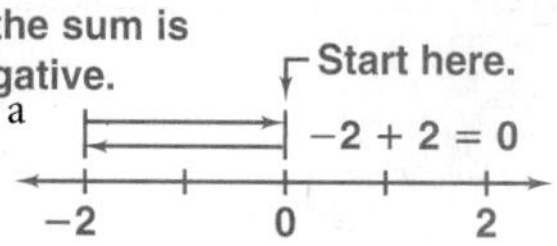

Write *positive* or *negative* to indicate the sign of the sum or difference.

3. $-6 + 13$ **positive** **4.** $6 - 13$ **negative** **5.** $-6 + (-13)$ **negative** **6.** $6 - (-13)$ **positive** **7.** $|-6 - 13|$ **positive**

CRITICAL THINKING Exercise 32 Point out that the word *opposite* has a meaning in mathematics related to its meaning in English.

CRITICAL THINKING Exercise 33 Have students test all four possible combinations (a positive, b negative; a positive, b positive; a negative, b negative; a negative, b positive) before making their conclusions.

CARTOON Have students write other silly expressions to simplify using addition and subtraction. Have students exchange their expressions with a partner and simplify, for example: (number of vegetables you can name) − (number of the house or apartment where you live) + your birth month.

MAKING CONNECTIONS Exercise 36 Point out that in geography the negative elevation does not mean that the Park Headquarters nor Death Valley are underground. Explain that elevation is based on sea level (the level of the ocean's surface). A place with a negative elevation is below the level of the ocean's surface.

WRITING Exercise 38 Have students compare their explanations with others. Help students realize that both in mathematics and in verbal language, there can be many ways of saying the same thing.

STANDARDIZED TEST TIP Exercise 55 Encourage students to determine mentally whether the answer could be negative before they actually calculate the answer. This helps them recognize the possibilities of an option that would obviously result in (e) none of these.

Evaluate each expression for $p = 4$ and $q = -3$.

8. $3 - q$ 6
9. $3p - q$ 15
10. $q - 7p$ −31
11. $|p - 9|$ 5
12. $|-12 + q|$ 15

Choose **Use a calculator, paper and pencil, or mental math to add or subtract.**

13. $-21 + (-14)$ −35
14. $13 + (-9)$ 4
15. $6 - (-5)$ 11
16. $-12 + 12$ 0
17. $-16 - 12$ −28
18. $|-16 - 12|$ 28
19. $|-16| - |12|$ 4
20. $15 - 19$ −4
21. $-18.6 - 25.3$ −43.9
22. $14.4 - 19.7$ −5.3
23. $|43.7 + (-45.2)|$ 1.5
24. $-25.7 - (-18.3)$ −7.4
25. $62.5 - 89.4$ −26.9
26. $-54.1 + 99.4$ 45.3
27. $|-28.2| + 17.5$ 45.7
28. $-65.7 - 98.9$ −164.6

Critical Thinking **Use the number line for Exercises 29 – 31.**

Q P R S T

29. If P is the opposite of T, what is the value of S? 0
30. If Q is the opposite of T, is R positive or negative? Why? Negative; R is to the left of the midpoint between Q and T.
31. If R is the opposite of T, which of the labeled points has the greatest absolute value? Why? Q; it is the farthest point from the midpoint between R and T.
32. **Critical Thinking** Explain what is wrong with the reasoning in the statement: *Since 20 is the opposite of −20, then 20°F must be very hot, because −20°F is very cold.* Both −20°F and 20°F are below freezing, so both are cold. Opposites are the numbers, not the measurements.
33. **Critical Thinking** Is $|a - b|$ always equal to $|b - a|$? Use examples to illustrate your answer. Yes; $3 - 4 = -1$, $4 - 3 = 1$; $|-1| = |1|$. So $a - b$ and $b - a$ are opposites. Absolute values of opposites are equal.
34. **Physics** Superconductors allow for very efficient passage of electric currents. For practical use, a superconductor must work above −196°C (the boiling point of nitrogen). In 1987, Paul Chu discovered a new class of materials that conduct electricity at −178°C.
    a. At how many degrees above the boiling point of nitrogen did the new materials conduct electricity? 18°C
    b. In 1990, researchers created a miniature transistor that conducts electricity at a temperature that is 48°C above the boiling point of nitrogen. What temperature is this? −148°C

Paul Ching-Wu Chu (b. 1941) is a physicist who heads a research group at the University of Houston.

35. **Critical Thinking** In the cartoon below, does the total "12 27" make sense? Explain. No, time and temperature are measures of different things. They cannot be added.

Frank & Ernest

Exercises MIXED REVIEW

OPEN-ENDED Exercise 57c Because the increases from year to year do not follow a predictable pattern, suggest that students base their predictions on the average of the three increases.

GETTING READY FOR LESSON 1-5 These problems prepare students for multiplication by reminding them of the patterns associated with multiplication.

Wrap Up

THE BIG IDEA Ask students to explain the process of subtracting integers.

RETEACHING ACTIVITY Students review addition and subtraction rules, then practice adding and subtracting integers and decimals with different signs. (Reteaching worksheet 1-4)

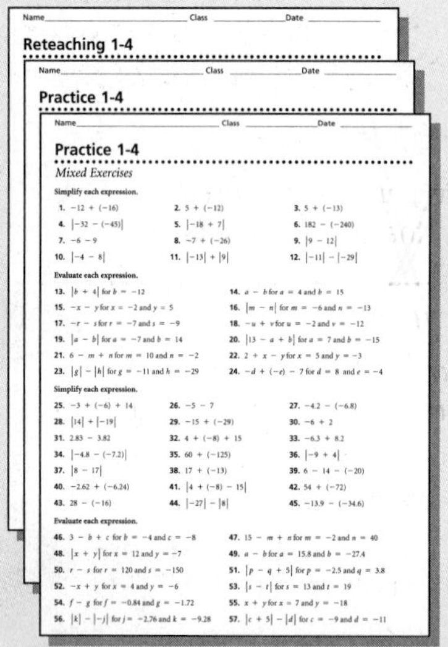
Name______ Class ______ Date ______

Reteaching 1-4

Name______ Class ______ Date ______

Practice 1-4

Name______ Class ______ Date ______

Practice 1-4

Mixed Exercises

Simplify each expression.

1. $-12 + (-16)$
2. $5 + (-12)$
3. $5 + (-13)$
4. $|-32 - (-45)|$
5. $|-18 + 7|$
6. $182 - (-240)$
7. $-6 - 9$
8. $-7 + (-26)$
9. $|9 - 12|$
10. $|-4 - 8|$
11. $|-13| + |9|$
12. $|-11| - |-29|$

Evaluate each expression.

13. $|b + 4|$ for $b = -12$
14. $a - b$ for $a = 4$ and $b = 15$
15. $-x - y$ for $x = -2$ and $y = 5$
16. $|m - n|$ for $m = -6$ and $n = -13$
17. $-r - s$ for $r = -7$ and $s = -9$
18. $-u + v$ for $u = -2$ and $v = -12$
19. $|a - b|$ for $a = -7$ and $b = 14$
20. $|13 - a + b|$ for $a = 7$ and $b = -15$
21. $6 - m + n$ for $m = 10$ and $n = -2$
22. $2 + x - y$ for $x = 5$ and $y = -3$
23. $|g| - |h|$ for $g = -11$ and $h = -29$
24. $-d + (-e) - 7$ for $d = 8$ and $e = -4$

Simplify each expression.

25. $-3 + (-6) + 14$
26. $-5 - 7$
27. $-4.2 - (-6.8)$
28. $|14| + |-19|$
29. $-15 + (-29)$
30. $-6 + 2$
31. $2.83 - 3.82$
32. $4 + (-8) + 15$
33. $-6.3 + 8.2$
34. $|-4.8 - (-7.2)|$
35. $60 + (-125)$
36. $|-9 + 4|$
37. $|8 - 17|$
38. $17 + (-13)$
39. $6 - 14 - (-20)$
40. $-2.62 + (-6.24)$
41. $|4 + (-8) - 15|$
42. $54 + (-72)$
43. $28 - (-16)$
44. $|-27| - |8|$
45. $-13.9 - (-34.6)$

Evaluate each expression.

46. $3 - b + c$ for $b = -4$ and $c = -8$
47. $15 - m + n$ for $m = -2$ and $n = 40$
48. $|x + y|$ for $x = 12$ and $y = -7$
49. $a - b$ for $a = 15.8$ and $b = -27.4$
50. $r - s$ for $r = 120$ and $s = -150$
51. $|p - q + 5|$ for $p = -2.5$ and $q = 3.8$
52. $-x + y$ for $x = 4$ and $y = -6$
53. $|s - t|$ for $s = 13$ and $t = 19$
54. $f - g$ for $f = -0.84$ and $g = -1.72$
55. $x + y$ for $x = 7$ and $y = -18$
56. $|k| - |-j|$ for $j = -2.76$ and $k = -9.28$
57. $|c + 5| - |d|$ for $c = -9$ and $d = -11$

Lesson Quiz

Lesson Quiz is also available in Transparencies.

In Exercises 1–4, evaluate each expression.

1. $12x - 51$ for $x = -4$ −99
2. $a - b + 3c$ for $a = 15$ $b = 12$ and $c = -5$ −12
3. $-p - 2r$ for $p = -8$ and $r = 3$ 2
4. $25 - 4x$ for $x = -5$ 45

36. *Geography* Suppose you travel from Park Headquarters to the floor of Death Valley.
    a. What is your change in elevation? −92 ft
    b. Why does it make sense to write your change in elevation with a negative number? It is a decline of 92 ft.

Telescope Peak to Death Valley

Site	Elevation
Telescope Peak	11,049 ft
Park Headquarters	−190 ft
Floor of Death Valley	−282 ft

37. Find the change in elevation from Telescope Peak to Death Valley. −11,331 ft

38. *Writing* Suppose a friend missed class today. Write an explanation for adding two numbers with different signs. See margin.

Evaluate each expression for $a = -2$, $b = 3$, and $c = -4$.

39. $6 - a$ 8
40. $a - b + c$ −9
41. $c - b + a$ −9
42. $|-a + 2|$ 4
43. $-|a|$ −2
44. $|a| + |b|$ 5
45. $|a + b|$ 1
46. $-|3 + a|$ −1
47. $a - b$ −5
48. $b - a$ 5
49. $4b - a$ 14
50. $4b - |a|$ 10
51. $a + 3b$ 7
52. $|a| + |3b|$ 11
53. $c + a + 5$ −1
54. $|c + a + 5|$ 1

55. *Standardized Test Prep* Evaluate $(a - b) - (c - d)$ for $a = -15$, $b = -8$, $c = 6$, and $d = 9$. B
    A. −20 B. −4 C. −10 D. −26 E. none of these

56. *Chemistry* A charged particle of magnesium has 12 protons and 10 electrons. Each proton has a charge of $^{+}1$ and each electron a charge of $^{-}1$. What is the total charge of the particle? +2

Exercises MIXED REVIEW

57. a. *Statistics* Draw a graph of the data in the table to show change in tuition from 1990 to 1993. See margin.
    b. Why did you choose the graph you drew? See below.
    c. *Open-ended* **Predict** what the tuition will be for a 2-year public college and a 4-year public college the year you graduate from high school. Explain. See margin.

57b. A line graph shows change over time.

Yearly College Tuition 1990–1993

Year	Public 2-Year Colleges	Public 4-Year Colleges
1990	$ 756	$2035
1991	$ 824	$2159
1992	$ 937	$2410
1993	$1018	$2610

Evaluate each expression.

58. $5t + 6$ for $t = 2$ 16
59. $18 - 7m$ for $m = 2$ 4

Simplify.

60. $\frac{3 + 5}{6 + 10}$ $\frac{1}{2}$
61. $\frac{7 - 2}{12 + 8}$ $\frac{1}{4}$
62. $\frac{1 + 4}{21 - 6}$ $\frac{1}{3}$
63. $\frac{8 - 5}{16 - 7}$ $\frac{1}{3}$

Getting Ready for Lesson 1-5

Patterns **Write the next three numbers in each pattern.**

64. −5, −1, 3, ■, ■, ■ 7, 11, 15
65. 9, 6, 3, 0, ■, ■, ■ −3, −6, −9
66. −15, −10, −5, ■, ■, ■ 0, 5, 10
67. 6, 4, 2, ■, ■, ■ 0, −2, −4

1-5 Teaching Notes

PROBLEM OF THE DAY

How many hair ribbons $7\frac{1}{2}$ in. long can be cut from a piece of ribbon 5 ft long? 8 ribbons

Problem of the Day is also available in Transparencies.

CONNECTING TO PRIOR KNOWLEDGE Ask students: *If you go to the store to buy two boxes of cereal that cost $3 each, and you have two 75-cent coupons, how do you calculate what the cereal will cost? What if you were buying 20 boxes?* Students' strategies may differ. Lead a class discussion on the strategies students could use.

WORK TOGETHER

Some students find the idea that the product of two negative numbers is positive to be extremely counter-intuitive. Students can be convinced that this idea makes sense by writing patterns such as those in Question 6. Ask students to explain each part of the pattern. Encourage students to create their own similar patterns.

For students who are having trouble remembering that two negatives make a positive, remind them of the example, *I am not unhealthy,* that was covered in Lesson 1-4 which showed how the *not* and the *un-* cancel each other.

Connections Space Flight . . . and more

1-5 Multiplying and Dividing Integers

What You'll Learn

- Multiplying and dividing with integers and decimals
- Simplifying expressions with exponents
- Evaluating expressions

...And Why

To compute with real-world data, like temperatures, that are negative as well as positive

WORK TOGETHER

Work with a partner.

Space Flight When the space shuttle is in the first stage of its return to Earth, it descends about 3.5 mi/min. You can write the rate as -3.5 mi/min.

To find how far the shuttle descends in 10 min, you need to find $(-3.5)(10)$. You know that $(3.5)(10) = 35$.

1. Did the shuttle's altitude *increase* or *decrease* during the 10 min? decrease
2. Which phrase better describes descending 35 mi? Why?
   A. a change in altitude of 35 mi **B.** a change in altitude of -35 mi
   B; because the shuttle is descending
3. Is the product $(-3.5)(10)$ equal to 35 or -35? Why?
   -35; the shuttle is descending, and the negative sign shows this.
4. Tell which result makes more sense and explain why. See below left.

	Situation	Expression	Result
a.	in football, the total of 2 penalties of 5 yd each	$2(-5)$	10 or -10?
b.	the average hourly decrease in temperature when it falls 20°F in a 4-h period	$\frac{-20}{4}$	5 or -5?
c.	the monthly income of a company with losses of $48,000 for one year	$\frac{-48{,}000}{12}$	4000 or -4000?

4a. -10; because it is a loss
4b. -5; because the temperature is falling
4c. -4000; because the company is losing money

5. a. **Generalize:** Is the product of a positive and a negative number *positive* or *negative*? negative
   b. **Generalize:** Is the quotient of a positive and a negative number *positive* or *negative*? negative
6. **Patterns** Use your knowledge of patterns to complete each column.

a.		b.	
$3 \cdot (-4) = \blacksquare$	-12	$15 \div (-5) = \blacksquare$	-3
$2 \cdot (-4) = \blacksquare$	-8	$10 \div (-5) = \blacksquare$	-2
$1 \cdot (-4) = \blacksquare$	-4	$5 \div (-5) = \blacksquare$	-1
$0 \cdot (-4) = \blacksquare$	0	$0 \div (-5) = \blacksquare$	0
$-1 \cdot (-4) = \blacksquare$	4	$-5 \div (-5) = \blacksquare$	1
$-2 \cdot (-4) = \blacksquare$	8	$-10 \div (-5) = \blacksquare$	2
$-3 \cdot (-4) = \blacksquare$	12	$-15 \div (-5) = \blacksquare$	3

7. Use the pattern you developed in Question 6. Is the product or quotient of two negative numbers positive or negative? positive
8. Compare your results with another pair of students. If you do not have the same results, explain your reasoning to each other. Check students' work.

Lesson Planning Options

Prerequisite Skills

- Finding number patterns
- Using the order of operations

Assignment Options

To provide flexible scheduling, this lesson can be subdivided into parts.

1 **Core** 1–8, 11, 13–15, 46
Extension 17, 18, 32–39

2 **Core** 4, 9, 10, 12, 16, 19–31
Extension 40–45, 47, 48

Use Mixed Review to maintain skills.

Resources

 **Student Edition**

Skills Handbook, pp. 569, 580
Extra Practice, p. 556
Glossary/Study Guide

Teaching Resources

Chapter Support File, Ch. 1
- Practice 1-5 (two worksheets)
- Reteaching 1-5

Classroom Manager 1-5
Glossary, Spanish Resources
Two-Year Algebra Handbook 1-5

 **Transparencies**

25, 34

THINK AND DISCUSS

Ask students to think of mnemonics (memory aids) they might use to remember the multiplication and division rules. Share these ideas with the class.

AUDITORY LEARNING Have students read aloud the multiplication and division rules. Ask students to state orally the multiplication and division rule used for each question and example.

ESL Have students write these rules in a short form that is meaningful for them, for example: $(+) \cdot (+) = (+)$ and $(-) \cdot (-) = (+)$.

Example 1

Remind students that $2yz$ is a short way of writing $2 \cdot y \cdot z$.

ERROR ALERT! Students may assume that the sign rules for products and quotients are different. **Remediation:** Remind students that products and quotients use the same rules for signs. Reinforce this by having students change the multiplication sentence $-4 \cdot 3 = -12$ into its related division sentences $\frac{-12}{-4} = 3$ and $\frac{-12}{3} = -4$.

Example 2 Relating to the Real World

Question 11a Students may need help remembering that a falling temperature is moving to the left on the number line. Encourage students to draw a number line from –20 to +6 to

Additional Examples

FOR EXAMPLE 1

Evaluate $\frac{x}{y} - 4x^2$ for $x = 4$, $y = 2$.

$$\frac{4}{2} - 4\left(4^2\right) = 2 - 4(16)$$
$$= 2 - 64$$
$$= -62$$

Discussion: *Which rules about the order of operations were used in this example?*

FOR EXAMPLE 2

Find the mean elevation of these locales: Death Valley –282 ft, Shoshone Peak 2151 ft, and Slocum Mountain 1562 ft.

$$\frac{-282 + 2151 + 1562}{3} = \frac{3431}{3} \approx 1143.67 \text{ ft}$$

Discussion: *How can elevation be negative?*

FOR EXAMPLE 3

Evaluate $2x^3$ for $x = -2$.

$$2(-2^3) = 2(-8)$$
$$= -16$$

Discussion: *Explain why $(-2)^3$ is –8 and not 8.*

THINK AND DISCUSS

Part 1

Multiplying and Dividing

The multiplication and division rules summarize what you discovered in the Work Together activity.

MULTIPLICATION AND DIVISION RULES

The product or quotient of two positive numbers is positive. The product or quotient of two negative numbers is positive.

Examples: $3 \cdot 5 = 15$ $\quad -3 \cdot (-5) = 15$

$\frac{15}{5} = 3$ $\quad \frac{-15}{-5} = 3$

The product or quotient of a positive number and a negative number is negative.

Examples: $-3 \cdot 5 = -15$ $\quad 3 \cdot (-5) = -15$

$\frac{-15}{5} = -3$ $\quad \frac{15}{-5} = -3$

QUICK REVIEW

You can indicate division using a fraction bar. $\frac{15}{5}$ means $15 \div 5$. Remember that division by 0 is undefined.

Example 1

Evaluate $\frac{-x}{-4} + 2yz$ for $x = -20$, $y = 6$, and $z = -1$.

$$\frac{-x}{-4} + 2yz = \frac{-(-20)}{-4} + 2(6)(-1) \quad \leftarrow \text{Substitute –20 for } x\text{, 6 for } y\text{, and –1 for } z.$$
$$= \frac{20}{-4} + 2(6)(-1) \quad \leftarrow \text{The opposite of –20 is 20.}$$
$$= -5 + (-12) \quad \leftarrow \text{Divide and multiply.}$$
$$= -17 \quad \leftarrow \text{Add.}$$

9. **Try This** Evaluate the expression in Example 1 for $x = 8$, $y = -7$, and $z = -3$.
44

When you divide, sometimes you need to round your result. The number line shows that -5.8 is closer to -6 than to -5. So, you round -5.8 to -6.

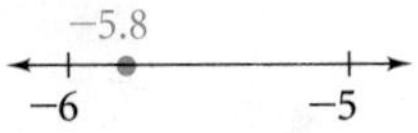

Example 2 Relating to the Real World

Math in the Media Find the mean high temperature. Round to the nearest degree.

$$\text{mean high temperature} = \frac{-17 + (-18) + (-9) + (-7) + (-6) + 4 + 1}{7}$$
$$= \frac{-52}{7}$$
$$\approx -7.428571429$$

The mean high temperature is about –7°F.

The Weekly Herald

Fairbanks, Alaska One Week in January

	High	Low
Sunday	–17°	–38°
Monday	–18°	–38°
Tuesday	–9°	–32°
Wednesday	–7°	–33°
Thursday	–6°	–31°
Friday	4°	–33°
Saturday	1°	–34°

Temperatures in Fahrenheit

represent the temperature. Have them write the time next to the corresponding temperatures.

Example 3

Tell students to think of the parentheses in part (b) as a pair of hands holding the – and the 3 together. They cannot be separated so the exponent 4 must apply to both of them. In part (a), there are no parentheses so they can be separated. The exponent applies only to the 3 in this situation.

Question 14 Have students look for a pattern that relates the exponents and the sign of the answer. For a negative number, odd exponents result in a negative; even exponents, in a positive.

Exercises ON YOUR OWN

ALTERNATIVE ASSESSMENT **Exercises 1–16** Assign Exercise 31 before 1 through 16. Then have students, taking turns, state the appropriate rule such as "positive times negative is negative" for each exercise. By concentrating only on the signs, students will exhibit their understanding of the multiplication and division rules.

Exercise 10 Have students first predict the sign of the result and then calculate.

Exercise 17 Remind students that *a* and *b* can represent negative numbers even though there is no negative sign.

10. Try This Find the mean low temperature to the nearest degree. $-34°F$
11. a. Suppose the temperature at 6 P.M. is $-8°$. Write an equation to find the temperature if the temperature falls 2°/h. Let n represent the number of hours since 6 P.M. t = temperature; $t = -8 - 2n$
    b. Find the temperature at 10 P.M. $-16°F$
12. Try This **Predict** whether the product of each expression is *positive* or *negative*. Then multiply.
    a. $(-1)(-2)$ positive; 2
    b. $(-1)(-2)(-3)$ negative; -6
    c. $(-1)(-2)(-3)(-4)$ positive; 24
    d. $(-1)(-2)(-3)(-4)(-5)$ negative; -120
13. Use your results in Question 12 to complete each statement.
    a. For an even number of negative factors, the product will be __?__. positive
    b. For an odd number of negative factors, the product will be __?__. negative
    c. For a product that includes negative and positive factors, do the positive factors affect the sign of the product? no

Part 2

Simplifying Expressions with Exponents

A term like -3^4 means the opposite of 3^4. Keep this in mind as you simplify expressions that have exponents and negatives.

Example 3

Use the order of operations to simplify each expression.

a. -3^4

$-3^4 = -(3 \cdot 3 \cdot 3 \cdot 3)$
$= -81$

b. $(-3)^4$

$(-3)^4 = (-3)(-3)(-3)(-3)$
$= 81$

QUICK REVIEW

1. Perform operations inside grouping symbols.
2. Simplify any terms with exponents.
3. Multiply and divide in order from left to right.
4. Add and subtract in order from left to right.

14. Evaluate each expression for $b = -2$.
    Sample: $b^2 = (-2)^2$
    $= 4$
    a. b^3 -8
    b. b^4 16
    c. b^5 -32
    d. b^6 64
15. Critical Thinking How can you use the exponent to predict whether an expression like $(-2)^{76}$ is positive or negative? If the exponent is even, the value is positive; if odd, the value is negative.

Exercises ON YOUR OWN

Simplify.

1. $10(-12)$ -120
2. $(-8)(-5)$ 40
3. -9^2 -81
4. $(-9)^2$ 81
5. $18 \div (-3)$ -6
6. $-18 \div (-3)$ 6
7. $\frac{-36}{9}$ -4
8. $(-20)(-5)$ 100
9. $(-1)^{23}$ -1
10. $-(-2)^3$ 8
11. $(-2)(5)(-3)$ 30
12. $3(-4)^3$ -192
13. $|4 + 8(-6)|$ 44
14. $13 - 3(6)$ -5
15. $\frac{(3 - 14)}{-2}$ $5\frac{1}{2}$
16. $(-5)(-1)^4$ -5

Technology Options

For Exercises 19–30, students may use spreadsheet software to evaluate each expression. Also for Exercise 32, students may use spreadsheet software to analyze the exercise.

Prentice Hall Technology

Calculator • Graphing Calculator Handbook: Procedure 1

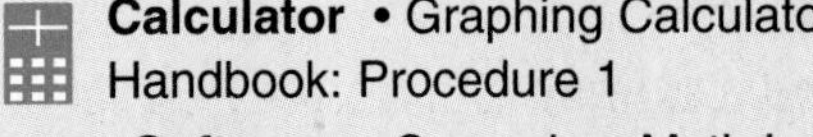

Software • Secondary Math Lab Toolkit • Computer Item List Generator 1-5

Internet • See the Prentice Hall site. (http://www.phschool.com)

Exercise 18 Suggest that students write out what happens, second by second, until they see a pattern emerging. The pattern begins at $t = 0$ when the parachute opens. After 1 s, $t = 1$ and the altitude is (5000 − 25).

ERROR ALERT! **Exercise 30** Students may begin by saying $n - n = 0$. **Remediation:** Remind students that, by the order of operations rules, the division is done first.

CRITICAL THINKING Exercise 32 Have students try several examples, using different values for *a* and *b*, until they feel confident enough to generalize.

OPEN-ENDED Exercises 41–45 Remind students that the method they choose may not be identical to that of their classmates because there are several ways to get the correct answer.

CRITICAL THINKING Exercise 48b Point out that the roller coaster has reached the bottom when $h = 0$. If students get a value of *h* which is less than 0, they have used a value of *t* that is too large.

Chapter Project **FIND OUT BY CALCULATING** Students are asked to calculate their own height using the given formulas. They are to make a supposition about an archaeological find. Students will choose one radius bone measurement to calculate height. This information is essential to their work on the Chapter Project. Have students add this task to work they have already completed for the project. You may wish to check students' progress on the project by having each group write a sentence or two describing what they have done for the project thus far.

17. Suppose *a* and *b* are integers.
a. When is the product ab positive? The product is positive when *a* and *b* have the same sign.
b. When is the product ab negative? The product is negative when *a* and *b* have different signs.

18. a. A parachutist opens her parachute at an altitude of 5000 feet. Her change in altitude is −25 ft/s. Write an equation to find her altitude at time *t* after she opens her parachute. a = altitude; $a = 5000 - 25t$
b. How far has she descended in 12.5 s? 312.5 ft
c. What is her altitude 12.5 s after she opens her parachute? 4687.5 ft

Evaluate each expression for $m = -4$, $n = 3$, and $p = -1$.

19. mn −12 **20.** mnp 12 **21.** $3m - n$ −15 **22.** $-5p^2$ −5

23. $2m$ −8 **24.** $7p - 2n$ −13 **25.** m^3 −64 **26.** $\frac{m}{p} - n$ 1

27. $\frac{m}{n}$ $-1\frac{1}{3}$ **28.** $8p \div (-2n)$ $1\frac{1}{3}$ **29.** $4n^3 \div m$ −27 **30.** $m \div n + (-n)$ $-4\frac{1}{3}$

31. *Writing* In your own words, write the rules for adding, subtracting, multiplying, and dividing a positive number and a negative number. Write an example for each rule. Check students' work. Make sure they understand the rules. Look for references to absolute value and for examples.

32. *Critical Thinking* Does $|ab|$ always equal $|a| \cdot |b|$? Explain. Yes; each product will be positive or zero. You can find absolute value before or after multiplying.

Is the value of each expression *positive* or *negative*? Explain.

33. $(-3)^4$ Pos.; even no. of neg. factors **34.** $(-4)^3$ Neg.; odd no. of neg. factors **35.** $(-3)^{103}$ Neg.; odd no. of neg. factors **36.** $(10)^5$ Pos.; the base is positive.

37. $-(-3)^2$ Neg.; $(-3)^2$ is pos. Opp of pos. num. is neg. **38.** $(-7)^2$ Pos.; even no. of neg. factors **39.** -5^{10} Neg.; 5^{10} is pos. Its opp. is neg. **40.** -5^{11} Neg.; 5^{11} is pos. Its opp. is neg.

Open-ended **Use $a = -3$, $b = 2$, and $c = -5$ to write an expression that has each value.** Answers may vary. Samples given.

41. 17 $ac + b$ **42.** 0 $b - a + c$ **43.** −1 $ab - c$ **44.** 1 $c - ab$ **45.** 7 $-bc + a$

46. *Oceanography* To map the features of the ocean floor, scientists take several sonar readings. Find the mean of these readings: −14,235 ft; −14,246 ft; and −14,230 ft. −14,237 ft

47. Complete each statement:
a. $(-7)^n < 0$ if n is _?_. odd
b. $(-7)^n > 0$ if n is _?_. even

48. *Entertainment* As riders plunge down the hill of the Mean Streak, you can find the approximate height, in feet, above the ground of their roller-coaster car. Use the formula $h = 155 - 16t^2$ where *t* is the number of seconds since the start of the descent.
a. How far is a rider from the bottom of the hill after 1 s? 2 s? 139 ft; 91 ft
b. *Critical Thinking* Does it take more than or less than 4 s to reach the bottom? Explain. See below.
c. Use your calculator and the problem solving strategy *Guess and Test* to find the time the roller coaster descends to the nearest tenth of a second. 3.1 s

48b. Less; for $t = 4$, $h = -101$. This would mean that the roller coaster is 101 ft below ground, which is not reasonable.

The Mean Streak, in Sandusky, Ohio, is one of the tallest and fastest wooden roller coasters in the world.

Exercises MIXED REVIEW

GETTING READY FOR LESSON 1-6 These problems give students practice in working with fractions to prepare them for manipulating rational numbers.

Wrap Up

THE BIG IDEA Ask students to tell in their own words the rules for multiplying and dividing integers. Then have students explain how to evaluate expressions by applying the order of operations.

RETEACHING ACTIVITY Students review multiplication and division rules, then practice multiplying and dividing integers and decimals with different signs. (Reteaching worksheet 1-5)

Exercises CHECKPOINT

In this Checkpoint, students will have an opportunity to assess their own progress on the topics covered in Lessons 1-1 through 1-5.

WRITING Exercise 10 Remind students that the exponent applies only to the 5 and not to the negative sign.

Chapter Project — *Find Out by Calculating*

Scientists use these formulas to approximate a person's height H, in inches, when they know the length of the tibia t, the humerus h, or the radius r.

- Use your tibia, humerus, and radius bone lengths to calculate your height. Are the calculated heights close to your actual height? Explain.
- An archaeologist found an 18-in. tibia on the site of an American colonial farm. Do you think it belonged to a man or a woman? Why?
- Choose one radius measurement from the data you collected for the Find Out question on page 9. Calculate the person's height. Can you tell whose height you have found? Explain.

Male
$H = 32.2 + 2.4t$
$H = 29.0 + 3.0h$
$H = 31.7 + 3.7r$
Female
$H = 28.6 + 2.5t$
$H = 25.6 + 3.1h$
$H = 28.9 + 3.9r$

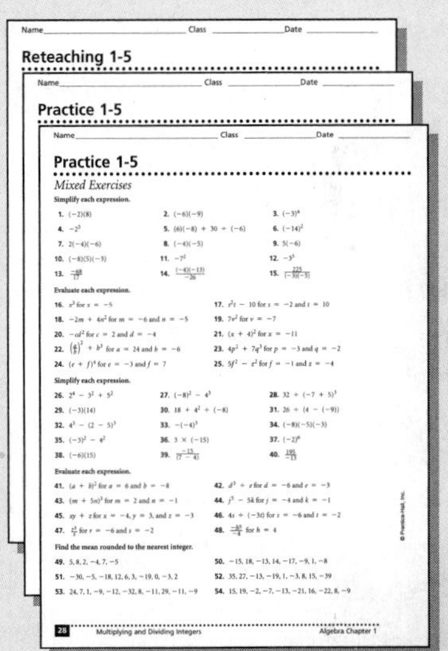
Reteaching 1-5

Practice 1-5

Practice 1-5

Mixed Exercises

Exercises MIXED REVIEW

Statistics **Use the line plot at the right.**

49. How many people watched more than three hours of television? 5 people

50. What other kind of graph could you use for the data? Explain. See below.

51. a. Find the mean, median, and mode of the data. 2.8; 2.5; 1

b. Is the *mean*, *median*, or *mode* the most useful to describe people's viewing habits? Why? Answers may vary. Sample: The median is the most useful because it is not affected by extreme values.

Hours Spent Watching TV Each Day

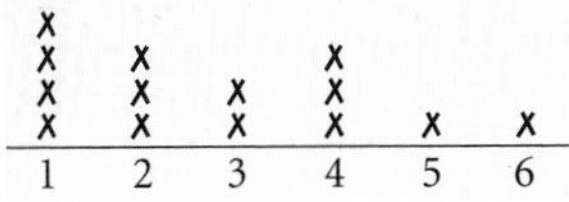

Getting Ready for Lesson 1-6

Find each sum or difference.

52. $\frac{3}{8} + \frac{1}{4}$ $\frac{5}{8}$ **53.** $\frac{7}{12} - \frac{1}{3}$ $\frac{1}{4}$ **54.** $\frac{5}{6} + \frac{2}{3}$ $1\frac{1}{2}$ **55.** $\frac{1}{2} - \frac{1}{6}$ $\frac{1}{3}$ **56.** $\frac{7}{8} - \frac{3}{5}$ $\frac{11}{40}$

50. Answers may vary. Sample: You could use a histogram since it also shows frequency.

Exercises CHECKPOINT

Simplify.

1. $8 + 4 \cdot 3$ 20 **2.** $(-3)^2 + (-5)$ 4 **3.** $\frac{7+5}{7-5}$ 6 **4.** $(9^2 - 60) \div 3$ 7

Evaluate each expression for $a = -7$ and $b = 4$.

5. $-a$ 7 **6.** a^2 49 **7.** $a + b$ −3 **8.** $|a - b|$ 11 **9.** $2b^2 + 3a$ 11

10. *Writing* Is -5^2 positive or negative? Explain. negative; because -5^2 is the opposite of 5^2

11. Find the mean, median, and mode. Round to the nearest tenth.
−12, 15, 8, −4, −6, 7, −4, 8, −5, −12, 0, −2, 5, −4 −0.4; −3; −4

12. Write an equation to describe the relationship shown at the right. $t = 175b$

boxes	tissues
1	175
2	350
3	525

Lesson Quiz

Lesson Quiz is also available in Transparencies.

1. Simplify $\frac{2^3}{-2}$. −4
2. Evaluate $\frac{5p - x}{3}$ for $p = 4$, $x = -1$. 7
3. Simplify $-4^2 + 2^3 + 5$. −3
4. Evaluate $n^3 + m$ for $n = -3$, $m = 8$. −19

1-6 Teaching Notes

PROBLEM OF THE DAY

Maria has eight pennies, seven nickels, seven dimes, and seven quarters. How can she divide this amount into three groups of equal value? **Answers may vary but must contain $.96 in each group.**

Problem of the Day is also available in Transparencies.

CONNECTING TO PRIOR KNOWLEDGE Ask the class to name as many categories of numbers, given an example of each, as they can. Ask questions to help students see how the categories they name are related.

THINK AND DISCUSS

ESL Make sure students do not confuse the mathematical meanings of *rational* and *irrational* with their non-mathematical meanings of "logical" and "illogical."

Question 4 Have a volunteer show the division on the board to convert $\frac{2}{11}$ to a decimal. Stop after six decimal places. Lead students to understand that the pattern shows that they could continue dividing forever, repeating the same set of digits. Now have students enter $\frac{2}{11}$ on their calculators. Lead students to understand that, since computers and calculators round answers, they can neither display an irrational number with complete accuracy, nor indicate whether the number is rational or irrational.

Lesson Planning Options

Prerequisite Skills

- Adding and subtracting fractions
- Using the order of operations

Assignment Options

To provide flexible scheduling, this lesson can be subdivided into parts.

 **Core** 1–12, 25–28
Extension 29, 30, 35

 **Core** 13–24, 31–34
Extension 36–45

Use Mixed Review to maintain skills.

Resources

 **Student Edition**

Skills Handbook, pp. 577, 579
Extra Practice, p. 556
Glossary/Study Guide

 **Teaching Resources**

Chapter Support File, Ch. 1
- Practice 1-6 (two worksheets)
- Reteaching 1-6

Classroom Manager 1-6
Glossary, Spanish Resources
Two-Year Algebra Handbook 1-6

 **Transparencies**
8, 11, 22, 34

Connections 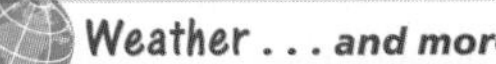 Weather . . . and more

1-6 Real Numbers and Rational Numbers

What You'll Learn
- Comparing and ordering rational numbers
- Evaluating expressions with rational numbers

...And Why

To solve real-world problems, such as converting temperatures

What You'll Need
- calculator

THINK AND DISCUSS

Part 1

Comparing Rational Numbers

The Venn diagram shows the relationship of sets of numbers.

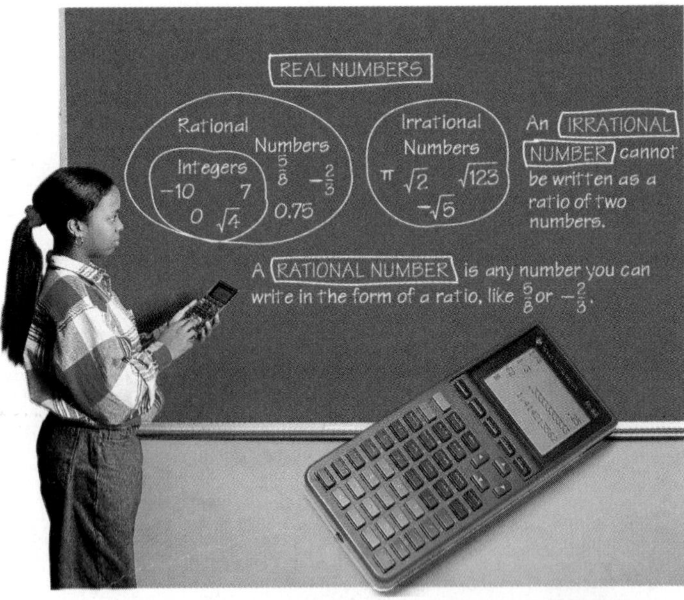

Integers are rational numbers because you can write them as ratios using 1 as the denominator: $7 = \frac{7}{1}$. Rational and irrational numbers make up the set of **real numbers.** Sample: $\frac{3}{4}$, 0.5, $-1\frac{5}{8}$

1. Write three numbers that are rational numbers but not integers. Choose numbers different from the ones in the diagram above.
2. Show that 0.75 is a rational number by writing it as a ratio. $0.75 = \frac{75}{100}$
3. *Open-ended* Where have you used irrational numbers? See left.
4. *Calculator* Use a calculator to find the square root of the numbers 1 through 10. Which do you think are rational? irrational? See left.

An irrational number expressed in decimal form is a nonterminating and nonrepeating decimal. If you use a calculator to express an irrational number as a decimal, the calculator gives you a decimal approximation.

QUICK REVIEW

$\frac{1}{4} = 0.25$, which is a terminating decimal.

$\frac{1}{3} = 0.\overline{3}$, which is a repeating decimal.

3. Answers may vary. Sample: π is used when finding circumference or area of a circle.
4. The square roots of 1, 4, and 9 are rational. The square roots of 2, 3, 5, 6, 7, 8, and 10 are irrational.

VISUAL LEARNING Have students write $a < b$ and $b > a$ in big letters on a sheet of paper. Explain that because the *greater than* and *less than* signs are so easily confused, students should design a visual reminder that will help them remember whether a or b is greater. Encourage students to put their visual aid on each page of their math work.

Example 1

Point out that because negative numbers are the opposites of positive numbers, students should expect comparisons of negative numbers to be the opposite of comparisons of positive numbers. For example, $\frac{1}{8} < \frac{1}{2}$ but $-\frac{1}{8} > -\frac{1}{2}$.

Example 2

Make sure students are aware of the correct calculator key to use for negative.

Stress that the bar over the 6 is a short way of writing a non-terminating decimal. Have students use their calculators to convert $\frac{3}{11}$ to a decimal, then tell you the result. Write $0.27272727\ldots = 0.\overline{27}$ on the board.

Example 3

Point out that you cannot subtract fractions until you have written both fractions with a common denominator.

When you compare two real numbers, only one of these can be true:

$a < b$	or	$a = b$	or	$a > b$
less than		**equal to**		**greater than**

Example 1

Use a number line to compare $-\frac{1}{8}$ and $-\frac{1}{2}$.

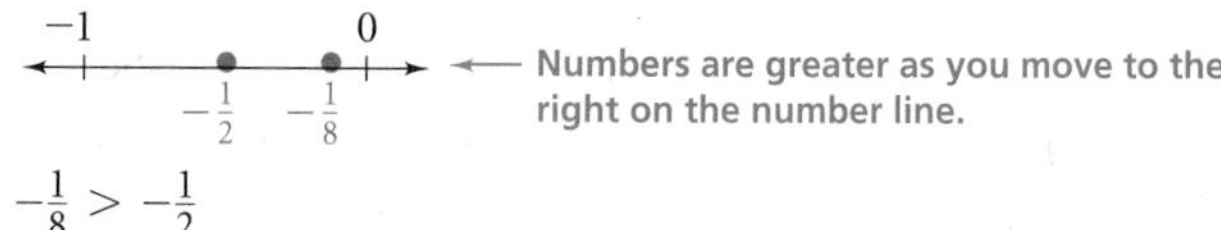

−1
0
$-\frac{1}{2}$
$-\frac{1}{8}$
Numbers are greater as you move to the right on the number line.

$-\frac{1}{8} > -\frac{1}{2}$

$-\frac{1}{2} < -\frac{1}{8}$ **5.** Rewrite the answer to Example 1 using the symbol for *less than*.

6. **Critical Thinking** Suppose you are comparing $-25\frac{1}{2}$ and $-25\frac{1}{8}$. Do you need a different number line from the one in Example 1 to determine which mixed number is greater? Explain. **No; since the integer part is the same for both numbers, you can compare the fractional parts.**

Many ratios are not easy to compare using a number line. Another method is to write the ratios as decimals and then compare.

Example 2

Write $-\frac{3}{8}$, $-\frac{1}{2}$, and $-\frac{5}{12}$ in order from least to greatest.

Use a calculator to write the rational numbers as decimals.

$-\frac{3}{8}$	(-) 3 ÷ 8 ENTER	−.375
$-\frac{1}{2}$	(-) 1 ÷ 2 ENTER	−.5
$-\frac{5}{12}$	(-) 5 ÷ 12 ENTER	−.4166666667

$-0.5 < -0.41\overline{6} < -0.375$

From least to greatest the ratios are $-\frac{1}{2}$, $-\frac{5}{12}$, and $-\frac{3}{8}$.

7. **Try This** Write $\frac{1}{12}$, $-\frac{2}{3}$, and $-\frac{5}{8}$ in order from least to greatest. $-\frac{2}{3}, -\frac{5}{8}, \frac{1}{12}$

Part 2

Evaluating Expressions

You evaluate expressions using rational numbers by substituting and performing the indicated operations.

Example 3

Evaluate $a + 2b$ where $a = \frac{2}{3}$ and $b = -\frac{5}{8}$.

$a + 2b = \frac{2}{3} + 2\left(-\frac{5}{8}\right)$ ← Substitute the values for a and b.

$= \frac{2}{3} - \frac{5}{4}$

$= -\frac{7}{12}$

CALCULATOR HINT

You can use this key sequence:

(2 ÷ 3) − (5 ÷ 4) ENTER

Additional Examples

FOR EXAMPLE 3

Evaluate $\frac{x}{y} - 2a$ for $a = \frac{3}{8}$, $x = \frac{5}{8}$, $y = \frac{2}{3}$.

$\left(\frac{5}{8} \div \frac{2}{3}\right) - 2\left(\frac{3}{8}\right) = \left(\frac{5}{8} \cdot \frac{3}{2}\right) - 2\left(\frac{3}{8}\right)$

$= \frac{15}{16} - \frac{3}{4}$

$= \frac{15 - 12}{16}$

$= \frac{3}{16}$

Discussion: *Why were you able to write $\frac{x}{y}$ as $\left(\frac{5}{8} \cdot \frac{3}{2}\right)$?*

FOR EXAMPLE 4

To change inches to centimeters, use the number of inches multiplied by 2.54, or $\frac{5}{2}$. About how many centimeters is 4 in.?

4 × 2.54 = 10.16 or

4 × $\frac{5}{2}$ = 10

Discussion: *Is the difference between the fraction and decimal values significant?*

Example 4 **Relating to the Real World** 

Make sure students understand that the wavy equal sign (≈) means *approximately equal to.*

KINESTHETIC LEARNING Use masking tape to make a large number line on the floor. Mark zero and use tick marks to indicate integers. Give students index cards containing numbers of all types: positive, negative, rational, irrational, absolute values. Have students go to the line with their numbers and stand in the correct place on the number line.

ERROR ALERT! Question 9 Students may confuse finding a reciprocal with finding an opposite number. **Remediation:** Help students discover for themselves that when adding opposites, the result is zero; when multiplying reciprocals, the result is one.

Example 5

As a warm-up for Question 11, have students evaluate $\frac{x}{y}$, using various combinations of positive and negative integers.

Exercises ON YOUR OWN

Exercises 1–8 Point out that it is difficult to compare fractions unless they have either the same denominator or the same numerator.

Technology Options

For Exercises 29 and 30, students may use spreadsheet software to analyze and graph data.

Prentice Hall Technology 

Software • Secondary Math Lab Toolkit • Computer Item List Generator 1-6

 **Internet** • See the Prentice Hall site. (http://www.phschool.com)

For practice with fractions, see Skills Handbook pages 575–579.

Example 4 **Relating to the Real World**

Weather Use the expression $\frac{5}{9}(F-32)$ to change from the Fahrenheit scale to the Celsius scale. What is 10°F in Celsius?

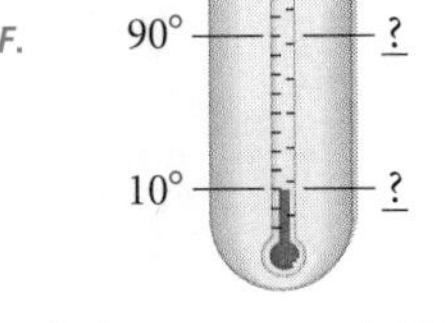

$\frac{5}{9}(F-32) = \frac{5}{9}(10-32)$ ← Substitute 10 for *F*.

$= \frac{5}{9}(-22)$

≈ -12

The temperature is about −12°C.

8. Find the Celsius temperature when the Fahrenheit temperature is 90°. about 32°C

The **reciprocal,** or *multiplicative inverse,* of a rational number $\frac{a}{b}$ is $\frac{b}{a}$. Zero does not have a reciprocal because division by zero is undefined.

9. a. **Patterns** Complete the chart.

b. What is the product of a number and its reciprocal? 1

Number	Reciprocal	Product	
3	$\frac{1}{3}$	$3 \cdot \frac{1}{3} = ■$	1
$\frac{1}{5}$	$\frac{5}{1}$ or 5	$\frac{1}{5} \cdot 5 = ■$	1
$-\frac{2}{3}$	$-\frac{3}{2}$	$-\frac{2}{3} \cdot \left(-\frac{3}{2}\right) = ■$	1

10. Is a number's reciprocal the same as its opposite? Explain. See left.

10. No; the reciprocal of a nonzero number $\frac{a}{b}$ is $\frac{b}{a}$. The opposite of $\frac{a}{b}$ is $-\frac{a}{b}$.

You know $12 \div 3 = 12 \cdot \frac{1}{3}$. This means that dividing by a nonzero number is the same as multiplying by its reciprocal.

$$a \div \frac{b}{c} = a \cdot \frac{c}{b} \text{ for } b \neq 0 \text{ and } c \neq 0$$

Example 5

Evaluate $\frac{x}{y}$ for $x = -\frac{3}{4}$ and $y = -\frac{5}{2}$.

$\frac{x}{y} = x \div y$

$= -\frac{3}{4} \div \left(-\frac{5}{2}\right)$

$= -\frac{3}{4} \cdot \left(-\frac{2}{5}\right)$ ← Multiply by $-\frac{2}{5}$, the reciprocal of $-\frac{5}{2}$.

$= \frac{3}{10}$

11. **Try This** Evaluate the expression in Example 5 for $x = 8$ and $y = -\frac{4}{5}$. −10

Exercises ON YOUR OWN

Use <, =, or > to compare.

1. $1\frac{2}{3}$ ■ $1\frac{1}{6}$ >

2. $-1\frac{2}{3}$ ■ $-1\frac{1}{6}$ <

3. $\frac{15}{8}$ ■ $1\frac{6}{8}$ >

4. $\frac{3}{5}$ ■ 0.6 =

5. $\frac{1}{2}$ ■ $\frac{1}{4}$ >

6. 0.14 ■ $\frac{1}{7}$ <

7. $-17\frac{1}{5}$ ■ $-17\frac{1}{4}$ >

8. −3.02 ■ −3.002 <

Exercises 9–12 Point out that a rational number may not appear in the form of a ratio. For example, 0.25 and $\frac{1}{4}$ are both rational numbers.

ALTERNATIVE ASSESSMENT **Exercises 25–28** Have students write all numbers as fractions for comparison. These exercises will help you assess whether students understand that they must use a common denominator to determine the order of the fractions.

Exercises 25–28 Encourage students to plot the numbers on a number line.

RESEARCH Exercise 29c. Students may find information on the height of mountains in state maps and almanacs or call their state or national park service.

Exercises 31–34 Remind students that they can replace the longer fraction bar with a division sign to write a problem like the one in Example 5. Also stress that students should compare the sizes of the fraction bars so that they do not, for example, mistake "5 divided by $\frac{4}{9}$" for "$\frac{5}{4}$ divided by 9." Encourage students to read the exercises to themselves as a fraction divided by a fraction, for example *"three eighths divided by negative two thirds."*

Complete with a rational number that makes each statement true.

9. $\frac{3}{5} \cdot ■ = 1$ $\frac{5}{3}$ 10. $-4 \cdot ■ = 1$ $-\frac{1}{4}$ 11. $■ \cdot \frac{7}{8} = 1$ $\frac{8}{7}$ 12. $■ \cdot \left(-\frac{8}{3}\right) = 1$ $-\frac{3}{8}$

Evaluate.

13. $x - \frac{3}{4}$ for $x = -1\frac{1}{4}$ -2
14. $a - b$ for $a = -\frac{1}{2}, b = \frac{1}{5}$ $-\frac{7}{10}$
15. rs for $r = \frac{2}{15}, s = 5$ $\frac{2}{3}$
16. $-p + t$ for $p = 1\frac{3}{8}, t = \frac{5}{8}$ $-\frac{3}{4}$
17. $2xy$ for $x = -\frac{2}{3}, y = -\frac{3}{4}$ 1
18. $\frac{v}{w}$ for $v = -\frac{5}{8}, w = -\frac{5}{6}$ $\frac{3}{4}$
19. ab for $a = 2\frac{1}{4}, b = -3$ $-6\frac{3}{4}$
20. $\frac{m}{n}$ for $m = -\frac{1}{2}, n = 5$ $-\frac{1}{10}$
21. $\frac{1}{2}(q - p)$ for $q = 4, p = -\frac{1}{3}$
22. $4rs$ for $r = -3\frac{1}{4}, s = -\frac{1}{5}$ $2\frac{3}{5}$
23. $b + c$ for $b = -2\frac{1}{3}, c = \frac{1}{4}$
24. $\frac{1}{5}z - y$ for $z = 1.5, y = 0.3$

21. $2\frac{1}{6}$ 23. $-2\frac{1}{12}$ 24. 0

Write each group of numbers in order from least to greatest.

25. $-\frac{1}{2}, -\frac{2}{3}, \frac{1}{4}$ 26. $-1.5, -\frac{4}{3}, -1\frac{1}{4}$ 27. $-9.7, -9\frac{7}{12}, -9\frac{3}{4}$ 28. $-4.12, -4.22, -4.05$

25–28. See below.

29. Science Air temperature drops as altitude increases. In the formula $t = (-5.5)\left(\frac{a}{1000}\right)$, t is the approximate change in Fahrenheit temperature, and a is the increase in altitude in feet.
   a. Find the change in temperature for the balloon. −44°F
   b. Suppose the temperature is 40°F at ground level. What is the approximate temperature at the balloon? −4°F
   c. Research Find the height of a mountain that is near you or that you are interested in. Suppose the temperature at the base of the mountain is 80°F. What is the approximate temperature at the top of the mountain? Check students' work.

30. Weather In the formula $w = -39 + \frac{3}{2}t$, w is the approximate windchill temperature when the wind speed is 20 mi/h, and t is the actual air temperature. Find the approximate windchill temperature when the actual air temperature is 10°F and the wind speed is 20 mi/h. −24°F

Rewrite each expression using the symbol ÷, then find each quotient.

Sample $\frac{-\frac{7}{12}}{4} = -\frac{7}{12} \div 4$ ← Rewrite as $-\frac{7}{12} \div 4$.

$= -\frac{7}{12} \cdot \left(\frac{1}{4}\right)$ ← Multiply by $\frac{1}{4}$, the reciprocal of 4.

$= -\frac{7}{48}$

25. $-\frac{2}{3}, -\frac{1}{2}, \frac{1}{4}$ 26. $-1.5, -\frac{4}{3}, -1\frac{1}{4}$ 27. $-9\frac{3}{4}, -9.7, -9\frac{7}{12}$ 28. $-4.22, -4.12, -4.05$

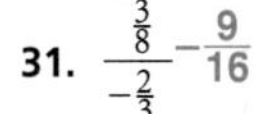

31. $\frac{\frac{3}{8}}{-\frac{2}{3}}$ $-\frac{9}{16}$ 32. $\frac{5}{\frac{4}{9}}$ $11\frac{1}{4}$ 33. $\frac{-\frac{5}{6}}{8}$ $-\frac{5}{48}$ 34. $\frac{-\frac{2}{5}}{-\frac{4}{5}}$ $\frac{1}{2}$

35. a. Open-ended Name a point between −2 and −3 on a number line. Sample: $-2\frac{1}{7}$
   b. Name a point between −2.8 and −2.9. Sample: −2.84
   c. Name a point between $-2\frac{1}{16}$ and $-2\frac{3}{8}$. Sample: $-2\frac{3}{16}$
   d. On a number line, is it possible to find a point between any two given points? Explain. Yes; for example, the average of any two numbers is between them.

Exercises MIXED REVIEW

GETTING READY FOR LESSON 1-7 This pattern prepares students for comparing the probabilities of different events.

Wrap Up

THE BIG IDEA Ask students: *What is a rational number?* any number that can be written in the form of the ratio of integers
Ask students: *How can you express an integer as a ratio?* Use 1 as the denominator.

RETEACHING ACTIVITY Students pick random rational numbers with a number cube, then substitute the rational numbers into variable equations and solve. (Reteaching worksheet 1-6)

A Point in Time

If you have block scheduling or an extended class period, you may wish to have students investigate these topics.

- What geometric calculations were perfected by the ancient Egyptians? area of squares, trapezoids, triangles, and circles; height and angles of pyramids
- What item from the Old Kingdom era played an essential part of the costume of men and women of the upper classes ? the woolen wig
- What musical instruments were popular before the New Kingdom era? harps and flutes

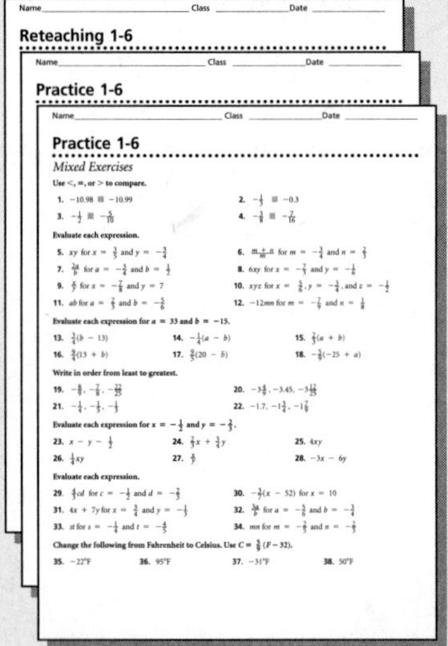
Name ______ Class ______ Date ______

Reteaching 1-6

Name ______ Class ______ Date ______

Practice 1-6

Name ______ Class ______ Date ______

Practice 1-6

Mixed Exercises

Use <, =, or > to compare.

1. $-10.98 \blacksquare -10.99$ 2. $-\frac{1}{3} \blacksquare -0.3$
3. $-\frac{1}{2} \blacksquare -\frac{5}{10}$ 4. $-\frac{3}{8} \blacksquare -\frac{7}{16}$

Evaluate each expression.

5. xy for $x = \frac{3}{5}$ and $y = -\frac{3}{4}$ 6. $\frac{m+n}{m}$ for $m = -\frac{3}{4}$ and $n = \frac{2}{3}$
7. $\frac{2a}{b}$ for $a = -\frac{3}{4}$ and $b = \frac{1}{2}$ 8. $6xy$ for $x = -\frac{2}{3}$ and $y = -\frac{1}{6}$
9. $\frac{x}{y}$ for $x = -\frac{7}{8}$ and $y = 7$ 10. xyz for $x = \frac{5}{6}, y = -\frac{3}{4}$, and $z = -\frac{1}{2}$
11. ab for $a = \frac{2}{3}$ and $b = -\frac{5}{6}$ 12. $-12mn$ for $m = -\frac{2}{9}$ and $n = \frac{1}{8}$

Evaluate each expression for $a = 33$ and $b = -15$.

13. $\frac{3}{4}(b - 13)$ 14. $-\frac{1}{4}(a - b)$ 15. $\frac{2}{3}(a + b)$
16. $\frac{9}{4}(13 + b)$ 17. $\frac{9}{5}(20 - b)$ 18. $-\frac{5}{9}(-25 + a)$

Write in order from least to greatest.

19. $-\frac{8}{9}, -\frac{7}{8}, -\frac{22}{25}$ 20. $-3\frac{4}{9}, -3.45, -3\frac{12}{25}$
21. $-\frac{1}{4}, -\frac{1}{5}, -\frac{1}{3}$ 22. $-1.7, -1\frac{3}{4}, -1\frac{7}{9}$

Evaluate each expression for $x = -\frac{1}{2}$ and $y = -\frac{2}{3}$.

23. $x - y - \frac{1}{2}$ 24. $\frac{2}{3}x + \frac{3}{4}y$ 25. $4xy$
26. $\frac{1}{4}xy$ 27. $\frac{x}{y}$ 28. $-3x - 6y$

Evaluate each expression.

29. $\frac{4}{3}cd$ for $c = -\frac{1}{2}$ and $d = -\frac{2}{3}$ 30. $-\frac{3}{7}(x - 52)$ for $x = 10$
31. $4x + 7y$ for $x = \frac{3}{4}$ and $y = -\frac{1}{3}$ 32. $\frac{3a}{b}$ for $a = -\frac{5}{6}$ and $b = -\frac{3}{4}$
33. st for $s = -\frac{1}{4}$ and $t = -\frac{4}{5}$ 34. mn for $m = -\frac{6}{5}$ and $n = -\frac{2}{3}$

Change the following from Fahrenheit to Celsius. Use $C = \frac{5}{9}(F - 32)$.

35. −22°F 36. 95°F 37. −31°F 38. 50°F

Lesson Quiz

Lesson Quiz is also available in Transparencies.

1. Evaluate $\frac{2}{3} \div \frac{2}{5}$. $1\frac{2}{3}$
2. Evaluate $2a + \frac{x}{y}$ for $a = -\frac{1}{8}$, $x = \frac{1}{2}, y = \frac{1}{4}$. $1\frac{3}{4}$
3. Use the comparison symbol < (less than) to order $\frac{3}{5}$ and $\frac{3}{8}$. $\frac{3}{8} < \frac{3}{5}$
4. Evaluate $\frac{2}{3}(a + b)$ for $a = \frac{3}{5}$, $b = -\frac{3}{4}$. $-\frac{1}{10}$

34

Evaluate.

36. $x + y$ for $x = \frac{13}{4}, y = -\frac{5}{2}$ $\frac{3}{4}$

37. $\frac{1}{5}(r - t)$ for $r = 6, t = -7$ $2\frac{3}{5}$

38. $\frac{3}{4}w - 7$ for $w = 1\frac{1}{3}$ -6

39. $\frac{x}{2y}$ for $x = 3.6, y = -0.4$ -4.5

40. $\frac{3a}{b} + c$ for $a = -2, b = -5, c = -1$ $\frac{1}{5}$

41. $\frac{n}{m}$ for $n = -\frac{4}{5}, m = 8$ $-\frac{1}{10}$

Writing **Decide if each statement is *true* or *false*. Justify your answer.**

42. All integers are rational numbers. T; any integer can be written as a ratio: $3 = \frac{3}{1}$.

43. All negative numbers are integers. F; $-\frac{3}{4}$

44. All rational numbers are integers. F; $-\frac{3}{4}$

45. Some real numbers are integers. T; since integers are real numbers, some real numbers are integers.

Exercises MIXED REVIEW

For Exercises 46–48, write an equation to model each situation.

46. the cost of several movie tickets that are $6.26 each c = cost, n = number of tickets; $c = 6.26n$

47. the change from a $10 bill after a purchase c = change, p = purchase; $c = 10 - p$

48. the total cost of an item with a shipping fee of $3.98. t = total cost, p = price of item; $t = p + 3.98$

49. Real Estate Boulder City, Nevada, bought 107,500 acres of land in Nevada's Eldorado Valley from the United States government. The price was $12/acre. How much did the city pay for the land? $1,290,000

Getting Ready for Lesson 1-7

50. Probability How much of the spinner is red? yellow? blue? $\frac{1}{2}, \frac{1}{8}, \frac{3}{8}$

A Point in Time

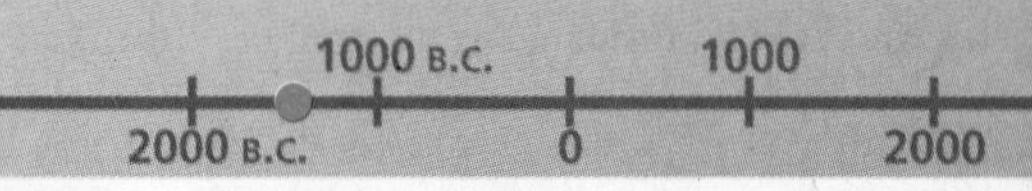

The Rhind Papyrus

A scroll discovered in Egypt shows that Egyptians were using symbols for plus, minus, equals, and an unknown quantity over 3500 years ago. Named for the British Egyptologist A. Henry Rhind, the papyrus is 18 ft long and 1 ft wide. It is a practical handbook containing 85 problems including work with rational numbers. In the Rhind Papyrus, rational numbers are written as unit fractions. A unit fraction has 1 as its numerator. Here are some examples.

$\frac{3}{4} = \frac{1}{2} + \frac{1}{4}$ $\frac{3}{8} = \frac{1}{4} + \frac{1}{8}$ $\frac{21}{30} = \frac{1}{6} + \frac{1}{5} + \frac{1}{3}$

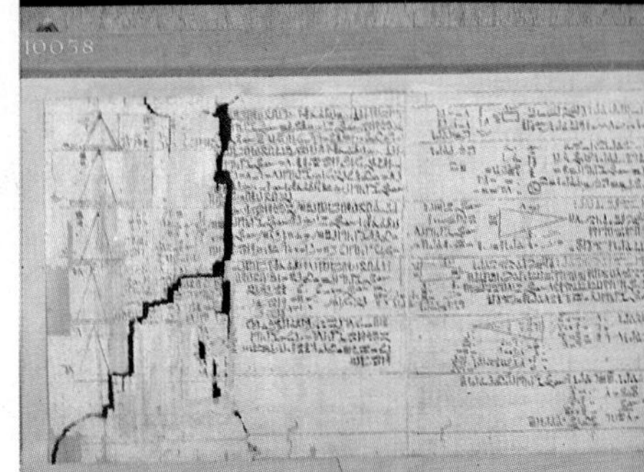

Math ToolboX

In Lesson 1-5, students practiced arithmetic operations with integers, decimals, and rational numbers. Students may have used the commutative, associative, and identity properties without knowing the name of the concept involved. Discuss these properties, using integers for examples, to help students understand what these properties mean. Make the names of the properties memorable by relating *commute* with commuting from home to work as in changing locations. Relate *associate* with associating with different groups of friends. Exercise 16 will help students see that the two expressions $(x - 2)$ and $(2 - x)$ are not equivalent.

ERROR ALERT! Students often misspell *commutative* by writing "communative." **Remediation:** Relate the property to the root word *commute* to help students understand the term and the spelling. Practice pronouncing *commutative* with the class, emphasizing each syllable.

WRITING Exercise 16 Students may need help justifying their answers. Have them write mathematical examples for each question.

ADDITIONAL PROBLEM Have students look up the words *commute*, *associate*, and *identity* in the dictionary and describe how each definition relates to its corresponding algebraic property.

Math ToolboX — Skills Review

Using Properties

After Lesson 1-6

These properties of mathematics help you to perform arithmetic and algebraic operations.

Commutative Properties of Addition and Multiplication

For all real numbers *a* and *b*:

$a + b = b + a$ $\qquad$ $a \cdot b = b \cdot a$

Examples: $2 + 3 = 3 + 2$ $\qquad$ $4 \cdot 5 = 5 \cdot 4$

Associative Properties of Addition and Multiplication

For all real numbers *a*, *b*, and *c*:

$(a + b) + c = a + (b + c)$ $\qquad$ $(a \cdot b) \cdot c = a \cdot (b \cdot c)$

Examples: $(5 + 6) + 7 = 5 + (6 + 7)$ $\qquad$ $(2 \cdot 3) \cdot 4 = 2 \cdot (3 \cdot 4)$

Identity Properties of Addition and Multiplication

For every real number *a*:

$a + 0 = a$ and $0 + a = a$ $\qquad$ $a \cdot 1 = a$ and $1 \cdot a = a$

Examples: $5 + 0 = 5$ and $0 + 5 = 5$ $\qquad$ $7 \cdot 1 = 7$ and $1 \cdot 7 = 7$

Name the property that each exercise illustrates.

1. $1m = m$ Id. Prop. of Mult.
2. $(-3 + 4) + 5 = -3 + (4 + 5)$ Assoc. Prop. of Add.
3. $3(8 \cdot 0) = (3 \cdot 8)0$ Assoc. Prop. of Mult.
4. $2 + 0 = 2$ Id. Prop. of Add.
5. $np = pn$ Comm. Prop. of Mult.
6. $f + g = g + f$ Comm. Prop. of Add.

Use the properties to simplify each expression.

7. $(5 \cdot 83) \cdot 2$ 830
8. $47 + 39 + 3 + 11$ 100
9. $25 \cdot 74 \cdot 2 \cdot 2$ 7400
10. $-5(7y)$ $-35y$
11. $8 + 9m + 7$ $9m + 15$
12. $4.75 + 2.95 + 1.25 + 6$ 14.95
13. $6\frac{1}{2} + 4\frac{1}{3} + 1\frac{1}{2} + \frac{2}{3}$ 13
14. $25 \cdot 1.7 \cdot 4$ 170
15. $(3p)(4q)(6r)$ $72pqr$
16. *Writing* Justify your answers to these questions: Is subtraction commutative? Is subtraction associative? Is division commutative? Is division associative? See margin.

page 35 Math Toolbox

16. subtraction commutative: no, $2 - 3 \neq 3 - 2$
    subtraction associative: no, $(2 - 3) - 4 \neq 2 - (3 - 4)$
    division commutative: no, $10 \div 5 \neq 5 \div 10$
    division associative: no, $(10 \div 5) \div 2 \neq 10 \div (5 \div 2)$

1-7 Teaching Notes

PROBLEM OF THE DAY

A couple would like to have three children. How many different possible outcomes of boys and girls could be born to this couple? List the different outcomes. eight; BBB, BBG, BGB, BGG, GBB, GBG, GGB, GGG

Problem of the Day is also available in Transparencies.

CONNECTING TO PRIOR KNOWLEDGE Ask: *What does the word "probability" mean?* how likely it is that an event will occur Ask students to describe situations in which they could use the word *probability*. Answers may vary. Sample: weather forecasting and sporting events

WORK TOGETHER

Organization and the division of labor needed to work in pairs help students develop team skills used in the workplace. Combine all the pairs' results into a larger group result. Have students compare these results to the results of their pair. Encourage students to comment on any differences.

THINK AND DISCUSS

ERROR ALERT! Students may think that a probability somehow guarantees that a specific event is going to occur. **Remediation:** Ask students to describe an event that was considered very unlikely, but occurred anyway. Help students understand that a probability measures the theoretical

Lesson Planning Options

Prerequisite Skills

- Reducing fractions
- Converting fractions to percents

Assignment Options

To provide flexible scheduling, this lesson can be subdivided into parts.

1 **Core** 1–5, 7–10
Extension 6, 11

2 **Core** 12–19
Extension 20–22

Use Mixed Review to maintain skills.

Resources

 **Student Edition**

Skills Handbook, pp. 575, 579
Extra Practice, p. 556
Glossary/Study Guide

 **Teaching Resources**

Chapter Support File, Ch. 1
- Practice 1-7 (two worksheets)
- Reteaching 1-7

Classroom Manager 1-7
Glossary, Spanish Resources
Two-Year Algebra Handbook 1-7

 **Transparencies**
35

1-7 Experimental Probability and Simulations

What You'll Learn
- Finding experimental probability
- Using simulations to find experimental probability

...And Why

To solve problems in business and sports

What You'll Need
- two number cubes

WORK TOGETHER

Work with a partner.

1. Suppose you roll two number cubes many times. Do you think that all of the sums will occur about the same number of times, or will some sums occur more often than others? Explain. Answers may vary.
2. Data Collection Roll two number cubes sixty times. On a line plot numbered from 2 to 12, mark the sum of the cubes each time you roll. Check students' work.
3. How well do your results of rolling the number cubes match your **prediction** in Question 1? Explain. Answers may vary.

THINK AND DISCUSS

Part 1

Finding Experimental Probability

Probability measures how likely something, usually called an **event,** is to happen. For **experimental probability** you gather data through observations or experiments. Use this ratio to find experimental probability.

You read *P* (event) as "the probability of an event."

$$P(\text{event}) = \frac{\text{number of times an event happens}}{\text{number of times experiment is done}}$$

Suppose that in the Work Together activity, the sum 3 occurred five times. Then, $P(3) = \frac{5}{60}$, or $\frac{1}{12}$. Each time you repeat an experiment, the experimental probability of an event may vary. The more data you collect, the more reliable your results will be.

Check students' work.

4. a. Use your Work Together data to find $P(2)$, $P(7)$, and $P(12)$.
   b. Which sum in part (a) has the greatest probability? Do you think this will be true if you repeat the experiment? Explain. See left.

4b. *P*(7) will have the greatest probability most of the time; yes; there are more ways to roll a 7 than to roll 2 or 12.

You can conduct an experiment by choosing items *at random*. This means that each item has an equal chance of being picked.

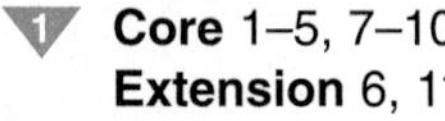

QUICK REVIEW

You can write $\frac{4}{5}$ as a fraction, decimal, or percent:
$\frac{4}{5} = 0.8 = 80\%$

Example 1 Relating to the Real World 

Business After receiving complaints, a retailer checked 100 light bulbs at random. Eighty of the bulbs worked. What is *P*(bulb works)?

$$P(\text{bulb works}) = \frac{\text{number of times an event happens}}{\text{number of times the experiment is done}}$$

$$= \frac{80}{100} \quad \leftarrow \text{The light bulbs worked 80 times.}$$

$$= \frac{4}{5} = 0.8 = 80\%$$

The probability that a light bulb works is 80%.

likelihood of the occurrence. For large numbers of events, the probability and the rate of occurrence should agree; for single events, a probability may not match the actual outcome.

DIVERSITY Have students brainstorm events that have a different probability of occurring in your city than in other places. **Answers may vary. Sample: the probability of some extreme weather conditions**

Example 1 Relating to the Real World

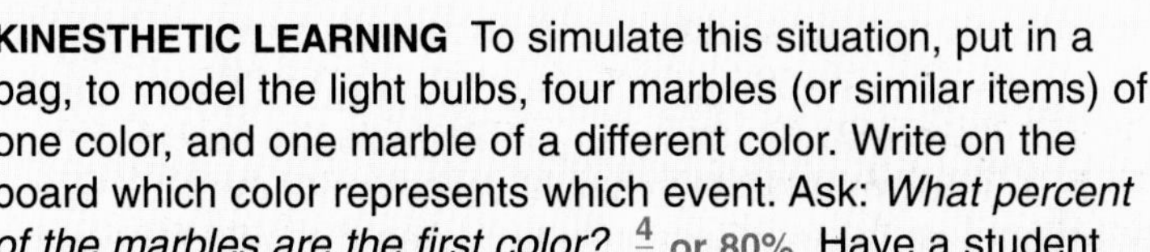

KINESTHETIC LEARNING To simulate this situation, put in a bag, to model the light bulbs, four marbles (or similar items) of one color, and one marble of a different color. Write on the board which color represents which event. Ask: *What percent of the marbles are the first color?* $\frac{4}{5}$ **or 80%** Have a student choose a marble blindly, see what color the marble is, replace the marble in the bag, and put a check mark on the board underneath the corresponding event. When all students have had a turn, have students compare the results with the probability they found experimentally in Example 1.

Example 2 Relating to the Real World 

CONNECTING TO THE STUDENTS' WORLD Explain to students that the table shows digits generated randomly then arbitrarily grouped in fives to make the digits easier to read. Point out that they should read across the table.

The alignment in the random number table may be confusing to students. The numbers are not five-digit numbers. They are one-digit numbers put in groups of five to make the table easier to read. Also, ensure that students understand that they need to read the table from left to right, and from top to bottom.

5. There are 500 light bulbs in stock. **Predict** how many of the light bulbs will work. (*Hint*: Multiply the probability found in Example 1 by the number of light bulbs in the shipment.) **400 bulbs**

Part 2 

Conducting a Simulation

A **simulation** is a model of a real-life situation. One way to do a simulation is to use a table of numbers that a computer has picked at random.

GRAPHING CALCULATOR HINT

You can generate random numbers using a graphing calculator or computer software.

Example 2 Relating to the Real World 

Medical Technician According to the American Red Cross, 40% of the people in the United States have type A blood. What is the probability that the next two people who donate blood have type A blood?

Define how the simulation will be done.

Use a random number table.

Random Number Table

23948 71470 12573 05954
65628 22310 09311 94864
41261 09943 34078 79481
34831 94510 71490 93312

Let the numbers 0–3 in the random number table represent having type A blood. Let 4–9 represent not having type A blood. Since the simulation is for two people, list the numbers in pairs. Look at twenty pairs.

← 40% of the numbers must represent having type A blood.

Conduct the simulation.

23 94 87 14 70 12 57 30 59 54 65 62 82 23 10 09 31 19 48 64

The arrows point to pairs in which both "people" have type A blood.

Interpret the simulation.

Six pairs represent "both people have type A blood" because these pairs contain only the numbers 0–3.

$$P(\text{both have type A blood}) = \frac{\text{number of times an event happens}}{\text{number of times the experiment is done}}$$
$$= \frac{6}{20}$$
$$= \frac{3}{10} = 0.3 = 30\%$$

The probability that both of the next two people have type A blood is 30%.

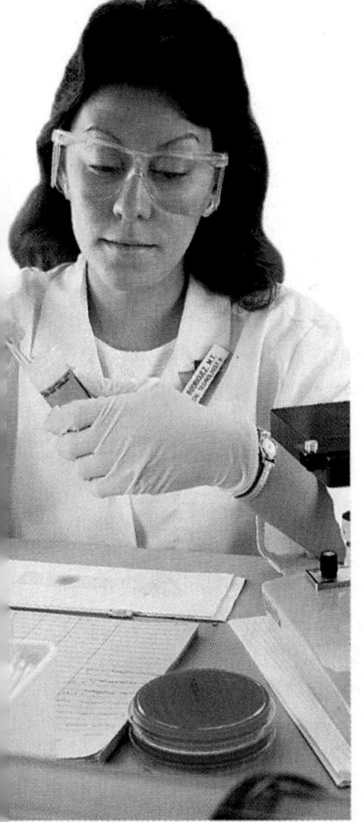

6. For the simulation in Example 2, could you use any four numbers from 0–9 to represent having type A blood? Explain. **Yes; any four numbers would represent 40% of the possibilities.**
7. **Try This** In the United States, about 50% of the people have type O blood. Use the random number table above to find the probability that the next two donors have type O blood. **Answers may vary. Using 20 pairs and 0-4 for type O blood, the probability is $\frac{7}{20}$ = 35%.**

Additional Examples

FOR EXAMPLE 1

a. Quality control checks 500 belts and finds 15 defective ones. What is the probability that a belt will pass quality control?

$$P(\text{belt passes}) = \frac{500 - 15}{500}$$
$$= \frac{485}{500}$$
$$= 0.97$$
$$= 97\%$$

b. About how many belts in a batch of 50,000 should pass quality control?

$P = 50{,}000 \times 0.97$
$= 48{,}500$

Discussion: *Why do you think the 97% is called "experimental probability"?*

FOR EXAMPLE 2

Use the random number table. What is the probability that of the next two people who donate blood, one is type A and one is type B? **For 0–3 to represent A and 4 to represent B, using all the data in the table, the probability is 12.5%. Theoretically it is 8%.**

Discussion: *In this simulation, how many digits will be needed to represent type B blood? Does it matter which digit is chosen to represent type B blood?*

Exercises ON YOUR OWN

GROUP ACTIVITY Exercise 1 Group activities help students develop team skills used in the workplace. Groups consisting of four students allow everyone in the group to have a partner. Partners should record each other's result. Then, the whole group can gather the final results in one table.

ESL Exercises 1, 10. and 11 Clarify the meaning of *heads* and *tails* of a coin. Make sure students understand any other terms used in these exercises.

ALTERNATIVE ASSESSMENT **Exercises 7–9** Have students discuss the way probability changes with the number of items involved. Change the given data to eight dimes. Have students discover and discuss the change in the results for these exercises. This will help you determine whether students understand the concept of probability.

Exercise 19 Point out to students that their results will vary depending on the random numbers selected and the number of experiments conducted. The greater the number of experiments, the more reliable the results. Have one group of students work with the first 30 random numbers and the other group the last 30. Compare the results.

CRITICAL THINKING Exercise 20 This exercise could be used to initiate a class discussion on sample bias.

Chapter Project FIND OUT BY ANALYZING Students are asked to review the data on height collected for the Find Out question on page 9 in Lesson 1-1. Students should separate the data collected into male and female, then organize and

Technology Options

For Exercise 22, students may use graphing calculators to create random numbers which simulate rolling a number cube.

Prentice Hall Technology

Calculator Graphing Calculator Handbook: Procedure 15

Software • Secondary Math Lab Toolkit • Computer Item Generator 1-7

CD-ROM Multimedia Algebra Lab 1

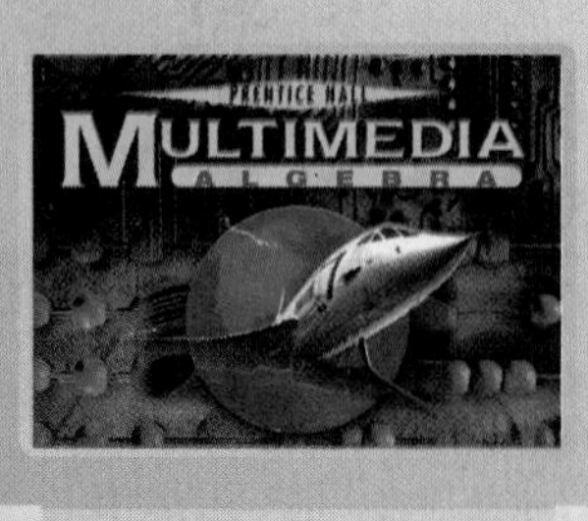

Internet • See the Prentice Hall site. (http://www.phschool.com)

Exercises ON YOUR OWN

1. a. *Group Activity* What do you expect the probability is for getting heads when you toss a coin? $\frac{1}{2}$
   b. Have each member in your group toss a coin 20 times, one person at a time. Use a table like the one below to record your data. After each person's turn, find the P(head) based on the total number of heads and the total number of tosses for your group. **1b–c. Check students' work.**

Team Member	Number of Heads	Number of Tosses	Total Number of Heads	Total Number of Tosses	P(head)
1	■	20	■	20	■
2	■	20	■	40	■

   c. As the total number of coin tosses increases, does the probability get closer to what you expect? Explain.

A driver collects data on the color of a certain traffic light. When she arrives at the light, it is green 16 times, yellow 5 times, and red 9 times. Find each probability.

2. P(red) $\frac{3}{10}$
3. P(green) $\frac{8}{15}$
4. P(yellow) $\frac{1}{6}$
5. P(not green) $\frac{7}{15}$

6. Suppose you bought a box of marbles. You select twelve marbles at random and record the color. The results are at the right.
   a. Find P(orange), P(green), P(blue), and P(yellow). $\frac{1}{2}$; $\frac{1}{12}$; $\frac{1}{4}$; $\frac{1}{6}$
   b. Suppose the box contains 160 marbles. About how many blue marbles do you expect? **40**
   c. Do you think there will be exactly 80 orange marbles? Explain. **No, the only way to know the exact number is to count all of the marbles.**

Suppose you grab a handful of coins from your change jar, and these coins are a random sample of the coins in your jar. You have 13 pennies, 3 quarters, 2 nickels, and 6 dimes. Find the probability that each is the next coin you select. Write each probability as a percent.

7. P(nickel) **8.3%**
8. P(penny) **54.2%**
9. P(coin worth less than 10¢) **62.5%**

10b. Now she has 8 hits in 26 times at bat. So the experimental probability is now $\frac{8}{26}$, not $\frac{7}{25}$.

10. a. A softball player is at bat 25 times. She gets a hit 7 times. What is the probability that she gets a hit the next time she comes to bat? $\frac{7}{25}$
    b. *Writing* Suppose the player gets a hit the 26th time at bat. Explain how getting a hit changes the experimental probability. **See above right.**

11. *Health* The table shows the results of the 1954 trials for the Salk polio vaccine. To test the vaccine, researchers gave a test group shots with vaccine and a control group shots without vaccine. Compare the probability of developing polio in the test group and in the control group. **P(test) ≈ 0.04%;**

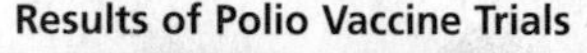

Results of Polio Vaccine Trials

Group	Number of Children	Number of Cases of Polio
Test	200,745	82
Control	201,229	162

display the data to see if there are differences between the sexes. This information is essential to their work on the Chapter Project. Have students add the results of this task to the information they have already collected. Check the students' progress at this point.

Exercises MIXED REVIEW

JOURNAL Remind students to keep their journal entries organized in one notebook or in one part of their math notebook.

GETTING READY FOR LESSON 1-8 These exercises prepare the students for adding and subtracting matrices.

Wrap Up

THE BIG IDEA Ask students to describe how to find the experimental probability of an event and how to conduct a simulation.

RETEACHING ACTIVITY Students use thumb tacks to find experimental probability. (Reteaching worksheet 1-7)

Use the data in the line plot to find each probability. 15. $\frac{7}{18}$

Students' Birthdays

Su	M	Tu	W	Th	F	Sa
X					X	
X	X				X	X
X	X	X		X	X	X
X	X	X		X	X	X

12. $P(\text{Sunday})$ $\frac{2}{9}$ 13. $P(\text{Tuesday})$ $\frac{1}{9}$ 14. $P(\text{Wednesday})$ 0
15. $P(\text{weekend day})$ 16. $P(\text{weekday})$ $\frac{11}{18}$ 17. $P(not\text{ Friday})$ $\frac{7}{9}$
18. Suppose you guess the answers to a two-question true/false test. Your probability of getting one answer correct is 50%. Do a simulation of this test by flipping a coin for each question. Let "heads" represent a correct answer and "tails" represent an incorrect answer. Simulate the test ten times and record the results. Find $P(\text{both answers are correct})$. **Check students' work.**
19. Suppose there are three stoplights on your way to school. Each one has a 30% chance of stopping you. Use the random number table on page 51 to do a simulation. Record your data and find $P(\text{stopped by all three lights})$. Write your probability as a decimal rounded to the nearest hundredth. Then write it as a percent. (*Hint*: Let the numbers 0–2 represent being stopped and 3–9 represent not being stopped.)
0.04, 4%
20. **Critical Thinking** Suppose you are taking a poll for student council elections. Which is the best method for choosing students at random? Explain your choice.
   A. You choose every 20th student on a list of all students.
   B. You choose students as they leave a band concert.
   C. You choose students in line at lunch.
   A; all students have an equal chance of being selected.
21. **Open-ended** Conduct a simulation to find the experimental probability that exactly two children in a family of five children will be girls. Record the results of your simulation.
21, 22 a–b. Check students' work.
22. a. **Data Collection** Roll a number cube and record how often each of the numbers 1 through 6 occurs. Roll the cube 50 times.
   b. Find the experimental probability for each number.
   c. Are the probabilities for each number the same? Explain.
   Answers may vary. Sample: No; the probabilities are close but not the same. Each number did not land exactly the same number of times.

Chapter Project *Find Out by Analyzing*

When predicting height, scientists use different formulas for men and women.

- Review the data collected for the Find Out by Question on page 9. Separate the data by male and female.
- Organize and display the data to see if there are differences between the sexes.

Exercises MIXED REVIEW

Simplify each expression.

23. $13 + 16(-2)$ **−19** 24. $36 \div (-4) - 2(-15)$ **21**
25. -7^2 **−49** 26. $(-7)^2$ **49**

Getting Ready for Lesson 1-8

27. What number is in the first row, second column? **53.6**
28. What number is in the second row, first column? **32**

17	53.6	42
32	52.5	−7
4	98	5.6

FOR YOUR JOURNAL

Explain why the experimental probability of an event can vary each time you perform the same experiment.

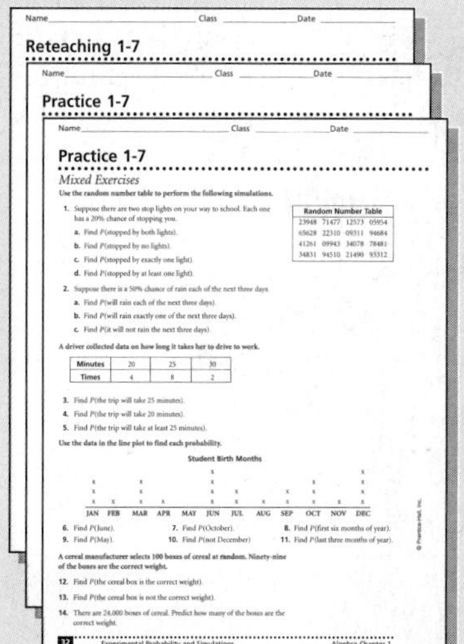

Lesson Quiz

Lesson Quiz is also available in Transparencies.

1. The socks in a drawer are unmatched and scattered randomly. You pull out eight socks and get 5 white, 2 blue, and 1 red. What is the probability that the next sock you pull is white?
$\frac{5}{8}$ = 62.5%
2. If you roll a wheel forward, when it slows down it will fall over, either to the left or right. What is the probability of it falling to the left?
$\frac{1}{2}$ = 0.5 = 50%

1-8 Teaching Notes

PROBLEM OF THE DAY

José made six three-point baskets and four two-point baskets at the basketball game last night. If Pete scored the same number of points as José and made ten two-point baskets, how many three-point baskets did Pete make? two

Problem of the Day is also available in Transparencies.

CONNECTING TO PRIOR KNOWLEDGE Have students brainstorm to list all the places they have seen data displayed in the form of a table or chart with rows and columns. **Answers may vary. Sample: bus timetables, newspaper reports, test grade result sheets**

THINK AND DISCUSS

ESL Make sure students understand the words *row* and *column*. Associate *row* with rows in a theater and *column* with columns in front of a building.

KINESTHETIC LEARNING If you have an extended class period or block scheduling, have students arrange their desks into a matrix of rows and columns. Tape row and column number labels on the walls in line with the desks. Call out random matrix entries and have the corresponding student stand. Alternatively, call students' names and have them stand and call out their row and column numbers.

Lesson Planning Options

Prerequisite Skills

- Adding and subtracting integers
- Understanding the terms row and column

Assignment Options

Core 1–18, 24–26
Extension 19, 20–23

Use Mixed Review to maintain skills.

Resources

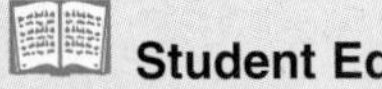

Student Edition

Skills Handbook, pp. 569, 579
Extra Practice, p. 556
Glossary/Study Guide

Teaching Resources

Chapter Support File, Ch. 1
- Practice 1-8 (two worksheets)
- Reteaching 1-8

Classroom Manager 1-8
Glossary, Spanish Resources
Two-Year Algebra Handbook 1-8

Transparencies
35, 38, 39

40

Connections Environment . . . and more

1-8 Organizing Data in Matrices

What You'll Learn
- Adding and subtracting matrices

...And Why

To investigate real-world data, such as school enrollment and employee wages

1992 School Enrollment in the United States (in millions)

Level	Public	Private
Elementary	27.1	3.1
High School	12.3	1.0
College	11.1	3.0

Source: *Statistical Abstract of the United States*

THINK AND DISCUSS

A **matrix** is a rectangular arrangement of numbers. The numbers are arranged in rows and columns and are usually written inside brackets. The matrix below displays the school enrollment data at the left.

$$\begin{array}{l} \\ \text{Elementary} \\ \text{High School} \\ \text{College} \end{array} \begin{array}{c} \text{Public} \quad \text{Private} \\ \begin{bmatrix} 27.1 & 3.1 \\ 12.3 & 1.0 \\ 11.1 & 3.0 \end{bmatrix} \end{array} \leftarrow \text{row}$$

column (arrow pointing to first column)

You identify the size of a matrix by the number of rows and the number of columns. This matrix has 3 rows and 2 columns, so it is a 3×2 matrix. Each item in a matrix is an **entry.**

1. What entry is in row 1, column 2? in row 2, column 1? 3.1; 12.3
2. *Open-ended* Make a 2×3 matrix. Sample: $\begin{bmatrix} 3 & 5 & 8 \\ 2 & 6 & -1 \end{bmatrix}$

Two matrices are equal if corresponding entries are equal.

corresponding entries

$$\begin{bmatrix} 0.5 & \frac{3}{8} \\ \frac{1}{4} & 0.4 \end{bmatrix} \quad \begin{bmatrix} \frac{1}{2} & 0.375 \\ 0.25 & \frac{2}{5} \end{bmatrix}$$

3. Are the two matrices equal? Explain. yes; $0.5 = \frac{1}{2}$, $\frac{3}{8} = 0.375$, $\frac{1}{4} = 0.25$, $0.4 = \frac{2}{5}$

You can add or subtract matrices if they are the same size. You do this by adding or subtracting corresponding entries.

Example 1

Subtract the two matrices. $\begin{bmatrix} 7 & -4 & 11 \\ 6 & 5 & -1 \end{bmatrix} - \begin{bmatrix} 5 & 9 & 7 \\ 8 & -7 & -3 \end{bmatrix}$

$$\begin{bmatrix} 7 & -4 & 11 \\ 6 & 5 & -1 \end{bmatrix} - \begin{bmatrix} 5 & 9 & 7 \\ 8 & -7 & -3 \end{bmatrix} = \begin{bmatrix} 7-5 & -4-9 & 11-7 \\ 6-8 & 5-(-7) & -1-(-3) \end{bmatrix}$$

$$= \begin{bmatrix} 2 & -13 & 4 \\ -2 & 12 & 2 \end{bmatrix}$$

4. What size are the matrices in Example 1? 2×3
5. **Try This** Add the two matrices that were subtracted in Example 1.

5. $\begin{bmatrix} 12 & 5 & 18 \\ 14 & -2 & -4 \end{bmatrix}$

Example 1 Relating to the Real World 

Point out that in the matrix which shows the combining of the entries, the expression in row 2, column 2 has a minus sign and a negative sign. Remind students to be careful about including all negative and minus signs when combining integers.

ERROR ALERT! **Question 5** Students may try to add the wrong entries. **Remediation:** Encourage students to draw a line through entries as they finish adding and also to validate their answers.

TACTILE LEARNING Let students write 3 × 2 matrix puzzles for each other to solve. Students should write clues for each entry, such as "row 2, column 2 is an integer between 1.3 and 2.7." Have students exchange their puzzles with a partner and solve. This activity would work well if you have an extended class period or block scheduling.

Example 2 Relating to the Real World 

VISUAL LEARNING In each row of the matrix, have students underline the first term, circle the second term, put a square around the third term, and put a triangle around the fourth term. This will help students identify the correct terms to add from each row.

Exercises ON YOUR OWN

Exercise 4 Ask students how they can tell that this is a matrix and not the absolute value of –8. The grouping symbols are not lines.

Matrices are a handy way to organize real-world data so that you can add or subtract the data.

 Relating to the Real World 

Environment Write each table as a matrix. Then add the matrices to find the total of endangered and threatened species.

Endangered Species

Species	United States Only	Both United States and Foreign	Foreign Only	Total
Mammals	36	20	251	307
Birds	57	16	153	226
Reptiles	8	8	63	79
Amphibians	6	0	8	14
Fishes	60	3	11	74

Threatened Species

Species	United States Only	Both United States and Foreign	Foreign Only	Total
Mammals	5	4	22	31
Birds	8	9	0	17
Reptiles	15	4	14	33
Amphibians	4	1	0	5
Fishes	32	6	0	38

$$\begin{bmatrix} 36 & 20 & 251 & 307 \\ 57 & 16 & 153 & 226 \\ 8 & 8 & 63 & 79 \\ 6 & 0 & 8 & 14 \\ 60 & 3 & 11 & 74 \end{bmatrix} + \begin{bmatrix} 5 & 4 & 22 & 31 \\ 8 & 9 & 0 & 17 \\ 15 & 4 & 14 & 33 \\ 4 & 1 & 0 & 5 \\ 32 & 6 & 0 & 38 \end{bmatrix} = \begin{bmatrix} 41 & 24 & 273 & 338 \\ 65 & 25 & 153 & 243 \\ 23 & 12 & 77 & 112 \\ 10 & 1 & 8 & 19 \\ 92 & 9 & 11 & 122 \end{bmatrix}$$

What? The Endangered Species Act of 1973 protects plant and animal species. *Endangered* species are close to extinction. The survival of *threatened* species is of great concern.

6. How many species of mammals are endangered or threatened? 338
7. *Critical Thinking* The grizzly bear is threatened in the United States but endangered in Mexico. So it is counted twice in the totals. There is a total of four mammal species counted twice. What is the actual total of endangered or threatened mammals? 334

Exercises ON YOUR OWN

What is the size of each matrix?

1. $\begin{bmatrix} 5 & 4 & 7 \\ 6 & 9 & 8 \end{bmatrix}$ 2 × 3
2. $\begin{bmatrix} 4 & 1 \end{bmatrix}$ 1 × 2
3. $\begin{bmatrix} 1 & 2 \\ 3 & 4 \\ 5 & 6 \end{bmatrix}$ 3 × 2
4. $\begin{bmatrix} -8 \end{bmatrix}$ 1 × 1
5. $\begin{bmatrix} 16 & -22 & 24 & 35 \\ 17 & -35 & 19 & 41 \\ 28 & -10 & 15 & 50 \end{bmatrix}$ 3 × 4

Additional Examples

FOR EXAMPLE 1

$$\begin{bmatrix} -6 & 8 & 11 \\ 2 & 5 & -3 \\ 1 & 4 & 7 \end{bmatrix} - \begin{bmatrix} 7 & -5 & -2 \\ 11 & 3 & -1 \\ 8 & 6 & 4 \end{bmatrix} =$$

$$\begin{bmatrix} -13 & 13 & 13 \\ -9 & 2 & -2 \\ -7 & -2 & 3 \end{bmatrix}$$

Discussion: *What would happen if the second matrix was a* 3 × 4 *matrix?*

FOR EXAMPLE 2

Write the tables as matrices, then add the matrices.

Week	Small	Medium	Large
Hamburgers	420	404	352
French fries	392	320	308
Shakes	187	159	152

Week	Small	Medium	Large
Hamburgers	385	315	367
French fries	439	406	302
Shakes	220	201	165

total:

	small	medium	large
hamburgers	805	719	719
french fries	831	726	610
shakes	407	360	317

Discussion: *How do matrices allow us to handle large amounts of data easily?*

Exercise 7 Reassure students that a matrix containing only zeros is quite acceptable. Students may think they have the wrong answer because the matrix looks strange.

Exercise 9 Remind students that, for two matrices to be equal, each pair of corresponding entries must be equal.

ALTERNATIVE ASSESSMENT **Exercises 10–17** Separate students into small groups with half the class solving the exercises by addition and the rest by subtraction. Have the answers written on index cards. Students can then solve the exercises using the opposite procedure and compare their results with other groups. This activity will help you assess students' understanding of the subtraction of negative numbers and the combining of matrices.

DIVERSITY **Exercise 18** Some students may be unfamiliar with the concept of investing in a company. Explain to the class how you invest in a company by buying shares of stock. Smart investors try to buy shares in companies that they believe will be highly successful. Have students brainstorm to decide which companies they think will grow in the near future.

Exercises MIXED REVIEW

Exercise 27–30 Have volunteers remind their classmates of the rules of the order of operations and of the rules for adding, subtracting, multiplying, and dividing integers.

GETTING READY FOR LESSON 1-9 These exercises give students practice substituting values into a given formula.

Technology Options

For Exercises 10–17, students may use graphing calculators to combine matrices.

Prentice Hall Technology

Calculator Graphing Calculator Handbook: Procedure 13

Software • Secondary Math Lab Toolkit • Computer Item Generator 1-8

CD-ROM Multimedia Algebra Lab 1

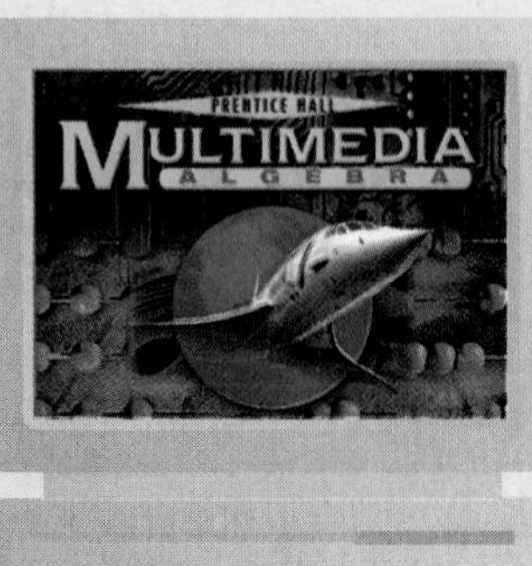

Internet • See the Prentice Hall site. (http://www.phschool.com)

Mental Math **Find each sum or difference.**

6. $\begin{bmatrix} 1 & 2 \\ 5 & 3 \\ -1 & -4 \end{bmatrix} + \begin{bmatrix} 2 & 0 \\ 9 & 1 \\ 3 & -8 \end{bmatrix}$
7. $\begin{bmatrix} 4 & 2 \\ 1 & -3 \end{bmatrix} + \begin{bmatrix} -4 & -2 \\ -1 & 3 \end{bmatrix}$
8. $\begin{bmatrix} -6 & -1 & 7 \\ 3 & -2 & -5 \end{bmatrix} - \begin{bmatrix} -8 & 6 & -2 \\ 14 & -3 & 1 \end{bmatrix}$

6. $\begin{bmatrix} 3 & 2 \\ 14 & 4 \\ 2 & -12 \end{bmatrix}$ 7. $\begin{bmatrix} 0 & 0 \\ 0 & 0 \end{bmatrix}$ 8. $\begin{bmatrix} 2 & -7 & 9 \\ -11 & 1 & -6 \end{bmatrix}$

9. What is the value of m, n, and p in the matrices?
$\begin{bmatrix} 5 & -1 \\ m & 0 \end{bmatrix} = \begin{bmatrix} n & -1 \\ -6 & p \end{bmatrix}$ −6; 5; 0

Add each pair of matrices. Then subtract the second matrix from the first matrix in each pair. See margin for 10–17.

10. $\begin{bmatrix} 16 & 11 & -20 \\ 22 & 8 & -10 \end{bmatrix}, \begin{bmatrix} 9 & 10 & 14 \\ -15 & -10 & 5 \end{bmatrix}$
11. $\begin{bmatrix} 42 & -36 \\ -10 & 54 \end{bmatrix}, \begin{bmatrix} 39 & 4 \\ 37 & 46 \end{bmatrix}$
12. $\begin{bmatrix} 3 & 2 & 1 \\ 4 & 2 & 0 \\ 0 & -2 & 5 \end{bmatrix}, \begin{bmatrix} 1 & -2 & 3 \\ -4 & 5 & -6 \\ 7 & -8 & 9 \end{bmatrix}$
13. $\begin{bmatrix} 3.4 & 2.1 \\ -6.4 & 5.7 \\ 8.8 & -9.3 \end{bmatrix}, \begin{bmatrix} 4.9 & -7.9 \\ 3.8 & -2.8 \\ 4.2 & 1.5 \end{bmatrix}$
14. $\begin{bmatrix} 5.0 & -1.7 \\ 1.2 & -3.8 \\ 1.5 & 2.4 \end{bmatrix}, \begin{bmatrix} 5.7 & 6.8 \\ -4.0 & -1.2 \\ 6.2 & 8.1 \end{bmatrix}$
15. $\begin{bmatrix} \frac{2}{3} & \frac{4}{5} \\ \frac{1}{8} & \frac{1}{6} \end{bmatrix}, \begin{bmatrix} \frac{1}{2} & \frac{2}{5} \\ \frac{3}{4} & 1 \end{bmatrix}$
16. $\begin{bmatrix} 356 & -190 & 171 \\ -256 & 321 & -150 \end{bmatrix}, \begin{bmatrix} 1 & 6 & -4 \\ -17 & -3 & 5 \end{bmatrix}$
17. $\begin{bmatrix} 3.5 & -0.2 \\ 0 & -7 \end{bmatrix}, \begin{bmatrix} -1 & 0.5 \\ -8.1 & 9.7 \end{bmatrix}$

18. a. *Sports* Write each table of data as a matrix. Check students' work.
b. Make a matrix that shows the change in participation from 1987 to 1992. See margin.
c. Did any categories lose participants? Which ones? See below.
d. In which category did the number of participants increase the most? basketball 7–11
e. *Writing* Suppose you have money to invest in sports equipment. In which sport would you invest? Use entries from your matrix to explain your choice. See margin.

Participation in Sports Activities (in millions)

	Sport	7–11 yr	12–17 yr	18–24 yr
1987	Basketball	3.7	8.3	5.1
	Tennis	1.0	3.4	3.9
	Soccer	3.6	4.1	0.9
	Volleyball	2.0	6.6	4.7

	Sport	7–11 yr	12–17 yr	18–24 yr
1992	Basketball	5.5	8.2	4.9
	Tennis	1.4	3.2	3.9
	Soccer	4.2	3.8	1.3
	Volleyball	1.6	5.2	5.1

19. *Critical Thinking* Suppose two matrices are equal. What will the matrix for their difference look like? Each entry will be zero.

20. a. *Open-ended* Make two 3×5 matrices.
b. Add your two matrices. 20a–c. See margin.
c. Subtract your two matrices.

18c. basketball 12–17 and 18–24, tennis 12–17, soccer 12–17, volleyball 7–11 and 12–17

Wrap Up

THE BIG IDEA Ask students to describe what a matrix is and to explain why matrices are useful.

RETEACHING ACTIVITY Students add and subtract corresponding matrix entries to add and subtract matrices. (Reteaching worksheet 1-8)

Exercises CHECKPOINT

In this Checkpoint, your students will have an opportunity to assess their own progress on the topics covered in Lessons 1-6 through 1-8.

Exercise 3 Remind students that dividing by a fraction is the same as multiplying by the reciprocal of the fraction.

Exercise 5 Emphasize that when students subtract matrices, they should check the positive and negative signs carefully to avoid mistakes.

Simplify each entry in the matrix.

21. $\begin{bmatrix} \frac{5}{6} \div \frac{2}{3} & -4 - 2\frac{1}{2} \\ \left(3\frac{1}{3}\right)\left(\frac{3}{4}\right) & \frac{1}{2} - \left(\frac{2}{5}\right)\left(\frac{5}{4}\right) \end{bmatrix}$ $\begin{bmatrix} 1\frac{1}{4} & -6\frac{1}{2} \\ 2\frac{1}{2} & 0 \end{bmatrix}$

22. $\begin{bmatrix} 8 + 2 \div 4 & 2^5 & -12 - 15 \\ 45 + 5(-13) & \frac{10 + 16}{4} & 4 - 2^2 \end{bmatrix}$ $\begin{bmatrix} 8.5 & 32 & -27 \\ -20 & 6.5 & 0 \end{bmatrix}$

23. a. *Jobs* Create a matrix to find the total number of workers in each pay category for each work shift. See margin.

b. How many weekend employees on the evening shift earn \$5.50/h? 4

c. *Critical Thinking* Suppose all employees work 8-h shifts both Saturday and Sunday. How could you use the matrix to find the total wages of the weekend employees? Multiply each entry times 8 and times the hourly wage. Then add all the products. See margin.

Number of Employees

Saturday Schedule

Shift	\$5.25	\$5.50	\$6.00	\$6.50
Day	8	3	5	1
Evening	10	2	2	1
Night	4	1	0	1

(Hourly Wage)

Sunday Schedule

Shift	\$5.25	\$5.50	\$6.00	\$6.50
Day	5	2	1	1
Evening	8	2	0	1
Night	2	1	0	1

(Hourly Wage)

Multiply each matrix by the given value.

Sample $2\begin{bmatrix} 2 & -3 \\ 1 & 5 \end{bmatrix} = \begin{bmatrix} 2(2) & 2(-3) \\ 2(1) & 2(5) \end{bmatrix} = \begin{bmatrix} 4 & -6 \\ 2 & 10 \end{bmatrix}$ ← Multiply each entry by the factor outside the matrix.

24. $3\begin{bmatrix} -9 & 2 \\ 6 & -12 \end{bmatrix}$ **25.** $-5\begin{bmatrix} 70 & 30 & -10 \\ 80 & 20 & -50 \end{bmatrix}$ **26.** $0\begin{bmatrix} -15 & 12 \\ 17 & 21 \end{bmatrix}$

24–26. See margin.

Exercises MIXED REVIEW

Evaluate each expression for $a = -5$, $b = 2.4$, $c = -0.5$, and $d = 3.7$.

27. $a + (-b) + c$ −7.9 **28.** $2b + (-d) + ac$ 3.6 **29.** $a \div c + 2d$ 17.4 **30.** $a^2 + bc$ 23.8

Getting Ready for Lesson 1-9

Geometry **Use the formula $A = \frac{1}{2}bh$ to find the area of each triangle.**

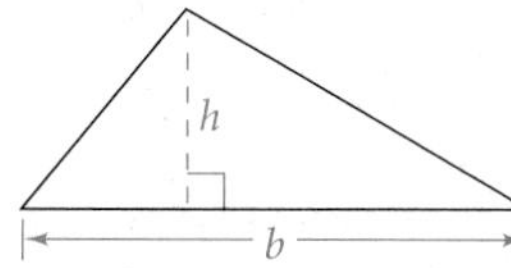

31. b = 12 in., h = 4 in. 24 in.^{2}

32. b = 15 cm, h = 7 cm 52.5 cm^2

33. b = 9.2 m, h = 14.8 m 68.08 m^2

Exercises CHECKPOINT

Evaluate each expression.

1. $2a - b$ for $a = \frac{3}{4}$, $b = \frac{7}{8}$ $\frac{5}{8}$ **2.** $3m - 2n$ for $m = 3.5$, $n = -0.8$ 12.1 **3.** $p \div q$ for $p = -\frac{5}{8}$, $q = -\frac{4}{5}$ $\frac{25}{32}$

Simplify.

4. $\begin{bmatrix} 3 & 5 \\ -4 & 2 \end{bmatrix} + \begin{bmatrix} \frac{1}{2} & -6 \\ 8 & -11 \end{bmatrix}$ $\begin{bmatrix} 3\frac{1}{2} & -1 \\ 4 & -9 \end{bmatrix}$

5. $\begin{bmatrix} 3.3 & 0.4 & 3.0 \\ -1.7 & 9.6 & -6.5 \end{bmatrix} - \begin{bmatrix} -2.3 & 7.2 & -8.5 \\ 4.1 & 6.3 & 9.2 \end{bmatrix}$ $\begin{bmatrix} 5.6 & -6.8 & 11.5 \\ -5.8 & 3.3 & -15.7 \end{bmatrix}$

6. *Open-ended* Write a problem you could solve by using a simulation. Describe how the simulation would be done. See margin.

Reteaching 1-8

Practice 1-8

Practice 1-8 *Mixed Exercises*

Lesson Quiz

Lesson Quiz is also available in Transparencies.

1. Subtract the following matrices.

$\begin{bmatrix} 5 & 2 \\ 3 & 6 \end{bmatrix} - \begin{bmatrix} 3 & -6 \\ 4 & -1 \end{bmatrix}$ $\begin{bmatrix} 2 & 8 \\ -1 & 7 \end{bmatrix}$

2. Put values in for each letter that will result in all numbers in the answer matrix being one.

$\begin{bmatrix} 5 & -7 & 4 \\ -2 & 15 & -6 \end{bmatrix} + \begin{bmatrix} -4 & a & b \\ c & -14 & d \end{bmatrix}$

$a = 8$, $b = -3$, $c = 3$, $d = 7$

pages 41–43 On Your Own

10–17, 18b, 18e, 20a–c, 23a, 24–26.

page 43 Checkpoint

6. See back of book.

Math ToolboX

In Lesson 1-8, students determined the size of matrices then added and subtracted them. This toolbox helps students see the importance of correctly determining matrix size. It will provide practice for those who have difficulty with the fine motor skills used in calculator math. Students will be immediately notified by an error message should they attempt to add matrices of unlike size. Once two matrices are entered into the calculator, students enjoy adding and then subtracting them without reentering the numbers. This lesson will also prepare students for solving simultaneous equations with matrices and the graphing calculator in Chapter 6.

ERROR ALERT! Students often incorrectly use the subtraction key instead of the negative key when entering negative numbers. **Remediation:** Briefly discuss the placement of the keys used in this toolbox, emphasizing the difference between the subtraction and negative keys. Have students work in pairs, one to press the keys and the other to watch and catch errors.

ADDITIONAL PROBLEM Find [A] − [B] and [A] + [B]

$$[A] = \begin{bmatrix} 3.1 & -4 & 5.2 \\ 5 & 2.2 & 4 \end{bmatrix} \quad [B] = \begin{bmatrix} 6.3 & 7 & -3 \\ -2.1 & 5.7 & 8 \end{bmatrix}$$

$$[A] - [B] = \begin{bmatrix} -3.2 & -11 & 8.2 \\ 7.1 & -3.5 & -4 \end{bmatrix}$$

$$[A] + [B] = \begin{bmatrix} 9.4 & 3 & 2.2 \\ 2.9 & 7.9 & 12 \end{bmatrix}$$

Materials and Manipulatives

- Graphing calculator

page 44 Math Toolbox

1. $\begin{bmatrix} 4.9 & 0.7 \\ 2.1 & 0.7 \\ -3.1 & 2.7 \end{bmatrix}; \begin{bmatrix} 2.7 & 3.5 \\ 0.9 & -2.3 \\ 0.3 & 1.1 \end{bmatrix}$

2. $\begin{bmatrix} 0 & 4 & 20 & -25 \\ 8 & 2 & -3 & 15 \\ -9 & -1 & -14 & 9 \end{bmatrix};$ $\begin{bmatrix} 10 & -12 & -6 & -1 \\ 0 & -20 & -17 & -5 \\ 21 & -13 & 2 & 15 \end{bmatrix}$

3. $\begin{bmatrix} 235 & 151 \\ 204 & 113 \end{bmatrix}; \begin{bmatrix} 11 & 23 \\ 20 & 13 \end{bmatrix}$

4. $\begin{bmatrix} 33 & 115 & -164 \\ 137 & 114 & 155 \\ -191 & 162 & 141 \\ -28 & 202 & 25 \end{bmatrix};$ $\begin{bmatrix} 211 & -19 & 38 \\ 53 & 220 & 1 \\ -29 & -44 & -51 \\ -146 & 26 & 239 \end{bmatrix}$

5a. $\begin{bmatrix} 4.9 & 0.7 \\ 2.1 & 0.7 \\ -3.1 & 2.7 \end{bmatrix}; \begin{bmatrix} -2.7 & -3.5 \\ -0.9 & 2.3 \\ -0.3 & -1.1 \end{bmatrix}$

Math ToolboX Technology

Matrices

After Lesson 1-8

You can use your graphing calculator to add or subtract matrices. In the calculator, a matrix is named using a variable. Some calculators will put brackets, [ ], around a variable to indicate a matrix. Before entering the values for a matrix, you must enter the size of the matrix.

$$[A] = \begin{bmatrix} 1 & 2 & 3 \\ 4 & 5 & 6 \end{bmatrix} \quad [B] = \begin{bmatrix} 6 & 9 & 2 \\ -1 & 4 & -7 \end{bmatrix}$$

← Remember that $[A]$ and $[B]$ can be added or subtracted because they are the same size, 2 × 3.

The key sequences below are for entering and adding two matrices. Do all the steps for matrix A, then for matrix B.

	Matrix A	Matrix B
To edit a matrix:	MATRX ▶ ▶ 1	MATRX ▶ ▶ 2
To enter matrix size:	2 ENTER 3 ENTER	2 ENTER 3 ENTER
To enter values for row 1:	1 ENTER 2 ENTER 3 ENTER	6 ENTER 9 ENTER 2 ENTER
To enter values for row 2:	4 ENTER 5 ENTER 6 ENTER	(−) 1 ENTER 4 ENTER (−) 7 ENTER
To go to the Home screen:	2nd QUIT	2nd QUIT

To add $[A]$ and $[B]$: MATRX 1 + MATRX 2 ENTER

Your calculator screen will display the sum.

```
[A]+[B]
      [[ 7  11  5 ]
       [ 3   9 -1 ]]
```

Use a graphing calculator. Find $[A] + [B]$ and $[A] - [B]$. 1–4. See margin.

1. $[A] = \begin{bmatrix} 3.8 & 2.1 \\ 1.5 & -0.8 \\ -1.4 & 1.9 \end{bmatrix}, [B] = \begin{bmatrix} 1.1 & -1.4 \\ 0.6 & 1.5 \\ -1.7 & 0.8 \end{bmatrix}$

2. $[A] = \begin{bmatrix} 123 & 87 \\ 112 & 63 \end{bmatrix}, [B] = \begin{bmatrix} 112 & 64 \\ 92 & 50 \end{bmatrix}$

3. $[A] = \begin{bmatrix} 5 & -4 & 7 & -13 \\ 4 & -9 & -10 & 5 \\ 6 & -7 & -6 & 12 \end{bmatrix}, [B] = \begin{bmatrix} -5 & 8 & 13 & -12 \\ 4 & 11 & 7 & 10 \\ -15 & 6 & -8 & -3 \end{bmatrix}$

4. $[A] = \begin{bmatrix} 122 & 48 & -63 \\ 95 & 167 & 78 \\ -110 & 59 & 45 \\ -87 & 114 & 132 \end{bmatrix}, [B] = \begin{bmatrix} -89 & 67 & -101 \\ 42 & -53 & 77 \\ -81 & 103 & 96 \\ 59 & 88 & -107 \end{bmatrix}$

5. **a.** Use the matrices in Exercise 1 to find $[B] + [A]$ and $[B] - [A]$. See margin.
   b. *Writing* Does changing the order of the matrices affect the result when you add matrices? subtract matrices? Explain. No; yes; addition and subtraction of matrices is like addition and subtraction of real numbers: $a + b = b + a$ but $a - b \neq b - a$.

1-9 Teaching Notes

PROBLEM OF THE DAY

At the park as I watched a group of ducks and squirrels, I counted 19 heads and 54 legs. How many ducks and how many squirrels were there? **11 ducks and 8 squirrels**

Problem of the Day is also available in Transparencies.

CONNECTING TO PRIOR KNOWLEDGE Bring to class an order form from a mail-order catalog. Fill the order blanks with fictitious items, quantities, and prices, but leave the subtotals and totals blank. Hand out copies to the class. Ask: *How many of you are familiar with ordering from catalogs? How long would it take you to calculate all the subtotals and totals? Do you see a pattern in how the subtotals and totals are calculated?* Have students brainstorm ways they could use this pattern to make the process quicker and easier.

THINK AND DISCUSS

Draw students' attention to the Technology Hint at the bottom of the page. Explain that, because of this feature of spreadsheets, it is not necessary to type in the formulas for cells D3 through D11. Once the formula for cell D2 is typed in, you can simply copy that formula and then paste it into cells D3 through D11. The software will automatically adjust each formula to refer to the correct cells. In other words, the formula C2 − B2 is changed by the computer to say C3 − B3. Help students understand how much time and effort this can save.

ESTIMATION Question 1a Ask students to describe how they could estimate their hours. Have students compare their

---

Connections Package Delivery . . . and more

What You'll Learn

- Using variables and formulas in spreadsheets

...And Why

To investigate real-world situations, such as television viewership

1-9 Variables and Formulas in Spreadsheets

THINK AND DISCUSS

How much time do you spend watching television? Is it about average for a teenager? You can use the spreadsheet below to find out.

Like a matrix, a **spreadsheet** organizes data in rows and columns. A **cell** is a box where a row and column meet.

Hours of Weekly TV Viewing

	A	B	C	D
1	Viewing Group	Nov. 91	Nov. 93	Change
2	Children 2–5	26.4	21.5	–4.9
3	Children 6–11	21.2	20.0	–1.2
4	Female Teens	22.2	20.8	–1.4
5	Male Teens	22.7	21.2	–1.5
6	Women 18–24	28.9	25.7	–3.2
7	Men 18–24	23.0	22.5	–0.5
8	Women 25–54	32.8	30.6	–2.2
9	Men 25–54	28.1	28.1	0
10	Women 55 and over	43.5	44.2	0.7
11	Men 55 and over	39.8	38.5	–1.3

column, cell B4, row

1. a. *Estimate* About how many hours do you watch TV in a week?
   b. Compare your weekly TV viewing hours with the November 1993 average for your group.
   1a–b. Check students' work.

You can use spreadsheet formulas to calculate the values in column D. The spreadsheet you see on a computer screen shows the value in the cell. The formula used to calculate the value is not shown in the spreadsheet.

Cell	Formula in Cell	Value Shown in Cell
D2	C2 − B2	−4.9
D3	C3 − B3	−1.2
D4	C4 − B4	−1.4
⋮	⋮	⋮

TECHNOLOGY HINT

A spreadsheet program allows you to put a formula in one cell and copy its format in other cells.

You can save time by using expressions to write formulas in cells. Suppose you change the value in cell B4. The computer will automatically evaluate the formula C4 − B4 and change the value in cell D4.

Lesson Planning Options

Prerequisite Skills

- Evaluating formulas
- Understanding the terms row and column

Assignment Options

Core 3–5, 7–15
Extension 1, 2, 6, 16–19

Use Mixed Review to maintain skills.

Resources

Student Edition

Skills Handbook, pp. 580, 582
Extra Practice, p. 556
Glossary/Study Guide

Teaching Resources

Chapter Support File, Ch. 1
- Practice 1-9 (two worksheets)
- Reteaching 1-9
- Alternative Activity

Classroom Manager 1-9
Glossary, Spanish Resources
Two-Year Algebra Handbook 1-9

Transparencies
3, 36, 40

methods. Discuss the advantages and disadvantages of each method.

DIVERSITY Some students will enjoy manipulating spreadsheets on the computer while others will be hesitant for fear of making mistakes. Encourage the more confident students to use their knowledge to help others by answering their questions without touching their computer's keys. Be sure all students have an opportunity to use the computer and the spreadsheet software.

Example 1

ERROR ALERT! Students may accidentally refer to the wrong cell when writing their formulas. **Remediation:** Encourage students to check each cell reference by tracing along the rows to the cell with one finger, and down the columns to the cell with another finger.

Question 2 and 3 Have students exchange formulas to check whether they are correct. Have them make sure they used the correct symbols for multiplication and exponents.

Question 4 Tell students that, although the spreadsheet software will follow the order of operations rules, students should test the spreadsheet's answer by calculating the value of the expression themselves for an example cell.

Example 2 **Relating to the Real World** 

VISUAL AND TACTILE LEARNING Have students name objects in the classroom that they estimate would have a delivery size of between 108 in. and 130 in. Measuring the objects will

Additional Examples

FOR EXAMPLE 1

	A	B	C	D
1	x	$0.5*x$	x^2	$x + 3$
2	0.5			
3	1.5			
4	2.5			

What are the spreadsheet formulas which evaluate cells B3, C2, and D4? **B3: 0.5*A3; C2: A2^2; D4: A4+3**

Discussion: *Describe ways in which computer spreadsheet cell formulas look different from mathematical formulas.*

FOR EXAMPLE 2

	A	B	C
1	height	area of base	volume
2	10	2	
3	15	4	
4	20	6	

Find the volume of each box given the heights and area of bases. Remember volume = height × area of base. **C2 = 20; C3 = 60; C4 = 120**

A computer spreadsheet program uses these operation symbols:

- To multiply: 3 * 5 means $3 \cdot 5$.
- To divide: 3/5 means $3 \div 5$.
- To raise to a power: 3^5 means 3^5.

Example 1

Cell A2 contains a value for the variable x. Write a spreadsheet formula for cell B2 to evaluate the expression in cell B1.

	A	B	C	D	E
1	x	x^3	x − 3	3x	■
2	−4	■	■	■	■
3	−1	■	■	■	■

Expression	Cell	Formula in Cell	Value Shown in Cell
x^3	B2	A2^3	−64

2. What formulas would the computer use to find the values in cells C2 and D2? **A2 − 3; 3 * A2**
3. Write the formulas and find the values for cells B3, C3, and D3. **A3 ^ 3; A3 − 3; 3 * A3; −1, −4, −3**
4. Suppose you use column E to evaluate $2x^3 - 5$. What spreadsheet formula would you use in cell E2? **2 * A2 ^ 3 − 5**

Example 2 **Relating to the Real World** 

Package Delivery Both the United States Postal Service (USPS) and the United Parcel Service (UPS) add the girth of a package to its length to find its delivery size. Both USPS and UPS measure dimensions in inches.

Write a formula for cell D2 that will find the delivery size of a package. Then find the delivery size of the package in row 2.

	A	B	C	D
1	width	height	length	delivery size
2	5	10	20	■
3	16	28	30	■
4	18.5	20.5	26	■

formula for cell D2: girth + length
$= 2w + 2h + l$
$= 2 * A2 + 2 * B2 + C2$

delivery size: $2 \cdot 5 + 2 \cdot 10 + 20$
$= 50$

The delivery size of the package in row 2 is 50 in.

girth = $2w + 2h$

HANDLE WITH CARE

width w

length l

height h

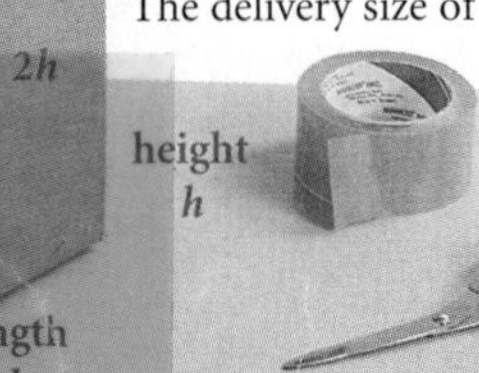

The length of a package is its longest dimension.

appeal to students with tactile learning styles, so let some volunteers measure the objects while the rest of the class calculates the actual delivery sizes. Compare the calculated delivery sizes to their estimates. Encourage students to comment on any differences.

CONNECTING TO THE STUDENTS' WORLD If you have an extended class period or block scheduling, use this opportunity to bring in newspaper advertisements for new cars which list different payment plan options. Collect advertisements for several different types of cars. Have students design a spreadsheet that would calculate the total purchase price for each payment plan. The spreadsheet formula should incorporate the down payment, monthly payment, number of months, and final payment.

Exercises ON YOUR OWN

CRITICAL THINKING Exercise 1 Students who are having difficulty getting started may need to write some formulas they have used and some spreadsheet formulas so they can compare and contrast them.

Exercise 2 Many students may not know how taxes and income after taxes are calculated. Ask students who understand to explain this process to the class.

ALTERNATIVE ASSESSMENT **Exercises 4–5** Before students calculate the value of each cell, ask them to write the math sentence for the expression by replacing the variables with their correct value. This procedure helps you assess whether

5. The maximum delivery size for packages sent by USPS is 108 in. For UPS the maximum delivery size is 130 in. Find the values in cells D3 and D4. Which package or packages qualify for UPS delivery but not for USPS delivery? 118; 104; package in row 3

6. *Critical Thinking* Suppose for the package in row 4 you used 26 in. as the height and 20.5 in. as the length. Is this helpful to the person mailing the package? Explain. No; this would make the delivery size 109.5 in. and would disqualify this package from USPS delivery.

Exercises ON YOUR OWN

1. *Critical Thinking* How are spreadsheet formulas similar to other kinds of formulas you have used? How are they different? Answers may vary. Sample: Both kinds of variables represent changing values. For spreadsheets, a variable has a letter and number for the cell it refers to.
2. Suppose you are an employer. You must find your employees' wages, taxes, and income after taxes. Can a spreadsheet program save you time? Explain. Yes; you need to do the same computations many times. The spreadsheet formulas can do this automatically.

Use the spreadsheet.

	A	B	C	D	E
1	x	4x^2	2x – 5	3x + 7	■
2	7	■	■	■	■
3	–2	■	■	■	■

3. Write the formulas you would use in cells B2, C2, and D2 to evaluate the expressions in cells B1, C1, and D1. 4 * A2 ^ 2; 2 * A2 – 5; 3 * A2 + 7
4. Find the values for cells B2, C2, and D2. 196, 9, 28
5. Find the values for cells B3, C3, and D3. 16, –9, 1
6. *Open-ended* Write an expression of your own for cell E1. Then write the formula for cell E2 that you would use to evaluate your expression. Sample: 2x ^ 3 – 2; 2 * A2 ^ 3 – 2

Evaluate each spreadsheet expression.

7. mean of three numbers: (A2 + B2 + C2)/3
   a. for A2 = –3, B2 = – 9, C2 = 3 –3
   b. for A2 = –10.8, B2 = 4.5, C2 = 0.6 –1.9
8. total of hourly wage minus deductions: A4 * B4 – C4
   a. for A4 = 6, B4 = 10, C4 = 20 40
   b. for A4 = 11.70, B4 = 40, C4 = 135 333
9. *Geometry* The formula for the volume of a rectangular prism is $V = lwh$. The spreadsheet gives the dimensions in feet for three rectangular prisms.
   a. Write a formula for cell D2 to find the volume of the rectangular prism in row 2. A2 * B2 * C2
   b. Find the volume of each rectangular prism. See table.

	A	B	C	D
1	length	width	height	volume
2	20	12	8	■ 1920
3	14.8	9.8	5	■ 725.2
4	30	25.4	7	■ 5334

10. *Writing* You can find the volume of a cube using the formula $V = s^3$, where s is the length of a side. Explain how to set up a spreadsheet to find the volumes of cubes with sides 4, 12, 3.8, and 6.25. See above.

10. Answers may vary. Sample: Make a list of entries in column A. In column B, enter the formulas to cube the column A entry from the same row. For example, in B2 enter A2^3.

Technology Options

For Exercises 12–15, students may use spreadsheet software to analyze data.

Prentice Hall Technology

Software • Secondary Math Lab Toolkit • Integrated Math Lab 2 • Computer Item List Generator 1-9

Internet • See the Prentice Hall site. (http://www.phschool.com)

students understand the use of variables and formulas in spreadsheets. It is important for students to realize the efficiency that spreadsheet formulas can bring to handling large amounts of data.

Exercises 7–8 Point out that the cells in these problems do not refer to any spreadsheets in the text.

GEOMETRY Exercise 9 Students are asked to find the volume of a rectangular prism using a spreadsheet.

Exercise 11 Ask students: *Will the values in column one be used in the formula?* no *Why or why not?* The year is not necessary to calculate the mean year wage.

STANDARDIZED TEST TIP **Exercise 19** Challenge students to think about the kinds of mistakes that someone could make when setting up this spreadsheet. Also have students brainstorm ways to find errors and the possible consequences of these errors.

Chapter Project FIND OUT BY CREATING Students are to measure the tibia, humerus, radius bones, and the heights of several adults. They will then create and organize a spreadsheet. Using the formulas on page 29, students will predict the heights and compare them with the measured heights. This information is essential to their work on the Chapter Project, which was introduced in the Chapter Opener. Have students add this task to work they have already completed for the project. You may wish to check students' progress on the project by having each group write a progress report telling what they have done so far and describing their plans for completion.

pages 47–49 On Your Own

17.

	A	B	C	D	E
1	Month	Number of Earrings	Buying Cost	Selling Cost	Monthly Profit
2	January	127	6.75 * B2	8.95 * B2	D2 − C2
3	February	174	6.75 * B3	8.95 * B3	D3 − C3
4	March	156	6.75 * B4	8.95 * B4	D4 − C4

11. a. *Business* Write a formula for cell D2 in the spreadsheet for Milo Construction Company. C2/B2

b. Find the values for cells D2, D3, D4, and D5. See below.

c. Suppose a worker earned the mean yearly wage in 1995. She worked 40 h each week and 50 weeks in a year. What was her hourly wage? $15.43

11b. $18,569.23; $24,259.50; $27,702; $30,861.03

Milo Construction Company

	A	B	C	D
1	Year	Number of Workers	Total Wages Paid to All Workers	Mean Yearly Wage
2	1980	13	$241,400	■
3	1985	20	$485,190	■
4	1990	35	$969,570	■
5	1995	39	$1,203,580	■

Geometry **Evaluate each spreadsheet expression.**

12. area of a parallelogram: B2 * C2
a. for B2 = 9, C2 = 12 108
b. for B2 = 15.3, C2 = 8.9 136.17

13. surface area of a sphere: 4/3 * 3.14 * A2^3
a. for A2 = $\frac{3}{4}$ 1.766 or 1.76625
b. for A2 = 7.2 1562.665

14. volume of a cylinder: 3.14 * B3^2 * C3
a. for B3 = 5, C3 = 8 628
b. for B3 = 3.5, C3 = 4.2 161.55

15. surface area of a rectangular prism:
2 * B2 * C2 + 2 * B2 * D2 + 2 * C2 * D2
a. for B2 = 2, C2 = 3, D2 = 4 52
b. for B2 = 3.5, C2 = 5.8, D2 = 4.2 118.72

16. a. How are the values in column G below related to values in columns A and D? They are the sum.

b. Write formulas to find the values of the cells in columns G and H.

c. *Critical Thinking* How does this spreadsheet model adding matrices? Corresponding entries are added.

G	H
A1 + D1	B1 + E1
A2 + D2	B2 + E2
A3 + D3	B3 + E3

	A	B	C	D	E	F	G	H
1	−3	4		−2	−4		−5	0
2	0	5		−1	7		−1	12
3	3	−2		4	−6		7	−8

17. *Computer* Eva Arturo owns Shine On, a small shop in the mall. She buys earrings for $6.75 and sells them for $8.95. The store sold 127 pairs of earrings in January, 174 in February, and 156 in March. Create a spreadsheet to calculate each amount. 17a–b. See margin.

a. the monthly earring sales **b.** the monthly profit from earring sales

18. *Critical Thinking* In Exercise 17, what computer formula could you use to find the total profit for all three months? E2 + E3 + E4 (in sample for Ex. 17)

Exercises MIXED REVIEW

Exercises 20–29 Have students first write the order of operations rules on a piece of paper. Then have them exchange their rules with a partner to verify that what they have written is correct and to discuss any differences

PORTFOLIO Share with students the criteria for assessing their work in portfolios. Students should also understand how the rubrics are used to assess their work, how each piece in the portfolio is scored, and how the scores they get in their portfolios will affect their overall evaluation.

Wrap Up

THE BIG IDEA Ask students to brainstorm how spreadsheets can help them evaluate formulas.

RETEACHING ACTIVITY Students use formulas and variables to calculate values and spreadsheets. (Reteaching worksheet 1-9)

Population of Four Cities in the United States

	A	B	C	D	E	F
1	City	1950 Population	1990 Population	Population change (1950 to 1990)	Area (mi^2)	1990 Density (people/mi^2)
2	Fort Wayne, IN	133,607	172,971	▪	62.7	▪
3	Albuquerque, NM	96,815	384,619	▪	132.2	▪
4	Houston, TX	596,163	1,629,902	▪	539.9	▪
5	Norfolk, VA	213,513	261,229	▪	53.8	▪

Source: *The World Almanac* and *The Statistical Abstract of the United States*

19. *Standardized Test Prep* For the spreadsheet, which statements are correct? C

I. The formula for cell D2 is C2 − B2.
II. In column D, negative entries indicate a population loss.
III. The formula for cell F5 is B5/E5.

A. I and III **B.** II and III **C.** I and II **D.** I, II, and III

Chapter Project *Find Out by Creating*

- Measure the tibia, humerus, and radius bones and the height of several adults to the nearest half-inch.
- Create a spreadsheet. Organize the measurements. Use the formulas from the Find Out question on page 29 in your spreadsheet to predict the heights of the adults.
- Compare the predicted heights with the measured heights. Does one of the formulas predict heights better than the other formulas? Explain.

Exercises MIXED REVIEW

Simplify each expression.

20. $8 - 14 \div 2$ 1 **21.** $(8 - 14) \div 2$ −3

22. $|8 - 32 \div 4| + 1$ 1 **23.** $(20 - 12) \div (3 + 1)$ 2

24. $9(3 + 5) + 6 \div 3$ 74 **25.** $[9(3 + 5) + 6] \div 3$ 26

26. $\frac{6 - 18}{2}$ −6 **27.** $\frac{-5 - 4^2}{-7}$ 3

28. $6 + 3^2 - 2 \cdot 55$ **29.** $10 - 4 \cdot 8 - 7^2$ −71

30. a. A notebook costs $1.95. Write an equation to find the cost of any number of notebooks. $c = 1.95n$

b. What is the cost of three notebooks? $5.85

PORTFOLIO

Select one or two items from your work for this chapter. Consider:
- corrected work
- diagrams, graphs, or charts
- a journal entry

Explain why you have included each selection that you make.

Name______ Class ______ Date ______

Reteaching 1-9

Name______ Class ______ Date ______

Practice 1-9

Name______ Class ______ Date ______

Practice 1-9

Mixed Exercises

1. Evaluate the spreadsheet expression (A3 + B3)/C3.
   a. for A3 = 6.8, B3 = 7.2, and C3 = 4
   b. for A3 = −12, B3 = 35 and C3 = −2

Use the spreadsheet at right for Exercises 2–4.

2. Write the spreadsheet formulas for cells B2 and C2.
3. Find the value for cells B2 and C2.
4. Find the value for cells B3 and C3.

	A	B	C
1	x	x^3	(x + 4)/5
2	5	▪	▪
3	−1	▪	▪

5. Evaluate the spreadsheet expression (B2 + C2)^2.
   a. for B2 = 8 and C2 = 14
   b. for B2 = −12.6 and C2 = 8.6
   c. for B2 = −7 and C2 = 3
6. A teacher wishes to use these percents to compute grades.
   Homework 20%
   Quiz 35%
   Test 45%
   Use the spreadsheet at right to answer **a–b.**
   a. Write the formula for cell E2.
   b. Find the values for cells E2 and E3.

	A	B	C	D	E
1	Student	Homework	Quiz	Test	Grade
2	A. Garcia	90	96	94	▪
3	C. Ho	94	92	88	▪

7. Evaluate the spreadsheet expression 0.5 * B2 * (C2 + D2).
   a. for B2 = 10, C2 = 4, and D2 = 8
   b. for B2 = 4, C2 = −4, and D2 = −6
8. Evaluate the spreadsheet expression (A2 + 2 * B2)/4.
   a. for A2 = 2 and B2 = 8
   b. for A2 = 1.6 and B2 = −2.5

Use the spreadsheet at right for Exercises 9–11.

9. Write the spreadsheet formulas for cells B2, C2, D2.
10. Find the values for cells B2, C2, and D2.
11. Find the values for cells B3, C3, and D3.

	A	B	C	D
1	x	x/4	(2x)^2	x^2 + x
2	2	▪	▪	▪
3	6	▪	▪	▪

Lesson Quiz

Lesson Quiz is also available in Transparencies.

1. Determine the cell values for D1, D2, and D3.

	A	B	C	D
1	37	390	15	B1/C1 = 26
2	42	294	14	B2/C2 = 21
3	79	323	17	B3/C3 = 19

2. Write the cell values for D1 and D2.

	A	B	C	D
1	5.80	30	18%	B1*C1+A1 = 11.2
2	7.50	25	19.5%	B2*C2+A2 = 12.375

Finishing the Chapter Project

PROJECT DAY You may wish to plan a project day on which students share their completed projects. Encourage groups to explain their process as well as their product.

PROJECT NOTEBOOK Have students review their project work and bring their notebooks up to date.

- Have students review their methods for finding, recording, and displaying the data they needed for the project.
- Ask groups to share insights they found for completing the project, such as shortcuts for doing graphs and spreadsheets. Also, ask if any mathematical ideas have become more obvious, and ask if there are areas about which they would like to learn more.

SCORING RUBRIC

3 Appropriate types of graphs and charts were chosen. The scales are accurate. The graphs and charts are well-labeled and contain accurate formulas and calculations. The spreadsheet is easy to follow. Your reasoning and explanations are correct and clearly expressed.

2 Student's spreadsheet, graphs, and formulas are correctly chosen and used. There are minor errors in scale or computation. Reasoning and explanations are essentially correct, but may contain awkward or unclear passages.

1 Some graphs or formulas are correct. Computations and scales contain errors. Explanations are barely adequate.

0 Major elements of the project are incomplete or missing.

CHAPTER PROJECT

Finishing the Chapter Project

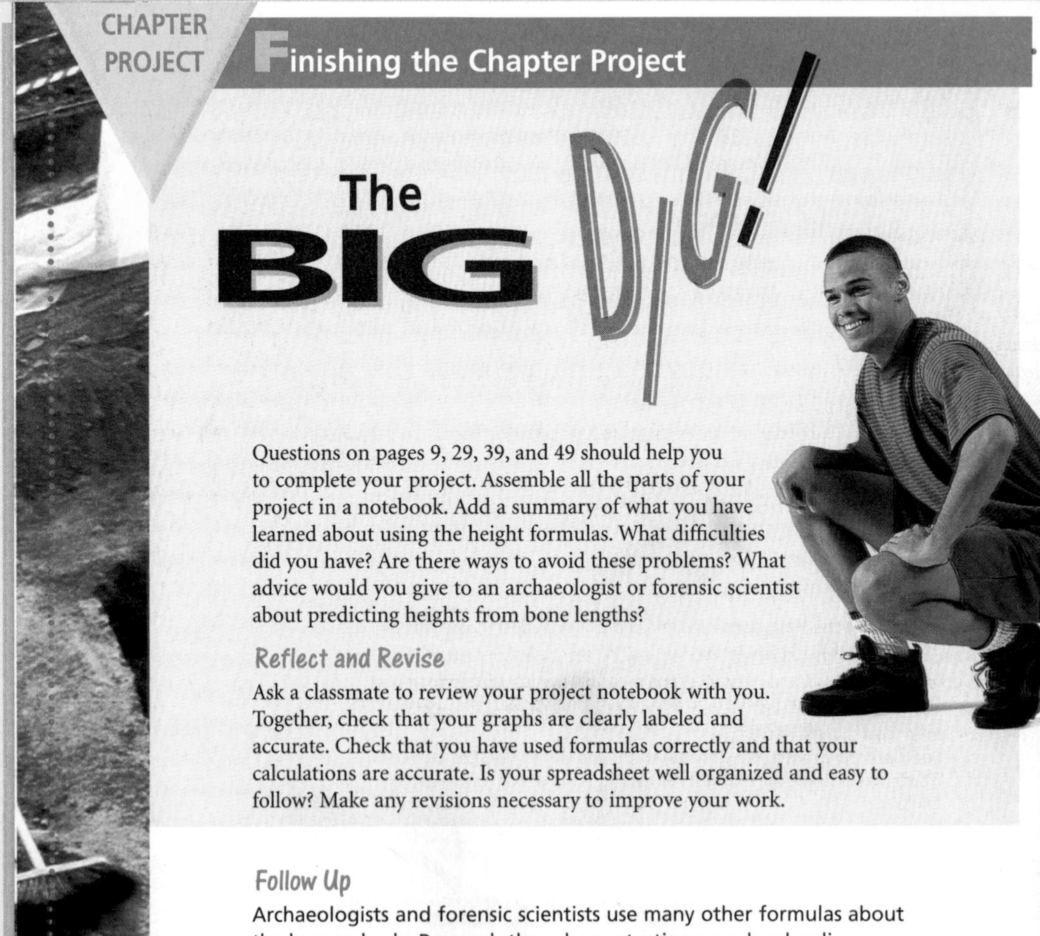

Questions on pages 9, 29, 39, and 49 should help you to complete your project. Assemble all the parts of your project in a notebook. Add a summary of what you have learned about using the height formulas. What difficulties did you have? Are there ways to avoid these problems? What advice would you give to an archaeologist or forensic scientist about predicting heights from bone lengths?

Reflect and Revise

Ask a classmate to review your project notebook with you. Together, check that your graphs are clearly labeled and accurate. Check that you have used formulas correctly and that your calculations are accurate. Is your spreadsheet well organized and easy to follow? Make any revisions necessary to improve your work.

Follow Up

Archaeologists and forensic scientists use many other formulas about the human body. Research these by contacting your local police department or by using one of the resources listed below.

For More Information

Avi-Yonah, Michael. *Dig This! How Archaeologists Uncover Our Past.* Minneapolis, Minnesota: Runestone, 1993.

Stones and Bones! How Archaeologists Trace Human Origins. Prepared by the Geography Department, Runestone Press. Minneapolis, Minnesota: Runestone, 1994.

"It's a Boy." *Scholastic Math Magazine* (April 1993):12–13.

"The Arm Bone's Connected to Math." *Scholastic Math Magazine* (October 1992): 8–9.

Wrap Up

HOW AM I DOING? Ask students to write their responses in journals and share their thoughts with a partner in small group sessions.

KEY TERMS The numbers in parentheses direct students to the pages where the terms (or symbols) are used or defined. Students should be able to (1) write a simple explanation of each term, (2) illustrate the term with a diagram, or (3) show an example that uses a term.

Exercises 1–3 Make sure students understand the difference between mean, median, and mode.

Exercise 11 Have students compare their variable expressions with those of the rest of the class to see how many different expressions were written.

Exercises 13–33 Remind students to be careful with positive and negative signs.

Exercises 35–37 Emphasize that students will have no trouble arranging data as long as everything is kept in an organized and systematic manner.

Remind students that the new mathematical terms in this chapter are defined in the Glossary/Study Guide in the back of the book.

1 Wrap Up

Key Terms

absolute value (p. 20)
bar graph (p. 6)
base (p. 15)
cell (p. 45)
entry (p. 40)
equation (p. 11)
evaluate (p. 16)
event (p. 36)
experimental probability (p. 36)
exponent (p. 15)
integers (p. 19)
irrational numbers (p. 30)
line graph (p. 7)
line plot (p. 4)
matrix (p. 40)
mean (p. 5)
median (p. 5)
mode (p. 4)
opposite (p. 19)
order of operations (p. 15)
rational numbers (p. 30)
real numbers (p. 30)
reciprocal (p. 32)
simulation (p. 37)
spreadsheet (p. 45)
term (p. 11)
variable (p. 11)
variable expression (p. 11)

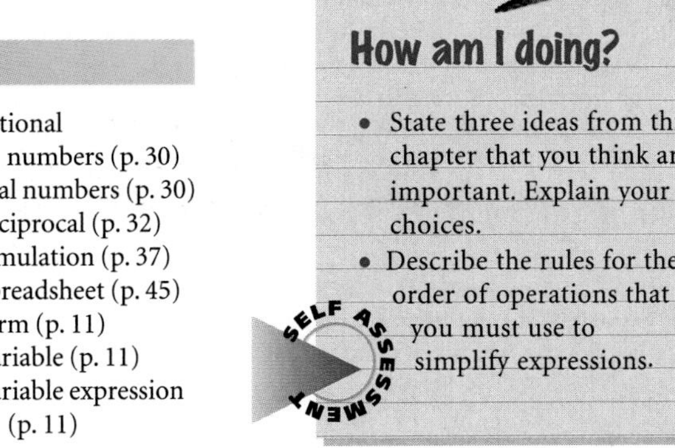

How am I doing?
• State three ideas from this chapter that you think are important. Explain your choices.
• Describe the rules for the order of operations that you must use to simplify expressions.
SELF ASSESSMENT

Displaying Data Relationships with Graphs — 1-1

You can use a **line plot** or **histogram** to show frequency, or the number of times a data item occurs. A **bar graph** compares amounts. A **line graph** shows how a set of values changes over time. **Mean, median,** and **mode** are three measures of central tendency.

Find the mean, median, and mode for each set of data.

1. 24, 45, 33, 27, 24 30.6; 27; 24
2. 80, 87, 81, 92, 87, 80, 83 84.3; 83; 80 and 87
3. 2.4, 2.3, 2.1, 2.5, 2.3, 2.2 2.3; 2.3; 2.3

Use the bar graph.

4. Find the difference between the 1996 and the 1997 sales for March. 4
5. Which month had the most sales in 1996? March
6. Find the average number of video recorders sold during the first three months for each year. 4.7 recorders; 8 recorders

Sales of Video Recorders

Number of Sales: 0, 2, 4, 6, 8, 10
1996, 1997
Jan. Feb. Mar.
Month

Variables, Order of Operations, Evaluating Expressions — 1-2, 1-3

A **variable** represents changing values. To **evaluate** a variable expression, you substitute a given number for each variable. Then you simplify the expression using the **order of operations.**

Evaluate each expression for $a = 3$, $b = 2$, and $c = 1$.

7. $4a - b^2$ 8
8. $9(a + 2b) + c$ 64
9. $\frac{2a + b}{2}$ 4
10. $2a^2 - (4b + c)$ 9

Resources

Student Edition

Extra Practice, p. 556
Glossary/Study Guide

Teaching Resources

Study Skills Handbook
Glossary, Spanish Resources

pages 51–53 Chapter 1 Wrap Up

35. $\begin{bmatrix} 7 & 7 \\ -2 & -10 \end{bmatrix}; \begin{bmatrix} -3 & -13 \\ 20 & 8 \end{bmatrix}$

36. $\begin{bmatrix} 4.3 & 0.6 \\ 1.0 & -0.3 \\ -1.7 & 1.2 \end{bmatrix}; \begin{bmatrix} -1.9 & -2.8 \\ 4.4 & 6.1 \\ -4.7 & -3.4 \end{bmatrix}$

37. $\begin{bmatrix} \frac{1}{8} \\ -\frac{7}{20} \end{bmatrix}; \begin{bmatrix} -\frac{9}{8} \\ 1\frac{23}{20} \end{bmatrix}$

38. Use a spreadsheet expression that finds the sum of the hours worked and multiplies the sum by the hourly wage.

11. *Open-ended* Use 3, –, 6, x, 5, and + to write a variable expression: Simplify your expression. Answers may vary. Sample: $3 - 6 + x - 5 = x - 8$

12. A data entry operator can input 140 records into a computer each day. The table shows the relationship between the number of days worked and the number of records entered.
   a. Write an equation to describe this relationship. $r = 140d$
   b. How many records will be entered after 21 days? after 30 days? 2940 records; 4200 records

Days	Records Entered
1	140
2	280
3	420
4	560

Adding and Subtracting Integers 1-4

Two numbers are **opposites** if they are the same distance from zero on the number line. Whole numbers and their opposites are **integers.** The **absolute value** of a number is its distance from zero on a number line.

To add two numbers with the same sign, *add* their absolute values. The sum has the same sign as the addends. To add two numbers with different signs, find the *difference* of their absolute values. The sum has the same sign as the number with the greater absolute value. To subtract a number, add its opposite.

Add or subtract.

13. $(-13) + (-4)$ -17
14. $5 - 17$ -12
15. $-12.4 + 22.3$ 9.9
16. $|54.3 - 29.4|$ 24.9
17. $-12 - (-7)$ -5

Evaluate each expression for $a = -5$, $b = 8$, and $c = -7$.

18. $a + c$ -12
19. $b - c$ 15
20. $-a + b$ 13
21. $|c - a|$ 2
22. $-|b| + |c|$ -1

Multiplying and Dividing Integers 1-5

The product or quotient of two integers with the same sign is *positive.* The product or quotient of two integers with different signs is *negative.*

Evaluate each variable expression for $x = 4$, $y = -2$, and $z = -3$.

23. $-3^2 + z^2$ 0
24. $y + (-4)x$ -18
25. $5xy$ -40
26. $\frac{x}{-2} + 2z$ -8
27. $\frac{x - y}{3}$ 2

28. Evaluate $a^2 - 2b^3$ for $a = -5$ and $b = -3$. Choose the correct answer. B
   A. -29 B. 79 C. 43 D. 241 E. -181

Real Numbers and Rational Numbers 1-6

A **rational number,** such as $\frac{5}{8}$ or $\frac{2}{3}$, is a ratio of two integers. An **irrational number** like π or $\sqrt{2}$, cannot be written as a ratio of integers. The product of a nonzero number and its **reciprocal** is 1.

Evaluate each expression for $a = \frac{5}{6}$ and $b = -\frac{2}{3}$.

29. $a - 3b$ $2\frac{5}{6}$
30. $3a - 4b$ $5\frac{1}{6}$
31. $6ab$ $-3\frac{1}{3}$
32. $a \div b$ $-1\frac{1}{4}$
33. $a + b^2$ $1\frac{5}{18}$

ALTERNATIVE ASSESSMENT Students may review the work of a partner after they have completed the exercises. They should evaluate which problems were found to be difficult. Students may explain to each other how they approached the solution. Allow students to summarize their findings in small groups.

Getting Ready for Chapter 2

Students may work these exercises independently or in small groups. The skills previewed will help prepare students to work with equations and functions.

Experimental Probability and Simulations 1-7

Experimental probability is based on data gathered through observations. The probability of an event, *P*(event), equals the number of times an event happens divided by the number of times the experiment is done.

34. The results of a blind taste test of fruit juices are at the right. Find each probability.
- **a.** *P* (preferred Brand A) $\frac{3}{14}$ **b.** *P* (preferred Brand B) $\frac{5}{14}$
- **c.** *P* (preferred Brand C) $\frac{3}{7}$ **d.** *P* (did *not* prefer Brand C) $\frac{4}{7}$

Brand A	Brand B	Brand C
卌 I	卌 卌	卌 卌 II

Organizing Data in Matrices 1-8

A **matrix** is a rectangular arrangement of numbers in rows and columns. You can add or subtract matrices if they are the same size. You add and subtract matrices by adding or subtracting corresponding entries.

Find the sum of each pair of matrices. Then find the difference of each pair of matrices by subtracting the second matrix from the first matrix. See margin.

35. $\begin{bmatrix} 2 & -3 \\ 9 & -1 \end{bmatrix}, \begin{bmatrix} 5 & 10 \\ -11 & -9 \end{bmatrix}$

36. $\begin{bmatrix} 1.2 & -1.1 \\ 2.7 & 2.9 \\ -3.2 & -1.1 \end{bmatrix}, \begin{bmatrix} 3.1 & 1.7 \\ -1.7 & -3.2 \\ 1.5 & 2.3 \end{bmatrix}$

37. $\begin{bmatrix} -\frac{1}{2} \\ \frac{2}{5} \end{bmatrix}, \begin{bmatrix} \frac{5}{8} \\ -\frac{3}{4} \end{bmatrix}$

Variables and Formulas in Spreadsheets 1-9

You can use a **spreadsheet** to analyze data. A spreadsheet organizes data in rows and columns. A **cell** is a box where a row and column meet. Spreadsheets use formulas to calculate values.

38. *Writing* A. Jones earns \$5.85/h, and M. Vasquez earns \$6.30/h. Explain how you could use a spreadsheet to find each person's total wage. See margin.

Employee	M	T	W	TH	F
A. Jones	7.8	7.3	7.9	7.2	7.3
M. Vasquez	7.5	7.6	7.4	7.8	7.8

Getting Ready for... CHAPTER 2

Evaluate each expression.

39. $3n + 2$ for $n = 1.2$ 5.6

40. $8(g + 1) - 3$ for $g = -2$ −11

41. $5.4b + 2.6b$ for $b = 10$ 80

Complete the spreadsheet at the right for each value of *x*.

		A	B	C	D	E
	1	x	x + 2	2x	x^2	x^3
42.	2	−2	0	−4	4	−8
43.	3	0	2	0	0	0
44.	4	4	6	8	16	64

Assessment

ENHANCED MULTIPLE CHOICE QUESTIONS are more complex than traditional multiple choice questions, which assess only one skill. Enhanced multiple choice questions assess the processes that students use as well as the end results. The questions are written so that students must use more than one strategy to solve the problem. Using multiple strategies is encouraged by the National Council of Teachers of Mathematics (NCTM). **Exercise 18** is an enhanced multiple choice question.

FREE-RESPONSE QUESTIONS do not give choices of answers. Some exercises have more than one possible answer, although students only need to give one correct answer.

Exercises 1–17 and 21–24 are free-response questions.

WRITING EXERCISES allow students to describe how they think about and understand the concepts they have learned. **Exercises 19 and 25** are writing exercises.

OPEN-ENDED PROBLEMS allow for more than one solution. Students must construct their own responses instead of choosing from possible answers. The student responses will help you determine the depth of understanding and any areas of difficulty experienced by students. **Exercise 20** is an open-ended problem. Make sure that students' responses contain the following elements:

- four rational numbers
- a correct number line, increasing from left to right

Resources

Teaching Resources

Chapter Support File, Ch. 1
- Chapter Assessment, Forms A and B
- Alternative Assessment

Chapter Assessment, Spanish Resources

Teacher's Edition

See also p. 2E for assessment options.

Software

Computer Item Generator

page 54 Chapter 1 Assessment

19a. Sometimes: $4 - 2 = 2$, but $2 - 4 = -2$

b. Never: $3 - (-2) = 3 + 2 = 5$

c. Sometimes: $-6 - (-3) = -6 + 3 = -3$, but $-2 - (-7) = -2 + 7 = 5$

d. Never: $-3 - 8 = -11$

24b. Estimates may vary. Sample:

	1995	1996
1st qtr.	22,000	20,000
2nd qtr.	31,000	42,000
3rd qtr.	29,000	23,000
4th qtr.	14,000	55,000
mean	24,000	35,000

25. Answers may vary. Sample: If the value for a is positive and the value for b is negative, the right hand side of the equation is negative while the left hand side is positive.

1 Assessment

Use an equation to model the relationship in each table.

1.

Number	Cost
1	\$2.30
2	\$4.60
3	\$6.90

$c = 2.30n$

2.

Payment	Change
\$1	\$9
\$2	\$8
\$3	\$7

$c = 10 - p$

Simplify each expression.

3. $3 + 5 - 4$ **4**

4. $8 - 2^4 \div 2$ **0**

5. $\frac{2 \cdot 3 - 1}{3^2}$ $\frac{5}{9}$

6. $36 - (4 + 5 \cdot 4)$ **12**

Evaluate each expression for $x = 3, y = -1$, and $z = 2$.

7. $2x + 3y + z$ **5**

8. $-xyz$ **6**

9. $-3x - 2z - 7$ **−20**

10. $-z^3 - 2z + z$ **−10**

11. $\frac{xy - 3z}{-5}$ $1\frac{4}{5}$

12. $x^2 + (-x)^2$ **18**

Find the sum or difference.

13. $\begin{bmatrix} 3 & 2 \\ -1 & 5 \end{bmatrix} + \begin{bmatrix} 8 & -5 \\ 3 & 0 \end{bmatrix}$ $\begin{bmatrix} 11 & -3 \\ 2 & 5 \end{bmatrix}$

14. $\begin{bmatrix} 1 & 9 & -4 \\ 5 & 2 & -1 \\ -6 & -2 & -1 \end{bmatrix} - \begin{bmatrix} 2 & -6 & 7 \\ -8 & 3 & -3 \\ 4 & -7 & 9 \end{bmatrix}$

See above right.

Explain why each statement is true or false.

15. All rational numbers are integers. **F; $\frac{1}{2}$ is not an integer**

16. The absolute value of a number is always positive. **F; $|0| = 0$, which is not positive**

17. A random survey of 60 students showed that 36 students used calculators for computation. What is the probability that a student chosen at random uses a calculator for computation? $\frac{3}{5}$

18. Complete the following: $3^2 \cdot ■ = 1$ **A**

A. $\frac{1}{9}$ B. -3^2 C. 0 D. 2^3 E. $-\frac{1}{3}$

14. $\begin{bmatrix} -1 & 15 & -11 \\ 13 & -1 & 2 \\ -10 & 5 & -10 \end{bmatrix}$

19. See margin.

19. *Writing* Tell if each of the subtraction sentences would *always*, *sometimes*, or *never* be true. Support your answer with two examples.

a. $(+) - (+) = (+)$ b. $(+) - (-) = (-)$

c. $(-) - (-) = (-)$ d. $(-) - (+) = (+)$

20. *Open-ended* Write four rational numbers. Then use a number line to order them from least to greatest. **Check students' work.**

Evaluate each spreadsheet expression.

21. (A3 + B1 + C2)/4 for A3 = −5.3, B1 = 7.5, C2 = 6.48 **2.17**

22. A1 − B1 * C2 for A1 = 16.09, B1 = 9.16, C2 = −5 **61.89**

23. A softball player made a hit 54 times in the last 171 times at bat. Find the probability that the softball player will get a hit the next time at bat. $\frac{6}{19}$

24. The double line graph shows the quarterly profit for a software company.

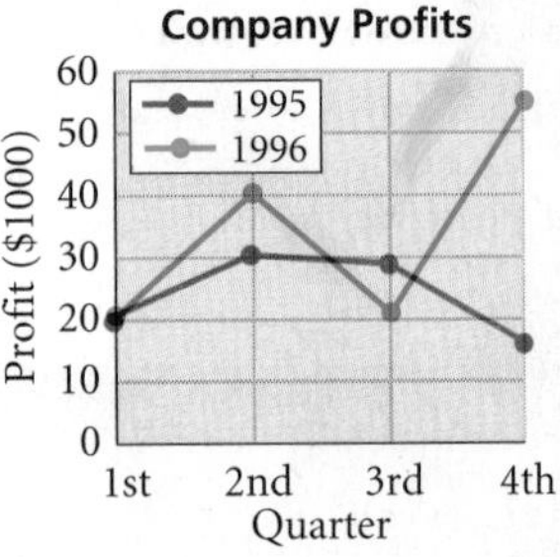

a. In which quarter did the company have its lowest quarterly profit? **4th quarter of 1995**

b. Estimate the company's profit for each quarter. Then find the mean quarterly profit for each year. **See margin.**

25. *Writing* Explain why the equation below is not always true.

$\left|\frac{a}{b}\right| = \frac{a}{b}$ **See margin.**

Preparing for Standardized Tests

Standardized tests, such as those administered for state assessment, the SAT, or the ACT, include regular math questions, quantitative comparison questions, open-ended problems, and free-response questions (which the SAT calls *grid-ins*).

MULTIPLE CHOICE QUESTIONS are followed by five answer choices, one of which is correct. **Exercises 1–9** are multiple choice questions.

QUANTITATIVE COMPARISON QUESTIONS ask students to compare two quantities. **Exercises 10 and 11** are quantitative comparison questions.

FREE-RESPONSE QUESTIONS do not give answer choices. Students must provide the one correct answer on their own. **Exercises 12–14 and 16** are free-response questions.

WRITING EXERCISES allow students to describe how they think about and understand the concepts they have learned. **Exercise 15** is a writing exercise.

STANDARDIZED TEST TIPS Encourage students to follow these four steps when taking standardized tests.

1. Read the entire question.
2. Decide whether to do the problem now or return to it later. If you skip it and want to come back to it later, circle the problem in your test booklet.
3. Look for the fastest way to solve the problem.
4. If you can eliminate one or more answer choices, make an educated guess.

1 Preparing for Standardized Tests

For Exercises 1–11, choose the correct letter.

1. Which of the following are true for the given data? 1, 2, 2, 2, 4, 4, 5, 6, 6, 7, 7 **D**
   I. mode > median II. mean < mode
   III. mean > median IV. median > mode
   A. I and III B. II and IV C. I and II
   D. III and IV E. I, II, and III

2. Simplify $2(5-3)^2 + 4 \div 2$. **C**
   A. 6 B. 8 C. 10
   D. 18 E. None of the above

3. Evaluate $\frac{3x-(-4)}{7}$ for $x = -6$. **A**
   A. –2 B. –3 C. 2
   D. 3 E. None of the above

4. You can find the distance d an object falls in feet for time t in seconds using the formula $d = 16t^2$. How far will a ball dropped out of the window of a high rise fall in 3 s? **C**
   A. 96 ft B. 160 ft C. 144 ft
   D. 16 ft E. 48 ft

5. The Fuller Book Company inspects a sample of 860 books and finds that 172 books have defective bindings. What is the probability that a book has a defective binding? **D**
   A. $\frac{1}{4}$ B. $\frac{1}{3}$ C. $\frac{1}{2}$
   D. $\frac{1}{5}$ E. None of the above

6. Which of the following expressions does *not* equal 27 when $x = -3$? **C**
   A. $-x^3$ B. $-9x$ C. $-(-x)^3$
   D. $3x^2$ E. $24 - x$

7. Tamara's teacher allows students to decide whether to use the mean, median, or mode for their test average. Tamara will receive the highest average if she uses the mean. Which set of test scores are Tamara's? **B**
   A. 95, 82, 76, 95, 96
   B. 79, 80, 91, 83, 80
   C. 65, 84, 75, 74, 65
   D. 100, 87, 94, 94, 81
   E. 89, 82, 84, 89, 79

8. Complete the following: $8^2 \cdot ■ = 4^2$ **E**
   A. 4 B. 2^4 C. $\frac{1}{8}$ D. 2^3 E. $\frac{1}{4}$

9. Which of the following lists the numbers in order from least to greatest? **B**
   A. $\frac{5}{6}, \frac{2}{5}, \frac{6}{7}$ B. $\frac{2}{5}, \frac{5}{6}, \frac{6}{7}$ C. $\frac{6}{7}, \frac{5}{6}, \frac{2}{5}$
   D. $\frac{2}{5}, \frac{6}{7}, \frac{5}{6}$ E. None of the above

Compare the boxed quantity in column A with the boxed quantity in column B. Choose the best answer.

A. The quantity in Column A is greater.
B. The quantity in Column B is greater.
C. The two quantities are equal.
D. The relationship cannot be determined on the basis of the information supplied.

	Column A	Column B
10.	-3^4	-4^3

B

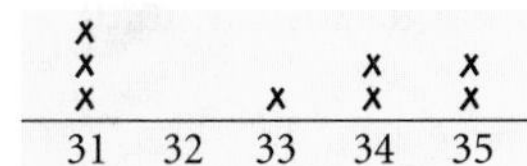

11.	median of data in line plot	mode of data in line plot

A

Find each answer.

12. $\begin{bmatrix} -2 & -1 \\ 3 & 0 \end{bmatrix} - \begin{bmatrix} -1 & 2 \\ 0 & 1 \end{bmatrix}$ $\begin{bmatrix} 1 & 3 \\ 3 & 1 \end{bmatrix}$

13. Evaluate $\frac{|x-2|}{|2x+10|}$ for $x = -4$. **3**

14. Find $\frac{3}{8} \div \frac{5}{16}$. $1\frac{1}{5}$

15. *Writing* Use the order of operations to explain why the statement $2(5+4) = 2(5) + 4$ is false. **See margin.**

16. Simplify $\frac{3^2 + 3^2 + 3^2}{4^2 + 4^2 + 4^2}$. $\frac{9}{16}$

Resources

Teaching Resources

Chapter Support File, Ch. 1
- Standardized Test Practice
- Cumulative Review

See also p. 2E for assessment options.

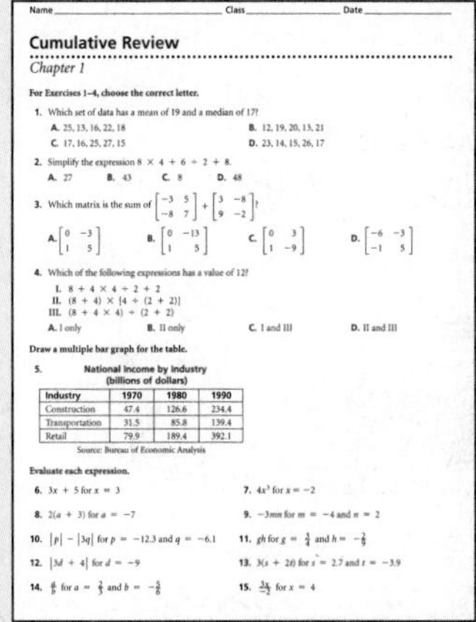
Name______ Class______ Date______

Cumulative Review

Chapter 1

For Exercises 1–4, choose the correct letter.

1. Which set of data has a mean of 19 and a median of 17?
   A. 25, 13, 16, 22, 18 B. 12, 19, 20, 13, 21
   C. 17, 16, 25, 27, 15 D. 23, 14, 15, 26, 17
2. Simplify the expression $8 \times 4 + 6 \div 2 + 8$.
   A. 27 B. 43 C. 8 D. 48
3. Which matrix is the sum of $\begin{bmatrix} -3 & 5 \\ -8 & 7 \end{bmatrix} + \begin{bmatrix} 3 & -8 \\ 9 & -2 \end{bmatrix}$?
   A. $\begin{bmatrix} 0 & -3 \\ 1 & 5 \end{bmatrix}$ B. $\begin{bmatrix} 0 & -13 \\ 1 & 5 \end{bmatrix}$ C. $\begin{bmatrix} 0 & 3 \\ 1 & -9 \end{bmatrix}$ D. $\begin{bmatrix} -6 & -3 \\ -1 & 5 \end{bmatrix}$
4. Which of the following expressions has a value of 12?
   I. $8 + 4 \times 4 \div 2 + 2$
   II. $(8 + 4) \times [4 \div (2 + 2)]$
   III. $(8 + 4 \times 4) \div (2 + 2)$
   A. I only B. II only C. I and III D. II and III

Draw a multiple bar graph for the table.

5. National Income by Industry (billions of dollars)

Industry	1970	1980	1990
Construction	47.4	126.6	234.4
Transportation	31.5	85.8	139.4
Retail	79.9	189.4	392.1

Source: Bureau of Economic Analysis

Evaluate each expression.

6. $3x + 5$ for $x = 3$
7. $4x^3$ for $x = -2$
8. $2(a + 3)$ for $a = -7$
9. $-3mn$ for $m = -4$ and $n = 2$
10. $|p| - |3q|$ for $p = -12.3$ and $q = -6.1$
11. gh for $g = \frac{3}{4}$ and $h = -\frac{2}{9}$
12. $|3d + 4|$ for $d = -9$
13. $3(s + 2t)$ for $s = 2.7$ and $t = -3.9$
14. $\frac{a}{b}$ for $a = \frac{2}{3}$ and $b = -\frac{5}{6}$
15. $\frac{3x}{-2}$ for $x = 4$

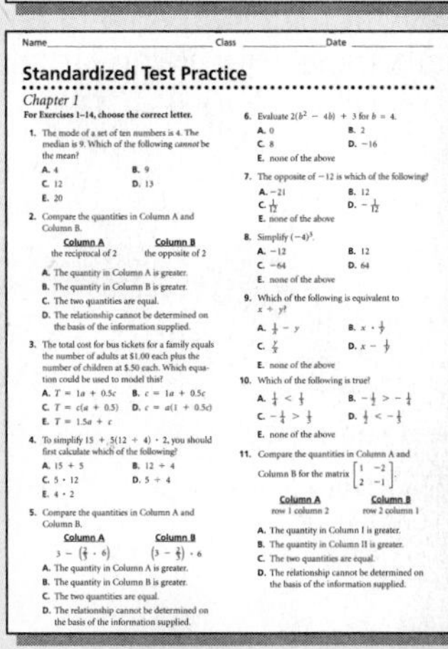
Name______ Class______ Date______

Standardized Test Practice

Chapter 1

For Exercises 1–14, choose the correct letter.

1. The mode of a set of ten numbers is 4. The median is 9. Which of the following *cannot* be the mean?
   A. 4 B. 9
   C. 12 D. 13
   E. 20
2. Compare the quantities in Column A and Column B.
   Column A: the reciprocal of 2 Column B: the opposite of 2
   A. The quantity in Column A is greater.
   B. The quantity in Column B is greater.
   C. The two quantities are equal.
   D. The relationship cannot be determined on the basis of the information supplied.
3. The total cost for bus tickets for a family equals the number of adults at $1.00 each plus the number of children at $.50 each. Which equation could be used to model this?
   A. $T = 1a + 0.5c$ B. $c = 1a + 0.5c$
   C. $T = c(a + 0.5)$ D. $c = a(1 + 0.5c)$
   E. $T = 1.5a + c$
4. To simplify $15 + 5(12 \div 4) \cdot 2$, you should first calculate which of the following?
   A. $15 + 5$ B. $12 \div 4$
   C. $5 \cdot 12$ D. $5 \div 4$
   E. $4 \cdot 2$
5. Compare the quantities in Column A and Column B.
   Column A: $3 - \left(\frac{2}{3} \cdot 6\right)$ Column B: $\left(3 - \frac{2}{3}\right) \cdot 6$
   A. The quantity in Column A is greater.
   B. The quantity in Column B is greater.
   C. The two quantities are equal.
   D. The relationship cannot be determined on the basis of the information supplied.
6. Evaluate $2(b^2 - 4b) + 3$ for $b = 4$.
   A. 0 B. 2
   C. 8 D. –16
   E. none of the above
7. The opposite of –12 is which of the following?
   A. –21 B. 12
   C. $\frac{1}{12}$ D. $-\frac{1}{12}$
   E. none of the above
8. Simplify $(-4)^3$.
   A. –12 B. 12
   C. –64 D. 64
   E. none of the above
9. Which of the following is equivalent to $x \div y$?
   A. $\frac{1}{x} - y$ B. $x \cdot \frac{1}{y}$
   C. $\frac{y}{x}$ D. $x - \frac{1}{y}$
   E. none of the above
10. Which of the following is true?
   A. $\frac{1}{4} < \frac{1}{5}$ B. $-\frac{1}{2} > -\frac{1}{4}$
   C. $-\frac{1}{4} > \frac{1}{3}$ D. $\frac{1}{2} < -\frac{1}{3}$
   E. none of the above
11. Compare the quantities in Column A and Column B for the matrix $\begin{bmatrix} 1 & -2 \\ 2 & -1 \end{bmatrix}$.
   Column A: row 1 column 2 Column B: row 2 column 1
   A. The quantity in Column I is greater.
   B. The quantity in Column II is greater.
   C. The two quantities are equal.
   D. The relationship cannot be determined on the basis of the information supplied.

15. For $2(5 + 4)$, add the values inside the parantheses first. For $2(5) + 4$, multiply first.

55

CHAPTER 2 Functions and Their Graphs

To accommodate flexible scheduling, some lessons are divided into parts. Assignment Options are given in the Lesson Planning Options for each lesson.

2-1 Analyzing Data Using Scatter Plots (pp. 59–63)
- Part 1 Drawing and Interpreting Scatter Plots
- Part 2 Sketching Graphs

Key Terms: scatter plot, trend line, positive correlation, negative correlation, no correlation

2-2 Relating Graphs to Events (pp. 64–68)
- Part 1 Interpreting Graphs
- Part 2 Sketching Graphs
- Part 3 Classifying Data

Key Terms: continuous data, discrete data

2-3 Linking Graphs to Tables (pp. 69–72)

Key Terms: dependent variable, independent variable

2-4 Functions (pp. 73—78)
- Part 1 Identifying Relations and Functions
- Part 2 Evaluating Functions
- Part 3 Analyzing Graphs

Key Terms: relation, function, function rule, domain, vertical-line test

2-5 Writing a Function Rule (pp. 79–83)
- Part 1 Understanding Function Notation
- Part 2 Using a Table of Values
- Part 3 Using Words to Write a Rule

Key Terms: function notation

2-6 The Three Views of a Function (pp. 84–88)

2-7 Families of Functions (pp. 90–94)
- Part 1 Identifying the Family of an Equation
- Part 2 Identifying the Family of a Graph

Key Terms: families of functions, linear functions, quadratic functions, absolute value functions

2-8 The Probability Formula (pp. 95–99)
- Part 1 Finding Theoretical Probability
- Part 2 Using a Tree Diagram to Find a Sample Space

Key Terms: outcomes, theoretical probability, impossible event, certain event, complement of an event, sample space

PACING OPTIONS

This chart suggests pacing only for the core lessons and their parts, and it is provided merely as a possible guide. It will help you determine how much time you have in your schedule to cover other features, such as the Chapter Project, Math Toolboxes, Wrap Up, and Assessment.

	1 Class Period	1 Class Period	1 Class Period	1 Class Period	1 Class Period	1 Class Period	1 Class Period	1 Class Period	1 Class Period
Traditional (40–45 min class periods)	2-1 Part 1	2-1 Part 2	2-2 Part 1, 2-2 Part 2, 2-2 Part 3	2-3	2-4 Part 1, 2-4 Part 2	2-4 Part 3	2-5 Part 1, 2-5 Part 2	2-5 Part 3	2-6
Two-Year Algebra (40–45 min class periods)	2-1 Part 1	2-1 Part 2	2-2 Part 1, 2-2 Part 2	2-2 Part 3	2-3	2-4 Part 1	2-4 Part 1	2-4 Part 2	2-4 Part 2
Block Scheduling (90 min class periods)	2-1 Part 1, 2-1 Part 2, 2-2 Part 1	2-2 Part 2, 2-2 Part 3, 2-3	2-4 Part 1, 2-4 Part 2, 2-4 Part 3	2-5 Part 1, 2-5 Part 2, 2-5 Part 3	2-6, 2-7 Part 1, 2-7 Part 2	2-8 Part 1, 2-8 Part 2			

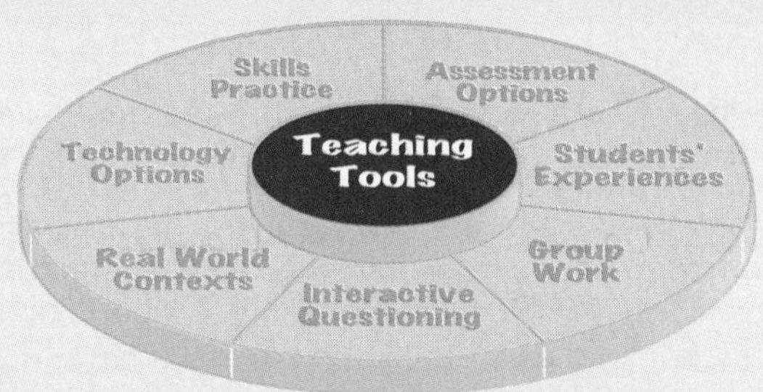

What Students Will Learn and Why

In this chapter, students will build on their knowledge of algebra skills learned in Chapter 1, by learning to predict trends, classify data as continuous or discreet, and identify dependent and independent variables in tables and graphs. They will learn to use the vertical-line test to analyze graphs. Students will write rules from tables and words, examine three views of a function, and study families of functions. Finally, they will use tree diagrams and sample space to learn about some probability concepts, and they will make predictions about real-world situations such as mall promotions.

Discussing the Chapter/Building on Experience

The concept map below relates chapter topics to real-world applications. You and your class may wish to add to the map or develop maps of your own. The center oval describes the topic of the chapter. The next level displays topics within the lessons. The outer ovals reflect applications of the content. As you and your class build a concept map, invite students to discuss applications with which they are familiar.

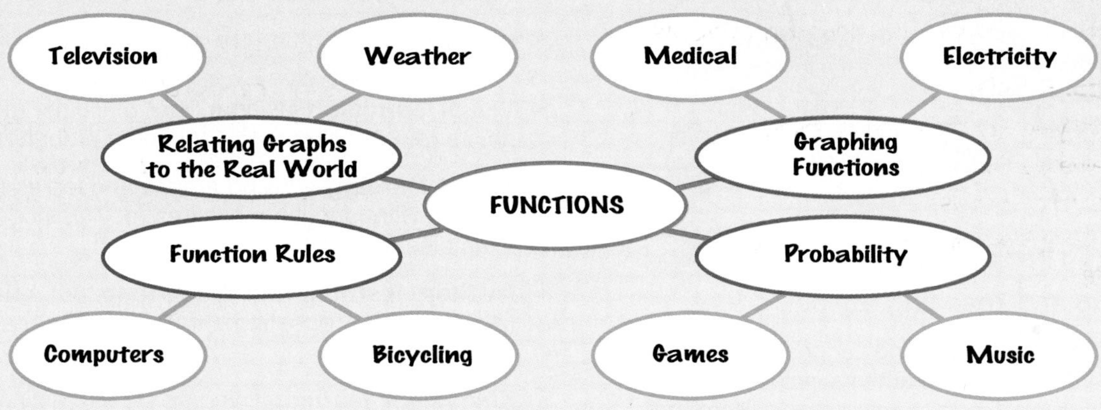

1 Class Period	1 Class Period	1 Class Period	1 Class Period	1 Class Period	1 Class Period	1 Class Period	1 Class Period	1 Class Period	1 Class Period	1 Class Period
2-7 ▼1	2-7 ▼2	2-8 ▼1 2-8 ▼2								
2-4 ▼3	2-5 ▼1	2-5 ▼2	2-5 ▼3	2-6	2-6	2-7 ▼1	2-7 ▼1	2-7 ▼2	2-8 ▼1	2-8 ▼2

Overview

Interactive Questioning Tips

A question is interactive when there is "give and take" between the questioner (teacher and student) and the respondent. When an open-ended question is asked, it is important that the questioner WAIT for a response. Question 6b in Lesson 2-2 asks students to give an example of each type of data (continuous and discreet). A student's response here is determined not only by their understanding of the concept but also their experience, which must be called up and considered.

Skills Practice

Every lesson provides skill practice with Exercises On Your Own and Exercises Mixed Review. The Student Edition includes Checkpoints (pp. 68 and 83) and a Cumulative Review (p. 105). In the Teacher's Edition, the Lesson Planning Options section for each lesson lists Prerequisite Skills students should know for that lesson. At the back of the Student Edition is the Skills Handbook—mini-lessons on math the students may need to review, The Chapter Support File for Chapter 2 in the Teaching Resources box includes two Practice worksheets per lesson, a worksheet for two Checkpoints, and worksheets for Cumulative Review and Standardized Test Preparation.

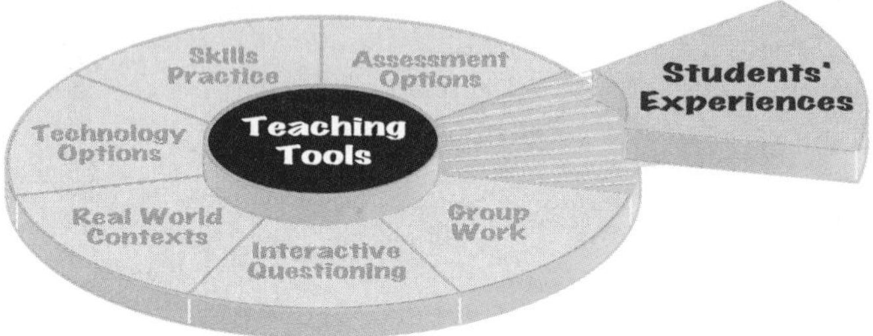

Diverse Learning and Teaching Styles

In your Teacher's Edition, you will find suggestions as to how you can help students complete mathematical tasks in Chapter 2 by reinforcing various learning styles. Here are some examples.

- **Visual learning** making a table of values (p. 81), plotting points (p. 85), inserting events on a probability line (p. 100).
- **Tactile learning** drawing graphs and tables (pp. 70, 91), working with function machines (p. 74), modeling probability examples (p. 96).
- **Auditory learning** writing a song (p. 60), questioning aloud (p. 69), quietly repeating phrases (p. 73), reading aloud (p. 96).
- **Kinesthetic learning** lining up (p. 59), acting out (p. 66), moving to appropriate zones (p. 74), flipping coins (p. 75), positioning on coordinate axes on the floor (pp. 86, 92).

Alternative Activity for Lesson 2-1

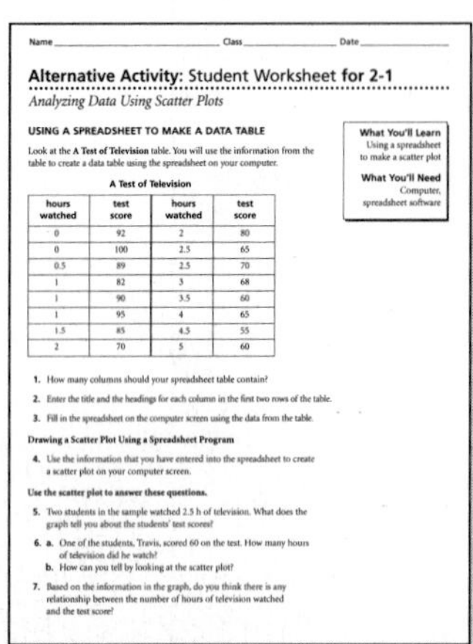

Name __________ Class ________ Date ________

Alternative Activity: Student Worksheet for 2-1

Analyzing Data Using Scatter Plots

What You'll Learn Using a spreadsheet to make a scatter plot

What You'll Need Computer, spreadsheet software

USING A SPREADSHEET TO MAKE A DATA TABLE

Look at the **A Test of Television** table. You will use the information from the table to create a data table using the spreadsheet on your computer.

A Test of Television

hours watched	test score	hours watched	test score
0	92	2	80
0	100	2.5	65
0.5	89	2.5	70
1	82	3	68
1	90	3.5	60
1	95	4	65
1.5	85	4.5	55
2	70	5	60

1. How many columns should your spreadsheet table contain?
2. Enter the title and the headings for each column in the first two rows of the table.
3. Fill in the spreadsheet on the computer screen using the data from the table.

Drawing a Scatter Plot Using a Spreadsheet Program

4. Use the information that you have entered into the spreadsheet to create a scatter plot on your computer screen.

Use the scatter plot to answer these questions.

5. Two students in the sample watched 2.5 h of television. What does the graph tell you about the students' test scores?
6. a. One of the students, Travis, scored 60 on the test. How many hours of television did he watch?
   b. How can you tell by looking at the scatter plot?
7. Based on the information in the graph, do you think there is any relationship between the number of hours of television watched and the test score?

for use with Example 1, addresses tactile and visual learning by using spreadsheet software on a computer to graph data from a table.

Alternative Activity for Lesson 2-3

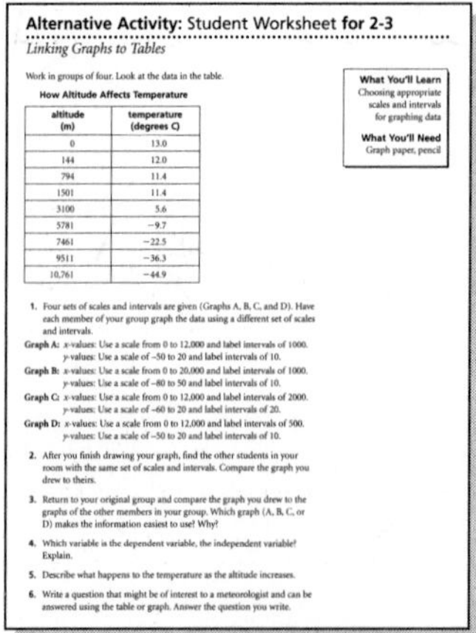

Alternative Activity: Student Worksheet for 2-3

Linking Graphs to Tables

What You'll Learn Choosing appropriate scales and intervals for graphing data

What You'll Need Graph paper, pencil

Work in groups of four. Look at the data in the table.

How Altitude Affects Temperature

altitude (m)	temperature (degrees C)
0	13.0
144	12.0
794	11.4
1501	11.4
3100	5.6
5781	−9.7
7461	−22.5
9511	−36.3
10,761	−44.9

1. Four sets of scales and intervals are given (Graphs A, B, C, and D). Have each member of your group graph the data using a different set of scales and intervals.

Graph A: x-values: Use a scale from 0 to 12,000 and label intervals of 1000. y-values: Use a scale of −50 to 20 and label intervals of 10.
Graph B: x-values: Use a scale from 0 to 20,000 and label intervals of 1000. y-values: Use a scale of −80 to 50 and label intervals of 10.
Graph C: x-values: Use a scale from 0 to 12,000 and label intervals of 2000. y-values: Use a scale of −60 to 20 and label intervals of 20.
Graph D: x-values: Use a scale from 0 to 12,000 and label intervals of 500. y-values: Use a scale of −50 to 20 and label intervals of 10.

2. After you finish drawing your graph, find the other students in your room with the same set of scales and intervals. Compare the graph you drew to theirs.
3. Return to your original group and compare the graph you drew to the graphs of the other members in your group. Which graph (A, B, C, or D) makes the information easiest to use? Why?
4. Which variable is the dependent variable, the independent variable? Explain.
5. Describe what happens to the temperature as the altitude increases.
6. Write a question that might be of interest to a meteorologist and can be answered using the table or graph. Answer the question you write.

for use with the Example, addresses visual learning by introducing the concept of choosing an appropriate scale display to data on a graph.

Alternative Activity for Lesson 2-7

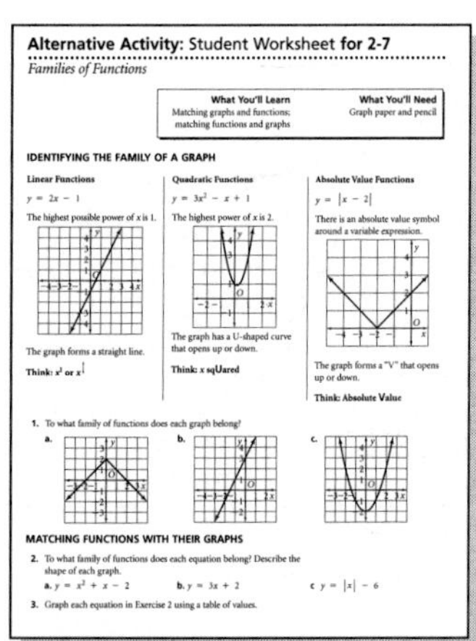

Alternative Activity: Student Worksheet for 2-7

Families of Functions

What You'll Learn Matching graphs and functions; matching functions and graphs

What You'll Need Graph paper and pencil

IDENTIFYING THE FAMILY OF A GRAPH

Linear Functions

$y = 2x - 1$

The highest possible power of x is 1.

The graph forms a straight line.

Think: x^1 or $x^{|}$

Quadratic Functions

$y = 3x^2 - x + 1$

The highest power of x is 2.

The graph has a U-shaped curve that opens up or down.

Think: x sqUared

Absolute Value Functions

$y = |x - 2|$

There is an absolute value symbol around a variable expression.

The graph forms a "V" that opens up or down.

Think: Absolute Value

1. To what family of functions does each graph belong?
   a. b. c.

MATCHING FUNCTIONS WITH THEIR GRAPHS

2. To what family of functions does each equation belong? Describe the shape of each graph.
   a. $y = x^2 + x - 2$ b. $y = 3x + 2$ c. $y = |x| - 6$
3. Graph each equation in Exercise 2 using a table of values.

for use with Example 2, addresses visual learning by showing ways to visualize graphs of different functions.

Cooperative Learning Tips

When used effectively, cooperative learning can help students develop interpersonal skills, learn to perform specific roles in a group, and learn to carry out specific responsibilities. The components of Chapter 2 provide a range of cooperative learning opportunities.

- In the Student Edition, the ***Work Together*** parts of lessons are specifically designed for cooperative learning activities. The Chapter 2 ***Chapter Project*** has students working cooperatively as they time each other to collect data about tongue twisters, draw and use scatter plots and graphs, and make predictions
- In the Teacher's Edition, you will find helpful hints for addressing diverse learning styles (see page C for Chapter 2). For every lesson, you will find a ***Reteaching Activity,*** which may involve cooperative learning.

Materials and Manipulatives

The authors expect all students to use scientific calculators. Calculator use is integrated throughout the course.

- **blocks or tiles (2-3)**
- **graphing calculator (2-7)**
- **graph paper (2-1)**
- **note cards (2-7)**
- **tape measure (2-1)**

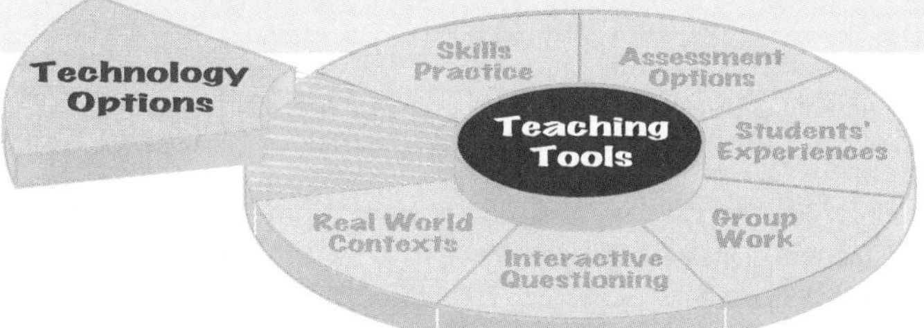

TECHNOLOGY OPTIONS

Technology Tools		Chapter Project	2-1	2-2	2-3	2-4	2-5	2-6	2-7	2-8
Calculator		Assumed that students have scientific calculators at any time.								
Graphing Calculator	Handbook		✔	✔	✔			✔		
	Student Edition		✔						✔	
Software	Secondary Math Lab Toolkit		✔	✔	✔	✔	✔	✔	✔	✔
	Integrated Math Lab									✔
	Computer Item Generator		✔	✔	✔	✔	✔	✔	✔	✔
	Student Edition		✔	✔	✔	✔	✔	✔	✔	✔
Video	Video Field Trip	✔								
CD-ROM	Multimedia Algebra Lab			✔	✔		✔	✔		
Internet		See the Prentice Hall site. (http://www.phschool.com)								

The Prentice Hall Algebra program offers you a rich variety of technology options. Be assured that all these options are provided as a means of enriching the program and are not essential for the successful completion of the course.

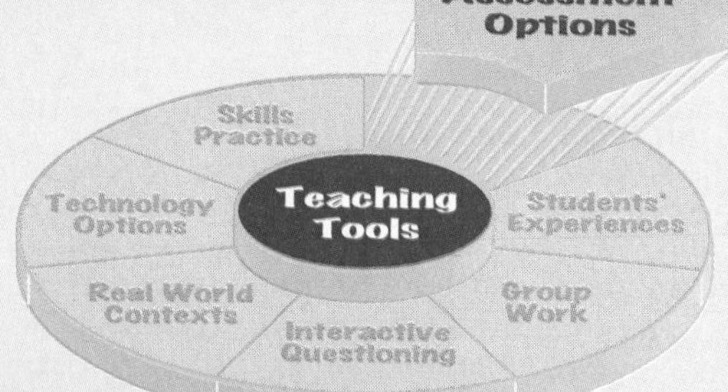

Assessment Options

The Prentice Hall Algebra Program provides you with many options. From these options, you may choose instructional materials and techniques appropriate for your students, or necessary to meet your district's curriculum requirements. As the chart indicates, the program also supports your teaching efforts by offering you many choices for assessment.

ASSESSMENT OPTIONS

Assessment Support Materials	Chapter Opener	2-1	2-2	2-3	2-4	2-5	2-6	2-7	2-8	Chapter End
Chapter Project	▲■●	▲■		▲■		▲■		▲■		▲■
Checkpoints			▲■			▲■				
Self-Assessment			▲■	▲■				▲■	▲■	
Writing Assignment		▲	▲●	▲	▲	▲	▲	▲	▲	●
Chapter Assessment										▲●
Alternative Assessment		■	■	■	■	■	■	■	■	●■
Cumulative Review										▲●
Standardized Test Prep					▲■		▲■	▲■	▲■	▲●
Computer Item Generator	Can be used to create custom-made practice or assessment at any time.									

▲ = Student Edition ■ = Teacher's Edition ● = Teaching Resources

Checkpoints

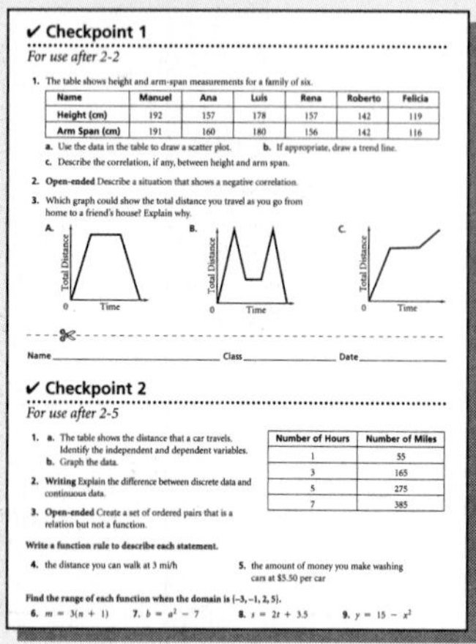

Alternative Assessment

Chapter Assessment

Available in both Form A and Form B

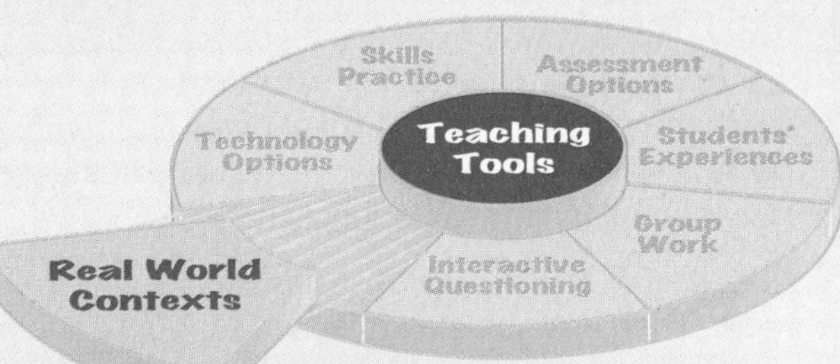

Making the Right Connections

Mathematics is imbedded in nearly every walk of life. The National Council of Teachers of Mathematics (NCTM) encourages educators to recognize these connections and to emphasize them for the purpose of better educating students for success in life and in a global economy. The **Connections** chart below highlights these connections for Chapter 2.

CONNECTIONS

Lesson	Interdiciplinary Connections	Career Prep	Other Real World Connections	Math Integration	NCTM Standards
Chapter Project			Language		Communication
2-1			Television Cars Nature Transportation		Algebra Communication Problem Solving
2-2		Meteorology	Transportation Travel Cars Cooking	Statistics	Algebra Communication Problem Solving
2-3	Health Science Physics		Weather Banking Baseball		Algebra Communication Problem Solving
2-4	Science	Business	Bicycling Communications Music		Algebra Communication Problem Solving Functions
2-5		Computers	Laundry Food		Algebra Communication Problem Solving Functions
2-6		Electrician Jobs Media Researcher	Electricity Conservation Communications		Algebra Communication Problem Solving Functions
2-7		Medical Careers			Algebra Communication Problem Solving Functions
2-8	Genetics		Shopping Games Music Cars		Algebra Communication Problem Solving Functions Statistics

CHAPTER 2 Functions and their Graphs

CONNECTING TO PRIOR LEARNING People use functions every day. Give the example: *Suppose you and your friends are going to the movies. The total cost of the tickets is a function of the cost of one ticket times the number of tickets bought.* Ask students to describe situations that could be represented in terms of a function.

CULTURAL CONNECTIONS Ask students why they listen to the radio. Then ask students to speculate about the importance of radio in cultures where there is no television. Some students may be familiar with the importance of radio communication during natural disasters or even government upheaval. Discuss why radio communication would be different in different parts of the world.

INTERDISCIPLINARY CONNECTIONS Discuss the importance of global and national communication. Elicit from students information about technology that allows us to communicate with the entire world. Relate the discussion to the development of high-speed, electronic communication, and how algebraic functions contribute to this technology.

ABOUT THE PROJECT Using tongue twisters provides students with a fun way to gather data to use functions and graphs. The Find Out questions in this chapter will give students practice at scatter plots, graphs, and tables as they investigate and display relationships in sets of data.

Technology Options

Prentice Hall Technology

Video Video Field Trip 2 "Auction Action," behind the scenes at an auction.

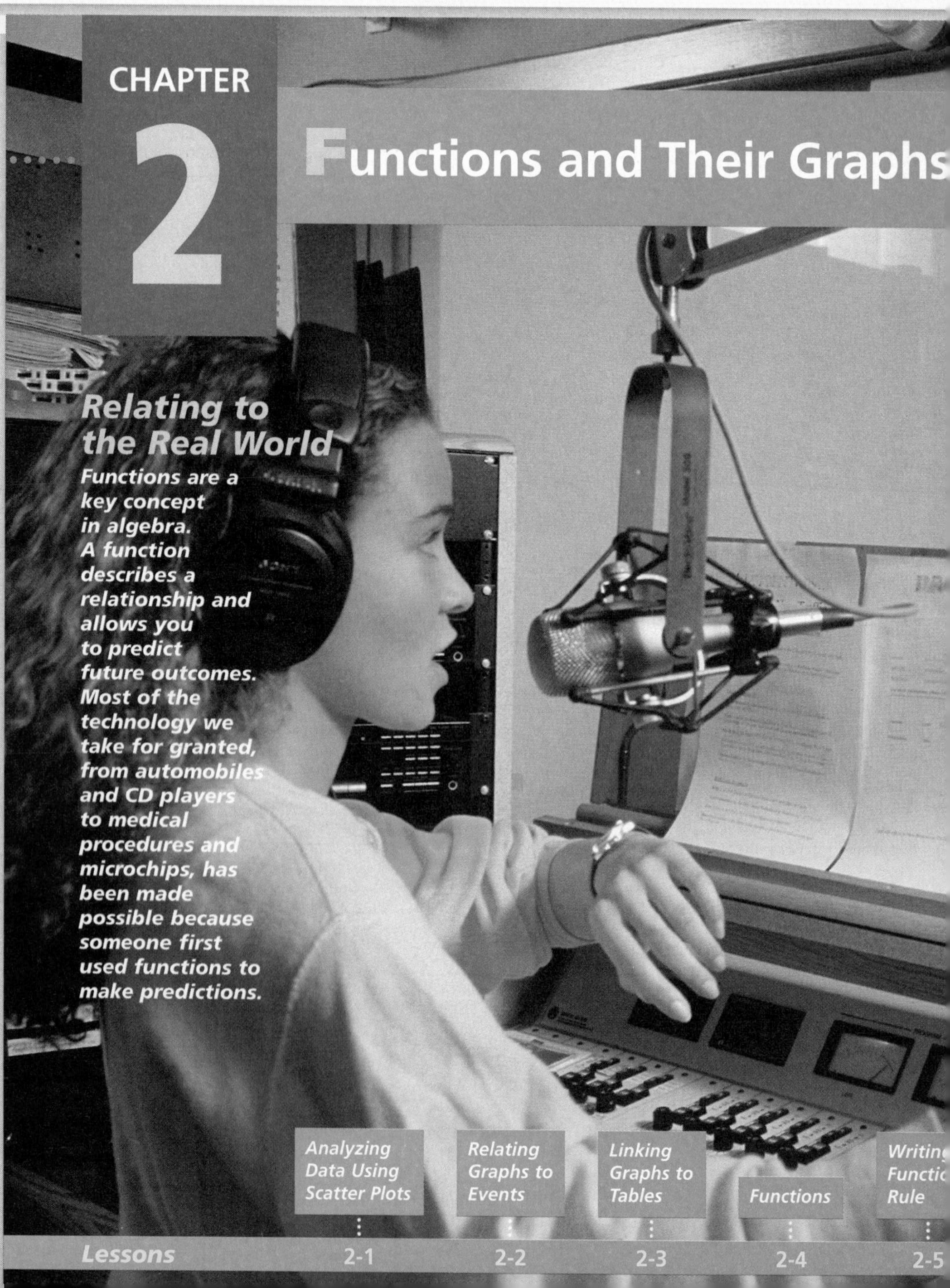

CHAPTER 2 Functions and Their Graphs

Relating to the Real World

Functions are a key concept in algebra. A function describes a relationship and allows you to predict future outcomes. Most of the technology we take for granted, from automobiles and CD players to medical procedures and microchips, has been made possible because someone first used functions to make predictions.

Launching the Project

PROJECT-NOTEBOOK Encourage students to keep all their project-related materials organized in a separate folder or notebook. See Chapter Project Manager and Scoring Rubric in the Chapter Support File.

- Assign students to work with a partner or in a small group. Select one of the tongue twisters on p. 57 (or one of your own choosing). Tell students that you are going to read a short sentence. Let them know how many words are in the sentence. Have each group predict how long it will take you to say the tongue twister.
- Read the tongue twister aloud and have the groups record the time.
- Ask the groups to compare their predicted time with the actual recorded time.
- Choose another tongue twister and repeat each step.
- Challenge students to create a table and graph to record the data they are gathering.
- Encourage the students to research more tongue twisters and investigate why some are "easier" to say than others.
- Suggest that students investigate the effect of practice on speed.

TRACKING THE PROJECT Have students read the Finishing the Chapter Project on p. 100 for an overview of the project. You may also want to set benchmark deadlines when students will show you their work in progress.

CHAPTER PROJECT

FAST TALKER

Radio announcers have to time their speech so that commericals and news updates are the correct length. Do you know how fast you talk? How about your friends? Try saying these tongue twisters: "The sunshade sheltered Sarah from the sunshine." — "Lavonne lingered, looking longingly for her lost laptop."

As you work through the chapter, you will time people as they say tongue twisters. You will use scatter plots and graphs to help investigate and display relationships in the data you collect. Then, using functions, you will summarize your findings and make predictions.

To help you complete the project:

- p. 63 *Find Out by Doing*
- p. 72 *Find Out by Graphing*
- p. 83 *Find Out by Writing*
- p. 94 *Find Out by Analyzing*
- p. 100 *Finishing the Project*

The Three Views of a Function 2-6
Families of Functions 2-7
The Probability Formula 2-8

Project Resources

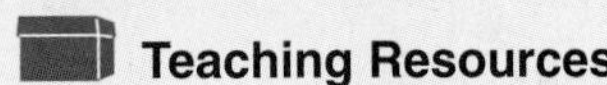

Teaching Resources

Chapter Support File, Ch. 2
- Chapter Project Manager and Scoring Rubric

Transparencies
41

Using the Rubric

Sharing the scoring rubric for the project with your students will alert them to your expectations before they begin work on the project.

As students complete each Find Out question in the chapter, you may wish to have them evaluate their own work or a partner's work, based on the scoring rubric. Students should have the opportunity to revise their work after it has been reviewed.

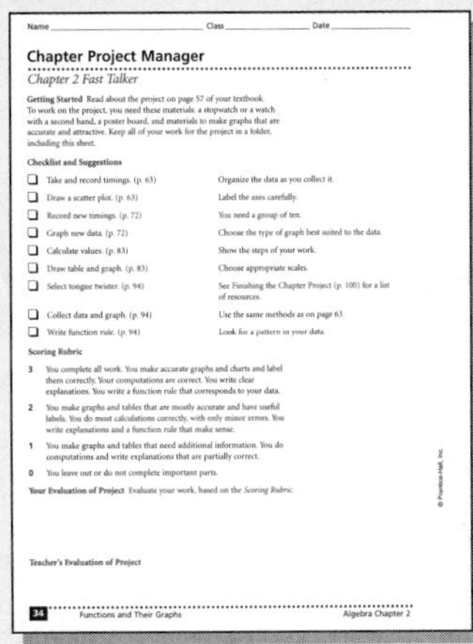

Name ______ Class ______ Date ______

Chapter Project Manager

Chapter 2 Fast Talker

Getting Started Read about the project on page 57 of your textbook. To work on the project, you need these materials: a stopwatch or a watch with a second hand, a poster board, and materials to make graphs that are accurate and attractive. Keep all of your work for the project in a folder, including this sheet.

Checklist and Suggestions

☐ Take and record timings. (p. 63)	Organize the data as you collect it.
☐ Draw a scatter plot. (p. 63)	Label the axes carefully.
☐ Record new timings. (p. 72)	You need a group of ten.
☐ Graph new data. (p. 72)	Choose the type of graph best suited to the data.
☐ Calculate values. (p. 83)	Show the steps of your work.
☐ Draw table and graph. (p. 83)	Choose appropriate scales.
☐ Select tongue twister. (p. 94)	See Finishing the Chapter Project (p. 100) for a list of resources.
☐ Collect data and graph. (p. 94)	Use the same methods as on page 63.
☐ Write function rule. (p. 94)	Look for a pattern in your data.

Scoring Rubric

3 You complete all work. You make accurate graphs and charts and label them correctly. Your computations are correct. You write clear explanations. You write a function rule that corresponds to your data.

2 You make graphs and tables that are mostly accurate and have useful labels. You do most calculations correctly, with only minor errors. You write explanations and a function rule that make sense.

1 You make graphs and tables that need additional information. You do computations and write explanations that are partially correct.

0 You leave out or do not complete important parts.

Your Evaluation of Project Evaluate your work, based on the *Scoring Rubric.*

Teacher's Evaluation of Project

34 Functions and Their Graphs Algebra Chapter 2

Math ToolboX

Throughout mathematics, the coordinate plane is used to solve problems and to represent data in a graphical form. This toolbox reviews the basic skills for graphing points and the vocabulary of graphs. Students also make generalizations about the coordinates in the different quadrants.

ERROR ALERT! Students may use the incorrect axis when graphing points. **Remediation:** Point out to students that the numbers in an ordered pair correspond to the axes in alphabetical order: first *x*, then *y*.

WRITING Exercise 18 Students may be divided into four groups with each group writing a sentence about a different quadrant.

ADDITIONAL PROBLEM Graph each point in the coordinate plane, and connect them in the order given. Describe the figure formed. (–5, –5), (–5, 5), (5, –5), (–5, –5), (5, 5), (–5, 5), (0, 11), (5, 5), (5, –5) **a square with two diagonals intersecting at the origin and a triangle on top of the square**

Materials and Manipulatives

- graph paper

page 58 Math Toolbox

9–12.

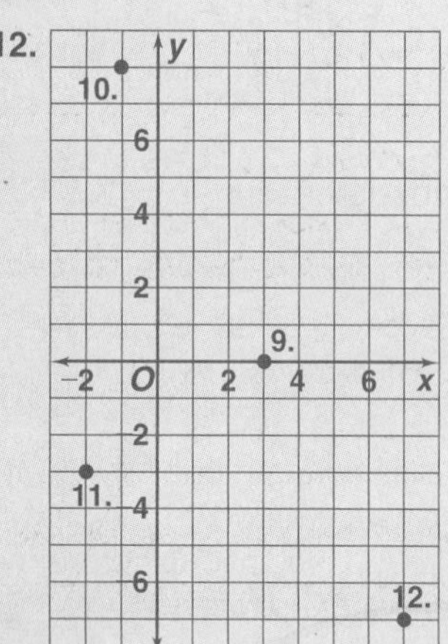

17a.

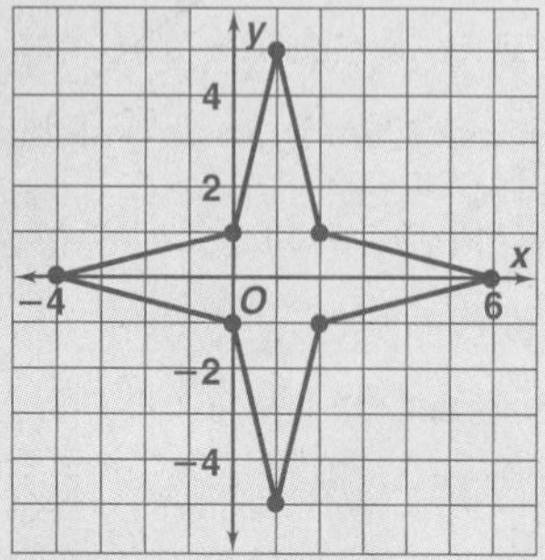

Math ToolboX — Skills Review

The Coordinate Plane

Before Lesson 2-1

The **coordinate plane** is formed when two number lines intersect at right angles. The horizontal axis is the **x-axis** and the vertical axis is the **y-axis.** The axes intersect at the **origin** and divide the coordinate plane into four sections called **quadrants.**

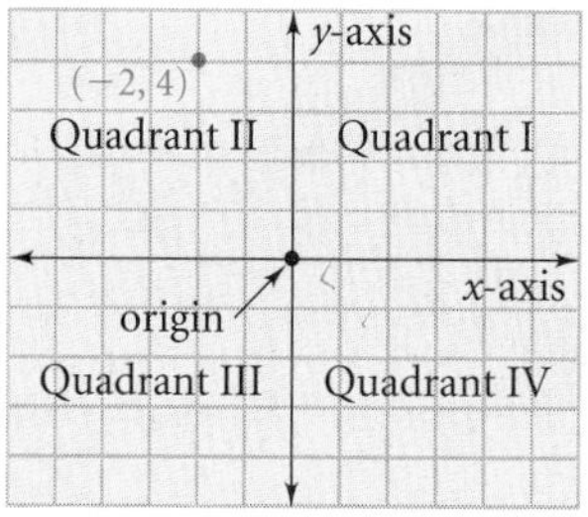

An **ordered pair** of numbers identifies the location of a point. These numbers are the **coordinates** of the point.

(–2, 4)
x-coordinate *y*-coordinate

(−2, 4) y-axis Quadrant II Quadrant I origin x-axis Quadrant III Quadrant IV

Name the point with the given coordinates in the graph at the right.

1. (2, 3) *H*
2. (−5, −3) *B*
3. (−3, 2) *E*
4. (0, −5) *G*

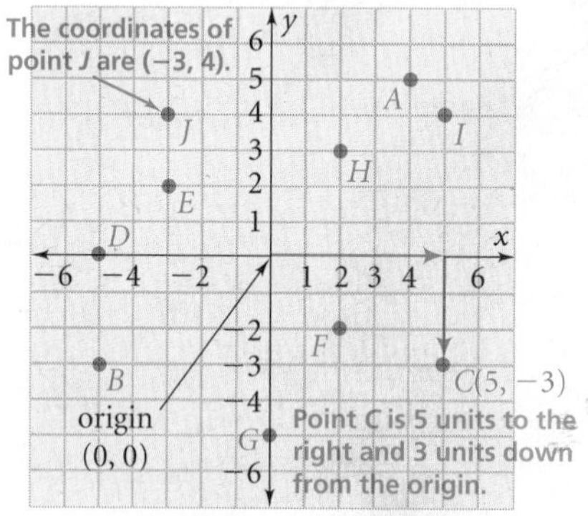

Name the coordinates of each point in the graph at the right.

5. *A* (4, 5)
6. *F* (2, −2)
7. *D* (−5, 0)
8. *I* (5, 4)

The coordinates of point *J* are (−3, 4).

Point *C* is 5 units to the right and 3 units down from the origin. *C*(5, −3) origin (0, 0)

Graph each point on a coordinate plane.

9. (3, 0)
10. (−1, 8)
11. (−2, −3)
12. (7, −7) See margin.

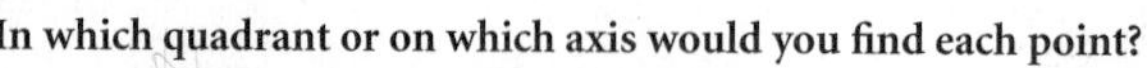

In which quadrant or on which axis would you find each point?

13. (−10, 6) II
14. (−12, 0) *x*-axis
15. (8, −18) IV
16. (0, 30) *y*-axis

17. a. Graph each point on a coordinate plane. See margin.
 (−4, 0), (0, 1), (1, 5), (2, 1), (6, 0), (2, −1), (1, −5), (0, −1), (−4, 0)
 b. Connect the points in order and describe the figure formed. It looks like a four-pointed star.

18. *Writing* Write a sentence describing the *x*- and *y*-coordinates of all points in Quadrant IV. The *x*-coordinates of all the points in Quadrant IV are positive; the *y*-coordinates are negative.

2-1 Teaching Notes

PROBLEM OF THE DAY

Five consecutive multiples of 6 have a sum of 270. What are they? 42, 48, 54, 60, 66

Problem of the Day is also available in Transparencies.

CONNECTING TO PRIOR KNOWLEDGE Have students brainstorm for pairs of items or events that seem to be related even though they are not. Answers may vary. Sample: washing the car and rainfall Discuss ways to test whether they are related.

WORK TOGETHER

DIVERSITY Because students may be sensitive about their body measurements, record the data without using names.

KINESTHETIC LEARNING Have students line up in the order of the lengths of their arm spans. Ask: *Will you be lined up in the same order if you line up in the order of your heights?*

THINK AND DISCUSS

Example 1 Relating to the Real World

Help students grasp the idea of negative correlation. Give examples of negative correlations. Encourage students to suggest other examples from their daily lives. Answers may

Connections Television . . . and more

2-1 Analyzing Data Using Scatter Plots

What You'll Learn

- Drawing and interpreting scatter plots
- Analyzing trends in scatter plots

...And Why

To predict trends related to television and transportation

What You'll Need

- graph paper
- tape measure

WORK TOGETHER

Work in groups. 1–2. Check students' work.

1. a. *Data Collection* Measure the heights and arm spans for each member of your group.
   b. Write the measurements as ordered pairs (height, arm span).
   c. Graph the ordered pairs for your data. Label the graph as shown.

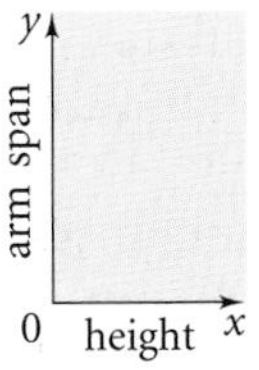

2. a. *Data Collection* Share your data with all other groups in your class.
   b. Graph the data for your entire class.
3. *Critical Thinking* Compare the graphs you drew in Questions 1(c) and 2(b). Why does the graph for the class show the relationship between height and arm span more clearly?
4. *Discussion* Describe the relationship between height and arm span.

3. Answers may vary. Sample: The trend is easier to see when you have a lot of data.
4. Answers may vary. Sample: The height and the arm span are approximately the same.

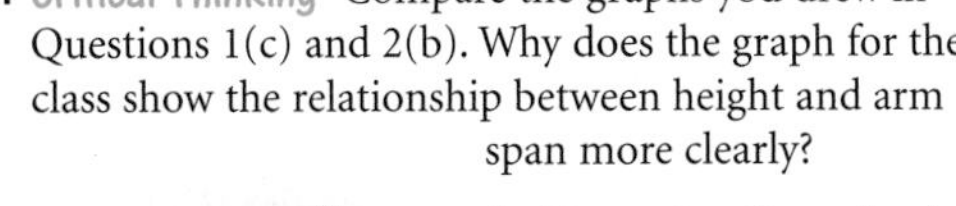

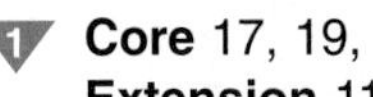

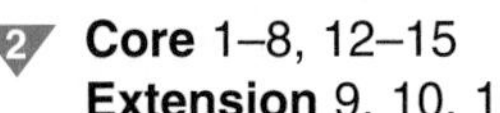

THINK AND DISCUSS

TECHNOLOGY HINT

You can use spreadsheet software or a graphing calculator to make a scatter plot.

Part 1 **Drawing and Interpreting Scatter Plots**

A **scatter plot** is a graph that relates data from two different sets. To make a scatter plot, the two sets of data are plotted as ordered pairs.

Lesson Planning Options

Prerequisite Skills

- Plotting ordered pairs

Assignment Options

To provide flexible scheduling, this lesson can be subdivided into parts.

1 **Core** 17, 19, 20, 22
Extension 11, 18, 21

2 **Core** 1–8, 12–15
Extension 9, 10, 16

Use Mixed Review to maintain skills.

Resources

Student Edition

Skills Handbook, p. 571, 580
Extra Practice, p. 557
Glossary/Study Guide

Teaching Resources

Chapter Support File, Ch. 2
- Practice 2-1 (two worksheets)
- Reteaching 2-1
- Alternative Activity 2-1

Classroom Manager 2-1
Glossary, Spanish Resources
Two-Year Algebra Handbook, 2-1

Transparencies
2, 41, 42, 46, 47

vary. Sample: As the bites taken from a sandwich increase, the amount of the sandwich remaining decreases.

ERROR ALERT! Students may not understand the concept of *trend* for data on scatter plots. **Remediation:** Tell students a *trend* means that the graph indicates any consistent relationship: the two sets of data increase together, decrease together, or change in opposite directions.

EXTENSION Question 5 Discuss the errors that can result from drawing conclusions based on too little data.

AUDITORY LEARNING Have pairs of students write a song that tells the difference between a positive and a negative correlation, with examples.

Example 2 *Relating to the Real World*

Point out that a trend line represents an approximation. There may be several ways to draw the line so that as many data points are above the line as below the line.

CONNECTING TO THE STUDENTS' WORLD Ask students to look and to listen for messages that include correlations. (for example, "Since I was elected, crime has dropped 12 percent.") Have the students share these messages with the

Additional Examples

FOR EXAMPLE 1

The table below shows the number of hours worked and the amount of money each person earned. Draw a scatter plot of the data.

Name	Hours worked	Amount earned
Janel	6	$25.50
Roscoe	12	$51.00
Victoria	11	$46.75
Alex	9	$38.25
Jordan	15	$63.75
Jennifer	10	$42.50

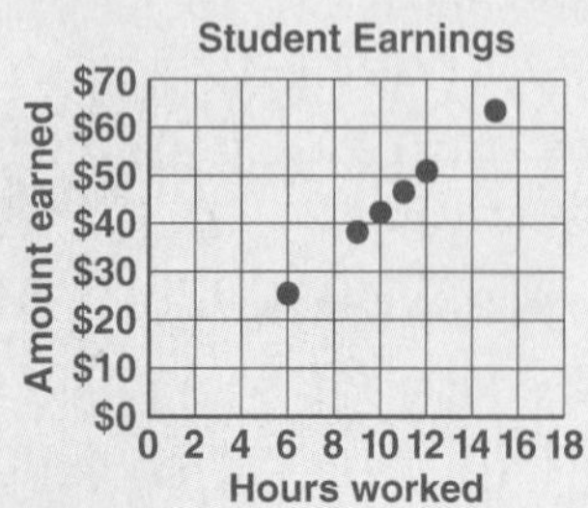

FOR EXAMPLE 2

Use the scatter plot in your answer for Example 1 to answer the following question. Is there a positive correlation, a negative correlation, or no correlation between the number of hours worked and the amount earned? Explain. **Positive correlation; both sets of data increase.**

Example 1 *Relating to the Real World* 

Television A group of students set out to see if the hours of television they watched yesterday relates to their scores on today's test. They polled each student in the group. Draw a scatter plot of the data.

A Test of Television			
Hours Watched	Test Score	Hours Watched	Test Score
0	92	2	80
0	100	2.5	65
0.5	89	2.5	70
1	82	3	68
1	90	3.5	60
1	95	4	65
1.5	85	4.5	55
2	70	5	60

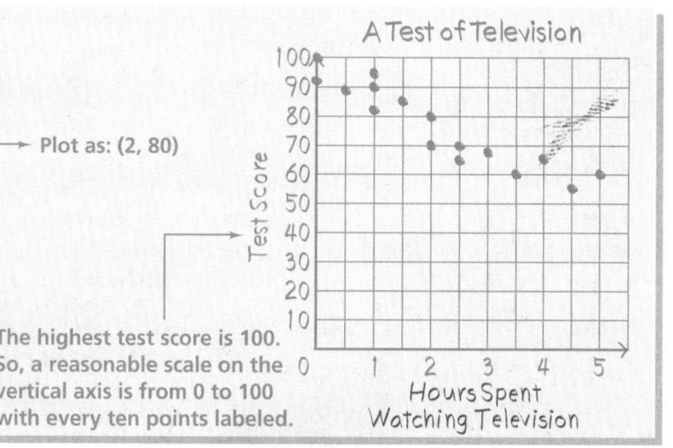

5. Two students in the sample watched 2 hours of television. What does the graph tell you about the students' test scores? **Their scores are 70 and 80.**
6. **a.** A student in the sample scores 85 on the test. How many hours of television did this student watch? $1\frac{1}{2}$ **h**
   b. How can you tell by looking at the scatter plot? **The data point with *y*-coordinate 85 has *x*-coordinate $1\frac{1}{2}$.**
7. **Try This** Each ordered pair represents (unit price, quantity sold) of various items for sale in a store. Draw a scatter plot of the data. ($5.00, 56), ($9.99, 4), ($8.49, 21), ($1.09, 71), ($.35, 92), ($5.00, 32) **See margin.**

Part 2

Analyzing Trends in Data

You use scatter plots to investigate trends relating two sets of data. These trends show positive, negative, or no correlation.

A **trend line** on a scatter plot shows a correlation more clearly.

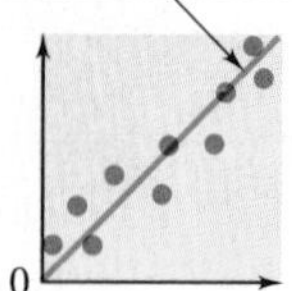

Positive correlation In general, both sets of data increase together.

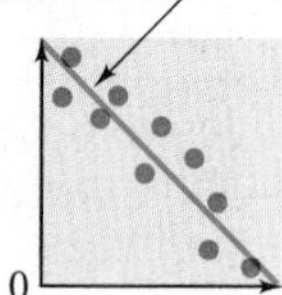

Negative correlation In general, one set of data decreases as the other set increases.

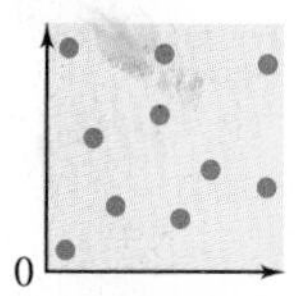

No correlation Sometimes data sets are not related.

class. As they analyze the messages, they can apply what they have learned about the vocabulary of correlations and cause-and-effect relationships.

MAKING CONNECTIONS Even though correlations do not necessarily show cause, correlations can be useful. Using correlations, public health investigators can sometimes describe the source of a disease years before the organism that causes it is found. For example, in 1854, John Snow concluded that a cholera epidemic in London was originating in the water from one well. It was years later that the organism causing the cholera was identified.

CRITICAL THINKING Question 8 Make sure students realize each data point represents a different car.

Exercises ON YOUR OWN

CRITICAL THINKING Exercise 1 Have students brainstorm for a list of other factors that might affect the amount of free time a student has.

ALTERNATIVE ASSESSMENT Exercises 1–8 These exercises will help you assess whether or not students understand the relationships within groups of data, and whether correlations are positive or negative. Have students describe the trend line for each exercise.

Example 2 Relating to the Real World 

Cars The scatter plot below shows information from newspaper advertisements. Is there a *positive correlation*, a *negative correlation*, or *no correlation* between the ages and asking prices of the cars? Explain.

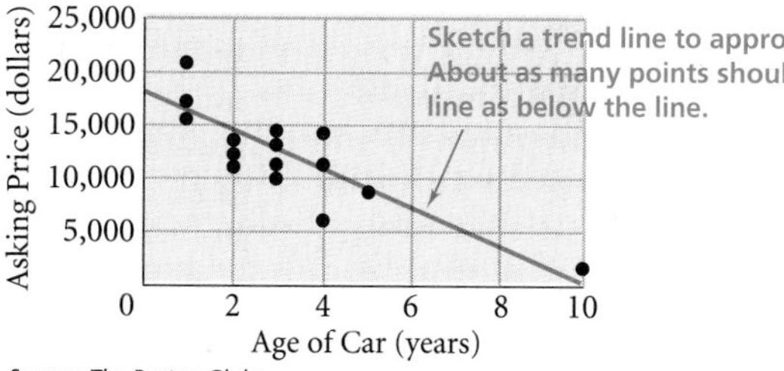
Prices of Ford Taurus Cars
Asking Price (dollars)
25,000
20,000
15,000
10,000
5,000
0
2
4
6
8
10
Age of Car (years)
Sketch a trend line to approximate the data. About as many points should be above the line as below the line.

Source: *The Boston Globe*

GRAPHING CALCULATOR HINT

You can use a graphing calculator to draw a trend line.

This trend line slants downward, so there is a negative correlation. As the age of the car increases, the price of the car decreases.

8. *Critical Thinking* What does each data point in Example 2 represent? Think of your answer in terms of (x–coordinate, y–coordinate).

8. The x-coordinate is the age of a car. The y-coordinate is the asking price of that car.

9. **Predict** the asking price of a Ford Taurus that is 7 years old.
about $5000

Technology Options

For Exercises 11 and 17, students may use spreadsheet software or a graphing calculator to make a scatter plot.

Prentice Hall Technology 

Calculator Graphing Calculator Handbook: Procedure 21

Software • Secondary Math Lab Toolkit • Integrated Math Lab 3 • Computer Item Generator 2-1

Internet • See the Prentice Hall site. (http://www.phschool.com)

Exercises ON YOUR OWN

Critical Thinking **Would you expect a *positive correlation*, a *negative correlation*, or *no correlation* between the two data sets? Why?**

1. Amount of Free Time vs. Number of Classes Taken Negative; with each class taken, you need to study more, so the amount of free time decreases.
2. Air Pollution Levels vs. Number of Cars on the Road Positive; each car on the road contributes to air pollution.
3. Snow Shovel Sales vs. Amount of Snowfall 3. Positive; you expect to sell more shovels as amount of snow increases.
4. Length at Birth vs. Month of Birth No correlation; the data are not related.
5. the Number of Gallons of Heating Oil Consumed vs. Average Daily Temperature Negative; as the temperature rises, people use less oil for heating.
6. Number of Cavities vs. Number of Times you Floss Your Teeth Negative; frequent flossing is claimed to prevent cavities.
7. Cost of a Pair of Sneakers vs. Sneaker Size No correlation; shoes are not priced by size.
8. Number of Calories Burned vs. Time Exercising 8. Positive; the longer you exercise the more calories you burn.
9. a. *Math in the Media* Think about the weather and its effect on voters. What correlation would you expect between inches of precipitation and voter turnout? Explain.
   b. Should candidates in the 11/4/97 election be concerned about the weather forecast? Why or why not?
   9a–b. See margin.

See special election-day supplement with complete guide to ballot questions.

Daily Tribune

RAIN, HEAVY AT TIMES
CHANCE OF SNOW
HIGH: 38 LOW: 30
FULL REPORT ON PAGE 20

CRITICAL THINKING Exercise 10b Use this problem to help students recognize that a correlation does not always prove a cause-and-effect relationship between the two data sets. You could graph the sample but that does not assure any correlation. If there is a correlation, there is a direct relationship; if no correlation, no relationship.

Exercise 14 Students may try to find a correlation in this plot even though none exists. Suggest this rule of thumb: If you can alter the trend line significantly by covering just one or two points, then there probably is no correlation.

OPEN-ENDED Exercise 16 Some students may have difficulty thinking up data sets on their own. Help them get started by suggesting that they first choose data involving a general topic such as sports, music, government, or health.

CRITICAL THINKING Exercise 18b Encourage students to learn more about this aspect of nutrition. Invite them to share their findings with the class.

OPEN-ENDED Exercise 21 For students who are not familiar with toll roads, explain that tolls are calculated in various ways. They can be based on the type of vehicle or on the number of passengers in the vehicle.

Chapter Project **FIND OUT BY DOING** Students choose a tongue twister and then time each other saying it. They combine other students' data with their own to create a table of values. This data will be displayed in a scatter plot. This information is essential to their work on the Chapter Project, introduced in the Chapter Opener. You may wish to

pages 59–61 Think and Discuss

7.

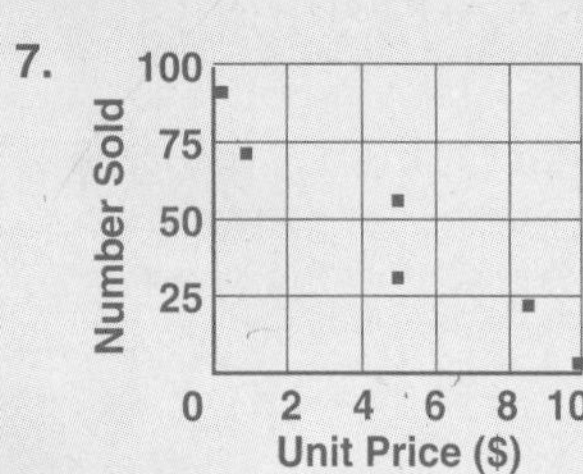

pages 61–63 On Your Own

9a. Negative correlation; everything else being equal, voters are more likely to stay home in bad weather.

b. Answers may vary. Sample: Yes; a low voter turnout may favor the candidates whose supporters feel strongly about the issues in the campaign and the turnout may be low because of bad weather.

10b. Answers may vary. Sample: No; the changes may be seasonal. In this case, the data are related, but one does not *cause* the change in the other.

11a–c, 16, 17a, 19, 31. See back of book.

62

10. During one month at a local deli, the number of pounds of ham sold decreased as the number of pounds of turkey sold increased.

a. Is this an example of a *positive correlation, negative correlation,* or *no correlation?* negative correlation

b. Critical Thinking Does this mean that fewer pounds of ham were sold *because* the deli sold more pounds of turkey? Could there be another reason? Explain. See margin.

11. a. Nature Use the data in the table to draw a scatter plot. 11a–c. See margin.

b. Draw a trend line on the scatter plot. Describe the correlation, if any, between length and mass of a bird egg.

c. Measure the length of the mallard duck egg at the right. **Predict** its mass.

How Big Is a Bird Egg?

Type of Bird Egg	Length (cm)	Mass (g)
Swallow	1.9	2
Swift	2.5	3.6
Turtledove	3.1	9
Partridge	3.6	14
Barn owl	3.9	20.7
Arctic tern	4.0	19
Louisiana egret	4.5	27.5
Grey heron	6.0	60
Chicken (small)	5.3	42.5
Chicken (extra large)	6.3	63.8

Is there a *positive correlation,* a *negative correlation,* or *no correlation* between the two data sets in each scatter plot?

12. negative

13.

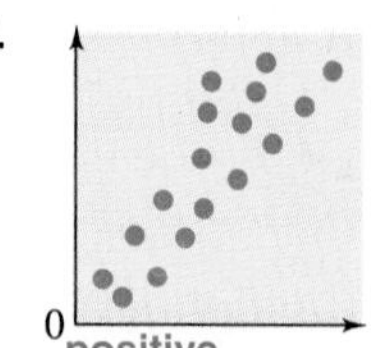

positive

14. no correlation

15. positive

16. Open-ended Describe three different situations: one that shows a positive correlation, one that shows a negative correlation, and one that shows no correlation. See margin.

Nutrition **Use the table below for Exercises 17–18.**

Calories and Fat in Some Common Foods

	Milk	Eggs	Chicken	Ham	Ice Cream	Corn	Ground Beef	Broccoli	Cheese
Grams of Fat*	8	6	4	19	14	1	10	1	9
Number of Calories*	150	80	90	245	270	70	185	45	115

*per serving

Source: *Home and Garden Bulletin No. 72*

17. a. Draw a scatter plot. Label the *x*-axis as Fat (g) and the *y*-axis as Number of Calories. See margin.

b. Is there a *positive correlation, negative correlation,* or *no correlation* between calories and grams of fat? positive correlation

18. a. Writing Write a statement to **generalize** the relationship between calories and grams of fat. As the amount of fat increases, the number of calories per serving tends to increase.

b. Critical Thinking Do you think this statement is always true? **Justify** your answer with an example. Sample: Yes; the more grams of fat there are in a piece of food, the more calories the food contains.

check progress by having each student write a sentence or two describing his or her duties within the group.

Exercises MIXED REVIEW

CRITICAL THINKING Exercise 31 Encourage students to make line plots with a set of random numbers in order to help them visualize the answer.

GETTING READY FOR LESSON 2-2 These exercises prepare students for interpreting information from a graph.

Wrap Up

THE BIG IDEA Ask students to explain how to test whether two sets of data are related by using the terms *scatter plot, trend, positive correlation,* and *negative correlation.*

RETEACHING ACTIVITY Students will learn to draw scatter plots and visually interpret trends. (Reteaching worksheet 2-1)

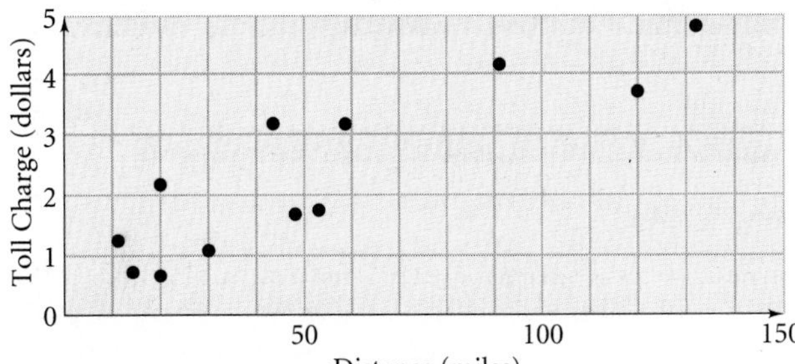

Source: *Indiana Department of Highways*

Transportation Use the scatter plot above.

19. What does each data point in the scatter plot represent? Think of your answer in terms of (x-coordinate, y-coordinate). **See margin.**

20. How can you tell which vehicles traveled the same distance? **The corresponding data points have the same x-coordinate.**

21. a. How can you tell which vehicles paid the same toll charge?

b. *Open-ended* Why might two vehicles that travel different distances pay the same toll charge? **21a–b. See above right.**

22. Is there a correlation between distance traveled and toll charges? Explain your reasoning. **Yes, there is a positive correlation; in general, the farther a vehicle travels, the more it is charged.**

21a. The corresponding points have the same y-coordinate.
b. Answers may vary. Sample: The vehicles could be in different toll categories.

Chapter Project — Find Out by Doing

Work in a group. Choose a tongue twister from p. 57.

- Time one person saying the tongue twister. Record the time to the nearest tenth of a second.
- Time two people saying the tongue twister. Be sure they speak one after the other.
- Add more people. Make a table for the data you collect.
- Collect data from other groups.
- Draw a scatter plot for all the data. Describe any correlation you see.

Exercises MIXED REVIEW

Evaluate each expression for $x = -3$, $y = 5$, and $z = 1$.

23. $3x + 2y - z$ **0**
24. $x^3 - 3z + y$ **−25**
25. $\frac{y^2 + 2z}{x}$ **−9**
26. $6x^2 - 4z^2$ **50**
27. $\frac{x + 6z}{2y}$ $\mathbf{\frac{3}{10}}$
28. $z(y - 8)$ **−3**
29. $x^2 + y^2 - z$ **33**
30. $z^3 - x^2$ **−8**

31. *Critical Thinking* Is it easiest to find the mean, median, or mode of data in a line plot? Explain. **See margin.**

Getting Ready for Lesson 2-2

Use the graph at the right.

Minimum Wages

Hourly Wage (dollars): 0, 2.50, 2.75, 3.00, 3.25, 3.50, 3.75, 4.00, 4.25, 4.50
Year: 1980, 1985, 1990, 1995

Source: *Statistical Abstract*

32. Describe the trend in minimum wage earnings. **The minimum wage earnings increase with time.**

33. Over which years did the minimum wage remain the same? **1981–1989 and 1991–1995**

34. Over which years did the greatest increase in minimum wage occur? How do you know? **1989–1991; the increase of $.45 per year was greater than in any other year.**

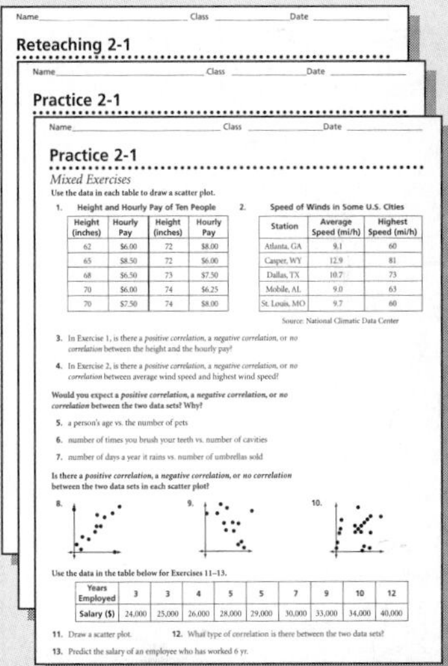
Name ______ Class ______ Date ______

Reteaching 2-1

Name ______ Class ______ Date ______

Practice 2-1

Name ______ Class ______ Date ______

Practice 2-1

Mixed Exercises

Use the data in each table to draw a scatter plot.

1. Height and Hourly Pay of Ten People

Height (inches)	Hourly Pay	Height (inches)	Hourly Pay
62	$6.00	72	$8.00
65	$8.50	72	$6.00
68	$6.50	73	$7.50
70	$6.00	74	$6.25
70	$7.50	74	$8.00

2. Speed of Winds in Some U.S. Cities

Station	Average Speed (mi/h)	Highest Speed (mi/h)
Atlanta, GA	9.1	60
Casper, WY	12.9	81
Dallas, TX	10.7	73
Mobile, AL	9.0	63
St. Louis, MO	9.7	60

Source: National Climatic Data Center

3. In Exercise 1, is there a *positive correlation*, a *negative correlation*, or *no correlation* between the height and the hourly pay?

4. In Exercise 2, is there a *positive correlation*, a *negative correlation*, or *no correlation* between average wind speed and highest wind speed?

Would you expect a *positive correlation*, a *negative correlation*, or *no correlation* between the two data sets? Why?

5. a person's age vs. the number of pets

6. number of times you brush your teeth vs. number of cavities

7. number of days a year it rains vs. number of umbrellas sold

Is there a *positive correlation*, a *negative correlation*, or *no correlation* between the two data sets in each scatter plot?

8. 9. 10.

Use the data in the table below for Exercises 11–13.

Years Employed	3	3	4	5	5	7	9	10	12
Salary ($)	24,000	25,000	26,000	28,000	29,000	30,000	33,000	34,000	40,000

11. Draw a scatter plot. 12. What type of correlation is there between the two data sets?

13. Predict the salary of an employee who has worked 6 yr.

Lesson Quiz

Lesson Quiz is also available in Transparencies.

1. Use the data to draw a scatter plot. **See back of book.**

Person	Words per Minute	Mistakes
Pam	25	8
Dena	65	11
Wai	53	12
Payton	40	10
Raul	32	9

2. Name the three types of correlation. **positive, negative, and no correlation**

3. If the trend line slants downward and to the right, what kind of correlation is this? **negative**

2-2 Teaching Notes

PROBLEM OF THE DAY

Pauline and Sarik took 7 other people deep-sea fishing off the coast of Matagorda Beach. Each person caught a fish. Pauline caught an 800-lb blue marlin. Each of the other people, in turn, caught a fish that weighed half as much as the previous fish caught. What was the total weight of the fish caught on this trip? **1596 lbs 14 oz**

Problem of the Day is also available in Transparencies.

CONNECTING TO PRIOR KNOWLEDGE Ask a student to perform the following moves as you count aloud and describe the move. *(1) take a step forward; (2) take a step forward; (3) jump up and down once on the spot; (4) take a step forward; (5) take a step backward.*

- *Use what you have learned in math to record on paper the series of moves, without using words.* **Answers may vary. Look for answers that use lines and symbols.**

THINK AND DISCUSS

Remind students that a variable is a symbol that can change or vary in value.

Example 1 **Relating to the Real World** 

CRITICAL THINKING Question 1 Ask questions such as the following to help students become familiar with the graph.

Lesson Planning Options

Prerequisite Skills

- Drawing and interpreting graphs

Assignment Options

To provide flexible scheduling, this lesson can be subdivided into parts.

1 **Core** 17–19
Extension 5, 6

2 **Core** 1–4
Extension 16

3 **Core** 7–14
Extension 15

Use Mixed Review to maintain skills.

Resources

Student Edition

Skills Handbook, p. 567
Extra Practice, p. 557
Glossary/Study Guide

Teaching Resources

Chapter Support File, Ch. 2
- Practice 2-2 (two worksheets)
- Reteaching 2-2
- Checkpoint

Classroom Manager 2-2
Glossary, Spanish Resources
Two-Year Algebra Handbook, 2-2

Transparencies
2, 42, 48

2-2 Relating Graphs to Events

What You'll Learn

- Interpreting and sketching graphs from stories
- Classifying data as discrete or continuous

...And Why

To solve problems relating to travel and transportation

THINK AND DISCUSS

Part 1

Interpreting Graphs

Graphs can describe real situations. A graph is not a picture like a photograph. A graph shows a relationship between two variables.

Example 1 **Relating to the Real World**

Transportation A commute home from school combines walking with taking the subway. The graph describes the trip by relating the variables time and total distance. Describe what the graph shows.

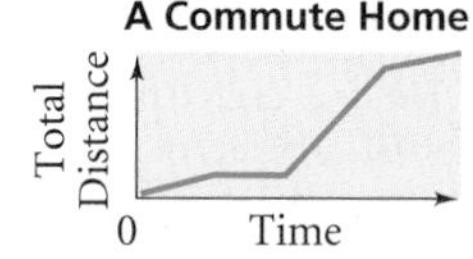

To describe a graph, you can label each part.

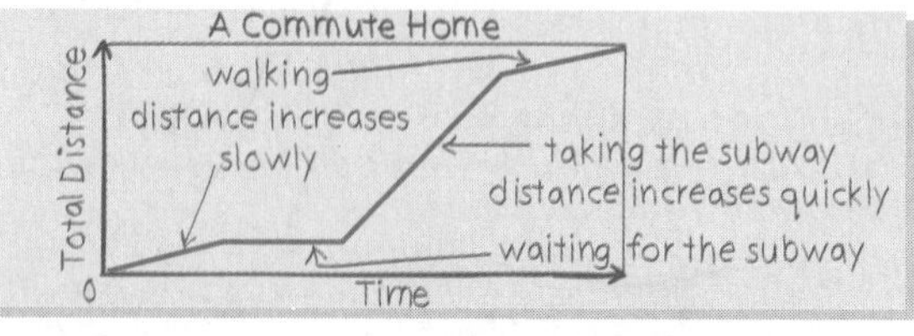

PROBLEM SOLVING

Look Back Another way to solve Example 1 is by writing a description. Write a short paragraph to describe the graph. **See margin.**

2. **You travel faster by subway than by walking.**

1. *Critical Thinking* Explain why the section of the graph showing the person waiting for the subway is flat. **The total distance traveled did not change while the person was waiting.**
2. Why is the subway section steeper than the walking sections? **See above.**

- *What does the Total Distance axis represent?* distance from school
- *What can you say about a point that is high on the Total Distance axis?* The point represents a location that is far from school.
- *How can you tell from the graph that the subway is faster than walking?* The student travels a greater distance in a shorter period of time when traveling on the subway.
- *Could this graph contain a vertical line?* No; a vertical line would mean that the person travels some distance in zero time.

ERROR ALERT! **Question 3** Students may have difficulty seeing that the *y*-axis represents speed rather than distance. **Remediation:** Draw the speed vs. time graph on the board. Underneath it draw the distance vs. time graph axes for the same event. Have students think and talk through how the distance graph will look second by second as you complete the graph on the board.

Question 3 Point out that a straight line indicates constant acceleration—constant change in speed—which is not realistic in real life. Encourage students to discuss how the first graph from Example 1 could be changed to show a more realistic representation of the subway ride.

Example 2 **Relating to the Real World** 

Point out that the Shermans end their trip at the same point from which they started. Explain that the graph cannot begin and end at the origin unless the Shermans travel backward in time.

This graph shows the relationship between time and speed on a bike ride.

Speed on a Bike Ride

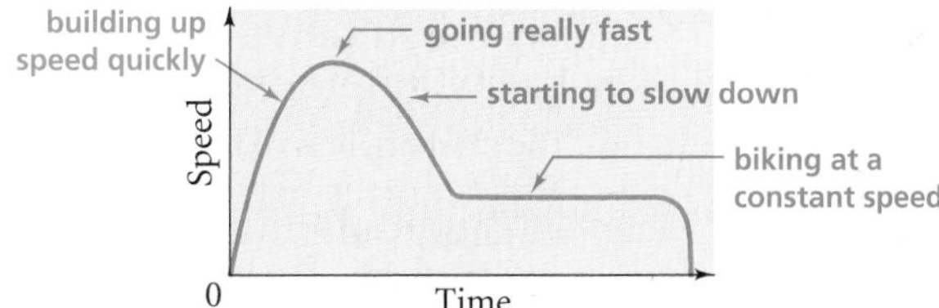

3. A friend mistakenly states that this graph describes a person bicycling up and then down a hill. What would you say to this friend to help clear up the misconception? Sample: This graph describes changes in speed, not elevation. Going fast, slowing down, and then finishing at a steady pace could have happened on a flat road.

Part 2

Sketching Graphs

When you draw a graph without actual data, the graph is called a *sketch.* A sketch is useful when you want to get an idea of what a graph looks like.

Example 2 **Relating to the Real World** 

Travel The Shermans are visiting friends several miles from home. On the way, they stop to buy juice drinks. After a few hours, they return home. Sketch a graph to describe the trip. Label the sections.

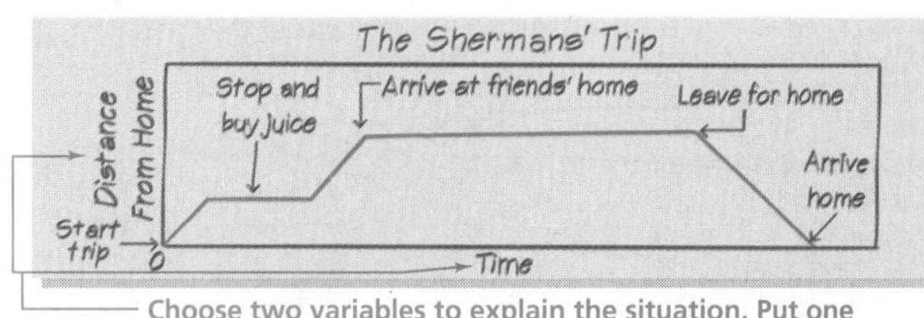

Choose two variables to explain the situation. Put one on the horizontal axis and the other on the vertical axis.

4. How is an increase in distance from home shown on the graph? a decrease? The corresponding line rises from left to right; the line declines.
5. Try This Sketch a graph to describe your height from the ground on a roller coaster ride. Explain the activity in each section of the graph. See margin.

Part 3

Classifying Data

In this lesson you have worked with variables such as time and distance. Variables represent data values that are *discrete* or *continuous.* **Continuous data** usually involve a measurement, such as temperature, length, or weight. **Discrete data** involve a count, such as a number of people or objects.

Additional Examples

FOR EXAMPLE 1

A bicycle ride from home to a friend's house combines riding on a flat surface with riding uphill and downhill. The graph describes the trip by relating the variables *time* and *distance.* Describe what the graph shows by labeling each part.

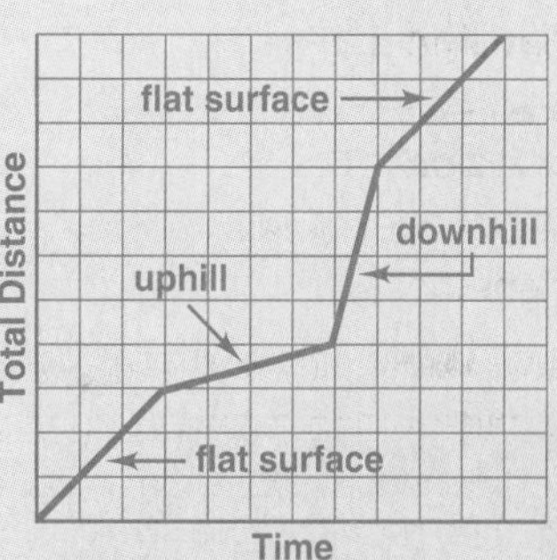

Discussion: *Describe how the graph represents the changes in sales throughout each year.*

FOR EXAMPLE 2

A train leaves the station. It travels for a short time, and then stops. It then climbs a hill before it stops again. On the way to the last stop, the train has a mechanical problem. It is delayed for two hours before it goes down a long decline. Sketch a graph to describe the train's trip. Label the axes.

See back of book.

Question 5 Have students sketch the side view of their roller coaster before they begin. Encourage roller coaster fans to describe their dream ride.

MAKING CONNECTIONS Students who are enthusiastic about roller coasters may be interested in learning more about how math and physics are used in roller coaster design. An article by Robert Speers includes charts and graphs, such as track elevation vs. time ("Physics and Roller Coasters: the Blue Streak at Cedar Point," *American Journal of Physics*, June 1991, v59, n6). Darlene Roy's article guides students in building model roller coasters ("Coaster Construction: Rolling Physics and Amusement Into One," *Science Teacher*, Sept. 1995, v62, n6).

Example 3 **Relating to the Real World** 

OPEN-ENDED Question 6b For students struggling with this question, ask: *What items or events use only whole numbers?* Have students think of a number situation in which a fraction would be meaningless.

WORK TOGETHER

KINESTHETIC LEARNING Have a student draw the Daily Trip to School graph on the board. Divide the time axis into five intervals. Have the students discuss the action performed in the first interval. Then ask a few volunteers to demonstrate the changes in speed depicted in the graph by moving around the room. Repeat for the remaining four intervals. If you have block scheduling or an extended class period, repeat this

Technology Options

For Exercises 1–4 and 7–14, students may use spreadsheet software to create a line plot using their own data.

Prentice Hall Technology 

Calculator Graphing Calculator Handbook: Procedure 4

Software • Secondary Math Lab Toolkit • Computer Item Generator 2-2

 **CD-ROM** Multimedia Algebra Lab 2

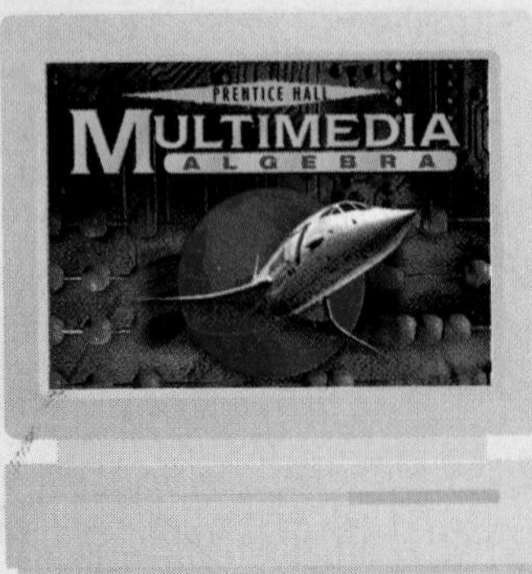

 **Internet** • See the Prentice Hall site. (http://www.phschool.com)

Who? In 1926, a Japanese plant pathologist named E. Kurosawa found a way to increase plant growth. He used a plant hormone called *gibberellin* to increase the growth rate of rice and corn stems.

Source: *Encyclopedia of Science and Technology*

6a. Graphs of continuous data are lines or curves. Graphs of discrete data are points that are not connected.
b. Samples: Continuous: changes in temperature over five days; discrete: daily high temperatures for the same five days.
7. Answers may vary. Sample: The student walked to a friend's house. Together they walked to the bus stop and waited for the bus. After picking them up, the bus made two more stops and then dropped them off at school.

Example 3

Classify the data as *continuous* or *discrete*. Explain your reasoning and sketch a graph of each situation.

a. height of a plant for five days **b.** class attendance for five days

a. The plant height is continuous. The plant continues to grow between the times its height is measured.

b. The attendance is discrete. Each day's count is distinct. For example, there is no meaning for values between Monday's and Tuesday's counts.

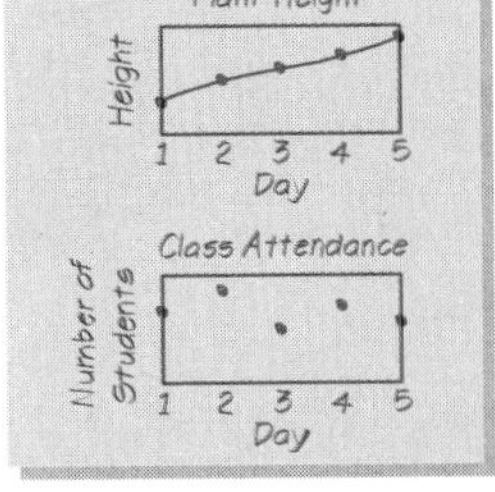

6. **a. Generalize** how graphs of continuous and discrete data differ.
b. Open-ended Give another example of each type of data.
See below left.

WORK TOGETHER

Students get to school in a variety of ways. They may walk, drive, take a bus or subway, or ride a bike. The graph shows one student's daily trip.

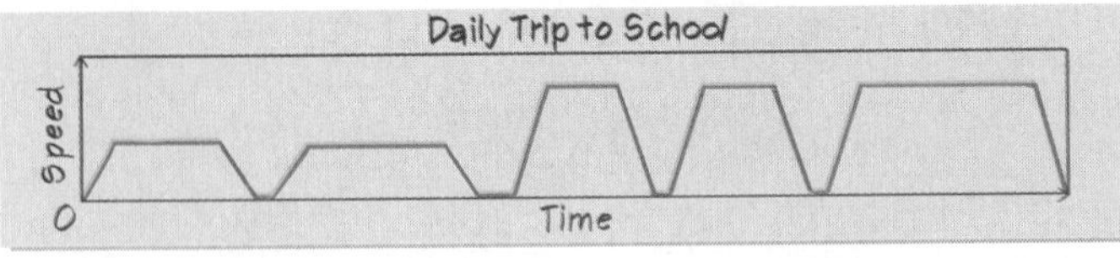

Work with a partner.

7. Write a story describing how this student traveled to school and the events that took place during the trip. See left.
8. Draw two new graphs to represent your daily trip and your partner's. Present your graphs to the class. Check students' work.

Exercises ON YOUR OWN

Sketch a graph to describe each situation. Explain the activity in each section of the graph. 1–4. See margin.

1. your distance from the ground as you jump rope
2. your blood pressure during one gym class
3. your pulse rate as you watch a scary movie
4. your speed as you skateboard downhill

exercise with different graphs. Alternatively, pair students so that each can draw a graph for their partner to act out.

EXTENSION Challenge students to draw a distance vs. time graph describing the science fiction journey of a time traveler. Encourage students to write an accompanying short story describing the traveler's adventures.

Exercises ON YOUR OWN

ALTERNATIVE ASSESSMENT **Exercises 7-14** Have students explain what makes the data continuous or discrete. These exercises will help you assess students' understanding of the differences between measuring and counting objects.

CARTOON Exercise 18 Encourage students to collect other cartoons that depict math for a bulletin board display. Challenge creative students to draw their own.

Exercises MIXED REVIEW

Exercise 23a Suggest that students substitute sample numbers into their expressions to test whether their expression matches the situation.

Exercise 23b Students need to note that this problem asks about each week, not each day.

5. *Critical Thinking* What is wrong with the graph at the right? Explain how you would fix this graph. The graph is continuous but data are discrete. To fix the graph, you must remove all lines connecting data points.

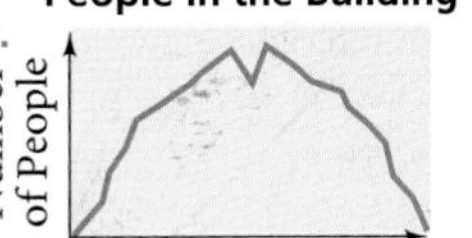

6. *Meteorology* The graph shows barometric pressure in Pittsburgh, Pennsylvania, during the blizzard of 1993. Describe what happened to pressure during the storm. See margin.

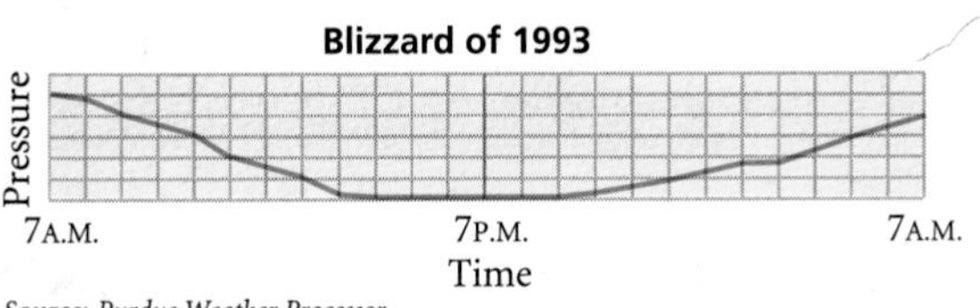

Source: *Purdue Weather Processor*

8. Disc.; there are no test grades between tests.
10. Cont.; between any two temperatures you measure, any value is possible.
12. Cont.; your hair is always growing.
14. Disc.; there is only one time total for each day

Classify the data as *discrete* or *continuous*. Explain your reasoning and sketch a graph of each situation. 7–14. See margin for sample graphs.

7. your weight from birth to age 14 Cont.; between any two weights you measure, any value is possible.
8. your algebra test grades for one term See above.
9. the time you get up each morning for one week Disc.; there is only one time for each morning.
10. temperatures throughout one week See above.
11. number of books you bring home each day for a week Disc.; you cannot have half a book.
12. length of your hair between haircuts See above.
13. your walking speed from this class to your locker Cont.; walking speed can be measured at all times.
14. time you spend reading each day for a week See above.
15. The graph at the right shows the weight of a baby and the weight of a puppy for their first two years.
    a. Which curve represents the puppy's weight? the baby's? blue; red
    b. Describe the growth patterns of the baby and the puppy. See margin.

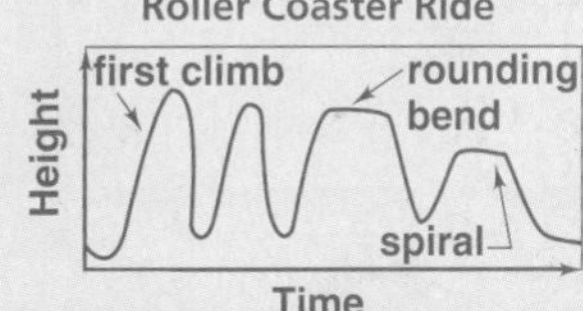

16. *Open-ended* Give an example of discrete data changing over time. Explain your reasoning and sketch a graph of the situation. See margin.
17. *Cars* On a highway, a driver sets a car's cruise control for a constant speed of 55 mi/h. Which graph shows the distance the car travels? Which graph shows the speed of the car? Explain your reasoning. Graph II; graph I; at constant positive speed, distance increases with time.

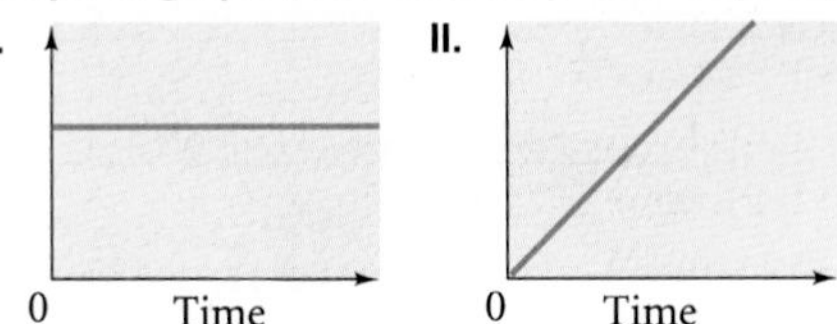

"As you can see, our profit picture is steady."

18. Use the cartoon at the right. See margin.
    a. What do you think the labels of the axes are?
    b. What is the presenter trying to imply with his graph?
    c. Why is his presentation misleading?
    d. Why might he want to do this?

page 64 Problem Solving

Answers may vary. Sample: The graph shows a slow increase as the commuter walks. While the commuter waits for the subway, the graph is flat. A steep line follows, for the short subway ride. Finally, there is another slow increase in distance as the commuter walks home.

pages 64–66 Think and Discuss

5. Graphs may vary. Sample:

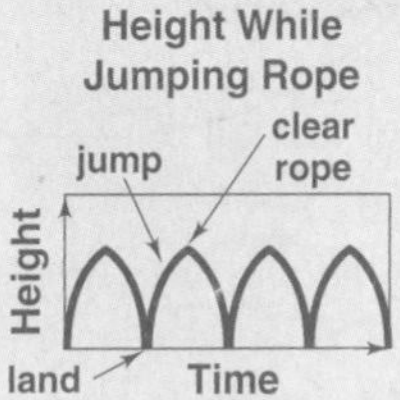

pages 66–68 On Your Own

1. Graphs may vary. Sample:

Height While Jumping Rope

2–4, 6–14, 15b, 16, 18a–d, 20–22, 23a, 24–27, Checkpoint 1a–b, 3. See back of book.

JOURNAL Students could also sketch and describe an alternative ending to the situation. For example, a student who describes a car accident delaying a trip could also draw a comparison graph describing the trip without the car accident.

GETTING READY FOR LESSON 2-3 These problems prepare students for plotting data from tables on the coordinate plane.

Wrap up

THE BIG IDEA Ask students to brainstorm for the advantages and disadvantages of using graphs to describe events.

RETEACHING ACTIVITY Students will learn to interpret and sketch graphs from stories. (Reteaching worksheet, 2-2)

Exercises CHECKPOINT

In this Checkpoint, students assess their own progress on the topics covered in Lessons 2-1 and 2-2.

Exercise 3 Encourage students to describe what the other three graphs could depict.

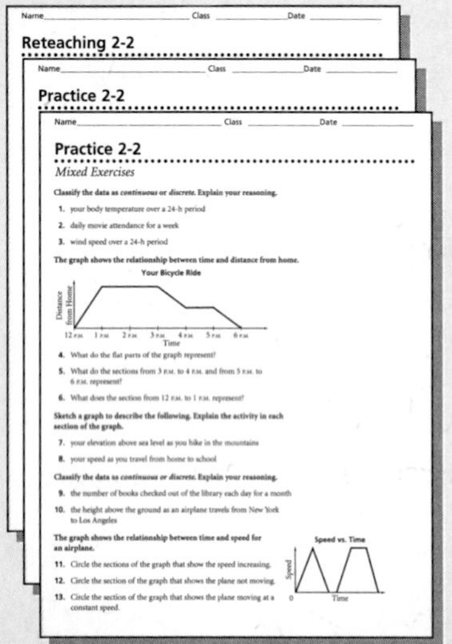
Reteaching 2-2

Practice 2-2

Practice 2-2
Mixed Exercises

Lesson Quiz

Lesson Quiz is also available in Transparencies.

Classify the data as *discrete* or *continuous*.

1. the amount of rainfall during a one month period continuous
2. the total number of plays during four football games discrete
3. the length of your hair over a one year period continuous

68

19. *Writing* Describe what the graph at the right shows about a student's in-line skating experience.

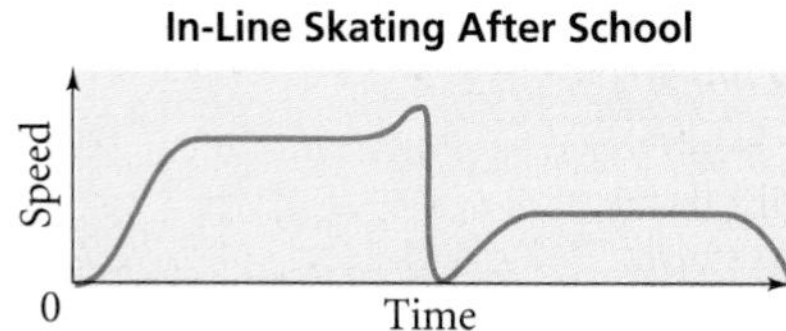

Answers may vary. Sample: The student started skating and got to cruising speed. After a while, he sped up going downhill, lost control, and crashed. After getting up, he decided not to go as fast as he did before the crash.

Exercises MIXED REVIEW

In Exercises 20–22, add or subtract. 20–22. See margin.

20. $\begin{bmatrix} 2 & -3 \\ 0 & 8 \end{bmatrix} + \begin{bmatrix} -5 & 9 \\ 7 & -6 \end{bmatrix}$ **21.** $\begin{bmatrix} 18 & 26 \\ 37 & 12 \end{bmatrix} - \begin{bmatrix} 32 & 15 \\ 39 & 63 \end{bmatrix}$ **22.** $\begin{bmatrix} 2.3 & 7.5 \\ 0.9 & 8.1 \end{bmatrix} - \begin{bmatrix} 6.8 & 9.2 \\ 1.3 & 0.9 \end{bmatrix}$

23. a. *Cooking* A hotel kitchen uses 200 lb more vegetables than fruit each day. Write an expression for how many pounds of vegetables the kitchen uses.

b. Suppose the kitchen uses 1000 lb of fruit daily. How many pounds of fruit and vegetables does the kitchen use each week?

a. f = amount of fruit used per day; $f + 200$ b. 15,400 lb

FOR YOUR JOURNAL

Sketch a graph to describe a situation that happened to you. Explain the activity in each section of the graph.

Getting Ready for Lesson 2-3

Graph each ordered pair on a coordinate plane.

24. (0, 4) **25.** (−9, 3) **26.** (−3, −2) **27.** (2.4, −6)

24–27. See margin.

Exercises CHECKPOINT

1. a. *Statistics* Use the data to draw a scatter plot.

b. Draw a trend line on the scatter plot. Describe the correlation, if any, between latitude and average low temperature. 1a–b. See margin.

c. What would you expect the average low temperature to be in a city at 20° north latitude? Answers may vary; about 70°F

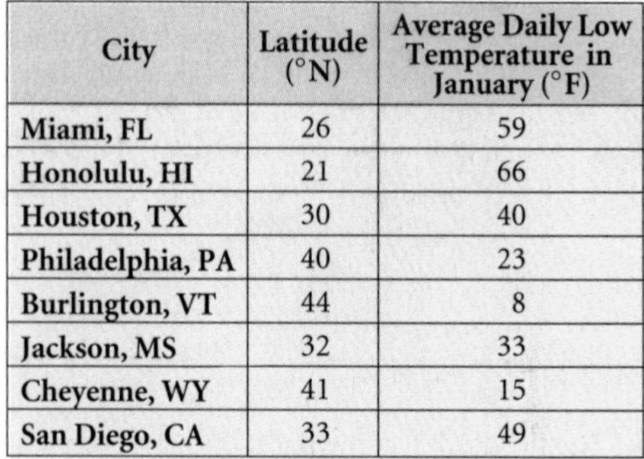

City	Latitude (°N)	Average Daily Low Temperature in January (°F)
Miami, FL	26	59
Honolulu, HI	21	66
Houston, TX	30	40
Philadelphia, PA	40	23
Burlington, VT	44	8
Jackson, MS	32	33
Cheyenne, WY	41	15
San Diego, CA	33	49

Source: *Statistical Abstract of the United States*

2. *Open-ended* Describe a situation that shows a positive correlation. Sample: shoe size vs. glove size

3. Which graph shows your distance from home as you walk to the library and back? Explain. See margin.

A.
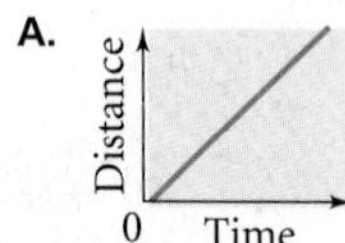

B.
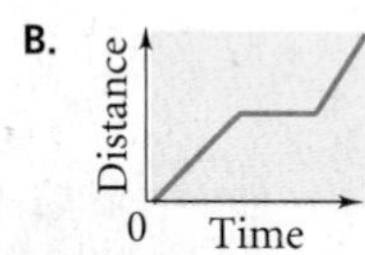

C.
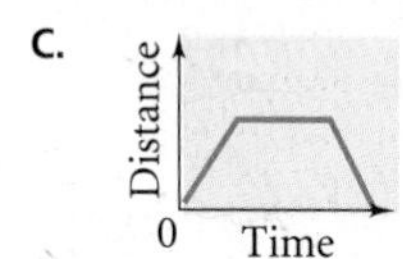

D.
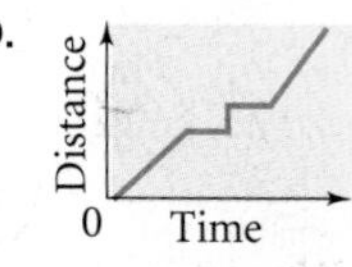

2-3 Teaching Notes

PROBLEM OF THE DAY

A South Haven shop bought 60 beach towels for $9 each. One third of the towels were sold for $20 each. The remaining towels were marked down to $15 and half of them sold. The rest of the towels were all sold for $12.50 each. How much profit did the shop make? **$410.00**

Problem of the Day is also available in Transparencies.

CONNECTING TO PRIOR KNOWLEDGE Ask the students: *What have you learned about graphs that you could use to describe the pattern in a table of numbers?*

WORK TOGETHER

Question 2 Make sure that students understand that levels are the same as layers or rows. Point out that the order (number of levels, number of blocks) tells you the levels are on the *x*-axis and the blocks are on the *y*-axis.

THINK AND DISCUSS

AUDITORY LEARNING Encourage students to ask themselves questions such as, "Does target heart rate depend on age?" as they identify the dependent and independent variables.

Connections Health . . . and more

2-3 Linking Graphs to Tables

What You'll Learn

- Choosing a scale and graphing data in tables
- Identifying independent and dependent variables

...And Why

To solve problems that involve health and weather

What You'll Need

- blocks or tiles

Number of Levels	Number of Blocks
1	1
2	3
3	■ 6
4	■ 10
5	■ 15
6	■ 21

Target Heart Rates

Age in Years	Beats per Minute
15	143.5
20	140.0
25	136.5
30	133.0
35	129.5
40	126.0
45	122.5

Source: *World Almanac and Book of Facts*

WORK TOGETHER

Patterns Work with a partner to construct or sketch stairs out of blocks.

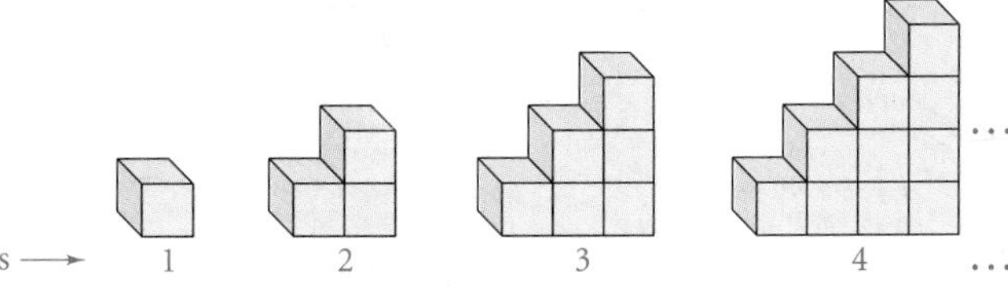

1. Copy and complete the table at the left. **See left.**
2. a. Graph the data in the table. Plot the points as the ordered pairs (number of levels, number of blocks). **See margin.**
   b. Should the points be connected? Why or why not? **No; the values are discrete.**
   c. Describe any trends you see between the number of levels and the number of blocks. **The increase in the number of blocks is equal to the number of levels.**

THINK AND DISCUSS

Health You use tables to organize data. Sometimes a graph shows patterns or trends in data better than a table does.

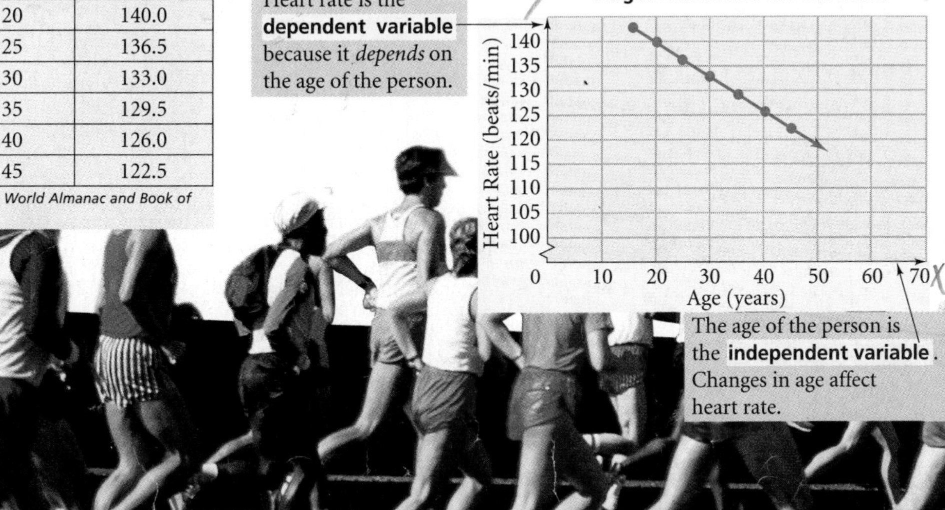

Heart rate is the **dependent variable** because it *depends* on the age of the person.

The age of the person is the **independent variable**. Changes in age affect heart rate.

Lesson Planning Options

Lesson Quiz is also available in Transparencies.

Prerequisite Skills

- Reading tables
- Plotting ordered pairs

Assignment Options

Core 1–9, 11, 12
Extension 10, 13, 14

Use Mixed Review to maintain skills.

Resources

Student Edition

Skills Handbook, p. 569, 571
Extra Practice, p. 557
Glossary/Study Guide

Teaching Resources

Chapter Support File, Ch. 2
- Practice 2-3 (two worksheets)
- Reteaching 2-3
- Alternative Activity 2-3

Classroom Manager 2-3
Glossary, Spanish Resources
Two-Year Algebra Handbook, 2-3

Transparencies
2, 43

page 69 Work Together

2a. See back of book.

pages 69–70 Think and Discuss

10a–b. See back of book.

Question 5 Tell students that the *x*-axis usually represents the independent variable. Ask students to suggest memory aids they might use to help them remember this.

Example 1 Relating to the Real World 

ERROR ALERT! Students may not immediately recognize that the graph and the table represent the same data because none of the points fall exactly on interval lines. **Remediation:** Have students practice finding points from the table on the graph. This will help students recognize how the location of each point in the graph has been estimated.

CONNECTING TO THE STUDENTS' WORLD Assign volunteers to research the altitudes of some nearby or famous mountains. Other volunteers can call or visit camping or sporting goods stores to find out the costs and temperature ratings of sleeping bags. Ask students to make a graph of sleeping bag cost vs. temperature rating. Have them use the two graphs to find out how much they would need to spend on a sleeping bag that was adequate for camping at one of the mountains researched by their classmates.

TACTILE LEARNING Students who need to feel and touch as they learn can preview the shape of a graph by placing counters or beans for the data points before drawing the line graph.

Exercises ON YOUR OWN

Exercise 2 Students might correctly suggest a scenario in which a person is paid first and then performs the task (for example, an author of best-selling novels). The amount paid would then be the independent variable.

Additional Examples

FOR EXAMPLE 1

When you are standing 25m (about 82 ft) from a jet aircraft, the sound level in dBA (decibels) is 140. The upper limit of comfort is 85dBA. A ringing alarm clock (1m away) measures 80dBA. This table shows standards, from the Occupational Safety and Health Administration for steady-state noise.

Permissible Daily Noise Exposure

Duration in Hours	Sound Level in dBA
8	90
6	92
4	95
3	97
2	100
1.5	102
1	105
0.5	110
0.25	115

Source: *Sizes*

This table shows that the permitted hours of exposure depend on the sound level. Graph the data in the table on a coordinate plane. See back of book.

Discussion: *How can you decide which axis to use for the sound level?*

70

3. Health What is the target heart rate for a 70-year-old? Explain how you got your answer. 105; you can find this by extending the line and finding the *y*-coordinate of the point with *x*-coordinate 70.
4. Critical Thinking Describe the relationship between the age of a person and the target heart rate. For every 5-year increase in age, the target rate decreases by 3.5 beats per minute.
5. Which axis relates to the independent variable? the dependent variable? Write a statement to **generalize** this. *x*-axis; *y*-axis; independent variable relates to *x*-axis and dependent variable relates to *y*-axis.
6. **Try This** Identify the independent and dependent variables in the Work Together activity. number of levels; number of blocks

TECHNOLOGY HINT

You can use spreadsheet software to graph data in tables. You can improve the look of your graph by changing the scales on the *x*- and *y*-axis.

You can use data in tables to make graphs. You need to choose an appropriate scale so the data are easy to graph and understand.

Example Relating to the Real World 

Weather Graph the data in the table on a coordinate plane.

How Altitude Affects Temperature

Altitude (meters)	Temperature (°C)
0	13.0
144	12.0
794	11.4
1501	11.4
3100	5.6
5781	−9.7
7461	−22.5
9511	−36.3
10,761	−44.9

Source: The University of Illinois at Urbana-Champaign's Web site *Weather World*

To draw a graph, first identify the least and greatest values for each set of data. Then, select a scale that includes these values and can easily be divided into equal sections.

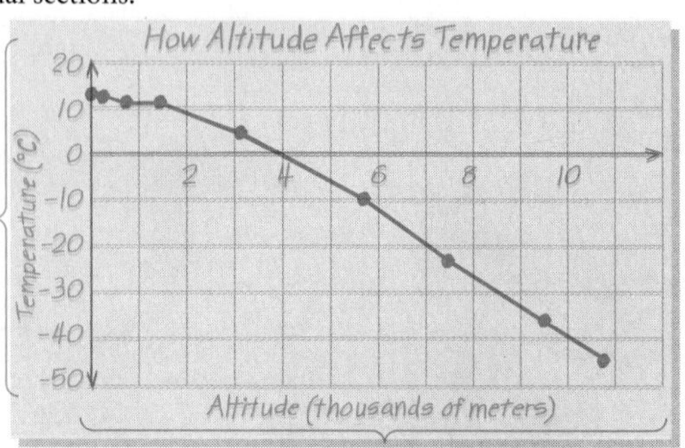

The *y*-values range from −44.9 to 13.0. So, choose a scale from −50 to 20 and label intervals of 10.

The *x*-values range from 0 to 10,761. So, choose a scale from 0 to 12,000 and label intervals of 2000. The *x*-axis is in thousands of meters to make graphing easier.

7. **Try This** Which variable is the dependent variable? the independent variable? Explain. Temperature; altitude; changes in altitude affect temperature.
8. Were all four quadrants needed to graph the data? Why or why not? No; there is no negative altitude in the data.
9. Describe what happens to the temperature as the altitude increases.
10. a. Open-ended Write a question that might be of interest to a meteorologist and can be answered using the table or graph.
    b. Answer the question you wrote in part (a). See margin.

9. As altitude increases temperature decreases, except from 794 to 1501 m, where the temperature stays level.

CRITICAL THINKING Exercise 10a Suggest that students multiply each monthly payment by the total number of months. These calculations help students understand that a lower monthly payment usually means that more money is being spent overall because of the interest.

ALTERNATIVE ASSESSMENT Exercise 12 Have students work in small groups to find the mathematical relationship between the temperature and the volume. This exercise helps you assess how well the students understand quadrants in graphing and relationships between variables.

DIVERSITY Exercise 14 Some students may not know the meaning of *home run.* Ask a student who knows baseball to explain this term by sketching a diamond and a stadium.

RESEARCH Exercise 14d Tell students: *Your school's baseball coach may know the dimensions of the high school's baseball stadium. You might call a little league to find the measurements of their fields. If you live in a city with a major league team, call the stadium to find the needed information.*

Chapter Project **FIND OUT BY GRAPHING** Have students time another tongue twister. First, they identify the dependent and independent variables. Then have students create a table and construct a graph. This is part of their work on the Chapter Project. Have students add this task to work they have already completed for the project. Check their progress by having each group write a sentence or two describing what they have accomplished on the project thus far.

Exercises ON YOUR OWN

1. a. *Science* What scales do you need to graph the data in the table?
   b. Draw the graph. 1a–c. See margin.
   c. How can you use the graph to approximate your own weight on the moon?

Weight on Earth (lb)	143	94	127	171
Weight on Moon (lb)	23.6	15.5	21	28.2

Identify the independent and dependent variables.

2. hours worked, amount paid indep.; dep.
3. area of a square, length of a side dep.; indep.
4. weight of book, number of pages dep.; indep.
5. weight of apples, cost of apples indep.; dep.
6. For Exercises 2–5, write a statement using the word *depends.* See margin.

Match each table of data to its graph.

7. B

x	y
−1	−1
0	0
1	−1
2	−4

8. A

x	y
−2	$\frac{1}{4}$
−1	$\frac{1}{2}$
1	2
2	4

9. C

x	y
−2	3
−1	1
0	−1
1	−3

A.

B.

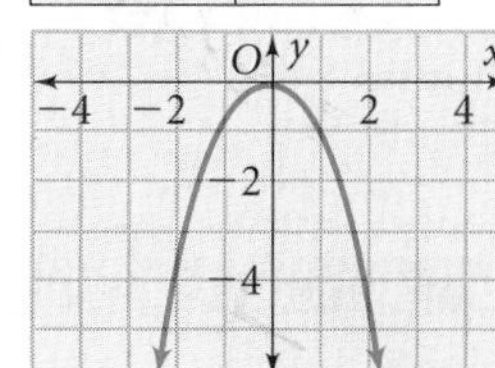

C.

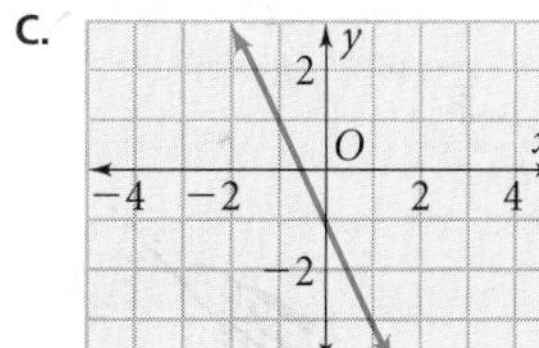

10. *Banking* Many people borrow money when they buy a car. The length of the loan might be anywhere from 2 to 6 years. The table shows the monthly payments on a loan of $10,000 at 9% interest. See margin.
   a. *Critical Thinking* What can you tell about the relationship between the length of the loan and the monthly payments?
   b. Identify the independent and dependent variables.
   c. Graph the data.
   d. Is it easier to see the relationship on a graph or in a table? **Justify** your answer.

Length of Loan (yr)	2	3	4	5	6
Monthly Payment ($)	456.95	318.08	248.92	207.64	180.31

11. a. Graph the data in the table at the right.
    b. What scales did you use and why? See margin.

High School Dance

Minutes Dancing	Admission
43	$5
12	$5
16	$5
111	$5
90	$5

Technology Options

For Exercises 10–12 and 14, students may use spreadsheet software or graphing calculators to graph the data.

Prentice Hall Technology

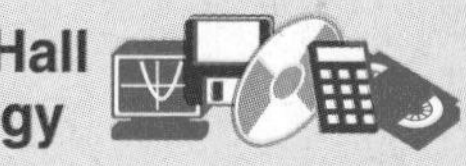

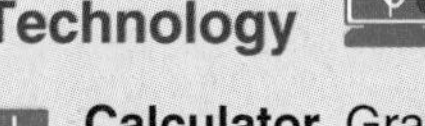

Calculator Graphing Calculator Handbook: Procedure 4, 6, 21

Software • Secondary Math Lab Toolkit • Computer Item Generator 2-3

CD-ROM Multimedia Algebra Lab 2

Internet • See the Prentice Hall site. (http://www.phschool.com)

pages 71–72 On Your Own

1a–c, 6, 10a–d, 11a–b, 12c–d, 13, 14a. See back of book.

page 72 Mixed Review

15, 16, 17a. See back of book.

ESL Group members who have limited English proficiency may not want to recite tongue twisters. Have these students use the practice as a way to improve their language skills. They may prefer to write a tongue twister in their original language and use this sentence for one of the experiments.

Exercises MIXED REVIEW

Exercise 16 Be aware that this question may bring up embarrassing contrasts in socioeconomic status.

SELF ASSESSMENT **JOURNAL** For students who had no difficulties with the lesson, suggest that they write about how they could apply what they learned today in their daily lives.

GETTING READY FOR LESSON 2-4 These exercises give students practice evaluating expressions to prepare them for evaluating functions.

Wrap Up

THE BIG IDEA Ask: *Is it easier to draw conclusions about data by looking at a table or at a graph? Why?*

RETEACHING ACTIVITY Students learn to select the proper scale so that differences in data can be recognized. (Reteaching worksheet 2-3)

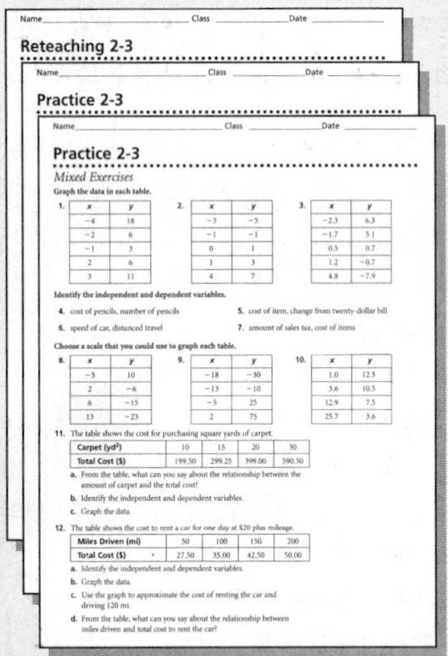

Name____ Class ____ Date ____

Reteaching 2-3

Name____ Class ____ Date ____

Practice 2-3

Name____ Class ____ Date ____

Practice 2-3

Mixed Exercises

Graph the data in each table.

1.

x	y
−4	18
−2	6
−1	3
2	6
3	11

2.

x	y
−3	−5
−1	−1
0	1
1	3
4	7

3.

x	y
−2.3	6.3
−1.7	5.1
0.5	0.7
1.2	−0.7
4.8	−7.9

Identify the independent and dependent variables.

4. cost of pencils, number of pencils
5. cost of item, change from twenty-dollar bill
6. speed of car, distanced travel
7. amount of sales tax, cost of items

Choose a scale that you could use to graph each table.

8.

x	y
−5	10
2	−6
6	−15
13	−23

9.

x	y
−18	−30
−13	−10
−5	25
2	75

10.

x	y
1.0	12.5
3.6	10.5
12.9	7.5
25.7	3.6

11. The table shows the cost for purchasing square yards of carpet.

Carpet (yd^2)	10	15	20	30
Total Cost ($)	199.50	299.25	399.00	590.50

a. From the table, what can you say about the relationship between the amount of carpet and the total cost?
b. Identify the independent and dependent variables.
c. Graph the data.

12. The table shows the cost to rent a car for one day at $20 plus mileage.

Miles Driven (mi)	50	100	150	200
Total Cost ($)	27.50	35.00	42.50	50.00

a. Identify the independent and dependent variables.
b. Graph the data.
c. Use the graph to approximate the cost of renting the car and driving 120 mi.
d. From the table, what can you say about the relationship between miles driven and total cost to rent the car?

Lesson Quiz

Lesson Quiz is also available in Transparencies.

Identify the dependent variable for each of the following pairs of items.

1. the circumference of a circle, the measure of the radius circumference
2. the price of a single compact disc, the total price of three compact discs total price
3. time spent studying for a test, score on the test score
4. number of hours it takes to type a paper, the length of the paper hours to type

12. Physics *Charles's Law* relates the volume of a gas to its temperature.
    a. What quadrant(s) do you need to draw to graph the data? I and II
    b. Identify the dependent and independent variables. volume; temperature
    c. What happens to the volume of a gas as temperature increases?
    d. Graph the data. c–d. See margin.

Charles's Law

Temperature (°C)	Volume (m^3)
−100	173
−50	223
0	273
50	323
100	373

13. Writing Define *dependent variable* and *independent variable* in your own words. Provide an example to illustrate your definitions.

13, 14a. See margin.

14. Baseball In 1919, Babe Ruth broke the record for the longest home run.
    a. Graph the data in the table to approximate the path of the ball.
    b. Use the graph to estimate how far Babe Ruth hit the ball. about 600 ft
    c. In 1960, Mickey Mantle of the New York Yankees hit a home run 643 ft. Did he break Babe Ruth's record? yes
    d. Research How far would a home run need to be hit to clear the baseball stadium nearest you?
    Check students' work.

Babe Ruth's Record Home Run

Distance	Height
0	3
100	83
200	132
300	147
400	128
500	75
575	13

BABE RUTH
LIFETIME BATTING STATISTICS

Chapter Project *Find Out by Graphing*

Sopchoppy, Florida, is the home of the Sopchoppy Shoe Shop. Time the tongue twister "The Sopchoppy Shoe Shop sells shoes" with a group of at least ten people. Use the same method you used in the Find Out question on page 63.

- Identify the dependent and independent variables.
- Record your data in a table.
- Display the data in an appropriate graph.

Exercises MIXED REVIEW

Sketch a graph to describe each situation. Explain the activity in each section of the graph. 15–16. See margin.

15. your speed as you march in a parade
16. the money you spend each day for a week
17. a. Statistics Make a line plot for the data: 33, 36, 32, 33, 35, 33, 36. See margin.
    b. What is the mode of the data? 33
    c. Open-ended What situation could the data represent?
    Sample: ages of mechanics working in a shop

Getting Ready for Lesson 2-4

Evaluate each expression.

18. $3a - 2$ for $a = -5$ -17
19. $\frac{x+3}{-6}$ for $x = 3$ -1
20. $8p - p$ for $p = -11$ -77
21. $3x^2$ for $x = 6$ 108
22. $\frac{x+5}{-2}$ for $x = 9$ -7
23. $b^3 - 4b$ for $b = -1$ 3
24. $7 - a$ for $a = -3$ 10
25. $\frac{3b}{4}$ for $b = 12$ 9

FOR YOUR JOURNAL

Record an example of anything you found difficult in today's lesson.

2-4 Teaching Notes

PROBLEM OF THE DAY

Kathi has a computer printer that can print in black, red, blue, or yellow. She has paper in white, tan, and lavender. How many different ways can she print a title page on her color printer? 12

Problem of the Day is also available in Transparencies.

CONNECTING TO PRIOR KNOWLEDGE Write the formula for the perimeter of a square on the board, P = 4s. Give several values for *s* and have students find the resulting values for *P*. Ask students how they could keep track of their results. Make a table with two columns on the board. Have a student label the columns *s* and *P* and fill in the values as they are calculated. Brainstorm with students for situations in which they have seen a similar table.

AUDITORY LEARNING Suggest that students quietly say to themselves phrases such as "pay depends on hours so pay is a function of hours," to help them become familiar with the vocabulary of functions.

THINK AND DISCUSS

Example 1

ERROR ALERT! The difference between functions and relations may be a difficult concept for students to grasp. They are likely to have difficulty remembering whether it is the *x* or *y* variable that must not be repeated. **Remediation:** Ask

Connections Bicycling . . . and more

2-4 Functions

What You'll Learn

- Defining relations, functions, domain, and range
- Evaluating functions
- Analyzing graphs

...And Why

To solve problems involving money and exercise

THINK AND DISCUSS

Part 1

Identifying Relations and Functions

Suppose your summer job pays \$4.25 an hour. Your pay *depends* on the number of hours you work. The number of hours is the independent variable and pay is the dependent variable. So, you can say that pay *is a function of* the number of hours you work.

1. Identify the independent and dependent variables for each situation.
   a. number of magazines sold and the profit made selling magazines
   b. amount of money in the yearbook committee treasury and the number of days the committee can afford to rent camera equipment

2. Write each situation in Question 1 using the words *is a function of.*

1a. number of magazines; profit
b. amount of money; number of days camera is rented
2. Profit is a function of the number of magazines sold; the amount of money in the yearbook committee's treasury is a function of the number of days the committee rents camera equipment.

Your Summer Job

Hours Worked	Pay
0	\$0
1	\$4.25
2	\$8.50
3	\$12.75

You can write the data in the table as a set of ordered pairs {(0, 0),(1, 4.25), (2, 8.50), (3, 12.75)}. Any set of ordered pairs is called a **relation.**

A **function** is a relation that assigns exactly one value of the dependent variable to each value of the independent variable. So, if *x* is the independent variable and *y* is the dependent variable, there can be only one *y*-value for each *x*-value.

Example 1

Determine if each relation is a function.

a.

x	y	
11	−2	→ (11, −2)
12	−1	→ (12, −1)
13	0	→ (13, 0)
20	7	→ (20, 7)

The relation is a function because exactly one *y*-value is assigned to each *x*-value.

b.

x	y	
−2	−1	→ (−2, −1)
−3	0	→ (−3, 0)
6	3	→ (6, 3)
−2	1	→ (−2, 1)

The relation is not a function because two *y*-values, −1 and 1, are assigned to one *x*-value, −2.

PROBLEM SOLVING

Look Back Graph(−2, −1) and (−2,1) on a coordinate plane. How are these points alike? How are they different? See margin.

Lesson Planning Options

Prerequisite Skills

- Using the order of operations

Assignment Options

To provide flexible scheduling, this lesson can be subdivided into parts.

1. **Core** 9–12, 18–21
   Extension 38, 51–53
2. **Core** 1–8, 25–37, 39–46
   Extension 47–50
3. **Core** 13–17, 24
   Extension 22, 23

Use Mixed Review to maintain skills.

Resources

Student Edition

Skills Handbook, p. 567, 569
Extra Practice, p. 557
Glossary/Study Guide

Teaching Resources

Chapter Support File, Ch. 2
- Practice 2-4 (two worksheets)
- Reteaching 2-4

Classroom Manager 2-4
Glossary, Spanish Resources
Two-Year Algebra Handbook, 2-4

Transparencies
12, 43

students to recall a mental picture of the vertical-line test. Have them create and share memory aids to help them remember these ideas:

- the difference between a relation and a function
- a function may have two *x*-values assigned to the same *y*-value, but may not have two *y*-values assigned to the same *x*-value.

To get ideas flowing, suggest that students imagine that *x* is a horse (the four parts of the *x* represent the four legs of the horse). A cowboy on the horse is represented by *y*. Encourage students to imagine that no cowboy *ever* lets another person (or two people at once) ride his horse. However, a cowboy may own and ride more than one horse. Invite an artistic student to write a large *x* and *y* on the board making the letters into a horse and cowboy.

Example 2

TACTILE LEARNING Encourage students to draw a function machine as a box. The *x*-values go in one end and the finished products, *y*-values, come out the other end. Students write various function rules on sticky notes and place one note at a time in the box. Have students label the drawings with some of the vocabulary from this lesson.

KINESTHETIC LEARNING Write the first six terms or words given in the table at the bottom of p. 74 on slips of paper, one for each student in the class. Mix up the papers and hand them out. Now divide the classroom floor into two zones—*x*-values and *y*-values. Tell students to look at the paper they have and move to the appropriate zone. Repeat this activity several times until students are familiar with the new terms.

Additional Examples

FOR EXAMPLE 1

Determine if the relations in Tables A and B are functions.

a.

x	y
3	2
5	−11
−7	13
15	4

b.

x	y
4	3
7	−9
−12	10
7	4

The relation in Table A is a function. The relation in Table B is not a function.

Discussion: *Explain why Table B does not represent a function.*

FOR EXAMPLE 2

Evaluate the function rule $y = 4x^2 + 2$ for $x = 3$. **38**

Discussion: *Explain why $y = 4x^2 + 2$ is a function.*

FOR EXAMPLE 3

The perimeter of a square is $p = 4s$. Find the range when the domain is {2, 5, 12}. **range = {8, 20, 48}**

Discussion: *What are two other ways to describe the domain and range?*

74

3a. Sample: (0, 27), (20, 22), (40, 21), (60, 13), (60, −3)
b. Sample: (1, 3.5), (2, 3.7), (4, 3.7), (5, 3.5)

QUICK REVIEW

dependent variable (y axis); independent variable (x axis); 0

3. a. *Open-ended* Create a data table or set of ordered pairs that is a relation but *not* a function. **See left.**

b. *Open-ended* Create a data table or set of ordered pairs that is a function. **See left.**

Part 2 **Evaluating Functions**

A **function rule** is an equation that describes a function. If you know the input values, you can use a function rule to find the output values.

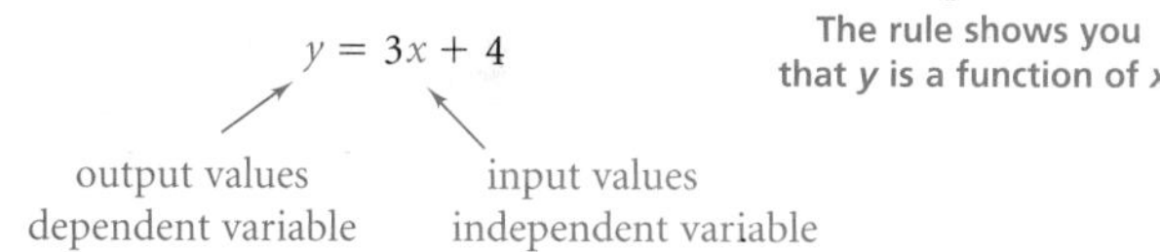

Example 2

Evaluate the function rule $y = 2x^2 - 7$ for $x = -4$.

$y = 2x^2 - 7$
$y = 2(-4)^2 - 7$ ← Substitute −4 for x.
$y = 2(16) - 7$ ← Find $(-4)^2$ first.
$y = 32 - 7$ ← Then multiply.
$y = 25$

When the *x*-value is −4, the *y*-value is 25.

QUICK REVIEW

Follow the Order of Operations.
1. Perform any operation(s) inside grouping symbols.
2. Simplify any terms with exponents.
3. Multiply and divide in order from left to right.
4. Add and subtract in order from left to right.

4. a. Evaluate the function rule in Example 2 for $x = 4$. **25**

b. *Critical Thinking* Explain why $y = 2x^2 - 7$ is a function. **For each value of *x*, the rule gives only one value of *y*.**

5. *Try This* Evaluate each function rule for $x = 6$.

a. $y = x - 11$ **−5**
b. $y = -3x^2 + 1$ **−107**
c. $y = \frac{1}{2}x^2$ **18**
d. $y = 1.5 + 2x$ **13.5**

The **domain** of a function is the set of all possible input values. The **range** of a function is the set of all possible output values. You can write both the domain and the range using braces, {}. The summary box shows some common names for the sets of values in a function.

Names of Values in a Function

independent variable	dependent variable
input	output
domain	range
x-values	*y*-values

Example 3 **Relating to the Real World** 

Point out that when a domain contains several values, the corresponding range is found by substituting each value of the domain in turn into the function.

Question 6 Have students suggest ways to help them remember the difference between *domain* and *range.* For example, if students are familiar with the folk song "Home on the Range," they might link the song to their mental image of the letter *y* being a cowboy. Invite an artistic student to draw a cartoon of a cowboy walking up the *y*-axis of the coordinate plane while singing the song.

KINESTHETIC LEARNING Flip a coin, assigning heads to be a function and tails to be a relation that is not a function. Announce *function* or *not a function* to the class at the end of each coin toss. Have students stand and use their arms or bodies to make a shape of the type of graph you called. Begin with straight-line graphs. Encourage students to correct each other's mistakes. Repeat this activity with U-shaped and V-shaped graphs until students quickly make the correct shapes.

Question 7 Check that students are clear on the difference between a horizontal line and a vertical line. Explain that a *horizontal* line is flat, like the *horizon*. A *vertical* line goes straight up and on the top you get *vertigo*.

Example 3 **Relating to the Real World** 

Bicycling The distance a wheel moves forward is a function of the number of rotations. The function rule $d = 7n$ describes the relationship between the distance d the wheel in the photo moves in feet and the number of rotations n. Find the range when the domain is $\{0, 2.5, 8\}$.

Substitute 0 for *n*.	Substitute 2.5 for *n*.	Substitute 8 for *n*.
$d = 7n$	$d = 7n$	$d = 7n$
$d = 7(0)$	$d = 7(2.5)$	$d = 7(8)$
$d = 0$	$d = 17.5$	$d = 56$

The range of the function is $\{0, 17.5, 56\}$.

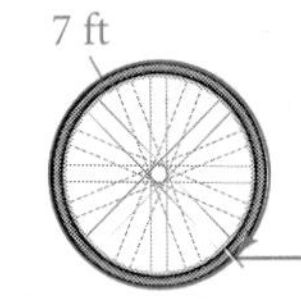

6. Describe what the domain and the range represent in Example 3. See left.

6. The domain is all the possible values of the number of rotations of the wheel. The range is all the possible values of the distance the wheel can move.

Part 3

Analyzing Graphs

The domain and range values can be written as ordered pairs (x, y). These ordered pairs are points on the graph of a function. You can tell if a relation is a function by analyzing its graph. One way is to use the **vertical-line test.** If a vertical line passes through a graph more than once, the graph is not the graph of a function.

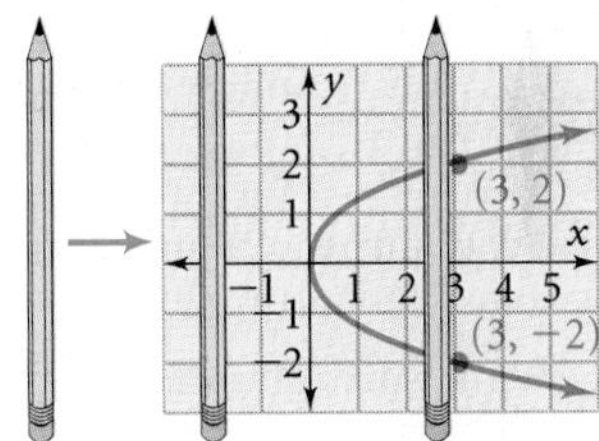

Pass a pencil across the graph as shown. Keep your pencil straight to represent a vertical line.

The pencil goes through more than one point on the graph. This graph is not the graph of a function because there are two *y*-values for the same *x*-value.

7. Why does the vertical-line test work? (*Hint:* How does the vertical-line test relate to the definition of a function?)

7. If a vertical line crosses the graph in two or more places, then there is more than one *y*-value for some *x*-value.

Technology Options

For Exercises 25–36 and 47–50, students may use spreadsheet software to find the range of each function.

Prentice Hall Technology 

 **Software** • Secondary Math Lab Toolkit • Computer Item List Generator 2-4

 **Internet** • See the Prentice Hall site. (http://www.phschool.com)

WORK TOGETHER

ERROR ALERT! **Question 8b** Students may think that this graph is a function where the graph touches the *x*-axis but not a function elsewhere. **Remediation:** Explain that if a vertical line passes through the graph more than once at *any* point on the graph, then the whole graph is not a function. Suggest that this misunderstanding is analogous to someone thinking that a flat tire is only deflated at the bottom.

Exercises ON YOUR OWN

MENTAL MATH Exercise 22a Suggest that students round 0.22 to 0.2 and then estimate how much the call should cost.

CONNECTING TO THE STUDENTS' WORLD Ask students to find out the typical costs of long-distance calls by asking their families or friends, or by calling phone companies. Students might compile a table showing, for each call, the city, length of time of the call, average cost per minute, and the total cost.

Exercises 25–36 Point out that, in these exercises, the dependent variable is isolated on the left side of the equation.

Exercise 37 For students who are not familiar with the term *analogy*, give a simpler example such as "glove is to hand as sock is to foot."

Exercise 38 Ask a student who plays the piano to share with the class how the instrument produces sound.

page 73 Problem Solving

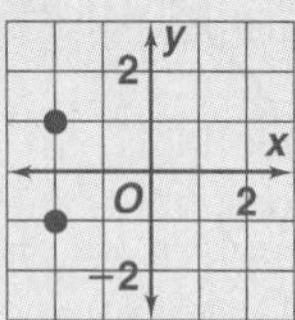

These points have the same x-coordinates but different y-coordinates.

pages 76–78 On Your Own

23a. Sample:

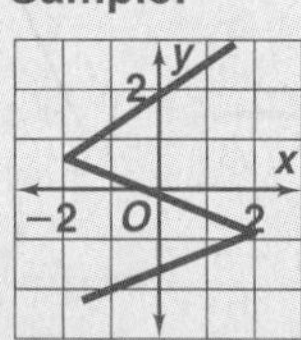

38a. Yes; to each input value there is only one output value.

b. For every increase of 7 in the key position, the frequency doubles.

52c. Answers may vary. Sample:

Time (s)	Distance (mi)
0	0
20	3,720,000
40	7,440,000
60	11,160,000
120	22,320,000
180	33,480,000

8a. Yes; no vertical line crosses the graph more than once.
b. No; many lines cross the graph twice.
c. Yes; no vertical line crosses the graph more than once.

WORK TOGETHER

Work with a partner.

8. Use the vertical-line test to determine if the graphs are graphs of functions. Explain why or why not.

a. 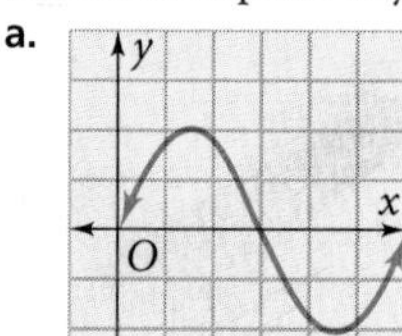 **b.** 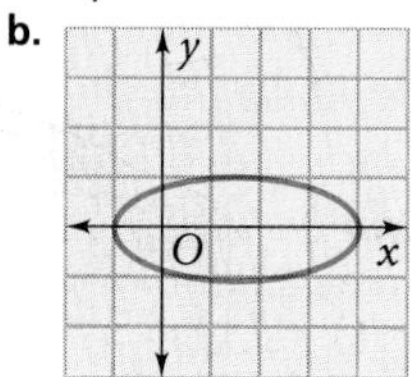 **c.** 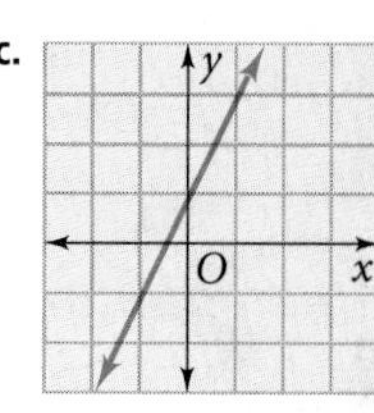

9. Can the graph of a function be a horizontal line? a vertical line? Why or why not? Yes, no; horizontal line graphs pass the vertical-line test, vertical line graphs fail the test.

Exercises ON YOUR OWN

Evaluate each function rule for $x = -3$.

1. $y = x + 7$ 4 **2.** $y = 11x - 1$ −34 **3.** $y = x^2$ 9 **4.** $y = -4x$ 12

5. $y = 15 - x$ 18 **6.** $y = 3x^2 + 2$ 29 **7.** $y = \frac{1}{4}x$ $-\frac{3}{4}$ **8.** $y = -x + 2$ 5

Does each graph represent a function? Why or why not? 9. No; graph fails the vertical-line test.

9. 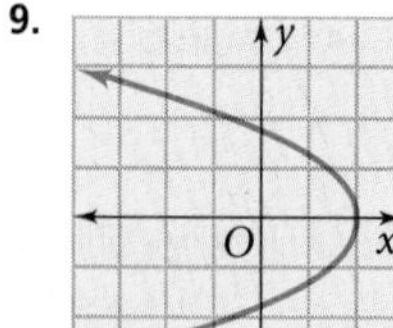 **10.** 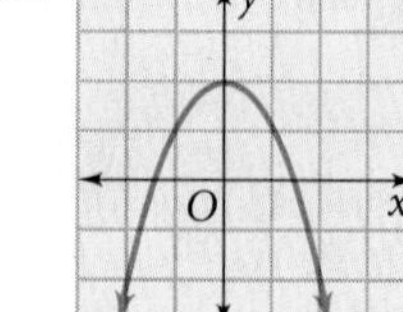 **11.** 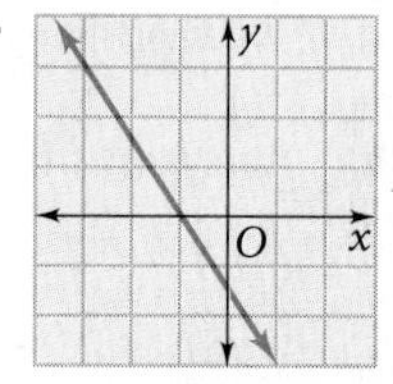 **12.** 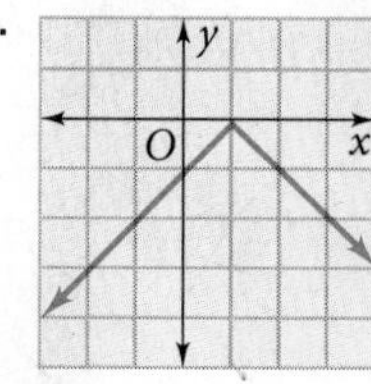

10–12. Yes; every graph passes the vertical-line test.

Which relations in Exercises 13–17 are functions? If the relation is a function, identify the dependent and independent variables.

13. number of tickets sold for a benefit play and amount of money made function; amount of money, number of tickets

14. students' heights and grade point averages not a function

15. amount of your monthly loan payment and the number of years you pay back the loan function; monthly payment, number of years

16. cost of electricity to run an air conditioner during peak usage hours and the number of hours it runs function; hours run, cost of electricity

17. time it takes to travel 50 miles and the speed of the vehicle function; speed, time

WRITING Exercise 38b Point out that the students only need to describe the vertical-line test; they do not need to graph the function. Students who wish to draw the graph, however, should be encouraged to do so. They may need help because of the large range of y-values.

ALTERNATIVE ASSESSMENT Exercises 39–46 Divide the class into small groups. Have half of the students solve the exercises using decimal values while the other half uses fractions. These exercises will help you assess whether students understand the procedures for evaluating functions.

Exercise 48 Make sure students read the first domain value as -1.2 and not $-1, 2$.

CRITICAL THINKING Exercise 51 Ask students to explain how they would correct the classmate's misunderstanding.

MAKING CONNECTIONS Exercise 52 A light-year is the distance light travels in one year. Assuming 31,557,600 seconds in a year (60 seconds · 60 minutes · 24 hours · 365.25 days) and the speed of light as 183,000 mi/s, a light-year is about $6 \cdot 10^{12}$ mi or 6,000,000,000,000 mi.

Another astronomical unit of distance, the parsec, is equal to 3.26 light-years. Alpha Centauri, our closest star, is 1.3 parsecs away.

Challenge students to use the speed of light in km/s to calculate the magnitude of a light-year. Remind them to use 365.25 days because of leap years.

Determine if each relation is a function.

18. no

x	y
1	−3
6	−2
9	−1
10	0
1	3

19. no

x	y
0	2
0	−2
3	1
3	−1
5	3

20. yes

x	y
2	12
1	5
0	0
−1	−3
−4	0

21. yes

x	y
−4	−4
−1	−4
0	−4
2	−4
3	−4

22. *Communications* The cost of a telephone call c is a function of the time spent talking t in minutes. Suppose the rule $c = 0.22t$ describes this function. A student calculates how much a 2-h phone call will cost. Use the student's work at the right.

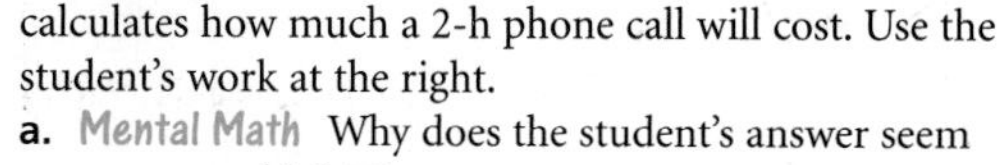

a. *Mental Math* Why does the student's answer seem unreasonable? The amount seems too low.
b. What mistake(s) did the student make? The student forgot to convert 2 h to minutes.
c. How much does it cost to make a 2-h phone call? \$26.40

23. a. *Open-ended* Sketch a graph of a relation that is not a function. See margin.
b. *Writing* Explain why the graph you drew is not a function graph. The graph fails the vertical line test.

24. a. *Business* A store bought disposable cameras for \$300. The store's profit p is a function of the number of cameras sold c. Find the range of $p = 6c - 300$ when the domain is {0, 15, 50, 62}. {−300, −210, 0, 72}
b. In this situation, what do the domain and range represent? Domain represents the number of cameras sold; the range represents the profit from selling these cameras.

Find the range of each function when the domain is {−1, 0, 6}.

25. $g = 5 + r$ {4, 5, 11}
26. $b = \frac{1}{2}a + 110$ {109.5, 110, 113}
27. $w = \frac{11(h - 40)}{2}$ $\{-225\frac{1}{2}, -220, -187\}$
28. $y = x^2 + 1$ {2, 1, 37}
29. $q = 4p - 7$ {−11, −7, 17}
30. $p = 2.4h$ {−2.4, 0, 14.4}
31. $t = 6s^2$ {6, 0, 216}
32. $d = 60r$ {−60, 0, 360}
33. $y = -x^3 + 6$ {7, 6, −210}
34. $d = 5t$ {−5, 0, 30}
35. $y = x^2 - x$ {2, 0, 30}
36. $y = -\frac{x}{14}$ $\{\frac{1}{14}, 0, -\frac{3}{7}\}$

37. a. *Language Arts* Copy and complete the analogy:
"Input value is to output value as independent variable is to __?__. dependent variable
b. Write another analogy that includes the words *input, output, domain,* and *range.* Input value is to output value as domain is to range.

38. *Music* There are 52 white keys on a piano. The frequency of each key is the number of vibrations per second the key's string makes.
a. Is this relation a function? Explain your reasoning. a–b. See margin.
b. *Writing* Describe the patterns in the table.

Key Position	1	8	15	22	29	36	43
Frequency	27.5	55	110	220	440	880	1760

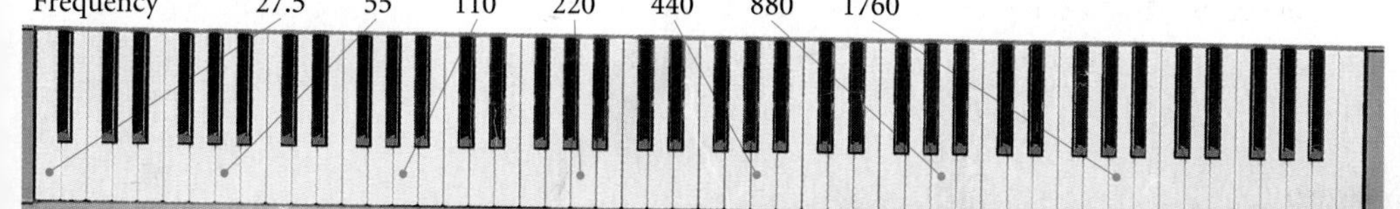

page 78 Mixed Review

54b.

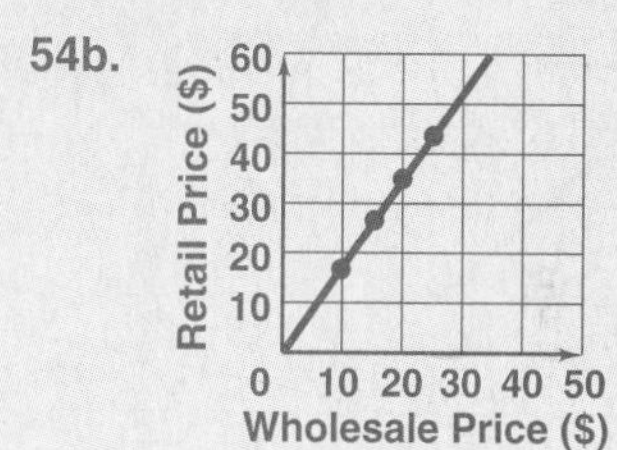

STANDARDIZED TEST TIP **Exercise 53** Point out to students that there are three answer choices beginning with 1. If you first substitute 2 into the function to get $y = 1$ you eliminate only two of the answer choices. You then have to spend time substituting a second value into the function. However, if you begin by substituting the last value, 6, into the function, you are more likely to get a y-value that allows you to eliminate more of the answer choices. There is only one answer choice containing 33. By studying the answer choices before you start, you can minimize the amount of calculation needed.

Exercises MIXED REVIEW

Exercise 54 Ask a volunteer, or a guest speaker, to explain the difference between wholesale and retail. Have students make conjectures about why the retail and wholesale prices are different.

Exercise 59 Encourage students to draw a number line to help them simplify this expression.

GETTING READY FOR LESSON 2-5 These exercises prepare students for writing function rules from data in tables.

Wrap Up

THE BIG IDEA Ask: *What is a function? How can you represent a function?*

RETEACHING ACTIVITY Students will use random numbers to evaluate functions. (Reteaching worksheet 2-4)

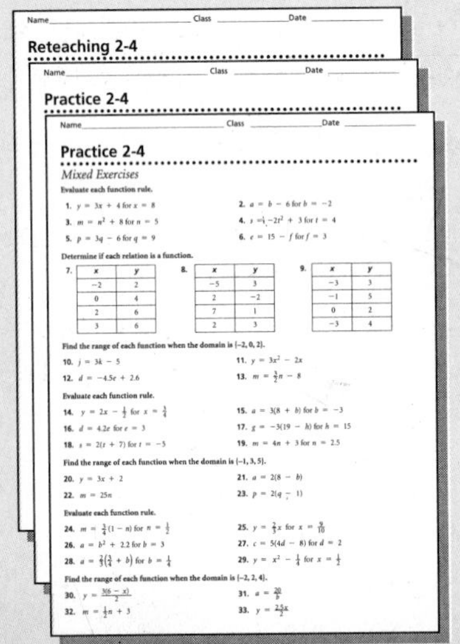

Lesson Quiz

Lesson Quiz is also available in Transparencies.

1. Evaluate for $x = -4$
   $y = 4x + 3$ **−13**
   $y = 2x^2 - 5$ **27**
2. Does the table represent a function? **yes**

Domain	Range
3	−1
7	4
2	11

3. Find the range of the following functions when the domain is $\{-1, 4, -2\}$.
   $t = \frac{2}{5}z + 2$ $\left\{\frac{8}{5}, \frac{18}{5}, \frac{6}{5}\right\}$
   $y = 4r - 3$ $\{-7, 13, -11\}$

78

Evaluate each function rule for $x = 0.6$.

39. $y = x + 1.53$ **2.13**
40. $y = 3x^2$ **1.08**
41. $y = \frac{1}{3}x - 4.5$ **−4.3**
42. $y = -4x$ **−2.4**
43. $y = \frac{x^2}{9}$ **0.04**
44. $y = 34 - x$ **33.4**
45. $y = -x^2 + 2x$ **0.84**
46. $y = \frac{4x - 2}{7}$ $\frac{2}{35}$

Find the range of the function rule $y = x^2$ for each domain.

47. $\{0.5, 11\}$ **{0.25, 121}**
48. $\{-1.2, 0, 4\}$ **{1.44, 0, 16}**
49. $\{-5, -1, 0, 2, 10\}$ **{25, 1, 0, 4, 100}**
50. $\left\{-\frac{1}{2}, \frac{1}{4}, \frac{2}{5}\right\}$ $\left\{\frac{1}{4}, \frac{1}{16}, \frac{4}{25}\right\}$

51. *Critical Thinking* Suppose a classmate looks at the graph at the right and says, "This graph is not a function because both 2 and −2 have a y-value of 4." Explain what is wrong with this student's statement. **The same y-value may correspond to any number of x-values.**
52. *Science* Light travels at a speed of approximately 186,000 miles per second. The function rule $d = 186{,}000t$ describes the relationship between distance d in miles and time t in seconds.
    a. How far does light travel in 20 seconds? **3,720,000 mi**
    b. How far does light travel in 1 minute? **11,160,000 mi**
    c. Make a table of values for this function rule. **See margin.**
53. *Standardized Test Prep* What is the range of the function $y = x^2 - 3$ when the domain is $\{2, 4, 6\}$? **D**
    A. $\{-1, 1, 3\}$ B. $\{1, 5, 9\}$ C. $\{1, 1, 9\}$ D. $\{1, 13, 33\}$ E. $\{6, 10, 14\}$

Exercises MIXED REVIEW

54. a. Which variable is the independent variable? the dependent variable? **wholesale price; retail price**
    b. Graph the data in the table. **See margin.**
    c. Use your graph to **predict** the retail price of an item with a \$30 wholesale price. **about \$53**

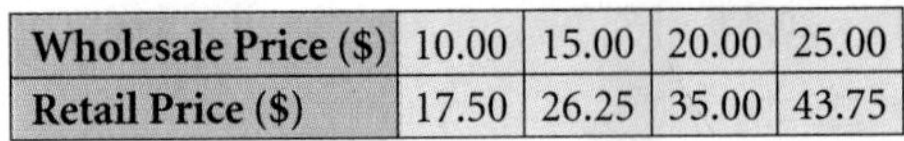

Wholesale Price (\$)	10.00	15.00	20.00	25.00
Retail Price (\$)	17.50	26.25	35.00	43.75

Simplify each expression.

55. $72 \div (8 + 4)$ **6**
56. $5^2 + 9.5$ **34.5**
57. $86 - 18 \div 3$ **80**
58. $\frac{6 + 2}{4}$ **2**
59. $\frac{2}{5} - \frac{1}{2}$ $-\frac{1}{10}$

Getting Ready for Lesson 2-5

Use an equation to model the relationship in each table.

60.

Number of People	Total Bill
1	\$3.00
2	\$6.00
3	\$9.00
4	\$12.00

$t = 3n$

61.

Amount Earned	Amount Spent
\$15	\$5
\$30	\$10
\$45	\$15
\$60	\$20

$s = \frac{1}{3}e$

62.

Number of Days	Supplies Remaining
0	12 lb
2	10 lb
4	8 lb
6	6 lb

$r = 12 - d$

2-5 Teaching Notes

PROBLEM OF THE DAY

If you wrote the integers from 1 to 100, how many times would you write the digit 3? 20

Problem of the Day is also available in Transparencies.

CONNECTING TO PRIOR KNOWLEDGE Ask students to share an experience in which they observed unrelated events that later formed a visible pattern. Ask students to recall experiences in math that involved finding and defining a pattern.

THINK AND DISCUSS

ERROR ALERT! Some students tend to misread function notation as "the variable f times the variable x." **Remediation:** Discuss this confusion and then write several functions on the board. Have the entire class, and then individual students, read them aloud using the words "f of."

Question 2b Suggest that students make a table of values for the domain and range.

Example 1

Point out that in place of $f(x) = x + 4$ or $f(x) = x^2$ students could also write $y = x + 4$ or $y = x^2$ Ask:

- *Why might you wish to use letters other than x and f(x)?*

Connections Computers . . . and more

2-5 Writing a Function Rule

What You'll Learn

- Writing rules for functions from tables and words

...And Why

To solve problems related to computers and jobs

THINK AND DISCUSS

Part 1 Understanding Function Notation

In the last lesson, you learned that a function rule is an equation that describes a function. Sometimes the equation is written using function notation. To write a rule in **function notation,** you use the symbol $f(x)$ in place of y. You read $f(x)$ as "f of x."

QUICK REVIEW

You can also write a function rule using "$y =$." For example, $f(x) = 2x + 5$ is equivalent to $y = 2x + 5$.

$$f(x) = 2x - 5$$

output values — dependent variable; input values — independent variable

The rule shows you that $f(x)$ is a function of x.

Function notation allows you to see the input value. Suppose the input value above is -2. Here is how to find $f(-2)$.

$$f(-2) = 2x - 5$$
$$f(-2) = 2(-2) - 5$$
$$f(-2) = -9$$

Be careful: $f(x)$ does not mean "f times x"!

When the input value is -2, the output value is -9.

1. Try This Find $f(7)$ and $f(0)$. 9; -5

2. a. Write an equation equivalent to $y = 7x + 6$ using function notation. $f(x) = 7x + 6$
   b. Find the range of the function when the domain is $\{-1, 0, 5\}$. $\{-1, 6, 41\}$

QUICK REVIEW

domain = input values
range = output values

Part 2 Using a Table of Values

You can write a rule for a function by analyzing a table of values.

Example 1

Patterns Write a function rule for each table.

Look for a pattern relating x and $f(x)$.

a.

x	$f(x)$
1	5
2	6
3	7
4	8

Ask yourself, "What can I do to 1 to get 5, 2 to get 6, . . . ?"

Relate $f(x)$ equals x plus four
Write $f(x) = x + 4$

b.

x	$f(x)$
0	0
3	9
6	36
9	81

Ask yourself, "What can I do to 3 to get 9, 6 to get 36, . . . ?"

$f(x)$ equals x times itself
$f(x) = x^2$

Lesson Planning Options

Prerequisite Skills

- Modeling equations

Assignment Options

To provide flexible scheduling, this lesson can be subdivided into parts.

1 **Core** 1–8, 18–25
Extension 10, 30

2 **Core** 12–14, 26–29, 32–34
Extension 11, 31

3 **Core** 15–17
Extension 9

Use Mixed Review to maintain skills.

Resources

Student Edition

Skills Handbook, p. 569, 571
Extra Practice, p. 557
Glossary/Study Guide

Teaching Resources

Chapter Support File, Ch. 2
- Practice 2-5 (two worksheets)
- Reteaching 2-5
- Checkpoint

Classroom Manager 2-5
Glossary, Spanish Resources
Two-Year Algebra Handbook 2-5

Transparencies
44

It may be easier to remember what each variable represents if you use variables other than *x* and *f(x)*.

- *Give an example.* Answers may vary. Sample: Using *c* to represent cost might be easier to remember than using *x*.

Example 2 **Relating to the Real World**

Explain that $L(n)$ means that L is a function of n. $L(n)$ means the same as $f(n)$ or $f(x)$. However, the letters L and n are being used because it is easier to remember that L represents the larger size and n represents the normal size.

CONNECTING TO THE STUDENTS' WORLD Challenge students to collect measurements of classroom items for an imaginary museum exhibit called The Walk-Through Classroom. Have students create and complete a table such as the following.

Item	Normal Length	Normal Width	Normal Depth	Larger Length	Larger Width	Larger Depth
Paper clip	1.75"	0.375"	0.0625"	35"	7.5"	1.25"

Example 3 **Relating to the Real World**

Students may expect the profit to be always a positive amount. Point out that, until you pay back your relative, your profit is going to be a negative amount; in other words, you will have a loss.

Additional Examples

FOR EXAMPLE 1

Write a function rule for each table.

A

x	$f(x)$
2	8
4	10
6	12
8	14

B

x	$f(x)$
1	1
4	7
7	13
10	19

a. $f(x) = x + 6$ b. $f(x) = 2x - 1$

FOR EXAMPLE 2

The journalism class makes $25 per page of advertising in the yearbook. If the class sells *n* pages, how much money will it earn? Write a function rule to describe this relationship.

Define n = number of pages sold
$P(n)$ = money earned
Relate money earned is 25 times number of pages sold
Write $P(n) = 25 \cdot n$
The rule $P(n) = 25$ describes the relationship.

FOR EXAMPLE 3

Suppose the choir is conducting a fundraiser selling ornaments. It cost the choir $75 for the ornaments and they sell them for $3.50 each. Write a function rule to show the profit as a function of the number of ornaments sold. $P(c) = 3.5c - 75$

Price	Tax
$5.00	$.25
$3.00	$.15
$2.00	$.10
$.60	$.03

3. a. Find $f(7)$ for parts (a) and (b) in Example 1. **11; 49**
   b. Did you use the function rule or the table in each case? Why? **The rule; 7 is not listed in the table as an *x*-value.**
4. a. Identify the independent variable and the dependent variable for the table at the left. **price; tax**
   b. **Try This** Let the independent variable be x and the dependent variable be $f(x)$. Write a function rule for the table. **$f(x) = 0.05x$**

Part 3

Using Words to Write a Rule

You can also write a rule from a description of a situation. When you define variables, you may wish to use letters other than x and $f(x)$.

Example 2 **Relating to the Real World**

Computers The Computer Museum in Boston, Massachusetts, features an exhibit called The Walk-Through Computer™ 2000. You can walk on and through this giant computer and experience it firsthand. The exhibit is about 20 times larger than a normal-sized desktop computer. Write a function rule to describe this relationship.

Define	n = normal size				
	$L(n)$ = larger size shown in museum exhibit				
Relate	larger size	is	20	times	normal size
Write	$L(n)$	=	20	$\cdot$	n

The rule $L(n) = 20n$ describes the relationship.

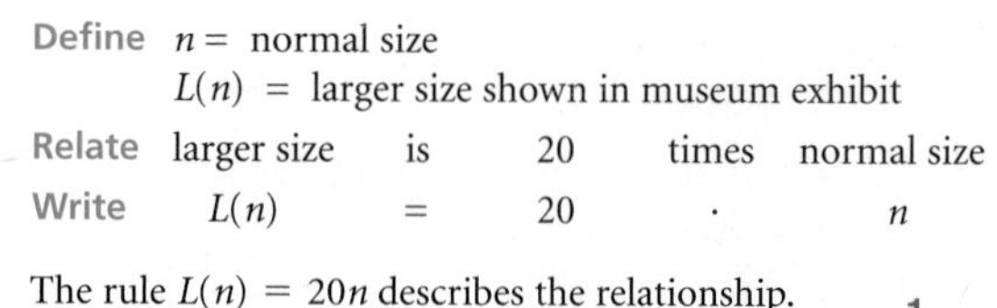

5. The space bar in the normal-sized computer measures $4\frac{3}{8}$ in. $\times$ $\frac{5}{8}$ in. Approximate the dimensions of the space bar in the museum exhibit. **5. $87\frac{1}{2}$ in. $\times$ $12\frac{1}{2}$ in.**

6. **Critical Thinking** The keyboard in the exhibit is 20 ft long and 10 ft wide. Find the approximate dimensions in inches of the normal-sized keyboard. **12 in. $\times$ 6 in.**

VISUAL LEARNING Question 8 Suggest that students make a table of values for l = 0, 1, 2, 3 and so on until they see the pattern emerging. Ask:

- *Before you mow any lawns, how much money do you owe?* **$245**
- *After you mow each lawn, what happens to the amount of money you owe?* **The amount owed is reduced by $20.**

Exercises ON YOUR OWN

MAKING CONNECTIONS Psychologists have found that people's ability to tell the difference between two weights depends on the amount of the weights. People can feel that a 1-oz envelope weighs less than a 2-oz envelope. But they cannot tell that a 50-lb box weighs less than a 50-lb, 1-oz box. A German physiologist named Ernst Weber wrote a function rule called Weber's Law. The law states that $f(w) = 0.02w$ where w is the weight of the lighter object, and $f(w)$ is the amount of weight that must be added before a person can detect the difference. For the 50-lb box, $f(50) = 1$, so 1 lb must be added to the box before a person would notice. Weber wrote other function rules for stimuli such as sound intensity and brightness.

ALTERNATIVE ASSESSMENT Exercises 1–8 Have students work the exercises using $f(4)$. Compare the results to those found when using $f(-4)$. These exercises will help you assess whether students understand the use of the notation $f(x)$. They also review operations with integers from previous lessons.

Writing a rule from words may involve more than one operation.

Example 3 *Relating to the Real World* 

Jobs Suppose you borrow money from a relative to buy a lawn mower that costs $245. You charge $18 to mow an average-size yard. Write a rule to describe your profit as a function of the number of lawns mowed.

Define l = number of lawns mowed
$P(l)$ = total profit

Relate	total profit	is	$18	times	lawns mowed	minus	cost of mower
Write	$P(l)$	=	18	·	l	−	245

The function rule $P(l) = 18l - 245$ describes your profit.

7. *Critical Thinking* Find $P(9)$. What does $P(9)$ mean?
−$83; $P(9)$ is negative, which means the mower is not yet paid for.
8. How many lawns would you need to mow to pay for the lawn mower?
14 lawns

Exercises ON YOUR OWN

Find $f(-4)$ for each function.

1. $f(x) = x^2$ **16**
2. $f(x) = 5 - x$ **9**
3. $f(x) = 7(x + 2)^2$ **28**
4. $f(x) = -x^3$ **64**
5. $f(x) = -x - 2$ **2**
6. $f(x) = x^3 + 6$ **−58**
7. $f(x) = x + 25$ **21**
8. $f(x) = 6x$ **−24**

9. *Laundry* Use the function in the table at the right.
   a. Identify the dependent and independent variables. **amount of water; number of loads**
   b. Write a rule to describe the function. **$w(n) = 42n$**
   c. How many gallons of water will you use for 7 loads of laundry? **294 gal**
   d. *Critical Thinking* In one month, you used 546 gal of water. How many loads did you wash? How did you find your answer? **13 loads; you divide 546 gal by 42 gal/load.**

Number of Loads	Gallons of Water Used
1	42
2	84
3	126
4	168

10. *Food* At a supermarket salad bar, the price of a salad depends on its weight. Salad costs $.19 per ounce.
    a. Write a rule to describe the function. **$s(w) = 0.19w$**
    b. How much will an 8-ounce salad cost? **$1.52**
11. *Writing* What advantage(s) can you see of having a function rule instead of a table of values for a function? **Answers may vary. Sample: The input values you need may not be in the table.**

Match each table of values with a rule at the right.

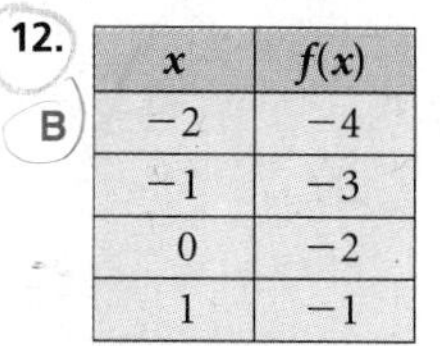

12. **B**

x	$f(x)$
−2	−4
−1	−3
0	−2
1	−1

13. **A**

x	$f(x)$
−1	−2
−2	−4
−3	−6
−4	−8

14. **C**

x	$f(x)$
−1	−3
0	−4
1	−5
2	−6

A. $f(x) = 2x$
B. $f(x) = x - 2$
C. $f(x) = -4 - x$

Technology Options

For Exercise 18-25, students may want to use spreadsheet software to find the range for each function.

Prentice Hall Technology 

Software • Secondary Math Lab Toolkit • Computer Item Generator 2-5

CD-ROM Multimedia Algebra Lab 2

Internet • See the Prentice Hall site. (http://www.phschool.com)

CRITICAL THINKING Exercise 9d Students may use the guess-and-check strategy to answer this question. Encourage students to confirm their answer by substituting the 546 gal into the function rule $W(n) = 42n$.

Exercise 16b Point out that many book clubs charge a shipping and handling fee for each shipment. There can be several books in one shipment. Using a shipping fee of \$3.50, have students calculate how many shipments are needed before the bookstore becomes more affordable. 4

RESEARCH Exercise 17 Supermarkets, bookstores, and libraries are excellent sources of magazines.

Exercises 18–25 Make sure students read the third domain value as 1.2 and not 1, 2.

STANDARDIZED TEST TIP Exercise 35 This is a typical quantitative comparison question. To do this type of question successfully, students need to be comfortable with what the four answer choices mean. Tell students that they can eliminate D if they can get a number for each column.

Chapter Project **FIND OUT BY WRITING** Students evaluate the given function and create a table and graph. Have students add this task to work they have already completed for the Chapter Project. Check students' progress on the project by having each group write a sentence or two about what they have done for the project and their plans for completion.

page 83 Mixed Review

40.

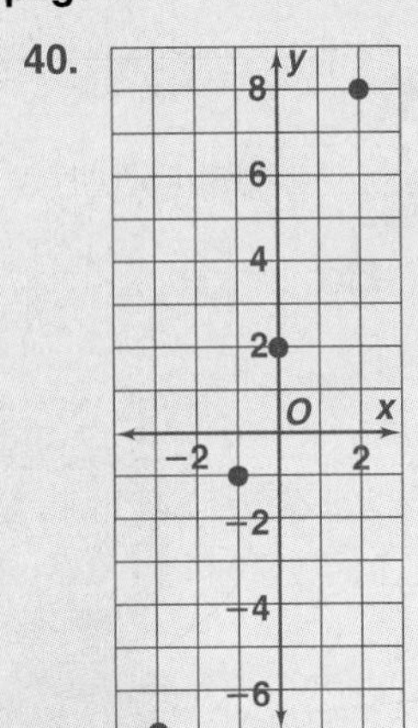

41.

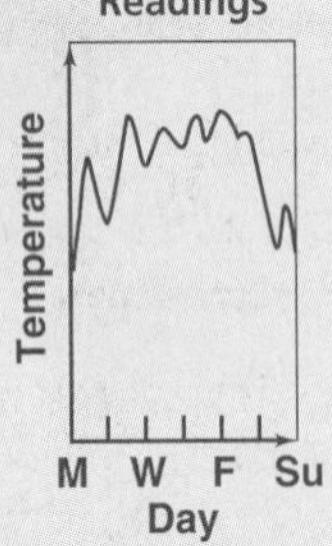

42.

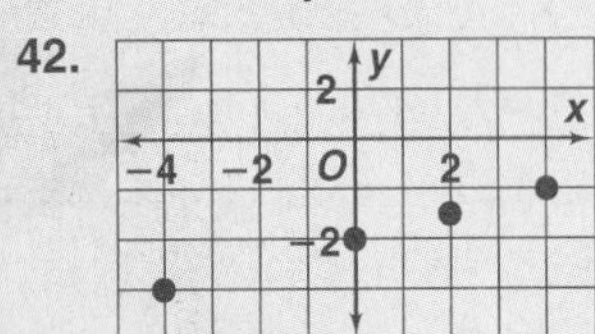

43, Checkpoint 1b, 2. See back of book.

Math in the Media **Use the advertisement for Exercises 15–17.**

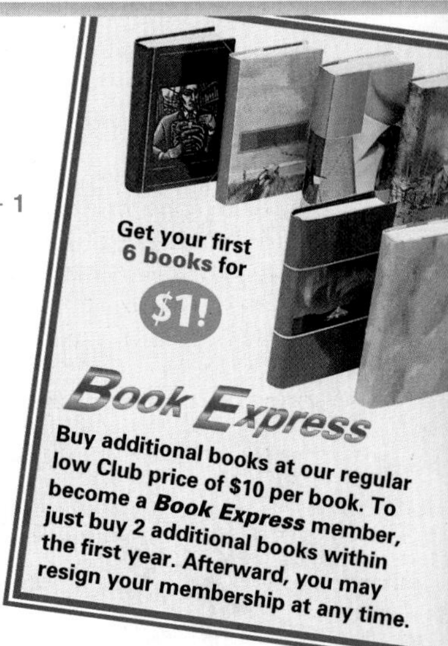

Get your first 6 books for
$1!
Book Express
Buy additional books at our regular low Club price of $10 per book. To become a Book Express member, just buy 2 additional books within the first year. Afterward, you may resign your membership at any time.

15. a. Write a rule to find the total cost $T(a)$ for all the books a person buys through Book Express. Let a represent the additional books bought (after the first 6 books). $T(a) = 10a + 1$

b. Suppose a person buys 9 books in all. Find the total cost. \$31

16. A bookstore sells the same books for an average price of \$6 each.

a. Suppose you plan to buy 12 books. What is the average cost per book as a member of Book Express? about \$5.08

b. Is it less expensive to buy 12 books at the club or at a bookstore? **Justify** your answer. The club saves you about \$.92 per book or \$11 in all.

17. a. *Research* Find an ad for another book club in a magazine.

b. Select 9 books you would be interested in buying. Compare the cost of the 9 books you selected to the cost at Book Express. a–b. Check students' work.

Find the range for each function when the domain is {–3, 0, 1.2, 5}.

18. $f(x) = x^3$ {–27, 0, 1.728, 125}
19. $C(g) = 15 - g$ {18, 15, 13.8, 10}
20. $y = 1.3x$ {–3.9, 0, 1.56, 6.5}
21. $R(s) = -s^2$ {–9, 0, –1.44, –25}
22. $y = -4x$ {12, 0, –4.8, 20}
23. $f(g) = g^2 + 1$ {10, 1, 2.44, 26}
24. $G(n) = (n - 5)^2$ {64, 25, 14.44, 0}
25. $f(x) = 6x - 4$ {–22, –4, 3.2, 26}

Patterns **Write a function rule for each table.**

26.

x	$f(x)$
1	0.5
2	1
3	1.5
4	2

$f(x) = 0.5x$

27.

x	y
–2	–8
–1	–4
0	0
1	4

$y = 4x$

28.

x	y
3	9
6	18
9	27
12	36

$y = 3x$

29.

x	$f(x)$
1	1
2	8
3	27
4	64

$f(x) = x^3$

30. Write another equivalent equation for each table in Exercises 26–29. $y = 0.5x$ $f(x) = 4x$ $f(x) = 3x$ $y = x^3$

31. *Food* Kathy and Rheta share an 8-slice pizza. The number of slices Rheta can eat depends on how many slices Kathy eats. Write a function rule to describe this relationship. What are the domain and range?
$R(K) = 8 - K$ {0, 1, 2, 3, 4, 5, 6, 7, 8}; {8, 7, 6, 5, 4, 3, 2, 1, 0}

For Exercises 32–33, write a function rule to describe each statement.

32. change from a one-dollar bill when buying pencils that cost 20¢ each
$f(p) = 1 - 0.2p$ or $f(p) = 100 - 20p$

33. a worker's earnings when the worker is paid \$4.25/h
$e(h) = 4.25h$

34. *Standardized Test Prep* Compare the quantities in Column A and Column B. B

Column A	Column B
$f(3)$ when $f(x) = 4x - 12$	$f(0)$ when $f(x) = x + 1$

A. The quantity in Column A is greater.
B. The quantity in Column B is greater.
C. The quantities are equal.
D. The relationship cannot be determined from the information given.

Exercises MIXED REVIEW

Exercise 36 Remind students that absolute value signs are not parentheses. Students may try to write $1.3 - |-2|$ as $1.3 + 2$ instead of $1.3 - 2$.

GETTING READY FOR LESSON 2-6 These exercises prepare students to graph functions.

Wrap Up

THE BIG IDEA Ask students to list the advantages of using function notation, instead of a table of values, to describe a pattern.

RETEACHING ACTIVITY Students learn to write rules for functions by interpreting the pattern of the data. (Reteaching worksheet 2-5)

Exercises CHECKPOINT

Students will assess their own progress on Lessons 2-3 to 2-5.

ESL **Exercise 2** Some students may need to give an example instead of writing their own definitions.

WRITING **Exercise 2** Encourage students to draw sketches to accompany their explanations.

Chapter Project ***Find Out by Writing***

The function $t(n) = 4.3n$ predicts the time t (in seconds) it takes n people in a row to say the tongue twister "A cricket critic cricked his neck at a critical cricket match."

- Find $t(5)$. Explain what it represents.
- What does $t(0) = 0$ mean?
- Suppose the output is 34.4 s. How can you find the input? What does the input represent?
- Make a table of values for the function and graph it.

Exercises MIXED REVIEW

Evaluate each expression for $t = -2$, $m = 5$, and $s = 1.3$.

35. $m^2 + t$ 23

36. $|s| - |t|$ −0.7

37. $-mt$ 10

38. $s^2 + t^2$ 5.69

39. $\frac{m}{t}$ −2.5

Getting Ready for Lesson 2-6

Graph the data in each table. 40–43. See margin.

40.

x	y
−3	−7
−1	−1
0	2
2	8

41.

x	y
−3	4
−2	0
0	−2
2	4

42.

x	y
−4	−3
0	−2
2	−1.5
4	−1

43.

x	y
−2	0.5
1	2
4	3.5
7	5

Exercises CHECKPOINT

1. a. The table at the right shows the number of words read within certain times. Identify the independent and dependent variables.
b. Graph the data. What scales did you use and why?
a. number of minutes; number of words b. See margin.

Number of Minutes	Number of Words Read
4	1100
8	2200
12	3300
16	4400

2. *Writing* Define *relation* and *function* in your own words. See margin.

3. Explain how you would find $f(3)$ for $f(x) = 4x - 12$.
You multiply 3 by 4, then subtract 12 from the result.

Write a function rule to describe each statement.

4. the price of rolls if they cost \$.30 each $p(r) = 0.3r$

5. the cost of your dinner if you leave a 15% tip $c(d) = 1.15d$

Find the range of each function when the domain is {−7, 0, 1, 3}.

6. $y = 12(x - 3)$ {−120, −36, −24, 0}

7. $v = 9 - u^2$ {−40, 9, 8, 0}

8. $y = 4x + 2.5$ {−25.5, 2.5, 6.5, 14.5}

9. $t = s + 4$ {−3, 4, 5, 7}

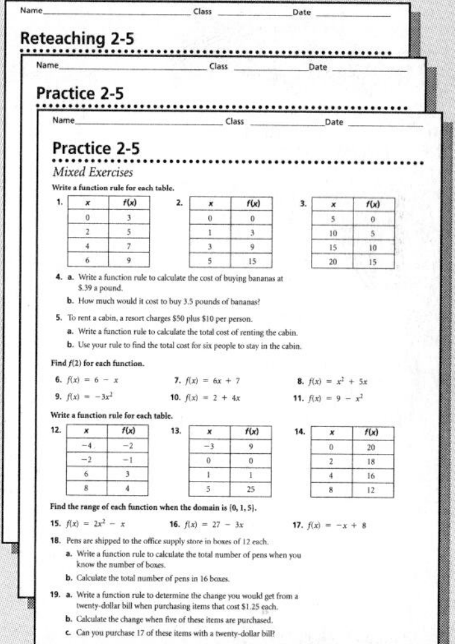

Name___ Class ___ Date ___

Reteaching 2-5

Name___ Class ___ Date ___

Practice 2-5

Name___ Class ___ Date ___

Practice 2-5

Mixed Exercises

Write a function rule for each table.

1.

x	$f(x)$
0	3
2	5
4	7
6	9

2.

x	$f(x)$
0	0
1	3
3	9
5	15

3.

x	$f(x)$
5	0
10	5
15	10
20	15

4. a. Write a function rule to calculate the cost of buying bananas at \$.39 a pound.
b. How much would it cost to buy 3.5 pounds of bananas?

5. To rent a cabin, a resort charges \$50 plus \$10 per person.
a. Write a function rule to calculate the total cost of renting the cabin.
b. Use your rule to find the total cost for six people to stay in the cabin.

Find $f(2)$ for each function.

6. $f(x) = 6 - x$ **7.** $f(x) = 6x + 7$ **8.** $f(x) = x^2 + 5x$

9. $f(x) = -3x^2$ **10.** $f(x) = 2 + 4x$ **11.** $f(x) = 9 - x^2$

Write a function rule for each table.

12.

x	$f(x)$
−4	−2
−2	−1
6	3
8	4

13.

x	$f(x)$
−3	9
0	0
1	1
5	25

14.

x	$f(x)$
0	20
2	18
4	16
8	12

Find the range of each function when the domain is {0, 1, 5}.

15. $f(x) = 2x^2 - x$ **16.** $f(x) = 27 - 3x$ **17.** $f(x) = -x + 8$

18. Pens are shipped to the office supply store in boxes of 12 each.
a. Write a function rule to calculate the total number of pens when you know the number of boxes.
b. Calculate the total number of pens in 16 boxes.

19. a. Write a function rule to determine the change you would get from a twenty-dollar bill when purchasing items that cost \$1.25 each.
b. Calculate the change when five of these items are purchased.
c. Can you purchase 17 of these items with a twenty-dollar bill?

Lesson Quiz

Lesson Quiz is also available in Transparencies.

1. Write a function rule for each table.

a.

x	$f(x)$
5	15
7	21
9	27

b.

x	$f(x)$
2	7
5	28
8	67

a. $f(x) = 3x$ b. $f(x) = x^2 + 3$

2. Find $f(3)$ for each function.
$f(x) = 4x + 3$ 15
$f(x) = 5(x - 1)^2$ 20
$f(x) = -x - 5$ −8

3. Write a function rule to describe the following statement: a model is 20% of the size of the real item.
$f(x) = 0.2x$

2-6 Teaching Notes

PROBLEM OF THE DAY

A soccer player is preparing for a tournament by running 3 mi every day for 6 wk. The first week her run took 21 min and 40 s. Each week she trimmed 25 s off her time. How long did the 3-mi run take her during the last week? 19 min 35 s

Problem of the Day is also available in Transparencies.

CONNECTING TO PRIOR KNOWLEDGE Write on the board 3, 6, 9, 12, . . . Ask students to name different ways to represent this pattern using words, sketches, symbols, or any other mathematical method.

THINK AND DISCUSS

EXTENSION Have students create a table of values and a graph for cars that get 20 mi/gal and 40 mi/gal. Have students conjecture about how the graphs will be different. Then have them compare the different graphs to see if their conjectures were correct.

CONNECTING TO THE STUDENTS' WORLD Ask students to write the name of a place, at least 100 miles away, that they would like to visit by car. Have students guess how much the gasoline for the trip might cost. Next, have them find the exact mileage to their destination on a map. Ask volunteers to find the average cost per gallon of gasoline in your city. Using the gasoline cost, and the assumption that gas consumption is 30 mi/gal, have each student calculate the cost to drive to the

Lesson Planning Options

Prerequisite Skills

- Using function notation
- Plotting ordered pairs

Assignment Options

Core 1–8, 12–30
Extension 9–11, 31–35

Use Mixed Review to maintain skills.

Resources

Student Edition

Skills Handbook, p. 569, 580
Extra Practice, p. 557
Glossary/Study Guide

Teaching Resources

Chapter Support File, Ch. 2
- Practice 2-6 (two worksheets)
- Reteaching 2-6

Classroom Manager 2-6
Glossary, Spanish Resources
Two-Year Algebra Handbook 2-6

Transparencies
2, 44

What You'll Learn
- Graphing a function
- Creating a table of values from a rule and a graph

...And Why

To solve problems involving electrician costs

1a. 420 mi
b–c. Possible answers: You can substitute 14 for g in the rule; you can extend the table to include 14 gal; you can extend the graph to find the y-coordinate of the point with x-coordinate 14.

Connections  Electrician . . . and more

2-6 The Three Views of a Function

THINK AND DISCUSS

You can model functions using rules, tables, and graphs. Suppose a car gets 30 mi/gal. Then the distance $d(g)$ that the car travels is a function of the number of gallons g.

Rule

$d(g) = 30g$

A function rule shows how the variables are related.

Table of Values

Gallons	Miles
g	$d(g)$
0	0
1	30
2	60
3	90

A table identifies specific values that satisfy the function.

Graph

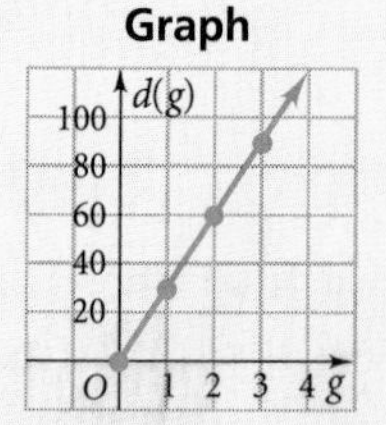

A graph gives a visual picture of the function.

1. a. Suppose the car used 14 gal of gasoline. How far did the car travel?
   b. How did you find your answer to part (a)? See left.
   c. Describe another method of solving the problem.

You can use a rule to model a function with a table of values and a graph.

Example 1 Relating to the Real World

Electrician Suppose you hire an electrician to install an electrical outlet in a wall. The electrician charges $68 for materials plus $40 an hour for service. The total cost $C(h)$ is a function of the number of hours it takes to do the job. Use the rule $C(h) = 68 + 40h$ to make a table of values and then a graph.

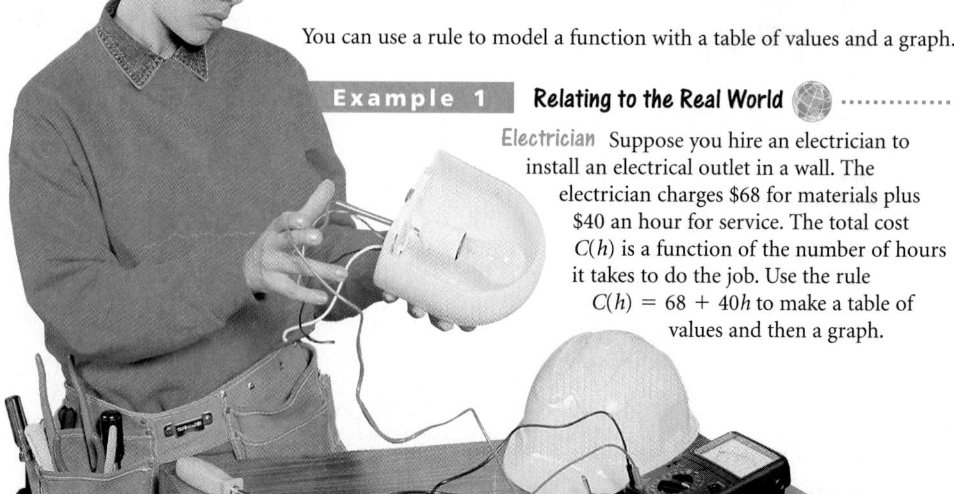

chosen destination. Ask them to compare the calculated cost to the guess.

Example 1 **Relating to the Real World** 

Question 2 Have students work in groups of three or four to discuss the advantages and disadvantages of using each the rule, the table, and the graph to answer part (a).

Question 3 The issue of whether or not to draw a line through a set of points provides an opportunity to assess students' number sense and ability to reason mathematically. Evaluating whether answers are reasonable and using mathematical reasoning are important skills in the world of work.

DIVERSITY Have students bring to class a piece of paper on which they have written an occupation and its associated hourly rate. Ask them to verify their information by talking to someone who has that job, perhaps in a phone call. Have them also write whether their job uses much, some, or only a little math. Post the information and discuss the patterns.

Example 2

VISUAL LEARNING Plot the points (2, –4) and (3, 0) on coordinate axes on the board. Draw a straight line through them. Ask students to explain why this graph of the given function is incorrect. Lead students to understand the importance of plotting a number of points before connecting the graph. Stress that students should include positive, negative, and zero values in their table to assure accuracy.

STEP 1: Choose values for h that seem reasonable, such as 1, 2.5, 4, and 7.

STEP 2: Input the values for h. Evaluate to find $C(h)$.

STEP 3: Plot the ordered pairs.

h	$C(h) = 68 + 40h$	$(h, C(h))$
1	$68 + 40(1) = 108$	(1, 108)
2.5	$68 + 40(2.5) = 168$	(2.5, 168)
4	$68 + 40(4) = 228$	(4, 228)
7	$68 + 40(7) = 348$	(7, 348)

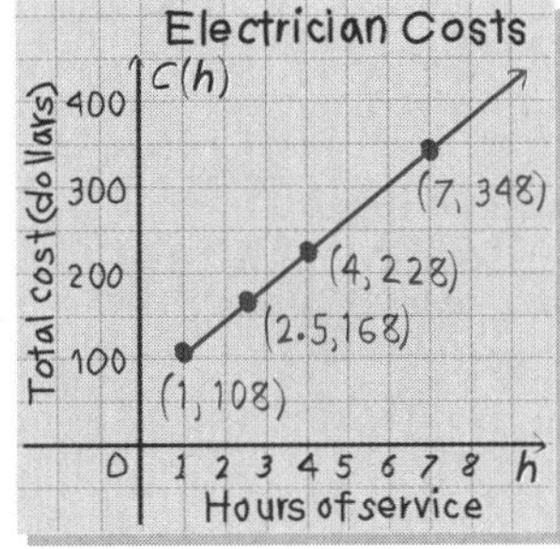

2. a. What is the total cost of the job if it takes $5\frac{1}{2}$ hours? $288
   b. Did you use the rule, the table, or the graph to answer part (a)? Answers may vary. Sample: rule
3. Why are the points on the graph connected by a line? The variable is continuous.
4. Are negative values for h reasonable? Explain. No; the electrician cannot work a negative number of hours.
5. **Electrician** Suppose this electrician tells you the job will take 2 to 6 hours to complete. What is the range of your costs? $148 to $308
6. **Try This** Model each rule with a table of values and a graph.
   a. $f(x) = -3x + 5$ b. $f(x) = \frac{1}{2}x$
   a–b. See margin.

QUICK REVIEW

You can also write a function rule using "$y =$ " form. For example, $f(x) = x^2 - x - 6$ is equivalent to $y = x^2 - x - 6$.

Some functions have graphs that are not straight lines. You can graph a function as long as you know its rule.

Example 2

Graph the function $y = x^2 - x - 6$.

First, make a table of values.

x	$y = x^2 - x - 6$	(x, y)
-2	$(-2)^2 - (-2) - 6 = 0$	(-2, 0)
-1	$(-1)^2 - (-1) - 6 = -4$	(-1, -4)
0	$(0)^2 - (0) - 6 = -6$	(0, -6)
1	$(1)^2 - (1) - 6 = -6$	(1, -6)
2	$(2)^2 - (2) - 6 = -4$	(2, -4)
3	$(3)^2 - (3) - 6 = 0$	(3, 0)

Then graph the data.

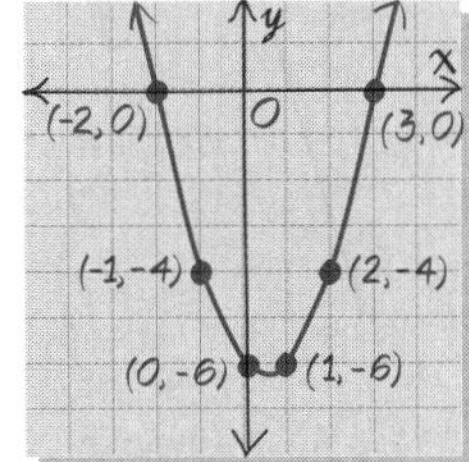

7. Answers may vary. Sample: You can find y for any value of x that is not in the table, and the values computed from the rule are more likely to be exact than the values estimated from a graph.

7. What advantage(s) does a rule have in describing a function? See left.
8. **Try This** Graph each function. a–b. See margin.
   a. $y = -3x^2 + 1$ b. $f(x) = x^2 - 4x - 5$

Additional Examples

FOR EXAMPLE 1

At the local video store you can rent a video game for $3. It costs you $5 a month to operate your video game player. The total monthly cost $C(v)$ is a function of the number of video games v you rent. Use the rule $C(v) = 5 + 3v$ to make a table of values and a graph.

v	$C(v) = 5 + 3v$	$(v, C(v))$
0	$5 + 3(0) = 5$	(0, 5)
1	$5 + 3(1) = 8$	(1, 8)
2	$5 + 3(2) = 11$	(2, 11)

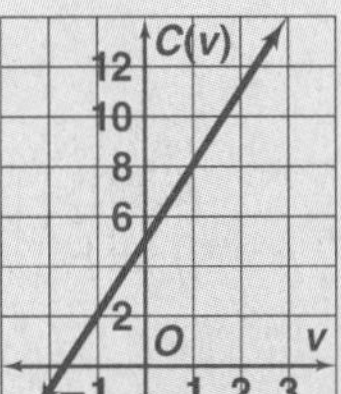

FOR EXAMPLE 2

Graph the function $y = x^2 + x - 2$.

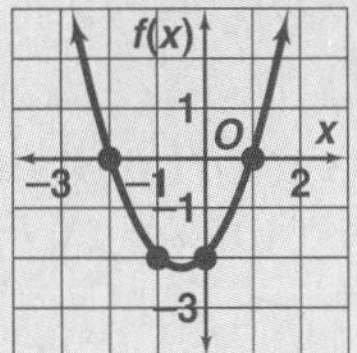

Example 3

KINESTHETIC LEARNING Use masking tape to make a large pair of coordinate axes on the floor in the center of the room with tick marks about 1 ft apart. Have students stand randomly around the room. As you call each student's name, have the student tell you the values of x and $f(x)$ that represent that location in the room. If you have block scheduling or extended class periods, extend this activity by directing individual students to move to new locations (for example, "Carlos go to (0, –3)").

Exercises ON YOUR OWN

ESTIMATION Exercise 10d This exercise helps students realize the importance of drawing clear graphs so that answers can be read from the graph. Encourage students to check their estimate by using the function rule.

CRITICAL THINKING Exercise 10e Lead a brainstorming session in which students list other examples of discrete and continuous data. Discuss how life might be different if some discrete data became continuous; for example, if people gave their age in decimals (such as 12.25 years old).

ESL Exercise 11 Because some students may not know what a shower head is, explain and point out that different shower heads use different amounts of water.

Exercises 31 and 32 Tell students that the cost of a call is based on rounding up the length of time.

Technology Options

For Exercises 1–8, 12–27, 31–32 and 34, students may use spreadsheet software or a graphing calculator to graph each function rule.

Prentice Hall Technology

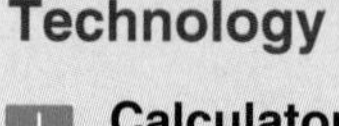

Calculator Graphing Calculator Handbook: Procedures 4, 6

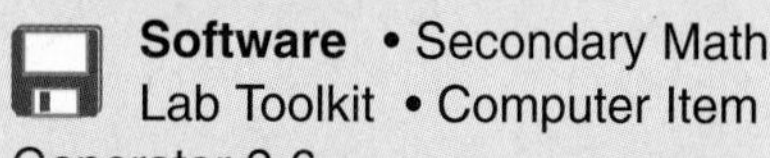

Software • Secondary Math Lab Toolkit • Computer Item Generator 2-6

CD-ROM Multimedia Algebra Lab 2

Internet • See the Prentice Hall site. (http://www.phschool.com)

86

You can create a table of values from a graph.

Example 3

Make a table of values for the graph below.

Use the scale of the graph to find ordered pairs.

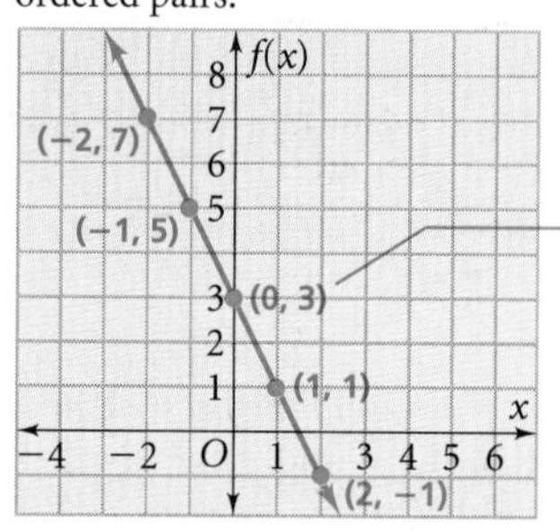

Put the ordered pairs in a table.

x	$f(x)$
−2	7
−1	5
0	3
1	1
2	−1

Exercises ON YOUR OWN

Model each rule with a table of values and a graph. 1–8. See margin.

1. $y = -3x$
2. $y = x^2 - 4x + 4$
3. $f(x) = 2x - 7$
4. $y = 6x^2$
5. $f(x) = \frac{1}{3}x$
6. $f(x) = 8 - x$
7. $f(x) = x^2 - 2$
8. $y = 5 + 4x$

9. *Patterns* The table of values below describes the perimeter of each figure in the pattern of blue tiles at the right. The perimeter P is a function of the number of tiles t.

Number of Tiles (t)	1	2	3	4
Perimeter (P)	4	6	8	10

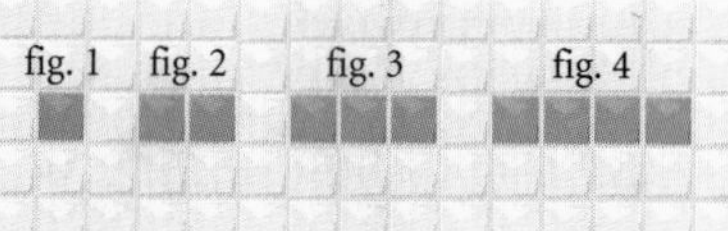

 a. Choose a rule to describe the function in the table. C
 A. $P = t + 3$ B. $P = 4t$ C. $P = 2t + 2$ D. $P = 6t - 2$
 b. How many tiles are in the figure if the perimeter is 20? 9
 c. Graph the function. See margin.

10. *Jobs* Juan charges \$3.50 per hour for baby-sitting.
 a. Write a rule to describe how the amount of money M earned is a function of the number of hours h spent baby-sitting. $M = 3.5h$
 b. Make a table of values. b–c. See margin.
 c. Graph the function.
 d. *Estimation* Use the graph to estimate how long it will take Juan to earn \$30. Answers may vary. Sample: about 8.5 hours
 e. *Critical Thinking* Do you think of baby-sitting data as discrete or continuous? Explain your reasoning. See margin.

QUICK REVIEW

Continuous data are usually measurements, such as temperatures and lengths. *Discrete data* are distinct counts, such as numbers of people or objects.

ERROR ALERT! Exercise 31 Students may draw a straight line connecting the points whenever the points appear collinear. **Remediation:** Explain that connecting the points is appropriate when both variables are continuous but not appropriate if either variable is discrete. To emphasize this, ask students to consider whether it is likely that a movie-rental store would charge you only half of the fee if you watched only half of a movie.

RESEARCH Exercise 32a Phone companies charge various rates. Students can find phone numbers in the yellow pages under "long distance services" or in the business section of the white pages.

STANDARDIZED TEST TIP Exercise 33 Suggest that first using $x = 0$ in each answer can easily eliminate some of the choices.

GEOMETRY Exercise 34 Students use a function rule rather than the standard formula to calculate the area of various isosceles triangles.

Exercises MIXED REVIEW

OPEN-ENDED Exercise 37 A scatter plot suggests situations that cannot be represented by linear functions.

ALTERNATIVE ASSESSMENT Exercises 38–42 Have students prepare a table of values and a graph for each exercise. This will help you assess whether students understand graphing functions.

GETTING READY FOR LESSON 2-7 These exercises give students practice in working with absolute values to prepare them to graph the absolute value function.

11. *Conservation* Use the data at the right. **a, b. See below**

a. Write a function rule for a standard shower head.

b. Write a function rule for a water-saving shower head.

c. Suppose you take a 6-min shower as recommended and use a water-saving head. How much water do you save compared to an average shower with a standard head? **73.2 gal; 36.6 gal**

d. Graph both functions on the same coordinate plane. What does the graph show you? **See margin.**

e. *Open-ended* How much water did you use during your last shower? How did you find your answer? **Check students' work.**

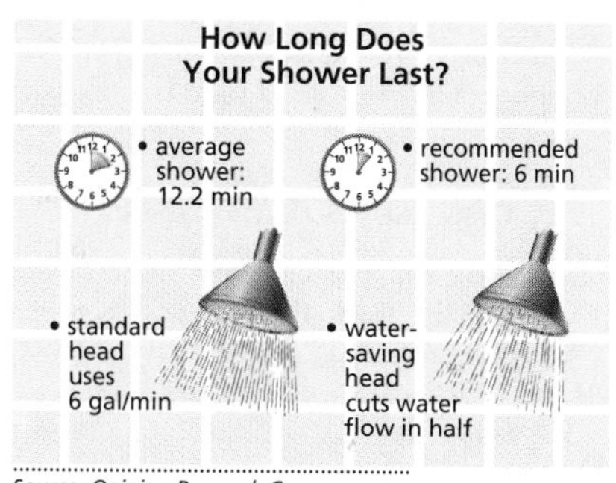

How Long Does Your Shower Last?
• average shower: 12.2 min
• recommended shower: 6 min
• standard head uses 6 gal/min
• water-saving head cuts water flow in half

Source: *Opinion Research Corp.*

Graph each function. **12–27. See margin.** **11a.** $s(t) = 6t$ **11b.** $w(t) = 3t$

12. $f(x) = x - 2$	**13.** $y = -10x$	**14.** $y = \frac{3}{4}x + 7$	**15.** $f(x) = x^2$
16. $y = x^2 - 3x + 2$	**17.** $f(x) = x$	**18.** $f(x) = x^2 - 9$	**19.** $y = 7 - 5x$
20. $f(x) = 6x + 1$	**21.** $f(x) = x - 3$	**22.** $y = x + \frac{1}{2}$	**23.** $y = 3.5x$
24. $f(x) = 4x$	**25.** $y = 1 - x^2$	**26.** $f(x) = 12 - x$	**27.** $f(x) = -5x^2$

Make a table of values for each graph. **28–30. See margin.**

28.

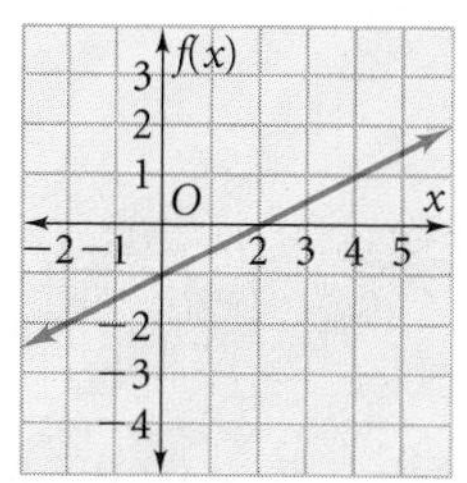

29.

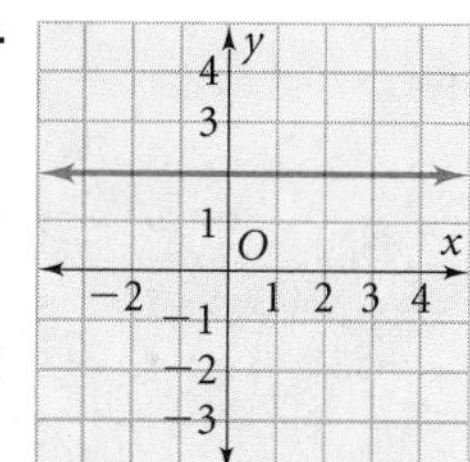

30.

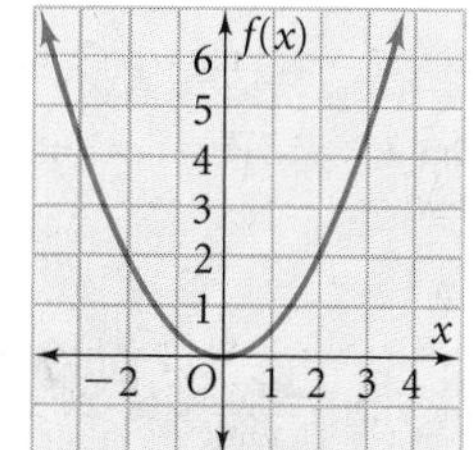

31. *Communications* The cost $C(a)$ of a call from Boston to Worcester, Massachusetts, is a function of the number of additional minutes a. The rule $C(a) = 0.27 + 0.11a$ closely models the cost.

a. How much will a 5-min call cost? (*Hint:* The number of additional minutes is one minute less than the length of the call.) **\$.71**

b. *Calculator* Make a table of values and graph the function. **See margin.**

c. *Critical Thinking* Suppose you don't want to spend any more than \$1.50 on this phone call. How many minutes can you talk? **at most 12 min**

32. a. *Research* Find the rate for weekday telephone calls in your area.

b. Write a rule to describe the cost of the call. Be sure to define your variables. **a–c. Check students' work.**

c. *Calculator* Make a table of values and graph the function.

33. *Standardized Test Prep* Which function is modeled by the table? **B**

A. $f(x) = x - 2$ **B.** $f(x) = 2x + 1$ **C.** $f(x) = 2x$

D. $f(x) = -x + 1$ **E.** $f(x) = \frac{1}{2}x - 1$

x	$f(x)$
−3	−5
0	1
2	5
5	11

pages 84–86 Think and Discuss

6a.

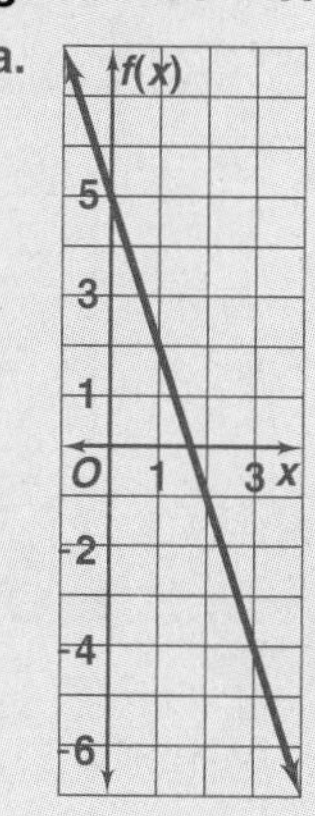

6b.

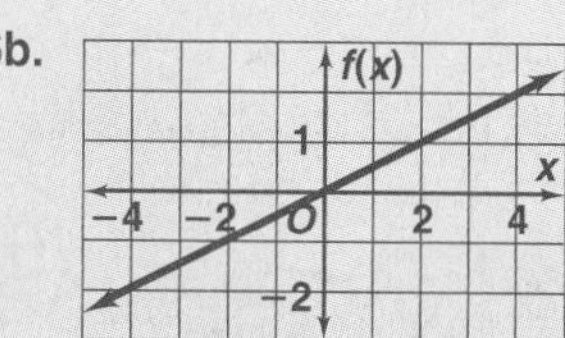

8a–b. See back of book.

pages 86–88 On Your Own

1–8, 9c, 10b–c, 10e, 11d, 12–30, 31b, 34a–b, 37a–b. See back of book.

Wrap Up

THE BIG IDEA Have students brainstorm for the advantages and disadvantages of representing functions using rules, tables, and graphs. For example, students might say that graphs make trends easier to see.

RETEACHING ACTIVITY Students create a data table and graph from randomly generated numbers. (Reteaching worksheet 2-6)

A Point in Time

If you have block scheduling or extended class periods, you may wish to have students investigate these topics which occurred in 1971.

- What constitutional amendment was ratified, giving 18-year-olds the vote? 26th
- On what planet did the U.S.S.R. soft-land a space capsule? Mars
- Which tennis star became the first woman athlete to win $100,000 in a single year? Billie Jean King
- Which NASA Apollo missions carried the third and fourth groups of astronauts to explore the moon's surface? Apollo 14 and 15

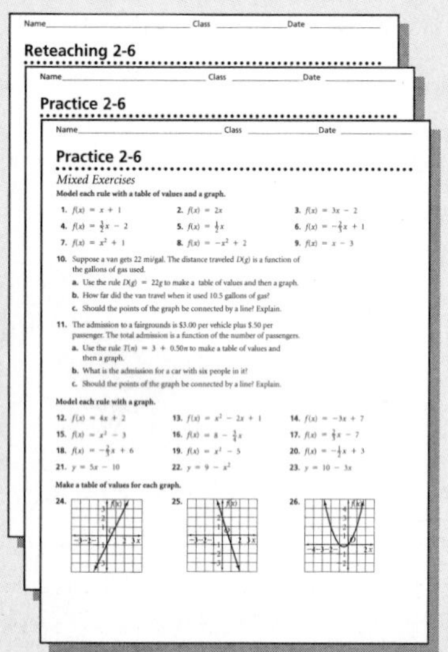
Reteaching 2-6

Practice 2-6

Practice 2-6

Mixed Exercises

Lesson Quiz

Lesson Quiz is also available in Transparencies.

1. Make a table of values for $y = 2x + 5$.

x	$y = 2x + 5$	(x, y)
0	$2(0) + 5 = 5$	(0, 5)
2	$2(2) + 5 = 9$	(2, 9)
4	$2(4) + 5 = 13$	(4, 13)

2. Graph $y = x^2 - 3$

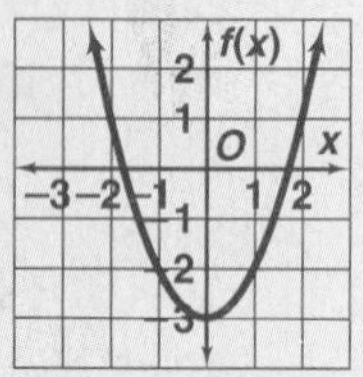

34. *Geometry* The function $A(l) = \frac{1}{2}l^2$ describes the area of an isosceles right triangle with leg l.
   a. Make a table of values for $l = 1, 2, 3, 4$.
   b. Graph the function. a–b. See margin.

35. *Writing* Suppose a student was not in class today. Describe to the student how to graph a function rule. Sample: Use the rule to make a table or a list of ordered pairs. Graph each ordered pair and then connect the points.

Exercises MIXED REVIEW

Use the scatter plot at the right.

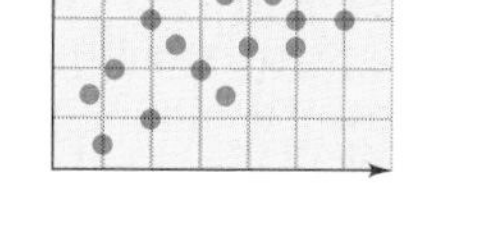

36. What kind of correlation does the scatter plot show? positive
37. a. *Open-ended* Describe a situation the scatter plot might represent.
    b. What would you label your axes? a–b. See margin.

Find the range of each function when the domain is {–2, 0, 1, 4}.

38. $f(x) = x - 1$ $\{-3, -1, 0, 3\}$
39. $f(c) = 2c + 7$ $\{3, 7, 9, 15\}$
40. $f(z) = \frac{1}{4}z$ $\{-\frac{1}{2}, 0, \frac{1}{4}, 1\}$
41. $f(w) = -w + 2$ $\{4, 2, 1, -2\}$
42. $f(t) = -2t$ $\{4, 0, -2, -8\}$

Getting Ready for Lesson 2-7

Simplify each expression.

43. $-4|5|$ -20
44. $|-5| + |-3.5|$ 8.5
45. $|-8 + 4|$ 4
46. $|-9.6|$ 9.6
47. $3 - |-4|$ -1

A Point in Time

Romana Acosta Bañuelos

In 1971, Romana Acosta Bañuelos became the first Mexican American woman to hold the office of United States Treasurer. Before her appointment to this post by President Nixon, she founded and managed her own multimillion-dollar food enterprise and established the Pan American National Bank of East Los Angeles. As a highly successful businesswoman, she had to work on a daily basis with interest rates, balance sheets, investments, and other functions used in the corporate world.

Math ToolboX

In Lesson 2-6, students looked at three views of a function: the function rule or equation, a table of values, and a graph. This toolbox focuses on how changes to the equation affect its graph. Because a graphing calculator can display multiple graphs on one screen, students can quickly see the similarities and the differences between the equations and the graphs. The graphing calculator allows students to focus on understanding which elements of the equation cause the changes in the graphs, rather than focusing on the calculations necessary to plot points.

ERROR ALERT! Students may not know how to enter exponents other than 2 into the calculator. **Remediation:** Have students practice entering different exponents.

WRITING Exercise 9 Students may wish to change the range to view the differences in the graph. Be sure to have students reset their calculators to the standard setting before completing the exercise.

ADDITIONAL PROBLEM Graph $y = x, y = x^2, y = x^3, y = x^4, y = x^5, y = x^6$. Make a generalization about the quadrants in which the graph of $y = x^n$ appears. **If n is even, the graph appears in the first and second quadrants. If n is odd, the graph appears in the first and third quadrants.**

Math ToolboX Technology

Graphing Functions

After Lesson 2-6

You can use a graphing calculator to graph a function. You can sketch a graph to record information about the function.

Example

Graph the function $y = x^2 - 9$. Use the standard setting for the range. Then sketch the graph on your paper.

STEP 1: Press ZOOM 6 to set the range at the **standard setting.** Press WINDOW to display the standard setting on the window screen.

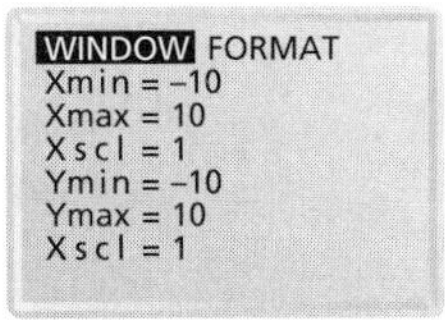

WINDOW FORMAT
Xmin = −10
Xmax = 10
Xscl = 1
Ymin = −10
Ymax = 10
Xscl = 1

STEP 2: Press Y= to go to the equation screen. Press CLEAR to get rid of any equation already in the calculator. Then press X,T,θ x^2 − 9 to enter the equation.

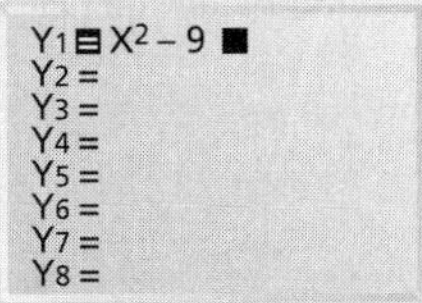

STEP 3: Press GRAPH to view the graph.

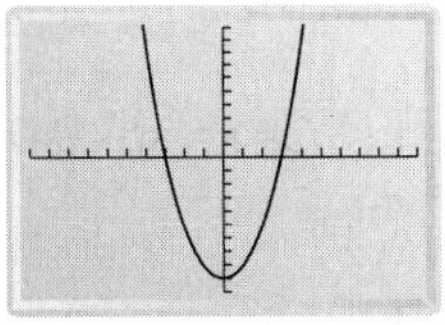

STEP 4: To sketch a graph that you see on your calculator screen, use the coordinates of at least three points. Plot these points on your paper and use them to make your sketch.

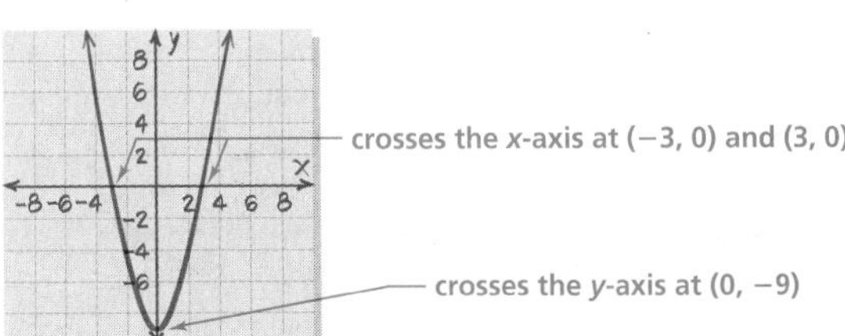

Graph each function on your graphing calculator. Use the standard setting for the range. Then sketch each graph. 1–8. See margin.

1. $y = -2x + 4$
2. $y = \frac{1}{2}x - 2$
3. $y = x^2 + 2$
4. $y = x^3 - 1$
5. $y = 3x - 6$
6. $y = \frac{1}{3}x$
7. $y = 4 - x^2$
8. $y = 3x^2 - 7$

9. *Writing* Examine the Window screen and the Graph screen in the Example. Explain what the values Xmin, Xmax, Ymin, and Ymax represent on the graph. **Xmin and Xmax are the least and the greatest values on the x-axis on the screen. Ymin and Ymax are the least and the greatest values on the y-axis on the screen.**

Materials and Manipulatives

- Graphing calculator

page 89 Math Toolbox

1.

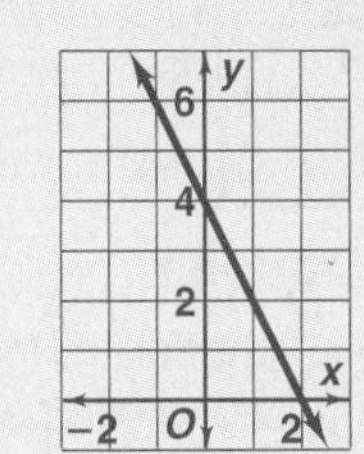

2.

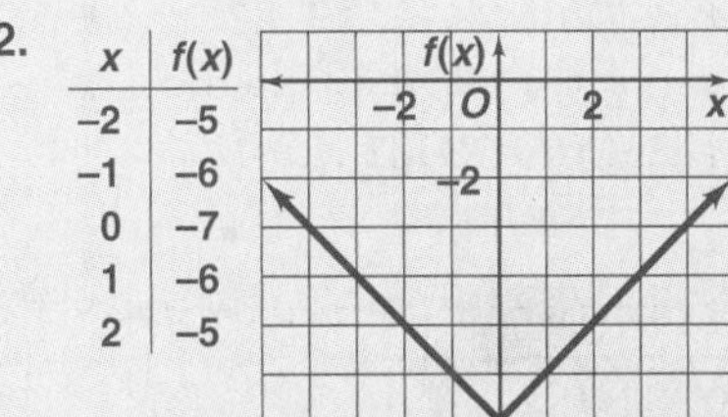

x	$f(x)$
−2	−5
−1	−6
0	−7
1	−6
2	−5

3.

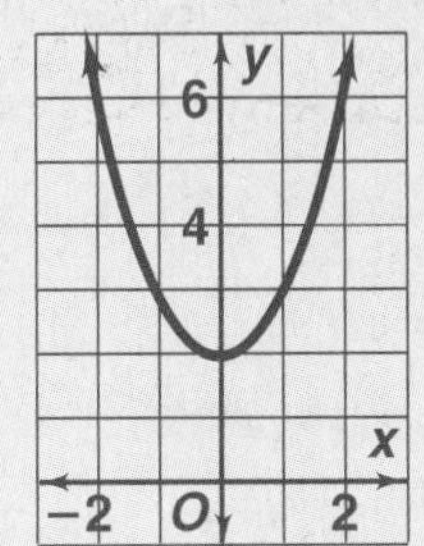

4–8. See back of book.

2-7 Teaching Notes

PROBLEM OF THE DAY

The average temperature in Mount Vernon for 7 days was 68°F. The temperatures were 72, 67, 63, 71, 75, and 61°F for the first 6 days. What must the temperature have been on the seventh day? 67°F

Problem of the Day is also available in Transparencies.

CONNECTING TO PRIOR KNOWLEDGE Ask students to conjecture about what a *family* of graphs might be.

WORK TOGETHER

If students use a computer:

- Make sure they know how to use the software to enter functions containing exponents and absolute values.
- Encourage students to make notes and sketches. A record of their work will help them find patterns and demonstrate their conclusions.
- If an equation produces no visible line suggest they try selecting a larger range.

If students use a graphing calculator:

- As with computers, encourage students to make notes and sketches.

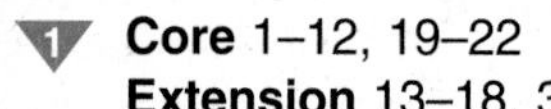

Lesson Planning Options

Prerequisite Skills

- Understanding the terms *power* and *absolute value*

Assignment Options

To provide flexible scheduling, this lesson can be subdivided into parts.

Core 1–12, 19–22
Extension 13–18, 38

Core 23–33
Extension 34–37, 39, 40

Use Mixed Review to maintain skills.

Resources

Student Edition

Skills Handbook, p. 578, 581
Extra Practice, p. 557
Glossary/Study Guide

Teaching Resources

Chapter Support File, Ch. 2
- Practice 2-7 (two worksheets)
- Reteaching 2-7
- Alternative Activity 2-7

Classroom Manager 2-7
Glossary, Spanish Resources
Two-Year Algebra Handbook, 2-7

Transparencies
1, 2, 45

What You'll Learn

- Identifying families of functions for equations and graphs

...And Why

To predict what the graph of an equation will look like

What You'll Need

- graphing calculator
- note cards

GRAPHING CALCULATOR HINT

Use the ABS key and the ( ) keys to enter the absolute value.

2a. $x^2 - 6$, $x^2 + 1$, $3x^2$; $x + 2$, $7x$, $-3x$; $|x| - 4$, $|x - 3|$
b. Graphs are U-shaped; graphs are straight lines; graphs are V-shaped.
c. Equations have x^2-term; equations have an x-term but no x^2-term or abs. value; equations have a variable expression inside the abs. value symbol.

4. Equation includes abs. value of a variable expression.

Connections Medicine . . . and more

2-7 Families of Functions

WORK TOGETHER

Work with a partner.

1. **Graphing Calculator** Graph each equation using the standard range setting. To help you keep track, sketch each graph on a note card. Label it with the correct equation. See margin.

$y = x^2 - 6$	$y = x + 2$
$y = \|x\| - 4$	$y = 7x$
$y = \|x - 3\|$	$y = x^2 + 1$
$y = 3x^2$	$y = -3x$

a–c. See left.

2. a. Sort the cards into three categories by grouping graphs that look alike.
   b. What similarities among the graphs in each category do you see?
   c. What similarities among the equations in each category do you see?

The categories you made can help you make predictions.

3. What does the graph of $y = 2x^2$ look like? It is U-shaped.

4. **Graphing Calculator** What can you say about the equation of this graph? See left.

THINK AND DISCUSS

Part 1
Identifying the Family of an Equation

You've already seen how grouping functions that are alike can help you make predictions. These groups are called *families of functions.* You can identify what family a function belongs to by looking at its equation.

- Explain the difference between the minus key and the negative (or change-sign) key.

DIVERSITY If you are using computers, make sure that all students know how to use them. Some students will have computers at home and be very familiar with their use. Others will have had little exposure to computers and may be anxious about using them. Have students who are familiar with computers help those who are not. Make sure that the inexperienced students get an equal opportunity to spend time on the computers.

TACTILE LEARNING Question 1 This question provides excellent practice in drawing different kinds of graphs.

THINK AND DISCUSS

Example 1

OPEN-ENDED Question 7 Point out that all quadratic functions are of the form $ax^2 + bx + c$. The highest power of x is 2.

Example 2

Note that in an absolute value function, the variable expression has x to the power of 1.

ERROR ALERT! Students may think that because an absolute value is never negative, the graph of an absolute value function will never contain points with negative coordinates.

QUICK REVIEW

Remember that $x = x^1$. So, the power of x is 1.

Example 1

To what family of functions does each equation belong? Explain.

a. $y = 2x - 6$

Its highest power of x is 1. So, $y = 2x - 6$ is a *linear function.*

b. $y = -8x^2$

Its highest power of x is 2. So, $y = -8x^2$ is a *quadratic function.*

5. The equation $y = |x + 7|$ is an *absolute value function.* What characteristic of the equation tells you this? the absolute value symbol
6. Try This To what family of functions does each equation belong? Explain.
   a. $y = 6x^2 + 1$ a, c. Quadratic; the highest power of x is 2.
   b. $y = 3|x|$ b. Abs. value; equation has x inside abs. value symbol.
   c. $y = x^2 + 3x + 2$
7. Open-ended Create three equations that belong to the quadratic family of functions. Samples: $y = 4x^2 - 2$; $y = -x^2$; $y = 2x^2 - x + 1$

Identifying the Family of a Graph

You can identify what family a function belongs to by looking at its graph.

Example 2

Graphing Calculator To what family of functions does each graph belong? Explain.

a.

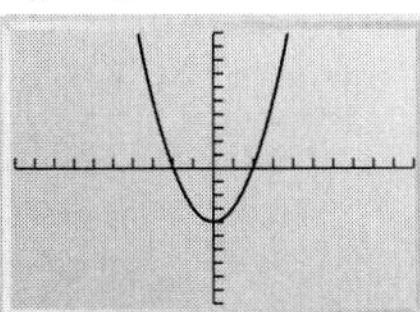

The graph is U-shaped. So, it is a *quadratic function.*

b.

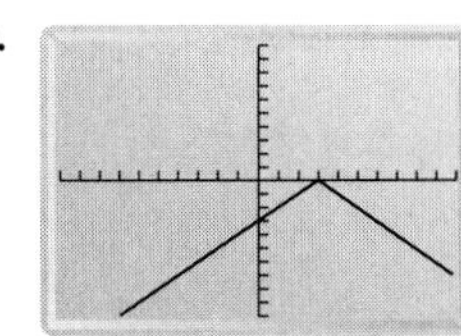

The graph forms a "V." So, it is an *absolute value function.*

8. a. The equation $y = -5x$ belongs to what family of functions? How do you know? Linear; the highest power of x is 1.
   b. Graphing Calculator Graph $y = -5x$. See margin.
   c. Look at your graph and **generalize.** What characteristic of a graph tells you it belongs to the *linear family of functions?* The graph is a straight line.
9. Open-ended Sketch three graphs that belong to the absolute value family of functions. See margin.

Additional Examples

FOR EXAMPLE 1

To what family of functions does each equation belong?

a. $y = 4x + 6$ linear

b. $y = 3x^2 + 1$ quadratic

c. $y = |x| + 7$ absolute value

Discussion: *Describe how you are able to tell one type of function from another.*

FOR EXAMPLE 2

To what family of functions does each graph belong?

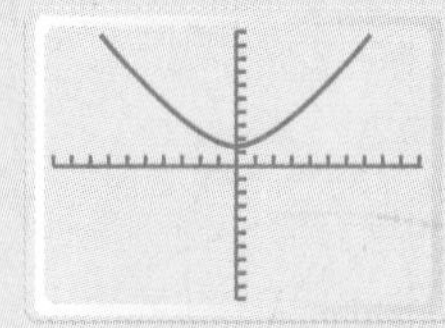

The graph is U-shaped so it is a *quadratic function.*

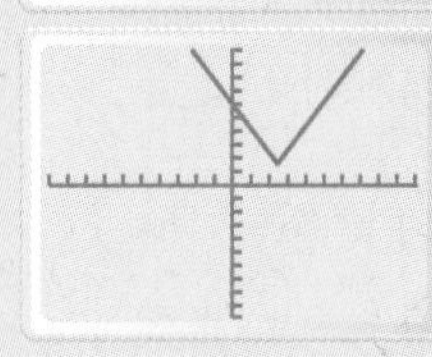

The graph is V-shaped so it is an *absolute value function.*

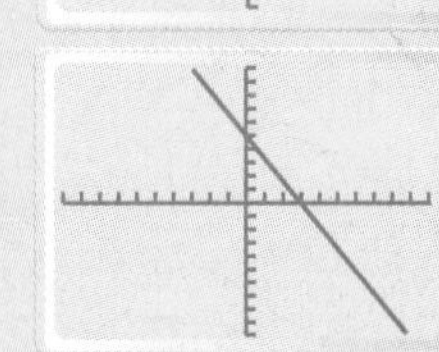

The graph forms a straight line so it is a *linear function.*

Remediation: Have students make a table of values for an equation such as $y = |x - 3|$ or $y = -|x - 3|$. Lead them to understand that x can be positive, negative, or zero. This error may occur again in On Your Own Exercises 8 and 38e.

KINESTHETIC LEARNING Clear the center of the classroom and use masking tape to make a large pair of coordinate axes on the floor. On the masking tape, make large tick marks about 1 ft apart. Give each group of 8 to 10 students a piece of string about 20 ft long and a simple equation from a function family. The group members will work together to make a table of values. Groups then take turns modeling their graph for the rest of the class by standing on points in the coordinate plane on the classroom floor and extending the string between them to show the shape of the graph. If you have block scheduling or extended class periods, challenge the other students to identify the equation of the graph.

EXTENSION Point out to students that only equations containing exponents of power 2, such as $y = x^2$ are considered quadratic equations. Equations such as $y = x^3$ or $y = x^4$ belong to other function families.

Exercises ON YOUR OWN

ALTERNATIVE ASSESSMENT **Exercises 1–9** Have students use the graphing calculator to graph each equation to verify their previous responses. These exercises help you assess whether the students have grasped the concept of analyzing equations to identify linear, quadratic, and absolute value functions.

Technology Options

For Exercises 25–33, students may use spreadsheet software or graphing calculators to verify their predictions by graphing each equation.

Prentice Hall Technology

Software • Secondary Math Lab Toolkit • Computer Item Generator 2-7

Internet • See the Prentice Hall site. (http://www.phschool.com)

The equations and graphs of functions that are in the same family are alike. Here are three families of functions.

Families of Functions

Linear Functions	Quadratic Functions	Absolute Value Functions
$y = 2x - 1$	$y = -3x^2 - x + 1$	$y = \|x + 2\|$
The highest power of x is 1.	The highest power of x is 2.	There is an absolute value symbol around a variable expression.
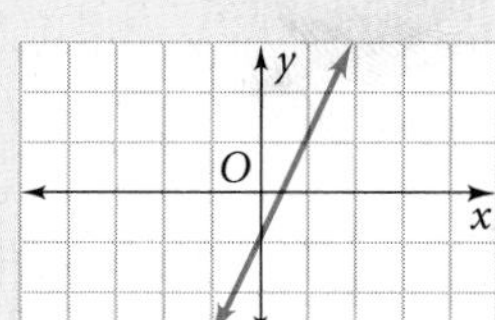	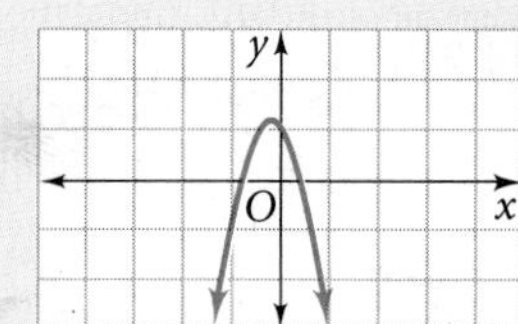	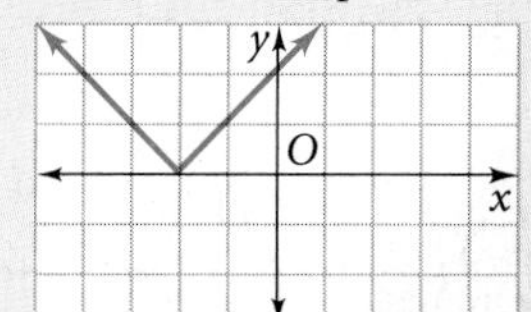
The graph forms a straight line.	The graph is a U-shaped curve that opens up or down.	The graph forms a "V" that opens up or down.

Exercises ON YOUR OWN

To what family of functions does each equation belong? Explain.

1. $y = \frac{4}{7}x + 7$
2. $y = 49 - x^2$
3. $y = |2 + 3x|$
4. $y = -x^2 + 11$
5. $y = -6x^2 + 13x - 5$
6. $y = \frac{1}{4}x$
7. $y = 0.5x^2$
8. $y = -2|x|$
9. $y = |x - 2|$

1–9. See margin.

Graphing Calculator **To what family of functions does each graph belong? Explain your reasoning.** 10–11. Abs. value; graph forms a "V" that opens up or down. 12. Quadratic; graph is a U-shaped curve that opens down.

10. 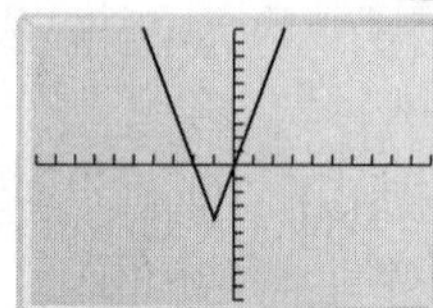

11. 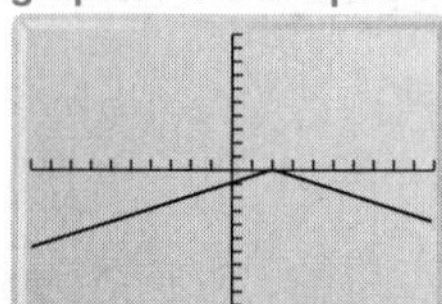

12. 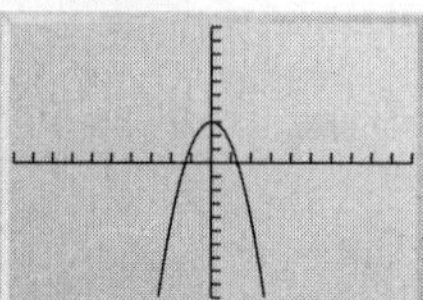

13. Medicine The recommended dosage D in milligrams of a certain medicine depends on a person's body weight w in kilograms. To what family of function does the formula $D = 0.1w^2 + 5w$ belong? Explain. Quadratic; the highest power of the variable is 2.

14. Writing How are linear functions and absolute value functions alike? How are they different? See margin.

15. Define a *family of functions* in your own words. Sample: A family of functions is a collection of functions whose graphs and equations are alike.

CRITICAL THINKING Exercise 16 Curious students may wish to make a table of values for the graph on the right. Have them see if they can find a pattern. Have students write an equation for the graph. This graph represents a function of the dependent variable, not the independent variable. $x = |y| - \frac{3}{2}$

CRITICAL THINKING Exercise 17 Help students find the equations of vertical and horizontal lines.

CRITICAL THINKING Exercises 23 and 24 Use these exercises to demonstrate that an equation or graph may model a real situation only for a certain range of values. Remind students that it is important to look at numerical answers, along with the situation being modeled, so that they don't come up with meaningless solutions. For example, in Exercise 23, the number of hours worked cannot be a negative number.

STANDARDIZED TEST TIP Exercise 40 Students can use the process of elimination to answer this. Since the graph is U-shaped, it represents a quadratic function. This eliminates answers A and C. Next, the students can substitute an x-value from the graph into the remaining functions. The $x = 0$ value is easiest, but does not eliminate an answer. Since the $x = 1$ value does not work for either B or D, the answer must be E.

Chapter Project **FIND OUT BY ANALYZING** Students select a new tongue twister. They use the method described in the Find Out question on p. 63 to record times. From the data, they construct a graph to determine if it is linear. The students write a function rule to predict the time taken to say the tongue twister. This information is essential to the Chapter Project. Have students add this task to work they have already

16. *Critical Thinking* Why are these graphs *not* quadratic or absolute value functions? They fail the vertical-line test.

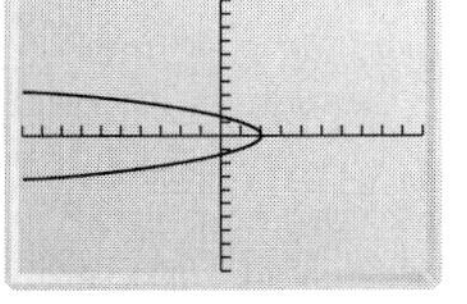

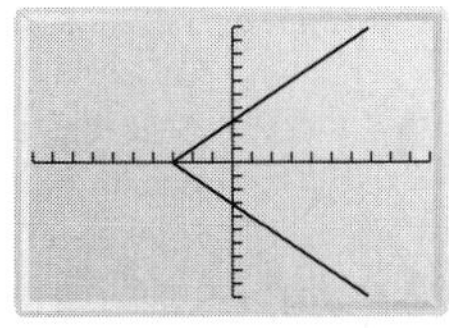

17. *Critical Thinking* Is a vertical line the graph of a linear function? Why or why not? No; it fails the vertical-line test so it is not a function.

18. *Open-ended* Write two linear, two quadratic, and two absolute value equations.
Samples: $y = x + 1$, $y = \frac{2}{3}x$; $y = x^2 + x$, $y = -3x^2 + 9$; $y = |x + 4|$, $y = -|x + 4| + 1$

What characteristic do you look for to identify the family of each?

19. graph of a quadratic function U-shaped curve

20. equation of a linear function highest power of the variable is 1

21. graph of an absolute value function V-shape

22. equation of an quadratic function highest power of the variable is 2

Critical Thinking **Determine to which family of functions each graph belongs. Then sketch a graph to model each situation.** 23–24. See margin for graphs.

23. Income is a function of of hours worked. linear

24. Height of a fly ball is a function of time. quadratic

Graphing Calculator **Predict the shape of the graph. Be as specific as you can. Graph each equation to verify your prediction.** 25–33. See margin for graphs.

25. $y = 7 - x$ line; slants down from left to right

26. $y = -4x^2$ U-shaped; opens down

27. $y = |1 - x|$ V-shaped; opens up

28. $y = 2x + 1$ line; slants up

29. $y = -|x| + 6$ V-shaped; opens down

30. $y = 2x^2 - 6$ U-shaped; opens up

31. $y = x^2 + 3$ U-shaped; opens up

32. $y = -x + 10$ line; slants down

33. $y = |-x| + 3$ V-shaped; opens up

To what family of functions does the function suggested by each picture belong? Explain your reasoning. 34–37. See margin.

34.

35.

36.

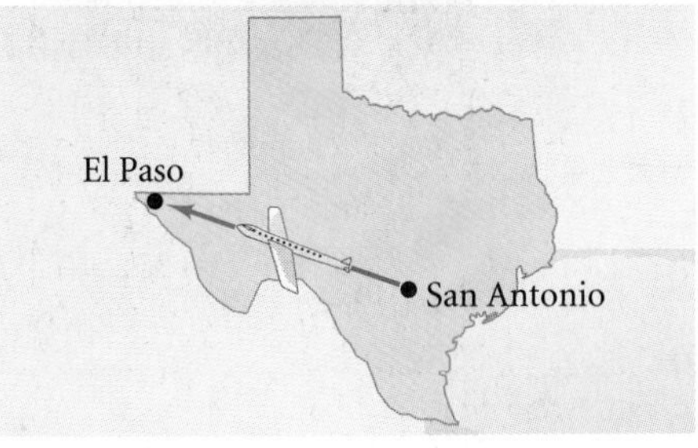

37.

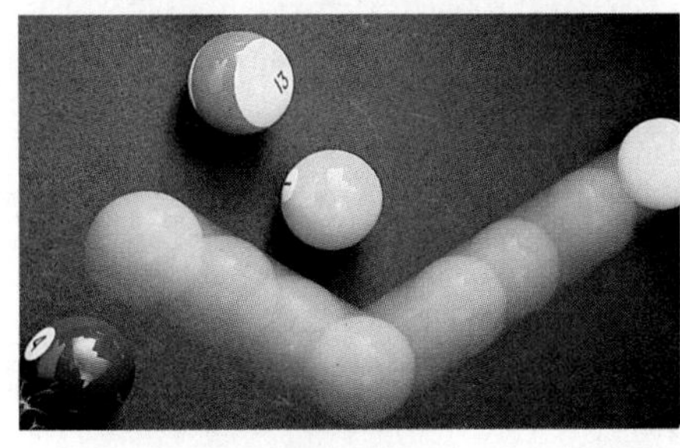

page 90 Work Together

1.

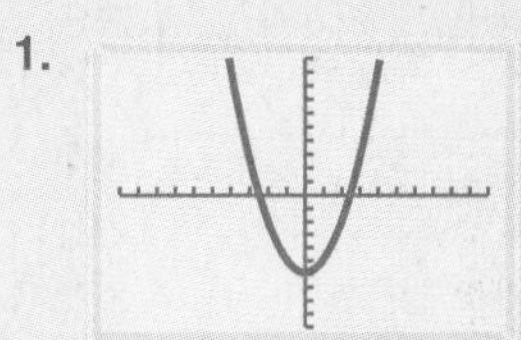

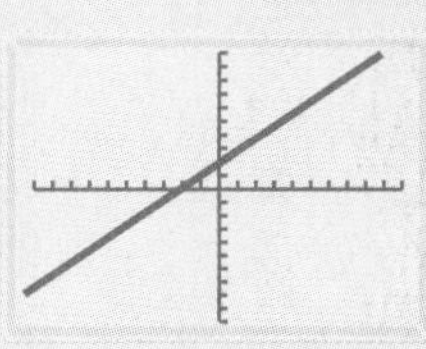

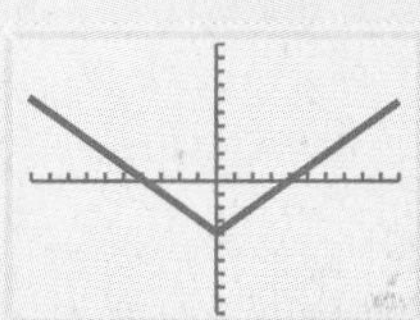

See back of book.

pages 91–92 Think and Discuss

8b, 9. See back of book.

pages 92–94 On Your Own

1–9, 14, 23–37, 38a–e, 39b, 41–48, 50a–b. See back of book.

completed for the project. You may wish to check students' progress on the project at this point.

Exercises MIXED REVIEW

GETTING READY FOR LESSON 2-8 These exercises prepare students for finding theoretical probability by asking them to find the number of favorable and possible outcomes.

SELF ASSESSMENT **JOURNAL** Encourage students to make sketches to illustrate their writing. Students may also want to write a brief answer to the question: *What graph in this lesson surprised you most? Why?*

Wrap Up

THE BIG IDEA Ask students how they can tell the shape of a graph by looking at the equation.

RETEACHING ACTIVITY Students learn to recognize the graphs of the different families of functions. (Reteaching worksheet 2-7)

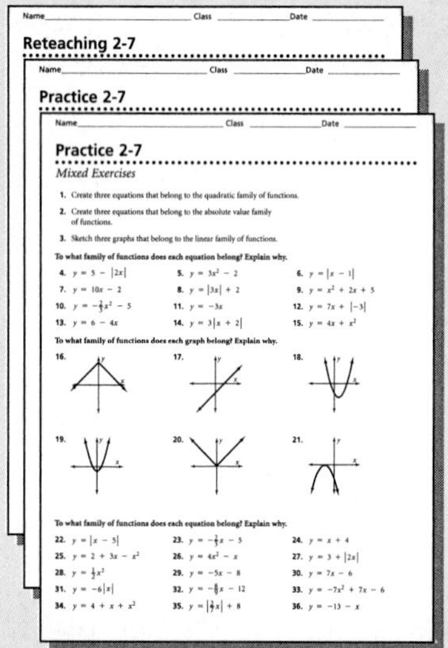
Name______ Class ______ Date ______

Reteaching 2-7

Name______ Class ______ Date ______

Practice 2-7

Name______ Class ______ Date ______

Practice 2-7

Mixed Exercises

1. Create three equations that belong to the quadratic family of functions.
2. Create three equations that belong to the absolute value family of functions.
3. Sketch three graphs that belong to the linear family of functions.

To what family of functions does each equation belong? Explain why.

4. $y = 5 - |2x|$ 5. $y = 3x^2 - 2$ 6. $y = |x - 1|$
7. $y = 10x - 2$ 8. $y = |3x| + 2$ 9. $y = x^2 + 2x + 5$
10. $y = -\frac{2}{3}x^2 - 5$ 11. $y = -3x$ 12. $y = 7x + |-3|$
13. $y = 6 - 4x$ 14. $y = 3|x + 2|$ 15. $y = 4x + x^2$

To what family of functions does each graph belong? Explain why.

16. 17. 18. 19. 20. 21.

To what family of functions does each equation belong? Explain why.

22. $y = |x - 5|$ 23. $y = -\frac{2}{3}x - 5$ 24. $y = x + 4$
25. $y = 2 + 3x - x^2$ 26. $y = 4x^2 - x$ 27. $y = 3 + |2x|$
28. $y = \frac{1}{2}x^2$ 29. $y = -5x - 8$ 30. $y = 7x - 6$
31. $y = -6|x|$ 32. $y = -\frac{8}{9}x - 12$ 33. $y = -7x^2 + 7x - 6$
34. $y = 4 + x + x^2$ 35. $y = |\frac{2}{7}x| + 8$ 36. $y = -13 - x$

Lesson Quiz

Lesson Quiz is also available in Transparencies.

To what family of functions does each equation or graph belong? Explain your reasoning. $y = |x| + 2$

1. $y = 24 - x^2$ **Quadratic; the highest power of *x* is 2.**
2. $y = |x| + 2$ **Absolute value; there is an absolute value symbol around a variable.**
3. $y = \frac{7}{8}x$ **Linear; the highest power of *x* is 1.**

38. a. Make a table of values for each equation. **a–d. See margin.**
 I. $y = |x|$ II. $y = -x^2$
 b. What do you notice about the signs of the *y*-values?
 c. What quadrant(s) do you expect each graph to be in?
 d. Graph each function. **See margin for graphs.**
 e. **Predict** whether each graph will open up or down. Graph each function to test your prediction.
 I. $y = -|x|$ **down** II. $y = x^2$ **up**

39. a. **Predict** the shape and opening of the graph $y = |x - 2|$.
 b. *Graphing Calculator* Test your prediction.
 a. V-shaped; opens up b. See margin.

40. *Standardized Test Prep* Which equation is shown in the graph? **C**

 A. $y = x + 2$
 B. $y = -x^2 + 2$
 C. $y = -|x| + 2$
 D. $y = x^2 + 2$
 E. None of these

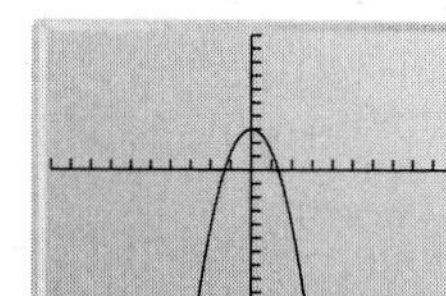

Chapter Project — Find Out by Analyzing

Think of a tongue twister not yet used in this chapter. (You may know one in a language other than English.)

- Use a group of at least ten people. Use the same method to record times as you did for the Find Out question on page 63.
- Graph the data.
- Do your data points form a line? Explain.
- Write a function rule that you could use to predict the time for *n* people to say your tongue twister.

Exercises MIXED REVIEW

Graph each function. **41–48. See margin.**

41. $f(x) = 5x - 1$ 42. $f(x) = -3x^2$ 43. $f(w) = 3w + 4.5$ 44. $q(p) = 3p - 1$
45. $f(x) = x^2 + 7$ 46. $f(t) = t + 2.5$ 47. $d(t) = 55t$ 48. $f(x) = \frac{1}{2}x + \frac{3}{2}$

49. a. Evaluate the function rule $f(x) = -2x^2 + 12$ for $x = -8$. **−116**
 b. Write the steps you used to evaluate the function. **Square −8; multiply by −2; add 12 to result.**

50. *Probability* Use the random number table.
 a. *Open-ended* A sock drawer has an equal number of white and blue socks. Simulate choosing a sock without looking in the drawer. Which numbers will represent white? blue?
 b. Conduct the simulation. What is the experimental probability of drawing a blue sock?
 50a–b. See margin.

Random Number Table

23948	71477	12573	05954
65628	22310	09311	94864
41261	09943	34078	70481

Getting Ready for Lesson 2-8

A bag of fruit has 5 apples, 6 bananas, 4 oranges, and one pomegranate. What fraction of the fruit does each represent?

51. apples $\frac{5}{16}$ 52. oranges $\frac{1}{4}$ 53. not pomegranates $\frac{15}{16}$
54. bananas $\frac{3}{8}$ 55. not apples $\frac{11}{16}$ 56. bananas and apples $\frac{11}{16}$

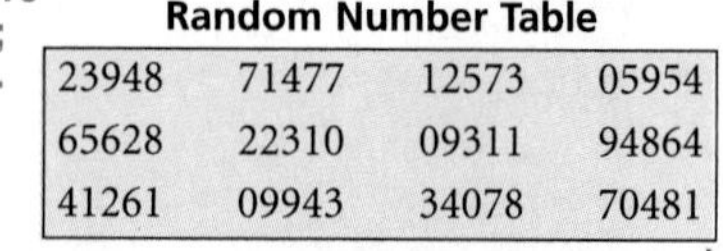

FOR YOUR JOURNAL

Summarize what you understand about functions. Give examples of how to write them and how to use them. Be specific about how they relate to graphs.

2-8 Teaching Notes

PROBLEM OF THE DAY

A square frame has sides 10 in. long. If the frame is 1 in. wide all around, what is the perimeter of the picture opening? **32 in.**

Problem of the Day is also available in Transparencies.

CONNECTING TO PRIOR KNOWLEDGE Ask students to discuss estimating the probability of an event occurring; for example, the probability that your favorite professional basketball player will make his next free throw. Ask:

- *What strategies did you use to estimate the probability?*
- *What have you learned that helps you estimate probability?*

THINK AND DISCUSS

ERROR ALERT! Students might think that favorable outcomes are only those they want to occur. **Remediation:** Explain that the term *favorable outcome* refers to the event you are specifically studying at that moment, whether it is good or bad. For example, if you calculate the probability that a hospital patient has a broken leg, you use the term *favorable outcome* for the number of patients with a broken leg.

Example 1 Relating to the Real World 

TACTILE LEARNING Distribute 170 equally-sized pieces of paper to students. Students will refer to the chart as they write the type of store on each piece of paper to be put into a box. They may choose to give names to the stores. Have students

---

Connections  Shopping . . . and more

2-8 The Probability Formula

What You'll Learn

- Finding theoretical probability
- Using a tree diagram and a sample space to find probability

...And Why

To make predictions about real-world situations such as mall promotions and game shows

THINK AND DISCUSS

Part 1

Finding Theoretical Probability

To promote the new Westside Mall, the mall manager gives away a store gift certificate. The winning shopper chooses the store by picking a piece of paper from a box. The box contains the names of 170 stores.

1. What type of store is the winner most likely to pick? Explain. **Clothing; there are more clothing stores than any other type.**

In Chapter 1, you found experimental probability by conducting experiments. The possible results of an experiment are called **outcomes.**

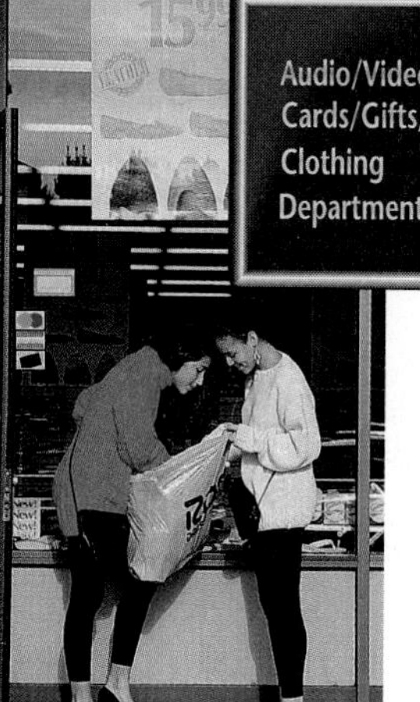

Stores at the Westside Mall

Audio/Video/Electronics	8	Food	25	Health/Beauty	9
Cards/Gifts/Books	11	Footwear	16	Specialty Shops	10
Clothing	47	Home Furnishings	7	Toys/Hobby/Sport	9
Department Store	4	Jewelry	14	Other	10

Total Number of Stores: 170

In the shopping mall giveaway, each store represents an outcome. If you want to pick a food store, the 25 food stores are *favorable outcomes.* If each outcome has an equal chance of happening, you can find **theoretical probability** using this formula.

$$P(\text{event}) = \frac{\text{number of favorable outcomes}}{\text{number of possible outcomes}}$$

Example 1 Relating to the Real World 

Shopping What is the probability of the winning shopper choosing a footwear store? Express your answer as a percent.

$$P(\text{footwear store}) = \frac{\text{number of favorable outcomes}}{\text{number of possible outcomes}}$$

$$= \frac{16}{170} \quad \leftarrow \text{There are 16 footwear stores in the mall and 170 stores in all.}$$

$$\approx 0.09 = 9\%$$

The probability of choosing a footwear store is about 0.09, or 9%.

For practice with percents, see Skills Handbook page 579.

2. *Calculator* Suppose you are the winning shopper. What is the probability you get a gift certificate to a jewelry store? **about 0.08 or 8%**

Lesson Planning Options

Prerequisite Skills

- Converting decimals to percents

Assignment Options

To provide flexible scheduling, this lesson can be subdivided into parts.

Part 1: **Core** 1–9, 12, 15–21, 23–30, 41, 42
Extension 10, 11, 22, 43

Part 2: **Core** 13, 31–38
Extension 14, 39, 40

Use Mixed Review to maintain skills.

Resources

 **Student Edition**

Skills Handbook, p. 570, 576
Extra Practice, p. 557
Glossary/Study Guide

Teaching Resources

Chapter Support File, Ch. 2
- Practice 2-8 (two worksheets)
- Reteaching 2-8

Classroom Manager 2-8
Glossary, Spanish Resources
Two-Year Algebra Handbook 2-8

 **Transparencies**

45

first calculate the theoretical probability of pulling out a paper representing the type of store they would like to visit. Then have each student in turn pull a paper at random from the box, tell whether the paper represents the type of store they want to visit, and then return the paper to the box. If you have block scheduling or extended class periods, have students choose, look at, and return papers repeatedly in order to test whether the experimental probability approximates the theoretical probability. This activity will help students appreciate the difference between the outcome of a single event and the probability of that event.

VISUAL LEARNING/OPEN-ENDED Question 4 Have a student draw a probability line from zero to one across the whole board. As students finish answering Question 4c, have them write the event they chose and the event's probability expressed as a percent in the appropriate place on the line.

ESL Avoid confusion with the *flattery* definition of compliment by explaining how a complement helps complete the set.

Example 2 **Relating to the Real World**

Ask students: *Can you think of a situation in which it would be easier to find the probability that an event does not happen?* **Answers may vary.**

AUDITORY LEARNING Working in pairs, have students write a description of an event and its opposite on separate sheets of paper. Separate students. Randomly hand out the sheets of paper. Have a student stand and read one sheet aloud. The class member with the opposite event should then read his or her sheet. Continue until all students have found their match.

Additional Examples

FOR EXAMPLE 1

The hair colors of students in a class are as follows: 8 brown, 10 black, 2 red, and 3 blond. If the teacher calls on one student at random, what is the probability of calling on a student with brown hair?

$P(\text{brown}) = \frac{8}{23} \approx$ **0.35 or 35%**

Discussion: *Does this mean that if the teacher chooses 100 times, 35 of the choices will always be students with brown hair?*

FOR EXAMPLE 2

Use the information from Example 1 to answer the following questions.

a. What is the probability that the teacher does not choose a student with black hair?
$P(\text{not black hair}) = \frac{13}{23} =$ **0.57 or 57%**

b. What would be the complement of this event? **43%**

Discussion: *Why is the sum of the probabilities of an event and its complement 1 or 100%?*

96

All probabilities range from 0 to 1.

Probability

← less likely — more likely →

0 **Impossible event** — 0.5 Equally likely as unlikely — 1 **Certain event**

When you roll a number cube marked with numbers from 1 to 6, rolling a 7 is an impossible event. Rolling a number less than 7 is a certain event.

3. **Try This** Find each probability when rolling a number cube.
   a. $P(\text{even})$ $\frac{1}{2}$ **b.** $P(0)$ **0** **c.** $P(5)$ $\frac{1}{6}$
4. ***Open-ended*** Suppose you chose a letter tile from a bag containing the letter tiles A, A, A, E, I, O, O, and U. **See left.**
   a. Name a certain event.
   b. Name an impossible event.
   c. Name an event that is more likely than choosing the letter I.

Answers may vary. Samples:
a. choosing a vowel
b. choosing a consonant
c. choosing an A

What? *Wheel of Fortune* was the first game show to be closed-captioned for the hearing impaired. The show is also seen in 52 overseas markets.

The **complement of an event** consists of all possible outcomes not in the event. For example, when you roll a number cube, the complement of "rolling a 4 or less" is "rolling a 5 or 6." The sum of the probabilities of an event and its complement is 1. You can use this formula to find the probability of the complement of an event.

$$P(\text{complement of event}) = 1 - P(\text{event})$$

You can write $P(\text{complement of event})$ as $P(\text{not event})$.

Example 2 **Relating to the Real World** 

Games On the television program *Wheel of Fortune,* the winning contestant must choose from five envelopes. Each envelope contains a grand prize, such as a car or a trip.

a. What is the probability of choosing a particular grand prize?
b. What is the probability of not choosing that grand prize?

a. $P(\text{grand prize}) = \frac{\text{number of favorable outcomes}}{\text{number of possible outcomes}} = \frac{1}{5}$

The probability the contestant chooses a specific grand prize is $\frac{1}{5}$.

b. $P(\text{not grand prize}) = 1 - P(\text{grand prize})$
$= 1 - \frac{1}{5}$
$= \frac{4}{5}$

The probability the contestant does not choose that grand prize is $\frac{4}{5}$.

PROBLEM SOLVING
Look Back Describe another method you could use to find $P(\text{not grand prize})$.
If the favorable event is not getting a specific grand prize, then there are 4 favorable outcomes out of a possible 5, so $P(\text{not grand prize}) = \frac{4}{5}$.

5. Suppose you choose from three colored envelopes: red, orange, blue.
   a. What is $P(\text{blue})$? $\frac{1}{3}$ **b.** What is $P(\text{not blue})$? $\frac{2}{3}$

Example 3 Relating to the Real World 

CONNECTING TO THE STUDENTS' WORLD Have students repeat the example to calculate the probability of the occurrence of the number of girls in their family. Invite students from very low probability families (for example, a family of four girls) to write this probability on the board.

MAKING CONNECTIONS The probabilities of a child being a boy or a girl are not exactly the same. Worldwide, 106 boys are born for every 100 girls. However, boys have a higher infant mortality rate, so girls soon outnumber boys.

EXTENSION Ask students to imagine a race of alien beings with three equally probable genders: agu, bluk, and cril. Ask:

- *What would you guess is the probability that an alien family with three children has at least two bluks?* **Answers may vary. Sample: 33%**
- *What answer do you get when you draw a tree diagram to calculate the probability?* $\frac{7}{27} \approx 26\%$

Exercises ON YOUR OWN

Exercise 12 Ask students whether they think the 20% chance of being picked *sounds* more likely to occur than the 80% chance of not being picked. Even though the two probabilities are identical in value, students may subjectively feel that one way of describing the probability makes it seem more likely. Encourage students to watch and listen for messages in advertisements that exploit this subjective effect. For example,

Part 2

Using a Tree Diagram to Find a Sample Space

The set of all possible outcomes is the **sample space.** Displaying the sample space with a *tree diagram* can help you find the probability of an event.

Example 3 Relating to the Real World 

Genetics What is the probability that there are at least two girls in a family of three children?

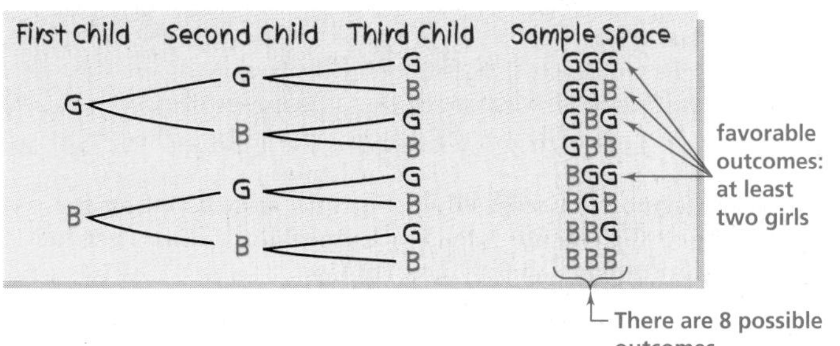

$$P(\text{at least two girls}) = \frac{\text{number of favorable outcomes}}{\text{total number of possible outcomes}} = \frac{4}{8} = \frac{1}{2}$$

The probability that there are at least two girls is $\frac{1}{2}$ or 50%.

6. **Try This** What is the probability that there are exactly two girls in a family of three children? $\frac{3}{8}$ **or 37.5%**

Exercises ON YOUR OWN

For Exercises 1–9, use the spinner at the right. Find each probability.

1. $P(\text{red})$ $\frac{1}{2}$
2. $P(\text{white})$ $\frac{1}{3}$
3. $P(5)$ $\frac{1}{6}$
4. $P(\text{even})$ $\frac{1}{2}$
5. $P(\text{red or blue})$ $\frac{2}{3}$
6. $P(\text{not } 2)$ $\frac{5}{6}$
7. $P(\text{number} < 4)$ $\frac{1}{2}$
8. $P(\text{not blue})$ $\frac{5}{6}$
9. $P(8)$ 0

10. **Open-ended** Suppose your teacher chooses a student at random from your algebra class. **a–b. Check students' work.**
   a. What is the probability that you are selected?
   b. What is the probability that a boy is not selected?
   c. How did you find *P*(not boy) in part (b)? Describe another way. **You can find the answer using either the complement formula or the probability formula.**
11. **Critical Thinking** What is the difference between theoretical probability and experimental probability? **See margin.**
12. Suppose you have a 20% chance of being picked for a committee at school. What is the probability that you will not be picked? Express your answer as a percent. **80%**

Technology Options

For Exercises 15–20, students may use graphing calculators or spreadsheet software to generate random numbers to test each theoretical probability.

Prentice Hall Technology 

 **Software** • Secondary Math Lab Toolkit • Integrated Math Lab 4 • Computer Item Generator 2-8

Internet • See the Prentice Hall site. (http://www.phschool.com)

an advertisement is more likely to state that "nine out of ten doctors like this product" than state "one out of ten doctors dislike this product."

Exercise 15 Have students work in pairs to develop a strategy for calculating the amount of odd numbers there are without counting them.

Exercises 15–20 The largest 3-digit positive number is 999. This is not saying that 999 positive 3-digit numbers exist. The first 99 numbers have 2-digits. To find how many positive 3-digit numbers there are, calculate 999 − 99.

ALTERNATIVE ASSESSMENT **Exercises 31–38** Have students work in small groups to find experimental probabilities for four tosses of a coin. Groups will make three sets of four tosses, recording the results each time. By having them compare the results of actual tosses to the theoretical probabilities, you can assess whether they understand variations from expected outcomes for probabilities.

STANDARDIZED TEST TIP **Exercise 39** Encourage students to think about what they are trying to evaluate in a probability exercise. It is important to understand what is being sought and what the conditions are for the choices.

Exercises 41–43 Ask students to describe the difference between the formula for probability and the formula for odds.

Exercises MIXED REVIEW

Exercise 44 According to the order of operations, students should simplify the absolute value before multiplying by 3.

pages 97–99 On Your Own

11. **Answers may vary. Sample: For theoretical probability, you know all the possible outcomes and you want to find out how likely a specific event is to happen. For experimental probability, you know the results of an experiment and you want to find out how frequently a specific event has happened.**

21a.

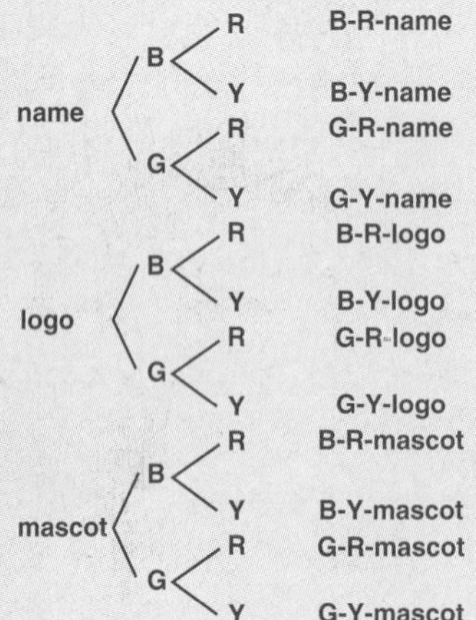

31–38. See back of book.

13. a. Suppose you roll a number cube and toss a coin at the same time. Copy and complete the tree diagram to list the possible outcomes in the sample space. **See right.**

b. What is P(4 and heads)? $\frac{1}{12}$

c. What is P(tails)? $\frac{1}{2}$

d. What is P(even number and tails)? $\frac{1}{4}$

Cube	Coin	Sample Space
1	H	1H
	T	1T
2	H	2H
	T	2T
3	H	3H
	T	3T
4	H	4H
	T	4T
5	H	5H
	T	5T
6	H	6H
	T	6T

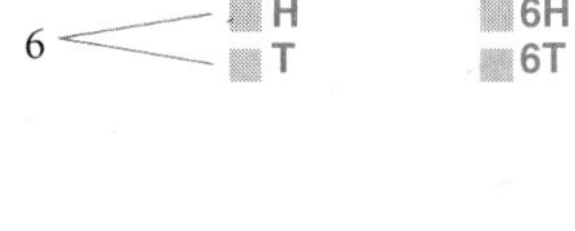

14. Music A disc jockey makes music selections for a radio program. For her first selection, she can choose from eight rock songs, three jazz pieces, five country-western ballads, and four rhythm and blues songs. Assume each selection has an equal chance of being chosen.

a. What is the probability that she chooses a rock song? $\frac{2}{5}$

b. What is the probability that she does not choose a ballad? $\frac{3}{4}$

c. What is the probability that she chooses a classical symphony? 0

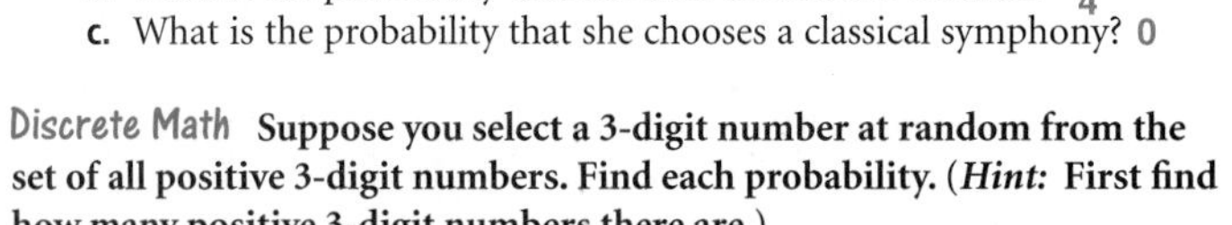

Discrete Math Suppose you select a 3-digit number at random from the set of all positive 3-digit numbers. Find each probability. (*Hint:* First find how many positive 3-digit numbers there are.)

15. P(an odd number) $\frac{1}{2}$ **16.** P(number > 900) $\frac{11}{100}$ **17.** P(number < 100) 0

18. P(number is a multiple of 30) $\frac{1}{30}$ **19.** P(number < 500) $\frac{4}{9}$ **20.** P(number = 243 or 244) $\frac{1}{450}$

21. On Cap Day each person who attends the baseball game receives a free cap. You have an equal chance of receiving any of the combinations at the right.

a. Use a tree diagram to show the sample space.

b. Find P(yellow cap, green visor, team mascot). $\frac{1}{12}$

c. Find P(red cap, blue or green visor, team logo). **21c.** $\frac{1}{6}$

21a. See margin.

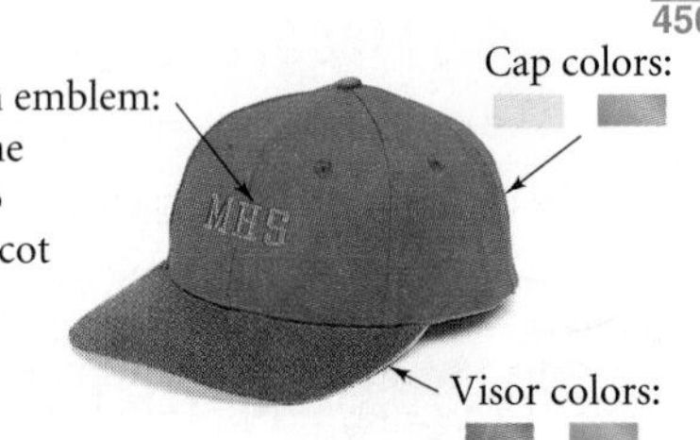

22. Calculator Each day in the United States, about 713,263 people have a birthday. About 9630 people celebrate their 16th birthday each day. What is the probability that someone celebrating a birthday today will turn 16? Express your answer as a percent. **about 1.4%**

Use the sample space at the right. Find each probability for the sum of two number cubes.

23. P(7) $\frac{1}{6}$ **24.** P(2) $\frac{1}{36}$

25. P(not 11) $\frac{17}{18}$ **26.** P(11 or 12) $\frac{1}{12}$

27. P(a multiple of 3) $\frac{1}{3}$ **28.** P(9) $\frac{1}{9}$

29. P(15) 0 **30.** P(even number) $\frac{1}{2}$

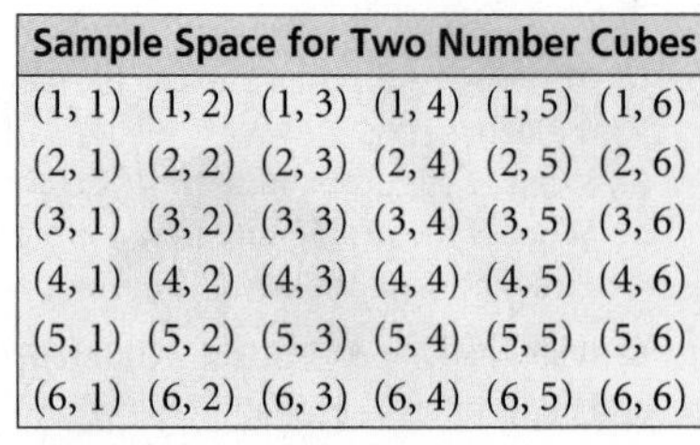

Sample Space for Two Number Cubes					
(1, 1)	(1, 2)	(1, 3)	(1, 4)	(1, 5)	(1, 6)
(2, 1)	(2, 2)	(2, 3)	(2, 4)	(2, 5)	(2, 6)
(3, 1)	(3, 2)	(3, 3)	(3, 4)	(3, 5)	(3, 6)
(4, 1)	(4, 2)	(4, 3)	(4, 4)	(4, 5)	(4, 6)
(5, 1)	(5, 2)	(5, 3)	(5, 4)	(5, 5)	(5, 6)
(6, 1)	(6, 2)	(6, 3)	(6, 4)	(6, 5)	(6, 6)

The outcome (2, 5) has a sum of 7.

Exercise 48 Ask students to explain the difference between -9^2 and $(-9)^2$.

SELF ASSESSMENT **PORTFOLIO** Share with students the criteria you will use to assess their work in portfolios, as well as how you plan to use the results. Students should understand how the rubrics are used to assess their work, how each piece in the portfolio counts, and how the scores they get on their portfolios will affect their overall evaluation.

Wrap Up

THE BIG IDEA Ask students to describe theoretical probability and tell why it is useful.

RETEACHING ACTIVITY Students will learn to calculate theoretical probability from a data table. (Reteaching worksheet 2-8)

Make a tree diagram to show the sample space for four tosses of a coin. Use the sample space to find each probability. 31–38. See margin for tree diagram.

31. P(exactly 2 tails) $\frac{3}{8}$ **32.** P(exactly 3 heads) $\frac{1}{4}$ **33.** P(at least 2 heads) $\frac{11}{16}$ **34.** P(4 tails) $\frac{1}{16}$

35. P(exactly 1 tail) $\frac{1}{4}$ **36.** P(no tails) $\frac{1}{16}$ **37.** P(5 tails) 0 **38.** P(1 or 2 heads) $\frac{5}{8}$

39. *Standardized Test Prep* You are going to roll a number cube once. Which has the same probability as P(not rolling a 4 or a 5)? B
A. P(rolling an even number) **B.** P(rolling a factor of 6) **C.** P(rolling a number less than 3)
D. P(rolling a 1 or a 2) **E.** P(not rolling an odd number)

40. *Cars* A car dealer has an equal chance of receiving any of the colors or styles in the diagram. What is the probability that the next car the dealer receives is a red or black convertible with a gray interior? $\frac{1}{12}$

Car Colors	Interior Colors	Style
		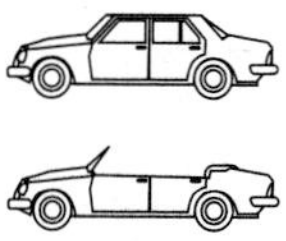

Use the formula below to find the *odds in favor* of an event.

$$\text{Odds} = \frac{\text{number of favorable outcomes}}{\text{number of unfavorable outcomes}}$$

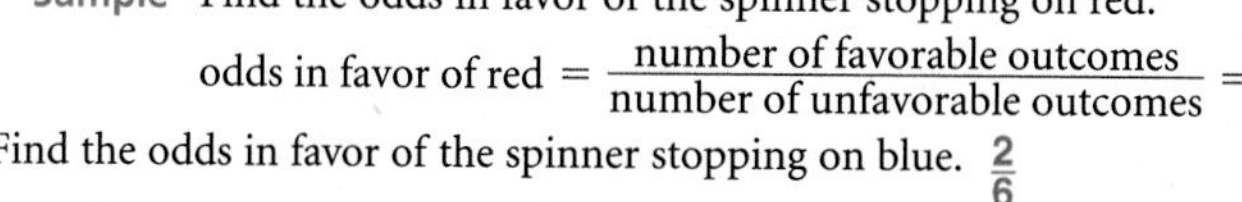

Sample Find the odds in favor of the spinner stopping on red.

$$\text{odds in favor of red} = \frac{\text{number of favorable outcomes}}{\text{number of unfavorable outcomes}} = \frac{3}{5}$$

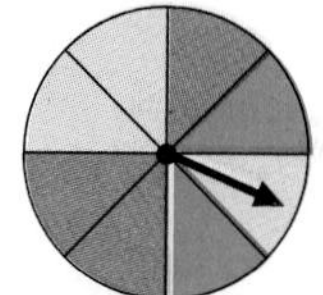

41. Find the odds in favor of the spinner stopping on blue. $\frac{2}{6}$

42. Find the odds in favor of the spinner not stopping on yellow. $\frac{5}{3}$

43. *Writing* Explain how you can use odds to find probability. Include an example.
Divide the numerator by the sum of numerator and denominator to find probability. For example, the probability of the spinner not stopping on yellow is $\frac{5}{5+3}$, which equals $\frac{5}{8}$.

Exercises MIXED REVIEW

Simplify each expression.

44. $|-8 - 3| \times 3$ 33 **45.** $(7^2 - 4) \div 9$ 5

46. $-3\frac{1}{2} - 5\frac{1}{6}$ $-8\frac{2}{3}$ **47.** $5.6 \div 7 - 0.2$ 0.6

48. $-9^2 + 5$ -76 **49.** $-\frac{2}{3} \div 6$ $-\frac{1}{9}$

50. a. Suppose you want a garden consisting of 8 tomato plants. Each plant costs \$1.50. How much will you spend on tomato plants? \$12
b. A gardening "rule of thumb" is that for every dollar you spend on plants, you will spend three dollars on supplies. How much should you budget for supplies? \$36
c. Estimate the total cost of your tomato garden. \$48

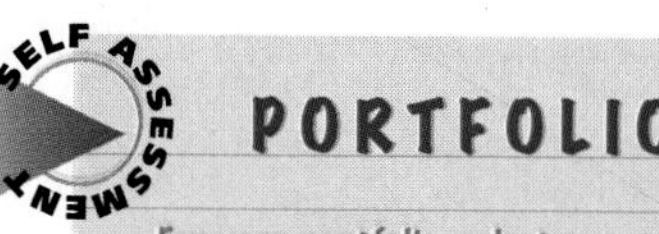

PORTFOLIO

For your portfolio, select one or two items from your work for this chapter. Here are some possibilities.
• corrected work
• most improved work
Explain why you have included each selection.

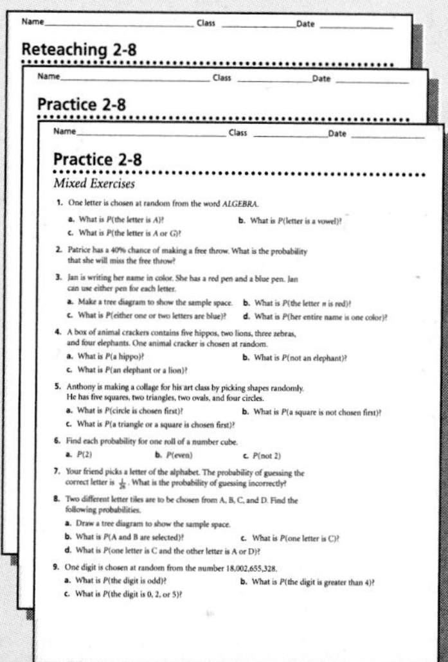
Name______ Class ______ Date ______

Reteaching 2-8

Name______ Class ______ Date ______

Practice 2-8

Name______ Class ______ Date ______

Practice 2-8

Mixed Exercises

1. One letter is chosen at random from the word *ALGEBRA*.
a. What is P(the letter is A)? b. What is P(letter is a vowel)?
c. What is P(the letter is A or G)?
2. Patrice has a 40% chance of making a free throw. What is the probability that she will miss the free throw?
3. Jan is writing her name in color. She has a red pen and a blue pen. Jan can use either pen for each letter.
a. Make a tree diagram to show the sample space. b. What is P(the letter n is red)?
c. What is P(either one or two letters are blue)? d. What is P(her entire name is one color)?
4. A box of animal crackers contains five hippos, two lions, three zebras, and four elephants. One animal cracker is chosen at random.
a. What is P(a hippo)? b. What is P(not an elephant)?
c. What is P(an elephant or a lion)?
5. Anthony is making a collage for his art class by picking shapes randomly. He has five squares, two triangles, two ovals, and four circles.
a. What is P(circle is chosen first)? b. What is P(a square is not chosen first)?
c. What is P(a triangle or a square is chosen first)?
6. Find each probability for one roll of a number cube.
a. P(2) b. P(even) c. P(not 2)
7. Your friend picks a letter of the alphabet. The probability of guessing the correct letter is $\frac{1}{26}$. What is the probability of guessing incorrectly?
8. Two different letter tiles are to be chosen from A, B, C, and D. Find the following probabilities.
a. Draw a tree diagram to show the sample space.
b. What is P(A and B are selected)? c. What is P(one letter is C)?
d. What is P(one letter is C and the other letter is A or D)?
9. One digit is chosen at random from the number 18,002,655,328.
a. What is P(the digit is odd)? b. What is P(the digit is greater than 4)?
c. What is P(the digit is 0, 2, or 5)?

Lesson Quiz

Lesson Quiz is also available in Transparencies.

Find each probability for one roll of a number cube with six sides.

1. P(even number) $\frac{3}{6}$ = 0.5 or 50%
2. P(not even) $\frac{3}{6}$ = 0.5 or 50%
3. P(number \$ 3) $\frac{4}{6}$ = $0.66\overline{6}$ or 67%
4. P(not 2) $\frac{5}{6}$ = $0.83\overline{3}$ or 83%

Finishing the Chapter Project

PROJECT DAY You may wish to plan a project day when students share their completed projects. Encourage groups to explain their process as well as their product.

PROJECT NOTEBOOK Ask students to review their project work and bring their notebooks up to date.

- Have students review their methods for finding, recording, and displaying the data they needed for the project.
- Ask groups to share strategies they found for completing the project, such as shortcuts for creating tables or making graphs. Ask the students if they can think of other ways language can be measured mathematically.

SCORING RUBRIC

3 All data is recorded and labeled accurately as well as plotted correctly. Written explanations are clear and easy to follow. Function rule fits the data. Project shows creativity and demonstrates understanding of the math.

2 Most data is recorded and labeled accurately and mostly plotted correctly, with minor errors. Written explanations are essentially correct but may contain awkward or unclear passages. Function rule fits the data fairly well.

1 Some data is recorded and labeled correctly and some plotting is correct. Room for improvement in written explanations. Function rule does not fit the data.

0 Major elements of the project are incomplete or missing.

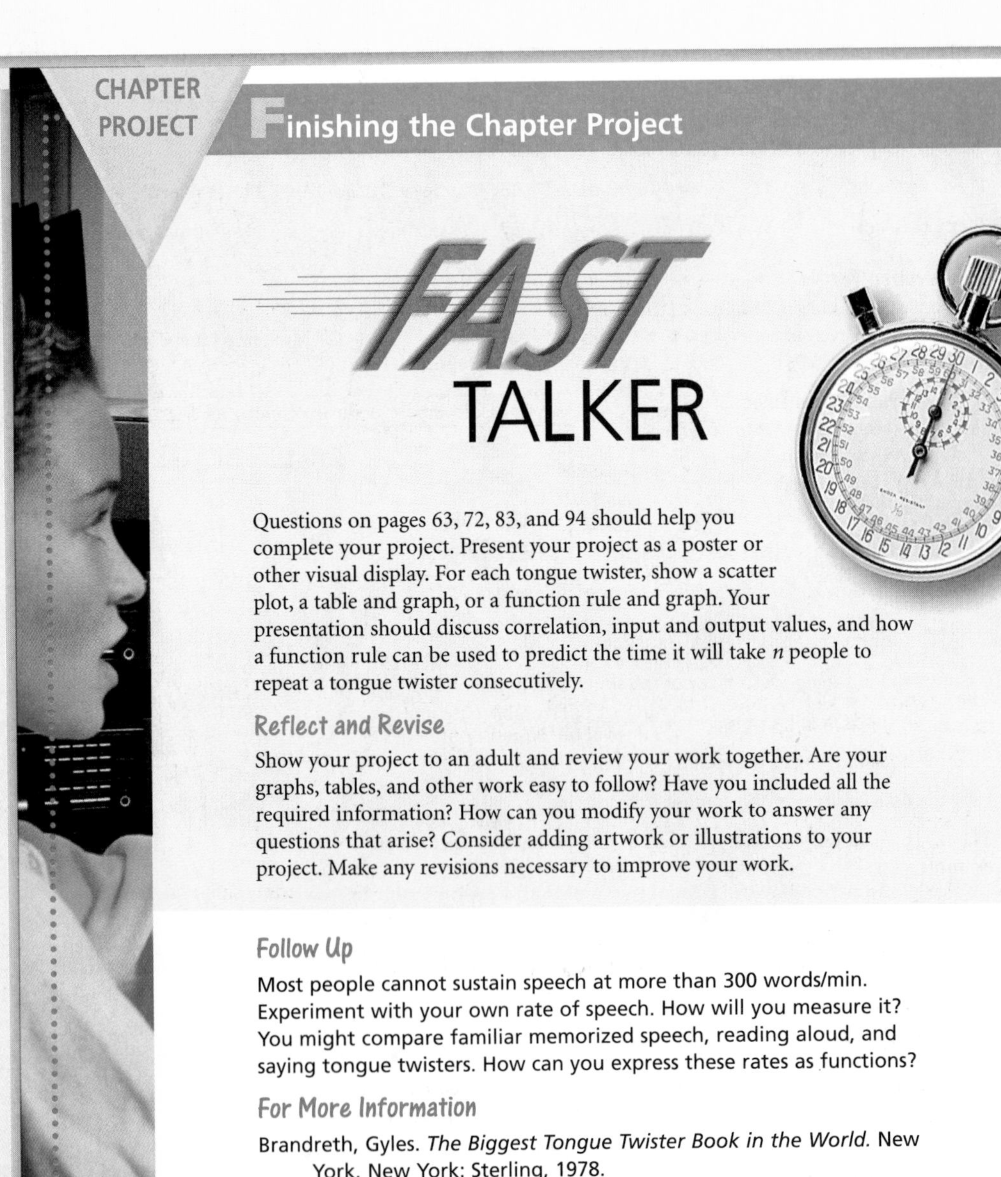

CHAPTER PROJECT

Finishing the Chapter Project

FAST TALKER

Questions on pages 63, 72, 83, and 94 should help you complete your project. Present your project as a poster or other visual display. For each tongue twister, show a scatter plot, a table and graph, or a function rule and graph. Your presentation should discuss correlation, input and output values, and how a function rule can be used to predict the time it will take n people to repeat a tongue twister consecutively.

Reflect and Revise

Show your project to an adult and review your work together. Are your graphs, tables, and other work easy to follow? Have you included all the required information? How can you modify your work to answer any questions that arise? Consider adding artwork or illustrations to your project. Make any revisions necessary to improve your work.

Follow Up

Most people cannot sustain speech at more than 300 words/min. Experiment with your own rate of speech. How will you measure it? You might compare familiar memorized speech, reading aloud, and saying tongue twisters. How can you express these rates as functions?

For More Information

Brandreth, Gyles. *The Biggest Tongue Twister Book in the World.* New York, New York: Sterling, 1978.

Schwartz, Alvin. *A Twister of Twists, a Tangler of Tongues.* Philadelphia, Pennsylvania: J. P. Lippincott, 1972.

Wrap Up

HOW AM I DOING? You may want to have students write their responses in journals. It may be helpful for students to share their thoughts with a math partner in a small group session.

KEY TERMS The numbers in parentheses direct students to the pages where the terms (or symbols) are used or defined. Students should be able to (1) write a simple explanation of each term, (2) illustrate the term with a diagram, or (3) show an example that uses a term.

Exercises 1–8 Have students discuss how visualizing graphs helps them understand information.

Exercises 4–8 Encourage students to use the notations they have practiced in this chapter.

Exercises 23–28 Encourage students to analyze the probabilities carefully, and then check that they understand how the outcomes are evaluated.

2 Wrap Up

Key Terms

certain event (p. 96)
complement of an event (p. 96)
continuous data (p. 65)
dependent variable (p. 69)
discrete data (p. 65)
domain (p. 74)
function (p. 73)
function notation (p. 79)
function rule (p. 74)
impossible event (p. 96)
independent variable (p. 69)
negative correlation (p. 60)
no correlation (p. 60)
outcomes (p. 95)
positive correlation (p. 60)
range (p. 74)
relation (p. 73)
sample space (p. 97)
scatter plot (p. 59)
theoretical probability (p. 95)
tree diagram (p. 97)
trend line (p. 60)
vertical-line test (p. 75)

How am I doing?

- State three ideas from this chapter that you think are important. Explain your choices.
- Describe the different ways you can model a function.

Analyzing Data Using Scatter Plots 2-1

You use **scatter plots** to find trends in data. Two data sets that increase together show a **positive correlation.** There is a **negative correlation** when one data set increases as another decreases. If there is no trend, there is **no correlation.**

Would you expect a *positive correlation, negative correlation,* or *no correlation* between the two data sets. Why?

1. daily temperature vs. the sales of juice drinks Positive correlation; people drink more cold drinks in warm weather.
2. hours a swimmer trains each day vs. the time it takes to swim 100 m Negative correlation; a better-trained swimmer takes less time to swim 100 m
3. *Open-ended* Describe two data sets that show a negative correlation. Samples: number of volunteers and length of time it takes to clean up the local park; bar weight and number of competitors able to lift the bar in a weight-lifting competition

Relating Graphs to Events 2-2

Graphs show a relationship between two variables. **Continuous data** usually involve measurements, such as temperatures, lengths, and weights. **Discrete data** involve a count, such as a number of people or objects.

Write a story to describe each graph. See margin.

4. money vs. time graph (0, money, time)

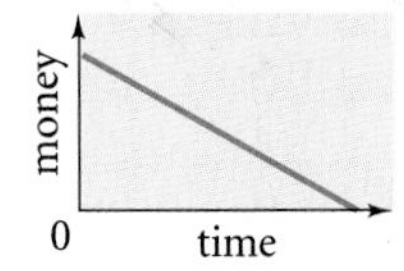

5. 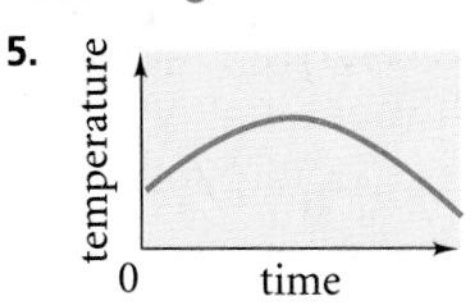

6. 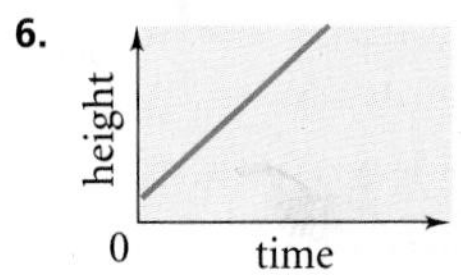

Resources

Student Edition
- Extra Practice, p. 557
- Glossary, Study Guide

Teaching Resources
- Study Skills Handbook
- Glossary, Spanish Resources

pages 101–103 Wrap Up

4. Answers may vary. Sample: A computer rental costs $2.50 per hour. If you start with a fixed amount of money, the longer you plan to work on the computer, the less money you will have left.
5. Answers may vary. Sample: A residential thermostat senses when the temperature in the room falls below the set level, turns the heater on until the temperature reaches 3°F above the set level, then turns the heater off. The graph shows the air temperature rising while the heater is working and then falling after the heater is turned off.
6. Answers may vary. Sample: An elevator is on the second floor. Someone gets in, goes to the eleventh floor, and gets off.

7. **Answers may vary. Sample:**

Height of a Sunflower Over a Summer

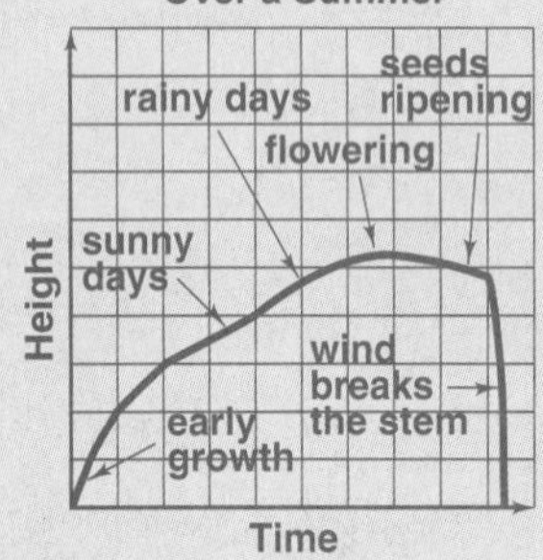

8. **Answers may vary. Sample:**

Number of People in a Restaurant

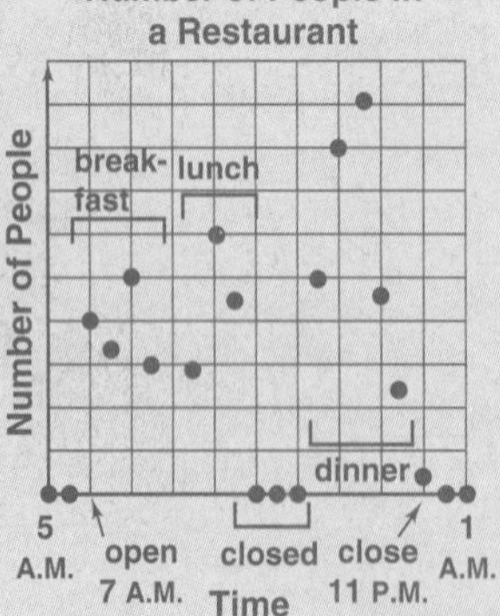

Sketch a graph to describe each situation. Explain the activity in each section of the graph. 7–8. See margin.

7. the height of a sunflower over a summer
8. the number of people in a restaurant each hour

Linking Graphs to Tables 2-3

When you graph data from a table, put the **independent variable** on the horizontal axis and the **dependent variable** on the vertical axis. The dependent variable *depends* on the independent variable.

Cost of a Median Priced House

Year	1985	1986	1987	1988	1989	1990	1991	1992	1993	1994
Price	\$75,500	\$80,300	\$85,600	\$90,600	\$93,100	\$97,500	\$99,700	\$100,900	\$106,100	\$107,600

9. a. Which variable in the table is independent? dependent? year; price
   b. Graph the data. b–c. See margin.
   c. What scales did you use for your axes? Explain why.

Functions 2-4

A set of ordered pairs is a **relation**. A **function** is a relation that has exactly one *y*-value for each *x*-value. A **function rule** is an equation that describes a function. The **domain** of a function is the set of all possible input values. The **range** of a function is the set of all possible output values. You can tell when a relation is a function by using the **vertical-line test.**

Find the range of each function when the domain is {–4, 0, 1, 5}.

10. $y = 4x - 7$ {−23, −7, −3, 13}
11. $m = 0.5n + 3$ {1, 3, 3.5, 5.5}
12. $p = q^2 + 1$ {17, 1, 2, 26}
13. Use the vertical-line test to determine if the graph is a function. not a function
14. Writing Explain when a relation is also a function. A relation is a function when exactly one output value corresponds to each input value.

Writing a Function Rule 2-5

A rule written using **function notation** allows you to see the input value. You can write a rule using a table of values or a description of a situation.

Write a function rule for each table of values.

15.

x	$f(x)$
2	3
4	5
6	7
8	9

$f(x) = x + 1$

16.

x	$f(x)$
−3	3
0	0
3	−3
6	−6

$f(x) = -x$

17.

x	$f(x)$
3.0	6.5
3.5	7.0
4.0	7.5
4.5	8.0

$f(x) = x + 3.5$

ALTERNATIVE ASSESSMENT Students may work these exercises independently or in small groups. For Exercises 23–28, students can make color-coded charts to depict the theoretical probability for each event. Have students summarize their findings in small groups.

Getting Ready for Chapter 3

Students may work these exercises independently or in small groups. The skills previewed will help prepare students for working with algebra concepts and solving simple equations.

The Three Views of a Function, Families of Functions — 2-6, 2-7

You can model functions using rules, tables, and graphs. Functions that are alike can be arranged into groups called *families of functions.* Some families are *quadratic, linear,* and *absolute value* functions.

Model each rule with a table of values and a graph. 18–21. See margin.

18. $f(x) = 4x^2 - 3$ **19.** $f(x) = |x + 5|$ **20.** $f(x) = 2x + 1$ **21.** $f(x) = |x| - 7$

22. For Exercises 18–21, determine whether each equation is a quadratic function, a linear function, or an absolute value function. Explain.
quadratic, highest power of x is 2; absolute value; linear, highest power of x is 1; absolute value

The Probability Formula — 2-8

You can find the **theoretical probability** of an event using this formula.

$$P(\text{event}) = \frac{\text{number of favorable outcomes}}{\text{number of possible outcomes}}$$

An **impossible event** has a probability of 0. A **certain event** has a probability of 1. The **complement** of an event consists of all outcomes not in the event. The set of all possible outcomes is the **sample space.**

Find each probability for one roll of a number cube.

23. $P(3)$ $\frac{1}{6}$ **24.** $P(\text{not } 5)$ $\frac{5}{6}$ **25.** $P(2 \text{ or } 6)$ $\frac{1}{3}$ **26.** $P(\text{number} \geq 7)$ 0

27. Suppose you toss a coin 4 times.
a. What is the probability that you toss exactly 3 heads? $\frac{1}{4}$
b. Explain what $P(\text{not 3 heads})$ means in your own words.
$P(\text{not 3 heads})$ means the chances of getting 1, 2, or 4 heads.

28. An astronomer calculates that the probability in favor of a visible meteor shower occurring in May is $\frac{3}{14}$. What is the probability that a visible meteor shower does not occur in May? E

A. 0 **B.** 1 **C.** $\frac{3}{14}$ **D.** $\frac{3}{17}$ **E.** $\frac{11}{14}$

Getting Ready for... CHAPTER 3

Simplify each expression.

29. $\frac{7 + 9}{2}$ 8 **30.** $3(2 + 1) - 4(2 + 1)$ −3 **31.** $\frac{3}{4}(3 + 5)$ 6

32. $\frac{7 - 4}{6}$ $\frac{1}{2}$ **33.** $5\left(\frac{6 + 1}{3 + 4}\right)$ 5 **34.** $(4 - 4^2) \div 10$ $-\frac{1}{2}$

Evaluate each expression for $m = 2$, $t = -3$, and $w = 1.5$.

35. $t + 2$ −1 **36.** $4m - 1$ 7 **37.** $w(m + 1)$ 4.5 **38.** $\frac{3}{2}w - t$ $5\frac{1}{4}$

9b.

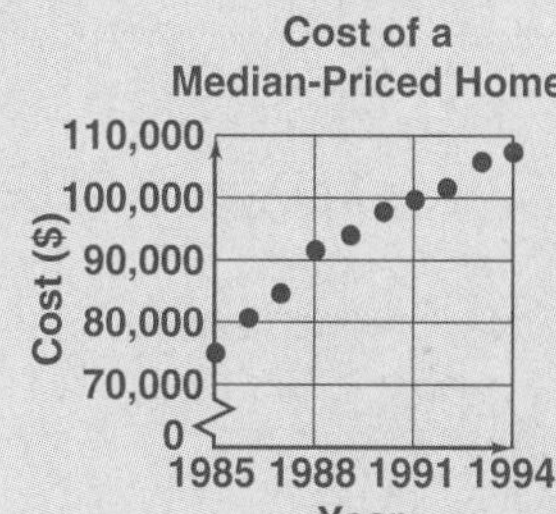

c. Answers may vary. Sample: Years from 1985 to 1994 on the x-axis with labels every 3 years; amounts from \$0 to \$110,000 with a break from \$0 to \$70,000 on the y-axis with labels every \$10,000; this graph most accurately represents all the data without taking too much space.

18. Tables may vary. Sample:

x	$f(x)$
−1	1
0	−3
1	1

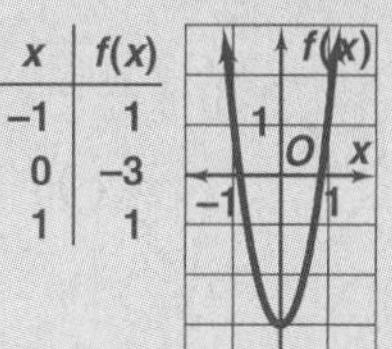

19–21. See back of book.

Assessment

ENHANCED MULTIPLE CHOICE QUESTIONS are more complex than traditional multiple choice questions, which assess only one skill. Enhanced multiple choice questions assess the processes that students use, as well as the end results. The questions are written so that students must use more than one strategy to solve the problem. Using multiple strategies is encouraged by the National Council of Teachers of Mathematics (NCTM). This Chapter Assessment does not contain an enhanced multiple choice question.

FREE RESPONSE QUESTIONS do not give answer choices. Some exercises have more than one possible answer. Students need to give only one correct response.

Exercises 1–11, 13–19, and 22 are free response questions.

WRITING EXERCISES allow students to describe how they think about and understand the concepts they have learned. **Exercise 12** is a writing exercise.

OPEN-ENDED PROBLEMS allow for more than one solution. Students must construct their own responses instead of choosing from possible answers. The student responses will help you determine the depth of their understanding and any possible areas of difficulty. **Exercise 20** is an open-ended problem. Make sure that the students' responses contain values for x and y and a written situation.

Resources

Teaching Resources

Chapter Support File, Ch. 2
- Chapter Assessment, Forms A and B
- Alternative Assessment

Chapter Assessments, Spanish Resources

Teacher's Edition

See also p. 56E for assessment options

Software • Computer Item Generator, Ch. 2

page 104 Chapter Assessment

8. Answers may vary. Sample:

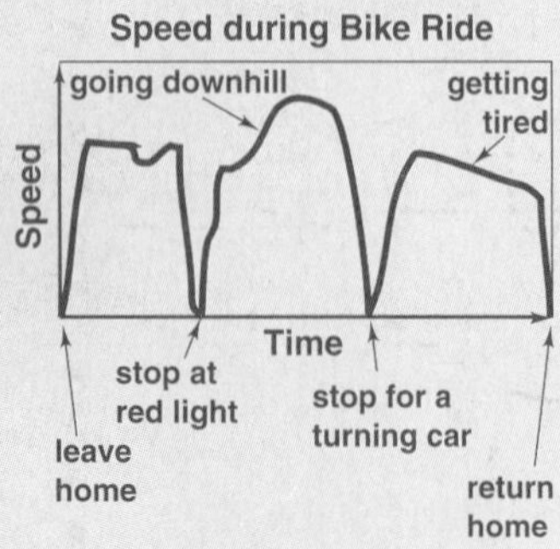

9, 17–20, 22. See back of book.

2 Assessment

Is there a *positive correlation,* a *negative correlation,* or *no correlation* for the data sets in each scatter plot?

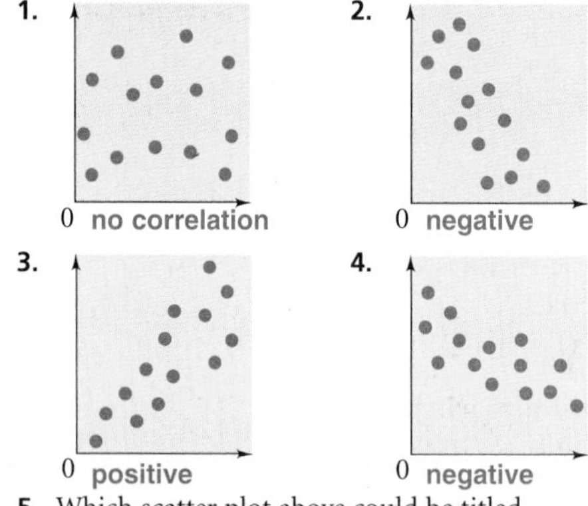

5. Which scatter plot above could be titled "Amount of Rain vs. Umbrella Sales"? 3

Classify the data as *discrete* or *continuous.*

6. your height from age 3 to age 14 continuous
7. the amount of money in a store's cash register during one day discrete

Sketch a graph to describe each situation. Explain the activity in each section of the graph.

8. the speed of a bicycle during an afternoon ride
9. the amount of milk in your container over one lunch period 8–9. See margin.

Determine if each relation is a function.

10.

x	y
−2	5
8	6
3	1.2
5	6

function

11.

x	y
9	6
3	8
4	9.5
9	2

not function

12. *Writing* Explain how to use the vertical-line test to determine if a graph is a function graph.

If any vertical line crosses the graph in two or more places the graph does not represent a function.

17. number of bills; cost of printing; number of lawns, money earned

Find the range of each function when the domain is $\{-3, -1.5, 0, 1, 4\}$.

13. $r = 4t^2 + 5$ $\{41, 14, 5, 9, 69\}$
14. $m = -3n - 2$ $\{7, 2.5, -2, -5, -14\}$

Write a function rule to describe each statement.

15. the cost of printing dollar bills when it costs 3.8¢ to make a dollar bill $c(d) = 0.038d$
16. the amount of money you earn mowing lawns at \$15 per lawn $m(n) = 15n$
17. For Exercises 15 and 16, identify the independent and dependent variables. 17. See above.

Model each rule with a table of values and a graph.

18. $f(x) = 1.5x - 3$
19. $f(x) = -x^2 + 4$

See margin.

20. *Open-ended* Describe a situation that could be modeled by the equation $y = 5x$. 20. See below.
21. Which graph is the graph of $y = -x + 2$? D

A.

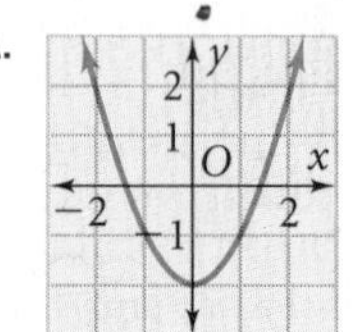

B.

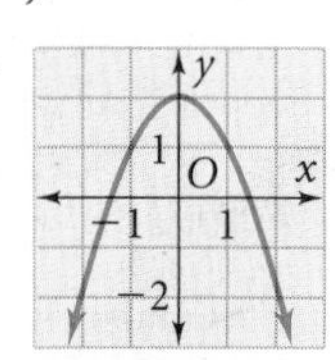

C.

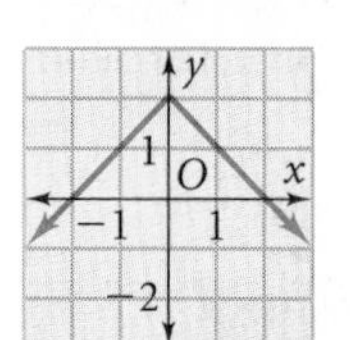

D.

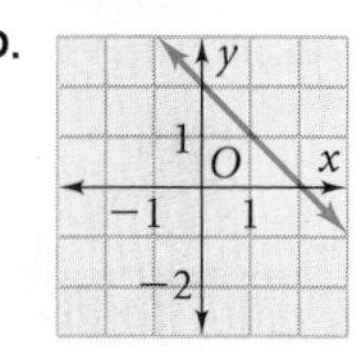

22. A vendor sells T-shirts in sizes small, medium, and large. The available colors are green, blue, red, purple, and black. All combinations of sizes and colors are equally likely.
    a. Draw a tree diagram showing all the possible combinations of sizes and colors. See margin.
    b. Find the probability of randomly selecting a black T-shirt that is size large. $\frac{1}{15}$

20. Sample: number of cars washed at \$5 each vs. money earned

Cumulative Review

Item	Review Topic	Chapter
1	data relationships	1
2	order of operations	1
3	multiplication/division of integers	1
4	formulas in spreadsheets	1
5	matrices	1
6	number patterns	pre-book
7, 11	functions	2
8	coordinate plane	2
9	rational numbers	1
10	families of functions	2
12	probability formulas	2

2 Cumulative Review

For Exercises 1–9, choose the correct letter.

1. Which line plot shows the following set of data? 0, 2, 4, 0, 5, 1, 2, 3, 5, 0, 1, 2 B

A.
```
X   X     X X
X X X X X X X
0 1 2 3 4 5
```
B.
```
X   X
X X X       X
X X X X X X X
0 1 2 3 4 5
```
C.
```
X X X X X X
X X X X X X
0 1 2 3 4 5
```
D.
```
X
X X   X
X X X X X X
0 1 2 3 4 5
```
E. None of the above

2. Which of the following expressions has a value of 48? B
   I. $8 + 2 \cdot 4 \cdot 3$
   II. $[8 + (2 \cdot 4)] \cdot 3$
   III. $(8 + 2) \cdot 4 \cdot 3$
   A. I only B. II only C. II and III
   D. I and III E. I, II, and III

3. Which of the following is a multiple of 4? D
   A. 3654 B. 8647 C. 19,354
   D. 28,640 E. 78,454

4. Which spreadsheet formula can you use to find the mean of four numbers? D
   A. A1 + B1 + C1 + D1/4
   B. A1 * B1 * C1 * D1
   C. (A1 + B1 + C1 + D1) * 4
   D. (A1 + B1 + C1 + D1)/4
   E. A1 + B1 + C1 + D1

5. Which matrix is the difference of $\begin{bmatrix} 8 & -7 \\ 3 & -5 \end{bmatrix} - \begin{bmatrix} 4 & 0 \\ -2 & -4 \end{bmatrix}$? C

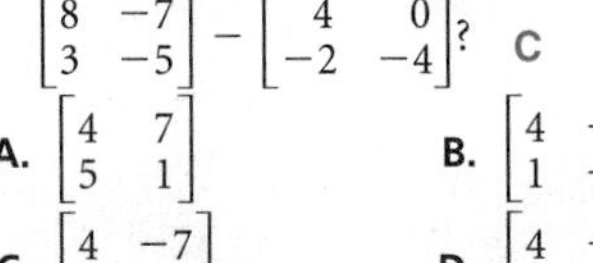

   A. $\begin{bmatrix} 4 & 7 \\ 5 & 1 \end{bmatrix}$ B. $\begin{bmatrix} 4 & -7 \\ 1 & -1 \end{bmatrix}$
   C. $\begin{bmatrix} 4 & -7 \\ 5 & -1 \end{bmatrix}$ D. $\begin{bmatrix} 4 & -7 \\ 1 & -9 \end{bmatrix}$
   E. None of the above

6. Find the sixth term of the pattern 14, 9, 4, ... B
   A. −13 B. −11 C. −1 D. 8 E. −14

7. Find $f(-2)$ when $f(x) = -3x^2 + 4x$. B
   A. 4 B. −20 C. −4 D. 12 E. 26

8. Choose the pair of coordinates that lie on the graph shown. E

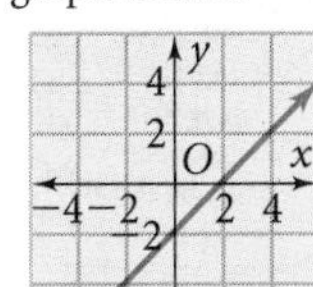

   A. (−1, 1), (1, 3)
   B. (−1, 2), (1, −2)
   C. (−1, 3), (1, 1)
   D. (−1, 2), (1, −2)
   E. (−1, −3), (1, −1)

9. Compare the quantities in Column A and Column B. A

Column A	Column B
$\frac{1}{2}$ of $\frac{3}{4}$	$\frac{1}{4}$ of $\frac{3}{5}$

   A. The quantity in Column A is greater.
   B. The quantity in Column B is greater.
   C. The two quantities are equal.
   D. The relationship cannot be determined on the basis of the information supplied.

For Exercises 10–12, write your answer.

10. The table shows the closing prices of a stock over a period of 5 days. Graph this relation and determine if it is a function. Is it a linear function? Explain why or why not. See margin.

Day	1	2	3	4	5
Price	12	$12\frac{1}{2}$	$12\frac{1}{4}$	13	$13\frac{1}{4}$

11. Find the range of $f(x) = |2x - 3|$ when the domain is {−5, −2, −1}. {13, 7, 5}

12. Suppose you have a bag containing 3 red, 4 blue, 5 white, and 2 black marbles. One marble is selected at random.
   a. Name a certain event. not green
   b. Name an impossible event. purple
   c. What is *P*(not selecting white)? $\frac{9}{14}$
   d. *Open-ended* Name an event more likely than choosing red. Sample: not red

Resources

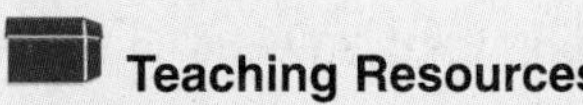

Teaching Resources

Chapter Support File, Ch. 2
- Cumulative Review
- Standardized Test Practice

Teacher's Edition

See also p. 56E for assessment options

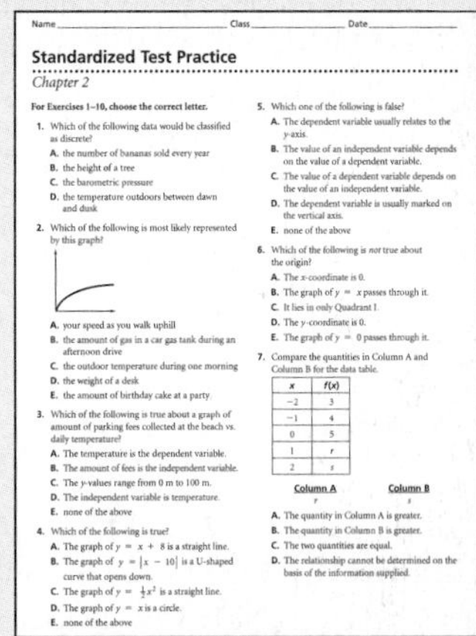

Standardized Test Practice
Chapter 2

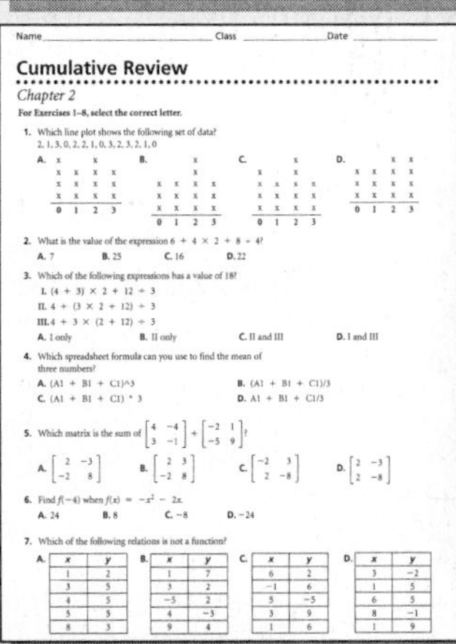

Cumulative Review
Chapter 2

page 105 Additional Answers

10. See back of book.

CHAPTER 3 Algebraic Concepts and Simple Equations

To accommodate flexible scheduling, some lessons are divided into parts. Assignment Options are given in the Lesson Planning Options for each lesson.

3-1 Modeling and Solving Equations (pp. 108–113)
- Part 1 Solving Addition and Subtraction Equations
- Part 2 Solving Multiplication and Division Equations
- Part 3 Modeling by Writing Equations

Key Terms: solutions, inverse operations, equivalent equations, properties of equality

3-2 Modeling and Solving Two-Step Equations (pp. 114–118)
- Part 1 Using Tiles
- Part 2 Using Properties

3-3 Combining Like Terms to Solve Equations (pp. 119–123)
- Part 1 Combining Like Terms
- Part 2 Solving Equations

Key Terms: coefficient, like terms, constant term

3-4 Using the Distributive Property (pp. 124–128)
- Part 1 Simplifying Variable Expressions
- Part 2 Solving and Modeling Equations

Key term: distributive property

3-5 Rational Numbers and Equations (pp. 129–133)
- Part 1 Multiplying by a Reciprocal
- Part 2 Multiplying by a Common Denominator

3-6 Using Probability (pp. 134–138)
- Part 1 Finding the Probability of Independent Events
- Part 2 Finding the Probability of Dependent Events
- Part 3 Finding Probability Using an Equation

Key terms: independent events, dependent events

3-7 Percent Equations (pp. 139–145)
- Part 1 Solving Percent Equations
- Part 2 Solving Real-World Problems with Percents
- Part 3 Simple Interest

Key term: simple interest

3-8 Percent of Change (pp. 146–149)
- Part 1 Percent of Change

Key terms: percent of change, percent of increase, percent of decrease

PACING OPTIONS

This chart suggests pacing only for the core lessons and their parts, and it is provided merely as a possible guide. It will help you determine how much time you have in your schedule to cover other features, such as the Chapter Project, Math Toolboxes, Wrap Up, and Assessment.

	1 Class Period	1 Class Period	1 Class Period	1 Class Period	1 Class Period	1 Class Period	1 Class Period	1 Class Period	1 Class Period
Traditional (40–45 min class periods)	3-1 Part 1, 3-1 Part 2	3-1 Part 1	3-2 Part 1	3-2 Part 2	3-3 Part 1, 3-3 Part 2	3-4 Part 1, 3-4 Part 2	3-5 Part 1	3-5 Part 2	3-6 Part 1, 3-6 Part 2
Two-Year Algebra (40–45 min class periods)	3-1 Part 1	3-1 Part 2	3-1 Part 3	3-2 Part 1	3-2 Part 2	3-2 Part 2	3-3 Part 1	3-3 Part 2	3-3 Part 2
Block Scheduling (90 min class periods)	3-1 Part 1, 3-1 Part 2, 3-1 Part 3	3-2 Part 1, 3-2 Part 2, 3-3 Part 1, 3-3 Part 2	3-4 Part 1, 3-4 Part 2	3-5 Part 1, 3-5 Part 2	3-6 Part 1, 3-6 Part 2, 3-6 Part 3	3-7 Part 1, 3-7 Part 2	3-8		

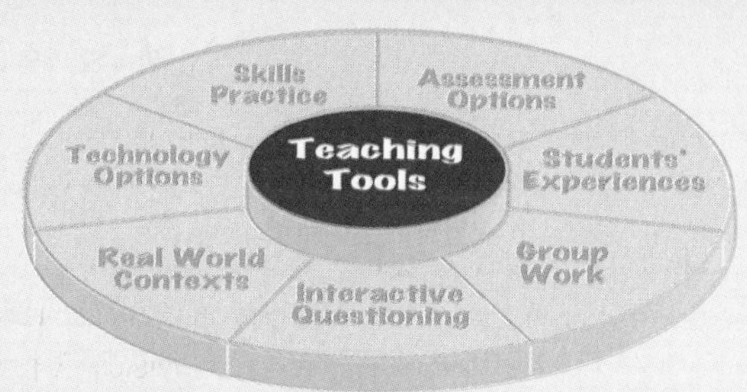

What Students Will Learn and Why

In this chapter, students will build on their knowledge of functions learned in Chapter 2, by learning to solve simple equations. They will learn to solve one- and two-step equations, and use these equations to solve real-world problems. They will learn to use the distributive property, and to combine like terms to solve equations. Students will solve equations involving rational numbers, and learn how to use equations to find the probability of dependent and independent events. Finally, students will use equations to find percents and percent of change.

Discussing the Chapter/Building on Experience

The concept map below relates chapter topics to real world applications. You and your class may wish to add to the map or develop maps of your own. The center oval describes the topic of the chapter. The next level displays topics within the lessons. The outer ovals reflect applications of the content. As you and your class build a concept map, invite students to discuss applications with which they are familiar.

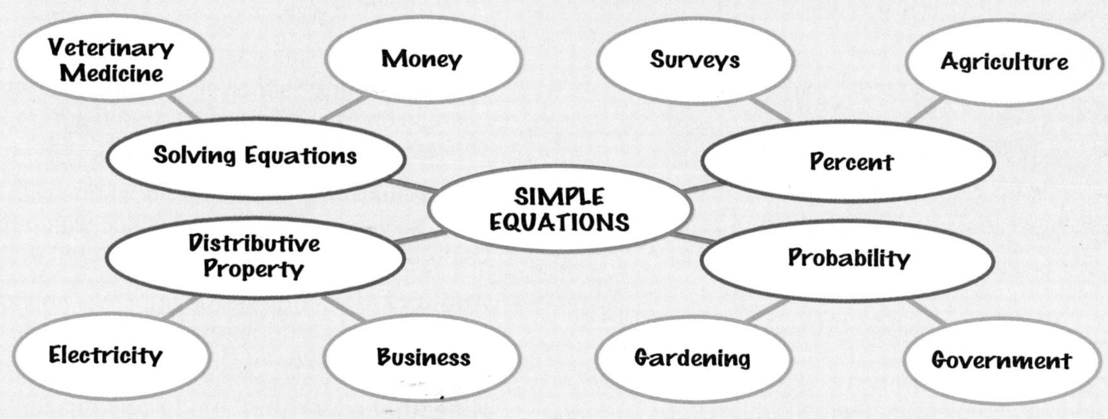

1 Class Period	1 Class Period	1 Class Period	1 Class Period	1 Class Period	1 Class Period	1 Class Period	1 Class Period	1 Class Period	1 Class Period	1 Class Period
3-6 (1)	3-7 (1) 3-7 (2)	3-8								
3-4 (1)	3-4 (2)	3-5 (1)	3-5 (2)	3-5 (2)	3-6 (1)	3-6 (2)	3-6 (3)	3-7 (1)	3-7 (2)	3-8

Interactive Questioning Tips

A question is interactive when there is "give and take" between the questioner (teacher and student) and the respondent. When a critical thinking question is asked, it is important for the questioner to understand what critical thinking skill is being tested, to elicit the correct response. For example, Question 3 in Lesson 3-4 asks students to find an error in a problem involving the distributive property. This is an analysis problem in which the questioner needs to lead the respondent to analyze the problem.

Skills Practice

Every lesson provides skill practice with Exercises On Your Own and Exercises Mixed Review. The Student Edition includes Checkpoints (pp. 128, 144) and a Preparing for Standardized Tests (p. 155). In the Teacher's Edition, the Lesson Planning Options section for each lesson lists Prerequisite Skills students should know for that lesson. At the back of the Student Edition is the Skills Handbook—mini-lessons on math the students may need to review. The Chapter Support File for Chapter 3 in the Teaching Resources box includes two Practice worksheets per lesson, a worksheet for two Checkpoints, and worksheets for Cumulative Review and Standardized Test Preparation.

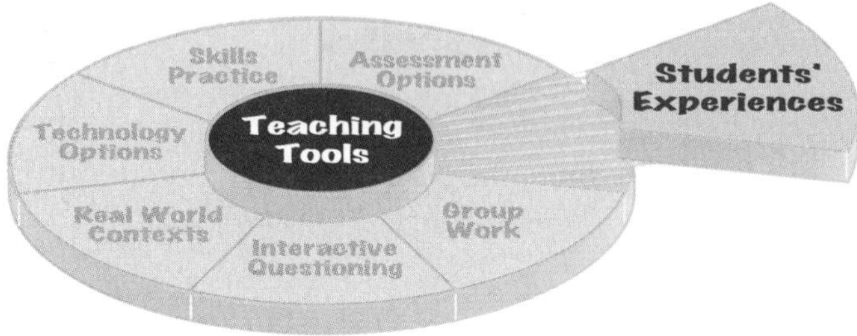

Diverse Learning and Teaching Styles

In your Teacher's Edition, you will find suggestions as to how you can help students complete mathematical tasks in Chapter 3 by reinforcing various learning styles. Here are some examples.

- **Visual learning** simplifying a long, disorganized expression (p. 120), using a tree diagram to show consecutive independent events (p. 135).
- **Tactile learning** modeling equations with manipulatives (p. 108), solving probability exercises with paper tiles (p. 134), using pieces of paper to solve percent problems (p. 140).
- **Auditory learning** assigning vowel sounds to variables (p. 120), pairs of students explaining steps to solve an equation (pp. 125, 147).
- **Kinesthetic learning** modeling equations by grouping students (p. 114), finding reciprocals by moving around the room (p. 130), grouping students by percents (p. 141).

Alternative Activity for Lesson 3-1

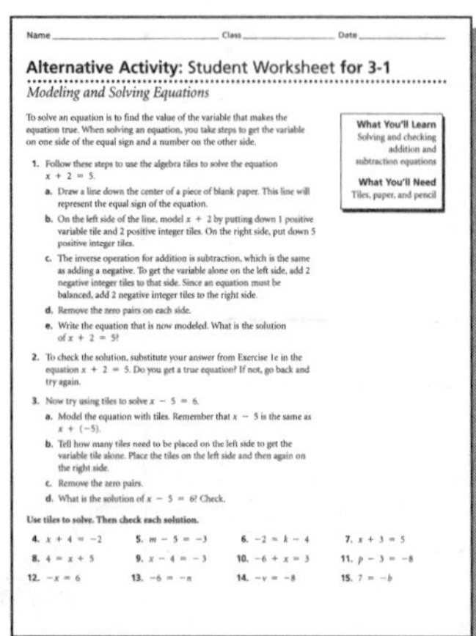
Alternative Activity: Student Worksheet for 3-1

Modeling and Solving Equations

for use with Example 1, addresses tactile learning by modeling and solving equations with tiles.

Alternative Activity for Lesson 3-2

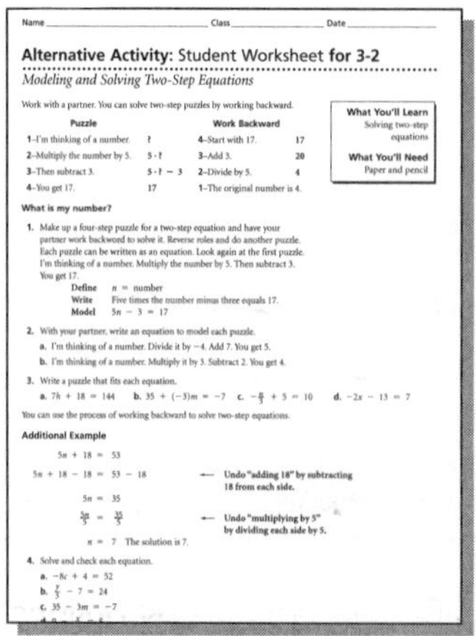
Alternative Activity: Student Worksheet for 3-2

Modeling and Solving Two-Step Equations

for use with Work Together and Example 1, addresses visual learning by working backwards to solve two step equations.

Alternative Activity for Lesson 3-7

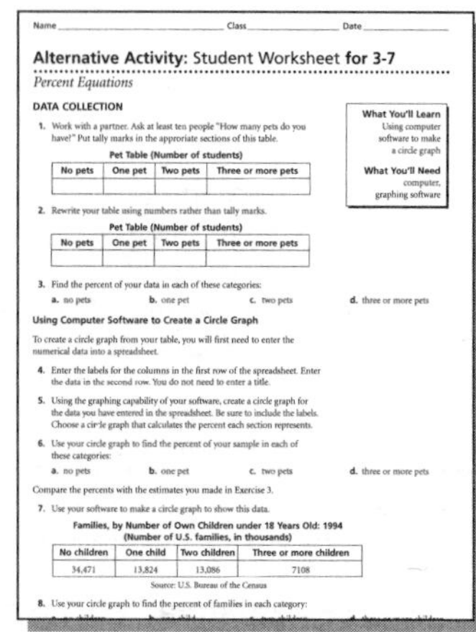
Alternative Activity: Student Worksheet for 3-7

Percent Equations

for use with Work Together, addresses tactile and visual learning by creating a circle graph on a computer.

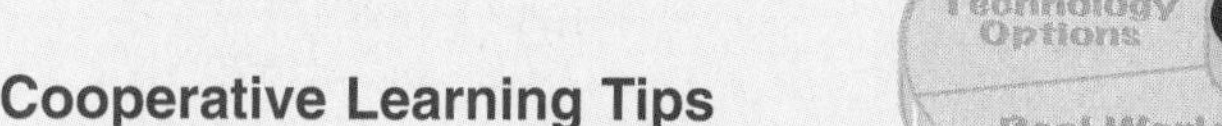

Cooperative Learning Tips

When used effectively, cooperative learning can help students develop interpersonal skills, learn to perform specific roles in a group, and learn to carry out specific responsibilities. The components of Chapter 3 provide a range of cooperative learning opportunities.

- In the Student Edition, the ***Work Together*** parts of lessons are specifically designed for cooperative learning activities.
- In the Teacher's Edition, you will find helpful hints for addressing diverse learning styles (see page C for Chapter 3). For every lesson, you will find a ***Reteaching Activity***, which may involve cooperative learning.

Materials and Manipulatives

The authors expect all students to use scientific calculators. Calculator use is integrated throughout the course.

- bag (3-6)
- colored cubes or chips (3-6)
- compass (3-7)
- protractor (3-8)
- tiles (3-2, 3-3)

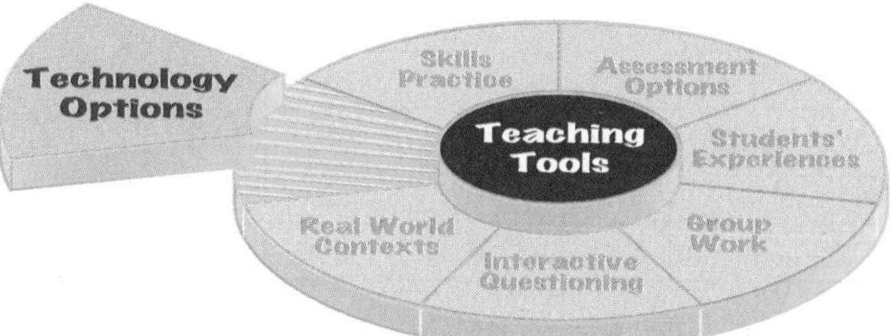

TECHNOLOGY OPTIONS

Technology Tools		Chapter Project	3-1	3-2	3-3	3-4	3-5	3-6	3-7	3-8
Calculator		Assumed that students have scientific calculators at any time.								
Graphing Calculator	Handbook									
	Student Edition		✔	✔			✔			
Software	Secondary Math Lab Toolkit		✔	✔	✔	✔	✔	✔	✔	✔
	Integrated Math Lab				✔		✔			
	Computer Item Generator		✔	✔	✔	✔	✔	✔	✔	✔
	Student Edition								✔	
Video	Video Field Trip	✔								
CD-ROM	Multimedia Algebra Lab		✔	✔				✔	✔	
Internet		See the Prentice Hall site. (http://www.phschool.com)								

The Prentice Hall Algebra program offers you a rich variety of technology options. Be assured that all these options are provided as a means of enriching the program and are not essential for the successful completion of the course.

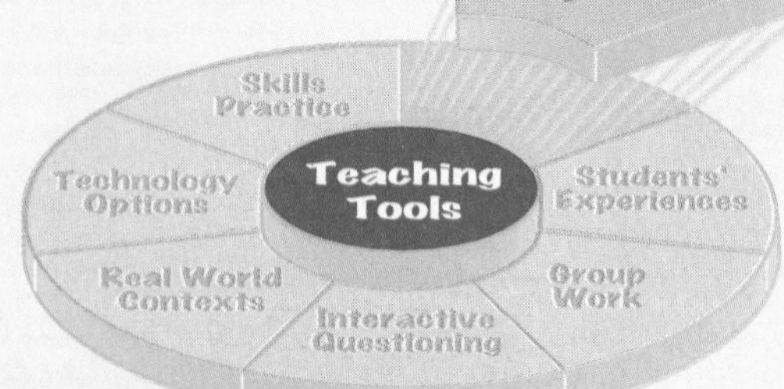

Assessment Options

The Prentice Hall Algebra Program provides you with many options. From these options, you may choose instructional materials and techniques appropriate for your students, or to meet you district's curriculum requirements. As the chart indicates, the program also supports your teaching efforts by offering you many choices for assessment.

ASSESSMENT OPTIONS

Assessment Support Materials	Chapter Opener	3-1	3-2	3-3	3-4	3-5	3-6	3-7	3-8	Chapter End
Chapter Project	▲■●	▲■	▲■	▲■				▲■		▲■
Checkpoints					▲●			▲●		
Self-Assessment		▲■				▲■			▲■	▲■
Writing Assignment		▲	▲■	▲	▲■●	▲	▲	▲	▲■	●
Chapter Assessment										▲●
Alternative Assessment										●
Cumulative Review										●
Standardized Test Prep		▲■			▲■	▲■			▲■	▲●
Computer Item Generator	Can be used to create custom-made practice or assessment at any time.									

▲ = Student Edition ■ = Teacher's Edition ● = Teaching Resources

Checkpoints

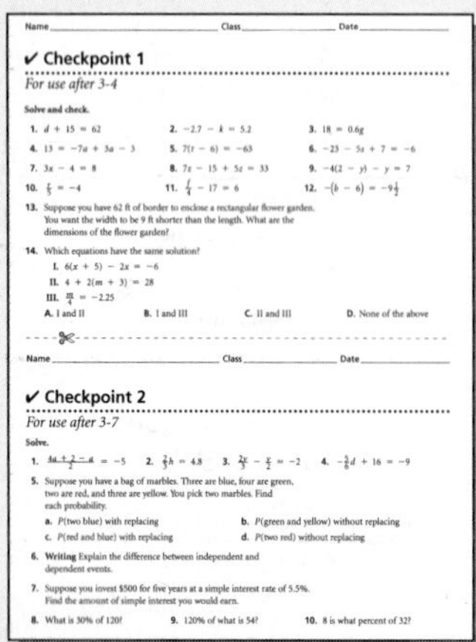
Checkpoint 1
For use after 3-4
Checkpoint 2
For use after 3-7

Alternative Assessment

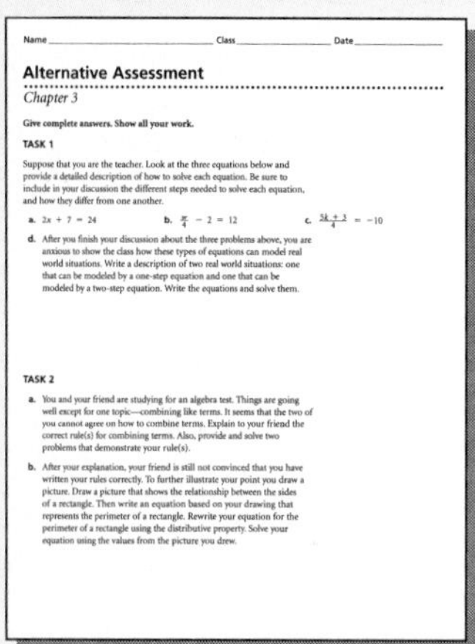
Alternative Assessment
Chapter 3

Chapter Assessment

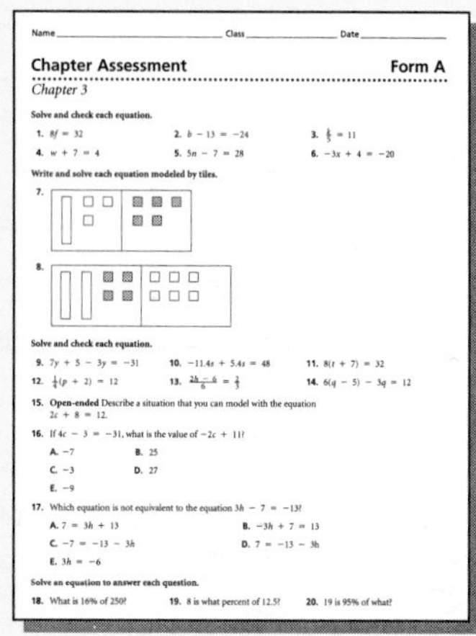
Chapter Assessment Form A
Chapter 3

Available in both Form A and Form B

Making the Right Connections

Mathematics is imbedded in nearly every walk of life. The National Council of Teachers of Mathematics (NCTM) encourages educators to recognize these connections and to emphasize them for the purpose of better educating students for success in life and in a global economy. The ***Connections*** chart below highlights these connections for Chapter 3.

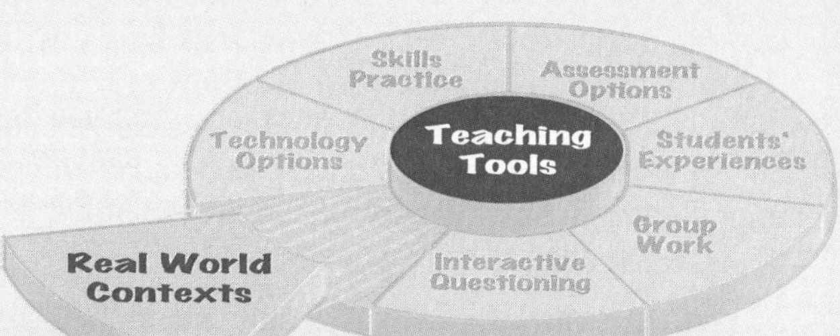

CONNECTIONS

Lesson	Interdiciplinary Connections	Career Prep	Other Real World Connections	Math Integration	NCTM Standards
Chapter Project	Economics		Budgeting		Problem Solving
3-1		Veterinary Medicine Physicians Assistant Car Repair	Weather Parking Recreation Sports Cities	Geometry	Algebra Communication Problem Solving
3-2	Writing	Farming Food Prep	Jobs Money Insurance Gardening		Algebra Communication Problem Solving
3-3	Art		Entertainment Money Shopping	Geometry	Algebra Communication Problem Solving
3-4	Geography	Business Sewing	Sports Shopping Electricity	Geometry	Algebra Communication Problem Solving
3-5	Geography		Nutrition Transportation Family Budget Space Exploration		Algebra Communication Problem Solving Mathematical Structure
3-6	Social Studies	Communications	Games Gardening		Algebra Communication Problem Solving Probability
3-7	Geography	Sales Community Service Finance	Surveys Data Collection Sales Tax		Algebra Communication Problem Solving Statistics
3-8	Health	Medical Careers Agriculture Physical Therapy	Environment Transportation		Algebra Communication Problem Solving Statistics

CHAPTER 3 Algebraic Concepts and Simple Equations

CONNECTING TO PRIOR LEARNING Ask students: *What does the term* balance *mean to you*? **Answers may vary. Sample: having weight distributed evenly** Have students think of a situation or setting that requires good balance. Have them share their ideas in a small group. Challenge the students to represent the situation mathematically.

CULTURAL CONNECTIONS Ask students to think about how much money they spend just on themselves. Is what they buy for themselves influenced by family needs? How many of them have jobs, get an allowance, or just ask for money when they need it? Teenagers from different cultures may have very different concepts about how money is to be used. Encourage students to share their viewpoints.

INTERDISCIPLINARY CONNECTIONS If a car payment is $150 a month, how many hours would a student have to work at $6.00 an hour just to make a car payment? Present information about how much income tax and social security would be deducted from each paycheck. Apply this information to the cost of gasoline, insurance, and repairs.

ABOUT THE PROJECT The Chapter Project gives students an opportunity to learn how to use equations to model a personal budget. In the Find Out questions, students write an equation

Technology Options

Prentice Hall Technology

Video Video Field Trip 3 "Walk on Wall Street," a look at commissions and how stock brokers earn their living.

CHAPTER 3

Algebraic Concepts and Simple Equations

Relating to the Real World

How do you solve a problem? Do you first decide what the problem really is, and then take a series of steps to improve the situation? In algebra, that's exactly what happens. To solve problems, many people, including pharmacists, marine biologists, and money managers, all use simple equations and step-by-step methods.

Lessons

- 3-1 Modeling and Solving Equations
- 3-2 Modeling and Solving Two-Step Equations
- 3-3 Combining Like Terms to Solve Equations
- 3-4 Using the Distributive Property
- 3-5 Rational Numbers and Equations

to model a budget plan. They also develop a spreadsheet to analyze their spending and saving habits. Students will display and present their budgets using a circle graph.

Launching the Project

PROJECT NOTEBOOK Encourage students to keep all project-related materials in a separate folder or notebook. **See Chapter Project and Scoring Rubric in Chapter Support File.**

- Assign students to work with a partner or in a small group. Ask groups to think of something important or expensive that they bought recently.
- Now ask them to consider what they would buy if they could spend \$150. **Answers may vary. Sample: clothes, musical instrument, stereo equipment, sporting goods**
- Have the groups brainstorm to list possible ways to find out how much the items they selected cost. **Answers may vary. Sample: catalogs, ads, visit stores, telephone stores**
- Ask each member of the group to select one item from the list as his or her goal to buy.
- Ask the students to explain why they chose that specific item.

TRACKING THE PROJECT You may wish to have students read Finishing the Chapter Project on page 154 an overview of the project. You may want to set benchmark deadlines for students to show you their work in progress.

CHAPTER PROJECT

checks and balances

When there is something you really want to buy, do you already have money saved for it? Or, do you put money aside each week until you can afford it? Maybe you just dream about it! A budget for your money can help you change dreams to reality.

As you work through the chapter, you will use equations to help model your personal finances. You will develop spreadsheets to analyze your weekly budget, including regular savings. You will use percents to create graphs. Then you will display and present your budget plan using the graphs and spreadsheets.

To help you complete the project:

- p. 113 *Find Out by Researching*
- p. 118 *Find Out by Modeling*
- p. 123 *Find Out by Organizing*
- p. 143 *Find Out by Graphing*
- p. 150 *Finishing the Project*

Project Resources

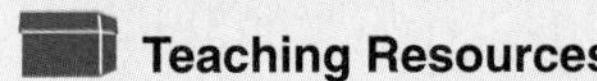

Teaching Resources

Chapter Support File, Ch. 3
- Chapter Project Manager and Scoring Rubric

Transparencies

50

Using the Rubric

Sharing the scoring rubric for the project with your students will alert them to your expectations before they begin work on the project.

As students complete each Find Out question in the chapter, you may wish to have them evaluate their own work or a partner's work, based on the scoring rubric. Students should have the opportunity to revise their work after it has been reviewed.

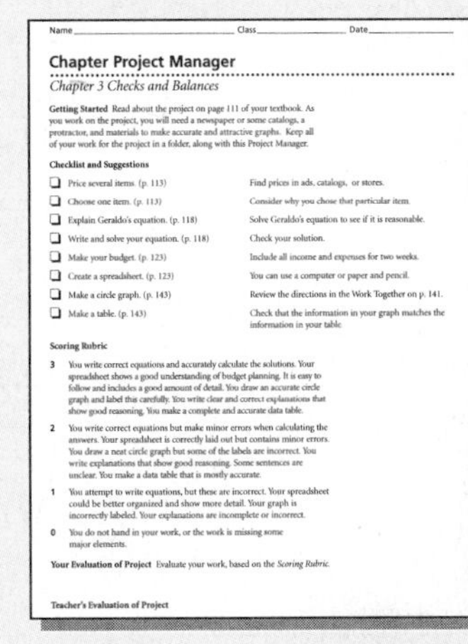

Name_____ Class_____ Date_____

Chapter Project Manager

Chapter 3 Checks and Balances

Getting Started Read about the project on page 111 of your textbook. As you work on the project, you will need a newspaper or some catalogs, a protractor, and materials to make accurate and attractive graphs. Keep all of your work for the project in a folder, along with this Project Manager.

Checklist and Suggestions

❑ Price several items. (p. 113)	Find prices in ads, catalogs, or stores.
❑ Choose one item. (p. 113)	Consider why you chose that particular item.
❑ Explain Geraldo's equation. (p. 118)	Solve Geraldo's equation to see if it is reasonable.
❑ Write and solve your equation. (p. 118)	Check your solution.
❑ Make your budget. (p. 123)	Include all income and expenses for two weeks.
❑ Create a spreadsheet. (p. 123)	You can use a computer or paper and pencil.
❑ Make a circle graph. (p. 143)	Review the directions in the Work Together on p. 141.
❑ Make a table. (p. 143)	Check that the information in your graph matches the information in your table.

Scoring Rubric

3 You write correct equations and accurately calculate the solutions. Your spreadsheet shows a good understanding of budget planning. It is easy to follow and includes a good amount of detail. You draw an accurate circle graph and label this carefully. You write clear and correct explanations that show good reasoning. You make a complete and accurate data table.

2 You write correct equations but make minor errors when calculating the answers. Your spreadsheet is correctly laid out but contains minor errors. You draw a neat circle graph but some of the labels are incorrect. You write explanations that show good reasoning. Some sentences are unclear. You make a data table that is mostly accurate.

1 You attempt to write equations, but these are incorrect. Your spreadsheet could be better organized and show more detail. Your graph is incorrectly labeled. Your explanations are incomplete or incorrect.

0 You do not hand in your work, or the work is missing some major elements.

Your Evaluation of Project Evaluate your work, based on the *Scoring Rubric.*

Teacher's Evaluation of Project

3-1 Teaching Notes

PROBLEM OF THE DAY

A triangle has an 8-in. base and a height of 4-in. A second triangle has three-fourths the area and half the height of the first triangle. What is the base of the second triangle? ($A = \frac{1}{2}bh$) 12 in.

Problem of the Day is also available in Transparencies.

CONNECTING TO PRIOR KNOWLEDGE Have students recall when they began to learn the basic arithmetic skills. Display the equation $3 + 8 = n$. Ask students to solve it. Then solve $n + 8 = 11$. Replace the n with an x. Explain that solving an equation containing a variable is simply finding the value of the variable that makes the equation true.

THINK AND DISCUSS

TACTILE LEARNING Encourage students to use manipulatives such as algebra tiles to model the equations throughout this lesson.

MENTAL MATH Question 1 Suggest that students use the guess-and-check strategy to solve each equation.

Example 1 Relating to the Real World 

Have students brainstorm other situations in which it is more practical to weigh something using this method than to weigh it by placing it on the scale.

Lesson Planning Options

Prerequisite Skills

- Performing operation with integers

Assignment Options

To provide flexible scheduling, this lesson can be subdivided into parts.

1 **Core** 4–5, 7, 12
Extension 10, 11, 28–29

2 **Core** 3, 6, 16–27, 39–50, 55–70
Extension 8–9, 30–31

3 **Core** 1–2, 14–15, 32–34, 35–38, 51–52, 54, 71
Extension 13, 53, 72

Use Mixed Review to maintain skills.

Resources

Student Edition

Skills Handbook, pp. 567, 576
Extra Practice, p. 558
Glossary/Study Guide

Teaching Resources

Chapter Support File, Ch. 3
- Practice 3-1 (two worksheets)
- Reteaching 3-1

Classroom Manager 3-1
Alternative Activity 3-1
Glossary, Spanish Resources
Two-Year Algebra Handbook 3-1

Transparencies
9, 10, 51

Connections Veterinary Medicine . . . and more

3-1 Modeling and Solving Equations

What You'll Learn
- Solving one-step equations
- Using equations to solve real-world problems

...And Why

To organize your thoughts and your work

THINK AND DISCUSS

Part 1

Solving Addition and Subtraction Equations

An equation is like a balance scale because it shows that two quantities are equal.

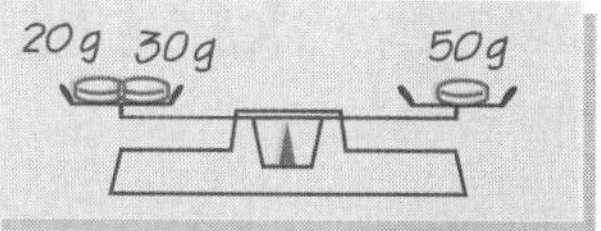

Equation: $20 + 30 = 50$

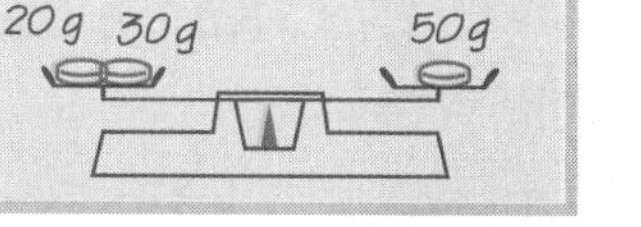

Equation: $x + 20 = 30$

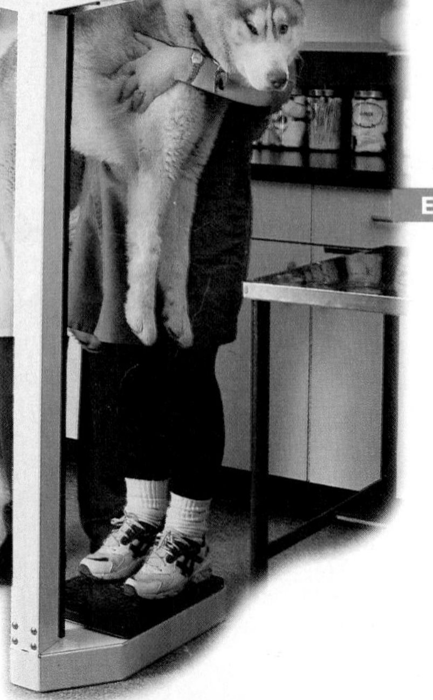

To solve an equation containing a variable, you find the value (or values) of the variable that make the equation true. These values are called **solutions.**

1. Mental Math Solve each equation.
   a. $x + 5 = 6$ 1 b. $y - 10 = 2$ 12
   c. $-4 = -2 + b$ −2

One way to solve an equation is to get the variable alone on one side of the equal sign. You can do this by using **inverse operations,** which are operations that undo one another.

Example 1 Relating to the Real World

Veterinary Medicine A veterinary assistant holds a dog and steps on a scale. The scale reads 193.7 lb. Alone, the assistant weighs 135 lb. To find the weight of the dog, solve the equation $x + 135 = 193.7$.

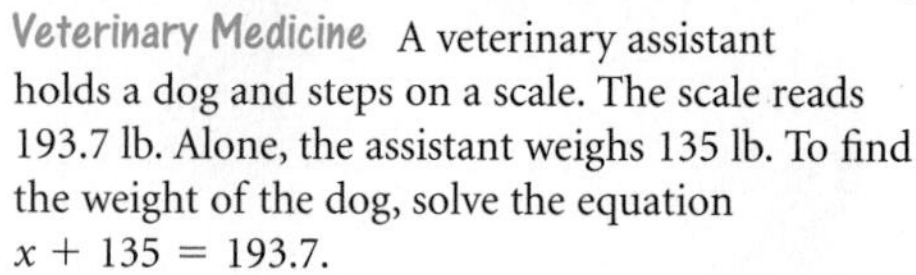

$$x + 135 = 193.7$$
$$x + 135 - 135 = 193.7 - 135$$
$$x = 58.7$$

← The inverse operation for addition is subtraction. Subtract 135 from each side.

The dog weighs 58.7 lb.

MENTAL MATH Question 3 Remind students that the solution to an equation does not depend on whether the variable is on the left side or on the right side of the equal sign.

ERROR ALERT! Students are sometimes confused about whether to add or subtract when a negative number is the first term in the variable side of an addition equation. Example: $-3 + r = 4$ **Remediation:** Before solving, have students rewrite the variable side of the equation so that the variable is first. The signs must stay in front of their original terms.

Example:
$$\begin{aligned} -3 + r &= 4 \\ r - 3 &= 4 \\ r - 3 + 3 &= 4 + 3 \\ r &= 7 \end{aligned}$$

Example 2

ESTIMATION Question 6 Encourage students to round Li's answer and substitute 400 into the equation. Then round the correct answer and substitute. Lead students to understand that even though estimation is not precise, it is a very useful skill. Estimation is used almost daily in personal and business situations.

CRITICAL THINKING Question 8 Have students imagine that a friend needs help understanding why division by zero is not defined. Ask students to tell how they would explain this concept to someone else.

Check $x + 135 = 193.7$

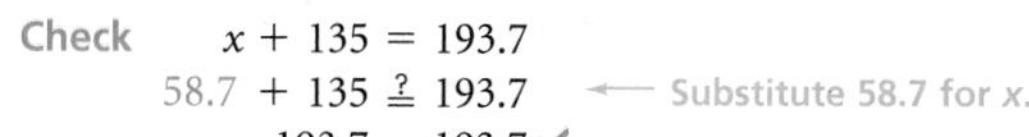

$58.7 + 135 \stackrel{?}{=} 193.7$ ← Substitute 58.7 for x.

$193.7 = 193.7$ ✓

QUICK REVIEW

The expression $-a$ means the opposite of a.

2. Try This What is the inverse operation for subtraction? Use this operation to solve $a - 14 = -26$. addition; -12

3. Mental Math Solve each equation by finding opposites.
   a. $-x = 7$ -7 b. $-y = -2$ 2 c. $5 = -z$ -5

Equivalent equations are equations that have the same solution. You can add or subtract the same number from each side of an equation to form an equivalent equation.

Addition Property of Equality

For any numbers a, b, and c, if $a = b$, then $a + c = b + c$.

Example: Since $\frac{8}{2} = 4$, then $\frac{8}{2} + 3 = 4 + 3$.

Subtraction Property of Equality

For any numbers a, b, and c, if $a = b$, then $a - c = b - c$.

Example: Since $\frac{15}{3} = 5$, then $\frac{15}{3} - 2 = 5 - 2$.

Part

2 Solving Multiplication and Division Equations

For simple multiplication and division equations, you can use mental math to find the solutions.

4. Mental Math Solve each equation.
   a. $4a = 12$ -3 b. $20 = -2x$ 10 c. $\frac{n}{6} = 5$ -30

5. Mental Math Is -16 a solution to the equation $-3x = -48$? Explain.
   No; $(-3) \cdot (-16)$ is not equal to -48.

Multiplication and division are inverse operations. You use multiplication to undo division and division to undo multiplication.

PROBLEM SOLVING

Look Back Show how you would check the solution.

Substitute 41.6 for r in the equation

$-\frac{41.6}{4} \stackrel{?}{=} -10.4;$

$-10.4 = -10.4$

Example 2

Solve $-\frac{r}{4} = -10.4$.

$-\frac{r}{4} = -10.4$ ← The operation is division. Multiply to undo.

$-4(-\frac{r}{4}) = -4(-10.4)$ ← Multiply each side of the equation by -4.

$r = 41.6$

The solution is 41.6.

Additional Examples

FOR EXAMPLE 1

James, who currently weighs 176 lb, wishes to compete as a middle-weight in a boxing tournament next month. To enter the tournament, he must weigh no more than 160 lb. To find out how much weight he must lose before next month, solve the equation $176 - x = 160$. 16 lb

Discussion: *Is James's weight-loss goal realistic? Explain.*

FOR EXAMPLE 2

Solve $6b = -15$. $b = -25$

Discussion: *How do you know that the solution must be negative?*

FOR EXAMPLE 3

Helen is planning to go to the grocery store and buy eggs for her bagel shop. If each carton of eggs contains 12 eggs, how many cartons must she buy to obtain 450 eggs? 37.5 cartons, or 38 whole cartons

Discussion: *Is it possible to write more than one equation that will help answer Helen's question?*

Example 3 **Relating to the Real World** 

Encourage students to estimate whether the answer is correct by rounding 8.8 to 9 and rounding 5.2 to 5. Ask students how many of them might have written the answer as 88 instead of 8.8. Stress the importance of using estimation to check calculations, especially those involving real numbers.

CRITICAL THINKING Question 9 Point out to students the phrase "about 5.2 times greater." Lead students to understand why an answer with more than one decimal place is sometimes not necessary.

ALTERNATIVE METHOD On the board, write two equations that result in $x = 8$. Ask students for more examples. Then, write $x = 7$ on the board. Tell students to write as many equations as they can that result in $x = 7$. Have student pairs exchange papers to check for correctness of each equation. Repeat the exercise with different values for *x*.

Exercises ON YOUR OWN

ALTERNATIVE ASSESSMENT Exercise 2 Ask students to draw a picture of a two-pan balance scale. Then have them use the picture to illustrate one of the four basic properties of equality. This exercise will help you assess students' perception of equality and balance in equations.

ESTIMATION Exercise 12 Once students conclude that 96.26 is not a reasonable answer, encourage students to estimate an answer before calculating the answer. Ask:

- *What equation would you use to estimate the answer?* $m - 60 = 160$.

Technology Options

For Exercise 13, students may want to use geometry software to check the measure of $\angle A$. For Exercises 38–49, students may want to use graphing calculators to check their answers.

Prentice Hall Technology 

 **Software** • Secondary Math Lab Toolkit • Computer Item Generator 3-1

CD-ROM Multimedia Algebra Lab 3

 **Internet** • See the Prentice Hall site. (http://www.phschool.com)

110

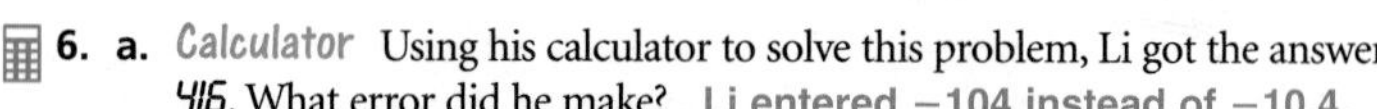

6. a. **Calculator** Using his calculator to solve this problem, Li got the answer 416. What error did he make? Li entered −104 instead of −10.4.
   b. **Estimation** How could Li use estimation to catch his error? −10.4 is close to −10 so the product should be close to (−4)(−10), or 40.

You can use multiplication and division to form equivalent equations.

> **Multiplication Property of Equality**
>
> For any numbers *a*, *b*, and *c*, if $a = b$, then $a \cdot c = b \cdot c$.
>
> **Example:** Since $\frac{6}{2} = 3$, then $\frac{6}{2} \cdot 2 = 3 \cdot 2$.

> **Division Property of Equality**
>
> For any numbers *a*, *b*, and *c*, with $c \neq 0$, if $a = b$, then $\frac{a}{c} = \frac{b}{c}$.
>
> **Example:** Since $3 + 1 = 4$, then $\frac{3+1}{2} = \frac{4}{2}$.

7. **Try This** Solve each equation.
   a. $3.2m = 5.44$ 1.7
   b. $\frac{b}{4} = -18$ −72
   c. $100.8 = -16y$ −6.3
8. **Critical Thinking** When you divide both sides of an equation by the same number, you cannot use the number zero. Explain what would result from the equation $3 \cdot 0 = 4 \cdot 0$ if you could divide both sides of the equation by zero. The equality is true because both sides are equal to 0. If you were allowed to divide both sides by 0, you would get 3 = 4, which is not true.

Part  3

Modeling by Writing Equations

A good strategy for solving a real-world problem is to model the problem with an equation. It organizes your thoughts and, of course, your work.

Example 3 **Relating to the Real World** 

Weather The average annual precipitation of Houston, Texas, is about 5.2 times that of El Paso, Texas. What is the average annual precipitation of El Paso?

Average Annual Precipitation

46 in. ? El Paso Houston

Define p = El Paso's precipitation

Relate Houston's precipitation is 5.2 times El Paso's precipitation

Write $46 = 5.2 \times p$

$46 = 5.2p$

$\frac{46}{5.2} = \frac{5.2p}{5.2}$ ← Divide each side by 5.2.

$8.8 \approx p$

The average annual precipitation of El Paso is about 8.8 in.

- *What is a reasonable estimate for the value of m?* 220
- *What is the exact value of m?* 221.74
- *Was estimation useful for predicting the answer?* yes

GEOMETRY Exercise 13 In this exercise, students use an equation to model a common geometry problem.

CARTOON Exercises 32–33 Have students brainstorm other situations in which the value of a variable changes from one event to another. For example, you may know several people named Ashley. The person you called Ashley yesterday may be different from the person you called Ashley today.

STANDARDIZED TEST TIP Exercise 34 In questions that ask students to compare several equations, suggest that students completely solve each equation before comparing them.

OPEN-ENDED Exercises 35–38 Suggest that pairs of students exchange papers to check that the equations correctly model the situations they describe.

DIVERSITY Exercise 51 Lead a discussion on the difficulties that people with disabilities face when their parking zones are taken by others. Have students note how such zones differ from other parking spaces. After students have gained an appreciation for the benefits of these parking zones, decide as a group what fine seems appropriate for people parking illegally.

CONNECTING TO THE STUDENTS' WORLD Exercise 53 Ask students to think of a type of car repair, and to guess the cost of the repair based on what they know about the cost of parts and labor. Have students research the actual costs of parts and labor. Then model the amounts using equations.

9a. Answers may vary. Sample: 8.846153846.
b. Abdul is right; the answer should not be more precise than the original data.

9. a. *Calculator* What does your calculator display for 46 ÷ 5.2 ENTER ?
b. *Critical Thinking* Akira thinks that 8.846153846 more accurately describes El Paso's precipitation than 8.8 does. Abdul thinks that because of the phrase "about 5.2 times greater," 8.8 is a better answer. Do you agree with Akira or Abdul? Explain. See left.

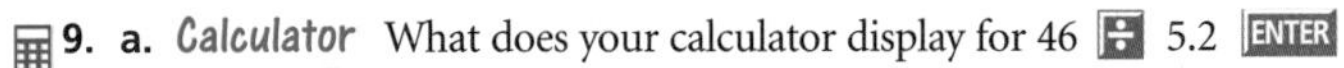

Exercises ON YOUR OWN

1. *Critical Thinking* To solve $c - 3 = 12$, Kendra added 3 to each side of the equation. Ted subtracted 12 from each side, then added 15. Which method is better? Why? Kendra's method is better; it needs only one operation and Ted's needs two.
2. Which property of equality do the balance scales model? addition property

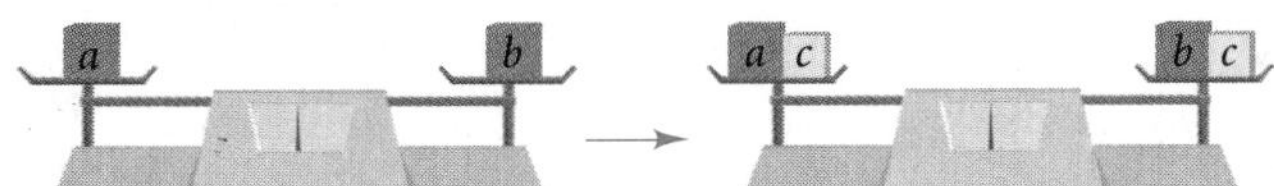

Mental Math **Solve each equation mentally.**

3. $4a = 24$ 6
4. $-6 + y = -11$ −5
5. $z - 8 = 0$ 8
6. $\frac{x}{7} = -3$ −21
7. $5 = -9 + v$ 14
8. $36 = -4x$ −9
9. $\frac{y}{5} = 100$ 500
10. $-5 = n - 2$ −3

11. Which equation is not equivalent to the others? C
A. $m + 7 = -4$ B. $7 + m = -4$ C. $-4 = m - 7$ D. $-4 = 7 + m$

12. *Estimation* Use estimation to check whether 96.26 is a reasonable solution for the equation $m - 62.74 = 159$. Explain your method. No; 96.26 is close to 100, 62.74 is close to 60, so the difference should be close to 40, and 159 is not.

Write an equation to model. Then solve each problem.

13. *Geometry* Suppose $\angle A$ and $\angle B$ are supplementary angles, and the measure of $\angle A$ is 109°. What is the measure of $\angle B$?
$B + 109 = 180$; 71°
14. *Wages* Suppose you work as a carpenter's apprentice. You earn \$93.50 for working 17 h. What is your hourly wage?
$17h = 93.5$; \$5.50/h
15. *Physician's Assistant* You measure a child and find that his height is $41\frac{1}{2}$ in. At his last visit, he was $38\frac{3}{4}$ in. tall. How much did he grow?
$38\frac{3}{4} + g = 41\frac{1}{2}$; $2\frac{3}{4}$ in.

QUICK REVIEW

Two angles are *supplementary* if the sum of their measures is 180°.

Solve and check.

16. $5p = -75$ −15
17. $2 = -2x$ −1
18. $y - \frac{2}{3} = \frac{1}{3}$ 1
19. $-8j = 12.56$ −1.57
20. $-7 + a = 28$ 35
21. $-\frac{p}{7} = 28$ −196
22. $-7n = 28$ −4
23. $y + 18 = 13.5$ −4.5
24. $-3.4 + x = 9.5$ 12.9
25. $10 = \frac{m}{-2}$ −20
26. $\frac{t}{2} = 98$ 196
27. $-5\frac{1}{3} = x + \frac{1}{2}$ $-5\frac{5}{6}$
28. $z - 0.35 = 1.65$ 2
29. $a + 3\frac{1}{4} = 7\frac{1}{2}$ $4\frac{1}{4}$
30. $6n = -120$ −20
31. $-\frac{y}{7} = 35$ −245

pages 111–113 On Your Own

32. Answers may vary. Sample: A letter only needs to represent the same value in related equations. If equations are not related, the values of the variable may be different.
35. Sample: Three friends shared the lunch bill at a restaurant and paid \$6 each. How much was the total bill?
36. Sample: You paid \$19.50 for a double CD after getting a 50 cent discount. What was the original price?
37. Sample: Today's high temperature is 16.1°C, which is 14°C higher than yesterday. What was the high temperature yesterday?
38. Sample: An amusement park charges a quarter for each ride. A family spent \$10 on all the rides over a weekend. How many times did they go on rides?

MAKING CONNECTIONS Exercise 71 The Pentagon is a five-sided building consisting of five concentric pentagons connected by spokelike corridors. It houses the U.S. Department of Defense, which includes the Army, Navy, and Air Force. The Pentagon was built during World War II and covers 34 acres.

MAKING CONNECTIONS Exercise 72 According to legend, a Greek soldier ran 36.2 km from the site of the battle at Marathon to Athens. Arriving in Athens, he announced the news of victory and then died. The modern marathon is 42.19 km long. Although all marathons cover the same distance, the terrain and weather conditions vary from race to race and region to region. Therefore, there is no single world record time for the running of this distance.

GEOMETRY Exercises 71–72 Have students compare the perimeter of the Pentagon to the length of a marathon. Ask:

- *We want to calculate the number of times a person must run around the Pentagon in order to run a length equal to a marathon. What equation could we use?* Answers may vary. Sample: $1.6t = 42.2$, where t = number of times around the Pentagon
- *How many times around must the person run?* about 26.4 times

Chapter Project **FIND OUT BY RESEARCHING** Ask groups of students to think of several items that they would like to buy for less than $150. Have each group price the items they select either using ads or by visiting stores. Each group then selects one item as a goal to buy. This task is essential to their work

112

Use the cartoon for Exercises 32 and 33.

"Just a darn minute! Yesterday you said X equals two!"

32. **Writing** Explain what the student does *not* understand about using letters in algebra. See margin.

33. What property of equality did the teacher use to solve the equation? subtraction property

34. **Standardized Test Prep** Which equations are equivalent? B

A. II and III	I. $x - 5 = 10$
B. II and IV	II. $4x = 20$
C. I and V	III. $-\frac{x}{8} = -40$
D. II, III, and IV	IV. $8 + x = 13$
E. I, II, and IV	V. $-30 = 2x$

Open-ended **Describe a situation that can be modeled by each equation.** 35–38. See margin.

35. $\frac{m}{3} = 6$
36. $p - 0.5 = 19.5$
37. $14 + t = 16.1$
38. $0.25q = 10$

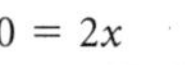
Choose **Use a calculator, paper and pencil, or mental math to solve each equation. Check your solutions.**

39. $z - 18 = 13.5$ 31.5
40. $\frac{x}{18} = 13.5$ 243
41. $-196 = m - 97$ -99
42. $9\frac{2}{3} = a + 19$ $-9\frac{1}{3}$
43. $0.9x = 11.7$ 13
44. $-6 = -3n$ 2
45. $q - 28\frac{1}{2} = -12$ $16\frac{1}{2}$
46. $-\frac{r}{3} = -101$ 303
47. $-1\frac{3}{4} = p - \frac{1}{2}$ $-1\frac{1}{4}$
48. $\frac{m}{4} = -80$ -320
49. $15x = 172.5$ 11.5
50. $15.9 = r + 27.3$ -11.4

Write an equation to model. Then solve each problem.

51. **Parking** In Los Angeles, the fine for illegally parking in a zone reserved for people with disabilities is $330. This is $280 higher than the fine in New York City. How much is the fine in New York City? $330 = 280 + x$; $50

52. **Recreation** The federal government spends $25.2 million each year to operate parks and historic sites in Washington, D.C. This is $7.8 million more than it spends to run Yellowstone National Park. How much does it cost to run Yellowstone National Park? $25.2 = y + 7.8$; $17.4 million

53. **Car Repair** Your bill for a car repair is $166.50.
    a. One third of the bill is for labor. How much is the charge for labor? $3L = 166.5$; $55.50
    b. The mechanic worked on your car for 1.5 h. What is the hourly charge for labor? $1.5r = 55.5$; $37

Baseball Diamond

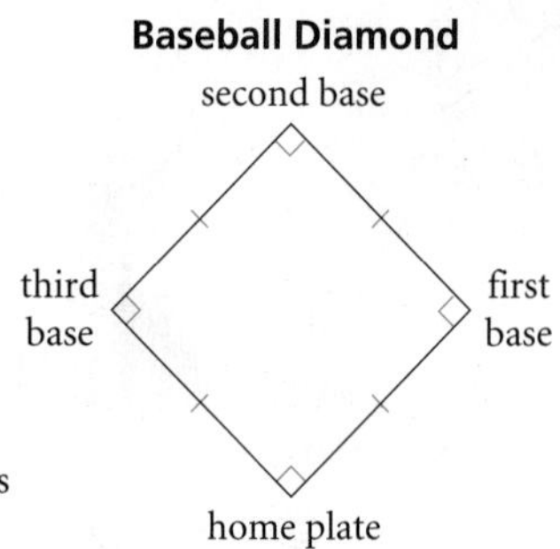

54. **Sports** The total distance around a baseball diamond is 360 ft. What is the distance from third base to home plate? $4b = 360$; 90 ft

on the chapter project. You may wish to check students' progress by having each student write a sentence or two describing his or her duties within the group.

Exercises MIXED REVIEW

GETTING READY FOR LESSON 3-2 These exercises prepare students to solve two-step equations.

SELF ASSESSMENT **JOURNAL** Students may also want to write a brief answer to the question: *What kinds of mistakes are most easily made when solving equations?*

Wrap Up

THE BIG IDEA Ask: *How is an equation useful for modeling a situation?*

RETEACHING ACTIVITY Students use tiles to solve one-step equations. They model the equation, the step, and the solution. (Reteaching worksheet 3-1)

Solve each equation.

55. $5 + v = -13$ −18
56. $\frac{x}{6} = -10$ −60
57. $8y = -64$ −8
58. $m - 1.2 = -2.5$ −1.3
59. $p + \frac{1}{4} = -3\frac{1}{2}$ $-3\frac{3}{4}$
60. $-\frac{a}{7} = 2.5$ −17.5
61. $314 = n + 576$ −262
62. $-1\frac{1}{3} = b - 5\frac{1}{2}$ $4\frac{1}{6}$
63. $-3.2x = 14.4$ −4.5
64. $32 = -0.8m$ −40
65. $b + 6\frac{1}{8} = 5\frac{3}{4}$ $-\frac{3}{8}$
66. $-\frac{p}{4} = 55$ −220
67. $-1.05 = t - 6.17$ 5.12
68. $-\frac{b}{3} = 107$ −321
69. $6x = -\frac{1}{4}$ $-\frac{1}{24}$
70. $-4.87 = x + 0.025$ −4.895

Write an equation to model. Then solve each problem.

71. *Geometry* The Pentagon is the headquarters of the United States Department of Defense. Its shape is a regular pentagon, and its perimeter is about 1.6 km. How long is one side of the Pentagon? $5s = 1.6$; 0.32 km

72. *Sports* In 1994, German Silva of Mexico won the New York Marathon with a time of 2 h 11 min 21 s. In the same year, Cosmas N'Deti of Kenya won the Boston Marathon with a time that was 4 min 6 s faster than Silva's time in New York. How long did it take N'Deti to run the Boston Marathon? $t + 246 = 7881$; 7635 s or 2 h 7 min 15 s

QUICK REVIEW

A *regular pentagon* has five sides of equal length.

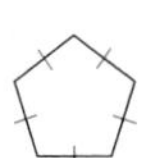

Reteaching 3-1

Practice 3-1

Practice 3-1
Mixed Exercises

Chapter Project *Find Out by Researching*

Think of several items you would like to buy for less than $150, such as a CD player, sports equipment, or some clothes.

- Price these items using ads or by visiting several stores.
- What factors other than price should you consider? Explain.
- After completing your research, choose one item as a goal to buy. Explain your decision.

Exercises MIXED REVIEW

73. Find $f(2)$ when $f(x) = 3x + 2.8$

74. Find the range of $f(x) = x^2 - 1$ when the domain is {−2, 1, 3}. {3, 0, 8}

75. *Cities* The population of New York City is about 15 times the population of New Orleans, Louisiana. The population of New York City is about 7,300,000. What is the population of New Orleans? about 500,000

Getting Ready for Lesson 3-2

Evaluate each expression.

76. $3n + 5$ for $n = -6$ −13
77. $2.8 + 5x$ for $x = 7.9$ 42.3
78. $3x - 7$ for $x = -12$ −43
79. $-4w - 17$ for $w = -12$ 31
80. $\frac{k}{5} + 3.8$ for $k = 14$ 6.6
81. $8 - \frac{x}{2}$ for $x = -40$ 28

SELF ASSESSMENT **FOR YOUR JOURNAL**

Describe some ways you can solve simple equations. Give a sample equation and solution for each method listed.

Lesson Quiz

Lesson Quiz is also available in Transparencies.

In Exercises 1–4, solve each equation, and check your solutions.

1. $m + 4.6 = 2.3$ $m = -2.3$

2. $14 - x = 8$ $x = 6$

3. $13y = 40$ $y \approx 3.1$

4. $\frac{c}{-2} = 15$ $c = -30$

5. A CD player that normally sells for $149.99 is on sale for $114.99. Write an equation and then solve it to find how much the price of the CD player is reduced. $149.99 - x = 114.99$; $x = 35$; The CD player was reduced by $35.

PROBLEM OF THE DAY

A long-distance phone company charges $.50 for the first minute and $.15 for each additional minute. If Mark budgets $4 to call his sister once a month, how long can they talk? How much will he be charged? **24 min; $3.95**

Problem of the Day is also available in Transparencies.

CONNECTING TO PRIOR KNOWLEDGE Ask students if they have heard or seen advertisements for cable television service. Ask students how they think the charge for cable service is determined. Usually the total cost is a combination of a base charge for several basic channels and a charge for special channels. Lead students to see that this situation can be represented by a two-step equation. Have students suggest equations and write them on the board.

WORK TOGETHER

KINESTHETIC LEARNING Divide the class into three groups. Give one group +1 tiles, the second group −1 tiles, and the third group *x* tiles. Have students walk around and arrange themselves in groups to model some of the equations in this lesson. Students can take turns volunteering to simplify the equations by using techniques such as removing zero pairs.

Lesson Planning Options

Prerequisite Skills

- Modeling equations with tiles
- Performing operations with integers

Assignment Options

To provide flexible scheduling, this lesson can be subdivided into parts.

1 **Core** 1–11
Extension 24–26, 50, 51, 60, 61

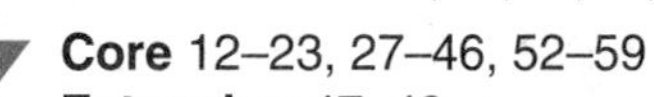

2 **Core** 12–23, 27–46, 52–59
Extension 47–49

Use Mixed Review to maintain skills.

Resources

Student Edition

Skills Handbook, p. 579
Extra Practice, p. 558
Glossary/Study Guide

Teaching Resources

Chapter Support File, Ch. 3
- Practice 3-2 (two worksheets)
- Reteaching 3-2
- Alternative Activity 3-2

Classroom Manager 3-2
Glossary, Spanish Resources
Two-Year Algebra Handbook 3-2

Transparencies
10, 17, 51

3-2 Modeling and Solving Two-Step Equations

What You'll Learn
- Solving two-step equations
- Using two-step equations to solve real-world problems

...And Why
To solve problems involving money

What You'll Need
- tiles

QUICK REVIEW
▫ represents 1.
▪ represents −1.
▫ ▪ is a zero pair.

WORK TOGETHER

Work in pairs. To model equations, use the ▮ tile to represent *x*.

$2x + 1 = -5$

3x − 2 is the same as 3x + (−2).

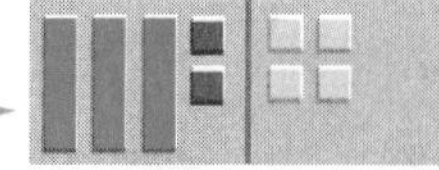

$3x - 2 = 4$

1. Write an equation for each model.
   a. 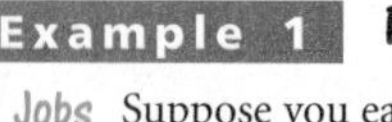 $2x + 3 = -4$
   b. 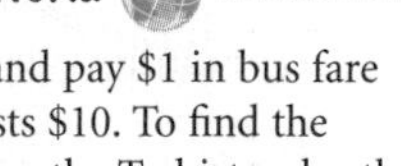 $3x - 1 = -7$
2. Use tiles to create a model for each equation. **See margin.**
   a. $2x = 8$ b. $x + 5 = -7$ c. $4x + 3 = -5$

THINK AND DISCUSS

Part 1 **Using Tiles**

A **two-step equation** is an equation that has two operations. You can use tiles to model and solve a two-step equation.

Example 1 **Relating to the Real World**

Jobs Suppose you earn $4/h baby-sitting and pay $1 in bus fare each way. You want to buy a T-shirt that costs $10. To find the number of hours you must work to buy the T-shirt, solve the equation $4x - 2 = 10$.

Model the equation with tiles.
$4x - 2 = 10$

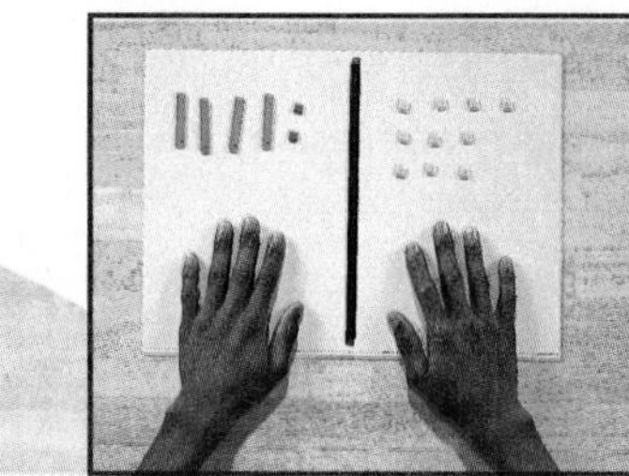

THINK AND DISCUSS

Example 1 **Relating to the Real World** 

ERROR ALERT! Students may not be sure whether to add first or divide first. **Remediation:** Remind students that the solution to an equation does not change if you perform the same operation on both sides of the equation. Show students that if you divide both sides by 4 first, you get $\frac{4x}{4} - \frac{2}{4} = \frac{10}{4}$.

The equation is still correct. Show students that simplifying the equation produces the correct solution.

$$x - \frac{1}{2} = \frac{5}{2}$$

$$x - \frac{1}{2} + \frac{1}{2} = \frac{5}{2} + \frac{1}{2}$$

$$x = \frac{6}{2}$$

$$x = 3$$

Solve the equation again by adding first. Then ask students whether the equation is easier to solve if you divide first or if you add first and give reasons. Lead students to understand that the order of operations is not necessary in solving two-step equations, but there is an order that makes the process easier: add or subtract before you multiply or divide.

1 Add 2 to each side of the equation. $4x - 2 + 2 = 10 + 2$

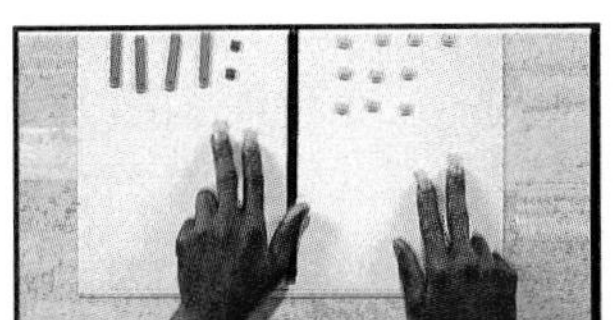

2 Simplify by removing zero pairs. $4x = 12$

3 Divide each side into four identical groups. $\frac{4x}{4} = \frac{12}{4}$

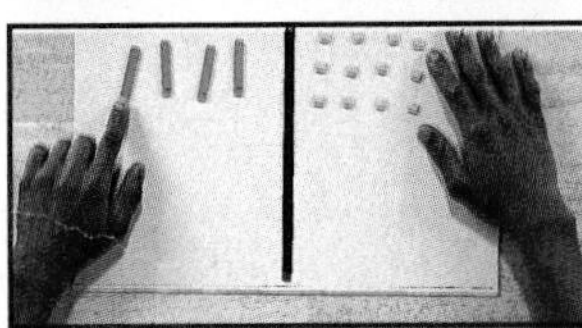

4 Solve for $1x$. $1x = 3$

$x = 3$

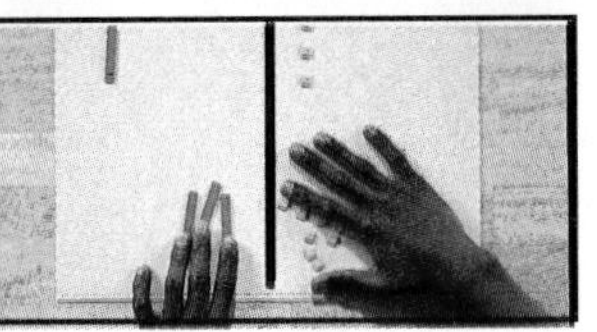

The solution is 3. You must baby-sit for 3 h to buy the T-shirt.

Check $4x - 2 = 10$

$4(3) - 2 \stackrel{?}{=} 10$ ← Replace x with 3.

$12 - 2 \stackrel{?}{=} 10$

$10 = 10$ ✓

3. Addition and division; they are inverse operations of subtraction and multiplication, respectively.

3. Which two operations were used to solve the equation? Why? **See left.**
4. **Try This** Use tiles to model and solve each equation.
   a. $2y - 3 = 7$ **5** b. $5 = 4m + 1$ **1** c. $3z - 2 = -8$ **−2**
   a–c. See margin for tile models.

5a. $-x + 7 - 7 = 12 - 7$
$-x = 5$
b. $(-1)(-x) = (-1)(5)$
$x = -5$
The solution is −5.

Part 2 Using Properties

To help you solve equations, you can write them in different ways. The variable x means $1x$. Similarly, $-x$ means $-1x$. Subtracting a variable is the same as adding its opposite. So, you can rewrite $4 - x$ as $4 + (-x)$.

5. It takes two operations to solve the equation $-x + 7 = 12$.
   a. Subtract 7 from each side of the equation. **See left.**
   b. Multiply each side of the equation by -1. What is the solution? **See left.**
6. Solve each equation.
   a. $-11 = -b + 6$ **17** b. $-9 - m = -2$ **−7** c. $15 = 3 - x$ **−12**

Additional Examples

FOR EXAMPLE 2

Solve the equation $10 - \frac{j}{6} = 5$.

$j = 30$

Discussion: *How would your answer change if $\frac{j}{6}$ was added to 10 rather than subtracted from 10? Explain the difference in the method used to solve the equation.*

FOR EXAMPLE 3

Frieda, a cab driver, rents her cab from Super Taxi for \$50 a day. If she makes an average of \$6 each time she picks up a passenger, how many passengers must she pick up to make a \$75 profit? Write an equation and solve.

$6x - 50 = 75$; $x \approx 20.8$; She must pick up 21 passengers.

Discussion: *If Frieda only picked up eight passengers, how much profit would she make? Explain the significance of your answer.*

Example 2

Point out to students that the negative sign can be eliminated during the multiplication or division step of solving an equation.

Example 3 Relating to the Real World 

Lead a discussion of whether students prefer buying individual ride tickets or paying a single, larger admission fee. Ask students who have been to amusement parks to share what they recall about the admission cost and the number of rides they could enjoy in one day. Compare the cost of buying individual ride tickets and the single admission fee.

MAKING CONNECTIONS The Ferris wheel was invented by an American engineer, G. W. Ferris, for the 1893 World's Columbian Exposition in Chicago. This first Ferris wheel was 250 ft high and could carry more than 2000 people. A year later, an even bigger Ferris wheel was built in London that was 328 ft high. Ferris wheels today are only about 45 ft high.

Exercises ON YOUR OWN

Point out that students can make their own algebra tiles by coloring pieces of paper. Students can also make sketches of tiles to help them solve equations.

ALTERNATIVE ASSESSMENT **Exercises 4–11** Have students work with a partner. Partner A uses tiles to set up Exercises 4–7; Partner B uses tiles to set up Exercises 8–11. After the setups are complete, ask partners to switch places and to

Technology Options

For Exercises 12–23 and 27–46, students may want to use graphing calculators to check their answers.

Prentice Hall Technology 

Software • Secondary Math Lab Toolkit • Computer Item Generator 3-2

CD-ROM Multimedia Algebra Lab 3

Internet • See the Prentice Hall site. (http://www.phschool.com)

116

Example 2

Solve the equation $1 = -\frac{k}{12} + 5$.

$$1 = -\frac{k}{12} + 5$$

$$1 - 5 = -\frac{k}{12} + 5 - 5 \quad \leftarrow \text{Subtract 5 from each side.}$$

$$-4 = -\frac{k}{12}$$

$$-12(-4) = -12\left(-\frac{k}{12}\right) \quad \leftarrow \text{Multiply each side by } -12.$$

$$48 = k$$

The solution is 48.

7. a. Which operations would you use to solve $-4n + 20 = 36$? Subtraction and division
   b. **Try This** Solve the equation and check your solution. -4

Example 3 Relating to the Real World 

Money Suppose you have \$18.75 to spend at Paradise Park. Admission is \$5.50. How many ride tickets can you buy?

Define	r = number of ride tickets you can buy				
Relate	entrance fee	plus	cost of ride tickets	equals	total cost
Write	5.5	+	$1.5r$	=	18.75

$$5.5 + 1.5r = 18.75$$

$$5.5 + 1.5r - 5.5 = 18.75 - 5.5 \quad \leftarrow \text{Subtract 5.5 from each side.}$$

$$1.5r = 13.25$$

$$\frac{1.5r}{1.5} = \frac{13.25}{1.5} \quad \leftarrow \text{Divide each side by 1.5.}$$

$$r \approx 8.8$$

You can buy 8 ride tickets.

PARADISE PARK GOOD FOR ONE RIDE RIDE TICKET \$1.50

CALCULATOR HINT

You can use a calculator to solve this equation. Write it in **calculator-ready form:** $x = (18.75 - 5.5) \div 1.5$. Use the sequence: (18.75 − 5.5) ÷ 1.5 ENTER.

8. Why should 8.8 be rounded to 8 in this situation? The number of tickets must be an integer and there is not enough money to buy 9 tickets.
9. If you buy 8 ride tickets, how much money do you have left? \$1.25

Exercises ON YOUR OWN

Write an equation for each model.

1.  $3x + 2 = 5$
2. 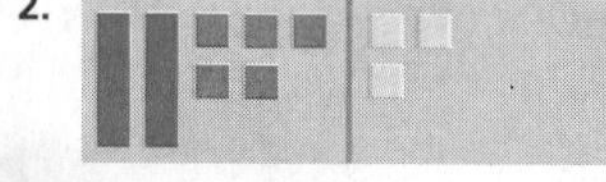 $2x - 5 = 3$
3. 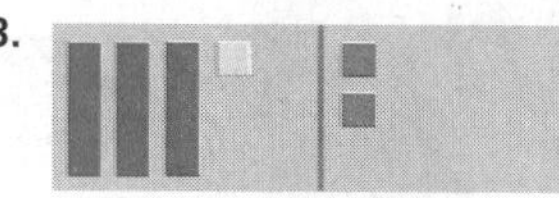 $3x + 1 = -2$

Use tiles to solve each equation.

4. $2n - 5 = 7$ 6
5. $3 + 4x = -1$ −1
6. $3b + 7 = -2$ −3
7. $-10 = -6 + 2y$ −2
8. $5y - 2 = -2$ 0
9. $0 = 3x - 3$ 1
10. $2z + 4 = -6$ −5
11. $4x + 9 = 1$ −2

solve their partner's equations. This activity will help assess whether students can set up equations using tiles. It will also assess their ability to solve equations with tiles.

ESTIMATION Exercise 47 Lead a discussion about how estimation can be used to evaluate whether a bill is reasonable. Ask a student to describe a situation in which a friend or family member found an error in a bill by comparing an estimate to the actual amount of the bill.

CONNECTING TO THE STUDENTS' WORLD Exercise 47 Have students research car-rental rates from some rental companies in your area. Invite students to write and solve equations based on the information they collect. Suggest that students estimate and then calculate the cost to rent a car to drive to some distant event.

ESL Exercise 48 Some students may be unfamiliar with the expressions *insurance policy* and *claim amount.* Ask: *Do you think an insurance company would pay a $3000 claim for a stolen computer if the insured held only a fire-damage insurance policy?* Lead a discussion of the meaning of these terms.

OPEN-ENDED Exercise 50 For students struggling to begin, suggest that they model a hobby or activity that involves counting.

CRITICAL THINKING Exercise 51 Suggest that students compare the results of multiplying each side of the equation by 10, by 100, and by 1000.

MENTAL MATH Exercise 60 Challenge students to devise an estimate rule for converting Fahrenheit to Celsius.

Solve each equation. Check your solutions.

12. $3x + 2 = 20$ 6
13. $-b + 5 = -16$ 21
14. $\frac{y}{2} + 5 = -12$ −34
15. $7 - 3k = -14$ 7
16. $1 + \frac{m}{4} = -1$ −8
17. $41 = 5 - 6h$ −6
18. $1.3n - 4 = 2.5$ 5
19. $\frac{x}{3} - 9 = 0$ 27
20. $-t - 4 = -3$ −1
21. $3.5 + 10m = 7.32$ 0.382
22. $7 = -2x + 7$ 0
23. $14 + \frac{h}{5} = 2$ −60

Use an equation to model and solve each problem.

24. **Farming** An orange grower ships oranges in boxes that weigh 2 kg. Each orange weighs 0.2 kg. The total weight of a box filled with oranges is 10 kg. How many oranges are packed in each box? $0.2n + 2 = 10$; 40 oranges

25. **Food Preparation** Suppose you are helping to prepare a large meal. You can peel 2 carrots/min. You need 60 peeled carrots. How long will it take you to finish if you have already peeled 18 carrots? $18 + 2t = 60$; 21 min

26. You can find the value of each variable in the matrices at the right by writing and solving equations. For example, solving the equation $2a + 1 = 11$, you get $a = 5$. Find the values of x, y, and k. −1; 6; −9

$$\begin{bmatrix} 2a+1 & -6 \\ -7 & -3k \end{bmatrix} = \begin{bmatrix} 11 & x-5 \\ 5-2y & 27 \end{bmatrix}$$

Choose **Use a calculator, paper and pencil, or mental math to solve each equation. Check your solutions.**

27. $-y + 7 = 13$ −6
28. $5 - b = 2$ 3
29. $10 = 2n + 1$ 4.5
30. $6 - 2p = 14$ −4
31. $3x - 15 = 33$ 16
32. $\frac{m}{3} - 9 = -21$ −36
33. $14 - \frac{y}{2} = -1$ 30
34. $\frac{x}{10} + 1.5 = 3.8$ 23
35. $-7 = 11 + 3b$ −6
36. $-6 + 6z = 0$ 1
37. $\frac{a}{5} + 15 = 30$ 75
38. $34 = 14 - 4p$ −5
39. $4 = 4 - \frac{a}{9}$ 0
40. $3x - 1 = 8$ 3
41. $-8 - c = 11$ −19
42. $\frac{m}{3} - 18 = 7$ 75
43. $3x - 2.1 = 4.5$ 2.2
44. $2 = -1 - \frac{k}{12}$ −36
45. $\frac{m}{2} - 1.002 = 0.93$ 3.864
46. $3 - 0.5c = 1.2$ 3.6

47. **Math in the Media** You rent a car for one day. Your total bill is $60.
    a. **Estimation** You estimate your mileage to be about 100 mi. Does the bill seem reasonable? Explain. See margin.
    b. Exactly how many miles did you drive? 125 mi

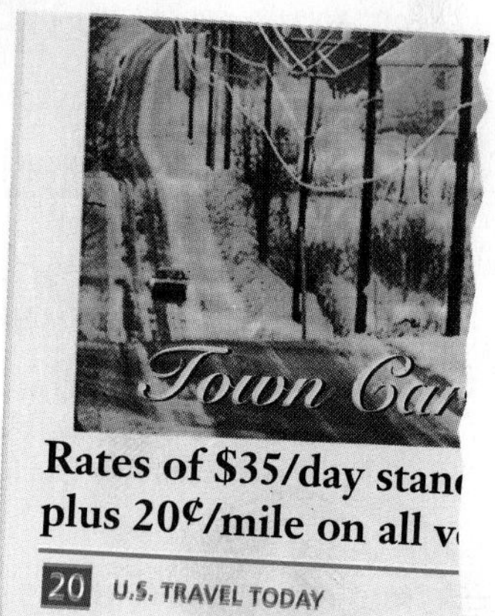

48. **Insurance** One insurance policy pays people for claims by multiplying the claim amount by 0.8 and then subtracting $500. If a person receives a check for $4650, how much was the claim amount? $6437.50

49. **Gardening** Tulip bulbs cost $.75 each plus $3.00 for shipping an entire order. You have $14.00. How many bulbs can you order? 14

50. **Open-ended** Write a problem that you can model with a two-step equation. Define the variable and show the solution to the problem. See margin.

51. **Critical Thinking** If you multiply each side of the equation $0.24r + 5.25 = -7.23$ by 100, the result is an equivalent equation. Explain why it might be helpful to do this. Multiplying by 100 changes all the numbers in the equation to integers and it might be easier to work with integers.

page 114 Work Together

2a.

b.

c.

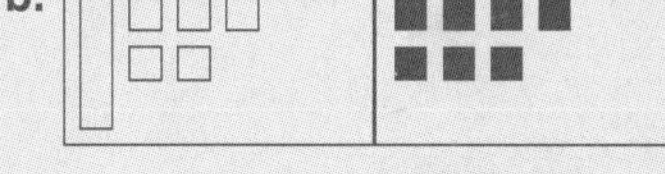

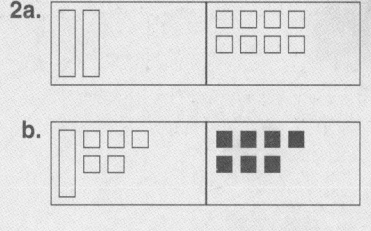

pages 114–116 Think and Discuss

4a.

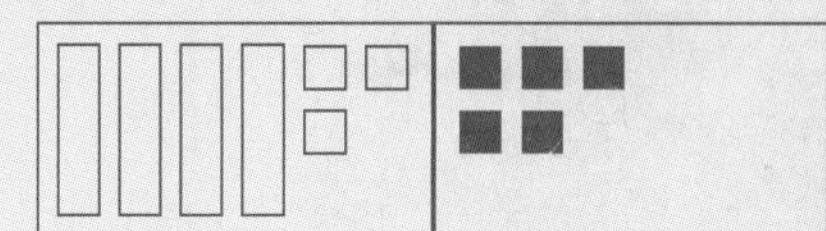

b.

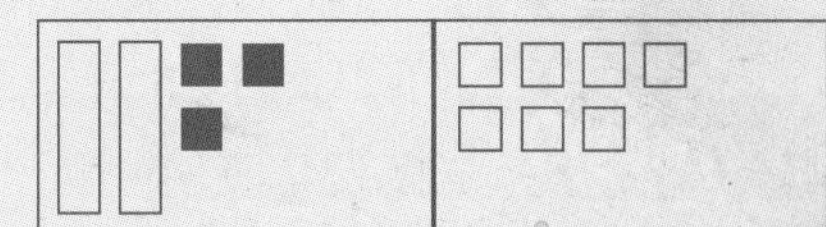

c.

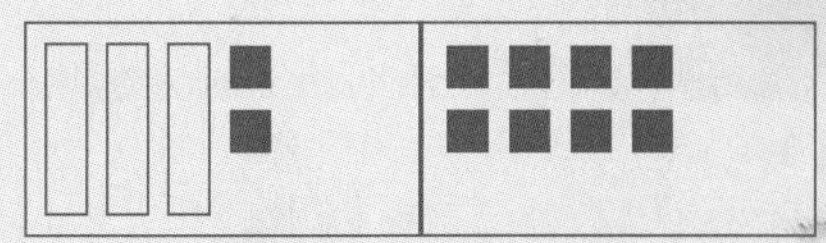

pages 116–118 On Your Own

47a, 50, 60a. See back of book.

WRITING Exercise 61 For students who are having difficulty, ask: *What items can you think of that come in sets of 15?*

Chapter Project **FIND OUT BY MODELING** Ask students to explain the example equation. Then, using the items they decided to save for, have them decide in how many weeks they wish to make their purchase. Ask students to write and to solve an equation to find out how much they need to save each week. Check students' progress on the project by having students describe what they have done for the project thus far.

Exercises MIXED REVIEW

Exercise 66 Prompt students to consider whether the solution represents the number of books in millions or just the number of books.

GETTING READY FOR LESSON 3-3 These exercises prepare students to combine and simplify like terms.

Wrap Up

THE BIG IDEA Ask: *How can a two-step equation be used to solve a real-world problem?*

RETEACHING ACTIVITY Students solve two-step equations. They fill in a chart by writing out each step needed to solve each equation and show each step mathematically. (Reteaching worksheet 3-2)

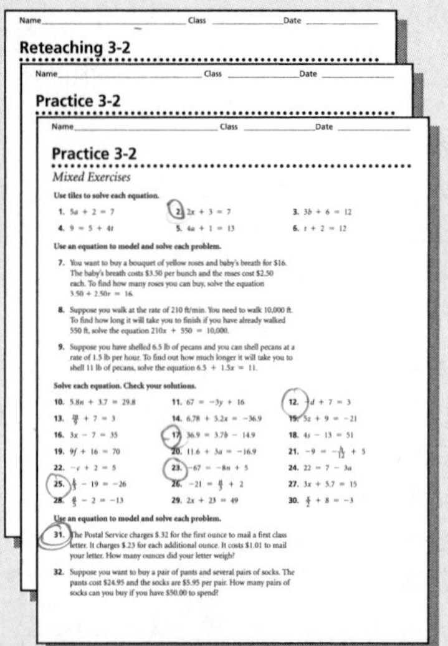
Reteaching 3-2

Practice 3-2

Practice 3-2

Mixed Exercises

Lesson Quiz

Lesson Quiz is also available in Transparencies.

In Exercises 1–3, solve each equation.

1. $2.7x + 3.4 = 19.6$ $x = 6$
2. $23 = 4c - 16$ $c = 9.75$
3. $8 + \frac{m}{2} = 6$ $m = -4$
4. A pizza delivery service promises to subtract \$.25 from the price of a pizza for each minute over the 20 minute delivery time limit they take getting the pizza to your house. You order a \$10 pizza and pay \$6 for it. Write and solve an equation showing how late the pizza was.
   $10 - 0.25x = 6$; $x = 16$

Calculator **Write each equation in calculator-ready form and solve.**

52. $1.2x + 0.6 = 32.4$ $x = (32.4 - 0.6) \div 1.2$; 26.5

53. $2.8 = 1.34 + \frac{r}{2}$ $r = (2.8 - 1.34) \cdot 2$; 2.92

54. $\frac{y}{4.5} - 9 = 3.5$ $y = (3.5 + 9) \cdot 4.5$; 56.25

55. $0.05n - 0.5 = 8$ $n = (8 + 0.5) \div 0.05$; 170

56. $9.007 - b = 8.32$ $b = (8.32 - 9.007) \cdot (-1)$; 0.687

57. $5.3 - 0.8n = 7$ $n = (7 - 5.3) \div (-0.8)$; −2.125

58. $0.3z + 0.91 = -0.08$ $z = (-0.08 - 0.91) \div 0.3$; −3.3

59. $3 - \frac{t}{2.5} = -7.06$ $t = (-7.06 - 3)(-2.5)$; 25.15

60. *Temperature* The formula for converting a temperature from Celsius, C, to Fahrenheit, F, is $F = 1.8C + 32$.

Fahrenheit	Celsius	Description of Temperature
212°	■	boiling point of water
■	37°	human body temperature
68°	■	room temperature
■	7°	average January high in Baltimore
■	0°	freezing point of water
19°	■	average January low in Chicago

a. Copy and complete the table. Round to the nearest degree. See margin.

b. *Mental Math* To convert from Celsius to Fahrenheit, you can get an estimate by using this rule: multiply the Celsius temperature by 2, then add 30. Use this strategy to convert 4°C, 15°C, and 50°C. 38°F; 60°F; 130°F

61. *Writing* Describe a situation that you can model with the equation $185 - 15n = 110$. Explain what the variable represents.

Sample: The neighborhood club collected \$185 in donations to enter a basketball team in the city tournament. The team needs to save \$110 for entry fees. How many practice balls can the team afford if the balls cost \$15 each?

Chapter Project **Find Out by Modeling**

To make a successful budget, you need to think about savings.

- Geraldo has already saved \$40 and wants to buy a CD player for \$129 in about four months. To find how much he should save each week, he wrote $40 + 16x = 129$. Explain his equation.
- On page 113, you chose one item as the goal for your project. How much does it cost? When do you want to buy this item?
- Write and solve an equation to find how much you should save per week.

Exercises MIXED REVIEW

Solve each equation.

62. $4s = 18$ 4.5   **63.** $x - 3 = 9$ 12   **64.** $\frac{m}{5} = 3$ 15   **65.** $-7 = n + 2$ −9

66. In 1995, the Library of Congress had 110 million books and other items. It is projected to have about 117.2 million items in 1999. Write and solve an equation to find how many items the Library of Congress adds each year. $110 + (1999 - 1995)b = 117.2$; 1.8 million items

Getting Ready for Lesson 3-3

Simplify.

67. $8 - (-1)$ 9   **68.** $9 - 11$ −2   **69.** $-15 + (-5)$ −20   **70.** $5 + (-12)$ −7   **71.** $-6 - (-9)$ 3

3-3 Teaching Notes

PROBLEM OF THE DAY

The geese in Algonquin Park left to migrate south. On the first day, five geese left. On the second day, ten geese left. On the third day, 20 geese left. If this pattern continued, how many geese left the park in a ten-day period? **5,115 geese**

Problem of the Day is also available in Transparencies.

CONNECTING TO PRIOR KNOWLEDGE Ask a student who is familiar with trading cards to briefly describe them to the class. Ask the student to explain how a disorganized collection of cards could be arranged. Lead students to understand that combining like objects can simplify a disorganized group. Ask students for other examples of grouping items in order to simplify a collection.

WORK TOGETHER

Explain that when you remove the zero pairs, the value of x is unchanged.

Question 1e Make sure students understand why this expression cannot be simplified.

THINK AND DISCUSS

The word *coefficient* can be difficult to remember. Ask students to think of and share mnemonic devices that will help them

---

Connections Money . . . and more

3-3 Combining Like Terms to Solve Equations

What You'll Learn
- Combining like terms
- Solving equations by combining like terms

...And Why

To find the price of an item by using the total bill

What You'll Need
- tiles

QUICK REVIEW

represents zero.

WORK TOGETHER

Work with a partner. When you use tiles to model an expression, you can combine variable tiles and integer tiles to create a simpler expression. You can use the tile to represent $-x$.

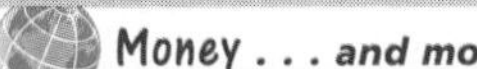

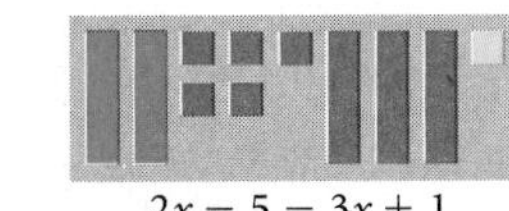

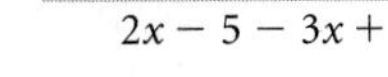

$2x - 5 - 3x + 1$

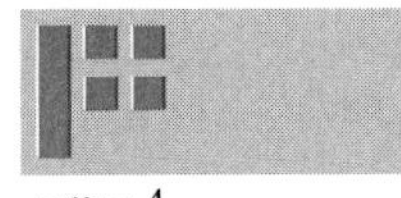

Remove zero pairs. → $-x - 4$

1. Model each expression. If possible, combine tiles and write a simpler expression. **See margin for models.**
   a. $4x - 2x$ **$2x$**
   b. $5 + 2x - 1$ **$2x + 4$**
   c. $3x - 6x + 4$ **$-3x + 4$**
   d. $8 + 3x - x - 6$ **$2x + 2$**
   e. $6 + 6x$ **$6x + 6$**
   f. $3x + 3x - x$ **$5x$**
   g. $2x - 1$ **$2x - 1$**
   h. $-5 - x + 4x - 4$ **$3x - 9$**
   i. $x + x + 1 + x$ **$3x + 1$**
2. Which expressions in Question 1 could not be simplified? Explain. **(e) and (g); each expression has only one term of each kind.**

THINK AND DISCUSS

Part 1

Combining Like Terms

QUICK REVIEW

A *term* is a number or a variable or the product or quotient of numbers and variables.

If a term has a variable, the numerical factor is called the **coefficient.** To identify the coefficients of each term in the expression $-x + 5y + 17$, you can write the expression as $-1x + 5y + 17$.

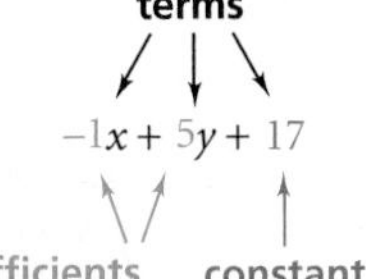

Terms are **like terms** if they have exactly the same variable factors. Terms like $5y$ and $2y^2$ are not like terms because their variable factors have different exponents. In the Work Together activity, you combined like terms using tiles. You can also combine like terms by adding coefficients.

Lesson Planning Options

Prerequisite Skills
- Modeling equations with tiles
- Adding integers

Assignment Options

To provide flexible scheduling, this lesson can be subdivided into parts.

1 **Core** 1, 3–13
Extension 2, 24, 25

2 **Core** 14–21, 26–39
Extension 22, 23, 40–49

Use Mixed Review to maintain skills.

Resources

Student Edition
Skills Handbook, p. 581, 582
Extra Practice, p. 558
Glossary/Study Guide

Teaching Resources
Chapter Support File, Ch. 3
- Practice 3-3 (two worksheets)
- Reteaching 3-3

Classroom Manager 3-3
Glossary, Spanish Resources
Two-Year Algebra Handbook 3-3

Transparencies
10, 17, 52, 55

remember the meaning of the word. Cooperative learning activities such as these provide opportunities to develop team skills students can use in the workplace.

Example 1

VISUAL LEARNING Write a long, disorganized expression containing several variables on the board. Sample: $12x + 5y - 2x + 3z - 7y + 4z + 2x + 2y + 4x$. Ask students to tell you how to group like terms to simplify the expression. Follow the students' suggestions for grouping the terms; then simplify the groups. This expression simplifies to $16x + 7z$. Make sure students understand that the y-terms cancel each other out.

AUDITORY LEARNING Question 4 Some students mistakenly combine unlike terms. Suggest that students assign a noun to each variable, such as *yam* for y and *mud pie* for m. Have students read the expressions aloud, substituting the nouns for each variable. Encourage students to make this activity more enjoyable by choosing nouns that are funny.

MAKING CONNECTIONS Invite students who are familiar with strategy games (Battleship, chess, Risk, and so on) to describe the aim of these games. Have them make a connection to how math skills such as combining like terms and removing zero pairs are used. For example, two opposing armies might have equal but opposite power. When the two armies meet, they cancel each other out and are removed from the game.

Example 2 **Relating to the Real World** 

Ask students whether this equation could be solved if the cost of a mystery book and the cost of a science-fiction book were

Additional Examples

FOR EXAMPLE 1

Simplify the following expression:
$4 - 2x + 3y + 7 + 2y - 5x$
$-7x + 5y + 11$

Discussion: *What would happen if you added –4y and 4y?*

FOR EXAMPLE 3

Three friends decide to order a box of egg rolls. Joe eats five egg rolls, Wanda eats two egg rolls, and Javier eats one egg roll. There is a $1 delivery charge for any order. If the total cost is $7, how much does each egg roll cost before the delivery charge is applied? Write an equation and solve.
$5s + 2s + s + 1 = 7; s = 0.75$; egg rolls cost 75¢ each.

Discussion: *What happens to the cost per egg roll if the friends decide to order more?*

Example 1

Simplify the expression $-2 + 6x + z - 2x + 8 - 4z$.

terms with x	terms with z	constant terms	← Group like terms.
$6x - 2x$	$1z - 4z$	$-2 + 8$	← Then add or subtract.
$4x$	$-3z$	6	

The simplified expression is $4x - 3z + 6$.

3. In Example 1, the term z was written as $1z$ before adding coefficients. Use tiles to verify that $z - 4z = -3z$. See margin.

4. **Try This** Combine like terms to simplify each expression.
   a. $9a + 10 - a + 3x - 5$ $8a + 3x + 5$
   b. $-4 + y - 9m - m + 2y$ $-10m + 3y - 4$

Part 2

Solving Equations

You can combine like terms to solve equations.

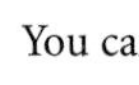

Example 2 **Relating to the Real World** 

Money Each used book at the library sale sold for the same amount. Suppose you bought 6 mysteries and 8 science-fiction books. Then you bought a poster for $2.40. You spent a total of $8.70. Solve the equation $6n + 8n + 2.40 = 8.70$ to find the cost of each book.

$$6n + 8n + 2.40 = 8.70$$
$$14n + 2.40 = 8.70 \quad \leftarrow \text{Combine like terms.}$$
$$14n + 2.40 - 2.40 = 8.70 - 2.40 \quad \leftarrow \text{Subtract 2.40 from each side.}$$
$$14n = 6.30$$
$$\frac{14n}{14} = \frac{6.30}{14} \quad \leftarrow \text{Divide each side by 14.}$$
$$n = 0.45$$

Each book costs $.45.

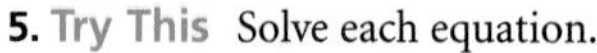
5. **Try This** Solve each equation.
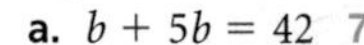
   a. $b + 5b = 42$ 7
   b. $3x - 4x + 6 = -2$ 8
   c. $7 = 4m - 2m + 1$ 3

different. Ask students what extra information would be needed to solve the problem.

MAKING CONNECTIONS Almost every day for two years, Steve Faloon, an undergraduate psychology student, spent an hour memorizing strings of digits. Most people, when asked to repeat a string of numbers they heard seconds before, can recall five to nine digits. Steve trained himself to recall up to 80 digits. He used a technique called *chunking*, which means grouping digits into meaningful groups. For example, a 10-digit number is easier to remember if you *chunk* it into a three-digit area code, a three-digit exchange code, and a four-digit number.

Example 3 **Relating to the Real World** 

Point out that you are buying only two concert tickets. The friend will pay for the express mail charge and the third ticket.

DIVERSITY Recognize that some students may think this is a lot to spend on entertainment. Remind these students that this is a hypothetical situation created for this problem.

ERROR ALERT! **Question 6** Students may try to subtract the value of the coupon from the total of $4.89. **Remediation:** Urge students to read the question carefully. Point out the phrase *After you use.* Ask students how this phrase affects the outcome.

Exercises ON YOUR OWN

ALTERNATIVE ASSESSMENT **Exercises 3–10** Have students draw lines connecting like terms. This will quickly assess their understanding of like terms.

Exercises 22–23 Students may misread these exercises. Urge them to check that their answers match the questions.

Example 3 **Relating to the Real World** 

What? The largest rock concert that has yet taken place was held on July 21, 1990, in Berlin, Germany, to celebrate German reunification.

Source: *The Guinness Book of Records*

Entertainment As a gift for your brother, you order two concert tickets. Two days before the concert a friend asks to join you. When you order the third ticket, there is a $12 charge for express delivery to be sure the ticket arrives on time. The total bill is $78. How much does each person owe?

Define c = charge for each of the original tickets
$c + 12$ = total charge for the late ticket

Relate: charge for the original tickets plus total charge for the late ticket is 78

Write: $2c + c + 12 = 78$

$$2c + c + 12 = 78$$
$$3c + 12 = 78 \quad \leftarrow \text{Combine like terms.}$$
$$3c + 12 - 12 = 78 - 12 \quad \leftarrow \text{Subtract 12 from both sides.}$$
$$3c = 66$$
$$\frac{3c}{3} = \frac{66}{3} \quad \leftarrow \text{Divide each side by 3.}$$
$$c = 22$$

You owe 2($22), or $44. Your friend owes $22 + $12, or $34.

Check $44 + $34 = $78 ✓

6. *Shopping* Suppose you buy 1.35 lb of apples and 3.55 lb of oranges. After you use a 50¢-off coupon, the total cost is $4.89. If apples and oranges sell for the same price, find their cost per pound. $1.10

Exercises ON YOUR OWN

1. Use tiles to simplify $-y + 5 + 3y - 4$. $2y + 1$; see margin for tiles.

2. *Critical Thinking* Suppose you want to solve the equation $-3m + 4 + 5m = -6$. As a first step, would you combine $-3m$ and $5m$ or subtract 4 from each side of the equation? Why? Answers may vary. Sample: Combine terms first so each side of the equation has the least possible number of terms before you change both sides.

Simplify each expression.

3. $2n - 3n$ $-n$
4. $28 - a + 4a$ $28 + 3a$
5. $-4 + 3b + 2 + 5b$ $8b - 2$
6. $x + 1.3 + 7x$ $8x + 1.3$
7. $-x + 7n - 3x + n$ $-4x + 8n$
8. $4 + 5z + z - 6\frac{1}{2}$ $6z - 2\frac{1}{2}$
9. $8 + x - 7x + m$ $-6x + m + 8$
10. $2k - 5b - b - k$ $k - 6b$

Write and solve each equation modeled by tiles.

11.  
$2x - x = 4$; 4

12.  
$2x - 1 + x = 5$; 2

13. 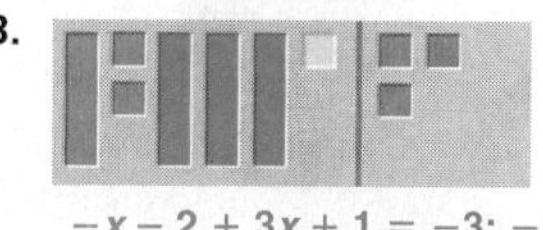
$-x - 2 + 3x + 1 = -3$; -1

Technology Options

For Exercises 14–21 and 26–39, students may want to use graphing calculators to check their answers.

Prentice Hall Technology 

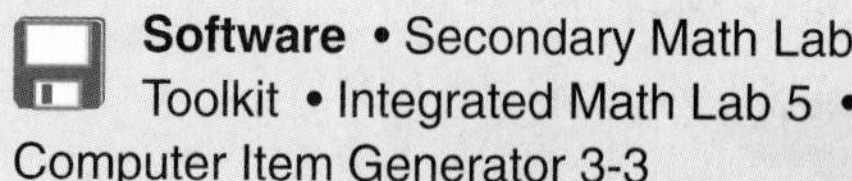
Software • Secondary Math Lab Toolkit • Integrated Math Lab 5 • Computer Item Generator 3-3

Internet • See the Prentice Hall site. (http://www.phschool.com)

OPEN-ENDED Exercise 25 Students may find this exercise easier if they begin with an expression with two terms and expand the expression to four terms.

Exercise 41 Suggest that students draw a diagram to help them understand the problem.

Exercise 44 Ask students:

- *Give an example of four consecutive even integers.* **Answers may vary. Sample: 4, 6, 8, 10**

Be sure students understand that a group of numbers such as 4, 6, 8, 10 are still considered consecutive even though the difference between each number is more than one.

GEOMETRY Exercise 45 Draw an equilateral triangle on the board similar to the triangle in this exercise. Write $x°$ in each angle. Ask students whether they can solve for x.

Chapter Project **FIND OUT BY ORGANIZING** Ask students to create a spreadsheet to help organize and analyze their budget plan. This task is essential to their work on the chapter project. Have students add this part of the project to work they have already completed. You may wish to check students' progress on the project by having them write a sentence or two describing what they have done for the project thus far and their plans for completion.

page 119 Work Together

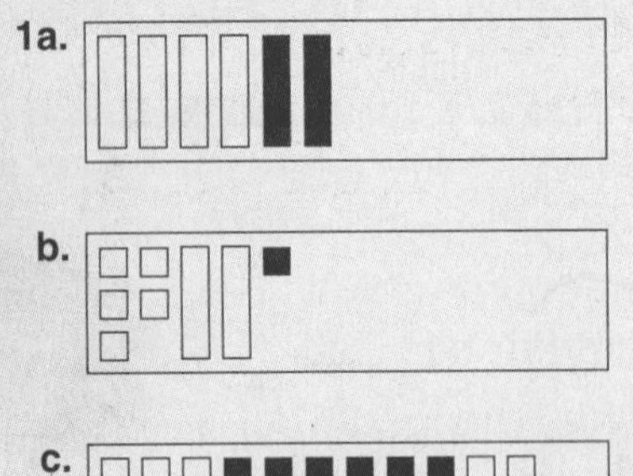

d–i. See back of book.

pages 119–121 Think and Discuss

3. See back of book.

pages 121–123 On Your Own

1. See back of book.

page 123 Mixed Review

53. See back of book.

Solve and check each equation.

14. $7x - 3x - 6 = 6$ **3**
15. $y + y + 2 = 18$ **8**
16. $-13 = 2b - b - 10$ **−3**
17. $5 - t - t = -1$ **3**
18. $9 = -3 + n + 2n$ **4**
19. $13 = 5 - 13 + 3x$ **7**
20. $1.6y + 3.2y = 96$ **20**
21. $0.5t - 3t + 5 = 0$ **2**

Use an equation to solve each problem.

22. *Geometry* A rectangle has perimeter 42 cm. The length is 3 cm greater than the width. Find the width and length of the rectangle. **9 cm; 12 cm**

23. *Entertainment* Movie tickets for an adult and three children cost \$20. An adult's ticket costs \$2 more than a child's ticket. Find the cost of an adult's ticket. **\$6.50**

24. *Writing* Three students simplify the expression $3p - 4p$. They get -1, $-p$, and $-1p$. State which answers are correct and explain why. **$-p$ and $-1p$; they are equivalent and both represent an accurate combination of like terms in this expression.**

25. *Open-ended* Write an expression with four terms that can be simplified to an expression with two terms. **Sample: $2a - 3k + 5a + k$**

Choose **Use a calculator, paper and pencil, or mental math to solve each equation. Check your answers.**

26. $4y - 2y = 18$ **9**
27. $-2b + 7b = 30$ **6**
28. $x + 3x - 7 = 29$ **9**
29. $6 - a + 4a = -6$ **−4**
30. $1 - 6t - 4t = 1$ **0**
31. $s + s - 6\frac{2}{3} = 4\frac{1}{3}$ **$5\frac{1}{2}$**
32. $72 + 4 - 14r = 36$ **$2\frac{6}{7}$**
33. $a + 6a - 9 = 30$ **$5\frac{4}{7}$**
34. $2 + 3n - n = -7$ **−4.5**
35. $4x + 3.6 + x = 1.2$ **−0.48**
36. $0 = -7n + 4 - 5n$ **$\frac{1}{3}$**
37. $3m + 4.5m = 18$ **2.4**
38. $x + x + 1 + x + 2 = -3.75$ **−2.25**
39. $x + x + 2 + 3x - 6 = 31$ **7**

Use an equation to solve each problem.

40. Suppose you are fencing a rectangular puppy kennel with 25 ft of fence. The side of the kennel next to your house does not need a fence. This side is 9 ft long. Find the dimensions of the kennel. **8 ft by 9 ft**

41. *Art* An art gallery owner wants to frame a 30 in.-wide portrait of poet Maya Angelou. He wants the width of the framed poster to be $38\frac{1}{2}$ in. How wide should each section of the frame be? **$4\frac{1}{4}$ in.**

Solve each consecutive integer problem.

Sample Find three consecutive integers with sum 219.

Consecutive integers are integers that differ by one.
Let the integers equal n, $n + 1$, and $n + 2$. Then solve this equation. $n + n + 1 + n + 2 = 219$ **72, 73, 74**

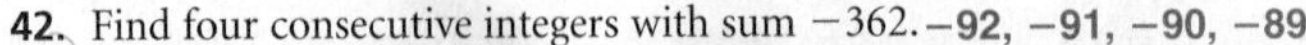

42. Find four consecutive integers with sum -362. **−92, −91, −90, −89**

43. Three friends were born in consecutive years. The sum of their birth years is 5961. Find the year in which each person was born. **1986, 1987, 1988**

44. The sum of four consecutive even integers is 308. Find the integers. **74, 76, 78, 80**

Exercises MIXED REVIEW

Exercise 52 Students may have difficulty getting started. Ask:

- *Are the x-values increasing or decreasing?* **increasing**
- *Are the y-values increasing or decreasing?* **increasing**
- *Are the x-values and y-values increasing by the same amount each time?* **yes**

GETTING READY FOR LESSON 3-4 These exercises provide practice in using the distributive property.

Wrap Up

THE BIG IDEA Ask: *How is the technique of combining like terms useful?*

RETEACHING ACTIVITY Students combine like terms by drawing circles, squares, and triangles around like terms. They group the like shapes and simplify the equations. (Reteaching worksheet 3-3)

Geometry **Use equations to find the value of each variable.**

45. 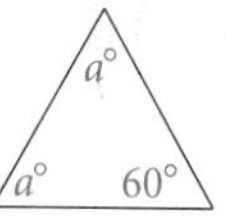

60

46. 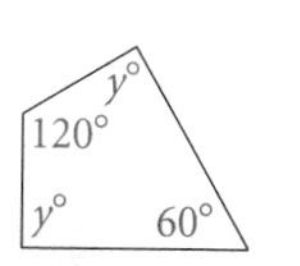

90

47. 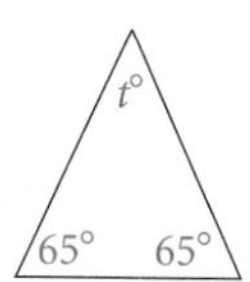

50

48. 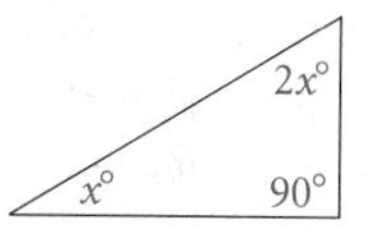

30

49. 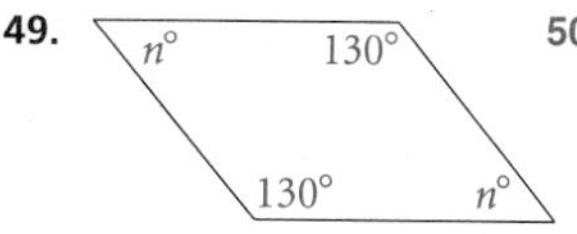

50

QUICK REVIEW

The sum of the measures of the angles of a triangle is 180°. The sum of the measures of the angles of a quadrilateral is 360°.

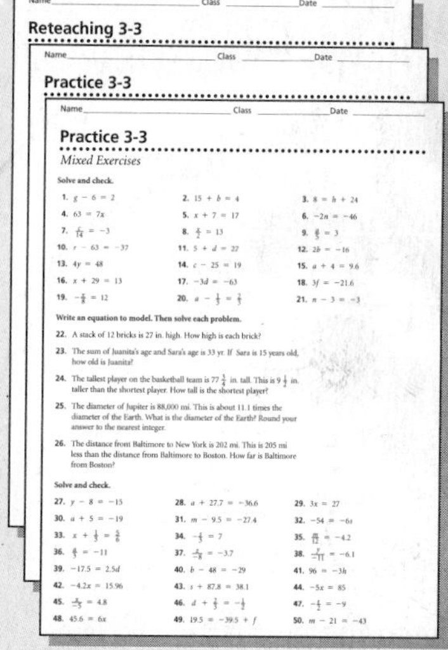
Reteaching 3-3
Practice 3-3
Practice 3-3
Mixed Exercises

Chapter Project ***Find Out by Organizing***

A spreadsheet can help you organize your information.

- Begin your budget by recording the amount of money you earn, the amount of money you save, and the amount of money you spend for two weeks.
- Analyze your expenses to plan how much you can spend each week, while still meeting your savings goal.
- Design a spreadsheet to show all the important categories in your budget plan. Include a column or row to show the total you will have saved after each week.
- Will you reach your savings goal when you hope to? Enter dollar amounts into your spreadsheet and verify that your budget works.

what I earn (chores, job)
what I save
what I spend each week (lunch, bus)
what I spend occasionally (movie, magazine)

Exercises MIXED REVIEW

Write a function rule for each table.

50.

x	y
3	7.5
4	10.0
5	12.5

$y = 2.5x$

51.

x	y
2	−4
4	−8
6	−12

$y = -2x$

52.

x	y
27	−3
28	−2
29	−1

$y = x - 30$

53. Sketch a graph that shows your speed as you traveled to school today. Explain the activity in each section of the graph. **See margin.**

Getting Ready for Lesson 3-4

Simplify.

54. $2(3 + 7)$ **20** **55.** $3(2 - 5)$ **−9** **56.** $-2(4 + 8)$ **−24** **57.** $-4(21 - 9)$ **−48**

Lesson Quiz

Lesson Quiz is also available in Transparencies.

1. Simplify the expression.
$3b + 4c - 8 + 6b - 4c + 2$
$9b - 6$

In Exercises 2 and 3, solve each equation.

2. $-7 = 5a + 3 - 2a$
$a \approx -3.33$

3. $0 = 4x + 7 + 2x - 3$
$x \approx -0.67$

4. Deborah is two years older than Chiu. The sum of their ages is 30. Write and solve an equation to find Chiu's age.
$(c + 2) + c = 30;\ c = 14$

3-4 Teaching Notes

PROBLEM OF THE DAY

Penny is 39 years old. She has lived in Texas for 16 years. How old will Penny be when she has lived in Texas for half her life? 46 years old

Problem of the Day is also available in Transparencies.

CONNECTING TO PRIOR KNOWLEDGE Have students brainstorm situations in which the word "distribute" has been used. Answers may vary. Sample answers: distribute papers to a class, distribute fliers to a neighborhood Ask students to make conjectures about what it might mean in math to use the distributive property.

WORK TOGETHER

Write $6\left(\frac{1}{2} - \frac{1}{3}\right)$ on the board. Ask: *What would be an easy way to simplify this expression?* Help students realize that it is easier to use the distributive property to simplify the expression to $3 - 2$ than to calculate $\frac{1}{2} - \frac{1}{3}$ and then multiply by 6.

Question 1b Ask students: *What does 50(84) represent?* area of a high school basketball court *What does 50(10) represent?* area of the difference in size between a high school basketball court and a college basketball court

Lesson Planning Options

Prerequisite Skills

- Multiplying integers
- Modeling equations with tiles

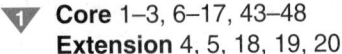

Assignment Options

To provide flexible scheduling, this lesson can be subdivided into parts.

1 **Core** 1–3, 6–17, 43–48
Extension 4, 5, 18, 19, 20

2 **Core** 21–36, 42
Extension 37–41

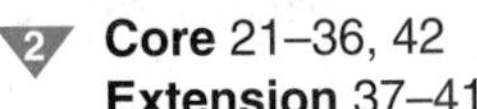

Use Mixed Review to maintain skills.

Resources

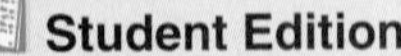

Student Edition
Skills Handbook, p. 582
Extra Practice, p. 558
Glossary/Study Guide

Teaching Resources
Chapter Support File, Ch. 3
- Practice 3-4 (two worksheets)
- Reteaching 3-4
- Checkpoint

Classroom Manager 3-4
Glossary, Spanish Resources
Two-Year Algebra Handbook 3-4

Transparencies
10, 14, 17, 52

Connections Sports . . . and more

What You'll Learn
- Using the distributive property
- Solving equations that involve the distributive property

...And Why
To solve problems involving electricity

3-4 Using the Distributive Property

WORK TOGETHER

Sports A high school basketball court is 84 ft long by 50 ft wide. A college basketball court is 10 ft longer than a high school basketball court.

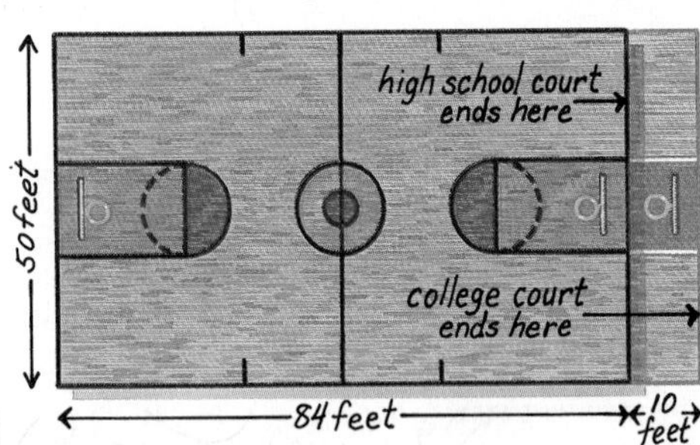

1. Work with a partner to evaluate each expression.
   a. $50(84 + 10)$ b. $50(84) + 50(10)$
2. Explain how each of the expressions in Question 1 represents the area of a college basketball court.

THINK AND DISCUSS

Simplifying Variable Expressions (Part 1)

Using two methods to find area illustrates the *distributive property.*

Distributive Property

For all real numbers a, b, and c:

$a(b + c) = ab + ac$ $(b + c)a = ba + ca$

$a(b - c) = ab - ac$ $(b - c)a = ba - ca$

Examples:

$5(20 + 6) = 5(20) + 5(6)$ $(20 + 6)5 = 20(5) + 6(5)$

$9(30 - 2) = 9(30) - 9(2)$ $(30 - 2)9 = 30(9) - 2(9)$

You can use the distributive property to simplify expressions.

$2(5x + 3) = 2(5x) + 2(3)$
$= 10x + 6$

$(3b - 2)\frac{1}{3} = 3b(\frac{1}{3}) - 2(\frac{1}{3})$
$= b - \frac{2}{3}$

THINK AND DISCUSS

ERROR ALERT! The negative signs in the expression $-(-2 - 9m)$ may confuse some students when they try to simplify this expression. **Remediation:** Have students distribute the first negative sign to the negative signs in the parentheses by drawing arrows as illustrated in the book. Remind the students that when multiplying two negatives, the answer is positive.

CRITICAL THINKING Question 3 Suggest that students describe other errors that students might make when simplifying this expression.

Question 4 Encourage students to move the fraction to the front of the expression if they find that form less confusing.

Example 1 **Relating to the Real World** 

AUDITORY LEARNING Have pairs of students take turns explaining the steps of this example to each other. Then have them work together on Question 5a. Examples of cooperative learning such as these help develop skills for working as part of a team.

Question 5b Encourage students to substitute the solution to Question 5a into the equation $3(2x - 1) = 5$ to see what happens when you try to combine like terms before using the distributive property. Make students aware that this is a common error students make.

To simplify an expression like $-(6x + 4)$, first rewrite it as $-1(6x + 4)$.

$$-(6x + 4) = -1(6x + 4) = -1(6x) + (-1)(4) = -6x - 4$$

$$-(-2 - 9m) = -1(-2 - 9m) = -1(-2) - (-1)(9m) = 2 + 9m$$

3. *Critical Thinking* A student rewrote $4(3x + 10)$ as $12x + 10$. Explain the student's error. The student did not use the distributive property correctly. 10 should also be multiplied by 4.
4. *Try This* Use the distributive property to simplify each expression.
   a. $-3(2x - 1)$ $-6x + 3$
   b. $-(7 - 5b)$ $-7 + 5b$
   c. $(3 - 8a)\frac{1}{4}$ $\frac{3}{4} - 2a$

Part 2
Solving and Modeling Equations

Example 1 **Relating to the Real World**

Shopping Posters of astronaut Sally Ride were on sale for \$3 off the regular price. Suppose you bought two posters and paid a total of \$8. Solve the equation $2(x - 3) = 8$ to find the regular price of each poster.

Model the equation with tiles.

You represent $2(x - 3)$ as two groups of tiles each containing $x - 3$.

$2(x - 3) = 8$

Rearrange tiles and use the distributive property.

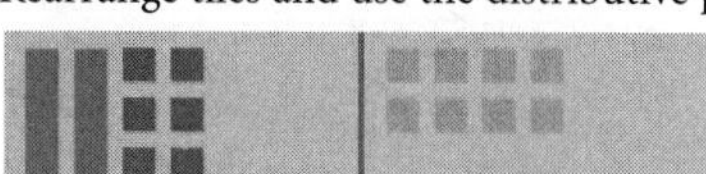

$2(x - 3) = 8$

$2x - 6 = 8$

Add six to each side.

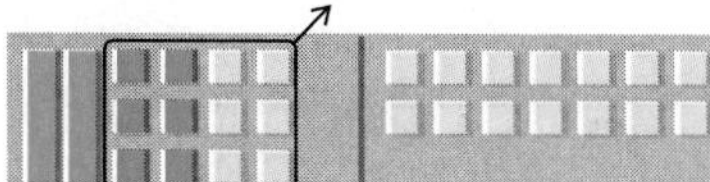

$2x - 6 + 6 = 8 + 6$

$2x = 14$

Divide each side into two identical groups.

$\frac{2x}{2} = \frac{14}{2}$

Solve for x.

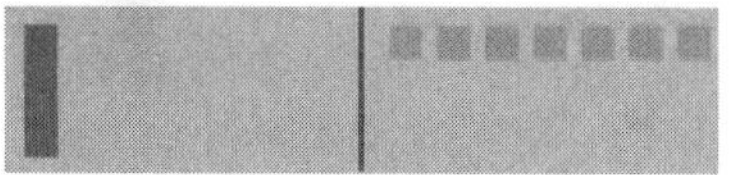

$x = 7$

Each poster regularly costs \$7.

Additional Examples

FOR EXAMPLE 1

Use algebra tiles to model and to solve the following equation.
$-2(y + 1) = -4$ $y = 1$

FOR EXAMPLE 2

Kwanda is driving down the highway and approaches a toll booth. She has five coins, including only dimes and nickels. If she has exactly 35¢, how many of the coins are dimes? Write an equation and solve.
$10d + 5(5 - d) = 35$; $d = 2$

Discussion: *What would your equation look like if you solved instead for the number of nickels?*

Example 2 **Relating to the Real World**

MAKING CONNECTIONS If you link several bulbs together in a series and connect them to a power supply, all the bulbs will light. If one bulb burns out, the circuit is broken, and all the other bulbs go out, too. Older strings of holiday lights are wired this way. The problem is solved by linking the bulbs together in parallel. At every bulb, there is an alternate route for the electricity to follow. If one bulb burns out, the other bulbs still light.

EXTENSION Challenge students to create and solve equations containing terms in parentheses that are within other terms in parentheses such as $4(x + 4(3x + 2)) = 10$. Stress that students must simplify the innermost term first and work outwards.

ESL To explain the meaning of *filament,* do the following activity. Ask students to draw a 60-watt light bulb. Ask: *Who remembered to draw the filament?* Choose a student who remembered, and ask him or her to draw the light bulb on the board.

Exercises ON YOUR OWN

ALTERNATIVE ASSESSMENT **Exercises 1–3** Group students in pairs. Have students write each equation and compare them with their partner's equations. If the equations differ, have students work together to determine the correct equation before solving it. This exercise will assess students' understanding of representing the distributive property with tiles.

WRITING Exercise 4 Encourage students to relate the meaning of the word *distribute* in math to the meaning of the word in other contexts.

Technology Options

For Exercises 21–36, students may want to use graphing calculators to check their answers.

Prentice Hall Technology

Software • Secondary Math Lab Toolkit • Computer Item Generator 3-4

Internet • See the Prentice Hall site. (http://www.phschool.com)

Who? Lewis Latimer (1848–1928) was an inventor who in 1882 received a patent for his method of making a light bulb filament. He was the author of the first book on electrical lighting.

Source: *Black Achievers in Science*

5. a. Use tiles to solve the equation $3(x - 1) + x = 5$. **2**
   b. In the equation $3(x - 1) + x = 5$, is it possible to combine like terms before you use the distributive property? Explain.
   No; you do not know what the coefficients are before you complete the multiplication.

You can use the distributive property to solve real-world problems.

Example 2 **Relating to the Real World** 

Electricity Several 6- and 12-volt batteries are wired so that the sum of their voltages produces a power supply of 84 volts. The total number of batteries is ten. How many of each type of battery are used?

Define $x =$ number of 6-volt batteries
$10 - x =$ number of 12-volt batteries

Relate	voltage from 6-volt batteries	plus	voltage from 12-volt batteries	equals	total voltage
Write	$6x$	$+$	$12(10 - x)$	$=$	84

$6x + 12(10 - x) = 84$

$6x + 120 - 12x = 84$ ← Use the distributive property.

$-6x + 120 = 84$ ← Combine like terms.

$-6x + 120 - 120 = 84 - 120$ ← Subtract 120 from each side.

$-6x = -36$

$\frac{-6x}{-6} = \frac{-36}{-6}$ ← Divide each side by -6.

$x = 6$

There are six 6-volt batteries.
$10 - x = 10 - 6 = 4$; there are four 12-volt batteries.

Check $6(6 \text{ volts}) + 4(12 \text{ volts}) \stackrel{?}{=} 84 \text{ volts}$
$36 \text{ volts} + 48 \text{ volts} = 84 \text{ volts}$ ✔

6. a. Suppose you let x represent the number of 12-volt batteries. Write and solve a new equation that models the problem in Example 2.
   b. Do you get a different answer to the problem? Explain.
   No; the equation has a different solution because the variable represents a different quantity, but the answer to the problem does not change.

6a. $12x + 6(10 - x) = 84$; 4

Exercises ON YOUR OWN

Write and solve each equation modeled by tiles.

1. 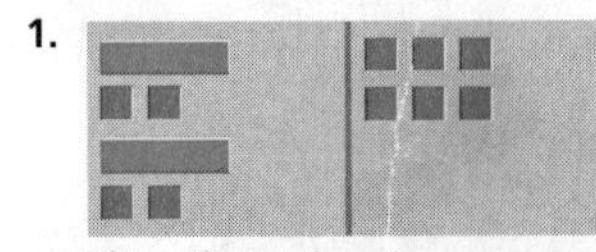
$2(t - 2) = -6$; -1

2. 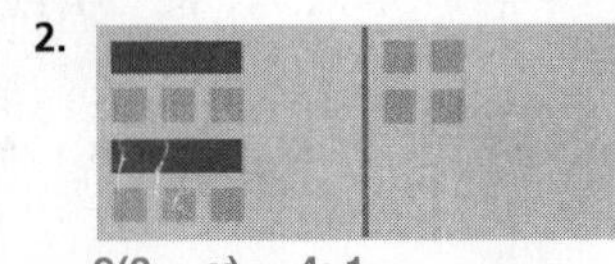
$2(3 - x) = 4$; 1

3. 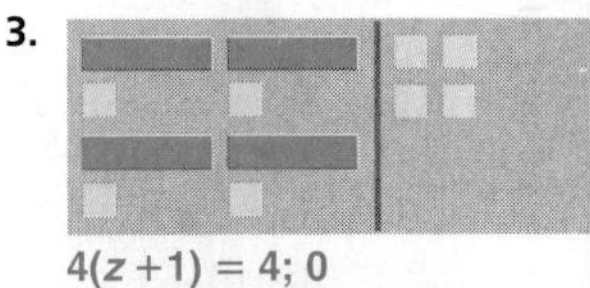
$4(z + 1) = 4$; 0

Exercise 18 Ask students: *If you didn't know that a pentagon was a five-sided figure, how could you find out from the question?* Solve the equation $180°(n - 2) = 540°$.

GEOMETRY Exercise 38 Make sure students don't confuse the subscripts with the exponents.

Exercise 39 For students struggling to start, suggest that they define a = number of aluminum bats bought. Lead students to understand that the number of wooden bats bought can be expressed as $(30 - a)$.

Exercise 40 Explain that equipment, such as an office copier, depreciates—loses value—over the years because of age, wear and tear, and so on.

STANDARDIZED TEST TIP Exercise 42 Standardized tests often contain questions that ask for the value of a variable multiplied or divided by a number, such as $\frac{y}{3}$ or $3y$. To find the new value, multiply or divide each term of the equation by this number. In this case, you would divide each term by 3.

MENTAL MATH Exercise 43 Lead students to understand that it is easier to subtract 4¢ from \$4 than to multiply 99¢ by 4.

CONNECTING TO THE STUDENTS' WORLD Many consumer items have prices that end in 9. Have students go to a store and note the price of one such item. Have them perform the mental math technique shown in Exercises 43–48 to calculate the cost of buying several of these items.

4. **Writing** In your own words, explain what the word *distribute* means. Sample: To *distribute* means "to spread" or "to share."
5. **Critical Thinking** Does $2ab = 2a \cdot 2b$? Explain your answer.
No; $2a \cdot 2b = 4ab$. You cannot use the distributive property because ab is a product, not a sum.

Simplify each expression.

6. $7(t - 4)$ $7t - 28$
7. $-2(n - 6)$ $-2n + 12$
8. $-(7x - 2)$ $-7x + 2$
9. $(5b - 4)\frac{2}{5}$ $2b - \frac{8}{5}$
10. $-2(x + 3)$ $-2x - 6$
11. $\frac{2}{3}(6y + 9)$ $4y + 6$
12. $(4 - z)(-1)$ $-4 + z$
13. $(3n - 7)(6)$ $18n - 42$
14. $-(2k + 5)$ $-2k - 5$
15. $-4.5(b - 3)$ $-4.5b + 13.5$
16. $\frac{2}{5}(5w + 10)$ $2w + 4$
17. $(9 - 4n)(-4)$ $-36 + 16n$

Geometry **If a polygon has n sides, the sum of the measures of its interior angles is $(n - 2)180°$. Use this for Exercises 18–20.**

18. The sum of the measures of the interior angles of a pentagon is 540°. What is the value of x in the figure to the right? 124°

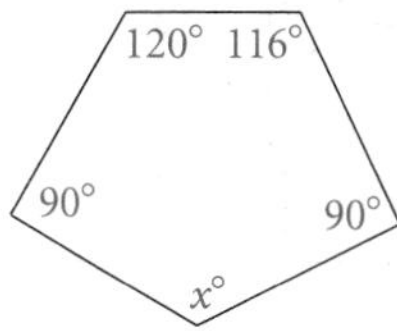

19. A polygon has seven sides. What is the sum of the measures of its interior angles? 900°
20. The sum of the measures of the interior angles of a polygon is 1440°. Use an equation to find the number of sides of the polygon.
$180(n - 2) = 1440$; 10

Solve and check each equation.

21. $2(8 + w) = 22$ 3
22. $m + 5(m - 1) = 11$ $2\frac{2}{3}$
23. $-(z + 5) = -14$ 9
24. $0.5(x - 12) = 4$ 20
25. $8y - (2y - 3) = 9$ 1
26. $\frac{1}{4}(m - 16) = 7$ 44
27. $15 = -3(x - 1) + 9$ -1
28. $\frac{3}{4}(8n - 4) = -2$ $\frac{1}{6}$
29. $5(a - 1) = 35$ 8
30. $6(x + 4) - 2x = -8$ -8
31. $-3(2t - 1) = 15$ -2
32. $-8 = -(3 + y)$ 5
33. $0 = \frac{1}{3}(6b + 9) + b$ -1
34. $2(1.5c + 4) = -1$ -3
35. $n - (3n + 4) = -6$ 1
36. $-\frac{1}{5}(10d - 5) = 9$ -4

In Exercises 37–40, use an equation to model and solve each problem.

37. **Geography** The shape of Colorado is nearly a rectangle. The length is 100 miles more than the width. The perimeter is about 1320 mi. Find the length and width of Colorado. 380 mi; 280 mi
38. **Geometry** The formula for the area of a trapezoid is $A = \frac{1}{2}h(b_1 + b_2)$. The area of $ABCD$ is 98 cm^2. Find the value of b_2. 11 cm

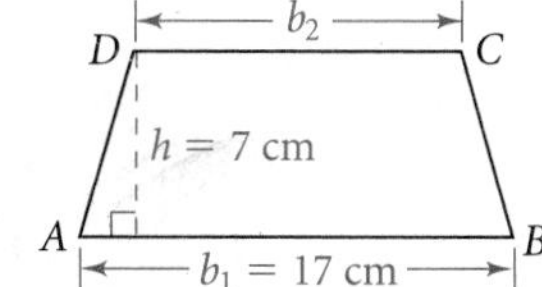

39. **Sports** A baseball team buys 15 bats for \$405. Aluminum bats cost \$25 and wooden bats cost \$30. How many of each type did they buy?
9 aluminum and 6 wooden bats
40. **Business** A company buys a copier for \$10,000. The Internal Revenue Service values the copier at $\$10{,}000(1 - \frac{n}{20})$ after n years. After how many years will the copier be valued at \$6500? 7 years
41. **Open-ended** Describe a situation where you would use the distributive property to solve a real-life problem. Sample: Your driveway is 9 ft long and 6 ft wide. You want to repave the driveway and make it 6 feet wider. Find the area of the new driveway so you can estimate the amount of materials needed.
42. **Standardized Test Prep** If $y = 3x - 10$, what is the value of $\frac{y}{3}$? C

A. $-x + 10$ B. $x + \frac{10}{3}$ C. $x - \frac{10}{3}$ D. $-x + \frac{10}{3}$ E. $x - 10$

Exercises MIXED REVIEW

Exercise 53c Have students calculate what each person spends on cranberry juice.

GETTING READY FOR LESSON 3-5 These exercises prepare students to solve equations with rational numbers.

Wrap Up

THE BIG IDEA Ask: *What is the distributive property and why is it useful?*

RETEACHING ACTIVITY Students will use the distributive property to solve equations. They will draw arrows to show distribution and show the steps to a solution. (Reteaching worksheet 3-4)

Exercises CHECKPOINT

In this Checkpoint, your students will have an opportunity to assess their own progress on the topics covered in Lessons 3-1 to 3-4.

Exercise 13 Explain to students that adding a border to a quilt is similar to adding a frame to a picture. Suggest that students make sketches.

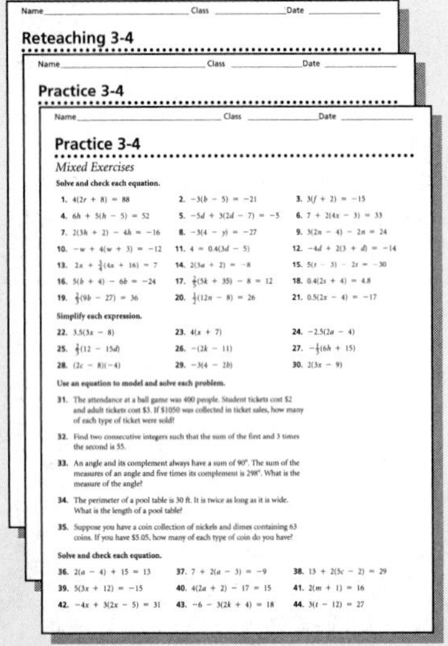
Reteaching 3-4

Practice 3-4

Practice 3-4

Mixed Exercises

Lesson Quiz

Lesson Quiz is also available in Transparencies.

1. Simplify the following expression. $(-3 + 2x)(-2)$ $6 - 4x$

In Exercises 2 and 3, solve each equation.

2. $3(4n - 2) = 7$ $n \approx 1.08$
3. $(4w + 6)\frac{3}{2} = 27$ $w = 3$
4. Kite rents five video tapes from Super Video at a total cost of $12. If new releases cost $3 a night, and older movies cost $2 a night, how many new releases did Kite rent? Write an equation and solve. $3x + 2(5 - x) = 12$; $x = 2$

Mental Math **You can use mental math and the distributive property to find prices quickly. Find each price.**

43. 4($.99) $3.96
44. 6($1.97) $11.82
45. 5($5.91) $29.55
46. 7($29.93) $209.51
47. 3 computer games at $32.99 each $98.97
48. 4 cans of fruit punch at $.69 each $2.76

$4(2.89) = 4(3.00 - 0.11)$
$= 4(3.00) - 4(0.11)$
$= 12 - 0.44$
$= 11.56$
$11.56

Exercises MIXED REVIEW

Solve and check.

49. $m - 4m = 2$ $-\frac{2}{3}$
50. $9 = -4y + 6y - 5$ 7
51. $2t - 8t + 1 = 43$ -7
52. $3.5 = 12s - 5s$ 0.5

53. *Technology* Use the spreadsheet at the right.
 a. Write a formula for cell C2 if there are 250 million people in the United States. B2/250
 b. Use your formula to find the missing values in the spreadsheet to one decimal place. 2.9; 0.9; 0.5
 c. Suppose a gallon of cranberry juice costs about $4. How much do people in the United States spend on cranberry juice in one year? $504 million

Annual U.S. Fruit Juice Sales

	A	B	C
1	juice	gal (millions)	gal/person
2	orange	734	■
3	apple	213	■
4	cranberry	126	■

Source: *Florida Dept. of Citrus*

Getting Ready for Lesson 3-5

Evaluate each expression.

54. $\frac{2}{3}n + 4$ for $n = -6$ 0
55. $\frac{k}{5} + \frac{2k}{7}$ for $k = -35$ -17
56. $\frac{3a + 1}{4}$ for $a = 7$ $5\frac{1}{2}$

Exercises CHECKPOINT

Solve and check.

1. $x - 7 = -6$ 1
2. $\frac{w}{3} = 11$ 33
3. $15 = 0.75v$ 20
4. $2t - 1 = 4$ $2\frac{1}{2}$
5. $\frac{b}{3} - 20 = 20$ 120
6. $-12 - 4x + 3 = -1$ -2
7. $10 = -5m - m - 2$ -2
8. $3n + n - 8 = 32$ 10
9. $-(z - 5) = -13\frac{1}{3}$ $18\frac{1}{3}$
10. $-0.8 - y = 1.9$ -2.7
11. $9(n + 7) = -81$ -16
12. $x + 2(3 - x) = 4$ 2
13. *Sewing* Suppose you are sewing a braid border on the edges of a quilt. The quilt is 20 in. longer than it is wide. You need 292 in. of braid to cover the edges of the quilt. What are the dimensions of the quilt? 63 in. by 83 in.
14. Which equation has the greatest solution? B
 A. $\frac{m}{4} = -12$
 B. $10 = 0.5(z + 3)$
 C. $4w - 5w + 9 = 8.4$
 D. $2c - 7 = 4$

3-5 Teaching Notes

PROBLEM OF THE DAY

A rectangular field is 120 yd long and 53 yd 1 ft wide. How much longer is the length of the field than the width? **66 yd 2 ft**

Problem of the Day is also available in Transparencies.

CONNECTING TO PRIOR KNOWLEDGE Write the equation $5(x + 3) = 2$ on the board. Ask students: *Do you think equations that solve real-world problems only contain integers? Do they ever contain rational numbers that are not integers?* **Answers may vary. Lead students to realize that rational numbers are common in real-world situations.**

Underneath the first equation on the board, write a second equation: $\frac{5}{4}\left(x + \frac{3}{4}\right) = 2$. Ask: *Do you think it is possible to solve this equation in the same way as the equation containing just integers?* **Answers may vary. Help students understand they can solve the equation in the same way.**

THINK AND DISCUSS

Example 1 Relating to the Real World 

ERROR ALERT! Students could easily make the mistake of multiplying $\frac{2}{3}$ times 200. **Remediation:** After students have read the question, ask: *Which has more calcium: the cheese or the milk?* **milk** *Will your answer be higher or lower than 200?* **higher**

Connections Nutrition . . . and more

3-5 Rational Numbers and Equations

What You'll Learn

- Solving equations involving rational numbers

...And Why

To solve nutrition and transportation problems

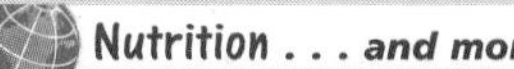

Part 1

Multiplying by a Reciprocal

In Chapter 1 you learned that a rational number can be represented as a ratio of two numbers. You can use reciprocals to solve equations involving rational numbers.

Example 1 Relating to the Real World 

Nutrition There are about 200 mg of calcium in 1 oz of cheddar cheese. How many milligrams of calcium are in 1 c of skim milk?

Define m = calcium (mg) in 1 c skim milk

Relate	two-thirds calcium in 1 c skim milk	equals	calcium in 1 oz cheddar cheese
Write	$\frac{2}{3}m$	$=$	200

$$\frac{2}{3}m = 200$$

$$\frac{3}{2}\left(\frac{2}{3}m\right) = \frac{3}{2}(200) \quad \leftarrow \text{Multiply each side by } \tfrac{3}{2}\text{, the reciprocal of } \tfrac{2}{3}.$$

$$m = 300$$

There are about 300 mg of calcium in 1 c of skim milk.

Check Two-thirds of the calcium in 1 c skim milk is 200.

$\frac{2}{3} \cdot 300 = 200$ ✔

MILK
The Calcium Winner
One ounce of cheddar cheese has only two-thirds the calcium of one cup of skim milk.

1. Solve the equation in Example 1 by multiplying each side by 3 and then dividing each side by 2. Which method do you prefer? Explain. **Check students' work.**

You can also use reciprocals to solve equations like $\frac{x}{5} = -7$. Write $\frac{x}{5}$ as $\frac{1}{5}x$ and multiply each side of the equation by 5, which is the reciprocal of $\frac{1}{5}$.

2. **Try This** Solve each equation.
   a. $\frac{2}{5}y = 1$ **$2\frac{1}{2}$** b. $-\frac{b}{8} = 2$ **−16** c. $-\frac{3}{4}x = 6$ **−8** d. $-2 = \frac{4c}{9}$ **$-4\frac{1}{2}$**

3. To solve $-\frac{1}{2}(3x - 5) = 7$, you can use the distributive property or you can multiply each side of the equation by −2, the reciprocal of $-\frac{1}{2}$. Explain why the second method is easier. Then solve the equation. **The second method involves fewer operations; −3.**

QUICK REVIEW

A *common denominator* of an equation is a multiple of all the denominators in the equation.

Part 2

Multiplying by a Common Denominator

To simplify an equation containing a fraction, you can multiply each side by the denominator of the fraction. The resulting equation is easier to solve.

Lesson Planning Options

Prerequisite Skills

- Performing operations with fractions
- Finding common denominators

Assignment Options

To provide flexible scheduling, this lesson can be subdivided into parts.

1. **Core** 1–8, 14, 15, 17, 19–27, 54
   Extension 10–13, 55
2. **Core** 16, 18, 28–53
   Extension 9, 56, 57

Use Mixed Review to maintain skills.

Resources

Student Edition

Skills Handbook, pp. 575–578
Extra Practice, p. 558
Glossary/Study Guide

Teaching Resources

Chapter Support File, Ch. 3
- Practice 3-5 (two worksheets)
- Reteaching 3-5

Classroom Manager 3-5
Glossary, Spanish Resources
Two-Year Algebra Handbook 3-5

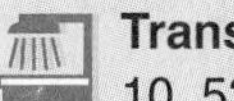

Transparencies
10, 53

Example 2 **Relating to the Real World** 

CONNECTING TO THE STUDENTS' WORLD Encourage students to repeat the example problem using their own grades.

KINESTHETIC LEARNING Write fractions on separate sheets of paper. Have enough for half the students in the class. Then, for each fraction written, write its reciprocal on a separate sheet for the other half of the class. Mix all the papers together and hand them out randomly. Challenge students to move around the room to find the partner holding their reciprocal fraction.

Example 3 **Relating to the Real World**

Be sure students don't confuse the meanings of $\frac{3}{4}$ empty and $\frac{3}{4}$ full. Have a student volunteer draw two cups on the board. Have the student shade the cups so that one appears $\frac{3}{4}$ empty and the other $\frac{3}{4}$ full.

DIVERSITY Many students may not know why it is important to monitor a car's gas consumption. Ask a student to explain the term *gas consumption* and how to calculate it. Have students brainstorm what might be wrong with a car if the rate of gas consumption begins to increase. **Answers may vary. Samples: tires under-inflated, leak in gas line, car needs tune-up, driver accelerates too hard and too often**

Additional Examples

FOR EXAMPLE 1

A neighborhood program rewards community service with free bowling. Each participant earns $\frac{3}{4}$ h of bowling for each hour of service performed. Write and solve an equation to determine how many hours of community service you must perform to receive six hours of bowling credit. $\frac{3}{4}h = 6;\ h = 8$

Discussion: *How could you solve this problem without using reciprocals?*

FOR EXAMPLE 2

Carl's semester grade in algebra is the average of his three 6-weeks' grades, and his final exam. If his 6-weeks' grades are 87, 93, and 91, what must he score on his final exam to make a 90 for the semester? Write an equation and solve.
$\frac{87 + 93 + 91 + x}{4} = 90;\ x = 89$

Discussion: *If you choose to exempt your final exam, your semester grade will be the average of your three 6-weeks' grades. What factors should you consider when making this decision?*

Score	Grade
90–100	A
80–89	B
70–79	C
60–69	D

CALCULATOR HINT

You can write the solution to this equation in calculator-ready form as $x = 3(90) - 92 - 75$.
To calculate, use the sequence
3 × 90 − 92 − 75 ENTER.

Example 2 **Relating to the Real World** 

School Your test scores are 92 and 75. Without extra credit, can you raise your test average to an A with your next test? Explain.

Define x = your next test score

Relate average of the scores equals lowest score for an A

Write $\frac{92 + 75 + x}{3} = 90$

$$\frac{92 + 75 + x}{3} = 90$$

$$3\left(\frac{92 + 75 + x}{3}\right) = 3(90) \quad \leftarrow \text{Multiply each side by 3.}$$

$$92 + 75 + x = 270$$

$$167 + x = 270 \quad \leftarrow \text{Simplify each side.}$$

$$167 + x - 167 = 270 - 167 \quad \leftarrow \text{Subtract 167 from each side.}$$

$$x = 103$$

Without extra credit the next test cannot bring your average to an A.

4. **Critical Thinking** How is multiplying by the denominator in Example 2 similar to multiplying by a reciprocal? **3 is the reciprocal of $\frac{1}{3}$.**

To solve an equation that has two or more fractions, multiply both sides of the equation by a common denominator.

Example 3 **Relating to the Real World** 

Cars You fill your car's gas tank when it is about $\frac{1}{2}$ empty. Later you fill the tank when it is about $\frac{3}{4}$ empty. You bought a total of $18\frac{1}{2}$ gal of gas on those two days. About how many gallons does the tank hold?

Define x = amount of gas (gal) the tank holds

Relate gallons from first fill plus gallons from second fill equals total bought

Write $\frac{1}{2} \cdot x + \frac{3}{4} \cdot x = 18\frac{1}{2}$

$$\frac{1}{2}x + \frac{3}{4}x = 18\frac{1}{2}$$

$$4\left(\frac{1}{2}x + \frac{3}{4}x\right) = 4\left(18\frac{1}{2}\right) \quad \leftarrow \text{Multiply each side by 4.}$$

$$4\left(\frac{1}{2}x\right) + 4\left(\frac{3}{4}x\right) = 74 \quad \leftarrow \text{Use the distributive property.}$$

$$2x + 3x = 74 \quad \leftarrow \text{Simplify each term.}$$

$$5x = 74 \quad \leftarrow \text{Combine like terms.}$$

$$\frac{5x}{5} = \frac{74}{5} \quad \leftarrow \text{Divide each side by 5.}$$

$$x = 14.8$$

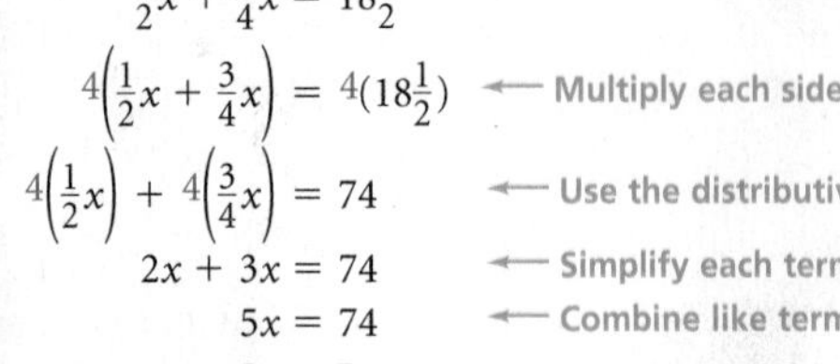

The gas tank holds about 14.8 gallons of gas.

OPEN-ENDED Question 6 Stress that students can use any common denominator to solve an equation. Suggest that students solve the equation in Part c. using a denominator of 6 and then using a denominator of 12. Compare the final answers to see that they are equal.

Exercises ON YOUR OWN

ALTERNATIVE ASSESSMENT Exercises 14, 15, 22, 24, and 27 Use these exercises to assess students' understanding of reciprocal algebra. Remind students that by using reciprocals they can solve algebraic equations more quickly.

MENTAL MATH Exercise 6 Remind students that any number multiplied by zero is zero.

CRITICAL THINKING Exercise 9 Ask students who are struggling to begin how $\frac{3}{8}x$ became $3x$.

Exercise 10 Ask a knowledgeable student to explain to the class what an overtime wage is.

Exercise 11 Challenge students to write their own equation before looking at the answer choices.

Exercise 12 Ask students: *Which is bigger: Kentucky or Wisconsin?* Wisconsin Urge students to keep this in mind when writing their equation and checking their answer.

Exercises 34–36 Explain that *d* represents the distance traveled in miles; *r* represents the rate of speed in mi/h; *t* represents the time in hours.

5. What other common denominator could have been used in Example 3? 8
6. *Open-ended* Name a common denominator that you would use to solve each equation. Samples given.
   a. $\frac{2}{3}x - \frac{5}{8}x = 26$ 24 b. $\frac{y}{8} + \frac{y}{12} = -4$ 48 c. $\frac{1}{2} = \frac{2}{3}b + \frac{1}{6}b$ 6

Exercises ON YOUR OWN

Mental Math **Solve each equation mentally.**

1. $\frac{3}{4}y = 9$ 12
2. $-\frac{2}{3}x = 6$ −9
3. $-4 = \frac{2}{5}a$ −10
4. $\frac{b}{10} = 5$ 50
5. $\frac{-7x}{8} = \frac{7}{8}$ −1
6. $\frac{3}{7}y = 0$ 0
7. $-\frac{n}{8} = 6$ −48
8. $\frac{2}{3}c = -18$ −27
9. *Critical Thinking* Explain the error in the student's work shown at the right. In using the distributive property, the student did not multiply 1 by 8.

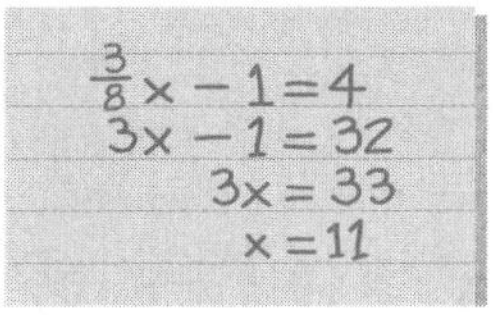

   $\frac{3}{8}x - 1 = 4$
   $3x - 1 = 32$
   $3x = 33$
   $x = 11$

10. *Jobs* Suppose you apply for a nurse's aide job that pays a \$12.90/h overtime wage. The overtime wage is $1\frac{1}{2}$ times the regular wage. What is the regular wage for the job? \$8.60/h
11. Suppose you buy $1\frac{1}{4}$ lb of roast beef for \$5. If p = price of the roast beef per pound, which equation models this situation? A
    A. $1\frac{1}{4}p = 5$ B. $p = 5 \cdot 1\frac{1}{4}$ C. $5p = 1\frac{1}{4}$ D. $p = 1\frac{1}{4} \div 5$
12. *Geography* The area of Kentucky is about 40,000 mi². This is about $\frac{5}{7}$ the area of Wisconsin. What is the area of Wisconsin? about 56,000 mi²
13. *Sewing* Suppose you buy $\frac{5}{8}$ of a yard of fabric for \$2.50. What is the price of the fabric per yard? \$4/yd

Choose **Use a calculator, paper and pencil, or mental math to solve each equation. Check your answers.**

14. $\frac{7}{8}x = 14$ 16
15. $-\frac{2}{9}y = 10$ −45
16. $\frac{x}{4} + \frac{3x}{5} = 17$ 20
17. $\frac{z}{3} = -1$ −3
18. $\frac{5a - 1}{8} = -5\frac{1}{4}$ $-8\frac{1}{5}$
19. $5 = -\frac{x}{6} + \frac{x}{2}$ 15
20. $5 = \frac{a}{6}$ 30
21. $\frac{y + 4}{3} = -1$ −7
22. $12 = \frac{7}{5}y$ $8\frac{4}{7}$
23. $5 = -\frac{1}{3}y + \frac{2}{7}$ $-14\frac{1}{7}$
24. $10 = -\frac{4}{3}d$ $-7\frac{1}{2}$
25. $\frac{-1 - 5x}{7} = 7$ −10
26. $\frac{2x - 1}{5} = 3$ 8
27. $-\frac{3x}{8} = -12$ 32
28. $\frac{6a + 4}{3} = -14$ $-7\frac{2}{3}$
29. $2y - \frac{3}{8}y = \frac{3}{4}$ $\frac{6}{13}$
30. $\frac{3}{4} = \frac{3 - b + 4b}{12}$ 2
31. $-\frac{n}{2} = 30$ −60
32. $\frac{2x}{3} + \frac{x}{2} = 7$ 6
33. $\frac{1}{3}x + \frac{1}{6}x = 27$ 54

Travel **Solve each problem using the formula $d = r \cdot t$.**

34. If you drive 65 mi/h for 3 h, how far do you drive? 195 mi
35. If you drive 200 mi in $3\frac{1}{4}$ h, how fast do you drive? about 61.5 mi/h
36. If you drive 210 mi at 60 mi/h, how long do you drive? $3\frac{1}{2}$ h

Technology Options

For Exercises 14–33 and 39–54, students may want to use graphing calculators to check their answers.

Prentice Hall Technology

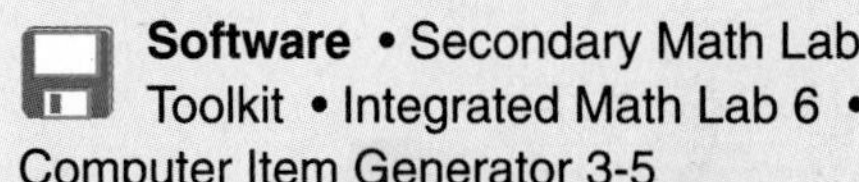

Software • Secondary Math Lab Toolkit • Integrated Math Lab 6 • Computer Item Generator 3-5

Internet • See the Prentice Hall site. (http://www.phschool.com)

pages 131–132 On Your Own

38. See back of book.

Exercise 56 Suggest that students begin by defining m to represent the monthly income. Lead them to create the equation $\frac{1}{3}m + \frac{1}{4}m = 1050$.

Exercises MIXED REVIEW

Exercise 66a Students may answer this question by repeatedly adding 8 to the starting time. Prompt them to answer the exercise in a different way by asking:

- *How long have the trains been running when you arrive at the station?* 44 min
- *What is the nearest number to 44 that is evenly divisible by 8?* 40 or 48
- *How long until the next train comes?* 4 min

GETTING READY FOR LESSON 3-6 These exercises give students practice with probability to prepare them for calculating the probability of dependent and independent events.

Wrap Up

THE BIG IDEA Ask: *When an equation contains rational numbers that are not integers, what do you do differently to solve that equation?*

RETEACHING ACTIVITY Students will solve equations involving rational numbers. They will use index cards to model rational numbers and their reciprocals. (Reteaching worksheet 3-5)

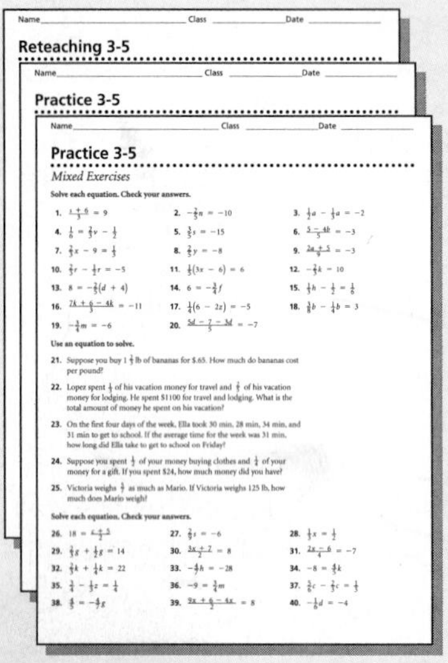
Reteaching 3-5

Practice 3-5

Mixed Exercises

Solve each equation. Check your answers.

Lesson Quiz

Lesson Quiz is also available in Transparencies.

In Exercises 1–3, solve and check each equation.

1. $-\frac{3}{5}z = 4$ $z = -6\frac{2}{3}$
2. $\frac{6a + a - 3}{2} = 9$ $a = 3$
3. $\frac{n}{4} + \frac{n}{3} = 7$ $n = 12$
4. Use the formula $d = r \cdot t$ to solve the following exercise. If you drive 90 miles in $2\frac{1}{2}$ hours, how fast do you drive? 36 miles per hour

37. **Work** What would the average number of hours per week have to be in the 1990s, to make the average for the three decades be 35 h/wk? See table.
38. **Writing** Explain how you can use common denominators to solve equations involving rational numbers. Give an example. See margin.

Average Work Week for Production Workers

Decade	Hours Per Week
1970s	36.4
1980s	34.9
1990s	■ 33.7

Choose **Use a calculator, paper and pencil, or mental math to solve each equation. Check your answers.**

39. $\frac{2x}{7} + \frac{x}{3} = -13$ -21
40. $-\frac{1}{5}(3x + 4) = 1$ -3
41. $\frac{2c - 1}{3} = 3$ 5
42. $1 = \frac{a}{7} - \frac{a}{3}$ $-5\frac{1}{4}$
43. $-\frac{8}{11}b = 24$ -33
44. $\frac{1}{2}(5d + 4) = 6$ $1\frac{3}{5}$
45. $-1 = \frac{2x - 1}{3}$ -1
46. $x - \frac{5x}{6} = -\frac{2}{3}$ -4
47. $\frac{3}{8}y + \frac{2}{3}y = \frac{5}{8}$ $\frac{3}{5}$
48. $\frac{n}{60} + \frac{n}{15} = -1$ -12
49. $\frac{7x}{3} = -21$ -9
50. $\frac{3t}{2} - \frac{3t}{4} = \frac{3}{2}$ 2
51. $\frac{3}{4}(2n - 5) = -1$ $1\frac{5}{6}$
52. $\frac{5c - 8}{18} = \frac{2}{3}$ 4
53. $\frac{b}{5} + \frac{5b}{3} = 2$ $1\frac{1}{14}$
54. $-\frac{3t}{8} = \frac{3}{2}$ -4
55. **Entertainment** As of 1994, the rental income for the movie *Back to the Future* was about \$105.5 million. This was about three fourths the rental income for *Home Alone.* Find the rental income for *Home Alone.* \$140 million
56. **Family Budget** A family allows $\frac{1}{3}$ of its monthly income for housing and $\frac{1}{4}$ of its monthly income for food. It budgets a total of \$1050 a month for housing and food. What is the family's monthly income? \$1800
57. **School** Suppose that on an average day, you spend $\frac{1}{5}$ of your homework time on math and $\frac{1}{2}$ of your homework time on literature. The time for these subjects totals $1\frac{3}{4}$ h. How much time do you spend on your homework? $2\frac{1}{2}$ h

Exercises MIXED REVIEW

Solve and check each equation.

58. $8h - 3 + 2h = 7$ 1
59. $-x + 6x = -35$ -7
60. $5k + 6 = -14$ -4
61. $m - 7m + 3 = 0$ $\frac{1}{2}$
62. $y + 15 + 2y = 0$ -5
63. $c - (5c - 1) = -47$ 12
64. $9(w - 1) = -27$ -2
65. $2j - 6j + 5 = 1$ 1
66. **Transportation** On Atlanta's rapid rail system, the MARTA, trains leave Airport Station every 8 min from 7:11 A.M. to 6:31 P.M.
    a. You get to Airport Station at 7:55 A.M. How long will you have to wait for a train? 4 min
    b. The ride to Doraville Station takes 41 min. When will you arrive? 8:40 A.M.

Getting Ready for Lesson 3-6

Probability **You have five quarters in your pocket. Their mint dates are 1987, 1991, 1989, 1994, and 1991. You pick one. Find each probability.**

67. P(mint date 1987) $\frac{1}{5}$
68. P(a quarter) 1
69. P(mint date 1991) $\frac{2}{5}$
70. P(a dime) 0

Math ToolboX

Throughout algebra, teachers should present students with opportunities to solve real-world problems. This toolbox adds logical reasoning and the use of a chart to the students' repertoire of problem-solving skills. Some students have difficulty translating words into equations, and a chart is a useful tool for organizing their work. The exercises also encourage students to read carefully. If time permits, encourage groups of students to write their own logic problems for other groups to solve.

ERROR ALERT! Students may not be able to see the additional eliminated possibilities when one part of the problem is solved. **Remediation:** Have students put Xs in the remaining rectangles of both the row and column that contain the check mark.

WRITING Exercise 4 Students may want to list the six possible solutions for the guess-and-test method. Check solutions by rereading the problem with that solution in mind.

ADDITIONAL PROBLEM John, Lucio, David, Cathy, and Marsha all go to Clinton High. Each has won an award for either volleyball, basketball, skiing, football, or karate. Football is an all male sport, while volleyball is an all female sport at Clinton High. David's and Marsha's sports do not require a ball but John's sport does. The karate student just won third place in her division. Lucio plays his sport indoors. Match each student with their after-school sport. John/football, Lucio/basketball, David/skiing, Cathy/volleyball, Marsha/karate

Math ToolboX — Problem Solving

Using Logical Reasoning

Before Lesson 3-6

You can use the strategy *Logical Reasoning* to solve problems. Organize what you know in a table. Use an X to eliminate an answer, and then draw conclusions.

Example

Space Exploration Astronauts Mae Carol Jemison, Ellen Ochoa, and Sidney M. Gutierrez flew separate space shuttle missions on the *Columbia*, the *Discovery*, and the *Endeavour*. Ellen Ochoa's mission occurred after the *Columbia's* mission. No crew member's initials were the same as the initial of his or her shuttle. On which shuttle did each person travel?

	M.C. Jemison	E. Ochoa	S.M. Gutierrez
Columbia	X	X	
Discovery			
Endeavour		X	

Ellen Ochoa's mission occurred after the *Columbia*'s, so she was not on the *Columbia*.
Eliminate possibilities with matching initials.

In both the first row and the second column, only one possibility remains. Put a check in each box.

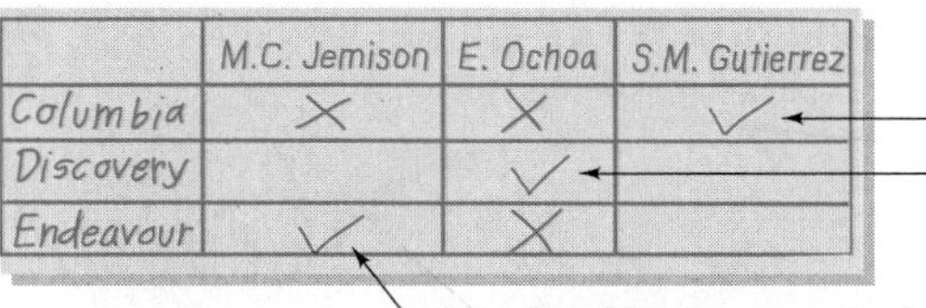

	M.C. Jemison	E. Ochoa	S.M. Gutierrez
Columbia	X	X	✓
Discovery		✓	
Endeavour	✓	X	

Sidney M. Gutierrez was on the *Columbia*.
Ellen Ochoa was on the *Discovery*.
Mae Carol Jemison was on the *Endeavour*.

Use the strategy *Logical Reasoning* to solve. Show your work.

1. The 1992 Olympic gold, silver, and bronze medals for soccer went to Ghana, Poland, and Spain, but not necessarily in that order. No country's medal began with the same letter as the country. If Ghana did not win the silver medal, what medal did each country win? Spain—gold; Poland—silver; Ghana—bronze
2. Latisha, Ken, and Ervin live on different floors of the same three-story building. The person on the third floor is the uncle of the person on the second floor. If Latisha lives on a higher floor than Ervin, where does each person live? Ervin lives on first floor, Latisha on second, Ken on third.
3. Coretta, David, Nando, and Helen live in Chicago, Dallas, New York, and Houston, but not necessarily in that order. No person's city begins with the same letter as his or her name. Coretta and Helen do not live in Texas. Where does each person live? Coretta: New York; David: Houston; Nando: Dallas; Helen: Chicago
4. **Writing** You could have solved the space shuttle problem by using the problem solving strategy *Guess and Test* instead of *Logical Reasoning*. Which method do you prefer? Explain your reasons. Sample: *Logical Reasoning* is more efficient since you do not make wrong guesses.

3-6 Teaching Notes

PROBLEM OF THE DAY

Jordan has a twenty-dollar bill, a ten-dollar bill, three one-dollar bills, two dimes, two nickels, and two pennies in his pocket. He wants his change for a $6.41 purchase to contain only a five-dollar bill and quarters. What is the least amount he should give the cashier? How much will his change be? **$12.16; $5.75**

Problem of the Day is also available in Transparencies.

CONNECTING TO PRIOR KNOWLEDGE Mediate a student discussion of the truthfulness of proverbs such as "Lightning never strikes twice in the same place."

WORK TOGETHER

Question 1 Urge students to make predictions before playing the game.

Question 4 Tell students that this is not a "fair" game. One player has a clear advantage. The point of the activity is not to win but to examine the advantage that the winner has.

THINK AND DISCUSS

Lead students to connect the term *independent event* to the discussion in the Connecting to Prior Knowledge activity.

TACTILE LEARNING Put students in groups of four or five. Have each group make one set of "paper tiles" to match those

Lesson Planning Options

Prerequisite Skills

- Finding the probability of single events

Assignment Options

To provide flexible scheduling, this lesson can be subdivided into parts.

1 **Core** 1–3, 5, 7–16, 30–37
Extension 4

2 **Core** 6, 17–22
Extension 23, 38

3 **Core** 26, 27, 29
Extension 24, 25, 28, 39

Use Mixed Review to maintain skills.

Resources

Student Edition

Skills Handbook, pp. 568, 578, 579
Extra Practice, p. 558
Glossary/Study Guide

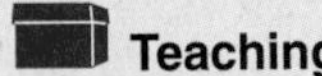

Teaching Resources

Chapter Support File, Ch. 3
- Practice 3-6 (two worksheets)
- Reteaching 3-6

Classroom Manager 3-6
Glossary, Spanish Resources
Two-Year Algebra Handbook 3-6

Transparencies
53

Connections Games . . . and more

What You'll Learn

- Finding the probability of independent and dependent events
- Using equations to solve probability problems

...And Why

To find probability in games of chance

What You'll Need

- colored cubes or chips
- bag

3-6 Using Probability

WORK TOGETHER

Work with a partner.
Play the game "Double Yellows." Put three yellow cubes and one red cube in a bag.

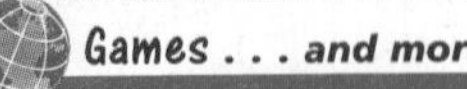

Partner X: Pick a cube. Put it back into the bag. Then pick again.

Partner Y: Pick a cube. Do not put it back into the bag. Then pick again.

Each partner takes turns choosing two cubes as described above. If a partner picks two yellow cubes, he or she scores a point.

1. Is the probability of choosing a yellow cube on the first pick the *same* or *different* for the two partners? Why? **Same; the sample spaces are the same.**
2. Is the probability of choosing a yellow cube on the second pick the *same* or *different* for the two partners? Why? **Different; X has 3 favorable outcomes out of 4 possible, Y has 2 favorable outcomes out of 3.**
3. **Predict** which partner is more likely to score points in "Double Yellows." Explain why. **X; they have the same chance of picking yellow on the first draw, but X has a better chance on the second draw.**
4. Play the game with your partner. Have each partner take ten turns. How do your results compare with the prediction you made in Question 1? **Check students' work.**

THINK AND DISCUSS

Part

1 Finding the Probability of Independent Events

In the game "Double Yellows" above, the two draws for Partner X are **independent events** because the result of the first draw does not affect the result of the second draw.

Probability of Two Independent Events

If A and B are independent events, you multiply the probabilities of the events to find the probability of both events occurring.

$P(A \text{ and } B) = P(A) \cdot P(B)$

in the text. Let students perform the activities in Examples 1 and 2. Have them compare their results with those in the text. To reinforce previously learned concepts, ask students which situation, the text problem or performing the activity, is theoretical probability and which is experimental probability. Have them explain the difference.

DIVERSITY Be sensitive to those students who feel negatively about probability because they associate it only with gambling. Stress that probability has many real-world applications such as statistics, scientific experiments, genetics, patient diagnosis, and the analysis of sports.

Example 1 **Relating to the Real World** 

VISUAL LEARNING Students may not understand why the two probabilities are multiplied rather than added. Draw a tree diagram on the board showing two consecutive independent events, J and K, each with a probability of $\frac{1}{2}$. Show how $P(J \text{ and } K) = \frac{1}{4}$ relates to $P(J) \cdot P(K)$.

ERROR ALERT! Some students may interpret $P(A \text{ and } E)$ as the probability that either event A or event E occurs. **Remediation:** Make sure students understand that $P(A \text{ and } E)$ means that event A happens and then event E happens.

Example 2 **Relating to the Real World** 

Have students compare the results of Examples 1 and 2. Ask them to explain why the results differ.

CONNECTING TO THE STUDENTS' WORLD Brainstorm ways people confuse dependent and independent events. Some examples to spark discussion are: bad luck runs in threes; if

Example 1 **Relating to the Real World** 

Games In a word game, you choose a tile from a bag containing the letter tiles shown. You *replace* the first tile in the bag and then choose again. What is the probability of choosing an A and then choosing an E?

Since you replace the first tile, the events are independent.

$P(A) = \frac{4}{15}$ ← There are 4 A's in the 15 tiles.

$P(E) = \frac{3}{15}$ ← There are 3 E's in the 15 tiles.

$$P(A \text{ and } E) = P(A) \cdot P(E) = \frac{4}{15} \cdot \frac{3}{15} = \frac{4}{75}$$

The probability of choosing an A, then an E is $\frac{4}{75}$.

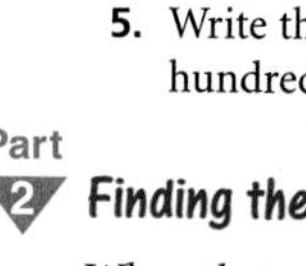

5. Write the answer to Example 1 as a decimal rounded to the nearest hundredth. Then write the decimal as a percent. 0.05; 5%

Part 2

Finding the Probability of Dependent Events

When the outcome of one event affects the outcome of a second event, the events are **dependent events.** In the game "Double Yellows," if you choose a second cube without replacing the first cube, the events are dependent.

> **Probability of Two Dependent Events**
>
> If A and B are dependent events, then
> $P(A \text{ and } B) = P(A) \cdot P(B \text{ after } A)$.

Example 2 **Relating to the Real World** 

Games Suppose you play the word game again. This time you *do not replace* the first tile before you choose the second tile. What is the probability of choosing an A and then choosing an E?

The events are dependent.

$P(A) = \frac{4}{15}$ ← There are 4 A's in the 15 tiles.

$P(E \text{ after } A) = \frac{3}{14}$ ← There are 3 E's in the 14 tiles left.

$$P(A \text{ and } E) = P(A) \cdot P(E \text{ after } A) = \frac{4}{15} \cdot \frac{3}{14} = \frac{2}{35}$$

The probability of choosing an A and then an E is $\frac{2}{35}$.

Additional Examples

FOR EXAMPLE 1

Absphemy has three quarters and five dimes in her pocket. She takes out one coin, then places it back in her pocket. Then she draws a second coin. What is the probability of drawing a dime and then a quarter? $\frac{15}{64}$

Discussion: *What are two other ways of displaying your answer?*

FOR EXAMPLE 2

Absphemy again draws two coins, but this time she does not replace the first coin after drawing it. What is the probability of first drawing a dime and then drawing a quarter? $\frac{15}{56}$

Discussion: *How would you calculate the probability of Absphemy drawing all three quarters?*

FOR EXAMPLE 3

A bag contains red and white buttons. Suppose you choose two buttons without replacement. The probability of choosing two red buttons is $\frac{3}{10}$. If the probability of choosing a red button first is $\frac{3}{5}$, what is the probability of choosing a second red button? $\frac{1}{2}$

you roll a six with the number cube, you are more likely to roll some other number next time.

Example 3

Point out that Kevin is grabbing socks at random without looking. A color-blind person has no difficulty telling white socks from black. Lead students to understand that $\frac{1}{3}$ represents the probability of both events happening. The unknown probability is that of choosing the second white sock. Students are "working backward" to find the unknown probability.

CRITICAL THINKING Question 8b After students have deduced that 18 of the socks are white, refer them to the $\frac{1}{3}$ probability. Ask students whether they are surprised that, with so many white socks, Kevin gets a white pair only one third of the time. Explain that the results of probability questions are frequently surprising and counterintuitive. Because of this, people often disagree in their answers.

DIVERSITY If there are color-blind students in the class, ask them to describe and compare their perceptions of color. As an alternative, challenge curious students to research colorblindness. Invite them to describe to the class how cones and ganglions in the eye work together to produce the sensation of color.

EXTENSION Mediate a student debate on the following opinion: "A state lottery is just a tax paid by people who don't understand probability." The ability to negotiate calmly with people whose views differ from your own is an important skill in the job market.

Technology Options

For Exercises 11–22, students may want to use spreadsheet software to create the sample space to calculate each probability.

Prentice Hall Technology 

Software • Secondary Math Lab Toolkit • Computer Item Generator 3-6

CD-ROM Multimedia Algebra Lab 3

Internet • See the Prentice Hall site. (http://www.phschool.com)

6. Write the answer to Example 2 as a decimal rounded to the nearest hundredth. Then write the decimal as a percent. 0.06; 6%
7. **Try This** What is the probability of choosing two *U's* if you pick a tile, do not replace it, and pick again? $\frac{1}{105}$

Part 3

Finding Probability Using an Equation

You can write an equation to solve some probability problems. When you are given two probabilities, solve the equation for the missing probability.

Example 3

Kevin is colorblind, so he buys only white socks and black socks. The probability that he picks out a pair of white socks is $\frac{1}{3}$. The probability that he reaches in and grabs one white sock is $\frac{3}{5}$. What is the probability that he chooses the second white sock?

Define $n = P(\text{white after choosing white})$

Relate $P(\text{two white}) = P(\text{white}) \cdot P(\text{white after choosing white})$

Write $\frac{1}{3} = \frac{3}{5} \cdot n$

$$\frac{1}{3} = \frac{3}{5}n$$

$$\frac{5}{3}\left(\frac{1}{3}\right) = \frac{5}{3}\left(\frac{3}{5}n\right) \quad \leftarrow \text{Multiply each side by } \tfrac{5}{3}.$$

$$\frac{5}{9} = n$$

The probability that Kevin will get the second white sock after choosing a white sock is $\frac{5}{9}$.

What? Most colorblind people have trouble recognizing red, brown, and green. About 8% of men and 0.5% of women are colorblind.

Source: *World Book Encyclopedia*

8. The probability of Kevin choosing a white sock on his first try is $\frac{3}{5}$.
   a. What does this tell you about the number of white socks he has?
   b. **Critical Thinking** If Kevin has 30 socks altogether, how many of them are white? 18 socks

8a. Three fifths of his socks are white.

Exercises ON YOUR OWN

Are the two events *dependent* or *independent*? Explain.

1. Toss a coin twice. indep.
2. Pick a vowel at random. Then pick a different vowel. dep.
3. Pick a ball from a basket of both yellow and pink balls. Pick again. dep.
4. **Gardening** A gardener's flat contains two plants with pink flowers, two with purple flowers, and two with white flowers. The plants have not yet flowered. You choose two plants from the flat. What is the probability that they both will have pink flowers? $\frac{1}{15}$

Exercises ON YOUR OWN

ALTERNATIVE ASSESSMENT **Exercises 11–16** Ask students to answer each question both with and without replacement of the marble. Then ask them to explain the effect of replacement on the probability of choosing two marbles. This exercise will assess whether students understand the difference between independent and dependent events.

OPEN-ENDED Exercise 25 Have students discuss the use of probability in games. Encourage students to talk about their favorite board games. Ask them to explain the probability of winning and the independence or dependence of events in the game.

RESEARCH Exercise 28 Information on federally owned acres and total acres in your state may be found in an atlas or almanac. The chamber of commerce and the internet may also be helpful.

STANDARDIZED TEST TIP **Exercise 39** Note that if you do not replace an object after withdrawing it, the probability of drawing the same type of object always decreases. Therefore the probability of Column A must be greater.

You have three $1 bills, two $5 bills, and a $20 bill in your pocket. You choose two bills. Find each probability.

5. $P(\$5 \text{ and } \$1)$ with replacing $\frac{1}{6}$

6. $P(\$1 \text{ and } \$20)$ without replacing $\frac{1}{10}$

7. $P(\$20 \text{ and } \$1)$ with replacing $\frac{1}{12}$

8. $P(\$1 \text{ and } \$1)$ without replacing $\frac{1}{5}$

9. A and B are independent. $P(A \text{ and } B) = \frac{1}{4}$ and $P(B) = \frac{3}{5}$. Find $P(A)$. $\frac{5}{12}$

10. *Writing* Use your own words to explain the difference between independent and dependent events. Give an original example of each. See margin.

You pick two marbles from the bag at the right. You replace the first one before you choose the second one. Find each probability.

11. $P(\text{red and green})$ $\frac{3}{50}$ **12.** $P(\text{two greens})$ $\frac{1}{25}$

13. $P(\text{blue and red})$ $\frac{3}{20}$ **14.** $P(\text{two yellows})$ 0

15. $P(\text{blue and green})$ $\frac{1}{10}$ **16.** $P(\text{two reds})$ $\frac{9}{100}$

You pick two marbles from the bag at the right. You pick the second one without replacing the first one. Find each probability.

17. $P(\text{red and blue})$ $\frac{1}{6}$ **18.** $P(\text{two blues})$ $\frac{2}{9}$

19. $P(\text{blue and green})$ $\frac{1}{9}$ **20.** $P(\text{two reds})$ $\frac{1}{15}$

21. $P(\text{green and yellow})$ 0 **22.** $P(\text{two greens})$ $\frac{1}{45}$

23. *School* Suppose you guess all the answers on a three-question "true or false" quiz. What is the probability you will guess them all correctly? D

A. $\frac{1}{2}$ B. $\frac{3}{2}$ C. $\frac{3}{8}$ D. $\frac{1}{8}$ E. $\frac{1}{3}$

24. A bag contains red cubes and blue cubes. You pick two cubes without replacing the first one. The probability of drawing two red cubes is $\frac{1}{15}$. The probability that your second cube is red if your first cube is red is $\frac{2}{9}$. Find the probability that the first cube you pick is red. $\frac{3}{10}$

25. *Open-ended* Make up a game like "Double Yellows." Give rules for playing the game. Explain the probability of winning and the independence or dependence of the events in your game. Check students' work.

Government **Use the data at the right.**

26. An acre of land in Idaho is chosen at random. What is the probability that the land is owned by the federal government? 62%

27. An acre of land is chosen at random from each of the three states listed. What is the probability that all three acres of land are owned by the federal government? about 1.2%

28. *Research* How many acres of land in your state are federally owned? What percent of the total land in your state is this? Check students' work.

State	Percent of Land Federally Owned
Idaho	62%
New Mexico	32%
Virginia	6%

pages 136–138 On Your Own

10. Answers may vary. Sample: Two events are *independent* if the outcome of one is not affected by the outcome of the other. For example, when you roll a number cube twice, the number on the first roll has no influence on the number on the second roll.
Two events are *dependent* if the outcome of one may affect the outcome of the other. For example, suppose there are 12 marbles in a bag and 3 of them are green. You take out one marble and then another. The chances of getting a green marble on the second trial will depend on the outcome of the first trial.

Exercises MIXED REVIEW

GETTING READY FOR LESSON 3-7 These exercises give students practice converting numbers to percents to prepare them for solving equations involving percents.

JOURNAL Students may also wish to write about how their new knowledge of probability has changed the way they think about some aspect of their lives.

Wrap Up

THE BIG IDEA Ask: *How do you find the probability of dependent and independent events?*

RETEACHING ACTIVITY Students use counters to find the probability of independent and dependent events. They count the counters, then find the probability of choosing one, then another, with and without replacement. (Reteaching worksheet 3-6)

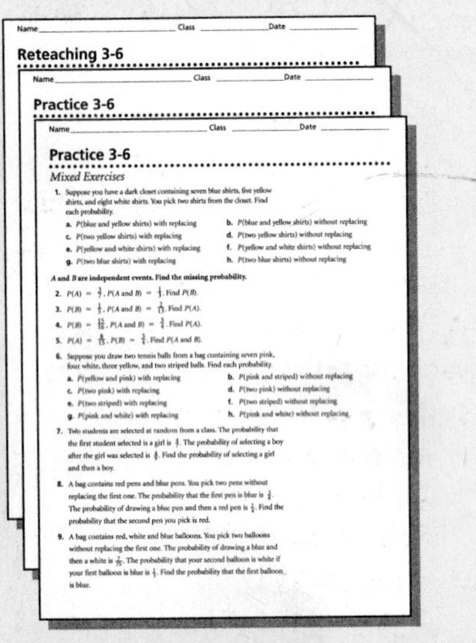

Name ______ Class ______ Date ______

Reteaching 3-6

Name ______ Class ______ Date ______

Practice 3-6

Name ______ Class ______ Date ______

Practice 3-6

Mixed Exercises

1. Suppose you have a dark closet containing seven blue shirts, five yellow shirts, and eight white shirts. You pick two shirts from the closet. Find each probability.
   a. P(blue and yellow shirts) with replacing
   b. P(blue and yellow shirts) without replacing
   c. P(two yellow shirts) with replacing
   d. P(two yellow shirts) without replacing
   e. P(yellow and white shirts) with replacing
   f. P(yellow and white shirts) without replacing
   g. P(two blue shirts) with replacing
   h. P(two blue shirts) without replacing

A and B are independent events. Find the missing probability.

2. P(A) = [illegible], P(A and B) = [illegible]. Find P(B).
3. P(B) = $\frac{1}{5}$, P(A and B) = $\frac{2}{15}$. Find P(A).
4. P(B) = $\frac{15}{16}$, P(A and B) = $\frac{3}{4}$. Find P(A).
5. P(A) = $\frac{8}{15}$, P(B) = $\frac{3}{4}$. Find P(A and B).
6. Suppose you draw two tennis balls from a bag containing seven pink, four white, three yellow, and two striped balls. Find each probability.
   a. P(yellow and pink) with replacing
   b. P(pink and striped) without replacing
   c. P(two pink) with replacing
   d. P(two pink) without replacing
   e. P(two striped) with replacing
   f. P(two striped) without replacing
   g. P(pink and white) with replacing
   h. P(pink and white) without replacing
7. Two students are selected at random from a class. The probability that the first student selected is a girl is [illegible]. The probability of selecting a boy after the girl was selected is [illegible]. Find the probability of selecting a girl and then a boy.
8. A bag contains red pens and blue pens. You pick two pens without replacing the first one. The probability that the first pen is blue is [illegible]. The probability of drawing a blue pen and then a red pen is $\frac{1}{4}$. Find the probability that the second pen you pick is red.
9. A bag contains red, white and blue balloons. You pick two balloons without replacing the first one. The probability of drawing a blue and then a white is [illegible]. The probability that your second balloon is white if your first balloon is blue is [illegible]. Find the probability that the first balloon is blue.

Lesson Quiz

Lesson Quiz is also available in Transparencies.

1. A bag contains two green and five blue marbles. What is the probability of drawing a green marble, replacing it, and then drawing another green marble? $\frac{4}{49}$
2. Using the same bag of marbles from Exercise 1, what is the probability of drawing two green marbles if the first one is not replaced? $\frac{1}{21}$
3. A and B are independent events. $P(A) = \frac{1}{3}$, $P(B) = \frac{2}{5}$, Find P(A and B) $\frac{2}{15}$

29. *Quality Control* The probability that a spark plug is defective is 0.06. You need two new spark plugs for a motorcycle. What is the probability that both spark plugs you buy are defective? 0.0036

A and B are independent events. Find the missing probability.

30. $P(A) = \frac{5}{8}$, $P(A \text{ and } B) = \frac{1}{4}$. Find $P(B)$. $\frac{2}{5}$
31. $P(A) = \frac{1}{3}$, $P(B) = \frac{3}{5}$. Find $P(A \text{ and } B)$. $\frac{1}{5}$
32. $P(A) = \frac{3}{20}$, $P(B) = \frac{5}{6}$. Find $P(A \text{ and } B)$. $\frac{1}{8}$
33. $P(A) = \frac{1}{4}$, $P(A \text{ and } B) = \frac{3}{20}$. Find $P(B)$. $\frac{3}{5}$
34. $P(A) = \frac{12}{35}$, $P(B) = \frac{7}{8}$. Find $P(A \text{ and } B)$. $\frac{3}{10}$
35. $P(A) = \frac{2}{3}$, $P(A \text{ and } B) = \frac{8}{15}$. Find $P(B)$. $\frac{4}{5}$
36. $P(A) = \frac{1}{20}$, $P(A \text{ and } B) = \frac{1}{50}$. Find $P(B)$. $\frac{2}{5}$
37. $P(A) = \frac{7}{9}$, $P(A \text{ and } B) = \frac{21}{36}$. Find $P(B)$. $\frac{3}{4}$

38. *School* A hat contains all the names of the students in a class that has 12 girls and 10 boys. To select representatives of the class, the teacher draws two names from the hat without replacing the first name.
    a. Find P(two girls). $\frac{2}{7}$
    b. Find P(two boys). $\frac{15}{77}$
    c. Find P(boy, then girl). $\frac{20}{77}$
    d. Find P(girl, then boy). $\frac{20}{77}$
    e. **Predict** the sum of the probabilities in parts (a)–(d). Check to see that the sum agrees with your prediction. 1

39. *Standardized Test Prep* You pick two balls from a bag containing balls of different colors. The bag contains at least two green balls. Compare the quantities in Column A and Column B. A

Column A	Column B
P(two greens) with replacement	P(two greens) without replacement

A. The quantity in Column A is greater.
B. The quantity in Column B is greater.
C. The quantities are equal.
D. The relationship cannot be determined from the given information.

Exercises MIXED REVIEW

Solve each equation.

40. $3.2(m + 5) = 16$ 0
41. $\frac{s}{5} - \frac{s}{3} = 8$ −60
42. $w - 13w = 3$ $-\frac{1}{4}$
43. $\frac{2c + 1}{7} = 3$ 10

44. a. *Communications* Of the 99 million households in the United States in 1995, 35 million were not listed in phone books. Six million of these households did not have phones. How many households had phones, but did not list their numbers? 29 million households
    b. Of the 35 million households not listed, $\frac{3}{7}$ had recently moved. How many households does this represent? 15 million households

Getting Ready for Lesson 3-7

Write each number as a percent.

45. 0.45 45%
46. $\frac{3}{4}$ 75%
47. 0.328 32.8%
48. 1.2 120%
49. $\frac{5}{5}$ 100%
50. 0.005 0.5%

FOR YOUR JOURNAL

Describe a situation in your life in which you could find the probability of independent and dependent events.

3-7 Teaching Notes

PROBLEM OF THE DAY

The average of three 6s and three 12s is also the average of three of what number? 9

Problem of the Day is also available in Transparencies.

CONNECTING TO PRIOR KNOWLEDGE Ask students how many of them have tried to calculate the sale price of an item marked down some percent. Have students share the mental math techniques they use to estimate the sale price.

THINK AND DISCUSS

Ask students to pay attention to the phrasing of the question because it determines which equation they need to use.

ESTIMATION Question 3 Encourage students to check their answers by using mental estimation. For example, in Question 3b, help students recognize that 75% means more than half. If 3 is more than half of the answer, then the answer must be less than 6.

Example 1 **Relating to the Real World** 

ERROR ALERT! Students may not know which equation to use for problems such as that in Example 1. **Remediation:** Percent problems use two numbers and a variable. Have stu-

---

Connections Community Service . . . and more

3-7 Percent Equations

What You'll Learn

- Using equations to solve problems involving percents
- Finding simple interest

...And Why

To solve problems involving commission, interest, and circle graphs

What You'll Need

- protractor
- compass

QUICK REVIEW

A *percent* is a ratio that compares a number to 100. You can write a percent with the percent symbol or as a fraction or decimal. For example, you can write 40% as $\frac{40}{100}$ or as 0.40.

For more practice with percents, see Skills Handbook, page 579.

THINK AND DISCUSS

Part 1

Solving Percent Equations

To model a percent problem with an equation, express each percent as a decimal. Three types of percent problems are modeled and solved below.

What is 40% of 8?	16 is 0.25% of what?	What percent of 40 is 5?
$n = 0.4 \times 8$ $n = 3.2$	$16 = 0.25 \times n$ $\frac{16}{0.25} = \frac{0.25n}{0.25}$ $64 = n$	$n \times 40 = 5$ $\frac{40n}{40} = \frac{5}{40}$ $n = 0.125$ $n = 12.5\%$

1. In the equations above, what symbol represents each of these words?
   a. is = b. of × c. what *n*
2. Match each problem with the equation that models it.
   A. 15 is what percent of 12? III I. $n = 0.15 \times 12$
   B. What is 15% of 12? I II. $12 = 0.15n$
   C. 12 is 15% of what? II III. $15 = n \times 12$
3. Write an equation to model each question. Solve each equation.
   a. What is 25% of 80? b. 3 is 75% of what?
   c. What percent of 44 is 11? d. 16% of what is 200? a–d. See left.

3. a. $n = 0.25 \times 80$; 20
b. $3 = 0.75 \times n$; 4
c. $n \times 44 = 11$; 25%
d. $0.16 \times n = 200$; 1250

Part 2

Writing Equations to Solve Percent Problems

Example 1 **Relating to the Real World**

Sales Suppose you work at a computer store. You earn 5% commission on everything you sell. How much do you earn on a computer you sell for $1900?

Define n = amount of your commission

Relate Your commission is 5% of $1900.

Write
$$n = 0.05 \times 1900$$
$$n = 0.05(1900)$$
$$n = 95$$

You earn $95 commission.

4. Tupi sells cars. He earns $145/wk in salary plus 8% commission. In one week he sells $36,000 worth of cars. How much does he earn? $3025

Lesson Planning Options

Prerequisite Skills

- Converting percents to decimals
- Multiplying and dividing with decimals

Assignment Options

To provide flexible scheduling, this lesson can be subdivided into parts.

1. **Core** 1–9, 15–26, 31–35
   Extension 14, 30
2. **Core** 13, 27, 28
   Extension 10, 12
3. **Core** 36–39
   Extension 11, 29

Use Mixed Review to maintain skills.

Resources

Student Edition

Skills Handbook, p. 579
Extra Practice, p. 558
Glossary/Study Guide

Teaching Resources

Chapter Support File, Ch. 3
- Practice 3-7 (two worksheets)
- Reteaching 3-7
- Checkpoint
- Alternative Activity 3-7

Classroom Manager 3-7
Glossary, Spanish Resources
Two-Year Algebra Handbook 3-7

Transparencies
54, 56, 57

dents find these. Then decide what the variable represents: the percent, the whole amount, or part of the whole amount. This will help them select the correct equation.

ESL **Examples 1 and 4** Find out if students know the difference between commission and interest. Ask: *Which would you earn—commission or interest—if you sold encyclopedias? clothes? houses?* Once they know the answer is "commission," ask which they would earn if they loaned someone money. Ensure students know they would earn commission from selling goods and earn interest from loaning money.

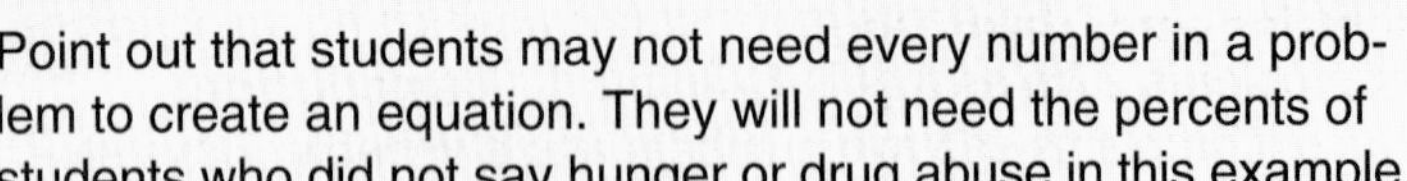

Example 2 **Relating to the Real World**

Point out that students may not need every number in a problem to create an equation. They will not need the percents of students who did not say hunger or drug abuse in this example.

TACTILE LEARNING Have students label three scraps of paper *part*, *whole*, and *percent*. After students read a problem, have them place the numerical information on the correct pieces of paper. The students can then rearrange the pieces of paper to form an equation before solving.

Example 3 **Relating to the Real World** 

Stress that, since 56 is larger than 36, you have volunteered more hours than are required. Lead students to understand that the percent calculated will therefore be more than 100%. Urge students to make mental estimates before calculating the answer. Knowing that the answer will be over 100% will help keep students from interpreting the answer 1.56 as 15.6%.

Additional Examples

FOR EXAMPLE 1

You plan to buy a radio that costs \$49.99. If the local sales tax is 7%, how much tax must you add to the price of the radio? \$3.50

Discussion: *After you write your equation, translate it back into words.*

FOR EXAMPLE 2

State Representative Garcia sent out survey forms to everyone in his district. After two weeks, 1450 people had returned their survey forms. If the returned survey forms represent 18% of the total sent out, how many survey forms were originally mailed? 8055

Discussion: *Is this an accurate way of finding out the opinions of a large group of people?*

FOR EXAMPLE 4

If you place \$1500 in a savings account drawing 7% simple interest, how much could you expect to earn in ten years? Use the formula $I = prt$. \$1050

You can use a circle graph to find data expressed in percents.

What Do You Think Is the Number One Problem in the World Today ?

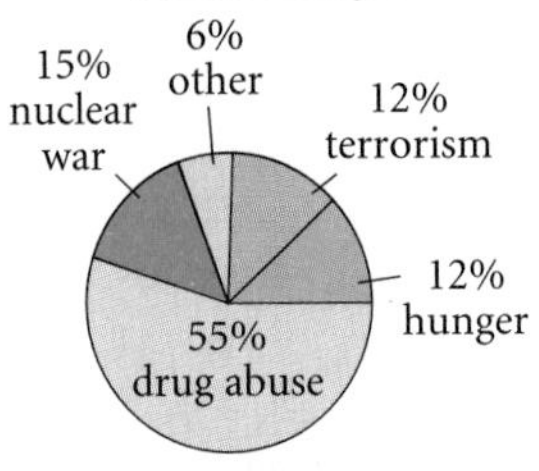

Source: *The Second Kids' World Almanac*

Example 2 **Relating to the Real World** 

Surveys In a survey by *The Second Kids' World Almanac*, 1650 students said that drug abuse is the number one problem in the world today. How many students were surveyed?

Define n = the number of students surveyed

Relate 55% of the students surveyed is 1650.

Write $0.55 \times n = 1650$

$0.55n = 1650$

$\frac{0.55n}{0.55} = \frac{1650}{0.55}$ ← Divide each side by 0.55.

$n = 3000$

There were 3000 students surveyed.

Check 0.55 × 3000 ENTER 1650 ✓

5. How many students chose hunger as the number one problem? 360 students

Percents can be numbers greater than 100. Suppose you agree to do some yardwork with your grandfather for \$20. You do such a good job that he pays you \$30. A payment of \$20 is 100% of what you expect. The payment of \$30 is more than 100% of what you expect.

6. **Open-ended** Think of two more situations in which you would use a percent greater than 100%. Sample: Retail cost is more than 100% of wholesale cost; world population in 1995 is more than 100% of world population in 1895.

MENTAL MATH Question 7 Make sure students understand that 200% is double 100%.

KINESTHETIC LEARNING Invite students to move around the room while you direct them with statements such as these:

- *50% of the girls go to the window*
- *20% of the boys go to the board*
- *30% of students wearing blue go to the door*
- *10% of the class go to the corner*

Use percents that are easy for students to manipulate mentally.

Example 4 Relating to the Real World 

CONNECTING TO THE STUDENTS' WORLD Have students find out the price of an item they would like to buy that costs more than $100. Assume students use a credit card charging a simple interest rate of 18% (1.5% per month). Have students calculate the interest that would be due after one month if they bought the item with a credit card.

WORK TOGETHER

Review with students how to use a protractor before beginning the Work Together activity.

Question 9 Students with access to computers may wish to redraw their circle graphs using computer software. Spreadsheet software and some word-processing software contains graphing options.

Example 3 Relating to the Real World

Community Service Suppose your school requires 36 hours of community service. You spent 56 hours volunteering at a hospital. What percent of the requirement have you fulfilled?

Define $n =$ the percent of the requirement you have fulfilled

Relate 56 is what percent of 36?

Write $56 = n \times 36$

$56 = 36n$

$\frac{56}{36} = \frac{36n}{36}$ ← Divide each side by 36.

$n = 1.555555555$

$n \approx 1.56$

You have fulfilled about 156% of the community-service requirement.

7. **Mental Math** How many hours would fulfill 200% of the requirement? 72 h

Part 3

Simple Interest

$I = prt$
I is the interest.
p is the principal.
r is the interest rate per year.
t is the time in years.

Today banks use *compound interest,* which you will study in Chapter 8. Compound interest is based on **simple interest.** Its formula is $I = prt$.

Example 4 Relating to the Real World 

Finance Suppose you invested $900 for two years. You earned $67.50 in simple interest. What was the annual rate of interest?

$I = prt$ ← Use the formula $I = prt$.

$67.50 = 900 \cdot r \cdot 2$ ← Substitute 67.50 for I, 900 for p, and 2 for t.

$\frac{67.50}{1800} = \frac{1800r}{1800}$ ← Divide each side by 1800.

$r = 0.0375$, or 3.75%

The annual rate of interest was 3.75%.

8. **Try This** Find the simple interest on $550 at 4.5% for one year. $24.75

QUICK REVIEW

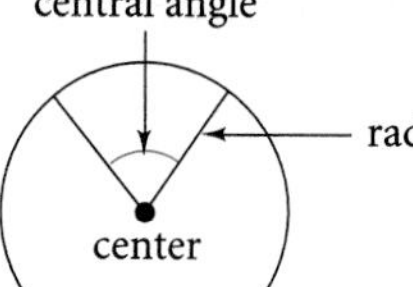

WORK TOGETHER

Data Collection Work with a partner. Ask at least 10 people "How many pets do you have?" Complete these steps to construct a circle graph.

a–c. Check students' work

9. a. Find the percent of your data in each of these categories: no pets, one pet, two pets, three or more pets.
   b. Express each percent as a decimal and multiply by 360°. This gives the number of degrees in the central angle for each category.
   c. Draw a circle and use a protractor to draw each central angle.

Technology Options

For Exercises 30, students may want to use spreadsheet software to find the percents.

Prentice Hall Technology 

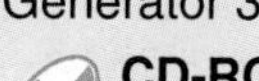

Software • Secondary Math Lab Toolkit • Computer Item Generator 3-7

CD-ROM Multimedia Algebra Lab 3

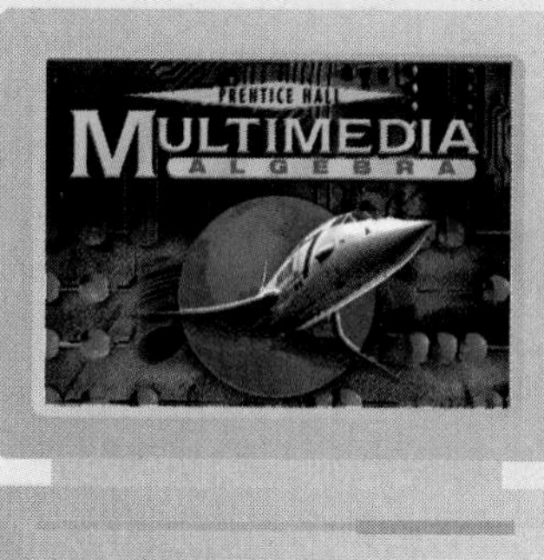

Internet • See the Prentice Hall site. (http://www.phschool.com)

Exercises ON YOUR OWN

ESTIMATION Exercise 12 Suggest that students estimate the answer by asking themselves the following:

- *What is 10% of $7.25?* 72.5¢
- *What do you get if you round that to the nearest ten cents?* 70¢
- *What's half of this amount, in other words, about 5% of the total?* 35¢
- *Since 70¢ is about 10%, and 35¢ is about 5%, if you add them together, you'll get 15%; how much is that?* $1.05
- *How much difference is there between this estimate and the actual amount?* 0.0375 or just under 4 cents

Encourage students to practice this mental estimation technique next time they go to a restaurant.

ALTERNATIVE ASSESSMENT Exercises 15–26 Have students refer to the Solving Percent Equations subsection on page 143. Assign the letter A to the first model of a percent equation, B to the second, and C to the third. Have the students work in pairs to decide which model – A, B, or C – should be used for each exercise. This activity will assess whether students understand how to model a percent problem as an equation.

MAKING CONNECTIONS Psychology and sociology books contain many tables of statistics describing percents of people with certain characteristics. By studying people's awareness of smell, for example, psychologists have found that almost everyone is incapable of sensing certain odors. About 33% of people cannot smell camphor.

pages 142–143 On Your Own

30c.

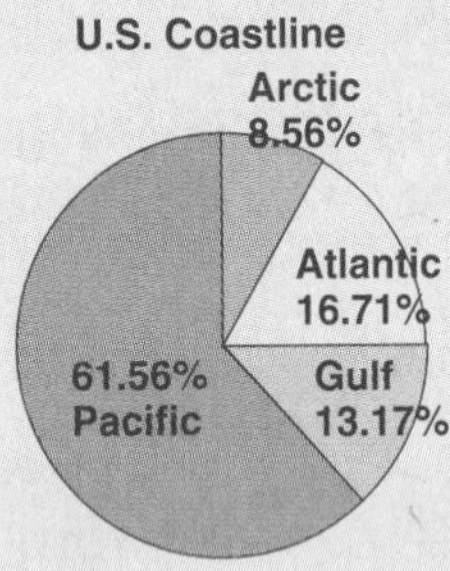

40. Tables may vary. Sample:

x	f(x)
−2	−4
−1	−1
0	2
1	5

41. Tables may vary. Sample:

x	f(x)
0	−2
2	1
3	2.5
4	5

Exercises ON YOUR OWN

Write an equation for each question. Do not solve the equation.

1. What is 4% of 150? $n = 0.04 \times 150$
2. 32 is what percent of 40? $32 = n \times 40$
3. 24 is 150% of what? $24 = 1.5 \times n$

Model with an equation. Then answer each question.

4. What percent of 51 is 17? $n \times 51 = 17$; about 33.3%
5. 10% of what is 8? $0.1 \times n = 8$; 80
6. $7\frac{1}{2}$% of $200 is what? $0.075 \times 200 = n$; $15
7. 6% of what is 36? $0.06 \times n = 36$; 600
8. 9% of 315 is what? $0.09 \times 315 = n$; 28.35
9. What percent of 45 is 18? $n \times 45 = 18$; 40%
10. *Communications* There were 96 million households in the United States in 1994. How many households had two telephones? about 29.8 million

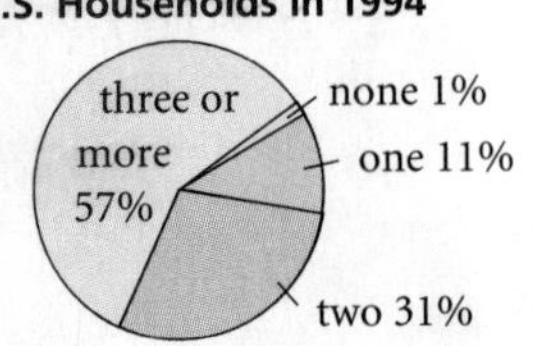

11. *Finance* A bank pays $3\frac{1}{4}$ % simple interest annually. How much interest does $640 earn in three years? $62.40
12. *Estimation* Estimate a 15% tip for a restaurant bill of $7.25. about $1
13. *Open-ended* What percent of your time do you spend sleeping? Describe the method you used to find this percent. Check students' work.
14. Choose the equation you would use to answer the question: What percent of 60 is 15? C
    A. $15n = 60$ B. $60(15) = n$ C. $60n = 15$ D. $n = \frac{60}{15}$

Choose **Use a calculator, paper and pencil, or mental math to answer each question.**

15. What is 25% of $80? $20
16. 1% of what is 7? 700
17. 5 is what percent of 10? 50%
18. What percent of 27.7 is 1.8? about 6.5%
19. 80% of 49 is what? 39.2
20. 120% of what is 60? 50
21. 9 is what percent of 200? 4.5%
22. 225% of 40 is what? 90
23. What is 55% of 600? 330
24. What percent of 150 is $7\frac{1}{2}$? 5%
25. 5.4 is what percent of 9? 60%
26. 95% of what is 83.125? 87.5
27. *Sales Tax* In Louisiana the state sales tax is 4%. If you buy a $2100 computer in Louisiana, how much tax will you pay? $84
28. *Geography* Alaska is the largest state in the United States, with an area of 570,374 mi^2. It accounts for about 15% of the country's area. Estimate the area of the United States. about 3,800,000 mi^2
29. *Finance* You received $41.60 in interest for a two-year investment at 6.5% simple interest. How much money did you invest? $320
30. *Geography* The information in the table gives the length of each of the four coastlines of the United States.
    a. How long is the coastline of the United States? 12,383 mi
    b. Find the percent of the coastline of the United States that each coast represents. Round to the nearest hundredth of a percent. See table.
    c. Draw and label a circle graph that represents the data. See margin.

U.S. Coastline Data

Coast	Miles	
Arctic	1060	8.56%
Atlantic	2069	16.71%
Gulf	1631	13.17%
Pacific	7623	61.56%

Chapter Project **FIND OUT BY GRAPHING** Ask students to make a circle graph for the budget that they created. Have them create a table showing dollar amounts, percents, and angle degrees they used to make the circle graph. This step is essential to their chapter project. Have students add this task to work they have already completed for the project. You may wish to check students' progress on the project at this point.

Exercises MIXED REVIEW

Exercise 45 Encourage students to review the absolute value function before attempting this exercise. Remind them to subtract the 3 after finding the absolute value of x.

GETTING READY FOR LESSON 3-8 These exercises give students practice simplifying the types of fractions they will be working with in Lesson 3-8.

Wrap Up

THE BIG IDEA Ask students how to use equations to solve problems containing percents.

RETEACHING ACTIVITY Students rewrite word sentences involving percents into equations. They draw triangles, rectangles and circles around words that represent *equals*, *multiplication*, and *the unknown*. Then write an equation using the same shape coding. (Reteaching worksheet 3-7)

Math in the Media **Use the news article.**

31. What percent of residents of the United States are 65 or older? about 12.6%
32. About how many elderly residents live below the poverty level? about 4.25 million people
33. What percent of elderly residents' income is from Social Security? 40%
34. *Estimation* The population of Florida is about 13,000,000. Estimate the number of Florida residents who are 65 or older. about 2.4 million people
35. *Writing* Explain how to use facts from the article to estimate the population of the United States in the year 2050. See below.

34 Million U.S. Residents are 65 or Older

Nearly 34 million of the 270 million residents of the United States have passed their 65th birthday. By the year 2050, roughly 20%, or 80 million people, will be 65 or older.

One of eight elderly residents lives below the poverty level. Two of every five dollars older residents earn are from Social Security.

Florida has the highest percent of elderly, with 18.4%. Alaska has the smallest share at 4.6%.

Finance **Use the formula for simple interest, $I = prt$. Find each missing value.**

36. $I = ■$, $p = \$340$, $r = 6\%$, $t = 3$ yr $61.20
37. $I = \$312.50$, $p = ■$, $r = 5\%$, $t = 5$ yr $1250
38. $I = \$392$, $p = \$1400$, $r = ■$, $t = 4$ yr 7%
39. $I = \$1540$, $p = \$22{,}000$, $r = 3.5\%$, $t = ■$ 2 yr

35. 20% of total population is 80 million. Solve $0.20 \times n = 80$ to get 400 million people in the U.S. in 2050.

Chapter Project ***Find Out by Graphing***

Make a circle graph for the personal budget you created in the Find Out question on page 123. In a table, show the dollar amounts, percents, and angle degrees you used to create the graph.

Exercises MIXED REVIEW

Make a table, then graph each function. 40–43. See margin.

40. $f(x) = 3x + 2$
41. $f(x) = 1.5x - 2$
42. $f(x) = \frac{4}{5}x - 0.5$
43. $f(x) = 4x + 1$

44. According to *USA TODAY*, children smile an average of 400 times a day. About how many times does a child smile in a week? in the month of October? 2800; 12,400

45. Find the range of the function $f(x) = |x| - 3$ when the domain is $\{-2, -1, 0, 1, 2\}$. $\{-1, -2, -3\}$

Getting Ready for Lesson 3-8

Simplify. Leave each answer in decimal form rounded to the nearest hundredth.

46. $\frac{45 - 35}{45}$ 0.22
47. $\frac{100 - 25}{100}$ 0.75
48. $\frac{80 - 21}{80}$ 0.74
49. $\frac{32 - 4}{32}$ 0.88

42. Tables may vary. Sample:

x	f(x)
−2	−2.1
−1	−1.3
0	−0.5
1	0.3
2	1.1

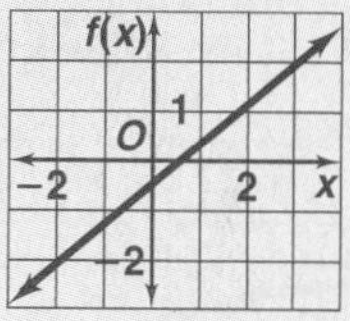

43. Tables may vary. Sample:

x	f(x)
−1	−3
0	1
1	5

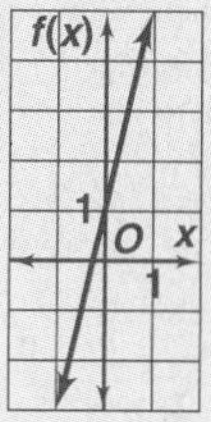

Exercises CHECKPOINT

In this Checkpoint, students will have an opportunity to assess their own progress on the topics covered in Lessons 3-5 to 3-7.

Exercise 3 Students who are used to exercises which simplify to common fractions may think that an answer of $3\frac{11}{15}$ is incorrect.

Exercises 6–10 Encourage students to make sketches to help them think about the exercises.

Algebra at Work

For further information about the training and skills necessary to be a *Media Researcher,* write to the following address:

- Nielsen Media Research
  299 Park Avenue
  New York, NY 10171

or research the company through the internet address:

- http://www.nielsenmedia.com/

Students can learn from discussions with experts. Here are some possible sources.

- Invite a general sales manager from a local TV station to speak to the class.
- Visit a company's marketing department and talk to the representatives about the work they do.

Encourage students to investigate these topics:

- How statistics is used on the job.
- The metering technology used to collect the data

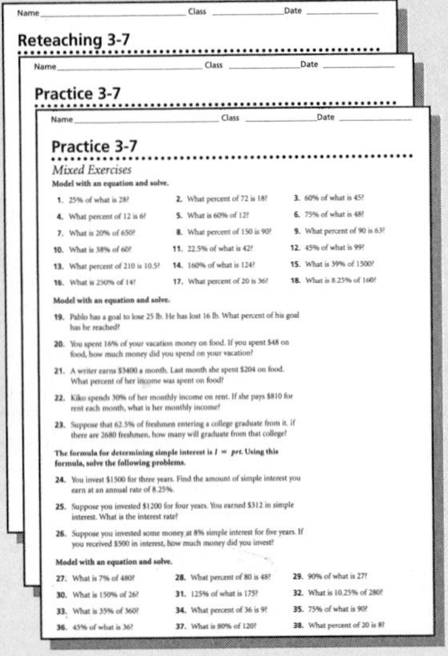

Name______ Class ______ Date ______

Reteaching 3-7

Name______ Class ______ Date ______

Practice 3-7

Name______ Class ______ Date ______

Practice 3-7

Mixed Exercises

Model with an equation and solve.

1. 25% of what is 28?
2. What percent of 72 is 18?
3. 60% of what is 45?
4. What percent of 12 is 6?
5. What is 60% of 12?
6. 75% of what is 48?
7. What is 20% of 650?
8. What percent of 150 is 90?
9. What percent of 90 is 63?
10. What is 38% of 60?
11. 22.5% of what is 42?
12. 45% of what is 99?
13. What percent of 210 is 10.5?
14. 160% of what is 124?
15. What is 39% of 1500?
16. What is 250% of 14?
17. What percent of 20 is 36?
18. What is 8.25% of 160?

Model with an equation and solve.

19. Pablo has a goal to lose 25 lb. He has lost 16 lb. What percent of his goal has he reached?
20. You spent 16% of your vacation money on food. If you spent \$48 on food, how much money did you spend on your vacation?
21. A writer earns \$3400 a month. Last month she spent \$204 on food. What percent of her income was spent on food?
22. Kiko spends 30% of her monthly income on rent. If she pays \$810 for rent each month, what is her monthly income?
23. Suppose that 62.5% of freshmen entering a college graduate from it. If there are 2680 freshmen, how many will graduate from that college?

The formula for determining simple interest is $I = prt$. Using this formula, solve the following problems.

24. You invest \$1500 for three years. Find the amount of simple interest you earn at an annual rate of 8.25%.
25. Suppose you invested \$1200 for four years. You earned \$312 in simple interest. What is the interest rate?
26. Suppose you invested some money at 8% simple interest for five years. If you received \$500 in interest, how much money did you invest?

Model with an equation and solve.

27. What is 7% of 480?
28. What percent of 80 is 48?
29. 90% of what is 27?
30. What is 150% of 26?
31. 125% of what is 175?
32. What is 10.25% of 280?
33. What is 35% of 360?
34. What percent of 36 is 9?
35. 75% of what is 90?
36. 45% of what is 36?
37. What is 80% of 120?
38. What percent of 20 is 8?

Lesson Quiz

Lesson Quiz is also available in Transparencies.

1. What percent of 6 is 19? about 317%
2. 15% of what is $5\frac{1}{4}$? 35
3. 13.5% of 45 is what? 6.075
4. The Earth's oceans cover approximately 70.8% of the planet's surface. If the surface area of the Earth is approximately 196 million square miles, what is the surface area of the Earth's oceans? about 139 million square miles

Exercises CHECKPOINT

Solve.

1. $\frac{5w + 2}{3} = -4$ $-\frac{14}{5}$
2. $\frac{3}{4}m = 7.2$ 9.6
3. $\frac{t}{8} + \frac{t}{7} = 1$ $3\frac{11}{15}$
4. $-\frac{4}{5}x - 2 = 7$ $-11\frac{1}{4}$
5. $6 = -\frac{1}{2}(3 - 5z)$ 3

Suppose you have marbles in your pocket. One is red, three are blue, two are green, and three are yellow. You pick two marbles. Find each probability.

6. *P*(two yellow) with replacing $\frac{1}{9}$
7. *P*(red and blue) without replacing $\frac{1}{24}$
8. *P*(red and blue) with replacing $\frac{1}{27}$
9. *P*(two green) without replacing $\frac{1}{36}$
10. **Writing** Explain how *P*(yellow and blue) changes if you switch from replacing to not replacing. The probability increases. The first draw is the same, but on the second draw, the number of marbles decreases and the number of favorable outcomes stays the same.
11. **Government** A majority is any number greater than 50% of the total. When the 435 voting members of the House of Representatives use majority rule, how many members are needed to make a majority? 218 members
12. **Open-ended** Give an example of a situation in which you would use percents. Sample: taking a survey as a poll of student opinion

Algebra at Work

Media Researcher

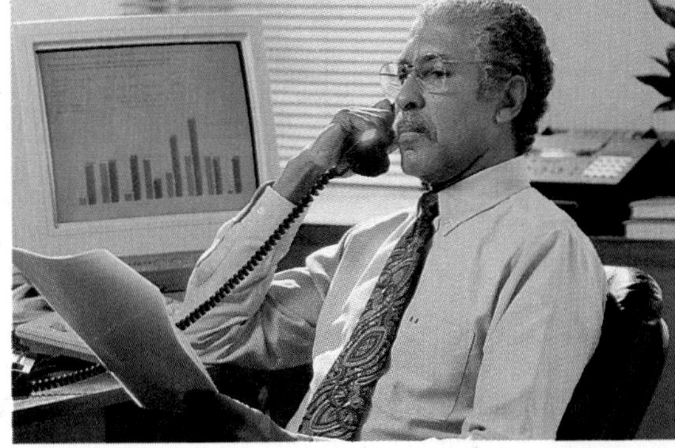

Nielsen Media Research estimates the number of people who watch each network television show. The company polls a random sample of viewers. For national surveys, Nielsen polls about 10,000 viewers. For local surveys, they poll about 300 to 500 viewers. The results are then applied to the larger population. Nielsen Media Research estimates a 2–4% possible error for its local surveys, but only 1–2% for national polls. As you can see, the estimated error decreases as the sampling size increases.

Mini Project: Describe how you would choose and contact a sample of 100 people to represent the entire population of your city.

Math ToolboX

This toolbox will help students see the many practical uses of their graphing skills. Because computers make graphing easy, students can concentrate on the appropriateness of their graph choice. Some graphing programs allow the scale to be changed so students can see how a different scale displays or distorts the data. Use sample graphs from a local newspaper to begin a discussion of the wide use of graphs. Sample graphs may also be found on most standardized tests.

ERROR ALERT! Some students may have difficulty entering the data into the correct rows or columns of a spreadsheet. **Remediation:** Have students work in pairs. Use the example data to guide students through data entry and graphing. Have students check each graph against the graph in the text.

WRITING Exercise 4 Students may refer to the comments below each example graph to help justify their answers.

ADDITIONAL PROBLEM Use a graphing program to make an appropriate graph for the data in the following table.

Student Grades on Chapter 3 Test

Grade	A	B	C	D	E
Number of students	10	7	8	3	1

A bar graph is appropriate.

Ask students to find the percentages necessary to make a circle graph of this data.

Math ToolboX Technology

Using a Graphing Program

After Lesson 3-7

In Chapter 1 you worked with graphs. In the last lesson you drew circle graphs. Computers can make graphing easier. To use a graphing program, you enter data in a spreadsheet. Then you choose the type of graph you want to draw. You must choose the type carefully because the computer will create the graph even if it is not appropriate for your data.

Instruments Played by Amateurs

	A	B
1	Instrument	People (millions)
2	Piano	24
3	Guitar	10
4	Flute	4
5	Drum	3

Source: *USA TODAY*

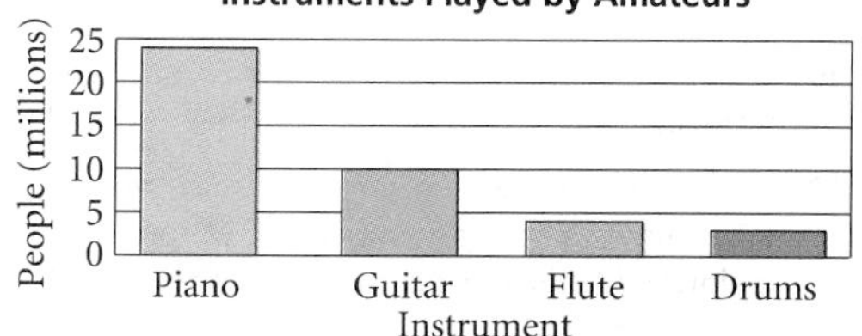

A bar graph is a good choice for this set of data. You can use a bar graph to compare amounts.

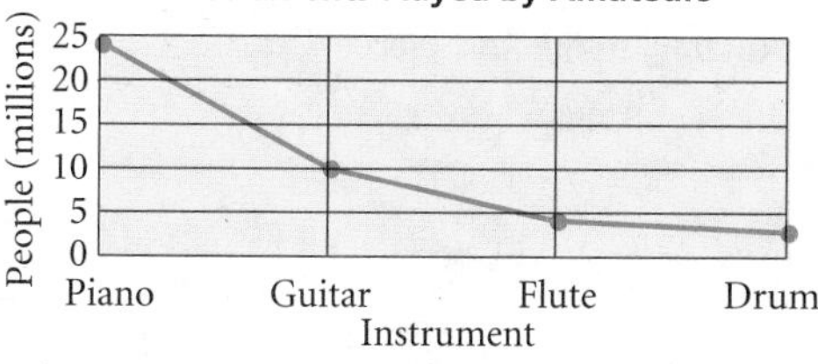

A line graph is *not* a good choice for this set of data. A line graph shows changes over time.

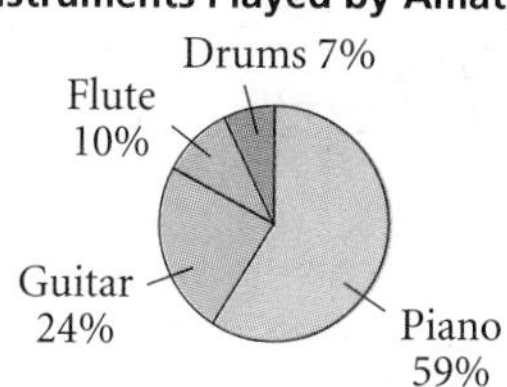

A circle graph is *not* a good choice for this set of data. A circle graph displays parts of a whole.

Use a graphing program to make an appropriate graph for the data. 1–4. See margin.

1. **Who Buys Take-Out Food?**

Age	18–29	30–44	45–59	60 plus
Percent	73%	64%	57%	39%

2. **People Who Telecommute (use a computer or a phone to work at home)**

Year	1990	1991	1992	1993
People (millions)	4.0	5.5	6.6	7.6

3. "How often do you lose your television remote control each week?" Here is how people answered. Never: 44%, 1–5 times: 38%, 5 or more times: 17%, Don't know: 1%. Make a graph of this set of data.

4. *Writing* For Exercises 1–3, explain why you chose the type of graph you did.

Materials and Manipulatives

- Computers with graphing programs

page 145 Math Toolbox

1.

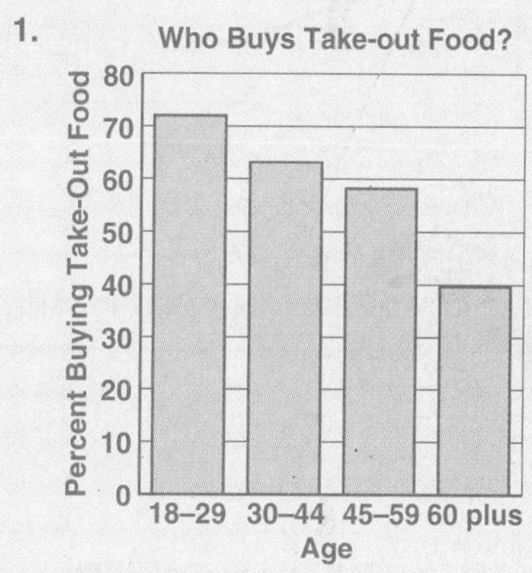

2.

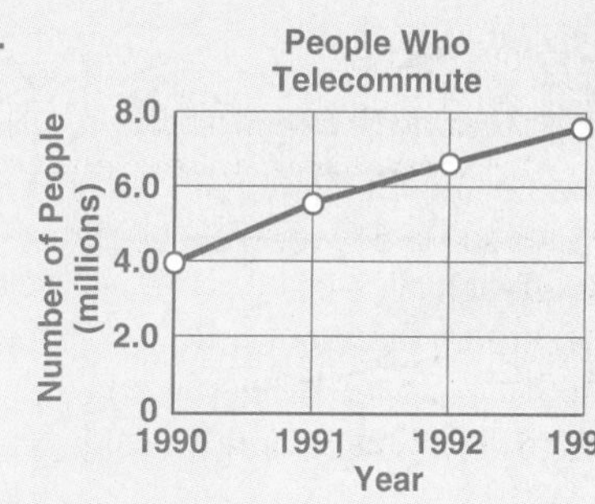

3, 4. See back of book.

3-8 Teaching Notes

PROBLEM OF THE DAY

A local music store is having a sale. A guitar that originally sold for $267 is now discounted by $26.70. A $30 tuner is now marked $3 off. An $80 harmonica is selling for $72. If the mark-down rate remains the same, for how much will a $365 drum be selling? $328.50

Problem of the Day is also available in Transparencies.

CONNECTING TO PRIOR KNOWLEDGE Ask students to explain what change is, and to describe changes around them. Brainstorm ways of describing change.

THINK AND DISCUSS

ESL Ensure students know that *quadruple* means to multiply by four. Ask how many children are in a set of *quadruplets.* Then make the analogy that *quadruple* is to *quadruplets* as *triple* is to *triplets.*

Example 1 Relating to the Real World 

Question 1 Students may feel that the word *quadrupled* makes the increase sound greater than it really is. Lead students to understand how the news media sometimes use this effect to make a story seem more dramatic.

MAKING CONNECTIONS Rheumatic diseases (one of which is arthritis) are diseases that affect the joints, connective tissue, and other supporting tissue. About 40 million Americans have

Lesson Planning Options

Prerequisite Skills

- Converting decimals to percents

Assignment Options

Core 1–20, 22–34, 39–44
Extension 21, 35–38, 45

Use Mixed Review to maintain skills.

Resources

 **Student Edition**

Skills Handbook, p. 579
Extra Practice, p. 558
Glossary/Study Guide

 **Teaching Resources**

Chapter Support File, Ch. 3
- Practice 3-8 (two worksheets)
- Reteaching 3-8

Classroom Manager 3-8
Glossary, Spanish Resources
Two-Year Algebra Handbook 3-8

 **Transparencies**
54

Connections 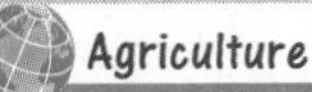 Agriculture . . . and more

3-8 Percent of Change

What You'll Learn

- Finding percent of change
- Solving problems involving percent of change

...And Why

To solve problems involving careers in medicine and agriculture

THINK AND DISCUSS

News reporters often use statistics to present information. Here are two different ways a reporter related this fact: The number of female physicians in the United States increased from 25,400 in 1970 to 104,200 in 1990.

A: From 1970 to 1990, the number of female physicians more than quadrupled.

B: From 1970 to 1990, the number of female physicians increased about 310%.

Statement A is easy to verify by estimating: 104,200 is about 4 times 25,400, since $4(25{,}000) = 100{,}000$.

To verify Statement B, you need to look at the percent of change in the data. The **percent of change** is the percent an amount changes from its original amount. The amount of change is the difference between these two values.

$$\text{percent of change} = \frac{\text{amount of change}}{\text{original amount}}$$

When a value increases from its original amount, you call the percent of change the **percent of increase.**

Example 1 Relating to the Real World 

Medical Careers Verify statement B above by finding the percent of increase in the number of female physicians.

$$\begin{aligned}\text{percent of change} &= \frac{\text{amount of change}}{\text{original amount}}\\ &= \frac{104{,}200 - 25{,}400}{25{,}400} && \leftarrow \text{Substitute.}\\ &= \frac{78{,}800}{25{,}400} && \leftarrow \text{Simplify the numerator.}\\ &\approx 3.10, \text{ or } 310\% && \leftarrow \text{Divide.}\end{aligned}$$

The percent of increase is about 310%.

Who? Dr. Graciela Solis Alarcón is a professor of medicine. Dr. Alarcón has won numerous awards for her research in rheumatology, which is the study of diseases of the muscles and joints.

1a. An increase of 300% raises the total to 400%, which is 4 times the original amount of 100%.

1. a. Explain why stating that an amount quadrupled is the same as stating that it increased 300%. See left.
   b. What would an increase of 100% mean? Increasing by 100% doubles the original amount.
2. When you find the percent of change, why is it important to know which amount is the original amount? It becomes the denominator.
3. Try This From 1970 to 1990, the total number of physicians in the United States increased from 334,000 to 615,400.
   a. Find the amount of change. 281,400 physicians
   b. Find the percent of increase. about 84%

arthritis, which is caused by infections, immune system abnormalities, injuries, aging, or genetics.

Example 2 **Relating to the Real World** 

ERROR ALERT! For problems such as the examples, students may lose track of whether a percent of change is an increase or a decrease. **Remediation:** Encourage students to make a note on the side of their papers as they calculate whether a percent change is an increase or a decrease.

EXTENSION Bring to class the business section of a newspaper. Show students how to read the stock market averages, such as the Dow Jones Industrial Average or the Standard and Poors 500. Have different groups track different averages for a few days. Ask each group to find the percent of change for a few days. Compare results with other groups.

AUDITORY LEARNING Ask pairs of students to explain to each other the meaning of "percent of change" and the steps to use when determining percent of increase or decrease.

Exercises ON YOUR OWN

Exercises 1–8 Encourage students to make rough estimates before calculating the answers.

MENTAL MATH Exercises 11–18 Encourage students to share with the class the mental math techniques they use.

OPEN-ENDED Exercise 19 As an alternative, challenge students to find store products with labels that describe percent of change. Ask students to check the manufacturers' math. If students find incorrect math, ask them to bring to class the product container or a description of the error.

When a value decreases from its original amount, you call the percent of change the **percent of decrease.**

Example 2 **Relating to the Real World** 

Agriculture Changes in the economy have affected the location and number of farms in the United States. Find the percent of decrease in the number of farms from 1940 to 1990.

Changes in Farms in the United States

Year	Number of Farms (thousands)	Acres of Farmland (millions)
1940	6102	1065
1990	2140	987

Computer technology is used to vary the rate and type of fertilizer applied to a field.

$$\text{percent of change} = \frac{\text{amount of change}}{\text{original amount}}$$

$= \frac{6102 - 2140}{6102}$ ← Substitute.

$= \frac{3962}{6102}$ ← Simplify the numerator.

≈ 0.65, or 65% ← Divide.

The percent of decrease in the number of farms is about 65%.

4. a. Find the percent of decrease in the number of acres of farmland from 1940 to 1990. about 7%
   b. Critical Thinking While the number of farms decreased by 65%, the number of acres of farmland decreased by a much smaller percent. What can you conclude about the size of the average farm?
   Answers may vary. Sample: The size of the average farm has increased.

Exercises ON YOUR OWN

Choose Use a calculator, pencil and paper, or mental math to find each percent of change. Describe the percent of change as a percent of increase or decrease. Round to the nearest percent.

1. \$12 to \$9 25% dec.
2. 19 in. to 25 in. 32% inc.
3. $5\frac{1}{2}$ ft to $5\frac{3}{4}$ ft 5% inc.
4. 180 lb to 150 lb 17% dec.
5. 18 to $17\frac{1}{2}$ 3% dec.
6. 15,000 to 12,000 20% dec.
7. \$1.75 to \$1.25 29% dec.
8. \$6/h to \$6.50/h 8% inc.

Additional Examples

FOR EXAMPLE 1

Anag's grade average for algebra changed from an 88 to a 94. What is the percent of change in his average? 6.8%

Discussion: *Would the percent of change be the same if his grade changed from 78 to 84? Explain.*

FOR EXAMPLE 2

Between 1940 and 1980, the federal budget increased from \$9.5 billion to \$725.3 billion. What percent of increase does this represent? 7535%

Discussion: *What other factors must you consider before drawing any conclusions about the federal government's spending policy?*

WRITING Exercise 21 For students who quickly decide that the net change is 0%, have them support their conclusion with an example showing the math calculations.

ALTERNATIVE ASSESSMENT Exercise 35 Randomly call on pairs of students to answer and to explain each subsection of the question. This exercise will assess students' understanding of percent of change and doubling or halving of the original amount.

Exercise 36 Direct students' attention to the phrase "grew by more than 40%" in the news article. Explain that the number of businesses is what grew. Students may interpret the phrase incorrectly to mean that existing businesses expanded.

STANDARDIZED TEST TIP Exercise 37 Notice that only response C reduces the original value. Since all other responses represent either a number larger than \$100 or an increase from \$100, C is the logical choice.

Exercise 45 Ask students to guess which percent of change will be larger before they calculate the answer.

Exercises MIXED REVIEW

Exercise 50 Point out that the exercise asks only for the percent. This is not a percent of change question.

Exercise 51 Urge students to read the exercise slowly and carefully.

SELF ASSESSMENT PORTFOLIO Share with students the criteria you will use to assess their work in portfolios, as well as how you plan to use the results. Help students understand how you

Technology Options

For Exercises 46–49, students may want to use graphing calculators to check their answers.

Prentice Hall Technology

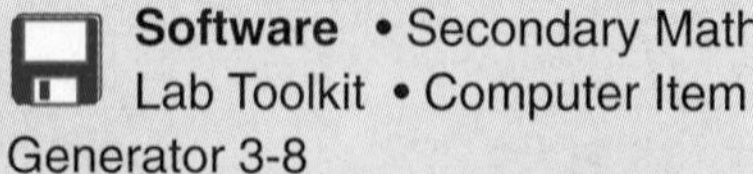

Software • Secondary Math Lab Toolkit • Computer Item Generator 3-8

Internet • See the Prentice Hall site. (http://www.phschool.com)

148

9. ***Physical Therapy*** Physical therapists measure strength on a dynamometer, which uses units called foot-pounds. Suppose you increase the strength in your elbow from 90 foot-pounds to 125 foot-pounds. Find the percent of increase. **about 39%**

10. ***Environment*** From 1987 to 1993, the number of days of unhealthy air quality in Atlanta dropped from 15 to 4. Find the percent of decrease in the number of days of unhealthy air. **about 73%**

Mental Math **Find each percent of change.** F = (A − B)/A

11. 5 cm to 10 cm **100%**
12. \$3.00 to \$1.50 **50%**
13. \$10 to \$4 **60%**
14. 12 in. to 6 in. **50%**
15. 3 ft to 4 ft **33%**
16. \$20,000 to \$25,000 **25%**
17. 2 m to 6 m **200%**
18. \$15 to \$5 **67%**

19. ***Open-ended*** Choose an item that you buy that has changed in price. Give its original price and its new price. Find the percent of change. **Check students' work.**

20. ***Sports*** In the 1960 Olympics, Wilma Rudolph of the United States won the women's 100-m run in 11.0 s. In 1988, Florence Griffith-Joyner, also of the United States, won with a time of 10.54 s. Find the percent of decrease in the winning time. **about 4%**

21. ***Writing*** Suppose you increase the price of a product by 20%, then reduce the price by 20%. Is the resulting price greater than, less than, or equal to the price you started with? Explain your answer. **Less; the 20% you subtracted is 20% of an amount greater than the original amount, so it is greater than the 20% you added.**

Calculator **Find each percent of change. Describe the percent of change as a percent of increase or decrease.**

22. \$26 to \$20 **23% dec.**
23. \$8/h to \$8.45/h **6% inc.**
24. 21 in. to 54 in. **157% inc.**
25. \$4.95 to \$3.87 **22% dec.**
26. 132 lb to 120 lb **9% dec.**
27. \$24,000 to \$25,000 **4% inc.**
28. \$8.99 to \$3.99 **56% dec.**
29. 25 mi/h to 55 mi/h **120% inc.**
30. \$42.69 to \$49.95 **17% inc.**
31. 2.3 cm to 2.8 cm **22% inc.**
32. 10.5 km to 4.2 km **60% dec.**
33. \$3.99/lb to \$2.89/lb **28% dec.**

34. ***Transportation*** The number of cars in China increased from 1986 to 1993. Use the data at the right to find the percent of increase. **about 245%**

Passenger Cars in China

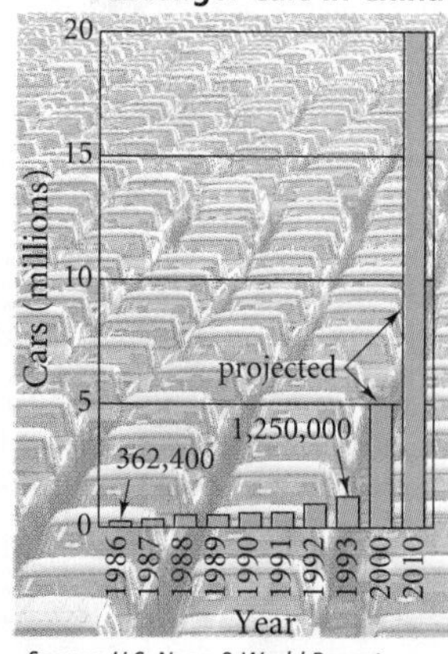

Source: *U.S. News & World Report*

35. ***Sales*** Suppose that you are selling sweatshirts for a class fund-raiser. The wholesaler charges you \$8 for each sweatshirt.
    a. You charge \$16 for each sweatshirt. Find the percent of increase. **100%**
    b. **Generalize** your answer to part (a). Doubling a price is the same as a __?__ percent increase. **100**
    c. After the fund-raiser is over, you reduce the price on the remaining sweatshirts to \$8. Find the percent of decrease. **50%**
    d. **Generalize** your answer to part (c). Cutting a price in half is the same as a __?__ percent decrease. **50**
    e. A few of the sale sweatshirts do not sell, so you give them away free to charity. Find the percent of decrease from the sale price. **100%**
    f. **Generalize** your answer to part (e). Reducing a price to zero is the same as a __?__ percent decrease. **100**

will use the rubrics to assess their work, how you will grade the portfolio, and how the scores they get in their portfolios will affect their overall evaluation.

Wrap Up

THE BIG IDEA Ask students to explain how to solve problems that involve percent of increase or decrease.

RETEACHING ACTIVITY Students use beans or paper clips to determine the percent of change. By adding or subtracting a number of beans or paper clips, students determine the percent of increase or decrease. (Reteaching worksheet 3-8)

36. **Math in the Media** Use the news article at the right. See below.
   a. How does the information in the second sentence verify the "more than 40%" claim in the first sentence?
   b. In 1994, what percent of the people employed in the United States worked for businesses owned by women? about 12%

36a. The increase from 5.4 to 7.7 million is about 43%.

37. **Standardized Test Prep** All of the following are equal *except:* C
   A. \$100 increased by 50% B. \$150 C. 50% of \$100
   D. 150% of \$100 E. \$100 increased by half

38. **Animal Studies** In 10 years, the African elephant population in Kenya decreased from 150,000 to 30,000.
   a. Find the percent of decrease. 80%
   b. On average, how many fewer elephants were there each year? 12,000 elephants

Find the percent of change in each price.

39. \$15 marked up to \$24 60%
40. \$750 discounted to \$600 20%
41. \$3500 marked up to \$4000 about 14%
42. \$1.80 marked up to \$2.20 about 22%
43. \$39.99 discounted to \$19.99 about 50%
44. \$24 marked up to \$36 50%

45. a. Find the percent of change from 20 min to 25 min. 25%
   b. Find the percent of change from 25 min to 30 min. 20%
   c. **Writing** Explain why the percent of change is different for parts (a) and (b) even though the amount of change is the same.
   Answers may vary. Sample: The percent of change is different because the original amount is different in each case.

Women-Owned Businesses on the Rise

The number of businesses owned by women grew more than 40% in the three years from 1991 to 1994. The National Foundation for Women Business Owners reported that 7.7 million companies were owned by women in 1994, up from 5.4 million in 1991. In 1994, these businesses employed about 15.5 million of the 131 million people employed in the United States.

Exercises MIXED REVIEW

Solve each equation.

46. $2 = \frac{c}{15}$ 30
47. $8w = 12$ 1.5
48. $z - 17 = -10$ 7
49. $m - 4 = 9$ 13

50. **Health** About 55 million Americans have foot problems. About 4.6 million of them have flat feet or fallen arches. What percent of people with foot problems have flat feet or fallen arches? about 8.4%

51. **Probability** You pick a colored chip from a bag, note its color, and replace it. You do this 20 times and get 7 red chips, 5 blue chips, and 8 green chips. What is the experimental probability that the next chip you choose will *not* be green? 60%

52. Match each scatter plot correlation with the line that describes it.
   II A. no correlation I. downward slanted line (\)
   III B. positive correlation II. no line
   I C. negative correlation III. upward slanted line (/)

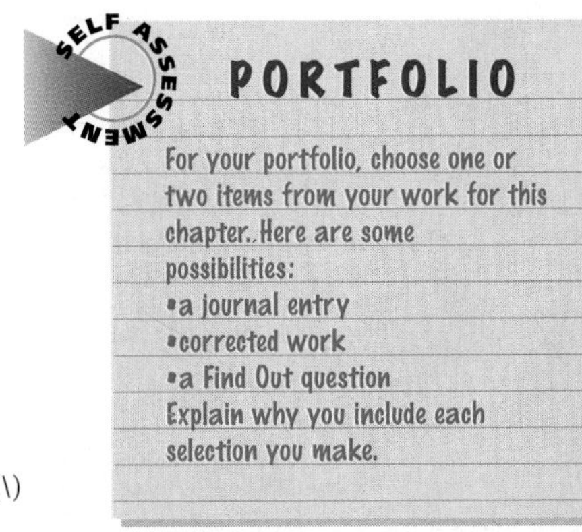

SELF ASSESSMENT
PORTFOLIO
For your portfolio, choose one or two items from your work for this chapter. Here are some possibilities:
•a journal entry
•corrected work
•a Find Out question
Explain why you include each selection you make.

Reteaching 3-8

Practice 3-8

Lesson Quiz

Lesson Quiz is also available in Transparencies.

In Exercises 1–3, find the percent of change.

1. \$26 to \$15 about –42%
2. 55 miles per hour to 65 miles per hour about 18%
3. 84 to 90 7.1%
4. A shirt originally costing \$20 was placed on sale at a discount of 20%. Later the price of the shirt was increased by 20%. What is the final price of the shirt? \$19.20

Finishing the Chapter Project

PROJECT DAY You may wish to plan a project day on which students share their completed projects. Encourage groups to explain their process as well as their product.

PROJECT NOTEBOOK Have students review their project work and bring their notebooks up to date.

- Have students review their methods for finding, recording, and displaying data they needed for the project.
- Ask groups to share any insights they found for completing the project, such as any shortcuts they found for creating circle graphs or spreadsheets or for analyzing budgets. Also, ask whether students feel they have gained insights about ways to manage their own money.

SCORING RUBRIC

3 Equations are correct and calculations are accurate. Spreadsheet shows detail, is easy to follow, and clearly shows a thorough understanding of maintaining a budget. The circle graph is accurate and well labeled. Reasoning and explanations are sound and clearly expressed.

2 The choice and usage of the formulas, spreadsheet, and graph are correct. There are minor errors in scale, computation, or accuracy. Reasoning and explanations are essentially correct, but may contain awkward or unclear passages.

1 Formulas are misused. Spreadsheet is not detailed or lacks organization. Graph could be more accurate. Explanations are incomplete or incorrect.

0 Major elements of the project are incomplete or missing.

CHAPTER PROJECT

Finishing the Chapter Project

checks and balances

Questions on pages 113, 118, 123, and 143 should help you to complete your project. Assemble all the parts of your project—the research on what you would like to buy, your expense record, your spreadsheet, and your circle graph. Are the expenses you recorded for two weeks typical for you? Does your budget support your purchase goal? Summarize what you think are the strengths and weaknesses of your budget.

Reflect and Revise

Present your budget and your purchase goal to a small group of classmates. Compare your decisions with theirs. Demonstrate to the group two equations of your own from the project. Check each other's work (including the circle graph) for reasonableness and accuracy. Use the comments and suggestions of your group to revise and improve your project.

Follow Up

Stick to your budget for several weeks. Are your savings on track? If not, what expenses can you reduce? Can you increase your income? Some of the resources listed below may give you ideas for a money-making project.

For More Information

Drew, Bonnie and Noel. *Fast Cash for Kids.* Hawthorne, New Jersey: Career Press, 1995.

Lamancusa, Joe. Kid Cash: *Creative Money-Making Ideas.* Blue Ridge Summit, Pennsylvania: TAB Books, 1993.

Spiselman, David. *A Teenager's Guide to Money, Banking, and Finance.* New York: J. Messner, 1987.

"The *Zillions* Allowance Survey." *Zillions* (April/May 1995): 14–17.

Sale! $99.99
Personal CD Player with Headphones
• Rich, full sound
• 16-program memory
Take it anywhere!
(runs on 2 AA batteries)

Wrap Up

HOW AM I DOING? Have students answer the two questions. Then organize a brainstorming session so that the whole class can discuss and refine the answers. Write the modified responses on the board for students to use as a study guide.

KEY TERMS The numbers in parentheses direct students to the pages where the terms (or symbols) are used or defined. Students should be able to (1) write a simple explanation of each term, or (2) illustrate the term with a diagram, or (3) show an example that uses a term.

Exercises 1–20 Ask students to write the symbols *A* (addition), *S* (subtraction), *M* (multiplication), or *D* (division) in order to identify the operations used to solve each equation. This helps students to apply inverse operations in a methodical way. Point out that, in the two-step exercises (9–20), an *A* or *S* symbol should precede an *M* or *D* symbol.

Exercises 30 and 31 Remind students that a negative sign in front of a set of parentheses is equivalent to multiplying all of the terms in the parentheses by –1.

Exercises 43–47 Remind students that the words *what, is,* and *of* are equivalent to the symbols *n*, =, and ×, respectively.

Remind students that the new mathematical terms in this chapter are defined in the Glossary/Study Guide in the back of the book.

3 Wrap Up

Key Terms

coefficient (p. 119)
constant term (p. 119)
dependent events (p. 135)
distributive property (p. 124)
equivalent equations (p. 109)
independent events (p. 134)
inverse operations (p. 108)
like terms (p. 119)
percent of change (p. 146)
percent of increase (p. 146)
percent of decrease (p. 147)
properties of equality (p. 109, 110)
simple interest (p. 141)
solutions (p. 108)

How am I doing?

- State three ideas from this chapter that you think are important. Explain your choices.
- Describe several important rules that you must apply in order to solve equations.

Modeling and Solving Equations 3-1

The value (or values) of a variable that make an equation true are called **solutions.** To solve an equation you can use **inverse operations,** which are operations that undo one another. To solve an addition or subtraction equation, subtract or add the same value to each side of the equation. To solve a multiplication or division equation, divide or multiply each side of the equation by the same nonzero value.

Solve and check.

1. $y - 7 = 9$ 16
2. $\frac{x}{12} = -3$ −36
3. $w + 23 = 54$ 31
4. $5d = 120$ 24
5. $-9 + t = 35$ 44
6. $c + 0.25 = 4.5$ 4.25
7. $7b = 84$ 12
8. $\frac{z}{4} = \frac{1}{2}$ 2

Modeling and Solving Two-Step Equations 3-2

A **two-step equation** is an equation that has two operations. You can use tiles to model and solve a two-step equation. To solve a two-step equation, first add or subtract. Then multiply or divide.

Write and solve each equation modeled by tiles.

9. 
$2x + 5 = 9; 2$

10. 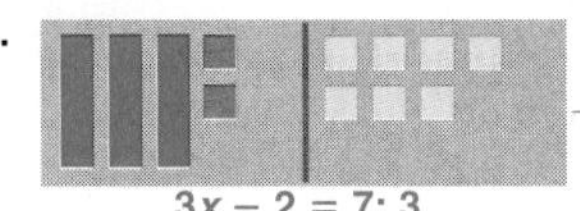
$3x - 2 = 7; 3$

11. 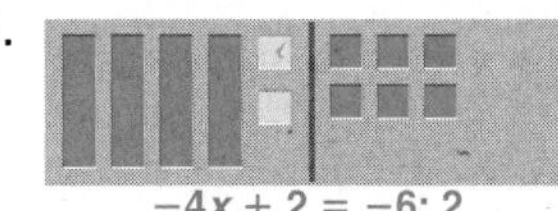
$-4x + 2 = -6; 2$

For Exercises 12–19, solve and check each equation.

12. $8u + 2 = 6$ $\frac{1}{2}$
13. $7t - 3 = 18$ 3
14. $-2q - 5 = -11$ 3
15. $11y + 9 = 130$ 11
16. $-z + 11 = -7$ 18
17. $5x - 8 = 12$ 4
18. $3m + 8 = 2$ −2
19. $10h - 4 = -94$ −9

20. **Writing** Describe a situation that you can model with $6m + 3 = 27$. Explain what the variable represents and solve the equation. Sample: A state park charges $6 per person admission plus a $3 parking fee. The Hsu family paid $27. How many people were in the car? 4 people

Materials

- tiles

Resources

Student Edition

Extra Practice, p. 558
Glossary/Study Guide

Teaching Resources

Student Study Survival Handbook
Glossary, Spanish Resources

pages 151–153 Wrap Up

42. Answers may vary. Sample: In mathematics, two events are *dependent* if the outcome of one influences the outcome of the other. A variable is *dependent* if its value is influenced by other variables. In social studies, something is *dependent* if it relies on another for financial, economic, or political support. In both cases, things that are *independent* do not rely on other things.

Combining Like Terms and Using the Distributive Property 3-3, 3-4

Terms with exactly the same variable factors are **like terms.** You can combine like terms and use the distributive property to simplify expressions and solve equations.

Simplify each expression.

21. $9m - 5m + 3$ $4m + 3$ **22.** $2b + 8 - b + 2$ $b + 6$ **23.** $-5(w - 4)$ $-5w + 20$ **24.** $9(4 - 3j)$ $36 - 27j$

Solve and check each equation.

25. $6b + 4b = -90$ -9 **26.** $-x + 7x = 24$ 4 **27.** $2(t + 5) = 9$ -0.5 **28.** $-2(r - \frac{1}{2}) = -2$ $1\frac{1}{2}$

29. $2b + 5(b + 1) = -9$ -2 **30.** $-(3 - 10y) = 12$ 1.5 **31.** $x - (4 - x) = 0$ 2 **32.** $1 = z + 3(z - 1)$ 1

33. Geometry The width of a rectangle is 6 cm less then the length. The perimeter is 72 cm. Write and solve an equation to find the dimensions of the rectangle. $2[\ell + (\ell - 6)] = 72$; 15 cm × 21 cm

Rational Numbers and Equations 3-5

You can simplify an equation that has fractional coefficients by multiplying each side of the equation by a reciprocal or by a common denominator.

34. Which operation will solve the equation $\frac{3}{4}w = \frac{9}{8}$? C

A. Divide each side by $\frac{4}{3}$.

B. Subtract $\frac{4}{3}$ from each side.

C. Multiply each side by $\frac{4}{3}$.

D. Add $\frac{4}{3}$ to each side.

Solve and check each equation.

35. $-\frac{2}{5}x = 18$ -45 **36.** $\frac{2y}{3} + \frac{y}{4} = 22$ 24 **37.** $\frac{3n - 2}{5} = -7$ -11 **38.** $-\frac{1}{3}(4b + 1) = 5$ -4

39. Suppose you spend $\frac{3}{5}$ of your monthly budget on food and $\frac{1}{4}$ on bus fare. Food and bus fare total \$34/mo. What is your monthly budget? \$40

Using Probability 3-6

Independent events do not affect one another. When the outcome of one event affects the outcome of a second event, the events are **dependent events.**

For independent events A and B:	For dependent events A and B:
$P(A \text{ and } B) = P(A) \cdot P(B)$	$P(A \text{ and } B) = P(A) \cdot P(B \text{ after } A)$

For Exercises 40 and 41, you choose two numbers from a box containing tiles numbered 1–10. State whether the two events are *independent* or *dependent* and then find each probability.

40. P(6 and an even number) without replacing dependent; $\frac{2}{45}$

41. P(1 and an odd number) with replacing independent; $\frac{1}{20}$

42. Writing Compare what the words *independent* and *dependent* mean in mathematics with what they mean in social studies. See margin.

ALTERNATIVE ASSESSMENT Ask each student to choose one problem from each review section which he or she found particularly difficult. Then have students work together in small peer-tutoring groups to solve these problems.

Getting Ready for Chapter 4

Students may work these exercises independently or in small groups. The skills previewed will help prepare students to work with absolute values and compound inequalities.

Percent Equations 3-7

You can use an equation to solve a percent problem.

The formula to calculate **simple interest** is $I = p \cdot r \cdot t$, where I is the interest, p is the principal, r is the annual interest rate, and t is the time in years.

Write and solve an equation to answer each question. Check your answer.

43. What is 15% of 86? $n = 0.15 \times 86$; 12.9

44. What percent of 5 is 40? $n \times 5 = 40$; 800%

45. 1.8 is 72% of what? $1.8 = 0.72 \times n$; 2.5

46. In one high school, 30 of the school's 800 students work on the school paper. What percent of the students work on the paper? 3.75%

47. *Finance* You invest $2000 in a bank account paying 5.5% simple interest annually. How much interest will you receive after 2 yr? $220

Percent of Change 3-8

The **percent of change** $= \frac{\text{amount of change}}{\text{original amount}}$. If a value increases from its original amount, the percent of change is called the percent of increase. If a value decreases from its original amount, the percent of change is called the **percent of decrease.**

For Exercises 48–50, find each percent of change. Describe the percent of change as a percent of increase or decrease.

48. $75,000 to $85,000 about 13% inc.

49. 20 ft to 15 ft 25% dec.

50. 60 h to 40 h about 33% dec.

51. In Irvine, California, parents donated 220,624 h of service to the schools in the 1993–1994 school year. This increased by about 60,000 hours the next year. Find the percent of increase of donated hours. 27.2%

52. *Open-ended* Describe a situation that involves a percent of increase that is more than 100%. Sample: It costs a coffee shop $.11 to make a cup of tea, which it sells for $.75. The markup is 582%.

Getting Ready for CHAPTER 4

Evaluate each expression.

53. $|-6| - |y|$ for $y = 8$ -2

54. $|t - 7|$ for $t = 5.6$ 1.4

55. $|c| + 9 = -10$ 19

56. $|x| + 3$ for $x = -4$ 7

57. $3z - 4.5$ for $z = 3$ 4.5

58. $13 - 5y$ for $y = -1$ 18

Solve and check.

59. $\frac{2}{3}k = \frac{1}{6}$ $\frac{1}{4}$

60. $\frac{1}{2}b = \frac{4}{7}$ $1\frac{1}{7}$

61. $\frac{1}{3} = \frac{1}{9}d$ 3

62. $\frac{3}{2}l = \frac{2}{3}$ $\frac{4}{9}$

63. $5t + 4 = -16$ -4

64. $7 - m = 2$ 5

65. $2(w - 7) = 90$ 52

66. $3y + 8 = -1$ -3

Assessment

ENHANCED MULTIPLE CHOICE QUESTIONS are more complex than traditional multiple choice questions, which assess only one skill. Enhanced multiple choice questions assess the processes that students use as well as the end results. The questions are written so that students must use more than one strategy to solve the problem. Using multiple strategies is encouraged by the National Council of Teachers of Mathematics (NCTM). **Exercises 15 and 16** are enhanced multiple choice questions.

FREE -RESPONSE QUESTIONS do not give choices of answers. Some exercises have more than one possible answer. Students only need to give one correct answer. **Exercises 1–13, 17–26a, 26c, and 27** are free response questions.

WRITING EXERCISES allow students to describe how they think about and understand the concepts they have learned. **Exercise 26b** is a writing exercise.

OPEN-ENDED PROBLEMS allow for more than one solution. Students must construct their own responses instead of choosing from possible answers. The student responses will help you determine the depth of understanding and any areas of difficulty experienced by students. **Exercise 14 and 28** are open-ended problems. Make sure that students' responses contain the following elements:

- a written description
- a probability problem with a solution

Resources

Teaching Resources

Chapter Support File, Ch. 3

- Chapter Assessment, Forms A and B
- Alternative Assessment

Chapter Assessment, Spanish Resources

Teacher's Edition

See also p. 106E for assessment options.

Software Computer Item Generator

page 154 **Assessment**

14. **Sample: Jeremy gets a commission on his paper route for one fifth of the amount collected. Last week he got \$4 in commission. How much money did he collect for the newspapers?**

26b. **Sample: Use the second row. Subtract the amount in the first column from the amount in the second column, divide by the first amount and multiply by 100.**

c. **Sample: Use the third row. Repeat the steps in part (b).**

28. **Sample: There are 15 gems in a sack. Exactly one is an onyx and the rest are opals. You take one gem at a time out of the sack. What is the probability that the first gem you pull out is an opal and the second the onyx?**

3 Assessment

Solve and check each equation.

1. $5n = -20$ **−4**
2. $t + 7 = 4$ **−3**
3. $\frac{r}{3} = 21$ **63**
4. $u - 8 = -15$ **−7**
5. $3q - 2 = 10$ **4**
6. $-2z + 1 = -9$ **5**

Write and solve each equation modeled by tiles.

7. 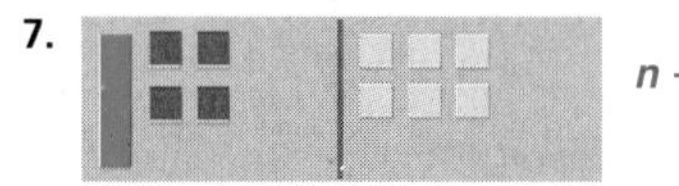 **n − 4 = 6; 10**

8. 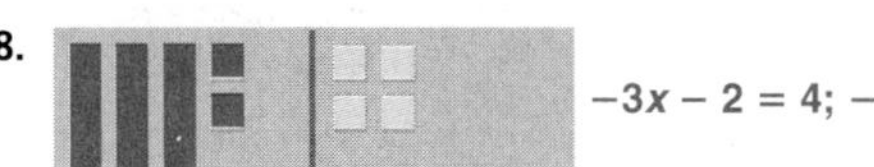 **−3x − 2 = 4; −2**

Solve and check each equation.

9. $3w + 2 - w = -4$ **−3**
10. $-9.5b + 4.5b = 25$ **−5**
11. $\frac{1}{4}(k - 1) = 10$ **41**
12. $6(y + 3) = 24$ **1**
13. $\frac{5n + 1}{8} = \frac{1}{2}$ **$\frac{3}{5}$**

14. *Open-ended* Describe a situation that you can model with the equation $\frac{m}{5} = 4$. **See margin.**

15. If $2t + 3 = -9$, what is the value of $-3t - 7$? **B**
    A. −6 B. 11
    C. −3 D. 2
    E. −9

16. Which equation is not equivalent to the equation $2x - 4 = -9$? **A**
    A. $4 = -7 - 2x$ B. $-4 = -7 - 2x$
    C. $4 = 2x + 7$ D. $2x = -3$
    E. $-2x + 4 = 7$

Solve an equation to answer each question.

17. What is 16% of 250? **40**
18. 8 is what percent of 12.5? **64%**
19. 19 is 95% of what? **20**

20. *Finance* You invest \$500 for three years and receive \$60 in simple interest. What is the annual interest rate? **4%**

21. Suppose a person contributes 6% of her salary to her retirement account. She works 20 h/wk at \$5.50/h. Find her weekly contribution. **\$6.60**

Calculate the percent of change. Describe each as a percent of increase or a percent of decrease.

22. \$4.50/h to \$5/h **11.1% inc.**
23. 60 km/h to 45 km/h **25% dec.**
24. 150 lb to 135 lb **10% dec.**
25. \$18 to \$24 **33% inc.**

26. The game Monopoly™ was introduced in 1935. The table shows how much some amounts in the game should have increased to have kept up with inflation.

	1935	1995
Total money in game	\$15,140	\$184,794
Amount each player starts the game with	1500	18,308
Park Place rent with no houses	35	428
Money collected when passing GO	200	2441

Source: Parker Brothers

a. Estimate the percent of inflation from 1935 to 1995 by finding the percent of increase in any one of the dollar amounts. **about 1120%**
b. *Writing* Describe the steps you used to calculate your answer to part (a).
c. Explain another way to get the same result.
b–c. See margin.

27. You have 8 red checkers and 8 black checkers in a bag. You choose two checkers. Find each probability.
    a. P(red and red) with replacing **$\frac{1}{4}$**
    b. P(red and black) without replacing **$\frac{4}{15}$**
    c. P(black and red) with replacing **$\frac{1}{4}$**

28. *Open-ended* Write and solve a probability problem involving dependent events. **See margin.**

Preparing for Standardized Tests

Standardized tests, such as those administered for state assessment, the SAT, or the ACT, include regular math questions, quantitative comparison questions, open-ended problems, and free-response questions (which the SAT calls *grid-ins*).

MULTIPLE CHOICE QUESTIONS are followed by five answer choices, one of which is correct. Exercises 1–4, 6, and 8 are multiple choice questions.

QUANTITATIVE COMPARISON QUESTIONS ask students to compare two quantities. Exercises 5 and 7 are quantitative comparison questions.

FREE-RESPONSE QUESTIONS do not give answer choices. Students must provide the one correct answer on their own. Exercises 9–11, 13, and 14 are free-response questions.

OPEN-ENDED PROBLEMS allow for more than one solution. Students must construct their own responses instead of choosing a single answer. The responses students give will help you determine the depth of their understanding and what difficulties, if any, they are experiencing. Exercise 12 is an open-ended problem.

STANDARDIZED TEST TIPS Standardized test questions usually appear in order of increasing difficulty. Therefore, if a question toward the beginning of the test has an easily identifiable answer, that choice is probably correct. However, if a problem toward the final part of a test seems to have an "obvious" answer, it is quite likely a "distracter." Make sure students know to reexamine this question very carefully!

3 Preparing for Standardized Tests

Choose the correct letter.

1. In which quadrants would the following points be graphed? (–2, 5) (3, –1) (5, –4) (–1, 4) **E**
   A. I and II B. II and III C. III and IV
   D. I and IV E. II and IV

2. Consider the function $f(x) = x^2 - 3$. Which of the following are true? **D**
   I. $f(1) > f(0)$ II. $f(2) > f(-3)$
   III. $f(2) = f(-2)$ IV. $f(-1) = f(3)$
   A. I only B. II only
   C. II and IV D. I and III

3. Match the graph with its equation. **D**

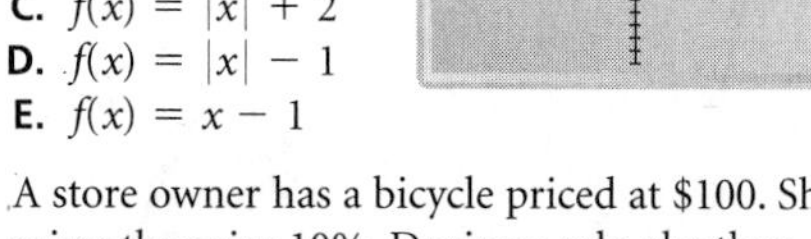

   A. $f(x) = |x| + 1$
   B. $f(x) = (x + 1)^2$
   C. $f(x) = |x| + 2$
   D. $f(x) = |x| - 1$
   E. $f(x) = x - 1$

4. A store owner has a bicycle priced at $100. She raises the price 10%. During a sale, she then lowers the price 10%. What is the new price of the bicycle? **C**
   A. $100 B. $101 C. $99
   D. $98 E. none of the above

5. Consider the function $f(x) = x^3 - x$. Compare the quantities in Column A and Column B. **A**

Column A	Column B
$f(0)$	$f(-2)$

   A. The quantity in Column A is greater.
   B. The quantity in Column B is greater.
   C. The two quantities are equal.
   D. The relationship cannot be determined on the basis of the information given.

6. If $\frac{2x}{3} = 5$, $\frac{2y - 2}{4} = 3$, and $\frac{z}{2} + \frac{z}{3} = 5$, which of the following is true? **A**
   A. $x > y$ B. $y < z$ C. $x = z$
   D. $z > x$ E. $x = y + z$

7. A number cube is rolled. Compare the quantities in Column A and Column B. **C**

Column A	Column B
the probability of rolling a number 5 or greater.	the probability of rolling a number 2 or less.

   A. The quantity in Column A is greater.
   B. The quantity in Column B is greater.
   C. The two quantities are equal.
   D. The relationship cannot be determined on the basis of the information given.

8. A bag contains 10 red marbles and 20 white marbles. You draw a marble, keep it, and draw another. What is the probability of drawing two red marbles? **C**
   A. $\frac{1}{3}$ B. $\frac{1}{9}$ C. $\frac{3}{29}$
   D. $\frac{1}{10}$ E. $\frac{1}{2}$

For Exercises 9–14, write your answer.

9. A table of values for the linear function $g(x)$ is given. Write an equation describing the function. **$g(x) = x - 3$**

x	1	2	3	4	5	6	7
$g(x)$	−2	−1	0	1	2	3	4

10. Solve $5(x - 7) - 2x = 4$. **13**

11. What is the median of the following values? −3, 7, 5, 5, −1, 0, 1, −4, 0, 1, 0 **0**

12. **Open-ended** Give an example of a real-life situation that you could model with the equation $3n + 5 = 68$. Define what the variable represents and then solve the equation. **See margin.**

13. Simplify $9a + 3b - 3 - 4a + 7 - 8b$. **$5a - 5b + 4$**

14. **Writing** Describe the shape of the graph of each type of function: linear, quadratic, and absolute value. **straight line; U-shaped; V-shaped**

Resources

Teaching Resources

Chapter Support File, Ch. 3
- Standardized Test Practice
- Cumulative Review

Teacher's Edition

See also p. 106E for assessment options.

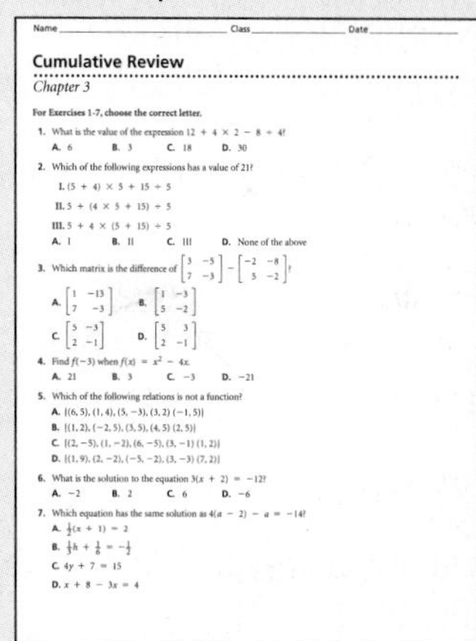

Name ______ Class ______ Date ______

Cumulative Review
Chapter 3

For Exercises 1-7, choose the correct letter.

1. What is the value of the expression $12 + 4 \times 2 - 8 \div 4$?
   A. 6 B. 3 C. 18 D. 30
2. Which of the following expressions has a value of 21?
   I. $(5 + 4) \times 5 + 15 \div 5$
   II. $5 + (4 \times 5 + 15) \div 5$
   III. $5 + 4 \times (5 + 15) \div 5$
   A. I B. II C. III D. None of the above
3. Which matrix is the difference of $\begin{bmatrix} 3 & -5 \\ 7 & -3 \end{bmatrix} - \begin{bmatrix} -2 & -8 \\ 5 & -2 \end{bmatrix}$?
   A. $\begin{bmatrix} 1 & -13 \\ 7 & -3 \end{bmatrix}$ B. $\begin{bmatrix} 1 & -3 \\ 5 & -2 \end{bmatrix}$
   C. $\begin{bmatrix} 5 & -3 \\ 2 & -1 \end{bmatrix}$ D. $\begin{bmatrix} 5 & 3 \\ 2 & -1 \end{bmatrix}$
4. Find $f(-3)$ when $f(x) = x^2 - 4x$.
   A. 21 B. 3 C. −3 D. −21
5. Which of the following relations is not a function?
   A. {(6, 5), (1, 4), (5, −3), (3, 2) (−1, 5)}
   B. {(1, 2), (−2, 5), (3, 5), (4, 5) (2, 5)}
   C. {(2, −5), (1, −2), (6, −5), (3, −1) (1, 2)}
   D. {(1, 9), (2, −2), (−5, −2), (3, −3) (7, 2)}
6. What is the solution to the equation $3(x + 2) = -12$?
   A. −2 B. 2 C. 6 D. −6
7. Which equation has the same solution as $4(a - 2) - a = -14$?
   A. $\frac{1}{2}(x + 1) = 2$
   B. $\frac{1}{3}h + \frac{1}{6} = -\frac{1}{2}$
   C. $4y + 7 = 15$
   D. $x + 8 - 3x = 4$

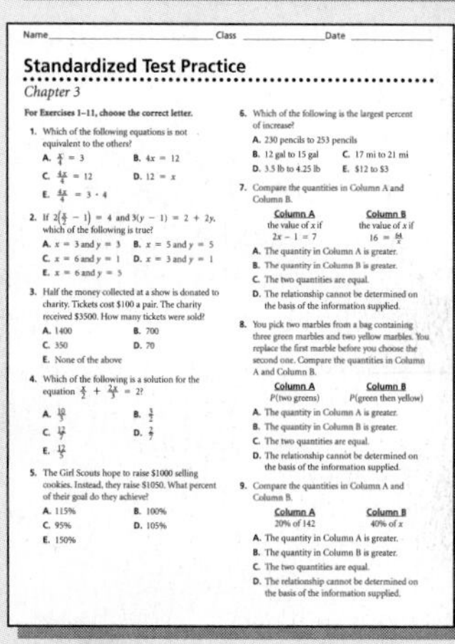

Name ______ Class ______ Date ______

Standardized Test Practice
Chapter 3

For Exercises 1–11, choose the correct letter.

1. Which of the following equations is not equivalent to the others?
   A. $\frac{x}{4} = 3$ B. $4x = 12$
   C. $\frac{4x}{4} = 12$ D. $12 = x$
   E. $\frac{4x}{4} = 3 \cdot 4$
2. If $2\left(\frac{x}{2} - 1\right) = 4$ and $3(y - 1) = 2 + 2y$, which of the following is true?
   A. $x = 3$ and $y = 3$ B. $x = 5$ and $y = 5$
   C. $x = 6$ and $y = 1$ D. $x = 3$ and $y = 1$
   E. $x = 6$ and $y = 5$
3. Half the money collected at a show is donated to charity. Tickets cost $100 a pair. The charity received $3500. How many tickets were sold?
   A. 1400 B. 700
   C. 350 D. 70
   E. None of the above
4. Which of the following is a solution for the equation $\frac{x}{2} + \frac{2x}{3} = 2$?
   A. $\frac{10}{3}$ B. $\frac{3}{2}$
   C. $\frac{12}{7}$ D. $\frac{2}{7}$
   E. $\frac{12}{5}$
5. The Girl Scouts hope to raise $1000 selling cookies. Instead, they raise $1050. What percent of their goal do they achieve?
   A. 115% B. 100%
   C. 95% D. 105%
   E. 150%
6. Which of the following is the largest percent of increase?
   A. 230 pencils to 253 pencils
   B. 12 gal to 15 gal C. 17 mi to 21 mi
   D. 3.5 lb to 4.25 lb E. $12 to $3
7. Compare the quantities in Column A and Column B.

Column A	Column B
the value of x if $2x - 1 = 7$	the value of x if $16 = \frac{64}{x}$

   A. The quantity in Column A is greater.
   B. The quantity in Column B is greater.
   C. The two quantities are equal.
   D. The relationship cannot be determined on the basis of the information supplied.
8. You pick two marbles from a bag containing three green marbles and two yellow marbles. You replace the first marble before you choose the second one. Compare the quantities in Column A and Column B.

Column A	Column B
P(two greens)	P(green then yellow)

   A. The quantity in Column A is greater.
   B. The quantity in Column B is greater.
   C. The two quantities are equal.
   D. The relationship cannot be determined on the basis of the information supplied.
9. Compare the quantities in Column A and Column B.

Column A	Column B
20% of 142	40% of x

   A. The quantity in Column A is greater.
   B. The quantity in Column B is greater.
   C. The two quantities are equal.
   D. The relationship cannot be determined on the basis of the information supplied.

page 155 Preparing for Standardized Tests

12. See back of book.

CHAPTER 4 Equations and Inequalities

To accommodate flexible scheduling, some lessons are divided into parts. Assignment Options are given in the Lesson Planning Options for each lesson.

4–1 Using Proportions (pp. 158–162)
- Part 1 Using Properties of Inequalities
- Part 2 Using Cross Products
- Part 3 Solving Percent Problems Using Proportions

Key Terms: ratio, scale, proportion, cross products, similar figures

4–2 Equations with Variables on Both Side (pp. 163–168)
- Part 1 Using Tiles to Solve Equations
- Part 2 Using Properties of Equality
- Part 3 Solving Special Types of Equations

Key Terms: identity

4–3 Solving Absolute Value Equations (pp. 170–174)
- Part 1 Solving Absolute Value Equations
- Part 2 Modeling by Writing Equations

4–4 Transforming Formulas (pp. 175–178)

Key Terms: literal equation

4–5 Solving Inequalities Using Addition and Subtraction (pp. 179–184)
- Part 1 Graphing and Writing Inequalities
- Part 2 Using Addition to Solve Inequalities
- Part 3 Using Subtraction to Solve Inequalities

Key Terms: solution of the inequality, equivalent inequalities

4–6 Solving Inequalities Using Multiplication and Division (pp. 185–189)
- Part 1 Solving Inequalities Using Multiplication
- Part 2 Solving inequalities Using Division

4–7 Solving Multi-Step Inequalities (pp. 190–194)
- Part 1 Solving with Variables on One Side
- Part 2 Solving with Variables on Both Sides

4–8 Compound Inequalities (pp. 195–200)
- Part 1 Solving Compound Inequalities Joined by *And*
- Part 2 Solving Compound Inequalites Joined by *Or*
- Part 3 Solving Absolute Value Inequalities

Key Terms: compound inequalities, absolute value

4–9 Interpreting Solutions (pp. 202–205)
- Part 1 Solving Inequalities Given a Replacement Set
- Part 2 Determining a Reasonable Answer

Key Terms: replacement set

PACING OPTIONS

This chart suggests pacing only for the core lessons and their parts, and it is provided merely as a possible guide. It will help you determine how much time you have in your schedule to cover other features, such as the Chapter Project, Math Toolboxes, Wrap Up, and Assessment.

	1 Class Period	1 Class Period	1 Class Period	1 Class Period	1 Class Period	1 Class Period	1 Class Period	1 Class Period	1 Class Period
Traditional (40–45 min class periods)	4-1 Part 1, 4-1 Part 2	4-1 Part 3	4-2 Part 1, 4-2 Part 2	4-2 Part 3	4-3 Part 1, 4-3 Part 2	4-4	4-5 Part 1, 4-5 Part 2	4-5 Part 3	4-6 Part 1, 4-6 Part 2
Two-Year Algebra (40–45 min class periods)	4-1 Part 1, 4-1 Part 2	4-1 Part 1	4-2 Part 1	4-2 Part 2	4-2 Part 3	4-3 Part 1	4-3 Part 2	4-4	4-5 Part 1
Block Scheduling (90 min class periods)	4-1 Part 1, 4-1 Part 2, 4-1 Part 3	4-2 Part 1, 4-2 Part 2, 4-2 Part 3	4-3 Part 1, 4-3 Part 2, 4-4	4-5 Part 1, 4-5 Part 2, 4-5 Part 3	4-6 Part 1, 4-6 Part 2, 4-7 Part 1, 4-7 Part 2	4-8 Part 1, 4-8 Part 2, 4-8 Part 3	4-9 Part 1, 4-9 Part 2		

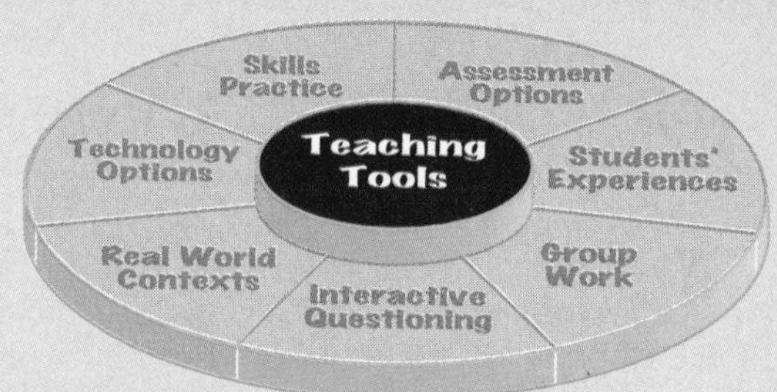

What Students Will Learn and Why

In this chapter, students will build on their knowledge of algebraic concepts and simple equations, learned in Chapter 3, by learning to solve equations and inequalities that involve proportions, variables on both sides, absolute values, and formulas.

Discussing the Chapter/Building on Experience

The concept map below relates chapter topics to real-world applications. You and your class may wish to add to the map or develop maps of your own. The center oval describes the topic of the chapter. The next level displays topics within the lessons. The outer ovals reflect applications of the content. As you and your class build a concept map, invite students to discuss applications with which they are familiar.

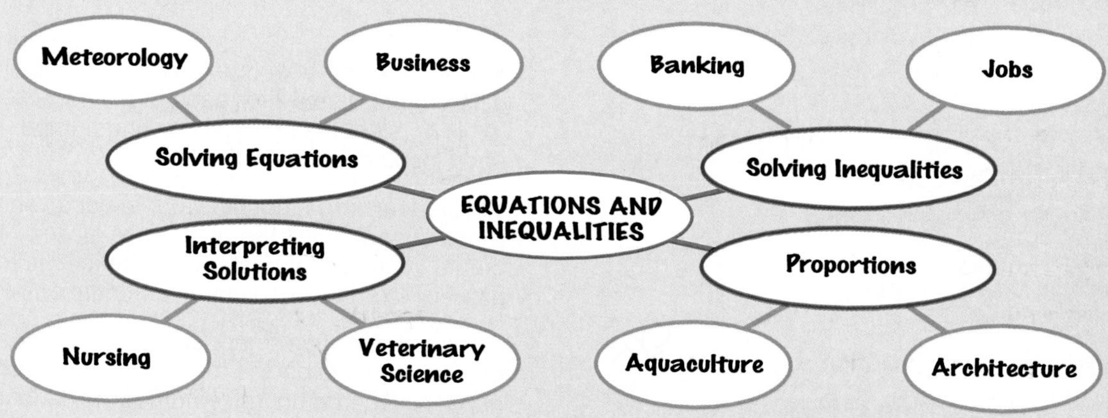

1 Class Period	1 Class Period	1 Class Period	1 Class Period	1 Class Period	1 Class Period	1 Class Period	1 Class Period	1 Class Period	1 Class Period	1 Class Period
4-7 (1) 4-7 (2)	4-8 (1) 4-8 (2)	4-8 (3)	4-9 (1) 4-9 (2)							
4-5 (2)	4-5 (3)	4-6 (1)	4-6 (2)	4-7 (1)	4-7 (2)	4-8 (1)	4-8 (2)	4-8 (3)	4-9 (1)	4-9 (2)

Interactive Questioning Tips

A question is interactive when there is "give and take" between the questioner (teacher and student) and the respondent. When asking critical thinking questions, it is important to encourage higher level thinking. For example, Question 6 in Lesson 4-3, Think and Discuss, asks students to determine if there is a solution for an equation. Students may be very quick with a yes or no answer but when they are asked to explain, they need time to organize an answer.

Skills Practice

Every lesson provides skill practice with Exercises On Your Own and Exercises Mixed Review. The Student Edition includes Checkpoints (pp. 182, 204) and a Cumulative Review (p. 215). In the Teacher's Edition, the Lesson Planning Options section for each lesson lists Prerequisite Skills students should know for that lesson. At the back of the Student Edition, is the Skills Handbook—mini-lessons on math the students may need to review. The Chapter Support File for Chapter 4 in the Teaching Resources box includes two Practice worksheets per lesson, a worksheet for two Checkpoints, and worksheets for Cumulative Review and Standardized Test Preparation.

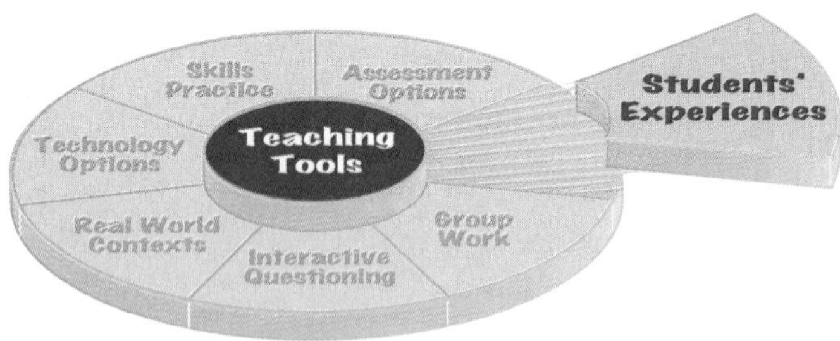

Diverse Learning and Teaching Styles

In your Teacher's Edition, you will find suggestions as to how you can help students complete mathematical tasks in Chapter 4 by reinforcing various learning styles. Here are some examples.

- **Visual learning** rewrite examples of problems using a different color pencil for each of the variables *a*, *b*, *c*, and *x* (p. 176), students practice drawing number lines with open and closed dots (p. 180)
- **Tactile learning** students work in pairs, while one works problems on paper, the other uses tiles (p. 164), design and create instructional posters that describes a step-by-step process to solve absolute value equations (p. 170)
- **Auditory learning** composing a rhyme, poem, song, or other mnemonic device to help them remember what happens when a negative number and inequality are multiplied (p. 187)
- **Kinesthetic learning** students are handed a paper with an integer, rational number, or mixed fraction and they arrange themselves on a number line (p. 196)

Alternative Activity for Lesson 4-2

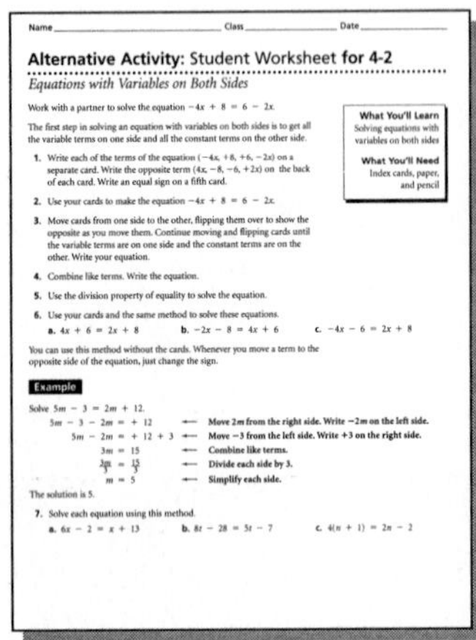
Alternative Activity: Student Worksheet for 4-2

Equations with Variables on Both Sides

for use with Example 1, addresses tactile learning by having students use tiles to solve problems.

Alternative Activity for Lesson 4-3

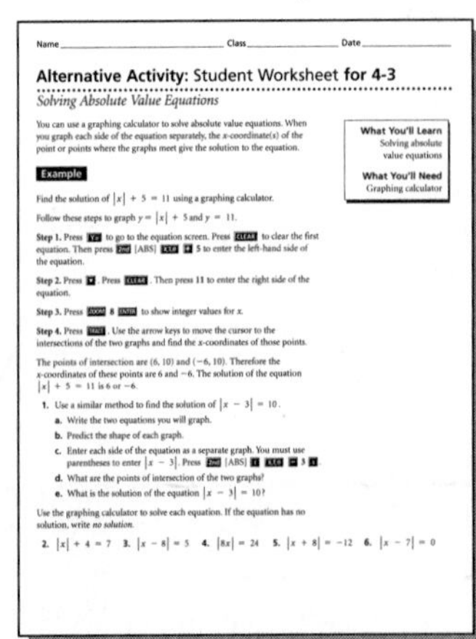
Alternative Activity: Student Worksheet for 4-3

Solving Absolute Value Equations

for use with Example 2, addresses visual learning by having students use a graphing calculator to demonstrate that the equation has two solutions.

Alternative Activity for Lesson 4-8

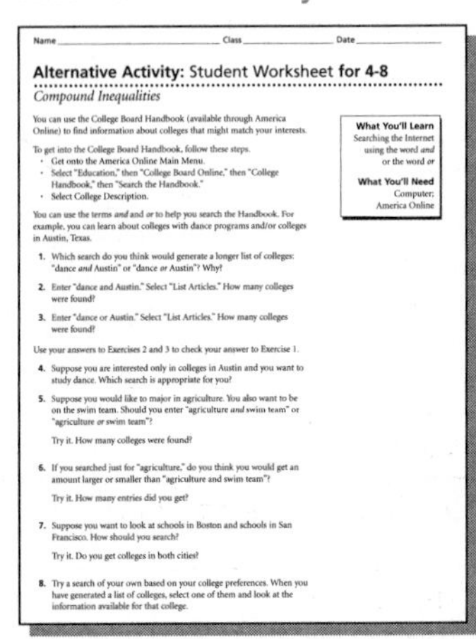
Alternative Activity: Student Worksheet for 4-8

Compound Inequalities

for use with Examples 1 and 3, address visual learning by having students see t results of using *and* and *or* in a search.

Cooperative Learning Tips

Cooperative learning is a dynamic activity. Used effectively, it can help students develop interpersonal skills, learn to perform specific roles in a group, and learn to carry out specific responsibilities. The components of Chapter 4 provide a range of cooperative learning opportunities.

- In the Student Edition, the ***Work Together*** parts of lessons are specifically designed for cooperative learning activities.
- In the Teacher's Edition, you will find helpful hints for addressing diverse learning styles (see page C for Chapter 4). For every lesson, you will find a ***Reteaching Activity,*** which may refer to cooperative learning.

Materials and Manipulatives

The authors expect all students to use scientific calculators. Calculator use is integrated throughout the course.

- **calculator (4-1, 4-2, 4-6, 4-9)**
- **centimeter ruler (4-1)**
- **tiles (4-2)**

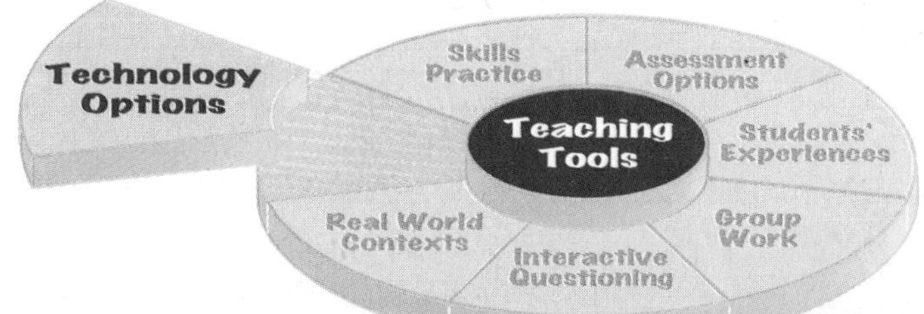

TECHNOLOGY OPTIONS

Technology Tools		Chapter Project	4-1	4-2	4-3	4-4	4-5	4-6	4-7	4-8	4-9
Calculator		Assumed that students have scientific calculators at any time.									
Graphing Calculator	Handbook				✔	✔					
	Student Edition				✔	✔					
Software	Secondary Math Lab Toolkit		✔	✔	✔	✔	✔	✔	✔	✔	✔
	Integrated Math Lab			✔	✔						
	Computer Item Generator		✔	✔	✔	✔	✔	✔	✔	✔	✔
	Student Edition			✔							
Video	Video Field Trip	✔									
CD-ROM	Multimedia Algebra Lab		✔			✔			✔	✔	
Internet		See the Prentice Hall site. (http://www.phschool.com)									

The Prentice Hall Algebra program offers you a rich variety of technology options. Be assured that all these options are provided as a means of enriching the program and are not essential for the successful completion of the course.

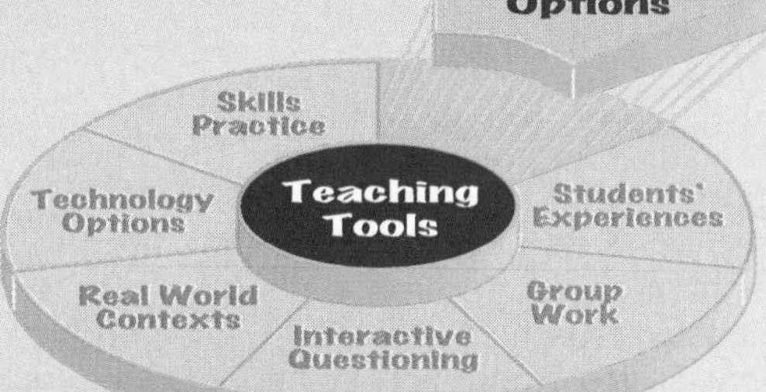

Assessment Options

The Prentice Hall Algebra Program provides you with many options. From these options, you may choose instructional materials and techniques appropriate for your students, or necessary to meet your district's curriculum requirements. As the ***Assessment Options*** chart indicates, the program also supports your teaching efforts by offering you many choices for assessment.

ASSESSMENT OPTIONS

Assessment Support Materials	Chapter Opener	4-1	4-2	4-3	4-4	4-5	4-6	4-7	4-8	4-9	Chapter End
Chapter Project	▲■●	▲■			▲■			▲■	▲■	▲■	▲■
Checkpoints					▲■				▲■		
Self-Assessment			▲■					▲■		▲■	▲■
Writing Assignment		▲	▲	▲	▲	▲	▲	▲	▲	▲	▲●
Chapter Assessment											▲●
Alternative Assessment		■	■	■	■	■	■	■	■	■	■●
Cumulative Review											▲●
Standardized Test Prep		▲■	▲■	▲■		▲■	▲■	▲■			●
Computer Item Generator	Can be used to create custom-made practice or assessment at any time.										

▲ = Student Edition ■ = Teacher's Edition ● = Teaching Resources

Checkpoints

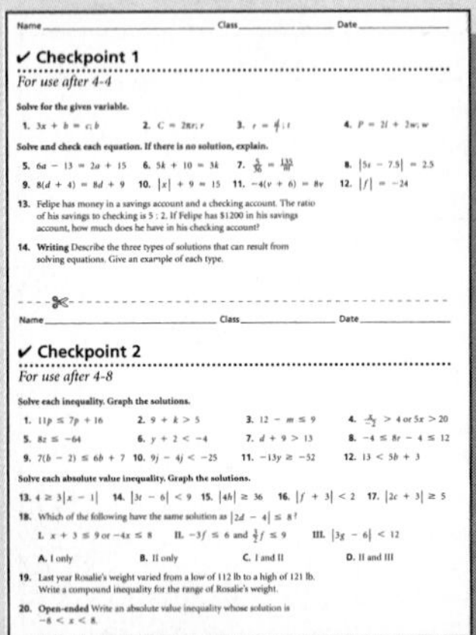
✔ Checkpoint 1

For use after 4-4

✔ Checkpoint 2

For use after 4-8

Alternative Assessment

Alternative Assessment

Chapter 4

Chapter Assessment

Chapter Assessment Form A

Chapter 4

Available in both Form A and Form B

Making the Right Connections

Mathematics is imbedded in nearly every walk of life. The National Council of Teachers of Mathematics (NCTM) encourages educators to recognize these connections and to emphasize them for the purpose of better educating students for success in life and in a global economy. The ***Connections*** chart below highlights these connections for Chapter 4.

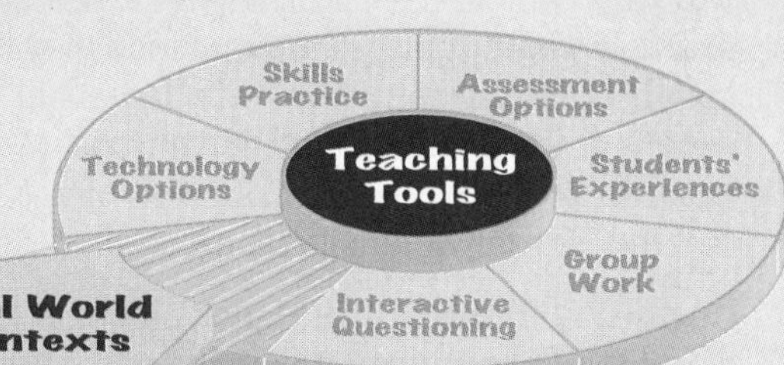

CONNECTIONS

Lesson	Interdisciplinary Connections	Career Prep	Other Real World Connections	Math Integration	NCTM Standards
Chapter Project	Health and Fitness		Health and Fitness		Connections
4-1		Aquaculture Architecture	Travel Hobbies	Geometry Statistics	Algebra Communication Problem Solving Connections
4-2			Transportation Business Recreation Technology	Geometry	Algebra Communication Problem Solving Connections
4-3		Meteorology	Manufacturing Polling Banking		Algebra Communication Problem Solving Connections
4-4	Science History		Travel Recreation Banking	Geometry	Algebra Communication Problem Solving Connections
4-5		Education	Banking		Algebra Communication Problem Solving Connections
4-6	Biology	Construction	Community Service		Algebra Communication Problem Solving Connections
4-7		Design Freight Handling Jobs	Business Recycling	Geometry	Algebra Communication Problem Solving Connections
4-8	Chemistry	Meteorology Medicine	Sports Manufacturing	Geometry	Algebra Communication Problem Solving Connections
4-9		Nursing Jobs Veterinary Science	Car Service Weather		Algebra Communication Problem Solving Connections

CHAPTER 4 Equations and Inequalities

CONNECTING TO PRIOR LEARNING Ask students to give examples of formulas and equations they have used, for instance, in geometry (area of a rectangle equals length times width), or for calculating simple interest ($I = prt$). Challenge students to think of other examples of equations and formulas that model real-world situations.

CULTURAL CONNECTIONS Ask students to share their ideas about the importance of exercise and whether it is addressed in different cultures.

INTERDISCIPLINARY CONNECTIONS Have students find out how many calories are contained in one gram of fat and how much exercise it would take to burn one gram of fat. Then ask students to estimate how much exercise it would take to work off the calories in their favorite dessert.

ABOUT THE PROJECT The Chapter Project gives students an opportunity to learn how mathematics, good health, and physical fitness are connected. In the Find Out questions found throughout the chapter, students use formulas, equations, and inequalities to calculate quantities that they can use to measure their own physical fitness. Students also design an exercise plan for themselves.

Technology Options

Prentice Hall Technology

Video Video Field Trip 4 "Focus on Fitness," a look at very specialized athletic training.

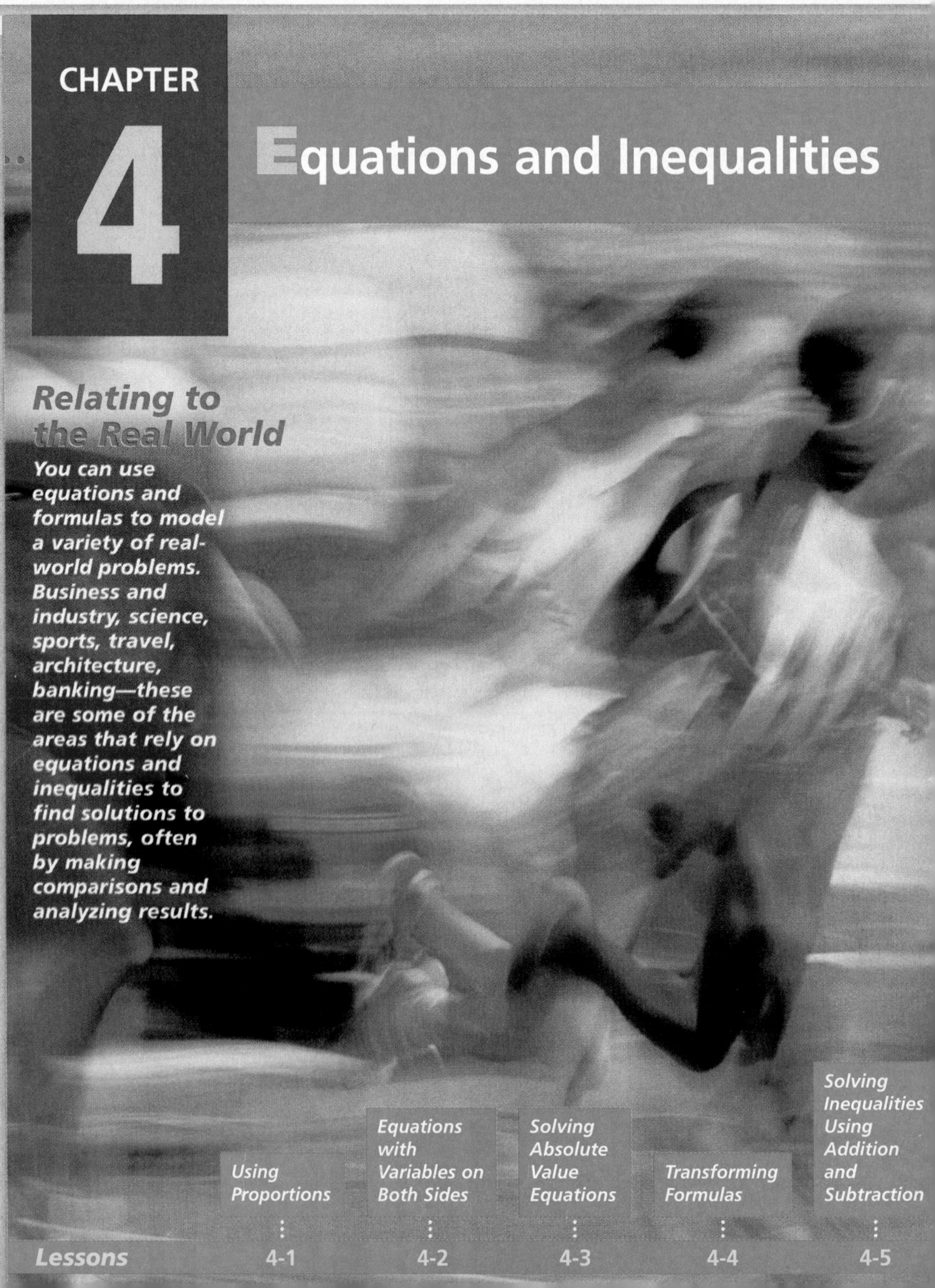

CHAPTER 4 Equations and Inequalities

Relating to the Real World

You can use equations and formulas to model a variety of real-world problems. Business and industry, science, sports, travel, architecture, banking—these are some of the areas that rely on equations and inequalities to find solutions to problems, often by making comparisons and analyzing results.

Launching the Project

PROJECT NOTEBOOK Encourage students to keep all project-related materials in a separate folder or notebook. **See Chapter Project and Scoring Rubric in Chapter Support File.**

- Ask students to estimate how many calories they burn when they exercise.
- Lead a discussion on the effects of exercise on the body. Ask students: *What are some of the metabolic rates that change when we exercise?* **Answers may vary. Samples: heart rate, lung capacity, blood pressure**
- Have the students look at the table on page 166. Explain that, when they begin the Chapter Project, they will choose an exercise from this table. Challenge students to research the calories burned during other activities and when resting.
- Have the students write and solve a proportion to find how many calories they would burn if they performed their chosen activity for 60 min.
- Challenge students to explain why some students burn more calories than others, and what that might mean.

TRACKING THE PROJECT You may wish to have students read the Finishing the Chapter Project on page 210 to help them get an overview of the project. Set benchmark deadlines for students to show you their work in progress.

CHAPTER PROJECT

NO SWEAT!

Your good health and physical fitness will enhance your quality of life for years to come. As you grow older, your needs will change. How much exercise should you get? What should your blood pressure be? You can use formulas and inequalities to describe many aspects of good health.

As you work through the chapter, you will use formulas for physical fitness and health. You will work with equations and inequalities that allow for differences in weight, height, and age. Finally, you will design an exercise plan for yourself.

To help you complete the project:

- p. 162 *Find Out by Calculating*
- p. 178 *Find Out by Solving*
- p. 194 *Find Out by Researching*
- p. 200 *Find Out by Writing*
- p. 205 *Find Out by Interviewing*
- p. 206 *Finishing the Project*

Project Resources

Teaching Resources
Chapter Support File, Ch. 4
- Chapter Project Manager and Scoring Rubric

Transparencies
58

Using the Rubric

Sharing the scoring rubric for the project with your students will alert them to your expectations before they begin work on the project.

As students complete each Find Out question in the chapter, you may wish to have them evaluate their own work or a partner's work, based on the scoring rubric. Students should have the opportunity to revise their work after it has been reviewed.

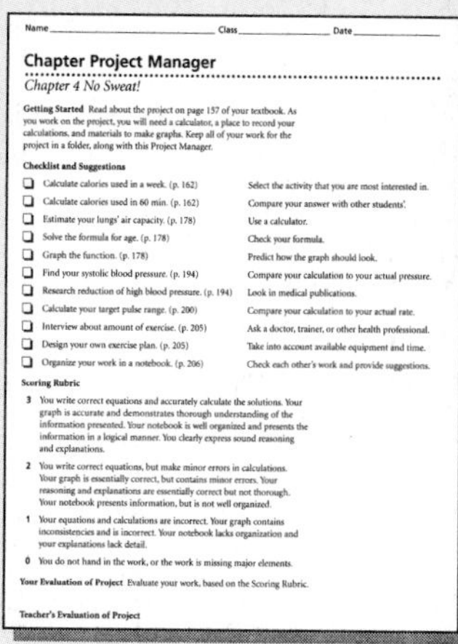

Name ________ Class ________ Date ________

Chapter Project Manager

Chapter 4 No Sweat!

Getting Started Read about the project on page 157 of your textbook. As you work on the project, you will need a calculator, a place to record your calculations, and materials to make graphs. Keep all of your work for the project in a folder, along with this Project Manager.

Checklist and Suggestions

❑ Calculate calories used in a week. (p. 162)	Select the activity that you are most interested in.
❑ Calculate calories used in 60 min. (p. 162)	Compare your answer with other students'.
❑ Estimate your lungs' air capacity. (p. 178)	Use a calculator.
❑ Solve the formula for age. (p. 178)	Check your formula.
❑ Graph the function. (p. 178)	Predict how the graph should look.
❑ Find your systolic blood pressure. (p. 194)	Compare your calculation to your actual pressure.
❑ Research reduction of high blood pressure. (p. 194)	Look in medical publications.
❑ Calculate your target pulse range. (p. 200)	Compare your calculation to your actual rate.
❑ Interview about amount of exercise. (p. 205)	Ask a doctor, trainer, or other health professional.
❑ Design your own exercise plan. (p. 205)	Take into account available equipment and time.
❑ Organize your work in a notebook. (p. 206)	Check each other's work and provide suggestions.

Scoring Rubric

3 You write correct equations and accurately calculate the solutions. Your graph is accurate and demonstrates thorough understanding of the information presented. Your notebook is well organized and presents the information in a logical manner. You clearly express sound reasoning and explanations.

2 You write correct equations, but make minor errors in calculations. Your graph is essentially correct, but contains minor errors. Your reasoning and explanations are essentially correct but not thorough. Your notebook presents information, but is not well organized.

1 Your equations and calculations are incorrect. Your graph contains inconsistencies and is incorrect. Your notebook lacks organization and your explanations lack detail.

0 You do not hand in the work, or the work is missing major elements.

Your Evaluation of Project Evaluate your work, based on the Scoring Rubric.

Teacher's Evaluation of Project

4-1 Teaching Notes

PROBLEM OF THE DAY

To calculate billing charges, a utility company uses the formula $C = 6 + a + b + f$ where a is the charge for the first 500 kWh calculated at $.031/kWh, b is the charge for the electricity used above 500 kWh calculated at $.060/kWh, and f is the fuel charge calculated at $.012/kWh. How much would you expect to pay for 850 kWh? $52.70

Problem of the Day is also available in Transparencies.

CONNECTING TO PRIOR KNOWLEDGE Ask students how many have seen or heard the term *scale model.* Lead a discussion of the meaning of the term *scale model.* Then ask students how the word *scale* affects or changes the meaning of the word *model.*

WORK TOGETHER

AUDITORY LEARNING Ask questions such as these to give students practice in talking and hearing about the concepts in this lesson:

- *What is the ratio of erasers to pieces of chalk in this room?*
- *What is the ratio of students who are left-handed to students who are right-handed?*
- *What is the ratio of desks to chairs?*

Discuss the definition of *ratio* before you organize the students in pairs. Explain that there are two different notations for ratio: a fraction and a comparison using a colon. Point out that we read 5 : 2 as "five to two."

Lesson Planning Options

Prerequisite Skills

- Understanding the terms *numerator* and *denominator*
- Finding common denominators

Assignment Options

To provide flexible scheduling, this lesson can be subdivided into parts.

1. **Core** 1–6, 24–27
   Extension 28–31, 38, 39
2. **Core** 7–8, 21
   Extension 23, 32–35, 40–42
3. **Core** 9–20
   Extension 22, 36, 37, 43

Use Mixed Review to maintain skills.

Resources

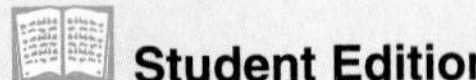

Student Edition

Skills Handbook, pp. 575, 577
Extra Practice, p. 559
Glossary/Study Guide

Teaching Resources

Chapter Support File, Ch. 4
- Practice 4-1 (two worksheets)
- Reteaching 4-1

Classroom Manager 4-1
Glossary, Spanish Resources
Two-Year Algebra Handbook 4-1

Transparencies
10, 59

158

Connections Aquaculture . . . and more

4-1 Using Proportions

What You'll Learn
- Solving proportions
- Using proportions to solve real-world problems

...And Why

To solve problems by comparing and evaluating quantities

What You'll Need
- centimeter ruler
- calculator

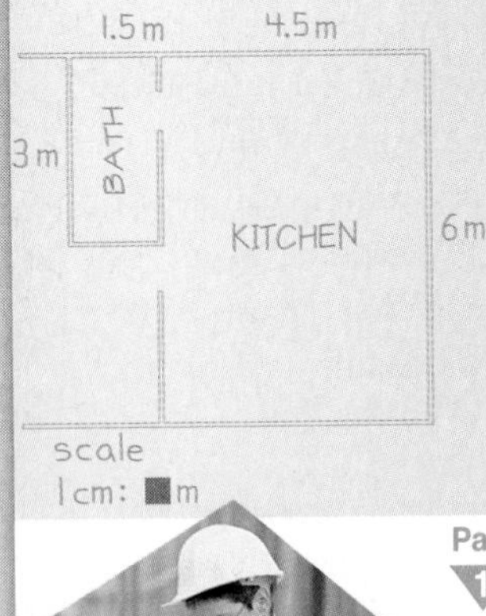

WORK TOGETHER

A **ratio** is a comparison of two numbers by division. For example, $\frac{3}{4}$ and 5 : 2 are ratios. The **scale** of a blueprint is the ratio of a length on the blueprint to the actual length it represents.

1. **a.** Work with a partner. Measure the length and width of the two rooms in the apartment blueprint in centimeters. a–b. See below.
   b. For the two rooms, record the measurements in a table like the one below. Complete the table by finding the ratios of the blueprint measurements to the actual measurements.

	Blueprint	Actual	Blueprint : Actual (cm : m)
Length	■ cm 2; 4	■ m 3; 6	■ : ■ 2 : 3; 4 : 6
Width	■ cm 1; 3	■ m 1.5; 4.5	■ : ■ 1 : 1.5; 3 : 4.5

2. Use the ratios in your table to complete the blueprint scale. 1 cm : 1.5 m

THINK AND DISCUSS

Using Properties of Equality

Part 1

A **proportion** is a statement that two ratios are equal. Another way to write the sample at the right is 3 : 4 = 12 : 16. Read this "3 is to 4 as 12 is to 16."

$$\frac{3}{4} = \frac{12}{16}$$

To solve a proportion with a variable, you can use the multiplication property of equality.

Example 1

Solve $\frac{t}{9} = \frac{5}{6}$.

$\frac{t}{9} \cdot 54 = \frac{5}{6} \cdot 54$ ← Multiply each side by a common denominator such as 54.

$6t = 45$

$\frac{6t}{6} = \frac{45}{6}$ ← Divide each side by 6.

$t = 7.5$

The solution is 7.5. The ratios $\frac{7.5}{9}$ and $\frac{5}{6}$ are equal.

3. Instead of 54, would another number have worked in Example 1? What number is the best choice? Why?
   Yes; 18; it is the least common denominator.

QUICK REVIEW

Multiplication Property of Equality

For any numbers a, b, and c, if $a = b$, then $ac = bc$.

THINK AND DISCUSS

ERROR ALERT! Some students may not understand the difference between a *ratio* and a *proportion.* **Remediation:** Write examples of each on the board. Have students tell which are ratios and which are proportions and explain why.

Example 1

Write on the board: $\frac{t}{9} = \frac{5}{6}$. Ask a student to write this proportion using ratio symbols. Then ask another student to read the proportion. Write on the board: $\frac{a}{b} = \frac{c}{d}$. Ask students:

- *Why can neither b nor d be equal to zero?* Division by zero is undefined.

MAKING CONNECTIONS The estimated ratio of students with engineering degrees to students with law degrees graduating each year in the United States is 1 : 50; the estimated ratio in Japan (each year) is 10 : 1.

Example 2

Question 5a Remind students that they can write any integer as a fraction. The numerator is the given integer, and the denominator is 1.

Example 3

GEOMETRY This example shows how to use proportions to solve problems about similar figures.

Part 2

Using Cross Products

The numerators and denominators of the ratios that form a proportion have a special relationship. The **cross products** of a proportion are equal.

$\frac{3}{4} = \frac{12}{16}$ 3 · 16 and 4 · 12 are cross products. 3 · 16 = 48 and 4 · 12 = 48

Cross Products of a Proportion

In a proportion, where $b \neq 0$ and $d \neq 0$; if $\frac{a}{b} = \frac{c}{d}$ then $ad = bc$.
Example: $\frac{2}{3} = \frac{8}{12}$ so $2 \cdot 12 = 3 \cdot 8$.

4. Use cross products to **justify** each statement.
a. $\frac{8}{3}$ and $\frac{4}{1.5}$ form a proportion. $8 \cdot 1.5 = 3 \cdot 4$
b. $\frac{5}{8}$ and $\frac{7}{10}$ do not form a proportion. $5 \cdot 10 \neq 8 \cdot 7$

Another way to solve a proportion with a variable is to use cross products.

Example 2

Solve $\frac{y}{2.5} = -\frac{3}{4}$.

$y(4) = (2.5)(-3)$ ← Use cross products.
$\frac{4y}{4} = \frac{-7.5}{4}$ ← Divide each side by 4.
$y = -1.875$ ← Simplify.

The solution is -1.875.

PROBLEM SOLVING
Look Back What steps would you take to solve this proportion using the multiplication property of equality?

5. a. Is $-4 = \frac{x}{16}$ a proportion? Why or why not?
b. Explain how to solve for x. Find x.
a. Answers may vary. Sample: No; -4 is not a ratio.
b. You need to multiply both sides by 16; -64.

Similar figures are figures that have the same shape but not necessarily the same size. The corresponding angles of similar figures are equal and the corresponding sides are in proportion.

Example 3

Geometry In the figure, $\triangle ABC$ is similar to $\triangle DFE$. Find length DE.

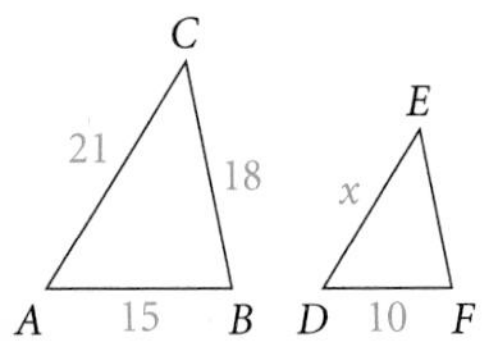

Define x = unknown length DE

Relate $\frac{\text{length } AB}{\text{length } DF} = \frac{\text{length } AC}{\text{length } DE}$

Write $\frac{15}{10} = \frac{21}{x}$ ← Write a proportion comparing the lengths of corresponding sides.

$15x = 10(21)$ ← Use cross products.

$\frac{15x}{15} = \frac{210}{15}$ ← Divide each side by 15.

$x = 14$

The length $DE = 14$. ← Simplify.

Additional Examples

FOR EXAMPLE 1

Solve $\frac{7}{3} = \frac{a}{6}$ using the multiplication property of equality. $a = 14$

FOR EXAMPLE 2

Solve $\frac{3}{x} = \frac{2}{1}$ using cross products. $x = 1.5$

Discussion: *What are the advantages and disadvantages of solving a proportion problem using cross products instead of the multiplication property of equality?*

FOR EXAMPLE 4

On a seven-question biology quiz, Clint answers six questions correctly. Use a proportion to find out the percent of questions he answered correctly. Round your answer to the nearest percent. $\frac{6}{7} = \frac{x}{100}$; $x = 86\%$

Discussion: *How could this method be used to find the percent of correct answers for a quiz of any length?*

Example 4 **Relating to the Real World** 

Point out that the solution of the equation is not the same as the answer to the word problem. To find the answer you must first round *n* to the nearest tenth. Since *n* represents the part per hundred, the answer to the problem is best expressed as a percent.

CRITICAL THINKING Question 8 Suggest that students begin by writing a proportion for each of the three questions. They can then identify differences among the three proportions.

KINESTHETIC LEARNING Challenge students to draw a scale model of the classroom floor. Suggest that each pair of students choose a different scale such as 1 cm = length of one foot, 1 in. = one meter stick, or 1 cm = one pace.

Exercises ON YOUR OWN

CONNECTING TO THE STUDENTS' WORLD Ask a student who is familiar with scale models to describe them to the class. Alternatively, have students visit hobby or toy stores in your area to find out the scales typically used on models and miniatures. Use their findings to develop questions.

OPEN-ENDED Exercise 8 Have volunteers share their problems with the class, or have students exchange problems with a partner.

GEOMETRY Exercise 22 This exercise can help you assess students' understanding of the relationship between similar triangles and proportions.

Technology Options

For exercises 22 and 36, students may want to use geometry software to find the length of the sides of the similar triangle and rectangle respectively.

Prentice Hall Technology 

 **Software** • Secondary Math Lab Toolkit • Computer Item Generator 4-1

CD-ROM Multimedia Algebra Lab 4

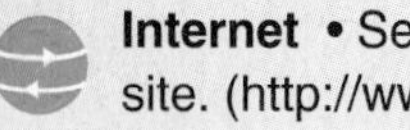 **Internet** • See the Prentice Hall site. (http://www.phschool.com)

QUICK REVIEW

The *perimeter* is the sum of the lengths of the sides of a figure.

6\. a. *Try This* Find length EF in the figure shown in Example 3. 12

b. What is the perimeter of $\triangle DFE$? 36

c. Write a ratio comparing the perimeters of $\triangle DFE$ and $\triangle ABC$. $\frac{36}{54}$

d. How does the ratio of the perimeters in part (c) compare with the ratio of length DE to length AC? The ratios are equal.

Part 3

Solving Percent Problems Using Proportions

QUICK REVIEW

$n\% = \frac{n}{100}$ ← part ← whole

In Chapter 3 you used equations to solve problems involving percents. Recall that percent is a ratio that compares a number to 100. You can use proportions to solve word problems that involve percent.

Example 4 **Relating to the Real World** 

Aquaculture In 1994, U.S. trout farms produced 52,100,000 lb of trout. Suppose a trout farmer raised 858,000 lb of trout. What percent of the 1994 U.S. trout production did the farmer raise? Round your answer to the nearest tenth of a percent.

Define n = the farmer's part of the trout production

Relate 858,000 is what percent of 52,100,000?

Write $\frac{858{,}000}{52{,}100{,}000} = \frac{n}{100}$ ← part ← whole

$858{,}000(100) = 52{,}100{,}000n$ ← Use cross products.

$\frac{858{,}000(100)}{52{,}100{,}000} = \frac{52{,}100{,}000n}{52{,}100{,}000}$ ← Divide each side by 52,100,000.

$\frac{858{,}000(100)}{52{,}100{,}000} = n$ ← Write in a calculator-ready form.

$1.646833013 = n$ ← Use a calculator.

$1.6 \approx n$ ← Round to the nearest tenth.

The farmer raised about 1.6% of the 1994 U.S. trout production.

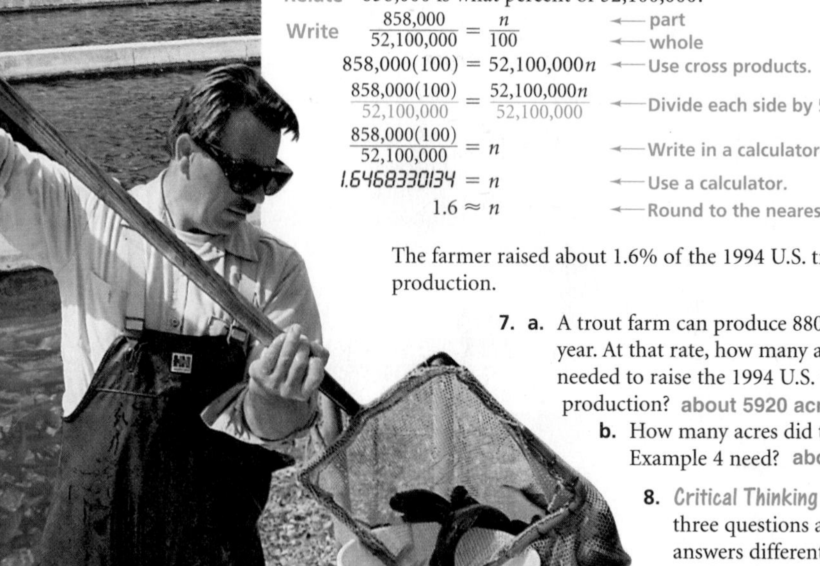

7\. a. A trout farm can produce 8800 lb/acre each year. At that rate, how many acres were needed to raise the 1994 U.S. trout production? about 5920 acres

b. How many acres did the farmer in Example 4 need? about 97.5 acres

8\. *Critical Thinking* How are these three questions and their answers different from each other?

- 10 is what percent of 25?
- What number is 10% of 25?
- 10% of what number is 25?

You write a different proportion to answer each question. Since the variable is in a different place in each proportion, the answers are different.

MENTAL MATH Exercises 24–27 Have students share the methods they used. Some may use cross products, while others may use equivalent fractions. Help students see that different methods produce the same result.

ALTERNATIVE ASSESSMENT Exercises 28–35 Ask students to express these ratio equations orally. This will help you assess the students' ability to translate word problems into symbols.

Exercises 28–35 Remind students to refer to Example 4 for help with writing in calculator-ready form.

Exercise 38b Some students may try to find the length of the segment (median) that connects the student's home to school. Have students trace the triangle on paper. Then use a colored pencil or pen to draw the route the student travels to school each day.

WRITING Exercise 40 Read students' responses to help you identify problem areas. Students may have difficulties in using cross products or solving proportions.

Exercises 42–43 Encourage students to estimate the answer first. After calculating the solution, they can use their estimate to assess whether the answer is reasonable.

DIVERSITY Ask a student who is mobility-impaired to describe some sporting activities popular with impaired athletes. For example, beepball is a form of baseball that uses bases and balls that beep and buzz for players with limited or no vision.

Exercises ON YOUR OWN

1. yes; $6 \cdot 20 = 8 \cdot 15$ 3. yes; $-0.12 \cdot 0.5 = -0.4 \cdot 0.15$ 5. no; $-3 \cdot 25 \neq -1 \cdot 100$

Which pairs of ratios could form a proportion? Justify each answer.

1. $\frac{6}{8}, \frac{15}{20}$ 2. $\frac{9}{12}, \frac{4}{5}$ 3. $-\frac{0.12}{0.15}, -\frac{0.4}{0.5}$ 4. $\frac{5}{6}, \frac{20}{24}$ 5. $-\frac{3}{100}, -\frac{1}{25}$ 6. $\frac{51}{3}, \frac{17}{1}$

2. no; $9 \cdot 5 \neq 4 \cdot 12$ 4. yes; $5 \cdot 24 = 6 \cdot 20$ 6. yes; $51 \cdot 1 = 17 \cdot 3$

7. A canary's heart beats 130 times in 12 s. Use a proportion to find how many times its heart beats in 40 s. about 433 times

8. *Open-ended* Write a problem that you can solve using the proportion $\frac{2}{5} = \frac{x}{9}$. Then show the solution to your problem. Sample: A recipe for bread uses 2 cups of liquid for every 5 cups of flour. You have 9 cups of flour. How many cups of liquid do you need? 3.6 cups

Use a proportion to answer Exercises 9–20. Round your answer to the nearest tenth or to the nearest tenth of a percent.

9. Find 45% of \$120. \$54
10. What percent of 200 is 25? 12.5
11. 17 is what percent of 85? 20%
12. 12.5 is 75% of what number? 16.7
13. What is $33\frac{1}{3}$% of 150? 50
14. 22 is 80% of what number? 27.5
15. What percent of 87 is 24? 27.6%
16. 15 is 35% of what number? 42.9
17. Find 89% of 345. 307.1
18. What number is 75% of 250? 187.5
19. 12.5 is 80% of what number? 15.6
20. What is 23% of 27? 6.21

21. *Architecture* A blueprint scale is 1.5 in. : 6 ft. On the plan, the den measures 2.5 in. by 3 in. What are the actual dimensions of the den? 10 ft by 12 ft
22. *Geometry* In the figure, $\triangle RST$ is similar to $\triangle XZY$. Find length ZY. 9.6

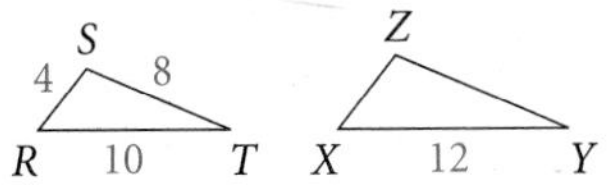

23. *Standardized Test Prep* Which equation does not have the same solution as $\frac{12}{x} = \frac{45}{60}$? E

A. $12 \cdot 60 = 45x$ B. $\frac{12}{45} = \frac{x}{60}$ C. $\frac{x}{12} = \frac{60}{45}$ D. $\frac{x}{60} = \frac{12}{45}$ E. $\frac{60}{x} = \frac{12}{45}$

Mental Math **Solve each proportion mentally.**

24. $\frac{x}{6} = \frac{12}{18}$ 4
25. $-\frac{12}{20} = -\frac{3}{y}$ 5
26. $\frac{8}{d} = \frac{40}{30}$ 6
27. $\frac{1}{2} = \frac{z}{25}$ 12.5

Choose **Use a calculator, paper and pencil, or mental math. Solve each proportion.**

28. $\frac{c}{6} = \frac{12}{15}$ 4.8
29. $\frac{21}{12} = \frac{7}{y}$ 4
30. $-\frac{37}{24} = \frac{k}{6}$ −9.25
31. $\frac{15}{n} = \frac{39}{13}$ 5
32. $\frac{17}{51} = \frac{n}{1}$ $\frac{1}{3}$
33. $-\frac{4.5}{x} = -\frac{1.8}{5}$ 12.5
34. $\frac{q}{56} = \frac{15}{14}$ 60
35. $\frac{2.5}{1.8} = -\frac{1.2}{r}$ −0.864

36. a. *Geometry* Rectangle $ABCD$ is 6 in. wide and 16 in. long. It is similar to rectangle $KLMN$, whose length is 24 in. What is the width of rectangle $KLMN$? 9 in.
b. Are the areas of the two rectangles proportional to the lengths of their corresponding sides? Explain. No; the ratio of the areas is not equal to the ratio of the lengths.

37. a. Do the two phrases in the picture offer the same discount? Explain.
b. *Open-ended* What would you expect to pay at the store's sale for a jacket regularly priced at \$67? Explain. a–b. See margin.

page 161–162 On Your Own

37a. No; for an item that originally costs \$12, the $\frac{1}{3}$-off sale results in the price \$8; 50% off results in the price \$6.

b. Answers may vary. Sample: You should expect to pay about \$45; $\frac{1}{3}$ off \$67 results in price \$44.67.

40. The cross products of a proportion are equal, so you can rewrite the equation $\frac{x}{2} = \frac{3}{4}$ as $4x = 2 \cdot 3$. The solution of this equation makes the proportion true: $x = 1.5$; $\frac{1.5}{2} = \frac{3}{4}$.

Chapter Project **FIND OUT BY CALCULATING** Ask students to estimate the number of calories they burn when they exercise. Then have students write and solve a proportion to calculate the number of calories they actually burn for a given activity. This task is essential to their work on the Chapter Project introduced in the Chapter Opener. You may wish to check students' progress by having each student write a sentence or two describing the objectives of this project.

Exercises MIXED REVIEW

GETTING READY FOR LESSON 4-2 These exercises give students practice in solving equations by combining like terms and using the distributive property. These skills are essential for solving the equations in the next lesson.

Wrap Up

THE BIG IDEA Ask students to define a proportion and to explain how to solve a proportion.

RETEACHING ACTIVITY Students use a cup and integer chips to represent money as they solve proportions. They will write ratios and use cross products to solve for the number of "coins" in the cup. (Reteaching worksheet 4-1)

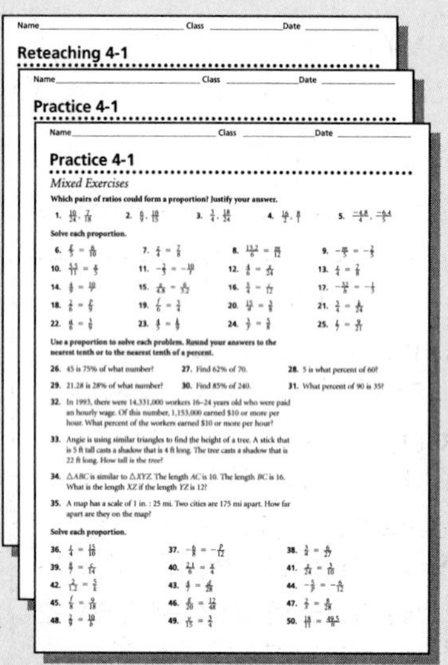

Lesson Quiz

Lesson Quiz is also available in Transparencies.

1. Write and solve a proportion to find the number of seconds in 60 min. $\frac{60}{1} = \frac{t}{60}$; $t = 3600$ s
2. What percent of 8 is 6.5? 81%
3. Solve $\frac{4}{-7} = \frac{2}{z}$ $z = -3.5$
4. Bianca is making a 30 in. model of the United States for her Geography class. How many miles must each inch represent if the United States is 2807 mi across? 94 mi

38. a. *Travel* Use a ruler and the map at the right. Find the distance from each town to the others. from D to L, 26 mi; from D to SP, 18 mi; from L to SP, 16 mi
    b. A student lives halfway between Lincoln and San Paulo and goes by the shortest route to school in Duncanville. How far does the student travel each day to and from school? 52 mi

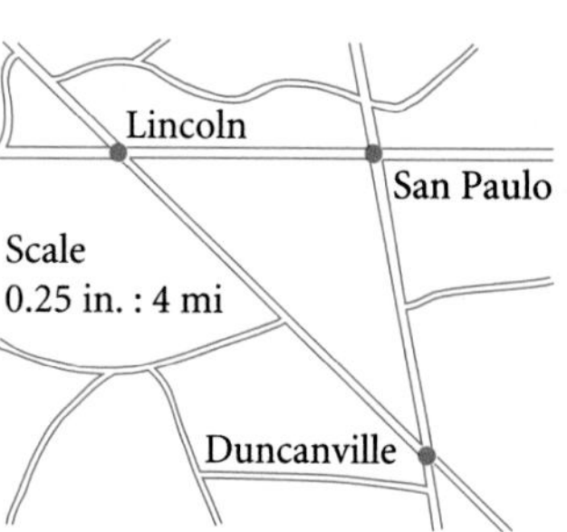

39. Suppose $\frac{12}{72} = \frac{x}{24}$ and $\frac{x}{36} = \frac{y}{81}$. Find y. 9
40. *Writing* Write an explanation telling an absent classmate how to use cross products to solve a proportion. Include an example. See margin.
41. *Hobbies* Some model trains are built to $\frac{1}{87}$ of actual size. Suppose an actual boxcar is 40 ft long. How many *inches* long is the model? 5.5 in.
42. *Statistics* In the southern United States, 68.6% of the population, or 58,656,000 people, live in urban areas. What is the total population of the southern United States? about 85,504,000 people
43. In 1992, there were 255,082,000 people living in the United States. Of these, 18,100,000 were 10–14 years old. What percent of the population was this? Round your answer to the nearest tenth of a percent. 7.1%

Chapter Project ***Find Out by Calculating***

When you exercise, the number of calories you burn is roughly proportional to your weight.

- Write and solve a proportion to find how many calories you would burn if you performed one activity in the table for 60 min.
- Write and solve a proportion to find how many calories you would burn if you did the activity for 60 min three times in a week.

Calories Burned During Exercise (150-lb person exercising for 60 min)

Exercise	Calories
Dancing	250
Biking, walking (3–5 mph)	300
Soccer, jogging (4–5 mph)	400
Swimming, skiing	500
Running (5–7 mph)	650

Exercises MIXED REVIEW

Add or subtract.

44. $\begin{bmatrix} 11 & -2 & 4 \\ 3.5 & 6.9 & 12.2 \end{bmatrix} + \begin{bmatrix} -5.4 & 3.2 & -8 \\ 7.6 & 2.3 & -9.7 \end{bmatrix}$ 44. $\begin{bmatrix} 5.6 & 1.2 & -4 \\ 11.1 & 9.2 & 2.5 \end{bmatrix}$

45. $\begin{bmatrix} -4.3 & 7.9 \\ 2.1 & -0.8 \end{bmatrix} - \begin{bmatrix} -3.1 & -6.7 \\ 3.5 & 4.9 \end{bmatrix}$ $\begin{bmatrix} -1.2 & 14.6 \\ -1.4 & -5.7 \end{bmatrix}$

Find the value of each function when $x = 3$.

46. $f(x) = 4x - 1$ 11
47. $f(x) = -1.5x - 7$ −11.5
48. $f(x) = 8x + 2$ 26
49. You have one foot of space left on a bookshelf, and your paperbacks are $\frac{3}{4}$ in. thick. How many more paperbacks can you fit on the shelf? 16

Getting Ready for Lesson 4-2

Solve and check each equation.

50. $7x + 4 - 15x = 36$ −4
51. $w + (w + 2) = -27$ −14.5
52. $6(7 - 2y) = 30$ 1

4-2 Teaching Notes

PROBLEM OF THE DAY

Find the values of a and b if $a^b = 81$ and $b^a = 64$.

$a = 3; b = 4$

Problem of the Day is also available in Transparencies.

CONNECTING TO PRIOR KNOWLEDGE On the board, write $x + 4 = 7$. Point out that students solve this by isolating the variable x on one side of the equation and combining the numerals on the other side of the equation. Now write $x + 4 = 2x + 7$. Lead students to understand that they can solve this equation the same way: combine the variables on one side and combine the numerals on the other side.

WORK TOGETHER

Question 7 Remind students to substitute the same number for x each time it appears in the equation.

THINK AND DISCUSS

ERROR ALERT! Some students may not understand the meaning of *sides of an equation*. **Remediation:** Explain that the equal sign separates the two sides of an equation.

Example 1

Point out that the goal is to get the variable x by itself on one side of the equation. Work through Example 1, step by step.

Connections Transportation . . . and more

4-2 Equations with Variables on Both Sides

What You'll Learn

- Solving equations with variables on both sides
- Identifying equations that have no solution or are identities

...And Why

To solve equations that model real-world situations, such as distance problems

What You'll Need

- tiles
- calculator

WORK TOGETHER

In Chapter 3 you used tiles and solved equations. Model each equation with tiles. Then solve. 1–4. See margin for models.

1. $x - 4 = -9$ -5 **2.** $-10 = 5y$ -2 **3.** $3a + 8 = 2$ -2 **4.** $-12 = 8 + 4b$ -5

Now look at an equation with variables on both sides: $6x + 3 = 4x + 9$.

5. Model this equation with tiles. See margin.

6. Discuss how you might use tiles to solve the equation. See below.

7. Use the problem solving strategy *Guess and Test* to solve the equation. 3

6. Remove 3 from each side. Next, remove $4x$ from each side. Then solve for x.

THINK AND DISCUSS

Part 1

Using Tiles to Solve Equations

Some equations cannot be solved easily using the strategy *Guess and Test*. In this lesson you will learn how to solve equations with variables on both sides. First you will use tiles to help you understand the process.

Example 1

Solve the equation

$$5x - 3 = 2x + 12.$$

Model the equation with tiles.

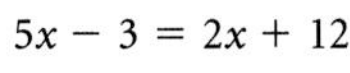

$$5x - 3 = 2x + 12$$

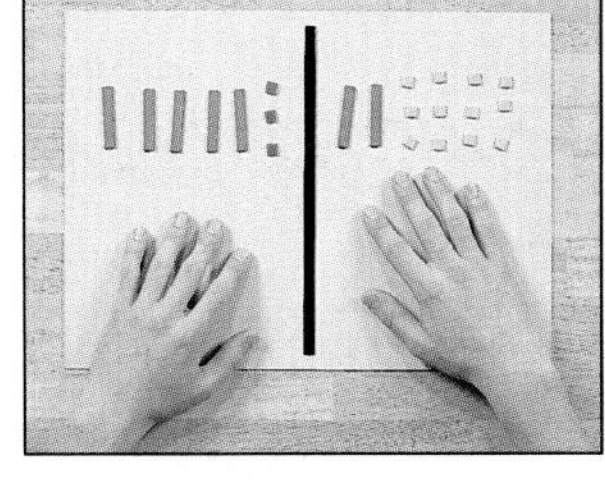

Add $-2x$ to each side, and simplify by removing zero pairs.

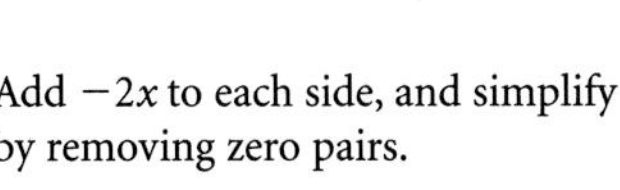

$$5x - 3 - 2x = 2x + 12 - 2x$$
$$3x - 3 = 12$$

Lesson Planning Options

Prerequisite Skills

- Performing operations with integers

Assignment Options

To provide flexible scheduling, this lesson can be subdivided into parts.

1 **Core** 1, 3–6, 22–24, 44–46
Extension 2, 7, 32

2 **Core** 14, 15, 25–27, 47–49
Extension 34, 35, 42, 43, 53

3 **Core** 8–13, 16–21, 28–30, 50–52
Extension 31, 33, 36–41

Use Mixed Review to maintain skills.

Resources

Student Edition

Skills Handbook, p. 567, 579
Extra Practice, p. 559
Glossary/Study Guide

Teaching Resources

Chapter Support File, Ch. 4

- Practice 4-2 (two worksheets)
- Reteaching 4-2
- Alternative Activity 4-2

Classroom Manager 4-2
Glossary, Spanish Resources
Two-Year Algebra Handbook 4-2

Transparencies
10, 17, 59

Ask for a volunteer to show what each side of the equation looks like after the zero pairs are removed. The volunteer may sketch tiles on the board or use tiles to model. Ask another volunteer to show the next step.

TACTILE LEARNING Have pairs of students work together throughout this lesson. One student will use pencil and paper while the other uses algebra tiles to model and to solve the equations.

ALTERNATIVE METHOD When a term is moved to the opposite side of the equal sign, it is performing its opposite operation. Use index cards to solve $5x - 3 = 2x + 12$ using this method. With a black marker, write each of the terms of the equation ($5x$, -3, $2x$, and $+12$) on a separate card. Write the opposite term ($-5x$, $+3$, $-2x$, and -12) in red on the back of each card. Write an equal sign on a fifth card. Tape the cards to the board to make the equation $5x - 3 = 2x + 12$. Now move cards from one side to the other, flipping them over to show their opposite as you move them. Continue moving and flipping cards until the variable terms are on one side and the constant terms are on the other. Have students simplify and solve. Challenge students to explain why this shortcut method works.

Additional Examples

FOR EXAMPLE 2

$(-5) + 3x = 5 + 2x$. $x = 10$

Discussion: *Which properties of equality did you use to arrive at your answer?*

FOR EXAMPLE 3

Chris and Maggie decide to meet after school at a nearby music store. Chris walks to the store at a speed of 2 mi/h. Maggie bicycles along the same path at 7 mi/h. Maggie gets to the store 30 min before Chris. How long did it take Maggie to get to the store? 12 min

FOR EXAMPLE 4

Prove that $(-2)(3z - 4) = 10 - 6z$ has no solution. $0 = 2$

FOR EXAMPLE 5

Prove that $(-2)(3z - 4) = 8 - 6z$ is an identity. $0 = 0$

164

QUICK REVIEW

■■ represents 0.

Add 3 to each side and simplify by removing zero pairs.

$3x - 3 + 3 = 12 + 3$
$3x = 15$

Divide each side into three identical groups.

$\frac{3x}{3} = \frac{15}{3}$

Solve for x.

$x = 5$

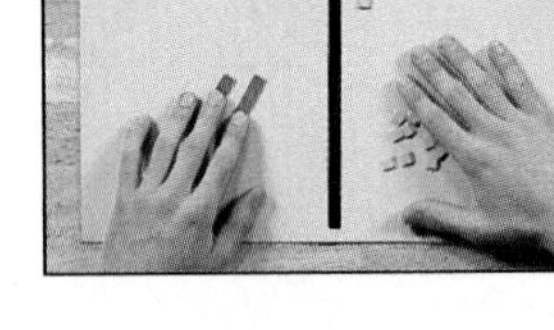

The solution is 5.

Check $5x - 3 = 2x + 12$
$5(5) - 3 \stackrel{?}{=} 2(5) + 12$ ← Substitute 5 for x.
$22 = 22$ ✓

8. Try This Use tiles to model and solve each equation.
a. $6x - 2 = x + 13$ 3
b. $4(x + 1) = 2x - 2$ −3
c. **Summarize** the steps you used to solve the equations. See margin.
8a,b. See margin for tiles.

Part 

Using Properties of Equality

You can use the properties of equality to get terms with variables on the same side of the equation.

Example 2

GRAPHING CALCULATOR HINT

You can check your solution to Example 2 by using the Solve feature to see if $\frac{1}{2}$ is the solution to the equation.

Solve $5t - 8 = 9t - 10$.

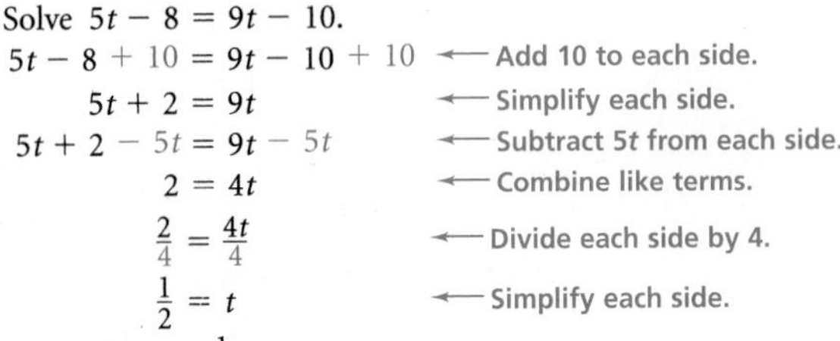

$5t - 8 + 10 = 9t - 10 + 10$ ← Add 10 to each side.
$5t + 2 = 9t$ ← Simplify each side.
$5t + 2 - 5t = 9t - 5t$ ← Subtract $5t$ from each side.
$2 = 4t$ ← Combine like terms.
$\frac{2}{4} = \frac{4t}{4}$ ← Divide each side by 4.
$\frac{1}{2} = t$ ← Simplify each side.

The solution is $\frac{1}{2}$.

Example 2

Question 10 Explain that it is often helpful to move the term with the lesser coefficient ($5t$ rather than $9t$) or the lesser constant (–10 rather than –8). The resulting equation often has fewer negative signs.

Example 3 **Relating to the Real World** 

DIVERSITY Be sensitive to those students who are surprised or confused by the sisters' bicycling home separately. Some families place greater importance than others on siblings being together.

Students may need to be reminded of how to use the distribution property.

CRITICAL THINKING Question 11 Suggest that students think about the units Ben used to define t.

EXTENSION Ask students:

- *If Ben defines t as time in minutes, what equation does he need to write and solve to answer the question?* $\frac{12}{60}t = \frac{9}{60}(t + 15)$
- *What if you multiply both sides of this equation by 60?* You get Ben's equation.

Example 4

Challenge students to demonstrate how to use the problem solving strategy of *Working Backwards* to write an equation that has variables on both sides and has no solution.

---

9. **Verify** the solution of Example 2. $5(\frac{1}{2}) - 8 \stackrel{?}{=} 9(\frac{1}{2}) - 10;\ -5\frac{1}{2} = -5\frac{1}{2}$ ✓
10. **a.** Suppose you began solving the equation in Example 2 by subtracting $9t$ from each side of the equation. Write the steps you would use to solve the equation. a–b. See margin.
    b. Compare your solution to the solution in Example 2. Does it matter that the variables are on different sides of the equal sign? Explain.

QUICK REVIEW

A formula for distance is distance = rate · time, or $d = rt$.

Equations are helpful when you solve distance problems.

Example 3 **Relating to the Real World** 

Transportation Mary and Jocelyn are sisters. They left school at 3:00 P.M. and bicycled home along the same bike path. Mary bicycled at a speed of 12 mi/h. Jocelyn bicycled at 9 mi/h. Mary got home 15 min before Jocelyn. How long did it take Mary to get home?

PROBLEM SOLVING HINT

Draw a diagram to help you visualize the conditions of the problem in Example 3.

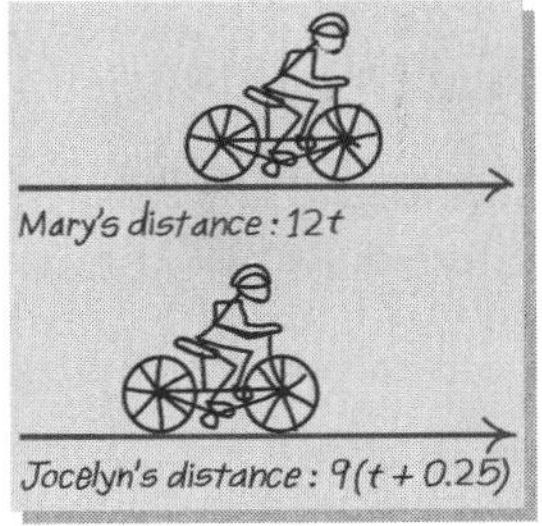

Define t = Mary's time in hours
$t + 0.25$ = Jocelyn's time in hours

Relate	Mary's distance (rate · time)	equals	Jocelyn's distance (rate · time)
Write	$12t$	$=$	$9(t + 0.25)$

$12t = 9(t + 0.25)$
$12t = 9t + 2.25$ ← Use the distributive property.
$12t - 9t = 9t + 2.25 - 9t$ ← Subtract $9t$ from each side.
$3t = 2.25$ ← Combine like terms.
$\frac{3t}{3} = \frac{2.25}{3}$ ← Divide each side by 3.
$t = 0.75$ ← Use a calculator.

It took Mary 0.75 h, or 45 min, to get home.

11. *Critical Thinking* To solve the problem in Example 3, Ben wrote the equation $12t = 9(t + 15)$. What mistake did he make?
    The units for t are hours and the units for 15 are minutes.

Part 3
Solving Special Types of Equations

An equation has **no solution** if no value makes the equation true.

Example 4

Solve $6m - 5 = 7m + 7 - m$.
$6m - 5 = 7m + 7 - m$
$6m - 5 = 6m + 7$ ← Combine like terms.
$6m - 5 - 6m = 6m + 7 - 6m$ ← Subtract $6m$ from each side.
$-5 = 7$ Not true for any m!

This equation has no solution.

Technology Options

For Exercises 16–21 and 44–52, students may want to use a graphing calculator to check their answers.

Prentice Hall Technology 

 **Software** • Secondary Math Lab Toolkit • Integrated Math Lab 7 • Computer Item Generator 4-2

Internet • See the Prentice Hall site. (http://www.phschool.com)

Example 5

Challenge students to use the problem solving strategy of *Working Backwards* to write an identity that has variables on both sides.

ESL Students might not understand the use of the word *identity* in this context. Explain that in this context, *identity* means sameness. Use the example that if you identify with someone, you recognize how you are the same as the other person. Thus, an equation can be an identity because it is the same for any value of the variable.

Question 14 Encourage students to share the mental math strategies they used.

Exercises ON YOUR OWN

CRITICAL THINKING Exercises 14 and 15 Have students check their own solutions to be sure they have not made errors.

page 163–166 Think and Discuss

1. 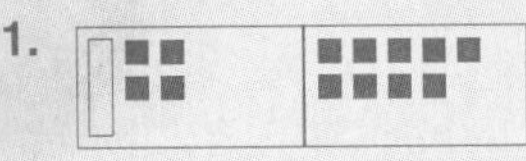

2. 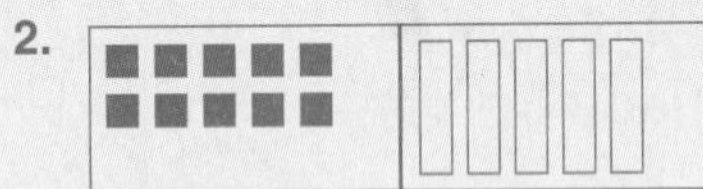

3. 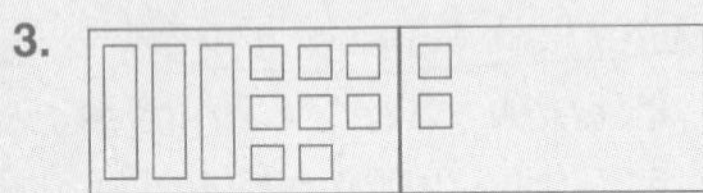

4. 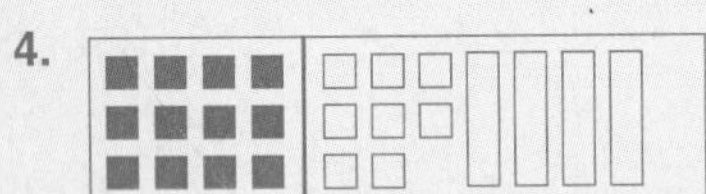

5. 

8a. 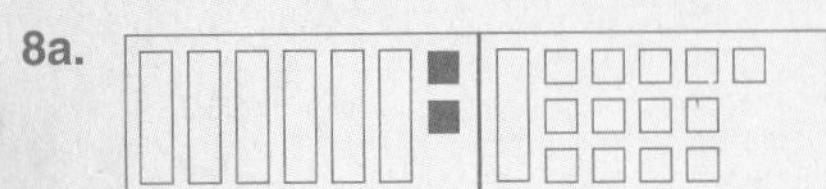

12. Is an equation that has 0 for a solution the same as an equation with no solution? Explain. No; 0 is a solution if the equation is true when the value of the variable is 0. An equation has no solutions if it is not true for any value of the variable.

An equation that is true for every value of the variable is an **identity.**

Example 5

Solve $10 - 8a = 2(5 - 4a)$.

$$10 - 8a = 2(5 - 4a)$$
$$10 - 8a = 10 - 8a \quad \text{← Use the distributive property.}$$
$$10 - 8a + 8a = 10 - 8a + 8a \quad \text{← Add 8a to each side.}$$
$$10 = 10 \quad \text{Always true!}$$

This equation is true for any value of a, so the equation is an identity.

13. Could you have stopped solving the equation when you saw that $10 - 8a = 10 - 8a$? Explain. Yes; the two sides are identical, so their values are equal for all values of a.

14. Mental Math Without writing the steps of a solution, tell whether the equation has *one solution*, *no solution*, or is an *identity*.

a. $9 + 5a = 5a - 1$ no solution
b. $5a + 9 = 2a$ one solution
c. $9 + 5a = 2a + 9$ one solution
d. $9 + 5a = 5a + 9$ identity

Exercises ON YOUR OWN

Write an equation for each model and solve.

1. 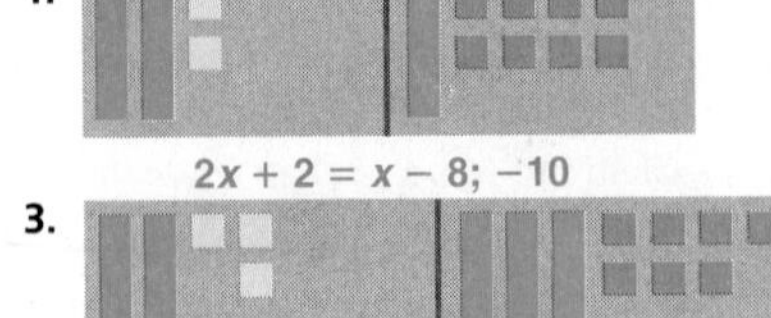

$2x + 2 = x - 8; -10$

2. 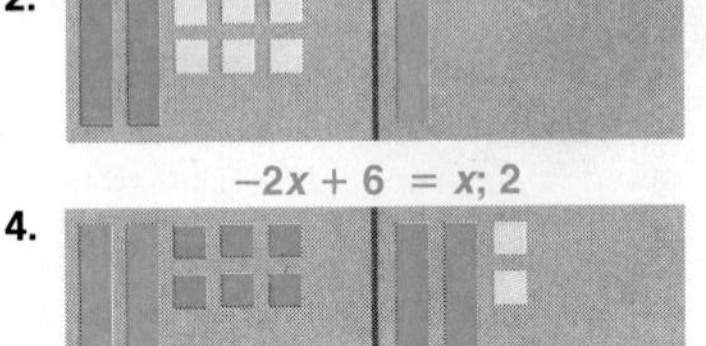

$-2x + 6 = x; 2$

3.

$2x + 3 = 3x - 7; 10$

4.

$2x - 6 = -2x + 2; 2$

Model each equation with tiles. Then solve. 5–7. See margin for models.

5. $4x - 3 = 3x + 4$ 7
6. $5x + 3 = 3x + 9$ 3
7. $8 - x = 2x - 1$ 3

Solve and check. If the equation is an identity or if it has no solution, write *identity* or *no solution*.

8. $3(x - 4) = 2x + 6$ 18
9. $4x - 7 = x + 3(4 + x)$ no solution
10. $5x = 3(x - 1) + (3 + 2x)$ identity
11. $0.5y + 2 = 0.8y - 0.3y$ no solution
12. $6 + 3m = -m - 6$ −3
13. $3t + 8 = 5t + 8 - 2t$ identity

MENTAL MATH Exercises 16–21 Students may use different techniques. Encourage students to share their methods with the class.

ALTERNATIVE ASSESSMENT Exercises 22–30 Ask students to explain why the solutions to some equations do not result in numerical answers. This will help you assess the students' understanding of the meanings of *identity* and *no solution.*

EXTENSION Exercise 28 Ask: *Which term could you eliminate from this equation to make the equation into an identity?* 2.4

WRITING Exercise 33 Tell the students that the exercise is describing an equation with only one variable on one side of the equal sign. Otherwise, they may use an equation such as $3x - 3x = 0$ (an identity) and not be able to achieve the correct answer.

Exercise 35 Have students discuss factors other than cost that they need to consider when deciding whether to buy the skates.

$\frac{1}{3}(2\cancel{4}6) - 50$ 125

Critical Thinking **Find the mistake in the solution of each equation. Explain the mistake and solve the equation correctly.**

14.
$$\begin{aligned} 2x &= 11x + 45 \\ 2x - 11x &= 11x - 11x + 45 \\ 9x &= 45 \\ \frac{9x}{9} &= \frac{45}{9} \\ x &= 5 \end{aligned}$$

$2x - 11x$ is $-9x$; -5

15.
$$\begin{aligned} 4.5 - y &= 2(y - 5.7) \\ 4.5 - y &= 2y - 11.4 \\ 4.5 - y - y &= 2y - y - 11.4 \\ 4.5 &= y - 11.4 \\ 4.5 + 11.4 &= y - 11.4 + 11.4 \\ 15.9 &= y \end{aligned}$$

You should add y, not subtract, on the third line; 5.3.

Mental Math **Solve and check each equation.**

16. $5y = y - 40$ -10
17. $7w = -7w$ 0
18. $r + 1 = 4r + 1$ 0
19. $6t + 1 = 6t - 8$ no solution
20. $2q + 4 = 4 - 2q$ 0
21. $3a + 1 = 9 - a$ 2

Choose **Use tiles, paper and pencil, calculator, or mental math to solve each equation. If appropriate, write *identity* or *no solution.***

22. $t + 1 = 3t - 5$ 3
23. $7y - 8 = 7y + 9$ no solution
24. $0.5k + 3.6 = 4.2 - 1.5k$ 0.3
25. $2r + 16 = r - 25$ -41
26. $\frac{3}{4}x = \frac{1}{2} + \frac{2}{3}x$ 6
27. $\frac{1}{3}(x - 7) = 5x$ $-\frac{1}{2}$
28. $0.7m = 0.9m + 2.4 - 0.2m$ no solution
29. $14 - (2q + 5) = -2q + 9$ identity
30. $0.3t + 1.4 = 4.2 - 0.1t$ 7

31. Find the value of each variable in the matrices. $\begin{bmatrix} 2x + 1 & a - 1 \\ w - 4 & 9y \end{bmatrix} = \begin{bmatrix} -5x - 6 & 5a \\ 3w + 4 & -3y \end{bmatrix}$
$a = -\frac{1}{4}$; $w = -4$; $x = -1$; $y = 0$

32. **Business** A toy company spends $1500 each day on plant costs plus $8 per toy for labor and materials. The toys sell for $12 each. How many toys must the company sell in one day to equal its daily costs? 375 toys

33. **Writing** Describe the two situations you learned about in this lesson that cannot occur when you are solving an equation with the variable on only one side of the equal sign. Give examples. See margin.

34. **Transportation** A truck traveling 45 mi/h and a train traveling 60 mi/h cover the same distance. The truck travels 2 h longer than the train. How many hours did each travel?
Train travels 6 h; truck travels 8 h.

35. **Recreation** You can buy used in-line skates from your cousin for $40, or you can rent them from the park. Either way you must rent safety equipment. How many hours must you skate at the park to justify buying your cousin's skates? 20 h

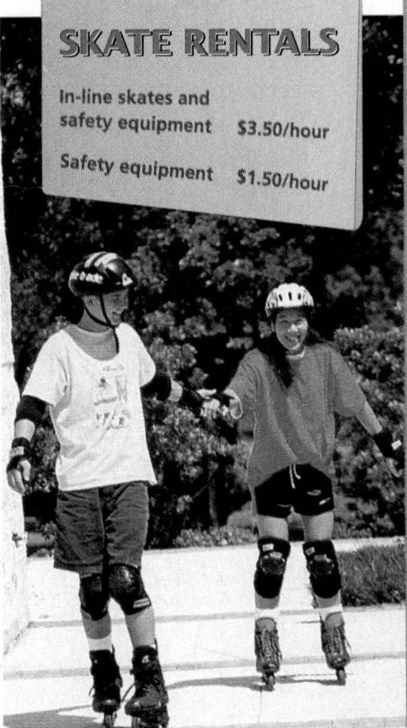

Open-ended **Write an equation with variables on both sides for each of the following solutions.** 36–41. Samples are given.

36. $x = 0$ $1 + 3x = x + 1$
37. x is a positive number. $2x = 5$
38. x is a negative number. $2.5x + 10 = 5$
39. All values of x are solutions. $-x + 3(x + 1) = 2x + 3$
40. $x = 1$ $3x - 2 = 1$
41. No values of x are solutions. $2x + 1 = 2x + 3$

b. 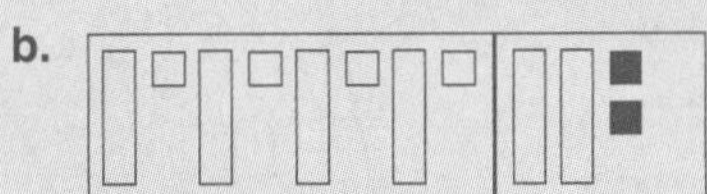

c. (a) Add 2 to each side and remove zero pairs. Remove x from each side. Divide each side into five identical groups.
(b) Rearrange tiles to simplify. Add -4 to each side and remove zero pairs. Remove $2x$ from each side. Divide each side into two identical groups.

10a. Subtract $9t$ from each side. Add 8 to each side. Divide each side by -4. $t = \frac{1}{2}$
b. Solutions are the same; no; two expressions in an equality are equal so it does not matter which is on the right and which is on the left.

pages 166–168 On Your Own

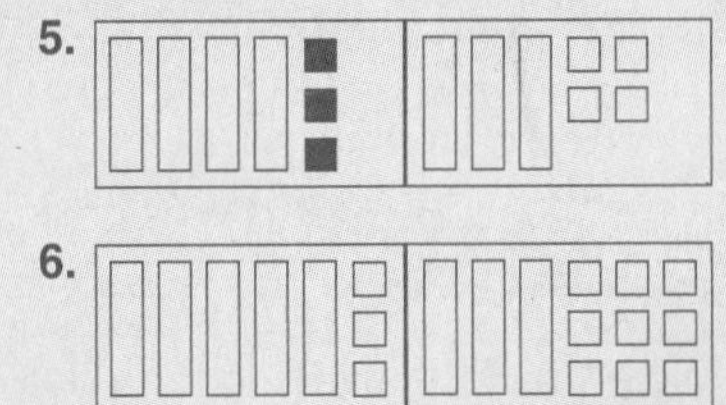

7, 33, 42b, 42e. See back of book.

OPEN-ENDED Exercises 36–41 Remind students to work backward to arrive at an equation.

GEOMETRY Exercise 43 Ask students to recall the meaning of *congruent*. If students need a hint to help them start, suggest that they first find the value of *x*.

STANDARDIZED TEST TIP **Exercise 53** Remind students to simplify each expression as much as possible, before attempting to compare two expressions.

Exercises MIXED REVIEW

GETTING READY FOR LESSON 4-3 Remind students to simplify expressions within absolute value symbols before finding the absolute value. These exercises prepare students for solving absolute value equations.

SELF ASSESSMENT **JOURNAL** Students may make up an example equation and show the steps needed to solve it. Encourage students to write a short explanation for each step.

Wrap Up

THE BIG IDEA Ask students to think about solving an equation with variables on only one side and solving an equation with variables on both sides. Ask: *What do you need to do differently to solve an equation with variables on both sides?*

RETEACHING ACTIVITY Students will solve equations with variables on both sides of the equal sign. They will fill in blanks for each step to solve and to check each inequality. (Reteaching worksheet, 4-2)

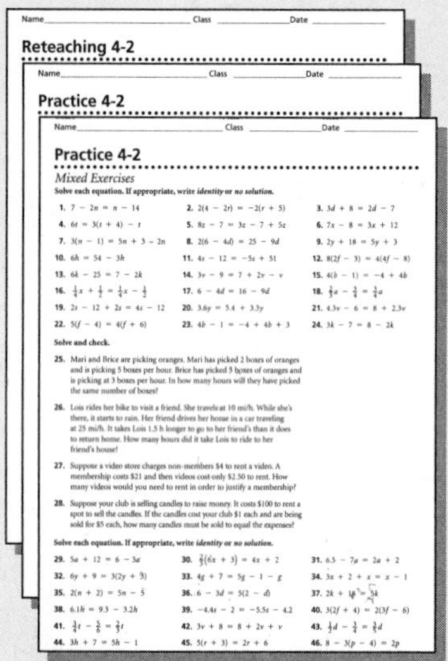
Reteaching 4-2

Practice 4-2

Practice 4-2

Lesson Quiz

Lesson Quiz is also available in Transparencies.

Solve and check. Write *identity* or *no solution* when appropriate.

1. $3 - 2t = 7t + 4$ $t = -\frac{1}{9}$
2. $4n = 2(n + 1) + 3(n - 1)$ $n = 1$
3. $3(1 - 2x) = 4 - 6x$ no solution
4. Kyle earns \$3/h working for a delivery service. He also receives a \$.50 bonus per delivery. If Kyle uses \$.20 worth of gas for each delivery, how many deliveries per hour must he make to earn \$6/h? ten deliveries

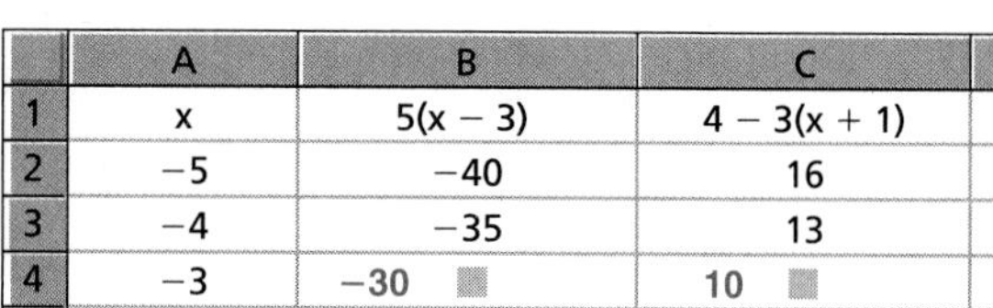

	A	B	C
1	x	5(x – 3)	4 – 3(x + 1)
2	–5	–40	16
3	–4	–35	13
4	–3	–30	10

42. a. *Technology* Write formulas for cells B2 and C2 to evaluate the expressions at the top of Columns B and C. See lower right.
    b. Enter the integers from –5 to 5 in Column A. Evaluate the expressions in Columns B and C using the values in Column A. See margin for spreadsheet.
    c. What is the value in Column A when the numbers in Columns B and C are equal? 2
    d. What equation have you solved? $5(x - 3) = 4 - 3(x + 1)$
    e. Use a spreadsheet to solve $5.2n - 9 = 11.2n + 3$. 2; see margin for spreadsheet.

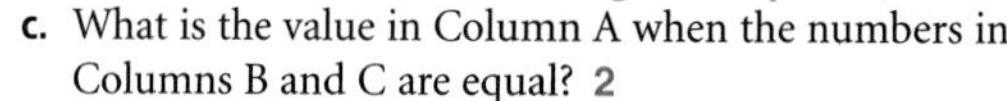

42a. formula for cell B2, 5* (A2 – 3); formula for cell C2, 4 – 3* (A2 + 1)

43. *Geometry* $\triangle ABC$ is congruent to $\triangle DEF$. Find the lengths of the sides of $\triangle DEF$. $DF = 7$, $DE = 10$, $EF = 6$

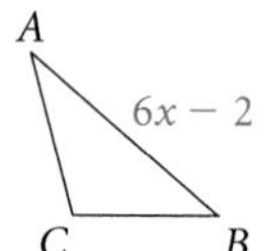

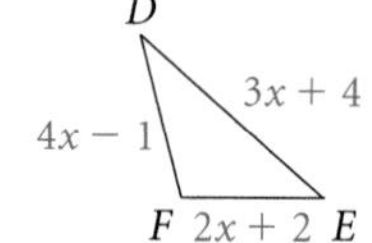

Solve each equation. Check your answers.

44. $2n - 5 = 8n + 7$ -2
45. $3x + 4 = x + 18$ 7
46. $3b + 5 - b = 4b$ -2.5
47. $5a - 14 = -5 + 8a$ -3
48. $4 - 6d = d + 4$ 0
49. $-10t + 6.25 = t + 11.75$ -0.5
50. $4x - 10 = x + 3x - 2x$ 5
51. $6x = 4(x + 5)$ 10
52. $\frac{3}{2}z - 2 = -\frac{5}{4}z - 4$ $-\frac{8}{11}$

53. *Standardized Test Prep* Compare the quantities in Column A and Column B. Which statement is true for all values of *x*? C

Column A	Column B
$5(x - 3)$	$7x - 12 - (2x + 3)$

A. The quantity in Column A is greater.
B. The quantity in Column B is greater.
C. The quantities are equal.
D. The relationship cannot be determined from the given information.

Exercises MIXED REVIEW

Solve and check each equation.

54. $-\frac{12}{14} = \frac{-9}{m}$ 10.5
55. $\frac{1}{3}(h - 5) = -11$ -28
56. $2x = 7x + 10$ -2
57. $\frac{w}{7} = \frac{11}{10}$ 7.7

58. *Money* The sales tax in Austin, Texas, is 8%. How much would you pay for three \$12 books and two \$15 books in Austin? \$71.28

Getting Ready for Lesson 4-3

Simplify.

59. $|15|$ 15
60. $|-12|$ 12
61. $|-34|$ 34
62. $|18 - 12|$ 6
63. $|9 + 2|$ 11
64. $|-12 - (-12)|$ 0
65. $-|-19|$ -19
66. $-|32|$ -32
67. $-|-10 + 8|$ -2

SELF ASSESSMENT **FOR YOUR JOURNAL**

Summarize what you know about solving equations with variables on both sides by writing a list of steps for solving this type of equation.

Math ToolboX

This toolbox uses the graphing calculator. The exercises help students see the relationship between solutions to equations and their graphs. First, work through the example with the class. Then have students work on their own or in small groups to complete the exercises. Encourage students to check their work by substituting their *x*-value into the equation.

ERROR ALERT! Some students may have difficulty checking a fractional answer. **Remediation:** Have students write their answer as a decimal rounded to the nearest hundredth rather than a fraction.

Use the ZBox function on the graphing calculator to enlarge the area around the point of intersection. This makes the integer marks on the axes easier to see. Be sure to have students include the origin in their zoom box as a point of reference.

WRITING Exercise 10 For students who have trouble starting, suggest that they first choose one of the equations they solved in Exercises 3–6. Then substitute the *x*-value of the solution back into their two equations. Compare the results with the original equation.

ADDITIONAL PROBLEM Use your graphing calculator to solve each equation.

1. $2x - 11 = x - 8$ 3
2. $4x - 7 = 0 - 3x$ 1
3. $x + 9 = 2(x + 7)$ −5

Math ToolboX Technology

Using Graphs to Solve or Check Equations

After Lesson 4-2

You can use a graphing calculator to solve equations with variables on both sides. When you graph each side of an equation separately, the *x*-coordinate of the point where the graphs meet gives the solution to the equation.

Example

Find the solution of $-\frac{1}{2}c = \frac{1}{2}c + 5$ using a graphing calculator.

STEP 1: Press [Y=] to go to the equation screen. Press [CLEAR] to clear the first equation. Then press [(-)] [.] [5] [X,T,θ] to enter the left-hand side of the equation.

STEP 2: Press the down arrow once. Press [CLEAR]. Then press [.] [5] [X,T,θ] [+] [5] to enter the right-hand side of the equation.

STEP 3: Press [GRAPH] to view the graphs of the equations.

X = −5 Y = 2.5

STEP 4: Press [2nd] [CALC] [5]. Move the cursor near the point of intersection. Press [ENTER] three times to find the coordinates of the intersection point.

The *x*-coordinate of the point of intersection is −5. The solution of the equation $-\frac{1}{2}c = \frac{1}{2}c + 5$ is −5.

1. The graphing calculator screen at the right shows the solution of $2x - 1 = x + 1$.
   a. What two equations were graphed? $y = 2x - 1$; $y = x + 1$
   b. What is the *x*-coordinate of the point where the graphs intersect? 2
   c. What is the solution of the equation? 2
2. David solved $3(a + 1) = 5a + 4$. His solution was $-\frac{3}{2}$. Graph $y = 3(x + 1)$ and $y = 5x + 4$. Use the [CALC] feature to find the *x*-coordinate of the intersection of the two lines. Is David's solution correct? Explain. No; the lines intersect at a point with *x*-coordinate $-\frac{1}{2}$, not $-\frac{3}{2}$.

10. The *y*-value is the value of expressions on both sides of the equation for the *x*-value that is the solution of the equation.

Use your graphing calculator to solve each equation. 3–9. See margin for graphs.

3. $3d - 9 = 0$ 3
4. $2n - 5 = 8n + 7$ −2
5. $2x - 15 = 3(x - 15)$ 30
6. $4 - 7q = -3$ 1

Use your graphing calculator to check each solution.

7. $5(n + 1) = n + 2; \frac{1}{4}$ no; $-\frac{3}{4}$
8. $2x - 9 = -3(x - 6); 5\frac{2}{5}$ yes
9. $b - 0 = -2(b + 1); 3\frac{2}{3}$ no; $-\frac{2}{3}$
10. Writing Explain the significance of the *y*-value of the point of intersection. (*Hint:* Use a pencil and paper to solve an equation and check the solution.) See above.

Materials and Manipulatives

- graph paper

Resources

Transparencies 11

page 169 Math Toolbox

3.

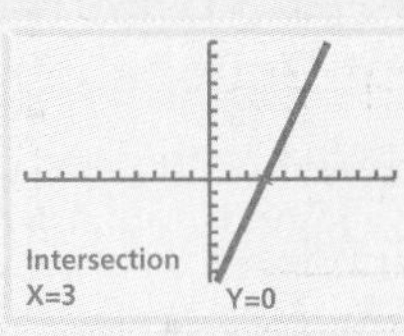

4.

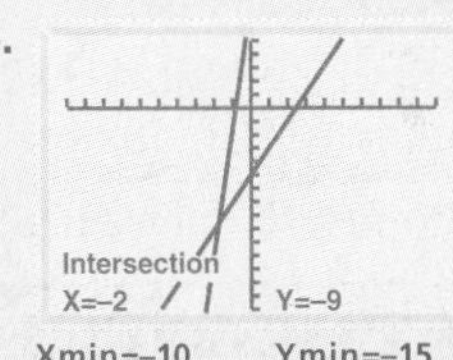

Xmin=−10 Ymin=−15
Xmax=10 Ymax=5
Xscl=1 Yscl=1

5.

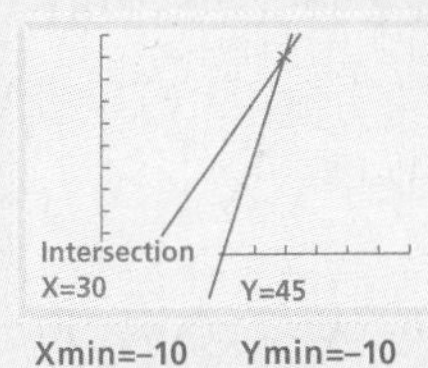

Xmin=−10 Ymin=−10
Xmax=50 Ymax=50
Xscl=5 Yscl=5

4-3 Teaching Notes

PROBLEM OF THE DAY

Three different blocks of paper pulp weigh 420 oz, 312 oz, and 180 oz. The recycling company cuts each of the blocks into pieces of identical weight for shipping. What is the maximum possible weight per piece? (Hint: Find the prime factors of each number.) 12 oz

Problem of the Day is also available in Transparencies.

CONNECTING TO PRIOR KNOWLEDGE Ask students how many of them have participated in surveys or polls. Encourage students to tell what kind of survey or poll it was and how the information was gathered.

WORK TOGETHER

TACTILE LEARNING Challenge students to design and create an instructional poster that describes a step-by-step process to solve absolute value equations.

Have students do this activity in pairs or in small groups. Cooperative learning experiences help students to develop team skills needed in the workplace. Have one student act as the recorder for the group. This encourages the students to work cooperatively.

Question 1 Students may not understand the terms "accurate to within 3%" and "results could vary from −3% to +3%." Point out that these terms mean that the actual results could be 3% more or 3% less than the poll results.

Lesson Planning Options

Prerequisite Skills

- Understanding the term *absolute value*
- Using number lines

Assignment Options

To provide flexible scheduling, this lesson can be subdivided into parts.

1 **Core** 1–8, 11–25
Extension 9, 33–40

2 **Core** 26, 27, 29–32, 41, 42
Extension 10, 28, 43, 44

Use Mixed Review to maintain skills.

Resources

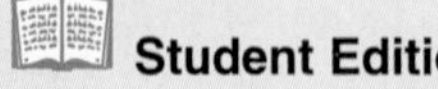

Student Edition

Skills Handbook, p. 582
Extra Practice, p. 559
Glossary/Study Guide

Teaching Resources

Chapter Support File, Ch. 4
- Practice 4-3 (two worksheets)
- Reteaching 4-3
- Alternative Activity 4-3

Classroom Manager 4-3
Glossary, Spanish Resources
Two-Year Algebra Handbook 4-3

Transparencies
8, 10, 60

170

4-3 Solving Absolute Value Equations

What You'll Learn
- Solving equations that involve absolute value
- Using absolute value equations to model real-world problems

...And Why

To solve problems involving opinion polls and quality control

WORK TOGETHER

Math in the Media Work with a partner to answer these questions.

Poll Shows Cortez Lead

A telephone poll of likely voters in the senate race was conducted by this newspaper and radio station KLRW. The poll shows that Maria Cortez has 49% of the vote, James Blake has 42%, and 9% of those polled are undecided. The results are accurate to within 3%.

1. The actual results could vary from the poll results by –3% to +3%.
   a. What is the greatest percent of voters who might vote for Blake? 54%
   b. What is the least percent of voters who might vote for Cortez? 46%
2. According to this poll, can Cortez get a majority of the votes? Explain.
3. *Open-ended* Write a short news article that includes numbers and a range around the numbers. (The article does not have to include percents and should not be about elections.) See margin.

2. Yes; the poll shows that Cortez has up to 52% of votes and may receive some "undecided" votes.

THINK AND DISCUSS

Part 1

Solving Absolute Value Equations

Situations involving a range of numbers can often be represented by absolute value equations. You can use a number line and the definition of absolute value to solve these equations.

QUICK REVIEW

The *absolute value* of a number is its distance from 0 on a number line. The symbol for the absolute value of a is $|a|$.

An absolute value is either zero or a positive number.

Example 1

Use a number line to solve $|m| = 2.5$.

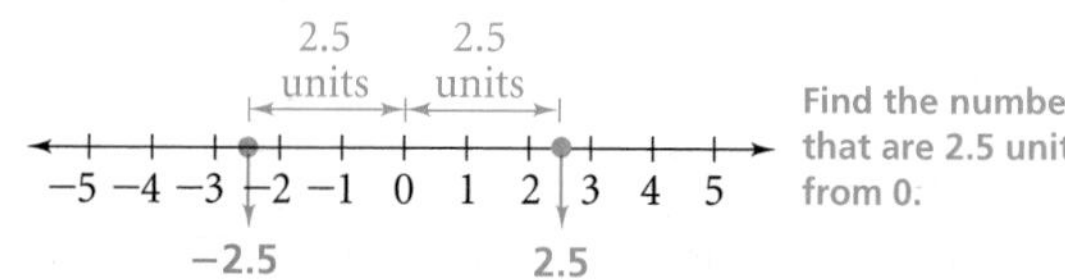

Find the numbers that are 2.5 units from 0.

The solution is −2.5 or 2.5.

4. **Try This** Use a number line to solve each equation.
   a. $|b| = 2$ −2, 2 b. $4 = |y|$ −4, 4 c. $|w| = 4.5$ −4.5, 4.5
5. Explain how the equations $|x| = 5$ and $x = |5|$ are different. The first equation has two solutions. The second has one solution.

Question 2 Ask: *What percent of votes must a candidate receive to get a majority of the votes?* more than 50%

OPEN-ENDED Question 3 Have students work in groups. Each group should submit only one news article. For groups struggling to begin, suggest topics such as an opinion poll or a report on quality control at a fast-food restaurant.

THINK AND DISCUSS

Example 1

CRITICAL THINKING Question 6 Suggest students try using the strategy *Guess and Check.*

Example 2

Question 9 Suggest that students solve each equation using mental math before solving on paper.

Example 3

Point out that the solutions of the equations differ from the answers to the problem. The answers must be expressed as percents.

MAKING CONNECTIONS Voter polls are much more accurate than they used to be. The science of random selection of poll participants began after a poll conducted in 1936 inaccurately predicted who would win the presidential election. The magazine *Literary Digest* predicted who would be elected the next president based on the responses received from two

---

6. Critical Thinking Is there a solution for $|m| = -3$? Explain.
No; abs. val. of a number cannot be neg.

7. For what value of a does the equation $|x| = a$ have just one solution?
0

You can use the properties of equality to solve an absolute value equation. Remember that you must get the absolute value by itself on one side of the equation.

Example 2

Solve $|x| + 5 = 11$.

$$|x| + 5 = 11$$
$$|x| + 5 - 5 = 11 - 5 \quad \text{← Subtract 5 from each side.}$$
$$|x| = 6$$
$$x = 6 \text{ or } x = -6 \quad \text{← The value of } x \text{ is either 6 or } -6.$$

The solution is 6 or −6.

Check both solutions.

$$|x| + 5 = 11$$
$$|-6| + 5 \stackrel{?}{=} 11 \quad \text{or} \quad |6| + 5 \stackrel{?}{=} 11 \quad \text{← Substitute 6 and } -6 \text{ for } x.$$
$$6 + 5 = 11 \checkmark \qquad 6 + 5 = 11 \checkmark$$

8\. This equation has no solution. If you subtract 11 from each side, the new equation is $|x| = -6$. Abs. val. cannot be negative.

8. What happens when you try to solve $|x| + 11 = 5$? Explain.
See left.

9. Try This Solve each equation.
a. $3 = |w| - 4$ −7, 7
b. $4|n| = 32$ −8, 8
c. $-6 = \frac{|x|}{-2}$ −12, 12
d. $|x| - 5 = -9$ no solution

Sometimes the absolute value expression has more than one term. You can use what you know about absolute value to write and solve two equations. The solutions of both equations are solutions of the absolute value equation.

Example 3

Read the article from the Work Together again. The equation $|p - 9| = 3$ represents the maximum and minimum percent of people in the election poll who are undecided. Solve the equation.

$$|p - 9| = 3 \quad \text{← The value of the expression } p - 9 \text{ is 3 or } -3.$$
$$p - 9 = 3 \quad \text{or} \quad p - 9 = -3 \quad \text{← Write two equations.}$$
$$p - 9 + 9 = 3 + 9 \qquad p - 9 + 9 = -3 + 9$$
$$p = 12 \qquad p = 6$$

Anywhere from 6% to 12% of the voters polled are undecided.

Check both solutions.

$$|p - 9| = 3$$
$$|12 - 9| \stackrel{?}{=} 3 \quad \text{or} \quad |6 - 9| \stackrel{?}{=} 3 \quad \text{← Substitute 12 and 6 for } p.$$
$$|3| = 3 \checkmark \qquad |-3| = 3 \checkmark$$

Additional Examples

FOR EXAMPLE 2

Solve $|x| - 5 = 11$.
$x = 16$ or -16

Discussion: *Why are there two solutions to this equation?*

FOR EXAMPLE 3

Solve $|i + 3| - 4 = 0$.
$i = 1$ or -7

FOR EXAMPLE 4

a. Geoff estimates his stride is 16 in. However, any given stride is likely to vary from his estimate by up to 2 in. Write and solve an expression to find Geoff's minimum and maximum stride length.
$|d - 16| = 2$; d is between 14 in. and 18 in.

b. If Geoff takes 130 strides to go from his Algebra class to his Biology class, what is the total distance between these classes? between 152 ft and 195 ft

Discussion: *When might estimates like Geoff's prove useful?*

million ballots the magazine sent to *Literary Digest* subscribers, owners of telephones, and owners of automobiles. Only wealthy people fit these criteria in 1936. The poll results indicated that the candidate favored by these respondents would win. However, the opposing candidate, Roosevelt, won the election by a large margin. *Literary Digest* lost credibility, and their sales declined.

Example 4 **Relating to the Real World** 

Have students use a calculator to check the answers.

Exercises ON YOUR OWN

OPEN-ENDED Exercise 9 Have students work in pairs and solve each other's equations.

ERROR ALERT! Students may try to combine a term inside the absolute value symbols with a term outside the symbols. **Remediation:** Write an equation such as $|m + 2| - 4 = 3$ on the board. Circle the absolute value expression and remind students that the first step in solving the equation is to get the absolute value expression by itself on one side of the equation.

Exercise 26 Have students check their percents to be sure they are reasonable.

WRITING Exercise 27 Encourage students to write solutions for the two equations side by side and to compare their answers.

Technology Options

For Exercise 10, students may want to use drawing software to verify their answers. For exercises 11–22, students may want to use graphing calculators to check their answers.

Prentice Hall Technology 

Calculator Graphing Calculator Handbook: Procedure 7, 9

Software • Secondary Math Lab Toolkit • Integrated Math Lab 8 • Computer Item Generator 4-3

 **Internet** • See the Prentice Hall site. (http://www.phschool.com)

172

10. Try This Solve and check. If there is no solution, explain why. **−3, −13**

a. $|c - 2| = 6$ **−4, 8** b. $-5.5 = |t + 2|$ c. $|x + 8| = 5$
d. $|7d| = 14$ **−2, 2** e. $12.9 = |3b|$ **−4.3, 4.3** f. $-6 = |x - 1|$
b, f. No solution; abs. val. cannot be neg.

Part 2

Modeling by Writing Equations

To maintain quality, a manufacturer sets limits for how much an item can vary from its specifications. You can use an absolute value equation to model a quality-control situation.

Example 4 **Relating to the Real World** 

Manufacturing The ideal diameter of a cylindrical machine part is 12.000 mm. At the factory that makes the parts, the quality control inspector is told that the actual diameter can vary from the ideal by at most 0.017 mm. Find the maximum and minimum diameters of the part.

12 − 0.017 mm; 12 mm; 12 + 0.017 mm

Define d = actual diameter in mm of the cylindrical part
Relate greatest difference between actual and ideal is 0.017 mm
Write $|d - 12.000| = 0.017$

$|d - 12.000| = 0.017$ ← The value of the expression $d - 12.000$ is 0.017 or −0.017.
$d - 12.000 = 0.017$ or $d - 12.000 = -0.017$ ← Write two equations.
$d = 12.017$ or $d = 11.983$

The maximum diameter is 12.017 mm and the minimum is 11.983 mm.

PROBLEM SOLVING
Look Back Show how you could find the maximum and minimum diameters by sketching the situation.
See above.

11a. Diameter cannot be neg.

11. a. Critical Thinking To solve the problem in Example 4, Mei wrote the equation $|d - 0.017| = 12.000$ and got the solution $d = 12.017$ or $d = -11.983$. Why is this solution not reasonable?
b. How would you explain why her original equation was incorrect? **The abs. val. should represent the amounts the diameter can vary, not the ideal diameter.**

Exercises ON YOUR OWN

Evaluate each expression.

1. $|x + 4|$ for $x = -1$ **3**
2. $|x - 7|$ for $x = -4$ **11**
3. $|8 - 2x|$ for $x = 3$ **2**
4. $|x| + 4$ for $x = 1$ **5**
5. $2|x|$ for $x = -3$ **6**
6. $|-3x|$ for $x = 2$ **6**
7. $6 - |x|$ for $x = -10$ **−4**
8. $-|x|$ for $x = 7$ **−7**

9. Open-ended Use each of the symbols $| \; |$, x, 5, 3, −12, +, and = to write an absolute value equation. Then solve your equation. **Sample: $|x + 5| - 12 = 3$; −20, 10**

10. a. Graphing Calculator Graph $y = 3|x| - 2$ and $y = 8$.
b. Use the CALC feature of the calculator to identify the x-coordinates of the points where the graphs intersect.
c. What absolute value equation have you solved? **$3|x| - 2 = 8$**

10a–b. 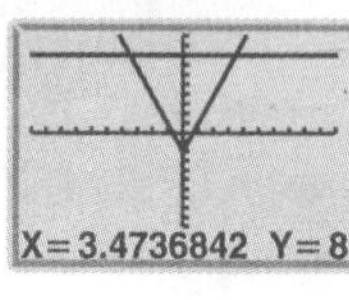

CRITICAL THINKING Exercise 28 Students may not understand what it means for an absolute value equation to have exactly one solution. Remind them that they want x to equal the same amount for both solutions. Emphasize that they should check their answer by substituting it into the equation.

ALTERNATIVE ASSESSMENT Exercises 33–40 Have students graph the solutions to these equations on a number line. This will help you assess student understanding of absolute values.

EXTENSION MENTAL MATH Exercises 33–40 Challenge students to write an absolute value equation that is an identity.

STANDARDIZED TEST TIP Exercise 44 To compare algebraic expressions such as these, choose values for the variable and try to find a pattern. Always pick at least three values: a positive number, a negative number, and zero.

Exercises MIXED REVIEW

GETTING READY FOR LESSON 4-4 These exercises prepare students for transforming formulas.

GEOMETRY Exercise 52 Ask: *What does the variable C stand for?* circumference of a circle

Solve and check each equation. If there is no solution, explain. 14, 16. No solution; abs. val. cannot be neg.

11. $|n| - 8 = -2$ −6, 6
12. $\frac{|v|}{-3} = -4.2$ −12.6, 12.6
13. $\left|a + \frac{1}{2}\right| = 3\frac{1}{2}$ −4, 3
14. $3 = |u| + 9$
15. $|r - 8| = 5$ 3, 13
16. $|t| + 7 = 4.5$
17. $9 = |c + 7|$ −16, 2
18. $|n - 4| = 0$ 4
19. $|m + 2| - 4 = 3$ −9, 5
20. $21 = |2d| + 3$ −9, 9
21. $|5p| = 3.6$ −0.72, 0.72
22. $3|v - 5| = 12$ 1, 9

Write an absolute value equation for each solution graphed below.

23. 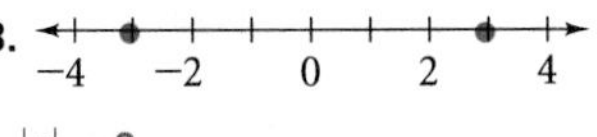
$|x| = 3$

24. 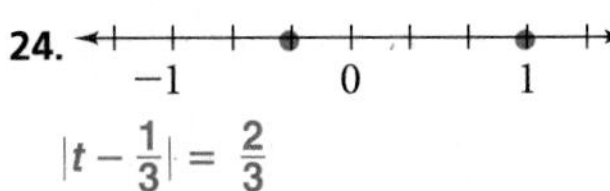
$\left|t - \frac{1}{3}\right| = \frac{2}{3}$

25. 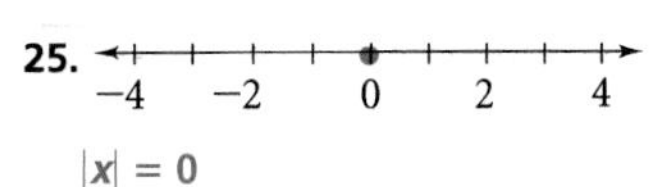
$|x| = 0$

26. Polling One poll reported that 53% of county residents favored building a recreation center. The polling service stated that this poll was accurate to within 4.5%. Use an absolute value equation to find the minimum and maximum percents of county residents who are in favor of building a recreation center. 48.5% to 57.5%

27. Writing Describe how solving $3|c| - 4 = 9$ is similar to solving $3c - 4 = 9$ and how it is different. See margin.

28. Critical Thinking For what values of a and b will the equation $|x - a| = b$ have exactly one solution? any value for a; b must be 0

PROBLEM SOLVING HINT
Use the strategy *Guess and Test* in Exercise 28.

Write *true* or *false* for each statement. Justify your response.

29. If $|q + 4| = -1$, there is no solution. True; abs. val. cannot be neg.
30. If $|h| + 9 = 4$, then $h = 5$ or $h = -5$. False; equation has no solution.
31. If $|3 + z| = 3$, then $z = -3$ or $z = 0$. False; $z = 0$ or $z = -6$.
32. If $|6 - s| = 0$, then $s = 0$. False; $6 - s$ must be 0.

Mental Math **Solve each equation. If there is no solution, explain.** 36. No solution; abs. val. cannot be neg.

33. $|b| = 12$ −12, 12
34. $|x| = 8$ −8, 8
35. $|m + 48| = 0$ −48
36. $-2 = |b|$
37. $|n| - 3 = 7$ −10, 10
38. $|z| + 5 = 1$
39. $|2m| = 18$ −9, 9
40. $|b| - 10 = -1$ −9, 9

38. No solution; abs. val. cannot be neg.

41. Meteorology A meteorologist reported that the previous day's temperatures varied 14°F from the normal temperature of 25°F. What were the maximum and minimum temperatures on the previous day? 11°F, 39°F

42. Manufacturing A box of crackers should weigh 454 g. The quality control inspector weighs every twentieth box. The inspector sends back any box that is not within 5 g of the ideal weight. $|w - 454| = 5$
    a. Write an absolute value equation for this situation.
    b. What are the minimum and maximum weights allowed? 449 g; 459 g

43. a. Banking To check that there are 40 nickels in a roll, a bank weighs the roll and allows for an error of 0.015 oz in the total weight. What are the maximum and minimum acceptable weights if the wrapper weighs 0.05 oz? 7.265 oz; 7.235 oz
    b. Critical Thinking If the roll weighs exactly 7.25 oz, can you be certain that all the coins are acceptable? Explain. See margin.

page 170 Work Together

3. Sample: The town-sponsored parade for the state champions started at the DuBois school and followed Main Street to Town Hall. Mayor Simpkins estimated that about 15,000 residents attended the parade. Other observers indicated that the official figure could be off by as much as 1000 people.

Page 172–174 On Your Own

27. All the computational steps are similar. For each equation you need to add 4 to each side and divide the results by 3. This solves $3c - 4 = 9$. You still need to re-write $|c| = \frac{5}{3}$ as two equations.

43b. No; one coin may weigh much more than 0.18 oz, while another may weigh much less.

Wrap Up

THE BIG IDEA Ask students to list the steps used in solving an absolute value equation.

RETEACHING ACTIVITY Students will solve equations that involve absolute values. They will write and show each step to solve and to check the absolute value equations. (Reteaching worksheet 4-3)

Algebra at Work

For further information about the training and skills necessary to be a cartographer, contact a human resource representative from one of the following companies:

- Rand McNally & Company
  P.O. Box 7600
  Chicago, IL 60680
- National Geographic Society
  1145 17th Street NW
  Washington, DC 20036

Students can learn from discussions with experts. Here are some possible sources.

- Invite a geography teacher to speak to the class.
- Research the internet and find expert cartographers through e-mail.

Encourage students to investigate these topics:

- the training necessary to qualify to become a cartographer.
- math and technology used on this job (topology, reading scales, understanding geographic information systems)
- how researchers are working to incorporate mapmaking experiences into the automated production of maps

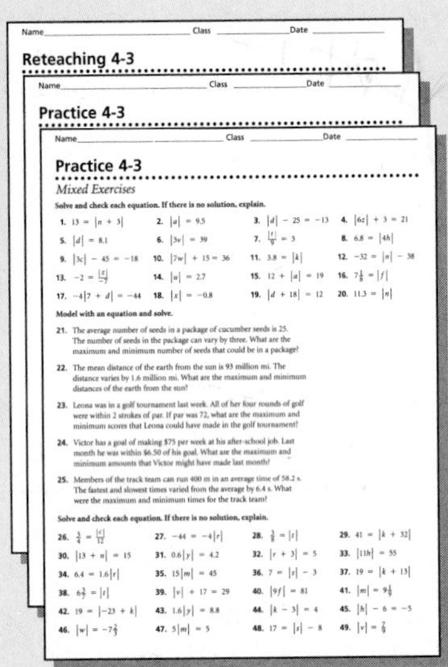

Name______ Class ______ Date ______

Reteaching 4-3

Name______ Class ______ Date ______

Practice 4-3

Name______ Class ______ Date ______

Practice 4-3

Mixed Exercises

Solve and check each equation. If there is no solution, explain.

1. $13 = |n + 3|$
2. $|a| = 9.5$
3. $|d| - 25 = -13$
4. $|6z| + 3 = 21$
5. $|d| = 8.1$
6. $|3v| = 39$
7. $\frac{|t|}{9} = 3$
8. $6.8 = |4h|$
9. $|3c| - 45 = -18$
10. $|7w| + 15 = 36$
11. $3.8 = |k|$
12. $-32 = |n| - 38$
13. $-2 = \frac{|z|}{-7}$
14. $|u| = 2.7$
15. $12 + |a| = 19$
16. $7\frac{1}{8} = |f|$
17. $-4|7 + d| = -44$
18. $|x| = -0.8$
19. $|d + 18| = 12$
20. $11.3 = |n|$

Model with an equation and solve.

21. The average number of seeds in a package of cucumber seeds is 25. The number of seeds in the package can vary by three. What are the maximum and minimum number of seeds that could be in a package?
22. The mean distance of the earth from the sun is 93 million mi. The distance varies by 1.6 million mi. What are the maximum and minimum distances of the earth from the sun?
23. Leona was in a golf tournament last week. All of her four rounds of golf were within 2 strokes of par. If par was 72, what are the maximum and minimum scores that Leona could have made in the golf tournament?
24. Victor has a goal of making $75 per week at his after-school job. Last month he was within $6.50 of his goal. What are the maximum and minimum amounts that Victor might have made last month?
25. Members of the track team can run 400 m in an average time of 58.2 s. The fastest and slowest times varied from the average by 6.4 s. What were the maximum and minimum times for the track team?

Solve and check each equation. If there is no solution, explain.

26. $\frac{3}{4} = \frac{|c|}{12}$
27. $-44 = -4|r|$
28. $\frac{1}{8} = |t|$
29. $41 = |k + 32|$
30. $|13 + n| = 15$
31. $0.6|y| = 4.2$
32. $|r + 3| = 5$
33. $|11h| = 55$
34. $6.4 = 1.6|r|$
35. $15|m| = 45$
36. $7 = |s| - 3$
37. $19 = |k + 13|$
38. $6\frac{3}{7} = |t|$
39. $|v| + 17 = 29$
40. $|9f| = 81$
41. $|m| = 9\frac{1}{9}$
42. $19 = |-23 + k|$
43. $1.6|y| = 8.8$
44. $|k - 3| = 4$
45. $|h| - 6 = -5$
46. $|w| = -7\frac{2}{3}$
47. $5|m| = 5$
48. $17 = |s| - 8$
49. $|v| = \frac{7}{9}$

Lesson Quiz

Lesson Quiz is also available in Transparencies.

Solve and check each equation. If there is no solution, explain.

1. $|x| - 6 = -2$ $x = 4$ or -4
2. $\left|b - \frac{2}{3}\right| = \frac{1}{3}$ $b = 1$ or $\frac{1}{3}$
3. $-4|3z - 2| = -16$
   $z = 2$ or $\frac{-2}{3}$
4. A certain stopwatch is accurate to within 0.1 s in each second. If the stopwatch reads 63.0 s, what are the minimum and maximum values of the true time?
   t is between 56.7 s and 69.3 s

44. *Standardized Test Prep* Compare the quantities in Column A and Column B. Which statement is true for all numbers t? **C**

Column A	Column B
$\lvert t - 4\rvert$	$\lvert 4 - t\rvert$

A. The quantity in Column A is greater.
B. The quantity in Column B is greater.
C. The two quantities are equal.
D. The relationship cannot be determined from the information given.

Exercises MIXED REVIEW

Solve and check each equation.

45. $-3c + 7 = 25$ **−6**
46. $\frac{1}{2}y = y - 6$ **12**
47. $9d + 3.5 = 2d$ **−0.5**
48. $4x - 5 = \frac{2}{3}(6x + 3)$ **no solution**
49. $9.2t = 6.3t - 7.25$ **−2.5**
50. $5.6m + 3.8 = 0.3$ **−0.625**

51. Elio earns $2 per hour more than Ted. They each worked for 5 h. Together they earned a total of $65. What is each person's pay rate?
Ted earns $5.50 per hour; Elio earns $7.50 per hour.

Getting Ready for Lesson 4-4

52. *Geometry* Explain how you can use the formula $C = \pi d$ to find the diameter of a tree without cutting down the tree. **You can measure the circumference of the tree trunk with a tape measure. Then divide the result by π.**

53. *Geometry* The formula for the area of a rectangle is $A = lw$. A rectangular room is 28 ft long and its area is 672 ft^2. What is its width? **24 ft**

Algebra at Work

Cartographer

A cartographer, or map maker, makes exact measurements of the area being mapped. The cartographer uses these dimensions to create the scale of the map, showing the ratio of map distance to actual distance. Knowing the scale of a map means that you can use a proportion to calculate any distance on the map.

Mini Project: Find two maps of the same area that use different scales. Choose several points that are on both maps. Calculate the actual distance between each pair of points on both maps. Are the distances the same? What could account for any differences you find? Write a paragraph detailing your conclusions. Include the maps and your calculations.

4-4 Teaching Notes

PROBLEM OF THE DAY

Subtract the smallest number of set A from the largest number of set A. Write your answer as a fraction.

Set A $= \left\{\frac{1}{2}, \frac{2}{5}, \frac{4}{7}, \frac{5}{6}, \frac{3}{10}, \frac{3}{8}\right\}$ $\frac{8}{15}$

Problem of the Day is also available in Transparencies.

CONNECTING TO PRIOR KNOWLEDGE Ask: *Savannah is Nora's aunt. What is Nora's relationship to Savannah?* Nora is Savannah's niece.

Lead students to understand that the two ways of describing this relationship are different but equivalent.

THINK AND DISCUSS

Example 1 Relating to the Real World 

ERROR ALERT! Some students may not be sure at what point they have completed transforming the equation or formula. **Remediation:** Emphasize that the transformation is complete when the variable *r* is alone on one side of the equation. Point out that all other variables are then on the other side.

Example 2 Relating to the Real World 

MAKING CONNECTIONS Steel, concrete, and other materials expand and contract due to changes in temperature. Engineers have to allow for the possible expansion and contraction when they build with these materials. The formula

Connections History . . . and more

4-4 Transforming Formulas

What You'll Learn

- Using formulas to solve real-world problems
- Solving a literal equation for one of its variables

...And Why

To transform formulas for real-world situations, such as sports, wage, and temperature problems

THINK AND DISCUSS

A formula shows the relationship between two or more variables. You can transform a formula to describe one quantity in terms of the others.

Example 1 Relating to the Real World

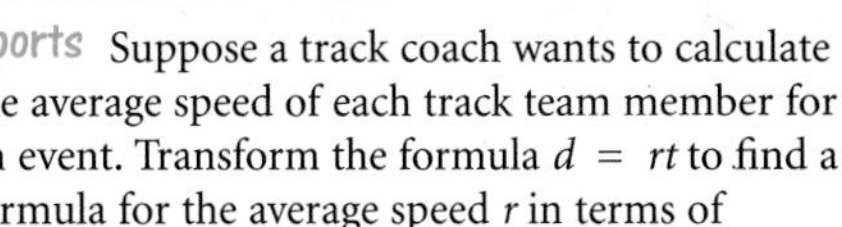

Sports Suppose a track coach wants to calculate the average speed of each track team member for an event. Transform the formula $d = rt$ to find a formula for the average speed r in terms of distance and time.

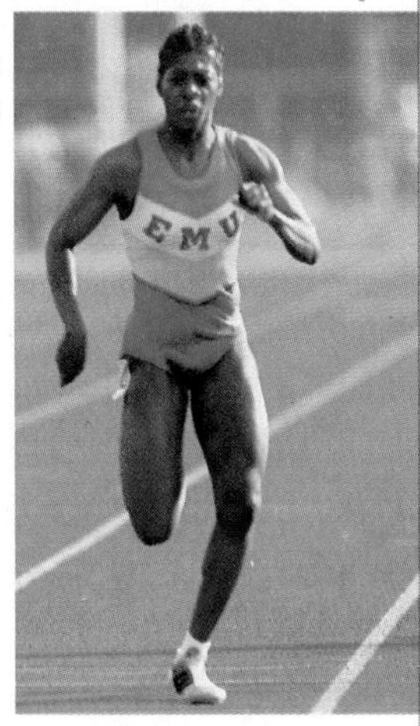

$$d = rt$$

$$\frac{d}{t} = \frac{rt}{t} \quad \leftarrow \text{Divide each side by } t,\ t \neq 0.$$

$$\frac{d}{t} = r \quad \leftarrow \text{Simplify.}$$

The formula for average speed is $r = \frac{d}{t}$.

PROBLEM SOLVING

Look Back What property of equality was used to transform $d = rt$ into $r = \frac{d}{t}$? division property of equality

1. In 1994, Sevetheda Fynes took first place at the New England Track and Field Championships. Shown at the right, she ran 200 m in 23.10 s. Find her average speed. about 8.66 m/s

Example 2 Relating to the Real World 

Science The formula $C = \frac{5}{9}(F - 32)$ gives the Celsius temperature C in terms of the Fahrenheit temperature F. Transform the formula to find Fahrenheit temperature in terms of Celsius temperature.

$$C = \frac{5}{9}(F - 32)$$

$$\frac{9}{5} \cdot C = \frac{9}{5} \cdot \frac{5}{9}(F - 32) \quad \leftarrow \text{Multiply each side by } \tfrac{9}{5}.$$

$$\frac{9}{5}C = F - 32 \quad \leftarrow \text{Simplify.}$$

$$\frac{9}{5}C + 32 = F - 32 + 32 \quad \leftarrow \text{Add 32 to each side.}$$

$$\frac{9}{5}C + 32 = F \quad \leftarrow \text{Simplify.}$$

The formula for Fahrenheit temperature is $F = \frac{9}{5}C + 32$.

PROBLEM SOLVING

Look Back Show how you can solve $C = \frac{5}{9}(F - 32)$ for F if you start by applying the distributive property to the right side of the formula. See margin.

2. The highest temperature ever recorded in the state of Oklahoma occurred in Tishomingo on July 26, 1943, when the temperature reached 49°C. Find the equivalent Fahrenheit temperature. 120.2°F

Lesson Planning Options

Prerequisite Skills

- Solving Equations

Assignment Options

Core 1–8, 10–23
Extension 9, 24–27

Use Mixed Review to maintain skills.

Resources

 **Student Edition**

Skills Handbook, pp. 578, 582
Extra Practice, p. 559
Glossary/Study Guide

Teaching Resources

Chapter Support File, Ch. 4
- Practice 4-4 (two worksheets)
- Reteaching 4-4
- Checkpoint

Classroom Manager 4-4
Glossary, Spanish Resources
Two-Year Algebra Handbook 4-4

 **Transparencies**
60

for the change in length of a 100-m bar of steel is $c = 1.1t$, where $c =$ the change in length in mm, and $t =$ the change in temperature in °C. For example, if the temperature increased 50°C, the 100-m bar would expand 55 mm. Ask students to explain how they would transform the formula to calculate the change in temperature in terms of the change in length.

DIVERSITY Students from other countries may not be familiar with the competitiveness of school sports in the United States. These students may need help understanding the purpose of techniques such as calculating the average speed.

Example 3 **Relating to the Real World**

Question 4 Suggest students use a calculator and round their answer to the nearest cent.

Example 4

VISUAL LEARNING Suggest that students rewrite Example 4 using a different color pencil for each of the variables *a, b, c,* and *x*.

CRITICAL THINKING Question 5 For each equation, have a volunteer go to the board and show the approach they would use to solve it. Then ask for another student to show a different approach. Emphasize that the different approaches lead to the same solution.

Exercises ON YOUR OWN

GEOMETRY Exercise 9a In this exercise, students use the formula for finding the midpoint of a segment. Suggest that

Additional Examples

FOR EXAMPLE 1

A motor's power rating can be found using the formula $P = \frac{W}{t}$. W is the work done by the motor in a certain amount of time t. Transform this formula to express the time t it takes for a motor to do a certain amount of work. $t = \frac{W}{P}$

Discussion: *Under what circumstances could this equation prove useful?*

FOR EXAMPLE 2

The formula to find the area of a trapezoid is $A = \frac{1}{2}(b_1 + b_2)$, where b_1 and b_2 represent the two bases. If $b_2 = 10$, transform the formula to find b_1 in terms of A.

$$A = \frac{1}{2}(b_1 + 10)$$
$$\frac{2}{1} \cdot A = \frac{2}{1} \cdot \frac{1}{2}(b_1 + 10)$$
$$2A = b_1 + 10$$
$$2A - 10 = b_1 + 10 - 10$$
$$2A - 10 = b_1$$

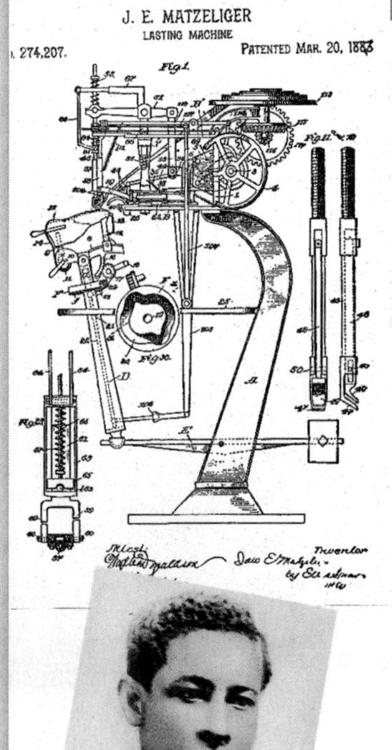

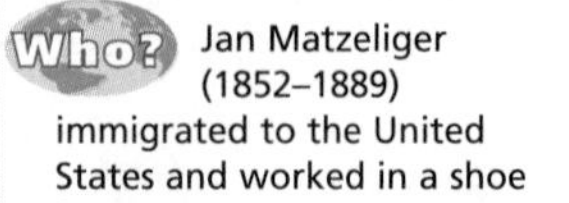

Jan Matzeliger (1852–1889) immigrated to the United States and worked in a shoe factory in Lynn, Massachusetts.

5. a. $m = -\frac{2}{7}n + 3$

b. $m = -18x + 5$

c. $m = \frac{27}{4}y - 2$

Example 3 **Relating to the Real World**

History In 1883, Jan Matzeliger invented the shoe-lasting machine to attach the upper part of a shoe to its sole. Before that, each shoe was assembled and sewn by hand. Wages were calculated using this formula.

hourly wage in \$ ⟶ $w = (0.34)(\frac{p}{t})$ ⟵ price in \$ of a pair of shoes; ⟵ time in minutes to assemble

Transform the formula to find the price of a pair of shoes in terms of the worker's hourly wage and the time needed to assemble the shoe.

$$w = (0.34)(\tfrac{p}{t})$$
$$w \cdot t = (0.34)(\tfrac{p}{t}) \cdot t \quad \leftarrow \text{Multiply each side by } t.$$
$$wt = 0.34p \quad \leftarrow \text{Simplify.}$$
$$\frac{wt}{0.34} = \frac{0.34p}{0.34} \quad \leftarrow \text{Divide each side by 0.34.}$$
$$\frac{wt}{0.34} = p \quad \leftarrow \text{Simplify.}$$

You can use the formula $p = \frac{wt}{0.34}$ to find the price of a pair of shoes.

3. **Try This** Transform the formula $w = (0.34)\frac{p}{t}$ to find a formula for time in terms of hourly wage and price. $t = (0.34)\frac{p}{w}$
4. In 1891 a worker who used the shoe-lasting machine could assemble one pair of shoes in two minutes and earned \$.16/h. What was the price of a pair of shoes in 1891? \$.94

A **literal equation** is an equation involving two or more variables. Formulas are special types of literal equations. You solve for one variable in terms of the others when you transform a literal equation.

Example 4

Solve $ax + b = c$ for x in terms of a, b, and c.

$$ax + b = c$$
$$ax + b - b = c - b \quad \leftarrow \text{Subtract } b \text{ from each side.}$$
$$ax = c - b \quad \leftarrow \text{Simplify.}$$
$$\frac{ax}{a} = \frac{c - b}{a} \quad \leftarrow \text{Divide each side by } a, a \neq 0.$$
$$x = \frac{c - b}{a} \quad \leftarrow \text{Simplify.}$$

5. **Critical Thinking** Describe the approach you would use to solve each equation for m. See margin for approach.
   a. $2n = 7(3 - m)$ b. $\frac{m - 5}{6} = -3x$ c. $\frac{2}{3}(2m + 4) = 9y$
6. Explain why you did or did not use the distributive property in each equation in Question 5. **Answers may vary. Sample: You do not use the distributive prop., because it is easier to use the mult. and div. props. first.**

students sketch the segment on a number line and estimate the midpoint of the segment before calculating.

ALTERNATIVE ASSESSMENT Exercises 10–21 After students have solved each equation for the given variable, have them solve the new equation for one of its other variables. This will help you assess student understanding of transformations.

Exercise 22b Help students to estimate. Ask: *Will Manuel need more than one hour?* yes *More than two hours?* yes *More than three hours?* no

ESL Exercise 23 Some students may not know what a *hang glider* is. To help these students understand the terms in this exercise, ask if anyone has ever been hang gliding. Ask what the sport involves and why it is called "hang" gliding. Challenge students to estimate the wingspan of a hang glider.

WRITING Exercise 24 Suggest that students compare solving $3x + 2 = y$ for x with solving $3x + 2 = 8$.

OPEN-ENDED Exercise 25 Students might use a formula that they remember from previous math or science lessons.

Chapter Project **FIND OUT BY SOLVING** Ask students to take a deep breath and to estimate the volume of air they think is in their lungs. Then have students calculate 90% of their air volume. After the students solve the given equation for age *a*, have them graph the function. This information is essential to their work on the Chapter Project. Have students add this task to work they have already completed for the project. You may wish to check students' progress on the project by having each one write a sentence or two describing what they have done for the project thus far.

Exercises ON YOUR OWN

Solve each equation for the given variable.

1. $3r + 4 = s; r$ $\quad r = \frac{s-4}{3}$
2. $c = \frac{d}{g}; g$ $\quad g = \frac{d}{c}$
3. $\frac{m}{n} = \frac{p}{q}; p$ $\quad p = \frac{mq}{n}$
4. $ax + by = c; y$ $\quad y = \frac{c - ax}{b}$
5. $C = 2\pi r; r$ $\quad r = \frac{C}{2\pi}$
6. $5x - 2y = -6; y$ $\quad y = \frac{5x+6}{2}$
7. $3m = 2(4 + t); t$ $\quad t = \frac{3m}{2} - 4$
8. $y = 3(t - s); s$ $\quad s = t - \frac{y}{3}$
9. **Geometry** Suppose *a* and *b* are the coordinates of the endpoints of a line segment. To find the midpoint of the segment, you can use the formula $m = \frac{a+b}{2}$.
   a. Find the midpoint of a segment with endpoints 8.2 and 3.5. 5.85
   b. Transform the formula to find *b* in terms of *a* and *m*. $b = 2m - a$
   c. Find the missing endpoint of a segment starting at –1.7 that has midpoint 2.1. Use the transformed formula from part (b). 5.9

a m b

Solve each equation for the given variable.

10. $x + y = 20; y$ $\quad y = 20 - x$
11. $4m - n = 6; n$ $\quad n = 4m - 6$
12. $2(t + r) = 5; t$ $\quad t = \frac{5}{2} - r$
13. $z - a = y; z$ $\quad z = a + y$
14. $\pi = \frac{C}{d}; d$ $\quad d = \frac{C}{\pi}$
15. $\frac{a}{2} = \frac{b}{7}; b$ $\quad b = \frac{7a}{2}$
16. $\frac{a+b}{c} = \frac{d}{4}; a$ $\quad a = \frac{cd}{4} - b$
17. $A = \frac{1}{2}bh; b$ $\quad b = \frac{2A}{h}$
18. $dx = c; x$ $\quad x = \frac{c}{d}$
19. $V = lwh; h$ $\quad h = \frac{V}{lw}$
20. $V = \frac{1}{3}\pi r^2 h; h$ $\quad h = \frac{3V}{\pi r^2}$
21. $|ap - b| = r; p$ $\quad p = \frac{b-r}{a}$ or $p = \frac{b+r}{a}$
22. a. **Travel** Manuel plans to drive to an amusement park. He will travel 121 mi on the highway at an average speed of 55 mi/h. Transform the formula $d = rt$ to find a formula for time. $t = \frac{d}{r}$
    b. How much time does Manuel need for this part of his trip? 2 h 12 min
23. a. **Recreation** The aspect ratio of a hang glider measures its ability to glide and soar. The formula $R = \frac{s^2}{A}$ gives the aspect ratio *R* for a glider with wingspan *s* and wing area *A*. Solve this formula for *A*. $A = \frac{s^2}{R}$
    b. Suppose you want to design a glider with a 9-ft wingspan and an aspect ratio of 3. Use the formula you found in part (a) to find the wing area. 27 ft²

s

24. **Writing** How is solving a literal equation similar to solving an equation with just one variable? How is it different? The same props. of equality are used, but the solution is a formula, not a number.
25. **Open-ended** Write an equation using more than one variable. Solve the equation for each variable. Show all your steps. Sample: $2x + 3y = 6$; $x = -\frac{3}{2}y + 3$; $y = -\frac{2}{3}x + 2$
26. a. **Banking** The formula $I = prt$ gives the amount of simple interest *I* earned by principal *p* at an annual interest rate *r* over *t* years. Solve this formula for *p*. $p = \frac{I}{rt}$
    b. Find *p* if $r = 0.035$, $t = 4$, and $I = \$420$. Write a sentence to explain the meaning of your answer. \$3000; \$3000 invested at 3.5% interest for 4 years will yield \$420 in interest.
27. a. Solve $2y + x = 4$ for *y*. $y = -\frac{1}{2}x + 2$
    b. **Graphing Calculator** Graph the equation you found in part (a).
    c. Use your graph and the TABLE feature of the calculator to find *y* when $x = 3$, $x = 4$, and $x = 5$. $\frac{1}{2}, 0, -\frac{1}{2}$

27b.

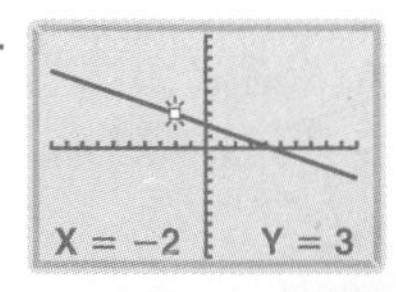

Technology Options

For Exercise 9, students may use geometry software to find the midpoint of the line segment. For Exercise 27, students may want to use drawing software to verify their answer.

Prentice Hall Technology

Calculator Graphing Calculator Handbook: Procedures 6, 7

Software • Secondary Math Lab Toolkit • Computer Item Generator 4-4

CD-ROM Multimedia Algebra Lab 4

Internet • See the Prentice Hall site. (http://www.phschool.com)

pages 175–176 Think and Discuss

Problem Solving, 5a–c, 27b. See back of book.

Exercises MIXED REVIEW

GETTING READY FOR LESSON 4-5 These exercises prepare students for solving inequalities.

OPEN-ENDED Exercise 31b Have students check their equation by using the data from part (a).

Wrap Up

THE BIG IDEA Ask students to explain what it means to transform a formula.

RETEACHING ACTIVITY Students use different shapes to solve literal equations for one value. They will draw shapes to model each equation and find the solution. (Reteaching worksheet 4-4)

Exercises CHECKPOINT

In this checkpoint, students have an opportunity to assess their own progress on the topics covered in Lessons 4-1 to 4-4.

Exercise 13 Encourage students to estimate first.

GEOMETRY Exercise 14 Students use the formula for the perimeter of a rectangle. Suggest that they begin by transforming the formula to find a formula for the width w in terms of the perimeter and the length.

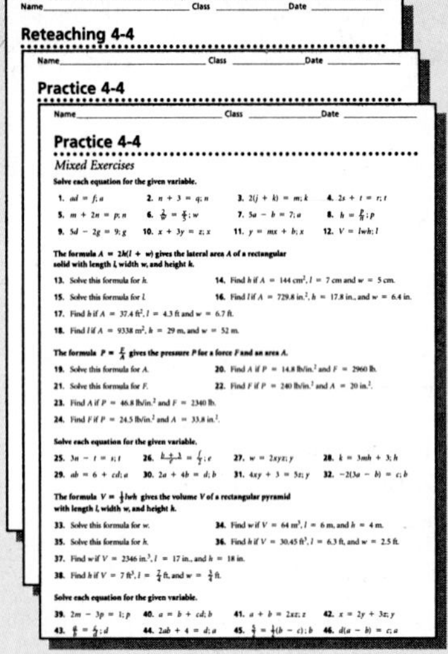

Lesson Quiz

Lesson Quiz is also available in Transparencies.

Solve each equation for the given variable.

1. $3x + 2y = z; y$ $y = \frac{1}{2}z - \frac{3}{2}x$
2. $\frac{a-b}{c} = \frac{d}{3}; a$ $a = \frac{cd}{3} + b$
3. $3x + 4 = 2(3 - y); y$ $y = 1 - \frac{3}{2}x$
4. A satellite's speed as it orbits the Earth is found using the formula $v^2 = \frac{Gm}{r}$. In this formula, m stands for the mass of the Earth. Transform this formula to find the mass of the Earth. $m = \frac{v^2r}{G}$

Chapter Project — Find Out by Solving

The volume of air an adult's lungs can hold decreases with age. The formula $V = 0.104h - 0.018a - 2.69$ estimates air volume V (in liters) of a person's lungs for someone of height h inches and age a years.

- Estimate your own lung volume. Then find 90% of your lung volume.
- Solve the formula for age a. Estimate how old you will be when your air volume is 90% of its current value. (Use your current height.)
- Substitute your current height for h in the original formula. Then graph the function to show how air volume changes as age increases.

Exercises MIXED REVIEW

28. 2.8 is what percent of 3.5? 80% **29.** 70 is $12\frac{1}{2}$% of what number? 560 **30.** What is 17% of 123? 20.91

31. a. Earnings Suppose you earn \$5.75/h and you work 12 h during one week. What are your earnings for the week? \$69

b. Write an equation that a worker paid hourly can use to find his or her earnings. Explain what each variable represents.
Answers may vary. Sample: p = total pay; r = hourly rate; h = number of hours; $p = rh$

Getting Ready for Lesson 4-5

Complete each statement with <, =, or >.

32. -3 ■ -5 > **33.** 4.8 ■ 4.29 > **34.** $-1 - 2$ ■ $6 - 9$ = **35.** $-\frac{3}{4}$ ■ $-\frac{4}{5}$ > **36.** $\frac{1}{2} + \frac{1}{2}$ ■ $\frac{1}{3} + \frac{1}{3}$ >

Exercises CHECKPOINT

Solve for the given variable.

1. $s = n - 90; n$ $n = s + 90$ **2.** $y = mx + b; b$ $b = y - mx$ **3.** $L = \frac{2b^2}{a}; a$ $a = \frac{2b^2}{L}$ **4.** $W = fd; f$ $f = \frac{W}{d}$

Solve and check each equation. If there is no solution, explain.

5. $\frac{6}{30} = \frac{m}{84}$ 16.8 **6.** $-4(t + 1) = 4 - 4t$ no solution; $-4 \neq 4$ **7.** $|x| = -7$ 7. No solution; abs. val. cannot be neg. **8.** $8w = 6w$ 0

9. $|n| + 2 = 10$ −8, 8 **10.** $y + 21 = 4y - 8$ $9\frac{2}{3}$ **11.** $5(2w - 4) = 6w$ 5 **12.** $|y - 4.5| = 3$ 1.5, 7.5

13. Maps The scale on a map is 3 cm : 10 km. On the map, the distance between two cities is 5.2 cm. What is the actual distance? $17\frac{1}{3}$ km

14. Geometry $P = 2l + 2w$ gives the perimeter P of a rectangle of length l and width w. Suppose a rectangle has perimeter 14.48 cm and length 4.16 cm. Which key sequence could you use to find the width? C

A. 2 × 14.48 + 2 × 4.16 ENTER
B. 14.48 − 2 × 4.16 ÷ 2 ENTER
C. (14.48 − 2 × 4.16) ÷ 2 ENTER
D. (2 × 4.16 − 14.48) ÷ 2 ENTER

4-5 Teaching Notes

PROBLEM OF THE DAY

Jackie has \$3.50 in quarters and nickels. If she has two more quarters than nickels, how many quarters and how many nickels does Jackie have? 12 quarters, 10 nickels

Problem of the Day is also available in Transparencies.

CONNECTING TO PRIOR KNOWLEDGE Ask students:

- *Kyle is older than Jaime. How would you write this using a greater than sign?* Kyle's age > Jaime's age
- *How would you write this using a less than sign?* Jaime's age < Kyle's age

Write these two answers on the board. Lead students to understand that they are equivalent in meaning.

WORK TOGETHER

Question 1b What would the inequality be if the express checkout sign read *Fewer Than 8 Items?* $x > 8$

OPEN-ENDED Question 3 For students struggling to begin, suggest examples such as vehicle capacities, weight limits, and number of people at an event.

THINK AND DISCUSS

Question 4 Ask: *How many solutions does the inequality* $x < 3$ *have?* an infinite number

Connections Banking . . . and more

4-5 Solving Inequalities Using Addition and Subtraction

What You'll Learn

- Using addition and subtraction to solve one-step inequalities
- Using inequalities to model real-world problems

...And Why

To solve inequalities that model real-world situations such as checking accounts

QUICK REVIEW

< is less than
≤ is less than or equal to
> is greater than
≥ is greater than or equal to
≠ is not equal to

WORK TOGETHER

Work with a partner.

1a. s = speed; $s \le 55$

1. Define a variable and write an inequality for each sign.

1b. n = number of items; $n \le 12$

1c. r = hourly rate; $r \ge 4.25$

2a. a = age; $a \ge 17$

2. Define a variable and write an inequality for each situation.
   a. Persons under 17 are not admitted.
   b. Visibility at the airport is less than two miles.
   c. You must be more than 36 in. tall to ride an amusement park ride.

b. v = visibility; $v < 2$ c. h = height of rider; $h > 36$

3. *Open-ended* Describe three other situations that involve inequalities. Write an inequality for each situation. Samples: The highest total rainfall for April is 2.8 in. $r \le 2.8$; Littering is subject to a fine of up to \$200 $f \le 200$; Over 2000 people participate in the charity race every year $p > 2000$.

THINK AND DISCUSS

Part 1 **Graphing and Writing Inequalities**

Some inequalities contain a variable. Any value of the variable that makes the inequality true is called a **solution of the inequality.** The solutions of the inequality $x < 3$ are all the numbers less than 3.

4. Tell whether each number is a solution of $x < 3$.
   a. 1 yes b. −7.3 yes c. 9.004 no d. 0 yes e. 3 no

Lesson Planning Options

Prerequisite Skills

- Understanding concepts of inequalities

Assignment Options

To provide flexible scheduling, this lesson can be subdivided into parts.

1 **Core** 1–6, 8–12 **Extension** 7, 16–21

2 **Core** 13, 15, 25–36 **Extension** 50, 51–56

3 **Core** 14, 23, 24, 37–48 **Extension** 22, 49, 57

Use Mixed Review to maintain skills.

Resources

Student Edition

Skills Handbook, p. 576
Extra Practice, p. 559
Glossary/Study Guide

Teaching Resources

Chapter Support File, Ch. 4
- Practice 4-5 (two worksheets)
- Reteaching 4-5

Classroom Manager 4-5
Glossary, Spanish Resources
Two-Year Algebra Handbook 4-5

Transparencies
8, 10, 61, 64

CRITICAL THINKING **Question 6** Have a volunteer display the solution to $x \neq 3$ on the board.

Question 7 Students can use the Problem Solving Hint to check the graphs of their solutions.

Have students compare the Addition Property of Inequality with the Addition Property of Equality on page 113.

VISUAL LEARNING Encourage students to practice drawing number lines with open and closed dots. Suggest that they decorate the open and closed dots in ways that will help them remember the different meanings of the dots.

Example 1

Use tiles to model solving the inequality. Write the inequality symbol on an index card. Place the card between the two sets of tiles. When the solution is complete, students can check by choosing a value for x and substituting that value in the original inequality.

ERROR ALERT! Some students may have difficulty deciding in which direction to draw the graph of the solution to the inequality. **Remediation:** Point out that when the solution has the variable on the left, the graph points in the same direction as the inequality symbol. When the solution has the variable on the right, the graph points in the opposite direction as the inequality symbol.

Additional Examples

FOR EXAMPLE 1

Solve $-3 + y < 2$. Graph the solutions on a number line. $y < 5$

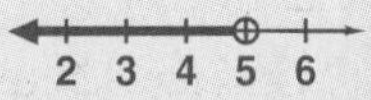

Discussion: *Why should we indicate the number 5 with an open dot?*

FOR EXAMPLE 2

In order to receive an A in your literature class, you must earn more than 350 points of reading credit. You have already earned 210 points of reading credit. Write an inequality and solve to determine the minimum number of points of reading credit you must still earn to receive an A.

$210 + p > 350$

$p > 140$

141 points of reading credit

180

The number lines show the solutions of $x < 3$ along with the graphs of three other inequalities comparing x and 3.

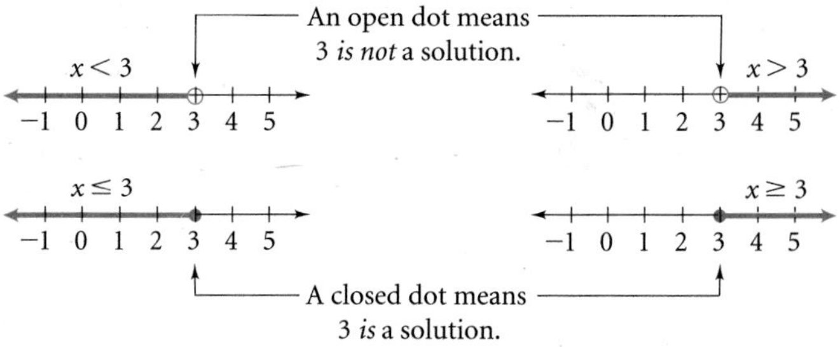

5. The solutions of $x < 3$ are all the numbers less than 3; the solution of $x = 3$ is only the number 3.

PROBLEM SOLVING HINT

When the variable is on the left, the inequality symbol points in the same direction as the graph.

7a.

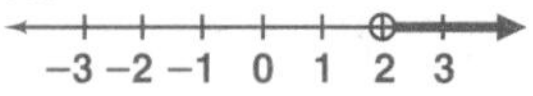

b.

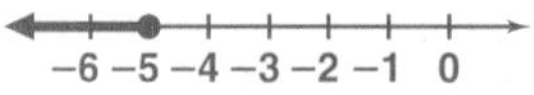

c.

5. Explain how the inequality $x < 3$ is different from the equation $x = 3$. **See left.**
6. *Critical Thinking* Describe how you can display the solutions of the inequality $x \neq 3$ on a number line. **You can put an open dot at 3 and color the rest of the number line.**
7. You can rewrite an inequality like $4 > n$ as $n < 4$. Rewrite each inequality so that the variable is on the left, then graph the solutions.
   a. $2 < x$ $x > 2$ **b.** $-5 \geq b$ $b \leq -5$ **c.** $0 \leq r$ $r \geq 0$
   See left for graphs.

Part 2

Using Addition to Solve Inequalities

The solutions of an inequality like $x < 3$ are easy to recognize. To solve some other inequalities, you may need to find a simpler, equivalent inequality. **Equivalent inequalities** have the same set of solutions.

Consider the inequality $-4 < 1$. The number line shows what happens when you add 2 to each side of the inequality.

$-4 < 1$

$-4 + 2 < 1 + 2$

$-2 < 3$

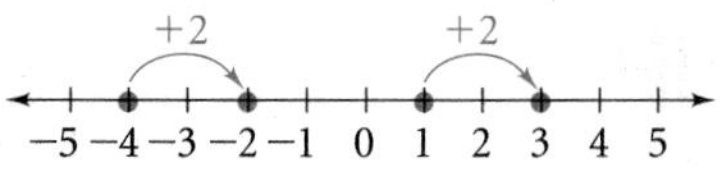

Notice that addition does not change the relationship between the numbers or the direction of the inequality symbol.

Addition Property of Inequality

For all real numbers a, b, and c, if $a > b$, then $a + c > b + c$.

Example: $5 > -1$, so $5 + 2 > -1 + 2$

For all real numbers a, b, and c, if $a < b$, then $a + c < b + c$.

Example: $-4 < 1$, so $-4 + 2 < 1 + 2$

8. What number would you add to both sides of each inequality to get a simpler, equivalent inequality?
   a. $x + 3 > -2$ **−3** **b.** $0 < -\frac{4}{3} + g$ $\frac{4}{3}$ **c.** $1.6 \geq z - 4.3$ **4.3**

Question 9d Have students use this generalization to check their solutions throughout the lesson.

Have students compare the Subtraction Property of Inequality with the Subtraction Property of Equality on page 113.

Question 11 When checking the solution, students may refer to the generalization they made in Question 9d.

CONNECTING TO THE STUDENTS' WORLD Remind students of the discussion you had about Kyle and Jaime at the beginning of the lesson. On the board, write Kyle's age > Jaime's age. Ask students:

- *Kyle is two years older than Jaime now. Five years from now, who will be older?* Kyle
- *How much older will Kyle be then?* two years older

Write Kyle's age + 5 > Jaime's age + 5 on the board. Ask students to think of friends or siblings who are older or younger. Then ask students to make up problems to illustrate the age difference in the future and in the past until students are confident that the age difference never changes. Help students make the connection between this discussion and the addition and subtraction properties of inequalities.

You can use the Addition Property of Inequality to solve inequalities.

Example 1

Solve $x - 3 < 5$. Graph the solutions on a number line.

$$x - 3 < 5$$
$$x - 3 + 3 < 5 + 3 \quad \leftarrow \text{Add 3 to each side.}$$
$$x < 8 \quad \leftarrow \text{Combine like terms.}$$

The solutions are all numbers less than 8.

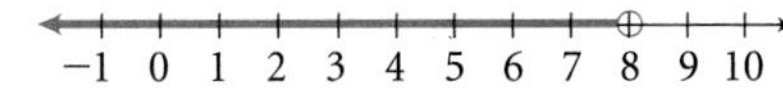

9d. Test a number to the right, to the left, and at the endpoint. The resulting inequality will only be true for numbers that are in the correct solution region.

9. a. Substitute a number greater than 8 for x in the inequality $x - 3 < 5$. Is your result true? no
   b. Substitute 8 for x in the inequality $x - 3 < 5$. Is your result true? no
   c. Substitute a number less than 8 for x in the inequality $x - 3 < 5$. Is your result true? yes
   d. Parts (a), (b), and (c) can serve as a check for Example 1. **Generalize** the steps you can use to check the solutions of an inequality. See left.

10a. $-3\ -2\ -1\ 0\ 1\ 2\ 3$

10. a. **Try This** Solve the inequality $f - 2 > -4$. Graph the solutions on a number line. $f > -2$
    b. Use the steps from Question 9, part (d), to check your work. $-1 - 2 > -4$ true; $-2 - 2 > -4$ false; $-3 - 2 > -4$ false

Part 3

Using Subtraction to Solve Inequalities

Just as you can add the same number to each side of an inequality, you can subtract the same number from each side. The order, or direction, of the inequality is not changed.

Subtraction Property of Inequality

For all real numbers a, b, and c, if $a > b$, then $a - c > b - c$.
Example: $3 > -1$, so $3 - 2 > -1 - 2$
For all real numbers a, b, and c, if $a < b$, then $a - c < b - c$.
Example: $-5 < 4$, so $-5 - 2 < 4 - 2$

11. 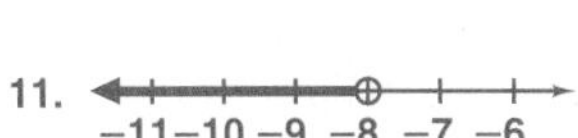

11. Solve and check the inequality $y + 2 < -6$. Graph the solutions on a number line. $y < -8$

12. What number would you subtract from both sides of each inequality to get a simpler, equivalent inequality?
    a. $w + 2 > -1$ 2
    b. $8 < \frac{5}{3} + r$ $\frac{5}{3}$
    c. $5.7 \geq k - 3.1$ -3.1

Technology Options

For Exercise 57, students may want to use spreadsheet software to sort and graph their research data. For Exercises 58–59, students may want to add and subtract the matrices with a graphing calculator.

Prentice Hall Technology

Software • Secondary Math Lab Toolkit • Computer Item Generator 4-5

Internet • See the Prentice Hall site. (http://www.phschool.com)

Example 2 **Relating to the Real World**

CONNECTING TO THE STUDENTS' WORLD Encourage students to research the minimum balance requirements of banks in your area. Use their data to create similar inequality problems.

CRITICAL THINKING Question 13b Help students realize that checking three random values does not guarantee that a solution is correct. They need to use the checking method described in Question 9.

Exercises ON YOUR OWN

Exercises 13–15 Have students write the second inequality under the first. Remind them to align the inequality symbols.

WRITING Exercise 22 Students might include a discussion of the difference between the solution of an addition or subtraction *equation* and the solution of an addition or subtraction *inequality.*

ESL Exercise 24 Students may be unfamiliar with the terms *free checking* and *service fees.* Briefly discuss how checking accounts work. Explain that banks often charge an annual fee unless you keep a minimum amount of money (for example, \$500) in the account.

ALTERNATIVE ASSESSMENT **Exercises 45–48** Randomly select a student and ask the student to state an appropriate first step for solving the first inequality. Then ask another student to suggest a subsequent step. Continue choosing students until all exercises are solved. This will help you assess the students' skill at solving inequalities.

pages 182–184 On Your Own

7d.

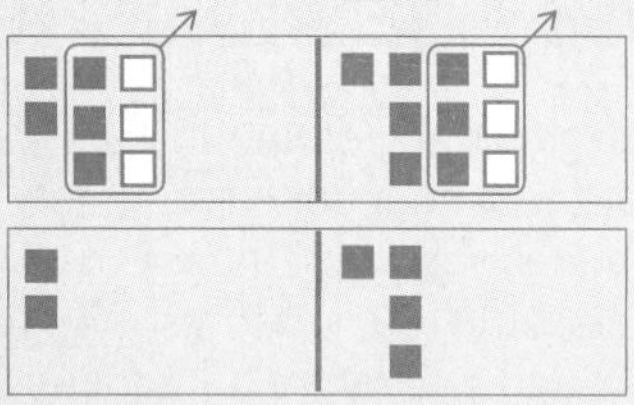

25.

26.

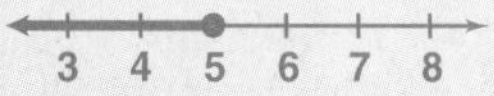

27.

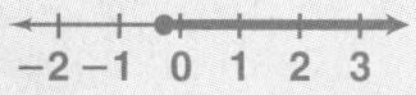

28.

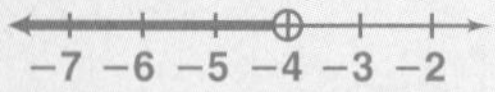

29.

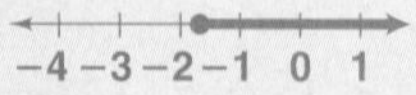

30.

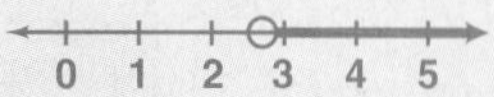

31.

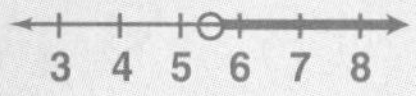

You can use inequalities to model real-world problems.

Example 2 **Relating to the Real World**

Banking The First National Bank offers free checking for accounts with a balance of at least \$500. Suppose you have a balance of \$516.46 and you write the check at the left. How much must you deposit to avoid being charged a service fee?

Define d = the amount you must deposit

Relate	current balance	minus	amount of check	plus	amount of deposit	is at least	\$500
Write	516.46	−	31.50	+	d	≥	500

$516.46 - 31.50 + d \geq 500$

$484.96 + d \geq 500$ ← Combine like terms.

$484.96 + d - 484.96 \geq 500 - 484.96$ ← Subtract 484.96 from each side.

$d \geq 15.04$ ← Simplify.

You must deposit at least \$15.04 in your account.

13. a. Check the answer to Example 2 by choosing three values and checking the values in the original inequality. Answers may vary.

b. *Critical Thinking* Explain why checking the values in the inequality does not guarantee that your solution is correct. The endpoint may be wrong, but the numbers check correctly.

14. Explain how the phrases "at least" and "at most" affect the inequality you write when you model a real-world problem. Both phrases imply that the endpoint is included. "At least" is translated as ≥. "At most" is translated as ≤.

Exercises ON YOUR OWN

Define a variable and write an inequality to model each situation.

1. A bus can seat 48 students or fewer. n = number of students; $n \leq 48$

2. There are over 20 species of crocodiles. c = number of species; $c > 20$

3. In many states, you must be at least 16 years old to obtain a driver's license. a = age of applicant; $a \geq 16$

4. At least 350 students attended the dance Friday night. s = number of students; $s \geq 350$

5. You may not use a light bulb of more than 60 watts in this light fixture. w = light bulb wattage ; $w \leq 60$

6. The Navy's flying team, the Blue Angels, makes more than 75 appearances each year. n = number of apperances in a year; $n > 75$

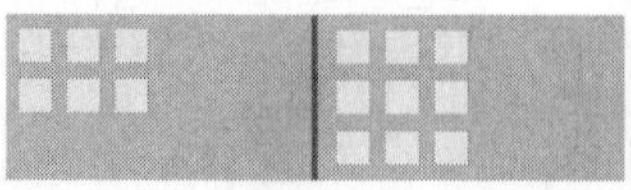

7. a. *Group Activity* Write an inequality for the tiles shown. $6 < 9$

b. Add four positive tiles to each side. Write an inequality for the tiles. $10 < 13$

c. Use the original set of tiles. Subtract seven positive tiles from each side. Write an inequality to represent the tiles. $-1 < 2$

d. Show how to use tiles to **verify** the inequality $-5 + 3 > -7 + 3$. See margin.

e. *Open-ended* Write an inequality you can solve with tiles. Solve the inequality. Show all the steps. Sample: $2x - 1 < x + 3$; $x < 4$

OPEN-ENDED Exercise 49a Encourage students to use numbers other than integers in their inequalities.

CRITICAL THINKING Exercise 50 Suggest students begin by solving the inequality.

STANDARDIZED TEST TIP Exercise 51 Since $x - y$ must be a positive quantity, $y - x$ must be a negative quantity. Therefore, substitute a small positive number such as +4 for $x - y$, and substitute a −4 for $y - x$ in I, II, and III.

DIVERSITY Exercises 52–56 Encourage a discussion of the implications of the graph on income. Invite students to make suggestions about what can be done to help reduce these inequalities. Help students understand the kinds of unfair business decisions that sometimes lead to wage gaps. As an example, part-time workers may be paid less than full-time workers for the same work. Remind students that women are more likely than men to work part-time as they balance career and family. Challenge students to think of other examples. Point out that the wage gap between men and women has decreased over the last few decades.

STATISTICS Exercises 52–56 Before students begin, ask:

- *What is a bachelor's degree?* a four-year college degree
- *What is an associate's degree?* a two-year college degree
- *What is meant by "some college"?* no college degree received, but some college courses taken

EXTENSION Exercises 52–56 Have students use the bar graph to write their own word problems. Students can exchange problems to solve them.

Write four numbers that are solutions of each inequality. 8–12. Answers may vary. Samples:

8. $v \geq -5$ −5, −3, 0, 12 **9.** $0.5 > c$ −20, −1, 0, 0.2 **10.** $2 + q < -7$ −13, −12, −10, −9.5 **11.** $f - 10 \leq 16$ 1, 2, 3, 26 **12.** $5 > 4 - z$ −0.5, 0, 1, 10

Tell what you must do to the first inequality in order to get the second.

13. $36 \leq -4 + y$; $40 \leq y$ add 4 to each side **14.** $9 + b > 24$; $b > 15$ subtract 9 from each side **15.** $m - \frac{1}{2} < \frac{3}{8}$; $m < \frac{7}{8}$ add $\frac{1}{2}$ to each side

Write each inequality in words. 18. x is at most 7.

16. $n < 5$ n is less than 5. **17.** $b > 0$ b is greater than 0. **18.** $x \leq 7$ **19.** $m \leq -1$ m is at most −1. **20.** $g - 2 < 7$ 2 less than g is less than 7. **21.** $z \geq -4$ 21. z is at least −4.

22. *Writing* How is solving an addition or subtraction inequality similar to solving an addition or subtraction equation? How is it different? You need to follow similar steps, but the solutions are different.

23. To earn an A in Ms. Orlando's math class, students must score a total of at least 135 points on the three tests. On the first two tests, Amy's scores were 47 and 48. What is the minimum score she must get on the third test in order to meet the requirement? 40

24. *Banking* Suppose you must maintain a balance of at least \$750 in your checking account in order to have free checking. The balance in your account is \$814.22 before you write a check for \$25. How much cash can you withdraw from the account and still have free checking? \$39.22

Solve each inequality. Graph the solutions on a number line. 25–48. See margin for graphs.

25. $x - 1 > 10$ $x > 11$ **26.** $w + 4 \leq 9$ $w \leq 5$ **27.** $h + \frac{3}{4} \geq \frac{1}{2}$ $h \geq -\frac{1}{4}$ **28.** $-5 > b - 1$ $b < -4$

29. $0 \leq x + 1.7$ $x \geq -1.7$ **30.** $n - 2\frac{1}{2} > \frac{1}{3}$ $n > 2\frac{5}{6}$ **31.** $3.5 < m - 2$ $m > 5.5$ **32.** $\frac{3}{2} + k \geq -45$ $k \geq -46\frac{1}{2}$

33. $y - 0.3 < 2.8$ $y < 3.1$ **34.** $2 < s - 8$ $s > 10$ **35.** $9.4 \leq t - 3.5$ $t \geq 12.9$ **36.** $h - \frac{1}{2} \geq -1$ $h \geq -\frac{1}{2}$

37. $-6 > n - \frac{1}{5}$ $n < -5\frac{4}{5}$ **38.** $8 + b < 1$ $b < -7$ **39.** $-7.7 \neq x - 2$ $x \neq -5.7$ **40.** $a + 3 \geq 2.7$ $a \geq -0.3$

41. $c + \frac{1}{2} \neq 3\frac{2}{7}$ $c \neq 2\frac{11}{14}$ **42.** $f - 2.3 \leq -1.21$ $f \leq 1.09$ **43.** $\frac{3}{2} + g \leq \frac{1}{3}$ $g \leq -1\frac{1}{6}$ **44.** $0 > k - 2\frac{3}{5}$ $k < 2\frac{3}{5}$

45. $7.5 + y < 13$ $y < 5.5$ **46.** $-1.4 + s + 2.1 > 11$ $s > 10.3$ **47.** $-7\frac{3}{4} + m + \frac{1}{2} \leq -2\frac{1}{4}$ $m \leq 5$ **48.** $\frac{2}{3} + t - \frac{5}{6} \neq 0$ $t \neq \frac{1}{6}$

49. a. *Open-ended* Use the inequality symbols <, ≤, >, and ≥ to write four addition or subtraction inequalities. Samples: $x + 4 < 3$; $b - 2 \leq 0$; $2 + m > 1.5$; $z - \frac{1}{2} \geq \frac{3}{4}$

b. Solve each of the inequalities in part (a) and graph the solutions on separate number lines. See margin.

50. *Critical Thinking* Sam says that he can solve $z - 8.6 \leq 5.2$ by replacing z with 13, 14, and 15. When $z = 13$, the inequality is false. When $z = 14$ and $z = 15$, the inequality is true. So Sam says that the solution is $z \geq 14$ Is his reasoning correct? **Justify** your answer. No; $z \geq 14$ is only a part of the solution. 13.8 is also a part of the solution but Sam missed it.

51. *Standardized Test Prep* If $x - y > 0$, which expression(s) must equal $|x - y|$? Choose A, B, C, D, or E. E

I. $x - y$ **II.** $-(x - y)$ **III.** $|y - x|$

A. I and II only **B.** II and III only **C.** III only **D.** I, II, and III **E.** I and III only

32. −49 −47 −45

33. 0 1 2 3 4 5

34. 8 9 10 11 12 13

35. 10 11 12 13 14 15

36. −3 −2 −1 0 1 2

37. −8 −7 −6 −5 −4 −3

38. −10 −9 −8 −7 −6 −5

39. −8 −7 −6 −5 −4 −3

40. −3 −2 −1 0 1 2

41. 0 1 2 3 4 5

42–48, 49b. See back of book.

RESEARCH Exercise 57 Your school's counseling office is a great source of information about various jobs. Also, the *Occupation Outlook Handbook*, which is found in many libraries, describes approximately 200 occupational categories, including the general nature of the occupation, training requirements, average salaries, employment outlook, and sources of additional information.

Exercises MIXED REVIEW

Exercise 60 Remind students that there are two ways of solving this equation: multiplying both sides by 24 and using cross products.

GETTING READY FOR LESSON 4-6 These exercises prepare students for solving inequalities using multiplication and division.

Wrap Up

THE BIG IDEA Ask students to compare solving addition and subtraction *inequalities* with solving addition and subtraction *equations*.

RETEACHING ACTIVITY Students use tiles to solve one-step inequalities using the Addition and Subtraction Properties. They will model the inequality, the steps, and the solution. (Reteaching worksheet 4-5)

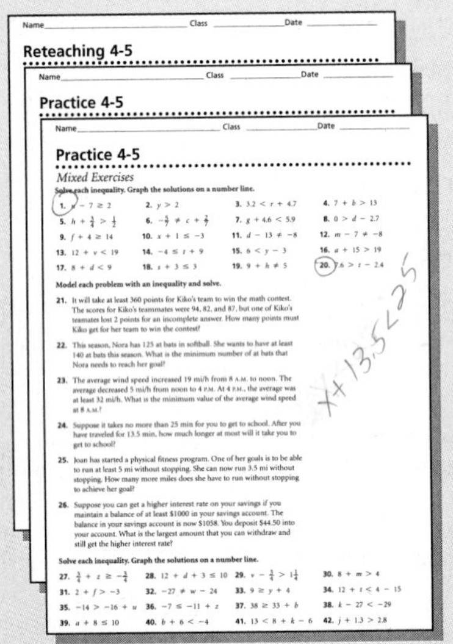
Reteaching 4-5

Practice 4-5

Practice 4-5

Mixed Exercises

Model each problem with an inequality and solve.

Solve each inequality. Graph the solutions on a number line.

Lesson Quiz

Lesson Quiz is also available in Transparencies.

1. Write four numbers that are solutions for the inequality $h - 4 < -2$. Answers may vary. Sample: 22, 21, 0, 1
2. Solve and graph $x + \frac{1}{2} > -2$. $x > -2\frac{1}{2}$
3. Solve and graph $w - 3 \neq -9$. $w \neq -6$

Statistics **Use the bar graph about education.**

52. Let w represent the average yearly income for women. For what education levels is $w > 22{,}000$ a true statement? associate or bachelor's degree or more
53. Write a single inequality that describes the average income for a man who does not graduate from high school. $s \leq 18{,}000$
54. On average, how much more do men who have at least a bachelor's degree earn than those who have a high school diploma? about $16,000
55. On average, how much more do women earn per year if they graduate from high school instead of dropping out before 9th grade? about $5000
56. At what education level does a woman's average income exceed the average income for a man with no high school education? high school graduate
57. *Research* Find the average salaries for men and women in a job of your choice. What level of education is necessary to get that job? Write inequalities to compare those salaries with the average salaries for that education level. Check students' work.

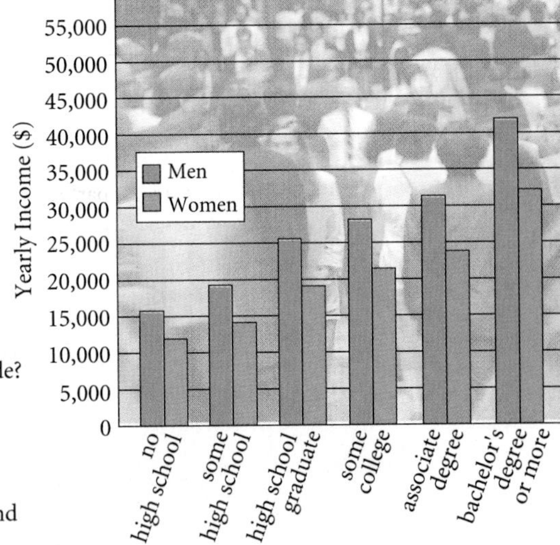

Exercises MIXED REVIEW

Find the sum or difference.

58. $\begin{bmatrix} 2 & 9 & 3 \\ 11 & 5 & 7 \end{bmatrix} + \begin{bmatrix} 18 & 0 & 15 \\ 6 & 12 & 4 \end{bmatrix}$ $\begin{bmatrix} 20 & 9 & 18 \\ 17 & 17 & 11 \end{bmatrix}$
59. $\begin{bmatrix} -4 & 13 \\ 8 & 1 \end{bmatrix} - \begin{bmatrix} 4 & -5 \\ 10 & -14 \end{bmatrix}$ $\begin{bmatrix} -8 & 18 \\ -2 & 15 \end{bmatrix}$

Solve.

60. $\frac{x}{3} = \frac{8}{24}$ 1
61. $|m + 6| = 5$ −11, −1
62. $\frac{3}{4}k = 15$ 20
63. $|p| - 2 = 10$ −12, 12
64. *Law Enforcement* In 1995, the 108-year-old California State Police became part of the California Highway Patrol. In what year was the California State Police founded? 1887

Getting Ready for Lesson 4-6

65. State the multiplication property of equality and the division property of equality in your own words. Two quant. are equal. If you multiply (divide) both of them by a number, the products (quotients) are also equal. You cannot divide by 0.

Solve and check each equation.

66. $8 = \frac{1}{2}t$ 16
67. $14 = -21x$ $-\frac{2}{3}$
68. $\frac{x}{6} = -1$ −6
69. $5d = 32$ 6.4
70. $0.5w = 4.5$ 9

4-6 Teaching Notes

PROBLEM OF THE DAY

A tree is 4.5 m tall and has a shadow 5.4 m long. How tall is a bush that has a shadow 1.5 m long? 1.25 m

Problem of the Day is also available in Transparencies.

CONNECTING TO PRIOR KNOWLEDGE Remind students of the discussion of Kyle and Jaime in previous lessons. Kyle is two years older than Jaime. Point out that if Kyle is now 14, Jaime is 12. On the board, write 14 yr $>$ 12 yr. Ask: *How would you convert the inequality shown on the board to months instead of years?* Multiply by 12 on both sides.

WORK TOGETHER

Question 2 Have students repeat the exploration used in Question 1 and Question 2 for the inequality $-3 < -1$. Then contrast and compare the answers to Question 2 for both inequalities.

THINK AND DISCUSS

Explain that *reverse the order* means *change the direction of the inequality symbol.*

Have students compare the Multiplication Property of Inequality with the Multiplication Property of Equality on page 114.

Connections Community Service . . . and more

What You'll Learn

- Using multiplication and division to solve one-step inequalities
- Using inequalities to model real-world problems

...And Why

To solve inequalities that model real-world situations such as fundraising

What You'll Need

- calculator

4-6 Solving Inequalities Using Multiplication and Division

WORK TOGETHER

Work with a partner to explore what happens to an inequality when you multiply each side by the same number. Consider the inequality $4 > 1$.

4 · 3	■ >	1 · 3
4 · 2	■ >	1 · 2
4 · 2	■ >	1 · 1
4 · 0	■ =	1 · 0
4 · −1	■ <	1 · −1
4 · −2	■ <	1 · −2
4 · −3	■ <	1 · −3

1. Complete each statement by replacing each ■ with $<$, $>$, or $=$. See right.
2. What happens to the inequality when you multiply it by a positive number? by zero? by a negative number? The inequal. remains true; both sides become equal to 0; the inequal. is reversed.

THINK AND DISCUSS

Part 

Solving Inequalities Using Multiplication

You can find an equivalent inequality by multiplying or dividing each side of an inequality by the same number. If the number is positive, the order remains the same. If the number is negative, you *reverse* the order.

Multiplication Property of Inequality

For all real numbers a and b, and for $c > 0$:

If $a > b$, then $ac > bc$. If $a < b$, then $ac < bc$.

Examples: $4 > -1$, so $4(5) > -1(5)$
$-6 < 3$, so $-6(5) < 3(5)$

For all real numbers a and b, and for $c < 0$:

If $a > b$, then $ac < bc$. If $a < b$, then $ac > bc$.

Examples: $4 > -1$, so $4(-2) < -1(-2)$
$-6 < 3$, so $-6(-2) > 3(-2)$

3. Consider the inequality $-2 \geq -5$. a. $-6 \geq -15$ b. $6 \leq 15$
   a. What inequality do you get when you multiply each side by 3?
   b. What inequality do you get when you multiply each side by -3?
   c. Explain why the Multiplication Property of Inequality applies to inequalities involving $\geq$ and $\leq$. See margin.

Lesson Planning Options

Prerequisite Skills

- Performing multiplication and division with inequalities
- Understanding concepts of inequalities

Assignment Options

To provide flexible scheduling, this lesson can be subdivided into parts.

1 **Core** 5–10, 19–22, 32–39
Extension 25, 30, 44–46

2 **Core** 1–4, 11–18, 31, 40–43
Extension 23, 24, 26–29

Use Mixed Review to maintain skills.

Resources

 **Student Edition**

Skills Handbook, p. 569, 578
Extra Practice, p. 559
Glossary/Study Guide

Teaching Resources

Chapter Support File, Ch. 4
- Practice 4-6 (two worksheets)
- Reteaching 4-6

Classroom Manager 4-6
Glossary, Spanish Resources
Two-Year Algebra Handbook 4-6

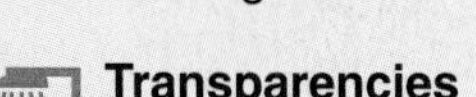

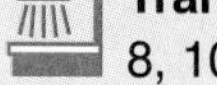 **Transparencies**
8, 10, 61

Example 1

Have students look at the graph for the solution to the inequality. Point out that the variable in the solution is on the left side of the inequality, therefore the graph points in the same direction as the inequality symbol. Since −2 is not part of the solution, there is an open circle at point −2.

Question 5 Ask: *What happens to the order of an inequality if the two sides are switched?* The inequality is reversed. If one inequality is written in the reverse order of another one, be sure the inequality symbol is "pointing" toward the same term to keep the statement correct.

ERROR ALERT! Some students reverse the order of the inequality when multiplying by a positive number.
Remediation: Work with tiles to show that multiplying by a positive number does not reverse the order of the inequality. As an example, model $4 < 7$ using tiles. Write the inequality symbol on an index card and place the card between the two sets of tiles. Double the number of tiles on each side to get $8 < 14$. Point out that it is not correct to switch the inequality symbol when multiplying by a positive number.

Additional Examples

FOR EXAMPLE 1

Solve $\frac{3}{2}z > -2$. Graph the solution on a number line. $z > -\frac{4}{3}$

−2 −1 0

Discussion: *Why does the negative sign not effect the inequality sign?*

FOR EXAMPLE 2

Solve $-\frac{3}{2}z > -2$. Graph the solution on a number line. $z < \frac{4}{3}$

0 1 2

Discussion: *Why does the negative sign in front of the $\frac{3}{2}$ change the inequality sign from $>$ to $<$?*

FOR EXAMPLE 3

Allison and Megan budgeted \$75 to spend on fuel for a cross-country trip. How many times can they fill the car's gas tank if it costs \$13 each time? Write an equation and solve.
$13f \leq 75$; $f \leq 5.8$; five times

Discussion: *Is this a reasonable budget for a 2000 mi trip?*

186

You can use the Multiplication Property of Inequality to write a simpler, equivalent inequality.

Example 1

Solve $\frac{x}{2} < -1$. Graph the solutions on a number line.

$\frac{x}{2} < -1$

$2\left(\frac{x}{2}\right) < 2(-1)$ ← Multiply each side by 2.

$x < -2$ ← Simplify each side.

The solutions are all numbers less than −2.

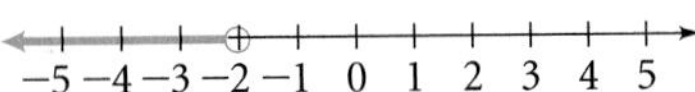

4. Describe the solutions for the inequality $\frac{x}{2} \leq -1$.
all the numbers not greater than −2
5. Jamal solved the inequality $-2 > \frac{y}{3}$ and got $-6 > y$. Erica solved the same inequality and got $y < -6$. Are they both correct? Explain.
Yes; both inequalities state that y must be less than −6.

Example 2

Solve $-\frac{2}{3}x \geq 2$. Graph the solutions on a number line.

$-\frac{2}{3}x \geq 2$

$\left(-\frac{3}{2}\right)\left(-\frac{2}{3}x\right) \leq \left(-\frac{3}{2}\right)(2)$ ← Multiply each side by $-\frac{3}{2}$. Reverse the order of the inequality.

$x \leq -3$ ← Simplify each side.

The solutions are all numbers less than or equal to −3.

−6 −5 −4 −3 −2 −1 0 1 2 3 4

PROBLEM SOLVING

Look Back How would the solution be different if the inequality were $\frac{2}{3}x \geq -2$?
The order of the inequal. would be reversed; $x \geq -3$.

6. How can you use substitution to explain to a classmate why the solution to Example 2 cannot be $x \geq -3$? You can substitute a number greater than −3 into the inequality. $-\frac{2}{3}(0) \geq 2$ is false.
7. a. Try This Solve the inequality $-t < \frac{1}{2}$.
   b. Graph the solution on a number line. 7a. $t > -\frac{1}{2}$ 7b. See left.
   c. Name four integers that are solutions of the inequality.
   Answers may vary. Samples: 0, 1, 2, 3

7b. −1 $-\frac{1}{2}$ 0 1 2

Part 2

Solving Inequalities Using Division

You can use division to simplify and solve inequalities.

8. Consider the inequality $12 > 6$. Make a table to show what happens when you divide each side by the same number. See margin.
9. What happens when an inequality is divided by a positive number? by a negative number? See left.

9. The order of the inequality is unchanged; the order of the inequality is reversed.

Example 2

Have students look at the graph for the solution of the inequality. Point out that, as in Example 1, the variable in the solution is on the left side of the inequality. The graph points in the same direction as the inequality symbol. Since -3 is part of the solution, there is a closed circle at point -3.

Have students compare the Division Property of Inequality with the Division Property of Equality on page 114.

AUDITORY LEARNING Challenge students to compose a rhyme, poem, song or other mnemonic device to help them remember what happens when you multiply an inequality by a negative number.

CONNECTING TO THE STUDENTS' WORLD Have students find out from student groups at school what types of fundraisers the groups have tried. Ask them to write down the amount of money raised. Use the students' notes to create inequality problems such as "How many coffee mugs must be sold to raise $300?"

CRITICAL THINKING Question 10 Tell students there are two answers to this question. Each property has a different reason why c cannot equal zero.

Example 3 **Relating to the Real World** 

Have students check the solution by substituting the values 15, 16, and 17 into the original inequality.

The pattern you see in your table can help you understand the Division Property of Inequality.

Division Property of Inequality

For all real numbers a and b, and for $c > 0$:

If $a > b$, then $\frac{a}{c} > \frac{b}{c}$. If $a < b$, then $\frac{a}{c} < \frac{b}{c}$.

Examples: $6 > -4$, so $\frac{6}{2} > \frac{-4}{2}$ $\quad -2 < 8$, so $\frac{-2}{2} < \frac{8}{2}$

For all real numbers a and b, and for $c < 0$:

If $a > b$, then $\frac{a}{c} < \frac{b}{c}$. If $a < b$, then $\frac{a}{c} > \frac{b}{c}$.

Examples: $6 > -4$, so $\frac{6}{-2} < \frac{-4}{-2}$ $\quad -2 < 8$, so $\frac{-2}{-2} > \frac{8}{-2}$

10. *Critical Thinking* Why can't $c = 0$ in the Multiplication and Division Properties of Inequality? **If you multiply an inequal. by 0, both sides become equal to 0. You cannot divide by 0 at all.**

Example 3 **Relating to the Real World** 

Community Service The student council is sponsoring a recycling drive to raise money to buy food for a food bank. A case of 12 jars of spaghetti sauce costs $13.50. What is the greatest number of cases of sauce the student council can buy if they collect $216?

Define	$c =$ the number of cases of spaghetti sauce.				
Relate	13.5	times	the number of cases	is less than or equal to	216
Write	13.5	$\cdot$	c	$\le$	216

$13.5c \le 216$

$\frac{13.5c}{13.5} \le \frac{216}{13.5}$ ← Divide each side by 13.5.

$c \le 16$ ← Use a calculator.

The council can buy at most 16 cases of sauce for the food bank. **11. The council can spend at most $216.**

11. Explain why you use the symbol $\le$ in the inequality in Example 3.

Technology Options

For Exercises 52–57, students may want to use a graphing calculator to check their answers for the equations.

Prentice Hall Technology 

Software • Secondary Math Lab Toolkit • Computer Item Generator 4-6

 **Internet** • See the Prentice Hall site. (http://www.phschool.com)

Exercises ON YOUR OWN

Exercise 1 Ask: *Does multiplying by −1 and dividing by −1 always produce the same result?* yes

Exercises 2–4 Have students write the second inequality under the first. Remind them to align the inequality symbols.

Exercises 5–10 Encourage students to use some numbers that are not integers.

STANDARDIZED TEST TIP **Exercise 25** The graph of a *less than* inequality will point to the *left* if the variable is on the *left* side of the solved inequality. Students might try mental math to decide which of the answer choices will not meet this description. Remind students that dividing by a negative number will reverse the order of the inequality.

ESTIMATION Exercises 26–29 Suggest that students round each number to the nearest integer to estimate the solution.

CRITICAL THINKING Exercise 31b Suggest that students begin by comparing the correct solution with Kia's solution.

ESL Exercise 44 Writing Some students may have difficulty writing an explanation for the Multiplication and Division Properties of Inequality. Suggest that these students make up examples to illustrate these properties. They can then write short phrases to describe the steps in their examples (for example, "divide both sides by 3" or "multiply both sides by 5").

WRITING Exercise 44 This exercise can help you assess whether students understand when to reverse the order of the inequality.

pages 185–187 Think and Discuss

3c. If two equal quantities are each multiplied by a number, the products are also equal. If the original quantities are not equal, you can use the Mult. Prop. of Inequal.

8.

12 ÷ 3	>	6 ÷ 3
12 ÷ 2	>	6 ÷ 2
12 ÷ 1	>	6 ÷ 1
12 ÷ 0	and	6 ÷ 0
	not defined	
12 ÷ −1	<	6 ÷ −1
12 ÷ −2	<	6 ÷ −2
12 ÷ −3	<	6 ÷ −3

pages 188–189 On Your Own

32. −10 −9 −8 −7 −6 −5

33. 32 33 34 35 36 37

34. −3 −2 −1 0 1 2

35. −32 −30 −28

36. 6 7 8 9 10 11

37. −5 −4 −3 −2 −1 0

38–44. See back of book.

Exercises ON YOUR OWN

1. a. By what number can you divide each side of the inequality $-t \leq 7$ to get $t \geq -7$? −1
   b. By what number can you multiply each side of the inequality $-m > -3$ to get $m < 3$? −1

Tell what you must do to the first inequality in order to get the second.

2. $5z > -25$; $z > -5$ div. by 5
3. $-b \geq 3.4$; $b \leq -3.4$ mult. by −1
4. $-\frac{7}{8}m > \frac{3}{4}$; $m < -\frac{6}{7}$ mult. by $-\frac{8}{7}$

Write four numbers that are solutions of each inequality. 5–10. Answers may vary. Samples:

5. $v \geq -5$ −5, −1, 0, 8
6. $0.5 > \frac{1}{2}c$ $\frac{1}{2}$, 0, −2, −8
7. $2q < -7$ −4, −7, −10, −20
8. $-3f \leq 16$ −5, −1, 0, 6
9. $\frac{r}{3} \geq -4$ −12, −3, −1, 10
10. $-3.5 < -m$ −6, −1, 0, 3

Replace each ■ with the number that makes the inequalities equivalent.

11. ■$s > 14$; $s < -7$ −2
12. $x +$ ■ ≥ 25; $x \geq -12$ 37
13. $8u \leq$ ■; $u \leq \frac{1}{2}$ 4
14. ■ $> -17 + a$; $a < -9$ −26
15. $36 <$ ■r; $r < -3.6$ −10
16. $-k \leq$ ■; $k \geq -7.5$ 7.5

Solve each inequality. Check your solutions. Graph the solutions on a number line. 17–22. See below for graphs.

17. $-4x \leq -16$ $x \geq 4$
18. $6 < -9g$ $-\frac{2}{3} > g$
19. $\frac{5}{6}x > -5$ $x > -6$
20. $-1 \geq \frac{t}{3}$ $-3 \geq t$
21. $-1.5d < -8$ $d > 5\frac{1}{3}$
22. $-\frac{5}{3}u > 1$ $u < -\frac{3}{5}$

17. 0 1 2 3 4 5 6
18. −2 −1 $-\frac{2}{3}$ 0
19. −7 −5 −3 −1
20. −5 −4 −3 −2 −1 0 1
21. 4 5 $5\frac{1}{3}$ 6
22. −2 −1 $-\frac{3}{5}$ 0

23. *Jobs* Suppose you earn \$5.85 per hour working part time at a dry cleaner. You need to earn at least \$100. How many full hours must you work to earn the money? 18 h

24. *Fund-raising* The science club is sponsoring a car wash. At \$2.50 per car, how many cars do they have to wash to earn at least \$300? 120 cars

25. *Standardized Test Prep* The number line shows the graph of all the solutions of an inequality. Which could *not* be that inequality? C

    −5 −4 −3 −2 −1 0 1 2

    A. $-2x > 4$ B. $-4 > 2x$ C. $-x < 2$
    D. $8 < -4x$ E. $-2 > x$

Estimation **Estimate the solution of each inequality.**

26. $-2.099r < 4$ $r > -2$
27. $3.87j > -24$ $j > -6$
28. $20.95 \geq \frac{1}{2}p$ $p \leq 42$
29. $\frac{20}{39}s \leq -14$ $s \leq -28$

30. *Construction* Swivel desks that each need 20 in. of space are being installed side by side in a new lecture hall. Each row of desks can be no longer than 25 ft. Find the maximum number of desks that will fit in each row. 15 desks

ALTERNATIVE ASSESSMENT Exercise 46 Group students into pairs. Have each pair complete Exercise 46. Then ask each pair to solve the inequalities written by one other group. This will help you assess the students' understanding of the Multiplication Property of Inequality.

Exercises MIXED REVIEW

GETTING READY FOR LESSON 4-7 These exercises prepare students for solving multi-step inequalities.

Wrap Up

THE BIG IDEA Have students state the main difference between solving multiplication and division *inequalities* and solving multiplication and division *equations.*

RETEACHING ACTIVITY Students use tiles and an index card, with "$>$" on one side and "$<$" on the other side, to solve one-step inequalities using the Multiplication and Division Properties. They model and solve inequalities, and explain whether the order of the inequality is reversed. (Reteaching worksheet 4-6)

31. a. Kia solved $-15q \le 135$ by adding 15 to each side of the inequality. What mistake did she make? **Kia should have divided each side by −15.**
    b. **Critical Thinking** Kia's solution was $q \le 150$. She checked her work by substituting 150 for q in the original problem. Why didn't her check let her know she'd made a mistake? **150 is a solution, but 151 is also a solution.**
    c. What substitution would have let her know she'd made a mistake? **Justify** your answer. **She could try any number less than −9; $-15(-10) \le 135$ is false.**

Solve each inequality. Graph the solutions on a number line. 32–43. See margin for graphs.

32. $4d \le -28$ $d \le -7$
33. $\frac{u}{7} > 5$ $u > 35$
34. $2 < -8s$ $s < \frac{-1}{4}$
35. $\frac{3}{2}k \ge -45$ $k \ge -30$
36. $0.3y < 2.7$ $y < 9$
37. $9.4 \le -4t$ $t \le -2.35$
38. $-h \ge 4$ $h \le -4$
39. $24 < -3x$ $x < -8$
40. $\frac{5}{2}x > 5$ $x > 2$
41. $0 < -7b$ $b < 0$
42. $\frac{5}{6} > -\frac{1}{3}p$ $p > -\frac{5}{2}$
43. $-0.2m \ge 9.4$ $m \le -47$

44. **Writing** Write a paragraph explaining the Multiplication and Division Properties of Inequality to a classmate. **See margin.**

45. **Remodeling** The Sumaris' den floor measures 18 ft by 15 ft. They want to cover the floor with tiles that cover $\frac{9}{16}$ ft^2.
    a. What is the least number of tiles they need to cover the floor? **480 tiles**
    b. Why might they need more tiles than the answer to part (a)? **Answers may vary. Sample: They may need to cut some tiles.**

46. **Open-ended** Write four different inequalities that you can solve using multiplication or division. Choose your inequalities so that the solution of each inequality is all numbers greater than 3. **Samples: $6x > 18$; $-\frac{1}{3}x < -1$; $2x > 6$; $-3x < -9$**

Exercises MIXED REVIEW

Solve for y in terms of x. 47. $y = \frac{6-2x}{3}$ 48. $y = 5x + 4$ 49. $y = \frac{15-4x}{-5}$ 50. $y = \frac{-2x-20}{3}$

47. $2x + 3y = 6$
48. $5x - y + 8 = 4$
49. $-5y + 4x = 15$
50. $2x + 3y = -20$

51. a. **Biology** There are appro 20 million bats in Brack Texas. They eat about 25 of insects every night. H many tons do they eat in a week? **1750 tons**
    b. About how much does a single bat eat each night? **about 0.4 oz**

Getting Ready for Lesson 4

Solve and check each equation.

52. $3(c + 4) = 5$ $-2\frac{1}{3}$
53. $2x - 7 + 3x + 4 = -25$ -4.4
54. $5p + 9 = 2p - 1$ $-3\frac{1}{3}$
55. $7x + 2(8 - x) = 4$ -2.4
56. $\frac{1}{2}k - \frac{2}{3} + k = \frac{7}{6}$ $\frac{11}{9}$
57. $2t - 32 = \frac{15}{4}t + 1$ $-18\frac{6}{7}$

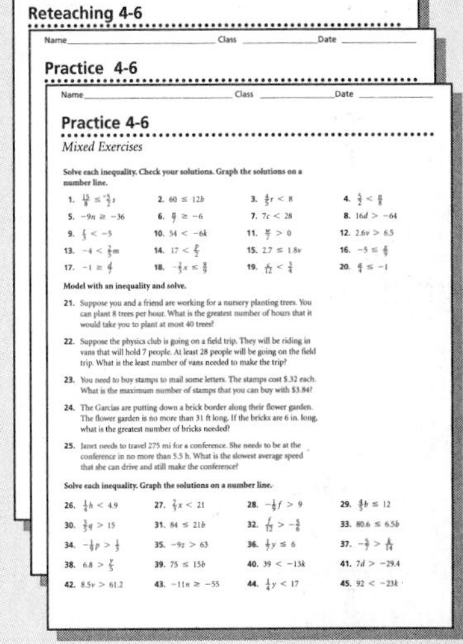
Name Class Date

Reteaching 4-6

Name Class Date

Practice 4-6

Name Class Date

Practice 4-6

Mixed Exercises

Solve each inequality. Check your solutions. Graph the solutions on a number line.

1. $\frac{15}{8} \le \frac{-5}{2}s$ 2. $60 \le 12b$ 3. $\frac{4}{5}r < 8$ 4. $\frac{5}{2} < \frac{n}{8}$
5. $-9n \ge -36$ 6. $\frac{y}{7} \ge -6$ 7. $7c < 28$ 8. $16d > -64$
9. $\frac{f}{3} < -5$ 10. $54 < -6k$ 11. $\frac{w}{7} > 0$ 12. $2.6v > 6.5$
13. $-4 < \frac{2}{5}m$ 14. $17 < \frac{p}{2}$ 15. $2.7 \le 1.8v$ 16. $-5 \le \frac{x}{9}$
17. $-1 \ge \frac{d}{7}$ 18. $-\frac{1}{3}x \le \frac{8}{9}$ 19. $\frac{c}{12} < \frac{1}{4}$ 20. $\frac{g}{4} \le -1$

Model with an inequality and solve.

21. Suppose you and a friend are working for a nursery planting trees. You can plant 8 trees per hour. What is the greatest number of hours that it would take you to plant at most 40 trees?
22. Suppose the physics club is going on a field trip. They will be riding in vans that will hold 7 people. At least 28 people will be going on the field trip. What is the least number of vans needed to make the trip?
23. You need to buy stamps to mail some letters. The stamps cost \$.32 each. What is the maximum number of stamps that you can buy with \$3.84?
24. The Garcias are putting down a brick border along their flower garden. The flower garden is no more than 31 ft long. If the bricks are 6 in. long, what is the greatest number of bricks needed?
25. Janet needs to travel 275 mi for a conference. She needs to be at the conference in no more than 5.5 h. What is the slowest average speed that she can drive and still make the conference?

Solve each inequality. Graph the solutions on a number line.

26. $\frac{1}{4}h < 4.9$ 27. $\frac{2}{3}x < 21$ 28. $-\frac{1}{5}f > 9$ 29. $\frac{4}{5}b \le 12$
30. $\frac{3}{5}q > 15$ 31. $84 \le 21b$ 32. $\frac{f}{12} > -\frac{5}{6}$ 33. $80.6 \le 6.5b$
34. $-\frac{1}{6}p > \frac{1}{3}$ 35. $-9z > 63$ 36. $\frac{1}{7}y \le 6$ 37. $-\frac{3}{7} > \frac{k}{14}$
38. $6.8 > \frac{y}{5}$ 39. $75 \le 15b$ 40. $39 < -13k$ 41. $7d > -29.4$
42. $8.5v > 61.2$ 43. $-11n \ge -55$ 44. $\frac{1}{4}y < 17$ 45. $92 < -23k$

Lesson Quiz

Lesson Quiz is also available in Transparencies.

1. Write four numbers that are solutions of the inequality $-6r \ge -2$. **Answers may vary. Sample: 0, −1, −2, −3**
2. Replace the ■ with the number that makes the inequalities equivalent.
   ■$a < -8$; $a > 4$
   ■ = −2
3. Solve and graph $\frac{1}{3} \ge \frac{3}{-2}a$.
   $a \ge -\frac{2}{9}$
   $-\frac{1}{3}$ 0 $\frac{1}{3}$

189

4-7 Teaching Notes

PROBLEM OF THE DAY

A three-digit number is divisible by 6. The sum of the digits is 9. Find the number. Must your answer be even or odd? **Answers may vary. Sample: 144; even**

Problem of the Day is also available in Transparencies.

CONNECTING TO PRIOR KNOWLEDGE Remind students about the multi-step equations they solved in Lesson 3-2. Have students make conjectures about the similarities and differences between solving multi-step inequalities and solving multi-step equations.

THINK AND DISCUSS

AUDITORY LEARNING As students work through the examples, encourage them to whisper quietly to themselves or to a partner the steps they are following to solve the inequality, for example, "Add 5 to each side," or "The order stays the same."

Example 1 **Relating to the Real World** 

Ask: *Would there be enough trim if the banner were 4 ft wide?* yes *5 ft wide?* no

Encourage students to draw a diagram such as the one under the Problem Solving Hint.

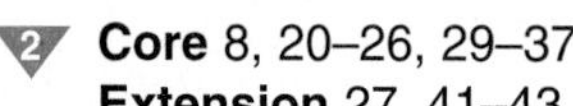

Lesson Planning Options

Prerequisite Skills

- Understanding number lines
- Solving one-step inequalities

Assignment Options

To provide flexible scheduling, this lesson can be subdivided into parts.

1 **Core** 1–7, 9–15, 17–19
Extension 16, 28, 38–40

2 **Core** 8, 20–26, 29–37
Extension 27, 41–43

Use Mixed Review to maintain skills.

Resources

 **Student Edition**

Skills Handbook, p. 567, 582
Extra Practice, p. 559
Glossary/Study Guide

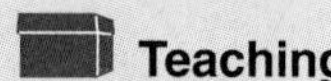

Teaching Resources

Chapter Support File, Ch. 4
- Practice 4-7 (two worksheets)
- Reteaching 4-7

Classroom Manager 4-7
Glossary, Spanish Resources
Two-Year Algebra Handbook 4-7

 **Transparencies**
8, 10, 62

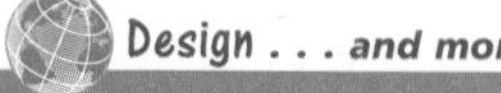

Connections Design . . . and more

4-7 Solving Multi-Step Inequalities

What You'll Learn

- Solving multi-step inequalities and graphing the solutions on a number line
- Using multi-step inequalities to model and solve real-world problems

...And Why

To solve inequalities that model real-world situations such as designing

THINK AND DISCUSS

Part 1 Solving with Variables on One Side

When you solve equations, sometimes you need to use more than one step. The same is true when you solve inequalities.

Example 1 **Relating to the Real World**

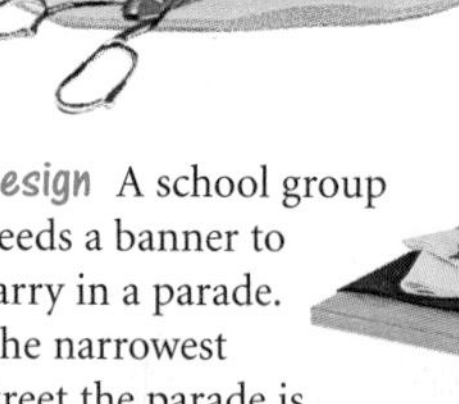

Design A school group needs a banner to carry in a parade. The narrowest street the parade is marching down measures 36 ft across, but some space is taken up by parked cars. The students have decided the length of the banner should be 18 ft. There are 45 ft of trim available to sew around the border of the banner. What is the greatest possible width for the banner?

PROBLEM SOLVING HINT

Draw a diagram.

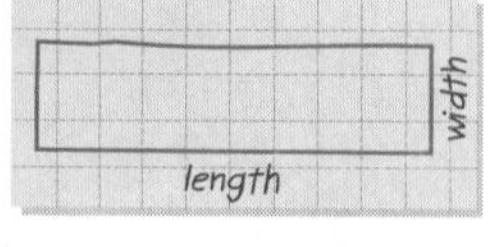

Define w = width of the banner

Relate Since the border goes around the edges of the banner, you can use the perimeter formula: $P = 2l + 2w$.

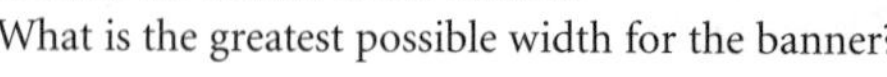

	twice the length	plus	twice the width	can be no more than	the border
Write	$2(18)$	$+$	$2w$	$\leq$	45

$$2(18) + 2w \leq 45$$
$$36 + 2w \leq 45 \quad \text{← Simplify the left side.}$$
$$36 + 2w - 36 \leq 45 - 36 \quad \text{← Subtract 36 from each side.}$$
$$2w \leq 9$$
$$\frac{2w}{2} \leq \frac{9}{2} \quad \text{← Divide each side by 2.}$$
$$w \leq 4.5$$

The greatest possible width for the banner is 4.5 ft.

Example 2

ERROR ALERT! When simplifying an expression, some students may apply the distributive property incorrectly. For example, they might simplify $2(w + 2) - 3w$ as $2w + 2 - 3w$ or $2w + 4 - 6w$. **Remediation:** Use tiles or draw a sketch to represent $2(w + 2) - 3w$. Show that this expression is equivalent to $2w + 4 - 3w$.

Example 3

Remind students that when solving equations with variables on both sides, it was easiest to get rid of the variable with the lowest coefficient first. This usually reduces the times you need to multiply or divide by negative numbers. Since multiplying and dividing by negative number in inequalities produces a chance for forgetting to reverse the order of the inequality, this is a good way to start the solving of inequalities with variables on both sides.

Question 3 Ask: *What other two numbers might you use to check the solution of the inequality?* **Answers may vary. Sample: 3 and 4**

CRITICAL THINKING Question 5 If students are having difficulty with this exercise, allow them to write down the steps.

1. What could the model $2l + 2w > 45$ mean in the situation described in Example 1? **A banner with these dimensions cannot be sewn.**

Sometimes solving an inequality involves the distributive property.

Example 2

Solve $2(w + 2) - 3w \geq -1$. Graph the solutions on a number line.

$2(w + 2) - 3w \geq -1$

$2w + 4 - 3w \geq -1$ ← Use the distributive property.

$-1w + 4 \geq -1$ ← Combine like terms.

$-1w + 4 - 4 \geq -1 - 4$ ← Subtract 4 from each side.

$-w \geq -5$ ← Simplify. Divide each side by -1.

$\frac{-w}{-1} \leq \frac{-5}{-1}$ ← Reverse the order of the inequality.

$w \leq 5$

All numbers less than or equal to 5 are solutions.

−1 0 1 2 3 4 5 6 7 8 9

PROBLEM SOLVING

Look Back What happens if you multiply each side of $-1w \geq -5$ by -1?

You reverse the order of the inequality.

2. You can check the solutions to Example 2 by substituting values into the inequality $2(w + 2) - 3w \geq -1$. **4 and 5**
 a. Use the values 4, 5, and 6. Which values make the inequality true?
 b. Explain how part (a) serves as a check on Example 2. **It shows that you selected the correct region.**

Part 2 Solving with Variables on Both Sides

Example 3

Solve $8z - 6 < 3z + 12$. Graph the solutions on a number line.

$8z - 6 < 3z + 12$

$8z - 6 - 3z < 3z + 12 - 3z$ ← Subtract 3z from each side.

$5z - 6 < 12$ ← Combine like terms.

$5z - 6 + 6 < 12 + 6$ ← Add 6 to each side.

$5z < 18$

$\frac{5z}{5} < \frac{18}{5}$ ← Divide each side by 5.

$z < 3.6$

All numbers less than 3.6 are solutions.

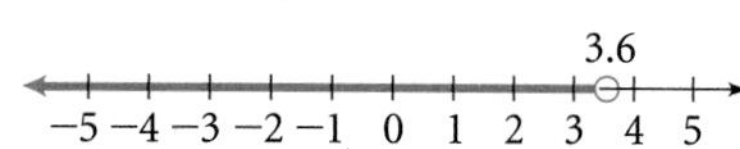

GRAPHING CALCULATOR HINT

You can check your solutions to Example 3 by using the TEST feature.

4b. 1 $1\frac{4}{5}$ 2 3

3. What happens when you replace z with 3.6 in $8z - 6 < 3z + 12$? **The two quant. are equal. The inequal. is false.**
4. a. **Try This** Solve $3b + 12 > 21 - 2b$. $b > \frac{9}{5}$
 b. Graph the solutions on a number line. **See left.**

Additional Examples

FOR EXAMPLE 1

A movie special-effects engineer wants to build a model car that will accelerate from 3 m/s to more than 10 m/s in only 4 s. Solve the equation $3 + 4a \geq 10$ to determine the required acceleration. $a \geq 1.75$ **m/s**

Discussion: *Why do the units of acceleration have to be m/s?*

FOR EXAMPLE 2

Solve $3x + 4(6 - x) < 2$. Graph the solution on a number line. $x > 2$

FOR EXAMPLE 3

Solve $-3w + 6 \geq 3(2w - 1)$. Graph the solution on a number line. $w \leq 1$

−1 0 1 2 3

Discussion: *Why is it necessary to sometimes reverse the order of an inequality?*

Exercises ON YOUR OWN

EXTENSION Exercises 1–9 Ask students:

- *Which exercise has an inequality that is true for any value of the variable?* Exercise 8
- *Which exercises show the solution of the inequality?* Exercises 1, 2, 4, 5, 7, 8, 9

Exercises 10–15 Have students first solve each inequality before attempting to match each inequality with its graph.

ALTERNATIVE ASSESSMENT **Exercise 27** Divide the class into small groups. Ask each member of the group to pick one equation from the exercises which they found difficult to solve. Then have the other members in the group solve this equation using the guidelines in Example 2. This will help you assess the students' understanding of multi-step inequalities.

WRITING Exercise 27 Students can also show and explain the graph of the solution and their method for checking.

CRITICAL THINKING Exercise 28 Suggest students begin by subtracting 4 from each side and then use the strategy *Guess and Test.*

STANDARDIZED TEST TIP **Exercise 38** Encourage students to simplify each expression as much as possible before substituting test values for the variable.

OPEN-ENDED Exercise 39 Suggest students use the problem solving strategy of *Working Backward.*

GEOMETRY Exercise 42 In this exercise students explore the *golden rectangle.* Draw an approximate golden rectangle on the board. Draw a line connecting the longer sides of the rectangle to divide it into a square and a smaller rectangle.

Technology Options

For Exercises 29–37 students may want to use a graphing calculator to check their solutions. For Exercises 46–49 students may want to graph the equations on a graphing calculator to help determine to which family each equation belongs.

Prentice Hall Technology

Software • Secondary Math Lab Toolkit • Computer Item Generator 4-7

CD-ROM Multimedia Algebra Lab 0

Internet • See the Prentice Hall site. (http://www.phschool.com)

QUICK REVIEW

Equations such as $1 + 3a = 3a + 1$ are *identities.* Equations such as $2 + a = a - 2$ have *no solution.*

Like equations, some inequalities are true for all values of the variable, and some inequalities are false for all values of the variable. When an inequality is false for all values of the variable, it has no solution.

5. *Critical Thinking* Without writing the steps of a solution, tell whether the inequality is *true* or *false* for all values of the variable. **Justify** your response. a–f. See margin.
   a. $4s - 5 < 4s - 7$ b. $4s - 5 < 3 + 4s$
   c. $4s + 6 \ge 6 + 4s$ d. $4s + 6 > 6 + 4s$
   e. $4s - 9 < 4s$ f. $4s \le 4s$

Exercises ON YOUR OWN

Tell what you must do to the first inequality in order to get the second. Be sure to list *all* the steps. 1–9. See margin.

1. $4j + 5 \ge 23; j \ge 4.5$
2. $2(q - 3) < 8; q < 7$
3. $8 - 4s > 16; -4s > 8$
4. $-8 > \frac{z}{-5} - 2; 30 < z$
5. $2y - 5 > 9 + y; y > 14$
6. $\frac{2}{3}g + 7 \ge 9; \frac{2}{3}g \ge 2$
7. $6 < 12 - s; s < 6$
8. $3 + 5t \ge 6(t - 1) - t; 3 \ge -6$
9. $6.2 < -r; -6.2 > r$

Match each inequality with its graph below.

10. $2x - 2 > 4$ D
11. $2 - 2x > 4$ E
12. $2x + 2 > 4$ F
13. $2x + 2 > 4x$ A
14. $-2x - 2 > 4$ B
15. $-2(x - 2) > 4$ C

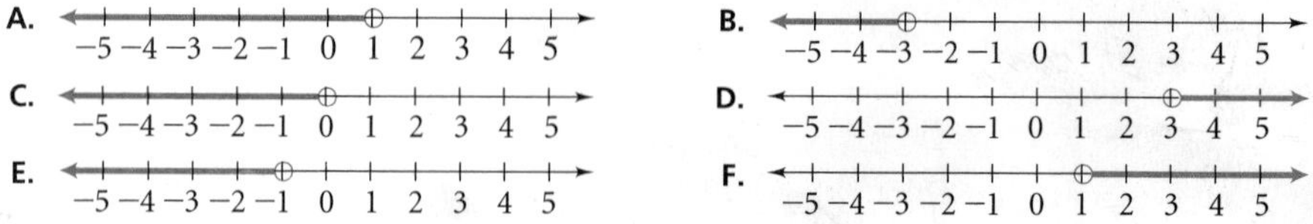

A. −5 −4 −3 −2 −1 0 1 2 3 4 5

B. −5 −4 −3 −2 −1 0 1 2 3 4 5

C. −5 −4 −3 −2 −1 0 1 2 3 4 5

D. −5 −4 −3 −2 −1 0 1 2 3 4 5

E. −5 −4 −3 −2 −1 0 1 2 3 4 5

F. −5 −4 −3 −2 −1 0 1 2 3 4 5

16. *Recreation* The sophomore class is planning a picnic. The cost of a permit to use the park is $250. To pay for the permit, there is a fee of $.75 for each sophomore and $1.25 for each guest who is not a sophomore. Two hundred sophomores plan to attend. How many guests must attend in order to pay for the permit? at least 80 guests

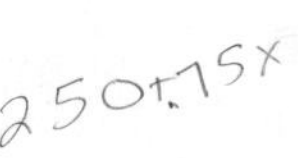

Solve each inequality. Graph the solutions on a number line. 17–25. See margin for graphs.

17. $5 \le 11 + 3h$ $h \ge -2$
18. $3(y - 5) > 6$ $y > 7$
19. $-4x - 2 < 8$ $x > -2\frac{1}{2}$
20. $r + 6 + 3r \ge 15 - 2r$ $r \ge 1\frac{1}{2}$
21. $5 - 2n \le 3 - n$ $n \ge 2$
22. $3(2v - 4) \le 2(3v - 6)$ all numbers
23. $2(m - 8) - 3m < -8$ $m > -8$
24. $-(6b - 2) > 0$ $b < \frac{1}{3}$
25. $7a - (9a + 1) > 5$ $a < -3$
26. *Writing* Suppose a friend is having difficulty solving the inequality $2.5(p - 4) > 3(p + 2)$. Explain how to solve the inequality, showing all necessary steps and identifying the properties you would use. See margin.

The smaller rectangle should also be an approximate golden rectangle. Challenge students to make a statement about the ratio width : length in the two rectangles.

EXTENSION Exercise 42 Challenge the students to compare the dimensions of a 3 in. × 5 in. index card with the dimensions of the golden rectangle. **The dimensions of the card are very close to the dimensions of the golden rectangle. If the dimensions of the card were 3 in. × 4.86 in., the card would be a golden rectangle.**

CONNECTING TO THE STUDENTS' WORLD Challenge students to think of part-time businesses that they could run from home. Examples are word processing, tutoring, window washing, house cleaning, lawn and garden care. Have students research the possible expenses associated with one such business and an hourly rate that is reasonable for your area. Using the data collected, students can create inequality problems similar to Exercise 43.

Exercise 43 Ask students: *Can you think of another way Carlos might charge his customers?* **Answers may vary. Sample: Carlos might charge his customer an amount per page.**

Chapter Project **FIND OUT BY RESEARCHING** Ask students to use the given equation to find their systolic blood pressure. Have them calculate the normal systolic blood pressure for a person twenty years old. Students will decide if this value is a maximum or a minimum systolic pressure and explain their reasoning. Have students further investigate lifestyle changes that can reduce high blood pressure. This task is essential to

---

27. *Freight Handling* The freight elevator of a building can safely carry a load of at most 4000 lb. A worker needs to move supplies in 50-lb boxes from the loading dock to the fourth floor of the building. The worker weighs 160 lb. The cart she uses weighs 95 lb. **74 boxes**

a. What is the greatest number of boxes she can move in one trip?

b. The worker must deliver 310 boxes to the fourth floor. How many trips must she make? **5 trips**

28. *Critical Thinking* Find a value of a such that the number line below shows all the solutions of $ax + 4 \leq -12$. **−8**

−5 −4 −3 −2 −1 0 1 2 3 4 5

Choose **Use a calculator or paper and pencil. Solve and check each inequality.** **33–34. no solutions**

29. $2 - 3k < 4 + 5k$ **$k > -\frac{1}{4}$**

30. $\frac{1}{2}n - \frac{1}{8} \geq \frac{3}{4} + \frac{5}{6}n$ **$n \leq -\frac{21}{8}$**

31. $-3(v - 3) \geq 5 - 3v$ **all numbers**

32. $8 \leq 5 - m + 1$ **$m \leq -2$**

33. $0.5(3 - 8t) > 20(1 - 0.2t)$

34. $\frac{2}{3}d - 4 > d + \frac{1}{8} - \frac{1}{3}d$

35. $38 - k \leq 5 - 2k$ **$k \leq -33$**

36. $\frac{4}{3}r - 3 < r + \frac{2}{3} - \frac{1}{3}r$ **$r < \frac{11}{2}$**

37. $-2(0.5 - 4s) \geq -3(4 + 3.5s)$ **$s \geq -\frac{22}{37}$**

38. *Standardized Test Prep* Which value of n is a solution of both $2(n + 5) \geq 4$ and $3(n - 1) < 3$? **B**

A. −7 **B.** −3 **C.** 2 **D.** 4 **E.** none of these

39. *Open-ended* Write two different inequalities that you can solve by adding 5 and multiplying by −3. Show how to solve each inequality. **Samples: $-\frac{1}{3}x - 5 < 2$, $x > -21$; $-\frac{1}{3}x - 5 \geq 1$, $x \leq -12$**

40. a. Generalize Solve $ax + b > c$ for x, where a is positive. **$x > \frac{c - b}{a}$**

b. Solve $ax + b > c$ for x, where a is negative. **$x < \frac{c - b}{a}$**

41. *Jobs* JoLeen is a sales associate in a clothing store. Each week she earns $250 plus a commission equal to 3% of her sales. This week she would like to earn no less than $460. What dollar amount of clothes must she sell? **at least $7000**

42. *Geometry* Artists often use the *golden rectangle* because it is considered to be pleasing to the eye. The length of a golden rectangle is about 1.62 times its width. Suppose you are making a picture frame in the shape of a golden rectangle. You have a 46-in. piece of wood. What are the length and width of the largest frame you can make? (Round your answers to the nearest tenth of an inch.) **14.2 in.; 8.8 in.**

Greece

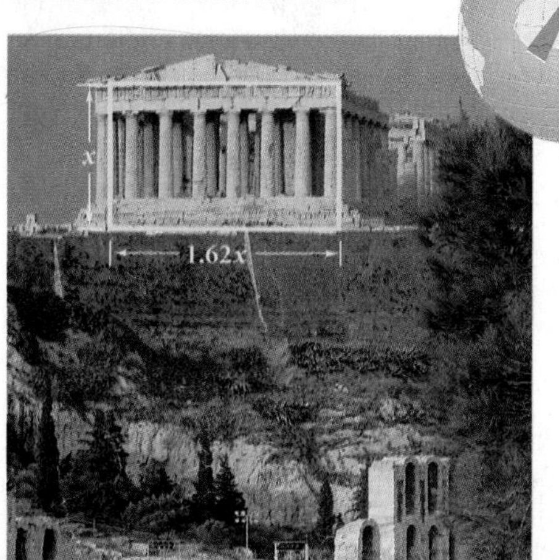

The Parthenon, an ancient Greek temple, was designed so that its dimensions form a golden rectangle.

page 190–192 Think and Discuss

5a. False; you subtract a smaller number from the quantity on the left than from the quantity on the right.

b. True; the quantity at the left is 5 less than 4*s* and the quantity on the right is 3 greater than 4*s*.

c. True; the two quantities are equal.

d. False; the two quantities are equal.

e. True; the quantity on the left is 9 smaller than the quantity on the right.

f. True; the two quantities are equal.

1. Subtract 5 from each side. Then divide each side by 4.
2. Divide each side by 2. Then add 3 to each side.
3. Subtract 8 from each side.
4. Add 2 to each side. Then multiply each side by −5.
5. Subtract *y* from each side. Then add 5 to each side.
6. Subtract 7 from each side.
7. Add *s* to each side. Then subtract 6 from each side.
8. Use the distributive property and collect like terms on the right. Then subtract 5*t* from each side.
9. Multiply or divide each side by −1. Reverse the order of the inequal.

page 194 Mixed Review

52–55. See back of book.

their work on the Chapter Project. Have students add this part of the project to work they have already completed. You may wish to check students' progress on the project by having each student write a short progress report telling what has been done so far and plans for completion.

Exercises MIXED REVIEW

Exercise 50 Ask: *How many grams (g) are there in one kilogram (kg)?* 1000

GETTING READY FOR LESSON 4-8 In this preview students graph pairs of inequalities on a single number line. This prepares them for graphing compound inequalities.

SELF ASSESSMENT **JOURNAL** Students should mention reversing the order of the inequality when multiplying or dividing both sides of the inequality by a negative number. They might also discuss differences in the solutions of equations and inequalities.

Wrap Up

THE BIG IDEA Ask students to tell one way in which solving multi-step *inequalities* differs from solving multi-step *equations*.

RETEACHING ACTIVITY Students use tiles to solve multi-step inequalities and then graph the solutions. They will model and solve each inequality, then draw the solution. (Reteaching worksheet 4-7)

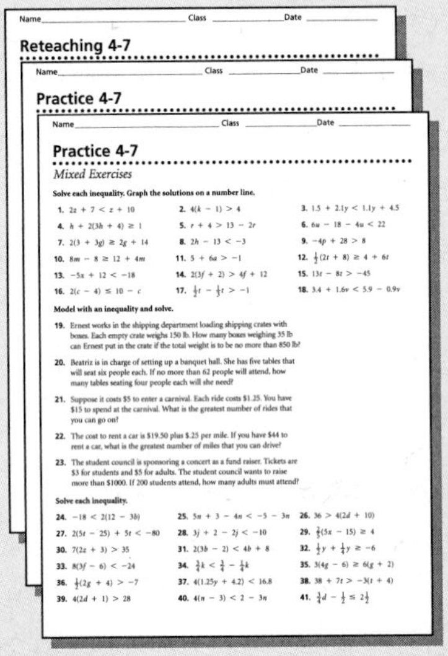

Name ________ Class ________ Date ________

Reteaching 4-7

Name ________ Class ________ Date ________

Practice 4-7

Name ________ Class ________ Date ________

Practice 4-7

Mixed Exercises

Solve each inequality. Graph the solutions on a number line.

1. $2z + 7 < z + 10$	2. $4(k - 1) > 4$	3. $1.5 + 2.1y < 1.1y + 4.5$
4. $h + 2(3h + 4) \geq 1$	5. $r + 4 > 13 - 2r$	6. $6u - 18 - 4u < 22$
7. $2(3 + 3g) \geq 2g + 14$	8. $2h - 13 < -3$	9. $-4p + 28 > 8$
10. $8m - 8 \geq 12 + 4m$	11. $5 + 6a > -1$	12. $\frac{1}{2}(2t + 8) \geq 4 + 6t$
13. $-5x + 12 < -18$	14. $2(3f + 2) > 4f + 12$	15. $13t - 8t > -45$
16. $2(c - 4) \leq 10 - c$	17. $\frac{1}{2}t - \frac{1}{3}t > -1$	18. $3.4 + 1.6v < 5.9 - 0.9v$

Model with an inequality and solve.

19. Ernest works in the shipping department loading shipping crates with boxes. Each empty crate weighs 150 lb. How many boxes weighing 35 lb can Ernest put in the crate if the total weight is to be no more than 850 lb?

20. Beatriz is in charge of setting up a banquet hall. She has five tables that will seat six people each. If no more than 62 people will attend, how many tables seating four people each will she need?

21. Suppose it costs $5 to enter a carnival. Each ride costs $1.25. You have $15 to spend at the carnival. What is the greatest number of rides that you can go on?

22. The cost to rent a car is $19.50 plus $.25 per mile. If you have $44 to rent a car, what is the greatest number of miles that you can drive?

23. The student council is sponsoring a concert as a fund raiser. Tickets are $3 for students and $5 for adults. The student council wants to raise more than $1000. If 200 students attend, how many adults must attend?

Solve each inequality.

24. $-18 < 2(12 - 3b)$	25. $5n + 3 - 4n < -5 - 3n$	26. $36 > 4(2d + 10)$
27. $2(5t - 25) + 5t < -80$	28. $3j + 2 - 2j < -10$	29. $\frac{2}{5}(5x - 15) \geq 4$
30. $7(2z + 3) > 35$	31. $2(3b - 2) < 4b + 8$	32. $\frac{1}{2}y + \frac{1}{4}y \geq -6$
33. $8(3f - 6) < -24$	34. $\frac{3}{4}k < \frac{3}{4} - \frac{1}{4}k$	35. $3(4g - 6) \geq 6(g + 2)$
36. $\frac{1}{2}(2g + 4) > -7$	37. $4(1.25y + 4.2) < 16.8$	38. $38 + 7t > -3(t + 4)$
39. $4(2d + 1) > 28$	40. $4(n - 3) < 2 - 3n$	41. $\frac{3}{4}d - \frac{1}{2} \leq 2\frac{1}{2}$

Lesson Quiz

Lesson Quiz is also available in Transparencies.

1. List the steps that you must do to the first inequality in order to get the second. $3(x - 4) > 4x + 7$; $x < -19$ **Distribute 3; add 12 to both sides; subtract $4x$ from both sides; multiply by -1; reverse the order of the inequality.**

2. Solve $-\frac{1}{3}p < \frac{1}{2}p - 6$. Graph the solution on a number line. $p > -7\frac{1}{5}$

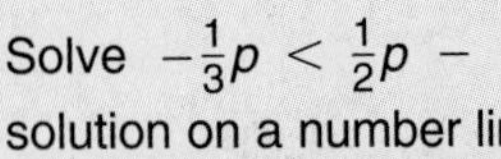

3. Solve and check the inequality $3(3c + 2) \leq 2(3c - 2)$. $c \leq -\frac{10}{3}$

43. *Business* Carlos plans to start a part-time word processing business out of his home. He is thinking of charging his customers $15 per hour. The table shows his expected monthly business expenses. Write an inequality to find the least number of hours he must work in a month to make a profit of at least $1200. $15n - (490 + 45 + 65) \geq 1200$

Monthly Expenses

Expense	Cost
Equipment Rental	$490
Materials	$45
Business Phone	$65

Chapter Project ***Find Out by Researching***

Systolic blood pressure, the higher number in a blood pressure reading, is measured as your heart muscle contracts. The formula $P \leq \frac{1}{2}a + 110$ gives the normal systolic blood pressure P based on age a.

- Find your normal systolic blood pressure.
- At age 20, does 120 represent a maximum or a minimum systolic pressure? Explain.
- A blood pressure reading higher than the normal value indicates a possible need for a change in lifestyle or for special medication. Research some lifestyle changes that can help reduce high blood pressure.

Exercises MIXED REVIEW

Write an inequality to model each situation.

44. An octopus can be up to 10 ft long. $p \leq 10$

45. A hummingbird migrates more than 1850 mi. $m \geq 1850$

To which family of functions does each graph belong? Explain your reasoning. 46. Quadratic; highest power of x is 2. 48. Linear; highest power of x is 1.

46. $y = x^2 - 3x$
47. $y = |x| - 2$
48. $y = 5x + 1$
49. $y = 9 - x^2$

47. Abs. val.; variable expression is inside abs. val. symbol. 49. Quadratic; highest power of x is 2.

50. It takes 4.5 million jasmine petals to make 450 g of jasmine oil. How many petals are needed to make 1 kg of jasmine oil? 10 million petals

51. *Recycling* Each year, 9.5 million vehicles are recycled.
    a. About 75% of each vehicle's mass is reused. The average vehicle weighs 1.5 tons. About how many tons of materials can be reused from one vehicle? about 1.125 tons
    b. How many tons of materials can be reused from all the recycled vehicles each year? about 10.7 million tons

SELF ASSESSMENT **FOR YOUR JOURNAL**

Summarize what you have learned about the similarities and differences between equations and inequalities, using specific examples.

Getting Ready for Lesson 4-8 52–55. See margin.

Graph each pair of inequalities on one number line.

52. $c < 8; c \geq 10$
53. $t \geq -2; t \leq -5$
54. $m \leq 7; m > 12$
55. $h > 1; h < 0$

4-8 Teaching Notes

PROBLEM OF THE DAY

Which is larger: x^5, y^4, or z^3, where $x = 2$, $y = 3$, and $z = 4$? y^4

Problem of the Day is also available in Transparencies.

CONNECTING TO PRIOR KNOWLEDGE Cut four different colored pieces of paper into small pieces. Include red and blue paper or change the wording accordingly in the following exercise. Randomly give each student two pieces. Ask students who have a red piece of paper to stand and then to sit. Ask students who have a blue piece of paper to stand and then to sit. Then ask students with red and blue pieces of paper to stand. Point out that, because you said *and,* only some members of each of the first two groups are now standing. Ask them all to sit. Then ask: *If I asked for the students with red or blue, who will stand?* **both groups**

THINK AND DISCUSS

Question 1 Ask: *How does the graph of $32 \leq t \leq 40$ differ from the graph of $32 \geq t \geq 40$?* **The graph of $32 \leq t \leq 40$ includes the points 32 and 40.**

Example 1

ALTERNATIVE METHOD You may wish to present this method as another way to graph $32 < t$ and $t < 40$. Draw a number line (from 30 to 42) on the board. Use colored chalk to show

---

What You'll Learn

- Solving compound inequalities and graphing the solutions on a number line
- Solving absolute value inequalities and graphing the solutions on a number line

...And Why

To solve problems involving the chemistry of a swimming pool

Meterologists use a variety of equipment to predict temperatures.

 You can search the Internet for more than one topic at once. If you join two topics with *or,* you get all the references for either topic. If you join the topics with *and,* you get only those that relate to both topics.

Source: America Online

Connections Chemistry . . . and more

4-8 Compound Inequalities

THINK AND DISCUSS

Part 1 Solving Compound Inequalities Joined by "And"

Today's temperatures will be above 32°F, but not as high as 40°F.

You can write this prediction as $32 < t$ and $t < 40$. Then you can combine the two inequalities into one, which you can read in two ways.

$$32 < t < 40$$

t is greater than 32, and less than 40. *t is between 32 and 40.*

The graph of $32 < t < 40$ is an *interval* on a number line.

30 31 32 33 34 35 36 37 38 39 40 41 42

In the same way, you can write an inequality with $>$, $\leq$, or $\geq$.

$$32 \leq t \leq 40$$

t is greater than or equal to 32, and less than or equal to 40. *t is between 32 and 40 inclusive.*

1. Graph $32 \leq t \leq 40$ on a number line.

32 34 36 38 40

Two inequalities that are joined by the word *and* or the word *or* are called **compound inequalities.** A solution of a compound inequality joined by *and* is any number that makes both inequalities true.

Example 1

Solve $-4 < r - 5 \leq -1$. Graph the solutions on a number line.

Write the compound inequality as two inequalities joined by *and.*

$-4 < r - 5$	and	$r - 5 \leq -1$	
$-4 + 5 < r - 5 + 5$	|	$r - 5 + 5 \leq -1 + 5$	← Add 5.
$1 < r$	and	$r \leq 4$	← Simplify.

The solutions are all numbers greater than 1 *and* less than or equal to 4.

−5 −4 −3 −2 −1 0 1 2 3 4 5

2. In Example 1, why is there an open circle at 1 and a closed circle at 4? 1 is not a solution, but 4 is.
3. To check the solution to Example 1, choose a value in the interval shown in the graph. Substitute your value in the original inequality to **verify** that the statement is true. **Answers may vary. Sample: $-4 < z - 5 \leq 21$ is true.**

Lesson Planning Options

Prerequisite Skills

- Understanding number lines
- Solving inequalities

Assignment Options

To provide flexible scheduling, this lesson can be subdivided into parts.

1. **Core** 1, 3, 5, 7, 13–14, 22–23 **Extension** 25–26, 45–48
2. **Core** 2, 4, 6, 8, 15–21, 24 **Extension** 27–28, 49–51
3. **Core** 29–44 **Extension** 9–12

Use Mixed Review to maintain skills.

Resources

Student Edition

Skills Handbook, pp. 577, 582
Extra Practice, p. 559
Glossary/Study Guide

Teaching Resources

Chapter Support File, Ch. 4
- Practice 4-8 (two worksheets)
- Reteaching 4-8
- Checkpoint
- Alternative Activity 4-8

Classroom Manager 4-8
Glossary, Spanish Resources
Two-Year Algebra Handbook 4-8

Transparencies
8, 10, 62, 65

the graph of $t < 40$. On the same number line, use a different color of chalk to show the graph of $t < 40$. Have students tell which points are marked in both red *and* blue. Draw a second number line with these points shaded.

Question 3 Students should also check points on either side of the interval to be sure those points are not solutions.

KINESTHETIC LEARNING Prepare one piece of paper per student. Write the integers from 1 to 10, as well as rational numbers and simple mixed fractions on the pieces of paper. Hand out the pieces of paper at random. Have students stand in the correct order to form these number lines:

- numbers greater than 5
- numbers greater than or equal to 7
- numbers greater than 2 and less than 6
- numbers less than 2 or greater than 8
- numbers less than or equal to 4 or greater than 9

Save the numbered pieces of paper for use in Lesson 4-9.

MAKING CONNECTIONS Invite students to guess the width of the world's narrowest and widest roads. Encourage students to state a guess and then to write it on the board in the form shown in the following example:

State: *The width is greater than 6 ft but less than 200 ft.*

Write: *6 ft < width < 200 ft*

After students record their guesses, share with them the following facts from the *Guinness Book of Records*. The

Additional Examples

FOR EXAMPLE 1

Solve $2 < 5 - f < 5$. Graph the solutions on a number line.

$0 < f < 3$

0 1 2 3

Discussion: *Why is it important to differentiate between open and closed dots on the number line?*

FOR EXAMPLE 2

Greg calculates his semester grade by averaging three grades he earned during the semester. So far he has scored 83 and 87. Set up and solve a compound inequality to determine the score Greg must make in order for his semester average to be between an 85 and a 95.

$85 < \frac{83 + 87 + g}{3} < 95$;

$85 < g < 115$

Discussion: *Why is it impossible for Greg to receive a 95 as his semester grade?*

196

A second way to solve a compound inequality involving *and* is to work on all three parts of the inequality at the same time. You work to get the variable alone between the inequality symbols.

Example 2 **Relating to the Real World** 

Chemistry The acidity of the water in a pool is considered normal if the average of three pH readings is between 7.2 and 7.8, inclusive. The first two readings for a pool are 7.4 and 7.9. What possible values for the third reading will make the pH normal?

Define p = value of third reading

Relate 7.2 is less than or equal to the average which is less than or equal to 7.8

Write $7.2 \leq \frac{7.4 + 7.9 + p}{3} \leq 7.8$

$$
\begin{aligned}
7.2 &\leq \frac{7.4 + 7.9 + p}{3} \leq 7.8 \\
3(7.2) &\leq 3\left(\frac{7.4 + 7.9 + p}{3}\right) \leq 3(7.8) && \leftarrow \text{Multiply by 3.} \\
21.6 &\leq 15.3 + p \leq 23.4 && \leftarrow \text{Simplify.} \\
21.6 - 15.3 &\leq 15.3 + p - 15.3 \leq 23.4 - 15.3 && \leftarrow \text{Subtract 15.3.} \\
6.3 &\leq p \leq 8.1 && \leftarrow \text{Simplify.}
\end{aligned}
$$

The value for the third reading must be between 6.3 and 8.1, inclusive.

4. Check the solution to Example 2 to **verify** that the solution gives an average pH in the normal range. See below left.

Solving Compound Inequalities Joined by "Or"

Part 2 

A solution of a compound inequality joined by *or* is any number that makes *either* inequality true.

Example 3

Solve $4v + 3 < -5$ or $-2v + 7 < 1$. Graph the solutions.

	or	
$4v + 3 < -5$		$-2v + 7 < 1$
$4v + 3 - 3 < -5 - 3$		$-2v + 7 - 7 < 1 - 7$
$4v < -8$		$-2v < -6$
$\frac{1}{4}(4v) < \frac{1}{4}(-8)$		$-\frac{1}{2}(-2v) > -\frac{1}{2}(-6)$
$v < -2$	or	$v > 3$

The solutions are all numbers that are less than –2 *or* are greater than 3.

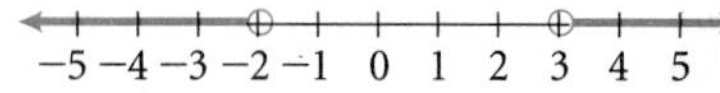

4. Answers may vary. Sample: $7.2 \leq \frac{7.4 + 7.9 + 6.3}{3} \leq 7.8$ is true.

narrowest road in the world is an alley in the village of Ripatransone, Italy, that is 16.9 in. wide. The widest road is a six-lane boulevard in Brasilia, Brazil, that is 820.2 ft wide.

Example 2 Relating to the Real World 

Question 4 Ask: *Would a third pH reading of 8 make the pH normal?* yes

Example 3

ERROR ALERT! Some students may try to write a combined inequality for inequalities joined by *or.* **Remediation:** Point out that combined inequalities are only used for inequalities joined by *and.*

EXTENSION Challenge students to solve $4v + 3 < -5$ *and* $-2v + 7 < 1$. This inequality has no solutions. Point out that the graphs of $v < -2$ and $v > 3$ do not have any points in common.

EXTENSION Question 9 Write an inequality that represents all numbers whose distance from three is less than six units. $|x - 3| < 6$

Example 4

Question 10 Encourage students to think about distances from a number on the number line. Students may need to go back and review Question 9 and the explanation above it.

5. **Try This** Graph each compound inequality on a separate line.
   a. $r < -1$ or $r > 3$
   b. $r > -1$ and $r < 3$
   5a–b. See margin.
6. Compare and contrast your graphs in Question 5.
   Each graph includes all the points not included in the other, except for endpoints. The endpoints are the same in both graphs.

Part 3
Solving Absolute Value Inequalities

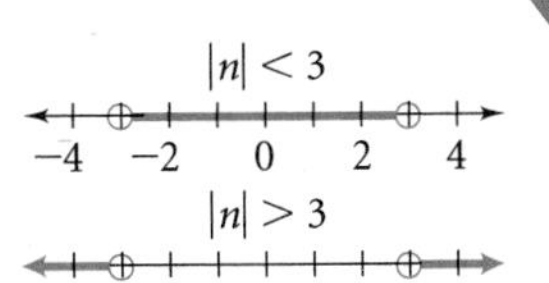

You can express the absolute value inequality $|n| < 3$ as the compound inequality $-3 < n < 3$. In a similar way, you can express the absolute value inequality $|n| > 3$ as the compound inequality $n > 3$ or $n < -3$. The graphs are at the left.

QUICK REVIEW
$|x| = 6$ means $x = 6$ or $x = -6$

7. Complete each statement with *less than* or *greater than.*
   a. For $|n| < 3$, the graph includes all points whose distance from zero is __?__ 3 units. less than
   b. For $|n| > 3$, the graph includes all points whose distance from zero is __?__ 3 units. greater than
8. Write an absolute value inequality to describe each graph.
   a. 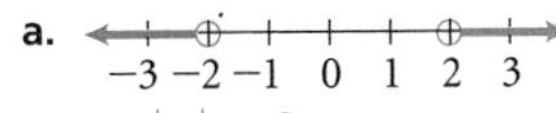
   $|m| > 2$
   b. 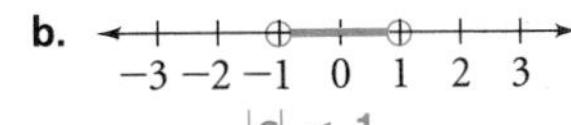
   $|a| < 1$

You can use absolute value inequalities to describe distances from numbers other than zero. The inequality $|n - 1| < 3$ represents all numbers whose distance from 1 is less than three units.

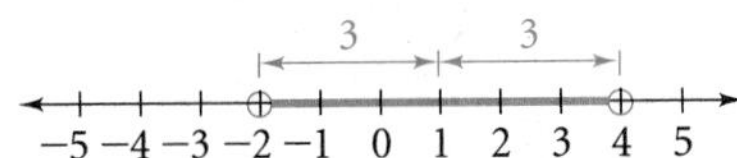

9. What does the inequality $|n - 1| > 3$ represent? These are all the numbers whose distance from 1 is more than 3 units.

You can also solve an absolute value inequality by first writing it as a compound inequality.

Example 4

Solve $|v - 3| \geq 4$. Graph the solutions on a number line.

Write $|v - 3| \geq 4$ as two inequalities joined by *or.*

$v - 3 \leq -4$	or	$v - 3 \geq 4$
$v - 3 + 3 \leq -4 + 3$		$v - 3 + 3 \geq 4 + 3$ ← Add 3.
$v \leq -1$	or	$v \geq 7$

The solutions are all numbers less than or equal to −1 *or* greater than or equal to 7.

−2 −1 0 1 2 3 4 5 6 7 8

PROBLEM SOLVING
Look Back How would the solution be different if the inequality were $|v - 3| > 4$?
−1 and 7 are not solutions of this inequality.

Technology Options

For Exercises 13–24 and 29–44, students may want to use graphing software to graph the inequalities on number lines.

Prentice Hall Technology 

Software • Secondary Math Lab Toolkit • Computer Item Generator 4-8

CD-ROM Multimedia Algebra Lab 4

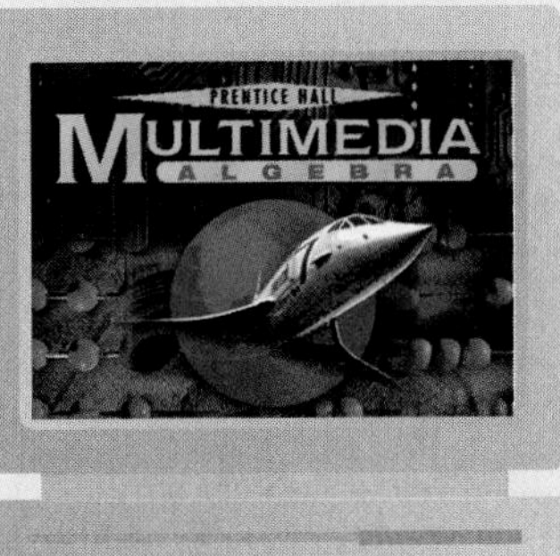

Internet • See the Prentice Hall site. (http://www.phschool.com)

Exercises ON YOUR OWN

EXTENSION Exercises 3–8 Ask students: *Which of these graphs represents the solution of* $\left|n - \frac{1}{2}\right| < 2\frac{1}{2}$? **the graph in Exercise 3**

OPEN-ENDED Exercise 25 Suggest topics, such as temperature or weight.

Exercise 26 Have students first write an inequality for the data in kilograms and then use the formula to rewrite the inequality with the data in pounds.

WRITING Exercise 28 If students have difficulty getting started, suggest they start with an example from Exercises 29–44.

GEOMETRY Exercises 45–48 Students explore the relationships among the lengths of the sides of a triangle. Have students look at the sample above the exercises. Point out that the length of the third side is between the sum and difference of the two given sides. Ask: *Can the length of the third side be 4 cm?* **no** *Why or why not?* **The sum of 4 and 4 is less than 10 and it must be greater than 10. Also, 4 is not between 4 and 10.**

pages 195–198 Think and Discuss

5a. 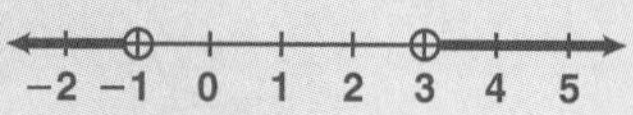

b. 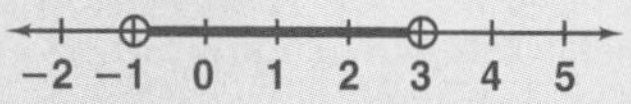

pages 198–200 On Your Own

11a. 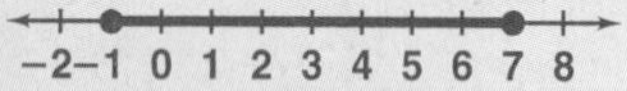

3. $-2 < x < 3$

4. $w < -3$ or $w \geq 2$

5. $-4 \leq x \leq 3$

6. $y \leq 0$ or $y > 2$

7. Answers may vary. Sample: $x < 1$ or $x > 0$

8. Answers may vary. Sample: $x < 3$ and $x > 4$

13. $-5 < j < 5$

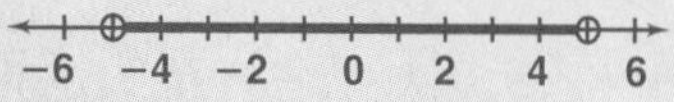

14. $-4 < q < 6$

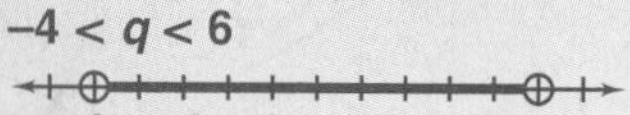

15. $k < -5$ or $k > -1$

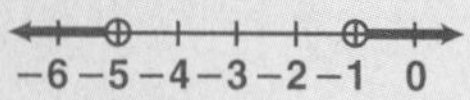

16. all numbers

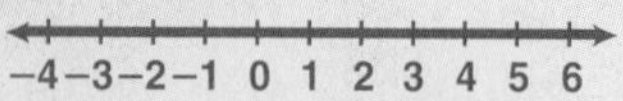

10. The solution is all the numbers whose distance from 3 is 4 or more, in either direction.
11a. $-1 \leq v \leq 7$; see margin for graph.

10. How could you solve $|v - 3| \geq 4$ using only a graph?

11. a. Try This Solve $|v - 3| \leq 4$. Graph the solutions on a number line.

b. Compare your solutions with the solutions in Example 4.
Each graph includes all the points not included in the other and the endpoints.

Exercises ON YOUR OWN

For each situation, define a variable and write a compound inequality.

1. The highest elevation in North America is 20,320 ft above sea level, at Mount McKinley, Alaska. The lowest elevation is 282 ft below sea level at Death Valley, California. e = elevation anywhere in North America; $-282 \leq e \leq 20{,}320$

2. Wind speeds of a *tropical storm* are at least 40 mi/h but no more than 74 mi/h. s = wind speed; $40 \leq s \leq 74$

Write a compound inequality that each graph could represent. 3–8. See margin.

3. 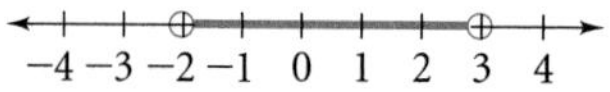

4. 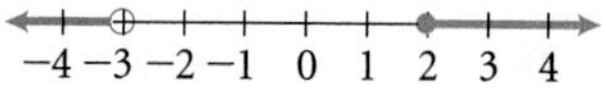

5. 

6. 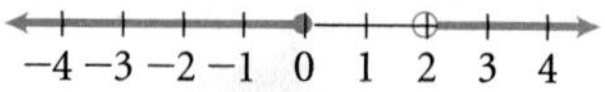

7. 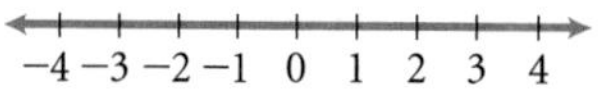

8. 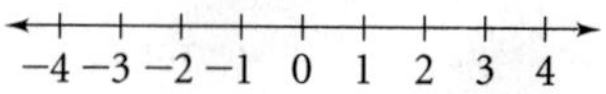

Choose a variable and write an absolute value inequality that represents each set of numbers on a number line.

9. all numbers less than 3 units from 0 $|x| < 3$

10. all numbers no less than 7.5 units from 0 $|x| \geq 7.5$

11. all numbers more than 2 units from 6 $|x - 6| > 2$

12. all numbers at least 3 units from −1 $|x + 1| \geq 3$

Solve each inequality and graph the solutions. 13–24. See margin.

13. $-3 < j + 2 < 7$

14. $12 > -3q$ and $-2q > -12$

15. $4 + k > 3$ or $6k < -30$

16. $x - 5 \leq 0$ or $x + 1 > -2$

17. $3 \geq 4r - 5 \geq -1$

18. $-1 \leq 3t - 2$ and $\frac{1}{2}t < -3$

19. $-2.8 \geq 2r + 0.2 > -3.8$

20. $6 - a < 1$ or $3a \leq 12$

21. $6.5 > w + 3 > 1.5$

22. $3a > -6$ and $7a < 14$

23. $0.25t \leq 3.5$ and $t \geq 4$

24. $25g < 400$ or $100 < 4g$

25. *Open-ended* Describe a real-life situation that you could represent with the inequality $-2 < x < 8$. Sample: elevation near a coastline varies between 2 m below and 8 m above sea level.

26. *Sports* A *welterweight* wrestler weighs at least 74 kg but no more than 82 kg. You can use the formula $k = 0.45p$ to find pounds p or kilograms k when one of the quantities is known. Write a compound inequality to describe the weight of a welterweight in pounds. $164.4 \leq p \leq 182.2$

Exercises 49–51 Be sure students understand how to read the graph. Ask:

- *What is the low temperature for Detroit in July?* about 60°F
- *What is the high temperature for Detroit in July?* about 82°F
- *What is the low temperature for Charlotte in July?* about 68°F
- *What is the high temperature for Charlotte in July?* about 90°F

ALTERNATIVE ASSESSMENT **Exercises 49–51** Ask students to write their own compound inequality expression using the data given by the graph. This will help you assess the students' understanding of compound inequality applications.

Chapter Project FIND OUT BY WRITING Ask students to solve the inequality to find the recommended pulse-rate range for their age. Have them solve the inequality for a person 25 yr old. Ask them to research and write a paragraph explaining why it is important to see a doctor before starting an exercise program. This task is essential to their work on the Chapter Project. Have students add this part of the project to work they have already completed. You may wish to check students' progress on the project by having each student write a sentence or two describing what they have done for the project thus far.

27. Manufacturing An electronics manufacturer needs a conveyer belt for its assembly plant. The completed conveyer belt may be 15 cm longer or shorter than shown. Find the possible lengths of the conveyer belt.
$7585 \text{ cm} \le l \le 7615 \text{ cm}$

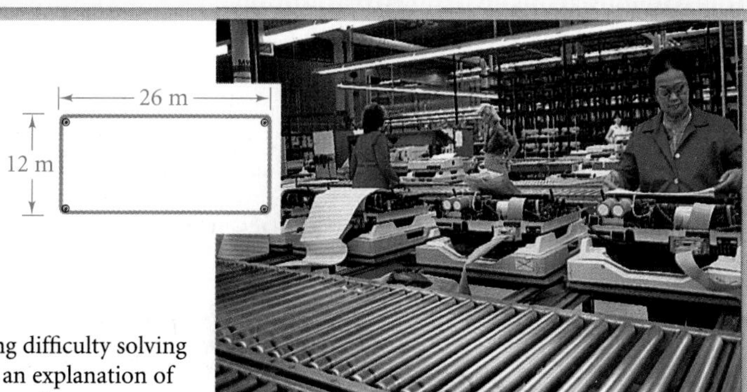

28. Writing Suppose a friend is having difficulty solving absolute value inequalities. Write an explanation of the process, with examples. See margin.

Express each absolute value inequality as a compound inequality. Solve and graph the solutions on a number line. 29–44. See margin.

29. $\lvert f\rvert > 2.5$	**30.** $\lvert x+3\rvert < 5$	**31.** $\lvert n+8\rvert \ge 5$	**32.** $\lvert 2y-3\rvert \ge 7$
33. $\lvert w\rvert > 2$	**34.** $\lvert n\rvert \le 5$	**35.** $\lvert 6.5x\rvert < 39$	**36.** $\lvert 3d\rvert \ge 6$
37. $\lvert y-2\rvert \ge 1$	**38.** $\lvert 2c-3\rvert < 9$	**39.** $\lvert 5t-4\rvert \ge 16$	**40.** $\lvert p-3\rvert < 5$
41. $\lvert 3t+1\rvert > 8$	**42.** $\lvert \frac{3}{4}x\rvert - 3 < -5$	**43.** $4.5 + \lvert 3m-2\rvert > 2$	**44.** $0 \le \lvert 3d-1\rvert$

Geometry **The sum of the lengths of any two sides of a triangle must be greater than the third side. The lengths of two sides of a triangle are given. Find the range of values for the possible lengths of the third side.**

Sample 3 cm, 7 cm

$x + 3 > 7$ and $x + 7 > 3$ and $3 + 7 > x$ ← Write inequalities.
$x > 4$ and $x > -4$ and $10 > x$ ← Solve each inequality.
$4 < x < 10$ ← Express as a compound inequality.

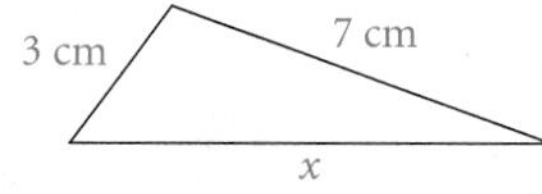

The length of the third side is greater than 4 cm and less than 10 cm.

45. 2.5 cm, 5 cm	**46.** 1 in., 4 in.	**47.** 28 mm, 21 mm	**48.** 12 ft, 18 ft
$2.5 < x < 7.5$	$3 < x < 5$	$7 < x < 49$	$6 < x < 30$

Meteorology **The *high–low graph* shows the average monthly high and low temperatures for Detroit, Michigan, and Charlotte, North Carolina.**

49. Write a compound inequality to represent the average temperature in Charlotte in June.
$66 \le C \le 88$

50. Write a compound inequality to represent the average temperature in Detroit in January. $15 \le D \le 30$

51. Use a compound inequality to describe the yearly temperature range for each city.
Charlotte: $29 \le C \le 90$
Detroit: $15 \le D \le 83$

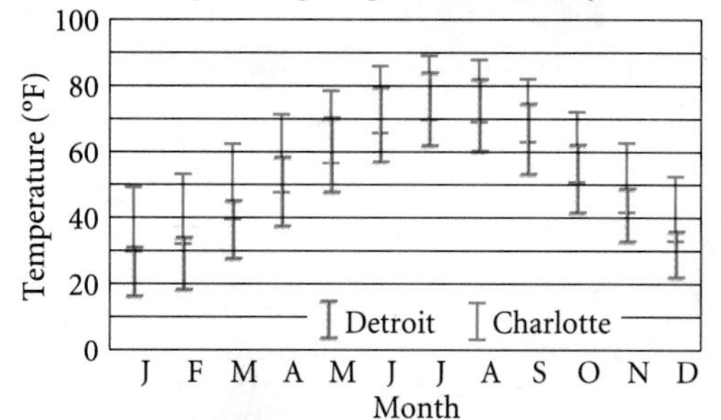

Source: *Statistical Abstract of the United States*

17. $1 \le r \le 2$
−1 0 1 2 3 4

18. no solution
−6 −5 −4 −3 −2 −1 0 1

19. $-2 < r \le -1.5$
−4 −3 −2 −1 0 1

20. $a \le 4$ or $a > 5$
2 3 4 5 6 7

21. $-1.5 < w < 3.5$
−2 −1 0 1 2 3 4

22. $-2 < a < 2$
−3 −2 −1 0 1 2 3

23. $4 \le t \le 14$
2 4 6 8 10 12 14 16

24. $g < 16$ or $g > 25$
15 16 17 18 19 20 21 22 23 24 25 26

28–44. See back of book.

page 200 Checkpoint

1–17, 20. See back of book.

Exercises MIXED REVIEW

GETTING READY FOR LESSON 4-9 These exercises prepare students for solving inequalities when the replacement set is the set of integers.

Wrap Up

THE BIG IDEA Have students discuss how the graph of a pair of inequalities joined by *and* differs from the graph of a pair of inequalities joined by *or*.

RETEACHING ACTIVITY Students solve compound inequalities, and then use pencils, paper hole-reinforcements, and pennies to graph the solution on a number line. They will use a pencil to model the solution on a given number line, then explain whether the endpoints should be open (reinforcements) or closed (pennies). (Reteaching worksheet 4-8)

Exercises CHECKPOINT

Students will assess their own progress on the topics covered in Lessons 4-5 to 4-8.

OPEN-ENDED Exercise 19 Suggest that students collect data from newspapers or television.

WRITING Exercise 20 Students might also refer to real-life situations.

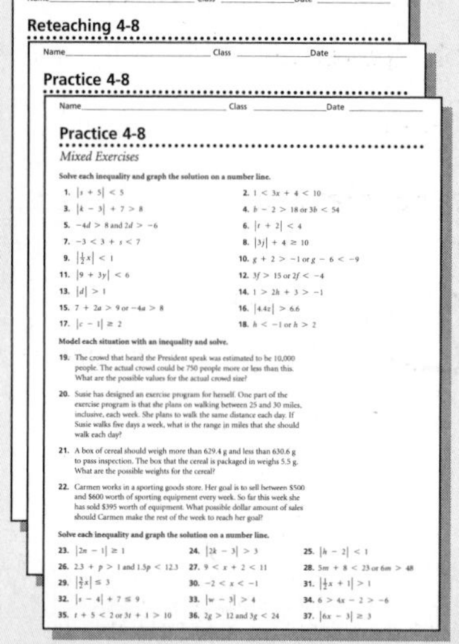
Reteaching 4-8
Practice 4-8
Practice 4-8
Mixed Exercises

Lesson Quiz

Lesson Quiz is also available in Transparencies.

1. Write a compound inequality for the graph. −2 −1 0 1 2
   $-2 < x < 2$

2. Solve the inequality $-0.75 \le 3t + 4 \le 0.75$ and graph its solution. Round your answer to the tenth's place.
   $-1.6 \le t \le -1.1$

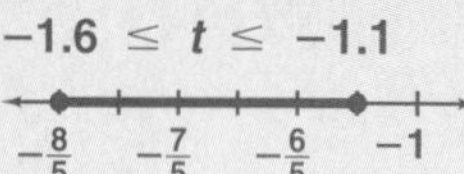

Chapter Project — Find Out by Writing

When you exercise, your pulse rate rises. Recommended pulse rates vary with age and physical condition. For vigorous exercise, such as jogging, the inequality $0.7(220 - a) \le R \le 0.85(220 - a)$ gives a target range for pulse rate R (in beats per minute), based on age a (in years).

- In what range should your pulse rate be when you are jogging?
- What is the target range for a person 25 years old?
- Why should you see a doctor before starting an exercise program?

Exercises MIXED REVIEW

Solve and check. State if the equation is an *identity* or has *no solution*.

52. $3y + 1 = 2.5$ 0.5
53. $-6m - 1 = 2m$ $-\frac{1}{8}$
54. $4(w + 8) = 10$ −5.5
55. $-2 = 7c - 5c$ −1
56. $8p - 4 = 4(2p - 1)$ identity
57. $3.8 = |4x| + 0.2$ −0.9, 0.9
58. $-5k + 5 = 5(-1 - k)$ no solution
59. $|t - \frac{1}{2}| = \frac{3}{4}$ $-\frac{1}{4}, 1\frac{1}{4}$
60. **Biology** Plasma makes up 55% of blood by volume. A 155-lb man has about 1.3 gal of blood. How much plasma does his blood contain? about 0.7 gal

Getting Ready for Lesson 4-9

Which of −3, −2, −1, 0, 1, 2, 3 are solutions of each inequality?

61. $m + 2 \le 3$ −3, −2, −1, 0, 1
62. $-4q > 0$ −3, −2, −1
63. $\frac{1}{2}z - 1 \le 0$ −3, −2, −1, 0, 1, 2
64. $3t + 5 < -4$ none

Exercises CHECKPOINT

Solve each inequality. Graph the solutions. 1–17. See margin.

1. $8 < c + 2$
2. $3x \le -24$
3. $-9m \ge 36$
4. $7 - c \le 12$
5. $5 < 6b + 3$
6. $12n \le 3n + 27$
7. $2 + 4r \ge 5(r - 1)$
8. $8w + 3w \le -22$
9. $6 + h \ge 2$
10. $0 \le 2t - 10t \le 4$
11. $\frac{x}{-3} > 9$ *or* $3x > 12$
12. $|g - 3| > 2$

Solve each absolute value inequality. Graph the solutions.

13. $|5d| \ge 15$
14. $|x + 3| < 7$
15. $|2x - 3| < 5$
16. $|4y - 2| \ge 18$
17. $\frac{4}{5} \ge |\frac{1}{2}x - 2|$
18. **Medicine** Normal body temperature is within 0.6°C of 36.6°C. Write a compound inequality for the range of normal body temperature. $36 < t < 37.2$
19. **Open-ended** Write a compound inequality to **predict** the temperature range for tomorrow. Check students' work.
20. **Writing** Explain the difference between the words *and* and *or* in a compound inequality. See margin.

Math ToolboX

Organizing and displaying data is an important skill in all fields of study and work. This toolbox adds the Venn diagram to the students' portfolio of problem-solving and data organization skills. A Venn diagram is a useful tool for organizing data. It is helpful to have students clearly label the sets in the diagram.

ERROR ALERT! Students may have difficulty working problems containing overlapping sets and entering this data into the Venn diagram. **Remediation:** Encourage students to first enter the data contained in the intersection of sets. Then students can work outward to find the number that is contained in the universal set.

WRITING Exercise 4 If time permits, groups of students may collect real data to display and present to the class. Possible topics include student music interests, participation in extracurricular activities, movies recently seen, or books read.

ADDITIONAL PROBLEM At Westwood High School, there are 219 boys. Two participate in football, basketball, and track. Twenty-one participate in track and football and three participate in basketball and track. Five participate in football and basketball. The football team has 62 members, the basketball team has 18 members, and the track team has 35 members. For each of these three sports, use a Venn diagram to determine the number of members that participate in that sport only. **football: 34; basketball: 8; track: 9**

Math ToolboX Problem Solving

Using a Venn Diagram

After Lesson 4-8

You can use a Venn diagram to illustrate relationships between sets and solve problems. First draw overlapping circles to represent the sets. Then draw a box around the circles to include any other information. Finally, fill in the diagram with information from the problem.

Example

An English teacher surveyed 48 ninth-grade students and found that 18 had read *Treasure Island*, 20 had read *Anne of Green Gables*, and 11 had read both books. How many students had not read either book?

After completing Steps 1 through 5, you can see that 21 students had not read either *Treasure Island* or *Anne of Green Gables*.

Step 1
number who read both books

Step 2
number who read only *Treasure Island*
$18 - 11 = 7$

Step 3
number who read only *Anne of Green Gables*
$20 - 11 = 9$

Step 4
number who read at least one book
$7 + 11 + 9 = 27$

Step 5
number who had not read either book
$48 - 27 = 21$

7 | 11 | 9 | 21

Use a Venn diagram to illustrate and solve each problem.
1–3. See margin for diagram.

1. Between 1933 and 1995 there were 11 presidents of the United States and 14 vice presidents. If 9 of the vice presidents were never president, how many of the presidents were never vice president?

2. Recently there were 118,519 female physicians in the United States. Of these, 16,573 were pediatricians, 40,431 were under the age of 35, and 6761 of the pediatricians were under the age of 35.
   a. How many female pediatricians were age 35 or older? 9812
   b. How many female physicians were 35 or older and not pediatricians? 68,276

3. In the 1994 Winter Olympics, 67 countries participated. The table shows how many countries won each possible medal.
   a. How many countries won gold, but not silver or bronze? 1
   b. How many countries won gold and silver, but not bronze? 2
   c. How many countries did not win any medals? 45

4. Writing Write a problem that you could solve using a Venn diagram. Then give your solution to the problem. See margin.

PROBLEM SOLVING HINT
For each exercise, first decide how many circles you need to draw.

1994 Winter Olympics
14 won gold.
12 won gold and silver.
11 won gold and bronze.
17 won silver.
14 won bronze and silver.
18 won bronze.
10 won gold, silver, and bronze.

Source: *The Guinness Book of Records*

Materials and Manipulatives

- Graphing calculator

Resources

Transparencies
5

page 201 Math Toolbox

1.

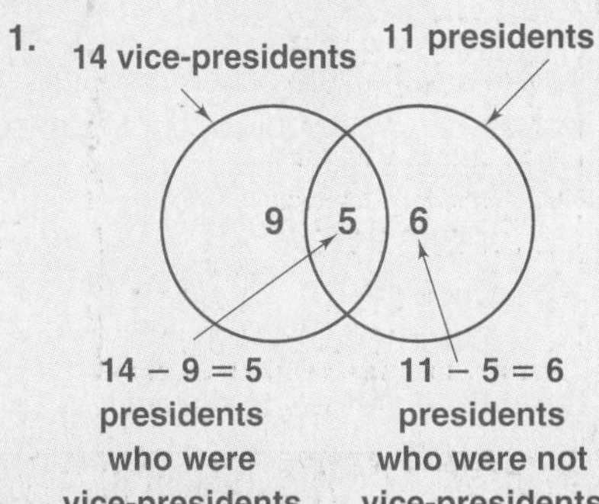

2–4. See back of book.

201

4-9 Teaching Notes

PROBLEM OF THE DAY

One hundred forty-two students joined the Kangaroo High School Band. Based on statistics from previous years, one-third of the students will drop out. Of the remaining members, only 6% will qualify for the state band. How many of the original members can be expected to qualify for the state band? **six members**

Problem of the Day is also available in Transparencies.

CONNECTING TO PRIOR KNOWLEDGE Ask students to list all the meanings they can think of for the word *replacement*. Lead them to understand that a replacement is an item that fits in the place of something else. When you have an inequality such as $f < 5$, a replacement set is all the numbers that can take the place of f and make the inequality true.

THINK AND DISCUSS

Example 1

Use the following steps to familiarize students with sets of numbers and the graph of each set.

- Draw a number line on the board. Shade the entire line to show that all numbers on the number line are real numbers. The set of real numbers can be defined as the set of all positive numbers, all negative numbers, and zero. Label the graph: Real Numbers.

Lesson Planning Options

Prerequisite Skills

- Understanding number lines
- Understanding the different types of numbers

Assignment Options

To provide flexible scheduling, this lesson can be subdivided into parts.

1 **Core** 1–16
Extension 24–29

2 **Core** 17–20
Extension 21–23

Use Mixed Review to maintain skills.

Resources

Student Edition

Skills Handbook, p. 569
Extra Practice, p. 559
Glossary/Study Guide

Teaching Resources

Chapter Support File, Ch. 4
- Practice 4-9 (two worksheets)
- Reteaching 4-9

Classroom Manager 4-9
Glossary, Spanish Resources
Two-Year Algebra Handbook 4-9

Transparencies
8,10, 22, 63

What You'll Learn

- Solving inequalities given a specific replacement set
- Checking the reasonableness of solutions

...And Why

To model real-world situations, like car inspection problems

What You'll Need

- calculator

Connections  Car Service . . . and more

4-9 Interpreting Solutions

THINK AND DISCUSS

Part 1

Solving Inequalities Given a Replacement Set

When you solve inequalities, the set of possible values for the variable, or **replacement set,** often is any real number, There are times, however, when the replacement set is limited to the set of integers or some other set. Your solution depends on the replacement set for the variable.

Example 1

Solve $-4 < 2k \leq 5$. Then graph the solutions on a separate number line for each replacement set.

a. the real numbers **b.** the integers **c.** $\{-5, -3, -2, 0, 3\}$

$$-4 < 2k \leq 5$$

$$\tfrac{1}{2}(-4) < \tfrac{1}{2}(2k) \leq \tfrac{1}{2}(5) \quad \leftarrow \text{Multiply by } \tfrac{1}{2}.$$

$$-2 < k \leq 2\tfrac{1}{2} \quad \leftarrow \text{Simplify.}$$

a. All real numbers greater than –2 and less than or equal to $2\frac{1}{2}$ satisfy the inequality.

−2 −1 0 1 2 3

b. The integers that satisfy the inequality are –1, 0, 1, and 2.

−2 −1 0 1 2 3

c. The value from {–5, –3, –2, 0, 3} that satisfies the inequality is 0.

−2 −1 0 1 2 3

1. **Try This** Match each replacement set with its graph of $|y| < 4$.
   A. the real numbers **IV** **I.** −5 −4 −3 −2 −1 0 1 2 3 4 5
   B. the integers **III** **II.** −5 −4 −3 −2 −1 0 1 2 3 4 5
   C. the positive integers **II** **III.** −5 −4 −3 −2 −1 0 1 2 3 4 5
   D. the positive real numbers **I** **IV.** −5 −4 −3 −2 −1 0 1 2 3 4 5

QUICK REVIEW

Three dots (• • •) indicate that a pattern continues.

2. **Open-ended** Write an inequality whose solutions are graphed on the number line below. Identify the replacement set for the variable.

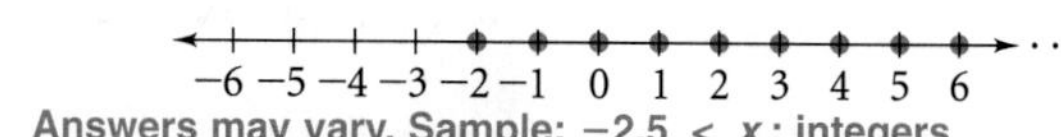

Answers may vary. Sample: $-2.5 < x$; integers

- Have a volunteer draw another number line to show the graph of the positive real numbers. Check to be sure the student does not include zero. Label the graph: Positive Real Numbers.
- Have another volunteer show the graph of the set of negative real numbers. Again, be sure the graph does not include zero. Label the graph.
- Next, draw another number line and use points to show the graph of all integers. Use two sets of three dots (. . .) to indicate that the pattern continues in both directions. Be sure to include zero. Label the graph.
- Have two more volunteers draw and label graphs for the positive integers and the negative integers. Be sure these graphs do not include zero.

Leave all graphs on the board so that students may refer to them throughout the lesson.

Example 2 **Relating to the Real World** 

Question 4 Ask students: *Other than the inspection fee, how might Winnie Johnson's Garage benefit from doing inspections?* **During the inspection, the mechanic might identify repairs that need to be done and then make a profit from doing the repairs.**

KINESTHETIC LEARNING If you have block scheduling or an extended class period, you may want students to participate in this exercise. Use the numbered pieces of paper from Lesson 4-8. Again use the numbers to have students form number lines in response to directions such as these:

- integers greater than 5

Part 2

Determining a Reasonable Answer

In many problems, you are given restrictions that indicate the replacement set for the variable. Other restrictions arise from common sense.

Example 2 **Relating to the Real World** 

Car Service Felix Ramiro's Garage has just purchased special equipment for the state emission inspection. The equipment cost \$1500. Each inspection costs the garage \$2.60 for labor and supplies. The garage gets to keep \$8.20 of each vehicle's inspection fee. How many inspections must the garage perform in order to make a profit?

Define	x = number of inspections				
Relate	income from inspections	is greater than	cost of equipment	plus	expense of inspections
Write	$8.2x$	$>$	1500	$+$	$2.6x$

$$8.2x > 1500 + 2.6x$$

$$8.2x - 2.6x > 1500 + 2.6x - 2.6x \quad \leftarrow \text{Subtract } 2.6x \text{ from each side.}$$

$$5.6x > 1500$$

$$\frac{5.6x}{5.6} > \frac{1500}{5.6} \quad \leftarrow \text{Divide each side by 5.6.}$$

$$x > 267.8571429$$

Since x represents the number of inspections, it makes sense to consider integer values only. So, the solutions are $x \geq 268$, where x is an integer. The garage must perform at least 268 inspections to make a profit.

266 267 268 269 270 271

3. Show how you would check Example 2. Be sure you check the words of the original problem, not simply the math of the inequality. See left.
4. a. *Car Service* Felix's garage is open six days a week. It has enough mechanics to do 11 inspections a day. What is the minimum number of weeks it will take to make a profit on inspections? Explain. a–b. See left.
   b. What is the maximum number of weeks it will take to make a profit on inspections? Explain.

3. Test 267 and 268: $8.2(267) > 1500 + 2.6(267)$ false; $8.2(268) > 1500 + 2.6(268)$ true. You do not need to test 267.8571429 because the situation demands integer solutions.

4a. 5 weeks; the garage can make 66 inspections in a week. To make 268 inspections it needs 4.06 weeks. This is more than 4 weeks.

4b. There is no maximum. If the number of inspections never reaches 268, the garage will never make a profit.

Additional Examples

FOR EXAMPLE 1

Solve $4 > -2d > -4$ Graph the solutions on a separate number line for the following replacement sets: real numbers, integers, negative integers, and positive real numbers.

$-2 < d < 2$

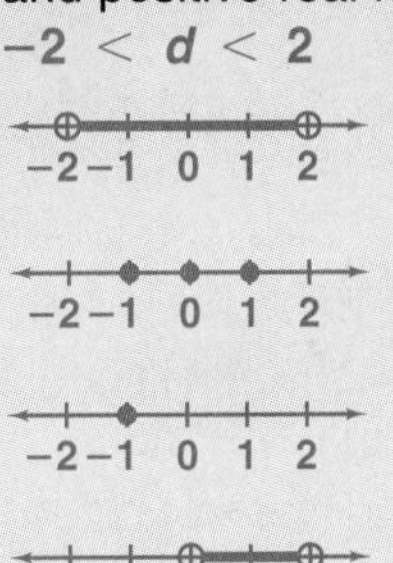

FOR EXAMPLE 2

Li plans to sell lemonade to raise \$40 for the science fair. She sells each cup of lemonade for \$.70 and spends \$.23 per cup for supplies. Set up and solve an inequality to find out the minimum number of cups she must sell.

$0.7c - 0.23c > 40$; $c > 5.3$; 86 cups

- real numbers greater than 5
- even integers greater than or equal to 7
- real numbers greater than 2 and less than 6
- odd integers less than 2 or greater than 8
- real numbers less than or equal to 4 or greater than 9

Exercises ON YOUR OWN

ERROR ALERT! Students may make the mistake of including zero in the set of positive integers. **Remediation:** Remind students that zero is neither positive nor negative.

OPEN-ENDED Exercise 22 Since the solution is any number between two positive numbers, students need to create a problem for which the solution is a continuous set of data.

ALTERNATIVE ASSESSMENT **Exercise 25** Randomly select students to answer this question orally. This will help you assess the students' understanding of replacement sets.

WRITING Exercise 25 Students might include a problem and a suitable replacement set for the solution.

Chapter Project FIND OUT BY INTERVIEWING Ask students to think of someone they could interview to find out more about exercising. Ask them to think of five questions they could ask. Have students conduct the interview and discuss their findings with their group or partner. Next, have the students design an exercise plan for themselves. This work is essential to their Chapter Project. Have students add this task to work they have already completed for the project. You may wish to check students' progress on the project at this point by inspecting all completed work in their project folder.

Technology Options

For Exercises 30–32, students may want to use spreadsheet software or a graphing calculator to manipulate the statistical data in these exercises.

Prentice Hall Technology

Software • Secondary Math Lab Toolkit • Computer Item Generator 4-9

Internet • See the Prentice Hall site. (http://www.phschool.com)

pages 204–205 On Your Own

1–2, 4, 6–7, 9–14, 16–22, 24a–b, 25–29. See back of book.

QUICK REVIEW

A set of data that involves measurements, such as length, weight, or temperature, is usually *continuous.*

A set of data that involves a count, such as numbers of people or objects, is *discrete.*

It is important to consider replacement sets when you use an inequality to model a real-world situation. For instance, consider this situation:

The temperature t ranges from 25°F to 35°F, inclusive.

A set of temperatures is a continuous set of data. You graph $25 \leq t \leq 35$ as an interval, with the real numbers as the replacement set for t.

24 25 26 27 28 29 30 31 32 33 34 35 36

Now suppose you want to model this situation:

The number n of students in a homeroom is between 25 and 35, inclusive.

The numbers of students is a discrete set of data. You graph $25 \leq n \leq 35$ as a set of points, with the positive integers as the replacement set for n.

24 25 26 27 28 29 30 31 32 33 34 35 36

Exercises ON YOUR OWN

Graph each inequality on a number line. Use the positive integers as a replacement set. If there are no solutions, write *no solutions.* 1–16. See margin for graphs.

1. $p \leq 6$
2. $|c - 2| > 3$ $c < -1$ or $c > 5$
3. $w \leq 0$ no solution
4. $-g > -8$ $g < 8$
5. $-3 \leq a < 1$ no solution
6. $r < 3$ or $r \geq 5$
7. $-4 < k + 1 < 7$ $-5 < k < 6$
8. $3(d + 2) < 6$ $d < 0$; no solution
9. $-y < -4.2$ $y < 4.2$
10. $|3 - r| > 2$ $r < 1$ or $r > 5$
11. $2 > u$ or $u \geq 3$
12. $w \geq 0$
13. $-4 \leq q < 2.25$
14. $3(d - 1) \leq 4.5$ $d \leq 2.5$
15. $4 < 2 - 2m \leq 8$ $-3 \leq m < -2$; no solution
16. $|2a + 1| < 9$ $-5 < a < 4$

Write an inequality that represents each situation. Graph the solutions on a number line. 17–20. See margin for graphs.

17. If you are 12 to 64 years old, you pay full price for admission to the Science Museum. $12 \leq a < 65$
18. A market researcher surveyed women who were at least 19 but less than 25 years old. $19 \leq a < 25$
19. The circumference of a baseball is between 23 cm and 23.5 cm. $23 < c < 23.5$
20. A box can hold from 15 to 20 books, inclusive. $15 \leq b \leq 20$

21. **Nursing** In nursing school, students learn temperature ranges for bath water. Tepid water is approximately 80°F to 93°F, warm water is approximately 94°F to 98°F, and hot water is approximately 110°F to 115°F. Model these ranges on one number line. Label each interval. See margin.
22. **Open-ended** Create a problem for which the solution is any number between two positive numbers a and b. See margin.

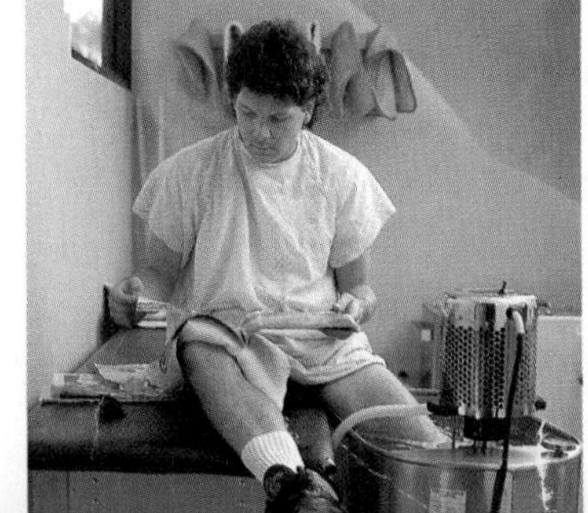

Exercises MIXED REVIEW

Exercises 30–32 Students who need a review of these terms can refer to the Playback sections on pages 4 and 5.

PORTFOLIO Share with students the criteria used to assess their work in portfolios, as well as how you plan to use the results. Students should understand how the rubrics are used to assess their work, how each piece in the portfolio counts, and how the scores they get in their portfolios affect their overall evaluation.

Wrap Up

THE BIG IDEA Ask students to choose a replacement set for x that is not the set of all real numbers. Then have them graph the solution of $x > 4$ for their replacement set.

RETEACHING ACTIVITY Students use tiles, pencils, and pennies to solve and to graph inequalities given a specific replacement set. After using tiles to solve a compound inequality, students will model the graph of the solutions with pennies, to show integer replacement, or with pencils, to show the real number replacement set. (Reteaching worksheet 4-9)

23. a. ***Jobs*** A summer employee at a store can work at most 8 h/day. How many hours must Nadine work on Friday to meet her goal of earning at least $175 for the week? 6 h

Employee	Hourly Wage	Mon.	Tues.	Wed.	Thurs.
Radam	$5.25	7	7	8	8
Nadine	$5.30	8	6	7	7

b. Summarize the steps you would use to find the number of hours each employee must work on Friday in order to meet that goal. Divide 175 by the hourly rate and subtract the hours worked so far. Round the result up to the next integer.

24. ***Veterinary Science*** The table lists respiratory rates for certain birds.

a. Model the respiratory rate for a parrot on a number line.

b. Draw two other number lines and align the 0 marks on them with the 0 mark on the number line in part (a). Model the respiratory rates for canaries and cockatiels on the number lines.

a–b. See margin.

At-Rest Respiratory Rates

Bird	Breaths/min
Canary	60–100
Cockatiel	100–125
Parrot	25–40

25. ***Writing*** Explain what a replacement set is and why you need to think about it when you are using an inequality to solve a problem.

See margin.

Graph each inequality for the given replacement set. 26–29. See margin for graphs.

26. $-8 < 2(c + 5) < 14$, for the set of negative integers $-9 < c < 2$

27. $-3 \leq 9v + 6 \leq 24$, for $\{-4, -2, 0, 2, 4\}$ $-1 \leq v \leq 2$

28. $12 \geq |6y - 3|$, for the set of positive real numbers $-1\frac{1}{2} \leq y \leq 2\frac{1}{2}$

29. $|2n + 6| < 12$, for the set of integers $-9 < n < 3$

Chapter Project — *Find Out by Interviewing*

How much exercise is enough? What kind of exercise is best for you?

- Interview a coach, a trainer, or a health professional to find out what type and amount of exercise they recommend for a teenager like you.
- Design your own exercise plan. Consider goals (muscle tone, heart workout) and amount and type of exercise.

Exercises MIXED REVIEW

Find each statistic for the data 1, 3, 4, 4, 5, 5, 7, 9, 9, 9, 13.

30. mean about 6.3 **31.** median 5 **32.** mode 9

Find the percent of change.

33. $24 to $36 50% **34.** 20 lb to 15 lb 25% **35.** 300 to 750 150%

36. Write an equation and solve. $3r = 5$; about 1.67 in./h

a. ***Weather*** In August 1995, a storm in Stillwater, Oklahoma, dropped 5 in. of rain in 3 h. What was the hourly rate?

b. Suppose it rained all day at this rate. How many inches of rain would fall in a 24-h period? $\frac{5}{3} = \frac{f}{24}$; 40 in.

SELF ASSESSMENT — PORTFOLIO

Select one or two items from your work for this chapter. Consider
- corrected work
- work based on manipulatives
- open-ended questions

Explain why you have included each selection you make.

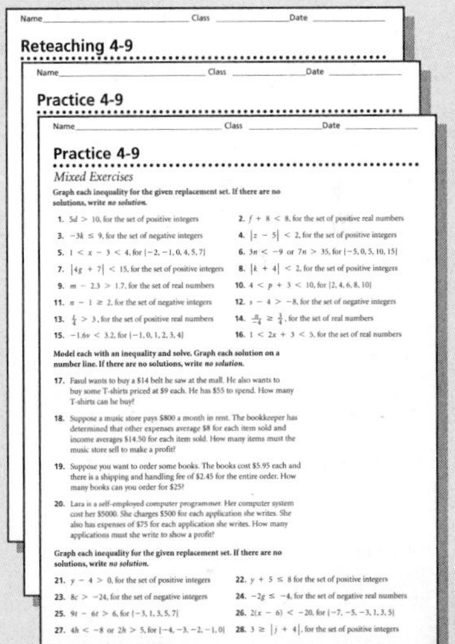
Name Class Date

Reteaching 4-9

Name Class Date

Practice 4-9

Name Class Date

Practice 4-9

Mixed Exercises

Graph each inequality for the given replacement set. If there are no solutions, write *no solution.*

1. $5d > 10$, for the set of positive integers
2. $f + 8 < 8$, for the set of positive real numbers
3. $-3k \leq 9$, for the set of negative integers
4. $|z - 5| < 2$, for the set of positive integers
5. $1 < x - 3 < 4$, for $\{-2, -1, 0, 4, 5, 7\}$
6. $3n < -9$ or $7n > 35$, for $\{-5, 0, 5, 10, 15\}$
7. $|4g + 7| < 15$, for the set of positive integers
8. $|k + 4| < 2$, for the set of positive integers
9. $m - 2.3 > 1.7$, for the set of real numbers
10. $4 < p + 3 < 10$, for $\{2, 4, 6, 8, 10\}$
11. $n - 1 \geq 2$, for the set of negative integers
12. $s - 4 > -8$, for the set of negative integers
13. $\frac{t}{4} > 3$, for the set of positive real numbers
14. $\frac{n}{-4} \geq \frac{3}{4}$, for the set of real numbers
15. $-1.6v < 3.2$, for $\{-1, 0, 1, 2, 3, 4\}$
16. $1 < 2x + 3 < 5$, for the set of real numbers

Model each with an inequality and solve. Graph each solution on a number line. If there are no solutions, write *no solution.*

17. Fasul wants to buy a $14 belt he saw at the mall. He also wants to buy some T-shirts priced at $9 each. He has $55 to spend. How many T-shirts can he buy?
18. Suppose a music store pays $800 a month in rent. The bookkeeper has determined that other expenses average $8 for each item sold and income averages $14.50 for each item sold. How many items must the music store sell to make a profit?
19. Suppose you want to order some books. The books cost $5.95 each and there is a shipping and handling fee of $2.45 for the entire order. How many books can you order for $25?
20. Lara is a self-employed computer programmer. Her computer system cost her $5000. She charges $500 for each application she writes. She also has expenses of $75 for each application she writes. How many applications must she write to show a profit?

Graph each inequality for the given replacement set. If there are no solutions, write *no solution.*

21. $y - 4 > 0$, for the set of positive integers
22. $y + 5 \leq 8$ for the set of positive integers
23. $8c > -24$, for the set of negative integers
24. $-2g \leq -4$, for the set of negative real numbers
25. $9t - 6t > 6$, for $\{-3, 1, 3, 5, 7\}$
26. $2(x - 6) < -20$, for $\{-7, -5, -3, 1, 3, 5\}$
27. $4h < -8$ or $2h > 5$, for $\{-4, -3, -2, -1, 0\}$
28. $3 \geq |j + 4|$, for the set of positive integers

Lesson Quiz

Lesson Quiz is also available in Transparencies.

Graph each inequality on a number line. Use the positive integers as the replacement set. If there are no solutions, write *no solutions.*

1. $-x \leq -4$ (graph: 3 4 5 6 7)

2. $2(a + 3) < 4$ no solutions

3. $|3 - r| < 2$ (graph: 1 2 3 4 5)

Finishing the Chapter Project

PROJECT DAY You may wish to plan a project day on which students share their completed projects. Encourage groups to explain their process as well as their product.

PROJECT NOTEBOOK Have students review their project work and bring their notebooks up to date.

- Have students review their methods for finding, recording, and solving equations and formulas used for the project.
- Ask groups to share any insights they found for completing the project, such as any shortcuts they found for solving formulas or equations. Also, find out if any students developed insights about their own level of physical fitness or how to improve it.

SCORING RUBRIC

3 Equations are correct, and calculations are accurate. Notebook is well organized and presents the information in a useful manner. Reasoning and explanations are sound and clearly expressed.

2 Calculations are mostly correct, but minor errors do exist. Reasoning and explanations are essentially correct but could be more thorough. Notebook presents information clearly but needs better organization.

1 Equations are inaccurate, and notebook lacks organization. Explanations lack detail.

0 Major elements of the project are incomplete or missing.

CHAPTER PROJECT

Finishing the Chapter Project

NO SWEAT!

Questions on pages 162, 178, 194, 200, and 205 should help you to complete your project. Present all your information for the project in a notebook. Is your exercise plan realistic for your available time and resources? How close is it to what you already do? (If it is very different, you should begin only with an adult's guidance.) Be sure to include what you have learned about blood pressure and about personal exercise.

Reflect and Revise

Organize your notebook around your use of equations and inequalities (one section for each). Share your notebook with an adult. Check your work for accuracy and look for any items that are not clear. Make any changes necessary in your work.

Follow Up

Find out about cross-training techniques or the training programs used in an Olympic sport.

For More Information

Kettelkamp, Larry. *Modern Sport Science.* New York: William Morrow, 1986.

"Physical Fitness." *Marshall Cavendish Encyclopedia of Health* (Vol. 10: 568–574). North Bellmore, New York: Marshall Cavendish, 1991.

Schwarzenegger, Arnold. *Arnold's Fitness for Kids Ages 11–14.* New York: Doubleday, 1993.

Virtue, Doreen. "What's Your Fitness Personality? Find an Exercise Plan that Really Fits." *Vegetarian Times* (January 1995): 55.

Wrap Up

HOW AM I DOING? Separate the class into small groups. Ask each group to put together a series of problems to illustrate the central ideas and rules of the chapter. Each group will solve the problems created by a different group. Solutions can then be checked and corrected by the group who designed the problems.

KEY TERMS The numbers in parentheses direct students to the pages where the terms (or symbols) are used or defined. Students should be able to (1) write a simple explanation of each term, (2) illustrate the term with a diagram, or (3) show an example that uses the term.

Exercises 1–6 Remind students that they can use either the *Multiplication Property of Equality* or *cross products* to solve proportion problems.

Exercise 14–17, 33, 40, 42 Emphasize that students need to set up two relationships before solving problems which include absolute value signs. For example, the equation $|x - 1| = 2$ has the solutions $x - 1 = 2$ and $x - 1 = -2$.

Exercises 27, 36, 46 Remind students that dividing by a negative number reverses the order of an inequality.

Remind students that the new mathematical terms in this chapter are defined in the Glossary/Study Guide in the back of the book.

4 Wrap Up

Key Terms

compound inequalities (p. 195)
cross products (p. 159)
equivalent inequalities (p. 180)
identity (p. 166)
no solution (p. 165)
literal equation (p. 176)
proportion (p. 158)
ratio (p. 158)
replacement set (p. 202)
scale (p. 158)
similar figures (p. 159)
solution of the inequality (p. 179)

How am I doing?

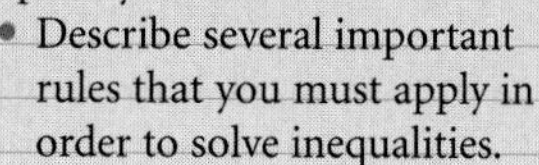

- State three ideas from this chapter that you think are important. Explain your choices.
- Describe several important rules that you must apply in order to solve inequalities.

Using Proportions 4-1

A **ratio** is a comparison of two numbers by division. A **proportion** is a statement that two ratios are equal. You can solve a proportion with a variable by using the **cross products**.

Solve each proportion.

1. $\frac{4}{12} = \frac{c}{6}$ 2 **2.** $\frac{t}{5} = \frac{23}{50}$ 2.3 **3.** $-\frac{9}{m} = \frac{3}{2}$ −6 **4.** $\frac{5}{4} = \frac{12.5}{z}$ 10 **5.** $\frac{x}{4} = -\frac{1}{10}$ −0.4 **6.** $\frac{8}{b} = \frac{16}{3}$ 1.5

7. *Hobbies* A model airplane can be built to $\frac{1}{48}$ of actual size. Suppose a wing of the model airplane is $\frac{3}{4}$ ft long. How long is the wing of the full-size airplane? 36 ft

Equations with Variables on Both Sides 4-2

You can use the properties of equality to solve an equation. An equation has no solution if no value of the variable makes the equation true. An equation is an **identity** if every value of the variable makes the equation true.

Write an equation for each model and solve.

8.

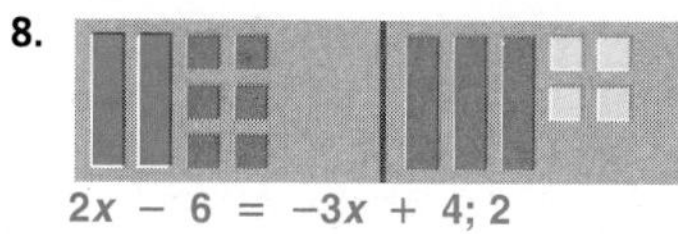

$2x - 6 = -3x + 4$; 2

9.

$-x + 8 = -2x - 9$; −17

Solve and check. If the equation is an identity or if it has no solution, write *identity* or *no solution*.

10. $5w = 6w + 11$ −11 **11.** $3(2t - 6) = 2(3t - 9)$ identity **12.** $4n - 6n = 2n$ 0 **13.** $9c + 4 = 3c - 8$ −2

Resources

Student Edition

Extra Practice, p. 559
Glossary/Study Guide

Teaching Resources

Study Skills Handbook
Glossary, Spanish Resources

page 207–209 Wrap Up

24. $h > -1$ (−3 −2 −1 0 1 2)

25. $k \leq -\frac{1}{2}$ (−3 −2 −1 0 1 2)

26. $b < 40$ (37 38 39 40 41 42)

27. $y \leq -168$ (−170 −169 −168 −167 −166 −165)

28. $c \leq -2$ (−5 −4 −3 −2 −1 0)

29. $m < -6$ (−9 −8 −7 −6 −5 −4)

30. $t < -5$ (−8 −7 −6 −5 −4 −3)

31. $x \geq 9$ (7 8 9 10 11 12)

33. $n \leq -6$ or $n \geq 2$

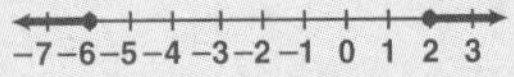

−7 −6 −5 −4 −3 −2 −1 0 1 2 3

34. $-2 \leq z < 7$

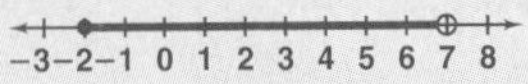

−3 −2 −1 0 1 2 3 4 5 6 7 8

35. $t \leq -2$ or $t \geq 7$

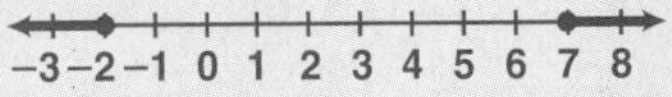

−3 −2 −1 0 1 2 3 4 5 6 7 8

36. $-1\frac{1}{2} \leq b < 0$

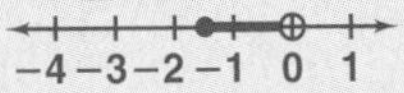

−4 −3 −2 −1 0 1

37. $2 \leq a < 4$

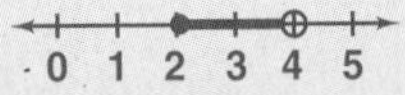

0 1 2 3 4 5

38. $-6\frac{1}{2} < d < 4$

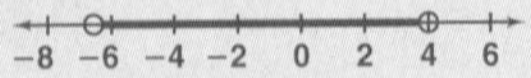

−8 −6 −4 −2 0 2 4 6

39. $2 \leq a < 5$

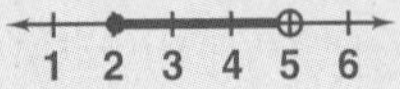

1 2 3 4 5 6

40. all numbers

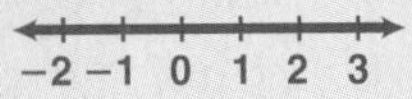

−2 −1 0 1 2 3

42.

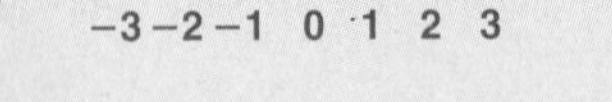

−3 −2 −1 0 1 2 3

43.

−1 0 1 2 3 4 5

208

Solving Absolute Value Equations 4-3

You can solve an **absolute value equation** by getting the absolute value by itself on one side of the equation. When the expression within the absolute value symbol contains more than one term, you must write and solve two equations.

Solve and check each equation. If there is no solution, explain.

14. $|y| = 5$ −5, 5 **15.** $|p + 3| = 9.5$ −12.5, 6.5 **16.** $|6 - b| = -1$ No solution; abs. val. cannot be neg. **17.** $|k - 8| = 0$ 8

18. *Open-ended* Write an absolute value equation that has no solution and another that has one solution. Solve the one that has one solution.
Samples: $|2x - 5| = -3$; $|3x - 1| = 0$; $\frac{1}{3}$

Transforming Formulas 4-4

A formula shows the relationship between two or more variables. When you express one variable in terms of the others, you are solving the equation for that variable.

Solve each equation for the given variable.

19. $m = \frac{a+b+c}{3}$; c $c = 3m - a - b$ **20.** $C = \pi d$; d $d = \frac{C}{\pi}$ **21.** $y = mx + b$; x $x = \frac{y-b}{m}$ **22.** $A = \frac{1}{2}bh$; h $h = \frac{2A}{b}$

23. *Science* Ohm's Law states that in an electrical circuit $E = IR$, where E represents the potential (in volts), I represents the current (in amperes), and R represents the resistance (in ohms).
a. Solve this formula for I. $I = \frac{E}{R}$
b. Find I if $E = 6$ volts and $R = 0.15$ ohms of resistance.
40 amperes

Solving Inequalities 4-5, 4-6, 4-7

You can add or subtract a number from both sides of an inequality to find a simpler **equivalent inequality.** You can multiply or divide both sides of an inequality by the same number to find a simpler equivalent inequality. If you multiply or divide by a positive number, the order of the inequality stays the same. If the number is negative, the order is *reversed.*

24. $h > -1$ 25. $k \leq -\frac{1}{2}$
26. $b < 40$ 27. $y \leq -168$
28. $c \leq -2$ 29. $m < -6$
30. $t < -5$ 31. $x \geq 9$

Solve each inequality. Graph the solutions on a number line. 24–31. See margin for graphs.

24. $h + 3 > 2$ **25.** $4k - 1 \leq -3$ **26.** $\frac{5}{8}b < 25$ **27.** $-\frac{2}{7}y - 6 \geq 42$

28. $6(c - 1) \leq -18$ **29.** $3m > 5m + 12$ **30.** $t - 4 < -9$ **31.** $5x - 2 \geq 4x + 7$

32. *Critical Thinking* Without writing the steps of the solution, describe the solutions of the inequality $-2x + \frac{3}{4} \leq -2x + \frac{1}{4}$. **Justify** your answer. There is no solution because $\frac{3}{4} \leq \frac{1}{4}$ is false.

ALTERNATIVE ASSESSMENT Group students into pairs. While partners look the other way, have one student from each pair choose a problem from the first review section and solve it on paper. With partners still looking away, have students explain the procedure while their partners try to follow the explanation and write down the exact procedure. Repeat this process until all review sections are complete.

Getting Ready for Chapter 5

Students may work these exercises independently or in small groups. The skills previewed will help prepare students to work with the Cartesian coordinate system and with function statements.

Compound Inequalities 4-8

Compound inequalities are two inequalities that are joined by the word *and* or the word *or*. A solution of a compound inequality joined by *and* makes *both* inequalities true. A solution of a compound inequality joined by *or* makes *either* inequality true.

Solve each inequality and graph the solutions. 33–40. See margin.

33. $|n + 2| \geq 4$ **34.** $-3 \leq z - 1 < 6$ **35.** $7t \geq 49$ or $2t \leq -4$ **36.** $0 < -8b \leq 12$

37. $-2 \leq 3a - 8 < 4$ **38.** $-6 < d + \frac{1}{2} < 4\frac{1}{2}$ **39.** $-1 \leq a - 3 < 2$ **40.** $|3x + 5| > -2$

41. Standardized Test Prep Which number is *not* a solution of the compound inequality $5w - 2 > 8$ or $-3w + 1 \geq 10$? C

A. –3 B. 5 C. 2 D. 3 E. –5

Interpreting Solutions 4-9

When you solve an inequality, the set of possible values for the variable is called the **replacement set.**

Graph each inequality on a number line. Use all integers as the replacement set. If there are no solutions, write *no solution*. 42–46. See margin for graphs.

42. $|4h - 1| \geq 7$ $h \leq -1.5$ and $h \geq 2$ **43.** $3t \leq 10$ $t \leq 3\frac{1}{3}$ **44.** $2 < y + 1 < 3$ no solution **45.** $-3 \leq m < 0$ **46.** $-2q > -q$ $q < 0$

Write and graph an inequality for each statement. 47–48. See margin for graphs.

47. Meteorology Cumulus clouds form $\frac{1}{4}$ mi to 4 mi above the Earth's surface. $\frac{1}{4} \leq c \leq 4$

48. Languages There are more than 1000 different languages spoken on the continent of Africa. $l > 1000$

49. Writing Suppose you use an inequality to model a real-world situation. Explain why you may need to specify a replacement set. Answers may vary. Sample: Real-world situations sometimes demand the solution to be an integer (number of people, amount of money) or a positive number (distance, speed). These conditions specify the replacement set.

Getting Ready for CHAPTER 5

Graph each set of points on a coordinate plane. 50–52. See margin.

50. (1, 2), (0, 1), (–2, –1), (4, 5) **51.** (4, –2), (2, 1), (–6, 3), (2, –1) **52.** (3, 1), (3, 4), (3, –2), (3, 3)

Write a function rule for each table.

53.

x	$f(x)$
0	–5
1	–4
2	–3
3	–2

$f(x) = x - 5$

54.

x	$f(x)$
–2	7
5	14
12	21
19	28

$f(x) = x + 9$

55.

x	$f(x)$
4	2
10	5
16	8
22	11

$f(x) = \frac{x}{2}$

56.

x	$f(x)$
–2	–6
1	3
4	12
7	21

$f(x) = 3x$

45. –4 –3 –2 –1 0 1

46. –3 –2 –1 0 1 2

47. 0 1 2 3 4 5

48. 0 200 400 600 800 1000 1200

50.

51.

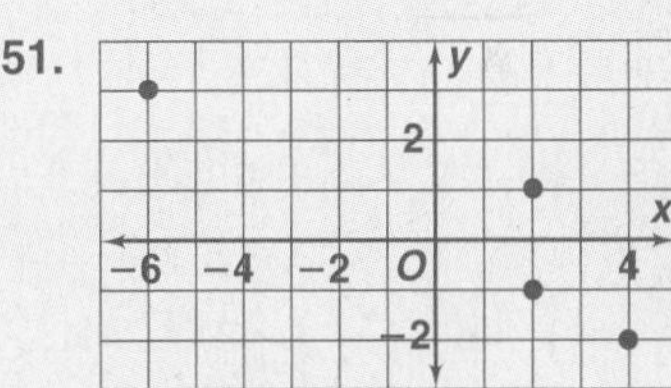

52. See back of book.

Assessment

ENHANCED MULTIPLE CHOICE QUESTIONS are more complex than traditional multiple choice questions, which assess only one skill. Enhanced multiple choice questions assess the processes that students use, as well as the end results. The questions are written so that students must use more than one strategy to solve the problem. Using multiple strategies is encouraged by the National Council of Teachers of Mathematics (NCTM). This Chapter Assessment does not contain an enhanced multiple choice question.

FREE RESPONSE QUESTIONS do not give answer choices. Some exercises have more than one possible answer. Students need to give only one correct response. **Exercises 1–12, 14–19, and 22–40** are free response questions.

WRITING EXERCISES allow students to describe how they think about and understand the concepts they have learned. **Exercise 13** is a writing exercise.

OPEN-ENDED PROBLEMS allow for more than one solution. Students must construct their own responses instead of choosing from possible answers. The student responses will help you determine the depth of their understanding and any possible areas of difficulty. **Exercise 21** is an open-ended problem. Make sure that the students' responses contain values for *x* and *y* and a written situation.

Resources

Teaching Resources

Chapter Support File, Ch. 4
- Chapter Assessment, Forms A and B
- Alternative Assessment

Chapter Assessment, Spanish Resources

Teacher's Edition

See also p.156E for assessment options.

Software • Computer Item Generator

page 210 Assessment

13. No; solutions of the equations are different. The first equation is true for *x*-values 3 and −11. The second equation is true for 3 and −3.

16. $m = 3t - v$

17. $c = \frac{8b - 2}{5}$

18. $b = \frac{2A}{h}$

19. $h = \frac{V}{2\pi r^2}$

28. $u > -4$

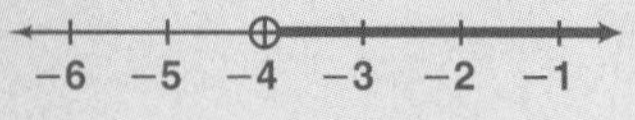

29. $t \le 2$

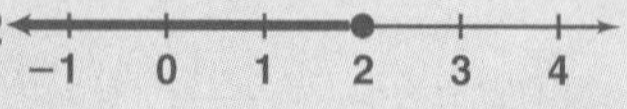

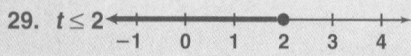

30. $w \ge 1$

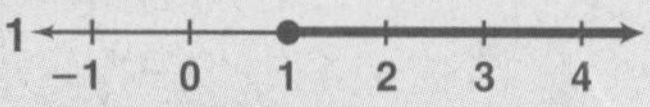

31–37, 39a–b. See back of book.

4 Assessment

Solve. If the equation is an identity or if it has no solution, write *identity* or *no solution*.

1. $\frac{3}{4} = \frac{c}{20}$ 15
2. $\frac{8}{15} = \frac{4}{w}$ 7.5
3. $\frac{w}{6} = \frac{6}{15}$ 2.4
4. $\frac{5}{t} = \frac{21}{100}$ $\frac{500}{21}$
5. $9j + 3 = 3(3j + 1)$
6. $8n = 5 + 3n$ 1
7. $4v - 9 = 6v + 7$ −8
8. $|8b - 3| = -21$
9. $|t - 6| = 4$ 2, 10
10. $|m + 1| = 11$ −12, 10

5. identity 8. no solution

11. Find the value of each variable in the matrices.
$$\begin{bmatrix} w+5 & 3x-1 \\ 2y & z+6 \end{bmatrix} = \begin{bmatrix} 2w-6 & 8x \\ 2-4y & 12 \end{bmatrix}$$
$w = 11$; $x = -\frac{1}{5}$; $y = \frac{1}{3}$; $z = 6$

12. The ratio of the length of a side of one square to that of another square is 3 : 4. A side of the smaller square is 9 cm. Find the length of a side of the larger square. 12 cm

13. **Writing** Are $|x + 4| = 7$ and $|x| + 4 = 7$ equivalent equations? Explain. See margin.

14. Suppose you score 9.1, 9.6, 9.7, 9.3, and 9.4 in a diving competition. The diver who is in first place has a final score of 56.4. What is the lowest score you can get on your last dive to win the competition? 9.4

15. A taxicab company charges a flat fee of $1.85 plus an additional $.40 per quarter-mile.
    a. Write a formula to find the total cost for each fare. $f = 0.40q + 1.85$
    b. Use this formula to find the total cost for traveling 8 mi. $14.65

Solve each equation for the given variable.

16. $t = \frac{v + m}{3}$; m
17. $8b - 5c = 2$; c
18. $A = \frac{1}{2}bh$; b
19. $V = 2\pi r^2 h$; h

16–19. See margin.

20. **Standardized Test Prep** If $6m + 3t = 8w$, then $m =$ ■. B
    A. $\frac{4}{3}w - 3t$
    B. $\frac{8w - 3t}{6}$
    C. $(8 - 3t) \div 6$
    D. $\frac{8w + 3t}{6}$
    E. $\frac{1}{2}t + 8w$

Samples: $-\frac{1}{6}w + 7 < 5$, $w > 12$; $-\frac{1}{6}x + 7 \ge 9$, $x \le -12$

21. **Open-ended** Write two different inequalities that you can solve by first subtracting 7, then multiplying by −6. Solve each inequality.

Define a variable and write an inequality to model each situation.

22. A student can take at most 7 classes.
    n = number of classes; $n \le 7$
23. The school needs at least 5 runners.
    r = number of runners; $n \ge 5$
24. Elephants can drink up to 40 gal at a time.
    w = amount of water; $w \le 40$
25. The paper route has more than 32 homes.
    h = number of homes; $h > 32$

Write a compound inequality that each graph could represent. $x < -1$ or $x > 1$

26. 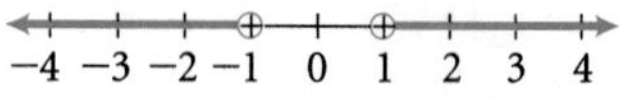

27. 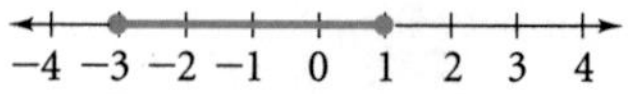

$-3 \le p \le 1$

Solve each inequality. Graph the solutions on a number line. 28–37. See margin.

28. $8 + u > 4$
29. $-5 + 4t \le 3$
30. $5w \ge -6w + 11$
31. $-\frac{7}{2}m < 13$
32. $|x - 5| \ge 10$
33. $|2h + 1| < 5$
34. $9 \le 6 - b < 12$
35. $-10 < 4q < 12$
36. $4 + 3n \ge 1$ or $-5n > 25$
37. $10k < 75$ and $4 - k \ge 0$

38. **Jobs** A clerk at the Radio Barn makes $300 a week plus 4% commission. How much does she have to sell to make at least $500 this week? at least $5000

39. Solve $-13 \le 5g + 7 < 20$ for each replacement set. See margin.
    a. positive integers
    b. all integers

40. Solve the equation modeled by tiles.

$3\frac{1}{3}$

Cumulative Review

Item	Review Topic	Chapter
1	Using proportions	4
2, 11	Modeling and solving equations	3
3, 9, 10	Solving absolute value equations	4
4	Solving one-step inequalities	4
5, 6, 13	Percent equations	3
7, 10	Using probability	3
8	Graphing inequalities	4
12	Matrices	1

4 Cumulative Review

For Exercises 1–9, choose the correct letter.

1. Which equation does *not* have the same solution as $\frac{7}{y} = \frac{31}{36}$? D

   A. $\frac{7}{31} = \frac{y}{36}$ **B.** $7 \cdot 36 = 31y$ **C.** $\frac{y}{36} = \frac{7}{31}$ **D.** $\frac{36}{31} = \frac{7}{y}$ **E.** $\frac{y}{7} = \frac{36}{31}$

2. Which of the following formulas correctly represent(s) the perimeter of the rectangle? D

   I. $p = x + x + y + y$
   II. $p = xy$
   III. $p = 2x + 2y$

   (rectangle with sides x and y)

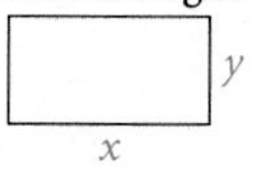

   A. I only **B.** II only **C.** II and III **D.** I and III **E.** I, II, and III

3. Solve $|a - 5| = 12$. C

   A. 18 **B.** -12 and 12 **C.** 17 and -7 **D.** -7 **E.** -17 and 17

4. Solve the inequality $4x + 2 < x - 5$. D

   A. $x \leq -\frac{1}{7}$ **B.** $x > \frac{7}{3}$ **C.** $x \geq -\frac{7}{3}$ **D.** $x < -\frac{7}{3}$ **E.** $x < 1$

5. Of 355 people surveyed, 62% agreed with the school committee's decision. About how many people did *not* agree with the committee? E

   A. 60 **B.** 140 **C.** 220 **D.** 300 **E.** None of the above

6. Students were asked to name their favorite type of motor vehicle. Seven preferred sport-utility vehicles, nine preferred sports cars, and five preferred luxury cars. What is the probability that a randomly selected student preferred a luxury car? B

   A. $\frac{1}{3}$ **B.** $\frac{5}{21}$ **C.** $\frac{5}{16}$ **D.** $\frac{3}{7}$ **E.** $\frac{2}{3}$

7. The number of subscribers to a magazine fell from 210,000 to 190,000. Find the approximate percent of decrease that this drop represents. B

   A. 5% **B.** 10% **C.** 20% **D.** 90% **E.** None of the above

8. Match the graph with its absolute value inequality. B

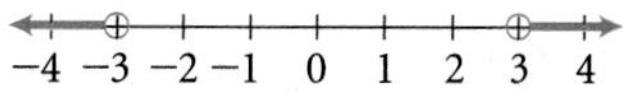

   (number line: −4 −3 −2 −1 0 1 2 3 4)

   A. $|s| \leq 3$ **B.** $|s| > 3$ **C.** $|s| \geq 3$ **D.** $|s| = 3$ **E.** $|s| < 3$

9. Compare the quantities in Column A and Column B for $x \neq 0$. B

Column A	Column B
$-\lvert x \rvert$	$\lvert -x \rvert$

   A. The quantity in Column A is greater.
   B. The quantity in Column B is greater.
   C. The two quantities are equal.
   D. The relationship cannot be determined from the information given.

For Exercises 10–13, write your answer.

10. A spinner numbered from 1 to 6 is spun. Each outcome is equally likely. Find the probability of getting an odd number. Then write this probability as a fraction, a decimal, and a percent. $\frac{1}{2}$, 0.5, 50%

11. Translate the following mathematical sentence into an equation and then solve it. Seventeen more than three times a number is 32. $3x + 17 = 32$ $x = 5$

12. *Open-ended* Use the numbers 5, –4, 9, 2, –1, and 3 to create two matrices of different sizes. See margin.

13. A CD player that normally costs \$225 would cost an employee \$180. What is the percent of the employee's discount? 20%

Resources

Teaching Resources

Chapter Support File, Ch. 4
- Cumulative Review
- Standardized Test Practice

Teacher's Edition

See also pp. 156E for assessment options.

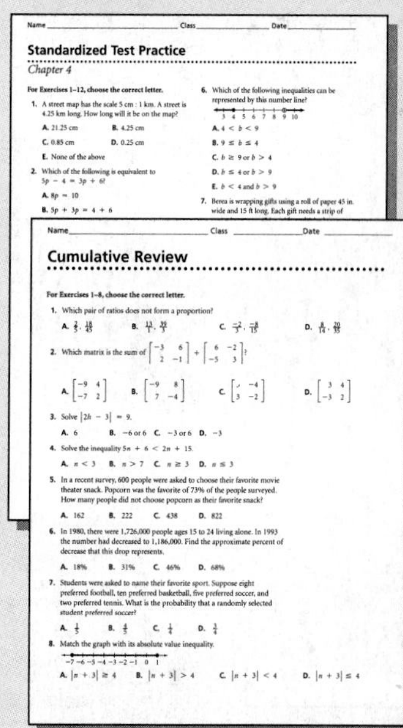

Cumulative Review

Name ______ Class ______ Date ______

Standardized Test Practice

Chapter 4

For Exercises 1–12, choose the correct letter.

1. A street map has the scale 5 cm : 1 km. A street is 4.25 km long. How long will it be on the map?

   A. 21.25 cm B. 4.25 cm C. 0.85 cm D. 0.25 cm E. None of the above

2. Which of the following is equivalent to $5p - 4 = 3p + 6$?

   A. $8p = 10$ B. $5p + 3p = 4 + 6$

6. Which of the following inequalities can be represented by this number line?

   (number line: 3 4 5 6 7 8 9 10)

   A. $4 < b < 9$ B. $9 \leq b \leq 4$ C. $b \geq 9$ or $b > 4$ D. $b \leq 4$ or $b > 9$ E. $b < 4$ and $b > 9$

7. Berea is wrapping gifts using a roll of paper 45 in. wide and 15 ft long. Each gift needs a strip of

Name ______ Class ______ Date ______

Cumulative Review

For Exercises 1–8, choose the correct letter.

1. Which pair of ratios does not form a proportion?

   A. $\frac{2}{5}, \frac{18}{45}$ B. $\frac{13}{1}, \frac{39}{3}$ C. $\frac{-2}{3}, \frac{-8}{15}$ D. $\frac{8}{14}, \frac{20}{35}$

2. Which matrix is the sum of $\begin{bmatrix} -3 & 6 \\ 2 & -1 \end{bmatrix} + \begin{bmatrix} 6 & -2 \\ -5 & 3 \end{bmatrix}$?

   A. $\begin{bmatrix} -9 & 4 \\ -7 & 2 \end{bmatrix}$ B. $\begin{bmatrix} -9 & 8 \\ 7 & -4 \end{bmatrix}$ C. $\begin{bmatrix} [illegible] & -4 \\ 3 & -2 \end{bmatrix}$ D. $\begin{bmatrix} 3 & 4 \\ -3 & 2 \end{bmatrix}$

3. Solve $|2h - 3| = 9$.

   A. 6 B. -6 or 6 C. -3 or 6 D. -3

4. Solve the inequality $5n + 6 < 2n + 15$.

   A. $n < 3$ B. $n > 7$ C. $n \geq 3$ D. $n \leq 3$

5. In a recent survey, 600 people were asked to choose their favorite movie theater snack. Popcorn was the favorite of 73% of the people surveyed. How many people did not choose popcorn as their favorite snack?

   A. 162 B. 222 C. 438 D. 822

6. In 1980, there were 1,726,000 people ages 15 to 24 living alone. In 1993 the number had decreased to 1,186,000. Find the approximate percent of decrease that this drop represents.

   A. 18% B. 31% C. 46% D. 68%

7. Students were asked to name their favorite sport. Suppose eight preferred football, ten preferred basketball, five preferred soccer, and two preferred tennis. What is the probability that a randomly selected student preferred soccer?

   A. $\frac{1}{5}$ B. $\frac{4}{5}$ C. $\frac{1}{4}$ D. $\frac{3}{4}$

8. Match the graph with its absolute value inequality.

   (number line: −7 −6 −5 −4 −3 −2 −1 0 1)

   A. $|n + 3| \geq 4$ B. $|n + 3| > 4$ C. $|n + 3| < 4$ D. $|n + 3| \leq 4$

page 211 Cumulative Review

12. $\begin{bmatrix} 5 & -4 \\ 9 & 2 \\ -1 & 3 \end{bmatrix}$, $\begin{bmatrix} 5 & -4 & 9 \\ 2 & -1 & 3 \end{bmatrix}$

CHAPTER 5 Graphing and Writing Linear Equations

To accommodate flexible scheduling, some lessons are divided into parts. Assignment Options are given in the Lesson Planning Options for each lesson.

5-1 Slope (pp. 214–218)
- Part 1 Counting Units to find Slope
- Part 2 Using Coordinates to Find Slope
- Part 3 Graphing Lines Given a Point and Its Slope

Key Terms: slope

5-2 Rate of Change (pp. 220–224)
- Part 1 Finding the Rate of Change
- Part 2 Using a Table
- Part 3 Linear Functions

Key Terms: rate of change

5-3 Direct Variation (pp. 225–229)
- Part 1 Direct Variation
- Part 2 Using the Constant of Variation to Write Equations
- Part 3 Using Proportions

Key Terms: constant of variation, direct variation

5-4 Slope-Intercept Form (pp. 230–234)
- Part 1 Defining the Slope-Intercept Form
- Part 2 Writing Equations

Key Terms: y-intercept, slope-intercept form

5-5 Writing the Equations of a Line (pp. 236–239)

5-6 Scatter Plots and Equations of Lines (pp. 241–245)
- Part 1 Trend Line
- Part 2 Line of Best Fit

Key Terms: line of best fit, correlation coefficient

5-7 $Ax + By = C$ Form (pp. 246–249)
- Part 1 Graphing Equations
- Part 2 Writing Equations

Key Terms: x-intercept, $Ax + By = C$ form

5-8 Parallel and Perpendicular Lines (pp. 250–255)
- Part 1 Slopes of Parallel Lines
- Part 2 Slopes of Perpendicular Lines

Key Terms: parallel lines, perpendicular lines

5-9 Using the *x*-intercept (pp. 256–259)

Key Terms: break-even point

PACING OPTIONS

This chart suggests pacing only for the core lessons and their parts, and it is provided merely as a possible guide. It will help you determine how much time you have in your schedule to cover other features, such as the Chapter Project, Math Toolboxes, Wrap Up, and Assessment.

	1 Class Period	1 Class Period	1 Class Period	1 Class Period	1 Class Period	1 Class Period	1 Class Period	1 Class Period	1 Class Period
Traditional (40–45 min class periods)	5-1 (1) 5-1 (2)	5-1 (3)	5-2 (1) 5-2 (2)	5-2 (3)	5-3 (1) 5-3 (2)	5-3 (3)	5-4 (1) 5-4 (2)	5-5	5-6 (1) 5-6 (2)
Two-Year Algebra (40–45 min class periods)	5-1 (1)	5-1 (2)	5-1 (3)	5-2 (1)	5-2 (2)	5-2 (3)	5-3 (1)	5-3 (2)	5-3 (3)
Block Scheduling (90 min class periods)	5-1 (1) 5-1 (2) 5-1 (3)	5-2 (1) 5-2 (2) 5-2 (3)	5-3 (1) 5-3 (2) 5-3 (3)	5-4 (1) 5-4 (2)	5-5 5-6 (1) 5-6 (2)	5-7 (1) 5-7 (2)	5-8 (1) 5-8 (2) 5-9		

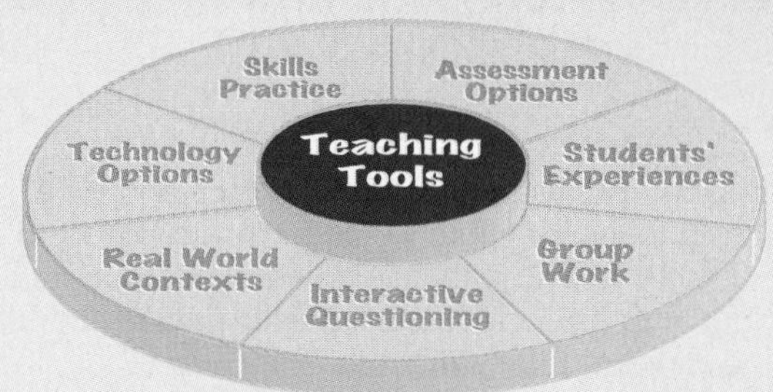

What Students Will Learn and Why

In this chapter, students will build on their knowledge of equations learned in Chapter 4, by learning to graph and write linear equations. They will learn to calculate rate of change, and they will learn to measure direct variation. They will learn the slope-intercept form and the $Ax + By = C$ form of a linear equation. Students will learn to write equations of lines and apply lines to scatter plots as well as use the x-intercept. Finally, students will learn about the relationship between the equations of parallel and perpendicular lines.

Discussing the Chapter/Building on Experience

The concept map below relates chapter topics to real-world applications. You and your class may wish to add to the map or develop maps of your own. The center oval describes the topic of the chapter. The next level displays topics within the lessons. The outer ovals reflect applications of the content. As you and your class build a concept map, invite students to discuss applications with which they are familiar.

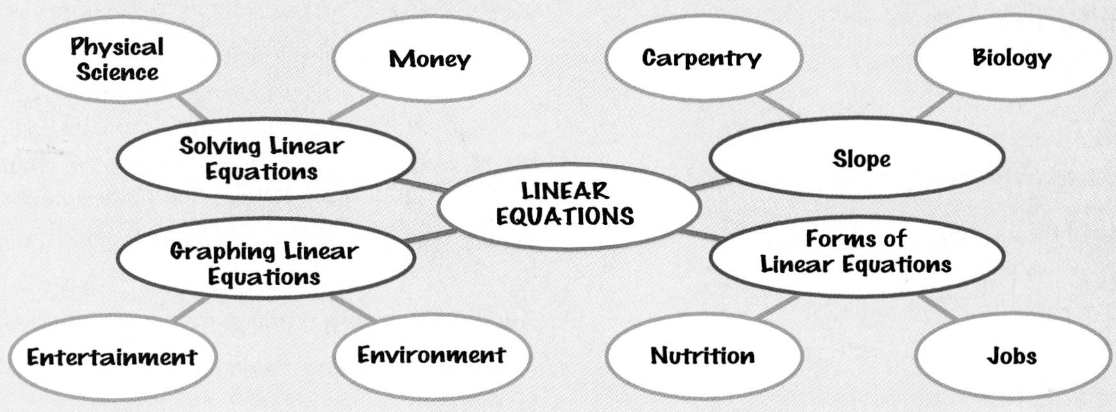

1 Class Period	1 Class Period	1 Class Period	1 Class Period	1 Class Period	1 Class Period	1 Class Period	1 Class Period	1 Class Period	1 Class Period	1 Class Period
5-7 ▽1 5-7 ▽2	5-8 ▽1 5-8 ▽2	5-9								
5-4 ▽1	5-4 ▽2	5-5	5-6 ▽1	5-6 ▽2	5-6 ▽2	5-7 ▽1	5-7 ▽2	5-8 ▽1	5-8 ▽2	5-9

Interactive Questioning Tips

A question is interactive when there is "give and take" between the questioner (teacher and student) and the respondent. One way to effectively use questioning as a teaching tool is to ask questions logically and sequentially. By avoiding random questions that lack clear focus, learning can be enhanced as students are led through difficult topics in small, sequential increments. For example, in the Skills Review on Dimensional Analysis, students first learn to convert hours to minutes. Then they are given a problem in which they must convert a distance and time into a speed and then convert the speed to different units.

Skills Practice

Every lesson provides skill practice with Exercises On Your Own and Exercises Mixed Review. The Student Edition includes Checkpoints (pp. 234, 255) and Preparing for Standardized Tests (p. 265). In the Teacher's Edition, the Lesson Planning Options section for each lesson lists Prerequisite Skills students should know for that lesson. At the back of the Student Edition is the Skills Handbook—mini-lessons on math the students may need to review. The Chapter Support File for Chapter 5 in the Teaching Resources box includes two Practice worksheets per lesson, a worksheet for two Checkpoints, and worksheets for Cumulative Review and Standardized Test Preparation.

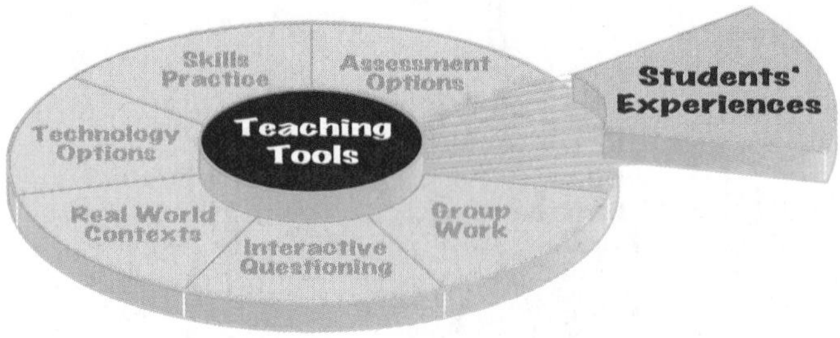

Diverse Learning and Teaching Styles

In your Teacher's Edition, you will find suggestions such as:

- **Visual learning** graphing data to see the shape of the graph (p. 222), discovering patterns of equations (p. 231)
- **Tactile learning** making coordinate planes by cutting yarn (p. 231)
- **Auditory learning** creating mnemonic devices (p. 231)
- **Kinesthetic learning** making a coordinate plane on the floor on which students arrange themselves to make graphs (p. 232)

Alternative Activity for Lesson 5–1

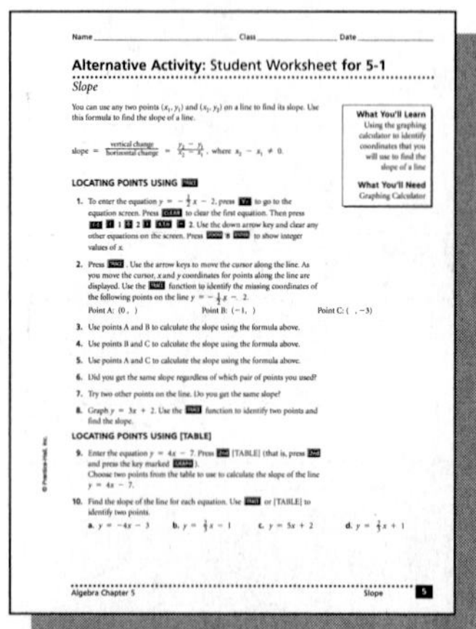

Name ______ Class ______ Date ______

Alternative Activity: Student Worksheet for 5-1

Slope

You can use any two points (x_1, y_1) and (x_2, y_2) on a line to find its slope. Use this formula to find the slope of a line.

$$\text{slope} = \frac{\text{vertical change}}{\text{horizontal change}} = \frac{y_2 - y_1}{x_2 - x_1}, \text{ where } x_2 - x_1 \neq 0.$$

What You'll Learn Using the graphing calculator to identify coordinates that you will use to find the slope of a line

What You'll Need Graphing Calculator

LOCATING POINTS USING [TRACE]

1. To enter the equation $y = -\frac{1}{2}x - 2$, press [Y=] to go to the equation screen. Press [CLEAR] to clear the first equation. Then press [(-)] [(] 1 [÷] 2 [)] [X,T,θ] [−] 2. Use the down arrow key and clear any other equations on the screen. Press [ZOOM] 8 [ENTER] to show integer values of x.
2. Press [TRACE]. Use the arrow keys to move the cursor along the line. As you move the cursor, x and y coordinates for points along the line are displayed. Use the [TRACE] function to identify the missing coordinates of the following points on the line $y = -\frac{1}{2}x - 2$.
   Point A: (0,) Point B: (−1,) Point C: (, −3)
3. Use points A and B to calculate the slope using the formula above.
4. Use points B and C to calculate the slope using the formula above.
5. Use points A and C to calculate the slope using the formula above.
6. Did you get the same slope regardless of which pair of points you used?
7. Try two other points on the line. Do you get the same slope?
8. Graph $y = 3x + 2$. Use the [TRACE] function to identify two points and find the slope.

LOCATING POINTS USING [TABLE]

9. Enter the equation $y = 4x - 7$. Press [2nd] [TABLE] (that is, press [2nd] and press the key marked [GRAPH]).
   Choose two points from the table to use to calculate the slope of the line $y = 4x - 7$.
10. Find the slope of the line for each equation. Use [TRACE] or [TABLE] to identify two points.
    a. $y = -4x - 3$ b. $y = \frac{2}{3}x - 1$ c. $y = 5x + 2$ d. $y = \frac{2}{5}x + 1$

Algebra Chapter 5 Slope 5

for use with Part 2, addresses visual learning by having students graph lines on a graphing calculator

Alternative Activity for Lesson 5–8

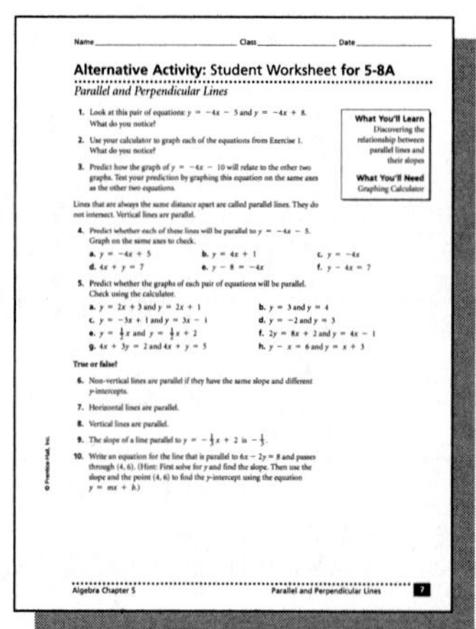

Name ______ Class ______ Date ______

Alternative Activity: Student Worksheet for 5-8A

Parallel and Perpendicular Lines

What You'll Learn Discovering the relationship between parallel lines and their slopes

What You'll Need Graphing Calculator

1. Look at this pair of equations: $y = -4x - 5$ and $y = -4x + 8$. What do you notice?
2. Use your calculator to graph each of the equations from Exercise 1. What do you notice?
3. Predict how the graph of $y = -4x - 10$ will relate to the other two graphs. Test your prediction by graphing this equation on the same axes as the other two equations.

Lines that are always the same distance apart are called parallel lines. They do not intersect. Vertical lines are parallel.

4. Predict whether each of these lines will be parallel to $y = -4x - 5$. Graph on the same axes to check.
   a. $y = -4x + 5$ b. $y = 4x + 1$ c. $y = -4x$
   d. $4x + y = 7$ e. $y - 8 = -4x$ f. $y - 4x = 7$
5. Predict whether the graphs of each pair of equations will be parallel. Check using the calculator.
   a. $y = 2x + 3$ and $y = 2x + 1$ b. $y = 3$ and $y = 4$
   c. $y = -3x + 1$ and $y = 3x - 1$ d. $y = -2$ and $y = 3$
   e. $y = \frac{1}{2}x$ and $y = \frac{1}{2}x + 2$ f. $2y = 8x + 2$ and $y = 4x - 1$
   g. $4x + 3y = 2$ and $4x + y = 5$ h. $y - x = 6$ and $y = x + 3$

True or false?

6. Non-vertical lines are parallel if they have the same slope and different y-intercepts.
7. Horizontal lines are parallel.
8. Vertical lines are parallel.
9. The slope of a line parallel to $y = -\frac{1}{3}x + 2$ is $-\frac{1}{3}$.
10. Write an equation for the line that is parallel to $6x - 2y = 8$ and passes through (4, 6). (Hint: First solve for y and find the slope. Then use the slope and the point (4, 6) to find the y-intercept using the equation $y = mx + b$.)

Algebra Chapter 5 Parallel and Perpendicular Lines 7

for use with Part 1, addresses visual learning by having students graph parallel lines

Alternative Activity for Lesson 5–8

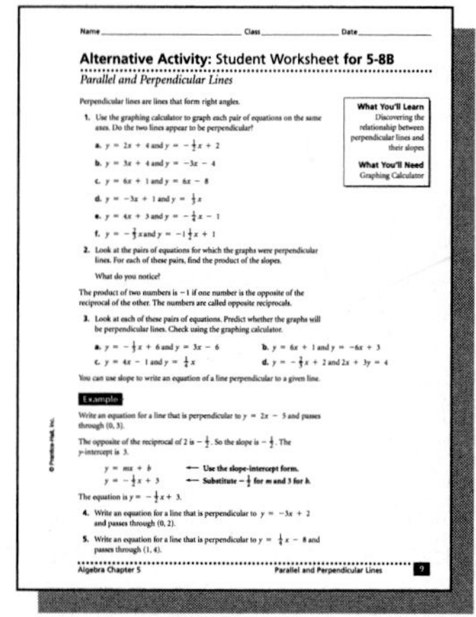

Name ______ Class ______ Date ______

Alternative Activity: Student Worksheet for 5-8B

Parallel and Perpendicular Lines

What You'll Learn Discovering the relationship between perpendicular lines and their slopes

What You'll Need Graphing Calculator

Perpendicular lines are lines that form right angles.

1. Use the graphing calculator to graph each pair of equations on the same axes. Do the two lines appear to be perpendicular?
   a. $y = 2x + 4$ and $y = -\frac{1}{2}x + 2$
   b. $y = 3x + 4$ and $y = -3x - 4$
   c. $y = 6x + 1$ and $y = 6x - 8$
   d. $y = -3x + 1$ and $y = \frac{1}{3}x$
   e. $y = 4x + 3$ and $y = -\frac{1}{4}x - 1$
   f. $y = -\frac{2}{3}x$ and $y = -1\frac{1}{2}x + 1$
2. Look at the pairs of equations for which the graphs were perpendicular lines. For each of these pairs, find the product of the slopes.
   What do you notice?

The product of two numbers is −1 if one number is the opposite of the reciprocal of the other. The numbers are called opposite reciprocals.

3. Look at each of these pairs of equations. Predict whether the graphs will be perpendicular lines. Check using the graphing calculator.
   a. $y = -\frac{1}{3}x + 6$ and $y = 3x - 6$ b. $y = 6x + 1$ and $y = -6x + 3$
   c. $y = 4x - 1$ and $y = \frac{1}{4}x$ d. $y = -\frac{2}{3}x + 2$ and $2x + 3y = 4$

You can use slope to write an equation of a line perpendicular to a given line.

Example

Write an equation for a line that is perpendicular to $y = 2x - 5$ and passes through (0, 3).

The opposite of the reciprocal of 2 is $-\frac{1}{2}$. So the slope is $-\frac{1}{2}$. The y-intercept is 3.

$y = mx + b$ ← **Use the slope-intercept form.**
$y = -\frac{1}{2}x + 3$ ← **Substitute $-\frac{1}{2}$ for m and 3 for b.**

The equation is $y = -\frac{1}{2}x + 3$.

4. Write an equation for a line that is perpendicular to $y = -3x + 2$ and passes through (0, 2).
5. Write an equation for a line that is perpendicular to $y = \frac{1}{4}x - 8$ and passes through (1, 4).

Algebra Chapter 5 Parallel and Perpendicular Lines 9

for use with Part 2, addresses visual learning by having students graph perpendicular lines

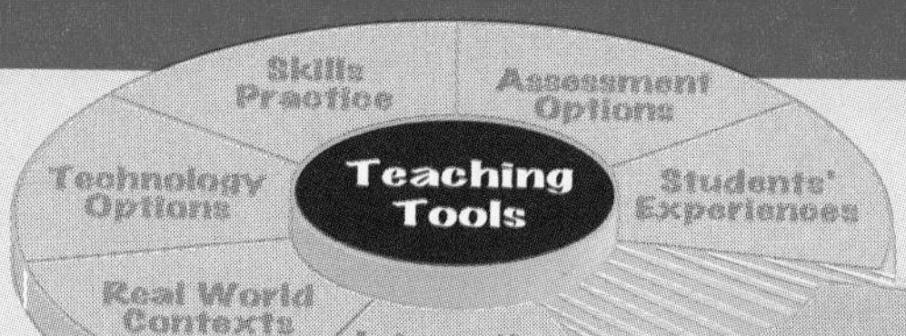

Cooperative Learning Tips

When used effectively, cooperative learning can help students develop interpersonal skills, learn to perform specific roles in a group, and learn to carry out specific responsibilities. The components of Chapter 5 provide a range of cooperative learning opportunities.

- In the Student Edition, the ***Work Together*** parts of lessons are specifically designed for cooperative learning activities.
- In the Teacher's Edition, you will find helpful hints for addressing diverse learning styles (see page C for Chapter 5). For every lesson, you will find a ***Reteaching Activity,*** which may involve cooperative learning.

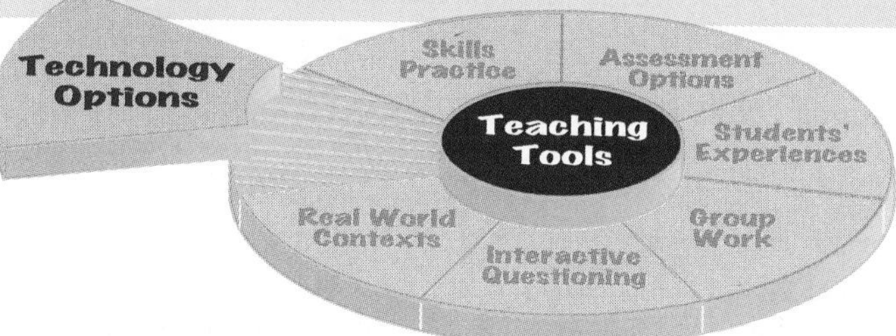

Materials and Manipulatives

The authors expect all students to use scientific calculators. Calculator use is integrated throughout the course.

- **clock or watch with second hand (5-6)**
- **graphing calculator (5-4, 5-6, 5-9)**
- **graph paper (5-1, 5-2, 5-6, 5-9)**

TECHNOLOGY OPTIONS

Technology Tools		Chapter Project	5-1	5-2	5-3	5-4	5-5	5-6	5-7	5-8	5-9
Calculator		Assumed that students have scientific calculators at any time.									
Graphing Calculator	Handbook					✔		✔	✔	✔	✔
	Student Edition					✔		✔	✔	✔	✔
Software	Secondary Math Lab Toolkit		✔	✔	✔	✔	✔	✔	✔	✔	✔
	Integrated Math Lab									✔	✔
	Computer Item Generator		✔	✔	✔	✔	✔	✔	✔	✔	✔
	Student Edition										
Video	Video Field Trip	✔									
CD-ROM	Multimedia Algebra Lab		✔		✔	✔			✔		
Internet		See the Prentice Hall site. (http://www.phschool.com)									

The Prentice Hall Algebra program offers you a rich variety of technology options. Be assured that all these options are provided as a means of enriching the program and are not essential for the successful completion of the course.

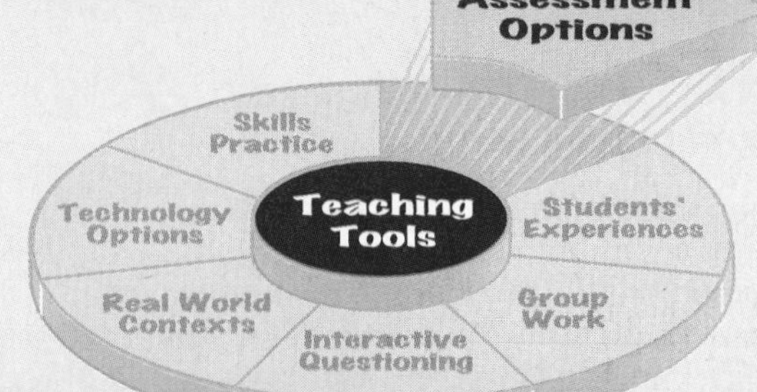

Assessment Options

The Prentice Hall Algebra Program provides you with many options. From these options, you may choose instructional materials and techniques appropriate for your students or those necessary to meet your district's curriculum requirements. As the chart indicates, the program also supports your teaching efforts by offering you many choices for assessment.

ASSESSMENT OPTIONS

Assessment Support Materials	Chapter Opener	5-1	5-2	5-3	5-4	5-5	5-6	5-7	5-8	5-9	Chapter End
Chapter Project	▲■●		▲■			▲■			▲■		▲■
Checkpoints					▲■				▲■		
Self-Assessment				▲■	▲■	▲■			▲■	▲■	▲■
Writing Assignment		▲	▲	▲	▲		▲	▲	▲	▲	●■
Chapter Assessment											▲●
Alternative Assessment		■	■	■	■	■	■	■	■	■	■●
Cumulative Review											●
Standardized Test Prep		▲■			▲■			▲■	▲■		▲●
Chapters 1-5 Assessment											●
Computer Item Generator	Can be used to create custom-made practice or assessment at any time.										

▲ = Student Edition ■ = Teacher's Edition ● = Teaching Resources

Checkpoints

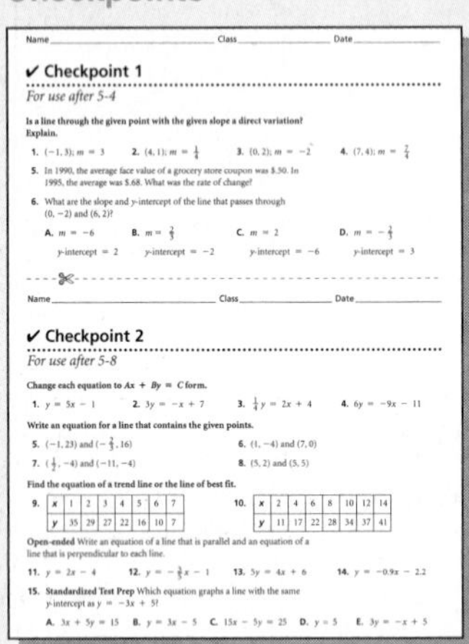
Checkpoint 1
For use after 5-4

Checkpoint 2
For use after 5-8

Alternative Assessment

Alternative Assessment
Chapter 5

Chapter Assessment

Chapter Assessment Form A
Chapter 5

Available in both Form A and Form B

Making the Right Connections

Mathematics is imbedded in nearly every walk of life. The National Council of Teachers of Mathematics (NCTM) encourages educators to recognize these connections and to emphasize them for the purpose of better educating students for success in life and in a global economy. The ***Connections*** chart below highlights these connections for Chapter 5.

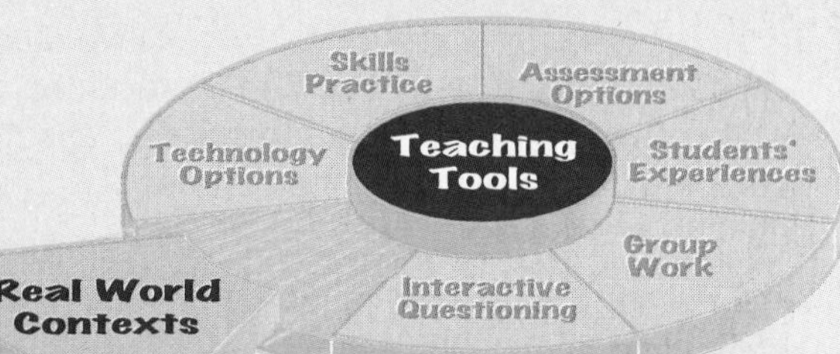

CONNECTIONS

Lesson	Interdiciplinary Connections	Career Prep	Other Real World Connections	Math Integration	NCTM Standards
Chapter Project	Economics	Jobs			Algebra Problem Solving
5-1		Carpentry	Fairs Environment		Algebra Communication Problem Solving
5-2	Science Biology	Rental Business	Parachuting Video Recording Agriculture		Algebra Communication Problem Solving
5-3	Physics Biology		Movies Weather Bicycling		Algebra Communication Problem Solving
5-4			Recreation Money		Algebra Communication Problem Solving
5-5	Physics	Business	Entertainment National Parks Real Estate		Algebra Communication Problem Solving
5-6		Business	Entertainment Technology Data Collection	Statistics	Algebra Communication Problem Solving Statistics
5-7			Fitness Fundraising Nutrition		Algebra Communication Problem Solving
5-8	Social Studies		Urban Planning Maps	Geometry	Algebra Communication Problem Solving Geometry from an Algebraic Perspective
5-9	Physical Science Zoology	Business	Money		Algebra Communication Problem Solving

CHAPTER 5 Graphing and Writing Linear Equations

CONNECTING TO PRIOR LEARNING Have students think of a situation or example that can be represented by a relationship where changing one of the components affects the other component. For example, the number of hours worked and amount earned.

CULTURAL CONNECTIONS Many teenagers expect to work some time during high school. Ask students who work how they spend the money they earn. Some students may be expected to contribute to the family budget, others may be saving for a car, and still others may not have any restrictions on how they spend their earnings. Ask students if there are cultural restrictions on the types of jobs some teenagers are allowed to do?

INTERDISCIPLINARY CONNECTIONS As students enter the work force, they are going to be faced with concepts such as minimum wage, social security, income tax, overtime, benefits and so on. Ask students to explain what they understand about these deductions.

ABOUT THE PROJECT This project will have students make predictions and comparisons about their first jobs. They will be asked to write, model, solve, and graph equations. Students will also interview adults about their job experiences as teenagers.

Technology Options

Prentice Hall Technology

Video Video Field Trip 5 "Rungs of the Ladder," how a carpenter makes his living.

CHAPTER 5 Graphing and Writing Linear Equations

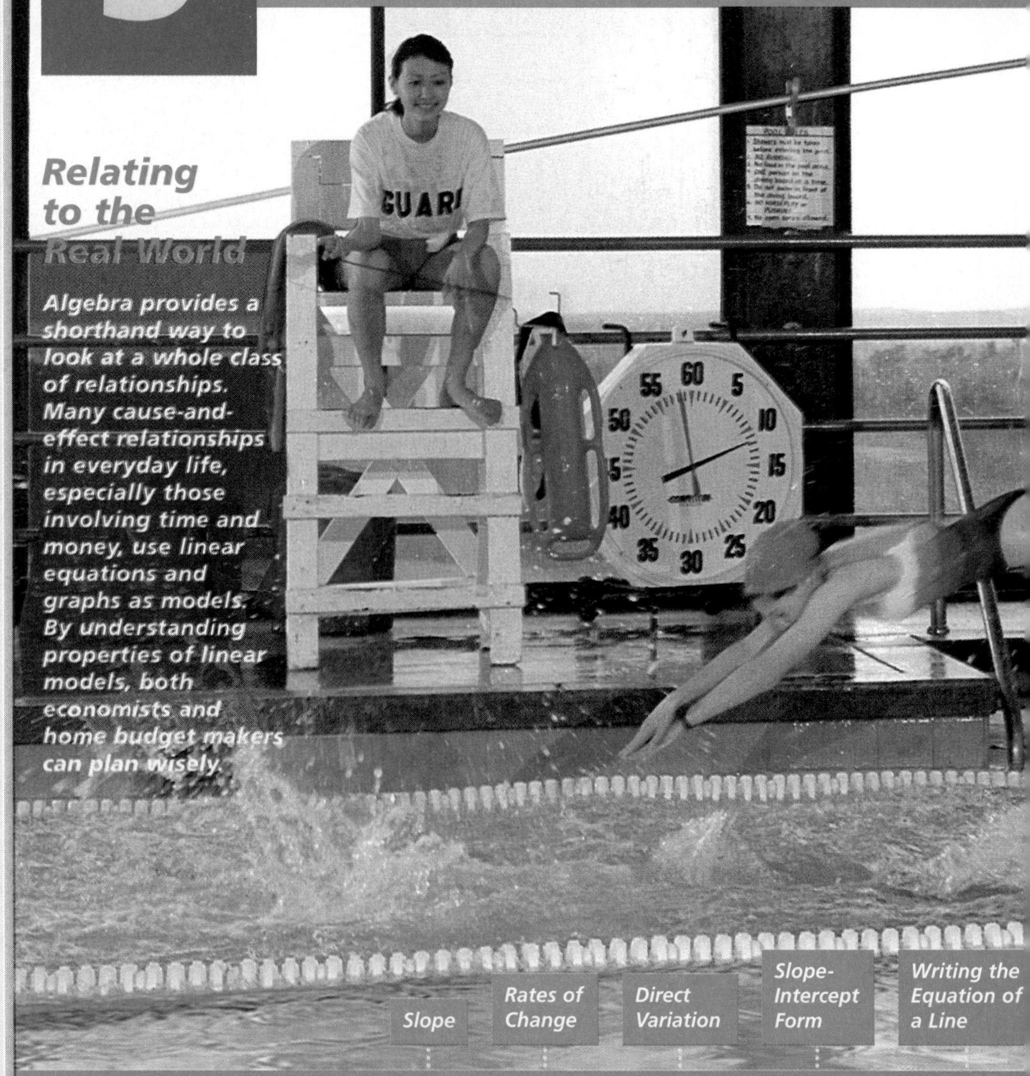

Relating to the Real World

Algebra provides a shorthand way to look at a whole class of relationships. Many cause-and-effect relationships in everyday life, especially those involving time and money, use linear equations and graphs as models. By understanding properties of linear models, both economists and home budget makers can plan wisely.

Launching the Project

PROJECT NOTEBOOK Encourage students to keep all project-related materials in a separate folder or notebook. **See Chapter Project and Scoring Rubric in Chapter Support File.**

- Have students brainstorm about jobs they might like as teenagers.
- From their brainstorming, have students select two jobs that interest them.
- Students will research these jobs and then construct a graph that shows income for each job.
- Elicit from the students possible scales to be used for the graphs.
- After they complete their graphs, have students compare the difference in income for 8 hours worked.

TRACKING THE PROJECT Have students read Finishing the Chapter Project on page 260 to help them get an overview of the project. Set benchmark deadlines for students to show you their work throughout the course of the chapter.

CHAPTER PROJECT

Taking THE PLUNGE

Do you have a job? If not, what will your first job be? What expenses will you have? How much money will you actually earn? How can you compare earnings between two jobs? Linear equations can help to answer all these questions.

As you work through the chapter, you will make graphs and write equations that model different jobs. You will use these models to predict income. After interviewing someone about their first job, you will choose a job that you might like to have and explain why.

To help you complete the project:

- **p. 224** *Find Out by Graphing*
- **p. 239** *Find Out by Modeling*
- **p. 255** *Find Out by Interviewing*
- **p. 260** *Finishing the Project*

Project Resources

Teaching Resources
Chapter Support File, Ch. 5
- Chapter Project Manager and Scoring Rubric

Transparencies
66

Using the Rubric

Sharing the scoring rubric for the project with your students will alert them to your expectations before they begin work on the project.

As students complete each Find Out question in the chapter, you may wish to have them evaluate their own work or a partner's work, based on the scoring rubric. Students should have the opportunity to revise their work after it has been reviewed.

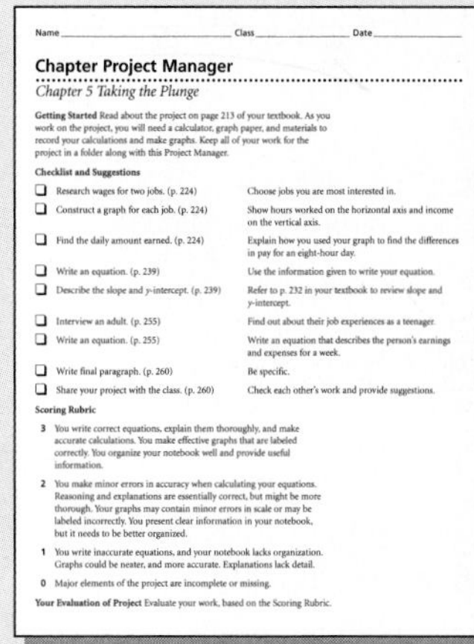

Name ______ Class ______ Date ______

Chapter Project Manager

Chapter 5 Taking the Plunge

Getting Started Read about the project on page 213 of your textbook. As you work on the project, you will need a calculator, graph paper, and materials to record your calculations and make graphs. Keep all of your work for the project in a folder along with this Project Manager.

Checklist and Suggestions

❑ Research wages for two jobs. (p. 224)	Choose jobs you are most interested in.
❑ Construct a graph for each job. (p. 224)	Show hours worked on the horizontal axis and income on the vertical axis.
❑ Find the daily amount earned. (p. 224)	Explain how you used your graph to find the differences in pay for an eight-hour day.
❑ Write an equation. (p. 239)	Use the information given to write your equation.
❑ Describe the slope and y-intercept. (p. 239)	Refer to p. 232 in your textbook to review slope and y-intercept.
❑ Interview an adult. (p. 255)	Find out about their job experiences as a teenager.
❑ Write an equation. (p. 255)	Write an equation that describes the person's earnings and expenses for a week.
❑ Write final paragraph. (p. 260)	Be specific.
❑ Share your project with the class. (p. 260)	Check each other's work and provide suggestions.

Scoring Rubric

3 You write correct equations, explain them thoroughly, and make accurate calculations. You make effective graphs that are labeled correctly. You organize your notebook well and provide useful information.

2 You make minor errors in accuracy when calculating your equations. Reasoning and explanations are essentially correct, but might be more thorough. Your graphs may contain minor errors in scale or may be labeled incorrectly. You present clear information in your notebook, but it needs to be better organized.

1 You write inaccurate equations, and your notebook lacks organization. Graphs could be neater, and more accurate. Explanations lack detail.

0 Major elements of the project are incomplete or missing.

Your Evaluation of Project Evaluate your work, based on the Scoring Rubric.

5-1 Teaching Notes

PROBLEM OF THE DAY

There are two natural numbers between zero and 100 that are both perfect squares and perfect cubes. What are they? 1 and 64

Problem of the Day is also available in Transparencies.

CONNECTING TO PRIOR KNOWLEDGE Ask students to explain what they think the word *slope* means. List some of their ideas on the board. Have students brainstorm a list of things that vary in slope. Suggestions might include roads, walkways, ramps, roofs, stairs, or slides. Ask students to describe ways in which the steepness of a slope might be measured.

WORK TOGETHER

Draw two slopes that have the same rise but different runs on the board. Lead students to understand that slope depends not just on the rise but on the ratio of rise and run.

MAKING CONNECTIONS The steepest street in the world is Baldwin Street in Dunedin, New Zealand, with a slope of 1.266. Invite students to model this slope using string or meter sticks.

THINK AND DISCUSS

Review with students what the words *vertical* and *horizontal* mean. Ask students to give definitions and examples.

Lesson Planning Options

Prerequisite Skills

- Graphing ordered pairs

Assignment Options

To provide flexible scheduling, this lesson can be subdivided into parts.

1. **Core** 1, 21–23
   Extension 2, 24, 31
2. **Core** 3–8, 25–27
   Extension 28–30, 32, 33
3. **Core** 9–16
   Extension 17–20

Use Mixed Review to maintain skills.

Resources

Student Edition

Skills Handbook, p. 579
Extra Practice, p. 560
Glossary/Study Guide

Teaching Resources

Chapter Support File, Ch. 5
- Practice 5-1 (two worksheets)
- Reteaching 5-1
- Alternative Activity 5-1

Classroom Manager 5-1
Glossary, Spanish Resources
Two-Year Algebra Handbook 5-1

Transparencies
2, 26, 67, 72, 73

Connections  Airplanes . . . and more

5-1 Slope

What You'll Learn
- Calculating the slope of a line
- Drawing a line through a point with a given slope

...And Why

To find slope in real-world situations, such as the landing of an airplane, and interpret its meaning

What You'll Need
- graph paper

WORK TOGETHER

Carpentry Carpenters use the terms rise and run to describe the steepness of a stairway or a roof line. You can use rise and run to describe the steepness of a hill.

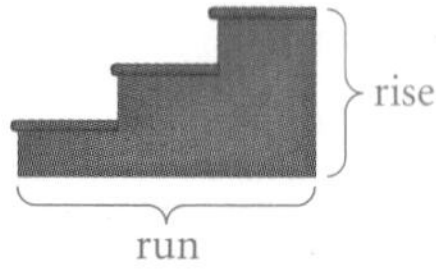

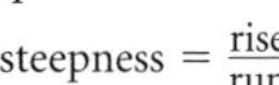

$$\text{steepness} = \frac{\text{rise}}{\text{run}}$$

1. Which hill appears to be steeper in the photos? A

A.

B.

2. a. Find a ratio for the steepness of each hill. A: $\frac{1}{3}$; B: $\frac{1}{5}$
   b. For which hill is the ratio greater? A

THINK AND DISCUSS

Part 1 **Counting Units to Find Slope**

The mathematical term to describe steepness is slope.

$$\textbf{slope} = \frac{\text{vertical change (rise)}}{\text{hortizontal change (run)}}$$

Example 1

Find the slope of each line.

a.

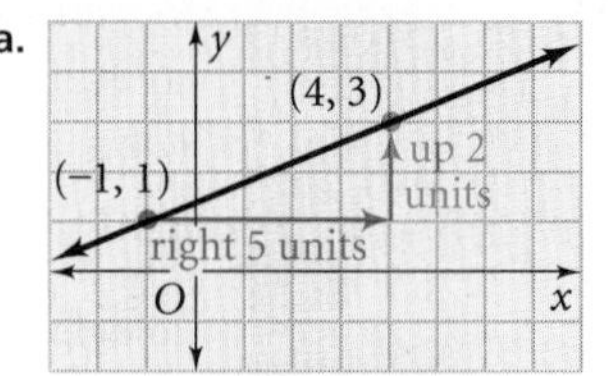

$$\text{slope} = \frac{\text{vertical change}}{\text{horizontal change}}$$
$$= \frac{2}{5}$$

The slope of the line is $\frac{2}{5}$.

b.

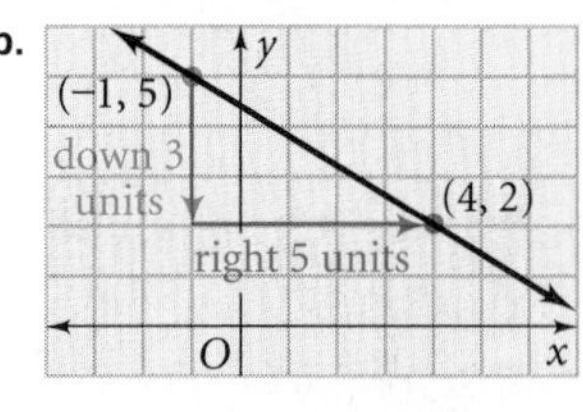

$$\text{slope} = \frac{\text{vertical change}}{\text{horizontal change}}$$
$$= -\frac{3}{5}$$

The slope of the line is $-\frac{3}{5}$.

Example 1

Point out that by convention we "read" slopes from left to right. Suggest that students remind themselves of this fact by remembering that we read text from left to right. When a line rises from left to right, its slope is positive. When a line falls from left to right, its slope is negative. Invite students to invent mnemonic devices that they could use to remember these facts.

Example 2 Relating to the Real World 

KINESTHETIC LEARNING Extend this example by having students make and fly paper airplanes, and measure the time it takes them to reach the ground. Divide students into groups of four. Have one group member time the flights while the other group members launch their planes from the same height. Then have students graph their data in the form of sloping lines as in Example 2. If time allows, students can transfer their graphs to a poster to compare the flight times of differently designed airplanes.

CRITICAL THINKING Question 4 Because students may expect slopes to be written as fractions, they may not recognize that a line can have a slope of 12. Point out that since 12 can be written as $\frac{12}{1}$, a line with slope 12 is the same as a line with slope $\frac{12}{1}$.

CRITICAL THINKING Question 5 To remind students that a number cannot be divided by zero, have them try the operation on a calculator. They will see that the calculator gives an error message.

---

3. Complete each statement with *upward* or *downward*.

a. A line with positive slope goes ■ from left to right. upward

b. A line with negative slope goes ■ from left to right. downward

You use the units associated with the axes to explain the meaning of the slope in a real-world situation.

Example 2 Relating to the Real World 

Airplanes The graph models the altitude of an airplane from the time the wheels are lowered (time = 0 s) to when the plane lands. Find the slope of the line. Explain what the slope means in this situation.

Find any two points on the graph. Use the points to find the slope.

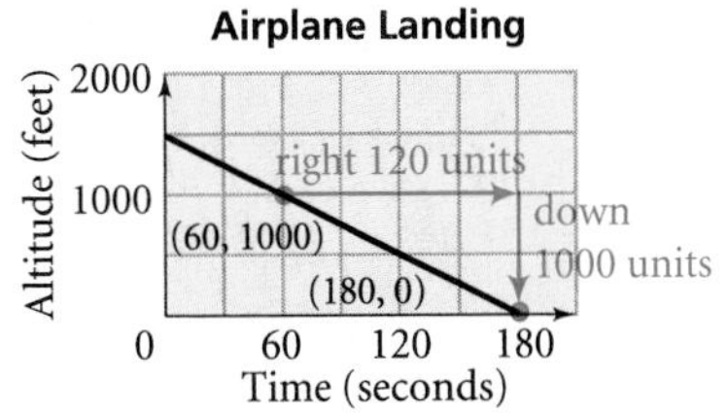

$$\text{slope} = \frac{\text{vertical change}}{\text{horizontal change}} = \frac{-1000}{120} = -8\tfrac{1}{3}$$

The slope of the line is $-8\frac{1}{3}$. The plane is descending $8\frac{1}{3}$ ft/s. ■

4. Critical Thinking Suppose a graph of a line with slope 12 indicates the relationship between altitude and time for another airplane. What would the slope mean in this situation? The airplane ascends at 12 ft/s.

Part 2 Using Coordinates to Find Slope

You read the coordinates (x_1, y_1) as "x sub 1, y sub 1."

You can use any two points on a line to find its slope. To find the slope of a line PQ (written $\overleftrightarrow{PQ}$), you can use this formula:

$$\text{slope} = \frac{\text{vertical change}}{\text{horizontal change}} = \frac{y_2 - y_1}{x_2 - x_1}, \text{ where } x_2 - x_1 \neq 0$$

5. Critical Thinking Why does the formula for the slope include the statement "where $x_2 - x_1 \neq 0$"? You cannot divide by 0.

Additional Examples

FOR EXAMPLE 2

The graph shows how the position of a motorcycle changes as the time passes.

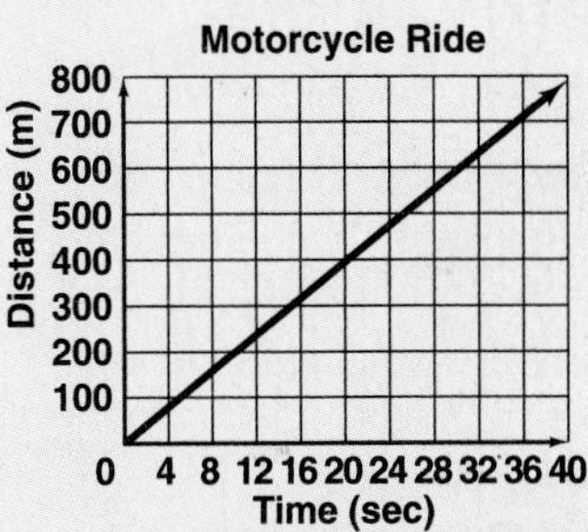

Find the slope of the line. What does the slope mean in this situation? 20, the motorcycle's velocity in m/s

Discussion: *What other physical properties can the slope of the line represent?*

FOR EXAMPLE 3

Find the slope of the line through $E(3, -2)$ and $F(-2, -1)$.

The slope of $\overleftrightarrow{EF}$ is $-\frac{1}{5}$.

Discussion: *How can you use the point-slope expression $\frac{y_2 - y_1}{x_2 - x_1}$ to prove that a number divided by zero is undefined?*

Example 3

ERROR ALERT! Students may think they must substitute the points in a specific order. **Remediation:** Encourage students to do the example again, but substitute (5, 7) for (x_1, y_1) and (−2, 1) for (x_2, y_2). Lead students to understand that either pair of coordinates may be (x_2, y_2).

Example 4

Invite students to create mnemonic devices to help them remember that a horizontal line has a slope of zero and a vertical line has an undefined slope. For example, a bicyclist on a flat road would have no (zero) problem, but a bicyclist riding into a vertical wall might lose consciousness (become undefined).

Example 5

ALTERNATIVE METHOD Invite students to substitute the known values into the slope formula as an alternative way of finding another point on the line.

Exercises ON YOUR OWN

OPEN-ENDED Exercise 2 Encourage students to draw conclusions about the range of values that a slope can have in each different colored region.

TACTILE LEARNING Exercises 3-8 Divide students into pairs. Have one student use the slope formula to find each slope while the other student finds the slope by plotting the points on graph paper. Then have students compare their answers.

Technology Options

For Exercises 3–8, students may use graphing software or a graphing calculator to find slope by graphing the lines.

Prentice Hall Technology 

 **Software** • Secondary Math Lab Toolkit • Computer Item Generator 5-1

CD-ROM Multimedia Algebra Lab 5

 **Internet** • See the Prentice Hall site. (http://www.phschool.com)

Example 3

Find the slope of a line through $A(-2, 1)$ and $B(5, 7)$.

$$\text{slope} = \frac{y_2 - y_1}{x_2 - x_1}$$
$$= \frac{7 - 1}{5 - (-2)} \quad \leftarrow \text{Substitute (5, 7) for } (x_2, y_2) \text{ and } (-2, 1) \text{ for } (x_1, y_1).$$
$$= \frac{6}{7}$$

The slope of $\overleftrightarrow{AB}$ is $\frac{6}{7}$.

6. **Try This** Find the slope of the line through $C(4, 0)$ and $D(-1, 5)$. −1

Example 4

Find the slope of each line using the points shown.

a. 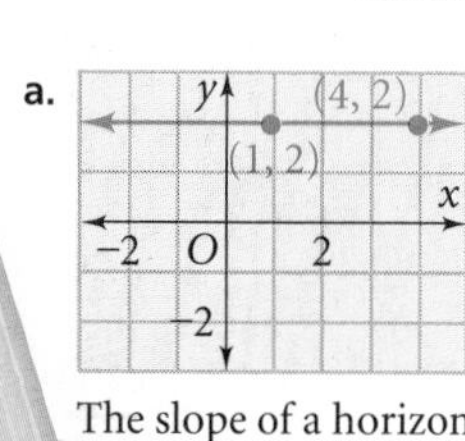

$$\text{slope} = \frac{y_2 - y_1}{x_2 - x_1} = \frac{2 - 2}{4 - 1} = \frac{0}{3} = 0$$

The slope of a horizontal line is 0.

b. 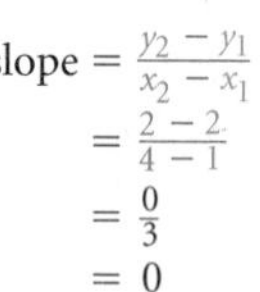

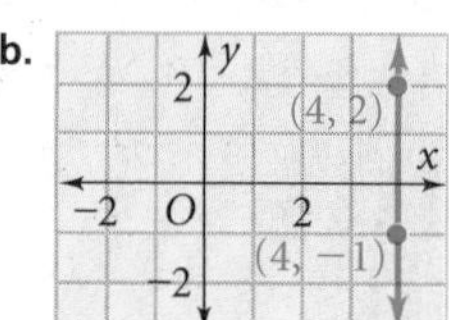

$$\text{slope} = \frac{y_2 - y_1}{x_2 - x_1} = \frac{2 - (-1)}{4 - 4} = \frac{3}{0},$$ undefined

The slope of a vertical line is undefined.

7. Complete this statement for (a) a horizontal line and (b) a vertical line. The ■-values change while the ■-values stay the same.
a. x; y b. y; x

Part 3 **Graphing a Line Given Its Slope and a Point**

Example 5

Draw a line through the point (1, 2) with the slope $-\frac{3}{2}$.

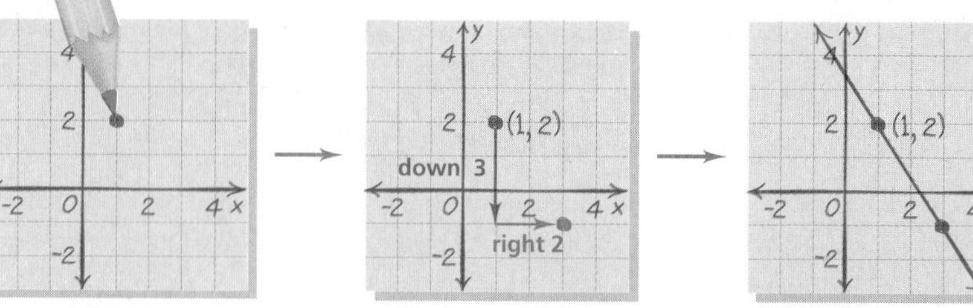

Plot (1, 2). → Find another point using the slope $-\frac{3}{2}$. → Draw a line through the points.

8. What are the coordinates of the second point in Example 5? (3, −1)

9. **Try This** Graph the line through the point (2, −3) with slope $\frac{5}{4}$. See margin.

ALTERNATIVE ASSESSMENT Exercises 9, 12, and 16 Ask students to graph these lines using the negative reciprocal of the given slope (-2, $-\frac{4}{3}$, $\frac{3}{5}$). This will help you assess the students' understanding of slope and help prepare them for Lesson 5-8.

DIVERSITY Exercises 21–24 Have teams of students measure the slopes of staircases and ramps at school or other public buildings. After students have shared their measurements with other teams, lead a discussion of how the slope of a staircase or ramp affects ease of use and the likelihood of falls. Discuss how the physical needs of people influence the steepness of stairs and ramps.

GEOMETRY Exercises 25–27 Encourage students to estimate each slope before finding the exact value.

Exercises 28–30 Make sure students understand that all parts of a line have the same slope. Lead students to understand that there is only one possible line with a specific slope that can pass through a specific point.

WRITING Exercise 32 Encourage students to describe in writing how they would help the friend correct the error.

STANDARDIZED TEST TIP Exercise 33 Have students create a sketch of figure-based problems such as these. With sketches, students can visualize the problem and check their answers.

Slope of Lines

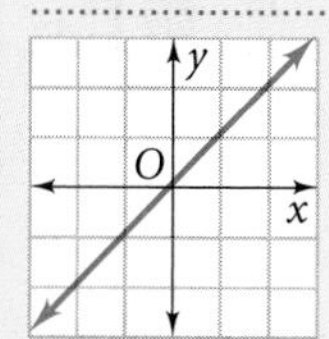

A line with positive slope goes upward from left to right.

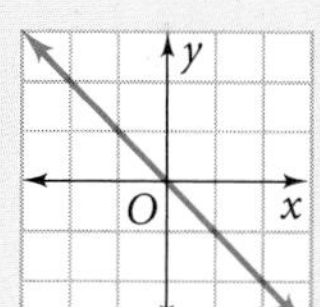

A line with negative slope goes downward from left to right.

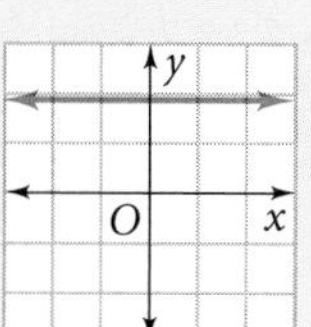

The slope of a horizontal line is 0.

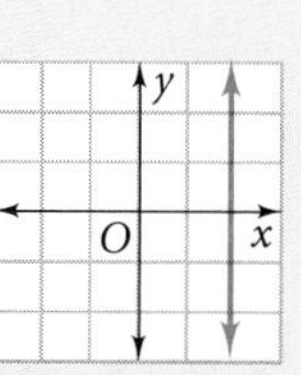

The slope of a vertical line is undefined.

Exercises ON YOUR OWN

Use the diagram at the right for Exercises 1 and 2.

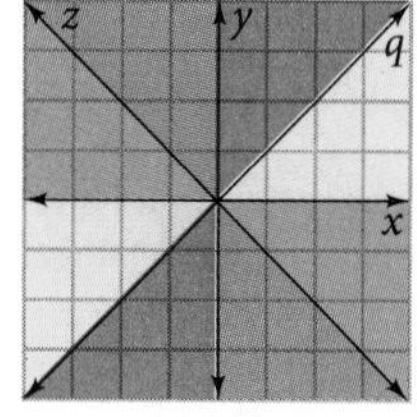

1. What is the slope of line q? of line z? 1; –1
2. *Open-ended* What is the slope of a line through the origin that is in the red region? the yellow region? the green region? the blue region? Answers may vary. Samples: 2; $\frac{1}{2}$; $-\frac{1}{2}$; -2

Find the slope of the line passing through each pair of points.

3. $(-8, 0), (1, 5)$ $\frac{5}{9}$
4. $(8, 3), (-4, 3)$ 0
5. $(-4, -5), (-9, 1)$ $-\frac{6}{5}$
6. $(\frac{1}{2}, 8), (1, -2)$ -20
7. $(4, -1), (4, 7)$ undefined
8. $(9, -2), (3, 4)$ -1

Through the given point, graph a line with the given slope. 9–16. See margin.

9. $(3, 4)$; slope $= \frac{1}{2}$
10. $(-2, 1)$; slope $= -2$
11. $(0, 3)$; slope $= 0$
12. $(-5, -2)$; slope $= \frac{3}{4}$
13. $(1, -3)$; slope $= -\frac{2}{3}$
14. $(2, 5)$; slope $= -\frac{4}{3}$
15. $(-1, 5)$; undefined slope
16. $(-2, 3)$; slope $= -\frac{5}{3}$

Critical Thinking Tell whether each statement is *true* or *false*. Explain.

17. All horizontal lines have the same slope. T; all horizontal lines have zero slope.
18. A line with slope 1 passes through the origin. F; for example, a line with slope 1 that passes through (0, 1) does not pass through the origin.
19. Two lines may have the same slope. T; two lines can have the same slope but will pass through different points.
20. The slope of a line in Quadrant III must be negative. F; a line through any point in Q III can have positive, negative, zero, or undefined slope.

Carpentry Tell whether the slope would *increase, decrease,* or *remain the same.*

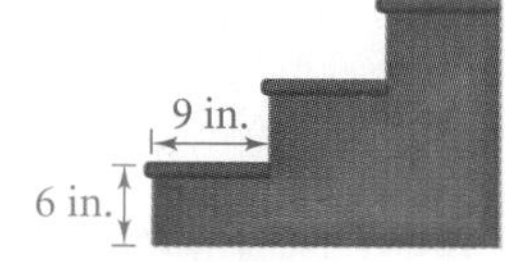

21. The rise of each step *increases* 1 in. increase
22. The run of each step *decreases* 1 in. increase
23. The rise and run both *increase* 1 in. increase
24. *Critical Thinking* How could the rise and run of each step change without changing the slope of the stairway? If you multiply the rise and run by the same number, the slope does not change.

pages 214–217 Think and Discuss

9.

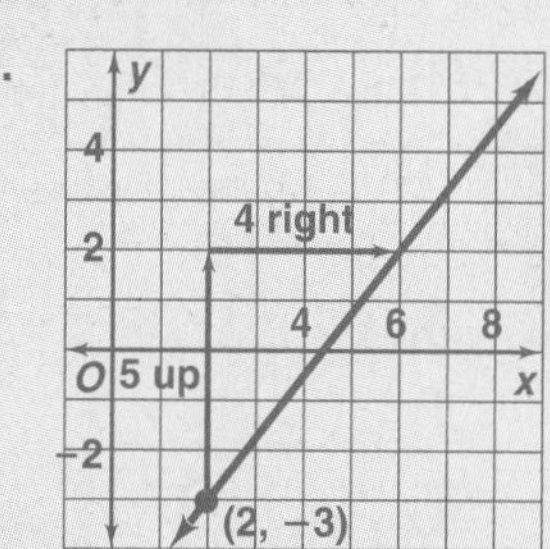

pages 217–218 On Your Own

9.

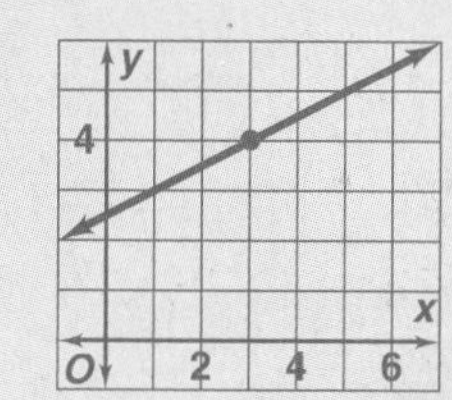

10.

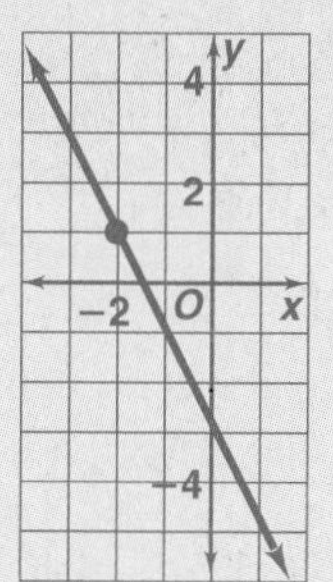

11–16. See back of book.

Exercises MIXED REVIEW

Exercises 34–38 Suggest that students quietly read aloud each probability to themselves. For example, for Exercise 34, say "the probability that I roll a one."

GETTING READY FOR LESSON 5-2 These exercises prepare students for finding the rate of change between independent and dependent variables.

Wrap Up

THE BIG IDEA Ask: *What is a slope? How do you describe the steepness of a slope?*

RETEACHING ACTIVITY Students use graph paper in order to calculate the slope of a line. (Reteaching worksheet 5-1)

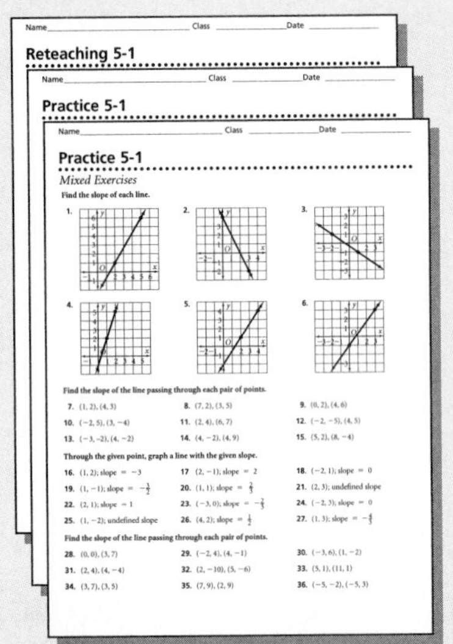
Reteaching 5-1

Practice 5-1

Practice 5-1
Mixed Exercises

Lesson Quiz

Lesson Quiz is also available in Transparencies.

1. What is the slope of line r?

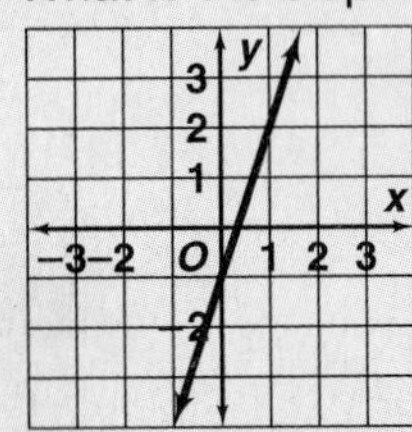

slope $= 3$

2. Find the slope of the line passing through $(-2, -3)$ and $(3, -\frac{1}{2})$. $\frac{1}{2}$

3. Graph a line with a slope of $-\frac{1}{2}$ which passes through the point $(2, 1)$. See back of book.

Geometry **Find the slope of the sides of each figure.**

25.

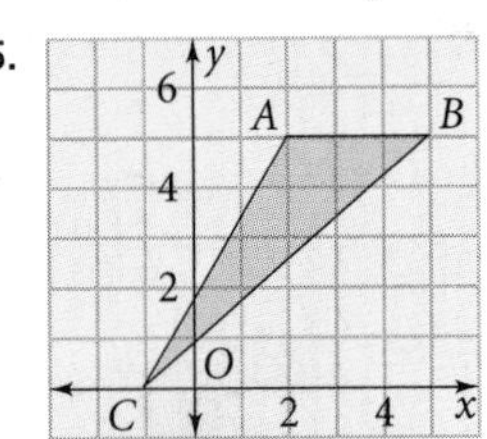

$\overleftrightarrow{AB}$: 0; $\overleftrightarrow{BC}$: $\frac{5}{6}$; $\overleftrightarrow{AC}$: $\frac{5}{3}$

26. $\overleftrightarrow{PQ}$: $\frac{1}{2}$; $\overleftrightarrow{QR}$: 0; $\overleftrightarrow{RS}$: 5; $\overleftrightarrow{PS}$: −1

27. $\overleftrightarrow{JK}$: $-\frac{1}{2}$; $\overleftrightarrow{KL}$: 2; $\overleftrightarrow{LM}$: $-\frac{1}{2}$; $\overleftrightarrow{JM}$: 2

Do the points lie on the same line?

Sample $A(1, 3), B(4, 2), C(-2, 4)$

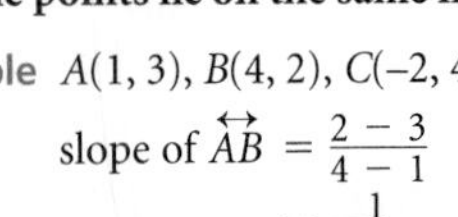

$$\text{slope of } \overleftrightarrow{AB} = \frac{2-3}{4-1} = -\frac{1}{3} \qquad \text{slope of } \overleftrightarrow{BC} = \frac{4-2}{-2-4} = \frac{2}{-6} = -\frac{1}{3}$$

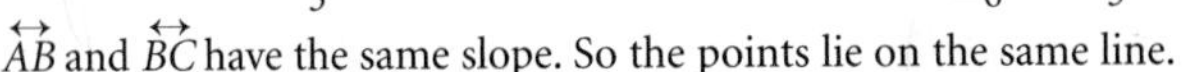

$\overleftrightarrow{AB}$ and $\overleftrightarrow{BC}$ have the same slope. So the points lie on the same line.

28. $A(3, 5), B(-1, 3), C(7, 7)$ yes
29. $P(4, 1), Q(-1, 5), R(1, 2)$ no
30. $L(6, 4), M(3, 2), N(0, 0)$ yes

31. *Fairs* The graph shows how much it costs to rent carousel equipment for a fair. Rental includes the cost of an operator.
   a. Estimate the slope of the line. What does that number mean?
   b. Customers pay \$2 for a ride. What is the number of customers needed per hour to cover the rental cost?
   a. About 110; the rent is \$110 per hour. b. 55 customers per hour

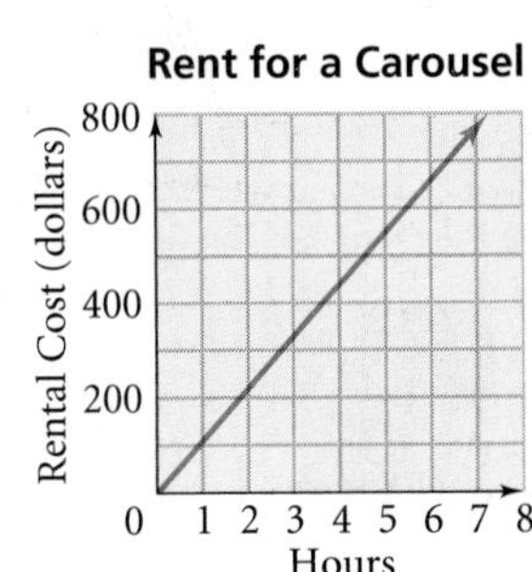

32. *Writing* A friend says the slope of the line passing through (1, 7) and (3, 9) is equal to the ratio $\frac{1-9}{7-3}$. Is this correct? Explain. See below.

33. *Standardized Test Prep* A line has a slope of $\frac{4}{3}$. Through which two points could this line pass? E
   A. $(-4, 10), (8, 19)$ B. $(8, 10), (0, 16)$ C. $(-8, 10), (-2, 2)$
   D. $(4, -4), (10, 2)$ E. none of the above

32. No; the slope is equal to the ratio $\frac{7-9}{1-3}$.

Exercises MIXED REVIEW

Find each probability based on one roll of a number cube.

34. $P(1)$ $\frac{1}{6}$
35. $P(\text{even number})$ $\frac{1}{2}$
36. $P(3 \text{ or } 5)$ $\frac{1}{3}$
37. $P(\text{integer})$ 1
38. $P(10)$ 0

39. *Environment* In 1994, biologists released 200,000 salmon into Alaska's Ninilchik River. In 1995, they released only 50,000 because the native salmon population had grown. Find the percent of change. 75% decrease

Getting Ready for Lesson 5-2

Identify the independent and dependent variable for each function.

40. distance an airplane flies and the time a flight takes indep.: distance; dep.: time

41. the amount of home heating fuel used and the outside temperature indep.: temperature; dep.: amount of fuel used

Math ToolboX

Many math and science problems involve conversions between different units. In fact, many standardized tests include problems that require unit changes. In this toolbox, students will learn to use cancellation (identity properties of multiplication and division) with units in unit conversion problems.

Have students multiply two simple fractions such as $\frac{3}{4}$ and $\frac{8}{9}$, and explain their work. Then, introduce the term *dimensional analysis*, and show students how it is related to the cancelling they used to solve the simple multiplication problem above.

VISUAL LEARNING To demonstrate the differences between units, show containers of different sizes (quart, gallon, pint, etc.) to the class.

ERROR ALERT! Students may have difficulty choosing the correct conversion factors. **Remediation**: Have students first write the proportion to be changed. They can then determine how to write the conversion factor so that the necessary units will cancel.

WRITING Exercise 12 Have students work through the problem to describe their thinking process.

ADDITIONAL PROBLEM The fastest recorded speed for an eastern gray kangaroo is 40 mi/h. What is the kangaroo's speed in feet per second? $58\frac{2}{3}$ ft/s

Math ToolboX Skills Review

Dimensional Analysis

Before Lesson 5-2

You can use conversion factors to change from one unit of measure to another. The process of analyzing units to decide which conversion factors to use is called **dimensional analysis.**

Since 60 min = 1 h, $\frac{60\text{ min}}{1\text{ h}}$ equals 1. You can use $\frac{60\text{ min}}{1\text{ h}}$ to convert hours to minutes.

$7\cancel{\text{h}} \cdot \frac{60\text{ min}}{1\cancel{\text{h}}} = 420\text{ min}$

The hour units cancel, and the result is minutes.

Sometimes you need to use more than one conversion factor.

Example

Animals A cheetah ran 300 ft in 2.92 s. What was the cheetah's speed in miles per hour?

You need to convert feet to miles and seconds to hours.

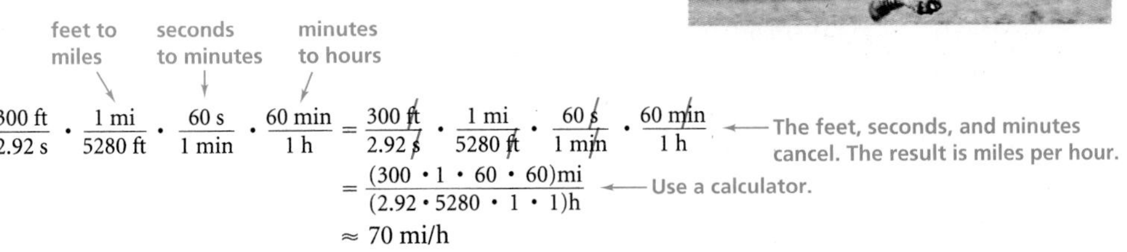

$$\frac{300\text{ ft}}{2.92\text{ s}} \cdot \frac{1\text{ mi}}{5280\text{ ft}} \cdot \frac{60\text{ s}}{1\text{ min}} \cdot \frac{60\text{ min}}{1\text{ h}} = \frac{300\cancel{\text{ft}}}{2.92\cancel{\text{s}}} \cdot \frac{1\text{ mi}}{5280\cancel{\text{ft}}} \cdot \frac{60\cancel{\text{s}}}{1\cancel{\text{min}}} \cdot \frac{60\cancel{\text{min}}}{1\text{ h}}$$

$$= \frac{(300 \cdot 1 \cdot 60 \cdot 60)\text{mi}}{(2.92 \cdot 5280 \cdot 1 \cdot 1)\text{h}}$$

$$\approx 70\text{ mi/h}$$

The cheetah's speed was about 70 mi/h.

Choose A or B for the correct conversion factor for changing the units.

1. quarts to gallons A
   A. $\frac{1\text{ gal}}{4\text{ qt}}$ **B.** $\frac{4\text{ qt}}{1\text{ gal}}$
2. ounces to pounds A
   A. $\frac{1\text{ lb}}{16\text{ oz}}$ **B.** $\frac{16\text{ oz}}{1\text{ lb}}$
3. inches to yards B
   A. $\frac{16\text{ in.}}{1\text{ yd}}$ **B.** $\frac{1\text{ yd}}{36\text{ in.}}$

Write each in the given unit or units.

4. 8 h = ■ s 28,800
5. 120 in. = ■ yd $3\frac{1}{3}$
6. \$1.85/3.25 lb = ■ ¢/oz about 4
7. 18 qt/s = ■ gal/min 270

Express each in miles per hour.

8. 300 yd in 10.9 min about 0.94 mi/h
9. 1 mi in 3.79 min about 15.83 mi/h
10. 120 ft in 30 s about 2.73 mi/h
11. 250 mi in 45 sec 20,000 mi/h
12. *Writing* Explain how you determine which conversion factors to use when changing 3 in./s to feet per minute. Use the conversion factors $\frac{1\text{ ft}}{12}$ and $\frac{60\text{ s}}{1\text{ min}}$ so that the inch and second units cancel.

219

5-2 Teaching Notes

PROBLEM OF THE DAY

The Hawks scored 40 points during last week's football game. Each quarter they scored more points than they did the previous quarter. How many points did they score during each quarter? 6, 7, 10, and 17

Problem of the Day is also available in Transparencies.

CONNECTING TO PRIOR KNOWLEDGE Ask students: *What terms have you used in the past to describe how quickly or how slowly something changes?*

THINK AND DISCUSS

Invite students to invent mnemonic devices to help them remember that the dependent variable goes in the numerator, and the independent variable goes in the denominator. For example, you might suggest that the variable on top is dependent on the variable underneath it; if the variable on the bottom goes away, the variable on top falls over.

Using the same mnemonic device, suggest that students think of the dependent variable climbing above the independent variable. This might help them remember to graph the dependent variable on the vertical axis.

Lesson Planning Options

Prerequisite Skills

- Understanding dependent and independent variables
- Understanding vertical and horizontal

Assignment Options

To provide flexible scheduling, this lesson can be subdivided into parts.

1 **Core** 1–6
Extension 22, 24

2 **Core** 7, 12–17
Extension 18

3 **Core** 8–11, 19, 20
Extension 21, 23

Use Mixed Review to maintain skills.

Resources

Student Edition

Skills Handbook, pp. 569, 570, 579
Extra Practice, p. 560
Glossary/Study Guide

Teaching Resources

Chapter Support File, Ch. 5
- Practice 5-2 (two worksheets)
- Reteaching 5-2

Classroom Manager 5-2
Glossary, Spanish Resources
Two-Year Algebra Handbook 5-2

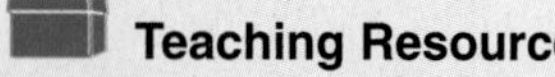

Transparencies
2, 67

Connections Parachuting . . . and more

5-2 Rates of Change

What You'll Learn
- Finding rates of change from tables and graphs

...And Why

To find rates of change in real-world situations, such as the rate of descent for a parachute or the cost of renting a computer

What You'll Need
- graph paper

QUICK REVIEW

A *rate* is a ratio that compares two quantities measured in different units.

THINK AND DISCUSS

Part 1 **Finding Rate of Change**

Suppose you type 140 words in 4 minutes. What is your typing rate?

$$\frac{140 \text{ words}}{4 \text{ min}} \text{ or } \frac{35 \text{ words}}{1 \text{ min}} \quad \begin{matrix}\leftarrow \text{dependent variable} \\ \leftarrow \text{independent variable}\end{matrix}$$

The number of words depends on the number of minutes you type. So, the number of words is the dependent variable.

1. Write each as a rate.
   a. You buy 5 yards of fabric for \$19.95. \$3.99/yd
   b. You travel 268.8 mi on 12 gal of gasoline. 22.4 mi/gal

You use a rate to find the amount of one quantity per one unit of another, such as typing 35 words in 1 min. The **rate of change** allows you to see the relationship between two quantities that are changing.

$$\text{rate of change} = \frac{\text{change in the dependent variable}}{\text{change in the independent variable}}$$

On a graph, you show the dependent variable on the vertical axis and the independent variable on the horizontal axis. The slope of a line is the rate of change relating the variables.

$$\text{slope} = \frac{\text{vertical change}}{\text{horizontal change}} = \frac{\text{change in the dependent variable}}{\text{change in the independent variable}}$$

Example 1 **Relating to the Real World**

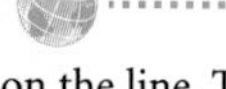

Parachuting Find the rate of change for data graphed on the line. Then explain what the rate of change means in this situation.

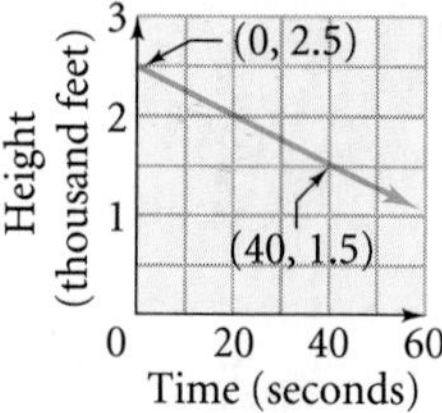

Find two points on the graph. Use the points to find the slope, which is also the rate of change.

$$\text{slope} = \frac{\text{vertical change}}{\text{horizontal change}} = \frac{1.5 - 2.5}{40 - 0} = -\frac{1}{40}$$

The rate of change is $-\frac{1}{40}$, which means that the parachute descends 1000 ft every 40 seconds.

MAKING CONNECTIONS Gregory Arakelian, one of 10,000 entrants at the Key Tronic World International Type-off, typed 158 words per minute on a personal computer. He made two mistakes. The record for a manual typewriter is 176 words per minute. This record has not been broken since 1959.

Example 1 *Relating to the Real World*

ERROR ALERT! Students may interpret the rate of change of $-\frac{1}{40}$ to mean that the parachute descends 1 ft for every 40 s. **Remediation:** Remind students to look at the axis labels before interpreting data from a graph.

MAKING CONNECTIONS The earliest recorded mid-air parachute rescue was reported in 1908. Dolly Shepherd rescued Louie May at 11,000 ft. The two women landed safely using a single parachute.

Example 2 *Relating to the Real World* 

Ask students to look at the table and describe the pattern before they work through the example.

ERROR ALERT! Students may interpret the increasing amounts in the table to mean that the per-day rate increases with each additional day. **Remediation:** Explain that the amounts in the table are the total rental cost for the given number of days.

2. In Example 1, why is height the dependent variable? The height of the parachute depends on the time it has descended.
3. How many feet does the parachute descend in a second? 25 ft

Part 2 Using a Table

Example 2 *Relating to the Real World*

Rental Business You can rent a computer from Hraibe's Rental Company. The first day's rent is \$60. Find the rate of change for renting a computer after the first day.

Cost of Renting a Computer

Number of Days	Hraibe's Rental Company
1	\$60
2	\$75
3	\$90
4	\$105
5	\$120

$$\text{rate of change} = \frac{\text{change in the dependent variable}}{\text{change in the independent variable}}$$

$$= \frac{\text{change in cost}}{\text{change in number of days}}$$ ← Cost depends on the number of days.

$$= \frac{105 - 60}{4 - 1}$$ ← Use any two pairs of data.

$$= \frac{45}{3}$$

$$= \frac{15}{1}$$

The rate of change is $\frac{15}{1}$, which means that it costs \$15 for each day a computer is rented after the first day.

4. Will the rate of change for the data in the table be the same for any pair of data items? Explain. Yes; between any two pairs of values in the table, the rate of change is \$15 per day.

Part 3 Linear Functions

The graph of a **linear function** is a line. You can also use a table to tell whether the relationship between sets of data is linear.

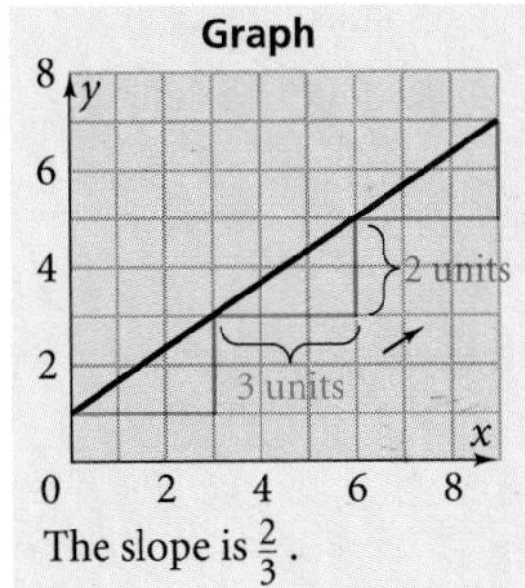

Table

x	y
0	1
3	3
6	5
9	7

3 units → 2 units

Rate of Change

(0, 1) to (3, 3) $\frac{2}{3}$

(3, 3) to (6, 5) $\frac{2}{3}$

(6, 5) to (9, 7) $\frac{2}{3}$

As you can see from the table for the graph, the rate of change between consecutive pairs of data is constant. So, the relationship between the x-values and the y-values is linear.

Additional Examples

FOR EXAMPLE 2

A car rental company charges its customers a base rate of \$50/day. In addition, the rental company charges customers an additional fee for every mile over 100 miles they drive. Using the table below, find the rate of change charged for the extra mileage.

Price per Mile

Miles	Fee
100	\$30.00
150	\$42.50
200	\$60.00
250	\$72.50

$\frac{\$0.25}{1 \text{ mi}}$

Discussion: *How reasonable is the extra charge/mi?*

FOR EXAMPLE 3

Laura wished to measure the effectiveness of her new exercise program. At the end of every week, she counts the number of sit-ups that she can do in one minute. If she does 30 sit-ups the first week, 42 sit-ups the second week, and 54 sit-ups the third week, is her progress linear? Yes; rate of change is 12.

Example 3 **Relating to the Real World** 

CRITICAL THINKING Question 6 Encourage students to estimate the answer before calculating the exact answer. Remind them to compare their answers to their estimates to help decide whether the answer they calculated is reasonable.

VISUAL LEARNING Question 7 If students have difficulty with this exercise, invite them to graph the data to see the shape of the graph. Point out that linear relationships are so named because their graphs form straight lines.

Exercises ON YOUR OWN

ALTERNATIVE ASSESSMENT **Exercises 1–3** Divide students into small groups and ask each group to write a "real world" rate-of-change expression. Then, randomly call on students to write an expression and graph it on the board. This will help you assess the students' understanding of linear rates of change.

Exercise 7 Explain that the body burns more calories at colder temperatures, because it has to work harder to maintain normal body temperature.

CRITICAL THINKING Exercise 7b Ask students to share with the class the shortcuts they used to answer this question.

ESL Exercise 7 Explain that *burn* means "used up."

CRITICAL THINKING Exercise 8 Refer students who say that the statement in Exercise 8 is true back to the sketch they drew for Question 7. Ask students to give examples of changes they are familiar with that do not follow linear patterns.

Technology Options

For Exercises 12–18, students may want to see if the relationships are linear by using graphing software to graph the pairs of points.

Prentice Hall Technology 

 **Software** • Secondary Math Lab Toolkit • Computer Item Generator 5-2

Internet • See the Prentice Hall site. (http://www.phschool.com)

Example 3 **Relating to the Real World** 

Science Tell whether the relationship shown by the data is linear.

Boiling Temperature of Water

	Altitude	Temperature	
	0 ft	212° F	
4000	4,000 ft	204.6° F	−7.4
4000	8,000 ft	197.2° F	−7.4
4000	12,000 ft	189.8° F	−7.4

Each rise of 4000 ft in altitude results in a 7.4° decline in the boiling temperature of water. So the relationship is linear.

5. *Critical Thinking* How much does the boiling temperature of water change for each 1000-ft rise in altitude? Temp. declines by 1.85°F.
6. *Critical Thinking* How could you find the boiling temperature of water in Denver, which has an altitude of 5300 ft? See left.
7. *Try This* Is the relationship shown by the data linear? no

x	0	1	3	5	7	9
y	0	1	4	9	16	25

6. Subtract 4000 ft from 5300 ft. Multiply the difference by the rate of change. Then add the result to 204.6°F. The boiling temperature in Denver is 202.2°F.

Exercises ON YOUR OWN

Find the rate of change for each situation.

1. A baby is 18 in. long at birth and 27 in. at ten months. 0.9 in./mo
2. The cost of group tickets for a museum is $48 for four people and $78 for ten people. $5 per person
3. You drive 30 mi in one hour and 120 mi in four hours. 30 mi/h

Find the rate of change.

4. **A Tank of Gas**

Fuel in Tank (gallons): 0, 5, 10, 15; Miles Traveled: 100, 200, 300

$-\frac{1}{15}$; about 1 gal used every 15 mi.

5. **Price of Oregano**

Cost (dollars): 1, 2; Weight (ounces): 0, 1, 2, 3, 4, 5, 6

$\frac{1}{4}$; $1 buys 4 oz of oregano.

6. 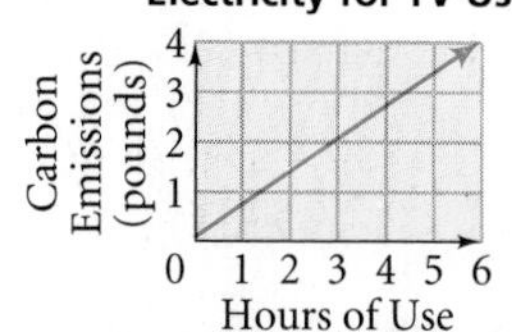

$\frac{2}{3}$; 2 lb of carbon emitted during 3 h of TV use.

CRITICAL THINKING Exercise 10 Suggest that students draw sketches to help them find the answer to this exercise.

Exercises 12–17 Encourage students to plot the coordinate pairs and draw the line connecting them to verify whether the data reflects a linear relationship.

CRITICAL THINKING Exercise 18a Students may see that the change in x is not consistent and thus conclude that the rate of change is not linear. Remind them that the rate of change depends on changes in both the dependent variable and the independent variable.

DIVERSITY Exercise 19 Students who are not familiar with videocassette recorders may not be aware that a VCR can record at several different speeds. Ask a knowledgeable student to explain the speed settings on a VCR and how they are used.

CRITICAL THINKING Exercise 19b Students may misread the question and answer which setting uses the longest amount of tape. Remind them to read questions slowly and carefully before calculating answers.

ESL Exercise 23 Paraphrase "monitor" as, "Lanai is checking or observing her distance." Relate it to the noun form with "just as a hall monitor checks or observes student behavior."

OPEN-ENDED Exercise 23 Prompt students' thinking by asking them whether a mountain hiking trail would be equally steep along its entire length.

CRITICAL THINKING Exercise 24c Challenge students to answer the same question for a vertical line.

7. a. At colder temperatures, people burn more calories. Is the relationship between the two quantities linear? Explain. See below.
   b. *Critical Thinking* How can you find the number of calories burned at 20°F? Subtract 20°F from 50°F. Multiply the diff. by the rate of change. Then add the result to 3330 cal. Answer: 3830 calories.

Working Outdoors

Temperature	Calories Burned Per Day
68°F	3030
62°F	3130
56°F	3230
50°F	3330

Critical Thinking **Tell whether each statement is *true* or *false*. If true, explain why. If false, give an example.**

8. Rates of change are always constant. F; you can ride a bicycle at different speeds.
9. The rate of change for a linear function is constant. T; rate of change is same as slope of graph, which is constant.
10. Two linear functions with positive slope are graphed on the same coordinate plane. The function with the greater rate of change will be steeper. T; line with greater slope has greater rate of change, so is steeper
11. A rate of change must be either positive or zero. F; the boiling temp. of water decreases as altitude increases.

Is the relationship shown by the data in each table linear?

12.

x	y
3	1
4	−2
5	−5
6	−8

yes

13.

x	y
2	0
4	1
6	4
8	9

no

14.

x	y
0	3
1	6
2	9
3	12

yes

15.

x	y
−3	5
−1	8
1	12
3	17

no

16.

x	y
1	0
8	1
15	2
22	3

yes

17.

x	y
9	−4
5	1
1	6
−3	11

yes

18. a. *Critical Thinking* Find the rate of change between consecutive pairs of data. 2

x	1	3	4	7
y	3	7	9	15

   b. Is the relationship shown by the data linear? Explain. Yes; the ratio between the change in y-value and the change in x-value remains the same for all consecutive pairs.

19. *Video Recording* The graph shows the lengths of videotape used in 30 min at three different settings of a VCR. See below.
   a. At which setting does the tape move fastest? slowest? Explain.
   b. *Critical Thinking* At which setting can you record for the longest time? Explain. C; you can record more if you use less tape per unit of time.

Video Recording

Length of Tape Used; Time; A; B; C

20. *Writing* How can you tell if the relationship between two sets of data is linear? Sample: The relationship is linear if the rate of change for all pairs of data is constant.
21. a. *Patterns* Draw the next two figures for the pattern at the right.
   b. Make a table for the relationship between the number of squares and the perimeter of each figure. a–b. See margin.
   c. Is the relationship between the number of squares and the perimeter linear? Explain. No; the number of squares increases by 1, 2, 3, etc., but the perimeter increases 4 units with each figure.

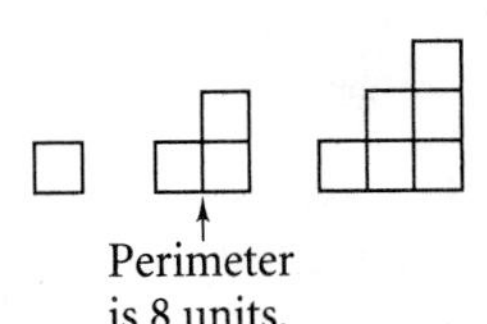

7a. Yes; for every 6°F decline in temp. the no. of calories burned increases by 100.

19a. A; more tape used in same amount of time.
C; less tape used in same amount of time.

pages 222–224 On Your Own

20. Answers may vary. Sample: The data are linear if the rate of change between consecutive pairs of data stays constant.

21a.

b.

No. of squares	Perimeter
1	4
3	8
6	12
10	16
15	20

page 224 Mixed Review

30. Tables may vary. Sample:

x	y
−2	−6
−1	−3
0	0
1	3
2	6

Chapter Project **FIND OUT BY GRAPHING** Ask students to research the starting hourly wage for two jobs that interest them. Then graph and compare the results. This task is essential to their work on the chapter project introduced in the Chapter Opener. Check students' understanding of the project by having each student write a sentence or two describing the objectives for this project.

Exercises MIXED REVIEW

Exercise 29 Suggest that students estimate the answer before solving.

GETTING READY FOR LESSON 5-3 These exercises prepare students for relating the slope of a line to a constant of variation.

Wrap Up

THE BIG IDEA Ask: *How could you describe a rate of change using a table or a graph?*

RETEACHING ACTIVITY Students use slope formula to find rate of change. They plot the data on graph paper to find each rate of change. (Reteaching worksheet 5-2)

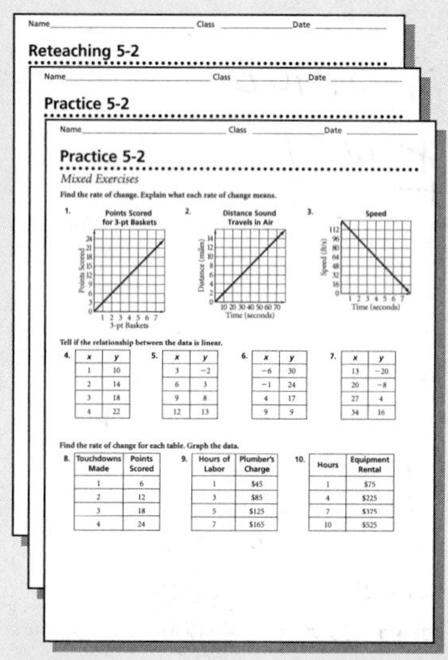

Lesson Quiz

Lesson Quiz is also available in Transparencies.

1. Find the rate of change. 0.075

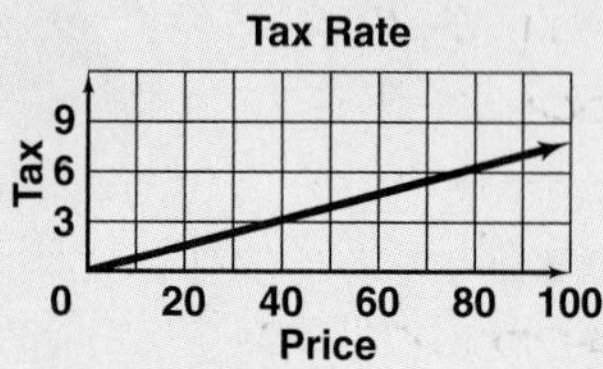

2. Tell if the relationship between the data is linear. no

x	y
2	0
4	3
6	7
8	12

22. a. *Biology* Which line in the graph at the right is the steepest?
b. During the 5-week period, which plant had the greatest rate of change? the least rate of change? How do you know? C; A; the rate of change for C is $\frac{2}{5}$ and the rate of change for A is about $\frac{1}{10}$.

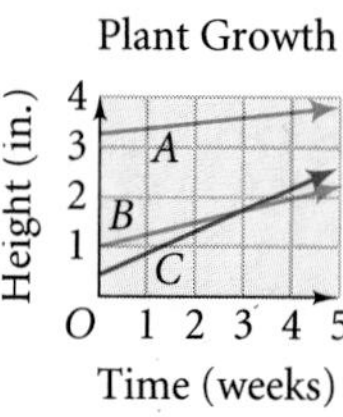

23. *Open-ended* Lanai is hiking up a mountain. She monitors and records her distance every half hour. Do you think the rates of change for every half hour are constant? Explain. Answers may vary. Sample: No; Lanai's speed changes as she gets tired.

24. a. Find the rate of change for the line. 0

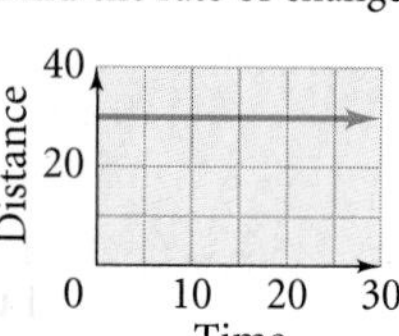

Sample: A car is parked 30 in. from the curb over 30 min.

b. *Writing* What situation could be modeled by the graph?
c. *Critical Thinking* What do you think is true about the rate of change for a horizontal line? Explain. The rate of change is 0; for any change in x-value, there is no change in y-value.

Chapter Project ***Find Out by Graphing***

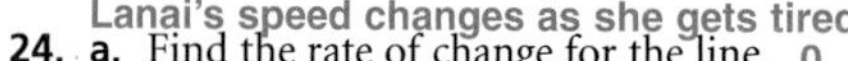

Find the starting hourly wage for two jobs that interest you.

- Make a graph that shows the income for each job. Show hours worked (0 to 10) on the horizontal axis and income on the vertical axis.
- Suppose you work eight hours. Explain how your graph shows the difference in income from the two jobs.

Exercises MIXED REVIEW

Solve.

25. $x + 3 = 2x - 1$ 4
26. $3(2t + 5) = -9$ −4
27. $2.8m - 1.3m = 4.5$ 3
28. $\frac{6s + 4}{8} = \frac{1}{2}$ 0

29. *Food* In the United States, people eat about 340 million pounds of cranberries per year. About 73 million pounds are eaten during Thanksgiving week alone. About what percent of the cranberries is consumed during the week of Thanksgiving? about 21%

Getting Ready for Lesson 5-3

Model each rule with a table of values and a graph. 30–33. See margin.

30. $y = 3x$
31. $y = 0.8x$
32. $y = \frac{3}{4}x$
33. $y = 7.2x$

5-3 Teaching Notes

PROBLEM OF THE DAY

Gary hung seven strings of holiday lights on his house. Some strings had 75 lights and the rest had 100 lights per string. When a fuse blew, one-third of the lights went out but 400 lights still worked. How many of the seven strings had 75 lights? four strings

Problem of the Day is also available in Transparencies.

CONNECTING TO PRIOR KNOWLEDGE Ask students to describe patterns that contain quantities that double. For example, the number of shoes in five pairs or the number of wings on eight birds. Ask students to express these patterns as equations with two variables.

THINK AND DISCUSS

VISUAL LEARNING Students may need help understanding that the table, graph, and function rule all represent the same relationship. Have students point to a row in the table, find that point on the graph, and then substitute the *x*- and *y*-values into the function rule. Repeat for each row in the table until students understand the relationship between the three forms.

Question 1c Make sure students can recall the meaning of the term *coefficient*.

MAKING CONNECTIONS For the animated Disney movie, *101 Dalmatians*, animators drew 6,469,952 Dalmatian spots.

5-3 Direct Variation

What You'll Learn

- Relating slope to constant of variation
- Using constant of variation to solve problems

...And Why

To solve real-world problems, such as finding the force needed to lift a given weight

Movie cameras project at 24 frames/s. Video cameras project at 30 frames/s.

THINK AND DISCUSS

Part 1

Direct Variation

Movies As you watch a movie, 24 individual pictures, or frames, flash on the screen each second. Here are three ways you can model the relationship between the number of frames and the number of seconds.

Table

x number of seconds	y number of frames
1	24
2	48
3	72
4	96
5	120

Graph

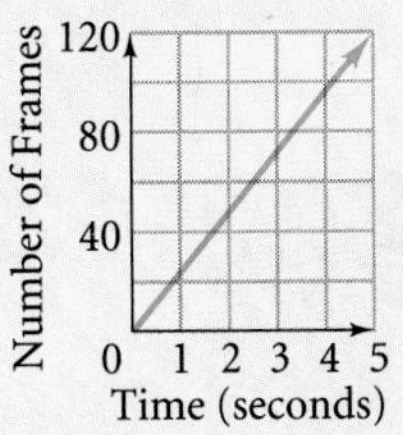

Number of Frames: 0, 40, 80, 120
Time (seconds): 0, 1, 2, 3, 4, 5

Function Rule

$y = 24x$

1. a. What is the rate of change for the data in the table? 24 frames/s
   b. What is the slope of the line shown in the graph? 24
   c. What is the relationship between the rate of change, the slope, and the coefficient of x in the function rule? The number is the same for the rate of change, the slope, and the coefficient of x.

The number of frames *varies directly* with the number of seconds the movie has been shown. This relationship is called a direct variation.

Direct Variation

A **direct variation** is a linear function that can be written in the form $y = kx$, where $k \neq 0$.

↑ constant of variation

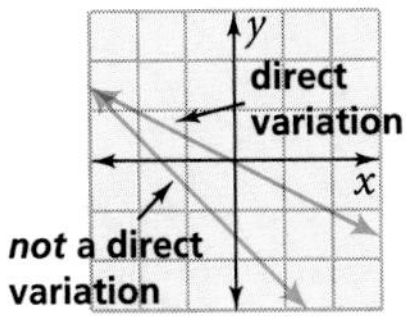

The function rule for a linear function is called a **linear equation.** Since a direct variation can be written as the linear equation $y = kx$, then when $x = 0$, $y = 0$. So the graph of a direct variation will pass through the origin, (0, 0).

y, *x*, direct variation, *not* a direct variation

You can write the linear equation $y = kx$ as $k = \frac{y}{x}$, where $x \neq 0$. The **constant of variation** k equals the rate of change for data that describe the variation. The value of k also equals the slope of the line that graphs the equation.

2. The direct variation graphed at the left includes the point (2, −1). What is its constant of variation? $-\frac{1}{2}$

Lesson Planning Options

Prerequisite Skills

- Isolating a variable
- Graphing slope

Assignment Options

To provide flexible scheduling, this lesson can be subdivided into parts.

1. **Core** 2–7, 25–32
   Extension 1, 8, 33
2. **Core** 9–12, 22
   Extension 19–21, 23
3. **Core** 13–17
   Extension 18, 24

Use Mixed Review to maintain skills.

Resources

Student Edition

Skills Handbook, p. 579
Extra Practice, p. 560
Glossary/Study Guide

Teaching Resources

Chapter Support File, Ch. 5
- Practice 5-3 (two worksheets)
- Reteaching 5-3

Classroom Manager 5-3
Glossary, Spanish Resources
Two-Year Algebra Handbook 5-3

Transparencies

2, 68

CONNECTING TO THE STUDENTS' WORLD Challenge students to find out the length of their favorite movies. Then have them use the function rule $y = 24x$ to calculate how many frames are in their favorite movie.

Help students understand that the constant of variation cannot be zero because y would have a value of zero regardless of the value of x.

Lead students to understand why the graph of a direct variation always passes through the origin.

Question 2 Make sure students understand why the graph of the second line is labeled "*not* a direct variation." On the board, draw several linear graphs that are direct variations, and some that are not. Have students identify the graphs of direct variations. Ask: *How can you quickly tell that a graph represents a direct variation?* Invite students to describe in their own words what all graphs of direct variations have in common.

Example 1 **Relating to the Real World** 

CRITICAL THINKING **Question 3** Students who tested Diane's method using only linear equations may conclude that the method works. Have them test the method again using simple quadratic equations such as $y = x^2$ or $y = 2x^2$. Invite students to graph these parabolas to see the difference between parabolas and lines.

Question 4 After students have solved each equation for y, ask them if there is a shortcut way to decide if an equation is a direct variation. **Answers may vary. Sample: There will be no constant term, except for zero.**

Additional Examples

FOR EXAMPLE 1

Decide if the equation is a direct variation. If it is, find the constant of variation.

a. $2x - 3y = 1$ **no**

b. $2x - 3y = 0$ **yes, $k = \frac{2}{3}$**

Discussion: *Which properties of equation (a) prevent it from being a direct variation?*

FOR EXAMPLE 2

The weight that an object exerts on a scale varies directly with the mass of the object. If a bowling ball has a mass of 6 kg, the scale reads 59 N. Write an equation for the relationship between weight and mass.
$y = 9.8x$, where x = mass and y = weight

FOR EXAMPLE 3

Write a function for the relationship between the data.

x	y
−2	1
0	0
2	−1
4	−2

$y = -0.5x$

226

To tell if a linear equation is a direct variation, you can transform the equation by solving for y.

Example 1

Is each equation a direct variation? If it is, find the constant of variation.

a. $5x + 2y = 0$
$2y = -5x$ ← Solve for y. →
$y = -\frac{5}{2}x$

Yes, $y = -\frac{5}{2}x$ is in the form $y = kx$.
The constant of variation is $-\frac{5}{2}$.

b. $5x + 2y = 9$
$2y = -5x + 9$
$y = -\frac{5}{2}x + \frac{9}{2}$

No, the equation is *not* in the form $y = kx$.

3\. Answers may vary. Sample: Yes; if the equality is true, the equation is a direct variation.

3. *Critical Thinking* Diane tries substituting (0, 0) for (x, y) to see if an equation is a direct variation. Does her method work? Explain. See left.

4. *Try This* Is each equation a direct variation? If it is, find the constant of variation.
a. $7y = 2x$ yes; $\frac{2}{7}$ **b.** $3y + 4x = 8$ no **c.** $y - 7.5x = 0$ yes; 7.5

Using the Constant of Variation to Write Equations (Part 2)

Example 2 **Relating to the Real World** 

Weather The time it takes you to hear thunder varies directly with your distance from the lightning. If you are 2 mi from where lightning strikes, you will hear thunder about 10 s after you see the lightning. Write an equation for the relationship between time and distance.

Define x = your distance in miles from the lightning
y = the number of seconds between seeing lightning and hearing thunder

Relate The time varies directly with the distance.
When $x = 2$, $y = 10$.

Write
$y = kx$ ← Use the general form of a direct variation.
$10 = k(2)$ ← Substitute 2 for x and 10 for y.
$\frac{10}{2} = \frac{2k}{2}$ ← Solve for k.
$5 = k$
$y = 5x$ ← Substitute 5 for k to write an equation.

The equation $y = 5x$ relates the distance x in miles you are from lightning to the time y in seconds it takes you to hear the thunder.

What? A bolt of lightning that you see during a thunderstorm has a speed that varies from 100 to 1000 mi/s.

Source: *The Guinness Book of Records*

5. Use the equation in Example 2 to find about how far you are from lightning if you hear thunder 7 s after you see lightning. 1.4 mi

Example 2 **Relating to the Real World** 

ERROR ALERT! **Question 5** Students may substitute the 7s into the wrong variable. **Remediation:** Remind students to check how the variables are defined before substituting any values.

Example 3

Encourage students to plot the points for each table to see how the graphs vary.

CRITICAL THINKING Question 6c Prompt students' thinking by asking questions such as these:

- *What point does the graph of a direct variation pass through?* (0, 0)
- *Do all linear functions pass through (0, 0)?* no

Example 4 **Relating to the Real World** 

ESTIMATION Encourage students to estimate an answer before answering the question. Then, have them compare their estimates to their answers to help them evaluate whether the answer is reasonable. Lead students to understand that even though estimation is not precise, it is a very useful skill. People use estimation almost daily in personal and business situations to evaluate whether an amount is reasonable.

KINESTHETIC LEARNING Clear a large area in the room and make a coordinate grid on the floor using string or masking tape. Direct students to arrange themselves in a line on the grid to model a simple direct variation graph such as $y = 2x$. Repeat for other direct variations such as $y = -2x$, $y = 3x$, $y = \frac{1}{2}x$, or $y = -\frac{2}{3}x$.

You can use the ratio $\frac{y}{x}$ to tell if two sets of data vary directly.

Example 3

For each table, tell whether y varies directly with x. If it does, write a function rule for the relationship shown by the data.

Find the ratio $\frac{y}{x}$ for each pair of data.

a.

x	y	$\frac{y}{x}$
−3	2.25	$\frac{2.25}{-3} = -0.75$
1	−0.75	$\frac{-0.75}{1} = -0.75$
4	−3	$\frac{-3}{4} = -0.75$

Yes, the constant of variation is −0.75. The function is $y = -0.75x$.

b.

x	y	$\frac{y}{x}$
2	−1	$\frac{-1}{2}$
4	1	$\frac{1}{4}$
6	3	$\frac{3}{6} = \frac{1}{2}$

No, the ratio $\frac{y}{x}$ is not the same for each pair of data.

6a. −0.75; 1
b. Yes; the rate of change in each table is constant between consecutive pairs of data.
c. No; a linear function that does not go through (0, 0) is not a direct variation.

6. **a.** Find the rate of change for the data in each table in Example 3.
   b. Is the relationship shown by the data in each table linear? Explain.
   c. *Critical Thinking* A direct variation is a linear function. Are all linear functions direct variations? Explain.

Part 3

Using Proportions

In a direct variation, the ratio $\frac{y}{x}$ is the same for any pair of data where $x \neq 0$. So the proportion $\frac{y_1}{x_1} = \frac{y_2}{x_2}$ is true for the ordered pairs (x_1, y_1) and (x_2, y_2) where $x_1 \neq 0$ and $x_2 \neq 0$.

Example 4 **Relating to the Real World** 

Physics The force you must apply to lift an object varies directly with the object's weight. You need to apply 0.625 lb of force to the windlass to lift a 28-lb weight. How much force would you need to lift 100 lb?

Relate The force of 0.625 lb lifts 28 lb. The force of n lb lifts 100 lb.

Define $\text{force}_1 = 0.625$ lb, $\text{weight}_1 = 28$ lb; $\text{force}_2 = n$, $\text{weight}_2 = 100$ lb

Write $\frac{\text{force}_1}{\text{weight}_1} = \frac{\text{force}_2}{\text{weight}_2}$ ← Use a proportion.

$\frac{0.625}{28} = \frac{n}{100}$ ← Substitute.

$100(0.625) = 28n$ ← Cross-multiply.

$\frac{100(0.625)}{28} = n$ ← Divide each side by 28.

$n \approx 2.232142857$ ← Use a calculator.

You need about 2.2 lb of force to lift a 100-lb object.

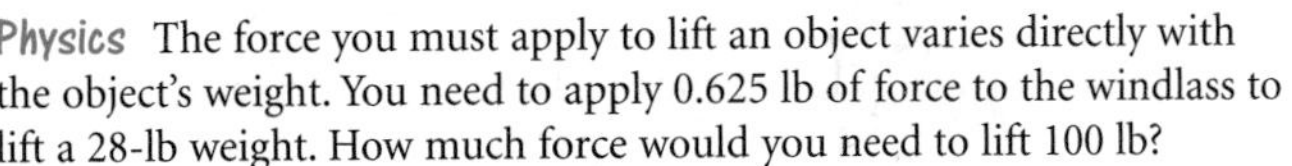

Technology Options

For Exercises 25–32, students may want to find the missing values by using graphing software or a graphing calculator to draw lines.

Prentice Hall Technology

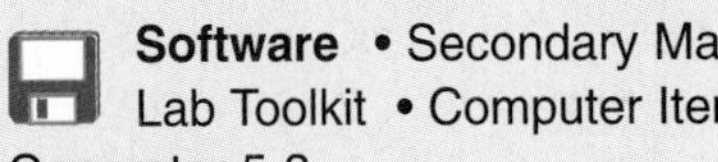

Software • Secondary Math Lab Toolkit • Computer Item Generator 5-3

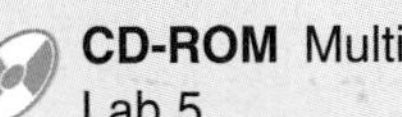

CD-ROM Multimedia Algebra Lab 5

Internet • See the Prentice Hall site. (http://www.phschool.com)

Exercises ON YOUR OWN

CRITICAL THINKING Exercise 8 Encourage students to plot the points and draw a graph to confirm their answers.

ALTERNATIVE ASSESSMENT Exercise 9 To assess the students' understanding of direct variation, have them graph the data in the table. Also have students use the resulting line to predict three additional data points that follow the same direct variation.

EXTENSION Exercise 17 Invite interested students to research how levers can be used to lift heavy objects. Have them share their findings with the class. If you have block scheduling or extended class periods, invite a science teacher to explain the physics involved. Have students apply their new math knowledge to calculate the length of the lever they would need to lift a heavy object.

WRITING Exercise 22a Encourage students to give examples of data that vary directly and data that do not vary directly.

MAKING CONNECTIONS Exercise 23c Born in Hawaii, Chad Rowan was the first foreign-born sumo wrestler to achieve the rank of *yokuzuna*—Grand Champion. He is the tallest and heaviest *yokuzuna* in Sumo history. When he became Grand Champion in 1993, he weighed 467.25 lb and was 6 ft 8 in. tall.

DIVERSITY Exercise 23c Students who are self-conscious about weight may be uncomfortable with this exercise. As an alternative, have students estimate the number of quarts of blood in the body of the sumo wrestler in the Making Connections.

pages 228–229 On Your Own

13. 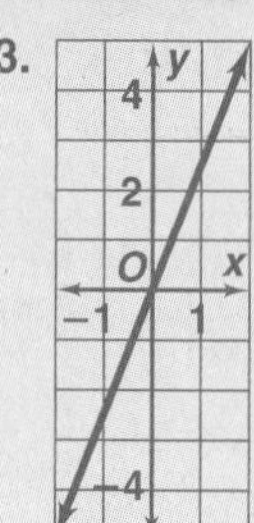

14. 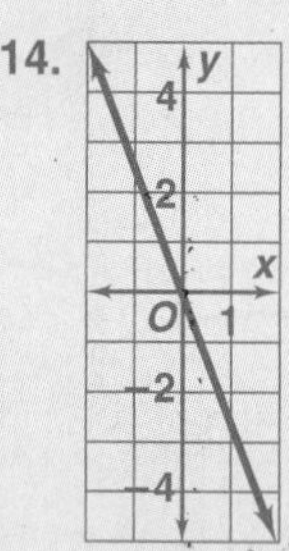

15. 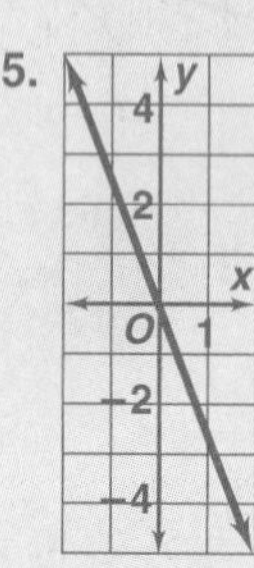

16, 23b. See back of book.

page 229 Mixed Review

34–36, 38–39. See back of book.

8a. using equation: about 195.5 lb

7. Write an equation for the direct variation in Example 4. $f = 0.022w$
8. a. Try This Use your equation in Question 7 or proportions to find how much weight you can lift with a force of 4.3 lb. See left.
   b. Why did you choose the method you used for part (a)? Answers may vary. Sample: the equation because it is easy to substitute for f and solve for w.

Exercises ON YOUR OWN

1. Which of the lines at the right are graphs of direct variations? Explain. A, B; the graph of a direct variation must pass through the origin.

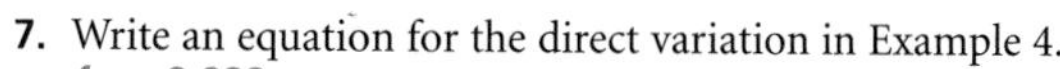

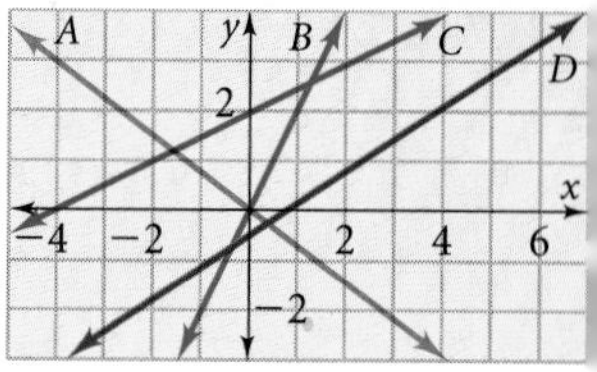

Is each equation a direct variation? If it is, what is the constant of variation?

2. $y = \frac{5}{3}x$ yes; $\frac{5}{3}$
3. $y = 2x + 4$ no
4. $y^2 = 2x$ no
5. $y = \frac{x}{4}$ yes; $\frac{1}{4}$
6. $3y = 8x$ yes; $\frac{8}{3}$
7. $x + 5y = 10$ no

8. Critical Thinking Can the points (2, 3) and (4, 6) be on the graph of the same direct variation? Explain. Yes; the ratios $\frac{3}{2}$ and $\frac{6}{4}$ are equal.

For each table, tell whether y varies directly with x. If it does, write a function rule for the relationship shown by the data.

9.

x	y
3	5.4
7	12.6
12	21.6

Yes; $y = 1.8x$

10.

x	y
−2	1
3	6
8	11

no

11.

x	y
−6	9
1	−1.5
8	−12

yes; $y = -1.5x$

12.

x	y
−5	−13
3	7.8
9	21.6

no

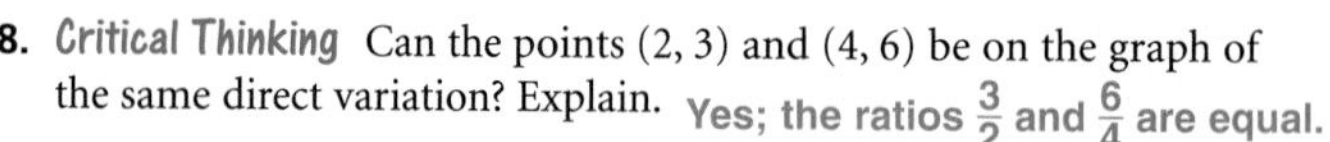

Draw the graph of a direct variation that includes the given point. Write an equation of the line. 13–16. See margin for graphs.

13. (2, 5) $y = \frac{5}{2}x$
14. (−2, 5) $y = -\frac{5}{2}x$
15. (2, −5) $y = -\frac{5}{2}x$
16. (−2, −5) $y = \frac{5}{2}x$

17. Physics The maximum weight you can lift with the lever in the diagram varies directly with the amount of force you apply. What force do you need to lift a friend who weighs 130 lb? 52 lb

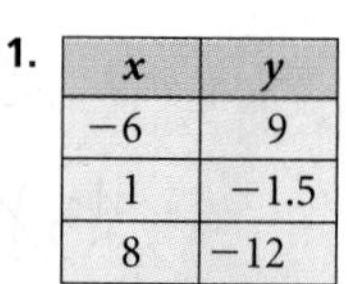

18. Critical Thinking For what value of c is $ax - by = c$ a direct variation? 0

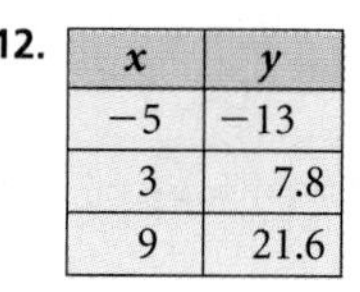

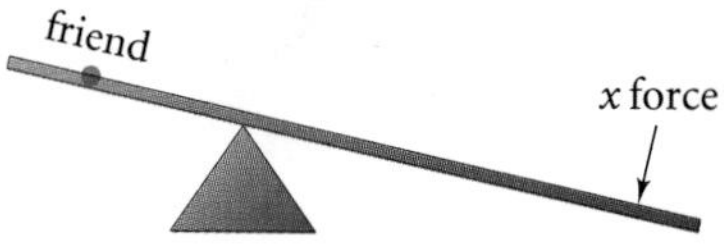

Critical Thinking **Is each statement true? Explain.**

19. The graph of a direct variation may pass through (0, 3). False; the graph must pass through (0, 0).
20. If you double an x-value of a direct variation, the y-value also doubles. True; if you double both values, the ratio does not change.
21. The graph of a direct variation can be a vertical line. False; a direct variation is a function. Vertical lines have more than one y-value for an x-value.
22. a. Writing How can you tell whether two sets of data vary directly? The ratio $\frac{y}{x}$ is the same for each pair of values.
   b. How can you tell if a line is the graph of a direct variation? b. A line through the origin that is neither vertical nor horizontal is the graph of a direct variation.

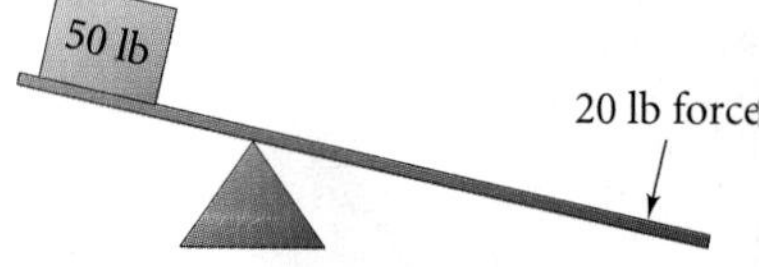

Exercises MIXED REVIEW

JOURNAL Suggest that students provide examples to show the relationships.

Exercises 34–36 Make sure students remember how the graph of an inequality containing $\geq$ differs from the graph of an inequality containing $>$.

Exercise 37 Point out that students could easily misread the sales choices. Remind students to read each choice slowly and carefully.

GETTING READY FOR LESSON 5-4 These exercises prepare students for manipulating equations that are written in slope-intercept form.

Wrap Up

THE BIG IDEA Ask: *What is direct variation? What relationship does direct variation describe?*

RETEACHING ACTIVITY Students work with a partner to draw and label graphs. Then one student uses a piece of wire or pipe cleaner to represent the equation of the line. The other student then answers questions about the line. (Reteaching worksheet 5-3)

23. Biology The amount of blood in a person's body varies directly with body weight. Someone weighing 160 lb has about 5 qt of blood.

a. Find the constant of variation and write an equation relating quarts of blood to weight. $\frac{1}{32}$; $b = \frac{1}{32}w$

b. Graph your equation. See margin.

c. Open-ended Estimate the number of quarts of blood in your body. Check students' work.

24. Bicycling A bicyclist traveled at a constant speed during a timed practice period. Use an equation or proportions to find the distance the cyclist will travel in 30 min. 9 mi

Bicyclist's Practice

Elapsed Time	Distance
10 min	3 mi
25 min	7.5 mi

Suppose the ordered pairs in each exercise are for the same direct variation. Find each missing value.

25. (3, 4) and (9, y) 12 **26.** (−1, 2) and (4, y) −8 **27.** (−5, 3) and (x, −4.8) 8 **28.** (1, y) and $(\frac{3}{2}, -9)$ −6

29. (2, 5) and (x, 12.5) 5 **30.** (−2, 5) and (x, −5) 2 **31.** (−6, −3) and (−4, y) −2 **32.** (x, −4) and (−9, −12) −3

33. a. Write an equation relating the data in the cartoon. $d = 7p$

b. How many dog years are 12 human years? 84 dog years

from the cartoon *Mother Goose and Grimm*

Exercises MIXED REVIEW

Solve each equation. Graph the solutions on a number line. 34–36. See margin for graphs.

34. $4m + 3 < 2$ $m < -\frac{1}{4}$ **35.** $9 > 5 - 3t$ $t > -1\frac{1}{3}$ **36.** $-z + 8 \geq 5$ $z \leq 3$

37. Jeans are on sale at four different stores. The original price of the jeans at each store is \$35. Which of the following sales will give you the best price on the jeans? B

A. 25% off the original price
B. $\frac{1}{3}$ off the original price
C. pay 70% of the original price
D. \$10 off the original price

FOR YOUR JOURNAL

Explain how rate of change, slope, and constant of variation are related.

Getting Ready for Lesson 5-4

Draw a line with the given slope that passes through the given point. 38–39. See margin.

38. slope = 3, passes through (0, −5)

39. slope = $-\frac{7}{8}$, passes through (0, 4)

Reteaching 5-3

Practice 5-3

Practice 5-3 *Mixed Exercises*

Lesson Quiz

Lesson Quiz is also available in Transparencies.

Is each equation a direct variation? If it is, what is the constant of variation?

1. $3x - 2y = 0$ yes, $\frac{3}{2}$

2. $3x - 2y = 3y$ yes, $\frac{3}{5}$

3. Tell whether y varies directly with x. If it does, write a function rule for the relationship between the data.

x	y
−2	4
−1	2
3	−6
4	−8

yes; $y = -2x$

5-4 Teaching Notes

PROBLEM OF THE DAY

Toni Rios earns an annual salary of $135,000 as a computer chip designer. If she usually works 40 hours per week for 50 weeks per year, how much does Toni earn per hour? per minute? $67.50; $1.13

Problem of the Day is also available in Transparencies.

CONNECTING TO PRIOR KNOWLEDGE On a coordinate plane, have students sketch a line passing through (0, 2). Have them compare their lines with those of students around them. Ask: *Did everybody draw the same line?* no *How do your lines differ?* Answers may vary. Encourage students to use terms such as slope, rise, fall, or steepness. *What information do I need to give you so that you all draw the same line?* the slope of the line

Next, have students sketch a line passing through (0, 4), with a slope of 1. Have them compare their lines with those of other students to confirm that they all drew the same line.

WORK TOGETHER

Question 1a The display on a graphing calculator may distort the graph so that the line for $y = x$ forms an angle of less than 45° with the x-axis. On some graphing calculators, you may be able to adjust the screen by choosing the square view of the graphs instead of the standard view.

Lesson Planning Options

Prerequisite Skills

- Graphing slope
- Drawing a line through a point with a given slope

Assignment Options

To provide flexible scheduling, this lesson can be subdivided into parts.

1 **Core** 1–6, 12–20
Extension 10, 11

2 **Core** 7–9, 21–23, 27–30
Extension 24–26

Use Mixed Review to maintain skills.

Resources

Student Edition

Skills Handbook, p. 579
Extra Practice, p. 560
Glossary/Study Guide

Teaching Resources

Chapter Support File, Ch. 5
- Practice 5-4 (two worksheets)
- Reteaching 5-4

Classroom Manager 5-4
Glossary, Spanish Resources
Two-Year Algebra Handbook 5-4

Transparencies

2, 27, 68, 74, 75, 76

5-4 Slope-Intercept Form

What You'll Learn

- Using the slope and y-intercept to draw graphs and write equations

...And Why

To investigate flag designs and real-world situations, such as salary plus commission

What You'll Need

- graphing calculator

WORK TOGETHER

1a. See margin.

1. a. Graphing Calculator Graph these equations on the same screen.
$y = \frac{1}{2}x$ $y = x$ $y = 5x$
b. **Generalize** How does the coefficient of x affect the graph of an equation? The greater the coefficient, the steeper the line.

2. a. Graphing Calculator Many national flags include designs formed from straight lines. You can use equations to model these designs. Choose values for k in the equation $y = kx$ to create a display like the one for the flag of Jamaica. Use the standard settings.

2a. Answers may vary. Sample: 0.9 and −0.9

Jamaica

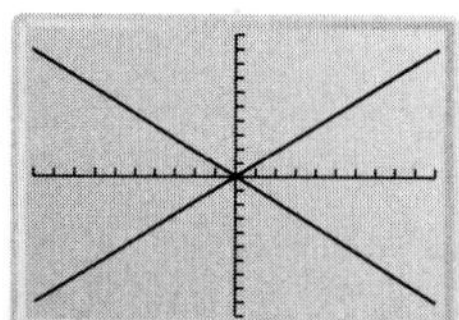

b. What do you notice about the values of k for the lines you graphed? The values of k are opposites of each other.

3. a. Graphing Calculator Graph these equations on the same screen.
$y = 2x$ $y = 2x + 3$ $y = 2x - 4$ See margin.
b. Where does each line cross the y-axis? 0; 3; −4

Tanzania

4. Choose values for b in the equation $y = 1.1x + b$. Create a display that resembles the Tanzanian flag. Write the equations you used for your display. Sample: $y = 1.1x - 4$, $y = 1.1x + 4$
5. **Generalize** What effect does the value of b have on the graph of an equation? The value of b tells you where the graph crosses the y-axis.
6. Open-ended Make a flag design of your own. Write the equations you use for your design.

Check students' work.

THINK AND DISCUSS

Part 1

Defining Slope-Intercept Form

The point where a line crosses the y-axis is the **y-intercept.**

$y = -\frac{1}{3}x + 2$

$y = -\frac{1}{3}x$

$y = -\frac{1}{3}x - \frac{4}{3}$

7. a. What is the y-intercept of each line at the left? $2, 0, -\frac{4}{3}$
b. **Generalize** What is the connection between a line's equation and its y-intercept? The constant term is equal to the y-intercept.

Question 1b Remind students that the coefficient of x is 1 in the equation $y = x$.

TACTILE LEARNING For Question 3, have students cut three pieces of different-colored yarn. Then use rubber cement or removable tape to place the three pieces of yarn on a coordinate plane to match the lines of the equations.

THINK AND DISCUSS

EXTENSION Invite interested students to research the history of mathematics to discover why the variables m and b were chosen to represent slope and y-intercept. Ask them to share their findings with the class.

AUDITORY LEARNING Have students create a mnemonic device for remembering how changes to m and b affect a graph. For example, increasing m makes the mountain harder to climb; b bumps the line up and down the y-axis.

ERROR ALERT! **Question 9** Some students may think that b cannot be negative because that changes the form of the equation to $y = x - b$. **Remediation:** Remind students that $y = x - b$ is the same as $y = x + (-b)$.

VISUAL LEARNING Question 9 Have students work in pairs. To help students discover the pattern of the slope-intercept form of a linear equation, ask them to draw a table with y, m, x, and b as the column headings. Have the pairs place the four parts of each equation into the four columns.

Example 1 **Relating to the Real World** 

Make sure students understand why the x-value of a y-intercept is always zero.

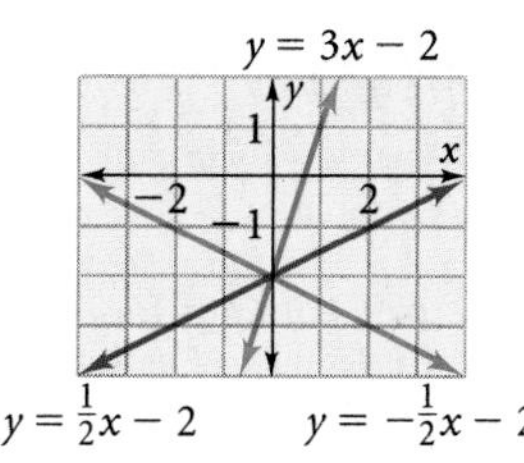

8. a. What is the slope of each line at the left? $3; \frac{1}{2}; -\frac{1}{2}$
   b. **Generalize** What is the connection between a line's equation and its slope? The coefficient of x is equal to the slope.

In the last lesson you learned that the letter k indicates the constant of variation of a direct variation. For linear equations in general, the letter m indicates the slope and the letter b indicates the y-intercept.

Slope-Intercept Form of a Linear Equation

The **slope-intercept form** of a linear equation is $y = mx + b$.
↑ slope ↑ y-intercept

9. Try This What are the slope and y-intercept of the line for each equation?
   a. $y = 3x - 5$ $3; -5$ b. $y = \frac{7}{6}x + \frac{3}{4}$ $\frac{7}{6}, \frac{3}{4}$ c. $y = -\frac{4}{5}x$ $-\frac{4}{5}; 0$
10. Try This Write an equation of a line with the given slope and y-intercept.
   a. $m = \frac{2}{3}, b = -5$ $y = \frac{2}{3}x - 5$ b. $m = -\frac{1}{2}, b = 0$ $y = -\frac{1}{2}x$ c. $m = 0, b = -2$ $y = -2$

You can use the slope and y-intercept to graph an equation.

Example 1

Graph the equation $y = 3x - 1$.

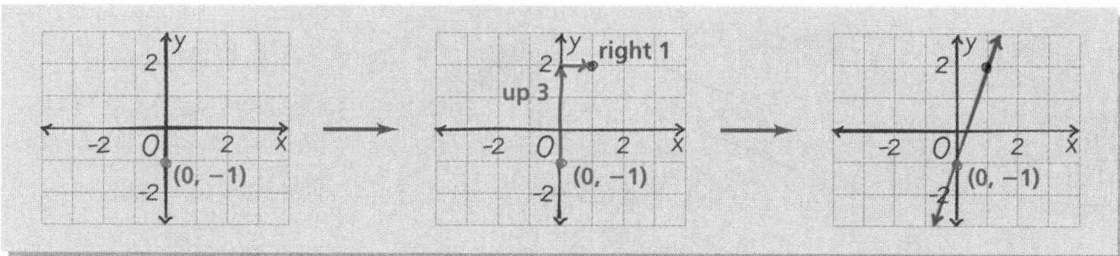

The y-intercept is -1. So plot a point at $(0, -1)$. The slope is 3, or $\frac{3}{1}$. Use the slope to plot a second point. Draw a line through the points.

11. Critical Thinking Could you find a second point by going down 3 and to the left 1? Explain. Yes; the slope is the same, $\frac{3}{1} = \frac{-3}{-1}$.
12. Try This Graph $y = -\frac{3}{2}x + 2$. See margin.

You may need to rewrite a linear equation to express it in slope-intercept form.

Additional Examples

FOR EXAMPLE 1

Graph the equation $y = -\frac{1}{2}x + 3$.

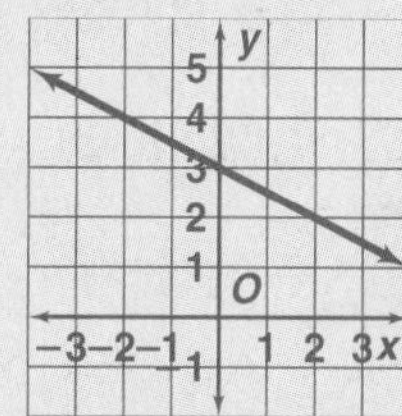

Discussion: *What would happen to the graph of this equation if you transposed the x and y variables?*

FOR EXAMPLE 3

Write an equation for the line shown below.

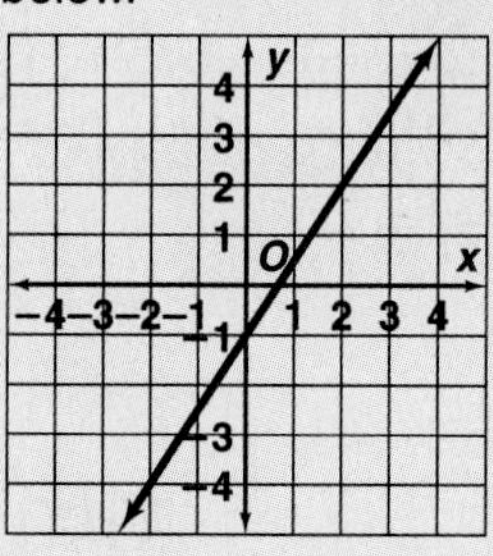

$y = \frac{3}{2}x - 1$

CRITICAL THINKING Question 10 Have students find this second point by graphing. Ask: *What integer rule makes this process the same as going up 3 and to the right 1?* The quotient of two negative numbers is positive.

KINESTHETIC LEARNING Question 11 Create a coordinate plane on a bulletin board in your classroom. Have students use yarn and thumb tacks to graph this and other slope-intercept equations.

Example 2 **Relating to the Real World** 

Invite students who are familiar with the concept of earning commissions to explain more about it to the class.

Remind students that the Commutative Property allows you to change the order of terms that are being added without changing the value of the equation.

Example 3 

Remind students that the formula for slope is $\frac{y_2 - y_1}{x_2 - x_1}$.

Exercises ON YOUR OWN

Exercise 2 Suggest that students rewrite the equation as $y = \frac{1}{2}x + 0$.

Exercise 5 Suggest that students rewrite the equation as $y = 0x + 3$.

Technology Options

For Exercises 12–20, students may use a graphing calculator to graph the equations.

Prentice Hall Technology 

Calculator Graphing Calculator Handbook: Procedure 4

Software • Secondary Math Lab Toolkit • Computer Item Generator 5-4

CD-ROM Multimedia Algebra Lab 5

Internet • See the Prentice Hall site. (http://www.phschool.com)

232

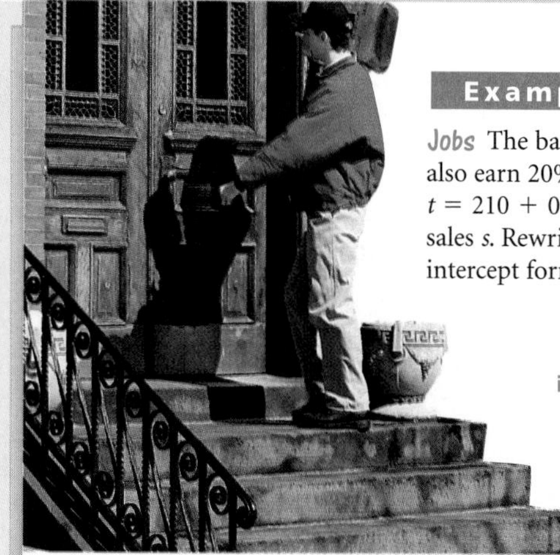

Example 2 **Relating to the Real World**

Jobs The base pay of a water delivery person is \$210 per week. He can also earn 20% commission on any sales he makes. The equation $t = 210 + 0.2s$ relates total earnings t to sales s. Rewrite the equation in slope-intercept form. Then graph the equation.

$t = 210 + 0.2s$ (210: intercept; 0.2: slope)

$t = 0.2s + 210$ ← slope-intercept form

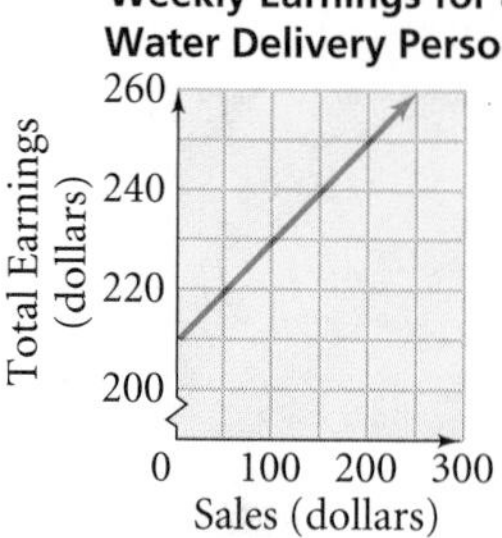

13. $y = 0.2x + 210$; y corresponds to the dependent variable t, and x corresponds to the independent variable c.

13. How would you express $t = 0.2s + 210$ using x and y in order to graph the equation on a graphing calculator? Explain. See left.

14. Would you use the equation or the graph to find a delivery person's total earnings with sales of \$225 in one week? Why? Answers may vary. Sample: Equation; you can compute the exact value using the equation.

15. **Try This** Rewrite $y + 5 = 4x$ in slope-intercept form. $y = 4x - 5$

Part 2 

Writing Equations

Example 3

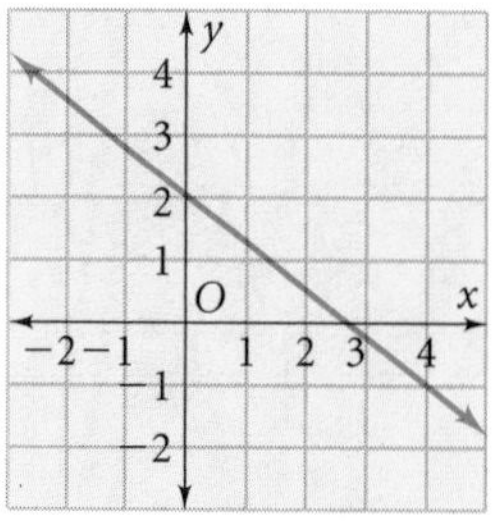

Write an equation for the line at the left.

Step 1 Find the y-intercept and another point.

The y-intercept is 2; (0, 2) and (4, −1) lie on the line.

Step 2 Find the slope.
Use (0, 2) and (4, −1).

$$\text{slope} = \frac{-1 - 2}{4 - 0} = -\frac{3}{4}$$

Step 3 Write an equation in slope-intercept form.
Substitute $-\frac{3}{4}$ for m and 2 for b.

$$y = mx + b$$
$$y = -\frac{3}{4}x + 2$$

Exercises ON YOUR OWN

Find the slope and y-intercept of each equation.

1. $y = -\frac{3}{4}x - 5$ $-\frac{3}{4}$; −5
2. $y = \frac{1}{2}x$ $\frac{1}{2}$; 0
3. $3x - 9 = y$ 3; −9
4. $2x = y + 7$ 2; −7
5. $y = 3$ 0; 3

STANDARDIZED TEST TIP Exercise 6 Explain that a negative total cost does not make sense (the store would be paying customers to buy CDs). For some values of *n*, you would be subtracting a larger amount from a smaller amount in answer choices B and E. Therefore, students can immediately discard these answers.

Exercise 7 Point out to students that this is the only equation in the set with a positive slope. Its graph should be obvious.

ALTERNATIVE ASSESSMENT Exercises 12–20 Assess students' knowledge of slope-intercept form by asking them to rewrite each equation in slope-intercept form. Then ask them to arrange the equations in order of increasing slope.

ERROR ALERT! Exercise 16 Many students confuse the graphing of horizontal and vertical lines. For example, they graph $x = 3$ parallel to the *x*-axis. **Remediation:** Since $x = 3$ means that the *x*-coordinate of all the points on the line is 3, have students write two points that fit this situation. (3, 1) and (3, –2) Then have students draw their line through these points.

OPEN-ENDED Exercise 25 For students having difficulty getting started, have them refer back to their flag designs in Question 6 for a line they drew.

WRITING Exercise 26 A mnemonic device that is helpful when graphing equations of lines is "*b*egin at *b*."

6. *Standardized Test Prep* A music store sells CDs for \$12 each. Customers may use one coupon good for \$4 off the total purchase. Suppose a customer buys *n* number of CDs using a coupon. Which equation models the relationship between the total cost *t* and the number of CDs a customer buys? A

   A. $t = 12n - 4$ **B.** $t = 4n - 12$ **C.** $4t = 12n$ **D.** $t = 12 - 4n$ **E.** $t = 4 - 12n$

Match the graph with the correct equation.

7. $y = x + 5$ III
8. $y = -\frac{5}{2}x + 5$ I
9. $y = -\frac{1}{2}x + 5$ II

I.

II.
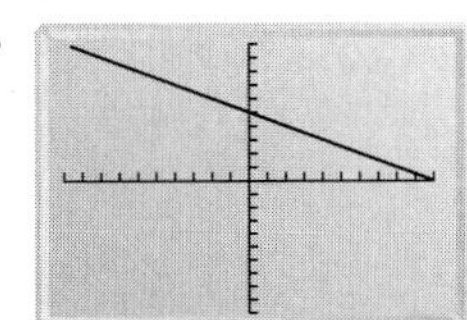

III.
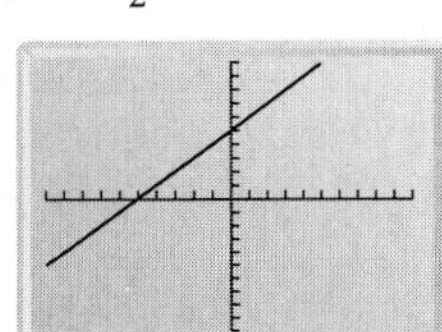

10. *Graphing Calculator* Suppose you want to graph the equation $y = \frac{5}{4}x - 3$. Enter each key sequence and display the graphs. a–b. See margin.

    a. Y= 5 ÷ 4 X,T,θ − 3

    b. Y= (5 ÷ 4) X,T,θ − 3

    c. Which equation gives you the graph of $y = \frac{5}{4}x - 3$? Explain. (b); the graph must be a line with *y*-intercept at –3.

11. A candle begins burning at time $t = 0$. Its height is measured over a period of 30 min. The data are graphed at the right.

    a. Use the graph to find the original height of the candle. 12 in.

    b. Write an equation that relates the height of the candle to the time it has been burning. $h = -\frac{2}{15}t + 12$

    c. How many minutes after the candle is lit will it burn out? 90 min

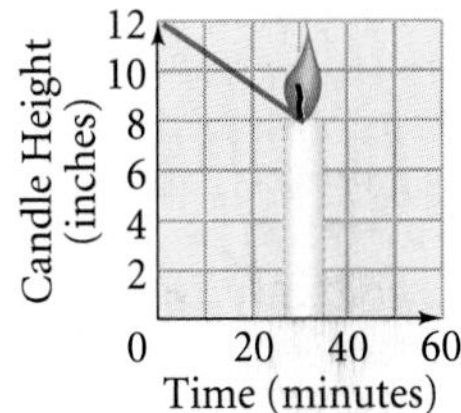

Graph each equation. 12–20. See margin.

12. $y = 2x - 1$
13. $y = 5 + 2x$
14. $y - 4 = x$
15. $y + \frac{3}{4}x = 0$
16. $y = 7$
17. $y = -\frac{1}{2}x + \frac{3}{2}$
18. $y = -\frac{2}{3}x + 0$
19. $y + 3 = \frac{7}{4}x$
20. $y + x = 3$

Find the slope and *y*-intercept. Write an equation of each line.

21.
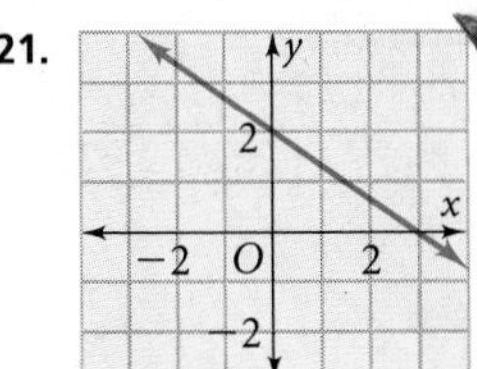

$-\frac{2}{3}$; 2; $y = -\frac{2}{3}x + 2$

22.
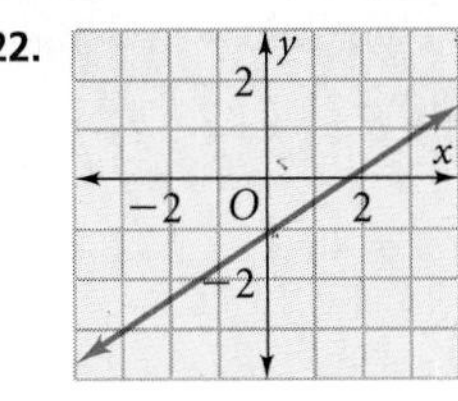

$\frac{2}{3}$; –1; $y = \frac{2}{3}x - 1$

23.
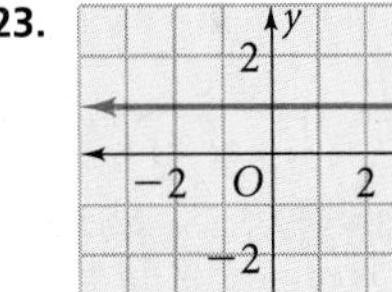

0; 1; $y = 1$

page 230 Work Together

1a.
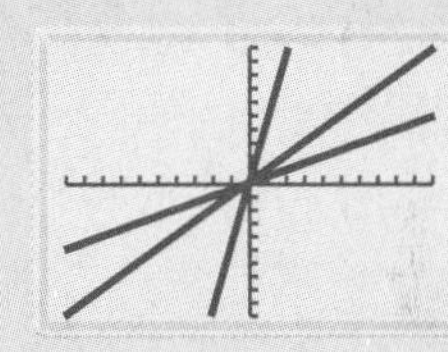

3a.
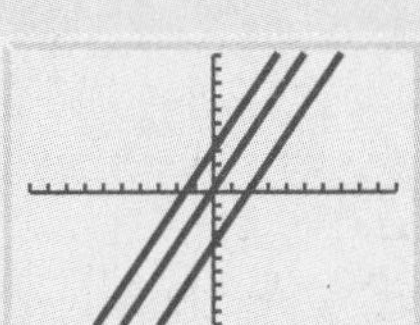

pages 230–232 Think and Discuss

11.
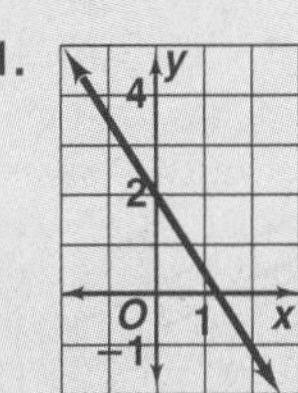

pages 233–234 On Your Own

10a–b, 12–20, 24b, 25–26. See back of book.

Exercises MIXED REVIEW

For students who could give short steps and picture examples, but might have difficulty with a paragraph explanation, reword the Journal directions: "Write each separate step needed to draw a graph with the slope and y-intercept. Draw an example for each step."

SELF ASSESSMENT **JOURNAL** Encourage students to illustrate their paragraphs with sample graphs.

GETTING READY FOR LESSON 5-5 These exercises prepare students to write the equation of a line in slope-intercept form given data about the line.

Wrap Up

THE BIG IDEA Ask: *Can you explain how changing the values of m and b affect the graph of the equation* $y = mx + b$?

RETEACHING ACTIVITY Students use counters to represent integers on the (x, y) axes. Slopes are represented on index cards. Partners take turns writing equations of the lines and drawing graphs. (Reteaching worksheet 5-4)

Exercises CHECKPOINT

In this Checkpoint, students will have an opportunity to assess their own progress on the topics covered in Lessons 5-1 to 5-4.

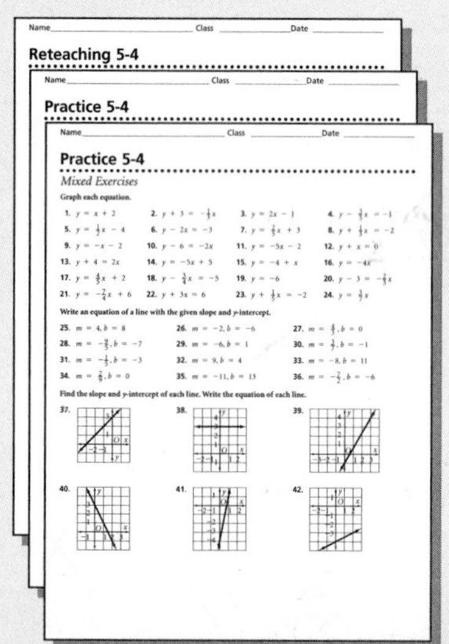

Name ______ Class ______ Date ______

Reteaching 5-4

Name ______ Class ______ Date ______

Practice 5-4

Name ______ Class ______ Date ______

Practice 5-4

Mixed Exercises

Graph each equation.

1. $y = x + 2$ 2. $y + 3 = -\frac{1}{3}x$ 3. $y = 2x - 1$ 4. $y - \frac{3}{5}x = -1$
5. $y = \frac{1}{2}x - 4$ 6. $y - 2x = -3$ 7. $y = \frac{2}{3}x + 3$ 8. $y + \frac{1}{3}x = -2$
9. $y = -x - 2$ 10. $y - 6 = -2x$ 11. $y = -5x - 2$ 12. $y + x = 0$
13. $y + 4 = 2x$ 14. $y = -5x + 5$ 15. $y = -4 + x$ 16. $y = -4x$
17. $y = \frac{4}{5}x + 2$ 18. $y - \frac{3}{4}x = -5$ 19. $y = -6$ 20. $y - 3 = -\frac{2}{3}x$
21. $y = -\frac{7}{4}x + 6$ 22. $y + 3x = 6$ 23. $y + \frac{1}{5}x = -2$ 24. $y = \frac{3}{7}x$

Write an equation of a line with the given slope and y-intercept.

25. $m = 4, b = 8$ 26. $m = -2, b = -6$ 27. $m = \frac{4}{3}, b = 0$
28. $m = -\frac{9}{5}, b = -7$ 29. $m = -6, b = 1$ 30. $m = \frac{3}{7}, b = -1$
31. $m = -\frac{1}{5}, b = -3$ 32. $m = 9, b = 4$ 33. $m = -8, b = 11$
34. $m = \frac{2}{9}, b = 0$ 35. $m = -11, b = 13$ 36. $m = -\frac{7}{2}, b = -6$

Find the slope and y-intercept of each line. Write the equation of each line.

37. 38. 39. 40. 41. 42.

Lesson Quiz

Lesson Quiz is also available in Transparencies.

1. Find the slope and y-intercept of $5x + 3 = y$. $m = 5, b = 3$
2. Graph the equation $y + 2 = \frac{3}{2}x$. See back of book.
3. Write an equation for the line shown below. $y = \frac{1}{2}x + \frac{3}{2}$

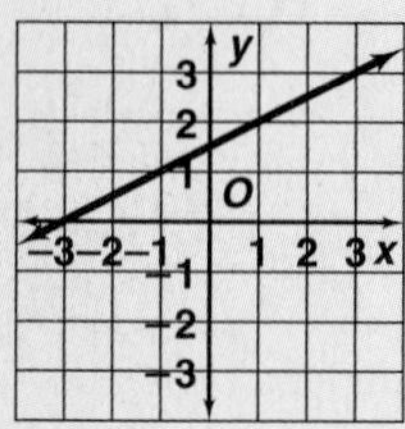

page 234 Mixed Review

35–38. See back of book.

24. *Recreation* A group of mountain climbers begins an expedition with 265 lb of food. They plan to eat a total of 15 lb of food per day. The equation $r = 265 - 15d$ relates the remaining food supply r to the number of days d. $r = -15d + 265$
    a. Write the equation in slope-intercept form.
    b. Graph your equation. See margin.
    c. The group plans to eat the last of their food the day their expedition ends. Use your graph to find how many days the expedition will last. about 18 days
25. *Open-Ended* Write an equation of your own. Identify the slope and y-intercept. Graph your equation.
26. *Writing* Explain how you would graph the line $y = \frac{3}{4}x + 5$. See margin.

25. Sample: $y = \frac{3}{2}x + 1$; $\frac{3}{2}$; 1; see margin for graph.

Write an equation of a line with the given slope and y-intercept.

27. $m = \frac{2}{9}, b = 3$ $y = \frac{2}{9}x + 3$
28. $m = 5, b = -\frac{2}{3}$ $y = 5x - \frac{2}{3}$
29. $m = -\frac{5}{4}, b = 0$ $y = -\frac{5}{4}x$
30. $m = 0, b = 1$ $y = 1$

Exercises MIXED REVIEW

Find the slope of the line through each set of points.

31. $(-2, 8), (5, -1)$ $-\frac{9}{7}$
32. $(4, 6), (2, -1)$ $\frac{7}{2}$
33. $(1, 2), (6, 1)$ $-\frac{1}{5}$
34. The United States Postal Service delivers 177 billion pieces of mail each year. This number represents 40% of the world's mail. How many pieces are sent world-wide each year? 442.5 billion pieces of mail

SELF ASSESSMENT **FOR YOUR JOURNAL** Write a paragraph explaining how to use the slope and y-intercept to write an equation and to draw a graph.

Getting Ready for Lesson 5-5

Through the given point, graph a line with the given slope. Then identify the y-intercept. 35–38. See margin for graphs.

35. $(2, 0)$; $m = -1$ 2
36. $(2, 3)$; $m = \frac{3}{2}$ 0
37. $(3, 2)$; $m = \frac{1}{3}$ 1
38. $(-2, 1)$; $m = -\frac{5}{2}$ −4

Exercises CHECKPOINT

Is a line through the given point with the given slope a direct variation? Explain.

1. $(4, 2)$; $m = \frac{1}{2}$ yes; $\frac{y}{x} = m$
2. $(-2, -2)$; $m = -1$ no; $\frac{y}{x} \neq m$
3. $(6, 9)$; m is undefined No; a vertical line cannot be a direct variation.
4. $(-3, 5)$; $m = -\frac{5}{3}$ yes; $\frac{y}{x} = m$
5. *Money* In 1990, people charged \$534 billion on the two most used types of credit cards. In 1994, people charged \$1.021 trillion on these same two credit cards. What was the rate of change? \$121.75 billion/yr
6. *Writing* How are the graphs of $y = 3x + 5$, $y = \frac{2}{3}x + 5$, and $y = \frac{3}{5}x + 5$ alike? How are they different? The three graphs have the same y-intercept. They have different slopes.

Math ToolboX

In Lessons 5-3 and 5-4, students related the slope of a line to rate of change and direct variation. This toolbox allows students to see the importance of range and scale when graphing linear functions.

Introduce this topic with a discussion on how linear graphs are used to convey information. Find newspaper articles that have linear graphs, and share them with the class. Ask students how changing the scale of the *x*- or *y*-axis of these graphs might affect how the information is perceived.

Direct students' attention to screens A and B. Ask students which screen a company president would want to use to show an increase in sales to the board of directors.

ERROR ALERT! Students may incorrectly choose a range that does not include the *x*-intercept. **Remediation**: Have students find the *x*-intercept for an equation by substituting zero for *y* and solving for *x*.

WRITING Exercise 11 To answer this question, have students refer to their work for Exercise 1.

ADDITIONAL PROBLEMS Find Xmin, Xmax, Ymin, and Ymax values that allow you to see where each line crosses both axes. Sketch the graph of each equation.

1. $y = \frac{1}{4}x - 75$ Answers may vary. The intercepts are (0, –75) and (300, 0)
2. $y = -20x + 30$ Answers may vary. The intercepts are (0, 30) and (1.5, 0)

Math ToolboX Technology

Exploring Range and Scale

After Lesson 5-4

The minimum and maximum values for an axis determine the range of values shown on that axis. Screens A and B show the graph of $y = 4x$.

Screen A

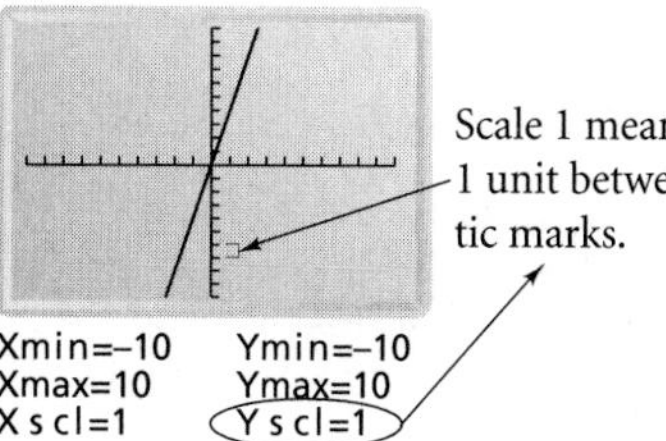

Scale 1 means 1 unit between tic marks.

Xmin=–10 Ymin=–10
Xmax=10 Ymax=10
Xscl=1 Yscl=1

Screen B

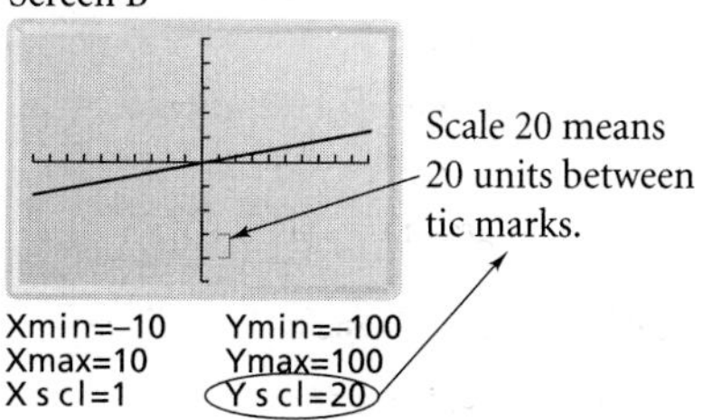

Scale 20 means 20 units between tic marks.

Xmin=–10 Ymin=–100
Xmax=10 Ymax=100
Xscl=1 Yscl=20

Screen A shows standard settings for a graphing calculator screen.

The range on the *y*-axis is 10 times greater for Screen B than Screen A. So the line in Screen B appears flatter than the line in Screen A.

1. a. Explain how you would find the Xscl and Yscl for the screen at the right. Divide Xmax and Ymax by the number of ticks between 0 and the edge; 5, 2.
   b. Write an equation for the line shown on the screen. $y = 2x + 5$

Xmin=–10 Ymin=–10
Xmax=10 Ymax=10
Xscl= Yscl=

To change the range, use the WINDOW feature. Then change the values of Xmin, Xmax, Ymin, or Ymax. For a negative number, use the (–) key. You can return to standard settings by pressing ZOOM 6.

2. a. *Open-ended* Graph the line $y = 2x + 30$ using standard settings. Then change the Ymin or Ymax value so that the *y*-intercept is on your screen. What Ymin or Ymax value did you choose? Answers may vary. Sample: Ymax = 40
   b. *Open-ended* Change the Xmin or Xmax value until you see where the line crosses the *x*-axis. Sketch your graph. State the minimum and maximum values you used for each axis. Answers may vary. Sample: Xmin = –20 See margin for sketch.

Find Xmin, Xmax, Ymin, and Ymax values that allow you to see where each line crosses both axes. Sketch the graph of each equation. 3–10. Answers may vary. See margin.

3. $y = 3x + 20$
4. $y = -x + 100$
5. $y = \frac{1}{5}x - 30$
6. $y = -\frac{1}{50}x + \frac{1}{10}$
7. $y = \frac{2}{3}x + 1500$
8. $y = -0.3x - 60$
9. $y = 55x + 55$
10. $y = -25x - \frac{1}{25}$
11. *Writing* How does changing the range affect how a graph looks? Increasing the range of values on the *y*-axis makes a line look flatter. Decreasing them makes a line look steeper. Increasing the range of values on the *x*-axis makes a line look steeper. Decreasing them makes a line look flatter.

Materials and Manipulatives

- Graphing calculator

Resources

Transparencies
2, 11, 29

page 235 Math Toolbox

2a.

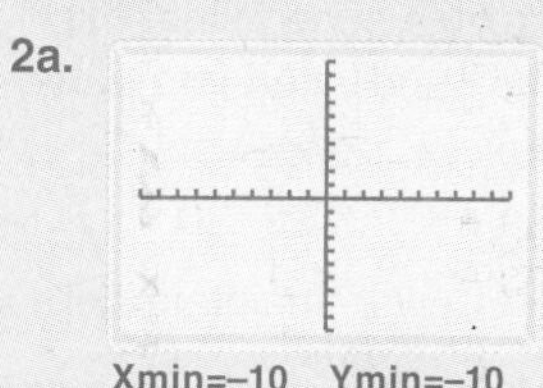

Xmin=–10 Ymin=–10
Xmax=10 Ymax=10
Xscl=1 Yscl=1

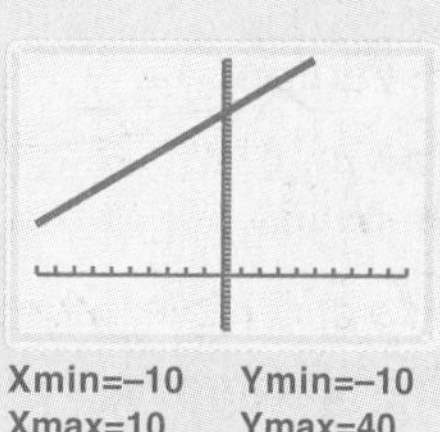

Xmin=–10 Ymin=–10
Xmax=10 Ymax=40
Xscl=1 Yscl=1

2b, 3–10. See back of book.

5-5 Teaching Notes

PROBLEM OF THE DAY

If $10^9 - 27$ is expressed as a single whole number, what is the sum of the digits? 73

Problem of the Day is also available in Transparencies.

CONNECTING TO PRIOR KNOWLEDGE Propose the following situation to students:

- *Your friend called you one night with a question about her algebra homework. You decide that the best way to help her is to have her draw the line of an equation on a graph. How would you describe the line without actually showing it to her?*

THINK AND DISCUSS

Example 1 **Relating to the Real World** 

Help students understand that when they are modeling real-world situations, a linear model is accurate for only a limited range of values. For example, x represents the year. Therefore, values of x less than zero correspond to years before 1900, and are not appropriate to this situation. Similarly, y, which represents the emissions in millions of metric tons, cannot have a value less than zero. Have students use the equation in Example 1 to estimate the emissions for 2025. The answer is -9.4, which is meaningless. Stress that students must consider the realistic range of values for any real-world situation that they are modeling with mathematics.

Lesson Planning Options

Prerequisite Skills

- Modeling equations
- Calculating slope

Assignment Options

To provide flexible scheduling, this lesson can be subdivided into parts.

Core 1–12, 14–22, 24–28
Extension 13, 23, 29, 30

Use Mixed Review to maintain skills.

Resources

Student Edition

Skills Handbook, p. 576
Extra Practice, p. 560
Glossary/Study Guide

Teaching Resources

Chapter Support File, Ch. 5
- Practice 5-5 (two worksheets)
- Reteaching 5-5

Classroom Manager 5-5
Glossary, Spanish Resources
Two-Year Algebra Handbook 5-5

Transparencies
2, 27, 69

Connections Environment . . . and more

5-5 Writing the Equation of a Line

What You'll Learn

- Writing an equation given the slope and a point
- Writing an equation given two points from a graph or a table

...And Why

To write equations to investigate real-world situations, such as carbon monoxide emissions

THINK AND DISCUSS

In some real-world situations you can identify the rate of change, or slope, and an ordered pair. Then you can use the slope and ordered pair to model the situation with a linear equation.

Example 1 **Relating to the Real World** 

Environment World-wide carbon monoxide emissions are decreasing about 2.6 million metric tons each year. In 1991, carbon monoxide emissions were 79 million metric tons. Use a linear equation to model the relationship between carbon monoxide emissions and time. Let $x = 91$ correspond to 1991.

Step 1 Use the data to write the slope and an ordered pair.

slope: -2.6; ordered pair: $(91, 79)$

Step 2 Find the y-intercept using the slope and the ordered pair.

$$y = mx + b$$
$$79 = -2.6(91) + b \quad \leftarrow \text{Substitute (91, 76) for } (x, y) \text{ and } -2.6 \text{ for } m.$$
$$79 = -236.6 + b$$
$$315.6 = b$$

Step 3 Substitute values for m and b to write an equation.

$$y = mx + b$$
$$y = -2.6x + 315.6 \quad \leftarrow \text{Substitute } -2.6 \text{ for } m \text{ and } 315.6 \text{ for } b.$$

The equation $y = -2.6x + 315.6$ models the relationship between carbon monoxide emissions and time.

1. a. Using the equation in Example 1, estimate the emissions for 1990. 81.6 million metric tons
   b. According to this model, what will the emissions be for 2000? 55.6 million metric tons

2. *Try This* Write an equation of a line with slope $\frac{2}{5}$ through the point $(4, -3)$. $y = \frac{2}{5}x - 4\frac{3}{5}$

You can use two points on a line to find an equation for the line. First find the slope of the line through the points. Then use the slope and one point to find the y-intercept and to write an equation of the line.

Question 2 Have pairs of students work together to answer this question. Have one student sketch the line while the other calculates the equation. Then have them compare their results to check each other's work. Cooperative learning opportunities help students develop the skills for working as part of a team.

MAKING CONNECTIONS In 1972, only 26 countries had environmental protection agencies. By 1991, that quantity had increased to 161.

Example 2

In Step 1, point out that students can substitute either (3, 4) or (−2, 1) for (x_2, y_2). In step 2, students can substitute either (3, 4) or (−2, 1) for (*x*, *y*). Students can substitute zero for *x* to find the value of the *y*-intercept. By comparing the value to the graph, they can assess whether the equation in Step 3 is correct.

Question 3 Have pairs of students work together as they did for Question 2.

AUDITORY LEARNING Invite students to compose songs or raps that describe the 3-step process for solving questions such as those in Example 2. Have students record their songs on tape or share them with the class.

Example 3

CRITICAL THINKING Question 4 If students are not confident about the answer, have them substitute other ordered pairs to find the *y*-intercept.

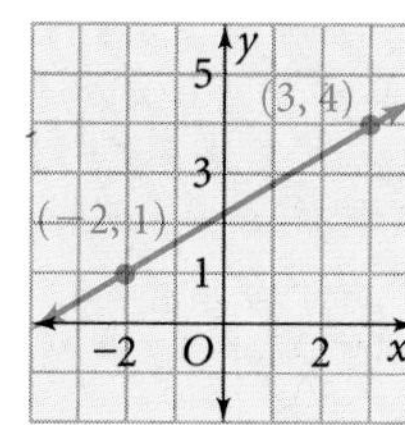

Example 2

Find an equation of the line at the left.

Step 1 Use the coordinates of two points to find the slope of the line.

$$\text{slope} = \frac{y_2 - y_1}{x_2 - x_1}$$

$$= \frac{4 - 1}{3 - (-2)} \quad \leftarrow \text{Substitute (3, 4) for } (x_2, y_2) \text{ and } (-2, 1) \text{ for } (x_1, y_1).$$

$$= \frac{3}{5}$$

Step 2 Find the *y*-intercept.

$$y = mx + b$$

$$4 = \frac{3}{5}(3) + b \quad \leftarrow \text{Substitute (3, 4) for } (x, y) \text{ and } \tfrac{3}{5} \text{ for } m. \text{ Solve for } b.$$

$$4 = \frac{9}{5} + b$$

$$2\tfrac{1}{5} = b$$

Step 3 Substitute values in $y = mx + b$.

$$y = \tfrac{3}{5}x + 2\tfrac{1}{5} \quad \leftarrow \text{Substitute } \tfrac{3}{5} \text{ for } m \text{ and } 2\tfrac{1}{5} \text{ for } b.$$

3. Try This Find an equation of the line through (−5, 3) and (−2, −4). $y = -\frac{7}{3}x - 8\frac{2}{3}$

You can also write a linear equation for data in tables. Two sets of data have a linear relationship if the rate of change between consecutive pairs of data is the same.

Example 3

x	y
−1	4
3	6
5	7
9	9

Is the relationship shown by the data linear? If it is, write an equation.

Step 1 Find the rate of change for consecutive ordered pairs.

(−1, 4) to (3, 6): $\frac{6-4}{3-(-1)} = \frac{2}{4} = \frac{1}{2}$

(3, 6) to (5, 7): $\frac{7-6}{5-3} = \frac{1}{2}$

(5, 7) to (9, 9): $\frac{9-7}{9-5} = \frac{2}{4} = \frac{1}{2}$

The relationship is linear. The rate of change equals the slope. The slope is $\frac{1}{2}$.

Step 2 Find the *y*-intercept and write an equation.

$$y = mx + b$$

$$4 = \tfrac{1}{2}(-1) + b \quad \leftarrow \text{Substitute } (-1, 4) \text{ for } (x, y) \text{ and } \tfrac{1}{2} \text{ for } m.$$

$$4 = -\tfrac{1}{2} + b$$

$$4\tfrac{1}{2} = b \quad \leftarrow \text{Add } \tfrac{1}{2} \text{ to each side of the equation.}$$

$$y = \tfrac{1}{2}x + 4\tfrac{1}{2} \quad \leftarrow \text{Substitute } \tfrac{1}{2} \text{ for } m \text{ and } 4\tfrac{1}{2} \text{ for } b \text{ in } y = mx + b.$$

PROBLEM SOLVING

Look Back Could you use a graph to find whether the relationship is linear? Explain. **Yes; the graph must be a straight line.**

4. *Critical Thinking* Is (−1, 4) the only ordered pair that you could use to find the *y*-intercept in Example 3? Explain. **No; you can use any of the four given ordered pairs because all four points lie on the same line.**

Additional Examples

FOR EXAMPLE 2

Find the equation of the line pictured below.

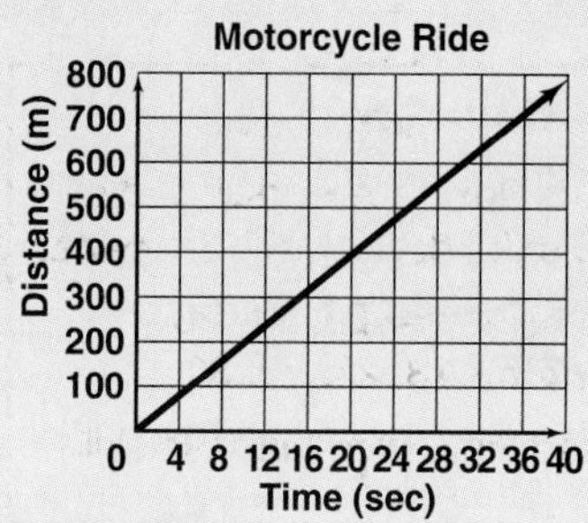

$y = \frac{2}{7}x - 1$

FOR EXAMPLE 3

Is the relationship listed below linear? If it is, write an equation for the relationship.

x	y
−2	−2
−1	−1
1	0
2	1

no

Discussion: *Why might this example be confusing?*

Exercises ON YOUR OWN

ALTERNATIVE ASSESSMENT **Exercises 14–19** Have students find a third point that the same line passes through. This will help you assess students' understanding of constant slope.

ERROR ALERT! **Exercise 13** Students may think that the linear equation they write will accurately model the situation for all possible values. **Remediation:** Point out that when you use an equation to model a real-world situation, there may be ranges of values for which the model does not work. The spring will only shorten or stretch to certain limits. Encourage students to write the ranges of values for which *x* and *y* realistically model the situation.

MAKING CONNECTIONS Exercise 13 If the 50-g mass was pulled down and then released, the mass would oscillate up and down for a while. Due to the effects of friction, however, the oscillations would get progressively smaller until the mass came to rest. If you could remove all effects of friction, the mass would oscillate up and down indefinitely. The speed of each oscillation depends on the tightness of the spring and on the size of the mass. The speed does not depend on how far the mass is pulled down before being released. Pulling the mass farther down results in bigger, but not faster, oscillations. If you have extended class periods, invite a science teacher to demonstrate the physics of springs.

Chapter Project FIND OUT BY MODELING The students are asked to correctly model a job situation by writing an equation. Have students solve their equations based on the parameters

Technology Options

For Exercises 14–19, students may use graphing software for drawing the lines through each point in order to help write the equations.

Prentice Hall Technology

Software • Secondary Math Lab Toolkit • Computer Item Generator 5-5

Internet • See the Prentice Hall site. (http://www.phschool.com)

Exercises ON YOUR OWN

Write an equation of a line through the given point with the given slope.

1. $(3, -5);\ m = 2$ $y = 2x - 9$
2. $(1, 2);\ m = -3$ $y = -3x + 5$
3. $(2, 6);\ m = \frac{4}{3}$ $y = \frac{4}{3}x + 3\frac{1}{3}$
4. $(-1, 5);\ m = -\frac{3}{5}$ $y = -\frac{3}{5}x + 4\frac{2}{5}$
5. $(0, 3);\ m = 1$ $y = x + 3$
6. $(3, 0);\ m = -1$ $y = -x + 3$
7. $(-5, 2);\ m = 0$ $y = 2$
8. $(6, 7);\ m$ undefined $x = 6$
9. $(3, 3);\ m = -\frac{1}{4}$ $y = -\frac{1}{4}x + 3\frac{3}{4}$
10. $(5, -2);\ m = \frac{7}{2}$ $y = \frac{7}{2}x - 19\frac{1}{2}$
11. $(-6, 1);\ m = -\frac{3}{4}$ $y = -\frac{3}{4}x - 3\frac{1}{2}$
12. $(2.8, 10.5);\ m = 0.25$ $y = 0.25x + 9.8$

13. a. Physics Each gram of mass stretches the spring 0.025 cm. Use $m = 0.025$ and the ordered pair (50, 8.5) to write a linear equation that models the relationship between the length of the spring and the mass. $L = 0.025M + 7.25$
   b. Critical Thinking What does the *y*-intercept mean in this situation?
   c. What is the length of the spring for a mass of 70 g? 9 cm
   b. The *y*-intercept is the length of the spring when no mass is attached.

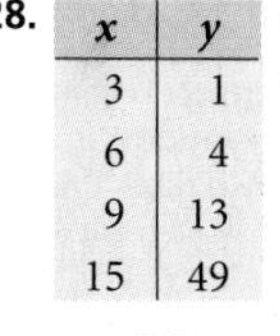

Write an equation of a line through the given points.

14. $(3, -3), (-3, 1)$ $y = -\frac{2}{3}x - 1$
15. $(7, 3), (2, 2)$ $y = \frac{1}{5}x + 1\frac{3}{5}$
16. $(3, 5), (5, 3)$ $y = -x + 8$
17. $(-8, 2), (1, 3)$ $y = \frac{1}{9}x + 2\frac{8}{9}$
18. $(-0.5, 2), (-22, 1.5)$ $y = \frac{1}{3}x + 2\frac{1}{6}$
19. $(25, 100), (15, 120)$ $y = -2x + 150$

Write an equation of each line.

20. 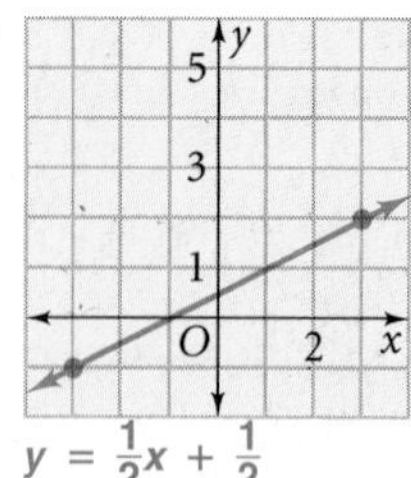

$y = \frac{1}{2}x + \frac{1}{2}$

21. 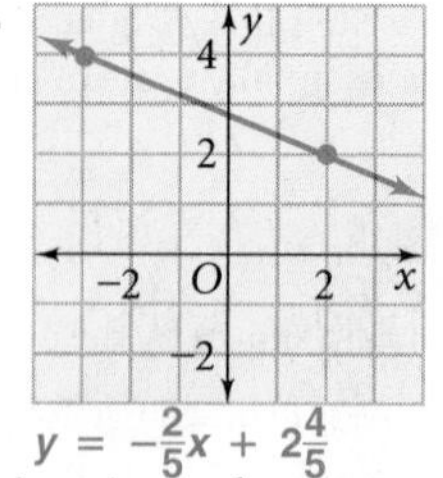

$y = -\frac{2}{5}x + 2\frac{4}{5}$

22. 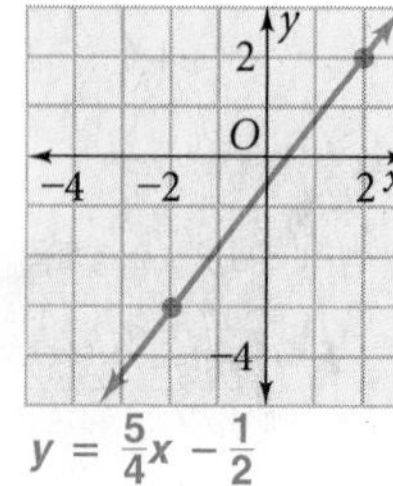

$y = \frac{5}{4}x - \frac{1}{2}$

23. Entertainment Total receipts for motion picture theaters were \$3.9 billion in 1986. Receipts were \$6.9 billion in 1992.
   a. Write an equation to model the relationship between receipts and time in years. Let 86 correspond to 1986. $r = 0.5t - 39.1$
   b. Use your equation to **predict** motion picture theater receipts in the year 2010. (*Hint:* Think about the number you will use for 2010.) \$15.9 billion

Tell whether the relationship shown by the data is linear. If it is, write an equation for the relationship.

24.

x	y
−10	−7
0	−3
5	−1
20	5

yes; $y = \frac{2}{5}x - 3$

25.

x	y
−4	9
2	−3
5	−9
9	−17

yes; $y = -2x + 1$

26.

x	y
1	7
2	8
3	10
4	13

no

27.

x	y
−10	−5
−2	19
5	40
11	58

yes; $y = 3x + 25$

28.

x	y
3	1
6	4
9	13
15	49

no

given. This information is essential to their work on the Chapter Project. Have students add this task to work they have already completed for the project. You may wish to check students' progress by having each one write a sentence or two describing what they have done for the project thus far.

Exercises MIXED REVIEW

JOURNAL Encourage students to make sketches that illustrate their explanations.

Exercises 34 and 36 Students may need to review absolute value inequalities before solving this problem.

CONNECTING TO THE STUDENTS' WORLD Exercise 37 Encourage students to find out more about how and why property is appraised from their friends or family. Have them find out what factors can change a property's value. Then share their findings with the class.

GETTING READY FOR LESSON 5-6 These exercises prepare students for relating linear equations to lines of best fit.

Wrap Up

THE BIG IDEA Ask: *Given two points on a line, how do you find the equation of that line?*

RETEACHING ACTIVITY Students write the equation of a line when given a slope and a specific point. (Reteaching worksheet 5-5)

---

29. a. Business A taxicab ride that is 2 mi long costs \$7. One that is 9 mi long costs \$24.50. Write an equation relating cost to length of ride. $c = 2.5\ell + 2$

b. What do the slope and y-intercept mean in this situation? slope: cost/mi; y-intercept: initial cost of ride

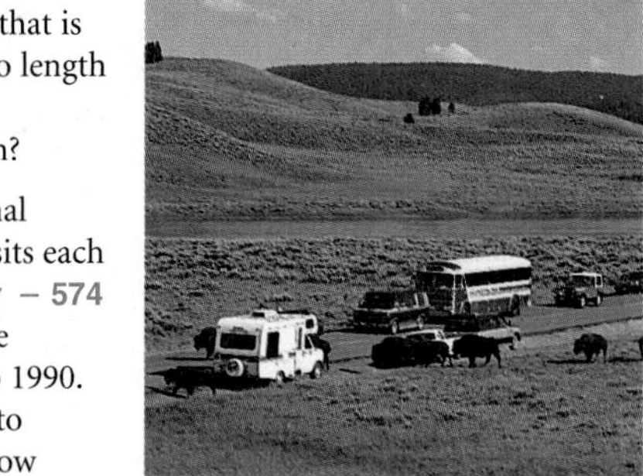

30. National Parks The number of recreational visits to National Parks in the United States increases by about 9.3 million visits each year. In 1990 there were about 263 million visits. $v = 9.3y - 574$

a. Write an equation to model the relationship between the number of visits and time in years. Let 90 correspond to 1990.

b. Open-ended Suppose the number of recreational visits to National Parks continues to increase at the same rate. How many visits will there be this year? Check students' work Samples: 337.4 million visits in 1998; 346.7 million visits in 1999; 356 million visits in 2000

Chapter Project — Find Out by Modeling

Suppose you earn \$5.50/h at a bakery. From your first paycheck you discover that \$1.15/h is withheld for taxes and benefits. You work x hours during a five-day week and you spend \$3.75 each day for lunch.

- Write an equation for your earnings for a week after taxes and expenses.
- In this situation what does the slope represent? the y-intercept?
- How many hours must you work to earn \$120 after taxes and expenses?

Exercises MIXED REVIEW

Solve.

31. $4c < 24 + c$ $c < 8$ **32.** $\frac{t}{2} + \frac{t}{3} = 5$ 6 **33.** $-4m \geq 7$ $m \leq -1\frac{3}{4}$

34. $|5c| < 16$ $-3\frac{1}{5} < c < 3\frac{1}{5}$ **35.** $4(3h + 2) = 5h - 3$ $-1\frac{4}{7}$ **36.** $|2n + 5| \geq 9$ $n \leq -7$ or $n \geq 2$

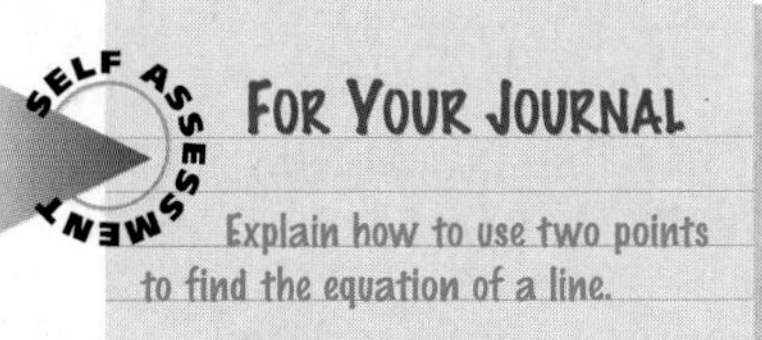

FOR YOUR JOURNAL

Explain how to use two points to find the equation of a line.

37. Real Estate In 1995, a new development was built near the home of Mary Davenport in Lawrence, Kansas. The appraised value of her property went from \$41,500 to \$426,870. What was the percent of change in the property's appraised value? 928.6% increase

Getting Ready for Lesson 5-6

Determine whether each scatter plot shows *positive, negative,* or *no correlation.*

38. 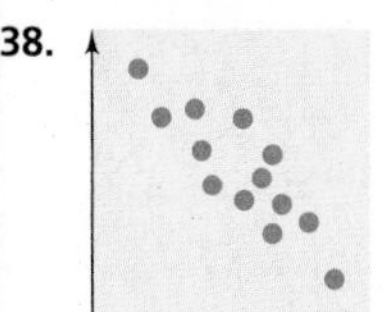 negative

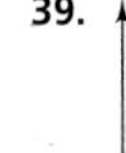

39. 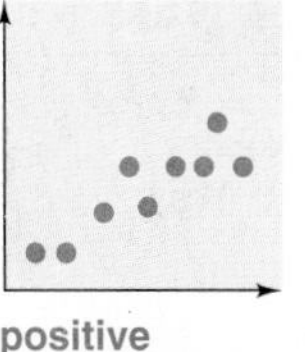 positive

40. 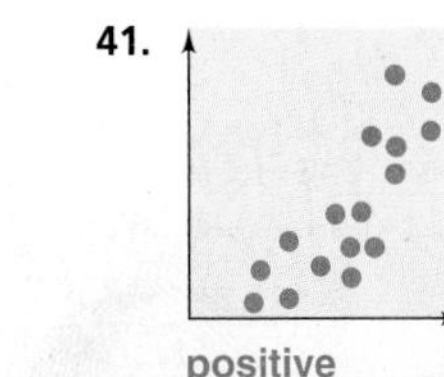 no correlation

41. positive

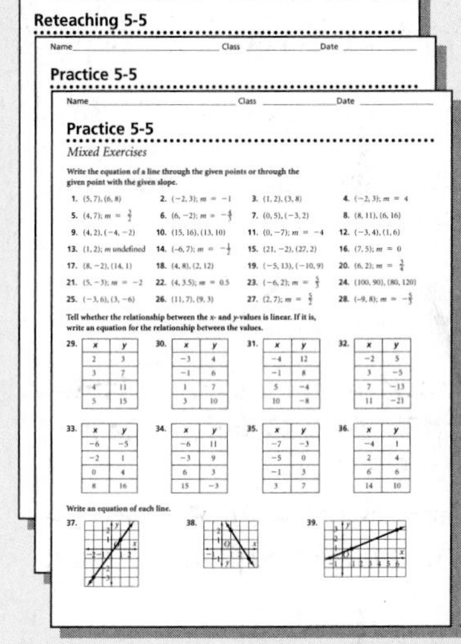

Reteaching 5-5

Practice 5-5

Practice 5-5
Mixed Exercises

Lesson Quiz

Lesson Quiz is also available in Transparencies.

1. Write an equation of a line through point (–3, –1) with a slope of –2. $y = -2x - 7$
2. Write an equation for the line that passes through points (–3, 2) and (3, 1). $y = -\frac{1}{6}x + 1\frac{1}{2}$
3. Decide if the following relationship between the x- and y-values is linear. If it is, write an equation for the relationship. $y = \frac{2}{6}x + 2$

x	y
–5	$\frac{1}{3}$
–1	$2\frac{1}{3}$
7	$4\frac{1}{3}$
12	6

Math Toolbox

In Lesson 5-6, students will use the graphing calculator to find the line of best fit. To ensure success, students must be proficient at entering data into the calculator and generating scatter plots.

Direct students' attention to the scatter plots on page 239 to introduce the concepts of positive, negative, and no correlation. Encourage students to check their scatter plots with a partner to quickly identify any data they may have entered incorrectly. To correct incorrectly-entered data, have them go back to the list, move the cursor to the incorrect number, press clear, and reenter the correct number.

ERROR ALERT! Some students may have difficulty changing between screens. **Remediation:** Have students press 2nd, then QUIT to return to the home screen.

WRITING Exercise 3 Have students list the advantages and disadvantages of both methods.

ADDITIONAL PROBLEM Display the data in a scatter plot. Tell whether the data has a *positive* correlation, *negative* correlation, or *no* correlation.

x	85	69	25	75	67	52	31	98
y	17	13.4	4.9	16	13.4	10.1	6.5	19.8

positive correlation

Materials and Manipulatives

- Graphing calculator

Resources

Transparencies
2, 11

page 240 Math Toolbox

1.

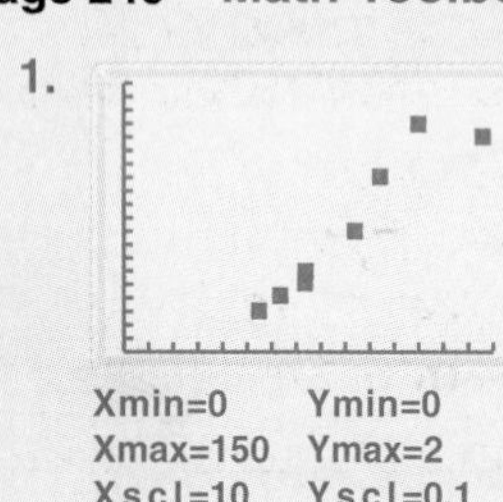

Xmin=0 Ymin=0
Xmax=150 Ymax=2
Xscl=10 Yscl=0.1

2.

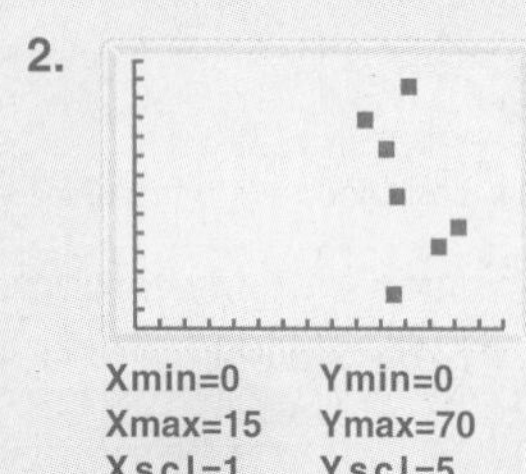

Xmin=0 Ymin=0
Xmax=15 Ymax=70
Xscl=1 Yscl=5

Math ToolboX Technology

Displaying Data

Before Lesson 5-6

The data in the table are displayed on the two graphing calculator screens.

Data

Year	Profit (in thousands of dollars)
1990 ↔ 90	125
91	205
92	296
93	510
94	620
95	904
96	1040

List Display

L1	L2	L3
90	125	-----
91	205	
92	296	
93	510	
94	620	
95	904	
96	1040	

L2 (7)=1040

Scatter Plot Display

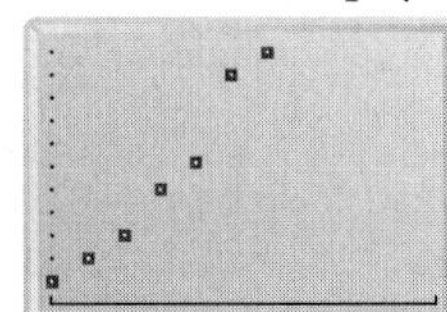

List Display To enter data into a list, press STAT 1. Clear old data in column L1, by pressing ▲ CLEAR ENTER. Enter the year data in the first list. After each data item press ENTER. Press ▶ to move to column L2. Then clear any old data. Enter the second set of data.

Scatter Plot Display You can make a scatter plot of data that are in lists.

WINDOW FORMAT
Xmin=90
Xmax=100
Xscl=10
Ymin=0
Ymax=1100
Yscl=100

Step 1 Set Up Variables Press STAT ▶ 3. Under 2-Var Stats, use the arrow keys to highlight L1 in the Xlist. Press ENTER. Highlight L2 in the Ylist.

Step 2 Choose Range The range for each variable should include the least and greatest number in the corresponding list.

Step 3 Make Graph Press 2nd Y= 1 ENTER. Move the cursor and press ENTER to select the options shown. Then press GRAPH.

Display the data in a scatter plot. For each exercise, tell whether the data have a *positive correlation, negative correlation,* or *no correlation.*

1–2. See margin for graphs.

1.

x	105	95	73	120	74	147	55	62
y	1.3	0.9	0.5	1.7	0.6	1.6	0.3	0.4

positive

2.

x	10.1	10.6	11.1	12.3	9.3	13.1	10.6
y	47	35	64	22	55	27	9

no correlation

3. *Writing* To view data to find correlations, do you prefer using a graphing calculator or graphing the data on graph paper? Why? Answers may vary. Sample: The calculator is easier to use because it graphs the points automatically.

5-6 Teaching Notes

PROBLEM OF THE DAY

Suppose two out of every 25 boys who join scouting will enter a profession first learned through merit badge programs. If a local district signs up 350 new scouts, how many will enter a profession first learned through merit badge programs? 28 scouts

Problem of the Day is also available in Transparencies.

CONNECTING TO PRIOR KNOWLEDGE On the board, draw a scatter plot showing a positive correlation. Ask students to make a prediction about a point not shown on the scatter plot. Have students describe how they made the prediction. Sketch a trend line on the scatter plot. Ask students to recall the name for this line that was introduced in Chapter 2. Remind students that a trend line is used to show the trend and make predictions.

THINK AND DISCUSS

Example 1 Relating to the Real World 

ERROR ALERT! Students may expect that their trend lines will be exactly like those of their neighbors. **Remediation:** Explain that trend lines are not precise. They are similar to estimates. Students may not all have drawn the same trend line. The equation that results in Step 2 may differ between students also. Students' predictions, like estimates, will be similar, but may differ slightly.

Contrast the trend line with the line of best fit. The trend line is

---

Connections Entertainment . . . and more

5-6 Scatter Plots and Equations of Lines

What You'll Learn
- Finding the equation of a trend line
- Using a calculator to find an equation for a line of best fit

...And Why

To analyze and make predictions based on real-world data

What You'll Need
- graphing calculator
- graph paper
- clock or watch with second hand

Ticket Sales for *Forrest Gump*	
Week	Sales (in millions)
1	$39.9
2	$39.2
3	$35.3
4	$28.6
5	$22.9
6	$21.5

Source: Variety

THINK AND DISCUSS

Part 1

Trend Line

Sometimes you can describe data that show a positive or negative correlation with a trend line. Then you can use the trend line to make predictions.

Example 1 Relating to the Real World 

Entertainment A film usually makes the most money in ticket sales during the first few weeks after its release. Find the equation of a trend line for the data about ticket sales for *Forrest Gump.*

Step 1 Draw a scatter plot. Then use a straightedge to draw a trend line. There should be about the same number of points above and below the trend line. Estimate two points on your trend line.

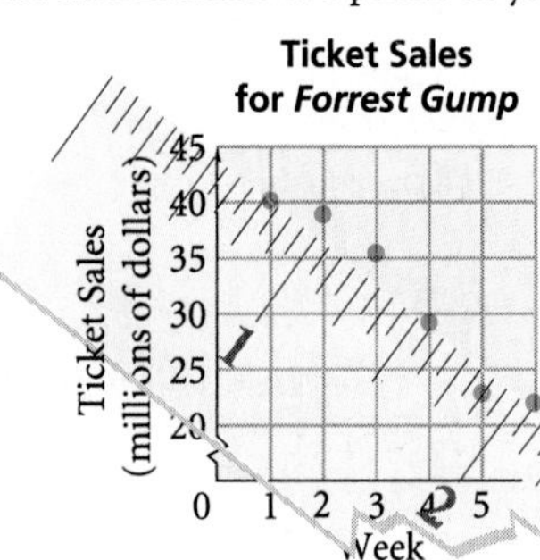

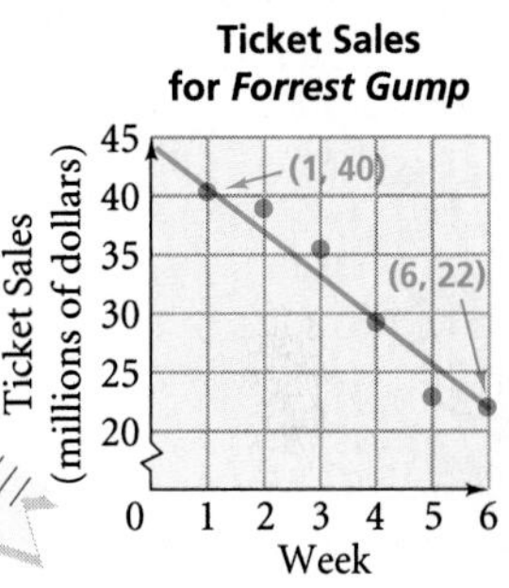

Step 2 Find the slope and y intercept.

$$\text{slope} = \frac{22 - 40}{6 - 1} \quad \leftarrow \frac{y_2 - y_1}{x_2 - x_1}$$

$$= -\frac{18}{5} = -3.6$$

$y = mx + b$ ← Use slope-intercept form to find the y-intercept.

$22 = -3.6(6) + b$ ← Substitute –3.6 for m and (6, 22) for (x, y). Then solve for b.

$22 = -21.6 + b$

$43.6 = b$ ← the y-intercept

Step 3 Write an equation of the line.

$y = -21.6x + 43.6$ ← Substitute –3.6 for m and 43.6 for b.

The equation $y = -3.6x + 43.6$ models the ticket sales.

1. **Predict** the ticket sales in the tenth week after the film's release. $7.6 million

Lesson Planning Options

Prerequisite Skills
- Graphing ordered pairs
- Understanding trend lines

Assignment Options

To provide flexible scheduling, this lesson can be subdivided into parts.

1 **Core** 1–6, 8–10
Extension 7, 18

2 **Core** 11–14, 16, 19
Extension 15, 17

Use Mixed Review to maintain skills.

Resources

 **Student Edition**

Skills Handbook, p. 579
Extra Practice, p. 560
Glossary/Study Guide

Teaching Resources

Chapter Support File, Ch. 5
- Practice 5-6 (two worksheets)
- Reteaching 5-6

Classroom Manager 5-6
Glossary, Spanish Resources
Two-Year Algebra Handbook 5-6

 **Transparencies**
2, 11, 69

an estimate, but the line of best fit should be identical for graphing calculators that use the same computing method.

EXTENSION Challenge students to research the method of linear regression. Most basic statistics textbooks will contain a description of this and other methods, such as the least squares method. Invite students to explain their findings to the class.

On the board, draw two scatter plots—one with a strong positive correlation and one with a weak positive correlation. Draw the scatter plots so they will have similar trend lines. Have students explain in their own words the differences between the two scatter plots. Then lead students to relate their descriptions to the value of the correlation coefficient.

ERROR ALERT! Students may be confused by the form of the slope-intercept equation on the graphing calculator, $y = ax + b$. **Remediation:** Point out to students that this equation ($y = ax + b$) is the same as ($y = mx + b$). The only difference is that a is used instead of m on the calculator.

Example 2 **Relating to the Real World** 

CRITICAL THINKING Question 2 Have students rewrite the slope in ratio form. Then they can use rise and run when describing the slope.

Question 3 Point out that making a prediction does not guarantee that your prediction will come true. For example, greeting card sales could drop in the future as more people send electronic mail.

Additional Examples

FOR EXAMPLE 1

In Mr. Arredondo's astronomy class, students keep a record of the sunset times on Mondays for a five-week period. Find the equation of the trend shown in the graph below.

Sunset Times

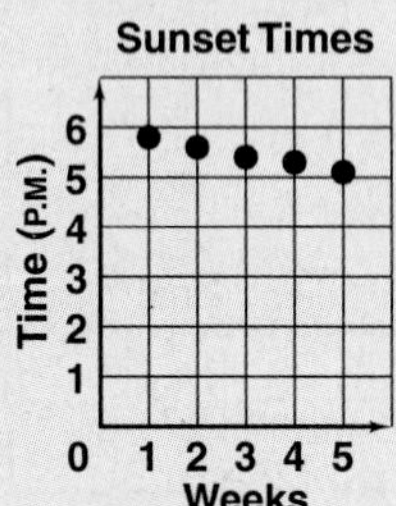

Answers may vary. Sample: $y = -0.16x + 6$

Discussion: *Would you expect this trend to continue indefinitely?*

FOR EXAMPLE 2

Plot the following points to find the equation of the line of best fit.

Students Receiving an "A"

Grading Period	Number of "A"s
1	5
2	8
3	11
4	11
5	14
6	16

$y = 2x + 4$

242

Part 2

Line of Best Fit

GRAPHING CALCULATOR HINT

The equation of the line of best fit will be in one of these forms:

$y = ax + b$ (a: slope, b: y-intercept)

$y = a + bx$ (a: y-intercept, b: slope)

The most accurate trend line showing the relationship between two sets of data is called the **line of best fit.** A graphing calculator computes the equation of a line of best fit using a method called linear regression.

The graphing calculator also gives you the **correlation coefficient** r, which will tell you how closely the equation models the data.

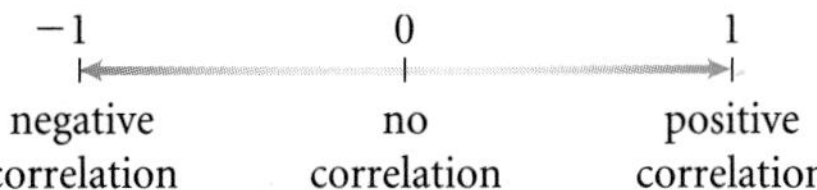

When the data points cluster around a line, there is a strong correlation between the line and the data. The nearer r is to -1 or 1, the more closely the data cluster around the line of best fit.

Example 2 **Relating to the Real World** 

Business Use a graphing calculator to find the equation of the line of best fit for the data at the left. Is there a strong correlation shown by the data?

Greeting Card Sales

Year	Sales (in billions)
1980	$2.05
1981	$2.30
1982	$2.45
1983	$2.70
1984	$3.20
1985	$3.45
1986	$3.65
1987	$3.75
1988	$3.90
1989	$4.20
1990	$4.60
1991	$5.00
1992	$5.35
1993	$5.60

Source: Greeting Card Association

Step 1 Let 80 correspond to 1980. Enter the data for years and then the data for sales in your graphing calculator.

Step 2 Find the equation for the line of best fit.

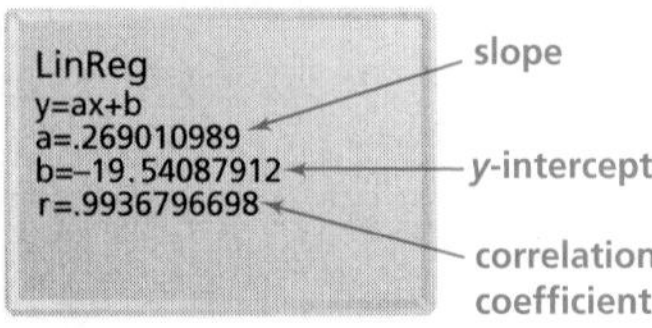

The equation for the line of best fit is $y = 0.269010989x - 19.54087912$. Since r is very close to 1, there is a strong correlation shown by the data.

2. *Critical Thinking* What is the meaning of the slope for the data in Example 2? Greeting card sales increased at the rate of about $.269 billion per year.
3. Use the equation in Example 2 to **predict** sales of greeting cards in the year 2010. $10.05 billion
4. *Try This* Find the equation of the line of best fit. Tell whether there is a strong correlation between the data.

a.

x	1	2	3	4	5
y	2	−3	8	9	−25

$y = -4.2x + 10.8$; no

b.

x	1	2	3	4	5
y	21	15	12	9	7

$y = -3.4x + 23$; yes

WORK TOGETHER

Students may have difficulty keeping time while counting their partner's beats. Help them by counting the seconds yourself and announcing the intervals of 10 s, 20 s, and so on.

ESL As an example for students unfamiliar with "to hum," "to tap out," and "count the beat," ask students to name a popular song. Ask them to hum and tap out the beat to the main chorus as the other students count the beats.

DIVERSITY Some students may feel embarrassed because they have difficulty keeping a steady beat. Explain that the ability to keep a steady beat is a learned skill just like the ability to do math, drive a car, or use a computer. Invite a music student to describe the devices that music teachers use, such as metronomes, to help students stay in rhythm.

KINESTHETIC LEARNING Students who have difficulty keeping rhythm while singing a song may find the task easier if they tap their feet, clap their hands, or snap their fingers.

CONNECTING TO THE STUDENTS' WORLD Encourage students who are in the school band to bring music they are working on to class. Have the class repeat the Work Together activity while band players practice their music.

Exercises ON YOUR OWN

ALTERNATIVE ASSESSMENT **Exercises 1–6** To assess students' understanding of correlation coefficients, randomly call on students and ask them to estimate the correlation coefficient of these scatter plots. Make sure they justify their answers. Remember that a correlation coefficient of 1

WORK TOGETHER

Music When you sing a song, your inner clock helps you keep the rhythm. Work with a partner or two to see if you keep a steady beat.

5–7. Check students' work.

5. *Data Collection* Choose a song. While you hum your song to yourself, tap out the beats. Have someone in your group count the number of beats you tap in 10 s, 20 s, 30 s, and 40 s.
6. Graph the data and find the equation of a trend line or use a graphing calculator to find the equation of the line of best fit.
7. Repeat Questions 5 and 6 for each person in your group.
8. Which song had the fastest beat? the slowest? How do you know? See left.
9. Which of you kept a steady beat the best? How do you know? Answers may vary. Sample: The line that is closest to the data points or the line with the greatest r-value models the steadiest beat.

8. Answers may vary. Sample: The line with the greatest slope modeled the fastest beat.

Exercises ON YOUR OWN

Decide if the data in each scatter plot follow a linear pattern. If they do, find the equation of a trend line. 1–6. Answers may vary. Samples are given.

1. **Calories and Fat for Fast Food Meals**

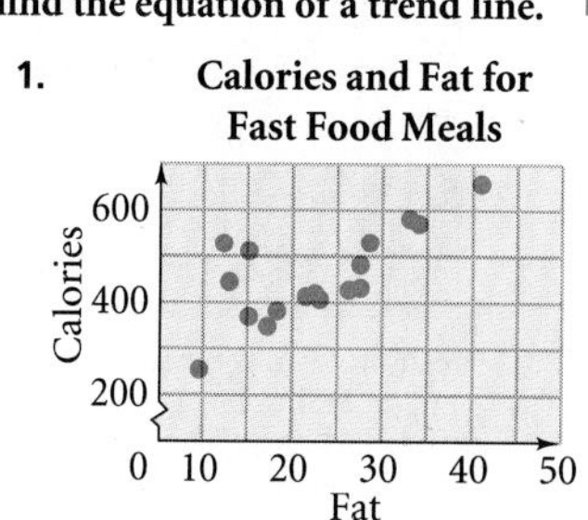

yes; $y = 11.25x + 200$

2. **Deer Population Bridger Mountains, Montana**

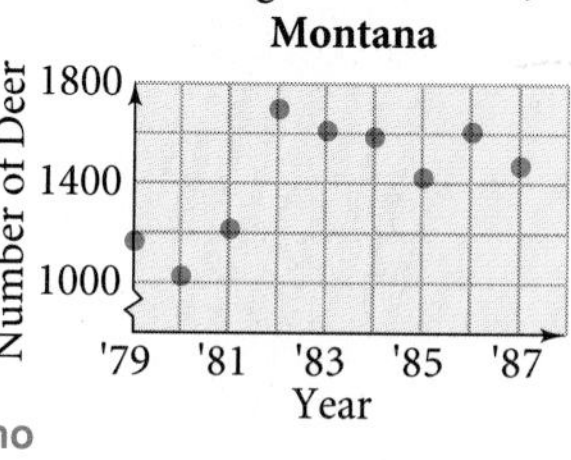

no

3. **Effect of Air Temperature on Speed of Sound**

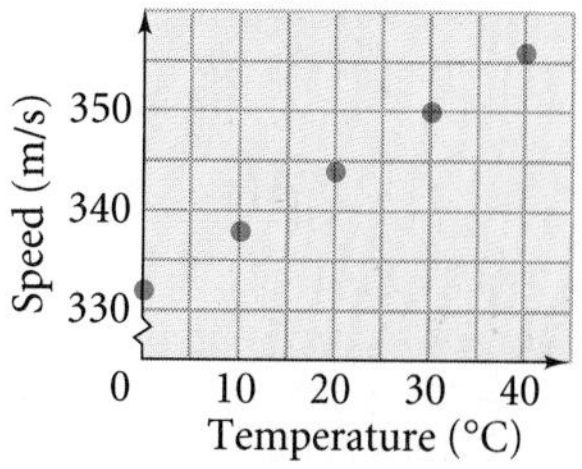

Speed (m/s): 330, 340, 350; Temperature (°C): 0, 10, 20, 30, 40

yes; $y = 0.6x + 330$

4. **Olympic 5000 Meter Men's Speed Skating**

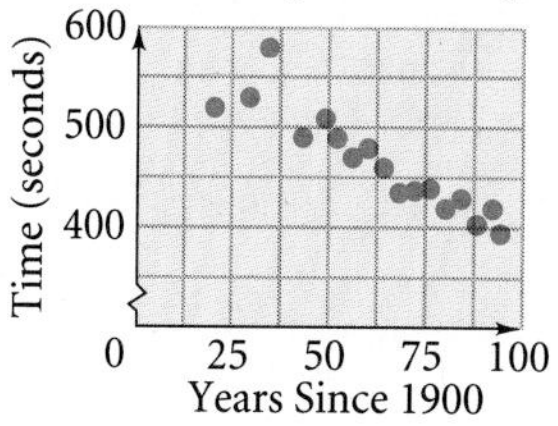

Time (seconds): 400, 500, 600; Years Since 1900: 0, 25, 50, 75, 100

yes; $y = -2.5x + 650$

5. **Animal Longevity and Gestation**

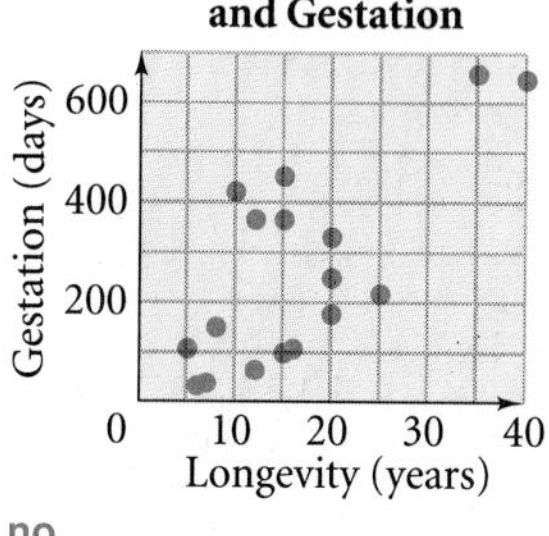

no

6. **Public College Enrollment**

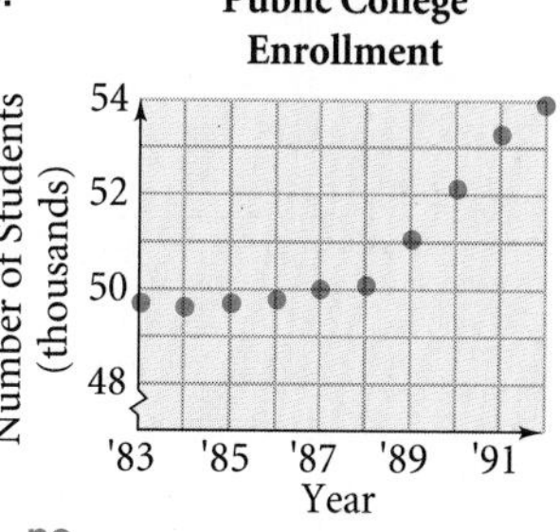

no

7. a. *Data Collection* Measure the wrist and neck sizes of five different people. Make a scatter plot with the data you collected.
   b. Write an equation for your data. Then **predict** the neck size of someone whose wrist measures 6.5 in. Check students' work.

Technology Options

For Exercises 7, and 11–18, students may use spreadsheet or graphing software to produce graphs to find the equation of a trend line.

Prentice Hall Technology

Calculator Graphing Calculator Handbook: Procedures 21, 23

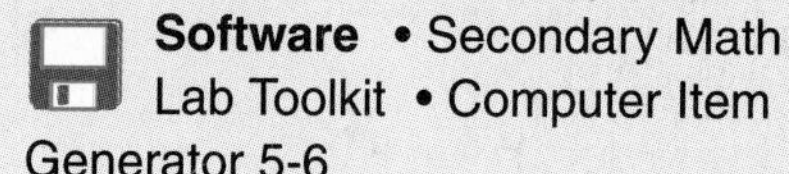

Software • Secondary Math Lab Toolkit • Computer Item Generator 5-6

Internet • See the Prentice Hall site. (http://www.phschool.com)

corresponds to a perfect correlation and a correlation coefficient of 0 means there is no correlation whatsoever.

Exercise 7 Advise students to conduct this exercise carefully, taking very loose measurements. Consider taking the measurements yourself and appoint a student to write them on the board for the entire class to use.

Exercises 11–14 Group the students in pairs. Have one student use paper and pencil to find a trend line while the partner uses the graphing calculator to find the line of best fit. Have students compare their answers.

Exercise 16c Remind students that a mathematical model is only an estimate. The model is only accurate for a limited range of values.

RESEARCH Exercise 18 You may also find population data on the World Wide Web at the U.S. Census Bureau website. Information can also be found on the web by searching for *census* on one of the search engines.

Exercise 18 Remind students that a correlation does not show cause and effect.

page 243-245 On Your Own

11a. 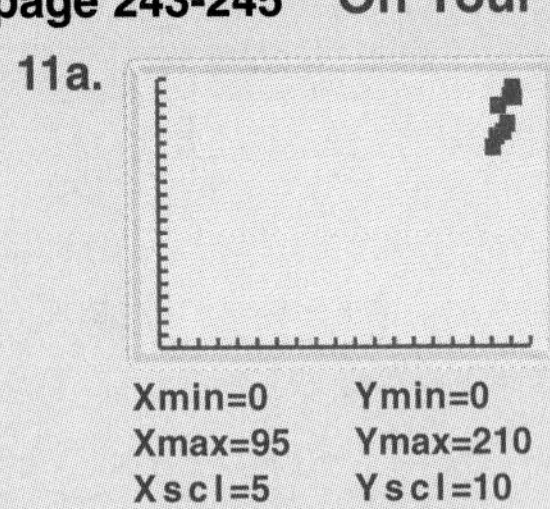

Xmin=0 Ymin=0
Xmax=95 Ymax=210
Xscl=5 Yscl=10

b. Answers may vary. Sample: $y = 6.404761905x - 392.7261905$

c. about 280 million TV sets

12a. 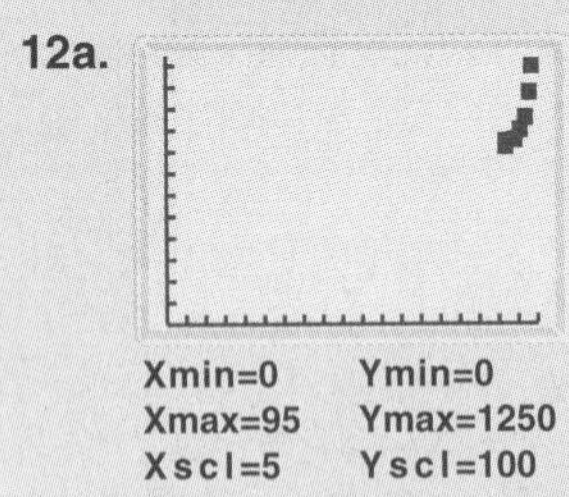

Xmin=0 Ymin=0
Xmax=95 Ymax=1250
Xscl=5 Yscl=100

b. Answers may vary. Sample: $y = 62.5x - 4676.428571$

c. about 1886 million gal

13a–c, 14a–c, 15a, 16a. See back of book.

page 245 Mixed Review

19a. See back of book.

Critical Thinking **Match each graph with its correlation coefficient.**

8. 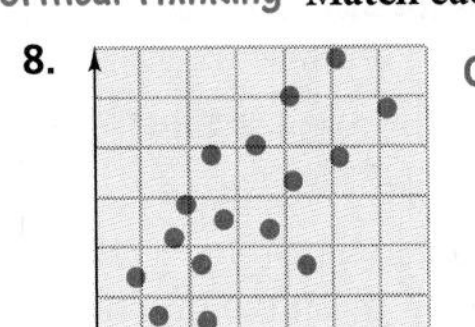 C

9. 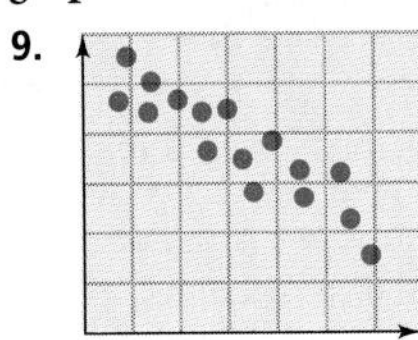 A

10. 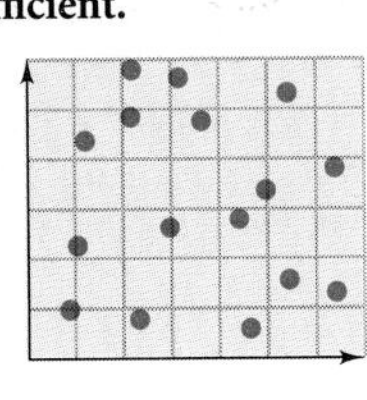 B

A. $r = -0.95$

B. $r = 0.13$

C. $r = 0.77$

For Exercises 11–14, follow these instructions. 11–14. See margin.

a. Make a scatter plot of the data.

b. *Choose* Use paper and pencil to find the equation of a trend line, or use a graphing calculator to find the equation of the line of best fit.

c. Use your equation to make a prediction for the year 2005.

11. **Households with Television sets (in millions)**

Year	Number
1986	158
1987	163
1988	168
1989	176
1990	193
1991	193
1992	192
1993	201

Source: *Statistical Abstract of the United States*

12. **Gallons of Ethanol Fuel Produced Annually (in millions)**

Year	Amount
1987	820
1988	830
1989	850
1990	900
1991	960
1992	1080
1993	1200

Source: National Corn Growers Association; Renewable Fuels Association

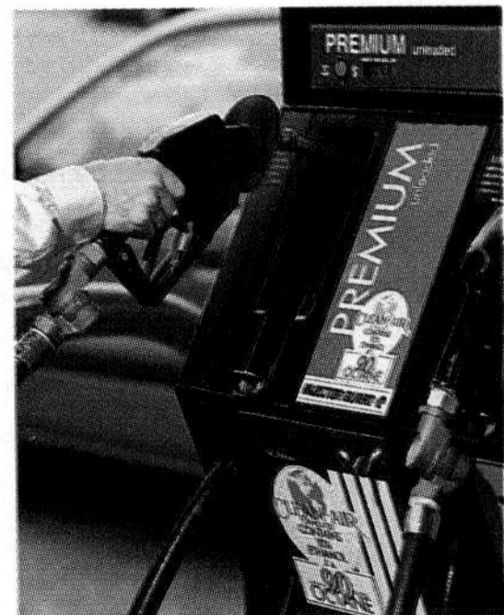

Ethanol is a renewable fuel made from plants. In the United States, it is usually made from corn. In Brazil, the largest producer of ethanol, it is made from sugar cane.

13. **Hispanic Population Growth in the United States (percent of total population)**

Year	Percent
1970	4.5
1980	6.4
1988	7.1
1990	9.0
1991	9.3
1992	9.5
1993	9.8

Source: United States Census Bureau

14. **Fast Food Sales (in billions)**

Year	Sales
1985	$46.4
1986	$49.4
1987	$57.6
1988	$65.0
1989	$68.3
1990	$74.2
1991	$77.0

Source: United States Department of Agriculture

15. a. *Open-ended* Make a table of data for a linear function. Use a graphing calculator to find the equation of the line of best fit. See margin for sample.

b. What is the correlation coefficient for your linear data?

for linear function with positive slope, 1; for linear function with negative slope, −1

Exercises MIXED REVIEW

Exercise 20 Remind students that once they have written the formula for D2, they can use the same formula in cells D3–D7. If students complete this spreadsheet on a computer, they can paste the formula entered in cell D2 into cells D3–D7.

GETTING READY FOR LESSON 5-7 These exercises prepare students for graphing equations using *x*- and *y*-intercepts.

Wrap Up

THE BIG IDEA Ask the following questions:

- *What is a trend line?*
- *What information can you get from a trend line?*
- *How is a trend line similar to and different from a line of best fit?*

RETEACHING ACTIVITY Students use toothpicks to determine slope of a trend line. One toothpick is placed vertically, and another is placed horizontally. (Reteaching worksheet 5-6)

16. a. *Business* Make a scatter plot of the data at the right. Find the equation of a trend line, or use a graphing calculator to find the equation of the line of best fit. **See margin.**

b. Use your equation to **predict** the percent of refillable bottles used in 1950 and the percent that will be used in the year 2000. **127%; −23%**

c. Are your answers to part(b) reasonable in the real-world situation? Explain. **No; the percent of refillable bottles can only be between 0 and 100.**

Refillable Soft-Drink Containers

Year	Percent of Total Sold
1960	96
1965	84
1970	65
1975	57
1980	34
1985	23
1990	7

17. *Writing* Decide whether the following statement is *true* or *false*. Then explain your answer: A trend line is a line that connects as many points in a set of data as possible. **False; a trend line need not pass through any data points.**

18. a. *Research* Choose a topic below or one of your own. You can find data in an almanac or the *Statistical Abstract of the United States*.
- population density of a state and its crime rate
- population of a state and the number of physicians who practice there

b. Find the equation of a trend line or use a graphing calculator to find the line of best fit for the data.

c. Is there a strong correlation shown by the data? Explain.

d. If the correlation is strong, use your equation to make a prediction.

18a–d. Check students' work.

19. a. Make a scatter plot of the data in the table below. **See margin.**

b. Use a graphing calculator to find the equation of the line of best fit. $y = 1.206586826x + 45.71556886$

c. Does the data show a strong correlation? Explain. **no; $r \approx 0.448$, which is not that close to 1.**

d. *Critical Thinking* Do you think that the correlation would be similar for other school choruses? Why or why not? **Sample: yes, there should be some positive correlation between age and height, since teens grow taller as they get older. However, there are many heights of people at all ages.**

Members of the School Chorus

Age (in years)	13	15	17	16	16	13	14	15	17	17	16	14	14
Height (in inches)	62	70	65	62	71	59	64	63	68	60	67	60	61

Exercises MIXED REVIEW

20. *Health* The spreadsheet shows the breathing rates for people of different ages.

a. Write a formula for cell D2. **C2/B2**

b. Find the values of cells D2 through D7. **See right.**

	A	B	C	C
1	Age	Breaths/min	Liters of oxygen/min	Liters of oxygen/breath
2	30–39	14.7	146.7	10.0
3	40–49	14.6	139.1	9.5
4	50–59	14.6	127.8	8.8
5	60–69	14.2	117.3	8.3
6	70–79	14.4	106.2	7.4
7	80–89	13.2	75.5	5.7

Source: *Prevention's Giant Book of Health Facts*

Solve each equation.

21. $3x + 4 = 73$ **23**
22. $-2.5 - 5z = -3z$ **−1.25**
23. $\frac{4}{x} = \frac{2}{3 + x}$ **−6**
24. $0.2t + 1.3 = -9.1$ **−52**
25. $4(3y - 5) = 16$ **3**
26. $\frac{1}{2}x - \frac{3}{4} = \frac{3}{2} + x$ $-4\frac{1}{2}$

Getting Ready for Lesson 5-7

Find the value of *x* when *y* = 0.

27. $y = 3x + 8$ $-\frac{8}{3}$
28. $4x - 5y = 7$ $\frac{7}{4}$
29. $y = \frac{1}{2}x - 10$ **20**
30. $y = 2x$ **0**

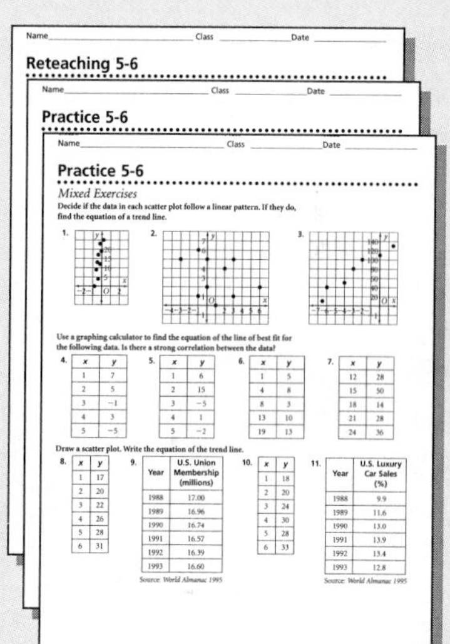

Lesson Quiz

Lesson Quiz is also available in Transparencies.

1. Decide if the data in the following scatter plot follows a linear pattern. If it does, find the equation of the trend line.

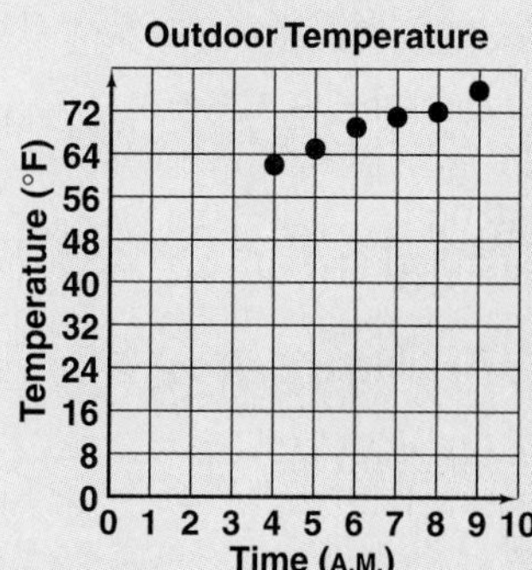

Yes; answers may vary. Sample: $y = 2.5x + 48$

5-7 Teaching Notes

PROBLEM OF THE DAY

The choir director offers to stop at a fast food restaurant with 60 students after a performance. A discussion reveals that 35 students like cheeseburgers, 37 students like pizza, 36 students like tacos, 22 students like cheeseburgers or pizza, 18 students like pizza or tacos, 23 students like tacos or cheeseburgers, and 15 students like all three choices. How many students like only one of the three fast food choices? 27 students

Problem of the Day is also available in Transparencies.

CONNECTING TO PRIOR KNOWLEDGE Ask: *How is the slope-intercept form of an equation like or unlike a recipe?* You add *A* amount of *x* to *B* amount of *y* for a result of *C*.

THINK AND DISCUSS

Help students understand why *A* and *B* in the equation cannot both be zero.

Lead students to understand that *y* is always zero at the *x*-intercept.

Lesson Planning Options

Prerequisite Skills

- Substituting numbers for variables in equations
- Solving equations

Assignment Options

To provide flexible scheduling, this lesson can be subdivided into parts.

1 **Core** 1–11, 15–22
Extension 12–14, 23

2 **Core** 24–31
Extension 32

Use Mixed Review to maintain skills.

Resources

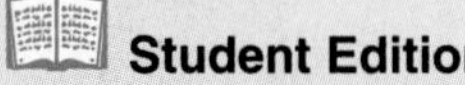

Student Edition
Skills Handbook, p. 576
Extra Practice, p. 560
Glossary/Study Guide

Teaching Resources
Chapter Support File, Ch. 5
- Practice 5-7 (two worksheets)
- Reteaching 5-7

Classroom Manager 5-7
Glossary, Spanish Resources
Two-Year Algebra Handbook 5-7

Transparencies
2, 27, 70

Connections Fitness . . . and more

5-7 $Ax + By = C$ Form

What You'll Learn
- Graphing equations using *x*- and *y*-intercepts
- Writing equations in $Ax + By = C$ form
- Modeling situations with equations in the form $Ax + By = C$

...And Why
To investigate real-world situations, such as burning calories when running and jogging

What You'll Need
- graph paper
- graphing calculator

THINK AND DISCUSS

Part 1

Graphing Equations

The slope-intercept form is just one form of a linear equation. Another form is $Ax + By = C$, which is useful in making quick graphs.

$Ax + By = C$ Form of a Linear Equation

$Ax + By = C$ is a linear equation, where *A* and *B* cannot both be zero.

$$\downarrow \quad \downarrow \quad \downarrow$$
$$3x + 4y = 8$$

To make a quick graph, you can use the *x*- and *y*-intercepts. The **x-intercept** is the *x*-coordinate of the point where a line crosses the *x*-axis.

Example 1

Graph $3x + 4y = 8$.

Step 1 To find the *x*-intercept, substitute 0 for *y* and solve for *x*.

$$3x + 4y = 8$$
$$3x + 4(0) = 8$$
$$3x = 8$$
$$x = \frac{8}{3}, \text{ or } 2\frac{2}{3}$$

The *x*-intercept is $2\frac{2}{3}$.

Step 2 To find the *y*-intercept, substitute 0 for *x* and solve for *y*.

$$3x + 4y = 8$$
$$3(0) + 4y = 8$$
$$4y = 8$$
$$y = 2$$

The *y*-intercept is 2.

Step 3 Plot $(2\frac{2}{3}, 0)$ and $(0, 2)$. Draw a line through the points.

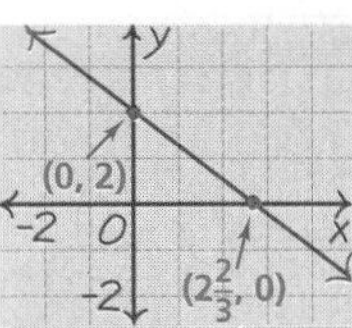

1. Mental Math Find the *x*- and *y*-intercept of each equation.
   a. $3x + 4y = 12$ 4; 3
   b. $5x + 2y = -10$ −2; −5
   c. $2x - y = 4$ 2; −4

To graph an equation on a graphing calculator, you must transform the equation to slope-intercept form. You can use the *x*- and *y*-intercepts to find an appropriate range for each axis.

Example 1

Students often enjoy learning these shortcuts as an alternative to plotting multiple points.

VISUAL LEARNING To find the *x*-intercept, have students cover the *y*-term with their fingers to concentrate on the equation that remains. Then cover the *x*-term to find the *y*-intercept.

ERROR ALERT! **Question 1** When finding the *x*- and *y*-intercepts, some students may confuse the *x*- and *y*- values. **Remediation:** Have students begin by writing down the ordered pair with a question mark in place of the intercept and a zero in place of the other variable.

Example 2

Point out that students could also find the *y*-intercept using the method shown in Step 2. Make sure students understand what range they must set on the calculators to include the *x*- and *y*-axes.

CRITICAL THINKING Question 4 Suggest that students test what happens by changing the Xmin and Xmax settings and redrawing the graph.

Example 3 **Relating to the Real World** 

OPEN-ENDED Question 6b Invite students to brainstorm for other examples of situations involving two interdependent variables.

Example 2

Graphing Calculator Graph $5x - 3y = 120$.

Step 1 Write the equation in slope-intercept form.

$5x - 3y = 120$ ← Solve for *y*.

$-3y = -5x + 120$ ← Subtract $5x$ from each side.

$y = \frac{5}{3}x - 40$ ← Divide each side by -3.

Step 2 Find the *x*- and *y*-intercepts.

$5x - 3y = 120$

$5x - 3(0) = 120$

$5x = 120$

$x = 24$ ← *x*-intercept

From Step 1, the *y*-intercept is –40.

Step 3 Set the ranges to include the *x*- and *y*-axes and the intercepts. Then graph $y = \frac{5}{3}x - 40$.

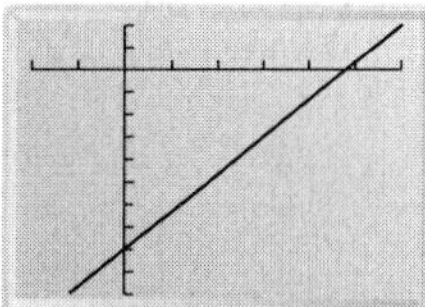

Xmin=–10 Ymin=–50
Xmax=30 Ymax=10
Xscl=5 Yscl=5

2. *Critical Thinking* What advantage is there in making Xmin less than 0 and Xmax greater than 24 for the graph on the calculator? You can see both axes and both intercepts on the screen.
3. *Graphing Calculator* Graph $4x - 12y = 54$. Sketch your graph. Include Xmin, Xmax, Ymin, Ymax, and the *x*- and *y*-intercepts. See margin.

Part 2 Writing Equations

You can write equations for real-world situations using the $Ax + By = C$ form.

Example 3 **Relating to the Real World** 

Fitness When you jog, you burn 7.3 calories/min. When you run, you burn 11.3 calories/min. Write an equation to find the times you would need to run and jog in order to burn 500 calories.

Define x = minutes spent jogging y = minutes spent running

Relate 7.3 × minutes jogging + 11.3 × minutes running = 500

Write $7.3x + 11.3y = 500$

4. **a.** Use the intercepts to graph the equation in Example 3, or graph the equation on a graphing calculator. See margin.
   b. *Open-ended* Use your graph to estimate three different running and jogging times needed to burn 500 calories. Sample: 5 min jog., 41 min run.; 22 min jog., 30 min run.; 39 min jog., 19 min run.

Additional Examples

FOR EXAMPLE 1

Graph $2x + 3y = -2$.

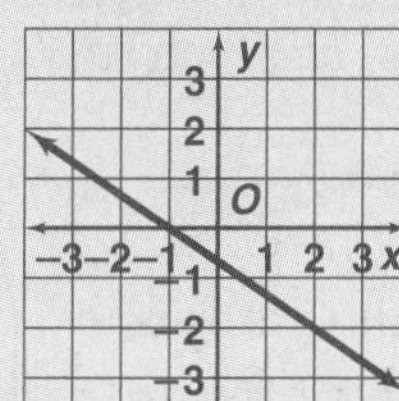

FOR EXAMPLE 3

Ryan has two part-time jobs. He can either make $12/hr mowing lawns or $5/hr delivering newspapers. Write and graph an equation to find the amount of time he must work at each job to make a total of $130.

$12x + 5y = 130$

Ryan's Choice

Hours delivering: 4, 8, 12, 16, 20, 24, 28
Hours mowing: 0, 1, 2, 3, 4, 5, 6, 7, 8, 9, 10

Discussion: *What possible reasons could Ryan have for not choosing to work only the highest paying job?*

Example 4 Relating to the Real World 

Suggest that students substitute the point (–1, 7) into the equation to check that the equation is correct.

Exercises ON YOUR OWN

ALTERNATIVE ASSESSMENT Exercises 4–11 Ask students to graph each equation using a graphing calculator. Follow the instructions for Exercises 15–22. This will help you assess students' understanding of the $Ax + Bx = C$ form.

OPEN-ENDED Exercise 12c Encourage students to think about a range of ticket prices that people would find most affordable.

STANDARDIZED TEST TIP Exercise 23 Remind students that memorizing the slope-intercept equivalent of the $Ax + By = C$ form ($y = -\frac{A}{B}x + \frac{C}{B}$) can save time during examinations.

Exercise 30 Remind students that the only way a ratio or fraction can equal zero is if the numerator is zero. The easiest slope ratio to use in this exercise is $\frac{0}{1}$.

CONNECTING TO THE STUDENTS' WORLD Exercise 32 Challenge students to research how much protein they need to eat every day. Then have them choose two sources of protein from among the following and write an equation modeling how much of each source they should eat each day. Sample sources of protein: Swiss cheese 7 g/oz, skinned

Technology Options

For Exercises 4–12 and 35–38, students may use graphing software or a graphing calculator to make graphs and then copy them on graph paper.

Prentice Hall Technology 

Calculator Graphing Calculator Handbook: Procedure 4

Software • Secondary Math Lab Toolkit • Computer Item Generator 5-7

CD-ROM Multimedia Algebra Lab 5

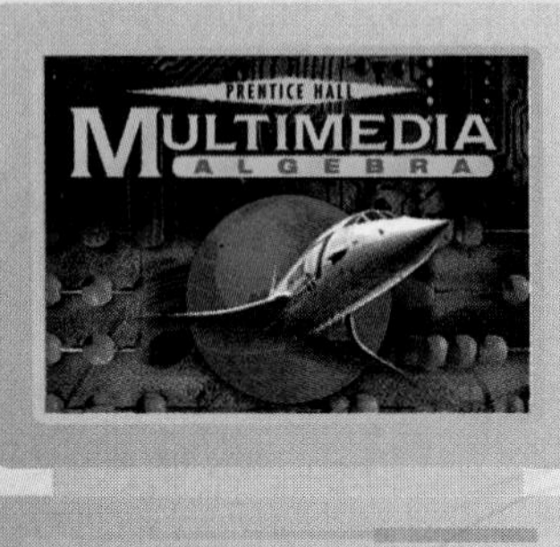

Internet • See the Prentice Hall site. (http://www.phschool.com)

page 246–248 Think and Discuss

5, 6a. See back of book.

pages 248–249 On Your Own

4–11, 12b, 13, 15–22, 32b. See back of book.

page 249 Mixed Review

35–38. See back of book.

You can write an equation in $Ax + By = C$ form if you know the slope and one point. For a line with slope m through point (x_1, y_1), this equation is true:

$$\frac{y - y_1}{x - x_1} = m$$

Example 4

Write an equation of the line with slope $-\frac{1}{2}$ through the point $(-1, 7)$.

$\frac{y - y_1}{x - x_1} = m$

$\frac{y - 7}{x - (-1)} = -\frac{1}{2}$ ← Substitute (–1, 7) for (x_1, y_1) and $-\frac{1}{2}$ for m.

$2(y - 7) = -1(x + 1)$ ← Simplify $x - (-1)$ and cross multiply.

$2y - 14 = -x - 1$ ← Use the distributive property.

$x + 2y = 13$ ← Add x and 14 to each side in order to write $Ax + By = C$ form.

5. *Critical Thinking* How is $\frac{y - y_1}{x - x_1}$ related to the ratio for finding the slope of a line through two points? The slope formula is $\frac{y_2 - y_1}{x_2 - x_1}$. You use (x, y) in place of (x_2, y_2) in $\frac{y - y_1}{x - x_1} = m$.

Exercises ON YOUR OWN

Match the equation with its graph.

1. $2x + 5y = 10$ A
2. $2x - 5y = 10$ B
3. $-2x + 5y = 10$ C

A. 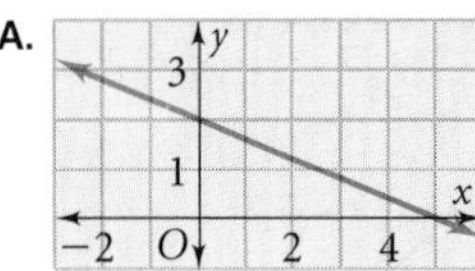

B. 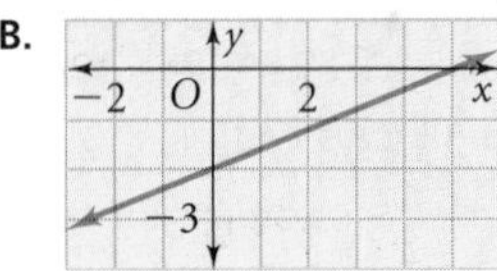

C. 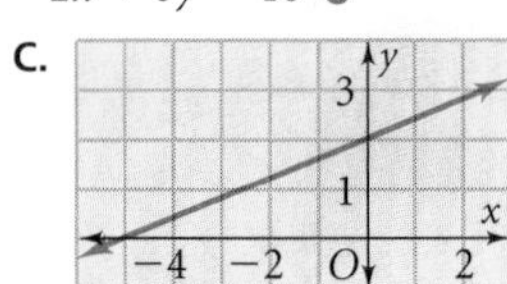

Graph each equation. 4–11. See margin.

4. $x + y = 2$
5. $5x - 12y = 30$
6. $-3x + y = 6$
7. $x - y = -7$
8. $-4x + y = -6$
9. $2x + 5y = 10$
10. $2x - y = 8$
11. $-3x + 4y = 12$

12. a. *Fund Raising* Suppose your school is having a dinner to raise money for new music and art supplies. You estimate that 200 children and 150 adults will attend. Write an equation to find what ticket prices you should set to raise \$900. $200c + 150a = 900$

    b. *Open-ended* Graph your equation. Choose three possible prices you could set for children's and adults' tickets. Explain which you think is the best choice. See margin.

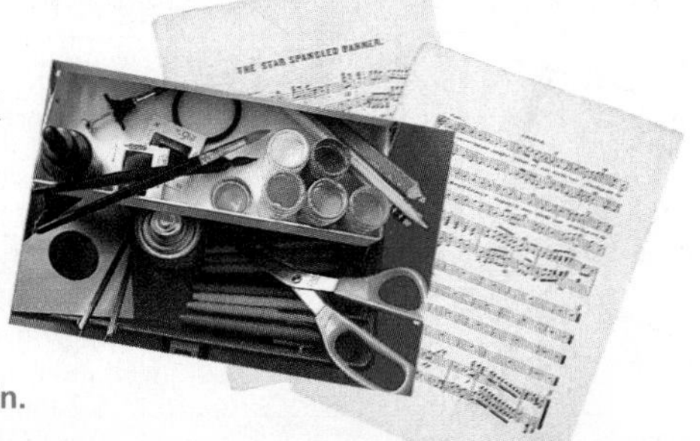

roasted chicken breast 9 g/oz, shredded wheat 3 g/oz, shelled sunflower seeds 6 g/oz, unsweetened lemonade $\frac{1}{6}$ g/oz.

Exercises MIXED REVIEW

PROBABILITY Exercises 33–34 Suggest that students review independent and dependent probabilities before answering these questions.

GETTING READY FOR LESSON 5-8 These exercises prepare students for examining the equations of parallel and perpendicular lines.

Wrap Up

THE BIG IDEA Ask: *Why is it useful to have two forms of a linear equation?*

RETEACHING ACTIVITY Students graph equations. Then they use self-stick notes to cover part of the equation. This enables them to find x- and y- intercepts with no confusion. (Reteaching worksheet 5-7)

3. **Writing** Two forms of a linear equation are the slope-intercept form and the $Ax + By = C$ form. Explain when each is most useful. **See margin.**

4. An equation is in standard form when A, B, and C are integers for $Ax + By = C$. Write each equation in standard form.

Sample: $Ax + By = C$ Form		Standard Form	
$3.5x + 7.2y = 12$	→	$35x + 72y = 120$	← Multiply each side by 10.

a. $3.8x + 7.2y = 5.4$ **$19x + 36y = 27$**
b. $0.5x - 0.75y = 1.25$ **$2x - 3y = 5$**
c. $\frac{2}{3}x + \frac{1}{6}y = 4$ **$4x + y = 24$**

Graphing Calculator **Graph each equation. Make a sketch of the graph. Include the x- and y-intercepts.** **15–22. See margin.**

5. $12x + 15y = -60$ 16. $8x - 10y = 100$ 17. $-5x + 11y = 120$ 18. $4x - 9y = -72$

9. $-3x + 7y = -42$ 20. $12x - 9y = 144$ 21. $9x + 7y = 210$ 22. $3x - 8y = 72$

3. **Standardized Test Prep**

a. Write $Ax + By = C$ in slope-intercept form by solving for y. **$y = -\frac{A}{B}x + \frac{C}{B}$**

b. Which expression equals the slope m? **C**

A. $-\frac{B}{A}$ B. $\frac{C}{A}$ C. $-\frac{A}{B}$ D. A E. $\frac{C}{B}$

c. Which expression equals the y-intercept b? **E**

A. $-\frac{B}{A}$ B. $\frac{C}{A}$ C. $-\frac{A}{B}$ D. A E. $\frac{C}{B}$

Write an equation in $Ax + By = C$ form for the line through the given point with the given slope.

4. $(3, -4)$; $m = 6$ **$-6x + y = -22$**
25. $(4, 2)$; $m = -\frac{5}{3}$ **$5x + 3y = 26$**
26. $(0, 2)$; $m = \frac{4}{5}$ **$-4x + 5y = 10$**
27. $(-2, -7)$; $m = -\frac{3}{2}$ **$3x + 2y = -20$**
8. $(4, 0)$; $m = 1$ **$-x + y = -4$**
29. $(5, -8)$; $m = -3$ **$3x + y = 7$**
30. $(-5, 2)$; $m = 0$ **$y = 2$**
31. $(1, -8)$; $m = -\frac{1}{5}$ **$x + 5y = -39$**

2. **Nutrition** Suppose you are preparing a snack mix. You want the total protein from peanuts and granola to equal 28 g.

a. Write an equation for the protein content of your mix.

b. Graph your equation. Use your graph to find how many ounces of granola you should use if you use one ounce of peanuts.

a. p = peanuts, g = granola; $7p + 3g = 28$
b. See margin for graph; 7 oz

Granola Protein: 3 g/oz
Peanuts Protein: 7 g/oz

Exercises MIXED REVIEW

Probability **Find each probability.**

3. P(rolling a 2, then a 4 on a number cube) **$\frac{1}{36}$**

34. P(getting heads on both coins when you toss two coins) **$\frac{1}{4}$**

Getting Ready for Lesson 5-8

Graph each pair of lines on one set of axes. **35–38. See margin.**

5. $y = 4x + 1$; $y = 4x - 3$
36. $y = 3x - 8$; $y = -\frac{1}{3}x - 2$
37. $3y = 2x + 6$; $y = \frac{2}{3}x + 4$
38. $4y = x - 8$; $y = -4x - 1$

Name____ Class ____ Date ____

Reteaching 5-7

Name____ Class ____ Date ____

Practice 5-7

Name____ Class ____ Date ____

Practice 5-7

Mixed Exercises

Graph each equation.

1. $x + y = 3$ 2. $x + 3y = -9$ 3. $-2x + 3y = 6$ 4. $5x - 4y = -20$
5. $3x + 4y = 12$ 6. $7x + 3y = 21$ 7. $3x - 5y = 15$ 8. $2x - 3y = 4$
9. $x + 4y = 4$ 10. $3x - 2y = -6$ 11. $5x + 2y = 5$ 12. $-7x + 2y = 14$
13. $3x + y = 3$ 14. $-3x + 5y = 15$ 15. $2x + y = 3$ 16. $8x - 3y = 24$

Graph each equation using a graphing calculator. Make a sketch of the graph. Include Xmin, Xmax, Ymin, Ymax, and the x- and y-intercepts.

17. $6x + 5y = 90$ 18. $4x + 7y = 84$ 19. $9x + 5y = 180$ 20. $3x + 8y = -120$
21. $7x - 10y = 140$ 22. $-6x + 11y = 132$ 23. $5x - 4y = -140$ 24. $-11x + 3y = 165$

Write an equation for a line through the given point with the given slope using the $Ax + By = C$ form.

25. $(3, 1)$; $m = 4$ 26. $(5, 4)$; $m = 2$ 27. $(-3, 3)$; $m = -2$ 28. $(6, -2)$; $m = 5$
29. $(2, -7)$; $m = \frac{2}{3}$ 30. $(9, 7)$; $m = -4$ 31. $(-1, 2)$; $m = -\frac{4}{5}$ 32. $(-5, 1)$; $m = -\frac{1}{5}$
33. $(6, -7)$; $m = \frac{5}{2}$ 34. $(-4, -1)$; $m = \frac{7}{3}$ 35. $(-4, 2)$; $m = -\frac{1}{3}$ 36. $(-8, 10)$; $m = -6$

37. The drama club sells 200 lb of fruit to raise money. They sell the fruit in 5-lb bags and 10-lb bags.
a. Write an equation to find the number of each type of bag that they should sell.
b. Graph your equation.
c. Use your graph to find two different combinations of types of bags.

38. The student council is sponsoring a carnival to raise money. Tickets cost \$5 for adults and \$3 for students. They want to raise \$450.
a. Write an equation to find the number of each type of ticket they should sell.
b. Graph your equation.
c. Use your graph to find two different combinations of tickets sold.

39. Anna goes to a store to buy \$70 worth of flour and sugar for her bakery. A bag of flour costs \$5 and a bag of sugar costs \$7.
a. Write an equation to find the number of bags of each type Anna can buy.
b. Graph your equation.

Lesson Quiz

Lesson Quiz is also available in Transparencies.

1. Graph the equation $-4x - 2y = 6$.

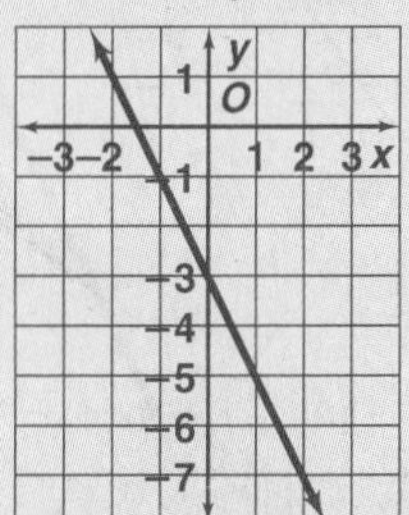

2. Write an equation for the line passing through point (3, 0) and having a slope of −2 using the $Ax + By = C$ form.
$2x + y = 6$

5-8 Teaching Notes

PROBLEM OF THE DAY

A 12 ft square patio is built in a 25 ft by 35 ft rectangular backyard. How many square feet of the backyard are not covered by the patio? 731 ft^2

Problem of the Day is also available in Transparencies.

CONNECTING TO PRIOR KNOWLEDGE Ask: *What is a perpendicular line? What math language have you learned that you could use to describe a line that is perpendicular to a line you just graphed?*

WORK TOGETHER

Make sure students understand which line represents each sister.

ESL Ask the class if they think it would be fair to allow one student to get a head start on the next test. Make sure that those unfamiliar with the expression *head start* understand that it means *to begin earlier.*

THINK AND DISCUSS

Explain that the definitions of vertical and nonvertical lines differ because the slopes of vertical lines are undefined.

Lesson Planning Options

Prerequisite Skills

- Understanding the terms parallel and perpendicular
- Isolating a variable

Assignment Options

To provide flexible scheduling, this lesson can be subdivided into parts.

1 **Core** 1–12, 25–30
Extension 32, 33, 37–40

2 **Core** 13–24, 41–50
Extension 31, 34–36, 51–53

Use Mixed Review to maintain skills.

Resources

Student Edition

Skills Handbook, p. 578, 581
Extra Practice, p. 560
Glossary/Study Guide

Teaching Resources

Chapter Support File, Ch. 5
- Practice 5-8 (two worksheets)
- Reteaching 5-8
- Checkpoint
- Alternative Activity 5-8 (2)

Classroom Manager 5-8
Glossary, Spanish Resources
Two-Year Algebra Handbook 5-8

Transparencies
2, 28, 70

What You'll Learn

- Writing equations for parallel and perpendicular lines
- Using slope to determine if lines are parallel, perpendicular, or neither

...And Why

To investigate real-world situations, such as urban planning

Connections Urban Planning . . . and more

5-8 Parallel and Perpendicular Lines

WORK TOGETHER

Work in pairs. See margin for reasoning.

1. Match the story with its graph. Explain your choices.
   a. Elena gives her younger sister Rosa a head start in a race. Elena runs faster than her sister. B
   b. Elena gives her younger sister Rosa a head start in a race. Elena is surprised that she and her sister run at the same rate. A

A. Distance (feet) 0–50; Time (seconds) 0–5

B. Distance (feet) 0–50; Time (seconds) 0–8

2. a. In graph A, how far apart are the two runners at 0 s? 1 s? 2 s? 3 s? a. 20 ft at all times
   b. Suppose the race were longer and the pattern for the distance between runners remained the same as in part (a). Would the lines intersect? Explain. No; Rosa would always be 20 ft ahead of Elena.
3. What is the slope of each line in graph A? 10

THINK AND DISCUSS

Part 1 **Parallel Lines**

In the Work Together activity, the lines in graph A are parallel. Two **parallel lines** are always the same distance apart. They do not intersect.

Slopes of Parallel Lines

Nonvertical lines are parallel if they have the same slope and different y-intercepts. Vertical lines are parallel.

Example The slope of $y = 2x + 3$ and $y = 2x - 5$ is 2. The graphs of these two equations are parallel.

4. Are horizontal lines parallel? Explain. Yes; all horizontal lines have 0 slope.
5. **Try This** What is the slope of a line parallel to $y = \frac{3}{5}x - 4$? $\frac{3}{5}$

Example 1 Relating to the Real World 

Point out that once students have found the slope of the equation $5x - 2y = 8$, they no longer need to work with that equation. To find the equation of the parallel line, they need only the slope of the given line because they already have a point on the parallel line.

Question 6 Ask students if they can tell the slope of $5x - 2y = 8$ by just glancing at the equation.

ERROR ALERT! **CRITICAL THINKING Question 9** Students may rearrange the equation to get $y = \frac{3}{4}x - \frac{9}{4}$ and then conclude that the lines are parallel. **Remediation:** Point out that the positive or negative sign is part of the coefficient. Slopes with opposite signs are not parallel. Encourage students to graph the equations to help them see this relationship.

Question 10 Point out that students need to find the slope of both lines in each graph.

AUDITORY LEARNING Students may be confused by the vocabulary in this lesson. Give students practice finding opposites, reciprocals, and opposite reciprocals by asking questions such as these.

- *What is the opposite of 2?* -2
- *What is the reciprocal of 2?* $\frac{1}{2}$
- *What is the opposite reciprocal of 2?* $-\frac{1}{2}$
- *What is the opposite of -4?* 4
- *What is the reciprocal of -4?* $-\frac{1}{4}$

You can use slope to write an equation of a line parallel to a given line.

Example 1 Relating to the Real World  

Write an equation for a line that contains $(-2, 3)$ and is parallel to the graph of $5x - 2y = 8$.

Find the slope of $5x - 2y = 8$.

$$5x - 2y = 8$$
$$-2y = -5x + 8$$
$$y = \frac{5}{2}x - 4 \quad \leftarrow \text{The slope is } \frac{5}{2}.$$

Use the slope-intercept form to find the y-intercept.

$$y = mx + b$$
$$3 = \frac{5}{2}(-2) + b \quad \leftarrow \text{Substitute the coordinates } (-2, 3) \text{ for } (x, y) \text{ and } \frac{5}{2} \text{ for } m.$$
$$3 = -5 + b$$
$$8 = b \quad \leftarrow \text{The } y\text{-intercept is 8.}$$

The equation is $y = \frac{5}{2}x + 8$.

6. Why was the equation $5x - 2y = 8$ changed to slope-intercept form? You can read the slope directly from the slope-intercept form.
7. Graph $5x - 2y = 8$ and $y = \frac{5}{2}x + 8$. Do the lines appear parallel? See margin for graphs; yes.
8. **Try This** Write an equation for the line that is parallel to $5x - 2y = 8$ and passes through the point $(4, 9)$. $y = \frac{5}{2}x - 1$
9. **Critical Thinking** Is the line graphed by the equation $3x - 4y = 9$ parallel to the line graphed by the equation $y = -\frac{3}{4}x + 5$? Explain. No; the lines have different slopes.

Part 2

Perpendicular Lines

Perpendicular lines are lines that form right angles. The slopes of perpendicular lines also have a special relationship.

10. Find the slope of the lines in each graph.

a. 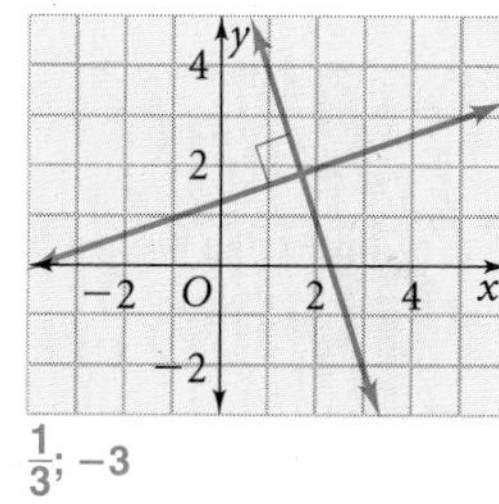
$\frac{1}{3}$; -3

b. 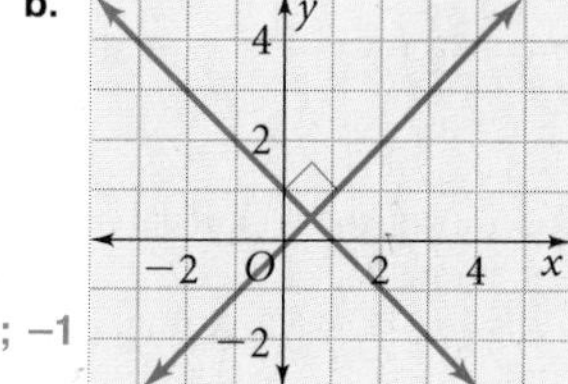
1; -1

c. 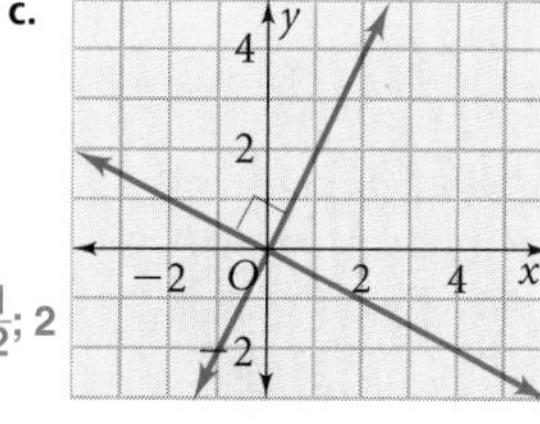
$-\frac{1}{2}$; 2

11. a. For each pair of lines, find the product of the slopes. -1; -1; -1
    b. What do you notice about the product of the slopes of the perpendicular lines? The product is always -1.

Additional Examples

FOR EXAMPLE 1

Write an equation for the line that contains $(-2, -1)$ and is parallel to the graph of $-3x + 2y = -3$.

$y = \frac{3}{2}x + 2$

Discussion: *How many other equations can you use to represent this line?*

FOR EXAMPLE 2

A sketch of a roof is drawn with one side represented by the equation $7x + 3y = 4$. Write an equation for the line representing the other side of the roof which goes through the origin and is perpendicular to the other side.

$y = \frac{3}{7}x$

- *What is the opposite reciprocal of –4?* $\frac{1}{4}$
- *What is the opposite of $\frac{1}{3}$?* $-\frac{1}{3}$
- *What is the reciprocal of $\frac{1}{3}$?* 3
- *What is the opposite reciprocal of $\frac{1}{3}$?* –3

Continue asking questions like these until students seem comfortable with the vocabulary.

Question 13d Remind students that a vertical line has an undefined slope.

Example 2 **Relating to the Real World** 

Students may be overwhelmed by the amount of data in this example. Point out that while they find the slope of Park Road, they can ignore the park entrance at (0, 4) because it is not relevant. Initially, students need only the two points on Park Road.

CRITICAL THINKING Question 14 Encourage students to draw two lines that are neither perpendicular nor parallel. Have them find the slopes of the lines. This will help reassure them that they do not need to graph the lines to be sure that the lines are perpendicular.

Technology Options

For Exercises 25–30, students may use a graphing calculator or graphing software to check their answers.

Prentice Hall Technology 

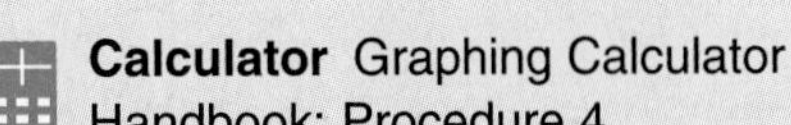

Calculator Graphing Calculator Handbook: Procedure 4

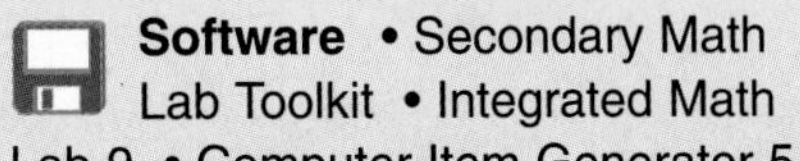

Software • Secondary Math Lab Toolkit • Integrated Math Lab 9 • Computer Item Generator 5-8

Internet • See the Prentice Hall site. (http://www.phschool.com)

Slopes of Perpendicular Lines

Two lines are perpendicular if the product of their slopes is -1. A vertical and a horizontal line are perpendicular.

Example The slope of $y = -4x + 7$ is -4. The slope of $y = \frac{1}{4}x + 2$ is $\frac{1}{4}$. Since $-4 \cdot \frac{1}{4} = -1$, the graphs of the equations are perpendicular.

The product of two numbers is -1 if one number is the opposite of the reciprocal of the other. These numbers are called opposite reciprocals.

fraction: $-\frac{3}{5}$ reciprocal: $-\frac{5}{3}$ opposite reciprocal: $\frac{5}{3}$

12. Is $-\frac{2}{3}$ the opposite reciprocal of $\frac{3}{2}$? Explain. **Yes; the product is –1.**

13. **Try This** Find the slope of a line perpendicular to a line with this slope.

a. -2 $\frac{1}{2}$ **b.** $\frac{2}{7}$ $-\frac{7}{2}$ **c.** $\frac{1}{5}$ -5 **d.** 0 **undefined**

You can use slope to write an equation of a line perpendicular to a given line.

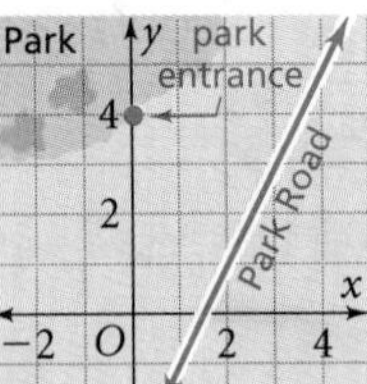

Example 2 **Relating to the Real World**

Urban Planning A bike path for a new city park will connect the park entrance to Park Road. The path will be perpendicular to Park Road. Write an equation for the line representing the bike path.

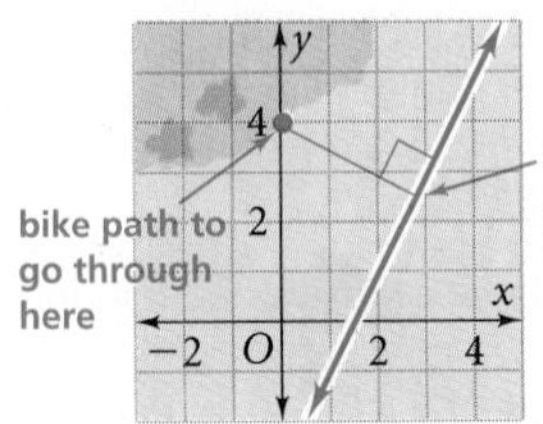

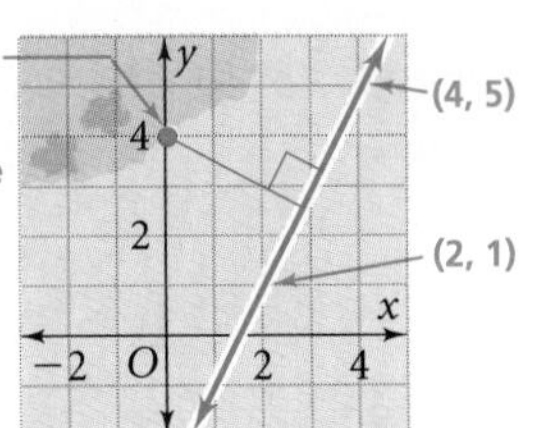

Find the slope m of Park Road.

$$m = \frac{y_2 - y_1}{x_2 - x_1} = \frac{5-1}{4-2} \quad \leftarrow \text{Points (2, 1) and (4, 5) are on Park Road.}$$
$$= \frac{4}{2} = 2$$

The opposite of the reciprocal of 2 is $-\frac{1}{2}$. So the slope of the bike path is $-\frac{1}{2}$. The y-intercept is 4.

$y = mx + b$ ← Use the y-intercept form.

$y = -\frac{1}{2}x + 4$ ← Substitute $-\frac{1}{2}$ for m and 4 for b.

The equation for the bike path is $y = -\frac{1}{2}x + 4$.

14. **Critical Thinking** How can you use slope to tell if two lines are parallel, perpendicular, or neither?

14. If slopes are the same, lines are parallel. If the product of slopes is –1, lines are perpendicular. If one line has slope 0 and the other has undefined slope, lines are perpendicular. Otherwise, lines are neither parallel nor perpendicular.

Exercises ON YOUR OWN

Exercises 5, 6, and 8 Suggest that students make sketches of these lines to help them solve the questions.

ALTERNATIVE ASSESSMENT **Exercises 13–24** To assess students' ability to differentiate between the slopes of parallel and perpendicular lines, ask them to find the slope of a line parallel to the graph of each equation.

Exercises 13–24 Have students check their answers by multiplying them by the slopes of the given lines. All results should be –1.

Exercises 25–30 Encourage students to share the techniques they use to make quick decisions.

Exercise 31 Point out that the appearance of the graphs is distorted on the calculator screen because the distance between points on the *x*-axis and *y*-axis are different.

Exercise 33a Lead students to recognize that after the two equations are rewritten in slope-intercept form, the coefficients of *x* will still be identical.

ERROR ALERT! **Exercise 35** Students may accidentally test the slopes of the two triangle sides that are not perpendicular. **Remediation:** Advise students to sketch the triangle before answering the question so that they test the slopes of the two sides that are most likely to be perpendicular.

GEOMETRY **Exercises 37–39** These exercises relate the sides of geometric figures to parallel and perpendicular lines. Review the definitions of quadrilaterals and parallelograms.

Exercises ON YOUR OWN

Find the slope of a line parallel to the graph of each equation.

1. $y = \frac{1}{2}x + 2.3$ $\frac{1}{2}$
2. $y = -\frac{2}{3}x - 1$ $-\frac{2}{3}$
3. $3x + 4y = 12$ $-\frac{3}{4}$
4. $2x - 3y = -2$ $\frac{2}{3}$
5. $y = 6$ 0
6. $x = 5$ undefined
7. $2x - y = 0$ 2
8. $x = 0$ undefined
9. $y = -3x + 2.3$ -3
10. $7x + 2y = 12$ $-\frac{7}{2}$
11. $15x - 12y = 7$ $\frac{5}{4}$
12. $y = 0.5x - 8$ 0.5

Find the slope of a line perpendicular to the graph of each equation.

13. $y = 2x$ $-\frac{1}{2}$
14. $y = -3x$ $\frac{1}{3}$
15. $y = \frac{7}{5}x - 2$ $-\frac{5}{7}$
16. $y = x$ -1
17. $2x + 3y = 5$ $\frac{3}{2}$
18. $y = \frac{x}{-5} - 7$ 5
19. $y = -8$ undefined
20. $x = 3$ 0
21. $4x - 2y = 9$ $-\frac{1}{2}$
22. $3x + 5y = 7$ $\frac{5}{3}$
23. $y = \frac{9}{2}x + 5$ $-\frac{2}{9}$
24. $y = 0.25x$ -4

Tell whether the lines for the pair of equations are *parallel, perpendicular,* or *neither*.

25. $y = 4x + \frac{3}{4}, y = -\frac{1}{4}x + 4$ perp.
26. $y = \frac{x}{3} - 4, y = \frac{1}{3}x + 2$ parallel
27. $y = -x + 5, y = x + 5$ perp.
28. $5x + y = 3, -5x + y = 8$ neither
29. $x = 2, y = 9$ perp.
30. $y = x, y = x + 2$ parallel

31. *Graphing Calculator* The graphs of $y = x$ and $y = -x$ are shown on the standard screen at the right. The product of the slopes is -1. Explain why the lines do not appear to be perpendicular. The vertical and the horizontal scales use units of different length.

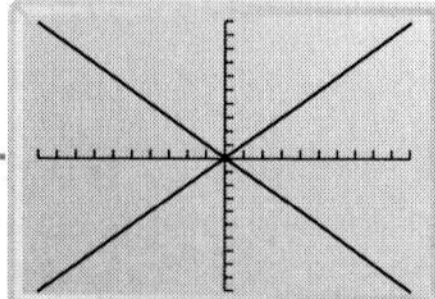

32. a. *Open-ended* Write an equation for a line parallel to the graph of $y = 4x - 1$. Sample: $y = 4x + 1$
    b. Can you write more than one equation for part (a)? Explain. Yes; the *y*-intercept of a parallel line can be any number other than -1.
33. a. Are the lines $3x + 5y = 6$ and $3x + 5y = 2$ parallel? Explain. Yes; they have the same slope, $-\frac{3}{5}$.
    b. Explain how you can tell that the lines $7x - 3y = 5$ and $7x - 3y = 8$ are parallel without finding slopes. Since the coefficients of *x* and *y* are the same, the ratio for the slope will be the same.
34. *Writing* Suppose a friend missed class today. Explain what you would tell your friend about using slope to find each equation.
    a. the equation of a line parallel to a given line The line has the same slope.
    b. the equation of a line perpendicular to a given line The line has the slope that is the opposite of the reciprocal of the slope of the given line.
35. *Geometry* Are the points (3, 8), (5, 4) and $(-3, 3)$ the vertices of a right triangle? Explain. No; the slopes are $\frac{5}{6}$, -2, and $\frac{1}{8}$. No pair of slopes has product -1.
36. a. *Maps* What is the slope of New Hampshire Avenue? about $\frac{5}{4}$
    b. Show that parts of Pennsylvania Avenue and Massachusetts Avenue near New Hampshire Avenue are parallel. See below.
    c. Show that New Hampshire Avenue is not perpendicular to Pennsylvania Avenue. See margin.

36b. Because the slope of both streets is $-\frac{1}{2}$, the streets are parallel.

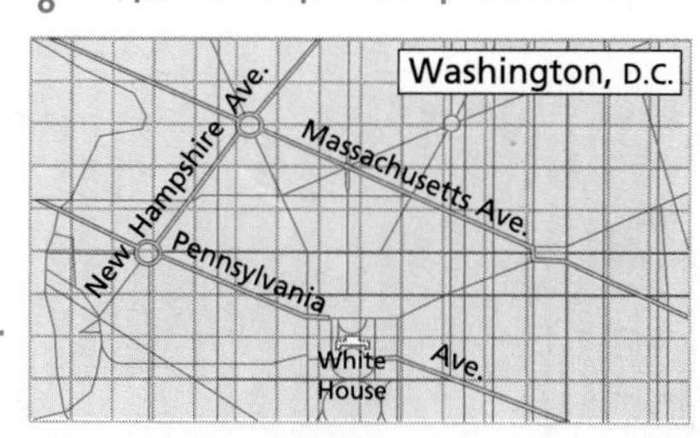

page 250 Work Together

1. I: The line representing Elena is steeper because Elena runs faster. So the lines will intersect.
   II: The lines are parallel because Rosa will always stay a constant distance ahead of Elena. So the lines never intersect

pages 250–252 Think and Discuss

7.

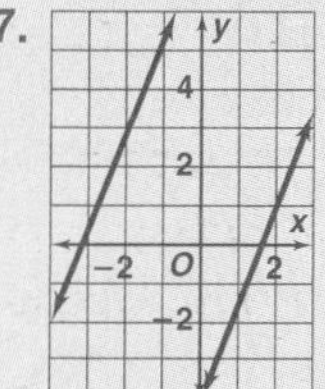

pages 253–255 On Your Own

36. The product of the slopes for New Hampshire Ave. and Pennsylvania Ave. is $\left(\frac{5}{4}\right)\left(-\frac{1}{2}\right) = -\frac{5}{8}$. Since the product is not -1, the streets are not perpendicular.
37. The slope of *AD* and *BC* is undefined, so they are parallel. The slope of *AB* and *CD* is $\frac{2}{5}$, so they are parallel. The quadrilateral is a parallelogram.

STANDARDIZED TEST TIP Exercise 40 When asked to compare the slopes of two lines, students should rewrite their equations in slope-intercept form.

Exercises 47–50 If students struggle with these problems, suggest that they experiment by using pencils to model lines on graph paper.

GEOMETRY/EXTENSION Exercises 51–53 Challenge students by asking: *Do you need to check all four angles to prove that the figure is a rectangle?* no Challenge pairs of students to explore what the minimum number of angles is that they must check.

Chapter Project **FIND OUT BY INTERVIEWING** Ask students to interview an adult regarding a job he or she had as a teenager. Have the students write an equation that describes the person's earnings and expenses for a week. This task is essential to their work on the Chapter Project. Have students add this part of the project to work they have already completed. Check students' progress on the project by having each student write a short progress report telling what has been done so far and their plans for completion.

Exercises MIXED REVIEW

JOURNAL Encourage students to make sketches to illustrate their writing.

Exercise 58 Encourage students to estimate an answer before solving. Then have them compare the estimate to the answer to check themselves.

pages 253–255 On Your Own

38. The slope of JK is $\frac{1}{5}$. The slope of KL is -2. The slope of LM is $\frac{1}{6}$. The slope of JM is -4. The quadrilateral is not a parallelogram.

39. The slope of PQ and RS is $-\frac{1}{2}$. The slope of QR and SP is $-\frac{3}{2}$. The quadrilateral is a parallelogram.

51. The slope of AB and CD is $\frac{2}{5}$. The slope of BC and AD is $-\frac{5}{2}$. The product is -1, so the quadrilateral is a rectangle.

52. The slope of KL and MN is $-\frac{1}{6}$. The slope of LM and KN is 5. The product is not -1, so the quadrilateral is not a rectangle.

53. The slope of PQ and RS is $\frac{1}{2}$. The slope of PS and QR is -2. The product is -1, so the quadrilateral is a rectangle.

Geometry **A quadrilateral with both pairs of opposite sides parallel is a parallelogram. Use slopes to show if each figure is a parallelogram.** 37–39. See margin.

37. 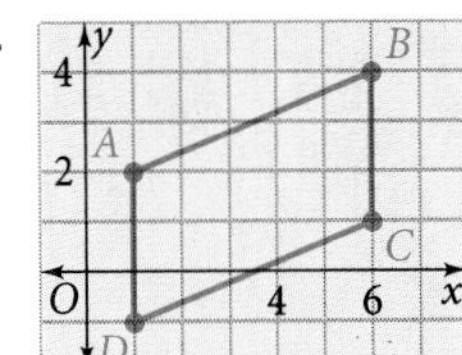

38. 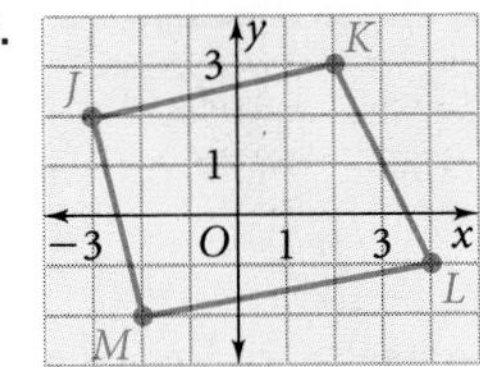

39. 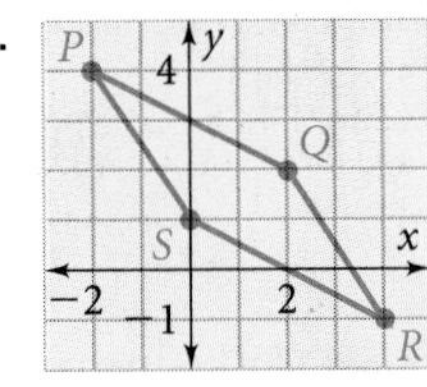

40. *Standardized Test Prep* Which of the following statements is true for the graphs of equations I and II? A

I: $2x + y = 15$ II: $3y = -6x + 30$

A. The graphs of I and II are parallel.
B. The graphs of I and II are perpendicular.
C. The graphs of I and II are the same line.
D. The graphs of I and II are neither parallel nor perpendicular.
E. none of the above

Write an equation that satisfies the given conditions.

41. parallel to $y = 6x - 2$, through $(0, 0)$ $y = 6x$
42. perpendicular to $y = 2x + 7$, through $(0, 0)$ $y = -\frac{1}{2}x$
43. perpendicular to $y = x - 3$, through $(4, 6)$ $y = -x + 10$
44. parallel to $y = -3x$, through $(3, 0)$ $y = -3x + 9$
45. parallel to $y = -\frac{2}{3}x + 12$, through $(5, -3)$ $y = -\frac{2}{3}x + \frac{1}{3}$
46. perpendicular to $y = -4x - \frac{5}{2}$, through $(1, 6)$ $y = \frac{1}{4}x + 5\frac{3}{4}$

Tell whether each statement is *true* or *false*. Explain your choice.

47. In a coordinate plane, every horizontal line is perpendicular to every vertical line. True; the definition of perpendicular lines says that horizontal and vertical lines are perpendicular.
48. Two lines with positive slopes can be perpendicular. False; the product of the slopes must be negative.
49. Two lines with positive slopes must be parallel. False; two positive numbers need not be equal.
50. The graphs of two direct variations can be parallel. False; graphs of all direct variations intersect at (0, 0).

Geometry **A quadrilateral with four right angles is a rectangle. Use slopes to show whether each figure is a rectangle.** 51–53. See margin.

51. 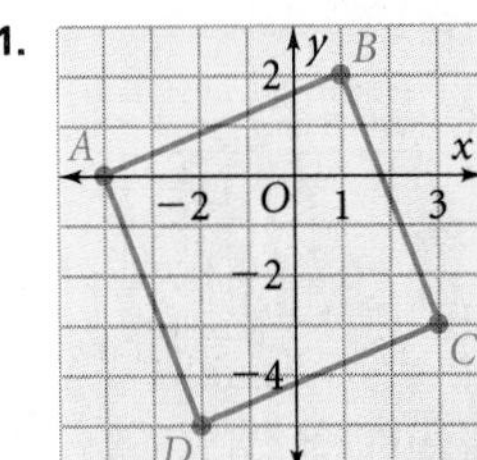

52. 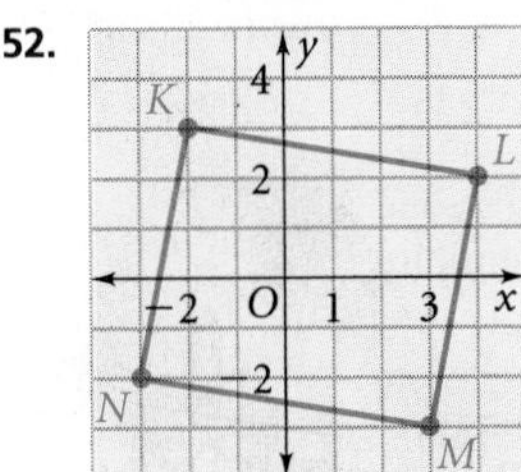

53. 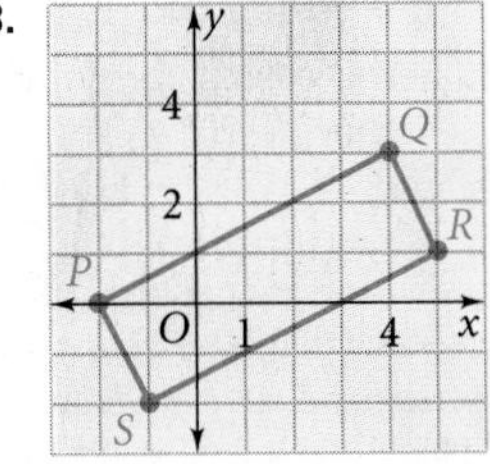

GETTING READY FOR LESSON 5-9 These exercises prepare students for finding x-intercepts.

Wrap Up

THE BIG IDEA Ask: *What are the all the differences you can name between perpendicular and parallel lines?*

RETEACHING ACTIVITY Students work in pairs using wire or coffee stirrers to represent given equations. One partner places the first wire, the other partner places the second. Then they take turns writing equations for the lines. (Reteaching worksheet 5-8)

Exercises CHECKPOINT

In this Checkpoint, students have an opportunity to assess their own progress on the topics covered in Lessons 5-5 to 5-8.

Exercise 7 Remind students to work slowly and carefully so that they do not confuse the negative signs with the minus signs.

STANDARDIZED TEST TIP **Exercise 15** Point out to students that some of the choices for this exercises are designed to look like the correct answer when they are actually incorrect. To avoid making a wrong choice, first change the equations into slope-intercept form.

Chapter Project *Find Out by Interviewing*

Interview an adult about a job he or she had as a teenager. Ask about positive and negative aspects of the job, salary, and expenses. Set up an equation that describes the person's earnings after expenses for a week.

Exercises MIXED REVIEW

Solve.

54. $\frac{x}{2} + \frac{x}{3} = 6$ $7\frac{1}{5}$

55. $3m - 4m > 2$ $m < -2$

56. $-\frac{x}{4} > 3$ $x < -12$

57. $9y - 3 > 5y + 1$ $y > 1$

58. *Social Studies* There were 132,300 African Americans among the troops serving in the Gulf War, or 24.5% of the total. How many troops served in the Gulf War? 540,000 troops

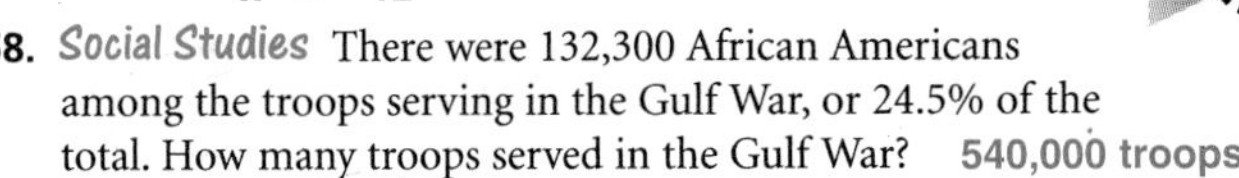

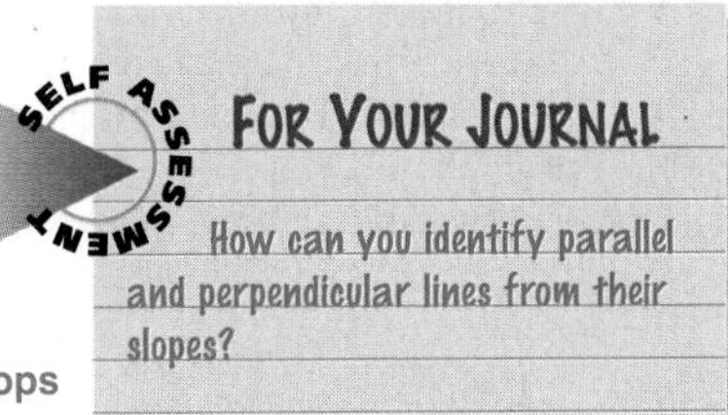

FOR YOUR JOURNAL

How can you identify parallel and perpendicular lines from their slopes?

Getting Ready for Lesson 5-9

Find the value of x when $y = 0$.

59. $y = 3x + 8$ $-\frac{8}{3}$

60. $4x - 5y = 7$ $\frac{7}{4}$

61. $y = \frac{1}{2}x - 10$ 20

62. $2x + 3y = 6$ 3

Exercises CHECKPOINT

Change each equation to $Ax + By = C$ form.

1. $3y = 2x + 6$ $-2x + 3y = 6$
2. $y = 4x - 7$ $-4x + y = -7$
3. $y = \frac{1}{3}x + 5$ $-\frac{1}{3}x + y = 5$
4. $15y = 9x - 12$ $-9x + 15y = -12$

Write an equation for a line that contains the given points.

5. $(3, 4)$ and $(-1, 5)$ $y = -\frac{1}{4}x + 4\frac{3}{4}$
6. $(0, -3)$ and $(\frac{1}{2}, 6)$ $y = 18x - 3$
7. $(-7, -5)$ and $(7, 5)$ $y = \frac{5}{7}x$
8. $(5, 9)$ and $(5, 10)$ $x = 5$

Find the equation of a trend line or the line of best fit. 9–10. Answers may vary. Samples are given.

9.

x	1	2	3	4	5	6	7
y	7	12	19	20	28	33	40

$y = 5.4x + 1.3$

10.

x	1	2	3	4	5	6	7
y	54	52	45	40	33	27	18

$y = -6.1x + 62.7$

Open-ended **Write an equation of a line that is parallel to each line and an equation of a line that is perpendicular to each line.** 11–14. Samples are given.

11. $y = 3x - 2$ $y = 3x$; $y = -\frac{1}{3}x$
12. $4x + 5y = 10$ $y = -\frac{4}{5}x - 1$; $y = \frac{5}{4}x + 3$
13. $y = \frac{2}{3}x + \frac{1}{6}$ $y = \frac{2}{3}x$; $y = -\frac{3}{2}x + 1$
14. $y = 1.6x - 0.4$ $y = 1.6x + 0.4$; $y = -\frac{5}{8}x$
15. *Standardized Test Prep* Which equation represents a line with the same y-intercept as $y = 4x - 3$? D

A. $y - 3 = x$ B. $y = 8x + 3$ C. $2y = 8x - 3$ D. $5x + 6y = -18$ E. $x = 3$

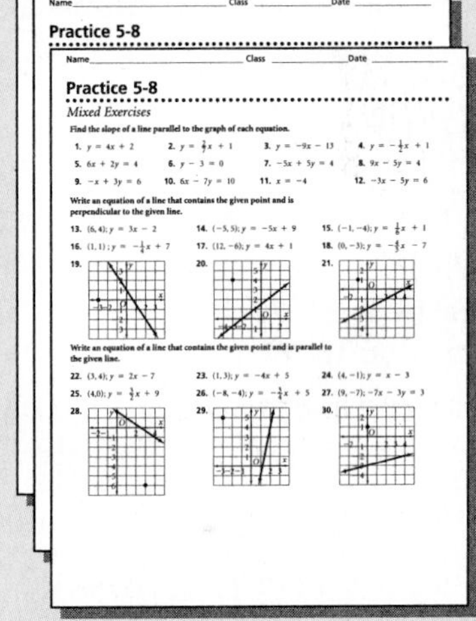

Lesson Quiz

Lesson Quiz is also available in Transparencies.

1. Find the slope of the line parallel to the equation $3x - 2y = 1$. $m = \frac{3}{2}$
2. Find the slope of the line perpendicular to the equation $y = -\frac{2}{3}x + 4$. $m = \frac{3}{2}$
3. Write an equation for the line that goes through point $(2, -1)$ and is parallel to the line $y = -\frac{1}{2}x + 3$. $y = -\frac{1}{2}x$
4. Write an equation for the line that goes through point $(2, -1)$ and is perpendicular to the line $y = -\frac{1}{2}x + 3$. $y = 2x - 5$

5-9 Teaching Notes

PROBLEM OF THE DAY

If a color printer can print five pages per minute, how long will it take to print a 63 page document? 12 min, 36 s

Problem of the Day is also available in Transparencies.

CONNECTING TO PRIOR KNOWLEDGE Propose the following scenario. *Think of a business that you would like to start. What would you spend to set up the business? How will you know when you have made a profit?*

WORK TOGETHER

Question 2 Remind students to read slowly and carefully. On the board, sketch a line crossing the x-axis at a point away from the origin. This will remind students that $y = 0$ on the x-axis, but x can equal any value.

THINK AND DISCUSS

Example 1

Explain that this is an alternative method of solving one-variable equations. This method is a more visual approach and incorporates the graphing calculator.

Lesson Planning Options

Prerequisite Skills

- Solving equations
- Graphing lines

Assignment Options

Core 1–17, 22–25
Extension 18–21, 26–29

Use Mixed Review to maintain skills.

Resources

 **Student Edition**

Skills Handbook, p. 579
Extra Practice, p. 560
Glossary/Study Guide

 **Teaching Resources**

Chapter Support File, Ch. 5
- Practice 5-9 (two worksheets)
- Reteaching 5-9
- Alternative Activity

Classroom Manager 5-9
Glossary, Spanish Resources
Two-Year Algebra Handbook 5-9

 **Transparencies**
2, 27, 71

5-9 Using the x-intercept

What You'll Learn

- Using the x-intercept of a linear equation to solve the related one-variable equation

...And Why

To use graphs to solve and check equations and to investigate business situations

What You'll Need

- graph paper
- graphing calculator

WORK TOGETHER

Work with a partner.

1. Solve the equation $\frac{2}{3}x + 4 = 0$. -6
2. a. Graph the function $y = \frac{2}{3}x + 4$. See margin.
   b. What is the value of x when the line crosses the x-axis? -6
   c. What is the value of y when the line crosses the x-axis? 0
3. Compare the results of Questions 1 and 2. What do you notice? The solution of the equation is equal to the x-value of the point where the line crosses the x-axis.

THINK AND DISCUSS

QUICK REVIEW

The *x-intercept* is the x-coordinate of the point where a line crosses the x-axis.

There is a special relationship between a one-variable equation and the related linear function. The solution of a one-variable equation equals the x-intercept of the graph of the related linear function.

One-Variable Equation	Related Linear Function
$\frac{2}{3}x + 4 = 0$	$y = \frac{2}{3}x + 4$

Notice that one side of the one-variable equation is 0. You replace 0 with y to write the related linear function.

Example 1

Solve $\frac{3}{4}x - 2 = -5$ by graphing.

Step 1 Write the equation so that one side is 0.

$\frac{3}{4}x - 2 = -5$

$\frac{3}{4}x + 3 = 0$ ← Add 5 to both sides.

Step 2 Replace 0 with y to write a function.

$y = \frac{3}{4}x + 3$

Step 3 Graph the function. Then find the x-intercept.

x-intercept is -4.

The solution of $\frac{3}{4}x - 2 = -5$ is -4.

Check $\frac{3}{4}x - 2 = -5$

$\frac{3}{4}(-4) - 2 \stackrel{?}{=} -5$ ← Substitute -4 for x.

$-3 - 2 \stackrel{?}{=} -5$

$-5 = -5$ ✔

Example 2

Point out that students need to graph $y = 0$ because the graphing calculator may not automatically display the x-axis and the X= and Y= display is only for graphed lines.

Example 3 **Relating to the Real World** 

Explain that when people start businesses, they typically spend money on equipment, an office, business supplies, and so on. During this set-up period, there is no income from the business. As long as expenses are greater than their business income, the business will lose money. The break-even point tells business owners how much business income they must generate so that expenses can be offset and the business will not continue to lose money.

AUDITORY LEARNING After students have worked Example 3, lead a discussion about business expenses. Have students brainstorm for all the expenses they might incur if they started a T-shirt-painting business. Ask a volunteer to record the responses on the board. Rework Example 3 on the board using the amounts provided by the class.

4. **Try This** Solve each equation by graphing. Check each solution.
   a. $2x + 4 = 0$ -2 **b.** $6x + 7 = 10$ $\frac{1}{2}$ **c.** $\frac{1}{2}x - 3 = 0$ 6

You can use a graphing calculator to solve a one-variable equation. Graph the related linear function. Next graph the equation for the x-axis, $y = 0$. To find the x-intercept, find the intersection point of the two lines.

Example 2

Graphing Calculator Find the solution of $5x - 12 = 0$.

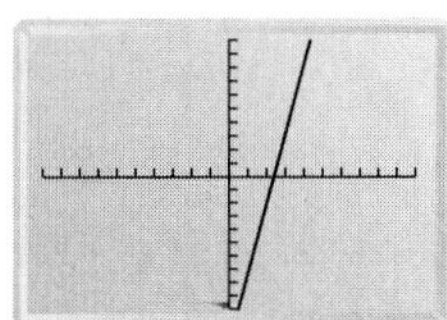

Graph $y = 5x - 12$ and $y = 0$.
The solution is 2.4.

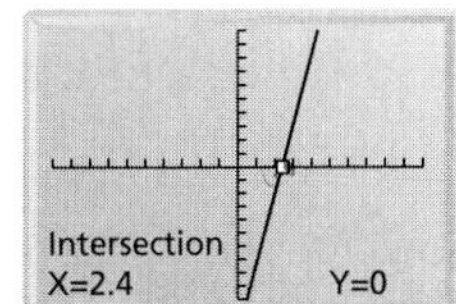

Use the CALC feature to find the coordinates of the intersection point.

PROBLEM SOLVING

Look Back Describe how to check the solution of Example 2. Then check that the solution is 2.4. **Substitute the solution into the original equation.**
$5(2.4) - 12 \stackrel{?}{=} 0$
$12 - 12 = 0$ ✔

When starting a business, people want to know the **break-even point,** the point at which their income equals their expenses.

Example 3 **Relating to the Real World** 

Business Suppose you invested \$140 to start a business selling painted T-shirts. You sell each shirt for \$7.50. Find the break-even point.

Relate income = expenses

Define x = the number of shirts you sell

Write $7.50x = 140$

Rewrite the equation as $7.50x - 140 = 0$. The related function is $y = 7.5x - 140$.

Select a range that includes the x-intercept.

WINDOW FORMAT
Xmin=-10
Xmax=50
Xscl=10
Ymin=-150
Ymax=50
Yscl=10

Graph $y = 7.5x - 140$ and $y = 0$. Find the intersection point of the lines.

Intersection
X=18.666667 Y=0

The break-even point is about 18.7.
You must sell 19 shirts to make a profit.

Additional Examples

FOR EXAMPLE 1

Solve $-\frac{1}{2}x + 4 = 6$ by graphing.

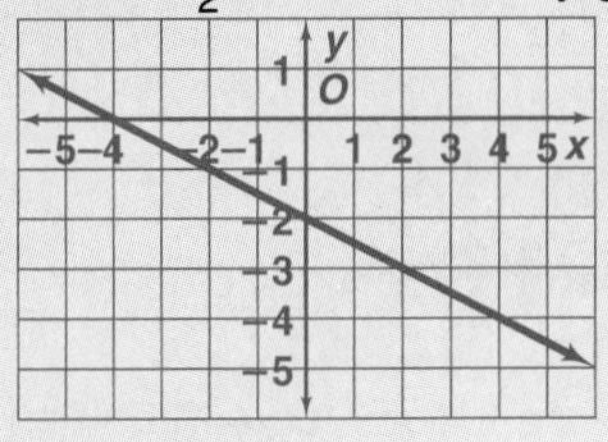

$x = -4$

FOR EXAMPLE 3

Your family is considering buying a washing machine. The machine costs \$400 and your family spends \$12 each week doing laundry. How many weeks must pass before your family reaches the break even point? Solve by using the x-intercept. 34

Exercises ON YOUR OWN

ERROR ALERT! Students may get wrong answers due to errors entering data. **Remediation:** Have students work in pairs, so they can check the data entered into the calculator.

Exercise 19 Point out that, according to Charles' theory, if a person inflated a balloon outside on a hot day and then put the balloon in a freezer, the balloon would visibly shrink. If you have block scheduling or extended class periods, try this experiment for yourself.

MAKING CONNECTIONS Exercise 19 Absolute zero is equal to −459.67°F. This temperature is only theoretical and has never been achieved. In 1993, however, Helsinki University in Finland announced that their Low Temperature Laboratory had achieved a temperature two billionth of a degree above absolute zero.

WRITING Exercise 21 Encourage students to think about all the things that could go wrong when they solve equations with calculators, and when they solve equations by hand.

ALTERNATIVE ASSESSMENT Exercises 22–25 Ask students to solve these equations using paper and pencil. Have students compare these solutions with those obtained using a graphing calculator. This will assess students' ability to use the x-intercept to solve single variable equations.

Technology Options

For Exercises 7–10, students may use a graphing calculator to check their answers.

Prentice Hall Technology

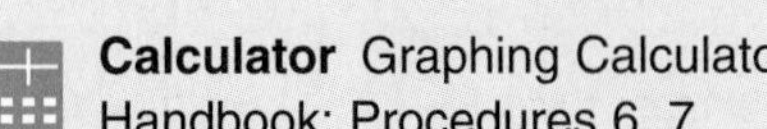
Calculator Graphing Calculator Handbook: Procedures 6, 7

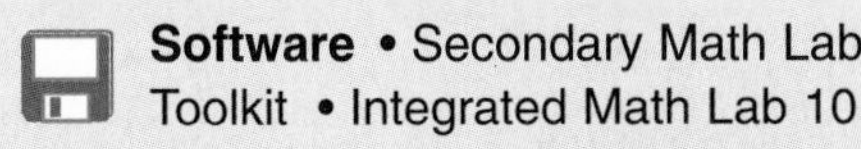
Software • Secondary Math Lab Toolkit • Integrated Math Lab 10 • Computer Item Generator 5-9

Internet • See the Prentice Hall site. (http://www.phschool.com)

page 256 Work Together

2a. See back of book.

pages 258–259 On Your Own

19b, 21. See back of book.

5. The y-values close to 0 are −5 for $x = 18$ and 2.5 for $x = 19$. This tells you that the line crosses the x-axis when x is between 18 and 19. Your range for the x-axis must include 18 and 19 to show the x-intercept in the calculator window.

5. *Graphing Calculator* Use the TABLE feature on your calculator to find values for y close to 0 in Example 3. How can the TABLE feature help you choose range values for the x-axis?

Exercises ON YOUR OWN

Write the function you would graph to find the solution of each one-variable equation.

1. $3x - 8 = 0$ $y = 3x - 8$
2. $7 = \frac{1}{2}x + 4$ $y = \frac{1}{2}x - 3$
3. $-x + 5 = 9$ $y = -x - 4$

Match each equation with its related graph.

4. $5 - 3x = 0$ C
5. $\frac{x}{3} - 2 = 0$ A
6. $x + 5 = 0$ B

A.

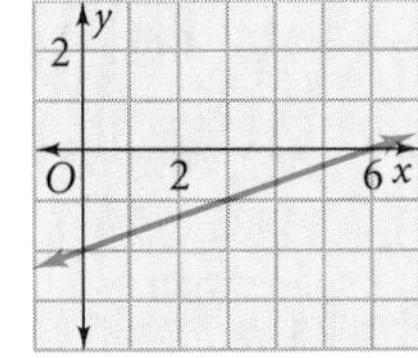

B.

C.

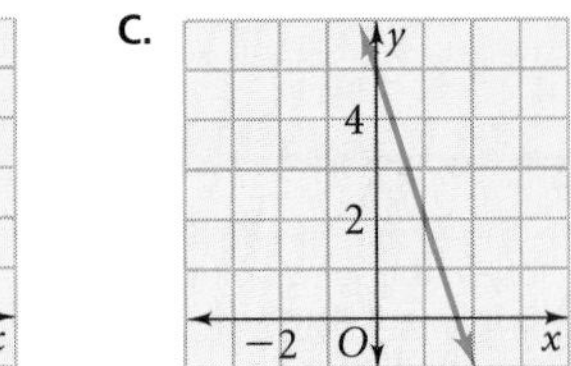

Solve each equation using a graph. Check each answer.

7. $0 = 2x + 3$ $-\frac{3}{2}$
8. $3x + 4 = 13$ 3
9. $4x - 3 = 9$ 3
10. $-x + 7 = 13$ −6

Choose **Solve each equation using a graphing calculator or with paper and pencil. Check each answer.**

11. $3x - 5 = 12$ $5\frac{2}{3}$
12. $\frac{2x}{5} = -4$ −10
13. $3x = 75$ 25
14. $5(x + 2) = 18$ $1\frac{3}{5}$
15. $27 = \frac{x}{3} + 15$ 36
16. $\frac{3}{4}x + 6 = -20$ $-34\frac{2}{3}$
17. $4(x - 6) = 5$ $7\frac{1}{4}$
18. $25x - 18 = 72$ $3\frac{3}{5}$

19. *Money* Suppose you build birdhouses. An investment of $100 allows you to build ten houses. You sell each house for $15.
    a. Write an equation relating the income to the expenses. $15x = 100$
    b. Graph a related linear function to find the break-even point. See below.
    c. Suppose you sell all the houses. How much profit will you make? $50
20. *Physical Science* In 1787, the French scientist Jacques Charles discovered that as a gas cooled at a constant pressure, the relationship between volume and temperature was linear. Suppose that at 60°C the volume of the gas is 555 cm³ and at 30°C the volume is 505 cm³.
    a. Write a linear function relating volume of gas to temperature. See below.
    b. Jacques Charles used a linear function to find *absolute zero*, the temperature at which a gas theoretically will have no volume. Graph your function to find absolute zero in degrees Celsius. b. −273°C

Jacques Charles

21. *Open-ended* Write a one-variable equation. Solve your equation using a graph. Sample: $2x + 7 = 6$; −0.5; see margin for graph.

20a. $v = \frac{5}{3}t + 455$

19b. See margin for graph; $6\frac{2}{3}$.

Exercises MIXED REVIEW

SELF ASSESSMENT PORTFOLIO Share with students the criteria you will use to assess their work in portfolios, as well as how you plan to use the results. Students should understand how the rubrics are used to assess their work, how each piece in the portfolio is valued, and how the scores the students receive in their portfolios will affect their overall evaluation.

Wrap Up

THE BIG IDEA Ask: *What is the x-intercept and what does it tell you?*

RETEACHING ACTIVITY Students use toothpicks to represent a line after finding the y-intercept. (Reteaching worksheet 5-9)

A Point in Time

If you have block scheduling or an extended class period, you may wish to have students investigate these topics:

- How old were you in 1984? Answers may vary.
- Who was the first American woman to fly in space? Dr. Sally K. Ride
- What year was the first space shuttle launch (Columbia)? 1981
- What was the cost of the Challenger shuttle? $1.2 billion

22. **Writing** What advantages and disadvantages do you see in using a calculator to solve equations? Answers may vary. Sample: If you enter information accurately, the calculator answer is correct. Using a calculator requires more steps than solving using paper and pencil.

Graphing Calculator **Solve each equation using a graphing calculator.**

23. $\frac{2x}{7} + 5 = 0$ -17.5
24. $2x + 10 = -15$ -12.5
25. $-8(x - 6) = 18$ 3.75
26. $\frac{4}{5}x - 18 = 15$ 41.25

Critical Thinking **Is each statement *true* or *false*? Explain.**

27. A vertical line has no x-intercept. False; a vertical line has no y-intercept.
28. The x-intercept of a direct variation is 0. True; the graph of a direct variation passes through (0, 0).
29. The x-intercept of $y = mx + b$ is the solution to $mx + b = 0$. True; the x-intercept is the x-value for y-value 0.

Exercises MIXED REVIEW

Evaluate each function for $x = -3.2$.

30. $f(x) = |5 - x|$ 8.2
31. $f(x) = 3x - 2$ -11.6
32. $f(x) = 4x^2 - 8$ 32.96
33. $f(x) = (x + 1)^2$ 4.84
34. **Zoology** In July 1995, three Siberian tigers were born in Garden City, Kansas. There were already about 100 Siberian tigers in the United States. By what percent did the population increase? about 3%

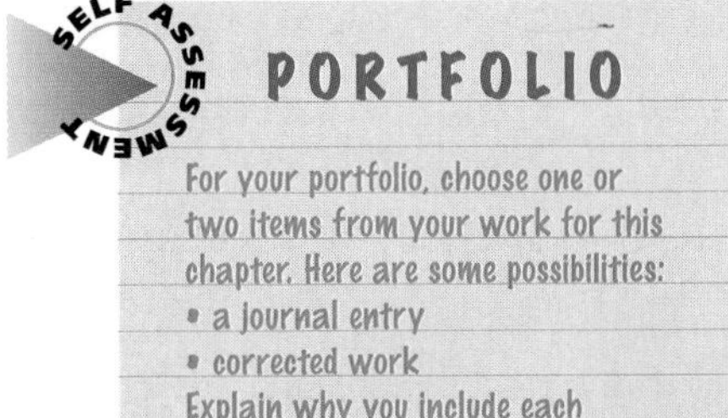

SELF ASSESSMENT
PORTFOLIO
For your portfolio, choose one or two items from your work for this chapter. Here are some possibilities:
• a journal entry
• corrected work
Explain why you include each selection that you make.

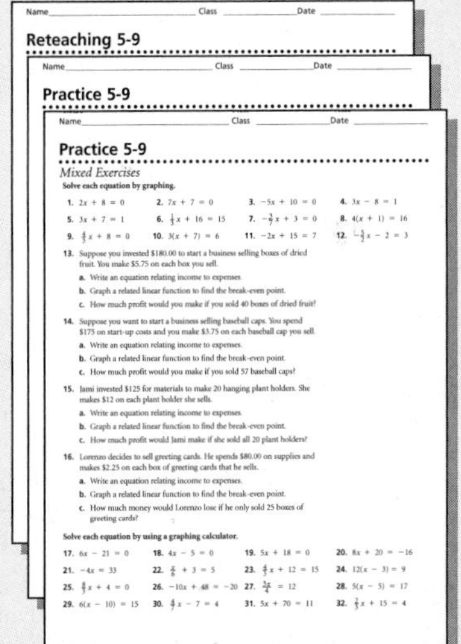
Reteaching 5-9

Practice 5-9

Practice 5-9

Mixed Exercises

Solve each equation by graphing.

Lesson Quiz

Lesson Quiz is also available in Transparencies.

1. Write the function you would use to find the solution of $-4 = 2x + 6$. $y = 2x + 10$
2. Find the solution to $5x - 2 = 4$ by graphing.

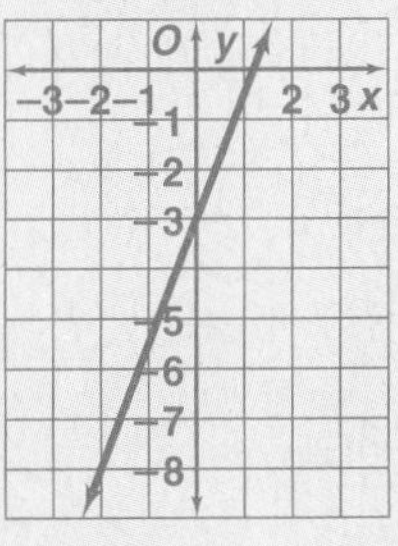

$x = \frac{6}{5}$

A Point in Time

1500 1600 1700 1800 1900 2000

Judith A. Resnik

On August 30, 1984, Astronaut Judith A. Resnik became the second American woman to fly in space, on the shuttle *Discovery's* first voyage. Resnik was an electrical engineer with a Ph.D. from the University of Maryland. On the ground she helped to design and develop a remote manipulator system, a task that required skill in linear equations. Her job during *Discovery's* six-day voyage was to manipulate a robotic arm and to extend and retract the shuttle's solar power array. Resnik died tragically in the shuttle *Challenger* disaster in 1986.

Finishing the Chapter Project

PROJECT DAY You may wish to plan a Project Day on which students share their completed projects. Encourage students to explain their process as well as their product.

PROJECT NOTEBOOK Have students review their project work and bring their notebooks up to date.

- Have students review the equations, graphs, and explanations that they needed for the project.
- Ask students to share any insights they found for completing the project, such as any shortcuts for writing equations or making graphs.

SCORING RUBRIC

3 Equations are written correctly, thoroughly explained, and accurate. Graphs are effective and labeled correctly. Notebook is well organized and provides useful information.

2 Equations are essentially correct with minor errors in calculations. Reasoning and explanations are essentially correct, but might be more thorough. Graphs may contain minor errors in scale, or may be labeled incorrectly. Notebook presents clear information, but needs to be better organized.

1 Equations are inaccurate, and notebook lacks organization. Graphs could be neater and more accurate. Explanations lack detail.

0 Major elements of the project are incomplete or missing.

CHAPTER PROJECT

Finishing the Chapter Project

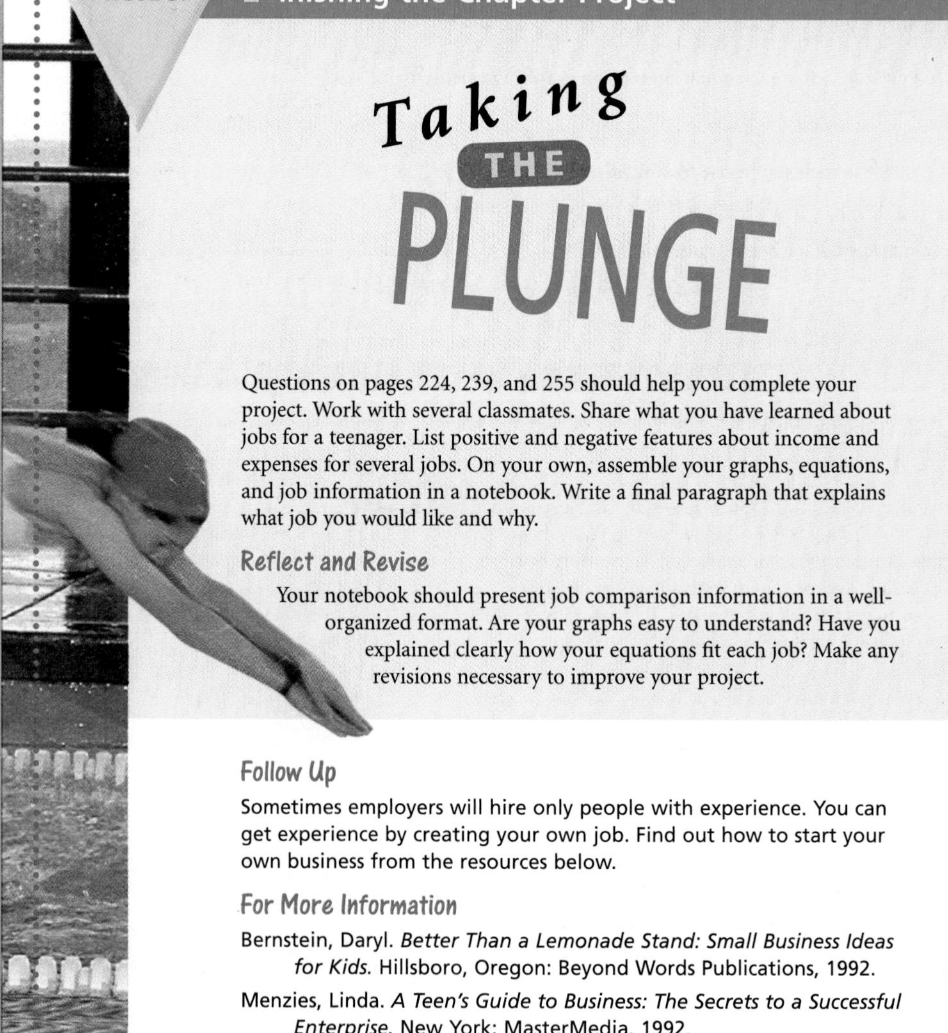

Taking THE PLUNGE

Questions on pages 224, 239, and 255 should help you complete your project. Work with several classmates. Share what you have learned about jobs for a teenager. List positive and negative features about income and expenses for several jobs. On your own, assemble your graphs, equations, and job information in a notebook. Write a final paragraph that explains what job you would like and why.

Reflect and Revise

Your notebook should present job comparison information in a well-organized format. Are your graphs easy to understand? Have you explained clearly how your equations fit each job? Make any revisions necessary to improve your project.

Follow Up

Sometimes employers will hire only people with experience. You can get experience by creating your own job. Find out how to start your own business from the resources below.

For More Information

Bernstein, Daryl. *Better Than a Lemonade Stand: Small Business Ideas for Kids.* Hillsboro, Oregon: Beyond Words Publications, 1992.

Menzies, Linda. *A Teen's Guide to Business: The Secrets to a Successful Enterprise.* New York: MasterMedia, 1992.

Wrap Up

HOW AM I DOING? Assess student understanding by holding a short competition. Divide the students into small groups and have each group respond to the questions on a large piece of paper. Award each group one point for every answer they have in common with another group. The group with the most points wins.

KEY TERMS The numbers in parentheses direct students to the pages where the terms (or symbols) are used or defined. Students should be able to (1) write a simple explanation of each term, (2) illustrate the term with a diagram, or (3) show an example that uses a term.

Exercises 1–3 Remind students that the slope of a horizontal line is zero and the slope of a vertical line is undefined.

Exercises 15–18 Emphasize the importance of the order of the coordinates when calculating slope $\left(\frac{y_2 - y_1}{x_2 - x_1}\right)$.

STANDARDIZED TEST TIP **Exercise 34** In an x-intercept, the y-coordinate is zero. So students may ignore the y-term altogether and just solve for $3x = 6$.

Remind students that the new mathematical terms in this chapter are defined in the Glossary/Study Guide in the back of the book.

5 Wrap Up

Key Terms

$Ax + By = C$ form (p. 246)
break-even point (p. 257)
constant of variation (p. 225)
correlation coefficient (p. 242)
dimensional analysis (p. 219)
direct variation (p. 225)
linear equation (p. 225)
linear function (p. 221)
line of best fit (p. 242)
parallel lines (p. 250)
perpendicular lines (p. 251)
rate of change (p. 220)
slope (p. 214)
slope-intercept form (p. 231)
x-intercept (p. 246)
y-intercept (p. 230)

How am I doing?

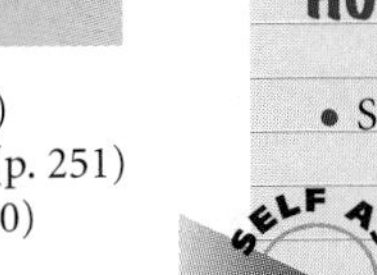

- State three ideas from this chapter that you think are important. Explain your choices.
- Describe two ways that you can write an equation of a line.

Slope and Rates of Change 5-1, 5-2

Rate of change allows you to look at how two quantities change relative to each other. **Slope** is the ratio of the vertical change to the horizontal change. The rate of change is also called slope.

$$\text{slope} = \frac{\text{vertical change}}{\text{horizontal change}} = \frac{\text{rise}}{\text{run}} \qquad \text{rate of change} = \frac{\text{change in the dependent variable}}{\text{change in the independent variable}}$$

Find two points on each graph. Then find the slope. 1–3. Answers may vary. Samples are given.

1.

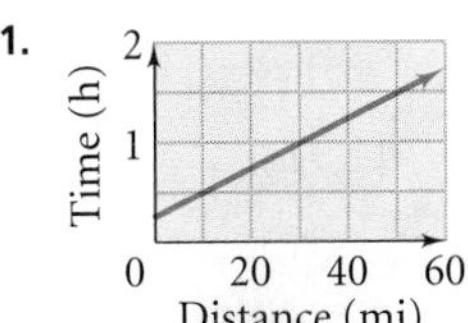

(10, 0.5), (30, 1); $\frac{1}{40}$

2. Height (ft) vs. Time (min)

(0, 150), (40, 150); 0

3. Gasoline in Tank (gal) vs. Hours of Driving

(1, 7.5), (3, 5); −1.25

Find the rate of change for each situation.

4. reading 5 pages in 6 min and 22 pages in 40 min $\frac{1}{2}$ page/min.

5. walking 1.5 mi in 25 min and 4 mi in 80 min $\frac{1}{22}$ mi/min

6. scoring 3 goals in 5 games and 9 goals in 7 games 3 goals/game

7. copying 4 pages in 1 min and 52 pages in 5 min 12 pages/min

Direct Variation 5-3

A function is a **direct variation** if it has the form $y = kx$, where $k \neq 0$. The coefficient k is the **constant of variation.**

Each set of ordered pairs is for a direct variation. Find each missing value.

8. (4, 8) and (3, y) 6

9. (−3, 9) and (x, −12) 4

10. (16, y) and (4, −1) −4

Resources

Student Edition

Extra Practice, p. 560
Glossary/Study Guide

Teaching Resources

Study Skills Handbook
Glossary, Spanish Resources

pages 262–264 Wrap Up

11.

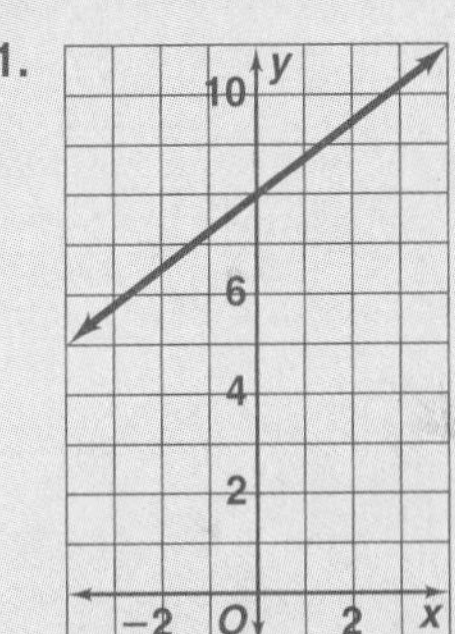

12.

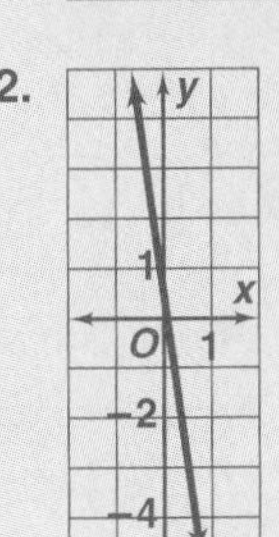

pages 262–264 Wrap Up

13.

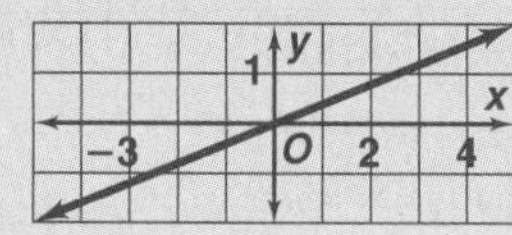

14.

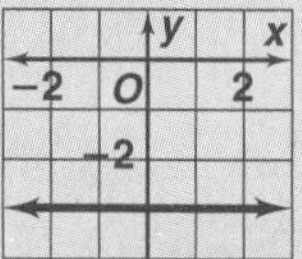

23.

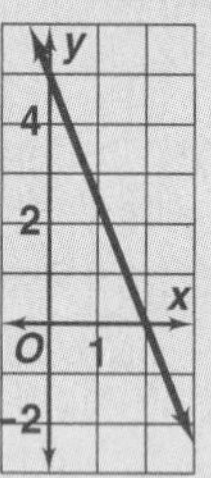

24.

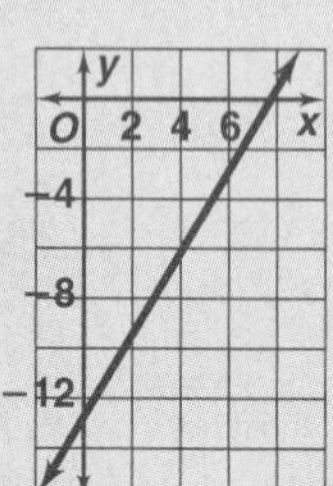

Slope-Intercept Form 5-4

The **y-intercept** of a line is the y-coordinate of the point where the line crosses the y-axis. The **slope-intercept form** of a linear equation is $y = mx + b$, where m is the slope and b is the y-intercept.

Write an equation of a line with the given slope and y-intercept. Then graph each equation. 11–14. See margin for graphs.

11. $m = \frac{3}{4}, b = 8$ $y = \frac{3}{4}x + 8$

12. $m = -7, b = \frac{1}{2}$ $y = -7x + \frac{1}{2}$

13. $m = \frac{2}{5}, b = 0$ $y = \frac{2}{5}x$

14. $m = 0, b = -3$ $y = -3$

Writing the Equation of a Line 5-5

When you know the slope of a line and a point on it, solve for the y-intercept. Then use the slope and the y-intercept to write an equation.

When you know two points on a line, first find the slope of the line through the points. Then find the y-intercept and write an equation.

Write an equation of a line through the given points.

15. $(4, 3), (-2, 1)$ $y = \frac{1}{3}x + 1\frac{2}{3}$

16. $(5, -4), (0, 2)$ $y = -\frac{6}{5}x + 2$

17. $(-1, 0), (-3, -1)$ $y = \frac{1}{2}x + \frac{1}{2}$

18. $(2, 7), (-8, 4)$ $y = \frac{3}{10}x + 6\frac{2}{5}$

Write an equation for each line.

19.

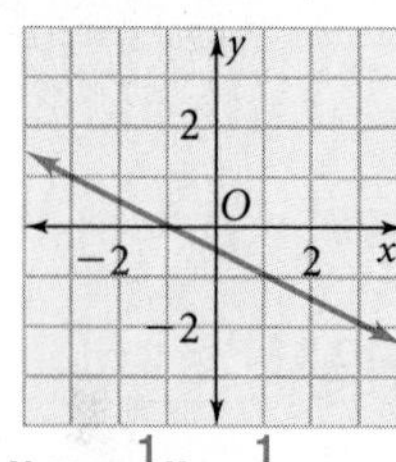

$y = -\frac{1}{2}x - \frac{1}{2}$

20.

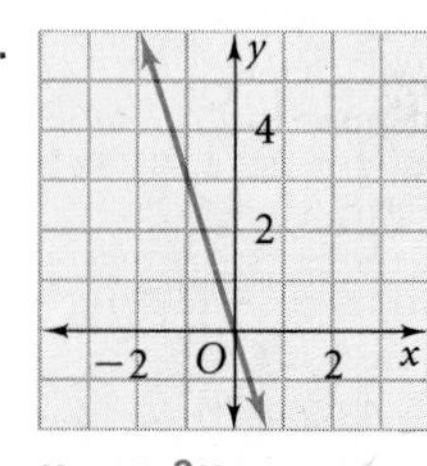

$y = -3x$

21.

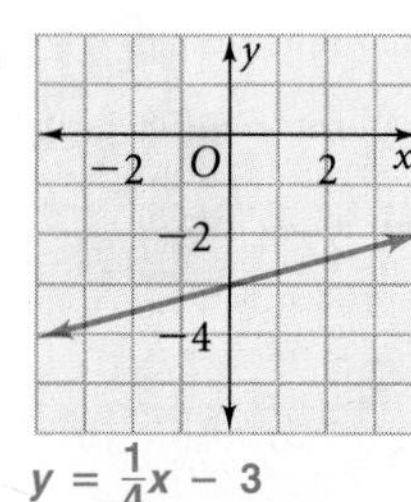

$y = \frac{1}{4}x - 3$

Scatter Plots and Equations of Lines 5-6

The **line of best fit** of a scatter plot is the most accurate trend line showing the relationship between the sets of data. You can find the equation of the line of best fit using a graphing calculator. The **correlation coefficient** tells how closely the equation models the data.

22. The table shows the average person's consumption of poultry from 1970 to 1993.

a. Find the equation of a trend line or use a graphing calculator to find the equation of the line of best fit. See below.

b. Use your equation to **predict** how much poultry the average person will eat in 2005. b. For sample in (a): 77.5 lb/person

22a. Sample: $y = 1.3x - 59$

Average Poultry Consumption (pounds/person)

Year	Pounds
1970	33.8
1975	32.9
1980	40.6
1985	45.2
1990	56.0
1993	61.1

ALTERNATIVE ASSESSMENT Divide students into small groups. Have each group write Exercises 11–21 on separate index cards. Then give the students the answers to these questions on a second set of cards. Ask students to match the question cards with the answer cards as quickly as possible.

Getting Ready for Chapter 6

Students may work these exercises independently or in small groups. The skills previewed will help prepare students to find the intersection points of two solution sets (equations).

$Ax + By = C$ Form 5-7

The **x-intercept** of a line is the x-coordinate of the point where the line crosses the x-axis. When a line is in $\boldsymbol{Ax + By = C}$ **form,** you can use the x- and y-intercepts to make a graph of the equation.

Find the x- and y- intercepts. Then graph each equation. 23–26. See margin for graphs.

23. $5x + 2y = 10$ 2; 5

24. $6.5x - 4y = 52$ 8; −13.

25. $-x + 3y = -15$ 15; −5

26. $8x - y = 104$ 13; −104

Parallel and Perpendicular Lines 5-8

Parallel lines have the same slope. Two lines are **perpendicular** if the product of their slopes is -1.

Write an equation that satisfies the given conditions.

27. parallel to $y = 5x - 2$, through $(2, -1)$ $y = 5x - 11$

28. perpendicular to $y = -3x + 7$, through $(3, 5)$ $y = \frac{1}{3}x + 4$

29. parallel to $y = 9x$, through $(0, -5)$ $y = 9x - 5$

30. perpendicular to $y = 8x - 1$, through $(4, 10)$ $y = -\frac{1}{8}x + 10\frac{1}{2}$

31. *Open-ended* Write the equations of four lines that form a rectangle. Sample: $y = x$, $y = -x + 5$, $y = x + 3$, $y = -x$

Using the x-intercept 5-9

You can use a linear function to solve the related one-variable equation. Graph the linear function. The x-intercept is the solution of the related one-variable equation.

32. You start a lawn-mowing business. You buy a used power mower for \$175. You plan to charge \$30 per lawn.

a. Write an equation relating your income to expenses. $30x = 175$

b. Graph the related linear function to find your break-even point. See margin for graph; $5\frac{5}{6}$ lawns, or 6 lawns to make a profit.

33. *Writing* How is the break-even point related to profit? The break-even point is the amount you need to produce to make \$0 profit.

34. *Standardized Test Prep* What is the x-intercept of $3x - 2y = 6$? C

A. -3 **B.** -2 **C.** 2 **D.** 3 **E.** −5

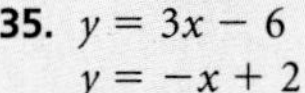

Getting Ready for CHAPTER 6

Graph each pair of equations on the same set of axes. 35–40. See margin.

35. $y = 3x - 6$
$y = -x + 2$

36. $y = -2x - 4$
$y = 0.5x + 9$

37. $y = 6x + 1$
$y = 6x - 4$

38. $y = x + 5$
$y = -3x + 5$

39. $y = x^2$
$y = 4x + 9$

40. $y = -4$
$y = 2x^2 + 1$

25.

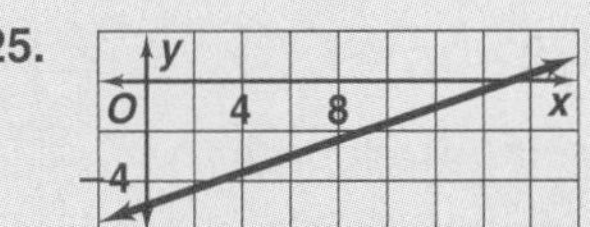

26.

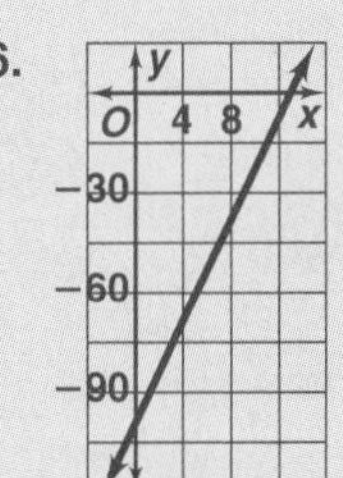

32b.

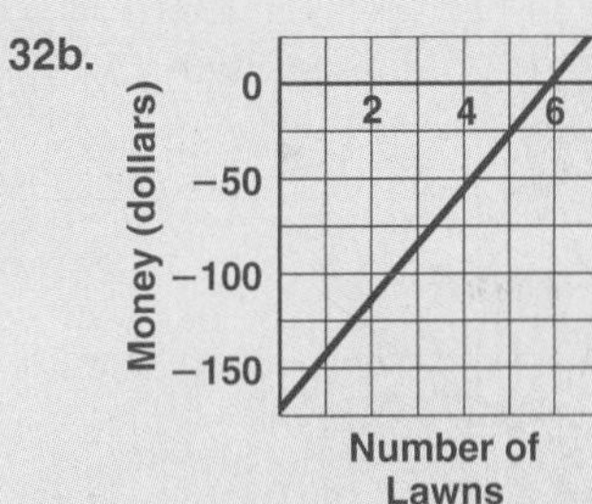

35.

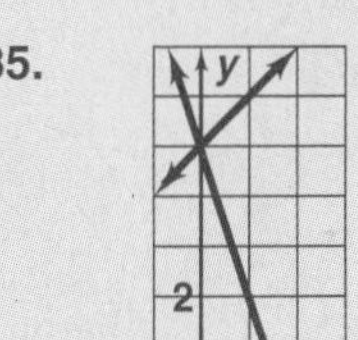

36–40. See back of book.

Assessment

ENHANCED MULTIPLE CHOICE QUESTIONS are more complex than traditional multiple choice questions, which assess only one skill. Enhanced multiple choice questions assess the processes that students use, as well as the end results. The questions are written so that students must use more than one strategy to solve the problem. Using multiple strategies is encouraged by the National Council of Teachers of Mathematics (NCTM). **Exercise 26** is an enhanced multiple choice question.

FREE RESPONSE QUESTIONS do not give answer choices. Some exercises have more than one possible answer. Students need to give only one correct response. **Exercises 3–16, 18–25, 27–30, and 32–35** are free response questions.

WRITING EXERCISES allow students to describe how they think about and understand the concepts they have learned. **Exercise 17** is a writing exercise.

OPEN-ENDED PROBLEMS allow for more than one solution. Students must construct their own responses instead of choosing from possible answers. The student responses will help you determine the depth of their understanding and any possible areas of difficulty. **Exercise 31** is an open-ended problem. Make sure that the students' responses contain values for x and y and a written situation.

Resources

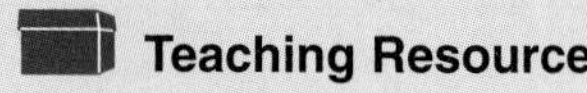

Teaching Resources

Chapter Support File, Ch. 5

- Chapter Assessment, Forms A and B
- Alternative Assessment

Chapter Assessment, Spanish Resources

Teacher's Edition

See also p. 212E for assessment options.

Software Computer Item Generator

page 264 Assessment

5.

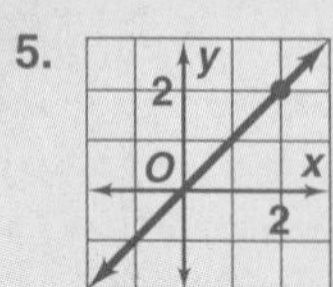

6.

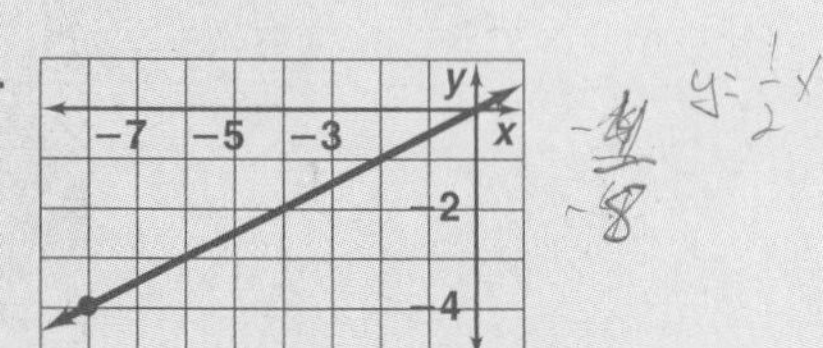

7.

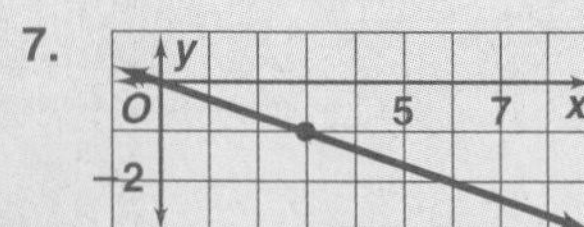

8, 17, 32b. See back of book.

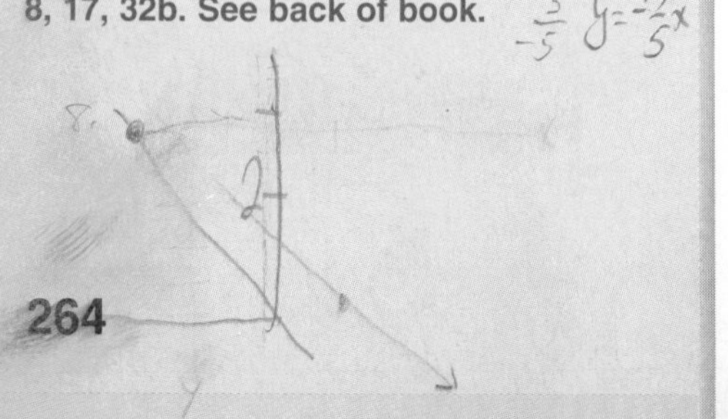

5 Assessment

Tell if each statement is *true* or *false*. Explain.

1. A rate of change must be positive. False; a rate of change could also be negative or 0.
2. The rate of change for a vertical line is 0. False; a vertical line has an undefined rate of change.

Find the slope of the line passing through each pair of points.

3. $(4, 3), (3, 8)$ -5
4. $(-2, 1), (6, -1)$ $-\frac{1}{4}$

Draw the graph of a direct variation that includes the given point. 5–8. See margin.

5. $(2, 2)$ 6. $(-8, -4)$ 7. $(3, -1)$ 8. $(-5, 3)$

Write each equation in $y = mx + b$ form.

9. $-7y = 8x - 3$ See below.
10. $x - 3y = 18$ $y = \frac{1}{3}x + 6$
11. $5x + 4y = 100$ $y = -\frac{5}{4}x + 25$
12. $9x = 2y + 13$ $y = \frac{9}{2}x - \frac{13}{2}$

Find the x- and y-intercepts of each line.

13. $3x + 4y = -24$ $-8; -6$
14. $2y = 3x - 8$ $\frac{8}{3}; -4$
15. $y = 0.3x + 6$ $-20; 6$
16. $y = -x + 1$ $1; 1$
17. **Writing** Explain why the x- and y-intercepts of a direct variation are always zero. See margin.

Write an equation of the line with the given slope that contains the given point. 18–21. Samples are given.

18. slope $= 3$, $(4, -8)$ $y = 3x - 20$
19. slope $= \frac{8}{3}$, $(-2, -7)$ $y = \frac{8}{3}x - \frac{5}{3}$
20. slope $= -\frac{1}{2}$, $(0, 3)$ $y = -\frac{1}{2}x + 3$
21. slope $= -5$, $(9, 0)$ $y = -5x - 45$

Write an equation of the line that contains the given points.

22. $(4, 9), (-2, -6)$ $y = \frac{5}{2}x - 1$
23. $(-1, 0), (3, 10)$ $y = \frac{5}{2}x + \frac{5}{2}$
24. $(5, -8), (-9, -8)$ $y = -8$
25. $(0, 7), (1, 5)$ $y = -2x + 7$
26. Which of the following lines is *not* perpendicular to $y = -2.5x + 13$? D
    - A. $y = 0.4x - 7$
    - B. $-2x + 5y = 8$
    - C. $y = \frac{2}{5}x + 4$
    - D. $2y = 5x + 1.5$

Write an equation of a line that satisfies the given conditions.

27. parallel to $y = 5x$, through $(2, -1)$ $y = 5x - 11$
28. perpendicular to $y = -2x$, through $(4, 0)$ $y = \frac{1}{2}x - 2$
29. parallel to $y = 5$, through $(-3, 6)$ $y = 6$
30. perpendicular to $x = -7$, through $(0, 2)$ $y = 2$
31. **Open-ended** Write the equation of a line parallel to $y = 0.5x - 10$. Sample: $y = 0.5x + 2$
32. You start a pet-washing service. You spend \$30 on supplies. You charge \$5 to wash each pet.
    a. Write an equation to relate your income y to the number of pets x you wash. $y = 5x - 30$
    b. Graph the equation. What are the x- and y-intercepts? See margin for graph; 6; -30.
    c. How many pets do you need to wash to break even? 6 pets

Use the data below for Exercises 33 and 34.

Local Governments in the United States (in thousands)

Year	Municipalities	School Districts
1962	18.0	34.7
1967	18.0	21.8
1972	18.5	15.8
1977	18.9	15.2
1982	19.1	14.9
1987	19.2	14.7
1992	19.3	14.6

33. a. **Graphing Calculator** Find the equation of a trend line or the line of best fit for the number of municipalities and the year.
    b. **Predict** the number of municipalities in 2005. about 20,000 municipalities

33a. $y = 0.0492857143x + 14.91928571$

34. a. **Graphing Calculator** Find the equation of a trend line or the line of best fit for the number of school districts and the year.
    b. Do the data show a strong correlation? Explain. No. The correlation coefficient is not near 1 or −1.

9. $y = -\frac{8}{7}x + \frac{3}{7}$

34a. $y = -0.5385714286x + 60.28428571$

Preparing for Standardized Tests

Standardized tests, such as those administered for state assessment, the SAT, or the ACT, include regular math questions, quantitative comparison questions, open-ended problems, and free response questions (which the SAT calls *grid-ins*).

MULTIPLE CHOICE QUESTIONS are followed by five answer choices, one of which is correct. **Exercises 1–9** are multiple choice questions.

QUANTITATIVE COMPARISON QUESTIONS ask students to compare two quantities. **Exercises 10 and 11** are quantitative comparison questions.

FREE-RESPONSE QUESTIONS do not give answer choices. Students must provide the one correct answer on their own. **Exercises 12–16** are free response questions.

STANDARDIZED TEST TIPS

1 Make sure to convert the problem's units into the same units as the given responses.

2 Memorize the standard choices for answers of questions involving quantitative comparisons to avoid having to keep rereading the instructions.

5 Preparing for Standardized Tests

For Exercises 1–11, choose the correct letter.

1. A horizontal line passes through $(5, -2)$. Which other point does it also pass through? B
   A. $(5, 2)$ B. $(-5, -2)$ C. $(-5, 2)$
   D. $(5, 0)$ E. none of the above

2. 180 is what percent of 60? A
   A. 300% B. 50% C. 3%
   D. 500% E. 30%

3. Match the graph with its equation. B

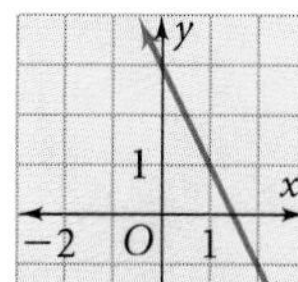

   A. $y = 2x + 3$
   B. $y = -2x + 3$
   C. $y = 2x - 3$
   D. $y = -2x - 3$
   E. none of the above

4. If $\frac{x}{3} = \frac{x+3}{5}$ and $\frac{y-7}{3} = \frac{y}{-7}$, then: B
   A. $x > y$ B. $y > x$ C. $x = y$
   D. $x = 2y$ E. $y > 2x$

5. Mariko runs 800 ft in one minute. What is her approximate speed in miles per hour? (Recall: 5280 ft = 1 mi) C
   A. 6 mi/h B. 8 mi/h C. 9 mi/h
   D. 12 mi/h E. 15 mi/h

6. If *a*, *b*, and *c* are three consecutive positive integers, which of the following could be true?
   I. $a + c < 2b$ II. $a + b < c$ B
   III. $a + c > 2b$ IV. $b + c > a$
   A. I only B. IV only C. I and II
   D. III and IV E. I and II

7. A line perpendicular to $y = 3x - 2$ passes through the point $(0, 6)$. Which other point lies on the line? A
   A. $(9, 3)$ B. $(-9, 3)$ C. $(-9, -3)$
   D. $(9, -3)$ E. none of the above

8. Which of the following is the best estimate of the fraction $\frac{63 \cdot 123 \cdot 0.49}{6.23 \cdot 11.93}$? A
   A. 50 B. 5 C. 20
   D. 250 E. 500

9. If $3x + 2 = 11$, then $5x + 1 = ■$. D
   A. 13 B. 14 C. 15 D. 16 E. 17

Compare the boxed quantity in Column A with the boxed quantity in Column B. Choose the best answer.

A. The quantity in Column A is greater.
B. The quantity in Column B is greater.
C. The two quantities are equal.
D. The relationship cannot be determined on the basis of the information supplied.

	Column A	Column B	
10.	the slope of $2x - 3y = 5$	the slope of $4y - 2 = 7x$	B

Use the equation $\frac{y-3}{x} = -3$.

	Column A	Column B	
11.	the slope	the *y* intercept	B

Find each answer.

12. Write and solve a compound inequality for the following statement. "The sum of a number and 15 is more than 27 but less than 32."
    $27 < n + 15 < 32$; $12 < n < 17$

13. Find the slope of the line that passes through $(-1, 3)$ and $(4, 6)$. $\frac{3}{5}$

14. Graphing Calculator Write the equation of a trend line or the line of best fit for the data.
    $y = 3.106382979x - 4.095744681$

x	1	1	2	3	4	4	5	6
y	0	−1	1	6	8	7	12	15

15. Find two consecutive integers such that the larger is seven less than three times the smaller. 4, 5

16. a. A new company employed 12 people. Two years later, it employed a total of 20 people. What was the percent of increase? about 67%
    b. If one of the 20 people retired, what would be the percent of decrease? 5%

Resources

Teaching Resources

Chapter Support File, Ch. 5
- Standardized Test Practice
- Cumulative Review

See also p. 212E for assessment options.

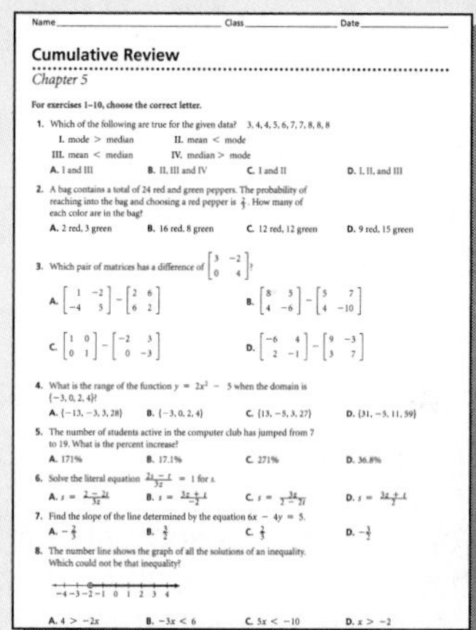

Name_____ Class_____ Date_____

Cumulative Review

Chapter 5

For exercises 1–10, choose the correct letter.

1. Which of the following are true for the given data? 3, 4, 4, 5, 6, 7, 7, 8, 8, 8
   I. mode > median II. mean < mode
   III. mean < median IV. median > mode
   A. I and III B. II, III and IV C. I and II D. I, II, and III
2. A bag contains a total of 24 red and green peppers. The probability of reaching into the bag and choosing a red pepper is $\frac{2}{3}$. How many of each color are in the bag?
   A. 2 red, 3 green B. 16 red, 8 green C. 12 red, 12 green D. 9 red, 15 green
3. Which pair of matrices has a difference of $\begin{bmatrix} 3 & -2 \\ 0 & 4 \end{bmatrix}$?
   A. $\begin{bmatrix} 1 & -2 \\ -4 & 5 \end{bmatrix} - \begin{bmatrix} 2 & 6 \\ 6 & 2 \end{bmatrix}$ B. $\begin{bmatrix} 8 & 5 \\ 4 & -6 \end{bmatrix} - \begin{bmatrix} 5 & 7 \\ 4 & -10 \end{bmatrix}$
   C. $\begin{bmatrix} 1 & 0 \\ 0 & 1 \end{bmatrix} - \begin{bmatrix} -2 & 3 \\ 0 & -3 \end{bmatrix}$ D. $\begin{bmatrix} -6 & 4 \\ 2 & -1 \end{bmatrix} - \begin{bmatrix} 9 & -3 \\ 3 & 7 \end{bmatrix}$
4. What is the range of the function $y = 2x^2 - 5$ when the domain is $\{-3, 0, 2, 4\}$?
   A. $\{-13, -3, 3, 28\}$ B. $\{-3, 0, 2, 4\}$ C. $\{13, -5, 3, 27\}$ D. $\{31, -5, 11, 59\}$
5. The number of students active in the computer club has jumped from 7 to 19. What is the percent increase?
   A. 171% B. 17.1% C. 271% D. 36.8%
6. Solve the literal equation $\frac{2s - t}{3z} = 1$ for *s*.
   A. $s = \frac{2 - 2t}{3z}$ B. $s = \frac{3z + t}{-2}$ C. $s = \frac{3z}{2 - 2t}$ D. $s = \frac{3z + t}{2}$
7. Find the slope of the line determined by the equation $6x - 4y = 5$.
   A. $-\frac{2}{3}$ B. $\frac{3}{2}$ C. $\frac{2}{3}$ D. $-\frac{3}{2}$
8. The number line shows the graph of all the solutions of an inequality. Which could not be that inequality?
   −4 −3 −2 −1 0 1 2 3 4
   A. $4 > -2x$ B. $-3x < 6$ C. $5x < -10$ D. $x > -2$

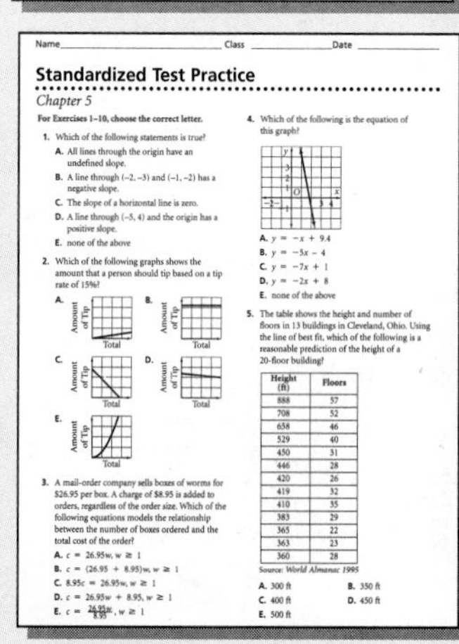

Name_____ Class_____ Date_____

Standardized Test Practice

Chapter 5

For Exercises 1–10, choose the correct letter.

1. Which of the following statements is true?
   A. All lines through the origin have an undefined slope.
   B. A line through $(-2, -3)$ and $(-1, -2)$ has a negative slope.
   C. The slope of a horizontal line is zero.
   D. A line through $(-5, 4)$ and the origin has a positive slope.
   E. none of the above
2. Which of the following graphs shows the amount that a person should tip based on a tip rate of 15%?
   A. B. C. D. E. (Amount of Tip / Total)
3. A mail-order company sells boxes of worms for \$26.95 per box. A charge of \$8.95 is added to orders, regardless of the order size. Which of the following equations models the relationship between the number of boxes ordered and the total cost of the order?
   A. $c = 26.95w, w \geq 1$
   B. $c = (26.95 + 8.95)w, w \geq 1$
   C. $8.95c = 26.95w, w \geq 1$
   D. $c = 26.95w + 8.95, w \geq 1$
   E. $c = \frac{26.95w}{8.95}, w \geq 1$
4. Which of the following is the equation of this graph?
   A. $y = -x + 9.4$
   B. $y = -5x - 4$
   C. $y = -7x + 1$
   D. $y = -2x + 8$
   E. none of the above
5. The table shows the height and number of floors in 13 buildings in Cleveland, Ohio. Using the line of best fit, which of the following is a reasonable prediction of the height of a 20-floor building?

Height (ft)	Floors
888	57
708	52
658	46
529	40
450	31
446	28
420	26
419	32
410	35
383	29
365	22
363	23
360	28

Source: *World Almanac 1995*

   A. 300 ft B. 350 ft
   C. 400 ft D. 450 ft
   E. 500 ft

CHAPTER 6

Systems of Equations and Inequalities

To accommodate flexible scheduling, some lessons are divided into parts. Assignment Options are given in the Lesson Planning Options for each lesson.

PACING OPTIONS

This chart suggests pacing only for the core lessons and their parts, and it is provided merely as a possible guide. It will help you determine how much time you have in your schedule to cover other features, such as the Chapter Project, Math Toolboxes, Wrap Up, and Assessment.

	1 Class Period	1 Class Period	1 Class Period	1 Class Period	1 Class Period	1 Class Period	1 Class Period	1 Class Period	1 Class Period
Traditional (40–45 min class periods)	6-1 Part 1, 6-1 Part 2	6-2 Part 1	6-2 Part 2	6-3 Part 1	6-3 Part 2	6-4	6-5	6-6	6-7
Algebra Two-Year (40–45 min class periods)	6-1 Part 1	6-1 Part 2	6-2 Part 1	6-2 Part 1	6-2 Part 2	6-2 Part 2	6-3 Part 1	6-3 Part 1	6-3 Part 2
Block Scheduling (90 min class periods)	6-1 Part 1, 6-1 Part 2	6-2 Part 1, 6-2 Part 2	6-3 Part 1, 6-3 Part 2	6-4, 6-5	6-6	6-7	6-8		

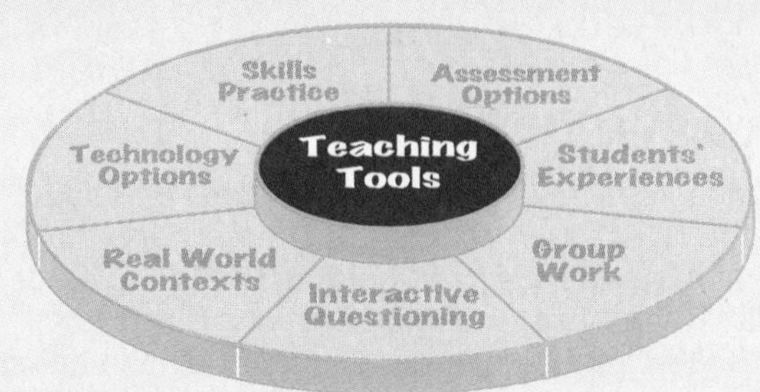

What Students Will Learn and Why

In this chapter, students will build on their knowledge of equations and inequalities learned in Chapter 5, by learning to solve systems of equations.

What Students Will Learn and Why

The concept map below relates chapter topics to real world applications. You and your class may wish to add to the map or develop maps of your own. The center oval describes the topic of the chapter. The next level displays topics within the lessons. The outer ovals reflect applications of the content. As you and your class build a concept map, invite students to discuss applications with which they are familiar.

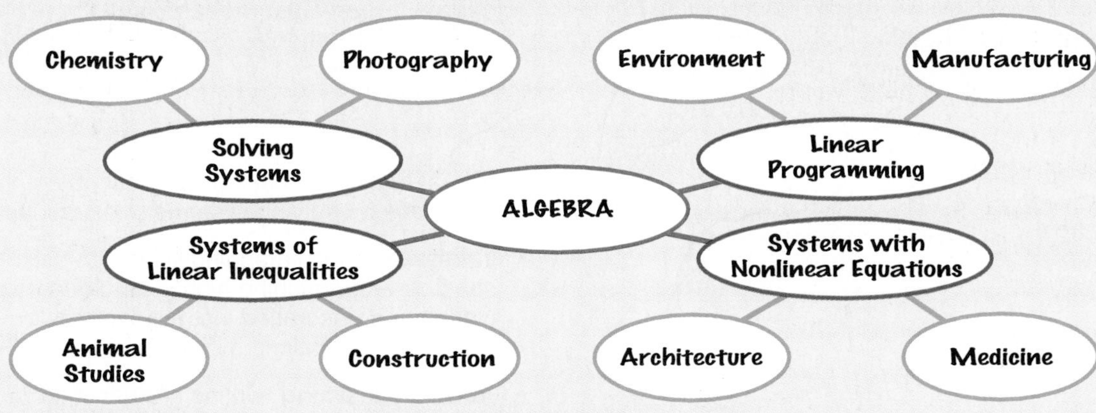

1 Class Period	1 Class Period	1 Class Period	1 Class Period	1 Class Period	1 Class Period	1 Class Period	1 Class Period	1 Class Period	1 Class Period	1 Class Period
6-8										
6-3 2	6-4	6-4	6-5	6-5	6-6	6-6	6-7	6-7	6-8	6-8

Skills Practice

Assessment Options | Technology Options | Teaching Tools | Students' Experiences | Real World Contexts | Group Work | Interactive Questioning

Interactive Questioning Tips

A question is interactive when there is "give and take" between the questioner (teacher or student) and the respondent. Questions should be adapted to the student's ability level. By asking students questions using terminology that they have already covered, understanding is enhanced and anxiety is reduced. For example, in Lesson 6-2, Example 3, students are questioned about parallel lines so that they may see how systems of parallel lines have no solution.

Skills Practice

Every lesson provides skill practice with Exercises On Your Own and Exercises Mixed Review. The Student Edition includes Checkpoints (pp. 284 and 299) and Cumulative Review (p. 315). In the Teacher's Edition, the Lesson Planning Options section for each lesson lists Prerequisite Skills students should know for that lesson. At the back of the Student Edition is the Skills Handbook—mini-lessons on math the students may need to review. The Chapter Support File for Chapter 6 in the Teaching Resources box includes two Practice worksheets per lesson, a worksheet for two Checkpoints, and worksheets for Cumulative Review and Standardized Test Preparation.

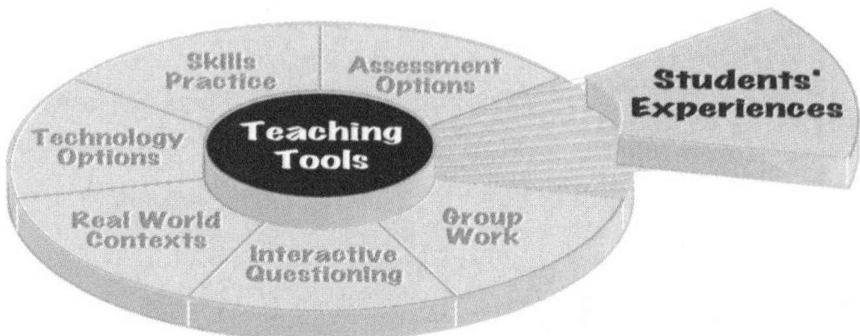

Diverse Learning and Teaching Styles

In your Teacher's Edition, you will find suggestions as to how you can help students complete mathematical tasks in Chapter 6 by reinforcing various learning styles. Here are some examples.

- **Visual learning** making and decorating charts of strategies for solving systems of equations (p. 282)
- **Tactile learning** graphing lines with color pencils and then shading with matching highlighting pen colors (p. 302), writing and decorating instructional posters for solving systems of equations by graphing (p. 306)
- **Auditory learning** writing a 30-second rap describing difference between solutions of systems (p. 277), identifying the quadrants in which an inequality is met (p. 296)
- **Kinesthetic learning** making a coordinate plane on the floor and represent lines between points with string (p. 271), moving to the correct side of an inequality on the coordinate plane on the floor (p. 290)

Alternative Activity for Lesson 6-1

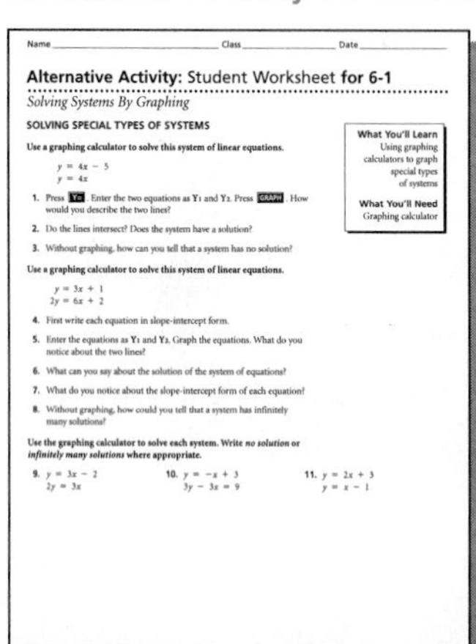

Name ____ Class ____ Date ____

Alternative Activity: Student Worksheet **for 6-1**

Solving Systems By Graphing

SOLVING SPECIAL TYPES OF SYSTEMS

What You'll Learn Using graphing calculators to graph special types of systems

What You'll Need Graphing calculator

Use a graphing calculator to solve this system of linear equations.

$y = 4x - 5$
$y = 4x$

1. Press Y=. Enter the two equations as Y1 and Y2. Press GRAPH. How would you describe the two lines?
2. Do the lines intersect? Does the system have a solution?
3. Without graphing, how can you tell that a system has no solution?

Use a graphing calculator to solve this system of linear equations.

$y = 3x + 1$
$2y = 6x + 2$

4. First write each equation in slope-intercept form.
5. Enter the equations as Y1 and Y2. Graph the equations. What do you notice about the two lines?
6. What can you say about the solution of the system of equations?
7. What do you notice about the slope-intercept form of each equation?
8. Without graphing, how could you tell that a system has infinitely many solutions?

Use the graphing calculator to solve each system. Write *no solution* or *infinitely many solutions* where appropriate.

9. $y = 3x - 2$, $2y = 3x$
10. $y = -x + 3$, $3y - 3x = 9$
11. $y = 2x + 3$, $y = x - 1$

for use with Part 2, addresses visual learning by using the graphing calculator to create graphs of systems of equations.

Alternative Activity for Lesson 6-7

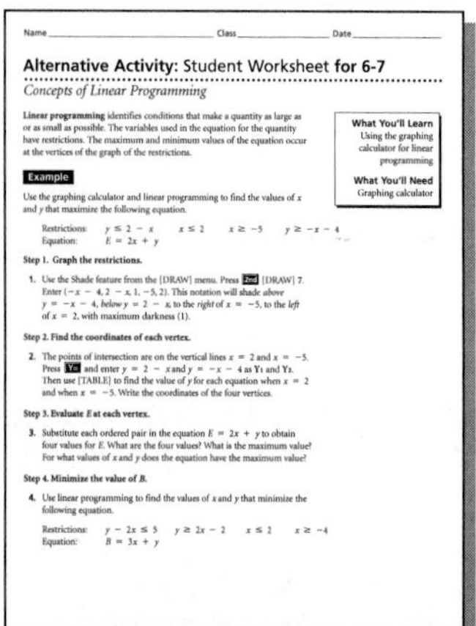

Name ____ Class ____ Date ____

Alternative Activity: Student Worksheet **for 6-7**

Concepts of Linear Programming

What You'll Learn Using the graphing calculator for linear programming

What You'll Need Graphing calculator

Linear programming identifies conditions that make a quantity as large as or as small as possible. The variables used in the equation for the quantity have restrictions. The maximum and minimum values of the equation occur at the vertices of the graph of the restrictions.

Example

Use the graphing calculator and linear programming to find the values of x and y that maximize the following equation.

Restrictions: $y \le 2 - x$ $x \le 2$ $x \ge -5$ $y \ge -x - 4$
Equation: $E = 2x + y$

Step 1. Graph the restrictions.

1. Use the Shade feature from the [DRAW] menu. Press 2nd [DRAW] 7. Enter $(-x - 4, 2 - x, 1, -5, 2)$. This notation will shade *above* $y = -x - 4$, *below* $y = 2 - x$, to the *right* of $x = -5$, to the *left* of $x = 2$, with maximum darkness (1).

Step 2. Find the coordinates of each vertex.

2. The points of intersection are on the vertical lines $x = 2$ and $x = -5$. Press Y= and enter $y = 2 - x$ and $y = -x - 4$ as Y1 and Y2. Then use [TABLE] to find the value of y for each equation when $x = 2$ and when $x = -5$. Write the coordinates of the four vertices.

Step 3. Evaluate *E* at each vertex.

3. Substitute each ordered pair in the equation $E = 2x + y$ to obtain four values for E. What are the four values? What is the maximum value? For what values of x and y does the equation have the maximum value?

Step 4. Minimize the value of *B*.

4. Use linear programming to find the values of x and y that minimize the following equation.

Restrictions: $y - 2x \le 5$ $y \ge 2x - 2$ $x \le 2$ $x \ge -4$
Equation: $B = 3x + y$

for use with Example 1, addresses tactile learning by using the graphing calculator for linear programming.

Alternative Activity for Lesson 6-8

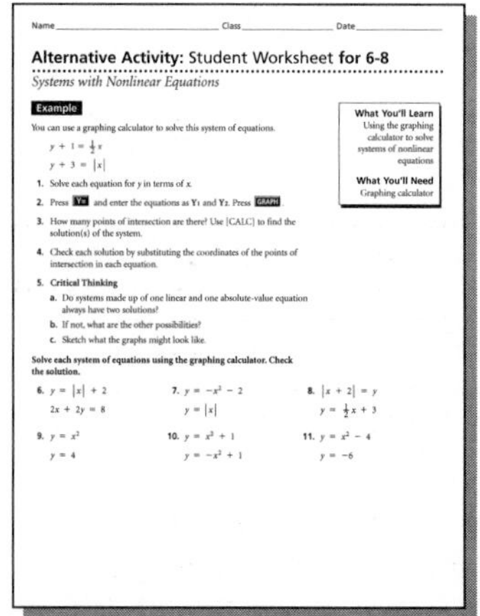

Name ____ Class ____ Date ____

Alternative Activity: Student Worksheet **for 6-8**

Systems with Nonlinear Equations

What You'll Learn Using the graphing calculator to solve systems of nonlinear equations

What You'll Need Graphing calculator

Example

You can use a graphing calculator to solve this system of equations.

$y + 1 = \frac{1}{2}x$
$y + 3 = |x|$

1. Solve each equation for y in terms of x.
2. Press Y= and enter the equations as Y1 and Y2. Press GRAPH.
3. How many points of intersection are there? Use [CALC] to find the solution(s) of the system.
4. Check each solution by substituting the coordinates of the points of intersection in each equation.
5. **Critical Thinking**
   a. Do systems made up of one linear and one absolute-value equation always have two solutions?
   b. If not, what are the other possibilities?
   c. Sketch what the graphs might look like.

Solve each system of equations using the graphing calculator. Check the solution.

6. $y = |x| + 2$, $2x + 2y = 8$
7. $y = -x^2 - 2$, $y = |x|$
8. $|x + 2| = y$, $y = \frac{1}{2}x + 3$
9. $y = x^2$, $y = 4$
10. $y = x^2 + 1$, $y = -x^2 + 1$
11. $y = x^2 - 4$, $y = -6$

for use with Example 1, addresses tact[ile] learning by using the graphing calculat[or to] solve the system of nonlinear equations

Cooperative Learning Tips

When used effectively, cooperative learning can help students develop interpersonal skills, learn to perform specific roles in a group, and learn to carry out specific responsibilities. The components of Chapter 6 provide a range of cooperative learning opportunities.

- In the Student Edition, the ***Work Together*** parts of lessons are specifically designed for cooperative learning activities.
- In the Teacher's Edition, you will find helpful hints for addressing diverse learning styles (see page C for Chapter 6). For every lesson, you will find a ***Reteaching Activity,*** which may involve cooperative learning.

Materials and Manipulatives

The authors expect all students to use scientific calculators. Calculator use is integrated throughout the course.

- **graph paper (6-1, 6-2, 6-3, 6-5, 6-6, 6-8)**
- **graphing calculator (6-1, 6-2, 6-6, 6-8)**

TECHNOLOGY OPTIONS

Technology Tools		Chapter Project	6-1	6-2	6-3	6-4	6-5	6-6	6-7	6-8
Calculator		Assumed that students have scientific calculators at any time.								
Graphing Calculator	Handbook		✔				✔	✔		✔
	Student Edition		✔	✔		✔	✔			✔
Software	Secondary Math Lab Toolkit		✔	✔	✔	✔	✔	✔	✔	✔
	Integrated Math Lab						✔			✔
	Computer Item Generator		✔	✔	✔	✔	✔	✔	✔	✔
	Student Edition								✔	
Video	Video Field Trip	✔								
CD-ROM	Multimedia Algebra Lab		✔	✔		✔		✔		
Internet		See the Prentice Hall site. (http://www.phschool.com)								

The Prentice Hall Algebra program offers you a rich variety of technology options. Be assured that all these options are provided as a means of enriching the program and are not essential for the successful completion of the course.

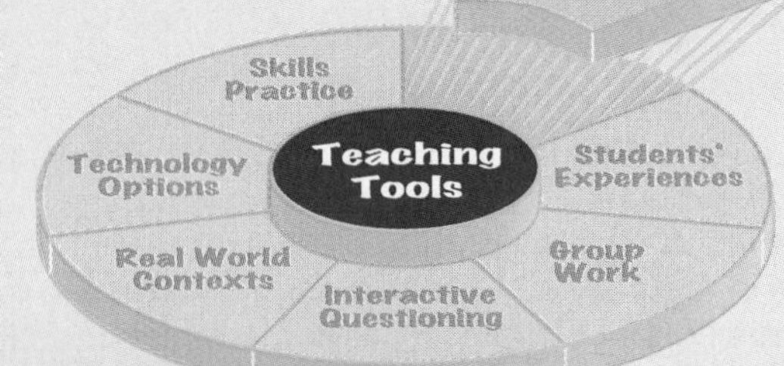

Assessment Options

The Prentice Hall Algebra Program provides you with many options. From these options, you may choose instructional materials and techniques appropriate for your students, or those necessary to meet your district's curriculum requirements. As the chart indicates, the program also supports your teaching efforts by offering you many choices for assessment.

ASSESSMENT OPTIONS

Assessment Support Materials	Chapter Opener	6-1	6-2	6-3	6-4	6-5	6-6	6-7	6-8	Chapter End
Chapter Project	●▲■	▲■		▲■		▲■	▲■			▲■
Checkpoints				▲●			▲●			
Self-Assessment				▲■			▲■		▲■	▲■
Writing Assignment		▲	▲	▲	▲	▲	▲	▲	▲	▲●
Chapter Assessment										▲●
Alternative Assessment		■	■	■	■	■	■	■	■	●■
Cumulative Review										▲●
Standardized Test Prep			▲■			▲■	▲■		▲■	●
Computer Item Generator	Can be used to create custom-made practice or assessment at any time.									

▲ = Student Edition ■ = Teacher's Edition ● = Teaching Resources

Checkpoints

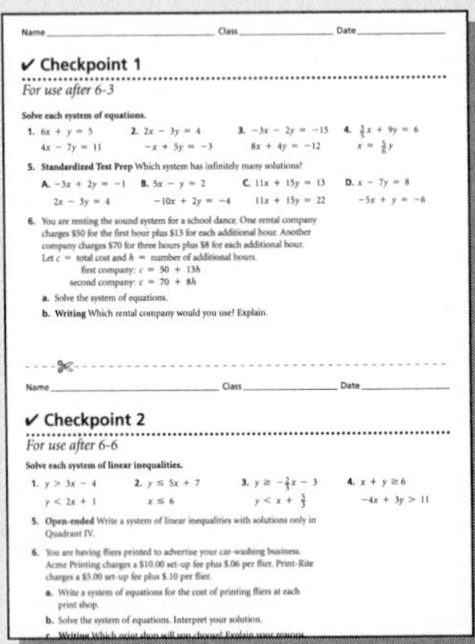
✔ Checkpoint 1

For use after 6-3

✔ Checkpoint 2

For use after 6-6

Alternative Assessment

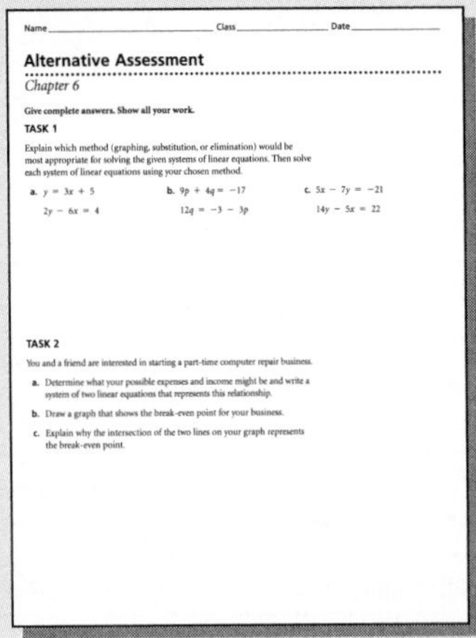
Alternative Assessment

Chapter 6

TASK 1

TASK 2

Chapter Assessment

Chapter Assessment Form A

Chapter 6

Available in both Form A and Form B

Making the Right Connections

Mathematics is imbedded in nearly every walk of life. The National Council of Teachers of Mathematics (NCTM) encourages educators to recognize these connections and to emphasize them for the purpose of better educating students for success in life and in a global economy. The Connections chart below highlights these connections for Chapter 6.

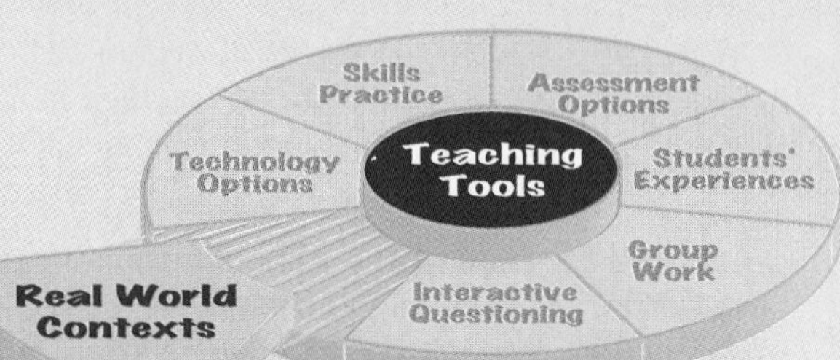

CONNECTIONS

Lesson	Interdiciplinary Connections	Career Prep	Other Real World Connections	Math Integration	NCTM Standards
Chapter Project			Cost Analysis		Algebra Problem Solving Reasoning
6-1			Entertainment Math in the Media Music		Algebra Communication Problem Solving
6-2	Biology	Agriculture	Transportation Internet	Geometry	Algebra Communication Problem Solving
6-3	Health	Sales Business Photography	Basketball Electricity Vacation		Algebra Communication Problem Solving
6-4	Chemistry	Publishing Business	School Musical Pets	Geometry	Algebra Communication Problem Solving
6-5		Manufacturing	Food Shopping Consumer Automobiles	Probability Geometry	Algebra Communication Problem Solving Probability
6-6		Construction	Animal Studies Shopping		Algebra Communication Problem Solving
6-7		Business Manufacturing	Time Management Environment		Algebra Communication Problem Solving
6-8		Architecture Medicine	Communications Native American Art		Algebra Communication Problem Solving

CONNECTING TO PRIOR LEARNING Have students reflect on the equations and inequalities they studied in Chapter 4. Challenge them to think of situations in which it would be useful to identify unknowns by more than one variable.

CULTURAL CONNECTIONS Although school dances have always been a part of high school socialization in America, other cultures may not permit boys and girls to even talk to each other. Ask students to discuss different customs involving dancing, dating, boys and girls attending the same school, and so on. Ask students to consider how they would react to a culture where girls are not encouraged to attend school.

INTERDISCIPLINARY CONNECTIONS Ask students about the costs involved in attending a dance. How do students figure the cost of going to a dance? If students work, how many hours do they have to work to make enough money to pay for a dance?

ABOUT THE PROJECT In this project, students assume the role of student council members planning for a dance. They use systems of equations to analyze costs and make decisions about dinner and bands. Then students write a report detailing the cost of catering and the band, and make a recommendation for ticket prices.

Technology Options

Prentice Hall Technology

Video Video Field Trip 6, "Thrill Ride," what it means to build an amusement park.

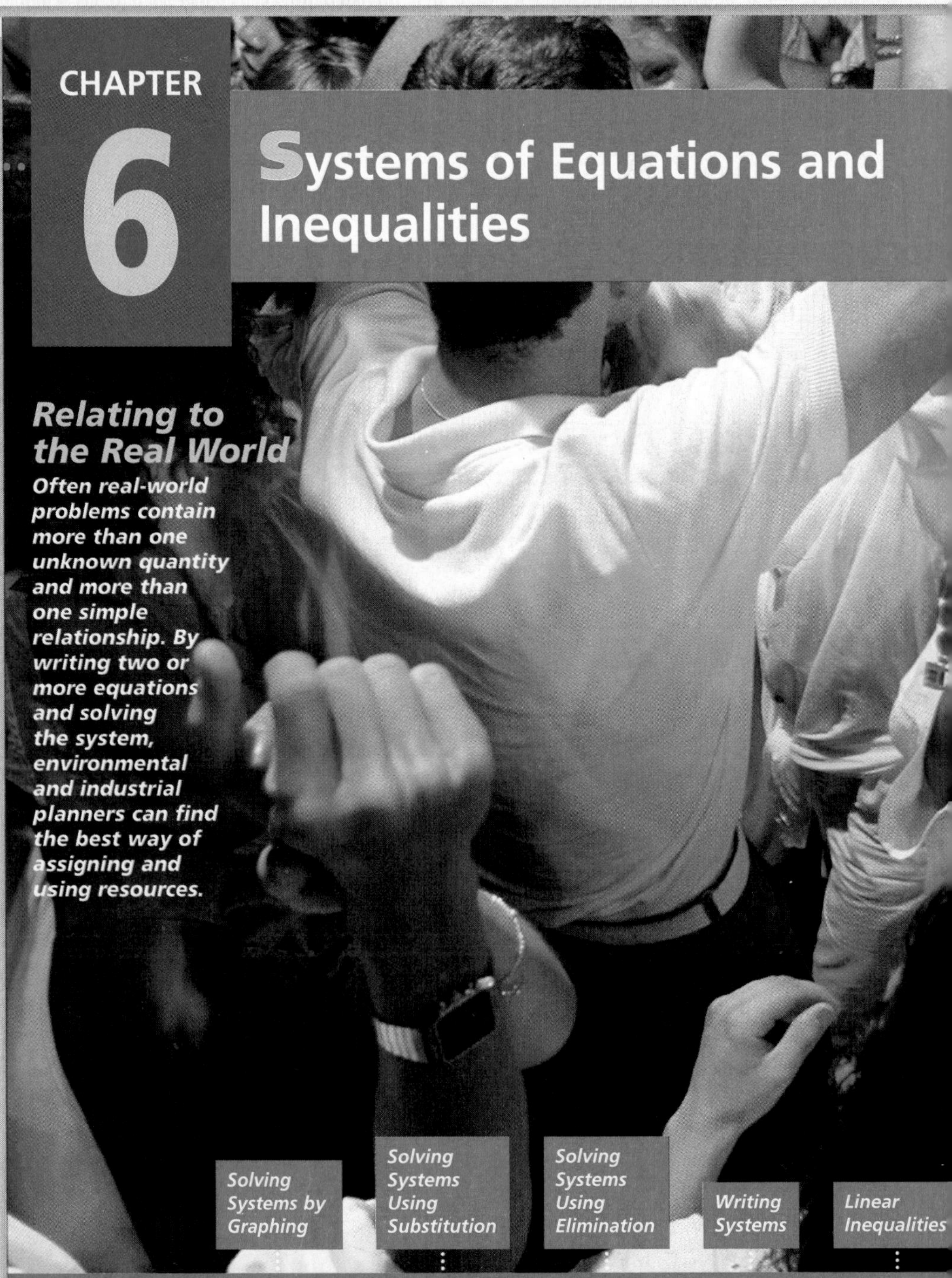

CHAPTER 6 Systems of Equations and Inequalities

Relating to the Real World

Often real-world problems contain more than one unknown quantity and more than one simple relationship. By writing two or more equations and solving the system, environmental and industrial planners can find the best way of assigning and using resources.

Lessons	6-1	6-2	6-3	6-4	6-5
	Solving Systems by Graphing	*Solving Systems Using Substitution*	*Solving Systems Using Elimination*	*Writing Systems*	*Linear Inequalities*

Launching the Project

PROJECT NOTEBOOK Encourage students to keep all project-related materials in a separate folder or notebook. See Chapter Project and Scoring Rubric in Chapter Support File.

- Assign students to work with a partner or in a small group. Have each group brainstorm items that would be necessary to put on a school dance. band, dinner, ticket prices, decorations, venue
- Ask the groups to consider how the cost of the band and dinner might affect the ticket price. The higher the cost of the band and dinner, the higher the cost of the tickets will be.
- Challenge the students to find out how much a typical band and dinner might cost. Have them create a spreadsheet to show their results.

TRACKING THE PROJECT Have students read Finishing the Chapter Project on page 310 to help them get an overview of the project. Also, set benchmark deadlines for students to show you their work in progress.

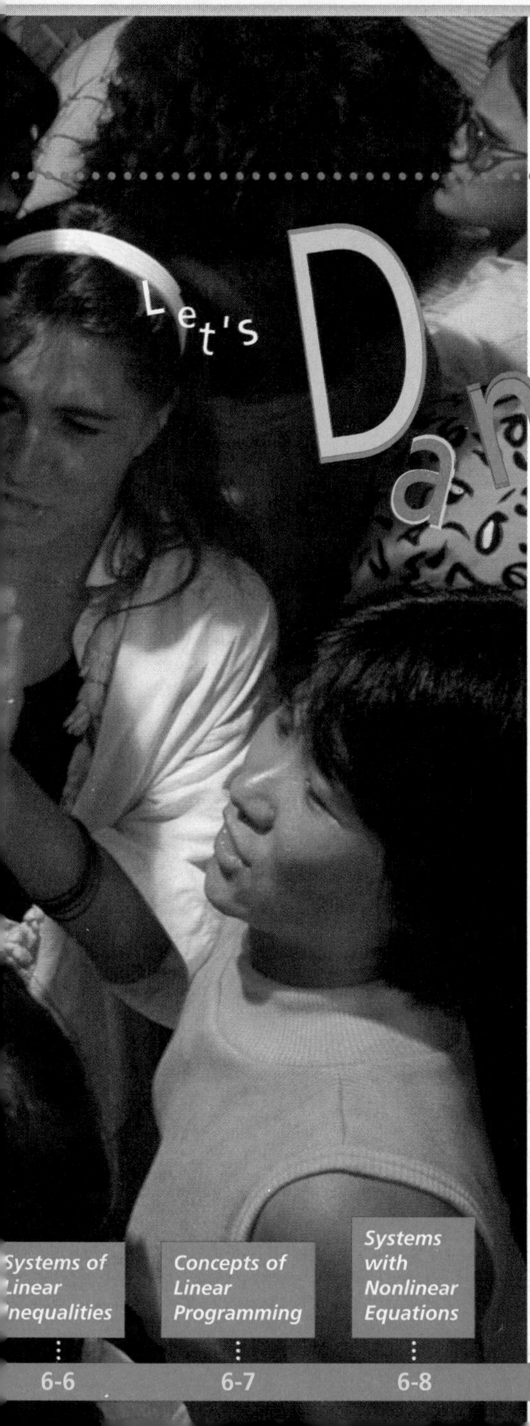

CHAPTER PROJECT

Let's Dance!

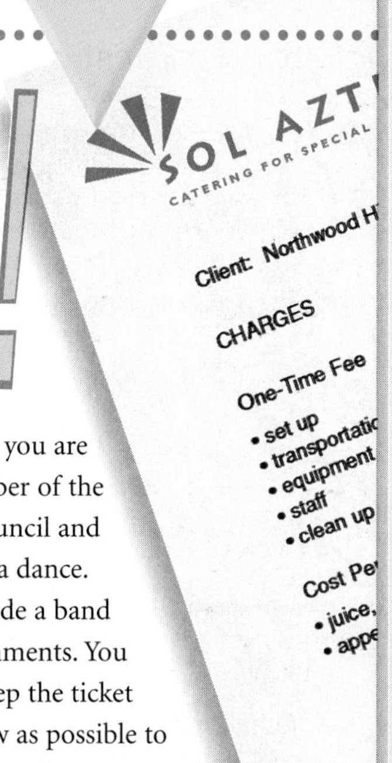

Suppose you are a member of the student council and must plan a dance. Plans include a band and refreshments. You want to keep the ticket price as low as possible to encourage students to attend.

As you work through the chapter, you will use systems of equations to analyze costs and make decisions. You will write a report detailing your choice of a band, the cost of a catering service, and what you would recommend as a ticket price.

To help you complete the project:

- **p. 274** ***Find Out by Graphing***
- **p. 284** ***Find Out by Calculating***
- **p. 293** ***Find Out by Writing***
- **p. 299** ***Find Out by Graphing***
- **p. 310** ***Finishing the Project***

Project Resources

Teaching Resources

Chapter Support File, Ch. 6
- Chapter Project Manager and Scoring Rubric

Transparencies

77

Using the Rubric

Sharing the scoring rubric for the project with your students will alert them to your expectations before they begin work on the project.

As students complete each Find Out question in the chapter, you may wish to have them evaluate their own work or a partner's work, based on the scoring rubric. Students should have the opportunity to revise their work after it has been reviewed.

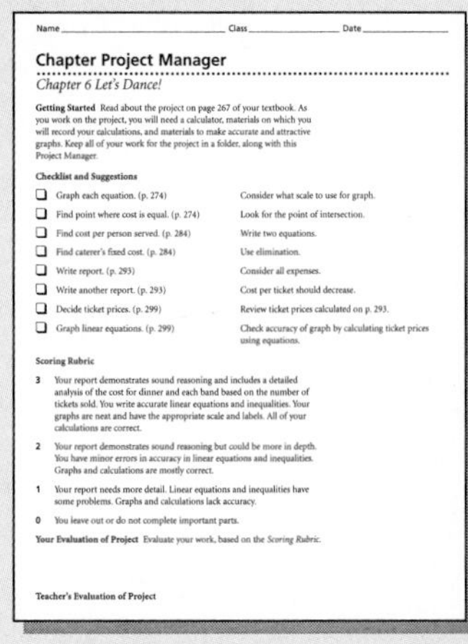

Name ________ Class ________ Date ________

Chapter Project Manager

Chapter 6 Let's Dance!

Getting Started Read about the project on page 267 of your textbook. As you work on the project, you will need a calculator, materials on which you will record your calculations, and materials to make accurate and attractive graphs. Keep all of your work for the project in a folder, along with this Project Manager.

Checklist and Suggestions

☐ Graph each equation. (p. 274)	Consider what scale to use for graph.
☐ Find point where cost is equal. (p. 274)	Look for the point of intersection.
☐ Find cost per person served. (p. 284)	Write two equations.
☐ Find caterer's fixed cost. (p. 284)	Use elimination.
☐ Write report. (p. 293)	Consider all expenses.
☐ Write another report. (p. 293)	Cost per ticket should decrease.
☐ Decide ticket prices. (p. 299)	Review ticket prices calculated on p. 293.
☐ Graph linear equations. (p. 299)	Check accuracy of graph by calculating ticket prices using equations.

Scoring Rubric

3 Your report demonstrates sound reasoning and includes a detailed analysis of the cost for dinner and each band based on the number of tickets sold. You write accurate linear equations and inequalities. Your graphs are neat and have the appropriate scale and labels. All of your calculations are correct.

2 Your report demonstrates sound reasoning but could be more in depth. You have minor errors in accuracy in linear equations and inequalities. Graphs and calculations are mostly correct.

1 Your report needs more detail. Linear equations and inequalities have some problems. Graphs and calculations lack accuracy.

0 You leave out or do not complete important parts.

Your Evaluation of Project Evaluate your work, based on the *Scoring Rubric.*

Teacher's Evaluation of Project

Math ToolboX

In the past, students may have been discouraged from using guessing as a strategy for problem solving. This toolbox demonstrates the value of guessing as a way to determine whether the actual answer is reasonable. Stress to students that the Guess and Test strategy requires a knowledge of the problem and the ability to revise the guess. Wild guessing is out of the question for this method. Point out the fact that tables, like the one in the example, make revising the guess easier. The goal of this toolbox is to help students develop the ability to approximate answers.

ERROR ALERT! Some students may have difficulty making the first guess. **Remediation**: Have students first try multiples of five or ten that may be easily computed. The guess may then be adjusted by comparing the solution to the conditions of the problem.

WRITING Exercise 5 Ask students what kinds of problems seem most suitable or least suitable for the guess and test method.

ADDITIONAL PROBLEM A mother is 2.8 times as old as her son. The sum of their ages is 57 yr. How old is the son? **15 yr old**

Math ToolboX Problem Solving

Guess and Test

Before Lesson 6-1

You can use the *Guess and Test* strategy to solve many types of problems. First, make a guess. Then test your guess against the conditions of the problem. Use the results from your first guess to make a more accurate guess. Continue to guess and test until you find the correct answer.

Example

The ratio of boys to girls in a ninth-grade class at a high school is about 3 to 2. There are about 600 ninth-graders. How many are boys? girls?

Guess the number of boys. — Subtract your guess from 600. — Write the ratio and compare.

	Number of Boys	Number of Girls	$\frac{\text{boys}}{\text{girls}} \overset{?}{=} \frac{3}{2} = 1.5$	
First guess	400	$600 - 400 = 200$	$\frac{400}{200} = 2$	← too high, so try a lower number of boys
Second guess	350	$600 - 350 = 250$	$\frac{350}{250} = 1.4$	← too low, so try a number between 350 and 400
Third guess	375	$600 - 375 = 225$	$\frac{375}{225} = 1.\overline{6}$	← too high, so try a number between 350 and 375
Fourth guess	360	$600 - 360 = 240$	$\frac{360}{240} = 1.5$	← correct

There are about 360 ninth-grade boys and 240 ninth-grade girls.

Solve each problem.

1. Find a pair of integers with a product of 32 and a sum of 12. **4 and 8**
2. Find a pair of integers with a sum of 114 and a difference of 2. **56 and 58**
3. Livingston is 25 mi east of Bozeman, Montana. Lisa left Bozeman at 2:00 P.M., driving east on I-90 at 65 mi/h. Jerome left Livingston at 2:00 P.M., driving west on I-90 at 55 mi/h.
   a. At what time will Lisa pass Jerome? **about 2:13 P.M.**
   b. How far will Lisa be from Bozeman when she passes Jerome? **about 13.5 mi**
4. Shigechiyo Izumi of Japan lived to be one of the oldest people in the world. Carrie White was the oldest known person in the United States. Carrie lived 4 years fewer than Izumi. The sum of their ages is 236 years. How many years did each person live? **Carrie White–116 yr; Shigechiyo Izumi–120 yr**
5. Writing Describe some advantages and disadvantages of using *Guess and Test* to solve a problem. **Answers may vary. Sample: In a simple situation, it is often easy to guess an answer. It is not as easy to guess multiple answers. Some answers, like $3\frac{3}{7}$, are difficult to guess.**

268

6-1 Teaching Notes

PROBLEM OF THE DAY

What operation signs must be placed in the blanks of 1__9__9__6 to get a result of 76? +, ×, −

Problem of the Day is also available in Transparencies.

CONNECTING TO PRIOR KNOWLEDGE Display an enlarged street map of your area or another city. Have students find roads that intersect once, roads that intersect more than once, and roads that never intersect. Help students describe the roads as intersecting lines, curves, or parallel lines.

THINK AND DISCUSS

Example 1

VISUAL LEARNING Have students graph the two lines in two different colors of highlighting markers. Then point out to them that the point of intersection is the point on the graph that contains both colors. Emphasize the difference between points that do and do not solve a system of equations.

To solve this system using the graphing calculator, students need to enter both equations on the same screen. Then, press ZOOM 8 ENTER. Press TRACE and use the arrow keys to move the cursor to the point of intersection of the two lines. Have students compare their solution with the solution in the text.

Connections Entertainment . . . and more

6-1 Solving Systems by Graphing

What You'll Learn

- Solving systems of linear equations by graphing

...And Why

To solve problems by comparing costs for services like television

What You'll Need

- graph paper
- graphing calculator

THINK AND DISCUSS

Part 1

Solving Systems with One Solution

How can you show all the solutions of the linear equation $y = 2x - 3$? Graph the line, of course! Each point on the line is a solution.

Linear Equation
$y = 2x - 3$

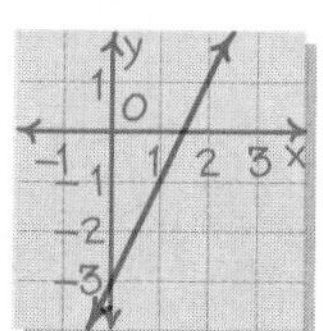

1. *Open-ended* Use the graph to write three different solutions of the equation $y = 2x - 3$. Then show that each ordered pair makes the equation true. Sample: (0, −3), (2, 1), (3, 3)

Two or more linear equations together form a **system of linear equations.** One way to solve a system of linear equations is by graphing. Any point common to all the lines is a **solution of the system.** So, any ordered pair that makes *all* the equations true is a solution of the system.

QUICK REVIEW

The slope-intercept form of a linear equation is $y = mx + b$, with m = slope and b = y-intercept.

Example 1

Solve the system of linear equations by graphing. $y = 2x - 3$, $y = x - 1$

Graph both equations on the same coordinate grid.

$y = 2x - 3$: slope is 2, y-intercept is −3.
$y = x - 1$: slope is 1, y-intercept is −1.

Find the point of intersection.

The lines intersect at (2, 1), so (2, 1) is the solution of the system.

Check See if (2, 1) makes both equations true.

$y = 2x - 3$		$y = x - 1$
$1 \stackrel{?}{=} 2(2) - 3$	← Substitute (2, 1) for (x, y). →	$1 \stackrel{?}{=} 2 - 1$
$1 \stackrel{?}{=} 4 - 3$		$1 = 1$ ✓
$1 = 1$ ✓		

It checks, so (2, 1) is the solution of the system of linear equations.

Lesson Planning Options

Prerequisite Skills

- Graphing linear equations
- Substituting numbers for variables in equations

Assignment Options

To provide flexible scheduling, this lesson can be subdivided into parts.

1. **Core** 1–5, 7–10; **Extension** 6, 11, 12
2. **Core** 13–24, 29–36, 41; **Extension** 25–28, 37–40, 42–46

Use Mixed Review to maintain skills.

Resources

Student Edition

Skills Handbook, pp. 567–569, 572
Extra Practice, p. 562
Glossary/Study Guide

Teaching Resources

Chapter Support File, Ch. 6
- Practice 6-1 (two worksheets)
- Reteaching 6-1
- Alternative Activity 6-1

Classroom Manager 6-1
Glossary, Spanish Resources
Two-Year Algebra Handbook 6-1

Transparencies
2, 11, 29, 78

Question 1 Have students use their graphing calculators to generate ordered pairs that are solutions to the equation. The first method is to explore the graph using TRACE. Have students enter the equation, press ZOOM **8** ENTER, press TRACE, and use the arrow keys to move the cursor along the line. Ordered pairs with integral values of *x* will be displayed. A second method is to enter the equation and then press 2nd [TABLE].

In this lesson, students will learn to solve systems of equations using both hand-drawn graphs and the graphing calculator. Detailed instructions for several calculators are available in the Graphing Calculator Handbook.

Have students use the graphing calculator to solve systems in which they are asked to draw a graph. The graphing calculator can be used to check the graphs drawn by the students.

Conversely, have students draw graphs for additional practice when the problems call for the graphing calculator. This will work particularly well if you have block scheduling or extended class periods.

Example 2 **Relating to the Real World**

Point out to students that [CALC] is the 2nd function of TRACE. To find the intersection point, have students press 2nd [CALC] **5** (intersect) ENTER. Then, have them respond to "First curve?" by placing the cursor somewhere along the first line (the curve number is indicated in the upper right corner of the screen) and pressing ENTER. Have students do the same for "Second curve?". To respond to "Guess?," have them move the cursor as close to the point of intersection as possible and press

Additional Examples

FOR EXAMPLE 1

Solve the system of equations by graphing.

$y = -\frac{2}{3}x + \frac{4}{9}$

$y = \frac{3}{2}x - 1$ $\left(\frac{12}{13}, \frac{5}{13}\right)$

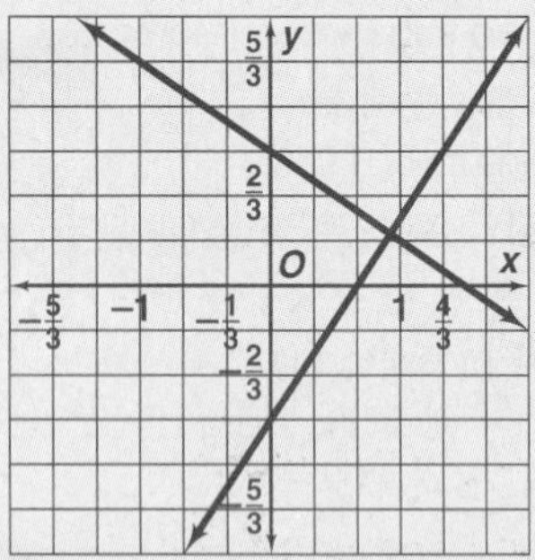

Discussion: *Is it possible to predict whether the lines for the two equations do, in fact, intersect?*

FOR EXAMPLE 2

You plan to start taking an aerobics class. Non-members pay $4.50 per class while members pay a $20.00 sign-up fee plus an additional $1.50 per class. Write and graph a system of equations to decide how many classes you will have to attend to make a membership purchase worthwhile. **(6.67, 30); seven classes**

3. Answers may vary. Sample: No; the solutions are points common to all the lines. If there is no such point, there is no solution. See margin for diagram.

2. Try This Solve the system $y = x + 5$. Check your solution.
$y = -4x$ **(−1, 4)**

3. Critical Thinking Do you think a system of linear equations always has exactly one solution? Draw diagrams to support your answer.

You can use the graph of a system of linear equations to solve problems.

Example 2 **Relating to the Real World**

Entertainment A cable company offers a "pay-per-view" club. Let c = the annual cost and n = the number of movies you watch in a year. Graph the system of equations below to decide whether to join the club.

Members: $c = 4n + 24$
Non-members: $c = 5.50n$

Step 1: Set an appropriate range.

```
WINDOW FORMAT
 Xmin=0
 Xmax=30
 Xscl=2
 Ymin=0
 Ymax=120
 Yscl=10
```

Step 2: Input the equations. Let $n = x$ and $c = y$.

```
Y1=4X+24
Y2=5.5X
Y3=
Y4=
Y5=
Y6=
Y7=
Y8=
```

Step 3: Use the CALC key to find the coordinates of the intersection point.

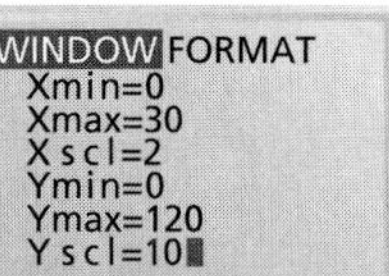

GRAPHING CALCULATOR HINT

You can also use the TABLE feature to find the intersection point. The ZOOM and TRACE keys will only estimate the intersection point.

The solution of the system is (16, 88) where $(x, y) = (n, c)$.

Check See if (16, 88) makes both equations true.

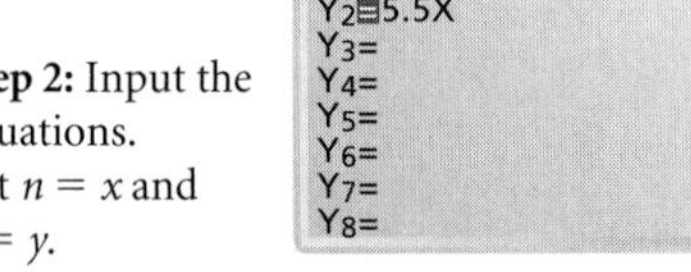

$c = 5.50n$
$88 \stackrel{?}{=} 5.50(16)$
$88 = 88$ ✔

You find that 16 movies in a year cost $88 for both members and nonmembers. If you plan to watch more than 16 movies in a year, join the club. If you plan to watch fewer than 16, do not join the club.

PROBLEM SOLVING HINT

For Question 4, you can use *Guess and Test* or *Draw a Graph.*

4. Suppose the annual fee is $15 instead of $24. What advice would you give a friend on whether or not to join the club? **Join the club if you plan to watch at least 10 movies in a year.**

ENTER. The point of intersection is then displayed. When students are finished, have them reset the screen to the standard viewing window by pressing ZOOM 6 ENTER.

CONNECTING TO THE STUDENTS' WORLD Have students research the pay-per-view rates of cable companies in your area. Repeat the example using data the students collected.

MAKING CONNECTIONS Economists use linear equations in quadrant I to represent supply and demand. The x-axis represents quantity and the y-axis represents price. A supply graph is a line with a positive slope; as the price rises, suppliers want to supply more of the product. A demand graph is a line with a negative slope; as the price rises, buyers are less willing to buy the product. The point of intersection of the supply and demand graph is called *equilibrium*.

Example 3

ERROR ALERT! Some students may interpret the identical equations as having zero solutions in common. **Remediation:** Have students test specific sets of coordinates in the original equations so they will discover that two different forms of the same equation have all solutions in common.

KINESTHETIC LEARNING Clear a large area in the room and make a large coordinate grid on the floor using string or masking tape. Choose two pairs of students. Give each pair a length of string long enough to reach across the room. Direct pairs of students to arrange themselves on the grid, stretching the string between them to model equations of lines, for example, direct the pairs to model $y = x$ and $y = x + 1$.

Part 2

Solving Special Types of Systems

A system of linear equations has **no solution** when the graphs of the equations are parallel. There are no points of intersection, so there is no solution.

$y = -x + 1$
$y = -x - 1$

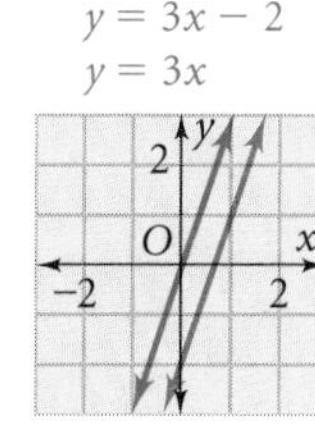

$y = 3x - 2$
$y = 3x$

y, 2, O, x, −2, 2

5. **Critical Thinking** Without graphing, how can you tell that a system has no solution? Give an example. From the equations, you can find the slopes and y-intercepts. When the slopes are the same and y-intercepts are different, the lines will not intersect. Sample: $y = -2x + 3$, $y = -2x - 6$

A system of linear equations has **infinitely many solutions** when the graphs of the equations are the same line. All points on the line are solutions of the system.

Example 3

Solve the system by graphing. $-4y = 4 + x$

$\frac{1}{4}x + y = -1$

GRAPHING CALCULATOR HINT

To enter an equation on a graphing calculator, you need to put it in slope-intercept form.

First, write each equation in slope-intercept form.

$-4y = 4 + x$ $\qquad$ $\frac{1}{4}x + y = -1$

$y = -\frac{1}{4}x - 1$ $\qquad$ $y = -\frac{1}{4}x - 1$

Then graph each equation on the same coordinate plane.

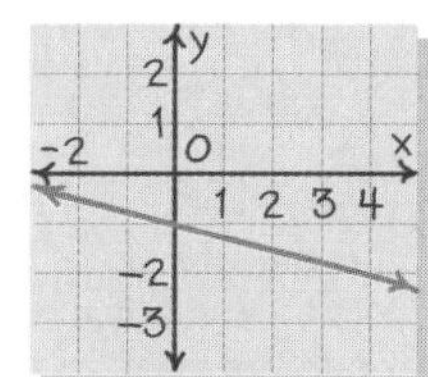

Since the graphs are the same line, the system has infinitely many solutions.

6. The equations are identical when written in the slope-intercept form. If the equations have the same slope-intercept form, the system has infinitely many solutions.

6. What do you notice about the slope-intercept form of each equation in Example 3? How could this help you solve a linear system?

7. **Try This** Solve each system by graphing. See margin for graphs.
   a. $y = x$, $y = x + 6$ no solution
   b. $2x + 2y = 1$, $y = -x + \frac{1}{2}$ inf. many solutions
   c. $x = 1$, $x = -2$ no solution

Technology Options

Students may find a graphing calculator useful for verifying their answers in Exercises 7–10.

Prentice Hall Technology

Calculator • Graphing Calculator Handbook: Procedure 9

Software • Secondary Math Lab Toolkit • Computer Item Generator 6-1

CD-ROM • Multimedia Algebra Lab 6

Internet • See the Prentice Hall site. (http://www.phschool.com)

As students take turns modeling lines, provide them with pairs of equations that demonstrate parallel and intersecting lines.

WORK TOGETHER

Have students work in pairs. Be sure they understand that each pair should prepare only one table. Some pairs may also choose to show the graphs of each of the three systems. Working cooperatively on a project like this helps students develop team skills that will be important when they have jobs. For example, teams of co-workers often work together to submit a solution to a problem, or to write reports or proposals.

Exercises ON YOUR OWN

Exercise 3 To solve this system using the calculator, students must enter $y = 6$ first. They can then use the [DRAW] function to graph $x = -1$ (a vertical line). To use the [DRAW] function, press 2nd and then PRGM. Select **4** (vertical). Press ENTER and press **4** (to graph $x = 4$) and then ENTER. Both graphs will then appear in the display.

Exercises 4–5 First, students need to solve these equations for *y*.

NUMBER THEORY Exercise 6 Students may wish to explore how this can be done using the graphing calculator with the [TABLE] and GRAPH functions.

pages 269–272 Think and Discuss

3. Answers may vary. Sample: No; the solutions are points in common to all the lines. If there is no such point, there is no solution.

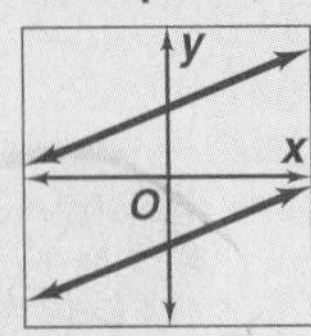

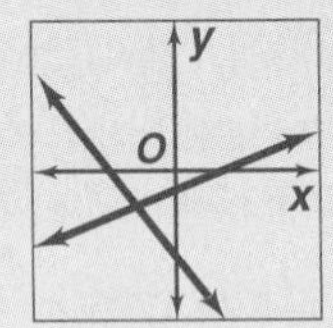

7a.

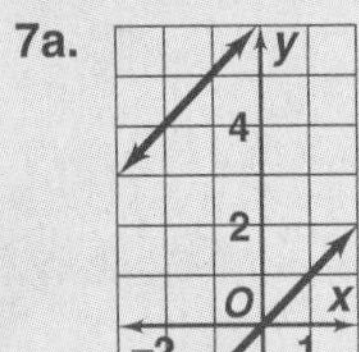

b.

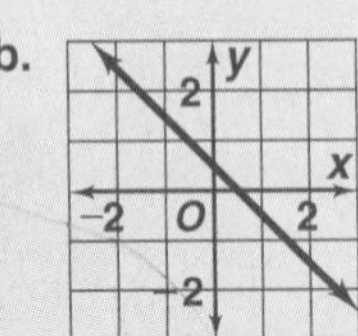

WORK TOGETHER

PROBLEM SOLVING HINT

Start with the known information and use the strategy *Work Backward* to write an appropriate system.

Work with a partner to copy and complete the table. Your goal is to create a system of equations that satisfies the conditions given.

	System of Equations	Description of Graph	One Solution of the System	Number of Solutions
8.	■	2 intersecting lines	$(1, -5)$	■
9.	■	2 non-intersecting lines	■	■
10.	■	■	■	infinitely many

8–10. See margin.

Exercises ON YOUR OWN

Graphing Calculator **Solve each system of linear equations by graphing. Sketch the graph on your paper.** 1–5. See margin for graphs.

1. $y = \frac{1}{3}x + 3$
   $y = \frac{1}{3}x - 3$
   no solution
2. $y = x$
   $y = 5x$
   (0, 0)
3. $y = 1$
   $y = x$
   (1, 1)
4. $2x + y = 3$
   $x - 2y = 4$
   (2, −1)
5. $3x - y = 7$
   $y = 3x - 7$
   infinitely many solutions

6. Number Theory You can represent the set of nonnegative even numbers by the expression $2n$, for $n = 0, 1, 2, \ldots$. You can represent the set of nonnegative odd numbers by $2n + 1$, for $n = 0, 1, 2, \ldots$.
   a. Copy and complete the table at the right.
   b. Graph the system. $y = 2n$ $y = 2n + 1$ See margin.
   c. Writing Why does it makes sense that this system has no solution? No number can be even and odd at the same time.

n	Even Numbers $2n$	Odd Numbers $2n + 1$
0	$2(0) = 0$	$2(0) + 1 = 1$
1	■ 2	■ 3
2	■ 4	■ 5
3	■ 6	■ 7
4	■ 8	■ 9
5	■ 10	■ 11

Is (−1, 5) a solution of each system? Verify your answer.

7. $x + y = 4$
   $x = -1$
   yes
8. $y = -x + 4$
   $y = -\frac{1}{5}x$
   no
9. $y = 5$
   $x = y - 6$
   yes
10. $y = 2x + 7$
    $y = x + 6$
    yes

11. a. Use the cartoon below. Are these systems of linear equations? Explain. No; the graphs are not straight lines.
    b. What does the solution of the system on the Marriage Chart mean? b–c. See margin.
    c. What does the solution of the system on Lily's Chart mean?
    d. What would you label the axes of Lily's chart? Answers may vary. Sample: feelings; time

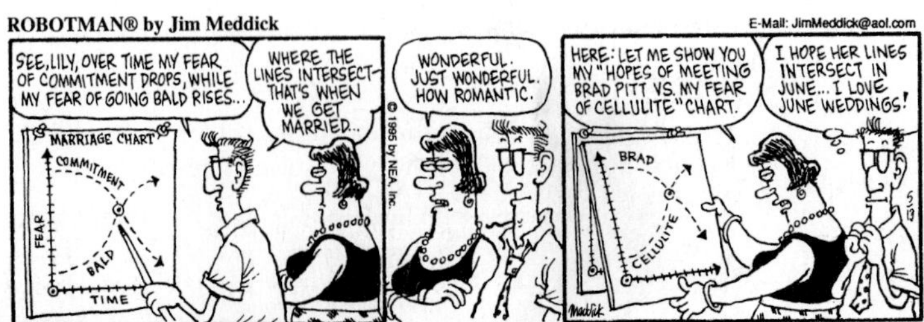

ALTERNATIVE ASSESSMENT Exercises 13–24 Ask students to predict—without graphing—whether the lines intersect, are parallel, or are the same line. This will help you assess students' understanding of systems of equations.

OPEN-ENDED Exercises 37–40 Have students share different ways to approach these exercises. For example, in Exercise 37, some students may recall that perpendicular lines have slopes that are opposite reciprocals. Others may draw two perpendicular lines and use the slopes and y-intercepts of the graphs to write the equations.

WRITING Exercise 42 Students might begin by graphing two equivalent equations.

EXTENSION AND ESTIMATION Exercise 43 Ask students:

- *Can you think of a way to solve this system without the graphs by using mental math?* Substitute 6.5 for y in the second equation and solve for x.
- *Can you use a similar method for Exercises 44–46?* The method will work for Exercise 44, but not for Exercises 45–46.

Chapter Project **FIND OUT BY GRAPHING** Students write linear equations to compare the costs of bands A and B. They then graph each equation to find out how many tickets must be sold for the cost of the bands to be equal. This task is an essential part of the Chapter Project which was introduced in the Chapter Opener. Check students' progress on the project by having each student write a sentence or two describing how this activity is related to the Chapter Project.

---

12. *Math in the Media* Suppose you see the two summer jobs advertised at the right. Let x = the amount of sales and y = money earned in a week.

Cellular Phone Sales: $y = 150 + 0.2x$
Stereo Sales: $y = 200 + 0.1x$

a. To earn the same amount of money at both jobs, how much will you need to sell in a week? $500

b. After talking with salespeople, you estimate weekly sales of about $600 with either job. At which job will you earn more money? Cellular Phone Sales

Sales Position
Salesperson Wanted
Knowledge of Cellular Phones
On-Site Sales
$150/week + 20% commission

CAREER OPPORTUNITY
Sell Stereo Equipment in
National Electronics Retail Chain!
$200/week + 10% commission

Solve each system by graphing. Write *no solution* or *infinitely many solutions* where appropriate. 13–24. See margin for graphs.

13. $y = -x + 4$; $y = 2x + 1$ (1, 3)
14. $x = 10$; $y = -7$ (10, –7)
15. $y = 3x$; $y = 5x$ (0, 0)
16. $y = 3x + 4$; $4y = 12x + 16$ infinitely many solutions
17. $3x + y = 5$; $x - y = 7$ (3, –4)
18. $2x - 2y = 4$; $y - x = 6$ no solution
19. $x + y = -1$; $x + y = 1$ no solution
20. $y = 1$; $3y + x = 9$ (6, 1)
21. $y = \frac{1}{2}x - 1$; $y = -\frac{1}{2}x - 1$ (0, –1)
22. $x + 2y = 3$; $-x = 2y - 3$ infinitely many solutions
23. $y = 4x - 3$; $y = 4x + 2$ no solution
24. $y = \frac{3}{4}x - 5$; $x = 4$ (4, –2)

Critical Thinking **Is each statement *true* or *false*? Explain your reasoning.**

25. A system of linear equations can have one solution, no solution, or infinitely many solutions. T; see margin for reasoning.

26. If a point is a solution of a system of linear equations, it is also a solution of each linear equation in the system. T; the point of intersection lies on both lines.

27. If a point is a solution of a linear equation, it is also a solution of any system containing that linear equation. F; the system may have no solutions.

28. If a system of linear equations has no solution, the graphs of the lines are parallel. T; if there is no solution, the lines do not intersect. So, they are parallel.

PROBLEM SOLVING HINT
For Exercises 25–28, you can *Draw a Graph.*

Without graphing, decide whether the lines in each system *intersect, are parallel,* or *are the same line.* Then write the number of solutions.

29. $y = 2x$; $y = 2x - 5$ parallel; no solution
30. $x + y = 4$; $2x + 2y = 8$ same line; inf. many solutions
31. $y = -3x + 1$; $y = 3x + 7$ intersect; one solution
32. $3x - 5y = 0$; $y = \frac{3}{5}x$ same line; inf. many solutions
33. $2y - 10x = 2$; $y = 5x + 1$ same line; inf. many solutions
34. $y = -4x + 4$; $y = -4x + 8$ parallel; no solution
35. $y = x + 1$; $y = \frac{1}{2}x$ intersect; one solution
36. $y = 2x + 3$; $y - 2x = 5$ parallel; no solution

Open-ended **Write a system of two linear equations with the given characteristics.** 37–40. Answers may vary. See margin for samples.

37. one solution; perpendicular lines
38. no solutions; one equation is $y = 2x + 5$
39. one solution; (0, –4)
40. infinitely many solutions; one equation is $y = 4x$

7c.

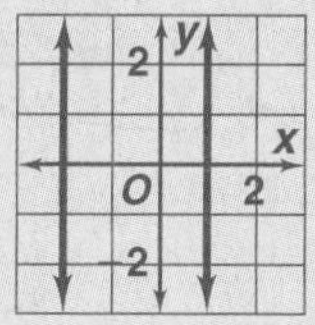

8. Sample: $y = x - 6$
$y = -5x$; 1

9. Sample: $y = 2x - 1$
$y = 2x + 3$;
no solution; 0

10. Sample: $2y - 4x = 3$
$y = \frac{1}{2}x - \frac{3}{4}$;
one line; infinitely many

pages 272–274 On Your Own

1.

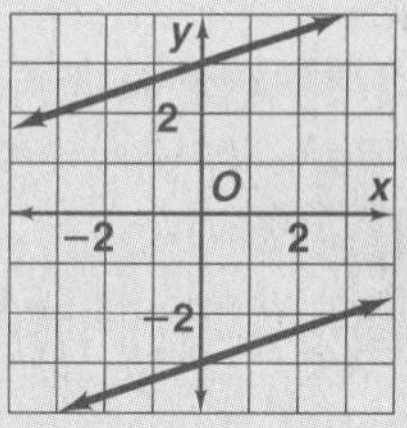

2.

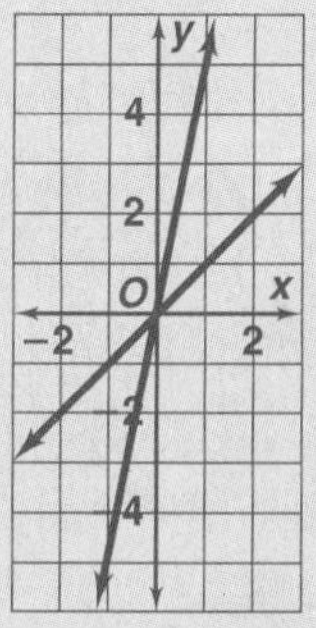

3–5, 6b, 11b, 11c, 13–25, 37–40. See back of book.

Exercises MIXED REVIEW

GETTING READY FOR LESSON 6-2 In this preview, students solve equations for one variable in terms of the other. This is a key skill for the next lesson on solving systems using substitution.

Wrap Up

THE BIG IDEA Ask students to tell what is meant by solving a system of linear equations. Then ask them to describe three different situations that can occur when graphing systems of linear equations.

RETEACHING ACTIVITY Students graph equations using toothpicks to represent linear equations. They find the point where the two lines intersect to find the solution. (Reteaching worksheet 6-1)

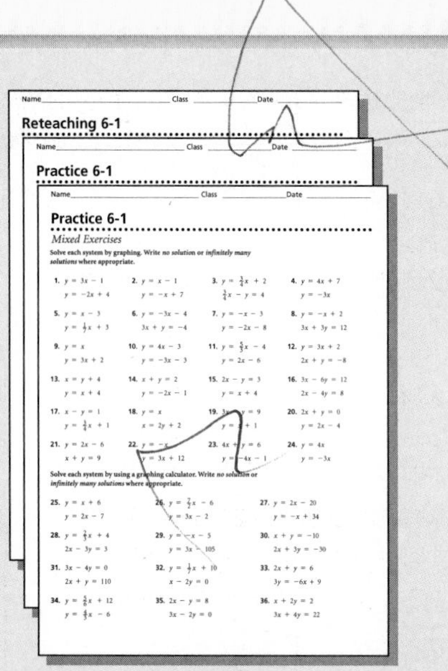
Name ______ Class ______ Date ______

Reteaching 6-1

Name ______ Class ______ Date ______

Practice 6-1

Name ______ Class ______ Date ______

Practice 6-1

Mixed Exercises

Solve each system by graphing. Write *no solution* or *infinitely many solutions* where appropriate.

1. $y = 3x - 1$, $y = -2x + 4$
2. $y = x - 1$, $y = -x + 7$
3. $y = \frac{3}{4}x + 2$, $\frac{3}{4}x - y = 4$
4. $y = 4x + 7$, $y = -3x$
5. $y = x - 3$, $y = \frac{1}{3}x + 3$
6. $y = -3x - 4$, $3x + y = -4$
7. $y = -x - 3$, $y = -2x - 8$
8. $y = -x + 2$, $3x + 3y = 12$
9. $y = x$, $y = 3x + 2$
10. $y = 4x - 3$, $y = -3x - 3$
11. $y = \frac{5}{3}x - 4$, $y = 2x - 6$
12. $y = 3x + 2$, $2x + y = -8$
13. $x = y + 4$, $y = x + 4$
14. $x + y = 2$, $y = -2x - 1$
15. $2x - y = 3$, $y = x + 4$
16. $3x - 6y = 12$, $2x - 4y = 8$
17. $x - y = 1$, $y = \frac{3}{4}x + 1$
18. $y = x$, $x = 2y + 2$
19. [illegible] $y = 9$, $y =$ [illegible] $+ 1$
20. $2x + y = 0$, $y = 2x - 4$
21. $y = 2x - 6$, $x + y = 9$
22. $y = -x$, $y = 3x + 12$
23. $4x + y = 6$, $y = -4x - 1$
24. $y = 4x$, $y = -3x$

Solve each system by using a graphing calculator. Write *no solution* or *infinitely many solutions* where appropriate.

25. $y = x + 6$, $y = 2x - 7$
26. $y = \frac{7}{2}x - 6$, $y = 3x - 2$
27. $y = 2x - 20$, $y = -x + 34$
28. $y = \frac{2}{3}x + 4$, $2x - 3y = 3$
29. $y = -x - 5$, $y = 3x - 105$
30. $x + y = -10$, $2x + 3y = -30$
31. $3x - 4y = 0$, $2x + y = 110$
32. $y = \frac{1}{2}x + 10$, $x - 2y = 0$
33. $2x + y = 6$, $3y = -6x + 9$
34. $y = \frac{5}{6}x + 12$, $y = \frac{4}{3}x - 6$
35. $2x - y = 8$, $3x - 2y = 0$
36. $x + 2y = 2$, $3x + 4y = 22$

Lesson Quiz

Lesson Quiz is also available in Transparencies.

Solve each system by graphing. Write *no solution* or *infinitely many solutions* where appropriate.

1. $x - y = 1$, $4x - 2y = 6$ (2, 1)

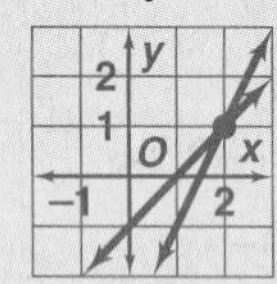

2. $y - 2x = 3$, $-2y + 4x = -6$
infinitely many solutions

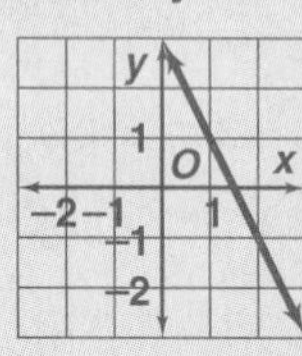

41. Music Suppose you and your friends form a band, and you want to record a demo tape. Studio A rents for \$100 plus \$50/h. Studio B rents for \$50 plus \$75/h. Let t = the number of hours and c = the cost.

Studio A: $c = 100 + 50t$
Studio B: $c = 50 + 75t$

a. Solve the system by graphing. (2, 200)

b. Explain what the solution of the system means in terms of your band renting a studio. Both studios charge \$200 rent for 2 h.

42. Writing When equivalent equations form a system, there are infinitely many solutions. Explain in your own words why this is true.
Equivalent equations have identical graphs. Each point on the graph is a solution of the system.

Chapter Project — Find Out by Graphing

Band A charges \$600 to play for the evening. Band B charges \$350 plus \$1.25 for each ticket sold. Write a linear equation for the cost of each band. Graph each equation and find the number of tickets for which the cost of the two bands will be equal.

Estimation **Estimate the solution of each system. Use the equations to test your estimate. Adjust your estimate until you find the exact solution.**

43. $y = 6.5$, $y = x + 2$ (4.5, 6.5)

44. $y = -4x - 10$, $y = -6$ (−1, −6)

45. $y = -0.75x - 3$, $y = 0.25x - 7$ (4, −6)

46. $y = -x - 1$, $y = x + 8$ (−4.5, 3.5)

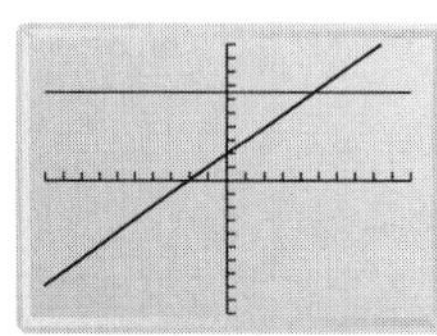
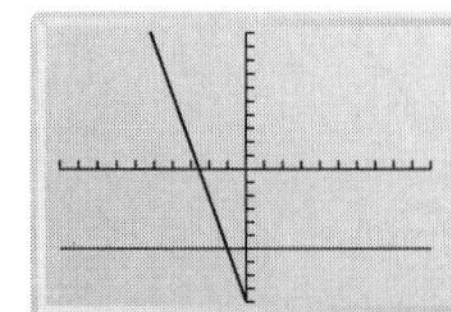
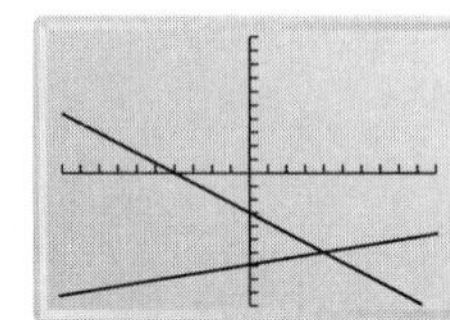
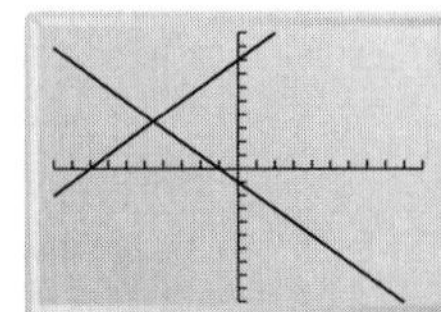
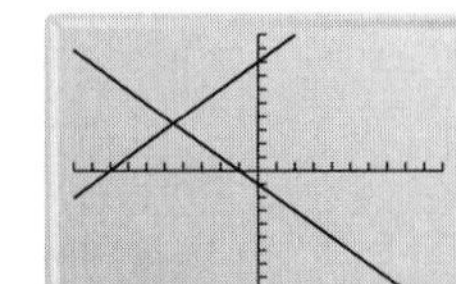

Exercises MIXED REVIEW

Probability **Find each probability for two rolls of a number cube.**

47. $P(3, \text{then } 4)$ $\frac{1}{36}$

48. $P(1, \text{then even})$ $\frac{1}{12}$

49. $P(\text{two integers})$ 1

50. $P(\text{at least one } 1)$ $\frac{11}{36}$

Write an inequality to model each situation.

51. Polar bears can swim as fast as 6 mi/h. $b \le 6$

52. Each eyelash is shed every 3 to 5 months. $3 \le e \le 5$

53. a. Geography Cairo, Egypt, has about 18 million residents. The average population density is 130,000 people/mi^2. What is Cairo's area? about 138.5 mi^2

b. Cairo has $\frac{1}{4}$ of Egypt's population. What is the population of Egypt? about 72 million

Egypt

Getting Ready for Lesson 6-2

Solve each equation for the given variable.

54. $x - y = 3; y$ $y = x - 3$

55. $\frac{1}{2}x = 4y; x$ $x = 8y$

56. $\frac{x}{2} = \frac{y}{4}; y$ $y = 2x$

57. $2x - 3y = 5; x$ $x = \frac{3}{2}y + \frac{5}{2}$

6-2 Teaching Notes

PROBLEM OF THE DAY

Chelsea got 33 hits out of 187 times at bat. Steve got 38 hits out of 224 times at bat. Who has the better batting average? Chelsea

Problem of the Day is also available in Transparencies.

CONNECTING TO PRIOR KNOWLEDGE On the board, write $x + 4 = 7$ and $x + 3 = 6$. Ask: *What value does x have in each of these equations?* 3 Write the second equation as $x = 6 - 3$. Substitute the expression $6 - 3$ for the variable in the first equation, resulting in $6 - 3 + 4 = 7$. Lead students to understand that substituting the expression does not change the equation because the values of x are equal.

WORK TOGETHER

Question 2 If students use the standard settings for [TABLE], the solution of the system will not appear in the table. Have pairs work together to use 2nd [TblSet] to discover how to change the interval for x-values on the table in order for the solution to appear.

THINK AND DISCUSS

Example 1

Point out that some systems of equations such as the system presented in the Work Together section are very difficult to solve accurately by drawing a graph. If a graphing calculator is

---

Connections Transportation . . . and more

6-2 Solving Systems Using Substitution

What You'll Learn

- Solving systems of linear equations by substitution

...And Why

To solve problems involving transportation

What You'll Need

- graph paper
- graphing calculator

System:
$y = x + 6.1$
$y = -2x - 1.4$

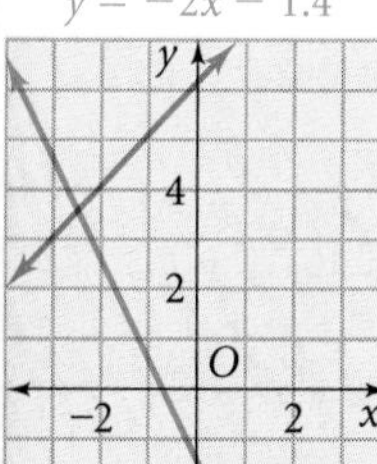

WORK TOGETHER

Work with a partner.

1. *Estimation* Use the graph at the left to estimate the solution of the system. Answers may vary. Sample: (−2.5, 3.5)
2. a. *Graphing Calculator* Graph the system at the left. See margin.
   b. Use the CALC or TABLE features to find the intersection point. (−2.5, 3.6)
3. Compare your estimate to your answer in Question 2(b). Check students' work.
4. *Choose* Use paper and pencil or a calculator to solve each system.
   a. $y = 3x - 6$, $y = -6x$ $\left(\frac{2}{3}, -4\right)$
   b. $y = x + 2.5$, $y = 2x - 0.5$ (3, 5.5)
   c. $2y - x = 3$, $y = x - 2$ (7, 5)
5. Explain why you chose the method(s) you used in Question 4. Check students' work.

THINK AND DISCUSS

Part 1

Solving Systems with One Solution

Sometimes you won't have a graphing calculator to use. Another way to solve a system is to use substitution. **Substitution** allows you to create a one-variable equation.

Example 1

Solve the system $y = x + 6.1$ using substitution.
$y = -2x - 1.4$

$y = -2x - 1.4$ ← Start with one equation.
$x + 6.1 = -2x - 1.4$ ← Substitute $x + 6.1$ for y in that equation.
$3x = -7.5$ ← Solve for x.
$x = -2.5$

Substitute −2.5 for x in either equation and solve for y.

$y = (-2.5) + 6.1$
$y = 3.6$

Since $x = -2.5$ and $y = 3.6$, the solution is (−2.5, 3.6).

Check See if (−2.5, 3.6) satisfies the other equation.

$3.6 \stackrel{?}{=} -2(-2.5) - 1.4$
$3.6 \stackrel{?}{=} 5 - 1.4$
$3.6 = 3.6$ ✓

PROBLEM SOLVING

Look Back Does the solution agree with the graph of the system in the Work Together? yes

Lesson Planning Options

Prerequisite Skills

- Graphing linear equations
- Substituting expressions and terms for variables in equations

Assignment Options

To provide flexible scheduling, this lesson can be subdivided into parts.

1 **Core** 9, 11–22, 29–33, 39
Extension 10, 28, 38

2 **Core** 1–8, 23–27, 34–37
Extension 40–48

Use Mixed Review to maintain skills.

Resources

Student Edition

Skills Handbook, p. 582
Extra Practice, p. 561
Glossary/Study Guide

Teaching Resources

Chapter Support File, Ch. 6
- Practice 6-2 (two worksheets)
- Reteaching 6-2

Classroom Manager 6-2
Glossary, Spanish Resources
Two-Year Algebra Handbook 6-2

Transparencies
11, 29, 78

not available, another method is needed. This example shows how to solve the system using the substitution method.

Have students graph the system to see how close their graph's solution is to the solution found by substituting. Ask them which method they think is more accurate. substitution

Example 2 Relating to the Real World 

ERROR ALERT! Some students may equate $6v$ with 6 vans. **Remediation:** Point out that v represents the number of vans and $6v$ represents the number of people each van can hold.

Question 7 Have students also check the solution in the second equation. Then have them go back to the original wording of the problem to see if the solution of the system of equations gives a correct solution to the word problem. This will ensure that the equations were written correctly.

EXTENSION Have students write four different one-variable equations that can be generated from this system of equations using the substitution method. $x + y = 4$; $x = 2y$
$2y + y = 4$; $x + \frac{1}{2}x = 4$; $4 - y = 2y$; $x = 2(4 - x)$

Question 9 Lead students in a discussion on how to decide which variable in which equation would be the least difficult to solve for in the first step.

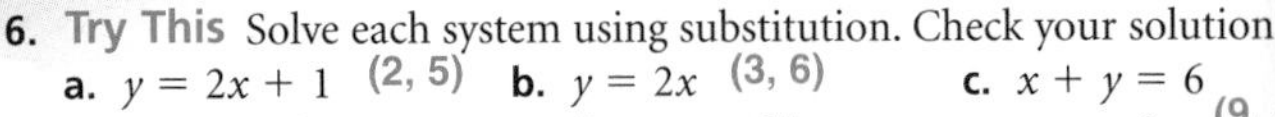

Example 3

This Example presents systems of equations for which using the substitution method produces a one-variable equation that is either an identity or has no solution. Students worked with such one-variable equations on pages 165–166.

Additional Examples

FOR EXAMPLE 1

Solve the following system using substitution.
$y = x + 2$
$y = -x + 4$ (1, 3)

FOR EXAMPLE 2

An economist determines the most profitable price for a product by finding where two different equations intersect. The equation of the supply curve, $n = 2p$, represents how many radios a local manufacturer is willing to make at a given price. The equation of the demand curve, $n = 100 - 3p$, represents the number of radios that local consumers are willing to buy at a given price. How many radios will the manufacturer produce for maximum profit? How much will the manufacturer charge if she wishes to sell all of the radios? 40 radios at $20 each

Discussion: *Why would selling either more or less than 40 radios result in a smaller profit for the radio manufacturer?*

6. **Try This** Solve each system using substitution. Check your solution.
   a. $y = 2x + 1$; $y = x + 3$ (2, 5)
   b. $y = 2x$; $7x - y = 15$ (3, 6)
   c. $x + y = 6$; $x = -3y$ (9, -

There is more than one way to solve a system using substitution. Solving a variable with a coefficient of 1 or –1 is a good place to start. No matter what variable you solve for first, you should always get the same answer.

Example 2 Relating to the Real World 

Transportation An art class is planning a trip to a museum. There are 22 people going on the trip. There are four drivers and two types of vehicles, vans and cars. The vans seat six people, and the cars seat four people, including drivers. How many vans and cars does the class need for the trip? Use the system below.

Let v = the number of vans and c = the number of cars.
Drivers: $v + c = 4$
People: $6v + 4c = 22$

You can solve the system by substitution.

$v + c = 4$ ← Solve the first equation for v.
$v = -c + 4$
$6(-c + 4) + 4c = 22$ ← Substitute $-c + 4$ for v in the second equation.
$-6c + 24 + 4c = 22$ ← Solve for c.
$-2c + 24 = 22$
$-2c = -2$
$c = 1$
$v + (1) = 4$ ← Substitute 1 for c in the first equation.
$v = 3$ ← Solve for v.

Since $c = 1$ and $v = 3$, the art class should use 1 car and 3 vans.

7. Check the solution in the first equation in Example 2. $1 + 3 = 4$

AUDITORY LEARNING Challenge groups of students to write a 30-second rap which describes the differences between systems of equations with one solution, no solution, or infinite solutions.

Question 12 Have students go back and look at the equations in the system solved in the example. Ask: *How could you determine that the system has no solution by looking at the equations?* Write each equation in slope-intercept form. Both have a slope of −1; each has a different *y*-intercept. Therefore, the lines are parallel and have no points in common. So, the system has no solution.

Question 14 Direct students to check their answers by using the substitution method. When students have completed their answers, have them answer these questions without actually drawing a graph.

- *How would you describe the graphs of the equations in system a?* The graphs are the same line.
- *How would you describe the graphs of the equations in system b?* The graphs are the same line.
- *How would you describe the graphs of the equations in system c?* The graphs are parallel lines.

8. a. Solve the system in Example 2 again. This time, start by solving the second equation for *c*. See margin.

b. *Critical Thinking* Why is this procedure more difficult? This procedure is longer and it involves computations with fractions.

9. Describe a possible first step for solving each system by substitution.

a. $3x - y = 17$ $2x + y = 8$	**b.** $x + 3y = 5$ $2x - 4y = -5$	**c.** $y = -2x - 3$ $y = x$

9a–c. See margin.

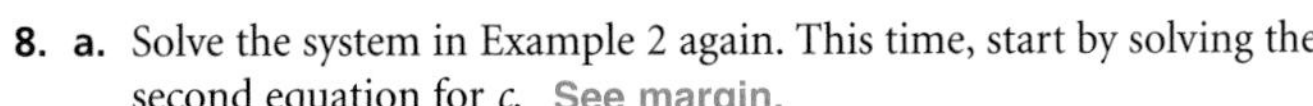

Part 2

Solving Special Types of Systems

You can use substitution to learn that systems have *no solution* or *infinitely many solutions.*

Example 3

Solve the system using substitution. $x + y = 6$
$5x + 5y = 10$

$x + y = 6$ ← Solve the first equation for *x*.
$x = 6 - y$
$5(6 - y) + 5y = 10$ ← Substitute 6 − *y* for *x* in the second equation.
$30 - 5y + 5y = 10$ ← Solve for *y*.
$30 = 10$ ← False!

Since $30 = 10$ is a false statement, the system has no solution.

10. Describe the graph of the system in Example 3. Graph the system to **verify** your answer. Two parallel lines; see margin for graph.

11. *Critical Thinking* How many solutions does a system of linear equations have if you get each result?

a. a true statement, such as $2 = 2$ infinitely many

b. a false statement, such as $10 = 1$ no solution

c. a statement such as $x = 4$ one solution

12. Graphs of systems with no solution are parallel lines. What do you know about the equations of parallel lines? The equations of parallel lines in slope-intercept form have the same slope and different intercepts.

13. Graphs of systems with infinitely many solutions are the same line. What do you know about the equations? The equations of these lines in slope-intercept form have the same slope and *y*-intercept.

14. Without using substitution, decide whether each system has *no solution* or *infinitely many solutions.* (*Hint:* Write each equation in slope-intercept form and compare.)

a. $3x - y = -2$ $y = 3x + 2$ infinitely many solutions	**b.** $y = 4x$ $2x - 0.5y = 0$ infinitely many solutions	**c.** $3x + y = 5$ $6x + 2y = 1$ no solution

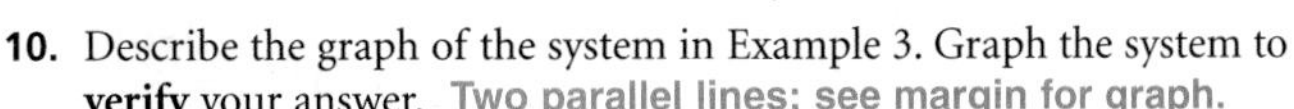

Technology Options

Students may want to estimate their answers for Exercises 23–26 by using a graphing calculator. Students may also want to use a graphing calculator for Exercises 41–48.

Prentice Hall Technology

Software • Secondary Math Lab Toolkit • Computer Item Generator 6-2

CD-ROM Multimedial Algebra Lab 6

Internet • See the Prentice Hall site (http://www.phschool.com)

Exercises ON YOUR OWN

ESL **Exercise 10** Use the term *on-line* in context by asking, "How many students spend at least half an hour per night on-line?" Ensure students know that *time on-line* means time the customer is connected to the Internet.

CONNECTING TO THE STUDENTS' WORLD Exercise 10 Invite students to find out the access rates for Internet providers in your area. Have students repeat the exercise using the data they collected.

ALTERNATIVE ASSESSMENT Exercises 11–22 Randomly call on students to estimate solutions to these exercises by graphing them on the board. This will help you assess students' understanding of the relationship between the graphical and algebraic methods of solving systems of equations.

ESTIMATION Exercises 23–27 Ask students why they think it is good to estimate before using substitution. There could be an error in their solving, and estimating will help them detect the error.

WRITING Exercise 28 Students might use one of the systems they have already solved in this lesson as an example.

STANDARDIZED TEST TIP Exercise 29 Explain that the only way a system of two linear equations can have no solution is if the two lines are parallel. Rewrite all the equations in the slope-intercept form, to verify which system has no solution.

pages 275–277 Think and Discuss

8a. $v + c = 4$

$6v + 4c = 18$

$4c = 18 - 6v$

$c = \frac{9}{2} - \frac{3}{2}v$

$v + (\frac{9}{2} - \frac{3}{2}v) = 4$

$2v + \frac{9}{2} = 4$

$-\frac{1}{2}v = -\frac{1}{2}$

$v = 1^2$

$6(1) + 4c = 18$

$4c = 12$

$c = 3$

9a. Answers may vary. Sample: Solve the first equation for *y*.

b. Answers may vary. Sample: Solve the first equation for *x*.

c. Answers may vary. Sample: Substitute *y* for *x* in the first equation.

10.

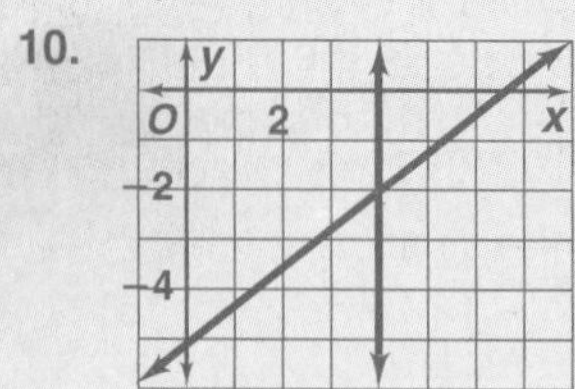

23–28. See back of book.

278

Exercises ON YOUR OWN

Critical Thinking **Suppose you try to solve a system of linear equations and get the following result. How many solutions does each system have?**

1. $x = 0$ one
2. $5 = 5$ infinitely many solutions
3. $-3 = 2$ no solution
4. $n = 10$ one
5. $0 = 0$ infinitely many solutions
6. $-8 = k$ one
7. $y = 6$ one
8. $0 = -9$ no solution

9. **Geometry** A rectangle is 4 times longer than it is wide ($l = 4w$). The perimeter of the rectangle is 30 cm ($2l + 2w = 30$). Find the dimensions of the rectangle. 12 cm by 3 cm

10. **Internet** Suppose you want access to the Internet. With a subscription to *Access,* you pay \$7.95 per month plus \$2.95 per on-line hour. With a subscription to *Network,* you pay \$12.95 per month plus \$1.95 per on-line hour. The system below models this situation. Let c = the monthly cost and h = the number of on-line hours.

    Access: $c = 7.95 + 2.95h$

    Network: $c = 12.95 + 1.95h$

    a. Use substitution to solve the system. (5, 22.70)

    b. Explain how to decide which subscription to buy. Subscribe to *Access* if you spend less than 5 h each month on-line. Subscribe to *Network* if you spend more than 5 h each month on-line.

Solve each system using substitution. Write *no solution* or *infinitely many solutions* where appropriate.

11. $y = 2x$, $6x - y = 8$ (2, 4)
12. $2x + y = 5$, $2y = 10 - 4x$ infinitely many solutions
13. $y = 3x + 1$, $x = 3y + 1$ $\left(-\frac{1}{2}, -\frac{1}{2}\right)$
14. $x - 3y = 14$, $x - 2 = 0$ (2, −4)
15. $2x + 2y = 5$, $y = \frac{1}{4}x$ $\left(2, \frac{1}{2}\right)$
16. $y = -3x$, $y + 3x = 2$ no solution
17. $4x + y = -2$, $-2x - 3y = 1$ $\left(-\frac{1}{2}, 0\right)$
18. $3x + 5y = 2$, $x + 4y = -4$ (4, −2)
19. $y = x + 2$, $y = 2x - 1$ (3, 5)
20. $y = 3$, $y = \frac{4}{3}x + 2$ $\left(\frac{3}{4}, 3\right)$
21. $x + 4y = -3$, $2x + 8y = -6$ infinitely many solutions
22. $2y = 0.2x + 7$, $3y - 2x = 2$ (5, 4)

Estimation **Graph each system to estimate the solution. Then use substitution to find the exact solution of the system.** 23–27. See margin for graphs.

23. $y = 2x$, $y = -6x + 4$ $\left(\frac{1}{2}, 1\right)$
24. $y = \frac{1}{2}x + 4$, $y = -4x - 5$ (−2, 3)
25. $x + y = 0$, $5x + 2y = -3$ (−1, 1)
26. $y = 0.7x + 3$, $y = -1.5x - 7$ $\left(-4\frac{6}{11}, -\frac{2}{11}\right)$
27. $y = 3x + 1$, $y = 3x - 2.5$ no solution

28. **Writing** Describe the advantages of using substitution to solve a system. **Justify** your answer with an example. See margin.

29. **Standardized Test Prep** Which system has no solution? D

    A. $y = x + 3$, $y + 4x = -2$

    B. $2x + 2y = 1$, $y = -x + \frac{1}{2}$

    C. $y = \frac{1}{2}x + 1$, $6y - 3x = 6$

    D. $y = x + 6$, $2y - 2x = 3$

    E. $y = 2x - 1$, $y = -2x$

DIVERSITY Exercise 39 Soybeans were such a staple of Chinese food that the Chinese considered them one of the five sacred grains. The United States Department of Agriculture learned about the soybean in the late 19th century and encouraged its production. Today, soybeans are used to make glycerin, paints, soaps, linoleum, rubber substitutes, and plastics. Soybean products are used in fertilizer and animal feed. People also eat soybeans in meat substitutes, vegetable oil, shortening, soy milk, cheese, and sauces.

MAKING CONNECTIONS Exercise 54 By age 5, Robert Wadlow—the world's tallest man—was already 5 ft 4 in. tall. When he died at age 22, he was 8 ft 11 in. tall. His femur measured 29.5 in.

Exercises MIXED REVIEW

GETTING READY FOR LESSON 6-3 Students review the distributive property which is needed for the next lesson on solving systems of equations using elimination.

Wrap Up

THE BIG IDEA Ask students to describe how to solve a system of equations using the substitution method.

RETEACHING ACTIVITY Students solve systems of linear equations using substitution. They rewrite equations in slope-intercept form, $y = mx + b$ and follow steps to solve. (Reteaching worksheet 6-2)

Mental Math **Match each system with its solution at the right.**

30. $y = x + 1$, $y = 2x - 1$ D
31. $2y - 8 = x$, $2y + 2x = 2$ C
32. $2y = x + 3$, $x = y$ B
33. $x - y = 1$, $x = \frac{1}{2}y + 2$ A

A. (3, 2)
B. (3, 3)
C. (−2, 3)
D. (2, 3)

How many solutions does each system have?

34. $3y + x = -1$, $x = -3y$ no solution
35. $2x + 4y = 0$, $y = -\frac{1}{2}x$ infinitely many solutions
36. $y = 6x$, $y = 3x$ one solution
37. $5x - y = 1$, $5x - y = 7$ no solution
38. If two linear equations have the same slope and different y-intercepts, their graphs are __?__ lines. Such a system has __?__ solution(s). parallel; no
39. ***Agriculture*** A farmer grows only soybeans and corn on his 240-acre farm ($s + c = 240$). This year he wants to plant 80 more acres of soybeans than of corn ($s = c + 80$). How many acres does the farmer need to plant of each crop? 160 acres of soybeans; 80 acres of corn
40. ***Open-ended*** Write a system of linear equations with exactly one solution. Use substitution to solve your system. Sample: $3x + 2y = 4$, $y - 2x = 9$; (−2, 5)

Choose **Solve each system by graphing or using substitution.**

41. $y = -2x + 3$, $y = x - 6$ (3, −3)
42. $y = \frac{1}{4}x$, $x + 2y = 12$ (8, 2)
43. $y = 0$, $4x - y = 1$ $\left(\frac{1}{4}, 0\right)$
44. $x - 3y = 1$, $2x - 6y = 2$ infinitely many solutions
45. $x - y = 20$, $2x + 3y = 0$ (12, −8)
46. $y = -x$, $x + y = 5$ no solution
47. $x = -2$, $3x - 2y = 4$ (−2, −5)
48. $0.4x + 0.5y = 1$, $x - y = 7$ (5, −2)

Exercises MIXED REVIEW

Find the slope and *y*-intercept of each line.

49. $y = 7x - 4$ 7; −4
50. $3x + 8y = 16$ $-\frac{3}{8}$; 2
51. $y = 9x$ 9; 0
52. $5y = 6x - 25$ $\frac{6}{5}$; −5
53. Write a linear function that passes through the points (2, 3) and (4, 6). $y = \frac{3}{2}x$
54. ***Human Biology*** The largest bone in the body is the femur. In a 5-ft tall woman, the femur is about 1.3 ft long. The smallest bone in the body, the stapes, is in the ear. It is only about 0.1 in. long. The femur is about how many times as long as the stapes? The femur is about 156 times as long as the stapes.

stapes

femur

Getting Ready for Lesson 6-3

Simplify each expression.

55. $(x + 4) - 4(2x + 1)$ $-7x$
56. $5(2x - 3) + (7x + 15)$ $17x$
57. $3(x - 2) + 6(2x + 1)$ $15x$

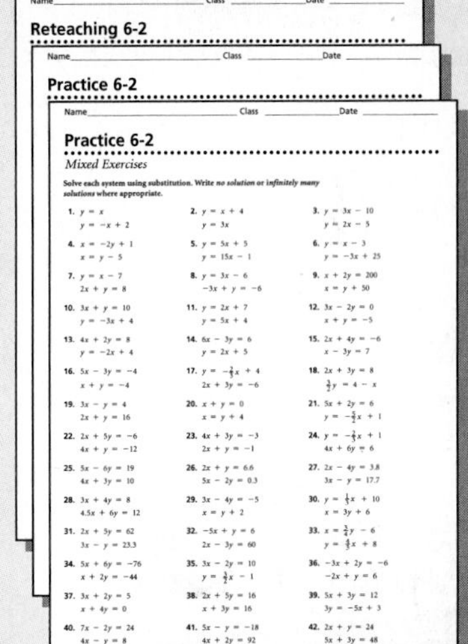

Name ______ Class ______ Date ______

Reteaching 6-2

Name ______ Class ______ Date ______

Practice 6-2

Name ______ Class ______ Date ______

Practice 6-2

Mixed Exercises

Solve each system using substitution. Write *no solution* or *infinitely many solutions* where appropriate.

1. $y = x$, $y = -x + 2$
2. $y = x + 4$, $y = 3x$
3. $y = 3x - 10$, $y = 2x - 5$
4. $x = -2y + 1$, $x = y - 5$
5. $y = 5x + 5$, $y = 15x - 1$
6. $y = x - 3$, $y = -3x + 25$
7. $y = x - 7$, $2x + y = 8$
8. $y = 3x - 6$, $-3x + y = -6$
9. $x + 2y = 200$, $x = y + 50$
10. $3x + y = 10$, $y = -3x + 4$
11. $y = 2x + 7$, $y = 5x + 4$
12. $3x - 2y = 0$, $x + y = -5$
13. $4x + 2y = 8$, $y = -2x + 4$
14. $6x - 3y = 6$, $y = 2x + 5$
15. $2x + 4y = -6$, $x - 3y = 7$
16. $5x - 3y = -4$, $x + y = -4$
17. $y = -\frac{2}{3}x + 4$, $2x + 3y = -6$
18. $2x + 3y = 8$, $\frac{3}{2}y = 4 - x$
19. $3x - y = 4$, $2x + y = 16$
20. $x + y = 0$, $x = y + 4$
21. $5x + 2y = 6$, $y = -\frac{5}{2}x + 1$
22. $2x + 5y = -6$, $4x + y = -12$
23. $4x + 3y = -3$, $2x + y = -1$
24. $y = -\frac{2}{3}x + 1$, $4x + 6y = 6$
25. $5x - 6y = 19$, $4x + 3y = 10$
26. $2x + y = 6.6$, $5x - 2y = 0.3$
27. $2x - 4y = 3.8$, $3x - y = 17.7$
28. $3x + 4y = 8$, $4.5x + 6y = 12$
29. $3x - 4y = -5$, $x = y + 2$
30. $y = \frac{1}{3}x + 10$, $x = 3y + 6$
31. $2x + 5y = 62$, $3x - y = 23.3$
32. $-5x + y = 6$, $2x - 3y = 60$
33. $x = \frac{3}{4}y - 6$, $y = \frac{4}{3}x + 8$
34. $5x + 6y = -76$, $x + 2y = -44$
35. $3x - 2y = 10$, $y = \frac{3}{2}x - 1$
36. $-3x + 2y = -6$, $-2x + y = 6$
37. $3x + 2y = 5$, $x + 4y = 0$
38. $2x + 5y = 16$, $x + 3y = 16$
39. $5x + 3y = 12$, $3y = -5x + 3$
40. $7x - 2y = 24$, $4x - y = 8$
41. $5x - y = -18$, $4x + 2y = 92$
42. $2x + y = 24$, $5x + 3y = 48$

Lesson Quiz

Lesson Quiz is also available in Transparencies.

Solve each system using substitution. Write *no solution* or *infinitely many solutions* where appropriate.

1. $x = 3\frac{1}{5} - \frac{4}{5}y$, $y = -1$ (4, −1)
2. $x = \frac{2}{3}y + 1\frac{1}{3}$, $y = \frac{3}{2}x + 1$ no solution
3. $x = -\frac{1}{2}y + 2$, $y = -2x + 4$ infinitely many solutions
4. $x = 4y - 4$, $y = -\frac{1}{2}x - 2$ (−4, 0)

6-3 Teaching Notes

PROBLEM OF THE DAY

Passengers boarded a plane in Phoenix. In Dallas, $\frac{1}{2}$ of the passengers got off and $\frac{4}{5}$ of the original number got on. In Memphis, $\frac{3}{4}$ of the passengers got off and twice the number of remaining passengers got on. In Chicago, $\frac{2}{3}$ of the passengers got off the plane, leaving 39 passengers. How many passengers originally boarded the plane in Phoenix? 120 passengers

Problem of the Day is also available in Transparencies.

CONNECTING TO PRIOR KNOWLEDGE Ask students to describe occasions in math when they use the strategy of elimination to simplify a problem.

THINK AND DISCUSS

Example 1

Ask: *What properties are reviewed in the Quick Review?* Addition Property of Equality and Subtraction Property of Equality Explain that the elimination method for solving systems of equations is based on these properties.

ALTERNATIVE METHOD Some students may prefer to add the equations as shown here. This is similar to the method

Lesson Planning Options

Prerequisite Skills

- Combining integers
- Substituting numbers for variables in equations

Assignment Options

To provide flexible scheduling, this lesson can be subdivided into parts.

1 **Core** 1–5, 10–17, 30–33
Extension 26, 27, 38

2 **Core** 6–8, 18–25, 34–37
Extension 9, 28, 29, 39

Use Mixed Review to maintain skills.

Resources

Student Edition

Skills Handbook, p. 574
Extra Practice, p. 561
Glossary/Study Guide

Teaching Resources

Chapter Support File, Ch. 6
- Practice 6-3 (two worksheets)
- Reteaching 6-3
- Checkpoint

Classroom Manager 6-3
Glossary, Spanish Resources
Two-Year Algebra Handbook 6-3

Transparencies
79

Connections 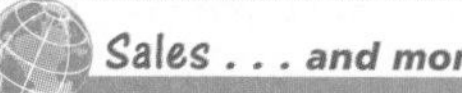 Sales . . . and more

6-3 Solving Systems Using Elimination

What You'll Learn

- Solving systems of linear equations using elimination

...And Why

To investigate real-world situations, such as sales

What You'll Need

- graph paper

QUICK REVIEW

If $a = b$ and $c = d$, then $a + c = b + d$ and $a - c = b - d$.

THINK AND DISCUSS

Part 1

Adding or Subtracting Equations

When both linear equations of a system are in the form $Ax + By = C$, you can solve the system using **elimination.** You can add or subtract the equations to eliminate a variable.

Example 1

Solve the system using elimination. Check your solution.

$$5x - 6y = -32$$
$$3x + 6y = 48$$

First, eliminate one variable.

$$\begin{array}{r} 5x - 6y = -32 \\ 3x + 6y = 48 \\ \hline 8x + 0 = 16 \\ x = 2 \end{array}$$

← Add the equations to eliminate y.
← Solve for x.

Then, find the value of the eliminated variable.

$3x + 6y = 48$ ← Pick one equation.
$3(2) + 6y = 48$ ← Substitute 2 for x.
$6 + 6y = 48$ ← Solve for y.
$6y = 42$
$y = 7$

Since $x = 2$ and $y = 7$, the solution is (2, 7).

Check See if (2, 7) makes the other equation true.

$5(2) - 6(7) \stackrel{?}{=} -32$
$10 - 42 \stackrel{?}{=} -32$
$-32 = -32$ ✓

PROBLEM SOLVING

Look Back Why was y eliminated? The coefficients of y are opposites.

Example 2  **Relating to the Real World** 

Basketball Altogether 292 tickets were sold for a high school basketball game. An adult ticket costs \$3. A student ticket costs \$1. Ticket sales were \$470. Use the system to find the number of each type of ticket sold.

number of tickets sold: $a + s = 292$
money collected: $3a + s = 470$

First, eliminate one variable.

$$\begin{array}{r} a + s = 292 \\ 3a + s = 470 \\ \hline -2a + 0 = -178 \\ a = 89 \end{array}$$

← Subtract the equations to eliminate s.
← Solve for a.

they used when applying the Addition Property of Equality to solving equations.

$5x - 6y + (3x + 6y) = -32 + 48$ ← Use the Addition Property of Equality.
$8x = 2$ ← Combine like terms.
$x = 2$ ← Solve for x.

$a + s - (3a + s) = 292 - 470$ ← Use the Subtraction Property of Equality.
$a + s - 3a - s = 292 - 470$ ← Use the Distributive Property.
$-2a = -178$ ← Combine like terms.
$a = 89$ ← Solve for a.

Example 2 **Relating to the Real World** 

ERROR ALERT! When subtracting the second equation, some students may fail to change the sign for *every* term. **Remediation:** Have them show the subtraction using the format shown in the following Alternative Method.

ALTERNATIVE METHOD Some students may prefer to subtract the equations in this way.

Example 3 **Relating to the Real World** 

Ask: *What property is reviewed in the Quick Review?* Multiplication Property of Equality Explain that in Example 3 both the Multiplication Property of Equality and the Subtraction Property of Equality are used to eliminate a variable.

Then, find the value of the eliminated variable.

$89 + s = 292$ ← Substitute 89 for a in the first equation.
$s = 203$ ← Solve for s.

There were 89 adult tickets sold and 203 student tickets sold.

1. Check the solution to Example 2. $89 + 203 = 292$ ✓ $3(89) + 203 = 470$ ✓
2. Would you *add* or *subtract* the equations to eliminate a variable?
   a. $a - b = 8$, $a + 2b = 5$ subtract
   b. $-3x + 2y = 1$, $4x - 2y = -3$ add
   c. $m + t = 6$, $5m + t = 14$ subtract
3. Try This Use elimination to solve the system in Question 2(a). (7, −1)

Multiplying First

Part 2

To eliminate a variable, you may need to multiply one or both equations in a system by a nonzero number. Then add or subtract the equations.

Example 3 **Relating to the Real World** 

Sales Suppose your class receives \$1084 for selling 205 packages of greeting cards and gift wrap. Let w = the number of packages of gift wrap sold and c = the number of packages of greeting cards sold. Use the system to find the number of each type of package sold.

total number of packages: $w + c = 205$
total amount of sales: $4w + 10c = 1084$

In the first equation, the coefficient of w is 1. In the second equation, the coefficient of w is 4. So multiply the first equation by 4. Then subtract to eliminate w.

$w + c = 205$: $4w + 4c = 820$ ← Multiply each side of the first equation by 4.
$4w + 10c = 1084$
$-6c = -264$ ← Subtract the two equations.
$c = 44$ ← Solve for c.

Find w.

$w + c = 205$ ← Use the first equation.
$w + 44 = 205$ ← Substitute 44 for c.
$w = 161$ ← Solve for w.

The class sold 161 packages of gift wrap and 44 packages of greeting cards.

4. Could you have multiplied the first equation by 10 rather than 4 and then solved the system? Why or why not? Yes; you could eliminate c first. Then you solve the first equation for w.

Additional Examples

FOR EXAMPLE 1

Solve the system using elimination. Check your solution.

$x + 3y = 7$
$x - 4y = 14$ (10, −1)

Discussion: *Why would subtracting these two equations be more effective than adding?*

FOR EXAMPLE 2

You are celebrating a victory with friends at a neighborhood sandwich shop. You plan to buy a total of 12 sandwiches with \$16.00. If a chicken sandwich costs \$1.50 and a tuna sandwich costs \$1.00, how many of each type should you buy? eight chicken and four tuna

FOR EXAMPLE 3

Solve the system using elimination. Check your solution.

$2x + 3y = 6$
$2x - 2y = 4$ $\left(\frac{12}{5}, \frac{2}{5}\right)$

Discussion: *How many different ways can you solve this problem?*

ERROR ALERT! Some students may forget to multiply both sides of the equation. **Remediation:** Encourage students to show the step $10(t + s) = 10(55)$.

VISUAL LEARNING Invite students to make and decorate charts showing the strategy used for solving systems of equations. Students may illustrate any of the three methods for solving that they have just learned.

WORK TOGETHER

ERROR ALERT! Question 5a Some students may think that they must multiply both equations by the same number. **Remediation:** Direct students' attention to the last step in the flowchart directly above the Work Together section.

Questions 6-8 Different students may try different methods, but the group should discuss and agree on the "best" method.

Exercises ON YOUR OWN

EXTENSION Exercises 10–12 Have students graph the system of equations in Exercise 10. Then on the same axes, have them graph the equation that resulted from adding the equations together. Ask: *How does the third graph relate to the other two?* **It passes through the point of intersection.** Have students follow a similar procedure for Exercises 11–12. Ask: *Why is the third line always horizontal or vertical?* **When one variable is eliminated, the result is a one-variable linear equation which, when graphed in the coordinate plane, is a horizontal or vertical line.**

Technology Options

A graphing calculator or graphing software may be used to solve the systems by graphing in Exercises 30–37.

Prentice Hall Technology

Software • Secondary Math Lab Toolkit • Computer Item Generator 6-3

Internet • See the Prentice Hall site. (http://www.phschool.com)

For systems with no solution or infinitely many solutions, look for the same results as you did when you used substitution.

When you solve systems using elimination, plan a strategy. A flowchart like this one may help you to decide how to eliminate a variable.

Can I eliminate a variable by adding or subtracting the given equations? — *yes* → Do it.

no →

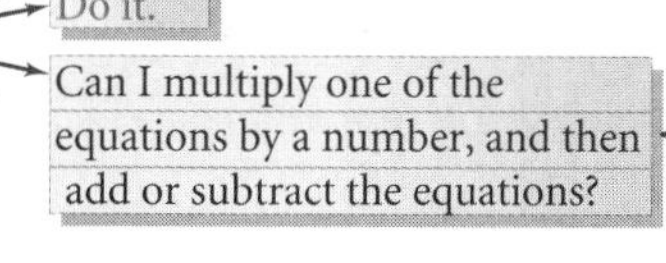

Can I multiply one of the equations by a number, and then add or subtract the equations? — *yes* → Do it.

no → Multiply both equations by different numbers. Then add or subtract the equations.

WORK TOGETHER

Work in groups.

5. Suppose you want to solve this system using elimination. $3x - 2y = 6$, $5x + 7y = 41$
   a. What would you multiply each equation by to eliminate x?
   b. What would you multiply each equation by to eliminate y?
   c. Solve the system using elimination. Be sure to check your solution.
   a–b. See margin. c. (4, 3)

Decide which method makes solving each system easier: graphing, substitution, or elimination. Then solve the system and explain the method you chose. **6–8. Answers may vary. See left for samples.**

6. $-5a + 14b = 13$, $9a = 72b$
7. $3p - 8q = 4$, $9p - 4q = 5$
8. $y = \frac{2}{3}x - 1$, $y = -x + 4$

6.Substitution; $\left(-4, -\frac{1}{2}\right)$; second equation is easy to solve for *a*.
7.Elimination; $\left(\frac{2}{5}, -\frac{7}{20}\right)$; you can multiply one equation to eliminate *p* or *q*.
8.Substitution; (3, 1); both equations are already solved for *y*.

9. Describe how to solve this system using elimination. **See margin.** $x - 2y = 8$, $y = x + 4$

Exercises ON YOUR OWN

Describe a first step for solving each system using elimination. Then solve each system. **1-8. Samples are given.**

1. $3x - y = 21$, $2x + y = 4$ **Add eqs.; (5, −6)**
2. $3x + 4y = -10$, $5x - 2y = 18$ **Mult. 2nd eq. by 2; (2, −4)**
3. $2x - y = 6$, $-3x + 4y = 1$ **Mult. 1st eq. by 4; (5, 4)**
4. $x - y = 12$, $x + y = 22$ **Add eqs.; (17, 5)**
5. $2r - 3n = 13$, $8r + 3n = 7$ **Add eqs.; (2, −3)**
6. $5a + 6b = 54$, $3a - 3b = 17$ **Mult. 2nd eq. by 2; $\left(8, \frac{7}{3}\right)$**
7. $x + y = 6$, $x + 3y = 10$ **Subtr. 1st eq. from 2nd; (4, 2)**
8. $2p - 5q = 6$, $4p + 3q = -1$ **Mult. 1st eq. by 2; $\left(\frac{1}{2}, -1\right)$**
9. *Business* A company orders two types of parts, brass *b* and steel *s*. One shipment contains 3 brass and 10 steel parts and costs \$48. A second shipment contains 7 brass and 4 steel parts and costs \$54. Solve the system to find the cost of each type of part. **Brass parts cost \$6; steel parts cost \$3.** $3b + 10s = 48$, $7b + 4s = 54$

ALTERNATIVE ASSESSMENT **Exercises 1–8** Divide the students into small groups. Have each group decide which first step would be most appropriate for each exercise, and record their choices. Then, hold a class discussion to compare each group's choice. Make sure each group defends its position. This will help you assess students' understanding of solving linear systems of equations by elimination.

Exercises 10–25 Remind students that when it is necessary to multiply one equation in a system to make elimination possible to use the equation that would be easiest to work with.

WRITING Exercise 26 Suggest to students that they use Exercises 10–25 as examples.

Exercise 39 Point out that students are to assume that the cost of one night or the cost of one meal is the same for weekends and weekdays.

Chapter Project FIND OUT BY CALCULATING Students calculate the fixed cost and the cost per person served based on the information supplied. This information is essential to their work on the Chapter Project. Have students add this task to work they have already completed for the project. Check students' progress on the project by having each one write a sentence or two describing what has been done for the project thus far.

Solve each system using elimination. Check your solution. 10–25. First number corresponds to the first variable in each equation.

10. $x + y = 12$, $x - y = 2$ (7, 5)
11. $-a + 2b = -1$, $a = 3b - 1$ (5, 2)
12. $3u + 4w = 9$, $-3u - 2w = -3$ (−1, 3)
13. $3x + 2y = 9$, $-x + 3y = 8$ (1, 3)
14. $r - 3p = 1$, $6r - p = 6$ (1, 0)
15. $5x + 3y = 1.5$, $-8x - 2y = 20$ (−4.5, 8)
16. $-2z + y = 3$, $z + 4y = 3$ (−1, 1)
17. $m - n = 0$, $m + n = 28$ (14, 14)
18. $2k - 3c = 6$, $6k - 9c = 9$ no solution
19. $3b + 4e = 6$, $-6b + e = 6$ $\left(-\frac{2}{3}, 2\right)$
20. $4x = -2y + 1$, $2x + y = 4$ no solution
21. $2p - 3t = 4$, $3p + 2t = 6$ (2, 0)
22. $3x + y = 8$, $x - y = -12$ (−1, 11)
23. $x + 4y = 1$, $3x + 12y = 3$ infinitely many solutions
24. $h = 2s - 1$, $2s - h = 1$ infinitely many solutions
25. $4x - 2y = 3$, $5x - 3y = 2$ $\left(2\frac{1}{2}, 3\frac{1}{2}\right)$
26. *Writing* Explain how to solve a system using elimination. Give examples of when you use addition, subtraction, and multiplication. See margin.
27. *Electricity* Two batteries produce a total voltage of 4.5 volts ($B_1 + B_2 = 4.5$). The difference in their voltages is 1.5 volts ($B_1 - B_2 = 1.5$). Determine the voltages of the two batteries. 3 V, 1.5 V

Critical Thinking **Do you *agree* or *disagree* with each statement? Explain.** 28–29. Answers may vary. Samples are given.

28. A system of linear equations written in the form $Ax + By = C$ is solved most easily by elimination. Agree; it is easy to multiply the equation in this form to match the coefficients of the other equation.
29. A system of linear equations written in slope-intercept form, $y = mx + b$, is solved most easily by substitution. Agree; you do not need to solve an equation for y before substituting the value.

Choose **Choose any method to solve each system.**

30. $y = x + 5$, $x + y = 1$ (−2, 3)
31. $4x - 2y = 6$, $-2x + y = -3$ infinitely many solutions
32. $y = 0.25x$, $y = -4$ (−16, −4)
33. $-5x + 2y = 14$, $-3x + y = -2$ (18, 52)
34. $y = -2x + 7$, $y = 4x - 5$ (2, 3)
35. $y = 4$, $3x - y = 5$ (3, 4)
36. $2x - 3y = 4$, $2x + y = -4$ (−1, −2)
37. $5x + 3y = 6$, $2x - 4y = 5$ $\left(1\frac{1}{2}, -\frac{1}{2}\right)$
38. *Open-ended* Write a system of linear equations that you would solve using elimination. Solve the system. Sample: $4x + 3y = 6$ $-2x - 5y = 4$; (3, −2)
39. *Vacation* A weekend at the Beach Bay Hotel in Florida includes two nights and four meals. A week includes seven nights and ten meals. The system of linear equations below models this situation. Let n = the cost of one night and m = the cost of one meal.

Weekend: $2n + 4m = 195$
Week: $7n + 10m = 650$

a. Use elimination to solve the system. (81.25; 8.125)
b. What does the solution mean in terms of the prices of the room and meals? Room for one night costs $81.25 per person; the average cost of one meal is about $8.13.

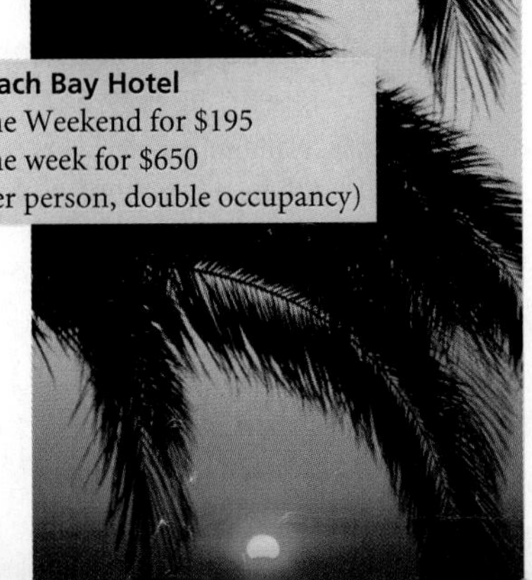
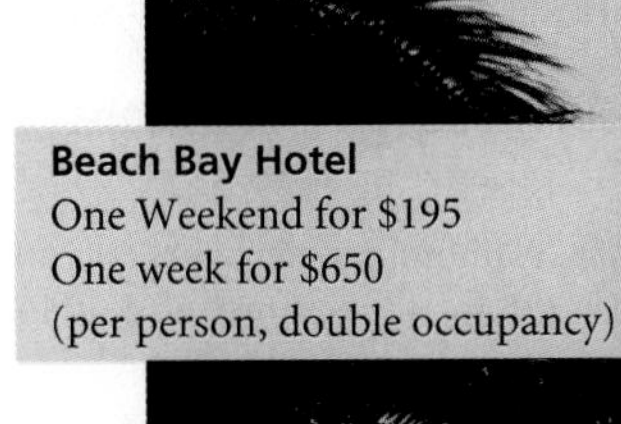

page 282 Work Together

5a. Answers may vary. Sample: Multiply the first equation by 5. Multiply the second equation by 3.

b. Answers may vary. Sample: Multiply the first equation by 7. Multiply the second equation by 2.

9. Answers may vary. Sample: Rewrite the first equation as $-2y = -x + 8$. Add the equations, $-y = 12$. Solve the equation, $y = -12$. Substitute the value in either equation and solve for x, $x = -16$. The solution is (−16, −12).

pages 282–284 On Your Own

26. Answers may vary. Sample: Rewrite each equation so that at least one pair of the corresponding coefficients are equal or opposite. Multiply each equation as necessary:
$2y - 4x = 3 \rightarrow 6y - 12x = 9$
$3y - 3x = 2 \quad 6y - 6x = 4$
If the coefficients are the same, subtract one equation from the other.
$6y - 12x = 9 \rightarrow -12x = 5$
$6y - 6x = 4$
If the coefficients are opposite, add the two equations.
$2y - 3x = 3 \rightarrow 6y = 5$
$4y + 3x = 2$
Then solve the new equation. Substitute the found value into one of the original equations and solve for the remaining variable.

Exercises MIXED REVIEW

MAKING CONNECTIONS Exercise 44 The poliomyelitis virus enters the bloodstream through the mouth. If the virus enters the central nervous system, it can attack the body's motor neurons and cause paralysis. Motor neurons make up the nerves that enable us to move our body parts. The vaccine against polio has dramatically reduced the number of polio cases.

GETTING READY FOR LESSON 6-4 Students write an equation to model a situation. In the next lesson they will write two equations to model a system of equations.

JOURNAL Encourage students to include methods that use the graphing calculator.

Wrap Up

THE BIG IDEA Ask students what is meant by solving a system of equations using elimination.

RETEACHING ACTIVITY Students will model equations using counters. Then they will solve each equation using elimination. (Reteaching worksheet 6-3)

Exercises CHECKPOINT

In this Checkpoint, your students will to assess their progress on the topics covered in Lessons 6-1 through 6-3.

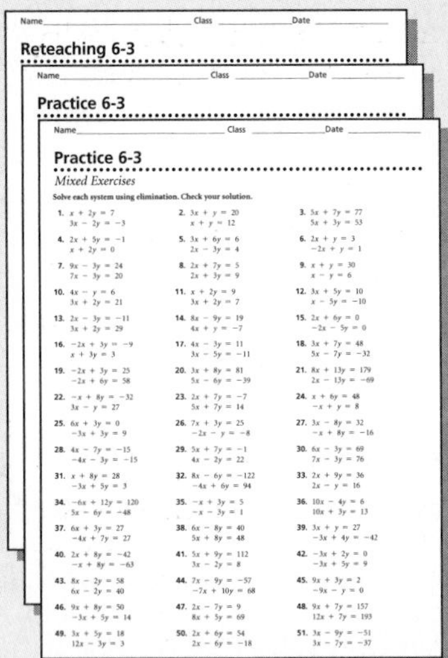
Name______ Class ______ Date ______

Reteaching 6-3

Name______ Class ______ Date ______

Practice 6-3

Name______ Class ______ Date ______

Practice 6-3

Mixed Exercises

Solve each system using elimination. Check your solution.

1. $x + 2y = 7$; $3x - 2y = -3$
2. $3x + y = 20$; $x + y = 12$
3. $5x + 7y = 77$; $5x + 3y = 53$
4. $2x + 5y = -1$; $x + 2y = 0$
5. $3x + 6y = 6$; $2x - 3y = 4$
6. $2x + y = 3$; $-2x + y = 1$
7. $9x - 3y = 24$; $7x - 3y = 20$
8. $2x + 7y = 5$; $2x + 3y = 9$
9. $x + y = 30$; $x - y = 6$
10. $4x - y = 6$; $3x + 2y = 21$
11. $x + 2y = 9$; $3x + 2y = 7$
12. $3x + 5y = 10$; $x - 5y = -10$
13. $2x - 3y = -11$; $3x + 2y = 29$
14. $8x - 9y = 19$; $4x + y = -7$
15. $2x + 6y = 0$; $-2x - 5y = 0$
16. $-2x + 3y = -9$; $x + 3y = 3$
17. $4x - 3y = 11$; $3x - 5y = -11$
18. $3x + 7y = 48$; $5x - 7y = -32$
19. $-2x + 3y = 25$; $-2x + 6y = 58$
20. $3x + 8y = 81$; $5x - 6y = -39$
21. $8x + 13y = 179$; $2x - 13y = -69$
22. $-x + 8y = -32$; $3x - y = 27$
23. $2x + 7y = -7$; $5x + 7y = 14$
24. $x + 6y = 48$; $-x + y = 8$
25. $6x + 3y = 0$; $-3x + 3y = 9$
26. $7x + 3y = 25$; $-2x - y = -8$
27. $3x - 8y = 32$; $-x + 8y = -16$
28. $4x - 7y = -15$; $-4x - 3y = -15$
29. $5x + 7y = -1$; $4x - 2y = 22$
30. $6x - 3y = 69$; $7x - 3y = 76$
31. $x + 8y = 28$; $-3x + 5y = 3$
32. $8x - 6y = -122$; $-4x + 6y = 94$
33. $2x + 9y = 36$; $2x - y = 16$
34. $-6x + 12y = 120$; $5x - 6y = -48$
35. $-x + 3y = 5$; $-x - 3y = 1$
36. $10x - 4y = 6$; $10x + 3y = 13$
37. $6x + 3y = 27$; $-4x + 7y = 27$
38. $6x - 8y = 40$; $5x + 8y = 48$
39. $3x + y = 27$; $-3x + 4y = -42$
40. $2x + 8y = -42$; $-x + 8y = -63$
41. $5x + 9y = 112$; $3x - 2y = 8$
42. $-3x + 2y = 0$; $-3x + 5y = 9$
43. $8x - 2y = 58$; $6x - 2y = 40$
44. $7x - 9y = -57$; $-7x + 10y = 68$
45. $9x + 3y = 2$; $-9x - y = 0$
46. $9x + 8y = 50$; $-3x + 5y = 14$
47. $2x - 7y = 9$; $8x + 5y = 69$
48. $9x + 7y = 157$; $12x + 7y = 193$
49. $3x + 5y = 18$; $12x - 3y = 3$
50. $2x + 6y = 54$; $2x - 6y = -18$
51. $3x - 9y = -51$; $3x - 7y = -37$

Lesson Quiz

Lesson Quiz is also available in Transparencies.

Solve each system using elimination. Check your solution.

1. $2a - 3b = 6$
   $4a + 3b = 6$ $\left(2, \frac{-2}{3}\right)$
2. $3z + 2y = 5$
   $-z + y = 3$ $\left(\frac{14}{5}, -\frac{1}{5}\right)$
3. $i - 3j = 2$
   $-2i + 6j = 4$
   infinitely many solutions
4. $-4x + 3y = 7$
   $4x + 6y = 2$ $(-1, 1)$

Chapter Project *Find Out by Calculating*

A caterer charges a fixed cost for preparing dinner plus a cost for each person served. You know that the cost for 100 people will be \$750 and the cost for 150 people will be \$1050. Find the caterer's fixed cost and the cost per person served.

Exercises MIXED REVIEW

Write an equation of the line passing through the given points.

40–43. Answers may vary. Samples are given.

40. $(3, -7)$ and $(-4, 1)$ $y = -\frac{8}{7}x - \frac{25}{7}$ **41.** $(0, 4)$ and $(0, -7)$ $x = 0$ **42.** $(5, -8)$ and $(9, 0)$ $2x - y = 18$ **43.** $(10, 2)$ and $(10, -2)$ $x = 10$

44. *Health* In 1980, there were 400,000 cases of polio reported worldwide. From 1980 to 1993, the number of cases declined 75%. How many cases of polio were reported in 1993? 100,000 cases

Getting Ready for Lesson 6-4

Write an equation to model each situation.

45. Two sandwiches and a drink cost \$6.50. $2s + d = 6.50$

46. A stack of ten paperbacks and three hardcover books is 18 in. high. $10p + 3h = 18$

47. Five pieces of plywood and two bags of nails weigh 32 lb. $5p + 2n = 32$

SELF ASSESSMENT — FOR YOUR JOURNAL

Make an outline or table titled "Methods for Solving Systems of Equations." Provide descriptions of procedures, helpful hints, and examples for each method.

Exercises CHECKPOINT

Solve each system of equations.

1. $x + 2y = 5$; $-4x + y = 8$ $\left(-1\frac{2}{9}, 3\frac{1}{9}\right)$ **2.** $5x - 3y = 27$; $5x + 4y = -1$ $(3, -4)$ **3.** $x + y = 19$; $10x - 7y = 20$ $(9, 10)$ **4.** $6x - \frac{3}{4}y = 16$; $y = -\frac{8}{3}x$ $\left(2, -\frac{5}{3}\right)$

5. *Open-ended* Write a system of equations where the solution is $x = 5$ and $y = 7$. Sample: $2x - y = 3$; $\frac{1}{2}x + \frac{1}{2}y = 6$

6. *Photography* A photographer offers two options for portraits. You can pay \$25 for 12 pictures and \$.40 for each extra print, or \$30 for 12 pictures and \$.15 for each extra print. Let c = the total cost and p = the number of extra prints.

first option: $c = 25 + 0.4p$
second option: $c = 30 + 0.15p$

a. Solve the system. $(20, 33)$

b. *Writing* Which option would you recommend to a friend? Explain. Choose the first option if you want fewer than 32 prints. If you want more than 32 prints, choose the second option.

6-4 Teaching Notes

PROBLEM OF THE DAY

A group of students are trying to evenly divide the cookies remaining from a bake sale. If each student takes three cookies, there are five cookies left. If each student takes four cookies, they are two cookies short. How many students and how many cookies are there?
7 students, 26 cookies

Problem of the Day is also available in Transparencies.

CONNECTING TO PRIOR KNOWLEDGE Ask students to share the techniques they use for translating word problems into the language of math.

THINK AND DISCUSS

AUDITORY LEARNING Encourage students to read the question quietly to themselves. Then have students ask themselves the following questions and write their answers in math language.

- *What do I know?*
- *What am I trying to find?*
- *What can I find using the facts I have?*

Example 1 **Relating to the Real World** 

ALTERNATIVE METHOD This problem can also be solved using one variable and one equation. Students solved similar problems in lesson 4-2.

Connections 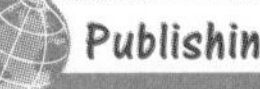 *Publishing . . . and more*

6-4 Writing Systems

What You'll Learn

- Writing and solving systems of linear equations
- Using systems to find the break-even point

...And Why

To model real-world situations, such as publishing a newsletter

THINK AND DISCUSS

In the Math Toolbox before Lesson 6-1, the following problem was solved using the strategy *Guess and Test.*

The ratio of boys to girls in a ninth-grade class at a high school is about 3 to 2. There are about 600 ninth-graders. How many are boys? How many are girls?

You can also solve this problem using a system of linear equations.

Example 1 **Relating to the Real World** 

Schools Use a system of linear equations to solve the problem stated above.

Define b = the number of boys
g = the number of girls

Relate The total number of ninth-graders is 600. The ratio of boys to girls is 3 to 2.

Write $b + g = 600$ $\qquad \frac{b}{g} = \frac{3}{2}$

You can solve the system by substitution.

$\frac{b}{g} = \frac{3}{2}$ ← Solve the second equation for b.

$b = \frac{3}{2}g$

$b + g = 600$ ← Use the first equation.

$\frac{3}{2}g + g = 600$ ← Substitute $\frac{3}{2}g$ for b.

$\frac{5}{2}g = 600$ ← Combine like terms.

$g = 240$ ← Multiply both sides by $\frac{2}{5}$.

$\frac{b}{g} = \frac{3}{2}$ ← Use the second equation.

$\frac{b}{240} = \frac{3}{2}$ ← Substitute 240 for g.

$b = 360$ ← Multiply both sides by 240.

There are about 360 ninth-grade boys and 240 ninth-grade girls.

PROBLEM SOLVING

Look Back Which method would you prefer to solve the system: substitution, graphing, or elimination? Why?
Check students' work.

1. *Graphing Calculator* Rewrite the system $b + g = 600$ and $\frac{b}{g} = \frac{3}{2}$ so that you could use a graphing calculator to find its solution.
$y = 600 - x; y = \frac{3}{2}x$

Lesson Planning Options

Prerequisite Skills

- Modeling equations
- Substituting expressions and terms for variables in equations

Assignment Options

Core 5–15, 19–24
Extension 1–4, 16–18, 25, 26

Use Mixed Review to maintain skills.

Resources

Student Edition

Skills Handbook, pp. 578, 582
Extra Practice, p. 561
Glossary/Study Guide

Teaching Resources

Chapter Support File, Ch. 6
- Practice 6-4 (two worksheets)
- Reteaching 6-4

Classroom Manager 6-4
Glossary, Spanish Resources
Two-Year Algebra Handbook 6-4

Transparencies
79

Define b = the number of boys

$600 - b$ = the number of girls

Relate The ratio of boys to girls is 3 to 2.

Write $\frac{b}{600 - b} = \frac{3}{2}$

$2b = 3(600 - b)$ ← Use cross products.

$2b = 1800 - 3b$ ← Use the distributive property.

$2b + 3b = 1800 - 3b + 3b$ ← Add $3b$ to both sides.

$5b = 1800$ ← Combine like terms.

$\frac{5b}{5} = \frac{1800}{5}$ ← Divide both sides by 5.

$b = 360$

There are about 360 ninth-grade boys. There are about 600 − 360 or 240 ninth-grade girls. (Substitute 360 for b in the expression 600 − b.)

Ask students which method was easier, one variable or two. Explain that both methods have their advantages. Solving the problem with one variable probably took less time, but with two variables, it is easier to see what part of the original question you are answering.

Question 1 Have students use the graphing calculator to check the solution for the system of equations. To do this, they will need to use  to change the variable ranges.

Example 2 **Relating to the Real World**

CONNECTING TO THE STUDENTS' WORLD Encourage students to research the costs involved in printing a school newsletter. Repeat the example using the data students collected.

Additional Examples

FOR EXAMPLE 1

A jar contains 130 coins. If the ratio of pennies to dimes is about 9 to 2, how many coins of each type are in the jar? What is the total value of the coins? Solve using a system of linear equations. **approximately 24 dimes and 106 pennies; $3.46**

Discussion: *Why is it impossible for the penny to dime ratio in the jar to be exactly 9 to 2?*

FOR EXAMPLE 2

You buy a personal computer that costs $1750.00. Your typing service charges $5.00 per page. Expenses are $.50 per page for ink, paper, electricity and other expenses. Use a system of equations to determine how many pages you must type to break even. **389 pages**

Discussion: *What other factors should you consider before making such a purchase?*

QUICK REVIEW

The *break-even point* is when income equals expenses.

In Chapter 5, you found the break-even point for a business using one equation and the x-intercept of its graph. The graph shows the break-even point of two equations.

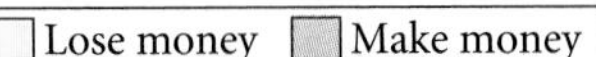

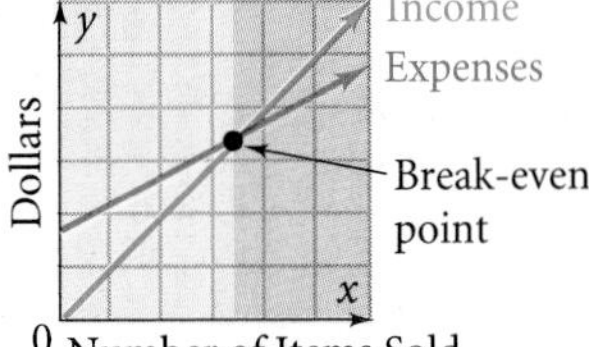

Example 2 **Relating to the Real World** 

Publishing Suppose a paper manufacturer publishes a newsletter. Expenses are $.90 for printing and mailing each copy, plus $600 for research and writing. The price of the newsletter is $1.50 per copy. How many copies of the newsletter must the company sell to break even?

Define x = the number of copies
y = the money for expenses or income

Relate Expenses: $.90 × copies printed + $600; Income: $1.50 × copies sold

Write $y = 0.9x + 600$; $y = 1.50x$

Choose a method to solve the system. Use substitution, since it is easy to substitute for y using these equations.

$$y = 0.9x + 600$$
$$y = 1.5x$$

$1.5x = 0.9x + 600$ ← Substitute $1.5x$ for y in the first equation.

$0.6x = 600$ ← Subtract $0.9x$ from each side.

$x = 1000$ ← Divide each side by 0.6.

To break even, the manufacturer must sell 1000 copies. ■

2. What are the expenses for 1000 copies? the income? **$1500; $1500**

3. *Critical Thinking* Can the company print more than it sells, but still earn a profit? If so, give an example. If not, explain why not. **See below.**

4. *Try This* Suppose printing and mailing expenses increase to $1.00 for each copy. How many copies of the newsletter must the company sell at $1.50 per copy to break even? **1200 copies**

3. Yes. Example: If the company prints 1500 copies and sells 1400, the profit is $150.

Exercises ON YOUR OWN

ERROR ALERT! When solving word problems, students might correctly solve a system of equations, but still get the wrong answer to the problem. **Remediation:** Students must check their answers to the problem with the original wording of the problem to make sure they have not made an error in writing the equations.

EXTENSION Exercise 2 Have students write a one-variable equation that could be used to solve this problem.
$0.25q + 0.10(12 - q) = 1.95$

ALTERNATIVE ASSESSMENT Exercises 5–12 Divide the students into eight groups so that each group is responsible for a different exercise. Have each member of the group solve their assigned system of equations using different methods. Then have them compare answers to check. This will help you assess students' understanding of systems of equations.

ESL Exercise 13 For the benefit of students unfamiliar with theater vocabulary, ask: *Why do you think the props and costumes are more expensive than the scripts?* As student volunteers answer, make sure they describe what they think *props*, *scripts*, and *costumes* are.

Exercise 16 Tell students that they must consider the distance to be driven.

Exercises 19–24 Help students determine when it is best to use which system. It might be easiest to use graphing if both equations are in slope-intercept form. If only one equation has a variable alone on one side of the equal sign, then substitution might be the best method.

Exercises ON YOUR OWN

1. *Geometry* The difference of the measures of two supplementary angles is 35°. Find both angle measures. 72.5° and 107.5°

QUICK REVIEW

The sum of the measures of supplementary angles is 180°.

2. Suppose you have just enough coins to pay for a loaf of bread priced at \$1.95. You have a total of 12 coins, with only quarters and dimes.
   a. Let $q =$ the number of quarters and $d =$ the number of dimes. Complete: ■ + ■ = 12 *q; d*
   b. Complete: 0.25■ + 0.10■ = ■ *q; d;* 1.95
   c. Use the equations you wrote for parts (a) and (b) to find how many of each coin you have. 5 quarters and 7 dimes
3. Suppose you have 10 coins that total \$.85. Some coins are dimes and some are nickels. How many of each coin do you have? 7 dimes and 3 nickels
4. *Open-ended* Write a problem for the total of two types of coins. Then give your problem to a classmate to solve. Sample: Suppose you have 8 coins totaling \$1. All the coins are quarters and nickels. How many of each type of coin do you have? 3 quarters, 5 nickels

Choose **Solve each linear system using any method. Tell why you chose the method you used.** 5–12. Check students' work for method.

5. $y = x + 2$, $y = -2x + 3$ $\left(\frac{1}{3}, 2\frac{1}{3}\right)$
6. $3x + 4y = -10$, $5x = 2y + 18$ $(2, -4)$
7. $5y = x$, $2x - 3y = 7$ $(5, 1)$
8. $3x - 2y = -12$, $5x + 4y = 2$ $(-2, 3)$
9. $4x = 5y$, $8x = 10y + 15$ no solution
10. $x = y - 3$, $x + 2y = 3$ $(-1, 2)$
11. $y = 3x$, $y = -\frac{1}{2}x$ $(0, 0)$
12. $y = 2x$, $7x - y = 35$ $(7, 14)$

13. *School Musical* Suppose you are the treasurer of the drama club. The cost for scripts for the spring musical is \$254. The cost of props and costumes is \$400. You must also pay royalty charges of \$1.20 per ticket to the play's publisher. You charge \$4.00 per ticket and expect to make \$150 on refreshments.
   a. Write an equation for the expenses. a. $m = 254 + 400 + 1.2n$
   b. Write an equation for the expected income. b. $m = 4n + 150$
   c. How many tickets must the drama club sell to break even? 180 tickets
   d. What method did you use to solve part (c)? Why? Sample: Substitution; both equations are already solved for *m*.
14. *Pets* The ratio of cats to dogs at your local animal shelter is about 5 to 2. The shelter accepts 40 cats and dogs. From about how many dogs would you have to choose? 11 or 12 dogs
15. *Business* Suppose you invest \$10,000 in equipment to manufacture a new board game. Each game costs \$2.65 to manufacture and sells for \$20. How many games must you make and sell before your business breaks even? 577 games
16. *Writing* Explain to a friend how to decide whether to rent a car from Auto-Rent or from Cars, Inc. Choose Cars, Inc., if you plan to drive more than 40 mi/d. Otherwise, rent from Auto-Rent.

AUTO-RENT: \$10/day + 50¢/mile

Cars, Inc.: \$20/day + 25¢/mile

Technology Options

Graphing calculators or graphing software may help students who want to solve Exercises 5–11 by graphing.

Prentice Hall Technology

Software • Secondary Math Lab Toolkit • Computer Item Generator 6-4

CD-ROM Multimedia Algebra Lab 6

Internet • See the Prentice Hall site (http://www.phschool.com)

pages 287–288 On Your Own

19–24, 32–35. See back of book.

Exercises MIXED REVIEW

GETTING READY FOR LESSON 6-5 In this preview, students review graphing linear *equations*. This will be the first step in graphing linear *inequalities* in the next lesson.

Wrap Up

THE BIG IDEA Ask students to tell what they found most difficult about writing a system of equations to solve a word problem.

RETEACHING ACTIVITY Students write and solve multistep systems of linear equations. They choose the strategy for solving based on the type of equation. (Reteaching worksheet 6-4)

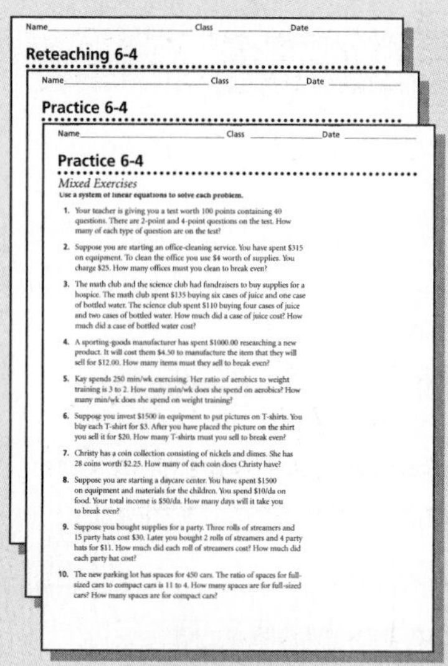

Name________ Class ________ Date ________

Reteaching 6-4

Name________ Class ________ Date ________

Practice 6-4

Name________ Class ________ Date ________

Practice 6-4

Mixed Exercises

Use a system of linear equations to solve each problem.

1. Your teacher is giving you a test worth 100 points containing 40 questions. There are 2-point and 4-point questions on the test. How many of each type of question are on the test?
2. Suppose you are starting an office-cleaning service. You have spent $315 on equipment. To clean the office you use $4 worth of supplies. You charge $25. How many offices must you clean to break even?
3. The math club and the science club had fundraisers to buy supplies for a hospice. The math club spent $135 buying six cases of juice and one case of bottled water. The science club spent $110 buying four cases of juice and two cases of bottled water. How much did a case of juice cost? How much did a case of bottled water cost?
4. A sporting-goods manufacturer has spent $1000.00 researching a new product. It will cost them $4.50 to manufacture the item that they will sell for $12.00. How many items must they sell to break even?
5. Kay spends 250 min/wk exercising. Her ratio of aerobics to weight training is 3 to 2. How many min/wk does she spend on aerobics? How many min/wk does she spend on weight training?
6. Suppose you invest $1500 in equipment to put pictures on T-shirts. You buy each T-shirt for $3. After you have placed the picture on the shirt you sell it for $20. How many T-shirts must you sell to break even?
7. Christy has a coin collection consisting of nickels and dimes. She has 28 coins worth $2.25. How many of each coin does Christy have?
8. Suppose you are starting a daycare center. You have spent $1500 on equipment and materials for the children. You spend $10/da on food. Your total income is $50/da. How many days will it take you to break even?
9. Suppose you bought supplies for a party. Three rolls of streamers and 15 party hats cost $30. Later you bought 2 rolls of streamers and 4 party hats for $11. How much did each roll of streamers cost? How much did each party hat cost?
10. The new parking lot has spaces for 450 cars. The ratio of spaces for full-sized cars to compact cars is 11 to 4. How many spaces are for full-sized cars? How many spaces are for compact cars?

Lesson Quiz

Lesson Quiz is also available in Transparencies.

1. Your band is planning to record a compact disk containing all of its greatest songs. It will cost $12,500.00 to record and produce a master copy, and an additional $2.50 to make each sale copy. If you plan to sell the final product for $7.99, how many disks must you sell to break even? 2277 disks
2. Suppose you wish to show a certain film at a local cinema. You must pay the distributor $600 and the cinema owner $2.50 per ticket sold. You charge $4.00 per ticket and expect to make an additional $2.00 per person from refreshment sales. How many tickets must you sell to break even? 172 tickets

17. **Chemistry** A piece of glass with an initial temperature of 99°C is cooled at a rate of 3.5°C/min. At the same time, a piece of copper with an initial temperature of 0°C is heated at a rate of 2.5°C/min. Let m = the number of minutes and t = the temperature in °C. Which system models the given information? C

   A. $t = 99 + 3.5m$, $t = 0 + 2.5m$ **B.** $t = 99 - 3.5m$, $t = 0 - 2.5m$

   C. $t = 99 - 3.5m$, $t = 0 + 2.5m$ **D.** $t = 99 + 3.5m$, $t = 0 - 2.5m$

18. Solve the system that models the situation in Exercise 17. Explain what the solution means in this situation. 16.5 min, 41.25°C; both objects will reach the temperature of 41.25°C in 16.5 min.

Glass can be drawn into optical fibers 16 km long. One fiber can carry 20 times as many phone calls as 500 copper wires.

Without solving, what method would you choose to solve each system: *graphing, substitution,* or *elimination*? Explain your reasoning. 19–24. Answers may vary. See margin for samples.

19. $4s - 3t = 8$, $t = -2s - 1$
20. $y = 3x - 1$, $y = 4x$
21. $3m - 4n = 1$, $3m - 2n = -1$
22. $y = -2x$, $y = -\frac{1}{2}x + 3$
23. $2x - y = 4$, $x + 3y = 16$
24. $u = 4v$, $3u - 2v = 7$

25. **Geometry** The perimeter of an isosceles triangle is 12 cm. The two sides s are each three times the length of the third side t.
    a. Write an equation for the perimeter of the triangle. $2s + t = 12$
    b. Write an equation that describes the relationship between one side s and side t. $s = 3t$
    c. Find the length of each side. $5\frac{1}{7}$, $5\frac{1}{7}$, $1\frac{5}{7}$

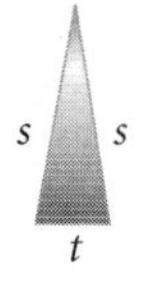

26. **Number Theory** Find two integers with a sum of 1244 and a difference of 90. 577 and 667

Exercises MIXED REVIEW

Solve each equation.

27. $7t + 4 = -10$ -2
28. $j + 9 = -j - 1$ -5
29. $|c - 4| = 21$ 25 or -17
30. $m - 5 = -6$ -1

31. In 1995, about $\frac{2}{3}$ of United States currency in circulation was outside the United States. There was $390 billion in circulation. How much was in use within the United States? $130 billion

Getting Ready for Lesson 6-5

Graph each equation. 32–35. See margin.

32. $y = 2x + 1$
33. $y - 4 = 0$
34. $y = -\frac{2}{3}x + 1$
35. $y + 2x = -5$

6-5 Teaching Notes

PROBLEM OF THE DAY

Each backpacker on the Lost Pines hike is allowed to carry only 20% of his or her weight. David weighs 120 lb and plans on bringing the following items: a 4 lb 1 oz tent, a 10 oz sleeping pad, a 3 lb 4 oz sleeping bag, 3 lb 15 oz of clothes, 2 lb 8 oz of camping supplies, and 6 lb of water. How much weight did David leave for food? 3 lb 10 oz

Problem of the Day is also available in Transparencies.

CONNECTING TO PRIOR KNOWLEDGE Have students brainstorm for real-world situations that can be expressed as inequalities. Some sample situations are "be home before midnight" and "more than 200 people attended the opening."

WORK TOGETHER

DIVERSITY Question 1 Girls, more than boys, tend to value getting the correct answer rather than developing higher order thinking skills. Encourage students to discuss strategies they can use instead of focusing on merely winning the game.

Question 1 The player who guesses the secret point can use a coordinate plane to keep a record of all points that are eliminated by each guess.

THINK AND DISCUSS

ERROR ALERT! Some students may be unable to remember when to use a *dashed* boundary line and when to use a *solid* boundary line. **Remediation:** Relate the *dashed* boundary

Connections Food Shopping . . . and more

6-5 Linear Inequalities

What You'll Learn

- Graphing linear inequalities

...And Why

To solve budget problems using linear inequalities

What You'll Need

- graph paper

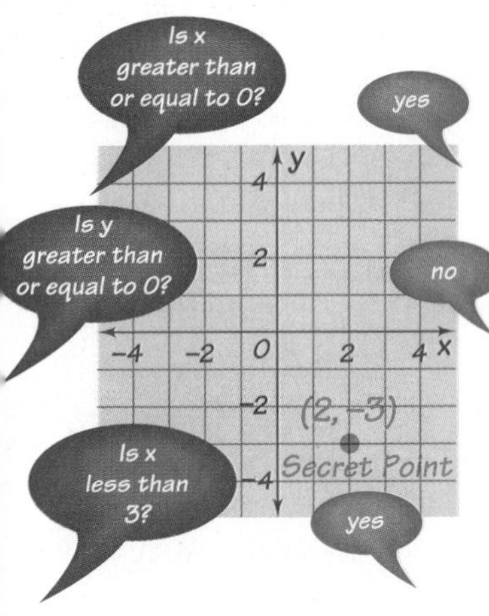

WORK TOGETHER

1. Play the game "What's the Point?" with a partner. Check students' work.

Object of the Game: To locate a secret point on the coordinate plane by asking as few questions as possible.

How to Play:

- Player A chooses a secret point on the coordinate plane. Each coordinate must be an integer from −10 to 10.
- Player B asks questions that contain the words *less than* or *greater than.* Player A answers each question with only *yes* or *no.* Count the number of questions asked until Player B names the secret point.
- The players switch roles to complete one round of the game.

How to Win: The player who names the point by asking fewer questions wins the round. The first player to win 3 rounds wins the game.

2. How many questions did you need to ask to locate the secret point? Check students' work.
3. If you were as lucky as possible, how many questions would you need to ask to locate the secret point? Explain with an example. Two; Is x greater than 9? yes. Is y greater than 9? yes. The point is (10, 10).
4. How do inequalities help you locate the secret point? Using an inequality eliminates many possible guesses with one question.
5. Describe a strategy for winning the game. Plan each question to come as close as you can to cutting the remaining possibilities in half.

QUICK REVIEW

$<$ "less than"
$\leq$ "less than or equal to"
$>$ "greater than"
$\geq$ "greater than or equal to"

THINK AND DISCUSS

Just as you have used inequalities to describe graphs on a number line, you can use inequalities to describe regions of a coordinate plane.

Number line
$x < 1$

−2 −1 0 1 2 3

Coordinate plane
$x < 1$

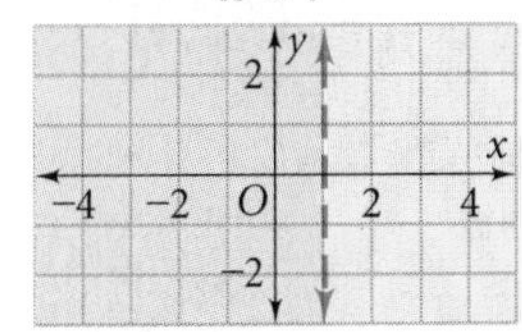

6. What do you think the graph of $y > 2$ looks like on a coordinate grid? a dashed horizontal line through (0, 2) with shading above the line

Lesson Planning Options

Prerequisite Skills

- Graphing linear equations
- Understanding inequalities

Assignment Options

Core 1–6, 8–24, 27–30, 36–39
Extension 7, 25, 26, 31–35, 40

Use Mixed Review to maintain skills.

Resources

 **Student Edition**

Skills Handbook, pp. 567, 582
Extra Practice, p. 561
Glossary/Study Guide

Teaching Resources

Chapter Support File, Ch. 6
- Practice 6-5 (two worksheets)
- Reteaching 6-5

Classroom Manager 6-5
Glossary, Spanish Resources
Two-Year Algebra Handbook 6-5

 **Transparencies**
2, 80, 82

line to the open circle used in graphing one-variable inequalities. Relate the *solid* boundary line to the closed circle used in graphing one-variable inequalities.

Example 1

Ask students: *Why is a dashed line rather than a solid line used?* **Since y is less than, but not equal to $2x + 3$, points on the line do not make the inequality true.**

Question 10 Suggest that students also test a point *not* on the graph and show that it does *not* satisfy the inequality.

If you have block scheduling or extended class periods, have students graph each of the example inequalities using the graphing calculator. Directions are given in the Math Toolbox on page 294.

Example 2

Ask students: *Why is the first step to write the inequality in slope-intercept form?* **Slope-intercept form makes it easy to graph the equation for the boundary line.** *Why is the inequality symbol reversed?* **Both sides of the inequality are divided by a negative number.**

KINESTHETIC LEARNING Clear a large area in the room and make a large coordinate grid on the floor using string or masking tape. Direct students to move to the correct regions on the grid as you call out simple inequalities such as $y \geq 3$, $x < 1$, $y \geq x$.

Additional Examples

FOR EXAMPLE 1

Graph $y > -2x + 1$.

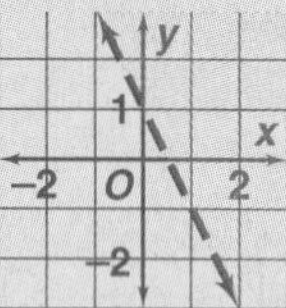

Discussion: *Why does the line of the equation need to be dotted?*

FOR EXAMPLE 2

Graph $\frac{2}{3}x - y \geq 1$.

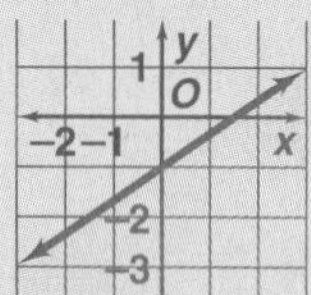

QUICK REVIEW

Use the slope m and the y-intercept b to graph an equation in the form $y = mx + b$.

PROBLEM SOLVING

Look Back Why did you test a point? Could you test the point (0, 3) to make your graph?

You test a point to check whether the inequality is satisfied above or below the line. (0, 3) is on the boundary line, so you cannot use it to test where the shading should be.

A **linear inequality** describes a region of the coordinate plane that has a boundary line. Every point in the region is a **solution of the inequality.**

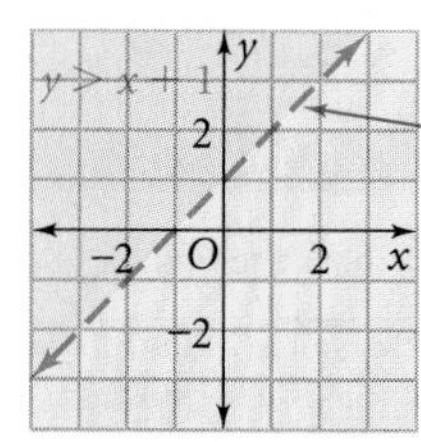

$y > x + 1$
Each point on a *dashed* boundary line is not a solution.

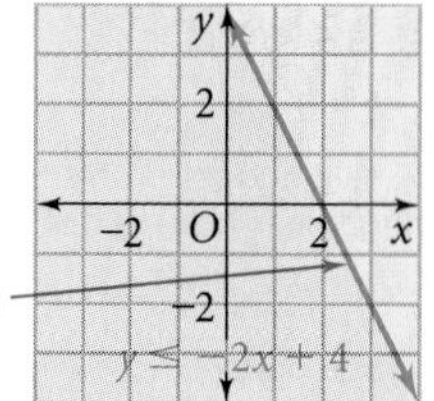

Each point on a *solid* boundary line is a solution.
$y \leq -2x + 4$

7. Is (1, 2) a solution for either inequality shown above? Explain. **Yes; (1, 2) is a solution for $y = -2x + 4$, so it is a solution for $y \leq -2x + 4$.**
8. *Open-ended* For each inequality above, name three solutions. **Sample: (0, 2), (3, 5), (−100, 100); (−50, 3), (−50, −3), (0, 0)**

Example 1

Graph $y < 2x + 3$.

First, graph the boundary line $y = 2x + 3$.

Points on the boundary line do *not* make the inequality true. Use a dashed line.

Next, test a point. Use (0, 0).

$y < 2x + 3$

$0 < 2(0) + 3$

$0 < 3$ True

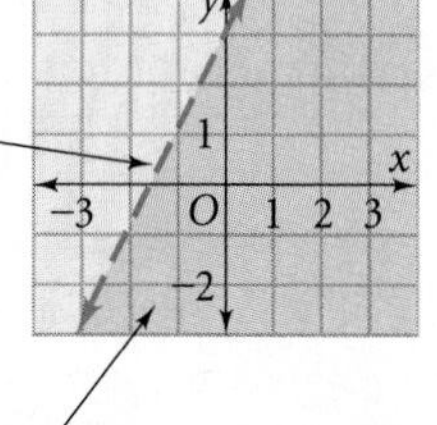

The inequality is true for (0, 0). Shade the region containing (0, 0).

9. **Try This** Graph the inequality $y \geq 2x + 3$. **See margin.**
10. You can test any point on the graph. Why is (0, 0) a good choice? **(0, 0) makes the multiplication and addition easy.**

Sometimes it helps to rewrite the inequality to find its solution.

Example 2

Graph $2x - 5y \leq 10$.

Write the inequality in slope-intercept form.

$2x - 5y \leq 10$

$-5y \leq -2x + 10$

$y \geq \frac{2}{5}x - 2$ ← Reverse the inequality symbol.

Graph $y = \frac{2}{5}x - 2$.

Test (0, 0) in $y \geq \frac{2}{5}x - 2$.

$0 \geq (0)x - 2$

$0 \geq -2$ True

Shade the region containing (0, 0).

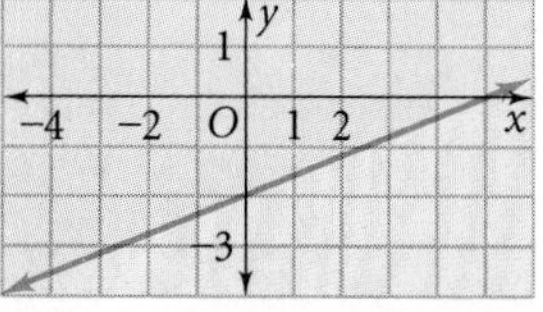

Example 3 **Relating to the Real World** 

CONNECTING TO STUDENTS' WORLD Have students find out the costs of alternatives less expensive than nuts, such as carrot sticks, celery sticks, or apple slices. Have students repeat the example using the costs they found.

The graph is shown only for quadrant 1 since only positive values for the number of pounds of peanuts and the number of pounds of cashews make sense. Students need to think about whether negative values for either variable make sense when they draw graphs to solve word problems.

Question 12 Students should check the answers by using the graph. Then they should check the answers by going back to the original word problem.

MAKING CONNECTIONS As described in Lesson 1 of this chapter, economists represent supply and demand using systems of linear equations. Economists also use inequalities in their models. The area of the supply-demand system above the point of intersection and between the two graphs represents the area of surplus: more product is being supplied than bought. The area of the supply-demand system below the point of intersection and between the two graphs represents the area of shortage: less product is being supplied than people want to buy.

11c. Shade above the line if the inequality uses > or ≥. Shade below the line if the inequality uses < or ≤.

11. a. When you graphed $y \geq \frac{2}{5}x - 2$ in Example 2, did you shade *above* or *below* the line? above

b. When you graphed $y < 2x + 3$ in Example 1, did you shade *above* or *below* the line? below

c. Critical Thinking Both inequalities are in slope-intercept form. Make a **conjecture** about the inequality symbol and the region shaded. See left.

d. Does your **conjecture** stay true for inequalities in $Ax + By < C$ form? Explain. No; $x + y < 0$ must be shaded below the line. $x - y < 0$ must be shaded above the line.

You can graph inequalities to solve real-world problems.

Example 3 **Relating to the Real World** 

Food Shopping Suppose you intend to spend no more than $12 on peanuts and cashews for a party. How many pounds of each can you buy?

Define x = the number of pounds of peanuts
y = the number of pounds of cashews

Relate cost of peanuts + cost of cashews ≤ maximum total cost

Write $2x + 4y \leq 12$

QUICK REVIEW

Use the intercepts to graph an equation in the form $Ax + By = C$.

Graph the boundary line $2x + 4y = 12$ using a solid line. Use only Quadrant I, since you cannot buy a negative amount of nuts.

Test (0, 0): $2(0) + 4(0) \leq 12$
$0 \leq 12$ True

Shade the region containing (0, 0).

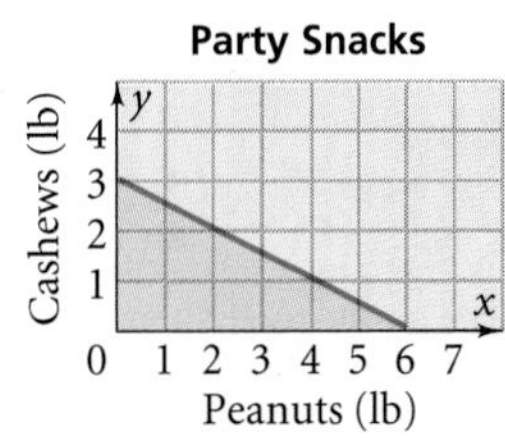

The graph shows all the possible solutions of the problem. For example, if you buy 2 lb of peanuts you can buy no more than 2 lb of cashews.

12. Which are solutions to Example 3? A, B, D

A. 2 lb peanuts and 1 lb cashews
B. 6 lb peanuts and no cashews
C. 1 lb peanuts and 3 lb cashews
D. 1.5 lb peanuts and 2 lb cashews

Technology Options

Students can graph the solutions to Exercises 8–23 and 31–35 with a graphing calculator or graphing software.

Prentice Hall Technology 

Calculator • Graphing Calculator Handbook: Procedure 8

Software • Secondary Math Lab Toolkit • Integrated Math Lab 11: • Computer Item Generator 6-5

Internet • See the Prentice Hall site. (http://www.phschool.com)

Exercises ON YOUR OWN

Exercise 6 Ask: *Which quadrant(s) should you use to show the graph of the inequality?* quadrant I

ALTERNATIVE ASSESSMENT **Exercises 8–23** Ask students to complete these exercises after reversing the sign of the inequality. For example, change $\leq$ to $\geq$. This will help you assess students' understanding of graphing linear inequalities.

STANDARDIZED TEST TIP **Exercise 24** Remind students that the only equations that will result in a shaded area bounded by a solid line are those which contain either a $\leq$ or $\geq$ sign. Therefore, they need only consider answer choices C and D.

CRITICAL THINKING **Exercise 25** This exercise helps students realize that the graph of a linear equation divides the coordinate plane into three sets of points: the points above (or to the right of) the line, the points on the line, and the points below (or to the left of) the line.

EXTENSION **Exercise 26** Ask students: *Suppose you win a prize if the sum of the number on the blue cube and the number on the red cube is less than 5. What is the probability that you win a prize?* $\frac{6}{36}$ or $\frac{1}{6}$

OPEN-ENDED **Exercise 35c** Ask: *How can you modify the graph of the inequality to show all possible sides for a square garden using 50 ft of fencing?* Draw the line for $x = y$. The points on this line represent all cases for which length $=$ width. The points both on the line and in the shaded area represent all possible sides for a square garden using 50 ft of fencing.

pages 289–291 Think and Discuss

9.

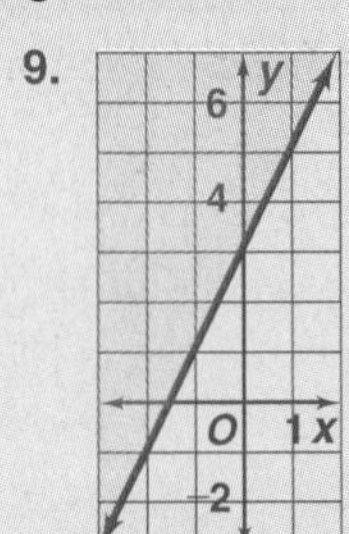

pages 292–293 On Your Own

6b.

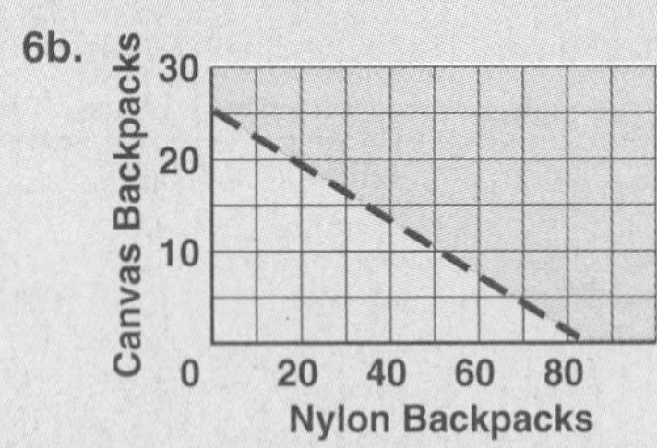

c. Sample: 50 nylon and 20 canvas backpacks, 100 nylon and no canvas backpacks, 11 nylon and 22 canvas backpacks

8.

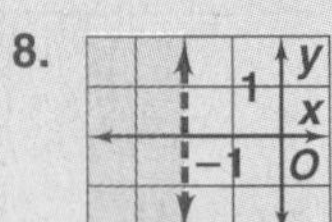

9–23, 31–34, 35b. See back of book.

Exercises ON YOUR OWN

Choose the linear inequality that describes each graph.

1.

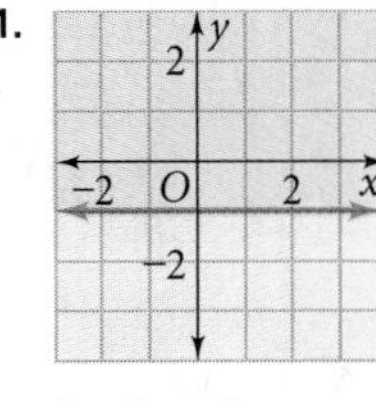

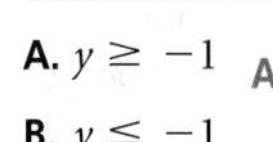

A. $y \geq -1$ A
B. $y \leq -1$

2.

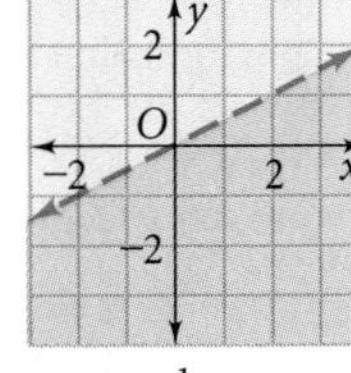

A. $y > \frac{1}{2}x$ B
B. $y < \frac{1}{2}x$

3.

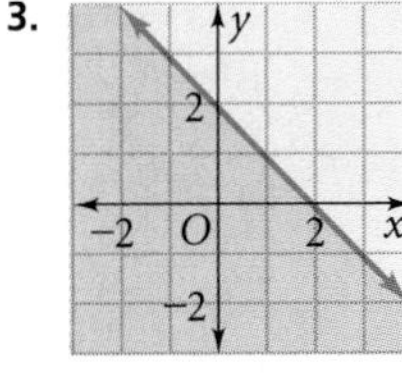

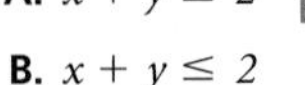

A. $x + y \geq 2$ B
B. $x + y \leq 2$

4.

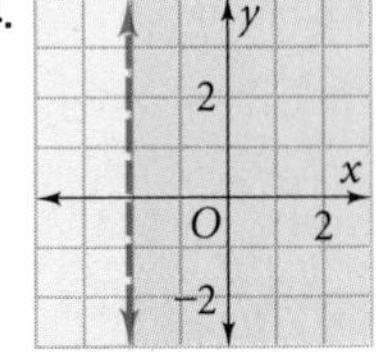

A. $x > -2$ A
B. $x < -2$

5. For which of the graphs in Exercises 1–4 is $(-2, -1)$ a solution? Exs. 1 and 3

6. **Manufacturing** A company makes backpacks. How many backpacks must the company sell to make a profit of more than \$250?
   a. Write a linear inequality that describes the situation. $3x + 10y > 250$
   b. Graph the linear inequality. b–c. See margin.
   c. Write three possible solutions to the problem.
   d. Why is "10 nylon packs and 22 canvas packs" *not* a solution? The point (10, 22) is on the boundary of the solution region.

7. **Writing** Explain why a linear inequality is useful when there are many solutions to a problem. A linear inequality allows you to describe a range of acceptable values.

Graph each linear inequality. 8–23. See margin.

8. $x < -2$	9. $y \geq 1$	10. $y < \frac{1}{4}x - 1$	11. $6y - 4x > 0$
12. $y < 5x - 5$	13. $y \leq 4x - 1$	14. $y > -3x + 4$	15. $y > -3x$
16. $x + y \geq 2$	17. $x + 3y \leq 6$	18. $\frac{1}{2}x + \frac{3}{2}y \geq \frac{3}{4}$	19. $y \geq \frac{1}{2}x$
20. $4y > 6x + 2$	21. $2x + 3y \leq 6$	22. $4x - 4y \geq 8$	23. $y - 2x < 2$

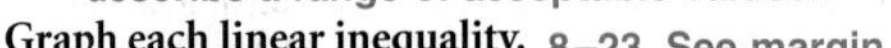

24. **Standardized Test Prep** Which statement describes the graph? D
   A. $y > x + 1$ B. $y < x + 1$ C. $y \leq x + 1$
   D. $y \geq x + 1$ E. $y = x + 1$

25. Write an inequality that describes the part of the coordinate plane *not* included in the graph of $y \geq x + 2$. $y < x + 2$

26. **Probability** Suppose you play a carnival game. You toss a blue and a red number cube. If the number on the blue cube is greater than the number on the red cube, you win a prize. The graph shows all the possible outcomes for tossing the cubes.
   a. Copy and shade the graph to show the winning outcomes. See right.
   b. Write an inequality that describes the shaded region. $y > x$
   c. What is the probability that you will win a prize? $\frac{5}{12}$

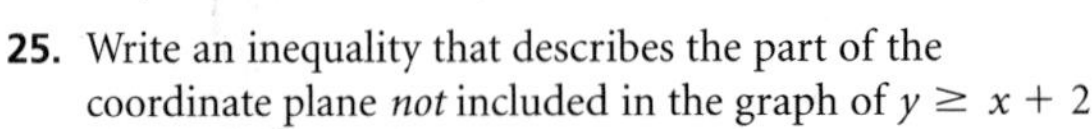

Comparing Cubes

Chapter Project **FIND OUT BY WRITING** Students use the information they previously calculated to write a report recommending a band and a ticket price. In their reports, they list their choices assuming both 200 and 300 attendees. Have students add this part of the project to work they have already completed as it is essential to their work on the chapter project. You may wish to check students' progress at this point by inspecting all completed work in their projects folders.

Wrap Up

THE BIG IDEA Ask students to compare graphing two-variable inequalities with graphing two-variable equations.

RETEACHING ACTIVITY Students graph inequalities and shade the appropriate region. They test a point from the shaded region to determine whether or not the equation is true. (Reteaching worksheet 6-5)

Exercises MIXED REVIEW

GETTING READY FOR LESSON 6-6 Students review solving systems of linear equations to prepare for solving systems of linear inequalities in the next lesson.

Write the inequality shown in each graph.

27. 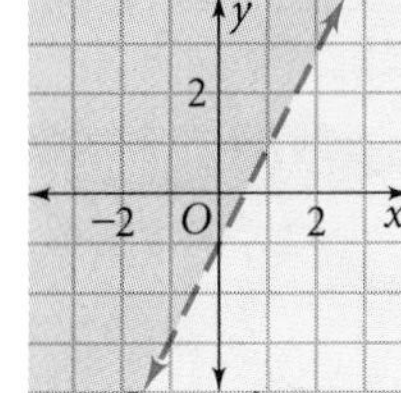
$y > 2x - 1$

28. 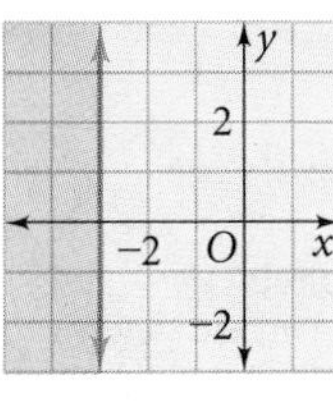
$x \leq -4$

29. 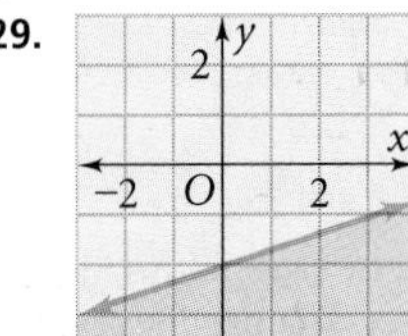
$y \leq \frac{1}{3}x - 2$

30. 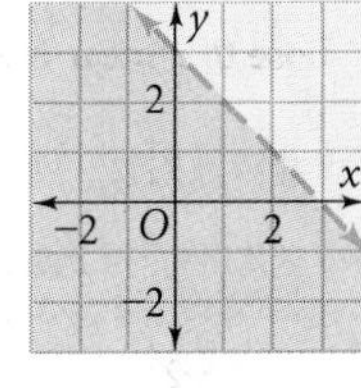
$y < -x + 3$

Write the linear inequality described. Then graph the inequality. 31–34. See margin for graphs.

31. x is positive. $x > 0$
32. y is negative. $y < 0$
33. y is not negative. $y \geq 0$
34. x is less than y. $x < y$

35. *Geometry* Suppose you have 50 ft of fencing. You want to fence a rectangular area of your yard for a garden.
   a. Use the formula for the perimeter of a rectangle to write a linear inequality that describes this situation. $2x + 2y \leq 50$
   b. Graph the inequality. See margin.
   c. *Open-ended* Give two possible sizes for a square garden.
   d. Can you make the garden 12 ft by 15 ft? **Justify** your answer, using both your graph and the inequality you wrote in part (a). No; $2(12) + 2(15) \leq 50$ is false. See margin for graph.

c. Samples: 11 ft by 11 ft; 12 ft by 12 ft

Is point *P* a solution of the linear inequality?

36. $y \leq -2x + 1$; $P(2, 2)$ no
37. $x < 2$; $P(1, 0)$ yes
38. $y \geq 3x - 2$; $P(0, 0)$ yes
39. $y > x - 1$; $P(0, 1)$ yes

40. *Consumer* Suppose you are shopping for crepe paper to decorate the gym for a school dance. Gold crepe paper costs \$5 per roll, and blue crepe paper costs \$3 per roll. You have at most \$48 to spend. How much gold and blue crepe paper can you buy? Explain your solution.
Answers may vary. Sample: There are many possibilities such as 6 rolls of blue and 6 rolls of gold paper. You can buy at most 16 rolls of blue or at most 9 rolls of gold crepe paper.

Chapter Project ***Find Out by Writing***

Use your information from the Find Out questions on pages 274 and 284. Assume that 200 people will come to the dance. Write a report listing which band you would choose and the cost per ticket that you need to charge to cover expenses. Then repeat the process assuming that 300 people will come.

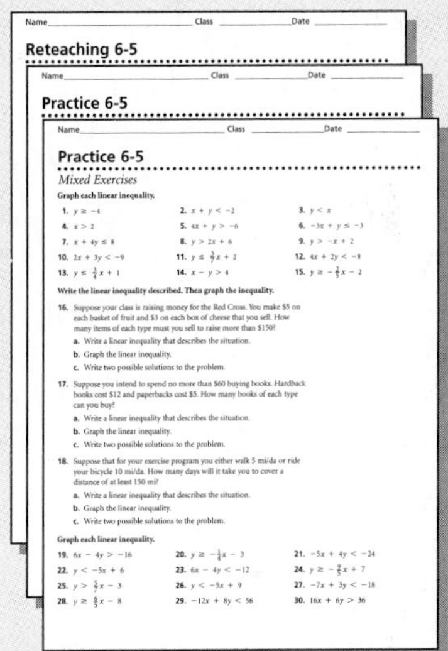
Name ______ Class ______ Date ______

Reteaching 6-5

Name ______ Class ______ Date ______

Practice 6-5

Name ______ Class ______ Date ______

Practice 6-5

Mixed Exercises

Graph each linear inequality.

1. $y \geq -4$
2. $x + y < -2$
3. $y < x$
4. $x > 2$
5. $4x + y > -6$
6. $-3x + y \leq -3$
7. $x + 4y \leq 8$
8. $y > 2x + 6$
9. $y > -x + 2$
10. $2x + 3y < -9$
11. $y \leq \frac{3}{7}x + 2$
12. $4x + 2y < -8$
13. $y \leq \frac{3}{4}x + 1$
14. $x - y > 4$
15. $y \geq -\frac{2}{5}x - 2$

Write the linear inequality described. Then graph the inequality.

16. Suppose your class is raising money for the Red Cross. You make \$5 on each basket of fruit and \$3 on each box of cheese that you sell. How many items of each type must you sell to raise more than \$150?
   a. Write a linear inequality that describes the situation.
   b. Graph the linear inequality.
   c. Write two possible solutions to the problem.
17. Suppose you intend to spend no more than \$60 buying books. Hardback books cost \$12 and paperbacks cost \$5. How many books of each type can you buy?
   a. Write a linear inequality that describes the situation.
   b. Graph the linear inequality.
   c. Write two possible solutions to the problem.
18. Suppose that for your exercise program you either walk 5 mi/da or ride your bicycle 10 mi/da. How many days will it take you to cover a distance of at least 150 mi?
   a. Write a linear inequality that describes the situation.
   b. Graph the linear inequality.
   c. Write two possible solutions to the problem.

Graph each linear inequality.

19. $6x - 4y > -16$
20. $y \geq -\frac{1}{4}x - 3$
21. $-5x + 4y < -24$
22. $y < -5x + 6$
23. $6x - 4y < -12$
24. $y \geq -\frac{9}{5}x + 7$
25. $y > \frac{5}{7}x - 3$
26. $y < -5x + 9$
27. $-7x + 3y < -18$
28. $y \geq \frac{6}{5}x - 8$
29. $-12x + 8y < 56$
30. $16x + 6y > 36$

Lesson Quiz

Lesson Quiz is also available in Transparencies.

Graph each linear inequality.

1. $y \leq -1$

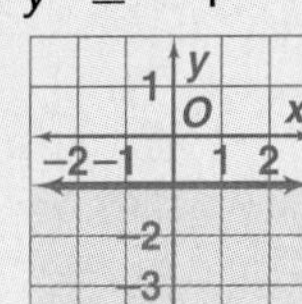

2. $\frac{3}{2}x - \frac{1}{2}y \leq 2.$

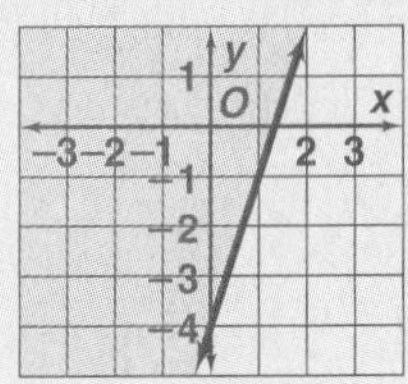

Exercises MIXED REVIEW

Find the slope of each line.

41. $y = 5x - 9$ 5
42. $7x + 4y = 20$ $-\frac{7}{4}$
43. $3y = 8x - 12$ $\frac{8}{3}$
44. $y = 8x + 1$ 8

45. *Automobiles* The average car is parked 95% of the time. How many hours is the average car on the road each day? 1.2 h

Getting Ready for Lesson 6-6

Solve each system of equations.

46. $y = 2x - 3$, $y = -4x$ $\left(\frac{1}{2}, -2\right)$
47. $4y = 3x + 11$, $y = 2x - 1$ $(3, 5)$
48. $y = 5x + 2$, $y = 4x - 6$ $(-8, -38)$
49. $y = 8x$, $y = 2x + 28$ $\left(4\frac{2}{3}, 37\frac{1}{3}\right)$

Math ToolboX

In Lesson 6-5, students determined by testing a point which portion of an inequality graph must be shaded. This toolbox allows students to determine shading by using the meaning of the inequality. Because the graphing calculator quickly displays the graphed inequality, students may focus their attention on which region needs to be shaded. Have students work in groups of four to compare their graphs. This toolbox will reinforce students' conjecture from Lesson 6-5 that the graphs of inequalities are shaded above the line if the inequality is greater than and below the line if the inequality is less than.

ERROR ALERT! Some students may try to adjust the level of shading after they have completed the graph. **Remediation:** Have students clear the equation from **Y1** and begin again. The level of shading must be determined before the graph is completed.

WRITING Exercise 9 Students may need to repeat one of the examples paying careful attention to the menus on the screen.

ADDITIONAL PROBLEM Use a graphing calculator to graph the inequality $3x + 4y \leq 4$. Sketch your graph.

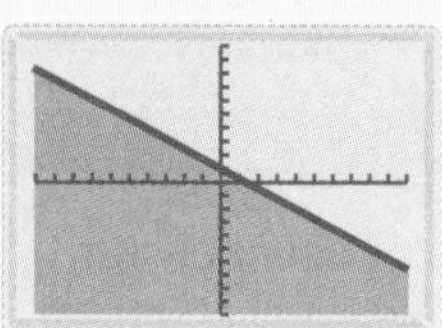

Materials and Manipulatives

- Graphing calculator

Resources

Transparencies
11, 29

page 294 Math Toolbox

1. 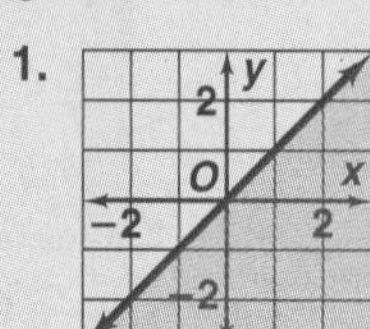

2. 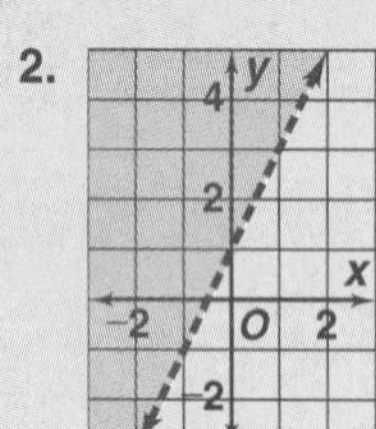

3. 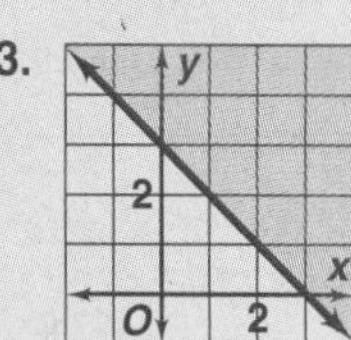

4–8. See back of book.

Math ToolboX Technology

Graphing Inequalities

After Lesson 6-5

You can use the DRAW feature of a graphing calculator to graph inequalities. The order you enter data depends on whether you are shading above or below a line. When shading below Y_1, use Shade (Ymin, Y_1). When shading above Y_1, use Shade (Y_1, Ymax). You do not have to use a close parenthesis before pressing ENTER.

Example

Graph each inequality.

a. $y < 2x + 3$

Shade below Y_1 for *less than.*

b. $y > 0.5x - 1$

Shade above Y_1 for *greater than.*

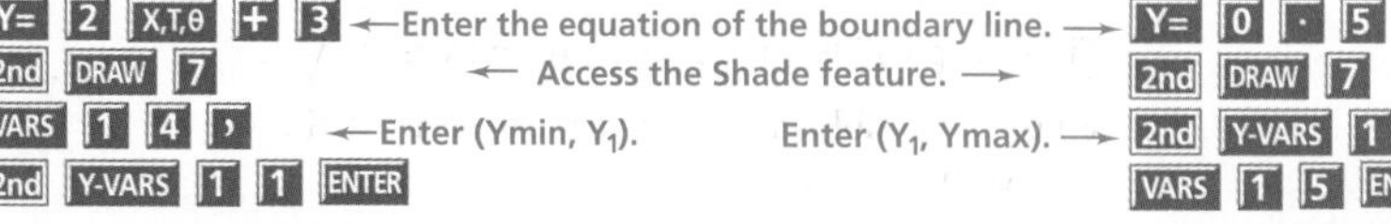

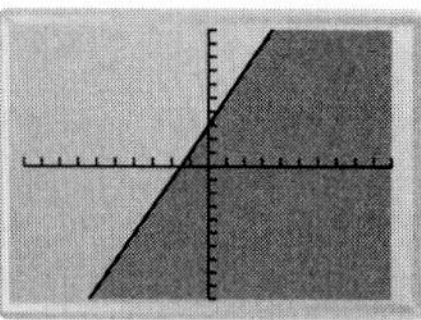

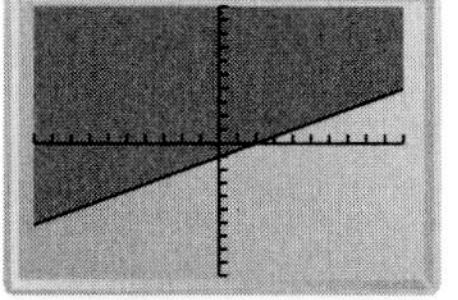

You can vary the darkness of the shading by entering an integer from 1 (light) to 8 (dark). Add a comma and the integer before pressing ENTER. The graph at the right has a darkness level of 2.

The graphing calculator does not make a distinction between a boundary line that is dotted (like *less than*) and a boundary line that is solid (like *less than or equal to*). You must decide if a boundary line should be solid or dotted when you sketch the inequality on your paper.

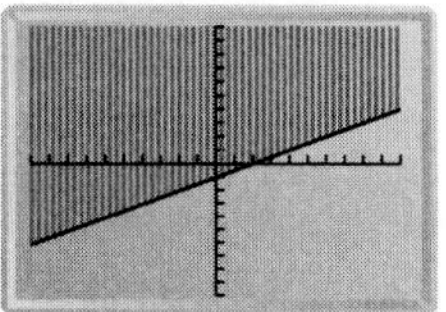

Shade(Y_1,Ymax,2)

Use a graphing calculator to graph each inequality. Sketch your graph. 1–8. See margin.

1. $y < x$
2. $y > 2x + 1$
3. $y \geq -x + 3$
4. $y \leq 5$
5. $x - y \geq 4$
6. $2x + 3y \leq 12$
7. $6x - 30y < 45$
8. $x - 2y \geq 50$

9. *Writing* What instructions would you need to change in Example (b) to graph $y > 0.5x - 1$ using Y_2 instead of Y_1?
Y= ▼ . 5 X,T,θ − 1 2nd DRAW 7 2nd Y-VARS 1 2) VARS 1 5 ENTER

6-6 Teaching Notes

PROBLEM OF THE DAY

A used car salesman buys a car for \$2500 and sells it for \$4327. He then buys a camper with two-thirds of his profit and sells the camper for twice what he paid for it. How much money did the salesman make during these transactions? \$3045

Problem of the Day is also available in Transparencies.

CONNECTING TO PRIOR KNOWLEDGE Invite students to suggest the minimum length and width of an envelope that they think a postal service would accept for delivery. Write their suggestions on the board. Summarize their suggestions by writing two linear inequalities such as $l \leq 10$ cm and $w \leq 6$ cm. Lead students to understand that real-world situations with many solutions can be modeled with a system of linear inequalities.

WORK TOGETHER

Have students work cooperatively in pairs or small groups, with one person as the recorder. Students may draw diagrams individually but have the group come to a consensus before recording the answer to a question.

THINK AND DISCUSS

In this chapter, students have learned several ways to solve a system of linear equations. In this lesson, they will learn the only way to solve systems of linear inequalities—by graphing.

Connections  Animal Studies . . . and more

6-6 Systems of Linear Inequalities

What You'll Learn

- Solving systems of linear inequalities by graphing

...And Why

To solve real-world problems that have many possible solutions, such as those in agriculture

What You'll Need

- graph paper
- graphing calculator

WORK TOGETHER

Work in pairs. Explore what happens when you graph lines or linear inequalities on the same coordinate plane.

1–2. See margin for graphs and reasoning.

1. Can two lines be drawn with the given intersection? Support each answer with a diagram or an explanation.
   a. a point yes b. a line yes c. a region no d. no intersection yes
2. Can the graphs of two linear inequalities be drawn with the given intersection? Support each answer with a diagram or an explanation.
   a. a point no b. a line yes c. a region yes d. no intersection yes
3. **Summarize** your findings in Questions 1 and 2. Compare the possible intersections of two lines with the possible intersections of the graphs of two linear inequalities. What do you notice? See margin.
4. Find the possible intersections for more than two lines. See margin.
5. Find the possible intersections for the graphs of more than two linear inequalities. See margin.

THINK AND DISCUSS

Two or more linear inequalities together form a **system of linear inequalities.** Here is an example of a system that describes the shaded region of the graph. Notice that there are two boundary lines.

System of Linear Inequalities

$x \geq 2$
$y < -1$

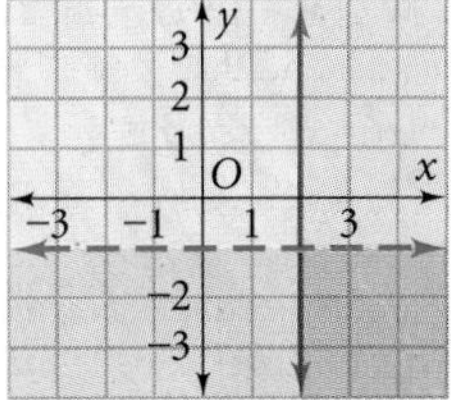

II | I | III | IV (quadrant diagrams with axes x, y, origin O)

6. You can describe all the points of a quadrant with a system of linear inequalities. Match each system with a quadrant.

   A. $x > 0$, $y > 0$ I **B.** $x > 0$, $y < 0$ IV **C.** $x < 0$, $y > 0$ II **D.** $x < 0$, $y < 0$ III

A **solution of a system of linear inequalities** makes each inequality in the system true. The graph of a system shows all of its solutions.

Lesson Planning Options

Prerequisite Skills

- Graphing linear inequalities

Assignment Options

Core 1–3, 5–16, 23–27
Extension 4, 17–22

Use Mixed Review to maintain skills.

Resources

Student Edition

Skills Handbook, pp. 567, 571
Extra Practice, p. 561
Glossary/Study Guide

Teaching Resources

Chapter Support File, Ch. 6
- Practice 6-6 (two worksheets)
- Reteaching 6-6
- Checkpoint

Classroom Manager 6-6
Glossary, Spanish Resources
Two-Year Algebra Handbook 6-6

Transparencies

2, 80, 83, 84, 85, 86

Question 6 Before beginning this question, have the students do the following.

- *Describe the graph of* $x = 0$. the y-axis
- *Describe the graph of* $y = 0$. the x-axis
- *Are the points on the axes part of any quadrant?* no

Be sure students understand that solving a system of inequalities means finding those points which are solutions to both inequalities.

VISUAL / AUDITORY LEARNING Draw a coordinate grid on the board. Label the four quadrants. Call out inequalities, and have students respond with the number(s) of the quadrant(s) in which the inequality is met. Samples:

- $x > 2$ quadrants I and IV
- $y < -3$ quadrants III and IV
- $x > 0$ *and* $y > 5$ quadrant I
- $x > -8$ *and* $y < 4$ all quadrants

Example 1 Relating to the Real World 

ALTERNATIVE METHOD Show the solution of this system of inequalities using the overhead projector. On one transparency draw the grid in black and the solution of $y \geq 80$ in red. On another transparency, draw the grid in black and the solution of $2x + 2y \leq 310$ in blue. Show each transparency separately and then show them super-imposed. The area that is both red and blue (purple) represents the graph of the solution. If an overhead projector is not available, show the solutions on the board using two colors.

Additional Examples

FOR EXAMPLE 1

Darell, who currently reads 50 words per min, is planning to enroll in a speed reading course. The course promises an increase in reading rate of between 7 and 12 words per min for each week of study. Graph the following equations and predict Darell's reading rate at the end of 6 weeks. $r \leq 50 + 12t$, $r \geq 50 + 7t$
between 92 and 122 words per min

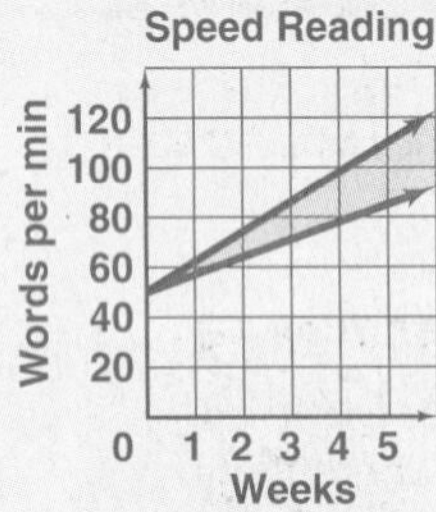

Discussion: *What would probably happen to this graph after many weeks pass?*

FOR EXAMPLE 2

Solve the system by graphing.
$-2y < 4x$
$2x + 3y \leq 4$

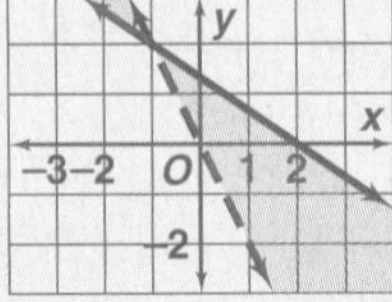

Example 1 Relating to the Real World 

Animal Studies A zoo keeper wants to fence a rectangular pen for goats. The length of the pen should be at least 80 ft, and the distance around it should be no more than 310 ft. What are the possible dimensions of the pen?

Define x = width of the pen
y = length of the pen

Relate	The length	is at least	80 ft.	The perimeter	is no more than	310 ft.
Write	y	$\geq$	80	$2x + 2y$	$\leq$	310

Use slope-intercept form to graph.

$y \geq 80$
$m = 0$
$b = 80$

Test (0, 1).
$y \geq 80$
$(1) \geq 80$ False

Shade above.

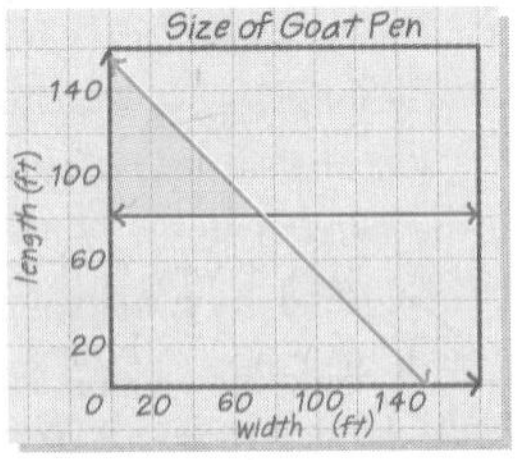

Size of Goat Pen
(length (ft): 20, 60, 100, 140; width (ft): 0, 20, 60, 100, 140)

Use intercepts to graph.

$2x + 2y \leq 310$
(155, 0)
(0, 155)

Test (0, 0).
$2(0) + 2(0) \leq 310$
$0 \leq 310$ True

Shade below.

The solutions are all the points in the shaded region above $y = 80$ but below $2x + 2y \leq 310$.

7. Give three possible dimensions (length and width) for the pen. How many solutions does this system have? **7. Samples: 20 ft by 80 ft, 20 ft by 120 ft, 70 ft by 85 ft; infinitely many solutions**
8. Why is the solution region shown only in Quadrant I? **Length and width cannot be negative.**
9. *Graphing Calculator* Use the *Shade* feature to graph the system of linear inequalities in Example 1. **See margin.**
10. *Try This* Solve each system by graphing.
    a. $y \leq 2x + 3$, $y \geq x - 1$
    b. $y > x$, $x \leq 3$
    c. $x + y > -2$, $x - y > 2$
    10. a–c. See margin.
11. *Critical Thinking* How can you decide whether or not points on boundary lines of a solution region are part of the solution of a system? **These points must be on the boundary line of a $\geq$ or $\leq$ inequality. They also must be within the shaded region for the second inequality.**

MAKING CONNECTIONS The alpaca is a domesticated llama and is bred mostly for its long, silky wool. The Incas made robes for royalty from alpaca hair. Today, about 10 million kg of alpaca hair is produced annually for wool textiles.

EXTENSION Question 7 Ask: *Is there any solution that is a square?* **No; the line $x = y$ does not intersect the region.**

Example 2

Question 13 Ask: *Without graphing, how could you determine that the lines are parallel?* **Write the equations for the two lines in slope-intercept form. They have the same slope and different y-intercepts.**

Exercises ON YOUR OWN

STANDARDIZED TEST TIP **Exercise 1** Instead of graphing both equations, use substitution to find the answer choice for which both equations produce a "true" result.

ERROR ALERT! **Exercises 5–16** Students may have difficulty deciding which region to shade. **Remediation:** Have them first show the graph of each inequality by using arrows on the boundary line pointing toward the region which represents the solution to that inequality. Then have them shade the appropriate region for the intersection.

Some systems of linear inequalities do not have a solution region. The graphs of the inequalities might intersect in a line or not at all.

Example 2

Solve the system by graphing. $4y \geq 6x$
$-3x + 2y \leq -6$

Use slope-intercept form to graph.

$4y \geq 6x$

$y \geq \frac{3}{2}x$

Test (0, 1).
Shade above.

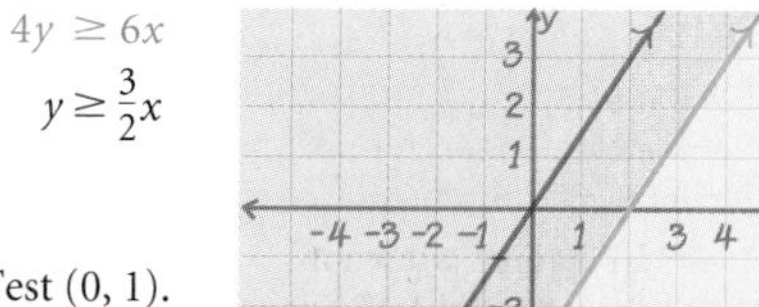

Use intercepts to graph.

$-3x - 2y \leq -6$

$(0, -3)$

$(2, 0)$

Test (0, 0).
Shade below.

Since the shaded regions do not overlap, the system has no solution.

12. Explain why (0, 0) was not tested when $4y \geq 6x$ was graphed. **(0, 0) is on the boundary line.**
13. How are the boundary lines in Example 2 related? **The lines are parallel.**
14. **a.** *Open-ended* Write another system of two linear inequalities that has no solution. **Sample: $y > 2x + 1$, $y < 2x - 3$**
    b. Must the boundary lines be parallel? Explain. **Yes; if the boundary lines are not parallel the solution regions overlap.**
15. *Critical Thinking* Write a system of linear inequalities in which the solution is a line. **Sample: $y \geq x$, $y \leq x$**

PROBLEM SOLVING HINT

For Questions 14 and 15, first draw a diagram. Then write the inequalities.

Exercises ON YOUR OWN

1. *Standardized Test Prep* Which of these points is a solution of the system $y \leq x + 5$ and $y + x > 3$? **C**
   A. (0, 0) **B.** (−1, 4) **C.** (3, 3) **D.** (−2, 6) **E.** (2, 0)
2. Which system is represented in the graph at the right? **D**
   A. $x + y \leq 3$, $y > x - 3$
   B. $x + y > 3$, $y \leq x - 3$
   C. $x + y \geq 3$, $y < x - 3$
   D. $x + y < 3$, $y \geq x - 3$

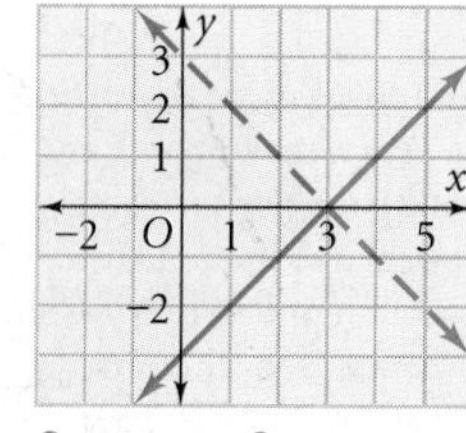

3. *Critical Thinking* Without graphing, explain why the point where the boundary lines intersect in the system $2x + y > 2$ and $x - y \geq 3$ is *not* a solution of the system. **This point is on the boundary line of $2x + y > 2$, so it is not a solution of $2x + y > 2$.**
4. *Writing* Write a problem that can be solved using the system of inequalities at the right. **Sample: You have 100 ft of fencing to build a pen for a goat. You have some space on a 30 ft long lawn between a road and a garden. What are possible dimensions of the pen?**
   $2l + 2w \leq 100$
   $l \leq 30$

Technology Options

Students can use a graphing calculator to find solutions for Exercises 5–16. Graphing calculators or geometry software may help students find the answers in Exercises 23–26.

Prentice Hall Technology

Calculator • Graphing Calculator Handbook: Procedure 8

Software • Secondary Math Lab Toolkit • Computer Item Generator 6-6

CD-ROM Multimedia Algebra Lab 6

Internet • See Prentice Hall site. (http://phschool.com)

ALTERNATIVE ASSESSMENT Exercises 5–16 Have students graph each equation on a separate overhead transparency. Then combine the two paired transparencies on the overhead projector. This will help you assess students' understanding of systems of linear inequalities.

DIVERSITY Exercise 17 Lead students to brainstorm for other ways that people could fasten materials together to build structures. Encourage students to research materials and methods that were used before metal nails were common. Samples: leather strips, wooden pegs, vines, dovetail joints.

GEOMETRY Exercise 25 The formula for the area of a trapezoid is $A = \frac{1}{2}h(b_1 + b_2)$.

Chapter Project **FIND OUT BY GRAPHING** Students use the ticket prices they calculated on page 293 to graph a system of linear inequalities that show the total amount received for the tickets. Check students' progress on the project by having each student write a short progress report telling what has been done so far and their plans for completion.

page 295 Work Together

1a.

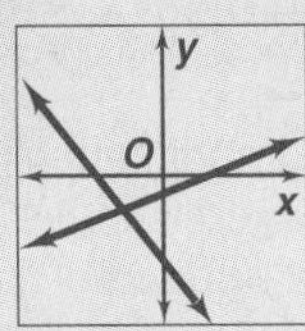

b.

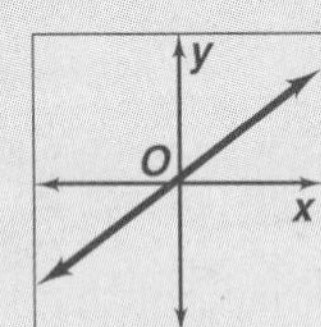

c. A line cannot contain a region.

d.

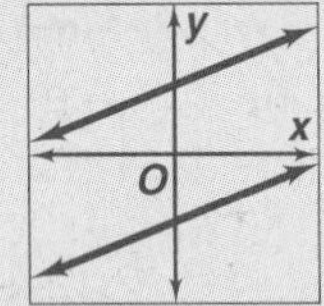

2a. Two regions cannot intersect in a point.

b.

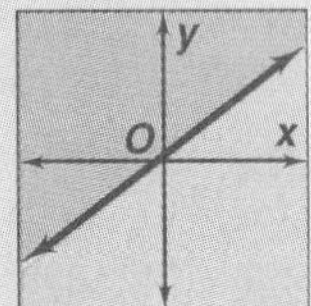

2c, 2d, 3–5, Think and Discuss, 9, 10a–c, On Your Own, 5–16, 17a, 17b, 18a, 23–26, 27a–c, Mixed Review, 36b, 36c, Checkpoint, 1–4, 6b. See back of book.

298

Solve each system of linear inequalities by graphing. 5–16. See margin for graphs.

5. $x \geq 2$, $y < 4$

6. $y > 3$, $x < -1$

7. $y \leq x$, $y \geq x + 1$ no solution

8. $y > 4x + 2$, $y \leq 4$

9. $y \leq 2x + 2$, $y < -x + 1$

10. $y \geq -2x + 1$, $y < x + 2$

11. $x + y \leq 6$, $x - y < 1$

12. $y - 3x < 6$, $y > 3x + 9$ no solution

13. $x > y$, $y > 0$

14. $x + y \leq 5$, $x \geq 1$

15. $y < \frac{1}{2}x$, $y > \frac{1}{2}x - 3$

16. $-x - y \leq 2$, $y - 2x > 1$

17. *Construction* A contractor has at most $33 to spend on nails for a project. The contractor needs at least 9 lb of finish nails and at least 12 lb of common nails. How many pounds of each type of nail should the contractor buy? a–b. See margin.
   a. Write a system of three inequalities that describes this situation.
   b. Graph the system to show all possible solutions.
   c. Name a point that is a solution of the system. Sample: (20, 20)
   d. Name a point that is *not* a solution of the system. Sample: (40, 25)

Open-ended **Write a system of linear inequalities with the given characteristics.** 18–21. Samples are given.

18. (0, 0) is a solution. $x \geq 0$, $y < 1$

19. Solutions are only in Quadrant II. $x < -1$, $y > 0$

20. There is no solution. $y > x$, $y < x - 1$

21. The solution region is triangular. $x + y < 6$, $y > 2x$, $y > -2x$

22. a. Solve the system of three inequalities at the right by graphing. See margin.
   b. **Verify** your solution by testing a point from the overlapping region in all three inequalities.
   (0, 0); $0 \leq 4$ ✓ $0 < 0 + 2$ ✓ $0 + 2(0) \geq -2$ ✓

   $x \leq 4$, $y < x + 2$, $x + 2y \geq -2$

Geometry **For the solution region of each system of linear inequalities, (a) describe the shape, (b) find the vertices, and (c) find the area.** 23–26. See margin for graphs.

23. $y \geq \frac{1}{2}x + 1$, $y \leq 2$, $x \geq -4$
   a. triangle b. (−4, −1), (−4, 2), (2, 2) c. 9 sq. units

24. $x \geq 1$, $x \leq 5$, $y \geq -1$, $y \leq 3$
   a. square b. (1, −1), (1, 3), (5, 3), (5, −1) c. 16 sq. units

25. $x \geq 0$, $x \leq 2$, $y \geq -4$, $y \leq -x + 2$
   a. trapezoid b. (0, −4), (0, 2), (2, 0), (2, −4) c. 10 sq. units

26. $x \geq 2$, $y \geq -3$, $x + y \leq 4$
   a. triangle b. (2, 2), (7, −3), (2, −3) c. 12.5 sq. units

27. *Shopping* Suppose you receive a $50 gift certificate to the Music and Books store. You want to buy some books and at least one CD. How can you spend your gift certificate on *x* paperbacks and *y* CDs?
   a. Write a system of linear inequalities that describes this situation.
   b. Graph the system to show possible solutions to this problem.
   c. What purchase does the ordered pair (2, 6) represent? Is it a solution to your system? Explain. d. 5 CDs, no paperbacks
   d. Find a solution in which you spend almost all of the gift certificate.
   e. What is the greatest number of paperbacks you can buy and still buy one CD? 6 paperbacks
   a–c. See margin.

Cityside Music and Books
All CDs $9.99
All books $5.99

Exercises MIXED REVIEW

GETTING READY FOR LESSON 6-7 In this preview students evaluate an equation for a given point. This will be used in linear programming.

JOURNAL Encourage students to include graphs with their journal entries.

Wrap Up

THE BIG IDEA Ask students how to describe the process for solving a system of linear inequalities.

RETEACHING ACTIVITY Students solve systems of linear inequalities by graphing. They use two colored markers to show the solutions for a system of equations. (Reteaching worksheet 6-6)

Exercises CHECKPOINT

In this Checkpoint, your students will have an opportunity to assess their own progress on the topics covered in Lessons 6-4 through 6-6.

Chapter Project — *Find Out by Graphing*

In the Find Out question on page 293, you found two ticket prices. Each price covers the cost of the dance under certain conditions. Decide what the ticket price should be. Plan for between 200 and 300 people. Graph a system of linear inequalities to show the total amount received from tickets.

Exercises MIXED REVIEW

Solve each equation.

28. $5m + 4 = 8m - 2$ 2 **29.** $5(t + 1) = 10$ 1 **30.** $4x = 2x + 5$ 2.5 **31.** $6h - 11 = 13$ 4

32. $3p = -\frac{3}{4}p + 5$ $\frac{4}{3}$ **33.** $-k - 7 = -3k + 1$ 4 **34.** $\frac{3c - 1}{4} = \frac{5}{2}$ $3\frac{2}{3}$ **35.** $\frac{1}{2}(t - 8) = 7$ 22

36. a. Identify the independent and dependent variables.
b. Graph the data.
c. What scales did you use and why?
a. price, sales tax
b–c. See margin.

Price	Sales Tax
$1.00	$.05
$3.00	$.15
$4.00	$.20
$7.00	$.35

FOR YOUR JOURNAL

Compare the processes of graphing a system of linear inequalities and graphing a system of linear equations. How are they similar? How are they different?

Getting Ready for Lesson 6-7

Evaluate each formula for the given point.

37. $B = 2x + 5y$; (6, 10) 62 **38.** $C = x + 6y$; (100, 550) 3400

39. $P = 6x + 2y$; (200, 75) 1350 **40.** $P = 2l + 2w$; (30, 18) 96

Exercises CHECKPOINT

Solve each system of linear inequalities. 1–4. See margin.

1. $y < -2x + 5$
 $y > 3x - 1$

2. $y \geq 2x - 1$
 $x \geq -5$

3. $y < 0.5x + 3$
 $y \geq -x + 2.5$

4. $3x + 2y \leq 12$
 $x - y < 10$

5. *Standardized Test Prep* Which of these points is a solution of the system $y \geq 4x - 1$ and $2x + 3y < 6$? D
A. (3, 0) **B.** (0, 4) **C.** (4, 1) **D.** (−3, 1) **E.** (3, −4)

6. You are going out for pizza! b. See margin.
a. Write a system of equations for the cost of a large pizza at each restaurant. $p = 7 + 0.75t$; $p = 8 + 0.5t$
b. Solve the system of equations. Interpret your solution.
c. *Open-ended* Where will you go for pizza? Explain your reasons. Sample: Tony's; unless you want more than 4 toppings, Tony's is a better deal.

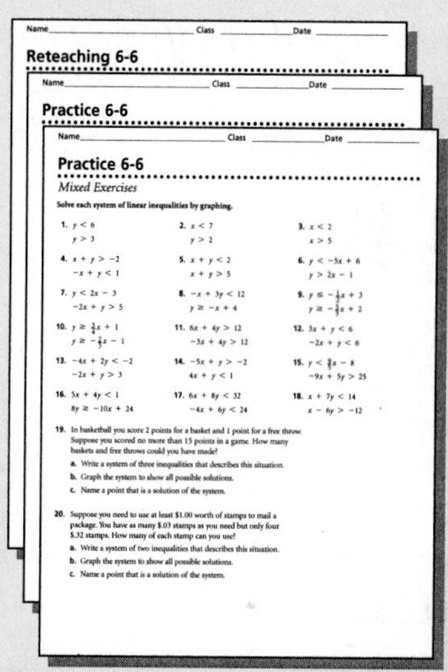

Name Class Date
Reteaching 6-6

Name Class Date
Practice 6-6

Name Class Date
Practice 6-6
Mixed Exercises

Solve each system of linear inequalities by graphing.

1. $y < 6$, $y > 3$ 2. $x < 7$, $y > 2$ 3. $x < 2$, $x > 5$
4. $x + y > -2$, $-x + y < 1$ 5. $x + y < 2$, $x + y > 5$ 6. $y < -5x + 6$, $y > 2x - 1$
7. $y < 2x - 3$, $-2x + y > 5$ 8. $-x + 3y < 12$, $y \geq -x + 4$ 9. $y \leq -\frac{1}{2}x + 3$, $y \geq -\frac{2}{3}x + 2$
10. $y \geq \frac{3}{4}x + 1$, $y \geq -\frac{2}{3}x - 1$ 11. $6x + 4y > 12$, $-3x + 4y > 12$ 12. $3x + y < 6$, $-2x + y < 6$
13. $-4x + 2y < -2$, $-2x + y > 3$ 14. $-5x + y > -2$, $4x + y < 1$ 15. $y < \frac{9}{5}x - 8$, $-9x + 5y > 25$
16. $5x + 4y < 1$, $8y \geq -10x + 24$ 17. $6x + 8y < 32$, $-4x + 6y < 24$ 18. $x + 7y < 14$, $x - 6y > -12$

19. In basketball you score 2 points for a basket and 1 point for a free throw. Suppose you scored no more than 15 points in a game. How many baskets and free throws could you have made?
a. Write a system of three inequalities that describes this situation.
b. Graph the system to show all possible solutions.
c. Name a point that is a solution of the system.

20. Suppose you need to use at least $1.00 worth of stamps to mail a package. You have as many $.03 stamps as you need but only four $.32 stamps. How many of each stamp can you use?
a. Write a system of two inequalities that describes this situation.
b. Graph the system to show all possible solutions.
c. Name a point that is a solution of the system.

Lesson Quiz

Lesson Quiz is also available in Transparencies.

Solve each system of linear inequalities by graphing.

1. $y \leq -1$
 $x \leq -1$

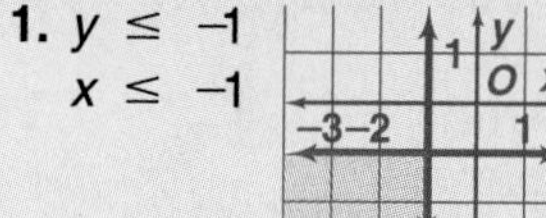

2. $y < 3x$
 $x \geq 2$

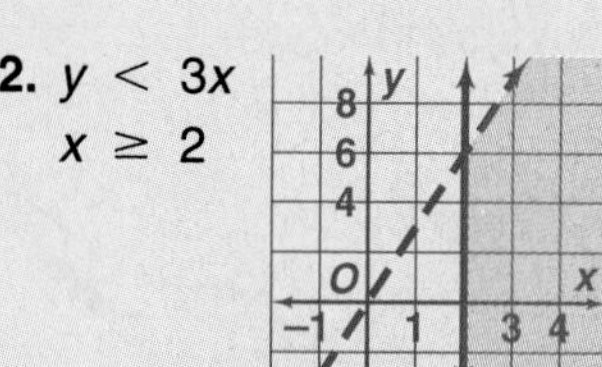

6-7 Teaching Notes

PROBLEM OF THE DAY

On a blueprint, the scale indicates that 2 cm represents 5 ft. What are the actual dimensions of a room that is 4.8 cm by 6 cm on the blueprint? 12 ft × 15 ft

Problem of the Day is also available in Transparencies.

CONNECTING TO PRIOR KNOWLEDGE An airline describes the maximum allowed size of a suitcase using the formula $l + w + h \leq 150$ in. Discuss what the formula means. Encourage students to measure various objects in the room to see whether the objects meet the airline requirements.

WORK TOGETHER

In this activity, students use job related skills such as time-management analysis and organization.

DIVERSITY Question 1 Some students may have no time available for a part-time job because of other responsibilities. Suggest that they make up a list for an imaginary person.

THINK AND DISCUSS

Example 1

Some students may ask for an explanation of why the maximum and minimum values occur at the vertices.

Lesson Planning Options

Prerequisite Skills

- Graphing linear inequalities
- Understanding maximum and minimum values

Assignment Options

Core 1–11
Extension 12–15

Use Mixed Review to maintain skills.

Resources

Student Edition

Skills Handbook, pp. 567, 571
Extra Practice, p. 561
Glossary/Study Guide

Teaching Resources

Chapter Support File, Ch. 6
- Practice 6-7 (two worksheets)
- Reteaching 6-7
- Alternative Activity 6-7

Classroom Manager 6-7
Glossary, Spanish Resources
Two-Year Algebra Handbook 6-7

Transparencies
2, 81

Connections Time Management . . . and more

6-7 Concepts of Linear Programming

What You'll Learn

- Solving linear programming problems

...And Why

To investigate real-world situations, such as time management

WORK TOGETHER

Suppose you are offered a part-time job. You wonder how much time you have available to work. You can use mathematics to help you organize your thoughts and make a good decision.

Work with a partner. 1–2. Check students' work.

1. a. Write the different ways you spend your time during a week.
   b. Organize your list into no more than ten categories.
2. Make a personal calendar for the last week.
   a. Assign time to the categories from Question 1.
   b. How much time do you have available to work at a part-time job?
   c. Discuss what you could or could not give up in your schedule.

THINK AND DISCUSS

You can answer questions like those above by using a process called linear programming. **Linear programming** identifies conditions that make a quantity as large as or as small as possible. The variables used in the equation for the quantity have restrictions. The maximum and minimum values of the quantity occur at vertices of the graph of the restrictions.

Example 1

Use linear programming. Find the values of x and y that maximize the quantity.

Restrictions $\begin{cases} x + y \leq 8 \\ x \geq 0 \\ y \geq 3 \end{cases}$

Equation $Q = 3x + 2y$

Step 1
Graph the restrictions.

Step 2
Find coordinates of each vertex.

Step 3
Evaluate Q at each vertex.

Vertex	$Q = 3x + 2y$	
$E(0, 3)$	$Q = 3(0) + 2(3) = 6$	
$F(5, 3)$	$Q = 3(5) + 2(3) = 21$	← maximum value of Q
$G(0, 8)$	$Q = 3(0) + 2(8) = 16$	

The maximum value 21 occurs when $x = 5$ and $y = 3$.

3. Find the values of x and y that minimize the quantity in Example 1. (0, 3)

On the same graph that shows the restrictions, show the graphs of the equation $Q = 3x + 2y$ for the values of Q that occur at the vertices.

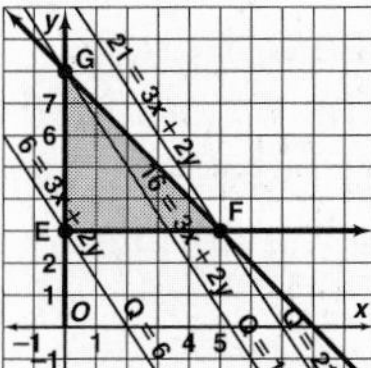

Notice that all points that satisfy the restrictions lie between the two outermost parallel lines, the maximum and the minimum.

Example 2 **Relating to the Real World** 

Marta's equation $B = 2x + y$ represents her own feelings about the relative importance of homework and work. Ask: *What equation would you write?* **Answers may vary. Sample: $B = x + 2y$ (I value work hours more than homework hours.)**

Example 3 **Relating to the Real World** 

AUDITORY LEARNING Have students read the question out loud to you one sentence at a time. After each sentence, repeat to students what they just read while writing the corresponding math inequality on the board. Help students make the connections between the sentences and the math language.

---

4. Find the value of the equation in Example 1 at each point.
a. (1, 3) 9 **b.** (4, 3) 18 **c.** (4, 4) 20 **d.** (1, 7) 17 **e.** (3, 4) 17

In the Work Together activity, you listed some restrictions you have on your time in relation to a part-time job. In the next example one student finds the best way to use her time.

Example 2 **Relating to the Real World** 

Time Management Marta plans to start a part-time job. Here are the restrictions she has found on her time for homework hours x and job hours y.

Restriction	Inequality
She has no more than 24 h/wk for homework and a job.	$x + y \leq 24$
The boss wants her to work at least 6 h/wk.	$y \geq 6$
She spends 10–15 h/wk on homework.	$x \geq 10$ $x \leq 15$

She decides that homework hours are twice as valuable as work hours. Find the best way B for Marta to split her time between homework and the job using the equation $B = 2x + y$.

Step 1
Graph the restrictions.

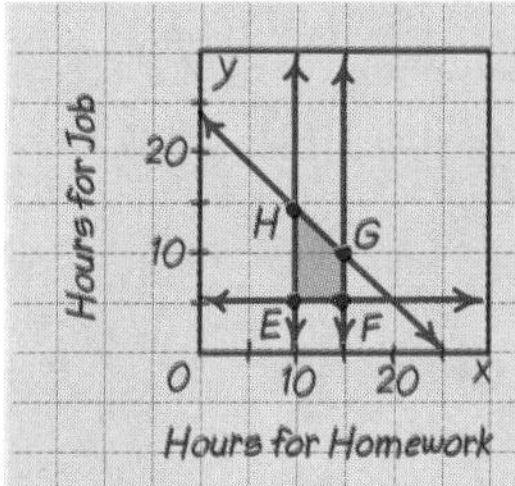

Step 2
Find the coordinates of each vertex.

Step 3
Evaluate B at each vertex.

Vertex	$B = 2x + y$	
$E(10, 6)$	$B = 2(10) + 6 = 26$	
$F(15, 6)$	$B = 2(15) + 6 = 36$	
$G(15, 9)$	$B = 2(15) + 9 = 39$	← maximum value of B
$H(10, 14)$	$B = 2(10) + 14 = 34$	

The best way for Marta to split her time each week is to spend 15 h on homework and 9 h at her job.

5. What does the point (10, 14) mean in terms of homework and job time? **Marta spends 10 h on homework and works for 14 h.**

Additional Examples

FOR EXAMPLE 1

Use linear programming to find the values of x and y that maximize the equation.

Restrictions: $y \geq 1$
$y \leq -2x$
$x \geq -2$

Equation: $Q = 2x - y$

Show your graph, vertex points, and test values. **max $(-\frac{1}{2}, 1)$**

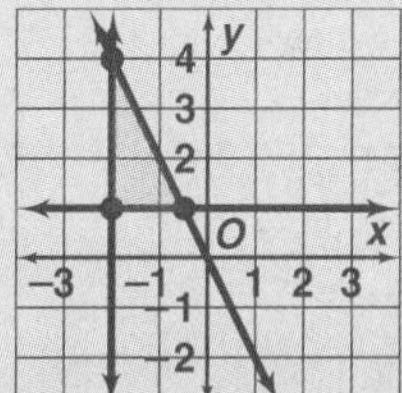

FOR EXAMPLE 3

You decide you will need at least 25 gal of paint for an upcoming project. At least 5 gal must be blue, and at least 10 gal must be white. If blue paint costs \$5/gal and white paint costs \$7/gal, how much of each type should you buy to minimize cost? **10 gal white, 15 gal blue**

STEP 3 Ask students:

- *What quadrant or quadrants will contain the graph?* only quadrant I
- *Why is this graph limited to quadrant I?* There cannot be a negative number of fish.

VISUAL / TACTILE LEARNING Exercises 4–7 Have students draw lines using colored pencils. They should then do the shading with highlighting pens that match the color of the pencils.

Exercises ON YOUR OWN

Exercises 4–7 Point out that the first step is to identify the vertices of the points of intersection of the boundary lines.

Exercises 8–11 Students must begin by graphing the inequalities that represent the restrictions.

ERROR ALERT! If students' graphs are not drawn accurately, they may incorrectly identify the vertices. **Remediation:** Have them check each vertex by substituting the ordered pair in the equations for the two boundary lines that intersect. If the point does not represent the solution, they might try another method (substitution or elimination) for solving the system of linear equations.

ALTERNATIVE ASSESSMENT **Exercise 12** After students have completed Exercise 12, ask them to find the maximum and minimum values for the equation $3x + 4y = C$. This will help you assess students' understanding of linear programming.

Technology Options

Students may use graphing software or graphing calculators to solve Exercises 4–11.

Prentice Hall Technology

Software • Secondary Math Lab Toolkit • Computer Item Generator 6-7

Internet • See the Prentice Hall site (http://www.phschool.com)

6a. Maximum; you want the greatest possible profit.
b. Minimum; you want the least expense.

6. Critical Thinking For each situation, would the best solution be a *maximum value* or a *minimum value*? Explain.
   a. You are selling tomatoes and beans from your garden. You want to determine how much of each to grow for the most profit.
   b. Suppose you manage a grocery store. You can buy tomatoes from two different farmers. You consider the price of the tomatoes and transportation costs. You must decide which supplier to use.

To solve linear programming problems, you must be able to write the inequalities for the restrictions and write the equation.

Example 3 Relating to the Real World

Halibut $4 each

Flounder $3 each

Business A seafood restaurant owner orders at least 50 fish. He cannot use more than 30 halibut or more than 35 flounder. How many of each fish should he use to minimize his cost?

Step 1 Write inequalities to describe the restrictions.

Define x = number of halibut used
y = number of flounder used

Relate	Write
He needs at least 50 fish.	$x + y \geq 50$
He cannot use more than 30 halibut.	$x \leq 30$
He cannot use more than 35 flounder.	$y \leq 35$

Step 2 Write the equation.

Define C = cost of fish

Relate cost is $4 for each halibut and $3 for each flounder

Write $C = 4x + 3y$

Step 3
Graph the restrictions.

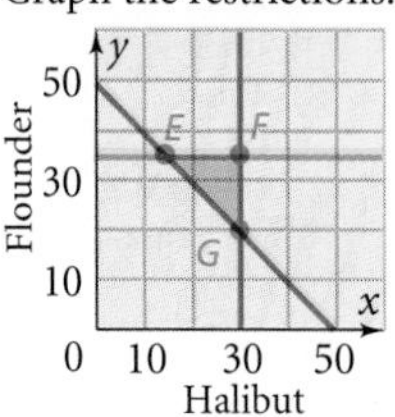

Step 4
Find the coordinates of each vertex.

Step 5
Evaluate C at each vertex.

Vertex	$C = 4x + 3y$	
$E(15, 35)$	$C = 4(15) + 3(35) = 165$	← minimum value of C
$F(30, 35)$	$C = 4(30) + 3(35) = 225$	
$G(30, 20)$	$C = 4(30) + 3(20) = 180$	

The seafood restaurant owner should buy 15 halibut and 35 flounder to minimize his cost.

7. Try This Suppose the restaurant owner changes his order to at least 40 fish. How many of each kind should he use to minimize his cost?
5 halibut; 35 flounder

OPEN-ENDED Exercise 12 A trapezoid is a four-sided figure with one pair of parallel sides. Have pairs of students check each other's systems.

Exercise 12 Have students think about this real-life situation. Ask students: *What kinds of factors might determine the restrictions for this problem?* **Answers may vary. Sample: Too much salt might affect the town's water and plant life. Too little salt might not be effective.**

ESL Exercise 14 Explain to students that the word *line* refers to a certain type or style of model airplane. Ask students to name the two lines of airplanes the toy manufacturer produces.

Exercises MIXED REVIEW

GETTING READY FOR LESSON 6-8 In this preview, students identify linear, quadratic, and absolute-value functions by looking at the equations. This skill will be useful in the next lesson on graphing systems of nonlinear equations.

Wrap Up

THE BIG IDEA Ask students to explain what is meant by linear programming.

RETEACHING ACTIVITY Students will solve linear programming problems. They will graph the restrictions of each equation, find the coordinates of each vertex and evaluate B at each vertex. (Reteaching worksheet 6-7)

Exercises ON YOUR OWN

Evaluate each equation at the points given. Which point gives the maximum value? the minimum value?

1. $Q = 3x + 5y$
   (8, 0), (4, 4), (3, 5), (0, 6)
   24, 32, 34, 30; (3, 5); (8, 0)
2. $B = 40x + 20y$
   (0, 0), (0, 40), (15, 10), (25, 0)
   0, 800, 800, 1000; (25, 0); (0, 0)
3. $A = 35x + 10y$
   (0, 0),(0, 10), (0, 40), (20, 10)
   0, 100, 400, 800; (20, 10); (0, 0)

Find the values of x and y that maximize or minimize each quantity for each graph.

4. Maximum for $P = x + y$ (4, 6)

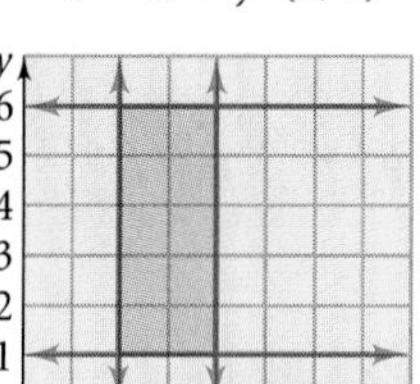

5. Maximum for $P = 3x + 2y$ (6, 0)

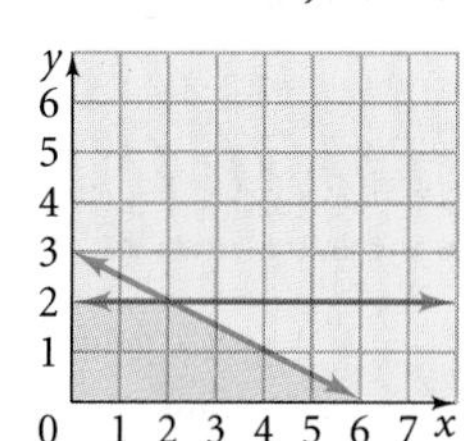

6. Minimum for $C = 2x + 5y$ (3, 0)

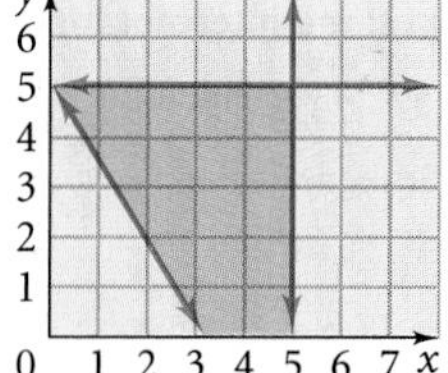

7. Minimum for $C = \frac{1}{2}x + y$ (4, 0)

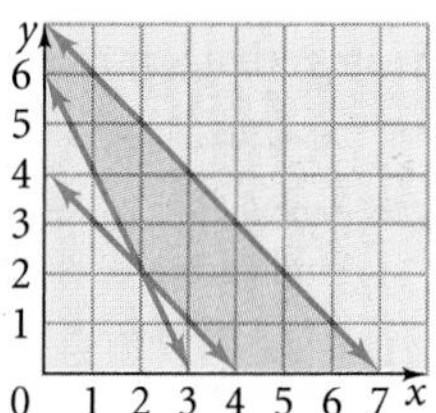

Evaluate the equation to find minimum and maximum values.

8. $x \geq 2$ 22; 49
   $x \leq 5$
   $y \geq 3$
   $y \leq 6$
   $A = 5x + 4y$
9. $x \geq 0$ 0; 36
   $y \geq 0$
   $x + y \leq 12$
   $D = 3x + y$
10. $y \geq -x + 4$ 8; 30
    $y \leq -2x + 10$
    $x \geq 0$
    $y \geq 0$
    $P = 2x + 3y$
11. $y \leq -x + 8$ 1; 8
    $y \geq 2x - 2$
    $x \geq 0$
    $y \geq 0$
    $Q = x + y$

12. **Open-ended** Write a system of restrictions that form a trapezoid. Sample: $x \leq 5, y \leq 5, x + y \geq 1, x + y \leq 8$
13. **Environment** A town is trying to find the best mix of sand and salt for treating icy roads. One consideration is cost, which they want to minimize. Sand costs $8 per ton, and salt costs $20 per ton. Write the equation for minimizing cost. $c = 8d + 20t$
14. **Manufacturing** A toy manufacturer wants to minimize her cost for producing two lines of toy airplanes. Because of the supply of materials, no more than 40 Flying Bats can be built each day, and no more than 60 Flying Falcons can be built each day. There are enough workers to build at least 70 toy airplanes each day.
    a. Write the inequalities for the restrictions. $b \leq 40; f \leq 60; f + b \geq 70; b \geq 0; f \geq 0$
    b. It costs $12 to manufacture a Flying Bat and $8 to build a Flying Falcon. Write an equation for the cost of manufacturing the toy airplanes. $c = 12b + 8f$
    c. Use linear programming to find how many of each toy airplane should be produced each day to minimize cost. 10 Flying Bats and 60 Flying Falcons
    d. **Critical Thinking** What else should the manufacturer consider before deciding how many of each toy to manufacture each day? Answers may vary. Sample: She needs to consider the profit she makes on each toy.

pages 303–304 On Your Own

15c. 400 ft^2 at Location A and 100 ft^2 at Location B

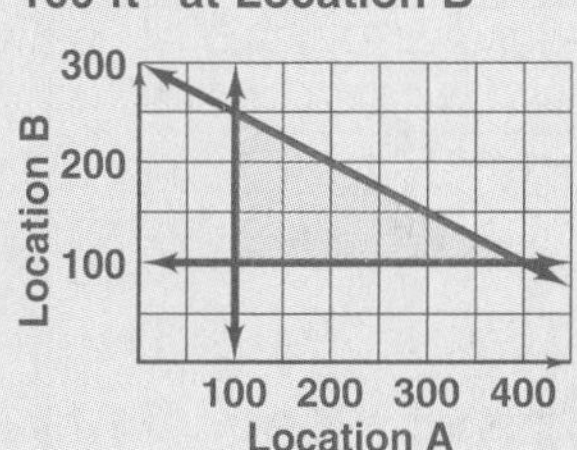

Algebra at Work

For further information about the training and skills necessary to manage a business students can:

- make an appointment and speak with the manager at any business establishment in your town.
- visit with a manager of a fast food restaurant .
- research small businesses through an internet address http://www.yahoo.com/business_and_economy/

Have students:

- invite a representative of the Small Business Association to speak to the class.
- go to the library to collect copies of The Wall Street Journal as a source of interesting articles on business.
- investigate ways in which math is used on the job—statistical process control, cycle time, use of charts and graphs.

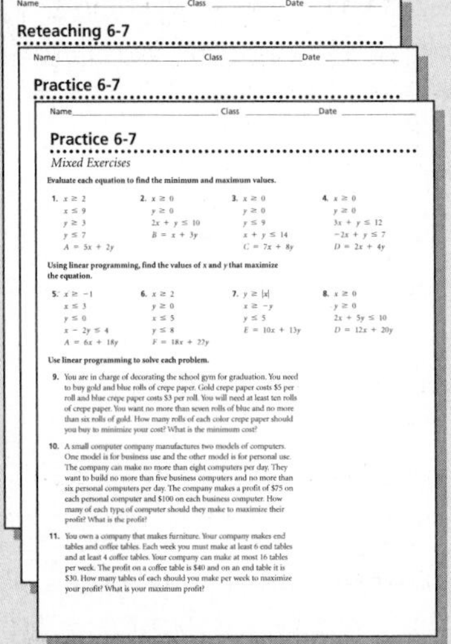

Name_______________ Class _________ Date _________

Reteaching 6-7

Name_______________ Class _________ Date _________

Practice 6-7

Name_______________ Class _________ Date _________

Practice 6-7

Mixed Exercises

Evaluate each equation to find the minimum and maximum values.

1. $x \geq 2$, $x \leq 9$, $y \geq 3$, $y \leq 7$, $A = 5x + 2y$
2. $x \geq 0$, $y \geq 0$, $2x + y \leq 10$, $B = x + 3y$
3. $x \geq 0$, $y \geq 0$, $y \leq 9$, $x + y \leq 14$, $C = 7x + 8y$
4. $x \geq 0$, $y \geq 0$, $3x + y \leq 12$, $-2x + y \leq 7$, $D = 2x + 4y$

Using linear programming, find the values of x and y that maximize the equation.

5. $x \geq -1$, $x \leq 3$, $y \leq 0$, $x - 2y \leq 4$, $A = 6x + 18y$
6. $x \geq 2$, $y \geq 0$, $x \leq 5$, $y \leq 8$, $F = 18x + 22y$
7. $y \geq |x|$, $x \geq -y$, $y \leq 5$, $E = 10x + 13y$
8. $x \geq 0$, $y \geq 0$, $2x + 5y \leq 10$, $D = 12x + 20y$

Use linear programming to solve each problem.

9. You are in charge of decorating the school gym for graduation. You need to buy gold and blue rolls of crepe paper. Gold crepe paper costs $5 per roll and blue crepe paper costs $3 per roll. You will need at least ten rolls of crepe paper. You want no more than seven rolls of blue and no more than six rolls of gold. How many rolls of each color crepe paper should you buy to minimize your cost? What is the minimum cost?
10. A small computer company manufactures two models of computers. One model is for business use and the other model is for personal use. The company can make no more than eight computers per day. They want to build no more than five business computers and no more than six personal computers per day. The company makes a profit of $75 on each personal computer and $100 on each business computer. How many of each type of computer should they make to maximize their profit? What is the profit?
11. You own a company that makes furniture. Your company makes end tables and coffee tables. Each week you must make at least 6 end tables and at least 4 coffee tables. Your company can make at most 16 tables per week. The profit on a coffee table is $40 and on an end table it is $30. How many tables of each should you make per week to maximize your profit? What is your maximum profit?

Lesson Quiz

Lesson Quiz is also available in Transparencies.

Using the given restrictions, evaluate the equation and find its maximum and minimum values.

1. Equation:
   $A = 3x + 2y$
   Restrictions:
   $y \leq 3$
   $x \leq -1$
   $y < 3x + 10$
   max 9, min −1

2. Equation:
   $B = x + 2y$
   Restrictions:
   $y \geq -1$
   $x \geq 1$
   $y < -2x + 4$
   max 5, min −1

15. ***Business*** A computer company has budgeted $6000 to rent display space at two locations for a new line of computers. Each location requires a minimum of 100 ft^2.
    a. ***Writing*** Explain what each inequality represents.
    Restrictions: $10x + 20y \leq 6000$, $x \geq 100$, $y \geq 100$
    b. Write an equation for the potential number of customers.
    c. Use linear programming to find the amount of space to rent at each location to maximize the number of potential customers.

Rental Locations

Location	Cost	Potential
A	$10/ft^2	30 customers/ft^2
B	$20/ft^2	40 customers/ft^2

a. Total cost must not exceed $6000; the company must rent at least 100 ft^2 at each location.
b. $N = 30x + 40y$ c. See margin.

Exercises MIXED REVIEW

Write each linear equation in slope-intercept form.

16. $5y = 6x - 3$ **$y = \frac{6}{5}x - \frac{3}{5}$**
17. $4x + 8y = 20$ **$y = -\frac{1}{2}x + 2\frac{1}{2}$**
18. $y = 3(x - 2)$ **$y = 3x - 6$**
19. $9y = 24x$ **$y = \frac{8}{3}x$**
20. ***Probability*** You roll a number cube and toss a coin at the same time. Find each probability.
    a. P(3 and T) **$\frac{1}{12}$**
    b. P(odd number and H) **$\frac{1}{4}$**
    c. P(prime and T) **$\frac{1}{4}$**
    d. $P(6)$ **$\frac{1}{6}$**

Getting Ready for Lesson 6-8

To which family of functions does each function belong?

21. $y = 3x^2 - x + 6$ **quadratic**
22. $y = |5x - 2|$ **abs. val.**
23. $y = 4x - 9$ **linear**
24. $y = -7x^2 - 1$ **quadratic**

Algebra at Work

Businessperson

Some of the goals of a business are to minimize costs and maximize profits. People in business use linear programming to analyze data in order to achieve these goals. The illustration lists some of the variables involved in operating a small manufacturing company. To solve a problem, a businessperson must identify the variables and restrictions and then search for the best possible solutions.

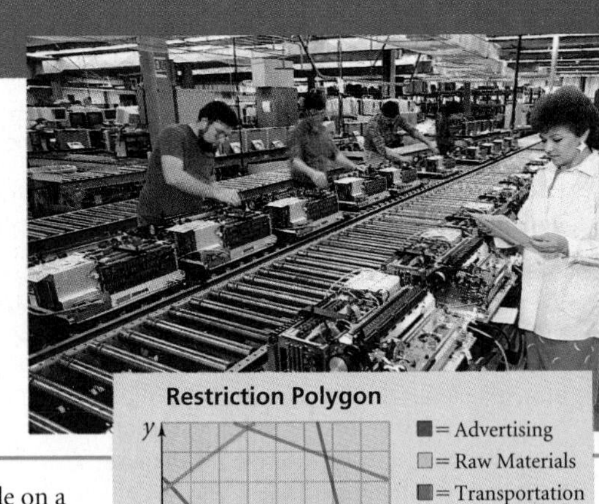

Mini Project: Work with a partner. Decide on a small business you could start. Identify the variables connected with your business and the restrictions on the variables.

6-8 Teaching Notes

PROBLEM OF THE DAY

Maria, Chantell, and Jon each brought a collection of rocks to school for a science project. The sum of the number of rocks in Maria's and Jon's collections is 32. The difference in the number of rocks between Chantell's and Maria's collections is 8. The total number of rocks brought by the three students is 42. How many rocks are in each student's collection?

Maria, 2 rocks; Chantell, 10 rocks; Jon, 30 rocks

Problem of the Day is also available in Transparencies.

CONNECTING TO PRIOR KNOWLEDGE On the board, draw a system of equations consisting of a horizontal linear equation and a downward-opening quadratic equation that intersect. Ask students to describe what the points of intersection of the two graphs represent.

THINK AND DISCUSS

Example 1

You may wish to have students follow these steps to work through the solution using the graphing calculator.

- Press [Y=] and enter the equations as **Y1** and **Y2.** Be sure students enter $|x - 1|$ correctly. The correct keystrokes are [2nd] [ABS] [(] [X,T,θ] [−] **1** [)]. Absolute value is the [2nd] function of [x^{-1}].
- Press [GRAPH].

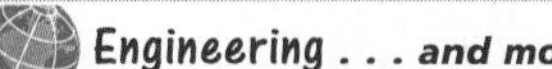

6-8 Systems with Nonlinear Equations

What You'll Learn

- Solving systems with linear, quadratic, and absolute value equations by graphing

...And Why

To solve problems involving engineering and architecture

What You'll Need

- graphing calculator
- graph paper

THINK AND DISCUSS

A system of equations can include equations that are not linear. The system shown consists of a linear equation and a quadratic equation.

$$y = x + 2$$
$$y = -x^2 + 4$$

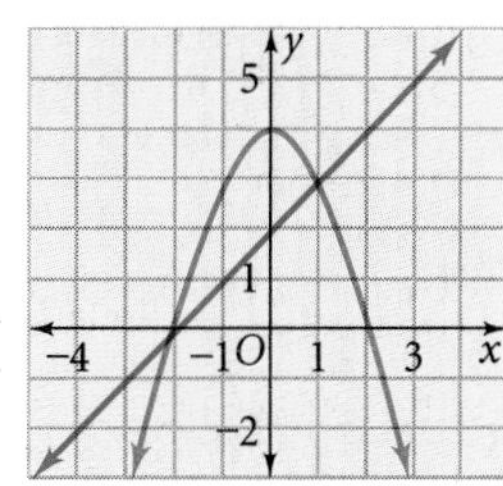

Notice that the graphs intersect at (–2, 0) and (1, 3). These two points are solutions of the system.

1. Check (–2, 0) and (1, 3) in both equations to **verify** that they are solutions of the system. $0 = -2 + 2$ ✓ $3 = 1 + 2$ ✓ $0 = -(-2)^2 + 4$ ✓ $3 = -(1)^2 + 4$ ✓

You can solve a system with a nonlinear equation by graphing.

Example 1

Solve the system of equations. $y = \frac{1}{2}x + 1$
$y = |x - 1|$

Graph each equation.

$y = \frac{1}{2}x + 1$

Use $y = mx + b$.

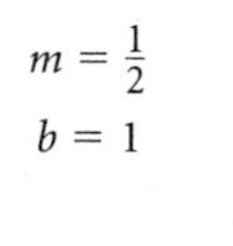

$m = \frac{1}{2}$

$b = 1$

$y = |x - 1|$

Make a table of values.

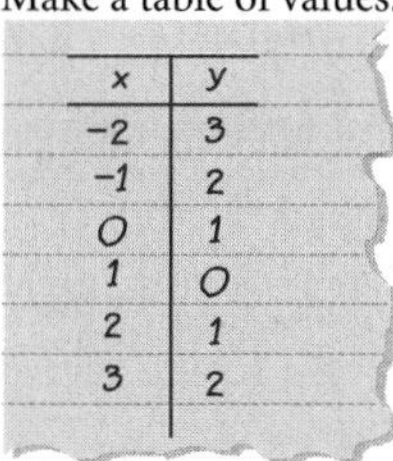

x	y
−2	3
−1	2
0	1
1	0
2	1
3	2

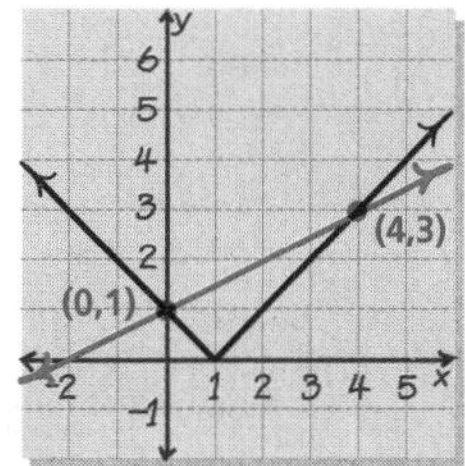

The graphs intersect at (0, 1) and (4, 3).

Check (0, 1):

$y = \frac{1}{2}x + 1$ | $y = |x - 1|$
$1 \stackrel{?}{=} \frac{1}{2}(0) + 1$ | $1 \stackrel{?}{=} |0 - 1|$
$1 \stackrel{?}{=} 0 + 1$ | $1 \stackrel{?}{=} |-1|$
$1 = 1$ ✓ | $1 = 1$ ✓

Check (4, 3):

$y = \frac{1}{2}x + 1$ | $y = |x - 1|$
$3 \stackrel{?}{=} \frac{1}{2}(4) + 1$ | $3 \stackrel{?}{=} |4 - 1|$
$3 \stackrel{?}{=} 2 + 1$ | $3 \stackrel{?}{=} |3|$
$3 = 3$ ✓ | $3 = 3$ ✓

Lesson Planning Options

Lesson Quiz is also available in Transparencies.

Prerequisite Skills

- Graphing linear equations
- Graphing quadratic equations

Assignment Options

Core 1–8, 10–13, 15, 20–31
Extension 9, 14, 16–19

Use Mixed Review to maintain skills.

Resources

Student Edition

Skills Handbook, pp. 567, 569, 579
Extra Practice, p. 561
Glossary/Study Guide

Teaching Resources

Chapter Support File, Ch. 6
- Practice 6-8 (two worksheets)
- Reteaching 6-8
- Alternative Activity 6-8

Classroom Manager 6-8
Glossary, Spanish Resources
Two-Year Algebra Handbook 6-8

Transparencies

2, 11, 29, 81

- Use [CALC] to identify the points of intersection. (For instruction on using [CALC], go back to the teaching notes for Example 2 in Lesson 6-1.)

TACTILE LEARNING Challenge students to write and decorate an instructional poster that describes steps to solve systems of equations by graphing.

CRITICAL THINKING Question 2 Have students sketch the different possibilities.

Example 2

You may also want to have students work through this example on the calculator. Point out that the correct keystrokes for x^2 are [X,T,θ] [x^2].

ERROR ALERT! **Question 3** If students graph using pencil and paper, they may not plot enough points to obtain an accurate graph. **Remediation:** Have students keep the function families in mind and check the shape of their graphs.

Example 3 **Relating to the Real World**

MAKING CONNECTIONS The curved shape of the bridge in Example 3 is similar to the curved flight of a projectile such as a model rocket. How far the rocket flies depends on the rocket's initial speed and the steepness of the rocket's launch. The horizontal distance that a projectile will travel—the range—can be calculated using the formula $R = \frac{v^2 \sin 2a}{g}$, where v = the initial velocity in m/s, a = the angle the projectile makes with the ground as it flies upward, and g = loss of acceleration due to gravity (approximately

Additional Examples

FOR EXAMPLE 1

Solve the system of equations by graphing.

$y = \frac{1}{2}x^3$

$y = 2x$ (2, 4), (0,0), (−2, −4)

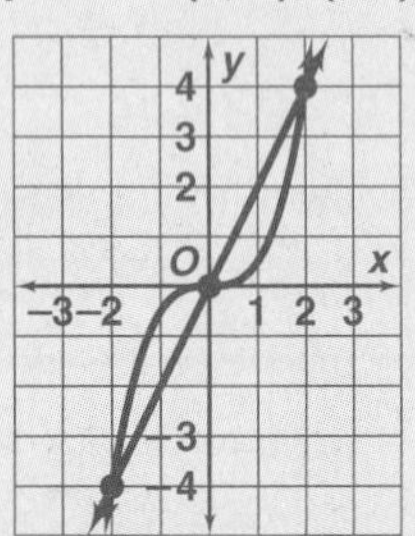

Discussion: *If you changed the exponent in the first equation from an odd to an even integer how would its graph change?*

FOR EXAMPLE 2

Solve the system of equations by graphing. Round your answers to the tenths place.

$y = 2x$

$y = x^2 + 1$ (1, 2)

Discussion: *How could you modify the second equation to result in no solutions? Two solutions?*

2. **Critical Thinking** Do systems made up of one linear equation and one absolute-value equation always have two solutions? If not, what are the other possibilities? Give examples. Answers may vary. Sample: No; no solution, $y = |x|$, $y = x - 1$; one solution, $y = |x|$, $y = x + 1$; infinitely many solutions, $y = |x|$, $y = x$.

Some systems of nonlinear equations have no solution.

Example 2

Solve the system of equations. $y = x^2 + 3$
$y = x$

Graph each equation.

$y = x^2 + 3$	$y = x$
Make a table of values.	Use $y = mx + b$. $m = 1$ $b = 0$

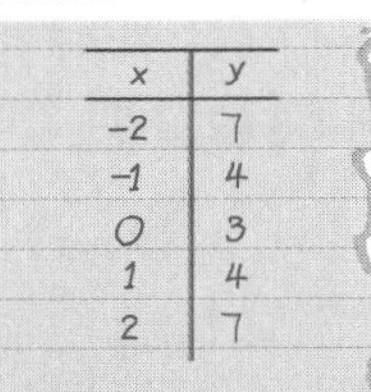

x	y
−2	7
−1	4
0	3
1	4
2	7

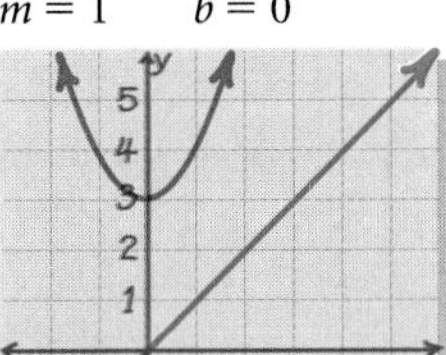

The graphs do not intersect, so the system has no solution.

3. **Try This** Solve each system.
   a. $y = |x| + 2$, $y = 4x - 1$ (1, 3)
   b. $y = -x^2$, $y = x^2 - 8$ (−2, −4), (2, −4)
   c. $y = |x - 2|$, $y = -\frac{1}{3}x + 2$ (0, 2), (3,1)

You can use a graphing calculator to solve systems with nonlinear equations.

Example 3 **Relating to the Real World**

Engineering Use the bridge diagram to find the coordinates of the points where the top arch intersects the road. Round answer to the nearest unit.

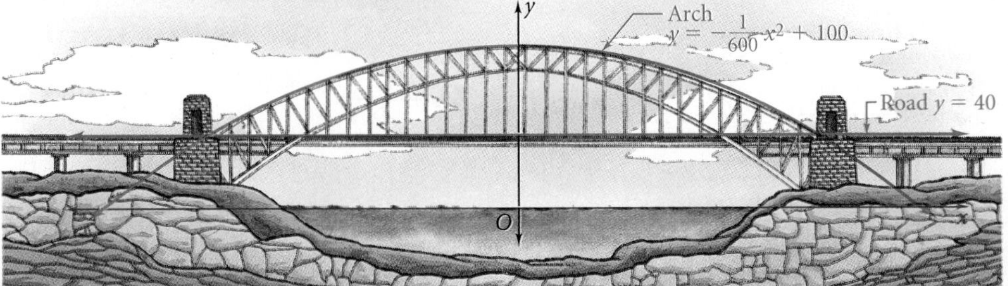

9.8 m/s^2). This formula does not take into account the slowing effects of air friction.

Have students work through this example using the calculator. Be sure they use [X,T,θ] [x^2] to enter x^2. (For instruction on using [CALC], go back to the teaching notes for Example 2 in Lesson 6-1.)

Have students also explore finding the solution using the [TABLE] and [TRACE] functions. When using [TABLE], find the x-values of the quadratic function for which the y-values are closest to 40 (190 and –190). To find the solution using [TRACE], press [TRACE], then [ZOOM] **8** [ENTER] to get integral values of x. Use the arrow keys to find the points of the quadratic function with the y-coordinates as close to 40 as possible.

Exercises ON YOUR OWN

ERROR ALERT! **Exercise 9** Students may obtain a syntax-error message when trying to enter the functions.

Remediation: Be sure to use the appropriate keystrokes. To enter $-|2.5x| + 5$, press [(-)] [2nd] [ABS] [(] 2.5 [X,T,θ] [)] [+] **5.** ([ABS] is the [2nd] function of [x^{-1}])To enter $0.04x^2$, press **0.04** [X,T,θ] [x^2].

Exercises 10–3 Students should be able to match the systems with their graphs by using what they know about families of functions.

GRAPHING CALCULATOR HINT

You can use the [TABLE] feature or the [ZOOM] and [TRACE] keys to estimate the intersection points.

Set an appropriate range.

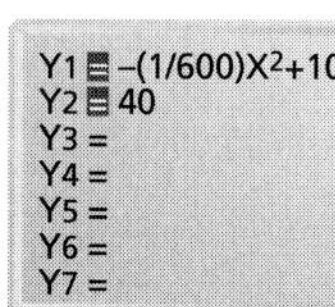
WINDOW FORMAT
Xmin=–300
Xmax=300
Xscl=20
Ymin=–40
Ymax=200
Yscl=20

Input the equations.

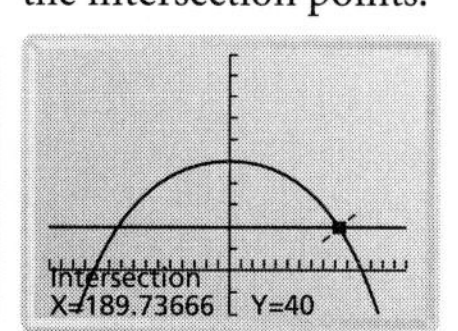
Y1 = –(1/600)X²+100
Y2 = 40
Y3 =
Y4 =
Y5 =
Y6 =
Y7 =

Use the [CALC] key to find the coordinates of the intersection points.

Intersection X=189.73666 Y=40

The solutions, rounded to the nearest unit, are (190, 40) and (–190, 40).

4. Each unit in Example 3 equals 1 ft. Find the length of the bridge. 380 ft

Exercises ON YOUR OWN

Write the solution(s) of each system of equations. Check that each solution makes both equations of the system true.

1. $y = 2x$ (0, 0)
$y = x^2$ (2, 4)

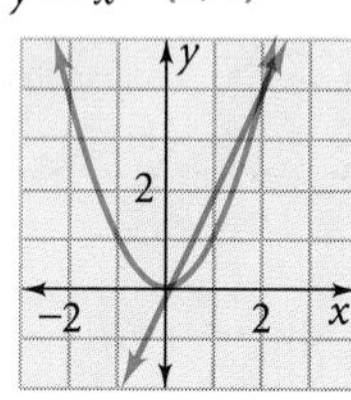

2. $y = 3$ (–3, 3)
$y = |x|$ (3, 3)

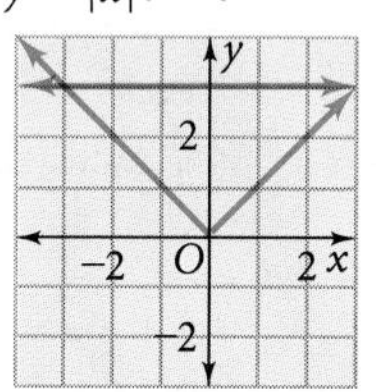

3. $y = -0.4x^2$ (0, 0)
$y = -4x^2$

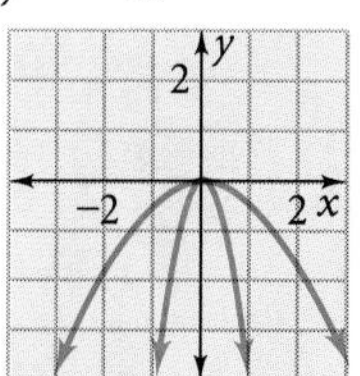

4. $y = |x|$ (2.5, 2.5)
$y = -x + 5$

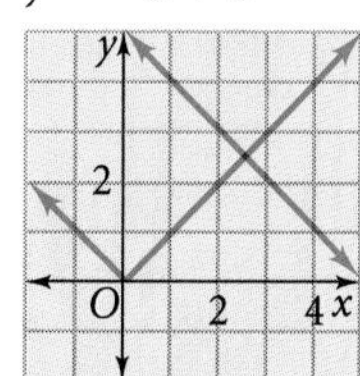

5. $y = -|x|$ (–2, –2)
$y = x^2 - 6$ (2, –2)

6. $y = x$ (0, 0)
$y = -x^2$ (–1, –1)

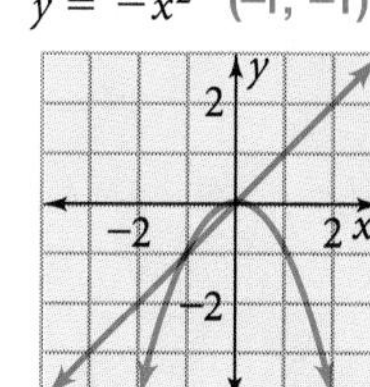

7. $y = |0.5x|$ (–2, 1)
$y = |3x| - 5$ (2, 1)

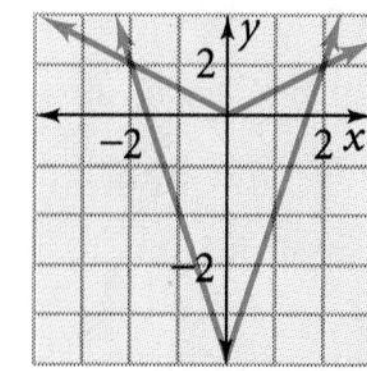

8. $y = x^2 + 2$ (0, 2)
$y = -x^2 + 2$

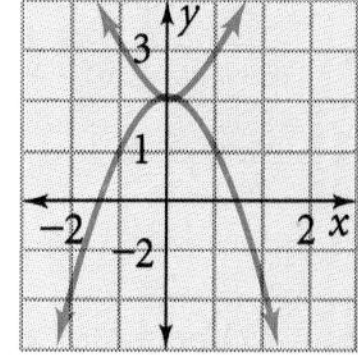

9. *Communications* Satellite dishes are used to receive television and radio signals. Use a graphing calculator to find the coordinates of S and T, the points at which the horn supports meet the dish. Round your answers to the nearest hundredth.
(–1.94, 0.15), (1.94, 0.15); See margin for graph.

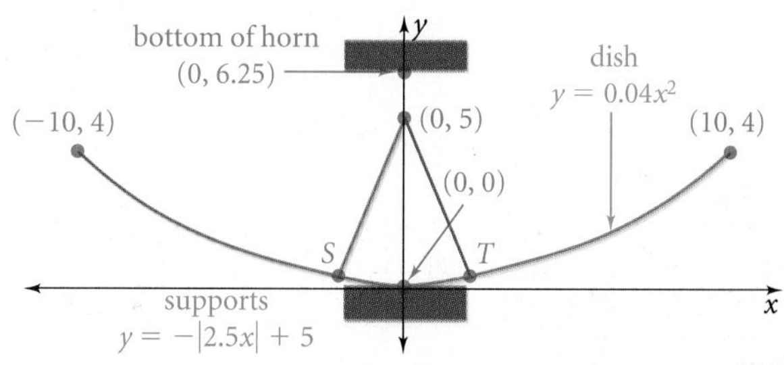

Technology Options

Students may want to check the answers they found in Exercises 1–8 using graphing calculators.

Prentice Hall Technology

Calculator • Graphing Calculator Handbook: Procedure 8

Software • Secondary Math Lab Toolkit • Integrated Math Lab 12 • Computer Item Generator 6-8

Internet • See the Prentice Hall site. (http://www.phschool.com)

CONNECTING TO THE STUDENTS' WORLD AND DIVERSITY
Exercise 14 Challenge students to think about designs featured in works of art from their cultural backgrounds. Encourage students to bring sketches or photographs of these designs to class. Have students model some of the lines in their design, using equations.

ALTERNATIVE ASSESSMENT **Exercises 16, 18, and 19** Ask students to complete these exercises by adding one more solution than is required to each system. This will help you assess students' understanding of nonlinear systems of equations.

STANDARDIZED TEST TIP **Exercise 15** If the two equations cannot be parallel, they also cannot have an infinite number of solutions.

Exercise 32 If students use a graphing calculator, this exercise provides them with a good opportunity to explore changing the maximum and minimum values for x and y using the WINDOW function. When the standard settings are used, no graphs will appear on the screen. For some settings only a portion of the graph of the system will appear. To solve the problem, both intersection points must appear on the screen. The following settings give a good view of both graphs and the points of intersection.

Xmin = −300 Xmax = 300
Ymin = 20 Ymax = 120

The Xscl and Yscl can be left as 1.

pages 305–307 Think and Discuss

3a.

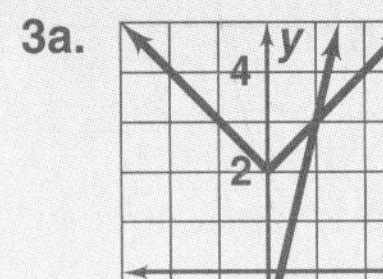

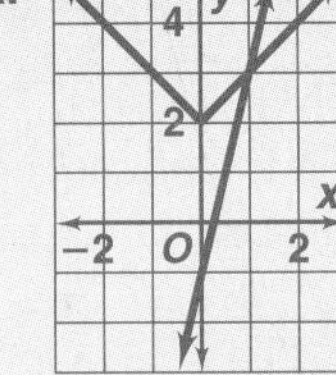

b.

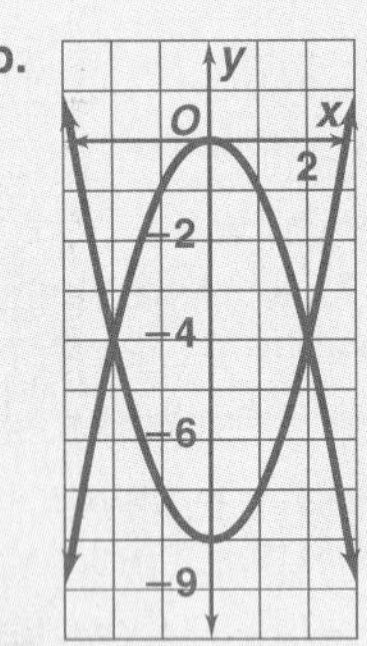

c.

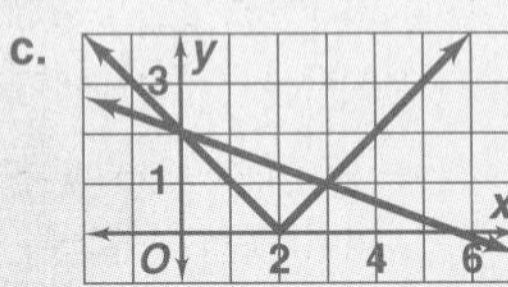

pages 307–309 On Your Own

9, 16–31, 32a, 42. See back of book.

Match each system of equations with its graph. Write the solution(s) of the system.

10. $y = |x + 2|$ B; $y = |x - 2|$ (0, 2)

11. $y = -2$ A; (−1, −2), $y = x^2 - 3$ (1, −2)

12. $y = 2.5x + 5$ D; $y = -|x - 2|$ (2, 0)

13. $y = x^2 + 2$ C; $y = -\frac{1}{2}x^2$ no solution

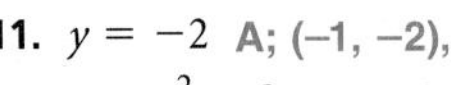

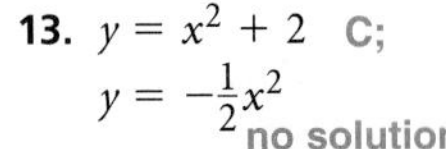

A.

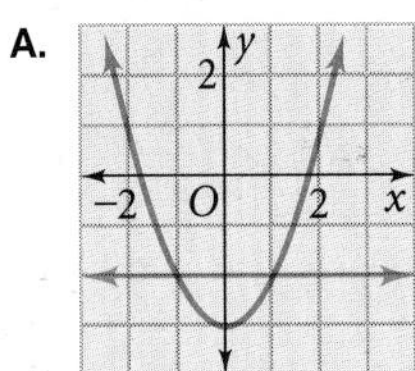

B.

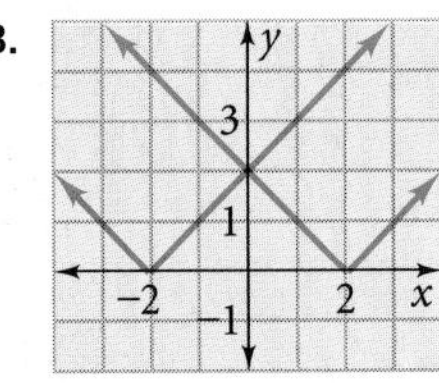

C.

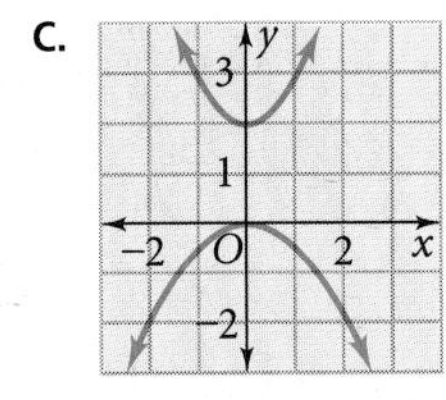

D.

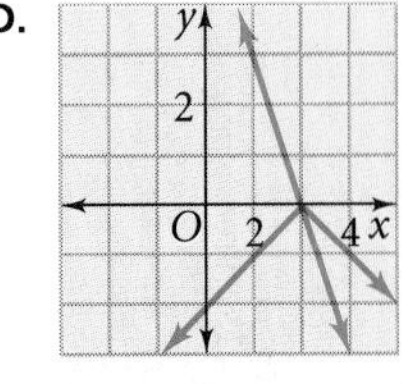

14. *Native American Art* This design was created by the Crow. The Crow are a people of the northern Great Plains of the United States. Patterns in the design can be described with a system of nonlinear equations. Find the coordinates of the points A and B. (−6, 9); (6, 9)

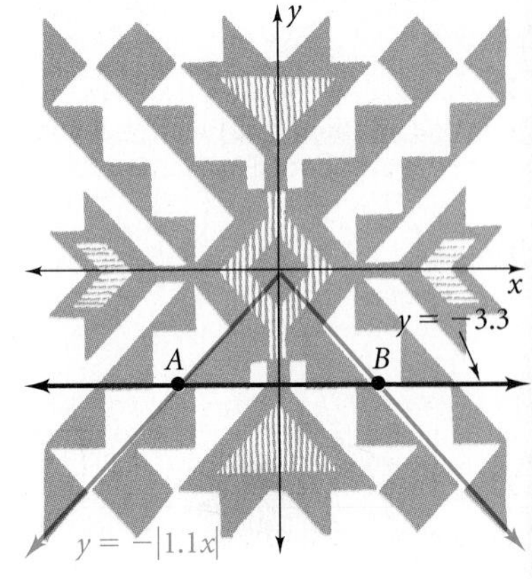

15. *Standardized Test Prep* How many solutions does this system of equations have? D

$$y = \frac{1}{2}x^2$$
$$y = -|x + 3|$$

A. one **B.** two **C.** three
D. none **E.** infinitely many

Open-ended **Sketch the graph of a system of two equations with the given characteristics.** 16–19. See margin for samples.

16. two quadratic equations, no solution

17. two absolute-value equations, two solutions

18. linear equation and absolute-value equation, one solution

19. linear equation and quadratic equation, no solution

Choose **Use paper and pencil or a graphing calculator to solve each system of equations.** 20-31. See margin for graphs.

20. $y = 3x - 4$; $y = -|x|$ (1, −1)

21. $y = |x + 3|$; $y = \frac{1}{2}x$ no solution

22. $y = \frac{3}{2}x + 1$ (0, 1); $y = |x| + 1$

23. $y = x^2 - 1$ (−2, 3), (1, 0); $y = -x + 1$

24. $y = |x|$ (0, 0); $y = -|x|$

25. $y = -x^2 + 2$; $y = x + 3$ no solution

26. $y = \frac{1}{2}x$ no solution; $y = |x + 3|$

27. $y = -2x^2$ (−1, −2), (1, −2); $y = |3x| - 5$

28. $y = -3$ no solution; $y = |x - 2|$

29. $y = x^2$ no solution; $y = x^2 - 2$

30. $y = |2x| - 3$ (3, 3), (−3, 3); $y = |x|$

31. $y = \frac{1}{4}x^2$ (−2, 1), (4, 4); $y = \frac{1}{2}x + 2$

Exercises MIXED REVIEW

Exercise 44 Point out that the answer should be given in doctors/year. Ask: If you were working in a spreadsheet program, what formula might you use to find the rate of change? (B3 – B2) / 95

SELF ASSESSMENT **PORTFOLIO** Share with students the criteria you will use to assess their work in portfolios, as well as how you plan to use the results. Students should understand how the rubrics are used to assess their work, how each piece in the portfolio counts, and how the scores they get in their portfolios will affect their overall evaluation.

Wrap Up

THE BIG IDEA Have students compare solving systems of nonlinear equations with solving systems of linear equations.

RETEACHING ACTIVITY Students solve linear, quadratic and absolute value equations by graphing. They use spaghetti or yarn to represent the equations on graph paper. (Reteaching worksheet 6-8)

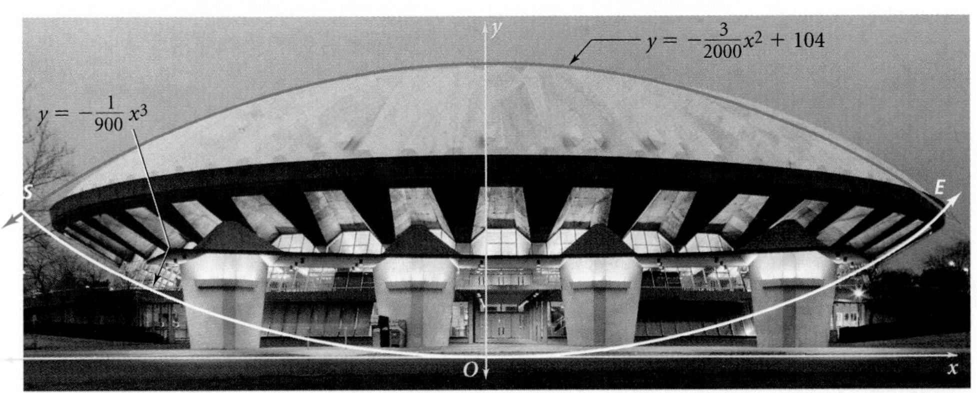

32. a. *Architecture* The University of Illinois Assembly Hall in Urbana can be described by a system with nonlinear equations. Use a graphing calculator to find the coordinates of points S and E. See margin for graph; about (−200, 44) and (200, 44).
    b. Find the length of $\overline{SE}$. about 400 ft

33. *Writing* How can you tell if a system with nonlinear equations has any solutions? A system of nonlinear equations has a solution if the graphs of the equations intersect.

Exercises MIXED REVIEW

34. What is 15% of 96? 14.4
35. 34 is what percent of 32? 106.25%
36. What percent of 40 is 27? 67.5%
37. 6 is what percent of 92? 6.5%
38. What percent of 6 is 2? $33\frac{1}{3}\%$
39. What is 85% of 108,000? 91,800

Find each sum or difference.

40. $\begin{bmatrix} -4 & a & 5 \\ 0 & b & 4c \end{bmatrix} + \begin{bmatrix} -12 & -a & 0 \\ d & 3b & -2c \end{bmatrix}$ $\begin{bmatrix} -16 & 0 & 5 \\ d & 4b & 2c \end{bmatrix}$

41. $\begin{bmatrix} 7 & x \\ 0 & -2y \\ 4z & -x \end{bmatrix} - \begin{bmatrix} x & 5 \\ 3x & -y \\ -4z & -x \end{bmatrix}$ $\begin{bmatrix} 7-x & x-5 \\ -3x & -y \\ 8z & 0 \end{bmatrix}$

42. *Writing* Explain the difference between the solution of an equation and the solution of a system of two equations. See margin.

43. a. *Medicine* Write a formula for cell D2 in the spreadsheet. 1000000 * B2 / C2
    b. Find the values in cells D2 and D3. 1576, 2423

44. What is the rate of change in the total number of doctors from 1900 to 1995? about 5457 doctors/yr

	A	B	C	D
1	Year	Number of Doctors	Population	Number of Doctors per Million People
2	1900	119,749	75,994,575	■
3	1995	638,200	263,434,000	■

PORTFOLIO

For your portfolio, choose one or two items from your work for this chapter. Here are some possibilities:
- a journal entry
- corrected work
- a Find Out question

Explain why you include each selection you make.

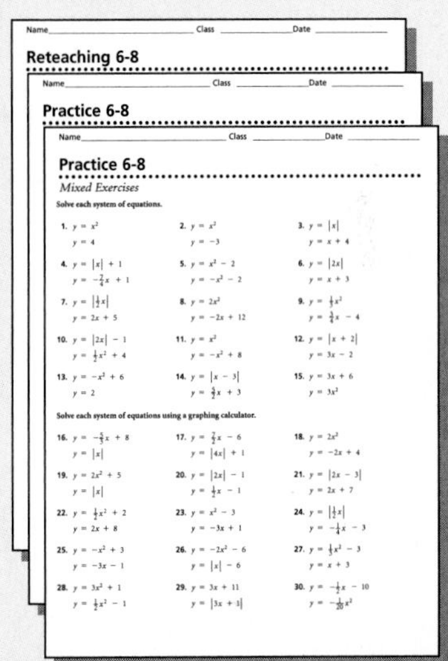

Name_____ Class _____ Date _____

Reteaching 6-8

Name_____ Class _____ Date _____

Practice 6-8

Name_____ Class _____ Date _____

Practice 6-8

Mixed Exercises

Solve each system of equations.

1. $y = x^2$; $y = 4$
2. $y = x^2$; $y = -3$
3. $y = |x|$; $y = x + 4$
4. $y = |x| + 1$; $y = -\frac{7}{4}x + 1$
5. $y = x^2 - 2$; $y = -x^2 - 2$
6. $y = |2x|$; $y = x + 3$
7. $y = |\frac{1}{2}x|$; $y = 2x + 5$
8. $y = 2x^2$; $y = -2x + 12$
9. $y = \frac{1}{3}x^2$; $y = \frac{3}{4}x - 4$
10. $y = |2x| - 1$; $y = \frac{1}{2}x^2 + 4$
11. $y = x^2$; $y = -x^2 + 8$
12. $y = |x + 2|$; $y = 3x - 2$
13. $y = -x^2 + 6$; $y = 2$
14. $y = |x - 3|$; $y = \frac{5}{2}x + 3$
15. $y = 3x + 6$; $y = 3x^2$

Solve each system of equations using a graphing calculator.

16. $y = -\frac{5}{3}x + 8$; $y = |x|$
17. $y = \frac{7}{2}x - 6$; $y = |4x| + 1$
18. $y = 2x^2$; $y = -2x + 4$
19. $y = 2x^2 + 5$; $y = |x|$
20. $y = |2x| - 1$; $y = \frac{1}{2}x - 1$
21. $y = |2x - 3|$; $y = 2x + 7$
22. $y = \frac{1}{2}x^2 + 2$; $y = 2x + 8$
23. $y = x^2 - 3$; $y = -3x + 1$
24. $y = |\frac{1}{2}x|$; $y = -\frac{1}{4}x - 3$
25. $y = -x^2 + 3$; $y = -3x - 1$
26. $y = -2x^2 - 6$; $y = |x| - 6$
27. $y = \frac{1}{3}x^2 - 3$; $y = x + 3$
28. $y = 3x^2 + 1$; $y = \frac{1}{2}x^2 - 1$
29. $y = 3x + 11$; $y = |3x + 1|$
30. $y = -\frac{1}{2}x - 10$; $y = -\frac{1}{20}x^2$

Lesson Quiz

Lesson Quiz is also available in Transparencies.

Use paper and pencil or a graphing calculator to solve each system of equations. Round your answers to the tenths place as appropriate.

1. $y = -|x| + 1$
   $y = -2$ (3, −2), (−3, −2)
2. $y = 3 + x^3$
   $y = 3 - x^4$ (0, 3) (−1, 2)
3. $y = -1 - 2x^2$
   $y = -3 + |x|$
   (0.8, −2.2), (−0.8, −2.2)
4. $y = |x - 2|$
   $y = \frac{1}{2}x^2$ (−3.2, 5.2), (1.2, .8)

Finishing the Chapter Project

PROJECT DAY Plan a project day on which students share their completed projects. Encourage groups to explain their process as well as their product.

PROJECT NOTEBOOK Have students review their project work and bring their notebooks up to date.

- Have students review their equations, graphs, and explanations that they needed for the project.
- Ask groups to share any insights they found for completing the project, such as any shortcuts they found for writing equations or making graphs.

SCORING RUBRIC

3 The report demonstrates sound reasoning and includes a detailed analysis of the cost of dinner and for each band based on the number of tickets sold. The student writes accurate linear equations and inequalities. The graphs are neat and have the appropriate scale and labels. All calculations are correct.

2 The report demonstrates sound reasoning but needs more detail. There are minor errors in the linear equations and inequalities. Graphs and calculations are mostly correct.

1 The report lacks essential details. Linear equations and inequalities have some problems. Graphs and calculations lack accuracy.

0 Major elements of the project are incomplete or missing.

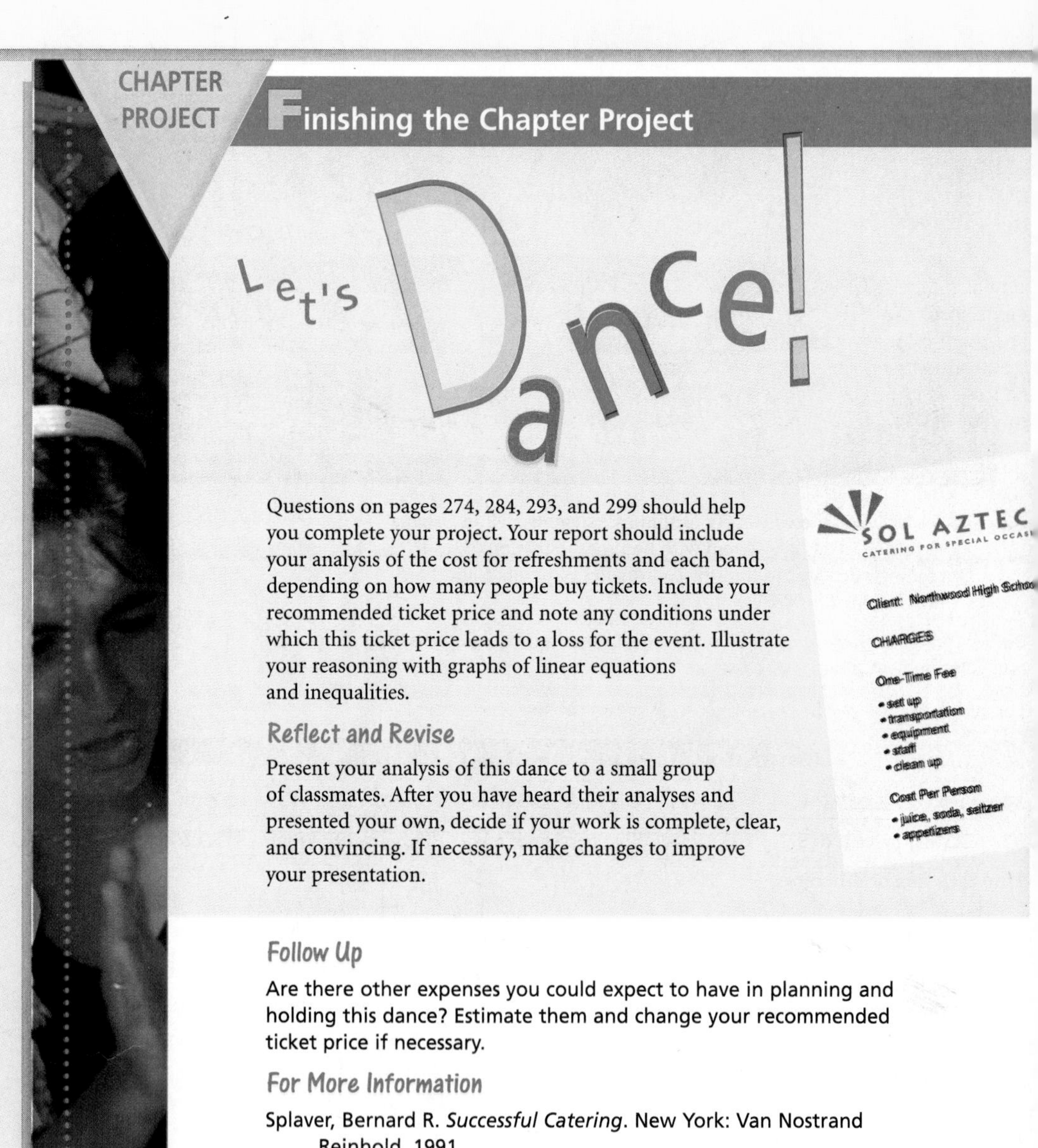

CHAPTER PROJECT

Finishing the Chapter Project

Let's Dance!

Questions on pages 274, 284, 293, and 299 should help you complete your project. Your report should include your analysis of the cost for refreshments and each band, depending on how many people buy tickets. Include your recommended ticket price and note any conditions under which this ticket price leads to a loss for the event. Illustrate your reasoning with graphs of linear equations and inequalities.

Reflect and Revise

Present your analysis of this dance to a small group of classmates. After you have heard their analyses and presented your own, decide if your work is complete, clear, and convincing. If necessary, make changes to improve your presentation.

Follow Up

Are there other expenses you could expect to have in planning and holding this dance? Estimate them and change your recommended ticket price if necessary.

For More Information

Splaver, Bernard R. *Successful Catering*. New York: Van Nostrand Reinhold, 1991.

Watkins, Andrea and Patricia Clarkson. *Dancing Longer, Dancing Stronger: A Dancer's Guide to Improving Technique and Preventing Injury*. Princeton, New Jersey: Princeton Book Company, 1990.

Wrap Up

HOW AM I DOING? Divide students into eight groups, and assign each group a section of Chapter 6. Have each group prepare a presentation that illustrates the main ideas of their section. Explain that the presentations should include both visual and verbal information.

KEY TERMS The numbers in parentheses direct students to the pages where the terms (or symbols) are used or defined. Students should be able to (1) write a simple explanation of each term, (2) illustrate the term with a diagram, or (3) show an example that uses a term.

Exercises 14–17 Remind students to multiply an expression by a factor to make elimination easier if necessary.

Exercises 20–23 Remind students that the boundary line of the graph of an inequality is dashed if the inequality symbol is $<$ or $>$. The boundary line of the graph of an inequality is solid, if the inequality symbol is $\leq$ or $\geq$.

Exercises 29–32 Encourage students to check every vertex of a linear programming solution set before deciding which ordered pair produces a minimum or maximum.

Remind students that the new mathematical terms in this chapter are defined in the Glossary/Study Guide in the back of the book.

6 Wrap Up

Key Terms

elimination (p. 280)
infinitely many solutions (p. 271)
linear inequality (p. 290)
linear progamming (p. 300)
no solution (p. 271)
solution of a system of linear inequalities (p. 295)
solution of the inequality (p. 290)
solution of the system (p. 269)
substitution (p. 275)
system of linear equations (p. 269)
system of linear inequalities (p. 295)

How am I doing?

- State three ideas from this chapter that you think are important. Explain your choices.
- Describe the different ways you can solve systems of equations and inequalities.

Resources

 **Student Edition**

Extra Practice, p. 561
Glossary/Study Guide

 **Teaching Resources**

Study Skills Handbook
Glossary, Spanish Resources

pages 311–313 Wrap Up

5.
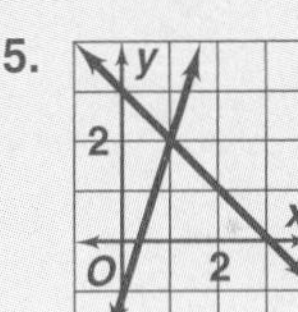

6.
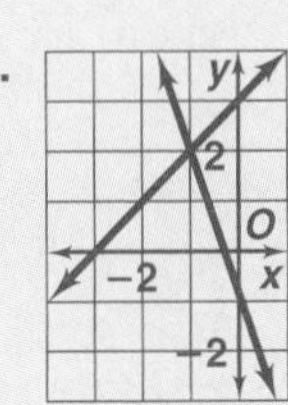

7.
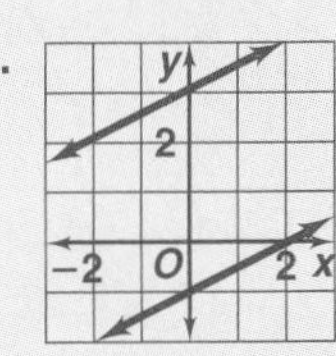

Solving Systems by Graphing — 6-1

Two or more linear equations form a **system of linear equations.** You can solve a system of linear equations by graphing. Any point where all the lines intersect is the **solution of the system.**

For each graph, write the system of linear equations and its solution.

1. 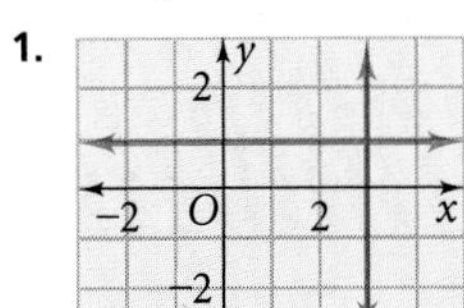
$y = 1$, $x = 3$; $(3, 1)$

2. 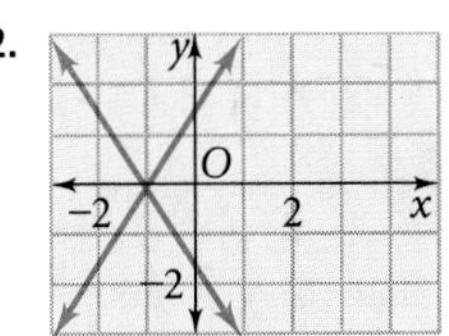
$y = -2x - 2$, $y = 2x + 2$; $(-1, 0)$

3. 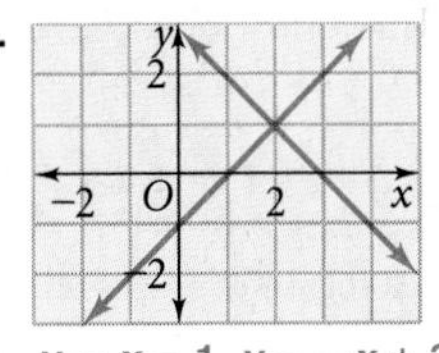
$y = x - 1$, $y = -x + 3$; $(2, 1)$

4. 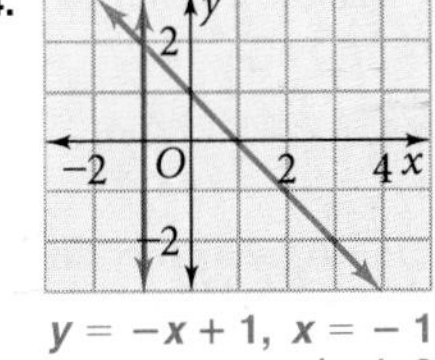
$y = -x + 1$, $x = -1$; $(-1, 2)$

Solve each system by graphing. 5-8. See margin for graphs.

5. $y = 3x - 1$
$y = -x + 3$ $(1, 2)$

6. $x - y = -3$
$3x + y = -1$ $(-1, 2)$

7. $-x + 2y = -2$
$y = \frac{1}{2}x + 3$ no solution

8. $y = -2x + 1$
$y = 2x - 3$ $(1, -1)$

Solving Systems Using Substitution — 6-2

You can also solve a system of linear equations using **substitution.** First, solve for one variable in terms of the other. Then substitute this result in either equation and solve.

Solve each system using substitution.

9. $y = 3x + 11$ $(-2, 5)$
$y = -2x + 1$

10. $4x - y = -12$ $\left(-4\frac{1}{2}, -6\right)$
$-6x + 5y = -3$

11. $y = 5x - 8$ $(2, 2)$
$5y = 2x + 6$

12. $8x = -2y - 10$
$2x = 4y$ $\left(-1\frac{1}{9}, -\frac{5}{9}\right)$

13. *Writing* Explain how you determine if a system has no solution or infinitely many solutions when you solve a system using substitution. See margin.

pages 311–313 Wrap Up

8. 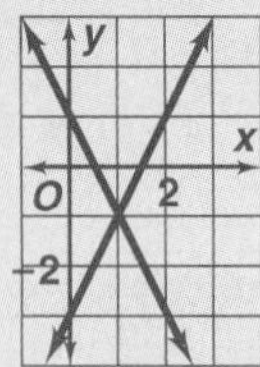

13. There is no solution when you get a false statement, such as 0 = 2. There are infinitely many solutions when you get a true statement, such as 5 = 5.

20. 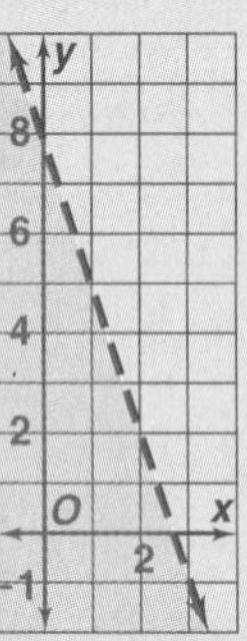

Solving Systems Using Elimination 6-3

You can solve a system of linear equations using **elimination.** You add or subtract the equations to eliminate one variable. You can multiply one or both of the equations by a nonzero number before adding or subtracting.

Solve each system using elimination. Check your solution.

14. $y = -3x + 5$
$y = -4x - 1$
(−6, 23)

15. $2x - 3y = 5$
$x + 2y = -1$
(1, −1)

16. $x + y = 10$
$x - y = 2$
(6, 4)

17. $-x + 4y = 12$
$2x - 3y = 6$
(12, 6)

Writing Systems 6-4

You can use systems of linear equations to solve word problems. First, define variables. Then model the situation with a system of linear equations.

Solve using any method. Check your solution.

18. A furniture finish consists of turpentine and linseed oil. It contains twice as much turpentine as linseed oil. If you need 16 fl oz of furniture finish, how much turpentine do you need? $10\frac{2}{3}$ fl oz

19. *Geometry* The difference between the measures of two complementary angles is 36°. Find both angle measures. (*Hint:* Two angles are complementary if the sum of their measures is 90°.) 63° and 27°

Linear Inequalities 6-5

A **linear inequality** describes a region of the coordinate plane. To graph the **solution of the inequality,** first graph the boundary line. Then test a point to shade the region that makes the inequality true.

Graph each linear inequality. 20–23. See margin.

20. $y < -3x + 8$ **21.** $y \geq 2x - 1$ **22.** $y \leq 0.5x + 6$ **23.** $y > -\frac{1}{4}x - 2$

Systems of Linear Inequalities 6-6

Two or more linear inequalities form a **system of linear inequalities.** To find the **solution of a system of linear inequalities,** graph each linear inequality. The solution region is where all the inequalities are true.

Solve each system of linear inequalities by graphing. 24–27. See margin.

24. $y \geq -4x + 1$
$y \leq 3x - 5$

25. $x - y < 10$
$x + y \leq 8$

26. $y \leq x - 3$
$y > x - 7$

27. $y < 5x$
$y \geq 0$

28. *Open-ended* Write a system of linear inequalities for which the solution region is a pentagon.
Sample: $x \geq -1$, $x \leq 5$, $y \geq -1$, $y \leq 5$, $x + y \leq 7$

ALTERNATIVE ASSESSMENT Exercises 9–12 and 14–17 Divide students into small groups. Have each group write the exercises on eight separate index cards. Then ask the students to graph the solution to each of these exercises on a second set of cards. Take up all of the cards, shuffle, and randomly distribute among the class. Ask students to work together to match the exercise cards with the correct solution cards.

Getting Ready for Chapter 7

Have students work these exercises independently or in small groups. The skills previewed will help prepare students to find the intersection points of two solution sets (equations).

Concepts of Linear Programming 6-7

You can use **linear programming** to minimize or maximize quantities. The variables used in the equation for the quantity have restrictions. The maximum and minimum values of the equation occur at the vertices of the graph of the restrictions.

Evaluate the equation to find minimum and maximum values. 29–32. See margin for graphs.

29. $x + y \le 7$, $x + 2y \le 8$, $x \ge 0$, $y \ge 0$, $P = 3x + y$ 0; 21

30. $x + 2y \le 5$, $x \ge 0$, $y \ge 1$, $P = 2x + 3y$ 3; 9

31. $x + y \le 6$, $2x + y \le 10$, $x \ge 0$, $y \ge 0$, $P = 4x + y$ 0; 20

32. $x \ge 1$, $x \le 4$, $y \ge 3$, $y \le 6$, $P = x + 2y$ 7; 16

33. *Standardized Test Prep* Which point minimizes the equation $C = 5x + 2y$? E

A. $(4, 1)$ **B.** $(5, 0)$ **C.** $(2, 6)$ **D.** $(3, 2)$ **E.** $(1, 5)$

Systems with Nonlinear Equations 6-8

You can solve systems that have linear, absolute value, and quadratic equations. Graph the equations on the same coordinate plane. Any point where the graphs intersect is a solution.

Solve each system of equations. 34–37. See margin for graphs.

34. $y = x^2 - 4$, $y = -x + 2$ (2, 0), (–3, 5)

35. $y = |2x| - 3$, $y = x - 1$ (2, 1), $\left(-\frac{2}{3}, -1\frac{2}{3}\right)$

36. $y = -x + 3$, $y = |2x - 3|$ (0, 3), (2, 1)

37. $y = 2x + 6$, $y = x^2 + 3$ (–1, 4), (3, 12)

38. *Critical Thinking* How many solutions can a system with an absolute-value equation and a quadratic equation have? Explain. 0, 1, 2, 3 or 4; See margin for reasoning.

Getting Ready for CHAPTER 7

Find the square of each number.

39. 3 9 **40.** –7 49 **41.** 4.5 20.25 **42.** $\frac{1}{2}$ $\frac{1}{4}$ **43.** 11 121 **44.** –8.2 67.24

Make a table of values to graph each function. 45–47. See margin.

45. $y = x^2 - 4x + 3$ **46.** $y = x^2 + x - 2$ **47.** $y = -x^2 + 4x + 5$

Evaluate the expression $b^2 - 4ac$ for the given values.

48. $a = 2, b = -5, c = 3$ 1 **49.** $a = -7, b = 9, c = 1$ 109 **50.** $a = 4.5, b = 8, c = 0$ 64

51. $a = 1, b = 3, c = -2$ 17 **52.** $a = 0.5, b = 4, c = 2.5$ 11 **53.** $a = -8, b = -3.2, c = 5$ 170.24

21.

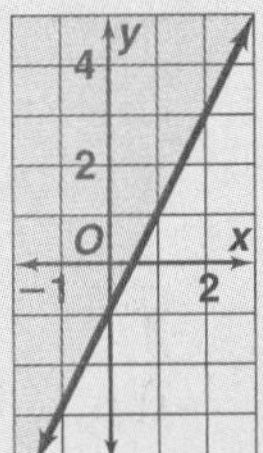

22.

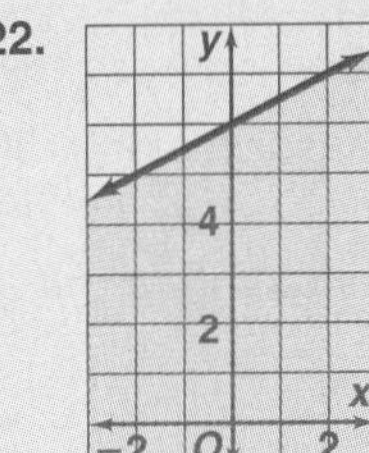

23.

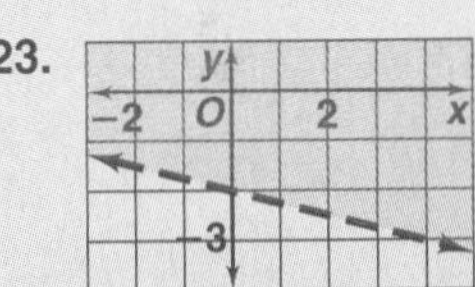

24–27, 29–32, 34–38, Getting Ready for Chapter 7, 45–47. See back of book.

Assessment

ENHANCED MULTIPLE CHOICE QUESTIONS are more complex than traditional multiple choice questions, which assess only one skill. Enhanced multiple choice questions assess the processes that students use, as well as the end results. The questions are written so that students must use more than one strategy to solve the problem. Using multiple strategies is encouraged by the National Council of Teachers of Mathematics (NCTM). **Exercise 19** is an enhanced multiple choice question.

FREE RESPONSE QUESTIONS do not give answer choices. Some exercises have more than one possible answer. Students need to give only one correct response. **Exercises 1–17, 20–23, and 27** are free response questions.

WRITING EXERCISES allow students to describe how they think about and understand the concepts they have learned. **Exercise 18** is a writing exercise.

OPEN-ENDED PROBLEMS allow for more than one solution. Students must construct their own responses instead of choosing from possible answers. The student responses will help you determine the depth of their understanding and any possible areas of difficulty. **Exercise 24–26** are open-ended problems. Make sure that the students' responses contain values for *x* and *y* and a written situation.

Resources

Teaching Resources

Chapter Support File, Ch. 6
- Chapter Assessment, Forms A and B
- Alternative Assessment

Chapter Assessment, Spanish Resources

Teacher's Edition

See also p. 266E for assessment options.

Software Computer Item Generator

page 314 Assessment

1.

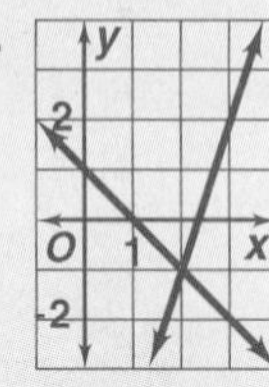

2.

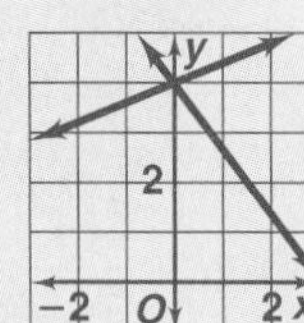

18, 20–23, 27a–c. See back of book.

6 Assessment

Solve each system of linear equations by graphing. 1–2. See margin for graphs.

1. $y = 3x - 7$ $y = -x + 1$ (2, –1)

2. $4x + 3y = 12$ $2x - 5y = -20$ (0, 4)

Critical Thinking **Suppose you try to solve a system of linear equations using substitution and get this result. How many solutions does each system have?**

3. $x = 8$ one

4. $5 = y$ one

5. $-7 = 4$ none

6. $x = -1$ one

7. $2 = y$ one

8. $9 = 9$ inf. many

Solve each system using substitution.

9. $y = 4x - 7$ $y = 2x + 9$ (8, 25)

10. $y = -2x - 1$ $y = 3x - 16$ (3, –7)

Solve each system using elimination.

11. $4x + y = 8$ $-3x - y = 0$ (8, –24)

12. $2x + 5y = 20$ $3x - 10y = 37$ $\left(11, -\frac{2}{5}\right)$

13. $x + y = 10$ $-x - 2y = -14$ (6, 4)

14. $3x + 2y = -19$ $x - 12y = 19$ (–5, –2)

Write a system of equations to model each situation. Then use your system to solve.

15. *Cable Service* Your local cable television company offers two plans: basic service with one movie channel for \$35 per month, or basic service with two movie channels for \$45 per month. What is the charge for the basic service and the charge for each movie channel?
$b + m = 35$ and $b + 2m = 45$; \$25; \$10

16. *Education* A writing workshop enrolls novelists and poets in a ratio of 5 to 3. There are 24 people at the workshop. How many novelists are there? How many poets?
15 novelists; 9 poets

17. You have 15 coins in your pocket that are either quarters or nickels. They total \$2.75. How many of each coin do you have?
10 quarters and 5 nickels

18. *Writing* Compare solving a linear equation with solving a linear inequality. What are the similarities? What are the differences?
See margin.

19. *Standardized Test Prep* Which point is *not* a solution of $y < 3x - 1$? C

A. (2, –4) **B.** (5, 7) **C.** (0, –1) **D.** (–3, –13) **E.** (4, –8)

Solve each system by graphing. 20–23. See margin.

20. $y > 4x - 1$ $y \leq -x + 4$

21. $y \geq 3x + 5$ $y > x - 2$

22. $y = x^2 + x + 1$ $y = -3x + 6$ (–5, 21), (1, 3)

23. $y = 4x - 5$ $y = |3x + 1|$ (6, 19)

Open-ended **Write a system of equations with the given characteristics.** 24–26. Samples are given.

24. two linear equations with no solution
$y = 3x + 1$; $y = 3x - 5$

25. a linear equation and a quadratic equation with two solutions $y = x$ $y = x^2$

26. three linear inequalities with a triangular solution region $x + y > 6$ $y < -5x + 12$ $y < 5x + 12$

27. *Fund Raising* You are making bread to sell at a holiday fair. A loaf of oatmeal bread takes 2 cups of flour and 2 eggs. A loaf of banana bread takes 3 cups of flour and 1 egg. You have 12 cups of flour and 8 eggs.

x = number of loaves of oatmeal bread
y = number of loaves of banana bread
Restrictions: $2x + 3y \leq 12$
$2x + y \leq 8$
$x \geq 0, y \geq 0$

a. Explain each restriction.

b. You will make \$1 profit for each loaf of oatmeal bread and \$2 profit for each loaf of banana bread. Write the equation.

c. Use linear programming to find how many loaves of each type you should make to maximize profits.

27a–c. See margin.

Cumulative Review

Item	Review Topic	Chapter
1, 3, 4	solving inequalities	4
2	proportions	4
5, 10–13	linear equations	5
6	families of functions	2
7, 16	solving systems of equations	6
8	line of best fit	5
9	solving equations	4
14	solving systems of inequalities	6
15	percent	3

6 Cumulative Review

For Exercises 1–11, choose the correct letter.

1 Which of the following points are solutions of $4y - 3x \le 8$? E
 I. $(0, 2)$ II. $(-3, \frac{1}{4})$
 III. $(5, -17.6)$ IV. $(-4, \frac{2}{5})$
 A. I only B. IV only C. I and II
 D. II and IV E. I and III

2. Suppose you earn \$74.25 for working 9 h. How much will you earn for working 15 h? D
 A. \$120 B. \$124.50 C. \$127.25
 D. \$123.75 E. none of the above

3. Which is *not* a solution of $5x - 4 < 12$? E
 A. −2 B. 3 C. 0 D. 1.8 E. 4

4. Which statement is true for every solution of the following system? B

$$2y > x + 4$$
$$3y + 3x > 13$$

 A. $x \le -3$ B. $y > 1$ C. $x > 4$
 D. $y < 5$ E. none of the above

5. Which of the following equations represents a vertical line through $(-7, -4)$? A
 A. $x = -7$ B. $x = 4$ C. $y = 7$
 D. $y = -4$ E. none of the above

6. The function $f(x) = |x - 5|$ belongs to which family of functions? D
 A. linear B. quadratic
 C. direct variation D. absolute value
 E. none of the above

7. What is true of the graphs of the two lines $3y - 8 = -5x$ and $3x = 2y - 18$? C
 A. no intersection B. intersect at $(2, -6)$
 C. intersect at $(-2, 6)$ D. are identical
 E. none of the above

8. A scatter plot shows a positive correlation. Which of the following could be an equation of the line of best fit? D
 A. $y = -5x + 1$ B. $2x + 3y = 6$
 C. $x = 16$ D. $y = 2x - 1$
 E. none of the above

9. Which of the following is the solution of $6(4x - 3) = -54$? E
 A. −2.5 B. −3 C. 2.5
 D. 3 E. none of the above

Compare the boxed quantity in Column A with the boxed quantity in Column B. Choose the best answer.

 A. The quantity in Column A is greater.
 B. The quantity in Column B is greater.
 C. The two quantities are equal.
 D. The relationship cannot be determined on the basis of the information supplied.

	Column A	Column B
10. A	the y-intercept of the graph of $6y - 5x = 2$	the y-intercept of the graph of $x + 9y = 2$
11. B	the slope of the line through $(2, -5)$ and $(-3, 1)$	the slope of the graph of $15y + 12x = 5$

Find each answer.

12. Write an equation of the line through (2, −1) and (3, 4).
 Answers may vary. Sample: $y = 5x - 11$

13. Write an equation of the line through (2, −3) that is perpendicular to the graph of $y = \frac{2}{5}x - \frac{7}{8}$. Answers may vary. Sample: $y = -\frac{5}{2}x + 2$

14. Graph $-3 \le 2x + 1 < 7$. See margin.

15. *Transportation* In 1996, the City Council of New York City voted to increase the number of taxis in the city from 11,787 to 12,187. What was the percent of increase? about 3.4%

16. Plumber A charges \$40 for a house call plus \$30 per hour for labor. Plumber B charges \$20 for a house call plus \$35 per hour for labor. How long must a job take before plumber A is the less expensive choice for the homeowner? more than 4 h

Resources

Teaching Resources

Chapter Support File, Ch. 6
- Cumulative Review
- Standardized Test Practice

Teacher's Edition

See also pp. 266E for assessment options.

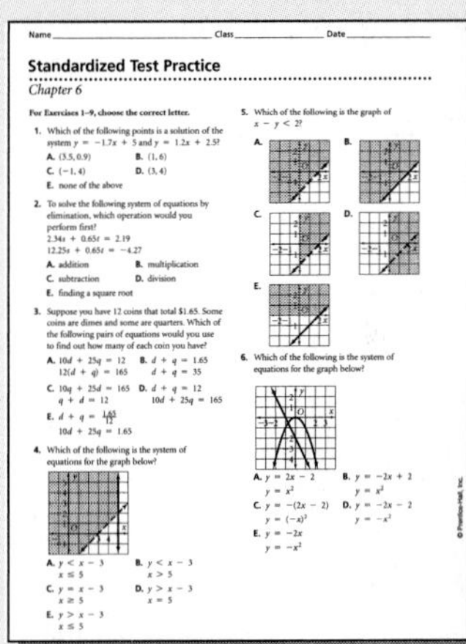

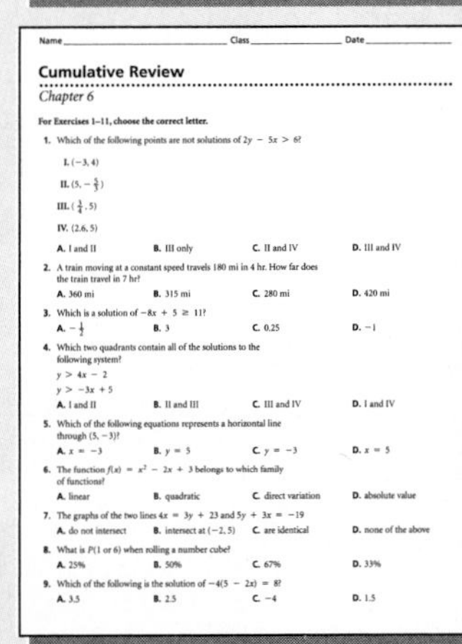

page 315 Cumulative Review

14. See back of book.

CHAPTER 7 Quadratic Equations and Functions

To accommodate flexible scheduling, some lessons are divided into parts. Assignment Options are given in the Lesson Planning Options for each lesson.

PACING OPTIONS

This chart suggests pacing only for the core lessons and their parts, and it is provided merely as a possible guide. It will help you determine how much time you have in your schedule to cover other features, such as the Chapter Project, Math Toolboxes, Wrap Up, and Assessment.

	1 Class Period	1 Class Period	1 Class Period	1 Class Period	1 Class Period	1 Class Period	1 Class Period	1 Class Period	1 Class Period
Traditional (40–45 min class periods)	7-1 Part 1	7-1 Part 2	7-2	7-3 Part 1	7-3 Part 2	7-4 Part 1	7-4 Part 2	7-4 Part 2	7-5 Parts 1, 2
Two-Year Algebra (40–45 min class periods)	7-1 Part 1	7-1 Part 1	7-1 Part 2	7-1 Part 2	7-2	7-2	7-3 Part 1	7-3 Part 1	7-3 Part 2
Block Scheduling (90 min class periods)	7-1 Parts 1, 2	7-2	7-3 Parts 1, 2	7-1 Parts 1, 2	7-5 Parts 1, 2	7-6	7-7		

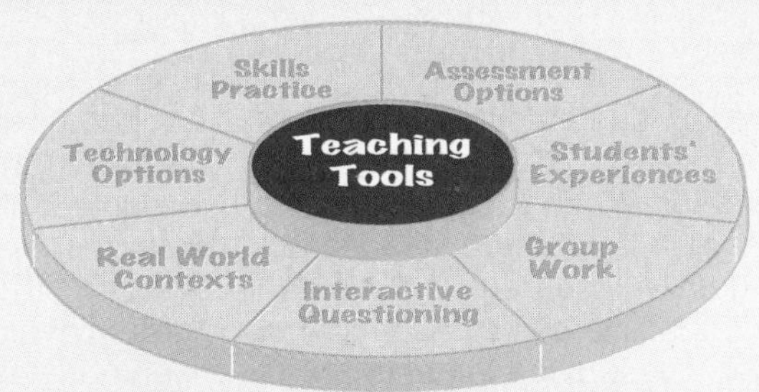

What Students Will Learn and Why

In this chapter, students will build on their knowledge of systems of equations and inequalities, learned in Chapter 6, by learning to solve quadratic equations and functions. They will learn to solve quadratic equations and use these equations to solve real-world problems. They will learn to find square roots and graph quadratic functions. Students will solve quadratic equations using the quadratic formula. Finally, students will use the discriminant to find the number of solutions of a quadratic equation.

Discussing the Chapter/Building on Experience

The concept map below relates chapter topics to real-world applications. You and your class may wish to add to the map or develop maps of your own. The center oval describes the topic of the chapter. The next level displays topics within the lessons. The outer ovals reflect applications of the content. As you and your class build a concept map, invite students to discuss applications with which they are familiar.

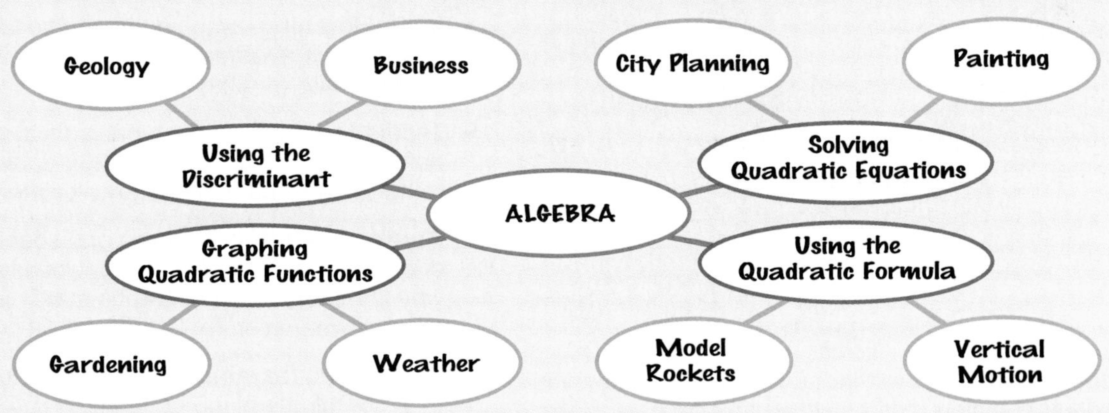

1 Class Period	1 Class Period	1 Class Period	1 Class Period	1 Class Period	1 Class Period	1 Class Period	1 Class Period	1 Class Period	1 Class Period	1 Class Period
7-6	7-7	7-1 ▼1, 7-1 ▼2, 7-1 ▼3	7-1 ▼1, 7-1 ▼2, 7-1 ▼3	7-1 ▼1, 7-1 ▼2, 7-1 ▼3	7-1 ▼1, 7-1 ▼2, 7-1 ▼3	7-1 ▼1, 7-1 ▼2, 7-1 ▼3	7-1 ▼1, 7-1 ▼2, 7-1 ▼3	7-1 ▼1, 7-1 ▼2, 7-1 ▼3	7-1 ▼1, 7-1 ▼2, 7-1 ▼3	7-1 ▼1, 7-1 ▼2, 7-1 ▼3
7-3 ▼2	7-4 ▼1	7-4 ▼1	7-4 ▼2	7-4 ▼2	7-5 ▼1	7-5 ▼1	7-5 ▼2	7-6	7-6	7-7

Skills Practice
Assessment Options
Technology Options
Teaching Tools
Students' Experiences
Real World Contexts
Group Work
Interactive Questioning

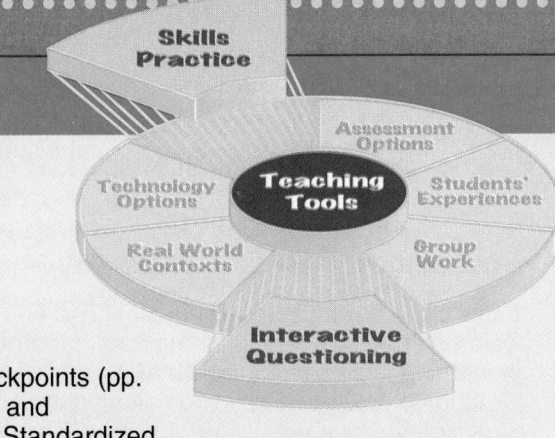

Interactive Questioning Tips

A question is interactive when there is "give and take" between the questioner (teacher or student) and the respondent. When a critical thinking question is asked, it is important to encourage students to explain their answer in more detail. Encouraging students to explain their answers promotes active participation and stimulates higher levels of thought. These two elements are essential in developing high level thinking skills. For example, in Lesson 7-4, Example 2, students are asked to estimate the maximum height of a rocket and then explain how they made their estimates.

Skills Practice

Every lesson provides skill practice with Exercises On Your Own and Exercises Mixed Review. The Student Edition includes Checkpoints (pp. 336 and 347) and Preparing for Standardized Tests (p. 359). In the Teacher's Edition, the Lesson Planning Options section for each lesson lists Prerequisite Skills students should know for that lesson. At the back of the Student Edition is the Skills Handbook—mini-lessons on math the students may need to review. The Chapter Support File for Chapter 7 in the Teaching Resources box includes two Practice worksheets per lesson, a worksheet for two Checkpoints, and worksheets for Cumulative Review and Standardized Test Preparation.

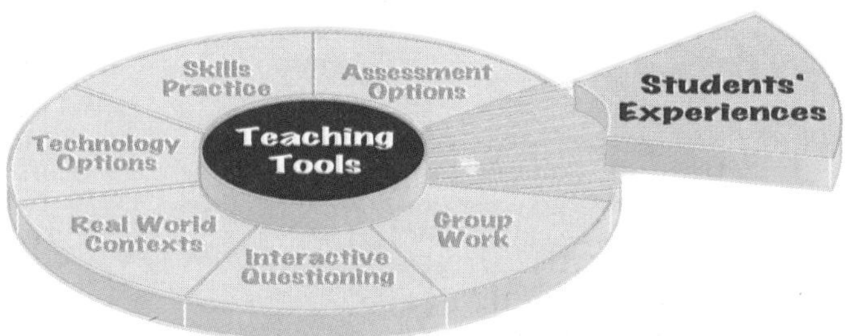

Diverse Learning and Teaching Styles

In your Teacher's Edition, you will find suggestions as to how you can help students complete mathematical tasks in Chapter 7 by reinforcing various learning styles. Here are some examples.

- **Visual learning** graphing quadratic functions as curves on the board or overhead (p. 318), observing parabola drawn on the board (p. 339)
- **Tactile learning** using paper clips to model graphs of quadratic functions (p. 324–325), drawing graphs by hand rather than using a graphing calculator (p. 329)
- **Auditory learning** listening and speaking about the graphs of quadratic functions (p. 320), working in pairs to restate why the intersection of graphs match the solutions for a quadratic equation (p. 345)
- **Kinesthetic learning** graphing simple quadratic functions on a coordinate plane on the floor (p. 320), making right triangles on the floor with string (p. 333)

Alternative Activity for Lesson 7-1

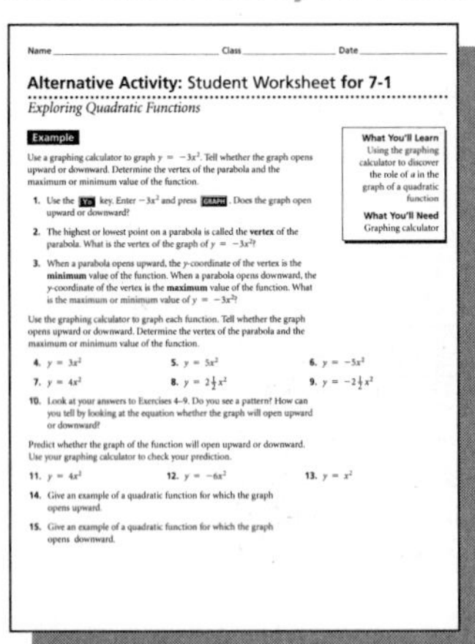

Name ____________ Class ____________ Date ____________

Alternative Activity: Student Worksheet for 7-1

Exploring Quadratic Functions

Example

Use a graphing calculator to graph $y = -3x^2$. Tell whether the graph opens upward or downward. Determine the vertex of the parabola and the maximum or minimum value of the function.

What You'll Learn Using the graphing calculator to discover the role of a in the graph of a quadratic function

What You'll Need Graphing calculator

1. Use the Y= key. Enter $-3x^2$ and press GRAPH. Does the graph open upward or downward?
2. The highest or lowest point on a parabola is called the **vertex** of the parabola. What is the vertex of the graph of $y = -3x^2$?
3. When a parabola opens upward, the y-coordinate of the vertex is the **minimum** value of the function. When a parabola opens downward, the y-coordinate of the vertex is the **maximum** value of the function. What is the maximum or minimum value of $y = -3x^2$?

Use the graphing calculator to graph each function. Tell whether the graph opens upward or downward. Determine the vertex of the parabola and the maximum or minimum value of the function.

4. $y = 3x^2$ 5. $y = 5x^2$ 6. $y = -5x^2$
7. $y = 4x^2$ 8. $y = 2\frac{1}{2}x^2$ 9. $y = -2\frac{1}{2}x^2$
10. Look at your answers to Exercises 4–9. Do you see a pattern? How can you tell by looking at the equation whether the graph will open upward or downward?

Predict whether the graph of the function will open upward or downward. Use your graphing calculator to check your prediction.

11. $y = 4x^2$ 12. $y = -6x^2$ 13. $y = x^2$
14. Give an example of a quadratic function for which the graph opens upward.
15. Give an example of a quadratic function for which the graph opens downward.

for use with Example 1, addresses visual and tactile learning by having students use a graphing calculator to discover the role of a in quadratic equations.

Alternative Activity for Lesson 7-5

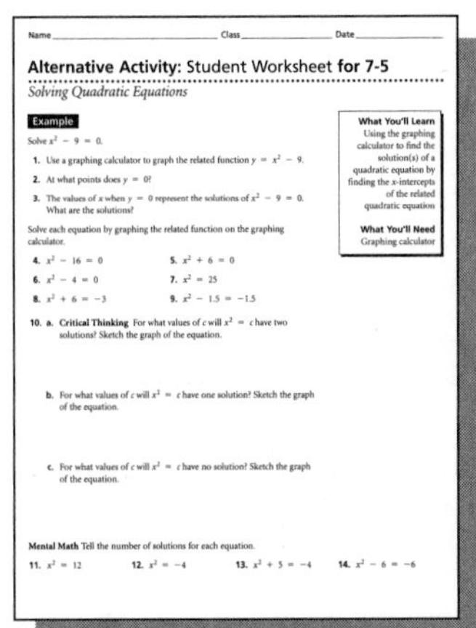

Name ____________ Class ____________ Date ____________

Alternative Activity: Student Worksheet for 7-5

Solving Quadratic Equations

Example

Solve $x^2 - 9 = 0$.

What You'll Learn Using the graphing calculator to find the solution(s) of a quadratic equation by finding the x-intercepts of the related quadratic equation

What You'll Need Graphing calculator

1. Use a graphing calculator to graph the related function $y = x^2 - 9$.
2. At what points does $y = 0$?
3. The values of x when $y = 0$ represent the solutions of $x^2 - 9 = 0$. What are the solutions?

Solve each equation by graphing the related function on the graphing calculator.

4. $x^2 - 16 = 0$ 5. $x^2 + 6 = 0$
6. $x^2 - 4 = 0$ 7. $x^2 = 25$
8. $x^2 + 6 = -3$ 9. $x^2 - 1.5 = -1.5$
10. a. **Critical Thinking** For what values of c will $x^2 = c$ have two solutions? Sketch the graph of the equation.
    b. For what values of c will $x^2 = c$ have one solution? Sketch the graph of the equation.
    c. For what values of c will $x^2 = c$ have no solution? Sketch the graph of the equation.

Mental Math Tell the number of solutions for each equation.

11. $x^2 = 12$ 12. $x^2 = -4$ 13. $x^2 + 5 = -4$ 14. $x^2 - 6 = -6$

for use with Example 3, addresses visual and tactile learning by using a graphing calculator to discover how to use a related equation to solve a quadratic equation by graphing.

Alternative Activity for Lesson 7-6

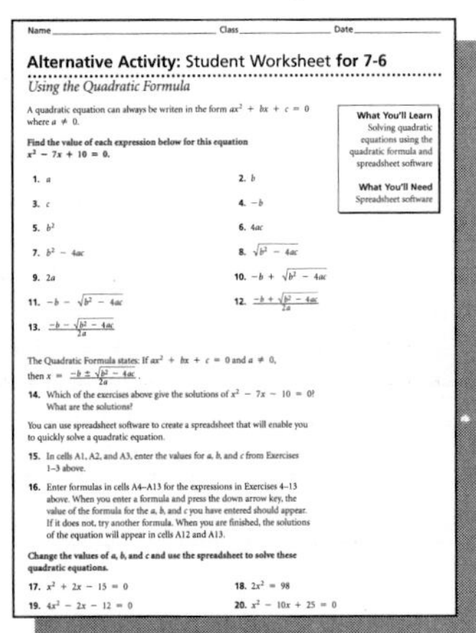

Name ____________ Class ____________ Date ____________

Alternative Activity: Student Worksheet for 7-6

Using the Quadratic Formula

A quadratic equation can always be writen in the form $ax^2 + bx + c = 0$ where $a \neq 0$.

What You'll Learn Solving quadratic equations using the quadratic formula and spreadsheet software

What You'll Need Spreadsheet software

Find the value of each expression below for this equation $x^2 - 7x + 10 = 0$.

1. a 2. b
3. c 4. $-b$
5. b^2 6. $4ac$
7. $b^2 - 4ac$ 8. $\sqrt{b^2 - 4ac}$
9. $2a$ 10. $-b + \sqrt{b^2 - 4ac}$
11. $-b - \sqrt{b^2 - 4ac}$ 12. $\frac{-b + \sqrt{b^2 - 4ac}}{2a}$
13. $\frac{-b - \sqrt{b^2 - 4ac}}{2a}$

The Quadratic Formula states: If $ax^2 + bx + c = 0$ and $a \neq 0$, then $x = \frac{-b \pm \sqrt{b^2 - 4ac}}{2a}$.

14. Which of the exercises above give the solutions of $x^2 - 7x - 10 = 0$? What are the solutions?

You can use spreadsheet software to create a spreadsheet that will enable you to quickly solve a quadratic equation.

15. In cells A1, A2, and A3, enter the values for a, b, and c from Exercises 1–3 above.
16. Enter formulas in cells A4–A13 for the expressions in Exercises 4–13 above. When you enter a formula and press the down arrow key, the value of the formula for the a, b, and c you have entered should appear. If it does not, try another formula. When you are finished, the solutions of the equation will appear in cells A12 and A13.

Change the values of a, b, and c and use the spreadsheet to solve these quadratic equations.

17. $x^2 + 2x - 15 = 0$ 18. $2x^2 = 98$
19. $4x^2 - 2x - 12 = 0$ 20. $x^2 - 10x + 25 = 0$

for use with Example 1, addresses visua and tactile learning by having students s quadratic equations using the quadratic formula and spreadsheet software.

Skills Practice / Assessment Options / Students' Experiences / Group Work / Interactive Questioning / Real World Contexts / Technology Options / Teaching Tools

Cooperative Learning Tips

When used effectively, cooperative learning can help students develop interpersonal skills, learn to perform specific roles in a group, and learn to carry out specific responsibilities. The components of Chapter 7 provide a range of cooperative learning opportunities.

- In the Student Edition, the ***Work Together*** parts of lessons are specifically designed for cooperative learning activities.
- In the Teacher's Edition, you will find helpful hints for addressing diverse learning styles (see page C for Chapter 7). For every lesson, you will find a ***Reteaching Activity***, which may involve cooperative learning.

Materials and Manipulatives

The authors expect all students to use scientific calculators. Calculator use is integrated throughout the course.

- calculator (7-4, 7-5, 7-6)
- graph paper (7-3)
- graphing calculator (7-1, 7-2, 7-7)

Technology Options / Skills Practice / Assessment Options / Students' Experiences / Group Work / Interactive Questioning / Real World Contexts / Teaching Tools

TECHNOLOGY OPTIONS

Technology Tools		Chapter Project	7-1	7-2	7-3	7-4	7-5	7-6	7-7
Calculator		Assumed that students have scientific calculators at any time.							
Graphing Calculator	Handbook			✔	✔				
	Student Edition		✔	✔	✔	✔	✔		✔
Software	Secondary Math Lab Toolkit		✔	✔	✔	✔	✔	✔	✔
	Integrated Math Lab		✔				✔		
	Computer Item Generator		✔	✔	✔	✔	✔	✔	✔
	Student Edition								
Video	Video Field Trip	✔							
CD-ROM	Multimedia Algebra Lab			✔	✔			✔	✔
Internet		See the Prentice Hall site. (http://www.phschool.com)							

The Prentice Hall Algebra program offers you a rich variety of technology options. Be assured that all these options are provided as a means of enriching the program and are not essential for the successful completion of the course.

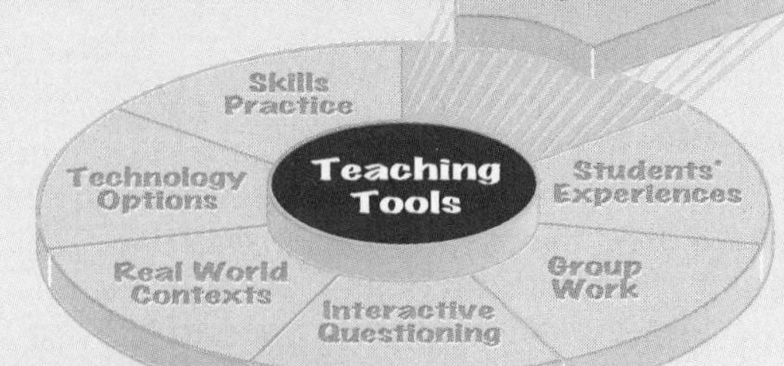

Assessment Options

The Prentice Hall Algebra Program provides you with many options. From these options, you may choose instructional materials and techniques appropriate for your students, or those necessary to meet your district's curriculum requirements. As the chart indicates, the program also supports your teaching efforts by offering you many choices for assessment.

ASSESSMENT OPTIONS

Assessment Support Materials	Chapter Opener	7-1	7-2	7-3	7-4	7-5	7-6	7-7	Chapter End
Chapter Project	▲■●			▲■	▲■		▲■	▲■	▲■
Checkpoints					▲●			▲●	
Self-Assessment		▲■		▲■			▲■	▲■	▲■
Writing Assignment		▲	▲	▲	▲	▲	▲	▲	▲●
Chapter Assessment									▲●
Alternative Assessment		■	■	■	■	■	■	■	■●
Cumulative Review									●
Standardized Test Prep					▲■	▲■	▲■	▲■	▲●
Computer Item Generator	Can be used to create custom-made practice or assessment at any time.								

▲ = Student Edition ■ = Teacher's Edition ● = Teaching Resources

Checkpoints

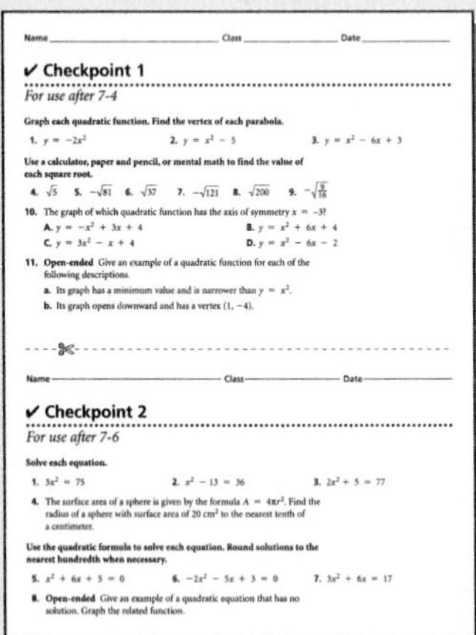
Name ______ Class ______ Date ______

✔ Checkpoint 1

For use after 7-4

Graph each quadratic function. Find the vertex of each parabola.

1. $y = -2x^2$ 2. $y = x^2 - 5$ 3. $y = x^2 - 6x + 3$

Use a calculator, paper and pencil, or mental math to find the value of each square root.

4. $\sqrt{5}$ 5. $-\sqrt{81}$ 6. $\sqrt{37}$ 7. $-\sqrt{121}$ 8. $\sqrt{200}$ 9. $-\sqrt{\frac{9}{16}}$

10. The graph of which quadratic function has the axis of symmetry $x = -3$?
 A. $y = -x^2 + 3x + 4$ B. $y = x^2 + 6x + 4$
 C. $y = 3x^2 - x + 4$ D. $y = x^2 - 6x - 2$

11. **Open-ended** Give an example of a quadratic function for each of the following descriptions.
 a. Its graph has a minimum value and is narrower than $y = x^2$.
 b. Its graph opens downward and has a vertex $(1, -4)$.

Name ______ Class ______ Date ______

✔ Checkpoint 2

For use after 7-6

Solve each equation.

1. $3x^2 = 75$ 2. $x^2 - 13 = 36$ 3. $2x^2 + 5 = 77$

4. The surface area of a sphere is given by the formula $A = 4\pi r^2$. Find the radius of a sphere with surface area of 20 cm^2 to the nearest tenth of a centimeter.

Use the quadratic formula to solve each equation. Round solutions to the nearest hundredth when necessary.

5. $x^2 + 6x + 5 = 0$ 6. $-2x^2 - 5x + 3 = 0$ 7. $3x^2 + 6x = 17$

8. **Open-ended** Give an example of a quadratic equation that has no solution. Graph the related function.

Alternative Assessment

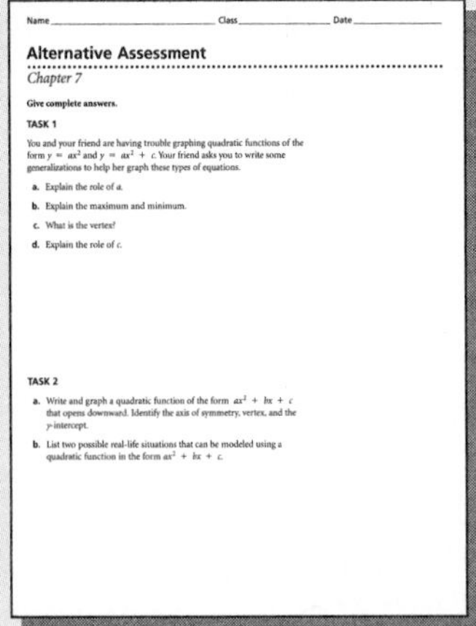
Name ______ Class ______ Date ______

Alternative Assessment

Chapter 7

Give complete answers.

TASK 1

You and your friend are having trouble graphing quadratic functions of the form $y = ax^2$ and $y = ax^2 + c$. Your friend asks you to write some generalizations to help her graph these types of equations.

a. Explain the role of a.
b. Explain the maximum and minimum.
c. What is the vertex?
d. Explain the role of c.

TASK 2

a. Write and graph a quadratic function of the form $ax^2 + bx + c$ that opens downward. Identify the axis of symmetry, vertex, and the y-intercept.
b. List two possible real-life situations that can be modeled using a quadratic function in the form $ax^2 + bx + c$.

Chapter Assessment

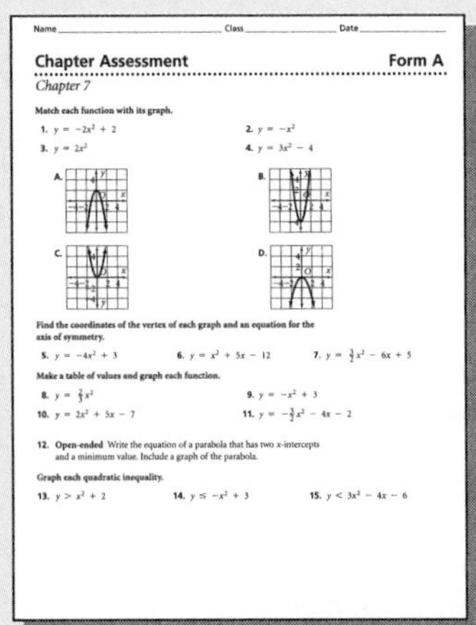
Name ______ Class ______ Date ______

Chapter Assessment **Form A**

Chapter 7

Match each function with its graph.

1. $y = -2x^2 + 2$ 2. $y = -x^2$
3. $y = 2x^2$ 4. $y = 3x^2 - 4$

A. B. C. D.

Find the coordinates of the vertex of each graph and an equation for the axis of symmetry.

5. $y = -4x^2 + 3$ 6. $y = x^2 + 5x - 12$ 7. $y = \frac{1}{2}x^2 - 6x + 5$

Make a table of values and graph each function.

8. $y = \frac{2}{3}x^2$ 9. $y = -x^2 + 3$
10. $y = 2x^2 + 5x - 7$ 11. $y = -\frac{3}{2}x^2 - 4x - 2$

12. **Open-ended** Write the equation of a parabola that has two x-intercepts and a minimum value. Include a graph of the parabola.

Graph each quadratic inequality.

13. $y > x^2 + 2$ 14. $y \le -x^2 + 3$ 15. $y < 3x^2 - 4x - 6$

Available in both Form A and Form B

Making the Right Connections

Mathematics is imbedded in nearly every walk of life. The National Council of Teachers of Mathematics (NCTM) encourages educators to recognize these connections and to emphasize them for the purpose of better educating students for success in life and in a global economy. The ***Connections*** chart below highlights these connections for Chapter 7.

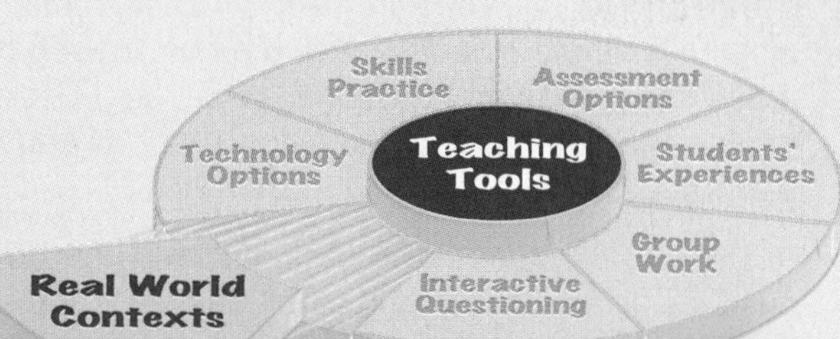

CONNECTIONS

Lesson	Interdiciplinary Connections	Career Prep	Other Real World Connections	Math Integration	NCTM Standards
Chapter Project		Transportation	Road Safety	Statistics	Problem Solving
7-1	Science	Construction	Toys Art Photography	Geometry	Algebra Communication Problem Solving
7-2	Economics Enviromental Science	Architecture Sales Business	Nature Landscaping Real Estate	Geometry	Algebra Communication Problem Solving
7-3	Physical Education Fitness	Meteorology	Weather Road Safety Fireworks Gardening Rowing		Algebra Communication Problem Solving
7-4	Physics History	Researcher	Space Towers Sports	Trigonometry	Algebra Communication Problem Solving
7-5	History Art Physics	Industry Scientist	City Planning Painting Space		Algebra Communication Problem Solving
7-6	Social Studies	Transportation Industry	Model Rockets Recreation Population	Statistics	Algebra Communication Problem Solving
7-7	Physics Geology Home Economics	Electrical Engineering Business	Home Improvements Business Computer		Algebra Communication Problem Solving

CHAPTER 7 Quadratic Equations and Functions

CONNECTING TO PRIOR LEARNING Elicit from students instances where objects travel in a non-linear path. (throwing a ball, roller coaster, traveling in a car, etc.) Explain that the motions can be represented using a type of equation called a *quadratic function.*

CULTURAL CONNECTIONS Ask students how driving laws and customs differ from those in other countries. Some students may be more accustomed to public transportation and bicycles and not really understand the American love of the automobile.

INTERDISCIPLINARY CONNECTIONS Ask students why it is important to understand the science and mathematics of driving. How does the weight and design of a car affect its overall safety? What do weather conditions have to do with driving? Ask students how mathematics and science are used to improve the safety and economy of cars.

ABOUT THE PROJECT The Chapter Project gives students an opportunity to apply mathematical models to real world driving situations. In the Find Out questions found throughout the chapter, students use formulas, equations, tables, and graphs to represent the relationship between speed, stopping distance, reaction time, and road type. Students also design a skit to present what they have learned about safe highway driving.

Technology Options

Prentice Hall Technology

Video Field Trip 7
"Downshift," a visit to the Bondurant School of Driving.

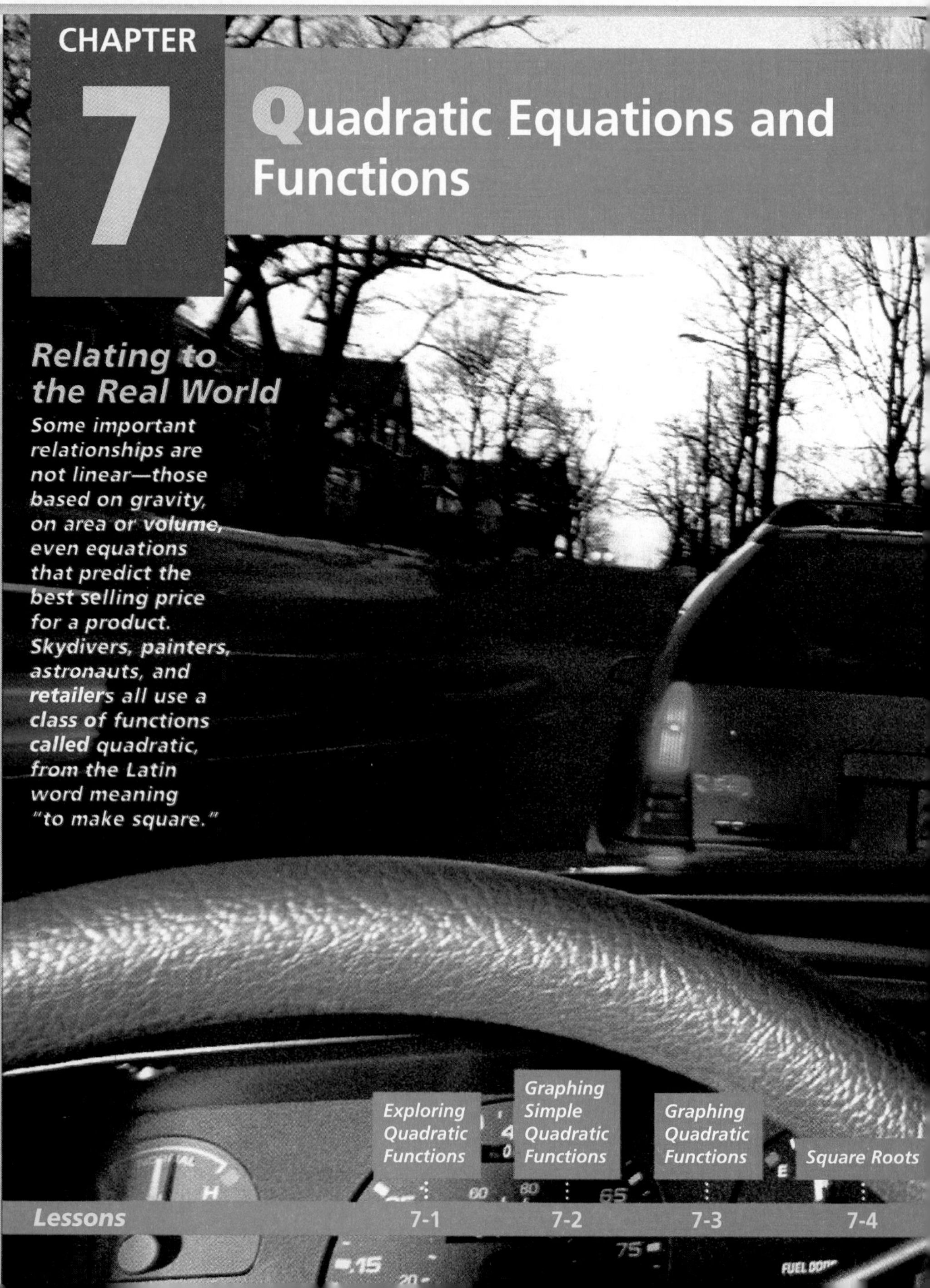

CHAPTER 7 Quadratic Equations and Functions

Relating to the Real World

Some important relationships are not linear—those based on gravity, on area or volume, even equations that predict the best selling price for a product. Skydivers, painters, astronauts, and retailers all use a class of functions called quadratic, from the Latin word meaning "to make square."

Launching the Project

PROJECT NOTEBOOK Encourage students to keep all project-related materials in a separate folder or notebook. See Chapter Project and Scoring Rubric in Chapter Support File.

- Ask students to estimate how long they think it takes to stop a car traveling on dry pavement. Encourage students to consider what factors affect stopping distance. speed, road conditions, reaction time
- Gather statistics to demonstrate actual stopping distances for vehicles. You may want to have the class measure the actual distances so that students can better visualize the data.
- Have students look at the Find Out by Graphing section on page 331 of their textbooks. Explain that, when they begin the Chapter Project, they will use the given formula to calculate safe stopping distances for different speeds.
- Have students graph the results of their calculations.
- Challenge students to compare their calculated results with their earlier estimations and draw a conclusion about the differences.

TRACKING THE PROJECT You may wish to have students read Finishing the Chapter Project on page 354 to help them get an overview of the project. Set benchmark deadlines for students to show you their work in progress.

CHAPTER PROJECT

Full STOP Ahead

What is a safe stopping distance for cars traveling on the highway? How do accident investigators determine whether cars involved in an accident were traveling at safe speeds? There are many variables that affect how quickly a car can stop. These include the car's speed, the driver's reaction time, the type of road, and the weather conditions.

As you work through the chapter, you will use formulas to estimate safe speeds under various conditions. You will make graphs to illustrate the relationships between speed, reaction time, and stopping distance. Then, with your classmates, you will plan a skit to present what you have learned about safe highway driving.

To help you complete the project:

- **p. 331** *Find Out by Graphing*
- **p. 336** *Find Out by Calculating*
- **p. 347** *Find Out by Reasoning*
- **p. 353** *Find Out by Communicating*
- **p. 354** *Finishing the Project*

Project Resources

Teaching Resources

Chapter Support File, Ch. 7

- Chapter Project Manager and Scoring Rubric

Transparencies

87

Using the Rubric

Sharing the scoring rubric for the project with your students will alert them to your expectations before they begin work on the project.

As students complete each Find Out question in the chapter, you may wish to have them evaluate their own work or a partner's work, based on the scoring rubric. Students should have the opportunity to revise their work after it has been reviewed.

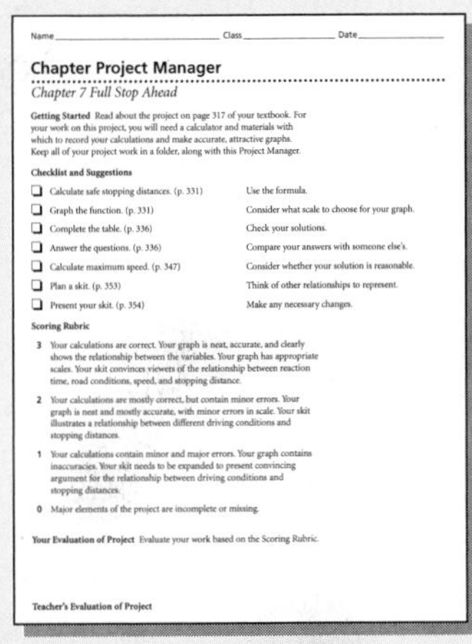

Name________ Class________ Date________

Chapter Project Manager

Chapter 7 Full Stop Ahead

Getting Started Read about the project on page 317 of your textbook. For your work on this project, you will need a calculator and materials with which to record your calculations and make accurate, attractive graphs. Keep all of your project work in a folder, along with this Project Manager.

Checklist and Suggestions

❑ Calculate safe stopping distances. (p. 331)	Use the formula.
❑ Graph the function. (p. 331)	Consider what scale to choose for your graph.
❑ Complete the table. (p. 336)	Check your solutions.
❑ Answer the questions. (p. 336)	Compare your answers with someone else's.
❑ Calculate maximum speed. (p. 347)	Consider whether your solution is reasonable.
❑ Plan a skit. (p. 353)	Think of other relationships to represent.
❑ Present your skit. (p. 354)	Make any necessary changes.

Scoring Rubric

3 Your calculations are correct. Your graph is neat, accurate, and clearly shows the relationship between the variables. Your graph has appropriate scales. Your skit convinces viewers of the relationship between reaction time, road conditions, speed, and stopping distance.

2 Your calculations are mostly correct, but contain minor errors. Your graph is neat and mostly accurate, with minor errors in scale. Your skit illustrates a relationship between different driving conditions and stopping distances.

1 Your calculations contain minor and major errors. Your graph contains inaccuracies. Your skit needs to be expanded to present convincing argument for the relationship between driving conditions and stopping distances.

0 Major elements of the project are incomplete or missing.

Your Evaluation of Project Evaluate your work based on the Scoring Rubric.

Teacher's Evaluation of Project

7-1 Teaching Notes

PROBLEM OF THE DAY

How many total squares of various sizes are there on an 8×8 grid? Hint: First find the number of 1×1 squares, then 2×2 squares, etc. Look for a pattern. 204

Problem of the Day is also available in Transparencies.

CONNECTING TO PRIOR KNOWLEDGE Have students write a short description of something with the same basic shape as the graphs in this lesson. Examples may include a mountain, a valley, a smile, a frown, or the path of a thrown ball.

WORK TOGETHER

Instruct students to put arrows on the ends of the graphs to show that the parabola continues. Have students use a calculator to evaluate the function for non-integer values of x, and locate the points on the graph to see that they are part of the curve.

VISUAL LEARNING Questions 1 and 2 Make sure students understand that quadratic functions are graphed as curves. Graph the points and draw the curve on the board or overhead projector. Make sure students do not connect points on a quadratic graph with straight lines.

Lesson Planning Options

Prerequisite Skills

- Graphing coordinate pairs
- Evaluating expressions

Assignment Options

To provide flexible scheduling, this lesson can be subdivided into parts.

1 **Core** 1–4, 17–20
Extension 21–24, 29–33

2 **Core** 5–16, 25–28
Extension 34–39

Use Mixed Review to maintain skills.

Resources

Student Edition

Skills Handbook, pp. 569, 580
Extra Practice, p. 562
Glossary/Study Guide

Teaching Resources

Chapter Support File, Ch. 7
- Practice 7-1 (two worksheets)
- Reteaching 7-1
- Alternative Activity 7-1

Classroom Manager 7-1
Glossary, Spanish Resources
Two-Year Algebra Handbook 7-1

Transparencies

2, 11, 29, 88, 92

Connections 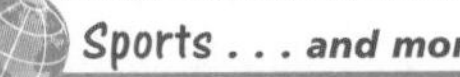 Sports . . . and more

7-1 Exploring Quadratic Functions

What You'll Learn

- Graphing quadratic functions of the form $y = ax^2$

...And Why

To understand how changing a affects the graph of a quadratic function

What You'll Need

- graphing calculator

WORK TOGETHER

Work with a partner. Complete a table of values and plot points, or use a graphing calculator set at the standard scale.

x	y
−2	■
−1	■
0	■
1	■
2	■

1. a. Graph the equations $y = x^2$ and $y = 3x^2$.
   b. How are the graphs in part (a) alike? different?
   1a–b. See margin.
2. a. Graph the equations $y = -x^2$ and $y = -3x^2$. 2a. See margin.
   b. How are the graphs in part (a) different from the graphs in Question 1? How are they like the graphs in Question 1? The graphs differ in the direction of their opening. $y = x^2$ and $y = -x^2$ are the same size; $y = 3x^2$ and $y = -3x^2$ are the same size.

THINK AND DISCUSS

Part 1

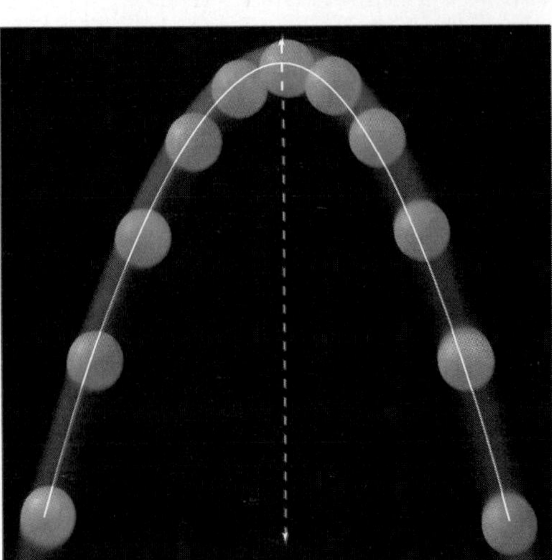

Quadratic Functions

The graphs you analyzed in the Work Together are all examples of **parabolas.** If you draw a parabola on a piece of paper, you can fold the paper down the middle of the parabola and the two sides will match exactly. The line down the middle of the parabola is the **axis of symmetry.**

3. **Try This** Trace each parabola on a sheet of paper and draw its axis of symmetry.

a.
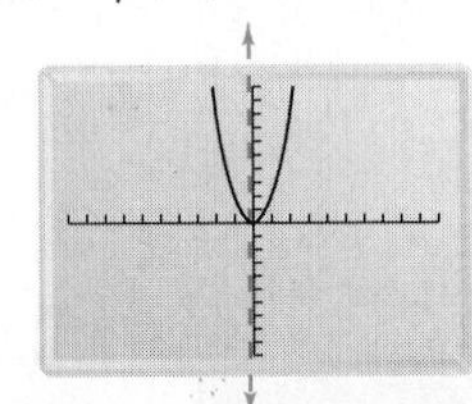

b.
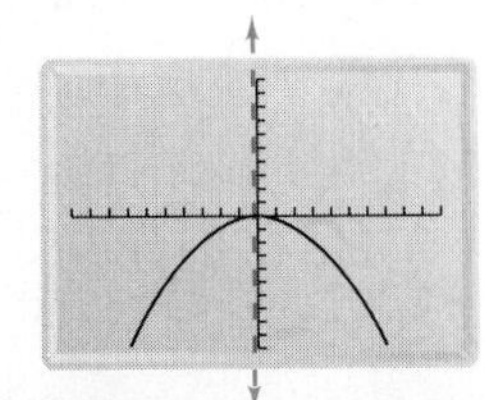

Question 2 Remind students to square the x-term before multiplying the x-term by **21** or **23**.

THINK AND DISCUSS

Have students substitute $a = 0$ into the quadratic function definition to help them understand that when $a = 0$, the function is not quadratic.

Question 4 Remind students that b and c can have values of zero within a quadratic function.

Example 1

ERROR ALERT! Students may think that the example sketch represents one graph instead of two. **Remediation:** Have students sketch the graph of $y = 2x^2$ on one coordinate plane and the graph of $y = -2x^2$ on another coordinate plane.

The term *minimum value* refers to the value of y. No point on a parabola can have a y-value lower than the minimum value.

Point out that we use the term *vertex* whether the point is a maximum or a minimum value.

Each parabola that you have seen is the graph of a *quadratic function.*

Quadratic Function

For $a \neq 0$, the function $y = ax^2 + bx + c$ is a **quadratic function.**

Examples: $y = 2x^2$, $y = x^2 + 2$, $y = -x^2 - x - 3$

When a quadratic function is written in the form $y = ax^2 + bx + c$, it is in **standard form.**

4. Name the values of a, b, and c for each quadratic function.
   a. $y = 3x^2 - 2x + 5$ 3; −2; 5
   b. $y = 3x^2$ 3; 0; 0
   c. $y = -0.5x^2 + 2x$ −0.5; 2; 0
5. Write each quadratic function in standard form.
   a. $y = 7x + 9x^2 - 4$ $y = 9x^2 + 7x - 4$
   b. $y = 3 - x^2$ $y = -x^2 + 3$

Part 

The Role of "a"

When you graph a quadratic function, if the value of a is positive, the parabola opens upward. If the value of a is negative, the parabola opens downward.

Example 1

Make a table of values and graph the quadratic functions $y = 2x^2$ and $y = -2x^2$.

x	$y=2x^2$	$y=-2x^2$
-3	18	-18
-2	8	-8
-1	2	-2
0	0	0
1	2	-2
2	8	-8
3	18	-18

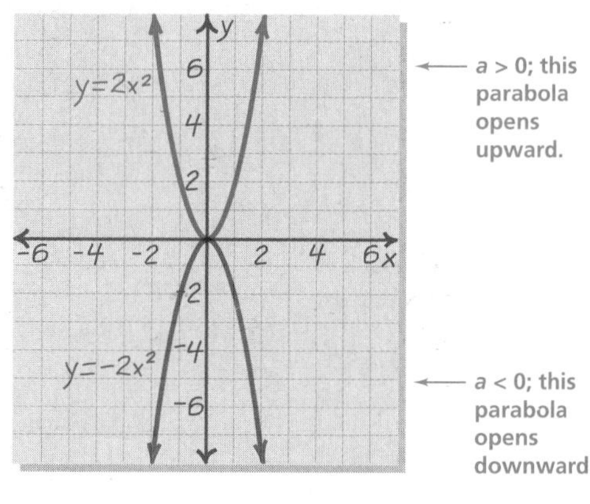

6. What is the axis of symmetry for the graphs in Example 1? y-axis
7. What would happen to the graph of $y = 2x^2$ if you could fold the graph over the x-axis? Explain. The graph matches the graph of $y = -2x^2$ because −2 is the opposite of 2.

Additional Examples

FOR EXAMPLE 1

Make a table of values and graph the quadratic functions $y = 3x^2$ and $y = -4x^2$.

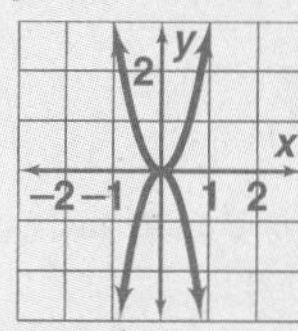

Discussion: *What effect does a negative sign have on the shape of a quadratic function's graph?*

FOR EXAMPLE 2

Graph $y = -\frac{1}{3}x^2$, $y = \frac{1}{2}x^2$, and $y = -x^2$. Compare the width of the graphs.

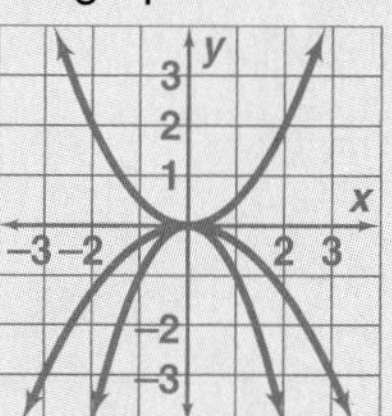

The order of equations from widest to narrowest are $y = -x^2$, $y = \frac{1}{2}x^2$, $y = -\frac{1}{3}x^2$.

Discussion: *Explain how you could quickly sketch the graph of a quadratic equation.*

Example 2

CRITICAL THINKING Question 10 Have students compare the graphs in Example 1 to see that a graph with $a = 2$ is the same width as a graph with $a = -2$.

AUDITORY LEARNING Give students practice hearing and speaking about graphs of quadratic equations by asking questions such as the following.

- *Which graph is wider: $y = 3x^2$ or $y = 5x^2$?* $y = 3x^2$
- *Does the graph of $y = -12x^2$ open up or down?* down
- *When a graph opens downward, is the vertex the minimum value or the maximum value?* maximum value

EXTENSION Challenge students to experiment with graphing equations that have an exponent of $x > 2$. Suggest that students make a table of values or use a graphing calculator to investigate the shape of graphs such as $y = x^3$ and $y = x^4$.

KINESTHETIC LEARNING Clear a large area in the room and make a large coordinate grid on the floor using string or masking tape. Direct students to model the graphs of simple quadratic functions by holding string or rope and standing on the grid. Begin by having the group model $y = x^2$. Then have them model equations that are wider and narrower. Lead students to understand how the value of a affects the width of the graph.

MAKING CONNECTIONS When you turn on a light bulb, the light radiates outward in all directions. However, when you turn on a flashlight or a car's headlights, the light comes out in a beam. Behind the bulb is a reflector in the shape of a

Technology Options

For Exercises 17–20, students may find graphing software or a graphing calculator helpful to order the quadratic functions.

Prentice Hall Technology 

Calculator • Graphing Calculator Handbook: Procedure 4

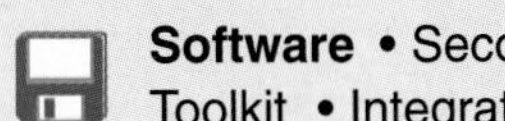

Software • Secondary Math Lab Toolkit • Integrated Math Lab 13 • Computer Item Generator 7-1

Internet • See the Prentice Hall site. (http://www.phschool.com)

The highest or lowest point on a parabola is called the **vertex** of the parabola.

When a parabola opens upward, the y-coordinate of the vertex is the **minimum value** of the function.

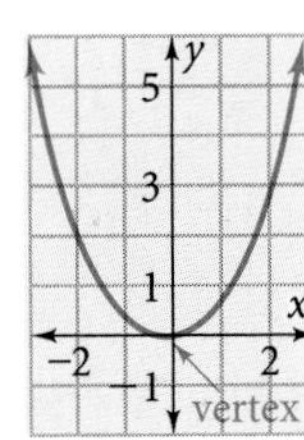

When a parabola opens downward, the y-coordinate of the vertex is the **maximum value** of the function.

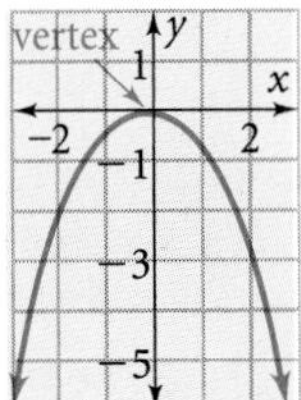

8. Answer these questions for $y = -3x^2$ and $y = 4x^2$.
   a. What is the value of a? −3; 4
   b. In which direction does each graph open? downward; upward
   c. Is the y-coordinate of the vertex a minimum or a maximum value of the function? maximum; minimum

9. **Summarize** what you know so far about how the value of a affects the parabola. When $a > 0$, the parabola opens upward. When $a < 0$, the parabola opens downward.

The value of a also affects the width of a parabola.

Example 2

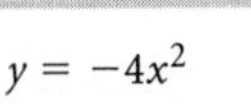

Graph $-4x^2$, $y = \frac{1}{4}x^2$, and $y = x^2$. Compare the widths of the graphs.

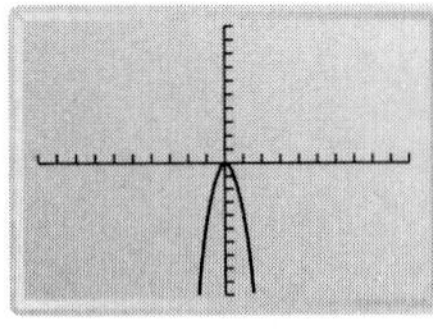

$y = -4x^2$

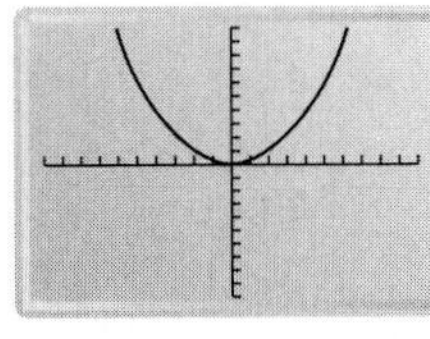

$y = \frac{1}{4}x^2$

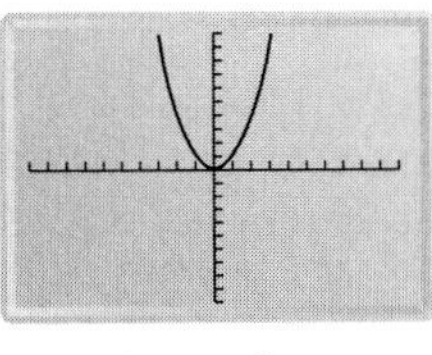

$y = x^2$

Of the three graphs, $y = \frac{1}{4}x^2$ is the widest, $y = x^2$ is narrower, and $y = -4x^2$ is the narrowest.

10. *Critical Thinking* What does the absolute value of a tell you about the width of a parabola? Use the graphs in Example 2 to explain your answer. The larger the abs. val. of a, the narrower the parabola.

11. *Try This* Order each group of quadratic functions from widest to narrowest graph.
    a. $y = 2x^2, y = 3x^2, y = -5x^2$ $\quad y = 2x^2, y = 3x^2, y = -5x^2$
    b. $y = \frac{3}{2}x^2, y = \frac{2}{3}x^2, y = \frac{1}{2}x^2$ $\quad y = \frac{1}{2}x^2, y = \frac{2}{3}x^2, y = \frac{3}{2}x^2$

parabola. The random light waves bounce off the reflector and are focused into a beam. Satellite receiver dishes work in a similar way. Signals from satellites orbiting Earth hit the back of the parabola-shaped dish and are focused onto a central receiver.

CONNECTING TO THE STUDENTS' WORLD Encourage students to find representations of parabolas in art, design, architecture, etc. Have students bring in a photograph or picture of a parabola and then have them estimate the value of *a* that best models the width of the parabola.

Exercises ON YOUR OWN

ALTERNATIVE ASSESSMENT **Exercises 9–16** To assess whether students understand the significance of the *a* coefficient in a quadratic function, have them order these functions from narrowest graph to widest graph.

Exercises 17–20 Remind students to use the absolute value of *a* when comparing graphs.

Exercise 25 Challenge students to use the points shown on the graph to find the equation of the quadratic function.

DIVERSITY Exercise 28 Some students may not be familiar with satellite-dish antennas. Have students who are familiar with them explain the most common use of satellite dishes—the reception of television broadcasts.

OPEN-ENDED Exercise 33 Have students write answers in the form of quadratic equations, not sketches.

Exercises ON YOUR OWN

Find the values of *a*, *b*, and *c* for each quadratic function.

1. $y = x^2 + 2x + 4$ 1; 2; 4
2. $y = 2x^2$ 2; 0; 0
3. $y = -x^2 - 3x - 9$ −1; −3; −9
4. $y = -2x^2 + 5$ −2; 0; 5

Tell whether each parabola opens *upward* or *downward* and whether the *y*-coordinate of the vertex is a *maximum* or a *minimum*.

5. $y = x^2$ upward; min.
6. $y = 9x^2$ upward; min.
7. $y = -\frac{2}{5}x^2$ downward; max.
8. $y = -6x^2$ downward; max.

Choose **Graph each quadratic function. Use either a table of values or a graphing calculator.** 9–16. See margin.

9. $y = \frac{1}{2}x^2$
10. $y = 1.5x^2$
11. $y = -4x^2$
12. $y = -\frac{1}{3}x^2$
13. $y = 4x^2$
14. $y = \frac{1}{3}x^2$
15. $y = -1.5x^2$
16. $y = -\frac{1}{2}x^2$

Order each group of quadratic functions from widest to narrowest graph. 17–20. See margin.

17. $y = 3x^2,\ y = x^2,\ y = 7x^2$
18. $y = 4x^2,\ y = \frac{1}{3}x^2,\ y = x^2$
19. $y = -2x^2,\ y = -\frac{2}{3}x^2,\ y = -4x^2$
20. $y = -\frac{1}{2}x^2,\ y = 5x^2,\ y = -\frac{1}{4}x^2$

Give the letter or letters of the graph(s) that make each statement true.

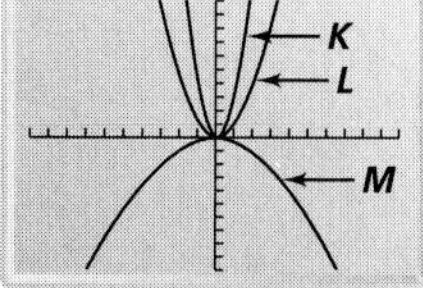

21. $a > 0$ *K, L*
22. $a < 0$ *M*
23. $|a|$ has the greatest value. *K*
24. $|a|$ has the least value. *M*

Trace each parabola on a sheet of paper and draw its axis of symmetry.

25. 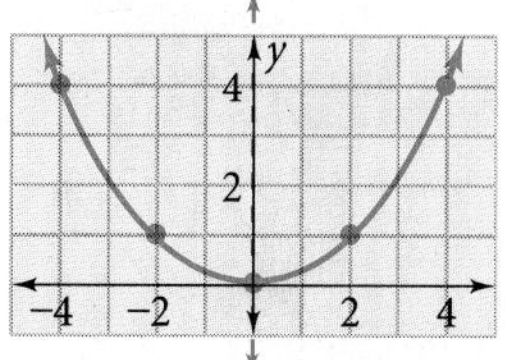

26.

27. 

28. 

page 318 Work Together

1a. 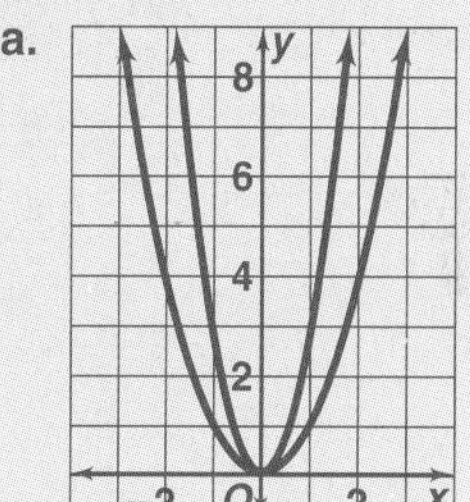

b. Each graph is a U-shaped curve opening upward. The graphs have different widths.

2a. 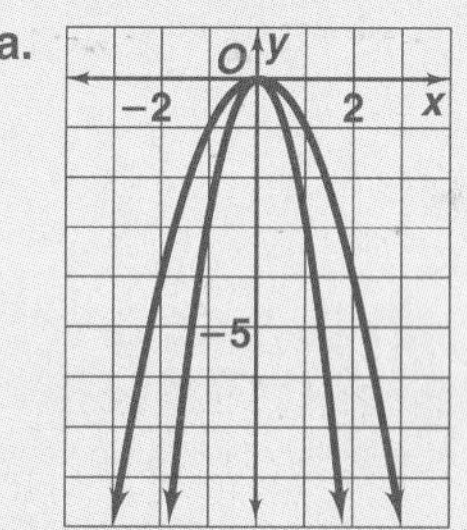

pages 321–322 On Your Own

9. 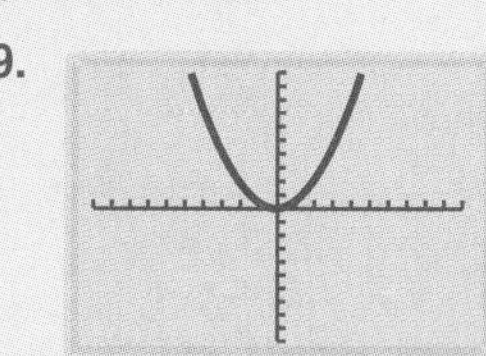

10–20, Getting Ready for Lesson 7-2, 44–46. See back of book.

Exercises MIXED REVIEW

JOURNAL Encourage students to make sketches showing the effects of *a* on the shape of graphs.

ESL Exercise 43 Many students may have seen a Slinky® but not know what it is called in English. Ask students to describe a Slinky®. Help students describe how it looks, what it does, and how to play with it. Try to have a Slinky® or facsimile to show students.

Exercise 43c Remind students that circumference = diameter × π.

CONNECTING TO THE STUDENTS' WORLD Ask students who have Slinkies® to verify that a Slinky® contains 80 ft of wire. Suggest that they use the formula for circumference to approximate the circumference of one coil of wire and then multiply that length by the number of coils.

GETTING READY FOR LESSON 7-2 These exercises prepare students for graphing quadratic equations of the form $y = ax^2 + c$.

Wrap Up

THE BIG IDEA Ask students: *What does the variable a tell you about the shape of the quadratic function graph?*

RETEACHING ACTIVITY Students graph quadratic functions of the form $y = ax^2$. They make a table of values, graph and label. (Reteaching worksheet 7-1)

Reteaching 7-1

Practice 7-1

Practice 7-1
Mixed Exercises

Lesson Quiz

Lesson Quiz is also available in Transparencies.

1. Find the values of *a*, *b*, and *c* for the quadratic equation $y = -2x^2 + \frac{1}{2}x - 3$.
   $a = -2, b = \frac{1}{2}, c = -3$

2. Graph the quadratic functions $y = \frac{2}{3}x^2$ and $y = -1.5x^2$.

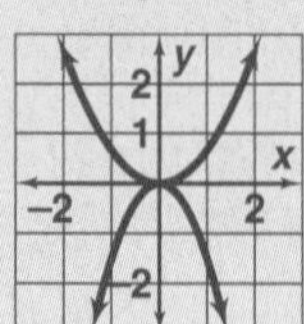

Writing **Without graphing, describe how each graph differs from the graph of $y = x^2$.**

29. $y = 2x^2$ narrower **30.** $y = -x^2$ opens downward **31.** $y = 1.5x^2$ narrower **32.** $y = \frac{1}{2}x^2$ wider

33. *Open-ended* Give an example of a quadratic function for each description. a–c. Answers may vary. Samples are given.
a. Its graph opens upward. $y = 2x^2$
b. Its graph has the same shape as the graph in part (a), but the graph opens downward. $y = -2x^2$
c. Its graph is wider than the graph in part (b). $y = x^2$

Match each function with its graph.

A. $y = x^2$ B. $y = -x^2$ C. $y = 3.5x^2$ D. $y = -3.5x^2$ E. $y = \frac{1}{4}x^2$ F. $y = -\frac{1}{4}x^2$

34. D
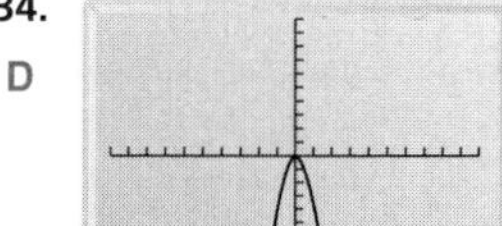

35. E
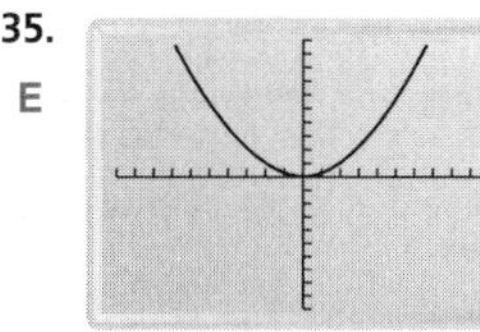

36. B
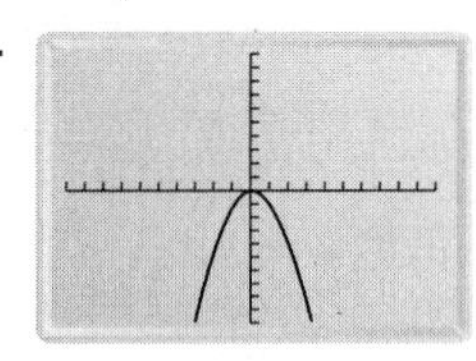

37. F
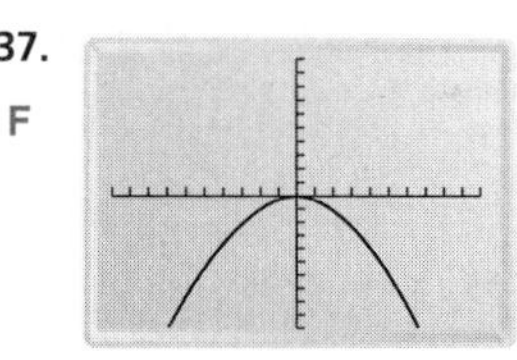

38. A
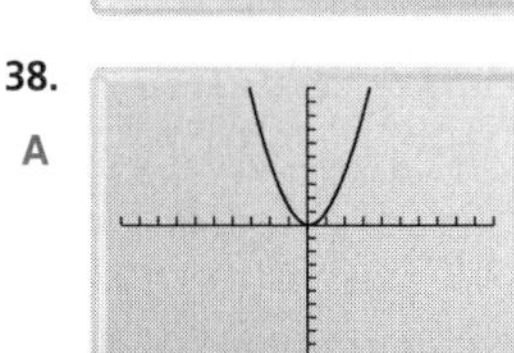

39. C
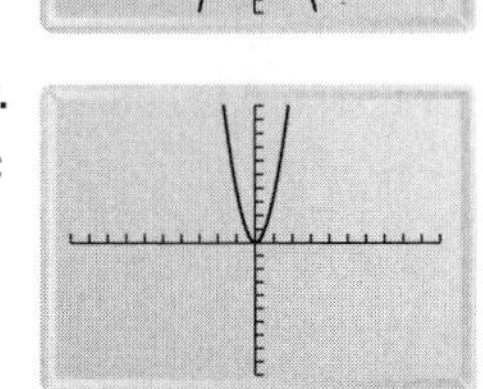

Exercises MIXED REVIEW

Find each percent of change.

40. 12 lb to 14 lb about 17% **41.** 5 ft to 7 ft 40% **42.** \$4.50 to \$2.25 50%

43. A Slinky® toy begins as 80 ft of wire. In 50 years of production, 3,030,000 mi of wire weighing 50,000 tons have been used.
a. How many Slinkies have been made? 199,980,000 Slinkies
b. How much does 1 mi of wire weigh? about 33 lb
c. The 3,030,000 mi of wire could go around the equator 126 times. What is the length of the equator? What is the diameter of Earth? about 24,048 mi; about 7655 mi

FOR YOUR JOURNAL

Describe a quadratic function and its graph. Explain how the value of "a" affects the graph. Give examples of quadratic functions to support your statements.

Getting Ready for Lesson 7-2

Graph each linear equation. 44–46. See margin.

44. $y = 2x$ **45.** $y = 2x - 3$ **46.** $y = 2x + 1$

7-2 Teaching Notes

PROBLEM OF THE DAY

Two positive numbers have a sum of 20. The difference of their squares is 40. Find the numbers. **9 and 11**

Problem of the Day is also available in Transparencies.

CONNECTING TO PRIOR KNOWLEDGE Sketch the graphs of $y = x$ and $y = x + 2$. Ask how the 2 affects the graph.

WORK TOGETHER

Questions 1 and 2 If students plot points from tables of values, have one student read aloud the ordered pairs while the partner plots the points.

THINK AND DISCUSS

Stress that the value of c changes only the *position* of the parabola, not the *shape* of the parabola.

Example 1

TACTILE LEARNING Question 3 After students have drawn the graph of $y = 2x^2$, have them bend a paper clip into the shape of the graph. Then have them move the paper clip up and down the y-axis. By tracing around the paper clip, they can check the points of the new graph and see that the shape of the graph has not been affected by its movement on the axis.

Question 4 Encourage students to use the term *vertex* in their summaries.

Connections Nature . . . and more

7-2 Graphing Simple Quadratic Functions

What You'll Learn

- Graphing quadratic functions of the form $y = ax^2 + c$
- Graphing quadratic functions that represent real-life situations

...And Why

To understand how changing c affects the graph of a quadratic function

What You'll Need

- graphing calculator

WORK TOGETHER

Work with a partner. Use a table of values or a graphing calculator set at the standard scale.

1. a. Graph the quadratic functions $y = x^2$, $y = x^2 - 5$, and $y = x^2 + 1$. **See margin.**
   b. How are the graphs in part (a) alike? How are they different? **The graphs are the same shape; they are shifted up or down.**
2. a. Graph the quadratic functions $y = -\frac{1}{2}x^2$, $y = -\frac{1}{2}x^2 + 4$, and $y = -\frac{1}{2}x^2 - 2$ **See margin.**
   b. How are the graphs in part (a) alike? How are they different? **The graphs are the same shape; they are shifted up or down.**

THINK AND DISCUSS

You have seen that changing the value of a in the function $y = ax^2$ affects whether the parabola opens upward or downward and how wide or narrow the parabola is. Changing the value of c in the function $y = ax^2 + c$ changes the position of the parabola.

Example 1

Graph the quadratic functions $y = -x^2 + 3$ and $y = -x^2 - 1$. Compare them to the graph of $y = -x^2$ at the left.

Make a table of values.

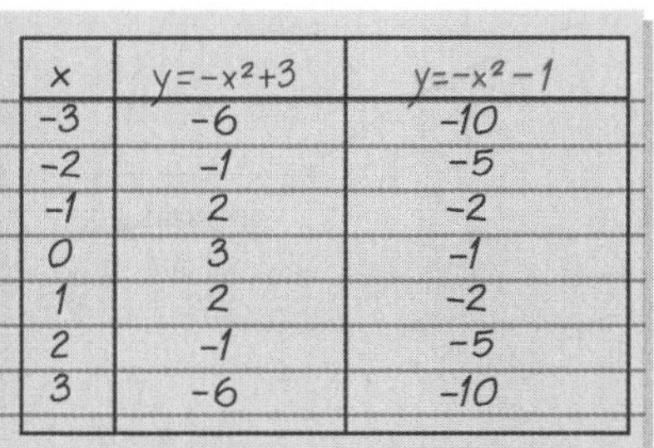

x	$y=-x^2+3$	$y=-x^2-1$
-3	-6	-10
-2	-1	-5
-1	2	-2
0	3	-1
1	2	-2
2	-1	-5
3	-6	-10

Plot the points. Connect the points to form smooth curves.

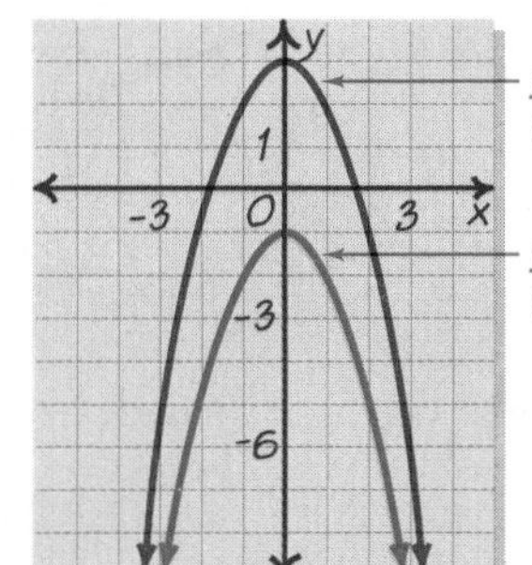

$y = -x^2 + 3$ shifts the parabola $y = -x^2$ *up* 3 units.

$y = -x^2 - 1$ shifts the parabola $y = -x^2$ *down* 1 unit.

3. **Try This** Graph $y = 2x^2$ and $y = 2x^2 - 4$. Compare the graphs. **See margin.**
4. **Summarize** how the graphs of $y = ax^2$ and $y = ax^2 + c$ are different and how they are alike. **The graphs are the same shape but are shifted up or down.**

Lesson Planning Options

Prerequisite Skills

- Making a table of values
- Graphing coordinate pairs

Assignment Options

Core 2–17, 21–33
Extension 1, 18–20, 34

Use Mixed Review to maintain skills.

Resources

Student Edition

Skills Handbook, pp. 567, 578
Extra Practice, p. 562
Glossary/Study Guide

Teaching Resources

Chapter Support File, Ch. 7
- Practice 7-2 (two worksheets)
- Reteaching 7-2

Classroom Manager 7-2
Glossary, Spanish Resources
Two-Year Algebra Handbook, 7-2

Transparencies
2, 11, 29, 88

Question 5 Point out that finding the maximum value of a function means the same thing as finding the value of y at the vertex.

Example 2 **Relating to the Real World** 

Question 6 Make sure students understand that t = time in seconds. Negative time in this situation does not make sense.

ERROR ALERT! **Question 8** Students may think that d can only be positive. **Remediation:** Point out that d represents the height above the edge of the canyon. If the stick drops below the edge of the canyon, then d has a negative value.

Exercises ON YOUR OWN

Exercises 2–5 Encourage students to share ideas they have on all the different ways they could answer these questions.

ALTERNATIVE ASSESSMENT **Exercises 6–17** Choose students at random and ask them to decide whether a given quadratic function has a maximum or a minimum. This will help you assess students' abilities to relate the shape of the graph of a quadratic equation to the sign of its coefficient.

Exercises 6–17 Encourage students to graph each function without using a table of values. After they have sketched a graph, have them check their sketch by testing a couple of points on the graph.

Additional Examples

FOR EXAMPLE 1

Graph the equations $y = 2x^2 - 1$ and $y = 2x^2 - 3$.

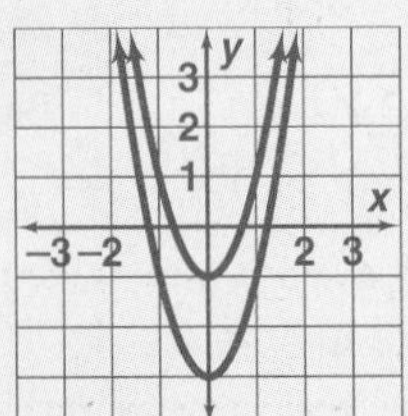

Discussion: *How do these graphs differ from the graph of* $y = 2x^2$? $y = -2x^2$? $y = \frac{1}{2}x^2$?

FOR EXAMPLE 2

The cross section of a swimming pool with a depth of 10 ft can be described by the quadratic equation $y = \frac{1}{4}x^2 - 10$. Graph the equation and estimate the width of the pool.

12.6 ft

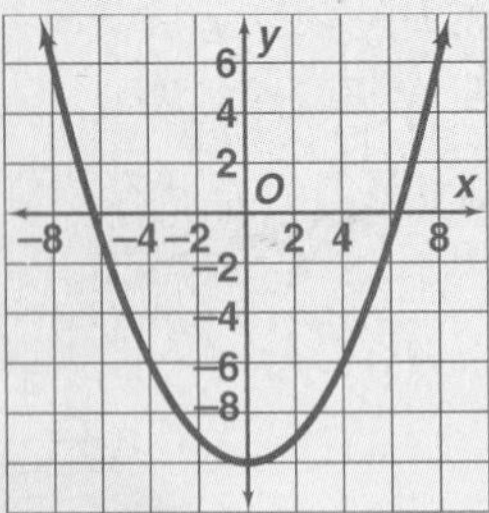

5. Since the parabolas in Example 1 open downward, the y-coordinate of each vertex is a maximum value. Find the maximum value of each function.
$y = -x^2 + 3$ has max. value 3; $y = -x^2 - 1$ has max. value −1.

When you graph a quadratic function that represents a real-life situation, you should limit the domain and range of your graph to x- and y-values that make sense in the situation.

Example 2 **Relating to the Real World** 

Nature Suppose you see an eagle flying over a canyon. The eagle is 30 ft above the level of the canyon's edge when it drops a stick from its claws. The function $d = -16t^2 + 30$ gives the height of the stick in feet after t seconds. Graph this quadratic function.

Graph t on the horizontal or x-axis. Graph d on the vertical or y-axis.

Choose nonnegative values of t that represent the first few seconds of the stick's fall.

Choose values for d that show the height of the stick as it falls.

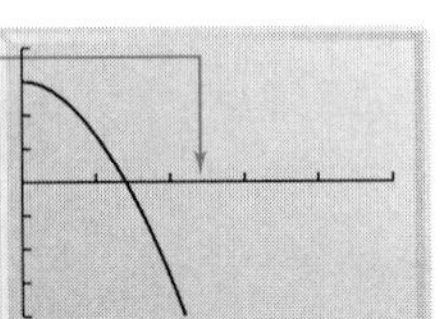

Xmin=0 Ymin=−40
Xmax=5 Ymax=40
Xscl=1 Yscl=10

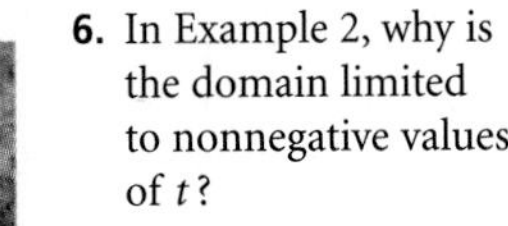

6. In Example 2, why is the domain limited to nonnegative values of t?
7. Explain why the range does not include any values greater than 30.
8. The height of the stick is represented by d. From what level is the height measured?

6. Time cannot be negative.
7. The stick starts falling from a height of 30 ft.
8. The height is measured from the canyon's edge.

ERROR ALERT! OPEN-ENDED Exercise 18 Students may think that if a graph has a vertex below the x-axis, then it must open downward. **Remediation:** Draw the graphs of equations such as $y = x^2 - 3$ and $y = -x^2 + 4$ to show that the location of the vertex has nothing to do with whether the graph opens upward or downward.

GEOMETRY Exercise 20 Students use a quadratic function to find the area of a circular pizza.

Exercise 20 Remind students of the difference between radius and diameter.

Exercise 20a Explain that the values of r that make sense are those values for which the pizza will fit inside the box.

EXTENSION Exercise 20 Challenge students to research why pizza boxes are square instead of round. Encourage students to brainstorm for a list of ideas and then call or visit pizza restaurants to find answers.

Exercises 21–26 Tell students to assume that each graph shown is drawn to the same scale.

Exercises 31 and 32 Ask:

- *Can a quadratic equation have neither a maximum nor a minimum?* no
- *Can a quadratic equation have both a maximum and a minimum?* no

Exercise 33 Encourage students to make sketches to help them understand the question. Ask: *Where did the value 240 come from in the function* $y = 240 - x^2$? 240 represents

Exercises ON YOUR OWN

1. *Writing* Describe how the graphs of the functions $y = 5x^2$, $y = 5x^2 + 50$, and $y = 5x^2 - 90$ are alike and how they are different.
The graphs of the functions have the same shape but are shifted up or down from each other.

Describe whether each quadratic function has a *maximum* or *minimum*.

2. $y = x^2 - 2$ minimum
3. $y = -x^2 + 6$ maximum
4. $y = 3x^2 + 1$ minimum
5. $y = -\frac{1}{2}x^2 - 9$ maximum

Choose **Graph each quadratic function. Use either a table of values or a graphing calculator.** 6–17. See margin.

6. $y = x^2 + 2$
7. $y = x^2 - 3$
8. $y = -x^2 + 4$
9. $y = -x^2 - 1$
10. $y = -2x^2 + 2$
11. $y = -2x^2 - 2$
12. $y = -\frac{1}{4}x^2$
13. $y = -\frac{1}{4}x^2 + 3$
14. $y = 4x^2$
15. $y = 4x^2 - 7$
16. $y = -1.5x^2 + 5$
17. $y = -1.5x^2 - 1$

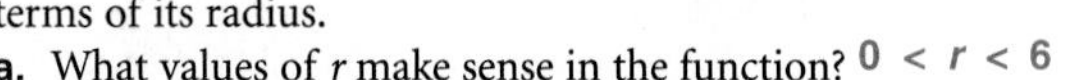

Open-ended **Give an example of a quadratic function for each description.** 18–19. Answers may vary. Samples are given.

18. It opens upward and its vertex is below the origin. $y = x^2 - 2$
19. It opens downward and its vertex is above the origin. $y = -x^2 + 2$
20. *Geometry* Suppose that a pizza must fit into a box with a base that is 12 in. long and 12 in. wide. You can use the quadratic function $A = \pi r^2$ to find the area of a pizza in terms of its radius.
    a. What values of r make sense in the function? $0 < r < 6$
    b. What values of A make sense in the function? $0 < A < 113$
    c. Graph the function. Use $\pi \approx 3.14$. See margin.

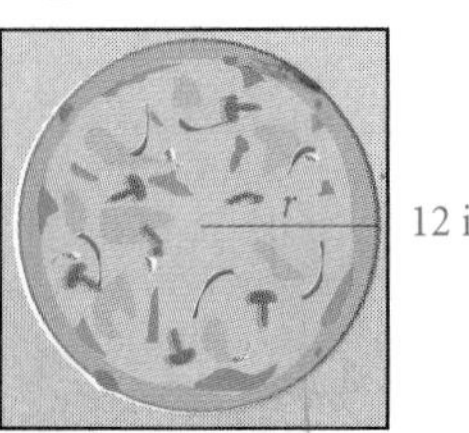

Match each function with its graph.

A. $y = x^2 - 1$
B. $y = x^2 + 4$
C. $y = -x^2 + 2$
D. $y = 3x^2 - 5$
E. $y = -3x^2 + 8$
F. $y = -0.2x^2 + 5$

21. 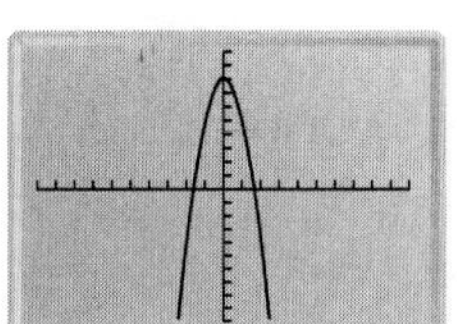 E
22. 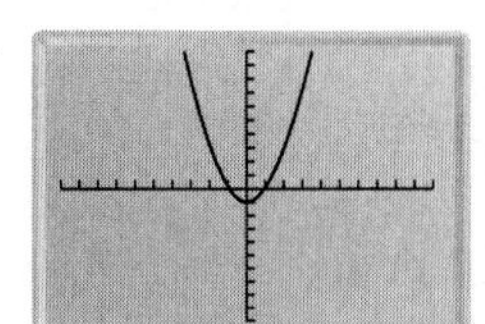 A
23. 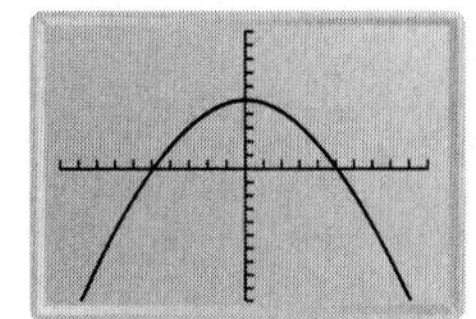 F
24. 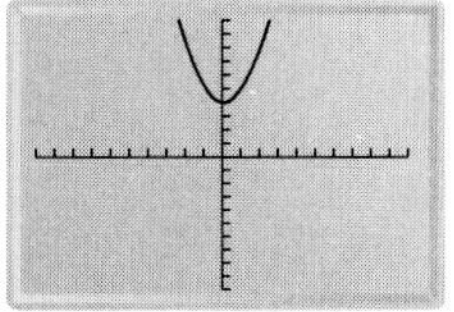 B
25. 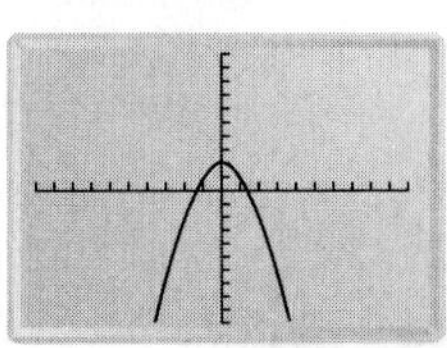 C
26. 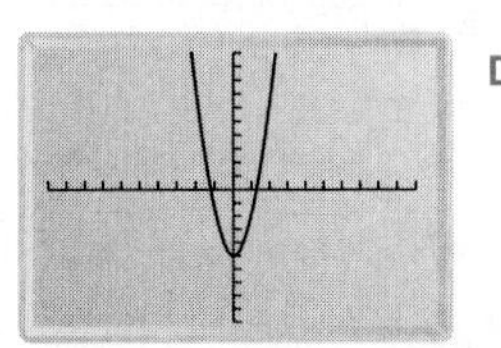 D

Technology Options

For Exercises 2–5 students may want to determine whether each function has a maximum or a minimum by using a graphing calculator.

Prentice Hall Technology

Calculator • Graphing Calculator Handbook: Procedure 4

Software • Secondary Math Lab Toolkit • Computer Item Generator 7-2

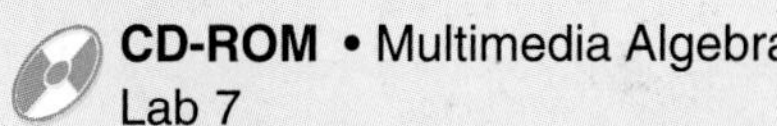

CD-ROM • Multimedia Algebra Lab 7

Internet • See the Prentice Hall site. (http://www.phschool.com)

Work Together, 1a, 2a, Think and Discuss, 3, On Your Own, 6–17, 20c, 33a, 34a, 34b. See back of book.

the area of the patio including the garden; 20 ft · 12 ft = 240 ft^2.

Exercise 34 Point out that the vertical sides should be measured perpendicular to the floor and not in a curve up the side of the arch. Also explain that the 6 ft width is measured at the floor.

RESEARCH Exercise 34c Students can find information in the library about how arches were used in ancient Rome or the Middle Ages. Look in the card catalog to find books about the history of architecture.

Exercises MIXED REVIEW

Exercises 38 and 39 Remind students to rewrite the equation in slope-intercept form before answering.

GETTING READY FOR LESSON 7-3 These exercises prepare students for finding the axis of symmetry of a quadratic equation of the form $y = ax^2 + bx + c$.

Wrap Up

THE BIG IDEA Ask students: *What does the variable c tell you about the graph of a quadratic function?*

RETEACHING ACTIVITY Students graph quadratic functions of the form $y = ax^2 + c$. Students make a table of values and graph. (Reteaching worksheet 7-2)

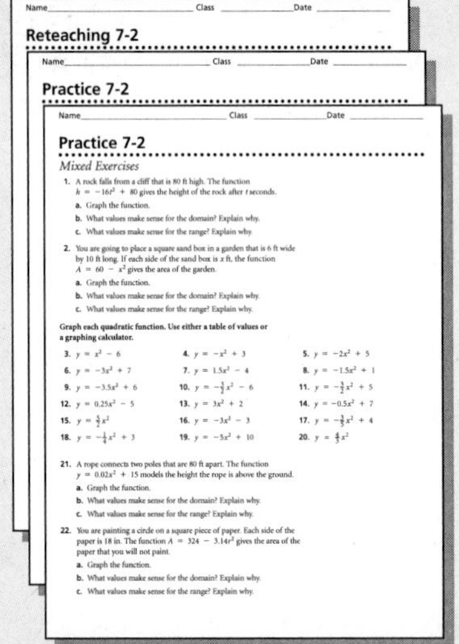

Name ______ Class ______ Date ______

Reteaching 7-2

Name ______ Class ______ Date ______

Practice 7-2

Name ______ Class ______ Date ______

Practice 7-2

Mixed Exercises

1. A rock falls from a cliff that is 80 ft high. The function $h = -16t^2 + 80$ gives the height of the rock after t seconds.
   a. Graph the function.
   b. What values make sense for the domain? Explain why.
   c. What values make sense for the range? Explain why.
2. You are going to place a square sand box in a garden that is 6 ft wide by 10 ft long. If each side of the sand box is x ft, the function $A = 60 - x^2$ gives the area of the garden.
   a. Graph the function.
   b. What values make sense for the domain? Explain why.
   c. What values make sense for the range? Explain why.

Graph each quadratic function. Use either a table of values or a graphing calculator.

3. $y = x^2 - 6$ 4. $y = -x^2 + 3$ 5. $y = -2x^2 + 5$
6. $y = -3x^2 + 7$ 7. $y = 1.5x^2 - 4$ 8. $y = -1.5x^2 + 1$
9. $y = -3.5x^2 + 6$ 10. $y = -\frac{3}{2}x^2 - 6$ 11. $y = -\frac{3}{2}x^2 + 5$
12. $y = 0.25x^2 - 5$ 13. $y = 3x^2 + 2$ 14. $y = -0.5x^2 + 7$
15. $y = \frac{5}{2}x^2$ 16. $y = -3x^2 - 3$ 17. $y = -\frac{3}{5}x^2 + 4$
18. $y = -\frac{1}{4}x^2 + 3$ 19. $y = -5x^2 + 10$ 20. $y = \frac{4}{3}x^2$

21. A rope connects two poles that are 80 ft apart. The function $y = 0.02x^2 + 15$ models the height the rope is above the ground.
   a. Graph the function.
   b. What values make sense for the domain? Explain why.
   c. What values make sense for the range? Explain why.
22. You are painting a circle on a square piece of paper. Each side of the paper is 18 in. The function $A = 324 - 3.14r^2$ gives the area of the paper that you will not paint.
   a. Graph the function.
   b. What values make sense for the domain? Explain why.
   c. What values make sense for the range? Explain why.

Lesson Quiz

Lesson Quiz is also available in Transparencies.

1. Does the function $y = -2x^2 - 1$ have a minimum or maximum? Explain. Maximum; the parabola opens downward.
2. Graph the quadratic function $y = 2x^2 - 4$.

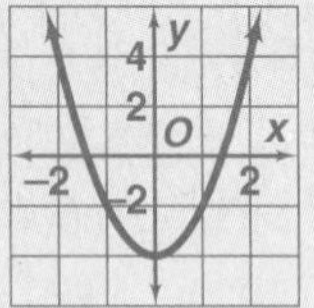

Give the letter of the parabola(s) that make each statement true.

27. $c > 0$ G
28. $c < 0$ F
29. $a > 0$ E, F
30. $a < 0$ G
31. The function has a maximum. G
32. The function has a minimum. E, F

33. a. Landscaping The plan for a rectangular patio has a square garden centered in the patio with sides parallel to the sides of the patio. The patio is 20 ft long by 12 ft wide. If each side of the garden is x ft, the function $y = 240 - x^2$ gives the area of the patio in ft^2. Graph this function. See margin.
   b. What values make sense for the domain? Explain why. $0 < x < 12$
   c. What is the range of the function? Explain why. $96 < y < 240$
   b–c. See margin for reasoning.
34. Architecture An architect wants to design an archway with these requirements.
   - The archway is 6 ft wide and has vertical sides 7 ft high.
   - The top of the archway is modeled by the function $y = -\frac{1}{3}x^2 + 10$. a–b. See margin.

   a. Sketch the architect's design by drawing vertical lines 7 units high at $x = -3$ and $x = 3$ and graphing the portion of the quadratic function that lies between $x = -3$ and $x = 3$.
   b. The plan for the archway is changed so that the top is modeled by the function $y = -0.5x^2 + 11.5$. Make a revised sketch of the archway.
   c. Research Find out how arches were used by the architects of ancient Rome or during the Middle Ages.
   Check students' work.

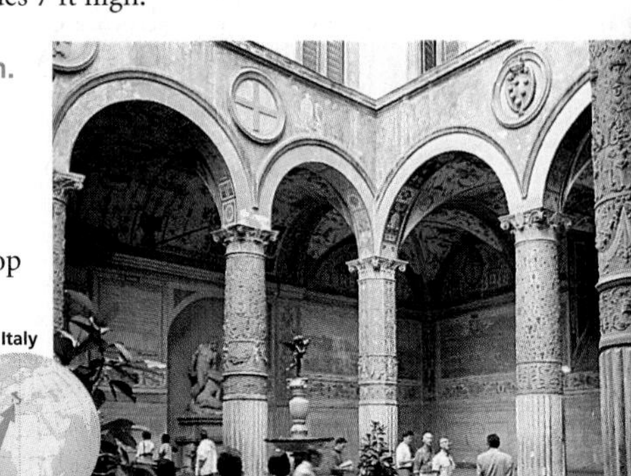

Exercises MIXED REVIEW

Write an equation of the line perpendicular to the given line through the given point.

35. $y = 5x - 2$; (7, 0) $y = -\frac{1}{5}x + \frac{7}{5}$
36. $y = -2x + 9$; (3, 5) $y = \frac{1}{2}x + \frac{7}{2}$
37. $y = 0.5x$; (−8, −2) $y = -2x - 18$
38. $x + y = 6$; (3, 1) $y = x - 2$
39. $y - 4x = 2$; (0, 3) $y = -\frac{1}{4}x + 3$
40. $y = -6x - 2$; (0, 0) $y = \frac{1}{6}x$
41. Real Estate Fox Island is a 4.5-acre island in Rhode Island. It has an assessed value of \$290,400. Use a proportion to find out how much a similar 2-acre island nearby might be worth. about \$129,067

Getting Ready for Lesson 7-3

Find $\frac{-b}{2a}$ for each quadratic function.

42. $y = x^2 + 4x - 2$ -2
43. $y = -8x^2 + x + 13$ $\frac{1}{16}$
44. $y = 6x^2 + x + 7$ $-\frac{1}{12}$

7-3 Teaching Notes

PROBLEM OF THE DAY

A recipe for strawberry salad can feed 12 children or 7 adults. Four recipes of strawberry salad are made for a banquet. 30 children eat a child's portion of strawberry salad. How many adult servings remain? 10.5 servings

Problem of the Day is also available in Transparencies.

CONNECTING TO PRIOR KNOWLEDGE Draw a parabola having an axis of symmetry on the y-axis. Then draw the parabola shifted two units to the right. Ask students: *What math language have you learned that you could apply to describe this shift?*

WORK TOGETHER

KINESTHETIC LEARNING Clear a large area in the room. Make a large coordinate grid on the floor using string or masking tape. Direct students to model the graph of a simple quadratic function, such as $y = x^2$ by holding string or rope and standing on the grid. Ask students to identify the axis of symmetry of their parabola. Have students name the coordinates of the point they are standing on. Then have each student's image partner name the coordinates of the image point.

THINK AND DISCUSS

Remind students that the y-intercept occurs when $x = 0$.

Connections Road Safety . . . and more

7-3 Graphing Quadratic Functions

What You'll Learn

- Graphing quadratic functions of the form $y = ax^2 + bx + c$
- Finding the axis of symmetry and vertex
- Graphing quadratic inequalities

...And Why

To solve problems involving weather and road safety

What You'll Need

- graph paper

WORK TOGETHER

To find the reflection of the point $(-1, 2)$ over the line $x = 2$, fold your paper on the line. Then mark the image of the original point.

Work with a partner to find the image of each point.

1. Find the reflection over the line $x = -2$.
   a. $(-4, 5)$ $(0, 5)$ b. $(-6, 1)$ $(2, 1)$ c. $(2, -2)$ $(-6, -2)$ d. $(-1, 0)$ $(-3, 0)$ e. $(0, 4)$ $(-4, 4)$

THINK AND DISCUSS

Part 

Graphing $y = ax^2 + bx + c$

In the quadratic functions you have graphed so far, $b = 0$. When $b \neq 0$, the parabola shifts right or left. The axis of symmetry is no longer the y-axis.

axis of symmetry $x = \frac{-b}{2a}$

vertex $\left(\frac{-b}{2a}, y\right)$

Graph of a Quadratic Function

The graph of $y = ax^2 + bx + c$, where $a \neq 0$, has the line $x = \frac{-b}{2a}$ as its axis of symmetry. The x-coordinate of the vertex is $\frac{-b}{2a}$.

The y-intercept of the graph of a quadratic function is c. This is because substituting $x = 0$ gives you the function value c. You can use the axis of symmetry and the y-intercept to help you graph the function.

Lesson Planning Options

Prerequisite Skills

- Graphing coordinate pairs
- Graphing linear inequalities

Assignment Options

To provide flexible scheduling, this lesson can be subdivided into parts.

1 **Core** 1–27 **Extension** 28–30

2 **Core** 31–39 **Extension** 40–42

Use Mixed Review to maintain skills.

Resources

 **Student Edition**

Skills Handbook, pp. 578, 582
Extra Practice, p. 562
Glossary/Study Guide

Teaching Resources

Chapter Support File, Ch. 7
- Practice 7-3 (two worksheets)
- Reteaching 7-3

Classroom Manager 7-3
Glossary, Spanish Resources
Two-Year Algebra Handbook 7-3

 **Transparencies**

1, 2, 11, 29, 89, 93

Example 1

Point out that the vertex is always on the axis of symmetry.

Example 2 **Relating to the Real World** 

Remind students that 12:00 A.M. is midnight and 12:00 P.M. is noon.

MAKING CONNECTIONS Evangelista Torricelli invented the mercury barometer in 1643. He filled a glass tube with mercury and stood the tube upside down in a bowl of mercury. He found that atmospheric pressure—the weight of the air—pushed the mercury down in the bowl and thus up inside the glass tube. The higher the atmospheric pressure, the higher the mercury was driven up the tube. Mercury is still used in barometers today because of its high density. Water, which is much less dense than mercury, is not practical for use in barometers. To measure atmospheric pressure using water, you would need to build a glass tube for the water that was at least 33 ft tall. Mercury columns need only be 30 in. tall.

ERROR ALERT! Students may interpret 11.3 h to mean 11 h and 30 min. **Remediation:** Remind students that the 0.3 means $\frac{3}{10}$ of an hour. To convert to minutes, multiply $\frac{3}{10}$ by 60 min to get 18 min.

DIVERSITY Have students brainstorm for a list of popular or folk methods that people use to predict the weather. For example, in the United States, a popular legend tells that if a groundhog emerges from hibernation on February 2 and sees his shadow cast by the sun, he will return to his burrow to sleep through six more weeks of winter. However, if the sky is cloudy, the groundhog will expect an early spring. In Europe,

Additional Examples

FOR EXAMPLE 1

Sketch the graph of the quadratic function $f(x) = -2 + 3x - 3x^2$ by finding its axis of symmetry, vertex, and y-intercept.

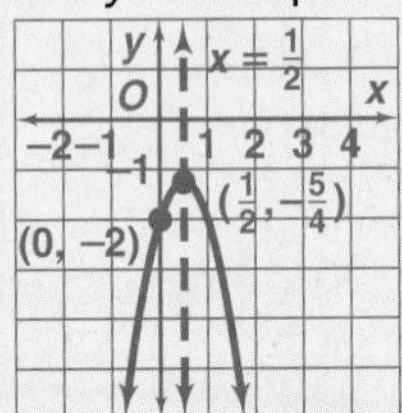

axis of symmetry: $x = \frac{1}{2}$
vertex: $(\frac{1}{2}, -1\frac{1}{4})$
y-intercept: -2

FOR EXAMPLE 2

The equation for the height of a ball thrown straight up at 9 m/s is $y = 9t - 4.9t^2$. In this equation, y represents the vertical position of the ball and t represents the time that the ball is in the air. At what time will the ball reach its highest point? How high will the ball be at this time?
0.92 s, 4.1 m

Discussion: *How would this graph change if the player threw the ball upward with a greater speed?*

328

Example 1

Graph the quadratic function $f(x) = 5 - 4x - x^2$.

Step 1 Find the y-intercept.

The value of c is 5, so the y-intercept of the graph is 5.

Step 2 Find the equation of the axis of symmetry and the coordinates of the vertex.

$f(x) = -x^2 - 4x + 5$ ← Write the function in standard form.

$x = -\frac{b}{2a} = -\frac{-4}{2(-1)} = -2$ ← Find the equation of the axis of symmetry.

The x-coordinate of the vertex is -2.

$f(x) = -(-2)^2 - 4(-2) + 5$ ← To find the y-coordinate of the vertex, substitute −2 for x.
$f(x) = 9$

The vertex is at $(-2, 9)$.

Step 3 Make a table of values and graph the function.

Pair each point on one side of the axis of symmetry with a point on the other side that will have the same $f(x)$ value.

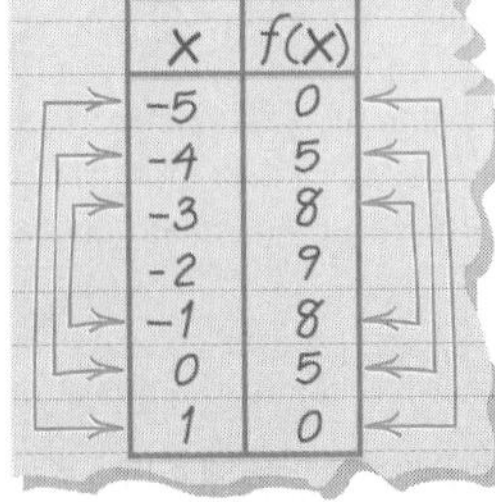

x	f(x)
-5	0
-4	5
-3	8
-2	9
-1	8
0	5
1	0

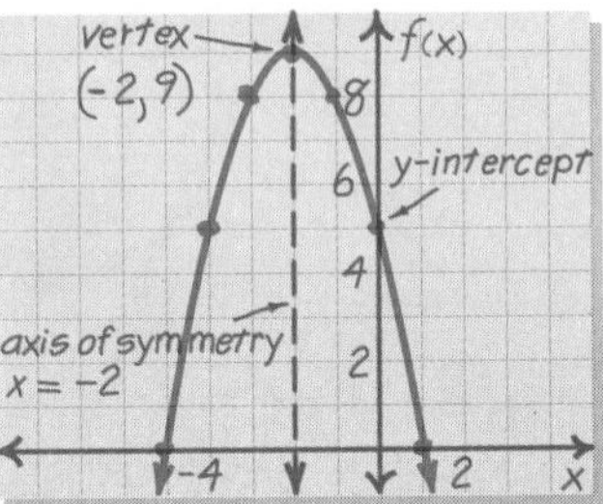

2. **Try This** Graph $y = x^2 - 6x + 9$. Find the equation of the axis of symmetry, the coordinates of the vertex, and the y-intercept.
See margin for graph; $x = 3$; $(3, 0)$; 9.

You can use what you know about quadratic functions to find maximum or minimum values in real-world problems.

Example 2 **Relating to the Real World** 

Weather Meteorologists use equations to model weather patterns. This function predicts atmospheric pressure over a certain 24-hour period.

$$y = 0.005x^2 - 0.113x + 30.22$$

In the equation, x represents the number of hours after 12:00 midnight and y represents the atmospheric pressure in inches of mercury. At what time will the pressure be lowest? What will the lowest pressure be?

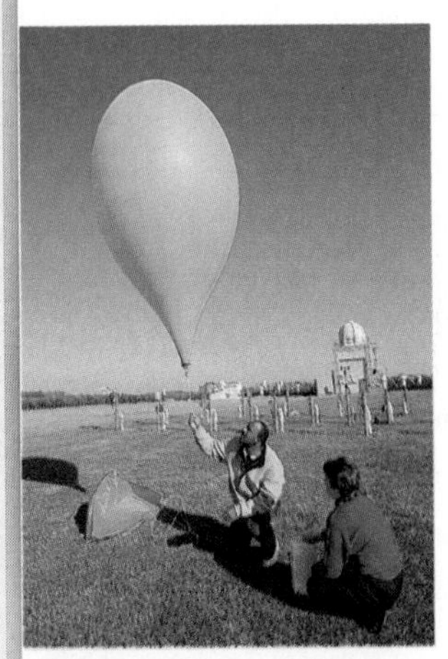

Weather balloons measure conditions in the atmosphere. About 1600 balloons a day are launched in the world.

some people believe that bears and badgers have similar weather-forecasting abilities.

Example 3 **Relating to the Real World** 

Help students set ranges and scales for the x-axis and y-axis so that the distance between the camper's top-right corner and the archway is clearly visible.

VISUAL LEARNING Encourage students to make sketches of the archway and the camper. Lead students to understand that the top-right corner of the camper at (6, 7) is the point most likely to scrape against the rock archway.

MAKING CONNECTIONS The largest camper ever made was built in 1990 for a sheik in the United Arab Emirates. The camper weighed 120 t and included eight bedrooms, eight bathrooms, and four garages. The camper was 66 ft long, 39 ft wide and 39 ft tall.

The archway in Example 3 is clearly too small for the sheik's camper to fit through. Challenge students to change the value of c in the equation $y = -0.1x^2 + c$ so that the arch would be big enough for the sheik's camper. Hint: Have students substitute the point (39, 39) into the equation and solve for c.

TACTILE LEARNING Some students may prefer drawing the graph by hand instead of using the graphing calculator. They can shade the different regions using different colored pencils.

Question 5 Remind students that inequalities containing < are shaded below the graph and inequalities containing > are shaded above the graph.

PROBLEM SOLVING

Look Back Use the graph of the function to check your solution.

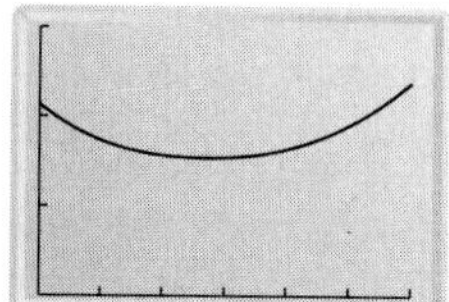

Xmin=0 Ymin=28
Xmax=24 Ymax=31
Xscl=4 Yscl=1

Since the coefficient of x^2 is positive, the curve opens upward and the y-coordinate of the vertex is a minimum.

$-\frac{b}{2a} = -\frac{-0.113}{2(0.005)} = 11.3$ ← Find the x-coordinate of the vertex.

After 11.3 hours (at 11:18 A.M.), the pressure will be at its lowest.

$y = 0.005(11.3)^2 - 0.113(11.3) + 30.22$ ← Substitute 11.3 for x.

$y = 29.58$

The minimum pressure will be 29.58 in. of mercury.

3. Why is it important to check the coefficient of the squared term when solving a real-world maximum or minimum problem? The coefficient shows whether there is a maximum ($a < 0$) or a minimum ($a > 0$).

Part 

Quadratic Inequalities

Graphing a quadratic inequality is similar to graphing a linear inequality. The curve is dashed if the inequality involves $<$ or $>$. The curve is solid if the inequality involves $\leq$ or $\geq$.

Example 3 **Relating to the Real World** 

Road Safety An archway over a road is cut out of rock. Its shape is modeled by the quadratic function $y = -0.1x^2 + 12$. Can a camper 6 ft wide and 7 ft high fit under the arch without crossing the median line?

The camper will fit if each point on it satisfies $y < -0.1x^2 + 12$. Graph the inequality $y < -0.1x^2 + 12$.

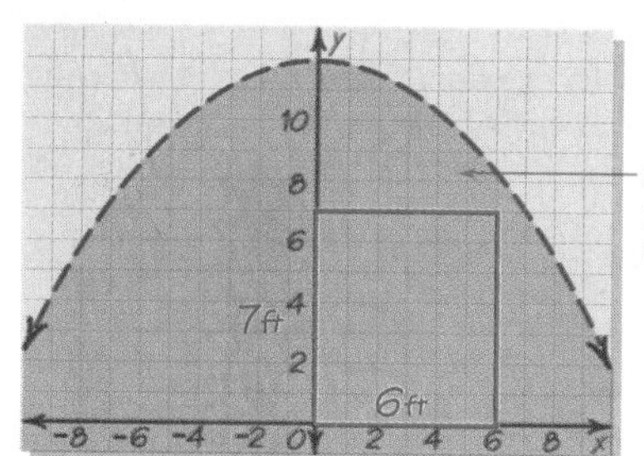

The camper should be able to fit since each point that represents the camper is in the shaded region.

4. Check the solution by showing that the values $x = 6$ and $y = 7$ satisfy the inequality $y < -0.1x^2 + 12$. $7 < -0.1(6)^2 + 12$ ✓

5. How does the graph of $y > -0.1x^2 + 12$ differ from the graph above? The region above the parabola is shaded.

6. **Try This** Graph $y \geq x^2 + 2x + 1$. See margin.

Technology Options

For Exercises 31–39, students may want to graph the inequalities with graphing software or a graphing calculator.

Prentice Hall Technology 

Calculator • Graphing Calculator Handbook: Procedure 4

Software • Secondary Math Lab Toolkit • Computer Item Generator 7-3

CD-ROM • Multimedia Algebra Lab 7

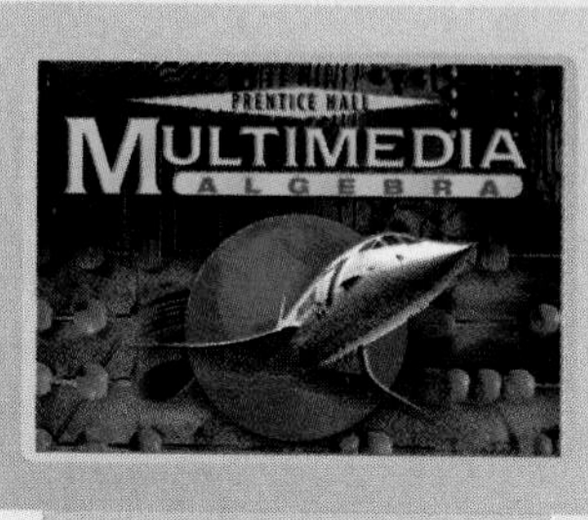

Internet • See the Prentice Hall site. (http://www.phschool.com)

Exercises ON YOUR OWN

Exercises 1–9 Remind students that the graph of an equation in the form of x = a constant is not parallel to the x-axis. It is perpendicular to the x-axis and intersects it at the constant that is in the equation. Since the formula for the axis of symmetry for a quadratic equation is $x = \frac{-b}{2a}$, every axis of symmetry will be perpendicular to the x-axis.

Exercises 10–15 Ask students to share the techniques they use to match the graphs to the functions. For example, by looking at the sign of a, you can determine whether the graph opens upward or downward. This single step eliminates three of the choices.

ALTERNATIVE ASSESSMENT **Exercises 16–21** Have students find the axis of symmetry and vertex coordinates of each equation using the $\frac{-b}{2a}$ method. This will help you assess the students' abilities to predict the shape of a quadratic equation. It will also serve as a self-check when students later complete these exercises on their own.

EXTENSION **Exercise 29** Challenge students to experiment with different garden shapes to see which shape provides the greatest area with only 80 ft of fence.

WRITING **Exercise 30** Encourage students to give examples and make sketches showing how the graphs change when the values of a, b, and c change.

Chapter Project FIND OUT BY GRAPHING Students evaluate a given equation to calculate safe stopping distances. They make a table and a graph to display the results. This task is essential to their work on the Chapter Project, which was

pages 327–329 Think and Discuss

2.

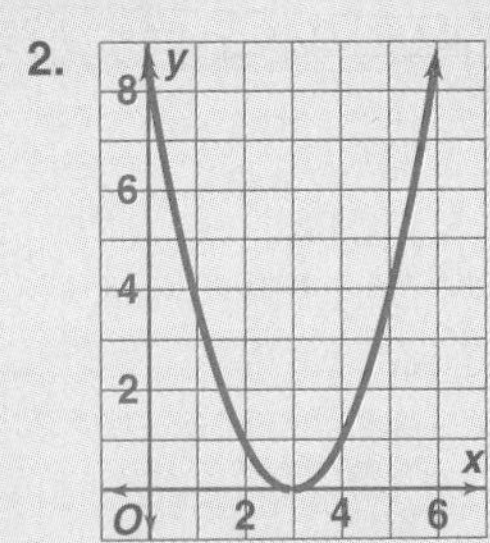

6.

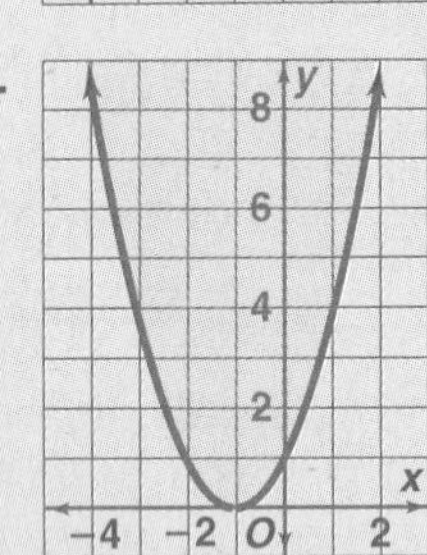

pages 330–331 On Your Own

16.

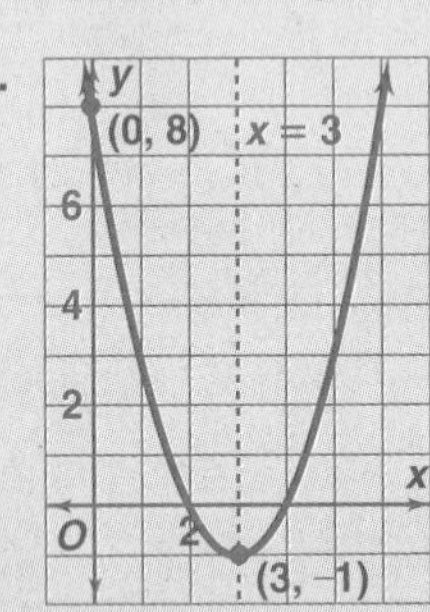

17–27, 30–39. See back of book.

Exercises ON YOUR OWN

Find the equation of the axis of symmetry and the coordinates of the vertex of the graph of each function.

1. $y = 0.2x^2 + 4$ $x = 0$; $(0, 4)$
2. $y = x^2 - 8x - 9$ $x = 4$; $(4, -25)$
3. $y = 2x^2 + 4x - 5$ $x = -1$; $(-1, -7)$
4. $y = 3x^2 - 9x + 5$ $x = 1\frac{1}{2}$; $\left(1\frac{1}{2}, -1\frac{3}{4}\right)$
5. $f(x) = 4x^2 - 3$ $x = 0$; $(0, -3)$
6. $y = 3x^2 - 9$ $x = 0$; $(0, -9)$
7. $f(x) = x^2 + 4x + 3$ $x = -2$; $(-2, -1)$
8. $y = 2x^2 - 6x$ $x = 1\frac{1}{2}$, $\left(1\frac{1}{2}, -4\frac{1}{2}\right)$
9. $y = 12 + x^2$ $x = 0$; $(0, 12)$

Match each graph with its function.

A. $y = x^2 - 6x$
B. $y = x^2 + 6x$
C. $y = -x^2 - 6x$
D. $y = -x^2 + 6x$
E. $y = -x^2 + 6$
F. $y = x^2 - 6$

10. 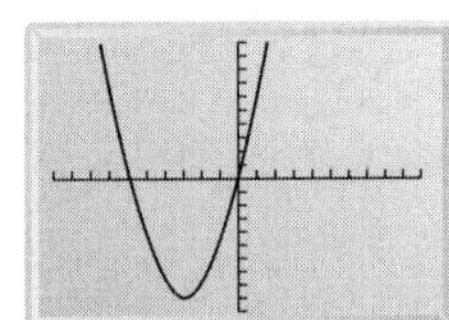 B

11. 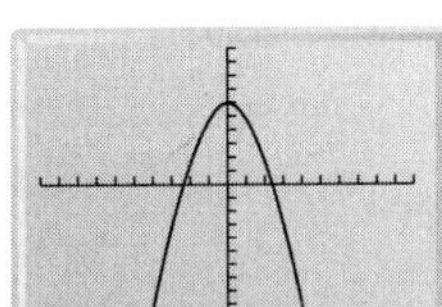 E

12. 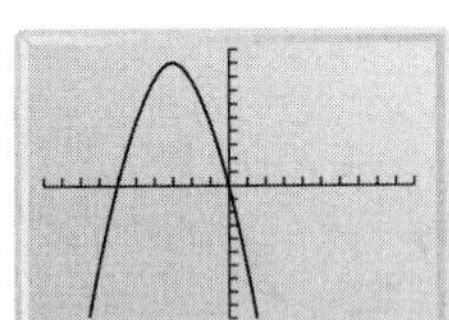

13. 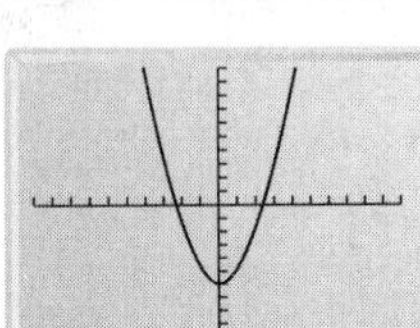 F

14. 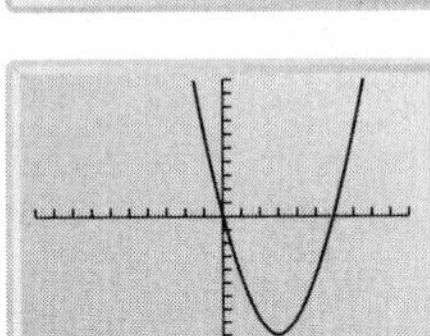 A

15. 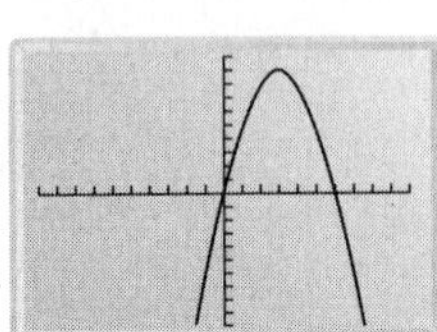

Choose **Graph each function. Use a graphing calculator or a table of values. Label the axis of symmetry, the vertex, and the y-intercept.** 16–27. See margin.

16. $f(x) = x^2 - 6x + 8$
17. $y = -x^2 + 4x - 4$
18. $y = x^2 + 1$
19. $y = -x^2 + 4x$
20. $y = -2x^2 + 6$
21. $f(x) = 3x^2 + 6x$
22. $y = x^2 - 4x + 3$
23. $f(x) = -x^2 - 4x - 6$
24. $y = x^2 - 2x + 1$
25. $f(x) = 2x^2 + x - 3$
26. $y = x^2 + 3x + 2$
27. $y = -x^2 + 4x - 7$

28. **Fireworks** A skyrocket is shot into the air. Its altitude h in feet after t seconds is given by the function $h = -16t^2 + 128t$.
    a. In how many seconds does the skyrocket reach maximum altitude? 4 s
    b. What is the skyrocket's maximum altitude? 256 ft

29. **Gardening** Suppose you have 80 ft of fence to enclose a rectangular garden. The function $A = 40x - x^2$ gives you the area of the garden in square feet where x is the width in feet. What width gives you the maximum gardening area? What is the maximum area? 20 ft; 400 ft^2

introduced in the Chapter Opener. Check students' progress on the project by having each student write a sentence or two describing the objectives of this project.

Exercises MIXED REVIEW

JOURNAL Invite students to describe a real-world problem that they can now solve with their new math knowledge.

GETTING READY FOR LESSON 7-4 These exercises give students practice squaring integers to prepare them for finding square roots.

Wrap Up

THE BIG IDEA Ask students: *What do you need to know before you can graph a quadratic equation?*

RETEACHING ACTIVITY Students graph quadratic functions for the form $y = ax^2 + bx + c$. They find the axis of symmetry and the vertex; then make a table of values and graph. (Reteaching worksheet 7-3)

30. *Writing* Explain how changing the values of a, b, and c in a quadratic function affects the graph of the function. See margin.

Graph each quadratic inequality. 31–39. See margin.

31. $y > x^2$ **32.** $y < -x^2$ **33.** $y \le x^2 + 3$

34. $y < -x^2 + 4$ **35.** $y \ge -2x^2 + 6$ **36.** $y > -x^2 + 4x - 4$

37. $y \le x^2 + 5x + 6$ **38.** $y < x^2 - x - 6$ **39.** $y \ge 3x^2 + 6x$

Open-ended **Give a quadratic function for each description.**

40. Its axis of symmetry is to the right of the y-axis. Sample: $y = x^2 - 4x$

41. Its graph opens downward and has vertex at (0, 0). Sample: $y = -3x^2$

42. Its graph lies entirely above the x-axis. Sample: $y = 2x^2 + 3$

Chapter Project — *Find Out by Graphing*

To avoid skidding, you want to know what a safe stopping distance is. Assume you are traveling on a dry road and have an average reaction time. The formula $f(x) = 0.044x^2 + 1.1x$ gives you a safe stopping distance in feet, where x is your speed in miles per hour. Make a table of values for speeds of 10, 20, 30, 40, 50, and 60 mi/h. Then graph the function.

Exercises MIXED REVIEW

Find each probability for two rolls of a number cube.

43. P(even and odd) $\frac{1}{4}$ **44.** P(7 and 5) 0

45. P(two odd numbers) $\frac{1}{4}$ **46.** P(6 and 2) $\frac{1}{36}$

47. a. *Rowing* The-Head-of-the-Charles Regatta is a 3 mi rowing race held annually in Cambridge, Massachusetts. In 1994, Xeno Muller won the men's singles title in 17 min, 47 s. What was his pace in mi/h? Round your answer to the nearest tenth. 10.1 mi/h

b. Muller rowed at 32 strokes/min. How many strokes did he row over the course of the race? 569 strokes

FOR YOUR JOURNAL

Summarize what you have learned about graphing quadratic functions. Include examples to support your statements.

Getting Ready for Lesson 7-4

Evaluate for $a = -1$, $b = 2$, $c = 3$, and $d = -4$.

48. a^2 1 **49.** $-a^2$ -1 **50.** $-b^2$ -4 **51.** dc^2 -36 **52.** $(dc)^2$ 144

Reteaching 7-3

Practice 7-3

Practice 7-3 *Mixed Exercises*

Lesson Quiz

Lesson Quiz is also available in Transparencies.

1. Graph $f(x) = 2x^2 + 4x - 1$. Label the axis of symmetry, the vertex and the y-intercept.

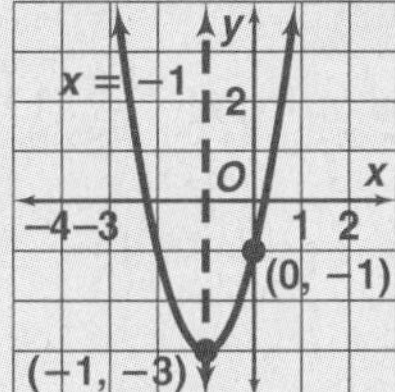

2. Graph the quadratic inequality $y < -3x^2 + 4x$. See back of book.

7-4 Teaching Notes

PROBLEM OF THE DAY

A square has an area of 144 cm^2. If each side is increased by 25%, what is the area of the new square? 225 cm^2

Problem of the Day is also available in Transparencies.

CONNECTING TO PRIOR KNOWLEDGE Have students compare the area of squares with the squares of numbers. Challenge students to explain why the word *square* is used.

THINK AND DISCUSS

Emphasize to students that although the problem "Find the square root of 16." has two answers, $\sqrt{16}$ has only one answer. Remind students that the radical sign, $\sqrt{\ }$, indicates only the principal (positive) square root.

Example 1

ERROR ALERT! **Question 1c** Students may assume that the square root is always smaller than the square. **Remediation:** Remind students that the square of a number is the number multiplied by itself. Therefore the square of a fraction like $\frac{1}{2}$ is the same as $\frac{1}{2} \cdot \frac{1}{2}$, which equals $\frac{1}{4}$. Since the square of a fraction results in a smaller number, the square root of a fraction will result in a larger number.

Make sure students understand the meaning of the bar over the 6 in the decimal $0.\overline{6}$ and the ellipses after the irrational numbers.

Lesson Planning Options

Prerequisite Skills

- Finding squares of numbers
- Using rational and irrational numbers

Assignment Options

To provide flexible scheduling, this lesson can be subdivided into parts.

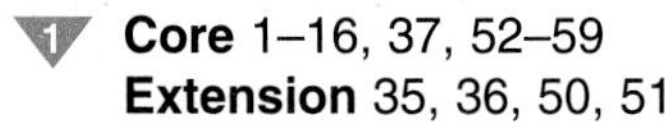

1 **Core** 1–16, 37, 52–59
Extension 35, 36, 50, 51

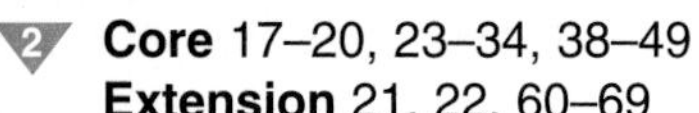

2 **Core** 17–20, 23–34, 38–49
Extension 21, 22, 60–69

Use Mixed Review to maintain skills.

Resources

Student Edition

Skills Handbook, pp. 579, 580
Extra Practice, p. 562
Glossary/Study Guide

Teaching Resources

Chapter Support File, Ch. 2
- Practice 7-4 (two worksheets)
- Reteaching 7-4
- Checkpoint

Classroom Manager 7-4
Glossary, Spanish Resources
Two-Year Algebra Handbook 7-4

Transparencies
11, 43, 89

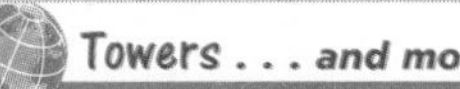

Connections Towers . . . and more

7-4 Square Roots

What You'll Learn
- Finding square roots
- Using square roots

...And Why

To use square roots in real-world situations, such as finding the distance from a satellite to the horizon

What You'll Need
- calculator

THINK AND DISCUSS

Part 1 Finding Square Roots

The diagram shows the relationship between squares and square roots. Every positive number has *two* square roots.

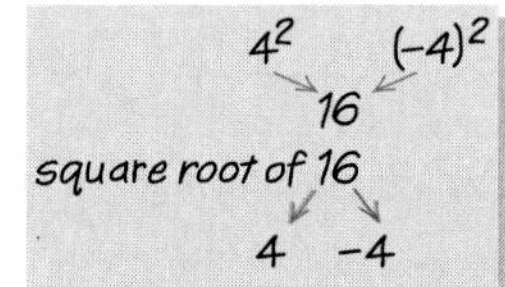

Square Root

If $a^2 = b$, then a is a **square root** of b.

Example: $4^2 = 16$ and $(-4)^2 = 16$, so 4 and -4 are square roots of 16.

A radical symbol $\sqrt{\ }$ indicates a square root. The expression $\sqrt{16}$ means the **principal** (or positive) **square root** of 16. The expression $-\sqrt{16}$ means the **negative square root** of 16. You can use the symbol $\pm$, read "plus or minus," to indicate both square roots.

Example 1

Simplify each expression.

a. $\sqrt{64} = 8$ ← positive square root

b. $-\sqrt{100} = -10$ ← negative square root

c. $\pm\sqrt{\frac{9}{16}} = \pm\frac{3}{4}$ ← The square roots are $\frac{3}{4}$ and $-\frac{3}{4}$.

d. $\pm\sqrt{0} = 0$ ← There is only one square root of 0.

e. $\sqrt{-16}$ is undefined ← For real numbers, the square root of a negative number is undefined.

1. **Try This** Simplify each expression.
 a. $\sqrt{49}$ 7 b. $\pm\sqrt{36}$ ±6 c. $-\sqrt{121}$ −11 d. $\sqrt{\frac{1}{25}}$ $\frac{1}{5}$

QUICK REVIEW

Rational and irrational numbers make up the set of *real numbers.* In decimal form, *rational* numbers terminate or repeat. *Irrational* numbers continue without repeating number patterns.

Some square roots are rational numbers and some are irrational numbers.

Rational: $\pm\sqrt{81} = \pm 9$, $-\sqrt{1.44} = -1.2$, $\sqrt{\frac{4}{9}} = \frac{2}{3} = 0.\overline{6}$

Irrational: $-\sqrt{5} = -2.23606797\ldots$, $\sqrt{\frac{1}{3}} = 0.57735026\ldots$

2. **Try This** Classify each expression as *rational* or *irrational.*
 a. $\sqrt{8}$ irrat. b. $\pm\sqrt{225}$ rat. c. $-\sqrt{75}$ irrat. d. $\sqrt{\frac{1}{4}}$ rat.

Example 2

ESTIMATION Because 14.52 is closer to 16 than 9, point out that students can expect the square root to be closer to 4 than to 3.

Example 3

Explain to students that estimation can help them spot errors when finding square roots with a calculator. In Example 2, they estimated the principal square root of 14.52 to be between 3 and 4. When they used a calculator to find $\sqrt{14.52}$, students found the principal square root rounded to the nearest hundredth to be 3.81, between 3 and 4 but closer to 4.

Example 4 Relating to the Real World 

KINESTHETIC LEARNING Group students into sets of four. Have students hold a length of string from a point on the floor to a point on the wall to form a right triangle with the floor and wall. The point on the wall should be twice the distance from the right angle as the point on the floor. While the third student measures the length of the three sides of the triangle, have the fourth student calculate the length of the string by using the formula in Example 4 and the length of the triangle legs formed by the wall and the floor. Then have students compare their answers.

ALTERNATIVE METHOD Students may not clearly understand the relationship between a square and a square root. Provide students with graph paper. Using different colored pens, have

Part 2

Estimating and Using Square Roots

The squares of integers are called **perfect squares.**

consecutive integers:	1	2	3	4	5	6	7
	↓	↓	↓	↓	↓	↓	↓
consecutive perfect squares:	1	4	9	16	25	36	49

You can estimate square roots by using perfect squares and by using a calculator.

Example 2

Estimation Between what two consecutive integers is $\sqrt{14.52}$?

$\sqrt{9} < \sqrt{14.52} < \sqrt{16}$ ← 14.52 is between the two consecutive square numbers 9 and 16.

$3 < \sqrt{14.52} < 4$

$\sqrt{14.52}$ is between 3 and 4.

3. **Try This** Between what two consecutive integers is $-\sqrt{105}$? -11 and -10

Example 3

Calculator Find $\sqrt{14.52}$ to the nearest hundredth.

√ 14.52 ENTER 3.81051177 ← Use a calculator.

$\sqrt{14.52} \approx 3.81$

4. **Critical Thinking** How can you use consecutive perfect squares to mentally check calculator answers for square roots? See margin.

Example 4 Relating to the Real World 

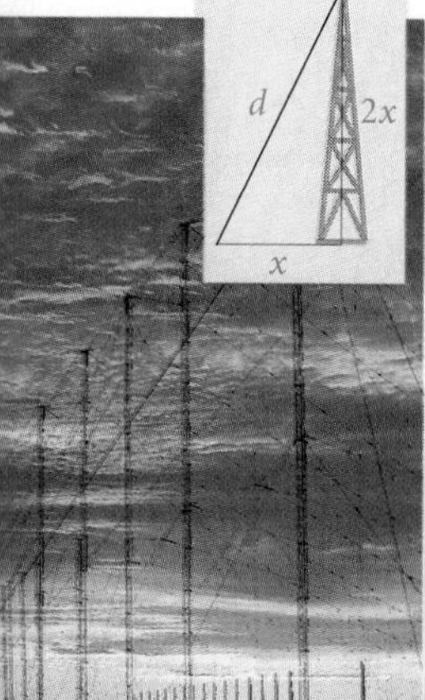

Towers A tower is supported with a wire. The formula $d = \sqrt{x^2 + (2x)^2}$ gives the length d of the wire for the tower at the left. Find the length of the wire if $x = 12$ ft.

$d = \sqrt{x^2 + (2x)^2}$

$d = \sqrt{12^2 + (2 \cdot 12)^2}$ ← Substitute 12 for x.

$d = \sqrt{144 + 576}$ ← Simplify.

$d = \sqrt{720}$

$d = 26.83281573$ ← Use a calculator.

$d \approx 26.8$

The wire is about 26.8 ft long.

5. **Try This** Suppose the tower is 140 ft tall. How long is the supporting wire? about 156.5 ft

Additional Examples

FOR EXAMPLE 1

Simplify $\sqrt{\frac{9}{25}}$. $\frac{3}{5}$

FOR EXAMPLE 2

Between what two consecutive integers is $\sqrt{19}$? $4 < \sqrt{19} < 5$

FOR EXAMPLE 3

Find $\sqrt{11.25}$ to the nearest hundredth. 3.35

FOR EXAMPLE 4

A ladder is leaning against a house to reach a second-story window. The formula $l = \sqrt{h^2 + d^2}$ gives the length l of the ladder for the height h of the window from the ground and the distance d between the bottom of the ladder and the house. Find the length of the ladder if $h = 12$ ft and $d = 5$ ft. 13 ft them outline squares that are one, two, three, four, and five units on a side. Make sure students label the side length of their squares. Then have students calculate how many unit squares are contained within each outline. Have them write the area of each outlined area underneath the length of the side. Lead students to understand that the area of each outlined square is the same as the length of one side of the outlined area squared.

MAKING CONNECTIONS A golden rectangle is a rectangle in which the ratio of the length *l* to the width *w* is equal to the ratio of (length + width) to length. In math notation, we write $\frac{l}{w} = \frac{l + w}{l}$. This ratio is called the *golden ratio.* If a square of side *w* is removed from the golden rectangle, the shape that remains is also a golden rectangle. This process can be repeated indefinitely. Ancient Greeks thought that a golden rectangle was aesthetically pleasing, and its shape has appeared in works of art and architecture.

We can write the equation $\frac{l}{w} = \frac{l + w}{l}$ as a quadratic equation and solve it.

$$l^2 = w^2 + wl$$

$$l^2 - wl - w^2 = 0$$

$$l = w\left(\frac{1 \pm \sqrt{5}}{2}\right)$$

Exercises ON YOUR OWN

EXTENSION Exercise 21 Challenge students to collect several different types of balls. Have them use the elasticity function to calculate each ball's elasticity coefficient. Have students compare the elasticity of different balls. Ask students

Technology Options

For Exercises 1–8, 23–34, and 61–69, some students may find it helpful to use a graphing calculator or math software to evaluate the square roots.

Prentice Hall Technology

Software • Secondary Math Lab Toolkit • Computer Item Generator 7-4

Internet • See Prentice Hall site. (http://www.phschool.com)

334

Exercises ON YOUR OWN

Find both the principal and negative square root of each number.

1. 169 ±13 **2.** 1.96 ±1.4 **3.** $\frac{1}{9}$ $\pm\frac{1}{3}$ **4.** 900 ±30

5. 0.25 ±0.5 **6.** $\frac{36}{49}$ $\pm\frac{6}{7}$ **7.** 1.21 ±1.1 **8.** 1681 ±41

Tell whether each expression is *rational* or *irrational*.

9. $\sqrt{37}$ irrat. **10.** $-\sqrt{0.04}$ rat. **11.** $\pm\sqrt{\frac{1}{5}}$ irrat. **12.** $-\sqrt{\frac{16}{121}}$ rat.

Between what two consecutive integers is each square root?

13. $\sqrt{35}$ 5 and 6 **14.** $\sqrt{27}$ 5 and 6 **15.** $-\sqrt{245}$ –16 and –15 **16.** $\sqrt{880}$ 29 and 30

Calculator **Use a calculator to simplify each expression. Round to the nearest hundredth.**

17. $\sqrt{12}$ 3.46 **18.** $-\sqrt{203}$ –14.25 **19.** $\sqrt{11{,}550}$ 107.47 **20.** $-\sqrt{150}$ –12.25

21. *Sports* The elasticity coefficient *e* of a ball relates the height *r* of its rebound to the height *h* from which it is dropped. You can find the elasticity coefficient using the function $e = \sqrt{\frac{r}{h}}$.

a. What is the elasticity coefficient for a tennis ball that rebounds 3 ft after it is dropped from a height of 3.5 ft? about 0.93

b. *Critical Thinking* Suppose that the elasticity coefficient of a basketball is 0.88. How high is its rebound if it is dropped 6 ft? about 4.6 ft

22. *Critical Thinking* What number other than 0 is its own square root? 1

Find the square roots of each number.

23. 0.36 ±0.6 **24.** 144 ±12 **25.** $\frac{25}{16}$ $\pm\frac{5}{4}$ **26.** 0.01 ±0.1

27. 400 ±20 **28.** 0 0 **29.** 625 ±25 **30.** $\frac{9}{49}$ $\pm\frac{3}{7}$

31. $\frac{1}{81}$ $\pm\frac{1}{9}$ **32.** 1.69 ±1.3 **33.** 729 ±27 **34.** 2.25 ±1.5

35. a. *Space* Find the distance to the horizon from a satellite 4200 km above Earth. The formula $d = \sqrt{12{,}800h + h^2}$ tells you the distance *d* in kilometers to the horizon from a satellite *h* kilometers above Earth. about 8500 km

b. Find the distance to the horizon from a satellite 3600 km above Earth. about 7700 km

36. *Standardized Test Prep* If $x^2 = 16$ and $y^2 = 25$, choose the least possible value for the expression $y - x$. D

A. −1 B. 0 C. 1
D. −9 E. −4

to make conjectures about why each type of ball has a different elasticity. Ask students to speculate about how some sports might be different if players used balls with higher or lower elasticities.

Exercises 23 and 34 Suggest that students convert decimals to fractions before finding square roots.

STANDARDIZED TEST TIP **Exercise 36** Remind students that the square root of a number can be both positive and negative. This means that the expression $y - x$ can have up to four possible values. Also remind students that if x has a negative value, the answer will be greater because subtracting a negative integer has the same result as adding a positive integer.

ESL **Exercise 37** Ask students to explain in what situation a golfer shouts, "Fore!".

CARTOON and DIVERSITY **Exercise 37** Challenge students to find out how and why the practice of shouting "Fore!" came about.

ALTERNATIVE ASSESSMENT **Exercises 38–49** To assess the students' understanding of square roots, ask them to find the two consecutive integers which lie on either side of each expression. This will also reinforce students' understanding of the relationship between a number and its square.

Exercises 61–69 Ask students to share the shortcuts they use to answer these questions quickly.

Chapter Project FIND OUT BY CALCULATING Students will evaluate the given formulas to complete a table. They will also come to conclusions about safe driving distances. This information is essential to their work on the Chapter Project.

37. In the cartoon, to what number is the golfer referring? **4**

from cartoon *Bound and Gagged*

Choose **Use a calculator, paper and pencil, or mental math. Find the value of each expression. If the value is irrational, round to the nearest hundredth.**

38. $\sqrt{441}$ **21**

39. $-\sqrt{\frac{4}{25}}$ **$-\frac{2}{5}$**

40. $\sqrt{2}$ **1.41**

41. $\sqrt{\frac{1}{36}}$ **$\frac{1}{6}$**

42. $-\sqrt{1.6}$ **−1.26**

43. $-\sqrt{157}$ **−12.53**

44. $\sqrt{200}$ **14.14**

45. $-\sqrt{13}$ **−3.61**

46. $\sqrt{30}$ **5.48**

47. $\sqrt{1089}$ **33**

48. $-\sqrt{0.64}$ **−0.8**

49. $\sqrt{41}$ **6.40**

50. **Writing** Explain the difference between $-\sqrt{1}$ and $\sqrt{-1}$. **$-\sqrt{1}$ is the negative of the square root of 1. $\sqrt{-1}$ is undefined.**

51. **Open-ended** Find two integers a and b, both between 1 and 25, such that $a^2 + b^2$ is a perfect square. **Answers may vary. Sample: 7 and 24**

Tell whether each expression is *rational, irrational,* or *undefined.*

52. $-\sqrt{3600}$ **rat.**

53. $\sqrt{8}$ **irrat.**

54. $\sqrt{-25}$ **undefined**

55. $\sqrt{6.25}$ **rat.**

56. $\sqrt{12.96}$ **rat.**

57. $\sqrt{129.6}$ **irrat.**

58. $-\sqrt{12.96}$ **rat.**

59. $\sqrt{-12.96}$ **undefined**

60. **Physics** If you drop an object, the time t in seconds that it takes to fall d feet is given by $t = \sqrt{\frac{d}{16}}$.

a. Find the time it takes an object to fall 400 ft. **5 sec**

b. Find the time it takes an object to fall 1600 ft. **10 sec**

c. **Critical Thinking** In part (b), the object falls four times as far as in part (a). Does it take four times as long to fall? Explain.

No; time varies as the square root of distance. It takes twice as long to fall.

Tell whether each statement is *true* or *false.* If the statement is false, rewrite it as a true statement.

61. $6 < \sqrt{38} < 7$ **true**

62. $-7 < -\sqrt{56} < -6$ **62. false; $-8 < -\sqrt{56} < -7$**

63. $-4 < -\sqrt{17} < -5$ **true**

64. $3.3 \le \sqrt{10.25} < 3.4$ **See above right.** **64. false; $3.2 \le \sqrt{10.25} < 3.3$**

65. $-16 < -\sqrt{280} < -15$ **See below.**

66. $21 < \sqrt{436} < 22$ **See below.**

67. $-38 < -\sqrt{1300} < -37$ **false; $-37 < -\sqrt{1300} < -36$**

68. $-9 \le -\sqrt{72} < -8$ **true**

69. $0.1 < \sqrt{0.03} < 0.2$ **true**

65. false; $-17 < -\sqrt{280} < -16$

66. $20 < \sqrt{436} < 21$ false;

pages 332–333 Think and Discuss

4. Square the integers on either side of the calculated values to see if the original number is between the squares.

page 336 Mixed Review

70. 2 3 4 5 6 7

71. −5 −4 −3 −2 −1 0

72. −2 −1 0 1 2 3

73. −9 −8 −7 −6 −5 −4

74. −5 −4 −3 −2 −1 0

75. −2 −1 0 1 2 3

76. 1 2 3 4 5 6

77. −1 0 1 2 3 4

Checkpoint, 1–3, 10. See back of book. s

Have students add this task to work they have already completed for the project. Check students' progress by having each student write a sentence or two describing what they have done for the project thus far.

Exercises MIXED REVIEW

Exercises 70–77 Remind students that when you multiply both sides of an inequality by a negative number, the inequality sign changes direction.

GETTING READY FOR LESSON 7-5 These exercises prepare students for solving quadratic equations in standard form.

Wrap Up

THE BIG IDEA Ask students to explain how to estimate and find the square root of a number.

RETEACHING ACTIVITY Students use a calculator to find square toots of specific numbers. They complete a table and identify numbers as rational or irrational. (Reteaching worksheet 7-4)

Exercises CHECKPOINT

Students will assess their own progress on Lessons 7-1 to 7-4.

Exercise 11 Make sure students recall what values of a would make the graph wider than the graph of $y = \frac{1}{2}x^2 - 3$.

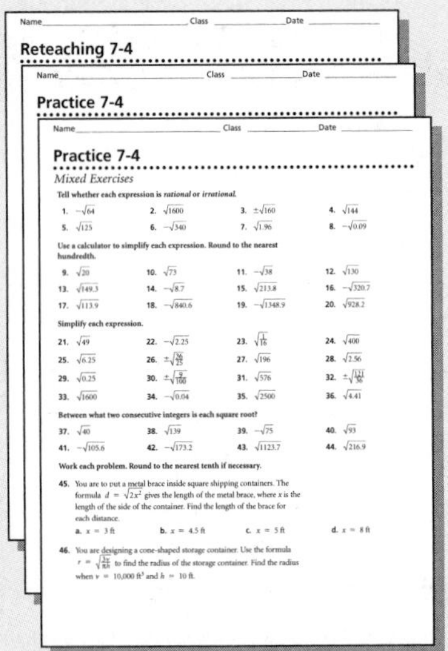
Name ______ Class ______ Date ______

Reteaching 7-4

Name ______ Class ______ Date ______

Practice 7-4

Name ______ Class ______ Date ______

Practice 7-4

Mixed Exercises

Tell whether each expression is *rational* or *irrational*.

1. $-\sqrt{64}$ 2. $\sqrt{1600}$ 3. $\pm\sqrt{160}$ 4. $\sqrt{144}$
5. $\sqrt{125}$ 6. $-\sqrt{340}$ 7. $\sqrt{1.96}$ 8. $-\sqrt{0.09}$

Use a calculator to simplify each expression. Round to the nearest hundredth.

9. $\sqrt{20}$ 10. $\sqrt{73}$ 11. $-\sqrt{38}$ 12. $\sqrt{130}$
13. $\sqrt{149.3}$ 14. $-\sqrt{8.7}$ 15. $\sqrt{213.8}$ 16. $-\sqrt{320.7}$
17. $\sqrt{113.9}$ 18. $-\sqrt{840.6}$ 19. $-\sqrt{1348.9}$ 20. $\sqrt{928.2}$

Simplify each expression.

21. $\sqrt{49}$ 22. $-\sqrt{2.25}$ 23. $\sqrt{\frac{1}{16}}$ 24. $\sqrt{400}$
25. $\sqrt{6.25}$ 26. $\pm\sqrt{\frac{36}{25}}$ 27. $\sqrt{196}$ 28. $\sqrt{2.56}$
29. $\sqrt{0.25}$ 30. $\pm\sqrt{\frac{9}{100}}$ 31. $\sqrt{576}$ 32. $\pm\sqrt{\frac{121}{36}}$
33. $\sqrt{1600}$ 34. $-\sqrt{0.04}$ 35. $\sqrt{2500}$ 36. $\sqrt{4.41}$

Between what two consecutive integers is each square root?

37. $\sqrt{40}$ 38. $\sqrt{139}$ 39. $-\sqrt{75}$ 40. $\sqrt{93}$
41. $-\sqrt{105.6}$ 42. $-\sqrt{173.2}$ 43. $\sqrt{1123.7}$ 44. $\sqrt{216.9}$

Work each problem. Round to the nearest tenth if necessary.

45. You are to put a metal brace inside square shipping containers. The formula $d = \sqrt{2x^2}$ gives the length of the metal brace, where x is the length of the side of the container. Find the length of the brace for each distance.
a. $x = 3$ ft b. $x = 4.5$ ft c. $x = 5$ ft d. $x = 8$ ft

46. You are designing a cone-shaped storage container. Use the formula $r = \sqrt{\frac{3v}{\pi h}}$ to find the radius of the storage container. Find the radius when $v = 10{,}000$ ft^3 and $h = 10$ ft.

Lesson Quiz

Lesson Quiz is also available in Transparencies.

Find the value of each expression. If the value is irrational, round to the nearest hundredth.

1. $\sqrt{3}$ 1.73
2. $\sqrt{\frac{1}{16}}$ $\frac{1}{4}$
3. $-\sqrt{25}$ -5

Tell whether each statement is *true* or *false*.

4. $-3 < -\sqrt{1} < -1$ false
5. $6 < \sqrt{65} < 12$ true

336

Chapter Project — *Find Out by Calculating*

Suppose a car left a skid mark d feet long. The formulas shown will estimate the speed s in miles per hour at which the car was traveling when the brakes were applied. Use the formulas to complete the table of estimated speeds.

- Why do you think the estimates of speed do not double when the skid marks double in length?
- Based on these results, what conclusions can you make about safe following distances?

Traveling Speed		
Dry Road	$s = \sqrt{27d}$	
Wet Road	$s = \sqrt{13.5d}$	
Skid Mark Length (d)	**Estimated Speed (s)**	
	Dry Road	Wet Road
60 ft	■	■
120 ft	■	■

Exercises MIXED REVIEW

Solve each inequality. Graph the solution on a number line. 70–77. See margin for graphs.

70. $2x > 8$ $x > 4$
71. $14 < -7s$ $s < -2$
72. $z - 1 \geq -1$ $z \geq 0$
73. $-2 - b > 4$ $b < -6$
74. $5b + 4 < -6$ $b < -2$
75. $-2c \geq 5c - 1$ $c \leq \frac{1}{7}$
76. $8(m - 3) \leq 4m$ $m \leq 3\frac{1}{2}$
77. $2 > 7 - 3t$ $t > 1\frac{2}{3}$

78. One in four residents of the United States has myopia, or nearsightedness. What is the probability that two people chosen at random are both nearsighted? $\frac{1}{16}$

Getting Ready for Lesson 7-5

Solve each equation.

79. $3x - 14 = 27$ $13\frac{2}{3}$
80. $4(y - 5) = 20$ 10
81. $-6b + 12 = 12$ 0
82. $3 - 9m = 0$ $\frac{1}{3}$

Exercises CHECKPOINT

Graph each quadratic function. Find the vertex of each parabola. 1–3. See margin for graphs.

1. $y = 4x^2$ (0, 0)
2. $y = 2x^2 + 7$ (0, 7)
3. $y = -x^2 - 2x + 10$ (−1, 11)

Choose **Use a calculator, paper and pencil, or mental math to find the value of each square root. Round to the nearest hundredth.**

4. $\sqrt{7}$ 2.65
5. $-\sqrt{100}$ −10
6. $\sqrt{23}$ 4.80
7. $\sqrt{144}$ 12
8. $-\sqrt{150}$ −12.25
9. $-\sqrt{\frac{1}{9}}$ $-\frac{1}{3}$

10. *Writing* What does the rule for a quadratic function tell you about how the graph of the function will look? See margin.

11. *Open-ended* Give an example of a quadratic function for each of the following descriptions. a–b. Samples are given.
a. Its graph opens downward. 11a. $y = -x^2 - 1$
b. Its graph is wider than the graph of $y = \frac{1}{2}x^2 - 3$. $y = \frac{1}{4}x^2$

7-5 Teaching Notes

PROBLEM OF THE DAY

A cockroach can travel at a speed of 29 cm/s. A centipede can travel at 30 m/min. Which can travel faster? Explain. centipede; 50 cm/s > 29 cm/s

Problem of the Day is also available in Transparencies.

CONNECTING TO PRIOR KNOWLEDGE Ask students to think of a question that has two possible, correct answers. For example: during one flight of the space shuttle, when is the shuttle exactly 1 km above the ground? The shuttle is 1 km above the ground twice: once during take-off and once during landing.

WORK TOGETHER

Question 1 Tell students to find the x-intercepts using the equations. Estimating the intercepts from the graphs will not be accurate.

Question 5 Lead students to understand that the values found in Question 5a satisfy the equation in Question 5b because $y = 0$ when the graph crosses the x-axis.

THINK AND DISCUSS

Have students describe in their own words what a quadratic equation looks like when written in standard form.

Make sure students recall the difference in notation between square roots and the principal square root.

Connections City Planning . . . and more

7-5 Solving Quadratic Equations

What You'll Learn

- Solving quadratic equations in $ax^2 = c$ form
- Finding if a quadratic equation has two solutions, one solution, or no solution

...And Why

To solve real-world problems, such as finding the radius of a pond

What You'll Need

- calculator

WORK TOGETHER

1. Work with a partner. Find the x-intercepts of each graph.
   a. $y = 2x - 3$ $1\frac{1}{2}$
   b. $y = -x^2 - 2x + 3$ $-3, 1$

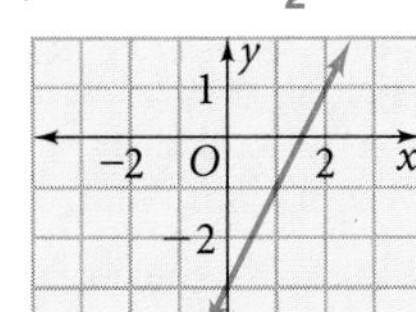

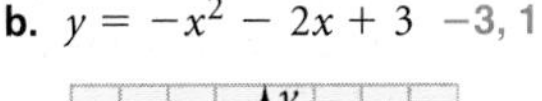

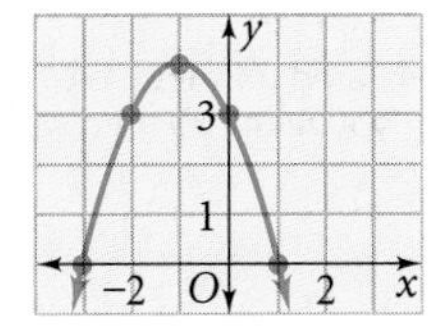

2. Use the graph in Question 1(a) to find the solution of $2x - 3 = 0$. $1\frac{1}{2}$
3. Are the x-intercepts that you found in Question 1(b) solutions of $-x^2 - 2x + 3 = 0$? Explain. Yes; -3 and 1 satisfy the equation.
4. Graph $y = x^2 + x - 6 = 0$. See margin.
5. a. Where does the graph of $y = x^2 + x - 6$ cross the x-axis? $x = -3, x = 2$
   b. Do the values you found in part (a) satisfy the equation $x^2 + x - 6 = 0$? yes

THINK AND DISCUSS

Part 1 **Using Square Roots to Solve Equations**

In the Work Together, you investigated a quadratic equation and its related quadratic function. Any values that make the equation true are solutions.

Standard Form of a Quadratic Equation

A **quadratic equation** is an equation that can be written in **standard form**:

$$ax^2 + bx + c = 0, \text{ where } a \neq 0$$

6. Write $6x^2 = 5x - 12$ in standard form. $6x^2 - 5x + 12 = 0$
7. Is $5x - 3 = 0$ a quadratic equation? Why or why not? No; the equation does not have an x^2 term.

There are two square roots for numbers greater than 0. So there are two solutions for an equation like $x^2 = 36$. You can solve equations in the form $x^2 = a$ by finding the square roots of each side.

$$x^2 = 36 \longrightarrow x = \pm 6$$

Lesson Planning Options

Prerequisite Skills

- Solving equations
- Finding square roots

Assignment Options

To provide flexible scheduling, this lesson can be subdivided into parts.

1 **Core** 13–19, 37–42
Extension 20, 21, 32–35

2 **Core** 1–12, 22–29
Extension 30, 31, 36

Use Mixed Review to maintain skills.

Resources

Student Edition

Skills Handbook, pp. 579, 582
Extra Practice, p. 562
Glossary/Study Guide

Teaching Resources

Chapter Support File, Ch. 7
- Practice 7-5 (two worksheets)
- Reteaching 7-5
- Alternative Activity 7-5

Classroom Manager 7-5
Glossary, Spanish Resources
Two-Year Algebra Handbook 7-5

Transparencies

1, 2, 11, 90, 94

Example 1

CRITICAL THINKING Question 9 Point out that solutions are the values of x that satisfy the equation. To find the value of c, substitute 11 and -11 for x.

Example 2 **Relating to the Real World** 

Make sure students understand the meaning of the word *volume.* Also, ensure that they understand why volume is stated in cubic units and has 3 as its exponent. If necessary, review with students the units for length, area, and volume.

CRITICAL THINKING Question 10 Lead students to understand that distance cannot be a negative quantity.

Question 13 Encourage students to guess before calculating the answer. Then have them compare their guesses to their answers.

Question 13b If students have difficulty understanding why the radius does not double, encourage them to experiment in the classroom or at home pouring water from narrow cans to wider cans.

CONNECTING TO THE STUDENTS' WORLD Ask students to measure the depth and radius of a circular pool, pond, or fountain. Then have them calculate the volume using the formula in Example 2. Explain that one cubic foot is equal to 7.48 gal. Have students calculate how many gallons are needed to fill the pool, pond, or fountain.

Additional Examples

FOR EXAMPLE 1

Solve $-2x^2 + 18 = 0$. $x = 3$

FOR EXAMPLE 2

Find the radius of a circle which has an area of 3.14 m^2. Use the equation $A = r^2$, where A is the area in m^2 and r is the radius of the circle in m. $r = 1$ m

Discussion: *If the radius of a circle doubles, does its area also double? Explain.*

FOR EXAMPLE 3

Solve the equation $2x^2 - 3 = 0$ by graphing the related function. 1.22

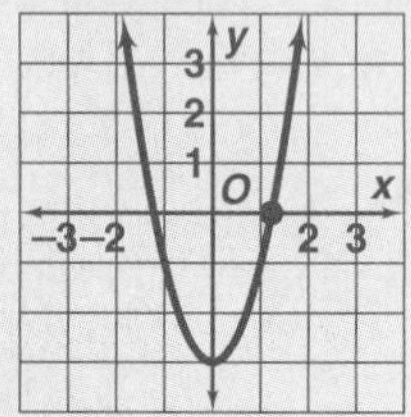

Example 1

Solve $2x^2 - 98 = 0$.

$$2x^2 - 98 = 0 + 98 \quad \text{Add 98 to each side.}$$
$$2x^2 = 98$$
$$x^2 = 49 \quad \text{Divide each side by 2.}$$
$$\sqrt{x^2} = \pm\sqrt{49} \quad \text{Find the square roots.}$$
$$x = \pm 7$$

8. **Try This** Use a calculator to solve $x^2 - 7 = 0$. Round your solutions to the nearest tenth. 2.6, −2.6

9. **Critical Thinking** Find a value for c so that the equation $x^2 - c = 0$ has 11 and -11 as solutions. 121

You can solve many geometric problems by finding square roots.

Example 2 **Relating to the Real World** 

City Planning A city is planning a circular duck pond for a new park. The depth of the pond will be 4 ft. Because of water resources, the maximum volume will be 20,000 ft^3. Find the radius of the pond. Use the equation $V = \pi r^2 h$, where V is the volume, r is the radius, and h is the depth.

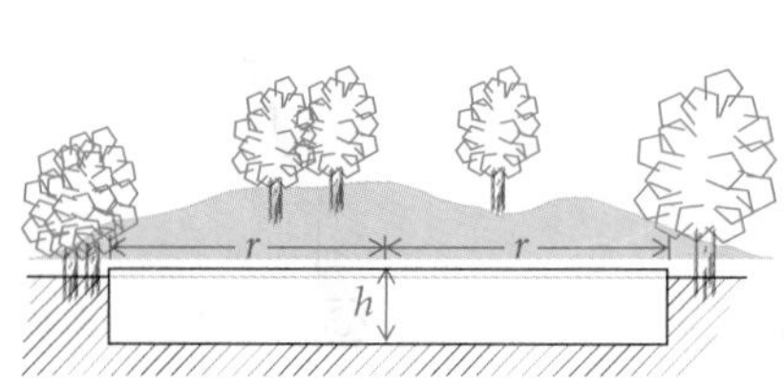

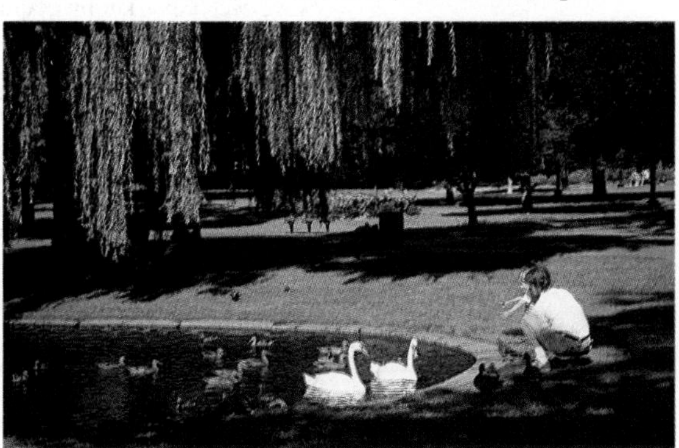

$$V = \pi r^2 h$$
$$20{,}000 = \pi r^2(4) \quad \text{Substitute 20,000 for } V \text{ and 4 for } h.$$
$$\frac{20{,}000}{\pi(4)} = r^2 \quad \text{Put in calculator-ready form.}$$
$$\sqrt{\frac{20{,}000}{\pi(4)}} = r \quad \text{Find the principal square root.}$$
$$39.89422804 = r \quad \text{Use a calculator.}$$

The pond will have a radius of about 39.9 ft.

Help students understand the size of a cubic foot and a gallon. Encourage students to make a box the size of a cubic foot from paper, tape, and other classroom materials. Bring in a one-gallon container so students can visualize the size of 7.48 gal.

MAKING CONNECTIONS The largest swimming pool in the world was the Fleishhacker Pool in San Francisco. The pool measured 1000 ft by 150 ft and was up to 14 ft deep. It contained 7.5 million gal of heated water. The pool was opened in 1925 but has since closed.

Have students calculate the capacity in gallons of a pool that has the same dimensions as Fleishhacker pool, but is 14 ft deep throughout. Then have them compare their answer to the 7.5 million gal provided. Encourage students to make conjectures about how much of the pool was 14 ft deep.

Example 3

Point out that equation (c) can be graphed, even though it has no solution. Because the graph does not touch the x-axis, y never has a value of zero, so there are no solutions.

CRITICAL THINKING Question 14 Encourage students to test their answers using values of c that are fractions or irrational numbers.

VISUAL LEARNING Question 14 On the board, draw a large coordinate grid with tick marks. Bend a pipe cleaner, coat hanger, or other piece of wire into the shape of a parabola. Hold the wire parabola over the coordinate grid to model equations such as $y = x^2 - 6$, $y = x^2$, and $y = x^2 + 6$. Have students identify the equation of the graph and the number of solutions.

10. Critical Thinking Why is the principal square root the only root that makes sense in Example 2? **The radius r will never be negative.**

11. **Justify** this step in Example 2:
If $20{,}000 = \pi r^2(4)$, then $\frac{20{,}000}{\pi(4)} = r^2$. **Divide each side by $\pi(4)$.**

12. Calculator Which keystrokes can you use to find r in Example 2? **A**
A. √ (20000 ÷ (π × 4)) ENTER
B. √ 20000 ÷ π × 4 ENTER

13. a. Try This Suppose the pond could have a volume of 40,000 ft^3. What will be the radius of the pond if the depth is not changed?
b. Critical Thinking Does the radius of the pond double when the volume doubles? Explain. **No; the radius is multiplied by $\sqrt{2}$ not 2.**
13a. about 56.4 ft

Part 2

Finding the Number of Solutions

You have seen quadratic equations that have two real numbers as solutions. For real numbers, a quadratic equation can have two solutions, one solution, or no solution.

You can use a graph to find the solution(s) of a quadratic equation by finding the x-intercepts of the related quadratic function.

Example 3

Solve each equation by graphing the related function.

a. $x^2 - 4 = 0$ b. $x^2 = 0$ c. $x^2 + 4 = 0$

Graph $y = x^2 - 4$. Graph $y = x^2$. Graph $y = x^2 + 4 = 0$.

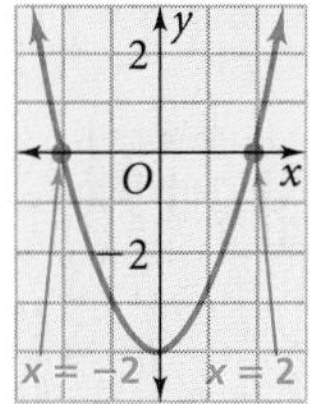

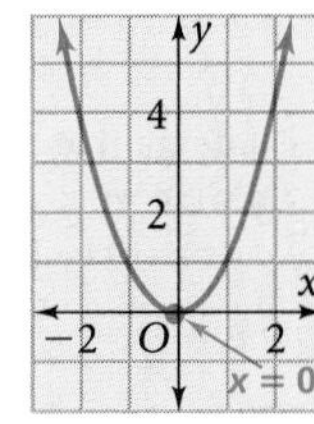

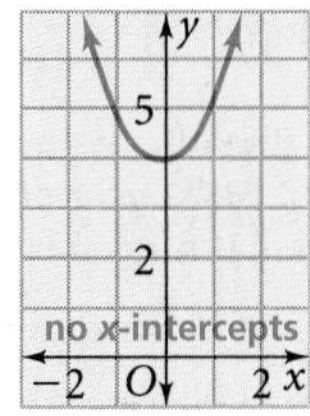

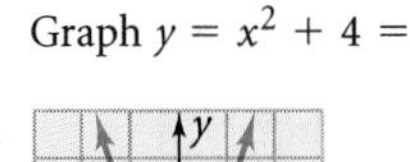

There are two solutions, $x = \pm 2$. There is one solution, $x = 0$. There is no solution.

14. a. Critical Thinking For what values of c will $x^2 = c$ have two solutions? **$c > 0$**
b. For what value of c will $x^2 = c$ have one solution? **0**
c. For what values of c will $x^2 = c$ have no solution? **$c < 0$**

15. Mental Math Tell the number of solutions for each equation.
a. $x^2 = -36$ **0** b. $x^2 - 12 = 6$ **2** c. $x^2 - 15 = -151$

Technology Options

For Exercises 1–12 and 37–42, students may want to solve the equations by graphing with a graphing calculator or with graphing software.

Prentice Hall Technology

Software • Secondary Math Lab Toolkit • Integrated Math Lab 14 • Computer Item Generator 7-5

Internet • See the Prentice Hall site. (http://www.phschool.com)

ERROR ALERT! **MENTAL MATH Question 15c** Students may quickly answer that there are no solutions because $-15 < 0$. **Remediation:** Remind students to eliminate everything but the x^2 from the left side of the equation before answering.

Exercises ON YOUR OWN

Exercise 3 Remind students to divide by 3 first.

ESL **Exercise 12** The term *keystroke* may be confusing. Ensure that students know the expression means pressing a calculator button.

ALTERNATIVE ASSESSMENT **Exercises 14–19** Have students predict the number of solutions that correspond to each expression by graphing its related function. This will help you assess students' understanding of quadratic equations.

Exercise 20 Tell students that the total area enclosed by the frame is the area enclosed by the outermost edge of the frame.

GEOMETRY Exercise 21 Students find the radius of a sphere given the surface area.

OPEN-ENDED Exercise 30 For students having difficulty getting started, explain that they are trying to find ways to make the sum of a number of perfect squares equal to 225. Suggest that they begin by writing all the squares that are less than 225.

KINESTHETIC LEARNING Exercise 31 If you have block scheduling or extended class periods, take students outside or to a large area such as the cafeteria or gymnasium. Have four students form a 20 ft by 20 ft square with string. They may need to use an 80 ft piece of string that has already been measured into 20 ft intervals. Then, have other students fill in

page 337 Work Together

4.

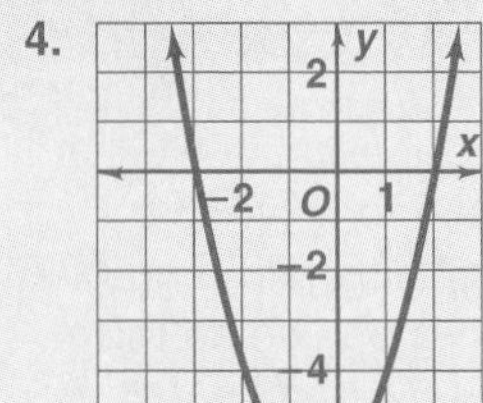

page 341 Mixed Review

46a. $58.5 \le x \le 76$

60 65 70 75

$64.5 \le x \le 72$

60 65 70 75

b. $58.5 \le x < 64.5$ or $72 < x \le 76$

60 65 70 75

Exercises ON YOUR OWN

Choose **Solve each equation by graphing, using mental math, or using paper and pencil. If the equation has no solution, write *no solution*.**

1. $x^2 = 4$ $2, -2$
2. $x^2 = 49$ $7, -7$
3. $3x^2 + 27 = 0$ no solution
4. $x^2 + 25 = 25$ 0
5. $3x^2 - 7 = -34$ no solution
6. $x^2 - 225 = 0$ $15, -15$
7. $49x^2 - 16 = -7$ $\frac{3}{7}, -\frac{3}{7}$
8. $x^2 - 9 = 16$ $5, -5$
9. $4x^2 = 25$ $\frac{5}{2}, -\frac{5}{2}$
10. $x^2 + 36 = 0$ no solution
11. $4x^2 - 100 = -100$ 0
12. $x^2 - 63 = 81$ $12, -12$
13. *Critical Thinking* Michael solved $x^2 + 25 = 0$ and found the solutions -5 and 5. Explain the mistake that Michael made. Answers may vary. Sample: Michael used -5 and 5 as square roots of -25.

Solve each equation. Round solutions to the nearest tenth.

14. $b^2 = 3$ $1.7, -1.7$
15. $8x^2 = 64$ $2.8, -2.8$
16. $n^2 - 5 = 16$ $4.6, -4.6$
17. $3m^2 + 7 = 13$ $1.4, -1.4$
18. $2x^2 - 179 = 0$ $9.5, -9.5$
19. $b^2 - 1 = 20$ $4.6, -4.6$

Model each problem with a quadratic equation. Then solve.

20. *Photography* Find the dimensions of the square picture that make the area of the picture equal to 75% of the total area enclosed by the frame. about 10.4 in. by 10.4 in.
21. *Geometry* Find the radius of a sphere with surface area 160 in.^{2}. Use the formula $A = 4\pi r^2$, where A is the surface area and r is the radius of the sphere. Round your answer to the nearest tenth of an inch. 3.6 in.

Susan La Flesche Picotte (1865–1915) was a physician and the leader of the Omaha people.

For each equation, you are given a statement about the number of solutions of each equation. If the claim is true, verify it by solving the equation. If the claim is false, write a correct statement.

22. $n^2 + 2 = 11$; there are two solutions. true; $3, -3$
23. $g^2 = -49$; there are two solutions. False; there are no solutions.
24. $x^2 + 9 = 25$; there is one solution. False; there are two solutions.
25. $4x^2 - 96 = 0$; there is one solution. False; there are two solutions.
26. $-4r^2 = -64$; there are two solutions. true; $4, -4$
27. $4n^2 - 256 = 0$; there are two solutions. true; $8, -8$
28. $4b^2 + 9 = 9$; there are two solutions. False; there is one solution.
29. $-x^2 - 15 = 0$; there is no solution. true
30. *Open-ended* Suppose you have 225 square tiles, all the same size. You can tile one surface using all the tiles. How could you tile more than one square surface using all the tiles? No surface can have only one tile. Answers may vary. Some possibilities for lengths of sides: 5, 6, 8, 10; nine 5's, 12, 9; 2, 5, 14; 2, 10, 11; 2, 4, 6, 13; 5, 10, 10; etc.
31. *Painting* Suppose you have a can of paint that will cover 400 ft^2. Find the radius of the largest circle you can paint. Round your answer to the nearest tenth of a foot. (*Hint:* Use the formula $A = \pi r^2$.) 11.3 ft

the sides of the square and pull the string gently outward until the students have formed a rough circle. Ask the students representing the middle of the square's sides how far outward they moved to make a circle. Add that distance to 10 ft (halfway across the square) to find the radius of the circle. Substitute this amount into the formula and see how close the result is to 400 ft^2.

STANDARDIZED TEST TIP **Exercise 36** Remind students that they should simplify a complex expression as much as possible before looking at the answer choices.

Exercises MIXED REVIEW

Exercise 43–45 Before students begin, have them review the different methods they can apply to solve these problems.

GETTING READY FOR LESSON 7-6 These exercises prepare students for using the quadratic formula.

Wrap Up

THE BIG IDEA Ask students: *How many solutions does a quadratic equation have? How can you tell before solving the equation how many solutions to expect?*

RETEACHING ACTIVITY Students solve quadratic equations in $ax^2 = c$ form. They complete a chart showing steps to solve each equation. (Reteaching worksheet 7-5)

Physics **Use this information for Exercises 32–35.** The time t it takes a pendulum to make a complete swing back and forth depends on the length of the pendulum. This formula relates the length of a pendulum ℓ in meters to the time t in seconds.

$$\ell = \frac{2.45t^2}{\pi^2}$$

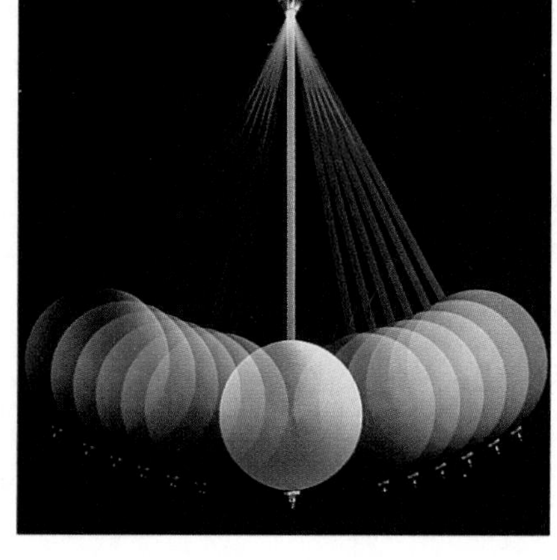

32. Find the length of the pendulum if $t = 1$ s. Round to the nearest tenth. 0.2 m
33. Find t if $\ell = 1.6$ m. Round to the nearest tenth. 2.5 s
34. Find t if $\ell = 2.2$ m. Round to the nearest tenth. 3.0 s
35. **Writing** You can adjust a clock with a pendulum by making the pendulum longer or shorter. If a clock is running slowly, would you lengthen or shorten the pendulum to make the clock run faster? Explain.
Shorten; decreasing ℓ will decrease t.

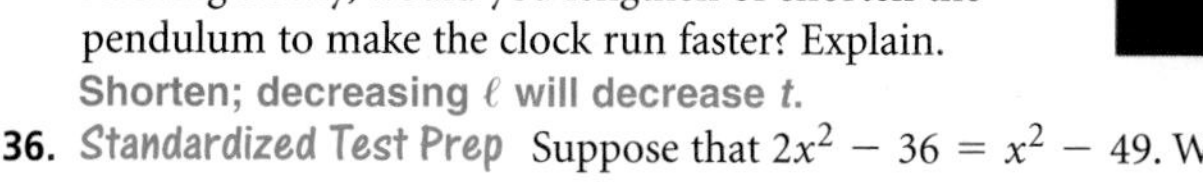

36. **Standardized Test Prep** Suppose that $2x^2 - 36 = x^2 - 49$. Which statement is correct? B
   - **A.** The equation has two real solutions.
   - **B.** The equation has no real solutions.
   - **C.** The equation has exactly one real solution.
   - **D.** You cannot determine the number of real solutions.

Solve each equation. If the equation has no solution, write *no solution*.

37. $5x^2 + 9 = 40$ 2.5, −2.5
38. $2x^2 + 8x^2 + 16 = 16$ 0
39. $8x^2 - 4 - 3x^2 + 6 = 30$ 2.4, −2.4
40. $x^2 + 3 - 2x^2 - 9 = 30$ no solution
41. $3x^2 - 10 + 5x^2 + 3 = 25$ 2, −2
42. $6x^2 + 22 + 2x^2 + 50 = 60$ no solution

Exercises MIXED REVIEW

Solve each system of equations.

43. $3x + 2y = 12$, $x - 2y = -4$ (2, 3)
44. $-x + 4y = 18$, $3x + y = -2$ (−2, 4)
45. $0.4x - 1.2y = 24$, $1.6x + 0.6y = 15$ (15, −15)

46. a. **Space** NASA astronauts must be between 4 ft $10\frac{1}{2}$ in. and 6 ft 4 in. tall. Russian cosmonauts must be between 5 ft $4\frac{1}{2}$ in. and 6 ft. Change these values to inches. Use a number line to graph an inequality that models each situation. 46a–b. See margin.
   b. Use a number line to graph the heights that are acceptable for an astronaut but not acceptable for a cosmonaut.

Getting Ready for Lesson 7-6

Evaluate each expression. Round to the nearest hundredth, where necessary.

47. $\sqrt{5^2 - 15}$ 3.16
48. $\sqrt{36 + 64}$ 10
49. $\sqrt{(-4)^2 - (2)(-3)}$ 4.69
50. $-3 - \sqrt{36}$ −9

Lesson Quiz

Lesson Quiz is also available in Transparencies.

Solve each equation by graphing, using mental math or pencil and paper. If the equation has no solution, write *no solution.*

1. $x^2 = 16$ 4
2. $2x^2 - 3 = 6$ 2.12
3. $2x^2 + 16 = 0$ no solution
4. $x^2 + 4 = 29$ 5
5. Find the speed s of a 4 kg bowling ball with a kinetic energy of 160 J (joules). Use the equation $E = \frac{1}{2}ms^2$ where m is the object's mass in kg and E is its kinetic energy. 8.94 m/s

Math ToolboX

In Lesson 7-5, students solved quadratic equations with $b = 0$. The graphing calculator allows students to solve quadratic equations when $b \neq 0$. First, emphasize that solutions to the equation are represented by the x-intercept in the graphs. Have students explain why the x-intercepts are the solutions. Then work through the example. Have students work in pairs to complete the exercises. Note: The [CALC] feature is located above the TRACE button.

ERROR ALERT! Some students may not be able to see their graphs or may get a graphing error display. **Remediation:** Have students press ZOOM **6** to display the standard graphing screen. Have students turn off all Stat Plots by pressing 2nd [STAT PLOT] **4** ENTER.

WRITING Exercise 8 Some students using the calculator method may need to be reminded to set one side of the equation equal to zero before entering it into the calculator.

ADDITIONAL PROBLEMS Use a graph to solve each equation.

1. $3x^2 + 4x - 8 = 0$ **−2.43, 1.10**
2. $x^2 - 7x + 10 = 0$ **2, 5**
3. $x^2 - 7x - 38 = 0$ **−3.59, 10.59**

Materials and Manipulatives

- Graphing calculator

Resources

Transparencies 11, 29

Math ToolboX Technology

Finding Roots

After Lesson 7-5

The solutions of a quadratic equation are the x-intercepts of the related quadratic function. The solutions of a quadratic equation and the related x-intercepts are often called *roots of the equation* or *roots of the function.*

Example

Use a graphing calculator to solve $x^2 - 6x + 3 = 0$.

Step 1

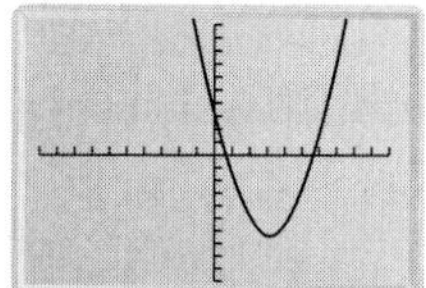

Enter $y = x^2 - 6x + 3$. Use the CALC feature. Select *2:ROOT*. The calculator will plot the graph.

Step 2

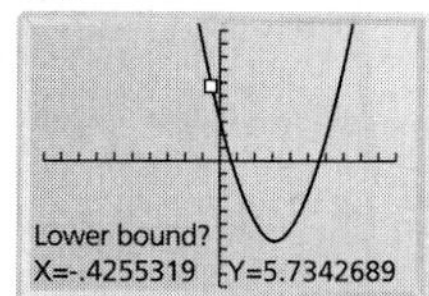

Move the cursor to the left of the first x-intercept. Press ENTER to set the lower bound.

Step 3

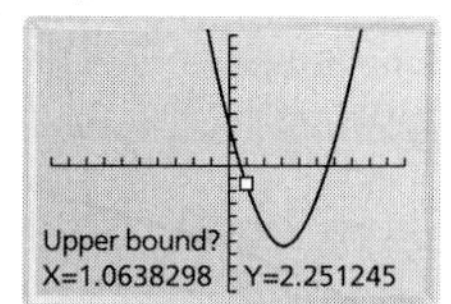

Move the cursor slightly to the right of the intercept. Press ENTER to set the upper bound.

Step 4

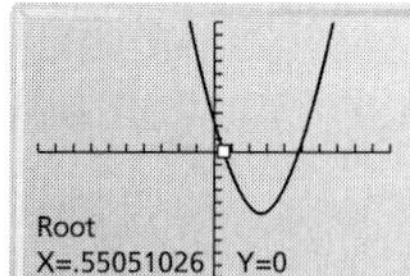

Press ENTER to display the first root, which is about 0.55.

Repeating the steps near the second intercept, you find the second root is about 5.45. So, the solutions are about 0.55 and 5.45.

Suppose you cannot see both of the x-intercepts on your graph. You can find the range for the x-axis by using the TABLE feature. The calculator screen at the right shows part of the table for $y = 2x^2 - 48x + 285$.

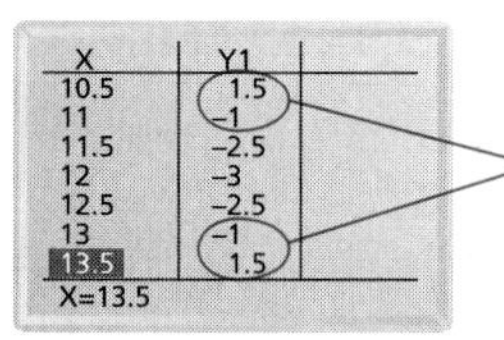

X	Y1
10.5	1.5
11	-1
11.5	-2.5
12	-3
12.5	-2.5
13	-1
13.5	1.5

X=13.5

The graph crosses the x-axis when the values for y change signs. So the range of values of x should include 10.5 and 13.5.

1. Find the x-intercepts of $y = 2x^2 - 48x + 285$. **about 10.78, 13.22**

Use a graph to solve each equation.

2. $x^2 - 6x - 16 = 0$ **−2, 8**
3. $2x^2 + x - 6 = 0$ **$-2, 1\frac{1}{2}$**
4. $\frac{1}{3}x^2 + 8x - 3 = 0$ **about −24.37, 0.37**
5. $x^2 - 18x + 5 = 0$ **about 0.28, 17.72**
6. $0.25x^2 - 8x - 45 = 0$ **about −4.88, 36.88**
7. $0.5x^2 + 3x - 36 = 0$ **−12, 6**
8. *Writing* Solve $3x^2 = 48$ using a calculator or paper and pencil. Explain why you chose the method you used. **4, −4; Check students' work.**

342

7-6 Teaching Notes

PROBLEM OF THE DAY

Mr. Sanchez bought 13 tickets to a movie and spent \$68. Children's tickets cost \$4 each and adult tickets cost \$6 each. How many adult tickets did Mr. Sanchez buy? 8

Problem of the Day is also available in Transparencies.

CONNECTING TO PRIOR KNOWLEDGE Ask students to think of formulas they can use to find answers to problems quickly. Samples: the Pythagorean Theorem, the formula for converting fractions to percents. Ask students to explain how these and other formulas work.

THINK AND DISCUSS

For now, the quadratic formula is presented without being derived. Tell students that the derivation of the formula will be presented in a later lesson.

Lead students to understand that $a \neq 0$ in the quadratic formula because division by 0 is undefined.

Example 1

ERROR ALERT! Students may become confused by all the minus signs and subtraction signs in the quadratic formula. **Remediation:** Suggest students begin substituting values by writing each value in parentheses, as shown in Example 1. Stress that if b is a negative number (such as -2 in Question 1), then $-b$ in the quadratic formula is $-(-2)$, which equals 2.

Connections Model Rockets . . . and more

7-6 Using the Quadratic Formula

What You'll Learn

- Using the quadratic formula to solve quadratic equations

...And Why

To investigate real-world situations, such as the vertical motion of model rockets

What You'll Need

- calculator

THINK AND DISCUSS

In Lesson 7-5, you solved some simple quadratic equations by finding square roots and by graphing. In this lesson, you will learn to solve any quadratic equation by using the **quadratic formula.** In Chapter 10 you will learn additional ways to solve a quadratic equation.

Quadratic Formula

If $ax^2 + bx + c = 0$ and $a \neq 0$, then $x = \frac{-b \pm \sqrt{b^2 - 4ac}}{2a}$.

Example 1

Solve $x^2 + 5x + 6 = 0$ by using the quadratic formula.

$x = \frac{-b \pm \sqrt{b^2 - 4ac}}{2a}$ ← Use the quadratic formula.

$x = \frac{-(5) \pm \sqrt{5^2 - (4)(1)(6)}}{2(1)}$ ← Substitute 1 for *a*, 5 for *b*, and 6 for *c*.

$x = \frac{-5 \pm \sqrt{1}}{2}$

$x = \frac{-5 + 1}{2}$ or $x = \frac{-5 - 1}{2}$ ← Write two solutions.

$x = -2$ or $x = -3$

The solutions are -2 and -3.

Check for $x = -2$

$(-2)^2 + 5(-2) + 6 \stackrel{?}{=} 0$

$4 - 10 + 6 \stackrel{?}{=} 0$

$0 = 0$ ✔

for $x = -3$

$(-3)^2 + 5(-3) + 6 \stackrel{?}{=} 0$

$9 - 15 + 6 \stackrel{?}{=} 0$

$0 = 0$ ✔

PROBLEM SOLVING

Look Back Could you check the solutions by graphing? Explain. Yes. Graph $y = x^2 + 5x + 6$ and find the x-intercepts.

1. **Try This** Use the quadratic formula to solve $x^2 - 2x - 8 = 0$. 4, −2
2. What values would you use for *a*, *b*, and *c* in the quadratic formula to solve $2x^2 = 140$? 2; 0; −140
3. How many solutions does $9x^2 - 24x + 16 = 0$ have? Explain. one; $b^2 - 4ac = (-24)^2 - 4(9)(16) = 0$

When the quantity under the radical sign in the quadratic formula is not a perfect square, you can use a calculator to approximate the solutions of an equation.

Lesson Planning Options

Prerequisite Skills

- Simplifying square root expressions
- Evaluating formulas

Assignment Options

Core 1–18, 21–32
Extension 19, 20, 33–35

Use Mixed Review to maintain skills.

Resources

Student Edition

Skills Handbook, pp. 572, 578, 579
Extra Practice, p. 562
Glossary/Study Guide

Teaching Resources

Chapter Support File, Ch. 7
- Practice 7-6 (two worksheets)
- Reteaching 7-6
- Alternative Activity 7-6
- Checkpoint

Classroom Manager 7-6
Glossary, Spanish Resources
Two-Year Algebra Handbook 7-6

Transparencies

11, 29, 90

Question 2 Point out to students that they cannot substitute values for *a*, *b*, and *c* until they have rewritten the equation in standard form.

Question 3 Ask students to tell how they realized that the equation had no solutions. Lead students to understand that as soon as they find that the value inside the radical is negative, they can stop working. Square roots of negative numbers are not real numbers, so there are no real roots for the corresponding equation.

Example 2 **Relating to the Real World** 

AUDITORY LEARNING Question 6 Group students in pairs. Have partners explain to each other why the intersection of the graphs matches the solutions for the quadratic equation.

Example 3 **Relating to the Real World** 

Explain to students that this example assumes the rocket will not accelerate after it leaves the ground. If the rocket continued to accelerate—from the thrust of its engine—this formula would not work.

Point out that students could multiply both sides of the equation $0 = -16t^2 + 96t - 128$ by -1 before changing it into the form of the quadratic formula. Multiplying by -1 eliminates one of the negatives. This step may help students who are easily confused by multiple negative signs.

CRITICAL THINKING Question 7b Encourage students to guess and then calculate the maximum height reached by the rocket.

KINESTHETIC LEARNING Divide students into small groups. Arrange students so that each group contains one student

Additional Examples

FOR EXAMPLE 1

Solve $2x^2 + 3x - 2 = 0$ by using the quadratic formula. 0.5, −2

Discussion: *Under what conditions will a quadratic equation have only one solution? No solutions?*

FOR EXAMPLE 3

Solve $-3x^2 + 4x + 3 = 0$. Round the solutions to the nearest hundredth. −0.54, 1.87

Example 2

Solve $2x^2 + 4x - 7 = 0$. Round the solutions to the nearest hundredth.

$x = \frac{-b \pm \sqrt{b^2 - 4ac}}{2a}$ ← Use the quadratic formula.

$x = \frac{-4 \pm \sqrt{4^2 - (4)(2)(-7)}}{2(2)}$ ← Substitute 2 for *a*, 4 for *b*, and −7 for *c*.

$x = \frac{-4 \pm \sqrt{72}}{4}$

$x = \frac{-4 + \sqrt{72}}{4}$ or $x = \frac{-4 - \sqrt{72}}{4}$ ← Write two solutions.

$x \approx \frac{-4 + 8.485281374}{4}$ or $x \approx \frac{-4 - 8.485281374}{4}$ ← Use a calculator.

$x \approx 1.12$ or $x \approx -3.12$

The solutions are approximately 1.12 and −3.12.

Check Graph the related function $y = 2x^2 + 4x - 7$. Use the *ROOT* option to find the *x*-intercept.

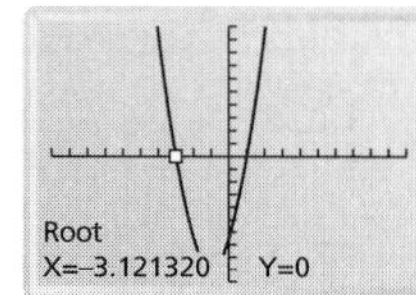

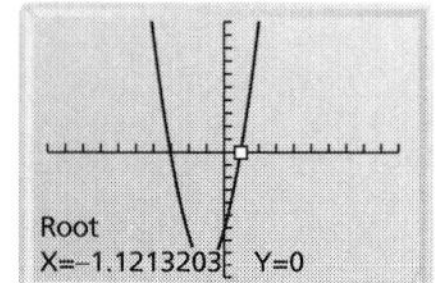

4. Do the graphing calculator screens indicate that the solutions in Example 2 check? Explain. Yes; solutions and *x*-intercepts match
5. **Try This** Find the solutions of $-3x^2 + 5x - 2 = 0$. Round to the nearest hundredth. 1, 0.67

The quadratic formula is important in physics when finding vertical motion. When an object is dropped, thrown, or launched either straight up or down, you can use the **vertical motion formula** to find the height of the object.

who has a watch with a timer that can measure in tenths of seconds. Have students drop a paper clip, table-tennis ball, or eraser from a carefully measured height. Then have students use the vertical motion formula and the time of the object's fall to calculate the distance the object fell. Advise students to decide how to time the fall before they begin the test. After students have compared the real distance with the distance calculated using the formula, have them make conjectures about why the two answers do not match exactly.

CONNECTING TO THE STUDENTS' WORLD Ask students who have built and flown model rockets to share their experiences and knowledge with the class. If possible, have them bring model rockets to school for a demonstration. Discuss safety concerns with students and administrators before you conduct the demonstration.

MAKING CONNECTIONS The record for throwing a heavier-than-air inert object is 1298 ft (433 yd). The object was a Skyro flying ring thrown by Tom McRann on May 21, 1986, at Great America Park, Santa Clara, California.

Exercises ON YOUR OWN

ALTERNATIVE ASSESSMENT Exercises 1–6 In order to assess whether students understand the quadratic formula, have them first complete these exercises on their own. Then, call on students to solve the expressions on the board, while they explain how they used the quadratic formula.

Exercise 19 Make sure students note what *t* and *P* represent. Invite students to use their knowledge of history to speculate about why the function models only the population after 1900.

Vertical motion formula: $h = -16t^2 + vt + s$
h is the height of the object in feet.
t is the time it takes an object to rise or fall to a given height.
v is the starting velocity in feet per second.
s is the starting height in feet.

Example 3 **Relating to the Real World** 

Model Rockets Members of the science club launch a model rocket from ground level with starting velocity of 96 ft/s. After how many seconds will the rocket have an altitude of 128 ft?

$h = -16t^2 + vt + s$ ← Use the vertical motion formula.
$128 = -16t^2 + 96t + 0$ ← Substitute 128 for h, 96 for v, and 0 for s.
$0 = -16t^2 + 96t - 128$ ← Subtract 128 from each side.

$x = \frac{-b \pm \sqrt{b^2 - 4ac}}{2a}$ ← Use the quadratic formula.

$t = \frac{-(96) \pm \sqrt{(96)^2 - (4)(-16)(-128)}}{2(-16)}$ ← Substitute −16 for a, 96 for b, and −128 for c.

$t = \frac{-96 \pm \sqrt{9216 - 8192}}{-32}$ ← Simplify.

$t = \frac{-96 \pm \sqrt{1024}}{-32}$

$t = \frac{-96 + 32}{-32}$ or $t = \frac{-96 - 32}{-32}$ ← Write two solutions and simplify.

$t = 2$ or $t = 4$

$s = 0$ because the rocket is launched from the ground.

The rocket is 128 ft above the ground after 2 s and after 4 s.

6. Critical Thinking Use a diagram to explain how the rocket could have an altitude of 128 ft at two different times. See margin.

7. a. Try This In Example 3, after how many seconds of flight does the rocket have an altitude of 80 ft? 1 s, 5 s
   b. Critical Thinking Estimate the number of seconds it will take the rocket to reach its maximum height. Explain how you made your estimate. At 3 s; this time is halfway between the time the rocket reached 80 ft going up and 80 ft going down.

Technology Options

For Exercises 21–32, students may be helped by using graphing software or a graphing calculator to solve the equations.

Prentice Hall Technology 

 **Software** • Secondary Math Lab Toolkit • Computer Item Generator 7-6

 **CD-ROM** • Multimedia Algebra Lab 7

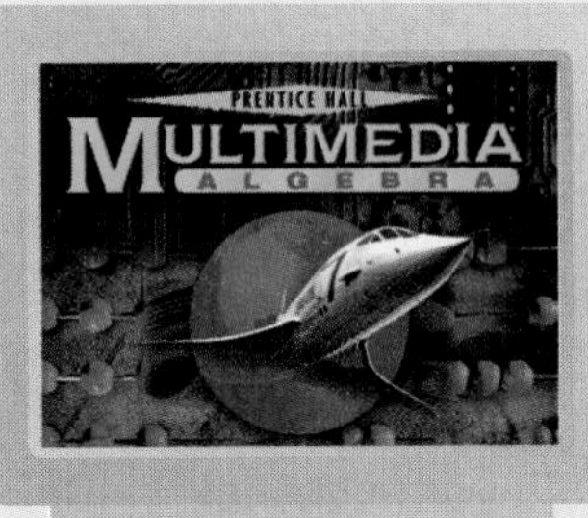

 **Internet** • See the Prentice Hall site. (http://www.phschool.com)

Think and Discuss, 6, On Your Own, 33. See back of book.

Exercises ON YOUR OWN

Write each equation in standard form.

1. $3x^2 - 10 = -13x$ $3x^2 + 13x - 10 = 0$
2. $4x^2 = 144x$ $4x^2 - 144x = 0$
3. $x^2 - 3x = 2x + 7$ $x^2 - 5x - 7 = 0$
4. $5x^2 = -7x - 8$ $5x^2 + 7x + 8 = 0$
5. $-12x^2 + 25x = 84$ $12x^2 - 25x + 84 = 0$
6. $-x^2 + 9x = 4x - 12$ $x^2 - 5x - 12 = 0$

Ask students to recall what they know about how the rates of immigration to the United States changed between the 19th and 20th centuries. Encourage interested students to make a graph of the United States' population from 1800 to the present. By studying the varying slope of the graph, lead students to understand that the graph is too complicated to be modeled by a simple function.

MAKING CONNECTIONS Exercise 35 The Sears Tower in Chicago has 110 floors that rise to a height of 1454 ft. Two TV antennae on top raise the height to 1707 ft. Suppose that a window washer on the 110th floor drops a cleaning brush. Using the vertical motion formula $h = -16t^2 + vt + s$, have students calculate how long it would take the brush to hit the ground.

Chapter Project **FIND OUT BY REASONING** Students use the given formula to calculate the maximum speed you should travel in order to stop in 150 ft. This task is essential to their work on the Chapter Project. Have students add this part of the project to work they have already completed. To check students' progress on the project, have them write a few sentences stating what they have done so far and their plans for completing the assignment.

Exercises MIXED REVIEW

SELF ASSESSMENT **JOURNAL** Encourage students to include sample quadratic equations and, if appropriate, graphs with their answers.

346

Use the quadratic formula to solve each equation. Round solutions to the nearest hundredth when necessary.

7. $6x^2 + 7x - 5 = 0$ 0.5, −1.67
8. $3x^2 - 3x - 1 = 0$ 1.26, −0.26
9. $6x^2 - 130 = 0$ 4.65, −4.65
10. $x^2 + 6x + 8 = -1$ −3
11. $5x^2 - 4x = 33$ 3, −2.2
12. $3x^2 = 6x + 4$ 2.53, −0.53
13. $9x^2 - 5x = 0$ 0.56, 0
14. $7x^2 = 13$ 1.36, −1.36
15. $2x^2 + 3x - 1 = 0$ 0.28, −1.78
16. $2x^2 - 12 = 11x$ 6.43, −0.93
17. $2x^2 + 5x + 3 = 0$ −1, −1.5
18. $4x^2 = 12x - 9$ 1.5

19. *Population* The function below models the United States population P in millions since 1900, where t is the number of years after 1900.

$$P = 0.0089t^2 + 1.1149t + 78.4491$$

   a. *Open-ended* Use the function to estimate the United States population the year you graduate from high school.
   b. Estimate the United States population in 2025. about 356.9 million
   c. Use the function to **predict** the year in which the population reaches 300 million. 2007

19a. Check students' work. Sample: 2001—281.8 million; 2002—284.8 million; 2003—287.7 million; 2004—290.7 million

20. *Recreation* Suppose you throw a ball in the air with a starting velocity of 30 ft/s. The ball is 5 ft high when it leaves your hand. After how many seconds will it hit the ground? Use the vertical motion formula $h = -16t^2 + vt + s$. about 2.0 s

Choose **Use any method you choose to solve each equation.**

21. $2t^2 = 72$ 6, −6
22. $3x^2 + 2x - 4 = 0$ 0.87, −1.54
23. $5b^2 - 10 = 0$ 1.41, −1.41
24. $3x^2 + 4x = 10$ 1.28, −2.61
25. $x^2 = -5x - 6$ −2, −3
26. $m^2 - 4m = -4$ 2
27. $d^2 - d - 6 = 0$ 3, −2
28. $13n^2 - 117 = 0$ 3, −3
29. $3s^2 - 4s = 2$ 1.72, −0.39
30. $5b^2 - 2b - 7 = 0$ 1.4, −1
31. $15x^2 - 12x - 48 = 0$ 2.23, −1.43
32. $4t^2 = 81$ 4.5, −4.5

33. *Writing* Compare how you solve the linear equation $mx + b = 0$ with how you solve the quadratic equation $ax^2 + bx + c = 0$. See margin.

34. *Open-ended* Give an example of a quadratic equation that is easier to solve by finding the square roots of each side than by using the quadratic formula or by graphing. Explain your choice. Answers may vary. Sample: $x^2 = 25$. Square roots of both sides are well known.

35. *Math in the Media* Use the data at the right. Suppose that a cleaner at the top of the Gateway Arch drops a cleaning brush. Use the vertical motion formula $h = -16t^2 + vt + s$.
   a. What is the value of s, the starting height? 630 ft
   b. What is the value of h when the brush hits the ground? 0 ft
   c. The starting velocity is 0. Find how many seconds it takes the brush to hit the ground. about 6.3 s

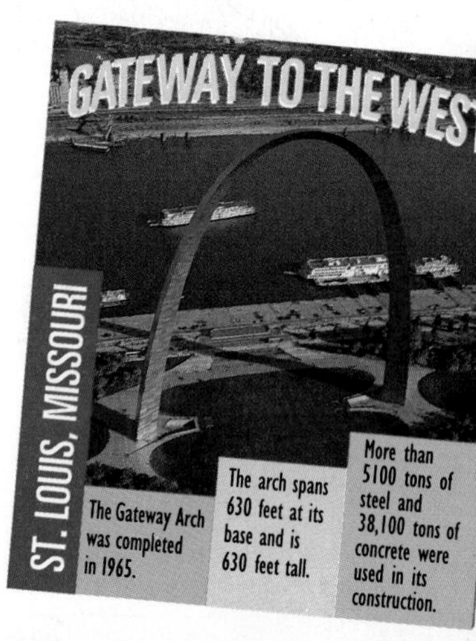

GATEWAY TO THE WEST
ST. LOUIS, MISSOURI
The Gateway Arch was completed in 1965.
The arch spans 630 feet at its base and is 630 feet tall.
More than 5100 tons of steel and 38,100 tons of concrete were used in its construction.

GETTING READY FOR LESSON 7-7 These exercises prepare students for using the discriminant to find the number of solutions of a quadratic equation.

Wrap Up

THE BIG IDEA Ask students: *What is the quadratic formula and why is it useful?*

RETEACHING ACTIVITY Students use the quadratic formula to solve quadratic equations. They find two solutions for each equation. (Reteaching worksheet 7-6)

Exercises CHECKPOINT

Students will assess their own progress on Lessons 7-5 to 7-6.

STANDARDIZED TEST TIP **Exercise 7** Remind students to always rewrite the equations in standard form before substituting into the quadratic formula.

Chapter Project — Find Out by Reasoning

The formula $d = 0.044s^2 + 1.1s$ relates the maximum speed s in miles per hour that you should travel in order to be able to stop in d feet. Suppose you have 150 ft (about 10 car lengths) between your car and the car in front of you. Find the maximum speed you should travel. about 47 mi/h

Exercises MIXED REVIEW

Find each probability for two rolls of a number cube.

36. P(a 2 and a 6) $\frac{1}{36}$ **37.** P(two odds) $\frac{1}{4}$

38. P(two 5's) $\frac{1}{36}$ **39.** P(two fractions) 0

40. a. *Transportation* A commuter train has a 150-ton locomotive and 70-ton double-decker passenger cars. Write a linear function for the weight of a train with c passenger cars. $w = 70c + 150$

b. How much does a train with 7 passenger cars weigh? 640 t

FOR YOUR JOURNAL

Write and answer three questions that review what you learned about quadratic equations in this chapter.

Getting Ready for Lesson 7-7

For each equation, find the value of $b^2 - 4ac$.

41. $2x^2 + 3x - 4 = 0$ 41 **42.** $3x^2 + 2x + 1 = 0$ −8 **43.** $x^2 + 2x - 5 = 0$ 24

44. $3x^2 - 6x = 5$ 96 **45.** $2x^2 - 5x = 3$ 49 **46.** $2x^2 + 3 = 7x$ 25

Exercises CHECKPOINT

Solve each equation.

1. $x^2 - 16 = 49$ 8.06, −8.06 **2.** $8x^2 + 1 = 33$ 2, −2 **3.** $x^2 - 120 = 1$ 11, −11

Use the quadratic formula to solve each equation.

4. $x^2 - 4x + 3 = 0$ 3, 1 **5.** $5x^2 + 3x - 2 = 0$ 0.4, −1 **6.** $4x^2 = 14x - 3$ 3.27, 0.23

7. *Standardized Test Prep* Which expression could you use to solve $2x^2 + 5 = 3x$? D

A. $\frac{-5 \pm \sqrt{5^2 - (4)(2)(3)}}{4}$ **B.** $\frac{-3 \pm \sqrt{3^2 - (4)(2)(5)}}{4}$

C. $\frac{-(-3) \pm \sqrt{(-3)^2 - (4)(2)(-3)}}{4}$ **D.** $\frac{-(-3) \pm \sqrt{(-3)^2 - (4)(2)(5)}}{4}$

Reteaching 7-6

Practice 7-6

Practice 7-6 Mixed Exercises

Lesson Quiz

Lesson Quiz is also available in Transparencies.

Write each equation in standard form.

1. $3x^2 - 7x = -12x + 3$
$3x^2 + 5x - 3 = 0$

2. $-2x^2 + 3x = -2x + 3$
$-2x^2 + 5x - 3 = 0$

Use the quadratic formula to solve each equation.

3. $5x^2 + 2x - 6 = 0$ 0.91, −1.31

4. $x^2 = -3x + 7$ 1.54, −4.54

Math ToolboX

This toolbox describes one method of writing a quadratic equation from its graph. It also provides students with additional practice solving systems of linear equations. To simplify the process, write the following steps on the board as you work through the example with students. **Step 1.** Find c using the y-intercept. **Step 2.** Substitute x-intercepts and c into $y = ax^2 + bx + c$ to find the system of equations. **Step 3.** Solve the system for a and b. **Step 4.** Use a, b, and c to write the equation. Have students work in small groups to complete the exercises.

ERROR ALERT! Some students may not put equations in proper form. **Remediation:** Have students put both equations in standard form before solving the system.

WRITING Exercise 4 Suggest students describe any problems they may have writing equations for the graphs on p. 330.

ADDITIONAL PROBLEM Work backward to write the equation of the quadratic function shown in the graph.

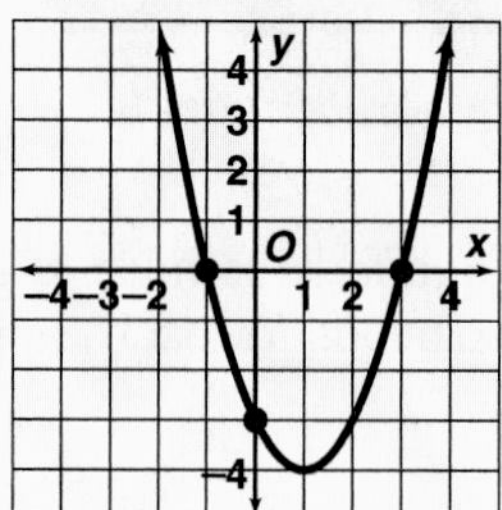

Resources

Transparencies
1,11, 29

Math ToolboX Problem Solving

Work Backward

After Lesson 7-6

Sometimes you can solve a problem by working backward. To write a quadratic function from its graph, you need to find the x- and y-intercepts.

Example

Work backward to write the equation of the quadratic function shown in the graph.

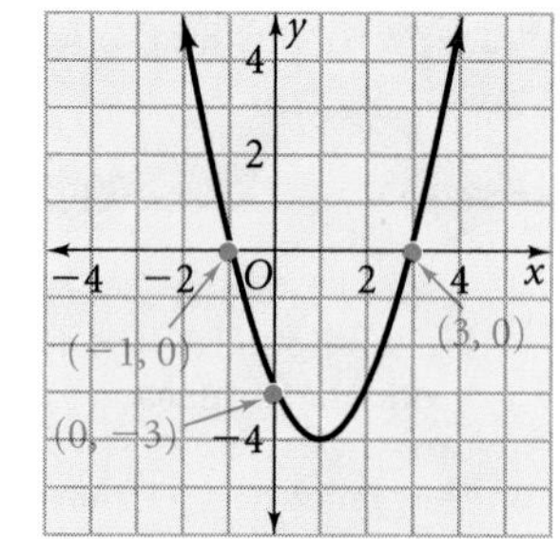

A quadratic function is in the form $y = ax^2 + bx + c$ $(a \neq 0)$. Since the y-intercept of the graph is -3, you know that $c = -3$. Notice that the graph has x-intercepts at -1 and 3. Substitute the points $(-1, 0)$ and $(3, 0)$ into $y = ax^2 + bx - 3$. The result is a system of equations.

$y = ax^2 + bx - 3$	$y = ax^2 + bx - 3$
$0 = a(-1)^2 + b(-1) - 3$	$0 = a(3)^2 + b(3) - 3$
$0 = a - b - 3$	$0 = 9a + 3b - 3$
$a - b = 3$	$9a + 3b = 3$
	$3a + b = 1$

When you solve the system of equations, you find that $a = 1$ and $b = -2$.

$$a - b = 3$$
$$3a + b = 1$$

The equation of the graph above is $y = x^2 - 2x - 3$.

Work backward to find the equation of each graph.

1. $y = x^2 + 6x + 8$

2. 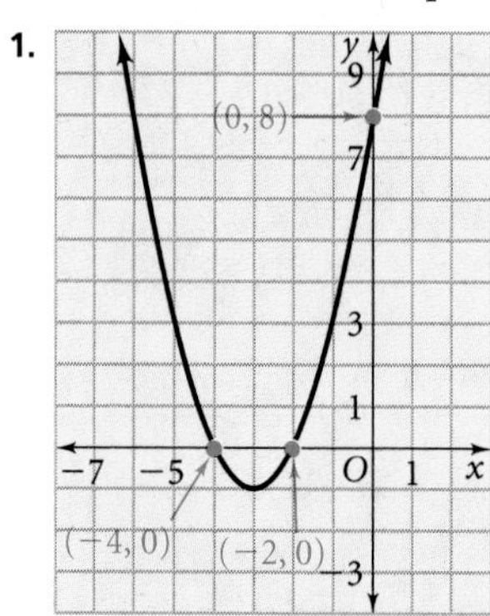

$y = -x^2 + x + 2$

3. $y = 2x^2 + 6x - 8$

4. *Writing* Explain why this method would not be ideal for all graphs.
Answers may vary. Sample: The x- and y-intercepts may be irrational numbers.

348

7-7 Teaching Notes

PROBLEM OF THE DAY

A horseback-riding camp must have two counselors for every nine campers and one trainer for every four horses. The camp must also have a nurse and an administrator. If 36 students sign up for camp, and each student gets his or her own horse, how many staff members will be needed? 19

Problem of the Day is also available in Transparencies.

CONNECTING TO PRIOR KNOWLEDGE Explain to students that they are going to learn to use the quadratic formula to determine whether there are two, one, or no solutions to a quadratic equation. Ask students to suggest scenarios where the value of the solutions is not important but the number of solutions is important.

WORK TOGETHER

Question 1 Stress that students do not need to solve each equation.

Question 2 Stress that students do not need to calculate what the x-intercepts are, just how many there are.

THINK AND DISCUSS

Make sure students understand that the discriminant is only what is inside the radical sign. The discriminant does not include the radical sign.

Connections Physics . . . and more

7-7 Using the Discriminant

What You'll Learn

- Using the discriminant to find the number of solutions of a quadratic equation

...And Why

To solve physics and home improvement problems

What You'll Need

- graphing calculator

WORK TOGETHER

Work with a partner.

1. Each equation is in the form of $ax^2 + bx + c = 0$. Find the value of the expression $b^2 - 4ac$ for each equation.
   a. $x^2 + 2x + 5 = 0$ -16
   b. $-3x^2 + 2x - 1 = 0$ -8
   c. $\frac{1}{2}x^2 + x + 4 = 0$ -7
2. a. Graphing Calculator Graph the related function for each equation in Question 1. See margin.
   b. How many x-intercepts do these graphs have? 0 each
   c. How many solutions do the equations in Question 1 have? 0 each
3. **Generalize** Based on Questions 1 and 2, complete this statement: For an equation in which $b^2 - 4ac$ ■ 0, the equation has ■ solutions. <; 0

THINK AND DISCUSS

In the Work Together activity, you investigated the discriminant of three quadratic equations. The quantity $b^2 - 4ac$ is called the **discriminant** of a quadratic equation. The discriminant is part of the quadratic formula.

$$x = \frac{-b \pm \sqrt{b^2 - 4ac}}{2a} \quad \leftarrow \text{the discriminant}$$

The graph of the related function of a quadratic equation gives you a picture of what happens when a discriminant is positive, 0, or negative.

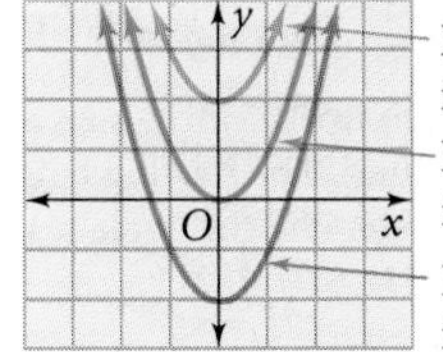

y
O
x
Discriminant is negative.
Discriminant is 0.
Discriminant is positive.

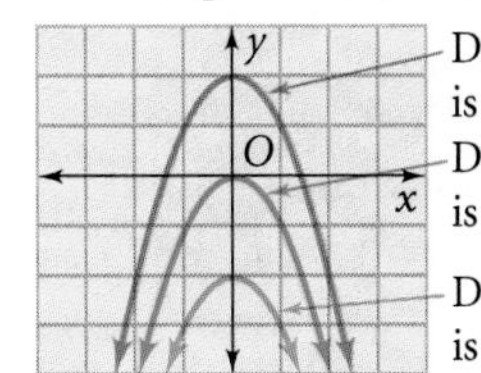

y
O
x
Discriminant is positive.
Discriminant is 0.
Discriminant is negative.

4. How many solutions will an equation have if the discriminant is positive? Explain. 2; there are 2 x-intercepts, that is, 2 values of x for which $y = 0$.
5. How many solutions will an equation have if the discriminant is 0? Explain. 1; there is just one x-intercept.
6. How many solutions will an equation have if the discriminant is negative? Explain. 0; there is no x-intercept.
7. Does the direction a graph opens affect the number of solutions found by using the discriminant? no

Lesson Planning Options

Prerequisite Skills

- Evaluating expressions
- Graphing quadratic functions

Assignment Options

Core 1–15, 22–30, 32–39
Extension 16–21, 31, 40, 41

Use Mixed Review to maintain skills.

Resources

Student Edition

Skills Handbook, pp. 580, 582
Extra Practice, p. 562
Glossary/Study Guide

Teaching Resources

Chapter Support File, Ch. 7
- Practice 7-7 (two worksheets)
- Reteaching 7-7

Classroom Manager 7-7
Glossary, Spanish Resources
Two-Year Algebra Handbook 7-7

Transparencies

11, 91, 95

Question 5 Point out that it is incorrect to think that if the discriminant is zero, there are zero solutions to an equation. Also point out that the sign of the discriminant does not predict the signs of the equation solutions.

ERROR ALERT! **Question 7** Students may think that the number of solutions indicated by the discriminant is dependent on the direction of the graph. **Remediation:** Help students understand that the direction a graph opens has no effect on the number of solutions found by using the discriminant. Point out that when you just want to know the number of solutions, you are actually asking whether the graph touches the x-axis, and, if so, how many times. The direction of the graph is irrelevant.

Some students may wish to calculate the discriminants and graph the related equations in order to prove to themselves that the diagrams on page 349 are correct. Suggest that they use the following equations for their proof: $y = x^2 + 4$, $y = x^2$, $y = x^2 - 4$, $y = -x^2 + 4$, $y = -x^2$, $y = -x^2 - 4$. Some students may have difficulty remembering which values of the discriminant result in which number of solutions. The least value ($b^2 - 4ac < 0$) results in the least amount of solutions (no solutions), while the greatest value ($b^2 - 4ac > 0$) results in the greatest amount of solutions (two solutions).

Example 1

Question 9 Remind students to rewrite the equation in standard form before calculating the discriminant.

Additional Examples

FOR EXAMPLE 1

Find the number of solutions for the equation $7x^2 - 3x = 4$. **two**

FOR EXAMPLE 2

In Example 2, we found that throwing an apple upward with a velocity of 30 ft/s would not cause it to reach a height of 25 ft. If the apple was instead thrown at 40 ft/s, would it then reach the required height? **yes**

Discussion: *How can the discriminant be used to find the minimum speed with which the apple can still reach the required height?*

Property of the Discriminant

For the quadratic equation $ax^2 + bx + c = 0$ where $a \neq 0$, the value of the discriminant tells you the number of solutions.

Discriminant	Number of Solutions
$b^2 - 4ac > 0$	two solutions
$b^2 - 4ac = 0$	one solution
$b^2 - 4ac < 0$	no solution

Example 1

Find the number of solutions of $3x^2 - 5x = 1$.

$3x^2 - 5x - 1 = 0$ ← Write in standard form.

$b^2 - 4ac = (-5)^2 - (4)(3)(-1)$ ← Substitute for a, b, and c.

$= 25 - (-12)$

$= 37$

Since $37 > 0$, the equation has two solutions.

8. **Graphing Calculator** Check the result in Example 1 by graphing the related function $y = 3x^2 - 5x - 1$. **See margin.**

9. **Try This** Find the number of solutions of $x^2 = 2x - 3$. **0**

10. **Critical Thinking** Kenji claimed that the discriminant of $2x^2 + 5x - 1 = 0$ had the value 17. What error did he make? **Kenji used $c = 1$ instead of $c = -1$.**

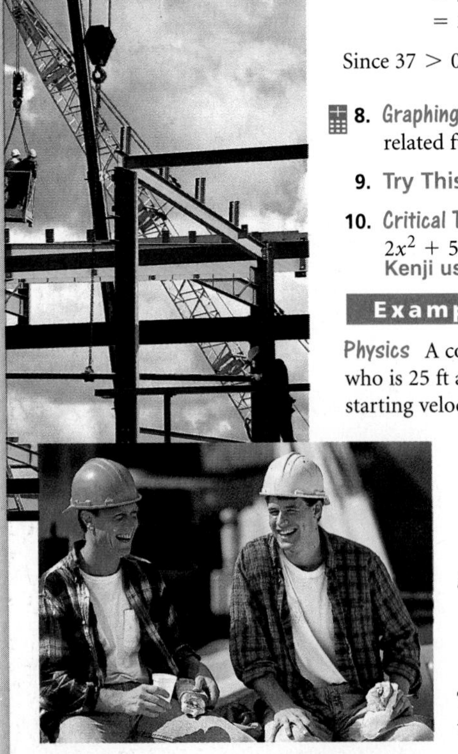

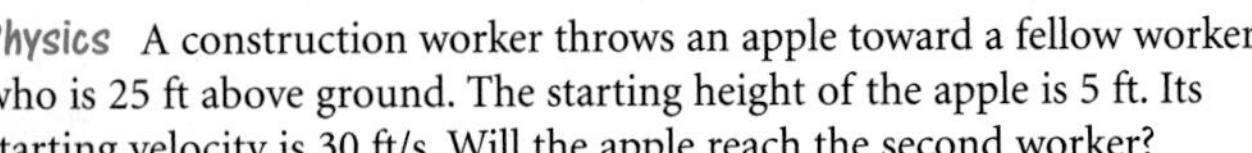

Example 2 **Relating to the Real World**

Physics A construction worker throws an apple toward a fellow worker who is 25 ft above ground. The starting height of the apple is 5 ft. Its starting velocity is 30 ft/s. Will the apple reach the second worker?

$h = -16t^2 + vt + s$ ← Use the vertical motion formula.

$25 = -16t^2 + 30t + 5$ ← Substitute 25 for h, 30 for v, and 5 for s.

$0 = -16t^2 + 30t - 20$ ← Write in standard form.

$b^2 - 4ac = (30)^2 - 4(-16)(-20)$ ← Evaluate the discriminant.

$= 900 - 1280$

$= -380$

The discriminant is negative. The apple will not reach the second worker.

CRITICAL THINKING Question 10 Have students calculate the correct value for the discriminant. As they find the value, they will find Kenji's error.

Example 2 **Relating to the Real World**

Challenge students to explain why the apple does not reach a height of 25 ft even though it is thrown with a speed of 30 ft/s.

AUDITORY LEARNING Challenge groups of students to write a poem or song they could recite or sing to remind themselves what the different values of the discriminant represent.

Exercises ON YOUR OWN

Have students first determine the number of solutions for each graph. Then, before applying the Property of the Discriminant, rewrite each answer choice in a form of less than, greater than, or equal to zero.

ALTERNATIVE ASSESSMENT **Exercises 4–7** Have students graph each equation to find the solutions. Then ask them to calculate the discriminant and write the value on their graphs. This will help you assess students' understandings of the discriminant.

OPEN-ENDED Exercise 16 Ask students to describe how they would go about writing such an equation. Ask: *What is the most efficient way to do this?*

Exercise 18 After students have written their answer to this exercise, have them write an equation based on their answer. Then have them find the discriminant to check their reasoning.

11. **Try This** Suppose the first construction worker in Example 2 goes up to the next floor. He then throws the apple at a starting height of 17 ft and starting velocity of 30 ft/s. Will the apple reach the second worker? **yes**

Exercises ON YOUR OWN

For which discriminant is each graph possible?

1. **C**
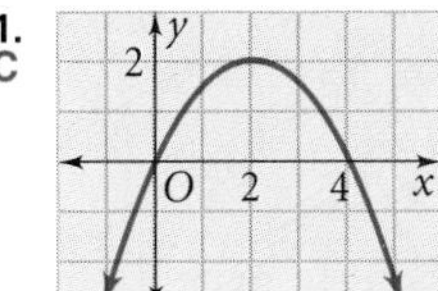

2. **B**
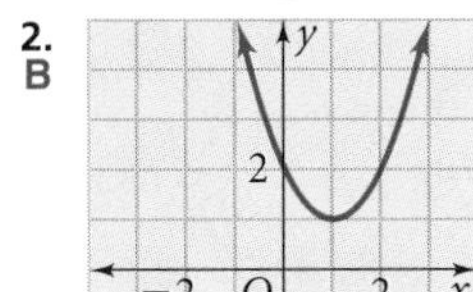

3. **A**
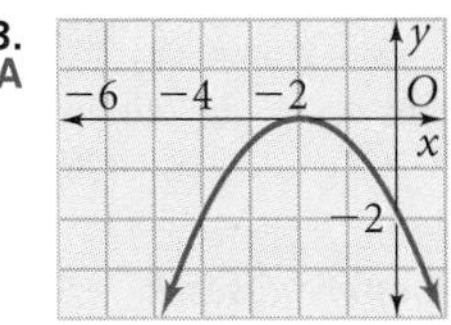

A. $b^2 - 4ac = 0$
B. $b^2 - 4ac = -2$
C. $b^2 - 4ac = 5$

Mental Math **Find the number of solutions of each equation.**

4. $x^2 - 3x + 4 = 0$ **0**
5. $x^2 - 6x + 9 = 0$ **1**
6. $x^2 + 4x - 2 = 0$ **2**
7. $x^2 - 1 = 0$ **2**
8. $x^2 - 2x - 3 = 0$ **2**
9. $x^2 + x = 0$ **2**
10. $2x^2 - 3x + 4 = 0$ **0**
11. $2x^2 + 4x = -15$ **0**
12. $x^2 - 7x + 6 = 0$ **2**
13. $x^2 + 2x + 1 = 0$ **1**
14. $4x^2 + 5x = -2$ **0**
15. $x^2 - 8x = -12$ **2**

16. **Open-ended** Write a quadratic equation that has no solution. **Answers may vary. Sample: $y = x^2 + 11$**

17. **Home Improvements** The Reeves family garden is 18 ft long and 15 ft wide. They want to modify it according to the diagram at the right. The new area is modeled by the equation $A = -x^2 + 3x + 270$.
   a. What value of x, if any, will give a new area of 280 ft^2? **no value of x**
   b. Is there any value of x for which the garden has an area of 266 ft^2? Explain. **Yes; $266 = -x^2 + 3x + 270$ transforms to $-x^2 + 3x + 4 = 0$. The discriminant is positive.**

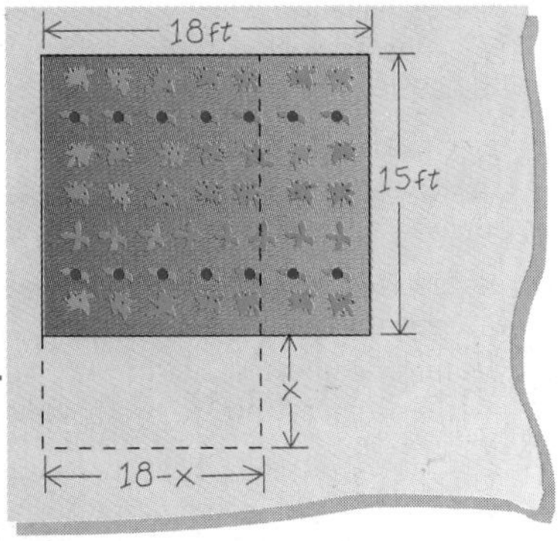

18. **Writing** How can you use the discriminant to write an equation that has two solutions? **Choose a, b, c, so that $b^2 - 4ac > 0$.**

19. **Physics** Suppose the equation $h = -16t^2 + 35t$ models the altitude a football will reach t seconds after it is kicked. Which of the following altitudes are possible? **A**
   A. $h = 16$ ft
   B. $h = 25$ ft
   C. $h = 30$ ft
   D. $h = 35$ ft

20. Find the value of the discriminant and the solutions of each equation.
   a. $x^2 - 6x + 5 = 0$ **16; 5, 1**
   b. $x^2 + x - 20 = 0$ **81; 4, –5**
   c. $2x^2 - 7x - 3 = 0$ **73; –0.39, 3.89**

21. **Critical Thinking** Use your results to Question 18. When the discriminant is a perfect square, are the solutions rational or irrational? Explain. **Rational; the square root of the discriminant is a positive integer.**

Technology Options

For Exercises 4–15 and 22–30, students may wish to use graphing software or a graphing calculator to find the number of solutions.

Prentice Hall Technology

Software • Secondary Math Lab Toolkit • Computer Item Generator 7-7

CD-ROM • Multimedia Algebra Lab 7

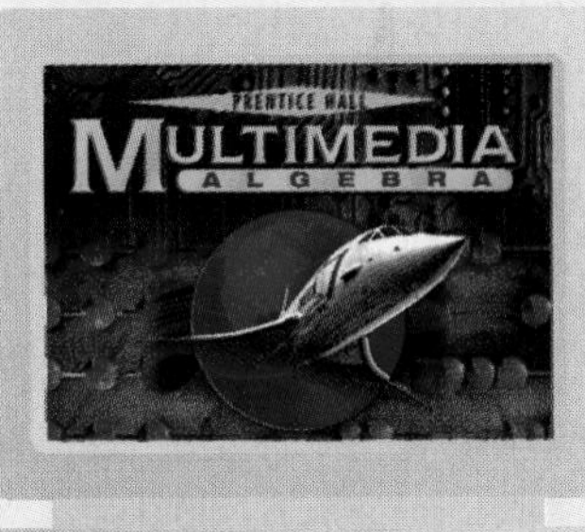

Internet • See the Prentice Hall site. (http://www.phschool.com)

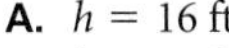

Exercises 22–30 Remind students that the x-intercepts are the solutions of each equation. They may apply the Property of Discriminant to make sure they found the correct amount of solutions.

Exercise 19 Ask students: *What is the value of t in the equation $h = 16t^2 + 35t$ for answer choice (C)?* 0

STANDARDIZED TEST TIP **Exercise 31** Note that there is no need to actually solve each expression. Simply use the discriminant of the equation in column A to predict the number of its solutions.

Chapter Project FIND OUT BY COMMUNICATING Students work as a group to plan a skit in which they present the information they learned about safe distances in driving. This task is the final part of their work on the Chapter Project. To check students progress on the project you may wish to inspect their Project Folders.

Exercises MIXED REVIEW

SELF ASSESSMENT PORTFOLIO Share with students the criteria you will use to assess their work in portfolios, as well as how you plan to use the results. Students should understand how the rubrics assess their work, how each piece in the portfolio counts, and how the scores they get in their portfolios will affect their overall evaluation.

Exercise 50 Challenge students to find out more about the power of volcanoes. Ask students to research at what speeds materials erupt from volcanoes.

page 349 Work Together

2a.

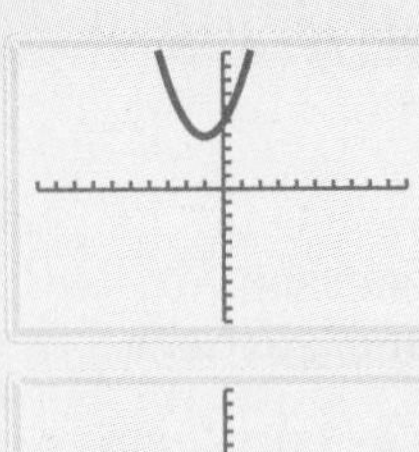

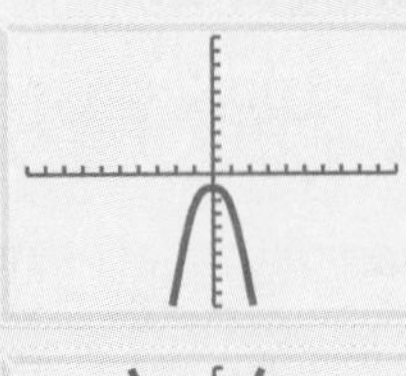

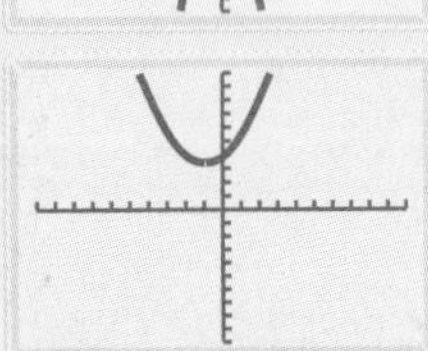

pages 349–351 Think and Discuss

8.

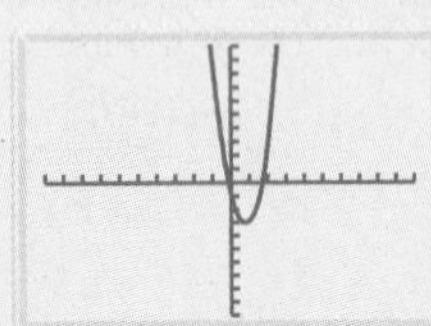

page 353 Mixed Review

42–45. See back of book.

352

Find the number of x-intercepts of each function.

22. $y = x^2 - 6x + 5$ 2
23. $y = 2x^2 + 4x - 3$ 2
24. $y = x^2 + 2x + 9$ 0
25. $y = -x^2 - 2x$ 2
26. $y = x^2$ 1
27. $y = 3x^2 - 2x + 5$ 0
28. $y = 8x^2 - 2x - 45$ 2
29. $y = -x^2 - 2$ 0
30. $y = -x^2$ 1

31. *Standardized Test Prep* Compare the quantities in Column A and Column B. B

Column A	Column B
the number of solutions of $35 = 20x^2 - 15x + 47$	the number of solutions of $15x + 7 = 0$

A. The quantity in Column A is greater.
B. The quantity in Column B is greater.
C. The quantities are equal.
D. The relationship cannot be determined from the information given.

For each function, decide if its graph crosses the x-axis. For those that do, find the coordinates of the points at which they cross.

32. $y = x^2 - 2x + 5$ no
33. $y = 2x^2 - 4x + 3$ no
34. $y = 4x^2 + x - 5$ yes; $\left(-1\frac{1}{4}, 0\right)$, $(1, 0)$
35. $y = -3x^2 - x + 2$ yes; $\left(\frac{2}{3}, 0\right)$, $(-1, 0)$
36. $y = x^2 - 5x + 7$ no
37. $y = 2x^2 - 3x - 5$ yes; $(-1, 0)$, $\left(2\frac{1}{2}, 0\right)$
38. For the equation $x^2 + 4x + k = 0$, find all values of k such that the equation has each number of solutions.
    a. none $k > 4$ b. one 4 c. two $k < 4$
39. *Electrical Engineering* The function $P = 3i^2 - 2i + 450$ models the power P in an electric circuit with a current i. Can the power in this circuit ever be zero? If so, at what value of i? no
40. *Business* An apartment rental agency uses the formula $I = 5400 + 300n - 50n^2$ to find its monthly income I based on renting n number of apartments. Will the agency's monthly income ever be $7000? Explain. No; the equation transforms to $-50n^2 + 300n - 1600 = 0$. The discriminant is negative.
41. *Computer* You can use a spreadsheet like the one at the right to find the discriminant for each value of b shown in column A.
    a. What spreadsheet formula would you use to find the value in cell B2? in cell C2?
    b. Describe the integer values of b for which $x^2 + bx + 1 = 0$ has solutions.
    c. Describe the integer values of b for which $x^2 + bx + 2 = 0$ has no solution. integer values ≥ -2 or ≤ 2

    a. A2ˆ2 − 4; A2ˆ2 − 8
    b. for integer values ≤ -2 or ≥ 2.

	A	B	C
1	b	x^2 + bx + 1 = 0	x^2 + bx + 2 = 0
2	−3	5	1
3	−2	0	−4
4	−1	−3	−7
5	0	−4	−8
6	1	−3	−7
7	2	0	−4
8	3	5	1

Wrap Up

THE BIG IDEA Ask students: *What is the discriminant and how is it useful?*

RETEACHING ACTIVITY Students find the value of the discriminant and the solutions for quadratic equations. They complete a chart showing the discriminant, solutions and x-intercepts. (Reteaching worksheet 7-7.)

A Point in Time

If you have block scheduling or an extended class period you may wish to have students investigate these topics:

- What well-noted game, introduced in the 1920s and of probable Chinese origin, became a U.S. craze In 1924? Mah Jong
- What publication, founded in 1923 in New York City by Henry Robinson Luce, was the first weekly news magazine? Time
- List five companies that manufacture parts for the aerospace industry. There are over 4,000 firms; examples may include: Boeing, General Electric, Texas Instruments.

Chapter Project **Find Out by Communicating**

Work with a group of your classmates to plan a skit that will present what you have learned about safe distances in driving. Illustrate the relationships among reaction time, road conditions, speed, and stopping distances.

Exercises MIXED REVIEW

Solve each system by graphing. 42–45. See margin.

42. $4x + y \leq 5$
$3x - 2y > 10$

43. $-6x - 3y \geq 8$
$y > -9$

44. $x < 7$
$-3x + 7y \geq 0$

45. $y = 3x - 2$
$y = -x + 6$

Evaluate each function for $x = 3$.

46. $f(x) = \frac{1}{2}x + 3$ $4\frac{1}{2}$

47. $g(x) = \frac{x+3}{2}$ 3

48. $h(x) = 3 - x$ 0

49. $f(x) = -x - 1$ -4

50. Geology Mount Shishalding is a 9372-ft tall volcano on Unimak Island, Alaska. In 1995 it erupted and sent up a 35,000-ft plume of ash. How far above sea level did the ash reach? 44,372 ft

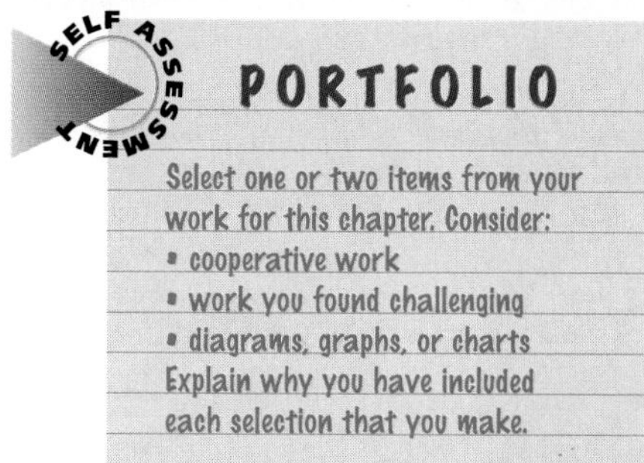

SELF ASSESSMENT
PORTFOLIO
Select one or two items from your work for this chapter. Consider:
• cooperative work
• work you found challenging
• diagrams, graphs, or charts
Explain why you have included each selection that you make.

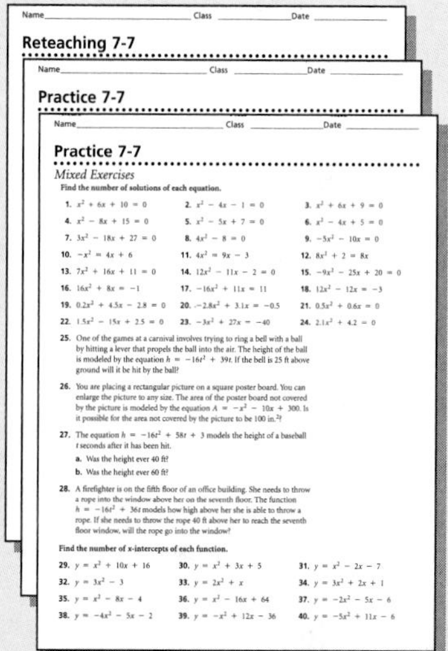

Lesson Quiz

Lesson Quiz is also available in Transparencies.

Find the number of solutions for each equation.

1. $3x^2 - 4x = 7$ two
2. $2x^2 = 3x + 10$ two
3. $2x^2 + 3 = 0$ none
4. $-3x^2 + 2x - 12 = 0$ none

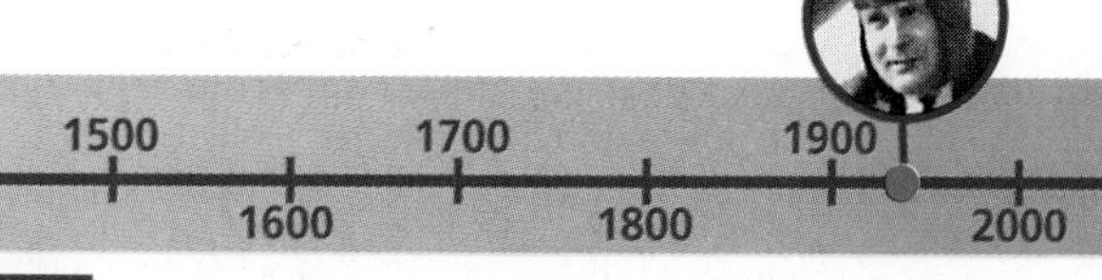

Juan de la Cierva

In 1923, Juan de la Cierva (1895–1936) designed the first successful autogyro, a rotor-based aircraft. The autogyro had rotating blades to give the aircraft lift, a propeller for forward thrust, and short, stubby wings for balance. Autogyros needed only short runways for takeoff and could descend almost vertically.

By hinging the rotor blades at the hub, de la Cierva allowed each blade to respond to aerodynamic forces. This was a significant contribution in the development of the modern helicopter.

De la Cierva's work on problems of lift and gravity, like the work of aeronautical engineers of today, involved quadratic functions.

Finishing the Chapter Project

PROJECT DAY You may wish to plan a project day on which students share their completed projects. Encourage students to explain their process as well as their product.

PROJECT NOTEBOOK Have students review their project work and bring their notebooks up to date.

- Have students review their methods for finding, recording, and solving equations and formulas used for the project.
- Ask groups to share any insights they found for completing the project, such as any shortcuts they found for solving formulas or making graphs. Also, find out if what students learned during the project might make them safer drivers.

SCORING RUBRIC

3 Calculations are correct. Graphs are neat, accurate, and clearly show the relationship between the variables. Scales are appropriate for your graphs. Your skit convinces viewers that there is a relationship between reaction time, road conditions, speed, and stopping distances.

2 Calculations are mostly correct, except for a few minor errors. Graphs are neat, and mostly accurate with minor errors in scale. Skit shows a relationship between different driving conditions and stopping distances.

1 Calculations contain minor and major errors. Graphs could be more accurate. Skit should be expanded to make convincing argument.

0 Major elements of the project are incomplete or missing.

CHAPTER PROJECT

Finishing the Chapter Project

Questions on pages 331, 336, 347, and 353 should help you to complete your project. Gather together all the data you compiled as you worked on the project. Include the equations you analyzed and your graphs. Discuss your conclusions about safe driving, stopping distance, road conditions, and so on with your classmates. Then, as a group, plan and rehearse your skit.

Reflect and Revise

Present your skit to a small group of classmates. After you have heard their comments, decide if your presentation is clear and convincing. If needed, make changes to improve your skit for the rest of the class.

Follow Up

If you have access to a commercial online service or the Internet, explore some of the forums and user groups that are related to driving and motor vehicles.

You may also want to contact highway patrol officers or registry of motor vehicle officials you know for information about the habits of drivers. Ask them what errors or violations are most common.

For More Information:

Highway Safety: Motorcycle helmet laws save lives and reduce costs to society. A Report to Congressional Requesters. Washington, D.C.: U.S. General Accounting Office, 1991.

Hewett, Joan. *Motorcycle on Patrol.* New York: Clarion Books, 1986.

Ross, Daniel Charles. "Ford F150." *Car and Driver* (January 1996): 134.

Saperstein, Robert. *Surviving an Auto Accident.* Ventura, California: Pathfinder Publishers, 1994.

Wrap Up

HOW AM I DOING? Divide students into small groups. Ask each group to put together a series of problems to illustrate the central ideas and rules of the chapter. Then, have groups exchange problems so that they solve the ones created by a different group. Finally, have the groups who wrote the problems check the answers.

KEY TERMS The numbers in parentheses direct students to the pages where the terms (or symbols) are used or defined. Students should be able to (1) write a simple explanation of each term, (2) illustrate the term with a diagram, or (3) show an example that uses the term.

Exercises 9–12 Remind students that they can predict whether a quadratic function curves upward or downward by noting the sign of the coefficient of the squared term.

STANDARDIZED TEST TIP **Exercise 31** The square root of a number greater than one is always smaller than the original number. Therefore, students can immediately eliminate answer choice (A).

Exercises 34–37 Remind students that a real number will always have a positive and a negative square root.

Exercises 39–42 Remind students that a quadratic equation can have zero, one, or two solutions.

Remind students that the new mathematical terms in this chapter are defined in the Glossary/Study Guide in the back of the book.

7 Wrap Up

Key Terms

axis of symmetry (p. 318)
discriminant (p. 349)
maximum value (p. 320)
minimum value (p. 320)
negative square root (p. 332)
parabola (p. 318)
perfect squares (p. 333)
principal square root (p. 332)
quadratic equation (p. 337)
quadratic formula (p. 343)
quadratic function (p. 319)
square root (p. 332)
standard form of a quadratic equation (p. 337)
standard form of a quadratic function (p. 319)
vertex (p. 320)
vertical motion formula (p. 344)

How am I doing?

- State three ideas from this chapter that you think are important. Explain your choices.
- Describe the different ways you can solve quadratic equations.

Resources

Student Edition
Extra Practice, p. 562
Glossary/Study Guide

Teaching Resources
Study Skills Handbook
Glossary, Spanish Resources

pages 355–357 Wrap Up

5. 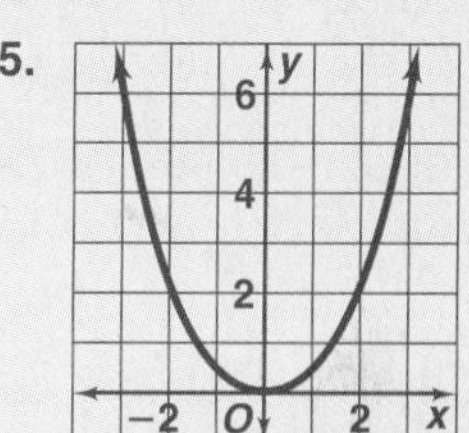

6. 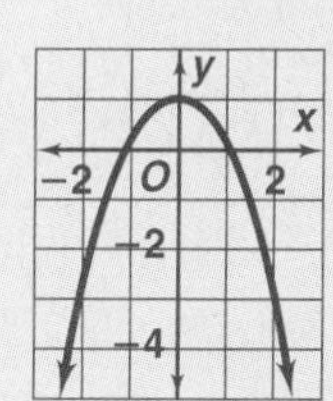

Exploring Quadratic Functions — 7-1

A function of the form $y = ax^2 + bx + c$ is a **quadratic function**. The shape of its graph is a **parabola.** The **axis of symmetry** of a parabola divides the parabola into two congruent halves.

The **vertex** of a parabola is where the axis of symmetry intersects the parabola. When a parabola opens downward, the y-coordinate of the vertex is a **maximum value** of the function. When a parabola opens upward, the y-coordinate of the vertex is a **minimum value** of the function.

The value of a determines whether the parabola opens upward or downward and how wide or narrow it is.

Open-ended Give an example of a quadratic function for each of the following descriptions.

1. Its graph opens downward. Sample: $y = -2x^2$
2. Its graph opens upward. Sample: $y = 2x^2$
3. Its vertex is at the origin. Sample: $y = x^2$
4. Its graph is wider than $y = x^2$. Sample: $y = \frac{1}{2}x^2$

Graphing Simple Quadratic Functions — 7-2

Changing the value of c in a quadratic function $y = ax^2 + c$ shifts the parabola up or down. The value of c is the y-intercept of the graph.

Graph each quadratic equation. 5–8. See margin.

5. $y = \frac{2}{3}x^2$
6. $y = -x^2 + 1$
7. $y = x^2 - 4$
8. $y = 5x^2 + 8$

State whether each function has a *maximum* or *minimum* value.

9. $y = 4x^2 + 1$ min.
10. $y = -3x^2 - 7$ max.
11. $y = \frac{1}{2}x^2 + 9$ min.
12. $y = -x^2 + 6$ max.

pages 355–357 Wrap Up

7. 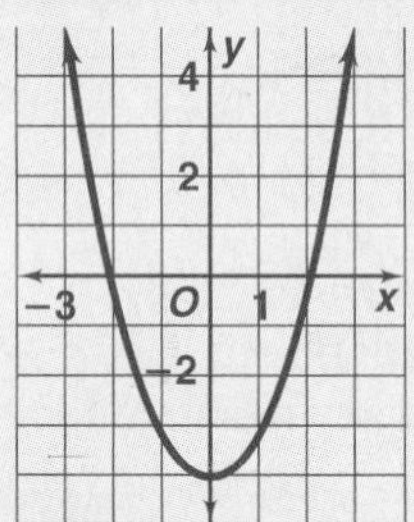

8. 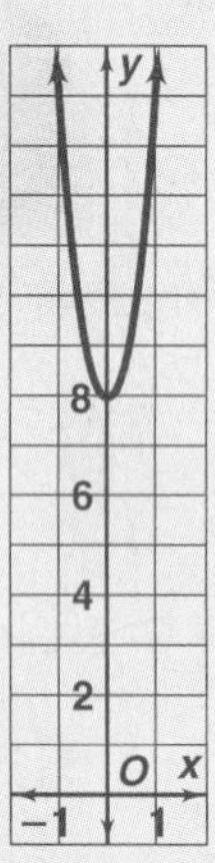

Graphing Quadratic Functions 7-3

The graph of $y = ax^2 + bx + c$, where $a \neq 0$, has the line $x = \frac{-b}{2a}$ as its axis of symmetry. The x-coordinate of the vertex is $\frac{-b}{2a}$.

Graph each function. Label the axis of symmetry and the vertex. 13–15. See margin.

13. $y = -\frac{1}{2}x^2 + 4x + 1$ **14.** $y = -2x^2 - 3x + 10$ **15.** $y = x^2 + 6x - 2$

Graph each quadratic inequality. 16–18. See margin.

16. $y \leq 3x^2 + x - 5$ **17.** $y > 2x^2 + 6x - 3$ **18.** $y \geq -x^2 - x - 8$

Square Roots 7-4

If $a^2 = b$, then a is a **square root** of b. The **principal** (or positive) **square root** of b is indicated by $\sqrt{b}$. The **negative square root** is indicated by $-\sqrt{b}$. The squares of integers are called **perfect squares.**

Tell whether each expression is *rational* or *irrational.*

19. $\sqrt{86}$ irrat. **20.** $-\sqrt{1.21}$ rat. **21.** $\pm\sqrt{\frac{1}{2}}$ irrat. **22.** $\sqrt{64}$ rat. **23.** $\sqrt{2.55}$ irrat. **24.** $-\sqrt{\frac{4}{25}}$ rat

Find the value of each expression. If the value is irrational, round your answer to the nearest hundredth.

25. $\sqrt{9}$ 3 **26.** $-\sqrt{47}$ −6.86 **27.** $\sqrt{0.36}$ 0.6 **28.** $\sqrt{140}$ 11.83 **29.** $-\sqrt{1}$ −1 **30.** $\sqrt{196}$ 14

31. *Standardized Test Prep* What is the principal square root of 2.25? B

A. 15 **B.** 1.5 **C.** −15 **D.** −1.5 **E.** 0.15

Solving Quadratic Equations 7-5

A **quadratic equation** can be written in the **standard form** $ax^2 + bx + c = 0$, where $a \neq 0$. Quadratic equations can have two, one, or no real solutions. You can solve some quadratic equations by taking the square root of each side, or by finding the x-intercepts of the related quadratic function.

If the statement is true, verify it by solving the equation. If it is false, write a true statement.

32. $x^2 - 10 = 3$; there is one solution.
False; there are two solutions.

33. $3x^2 = 27$; there are two solutions.
true; 3, −3

Solve each equation. If the equation has no solution, write *no solution.*

34. $6(x^2 - 2) = 12$ 2, −2 **35.** $-5m^2 = -125$ 5, −5 **36.** $9(w^2 + 1) = 9$ 0 **37.** $3r^2 + 27 = 0$ no solution

38. *Geometry* The area of a circle is given by the formula $A = \pi r^2$. Find the radius of a circle with area 16 in.^{2} to the nearest tenth of an inch. 2.3 in.

ALTERNATIVE ASSESSMENT Exercises 13–15, 34–37
Have students multiply the a coefficient of each term by $-\frac{1}{2}$ and solve. Then have them multiply the original equation by 2 and solve again. This will help you assess students' understanding of quadratic equations.

Getting Ready for Chapter 8

Students may complete these exercises independently or in small groups. The skills previewed will help prepare students for working with the Cartesian coordinate system and function statements.

Using the Quadratic Formula 7-6

You can solve a quadratic equation using the **quadratic formula**.

If $ax^2 + bx + c = 0$ and $a \neq 0$, then $x = \frac{-b \pm \sqrt{b^2 - 4ac}}{2a}$.

Use the quadratic formula to solve each equation. Round solutions to the nearest hundredth when necessary.

39. $4x^2 + 3x - 8 = 0$ 1.09, −1.84
40. $2x^2 - 7x = -3$ 3, 0.5
41. $-x^2 + 8x + 4 = 5$ 7.87, 0.13
42. $9x^2 - 270 = 0$ 5.48, −5.48

43. *Vertical Motion* Suppose you throw a ball in the air. The ball is 6 ft high when it leaves your hand. Use the equation $0 = -16t^2 + 20t + 6$ to find the number of seconds t that the ball is in the air. 1.5 s

Using the Discriminant 7-7

You can use the **discriminant** to find the number of real solutions of a quadratic equation. When a quadratic equation is in the form $ax^2 + bx + c = 0$ $(a \neq 0)$, the discriminant is $b^2 - 4ac$.

If $b^2 - 4ac > 0$, there are two solutions.
If $b^2 - 4ac = 0$, there is one solution.
If $b^2 - 4ac < 0$, there is no solution.

Evaluate the discriminant. Determine the number of real solutions of each equation.

44. $x^2 + 5x - 6 = 0$ 49; 2
45. $-3x^2 - 4x + 8 = 0$ 112; 2
46. $2x^2 + 7x + 11 = 0$ −39; 0

47. *Writing* Explain why a quadratic equation has one real solution if its discriminant equals zero. Answers may vary. Sample: If the discriminant is 0, both choices in the quadratic formula give the same value, $-\frac{b}{2a}$.

Getting Ready for... CHAPTER 8

Use exponents to write each expression.

48. $3 \cdot 3 \cdot 5 \cdot 5 \cdot 5$ $3^2 5^3$
49. $8 \cdot 8 \cdot 8 \cdot x \cdot x \cdot x \cdot x$ $8^3 x^4$
50. $h \cdot h \cdot h \cdot h \cdot h \cdot w \cdot w$ $h^5 w^2$

Simplify each expression.

51. $5 \cdot 10^3$ 5000
52. $-4^3 - (-4)^3$ 0
53. $8 \cdot 10^4$ 80,000
54. $\frac{3^4}{3^2}$ 9
55. $10(3^2 - 3^4)$ −720
56. $7 \cdot 10^2 \div 10^3$ 0.7

Evaluate each expression.

57. $d^2 \cdot g^2$ for $d = -2$ and $g = 3$ 36
58. $7n^8 - 5n^3$ for $n = -1$ 12
59. $\frac{m^3}{m^4}$ for $m = 3$ $\frac{1}{3}$

13.
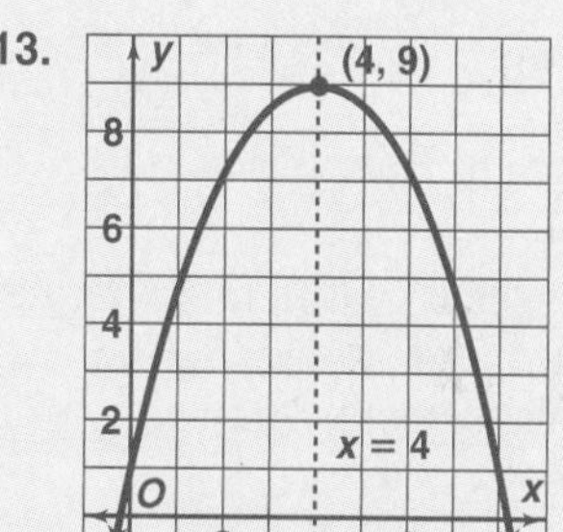

14.
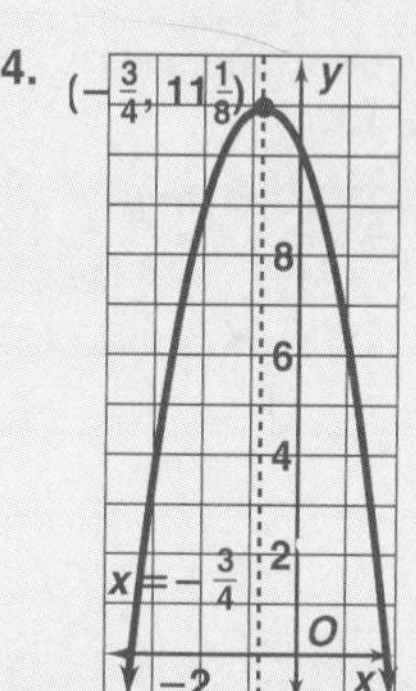

15–18. See back of book.

Assessment

ENHANCED MULTIPLE CHOICE QUESTIONS are more complex than traditional multiple choice questions, which assess only one skill. Enhanced multiple choice questions assess the processes that students use, as well as the end results. The questions are written so that students must use more than one strategy to solve the problem. Using multiple strategies is encouraged by the National Council of Teachers of Mathematics (NCTM). **Exercise 38** is an enhanced multiple choice question.

FREE RESPONSE QUESTIONS do not give answer choices. Some exercises have more than one possible answer. Students need to give only one correct response. **Exercises 5–12, 14–30, and 32–37** are free response questions.

WRITING EXERCISES allow students to describe how they think about and understand the concepts they have learned. **Exercise 13** is a writing exercise.

OPEN-ENDED PROBLEMS allow for more than one solution. Students must construct their own responses instead of choosing from possible answers. The student responses will help you determine the depth of their understanding and any possible areas of difficulty. **Exercise 31** is an open-ended problem. Make sure that the students' responses contain values for x and y and a written situation.

Resources

Teaching Resources

Chapter Support File, Ch. 7
Chapter Assessment, Forms A and B
Alternative Assessment
Chapter Assessment, Spanish Resources

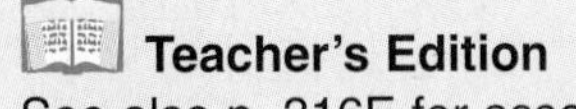

Teacher's Edition

See also p. 316E for assessment options.

Software Computer Item Generator

page 358 Chapter Assessment

9.

x	y
−3	5
−2	0
−1	−3
0	−4
1	−3

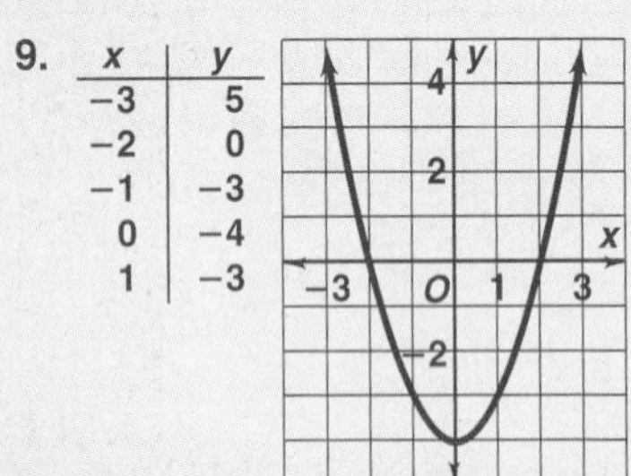

10.

x	y
−3	−8
−2	−3
−1	0
0	1
1	0

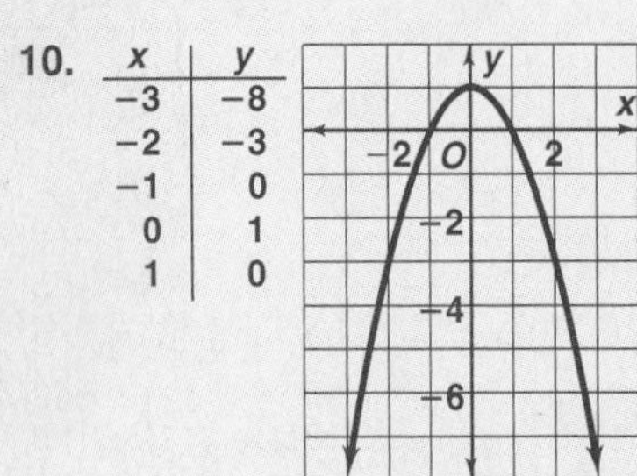

11–12, 18–20, 31. See back of book.

7 Assessment

Match each function with its graph.

1. $y = 3x^2$ **C**
2. $y = -3x^2 + 1$ **D**
3. $y = -2x^2$ **B**
4. $y = x^2 - 3$ **A**

A.

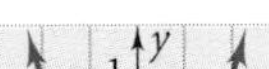

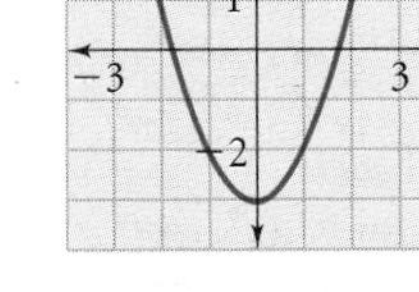

B.

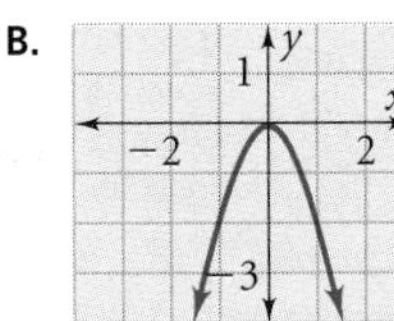

C.

D.

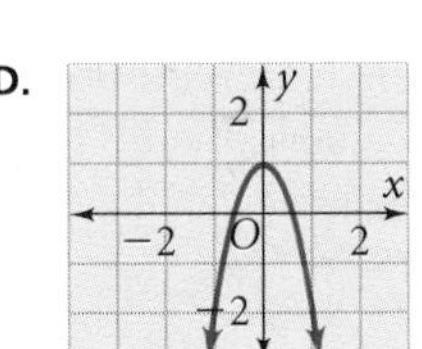

Find the coordinates of the vertex of the graph and an equation for the axis of symmetry.

5. $y = 3x^2 - 7$ **(0, −7); $x = 0$**
6. $y = x^2 - 3x + 2$ **(1.5, −0.25); $x = 1.5$**
7. $y = -2x^2 + 10x - 1$ **(2.5, 11.5); $x = 2.5$**
8. $y = \frac{1}{2}x^2 + 6x$ **(−6, −18); $x = -6$**

Make a table of values and graph each function.

9. $y = x^2 - 4$
10. $y = -x^2 + 1$
11. $y = 5x^2$
12. $y = \frac{1}{2}x^2 - 2$

9–12. See margin.

13. Writing Explain what you can determine about the shape of a parabola from its equation without graphing. **You can tell how wide it is and whether it opens upward or downward.**

Find the number of x-intercepts of each function.

14. $y = 5x^2$ **1**
15. $y = 3x^2 + 10$ **0**
16. $y = -2x^2 + x + 7$ **2**
17. $y = x^2 - 4x$ **2**

Graph each quadratic function. **18–20. See margin.**

18. $y = -3x^2 - x + 10$
19. $y = \frac{1}{2}x^2 + 2x + 4$
20. $y = x^2 - 3x + 5$

Find both the principal and negative square roots of each number.

21. 1.44 **1.2; −1.2**
22. 1600 **40; −40**
23. $\frac{4}{9}$ **$\frac{2}{3}$; $-\frac{2}{3}$**

Between what two consecutive integers is each square root?

24. $\sqrt{28}$ **5, 6**
25. $\sqrt{136}$ **11, 12**
26. $\sqrt{332}$ **18, 19**

Use any method to solve each quadratic equation. Round solutions to the nearest hundredth.

27. $2x^2 = 50$ **5, −5**
28. $-3x^2 + 7x = -10$ **3.33, −1**
29. $x^2 + 6x + 9 = 25$ **2, −8**
30. $-x^2 - x + 2 = 0$ **1, −2**

31. Open-ended Write the equation of a parabola that has two x-intercepts and a maximum value. Include a graph of your parabola. **Answers may vary Sample: $y = -x^2 + 4$; see margin for graph.**

Model each problem with a quadratic equation. Then solve the problem.

32. Geometry The volume of a cylinder is given by the formula $V = \pi r^2 h$, where r is the radius of the cylinder and h is the height. A cylinder with height of 10 ft has volume 140 ft³. What is the radius to the nearest tenth of a foot? **$140 = 10\pi r^2$; 2.1 ft**

33. Landscaping The area of a rectangular patio is 800 ft². The length of the patio is twice the width. Find the dimensions of the patio. (*Hint:* use the formula $A = l \cdot w$.) **$800 = 2w^2$; $w = 20$ ft, $\ell = 40$ ft**

Evaluate the discriminant. Determine the number of real solutions of each equation.

34. $x^2 + 4x = 5$ **36; 2**
35. $x^2 - 8 = 0$ **32; 2**
36. $2x^2 + x = 0$ **1; 2**
37. $3x^2 - 9x = -5$ **21; 2**

38. Standardized Test Prep Find the value of k for which the equation $kx^2 - 10x + 25k = 0$ has one real root. **A**

A. −1 B. 2 C. 3 D. 5 E. 10

Preparing for Standardized Tests

Standardized tests, such as those administered for state assessment, the SAT, or the ACT, include regular math questions, quantitative comparison questions, open-ended problems, and free response questions (which the SAT calls *grid-ins*).

MULTIPLE CHOICE QUESTIONS are followed by five answer choices, one of which is correct. **Exercises 1–10** are multiple choice questions.

QUANTITATIVE COMPARISON QUESTIONS ask students to compare two quantities. **Exercises 11 and 12** are quantitative comparison questions.

FREE-RESPONSE QUESTIONS do not give answer choices. Students must provide the one correct answer on their own. **Exercises 13–16** are free-response questions.

OPEN-ENDED PROBLEMS allow for more than one solution. Students must construct their own responses instead of choosing a single answer. The responses students give will help you determine the depth of their understanding and what difficulties, if any, they are experiencing. This standardized test does not have an open-ended-problem.

STANDARDIZED TEST TIPS **Exercise 3** Memorizing the squares of the first nine integers can be a substantial time saver when making comparisons between integers and square roots.

7 Preparing for Standardized Tests

For Exercises 1–12, choose the correct letter.

1. Choose the best approximation of the solutions of $3x^2 - 5x + 1 = 0$. E
   A. 2 and −3 B. 1.5 and −0.5
   C. 1.5 and 1.75 D. −3 and 2
   E. 1.5 and 0.25

2. What is the equation of the axis of symmetry of $y = 5x^2 - 2x + 3$? E
   A. $x = \frac{4}{5}$ B. $x = -\frac{4}{5}$ C. $y = \frac{4}{5}$
   D. $y = -\frac{4}{5}$ E. none of the above

3. Between what two consecutive integers is $\sqrt{52}$? C
   A. 5 and 6 B. 6 and 7 C. 7 and 8
   D. 8 and 9 E. 9 and 10

4. How many solutions are there for the system $y = |x - 3|$ and $6y - x = 24$? C
   A. 0 B. 1 C. 2 D. 3 E. 4

5. What are the solutions of the equation $x^2 - 6x - 11 = 0$? D
   A. −8 and 3 B. $3 \pm 4\sqrt{5}$ C. 8 and −3
   D. $3 \pm 2\sqrt{5}$ E. none of the above

6. What is the solution of the system $-2x - 3y = -13$ and $3x + 2y = 0$? E
   A. (−2, 3) B. (−2, −3) C. (2, 3)
   D. (2, −3) E. none of the above

7. What is the maximum value of y in $y = -3x^2 - 6x - 1$? A
   A. 2 B. −2 C. 1 D. −1 E. 0

8. If a line passes through (5, 2) and (−7, −1), its slope is between which two numbers? A
   A. 0 and 1 B. 1 and 100
   C. −1 and −100 D. 0 and −1
   E. Cannot be determined from the information given.

9. What is the value of the discriminant of $0 = 3x^2 - 4x - 3$? B
   A. −20 B. 52 C. 25
   D. 4 E. none of the above

10. If $x^2 + 4x + 4 = 49$, then
    I. $x = 5$ II. $x = -9$
    III. $x = \sqrt{47}$ IV. $x = -\sqrt{47}$
    A. I only B. II only C. III and IV
    D. I and II E. III only D

Compare the boxed quantity in Column A with the boxed quantity in Column B. Choose the best answer.

A. The quantity in Column A is greater.
B. The quantity in Column B is greater.
C. The two quantities are equal.
D. The relationship cannot be determined on the basis of the information supplied.

	Column A	Column B	
	Use $\begin{bmatrix} 6 & 1 \\ 0 & x \end{bmatrix} + \begin{bmatrix} 1 & y \\ -5 & 3 \end{bmatrix} = \begin{bmatrix} 7 & 9 \\ -5 & 6 \end{bmatrix}$.		
11.	x	y	B
	Use $x + y = 5$ and $2y - x = 4$.		
12.	2	x	C

Find each answer.

13. Find the vertex and the axis of symmetry of the graph of the equation $y = 4x^2 - 3x$. $\left(\frac{3}{8}, -\frac{9}{16}\right)$; $x = \frac{3}{8}$

14. Ben leaves his home 20 mi west of Boston at 10:00 A.M., traveling west by bicycle at 15 mi/h. Esmira leaves Boston by car at 12:00 noon, driving west at 55 mi/h. At what time will she pass Ben? 1:15 P.M.

15. Graph the inequality $|x - 2| \le 9$ on a number line. See margin.

16. A baker can form 2 loaves of bread in 5 min. How many loaves can the baker form in an hour? 24 loaves

Resources

Teaching Resources
Chapter Support File, Ch. 7
- Standardized Test Practice
- Cumulative Review

Teacher's Edition
See also pp. 316E for assessment options.

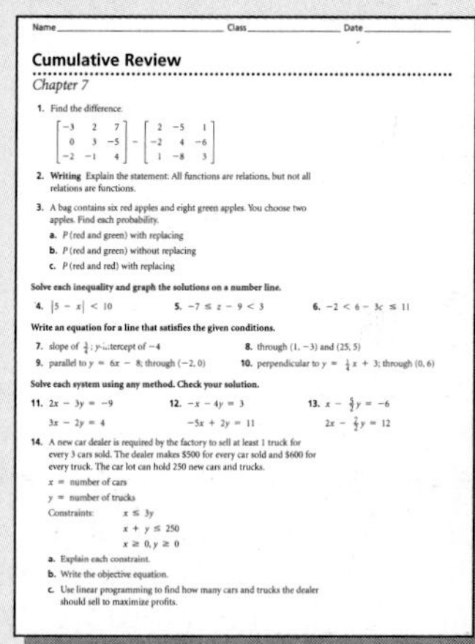
Name______ Class______ Date______

Cumulative Review

Chapter 7

1. Find the difference.
$\begin{bmatrix} -3 & 2 & 7 \\ 0 & 3 & -5 \\ -2 & -1 & 4 \end{bmatrix} - \begin{bmatrix} 2 & -5 & 1 \\ -2 & 4 & -6 \\ 1 & -8 & 3 \end{bmatrix}$
2. **Writing** Explain the statement: All functions are relations, but not all relations are functions.
3. A bag contains six red apples and eight green apples. You choose two apples. Find each probability.
   a. P(red and green) with replacing
   b. P(red and green) without replacing
   c. P(red and red) with replacing

Solve each inequality and graph the solutions on a number line.

4. $|5 - x| < 10$ 5. $-7 \le z - 9 < 3$ 6. $-2 < 6 - 3c \le 11$

Write an equation for a line that satisfies the given conditions.

7. slope of $\frac{1}{4}$; y-intercept of −4 8. through (1, −3) and (25, 5)
9. parallel to $y = 6x - 8$; through (−2, 0) 10. perpendicular to $y = \frac{1}{4}x + 3$; through (0, 6)

Solve each system using any method. Check your solution.

11. $2x - 3y = -9$, $3x - 2y = 4$ 12. $-x - 4y = 3$, $-5x + 2y = 11$ 13. $x - \frac{5}{2}y = -6$, $2x - \frac{7}{2}y = 12$

14. A new car dealer is required by the factory to sell at least 1 truck for every 3 cars sold. The dealer makes \$500 for every car sold and \$600 for every truck. The car lot can hold 250 new cars and trucks.
x = number of cars
y = number of trucks
Constraints: $x \le 3y$; $x + y \le 250$; $x \ge 0, y \ge 0$
   a. Explain each constraint.
   b. Write the objective equation.
   c. Use linear programming to find how many cars and trucks the dealer should sell to maximize profits.

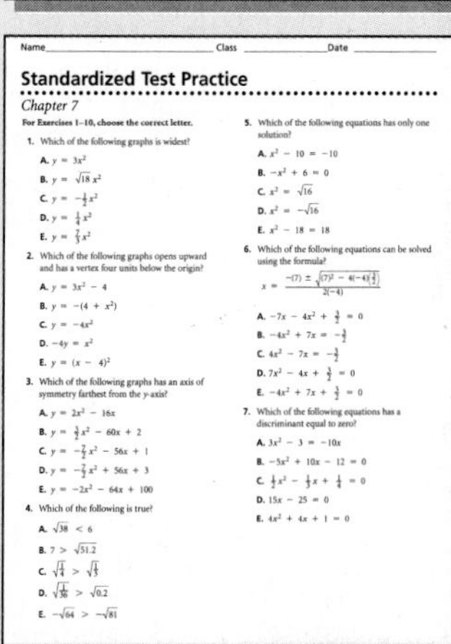
Name______ Class______ Date______

Standardized Test Practice

Chapter 7

For Exercises 1–10, choose the correct letter.

1. Which of the following graphs is widest?
   A. $y = 3x^2$ B. $y = \sqrt{18}x^2$ C. $y = -\frac{1}{2}x^2$ D. $y = \frac{1}{4}x^2$ E. $y = \frac{7}{3}x^2$
2. Which of the following graphs opens upward and has a vertex four units below the origin?
   A. $y = 3x^2 - 4$ B. $y = -(4 + x^2)$ C. $y = -4x^2$ D. $-4y = x^2$ E. $y = (x - 4)^2$
3. Which of the following graphs has an axis of symmetry farthest from the y-axis?
   A. $y = 2x^2 - 16x$ B. $y = \frac{3}{2}x^2 - 60x + 2$ C. $y = -\frac{7}{2}x^2 - 56x + 1$ D. $y = -\frac{7}{2}x^2 + 56x + 3$ E. $y = -2x^2 - 64x + 100$
4. Which of the following is true?
   A. $\sqrt{38} < 6$ B. $7 > \sqrt{51.2}$ C. $\sqrt{\frac{1}{4}} > \sqrt{\frac{1}{3}}$ D. $\sqrt{\frac{1}{36}} > \sqrt{0.2}$ E. $-\sqrt{64} > -\sqrt{81}$
5. Which of the following equations has only one solution?
   A. $x^2 - 10 = -10$ B. $-x^2 + 6 = 0$ C. $x^2 = \sqrt{16}$ D. $x^2 = -\sqrt{16}$ E. $x^2 - 18 = 18$
6. Which of the following equations can be solved using the formula?
$x = \frac{-(7) \pm \sqrt{(7)^2 - 4(-4)\left(\frac{3}{2}\right)}}{2(-4)}$
   A. $-7x - 4x^2 + \frac{3}{2} = 0$ B. $-4x^2 + 7x = -\frac{3}{2}$ C. $4x^2 - 7x = -\frac{3}{2}$ D. $7x^2 - 4x + \frac{3}{2} = 0$ E. $-4x^2 + 7x + \frac{3}{2} = 0$
7. Which of the following equations has a discriminant equal to zero?
   A. $3x^2 - 3 = -10x$ B. $-5x^2 + 10x - 12 = 0$ C. $\frac{1}{2}x^2 - \frac{1}{3}x + \frac{1}{4} = 0$ D. $15x - 25 = 0$ E. $4x^2 + 4x + 1 = 0$

page 359 Preparing for Standardized Test

15. See back of book.

CHAPTER 8 Exponents and Exponential Functions

To accommodate flexible scheduling, some lessons are divided into parts. Assignment Options are given in the Lesson Planning Options for each lesson.

PACING OPTIONS

This chart suggests pacing only for the core lessons and their parts, and it is provided merely as a possible guide. It will help you determine how much time you have in your schedule to cover other features, such as the Chapter Project, Math Toolboxes, Wrap Up, and Assessment.

	1 Class Period	1 Class Period	1 Class Period	1 Class Period	1 Class Period	1 Class Period	1 Class Period	1 Class Period	1 Class Period
Traditional (40–45 min class periods)	8-1 Part 1, 8-1 Part 2	8-1 Part 3	8-2 Part 1, 8-2 Part 2	8-3	8-4 Part 1	8-4 Part 2	8-5 Part 1	8-5 Part 2	8-6 Part 1, 8-6 Part 2
Two-Year Algebra (40–45 min class periods)	8-1 Part 2	8-1 Part 2	8-1 Part 3	8-2 Part 1	8-2 Part 1	8-2 Part 2	8-3	8-3	8-4 Part 1
Block Scheduling (90 min class periods)	8-1 Part 1, 8-1 Part 2, 8-1 Part 3	8-2 Part 1, 8-2 Part 2, 8-3	8-4 Part 1, 8-4 Part 2	8-5 Part 1, 8-5 Part 2	8-6 Part 1, 8-6 Part 2	8-7 Part 1, 8-7 Part 2	8-8 Part 1, 8-8 Part 2		

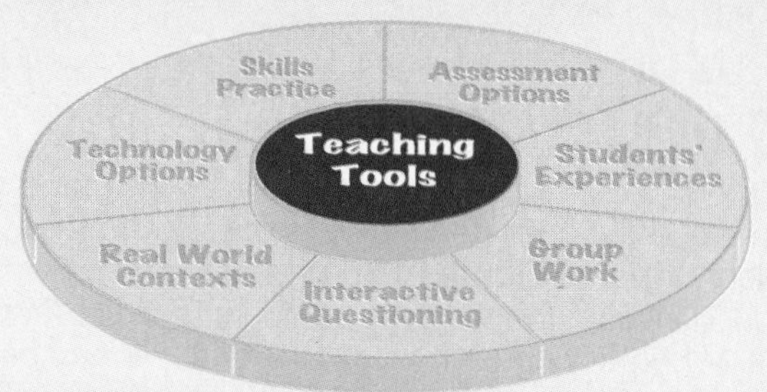

What Students Will Learn and Why

In this chapter, students will build on their knowledge of quadratic equations and functions, learned in Chapter 7, by learning about exponents and exponential functions. They will learn to find exponential growth and decay, and they will use these equations to solve real-world problems. They will learn to use zero and negative exponents. Students will use scientific notation to express numbers that are very large and very small. Finally, students will learn about the multiplication and division properties of exponents.

Discussing the Chapter/Building on Experience

The concept map below relates chapter topics to real-world applications. You and your class may wish to add to the map or develop maps of your own. The center oval describes the topic of the chapter. The next level displays topics within the lessons. The outer ovals reflect applications of the content. As you and your class build a concept map, invite students to discuss applications with which they are familiar.

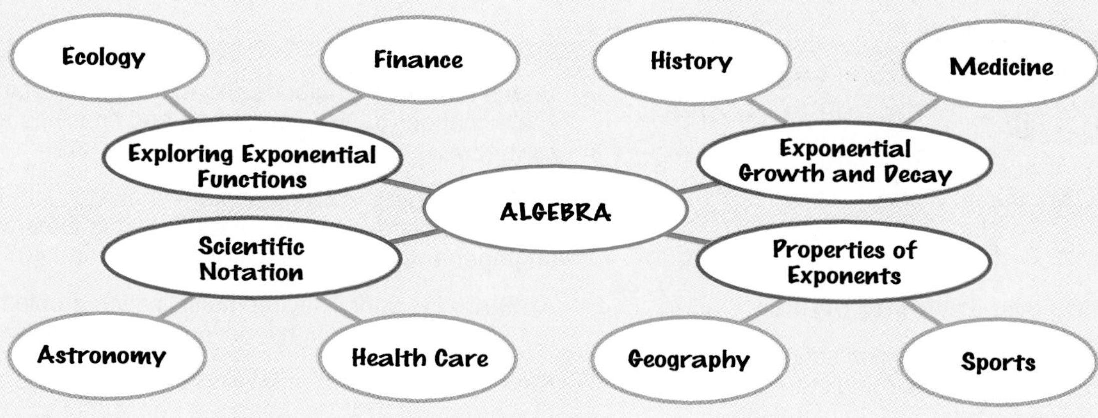

1 Class Period	1 Class Period	1 Class Period	1 Class Period	1 Class Period	1 Class Period	1 Class Period	1 Class Period	1 Class Period	1 Class Period	1 Class Period
8-7 1	8-7 2	8-8 1	8-8 2							
8-4 2	8-5 1	8-5 2	8-6 1	8-6 2	8-7 1	8-7 2	8-7 2	8-8 1	8-8 2	8-8 2

Interactive Questioning Tips

A question is interactive when there is "give and take" between the questioner (teacher or student) and the respondent. When a critical thinking question is asked, it is important to involve the majority of students in the learning activities. When calling on hesitant students, it may be helpful to use encouraging phrases such as "Keep going, you're doing fine."

Skills Practice

Every lesson provides skill practice with Exercises On Your Own and Exercises Mixed Review. The Student Edition includes Checkpoints (pp. 377 and 395) and Cumulative Review (p. 411). In the Teacher's Edition, the Lesson Planning Options section for each lesson lists Prerequisite Skills students should know for that lesson. At the back of the Student Edition is the Skills Handbook—mini-lessons on math the students may need to review. The Chapter Support File for Chapter 8 in the Teaching Resources box includes two Practice worksheets per lesson, a worksheet for two Checkpoints, and worksheets for Cumulative Review and Standardized Test Preparation.

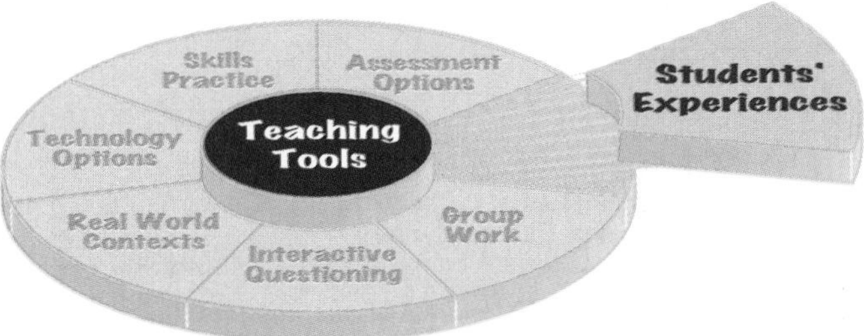

Diverse Learning and Teaching Styles

In your Teacher's Edition, you will find suggestions as to how you can help students complete mathematical tasks in Chapter 8 by reinforcing various learning styles. Here are some examples.

- **Visual learning** photocopying rulers on enlarging copiers to see scale change (p. 364), observing and checking each other's work as they work problems on the board (p. 369)
- **Tactile learning** making posters showing what they have learned about exponents (p. 397), taking turns working problems on paper and with a calculator (pp. 402 and 403)
- **Auditory learning** saying multiplication problems out loud (p. 386), talking through working interest problems (p. 369)
- **Kinesthetic learning** modeling exponential decay by moving out of a group (p. 374), learning the concept of zero or negative exponents by taking increasingly smaller steps (p. 381)

Alternative Activity for lesson 8-2 A

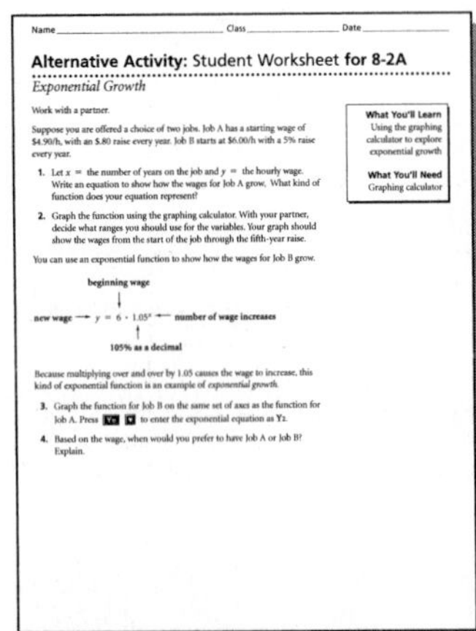

Name ______ Class ______ Date ______

Alternative Activity: Student Worksheet for 8-2A

Exponential Growth

Work with a partner.

Suppose you are offered a choice of two jobs. Job A has a starting wage of \$4.90/h, with an \$.80 raise every year. Job B starts at \$6.00/h with a 5% raise every year.

1. Let x = the number of years on the job and y = the hourly wage. Write an equation to show how the wages for Job A grow. What kind of function does your equation represent?
2. Graph the function using the graphing calculator. With your partner, decide what ranges you should use for the variables. Your graph should show the wages from the start of the job through the fifth-year raise.

You can use an exponential function to show how the wages for Job B grow.

beginning wage ↓
new wage → $y = 6 \cdot 1.05^x$ ← **number of wage increases**
↑ **105% as a decimal**

Because multiplying over and over by 1.05 causes the wage to increase, this kind of exponential function is an example of *exponential growth*.

3. Graph the function for Job B on the same set of axes as the function for Job A. Press [Y=] [▼] to enter the exponential equation as Y2.
4. Based on the wage, when would you prefer to have Job A or Job B? Explain.

What You'll Learn
Using the graphing calculator to explore exponential growth

What You'll Need
Graphing calculator

for use with Work Together, addresses visual and tactile learning by having students graph lines with a graphing calculator.

Alternative Activity for Lesson 8-2 B

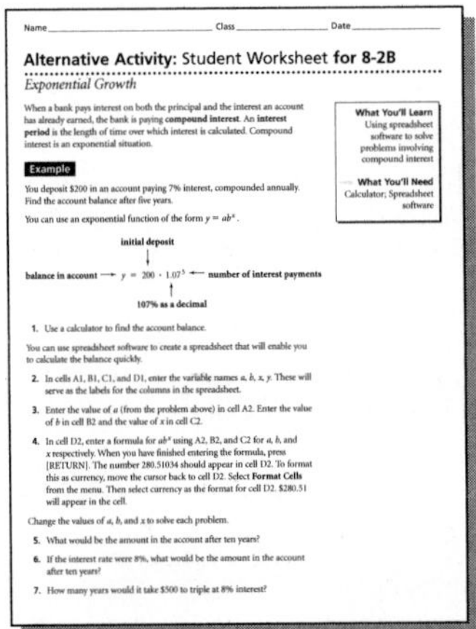

Name ______ Class ______ Date ______

Alternative Activity: Student Worksheet for 8-2B

Exponential Growth

When a bank pays interest on both the principal and the interest an account has already earned, the bank is paying **compound interest**. An **interest period** is the length of time over which interest is calculated. Compound interest is an exponential situation.

Example

You deposit \$200 in an account paying 7% interest, compounded annually. Find the account balance after five years.

You can use an exponential function of the form $y = ab^x$.

initial deposit ↓
balance in account → $y = 200 \cdot 1.07^5$ ← **number of interest payments**
↑ **107% as a decimal**

1. Use a calculator to find the account balance.

You can use spreadsheet software to create a spreadsheet that will enable you to calculate the balance quickly.

2. In cells A1, B1, C1, and D1, enter the variable names a, b, x, y. These will serve as the labels for the columns in the spreadsheet.
3. Enter the value of a (from the problem above) in cell A2. Enter the value of b in cell B2 and the value of x in cell C2.
4. In cell D2, enter a formula for ab^x using A2, B2, and C2 for a, b, and x respectively. When you have finished entering the formula, press [RETURN]. The number 280.51034 should appear in cell D2. To format this as currency, move the cursor back to cell D2. Select **Format Cells** from the menu. Then select currency as the format for cell D2. \$280.51 will appear in the cell.

Change the values of a, b, and x to solve each problem.

5. What would be the amount in the account after ten years?
6. If the interest rate were 8%, what would be the amount in the account after ten years?
7. How many years would it take \$500 to triple at 8% interest?

What You'll Learn
Using spreadsheet software to solve problems involving compound interest

What You'll Need
Calculator; Spreadsheet software

for use with page 369, addresses visual and tactile learning by having students work together on a computer to develop a spreadsheet for calculating interest.

Alternative Activity for Lesson 8-4

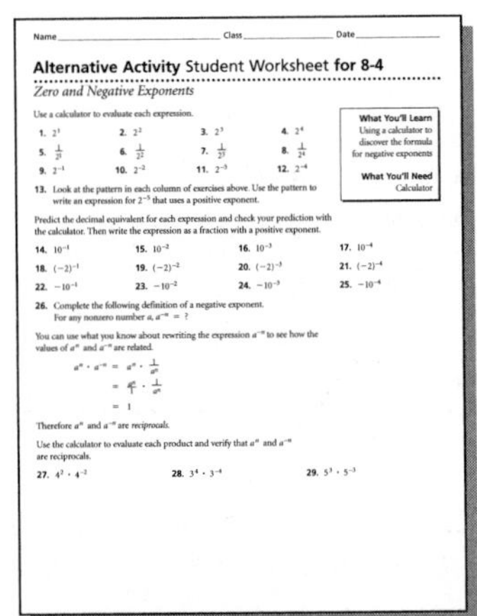

Name ______ Class ______ Date ______

Alternative Activity Student Worksheet for 8-4

Zero and Negative Exponents

Use a calculator to evaluate each expression.

1. 2^1	2. 2^2	3. 2^3	4. 2^4
5. $\frac{1}{2^1}$	6. $\frac{1}{2^2}$	7. $\frac{1}{2^3}$	8. $\frac{1}{2^4}$
9. 2^{-1}	10. 2^{-2}	11. 2^{-3}	12. 2^{-4}

13. Look at the pattern in each column of exercises above. Use the pattern to write an expression for 2^{-5} that uses a positive exponent.

Predict the decimal equivalent for each expression and check your prediction with the calculator. Then write the expression as a fraction with a positive exponent.

14. 10^{-1}	15. 10^{-2}	16. 10^{-3}	17. 10^{-4}
18. $(-2)^{-1}$	19. $(-2)^{-2}$	20. $(-2)^{-3}$	21. $(-2)^{-4}$
22. -10^{-1}	23. -10^{-2}	24. -10^{-3}	25. -10^{-4}

26. Complete the following definition of a negative exponent. For any nonzero number a, a^{-n} = ?

You can use what you know about rewriting the expression a^{-n} to see how the values of a^n and a^{-n} are related.

$$a^n \cdot a^{-n} = a^n \cdot \frac{1}{a^n}$$
$$= \frac{a^n}{1} \cdot \frac{1}{a^n}$$
$$= 1$$

Therefore a^n and a^{-n} are *reciprocals*.

Use the calculator to evaluate each product and verify that a^n and a^{-n} are reciprocals.

27. $4^2 \cdot 4^{-2}$ 28. $3^4 \cdot 3^{-4}$ 29. $5^3 \cdot 5^{-3}$

What You'll Learn
Using a calculator to discover the formula for negative exponents

What You'll Need
Calculator

for use with Example 2, addresses visual and tactile learning by having students use a calculator to find the values of numbers with negative exponents.

Cooperative Learning Tips

When used effectively, cooperative learning can help students develop interpersonal skills, learn to perform specific roles in a group, and learn to carry out specific responsibilities. The components of Chapter 8 provide a range of cooperative learning opportunities.

- In the Student Edition, the ***Work Together*** parts of lessons are specifically designed for cooperative learning activities.
- In the Teacher's Edition, you will find helpful hints for addressing diverse learning styles (see page C for Chapter 8). For every lesson, you will find a ***Reteaching Activity,*** which may involve cooperative learning.

Materials and Manipulatives

The authors expect all students to use scientific calculators. Calculator use is integrated throughout the course.

- **graphing calculator (8-4)**
- **notebook paper (8-1)**

TECHNOLOGY OPTIONS

Technology Tools		Chapter Project	8-1	8-2	8-3	8-4	8-5	8-6	8-7	8-8
Calculator		Assumed that students have scientific calculators at any time.								
Graphing Calculator	Handbook		✔	✔	✔		✔			
	Student Edition		✔	✔	✔	✔	✔			
Software	Secondary Math Lab Toolkit		✔	✔	✔	✔	✔	✔	✔	✔
	Integrated Math Lab					✔		✔		
	Computer Item Generator		✔	✔	✔	✔	✔	✔	✔	✔
	Student Edition									
Video	Video Field Trip	✔								
CD-ROM	Multimedia Algebra Lab		✔	✔	✔		✔			
Internet		See the Prentice Hall site. (http://www.phschool.com)								

The Prentice Hall Algebra program offers you a rich variety of technology options. Be assured that all these options are provided as a means of enriching the program and are not essential for the successful completion of the course.

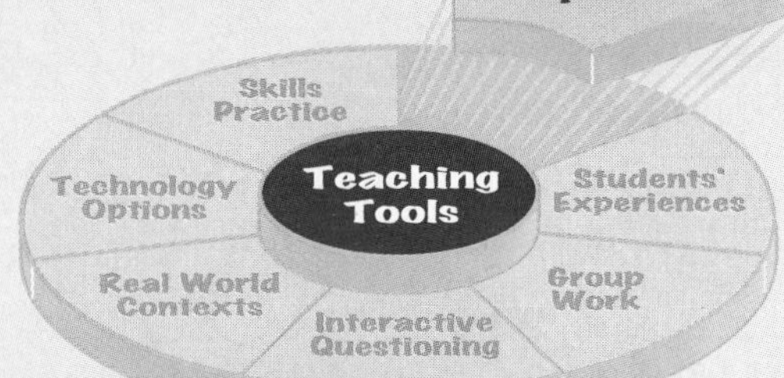

Assessment Options

The Prentice Hall Algebra Program provides you with many options. From these options, you may choose instructional materials and techniques appropriate for your students or those necessary to meet your district's curriculum requirements. As the chart indicates, the program also supports your teaching efforts by offering you many choices for assessment.

ASSESSMENT OPTIONS

Assessment Support Materials	Chapter Opener	8-1	8-2	8-3	8-4	8-5	8-6	8-7	8-8	Chapter End
Chapter Project	● ▲ ■	▲ ■	▲ ■			▲ ■	▲ ■	▲ ■		▲ ■
Checkpoints				▲ ●			▲ ●			
Self-Assessment				▲ ■				▲ ■	▲ ■	▲ ■
Writing Assignment		▲	▲	▲	▲	▲	▲	▲	▲	▲ ●
Chapter Assessment										▲ ●
Alternative Assessment		■	■	■	■	■	■	■	■	● ■
Cumulative Review										● ▲
Standardized Test Prep		▲ ■	▲ ■		▲ ■		▲ ■			●
Computer Item Generator	Can be used to create custom-made practice or assessment at any time.									

▲ = Student Edition ■ = Teacher's Edition ● = Teaching Resources

Checkpoints

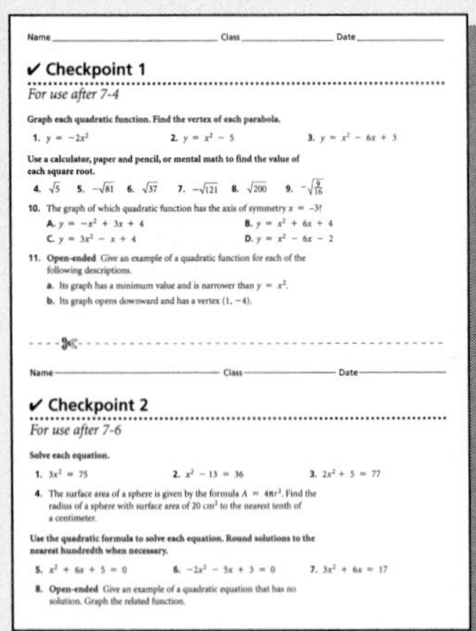
Name_____ Class_____ Date_____

✔ Checkpoint 1

For use after 7-4

Graph each quadratic function. Find the vertex of each parabola.

1. $y = -2x^2$ 2. $y = x^2 - 5$ 3. $y = x^2 - 6x + 3$

Use a calculator, paper and pencil, or mental math to find the value of each square root.

4. $\sqrt{5}$ 5. $-\sqrt{81}$ 6. $\sqrt{37}$ 7. $-\sqrt{121}$ 8. $\sqrt{200}$ 9. $-\sqrt{\frac{9}{16}}$

10. The graph of which quadratic function has the axis of symmetry $x = -3$?
A. $y = -x^2 + 3x + 4$ B. $y = x^2 + 6x + 4$
C. $y = 3x^2 - x + 4$ D. $y = x^2 - 6x - 2$

11. **Open-ended** Give an example of a quadratic function for each of the following descriptions.
a. Its graph has a minimum value and is narrower than $y = x^2$.
b. Its graph opens downward and has a vertex $(1, -4)$.

Name_____ Class_____ Date_____

✔ Checkpoint 2

For use after 7-6

Solve each equation.

1. $3x^2 = 75$ 2. $x^2 - 13 = 36$ 3. $2x^2 + 5 = 77$

4. The surface area of a sphere is given by the formula $A = 4\pi r^2$. Find the radius of a sphere with surface area of 20 cm² to the nearest tenth of a centimeter.

Use the quadratic formula to solve each equation. Round solutions to the nearest hundredth when necessary.

5. $x^2 + 6x + 5 = 0$ 6. $-2x^2 - 5x + 3 = 0$ 7. $3x^2 + 6x = 17$

8. **Open-ended** Give an example of a quadratic equation that has no solution. Graph the related function.

Alternative Assessment

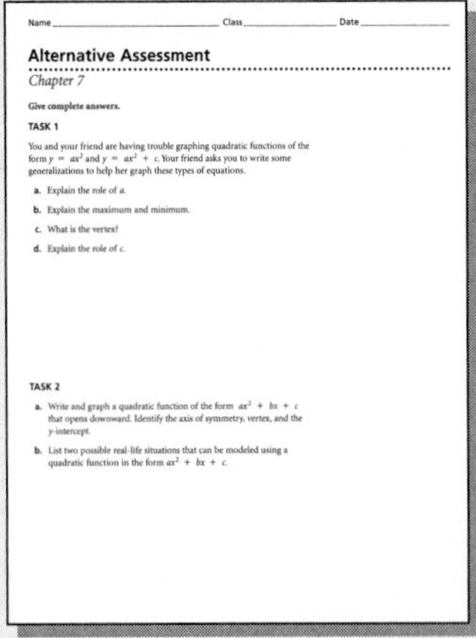
Name_____ Class_____ Date_____

Alternative Assessment

Chapter 7

Give complete answers.

TASK 1

You and your friend are having trouble graphing quadratic functions of the form $y = ax^2$ and $y = ax^2 + c$. Your friend asks you to write some generalizations to help her graph these types of equations.

a. Explain the role of a.
b. Explain the maximum and minimum.
c. What is the vertex?
d. Explain the role of c.

TASK 2

a. Write and graph a quadratic function of the form $ax^2 + bx + c$ that opens downward. Identify the axis of symmetry, vertex, and the y-intercept.
b. List two possible real-life situations that can be modeled using a quadratic function in the form $ax^2 + bx + c$.

Chapter Assessment

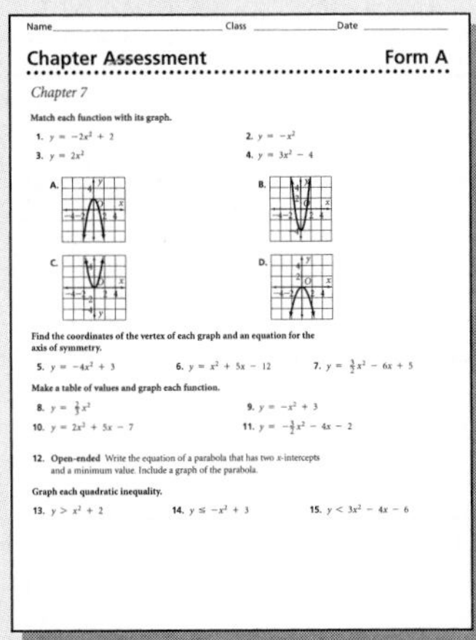
Name_____ Class_____ Date_____

Chapter Assessment **Form A**

Chapter 7

Match each function with its graph.

1. $y = -2x^2 + 2$ 2. $y = -x^2$
3. $y = 2x^2$ 4. $y = 3x^2 - 4$

A. B. C. D.

Find the coordinates of the vertex of each graph and an equation for the axis of symmetry.

5. $y = -4x^2 + 3$ 6. $y = x^2 + 5x - 12$ 7. $y = \frac{3}{2}x^2 - 6x + 5$

Make a table of values and graph each function.

8. $y = \frac{2}{3}x^2$ 9. $y = -x^2 + 3$
10. $y = 2x^2 + 5x - 7$ 11. $y = -\frac{3}{2}x^2 - 4x - 2$

12. **Open-ended** Write the equation of a parabola that has two x-intercepts and a minimum value. Include a graph of the parabola.

Graph each quadratic inequality.

13. $y > x^2 + 2$ 14. $y \leq -x^2 + 3$ 15. $y < 3x^2 - 4x - 6$

Available in both Form A and Form B

Making the Right Connections

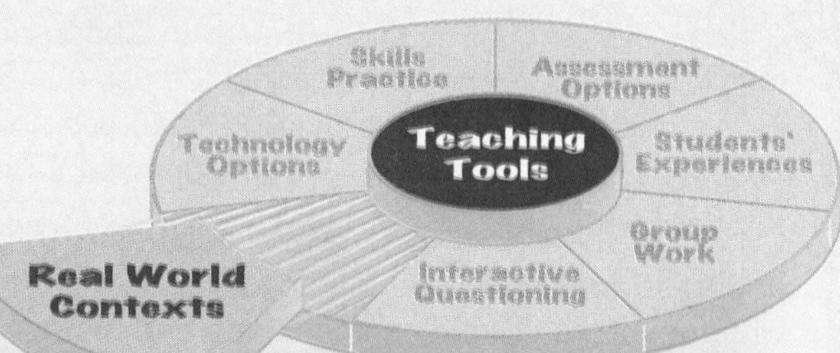

Mathematics is imbedded in nearly every walk of life. The National Council of Teachers of Mathematics (NCTM) encourages educators to recognize these connections and to emphasize them for the purpose of better educating students for success in life and in a global economy. The ***Connections*** chart below highlights these connections for Chapter 8.

CONNECTIONS

Lesson	Interdisciplinary Connections	Career Prep	Other Real World Connections	Math Integration	NCTM Standards
Chapter Project	Biology				Communication
8-1	Biology Ecology		Technology Finance Books		Algebra Communication Problem Solving
8-2	History	Wages	Medical Care Savings Education		Algebra Communication Problem Solving
8-3			Medicine Consumer Trends	Statistics	Algebra Communication Problem Solving
8-4	Botany		Population Growth Communications	Geometry	Algebra Communication Problem Solving
8-5	Physical Science Astronomy		Telecommunications Probability Precious Metals Health Care Sports		Algebra Communication Problem Solving
8-6	Biology Chemistry Entomology		Technology Medicine Sports		Algebra Communication Problem Solving
8-7	Physical Science Geography Biology		Technology Textiles	Geometry	Algebra Communication Problem Solving
8-8	Astronomy		Environment Television Finance Medicine		Algebra Communication Problem Solving

CHAPTER 8 Exponents and Exponential Functions

CONNECTING TO PRIOR LEARNING In Chapter 5, students learned to express direct relationships using linear functions. In this chapter, students will learn to express non-linear situations, such as growth, with exponential functions.

CULTURAL CONNECTIONS Many cultures emphasize science and math education for high school students. Ask students why they think education in science and math is considered important. Ask students to share information about what science has contributed to the advancement of humanity and how this information is shared cross-culturally.

INTERDISCIPLINARY CONNECTIONS Math and science are connected to many fields of study—medicine, computers, the space program, etc. Ask students how these different disciplines are related and how exponential functions could be applied across more than one of these fields of study.

ABOUT THE PROJECT The Chapter Project gives students an opportunity to measure the exponential growth of mold and to express the growth using exponential equations. Students grow their own mold, take quantitative measurements of the growth, graph the growth, and derive an exponential equation to represent the growth and predict further growth.

Technology Options

Prentice Hall Technology

Video Field Trip • "Breaking the Mold," an investigation of bacteria and how it is useful in the making of cheese.

CHAPTER 8 Exponents and Exponential Functions

Relating to the Real World

What do money in a savings account, the population of the world, and radioactive waste all have in common? You can use exponential functions to describe them and predict the future. Exponential relationships are widely used by scientists, business people, and even politicians trying to predict budget surpluses and deficits.

Lessons	8-1	8-2	8-3	8-4	8-5
	Exploring Exponential Functions	*Exponential Growth*	*Exponential Decay*	*Zero and Negative Exponents*	*Scientific Notation*

Launching the Project

PROJECT NOTEBOOK Encourage students to keep all project-related materials in a separate folder or notebook. **See Chapter Project and Scoring Rubric in Chapter Support File.**

Ask students if they have ever opened a container in the refrigerator and been surprised by moldy leftovers. Explain that mold grows at an exponential rate. Food that was edible one day may be covered with mold the following day. Ask students what they think a graph of mold growth might look like.

Have students research mold growth and find out what conditions are best for growing mold for this project.

- Direct student attention to page 366 and have students make a list of materials they will need to begin the project.
- Have students predict the length of time required for mold to start growing and to cover the dish.
- When the project is complete, challenge students to compare their predictions to the actual mold growth. Ask them to explain why the growth of the mold was exponential.

TRACKING THE PROJECT You may wish to have students read Finishing the Chapter Project on page 406 to help them get an overview of the project. Set benchmark deadlines for students to show you their work in progress.

CHAPTER PROJECT

MOLDY OLDIES

You take a piece of bread from the bread bag and find that there is green mold on it. The bread was fine two days ago! You open the refrigerator to look for a snack only to see that the cheese is covered with a fuzzy white mold. So, just how fast does mold grow, anyway?

As you work through the chapter, you will grow your own mold. You will gather data, create graphs, and make predictions. As part of your research, you will plan and complete an experiment to monitor growth.

To help you complete the project:

- p. 366 *Find Out by Doing*
- p. 372 *Find Out by Recording*
- p. 389 *Find Out by Graphing*
- p. 395 *Find Out by Analyzing*
- p. 400 *Find Out by Interpreting*
- p. 406 *Finishing the Project*

A Multiplication Property of Exponents 8-6

More Multiplication Properties of Exponents 8-7

Division Properties of Exponents 8-8

Project Resources

Teaching Resources

Chapter Support File, Ch. 8

- Chapter Project Manager and Scoring Rubric

Transparencies

96

Using the Rubric

Sharing the scoring rubric for the project with your students will alert them to your expectations before they begin work on the project.

As students complete each Find Out question in the chapter, you may wish to have them evaluate their own work or a partner's work, based on the scoring rubric. Students should have the opportunity to revise their work after it has been reviewed.

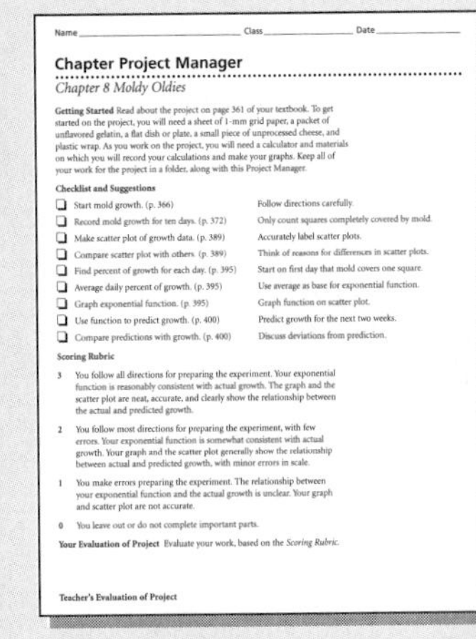

Name ______ Class ______ Date ______

Chapter Project Manager

Chapter 8 Moldy Oldies

Getting Started Read about the project on page 361 of your textbook. To get started on the project, you will need a sheet of 1-mm grid paper, a packet of unflavored gelatin, a flat dish or plate, a small piece of unprocessed cheese, and plastic wrap. As you work on the project, you will need a calculator and materials on which you will record your calculations and make your graphs. Keep all of your work for the project in a folder, along with this Project Manager.

Checklist and Suggestions

❑ Start mold growth. (p. 366)	Follow directions carefully.
❑ Record mold growth for ten days. (p. 372)	Only count squares completely covered by mold.
❑ Make scatter plot of growth data. (p. 389)	Accurately label scatter plots.
❑ Compare scatter plot with others. (p. 389)	Think of reasons for differences in scatter plots.
❑ Find percent of growth for each day. (p. 395)	Start on first day that mold covers one square.
❑ Average daily percent of growth. (p. 395)	Use average as base for exponential function.
❑ Graph exponential function. (p. 395)	Graph function on scatter plot.
❑ Use function to predict growth. (p. 400)	Predict growth for the next two weeks.
❑ Compare predictions with growth. (p. 400)	Discuss deviations from prediction.

Scoring Rubric

3 You follow all directions for preparing the experiment. Your exponential function is reasonably consistent with actual growth. The graph and the scatter plot are neat, accurate, and clearly show the relationship between the actual and predicted growth.

2 You follow most directions for preparing the experiment, with few errors. Your exponential function is somewhat consistent with actual growth. Your graph and the scatter plot generally show the relationship between actual and predicted growth, with minor errors in scale.

1 You make errors preparing the experiment. The relationship between your exponential function and the actual growth is unclear. Your graph and scatter plot are not accurate.

0 You leave out or do not complete important parts.

Your Evaluation of Project Evaluate your work, based on the *Scoring Rubric*.

Teacher's Evaluation of Project

8-1 Teaching Notes

PROBLEM OF THE DAY

Kevin has exactly $1.00 in change in his pocket consisting of pennies, nickels, dimes and quarters. If he has 13 coins, what are they? **5 pennies, 2 quarters, 3 dimes, 3 nickels**

Problem of the Day is also available in Transparencies.

CONNECTING TO PRIOR KNOWLEDGE Ask students who have seen the Disney movie *Fantasia* to describe the Sorcerer's Apprentice sequence. Mickey Mouse tries to stop the broom from flooding the castle so he chops the broom in half. Both halves come to life and continue pouring in water. He chops both in half again. This continues until there is an army of brooms. Ask students what math concepts and terminology could describe the way the number of brooms increased.

WORK TOGETHER

CONNECTING TO THE STUDENTS' WORLD Challenge students to experiment with different sizes and thicknesses of paper to discover the maximum number of folds they can make. Only by using very large sheets of very thin tissue paper, can a person make more than seven folds.

THINK AND DISCUSS

Have students extend the table for six and seven folds to be sure they understand the pattern. As an extension, have students study the patterns in the table and use them to

Lesson Planning Options

Prerequisite Skills

- Using exponents

Assignment Options

To provide flexible scheduling, this lesson can be subdivided into parts.

1 **Core** 1, 2, 23
Extension 17, 18, 23–25

2 **Core** 3–14, 19–21
Extension 15, 22

3 **Core** 26–31
Extension 16

Use Mixed Review to maintain skills.

Resources

 **Student Edition**

Skills Handbook, pp. 569, 580, 582
Extra Practice, p. 563
Glossary/Study Guide

Teaching Resources

Chapter Support File, Ch. 8
- Practice 8-1 (two worksheets)
- Reteaching 8-1

Classroom Manager 8-1
Glossary, Spanish Resources
Two-Year Algebra Handbook 8-1

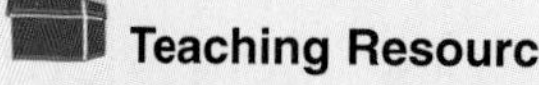

Transparencies
11, 23, 97

Connections Biology . . . and mo

8-1 Exploring Exponential Functions

What You'll Learn
- Examining patterns in exponential functions

...And Why
To model different patterns

What You'll Need
- notebook paper
- calculator

WORK TOGETHER

Work with a partner.

1. Fold a sheet of notebook paper in half. Notice that the fold line divide the paper into 2 rectangles. **Check students' work.**
2. Fold the paper in half again. Now how many rectangles are there? **4**
3. Continue folding the paper in half until you cannot make another fold. Keep track of your results in a table like the one at the right.
4. Patterns What pattern do you notice in the number of rectangles as the number of folds increases? Explain. **The number of rectangles doubles with each fold.**
5. Suppose you could continue to fold the paper. Extend your table to include 10 folds. How many rectangles would there be? **64, 128, 256, 512, 1024; 1024**

Number of Folds	Number of Rectangles
0	1
1	2
2	4
3	8 ■
4	16 ■
5	32 ■

THINK AND DISCUSS

Part 1 Exploring Exponential Patterns

The pattern that you explored in the Work Together involves repeated multiplication by 2. The table below uses exponents to show the pattern.

QUICK REVIEW

base → b^x ← exponent

Number of Folds	Number of Rectangles	Pattern	Written with Exponents
0	1		
1	2	$= 2$	$= 2^1$
2	4	$= 2 \cdot 2$	$= 2^2$
3	8	$= 2 \cdot 2 \cdot 2$	$= 2^3$
4	16	$= 2 \cdot 2 \cdot 2 \cdot 2$	$= 2^4$
5	32	$= 2 \cdot 2 \cdot 2 \cdot 2 \cdot 2$	$= 2^5$

6. Use an exponent to write each number.
   a. $3 \cdot 3 \cdot 3 \cdot 3$ **3^4** b. $(-2)(-2)(-2)$ **$(-2)^3$** c. 125 **5^3**

discover what the value of 2^0 might be. Explain that they will learn later in this chapter that any number raised to the zero power is equal to one.

Example 1 **Relating to the Real World** 

Time	Number of Rabbits	Written with Exponents
Initial	20	
$\frac{1}{2}$ yr	$20 \cdot 3 = 60$	$20 \cdot 3^1$
1 yr	$60 \cdot 3 = 180$	$20 \cdot 3^2$
$1\frac{1}{2}$ yr	$180 \cdot 3 = 540$	$20 \cdot 3^3$
2 yr	$540 \cdot 3 = 1620$	$20 \cdot 3^4$

Copy the table from Example 1 onto the board. Work with students to extend the table to include a third column entitled "Written with Exponents" like the one to the left.

Go over the section in the text on exponential functions. Then challenge students to write an exponential function for the rabbit problem in Example 1. Have them let *r* equal the number of rabbits and *t* equal time in years. $r = 20 \cdot 3^{2t}$

MAKING CONNECTIONS Rabbits can have seven litters a year, each litter consisting of up to eight babies; each rabbit can begin producing its own litters by four or five months of age. When rabbits were introduced to Australia, this exponential birthrate caused major destruction of grazing lands. In the 1950s, the Australian rabbits were deliberately given the viral disease myxomatosis. This disease reduced

Example 1 **Relating to the Real World** 

Biology Suppose there are 20 rabbits on an island and that the rabbit population can triple every half-year. How many rabbits would there be after 2 years?

Time	Number of Rabbits
Initial	20
$\frac{1}{2}$ year	$20 \cdot 3 = 60$
1 year	$60 \cdot 3 = 180$
$1\frac{1}{2}$ years	$180 \cdot 3 = 540$
2 years	$540 \cdot 3 = 1620$

Use the problem-solving strategy *Make a Table.*

To triple the amount, multiply the previous half-year's total by 3.

After two years, there would be 1620 rabbits.

Rabbits were brought from Europe to Australia around 1860. The number of rabbits increased exponentially, and by 1870 there were millions of rabbits.

Part 2

Evaluating Exponential Functions

CALCULATOR HINT

To evaluate 2^n for $n = 10$, press 2 [^] 10 [=] or press 2 [^] 10 [ENTER].

You can write the pattern you found in the Work Together as a function with a variable as an exponent. To find the number of rectangles *r* created by *n* folds, use the function $r = 2^n$. You read the expression 2^n as "2 to the *n*th power." The number of rectangles increases *exponentially* as the paper-folding continues.

7. a. Substitute 10 for *n* in the function $r = 2^n$. Use your calculator to find the value for *r*. 1024

b. How does your answer compare to your answer to Question 5? It is the same number.

The function $r = 2^n$ is an *exponential function.*

Exponential Function

For all numbers *a* and for $b > 0$ and $b \neq 1$, the function $y = a \cdot b^x$ is an **exponential function.**

Examples: $y = 0.5 \cdot 2^x$; $f(x) = -2 \cdot 0.5^x$

Additional Examples

FOR EXAMPLE 1

Suppose that two mice live in a farmhouse. If the number of mice quadruples every 3 mo, how many mice will be in the house after 2 yr? 131,072

Discussion: *Is this growth rate reasonable? What kinds of environmental factors would tend to limit the total number of mice in the farmhouse?*

FOR EXAMPLE 2

Evaluate $y = 2.5^x$ for the domain {2, 3, 4}. Round your answers to the nearest hundredths. 6.25, 15.63, 39.06

FOR EXAMPLE 3

Graph the functions $y = 2^x$ and $y = 4^x$ on the same graph.

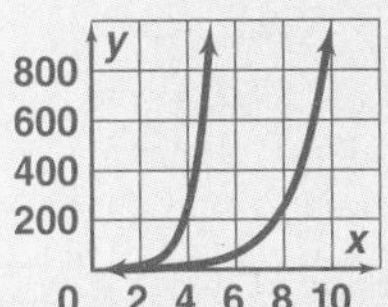

Discussion: *What effect does increasing the value of the base have on the shape of an exponential function?*

the rabbit population temporarily, but now rabbits have developed a resistance to the virus.

Example 2

Question 8 Students can visualize the different rates at which the functions increase by graphing them. To graph these functions on the graphing calculator, students should first press WINDOW to set the size of the viewing window. Use the settings below.

Xmin = 0 Xmax = 3 Xscl = 1
Ymin = 0 Ymax = 250 Yscl = 1

Example 3 **Relating to the Real World**

DIVERSITY Some students may not be familiar with photocopiers. If you have copies to make for your classes, consider inviting those students to accompany you while you make copies to see how the process works.

VISUAL LEARNING If copiers are available, have students photocopy a ruler at 150%. Then have them photocopy the enlargement at 150%. Have students measure (with a normal-sized ruler) the enlargements to compare the enlarged ruler to a normal-sized ruler.

Question 9 Point out that the ranges can be changed by using WINDOW. The correct keystrokes for 2^x are **2** ^ X,T,θ.

Exercises ON YOUR OWN

Exercise 1 Let x equal the number of 20-min time periods. Let y equal the number of bacteria cells. Have students write an equation which represents this problem. $y = 75 \cdot 2x$.

Technology Options

For Exercises 23 and 26–31 students may be helped by using graphing software or a graphing calculator.

Prentice Hall Technology

Calculator • Graphing Calculator Handbook: Procedure 18

Software • Secondary Math Lab Toolkit • Computer Item Generator 8-1

CD-ROM • Multimedia Algebra Lab 8

Internet • See the Prentice Hall site. (http://www.phschool.com)

QUICK REVIEW

The order of operations is parentheses, exponents, multiplication and division, addition and subtraction.

GRAPHING CALCULATOR HINT

These range values will give you a clear picture of $y = 2^x$.

Xmin = 0 Ymin = 0
Xmax = 10 Ymax = 100
Xscl = 1 Yscl = 10

Example 2

Evaluate each exponential function.

a. $y = 5^x$ for $x = 2, 3, 4$

x	$y = 5^x$	y
2	$5^2 = 25$	25
3	$5^3 = 125$	125
4	$5^4 = 625$	625

b. $t(n) = 4(3^n)$ for the domain {3, 6}

n	$t(n) = 4(3^n)$	$t(n)$
3	$4 \cdot 3^3 = 4 \cdot 27 = 108$	108
6	$4 \cdot 3^6 = 4 \cdot 729 = 2916$	2916

8. **Try This** Evaluate the functions $y = 6^x$ and $y = 3(2^x)$ for $x = 1, 2,$ and 3. Which function increases more quickly? Why? **6, 36, 216; 6, 12, 24; 6^x increases more quickly; each increase in x by 1 corresponds to multiplication by 6 in the first function and multiplication by 2 in the second function.**

Part 3

Graphing Exponential Functions

The graphs of many exponential functions look alike.

Example 3 **Relating to the Real World** 

Technology Some photocopiers allow you to choose how large you want an image to be. The function $f(x) = 1.5^x$ models the increase in size of a picture being copied over and over at 150%. Graph the function.

Make a table of values.

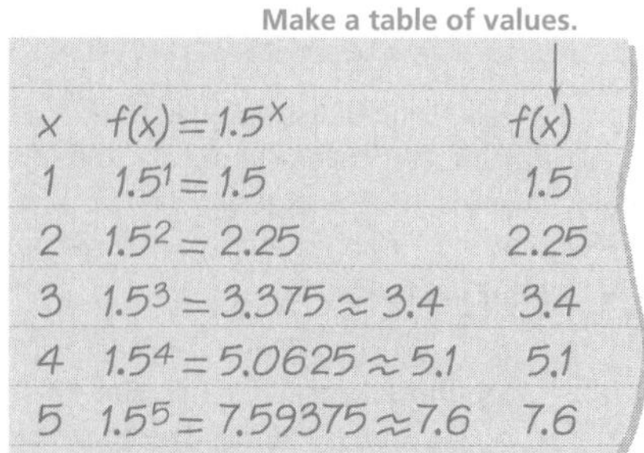

x	$f(x) = 1.5^x$	$f(x)$
1	$1.5^1 = 1.5$	1.5
2	$1.5^2 = 2.25$	2.25
3	$1.5^3 = 3.375 \approx 3.4$	3.4
4	$1.5^4 = 5.0625 \approx 5.1$	5.1
5	$1.5^5 = 7.59375 \approx 7.6$	7.6

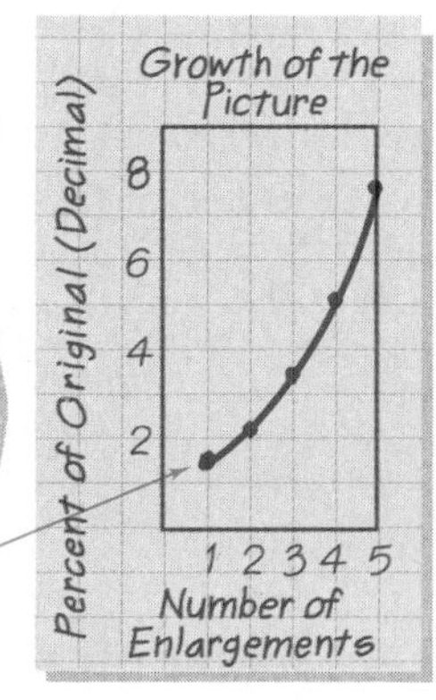

Plot the points. Connect the points with a smooth curve.

9. **a. Graphing Calculator** Use your graphing calculator to graph the function $y = 2^x$. **See margin.**

b. How is the graph of $y = 2^x$ similar to the graph of $f(x) = 1.5^x$? How is it different? **The shapes of $y = 2^x$ and $f(x) = 1.5^x$ are similar, but $y = 2^x$ is steeper.**

Students can then use this equation to check the answer to the problem by using the [TABLE] function on the graphing calculator.

ESL **Exercise 1** Many students may not understand this use of the word *culture*. Ask students how they would describe or define a laboratory culture. Ensure that they know a bacteria culture is a cultivation of bacteria living in a nutrient material.

ERROR ALERT! **Exercise 5** Some students may begin by multiplying 100 by 10 and then squaring the product. **Remediation:** Remind them that the order of operations is: parentheses, exponents, multiplication and division, addition and subtraction.

ALTERNATIVE ASSESSMENT **Exercises 7–14** To check students' understanding of exponential functions, have them graph each expression.

STANDARDIZED TEST TIP **Exercise 15** In the expression $y = a \cdot b^x$, *a* represents the number in the original population, and *b* represents the population's growth factor—2 to double the population, 3 to triple the population, etc. Therefore, the only possible answer choice is *B*.

OPEN-ENDED Exercise 22 Use this exercise for a class discussion. Make a list on the board that shows students' functions and tells whether the outputs of the function increase, decrease, or remain the same. Have students describe any patterns they see. Students should discover that if the number multiplied repeatedly is a number less than one, the outputs will decrease. If the number is one, the outputs will

Exercises ON YOUR OWN

1. Patterns Bacteria in a laboratory culture can double in number every 20 min. Suppose a culture starts with 75 cells. Copy, complete, and extend the table to find when there will be more than 30,000 bacteria cells. **See margin for table; after 3 h.**

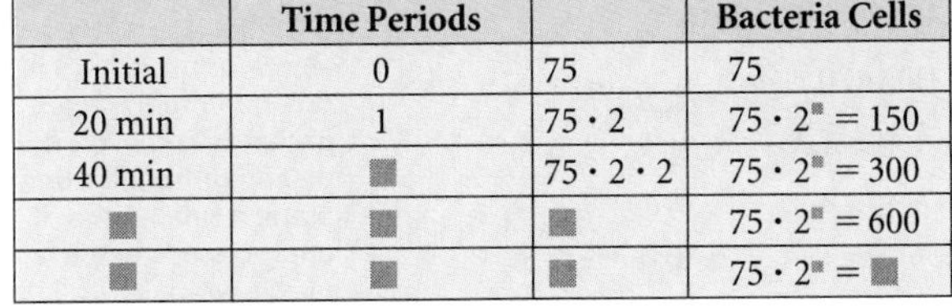

Time	Number of 20-min Time Periods	Pattern	Number of Bacteria Cells
Initial	0	75	75
20 min	1	$75 \cdot 2$	$75 \cdot 2^{■} = 150$
40 min	■	$75 \cdot 2 \cdot 2$	$75 \cdot 2^{■} = 300$
■	■	■	$75 \cdot 2^{■} = 600$
■	■	■	$75 \cdot 2^{■} =$ ■

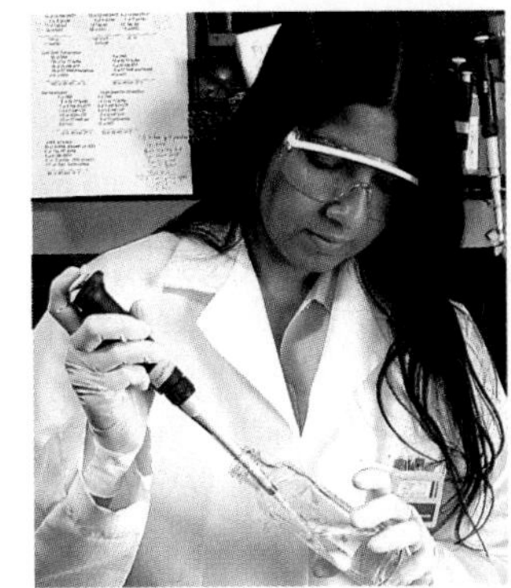

2. Finance An investment of \$10,000 doubles in value every 13 years. How much is the investment worth after 52 years? after 65 years? **\$160,000; \$320,000**

Which function is greater at the given value?

3. $y = 5^x$ and $y = x^5$ at $x = 5$ **equal**
4. $f = 10 \cdot 2^t$ and $f = 200 \cdot t^2$ at $t = 7$ **$200 \cdot t^2$**
5. $f(x) = 2^x$ and $f(x) = 100x^2$ at $x = 10$ **$100x^2$**
6. $y = 3^x$ and $y = x^3$ at $x = 4$ **3^x**

Evaluate each function for the domain {1, 2, 3, 4, 5}. Is the function *increasing, decreasing,* or *neither?*

7. $f(x) = 4^x$ **4, 16, 64, 256, 1024; incr.**
8. $c = a^3$ **1, 8, 27, 64, 125; incr.**
9. $h(x) = 1^x$ **1, 1, 1, 1, 1; neither**
10. $f(x) = 5 \cdot 4^x$ **20, 80, 320, 1280, 5120; incr.**
11. $y = 0.5^x$ **0.5, 0.25, 0.125, 0.0625, 0.03125; decr.**
12. $y = \left(\frac{2}{3}\right)^x$ **$\frac{2}{3}, \frac{4}{9}, \frac{8}{27}, \frac{16}{81}, \frac{32}{243}$; decr.**
13. $g(x) = 4 \cdot 10^x$ **40, 400, 4000, 40000, 400000; incr.**
14. $d = 100 \cdot 0.3^t$ **30, 9, 2.7, 0.81, 0.243; decr.**
15. Standardized Test Prep A population of 6000 doubles in size every 10 years. Which equation relates the size of the population *y* to the number of 10-year periods *x*? **B**
    A. $y = 6000 \cdot 10^x$ B. $y = 6000 \cdot 2^x$ C. $y = 10 \cdot 2^x$ D. $y = 2 \cdot 10^x$ E. $y = 2 \cdot 6000^x$
16. Graphing Calculator Graph the functions $y = x^2$ and $y = 2^x$ on the same set of axes.
    a. What happens to the graphs between $x = 1$ and $x = 3$? **See margin.**
    b. Critical Thinking How do you think the graph of $y = 6^x$ would compare to the graphs of $y = x^2$ and $y = 2^x$? **The graph of $y = 6^x$ is steeper than $y = x^2$ and $y = 2^x$.**
17. Writing **Analyze** the range for the function $f(x) = 500 \cdot 1^x$ using the domain {1, 2, 3, 4, 5}. Explain why the restriction $b \neq 1$ is included in the definition of an exponential function. **500, 500, 500, 500, 500; $b = 1$ produces a linear graph.**
18. Ecology In 50 days, a water hyacinth can generate 1000 offspring (the number of plants is multiplied by 1000). How many hyacinth plants could there be after 150 days? **1,000,000,000 plants**

pages 362–364 Think and Discuss

9a.

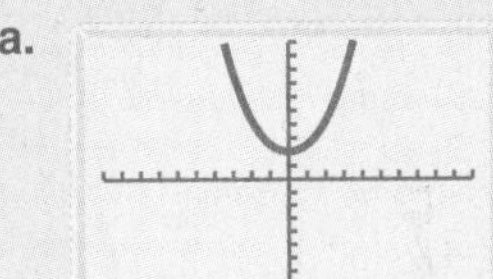

pages 365–366 On Your Own

1.

Time	Time Periods	Pattern	Number of Bacteria Cells
Initial	0	75	75
20 min	1	$75 \cdot 2$	$75 \cdot 2^1 = 150$
40 min	2	$75 \cdot 2 \cdot 2$	$75 \cdot 2^2 = 300$
60 min	3	$75 \cdot 2 \cdot 2 \cdot 2$	$75 \cdot 2^3 = 600$
80 min	4	$75 \cdot 2 \cdot 2 \cdot 2 \cdot 2$	$75 \cdot 2^4 = 1200$

16.

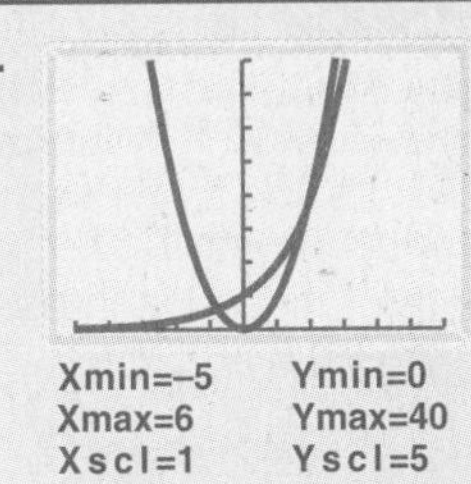

16a. Between $x = 1$ and $x = 3$, the graph of $y = x^2$ rises faster than the graph of $y = 2^x$. The graphs intersect at $x = 2$.

24, 25a–b, 26–31. See back of book.

stay the same. If the number is greater than one, the outputs will increase.

CRITICAL THINKING Exercise 25c Students should read the definition of an exponential function on page 363 and think about whether $y = (-2)^x$ can be written in the form $y = a \cdot b^x$ with $b > 0$.

Chapter Project **FIND OUT BY DOING** Students will begin the project by preparing an experiment to measure mold growth. Ask a science teacher to visit your classroom to explain how the experiment works. This task is essential to students' work on the Chapter Project, which was introduced in the Chapter Opener. Check students' progress on the project by having students write a sentence or two each to describe what steps they took to begin the mold growth.

Exercises MIXED REVIEW

Exercise 36 Students may enjoy drawing an actual-size diagram of the size of a page in *Old King Cole.*

GETTING READY FOR LESSON 8-2 In these exercises, students evaluate exponential functions to prepare them for using similar functions to model exponential growth.

Wrap Up

THE BIG IDEA Ask students to explain what an exponential function is.

RETEACHING ACTIVITY Students make data tables and write equations to represent exponential functions. (Reteaching worksheet 8-1)

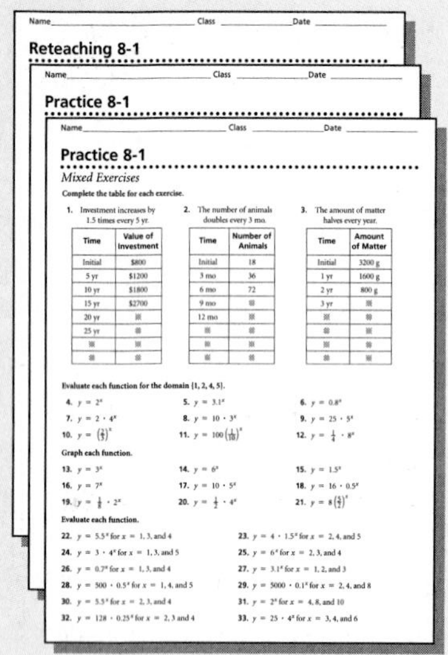

Name______ Class ______ Date ______

Reteaching 8-1

Name______ Class ______ Date ______

Practice 8-1

Name______ Class ______ Date ______

Practice 8-1

Mixed Exercises

Complete the table for each exercise.

1. Investment increases by 1.5 times every 5 yr.

Time	Value of Investment
Initial	\$800
5 yr	\$1200
10 yr	\$1800
15 yr	\$2700
20 yr	
25 yr	

2. The number of animals doubles every 3 mo.

Time	Number of Animals
Initial	18
3 mo	36
6 mo	72
9 mo	
12 mo	

3. The amount of matter halves every year.

Time	Amount of Matter
Initial	3200 g
1 yr	1600 g
2 yr	800 g
3 yr	

Evaluate each function for the domain {1, 2, 4, 5}.

4. $y = 2^x$
5. $y = 3.1^x$
6. $y = 0.8^x$
7. $y = 2 \cdot 4^x$
8. $y = 10 \cdot 3^x$
9. $y = 25 \cdot 5^x$
10. $y = \left(\frac{2}{3}\right)^x$
11. $y = 100\left(\frac{1}{10}\right)^x$
12. $y = \frac{1}{4} \cdot 8^x$

Graph each function.

13. $y = 3^x$
14. $y = 6^x$
15. $y = 1.5^x$
16. $y = 7^x$
17. $y = 10 \cdot 5^x$
18. $y = 16 \cdot 0.5^x$
19. $y = \frac{1}{8} \cdot 2^x$
20. $y = \frac{1}{2} \cdot 4^x$
21. $y = 8\left(\frac{5}{2}\right)^x$

Evaluate each function.

22. $y = 5.5^x$ for $x = 1, 3,$ and 4
23. $y = 4 \cdot 1.5^x$ for $x = 2, 4,$ and 5
24. $y = 3 \cdot 4^x$ for $x = 1, 3,$ and 5
25. $y = 6^x$ for $x = 2, 3,$ and 4
26. $y = 0.7^x$ for $x = 1, 3,$ and 4
27. $y = 3.1^x$ for $x = 1, 2,$ and 3
28. $y = 500 \cdot 0.5^x$ for $x = 1, 4,$ and 5
29. $y = 5000 \cdot 0.1^x$ for $x = 2, 4,$ and 8
30. $y = 5.5^x$ for $x = 2, 3,$ and 4
31. $y = 2^x$ for $x = 4, 8,$ and 10
32. $y = 128 \cdot 0.25^x$ for $x = 2, 3$ and 4
33. $y = 25 \cdot 4^x$ for $x = 3, 4,$ and 6

Lesson Quiz

Lesson Quiz is also available in Transparencies.

Evaluate each function for the domain {2, 4, 6}. Does the function increase, decrease, or neither?

1. $y = \left(\frac{4}{3}\right)^x$ {1.78, 3.16, 5.62} increases
2. $y = \left(\frac{3}{4}\right)^x$ {0.56, 0.32, 0.18} decreases
3. $y = 1^x$ neither
4. $y = 10 \cdot 0.4^x$ {1.6, 0.26, 0.04} decreases

Evaluate each expression.

19. $50 \cdot x^5$ for $x = 0.5$ 1.5625 **20.** $50{,}000 \cdot m^3$ for $m = 1.1$ 66,550 **21.** $0.0125 \cdot c^4$ for $c = 2$ 0.2

22. *Open-ended* Select one of the exponential functions from this lesson. 22a–b. Check students' work.
  a. What number is multiplied repeatedly in your example?
  b. As the exponent increases, tell whether the outputs of your function increase, decrease, or do neither.

23. Match each table with the function that models the data.

Table I A

x	y
1	3
2	6
3	9
4	12

Table II C

x	y
1	3
2	9
3	27
4	81

Table III B

x	y
1	1
2	8
3	27
4	64

Functions:
A. $y = 3x$
B. $y = x^3$
C. $y = 3^x$

24. *Critical Thinking* Why don't two 150% enlargements on a photocopier produce the same size picture as one 300% enlargement? See margin.

25. *Patterns* The base in the function $y = (-2)^x$ is a negative number.
  a. Make a table of values for the domain {1, 2, 3, 4, 5, 6}.
  b. What pattern do you see in the outputs? 25a–b. See margin.
  c. *Critical Thinking* Is $y = (-2)^x$ an exponential function? **Justify** your answer. No; the shape of the graph is not similar to the shape of the graph of an exponential function.

Graph each function. 26–31. See margin.

26. $y = 3^x$ **27.** $y = 3\left(\frac{3}{2}\right)^x$ **28.** $y = 1.5^x$

29. $y = \frac{1}{4} \cdot 2^x$ **30.** $y = 0.1 \cdot 2^x$ **31.** $y = 10^x$

Chapter Project *Find Out by Doing*

Gather the materials for your project: $\frac{1}{16}$ in. or 1 mm graph paper, a packet of unflavored gelatin, a flat dish or plate, a small piece of unprocessed cheese, and plastic wrap.

• Decide where you will keep your dish. A warm, humid place is best for growing mold.

• Cut the graph paper to cover the bottom of the dish. Follow the directions on the gelatin packet. Cover the graph paper with about $\frac{1}{8}$ in. of gelatin. Add the cheese to the gelatin. Leave the dish uncovered overnight; then cover tightly with the plastic wrap.

Exercises MIXED REVIEW

Solve each equation.

32. $5(g - 1) = \frac{1}{2}$ $1\frac{1}{10}$ **33.** $t^2 + 3t - 4 = 0$ 1, −4 **34.** $|m - 7| = 9$ −2, 16 **35.** $x^2 - 5x + 6 = 0$ 2, 3

36. *Books* The smallest book in the Library of Congress is *Old King Cole.* Each square page has area $\frac{1}{25}$ in.^{2}. How wide is one page? $\frac{1}{5}$ in.

Getting Ready for Lesson 8-2

Find the range of each function for the domain {1, 2, 3, 4, 5}.

37. $y = 2^x$ {2, 4, 8, 16, 32} **38.** $y = 4 \cdot 2^x$ {8, 16, 32, 64, 128} **39.** $y = 0.5 \cdot 2^x$ {1, 2, 4, 8, 16} **40.** $y = \frac{3}{2} \cdot 2^x$ {3, 6, 12, 24, 48} **41.** $y = 3 \cdot 2^x$ {6, 12, 24, 48, 96}

42. How are the functions in Exercises 37–41 alike? How are they different? All functions are increasing; the basic function 2^x is multiplied by different numbers.

8-2 Teaching Notes

PROBLEM OF THE DAY

Which has a larger area, a square with 11 in. sides or a triangle with a 16 in. base and a 15 in. height? the square

Problem of the Day is also available in Transparencies.

CONNECTING TO PRIOR KNOWLEDGE Ask students: *Let's say I give one student one bean. The next student gets double that, two beans. The third student gets double the previous amount, four beans. How many beans will the last student in the class receive?* 2^{n-1}, where n is the number of students Encourage students to make guesses before calculating the answer. Students will most likely not predict such a large number.

WORK TOGETHER

Have students work in pairs. One student in each pair should act as the recorder. This will help students to develop skills that they will need in the workplace.

Ask students: *If y represents the new wage and x represents the number of wage increases, what function could be written to show how the wages for Job A grow?* $y = 5 + 0.5x$ *What kind of equation is this?* linear

THINK AND DISCUSS

This section shows how an exponential function can be used to show how the wages for Job B grow.

Connections Savings. . . and more

8-2 Exponential Growth

What You'll Learn

- Modeling exponential growth
- Calculating compound interest

...And Why

To solve problems involving medical costs and finance

What You'll Need

- calculator

QUICK REVIEW

A 10% increase in an amount means you have 110% of the original amount. To find 110%, multiply by 1.1.

Time	Job A	Job B
Start	\$5.00	\$4.80
6 mo	■	■
1 yr	■	■

WORK TOGETHER

Jobs Suppose you are offered a choice of two jobs. Job A has a starting wage of \$5.00/h, with a \$.50 raise every 6 months. Job B starts at \$4.80/h, with a 10% raise every 6 months. Work with a partner.

1. How do you find the new wage after each raise for Job A? You add \$.50 to the previous wage.
2. In Job B, each new wage is 110% of the previous wage. How do you find each new wage for Job B? You multiply the previous wage by 1.1.
3. Organize the wages for each job in a table like the one at the left. Show wages from the start of the job through the raise at three years. Round each wage to the nearest cent. See margin.
4. a. Patterns Which wage pattern involves repeated multiplication?
   b. Which wage pattern results in a linear function? b. Job A a. Job B
5. Which graph represents the wages for Job A? for Job B? red; blue
6. When would you prefer to have Job A? Job B? Explain. The first 2 yr; after 2 yr; Job A pays better wages for the first two years, but Job B wages grow much faster after that.

THINK AND DISCUSS

Modeling Exponential Growth

Part 1

You can use an exponential function to show how the wages for Job B grow.

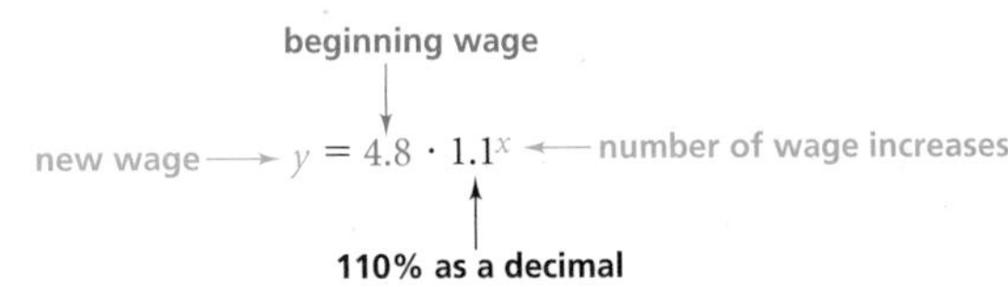

Because multiplying over and over by 1.1 causes the wage to increase, this kind of exponential function is an example of *exponential growth.*

Lesson Planning Options

Prerequisite Skills

- Finding the range of functions
- Evaluating expressions

Assignment Options

To provide flexible scheduling, this lesson can be subdivided into parts.

1 **Core** 1–14, 23–29 **Extension** 15, 16, 19, 22, 30

2 **Core** 17, 21, 31–34 **Extension** 18, 20

Use Mixed Review to maintain skills.

Resources

Student Edition

Skills Handbook, pp. 579, 580
Extra Practice, p. 563
Glossary/Study Guide

Teaching Resources

Chapter Support File, Ch. 8
- Practice 8-2 (two worksheets)
- Reteaching 8-2
- Alternative Activity 8-2, A and B

Classroom Manager 8-2
Glossary, Spanish Resources
Two-Year Algebra Handbook 8-2

Transparencies

11, 97, 101

After students have read the section on exponential growth, ask: *Does the function $y = 3 \cdot 4^x$ model exponential growth?* yes

CONNECTING TO THE STUDENTS' WORLD Question 7 Encourage students to find out the population and growth rate of your city or a nearby city that is experiencing a steady growth rate. Have students repeat Question 7 using the data they collected.

Example 1 Relating to the Real World 

Question 8 Be sure students understand that they are to make a prediction using the formula.

MAKING CONNECTIONS Read this famous riddle to students: As I was going to St. Ives, I met a man with seven wives. Every wife had seven sacks, every sack had seven cats, every cat had seven kits. Kits, cats, sacks, and wives, how many were going to St. Ives? Lead students to understand that this quantity can be expressed as $7 \cdot 7 \cdot 7 \cdot 7 = 7^4$. However, the answer to the riddle is one. The man with seven wives was going the other way.

Example 2 Relating to the Real World 

EXTENSION Have students solve the following using the [TABLE] function on the graphing calculator. Ask students: *At 6.5% interest, how long will it take the $500 deposit to reach a balance of more than $3000?* 29 yr

Additional Examples

FOR EXAMPLE 1

A certain town has a population of 13,000 people. The town grows 10% annually.

a. Write an equation to model the increases in population. $p = 13{,}000 \cdot 1.1^t$

b. Use your equation to find how many years it will take for the town's population to double. 8 yr

Discussion: *Why does it take less than 10 years for a 10% annual increase to total 100%?*

FOR EXAMPLE 3

Suppose you deposited a $1,000 scholarship in an account paying 7.9% interest compounded quarterly. How much will your scholarship be worth in 2 years? $1169.36

Exponential Growth

For $a > 0$ and $b > 1$, the function $y = a \cdot b^x$ models **exponential growth.**

$y = a \cdot b^x$ — a: starting amount; x: number of increases; b: the base, called the **growth factor**

Example: $y = 1000 \cdot 2^x$

When you use exponential functions to model real-world situations, you must identify the initial amount a and the growth factor b. To show growth, b must be greater than 1.

7. Suppose the population of a city is 50,000 and is growing 3% each year.
 a. The initial amount a is ■. 50,000
 b. The growth factor b is 100% + 3%, which is 1 + ■ = ■. 0.03; 1.0
 c. To find the population after one year, you multiply ■ · 1.03. 50,000
 d. Complete the equation to find the population after x years. $y = ■ \cdot ■^{■}$ 50,000; 1.03; x
 e. Use your equation to find the population after 25 years. about 104,689

Example 1 Relating to the Real World

Medical Care Since 1985, the daily cost of patient care to community hospitals in the United States has increased about 8.6% per year. In 1985, hospitals paid an average cost per day of $460.

a. Write an equation to model the cost of hospital care.
b. Use your equation to find the approximate cost per day in 1995.

a. Use an exponential function to model repeated percent increases.

Relate $y = a \cdot b^x$

Define x = the number of years since 1985
y = the cost of hospital care at various times
b = 100% plus 8.6% of the cost = 108.6% = 1.086
a = initial cost in 1985 = $460

Write $y = 460 \cdot 1.086^x$ ← Substitute values for the initial amount a and the growth factor b.

b. 1995 is 10 years after 1985, so solve the equation for $x = 10$.

$y = 460 \cdot 1.086^{10}$ ← Substitute.
$= 1049.677974$

The average cost per day in 1995 was about $1050.

CALCULATOR HINT

To evaluate $460 \cdot 1.086^{10}$ press 460 [×] 1.086 [^] 10 [ENTER].

8. **Try This** **Predict** the cost per day for the year 2000. about $1586

DIVERSITY Be sensitive to those students whose families may not understand or trust banks and choose not to use them. Point out that, as in any business, a customer's experience will vary from bank to bank.

VISUAL/AUDITORY LEARNING Help students see how the \$500 grows year by year. Write \$500 on the board. Have a student calculate and then tell you the interest earned and the new balance for the first year. interest \$32.50; new balance \$532.50 Write these answers on the board. Have a second student calculate the interest and balance for the second year. interest \$34.61; new balance \$567.11 Continue year by year until students see how the amount is growing. Have a student come to the board to graph how the amount grows.

Some students may not understand why the interest rate ***b*** is represented by 106.5% in the equation. Explain that the 100% represents the amount already in the account and you are adding 6.5% to that amount. If you only used 6.5% as ***b***, then you would only calculate the interest added to the account, not the account balance.

EXTENSION Question 11 Have students research current interest rates and calculate the account balance based on these rates. Suggest that students check the financial section of a newspaper or contact local banking institutions to find current interest rates for savings accounts. Ask students to find out whether the interest rate varies for different types of accounts such as certificates of deposit.

PROBLEM SOLVING HINT

For Questions 9 and 10, use the *Guess and Test* strategy.

9. Find the first year in which the predicted cost per day will be greater than \$2000. 2003

10. The cost per day doubled in less than 10 years between 1985 and 1995. Using the function from Example 1, about how long would it take to double again? less than 9 yr

Part 2

Finding Compound Interest

QUICK REVIEW

Principal is the amount of money you put in a bank account. *Interest* is the amount the bank pays you for letting it use your money.

When a bank pays interest on both the principal *and* the interest an account has already earned, the bank is paying **compound interest.** An **interest period** is the length of time over which interest is calculated. Compound interest is an exponential growth situation.

Example 2 **Relating to the Real World** 

Savings Suppose your parents deposited \$500 in an account paying 6.5% interest, compounded annually (once a year), when you were born. Find the account balance after 18 years.

Relate $y = a \cdot b^x$ ← Use an exponential function.

Define $x =$ the number of interest periods
$y =$ the balance at various times
$a = 500$ ← initial deposit
$b = 1.065$ ← 100% + 6.5% = 106.5% = 1.065

Write $y = 500 \cdot 1.065^x$
$= 500 \cdot 1.065^{18}$ ← Once a year for 18 years is 18 interest periods. Substitute 18 for *x*.
$= 1553.32719$

The balance after 18 years will be \$1553.33.

11. Try This Suppose the interest rate on the account was 8%. How much would be in the account after 18 years? \$1998.01

For Exercises 24–27, students may want to use a graphing calculator or spreadsheet software to determine which function each table represents.

Prentice Hall Technology

Calculator • Graphing Calculator Handbook: Procedure 18

Software • Secondary Math Lab Toolkit • Computer Item Generator 8-2

CD-ROM • Multimedia Algebra Lab 8

Internet • See the Prentice Hall site. (http://www.phschool.com)

Example 3

Be sure students understand that the annual rate of interest (6.5%) must be divided by 4 if the interest is compounded quarterly. Some students may think that 6.5% interest will be given each quarter. Point out that the number of years is multiplied by 4 to obtain the number of interest payments. Have students compare the balance in the account when the interest is compounded quarterly with the balance when the interest is compounded annually (Example 2). Have students discuss the significance of this difference.

Question 13b Have students use a calculator to solve this problem. To make sure that they have not made errors, they should check whether their answer is reasonable. Discuss how to decide if the answer is reasonable. Help students realize that the answer will be more than $1595.92, but not too much more.

Exercises ON YOUR OWN

EXTENSION Exercise 7 Ask: *Write a problem that can be modeled by this equation.* **Answers may vary. Sample: If you deposit $1000 in an account paying 4% interest, compounded annually, what will be the account balance after 3 yr?**

WRITING Exercise 18 Students may use the [TABLE] function on the graphing calculator to help them answer this question.

OPEN-ENDED Exercise 19d Point out that students should use the function to make their predictions.

page 367 Work Together

3.

Time	Job A	Job B
Start	$5.00	$4.80
6 mo	$5.50	$5.28
1 yr	$6.00	$5.81
1.5 yr	$6.50	$6.39
2 yr	$7.00	$7.03
2.5 yr	$7.50	$7.73
3 yr	$8.00	$8.50

pages 370–372 On Your Own

15. x = number of years; y = population; $y = 130{,}000 \cdot 1.01^x$
16. x = number of years; y = price; $y = 50 \cdot 1.06^x$
17. x = number of months; y = deposit with interest; $y = 3000 \cdot \left(1 + \frac{0.05}{12}\right)^x$
23.

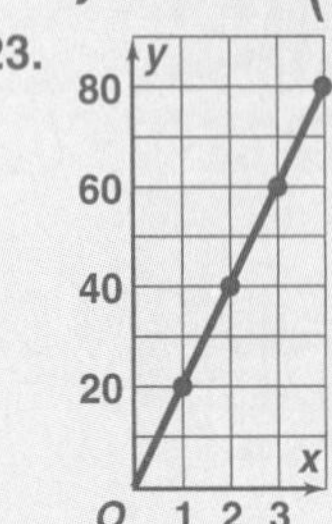

Banks sometimes pay compound interest more than once a year. When they use shorter interest periods, the interest rate for each period is also reduced.

Annual Interest Rate of 8%

Compounded	Periods per Year	Rate per Period
annually	1	8% every year
semi-annually	2	$\frac{8\%}{2}$ = 4% every 6 months
quarterly	4	$\frac{8\%}{4}$ = 2% every 3 months
monthly	12	$\frac{8\%}{12}$ = 0.6% every month

12. In an account that pays 6.5% interest, what is the interest rate if the interest is compounded quarterly? monthly? 1.625%; 0.542%

Example 3

Suppose the account in Example 2 paid interest compounded quarterly instead of annually. Find the account balance after 18 years.

Relate $y = a \cdot b^x$ ← Use an exponential function.

Define x = the number of interest periods (quarters)
y = the balance at various times
$a = 500$ ← initial deposit
$b = 1 + \frac{0.065}{4}$ ← There are 4 interest periods in 1 year, so divide the interest into four parts.
$= 1.01625$

Write $y = 500 \cdot 1.01625^x$
$= 500 \cdot 1.01625^{72}$ ← 18 • 4 = 72 interest periods in 18 years
$= 1595.916716$ ← Use a calculator.

The balance after 18 years will be $1595.92.

13. a. How many interest periods per year are there for an account with interest compounded daily? 365 periods
b. Try This Suppose the account above paid interest compounded daily. How much money would be in the account after 18 years? $1610.83

Exercises ON YOUR OWN

Identify the initial amount *a* and the growth factor *b* in each exponential function.

1. $g(x) = 20 \cdot 2^x$ 20; 2
2. $y = 200 \cdot 1.0875^x$ 200; 1.0875
3. $y = 10{,}000 \cdot 1.01^x$ 10,000; 1.01
4. $f(t) = 1.5^t$ 1; 1.5

What repeated percent of increase is modeled in each function?

5. $r = 70 \cdot 1.5^n$ 50%
6. $f(t) = 30 \cdot 1.095^t$ 9.5%
7. $y = 1000 \cdot 1.04^x$ 4%
8. $y = 2^x$ 100%

ALTERNATIVE ASSESSMENT **Exercise 20** Ask students to repeat this exercise using the same interest rate compounded monthly. This will help assess students' understanding of compound interest rates.

STANDARDIZED TEST TIP **Exercise 21** Remind students that the base for a percent increase exponential function is always one plus the decimal equivalent of the percent increase.

ESL **Exercise 27–29** Explain that the expression *justify your reasoning* means to explain why your answers are correct.

EXTENSION Exercise 31 Have students research, write, and solve a similar problem for their own state or another state that they choose. Have students compare their information and their predictions.

ERROR ALERT! **Exercise 31** Some students may interpret 3.5% as the amount of interest given each quarter. **Remediation:** Explain that the interest rate given always applies to the annual rate of interest. It may help students to read Example 3 on page 370.

Exercise 32 Have students use 365 days, not 365.25 days in a year when computing the interest compounded daily.

Chapter Project FIND OUT BY RECORDING Students will begin recording mold growth and continue recording growth each day for ten days. This information is essential to their work on the chapter project. Have students add their recorded data to work they have already completed for the project. Check student progress by having each student write a sentence or two describing what they have done for the project thus far.

Write the growth factor used to model each percent of increase in an exponential function.

9. 4% 1.04 **10.** 5% 1.05 **11.** 3.7% 1.037 **12.** 8.75% 1.0875 **13.** 0.5% 1.005 **14.** 15% 1.15

Write an exponential function to model each situation. Tell what each variable you use represents. 15–17. See margin.

15. A population of 130,000 grows 1% per year

16. A price of \$50 increases 6% each year

17. A deposit of \$3000 earns 5% annual interest compounded monthly.

18. *Writing* Would you rather have \$500 in an account paying 6% interest compounded quarterly or \$750 in an account paying 5.5% compounded annually? **Summarize** your reasoning. \$750; despite the lower interest rate the second account earns more money for up to 67 yr.

19. *Education* The function $y = 355 \cdot 1.08^x$ models the average annual cost y (in dollars) for tuition and fees at public two-year colleges. The variable x represents the number of years since 1980.

a. What was the average annual cost in 1980? \$355

b. What is the average percent increase in the annual cost? 8%

c. Find the average annual cost for 1990. about \$766

d. *Open-ended* **Predict** the average annual cost for the year you plan to graduate from high school. Check students' work. Sample: 2001—\$1787; 2002—\$1930; 2003—\$2084; 2004—\$2251

20. *History* The Dutch bought Manhattan Island in 1626 for \$24 worth of merchandise. Suppose the \$24 had been invested in 1626 in an account paying 4.5% interest compounded annually. Find the balance today. Check students' work. Sample: about \$300 million

21. *Standardized Test Prep* An investment of \$100 earns 5% interest compounded annually. Which expression represents the value of the investment after 10 years? E

A. $10 \cdot 100^5$ B. $100 \cdot 0.05^{10}$ C. $100 \cdot 10^{0.05}$
D. $10 \cdot 100^{1.05}$ E. $100 \cdot 1.05^{10}$

Workstations Replace Supercomputers

Workstations–sophisticated computers that sit on a desktop–are replacing larger mainframe computers in industry. Since 1987, sales of workstation computers in industry have increased about 30% per year. In 1987, sales totaled about \$3 billion.

22. a. *Math in the Media* Write an equation to model the sales of workstation computers since 1987. See below.

b. Use your model to find the total sales in 1995. about \$24.5 billion

a. s = sales in billions of dollars; x = years since 1987; $s = 3 \cdot 1.3^x$

Graph the function represented in each table. Then tell whether the table represents a *linear function* or an *exponential function*. 23–26. See margin for graphs.

23.

x	y
1	20
2	40
3	60
4	80

linear

24.

x	y
1	3
2	9
3	27
4	81

exponential

25.

x	y
1	6
2	12
3	24
4	48

exponential

26.

x	y
1	3
2	9
3	15
4	21

linear

24.

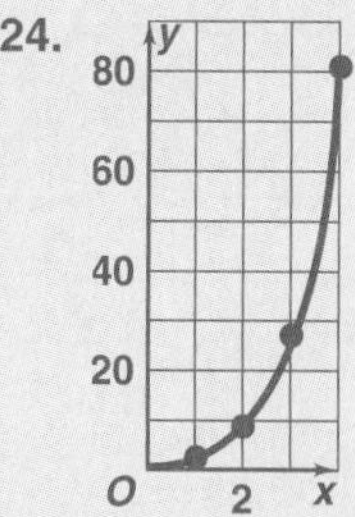

25.

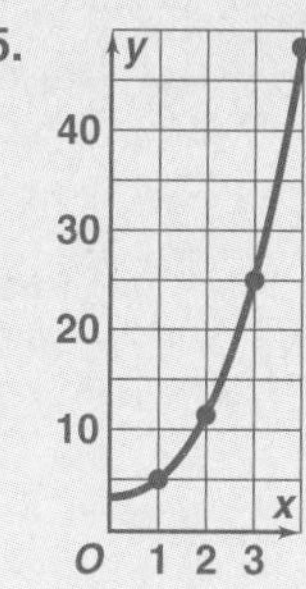

26, 30a. See back of book.

Exercises MIXED REVIEW

GETTING READY FOR LESSON 8-3 Students evaluate exponential expressions in which the number that is multiplied repeatedly is a fraction less than one. This will prepare students for working with expressions modeling exponential decay in the next lesson.

Wrap Up

THE BIG IDEA Ask students to explain what exponential growth is.

RETEACHING ACTIVITY Students model exponential growth using tables. Then write exponential functions and solve. (Reteaching worksheet 8-2)

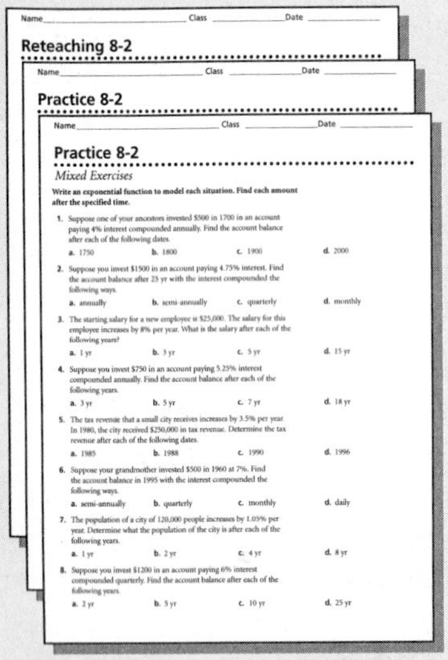

Reteaching 8-2

Practice 8-2

Practice 8-2
Mixed Exercises

Lesson Quiz

Lesson Quiz is also available in Transparencies.

1. Identify the original amount *a* and the growth factor *b* in the exponential function $y = 10 \cdot 1.036^x$. $a = 10, b = 1.036$

2. What repeated percent of increase is modeled in the function $y = 19 \cdot 1.9^n$? 90%

3. What growth factor would you use to model a 0.3% of increase in an exponential function? 1.003

4. A loan of $1000 is due at the end of 1 year. The loan's interest rate of 21% is compounded monthly. What is the total repayment amount? $1231.44

Tell whether each graph is a *linear function*, an *exponential function*, or *neither*. Justify your reasoning.

27. 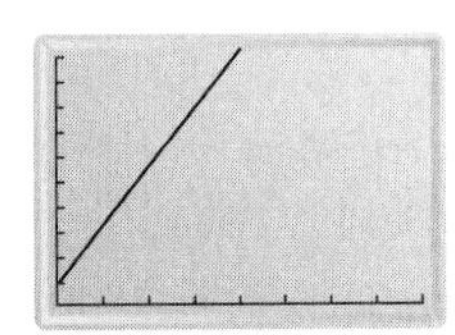
Linear; graph is a straight line.

28. 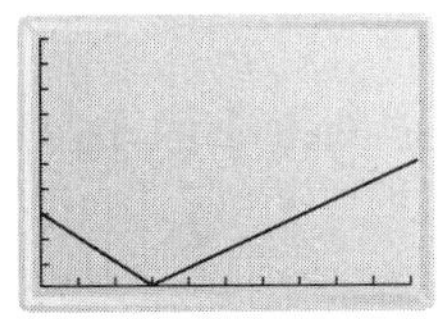
Neither; this is an abs. val. graph.

29. 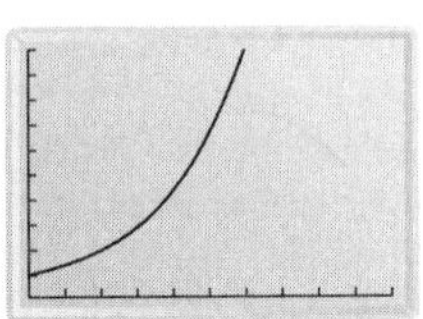
Exponential; graph curves upward.

30. *Statistics* Since 1970, the population of Virginia has grown at an average annual rate of about 1.015%. In 1970, the population was about 4,651,000.
   a. *Graphing Calculator* Write and graph a function to model population growth in Virginia since 1970. $y = 4{,}651{,}000 \cdot 1.01015^x$; see margin for graph.
   b. Estimate the population of Virginia in 1980 and 1990. about 5,100,000; about 5,700,000
   c. **Predict** the population of Virginia in the years 2000 and 2025. 6,296,836; 8,105,300

Write an exponential function to model each situation. Find each amount after the specified time.

31. $20,000 principal
3.5% compounded quarterly
10 years
$y = 20{,}000 \cdot \left(1 + \frac{0.035}{4}\right)^{4x}$; $22,092.44

32. $30 principal
4.5% compounded daily
2 years
$y = 30 \cdot \left(1 + \frac{0.045}{365}\right)^{365x}$; $73.24

33. $2400 principal
7% compounded annually
10 years
$y = 2400 \cdot (1 + 0.07)^x$; $4721.16

34. $2400 principal
7% compounded monthly
10 years
$y = 2400 \cdot \left(1 + \frac{0.07}{12}\right)^{12x}$; $4821.27

Chapter Project **Find Out by Recording**

Count the number of graph paper squares covered by mold once a day for 10 days. Record the mold growth. (Don't worry if you don't see any mold for the first few days.)

Exercises MIXED REVIEW

Find the vertex of each parabola. Identify it as a *maximum* or a *minimum*.

35. $y = 3x^2 + 2x - 8$ $\left(-\frac{1}{3}, -8\frac{1}{3}\right)$; min.

36. $y = \frac{1}{2}x^2 - 5x + 1$ $\left(5, -11\frac{1}{2}\right)$; min.

37. $y = 4x^2 - 11$ (0, −11); min.

38. a. *History* The boardwalk in Atlantic City was 1 mi long and 12 ft wide when it was built in 1870. What was its area in square feet?
   b. In 1995, the boardwalk was 4.5 mi long and up to 60 ft wide. What could be its maximum area in square feet?
   c. About how many times larger was the boardwalk in 1995 than in 1870?

a. 63,360 ft^2 b. 1,425,600 ft^2 c. 22.5 times larger

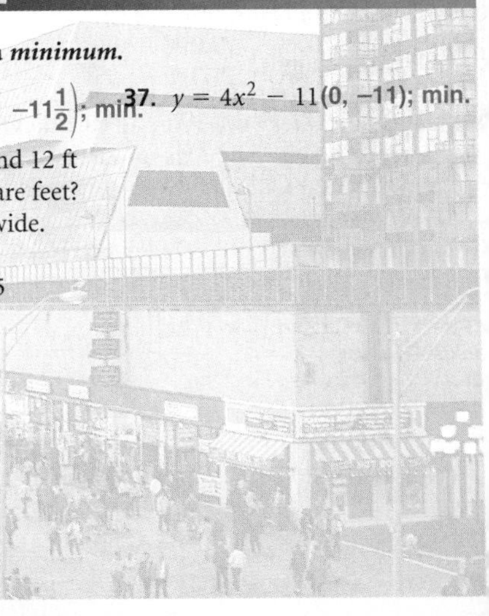

Getting Ready for Lesson 8-3

Simplify each expression.

39. $\left(\frac{1}{2}\right)^2$ $\frac{1}{4}$

40. 0.5^4 0.0625

41. $\left(\frac{3}{4}\right)^1$ $\frac{3}{4}$

42. 0.9^3 0.729

43. 0.98^2 0.9604

44. $\left(\frac{5}{6}\right)^2$ $\frac{25}{36}$

8-3 Teaching Notes

PROBLEM OF THE DAY

The fastest recorded speed for a bull killer whale is 50.6 ft/s. What is the whale's speed in mph? 34.5 mi/h

Problem of the Day is also available in Transparencies.

CONNECTING TO PRIOR KNOWLEDGE Remind students of the paper-folding activity in 8-1. Point out that after every fold, the top surface of the paper is half its previous size. Use this idea to introduce students to the concept of exponential decay.

THINK AND DISCUSS

Sketch graphs which model exponential decay or exponential growth on the board. Have students identify which is modeled. Ask volunteers to sketch their own graphs for each type of function.

Ask: *Does the function model exponential decay or growth?*

$y = 6 \cdot 4^x$ growth

$y = 6 \cdot \left(\frac{1}{4}\right)^x$ decay

Example 1 Relating to the Real World 

Be sure students understand that the half-life of a substance remains constant regardless of the size of the sample.

Connections Medical Care . . . and more

8-3 Exponential Decay

What You'll Learn

- Modeling exponential decay
- Using half-life models

...And Why

To solve problems involving population and radioactive decay

What You'll Need

- calculator

THINK AND DISCUSS

In Lesson 8-2 you used the exponential function $y = a \cdot b^x$ to model growth. You can also use it to model decay. The difference between growth and decay is the value of b, the base. With growth, b is greater than 1. With decay, b is between 0 and 1 and is called the **decay factor.**

Exponential Decay

For $a > 0$ and $0 < b < 1$, the function $y = a \cdot b^x$ models **exponential decay.**

Example: $y = 5 \cdot \left(\frac{1}{2}\right)^x$

Exponential Decay y-values decrease because of repeated multiplication by a number between 0 and 1.

$y = 5 \cdot 2^x$

$y = 5 \cdot \frac{1}{2}^x$

Xmin=0 Xmax=4 Xscl=0.5 Ymin=0 Ymax=15 Yscl=1

Exponential Growth y-values increase because of repeated multiplication by a number greater than 1.

You can use an exponential function to model the decay of a radioactive substance.

Who? Marie Curie (1867–1934) was born in Poland. She received Nobel Prizes in physics and chemistry for her pioneering work with radioactive elements. The *curie* is named for Marie Curie.

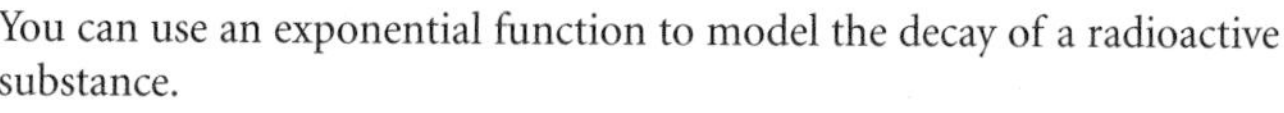

Example 1 Relating to the Real World 

Medicine To treat some forms of cancer, doctors use radioactive iodine. Use the graph to find how much iodine-131 is left in a patient eight days after the patient receives a dose of 20 mCi (millicuries).

Iodine-131 Decay

Radioactivity Level (mCi): 0, 10, 20

Time (days): 0, 10, 20, 30

The x-value 8 represents eight days. The y-value 10 represents the amount of iodine-131 remaining.

After eight days, there are 10 mCi of iodine-131 left.

Lesson Planning Options

Prerequisite Skills

- Graphing functions
- Converting decimals to percents

Assignment Options

Core 1–15, 26–28, 31–36
Extension 16–25, 29, 30, 37–39

Use Mixed Review to maintain skills.

Resources

 **Student Edition**

Skills Handbook, pp. 569, 579, 580
Extra Practice, p. 563
Glossary/Study Guide

Teaching Resources

Chapter Support File, Ch. 8
- Practice 8-3 (two worksheets)
- Reteaching 8-3
- Checkpoint

Classroom Manager 8-3
Glossary, Spanish Resources
Two-Year Algebra Handbook 8-3

 **Transparencies**

98, 101

DIVERSITY Consider that some students may feel uncomfortable discussing cancer because of personal or family experiences with the disease.

MAKING CONNECTIONS Nuclear medicine, a field of radiology, uses radioactive isotopes for diagnosing and treating cancer. The isotopes are added to the blood. As the isotopes travel through body organs, special cameras use the pattern of radiation being released to build up a picture of the organ. This process can help in diagnosing cancer. Isotopes can also be used to treat cancer when a dose of radiation is introduced into or next to a cancerous tumor in order to destroy it.

KINESTHETIC LEARNING Have students all move to one side of the classroom. Direct one-third of the students to move to the other side. Explain that as they move across the room, they are experiencing exponential decay. Then direct one-third of the remaining students to decay. Continue until fewer than three students remain. Direct students' attention to the graph of Iodine-131 decay on page 373. Help them recognize that, initially, a large number decayed, but as decay continued, progressively smaller numbers were affected. If you have block scheduling or extended class periods, repeat the activity using fractions other than one-third.

MAKING CONNECTIONS Radiocarbon dating is a method scientists use to determine the age of very old artifacts. All living things contain the element carbon. A very small percentage of carbon atoms are the radioactive isotope of carbon called carbon-14. Organisms take in carbon and carbon-14 by eating or breathing. When the organism dies, no more carbon-14 is brought into the body. To find the age of an ancient artifact such as a piece of bone, scientists first

Additional Examples

FOR EXAMPLE 2

The number of students remaining after school can be modeled by the exponential function $n = 1700 \cdot 0.2^t$. In this equation, t is the number of hours after the end of the school day, and n is the number of students remaining in the school building. Plot this expression to find out how long it takes for 1000 students to leave the school. How much time will pass before there are only 10 students left in the building?

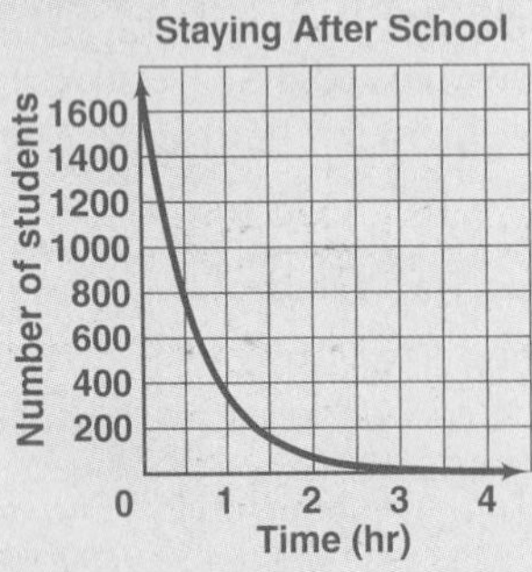

0.55 hr; 3.2 hr

Discussion: *What factors could explain the shape of the graph?*

FOR EXAMPLE 3

Use the equation $n = 1700 \cdot 0.2^t$ to find the hourly percent of decrease in the number of students who remain at school after classes are over. 80%/hr

1. **Try This** Use the graph in Example 1 to find the amount of iodine-131 left after 24 days. about 2.5 mCi

The *half-life* of a radioactive substance is the length of time it takes for one half of the substance to decay.

2. a. How long does it take for half of the 20 mCi dose in Example 1 to decay? How much is left? 8 da; 10 mCi
   b. Use your answer to part (a). How long does it take for half of that amount to decay? How much is left? 8 da; 5 mCi
   c. What is the half-life of iodine-131? 8 da
   d. How many half-lives of iodine-131 occur in 32 days? 4
   e. **Critical Thinking** Suppose you start with a 50 mCi sample of iodine-131. What is its half-life? How much iodine-131 is left after one half-life? after two half-lives? 8 da; 25 mCi; 12.5 mCi

You can use exponential decay to model other real-world situations.

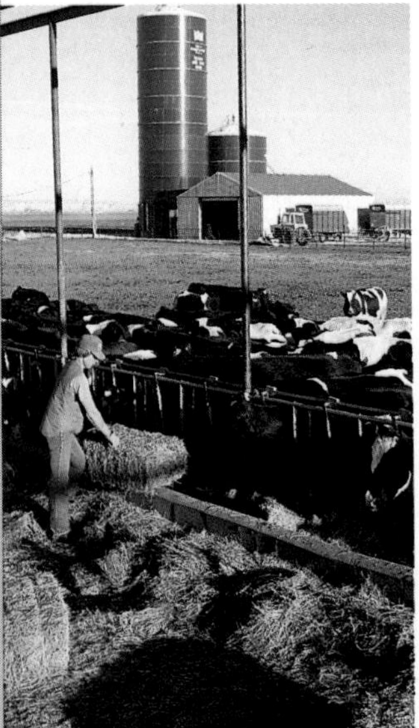

Example 2 **Relating to the Real World** 

Consumer Trends An exponential function models the amount of whole milk each person in the United States drinks in a year. Graph the function $y = 21.5 \cdot 0.955^x$, where y is the number of gallons of whole milk and x is the number of years since 1975.

Make a table of values.

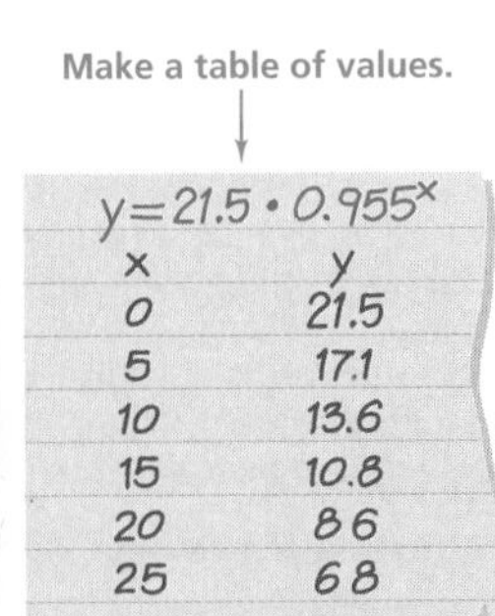

$y = 21.5 \cdot 0.955^x$

x	y
0	21.5
5	17.1
10	13.6
15	10.8
20	8.6
25	6.8

Graph the points. Draw a smooth curve through the points.

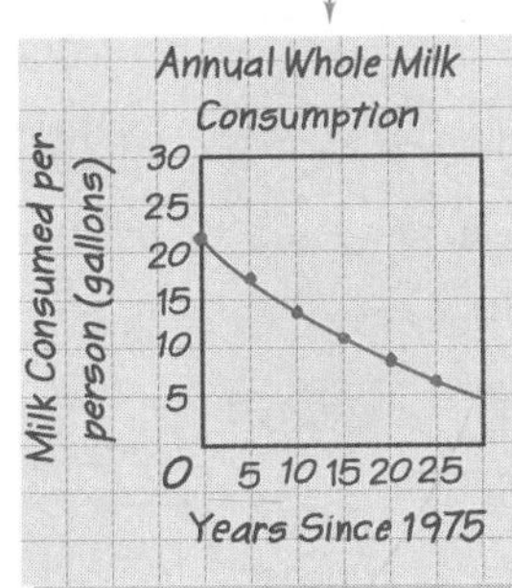

3. Which x-value corresponds to the year 1995? 20
4. In which year did whole milk consumption fall to about 10.8 gal/person? 1990
5. a. Use the function $y = 21.5 \cdot 0.955^x$ and your calculator to **predict** whole milk consumption for the year 2010. about 4.3 gal/person
   b. **Open-ended** Use the function to **predict** whole milk consumption 10 years from now. Check students' work. Sample: 2007—4.9 gal/person; 2008—4.7 gal/person; 2009—4.5 gal/person

measure the amount of carbon-14 in the sample. Since carbon-14 has a known half-life, the amount of carbon-14 left in the sample indicates the approximate age of the sample.

Example 2 **Relating to the Real World** 

Discuss with students why the annual consumption of whole milk may be decreasing. Point out that this data does not mean that all milk consumption is decreasing. People are becoming more health conscious and are switching from whole milk to milk with a lower percentage of fat.

Question 4 Point out that this question refers to whole milk consumption.

Example 3

Question 7 Students can check their answers by actually doing the calculation to see if the answer is reasonable.

Exercises ON YOUR OWN

ALTERNATIVE ASSESSMENT **Exercises 5–8** To assess students' understanding of exponential decay, ask them to graph these equations. Then have them use their graphs to find the half-life of each function.

Exercises 16–24 Have students look at each function to decide whether it models growth or decay. They can then use their graphs to check.

Example 3

Consumer Trends Use the equation $y = 21.5 \cdot 0.955^x$ to find the annual percent of decrease in whole milk consumption in the United States.

$0.955 = 95.5\%$ ← Change the decay factor to a percent.

$100\% - 95.5\% = 4.5\%$ ← Subtract.

Whole milk consumption is decreasing by 4.5% per year.

6. Try This What percent of decrease does each decay factor model?
   a. 0.75 25% b. 0.4 60% c. 0.135 86.5% d. 0.0074 99.26%
7. By what number would you multiply 15 to decrease it
   a. by 6%? 0.94 b. by 12%? 0.88 c. by 3.5%? 0.965 d. by 53.9%? 0.461
8. Suppose an initial population of 10,000 people decreases by 2.4% each year. Write an exponential function in the form $y = ■ \cdot ■^x$ to model the population y after x years have passed. $y = 10{,}000 \cdot 0.976^x$

Technology Options

For Exercises 13–15, students may find it helpful to check their answers by graphing with a graphing calculator or graphing software.

Prentice Hall Technology 

 **Calculator** • Graphing Calculator Handbook: Procedure 18

Software • Secondary Math Lab Toolkit • Computer Item Generator 8-3

CD-ROM • Multimedia Algebra Lab 8

 **Internet** • See the Prentice Hall site. (http://www.phschool.com)

Exercises ON YOUR OWN

Identify each function as *exponential growth* or *exponential decay*.

1. $y = 0.68 \cdot 2^x$ growth
2. $y = 2 \cdot 0.68^x$ decay
3. $y = 68 \cdot 2^x$ growth
4. $y = 68 \cdot 0.2^x$ decay

Identify the decay factor in each function.

5. $y = 5 \cdot 0.5^x$ 0.5
6. $f(x) = 10 \cdot 0.1^x$ 0.1
7. $g(x) = 100 \cdot \left(\frac{2}{3}\right)^x$ $\frac{2}{3}$
8. $y = 0.1 \cdot 0.9^x$ 0.9

Find the percent of decrease for each function.

9. $r = 70 \cdot 0.9^n$ 10%
10. $f(t) = 45 \cdot 0.998^t$ 0.2%
11. $r = 50 \cdot \left(\frac{1}{2}\right)^n$ 50%
12. $y = 1000 \cdot 0.75^x$ 25%

Use the graph to estimate the half-life of each radioactive substance.

13. 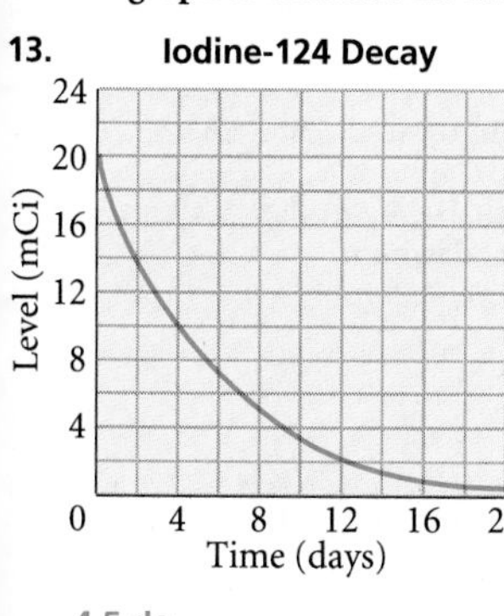

4.5 da

14. 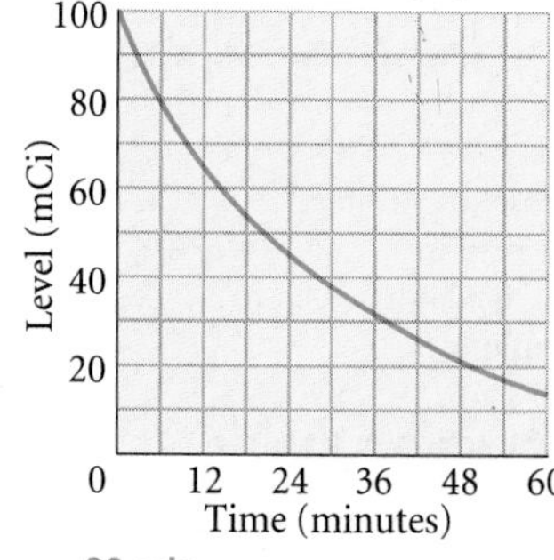

20 min

15. 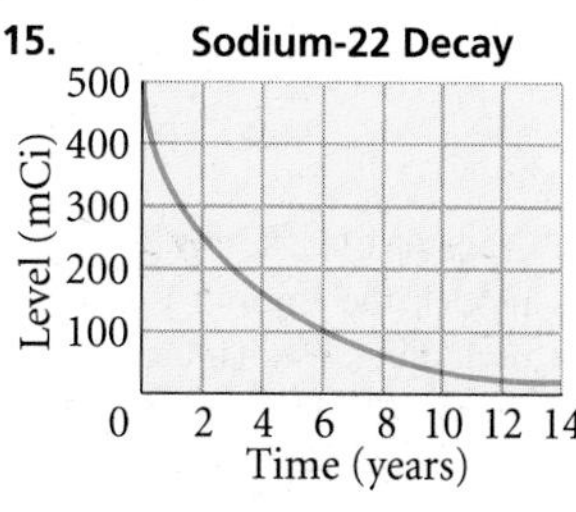

2.5 yr

WRITING Exercise 25 Have volunteers share their problems with the class.

Exercise 26–28 Be sure students use two variables to write the function and then substitute the number of years for the independent variable.

STATISTICS Exercise 29 Discuss factors that might cause the population of a city or town to decrease. Possible factors might include companies or factories moving out or downsizing, families having fewer children, etc.

OPEN-ENDED Exercise 30 Have students think about what the graph would look like for $b = 0$. The graph would be the graph of $y = 0$, the x-axis. Have students think about what the graph would look like for $b = 1$. The graph would be the graph of $y = a$, a horizontal line. Have students compare these graphs with the graphs they drew in the exercise.

ESTIMATION Exercise 37a Point out that the cursor in the diagram marks the point needed to make the estimate.

Exercises MIXED REVIEW

GETTING READY FOR LESSON 8-4 These exercises prepare students for working with negative exponents in the next lesson.

JOURNAL Encourage students to write equations and to show graphs.

pages 375–377 On Your Own

16.

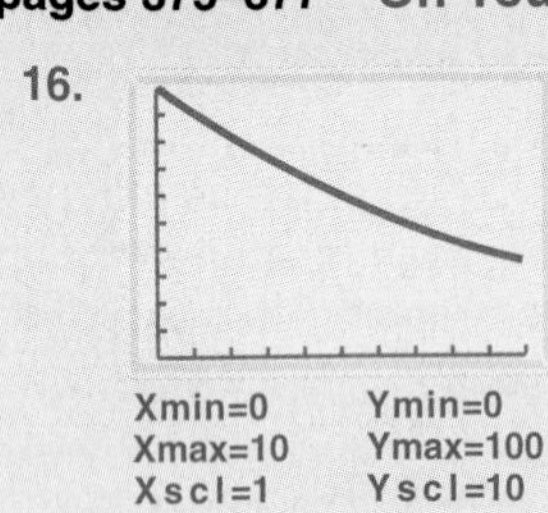

Xmin=0 Ymin=0
Xmax=10 Ymax=100
Xscl=1 Yscl=10

17.

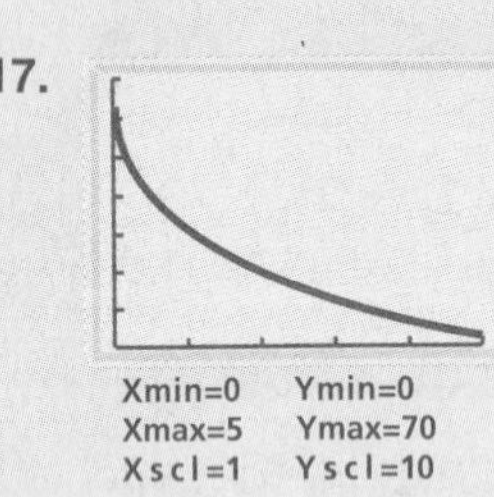

Xmin=0 Ymin=0
Xmax=5 Ymax=70
Xscl=1 Yscl=10

18.

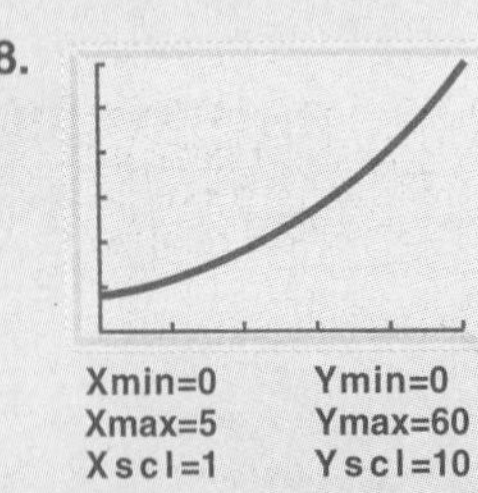

Xmin=0 Ymin=0
Xmax=5 Ymax=60
Xscl=1 Yscl=10

19–24, 30c. See back of book.

Checkpoint, 5–7, 8a. See back of book.

Choose **Use a graphing calculator or make a table of values to graph each function. Label each graph as *exponential growth* or *exponential decay*.** 16–24. See margin for tables and graphs.

16. $f(x) = 100 \cdot 0.9^x$ decay
17. $s = 64 \cdot \left(\frac{1}{2}\right)^n$ decay
18. $g = 8 \cdot 1.5^x$ growth
19. $y = 3.5 \cdot 0.01^x$ decay
20. $y = 2 \cdot 10^x$ growth
21. $g(x) = \left(\frac{1}{10}\right) \cdot 0.1^x$ decay
22. $f = 10 \cdot 0.1^x$ decay
23. $y = \frac{2}{5} \cdot \left(\frac{1}{2}\right)^x$ decay
24. $y = 0.5 \cdot 2^x$ growth

25. **Writing** Describe a situation that can be modeled by the equation $y = 100 \cdot 0.9^x$. Sample: A \$100 sewing machine depreciates at 10% per year.

Write an exponential function to model each situation. Find each amount after the specified time.

26. 3,000,000 initial population; 1.5% annual decrease; 10 years $y = 3{,}000{,}000 \cdot 0.985^x$; 2,579,191
27. \$900 purchase; 20% loss in value each year; 6 years $y = 900 \cdot 0.8^x$; \$235.93
28. \$10,000 investment; 12.5% loss each year; 7 years $y = 10{,}000 \cdot 0.875^x$; \$3926.96

29. **Statistics** In 1980, the population of Warren, Michigan, was about 161,000. Since then the population has decreased about 1% per year.
    a. Write an equation to model the population of Warren since 1980. $y = 161{,}000 \cdot 0.99^x$
    b. **Estimation** Estimate the population of Warren in 1990. 145,606
    c. Suppose the current trend continues. **Predict** the population of Warren in 2010. 119,092

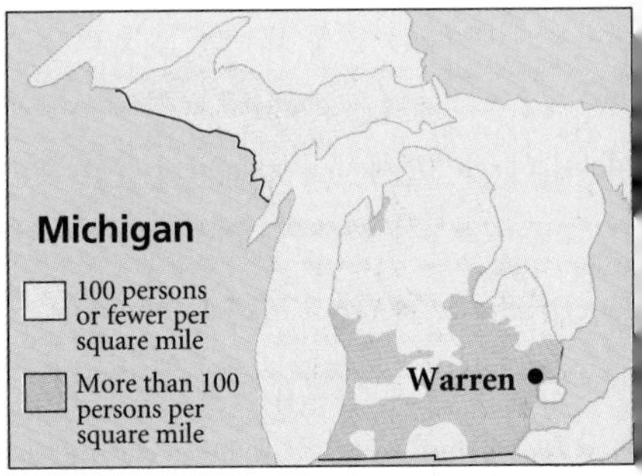

Michigan

100 persons or fewer per square mile

More than 100 persons per square mile

Warren

30. a. **Open-ended** Write two exponential decay functions, one with a base near 0 and one with a base near 1. Sample: $y = 0.1^x$; $y = 0.98^x$
    b. Find the range of each function using the domain {1, 2, 3, 4}.
    c. Graph each function. b. for sample in part (a): {0.1, 0.01, 0.001, 0.0001}; {0.98, 0.96, 0.94, 0.92} c. See margin.

Calculate the decay factor for each percent of decrease.

31. 3% 0.97
32. 70% 0.30
33. 2.6% 0.974
34. 4.75% 0.9525
35. 0.7% 0.993
36. 23.4% 0.766

37. **Graphing Calculator** The function $y = 15 \cdot 0.84^x$ models the amount y of a 15-mg dose of antibiotic remaining in the bloodstream after x hours.
    a. **Estimation** Study the graphing calculator screen to estimate the half-life of this antibiotic in the bloodstream. about 4 h
    b. Use your estimate to **predict** the fraction of the dose that will remain in the bloodstream after 8 hours. b. about $\frac{1}{4}$
    c. **Verify** your prediction by using the function to find the amount of antibiotic remaining after 8 hours. $\frac{15 \cdot 0.84^8}{15} = 0.247875891 \approx \frac{1}{4}$

Antibiotic Decay in the Bloodstream

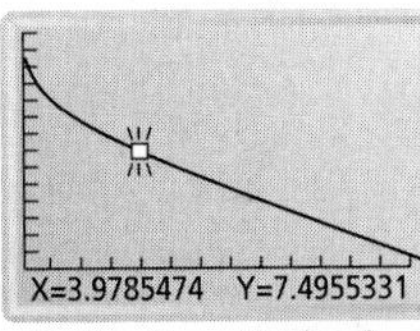

X=3.9785474 Y=7.4955331

Xmin=0 Ymin=0
Xmax=13 Ymax=15

Wrap Up

THE BIG IDEA Ask students to explain what is meant by *exponential decay.*

RETEACHING ACTIVITY Students write exponential functions modeling decay. The find amounts after specified time periods. (Reteaching worksheet 8-3)

Exercises CHECKPOINT

Students will assess their own progress on Lessons 8-1 to 8-3.

STANDARDIZED TEST TIP **Exercise 4** Remind students that when an exponential function is written to model growth the equation must show the initial amount times 100% + rate of growth. Answer B is the only answer which could be correct.

ERROR ALERT! **Exercise 7** Some students may glance at the fraction in this function and assume that the function represents decay. **Remediation:** Point out that the fraction $\frac{3}{2}$ is greater than 1.

How many half-lives occur in each length of time?

38. 2 days (1 half-life = 8 h) 6 half-lives

39. 300 years (1 half-life = 75 yr) 4 half-lives

Exercises MIXED REVIEW

Solve.

40. $3 = |x - 7|$ 4, 10

41. $2 - x > 5$ $x < -3$

42. $\frac{t}{3} - \frac{t}{6} = 15$ 90

43. $j^2 + 3 = 12$ −3, 3

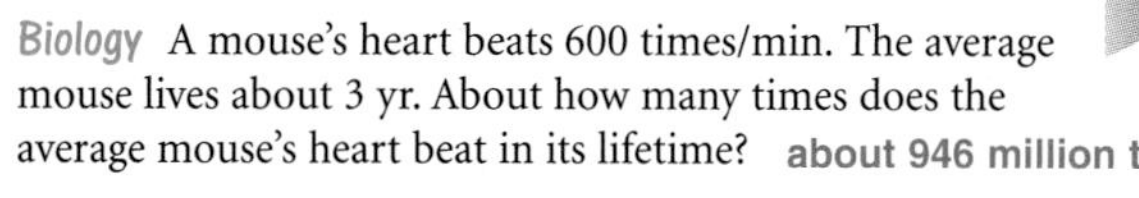

44. Biology A mouse's heart beats 600 times/min. The average mouse lives about 3 yr. About how many times does the average mouse's heart beat in its lifetime? about 946 million times

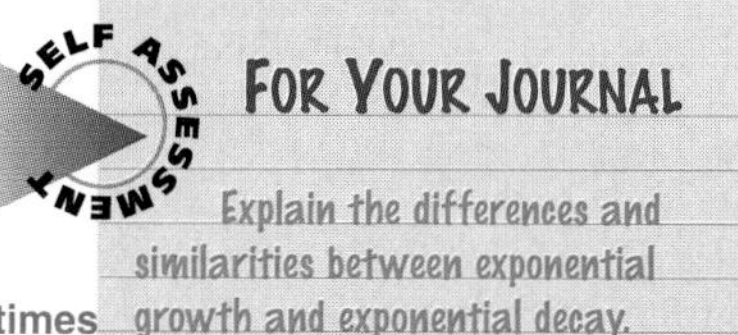

FOR YOUR JOURNAL

Explain the differences and similarities between exponential growth and exponential decay. Give an example of each.

Getting Ready for Lesson 8-4

Complete each expression.

45. $\frac{1}{2^2} = \frac{1}{\blacksquare}$ 4

46. $\frac{8}{27} = \frac{2^{\blacksquare}}{3^{\blacksquare}}$ 3; 3

47. $\frac{2^2}{4^2} = \frac{\blacksquare}{\blacksquare} = \blacksquare$ 4; 16; $\frac{1}{4}$

Exercises CHECKPOINT

Write an exponential function to model each situation. Find each amount after the specified time.

1. \$65,000 initial market value
3.2% annual increase
15 years $y = 65{,}000 \cdot 1.032^x$; \$104,257.86

2. 200,000 initial population
5% loss each year
20 years $y = 200{,}000 \cdot 0.95^x$; 71,697

3. \$300 initial value
doubles every 10 years
50 years $y = 300 \cdot 2^x$; x = ten-year period; \$9600

4. Standardized Test Prep From 1790 to 1860, the population of the United States grew slowly but steadily from 4 million. Which exponential function is the most likely model for the population p in the years after 1790? B

A. $p = 4 \cdot 2.6^x$ **B.** $p = 4 \cdot 1.03^x$ **C.** $p = 4 \cdot 0.98^x$ **D.** $p = 4 \cdot 3^x$ **E.** $p = 0.98 \cdot 4^x$

Graph each function. Label each graph as *exponential growth* or *exponential decay*. 5–7. See margin for graphs.

5. $y = \left(\frac{9}{10}\right) \cdot 2^x$ growth

6. $f(x) = 5 \cdot 0.5^x$ decay

7. $y = 2 \cdot \left(\frac{3}{2}\right)^x$ growth

8. Open-ended Suppose you have \$1000 to deposit into a savings account for your education. One account pays 6.7% compounded annually. Another account pays 5% compounded monthly.

a. Which account would you choose? Consider the length of time you would leave the money in the account. **Summarize** your reasoning. See margin.

b. Calculate how much will be in the account if you close it when you are 18. Check students' work.

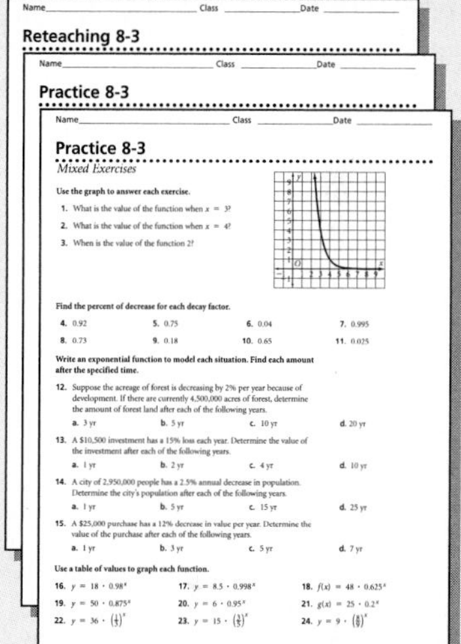

Name ______ Class ______ Date ______

Reteaching 8-3

Name ______ Class ______ Date ______

Practice 8-3

Name ______ Class ______ Date ______

Practice 8-3

Mixed Exercises

Use the graph to answer each exercise.

1. What is the value of the function when $x = 3$?

2. What is the value of the function when $x = 4$?

3. When is the value of the function 2?

Find the percent of decrease for each decay factor.

4. 0.92 **5.** 0.75 **6.** 0.04 **7.** 0.995

8. 0.73 **9.** 0.18 **10.** 0.65 **11.** 0.025

Write an exponential function to model each situation. Find each amount after the specified time.

12. Suppose the acreage of forest is decreasing by 2% per year because of development. If there are currently 4,500,000 acres of forest, determine the amount of forest land after each of the following years.

a. 3 yr **b.** 5 yr **c.** 10 yr **d.** 20 yr

13. A \$10,500 investment has a 15% loss each year. Determine the value of the investment after each of the following years.

a. 1 yr **b.** 2 yr **c.** 4 yr **d.** 10 yr

14. A city of 2,950,000 people has a 2.5% annual decrease in population. Determine the city's population after each of the following years.

a. 1 yr **b.** 5 yr **c.** 15 yr **d.** 25 yr

15. A \$25,000 purchase has a 12% decrease in value per year. Determine the value of the purchase after each of the following years.

a. 1 yr **b.** 3 yr **c.** 5 yr **d.** 7 yr

Use a table of values to graph each function.

16. $y = 18 \cdot 0.98^x$ **17.** $y = 8.5 \cdot 0.998^x$ **18.** $f(x) = 48 \cdot 0.625^x$

19. $y = 50 \cdot 0.875^x$ **20.** $y = 6 \cdot 0.95^x$ **21.** $g(x) = 25 \cdot 0.2^x$

22. $y = 36 \cdot \left(\frac{1}{3}\right)^x$ **23.** $y = 15 \cdot \left(\frac{3}{5}\right)^x$ **24.** $y = 9 \cdot \left(\frac{8}{9}\right)^x$

Lesson Quiz

Lesson Quiz is also available in Transparencies.

1. Identify the decay factor in the function $f(x) = 20 \cdot \left(\frac{1}{3}\right)^x$. $\frac{1}{3}$

2. Find the percent of decrease for the function $r = 3.0 \cdot 0.26^x$. 74%

3. A town with a population of 270,000 loses 2.3% of its population every year. What will its population be after 20 years? 169,533

4. A new car, worth \$16,000, loses 20% of its value each year. How much is the car worth after 10 years? \$1,718

Math ToolboX

This toolbox uses the graphing calculator to find the exponential function of best fit. Begin the lesson with a brief discussion of real-world exponential data sets. Check local newspapers, *USA Today*, and magazines such as *Consumer Reports* for current examples. Work through the example with students. Have students work in small groups to complete the exercises. Have each group write an exponential function. The table function of the graphing calculator will help students generate sets of points. Have groups exchange data and find the corresponding function.

ERROR ALERT! Some students may have difficulty setting the proper window. **Remediation:** Have students determine the lowest and highest values for x and y from the data. Those values must be between the minimum and maximum settings. Explain that scale is determined by the size of the numbers. Larger numbers require a larger scale.

WRITING Exercise 3b Refer students who are having difficulty to the discussion of r on page 242.

ADDITIONAL PROBLEM Use a graphing calculator to write the exponential function that fits the following data.

Tech High Students With Computers at Home

Year	1984	1986	1988	1990	1992	1994	1996
	17	27	48	82	140	237	403

$y = 5.693712584 \times 1.305109236^x$

Materials and Manipulatives

- Graphing calculator

Resources

Transparencies
11, 29

page 378 Math Toolbox

1. $y = 13.66 \cdot 1.11^x$

U.S. Movie Earnings

Dollars (billions); Year (after 1980)

2. $y = 34.23 \cdot 0.88^x$

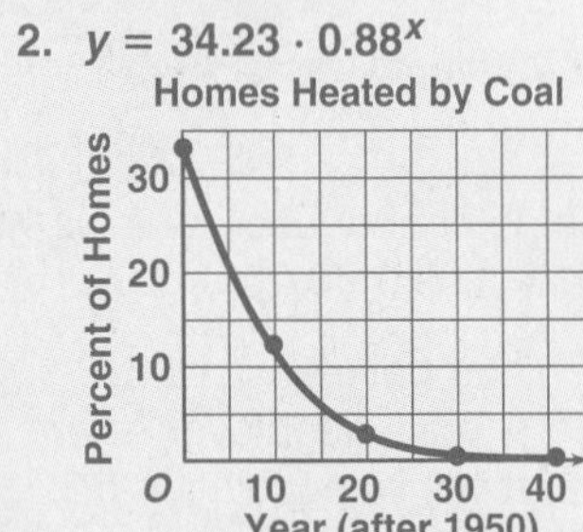

3a–b. See back of book.

Math ToolboX Technology

Fitting Exponential Curves to Data

After Lesson 8-3

In Chapter 5, you learned how to find a line of best fit for a set of data. Some data sets are better modeled by exponential functions. A graphing calculator makes it easy to graph exponential functions for a set of data.

U.S. Sales of Compact Discs

Year	Millions of CDs
1987 ↔ 7	102.1
8	149.7
9	207.2
10	286.5
11	333.3
12	407.5
13	495.4
1994 ↔ 14	662.1

Example

Use a graphing calculator to graph the data and find the best-fitting exponential function $y = a \cdot b^x$.

Step 1: Use the STAT feature to enter the data. Let 1980 correspond to $x = 0$.

Step 2: Use the STAT PLOT feature to graph the data in a scatter plot.

Step 3: Find the equation for the best-fitting exponential function. Press STAT ▶ ALPHA A ENTER to get the ExpReg equation.

```
ExpReg
y=a*b^x
a=19.82510651
b=1.287856937
r=.9908425236
```

Step 4: Graph the function. Press Y= CLEAR VARS 5 ▶ ▶ 7 to enter the ExpReg results. Press GRAPH to display the data and the function together.

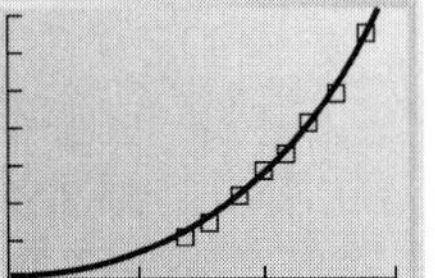

Xmin=0 Ymin=0
Xmax=15 Ymax=700
Xscl=5 Yscl=100

Use a graphing calculator to write the exponential function that fits each set of data. Sketch the graph of the function and show the data points. 1–2. See margin.

1. **U.S. Movie Earnings.** Let 1980 correspond to $x = 0$.

Year	1986	1987	1988	1989	1990	1991	1992
Billions of Dollars	23.8	27.8	31.2	35.0	38.1	41.1	43.8

2. **U.S. Homes Heated by Coal.** Let 1950 correspond to $x = 0$.

Year	1950	1960	1970	1980	1991
Percent of Homes	33.8	12.2	2.9	0.4	0.3

Population of New Mexico (millions)

1970	1.017
1980	1.303
1985	1.438
1990	1.515
1994	1.654

3. **a.** The table gives the population of New Mexico at various times. Use the best-fitting exponential function to **predict** the population of New Mexico in 2010.
   b. Writing Explain how the r-value affects your answer to part (a).
   a–b. See margin.

8-4 Teaching Notes

PROBLEM OF THE DAY

On the Warrior basketball team, the starting players have an average height of 6 ft. Kendall is 5 ft 9 in. Lorenzo is 5 ft 10 in. Malcolm is 6 ft 1 in. Juan is 5 ft 11 in. How tall is the remaining player? 6 ft 5 in.

Problem of the Day is also available in Transparencies.

CONNECTING TO PRIOR KNOWLEDGE Have students recall the solution to an expression with a zero exponent. If they are having difficulty, have them look back at the chart at the bottom of page 362.

WORK TOGETHER

If students have difficulty completing the table, have them look at Question 1 for a hint.

Question 2 Remind students to use WINDOW to change the range values.

Question 4c Students will need to extend the pattern in the table to answer this question.

THINK AND DISCUSS

Go over the section "Zero as an Exponent." Then, have students go back to the table at the bottom of page 362. Ask: *What is the number of rectangles for zero folds?* 1 *Look at*

Connections Population Growth . . . and more

8-4 Zero and Negative Exponents

What You'll Learn

Evaluating and simplifying expressions in which zero and negative numbers are used as exponents

...And Why

To analyze exponential functions over a broader domain

What You'll Need

graphing calculator

WORK TOGETHER

Work with a partner to copy and complete the table. Replace each box with a whole number or a fraction in lowest terms. See right.

$y = 2^x$	$y = 5^x$	$y = 10^x$
$2^2 = 4$	$5^2 = 25$	$10^2 = 100$
$2^1 = 2$	$5^1 = 5$	$10^1 = 10$
$2^0 = ■$ 1	$5^0 = ■$ 1	$10^0 = ■$ 1
$2^{-1} = ■$ $\frac{1}{2}$	$5^{-1} = ■$ $\frac{1}{5}$	$10^{-1} = ■$ $\frac{1}{10}$
$2^{-2} = ■$ $\frac{1}{4}$	$5^{-2} = ■$ $\frac{1}{25}$	$10^{-2} = ■$ $\frac{1}{100}$

1. You can describe what happens in the first column of the table as division by 2. What happens in the other columns of the table? division by 5; division by 10

Xmin=-3
Xmax=3
Xscl=1
Ymin=0
Ymax=100
Yscl=10

2. a. Graph the three functions on your calculator. Use the range values at the left. Sketch the graphs. See margin.
   b. At what point do the three graphs intersect? (0, 1)
3. a. What pattern do you notice in the row containing 0 as an exponent?
   b. Use your calculator to calculate other numbers to the zero power. What do you notice? a–b. When $x = 0$, $y = 1$.
4. Copy and complete each expression.
   a. $2^{-1} = \frac{1}{■} = \frac{1}{2^{■}}$ 2; 1 b. $2^{-2} = \frac{1}{■} = \frac{1}{2^{■}}$ 4; 2 c. $2^{-3} = \frac{1}{■} = \frac{1}{2^{■}}$ 8; 3
5. *Critical Thinking* Look for a pattern in your answers to Question 4. Does this pattern hold true for the other columns? Explain. Yes; for each column, $a^{-n} = \frac{1}{a^n}$.

THINK AND DISCUSS

Part 1

Using Zero and Negative Integers as Exponents

The pattern you saw in Question 3 is an important property of exponents.

Zero as an Exponent

For any nonzero number a, $a^0 = 1$.

Examples: $5^0 = 1$; $(-2)^0 = 1$; $\left(\frac{3}{8}\right)^0 = 1$; $1.02^0 = 1$

Notice that 0 is excluded as a base. The expression 0^0 is undefined, just as the expressions $\frac{2}{0}$ and $\frac{0}{0}$ are undefined.

Lesson Planning Options

Prerequisite Skills

- Simplifying exponents
- Graphing functions

Assignment Options

To provide flexible scheduling, this lesson can be subdivided into parts.

1 **Core** 1–18, 20–37, 52–65
Extension 19, 38–43

2 **Core** 44–49, 73–75
Extension 50, 51, 66–72

Use Mixed Review to maintain skills.

Resources

Student Edition

Skills Handbook, pp. 576, 580, 582
Extra Practice, p. 563
Glossary/Study Guide

Teaching Resources

Chapter Support File, Ch. 8
- Practice 8-4 (two worksheets)
- Reteaching 8-4
- Alternative Activity 8-4

Classroom Manager 8-4
Glossary, Spanish Resources
Two-Year Algebra Handbook 8-4

Transparencies
11, 29, 98, 102

the pattern of exponents. *How could the number of rectangles for zero folds be written as a power of 2?* 2^0 This should help give meaning to the definition of a zero exponent.

Example 1 **Relating to the Real World**

After students have worked through this example, have them go back to Example 3 on page 364. Have them evaluate the function for $x = 0$. $f(0) = 1$ Ask: *What does f(0) represent in this situation?* the percent of the original after zero enlargements *Does this make sense?* Yes. After zero enlargements the original is 100% (or 1 as a decimal) of the original. Again, this gives meaning to the definition of a zero exponent.

Example 2

ERROR ALERT! **Question 6e** Some students may evaluate this expression as $\frac{1}{(-3)^2}$ or $\frac{1}{9}$. **Remediation:** Point out that in the absence of parentheses the exponent operates on the 3 only. $-3^{-2} = -\frac{1}{3^2} = -\frac{1}{9}$

Example 3

Question 10 Have students use a calculator to verify their answers. They will need to choose a value for the variables. Then evaluate each expression to be sure the two expressions are equivalent.

Additional Examples

FOR EXAMPLE 1

Which of the following expressions has the greatest value when $x = 0$?

$y = 20.3^x$, $y = 20.2^x$, $y = 20 \cdot \left(\frac{1}{2}\right)^x$

They all have the same value, 20.

FOR EXAMPLE 2

Write each expression as a simple fraction.

a. 3^{-2} $\frac{1}{9}$

b. $(-2)^{-3}$ $-\frac{1}{8}$

Discussion: *What effect does the sign of the exponent have on the sign of the simplified fraction?*

FOR EXAMPLE 3

Rewrite each expression so that all exponents are positive.

a. $ab^{-2}z^3$ $\frac{az^3}{b^2}$

b. $\frac{x^{-2}}{y^3}$ $\frac{1}{x^2y^3}$

Example 1 **Relating to the Real World** 

Population Growth The function $f(t) = 1000 \cdot 2^t$ models an initial population of 1000 insects that doubles every time period t. Evaluate the function for $t = 0$. Then describe what $f(0)$ represents in the situation.

$$f(0) = 1000 \cdot 2^0 \quad \leftarrow \text{Substitute 0 for } t.$$
$$= 1000 \cdot 1 \quad \leftarrow 2^0 = 1$$
$$= 1000$$

The value of $f(0)$ represents the initial population of insects. This makes sense because when $t = 0$, no time has passed.

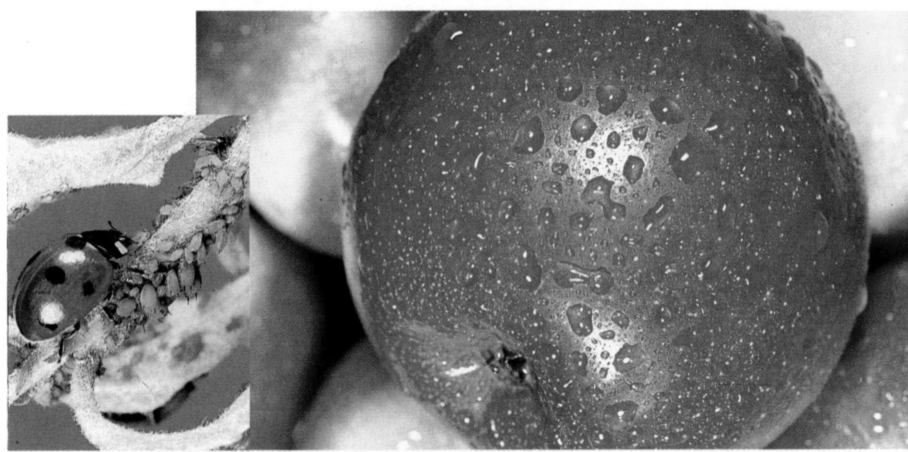

A large aphid population can destroy an apple orchard's produce. One way of controlling the aphids is to release ladybugs into the orchard, where they feed on the apple aphid colonies.

The pattern from Questions 4 and 5 illustrates another important property.

Negative Exponents

For any nonzero number a and any integer n, $a^{-n} = \frac{1}{a^n}$.

Examples: $6^{-4} = \frac{1}{6^4}$ and $7^{-1} = \frac{1}{7^1}$

QUICK REVIEW

Unless grouping symbols are used, exponents operate on only one factor.

$-4^2 = -(4 \cdot 4) = -16$

$(-4)^2 = -4 \cdot -4 = 16$

$2x^3 = 2 \cdot x \cdot x \cdot x$

$(2x)^3 = 2x \cdot 2x \cdot 2x$

Example 2

Write each expression as a simple fraction.

a. 4^{-3}

$4^{-3} = \frac{1}{4^3}$ ← definition of a negative exponent

$= \frac{1}{64}$

b. $(-3)^{-2}$

$(-3)^{-2} = \frac{1}{(-3)^2}$

$= \frac{1}{9}$

6. Try This Write each expression as a simple fraction.

a. 3^{-4} $\frac{1}{81}$ **b.** $(-7)^0$ 1 **c.** $(-4)^{-3}$ $-\frac{1}{64}$ **d.** 7^{-3} $\frac{1}{343}$ **e.** -3^{-2} $-\frac{1}{9}$

Example 4

Question 11 To review how the quadrants are numbered, students can look back to page 295.

CRITICAL THINKING Question 12 Have students use the graphing calculator to check the explanation. Remind them to use the same range values as shown on page 381.

KINESTHETIC LEARNING For students who are having trouble grasping the concept of zero or negative exponents, try the following activity.

Clear a large space in the room. Have a few students stand against a wall and take four paces forward. Ask: *What's one-half of four?* 2 Direct students to take two steps forward. Ask: *What's one-half of two?* 1 Direct students to take one step forward. Continue with one-quarter, one-eighth, and one-sixteenth. Ask students to describe the pattern using math language. Ask: *Are you ever going to start going backward?* no Help students make the connection between the actions they just performed and the values in the first column of the Work Together table. Be sure they understand that powers can be very large numbers or very small fractions but the values are always positive.

You can use what you know about rewriting the expression a^{-n} to see how the values of a^n and a^{-n} are related.

$$a^n \cdot a^{-n} = a^n \cdot \frac{1}{a^n}$$
$$= \frac{a^n}{1} \cdot \frac{1}{a^n} = 1$$

Therefore, a^n and a^{-n} are *reciprocals.*

7. **Verify** that a^n and a^{-n} are reciprocals by evaluating each product.
   a. $3^2 \cdot 3^{-2}$ $9 \cdot \frac{1}{9} = 1$ b. $2^4 \cdot 2^{-4}$ $16 \cdot \frac{1}{16} = 1$ c. $5^3 \cdot 5^{-3}$ $125 \cdot \frac{1}{125} = 1$
8. Write the reciprocal of each number in two ways: as a simple fraction and using a negative exponent.
   a. 10^1 $\frac{1}{10}, 10^{-1}$ b. 10^2 $\frac{1}{100}, 10^{-2}$ c. 1000 $\frac{1}{1000}, 10^{-3}$ d. $10{,}000$ $\frac{1}{10{,}000}, 10^{-4}$
9. Write each expression as a decimal.
   a. 10^{-3} 0.001 b. $3 \cdot 10^{-2}$ 0.03 c. $-5 \cdot 10^{-4}$ −0.0005 d. 10^{-6} 0.000 001

Example 3

Rewrite each expression so that all exponents are positive.

a. $4yx^{-3} = 4y\left(\frac{1}{x^3}\right)$ ← definition of negative exponent
$= \frac{4y}{x^3}$

b. $\frac{1}{w^{-4}} = 1 \div w^{-4}$ ← rewrite using a division symbol
$= 1 \div \frac{1}{w^4}$
$= 1 \cdot w^4$ ← multiply by the reciprocal
$= w^4$

10. **Try This** Complete each expression using only positive exponents.
    a. $\frac{1}{x^{-3}} = x^{\blacksquare}$ 3 b. $\frac{1}{v^{-2}} = v^{\blacksquare}$ 2 c. $w^{-3} = \frac{1}{w^{\blacksquare}}$ 3 d. $\frac{w^{-3}}{v^{-2}} = \frac{\blacksquare^{\blacksquare}}{\blacksquare^{\blacksquare}}$ $\frac{v^2}{w^3}$

Part 2

Relating the Properties to Exponential Functions

Zero and negative integer exponents allow you to understand the graph of an exponential function more completely.

Example 4

GRAPHING CALCULATOR HINT You can generate values for a function with the key on some graphing calculators.

Graphing Calculator Graph the functions $y = 2^x$ and $y = \left(\frac{1}{2}\right)^x$ on the same set of axes. Show the functions over the domain $\{-3 \le x \le 3\}$.

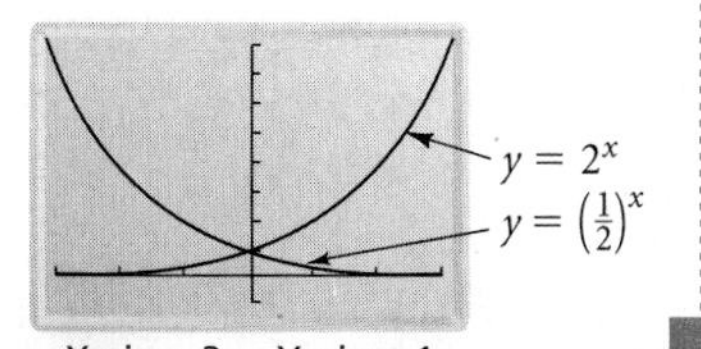

Xmin=-3 Ymin=-1
Xmax=3 Ymax=8

Technology Options

For Exercises 38–47, students can use a graphing calculator to help them evaluate the expressions and graph the functions.

Prentice Hall Technology

Software • Secondary Math Lab Toolkit • Integrated Math Lab 15 • Computer Item Generator 8-4

Internet • See the Prentice Hall site. (http://www.phschool.com)

Exercises ON YOUR OWN

ERROR ALERT! **Exercise 15** Students may try to multiply 5 · 10 before calculating 10^{-2}. **Remediation:** Remind students that in the order of operations, finding exponents comes before multiplication.

Exercises 20–37 Remind students that when an expression does not contain parentheses, the exponent relates only to the variable or number that is adjacent and to the left of the exponent.

ALTERNATIVE ASSESSMENT **Exercises 33–37** Write the inverse of each expression on the board. For example, the expression $\frac{7s^{-5}}{5t^{-3}}$ should be rewritten as $\frac{5t^{-3}}{7s^{-5}}$. Then randomly choose students to rewrite the expressions to contain only positive exponents. This will help you assess students' understanding of negative exponents.

ERROR ALERT! **Exercise 38** Some students may evaluate c^b as -4^2, rather than $(-4)^2$. **Remediation:** Point out that since -4 represents the value of c, the exponent operates on -4.

Exercises 48 and 49 Point out that the graphs are displayed using the standard settings for the variable ranges -10 to 10. Students may have difficulty determining whether function II or III is displayed in Exercise 48. Suggest that they find the values for each function when $x = 2$ and then check the graphs.

Exercises 52–57 Have students recall the rules for multiplying integers. Encourage them to generalize rules for multiplying negative integers with odd exponents and with even exponents.

page 379 Work Together

2a. $y = 2^x$

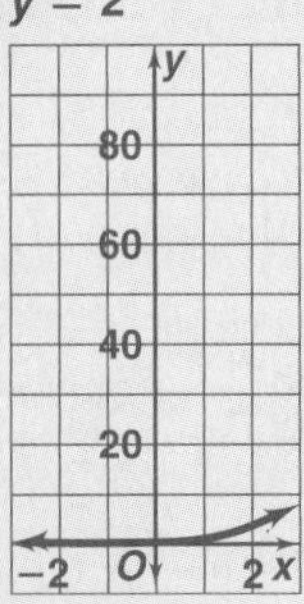

$y = 5^x$

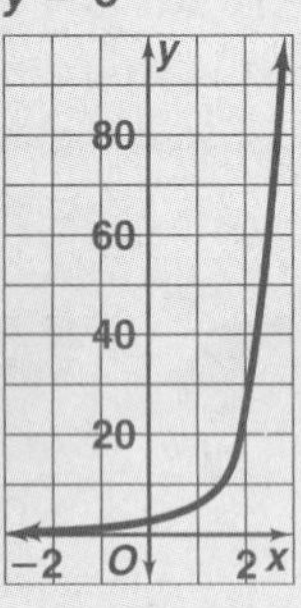

$y = 10^x$

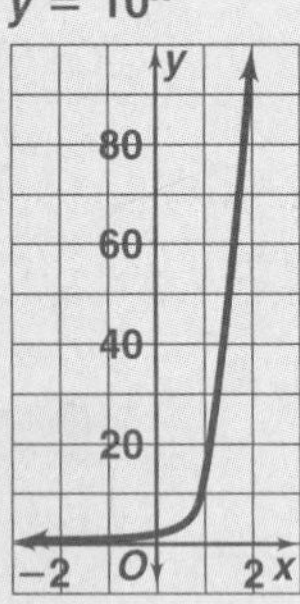

11a. Yes; as x gets smaller (moving left on the x-axis), 2^x gets closer and closer to, but never reaches, 0.

11. a. Is the value of 2^x always positive? Explain.

b. In what quadrants do the graphs of $y = 2^x$ and $y = \left(\frac{1}{2}\right)^x$ appear? Explain. I and II; the values of 2^x and $\left(\frac{1}{2}\right)^x$ are always positive

12. *Critical Thinking* The graph of $y = 2^{-x}$ is identical to one of the two graphs shown in Example 4. Use the definition of negative exponents to help decide which one. Explain. $y = \left(\frac{1}{2}\right)^x$; both functions show a decay.

Exercises ON YOUR OWN

Write each expression as an integer or simple fraction.

1. -2.57^0 -1
2. 4^{-2} $\frac{1}{16}$
3. $(-5)^{-1}$ $-\frac{1}{5}$
4. $\left(\frac{2}{3}\right)^{-1}$ $\frac{3}{2}$
5. $\frac{1}{2^{-3}}$ 8
6. 5^{-3} $\frac{1}{125}$
7. $\left(\frac{1}{3}\right)^{-2}$ 9
8. -3^{-4} $-\frac{1}{81}$
9. 2^{-6} $\frac{1}{64}$
10. $(-12)^{-1}$ $-\frac{1}{12}$
11. $45 \cdot (0.5)^0$ 45
12. $\left(\frac{1}{3}\right)^{-2}$ 9
13. $54 \cdot 3^{-2}$ 6
14. $\left(-\frac{1}{4}\right)^{-3}$ -64
15. $5 \cdot 10^{-2}$ $\frac{1}{20}$
16. $\frac{5^{-2}}{7^{-3}}$ $\frac{343}{25}$
17. $\frac{4^{-1}}{9^0}$ $\frac{1}{4}$
18. $\frac{(-3)^{-1}}{-2^{-1}}$

19. a. *Patterns* Complete the pattern.

$\frac{1}{5^2} = \blacksquare$ 5^{-2} $\quad \frac{1}{5^1} = \blacksquare$ 5^{-1} $\quad \frac{1}{5^0} = \blacksquare$ 5^0 or 1 $\quad \frac{1}{5^{-1}} = \blacksquare$ 5^1 $\quad \frac{1}{5^{-2}} = \blacksquare$ 5^2

b. Write $\frac{1}{5^{-4}}$ using a positive exponent. 5^4

c. *Generalize* Rewrite $\frac{1}{a^{-n}}$ so that the power of a is in the numerator. a^n

Write each expression so that it contains only positive exponents.

20. $\frac{1}{c^{-1}}$ c
21. $\frac{1}{x^{-7}}$ x^7
22. $3ab^0$ $3a$
23. $(5x)^{-4}$ $\frac{1}{625x^4}$
24. $\frac{5^{-2}}{p}$ $\frac{1}{25p}$
25. $a^{-4}b^0$ $\frac{1}{a^4}$
26. $\frac{3x^{-2}}{y}$ $\frac{3}{x^2y}$
27. $12xy^{-3}$ $\frac{12x}{y^3}$
28. $\frac{7ab^{-2}}{3w}$ $\frac{7a}{3b^2w}$
29. $5ac^{-5}$ $\frac{5a}{c^5}$
30. $x^{-5}y^{-7}$ $\frac{1}{x^5y^7}$
31. $\frac{8a^{-5}}{c^{-3}d^3}$ $\frac{8c^3}{a^5d}$
32. $x^{-5}y^7$ $\frac{y^7}{x^5}$
33. $\frac{7s^{-5}}{5t^{-3}}$ $\frac{7t^3}{5s^5}$
34. $\frac{6a^{-1}c^{-3}}{b^0}$ $\frac{6}{ac^3}$
35. $\frac{1}{a^{-3}b^3}$ $\frac{a^3}{b^3}$
36. $\frac{c^4}{x^2y^{-1}}$ $\frac{c^4y}{x^2}$
37. $\frac{mn^{-4}}{p^0q^{-2}}$ $\frac{mq}{n^4}$

Evaluate each expression for $a = 3$, $b = 2$, and $c = -4$.

38. c^b 16
39. $a^{-b}b$ $\frac{2}{9}$
40. b^{-a} $\frac{1}{8}$
41. b^c $\frac{1}{16}$
42. $c^{-a}b^{ab}$ -1
43. c^a -64

Graph each function over the domain $\{-3 \le x \le 3\}$. 44–47. See margin.

44. $y = 2^x$
45. $f(x) = (2)^{-x}$
46. $g(x) = 0.5 \cdot 3^x$
47. $y = 1.5 \cdot (1.5)^{-x}$

Match each graphing calculator screen with the functions it displays.

48. 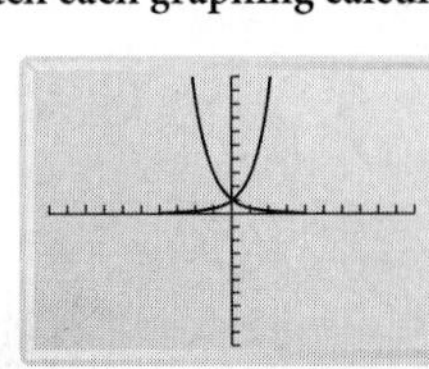 II

49. 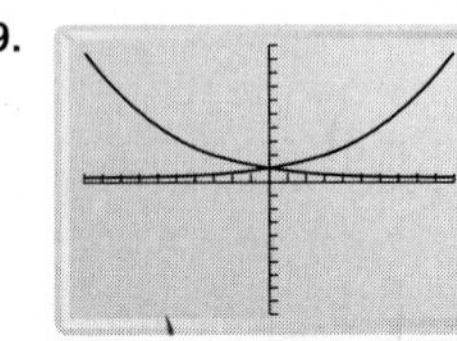 I

I. $y = \left(\frac{5}{4}\right)^x$; $y = \left(\frac{4}{5}\right)^x$

II. $y = 3^x$; $y = \left(\frac{1}{3}\right)^x$

III. $y = 15^x$; $y = \left(\frac{1}{15}\right)^x$

Exercise 58 Suggest students rewrite 0.5 as a fraction.

STANDARDIZED TEST TIP **Exercise 66** These two expressions are inverse functions, which suggests that their graphs are mirror images of one another. One function will have a greater value for $x < 0$ and the other will have a greater value for $x > 0$. Therefore, the only possible answer choice is *D*.

OPEN-ENDED **Exercise 68** Point out that the fraction may be less than or greater than 1, positive, or negative.

RESEARCH **Exercise 69** Students may find information on the size of certain small metric units by finding definitions in dictionaries or in tables in scientific reference books.

Exercise 70a Students should use the function instead of the graph to find the best answer.

Exercise 70c Students can also use the [TABLE] function to find the answer. Then they can discuss which was easier to use, TRACE or [TABLE].

KINESTHETIC LEARNING **Exercise 71** Have students act out this exercise. Instead of telling news, direct a student to begin by standing in a corner of the room. Have the student call the names of two neighboring students. They, in turn, stand and each call the names of two nearby students. Continue until all students are standing. If you have block scheduling or extended class periods, have students repeat the exercise with 30-second intervals between students' standing and calling on others. Record on the board how many students are standing at each interval.

CRITICAL THINKING **Exercise 72** Students can begin by writing the reciprocal of $3x^{-2}$ and the reciprocal of $3x^2$.

50. Writing Explain why the value of -3^0 is negative and the value of $(-3)^0$ is positive. -3^0 is the opposite of 3^0 and $3^0 = 1$. $(-3)^0 = 1$, by definition.

51. Copy and complete the table.

a	4	0.2	■ $\frac{1}{3}$	■ 6	$\frac{7}{8}$	■ 2
a^{-1}	■ $\frac{1}{4}$	■ 5	3	$\frac{1}{6}$	■ $\frac{8}{7}$	0.5

Determine whether the value of each expression is *positive* or *negative*.

52. -2^2 negative
53. $(-2)^2$ positive
54. 2^{-2} positive
55. $(-2)^3$ negative
56. $(-2)^{-3}$ negative
57. 4^{-1} positive

Copy and complete each equation.

58. $2^{■} = 0.5$ -1
59. $3xy^{■} = \frac{3x}{y^3}$ -3
60. $\frac{x^{■}}{2y^{■}} = \frac{1}{2x^{-3}y^4}$ 3; 4
61. $\frac{a^{■}}{3b^{■}} = \frac{b^3}{3}$ 0; -3
62. $(-5)^{■} = -\frac{1}{125}$ -3
63. $\frac{4n^{■}}{m^{■}} = \frac{4m^2}{n^3}$ -3; -2
64. $\frac{5x^{■}}{y^{■}} = \frac{5}{xy^2}$ -1; 2
65. $\frac{x^{■}}{y^{■}} = x^{-2}y^3$ -2; -3

66. Standardized Test Prep Compare the quantities in Column A and Column B. D

Column A	Column B
3^x	3^{-x}

A. The quantity in Column A is greater.
B. The quantity in Column B is greater.
C. The quantities are equal.
D. The relationship cannot be determined from the information given.

67. Which expressions equal $\frac{1}{4}$? A, B, D

A. 4^{-1} B. 2^{-2} C. -4^1
D. $\frac{1}{2^2}$ E. 1^4 F. -2^{-2}

68. Open-ended Choose a fraction to use as a value for the variable a. Find the values of a^{-1}, a^2, and a^{-2}. Sample: $a = \frac{2}{3}; \frac{3}{2}; \frac{4}{9}; \frac{9}{4}$

69. Research Certain small units of length have special names. For each unit, give its length in the form $10^{■}$ meters.

a. fermi 10^{-15} m b. micron 10^{-6} m c. angstrom 10^{-10} m

70. Botany A botanist studying plant life on a remote island in the Pacific Ocean discovers that the number of plants of a particular species is increasing at a high rate. Each month for the next eight months, she counts the number of these plants in an acre plot of land. She then fits an exponential function to the data.

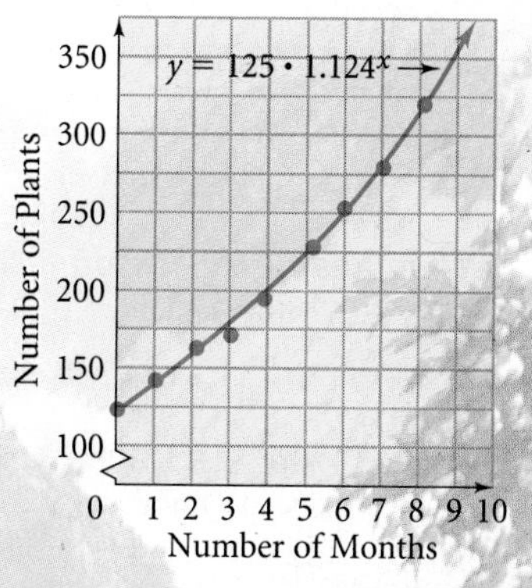

a. By what percent is the number of plants increasing each month? 12.4%

b. If $x = 0$ represents the time of her initial count of the plants, what does $x = -3$ represent? 3 mo before the initial count

c. Graphing Calculator Graph the function on a graphing calculator. Use the TRACE feature to estimate when the plant was first introduced to the island. (*Hint:* Find the value of x when $y = 1$.) about 41 mo before the initial count

pages 382–384 On Your Own

44.

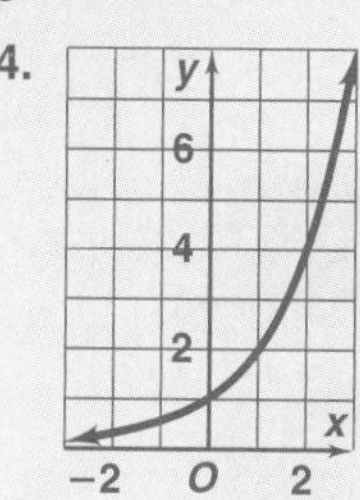

45.

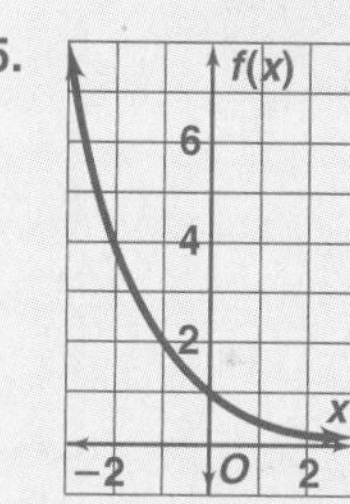

46.

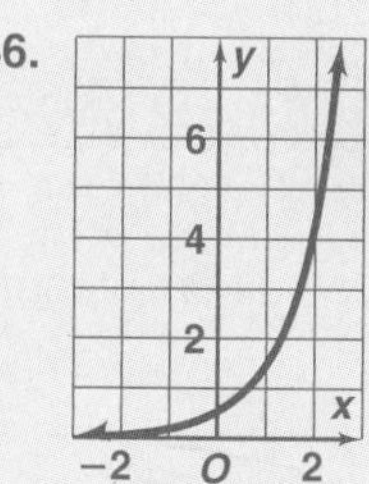

47, 73, 74. See back of book.

Mixed Review, 77. See back of book.

Exercise 75c Be sure students understand that *r* represents the shaded fraction of the whole rectangle.

Exercise 75d Have students try to generate all possible forms for expressing this answer: for example, 2^{-10}, $\frac{1}{1024}$, etc.

Exercises MIXED REVIEW

Exercise 79 Discuss possible ways this average amount might be determined. For example: surveys of a sample of people, daily or weekly sales figures, etc.

GETTING READY FOR LESSON 8-5 These exercises help students make the connection between powers of 10 and standard numerals to prepare students for writing numbers in scientific notation.

Wrap Up

THE BIG IDEA Ask students to explain the meaning of a zero exponent and the meaning of a negative exponent.

RETEACHING ACTIVITY Students simplify expressions with exponents. (Reteaching worksheet 8-4)

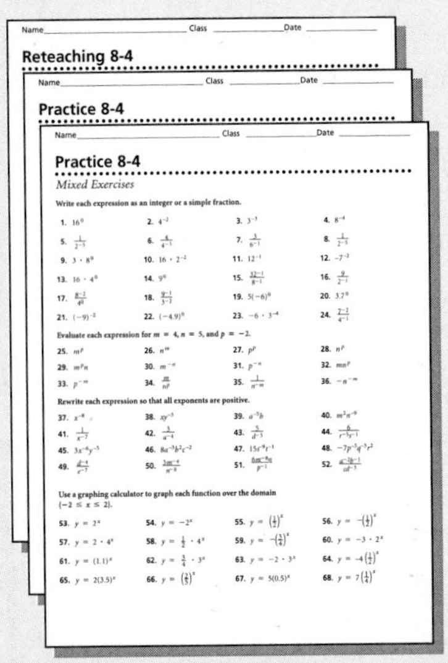
Name______ Class ______ Date ______

Reteaching 8-4

Name______ Class ______ Date ______

Practice 8-4

Name______ Class ______ Date ______

Practice 8-4

Mixed Exercises

Write each expression as an integer or a simple fraction.

1. 16^0 2. 4^{-2} 3. 3^{-3} 4. 8^{-4}
5. $\frac{1}{2^{-5}}$ 6. $\frac{4}{4^{-3}}$ 7. $\frac{3}{6^{-1}}$ 8. $\frac{1}{2^{-3}}$
9. $3 \cdot 8^0$ 10. $16 \cdot 2^{-2}$ 11. 12^{-1} 12. -7^{-2}
13. $16 \cdot 4^0$ 14. 9^0 15. $\frac{32^{-1}}{8^{-1}}$ 16. $\frac{9}{2^{-1}}$
17. $\frac{8^{-2}}{4^0}$ 18. $\frac{9^{-1}}{3^{-2}}$ 19. $5(-6)^0$ 20. 3.7^0
21. $(-9)^{-2}$ 22. $(-4.9)^0$ 23. $-6 \cdot 3^{-4}$ 24. $\frac{7^{-2}}{4^{-1}}$

Evaluate each expression for $m = 4$, $n = 5$, and $p = -2$.

25. m^p 26. n^m 27. p^p 28. n^p
29. m^pn 30. m^{-n} 31. p^{-n} 32. mn^p
33. p^{-m} 34. $\frac{m}{n^p}$ 35. $\frac{1}{n^{-m}}$ 36. $-n^{-m}$

Rewrite each expression so that all exponents are positive.

37. x^{-8} 38. xy^{-3} 39. $a^{-5}b$ 40. m^2n^{-9}
41. $\frac{1}{x^{-7}}$ 42. $\frac{3}{a^{-4}}$ 43. $\frac{5}{d^{-3}}$ 44. $\frac{6}{r^{-5}t^{-1}}$
45. $3x^{-6}y^{-5}$ 46. $8a^{-3}b^2c^{-2}$ 47. $15s^{-9}t^{-1}$ 48. $-7p^{-5}q^{-3}r^2$
49. $\frac{d^{-4}}{c^{-7}}$ 50. $\frac{3m^{-4}}{n^{-8}}$ 51. $\frac{6m^{-8}n}{p^{-1}}$ 52. $\frac{a^{-2}b^{-1}}{cd^{-3}}$

Use a graphing calculator to graph each function over the domain $\{-2 \le x \le 2\}$.

53. $y = 2^x$ 54. $y = -2^x$ 55. $y = \left(\frac{1}{2}\right)^x$ 56. $y = -\left(\frac{1}{2}\right)^x$
57. $y = 2 \cdot 4^x$ 58. $y = \frac{1}{2} \cdot 4^x$ 59. $y = -\left(\frac{3}{4}\right)^x$ 60. $y = -3 \cdot 2^x$
61. $y = (1.1)^x$ 62. $y = \frac{3}{4} \cdot 3^x$ 63. $y = -2 \cdot 3^x$ 64. $y = -4\left(\frac{1}{2}\right)^x$
65. $y = 2(3.5)^x$ 66. $y = \left(\frac{2}{5}\right)^x$ 67. $y = 5(0.5)^x$ 68. $y = 7\left(\frac{1}{4}\right)^x$

Lesson Quiz

Lesson Quiz is also available in Transparencies.

Write each expression as an integer or simple fraction.

1. $3 \cdot 4^{-2}$ $\frac{3}{16}$

2. $-2 \cdot 10^0$ -2

Write each expression so that it contains only positive exponents.

3. $-2a^0b^{-2}$ $\frac{-2}{b^2}$

4. $\frac{i^{-3}j^2}{k^{-3}j^{-2}}$ $\frac{j^{2k}j^2}{i^3}$

71. Communications Suppose you are the only person in your class who knows a story. After a minute you tell a classmate. Each minute after that, every student who knows the news tells another student (sometimes the person being told will already have heard it). In a class of 30 students, the formula $N = \frac{30}{1 + 29 \cdot 2^{-t}}$ predicts the number of people N who will have heard the news after t minutes. Find how many students will have heard your news after 2 minutes, 5 minutes, and 10 minutes. **about 2 students; about 16 students; about 29 students**

72. Critical Thinking Are $3x^{-2}$ and $3x^2$ reciprocals? Explain. **no; $3x^{-2} \cdot 3x^2 \neq 1$**

Graph each pair of functions on the same set of axes over the domain $\{-3 \le x \le 3\}$. **73–74. See margin.**

73. $y = 10 \cdot 2^x$; $y = 20 \cdot 2^x$

74. $y = 1.2^x$; $y = 1.8^x$

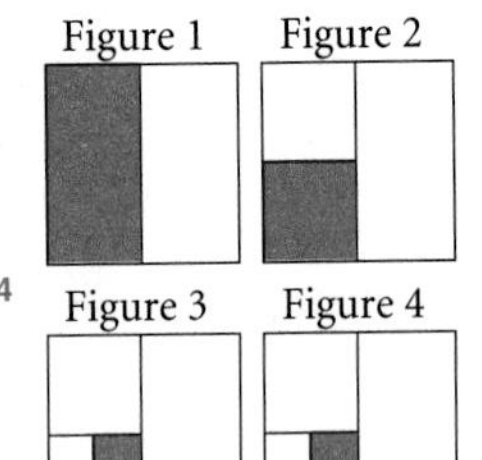

75. a. Geometry What fraction of each figure is shaded? **$\frac{1}{2}$; $\frac{1}{4}$; $\frac{1}{8}$; $\frac{1}{16}$**
    b. Rewrite each fraction from part (a) in the form $2^{\blacksquare}$. **2^{-1}; 2^{-2}; 2^{-3}; 2^{-4}**
    c. Patterns Look for a pattern in your answers to part (b). Write a function that relates the figure number n to the shaded rectangle r. **$r = 2^{-n}$**
    d. What portion of the square would be shaded in Figure 10? **2^{-10} or $\frac{1}{2^{10}}$**

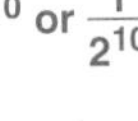

Exercises MIXED REVIEW

Solve each system.

76. $2x + 2y = 3$; $4x + 3y = 5$ **$\left(\frac{1}{2}, 1\right)$**

77. $4x + y < 8$; $2x \ge 3$ **See margin.**

78. $3y = \frac{1}{3}|x| - 2$; $|x| + 9y = 6$ **(−6, 0), (6, 0)**

79. Each person in Colorado spends an average of \$31.17/yr on books.
    a. Write a function to represent the amount b a group of p people spend on books in a year. **$b = 31.17p$**
    b. On average, how much does a family of four living in Denver spend on books in a year? **\$124.68**
    c. Writing Does this mean every person in Colorado spends exactly \$31.17 on books? Explain. **No; some people may spend more than the average amount and some may spend less.**

Getting Ready for Lesson 8-5

Simplify each expression.

80. $3.4 \cdot 10^1$ **34**

81. $7 \cdot 10^{-2}$ **0.07**

82. $8.2 \cdot 10^5$ **820,000**

83. $3 \cdot 10^{-3}$ **0.003**

84. $6 \cdot 10^4$ **60,000**

85. Write $3 \times 10^3 + 6 \times 10^1 + 7 \times 10^0 + 8 \times 10^{-1} + 5 \times 10^{-2}$ as a standard decimal number. **3067.85**

8-5 Teaching Notes

PROBLEM OF THE DAY

A waitress at a Vietnamese restaurant must pay 10% of her tips to the hostess, 25% to the busperson, and 5% to the cashier. If the waitress earns $65.50 in tips on Friday night, how much money in tips does she get to take home? $39.30

Problem of the Day is also available in Transparencies.

CONNECTING TO PRIOR KNOWLEDGE On the board, write 1,000,000; 10,000,000; and 100,000,000. Ask students if they would get tired of writing so many digits. Then ask: *What math language have you learned in this chapter that you could apply to write these numbers in a shorter way?*

THINK AND DISCUSS

As an extra project, have students research science and news magazines for facts that contain very large or small numbers. After the lesson, have students record these facts, along with the numbers written in standard and scientific notation. Have students report their findings to the class and record the facts and numbers on a class poster.

Explain that a number is in scientific notation if it is in the form $a \times 10^n$, where $1 \leq a < 10$. Be sure students understand that this means that the value of a must be greater than or equal to 1 and less than 10. Ask: *Where would you place the decimal point in each of the following examples to get a number that could represent a?*

Connections Physical Science . . . and more

8-5 Scientific Notation

What You'll Learn
- Writing numbers in scientific notation
- Using scientific notation

...And Why

To calculate with very large or very small numbers

What You'll Need
- calculator

THINK AND DISCUSS

Part 1

Writing Numbers in Scientific Notation

Jupiter has an average radius of 69,075 km. What is Jupiter's volume?

To answer this question, you probably want to find the formula for the volume of a sphere and use a calculator.

$V = \frac{4}{3}\pi r^3$ ← formula for the volume of a sphere

$= \frac{4}{3}\pi(69{,}075)^3$ ← Substitute 69,075 for r.

When you use a calculator to find the answer, the display looks something like 1.380547297E15.

The calculator displays the answer in this form, called *scientific notation*, because the answer contains more digits than the calculator can display. Scientific notation is a kind of shorthand for very large or very small numbers.

Scientific Notation

A number is in **scientific notation** if it is written in the form

$$a \times 10^n,$$

where n is an integer and $1 \leq |a| < 10$.

Examples: 3.4×10^6, 5.43×10^{13}, 9×10^{-10}

You can change a number from scientific notation into standard notation.

1.380547297E15

$\approx 1.38 \times 10^{15}$ ← scientific notation

$= 1.38 \times 1{,}000{,}000{,}000{,}000{,}000$ ← 10^{15} has 15 zeros.

$= 1{,}380{,}000{,}000{,}000{,}000$ ← standard notation

Jupiter has a volume slightly greater than 1 quintillion km^3.

1. a. Try This Evaluate the expression $\frac{4}{3}\pi(6000)^3$ on your calculator. What does your calculator display? 9.047786842E11
   b. Write the number in standard notation. 904,778,684,200

Lesson Planning Options

Prerequisite Skills
- Simplifying expressions
- Using positive and negative exponents

Assignment Options

To provide flexible scheduling, this lesson can be subdivided into parts.

1 **Core** 1–4, 10–30
Extension 31, 47–53

2 **Core** 5–9, 32–36, 37–42
Extension 43–46

Use Mixed Review to maintain skills.

Resources

Student Edition

Skills Handbook, pp. 569, 580, 582
Extra Practice, p. 563
Glossary/Study Guide

Teaching Resources

Chapter Support File, Ch. 8
- Practice 8-5 (two worksheets)
- Reteaching 8-5

Classroom Manager 8-5
Glossary, Spanish Resources
Two-Year Algebra Handbook 8-5

Transparencies
99

0.006745 between 6 and 7
127,678,754 between 1 and 2
0.00567503 between 5 and 6
1,000,000 between 1 and 0

EXTENSION Question 1 *Give an example of a problem you might solve by evaluating this expression.* Answers may vary. Sample: What is the volume of a sphere with a radius of 6000 ft?

Question 3 Be sure students realize that this question refers to the table below the picture. Stress to students that in ordering numbers in scientific notation, you compare the exponents first, then numbers with the same exponent.

Question 4 If students enter the standard number in their calculator, the display will show the number in scientific notation. Students can then use the calculator to check their work.

Example 1 **Relating to the Real World** 

ERROR ALERT! **Question 5** Some students may forget which way to move the decimal point. **Remediation:** Avoid the analogy between sign and decimal movement. Explain that if the exponent is positive, the standard form of the number is greater than one (large). If the exponent is negative, the standard form of the number is less than one (small).

AUDITORY LEARNING Question 8 Ask questions such as the following to give students practice using the powers of ten.

- *What is 10^2 times 10?* 10^3
- *What is 10^3 divided by 10?* 10^2
- *What is 10^2 described without powers of ten?* 100
- *What is 100 times 100 in powers of ten?* 10^4
- *What is 10^4 times 1000?* 10^7

Additional Examples

FOR EXAMPLE 1

Write each number in standard notation.

a. An elephant has a mass of 8.8×10^4 kg. 88,000 kg

b. An ant has a mass of 7.3×10^{-5} kg. 0.000 073 kg

FOR EXAMPLE 2

Simplify $8 \times (3 \times 10^3)$. Give your answer in scientific notation.
2.4×10^4

FOR EXAMPLE 3

A certain container of water contains 6.022×10^{23} water molecules. If these water molecules have a mass of 18 g, how much mass does each water molecule have? 2.99×10^{-23} g

Discussion: *What are the advantages of using scientific notation to study things on the molecular scale?*

To write a number in scientific notation, you use a *power of 10*. A **power of 10** is an expression in the form $10^{■}$.

2. a. Copy and complete the table. See margin.

Power of 10	$10^{■}$	10^3	$10^{■}$	$10^{■}$	$10^{■}$
Standard Notation	1,000,000	■	1	0.001	■
Unit Name	millions	■	ones	■	millionths

b. *Patterns* What pattern did you notice in part (a)? Exponents decrease by 3 from left to right.

3. a. Order the data in the table from least to greatest mass. 8.7×10^{25}, 1.0×10^{26}, 5.7×10^{26}, 1.9×10^{27}

computer-generated image of the solar system

Masses of Planets (kg)

Jupiter	Saturn	Uranus	Neptune
1.9×10^{27}	5.7×10^{26}	8.7×10^{25}	1.0×10^{26}

b. Summarize the reasoning you used and the steps you took to order the data. Answers may vary. Sample: Write each number in the form $a \times 10^{25}$. Then arrange the data in order of increasing values of a.

4. *Physical Science* Write each number in scientific notation.

a. mass of Earth's atmosphere: 5,700,000,000,000,000 tons 5.7×10^{15}

b. mass of the smallest insect, a parasitic wasp: 0.000 004 92 g 4.92×10^{-6}

You can use what you know about place value and multiplication by 10 to convert from scientific notation to standard notation.

Example 1 **Relating to the Real World** 

Physical Science Write each number in standard notation.

a. temperature at the Sun's core: 1.55×10^6 K

b. lowest temperature ever in a lab: 2×10^{-11} K

a. $1.55 \times 10^6 = 1.550000.$
$= 1,550,000$

A positive exponent indicates a large number. Move the decimal point 6 places to the right.

b. $2 \times 10^{-11} = 0.00000000002.$
$= 0.000\ 000\ 000\ 02$

A negative exponent indicates a small number. Move the decimal point 11 places to the left.

What? One kelvin (1 K) is equal to 1°C, but the kelvin temperature scale starts at absolute zero (–273.15°C). Because nothing can be colder than absolute zero, there are no negative temperatures on the kelvin scale.

5. *Try This* Write each number in standard notation.

a. distance light travels in one year (one light-year): 5.88×10^{12} mi 5,880,000,000,000

b. highest elevation in Florida: 6.53×10^{-2} mi 0.0653

6. Which numbers are *not* in scientific notation? Explain. Write each number in scientific notation.

a. 11.24×10^4 **b.** 2.004×10^{-23} **c.** -12×10^{-2}

6. a, c; in $a \times 10^n$, abs. val. of a must be between 1 and 10; 1.124×10^5; -1.2×10^{-1}

Continue asking similar questions until students are comfortable with the terminology and with multiplying and dividing powers of ten.

Example 2

Question 9 Explain to students that the exponent of 10 did not change in all cases because the numbers that are being multiplied by the power of 10 are all less than 10.

Example 3 **Relating to the Real World** 

Help make the data in the problem understandable to the students. Discuss the term *subscribers.* Be sure students understand that subscribers may be businesses or families. Ask students to use the data in the example to estimate the number of calls per day per subscriber. **about 10** Discuss whether this seems reasonable.

CONNECTING TO THE STUDENTS' WORLD Encourage students to collect data from their friends and families about how many telephone calls each person makes every day. Ask students to collect data on the number of calls made from businesses as well as calls from home. Advise students to include the number of calls made using a fax machine or a modem, and calls placed that were not answered or that were busy. Have students compare the data they collect to the national average.

Part 2

Calculating with Scientific Notation

Standard Notation	Scientific Notation	
0.0617	$6.17 \times 10^{■}$	−2
0.617	$6.17 \times 10^{■}$	−1
6.17	$6.17 \times 10^{■}$	0
61.7	$6.17 \times 10^{■}$	1
617	$6.17 \times 10^{■}$	2
6170	$6.17 \times 10^{■}$	3

7. a. Copy and complete the table at the left.
   b. *Patterns* Look for a pattern in the exponents as you scan down the table. As you multiply a number by 10 repeatedly, the exponent in the power of 10 _?_ by ■ repeatedly. **increases; 1**
8. Write each expression as a single power of ten.
   a. $10^{12} \times 10$ 10^{13} b. 10×10^{-8} 10^{-7} c. $10^{-7} \times 10$ 10^{-6}

Example 2

Mental Math Simplify $7 \times (4 \times 10^5)$. Give your answer in scientific notation.

$7 \times (4 \times 10^5) = 28 \times 10^5$ ← Multiply whole numbers.
$= 2.8 \times 10 \times 10^5$ ← Write 28 as 2.8 × 10.
$= 2.8 \times 10^6$ ← Combine powers of 10.

9. *Try This* Simplify. Give each answer in scientific notation.
   a. $2.5 \times (6 \times 10^3)$ **1.5×10^4** b. $1.5 \times (3 \times 10^4)$ **4.5×10^4** c. $9 \times (7 \times 10^{-9})$ **6.3×10^{-8}**
10. *Mental Math* Double each number. Give your answers in scientific notation.
    a. 4×10^5 **8×10^5** b. \$6.3 million **\$1.26 × 10^7** c. 1.2×10^{-3} **2.4×10^{-3}**
11. You express 1 billion as 10^9. Explain why you express 436 billion as 4.36×10^{11}. **436 billion is $436 \times 10^9 = 4.36 \times 10^{11}$.**

Most calculators have a key labeled EE or EXP that allows you to enter a number in scientific notation.

Example 3 **Relating to the Real World** 

Telecommunications In 1993, 436 billion telephone calls were placed by 130 million United States telephone subscribers. What was the average number of calls placed per subscriber?

$\frac{436 \text{ billion calls}}{130 \text{ million subscribers}} = \frac{4.36 \times 10^{11}}{1.3 \times 10^8}$ ← Write in scientific notation.

= 4.36 EE 11 ÷ 1.3 EE 8 ENTER

= 3353.846154

Each subscriber made an average of 3354 calls in 1993.

12. *Try This* The closest star to Earth (other than the sun) is Alpha Centauri, 4.35 light-years from Earth. How many miles from Earth is Alpha Centauri? (*Hint:* See Question 5(a).)
    about 2.56×10^{13} miles

Technology Options

For Exercises 5–30, students may find it helpful to use a graphing calculator or spreadsheet software to write the correct scientific notation for each number.

Prentice Hall Technology 

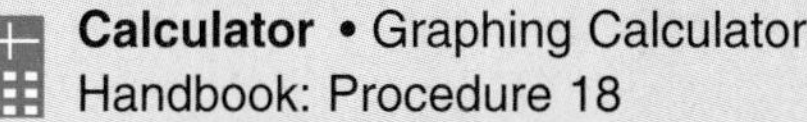

Calculator • Graphing Calculator Handbook: Procedure 18

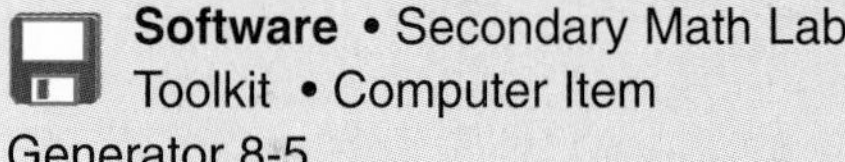

Software • Secondary Math Lab Toolkit • Computer Item Generator 8-5

CD-ROM • Multimedia Algebra Lab 8

Internet • See the Prentice Hall site. (http://www.phschool.com)

Exercises ON YOUR OWN

ALTERNATIVE ASSESSMENT **Exercises 10–30** Have students write all the exercises on separate index cards. Then exchange cards and rewrite the expressions in scientific notation. Finally have students compare numbers and arrange the cards in a line on the floor from least to greatest.

ERROR ALERT! **Exercise 30** Some students may not recall how to write 25 trillion as a standard numeral. **Remediation:** Point out that the trillions comma comes after the billions comma, and that 25 trillion is written as 25,000,000,000,000.

CONNECTING TO THE STUDENTS' WORLD Exercise 43 Based on the statistics provided in Exercise 43, in 1990, there was one centenarian, on average, in every 6675 people. Ask students if they know any centenarians. Suggest that students conduct a brief survey of other math classes to see how many other students know centenarians. Have students compare their statistics with the national figures. Interested students may wish to explore how the number of centenarians and number of United States residents has changed since 1990.

MAKING CONNECTIONS A centillion equals $(10^6)^{10C}$ or 1 followed by 600 zeros. The term *googol*, to represent a number equal to 10^{100}, was suggested by a 9-year-old. A googolplex equals 10 to the power googol, or $10^{(10^{100})}$. A number called *asankhyeya*, equal to 10^{140}, is mentioned in a Buddhist work dating from 100 B.C.

ESL **CARTOON Exercise 46** The expression *shot 240 frames* may be confusing to some students. Ask students what they would use to do the shooting in this case and what the result would be.

pages 385–387 Think and Discuss

2a.

Power of 10	Standard Notation	Unit Name
10^6	1,000,000	millions
10^3	1000	thousands
10^0	1	ones
10^{-3}	0.001	thousandths
10^{-6}	0.000 001	millionths

page 389 Mixed Review

54. $\begin{bmatrix} 7 & 11 & 14 \\ 6 & 10 & 8 \end{bmatrix}$

55. $\begin{bmatrix} -3.2 & 2.4 \\ 4.1 & -1.2 \end{bmatrix}$

56. Hockey Standings
Central Division

Goals	FOR	OPP	DIFF
Detroit	356	275	+ 81
Toronto	280	243	+ 37
Dallas	286	265	+ 21
St. Louis	270	283	– 13
Chicago	254	240	+ 14
Winnipeg	245	344	– 99

Exercises ON YOUR OWN

Order each set of numbers from least to greatest.

1. $10^5, 10^{-3}, 10^0, 10^{-1}, 10^1$ $10^{-3}, 10^{-1}, 10^0, 10^1, 10^5$
2. $6.2 \times 10^7, 5.1 \times 10^7, 8 \times 10^7, 1.02 \times 10^7$ $1.02 \times 10^7, 5.1 \times 10^7, 6.2 \times 10^7, 8 \times 10^7$
3. $4.02 \times 10^5, 4.1 \times 10^4, 4.1 \times 10^5, 4 \times 10^5$ $4.1 \times 10^4, 4 \times 10^5, 4.02 \times 10^5, 4.1 \times 10^5$
4. $5.1 \times 10^{-3}, 4.8 \times 10^{-1}, 5.2 \times 10^{-3}, 5.6 \times 10^{-}$ $5.1 \times 10^{-3}, 5.2 \times 10^{-3}, 5.6 \times 10^{-2}, 4.8 \times 10^{-1}$

Determine whether each number is in scientific notation. If it is not, write it in scientific notation.

5. 23×10^7 2.3×10^8
6. 385×10^{-6} 3.85×10^{-4}
7. 0.0027×10^{-4} 2.7×10^{-7}
8. 9.37×10^{-8} yes
9. $25.79 \times 10^{-}$ 2.579×10^{-4}

Write each number in scientific notation.

10. 0.00325 3.25×10^{-3}
11. 9,040,000,000 9.04×10^9
12. 13,030,000 1.303×10^7
13. 0.00092 9.2×10^{-4}
14. 0.001 002 1.002×10^{-3}
15. 370 billion 3.7×10^{11}
16. 9.3 million 9.3×10^6
17. 41.8 billion 4.18×10^{10}
18. 60.7×10^{22} 6.07×10^{23}
19. 62.9×10^{15} 6.29×10^{16}
20. 2 thousandths 2×10^{-3}
21. 33 billionths 3.3×10^{-8}
22. 950 millionths 9.5×10^{-4}
23. 83.5×10^{-6} 8.35×10^{-5}
24. 350×10^{-9} 3.5×10^{-7}
25. lightest blue whale: 418,000 lb 4.18×10^5
26. thinnest glass: 0.00098 in. 9.8×10^{-4}
27. lightest bird egg: 0.0128 oz 1.28×10^{-2}
28. diameter of thinnest copper wire: 0.000 5 in. 5×10^{-4}
29. diameter of smallest bacteria cells: 4 millionths in. 4×10^{-6}
30. closest star (other than the Sun): 25.6 trillion m 2.56×10^{13}
31. **Writing** Explain how to convert numbers like 350 billion, 7.2 trillion, or 48 millionths quickly to scientific notation. Sample: Write 1 billion as 10^9, then 350 billion is $350 \times 10^9 = 3.5 \times 10^{11}$.

Write each number in standard notation.

32. 7.042×10^9 7,042,000,000
33. 4.69×10^{-6} 0.000 004 69
34. 1.7×10^{-13} 0.000 000 000 000 17
35. 5×10^{10} 50,000,000,000
36. 1.097×10^8 109,700,000

Mental Math **Calculate each product or quotient. Give your answers in scientific notation.**

37. $8 \times (7 \times 10^{-3})$ 5.6×10^{-2}
38. $8 \times (3 \times 10^{14})$ 2.4×10^{15}
39. $2 \times (3 \times 10^2)$ 6×10^2
40. $(28 \times 10^5) \div 7$ 4×10^5
41. $(8 \times 10^{-8}) \div 4$ 2×10^{-8}
42. $(8 \times 10^{12}) \div 4$ 2×10^{12}
43. **Probability** In 1990, there were approximately 249 million residents of the United States. The census counted 37,306 centenarians (age 100 or greater). What is the probability that a randomly selected U.S. resident is a centenarian? Give your answer in scientific notation. about 1.5×10^{-4}
44. **Precious Metals** Earth's crust contains approximately 120 trillion metric tons of gold. One metric ton of gold is worth about \$11.5 million. What is the approximate value of the gold in Earth's crust? $\$1.38 \times 10^{21}$
45. **Astronomy** Light travels through space at a constant speed of about 3×10^5 km/s. Earth is about 1.5×10^8 km from the Sun. How long does it take for light from the sun to reach Earth? about 5×10^2 s

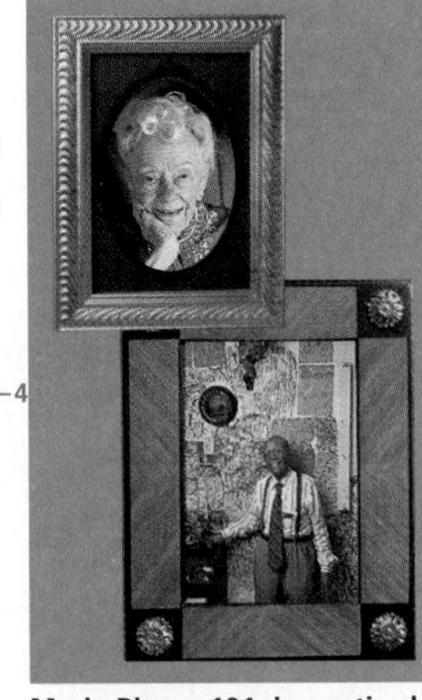

Marie Rinne, 101, is a retired teacher. Ezekiel Gibbs, 102, is a folk artist.

CARTOON Exercise 46 Ask students to think about the number of pieces of artwork needed to make a full-length animated cartoon. Discuss the impact that computer graphics has made on this kind of film making.

OPEN-ENDED Exercise 46c Students can find the length of their favorite movie by looking at the label on the videotape of the movie. Students will enjoy sharing their answers to this question.

Chapter Project **FIND OUT BY GRAPHING** Students will make a scatter plot from the mold growth data they have recorded and compare their plots. Have students add their scatter plots to the other project work they have already completed.

Exercises MIXED REVIEW

GETTING READY FOR LESSON 8-6 These exercises prepare students for learning about the multiplication property of exponents.

Wrap Up

THE BIG IDEA Ask students to explain how to write a number using scientific notation.

RETEACHING ACTIVITY Students write numbers with positive and negative exponents in scientific notation. They also write numbers shown in scientific notation in standard notation. (Reteaching worksheet 8-5)

FOX TROT by Bill Amend

46. a. Write 500 trillion in scientific notation. 5×10^{14}
    b. Since the 10-second length of the movie is off by a factor of 500 trillion, what time span does the movie actually represent? about 1.6×10^8 yr
    c. *Open-ended* Find out the length of one of your favorite movies. Calculate how many frames it has, and express the number of frames in scientific notation. Answers may vary. Sample: A 90-min movie has 1.296×10^5 frames.

Mental Math **Double each number. Give your answers in scientific notation.**

47. 3.5×10^{-3} 7×10^{-3}
48. \$75 million \$$1.5 \times 10^8$
49. 450×10^{-1} 9×10^1
50. 3550 7.1×10^3
51. 250×10^5 5×10^7
52. 790 1.58×10^3

53. *Health Care* The total amount spent for health care in the United States in 1993 was \$884.2 billion. The U.S. population in 1993 was 258.1 million. What was the average amount spent per person on health care? about \$3426

Chapter Project **Find Out by Graphing**

Make a scatter plot of the growth data. Compare your graph with graphs made by other students in your group. Do the data plots look exponential? Are there differences that seem to be related to where and how long the dish was exposed?

Name ______ Class ______ Date ______

Reteaching 8-5

Practice 8-5

Practice 8-5

Mixed Exercises

Lesson Quiz

Lesson Quiz is also available in Transparencies.

1. Order the following numbers from least to greatest.
   9.8×10^{-2}, 1.6×10^3, 2.4×10^{-1}, 1.1×10^{-1}
   9.8×10^{-2}, 1.1×10^{-1}, 2.4×10^{-1}, 1.6×10^3
2. Write 0.006 27 in scientific notation. 6.27×10^{-3}
3. Write 9.4×10^4 in standard notation. 94,000
4. Calculate the product $7 \times (6 \times 10^{-2})$. Give your answer in scientific notation. 4.2×10^{-1}

Exercises MIXED REVIEW

Find each sum or difference. 54–55. See margin.

54. $\begin{bmatrix} 2 & 9 & 5 \\ 3 & 6 & 1 \end{bmatrix} + \begin{bmatrix} 5 & 2 & 9 \\ 3 & 4 & 7 \end{bmatrix}$

55. $\begin{bmatrix} 2.6 & 4 \\ 8.1 & 6.7 \end{bmatrix} - \begin{bmatrix} 5.8 & 1.6 \\ 4 & 7.9 \end{bmatrix}$

56. *Sports* In the hockey statistics, the FOR column tells the number of goals made by each team. The OPP column tells the number of goals made by a team's opponents. The DIF column reports the difference. Find the errors in the DIF column. See margin.

Hockey Standings Central Division

Goals	FOR	OPP	DIF
Detroit	356	275	+81
Toronto	280	243	−37
Dallas	286	265	+21
St. Louis	270	283	−13
Chicago	254	240	−14
Winnipeg	245	344	+99

Getting Ready for Lesson 8-6

Rewrite each expression using exponents.

57. $t \cdot t \cdot t \cdot t \cdot t \cdot t \cdot t$ t^7
58. $(6 - m)(6 - m)(6 - m)$ $(6 - m)^3$
59. $5 \cdot 5 \cdot 5 \cdot s \cdot s \cdot s$ 5^3s^3

Math ToolboX

In this toolbox, students will combine the skills of scientific notation and rounding with the concept of significant digits. Begin by asking students to define the terms *significant* and *digit*. Then work through the examples with the class. Point out to students that significant digits may be located on either side of the decimal point. Then, have students work through the exercises on their own or in small groups.

ERROR ALERT! Some students may have difficulty determining the correct sign of the exponent. **Remediation:** Have students refer to their work in Lesson 8-4. Remind students that decimals may be written as fractions. Students can then determine the correct sign of the exponent by deciding if the answer should be multiplied by a power of ten or by a power of the reciprocal of ten.

WRITING Exercise 10 For students who have trouble starting, ask them to think of situations where very accurate measurements might be necessary. Explain that in these situations, very accurate instruments can increase the number of significant digits in a measurement.

ADDITIONAL PROBLEM Express each number in scientific notation to three significant digits.

1. 0.000 031 78 3.18×10^{-5}
2. 58,416,327 5.84×10^{7}
3. 34,002 3.40×10^{4}
4. 0.005 555 5.56×10^{-3}

Math ToolboX **Skills Review**

Significant Digits

After Lesson 8-5

Significant digits tell scientists how precise a measurement is. The more significant digits there are, the more precise the measurement is. Scientists consider all the significant digits in a measurement to be exact except for the final digit, which is considered to be rounded.

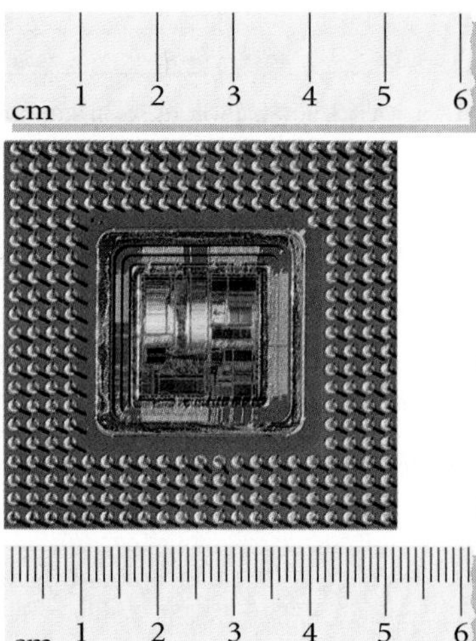

Example 1

Express the length of the computer chip in centimeters to (a) one significant digit and (b) two significant digits.

a. The length of the chip is 5 cm.
b. The length of the chip is 5.2 cm.

Example 2

The moon can come within 221,463 mi of Earth. Express this distance in scientific notation to three significant digits.

$221{,}463 = 2.21463 \times 10^5$ ← Write in scientific notation.

$\approx 2.21 \times 10^5$ ← Round 2.21463 to the hundredths' place.

To three significant digits, the distance is 2.21×10^5 mi.

Tell how many significant digits are in each measurement.

1. Mercury's period of revolution: 87.9686 da 6
2. largest known galaxy: 3.3×10^{19} mi in diameter 2
3. tallest sand castle: 19.5 ft high 3

Express each number in scientific notation to three significant digits.

4. the surface area of Earth: 196,949,970 mi^2 1.97×10^{8}
5. smallest diamond: 0.000 102 2 carat 1.02×10^{-4}
6. thinnest commercial glass: 0.000 984 in. 9.84×10^{-4}
7. the farthest the moon can be from Earth: 252,710 mi 2.53×10^{5}
8. average weight of the smallest bone in the inner ear: 0.010 853 75 oz 1.09×10^{-2}
9. 1991 U.S. deaths from heart disease: 725,010 7.25×10^{5}
10. *Writing* When would it be important to measure something to three or more significant digits? Explain. Sample: Precise measurements, such as those made on machined parts for the space shuttle, may need more than two significant digits.

8-6 Teaching Notes

PROBLEM OF THE DAY

A tree casts a shadow that is 63.75 ft long at the same time a 4 ft boy casts a 3 ft shadow. How tall is the tree?
85 ft

Problem of the Day is also available in Transparencies.

CONNECTING TO PRIOR KNOWLEDGE On the board, write $2^3 \cdot 2^4 = ?$. Ask students: *What math skills have you learned that you could apply to answer this question?* Students may suggest the guess-and-check strategy or suggest writing each power without exponents to see what pattern develops.

WORK TOGETHER

Organize students in groups of three or four. One student can act as the recorder while the others use calculators and compare their calculations.

THINK AND DISCUSS

Take time to fully explain how 8^6 can be written as a product of two powers of 8. Have students help you list other possibilities. Students are more likely to remember the rules of exponents if they understand how the rules work.

Question 8 This question highlights a common error that students make. Emphasize that all factors must have the same base in order to apply the rule for addition of exponents.

Connections Biology . . . and more

8-6 A Multiplication Property of Exponents

What You'll Learn
Multiplying powers with the same base

...And Why
To solve problems that involve the multiplication of numbers in scientific notation

What You'll Need
calculator

WORK TOGETHER

To evaluate $5^3 \cdot 5^5$, you could multiply the value of 5^3 by the value of 5^5. But is there a shortcut for finding the value of expressions like this? Work in your group to find one.

1. Calculator Find the value of each expression.
   a. $5^3 \cdot 5^5$ 390,625
   b. $5^6 \cdot 5^2$ 390,625
   c. $5^1 \cdot 5^7$ 390,625
   d. $5^4 \cdot 5^4$ 390,625
2. Patterns What pattern do you notice in your answers to Question 1? All the numbers are the same.
3. Write each expression from Question 1 in the form shown below.
   $$5^3 \cdot 5^5 = (5 \cdot 5 \cdot 5) \cdot (5 \cdot 5 \cdot 5 \cdot 5 \cdot 5) = 5^8$$ See margin.
4. Look for a pattern in the expressions you wrote in Question 3. Write a shortcut for finding the value of expressions such as $5^3 \cdot 5^5$. See margin.
5. Use your shortcut to find the missing value in each expression. **Verify** your answers by using a calculator.
   a. $5^3 \cdot 5^4 = 5^{■}$ 7; 78,125
   b. $3^2 \cdot 3^5 = 3^{■}$ 7; 2187
   c. $1.2^3 \cdot 1.2^3 = 1.2^{■}$ 6; 2.985 984
   d. $7^3 \cdot 7^2 = 7^{■}$ 5; 16,807

CALCULATOR HINT
Use the y^x or ^ key to evaluate expressions with exponent.

THINK AND DISCUSS

Part 1

Multiplying Powers

QUICK REVIEW
Read 5^4 as "5 to the 4th power."

Any expression in the form a^n, such as 5^4, is called a **power.** In the Work Together, you discovered a shortcut for multiplying powers with the same base. This shortcut works because the factors of a power such as 8^6 can be combined in different ways. Here are two examples:

$$8^6 = \underbrace{8 \cdot 8 \cdot 8 \cdot 8}_{8^4} \cdot \underbrace{8 \cdot 8}_{8^2} = 8^4 \cdot 8^2 \qquad 8^6 = \underbrace{8 \cdot 8 \cdot 8}_{8^3} \cdot \underbrace{8 \cdot 8 \cdot 8}_{8^3} = 8^3 \cdot 8^3$$

Notice that $8^4 \cdot 8^2 = 8^3 \cdot 8^3 = 8^6$. When you multiply powers with the same base, you add the exponents.

Multiplying Powers with the Same Base

For any nonzero number a and any integers m and n, $a^m \cdot a^n = a^{m+n}$.

Example: $3^5 \cdot 3^4 = 3^{5+4} = 3^9$

Lesson Planning Options

Prerequisite Skills

- Simplifying exponents
- Simplifying expressions

Assignment Options

To provide flexible scheduling, this lesson can be subdivided into parts.

1 **Core** 1–11, 20–47
Extension 19, 48–50

2 **Core** 12–17, 55–57
Extension 18, 51–54, 58, 59

Use Mixed Review to maintain skills.

Resources

Student Edition

Skills Handbook, pp. 574, 580, 582
Extra Practice, p. 563
Glossary/Study Guide

Teaching Resources

Chapter Support File, Ch. 8
- Practice 8-6 (two worksheets)
- Reteaching 8-6
- Checkpoint

Classroom Manager 8-6
Glossary, Spanish Resources
Two-Year Algebra Handbook 8-6

Transparencies
90

Example 1

Some students may benefit from seeing these examples showing all the factors worked out.

$$c^4 \cdot d^3 \cdot c^2 = c \cdot c \cdot c \cdot c \cdot d \cdot d \cdot d \cdot c \cdot c$$
$$= c \cdot c \cdot c \cdot c \cdot c \cdot c \cdot d \cdot d \cdot d$$
$$= c^6d^3$$

Question 9a

ERROR ALERT! Some students may think of a as a^0. **Remediation:** Remind them that $a^0 = 1$ and that $a^1 = a$.

AUDITORY LEARNING Have a student raise 4 to the power of a number between 1 and 9 and say it aloud. Have a second student do the same. Then, have a third student combine their powers of 4 and say the answer aloud. For example, if two students combine 4^3 and 4^9, their result would be 4^{12}. Repeat this process until you are comfortable that students understand the rule for multiplying exponents with the same base.

Example 2 **Relating to the Real World**

To review dimensional analysis, students can go back to the Math Toolbox on page 219 in Chapter 5.

After you have worked through the steps of the Example, ask students: *Why do you think 2000 was written as 2×10^3?* The first number in scientific notation must fit this description: $2 \leq x < 10$. To change to this format, you must first write 2000 in its own scientific notation and then apply this result to the rest of the problem.

Additional Examples

FOR EXAMPLE 1

Simplify each expression.

a. $a^2 \cdot b^3 \cdot a^2$ a^4b^3

b. $2x \cdot 3x^3 \cdot 4x^{-2}$ $24x^2$

Discussion: *What is the procedure for multiplying positive and negative exponents having the same base?*

FOR EXAMPLE 2

The speed of light is 3×10^8 m/s. If there are 1×10^3 m in 1 km, and 3.6×10^3 s in 1 h, find the speed of light in km/hr. 1.08×10^9 km/hr

Discussion: *How can you use dimensional analysis to help solve this problem?*

FOR EXAMPLE 3

Simplify $(3.1 \times 10^{-3})(1.2 \times 10^{-5})$. 3.72×10^{-8}

6. Write 5^7 as a product of powers in three different ways. $5^1 \cdot 5^6; 5^2 \cdot 5^5; 5^3 \cdot 5^4$

7. Does $3^2 \cdot 3^5 \cdot 3^1 = 3^{2+5+1} = 3^8$? Explain. Yes; you can use the multiplication rules to multiply several powers at the same time.

8. Does $3^4 \cdot 2^2 = 6^{4+2}$? Explain. No; the rule applies only to powers with the same base.

QUICK REVIEW

The coefficient in $3x^8$ is the numerical factor 3.

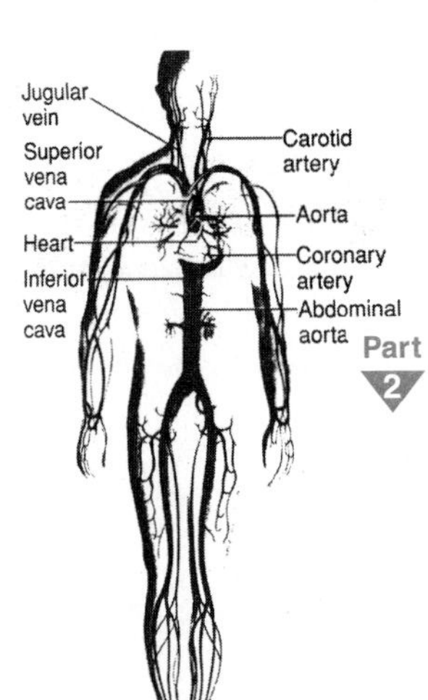

Example 1

Simplify each expression.

a. $c^4 \cdot d^3 \cdot c^2$

$= c^4 \cdot c^2 \cdot d^3$ ← Rearrange factors.

$= c^{4+2} \cdot d^3$ ← Add exponents of powers with the same base.

$= c^6d^3$

b. $5x \cdot 2y^4 \cdot 3x^8$

Rearrange factors. → $= (5 \cdot 2 \cdot 3)(x \cdot x^8)(y^4)$

Multiply coefficients. → $= (30)(x^1 \cdot x^8)(y^4)$

Add exponents of powers with the same base. → $= (30)(x^{1+8})(y^4)$

$= 30x^9y^4$

9. Try This Simplify each expression.

a. $a \cdot b^2 \cdot a^5$ a^6b^2

b. $6x^2 \cdot 3y^3 \cdot 2y^4$ $36x^2y^7$

c. $m^2 \cdot n^2 \cdot 7m$ $7m^3n^2$

Part 2 Working with Scientific Notation

Example 2 **Relating to the Real World** 

Biology A human body contains about 3.2×10^4 µL (microliters) of blood for each pound of body weight. Each microliter of blood contains about 5×10^6 red blood cells. Find the approximate number of red blood cells in the body of a 125-lb person.

$\text{pounds} \cdot \frac{\text{microliters}}{\text{pound}} \cdot \frac{\text{cells}}{\text{microliter}} = \text{cells}$ ← Use dimensional analysis.

$125 \text{ lb} \cdot (3.2 \times 10^4)\frac{\mu L}{lb} \cdot (5 \times 10^6)\frac{\text{cells}}{\mu L}$

$= (125 \cdot 3.2 \cdot 5) \cdot (10^4 \cdot 10^6)$ ← Rearrange factors.

$= (125 \cdot 3.2 \cdot 5) \cdot 10^{10}$ ← Add exponents.

$= 2000 \cdot 10^{10}$ ← Simplify.

$= 2 \times 10^3 \cdot 10^{10}$ ← Write 2000 as 2×10^3.

$= 2 \times 10^{13}$ ← Use scientific notation.

There are about 2×10^{13} (20 trillion!) red blood cells in a 125-lb person.

10. Try This About how many red blood cells are in your body? Check students' work.

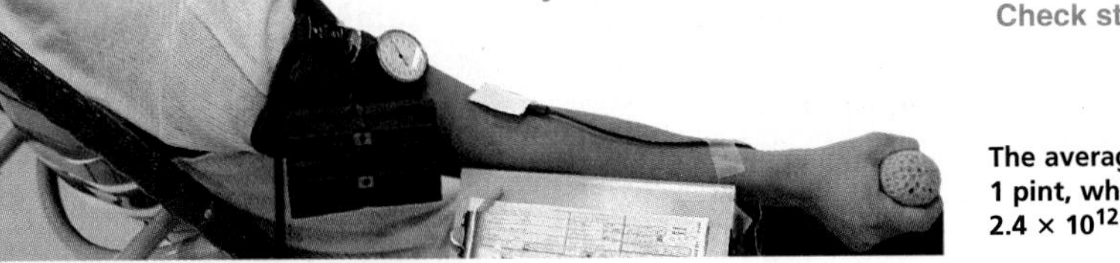

The average blood donation is 1 pint, which contains about 2.4×10^{12} red blood cells.

DIVERSITY Question 10 Students who are self-conscious about weight may be uncomfortable with this question. As an alternative, have students estimate the number of red blood cells in a 10-year-old child. According to the 1990 *Physician's Handbook*, average 10-year-olds weigh 69 lb.

MAKING CONNECTIONS Red blood cells are tiny circular disks about 0.003 in. in diameter. Their main functions are to transport the waste product carbon dioxide to the lungs for release from the body and then to pick up oxygen from the lungs to take to the body's tissues. Each red blood cell has a life span of about 120 days. Each day, the body destroys and replaces more than 200 billion (2×10^{11}) cells.

Example 3

This is students' first introduction to multiplying powers with negative exponents. Help strengthen students' understanding by working through the following example.

$$10^3 \cdot 10^{-5} = 10 \cdot 10 \cdot 10 \cdot \frac{1}{10 \cdot 10 \cdot 10 \cdot 10 \cdot 10}$$
$$= \frac{1}{10 \cdot 10}$$
$$= \frac{1}{10^2} = 10^{-2}$$

Exercises ON YOUR OWN

Exercise 9 If a student has difficulty with this problem, ask: *How can you write 1 as a power of a?* a^0

The property of multiplying powers with the same base works with negative exponents, also.

Example 3

Simplify $(0.2 \times 10^5)(4 \times 10^{-12})$. Give the answer in scientific notation.

$(0.2 \times 10^5)(4 \times 10^{-12})$
$= (0.2 \times 4)(10^5 \times 10^{-12})$ ← Rearrange factors.
$= 0.8 \times 10^{5+(-12)}$ ← Multiply.
$= 0.8 \times 10^{-7}$ ← Add exponents.
$= 8 \times 10^{-1} \times 10^{-7}$ ← Write 0.8 as 8×10^{-1}.
$= 8 \times 10^{-8}$ ← Add exponents.

11. Try This Simplify each expression. Give your answers in scientific notation.
    a. $(0.5 \times 10^{13})(0.3 \times 10^{-4})$ 1.5×10^8
    b. $(0.7 \times 10^{-9})(0.03 \times 10^8)$ 2.1×10^{-3}
12. **Verify** that $2^5 \cdot 2^{-2} = 2^3$ by writing 2^5 as an integer and 2^{-2} as a fraction, then multiplying. $32 \cdot \frac{1}{4} = 8$ ✓

Technology Options

For Exercises 12–17, students may want to use a graphing calculator to write their answers in scientific notation.

Prentice Hall Technology

Software • Secondary Math Lab Toolkit • Integrated Math Lab 16 • Computer Item Generator 8-6

Internet • See the Prentice Hall site. (http://www.phschool.com)

Exercises ON YOUR OWN

Write each expression as a product of powers, then simplify.

1. $(x \cdot x \cdot x \cdot x)(x \cdot x \cdot x)$ $x^4 \cdot x^3 = x^7$
2. $(x \cdot x)\left(\frac{1}{x \cdot x \cdot x}\right)$ $x^2 \cdot x^{-3} = x^{-1}$
3. $\left(\frac{1}{x \cdot x}\right)\left(\frac{1}{x \cdot x \cdot x \cdot x \cdot x}\right)$ $x^{-2} \cdot x^{-5} = x^{-7}$

Complete each equation.

4. $5^2 \cdot 5^{■} = 5^{11}$ 9
5. $5^7 \cdot 5^{■} = 5^3$ −4
6. $a^{12} \cdot a^{■} = a^{12}$ 0
7. $a \cdot a \cdot a^3 = a^{■}$ 5
8. $2^{■} \cdot 2^4 = 2^1$ −3
9. $a^{■} \cdot a^4 = 1$ −4
10. $c^{-5} \cdot c^{■} = c^6$ 11
11. $x^3y^{■} \cdot x^{■} = y^2$ 2; −3

Multiply. Give your answers in scientific notation.

12. $(9 \times 10^7)(3 \times 10^{-16})$ 2.7×10^{-8}
13. $(8 \times 10^{-3})(0.1 \times 10^9)$ 8×10^5
14. $(0.7 \times 10^{-12})(0.3 \times 10^8)$ 2.1×10^{-5}
15. $(2 \times 10^6)(3 \times 10^3)$ 6×10^9
16. $1 \times 10^3 \cdot 10^{-8}$ 1×10^{-5}
17. $(4 \times 10^6) \times 10^{-3}$ 4×10^3
18. Standardized Test Prep Simplify $p^8 \cdot 9q^2 \cdot q^4 \cdot p^3 \cdot 2q$. D
    A. $-11p^{11}q^7$ B. $11p^{14}q^8$ C. $18p^{11}q^6$ D. $18p^{11}q^7$ E. $18pq^{17}$
19. a. Open-ended Write y^8 in four different ways as the product of two powers with the same base. Use only positive exponents. $y^1 \cdot y^7$; $y^2 \cdot y^6$; $y^3 \cdot y^5$; $y^4 \cdot y^4$
    b. Write y^8 in four different ways as the product of two powers with the same base, using negative or zero exponents in each. Sample: $y^0 \cdot y^8$; $y^{-2} \cdot y^{10}$; $y^{-3} \cdot y^{11}$; $y^{-8} \cdot y^{16}$
    c. How many ways are there to write y^8 as the product of two powers? **Summarize** your reasoning.
    Infinitely many ways; there are infinitely many ways to write 8 as a sum of two integers.

ERROR ALERT! Exercise 12 Students may give the answer as 27×10^{-9}. **Remediation:** Point out that in scientific notation, one factor must be a number greater than or equal to one and less than ten.

ALTERNATIVE ASSESSMENT Exercise 18 Have students rewrite each answer choice two different ways. For example, $-11p^{11}q^{7}$ can be rewritten as $-\frac{11p^{11}}{q^{-7}}$ or as $(-11p^{5})(q^{7})(p^{6})$. This will help you assess student understanding of the multiplication property of exponents.

STANDARDIZED TEST TIP Exercise 18 Remind students that a variable exponent has no effect on the coefficient unless both the variable and the coefficient are in the same set of parentheses. Therefore, the only answer choices that need be considered are *C*, *D*, and *E*.

GEOMETRY Exercises 51–54 Review area formulas if necessary.

Rectangle: $A = lw$

Triangle: $A = \frac{1}{2}bh$

Parallelogram: $A = bh$

Square: $A = s^2$

CRITICAL THINKING Exercises 55–57 Have students explain what common error was made in each exercise.

ESL Exercise 58 Use a gesture to demonstrate the meaning of the term *spiral* or draw a spiral on the board.

Chapter Project **FIND OUT BY ANALYZING** Students will find the percent growth from each day to the next and then find the average daily percent growth. Finally, they will create

page 391 Work Together

3. $5^6 \cdot 5^2 =$
$(5 \cdot 5 \cdot 5 \cdot 5 \cdot 5 \cdot 5) \cdot (5 \cdot 5) = 5^8$
$5^1 \cdot 5^7 =$
$(5) \cdot (5 \cdot 5 \cdot 5 \cdot 5 \cdot 5 \cdot 5 \cdot 5) = 5^8$
$5^4 \cdot 5^4 =$
$(5 \cdot 5 \cdot 5 \cdot 5) \cdot (5 \cdot 5 \cdot 5 \cdot 5) = 5^8$
4. Answers may vary. Sample: The exponent of the product is equal to the sum of the exponents of the factors. $5^3 \cdot 5^5 = 5^{3+5}$

page 395 Checkpoint

18.

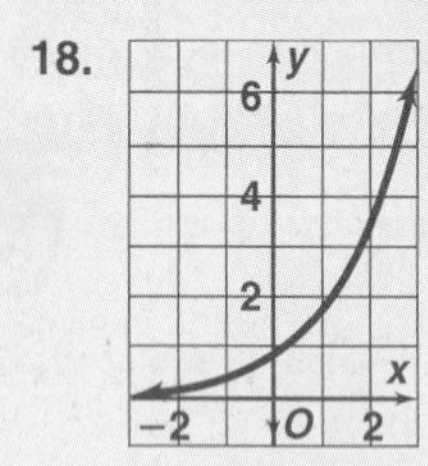

19.

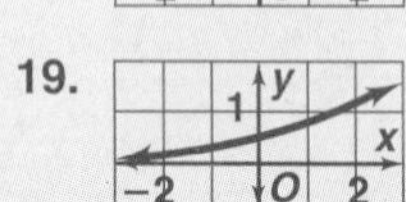

20. See back of book.

Simplify each expression. Use only positive exponents.

20. $(0.99^2)(0.99^4)$ 0.99^6
21. $(1.025)^3(1.025)^{-3}$ 1
22. $c^{-2}c^7$ c^5
23. $3r \cdot r^4$ $3r^5$
24. $5t^{-2} \cdot 2t^{-5}$ $\frac{10}{t^7}$
25. $(a^2b^3)(a^6)$ a^8b^3
26. $(x^5y^2)(x^{-6}y)$ $\frac{y^3}{x}$
27. $3x^2 \cdot x^2$ $3x^4$
28. $(1.03^8)(1.03^4)$ 1.03^{12}
29. $-(0.99^3)(0.99^0)$ -0.99^3
30. $a \cdot a^{-7}$ $\frac{1}{a^6}$
31. $b^{-2} \cdot b^4 \cdot b$ b^3
32. $10^{-13} \cdot 10^5$ $\frac{1}{10^8}$
33. $(-2m^3)(3.5m^{-3})$ -7
34. $(7x^5)(8x)$ $56x^6$
35. $(15a^3)(-3a)$ $-45a^4$
36. $(-2.4n^4)(2n^{-1})$ $-4.8n^3$
37. $bc^{-6} \cdot b$ $\frac{b^2}{c^6}$
38. $(5x^5)(3y^6)(3x^2)$ $45x^7y^6$
39. $(4c^4)(ac^3)(3a^5c)$ $12a^6c^8$
40. $x^6 \cdot y^2 \cdot x^4$ $x^{10}y^2$
41. $a^6b^3 \cdot a^2b^{-2}$ a^8b
42. $\frac{1}{x^2 \cdot x^{-5}}$ x^3
43. $\frac{1}{a^3 \cdot a^{-2}}$ a^5
44. $\frac{5}{c \cdot c^{-4}}$ $5c^3$
45. $2a^2(3a + 5)$ $6a^3 + 10a^2$
46. $8m^3(m^2 + 7)$ $8m^5 + 56m^3$
47. $-4x^3(2x^2 - 9x)$ $-8x^5 + 36x^4$

48. **Chemistry** The term *mole* is used in chemistry to refer to a specific number of atoms or molecules. One mole is equal to 6.02×10^{23}. The mass of a single hydrogen atom is approximately 1.67×10^{-24} gram. What is the mass of 1 mole of hydrogen atoms? about 1.01 g

49. a. Jerome wrote $a^3b^2b^4 = (a \cdot a \cdot a)(b \cdot b)(b \cdot b \cdot b \cdot b)$. Jeremy wrote $a^3b^2b^4 = ab^9$. Whose work is correct? Explain. Jerome's work; Jeremy added exponents having different bases.
b. Use the correct work to simplify $a^3b^2b^4$. a^3b^6

50. **Writing** Explain why $x^3 \cdot y^5$ cannot be simplified. x^3 and y^5 have different bases.

Geometry Find the area of each figure.

51.

52.

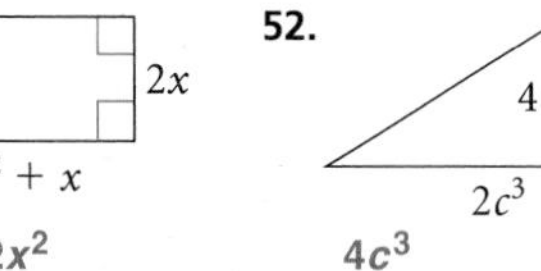

$6x^3 + 2x^2$ (51); $4c^3$ (52)

53.

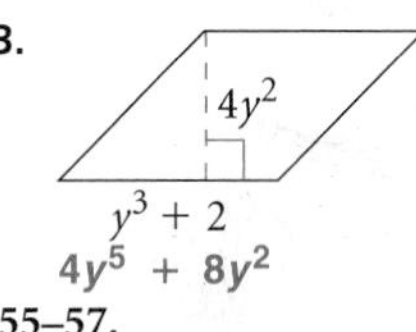

$4y^5 + 8y^2$

54.

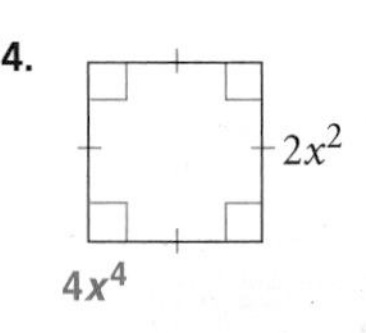

$4x^4$

Critical Thinking Find and correct the errors in Exercises 55–57.

55. $(3x^2)(-2x^4) = 3(-2)x^{2 \cdot 4}$
$= -6x^8$
x^{2+4}; $-6x^6$

56. $4a^2 \cdot 3a^5 = (4 + 3)a^{2+5}$
$= 7a^7$
$4 \cdot 3$; $12a^7$

57. $x^6 \cdot x \cdot x^3 = x^{6+3}$
$= x^9$
x^{6+1+3}; x^{10}

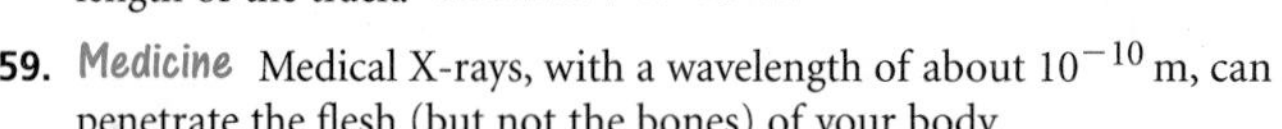

58. **Technology** A CD-ROM stores about 600 megabytes (6×10^8 bytes) of information along a spiral track. Each byte uses about 9 micrometers (9×10^{-6} m) of space along the track. Find the length of the track. about 5.4×10^3 m

The music on a compact disc comes from a series of notches that can be read by a laser. The disc is enclosed in two layers of plastic. The plastic surface of this CD is cracked to show the notched layer underneath.

59. **Medicine** Medical X-rays, with a wavelength of about 10^{-10} m, can penetrate the flesh (but not the bones) of your body.
a. Ultraviolet rays, which cause sunburn by penetrating only the top layers of skin, have a wavelength about 1000 times as long as X-rays. Find the wavelength of ultraviolet rays. about 10^{-7} m
b. **Critical Thinking** The wavelengths of visible light are between 4×10^{-7}m and 7.5×10^{-7}m. Are these wavelengths longer or shorter than those of ultraviolet rays? longer

an exponential function from the average growth. Have students keep this task with their other project tasks in their project notebook.

Exercises MIXED REVIEW

GETTING READY FOR LESSON 8-7 These exercises prepare students for another multiplication property of exponents that will be covered in the next lesson.

Wrap Up

THE BIG IDEA Ask students to give an example that illustrates the multiplication property of exponents used in this lesson.

RETEACHING ACTIVITY Students multiply powers with the same base. (Reteaching worksheet 8-6)

Exercises CHECKPOINT

Students will assess their own progress on Lessons 8-4 to 8-6.

WRITING Exercise 17 Students should recognize that this exercise involves combining like terms.

Chapter Project *Find Out by Analyzing*

- Starting with the first day you see at least one square of mold, find the percent of growth from each day to the next.
- Use the average of these percents as an estimate for the base of an exponential function.
- Write an exponential function to fit the data. Graph it on your data plot.

Exercises MIXED REVIEW

Identify each square root as *rational* or *irrational*. Simplify if possible.

60. $\sqrt{121}$ rat.; 11 **61.** $\sqrt{67}$ irrat. **62.** $-\sqrt{49}$ rat.; -7 **63.** $\sqrt{\frac{1}{4}}$ rat.; $\frac{1}{2}$ **64.** $-\sqrt{12}$ irrat. **65.** $\sqrt{\frac{9}{16}}$ rat.; $\frac{3}{4}$

66. Sports In 1989, United States sales of in-line skates were \$21 million. In 1994, sales were \$369 million. Find the percent of increase. about 1657%

Getting Ready for Lesson 8-7

Rewrite each expression with one exponent.

67. $3^2 \cdot 3^2 \cdot 3^2$ 3^6 **68.** $2^3 \cdot 2^3 \cdot 2^3 \cdot 2^3$ 2^{12} **69.** $5^7 \cdot 5^7 \cdot 5^7 \cdot 5^7$ 5^{28} **70.** $7 \cdot 7 \cdot 7$ 7^3

Exercises CHECKPOINT

Write each expression as an integer or simple fraction.

1. $(-7.3)^0$ 1 **2.** 3^{-2} $\frac{1}{9}$ **3.** -8^{-1} $-\frac{1}{8}$ **4.** $\left(\frac{1}{5}\right)^{-3}$ 125 **5.** -4^2 -16 **6.** $(-2)^{-3}$ $-\frac{1}{8}$

Simplify each expression. Use only positive exponents.

7. $s^{-2} \cdot s^4 \cdot s$ s^3 **8.** $a^7b^2 \cdot 21a^{-6}$ $21ab^2$ **9.** $(2a^3)(-3a^{-3})(\frac{1}{6}a^0)$ -1 **10.** g^3h^{-4} $\frac{g^3}{h^4}$ **11.** $\frac{x^{-5}}{y^2y^{-8}}$ $\frac{y^6}{x^5}$

Mental Math **Simplify. Give your answers in scientific notation.**

12. $(5 \times 10^6) \times 3$ 1.5×10^7 **13.** $0.4 \times (2 \times 10^{-7})$ 8×10^{-8} **14.** $(9 \times 10^{11}) \div 3$ 3×10^{11} **15.** $6 \times (4 \times 10^3)$ 2.4×10^4

16. a. Biology Georg Frey collected beetles. When he died, his collection contained 3 million beetles of 90,000 different species. Write each number in scientific notation. 3×10^6; 9×10^4

b. About how many beetles fit into each of his 6500 packing cases? about 462

17. Writing LaWanda wrote $a^5 + a^5 = 2a^5$. Amanda wrote $a^5 + a^5 = a^{10}$. Whose work is correct? Explain. LaWanda; Amanda used the multiplication rule to add powers.

Graph each function over the domain $\{-3 \leq x \leq 3\}$. 18–20. See margin.

18. $f(x) = 0.8 \cdot 2^x$ **19.** $y = 0.5 \cdot 1.5^x$ **20.** $g(x) = 2 \cdot 3^x$

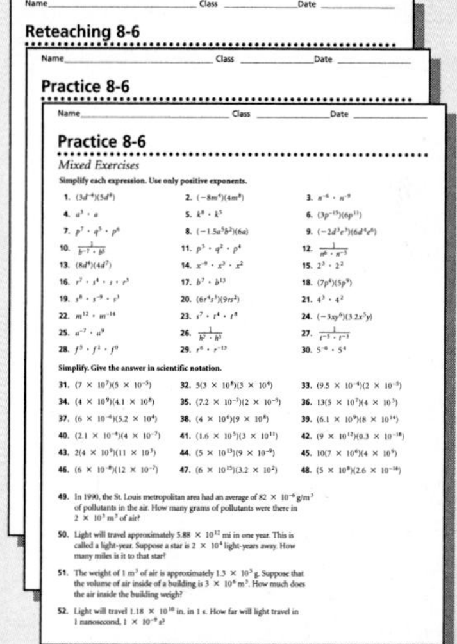
Reteaching 8-6

Practice 8-6

Practice 8-6

Mixed Exercises

Lesson Quiz

Lesson Quiz is also available in Transparencies.

Multiply. Give your answers in scientific notation.

1. $(3 \times 10^4)(5 \times 10^2)$ 1.5×10^7

2. $(7 \times 10^{-4})(1 \times 10^5)$ 7×10^1

Simplify each expression. Use only positive exponents.

3. $(2.5z^{-2})(-3z^4)$ $-7.5z^2$

4. $(-2i^{-2})(-3i^2j^{-2})(-5j^{-3})$ $-\frac{30}{j^5}$

8-7 Teaching Notes

PROBLEM OF THE DAY

The expected enrollment at Jordan High School is as follows: 9th grade, 693; 10th grade, 652; 11th grade, 594; 12th grade, 523. The current building capacity is 1750 students. The new wing under construction has a capacity of 450 students. How many portable classrooms must be built at Jordan High School if each one has a capacity of 28 students? ten portables

Problem of the Day is also available in Transparencies.

CONNECTING TO PRIOR KNOWLEDGE Ask students how 8 can be written as a power of 2. 2^3 On the board, write $2^3 = 8$. Then write $2^8 = 2^{2^3}$. Ask students whether they think it is correct to replace the 8 with 2^3. Ask: *What math concepts have you learned in this chapter that can help you decide whether this expression is correct?*

THINK AND DISCUSS

Questions 1 and 2 These questions will help students understand the rule for raising a power to a power. It is essential that students understand why the rules of exponents work. Without this understanding, the rules can be confusing and difficult for students to remember.

ERROR ALERT! **Question 4** Some students may evaluate 25^4 as 100. **Remediation:** Remind them that $25^4 = 25 \cdot 25 \cdot 25 \cdot 25$.

Lesson Planning Options

Prerequisite Skills

- Using positive and negative exponents
- Simplifying expressions

Assignment Options

To provide flexible scheduling, this lesson can be subdivided into parts.

1 **Core** 1–28, 41–46
Extension 29, 47–50

2 **Core** 30–36, 52–55
Extension 37–40, 51

Use Mixed Review to maintain skills.

Resources

Student Edition

Skills Handbook, pp. 580, 582
Extra Practice, p. 563
Glossary/Study Guide

Teaching Resources

Chapter Support File, Ch. 8
- Practice 8-7 (two worksheets)
- Reteaching 8-7

Classroom Manager 8-7
Glossary, Spanish Resources
Two-Year Algebra Handbook 8-7

Transparencies
11, 100

Connections  Physical Science . . . and more

8-7 More Multiplication Properties of Exponents

What You'll Learn

- Using two more multiplication properties of exponents

...And Why

To solve problems that involve raising numbers in scientific notation to a power

What You'll Need

- calculator

THINK AND DISCUSS

Part 1

Raising a Power to a Power

In Lesson 8-6, you used patterns to discover how to multiply powers with the same base. You can use what you discovered there to find a shortcut for simplifying expressions such as $(8^6)^3$.

1. Copy and complete each statement.
   a. $(a^3)^2 = a^3 \cdot a^3 = a^{3 \cdot \blacksquare} = a^{\blacksquare}$ 2; 6
   b. $(4^5)^3 = 4^5 \cdot 4^5 \cdot 4^5 = 4^{\blacksquare + \blacksquare + \blacksquare} = 4^{5 \cdot \blacksquare} = 4^{\blacksquare}$ 5; 5; 5; 3; 15
   c. $(2^7)^4 = 2^7 \cdot 2^7 \cdot 2^7 \cdot 2^7 = 2^{7 \cdot \blacksquare} = 2^{\blacksquare}$ 4; 28
   d. $(8^6) = 8^6 \cdot 8^6 = 8^{\blacksquare + \blacksquare} = 8^{6 \cdot \blacksquare} = 8^{\blacksquare}$ 6; 6; 2; 12
   e. $(g^4)^3 = g^4 \cdot g^4 \cdot g^4 = g^{4 \cdot \blacksquare} = g^{\blacksquare}$ 3; 12

2. *Patterns* Look for a pattern in your answers to Question 1. Write a shortcut for simplifying an expression such as $(8^6)^3$. See left.

2. The power of the result is the product of the powers in the original expression; $(8^6)^3 = 8^{6 \cdot 3} = 8^{18}$.

You can use the property below to simplify some exponential expressions.

Raising a Power to a Power

For any nonzero number a and any integers m and n, $(a^m)^n = a^{mn}$.

Example: $(5^4)^2 = 5^{4 \cdot 2} = 5^8$

3. The work you did in Question 1 involves repeated multiplication. Simplify each expression below by using repeated multiplication. Then check your work by using the property.
   a. $(2^3)^2$ 2^6
   b. $(h^2)^4$ h^8
   c. $(3^3)^4$ 3^{12}

4. Simplify each expression. What do you notice?
   a. $(5^2)^4 = 25^4 = \blacksquare$ 390,625
   b. $(5^2)^4 = 5^8 = \blacksquare$ 390,625

   The result is the same using either procedure.

5. Use the property to find each missing value. Then use a calculator to evaluate each expression and verify your answers.
   a. $(3^{\blacksquare})^4 = 9^4$ 2; 6561
   b. $(5^2)^{\blacksquare} = 5^8$ 4; 390,625
   c. $(2^{\blacksquare})^{\blacksquare} = 4^2$ 2; 2; 16
   d. $(3^4)^2 = 3^{\blacksquare}$ 8; 6561
   e. $(1.1^5)^7 = 1.1^{\blacksquare}$ 35; 28.1
   f. $(123^0)^{87} = 123^{\blacksquare}$ 0; 1

CALCULATOR HINT

When evaluating an expression such as $(3^4)^2$, you do not need to use parentheses because the calculator evaluates powers from left to right. Pressing 3 ^ 4 ^ 2 ENTER will give you 6561.

Example 1

Simplify each expression.

a. $(x^3)^6 = x^{3 \cdot 6}$ ← Multiply exponents when raising a power to a power.
$= x^{18}$ ← Simplify.

Example 1

Part (b) allows students to use the multiplication property from this lesson and the multiplication property from the previous lesson. Some students may confuse these two properties. It may help to show the following example using repeated factors of c.

$c^5 (c^3)^2 = (c \cdot c \cdot c \cdot c \cdot c)(c \cdot c \cdot c)(c \cdot c \cdot c)$

VISUAL/TACTILE LEARNING Encourage small groups of students to make class posters that describe what they have learned about using exponents. Have them include examples of each rule they describe in the poster. Suggest that students leave space on the poster to add more material after they have completed Lesson 8-8.

Example 2

You may wish to have students verify each example using repeated multiplication. For example,

$$\begin{aligned}(2x^2)^4 &= (2x^2)(2x^2)(2x^2)(2x^2)\\ &= (2 \cdot x \cdot x)(2 \cdot x \cdot x)(2 \cdot x \cdot x)(2 \cdot x \cdot x)\\ &= (2 \cdot 2 \cdot 2 \cdot 2)(x \cdot x \cdot x \cdot x \cdot x \cdot x \cdot x \cdot x)\\ &= 2^4x^8\end{aligned}$$

To help students who have difficulty evaluating 2^4 mentally, have them think of $2 \cdot 2 \cdot 2 \cdot 2$ as $4 \cdot 4$. In a similar fashion, they can think of 3^4 as $3 \cdot 3 \cdot 3 \cdot 3$ and as $9 \cdot 9$.

b.
$$\begin{aligned}c^5(c^3)^2 &= c^5 \cdot c^{3 \cdot 2} && \leftarrow \text{Multiply exponents in } (c^3)^2.\\ &= c^5 \cdot c^6 && \leftarrow \text{Simplify.}\\ &= c^{5+6} && \leftarrow \text{Add exponents when multiplying powers with the same base.}\\ &= c^{11}\end{aligned}$$

6. Try This Simplify $(a^4)^2 \cdot (a^2)^5$. a^{18}

Part 2

Raising a Product to a Power

You can use repeated multiplication to simplify expressions like $(5y)^3$.

$$\begin{aligned}(5y)^3 &= 5y \cdot 5y \cdot 5y\\ &= 5 \cdot 5 \cdot 5 \cdot y \cdot y \cdot y\\ &= 5^3 \cdot y^3\\ &= 125y^3\end{aligned}$$

Notice from the steps above that $(5y)^3 = 5^3y^3$. This illustrates another property of exponents.

Raising a Product to a Power

For any nonzero numbers a and b and any integer n, $(ab)^n = a^nb^n$.

Example: $(3x)^4 = 3^4x^4 = 81x^4$

7. Calculator **Verify** each equation for two values of the variable (other than 0 and 1). Answers may vary. See left for samples.

a. $(5y)^3 = 125y^3$ b. $(3x)^4 = 3^4x^4$ c. $(7c)^2 = 49c^2$

7a. $(5 \cdot 2)^3 = 1000$; $125 \cdot 3^3 = 1000$; $(5 \cdot 3)^3 = 3375$; $125 \cdot 3^3 = 3375$
7b. $(3 \cdot 2)^4 = 1296$; $3^4 2^4 = 1296$; $(3 \cdot 5)^4 = 50{,}625$; $3^4 5^4 = 50{,}625$
7c. $(7 \cdot 2)^2 = 196$; $49 \cdot 2^2 = 196$; $(7 \cdot 4)^2 = 784$; $49 \cdot 4^2 = 784$

Example 2

Simplify each expression.

a.
$$\begin{aligned}(2x^2)^4 &= 2^4(x^2)^4 && \leftarrow \text{Raise each factor to the 4th power.}\\ &= 2^4x^8 && \leftarrow \text{Multiply exponents.}\\ &= 16x^8 && \leftarrow \text{Simplify.}\end{aligned}$$

b.
$$\begin{aligned}(x^{-2})^2(3xy^2)^4 &= (x^{-2})^2 \cdot 3^4x^4(y^2)^4 && \leftarrow \text{Raise three factors to the 4th power.}\\ &= x^{-4} \cdot 3^4x^4y^8 && \leftarrow \text{Multiply exponents.}\\ &= 3^4 \cdot x^{-4} \cdot x^4 \cdot y^8 && \leftarrow \text{Rearrange factors.}\\ &= 3^4x^0y^8 && \leftarrow \text{Add exponents.}\\ &= 81y^8 && \leftarrow \text{Simplify.}\end{aligned}$$

8. Try This Simplify each expression.

a. $(15z)^3$ $3375z^3$ b. $(4g^5)^2$ $16g^{10}$ c. $(6mn)^3(5m^{-3})^4$ $\frac{135{,}000n^3}{m^9}$

Additional Examples

FOR EXAMPLE 1

Simplify each expression.

a. $(a^3)^4$ a^{12}

b. $(b^2)^3b^2$ b^8

FOR EXAMPLE 2

Simplify each expression.

a. $(3x^3)^2$ $9x^6$

b. $(4x^2y^3)^2(x^3)^{-2}$ $\frac{16y^6}{x^2}$

FOR EXAMPLE 3

Find the kinetic energy in joules of a spacecraft using the formula $E = \frac{1}{2}mv^2$. The spacecraft has a mass m of 1.7×10^4 kg and a velocity v of 6.2×10^4 m/s. 3.27×10^{18} joules

Example 3 Relating to the Real World 

Be sure students see the connection between Einstein's famous formula $e = mc^2$ (See margin of student page.) and the expression $10^{-3} \cdot (3 \times 10^8)^2$.

MAKING CONNECTIONS In 1939, scientists under Enrico Fermi succeeded in firing neutrons—sub-atomic particles—at uranium atoms. The uranium atoms split into smaller fragments, and part of the uranium's mass was converted to energy, proving Einstein's theory of relativity ($e = mc^2$).

Exercises ON YOUR OWN

ALTERNATIVE ASSESSMENT **Exercises 1–28** Assign one exercise to each student. Ask students to rewrite their assigned expression three ways by using carefully chosen parentheses and exponents. Then simplify each new expression and compare the results.

Exercises 21, 23, 26 This lesson stressed the rule $(ab)^n = a^n b^n$. Help students discover that to simplify the expressions in these exercises, they must work backward and use the rule $a^n b^n = (ab)^n$.

ERROR ALERT! **Exercise 22** Some students may multiply $25x^2$ and $5x^2$. **Remediation:** Point out that this exercise involves addition and requires use of the rule for combining like terms.

Exercise 29c Encourage students to talk about how they visualize solving this problem. For instance, they might imagine all the water poured into a box for which the area of the bottom equals the surface area of the Earth that is water.

Technology Options

For Exercises 30–35, students find it helpful if they use a graphing calculator to multiply the numbers in scientific notation.

Prentice Hall Technology 

 **Software** • Secondary Math Lab Toolkit • Computer Item Generator 8-7

Internet • See the Prentice Hall site. (http://www.phschool.com)

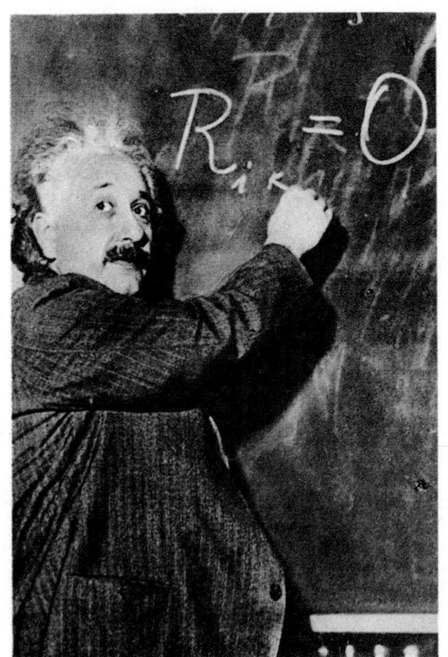

Who? Albert Einstein is famous for discovering the relationship $e = mc^2$, where e is energy (in joules), m is mass (in kg), and c is the speed of light (about 3×10^8 meters per second).

You can use the properties of exponents to solve real-world problems.

Example 3 Relating to the Real World 

Physical Science All objects, even resting ones, contain energy. The expression $10^{-3} \cdot (3 \times 10^8)^2$ describes the amount of resting energy a raisin contains, where 10^{-3} kg is the mass of the raisin. Simplify the expression.

$10^{-3} \cdot (3 \times 10^8)^2$
$= 10^{-3} \cdot 3^2 \cdot (10^8)^2$ ← Raise two factors to the 2nd power.
$= 10^{-3} \cdot 3^2 \cdot 10^{16}$ ← Multiply exponents.
$= 3^2 \cdot 10^{-3} \cdot 10^{16}$ ← Rearrange factors.
$= 3^2 \cdot 10^{-3+16}$ ← Add exponents.
$= 9 \times 10^{13}$ ← Simplify.

One raisin contains about 9×10^{13} joules of resting energy.

9. Multiply. Give your answers in scientific notation.
   a. $(4 \times 10^5)^2$ 1.6×10^{11}
   b. $(2 \times 10^{-10})^3$ 8×10^{-30}
   c. $(10^3)^4(4.3 \times 10^{-8})$ 4.3×10^4

10. Energy An hour of television use consumes 1.45×10^{-1} kWh of electricity. Each kilowatt-hour (kWh) of electric use is equivalent to 3.6×10^6 joules of energy.
    a. How many joules does a television use in 1 h? (Hint: Use a proportion.) 522,000
    b. Suppose you could release the resting energy in a raisin. About how many hours of television use could be powered by that energy? 172,413,793.1 h

Exercises ON YOUR OWN

Simplify each expression. Use positive exponents.

1. $(xy)^9$ x^9y^9
2. $(c^5)^2$ c^{10}
3. $(a^2b^4)^3$ a^6b^{12}
4. $(3m^3)^4$ $81m^{12}$
5. $(g^{10})^{-4}$ $\frac{1}{g^{40}}$
6. $g^{10} \cdot g^{-4}$ g^6
7. $(x^3y)^4$ $x^{12}y^4$
8. $(x^{-2})^3x^{-12}$ $\frac{1}{x^{18}}$
9. $(0.5^2)^{-2}$ 16
10. $(2xy)^3x^2$ $8x^5y^3$
11. $(g \cdot g^4)^2$ g^{10}
12. $(c^2)^{-2}(c^3)^4$ c^8
13. $s^3(s^2)^4$ s^{11}
14. $(mg^4)^{-1}$ $\frac{1}{mg^4}$
15. $m(g^4)^{-1}$ $\frac{m}{g^4}$
16. $(7cd^4)^2$ $49c^2d^8$
17. $(2p^6)^0$ 1
18. $(5ac^3)^{-2}$ $\frac{1}{25a^2c^6}$
19. $(4a^2b)^3(ab)^3$ $64a^9b^6$
20. $(4xy^2)^4(2y)^{-3}$ $32x^4y^5$
21. $3^7 \cdot \left(\frac{1}{3}\right)^7$ 1
22. $(5x)^2+5x^2$ $30x^2$
23. $2^4 \cdot 5^4$ 10,000
24. $(64.1^{-3})^0$ 1
25. $(b^n)^3b^2$ b^{3n+2}
26. $15^2 \cdot (0.2)^2$ 9
27. $(4.1)^5 \cdot (4.1)^{-5}$ 1
28. $3^2 \cdot (3x)^3$ $243x^3$

Exercise 36 Suggest to students that they begin by writing 6^4 as a power of 2. 2^6

WRITING Exercise 37 Students should include properties from this lesson and the previous lesson.

GEOMETRY Exercise 38 Ask: How many surfaces does a cube have? 6 Make a model or have groups of students make a model of one or both cubes. Choose a value for x and construct the cubes out of cardboard. The models can be used to check the answer for parts (b) and (d).

EXTENSION/OPEN-ENDED Exercise 39 Ask students: *List all the different possibilities for m and n.* $m = 1, n = 24$; $m = 2, n = 12$; $m = 3, n = 8$; $m = 4, n = 6$; $m = 6, n = 4$; $m = 8, n = 3$; $m = 2, n = 2$; $m = 24, n = 1$

How many possibilities are there? 8

Exercise 40 Sometimes students will solve a problem without giving much thought to the content. Encourage them to think about how small the chip really is.

ERROR ALERT! Exercise 47 Some students may give the answer as 2^6. **Remediation:** Point out that the answer is 6. They have been asked to find the value of x that makes the equation true.

Chapter Project **FIND OUT BY INTERPRETING** Students will use the exponential functions they created to predict mold growth for the next two weeks. After the two weeks, they will discuss the factors that affected the accuracy of their predictions. After the experiment is finished, have them complete Finishing the Project on page 406.

29. a. **Geography** Earth has a radius of about 6.4×10^6 m. Approximate the surface area of Earth by using the formula for the surface area of a sphere, $S = 4\pi r^2$. about 5.15×10^{14} m^2
    b. Earth's surface is about 70% water, almost all of it in oceans. About how many square meters of Earth's surface are covered with water? See below.
    c. The oceans have an average depth of 3795 m. What is the approximate volume of water on Earth?

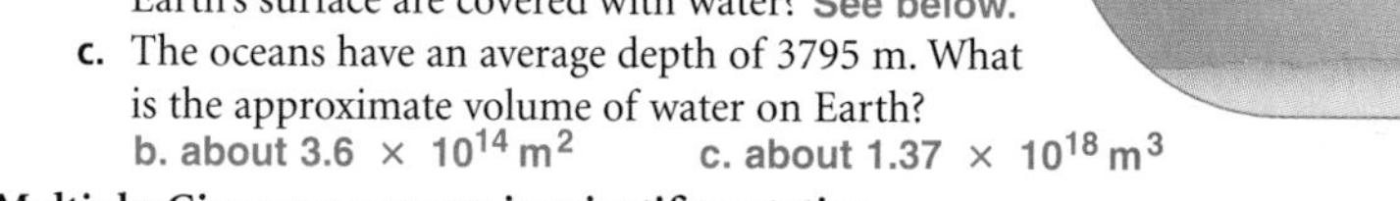

b. about 3.6×10^{14} m^2 c. about 1.37×10^{18} m^3

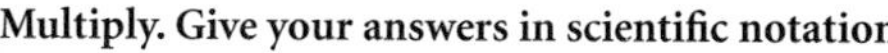

Multiply. Give your answers in scientific notation.

30. $(3 \times 10^5)^2$ 9×10^{10}
31. $(2 \times 10^{-3})^3$ 8×10^{-9}
32. $(7 \times 10^4)^2$ 4.9×10^9
33. $(6 \times 10^{12})^2$ 3.6×10^{25}
34. $(3 \times 10^{-4})^3$ 2.7×10^{-11}
35. $(4 \times 10^8)^{-2}$ 6.25×10^{-18}

36. Which expression or expressions do *not* equal 64? C
    A. $2^5 \cdot 2$ B. 2^6 C. $2^2 \cdot 2^3$ D. $(2^3)^2$ E. $(2^2)(2^2)^2$

37. **Writing** Explain how the properties of exponents help you simplify algebraic expressions. Answers may vary. Sample: Products of powers with the same base can be combined by adding exponents.

38. a. **Geometry** Write an expression for the surface area of each cube. $24x^2$, $96x^2$
    b. How many times greater than the surface area of the small cube is the surface area of the large cube? 4
    c. Write an expression for the volume of each cube. $8x^3$, $64x^3$
    d. How many times greater than the volume of the small cube is the volume of the large cube? 8

2x 4x

39. **Open-ended** Write a^{24} as a product of the form $(a^m)^n$ in four different ways. Use only positive exponents.
Answers may vary. Sample: $(a^1)^{24}$, $(a^2)^{12}$, $(a^3)^8$, $(a^8)^3$

40. **Technology** In computer memory chips currently being developed, a square piece of chip is one thousandth of an inch (10^{-3} in.) on each side. It will hold 3000 bits of data. Find the area of the piece of chip. 10^{-6} in.^{2}

Complete each equation.

41. $(x^2)^{■} = x^6$ 3
42. $(x^{■})^3 = x^{-12}$ −4
43. $(ab^2)^{■} = a^4b^8$ 4
44. $(5x^{■})^2 = 25x^4$ 2
45. $(3x^3y^{■})^3 = 27x^6$ 0
46. $(m^2n^3)^{■} = \frac{1}{m^6n^9}$ −3

Solve each equation. Use the sample below.

Sample:
$25^3 = 5^x$
$(5^2)^3 = 5^x$ ← Write 25 as a power of 5.
$5^6 = 5^x$ ← Multiply exponents.
$x = 6$

The solution is 6.

47. $8^2 = 2^x$ 6
48. $3^x = 27^4$ 12
49. $4^x = 2^6$ 3
50. $5^6 = 25^x$ 3

Exercises MIXED REVIEW

GETTING READY FOR LESSON 8-8 These exercises prepare students for the division property of exponents which will be taught in the next lesson.

JOURNAL Encourage students to use both numerals and variables in their examples.

Wrap Up

THE BIG IDEA Ask students to tell what is meant by "raising a power to a power" and "raising a product to a power."

RETEACHING ACTIVITY Students simplify expressions by multiplying exponents. (Reteaching worksheet 8-7)

A Point in Time

If you have block scheduling or an extended class period, you may wish to have students investigate these topics:

- *To cover the head of a pin would require 10,000 human cells. How many cells are in each human being?* **10,000,000,000,000 cells**
- *October 15, 1924, the longest airship flight ended in Lakehurst, New Jersey. Name the German dirigible and the amount of miles for the flight.* **ZR-3, 5060 mi**
- *The development of the skyscraper was attributed to Louis Henri Sullivan, who died in 1924. What structures in Chicago were designed by this architect?* **Chicago Auditorium, Schiller Building, Schlesinger and Meyer Building**

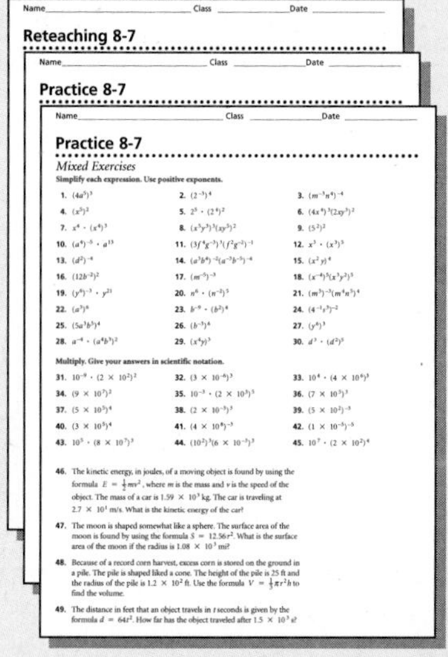

Reteaching 8-7

Practice 8-7

Practice 8-7

Mixed Exercises

Lesson Quiz

Lesson Quiz is also available in Transparencies.

1. Simplify $(x^3y^2)^3$ using only positive exponents. x^9y^6
2. Simplify $(2x^2)^{-2}(3x^2y)^3$ using only positive exponents. $\frac{27}{4}x^2y^3$
3. Multiply $(3 \times 10^{12})^3$. Express your answer in scientific notation. 2.7×10^{37}
4. Rewrite the expression $4x^2y^{-2}$ using only one exponent. Use parentheses. $\left(\frac{2x}{y}\right)^2$

51. *Biology* There are an estimated 200 million insects for each person on Earth. The world population is about 5.5 billion. About how many insects are there on Earth? 1.1×10^{18} **insects**

Write each expression with only one exponent. Use parentheses.

52. $m^4 \cdot n^4$ $(mn)^4$

53. $(a^5)(b^5)(a^0)$ $(ab)^5$

54. $49x^2y^2z^2$ $(7xyz)^2$

55. $\frac{12x^{-2}}{3y^2}$ $\left(\frac{2}{xy}\right)^2$

Chapter Project *Find Out by Interpreting*

Use your exponential function to predict the growth of your mold at two weeks. Compare your predictions to the actual growth of the mold. Discuss any factors that may influence the accuracy of your exponential model.

Exercises MIXED REVIEW

56. Solve $3x^2 - 5x - 2 = 0$. $2, -\frac{1}{3}$

57. Solve $x^2 + 2x + 1 = 0$. -1

58. *Textiles* To make felt, thin layers of cotton are built up $\frac{1}{32}$ in. at a time. How many layers does it take to make felt $\frac{1}{4}$ in. thick? **8 layers**

SELF ASSESSMENT — FOR YOUR JOURNAL

Summarize the three multiplication properties of exponents that you have learned in the last two lessons. Give an example of each.

Getting Ready for Lesson 8-8

Simplify each expression.

59. $x^2 \cdot x^{-3}$ $\frac{1}{x}$

60. $u^8 \cdot u^{-4}$ u^4

61. $t^{-6} \cdot t$ $\frac{1}{t^5}$

62. $h^{-3} \cdot h^0$ $\frac{1}{h^3}$

A Point in Time

1500 1600 1700 1800 1900 2000

Dr. Jewel Plummer Cobb

Dr. Jewel Plummer Cobb was born in 1924 and obtained her master's and doctorate degrees in cell physiology from New York University. Dr. Cobb has concentrated on the study of normal and malignant skin cells and has published nearly fifty books, articles, and reports.

Because the number of cancer cells grows exponentially, cell biologists often write cancer cell data in scientific notation.

8-8 Teaching Notes

PROBLEM OF THE DAY

A furniture builder receives materials to build three-legged stools and four-legged stools. He received 32 stool tops and 108 legs. How many four-legged stools should he build? 12

Problem of the Day is also available in Transparencies.

CONNECTING TO PRIOR KNOWLEDGE Ask students: *What have you learned recently about multiplying two powers with the same base, such as 2^3 times 2^4?* To multiply two powers with the same base, add the exponents. Encourage students to make conjectures about how they might divide two powers with the same base.

THINK AND DISCUSS

Help students see the connection between the rule for dividing powers with the same base and the rule for multiplying powers with the same base.

$$\frac{5^6}{5^2} = 5^6 \div 5^2 = 5^6 \times \frac{1}{5^2} = 5^6 \times 5^{-2} = 5^4$$

Example 1

You can verify the answer to part (b) by using the rule for multiplying powers with the same base.

Connections Environment . . . and more

8-8 Division Properties of Exponents

What You'll Learn

Applying division properties of exponents

...And Why

To solve problems that involve dividing numbers in scientific notation

What You'll Need

calculator

THINK AND DISCUSS

Part 1 Dividing Powers with the Same Base

In Lessons 8-6 and 8-7 you studied patterns that occur when you multiply powers with the same base. You can see a similar pattern when you divide powers with the same base.

You can use repeated multiplication to simplify fractions. Expand the numerator and denominator using repeated multiplication. Then cancel like terms.

$$\frac{5^6}{5^2} = \frac{5 \cdot 5 \cdot 5 \cdot 5 \cdot 5 \cdot 5}{5 \cdot 5} = 5^4$$

The pattern is that the exponents are subtracted.

1. Simplify each expression by expanding the numerator and the denominator and then canceling. Describe any patterns you see.
   a. $\frac{5^7}{5^3}$ 5^4 b. $\frac{5^{12}}{5^8}$ 5^4 c. $\frac{5^8}{5^4}$ 5^4 d. $\frac{5^5}{5^1}$ 5^4 e. $\frac{5^5}{5^8}$ $\frac{1}{5^3}$

Dividing Powers with the Same Base

For any nonzero number a and any integers m and n, $\frac{a^m}{a^n} = a^{m-n}$.

Example: $\frac{3^7}{3^2} = 3^{7-2} = 3^5$

2. **Calculator** Find each missing value. Use a calculator to check.
   a. $\frac{5^9}{5^2} = 5^{■}$ 7 b. $\frac{2^4}{2^3} = 2^{■}$ 1 c. $\frac{3^2}{3^5} = 3^{■}$ −3 d. $\frac{5^3}{5^2} = 5^{■}$ 1

Example 1

Simplify each expression. Use only positive exponents.

a. $\frac{a^6}{a^{14}} = a^{6-14}$ ← Subtract exponents when dividing powers with the same base.
$= a^{-8}$
$= \frac{1}{a^8}$ ← Rewrite using positive exponents.

b. $\frac{c^{-1}d^3}{c^5d^{-4}} = c^{-1-5} \cdot d^{3-(-4)}$ ← Subtract exponents.
$= c^{-6}d^7$
$= \frac{d^7}{c^6}$ ← Use positive exponents.

3. **Try This** Simplify each expression. Use positive exponents.
   a. $\frac{b^4}{b^9}$ $\frac{1}{b^5}$ b. $\frac{z^{10}}{z^5}$ z^5 c. $\frac{a^2b}{a^4b^3}$ $\frac{1}{a^2b^2}$ d. $\frac{m^{-1}n^2}{m^3n}$ $\frac{n}{m^4}$ e. $\frac{x^2yz^4}{xy^4z^{-3}}$ $\frac{xz^7}{y^3}$

Lesson Planning Options

Prerequisite Skills

- Using positive and negative exponents
- Simplifying expressions

Assignment Options

To provide flexible scheduling, this lesson can be subdivided into parts.

Core 1–12, 19–23
Extension 13–18, 24–26

Core 27–50
Extension 51–58

Use Mixed Review to maintain skills.

Resources

Student Edition

Skills Handbook, p. 580
Extra Practice, p. 563
Glossary/Study Guide

Teaching Resources

Chapter Support File, Ch. 8
- Practice 8-8 (two worksheets)
- Reteaching 8-8

Classroom Manager 8-8
Glossary, Spanish Resources
Two-Year Algebra Handbook 8-8

Transparencies

100

$$\begin{aligned}\frac{c^{-1}d^3}{c^5d^{-4}} &= c^{-1}d^3 \div c^5d^{-4}\\ &= c^{-1}d^3 \times \frac{1}{c^5d^{-4}}\\ &= c^{-1}d^3 \times c^{-5}d^4\\ &= c^{-1}c^{-5}d^3d^4\\ &= c^{-6}d^7\\ &= \frac{d^7}{c^6}\end{aligned}$$

ALTERNATIVE METHOD As students work through examples, some may notice that when a number or a variable in the numerator or denominator is raised to a power, it can be moved to the denominator or numerator respectively by changing the sign of the power. $\frac{c^{-1}d^3}{c^5d^{-4}} = \frac{d^3d^4}{c^5c^1} = \frac{d^7}{c^6}$

Example 2 **Relating to the Real World**

CONNECTING TO THE STUDENTS' WORLD Challenge students to collect and weigh all the recyclable and unrecyclable waste paper that their household generates in one week. Have students bring their data to class to combine with other students' data. Have students calculate the weight of the paper they recycled and compare the weight to the national average.

Question 5 Ask students: *How can you check that your answer is in scientific notation?* **One factor must be less than or equal to one and less than ten. The other factor must be written as a power of ten.**

TACTILE LEARNING Question 6 Have pairs of students work together. Have one student perform the calculation by hand

Additional Examples

FOR EXAMPLE 1

Simplify each expression. Use only positive exponents.

a. $\frac{x^4}{x^5}$ $\frac{1}{x}$

b. $\frac{i^3j^{-4}}{i^{-3}j^{-6}}$ i^6j^2

FOR EXAMPLE 2

James' heart beats approximately 4.6×10^7 beats in a year. If there are 3.8×10^5 min in a year, what is James' average heart rate (beats per min)? **121 beats per min**

Discussion: *Why is your answer only an approximation?*

FOR EXAMPLE 3

Simplify each expression.

a. $\left(\frac{2}{y^3}\right)^2$ $\frac{4}{y^6}$

b. $\left(\frac{2x}{3}\right)^{-2}$ $\frac{9}{4x^2}$

Example 2 **Relating to the Real World** 

Environment In 1993 the total amount of waste paper and cardboard recycled in the United States was 77.8 million tons. The population in 1993 was 258.1 million. How much paper was recycled per person?

$$\begin{aligned}\frac{77.8 \text{ million tons}}{258.1 \text{ million people}} &= \frac{77.8 \times 10^7 \text{ tons}}{2.581 \times 10^8 \text{ people}} && \leftarrow \text{Write in scientific notation.}\\ &= \frac{7.78}{2.581} \times 10^{7-8} && \leftarrow \text{Subtract exponents when dividing powers with the same base.}\\ &= \frac{7.78}{2.581} \times 10^{-1}\\ &\approx 3.01 \times 10^{-1} && \leftarrow \text{Simplify.}\\ &\approx 0.3\end{aligned}$$

In 1993, 0.3 ton (600 lb!) of waste paper was recycled per person.

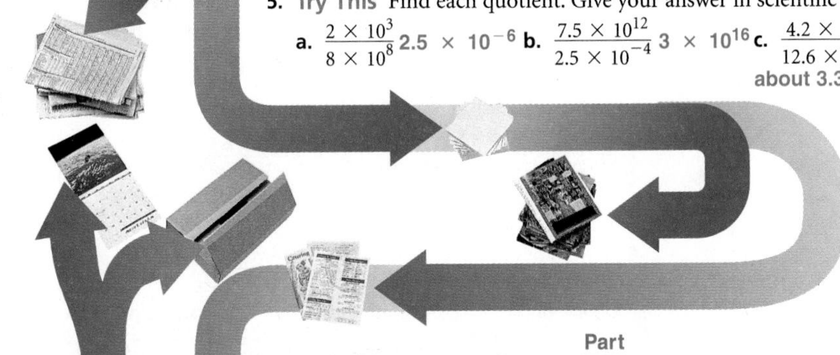

4. Only 34% of the waste paper generated in 1993 was recycled. How many pounds of waste paper were generated per person in 1993? **Answers may vary. Sample: about 1800 lb**
5. *Try This* Find each quotient. Give your answer in scientific notation.
   a. $\frac{2 \times 10^3}{8 \times 10^8}$ **2.5×10^{-6}** **b.** $\frac{7.5 \times 10^{12}}{2.5 \times 10^{-4}}$ **3×10^{16}** **c.** $\frac{4.2 \times 10^5}{12.6 \times 10^2}$ **about 3.33×10^2**

Part 2

Raising a Quotient to a Power

You can use repeated multiplication to simplify the expression $\left(\frac{x}{y}\right)^3$.

$$\begin{aligned}\left(\frac{x}{y}\right)^3 &= \frac{x}{y} \cdot \frac{x}{y} \cdot \frac{x}{y}\\ &= \frac{x \cdot x \cdot x}{y \cdot y \cdot y}\\ &= \frac{x^3}{y^3}\end{aligned}$$

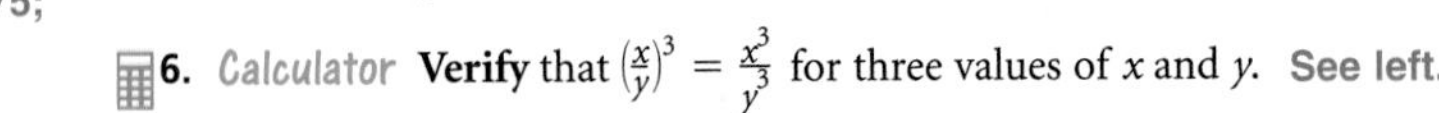

6. *Calculator* **Verify** that $\left(\frac{x}{y}\right)^3 = \frac{x^3}{y^3}$ for three values of x and y. **See left.**

6. Sample:
$\left(\frac{3}{2}\right)^3 = 3.375, \frac{3^3}{2^3} = 3.375;$
$\left(\frac{6}{3}\right)^3 = 8, \frac{6^3}{3^3} = 8$
$\left(\frac{4}{5}\right)^3 = 0.512, \frac{4^3}{5^3} = 0.512$

Raising a Quotient to a Power

For any nonzero numbers a and b, and any integer n, $\left(\frac{a}{b}\right)^n = \frac{a^n}{b^n}$.

Example: $\left(\frac{4}{5}\right)^3 = \frac{4^3}{5^3} = \frac{64}{125}$

while the other uses a calculator. Then have students compare answers to check each other's work.

Question 6 Ask: *What value cannot be used for y?* zero *Why?* Dividing by zero is undefined.

Example 3

Students have had experience writing fractions in simplest form. The concept here is similar, except that in this case they must also check that all exponents are positive.

Exercises ON YOUR OWN

Exercises 19–23 Ask: *For which exercises did you use mental math?*

OPEN-ENDED Exercise 24b Ask volunteers to share their examples.

Exercise 25 Encourage students to use scientific notation to solve this problem.

ESL Exercise 25b The term *projection* may be confusing since it can mean "to plan" in some other languages. Ensure students know that in this context, *projections* are predictions or educated guesses based on past experience or evidence.

ERROR ALERT! **Exercise 41** Some students may try to put the variable *b* in the simplified expression. **Remediation:** Remind them that any number raised to the zero power equals one, so no variable will appear when this expression is simplified.

Example 3

Simplify each expression.

a. $\left(\frac{4}{x^2}\right)^3 = \frac{4^3}{(x^2)^3}$ ← Raise the numerator and denominator to the 3rd power.

$= \frac{4^3}{x^6}$ ← Multiply exponents.

$= \frac{64}{x^6}$ ← Simplify.

b. $\left(-\frac{3}{5}\right)^{-2} = \left(\frac{-3}{5}\right)^{-2}$ ← Write the fraction with a negative numerator.

$= \frac{(-3)^{-2}}{5^{-2}}$ ← Raise the numerator and denominator to the −2 power.

$= \frac{5^2}{(-3)^2}$ ← Apply the definition of negative exponents.

$= \frac{25}{9}$ ← Simplify.

7. In Example 3, part (b), the first step could have been to rewrite the fraction as $\left(\frac{3}{-5}\right)^{-2}$. Explain how this step would affect the rest of your work. The last step would be $\frac{(-5)^2}{3^2} = \frac{25}{9}$.

8. *Open-ended* Suppose you wanted to simplify $\left(\frac{5}{3}\right)^3$. What would be your first step? Answers may vary. Sample: $\left(\frac{5}{3}\right)^3 = \frac{5^3}{3^3}$

When you have reduced an expression as far as possible, and when all your exponents are positive, the expression is in **simplest form.**

9. Is each expression in simplest form? If not, simplify it.

a. $\frac{(x^3)^2}{y^7}$ $\frac{x^6}{y^7}$ b. $x^{-3}y^2$ $\frac{y^2}{x^3}$ c. $\frac{a^5}{ab}$ $\frac{a^4}{b}$ d. $\frac{(2x)^2}{2x^2}$ 2

Sometimes you can combine steps to shorten your work.

10. Copy and complete each statement.

a. $\left(\frac{4a^3}{b}\right)^2 = \frac{\blacksquare a^{\blacksquare}}{b^{\blacksquare}}$ 16; 6; 2 ← Raising products, quotients, and powers to a power.

b. $\frac{x^{-3}x^8}{x^2} = x^{\blacksquare + \blacksquare - \blacksquare} = x^{\blacksquare}$ ← Multiplying and dividing powers with the same base.
−3; 8; 2; 3

11. *Try This* Simplify each expression.

a. $\left(\frac{2m^5}{m^2}\right)^4$ $16m^{12}$ b. $\left(\frac{2n^3t}{t^2}\right)^2$ $\frac{8n^9}{t^3}$ c. $\left(\frac{5x^7y^4}{x^{-2}}\right)^2$ $25x^{18}y^8$

Exercises ON YOUR OWN

Simplify each expression. Use only positive exponents.

1. $\frac{2^5}{2^7}$ $\frac{1}{4}$
2. $\left(\frac{2^2}{5}\right)^2$ $\frac{16}{25}$
3. $\left(\frac{3}{4}\right)^2$ $\frac{9}{16}$
4. $\left(\frac{2}{3}\right)^{-2}$ $\frac{9}{4}$
5. $\left(\frac{3^3}{3^4}\right)^2$ $\frac{1}{9}$
6. $\frac{c^{12}}{c^{15}}$ $\frac{1}{c^3}$
7. $\frac{x^{13}y^2}{x^{13}y}$ y
8. $\frac{m^{-2}}{m^{-5}}$ m^3
9. $\frac{(2a^7)(3a^2)}{6a^3}$ a^6
10. $\left(\frac{3b^2}{5}\right)^0$ 1
11. $\left(\frac{-2}{3}\right)^{-3}$ $-\frac{27}{8}$
12. $\frac{3^2 \cdot 5^0}{2^3}$ $\frac{9}{8}$

Technology Options

For Exercises 19–23 and 57, students may wish to use spreadsheet software or a graphing calculator to help handle the scientific notation.

Prentice Hall Technology

Software • Secondary Math Lab Toolkit • Computer Item Generator 8-8

Internet • See the Prentice Hall site. (http://www.phschool.com)

ALTERNATIVE ASSESSMENT **Exercises 38–41** To assess students' understanding of the division properties of exponents, ask them to solve these problems after changing the outside exponent to its corresponding negative number. For example, $\left(\frac{c^5}{c^9}\right)^3$ would become $\left(\frac{c^5}{c^9}\right)^{-3}$.

Exercises 52–56 Before students begin, have them generate the list of properties of exponents. Definitions of the properties of exponents can be found on the following pages.

Definition of Zero as an Exponent (page 379)

Definition of Negative Exponent (page 380)

Multiplying Powers with the Same Base (page 391)

Raising a Power to a Power (page 396)

Raising a Product to a Power (page 397)

Dividing Powers with the Same Base (page 401)

Exercises 55–56 Students should simplify the expressions, and write down each property as it is used.

Exercise 57 Ask: *What would the ratio be if the planet had a perfectly circular orbit?* 1 Have students discuss how they might predict the answer to part (b) before they calculate the ratios. Help students realize that they can look at just the first factor of each maximum and minimum and find the pair which are closest together and farthest apart.

MAKING CONNECTIONS **Exercise 58** Which blood type is the most common depends on where you are in the world. Type O is the most common worldwide; however, other types may be more common in specific regions. In the United States, subtype O+ is found in 38.4% of the population and is the most common; type AB—is the rarest and is found in only 0.7% of the population. The rarest type

pages 403–405 On Your Own

24a. Answers may vary. Sample: The quotient can be written as a product:

$$\frac{c^4}{c^6} = c^4c^{-6} = c^{4-6} = \frac{1}{c^{-(4-6)}} = \frac{1}{c^{6-4}}$$

52. $\frac{1}{2^3}$; Neg. Exp. Prop.
53. $\frac{1}{2^3}$; Div. and Neg. Exp. Prop.
54. $\frac{1}{2^3}$; Raising a Quotient to a Power Prop.
55. $\frac{1}{2^3}$; Mult. Prop.
56. $\frac{1}{2^3}$; Raising a Power to a Power, Div. and Neg. Exp. Prop.
57. 6.97×10^7; 4.59×10^7
1.075×10^8; 1.01
1.521×10^8; 1.03
2.491×10^8; 2.067×10^8; 1.21
8.157×10^8; 7.409×10^8; 1.10
1.507×10^9; 1.347×10^9; 1.12
3.004×10^9; 2.735×10^9; 1.10
4.537×10^9; 4.457×10^9; 1.08
7.375×10^9; 4.425×10^9; 1.67

pages 405 Mixed Review

62.
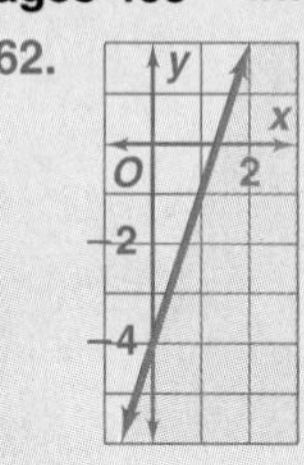

63–64. See back of book.

Explain why each expression is *not* in simplest form.

13. 5^3m^3 $5^3 = 125$
14. x^5y^{-2} $y^{-2} = \frac{1}{y^2}$
15. $(2c)^4$ $(2c)^4 = 16c^4$
16. $\frac{d^7}{d}$ $\frac{d^7}{d} = d^6$
17. x^0y $x^0 = 1$
18. $(3z^2)^3$ $(3z^2)^3 = 27z^6$

Choose **Use a calculator, paper and pencil, or mental math to simplify each quotient. Give your answers in scientific notation.**

19. $\frac{6.5 \times 10^{15}}{1.3 \times 10^8}$ 5×10^7
20. $\frac{2.7 \times 10^8}{0.9 \times 10^3}$ 3×10^5
21. $\frac{2.7 \times 10^8}{3 \times 10^5}$ 9×10^2
22. $\frac{8.4 \times 10^{-5}}{2 \times 10^{-8}}$ 4.2×10^3
23. $\frac{4.7 \times 10^{-4}}{3.1 \times 10^2}$ about 1.52×10^{-6}

24. a. **Writing** While simplifying the expression $\frac{c^4}{c^6}$, Kneale said, "I've found a property of exponents that's not in my algebra book!" Write an explanation of why Kneale's method works. See margin.
b. **Open-ended** Apply Kneale's method to an example you create.

Kneale
$\frac{c^4}{c^6} = \frac{1}{c^{6-4}} = \frac{1}{c^2}$

24b. Sample: $\frac{2^5}{2^8} = \frac{1}{2^{8-5}} = \frac{1}{2^3}$

25. a. **Television** In 1999 people in the United States will watch television a total of 450 billion hours, according to industry projections. The population will be about 274 million people. Find the number of hours of TV viewing per person for 1999. about 1642 h
b. According to these projections, about how many hours per day will people watch television in 1999? about 4.5 h/da

26. Lena and Jared used different methods to simplify $\left(\frac{b^7}{b^3}\right)^2$. Are both methods correct? Explain.

Lena
$\left(\frac{b^7}{b^3}\right)^2 = \frac{b^{14}}{b^6} = b^8$

Jared
$\left(\frac{b^7}{b^3}\right)^2 = (b^4)^2 = b^8$

Yes; a quotient may be simplified before or after it is raised to a power.

Simplify each expression. Use only positive exponents.

27. $\frac{2^7}{2^5}$ 4
28. $2^5 \cdot 2^{-7}$ $\frac{1}{4}$
29. $\left(\frac{3^5}{3^2}\right)^2$ 729
30. $\frac{a^7}{a^9}$ $\frac{1}{a^2}$
31. $\frac{6c^7}{3c}$ $2c^6$
32. $\frac{5x^5}{15x^3}$ $\frac{x^2}{3}$
33. $\frac{a^7b^3c^2}{a^2b^6c^2}$ $\frac{a^5}{b^3}$
34. $\frac{a^{-21}a^{15}}{a^3}$ $\frac{1}{a^9}$
35. $\frac{c^3d^7}{c^8d^{-1}}$ $\frac{d^8}{c^5}$
36. $\frac{p^7qr^{-1}}{pq^{-2}r^5}$ $\frac{p^6q^3}{r^6}$
37. $\frac{5x^3}{(5x)^3}$ $\frac{1}{25}$
38. $\left(\frac{5x^3}{20x}\right)^3$ $\frac{x^6}{64}$
39. $\left(\frac{c^5}{c^9}\right)^3$ $\frac{1}{c^{12}}$
40. $\left(\frac{7a}{b^3}\right)^2$ $\frac{49a^2}{b^6}$
41. $\left(\frac{3b^0}{5}\right)^2$ $\frac{9}{25}$
42. $\frac{(3a^3)^2}{10a^{-1}}$ $\frac{9a^7}{10}$
43. $\left(\frac{5m^3n}{m^5}\right)^3$ $\frac{125n^3}{m^6}$
44. $\left(\frac{x^4x}{x^{-2}}\right)^{-4}$ $\frac{1}{x^{28}}$

Write each expression with only one exponent. You may need to use parentheses.

45. $\frac{3^5}{5^5}$ $\left(\frac{3}{5}\right)^5$
46. $\frac{m^7}{n^7}$ $\left(\frac{m}{n}\right)^7$
47. $\frac{d^8}{d^5}$ d^3
48. $\frac{10^7 \cdot 10^0}{10^{-3}}$ 10^{10}
49. $\frac{27x^3}{8y^3}$ $\left(\frac{3x}{2y}\right)^3$
50. $\frac{4m^2}{169m^4}$ $\left(\frac{2}{13m}\right)^2$

51. a. **Finance** In 1980, the U.S. Government owed $909 billion to its creditors. The population of the United States was 226.5 million people. How much did the government owe per person in 1980? about $4013
b. In 1994 the debt had grown to $4.64 trillion, with a population of 260 million. How much did the government owe per person? about $17,846
c. What was the percent of increase in the amount owed per person from 1980 to 1994? about 345%

PROBLEM SOLVING HINT
Use a proportion.

of blood found worldwide is subtype h-h. It has been found in only three people.

Exercises MIXED REVIEW

Exercise 65 Have students express the answer in words and in scientific notation. 326 trillion and 3.26×10^{11}

SELF ASSESSMENT **PORTFOLIO** Share with students the criteria you will use to assess their work in portfolios as well as how you plan to use the results. Students should understand how the rubrics are used to assess their work, how each piece in the portfolio counts, and how the scores they get in their portfolios will affect their overall evaluation.

Wrap Up

THE BIG IDEA Ask students to explain the rule for dividing powers with the same base.

RETEACHING ACTIVITY Students simplify expressions with exponents using division properties. (Reteaching worksheet 8-8)

Which property(ies) of exponents would you use to simplify each expression? 52–56. See margin.

52. 2^{-3} **53.** $\frac{2^2}{2^5}$ **54.** $\left(\frac{1}{2}\right)^3$ **55.** $\frac{1}{2^{-4} \cdot 2^7}$ **56.** $\frac{(2^4)^3}{2^{15}}$

57. *Astronomy* The ratio of a planet's maximum to minimum distance from the Sun is a measure of how circular its orbit is.

a. Copy and complete the table below. See margin.

b. Which planet has the least circular orbit? the most circular orbit? Explain your reasoning. Pluto; its ratio is farthest from 1; Venus; its ratio is closest to 1.

Distances from the Sun (km)

Planet	Maximum	Minimum	Maximum : Minimum
Mercury	6.97×10^7	4.59×10^7	■ : ■ $= \frac{6.97 \times 10^7}{4.59 \times 10^7} = \frac{6.97}{4.59} \approx 1.52$
Venus	1.089×10^8	1.075×10^8	1.089×10^8 : ■ ≈ ■
Earth	1.521×10^8	1.471×10^8	■ : 1.471×10^8 ≈ ■
Mars	2.491×10^8	2.067×10^8	■ : ■ ≈ ■
Jupiter	8.157×10^8	7.409×10^8	■ : ■ ≈ ■
Saturn	1.507×10^9	1.347×10^9	■ : ■ ≈ ■
Uranus	3.004×10^9	2.735×10^9	■ : ■ ≈ ■
Neptune	4.537×10^9	4.457×10^9	■ : ■ ≈ ■
Pluto	7.375×10^9	4.425×10^9	■ : ■ ≈ ■

min.

max.

58. *Medicine* If you donate blood regularly, the American Red Cross recommends a 56-day waiting period between donations. One pint of blood contains about 2.4×10^{12} red blood cells. Your body normally produces about 2×10^6 red blood cells per second.

a. At its normal rate, in how many seconds will your body replace the red blood cells lost by giving one pint of blood? 1.2×10^6 s

b. Convert your answer from part (a) to days. about 14 da

Exercises MIXED REVIEW

Solve each equation.

59. $x^2 = 0.36$ 0.6, −0.6 **60.** $t^2 - 13 = 0$ 3.6, −3.6 **61.** $m + 4 = -2$ −6

Graph each inequality. 62–64. See margin.

62. $y \leq 3x - 4$ **63.** $y > -2x$ **64.** $y \geq -x + 1$

65. The Folsom Dam in California holds 1 million acre-feet of water in a reservoir. An acre-foot of water is the amount of water that covers an acre to the depth of one foot, or 326,000 gal. How many gallons are in the reservoir?

3.26×10^{11} gal

PORTFOLIO

For your portfolio, select one or two items from your work for this chapter. Here are some possibilities:
- best work
- work you found challenging
- part of your project

Explain why you have included each selection that you make.

Reteaching 8-8

Practice 8-8

Practice 8-8
Mixed Exercises

Lesson Quiz

Lesson Quiz is also available in Transparencies.

Simplify each expression. Use only positive exponents.

1. $\left(-\frac{1}{3}\right)^{-3}$ −27
2. $\frac{(3a^4)(2a^{-2})}{6a^2}$ 1
3. $\frac{1.6 \times 10^3}{4 \times 10^{-2}}$ 4×10^4
4. Write the expression $\frac{9x^2}{16y^2}$ using only one exponent. $\left(\frac{3x}{4y}\right)^2$

Finishing the Chapter Project

PROJECT DAY You may wish to plan a project day on which students share their completed projects. Encourage students to explain their processes as well as their products.

PROJECT NOTEBOOK Have students review their project work and bring their notebooks up to date.

- Have students make sure all elements of the project have been completed and are in their notebooks.
- Ask groups to share what they learned during the project, such as what they learned about exponential growth.

SCORING RUBRIC

3 Student followed all directions for preparing the experiment. Exponential function is reasonably consistent with actual growth. The graph and scatter plot are neat, accurate, and clearly show the relationship between the actual and predicted growth.

2 Student followed most directions for preparing the experiment, with few errors. Exponential function is somewhat consistent with actual growth. The graph and scatter plot generally show the relationship between the actual and predicted growth, with minor errors.

1 Student made errors preparing the experiment. The relationship between the function and the actual growth is unclear. Graph and scatter plot are not accurate.

0 Major elements of the project are incomplete or missing.

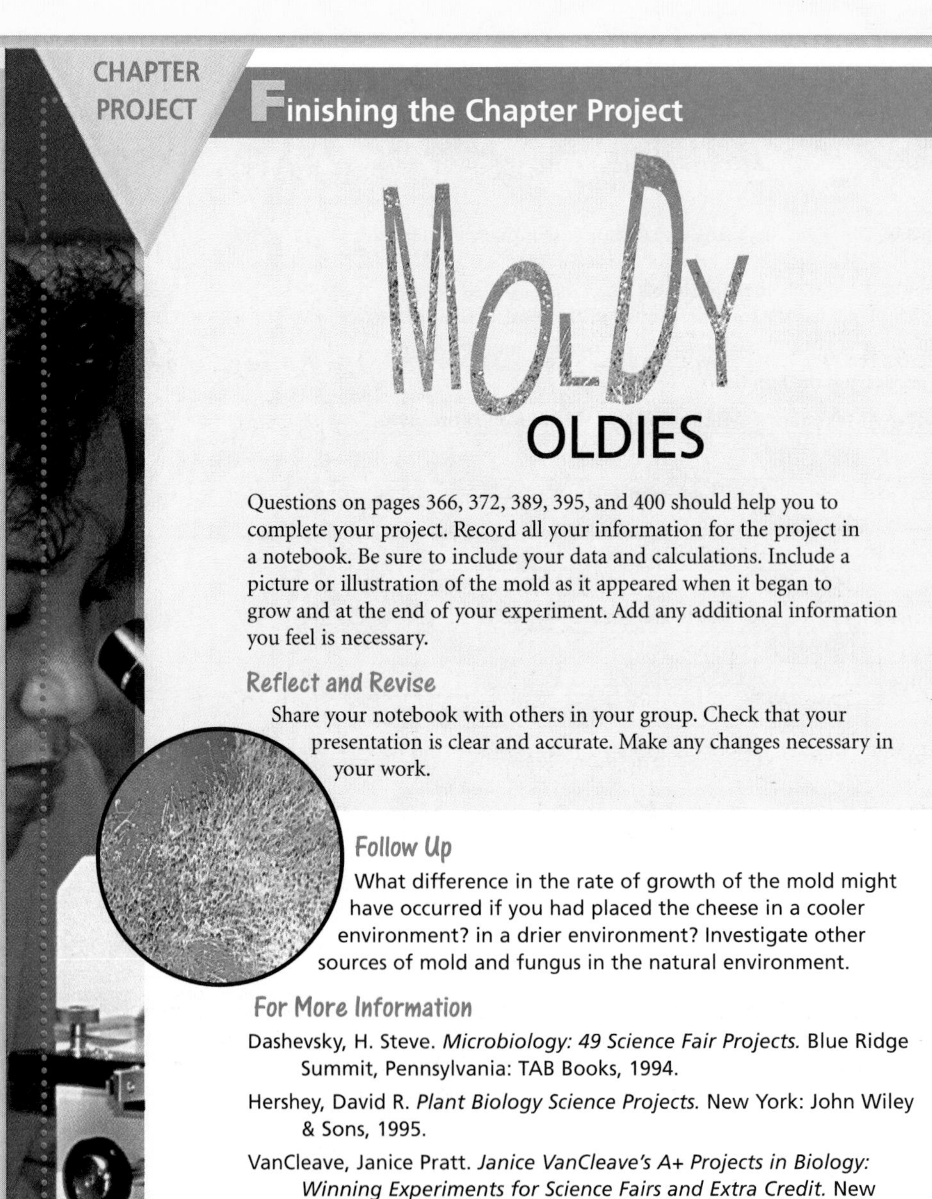

CHAPTER PROJECT

Finishing the Chapter Project

MOLDY OLDIES

Questions on pages 366, 372, 389, 395, and 400 should help you to complete your project. Record all your information for the project in a notebook. Be sure to include your data and calculations. Include a picture or illustration of the mold as it appeared when it began to grow and at the end of your experiment. Add any additional information you feel is necessary.

Reflect and Revise

Share your notebook with others in your group. Check that your presentation is clear and accurate. Make any changes necessary in your work.

Follow Up

What difference in the rate of growth of the mold might have occurred if you had placed the cheese in a cooler environment? in a drier environment? Investigate other sources of mold and fungus in the natural environment.

For More Information

Dashevsky, H. Steve. *Microbiology: 49 Science Fair Projects.* Blue Ridge Summit, Pennsylvania: TAB Books, 1994.

Hershey, David R. *Plant Biology Science Projects.* New York: John Wiley & Sons, 1995.

VanCleave, Janice Pratt. *Janice VanCleave's A+ Projects in Biology: Winning Experiments for Science Fairs and Extra Credit.* New York: John Wiley & Sons, 1993.

Wrap Up

Chapter Project HOW AM I DOING? Divide students into six groups. Ask each group to solve the review problems in one of the chapter wrap up exercise sections. Then, have each group prepare a brief presentation to summarize the central ideas of their assigned section. Have students discuss any exercises they found challenging.

KEY TERMS The numbers in parentheses direct students to the pages where the terms (or symbols) are used or defined. Students should be able to (1) write a simple explanation of each term, (2) illustrate the term with a diagram, or (3) show an example that uses the term.

Exercises 1–2 Remind students that the domain of a function is the set of numbers used to evaluate that function.

Exercises 4–11 Remind students that if the base of an exponential function is greater than one, the exponential function will increase. Conversely, if the base of an exponential function is less than one, the function will decrease.

Exercises 12–15 Emphasize that students should convert a percent of change into its decimal equivalent before using it as the base of an exponential function.

Exercise 17 Make sure that students understand the difference between interest compounded quarterly and interest compounded annually.

8 Wrap Up

Key Terms

compound interest (p. 369)
decay factor (p. 373)
exponential decay (p. 373)
exponential function (p. 363)
exponential growth (p. 368)
growth factor (p. 368)
interest period (p. 369)
power (p. 391)
power of ten (p. 386)
scientific notation (p. 385)
simplest form (p. 403)

How am I doing?

SELF ASSESSMENT

- State three ideas from this chapter that you think are important. Explain your choices.
- Describe several rules that you can use to simplify expressions with exponents.

Exploring Exponential Functions 8-1

You can use exponents to show repeated multiplication. An **exponential function** repeatedly multiplies an amount by the same positive number.

Evaluate each function for the given values.

1. $f(x) = 3 \cdot 2^x$ for the domain $\{1, 2, 3, 4\}$ 6, 12, 24, 48
2. $y = 10 \cdot (0.75)^x$ for the domain $\{1, 2, 3\}$ 7.5, 5.625, 4.21875
3. a. One kind of bacterium in a laboratory culture triples in number every 30 minutes. Suppose a culture is started with 30 bacteria cells. How many bacteria will there be after 2 hours? 2430 bacteria
   b. After how many minutes will there be more than 20,000 bacteria? about 178 min

Exponential Growth and Decay 8-2, 8-3

The general form of an exponential function is $y = a \cdot b^x$.

When $b > 1$, the function increases, and the function shows **exponential growth.** The base, b, of the exponent is called the **growth factor.** An example of exponential growth is **compound interest.**

When $0 < b < 1$, the function decreases, and the function shows **exponential decay.** Then b is called the **decay factor.** An example of exponential decay is half-life.

Identify each exponential function as *exponential growth* or *exponential decay.* Then identify the growth or decay factor.

4. $y = 5.2 \cdot 3^x$ growth; 3
5. $y = 0.15 \cdot \left(\frac{3}{2}\right)^x$ growth; $\frac{3}{2}$
6. $y = 7 \cdot 0.32^x$ decay; 0.32
7. $y = 1.3 \cdot \left(\frac{1}{4}\right)^x$ decay; $\frac{1}{4}$

Graph each function. 8–11. See margin.

8. $f(x) = 2.5^x$
9. $y = 0.5 \cdot (0.5)^x$
10. $f(x) = \left(\frac{1}{2}\right) \cdot 3^x$
11. $y = 0.1^x$

What percent increase or decrease is modeled in each function?

12. $y = 100 \cdot 1.025^x$ 2.5% incr.
13. $y = 32 \cdot 0.75^x$ 25% decr.
14. $y = 0.4 \cdot 2^x$ 100% incr.
15. $y = 1.01 \cdot 0.9^x$ 10% decr.

Resources

Student Edition

Extra Practice, p. 563
Glossary/Study Guide

Teaching Resources

Study Skills Handbook
Glossary, Spanish Resources

pages 407–409 Wrap Up

8.

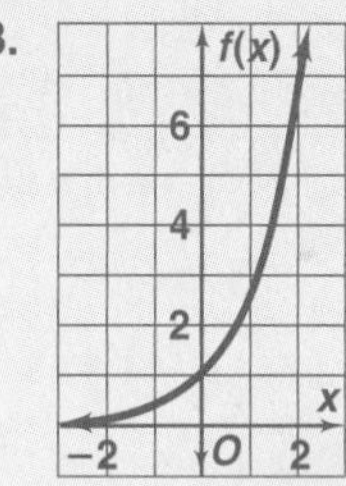

9.

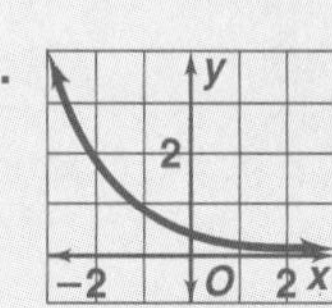

10.

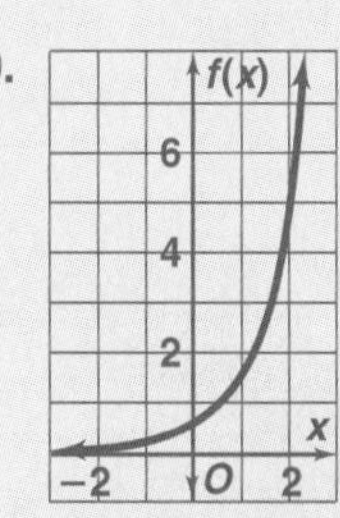

Exercises 19–25 Remind students that a negative exponent in the denominator is equivalent to a positive exponent in the numerator.

Exercises 26–30 Emphasize that students should always express scientific notation in the correct form.

Exercises 37–41, 44–49 Emphasize that the same variable should not appear more than once in a simplified expression.

Remind students that the new mathematical terms in this chapter are defined in the Student Study Guide/Glossary in the back of the book.

pages 407–409 Wrap Up

11.

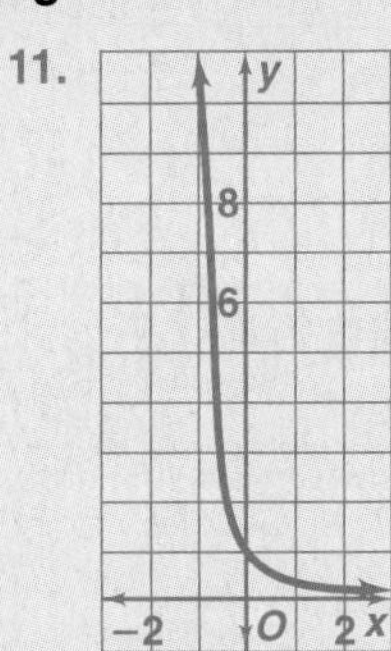

54. Answers may vary. Sample: Use the Div. Prop.: $\left(\frac{5a^2}{10}\right)^{-3}$; simplify the fraction: $\left(\frac{a^2}{2}\right)^{-3}$; use Raising a Quotient to a Power Prop.: $\frac{(a^2)^{-3}}{2^{-3}}$; use Raising a Power to a Power Prop.: $\frac{a^{-6}}{2^{-3}}$; use the Neg. Exp. Prop.: $\frac{2^3}{a^6}$ or $\frac{8}{a^6}$.

16. The population of a city is 100,000 and is growing 7% each year.
   a. Write an equation to model the population of the city after any number of years. $y = 100{,}000 \cdot 1.07^x$
   b. Use your equation to find the population after 25 years. 542,743

17. a. *Finance* Suppose you earned \$1200 last summer and you put it into a savings account that pays 5.5% interest compounded quarterly. Find the balance after 9 months and after 12 months. \$1250.18; \$1267.37
   b. *Writing* Would you rather have an account that pays 5.5% interest compounded quarterly or 6% interest compounded annually? Explain. 6%; it pays more interest.

18. *Chemistry* The half-life of radioactive carbon-11 is 20 min. You start an experiment with 160 mCi of carbon-11. After 2 h, how much radioactivity remains? 2.5 mCi

Zero and Negative Exponents — 8-4

You can use zero and negative numbers as exponents. For any nonzero number a, $a^0 = 1$. For any nonzero number a and any integer n, $a^{-n} = \frac{1}{a^n}$.

Write each expression so that all exponents are positive.

19. $b^{-4}c^0d^6$ $\frac{d^6}{b^4}$
20. $\frac{x^{-2}}{y^{-8}}$ $\frac{y^8}{x^2}$
21. $7k^{-8}h^3$ $\frac{7h^3}{k^8}$
22. $\frac{1}{p^2q^{-4}r^0}$ $\frac{q^4}{p^2}$
23. $\left(\frac{2}{3}\right)^{-4}$ $\frac{5^4}{2^4}$

24. *Critical Thinking* Is $(-3b)^4 = -12b^4$? Why or why not? Yes if $b = 0$, no if $b \neq 0$. $(-3b)^4$ is positive and $-12b^4$ is negative.

25. *Standardized Test Prep* If $a = 4$, $b = -3$, and $c = 0$, which expression has the greatest value? C
   A. a^b B. b^c C. $\frac{1}{b^{-a}}$ D. $\frac{a^c}{b^c}$ E. $\frac{c}{a^{-b}}$

Scientific Notation — 8-5

You can use **scientific notation** to express very large or very small numbers. A number is in scientific notation if it is in the form $a \times 10^n$, where $1 \leq |a| < 10$ and n is an integer.

Determine whether each number is in scientific notation. If it is not, write it in scientific notation.

26. 950×10^5 9.5×10^7
27. 72.35×10^8 7.235×10^9
28. 1.6×10^{-6} yes
29. 84×10^{-5} 8.4×10^{-4}
30. 0.26×10^{-3} 2.6×10^{-4}

Write each number in scientific notation.

31. The space probe *Voyager 2* traveled 2,793,000 miles. 2.793×10^6 mi

32. There are 189 million passenger cars and trucks in use in the United States. 1.89×10^8 cars and trucks

Double each number. Give your answers in scientific notation.

33. 8.03×10^7 1.606×10^8
34. 2.3×10^{-9} 4.6×10^{-9}
35. 7.084×10^6 1.4168×10^7
36. 5×10^{-13} 1×10^{-12}

ALTERNATIVE ASSESSMENT Ask students to pick one problem from each review section they feel illustrates the central idea of that section. Then have them explain, in writing, why their chosen problem is an appropriate example.

Getting Ready for Chapter 9

Students may work these exercises independently or in small groups. The skills previewed will help prepare students to work with expressions containing radicals and with the analysis of statistical data.

Multiplication Properties of Exponents — 8-6, 8-7

To multiply powers with the same base, add the exponents. $a^m \cdot a^n = a^{m+n}$

To raise a power to a power, multiply the exponents. $(a^m)^n = a^{mn}$

To raise a product to a power, raise each factor in the product to the power. $(ab)^n = a^nb^n$

Simplify each expression. Use only positive exponents.

37. $2d^2d^3$ $2d^5$ **38.** $(q^3r)^4$ $q^{12}r^4$ **39.** $(5c^{-4})(-4m^2c^8)$ $-20c^4m^2$ **40.** $(1.34^2)^5(1.34)^{-8}$ 1.7956 **41.** $(12x^2y^{-2})^5(4xy^{-3})^{-8}$ $\frac{243}{64}x^2y^{14}$

42. *Estimation* Each square inch of your body has about 6.5×10^2 pores. The back of your hand has area about 0.12×10^2 in.^{2}. About how many pores are on the back of one hand? Estimates may vary: about 7×10^3 pores

43. *Open-ended* Write and solve a problem that involves multiplying exponents. Sample: Simplify $(2a^2)^2$; $4a^4$

Division Properties of Exponents — 8-8

To divide powers with the same base, subtract the exponents. $\frac{a^m}{a^n} = a^{m-n}$

To raise a quotient to a power, raise the dividend and the divisor to the power. $\left(\frac{a}{b}\right)^n = \frac{a^n}{b^n}$

Determine whether each expression is in simplest form. If it is not, simplify it.

44. $\frac{w^2}{w^5}$ $\frac{1}{w^3}$ **45.** $(8^3) \cdot 8^{-5}$ $\frac{1}{64}$ **46.** $\left(\frac{21x^3}{5y^2}\right)$ yes **47.** $\left(\frac{n^5}{v^3}\right)^7$ $\frac{n^{35}}{v^{21}}$ **48.** $\frac{e^{-6}c^3}{e^5}$ $\frac{c^3}{e^{11}}$ **49.** $\left(\frac{x^9}{s^{-3}}\right)^5$ $s^{15}x^{45}$

Find each quotient. Give your answer in scientific notation.

50. $\frac{4.2 \times 10^8}{2.1 \times 10^{11}}$ 2×10^{-3} **51.** $\frac{3.1 \times 10^4}{12.4 \times 10^2}$ 2.5×10^1 **52.** $\frac{4.5 \times 10^3}{9 \times 10^7}$ 5×10^{-5} **53.** $\frac{5.1 \times 10^5}{1.7 \times 10^2}$ 3×10^3

54. *Writing* List the steps that you would follow to simplify $\left(\frac{5a^8}{10a^6}\right)^{-3}$. See margin.

Getting Ready for... CHAPTER 9

Find the distance between the numbers on a number line.

55. 5 and 3 2 **56.** –7 and 4 11 **57.** –5 and –11 6

Simplify each expression.

58. $7r(11 + 4x)$ $77r + 28rx$ **59.** $8m(3 - 2t)$ $24m - 16mt$ **60.** $b(8 + 2b)$ $8b + 2b^2$

61. $8p + 6d - 3p$ $5p + 6d$ **62.** $-5n - 4n + 10nn$ **63.** $5^2 + 6^2$ 61

64. $9^2 - 4^2$ 65 **65.** $(3t)^2 + (2t)^2$ $13t^2$ **66.** $20y^2 - (4y)^2$ $4y^2$

Find the range and mean of each set of data.

67. \$40, \$58, \$44, \$47, \$39, \$58, \$56 \$19; about \$48.86

68. 4 kg, 3 kg, 6 kg, 3 kg, 5 kg, 8 kg, 4 kg, 3 kg 5 kg; 4.5 kg

69. 1.3 min, 1.4 min, 1.1 min, 1.0 min, 1.4 min 0.4 min; 1.24 min

Assessment

ENHANCED MULTIPLE CHOICE QUESTIONS are more complex than traditional multiple choice questions, which assess only one skill. Enhanced multiple choice questions assess the processes that students use, as well as the end results. The questions are written so that students must use more than one strategy to solve the problem. Using multiple strategies is encouraged by the National Council of Teachers of Mathematics (NCTM). **Exercise 9** is an enhanced multiple choice question.

FREE RESPONSE QUESTIONS do not give answer choices. Some exercises have more than one possible answer. Students need to give only one correct response. **Exercises 1–6 and 10–32** are free response questions.

WRITING EXERCISES allow students to describe how they think about and understand the concepts they have learned. **Exercise 8** is a writing exercise.

OPEN-ENDED PROBLEMS allow for more than one solution. Students must construct their own responses instead of choosing from possible answers. The student responses will help you determine the depth of their understanding and any possible areas of difficulty. **Exercise 7** is an open-ended problem. Make sure that the students' responses contain values for *x* and *y* and a written situation.

Resources

Teaching Resources

Chapter Support File, Ch. 8

- Chapter Assessment, Forms A and B
- Alternative assessment

Chapter Assessment, Spanish Resources

Teacher's Edition

See also p. 360E for assessment options.

Software Computer Item Generator

page 410 Assessment

5a.

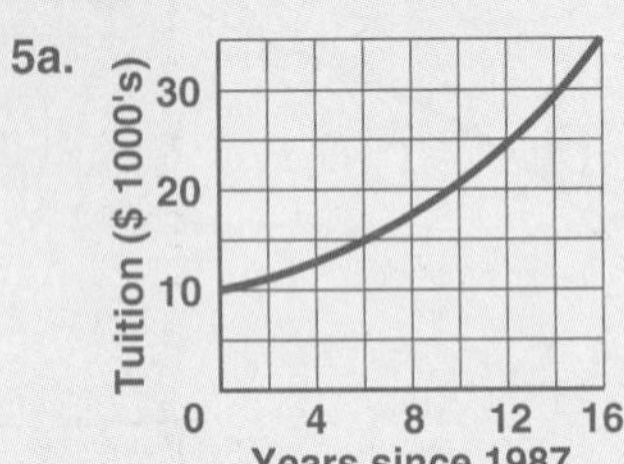

6b. **Check students' work. Samples:**

1998: $n = 33, y = 12.12$;

1999: $n = 34, y = 12.97$;

2000: $n = 35, y = 13.88$;

2001: $n = 36, y = 14.85$

c. **Check students' work. Samples:**

2008—23.85 billion kW•h; 2009—25.52 billion kW•h; 2010—27.30 billion kW•h; 2011—29.21 billion kW•h

6d, 7, 10–12. See back of book.

8 Assessment

Evaluate each function for $x = 1, 2,$ and 3.

1. $y = 3 \cdot 5^x$ 15, 75, 375
2. $f(x) = \frac{1}{2} \cdot 4^x$ 2, 8, 32
3. $f(x) = 4 \cdot (0.95)^x$ 3.8, 3.61, 3.4295
4. $g(x) = 5 \cdot \left(\frac{3}{4}\right)^x$ 3.75, 2.8125, 2.109375
5. The function $y = 10 \cdot 1.08^x$ models the cost of annual tuition (in thousands of dollars) at a local college in the years since 1987.
   a. Graph the function. See margin.
   b. What is the annual percent increase? 8%
   c. How much was tuition in 1987? in 1992?
   d. How much will tuition be the year you plan to graduate from high school?
   c. \$10,000; \$14,693 d. Check students' work.
6. The function $y = 1.3 \cdot (1.07)^n$ models a city's annual electrical consumption for the n years since 1965, where y is billions of kilowatt-hours.
   a. Determine whether the function models exponential growth or decay, and find the growth or decay factor. growth; 1.07
   b. What value of n should be substituted to find the value of y now? Use this value for n to find y. b–d. See margin.
   c. What will be the annual electrical usage in 10 years?
   d. What was the annual electrical usage 10 years ago?
7. *Open-ended* Write and solve a problem involving exponential decay. See margin.
8. *Writing* Explain when the function $y = a \cdot b^x$ shows exponential growth and when it shows exponential decay. The function shows growth when $b > 1$ and decay when $0 < b < 1$.
9. *Standardized Test Prep* If $a = -3$, which expression has the least value? C

   A. a^2a^0 B. a^a C. a^8a^{-5}
   D. $-a^aa^{-4}$ E. $(a^3a^{-4})^2$

Graph each function. 10–12. See margin.

10. $y = \frac{1}{2} \cdot 2^x$
11. $y = 2 \cdot \left(\frac{1}{2}\right)^x$
12. $f(x) = 3^x$
13. *Critical Thinking* Is there a solution to the equation $3^x = 5^x$? Explain. Yes; if $x = 0$, $3^x = 1$ and $5^x = 1$.

Write each number in scientific notation.

14. There were 44,909,000 votes cast for Bill Clinton in the 1992 presidential election. 4.4909×10^7 votes
15. More than 450,000 households in the United States have reptiles as pets. 4.5×10^5 households

Determine whether each number is in scientific notation. If it is not, write it in scientific notation.

16. 76×10^{-9} 7.6×10^{-8}
17. 7.3×10^5 yes
18. $4.05 \times 10 \times 10^{-8}$ 4.05×10^{-7}
19. 32.5×10^{13} 3.25×10^{14}

Simplify each expression. Use positive exponents.

20. $\frac{r^3t^{-7}}{t^5}$ $\frac{r^3}{t^{12}}$
21. $\left(\frac{a^3}{m}\right)^{-4}$ $\frac{m^4}{a^{12}}$
22. $\frac{t^{-8}m^2}{m^{-3}}$ $\frac{m^5}{t^8}$
23. $c^3v^9c^{-1}c^0$ c^2v^9
24. $h^2k^{-5}d^3k^2$ $\frac{h^2d^3}{k^3}$
25. $9y^4j^2y^{-9}$
26. $(w^2k^0p^{-5})^{-7}$ $\frac{p^{35}}{w^{14}}$
27. $2y^{-9}h^2(2y^0h^{-4})^{-6}$
28. $(1.2)^5(1.2)^{-2}$ 1.728
29. $(-3q^{-1})^3q^2$ $-\frac{27}{q}$

25. $\frac{9j^2}{y^5}$ 27. $\frac{h^{26}}{32y^9}$

30. a. *Astronomy* The speed of light in a vacuum is 186,300 mi/s. Use scientific notation to express how far light travels in one hour.
    b. How long does light take to travel to Saturn, about 2.3×10^9 mi away from Earth?
    a. 6.7×10^8 mi b. about 3.4 h
31. *Banking* A customer deposits \$1000 in a savings account that pays 4% interest compounded quarterly. How much money will the customer have in the account after 2 years? after 5 years? \$1082.86; \$1220.19
32. *Automobiles* Suppose a new car is worth \$14,000. Its value decreases by one fifth each year. The function $y = 14{,}000(0.8)^x$ models the car's value after x years. \$11,200
    a. Find the value of the car after one year.
    b. Find the value of the car after four years. \$5734.40

Cumulative Review

Item	Review Topic	Chapter
1	absolute value	4
2, 11	quadratic functions	7
3	solving equations	3
4	median	1
5, 13	quadratic formula	7
6, 17	linear equations	5
7	linear inequalities	6
8	solving systems of equations	6
9	percent	3
10	probability	3
12	scientific notations	8
14	exponential functions	8
15	graphing quadratic equations	7
16	modeling equations	3

8 Cumulative Review

For Exercises 1–14, choose the correct letter.

1. If a is positive and b is negative, which of the following is negative? D
   A. $|ab|$ B. $a + |b|$ C. $a|b|$
   D. $|a|b$ E. $|a| - b$

2. What is the value of the function $f(x) = 2x^2 + x + 3$ when $x = -3$? E
   A. 42 B. -18 C. 24 D. 36 E. 18

3. Which equation does *not* have the solution -1.5? D
   A. $\frac{9}{w} = -6$ B. $-10w = 15$
   C. $4 - 3w = 8.5$ D. $-1 - 2w = -4$
   E. $w^2 = 2.25$

4. Your test scores are 88, 78, 81, 83, and 90. What score do you need on your next test to raise your median to 85? B
   A. 90 B. 87 C. 85 D. 86 E. 92

5. Use the quadratic formula to find the solutions of $2x^2 + 5x + 3 = 0$. A
   A. $\frac{-3}{2}, -1$ B. $\frac{3}{2}, -1$ C. $\frac{3}{2}, 1$
   D. $-3, -1$ E. $-3, -2$

6. Identify the function. B
   A. $2y + x = 2$
   B. $y + 2x = -2$
   C. $2y - x = 1$
   D. $y + 2x = 1$
   E. $y - 2x = 1$

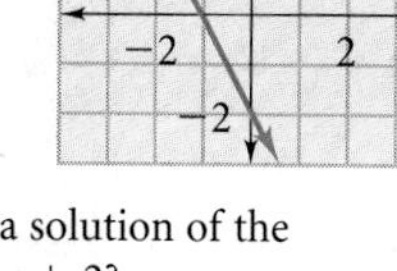

7. Which value of x is *not* a solution of the inequality $5 - 6x < -x + 2$? A
   A. -1 B. 5 C. 1 D. $\frac{3}{4}$ E. 3

8. Find the solution of the system of equations. C
   $\frac{1}{3}x - y = 4$ $\quad x + 3y = 0$
   A. $(9, -1)$ B. $(-6, 2)$ C. $(6, -2)$
   D. $(-3, 1)$ E. $(-9, 1)$

9. You earn a commission of 6% on your first \$500 of sales and 10% on all sales above \$500. If you earn \$130 in commissions, what are your total sales? D
   A. \$800 B. \$1000 C. \$1300
   D. \$1500 E. \$2000

10. You flip a coin and roll a number cube. Find the probability of getting a head and a multiple of 3. B
    A. $\frac{1}{4}$ B. $\frac{1}{6}$ C. $\frac{1}{12}$ D. $\frac{3}{2}$ E. $\frac{5}{6}$

11. Find the minimum value of the function $f(x) = x^2 + x + 4$. A
    A. $\frac{15}{4}$ B. 4 C. $\frac{17}{4}$ D. -2 E. -1

12. Which number has the least value? D
    A. $2.8 \cdot 10^{-5}$ B. $5.3 \cdot 10^{-4}$
    C. $8.3 \cdot 10^{-7}$ D. $1.6 \cdot 10^{-8}$
    E. $7.04 \cdot 10^{-8}$

Compare the boxed quantity in Column A with the boxed quantity in Column B. Choose the best answer.

A. The quantity in Column A is greater.
B. The quantity in Column B is greater.
C. The two quantities are equal.
D. The relationship cannot be determined on the basis of the information supplied.

Find each answer.

	Column A	Column B	
13. $x^2 + x - 20 = 0$	the value of the discriminant of the equation	the sum of the solutions of the equation	A
14.	the growth factor of an exponential function	the decay factor of an exponential function	A

15. Graph $y = -2x^2 + 3x + 8$. See margin.

16. *Open-ended* Describe a situation that you could model using the equation $17 - 3x = 5$. See margin.

17. *Writing* The slopes of four lines are
    a: $\frac{3}{5}$ $\quad b$: $-\frac{10}{6}$ $\quad c$: $-\frac{5}{3}$ $\quad d$: $\frac{9}{15}$
    Do these lines determine a rectangle? Explain why or why not. Yes; sides are parallel and at right angles.

Resources

Teaching Resources

Chapter Support File, Ch. 8
- Cumulative Review
- Standardized Test Practice

Teacher's Edition

See also pp. 360E for assessment options.

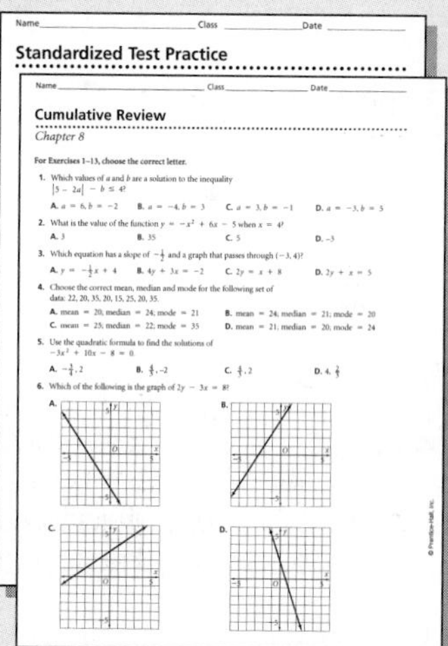

15.

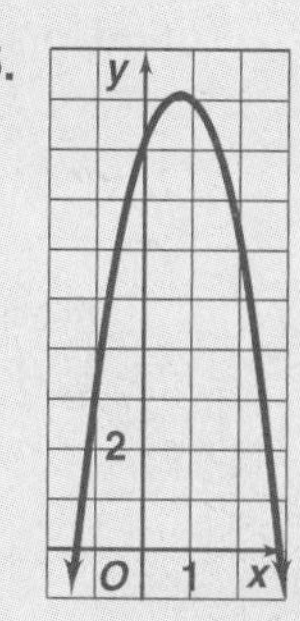

16. Sample: Kalafia had \$17. He bought several melons at \$3 each. He has \$5 left.

CHAPTER 9 Right Triangles and Radical Expressions

To accommodate flexible scheduling, some lessons are divided into parts. Assignment Options are given in the Lesson Planning Options for each lesson.

PACING OPTIONS

This chart suggests pacing only for the core lessons and their parts, and it is provided merely as a possible guide. It will help you determine how much time you have in your schedule to cover other features, such as the Chapter Project, Math Toolboxes, Wrap Up, and Assessment.

	1 Class Period	1 Class Period	1 Class Period	1 Class Period	1 Class Period	1 Class Period	1 Class Period	1 Class Period	1 Class Period
Traditional (40–45 min class periods)	9-1 Part 1	9-1 Part 2	9-2 Part 1	9-2 Part 2	9-3 Part 1, 9-3 Part 2	9-4 Part 1, 9-4 Part 2	9-5 Part 1	9-5 Part 2	9-6 Part 1
Algebra Over 2 Years (40–45 min class periods)	9-1 Part 1	9-1 Part 2	9-1 Part 2	9-2 Part 1	9-2 Part 1	9-2 Part 2	9-3 Part 1	9-3 Part 2	9-4 Part 1
Block Scheduling (90 min class periods)	9-1 Part 1	9-1 Part 2	9-2 Part 1	9-2 Part 2	9-3 Part 1, 9-3 Part 2	9-4 Part 1, 9-4 Part 2	9-5 Part 1	9-5 Part 2	9-6 Part 1

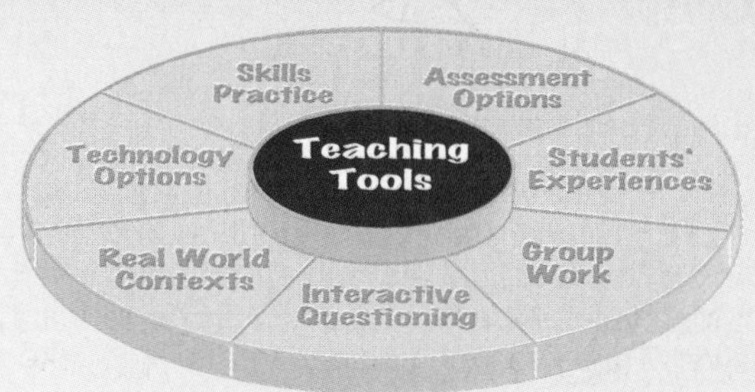

What Students Will Learn and Why

In this chapter, students will build on their knowledge of exponents and exponential functions, learned in Chapter 8, by applying exponential functions to the sides of right triangles. They will learn to use the Pythagorean theorem, and learn about the basic trigonometric ratios. They will learn to add and subtract radicals. Students will solve radical equations and graph square root functions. Finally, students will analyze real-world data using standard deviation.

Discussing the Chapter/Building on Experience

The concept map below relates chapter topics to real-world applications. You and your class may wish to add to the map or develop maps of your own. The center oval describes the topic of the chapter. The next level displays topics within the lessons. The outer ovals reflect applications of the content. As you and your class build a concept map, invite students to discuss applications with which they are familiar.

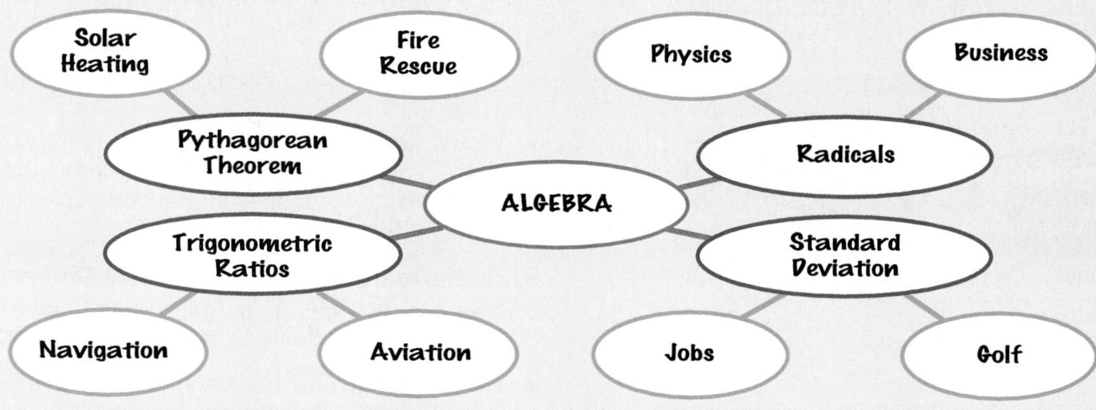

1 Class Period	1 Class Period	1 Class Period	1 Class Period	1 Class Period	1 Class Period	1 Class Period	1 Class Period	1 Class Period	1 Class Period	1 Class Period
9-6 ▽2	9-7	9-8								
9-4 ▽2	9-5 ▽1	9-5 ▽2	9-5 ▽2	9-6 ▽1	9-6 ▽2	9-6 ▽2	9-7	9-7	9-8	9-8
9-6 ▽2	9-7	9-8								

Skills Practice
Teaching Tools
Assessment Options
Students' Experiences
Group Work
Interactive Questioning
Real World Contexts
Technology Options

Interactive Questioning Tips

A question is interactive when there is "give and take" between the questioner (teacher or student) and the respondent. When a critical thinking question is asked, it is important to develop a pattern by which students clarify their initial answer. This gives the student a chance to take their answer to a higher thought level. When students become used to answering questions in this manner, higher level thinking will be stimulated on every question. In the textbook, this technique is often used in critical thinking questions by asking students to explain their answers.

Skills Practice

Every lesson provides skill practice with Exercises On Your Own and Exercises Mixed Review. The Student Edition includes Checkpoints (pp. 429, 449) and Preparing for Standardized Tests (p. 461). In the Teacher's Edition, the Lesson Planning Options section for each lesson lists Prerequisite Skills students should know for that lesson. At the back of the Student Edition is the Skills Handbook—mini-lessons on math the students may need to review, The Chapter Support File for Chapter 9 in the Teaching Resources box includes two Practice worksheets per lesson, a worksheet for two Checkpoints, and worksheets for Cumulative Review and Standardized Test Preparation.

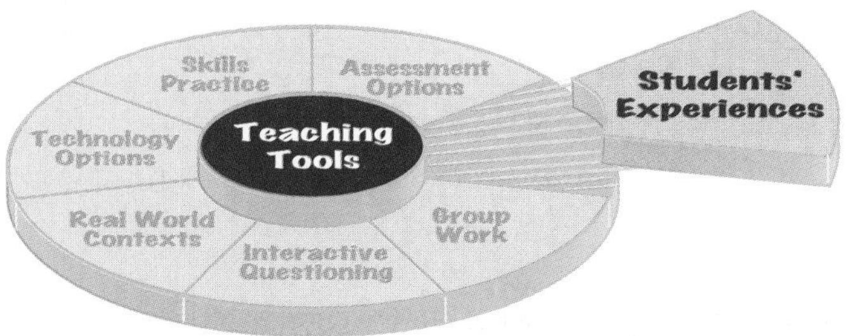

Diverse Learning and Teaching Styles

In your Teacher's Edition, you will find suggestions as to how you can help students complete mathematical tasks in Chapter 9 by reinforcing various learning styles. Here are some examples.

- **Visual learning** reinforcing the term *legs* by drawing a cartoon character (p. 415), writing *x*-coordinates in one color and *y*-coordinates in a second color to clarify the distance formula (p. 421)
- **Tactile learning** using a model ladder to make a triangle against a book (p. 415), building a model roller coaster to use radical equations to solve real-world problems (p. 441)
- **Auditory learning** reinforcing understanding of the distance formula with verbal explanations (p. 420), verbally repeating the ratios of a triangle to become familiar with the terms (p.425)
- **Kinesthetic learning** using string to model a triangle and measure the hypotenuse (p. 416), using radicals on paper to help students simplify different radicals (p. 436)

Alternative Activity for Lesson 9-1

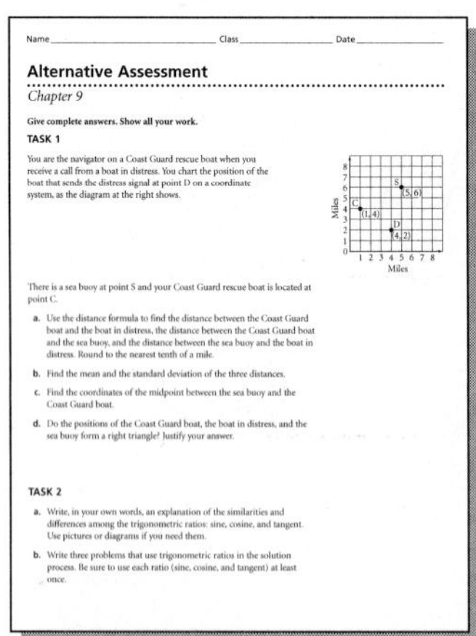

Name ______ Class ______ Date ______

Alternative Assessment

Chapter 9

Give complete answers. Show all your work.

TASK 1

You are the navigator on a Coast Guard rescue boat when you receive a call from a boat in distress. You chart the position of the boat that sends the distress signal at point D on a coordinate system, as the diagram at the right shows.

There is a sea buoy at point S and your Coast Guard rescue boat is located at point C.

a. Use the distance formula to find the distance between the Coast Guard boat and the boat in distress, the distance between the Coast Guard boat and the sea buoy, and the distance between the sea buoy and the boat in distress. Round to the nearest tenth of a mile.

b. Find the mean and the standard deviation of the three distances.

c. Find the coordinates of the midpoint between the sea buoy and the Coast Guard boat.

d. Do the positions of the Coast Guard boat, the boat in distress, and the sea buoy form a right triangle? Justify your answer.

TASK 2

a. Write, in your own words, an explanation of the similarities and differences among the trigonometric ratios: sine, cosine, and tangent. Use pictures or diagrams if you need them.

b. Write three problems that use trigonometric ratios in the solution process. Be sure to use each ratio (sine, cosine, and tangent) at least once.

for use with Work Together, addresses tactile learning by making a puzzle whose solution is a right triangle.

Alternative Activity for Lesson 9-2

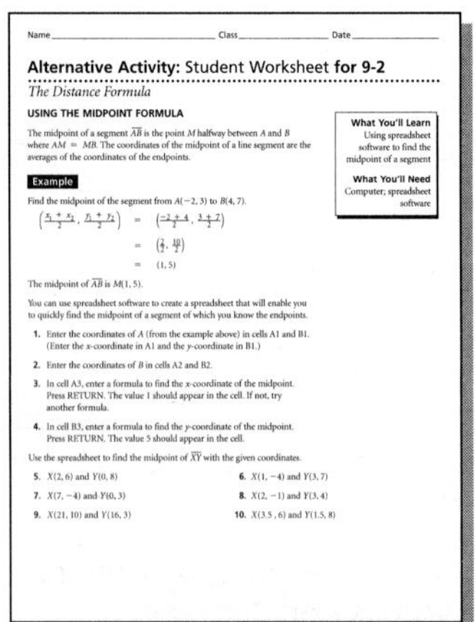

Name ______ Class ______ Date ______

Alternative Activity: Student Worksheet **for 9-2**

The Distance Formula

USING THE MIDPOINT FORMULA

The midpoint of a segment $\overline{AB}$ is the point M halfway between A and B where $AM = MB$. The coordinates of the midpoint of a line segment are the averages of the coordinates of the endpoints.

What You'll Learn Using spreadsheet software to find the midpoint of a segment

What You'll Need Computer; spreadsheet software

Example

Find the midpoint of the segment from $A(-2, 3)$ to $B(4, 7)$.

$$\left(\frac{x_1 + x_2}{2}, \frac{y_1 + y_2}{2}\right) = \left(\frac{-2 + 4}{2}, \frac{3 + 7}{2}\right) = \left(\frac{2}{2}, \frac{10}{2}\right) = (1, 5)$$

The midpoint of $\overline{AB}$ is $M(1, 5)$.

You can use spreadsheet software to create a spreadsheet that will enable you to quickly find the midpoint of a segment of which you know the endpoints.

1. Enter the coordinates of A (from the example above) in cells A1 and B1. (Enter the x-coordinate in A1 and the y-coordinate in B1.)
2. Enter the coordinates of B in cells A2 and B2.
3. In cell A3, enter a formula to find the x-coordinate of the midpoint. Press RETURN. The value 1 should appear in the cell. If not, try another formula.
4. In cell B3, enter a formula to find the y-coordinate of the midpoint. Press RETURN. The value 5 should appear in the cell.

Use the spreadsheet to find the midpoint of $\overline{XY}$ with the given coordinates.

5. $X(2, 6)$ and $Y(0, 8)$
6. $X(1, -4)$ and $Y(3, 7)$
7. $X(7, -4)$ and $Y(0, 3)$
8. $X(2, -1)$ and $Y(3, 4)$
9. $X(21, 10)$ and $Y(16, 3)$
10. $X(3.5, 6)$ and $Y(1.5, 8)$

for use with Example 3, addresses tactile and visual learning by finding the midpoint of a segment using a spreadsheet.

Alternative Activity for Lesson 9-8

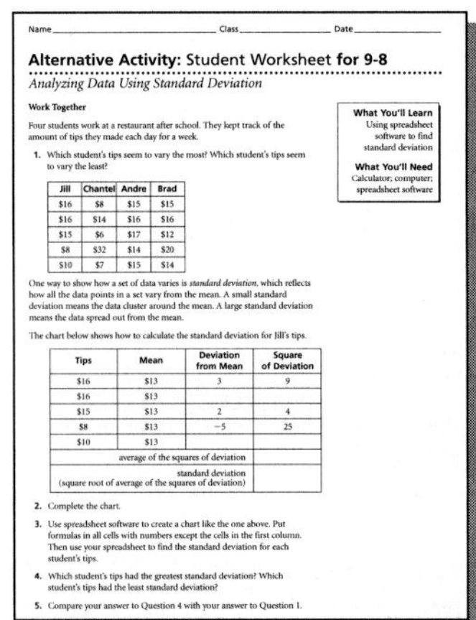

Name ______ Class ______ Date ______

Alternative Activity: Student Worksheet **for 9-8**

Analyzing Data Using Standard Deviation

Work Together

Four students work at a restaurant after school. They kept track of the amount of tips they made each day for a week.

What You'll Learn Using spreadsheet software to find standard deviation

What You'll Need Calculator; computer; spreadsheet software

1. Which student's tips seem to vary the most? Which student's tips seem to vary the least?

Jill	Chantel	Andre	Brad
\$16	\$8	\$15	\$15
\$16	\$14	\$16	\$16
\$15	\$6	\$17	\$12
\$8	\$32	\$14	\$20
\$10	\$7	\$15	\$14

One way to show how a set of data varies is *standard deviation*, which reflects how all the data points in a set vary from the mean. A small standard deviation means the data cluster around the mean. A large standard deviation means the data spread out from the mean.

The chart below shows how to calculate the standard deviation for Jill's tips.

Tips	Mean	Deviation from Mean	Square of Deviation
\$16	\$13	3	9
\$16	\$13		
\$15	\$13	2	4
\$8	\$13	−5	25
\$10	\$13		
average of the squares of deviation			
standard deviation (square root of average of the squares of deviation)			

2. Complete the chart.
3. Use spreadsheet software to create a chart like the one above. Put formulas in all cells with numbers except the cells in the first column. Then use your spreadsheet to find the standard deviation for each student's tips.
4. Which student's tips had the greatest standard deviation? Which student's tips had the least standard deviation?
5. Compare your answer to Question 4 with your answer to Question 1.

for use with Work Together, addresses tactile and visual learning by calculating standard deviations using a spreadshe

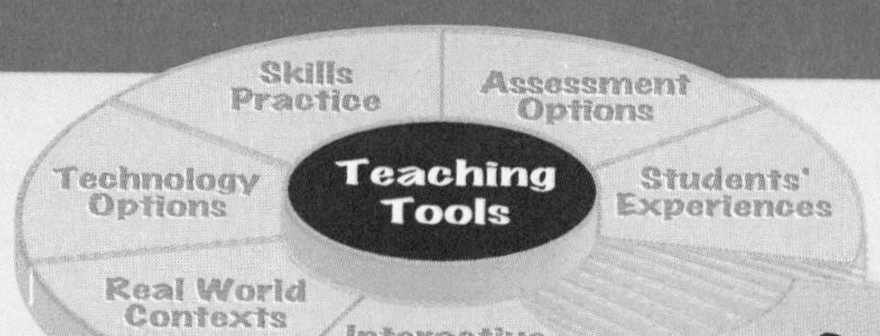

Cooperative Learning Tips

When used effectively, cooperative learning can help students develop interpersonal skills, learn to perform specific roles in a group, and learn to carry out specific responsibilities. The components of Chapter 9 provide a range of cooperative learning opportunities.

- In the Student Edition, the ***Work Together*** parts of lessons are specifically designed for cooperative learning activities.
- In the Teacher's Edition, you will find helpful hints for addressing diverse learning styles (see page C for Chapter 9). For every lesson, you will find a ***Reteaching Activity***, which may involve cooperative learning.

Materials and Manipulatives

The authors expect all students to use scientific calculators. Calculator use is integrated throughout the course.

- **graph paper (9-1, 9-2, 9-3)**
- **graphing calculator (9-7)**
- **rectangular objects (9-1)**
- **ruler (9-1, 9-2, 9-3)**

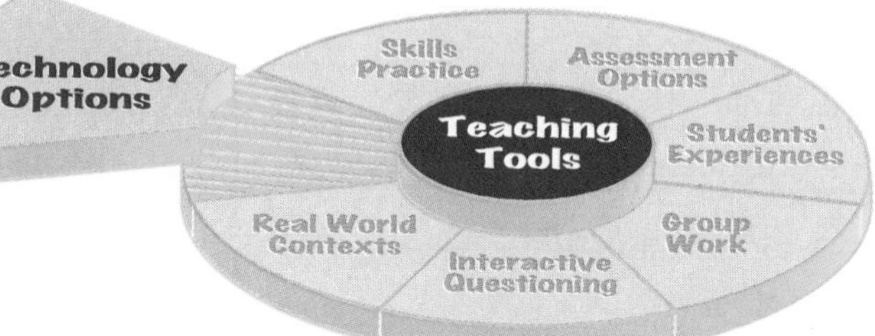

TECHNOLOGY OPTIONS

Technology Tools		Chapter Project	9-1	9-2	9-3	9-4	9-5	9-6	9-7	9-8
Calculator		Assumed that students have scientific calculators at any time.								
Graphing Calculator	Handbook		✔		✔			✔	✔	✔
	Student Edition		✔					✔	✔	✔
Software	Secondary Math Lab Toolkit™		✔	✔	✔	✔	✔	✔	✔	✔
	Integrated Math Lab						✔		✔	
	Computer Item Generator		✔	✔	✔	✔	✔	✔	✔	✔
	Student Edition									
Video	Video Field Trip	✔								
CD-ROM	Multimedia Algebra Lab		✔	✔	✔					✔
Internet		See the Prentice Hall site. (http://www.phschool.com)								

The Prentice Hall Algebra program offers you a rich variety of technology options. Be assured that all these options are provided as a means of enriching the program and are not essential for the successful completion of the course.

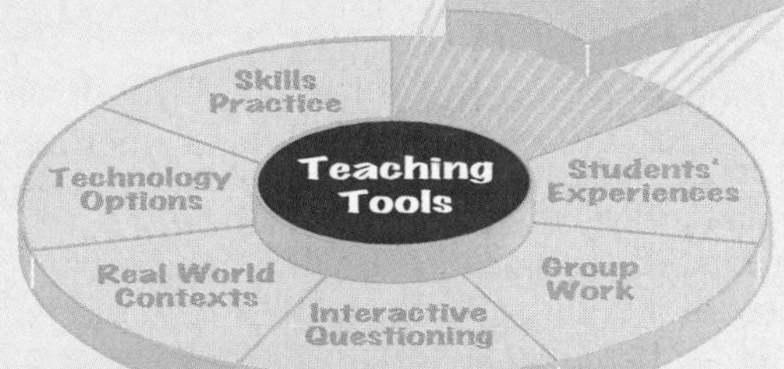

Assessment Options

The Prentice Hall Algebra Program provides you with many options. From these options, you may choose instructional materials and techniques appropriate for your students, or necessary to meet your district's curriculum requirements. As the chart indicates, the program also supports your teaching efforts by offering you many choices for assessment.

ASSESSMENT OPTIONS

Assessment Support Materials	Chapter Opener	9-1	9-2	9-3	9-4	9-5	9-6	9-7	9-8	Chapter End
Chapter Project	● ▲ ■	▲ ■	▲ ■	▲ ■			▲ ■	▲ ■		▲ ■
Checkpoints				▲ ●				▲ ●		
Self-Assessment				▲ ■				▲ ■	▲ ■	
Writing Assignment		▲	▲	▲	▲	▲	▲	▲	▲	● ▲
Chapter Assessment										▲ ●
Alternative Assessment		■	■	■	■	■	■	■	■	● ■
Cumulative Review										●
Standardized Test Prep				▲ ■	▲ ■	▲ ■	▲ ■	▲ ■		▲ ●
Computer Item Generator	Can be used to create custom-made practice or assessment at any time.									

▲ = Student Edition ■ = Teacher's Edition ● = Teaching Resources

Checkpoints

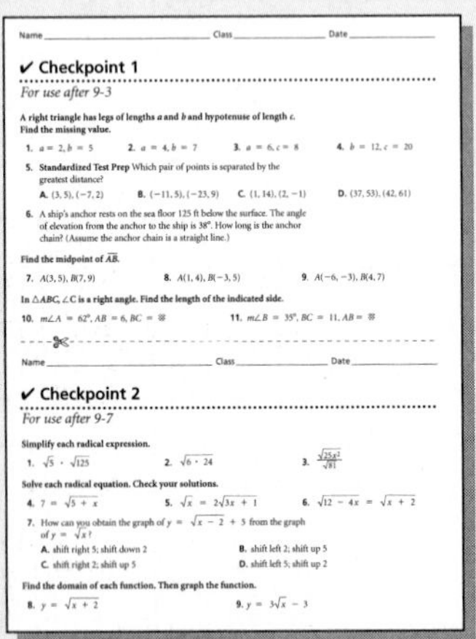

Name ________ Class ________ Date ________

✔ Checkpoint 1

For use after 9-3

A right triangle has legs of lengths a and b and hypotenuse of length c. Find the missing value.

1. $a = 2, b = 5$ 2. $a = 4, b = 7$ 3. $a = 6, c = 8$ 4. $b = 12, c = 20$

5. **Standardized Test Prep** Which pair of points is separated by the greatest distance?
A. (3, 5), (−7, 2) B. (−11, 5), (−23, 9) C. (1, 14), (2, −1) D. (37, 53), (42, 61)

6. A ship's anchor rests on the sea floor 125 ft below the surface. The angle of elevation from the anchor to the ship is 38°. How long is the anchor chain? (Assume the anchor chain is a straight line.)

Find the midpoint of $\overline{AB}$.

7. $A(3, 5), B(7, 9)$ 8. $A(1, 4), B(-3, 5)$ 9. $A(-6, -3), B(4, 7)$

In $\triangle ABC$, $\angle C$ is a right angle. Find the length of the indicated side.

10. $m\angle A = 62°, AB = 6, BC = ▒$ 11. $m\angle B = 35°, BC = 11, AB = ▒$

Name ________ Class ________ Date ________

✔ Checkpoint 2

For use after 9-7

Simplify each radical expression.

1. $\sqrt{5} \cdot \sqrt{125}$ 2. $\sqrt{6 \cdot 24}$ 3. $\frac{\sqrt{25x^2}}{\sqrt{81}}$

Solve each radical equation. Check your solutions.

4. $7 = \sqrt{5 + x}$ 5. $\sqrt{x} = 2\sqrt{3x + 1}$ 6. $\sqrt{12 - 4x} = \sqrt{x + 2}$

7. How can you obtain the graph of $y = \sqrt{x - 2} + 5$ from the graph of $y = \sqrt{x}$?
A. shift right 5; shift down 2 B. shift left 2; shift up 5
C. shift right 2; shift up 5 D. shift left 5; shift up 2

Find the domain of each function. Then graph the function.

8. $y = \sqrt{x + 2}$ 9. $y = 3\sqrt{x} - 3$

Alternative Assessment

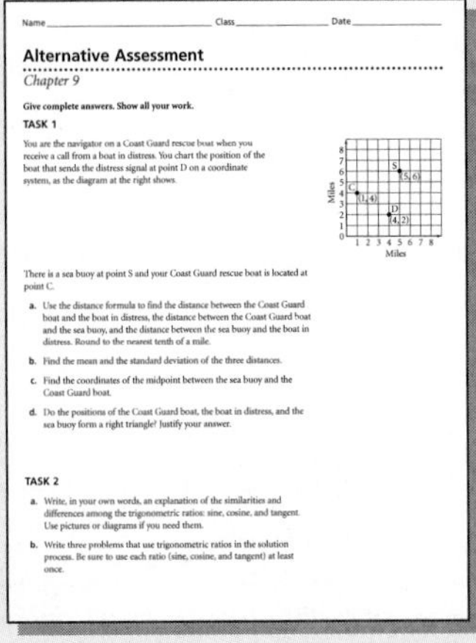

Name ________ Class ________ Date ________

Alternative Assessment

Chapter 9

Give complete answers. Show all your work.

TASK 1

You are the navigator on a Coast Guard rescue boat when you receive a call from a boat in distress. You chart the position of the boat that sends the distress signal at point D on a coordinate system, as the diagram at the right shows.

There is a sea buoy at point S and your Coast Guard rescue boat is located at point C.

a. Use the distance formula to find the distance between the Coast Guard boat and the boat in distress, the distance between the Coast Guard boat and the sea buoy, and the distance between the sea buoy and the boat in distress. Round to the nearest tenth of a mile.

b. Find the mean and the standard deviation of the three distances.

c. Find the coordinates of the midpoint between the sea buoy and the Coast Guard boat.

d. Do the positions of the Coast Guard boat, the boat in distress, and the sea buoy form a right triangle? Justify your answer.

TASK 2

a. Write, in your own words, an explanation of the similarities and differences among the trigonometric ratios: sine, cosine, and tangent. Use pictures or diagrams if you need them.

b. Write three problems that use trigonometric ratios in the solution process. Be sure to use each ratio (sine, cosine, and tangent) at least once.

Chapter Assessment

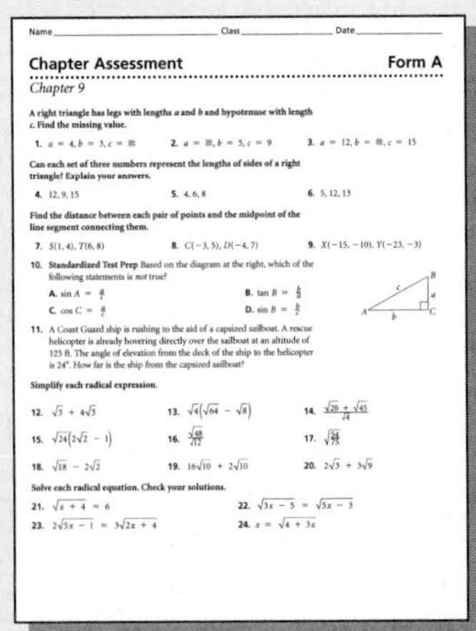

Name ________ Class ________ Date ________

Chapter Assessment **Form A**

Chapter 9

A right triangle has legs with lengths a and b and hypotenuse with length c. Find the missing value.

1. $a = 4, b = 3, c = ▒$ 2. $a = ▒, b = 5, c = 9$ 3. $a = 12, b = ▒, c = 15$

Can each set of three numbers represent the lengths of sides of a right triangle? Explain your answers.

4. 12, 9, 15 5. 4, 6, 8 6. 5, 12, 13

Find the distance between each pair of points and the midpoint of the line segment connecting them.

7. $S(1, 4), T(6, 8)$ 8. $C(-3, 5), D(-4, 7)$ 9. $X(-15, -10), Y(-23, -3)$

10. **Standardized Test Prep** Based on the diagram at the right, which of the following statements is *not* true?
A. $\sin A = \frac{a}{c}$ B. $\tan B = \frac{b}{a}$
C. $\cos C = \frac{a}{c}$ D. $\sin B = \frac{b}{c}$

11. A Coast Guard ship is rushing to the aid of a capsized sailboat. A rescue helicopter is already hovering directly over the sailboat at an altitude of 125 ft. The angle of elevation from the deck of the ship to the helicopter is 24°. How far is the ship from the capsized sailboat?

Simplify each radical expression.

12. $\sqrt{3} + 4\sqrt{3}$ 13. $\sqrt{4}(\sqrt{64} - \sqrt{8})$ 14. $\frac{\sqrt{20} + \sqrt{45}}{\sqrt{4}}$

15. $\sqrt{24}(2\sqrt{2} - 1)$ 16. $\frac{\sqrt{48}}{\sqrt{12}}$ 17. $\sqrt{\frac{54}{75}}$

18. $\sqrt{18} - 2\sqrt{2}$ 19. $16\sqrt{10} + 2\sqrt{10}$ 20. $2\sqrt{3} + 3\sqrt{9}$

Solve each radical equation. Check your solutions.

21. $\sqrt{x + 4} = 6$ 22. $\sqrt{3x - 5} = \sqrt{5x - 3}$

23. $2\sqrt{5x - 1} = 3\sqrt{2x + 4}$ 24. $x = \sqrt{4 + 3x}$

Available in both Form A and Form B

Making the Right Connections

Mathematics is imbedded in nearly every walk of life. The National Council of Teachers of Mathematics (NCTM) encourages educators to recognize these connections and to emphasize them for the purpose of better educating students for success in life and in a global economy. The ***Connections*** chart below highlights these connections for Chapter 9.

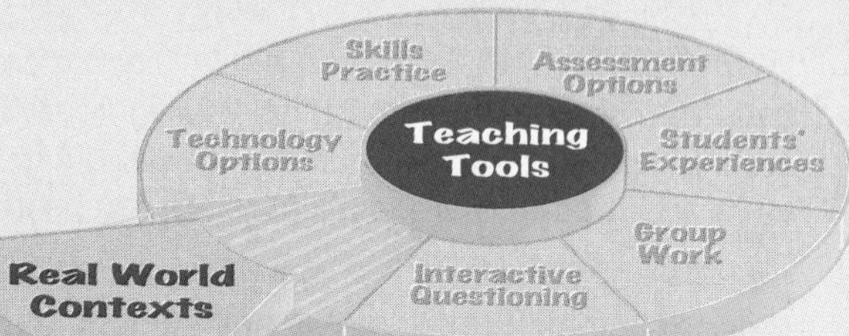

CONNECTIONS

Lesson	Interdisciplinary Connections	Career Prep	Other Real World Connections	Math Integration	NCTM Standards
Chapter Project	Geography Science		Space Observation	Trigonometry Geometry	Communication Trigonometry
9-1	Physical Science History	Solar heating Technician	Fire Rescue Birds Sightseeing Architecture	Geometry	Algebra Trigonometry Communication Problem Solving
9-2	Fitness Social Studies	Electrical Line Technician	Hiking News Coverage Transportation Finance	Geometry	Algebra Trigonometry Communication Problem Solving
9-3	Language Arts Environmental Science Physical Education	Aeronautics Mariner Industry	Navigation Nature Recreation Aviation Transportation	Trigonometry	Algebra Trigonometry Communication Problem Solving
9-4	Accounting Home Economics	Marketing Sales	Commuting Packaging Sailing Hobbies Consumer	Geometry	Algebra Trigonometry Communication Problem Solving
9-5	Art Physical Science	Construction Technician	Art Recreation Architecture Travel	Geometry	Algebra Trigonometry Communication Problem Solving
9-6	Physics Economics	Business Shipping	Amusement Parks Packaging Business	Geometry	Algebra Trigonometry Communication Problem Solving
9-7	Science Accounting	Firefighters Sales	Business Finance	Mathmatical Finance	Algebra Trigonometry Communication Problem Solving
9-8	Computer Science	Industry Business Research	Jobs Golf Automobiles	Statistics	Algebra Trigonometry Communication Problem Solving

CHAPTER 9 Right Triangles and Radical Expressions

CONNECTING TO PRIOR LEARNING Since most students will have had some exposure to the Pythagorean theorem, have pairs of students write what they know about it. Elicit ideas they have about other mathematical relationships in a right triangle.

CULTURAL CONNECTIONS Not all countries have advanced radar to guide air traffic. Ask if any students have seen an air-traffic control tower. Ask: *What happens when conditions are poor and there is no advanced technology to guide pilots?*

INTERDISCIPLINARY CONNECTIONS Even before knowledge of the Pythagorean theorem, ancient peoples used mathematical concepts in construction. Ask students to think of massive structures built throughout the history of the world and speculate about the math concepts used in building them.

ABOUT THE PROJECT The Chapter Project has students use the Pythagorean theorem in real-life situations. Students use formulas, equations, tables, and graphs to determine and compare the distances they would be able to see to the horizon if they were standing on other planets.

Technology Options

Prentice Hall Technology

Video Video Field Trip 9, "Expanding Horizons" a visit with astronomers at Kitt Peak Observatory

CHAPTER 9 Right Triangles and Radical Expressions

Relating to the Real World

Right triangles have been important in surveying and construction since early civilization. Since then, special relationships in right triangles have led to improved navigation for ships and planes, to formulas for a pendulum and for distance, and to other areas of mathematics that use radical expressions.

Lessons	9-1	9-2	9-3	9-4	9-5
	The Pythagorean Theorem	*The Distance Formula*	*Trigonometric Ratios*	*Simplifying Radicals*	*Adding and Subtracting Radicals*

Launching the Project

PROJECT NOTEBOOK Encourage students to keep all project-related materials in a separate folder or notebook. See the Chapter Project and Scoring Rubric in the Chapter Support File.

- Ask students to think of a familiar distance measured in feet. depth of a swimming pool, distance between softball bases, height of a volleyball net
- Have them estimate how many of these distances would fit into a mile.
- Challenge students to convert their measurement into miles or to list the steps that would be necessary to make the conversion.
- Have the students look at the Find Out by Writing section on page 418 of their textbooks. Explain that when students begin the Chapter Project, they will learn how to convert feet to miles and to use the Pythagorean theorem to write an equation relating r, d, and h.

TRACKING THE PROJECT You may wish to have students read the Finishing the Chapter Project on page 456 to help them get an overview of the project. Set benchmark deadlines for students to show you their work in progress.

CHAPTER PROJECT

On a CLEAR Day...

Suppose it's a clear day and you have a view with no obstructions — maybe not as clear as from an air traffic control tower, but fairly clear. How far would you be able to see to the horizon? You can use the Pythagorean theorem and other concepts in this chapter to find this distance.

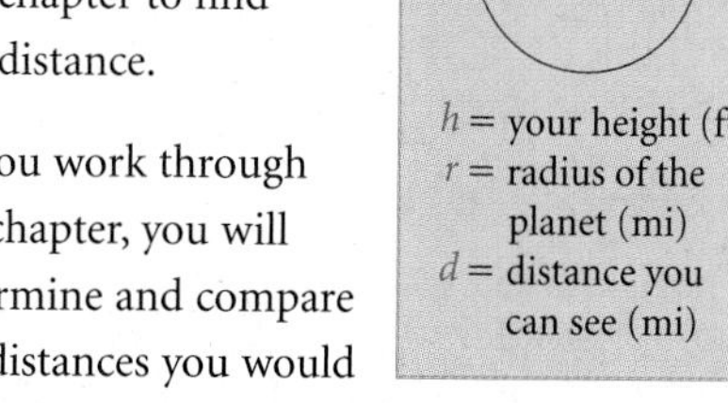

h = your height (ft)
r = radius of the planet (mi)
d = distance you can see (mi)

As you work through the chapter, you will determine and compare the distances you would be able to see to the horizon if you could stand on any planet, including Earth. Your project should include diagrams of the planets, formulas for the distances visible, and graphs of these formulas.

To help you complete the project:

- **p. 418** *Find Out by Writing*
- **p. 424** *Find Out by Calculating*
- **p. 429** *Find Out by Researching*
- **p. 444** *Find Out by Communicating*
- **p. 449** *Find Out by Organizing*
- **p. 456** *Finishing the Project*

Project Resources

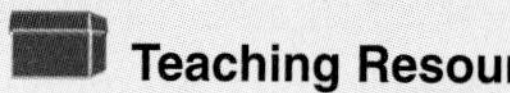

Teaching Resources

Chapter Support File, Ch. 9
- Chapter Project Manager and Scoring Rubric

Transparencies
104

Using the Rubric

Sharing the scoring rubric for the project with your students will alert them to your expectations before they begin work on the project.

As students complete each Find Out question in the chapter, you may wish to have them evaluate their own work or a partner's work, based on the scoring rubric. Students should have the opportunity to revise their work after it has been reviewed.

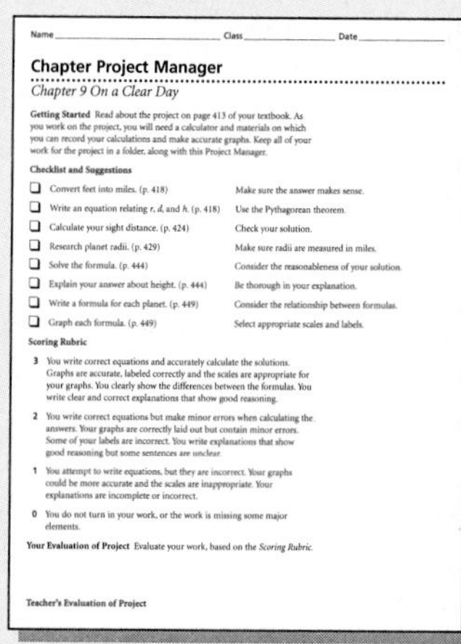

Name ____________ Class ____________ Date ____________

Chapter Project Manager

Chapter 9 On a Clear Day

Getting Started Read about the project on page 413 of your textbook. As you work on the project, you will need a calculator and materials on which you can record your calculations and make accurate graphs. Keep all of your work for the project in a folder, along with this Project Manager.

Checklist and Suggestions

- ☐ Convert feet into miles. (p. 418) — Make sure the answer makes sense.
- ☐ Write an equation relating r, d, and h. (p. 418) — Use the Pythagorean theorem.
- ☐ Calculate your sight distance. (p. 424) — Check your solution.
- ☐ Research planet radii. (p. 429) — Make sure radii are measured in miles.
- ☐ Solve the formula. (p. 444) — Consider the reasonableness of your solution.
- ☐ Explain your answer about height. (p. 444) — Be thorough in your explanation.
- ☐ Write a formula for each planet. (p. 449) — Consider the relationship between formulas.
- ☐ Graph each formula. (p. 449) — Select appropriate scales and labels.

Scoring Rubric

3 You write correct equations and accurately calculate the solutions. Graphs are accurate, labeled correctly and the scales are appropriate for your graphs. You clearly show the differences between the formulas. You write clear and correct explanations that show good reasoning.

2 You write correct equations but make minor errors when calculating the answers. Your graphs are correctly laid out but contain minor errors. Some of your labels are incorrect. You write explanations that show good reasoning but some sentences are unclear.

1 You attempt to write equations, but they are incorrect. Your graphs could be more accurate and the scales are inappropriate. Your explanations are incomplete or incorrect.

0 You do not turn in your work, or the work is missing some major elements.

Your Evaluation of Project Evaluate your work, based on the *Scoring Rubric.*

Teacher's Evaluation of Project

9-1 Teaching Notes

PROBLEM OF THE DAY

Suyin paid $7 for a share of stock that doubles in value every year. How much will the share of stock be worth in 6 yr? $448

Problem of the Day is also available in Transparencies.

CONNECTING TO PRIOR KNOWLEDGE Have students describe a right triangle. Then ask them what they know about the angles of a right triangle. One is a right angle, or 90°; the other two angles are acute; together the angles equal 180°.

WORK TOGETHER

Question 3 Make sure students understand what is meant by the diagonal of the cover.

THINK AND DISCUSS

Make sure students understand the definition of a right angle and can recognize right angles.

ERROR ALERT! Students may think that the hypotenuse is always the side on top. **Remediation:** On the board, draw several triangles oriented different ways and have students label the hypotenuse of each triangle. Ask students: *Is one side of a right triangle always the longest side?* yes *Which one?* the hypotenuse Have students who are having difficulty

Lesson Planning Options

Prerequisite Skills

- Simplifying square roots
- Knowing parts of right triangles

Assignment Options for Exercises On Your Own

To provide flexible scheduling, this lesson can be subdivided into parts.

Core 1–12, 14–17
Extension 13, 18–20

Core 21–36
Extension 37–39

Use Mixed Review to maintain skills.

Resources

Student Edition

Skills Handbook, pp. 581, 582
Extra Practice, p. 564
Glossary/Study Guide

Teaching Resources

Chapter Support File, Ch. 9
- Practice 9-1 (two worksheets)
- Reteaching 9-1
- Alternative Activity 9-1

Classroom Manager 9-1
Glossary, Spanish Resources
Two-Year Algebra Handbook 9-1

Transparencies

105, 109

9-1 The Pythagorean Theorem

What You'll Learn
- Finding lengths of sides of a right triangle
- Deciding if a triangle is a right triangle

...And Why

To calculate distances that cannot be measured directly

What You'll Need
- rectangular objects
- ruler
- graph paper

WORK TOGETHER

Work with a partner.

1. Measure the length l and width w of the cover of your mathematics textbook. Be as precise in your measurements as you can. Answers may vary. Sample: ℓ = 8.75 in.; w = 10.25 in.
2. Use a calculator to find the value of the expression $l^2 + w^2$. 181.625 (for sample given)
3. Measure the length d of one of the diagonals of the cover of your book. Calculate the value of d^2. d = 13.5 in.; d^2 = 182.25
4. What can you say about the values of $l^2 + w^2$ and d^2? The values are about equal.
5. Choose rectangular objects or draw rectangles on graph paper. Repeat the same steps. What can you say about the length of a diagonal? The square of the length of the diagonal is about equal to the sum of the square of the length and the square of the width.

THINK AND DISCUSS

Part 1

Solving Equations Using the Pythagorean Theorem

When you draw a diagonal of a rectangle, you separate the rectangle into two right triangles. In a right triangle, the side opposite the right angle is the longest. It is called the **hypotenuse.** The other two sides are called **legs.** There is a relationship among the lengths of the sides of a right triangle.

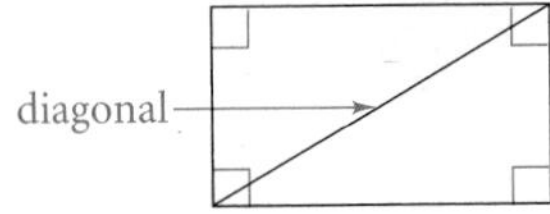

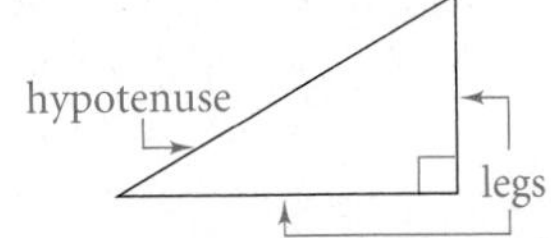

QUICK REVIEW

A *right triangle* is a triangle with one right angle.

The Pythagorean Theorem

In a right triangle, the sum of the squares of the lengths of the legs is equal to the square of the length of the hypotenuse.

$a^2 + b^2 = c^2$

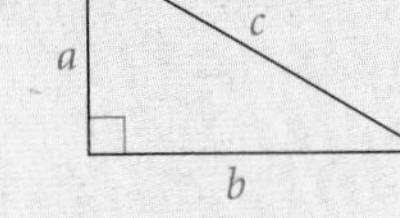

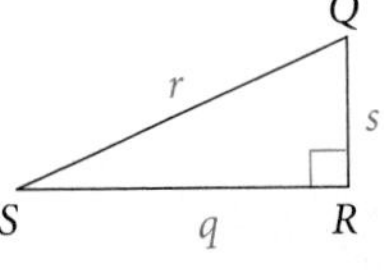

6. Restate the Pythagorean theorem for $\triangle QRS$ to the left. $s^2 + q^2 = r^2$
7. A right triangle has legs of length 5 in. and 12 in. The length of the hypotenuse is 13 in. **Verify** that the Pythagorean theorem is true for this triangle. $5^2 + 12^2 = 169$; $13^2 = 169$; $5^2 + 12^2 = 13^2$ ✓

understanding why the hypotenuse is always the longest side of a right triangle draw a variety of right triangles on the board. Then, have them measure the sides to validate this fact.

Make sure students recognize that in the Pythagorean theorem, the hypotenuse, c^2, is always written alone on one side of the equal sign. The sum of the squares of the two smaller sides equals the square of the longest side.

Example 1

VISUAL LEARNING To help students learn the vocabulary of this lesson, challenge them to draw at each end of a line (labeled *hypotenuse*) long-legged cartoon characters who are each standing on one foot. Point out that the term *legs* refers to the two sides of the right triangle that are not the hypotenuse.

Example 2 **Relating to the Real World** 

ESTIMATION Before students calculate the answer, have them estimate how far up the building the ladder can reach. Ask: *Can it be more than 100 ft?* **No; the ladder is only 100 ft long.**

TACTILE LEARNING Encourage students to cut a strip of stiff paper or cardboard 10 cm long to model the ladder. Have them lean the strip against a book 5 cm away and measure how far up the book the model leans.

Example 1

A right triangle has legs of lengths 9 cm and 12 cm. What is the length of the hypotenuse?

$a^2 + b^2 = c^2$ ← Use the Pythagorean theorem.
$9^2 + 12^2 = c^2$ ← Substitute 9 for *a* and 12 for *b*.
$81 + 144 = c^2$ ← Simplify.
$\sqrt{225} = \sqrt{c^2}$ ← Take the square root of each side.
$c = 15$

The hypotenuse has length 15 cm.

PROBLEM SOLVING

Look Back Why did we consider only the principal square root of 225? **The length of a side of a triangle cannot be negative.**

8. Try This What is the length of the hypotenuse of a right triangle with legs of lengths 7 and 24? **25**

You can use the Pythagorean theorem to find the length of a leg of a right triangle when you know the lengths of the other sides.

Example 2 **Relating to the Real World**

Fire Rescue A fire truck parks 16 ft away from a building. The fire truck extends its ladder 30 ft . How far up the building from the truck's roof does the extension ladder reach?

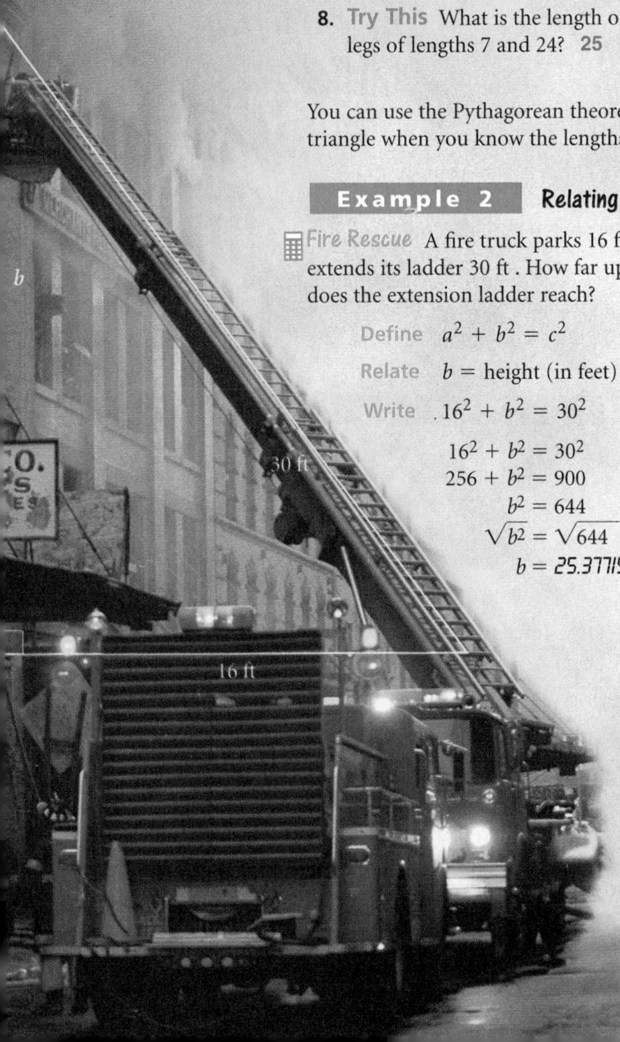

Define $a^2 + b^2 = c^2$ ← Use the Pythagorean theorem.

Relate b = height (in feet) the ladder reaches

Write $16^2 + b^2 = 30^2$ ← Substitute 16 for *a* and 30 for *c*.

$16^2 + b^2 = 30^2$
$256 + b^2 = 900$ ← Simplify.
$b^2 = 644$ ← Subtract 256 from each side.
$\sqrt{b^2} = \sqrt{644}$ ← Take the square root of each side.
$b = 25.37715508$ ← Use a calculator.

The ladder reaches about 25 ft up the building.

9. Calculator A right triangle has a 47-in. hypotenuse and a 19-in. leg. What is the length of the other leg? **about 43 in.**

Additional Examples

FOR EXAMPLE 1

A right triangle has legs of lengths 4 m and 8 m. What is the length of the hypotenuse? **8.94 m**

FOR EXAMPLE 2

A water tower is supported by guy wires. A 120 m wire leads from the top of the tower to the ground. If the wire meets the ground 35 m from the base of the tower, how high is the water tower? **115 m**

Discussion: *If the guy wire reaches the ground farther from the tower's base, how would the value of the wire's angle with the ground change? How would the value of the angle of the wire with the tower change?*

FOR EXAMPLE 3

A triangle has sides of lengths 7, 5, and 5. Is it a right triangle? **no**

Example 3

ERROR ALERT! Students may not substitute the correct values in the correct places. **Remediation:** First have students find the greatest number. Substitute this value for *c*. Then substitute the two smaller numbers for *a* and *b.* It does not matter which of the two smaller values replace *a* or *b* because addition is commutative.

KINESTHETIC LEARNING To help students develop the teamwork skills needed in the workplace, have them collaborate in the following activity. Give each group of students three lengths of string, each at least 12 ft long. Have students cut two pieces so that they are exactly 7 ft and 9 ft. Then have them model the triangle in Example 3 to determine the length of the hypotenuse. Have each group choose a person to measure the hypotenuse while three other group members hold the strings. The hypotenuse is 11.4 ft long.

MAKING CONNECTIONS Pythagoras was a Greek philosopher and religious leader. He founded a philosophical and religious school in Croton. Because the doctrine of the school included secrecy and communal participation, we do not actually know whether Pythagoras himself or one of his followers developed the Pythagorean theorem. The Pythagoreans were vegetarians, believed in reincarnation and numerology, and thought that the structure of the cosmos was based on the ratios 4 : 3, 3 : 2, and 2 : 1. Later, Pythagoreans came to believe that odd and even numbers were related to male and female qualities and the virtues of good and bad.

Technology Options

For Exercises 21–36, students may find it helpful to use geometry software to determine if the numbers represent right triangles.

Prentice Hall Technology

Calculator
- Graphing Calculator Handbook: Procedures 19, 20

Software
- Secondary Math Lab Toolkit™
- Computer Item Generator 9-1

CD-ROM
- Multimedia Algebra Lab 9

Internet
- See the Prentice Hall site. (http://www.phschool.com)

416

Part 2

Using the Converse

You can find out whether a triangle is a right triangle by using the converse of the Pythagorean theorem.

The Converse of the Pythagorean Theorem

If a triangle has sides of lengths *a*, *b*, and *c*, and $a^2 + b^2 = c^2$, then the triangle is a right triangle with hypotenuse of length *c*.

Example 1

A triangle has sides of lengths 7, 9, and 12. Is it a right triangle?

Apply the converse of the Pythagorean theorem.

$a^2 + b^2 \stackrel{?}{=} c^2$

$7^2 + 9^2 \stackrel{?}{=} 12^2$ ← Substitute 12 for *c*, since 12 is the length of the longest side. Substitute 7 and 9 for *a* and *b*.

$49 + 81 \stackrel{?}{=} 144$

$130 \neq 144$

The triangle is not a right triangle.

10. Try This A triangle has sides of lengths 10, 24, and 26. Is the triangle a right triangle? Explain. Yes; $10^2 + 24^2 = 676$ and $26^2 = 676$.

Exercises ON YOUR OWN

Calculator **Use the triangle at the right. Find the length of the missing side to the nearest tenth.**

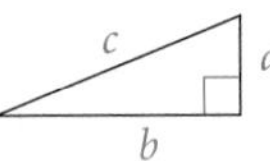

1. $a = 6, b = 8, c = ■$ 10
2. $a = 8, b = 24, c = ■$ 25.3
3. $a = 1, b = 2, c = ■$ 2.2
4. $a = 3, b = ■, c = 5$ 4
5. $a = ■, b = 7, c = 10$ 7.1
6. $a = ■, b = 12, c = 13$ 5
7. $a = 4, b = 4, c = ■$ 5.7
8. $a = 13, b = 2, c = ■$ 13.2
9. $a = 15, b = ■, c = 25$ 20
10. $a = ■, b = 10, c = 15$ 11.2
11. $a = 75, b = ■, c = 100$ 66.1
12. $a = ■, b = 12, c = 18$ 13.4

13. Any set of three positive integers that satisfies the relationship $a^2 + b^2 = c^2$ is called a *Pythagorean triple.*
    a. **Verify** that the numbers 6, 8, and 10 form a Pythagorean triple. $6^2 + 8^2 = 100$ and $10^2 = 100$
    b. Copy the table at the right. Complete the table so that the values in each row form a Pythagorean triple. 5; 12; 7; 41
    c. Group Activity Find a Pythagorean triple that does not appear in the table. Answers may vary. Sample: 8, 15, 17

a	b	c
3	4	■
5	■	13
■	24	25
9	40	■

Exercises ON YOUR OWN

ESL Exercise 18 Explain that *due* east and *due* north mean *exactly* east or *exactly* north.

WRITING Exercise 19a Point out that the term *sides* can mean *legs* or *hypotenuse*. Make sure students clearly understand the difference in meaning.

GEOMETRY Exercise 20 Students apply their knowledge of squares to solve a problem involving the Pythagorean theorem.

ALTERNATIVE ASSESSMENT **Exercises 21–36** For any exercises where students think the lengths do not form a right triangle, have them change the given lengths to ones that will form a right triangle. This will assess understanding of the Pythagorean theorem.

Exercises 21–36 Remind students that the longest side in a right triangle is the hypotenuse.

DIVERSITY Exercise 37 Some students may not be familiar with gondolas. Others may think the word *gondola* applies only to boats in Venice, Italy. Clarify to which type of gondola the problem refers. Ask students who have ridden in a gondola to share with the class their knowledge of how it works and what it feels like to ride in one. Ask students to bring in photographs of gondolas.

Exercise 39 Encourage students to make sketches or model the problem by making a rod with paper and tape. Point out that the length of the rod—10 cm—is irrelevant in this question. Help students develop the skill of identifying and ignoring extraneous information.

Find the missing length to the nearest tenth.

14. 23.3

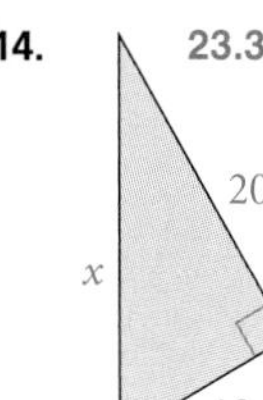

15. 9.7

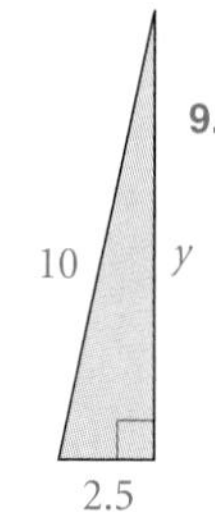

16. 9

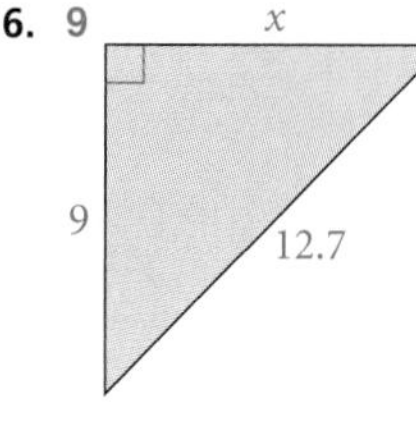

17. 559.9

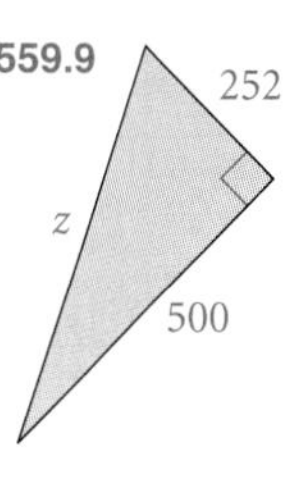

18. Birds A pigeon leaves its nest and flies 5 km due east. The pigeon then flies 3 km due north. How far is the pigeon from its nest? about 5.8 km

19. You know that two sides of a right triangle measure 10 in. and 8 in.

a. Writing Why is this not enough information to be sure of finding the length of the third side? See margin.

b. Give two different possible values for the length of the third side. Explain how you found your answers. 6 in.; $\sqrt{10^2 - 8^2}$; about 12.8 in.; $\sqrt{10^2 + 8^2}$

20. Geometry The yellow, green, and blue figures are squares. Use the Pythagorean theorem and your knowledge of lengths to find the length of one side of the blue square. about 7.2

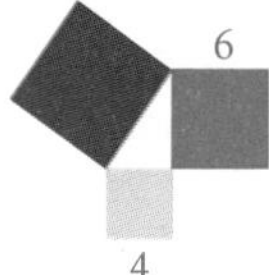

Can each set of three numbers represent the lengths of the sides of a right triangle? Explain your answers.

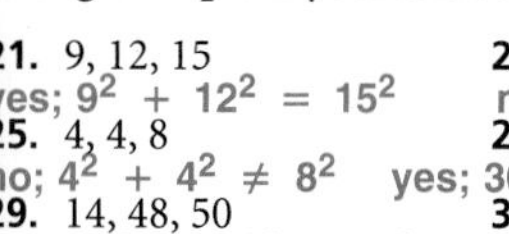

21. 9, 12, 15 yes; $9^2 + 12^2 = 15^2$

22. 1, 2, 3 no; $1^2 + 2^2 \neq 3^2$

23. 2, 4, 5 no; $2^2 + 4^2 \neq 5^2$

24. 34, 16, 30 yes; $16^2 + 30^2 = 34^2$

25. 4, 4, 8 no; $4^2 + 4^2 \neq 8^2$

26. 5000, 4000, 3000 yes; $3000^2 + 4000^2 = 5000^2$

27. 1.25, 3, 3.25 yes; $1.25^2 + 3^2 = 3.25^2$

28. 19, 21, 23 no; $19^2 + 21^2 \neq 23^2$

29. 14, 48, 50 yes; $14^2 + 48^2 = 50^2$

30. $\frac{1}{3}, \frac{1}{4}, \frac{1}{5}$ 30. no; $\left(\frac{1}{4}\right)^2 + \left(\frac{1}{5}\right)^2 \neq \left(\frac{1}{3}\right)^2$

31. 2, 1.5, 1 no; $1^2 + 1.5^2 \neq 2^2$

32. 10, 24, 26 yes; $10^2 + 24^2 = 26^2$

33. 4, 5, 6 no; $4^2 + 5^2 \neq 6^2$

34. 15, 20, 25 yes; $15^2 + 20^2 = 25^2$

35. 18, 80, 82 yes; $18^2 + 80^2 = 82^2$

36. $\frac{3}{4}, 1, 1\frac{1}{4}$ 36. yes; $\left(\frac{3}{4}\right)^2 + 1^2 = \left(1\frac{1}{4}\right)^2$

37. a. Sightseeing A gondola travels between two elevations along a cable. What is the distance the gondola travels from the bottom of the hill to the top? about 1340 ft

b. The gondola travels from the bottom of the hill to the top of the hill in 20 min. What is the average speed of the gondola in feet per minute? in miles per hour?

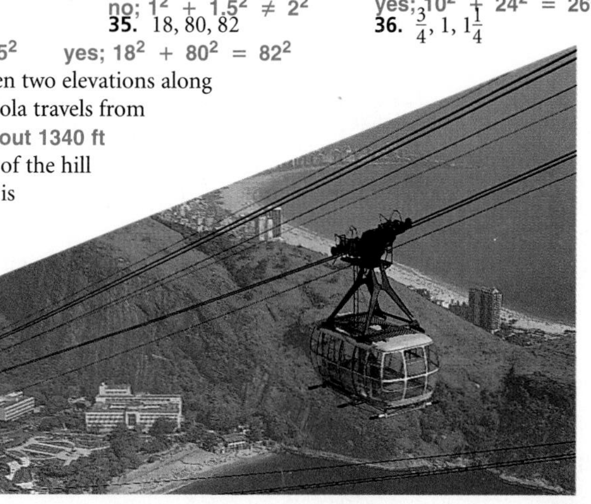

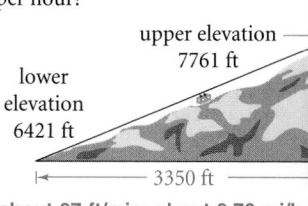

b. about 67 ft/min; about 0.76 mi/h

pages 416–418 On Your Own

19a. The third side can be the hypotenuse, the longest of the three sides, or it could be the shortest leg.

Chapter Project FIND OUT BY WRITING Students write responses and use the Pythagorean theorem to answer the given questions. This task is essential to their work on the Chapter Project which is introduced in the Chapter Opener. Check students' progress on the project by having each student write a sentence or two explaining how this task is related to the project as a whole.

Exercises MIXED REVIEW

Exercise 42 Encourage students to make sketches to help them solve the problem.

GETTING READY FOR LESSON 9-2 These exercises prepare students for using the distance formula.

Wrap Up

THE BIG IDEA Ask students to describe the Pythagorean theorem and tell how they can use it to determine whether a triangle is a right triangle.

RETEACHING ACTIVITY Students find lengths of sides of a right triangle using Pythagorean theorem. (Reteaching worksheet 9-1)

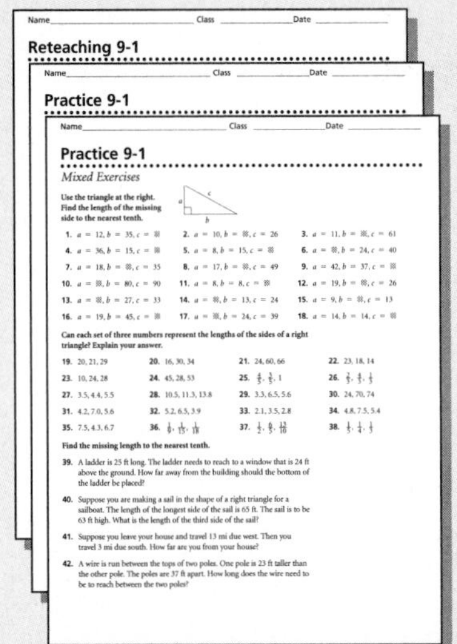

Name______ Class ______ Date ______

Reteaching 9-1

Name______ Class ______ Date ______

Practice 9-1

Name______ Class ______ Date ______

Practice 9-1

Mixed Exercises

Use the triangle at the right. Find the length of the missing side to the nearest tenth.

1. $a = 12, b = 35, c = ■$
2. $a = 10, b = ■, c = 26$
3. $a = 11, b = ■, c = 61$
4. $a = 36, b = 15, c = ■$
5. $a = 8, b = 15, c = ■$
6. $a = ■, b = 24, c = 40$
7. $a = 18, b = ■, c = 35$
8. $a = 17, b = ■, c = 49$
9. $a = 42, b = 37, c = ■$
10. $a = ■, b = 80, c = 90$
11. $a = 8, b = 8, c = ■$
12. $a = 19, b = ■, c = 26$
13. $a = ■, b = 27, c = 33$
14. $a = ■, b = 13, c = 24$
15. $a = 9, b = ■, c = 13$
16. $a = 19, b = 45, c = ■$
17. $a = ■, b = 24, c = 39$
18. $a = 14, b = 14, c = ■$

Can each set of three numbers represent the lengths of the sides of a right triangle? Explain your answer.

19. 20, 21, 29
20. 16, 30, 34
21. 24, 60, 66
22. 23, 18, 14
23. 10, 24, 28
24. 45, 28, 53
25. $\frac{4}{5}, \frac{3}{5}, 1$
26. [illegible]
27. 3.5, 4.4, 5.5
28. 10.5, 11.3, 13.8
29. 3.3, 6.5, 5.6
30. 24, 70, 74
31. 4.2, 7.0, 5.6
32. 5.2, 6.5, 3.9
33. 2.1, 3.5, 2.8
34. 4.8, 7.5, 5.4
35. 7.5, 4.3, 6.7
36. $\frac{1}{9}, \frac{1}{15}, \frac{1}{18}$
37. $\frac{1}{2}, \frac{6}{5}, \frac{13}{10}$
38. $\frac{1}{5}, \frac{1}{4}, \frac{1}{3}$

Find the missing length to the nearest tenth.

39. A ladder is 25 ft long. The ladder needs to reach to a window that is 24 ft above the ground. How far away from the building should the bottom of the ladder be placed?
40. Suppose you are making a sail in the shape of a right triangle for a sailboat. The length of the longest side of the sail is 65 ft. The sail is to be 63 ft high. What is the length of the third side of the sail?
41. Suppose you leave your house and travel 13 mi due west. Then you travel 3 mi due south. How far are you from your house?
42. A wire is run between the tops of two poles. One pole is 23 ft taller than the other pole. The poles are 37 ft apart. How long does the wire need to be to reach between the two poles?

Lesson Quiz

Lesson Quiz is also available in Transparencies.

1. Find the missing length to the nearest tenth. 16.6

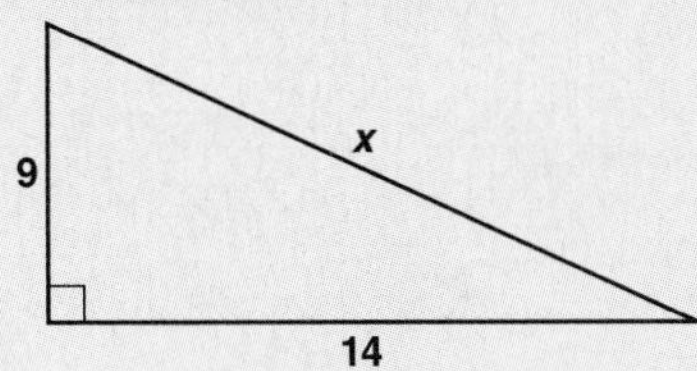

2. Find the missing length to the nearest tenth. 8.1

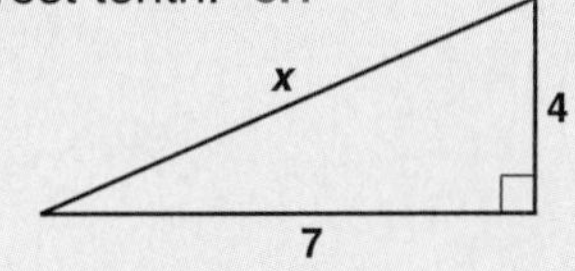

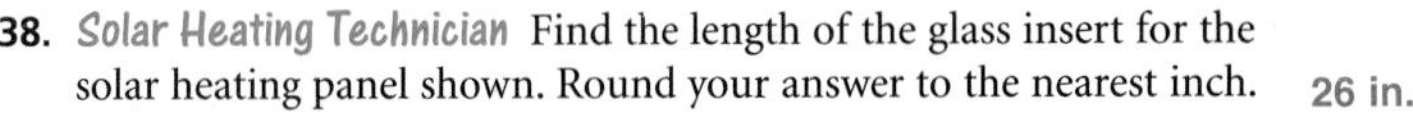

38. *Solar Heating Technician* Find the length of the glass insert for the solar heating panel shown. Round your answer to the nearest inch. 26 in.

39. What is the diameter of the smallest circular opening that the rectangular rod shown will fit through? Round your answer to the nearest tenth. 4.3 cm

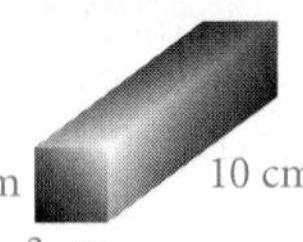

Chapter Project ***Find Out by Writing***

- How many feet are in a mile?
- How would you convert 15 ft into miles?
- How would you represent the quantity *h* feet in miles?
- In the diagram on page 413, replace *h* with the expression that represents *h* feet in miles. Use the Pythagorean theorem to write an equation relating *r*, *d*, and *h*. Do not simplify the equation.

Exercises MIXED REVIEW

Solve each system of equations by substitution.

40. $y = 3x + 5,\ x + y = 4$ $-\frac{1}{4}, 4\frac{1}{4}$

41. $y = -3x + 2,\ x - y = 0$ $\frac{1}{2}, \frac{1}{2}$

42. *Architecture* A model of a house is $\frac{1}{25}$ of its actual size. One side of the model house is 48 in. long. How long is the corresponding side in the actual house? 100 ft

Getting Ready for Lesson 9-2

Let $c = \sqrt{a^2 + b^2}$. For each set of values, calculate c to the nearest tenth.

43. $a = 3,\ b = 4$ 5

44. $a = -2,\ b = 5$ 5.4

45. $a = -3,\ b = 8$ 8.5

46. $a = 7,\ b = -5$ 8.6

Math ToolboX

In Lesson 9-1, students learned the Pythagorean theorem. This toolbox allows students to test for a right triangle using a calculator program. First demonstrate the use of the program on an overhead calculator, if available. Then have students work in groups of four to enter the program and complete the exercises. This toolbox will reinforce the fact that *c* represents the longest side of a triangle, the hypotenuse.

If you want each class to experience entering the program, be sure to have students delete the program before they return the calculators by pressing 2nd [MEM] **2** and then the program number. Some types of calculators can be linked together to quickly enter entire programs for students having difficulty.

ERROR ALERT! Some students may have difficulty locating some commands to enter the complete program.
Remediation: List directions for finding these commands on the chalkboard. For If, Goto, and Lbl, press PRGM; for Prompt and Disp, PRGM ▶; for =, 2nd [TEST]. Students must use the down cursor to view the entire menu.

WRITING Exercise 17 For students who have trouble starting, suggest that they examine the program directions and work the problem using paper and pencil.

ADDITIONAL PROBLEMS Use the program to tell if each set of numbers can represent the sides of a right triangle.

1. $\{1, 1, \sqrt{2}\}$ yes **2.** $\{4, 5, 6\}$ no
3. $\{7, 14, 7\sqrt{3}\}$ yes **4.** $\{12, 16, 20\}$ yes
5. $\{11, 17, 28\}$ no

Math ToolboX — Technology

Testing for a Right Triangle

After Lesson 9-1

With a graphing calculator, you can create a program and save it to use in the future. This program tests three numbers to see if they can be the lengths of the sides of a right triangle.

To input the program, choose PRGM, then *NEW*. To name the program, press 1, type PYTHAG, and then press ENTER.

To enter the program you will find commands in the PRGM feature under *CTL* and *I/O*. After each line press ENTER to go to the next line.

To type the variables *A*, *B*, and *C*, use ALPHA before typing each letter. You can use *A-LOCK* to type a string of letters such as the information in quotes. The equal sign is in the TEST feature.

After entering your program, press 2nd QUIT. To run your program, press PRGM, select PYTHAG, and choose *EXEC*.

```
PROGRAM:PYTHAG
:Prompt A,B,C
:If A²+B²=C²
:Goto 1
:Disp "NOT A RIGHT
TRIANGLE"
:Goto 2
:Lbl 1
:Disp "RIGHT TRIANGLE"
:Lbl 2
```

Use the program to tell if each set of numbers can represent the sides of a right triangle. Always input the longest side as *c*.

1. {3, 4, 5} yes **2.** {7, 9, 12} no **3.** {9, 12, 15} yes **4.** {10, 11, 12} no
5. {8, 15, 17} yes **6.** {17, 17, 17} no **7.** {1.4, 4.8, 5} yes **8.** {6, 8, 10} yes
9. {3, 7, 8} no **10.** {7, 24, 25 } yes **11.** {20, 21, 29} yes **12.** {12, 13, 20} no

Use the program to see if each triangle is a right triangle.

13. yes
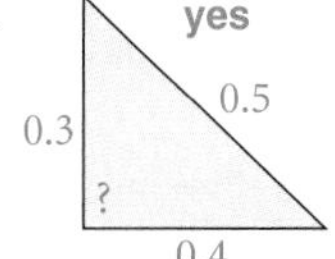

14. no
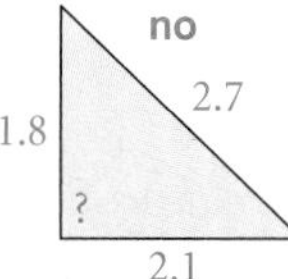

15. yes
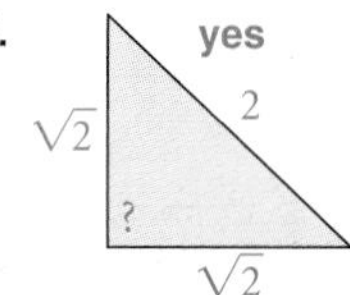

16. Any three integers *a*, *b*, and *c* form a Pythagorean triple if $a^2 + b^2 = c^2$. The following formulas generate Pythagorean triples using positive integers *x* and *y* where $x > y$.

$$a = x^2 - y^2$$
$$b = 2xy$$
$$c = x^2 + y^2$$

a. Find the Pythagorean triple that is generated by the values $x = 3$ and $y = 2$. 5, 12, 13

b. Find the values of *x* and *y* that generate the set {9, 40, 41}. 5, 4

17. *Writing* The set {–3, –4, 5} could not represent the sides of a right triangle because the sides of triangles cannot have negative lengths. Explain why the calculator display indicates a right triangle when you input the values –3, –4, and 5. $(-3)^2 + (-4)^2 = 5^2$; the program does not check for positive numbers.

```
prgmPYTHAG
A=?-3
B=?-4
C=?5
RIGHT TRIANGLE
              Done
```

Materials and Manipulatives

- Graphing calculator

Resources

Transparencies
11

419

9-2 Teaching Notes

PROBLEM OF THE DAY

Michael can run the 100-yd dash in 10 s. If he could keep up this pace, how fast could he run the mile?
2 min 56 s

Problem of the Day is also available in Transparencies.

CONNECTING TO PRIOR KNOWLEDGE Draw a set of coordinate axes on the board. Draw two points that do not line up either horizontally or vertically. Ask: *What theorem or formula have you learned that could be used to find the distance between these two points?* Pythagorean theorem

WORK TOGETHER

Question 2 Point out that *AC* is the distance from point *A* to point *C*.

ERROR ALERT! Students may become confused by the variables used in the Pythagorean theorem and the names of the sides of ΔABC. **Remediation:** Encourage students to rewrite the Pythagorean theorem using the names of the sides of ΔABC.

THINK AND DISCUSS

AUDITORY LEARNING Help students make the connection between the distance formula and the process they used in the Work Together activity. Ask students to explain out loud

Lesson Planning Options

Prerequisite Skills

- Simplifying square roots
- Evaluating formulas

Assignment Options for Exercises On Your Own

To provide flexible scheduling, this lesson can be subdivided into parts.

1 **Core** 1–15
Extension 16, 17, 31–34

2 **Core** 18–29
Extension 30, 35, 36

Use Mixed Review to maintain skills.

Resources

 **Student Edition**

Skills Handbook, pp. 581, 582
Extra Practice, p. 564
Glossary/Study Guide

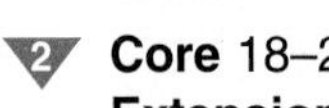

Teaching Resources

Chapter Support File, Ch. 9
- Practice 9-2 (two worksheets)
- Reteaching 9-2
- Alternative Activity 9-2

Classroom Manager 9-2
Glossary, Spanish Resources
Two-Year Algebra Handbook 9-2

 **Transparencies**
1, 2, 105, 110

9-2 The Distance Formula

What You'll Learn

- Finding the distance between two points in a coordinate plane
- Finding the coordinates of the midpoint of two points

...And Why

To find the distance between two groups of hikers

What You'll Need

- graph paper
- ruler

QUICK REVIEW

AB is the distance between points *A* and *B* and the length of $\overline{AB}$.

WORK TOGETHER

Work with a partner as you apply the Pythagorean theorem to find the distance between two points.

1. On graph paper, graph points $A(-3, 4)$, $B(1, 1)$, and $C(-3, 1)$. Then connect the points. See margin.
2. Find *AC* and *BC*. 3, 4
3. $\triangle ABC$ is a right triangle. Use the Pythagorean theorem to write an equation relating *AC*, *BC*, and *AB*. $AC^2 + BC^2 = AB^2$
4. Substitute values for *AC* and *BC* in the equation you wrote in Question 3. Solve the equation to find *AB*. 5

THINK AND DISCUSS

Part 1 **Finding the Distance**

In the Work Together activity, you found the distance between two particular points. If you have any two points $P(x_1, y_1)$ and $Q(x_2, y_2)$, you can graph them and form a right triangle as shown in the diagram. You can then use the Pythagorean theorem to find the distance between the points.

$(PQ)^2 = (PR)^2 + (RQ)^2$ ← Use the Pythagorean theorem.

$(PQ)^2 = (x_2 - x_1)^2 + (y_2 - y_1)^2$ ← Substitute the lengths you know.

$PQ = \sqrt{(x_2 - x_1)^2 + (y_2 - y_1)^2}$ ← Take the principal square root of each side.

[Graph: y, Q(x₂, y₂), P(x₁, y₁), R(x₂, y₁), O, x]

This method for finding the distance between two points is summarized in the *distance formula*.

The Distance Formula

The distance *d* between any two points (x_1, y_1) and (x_2, y_2) is

$d = \sqrt{(x_2 - x_1)^2 + (y_2 - y_1)^2}$.

Example 1 **Relating to the Real World**  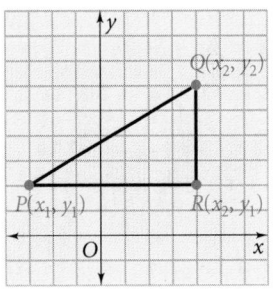

Hiking The Gato and Wilson families are staying at a campground. The Gatos leave camp and hike 2 km west and 5 km south. The Wilsons leave camp and hike 1 km east and 4 km north. How far apart are the families?

how they found AC. Lead them to understand that the equation they used to find AC, $4 - 1 = 3$, corresponds to the $(x_2 - x_1)$ section of the distance formula.

ERROR ALERT! **VISUAL LEARNING** Students may calculate $(x_1 - y_1)$ and $(x_2 - y_2)$ instead of $(x_2 - x_1)$ and $(y_2 - y_1)$. **Remediation:** Have students write the x-coordinates in one color and the y-coordinates in a second color. Then have them write the distance formula using the corresponding color for each coordinate.

ERROR ALERT! Students may be concerned about which of the two points to call (x_1, y_1). **Remediation:** Have students calculate a distance twice, changing the point they use as $(x_1 - y_1)$. Since the distances are squared, the order in which they choose points does not matter.

Example 1 ***Relating to the Real World*** 

Point out that when students enter points with negative coordinates into the distance formula, they may easily overlook a negative sign or minus sign. Stress that students must work slowly and carefully to avoid overlooking important details.

Example 2

GEOMETRY Students use the distance formula to determine that opposite sides of a quadrilateral are of equal length.

Question 6 Remind students that the formula for slope is $\frac{y_2 - y_1}{x_2 - x_1}$.

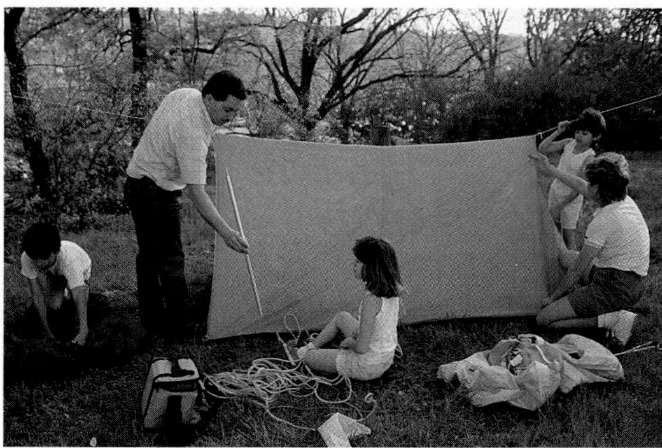

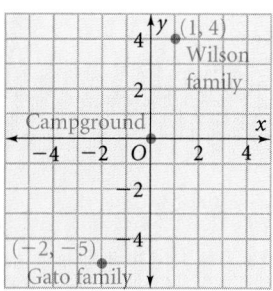

Let the campground be (0, 0). Then the Gato family's location is (−2, −5) and the Wilson family's location is (1, 4).

$d = \sqrt{(x_2 - x_1)^2 + (y_2 - y_1)^2}$ ← Use the distance formula.

$d = \sqrt{(1 - (-2))^2 + (4 - (-5))^2}$ ← Substitute (1, 4) for (x_2, y_2) and (−2, −5) for (x_1, y_1).

$d = \sqrt{3^2 + 9^2}$ ← Simplify.

$d = \sqrt{90}$

$d = 9.486832981$ ← Use a calculator.

The two families are about 9.5 km apart.

PROBLEM SOLVING

Look Back Would the result in Example 1 be different if you used (1, 4) for (x_1, y_1) and (−2, −5) for (x_2, y_2)? Explain your answer.

No; 1 − (−2) and −2 − 1 are opposites, but their squares are the same.

5. Try This One hiker is 4 mi west and 3 mi north of the campground. Another is 6 mi east and 3 mi south of the campground. How far apart are the hikers? **about 11.7 mi**

You can also use the distance formula to determine if lengths of opposite sides in a figure are equal.

Example 2

Geometry Quadrilateral $ORST$ is shown here in a coordinate plane. Use the distance formula to show that $\overline{OR}$ and $\overline{ST}$ are equal in length.

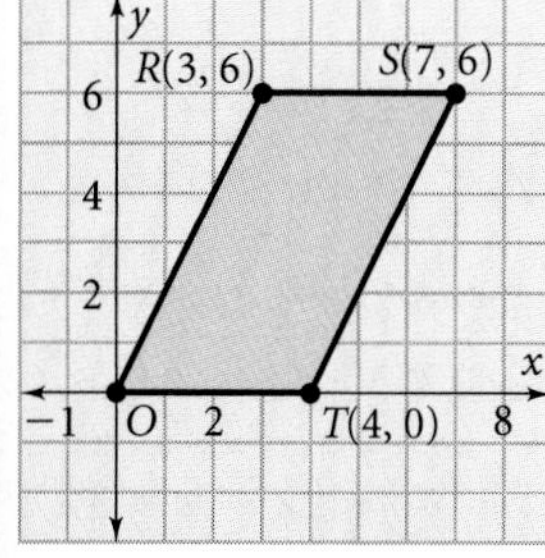

$OR = \sqrt{(3-0)^2 + (6-0)^2}$ $\quad ST = \sqrt{(7-4)^2 + (6-0)^2}$

$OR = \sqrt{3^2 + 6^2}$ $\quad ST = \sqrt{3^2 + 6^2}$

$OR = \sqrt{45}$ $\quad ST = \sqrt{45}$

$OR = ST$. So, $\overline{OR}$ and $\overline{ST}$ are equal in length.

6. Use slopes to show that $\overline{OR}$ and $\overline{ST}$ are parallel.

Slope of $\overline{OR}$ = 2; slope of $\overline{ST}$ = 2; slopes are equal, so segments are parallel.

Additional Examples

FOR EXAMPLE 1

Mickey and Jamie are playing baseball. Mickey runs 20 m north and 15 m west. Jamie runs 10 m south and 5 m east. How far apart are the two players? **36 m**

Discussion: *If you placed the origin at 40 m south and 40 m east, would the distance between the two players change? Explain.*

FOR EXAMPLE 2

What is the distance between a point at (−3, 2) and a point at (3, −2)? **7.2**

FOR EXAMPLE 3

Find the midpoint of a the segment from $A(2, -1)$ to $B(4, -3)$. **(3, −2)**

CRITICAL THINKING Question 7 Remind students of the difference between a parallelogram and a quadrilateral.

ALTERNATIVE METHOD Draw the coordinate axes on the board. Draw and label points $A(1, 8)$, $B(7, 4)$, and $C(1, 4)$. Draw the line connecting points A and B. Using a different color chalk, draw the vertical and horizontal legs of the right triangle. Ask:

- *What is the length of the vertical leg?* **4 units**
- *Where is the midpoint of the vertical leg?* **halfway down the leg at (1, 6)**
- *What is the length of the horizontal leg?* **6 units**
- *Where is the midpoint of the horizontal leg?* **halfway along the leg at (4, 4)**

On the board, draw and label points (1, 6) and (4, 4). Then draw dotted horizontal and vertical lines connecting these points to the line between A and B. Label the point (4, 6) and point out that this is the midpoint of A and B.

Help students make the connection between the method you used and the midpoint formula.

Example 3

GEOMETRY In this example, students find the midpoint of a line segment.

Technology Options

For Exercises 4–15, students may measure the distance between pairs of points by using geometry software.

Prentice Hall Technology

Software
- Secondary Math Lab Toolkit™
- Computer Item Generator 9-2

CD-ROM
- Multimedia Algebra Lab 9

Internet
- See the Prentice Hall site. (http://www.phschool.com)

7a. $LK = \sqrt{68}$; $MN = \sqrt{68}$

7. a. **Try This** Quadrilateral $KLMN$ has vertices with coordinates $K(-3, -2)$, $L(-5, 6)$, $M(2, 6)$, and $N(4, -2)$. Show that $LK = MN$.
   b. Use slopes to show that $\overline{LK}$ and $\overline{MN}$ are parallel. **Each slope is −4.**
   c. *Critical Thinking* Describe quadrilateral $KLMN$ in as much detail as you can. **Justify** your descriptions and conclusions. **$KLMN$ is a parallelogram. Opposite sides are parallel. The product of slopes is not −1.**

Part 

2 Using the Midpoint Formula

The **midpoint** of a segment $\overline{AB}$ is the point M halfway between A and B where $AM = MB$. The coordinates of the midpoint of a line segment are the averages of the coordinates of the endpoints.

The Midpoint Formula

The midpoint M of a line segment with endpoints $A(x_1, y_1)$ and $B(x_2, y_2)$ is $\left(\frac{x_1 + x_2}{2}, \frac{y_1 + y_2}{2}\right)$.

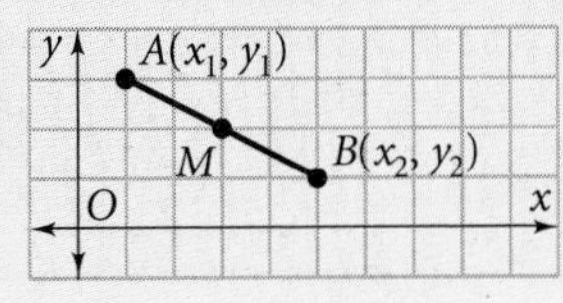

Example 3

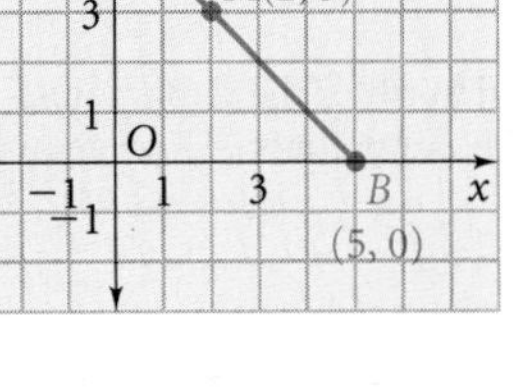

Geometry Find the midpoint of the segment from $A(-1, 6)$ to $B(5, 0)$.

$$\left(\frac{x_1 + x_2}{2}, \frac{y_1 + y_2}{2}\right) = \left(\frac{-1 + 5}{2}, \frac{6 + 0}{2}\right)$$
$$= \left(\frac{4}{2}, \frac{6}{2}\right)$$
$$= (2, 3)$$

The midpoint of $\overline{AB}$ is $M(2, 3)$.

8. In Example 3, use the distance formula to **verify** that $AM = MB$. **$AM = \sqrt{18}$; $MB = \sqrt{18}$**
9. **Try This** Find the coordinates of the midpoint of the line segment with endpoints $X(-7.2, 2)$ and $Y(4.5, 7.5)$. **(−1.35, 4.75)**

Exercises ON YOUR OWN

Calculator **Approximate AB to the nearest tenth of a unit.**

1. 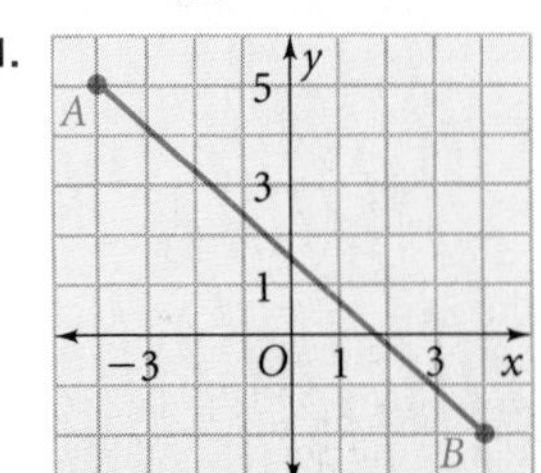

10.6 2. 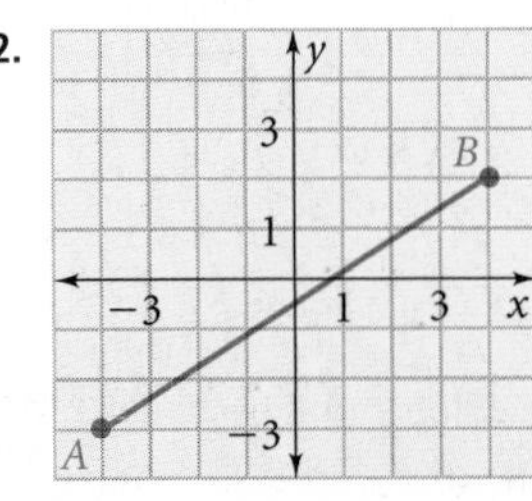

9.4 3. 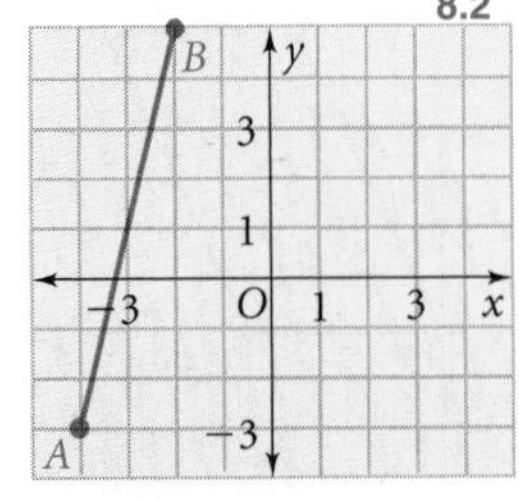

8.2

Exercises ON YOUR OWN

Exercise 16 Encourage students to draw a set of coordinate axes. Have them place the political rally at (0, 0). Then, add the details of the problem to the coordinate grid.

ESL Exercise 17 Ask a student to explain the meaning of "Copy that," spoken by Mr. Tanaka. Explain that Ms. Elisa and Mr. Tanaka are speaking over two-way radios. If students are unsure of the meaning, explain that "copy that" is radio shorthand for "I understand what you are saying."

ALTERNATIVE ASSESSMENT **Exercise 18** Ask students to find the midpoint between the first given point and the midpoint found in the exercise. This will help you assess student understanding of the midpoint formula.

CONNECTING TO THE STUDENTS' WORLD Exercise 30 Have pairs of students study a city map. Using the scale of the map, have them apply the distance formula to calculate the distance between various landmarks in the city.

GEOMETRY Exercises 31–33 Students use the distance formula to find the perimeter of geometric figures.

OPEN-ENDED Exercise 34 Suggest that students use their thumb and forefinger to measure a segment between 10 and 20 units on the coordinate plane. Then, keeping their fingers set at this distance, move them to a new location on the plane to plot two points that meet the criteria of the exercise.

CRITICAL THINKING Exercise 35 Encourage students to make sketches to answer the question.

Find the distance between each pair of points to the nearest tenth.

4. $(3, -2), (-1, 5)$ **8.1** **5.** $(-4, -4), (4, 4)$ **11.3** **6.** $(0, 3), (13, -6)$ **15.8**

7. $(4, 0), (2, -1)$ **2.2** **8.** $(7, -2), (-8, -2)$ **15** **9.** $(5, 9), (10, -1)$ **11.2**

10. $(0, 0), (6, -9)$ **10.8** **11.** $(-3, 7), (-11, 9)$ **8.2** **12.** $(11, 0), (-3, -7)$ **15.7**

13. $(-2, 7), (-2, -7)$ **14** **14.** $(9, 10), (11, 12)$ **2.8** **15.** $(-3.5, 4.5), (-4.5, 5.5)$ **1.4**

16. a. *News Coverage* Two news helicopters are on their way to a political rally, flying at the same altitude. One helicopter is 20 mi due west of the rally. The other is 15 mi south and 15 mi east of the rally. How far apart are they? **about 38.1 mi**

b. How far from the rally are they? **20 mi; about 21.2 mi**

c. Each helicopter is flying at 80 mi/h. How many minutes will it take each of them to arrive at the scene? **15 min; 15.9 min**

17. a. *Electrical Line Technician* Graph Mr. Tanaka's and Ms. Elisa's locations on a coordinate grid with the substation at the origin. **See margin.**

b. What are the coordinates of the point where they will meet? **(0, −0.5)**

c. Describe their meeting point in miles north/south and east/west of the substation. **The meeting point is 0.5 mile south of the substation.**

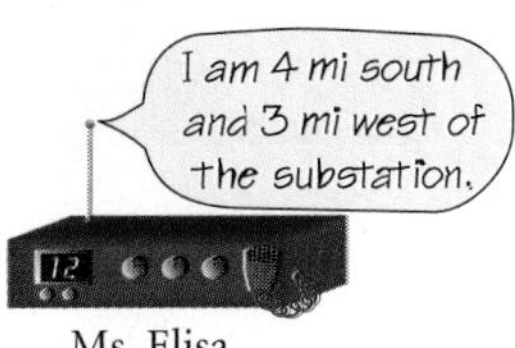

I am 4 mi south and 3 mi west of the substation.

Ms. Elisa

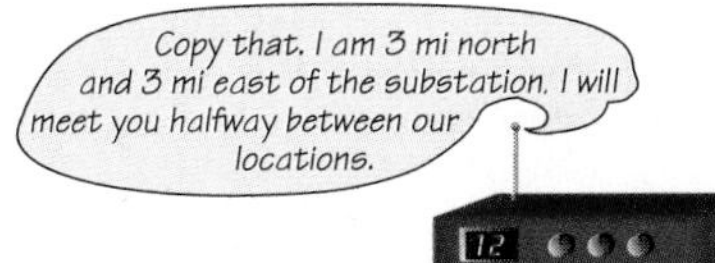

Copy that. I am 3 mi north and 3 mi east of the substation. I will meet you halfway between our locations.

Mr. Tanaka

Find the midpoint of $\overline{XY}$.

18. $X(2, 5)$ and $Y(0, 7)$ **(1, 6)** **19.** $X(-3, 14)$ and $Y(6, 1)$ $\left(1\frac{1}{2}, 7\frac{1}{2}\right)$ **20.** $X(8, -5)$ and $Y(-4, 5.5)$ $\left(2, \frac{1}{4}\right)$

21. $X(4, 3)$ and $Y(-9, 3)$ $\left(-2\frac{1}{2}, 3\right)$ **22.** $X(0, 6)$ and $Y(-5, -8)$ $\left(-2\frac{1}{2}, -1\right)$ **23.** $X(-1, 8)$ and $Y(-7, 0)$ $(-4, 4)$

24. $X(4, 1)$ and $Y(1, 4)$ **25.** $X(5, -11)$ and $Y(12, -7)$ **26.** $X(2, 9)$ and $Y(-2, -9)$ **(0, 0)**

27. $X(3, 11)$ and $Y(11, 3)$ **(7, 7)** **28.** $X\left(4\frac{1}{2}, -2\right)$ and $Y\left(-1\frac{1}{2}, 5\right)$ **29.** $X(9, 7)$ and $Y(-9, -7)$ **(0, 0)**

24. $\left(2\frac{1}{2}, 2\frac{1}{2}\right)$ **25.** $\left(8\frac{1}{2}, -9\right)$ **28.** $\left(1\frac{1}{2}, 1\frac{1}{2}\right)$

30. a. *Transportation* On the map, each unit represents one mile. A van breaks down on its way to a factory. The driver calls a garage for a tow truck. There is a bridge halfway between the garage and the van. How far is the bridge from the van?

b. The van is towed to the factory and then to the garage. How many miles does the tow truck tow the van? **a. about 4.3 mi b. about 17.4 mi**

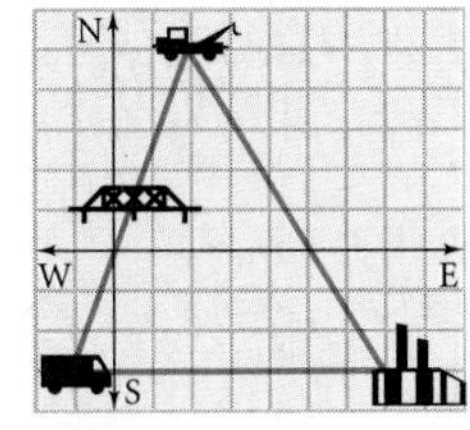

page 420 Work Together

1.

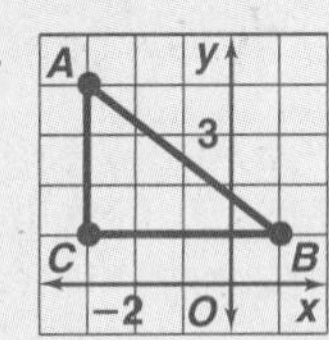

page 422–424 On Your Own

17a.

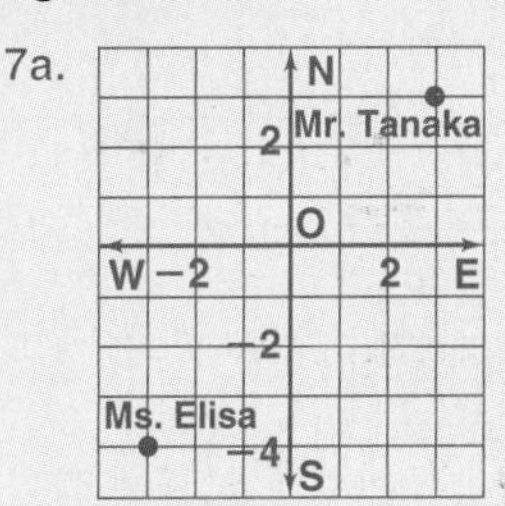

34b.

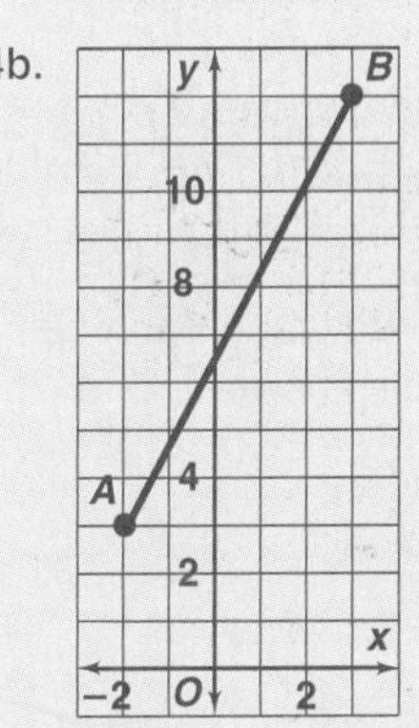

36. See back of book.

WRITING Exercise 36 Have students write each formula. Then ask them to explain in their own words what each variable in the formula represents.

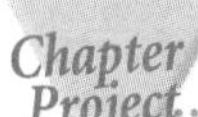

FIND OUT BY CALCULATING Students evaluate the given formula to calculate the distance they can see to the horizon. This is essential to their work on the Chapter Project. Have students add this task to work they have already completed for the project. Check students' progress by having them write a sentence or two describing what they have done for the project thus far.

Exercises MIXED REVIEW

Exercise 41 Point out that the problem asks students to calculate the amount of interest that has accumulated after four years, not how much interest the bank pays after the fourth year.

GETTING READY FOR LESSON 9-3 These exercises prepare students for using trigonometric ratios.

Wrap Up

THE BIG IDEA Ask: *What is the distance formula? How do you use it? Why is the formula useful? What is the midpoint formula? How do you use it? Why is the formula useful?*

RETEACHING ACTIVITY Students underline *x*-coordinates and circle *y*-coordinates to help them apply the distance formula. Then they find the distance between pairs of points. (Reteaching worksheet 9-2)

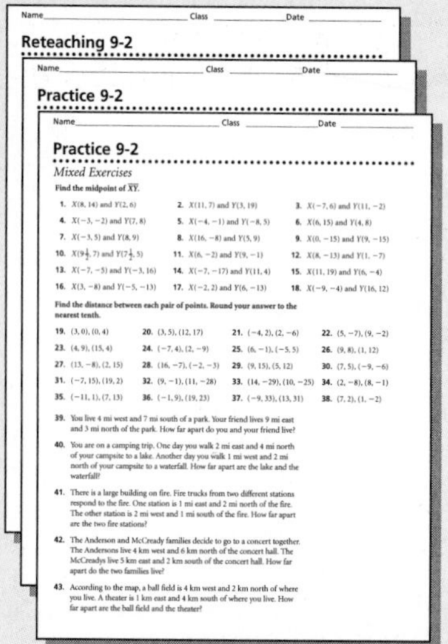

Name________ Class ________ Date ________

Reteaching 9-2

Name________ Class ________ Date ________

Practice 9-2

Name________ Class ________ Date ________

Practice 9-2

Mixed Exercises

Find the midpoint of $\overline{XY}$.

1. $X(8, 14)$ and $Y(2, 6)$
2. $X(11, 7)$ and $Y(3, 19)$
3. $X(-7, 6)$ and $Y(11, -2)$
4. $X(-3, -2)$ and $Y(7, 8)$
5. $X(-4, -1)$ and $Y(-8, 5)$
6. $X(6, 15)$ and $Y(4, 8)$
7. $X(-3, 5)$ and $Y(8, 9)$
8. $X(16, -8)$ and $Y(5, 9)$
9. $X(0, -15)$ and $Y(9, -15)$
10. $X(9\frac{1}{2}, 7)$ and $Y(7\frac{1}{2}, 5)$
11. $X(6, -2)$ and $Y(9, -1)$
12. $X(8, -13)$ and $Y(1, -7)$
13. $X(-7, -5)$ and $Y(-3, 16)$
14. $X(-7, -17)$ and $Y(11, 4)$
15. $X(11, 19)$ and $Y(6, -4)$
16. $X(3, -8)$ and $Y(-5, -13)$
17. $X(-2, 2)$ and $Y(6, -13)$
18. $X(-9, -4)$ and $Y(16, 12)$

Find the distance between each pair of points. Round your answer to the nearest tenth.

19. (3, 0), (0, 4)
20. (3, 5), (12, 17)
21. (−4, 2), (2, −6)
22. (5, −7), (9, −2)
23. (4, 9), (15, 4)
24. (−7, 4), (2, −9)
25. (6, −1), (−5, 5)
26. (9, 8), (1, 12)
27. (13, −8), (2, 15)
28. (16, −7), (−2, −3)
29. (9, 15), (5, 12)
30. (7, 5), (−9, −6)
31. (−7, 15), (19, 2)
32. (9, −1), (11, −28)
33. (14, −29), (10, −25)
34. (2, −8), (8, −1)
35. (−11, 1), (7, 13)
36. (−1, 9), (19, 23)
37. (−9, 33), (13, 31)
38. (7, 2), (1, −2)
39. You live 4 mi west and 7 mi south of a park. Your friend lives 9 mi east and 3 mi north of the park. How far apart do you and your friend live?
40. You are on a camping trip. One day you walk 2 mi east and 4 mi north of your campsite to a lake. Another day you walk 1 mi west and 2 mi north of your campsite to a waterfall. How far apart are the lake and the waterfall?
41. There is a large building on fire. Fire trucks from two different stations respond to the fire. One station is 1 mi east and 2 mi north of the fire. The other station is 2 mi west and 1 mi south of the fire. How far apart are the two fire stations?
42. The Anderson and McCready families decide to go to a concert together. The Andersons live 4 km west and 6 km north of the concert hall. The McCreadys live 5 km east and 2 km south of the concert hall. How far apart do the two families live?
43. According to the map, a ball field is 4 km west and 2 km north of where you live. A theater is 1 km east and 4 km south of where you live. How far apart are the ball field and the theater?

Lesson Quiz

Lesson Quiz is also available in Transparencies.

Find the distance between each pair of points. Round your answer to the nearest tenth.

1. (2, −1), (−4, 3) 7.2

2. (−2.5, 3.5), (−7.5, 8.5) 7.1

Find the midpoint of $\overline{XY}$.

3. $X(3, 6)$ and $Y(0, 2)$ (1.5, 4)

4. $X(6, -4)$ and $Y(12, -2)$ (9, −3)

424

Geometry **Find the perimeter of each figure to the nearest tenth.**

31.

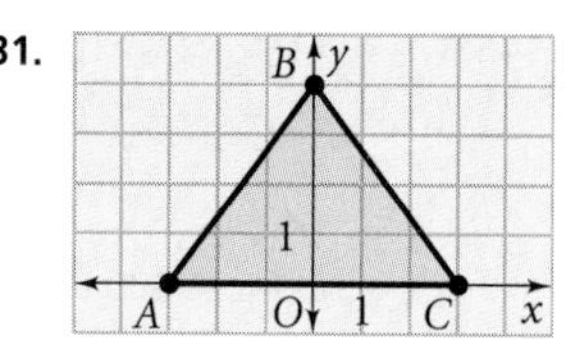

16

32.

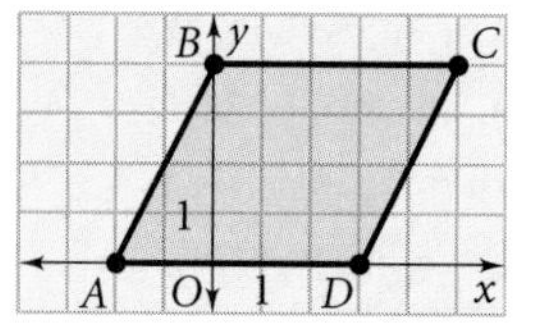

18.9

33.

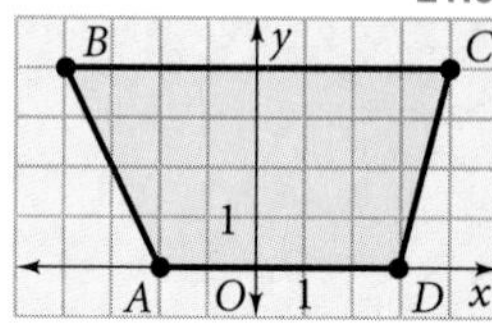

21.6

34. a. *Open-ended* Suppose the distance between two points on a coordinate plane is between 10 and 13 units. Identify two points A and B not directly above or across from one another that meet this requirement. Answers may vary. Sample: $A(-2, 3)$, $B(3, 12)$

b. Plot your points on graph paper. See margin.

c. **Verify** that your points satisfy the requirement by finding AB. For sample in part (a): about 10.3

35. *Critical Thinking* If the midpoint of a line segment is the origin, what must be true of the coordinates of the endpoints of the segment? *x*-coordinates are opposites; *y*-coordinates are opposites.

36. *Writing* **Summarize** the distance formula and the midpoint formula. Use examples of your own to show the use of each formula. See margin.

Chapter Project

Find Out by Calculating

The radius of Earth is about 3960 mi. The formula for the distance you can see to the horizon on Earth is $d = 1.225\sqrt{h}$ where d is the distance visible in miles and h is your height in feet. How far can you see on a clear day with no obstructions?

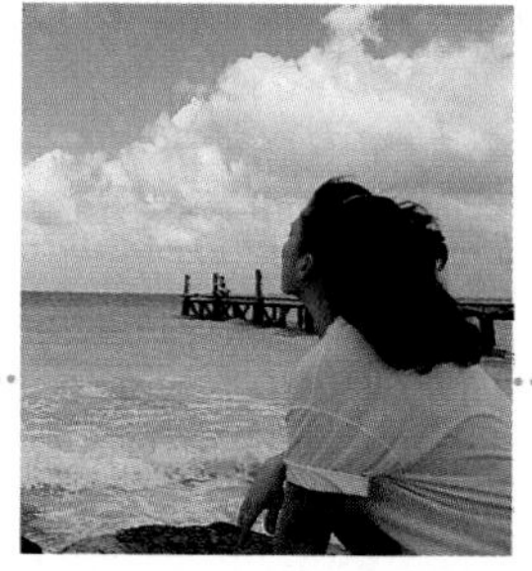

Exercises MIXED REVIEW

Solve each equation for the given variable.

37. $2x + 7y = 4$; x $x = 2 - \frac{7}{2}y$

38. $3x - 5y = 7$; y $y = \frac{3}{5}x - \frac{7}{5}$

39. $V = \pi r^2 h$; h $h = \frac{V}{\pi r^2}$

40. $S = 2\pi rh$; r $r = \frac{S}{2\pi h}$

41. *Finance* You deposit $3000 into a bank account paying 6% simple interest annually. How much interest will you receive after 4 years? $720

Getting Ready for Lesson 9-3

Let $c = \frac{A}{H}$, $s = \frac{O}{H}$, $t = \frac{O}{A}$. **Calculate c, s, and t for each of the following.**

42. $A = 3, O = 4, H = 5$ $\frac{3}{5}, \frac{4}{5}, \frac{4}{3}$

43. $A = 5, O = 12, H = 13$ $\frac{5}{13}, \frac{12}{13}, \frac{12}{5}$

Solve each equation.

44. $\frac{15}{x} = 0.75$ 20

45. $\frac{x}{20} = 0.34$ 6.8

46. $0.82x = 25$ about 30.5

47. $\frac{x}{0.52} = 14$ 7.28

9-3 Teaching Notes

PROBLEM OF THE DAY

If water weighs 8 lb/gal, how much does 5 gal, plus 10 qt, plus 7 pt, plus 2 qt weigh? 78 lb

Problem of the Day is also available in Transparencies.

CONNECTING TO PRIOR KNOWLEDGE Ask students to discuss what they have discovered about right triangles and the Pythagorean theorem. Have students brainstorm about applications of the Pythagorean theorem and how they might use their knowledge of right triangles to solve problems.

WORK TOGETHER

This activity demonstrates that the ratios are constant for a given angle. Students also learn that the larger the angle, the larger the ratio.

THINK AND DISCUSS

AUDITORY LEARNING Stress the parts of a triangle used to make up each ratio. Have the students repeat the ratios to you several times to become familiar with the terms.

Example 1

ERROR ALERT! Students may have trouble remembering the trigonometric ratios. **Remediation:** Have students write the

---

Connections Navigation . . . and more

9-3 Trigonometric Ratios

What You'll Learn

- Exploring and calculating trigonometric ratios
- Using sine, cosine, and tangent to solve problems

...And Why

To find distances indirectly

What You'll Need

- graph paper
- ruler

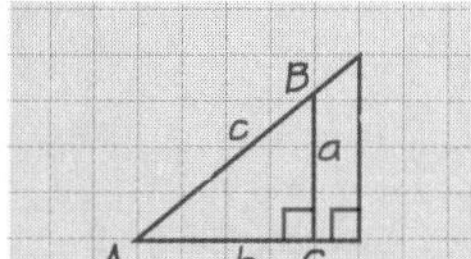

WORK TOGETHER

Work with a partner.

1. On graph paper, draw a right triangle like the one at the left. Extend sides $\overline{AB}$ and $\overline{AC}$ to form a second triangle similar to the first triangle. One example is shown at the left. Check students' work.
2. a. Copy the table below. Measure and record the lengths of the legs of each triangle. a–c. Answers may vary. Check students' work.

Triangle	a	b	c	$\frac{a}{b}$	$\frac{a}{c}$	$\frac{b}{c}$
first	■	■	■	■	■	■
second	■	■	■	■	■	■

   b. Calculate and record c, the length of each hypotenuse.
   c. Calculate and record the ratios $\frac{a}{b}$, $\frac{a}{c}$, and $\frac{b}{c}$ for each triangle.
3. How do corresponding ratios in the two triangles compare? The corresponding ratios are equal.

THINK AND DISCUSS

Part 1

Finding Trigonometric Ratios

The ratios you explored in the Work Together are called **trigonometric ratios,** meaning triangle measurement ratios. In $\triangle ABC$, $\overline{BC}$ is the side opposite $\angle A$ and $\overline{AC}$ is the side adjacent to $\angle A$. The hypotenuse is $\overline{AB}$.

You can use these relationships to express trigonometric ratios.

sine of $\angle A = \dfrac{\text{length of side opposite } \angle A}{\text{length of hypotenuse}}$ or $\sin A = \frac{a}{c}$

cosine of $\angle A = \dfrac{\text{length of side adjacent to } \angle A}{\text{length of hypotenuse}}$ or $\cos A = \frac{b}{c}$

tangent of $\angle A = \dfrac{\text{length of side opposite } \angle A}{\text{length of side adjacent to } \angle A}$ or $\tan A = \frac{a}{b}$

Example 1

Use the diagram at the left. Find sin A, cos A, and tan A.

$\sin A = \frac{5}{13}$ $\quad \cos A = \frac{12}{13}$ $\quad \tan A = \frac{5}{12}$

4. **Try This** Use the diagram to find sin B, cos B, and tan B. $\frac{12}{13}, \frac{5}{13}, \frac{12}{5}$

Lesson Planning Options

Prerequisite Skills

- Simplifying ratios
- Knowing parts of right triangles

Assignment Options for Exercises On Your Own

To provide flexible scheduling, this lesson can be subdivided into parts.

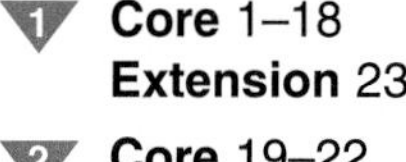

1 **Core** 1–18
Extension 23–25

2 **Core** 19–22
Extension 26–28

Use Mixed Review to maintain skills.

Resources

Student Edition

Skills Handbook, pp. 581, 582
Extra Practice, p. 564
Glossary/Study Guide

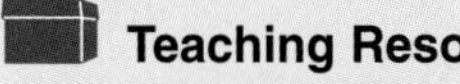

Teaching Resources

Chapter Support File, Ch. 9
- Practice 9-3 (two worksheets)
- Reteaching 9-3
- Checkpoint

Classroom Manager 9-3
Glossary, Spanish Resources
Two-Year Algebra Handbook 9-3

Transparencies

1, 106, 111

ratios on an index card. Encourage them to create their own mnemonic devices to remember the ratios, making acronyms from the letters O (opposite side), A (adjacent side), H (hypotenuse), S (sine), C (cosine), and T (tangent). One example is: Some Old Hags Can't Always Hide Their Old Age.

VISUAL LEARNING Encourage students to use color coding to keep track of which length is in the numerator and which length is in the denominator.

TACTILE LEARNING Ask students to construct a right triangle on a geoboard. Have them measure and record the opposite side and then the adjacent side. Have them do the same for the other acute angle, then calculate the ratios.

Example 2

ERROR ALERT! In problems such as Example 2, students may become confused finding the opposite and adjacent sides if the angles are switched or the triangle is rotated to other positions. **Remediation:** Have students practice naming the sides that are opposite and adjacent to various angles. Make sure the triangles are rotated or reflected when students practice so they are not just memorizing positions.

KINESTHETIC LEARNING To help students decide which trigonometric ratio to use, have them draw the triangle described in the example. Then label the measurements of the given parts of the triangle with colored pencils or markers.

Additional Examples

FOR EXAMPLE 1

Use the diagram to find sin A, cos A, and tan A. $\frac{6}{10}, \frac{8}{10}, \frac{6}{8}$

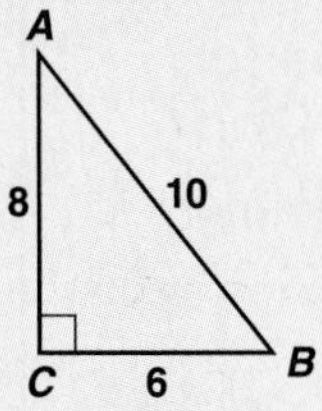

Discussion: *Would your answers change if the triangle was rotated 180°? Explain.*

FOR EXAMPLE 2

Find the value of x in the triangle below. 5

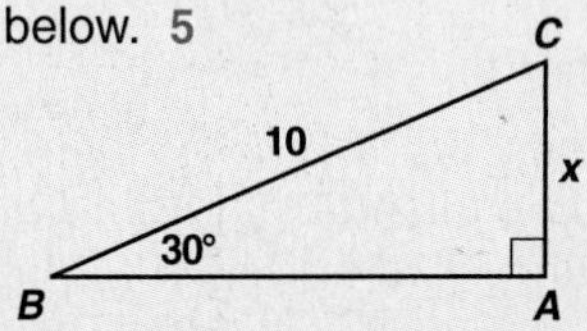

FOR EXAMPLE 3

Suppose the angle of elevation between the ground and the top of a cliff is 70°. If the base of the cliff is 25 m from your current position, how high is the cliff? 69 m

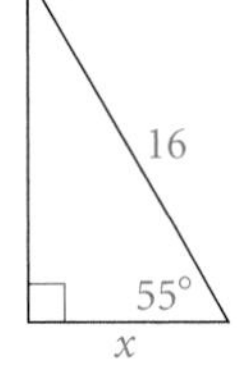

You can use trigonometry to find missing lengths in a triangle.

Example 2

Find the value of x in the triangle at the left.

First, decide which trigonometric ratio to use. You know the angle and hypotenuse and are trying to find the adjacent side. So use cosine.

$$\cos 55° = \frac{\text{side adjacent}}{\text{hypotenuse}}$$ ← This is a short form of the definition.

$$\cos 55° = \frac{x}{16}$$

$$0.573576436 = \frac{x}{16}$$ ← Use a calculator.

$$16(0.573576436) = x$$ ← Cross multiply.

$$9.177222982 = x$$ ← Use a calculator.

The value of x is about 9.2.

CALCULATOR HINT

Use degree mode when finding trigonometric ratios. Use the key sequence  55 ENTER.

5. You know the lengths of two sides of the triangle in Example 2. You can use different methods to find the length of the third side.
   a. Use the Pythagorean theorem to find the length of the third side. about 13.1
   b. Use either the sine or tangent ratio to find the length of the third side. about 13.1
   c. Compare your answers to parts (a) and (b). The answers are the same.

Part 2 Solving Problems Using Trigonometric Ratios

You can use trigonometric ratios to measure distances indirectly when you know an angle of elevation. An **angle of elevation** is an angle from the horizontal up to a line of sight.

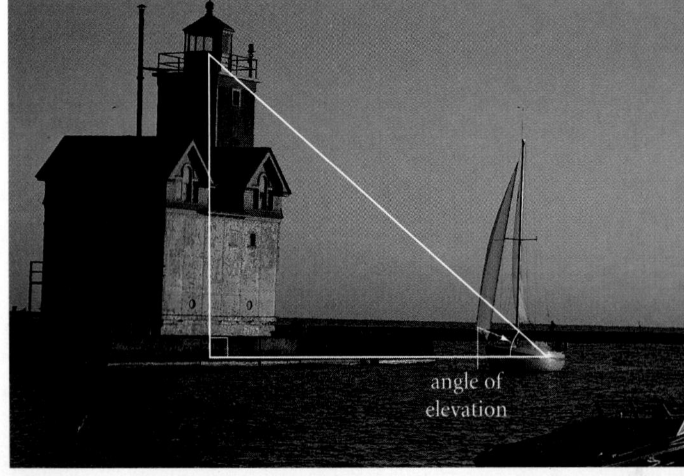

Example 3 Relating to the Real World

Discuss whether it would be practical to try to solve this problem by directly measuring the distance. Ask students how they could measure a distance of 137 ft. Develop the conclusion that indirect measurement can save enormous amounts of time, money, and labor.

EXTENSION Challenge interested students to find out more about trigonometric ratios. Encourage them to learn the inverse ratios (arc sin, arc cosin, and arc tan) and explore some of the trigonometric formulas, such as $\sin^2 x + \cos^2 x = 1$.

Exercises ON YOUR OWN

Exercises 8–12 Explain that rounding to four decimal places is traditional because tables of trigonometric ratios are written to four decimal places. Today, calculators are used more often than trigonometric value tables to find values.

Exercises 13–15 Tell students that it makes no sense to write $\sin = \frac{3}{5}$. This is similar to writing "half is 7." Half of what? The correct expression is "half of 14 is 7" and "$\sin A = \frac{3}{5}$"

ALTERNATIVE ASSESSMENT **Exercises 16, 17, and 18** To assess understanding of trigonometric ratios, ask students to find the third side of each triangle.

Example 3 Relating to the Real World 

Navigation Suppose that an angle of elevation from a rowboat to a lighthouse is 35°. You know that the lighthouse is 96 ft tall. How far from the lighthouse is the rowboat?

Make a diagram.

Define x = the distance from the lighthouse to the boat

Relate You know the angle and the opposite side and you are trying to find the adjacent side. So use tangent.

Write $\tan A = \frac{\text{side opposite}}{\text{side adjacent}}$

$\tan 35° = \frac{96}{x}$ ← Substitute for the angle and sides.

$x(\tan 35°) = 96$ ← Cross multiply.

$x = \frac{96}{\tan 35°}$ ← Divide to put in calculator-ready form.

$x = 137.1022086$ ← Use a calculator.

$x \approx 137$

The rowboat is about 137 ft from the lighthouse.

6. Could you use the sine or cosine to solve Example 3? Explain. **Answers may vary. Sample: Yes; use the sine to solve for the hypotenuse. Then use the cosine to solve for x.**

Exercises ON YOUR OWN

1. Language Arts Write a nonmathematical sentence using the word *adjacent*. **Answers may vary. Sample: Kentucky and Virginia are adjacent states.**

Use △RST to evaluate each expression.

2. $\sin R$ $\frac{3}{5}$
3. $\cos R$ $\frac{4}{5}$
4. $\tan R$ $\frac{3}{4}$
5. $\sin S$ $\frac{4}{5}$
6. $\cos S$ $\frac{3}{5}$
7. $\tan S$ $\frac{4}{3}$

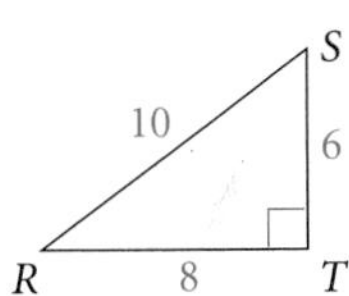

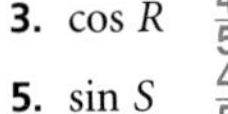

Calculator **Evaluate each expression. Round to four decimal places.**

8. sin 32° **0.5299**
9. cos 40° **0.7660**
10. tan 52° **1.2799**
11. sin 85° **0.9962**
12. tan 7° **0.1228**

For each figure, find sin *A*, cos *A*, and tan *A*.

13. $\frac{21}{29}, \frac{20}{29}, \frac{21}{20}$

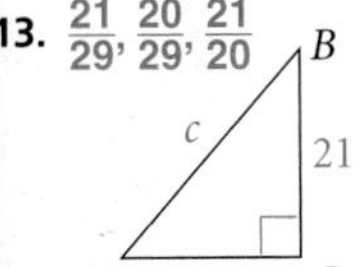

14. $\frac{8}{17}, \frac{15}{17}, \frac{8}{15}$

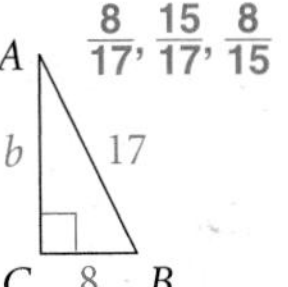

15. **0.9231, 0.3846, 2.4**

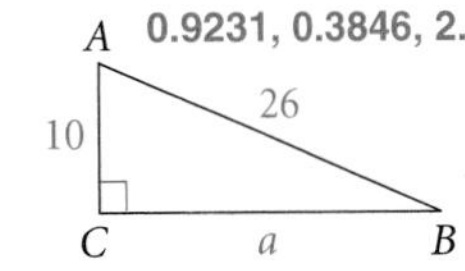

Technology Options

For Exercises 13–18, students may wish to find the missing information with geometry software or a graphing calculator.

Prentice Hall Technology

Calculator
- Graphing Calculator Handbook: Procedure 11

Software
- Secondary Math Lab Toolkit™
- Computer Item Generator 9-3

CD-ROM
- Multimedia Algebra Lab 9

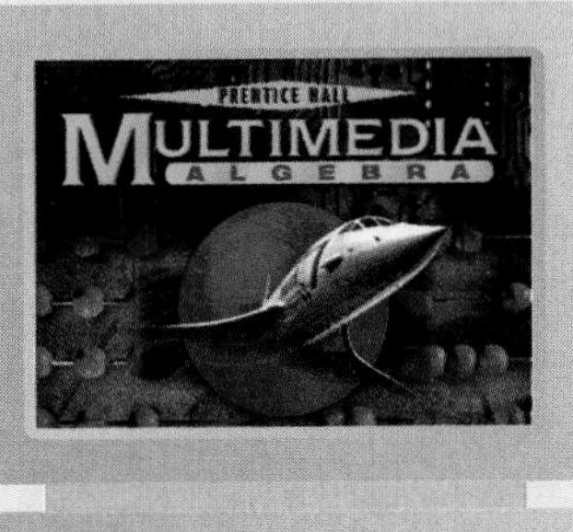

Internet
- See the Prentice Hall site. (http://www.phschool.com)

ESL Exercise 19 Unfamiliar vocabulary or the use of certain words in different contexts may make this first sentence difficult for students to visualize. On the board, sketch the *cabin*, the *canyon*, the *cliff*, and the *vertical wall of rock* to illustrate their physical relationship.

WRITING Exercise 22 Encourage students to draw the figure and label the known values when starting a problem such as this one. By describing the known information, students can often use a trigonometric ratio to find the unknown.

STANDARDIZED TEST TIP Exercise 26 Have students draw the figure described. Explain that drawing the figure will help students understand the problem and avoid mistakes.

OPEN-ENDED Exercise 27 Have students estimate the potential time- and money-savings achieved in Example 3 by measuring indirectly instead of using 137 ft of measuring tape.

Chapter Project **FIND OUT BY RESEARCHING** Students research to find the radii of all the planets in our solar system. Have students add this part of the Chapter Project to work they have already completed.

Exercises MIXED REVIEW

SELF ASSESSMENT **JOURNAL** Ask students if the angles of triangles are equal, will the size of the triangle make any difference in the size of the ratios. Ask them to explain with examples.

GETTING READY FOR LESSON 9-4 These exercises prepare students for simplifying radicals that contain products and quotients.

pages 427–429 On Your Own

22. Multiply the known length by tan 30°, or $\frac{\sqrt{3}}{3}$.
27. Answers may vary. Sample: You can use trigonometric ratios to find the height of a tree or a building; you can find the distance between landmarks on the opposite bank of a river or a pond.

page 429 Mixed Review

29. [graph]

30. 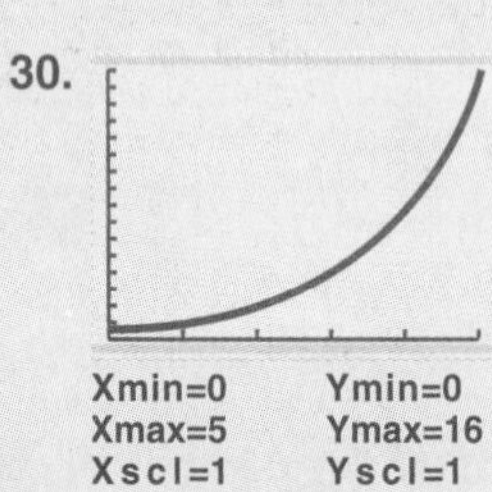

Xmin=0 Ymin=0
Xmax=5 Ymax=16
Xscl=1 Yscl=1

31–32. See back of book.

page 429 Checkpoint

9. See back of book.

428

Find the value of x to the nearest tenth.

16. 5.5 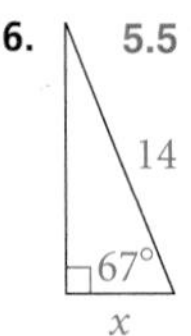

17. 10.4 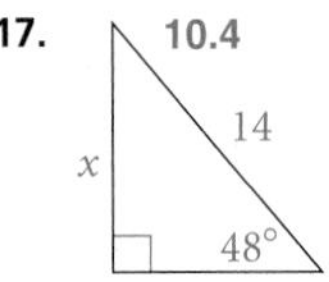

18. 19.2 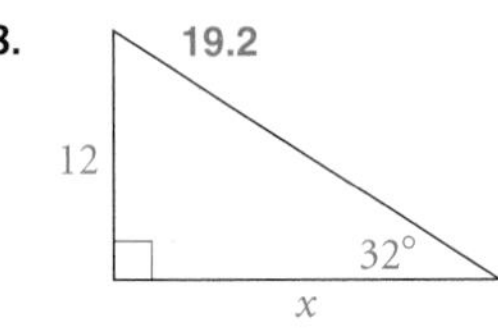

19. **Nature** Suppose you look up from a cabin on the floor of a canyon to a cliff at the top of a vertical wall of rock. The angle of elevation from your location to the cliff is 32°. The cabin is 400 ft from the base of the canyon wall. How high is the canyon wall? about 250 ft

20. **Recreation** A Ferris wheel is shown at the right. Find the distance x that the seat is above the horizontal line through the center of the wheel. Then find the seat's height above the ground. 26.2 ft, 67.4 ft

21. **Aviation** Suppose you live about 5 mi from a tower. From your home, you see a plane directly above the tower. Your angle of elevation to the plane is 21°. What is the plane's altitude? about 10,000 ft

22. **Writing** Suppose you know that a right triangle has a 30° angle and you know the length of the leg adjacent to the 30° angle. Describe how you would find the length of the leg opposite the 30° angle. See margin.

Find the indicated length to the nearest tenth.

23. 59.2 78.4

24. 4.5 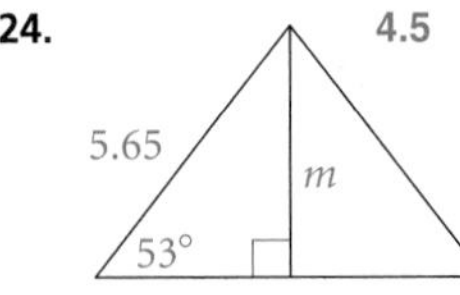

25. 514.3 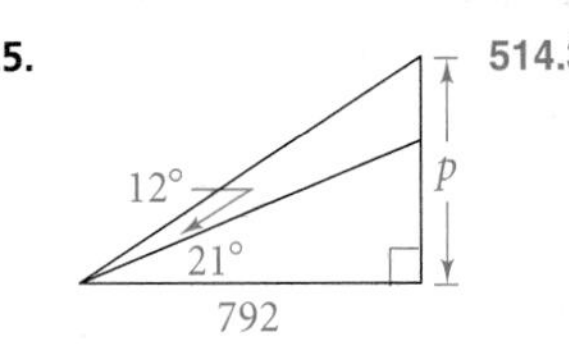

26. **Standardized Test Prep** $\triangle KLM$ is a right triangle with a right angle at M. Which of the following is false? B

A. $\sin K = \frac{LM}{KL}$ B. $\cos K = \frac{KL}{KM}$ C. $\tan K = \frac{LM}{KM}$ D. $\cos L = \frac{LM}{KL}$ E. $\sin L = \frac{KM}{KL}$

27. **Open-ended** Name some ways you might use a trigonometric ratio to calculate a distance instead of measuring it directly. See margin.

28. **Nature** Suppose you are lying on the ground looking up at a California redwood tree. Your angle of elevation to the top of the tree is 42°. You are 280 ft from the base of the tree.
    a. How tall is the tree? about 252 ft
    b. How far would a bird have to fly to get from the top of the tree to your location? about 377 ft

Wrap Up

THE BIG IDEA Ask students: *Why are trigonometric ratios useful in solving problems? What kinds of problems can be solved using trigonometric ratios?*

RETEACHING ACTIVITY Students use trigonometric ratios for sine, cos and tan to evaluate expressions. (Reteaching worksheet 9-3)

Exercises CHECKPOINT

Students assess their own progress on Lessons 9-1 to 9-3.

WRITING Exercise 9 Ask students what methods they know for finding the distance. Ask: *Is it better to use the distance formula?*

OPEN-ENDED Exercise 15 Encourage students to use large numbers in the coordinates.

Chapter Project *Find Out by Researching*

Do research to find the radii of all the planets in our solar system. Express your answers in miles.

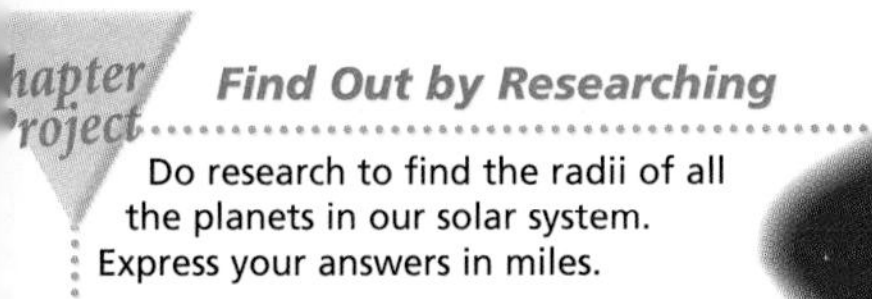

Exercises MIXED REVIEW

Graph each equation. 29–32. See margin.

29. $y = x + 2$ 30. $y = |x| + 2$

31. $y = x^2 + 2$ 32. $y = 2^x$

33. *Transportation* A car is traveling at a constant speed. After 3 h, the car has gone 150 mi. How many miles will the car have gone after 5 h? 250 mi

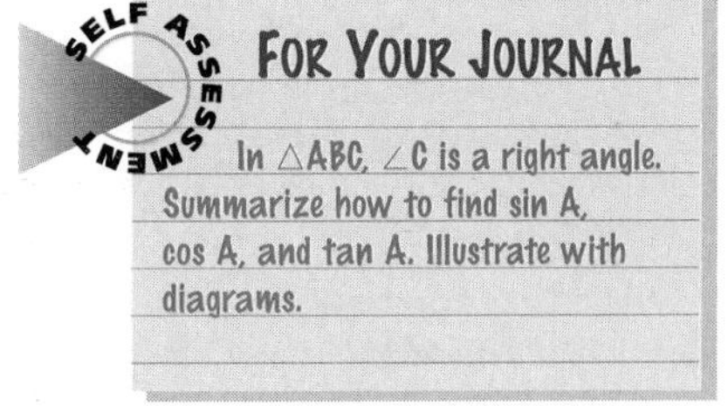

FOR YOUR JOURNAL

In $\triangle ABC$, $\angle C$ is a right angle. Summarize how to find sin *A*, cos *A*, and tan *A*. Illustrate with diagrams.

Getting Ready for Lesson 9-4

Find each value.

34. $\sqrt{4}$ 2 35. $\sqrt{169}$ 13 36. $3\sqrt{25}$ 15 37. $-2\sqrt{9}$ −6 38. $-3\sqrt{49}$ −21

Exercises CHECKPOINT

A right triangle has legs of lengths *a* and *b* and hypotenuse of length *c*. Find the missing value.

1. $a = 2, b = 4$ $c \approx 4.5$ 2. $a = 3, b = 5$ $c \approx 5.8$ 3. $a = 1, b = 7$ $c \approx 7.1$ 4. $a = 4, b = 3$ $c = 5$

5. $a = 10, c = 26$ $b = 24$ 6. $b = 3, c = 9$ $a \approx 8.5$ 7. $a = 8, c = 12$ $b \approx 8.9$ 8. $b = 20, c = 25$ $a = 15$

9. *Writing* Describe how to find the distance between points (x_1, y_1) and (x_2, y_2). Include examples. See margin.

10. The angle of elevation from a point on the ground 300 ft from a tower is 42°. How tall is the tower? about 270 ft

Approximate *AB* to the nearest tenth.

11. $A(1, 4), B(2, 7)$ 3.2 12. $A(-2, -1), B(4, 2)$ 6.7 13. $A(-3, 5), B(6, 1)$ 9.8 14. $A(2, 3), B(5, 4)$ 3.2

15. *Open-ended* Write the coordinates of a pair of points *A* and *B*. Then find the midpoint of $\overline{AB}$. Sample: $A(-5, -2)$, $B(8, -4)$, $M(1.5, -3)$

In $\triangle ABC$, $\angle C$ is a right angle. Find the length of the indicated side.

16. $m\angle B = 43°, AB = 4, AC =$ ■ about 2.7 17. $m\angle A = 33°, AC = 7, BC =$ ■ about 4.5

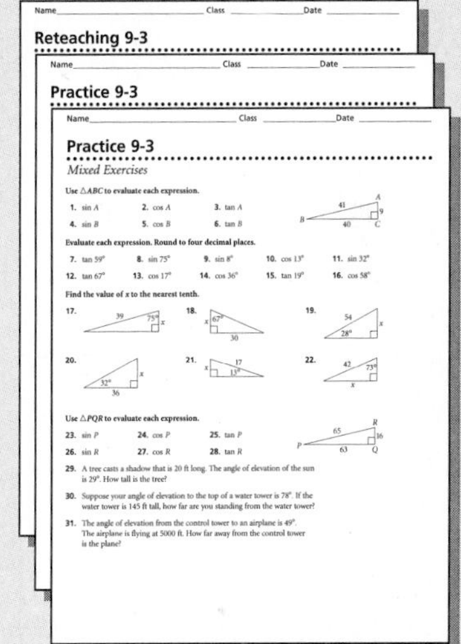

Lesson Quiz

Lesson Quiz is also available in Transparencies.

1. Use the figure below to find sin *A*, cos *A*, and tan *A*. 0.82, 0.57, 1.4

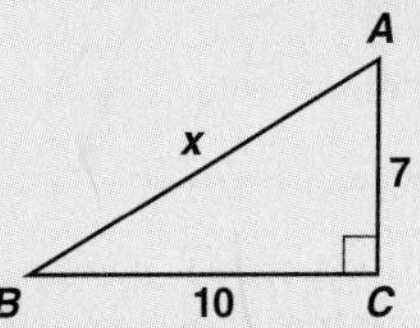

2. Find the value of *x* to the nearest tenth. $4\sqrt{3}$

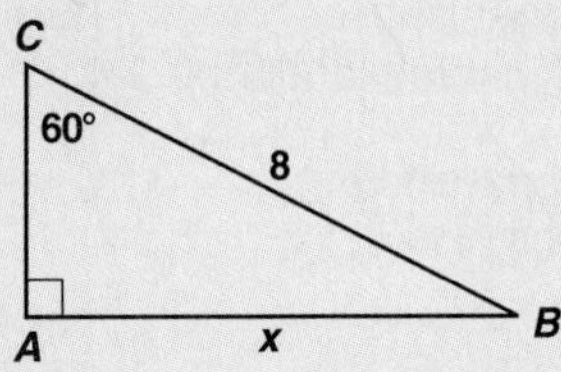

9-4 Teaching Notes

PROBLEM OF THE DAY

Find four consecutive multiples of 15 whose sum is three times the LCM of 10 and 22. 60, 75, 90, 105

Problem of the Day is also available in Transparencies.

CONNECTING TO PRIOR KNOWLEDGE Write $4^2 \cdot 4^3$ on the board. Ask students how to simplify the expression. Add the exponents. Now write $\sqrt{4} \cdot \sqrt{5}$. Ask students to make conjectures about how they could simplify this expression.

THINK AND DISCUSS

Question 1 Have students practice writing and identifying other expressions that *are* or *are not* in simplest form.

Remind students that the radical sign refers only to the principal square root.

Example 1

ALTERNATIVE METHOD As an alternative to factoring expressions into perfect squares, remind students how to factor expressions. On the board, divide 192 repeatedly by 2 until only 3 remains. Then write $192 = 2 \cdot 2 \cdot 2 \cdot 2 \cdot 2 \cdot 2 \cdot 3$. Group each pair of 2's. Point out that a pair of identical factors makes a perfect square. A pair of 2's is 4, which is a

Lesson Planning Options

Prerequisite Skills

- Simplifying square roots
- Using the Pythagorean theorem

Assignment Options for Exercises On Your Own

To provide flexible scheduling, this lesson can be subdivided into parts.

 **Core** 1–20, 40–47
Extension 21–24, 56

 **Core** 25–34, 48–55
Extension 35–39

Use Mixed Review to maintain skills.

Resources

 **Student Edition**

Skills Handbook, pp. 578, 581, 582
Extra Practice, p. 564
Glossary/Study Guide

 **Teaching Resources**

Chapter Support File, Ch. 9
- Practice 9-4 (two worksheets)
- Reteaching 9-4

Classroom Manager 9-4
Glossary, Spanish Resources
Two-Year Algebra Handbook 9-4

 **Transparencies**

19, 106, 112

Connections  Commuting . . . and more

9-4 Simplifying Radicals

What You'll Learn
- Simplifying radicals involving products and quotients
- Solving problems involving radicals

...And Why

To solve distance problems

THINK AND DISCUSS

Part 1 **Multiplication with Radicals**

A radical expression is in **simplest form** when *all three* statements are true.

- The expression under the radical sign has no perfect square factors other than 1.
- The expression under the radical sign does not contain a fraction.
- The denominator does not contain a radical expression.

1. Explain why each expression *is* or *is not* in simplest form.

a. $\sqrt{20}$ **b.** $4\sqrt{5}$ **c.** $\frac{1}{\sqrt{3}}$ **d.** $\sqrt{\frac{2}{5}}$

a–d. See left.

1a. Not simplest form; 20 has 2^2 as a factor.
b. Simplest form; all three statements are true.
c. Not simplest form; denominator has radical expression.
d. Not simplest form; the radical sign contains a fraction.

Multiplication Property of Square Roots

For any numbers $a \geq 0$ and $b \geq 0$, $\sqrt{ab} = \sqrt{a} \cdot \sqrt{b}$.

Example: $\sqrt{54} = \sqrt{9} \cdot \sqrt{6} = 3 \cdot \sqrt{6} = 3\sqrt{6}$

You can simplify radical expressions that contain numbers and variables. Assume that all variables under radicals represent positive numbers.

Example 1

Simplify each radical expression.

a. $\sqrt{192} = \sqrt{64 \cdot 3}$ ← 64 is a perfect square and a factor of 192.
$= \sqrt{64} \cdot \sqrt{3}$ ← Use the multiplication property.
$= 8\sqrt{3}$ ← Simplify $\sqrt{64}$.

b. $\sqrt{16a^3} = \sqrt{16} \cdot \sqrt{a^2} \cdot \sqrt{a}$ ← Use the multiplication property.
$= 4a\sqrt{a}$ ← Simplify.

Example 2

Simplify the radical expression $\sqrt{6} \cdot \sqrt{15}$.

$\sqrt{6} \cdot \sqrt{15} = \sqrt{90}$ ← Combine factors under one radical.
$= \sqrt{9 \cdot 10}$ ← 9 is a perfect square and a factor of 90.
$= \sqrt{9} \cdot \sqrt{10}$ ← Use the multiplication property.
$= 3\sqrt{10}$ ← Simplify $\sqrt{9}$.

2. Try This Simplify each expression.

a. $5\sqrt{300}$ $50\sqrt{3}$ **b.** $\sqrt{13} \cdot \sqrt{52}$ 26 **c.** $\sqrt{x^2y^5}$ $xy^2\sqrt{y}$ **d.** $(3\sqrt{5})^2$ 45 perfect square. Each perfect square can be removed from the radical sign and written as a square root. The 3 cannot be removed from the radical sign since it is not paired with another 3. On the board, write

$$\sqrt{192} = \sqrt{(2\cdot 2)\cdot(2\cdot 2)\cdot(2\cdot 2)\cdot 3}$$
$$= 2\cdot 2\cdot 2\sqrt{3}$$
$$= 8\sqrt{3}$$

Example 2

VISUAL AND TACTILE LEARNING Suggest that students make a poster showing all the perfect squares from 0 to 200. Making the poster will help them memorize and recognize perfect squares.

Example 3 **Relating to the Real World** 

If calculators are not available, encourage students to leave their answers in simplified radical form. The simplified form for Example 3 is $4\sqrt{3}$.

AUDITORY LEARNING Encourage students to create and tell each other numerical facts expressed as radicals; for example, my sister is $\sqrt{81}$ yr old.

Question 4 Explain that the square roots of negative numbers, such as $\sqrt{-8}$, are not real numbers. They are called *imaginary* numbers because they cannot be graphed on the real number line. Imaginary numbers are explored in Advanced Algebra. However, interested students may wish to consult an Advanced Algebra book to find out more about imaginary numbers.

When you use radical expressions to solve real-world problems, you need to evaluate any radicals to get a numerical answer.

Example 3 **Relating to the Real World** 

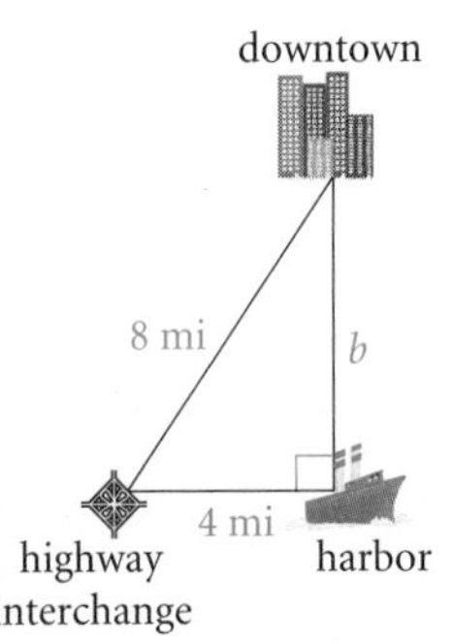

Commuting Use the figure at the left. About how many miles is it from downtown to the harbor? Round to the nearest tenth of a mile.

$a^2 + b^2 = c^2$ ← Use the Pythagorean theorem.
$4^2 + b^2 = 8^2$ ← Substitute. Remember that the hypotenuse is the longest side.
$b^2 = 8^2 - 4^2$
$b = \sqrt{8^2 - 4^2}$ ← Solve for b.
$= \sqrt{48}$ ← Use a calculator.
$= 6.92820323$

It is about 6.9 miles from downtown to the harbor.

PROBLEM SOLVING

Look Back Explain how you can use $\sqrt{49}$ to check the reasonableness of the answer in Example 3.
48 is close to 49, so $\sqrt{48}$ is close to $\sqrt{49}$. $\sqrt{48}$ is close to 7.

3. *Calculator* Suppose a classmate did Example 3 and got the answer $4\sqrt{3}$. Use a calculator to check if this answer is correct.
$4\sqrt{3} = 6.92820323$ ✓

Part 2 Division with Radicals

You can use the division property of square roots to simplify expressions.

Division Property of Square Roots

For any numbers $a \geq 0$ and $b > 0$, $\sqrt{\frac{a}{b}} = \frac{\sqrt{a}}{\sqrt{b}}$.

Example: $\sqrt{\frac{4}{9}} = \frac{\sqrt{4}}{\sqrt{9}} = \frac{2}{3}$

4. Explain why there are the restrictions $a \geq 0$ and $b > 0$ in the properties of square roots. **You can compute the sq. root of only a nonnegative number. Also, the value under the radical in the denominator must be positive.**

Additional Examples

FOR EXAMPLE 1

Simplify the radical expression $\sqrt{8x^5}$.
$2x^2\sqrt{2x}$

FOR EXAMPLE 2

Simplify the radical expression $\sqrt{20}\cdot\sqrt{5}$. **10**

FOR EXAMPLE 4

Simplify the radical expression $\sqrt{\frac{8x^2}{81}}$.
$\frac{2x\sqrt{2}}{9}$

FOR EXAMPLE 5

Simplify the radical expression $\sqrt{\frac{150x^5}{3x^2}}$. **$5x\sqrt{2x}$**

FOR EXAMPLE 6

Simplify the radical expression $\frac{3x}{\sqrt{8}}$.
$\frac{3x\sqrt{2}}{4}$

Example 4

Remind students that $b^4 = b^2 \cdot b^2$. So $\sqrt{b^4} = b^2$.

Example 5

Write the following on the board. Have students use their calculators to find and dictate to you the square roots of 12 and 3.

$$\sqrt{\frac{12}{3}} = \frac{\sqrt{12}}{\sqrt{3}} = \frac{3.464101615}{1.732050808} = 2$$

Students may be surprised by the simple answer. Using this example, lead students to understand that they must study an expression carefully to decide whether to divide first or to evaluate the numerator and denominator separately.

ERROR ALERT! **Question 5c** Students often forget to move a square root outside the radical sign as they simplify. They might write, $\sqrt{25} = \sqrt{5}$ instead of 5. **Remediation:** Make students aware of this common error. Remind them to work carefully, and check their work after each problem.

Example 6

ERROR ALERT! Students may think that $\frac{2}{\sqrt{5}}$ is in a simpler form than $\frac{2\sqrt{5}}{5}$ because $\frac{2}{\sqrt{5}}$ contains fewer digits.

Remediation: Have students reread the definition of rationalization. Encourage them to pay attention to the denominator when they are deciding whether the expression is in its simplest form.

Technology Options

For Exercises 6–20 and 25–34, students may find it helpful to simplify the radical expressions with a graphing calculator.

Prentice Hall Technology

Software
- Secondary Math Lab Toolkit™
- Computer Item Generator 9-4

Internet
- See the Prentice Hall site. (http://www.phschool.com)

When you simplify a radical expression involving division, sometimes it is easier to simplify the numerator and denominator separately.

Example 4

Simplify each radical expression.

a. $\sqrt{\frac{11}{49}} = \frac{\sqrt{11}}{\sqrt{49}}$ ← Use the division property.

$= \frac{\sqrt{11}}{7}$ ← Simplify $\sqrt{49}$.

b. $\sqrt{\frac{25}{b^4}} = \frac{\sqrt{25}}{\sqrt{b^4}}$ ← Use the division property.

$= \frac{5}{b^2}$ ← Simplify $\sqrt{25}$ and $\sqrt{b^4}$.

When you simplify a radical expression involving division, sometimes it is easier to divide first and then simplify the radical expression.

Example 5

Simplify the radical expression $\frac{\sqrt{96}}{\sqrt{12}}$.

$\frac{\sqrt{96}}{\sqrt{12}} = \sqrt{\frac{96}{12}}$ ← Use the division property.

$= \sqrt{8}$ ← Divide.

$= \sqrt{4 \cdot 2}$ ← 4 is a perfect square and a factor of 8.

$= \sqrt{4} \cdot \sqrt{2}$ ← Use the multiplication property.

$= 2\sqrt{2}$ ← Simplify $\sqrt{4}$.

5. **Try This** Simplify each expression.

a. $\sqrt{\frac{144}{9}}$ 4   b. $\frac{\sqrt{24}}{\sqrt{8}}$ $\sqrt{3}$ c. $\sqrt{\frac{25c^3}{b^2}}$ $\frac{5c\sqrt{c}}{b}$

When you have a square root in a denominator that is not a perfect square, you should **rationalize** the denominator. To do this, make the denominator a rational number without changing the value of the expression.

Example 6

Simplify $\frac{2}{\sqrt{5}}$.

$\frac{2}{\sqrt{5}} = \frac{2}{\sqrt{5}} \cdot \frac{\sqrt{5}}{\sqrt{5}}$ ← Multiply by $\frac{\sqrt{5}}{\sqrt{5}} = 1$.

$= \frac{2\sqrt{5}}{\sqrt{25}}$ ← Use the multiplication property.

$= \frac{2\sqrt{5}}{5}$ ← Simplify.

PROBLEM SOLVING

Look Back Why did you choose to multiply by $\frac{\sqrt{5}}{\sqrt{5}}$ to rationalize the expression? The denominator in simplest form cannot have radical expressions. $\sqrt{5} \cdot \sqrt{5} = 5$

6. **Try This** Simplify each expression by rationalizing the denominator.

a. $\frac{3}{\sqrt{3}}$ $\sqrt{3}$ b. $\frac{14}{\sqrt{7}}$ $2\sqrt{7}$ c. $\frac{9}{\sqrt{10}}$ $\frac{9\sqrt{10}}{10}$

EXTENSION Point out that the process of finding a square root is the opposite of the process of finding a square. A square root is written with a radical sign while a square is written with an exponent. Challenge students to explore whether a square root can be expressed using an exponent. In Advanced Algebra, students learn that square roots can be expressed as numbers with exponents of $\frac{1}{2}$.

Exercises ON YOUR OWN

Exercise 13 Point out that the exponent applies to everything inside the parentheses. Write the following:

$$(5\sqrt{7})^3 = 5 \cdot 5 \cdot 5 \cdot \sqrt{7} \cdot \sqrt{7} \cdot \sqrt{7} = 125 \cdot 7 \cdot \sqrt{7} = 875\sqrt{7}$$

CRITICAL THINKING Exercise 24d Point out that only the height changes, not the horizontal distance to the raft.

ALTERNATIVE ASSESSMENT Exercises 6 and 8 Ask students to simplify the radical expressions before multiplying. This will help you assess their understanding of radical expressions.

STANDARDIZED TEST TIP Exercise 39 Point out that it is often useful to use squared variables as a guide in eliminating answer choices. The x^2 term under the radical tells us that the final response must contain either an x outside of a radical or a $\sqrt{x}\sqrt{x}$ term. Thus, answer choice D is the only possible solution.

Exercises ON YOUR OWN

Tell whether each expression is in simplest form.

1. $\frac{13}{\sqrt{4}}$ no
2. $2\sqrt{12}$ no
3. $\frac{3}{\sqrt{13}}$ no
4. $5\sqrt{30}$ yes
5. $\sqrt{\frac{2}{5}}$ no

Simplify each radical expression.

6. $\sqrt{25} \cdot \sqrt{100}$ 50
7. $\sqrt{8} \cdot \sqrt{32}$ 16
8. $3\sqrt{81} \cdot \sqrt{81}$ 243
9. $\sqrt{10} \cdot \sqrt{40}$ 20
10. $\sqrt{10^4}$ 100
11. $\sqrt{200}$ $10\sqrt{2}$
12. $5\sqrt{320}$ $40\sqrt{5}$
13. $(5\sqrt{7})^3$ $875\sqrt{7}$
14. $\sqrt{96}$ $4\sqrt{6}$
15. $(2\sqrt{18})^2$ 72
16. $\sqrt{12} \cdot \sqrt{75}$ 30
17. $\sqrt{3} \cdot \sqrt{51}$ $3\sqrt{17}$
18. $\sqrt{26} \cdot 2$ $2\sqrt{13}$
19. $2\sqrt{6} \cdot 4$ $4\sqrt{6}$
20. $\sqrt{8} \cdot \sqrt{26}$ $4\sqrt{13}$

Use the Pythagorean theorem to find *s*. Express *s* as a radical in simplest form.

21. 10

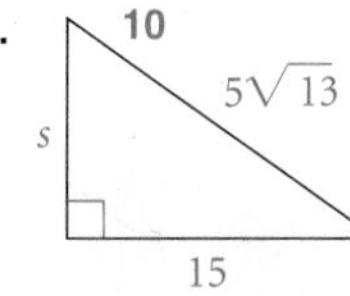

22. $4\sqrt{3}$

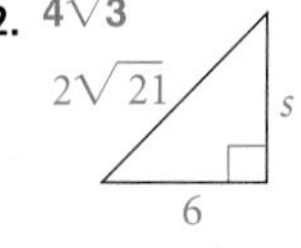

23. $5\sqrt{10}$

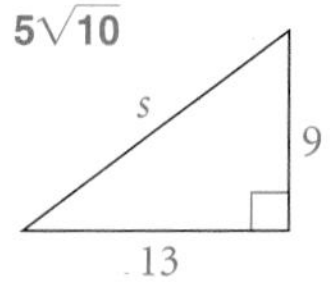

24. **Diving** Suppose you are standing at the top of a diving platform *h* feet tall. Looking down, you can see a raft on the water 8 feet from the bottom of the diving platform.
    a. Find the distance *d* from you to the raft if $h = 6$ ft. 10 ft
    b. Find the distance *d* from you to the raft if $h = 12$ ft.
    c. Suppose you know the distance *d* from you to the raft is 16 ft. About how tall is the diving platform? about 13.9 ft
    d. **Critical Thinking** Could the distance *d* from you to the raft be 7 ft? Why or why not? No, *d* must be greater than 8 ft.

    b. about 14.4 ft

h
d
8 ft

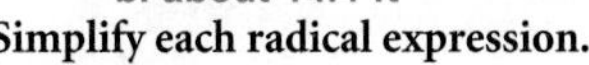

Simplify each radical expression.

25. $3\sqrt{\frac{1}{4}}$ $\frac{3}{2}$
26. $\sqrt{\frac{21}{49}}$ $\frac{\sqrt{21}}{7}$
27. $\sqrt{\frac{625}{100}}$ $\frac{5}{2}$
28. $\frac{\sqrt{96}}{\sqrt{9}}$ $\frac{4\sqrt{6}}{3}$
29. $\sqrt{\frac{120}{121}}$ $\frac{2\sqrt{30}}{11}$
30. $\frac{\sqrt{15}}{\sqrt{5}}$ $\sqrt{3}$
31. $\frac{\sqrt{72}}{\sqrt{64}}$ $\frac{3\sqrt{2}}{4}$
32. $\sqrt{\frac{48}{24}}$ $\sqrt{2}$
33. $\frac{\sqrt{169}}{\sqrt{144}}$ $\frac{13}{12}$
34. $\frac{\sqrt{400}}{\sqrt{121}}$ $\frac{20}{11}$

35. **Packaging** Use the diagram at right. Find the width *w* that the box needs to be to fit the fishing rod. 1.2 m
36. **Writing** How do you know when to rationalize a radical expression?
    Rationalize when a denominator contains a radical.
37. Suppose *a* and *b* are positive integers.
    a. **Verify** that if $a = 18$ and $b = 10$, then $\sqrt{a} \cdot \sqrt{b} = 6\sqrt{5}$. See margin.
    b. **Open-ended** Find several other pairs of positive integers *a* and *b* such that $\sqrt{a} \cdot \sqrt{b} = 6\sqrt{5}$.
    Samples: 9 and 20; 3 and 60; 45 and 4

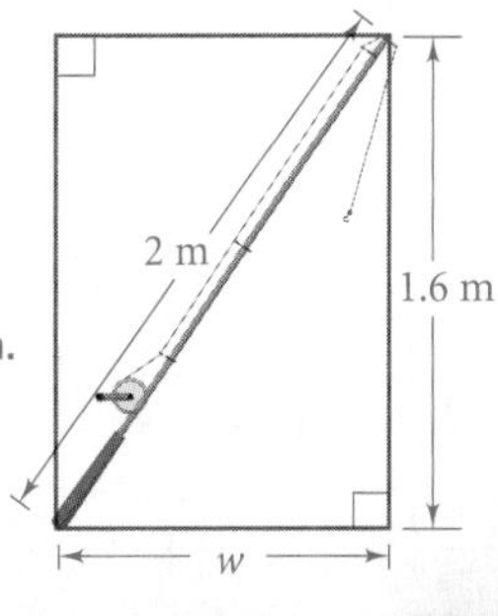

pages 433–434 On Your Own

37a. $\sqrt{18} \cdot \sqrt{10} = \sqrt{180} = \sqrt{36 \cdot 5} = 6\sqrt{5}$

page 434 Mixed Review

61. $5c + 3t \geq 30$; you can buy as many as 6 crab and 0 turkey, 5 crab and 1 turkey, 3 crab and 5 turkey, 1 crab and 8 turkey, or 0 crab and 10 turkey sandwiches.

Exercises MIXED REVIEW

DIVERSITY Exercise 61 Some students may not want to serve only sandwiches that contain meat. Encourage them to repeat the problem using some vegetarian sandwich fillings as peanut butter, tomato, avocado, cheese, lettuce, sprouts, etc.

GETTING READY FOR LESSON 9-5 These exercises prepare students for simplifying radicals involving addition and subtraction.

Wrap Up

THE BIG IDEA Ask students to describe what they have learned about simplifying radicals.

RETEACHING ACTIVITY Students simplify radicals using multiplication and division properties of radicals. (Reteaching worksheet 9-4)

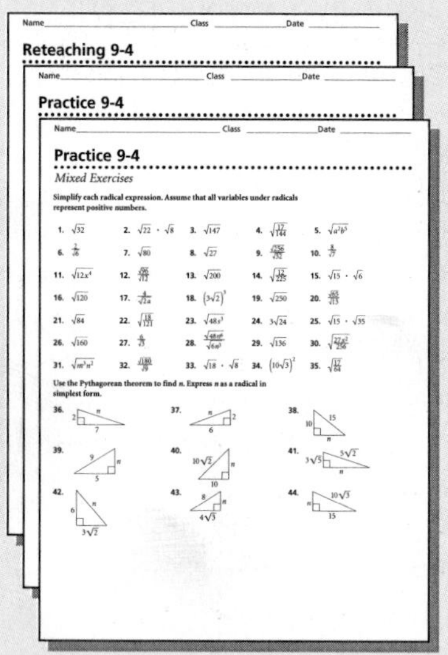

Lesson Quiz

Lesson Quiz is also available in Transparencies.

Simplify each radical expression.

1. $\sqrt{16} \cdot \sqrt{8}$ $8\sqrt{2}$
2. $\sqrt{\frac{12}{36}}$ $\frac{\sqrt{3}}{3}$
3. $\frac{2}{\sqrt{a^5}}$ $\frac{2\sqrt{a}}{a^3}$
4. $\frac{2\sqrt{8x^3}}{\sqrt{x^5}}$ $\frac{4\sqrt{2}}{x}$

434

38. *Sailing* The diagram at the right shows a sailboat.
    a. Use the Pythagorean theorem to find the height of the sail in simplest radical form. $6\sqrt{5}$ ft
    b. Use the result of part (a) and the formula for the area of a triangle to find the area of the sail. Round your answer to the nearest tenth. 80.5 ft^2

18 ft
12 ft

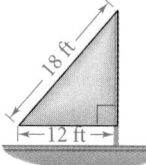

39. *Standardized Test Prep* Suppose that x and y are positive numbers. Which of the following is *not* equivalent to $\sqrt{24x^2y}$? D

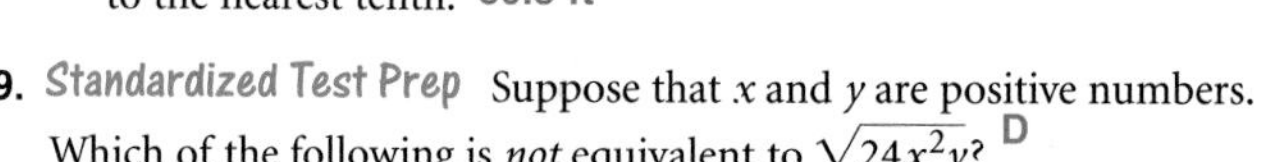

A. $2x\sqrt{6y}$ B. $\sqrt{24x}\sqrt{xy}$ C. $\sqrt{4xy}\sqrt{6x}$
D. $2x^2\sqrt{6y}$ E. $x\sqrt{24y}$

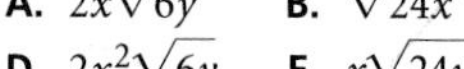

Simplify each radical expression. Assume that all variables represent positive numbers.

40. $\sqrt{v^6}$ v^3
41. $\frac{2}{\sqrt{a^3}}$ $\frac{2\sqrt{a}}{a^2}$
42. $\sqrt{20a^2b^3}$ $2ab\sqrt{5b}$
43. $\sqrt{a^3b^5c^3}$ $ab^2c\sqrt{abc}$
44. $\sqrt{12x^3y^2}$ $2xy\sqrt{3x}$
45. $\sqrt{4y^4}$ $2y^2$
46. $\frac{\sqrt{x^2}}{\sqrt{y^3}}$ $\frac{x\sqrt{y}}{y^2}$
47. $\sqrt{\frac{18x}{81}}$ $\frac{\sqrt{2x}}{3}$

Simplify each expression by rationalizing the denominator.

48. $\frac{3}{\sqrt{7}}$ $\frac{3\sqrt{7}}{7}$
49. $\frac{12}{\sqrt{12}}$ $2\sqrt{3}$
50. $\frac{2\sqrt{2}}{\sqrt{5}}$ $\frac{2\sqrt{10}}{5}$
51. $\frac{9}{\sqrt{8}}$ $\frac{9\sqrt{2}}{4}$
52. $\frac{3\sqrt{2}}{\sqrt{6}}$ $\sqrt{3}$
53. $\frac{25}{\sqrt{5}}$ $5\sqrt{5}$
54. $\frac{2\sqrt{5}}{\sqrt{12}}$ $\frac{\sqrt{15}}{3}$
55. $\frac{16}{\sqrt{6}}$ $\frac{8\sqrt{6}}{3}$

56. *Hobbies* Use the Pythagorean theorem to find the missing dimensions a, b, and c of each triangle in the quilt square. 6 in.; $6\sqrt{2}$ in.; $3\sqrt{2}$ in.

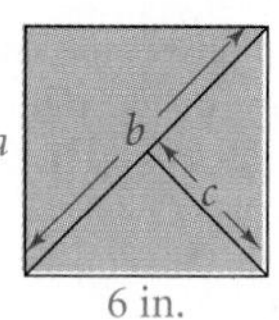

Exercises MIXED REVIEW

Find the x- and y-intercepts of each equation.

57. $2x + 3y = 6$ 3; 2
58. $4x + 2y = 8$ 2; 4
59. $-3x + 5y = 15$ −5; 3
60. $-2x - 4y = 12$ −6; −3

61. *Consumer* You are shopping for food to serve for a lunch party. Crab sandwiches cost \$5 per serving and turkey sandwiches cost \$3 per serving. You have \$30.00 total to spend on sandwiches. Write an inequality to model the situation. How many of each kind can you buy? See margin.

Getting Ready for Lesson 9-5

Simplify each square root. Then add or subtract.

62. $\sqrt{16} + \sqrt{36}$ 10
63. $\sqrt{49} - \sqrt{64}$ −1
64. $\sqrt{121} + \sqrt{81}$ 20
65. $\sqrt{400} - \sqrt{100}$ 10

9-5 Teaching Notes

PROBLEM OF THE DAY

Julie worked 24 h and was paid \$5.50/h, after taxes. How many more hours must she work to pay her car insurance bill of \$231.00? 18 h

Problem of the Day is also available in Transparencies.

CONNECTING TO PRIOR KNOWLEDGE Write $2x + 3x$ on the board. Ask: *Can this be simplified?* yes Then write $2x + 3y$. Ask: *Can this be simplified?* no Now write $2\sqrt{x} + 3\sqrt{x}$. Ask: *Can this be simplified?* yes Now write $2\sqrt{x} + 3\sqrt{y}$. Ask: *Do you think this can be simplified?* no Lead students to understand that because the radicals are different, the expression cannot be simplified.

WORK TOGETHER

ERROR ALERT! Students may confuse $\sqrt{a} + \sqrt{b}$ with $\sqrt{a + b}$. **Remediation:** For $\sqrt{a} + \sqrt{b}$, the radical sign does not cover the plus sign. For $\sqrt{a + b}$, the radical sign covers the plus sign. Stress that students must pay careful attention to whether the radical covers the operation symbol.

THINK AND DISCUSS

Example 1

Encourage students to use a calculator to verify that combining like terms is correct.

Connections  Art . . . and more

What You'll Learn

- Simplifying radicals involving addition and subtraction
- Solving problems involving sums and differences of radicals

...And Why

To solve geometric problems involving art

9-5 Adding and Subtracting Radicals

WORK TOGETHER

Work in pairs.

1. Copy and complete the table. See margin.

a	b	$\sqrt{a}$	$\sqrt{b}$	$\sqrt{a} + \sqrt{b}$	$\sqrt{a+b}$	$\sqrt{a} - \sqrt{b}$	$\sqrt{a-b}$
9	16	■	■	■	■	■	■
25	100	■	■	■	■	■	■
64	36	■	■	■	■	■	■
4	121	■	■	■	■	■	■
49	1	■	■	■	■	■	■

2. a. Compare the values in the addition columns. What do you notice? $\sqrt{a} + \sqrt{b}$ entries are not equal to the corresponding $\sqrt{a + b}$ entries.

b. In general, does $\sqrt{a} + \sqrt{b} = \sqrt{a + b}$? Explain. See margin.

3. a. Compare the values in the subtraction columns. What do you notice? $\sqrt{a} - \sqrt{b}$ entries are not equal to the corresponding $\sqrt{a - b}$ entries.

b. Does $\sqrt{a} - \sqrt{b} = \sqrt{a - b}$? Explain. No; the distributive prop. applies only to multiplication and division.

4. Is there a rule for adding and subtracting radicals? **Justify** your reasoning. Answers may vary. Sample: No; you cannot find a pattern that works for all numbers.

THINK AND DISCUSS

Part 1

Simplifying Sums and Differences

You can simplify radical expressions by combining like terms. Like terms have the same radical part.

like terms	unlike terms
$4\sqrt{7}$ and $-12\sqrt{7}$	$3\sqrt{11}$ and $2\sqrt{5}$

5. Identify each pair of expressions as like or unlike terms. **Justify** your answers.

a. $\sqrt{8}$ and $2\sqrt{8}$ like; same radical factors
b. $2\sqrt{3}$ and $3\sqrt{2}$ unlike; different radical factors
c. $2\sqrt{7}$ and $-3\sqrt{7}$ like; same radical factors

Example 1

Simplify the expression $\sqrt{2} + 3\sqrt{2}$.

$\sqrt{2} + 3\sqrt{2} = 1\sqrt{2} + 3\sqrt{2}$ ← Both $1\sqrt{2}$ and $3\sqrt{2}$ have $\sqrt{2}$.
$= 4\sqrt{2}$ ← Combine like terms.

Lesson Planning Options

Prerequisite Skills

- Simplifying square roots
- Combining like terms

Assignment Options for Exercises On Your Own

To provide flexible scheduling, this lesson can be subdivided into parts.

1. **Core** 1–15, 17–25
   Extension 16, 39–41
2. **Core** 26–34
   Extension 35–38, 42, 43

Use Mixed Review to maintain skills.

Resources

Student Edition

Skills Handbook, p. 569
Extra Practice, p. 564
Glossary/Study Guide

Teaching Resources

Chapter Support File, Ch. 9
- Practice 9-5 (two worksheets)
- Reteaching 9-5

Classroom Manager 9-5
Glossary, Spanish Resources
Two-Year Algebra Handbook 9-5

Transparencies

107, 113

Example 2

KINESTHETIC LEARNING Write each of the following radicals on separate pieces of paper: $\sqrt{3}$, $\sqrt{5}$, $\sqrt{12}$, and $\sqrt{125}$. Make enough copies so each student has one radical. Hand out the radicals at random. Instruct students to simplify their radicals, if possible. Lead students to recognize that they can simplify $\sqrt{12}$ to $2\sqrt{3}$ which makes it a like term with $\sqrt{3}$. Ask students how they can simplify $\sqrt{125}$. Direct students to compare radicals until they find a partner with a like term. Have students with like terms form groups. Challenge students to write an expression describing the addition of all the radical terms in their group.

ERROR ALERT! Students may think $\sqrt{12}$ cannot be simplified because they chose only 2 and 6 as factors. **Remediation:** remind students that they can remove only perfect squares from within radical signs. Encourage students to test whether perfect squares are factors of the number within the radical: 4, a perfect square, is a factor of 12, so factoring 12 into 4 and 3 leads to simplifying the radical.

Example 3 **Relating to the Real World**

Remind students of the meaning of *ratio.* Make sure they do not interpret the question to mean that the length of the painting is $1 + \sqrt{5}$.

Example 4

Remind students to write clearly and carefully when solving problems such as this. Students make numerous mistakes because they fail to read their own handwriting correctly.

Additional Examples

FOR EXAMPLE 1

Simplify the expression

$4\sqrt{3} + 2\sqrt{3}$ $6\sqrt{3}$

FOR EXAMPLE 2

Simplify the expression

$3\sqrt{5} - 5\sqrt{5}$ $-2\sqrt{5}$

FOR EXAMPLE 4

Simplify $\sqrt{3}(2 - \sqrt{3})$ $2\sqrt{3} - 3$

436

Example 2

Simplify the expression $4\sqrt{3} - \sqrt{12}$.

$$4\sqrt{3} - \sqrt{12} = 4\sqrt{3} - \sqrt{4 \cdot 3}$$ ← 4 is a perfect square and a factor of 12.

$$= 4\sqrt{3} - \sqrt{4} \cdot \sqrt{3}$$ ← Use the multiplication property.

$$= 4\sqrt{3} - 2\sqrt{3}$$ ← Simplify $\sqrt{4}$.

$$= 2\sqrt{3}$$ ← Combine like terms.

6. Try This Simplify each expression.

a. $2\sqrt{5} + \sqrt{5}$ $3\sqrt{5}$ **b.** $3\sqrt{45} + 2\sqrt{5}$ $11\sqrt{5}$ **c.** $3\sqrt{3} - 2\sqrt{12}$ $-\sqrt{3}$

Part 2

Simplifying Products, Sums, and Differences

QUICK REVIEW

The distributive property says $a(b + c) = ab + ac$.

Sometimes you need to use the distributive property and what you have learned in lessons 9-4 and 9-5 to simplify expressions.

Example 3 **Relating to the Real World**

Art The ratio length : width of this painting by Mondrian is approximately equal to the *golden ratio* $(1 + \sqrt{5}) : 2$. The width of the painting is 50 in. Find the length of the painting. Express your answer in simplest radical form. Then estimate the length in inches.

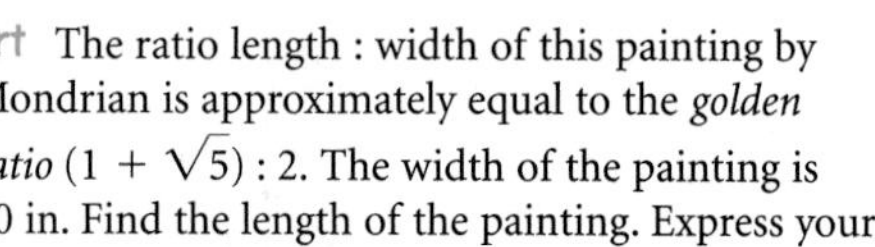

Define 50 = width of painting
x = length of painting

Relate $(1 + \sqrt{5}) : 2 =$ length : width

Write
$$\frac{1 + \sqrt{5}}{2} = \frac{x}{50}$$
$$50(1 + \sqrt{5}) = 2x$$ ← Cross multiply.
$$\frac{50(1 + \sqrt{5})}{2} = x$$ ← Solve for x.
$$25(1 + \sqrt{5}) = x$$
$$25 + 25\sqrt{5} = x$$ ← Use the distributive property.
$$80.90169944 = x$$ ← Use a calculator.
$$81 \approx x$$

The length of Mondrian's painting is about 81 in.

7. a. Calculator Write the golden ratio as a decimal to the nearest hundredth. 1.62

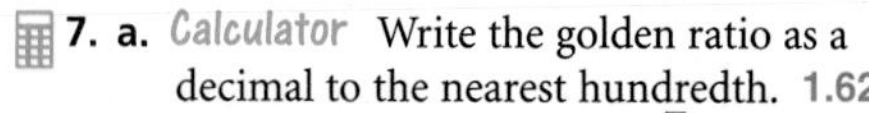

b. Calculator Write $\frac{25 + 25\sqrt{5}}{50}$ as a decimal to the nearest hundredth. 1.62

c. Compare your answers to parts (a) and (b). What do you notice about the two values? The numbers are the same.

Exercises ON YOUR OWN

MAKING CONNECTIONS Exercise 16 The greatest number of kites flown on a single line is 11,284. Sadao Harada and a team of assistants achieved this feat in Kagoshima, Japan in October, 1990.

ALTERNATIVE ASSESSMENT **Exercises 26–28** Have students work in pairs. Each student in the pair should sit facing away from the other. Have one student in the pair explain the steps required to simplify each expression while the other student follows the instructions. Then, ask the students to switch roles. This will help you assess their knowledge of adding and subtracting radicals.

CARTOON Exercise 36 Professor Hinkle's age (about 251) is bizarre. Challenge students to use parentheses to make a more reasonable age. **parentheses around the fraction; about 50** Discuss what might have gone wrong with the cartoonist's original idea. **Answers may vary. Sample: thought putting a square on the numerator affected the denominator**

CONNECTING TO THE STUDENTS' WORLD Exercise 36 Suggest that students make a class poster displaying the numbers of their houses or apartments. Have them write the numbers as expressions containing radicals.

DIVERSITY Exercise 37b Remind students to be aware of others' feelings when correcting each other's work. Help them understand the value of helping and correcting one another in a positive way.

Example 4

Simplify $\sqrt{2}(5 - \sqrt{8})$.

$\sqrt{2}(5 - \sqrt{8}) = \sqrt{2}(5) - \sqrt{2}(\sqrt{8})$ ← Use the distributive property.
$= 5\sqrt{2} - \sqrt{2 \cdot 8}$ ← Use the multiplication property.
$= 5\sqrt{2} - \sqrt{16}$ ← Multiply.
$= 5\sqrt{2} - 4$ ← Simplify.

8. Try This Simplify each expression.
   a. $2(2 + \sqrt{3})$ $4 + 2\sqrt{3}$
   b. $\sqrt{2}(1 - 2\sqrt{10})$ $\sqrt{2} - 4\sqrt{5}$
   c. $\sqrt{3}(5\sqrt{2} - 2\sqrt{6})$ $5\sqrt{6} - 6\sqrt{2}$
9. Simplify $3\sqrt{2}(\sqrt{24} + 2\sqrt{6})$ by first simplifying $\sqrt{24}$. $24\sqrt{3}$

Exercises ON YOUR OWN

Simplify each expression.

1. $15\sqrt{9} - \sqrt{9}$ 42
2. $3(\sqrt{27} + 1)$ $9\sqrt{3} + 3$
3. $\sqrt{18} + \sqrt{3}$ $3\sqrt{2} + \sqrt{3}$
4. $2\sqrt{12} - 7\sqrt{3}$ $-3\sqrt{3}$
5. $2\sqrt{3}(\sqrt{3} - 1)$ $6 - 2\sqrt{3}$
6. $\sqrt{8} + 2\sqrt{2}$ $4\sqrt{2}$
7. $\sqrt{27} - \sqrt{18}$ $3\sqrt{3} - 3\sqrt{2}$
8. $-3\sqrt{6} + 8\sqrt{6}$ $5\sqrt{6}$
9. $3\sqrt{7} - \sqrt{28}$ $\sqrt{7}$
10. $16\sqrt{10} + 2\sqrt{10}$ $18\sqrt{10}$
11. $\sqrt{3}(\sqrt{15} + \sqrt{4})$ $3\sqrt{5} + 2\sqrt{3}$
12. $\sqrt{5} - 3\sqrt{5}$ $-2\sqrt{5}$
13. $\sqrt{12} + \sqrt{24} - \sqrt{36}$ $2\sqrt{3} + 2\sqrt{6} - 6$
14. $\sqrt{3}(\sqrt{2} + 2\sqrt{3})$ $\sqrt{6} + 6$
15. $\sqrt{2}(\sqrt{8} - 4)$ $4 - 4\sqrt{2}$
16. Recreation You can make a box kite in the shape of a rectangular solid. The opening at each end of the kite is a square.
   a. Suppose the sides of the square are 2 ft long. How long are the diagonal struts used for bracing? **about 2.8 ft**
   b. Suppose each side of the square has length s. Find the length of the diagonal struts in terms of s. Write your answer in simplest form. $s\sqrt{2}$

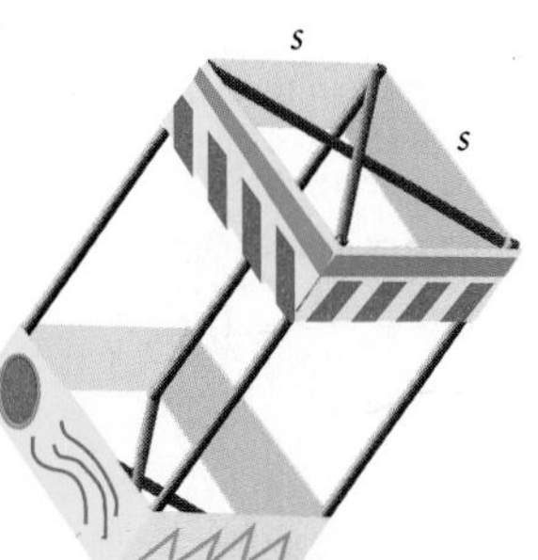

Calculator **Use a calculator to evaluate each expression. Round your answers to the nearest tenth.**

17. $\sqrt{2} + \sqrt{3}$ 3.1
18. $4\sqrt{5} + \sqrt{10}$ 12.1
19. $\sqrt{40} + \sqrt{90}$ 15.8
20. $6\sqrt{8} - 8\sqrt{6}$ −2.6
21. $\sqrt{3}(\sqrt{6} + 1)$ 6.0
22. $\sqrt{18} + \sqrt{2}$ 5.7
23. $5\sqrt{5} - \sqrt{7}$ 8.5
24. $\sqrt{13} - 2\sqrt{2}$ 0.8
25. $3\sqrt{2} - 3\sqrt{3}$ −1.0

Technology Options

For Exercises 18–25, students may want to simplify the radical expressions by using algebra software or a graphing calculator.

Prentice Hall Technology

Software
- Secondary Math Lab Toolkit™
- Integrated Math Lab 17
- Computer Item Generator 9-5

Internet • See the Prentice Hall site. (http://www.phschool.com)

GEOMETRY Exercises 39–41 Students use radical expressions to find the perimeters of figures.

CONNECTING TO THE STUDENTS' WORLD Exercise 42 Encourage students to find out more about the *golden ratio* and how it was used in Renaissance architecture. Invite interested students to make a presentation for the class, describing the *golden ratio* and showing pictures of buildings incorporating the *golden ratio* proportions.

STANDARDIZED TEST TIP **Exercise 43** Remind students that it is helpful to divide each term in the numerator by the common denominator. This will simplify the given expression to $\sqrt{25} + \sqrt{16}$, or $5 + 4$.

Exercises MIXED REVIEW

ESTIMATION Exercise 44 Suggest that students round 75 to 80 and then find $\frac{1}{4}$ of 80. Students can then use their estimate of 20 to check their answers.

GETTING READY FOR LESSON 9-6 These exercises prepare students for solving equations that contain radicals.

Wrap Up

THE BIG IDEA Ask students to describe how to add and subtract radicals.

RETEACHING ACTIVITY Students simplify radicals involving addition and subtraction. (Reteaching worksheet 9-5)

page 435 Work Together

a	b	$\sqrt{a}$	$\sqrt{b}$	$\sqrt{a}+\sqrt{b}$
9	16	3	4	7
25	100	5	10	15
64	36	8	6	14
4	121	2	11	13
49	1	7	1	8

a	b	$\sqrt{a+b}$	$\sqrt{a}-\sqrt{b}$	$\sqrt{a-b}$
9	16	5	−1	undef.
25	100	11.18	−5	undef.
64	36	10	2	5.29
4	121	11.18	−9	undef.
49	1	7.07	6	6.93

2b. No; you cannot add numbers under different radicals.

Simplify each expression.

26. $3\sqrt{3}(4\sqrt{27} - \sqrt{3})$ 99
27. $\sqrt{12} + \sqrt{32}$ $2\sqrt{3} + 4\sqrt{2}$
28. $\sqrt{5} + 2\sqrt{50}$ $\sqrt{5} + 10\sqrt{2}$
29. $2\sqrt{2}(-2\sqrt{32} + \sqrt{8})$ −24
30. $3\sqrt{10}(\sqrt{10} + 4\sqrt{5})$ $30 + 60\sqrt{2}$
31. $3\sqrt{2}(2 + \sqrt{6})$ $6\sqrt{2} + 6\sqrt{3}$
32. $\sqrt{18} - 2\sqrt{2}$ $\sqrt{2}$
33. $\sqrt{68} + 17$ $2\sqrt{17} + 17$
34. $-\sqrt{6}(\sqrt{6} - 5)$ $-6 + 5\sqrt{6}$

35. *Open-ended* Make up three sums that are less than or equal to 50. Use the square roots of 2, 3, 5, or 7 and the whole numbers less than 10. For example, $8\sqrt{5} + 9\sqrt{7} \le 50$. Samples: $9\sqrt{2} + 9\sqrt{3}$; $5\sqrt{7} + \sqrt{2}$; $3\sqrt{5} + 9\sqrt{3}$

36. *Choose* Use a calculator, paper and pencil, or estimation to find the value of the numerical expression in the cartoon. about 251

from the cartoon *Bound & Gagged*

37. Bill wrote $\sqrt{24} + \sqrt{48} = 3\sqrt{24} = 6\sqrt{6}$.
    a. Simplify $\sqrt{24} + \sqrt{48}$. $2\sqrt{6} + 4\sqrt{3}$
    b. *Critical Thinking* What error did Bill make? Explain. Bill simplified $\sqrt{48}$ as $2\sqrt{24}$ instead of $2\sqrt{12}$ or $4\sqrt{3}$.

38. *Writing* Explain the errors in the work below.
    a. $\sqrt{5} + \sqrt{11} = \sqrt{16} = 4$ You cannot add numbers under different radicals.
    b. $\sqrt{41} = \sqrt{16 + 25} = \sqrt{16} + \sqrt{25} = 9$ You cannot express the square root of a sum as the sum of the square roots.

Geometry **Find the perimeter of each figure below.**

39.
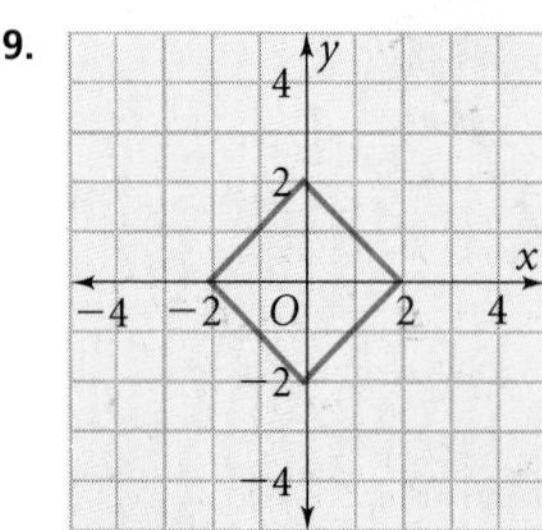

$8\sqrt{2}$

40.
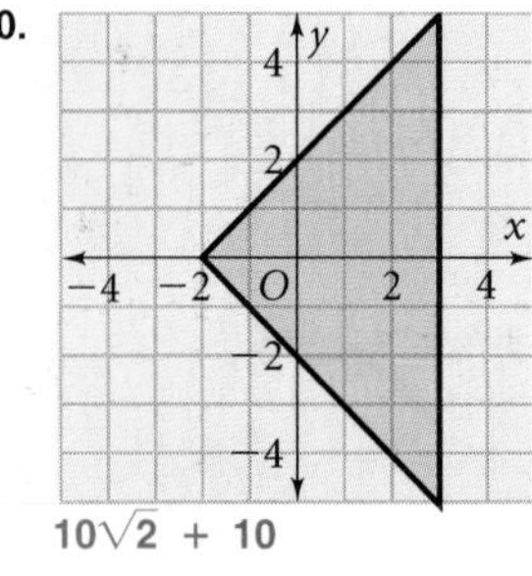

$10\sqrt{2} + 10$

41.
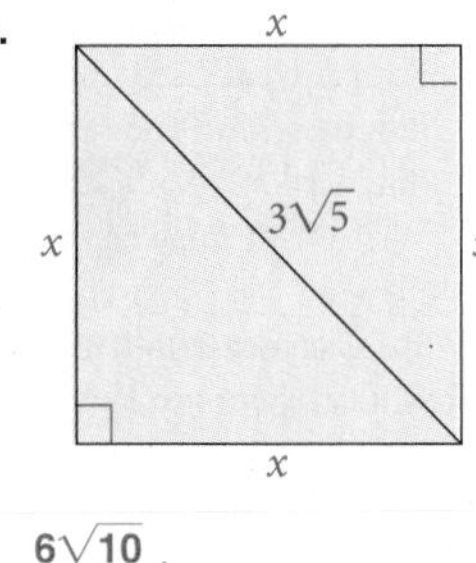

$6\sqrt{10}$

42. *Architecture* The ratio width : height of the front face of a building is equal to the *golden ratio* $(1 + \sqrt{5}) : 2$. The height of the front face of the building is 24 ft. Find the width of the building. Express your answer in simplest radical form. Then estimate in feet. $12 + 12\sqrt{5}$; about 39 ft

Algebra at Work

For further information about the comprehensive national program that is the recognized leader in automotive technician testing and certification, write to the following address:

- ASE National Institute for Automotive Service Excellence
  13505 Dulles Technology Drive
  Herndon, VA 22071-3415
  Tel: (703) 713-3800

or link to various websites that are of interest to auto mechanics through the Valvoline Motor Oil Homepage http://149.55.30.56, Valvoline's Mechanic's Only Homepage http://149.55.30.56/mech_only/.

Students learn from discussions with experts. Here are some possible sources.

- Invite an instructor of auto mechanics from a local community college to speak to the class.
- Plan a field trip to an automobile dealership.

Encourage students to investigate these topics:

- the training necessary to become an auto mechanic.
- how an auto mechanic uses math on the job: selecting appropriate tools, interpreting computer diagnostics, evaluating engine performance

43. *Standardized Test Prep* Simplify $\frac{\sqrt{50} + \sqrt{32}}{\sqrt{2}}$. E

A. $\sqrt{41}$ **B.** $5 + 4\sqrt{2}$ **C.** $4 + 5\sqrt{2}$ **D.** $9\sqrt{2}$ **E.** 9

Exercises MIXED REVIEW

44. **a.** Travel Aruba is changing $\frac{1}{4}$ of its 75 mi^2 into protected parkland. How many square miles will the protected area be? $18\frac{3}{4}$ mi^2
b. The island has 88,000 residents. What is its population density (people/mi^2) for the whole island? about 1173 people/mi^2
c. What is the population density excluding the protected parkland? about 1564 people/mi^2

Aruba

Solve using the quadratic formula. Tell whether each solution is *rational* or *irrational*.

45. $x^2 - 2x - 15 = 0$ −3, 5; rat.

46. $x^2 + 6x + 2 = 0$ −5.65, −0.35; irrat.

Getting Ready for Lesson 9-6

Evaluate each expression for the given value.

47. $\sqrt{x} - 3$ for $x = 16$ 1

48. $\sqrt{x} + 7$ for $x = 9$ 4

49. $2\sqrt{x} + 3 - 4$ for $x = 1$ 0

Algebra at Work

Auto Mechanic

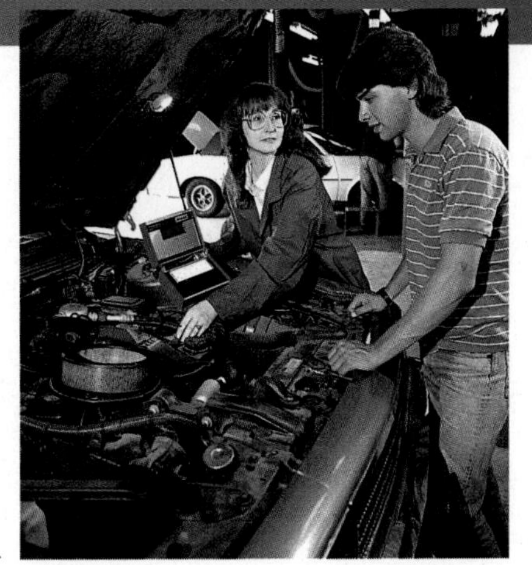

An auto mechanic's job is to see that car engines get the most out of every gallon of gas. Formulas used by mechanics often involve radicals. For example, a car gets its power when gas and air in each cylinder are compressed and ignited by a spark plug. An engine's efficiency (e) is given by the formula $e = \frac{c - \sqrt{c}}{c}$, where c is the compression ratio.

Because of the complexity of such formulas and of modern high-performance engines, today's auto mechanic must be a highly trained and educated professional who understands algebra, graph reading, and the operation of computerized equipment.

Mini Project: Evaluate the formula to find the engine efficiency for even-numbered compression ratios from 2 to 20.

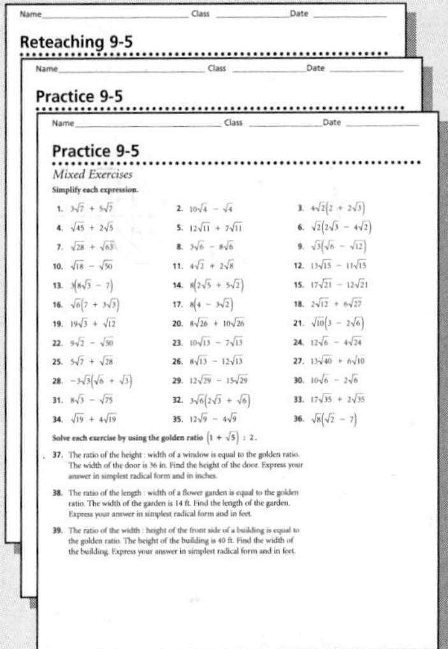

Name ______ Class ______ Date ______

Reteaching 9-5

Name ______ Class ______ Date ______

Practice 9-5

Name ______ Class ______ Date ______

Practice 9-5

Mixed Exercises

Simplify each expression.

1. $3\sqrt{7} + 5\sqrt{7}$
2. $10\sqrt{4} - \sqrt{4}$
3. $4\sqrt{2}(2 + 2\sqrt{3})$
4. $\sqrt{45} + 2\sqrt{5}$
5. $12\sqrt{11} + 7\sqrt{11}$
6. $\sqrt{2}(2\sqrt{3} - 4\sqrt{2})$
7. $\sqrt{28} + \sqrt{63}$
8. $3\sqrt{6} - 8\sqrt{6}$
9. $\sqrt{3}(\sqrt{6} - \sqrt{12})$
10. $\sqrt{18} - \sqrt{50}$
11. $4\sqrt{2} + 2\sqrt{8}$
12. $13\sqrt{15} - 11\sqrt{15}$
13. $3(8\sqrt{3} - 7)$
14. $8(2\sqrt{5} + 5\sqrt{2})$
15. $17\sqrt{21} - 12\sqrt{21}$
16. $\sqrt{6}(7 + 3\sqrt{3})$
17. $8(4 - 3\sqrt{2})$
18. $2\sqrt{12} + 6\sqrt{27}$
19. $19\sqrt{3} + \sqrt{12}$
20. $8\sqrt{26} + 10\sqrt{26}$
21. $\sqrt{10}(3 - 2\sqrt{6})$
22. $9\sqrt{2} - \sqrt{50}$
23. $10\sqrt{13} - 7\sqrt{13}$
24. $12\sqrt{6} - 4\sqrt{24}$
25. $5\sqrt{7} + \sqrt{28}$
26. $8\sqrt{13} - 12\sqrt{13}$
27. $13\sqrt{40} + 6\sqrt{10}$
28. $-3\sqrt{3}(\sqrt{6} + \sqrt{3})$
29. $12\sqrt{29} - 15\sqrt{29}$
30. $10\sqrt{6} - 2\sqrt{6}$
31. $8\sqrt{3} - \sqrt{75}$
32. $3\sqrt{6}(2\sqrt{3} + \sqrt{6})$
33. $17\sqrt{35} + 2\sqrt{35}$
34. $\sqrt{19} + 4\sqrt{19}$
35. $12\sqrt{9} - 4\sqrt{9}$
36. $\sqrt{8}(\sqrt{2} - 7)$

Solve each exercise by using the golden ratio $(1 + \sqrt{5}) : 2$.

37. The ratio of the height : width of a window is equal to the golden ratio. The width of the door is 36 in. Find the height of the door. Express your answer in simplest radical form and in inches.

38. The ratio of the length : width of a flower garden is equal to the golden ratio. The width of the garden is 14 ft. Find the length of the garden. Express your answer in simplest radical form and in feet.

39. The ratio of the width : height of the front side of a building is equal to the golden ratio. The height of the building is 40 ft. Find the width of the building. Express your answer in simplest radical form and in feet.

Lesson Quiz

Lesson Quiz is also available in Transparencies.

Simplify each expression.

1. $12\sqrt{16} - 2\sqrt{16}$ 40
2. $\sqrt{2}(\sqrt{2} + 3\sqrt{3})$ $2 + 3\sqrt{6}$
3. $\sqrt{3}(\sqrt{12} - 2)$ $6 - 2\sqrt{3}$
4. $\sqrt{20} - 2\sqrt{5}$ 0

9-6 Teaching Notes

PROBLEM OF THE DAY

A farmer needs 325 ft of burlap to make bags for grain. His foreman bought 125 yd of burlap for the project. How many extra feet were bought? **50 ft**

Problem of the Day is also available in Transparencies.

CONNECTING TO PRIOR KNOWLEDGE Write $x^2 - 3 = 6$ on the board. Ask students: *How do you solve this?* **Add 3 to both sides, then take the square root of both sides; $x = +3$ or -3.** Write $\sqrt{x^2 - 3} = \sqrt{6}$ on the board. Ask students: *Why isn't this the right way to solve the equation?* **Answers may vary. Sample: You cannot simplify the left side.** Lead students to understand that solving equations containing radicals uses the same processes as solving equations containing exponents: isolate the radical that contains the variable; then square both sides to eliminate the radical sign.

THINK AND DISCUSS

Point out that x cannot be negative because the square roots of negative numbers—*imaginary* numbers—are not real numbers.

Example 1

ESTIMATION Question 1c Lead students to understand that if $\sqrt{64} - 3 = 5$, then $\sqrt{65} - 3$ will be slightly more than

Lesson Planning Options

Prerequisite Skills

- Solving equations
- Calculating squares

Assignment Options for Exercises On Your Own

To provide flexible scheduling, this lesson can be subdivided into parts.

Core 1–6, 9–17
Extension 7, 8, 18–20

Core 21–29, 32–37
Extension 30, 31

Use Mixed Review to maintain skills.

Resources

Student Edition

Skills Handbook, pp. 576, 582
Extra Practice, p. 564
Glossary/Study Guide

Teaching Resources

Chapter Support File, Ch. 9
- Practice 9-6 (two worksheets)
- Reteaching 9-6
- Checkpoint

Classroom Manager 9-6
Glossary, Spanish Resources
Two-Year Algebra Handbook 9-6

Transparencies

11, 107

What You'll Learn

- Solving equations that contain radicals
- Identifying extraneous solutions

...And Why

To design amusement park rides

9-6 Solving Radical Equations

THINK AND DISCUSS

Part 

Solving a Radical Equation

An equation that has a variable under a radical is a **radical equation.** You can often solve a radical equation by squaring both sides. To do this, first get the radical by itself on one side of the equation. Remember that the expression under the radical must be positive.

When $x \geq 0$, $(\sqrt{x})^2 = x$.

Example 1

Solve $\sqrt{x} - 3 = 4$. Check your solution.

$\sqrt{x} - 3 = 4$
$\sqrt{x} - 3 + 3 = 4 + 3$ ← Add 3 to both sides.
$\sqrt{x} = 7$ ← Simplify.
$(\sqrt{x})^2 = 7^2$ ← Square both sides.
$x = 49$

Check
$\sqrt{x} - 3 = 4$
$\sqrt{49} - 3 \stackrel{?}{=} 4$ ← Substitute 49 for x.
$7 - 3 = 4$ ✔

The solution of $\sqrt{x} - 3 = 4$ is 49.

1. a. Find the values of $\sqrt{36} - 3$ and $\sqrt{64} - 3$. **3; 5**
   b. *Critical Thinking* Do the results in part (a) suggest that $x = 49$ is a reasonable solution to $\sqrt{x} - 3 = 4$? Explain.
   c. Explain how parts (a) and (b) are a form of estimation.
   b–c. See margin.
2. *Try This* Solve and check.
   a. $\sqrt{x} + 5 = 12$ **49** b. $2\sqrt{x} + 7 = 19$ **36** c. $\sqrt{x + 3} = 5$ **2**
3. a. Solve $x^2 - 3 = 4$. **2.65, −2.65**
   b. Compare and contrast how you solved the equation in part (a) with how you solved $\sqrt{x} - 3 = 4$. **See left.**

3b. In each solution the first step is to add 3 to each side. Then you compute a square in one equation and a square root in the other.

4. a. *Graphing Calculator* Graph $y = \sqrt{x} - 3$ and $y = 4$ together. Use the range shown on the screen at the left. **See margin.**
   b. How many intersection points are there? **1**
   c. What are the coordinates of the intersection point(s)? **(49, 4)**
   d. *Critical Thinking* How does your answer to part (c) confirm the solution of $\sqrt{x} - 3 = 4$ in Example 1? **The x-coordinate of the intersection point is the solution of the equation.**

```
WINDOW FORMAT
Xmin=0
Xmax=64
Xscl=1
Ymin=-10
Ymax=10
Yscl=1
```

5. Have students estimate $\sqrt{65} - 3$ and then check their estimates with a calculator. 5.06

Question 3 Point out that, because of the exponent, students can expect two possible solutions.

CRITICAL THINKING Question 4d Ask students: *If you graphed the two equations, and found that there was no point of intersection, what would that tell you?* **The corresponding equation has no solution.**

Example 2 Relating to the Real World

Students can assume that the roller coaster car has a velocity $v = 0$ at the top of the loop.

DIVERSITY Some students will not have ridden a roller coaster or even know what one looks like. Try to bring pictures or information about roller coasters to class. Students may not know that roller coasters rely on the acceleration of gravity to provide most of their speed. Invite students to speculate about how roller coasters work and why the height of the first drop is important. Encourage them to sketch the side view of a typical roller coaster and then decide where the roller coaster requires electrical power to move upward.

TACTILE LEARNING Challenge interested students to build a model roller coaster. Have them build tracks wide enough for a marble using heavy construction paper and tape. Encourage them to share what they learn with the class. Have them explain how high a drop should be to provide the marble enough speed but not so high that the marble flies off the tracks.

MAKING CONNECTIONS The longest vertical drop of a roller coaster is 225 ft, producing a speed of 80 mi/h. This roller

Example 2 Relating to the Real World 

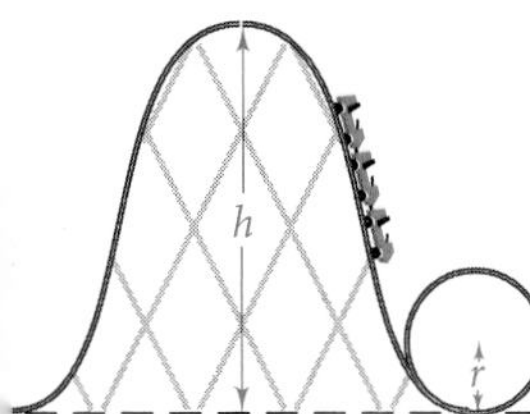

Amusement Parks On a roller coaster ride, your speed in a loop depends on the height of the hill you have just come down and the radius of the loop in feet. The equation $v = 8\sqrt{h - 2r}$ gives the velocity v in feet per second of a car at the top of the loop. Suppose the loop has a radius of 18 ft. You want the car to have a velocity of 30 ft/s at the top of the loop. How high should the hill be?

Solve $v = 8\sqrt{h - 2r}$ for $v = 30$ and $r = 18$.

$$30 = 8\sqrt{h - 2(18)} \quad \leftarrow \text{Substitute for } r \text{ and } v.$$
$$\frac{30}{8} = \sqrt{h - 2(18)} \quad \leftarrow \text{Divide each side by 8 to get the radical alone.}$$
$$3.75 = \sqrt{h - 36}$$
$$(3.75)^2 = (\sqrt{h - 36})^2 \quad \leftarrow \text{Square both sides.}$$
$$14.0625 = h - 36$$
$$50.0625 = h$$

The hill should be about 50 ft high.

5. **Try This** Find the height of the hill when the velocity at the top of the loop is 35 ft/s and radius of the loop is 24 ft. **about 67.14 ft**

6. **Try This** Find the radius of the loop when the hill is 150 ft high and the velocity of the car is 30 ft/s. **about 67.97 ft**

7. About how many miles per hour is 30 ft/s? (*Hint:* 1 mi = 5280 ft) **about 20.5 mi/h**

8. **Critical Thinking** Would you expect the velocity of the car to increase or decrease in each situation? Explain your reasoning.
   a. as the radius of the loop increases **Decrease; $h-2r$ decreases as r increases.**
   b. as the height of the hill decreases **Decrease; $h-2r$ decreases as h decreases.**

Additional Examples

FOR EXAMPLE 1

Solve $\sqrt{x} + 2 = 3$. **1**

FOR EXAMPLE 2

How high must the hill in Example 2 be if the roller coaster car is to have a velocity of 120 ft/s at the top of the loop? Note: r will still be 18 ft, but v will now be 120 ft/s. **261 ft**

Discussion: *Why does the height of the hill have to be more than doubled to produce a doubling of the speed?*

FOR EXAMPLE 3

Solve $\sqrt{2x + 3} = \sqrt{3x - 4}$. **7**

FOR EXAMPLE 4

Solve $x = \sqrt{x + 4}$. $\frac{1 \pm \sqrt{17}}{2}$

coaster—*Steel Phantom*—opened in 1991 at Kennywood Amusement Park in West Mifflin, PA.

CRITICAL THINKING Question 8 For students having difficulty answering these questions, suggest that they imagine themselves on roller coasters with differing dimensions.

Example 3

After students have worked through this example, write $\sqrt{x+4} = \sqrt{x+5}$ on the board. Ask students to make conjectures about the solution of this equation without actually solving it. Lead them to recognize that this equation will have no solution. Help students to understand that they can sometimes save time and effort by thinking about a problem before solving it.

Example 4

ERROR ALERT! Students may square $(x + 6)$ instead of $\sqrt{x+6}$. **Remediation:** Because of the exponent outside the parentheses, they may be expecting to see an x^2 term. On the board, write $(\sqrt{x})^2 = x$ and $(\sqrt{3})^2 = 3$. Help students understand that when they square a square root, they will not get an answer that is squared.

AUDITORY LEARNING Question 11 Ask students to discuss this question and then make a broader generalization. Lead them to understand that if x is the square root of any expression, x cannot be a negative number. Any negative number squared is a positive number. Have students repeat the generalization out loud.

Technology Options

For Exercises 1–6, 9–17, and 21–29, students may find it helpful to check their answers by graphing the equations on a graphing calculator.

Prentice Hall Technology

Calculator
- Graphing Calculator Handbook: Procedure 7

Software
- Secondary Math Lab Toolkit™
- Computer Item Generator 9-6

Internet
- See the Prentice Hall site. (http://www.phschool.com)

Squaring both sides of an equation also works when each side of the equation is a radical expression.

Example 3

Solve $\sqrt{3x-2} = \sqrt{x+6}$.

$$(\sqrt{3x-2})^2 = (\sqrt{x+6})^2 \quad \leftarrow \text{Square both sides.}$$
$$3x - 2 = x + 6$$
$$3x = x + 8$$
$$2x = 8$$
$$x = 4$$

The solution is 4.

PROBLEM SOLVING

Look Back Show how to check that 4 is the solution of $\sqrt{3x-2} = \sqrt{x+6}$.

$\sqrt{3(4)-2} \stackrel{?}{=} \sqrt{4+6}$
$\sqrt{10} = \sqrt{10}$ ✔

9. Graphing Calculator Graph the equations $y = \sqrt{3x-2}$ and $y = \sqrt{x+6}$ together. How does the display confirm the solution of Example 3? See margin.

10. Try This Solve $\sqrt{5x-6} = \sqrt{3x+5}$. Check your solution. 5.5

Part 2 Solving Equations with Extraneous Solutions

When you solve equations by squaring both sides, you sometimes find two possible solutions. You need to determine which solution actually satisfies the original equation.

Example 4

Solve $x = \sqrt{x+6}$.

$$(x)^2 = (\sqrt{x+6})^2 \quad \leftarrow \text{Square both sides.}$$
$$x^2 = x + 6$$
$$x^2 - x - 6 = 0 \quad \leftarrow \text{Subtract } x \text{ and 6 from both sides.}$$
$$x = \frac{-(-1) \pm \sqrt{(-1)^2 - 4(1)(-6)}}{2(1)} \quad \leftarrow \text{Use the quadratic formula to solve for } x.$$
$$x = \frac{1 \pm \sqrt{1-(-24)}}{2}$$
$$x = \frac{1 \pm \sqrt{25}}{2}$$
$$x = \frac{1 \pm 5}{2} \quad \leftarrow \text{Simplify } \sqrt{25}.$$
$$x = \frac{1+5}{2} \text{ or } x = \frac{1-5}{2}$$
$$x = 3 \text{ or } x = -2$$

QUICK REVIEW

The quadratic formula states that for an equation of the form $ax^2 + bx + c = 0$, if $a \neq 0$, then $x = \frac{-b \pm \sqrt{b^2 - 4ac}}{2a}$.

Check $x = \sqrt{x+6}$

$3 \stackrel{?}{=} \sqrt{3+6}$ $\qquad$ $-2 \stackrel{?}{=} \sqrt{-2+6}$ $\leftarrow \sqrt{4} = 2$

$3 = 3$ ✔ $\qquad$ $-2 \neq 2$

The only solution is 3.

ERROR ALERT! At the end of the lesson, write $x = \sqrt{9} + 4$ on the board. Ask students how they would solve this equation. Expect some students to automatically reply, "You isolate the square root on one side of the equation and then square both sides." **Remediation:** Suggest that students review the earlier lessons of the chapter. Point out that they need only use the "isolate and square both sides" strategy when there is a variable inside the radical. Provide extra practice if students need help deciding when to use this strategy.

Exercises ON YOUR OWN

GEOMETRY **Exercise 7** Encourage students to draw the diagram carefully, using the measurements provided to check their calculations.

STANDARDIZED TEST TIP **Exercise 18** Since the square root of an expression must be positive, answer choice D is the only possible choice.

CONNECTING TO THE STUDENTS' WORLD Exercise 19 Encourage students to measure the height and radius of cans at home. Have students bring their data to class. Students might wish to display cans of different sizes in the classroom and to label each can with its dimension and volume.

WRITING Exercise 20 Students can assume that this question refers to radicals that contain variables. Radicals that do *not* contain variables do not need to be isolated.

ALTERNATIVE ASSESSMENT **Exercise 20** To assess understanding of solving radical expressions, have students give an example of getting the radical alone on one side of the

The value -2 is called an **extraneous solution** of Example 4. It is a solution of the derived equation ($x^2 = x + 6$), but not of the original equation ($x = \sqrt{x + 6}$).

11. How could you have determined that -2 was not a solution of $x = \sqrt{x + 6}$ without going through all the steps of the check? **The expression on the right is nonnegative because it is a radical expression. So x cannot be negative.**

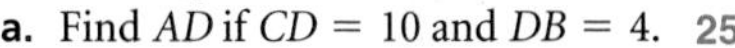

Exercises ON YOUR OWN

Solve each radical equation. Check your solutions.

1. $\sqrt{5x + 10} = 5$ 3
2. $7 = \sqrt{x + 5}$ 44
3. $6 - \sqrt{3x} = -3$ 27
4. $\sqrt{n} = \frac{7}{8}$ $\frac{49}{64}$
5. $20 = \sqrt{x} - 5$ 625
6. $\sqrt{x - 10} = 1$ 11

7. *Geometry* In the right triangle $\triangle ABC$, the altitude $\overline{CD}$ is at a right angle to the hypotenuse. You can use $CD = \sqrt{(AD)(DB)}$ to find certain lengths.
   a. Find AD if $CD = 10$ and $DB = 4$. 25
   b. Find DB if $AD = 20$ and $CD = 15$. $11\frac{1}{4}$

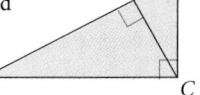

8. *Physics* The equation $T = \sqrt{\frac{2\pi^2 r}{F}}$ gives the time T in seconds it takes a body with mass 0.5 kg to complete one orbit of radius r meters. The force F in newtons pulls the body toward the center of the orbit.
   a. It takes 2 s for an object to make one revolution with a force of 10 N (newtons). Find the radius of the orbit. about 2 m
   b. Find the radius of the orbit if the force is 160 N and $T = 2$. about 32.4 m

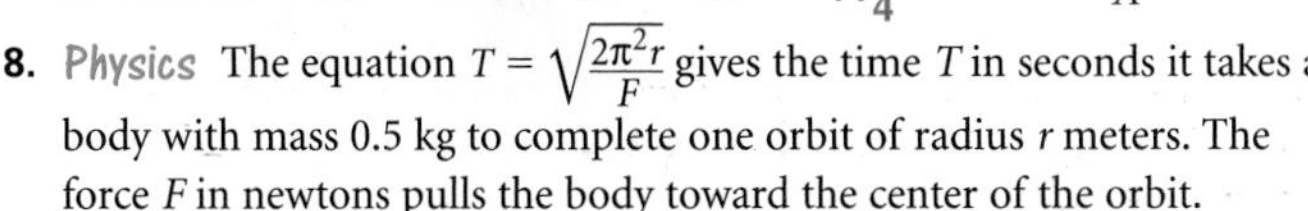

Solve each radical equation. Check your solutions.

9. $\sqrt{3x + 1} = \sqrt{x}$ no solution
10. $\sqrt{x + 5} = \sqrt{3x + 6}$ $-\frac{1}{2}$
11. $\sqrt{3x + 10} = \sqrt{9 - x}$ $-\frac{1}{4}$
12. $\sqrt{7x + 5} = \sqrt{x - 3}$ no solution
13. $\frac{x}{2} = \sqrt{3x}$ 0, 12
14. $\sqrt{x + 12} = 3\sqrt{x}$ $\frac{3}{2}$
15. $\sqrt{5x - 4} = \sqrt{3x + 10}$ 7
16. $\sqrt{2x} = \sqrt{9 - x}$ 3
17. $x = \sqrt{2x + 3}$ 3

18. *Standardized Test Prep* Which of the following radical equations has no *real* solution? D
   A. $-\sqrt{x} = -25x$
   B. $\sqrt{2x + 1} = \sqrt{3x - 5}$
   C. $-3\sqrt{3x} = -5$
   D. $\sqrt{3x + 1} = -10$
   E. $\frac{x}{2} = \sqrt{x - 1}$

19. *Packaging* The volume V in cubic units of a cylindrical can is given by the formula $V = \pi r^2 h$, where r is the radius of the can and h is its height. The radius of a can is 2.5 in. and the height of the can is 5 in. Find the can's volume in cubic inches. about 98.2 in.^{3}

20. *Writing* Suppose an equation has only one radical. Why must you get the radical alone on one side of an equation before squaring both sides? Support your answer by giving an example. **If you don't, the new expression also will have a radical. Sample: $(x + \sqrt{x - 1})^2 = x^2 + x - 1 + 2x\sqrt{x - 1}$**

Storage tanks in Knoxville, Tennessee

pages 440–443 Think and Discuss

9.

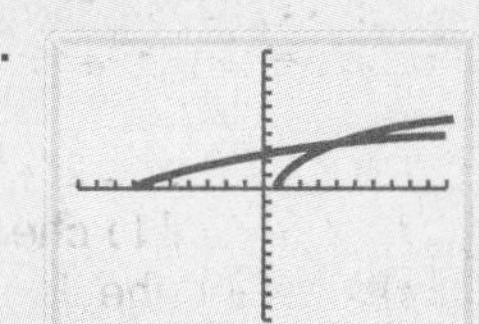

The graphs intersect at $x = 4$.

1b. Yes; the solution is between 36 and 64.

c. The values of $\sqrt{36} - 3$ and of $\sqrt{64} - 3$ are colose to 4, so you can estimate that the solution is between 36 and 64.

4a.

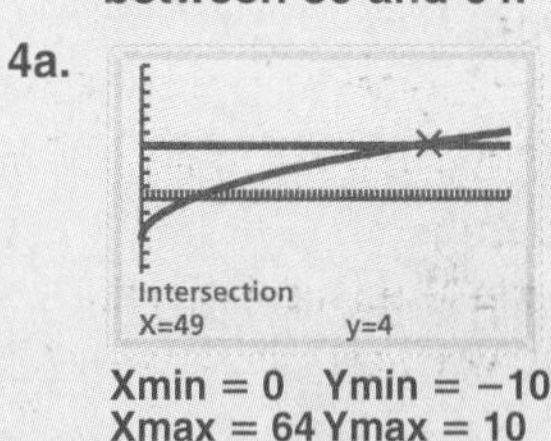

Xmin = 0 Ymin = −10
Xmax = 64 Ymax = 10
Xscl = 1 Yscl = 1

9.

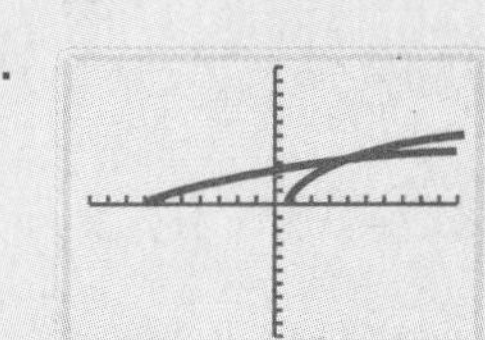

The graphs intersect at $x = 4$.

equation before squaring both sides and an example of squaring first.

TACTILE LEARNING Exercise 31 Some students may not recognize that the diagram represents an unfolded box. Encourage students to find a box and unfold it, or to copy the diagram and cut and fold it to make their own box.

Chapter Project **FIND OUT BY COMMUNICATING** Students use the given formula to answer questions. Check students' progress by having them write a short report telling what they have done so far and what their plans are for completing the project.

Exercises MIXED REVIEW

Exercise 42 Point out that the 8% commission is earned on sales, not on the $750 base pay.

GETTING READY FOR LESSON 9-7 These exercises prepare students for graphing square root functions.

Wrap Up

THE BIG IDEA Ask students: *What are radical equations? How do you solve radical equations?*

RETEACHING ACTIVITY Students use index cards or similar size pieces of paper to show steps for solving radical equations. (Reteaching worksheet 9-6)

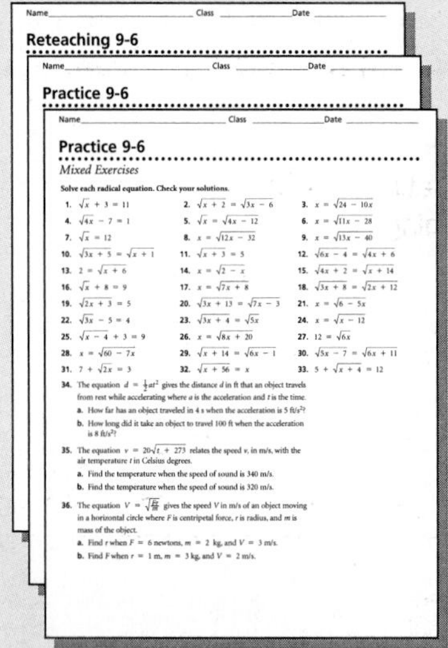

Name______ Class ______ Date ______

Reteaching 9-6

Name______ Class ______ Date ______

Practice 9-6

Name______ Class ______ Date ______

Practice 9-6

Mixed Exercises

Solve each radical equation. Check your solutions.

1. $\sqrt{x} + 3 = 11$	2. $\sqrt{x + 2} = \sqrt{3x - 6}$	3. $x = \sqrt{24 - 10x}$
4. $\sqrt{4x} - 7 = 1$	5. $\sqrt{x} = \sqrt{4x - 12}$	6. $x = \sqrt{11x - 28}$
7. $\sqrt{x} = 12$	8. $x = \sqrt{12x - 32}$	9. $x = \sqrt{13x - 40}$
10. $\sqrt{3x + 5} = \sqrt{x + 1}$	11. $\sqrt{x + 3} = 5$	12. $\sqrt{6x - 4} = \sqrt{4x + 6}$
13. $2 = \sqrt{x + 6}$	14. $x = \sqrt{2 - x}$	15. $\sqrt{4x + 2} = \sqrt{x + 14}$
16. $\sqrt{x} + 8 = 9$	17. $x = \sqrt{7x + 8}$	18. $\sqrt{3x + 8} = \sqrt{2x + 12}$
19. $\sqrt{2x + 3} = 5$	20. $\sqrt{3x + 13} = \sqrt{7x - 3}$	21. $x = \sqrt{6 - 5x}$
22. $\sqrt{3x} - 5 = 4$	23. $\sqrt{3x + 4} = \sqrt{5x}$	24. $x = \sqrt{x - 12}$
25. $\sqrt{x - 4} + 3 = 9$	26. $x = \sqrt{8x + 20}$	27. $12 = \sqrt{6x}$
28. $x = \sqrt{60 - 7x}$	29. $\sqrt{x + 14} = \sqrt{6x - 1}$	30. $\sqrt{5x - 7} = \sqrt{6x + 11}$
31. $7 + \sqrt{2x} = 3$	32. $\sqrt{x + 56} = x$	33. $5 + \sqrt{x + 4} = 12$

34. The equation $d = \frac{1}{2}at^2$ gives the distance d in ft that an object travels from rest while accelerating where a is the acceleration and t is the time.
   a. How far has an object traveled in 4 s when the acceleration is 5 ft/s²?
   b. How long did it take an object to travel 100 ft when the acceleration is 8 ft/s²?

35. The equation $v = 20\sqrt{t + 273}$ relates the speed v, in m/s, with the air temperature t in Celsius degrees.
   a. Find the temperature when the speed of sound is 340 m/s.
   b. Find the temperature when the speed of sound is 320 m/s.

36. The equation $V = \sqrt{\frac{Fr}{m}}$ gives the speed V in m/s of an object moving in a horizontal circle where F is centripetal force, r is radius, and m is mass of the object.
   a. Find r when $F = 6$ newtons, $m = 2$ kg, and $V = 3$ m/s.
   b. Find F when $r = 1$ m, $m = 3$ kg, and $V = 2$ m/s.

Lesson Quiz

Lesson Quiz is also available in Transparencies.

Solve each radical equation. Check your solutions.

1. $\sqrt{7x - 3} = 4$ $2\frac{5}{7}$
2. $\sqrt{3x + 2} = \sqrt{x}$ -1
3. $\sqrt{2x + 7} = \sqrt{5x - 8}$ 5
4. $4x = \sqrt{8x + 3}$ 0.75, −0.25

Solve each radical equation. Check your solutions.

21. $x = \sqrt{x + 2}$ 2
22. $2x = \sqrt{10x + 6}$ 3
23. $\frac{x}{3} = \sqrt{x - 2}$ 3, 6
24. $\sqrt{x + 5} = 2x$ $1\frac{1}{4}$
25. $2x = 2\sqrt{x - 5}$ no solution
26. $\sqrt{2x - 15} = \frac{x}{4}$ 12, 20
27. $\sqrt{x + 12} = x$ 4
28. $\sqrt{2x - 1} = \frac{x + 8}{2}$ no solution
29. $x = \sqrt{7x - 6}$ 1, 6
30. *Writing* Explain in your own words what an extraneous solution is. Answers may vary. Sample: A solution is extraneous when it solves a derived equation but not the original equation.
31. The diagram shows a piece of cardboard that makes a box when sections of it are folded and taped. The ends of the box are x in. long and the box is 10 in. long.

    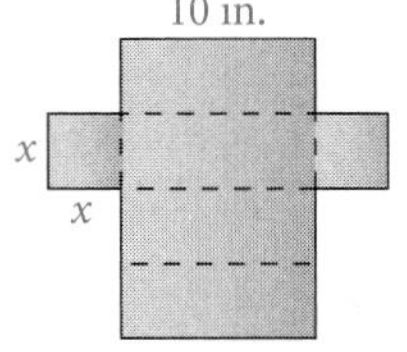

    a. Write the formula for the volume V of the box. $V = 10x^2$

    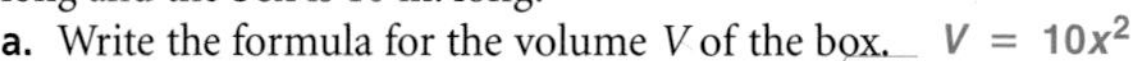

    b. Solve the equation in part (a) for x. $x = \frac{\sqrt{10V}}{10}$
    c. *Open-ended* Find some integer values of x that make the box have a volume between 40 in.³ and 490 in.³. Complete list: 2 in., 3 in., 4 in., 5 in., 6 in., 7 in.

Tell which solution(s), if any, are extraneous for the given equation.

32. $-x = \sqrt{-x + 6}$; $x = -3, x = 2$ 2
33. $\sqrt{12 - x} = x$; $x = -4, x = 3$ −4
34. $x = \sqrt{2x}$; $x = 0, x = 2$ none
35. $2x = \sqrt{4x + 3}$; $x = \frac{3}{2}, x = -\frac{1}{2}$ $-\frac{1}{2}$
36. $x = \sqrt{28 - 3x}$; $x = 4, x = -7$ −7
37. $-x = \sqrt{-6x - 5}$; $x = -5, x = -1$ none

Chapter Project **Find Out by Communicating**

- Solve the formula $d = 1.225\sqrt{h}$ for h where d is in miles and h is in feet. (*Hint:* See page 424.)
- How tall would a person on Earth have to be to see 7 mi to the horizon?
- Could a person possibly see such a distance? Why or why not?

Exercises MIXED REVIEW

Simplify each expression. Assume that no denominator is equal to zero.

38. $\frac{x^2y^7}{x^4y^5}$ $\frac{y^2}{x^2}$
39. $\frac{a^2b^5}{a^7b^{10}}$ $\frac{1}{a^5b^5}$
40. $\frac{15x^4y^5}{45x^2y^3}$ $\frac{x^2y^2}{3}$
41. $\frac{14x^2y^3}{7x^3y^5}$ $\frac{2}{xy^2}$
42. *Business* A salesperson earns $750 per month plus 8% commission on the amount she sells. Write an equation to show how her monthly income relates to her sales in dollars. $m = 750 + 0.08s$

Getting Ready for Lesson 9-7

Find the value of y for the given value of x.

43. $y = \sqrt{x + 7} - 3$ for $x = 2$ 0
44. $y = 3\sqrt{x} + 2$ for $x = 9$ 11
45. $y = -2\sqrt{x - 1} + 2$ for $x = 1$ 2

9-7 Teaching Notes

PROBLEM OF THE DAY

The ratio of girls to boys at Canyon Vista High is 5 to 4 for every class. The freshman and sophomore classes have 235 and 195 girls, respectively. How many fewer boys are in the sophomore class than in the freshman class? **32 boys**

Problem of the Day is also available in Transparencies.

CONNECTING TO PRIOR KNOWLEDGE Ask students the following questions:

- *What does the graph of $y = x^2$ look like?* **It is U-shaped.** *How does it differ from the graph of $y = x$?* **The graph of $y = x$ is linear.**
- *What do you think the graph of $y = \sqrt{x}$ looks like?* **The graph is the top half of a sideways U that opens to the right.**

THINK AND DISCUSS

ESL Ask a student to explain what the *nozzle* of a hose is and where on the hose the nozzle is located. To understand how the expression *nozzle pressure* is being used, students should know a *nozzle* is the attachment at the end of the hose that directs the flow of water.

EXTENSION Encourage interested students to find out more about how and why the flow rate of water is important to firefighters. Have students research the different kinds of hoses and nozzles that firefighters use for different tasks. Ask students to share their findings with the class.

Connections Firefighting . . . and more

What You'll Learn

- Graphing and exploring square root functions
- Solving real-world problems using square root functions

...And Why

To solve problems involving firefighting

What You'll Need

- graphing calculator

9-7 Graphing Square Root Functions

THINK AND DISCUSS

Firefighters When firefighters are trying to put out a fire, the rate at which they can spray water on the fire is very important to them. For a hose with a 2 in. nozzle diameter, the flow rate, f, in gal/min is given by this formula.

$$f = 120\sqrt{p}, \text{ where } p \text{ is the nozzle pressure in lb/in.}^2$$

The flow-rate function is an example of a square root function. The simplest **square root function** is $y = \sqrt{x}$.

x	y
0	0
1	1
4	2
9	3
16	4

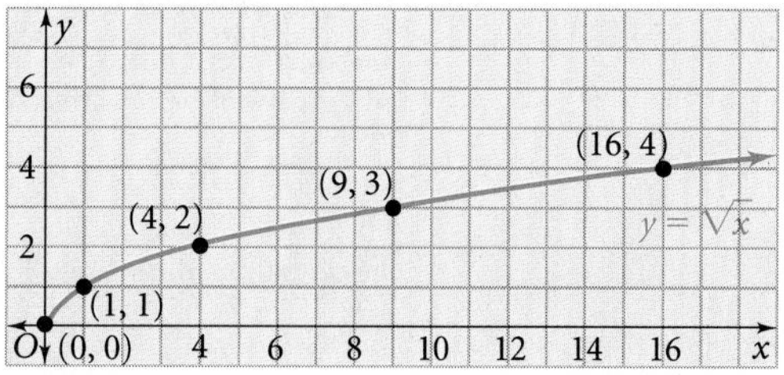

1. Why do you think the x-values in the table were chosen? **The x-values were chosen to find points with integer coordinates.**
2. Find an approximate value of $\sqrt{3}$ and see if it seems to fit the graph. **about 1.7; yes**
3. Why is the graph not continued to the left of the y-axis? **You cannot take the square root of a negative number.**

Lesson Planning Options

Prerequisite Skills

- Evaluating radical expressions
- Graphing functions

Assignment Options for Exercises On Your Own

Core 1–9, 11–13, 15–26
Extension 10, 14, 27–31

Use Mixed Review to maintain skills.

Resources

 **Student Edition**

Skills Handbook, pp. 567, 576, 582
Extra Practice, p. 564
Glossary/Study Guide

 **Teaching Resources**

Chapter Support File, Ch. 9
- Practice 9-7 (two worksheets)
- Reteaching 9-7

Checkpoint
Classroom Manager 9-7
Glossary, Spanish Resources
Two-Year Algebra Handbook 9-7

 **Transparencies**
2, 6, 11, 108

VISUAL LEARNING On an overhead transparency, draw the graph of $y = x^2$. Flip the transparency over backward and turn it sideways to produce the graph of $y = \sqrt{x}$. Lead students to understand that square roots of negative numbers are not real numbers, so neither y nor x can ever be negative. Thus, the graph of $y = \sqrt{x}$ is only in the first quadrant. Erase the section of the $y = x^2$ graph that is in the fourth quadrant.

Example 1

ERROR ALERT! Students may think that all equations containing square roots have graphs only in the first quadrant. **Remediation:** Have students make a table of values for $y = \sqrt{x}$ and $y = \sqrt{x} - 1$.

Example 2

Help students understand how the domain of a function containing a square root can have negative values. Encourage students to substitute the values –3, –2, and –1 for x into the equation $y = \sqrt{x + 3}$ to explore what y-values they get.

Example 3 **Relating to the Real World** 

MAKING CONNECTIONS The Oshkosh firetruck, which is used for aircraft and runway fires, can discharge 50,200 gal of foam through its two turrets in exactly 2.5 min. Encourage students to convert this flow rate to gal/min and compare it to the flow rate in Example 3.

Additional Examples

FOR EXAMPLE 1

Analyze how the graph of $y = \sqrt{x} + 2$ compares to the graph of $y = \sqrt{x}$. Use a graphing calculator.

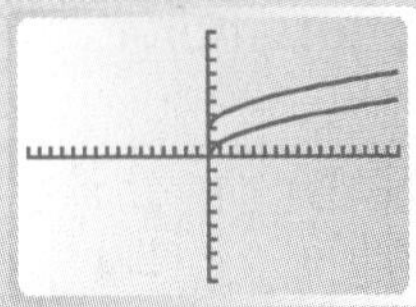

The graph of $y = \sqrt{x} + 2$ is two units higher than the graph of $y = \sqrt{x}$.

Discussion: *What would happen if you replaced the 2 with a $\frac{1}{2}$?*

FOR EXAMPLE 2

Find the domain of $y = \sqrt{x + 5}$. Then graph the function. $x \leq -5$

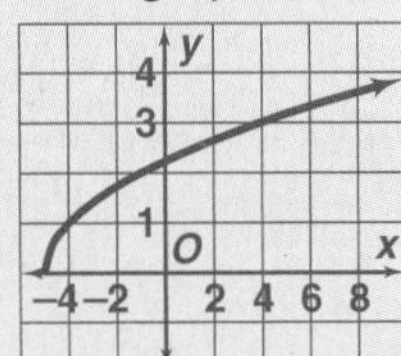

Discussion: *Why can x not have a value less than* 6*?*

446

Example 1

Compare the graph of $y = \sqrt{x} - 1$ to the graph of $y = \sqrt{x}$.

Method 1 Use a table.

x	$\sqrt{x}$	$\sqrt{x} - 1$
0	$\sqrt{0} = 0$	$\sqrt{0} - 1 = -1$
1	$\sqrt{1} = 1$	$\sqrt{1} - 1 = 0$
4	$\sqrt{4} = 2$	$\sqrt{4} - 1 = 1$
9	$\sqrt{9} = 3$	$\sqrt{9} - 1 = 2$
16	$\sqrt{16} = 4$	$\sqrt{16} - 1 = 3$

Method 2 Use a graphing calculator.

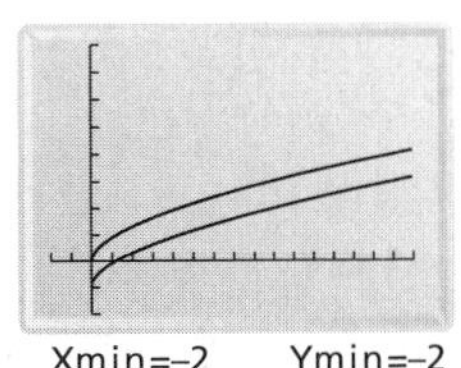

Xmin=–2 Ymin=–2
Xmax=16 Ymax=8
Xscl=1 Yscl=1

For each value of x, the value of $\sqrt{x} - 1$ is one less than the value of $\sqrt{x}$. The graph of $y = \sqrt{x} - 1$ is one unit lower than the graph of $y = \sqrt{x}$.

4. Try This **Analyze** the graph of $y = \sqrt{x} + 1$ by comparing it to the graph of $y = \sqrt{x}$. **The graphs have the same shape. The graph of $y = \sqrt{x} + 1$ is 1 unit above the graph of $y = \sqrt{x}$.**
5. a. **Predict** how the graph of $y = \sqrt{x} + k$ will compare to the graph of $y = \sqrt{x}$ when k is a negative number. **The graph is k units lower.**
   b. Graphing Calculator Confirm your prediction by graphing $y = \sqrt{x} - 3$ and $y = \sqrt{x}$. **See margin.**
6. The calculator display at the left shows the graph of a square root function of the form $y = \sqrt{x} + k$. What is the value of k? **3**

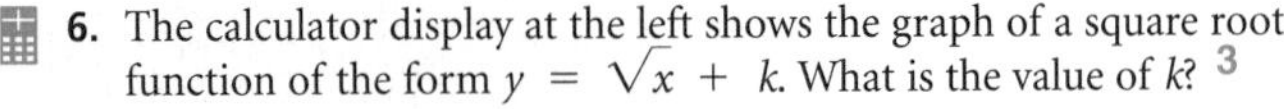

Xmin=–2 Ymin=–2
Xmax=15 Ymax=8
Xscl=1 Yscl=1

You can solve an inequality to find the domain of a square root function.

Example 2

Find the domain of $y = \sqrt{x + 3}$. Then graph the function.

The square root limits the domain because the expression under the radical cannot be negative. To find the domain, solve $x + 3 \geq 0$.

$$x + 3 \geq 0$$
$$x \geq -3$$

The domain is the set of all real numbers greater than or equal to –3.

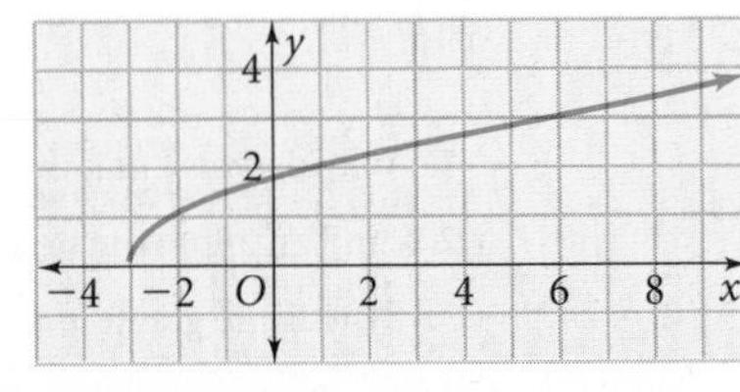

TACTILE LEARNING Question 12 Encourage students to draw and compare the graphs of functions such as $y = 2\sqrt{x}$, $y = 3\sqrt{x}$, $y = \frac{1}{2}\sqrt{x}$, and $y = \frac{1}{4}\sqrt{x}$. Suggest that students draw each graph using a different colored pencil. After they have drawn and studied the graphs, have them make generalizations about how the value of *k* affects the appearance of the graph.

MAKING CONNECTIONS The *period* of a pendulum—the amount of time in seconds that the pendulum takes to swing back and forth once—depends only on the length of the pendulum. The equation for the period of a pendulum is $T = 2\pi\sqrt{\frac{l}{g}}$, where *T* is the period in seconds, *l* is the length of the pendulum in meters, and *g* is the acceleration due to gravity: 9.8 m/s^2. Point out that the period of a pendulum is the same no matter how far it swings back and forth.

KINESTHETIC LEARNING Have students make pendulums out of different lengths of string and erasers. Have students compare their measurements to the values they calculate using the formula. Encourage students to draw the graph of the pendulum function; then, on their graphs, plot the points they confirmed experimentally.

Exercises ON YOUR OWN

ALTERNATIVE ASSESSMENT **Exercises 1–9** Have students graph each of the functions. This will help you assess their understanding of graphing square-root functions.

7. The graph is identical to the graph of $y = \sqrt{x}$ shifted left 3 units.

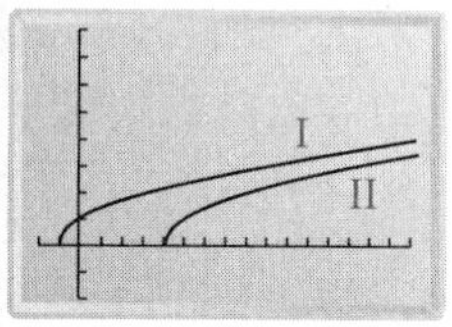

Xmin=-2 Ymin=-2
Xmax=16 Ymax=8
Xscl=1 Yscl=1

7. Compare the graph in Example 2 to the graph of $y = \sqrt{x}$.

8. **Try This** Find the domain of $y = \sqrt{x - 4}$. Then graph the function. $x \geq 4$; see margin for graph

9. a. The diagram at the left shows the graphs of two functions of the form $y = \sqrt{x + k}$. Describe the domain of each function.
   b. Which graph is the graph of $y = \sqrt{x + 1}$? **Justify** your response.
   a. $x \geq -1$; $x \geq 4$ b. I; the function can be computed for any $x \geq -1$.

10. **Summarize** how the graph of $y = \sqrt{x + k}$ compares to the graph of $y = \sqrt{x}$. For positive *k*, the graph of $y = \sqrt{x + k}$ is *k* units to the left. For negative *k*, the graph of $y = \sqrt{x + k}$ is $-k$ units to the right.

You can use square root functions to describe real-life situations.

Example 3 Relating to the Real World

Firefighting Graph the flow-rate function $f = 120\sqrt{p}$ introduced on page 445. Evaluate the function when the nozzle pressure is 40 lb/in.^{2}.

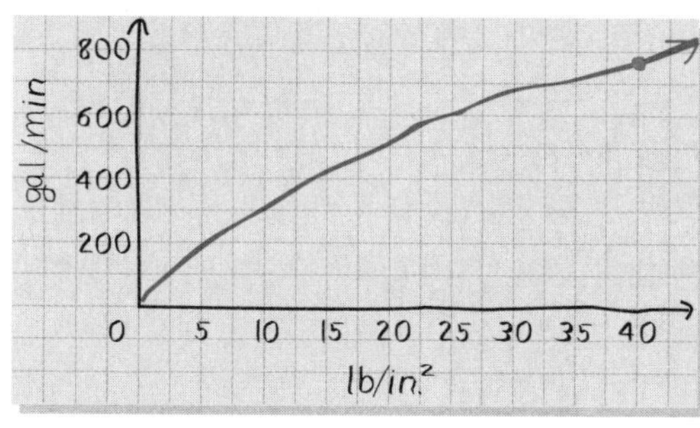

$f = 120\sqrt{40}$ ← Substitute 40 for *p*.
$= 758.9466384$ ← Use a calculator.

The flow rate is about 759 gal/min when the nozzle pressure is 40 lb/in.^{2}.

11. In Example 3, why are there different scales on the axes? The graph would not fit on the page if the scales were the same.

12. Compare the graph in Example 3 to the graph of $y = \sqrt{x}$. The graphs have the same shape, but $f = 120\sqrt{p}$ is wider.

13. **Summarize** how the graph of $y = k\sqrt{x}$ compares to the graph of $y = \sqrt{x}$. The graphs have the same shape, but the graph of $y = k\sqrt{x}$ is wider when $k > 1$ and narrower when $0 < k < 1$.

Exercises ON YOUR OWN

Find the domain of each function.

1. $y = \sqrt{x - 2}$ $x \geq 2$
2. $f(x) = \sqrt{4x + 3}$ $x \geq -\frac{3}{4}$
3. $y = \sqrt{1.5x}$ $x \geq 0$
4. $y = \sqrt{3x + 5}$ $x \geq -\frac{5}{3}$
5. $y = \sqrt{7 + x}$ $x \geq -7$
6. $f(x) = \sqrt{2 + x}$ $x \geq -2$
7. $f(x) = \sqrt{3 + x}$ $x \geq -3$
8. $y = \sqrt{x + 3} - 1$ $x \geq -3$
9. $y = \sqrt{x - 5} + 1$ $x \geq 5$

Technology Options

For Exercises 1–9, students may wish to find the domains by graphing the equations on a graphing calculator.

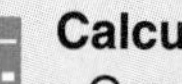

Calculator
- Graphing Calculator Handbook: Procedure 4

Software
- Secondary Math Lab Toolkit™
- Integrated Math Lab 18
- Computer Item Generator 9-7

Internet
- See the Prentice Hall site. (http://www.phschool.com)

Exercises 8–9 Make sure students understand that the numeral 1 that is outside the radical sign will affect the range but not the domain of each function.

STANDARDIZED TEST TIP **Exercise 14** Remind students that the square root of an expression can never have a negative value. Since y is always positive, answer choices A, B, and D can be eliminated since they contain quadrants III and IV.

OPEN-ENDED Exercise 31 Point out that students must use the same value of k for each of the three functions.

Chapter Project FIND OUT BY ORGANIZING Students write sight-distance formulas for each planet. This task is essential to their work on the Chapter Project. Have students add this part of the project to work they have already completed.

Exercises MIXED REVIEW

SELF ASSESSMENT **JOURNAL** As an alternative, suggest that students compare how square-root function graphs are shifted depending on whether the k-value is inside or outside the radical. For example, have students compare how the graph of $y = \sqrt{x + 3}$ is shifted differently than the graph of $y = \sqrt{x} + 3$.

Exercises 32–33 Remind students to rewrite the graphs in slope-intercept form before graphing.

GETTING READY FOR LESSON 9-8 These exercises prepare students for calculating and applying the standard deviation.

pages 445–447 Think and Discuss

5b.

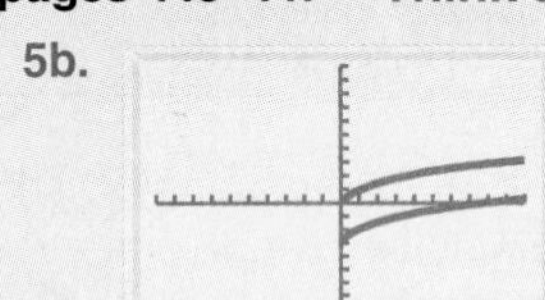

pages 447–449 On Your Own

15.

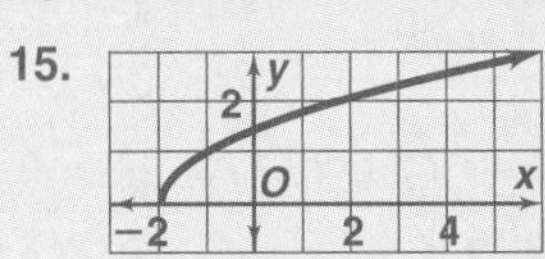

16.

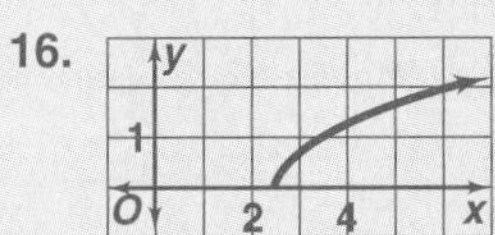

17.

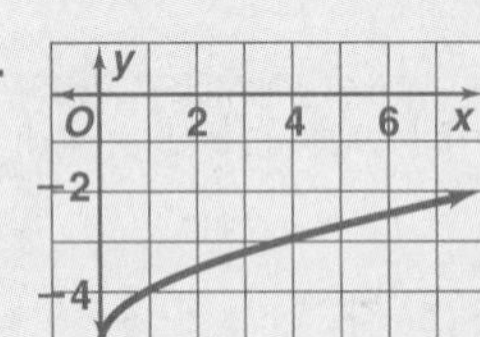

18–26, 31b, 32, 33. See back of book.

page 449 Checkpoint

13–17. See back of book.

10. *Business* Last year a store had an advertising campaign. Sales figures for disposable cameras are shown. The function $n = 27\sqrt{5t} + 53$ models sales volume n of the cameras as a function of time t, the number of months after the start of the campaign.
    a. Evaluate the function to find how many disposable cameras the store sold in the seventh month. about 213
    b. Solve an equation to find the month in which the number of disposable cameras sold was about 175. 4th month

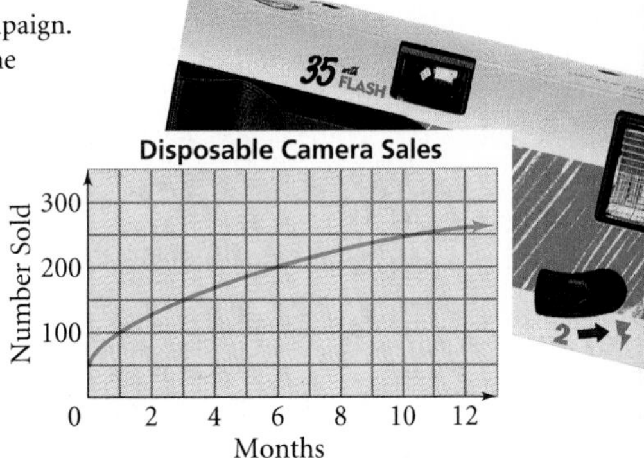

Match each function with its graph.

11. $y = \sqrt{x + 4}$ B
12. $y = \sqrt{x} - 2$ C
13. $y = 3\sqrt{x}$ A

A.

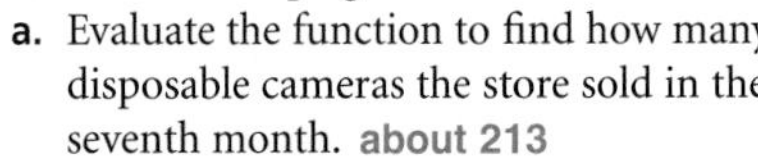

B.

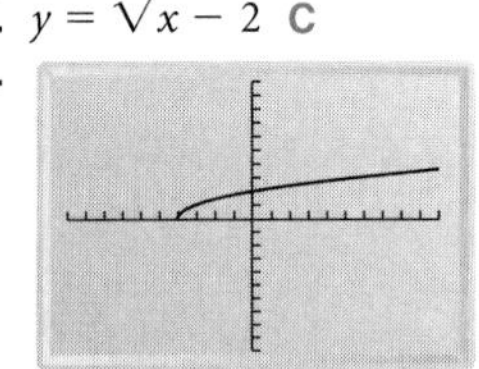

C.

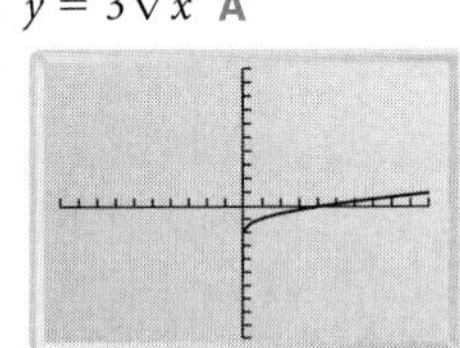

14. *Standardized Test Prep* In which of the following quadrants will your calculator display the graph of $y = \sqrt{x + 7}$? C
    A. I, II, and III B. I and IV C. I D. IV E. I and II

Graph each function. 15–26. See margin.

15. $y = \sqrt{x + 2}$
16. $y = \sqrt{x - 2.5}$
17. $f(x) = \sqrt{x} - 5$
18. $y = 4\sqrt{x}$
19. $f(x) = \sqrt{x} + 2$
20. $y = \sqrt{x + 3}$
21. $y = \sqrt{x - 6}$
22. $f(x) = 5\sqrt{x}$
23. $y = \sqrt{x - 2} + 3$
24. $f(x) = \sqrt{x - 3} - 1$
25. $y = \sqrt{x + 5} + 2$
26. $f(x) = \sqrt{x + 2} - 4$

Writing **Using words like "shift up," "shift down," "shift left," and "shift right," describe to a friend how to use the graph of $y = \sqrt{x}$ to obtain the graph of each function.**

27. $f(x) = \sqrt{x + 8}$ shift left 8
28. $y = \sqrt{x} - 10$ shift down 10
29. $y = \sqrt{x} + 12$ shift up 12
30. $f(x) = \sqrt{x - 9}$ shift right 9

31. a. *Open-ended* Give an example of a square root function in each form. Assume $k > 0$. Sample: $y = \sqrt{x + 2}$, $y = \sqrt{x} + 2$, $y = 2\sqrt{x}$
    $y = \sqrt{x + k}$,
    $y = \sqrt{x} + k$,
    $y = k\sqrt{x}$
    b. Graph each function on the same coordinate grid. See margin.

Wrap Up

THE BIG IDEA Ask students to explain what the graph of the square root function looks like compared to other graphs they have studied.

RETEACHING ACTIVITY Students make a table of values and then graph square root functions on graph paper. (Reteaching worksheet 9-7)

Exercises CHECKPOINT

Students will assess their progress on Lessons 9-4 to 9-7.

Exercise 12 Suggest that students first multiply both sides of the equation by -1.

STANDARDIZED TEST TIP **Exercise 18** Students should be reminded that the formula for the perimeter of a rectangle is $2l + 2w$. Therefore, the answer choices A, B, and C can all be eliminated. Answer choice D is the only choice showing the correct formula.

Chapter Project — Find Out by Organizing

- The formula for the distance you can see on any planet is $d = \sqrt{\frac{rh}{2640}}$, where h is your height in feet and r is the radius of the planet in miles. Use the planet radii you researched on page 429 to write a separate formula for each planet.
- Graph each formula.

Exercises MIXED REVIEW

Solve each system of equations by graphing. 32–33. See margin for graphs.

32. $x + y = 5$
$2x - 3y = -5$ (2, 3)

33. $3x - y = 4$
$4x + 2y = 2$ (1, −1)

34. *Finance* You deposit \$1000 into a bank account that pays 6% compounded annually. How much will be in the bank account at the end of 3 years? \$1191.02

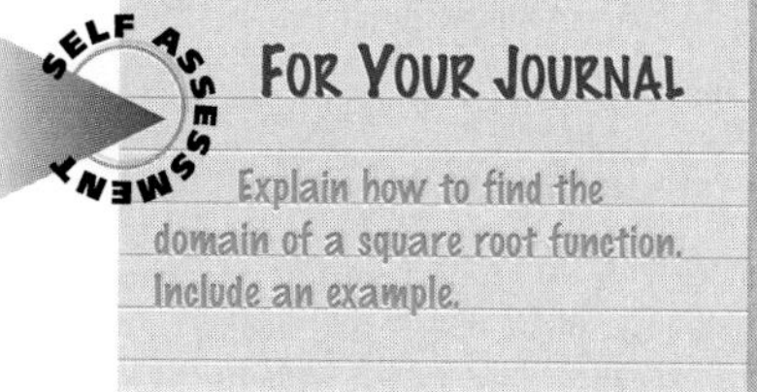

SELF ASSESSMENT
FOR YOUR JOURNAL
Explain how to find the domain of a square root function. Include an example.

Getting Ready for Lesson 9-8

Find the mean of each set of data.

35. 4, 6, 8, 10, 11 7.8
36. 3, 4, 5, 7, 9 5.6
37. −8, −6, 0, 2, 3 −1.8
38. −3, −2, 0, 2, 9 1.2

Exercises CHECKPOINT

Simplify each radical expression.

1. $\sqrt{3} \cdot \sqrt{27}$ 9
2. $\sqrt{7 \cdot 28}$ 14
3. $\sqrt{64b^5}$ $8b^2\sqrt{b}$
4. $\sqrt{18} - \sqrt{8}$ $\sqrt{2}$
5. $\sqrt{3}(\sqrt{3} + 4)$ $3 + 4\sqrt{3}$
6. $\frac{\sqrt{24}}{\sqrt{3}}$ $2\sqrt{2}$
7. $\sqrt{\frac{x^3}{4}}$ $\frac{x\sqrt{x}}{2}$
8. $\frac{6}{\sqrt{3}}$ $2\sqrt{3}$

Solve each radical equation. Check your solutions.

9. $8 = \sqrt{x - 4}$ 68
10. $\sqrt{x} = \sqrt{3x - 12}$ 6
11. $\sqrt{2x - 5} = \sqrt{11}$ 8
12. $7 - \sqrt{2x} = 1$ 18

Graph each function. 13–16. See margin.

13. $y = \sqrt{x} + 4$
14. $f(x) = \sqrt{x - 5}$
15. $f(x) = 2\sqrt{x}$
16. $y = \sqrt{x} - 3$

17. *Writing* How is combining like terms with radicals similar to combining like terms with variables? See margin.

18. *Standardized Test Prep* What is the perimeter of the figure? D

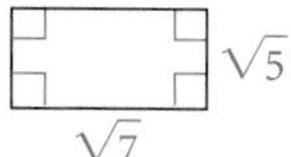

A. $2\sqrt{7} + \sqrt{5}$
B. $\sqrt{5} + \sqrt{7}$
C. $\sqrt{35}$
D. $2\sqrt{7} + 2\sqrt{5}$
E. none of the above

Reteaching 9-7

Practice 9-7

Practice 9-7 *Mixed Exercises*

Lesson Quiz

Lesson Quiz is also available in Transparencies.

1. Find the domain of $f(x) = \sqrt{2x - 4}$. $x \geq 2$

Graph each function.

2. $y = 3\sqrt{x}$.

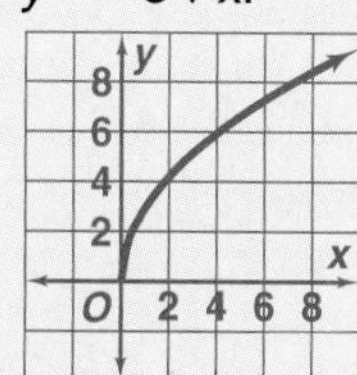

3. $f(x) = \sqrt{x - 2} + 3$
See back of book.

Math ToolboX

Both the NCTM standards and the business community promote an increase in students' abilities to understand statistics. In this toolbox, students learn another graphical method to display data, the box-and-whisker plot. Begin by briefly reviewing the concept of median and introduce quartiles. Then work though the example with students. Emphasize the proper position of the box. Point out to students that the box represents the middle 50% of the data. Rulers and graph paper will help students make more accurate box-and-whisker plots.

ERROR ALERT! Some students mistakenly place the vertical line at the center of the box. **Remediation:** Remind students that the vertical line segment represents the median of the data set.

WRITING Exercise 9 For students who have trouble starting, suggest that they consider how histograms and box-and whisker-plots are constructed and what information can be learned from viewing each display.

ADDITIONAL PROBLEM Bug Gone Exterminators test a new fire-ant pesticide by treating 11 containers with the product and placing 50 ants in each container. They record the number of survivors after 12 h. Create a box-and-whisker plot for the data. {12, 2, 7, 6, 18, 1, 5, 19, 5, 4}

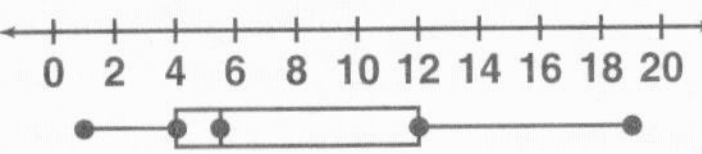

Resources

Transparencies
8

page 450 Math Toolbox

1.
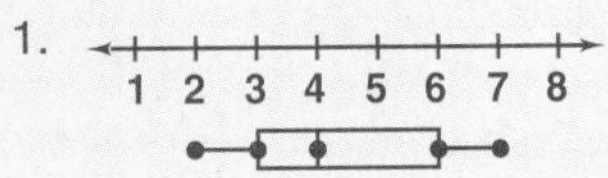

2.
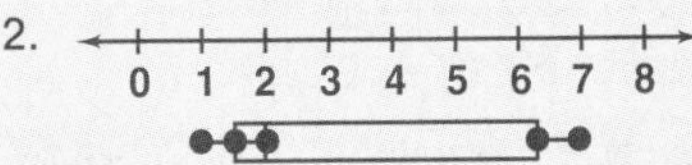

3.
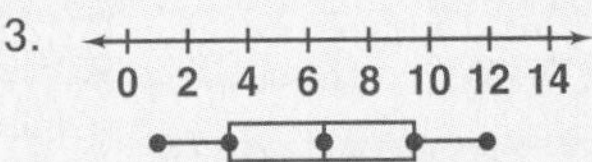

4.
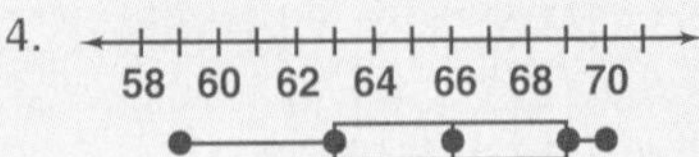

5.
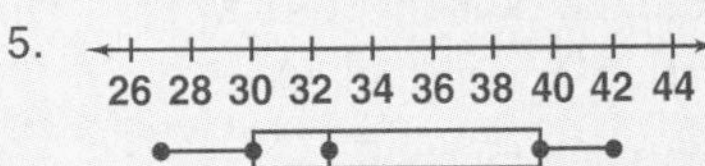

6.
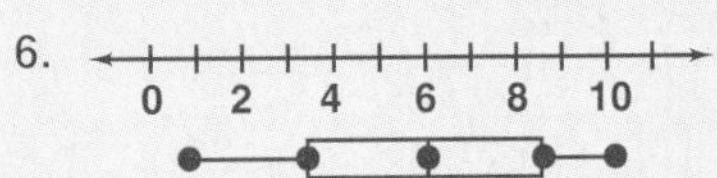

7–8, 9a. See back of book.

Math ToolboX — Skills Review

Box-and-Whisker Plots

Before Lesson 9-8

To show how data items are spread out, you can arrange a set of data in order from least to greatest. The maximum, minimum, and median give you some information about the data. You can better describe the data by dividing it into fourths. The **lower quartile** is the median of the lower half of the data. The **upper quartile** is the median of the upper half of the data.

The data below describes the highway gas mileage (mi/gal) for several brands of cars.

minimum ↓ median = 38.5 ↓ maximum ↓

17 19 27 37 40 42 52 58

lower quartile $\frac{19+27}{2} = 23$ ↑ ↑ upper quartile $\frac{42+52}{2} = 47$

A **box-and-whisker plot** is a visual representation of data. The box-and-whisker plot at the right displays the gas mileage information. The box represents the data from the lower quartile to the upper quartile. The vertical line segment represents the median. Horizontal line segments called whiskers show the spread of the data to the minimum and to the maximum.

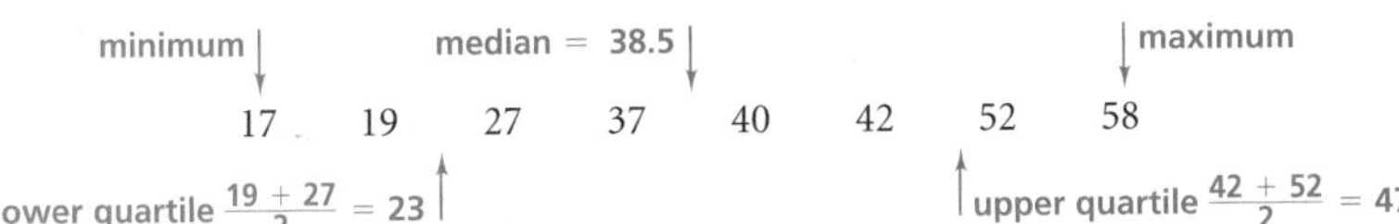

Create a box-and-whisker plot for each data set. 1–8. See margin.

1. {3, 2, 3, 4, 6, 6, 7}
2. {1, 1.5, 1.7, 2, 6.1, 6.2, 7}
3. {1, 2, 5, 6, 9, 12, 7, 10}
4. {65, 66, 59, 61, 67, 70, 67, 66, 69, 70, 63}
5. {29, 32, 40, 31, 33, 39, 27, 42}
6. {3, 3, 5, 7, 1, 10, 10, 4, 4, 7, 9, 8, 6}
7. {1, 1.2, 1.3, 4, 4.1, 4.2, 7}
8. {1, 3.8, 3.9, 4, 4.3, 4.4, 7, 5}

9. *Jobs* Below are the number of hours a student worked each week at her summer job. When she applied for the job, she was told the typical work week was 29 hours.
   29, 23, 21, 20, 17, 16, 15, 33, 33, 32, 15
   a. Make a box-and-whisker plot for the data. **See margin.**
   b. How many weeks are above the upper quartile? What are the number of hours worked? **2; 33, 33**
   c. What is the median number of hours she worked? What is the mean? Compare these to the typical work week. **21 h; 23 h; both are much less than the 29-h typical week.**

10. *Writing* In what ways are histograms and box-and-whisker plots alike and in what ways are they different? **Each diagram shows how the data are distributed. A histogram shows how many data points are in each part of the range. A box-and-whisker plot gives information about the maximum, the minimum, and the median.**

9-8 Teaching Notes

PROBLEM OF THE DAY

A local hardware store displays screws in order of size from least to greatest. If a shipment comes in with screws of lengths $\frac{1}{4}$ in., $\frac{5}{6}$ in., $\frac{1}{36}$ in., $\frac{1}{2}$ in., $\frac{3}{4}$ in., and $\frac{1}{6}$ in., in what order should they be displayed?

$\frac{1}{36}, \frac{1}{6}, \frac{1}{4}, \frac{1}{2}, \frac{3}{4}, \frac{5}{6}$

Problem of the Day is also available in Transparencies.

CONNECTING TO PRIOR KNOWLEDGE Write 3, 4, 5 on the board. Ask: *What is the mean of these three numbers?* 4 Now write −47, 4, 55. Ask: *What is the mean of these three numbers?* 4 Point out that both sets of numbers have the same mean. Ask: *What is different about these sets of numbers?* Answers may vary. Sample: The second set is more spread out. Invite students to make conjectures about how to describe the spread of a set of numbers using math terms.

WORK TOGETHER

Before students begin the Work Together activity, invite them to guess whether students working at the camp or the nursing home have higher average salaries.

Question 4 Point out that because students later square the deviation, it doesn't matter whether students subtract the mean from the salary or the salary from the mean. The important part of the deviation value is the size—how far from

Connections Jobs . . . and more

9-8 Analyzing Data Using Standard Deviation

What You'll Learn

- Exploring standard deviation
- Calculating and using standard deviation

...And Why

To analyze job opportunities

WORK TOGETHER

Jobs Work with a partner.

Four students had summer jobs at a camp last year. Four other students had jobs at a nursing home. Their weekly pay is listed below.

Salaries

Camp	Nursing Home
$150	$140
$160	$190
$220	$210
$270	$260

1. a. Calculate the mean for the pay in each location. $200; $200
   b. What does the mean represent? the average weekly pay
2. a What is the range of weekly pay for each location? $120; $120
   b. Use the ranges to describe how the pay varies at the two locations. Answers may vary. Sample: The weekly pay at each job varies by as much as $120.
3. Copy the table below.

QUICK REVIEW

The *range* of a set of data is the difference between the greatest number and the least number in the set.

Salary	Mean	Deviation from Mean	Square of Deviation
$150	$200	−50	2500
$160	$200	−40	1600
$220	$200	20	400
$270	$200	70	4900
sum of the squares of the deviations →			9400

4. You find deviation from the mean by subtracting the mean from each salary value. Fill in the missing values in column 3. See above.
5. Complete column 4 by squaring each value in column 3. See above.
6. Complete a table like the one above for the nursing home workers. See margin.
7. a. Compare the deviations from the mean at the two locations. At which location does the pay vary most from the mean? camp
   b. Compare the sum of the squares of the deviations for the two locations. How does this support your answer to part (a)? 9400 > 7400; the sum of squares of deviations is greater for the camp.

Lesson Planning Options

Prerequisite Skills

- Calculating the mean
- Finding the range of a set of data

Assignment Options for Exercises On Your Own

Core 1–6, 11–14
Extension 7–10, 15, 16

Use Mixed Review to maintain skills.

Resources

Student Edition

Skills Handbook, pp. 569, 582
Extra Practice, p. 564
Glossary/Study Guide

Teaching Resources

Chapter Support File, Ch. 9

- Practice 9-8 (two worksheets)
- Reteaching 9-8
- Alternative Activity 9-8

Classroom Manager 9-8
Glossary, Spanish Resources
Two-Year Algebra Handbook 9-8

Transparencies
108

the mean each value is. Whether the distance is positive or negative is irrelevant.

THINK AND DISCUSS

MAKING CONNECTIONS The table shows the median weekly earnings of full-time workers in 1993 categorized by age and sex. Have students follow the steps to calculate and to compare the mean and standard deviation of weekly earnings for male workers, female workers, and both combined.

Age Range	Men	Women
16–24	\$289	\$274
25–34	\$478	\$396
35–44	\$598	\$437
45–54	\$656	\$441
55–64	\$586	\$396
65+	\$453	\$335

Source: *World Almanac 1995*

VISUAL LEARNING Invite students to make a poster for the classroom describing the steps used to calculate the standard deviation. Suggest that students illustrate each step with an example to make the explanation more clear.

Example 1

KINESTHETIC LEARNING Use masking tape to make a number line, numbered from 67 to 90, on the classroom floor . Write the numbers as far apart as possible for the size of your classroom. Have a group of five students stand on the points given in Data Set 1. Then, have them move to the points given in Data Set 2. Encourage the rest of the class to describe the difference between the two sets in terms of how spread out the students are on the number line. Point out that the means of both data sets are the same.

Additional Examples

FOR EXAMPLE 1

Find the standard deviation of the two data sets below.

a. Data Set 1 {12, 6, 14, 8, 12} **2.94**

b. Data Set 2 {8, 8, 9, 20, 10} **4.56**

Discussion: *How does a data set's standard deviation differ from its mean?*

FOR EXAMPLE 2

The scores for a seven-game softball series are listed below.

Team A {4, 7, 2, 3, 5, 6, 9}.

Team B {5, 8, 1, 4, 4, 5, 8).

a. Which team won the series? **Team A**

b. What was the average score per game for each team? **Team A: 5.14, Team B: 5**

c. Which team has the smallest standard deviation? **Team A: 2.23, Team B: 2.27**

Discussion: *If you were the manager of a softball team, would you rather your team has a smaller standard deviation or a larger standard deviation? Explain.*

THINK AND DISCUSS

A way to show how a set of data is spread out is the **standard deviation.** It reflects how all the data points in a set vary from the mean. The symbol for standard deviation is σ, pronounced "sigma."

small standard deviation ⟶ data cluster around the mean
large standard deviation ⟶ data spread out from the mean

You can calculate the standard deviation by following these six steps.

- Find the mean of the data set. The expression $\bar{x}$ represents the mean.
- Find the difference of each data value from the mean.
- Calculate the square of each difference.
- Find the sum of the squares of the differences.
- Divide the sum by the number of data values.
- Take the square root of the quotient just calculated.

Example 1

Find the standard deviation of the two data sets below.
a. Data Set 1 {83, 88, 75, 69, 70} **b.** Data Set 2 {69, 75, 76, 77, 88}

a. Find the mean.

mean $= \frac{83 + 88 + 75 + 69 + 70}{5}$
$= \frac{385}{5}$
$= 77$

Calculate the standard deviation.

x	$\bar{x}$	$x - \bar{x}$	$(x - \bar{x})^2$
83	77	6	36
88	77	11	121
75	77	−2	4
70	77	−7	49
69	77	−8	64
		Sum:	274

sum of squares = 274
$\frac{\text{sum of squares}}{5} = \frac{274}{5}$
$= 54.8$
$\sqrt{54.8} = 7.402702209$

The standard deviation is about 7.4.

b. Find the mean.

mean $= \frac{69 + 75 + 76 + 77 + 88}{5}$
$= \frac{385}{5}$
$= 77$

Calculate the standard deviation.

x	$\bar{x}$	$x - \bar{x}$	$(x - \bar{x})^2$
69	77	−8	64
75	77	−2	4
76	77	−1	1
77	77	0	0
88	77	11	121
		Sum:	190

sum of squares = 190
$\frac{\text{sum of squares}}{5} = \frac{190}{5}$
$= 38$
$\sqrt{38} = 6.164414003$

The standard deviation is about 6.2.

8. In which data set do the values cluster closer to the mean? **Data Set 2**

Students may need further examples to help them make the connection between how spread out the students are and the value of the standard deviation. Here are more data set values and their standard deviations. All sets have a mean of 77.

Data Set	Standard Deviation
{67, 68, 75, 85, 90}	9.143 304
{67, 67, 71, 90, 90}	10.714 48
{75, 76, 77, 78, 79}	1.414 214

Example 2 Relating to the Real World 

Encourage students with access to computer spreadsheet programs to learn how to use the software' s standard deviation feature. They can either use the program's **Help** feature or read the software manuals. For *Microsoft Excel*, the function STDEVP (range of cells) calculates the standard deviation of the cells indicated in the range. Be sure students use the command STDEVP and not the commands DSTDEV or STDEV, which calculate something else.

ERROR ALERT! Students may get incorrect answers because of calculator keystroke errors. **Remediation:** Have students work in pairs, one entering the data and the other dictating which keys to use.

TACTILE LEARNING Divide students into pairs. Have one student use a calculator while the other calculates the standard deviation by hand. Have students compare their answers.

ESTIMATION Question 10c Suggest that students write Player 3's scores in numerical order in a line. Then have them

You can use a calculator to find the standard deviation of a set of data.

Example 2 Relating to the Real World

Golf Three friends play golf. Their scores on six holes of golf are below. Calculate the mean and standard deviation for Player 1's scores.

Player 1	5	4	2	4	10	5
Player 2	5	5	5	5	5	5
Player 3	3	10	4	4	7	2

You can use a calculator to calculate the mean and standard deviation.

- Choose STAT and *EDIT* to enter your data.
- Choose STAT, *CALC*, and *1-VAR STATS* to calculate.
- The screen displays information about your data including the mean, $\bar{x}$, the standard deviation, σ_x, and the number of entries, n.

The mean of Player 1's scores is 5. The standard deviation is about 2.4.

9. a. **Predict** whether the standard deviations of Player 2's and Player 3's scores will be greater or less than the standard deviation of Player 1's scores. Player 2 has the least st. dev.; Player 3 has the greatest st. dev.
   b. Explain how you made your predictions in part (a). See margin.

10. a. Try This Use a calculator to find the mean and standard deviation for Player 2's and Player 3's golf scores. 5; 0; 5; 2.7
    b. What does a standard deviation of zero mean? All data are the same.
    c. Estimation Suppose a friend got the answer 27.5 when using a calculator to find the standard deviation of Player 3's golf scores. How do you know that your friend made an error? The standard deviation cannot be greater than the maximum deviation for data.

At age 20, Stanford University sophomore Tiger Woods is a two-time United States Amateur champion.

Technology Options

For Exercises 1–4 and 11–14, students may find it helpful to use spreadsheet software to find the standard deviation or each set of data.

Prentice Hall Technology 

 **Calculator**
- Graphing Calculator Handbook: Procedure 21, 22

 **Software**
- Secondary Math Lab Toolkit™
- Computer Item Generator 9-8

 **CD-ROM**
- Multimedia Algebra Lab 9

 **Internet**
- See the Prentice Hall site. (http://www.phschool.com)

estimate the mean. If they choose an estimate mean of 6, point out that the scores 2 and 10 are both 4 away from the mean, so the standard deviation cannot be more than 4.

Exercises ON YOUR OWN

ALTERNATIVE ASSESSMENT **Exercise 6** To assess understanding of standard deviation, ask students to repeat this exercise using the high temperatures in their area for 14 days.

WRITING Exercise 7 Encourage students to make up test sets of data before drawing conclusions.

DIVERSITY Exercise 9 Use this exercise as a starting point for a discussion on ethics. Explain that all figures in the exercise are correct. Ask: *If 17 out of 20 employees earn \$265/wk and the average salary is \$400/wk, what must be true of the remaining 3 employees?* Their salaries must be much higher. If the employer had advertised the average weekly salary as \$400 but only planned to pay \$265/wk, would this have been an ethical business practice?

ESL Exercise 15 The definition of *lift* as a device used to carry skiers to the top of a ski slope may be unfamiliar to some students. Ask a volunteer to explain the meaning of a *lift ticket.*

RESEARCH Exercise 16 Students can find people to survey from among their friends, classmates, or other peer groups. Remind students that they need to collect some type of numerical data.

CONNECTING TO THE STUDENTS' WORLD Exercise 16 Before students begin their research, have them choose an interesting topic for their data collection. Guide students away

page 451 Work Together

6.

Salary	Mean	Deviation	Sqr. of Dev.
\$140	\$200	−60	3600
\$190	\$200	−10	100
\$210	\$200	10	100
\$260	\$200	60	3600
sum of sqrs. of deviations			7400

page 452–453 Think and Discuss

9b. Answers may vary. Sample: Each player has the average score of 5. Player 2's scores are all the same so the standard deviation is very low. Player 3's and Player 1's scores are almost the same, but Player 1 has a 5 and a 5, while Player 3 has a 3 and a 7.

pages 454–455 On Your Own

1.

x	$\bar{x}$	$x - \bar{x}$	$(x - \bar{x})^2$
5	5	0	0
3	5	−2	4
2	5	−3	9
5	5	0	0
10	5	5	25
		Sum:	38

2–4, 7, 9a. See back of book.

Exercises ON YOUR OWN

Calculator **Make a table like the one in Example 1 to find the standard deviation for each data set. Use a calculator to find the square root.** 1–4. See margin for tables.

1. {5, 3, 2, 5, 10} 2.8
2. {10, 9, 10, 12, 11, 14} 1.6
3. {3.5, 4.5, 6.0, 4.0, 2.5, 2.5, 5.0} 1.1
4. {11, 11, 17, 17, 10, 11, 12, 19} 3.3

5. Tell what you know about the following data sets.
   a. Data Set 1: The mean is 25 and the standard deviation is 100. The data are spread far apart.
   b. Data Set 2: The mean is 25 and the standard deviation is 2. Most data are near 25.

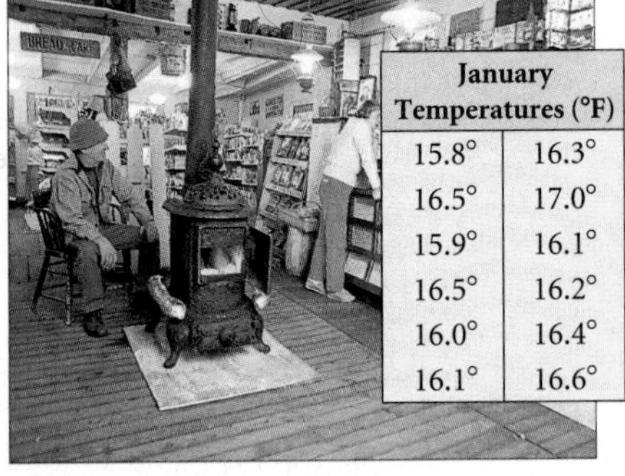

January Temperatures (°F)	
15.8°	16.3°
16.5°	17.0°
15.9°	16.1°
16.5°	16.2°
16.0°	16.4°
16.1°	16.6°

6. The table at the right records the outdoor temperatures (°F) reported to a local meteorologist by twelve weather watchers near Denver, Colorado, in January. Find the standard deviation of the temperature data. 0.3

7. Writing Explain the effects of a very large or very small data value on the mean and standard deviation of a data set. See margin.

8. Open-ended Create data sets with at least 6 entries as follows.
   a. The mean is 12 and the standard deviation is 0. Sample: 12, 12, 12, 12, 12, 12, 12
   b. The mean is 7 and the range is 12. Sample: 0, 2, 12, 12, 9, 7, 7

9. Jobs When Lyman interviewed for a summer job, he asked about the average salary of summer employees. He was excited that the average weekly wage was \$400. When his employer offered him a job at \$265 a week, he was confused. His employer stated: "Of my 20 summer employees, 17 make \$265 per week."
   a. During the job interview, what other statistics should Lyman have asked about in order to find out more information about salaries? Explain your choice(s). See margin.
   b. The three highly-paid summer employees each earn the same amount. How much do they earn? \$1165
   c. Estimate Do you expect the standard deviation for the company's summer salaries to be large or small? Why? c. Large; all the data are far from the mean.
   d. Verify your prediction by calculating the standard deviation. d. 321.36

10. The graph at the right shows the differences from the mean in a set of data containing eight data values.
   a. Find the square of each difference from the mean. 25, 4, 1, 1, 16, 9, 4, 4
   b. Find the sum of the squares calculated in part (a). 64
   c. Use the result of part (b) to find the standard deviation of the data set. 2.83

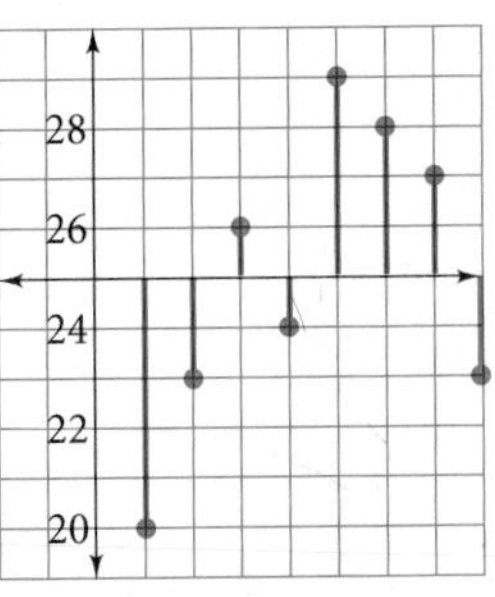

from topics that may be inappropriate or may cause discomfort or embarrassment to some students.

Exercises MIXED REVIEW

PORTFOLIO Share with students the criteria you will use to assess their work in portfolios, as well as how you plan to use the results. Students should understand how the rubrics are used to assess their work, how each piece in the portfolio counts, and how the scores they get in their portfolios will affect their overall evaluation.

Exercise 21 If students are unsure what to do, suggest that they multiply the first equation by 2 and the second equation by 3.

Exercise 26 Point out that the 5 has no effect on the domain.

Wrap Up

THE BIG IDEA Ask students: *What is a standard deviation and how is it useful?*

RETEACHING ACTIVITY Students make a table showing steps to find the standard deviation of a data set. (Reteaching worksheet 9-8)

Graphing Calculator **Use a calculator to find the standard deviation of each set of data.**

11. {11.4, 10.1, 9.5, 9.9, 10.1, 11.2, 12.0, 12.3} 0.98
12. {102.4, 100.8, 99.5, 103.4, 105.6, 111.5, 120.5} 6.86
13. {13, 19, 23, 50, 43, 44, 50, 52, 74, 83, 88, 90} 25.47
14. {−3.2, 0, 1.3, −2.0, −3.5, 0, 3.2, 2.3, 1.1, 0.3} 2.12

15. **Math in the Media** You are writing a magazine article on ski resorts in the Sierras. The information at the right is part of your research.
    a. What is the range for the price of a lift ticket? $20
    b. Find the standard deviation of the ticket prices. 5.83
    c. To give your readers a feel for the variability of ticket prices, you include the following sentence in your article. Fill in the blanks.

       Lift ticket prices vary about $ ■ around an average of $ ■ for an all-day adult ticket. 6; 39

    d. **Writing** Explain why your sentence in part (c) is more helpful than simply telling your readers the range, as in "The prices vary from $26 to $46." Answers may vary. Sample: Only one ticket cost less than $34. Your sentence gives a better description of the cost.

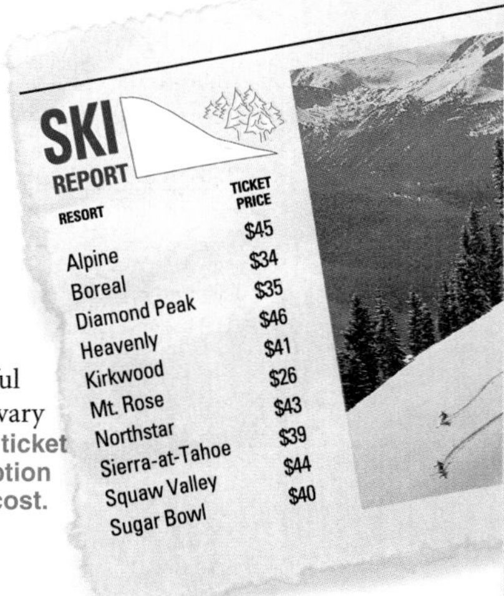

SKI
REPORT
RESORT
TICKET PRICE
Alpine $45
Boreal $34
Diamond Peak $35
Heavenly $46
Kirkwood $41
Mt. Rose $26
Northstar $43
Sierra-at-Tahoe $39
Squaw Valley $44
Sugar Bowl $40

16. **Research** Collect data on a topic of interest such as classmates' batting averages or students' curfew times. Gather at least 15 measurements and calculate the mean and standard deviation. Write a paragraph describing your data. Check students' work.

Exercises MIXED REVIEW

Write each number in scientific notation.

17. 0.00347 3.47×10^{-3}
18. 3,112,200 3.1122×10^{6}
19. 0.000825 8.25×10^{-4}
20. 50,147,235 5.0147235×10^{7}

Solve each system of equations.

21. $3x + 2y = 7$, $2x + 3y = 8$ (1, 2)
22. $5x - 2y = 3$, $4x + 2y = 6$ (1, 1)
23. $x + y = 7$, $2x - y = 8$ (5, 2)
24. $3x - 4y = 5$, $3x + 2y = 11$ (3, 1)

Find the domain of each function.

25. $y = \sqrt{x - 5}$ $x \geq 5$
26. $f(x) = 5 + \sqrt{6 + 3x}$ $x \geq -2$

27. **Automobiles** A new car costs $20,000. It loses value at a rate of 7% per year. How much is the car worth in four years? $14,961.04

PORTFOLIO

For your portfolio, select one or two items from your work for this chapter. Here are possibilities:
- corrected work
- diagrams, graphs, or charts
- a journal entry

Explain why you have included each selection you make.

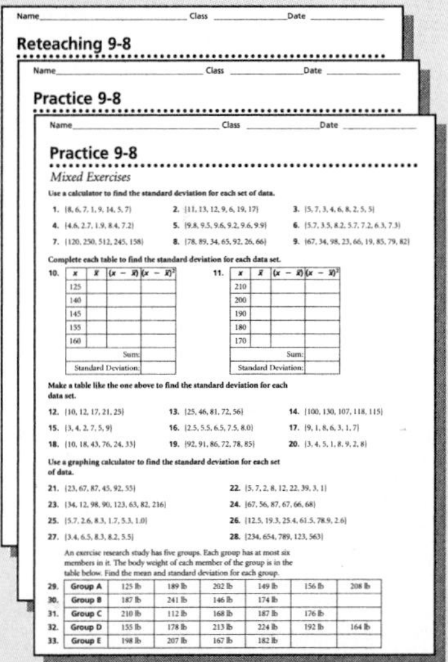
Reteaching 9-8
Practice 9-8
Practice 9-8
Mixed Exercises

Lesson Quiz

Lesson Quiz is also available in Transparencies.

Find the standard deviation of each data set.

1. temperatures (°F) during a one-week period {72, 76, 81, 85, 80, 89, 100} 8.55
2. scores of five algebra tests {95, 92, 99, 90, 88} 3.87
3. weight of six students (lb) {145, 120, 160, 130, 119, 120} 15.35

Finishing the Chapter Project

PROJECT DAY You may wish to plan a project day on which students share their completed projects. Encourage students to explain their process as well as their product.

PROJECT NOTEBOOK Ask students to review their project work and bring their notebooks up to date.

- Have students review their methods for finding, recording, and solving equations and formulas used for the project.
- Ask groups to share any insights they found for completing the project, such as any shortcuts they found for solving formulas or making graphs.

SCORING RUBRIC

3 Calculations are correct. Graphs are neat, accurate, and clearly show the differences between the formulas. Scales are appropriate for the graphs. Explanations are thorough and well thought out.

2 Calculations are mostly correct, with some minor errors. Graphs are neat and mostly accurate with minor errors in scale.

1 Calculations contain errors. Graphs could be more accurate.

0 Major elements of the project are incomplete or missing.

CHAPTER PROJECT

Finishing the Chapter Project

On a CLEAR Day...

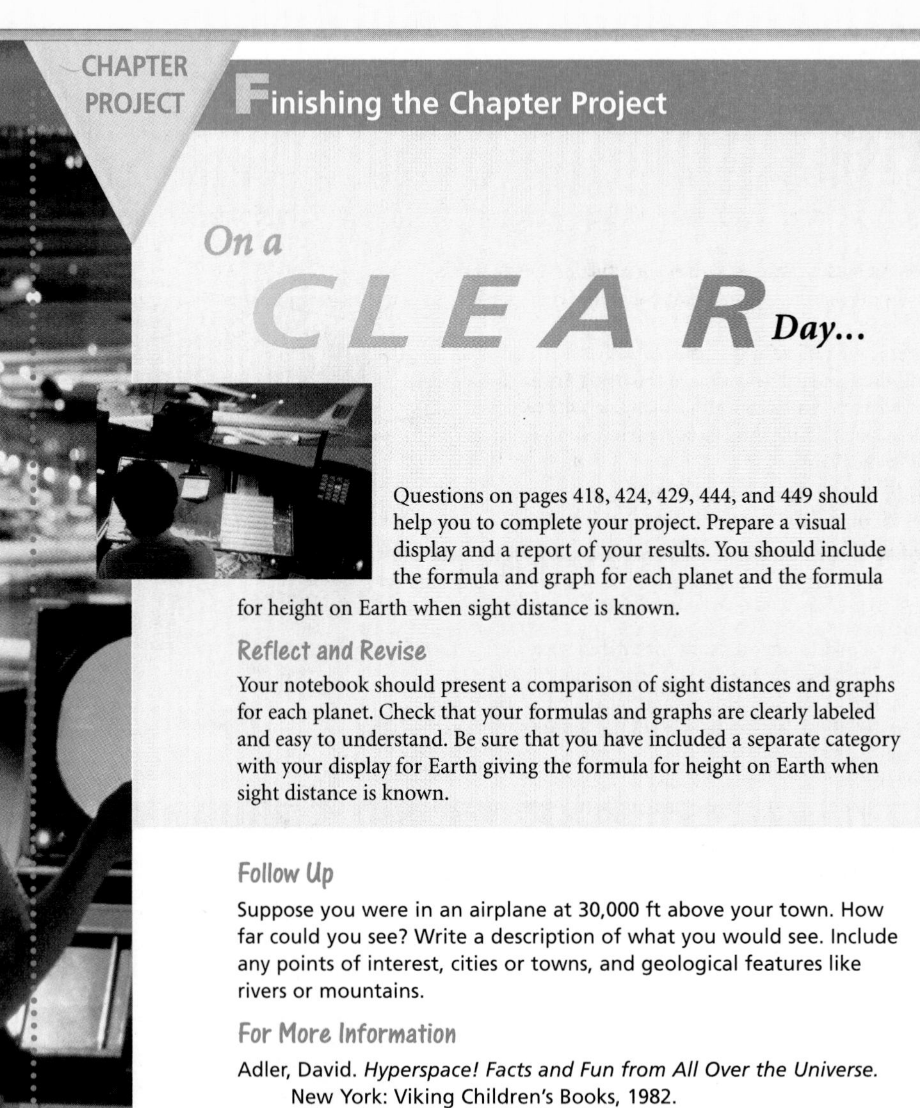

Questions on pages 418, 424, 429, 444, and 449 should help you to complete your project. Prepare a visual display and a report of your results. You should include the formula and graph for each planet and the formula for height on Earth when sight distance is known.

Reflect and Revise

Your notebook should present a comparison of sight distances and graphs for each planet. Check that your formulas and graphs are clearly labeled and easy to understand. Be sure that you have included a separate category with your display for Earth giving the formula for height on Earth when sight distance is known.

Follow Up

Suppose you were in an airplane at 30,000 ft above your town. How far could you see? Write a description of what you would see. Include any points of interest, cities or towns, and geological features like rivers or mountains.

For More Information

Adler, David. *Hyperspace! Facts and Fun from All Over the Universe.* New York: Viking Children's Books, 1982.

"Living and Working in Space: The Countdown Has Begun." PBS Video, 1320 Braddock Place, Alexandria, Virginia 22314. (800) 344-3337.

Free videotapes, slides, books, pamphlets: National Aeronautics and Space Administration, Education Programs Officer, NASA Headquarters, Code XEE, Washington, D.C. 20546. (202) 453-8396.

Wrap Up

Chapter Project **HOW AM I DOING?** After students have written their individual lists of important ideas, ask each student to share his or her list with the class. Compile students' answers on the board, then use the composite list as a focus for discussion and review.

KEY TERMS The numbers in parentheses direct students to the pages where the terms (or symbols) are used or defined. Students should be able to (1) write a simple explanation of each term, (2) illustrate the term with a diagram, or (3) show an example that uses the term.

Exercises 6–9 Suggest that students check their answers by graphing their solutions on a coordinate plane.

Exercises 11–13 Point out that the sine of an angle is the same as the cosine of 90° minus the angle.

Exercises 20–27 Suggest that students check their answers by finding the decimal equivalent of each expression with a calculator.

Exercises 34–37 Have students check their answers by sketching each function.

Exercises 38–41 Remind students that some calculators are able to calculate the standard deviation automatically .

Remind students that the new mathematical terms in this chapter are defined in the Glossary/Study Guide in the back of the book.

9 Wrap Up

Key Terms

angle of elevation (p. 426)
box-and-whisker plot (p. 450)
converse of the Pythagorean theorem (p. 416)
cosine (p. 425)
distance formula (p. 420)
division property of square roots (p. 431)
extraneous solution (p. 443)
hypotenuse (p. 414)
legs (p. 414)
like terms (p. 435)
lower quartile (p. 450)
midpoint (p. 422)
midpoint formula (p. 422)
multiplication property of square roots (p. 430)
Pythagorean theorem (p. 414)
radical equation (p. 440)
rationalize (p. 432)
simplest form (p. 430)
sine (p. 425)
square root function (p. 445)
standard deviation (p. 452)
tangent (p. 425)
trigonometric ratios (p. 425)
unlike terms (p. 435)
upper quartile (p. 450)

How am I doing?

SELF ASSESSMENT

- State three ideas from this chapter that you think are important. Explain your choices.
- Describe the method for finding the distance between two points *A* and *B*.

Resources

Teaching Resources

Extra Practice, p. 564
Glossary/Study Guide

Student Edition

Study Skills Handbook
Glossary, Spanish Resources

The Pythagorean Theorem 9-1

For a right triangle with **legs** a and b and **hypotenuse** c, the **Pythagorean theorem** states that $a^2 + b^2 = c^2$. The **converse of the Pythagorean theorem** states that if a triangle has sides of lengths a, b, and c, and if $a^2 + b^2 = c^2$, then it is a right triangle with hypotenuse of length c.

A rectangle has sides with lengths *a* and *b*. Find the length of the diagonal to the nearest hundredth.

1. $a = 3, b = 5$ 5.83
2. $a = 11, b = 14$ 17.80
3. $a = 7, b = 13$ 14.76
4. $a = 4, b = 9$ 9.85
5. *Writing* The hypotenuse of a right triangle is 26 cm. The length of one leg is 10 cm. Describe how to find the length of the other leg. **Square the length of the hypotenuse. Subtract the square of the known leg length. Take the square root of the difference.**

The Distance Formula 9-2

The **distance formula** $d = \sqrt{(x_2 - x_1)^2 + (y_2 - y_1)^2}$ gives the distance between two points (x_1, y_1) and (x_2, y_2). The **midpoint formula** $M = \left(\frac{x_1 + x_2}{2}, \frac{y_1 + y_2}{2}\right)$ gives the coordinates of their midpoint.

Find the midpoint of $\overline{AB}$.

6. $A(3, 7), B(-2, 4)$ (0.5, 5.5)
7. $A(5, -2), B(6, 14)$ (5.5, 6)
8. $A(3, -9), B(14, 16)$ (8.5, 3.5)
9. $A(12, 17), B(-7, 9)$ (2.5, 13)
10. *Open-ended* Find two points with integer coordinates $\sqrt{74}$ units apart. **Answers may vary. Sample: (–2, 3) and (5, 8)**

pages 457–459 Wrap Up

34.

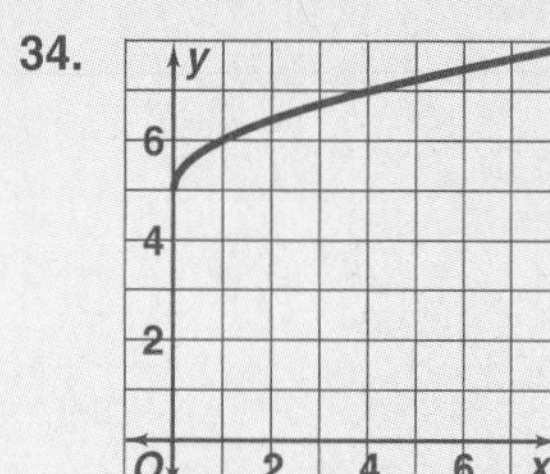

35.

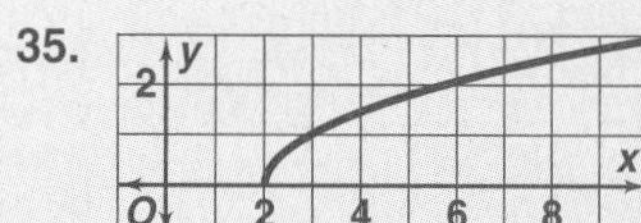

Trigonometric Ratios 9-3

Trigonometric ratios are triangle measurement ratios. For a right triangle of a given shape, they do not change no matter how large or small the triangle is. Three trigonometric ratios are **sine** (sin), **cosine** (cos), and **tangent** (tan), shown at the right.

$\sin A = \frac{\text{length of side of opposite } \angle A}{\text{length of hypotenuse}}$

$\cos A = \frac{\text{length of side of adjacent to } \angle A}{\text{length of hypotenuse}}$

$\tan A = \frac{\text{length of side opposite } \angle A}{\text{length of side adjacent to } \angle A}$

You can use trigonometric ratios to measure distances indirectly. You can use an **angle of elevation** to measure heights indirectly.

In $\triangle ABC$, $\angle C$ is a right angle. Find the lengths of the missing sides.

11. $AB = 12$, $m\angle A = 34°$ $BC \approx 6.71$; $AC \approx 9.95$

12. $BC = 9$, $m\angle B = 72°$ $AC \approx 27.7$; $AB \approx 29.12$

13. $AC = 8$, $m\angle B = 52°$ $BC \approx 6.25$; $AB \approx 10.15$

14. *Standardized Test Prep* In $\triangle ABC$, $\angle C$ is a right angle. Which of the following is (are) true? C

I. $\cos B = \frac{AC}{AB}$ **II.** $\tan B = \frac{BC}{AC}$ **III.** $\sin A = \frac{BC}{AB}$

A. I only **B.** II only **C.** III only **D.** I and II **E.** II and III

Simplifying Radicals 9-4

You can simplify some radical expressions by using products or quotients. The **multiplication property of square roots** states that for any two nonnegative numbers $\sqrt{ab} = \sqrt{a} \cdot \sqrt{b}$. The **division property of square roots** is at the right.

$\sqrt{\frac{a}{b}} = \frac{\sqrt{a}}{\sqrt{b}}$

$a \geq 0, b > 0$
division property

Simplify each radical expression.

15. $\sqrt{32} \cdot \sqrt{144}$ $48\sqrt{2}$ **16.** $\frac{\sqrt{84}}{\sqrt{121}}$ $\frac{2\sqrt{21}}{11}$ **17.** $\sqrt{96} \cdot \sqrt{25}$ $20\sqrt{6}$ **18.** $\sqrt{\frac{100}{169}}$ $\frac{10}{13}$

19. A rectangle is 7 times as long as it is wide. Its area is 1400 cm^2. Find the dimensions of the rectangle in simplified form. $10\sqrt{2}$ cm by $70\sqrt{2}$ cm

Adding and Subtracting Radicals 9-5

You can use the distributive property to simplify expressions with sums and differences of radicals. First, simplify the radicals to have like terms.

Simplify each radical expression.

20. $6\sqrt{7} - 2\sqrt{28}$ $2\sqrt{7}$ **21.** $5(\sqrt{20} + \sqrt{80})$ $30\sqrt{5}$ **22.** $\sqrt{54} - 2\sqrt{6}$ $\sqrt{6}$ **23.** $\sqrt{125} - 3\sqrt{5}$ $2\sqrt{5}$

24. $\sqrt{10}(\sqrt{10} - \sqrt{20})$ $10 - 10\sqrt{2}$ **25.** $7\sqrt{90} + \sqrt{160}$ $25\sqrt{10}$ **26.** $\sqrt{72} + 3\sqrt{32}$ $18\sqrt{2}$ **27.** $\sqrt{28} + 5\sqrt{63}$ $17\sqrt{7}$

28. A box is 2 in. long, 3 in. wide, and 4 in. tall. What is the length of the longest distance between corners of the box? Express your answer in simplified form. $\sqrt{29}$ in.

ALTERNATIVE ASSESSMENT Ask pairs of students to create a problem to illustrate the principle idea of each review section. Then have students exchange their work, and solve each other's sample problem set.

Getting Ready for Chapter 10

Students may work these exercises independently or in small groups. The skills previewed will help prepare students for factoring quadratic expressions.

Solving Radical Equations 9-6

A **radical equation** has a variable under a radical. You can often solve a radical equation by solving for the square root, then squaring both sides of the equation. Squaring each side of an equation also works when both sides are square roots. When you square both sides of a radical equation, you may produce an **extraneous solution**. It is not a solution of the original equation.

Solve each radical equation.

29. $\sqrt{x+7}=3$ **2** 30. $\sqrt{x}+3\sqrt{x}=16$ **16** 31. $\sqrt{x+7}=\sqrt{2x-1}$ **8** 32. $\sqrt{x}-5=4$ **81**

33. The volume V of a cylinder is given by $V=\pi r^2h$, where r is the radius of the cylinder and h is its height. If the volume of a cylinder is 54 in.^{2}, and its height is 2 in., what is its radius to the nearest 0.01 in.? **2.93 in.**

Graphing Square Root Functions 9-7

The simplest **square root function** is $y=\sqrt{x}$. To find the domain of a square root function, solve the inequality where the expression under the radical is greater than or equal to zero.

Find the domain of each function. Then graph the function. **34–37. See margin for graphs.**

34. $y=\sqrt{x}+5$ $x\geq 0$
35. $y=\sqrt{x-2}$ $x\geq 2$
36. $y=\sqrt{x+1}$ $x\geq -1$
37. $y=2\sqrt{x}$ $x\geq 0$

36.

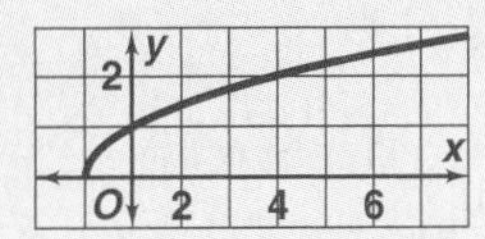

37.

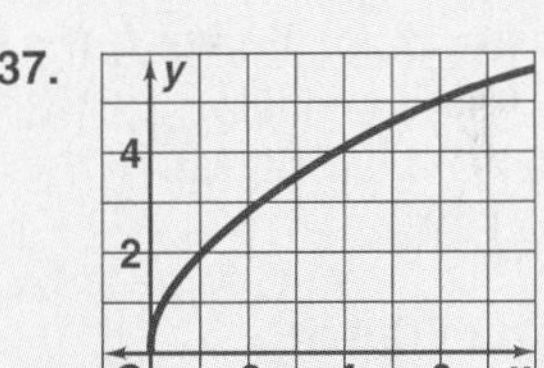

Analyzing Data Using Standard Deviation 9-8

The **standard deviation** shows how spread out a set of data is from the mean. You can calculate the standard deviation by following these six steps. Find the mean, $\bar{x}$, of the data set. Find the difference of each data value from the mean. Square each difference. Find the sum of the squares of the differences. Divide the sum by the number of data values. The square root of the quotient just calculated is the standard deviation.

Find the standard deviation of each set of data.

38. 2, 7, 9, 15, 17 **5.44**
39. 5, 8, 11, 12, 19 **4.69**
40. 4, 10, 12, 13, 23 **6.15**
41. 7, 15, 20, 23, 29 **7.44**

Getting Ready for... CHAPTER 10

Use the quadratic formula to solve these equations to the nearest hundredth. If the equation has no real solution, write *no solution*.

42. $2x^2+5x-4=0$ **0.64, −3.14**
43. $3x^2-2x-7=0$ **1.90, −1.23**
44. $x^2-3x-8=0$ **4.70, −1.70**
45. $2x^2-5x+15=0$ **no solution**
46. $5x^2-4x-12=0$ **2, −1.2**
47. $7x^2+2x-18=0$ **1.47, −1.75**

Assessment

ENHANCED MULTIPLE CHOICE QUESTIONS are more complex than traditional multiple choice questions, which assess only one skill. Enhanced multiple choice questions assess the processes that students use, as well as the end results. The questions are written so that students must use more than one strategy to solve the problem. Using multiple strategies is encouraged by the National Council of Teachers of Mathematics (NCTM). **Exercise 25** is an enhanced multiple choice question.

FREE RESPONSE QUESTIONS do not give answer choices. Some exercises have more than one possible answer. Students need to give only one correct response. **Exercises 1–23, 27–41, and 43–44** are free response questions.

WRITING EXERCISES allow students to describe how they think about and understand the concepts they have learned. **Exercise 42** is a writing exercise.

OPEN-ENDED PROBLEMS allow for more than one solution. Students must construct their own responses instead of choosing from possible answers. The student responses will help you determine the depth of their understanding and any possible areas of difficulty. **Exercise 26** is an open-ended problem. Make sure that the students' responses contain values for x and y and a written situation.

Resources

Teaching Resources

Chapter Support File, Ch. 9
- Chapter Assessment, Forms A and B
- Alternative Assessment

Chapter Assessment, Spanish Resources

Teacher's Edition

See also p. 412E for assessment options

Software • Computer Item Generator

page 460 Assessment

33.

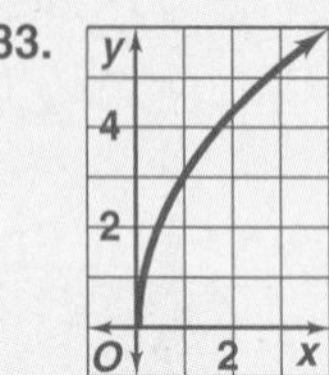

34.

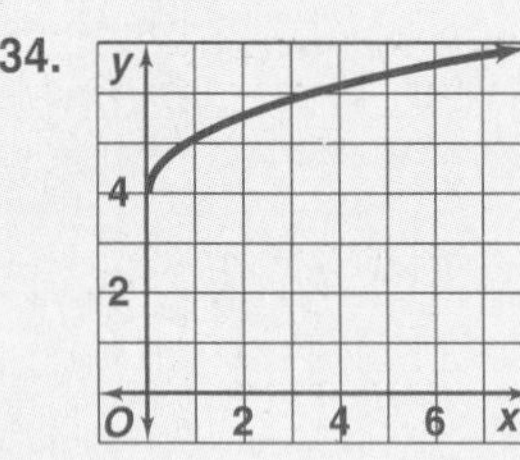

9 Assessment

Find whether the following sets of numbers determine a right triangle.

1. 6, 8, 10 yes
2. 6, 7, 9 no
3. 4, 5, 11 no
4. 10, 24, 26 yes

Approximate *AB* to the nearest hundredth.

5. $A(1, -2), B(5, 7)$ 9.85
6. $A(3, 5), B(7, 4)$ 4.12
7. $A(4, 7), B(-11, -6)$ 19.85
8. $A(0, -5), B(3, 2)$ 7.62

Find the coordinates of the midpoint of $\overline{AB}$.

9. $A(4, 9), B(1, -5)$ (2.5, 2)
10. $A(-2, -7), B(3, 0)$ (0.5, −3.5)
11. $A(3, -10), B(-4, 6)$ (−0.5, −2)
12. $A(0, 8), B(-1, 1)$ (−0.5, 4.5)

Find the lengths to the nearest hundredth.

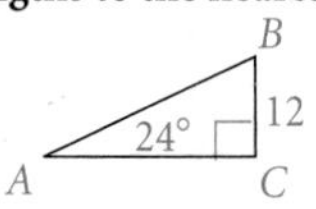

13. AB 29.50
14. AC 26.95

(Triangle ABC with angle 24° at A, right angle at C, $BC = 12$)

15. The distance between consecutive bases in a baseball diamond is 90 ft. How far is it from first base to third base? about 127.3 ft

16. One house is 12 mi east of a school. Another house is 9 mi north of the school. How far apart from each other are the houses? 15 mi

Simplify each radical expression.

17. $\sqrt{\frac{128}{64}}$ $\sqrt{2}$
18. $\frac{\sqrt{27}}{\sqrt{75}}$ $\frac{3}{5}$
19. $\sqrt{48}$ $4\sqrt{3}$
20. $\sqrt{12} \cdot \sqrt{8}$ $4\sqrt{6}$
21. $3\sqrt{32} + 5\sqrt{2}$ $17\sqrt{2}$
22. $2\sqrt{27} + 5\sqrt{3}$ $11\sqrt{3}$
23. $7\sqrt{125} - 3\sqrt{175}$ $35\sqrt{5} - 15\sqrt{7}$
24. $\sqrt{128} - \sqrt{192}$ $8\sqrt{2} - 8\sqrt{3}$

25. Standardized Test Prep If x and y are positive, which expression(s) is (are) equivalent to $\sqrt{24x^2y^3}$? E

I. $2xy\sqrt{12xy^2}$ II. $2xy\sqrt{6y}$ III. $xy\sqrt{24y}$

A. I only B. II only C. III only
D. I and II E. II and III

26. Open-ended Write a problem involving addition of two like terms with radical expressions. Simplify the sum. Sample: $2\sqrt{5} + 4\sqrt{5} = 6\sqrt{5}$

Solve each radical equation.

27. $3\sqrt{x} + 2\sqrt{x} = 10$ 4
28. $8 = \sqrt{5x - 1}$ 13
29. $5\sqrt{x} = \sqrt{15x + 60}$ 6
30. $\sqrt{x} = \sqrt{2x - 7}$ 7
31. $3\sqrt{x + 3} = 2\sqrt{x + 9}$ $\frac{9}{5}$

32. A rectangle is 5 times as long as it is wide. The area of the rectangle is 100 ft². How wide is the rectangle? Express your answer in simplified form. $2\sqrt{5}$ ft

Find the domain of each function. Then graph the function. 33–36. See margin for graphs.

33. $y = 3\sqrt{x}$ $x \geq 0$
34. $y = \sqrt{x} + 4$ $x \geq 0$
35. $y = \sqrt{x - 4}$ $x \geq 4$
36. $y = \sqrt{x + 9}$ $x \geq -9$

37. A cube has 3-in. sides. Find the longest distance between a pair of corners. Express your answer in simplified form. $3\sqrt{3}$ in.

Find the standard deviation of each set of data.

38. 5, 7, 9, 11, 12 2.56
39. 2, 4, 7, 11, 15 4.71
40. 4, 12, 13, 15, 17 4.45
41. 1, 2, 4, 6, 7 2.28

42. Writing Explain how these two data sets differ based on the following information.
Set A: mean 20; range 10; standard deviation 4
Set B: mean 20; range 20; standard deviation 2
See margin.

43. Geometry The formula for the volume V of a cylinder of height h and radius r is $V = \pi r^2 h$. Solve the formula for r in terms of V and h. $r = \sqrt{\frac{V}{\pi h}}$

44. From the ground you can see a satellite dish on the roof of a building 60 feet high. The angle of elevation is 62°. How far away is the building from you? about 31.9 ft

Preparing for Standardized Tests

Standardized tests, such as those administered for state assessment, the SAT, or the ACT, include regular math questions, quantitative comparison questions, open-ended problems, and free response questions (which the SAT calls *grid-ins*).

MULTIPLE CHOICE QUESTIONS are followed by five answer choices, one of which is correct. **Exercises 1–13** are multiple choice questions.

QUANTITATIVE COMPARISON QUESTIONS ask students to compare two quantities. **Exercises 12 and 13** are quantitative comparison questions.

FREE-RESPONSE QUESTIONS do not give answer choices. Students must provide the one correct answer on their own. **Exercises 14–18** are free-response questions.

OPEN-ENDED PROBLEMS allow for more than one solution. Students must construct their own responses instead of choosing a single answer. The responses students give will help you determine the depth of their understanding and what difficulties, if any, they are experiencing. **Exercise 14** is an open-ended problem.

STANDARDIZED TEST TIPS **Exercise 13** Before comparing two expressions, it is often useful to simplify them. In this case, adding an x and subtracting a 1 from both expressions makes the relationship between the two expressions much easier to see.

9 Preparing for Standardized Tests

For Exercises 1–13, choose the correct letter.

1. At a supermarket, salad costs \$.40 per ounce. Which rule represents the cost in dollars of buying x ounces of salad? B
   A. $y = -0.4x$ **B.** $y = 0.4x$
   C. $y = 0.4$ **D.** $y = x + 0.4$
   E. $y = x - 0.4$

2. Which function is modeled by the table? E

x	-1	1	3	5
y	-5	-1	3	7

   A. $y = 2x$ **B.** $y = 2x + 3$
   C. $y = \frac{1}{2}x$ **D.** $y = \frac{1}{2}x + 3$
   E. $y = 2x - 3$

3. Find the value of n if $3n - 5 = 7$. B
   A. 3 **B.** 4 **C.** 5 **D.** 6 **E.** 7

4. The sum of four consecutive integers is 190. What is the third integer? E
   A. 44 **B.** 45 **C.** 46 **D.** 47 **E.** 48

5. Which are solutions of $3(x - 4) \leq 18$ and $2(x - 1) \geq 6$? A
   I. 9 **II.** 12 **III.** 15
   A. I only **B.** II only **C.** III only
   D. I and III **E.** II and III

6. What is the slope of a line perpendicular to $3x + 2y = 7$? C
   A. $-\frac{3}{2}$ **B.** $-\frac{2}{3}$ **C.** $\frac{2}{3}$ **D.** $\frac{3}{2}$ **E.** 7

7. How many solutions are there to the quadratic equation $2x^2 + 5x + 1 = 0$? C
   A. 0 **B.** 1 **C.** 2 **D.** 3 **E.** 4

8. Which of the functions is quadratic? D
   I. $y = 2x^2 - 7$ **II.** $y = 2x + 3$
   III. $y = 3x^2 + 2x$
   A. I only **B.** II only **C.** III only
   D. I and III **E.** I, II, and III

9. If $y - x > 0$, what equals $|x - y|$? A
   I. $y - x$ **II.** $|y - x|$ **III.** $x - y$
   A. I and II **B.** II and III **C.** I and III
   D. II only **E.** I, II, and III

10. What is the probability of *not* rolling a 1 or 2 on a number cube? D
   A. $\frac{1}{6}$ **B.** $\frac{1}{3}$ **C.** $\frac{1}{2}$ **D.** $\frac{2}{3}$ **E.** $\frac{5}{6}$

11. A box is 2 cm wide, 5 cm long, and 4 cm high. What is the distance in centimeters between the most distant pair of corners? B
   A. 11 **B.** $3\sqrt{5}$ **C.** $\sqrt{41}$ **D.** $\sqrt{29}$ **E.** $2\sqrt{5}$

Compare the boxed quantity in Column A with the boxed quantity in Column B. Choose the best answer.

A. The quantity in Column A is greater.
B. The quantity in Column B is greater.
C. The two quantities are equal.
D. The relationship cannot be determined on the basis of the information supplied.

	Column A	Column B	
12.	$5\sqrt{3} + 1$	$6\sqrt{3} - 1$	A
13.	$-x - 1$	$x + 1$	D

Find each answer.

14. **Open-ended** Find two points with integer coordinates that are $\sqrt{41}$ units apart.
   Sample: (−3, 4) and (2, 8)

15. The test scores of one student are 79, 80, 85, 87, and 94. Find the mean and standard deviation of these scores. 85; 5.40

16. **Geometry** In $\triangle ABC$, $\angle C$ is a right angle, $AB = 7$, and $m\angle B = 28°$. What are the lengths of $\overline{BC}$ and $\overline{AC}$ to the nearest hundredth?
   6.18; 3.29

17. **Art** In 1996, the National Gallery in Washington, DC, exhibited 21 paintings by Johannes Vermeer. This is about $\frac{2}{3}$ of his paintings that are known to exist. How many of his paintings do we know of today? 31 or 32

18. **Geometry** What is the length of the diagonal of a rectangle with sides 6 cm and 10 cm?
   about 11.66 cm

Resources

Teaching Resources

Chapter Support File, Ch. 9
- Standardized Test Practice
- Cumulative Review

Teacher's Edition

See also p. 412E for assessment options

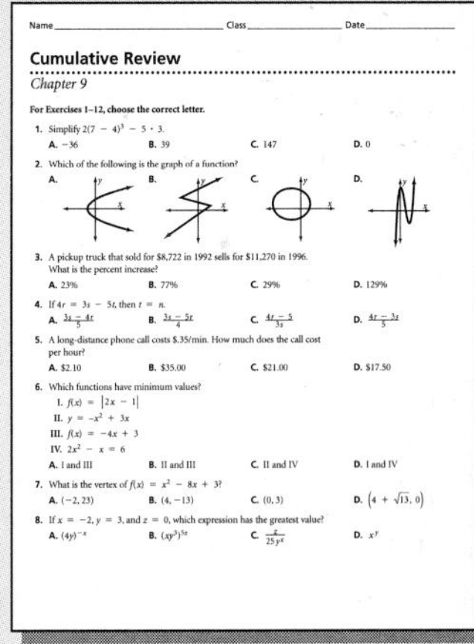

Name______ Class______ Date______

Cumulative Review

Chapter 9

For Exercises 1–12, choose the correct letter.

1. Simplify $2(7 - 4)^3 - 5 \cdot 3$.
   A. −36 B. 39 C. 147 D. 0
2. Which of the following is the graph of a function?
   A. B. C. D.
3. A pickup truck that sold for \$8,722 in 1992 sells for \$11,270 in 1996. What is the percent increase?
   A. 23% B. 77% C. 29% D. 129%
4. If $4r = 3s - 5t$, then $t = n$.
   A. $\frac{3s - 4r}{5}$ B. $\frac{3s - 5r}{4}$ C. $\frac{4r - 5}{3s}$ D. $\frac{4r - 3s}{5}$
5. A long-distance phone call costs \$.35/min. How much does the call cost per hour?
   A. \$2.10 B. \$35.00 C. \$21.00 D. \$17.50
6. Which functions have minimum values?
   I. $f(x) = |2x - 1|$
   II. $y = -x^2 + 3x$
   III. $f(x) = -4x + 3$
   IV. $2x^2 - x = 6$
   A. I and III B. II and III C. II and IV D. I and IV
7. What is the vertex of $f(x) = x^2 - 8x + 3$?
   A. $(-2, 23)$ B. $(4, -13)$ C. $(0, 3)$ D. $(4 + \sqrt{13}, 0)$
8. If $x = -2$, $y = 3$, and $z = 0$, which expression has the greatest value?
   A. $(4y)^{-x}$ B. $(xy^3)^{5z}$ C. $\frac{z}{25y^x}$ D. x^y

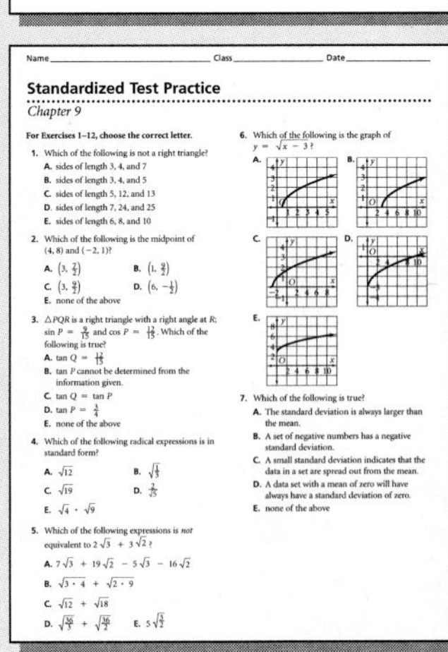

Name______ Class______ Date______

Standardized Test Practice

Chapter 9

For Exercises 1–12, choose the correct letter.

1. Which of the following is not a right triangle?
   A. sides of length 3, 4, and 7
   B. sides of length 3, 4, and 5
   C. sides of length 5, 12, and 13
   D. sides of length 7, 24, and 25
   E. sides of length 6, 8, and 10
2. Which of the following is the midpoint of (4, 8) and (−2, 1)?
   A. $\left(3, \frac{7}{2}\right)$ B. $\left(1, \frac{9}{2}\right)$
   C. $\left(3, \frac{9}{2}\right)$ D. $\left(6, -\frac{1}{2}\right)$
   E. none of the above
3. $\triangle PQR$ is a right triangle with a right angle at R; $\sin P = \frac{9}{15}$ and $\cos P = \frac{12}{15}$. Which of the following is true?
   A. $\tan Q = \frac{12}{15}$
   B. $\tan P$ cannot be determined from the information given.
   C. $\tan Q = \tan P$
   D. $\tan P = \frac{3}{4}$
   E. none of the above
4. Which of the following radical expressions is in standard form?
   A. $\sqrt{12}$ B. $\sqrt{\frac{1}{3}}$
   C. $\sqrt{19}$ D. $\frac{2}{\sqrt{5}}$
   E. $\sqrt{4} \cdot \sqrt{9}$
5. Which of the following expressions is *not* equivalent to $2\sqrt{3} + 3\sqrt{2}$?
   A. $7\sqrt{3} + 19\sqrt{2} - 5\sqrt{3} - 16\sqrt{2}$
   B. $\sqrt{3 \cdot 4} + \sqrt{2 \cdot 9}$
   C. $\sqrt{12} + \sqrt{18}$
   D. $\sqrt{\frac{36}{3}} + \sqrt{\frac{36}{2}}$ E. $5\sqrt{\frac{3}{2}}$
6. Which of the following is the graph of $y = \sqrt{x - 3}$?
   A. B. C. D. E.
7. Which of the following is true?
   A. The standard deviation is always larger than the mean.
   B. A set of negative numbers has a negative standard deviation.
   C. A small standard deviation indicates that the data in a set are spread out from the mean.
   D. A data set with a mean of zero will have always have a standard deviation of zero.
   E. none of the above

CHAPTER 10 Polynomials and Polynomial Functions

To accommodate flexible scheduling, some lessons are divided into parts. Assignment Options are given in the Lesson Planning Options for each lesson.

PACING OPTIONS

This chart suggests pacing only for the core lessons and their parts, and it is provided merely as a possible guide. It will help you determine how much time you have in your schedule to cover other features, such as the Chapter Project, Math Toolboxes, Wrap Up, and Assessment.

	1 Class Period	1 Class Period	1 Class Period	1 Class Period	1 Class Period	1 Class Period	1 Class Period	1 Class Period	1 Class Period
Traditional (40–45 min class periods)	10-1 (1) 10-1 (2)	10-1 (3)	10-2 (1)	10-2 (2)	10-3 (1) 10-3 (2)	10-3 (3)	10-4 (1) 10-4 (2)	10-4 (3)	10-5 (1) 10-5 (2)
Algebra Over 2 Years (40–45 min class periods)	10-1 (1)	10-1 (2)	10-1 (3)	10-2 (1)	10-2 (2)	10-2 (2)	10-3 (1)	10-3 (2)	10-3 (3)
Block Scheduling (90 min class periods)	10-1 (1) 10-2 (2) 10-3 (3)	10-2 (1) 10-2 (2)	10-3 (1) 10-3 (2) 10-3 (3)	10-4 (1) 10-4 (2)	10-4 (3)	10-5 (1) 10-5 (2)	10-6 10-7		

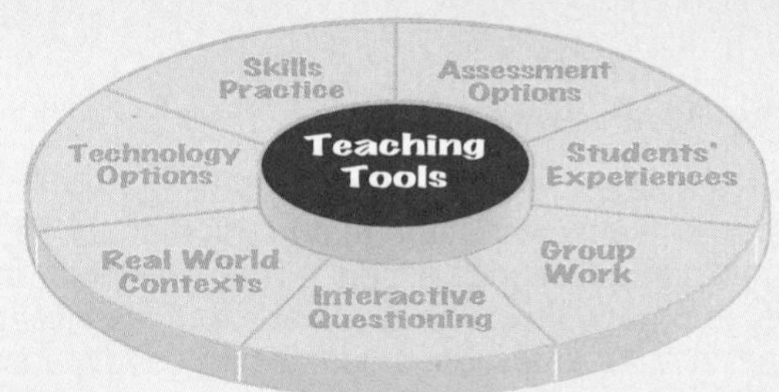

What Students Will Learn and Why

In this chapter, students will build on their knowledge of exponents, learned in Chapter 9, by learning to solve polynomials. They will learn to add and subtract polynomials and to use polynomials to solve real-world problems. They will learn to multiply and factor polynomials and to factor special cases. Students will factor trinomials and will learn how to solve equations by factoring. Finally, students will learn how to choose the appropriate method for solving polynomial functions.

Discussing the Chapter/Building on Experience

The concept map below relates chapter topics to real-world applications. You and your class may wish to add to the map or develop maps of your own. The center oval describes the topic of the chapter. The next level displays topics within the lessons. The outer ovals reflect applications of the content. As you and your class build a concept map, invite students to discuss applications with which they are familiar.

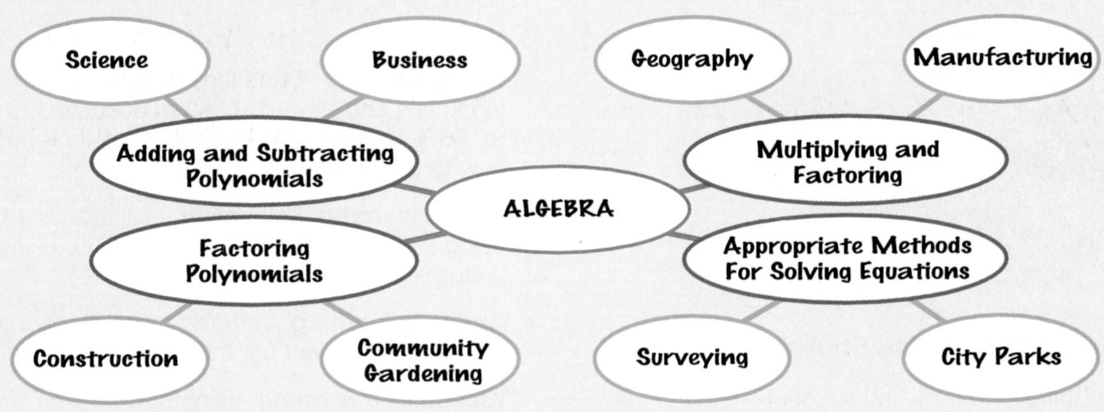

1 Class Period	1 Class Period	1 Class Period	1 Class Period	1 Class Period	1 Class Period	1 Class Period	1 Class Period	1 Class Period	1 Class Period	1 Class Period
10-6	10-7									
10-4 ▽1	10-4 ▽2	10-4 ▽3	10-4 ▽3	10-5 ▽1	10-5 ▽2	10-5 ▽2	10-6	10-6	10-7	10-7

Interactive Questioning Tips

A question is interactive when there is "give and take" between the questioner (teacher or student) and the respondent. When a critical thinking question is asked, it is important for the questioner to phrase the question clearly and specifically. When given a direct question, students can use time efficiently to formulate responses rather than taking time to interpret the question.

Skills Practice

Every lesson provides skill practice with Exercises On Your Own and Exercises Mixed Review. The Student Edition includes Checkpoints (pp. 128, 144) and Preparing for Standardized Tests (p. 155). In the Teacher's Edition, the Lesson Planning Options section for each lesson lists Prerequisite Skills students should know for that lesson. At the back of the Student Edition is the Skills Handbook—mini-lessons on math the students may need to review. The Chapter Support File for Chapter 10 in the Teaching Resources box includes two Practice worksheets per lesson, a worksheet for two Checkpoints, and worksheets for Cumulative Review and Standardized Test Preparation.

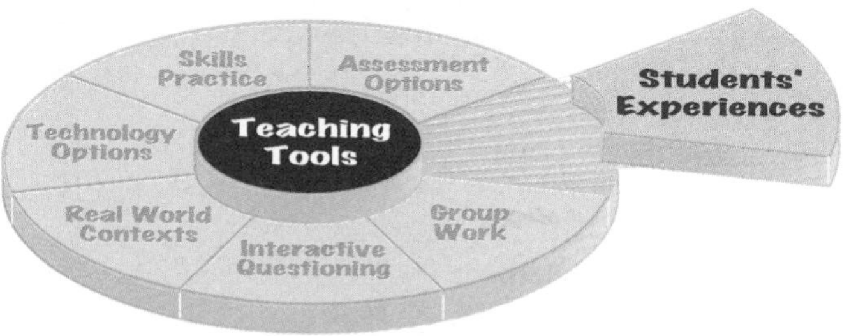

Diverse Learning and Teaching Styles

In your Teacher's Edition, you will find suggestions as to how you can help students complete mathematical tasks in Chapter 10 by reinforcing various learning styles. Here are some examples.

- **Visual learning** using tiles to factor perfect square trinomials and having students describe the patterns used (p. 487), making posters describing the special cases and how to factor them (p. 488)
- **Tactile learning** comparing using pencil and paper with using tiles to factor trinomials (p. 482), using tiles to see the patterns in the Work Together (p. 486)
- **Auditory learning** verbalizing the steps in the FOIL method as they act out the method (p. 476)
- **Kinesthetic learning** using polynomials written on paper to help students identify different polynomials (p. 466), using four students to act out the FOIL method by shaking hands (p. 476)

Alternative Activity for Lesson 10-2

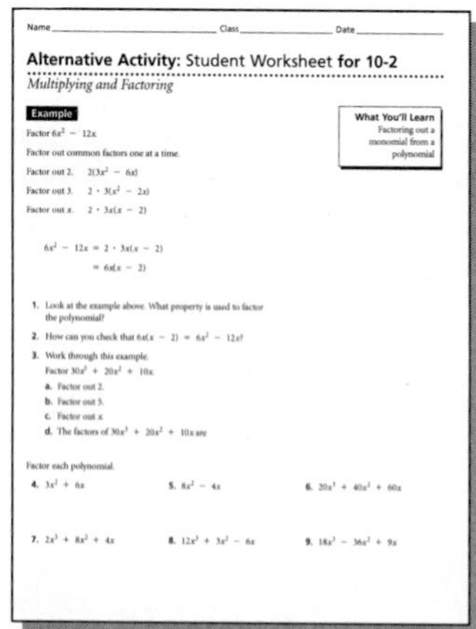
Alternative Activity: Student Worksheet **for 10-2**

Multiplying and Factoring

for use with Example 2, addresses auditory learning by reviewing the meaning of key terms and working through Example 2.

Alternative Activity for Lesson 10-3

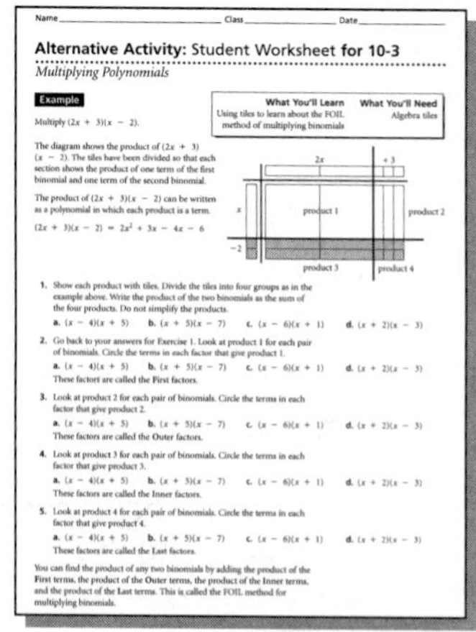
Alternative Activity: Student Worksheet **for 10-3**

Multiplying Polynomials

for use with Example 2, addresses visual and tactile learning by teaching the FOIL method with tiles.

Alternative Activity for Lesson 10-4

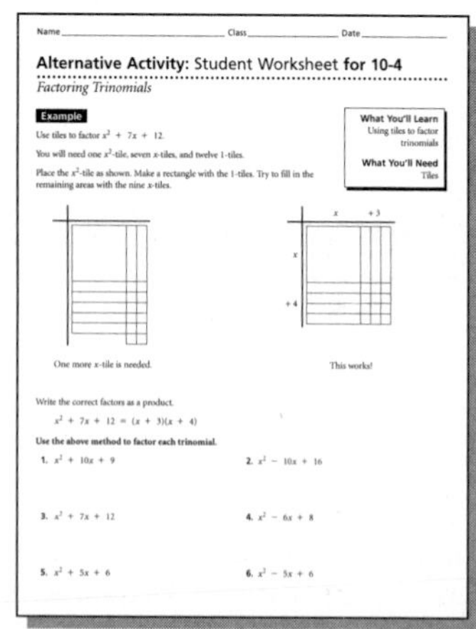
Alternative Activity: Student Worksheet **for 10-4**

Factoring Trinomials

for use with Example 2, addresses tacti learning by using tiles to factor trinomia

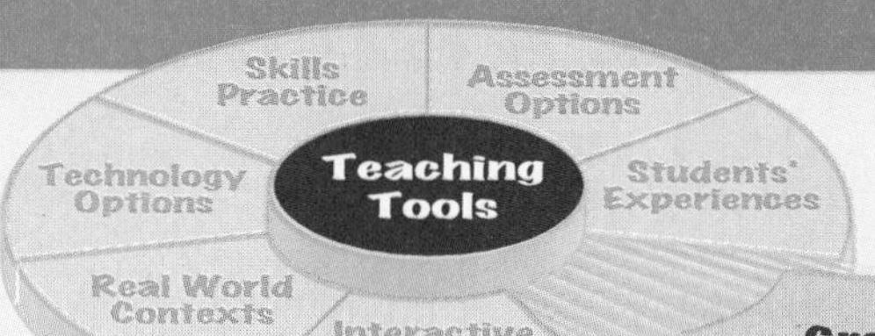

Cooperative Learning Tips

When used effectively, cooperative learning can help students develop interpersonal skills, learn to perform specific roles in a group, and learn to carry out specific responsibilities. The components of Chapter 10 provide a range of cooperative learning opportunities.

- In the Student Edition, the ***Work Together*** parts of lessons are specifically designed for cooperative learning activities.
- In the Teacher's Edition, you will find helpful hints for addressing diverse learning styles (see page C for Chapter 10). For every lesson, you will find a ***Reteaching Activity***, which may involve cooperative learning.

Materials and Manipulatives

The authors expect all students to use scientific calculators. Calculator use is integrated throughout the course.

- **graph paper (10-4)**
- **graphing calculator (10-3, 10-5, 10-6, 10-7)**
- **tiles (10-1, 10-2, 10-3, 10-4, 10-5)**

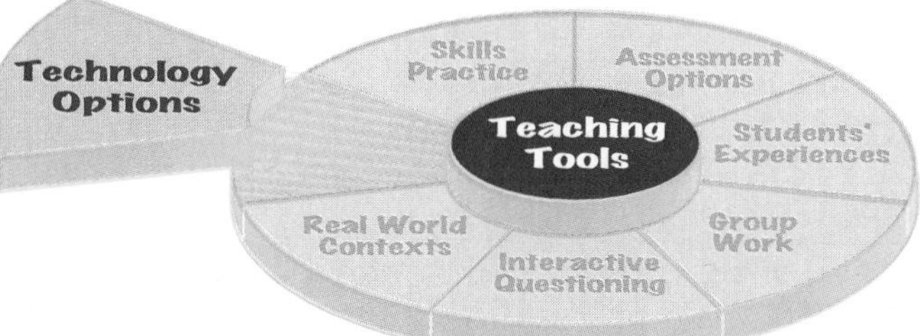

TECHNOLOGY OPTIONS

Technology Tools		Chapter Project	10-1	10-2	10-3	10-4	10-5	10-6	10-7
Calculator		Assumed that students have scientific calculators at any time.							
Graphing Calculator	Handbook				✔		✔	✔	✔
	Student Edition				✔		✔	✔	✔
Software	Secondary Math Lab Toolkit™		✔	✔	✔	✔	✔	✔	✔
	Integrated Math Lab		✔			✔			
	Computer Item Generator		✔	✔	✔	✔	✔	✔	✔
	Student Edition								
Video	Video Field Trip	✔							
CD-ROM	Multimedia Algebra Lab		✔	✔	✔			✔	
Internet		See the Prentice Hall site. (http://www.phschool.com)							

The Prentice Hall Algebra program offers you a rich variety of technology options. Be assured that all these options are provided as a means of enriching the program and are not essential for the successful completion of the course.

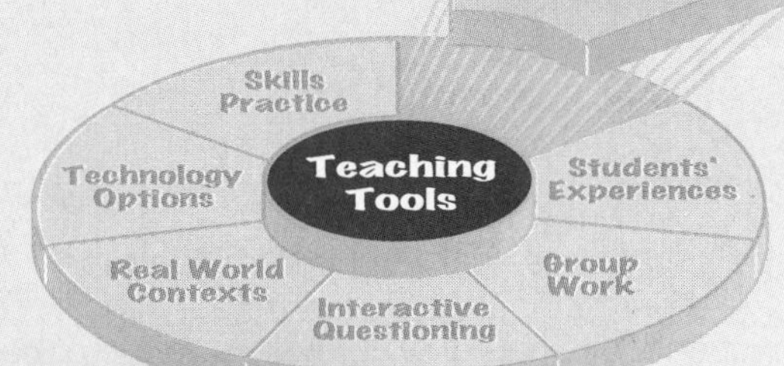

Assessment Options

The Prentice Hall Algebra Program provides you with many options. From these options, you may choose instructional materials and techniques appropriate for your students or those necessary to meet your district's curriculum requirements. As the chart indicates, the program also supports your teaching efforts by offering you many choices for assessment.

ASSESSMENT OPTIONS

Assessment Support Materials	Chapter Project	10-1	10-2	10-3	10-4	10-5	10-6	10-7	Chapter End
Chapter Project	● ▲ ■		▲ ■	▲ ■	▲ ■		▲ ■		▲ ■
Checkpoints				▲ ●			▲ ●		
Self-Assessment				▲ ■	▲ ■		▲ ■	▲ ■	▲ ■
Writing Assignment		▲	▲	▲	▲	▲	▲	▲	● ▲
Chapter Assessment									▲ ●
Alternative Assessment		■	■	■	■	■	■	■	● ■
Cumulative Review									▲ ●
Standardized Test Prep						▲ ■	▲ ■	▲ ■	●
Computer Item Generator	Can be used to create custom-made practice or assessment at any time.								

▲ = Student Edition ■ = Teacher's Edition ● = Teaching Resources

Checkpoints

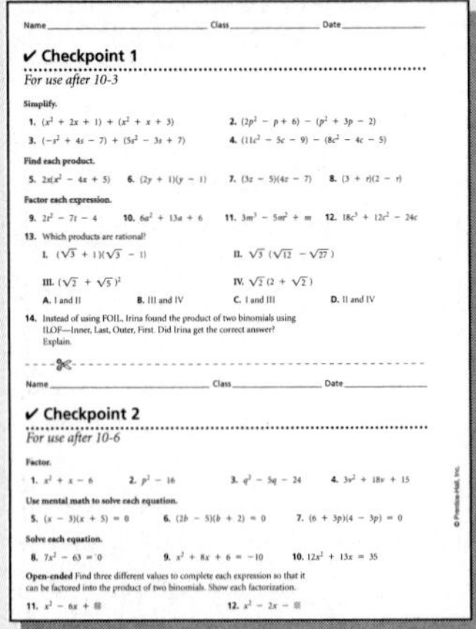

✔ Checkpoint 1
For use after 10-3

✔ Checkpoint 2
For use after 10-6

Alternative Assessment

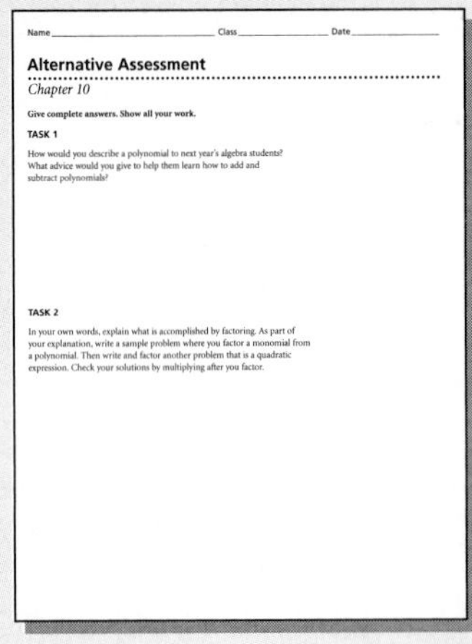

Alternative Assessment
Chapter 10

Chapter Assessment

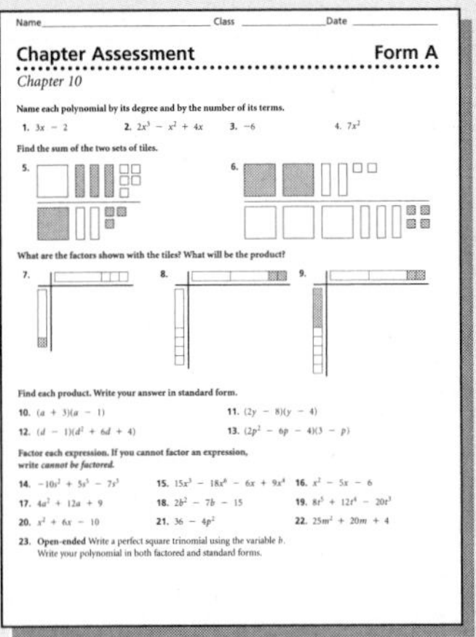

Chapter Assessment Form A
Chapter 10

Available in both Form A and Form B

Making the Right Connections

Mathematics is imbedded in nearly every walk of life. The National Council of Teachers of Mathematics (NCTM) encourages educators to recognize these connections and to emphasize them for the purpose of better educating students for success in life and in a global economy. The ***Connections*** chart below highlights these connections for Chapter 10.

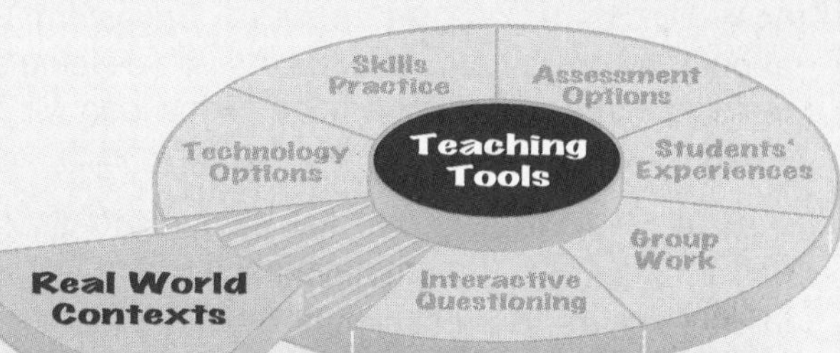

CONNECTIONS

Lesson	Interdisciplinary Connections	Career Prep	Other Real World Connections	Math Integration	NCTM Standards
Chapter Project	Envirnmental Science	Production Carpentry	Ecology Conservation	Trigonometry	Problem Solving
10-1	Language Science Accounting	Business Industry	Language Arts Other Planets	Geometry	Algebra Communication Problem Solving
10-2	Geography Social Studies	Manufacturing	Building Models Geography	Geometry	Algebra Communication Problem Solving Geometry
10-3	Language Writing	Construction Investments	Financial Planning Banking Building	Geometry Mathematical Finance	Algebra Communication Problem Solving
10-4	Writing Sociology	Statistical Analysis	Construction Community Gardening	Statistics	Algebra Communication Problem Solving Statistics
10-5	Writing Economics	Math in the Media Industry Sales	Crane Operators Hardware Stores	Geometry	Algebra Communication Problem Solving Geometry
10-6	Physical Education Fitness	Manufacturing Business	Baseball Sailing Construction	Geometry	Algebra Communication Problem Solving Geometry
10-7	Writing History	Surveying Marketing	City Parks	Geometry	Algebra Communication Problem Solving Geometry

CHAPTER 10 Polynomials and Polynomial Functions

CONNECTING TO PRIOR LEARNING Brainstorm with students about the concept of *like terms*. As you develop a working definition, stress that knowledge of this idea will be key to students' understanding of the problems in the chapter.

CULTURAL CONNECTIONS Ask students to discuss Arbor Day, and whether any of them have participated in a tree-planting project. Ask what students know about tree conservation efforts in other countries. Discuss also the impact of natural and man-made disasters on the forest resources in this country and other countries.

INTERDISCIPLINARY CONNECTIONS Have students brainstorm for a list of reasons for cutting trees and reasons for preserving them. Have students investigate how many trees it takes to produce a toothpick, a sheet of paper, or a textbook. Ask them to find out whether recycling has made a difference in the number of trees used by the paper industry.

ABOUT THE PROJECT The Chapter Project gives students an opportunity to study and learn more about the uses of trees and wood. In the Find Out questions found throughout the chapter, students use formulas, equations, and graphs to analyze data and predict the production of wood and fruit.

Technology Options

Prentice Hall Technology

Video • Video Field Trip 10, "Silent Sentinels," a look at some of the giant trees of the Redwood National Forest.

CHAPTER 10 Polynomials

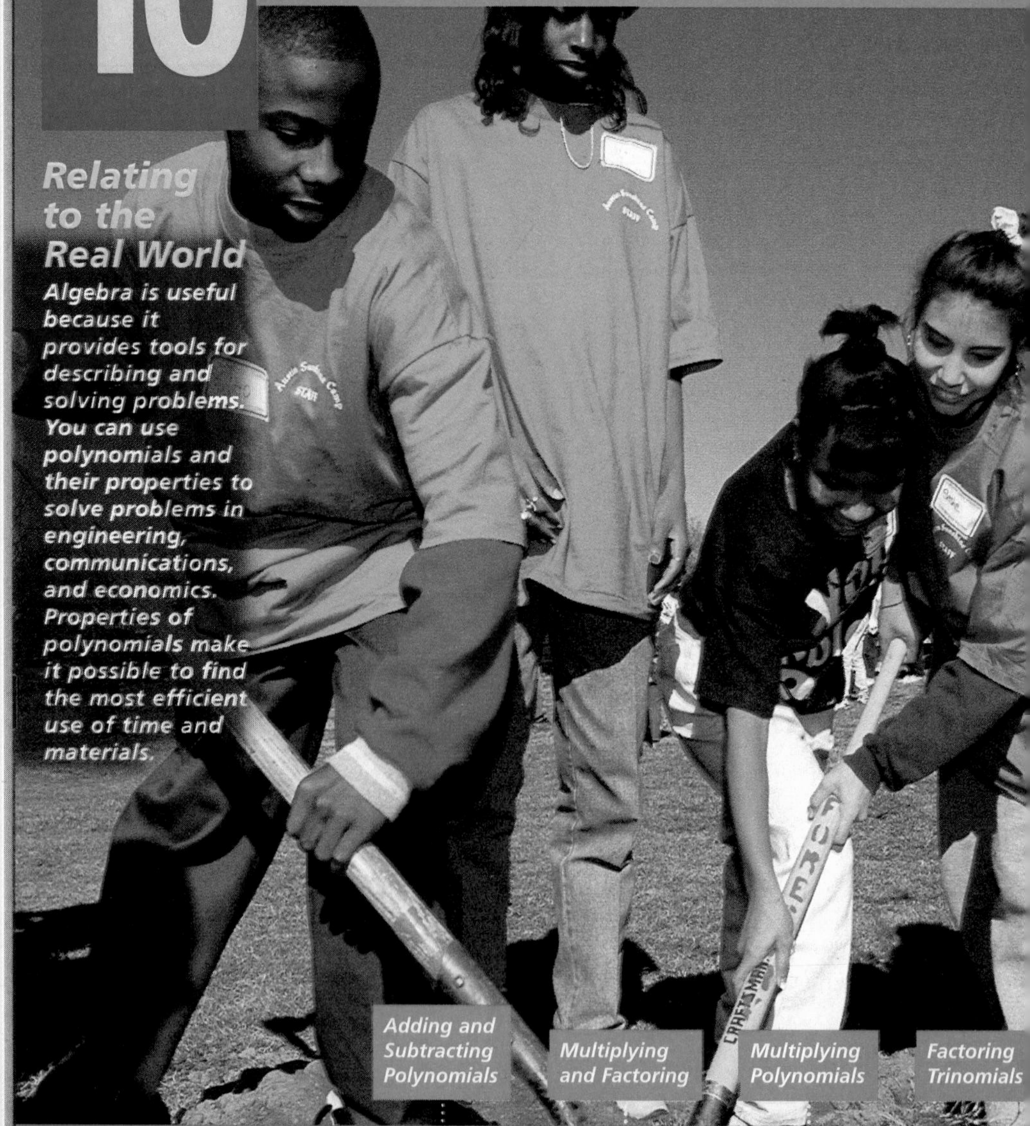

Relating to the Real World

Algebra is useful because it provides tools for describing and solving problems. You can use polynomials and their properties to solve problems in engineering, communications, and economics. Properties of polynomials make it possible to find the most efficient use of time and materials.

Launching the Project

PROJECT NOTEBOOK Encourage students to keep all project-related materials in a separate folder or notebook. See Chapter Project and Scoring Rubric in Chapter Support File.

- Ask students to think of something flat that is made of wood, such as a table top or door.
- Have them estimate how many pieces of wood, each 1 ft^2, make up their object.
- Have students compare results with partners.
- Have students look at the Find Out by Researching section on page 474 of their textbooks. Explain that they will research types of wood and the tools carpenters use to work with wood.

TRACKING THE PROJECT You may wish to have students read Finishing the Chapter Project on page 502 to help them get an overview of the project. Set benchmark deadlines for students to show you their work in progress.

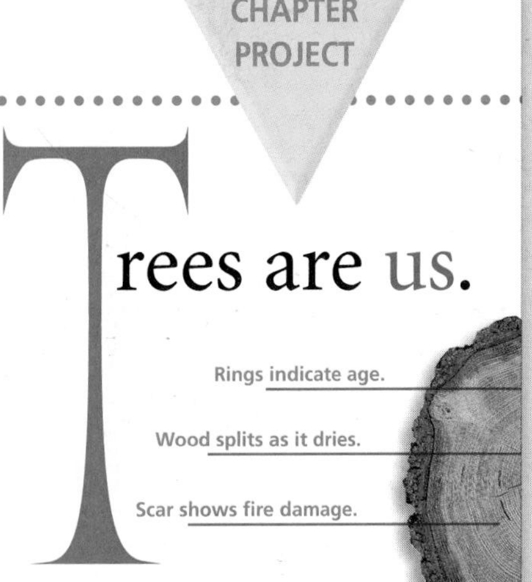

Trees are us.

Many schools celebrate Arbor Day by planting young trees to replenish our ecosystem. Trees use the carbon dioxide that humans and animals exhale to make oxygen. Trees anchor the soil and prevent erosion. They also produce fruit. Wood from trees is used for the construction of everything from pencils to houses.

As you work through the chapter, you will learn more about the uses of trees. You will use formulas to analyze data and predict the production of wood and fruit. Then you will decide how to organize and display your results.

To help you complete the project:

Resources

Teaching Resources

Chapter Support File, Ch. 10
- Chapter Project Manager and Scoring Rubric

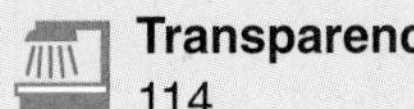

Transparencies

114

Using the Rubric

Sharing the scoring rubric for the project with your students will alert them to your expectations before they begin work on the project.

As students complete each Find Out question in the chapter, you may wish to have them evaluate their own work or a partner's work, based on the scoring rubric. Students should have the opportunity to revise their work after it has been reviewed.

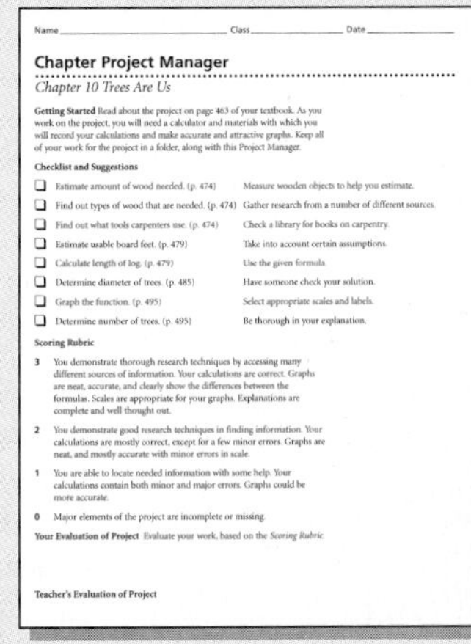

Name ______ Class ______ Date ______

Chapter Project Manager

Chapter 10 Trees Are Us

Getting Started Read about the project on page 463 of your textbook. As you work on the project, you will need a calculator and materials with which you will record your calculations and make accurate and attractive graphs. Keep all of your work for the project in a folder, along with this Project Manager.

Checklist and Suggestions

	Checklist	Suggestions
❑	Estimate amount of wood needed. (p. 474)	Measure wooden objects to help you estimate.
❑	Find out types of wood that are needed. (p. 474)	Gather research from a number of different sources.
❑	Find out what tools carpenters use. (p. 474)	Check a library for books on carpentry.
❑	Estimate usable board feet. (p. 479)	Take into account certain assumptions.
❑	Calculate length of log. (p. 479)	Use the given formula.
❑	Determine diameter of trees. (p. 485)	Have someone check your solution.
❑	Graph the function. (p. 495)	Select appropriate scales and labels.
❑	Determine number of trees. (p. 495)	Be thorough in your explanation.

Scoring Rubric

3 You demonstrate thorough research techniques by accessing many different sources of information. Your calculations are correct. Graphs are neat, accurate, and clearly show the differences between the formulas. Scales are appropriate for your graphs. Explanations are complete and well thought out.

2 You demonstrate good research techniques in finding information. Your calculations are mostly correct, except for a few minor errors. Graphs are neat, and mostly accurate with minor errors in scale.

1 You are able to locate needed information with some help. Your calculations contain both minor and major errors. Graphs could be more accurate.

0 Major elements of the project are incomplete or missing.

Your Evaluation of Project Evaluate your work, based on the *Scoring Rubric*.

Teacher's Evaluation of Project

10-1 Teaching Notes

PROBLEM OF THE DAY

A triangular pennant has an area of 40 ft^2. The base is 2 ft more than the height. Find the base and the height of the pennant. 8 ft and 10 ft

Problem of the Day is also available in Transparencies.

CONNECTING TO PRIOR KNOWLEDGE On the board, write $2a + 3b + 4c + 6a + 2b + 12c$. Ask students: *How would you simplify this expression?* combine like terms Help students recognize that $2a + 6a$ can be written $a(2 + 6)$. Explain that when you add and subtract polynomials, you combine like terms.

WORK TOGETHER

DIVERSITY Some students may not know the purpose of all the supplies for birds that are carried in the pet store. Ask knowledgeable students to describe the various items.

CONNECTING TO THE STUDENTS' WORLD Have students each choose an animal they would like to own as a pet. Then have students research the supplies needed per month and the cost of each item. Once students have their data, encourage them to create problems like the Work Together question for other students to solve.

MAKING CONNECTIONS *Harper's Index* estimates that it costs \$6400 to raise a medium-size dog to the age of 11 years.

Lesson Planning Options

Prerequisite Skills

- Combining like terms
- Using tiles

Assignment Options for Exercises On Your Own

To provide flexible scheduling, this lesson can be subdivided into parts.

1. **Core** 1–13
   Extension 23, 24, 29, 30
2. **Core** 14, 15, 25–28
   Extension 31–33
3. **Core** 16–21, 34–39
   Extension 22, 40

Use Mixed Review to maintain skills.

Resources

Student Edition

Skills Handbook, pp. 580, 582
Extra Practice, p. 565
Glossary/Study Guide

Teaching Resources

Chapter Support File, Ch. 10
- Practice 10-1 (two worksheets)
- Reteaching 10-1

Classroom Manager 10-1
Glossary, Spanish Resources
Two-Year Algebra Handbook 10-1

Transparencies

17, 24, 115, 119

What You'll Learn
- Describing polynomials
- Adding and subtracting polynomials

...And Why

To use polynomials in real-world situations, such as working in a store

What You'll Need
- tiles

10-1 Adding and Subtracting Polynomials

WORK TOGETHER

Business Work in groups. Suppose you work at a pet store. The spreadsheet shows the contents of several customers' orders.

	A	B	C	D	E	F
1	Customer	Seed	Cuttlebone	Millet	G. Paper	Perches
2	Davis			✔		
3	Brooks	✔	✔			
4	Casic	✔		✔		
5	Martino	✔			✔	✔

ROCKY'S FRIENDS
Bird Supplies
- bird seed (5 lb) \$3.99
- cuttlebone (2 ct) 2.00
- spray millet (5 lb) 24.00
- gravel paper (1 pkg) 2.29
- perches (2 ct) 1.89

The variables represent the number of each item ordered.
s = bags of birdseed m = bags of millet p = packages of perches
c = packages of cuttlebone g = packages of gravel paper

1. Which expression represents the cost of Casic's order? B
   A. $27.99(s + m)$ **B.** $3.99s + 24m$ **C.** $27.99sm$
2. Write expressions to represent each of the other customers' orders. Davis $24m$; Brooks $3.99s + 2c$; Martino $3.99s + 2.29g + 1.89p$
3. Martino buys 10 bags of birdseed, 4 packages of gravel paper, and 2 packages of perches. What is the total cost of his order? \$52.84
4. *Open-ended* Make up several orders and write expressions for these orders. Have other members of your group find the cost of your orders. Check each other's work. Answers may vary. Sample: 2 bags birdseed, 2 pkgs cuttlebone, 1 pkg perches; $2s + 2c + p$; \$13.87

Question 4 Students will need to give values for each variable. The values will represent the number of each item purchased.

Suggest the following as an independent project or use as a class project if you have block scheduling or an extended class period. Have students use computer software to replicate this spreadsheet . They can then use the spreadsheet to calculate the total cost of a sample order.

THINK AND DISCUSS

To check students' understanding of the section entitled "Describing Polynomials," ask the following questions:

- *Is each of the following expressions a polynomial?* $x^3 + 2x^2 + 1$ yes $2x^2 + \frac{1}{x}$ no $4x^{-2}$ no $-3x + 5$ yes
- *Why does the polynomial* $7x + 4$ *have a degree of* 1? The exponent of x is 1.
- *Is the polynomial* $3x^2 + 2x + 3$ *in standard form?* yes
- *Is the polynomial* $3x^2 + 2x^2 + 3$ *in standard form?* no *Why not?* Two terms have the same degree.

Example 1

Question 5 Point out that for a polynomial to be in standard form, the terms must decrease in degree from left to right. Check students' answers to be sure they understand exactly what standard form means.

THINK AND DISCUSS

Part 1 Describing Polynomials

QUICK REVIEW

A *term* is a number or the product or quotient of a number and a variable.

A **polynomial** is one term or the sum or difference of two or more terms. A polynomial has no variables in a denominator. For a term that has only one variable, the **degree of a term** is the exponent of the variable.

$$x^3 - 4x + 5x^2 + 7$$

The degree of a constant is 0.

degree → 3 1 2 0

The **degree of a polynomial** is the same as the degree of the term with the highest degree. You can name a polynomial by its degree or by the number of its terms.

QUICK REVIEW

$x = x^1$

Polynomial	Degree	Name Using Degree	Number of Terms	Name Using Number of Terms
$7x + 4$	1	linear	2	**binomial**
$3x^2 + 2x + 1$	2	quadratic	3	**trinomial**
$4x^3$	3	cubic	1	**monomial**
5	0	constant	1	monomial

The polynomials in the chart are in **standard form,** which means the terms decrease in degree from left to right and no terms have the same degree.

Example 1

Write each polynomial in standard form. Then name each polynomial by its degree and the number of its terms.

a. $5 - 2x$
$-2x + 5$
linear binomial

b. $3x^4 - 4 + 2x^2$
$3x^4 + 2x^2 - 4$
fourth degree trinomial

c. $-2x + 5 - 4x^2 + x^3$
$x^3 - 4x^2 - 2x + 5$
cubic polynomial with four terms

5a. $-9x^4 + 6x^2 + 7$; fourth degree trinomial

b. $-y^3 + 3y - 9$; cubic trinomial

c. $7v + 9$; linear binomial

5. Try This Write each polynomial in standard form. Then name each polynomial by its degree and the number of its terms. See left.

a. $6x^2 + 7 - 9x^4$ **b.** $3y - 9 - y^3$ **c.** $9 + 7v$

Part 2 Adding Polynomials

You can use tiles to add and subtract polynomials.

Opposite terms form zero pairs. → x^2 $-x^2$ = 0 x $-x$ = 0 1 -1 = 0

Additional Examples

FOR EXAMPLE 1

Write each polynomial in standard form. Then name each polynomial by its degree and the number of its terms.

a. $-2 + 7x$
$7x - 2$ linear binomial

b. $2x^3 - x^4 + 3$ $-x^4 + 2x^3 + 3$
fourth degree trinomial

c. $3x^4 - 2 - 2x^4 + 7x$
$x^4 + 7x - 2$
fourth degree trinomial

Discussion: *Why is the expression* $3x^4 - 2 - 2x^4 + 7x$ *considered a fourth degree polynomial with three terms?*

FOR EXAMPLE 2

Find $(5x^2 - 3x + 7) + (2x^2 + 5x - 7)$. $7x^2 + 2x$

FOR EXAMPLE 3

Find $(2x^3 + 4x^2 - 6) - (3x^3 + 2x - 2)$
$-x^3 + 4x^2 - 2x - 4$

If tiles are available, introduce the x^2-tile. Point out that each side is the same length as the x-tile. Have students use tiles to make polynomials such as $2x^2 + 3x - 1$ and $-2x^2 + 4x + 3$.

KINESTHETIC LEARNING Have students write on pieces of paper, polynomials of different degrees with different numbers of terms. Distribute the papers randomly. Then direct students to move to locations in the room with instructions such as the following.

- *All linear polynomials go to the front.*
- *All cubic polynomials go to the windows.*
- *All trinomial polynomials go to the back.*
- *All binomial cubic polynomials go to the front.*
- *All monomial linear polynomials go to the door.*

After each move, have students check with each other to make sure that they are in the correct location.

Example 2

You can relate adding polynomials vertically and horizontally to adding whole numbers vertically and horizontally.

Ask: *Will the tile method work for adding any two polynomials?* no *Why not?* It will only work for polynomials for which the degree is 2 or less. The tiles do not contain a model for x^3.

Question 7 Remind students to give their answers in standard form.

Technology Options

For Exercises 16–21 and 34–39, students may wish to use algebra software to simplify the expressions.

Prentice Hall Technology

Software
- Secondary Math Lab Toolkit™
- Integrated Math Lab 19
- Computer Item Generator 10-1

CD-ROM
- Multimedia Algebra Lab 10

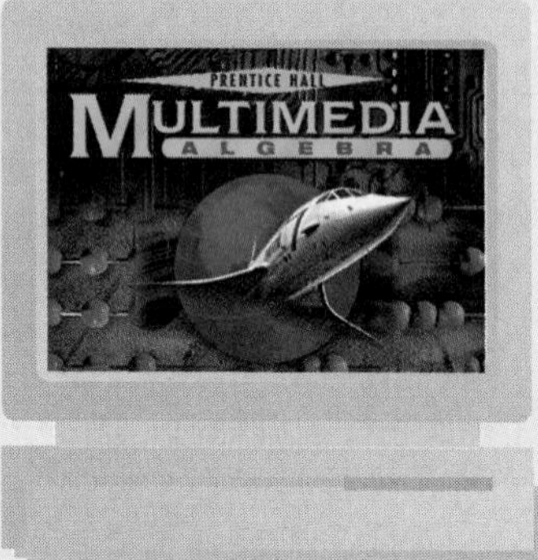

Internet
- See the Prentice Hall site. (http://www.phschool.com)

6. What polynomial is shown using each set of tiles?

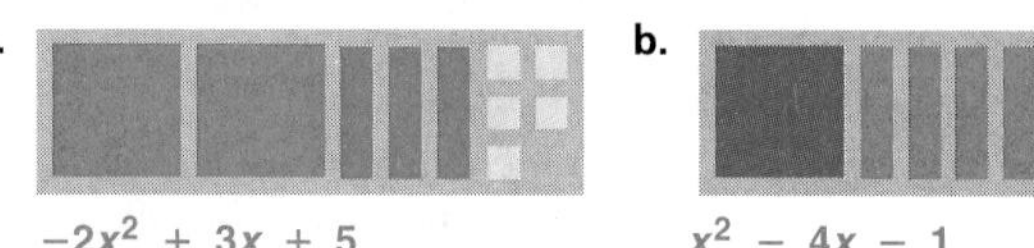

a. $-2x^2 + 3x + 5$ b. $x^2 - 4x - 1$

Example 2

Find $(2x^2 - 3x + 4) + (3x^2 + 2x - 3)$.

Method 1 Add using tiles.

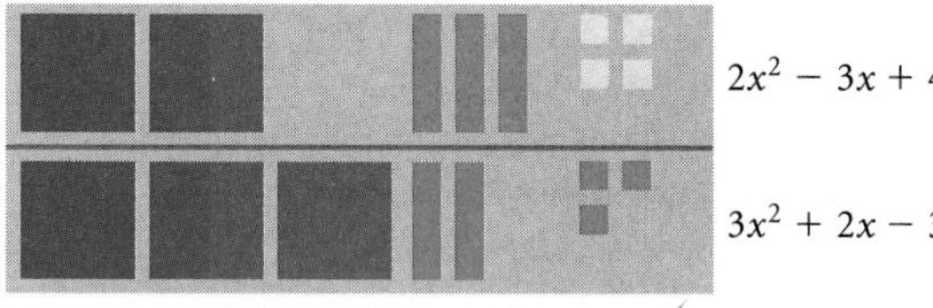

Group like tiles together. Remove zero pairs. Write an expression for the remaining tiles.

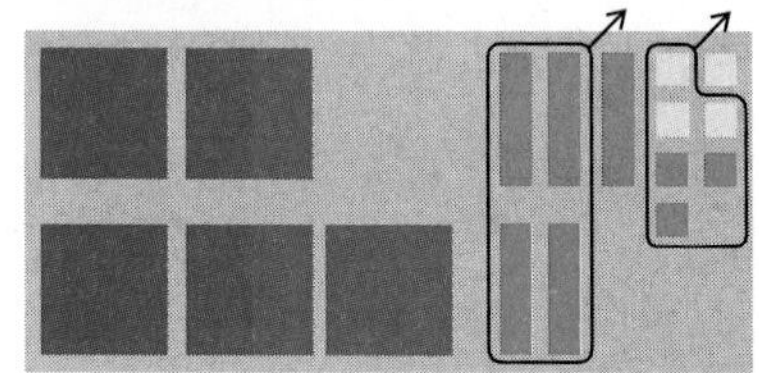

$5x^2 - x + 1$

QUICK REVIEW

Like terms have the same variable and the same number of variable factors, or the same degree.

Method 2 Add vertically.

Line up like terms.
Then add the coefficients.

$$\begin{array}{r} 2x^2 - 3x + 4 \\ \underline{3x^2 + 2x - 3} \\ 5x^2 - x + 1 \end{array}$$

The sum is $5x^2 - x + 1$.

Method 3 Add horizontally.

Group like terms. Then add the coefficients.

$(2x^2 - 3x + 4) + (3x^2 + 2x - 3)$
$= (2x^2 + 3x^2) + (-3x + 2x) + (4 - 3)$
$= 5x^2 - x + 1$

7. a. **Try This** Find $(5x^2 + 4x - 7) + (-4x^2 + 8x - 1)$ using any method you choose. $x^2 + 12x - 8$
   b. Why did you choose the method you used? Answers may vary. Sample: adding vertically; it helps to keep like terms in order.

Part

Subtracting Polynomials

In Chapter 1, you learned that subtraction means to add the opposite. So when you subtract a polynomial, change each of its terms to the opposite. Then add the coefficients.

Example 3

While discussing the **Check** section, ask: *Would zero be a good value to substitute for x?* no *Why not?* All the variable terms would become zero and you wouldn't know if one of the coefficients was incorrect.

ERROR ALERT! **Question 8** Some students may use zero for the coefficient of *x*. **Remediation:** Remind them that *x* means 1*x*.

EXTENSION Question 8 Have students verify their answers by using tiles. They can do this by adding the opposite of each term being subtracted.

Exercises ON YOUR OWN

ALTERNATIVE ASSESSMENT **Exercises 1–5** To assess students' understanding of polynomial nomenclature, ask them to write their own examples for A–E.

ERROR ALERT! **Exercises 6–13** Many students forget that the sign in front of a term must move with that term. For example, when rewriting $3a - 2a^3$, many students will write $2a^3 - 3a$. The correct way is $-2a^3 + 3a$. **Remediation:** Have the students circle all of the terms with signs that are in front of each term. Students must move everything in the circle together.

Exercises 16–21 Have students discuss which methods they used to find each sum or difference. Ask: *Which exercises*

Example 3

Find $(7x^3 - 3x + 1) - (x^3 + 4x^2 - 2)$.

Method 1 Subtract vertically.

Line up like terms.		Add the opposite.
$7x^3 \qquad - 3x + 1$		$7x^3 \qquad - 3x + 1$
$-(x^3 + 4x^2 \qquad - 2)$	→	$-x^3 - 4x^2 \qquad + 2$
		$6x^3 - 4x^2 - 3x + 3$

Method 2 Subtract horizontally.

Write the opposite of each term in the polynomial being subtracted. Group like terms. Then add the coefficients of like terms.

$(7x^3 - 3x + 1) - (x^3 + 4x^2 - 2)$

$= 7x^3 - 3x + 1 - x^3 - 4x^2 + 2$

$= (7x^3 - x^3) - 4x^2 - 3x + (1 + 2)$ ← The coefficients of $7x^3$ and $-x^3$ are 7 and −1. $7x^3 - 1x^3 = 6x^3$.

$= 6x^3 - 4x^2 - 3x + 3$

The difference is $6x^3 - 4x^2 - 3x + 3$.

Check Substitute a value for *x* to check. Here 2 is substituted for *x*.

$$(7x^3 - 3x + 1) - (x^3 + 4x^2 - 2) \stackrel{?}{=} 6x^3 - 4x^2 - 3x + 3$$

$$7(2^3) - 3(2) + 1 - [2^3 + 4(2^2) - 2] \stackrel{?}{=} 6(2^3) - 4(2^2) - 3(2) + 3$$

$$(56 - 6 + 1) - (8 + 16 - 2) \stackrel{?}{=} 48 - 16 - 6 + 3$$

$$51 - 22 \stackrel{?}{=} 29$$

$$29 = 29 ✓$$

8. Try This Subtract $(3x^2 + 4x - 1) - (x^2 - x - 2)$. $2x^2 + 5x + 1$

9. Critical Thinking How are subtracting vertically and subtracting horizontally alike? Answers may vary. Sample: Each term being subtracted is changed to its opposite.

Exercises ON YOUR OWN

Match each expression with its name.

1. $5x^2 - 2x + 3$ C
2. $\frac{3}{4}z + 5$ D
3. $7a^3 + 4a - 12$ E
4. $\frac{3}{x} + 5$ B
5. -15 A

A. constant monomial
B. *not* a polynomial
C. quadratic trinomial
D. linear binomial
E. cubic trinomial

could not be done with tiles? **Exercises 17–20** *Why?* **There is no x^3-model included with the tiles.**

Exercises 20–21 These exercises extend the concept of adding and subtracting polynomials to include three polynomials.

Exercise 22 Ask: *What should Kwan have written?* **$5x^2 - 3x + 1 - 2x^2 + 4x + 2$**

WRITING Exercise 23 Encourage students to relate the word to the meaning of the prefix: mono (one); bi (two); tri (three); poly (many). For example, a monogram is a design made from *one* letter of a name: the initial letter. A binocular device is designed for use by *two* eyes at once. A tricuspid is a tooth having *three* points. A polyglot is a person who speaks *many* languages.

Exercise 24b Have students name mathematical terms that begin with these prefixes. Have students describe each term.

GEOMETRY Exercises 25–28 Remind students that to find the perimeter, they must add the lengths of all the sides for each figure, not just the labeled sides.

Exercise 27 Point out that sides with the same number of hash marks on them are equal in length.

CRITICAL THINKING Exercise 30 Have students look back at the chart on Naming Polynomials on page 465 to help them answer this question.

GEOMETRY Exercises 31–33 Students must add the known sides first. Then subtract that sum from each perimeter to find the lengths of the missing sides.

pages 467–469 On Your Own

23a. monogram: as a noun, a single design made up of two or more letters, often initials of a name

b. binocular: as an adjective, for the use of, or using, both eyes at the same time

c. tricuspid: as a noun, a tooth with three cusps; as an adjective, the tricuspid valve in the heart has three flaps

d. polyglot: as a noun, a person who uses several languages

Write each polynomial in standard form. Then name each polynomial by its degree and number of terms.

6. $4x - 3x^2$ **$-3x^2 + 4x$; quad. bin.**

7. $4x + 9$ **$4x + 9$; linear bin.**

8. $6 - 3x - 7x^2$ **$-7x^2 - 3x + 6$; quad. tri.**

9. $9z^2 - 11z^3 + 5z - 5$ **$-11z^3 + 9z^2 + 5z - 5$ cubic with 4 terms**

10. $y - 7y^3 + 15y^8$ **$15y^8 - 7y^3 + y$; 8th deg. tri.**

11. $c^2 - 2 + 4c$ **$c^2 + 4c - 2$; quad. tri.**

12. $7 + 5b^2$ **$5b^2 + 7$; quad. bin.**

13. $-10 + 4q^4 - 8q + 3q^2$ **$4q^4 + 3q^2 - 8q - 10$; 4th deg. poly. with 4 terms**

Find the sum of the two sets of tiles.

14. 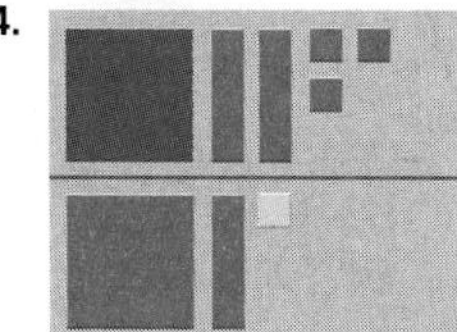 **$-3x - 2$**

15.  **$x^2 - 1$**

Find each sum or difference.

16. $(7y^2 - 3y + 4y) + (8y^2 + 3y^2 + 4y)$ **$18y^2 + 5y$**

17. $(2x^3 - 5x^2 + 3x - 1) - (8x^3 - 8x^2 + 4x + 3)$ **$-6x^3 + 3x^2 - x - 4$**

18. $(-7z^3 + 3z - 1) - (-6z^2 + z + 4)$ **$-7z^3 + 6z^2 + 2z - 5$**

19. $(7a^3 + 3a^2 - a + 2) + (8a^2 - 3a - 4)$ **$7a^3 + 11a^2 - 4a - 2$**

20. $(5y^3 + 7y) - (3y^3 + 9y^2) + (7y^3 + 2y)$ **$9y^3 - 9y^2 + 9y$**

21. $(2x^2 - 4) + (3x^2 - 6) - (-x^2 + 2)$ **$6x^2 - 12$**

22. Critical Thinking Kwan rewrote $(5x^2 - 3x + 1) - (2x^2 - 4x - 2)$ as $5x^2 - 3x + 1 - 2x^2 - 4x - 2$. What mistake did he make? **Answers may vary. Sample: Kwan did not change two of the terms being subtracted to their opposites.**

Language Arts **Use a dictionary if necessary.**

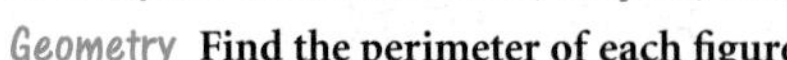

23. Writing Write the definition of each word. **See margin.**
   a. monogram b. binocular c. tricuspid d. polyglot

24. a. Open-ended Find other words that begin with *mono, bi, tri,* or *poly.*
   b. Do these prefixes have meanings similar to those in mathematics?
   a. Samples: mononucleosis, bicycle, triangle, polyunsaturated b. yes

Geometry **Find the perimeter of each figure.**

25. 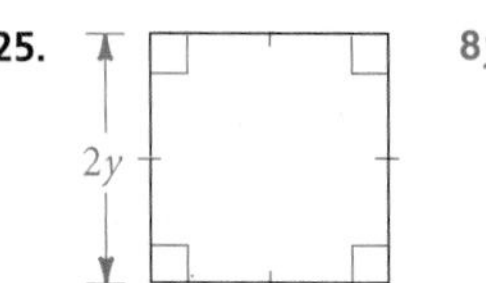

$8y$

26. 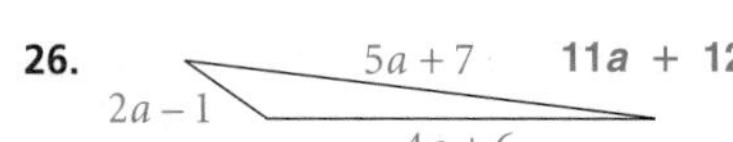

$11a + 12$

27. 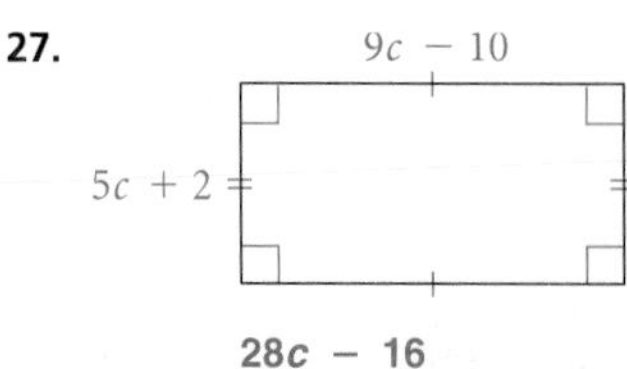

$28c - 16$

28. 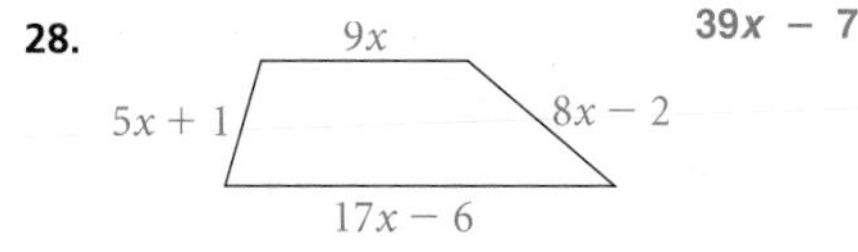

$39x - 7$

CRITICAL THINKING Exercise 40 Have students look back to the description of polynomials given on page 465.

Exercises MIXED REVIEW

ERROR ALERT! Exercise 43 Some students may omit the parentheses when writing the expression in exponential notation. **Remediation:** Point out that if the parentheses are omitted, it appears that the exponent relates only to the numerator and not to the whole fraction.

GETTING READY FOR LESSON 10-2 In these exercises students multiply a polynomial by a constant. This prepares them for multiplying a polynomial by a monomial in the next lesson.

Wrap Up

THE BIG IDEA Ask students to explain a method for adding and subtracting polynomials.

RETEACHING ACTIVITY Students use tiles to add and subtract polynomials. (Reteaching worksheet 10-1)

---

29. *Open-ended* In his will, Mr. McAdoo is leaving equal shares of the land shown at the right to his two brothers. Write a polynomial expression for the land that each brother should inherit. $3x^2 + 4x + 8$

30. *Critical Thinking* Is it possible to write a binomial with degree 0? Explain. No; both terms of a binomial cannot be constants.

Geometry **Find each missing length.**

31. Perimeter $= 25x + 8$

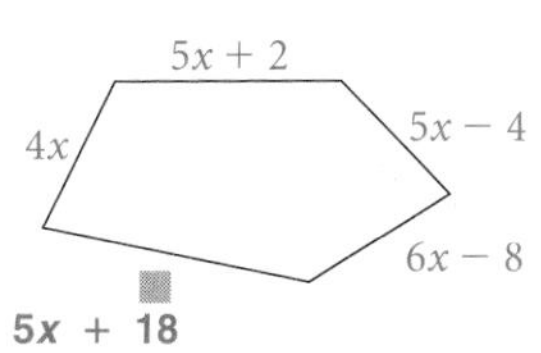

$5x + 18$

32. Perimeter $= 23a - 7$

$9a - 6$

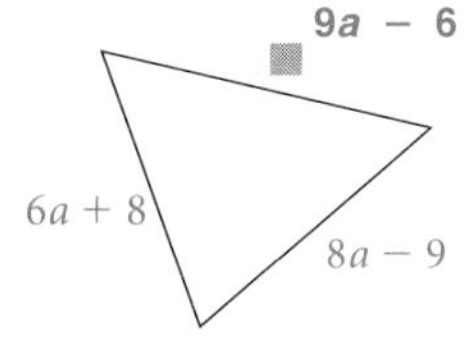

33. Perimeter $= 38y + 2$

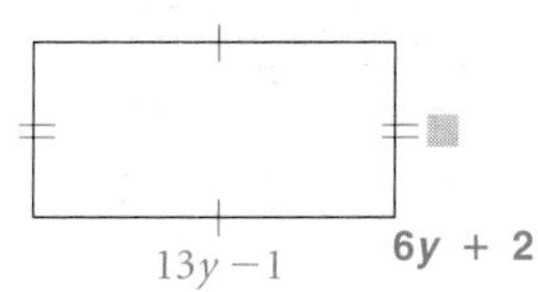

$6y + 2$

Simplify. Write each answer in standard form.

34. $(3 - 2x + 3x^2) + (7 + 6x - 2x^2)$ $x^2 + 4x + 10$
35. $(4x^3 - 2x^2 - 13) + (-4x^3 + 2 - 7x)$ $-2x^2 - 7x - 11$
36. $(3x^3 - 3x^2 - x - 1) - (3x^2 - 6x)$ $3x^3 - 6x^2 + 5x - 1$
37. $(-2r^2 + r - 6) - (1 + 2r - 4r^2)$ $2r^2 - r - 7$
38. $(9c^3 - c + 8 + 6c^2) - (3c^3 + 3c - 4)$ $6c^3 + 6c^2 - 4c + 12$
39. $(b^4 - 6 + 5b) + (8b^4 + 2b - 3b^2)$ $9b^4 - 3b^2 + 7b - 6$

40. *Critical Thinking* Why is $3x^2$ a monomial and $3x^{-2}$ *not* a monomial? $3x^{-2} = \frac{3}{x^2}$, which has a variable in the denominator

Exercises MIXED REVIEW

Use exponential notation to write each expression.

41. $4 \cdot 4 \cdot 4 \cdot 4$ 4^4
42. $(0.5)(0.5)(0.5)$ $(0.5)^3$
43. $\left(\frac{2}{3}\right)\left(\frac{2}{3}\right)\left(\frac{2}{3}\right)\left(\frac{2}{3}\right)\left(\frac{2}{3}\right)$ $\left(\frac{2}{3}\right)^5$
44. $28 \cdot 28 \cdot 28$ 28^3

Find the standard deviation of each set of numbers to the nearest tenth.

45. 10, 12, 16, 5, 2 5.0
46. 11, 7, 10, 12 1.9
47. 5, 3, 7, 8, 3, 4 1.9

48. *Science* The surface area of a sphere is found using the formula $S = 4\pi r^2$. The approximate radius of Jupiter is 4.4×10^4 mi. Find the approximate surface area of Jupiter. 2.43×10^{10} mi^2

Getting Ready for Lesson 10-2

Multiply.

49. $2(x - 3)$ $2x - 6$
50. $-6(3x - 2)$ $-18x + 12$
51. $(-3)(-5a + 7)$ $15a - 21$
52. $(-7)(c^2 - 8c)$ $-7c^2 + 56c$

Lesson Quiz

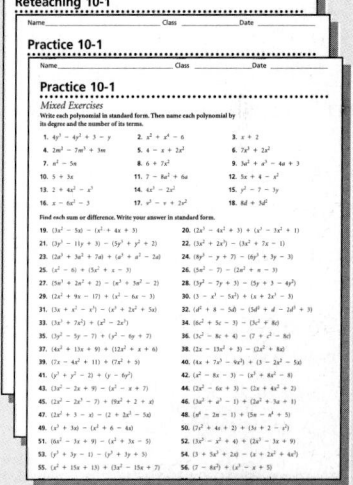

Lesson Quiz is also available in Transparencies.

Simplify each expression. Write each answer in standard form and classify each polynomial by its degree and number of terms.

1. $-4 + 3x - 2x^2$
   $-2x^2 + 3x - 4$
   quadratic trinomial
2. $2b^2 + 3b - 4b^3 + 6$
   $-4b^3 + 2b^2 + 3b + 6$; cubic polynomial with four terms
3. $(2x^3 + 3x - 4) + (-2x^2 + 3x + 5)$
   $2x^3 - 2x^2 + 6x + 1$; cubic polynomial with four terms
4. $(-3r + 4r^2 - 3) - (4r^2 + 6r - 2)$
   $-9r - 1$; linear binomial

10-2 Teaching Notes

PROBLEM OF THE DAY

Tyrone earned $10.00/h at a local grocery store. Due to cutbacks, Tyrone had to take a 10% pay cut. Later, all employees were given a 10% raise. How much is Tyrone now earning? $9.90/hr.

Problem of the Day is also available in Transparencies.

CONNECTING TO PRIOR KNOWLEDGE Write $3 \cdot 9$ on the board. Ask: *Is* 9 *the same as* $4 + 5$? yes Write $3 \cdot (4 + 5)$. Point out that you can use the distributive property to multiply this expression. Now write $(3 \cdot 4) + (3 \cdot 5)$. Lead students to understand you are using the distributive property.

THINK AND DISCUSS

ERROR ALERT! Some students may conclude that on tiles the length of x is equivalent to 5 units. **Remediation:** Have students line up one x-tile with five 1-tiles. They should find that the x-tile is a little longer (0.3 cm longer). The tiles are designed this way so the x-tile can represent *any* number.

Example 1

ERROR ALERT! **Question 2** When counting tiles for the product, some students may include the tiles that indicate the factors. **Remediation:** Have them look at the diagram in Example 1. Point out that only the tiles below and to the right of the dark black lines represent the product.

Lesson Planning Options

Prerequisite Skills

- Multiplying and dividing exponents
- Using the distributive property

Assignment Options for Exercises On Your Own

To provide flexible scheduling, this lesson can be subdivided into parts.

1 **Core** 1–5, 9–20
Extension 33, 51

2 **Core** 6–8, 21–32, 36–47
Extension 34, 35, 48–50

Use Mixed Review to maintain skills.

Resources

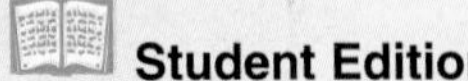

Student Edition

Skills Handbook, pp. 573–575, 582
Extra Practice, p. 565
Glossary/Study Guide

Teaching Resources

Chapter Support File, Ch. 10
- Practice 10-2 (two worksheets)
- Reteaching 10-2
- Alternative Activity 10-2

Classroom Manager 10-2
Glossary, Spanish Resources
Two-Year Algebra Handbook 10-2

Transparencies
17, 25, 115

Connections Building Models . . . and more

What You'll Learn
- Multiplying a polynomial by a monomial
- Factoring a monomial from a polynomial

...And Why
To explore formulas for area

What You'll Need
- tiles

10-2 Multiplying and Factoring

THINK AND DISCUSS

Part 1

Multiplying by a Monomial

You can use the distributive property to multiply polynomials. You can also use tiles to multiply polynomials.

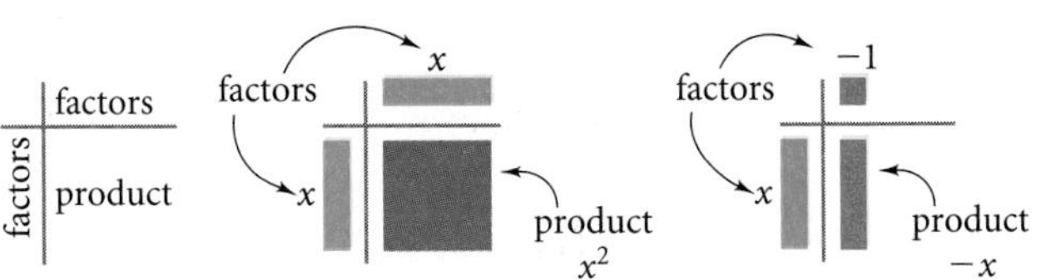

Example 1

Multiply $3x$ and $(2x + 1)$.

Method 1 Use tiles.

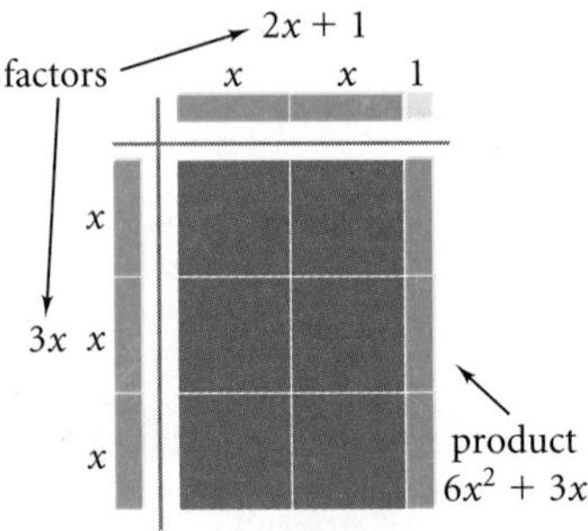

Method 2 Use the distributive property.

$$3x(2x + 1) = 3x(2x) + 3x(1)$$
$$= 6x^2 + 3x$$

1. **Try This** Use the distributive property to multiply $2x(x - 2)$. Then check your work by using tiles. $2x^2 - 4x$
2. a. Use tiles to find each product: $2x(4x - 3)$ and $(4x - 3)(2x)$. Each is $8x^2 - 6x$
   b. Do both models represent the same area? Explain. Yes; see margin for explanation.
3. a. *Critical Thinking* Kevin said that $-2x(4x - 3) = -8x^2 - 6x$. Karla said that $-2x(4x - 3) = -8x^2 + 6x$. Who is correct? Karla
   b. Explain the error that Kevin or Karla made. $(-2x)(-3) = 6x$, not $-6x$.

Example 2

ERROR ALERT! Students may place the tiles incorrectly, as shown here. **Remediation:** When making rectangles, remind students to line up like tiles, such as an x-length next to another x-length. A diagram such as the one shown here would not be helpful in finding factors.

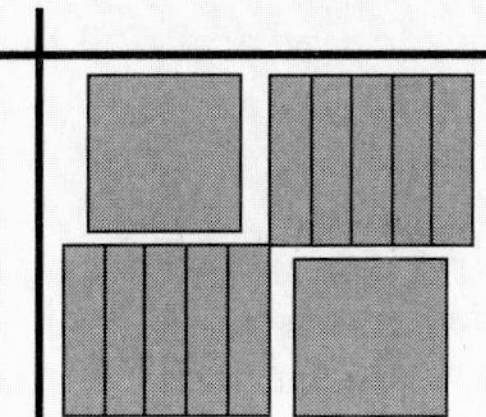

Example 3

Review briefly how to find the greatest common factor (GCF) of a set of whole numbers.

To find the GCF of 4, 8, and 12, first list the prime factorization of each number.

$$4 = 2 \cdot 2$$
$$8 = 2 \cdot 2 \cdot 2$$
$$12 = 2 \cdot 2 \cdot 3$$

Find all the common factors of the numbers. The GCF is $2 \cdot 2$, or 4.

Part 2

Factoring Out a Monomial

Factoring a polynomial reverses the multiplication process. You can use tiles to make a rectangle to find the factors of a polynomial.

Example 2

Factor $2x^2 - 10x$.

← Model using tiles.

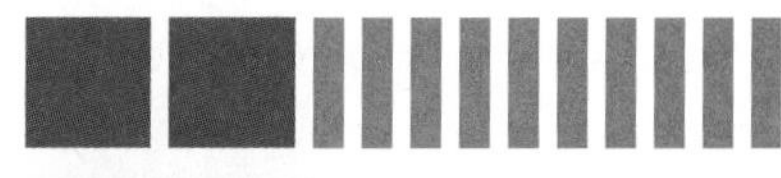

← Make a rectangle that is as close to square as possible.

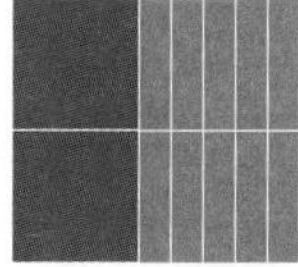

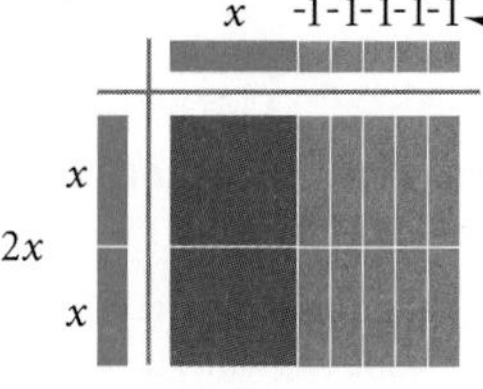

← Find the factors.

$2x^2 - 10x = 2x(x - 5)$

4. **Try This** Use tiles to factor each polynomial.
 a. $x^2 + 8x$ $x(x + 8)$ b. $3x^2 - 12x$ $3x(x - 4)$ c. $-3x^2 + 6x$ $-3x(x - 2)$

QUICK REVIEW

The *greatest common factor* (GCF) is the greatest factor that divides evenly into each term.

For practice with greatest common factors, see Skills Handbook page 574.

To factor out a monomial using the distributive property, it is helpful to find the greatest common factor (GCF).

Example 3

Find the GCF of the terms of the polynomial $4x^3 + 12x^2 - 8x$.

List the factors of each term. Identify the factors common to all terms.

$$4x^3 = 2 \cdot 2 \cdot x \cdot x \cdot x$$
$$12x^2 = 2 \cdot 2 \cdot 3 \cdot x \cdot x$$
$$8x = 2 \cdot 2 \cdot 2 \cdot x$$

The GCF is $2 \cdot 2 \cdot x$ or $4x$.

5. **Try This** Find the GCF of the terms of each polynomial.
 a. $4x^3 - 2x^2 - 6x$ $2x$ b. $5x^5 + 10x^3$ $5x^3$ c. $3x^2 - 18$ 3

Additional Examples

FOR EXAMPLE 1

Multiply $3x^2$ and $x + 3$ $3x^3 + 9x^2$

FOR EXAMPLE 4

Factor $2x^3 + 8x^2 - 16x$
$2x(x^2 + 4x - 8)$

FOR EXAMPLE 5

Find the area of the shaded portion of the figure below. The area of a triangle is $\frac{1}{2}bh$, where b is the width of the base of the triangle and h is the height of the triangle. The area of a circle is πr^2, where r is the radius of the circle. $\pi r^2 - r^2$

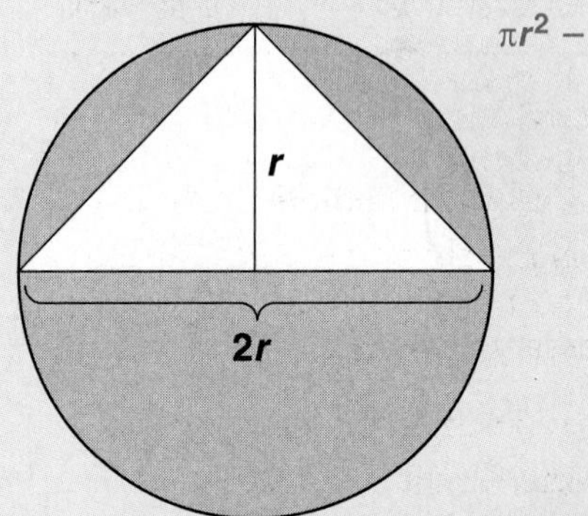

Discussion: *What would happen to the shaded area if the figure were twice as large? Explain.*

VISUAL LEARNING On the board, write Becki Martinez, Selena Martinez, and Miguel Martinez. Point out that this is a lot of writing. Then write (Becki, Selena, and Miguel) Martinez. Ask students to describe in their own words what you just did. Lead them to understand that when you factor an expression, you write the factor that each term has in common outside the parentheses.

Example 4

Have students use the GCF method to factor $2x^2 - 10x$. Then have them go back and compare the factors with the factors found by using tiles in Example 2.

Example 5 Relating to the Real World  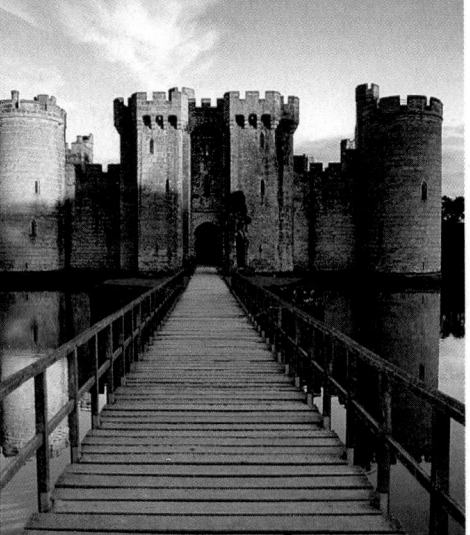

ESL Ask volunteers to describe a *moat* and discuss its purpose.

ERROR ALERT! **Question 8** Students might divide only the first term by the GCF. **Remediation:** Point out that the whole polynomial must be divided by the GCF. This means every term of the polynomial must be divided by the GCF.

Exercises ON YOUR OWN

OPEN-ENDED Exercise 33 As a class, try to generate all the possibilities. $x(6x + 12)$; $2x(3x + 6)$; $3x(2x + 4)$; $6x(x + 2)$

WRITING Exercise 34b If students have difficulty getting started, have them substitute first an even integer and then an odd integer into the expression to see what kinds of results they will get.

Technology Options

For Exercises 9–20, students may find it helpful if they use algebra software to simplify the expressions.

Prentice Hall Technology 

 **Software**
- Secondary Math Lab Toolkit™
- Computer Item Generator 10-2

 **CD-ROM**
- Multimedia Algebra Lab 10

 **Internet**
- See the Prentice Hall site. (http://www.phschool.com)

472

Example 4

Factor $3x^3 - 9x^2 + 15x$.

Step 1 Find the GCF.

$$3x^3 = 3 \cdot x \cdot x \cdot x$$
$$9x^2 = 3 \cdot 3 \cdot x \cdot x$$
$$15x = 3 \cdot 5 \cdot x$$

The GCF is $3 \cdot x$ or $3x$.

Step 2 Factor out the GCF.

$$3x^3 - 9x^2 + 15x = 3x(x^2) - 3x(3x) + 3x(5) = 3x(x^2 - 3x + 5)$$

6. Use the distributive property to check the factoring in Example 4. $3x(x^2 - 3x + 5) = 3x^3 - 9x^2 + 15x$ ✓
7. **Try This** Use the GCF to factor each polynomial.
   a. $8x^2 - 12x$ $4x(2x - 3)$
   b. $5x^3 + 10x$ $5x(x^2 + 2)$
   c. $6x^3 - 12x^2 - 24x$ $6x(x^2 - 2x - 4)$

Example 5 Relating to the Real World

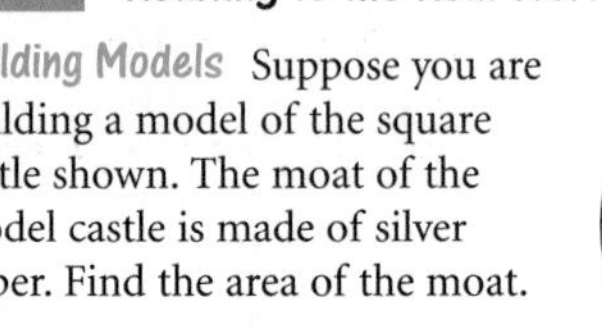

Building Models Suppose you are building a model of the square castle shown. The moat of the model castle is made of silver paper. Find the area of the moat.

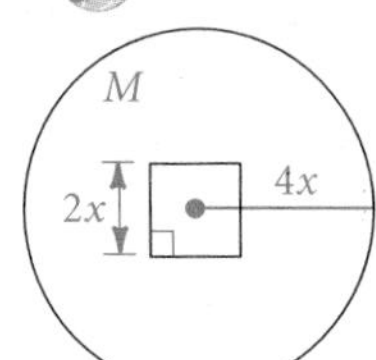

Define M = area of the moat
$2x$ = length of the side of the castle
$4x$ = radius of the moat
$A = \pi r^2$ ← formula for the area of a circle

Relate area of moat is area of circle minus area of square

Write $M = \pi(4x)^2 - (2x)^2$

$M = 16\pi x^2 - 4x^2$ ← Simplify $(4x)^2$ and $(2x)^2$.
$= 4x^2(4\pi - 1)$ ← The GCF is $4x^2$.

The area of the moat is $4x^2(4\pi - 1)$.

8. **Try This** Use the GCF to factor each polynomial.
   a. $2g^2 - 4$ $2(g^2 - 2)$
   b. $2x^3 - 4x^2 + 6x$ $2x(x^2 - 2x + 3)$
   c. $6x^3 + 24x^2 + 6x$ $6x(x^2 + 4x + 1)$

ALTERNATIVE ASSESSMENT **Exercises 36–47** Write each exercise, along with its corresponding factors, on separate index cards. For example, write the expression $6x - 4$ on one card, and the factors 2 and $(3x - 2)$ on two additional cards. Shuffle the cards from all the exercises and randomly distribute them among the students. Ask each student whose card contains an unfactored expression to find the classmates that have the corresponding factor cards.

GEOMETRY Exercise 51c Ask: *Why is $(n - 3)$ multiplied by $\frac{n}{2}$ rather than by n?* If you multiplied by n, each diagonal would be counted twice, since each diagonal connects two points.

EXTENSION Exercise 51 Ask students: *If n people all write letters to each other, how many letters will be written? Give your answer as a polynomial in standard form.*
$n(n - 1) = n^2 - n$

Exercises ON YOUR OWN

Use tiles to find each product.

1. $3(x + 4)$ $3x + 12$
2. $x(x - 3)$ $x^2 - 3x$
3. $2x(2x + 1)$ $4x^2 + 2x$
4. $4x(5x - 8)$ $20x^2 - 32x$
5. $4x(2x + 3)$ $8x^2 + 12x$

For each set of tiles, find the missing factors or product. Then write the factors and product as variable expressions. See margin for diagrams.

6. 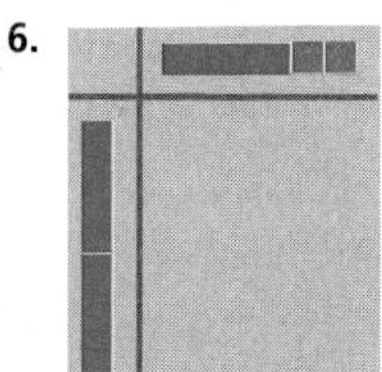
$(2x)(x - 2) = 2x^2 - 4x$

7. 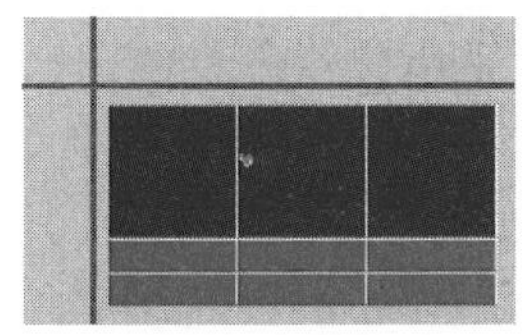
$(x + 2)(3x) = 3x^2 + 6x$

8. 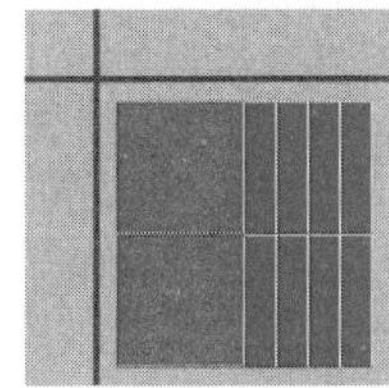
$-2x^2 + 8x = (-2x)(x - 4)$ or $= (2x)(-x + 4)$

Find each product.

9. $4(2x + 7)$ $8x + 28$
10. $t(5t^2 + 6t)$ $5t^3 + 6t^2$
11. $6x(-9x^3 + 6x - 8)$ $-54x^4 + 36x^2 - 48x$
12. $2g^2(g^2 + 6g + 5)$ $2g^4 + 12g^3 + 10g^2$
13. $-3a(4a^2 - 5a + 9)$ $-12a^3 + 15a^2 - 27a$
14. $7x^2(5x^2 - 3)$ $35x^4 - 21x^2$
15. $-3p^2(-2p^3 + 5p)$ $6p^5 - 15p^3$
16. $4n^2(2n^2 + 4n)$ $8n^4 + 16n^3$
17. $x(x + 3) - 5x(x - 2)$ $-4x^2 + 13x$
18. $12c(-5c^2 + 3c - 4)$ $-60c^3 + 36c^2 - 48c$
19. $x^2(x + 1) - x(x^2 - 1)$ $x^2 + x$
20. $-4j(3j^2 - 4j + 3)$ $-12j^3 + 16j^2 - 12j$

Find the greatest common factor (GCF) for each polynomial.

21. $15x + 21$ 3
22. $6a^2 - 8a$ $2a$
23. $36s + 24$ 12
24. $x^3 + 7x^2 - 5x$ x
25. $5b^3 - 30$ 5
26. $w^4 - 9w^2$ w^2
27. $9x^3 - 6x^2 + 12x$ $3x$
28. $5r^5 - 3r^2 + 4r$ r
29. $25s^2 + 5s - 15s^3$ $5s$
30. $8p^3 - 24p^2 + 16p$ $8p$
31. $56x^4 - 32x^3 - 72x^2$ $8x^2$
32. $2x + 3x^2$ x

33. a. *Open-ended* Draw two different tile diagrams to represent $6x^2 + 12x$ as a product. Place the x^2 tiles in the upper left area.
 b. Write the factored form of $6x^2 + 12x$ for each diagram in part (a).
 a–b. See margin.
34. a. Factor $n^2 - n$. $n(n - 1)$
 b. *Writing* Suppose n is an integer. Is $n^2 - n$ *always, sometimes,* or *never* even? **Justify** your answer. See margin.

35. *Manufacturing* The diagram shows a solid block of metal with a cylinder cut out of it. The formula for the volume of a cylinder is $V = \pi r^2 h$, where r is the radius and h is the height.
 a. Write a formula for the volume of the cube in terms of s. $V = 64s^3$
 b. Write a formula for the volume of the cylinder in terms of s. $V = 48\pi s^2$
 c. Write a formula in terms of s for the volume of the metal left after the cylinder has been removed. $V = 64s^3 - 48\pi s^2$
 d. Factor your formula from part (c). $V = 16s^2(4s - 3\pi)$
 e. What is the volume of the block of metal after the cylinder has been removed if $s = 15$ in? about 182,071 in.^{3}

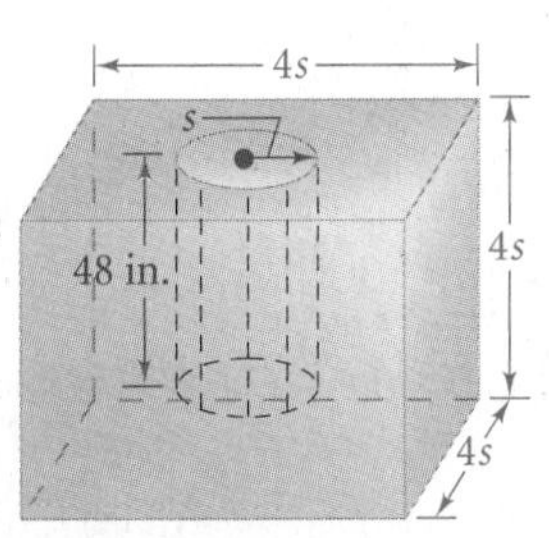

pages 470–472 Think and Discuss

2a. $2x(4x - 3)$:

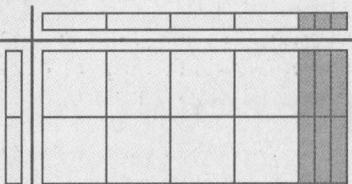

$2x(4x - 3) = 8x^2 - 6x$

$(4x - 3)(2x)$:

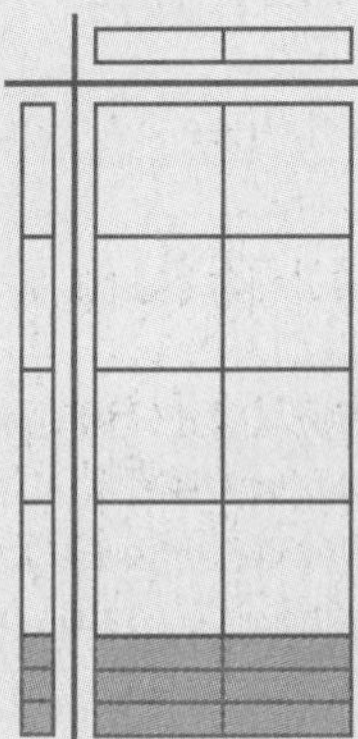

$(4x - 3)(2x) = 8x^2 - 6x$

b. Yes; the models have an equal number of x^2-tiles, 8, an equal number of x-tiles, 6, and an equal number of 1-tiles, 0. They are arranged differently.

pages 473–474 On Your Own

6–8, 33a–b, 34b. See back of book.

Chapter Project **FIND OUT BY RESEARCHING** Students research lumber and tool requirements for the construction of a house. This task is essential to their work on the Chapter Project, introduced in the Chapter Opener. Check students' understanding of the project by having each student write a sentence or two describing the objectives of this project.

Exercises MIXED REVIEW

GETTING READY FOR LESSON 10-3 Students simplify polynomials for which each term is the product of two monomials. This skill will be used in multiplying two polynomials.

Wrap Up

THE BIG IDEA Ask students to explain one way to factor a polynomial and how they would check the factors.

RETEACHING ACTIVITY Students factor polynomials by finding the GCF. (Reteaching worksheet 10-2)

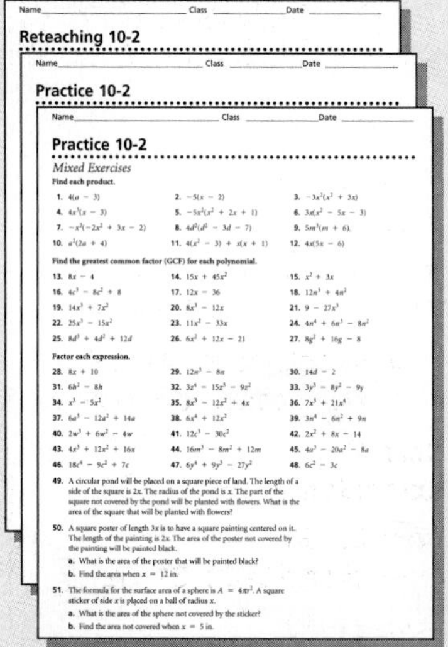
Name Class Date

Reteaching 10-2

Practice 10-2

Mixed Exercises

Find each product.

Find the greatest common factor (GCF) for each polynomial.

Factor each expression.

Lesson Quiz

Lesson Quiz is also available in Transparencies.

1. Simplify $-2x^2(-3x^2 + 8 - 2x)$. $6x^4 + 4x^3 - 16x^2$
2. Find the GCF of the polynomial $16x^4 - 4x^3 + 8x^2$. $4x^2$
3. Factor $3x^3 + 9x^2$. $3x^2(x + 3)$
4. Factor $12x - 4x^3 + 8x^2$. $4x(3 - x^2 + 2x)$

Factor each expression.

36. $6x - 4$ $2(3x - 2)$
37. $s^4 + 4s^3 - 2s$ $s(s^3 + 4s^2 - 2)$
38. $10r^2 - 25r + 20$ $5(2r^2 - 5r + 4)$
39. $2x^2 - 4x^4$ $2x^2(1 - 2x^2)$
40. $12p^3 + 4p^2 - 2p$ $2p(6p^2 + 2p - 1)$
41. $7k^3 - 35k^2 + 70k$ $7k(k^2 - 5k + 10)$
42. $15n^3 + 3n^2 - 12n$ $3n(5n^2 + n - 4)$
43. $9x + 12x^2$ $3x(3 + 4x)$
44. $24n^3 - 12n^2 + 12n$ $12n(2n^2 - n + 1)$
45. $6m^6 - 24m^4 + 6m^2$ $6m^2(m^4 - 4m^2 + 1)$
46. $15k^3 + 3k^2 - 12k$ $3k(5k^2 + k - 4)$
47. $5m^3 - 7m^2$ $m^2(5m - 7)$

Factor by grouping like terms.

Sample		
	$2x^3 + 2x + 3x^2 + 3$	← Group the terms with common factors together.
	$2x(x^2 + 1) + 3(x^2 + 1)$	← Factor the GCF from each group.
	$(2x + 3)(x^2 + 1)$	← Factor out the common polynomial.

48. $3v^3 + 18v^2 - 4v - 24$ $(3v^2 - 4)(v + 6)$
49. $2x^3 + x^2 - 14x - 7$ $(x^2 - 7)(2x + 1)$
50. $2x^3 + 3x^2 + 4x + 6$ $(x^2 + 2)(2x + 3)$
51. a. *Geometry* How many sides does the polygon at the right have? How many diagonals does it have from one vertex? 7; 4
    b. Suppose a polygon, like the one at the right, has n sides. How many diagonals will it have from one vertex? $n - 3$
    c. The number of diagonals that can be drawn from all the vertices is $\frac{n}{2}(n - 3)$. Multiply the two factors. $\frac{1}{2}n^2 - \frac{3}{2}n$

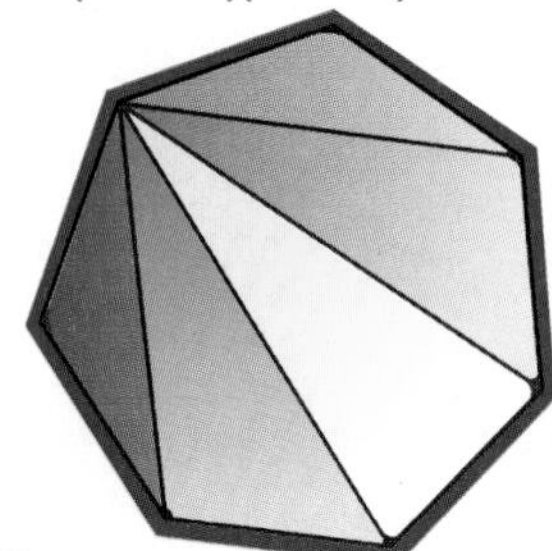

Chapter Project *Find Out by Researching*

A board foot is a linear measure of lumber equal to a square foot of wood 1 in. thick. What can you make from 10 board feet? 100 board feet? 1000 board feet? How is the size of a house related to the amount of wood used to build it? What different types of wood are needed for cabinets, floors, and roofs? What tools do carpenters use to make these items?

Exercises MIXED REVIEW

Use $\triangle ABC$ to find each trigonometric ratio.

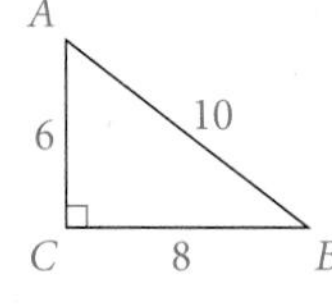

52. $\sin A$ $\frac{4}{5}$
53. $\cos B$ $\frac{4}{5}$
54. $\tan A$ $\frac{4}{3}$
55. $\sin B$ $\frac{3}{5}$
56. $\cos A$ $\frac{3}{5}$
57. $\tan B$ $\frac{3}{4}$
58. *Geography* Two buses are on their way to the same bus terminal. One bus is 1.5 mi due north of the terminal. The other bus is 2.0 mi due east of the terminal. How far from each other are the buses? 2.5 mi

Getting Ready for Lesson 10-3

Simplify and write in standard form.

59. $(4x)(32) - (4x)(12x) + (4x)(7x^2)$ $28x^3 - 48x^2 + 128x$
60. $(3y)(7) - (3y)(2y) + (3)(5y^2) - (7y)(6) - (7y)(8y)$ $-47y^2 - 21y$

10-3 Teaching Notes

PROBLEM OF THE DAY

To be exempt from the last test in algebra class, a student must have a test average of 85. To be exempt, what must a student make on the fifth test if the first four test grades are 92, 75, 88, and 74? 96

Problem of the Day is also available in Transparencies.

CONNECTING TO PRIOR KNOWLEDGE Write these steps on the board. After each step, make sure students understand the changes you've made before you write the next step.

$$8 \cdot 12 = (6 + 2) \cdot (9 + 3)$$
$$= 6(9 + 3) + 2(9 + 3)$$
$$= (6 \cdot 9) + (6 \cdot 3) + (2 \cdot 9) + (2 \cdot 3)$$
$$= 54 + 18 + 18 + 6$$
$$= 96$$

Show students how the distributive property is used to multiply the terms in this example. Explain that multiplying polynomials uses the same process.

THINK AND DISCUSS

Example 1

Some students may find it helpful to use the following as a first step. $(2x + 1)(x - 5) = 2x(x - 5) + 1(x - 5)$

KINESTHETIC and AUDITORY LEARNING Have four students stand facing the class with their backs to the board. The

Connections Savings . . . and more

10-3 Multiplying Polynomials

What You'll Learn

- Multiplying two binomials
- Multiplying a trinomial and a binomial

...And Why

To investigate real-world situations, such as savings accounts

What You'll Need

- tiles
- graphing calculator

THINK AND DISCUSS

Part 1

Multiplying Two Binomials

You can use tiles or the distributive property to multiply two binomials.

Example 1

Find the product $(2x + 1)(x - 5)$.

Method 1 Use tiles.

Step 1 Show the factors.

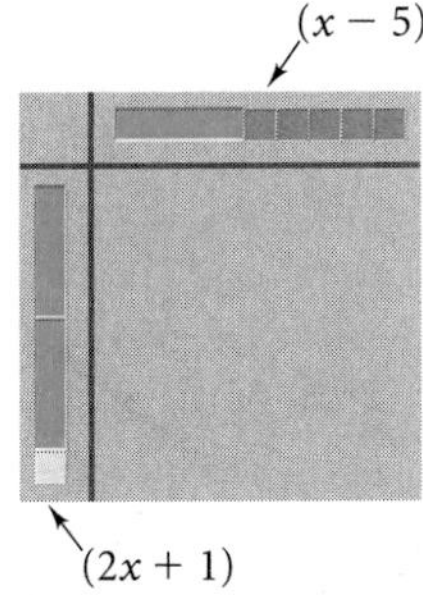

Step 2 Find the product.

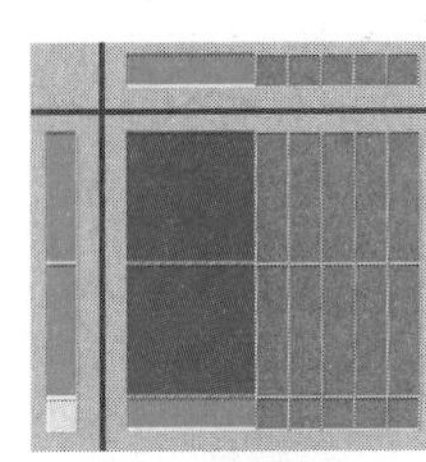

$2x^2 - 10x + x - 5$ ← Add coefficients of like terms.
$2x^2 - 9x - 5$

Method 2 Use the distributive property.

$$(2x + 1)(x - 5) = 2x(x) + 2x(-5) + 1(x) + 1(-5)$$
$$= 2x^2 - 10x + x - 5$$
$$= 2x^2 - 9x - 5$$

The product is $2x^2 - 9x - 5$.

1. Rework Example 1 using tiles. Put $2x + 1$ on the horizontal line and $x - 5$ on the vertical line. Do you get the same result as in Example 1? yes
2. **Try This** Find the product $(3x + 1)(2x + 3)$ using tiles. $6x^2 + 11x + 3$
3. **Try This** Find the product $(6x - 7)(3x + 5)$ using the distributive property. $18x^2 + 9x - 35$
4. a. **Graphing Calculator** Graph $y = (2x + 1)(x - 5)$ and $y = 2x^2 - 9x - 5$ on the same calculator screen. What appears to be true of the two graphs? The graphs appear to be the same.
   b. Use graphs to check your answers to Questions 2 and 3. See margin.

Lesson Planning Options

Prerequisite Skills

- Simplifying expressions

Assignment Options for Exercises On Your Own

To provide flexible scheduling, this lesson can be subdivided into parts.

1. **Core** 1–6; **Extension** 7–10
2. **Core** 23–30; **Extension** 20–22, 31, 32
3. **Core** 11–19; **Extension** 33–36

Use Mixed Review to maintain skills.

Resources

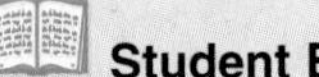

Student Edition

Skills Handbook, pp. 573, 574, 579, 582
Extra Practice, p. 565
Glossary/Study Guide

Teaching Resources

Chapter Support File, Ch. 10
- Practice 10-3 (two worksheets)
- Reteaching 10-3
- Checkpoint
- Alternative Activity 10-3

Classroom Manager 10-3
Glossary, Spanish Resources
Two-Year Algebra Handbook 10-3

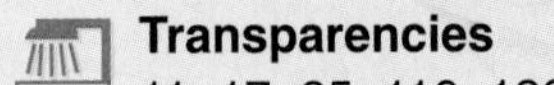

Transparencies

11, 17, 25, 116, 120

students represent the four terms in two binomials. Draw huge parentheses on the board around each pair of students who are representing binomials. Now demonstrate FOIL by having the First pair of students shake hands. Then have the Outer pair of students shake hands. Continue for the Inner and the Last pairs. Have students say aloud the FOIL step they are performing.

If you have block scheduling or extended class periods, put the students in groups of four. Have students perform FOIL within their group. Instead of shaking hands, students work in pairs to calculate the product of their terms. Have each group check their work when they are finished.

Example 2

On an overhead projector, demonstrate the FOIL method using tiles. Use the tiles to show $(3x - 5)(2x + 7)$.

Separate the tiles so that one set shows the product of the FIRST terms, one set shows the product of the OUTER terms, one set shows the product of the INNER terms, and one set shows the product of the LAST terms.

Example 3 Relating to the Real World 

CONNECTING TO THE STUDENTS' WORLD Have students research financial institutions in your city to find out the various interest rates paid on \$500 accounts. Encourage students to research how interest rates increase for higher balances. Have students repeat Example 3 using the data they collected.

EXTENSION Challenge students to show that the expression used to find the amount in the account is equivalent to the expression $500(1 + r) + 500(1 + r)^2$. This latter expression

Additional Examples

FOR EXAMPLE 1

Find the product $(2y - 3)(-y + 1)$.
$-2y^2 + 5y - 3$

FOR EXAMPLE 2

Find the product $(2x + 3)(2x - 3)$.
$4x^2 - 9$

Discussion: *Why is the product of these two binomials another binomial?*

FOR EXAMPLE 4

Find the product $(3x^2 - 2x + 3)(2x + 7)$. $6x^3 + 17x^2 - 8x + 21$

Discussion: *Describe two methods you could use to find the product of this expression. Which of the these methods do you prefer? Explain.*

Part 2

Multiplying Using FOIL

One way to organize how you multiply two binomials is to use *FOIL*, which stands for "First, Outer, Inner, Last." The term FOIL is a memory device for applying the distributive property.

Example 2

Find the product $(3x - 5)(2x + 7)$.

First, Last, Inner, Outer

$$\begin{aligned}(3x - 5)(2x + 7) &= \overset{\text{First}}{(3x)(2x)} + \overset{\text{Outer}}{(3x)(7)} - \overset{\text{Inner}}{(5)(2x)} - \overset{\text{Last}}{(5)(7)} \\ &= 6x^2 + 21x - 10x - 35 \\ &= 6x^2 + 11x - 35\end{aligned}$$

middle term

The product is $6x^2 + 11x - 35$.

5. **Mental Math** What is the middle term of each product?
   a. $(2x + 3)(x + 1)$ $5x$
   b. $(2x - 3)(x + 1)$ $-x$
   c. $(2x + 3)(x - 1)$ x
   d. $(2x - 3)(x - 1)$ $-5x$
6. **Try This** Find each product.
   a. $(3x + 4)(2x + 5)$ $6x^2 + 23x + 20$
   b. $(3x - 4)(2x + 5)$ $6x^2 + 7x - 20$
   c. $(3x + 4)(2x - 5)$ $6x^2 - 7x - 20$
   d. $(3x - 4)(2x - 5)$ $6x^2 - 23x + 20$
7. **Graphing Calculator** Use a graphing calculator to check your answers in Question 6. See margin.

Example 3 Relating to the Real World 

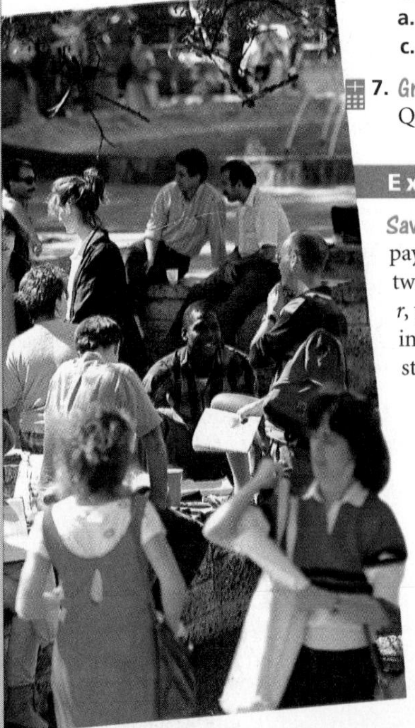

Savings Many students and their families start saving money early to pay for college. Suppose you deposit \$500 at the beginning of each of two consecutive years. If your bank pays interest annually at the rate r, you can use the expression $(1 + r)(2 + r)500$ to find the amount in your account at the end of the two years. Write the expression in standard form.

$(1 + r)(2 + r)500$

$= (1 + r)(2 + r)500$ ← Use FOIL to simplify $(1 + r)(2 + r)$.

$= (2 + r + 2r + r^2)(500)$

$= (2 + 3r + r^2)(500)$ ← Add like terms.

$= 1000 + 1500r + 500r^2$ ← Use the distributive property.

$= 500r^2 + 1500r + 1000$ ← Write in standard form.

8. How much money is in your account if the interest rate in Example 3 is 4%? (*Hint*: Write the interest rate as a decimal.) \$1060.80

is obtained by using the formula presented in Example 2 on page 369.

$$\begin{aligned} 500(1 + r) + 500(1 + r)^2 &= 500[(1 + r) + (1 + r)^2] \\ &= 500[(1 + r) + (1 + r)(1 + r)] \\ &= 500[(1 + r) + (1 + 2r + r^2)] \\ &= 500(1 + r + 1 + 2r + r^2) \\ &= 500(2 + 3r + r^2) \\ &= (2 + 3r + r^2)(500) \end{aligned}$$

This last expression is equivalent to the fourth line in the solution of Example 3.

Example 4

Method 1 shows how to multiply the factors vertically. Relate this to multiplication of whole numbers.

$$\begin{array}{r} 316 \\ \times 23 \\ \hline 948 \\ 6320 \\ \hline 7268 \end{array}$$

When multiplying whole numbers, the partial products are lined up according to place value. When multiplying polynomials, they are lined up by like terms.

Exercises ON YOUR OWN

ERROR ALERT! **Exercises 1–3** Some students may include the tiles that represent the factors when determining the product. **Remediation:** Remind students that the product is represented by the tiles below and to the right of the lines. You might relate this diagram to a multiplication table.

Part  3

Multiplying a Trinomial and a Binomial

FOIL works when you multiply two binomials, but it is not helpful when multiplying a trinomial and a binomial. You can use the vertical method or the horizontal method to distribute each term in a factor.

Example 4

Find the product $(3x^2 + x - 6)(2x - 3)$.

Method 1 Multiply vertically.

$$\begin{array}{rl} 3x^2 + x - 6 & \\ 2x - 3 & \\ \hline -9x^2 - 3x + 18 & \leftarrow \text{Multiply by } -3. \\ 6x^3 + 2x^2 - 12x \quad\quad & \leftarrow \text{Multiply by } 2x. \\ \hline 6x^3 - 7x^2 - 15x + 18 & \leftarrow \text{Add like terms.} \end{array}$$

Method 2 Multiply horizontally.

$$(2x - 3)(3x^2 + x - 6)$$

$$= 2x(3x^2) + 2x(x) + 2x(-6) - 3(3x^2) - 3(x) - 3(-6)$$

$$= 6x^3 + 2x^2 - 12x - 9x^2 - 3x + 18$$

$$= 6x^3 - 7x^2 - 15x + 18 \quad \leftarrow \text{Add like terms.}$$

The product is $6x^3 - 7x^2 - 15x + 18$.

9. a. Try This Find the product $(3a + 4)(5a^2 + 2a - 3)$ using both methods shown in Example 4. $15a^3 + 26a^2 - a - 12$
   b. Do you prefer the vertical or the horizontal method? Why? Answers may vary. Sample: Vertical method; it is easier to keep track of like terms.

Exercises ON YOUR OWN

What are the factors shown with the tiles? What will be the product?

1. 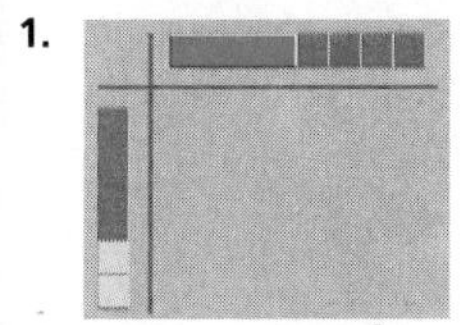  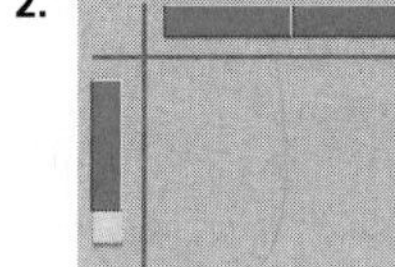
$-x + 2, x - 4; -x^2 + 6x - 8$

2.  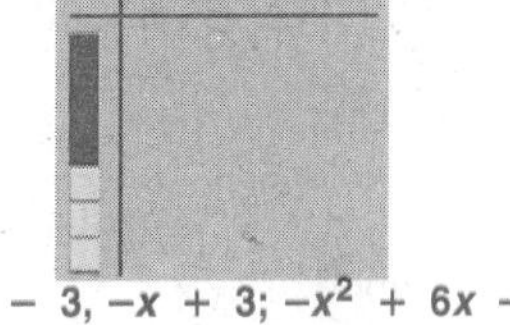
$x + 1, 2x - 3; 2x^2 - x - 3$

3. $x - 3, -x + 3; -x^2 + 6x - 9$

Use tiles to find each product. See margin for models.

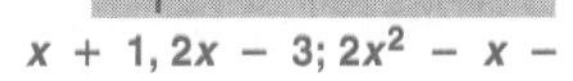
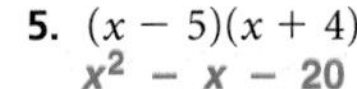

4. $(x + 2)(x + 5)$ $x^2 + 7x + 10$
5. $(x - 5)(x + 4)$ $x^2 - x - 20$
6. $(2x - 1)(x + 2)$ $2x^2 + 3x - 2$

Technology Options

For Exercises 11–19, students may want to simplify the products using algebra software.

Prentice Hall Technology

Calculator
- Graphing Calculator Handbook: Procedure 4

Software
- Secondary Math Lab Toolkit™
- Integrated Math Lab 20
- Computer Item Generator 10-3

CD-ROM
- Multimedia Algebra Lab 10

Internet
- See the Prentice Hall site. (http://www.phschool.com)

OPEN-ENDED Exercise 21 Ask: *Will the product of a binomial and a trinomial always have four terms?* yes To help convince students that this is true, challenge them to come up with an exception.

Exercise 22 Have a volunteer draw a diagram to represent the problem. Have the other students decide whether the diagram meets the conditions of the problem.

ALTERNATIVE ASSESSMENT **Exercises 23, 25, 27, 29** To evaluate student understanding of the FOIL method, ask students to multiply the first binomial factor of Exercise 23 and the first binomial factor of Exercise 25 (example: $(x + 3)(a - 1)$). Repeat for Exercises 27 and 29. Emphasize that the FOIL method can be used to multiply two binomial expressions that contain different variables.

Chapter Project **FIND OUT BY CALCULATING** Students evaluate the given expression to calculate the usable board feet of a log. This information is essential to their work on the Chapter Project. Have students add this task to work they have already completed for the project. Check students' progress by having each student write a sentence or two describing what they have done for the project thus far.

Exercises MIXED REVIEW

GETTING READY FOR LESSON 10-4 In these exercises, students factor out a common polynomial. This prepares students for learning other ways of factoring trinomials.

pages 475–477 Think and Discuss

4a.

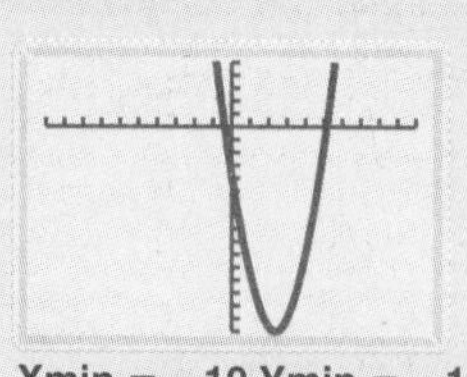

Xmin = −10 Ymin = −16
Xmax = 10 Ymax = 5
Xscl = 1 Yscl = 1

b. $y = (3x + 1)(2x + 3)$ and $y = 6x^2 + 11x + 3$;

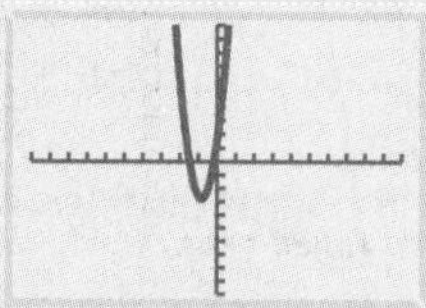

$y = 18x^2 + 9x - 35$ and $y = (6x - 7)(3x + 5)$

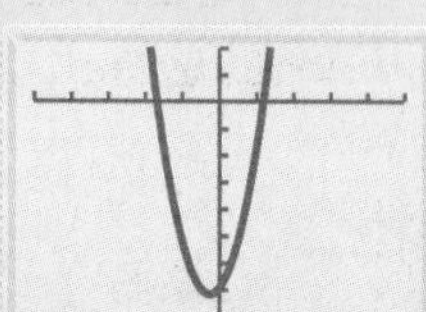

Xmin = −5 Ymin = −40
Xmax = 5 Ymax = 10
Xscl = 1 Yscl = 5

pages 477–479

Think and Discuss 7, On Your Own 4–6, Checkpoint 14. See back of book.

Copy and fill in each blank.

7. $(5a + 2)(6a - 1) = \blacksquare a^2 + 7a - 2$ 30
8. $(3c - 7)(2c - 5) = 6c^2 - 29c + \blacksquare$ 35
9. $(z - 4)(2z + 1) = 2z^2 - \blacksquare z - 4$ 7
10. $(2x + 9)(x + 2) = 2x^2 + \blacksquare x + 18$ 13

Choose **Use any method you choose to find each product.**

11. $(x + 7)(x - 6)$ $x^2 + x - 42$
12. $(a - 8)(a - 9)$ $a^2 - 17a + 72$
13. $(2y + 5)(y - 3)$ $2y^2 - y - 15$
14. $(r + 6)(r - 4)$ $r^2 + 2r - 24$
15. $(y + 4)(5y - 8)$ $5y^2 + 12y - 32$
16. $(x + 9)(x^2 - 4x + 1)$ $x^3 + 5x^2 - 35x + 9$
17. $(a - 4)(a^2 - 2a + 1)$ $a^3 - 6a^2 + 9a - 4$
18. $(x - 3)(2x^2 + 3x + 3)$ $2x^3 - 3x^2 - 6x - 9$
19. $(2t^2 - 6t + 3)(2t - 5)$ $4t^3 - 22t^2 + 36t - 15$
20. *Geometry* Use the formula $V = lwh$ to write a polynomial in standard form for the volume of the box shown at the right. $n^3 + 15n^2 + 56n$

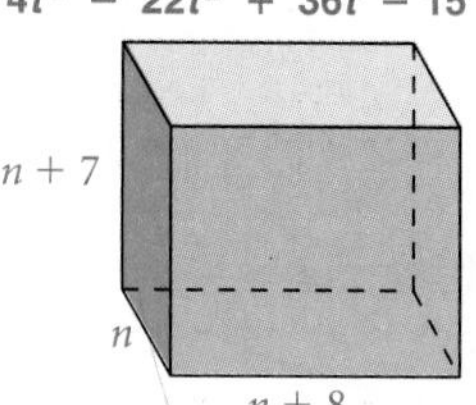

21. *Open-ended* Write a binomial and a trinomial. Find their product. Sample: $(2x - 1)(x^2 - 5x + 3) = 2x^3 - 11x^2 + 11x - 3$
22. *Construction* You are planning a rectangular garden. Its length is 4 ft more than twice its width. You want a walkway 2 ft wide around the garden. Write an expression for the area of the garden and walk. (*Hint:* Draw a diagram.) w = width; $2w^2 + 16w + 32$

Find each product using FOIL.

23. $(x + 3)(x + 5)$ $x^2 + 8x + 15$
24. $(2y + 1)(3y + 4)$ $6y^2 + 11y + 4$
25. $(a - 1)(a - 7)$ $a^2 - 8a + 7$
26. $(4x + 3)(4x - 3)$ $16x^2 - 9$
27. $(5a - 2)(a + 3)$ $5a^2 + 13a - 6$
28. $(6x + 1)(2x - 3)$ $12x^2 - 16x - 3$
29. $(3y - 7)(-2y + 2)$ $-6y^2 + 20y - 14$
30. $(8 - 6x)(5 + 2x)$ $40 - 14x - 12x^2$
31. *Writing* Which method do you prefer for multiplying two binomials? Why? Answers may vary. Sample: FOIL; many problems can be done mentally using FOIL.
32. *Financial Planning* Suppose you deposit \$2000 in a savings account for college that has an annual interest rate r. At the end of three years, the value of your account will be $2000(1 + r)^3$ dollars.
    a. Simplify $2000(1 + r)^3$ by finding the product of $2000(1 + r)(1 + r)(1 + r)$. Write your answer in standard form.
    b. Find the amount of money in the account if the interest rate is 3%.

    a. $2000r^3 + 6000r^2 + 6000r + 2000$ b. \$2185.45

Find each product. Is the product *rational* or *irrational?*

33. $(\sqrt{3} + \sqrt{2})(\sqrt{3} - \sqrt{2})$ 1; rational
34. $(\sqrt{3} + 2)^2(\sqrt{3} - 2)^2$ 1; rational
35. $(\sqrt{5} + 3)(\sqrt{5} + 7)$ $26 + 10\sqrt{5}$; irrational
36. a. Find $(x + 1)(x + 1)$. $x^2 + 2x + 1$
    b. Find $(x + 1)(x^2 + x + 1)$. $x^3 + 2x^2 + 2x + 1$
    c. Find $(x + 1)(x^3 + x^2 + x + 1)$. $x^4 + 2x^3 + 2x^2 + 2x + 1$
    d. *Patterns* Use the pattern you see in parts (a) – (c) to **predict** the product of $(x + 1)(x^7 + x^6 + x^5 + x^4 + x^3 + x^2 + x + 1)$. $x^8 + 2x^7 + 2x^6 + 2x^5 + 2x^4 + 2x^3 + 2x^2 + 2x + 1$

JOURNAL Students should describe each step as well as show each step.

Wrap Up

THE BIG IDEA Ask students to explain how to multiply two binomials.

RETEACHING ACTIVITY Students draw arrows to show multiplication of terms in two binomials. (Reteaching worksheet 10-3)

Exercises CHECKPOINT

Students will assess their own progress on Lessons 10-1 to 10-3.

Exercises 2 and 3 Remind students to distribute the minus (or negative) sign to all of the terms in the second parentheses.

Chapter Project — *Find Out by Calculating*

You can use the expression $0.0655\ell(1 - p)(d - s)^2$ to find the number of usable board feet in a log.

- Estimate the usable board feet in a 35-ft log if its diameter is 20 in. Assume the log loses 10% (0.10) of its volume from the saw cuts and a total of 2 in. is trimmed off the ends.
- The diameter of a log is 25 in. A total of 2 in. will be trimmed off the ends. The estimated volume loss due to saw cuts is 10%. How long must the log be to yield 600 board feet of lumber?

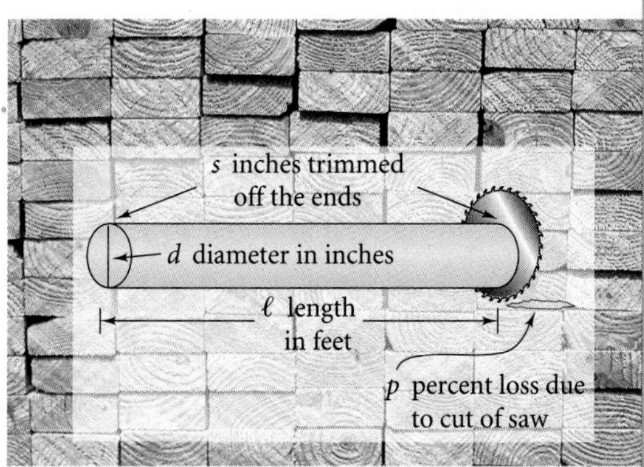

Exercises MIXED REVIEW

Solve each equation.

37. $6x^2 = 24$ 2, −2 **38.** $8x - 23 = 41$ 8 **39.** $-x^2 + x + 2 = 0$ 2, −1

40. *Language* The *Oxford English Dictionary* defines 616,500 words. Write this number in scientific notation. 6.165×10^5

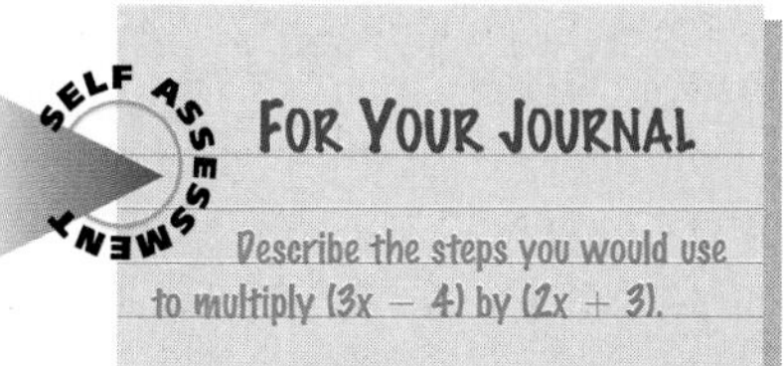

FOR YOUR JOURNAL
Describe the steps you would use to multiply (3x − 4) by (2x + 3).

Getting Ready for Lesson 10-4

Factor by grouping.

41. $2x(3x + 5) + 4(3x + 5)$ $(2x + 4)(3x + 5)$ **42.** $5x(2x - 7) + 3(2x - 7)$ $(5x + 3)(2x - 7)$ **43.** $-3x(4x + 1) + 5(4x + 1)$ $(-3x + 5)(4x + 1)$

Exercises CHECKPOINT

Simplify.

1. $(x^2 + x - 3) + (2x^2 + 4x - 1)$ $3x^2 + 5x - 4$ **2.** $(3a^3 + 2a^2 - 5) - (a^3 + a^2 + 2)$ $2a^3 + a^2 - 7$

3. $(m^2 - m + 7) - (3m^2 + 4m - 1)$ $-2m^2 - 5m + 8$ **4.** $(6x^2 - 2x - 5) + (3x^2 + x - 3)$ $9x^2 - x - 8$

Find each product.

5. $(-w + 3)(w + 3)$ $-w^2 + 9$ **6.** $2t(-3t^2 - 2t + 6)$ $-6t^3 - 4t^2 + 12t$ **7.** $(m + 4)(m - 3)$ $m^2 + m - 12$ **8.** $(b - 6)(b - 3)$ $b^2 - 9b + 18$

Factor each expression.

9. $-3c^3 + 15c^2 - 3c$ $-3c(c^2 - 5c + 1)$ **10.** $10a^3 + 5a^2 + 5a$ $5a(2a^2 + a + 1)$ **11.** $8p^3 - 20p^2 - 24p$ $4p(2p^2 - 5p - 6)$ **12.** $x^3 + 4x^2 + 7x$ $x(x^2 + 4x + 7)$

13. a. *Open-ended* Write two binomials using the variable *z*. Sample: $(z - 2), (3z + 1)$
b. Find the sum and product of the two binomials. $4z - 1;\ 3z^2 - 5z - 2$

Reteaching 10-3

Practice 10-3

Practice 10-3
Mixed Exercises

Lesson Quiz

Lesson Quiz is also available in Transparencies.

Use any method you choose to find each product.

1. $(x + 3)(x - 6)$ $x^2 - 3x - 18$

2. $(b - 3)(b - 4)$ $b^2 - 7b + 12$

3. $(x + 1)(x^2 + x + 1)$ $x^3 + 2x^2 + 2x + 1$

4. $(-3y^2 - 2y + 3)(2y + 2)$ $-6y^3 - 10y^2 + 2y + 6$

Math ToolboX

In Lesson 10-6, students will learn to solve quadratic equations by factoring. This toolbox uses the graphing calculator to show the relationship between a quadratic function and its linear factors. Use this toolbox with a discussion of the x-intercept to give students a visual preview of the solutions to quadratic equations. The graphing skills learned in this toolbox will give students a method to check their work in Lesson 10-6. Encourage students to change the viewing window so that the entire parabola can be seen. Have students press ZOOM **6** to return to the standard screen before beginning each problem.

ERROR ALERT! Some students may have forgotten how to write a linear equation from its graph. **Remediation:** Briefly review writing equations by sketching the graph of $y = 2x + 4$ on the board and working through the process of finding the corresponding equation.

WRITING Exercise 12 Students who have difficulty may use one of the exercises in this lesson as their example.

ADDITIONAL PROBLEM Graph the functions on the same calculator screen. Sketch the graph.

$y = 3x - 1 \qquad y = x + 2 \qquad y = 3x^2 + 5x - 2$

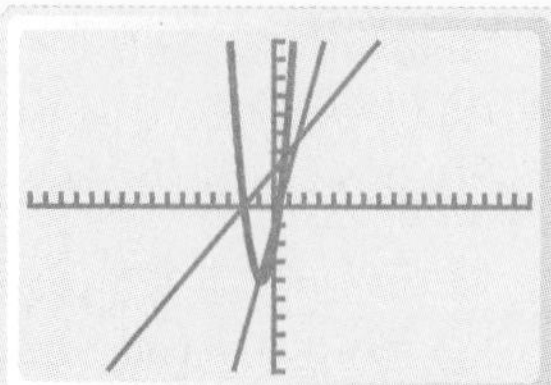

Materials and Manipulatives

- Graphing calculator

Resources

Transparencies
2, 11

page 480 Math Toolbox

1.

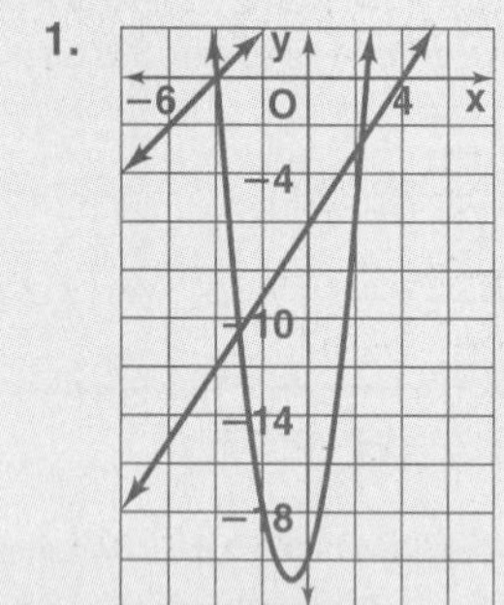

2.

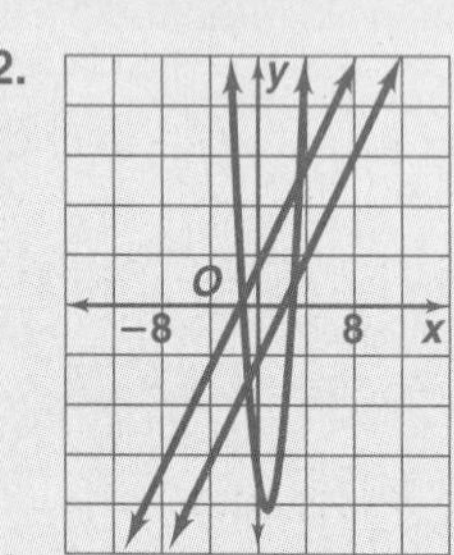

3–6, 8d. See back of book.

Math ToolboX — Technology

Exploring Factors and Products

After Lesson 10-3

You can often write a quadratic expression using factored form as well as standard form.

Factored Form	Standard Form
$(2x - 5)(x + 4)$	$2x^2 + 3x - 20$

The product of the linear expressions $2x - 5$ and $x + 4$ is the quadratic expression $2x^2 + 3x - 20$. You can explore the linear factors and their quadratic product using a graphing calculator. The related function of the expression $2x - 5$ is $y = 2x - 5$.

Graphing Calculator **Graph the functions in each exercise on the same calculator screen. Make sketches of each display.** 1–6. See margin.

1. $y = 2x - 5$
   $y = x + 4$
   $y = 2x^2 + 3x - 20$
2. $y = 2x - 5$
   $y = 2x + 3$
   $y = 4x^2 - 4x - 15$
3. $y = -x + 1$
   $y = x - 5$
   $y = -x^2 + 6x - 5$
4. $y = -\frac{1}{2}x + 3$
   $y = x + 2$
   $y = -\frac{1}{2}x^2 + 2x + 6$
5. $y = x + 3$
   $y = x - 3$
   $y = x^2 - 9$
6. $y = x + 3$
   $y = -2x - 1$
   $y = -2x^2 - 7x - 3$

7. Use your graphs in Exercises 1–6. How are the x-intercepts of the linear functions related to the x-intercepts of the quadratic function? The x-intercepts are the same.
8. **a.** Write equations in slope-intercept form of the lines at the right.
   b. What linear expressions are related to the equations in part (a)?
   c. Write the product of the linear expressions. Use standard form.
   d. Check your work by graphing the linear equations in part (a) and the quadratic function related to the expression in part (c). See margin.
   8a. $y = -x + 2$; $y = x + 3$ b. $-x + 2$; $x + 3$ c. $-x^2 - x + 6$

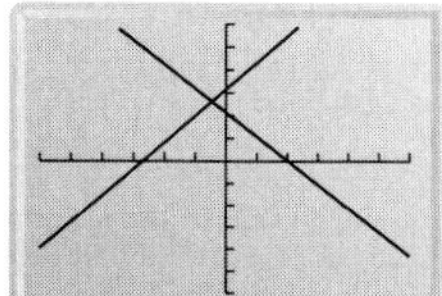

Xmin=-6 Ymin=-6
Xmax=6 Ymax=6
Xscl=1 Yscl=1

Write the expressions related to each linear graph. Then write the quadratic product of the expressions in standard form.

9. $-x + 2$; $x - 2$; $-x^2 + 4x - 4$
10. 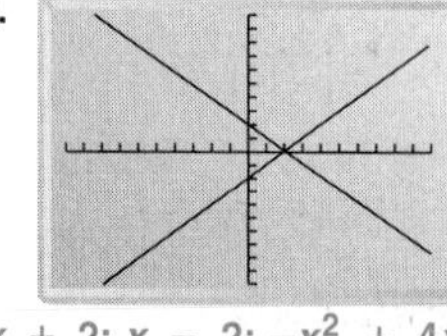
   $2x$; $\frac{1}{2}x - 2$; $x^2 - 4x$
11. 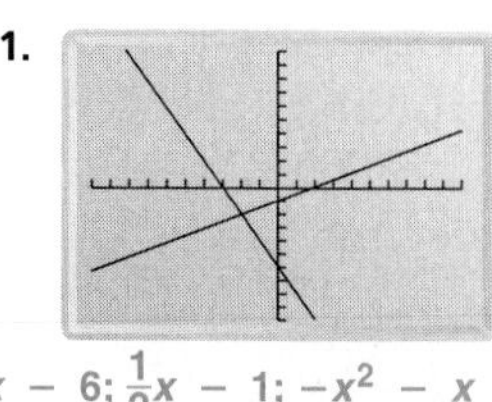
   $-2x - 6$; $\frac{1}{2}x - 1$; $-x^2 - x + 6$

12. *Writing* **Summarize** the relationship between a quadratic expression and its linear factors. Include an example. Answers may vary. Sample: A quadratic expression has the same x-intercepts as each of its linear factors. $x^2 + x - 20$ and $(x + 5)(x - 4)$ have x-intercepts -5, 4.

480

10-4 Teaching Notes

PROBLEM OF THE DAY

Mr. Cruz needs to paint a rectangular wooden trailer that is 9 ft wide, 18 ft long, and 7 ft high. If he doesn't paint the bottom, how many square feet will he paint? 540 sq ft

Problem of the Day is also available in Transparencies.

CONNECTING TO PRIOR KNOWLEDGE On the board, write $x^2 + 10x + 16 = 0$. Have students make conjectures about how they could solve the equation by extracting the factors, without using either the graphing technique or the quadratic formula.

WORK TOGETHER

MAKING CONNECTIONS Concrete is a mixture of sand, gravel, Portland cement, and water. For driveways and sidewalks, a cement : sand : gravel ratio of $1 : 2\frac{1}{4} : 3$ is recommended.

THINK AND DISCUSS

Example 1

Point out that in making the rectangles, the x-dimension of a tile must be placed next to the x-dimension of another tile; the 1-dimension must be placed next to the 1-dimension.

Connections Construction . . . and more

What You'll Learn

- Factoring quadratic expressions
- Identifying quadratic expressions that cannot be factored

...And Why

To solve civil engineering and landscaping problems

What You'll Need

- tiles
- graph paper

10-4 Factoring Trinomials

WORK TOGETHER

Construction A contractor has agreed to pour the concrete for the floor of a garage. He knows the area of the floor is 221 ft^2 but cannot remember its dimensions. He does remember that the dimensions are prime numbers.

Work with a partner to find the dimensions of the garage floor.

1. Explain how you know that 221 is not the product of two 1-digit numbers. Answers may vary. Sample: Products of 1-digit numbers are less than 100.
2. Explain how you could decide if 221 has a factor greater than 20 other than itself. See margin.
3. Use graph paper. Cut out several 10×10 squares, some 1×10 rectangles, and some 1×1 squares. Use these pieces to make a rectangle representing the garage floor. What are its dimensions? 13×17
4. Repeat this process to find a pair of prime numbers with each product.
   a. 133 7, 19 **b.** 161 7, 23 **c.** 209 11, 19

THINK AND DISCUSS

Part 1 **Using Tiles**

Some quadratic trinomials are the product of two **binomial factors.**

quadratic trinomial		binomial factors
$x^2 + 10x + 16$	→	$(x + 2)(x + 8)$

The diagram at the left shows how $x^2 + 10x + 16$ can be displayed as a rectangle with sides of length $x + 2$ and $x + 8$.

Lesson Planning Options

Prerequisite Skills

- Finding factors of numbers

Assignment Options for Exercises On Your Own

To provide flexible scheduling, this lesson can be subdivided into parts.

1 **Core** 1–15
Extension 22, 23

2 **Core** 24–39
Extension 16–21

3 **Core** 43–54
Extension 40–42

Use Mixed Review to maintain skills.

Resources

Student Edition

Skills Handbook, pp. 573, 574, 582
Extra Practice, p. 565
Glossary/Study Guide

Teaching Resources

Chapter Support File, Ch. 10
- Practice 10-4 (two worksheets)
- Reteaching 10-4
- Alternative Activity 10-4

Classroom Manager 10-4
Glossary, Spanish Resources
Two-Year Algebra Handbook 10-4

Transparencies
1, 17, 116, 121

Have students verify their answers using the FOIL method of multiplying binomials.

Example 2

To parallel this example with tiles, have students make rectangles with the twenty 1-tiles until they obtain a rectangle that allows them to fit in the nine negative x-tiles. This should confirm why they need to find two numbers with a product of 20 and a sum of -9.

Have tiles available so that students can experience working through the same examples with and without tiles. This will work particularly well if you have extended class periods or block scheduling.

TACTILE LEARNING Divide students into pairs. Have one student factor using pencil and paper while the other uses tiles. Have students compare their answers and resolve any differences.

Example 3

Ask: *How is this trinomial different from the trinomials in Examples 1 and 2?* The constant term is negative.

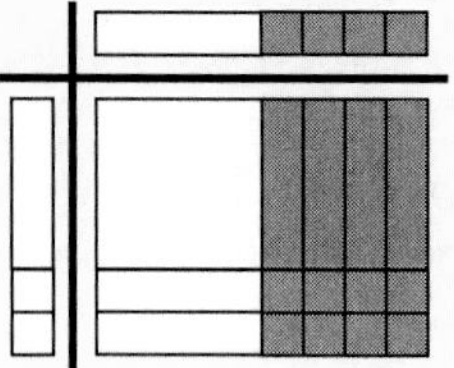

Additional Examples

FOR EXAMPLE 1

Use tiles to factor $x^2 + 4x + 3$.

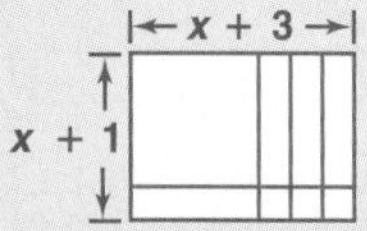

Discussion: *How do you know that both factors must be positive?*

FOR EXAMPLE 3

Factor $x^2 - x - 12$. $(x + 3)(x - 4)$

Discussion: *How do you know that one factor must be positive and the other factor must be negative? Can you predict which factor (positive or negative) will have the greatest absolute value? Explain.*

FOR EXAMPLE 4

Factor $x^2 - 5x + 6$. $(x - 2)(x - 3)$

Discussion: *How do you know that both factors must be negative?*

You can use tiles to factor quadratic trinomials.

Example 1

Use tiles to factor $x^2 + 7x + 12$.

Choose one x^2-tile, seven x-tiles and twelve 1-tiles. Use the strategy *Guess and Test* to form a rectangle using all the tiles.

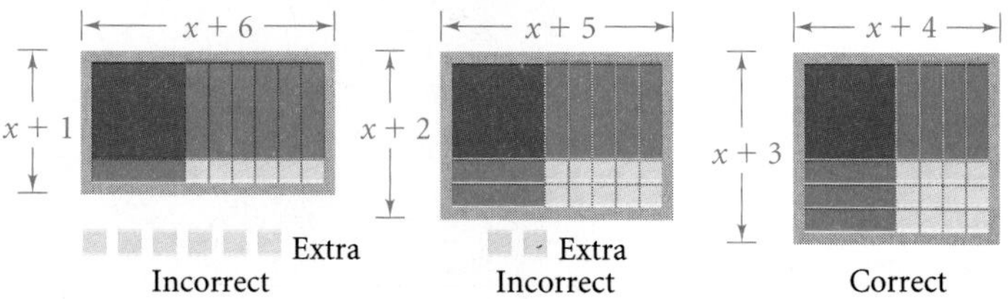

Write the correct factors as a product.

$x^2 + 7x + 12 = (x + 3)(x + 4)$

5. **Try This** Use tiles to factor these trinomials. **Verify** your answers using FOIL.
   a. $x^2 + 6x + 8$ $(x + 2)(x + 4)$ b. $x^2 + 11x + 10$ $(x + 10)(x + 1)$

Part 2

Testing Possible Factors

To factor trinomials of the form $x^2 + bx + c$, you can use FOIL with the strategy *Guess and Test.*

The sum of the numbers you use here must equal b.

$$x^2 + bx + c = (x + \square)(x + \square)$$

The product of the numbers you use here must equal c.

Example 2

Factor $x^2 - 9x + 20$.

Choose numbers that are factors of 20. Look for a pair with sum -9.

Factors of 20	Sum of Factors
-1 and -20	$-1 + (-20) = -21$
-2 and -10	$-2 + (-10) = -12$
-4 and -5	$-4 + (-5) = -9$

List only negative factors because you are looking for a sum of -9. Two positive numbers cannot have a negative sum.

The numbers -4 and -5 have a product of 20 and a sum of -9.
The correct factors are $(x - 4)$ and $(x - 5)$.
So, $x^2 - 9x + 20 = (x - 4)(x - 5)$.

To factor this trinomial with tiles, students will need to use a combination of positive x-tiles and negative x-tiles that have a sum of $-2x$.

Example 4

Ask: *How is this trinomial different from the trinomials in Examples 1–3?* **The coefficient of x^2 is not equal to 1.**

Have students verify the solution with tiles.

Check $x^2 - 9x + 20 \stackrel{?}{=} (x-4)(x-5)$ ← Find the product of the right side.

$x^2 - 9x + 20 \stackrel{?}{=} x^2 - 5x - 4x + 20$

$x^2 - 9x + 20 = x^2 - 9x + 20$ ✔

6. Critical Thinking Is $x^2 - 6x - 16 = (x-8)(x+2)$ factored correctly? Explain. **Yes; $-8 + 2 = -6$ for the middle term and $(-8)(2) = -16$ for the last term.**

Example 3

Factor $x^2 - 2x - 8$.

Choose numbers that are factors of -8. Look for a pair with sum -2.

Factors of -8	Sum of Factors
-1 and 8	$-1 + 8 = 7$
-8 and 1	$-8 + 1 = -7$
-2 and 4	$-2 + 4 = 2$
-4 and 2	$-4 + 2 = -2$

← -4 and 2 have a sum of -2.

$x^2 - 2x - 8 = (x-4)(x+2)$

7. Try This Factor $x^2 - 4x - 12$. **$(x-6)(x+2)$**

Part 3

Factoring $ax^2 + bx + c$

To factor quadratic trinomials where $a \neq 1$, list factors of a and c. Use these factors to write binomials. Test for the correct value for b.

Example 4

Factor $3x^2 - 7x - 6$.

List factors of 3: 1 and 3; -1 and -3.

List factors of -6: 1 and -6; -1 and 6; 2 and -3; -2 and 3.

Use the factors to write binomials. Look for -7 as the middle term.

$(\boxed{1}x + \boxed{1})(\boxed{3}x + \boxed{-6})$ $-6x + 3x = -3x$

$(\boxed{1}x + \boxed{-6})(\boxed{3}x + \boxed{1})$ $1x - 18x = -17x$

$(\boxed{1}x + \boxed{-1})(\boxed{3}x + \boxed{6})$ $6x - 3x = 3x$

$(\boxed{1}x + \boxed{6})(\boxed{3}x + \boxed{-1})$ $-1x + 18x = 17x$

$(\boxed{1}x + \boxed{2})(\boxed{3}x + \boxed{-3})$ $-3x + 6x = 3x$

$(\boxed{1}x + \boxed{-3})(\boxed{3}x + \boxed{2})$ $2x - 9x = -7x$ Correct!

$3x^2 - 7x - 6 = (x-3)(3x+2)$.

8. Try This Factor $2x^2 - 3x - 5$. **$(2x-5)(x+1)$**

Technology Options

For Exercises 8–15, students may want to use tiles found in algebra software.

Prentice Hall Technology 

Software

- Secondary Math Lab Toolkit™
- Computer Item Generator 10-4

Internet

- See the Prentice Hall site. (http://www.phschool.com)

Exercises ON YOUR OWN

ERROR ALERT! **Exercise 16** Some students may not realize that the missing number can be a negative number. **Remediation:** Point out that the value of the missing number must be negative in order to obtain the given trinomial. Students can rewrite $x + (-7)$ as $x - 7$.

WRITING **Exercise 22** To answer this exercise, have students go back and analyze the trinomials in Examples 1–3.

DIVERSITY **Exercise 23** Some students may not be familiar with community gardens. Ask a student who is familiar with them to describe them to the class. If there is a community garden in your community, recommend that students visit the garden.

ESL **Exercise 23** To avoid confusion with plotted points, such as those on a graph, explain that *plot* in this instance means a piece of land. Ask students what kinds of vegetables they would plant if the class worked together on a plot in a community garden.

ALTERNATIVE ASSESSMENT **Exercises 24–31** Ask students to solve these exercises by factoring and then justify their results using drawings or tiles. This will help you assess student understanding of the factors of a quadratic equation.

EXTENSION **Exercise 40** Ask: *How many possible values are there?* an infinite number *How could you use algebra to describe all the possible values?* $n(n + 3)$

Chapter Project FIND OUT BY CALCULATING Students use the given expression to determine the diameter of trees. This

page 481 Work Together

2. Answers may vary. Sample: If one prime factor of 221 is greater than 20, it must also be greater than or equal to 23. Since $23 \cdot 10 > 221$, the other prime factor of 221 would have to be 3, 5, or 7, which are easily checked.

pages 484–485 On Your Own

8.

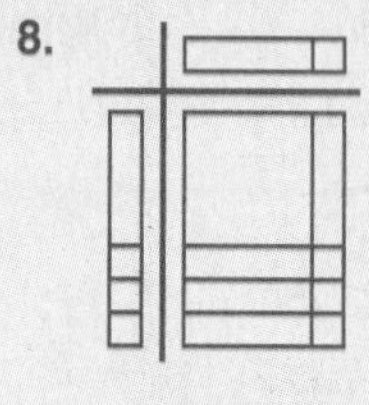

9.

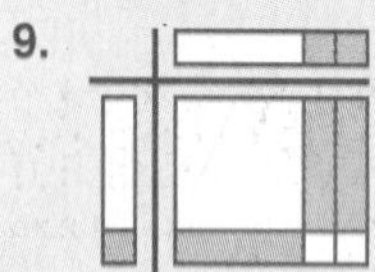

10–15, 22a–b, 40–42. See back of book.

page 485 Mixed Review

59–60. See back of book.

484

Exercises ON YOUR OWN

Write the length and width of each rectangle as a binomial. Then write an expression for the area of each rectangle.

1. 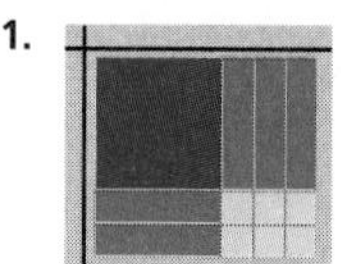
$x + 3$; $x + 2$; $x^2 + 5x + 6$

2. 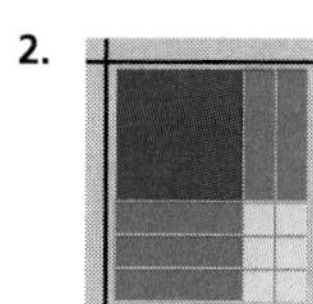
$x + 2$; $x + 3$; $x^2 + 5x + 6$

3. 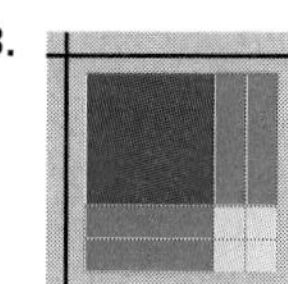
$x + 2$; $x + 2$; $x^2 + 4x + 4$

Can you form a rectangle using all the pieces in each set? Explain.

4. one x^2-tile, two x-tiles, and one 1-tile
Yes; $(x + 1)(x + 1) = x^2 + 2x + 1$
5. one x^2-tile, five x-tiles, and eight 1-tiles
No; no two numbers with product 8 have sum 5.
6. one x^2-tile, six x-tiles, and six 1-tiles
No; no two numbers with product 6 have sum 6.
7. one x^2-tile, nine x-tiles, and eight 1-tiles
Yes; $(x + 8)(x + 1) = x^2 + 9x + 8$

Use tiles or make drawings to represent each expression as a rectangle. Then write the area as the product of two binomials. 8–15. See margin for diagrams.

8. $x^2 + 4x + 3$ $(x + 3)(x + 1)$
9. $x^2 - 3x + 2$ $(x - 2)(x - 1)$
10. $x^2 + 3x - 4$ $(x + 4)(x - 1)$
11. $x^2 - 2x - 8$ $(x - 4)(x + 2)$
12. $x^2 + 5x + 6$ $(x + 2)(x + 3)$
13. $x^2 - 3x - 4$ $(x - 4)(x + 1)$
14. $x^2 + x - 6$ $(x + 3)(x - 2)$
15. $x^2 - 2x + 1$ $(x - 1)(x - 1)$

Complete.

16. $x^2 - 6x - 7 = (x + 1)(x + ■)$ −7
17. $k^2 - 4k - 12 = (k - 6)(k + ■)$ 2
18. $t^2 + 7t + 10 = (t + 2)(t + ■)$ 5
19. $c^2 + c - 2 = (c + 2)(c + ■)$ −1
20. $y^2 - 13y + 36 = (y - 4)(y + ■)$ −9
21. $x^2 + 3x - 18 = (x + 6)(x + ■)$ −3

22. *Writing* Suppose you can factor $x^2 + bx + c$ into the product of two binomials. a–b. See margin.
    a. Explain what you know about the factors if $c > 0$.
    b. Explain what you know about the factors if $c < 0$.

23. *Community Gardening* The diagram at the right shows 72 plots in a community garden.
    a. Write a quadratic expression that represents the area of the garden. $x^2 + 15x + 56$
    b. Write the factors of the expression you wrote in part (a). $(x + 7)(x + 8)$

Factor each quadratic trinomial.

24. $x^2 + 6x + 8$ $(x + 2)(x + 4)$
25. $a^2 - 5a + 6$ $(a - 2)(a - 3)$
26. $d^2 - 7d + 12$ $(d - 4)(d - 3)$
27. $k^2 + 9k + 8$ $(k + 1)(k + 8)$
28. $y^2 - 4y - 45$ $(y - 9)(y + 5)$
29. $r^2 - 10r - 11$ $(r - 11)(r + 1)$
30. $c^2 + 2c + 1$ $(c + 1)(c + 1)$
31. $x^2 + 2x - 15$ $(x + 5)(x - 3)$
32. $t^2 + 7t - 18$ $(t + 9)(t - 2)$
33. $x^2 + 12x + 35$ $(x + 5)(x + 7)$
34. $y^2 - 10y + 16$ $(y - 8)(y - 2)$
35. $a^2 - 9a + 14$ $(a - 7)(a - 2)$
36. $r^2 + 6r - 16$ $(r + 8)(r - 2)$
37. $y^2 + 13y - 48$ $(y + 16)(y - 3)$
38. $x^2 + 10x + 25$ $(x + 5)(x + 5)$
39. $w^2 - 2w - 24$ $(w - 6)(w + 4)$

task is essential to their work on the Chapter Project. Have students add this part of the project to work they have already completed. This is a good time to check students' progress by examining their project folders.

Exercises MIXED REVIEW

GETTING READY FOR LESSON 10-5 These exercises prepare students for factoring perfect square trinomials and the difference of two squares.

SELF ASSESSMENT **JOURNAL** Students' journal entries may give you clues to possible misunderstandings they may have about tiles.

Wrap Up

THE BIG IDEA Ask students to tell what they have learned about finding the binomial factors of a trinomial.

RETEACHING ACTIVITY Students use tiles to factor quadratic expressions. Then they factor trinomials and check their answers using tiles. (Reteaching worksheet 10-4)

Open-ended **Find three different values to complete each expression so that it can be factored into the product of two binomials. Show each factorization.** 40–42. See margin.

40. $x^2 - 3x - \blacksquare$ 41. $x^2 + x - \blacksquare$ 42. $x^2 + \blacksquare x + 12$

Factor each expression.

43. $2x^2 - 15x + 7$ $(2x - 1)(x - 7)$
44. $5x^2 - 2x - 7$ $(5x - 7)(x + 1)$
45. $2x^2 - x - 3$ $(2x - 3)(x + 1)$
46. $8x^2 - 14x + 3$ $(4x - 1)(2x - 3)$
47. $2x^2 - 11x - 21$ $(2x + 3)(x - 7)$
48. $3x^2 + 13x - 10$ $(3x - 2)(x + 5)$
49. $2x^2 - 7x + 3$ $(2x - 1)(x - 3)$
50. $6t^2 + 13t - 5$ $(3t - 1)(2t + 5)$
51. $7x^2 - 20x - 3$ $(7x + 1)(x - 3)$
52. $2x^2 + x - 3$ $(2x + 3)(x - 1)$
53. $3x^2 + 17x + 20$ $(3x + 5)(x + 4)$
54. $2x^2 + 3x - 20$ $(2x - 5)(x + 4)$

Chapter Project — Find Out by Calculating

20 ft

With aerial photography, you can study a forest of ponderosa pines without ever walking through it.

To find the diameter in inches of trees in the forest, use this expression:

$3.76 + (1.35 \times 10^{-2})hv - (2.45 \times 10^{-6})hv^2 + (2.44 \times 10^{-10})hv^3$

The variable h is the height of the tree in feet, and v is the crown diameter visible in feet (from a photograph).

- Determine the diameter of 100-ft trees that have a visible crown diameter of 20 ft.

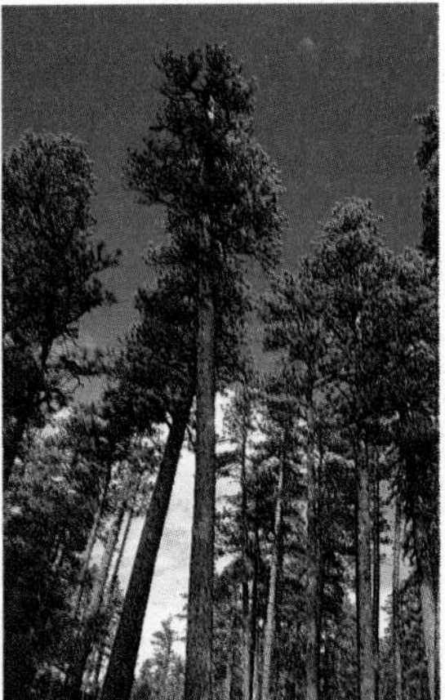

Exercises MIXED REVIEW

Find the distance between the given points.

55. (3, 5); (6, 1) 5
56. (−2, 8); (4, −1) about 10.8
57. (−7, −2); (−3, −9) about 8.1
58. (8, 0); (−3, −4) about 11.7

Graph each inequality. 59–60. See margin.

59. $y \leq 4x - 9$
60. $y > x^2 + 3x + 1$

61. *Statistics* Suppose you have taken four history tests. Your test scores are 84, 78, 75, and 79. What must you score on your next test to average at least 80 points on all five tests? 84

SELF ASSESSMENT **FOR YOUR JOURNAL**

Describe several things you like and/or do not like about using tiles.

Getting Ready for Lesson 10-5

Find each product.

62. $(x + 9)^2$ $x^2 + 18x + 81$
63. $(2x + 3)^2$ $4x^2 + 12x + 9$
64. $(5x - 4)^2$ $25x^2 - 40x + 16$
65. $(x + 8)(x - 8)$ $x^2 - 64$
66. $(x + 4)(x - 4)$ $x^2 - 16$
67. $(2x + 7)(2x - 7)$ $4x^2 - 49$
68. $(3x + 5)(3x - 5)$ $9x^2 - 25$
69. $(2x - 1)^2$ $4x^2 - 4x + 1$

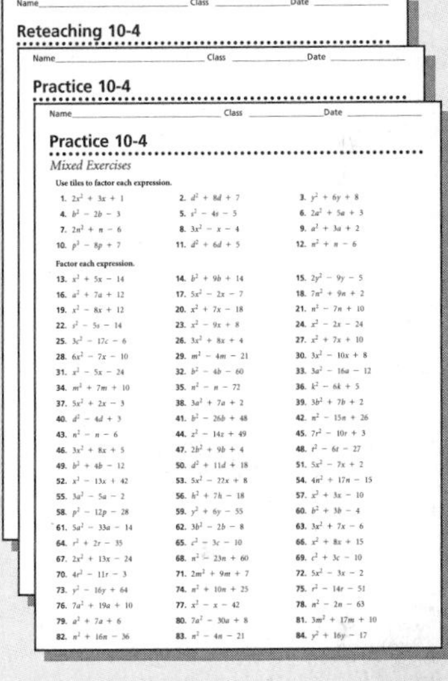

Name______ Class ______ Date ______

Reteaching 10-4

Name______ Class ______ Date ______

Practice 10-4

Name______ Class ______ Date ______

Practice 10-4

Mixed Exercises

Use tiles to factor each expression.

1. $2x^2 + 3x + 1$ 2. $d^2 + 8d + 7$ 3. $y^2 + 6y + 8$
4. $b^2 - 2b - 3$ 5. $s^2 - 4s - 5$ 6. $2a^2 + 5a + 3$
7. $2n^2 + n - 6$ 8. $3x^2 - x - 4$ 9. $a^2 + 3a + 2$
10. $p^2 - 8p + 7$ 11. $d^2 + 6d + 5$ 12. $n^2 + n - 6$

Factor each expression.

13. $x^2 + 5x - 14$ 14. $b^2 + 9b + 14$ 15. $2y^2 - 9y - 5$
16. $a^2 + 7a + 12$ 17. $5x^2 - 2x - 7$ 18. $7n^2 + 9n + 2$
19. $x^2 - 8x + 12$ 20. $x^2 + 7x - 18$ 21. $n^2 - 7n + 10$
22. $s^2 - 5s - 14$ 23. $x^2 - 9x + 8$ 24. $x^2 - 2x - 24$
25. $3c^2 - 17c - 6$ 26. $3x^2 + 8x + 4$ 27. $x^2 + 7x + 10$
28. $6x^2 - 7x - 10$ 29. $m^2 - 4m - 21$ 30. $3x^2 - 10x + 8$
31. $x^2 - 5x - 24$ 32. $b^2 - 4b - 60$ 33. $3a^2 - 16a - 12$
34. $m^2 + 7m + 10$ 35. $n^2 - n - 72$ 36. $k^2 - 6k + 5$
37. $5x^2 + 2x - 3$ 38. $3a^2 + 7a + 2$ 39. $3b^2 + 7b + 2$
40. $d^2 - 4d + 3$ 41. $b^2 - 26b + 48$ 42. $n^2 - 15n + 26$
43. $n^2 - n - 6$ 44. $z^2 - 14z + 49$ 45. $7r^2 - 10r + 3$
46. $3x^2 + 8x + 5$ 47. $2b^2 + 9b + 4$ 48. $t^2 - 6t - 27$
49. $b^2 + 4b - 12$ 50. $d^2 + 11d + 18$ 51. $5x^2 - 7x + 2$
52. $x^2 - 13x + 42$ 53. $5x^2 - 22x + 8$ 54. $4n^2 + 17n - 15$
55. $3a^2 - 5a - 2$ 56. $h^2 + 7h - 18$ 57. $x^2 + 3x - 10$
58. $p^2 - 12p - 28$ 59. $y^2 + 6y - 55$ 60. $b^2 + 3b - 4$
61. $5a^2 - 33a - 14$ 62. $3b^2 - 2b - 8$ 63. $3x^2 + 7x - 6$
64. $r^2 + 2r - 35$ 65. $c^2 - 3c - 10$ 66. $x^2 + 8x + 15$
67. $2x^2 + 13x - 24$ 68. $n^2 - 23n + 60$ 69. $c^2 + 3c - 10$
70. $4r^2 - 11r - 3$ 71. $2m^2 + 9m + 7$ 72. $5x^2 - 3x - 2$
73. $y^2 - 16y + 64$ 74. $n^2 + 10n + 25$ 75. $r^2 - 14r - 51$
76. $7a^2 + 19a + 10$ 77. $x^2 - x - 42$ 78. $n^2 - 2n - 63$
79. $a^2 + 7a + 6$ 80. $7a^2 - 30a + 8$ 81. $3m^2 + 17m + 10$
82. $n^2 + 16n - 36$ 83. $n^2 - 4n - 21$ 84. $y^2 + 16y - 17$

Lesson Quiz

Lesson Quiz is also available in Transparencies.

Factor each quadratic trinomial.

1. $x^2 - x - 2$ $(x + 1)(x - 2)$
2. $x^2 - 5x + 6$ $(x - 3)(x - 2)$
3. $x^2 + 7x + 6$ $(x + 1)(x + 6)$
4. $a^2 + 2a - 8$ $(a + 4)(a - 2)$

10-5 Teaching Notes

PROBLEM OF THE DAY

Chef Gloria Giovonni needs 6 c of sugar but has only a $\frac{3}{4}$ c scoop with her. How many full scoops of sugar should she add? 8

Problem of the Day is also available in Transparencies.

CONNECTING TO PRIOR KNOWLEDGE Ask students: *Can you give an example of a situation in math where you saw a pattern and used that knowledge to save yourself time in solving problems?* **Answers may vary. Sample: If you notice that the k in $y = x + k$ moves the graph of $y = x$ up and down, you can save time by not having to make a table of values.**

Lead students to understand that, in this lesson, they will be learning about patterns which will save time and effort.

WORK TOGETHER

Pattern recognition activities such as these are excellent for cooperative learning. This type of team work prepares students for teamwork in the work place. For example, many scientists work in teams to analyze data and make predictions based on patterns.

TACTILE LEARNING Encourage students to use tiles to help them see the pattern in each group.

Question 2 Some students may choose to use tiles to find the products.

Lesson Planning Options

Prerequisite Skills

- Multiplying trinomials
- Factoring trinomials

Assignment Options for Exercises On Your Own

To provide flexible scheduling, this lesson can be subdivided into parts.

1 **Core** 25–32
Extension 1–4, 37, 38

2 **Core** 5–20
Extension 21–24, 33–36

Use Mixed Review to maintain skills.

Resources

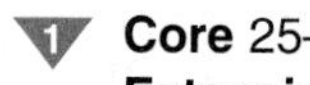

Student Edition

Skills Handbook, pp. 573, 574, 579
Extra Practice, p. 565
Glossary/Study Guide

Teaching Resources

Chapter Support File, Ch. 10
- Practice 10-5 (two worksheets)
- Reteaching 10-5

Classroom Manager 10-5
Glossary, Spanish Resources
Two-Year Algebra Handbook 10-5

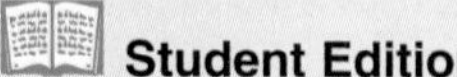

Transparencies
11, 17, 117, 122

10-5 Factoring Special Cases

What You'll Learn

- Factoring the difference of two squares
- Factoring perfect square trinomials

...And Why

To solve problems related to geometry and construction

What You'll Need

- tiles
- graphing calculator

WORK TOGETHER

Work with a partner. Answer each question for Groups A, B, and C.

Group A	Group B	Group C
$(x + 7)(x - 7)$	$(x + 7)(x + 7)$	$(x - 7)(x - 7)$
$(k + 3)(k - 3)$	$(k + 3)(k + 3)$	$(k - 3)(k - 3)$
$(w + 5)(w - 5)$	$(w + 5)(w + 5)$	$(w - 5)(w - 5)$
$(3x + 1)(3x - 1)$	$(3x + 1)(3x + 1)$	$(3x - 1)(3x - 1)$

1. Describe the pattern in each of the pairs of factors. See margin.
2. Find each product. See margin.
3. How can you use mental math to quickly multiply binomials like those in each group? Explain using examples. See margin.

THINK AND DISCUSS

Factoring a Difference of Two Squares

Part 1

As you saw in the Work Together activity, sometimes when you multiply two binomials, the *middle* term in the product is 0.

4. What polynomial is modeled by each set of tiles? What are the factors of each polynomial?

a. 

$x^2 - 4$
$(x + 2)(x - 2)$

b. $4x^2 - 9, (2x + 3)(2x - 3)$

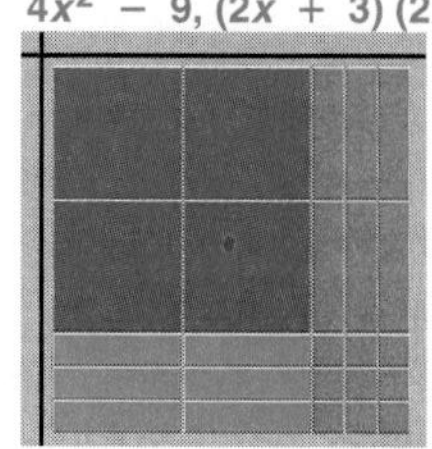

Each product represents either a perfect square trinomial or the difference of two squares. In this lesson, students will learn special methods for factoring these types of trinomials.

THINK AND DISCUSS

Example 1

Have students generate a list of binomials of this type. Any binomial $x^2 - n$, for which n is a perfect square is acceptable.

Example 2

Have students generate a list of binomials of this type. Any binomial $ax^2 - n$, for which a and n are both perfect squares is acceptable.

VISUAL LEARNING Use tiles to demonstrate how to factor perfect square trinomials. Encourage students to describe the visual patterns and relate the patterns to the math.

Direct students' attention to the section Perfect Square Trinomial at the bottom of the page. On the board, write $(x + 4)(x + 4) = x^2 + 8x + 6$. Ask: *What is the value of a?* x *What is the value of b?* 4

Illustrate $(x + 4)(x + 4) = x^2 + 8x + 16$ with tiles. Point out the tile that represents $a^2(x^2)$, the two sets of tiles that represent $2ab$ ($4x$ and $4x$), and the tiles that represent b^2 (16).

MAKING CONNECTIONS The formula for the difference of two squares can be used to calculate mentally the squares of difficult numbers. For example, 99^2 is very difficult to evaluate mentally. However, 99^2 can be written as the difference of two

When you factor a difference of two squares, the result is two binomial factors that are the same except for the signs between the terms.

Difference of Two Squares

For all real numbers a and b, $a^2 - b^2 = (a + b)(a - b)$.

Example: $x^2 - 16 = (x + 4)(x - 4)$

Example 1

Factor $x^2 - 64$.

$x^2 - 64 = x^2 - 8^2$ ← Rewrite 64 as 8^2.

$= (x + 8)(x - 8)$ ← Factor.

Check Use FOIL to multiply.

$(x + 8)(x - 8)$

$x^2 - 8x + 8x - 64$

$x^2 - 64$ ✔

5. Mental Math Factor $x^2 - 36$. $(x + 6)(x - 6)$

Example 2

Factor $4x^2 - 121$.

$4x^2 - 121 = 4x^2 - 11^2$ ← Rewrite 121 as 11^2.

$= (2x)^2 - (11)^2$ ← Rewrite $4x^2$ as $(2x)^2$.

$= (2x + 11)(2x - 11)$ ← Factor.

QUICK REVIEW

For $a \neq 0$ and $b \neq 0$ and all integers n, $(ab)^n = a^n b^n$.

6. Critical Thinking Suppose a classmate factored $4x^2 - 121$ and got $(4x + 11)(4x - 11)$. What mistake did this classmate make? $4x^2$ was not factored into $(2x)^2$.

7. Mental Math Factor $9x^2 - 25$. $(3x + 5)(3x - 5)$

Part 2 Factoring a Perfect Square Trinomial

In the Work Together activity you multiplied a binomial by itself. This is called squaring a binomial. The result is a **perfect square trinomial.** When you factor a perfect square trinomial, the two binomial factors are the same.

Perfect Square Trinomial

For all real numbers a and b:

$$a^2 + 2ab + b^2 = (a + b)(a + b) = (a + b)^2$$
$$a^2 - 2ab + b^2 = (a - b)(a - b) = (a - b)^2$$

Examples: $x^2 + 10x + 25 = (x + 5)(x + 5) = (x + 5)^2$

$x^2 - 10x + 25 = (x - 5)(x - 5) = (x - 5)^2$

Additional Examples

FOR EXAMPLE 1

Factor $a^2 - 16$. $(a + 4)(a - 4)$

FOR EXAMPLE 3

Factor $x^2 - 6x + 9$. $(x - 3)^2$

FOR EXAMPLE 4

Factor $4x^2 + 12x + 9$ $(2x + 3)^2$

FOR EXAMPLE 5

Factor $18x^3 - 32x$ $2x(3x - 4)(3x + 4)$

Discussion: *What are some guidelines that should be followed when trying to factor a quadratic expression?*

squares: $(100 - 1)^2$. Show the following steps:

$$(100 - 1)^2 = 100^2 - 2(100)(1) + 1^2$$
$$= 10,000 - 200 + 1$$
$$= 9801$$

Example 3

Question 8 *Ask: What is the axis of symmetry?* $x = 4$

Check students' abilities to identify perfect square trinomials. Write the following trinomials on the board. Ask: *Is the trinomial a perfect square trinomial?*

$x^2 - 6x + 9$ yes
$x^2 - 8x + 25$ no
$x^2 - 4x + 4$ yes
$x^2 - 10x + 36$ no

Example 4

Have students verify their answers using tiles.

Question 11 Have students illustrate their trinomial in an algebra-tile diagram. Discuss students' strategies for writing their trinomials.

Example 5

Have students give examples of other binomials that can be factored this way.

TACTILE and VISUAL LEARNING Encourage students to make a poster listing the special cases and describing how to factor them. Have students provide examples for each special case. Have students write the examples using the variables *a* and *b*,

Technology Options

For Exercises 5–20, students may find it helpful to factor the expressions using algebra software.

Prentice Hall Technology

Calculator
- Graphing Calculator Handbook: Procedure 7

Software
- Secondary Math Lab Toolkit™
- Computer Item Generator 10-5

Internet
- See the Prentice Hall site. (http://www.phschool.com)

488

Example 3

Factor $x^2 - 8x + 16$.

$x^2 - 8x + 16 = x^2 - 8x + 4^2$ ← Rewrite 16 as 4^2.

$= x^2 - 2(x)(4) + 4^2$ ← Does the middle term equal $2ab$? $8x = 2(x)(4)$ ✓

$= (x - 4)^2$ ← Factor as a squared binomial.

So, $x^2 - 8x + 16 = (x - 4)^2$.

8. a. **Graphing Calculator** Graph $y = x^2 - 8x + 16$. See margin.
   b. Find the x-intercept(s) of the graph. 4
   c. **Critical Thinking** What information about the factorization does the x-intercept(s) give you? The x-intercept 4 indicates that $x - 4$ is a factor of $x^2 - 8x + 16$.
9. The expression $(3x + 4)^2$ equals $9x^2 + ■ + 16$. What is the middle term? $24x$

Example 4

Factor $9x + 12x + 4$.

$9x^2 + 12x + 4$

$= (3x)^2 + 12x + 2^2$ ← Rewrite $9x^2$ as $(3x)^2$ and 4 as 2^2.

$= (3x)^2 + 2(3x)(2) + 2^2$ ← Does the middle term equal $2ab$? $12x = 2(3x)(2)$ ✓

$= (3x + 2)^2$ ← Factor as a squared binomial.

So, $9x + 12x + 4 = (3x + 2)^2$.

PROBLEM SOLVING
Look Back Multiply the factors to check the factorization.

10. **Try This** Factor each trinomial.
    a. $x^2 - 14x + 49$ $(x - 7)^2$ b. $x^2 + 18x + 81$ $(x + 9)^2$ c. $4x^2 - 12x + 9$ $(2x - 3)^2$
11. a. **Open-ended** Write a quadratic expression of your own that is a perfect square trinomial. Sample: $9x^2 + 24x + 16$
    b. Explain how you know your trinomial is a perfect square trinomial. The first and last terms are perfect squares and the middle term is twice the product of the square roots of the first and last terms.

Sometimes a quadratic expression looks like it can't be factored when actually it can. Take out any common factors. Then see if you can factor further.

Example 5

Factor $10x^2 - 40$.

$10x^2 - 40 = 10(x^2 - 4)$ ← Factor out the GCF: 10.

$= 10(x - 2)(x + 2)$ ← Factor $(x^2 - 4)$.

So, $10x^2 - 40 = 10(x - 2)(x + 2)$.

12. **Try This** Factor $8x^2 - 50$. $2(2x + 5)(2x - 5)$

as are used in their texts. Suggest that students use one color pen for a and a different color for b.

Exercises ON YOUR OWN

OPEN-ENDED Exercise 4 Have volunteers use the overhead projector to show their models.

ALTERNATIVE ASSESSMENT **Exercises 5–20** In order to assess student understanding of factoring special cases, ask them to represent the factors of these two expressions using illustrations or tiles.

ERROR ALERT! **Exercise 20** Some students may factor out 2 as the GCF. **Remediation:** Point out that the GCF of the expression is $2x$.

WRITING Exercise 21 Suggest one example be representative of $(a + b)(a + b)$ and one representative of $(a - b)(a - b)$.

DIVERSITY Exercise 22 Most students will not be familiar with cranes or the kinds of decisions that crane operators make. Have someone who is familiar with cranes and the physics of their operation speak to the class. Since construction has traditionally been a male-dominated field, try to have a female speaker address the class.

CONNECTING TO THE STUDENTS' WORLD Exercise 22 Encourage students to visit hardware stores to find out the difference in strength between nylon rope, cable, and chain. Also have them find out the diameters of cable typically used in industrial cranes. Using the data they collect, have them repeat the exercise to calculate the weight of load they can lift with each diameter cable they research.

Exercises ON YOUR OWN

What polynomial is modeled by each set of tiles and what are the factors of each polynomial?

1. $x^2 + 10x + 25$; $x + 5$, $x + 5$

2. $x^2 - 9$; $x + 3$; $x - 3$

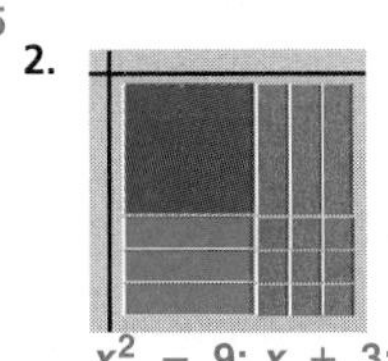

3. $9x^2 + 6x + 1$; $3x + 1$, $3x + 1$

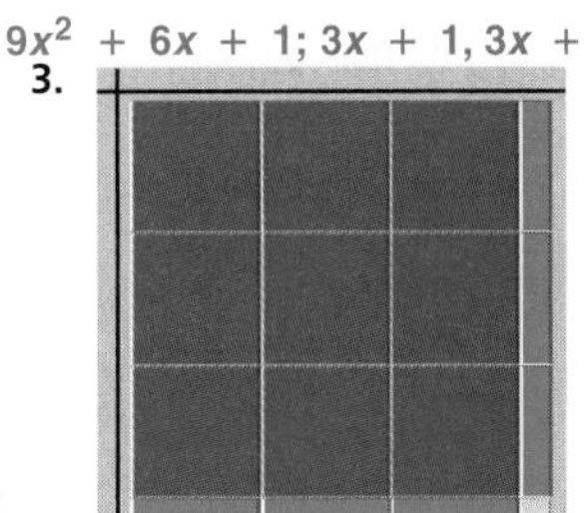

4. a. *Open-ended* Use a minimum of three x^2-tiles, one x-tile, and one 1-tile. Draw the model of the difference of two squares and a model of a perfect square trinomial. a–b. See margin.
   b. Represent each model from part (a) as a polynomial in both factored and unfactored form.

Factor each expression.

5. $x^2 + 2x + 1$ $(x + 1)^2$
6. $t^2 - 144$ $(t + 12)(t - 12)$
7. $x^2 - 18x + 81$ $(x - 9)^2$
8. $15t^2 - 15$ $15(t + 1)(t - 1)$
9. $3x^2 - 6x + 3$ $3(x - 1)^2$
10. $9w^2 - 16$ $(3w + 4)(3w - 4)$
11. $6x^2 - 150$ $6(x + 5)(x - 5)$
12. $k^2 - 6k + 9$ $(k - 3)^2$
13. $x^2 - 49$ $(x + 7)(x - 7)$
14. $a^2 + 12a + 36$ $(a + 6)^2$
15. $4x^2 - 4x + 1$ $(2x - 1)^2$
16. $16n^2 - 56n + 49$ $(4n - 7)^2$
17. $9x^2 + 6x + 1$ $(3x + 1)^2$
18. $2g^2 + 24g + 72$ $2(g + 6)^2$
19. $x^2 - 400$ $(x + 20)(x - 20)$
20. $2x^3 - 18x$ $2x(x + 3)(x - 3)$

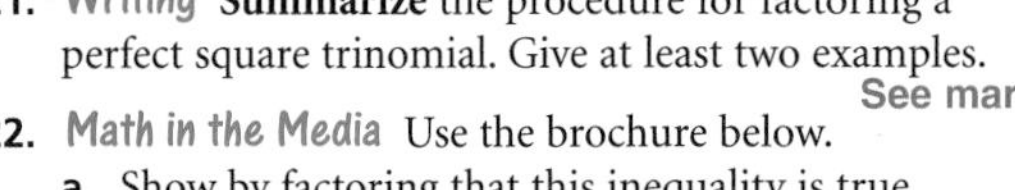

21. *Writing* **Summarize** the procedure for factoring a perfect square trinomial. Give at least two examples. See margin.

22. *Math in the Media* Use the brochure below.
    a. Show by factoring that this inequality is true.
    $(\pi d - \sqrt{15w})(\pi d + \sqrt{15w}) \geq 0$

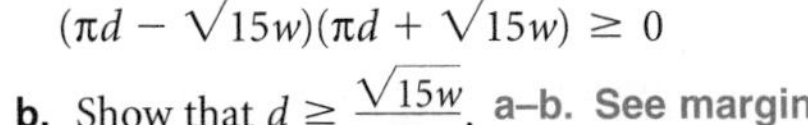

    b. Show that $d \geq \frac{\sqrt{15w}}{\pi}$. a–b. See margin.
    c. *Calculator* Is a cable 3 in. in diameter sufficient to lift an object weighing 5 tons? **Justify** your response. Yes; $(\pi 3)^2 - (\sqrt{15 \cdot 5})^2 \approx 88.83 - 75 = 13.83$; $13.83 > 0$

Load Requirements For Crane Operators
The weight of a load is limited by the diameter of the fiber cable. As the load gets heavier, the diameter of the cable must increase. To raise a load weighing w tons, a fiber cable having diameter d in inches must satisfy the inequality (πd)² – (√15w)² ≥ 0.

page 486 Work Together

1. Group A: One factor is a sum of two terms and the other is a difference of the same terms.

   Group B: The factors are the same. Each factor is the sum of the two terms.

   Group C: The factors are the same. Each factor is a difference of two terms.

2. $x^2 - 49$ $x^2 + 14x + 49$
   $x^2 - 14x + 49$
   $k^2 - 9$ $k^2 + 6k + 9$
   $k^2 - 6k + 9$
   $w^2 - 25$ $w^2 + 10w + 25$
   $w^2 - 10w + 25$
   $9x^2 - 1$ $9x^2 + 6x + 1$
   $9x^2 - 6x + 1$

3. Group A: Square the first term, subtract the square of the second term.

   Group B: Square the first term, add twice the product of the first and second terms, add the square of the second term.

   Group C: Square the first term, subtract twice the product of the first and second terms, add the square of the second term.

pages 486–488 Think and Discuss

8a. See back of book.

pages 489–490 On Your Own

4a–b, 21, 22a–b, 38a. See back of book.

STANDARDIZED TEST TIP Exercise 24 Standardized tests are often designed so that the average student will run out of time before completing the test. The ability to recognize special factoring cases can help students overcome this constraint.

MENTAL MATH Exercises 25–32 Challenge students to make up similar problems.

EXTENSION and CRITICAL THINKING Exercise 37 Ask: *What are the factors of* $(t - 1)^2 - 25$? $(t + 4)(t - 6)$

Exercises MIXED REVIEW

GETTING READY FOR LESSON 10-6 Students solve equations for which one side of the equation is zero. This prepares students for solving equations by factoring.

Wrap Up

THE BIG IDEA Ask students to tell what special polynomials they learned to factor in this lesson.

RETEACHING ACTIVITY Students factor the difference of two squares. (Reteaching worksheet 10-5)

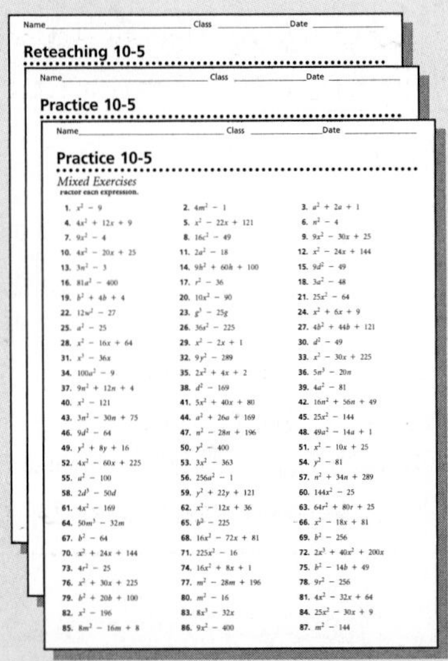
Name Class Date
Reteaching 10-5
Practice 10-5
Practice 10-5
Mixed Exercises

Lesson Quiz

Lesson Quiz is also available in Transparencies.

1. Factor $4y^2 - 25$. $(2y + 5)(2y - 5)$
2. Factor $9a^2 - 24a + 16$. $(3a - 4)^2$
3. Find the product of (49)(51) using the difference of two squares. Show all your work.
$(50 - 1)(50 + 1) = 2499$
4. Factor $\frac{1}{9}x^2 - 2x + 9$.
$\left(\frac{1}{3}x - 3\right)\left(\frac{1}{3}x + 3\right)$

23. a. *Geometry* Write two expressions in terms of n and m for the area of the solid region at the right. One expression should be in factored form. $3.14n^2 - 3.14m^2 = 3.14(n + m)(n - m)$

b. Use either form of your answer to part (a). Find the area of the solid region if $n = 10$ in. and $m = 3$ in. 285.74 in.^{2}

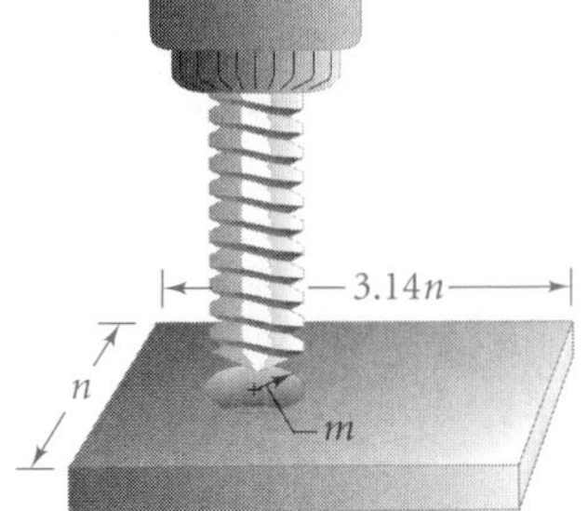

24. *Standardized Test Prep* Which of the following expressions is the factorization of $100x^2 + 220x + 121$? C

A. $(10x + 1)(10x - 1)$ B. $(10x - 11)(10x - 11)$
C. $(10x + 11)(10x + 11)$ D. $(10x - 10)(11x + 11)$

Mental Math **Find each product using the difference of two squares.**

Sample: $(17)(23) = (20 - 3)(20 + 3)$ ← Write the factors in the form $(a - b)(a + b)$.
$= 400 - 9$ ← Multiply.
$= 391$ ← Subtract.

25. (27)(33) 891	**26.** (19)(21) 399	**27.** (43)(37) 1591	**28.** (29)(31) 899
29. (16)(24) 384	**30.** (51)(49) 2499	**31.** (18)(22) 396	**32.** (98)(102) 9996

Factor each expression using rational numbers in your factors.

33. $\frac{1}{4}m^2 - \frac{1}{9}$ $\left(\frac{1}{2}m + \frac{1}{3}\right)\left(\frac{1}{2}m - \frac{1}{3}\right)$
34. $\frac{1}{4}p^2 - 2p + 4$ $\left(\frac{1}{2}p - 2\right)^2$
35. $\frac{1}{9}n^2 - \frac{1}{25}$ $\left(\frac{1}{3}n + \frac{1}{5}\right)\left(\frac{1}{3}n - \frac{1}{5}\right)$
36. $\frac{1}{25}k^2 + \frac{6}{5}k + 9$ $\left(\frac{1}{5}k + 3\right)^2$

37. a. *Critical Thinking* The expression $(t - 3)^2 - 16$ is a difference of two squares. Using the expression $a^2 - b^2$, identify a and b. $t - 3$; 4

b. Factor $(t - 3)^2 - 16$. $(t + 1)(t - 7)$

38. a. *Graphing Calculator* Graph $y = 4x^2 - 12x + 9$. See margin.

b. Use the x-intercept(s) of the graph to write $4x^2 - 12x + 9$ in factored form. $(2x - 3)^2$

Exercises MIXED REVIEW

Simplify each expression.

39. $\sqrt{25} + \sqrt{72}$ $5 + 6\sqrt{2}$	**40.** $8^2 - (4 + \sqrt{8})$ $60 - 2\sqrt{2}$	**41.** $\sqrt{5^2 + 4^2}$ $\sqrt{41}$	**42.** $\sqrt{100} - \sqrt{4}$ 8

43. *Sales* Suppose you buy a watch for \$35 that regularly sells for \$50. Find the percent of decrease in the price of the watch. 30%

Getting Ready for Lesson 10-6

Solve each equation.

44. $2x + 3 = 0$ $-\frac{3}{2}$ **45.** $-3x - 4 = 0$ $-\frac{4}{3}$

46. $8x - 9 = 0$ $\frac{9}{8}$ **47.** $-3x + 5 = 0$ $\frac{5}{3}$

10-6 Teaching Notes

PROBLEM OF THE DAY

How many 2 in. cubes will fit in a rectangular box that measures 7 in. by 13 in. by 9 in.? 72

Problem of the Day is also available in Transparencies.

CONNECTING TO PRIOR KNOWLEDGE Write $fg = 0$. Ask students: *What can you tell me about f and g based on this piece of information?* Either f is zero or g is zero.

THINK AND DISCUSS

VISUAL LEARNING Display $(x + 3)(x + 2) = 0$. Explain that one of the factors equals zero. Let the first factor be zero. Write $x + 3 = 0$. Lead students to understand that the value of the second factor is irrelevant because anything multiplied by zero is zero. Ask: *What is x?* −3 Now tell students to assume that the second factor is zero. Write $x + 2 = 0$. Ask: *What is x?* −2 Conclude by writing $x = -3$ or $x = -2$.

Example 1

Have students write the equation in this example as a quadratic equation in standard form. $x^2 + 11x + 30 = 0$ Making this connection will help students see how Examples 1 and 2 are related.

Example 2

Question 2 Ask: *If a graphing calculator is not available, how else might you check the solution?* Substitute −5.5 and 8 into

What You'll Learn

- Solving quadratic equations by factoring

...And Why

To find the dimensions of a box that can be manufactured from a given amount of material

What You'll Need

- graphing calculator

Connections  Manufacturing . . . and more

10-6 Solving Equations by Factoring

THINK AND DISCUSS

When you solve a quadratic equation by factoring, you use the zero-product property.

Zero-Product Property

For all real numbers a and b, if $ab = 0$, then $a = 0$ or $b = 0$.

Example: If $(x + 3)(x + 2) = 0$, then $x + 3 = 0$ or $x + 2 = 0$.

Example 1

Solve $(x + 5)(x + 6) = 0$.

$(x + 5)(x + 6) = 0$

$x + 5 = 0$ or $x + 6 = 0$ ← Use the zero-product property.

$x = -5$ or $x = -6$ ← Solve for x.

The solutions are −5 and −6.

Check Substitute −5 for x. $(-5 + 5)(-5 + 6) \stackrel{?}{=} 0$, $(0)(1) = 0$ ✓

Substitute −6 for x. $(-6 + 5)(-6 + 6) \stackrel{?}{=} 0$, $(-1)(0) = 0$ ✓

PROBLEM SOLVING

Look Back How could you solve $(x + 5)(x + 6) = 0$ by graphing?

Graph $y = (x + 5)(x + 6)$ and find the x-intercepts.

1. Try This Solve each equation.
   a. $(x + 7)(x - 4) = 0$ −7, 4
   b. $(3y - 5)(y - 2) = 0$ $1\frac{2}{3}$, 2

You can sometimes solve a quadratic equation by factoring. Write the equation in standard form. Factor the quadratic expression. Then use the zero-product property.

Example 2

Solve $2x^2 - 5x = 88$ by factoring.

$2x^2 - 5x = 88$

$2x^2 - 5x - 88 = 0$ ← Subtract 88 from each side.

$(2x + 11)(x - 8) = 0$ ← Factor $2x^2 - 5x - 88$.

$2x + 11 = 0$ or $x - 8 = 0$ ← Use the zero-product property.

$x = -5.5$ or $x = 8$ ← Solve for x.

The solutions are −5.5 and 8.

Lesson Planning Options

Prerequisite Skills

- Solving equations
- Factoring equations

Assignment Options for Exercises On Your Own

Core 1–15, 18–29, 35–40

Extension 16, 17, 30–34, 41–45

Use Mixed Review to maintain skills.

Resources

 **Student Edition**

Skills Handbook, pp. 573, 574, 582
Extra Practice, p. 565
Glossary/Study Guide

Teaching Resources

Chapter Support File, Ch. 10
- Practice 10-6 (two worksheets)
- Reteaching 10-6
- Checkpoint

Classroom Manager 10-6
Glossary, Spanish Resources
Two-Year Algebra Handbook 10-6

 **Transparencies**

11, 117

the original equation to see if these values make the equation true.

Example 3 **Relating to the Real World** 

Check students' understanding of the problem. Ask: *Does* 144 in.^{2} *include the waste material?* **yes**

If students have difficulty with this problem, have them make a model of the box described in this problem.

EXTENSION Question 4 Have students make up a similar problem. Choose one or two problems for the class to solve.

Exercises ON YOUR OWN

Exercises 1–3 Students may need to look back to Lesson 10-3 to reexamine the methods for multiplying polynomials.

Exercises 5–15 Remind students to check their answers by either graphing or substituting.

Exercises 13–15 Be sure students realize that to solve these equations, they must first get zero on one side of the equation.

ERROR ALERT! GEOMETRY Exercise 17 Many students will use 3 cm as the lengths of the sides of the square. **Remediation:** Have the students draw two different-sized squares. Write an x on each side of the smaller square. The x represents the unknown length of the original square. Have

Additional Examples

FOR EXAMPLE 1

Solve $(x - 2)(x + 3) = 0$ **2, −3**

FOR EXAMPLE 2

Solve $3x^2 + x = 10$ $-2, \frac{5}{3}$

Discussion: *Will there always be exactly two solutions for every quadratic equation? Explain.*

FOR EXAMPLE 3

The length of a rectangle is 2 cm longer than the width. The sides of the rectangle are increased by 2 cm. The area of the new rectangle is 80 cm^2. Find the length and width of the original rectangle. **length = 8 cm and width = 6 cm**

492

2. a. *Graphing Calculator* Graph $y = 2x^2 - 5x - 88$. **See margin.**
   b. In the CALC feature, use *ROOT* to find the x-intercepts of the graph. **−5.5, 8**
   c. Explain how your answers to part (b) are related to the solutions found in Example 2. **The x-intercepts of $y = 2x^2 - 5x - 88$ are the solutions of $2x^2 - 5x - 88 = 0$.**
3. a. Solve $x^2 - 12x + 36 = 0$ by factoring. **6**
   b. *Critical Thinking* Why does the equation in part (a) have only one solution? **The factors of $x^2 - 12x + 36$ are identical.**

You can solve some real-world problems by factoring and using the zero-product property.

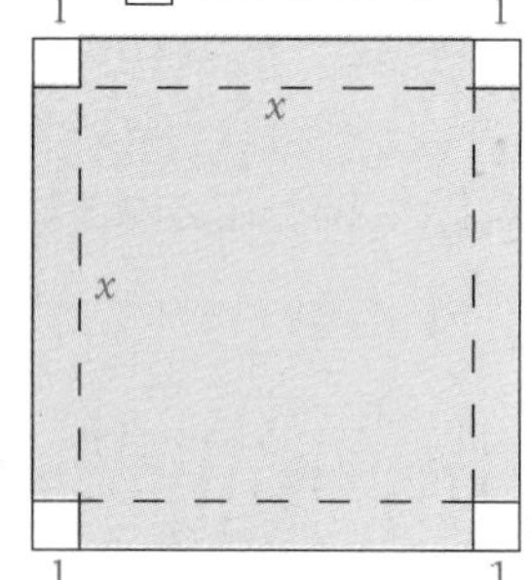

Example 3 **Relating to the Real World** 

Manufacturing The diagram shows a pattern for an open-top box. The total area of the sheet of material used to manufacture the box is 144 in.^{2}. The height of the box is 1 in. Therefore 1 in. × 1 in. squares are cut from each corner. Find the dimensions of the box.

Define width $= x + 1 + 1 = x + 2$
length $= x + 1 + 1 = x + 2$

Relate length × width = area

Write $(x + 2)(x + 2) = 144$

$$(x + 2)(x + 2) = 144$$
$$x^2 + 4x + 4 = 144 \quad \leftarrow \text{Find the product } (x + 2)(x + 2).$$
$$x^2 + 4x - 140 = 0 \quad \leftarrow \text{Subtract 144 from each side.}$$
$$(x + 14)(x - 10) = 0 \quad \leftarrow \text{Factor } x^2 + 4x - 140.$$
$$x + 14 = 0 \text{ or } x - 10 = 0 \quad \leftarrow \text{Use the zero-product property.}$$
$$x = -14 \text{ or } x = 10$$

Since the length must be positive, the solution is 10. The dimensions of the box are 10 in. × 10 in. × 1 in.

4. Suppose that a box with a square base has height 2 in. It is cut from a square sheet of material with area 121 in.^{2}. Find the dimensions of the box. **7 in. × 7 in. × 2 in.**

students write $x + 3$ on each side of the larger square to represent the increase of 3 cm.

STANDARDIZED TEST TIP **Exercise 30** Point out that if the expression $(a + b)^2$ is expanded, the other expressions may be substituted. This gives us

$$\begin{aligned} &a^2 + 2ab + b^2 \\ &= (a^2 + b^2) + (2ab) \\ &= (9) + (12) \\ &= 21. \end{aligned}$$

Exercise 32a Ask: *What is the value of h when the ball hits the ground?* 0

Exercise 32b Have students compare their estimate with their answer for part (a).

Exercises 33–34 Have students illustrate the problems using the appropriate geometric figure. Check students' diagrams to assess their understanding of the problems.

ERROR ALERT! **Exercises 41 and 42** Some students may give the value of *n* as the answer. **Remediation:** Point out that the solution to the problem is obtained by substituting the value of *n* in the three expressions.

ALTERNATIVE ASSESSMENT **Exercise 42** To assess understanding of the Pythagorean theorem, have students find three sets of integers which will always represent the sides of a right triangle.

Exercises ON YOUR OWN

Mental Math **Use mental math to solve each equation.**

1. $(x - 3)(x - 7) = 0$ 3, 7
2. $(x + 4)(2x - 9) = 0$ $-4, 4\frac{1}{2}$
3. $(7x + 2)(5x + 4) = 0$ $-\frac{2}{7}, -\frac{4}{5}$

Solve each equation by factoring.

4. $b^2 + 3b - 4 = 0$ $-4, 1$
5. $m^2 - 5m - 14 = 0$ $-2, 7$
6. $w^2 - 8w = 0$ 0, 8
7. $x^2 - 16x + 55 = 0$ 5, 11
8. $x^2 - 3x - 10 = 0$ $-2, 5$
9. $n^2 + n - 12 = 0$ $-4, 3$
10. $2x^2 - 7x + 5 = 0$ $1, 2\frac{1}{2}$
11. $x^2 - 10x = 0$ 0, 10
12. $4x^2 - 25 = 0$ $-2\frac{1}{2}, 2\frac{1}{2}$
13. $5q^2 + 18q = 8$ $-4, \frac{2}{5}$
14. $z^2 - 5z = -6$ 2, 3
15. $20 = x^2 - 9x$ $-1, 10$
16. Writing **Summarize** the procedure for solving a quadratic equation by factoring. Include an example. See margin.
17. Geometry The sides of a square are each increased by 3 cm. The area of the new square is 64 cm^2. Find the length of a side of the original square. 5 cm

Simplify each equation and write it in standard form. Then solve each equation.

18. $3a^2 + 4a = 2a^2 - 2a - 9$ $a^2 + 6a + 9 = 0$; -3
19. $4x^2 + 20 = 10x + 3x^2 - 4$ $x^2 - 10x + 24 = 0$; 4, 6
20. $6y^2 + 12y + 13 = 2y^2 + 4$ $4y^2 + 12y + 9 = 0$; $-1\frac{1}{2}$
21. $2q^2 + 22q = -60$ $2q^3 + 22q + 60 = 0$; $-6, -5$
22. $3t^2 + 8t = t^2 - 3t - 12$ $2t^2 + 11t + 12 = 0$; $-1\frac{1}{2}, -4$
23. $4 = -5n + 6n^2$ $6n^2 - 5n - 4 = 0$; $-\frac{1}{2}, \frac{4}{3}$
24. $3x^2 = 9x + 30$ $3x^2 - 9x - 30 = 0$; $-2, 5$
25. $20p^2 - 74 = 6$ $20p^3 - 80 = 0$; $-2, 2$
26. $2x^2 + 5x^2 = 3x$ $7x^2 - 3x = 0$; $0, \frac{3}{7}$
27. $7n^2 = 3n^2 + 100$ $4n^2 - 100 = 0$; 5, -5
28. $12x^2 + 8x = -4x^2 + 8x$ $16x^2 = 0$; 0
29. $9c^2 = 36$ $9c^2 - 36 = 0$; 2, -2
30. Standardized Test Prep If $a^2 + b^2 = 9$ and $ab = 6$, what does $(a + b)^2$ equal? C

    A. 3 B. 15 C. 21 D. 30 E. 36
31. Geometry A rectangular box has volume 280 $in.^3$. Its dimensions are 4 in. × $(x + 2)$ in. × $(x + 5)$ in. Find x. Use the formula $V = lwh$. 5 in.
32. Baseball Suppose you throw a baseball into the air from a starting height *s*. You toss the ball with an upward starting velocity of *v* ft/s. You can use the equation $h = -16t^2 + vt + s$ to find the ball's height *h* in feet *t* seconds after it is thrown.
    a. Suppose you toss a baseball directly upward with an starting velocity of 46 ft/s from a starting height of 6 ft. When will the ball hit the ground? 3 s
    b. Graphing Calculator Graph the related function for the equation you wrote in part (a). Use your graph to estimate how high the ball is tossed.

    See margin for graph; about 39.1 ft.

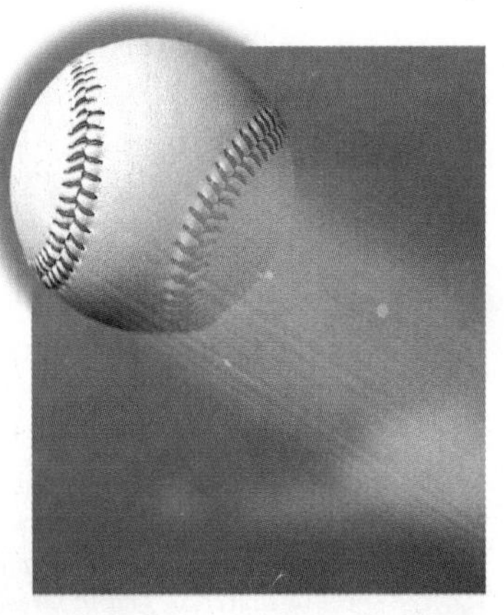

Technology Options

For Exercises 18–29, students may want to simplify each equation with algebra software.

Prentice Hall Technology

Calculator
- Graphing Calculator Handbook: Procedure 7

Software
- Secondary Math Lab Toolkit™
- Computer Item Generator 10-6

CD-ROM
- Multimedia Algebra Lab 10

Internet
- See the Prentice Hall site. (http://www.phschool.com)

CRITICAL THINKING Exercise 43 Choose students to explain to the class how they solved the problem. Some may use Guess and Test. Others may write a quadratic equation.

OPEN-ENDED Exercise 44 Have volunteers tell how they determined their values for *x*, *a*, and *b*.

Chapter Project **FIND OUT BY GRAPHING** Students use the given function to calculate and graph the number of bushels of walnuts produced on an acre of land. Check students' progress on the Chapter Project by having each student write a short progress report on what has been done to date and what the plans are for completion.

Exercises MIXED REVIEW

DIVERSITY Exercise 53 Some students may not participate in sports or be interested in sports. Encourage students to work together to write similar problems that reflect other activities.

Exercise 53 Students may choose to make a table to represent the data.

GETTING READY FOR LESSON 10-7 These exercises prepare students for choosing the best method for solving quadratic equations.

pages 491–492 Think and Discuss

2a.

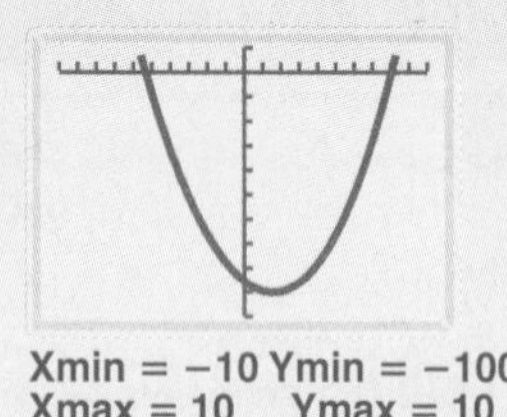

Xmin = −10 Ymin = −100
Xmax = 10 Ymax = 10
Xscl = 1 Yscl = 10

pages 493–495 On Your Own

16. Rewrite the equation in standard form. Factor the quadratic expression into linear factors. Equate each factor to 0 and solve the related linear equations. Each solution of the linear equations is a solution of the quadratic equation. Sample:
$x^2 + 3x - 10 = 0$
$(x - 2)(x + 5) = 0$
$x - 2 = 0$ or $x + 5 = 0$
$x = 2$ or $x = -5$
The solutions are 2 and −5.
If the quadratic expression does not factor, such as $x^2 + x + 1$, the corresponding equation, $x^2 + x + 1 = 0$, has no real solution.

32b. See back of book.

33. *Sailing* The height of a right-triangular sail on a boat is 2 ft greater than twice the base of the sail. Suppose the area of the sail is 110 ft^2.
   a. Find the dimensions of the sail. 10 ft base; 22 ft height
   b. Find the approximate length of the hypotenuse of the sail. (*Hint:* Use the Pythagorean theorem ($c^2 = a^2 + b^2$).) about 24.2 ft

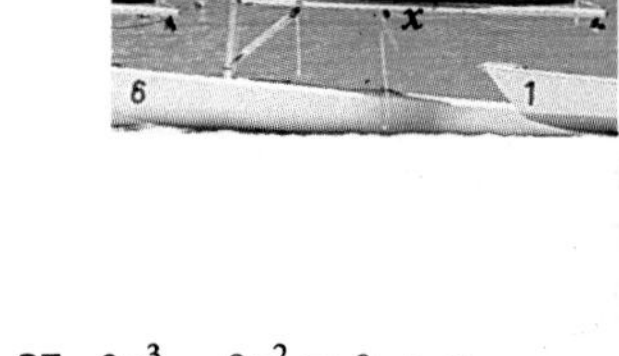

34. *Construction* You are building a rectangular wading pool. You want the area of the bottom to be 90 ft^2. You want the length of the pool to be 3 ft longer than twice its width. What will be the dimensions of the pool? 6 ft by 15 ft

Solve each cubic equation.

Sample: $x^3 + 7x^2 + 12x = 0$ ← The highest degree of a term is three. This is a cubic equation.

$x^3 + 7x^2 + 12x = 0$
$x(x^2 + 7x + 12) = 0$ ← Factor out the GCF.
$x(x + 3)(x + 4) = 0$ ← Factor the quadratic trinomial.
$x = 0,\ x + 3 = 0,$ or $x + 4 = 0$ ← Use the zero-product property.
$x = 0,\quad x = -3,$ or $x = -4$ ← Solve for *x*.

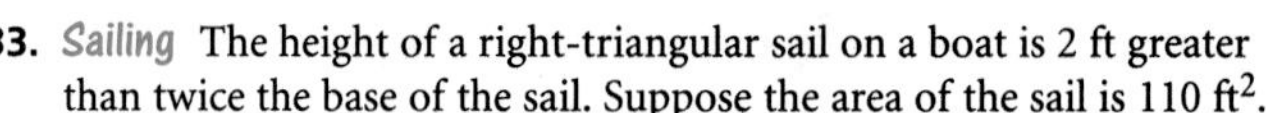

The solutions are 0, −3, and −4.

35. $x^3 - 10x^2 + 24x = 0$ 0, 4, 6
36. $x^3 - 5x^2 + 4x = 0$ 0, 1, 4
37. $3x^3 - 9x^2 = 0$ 0, 3
38. $x^3 + 3x^2 - 70x = 0$ 0, 7, −10
39. $3x^3 - 30x^2 + 27x = 0$ 0, 1, 9
40. $2x^3 = -2x^2 + 40x$ 0, 4, −5

In each diagram, you are given a right triangle that has special characteristics. Use the Pythagorean theorem ($c^2 = a^2 + b^2$) to find possible lengths of the sides of each right triangle.

41. three consecutive integers
3, 4, 5

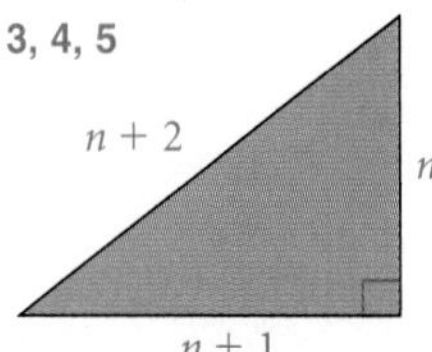

42. three consecutive even integers
6, 8, 10

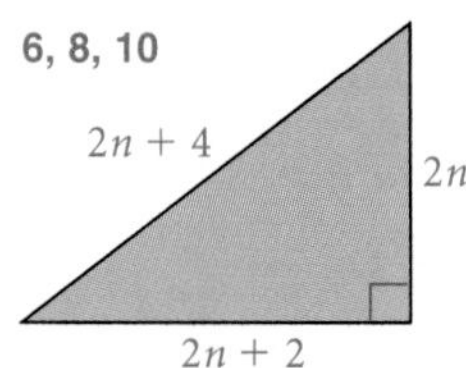

43. *Critical Thinking* A number plus its square equals zero. For which two numbers is this true? 0, −1

44. *Open-ended* In the diagram at the right, *x* is a positive integer and *a* and *b* are integers. List several possible values for *x*, *a*, and *b* so that the large rectangle will have an area of 56 square units.
Samples: 1, 27, 1; 2, 12, 2; 6, 2, 1

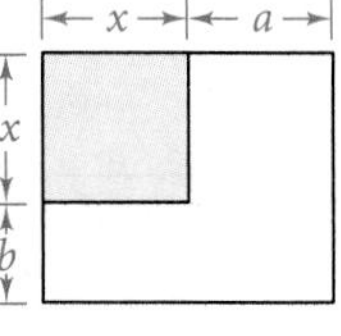

45. *Manufacturing* An open box with height 1 in. has a length that is 2 in. greater than its width. The box was made with minimum waste from an 80 in.^{2} rectangular sheet of material. What were the dimensions of the sheet of material? (*Hint:* Draw a diagram.)
8 in. × 10 in.

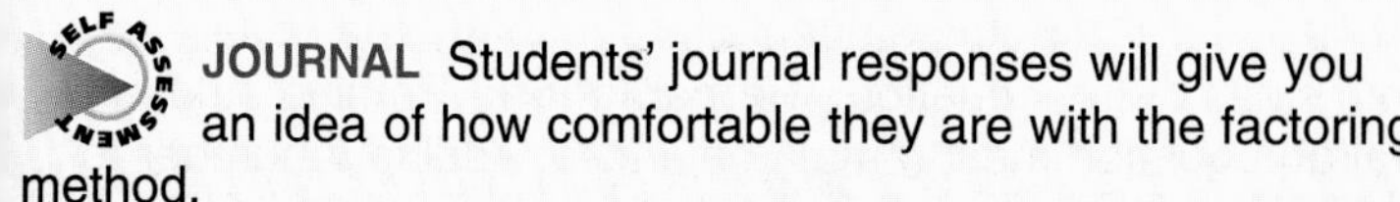

JOURNAL Students' journal responses will give you an idea of how comfortable they are with the factoring method.

Wrap Up

THE BIG IDEA Ask students to explain how factoring can be used to solve a quadratic equation.

RETEACHING ACTIVITY Students solve quadratic equations by factoring. (Reteaching worksheet 10-6)

Exercises CHECKPOINT

Students will assess their own progress on Lessons 10-4 to 10-6.

STANDARDIZED TEST TIP **Exercise 11** After moving –4 to the left side of the equal sign, it becomes obvious the x will be a positive integer. This eliminates answer choices A and B. Answer choice D can also be eliminated since x^3 cannot equal 16 for any positive integer.

Chapter Project — *Find Out by Graphing*

You can use the function $b = -0.01t^2 + 0.8t$ to find the number of bushels b of walnuts produced on an acre of land. The variable t represents the number of trees per acre.

- Graph this function. What number of trees per acre gives the greatest yield?
- How many walnut trees would you advise a farmer to plant on 5 acres of land? Explain your reasoning.

Exercises MIXED REVIEW

Simplify.

46. $(x^2)^4$ x^8 **47.** $(3y^3)^{-1}$ $\frac{1}{3y^3}$ **48.** $4(n^3)^0$ 4 **49.** $(x^3)(x^{-3})$ 1

Solve each equation. If there is no real solution, write *no real solution*.

50. $\sqrt{a} = 5$ 25 **51.** $\sqrt{3a} = 5$ $\frac{25}{3}$ **52.** $3\sqrt{a} = \sqrt{a} + 18$ 81

53. Out of 100 students, 30 students play football, 25 students play baseball, and 15 play both sports. How many play neither sport? 60 students

FOR YOUR JOURNAL

Suppose you are to solve $2x^2 - 11x + 5 = 0$. Would you use the quadratic formula or factoring? Explain why.

Getting Ready for Lesson 10-7

Use the quadratic formula to solve each equation.

54. $x^2 + x - 12 = 0$ 3, –4

55. $10x^2 + 13x - 3 = 0$ $\frac{1}{5}, -\frac{3}{2}$

56. $4x^2 + 4x + 3 = 0$ no real solution

Exercises CHECKPOINT

Factor.

1. $x^2 - 5x + 6$ $(x - 3)(x - 2)$

2. $n^2 - 6n + 9$ $(n - 3)^2$

3. $g^2 - 8g - 20$ $(g - 10)(g + 2)$

4. $9x^2 - 49$ $(3x + 7)(3x - 7)$

Mental Math **Use mental math to solve each equation.**

5. $(a + 3)(3a - 2) = 0$ $-3, \frac{2}{3}$

6. $(m - 6)(2m + 3) = 0$ $6, -\frac{3}{2}$

7. $(x + 7)(x + 2) = 0$ –7, –2

Solve each equation.

8. $25x^2 - 100 = 0$ 2, –2

9. $9x^2 + 24x = -16$ $-\frac{4}{3}$

10. $10x^2 - 11x - 6 = 0$ $\frac{3}{2}, -\frac{2}{5}$

11. Standardized Test Prep If $x^2 + 4x = -4$, what is the value of x^3? A

A. –8 B. 0 C. 8 D. 16

E. It cannot be determined from the information given.

Reteaching 10-6

Practice 10-6

Practice 10-6

Mixed Exercises

Lesson Quiz

Simplify each equation and write in standard form. Then solve the equation.

1. $6 = a^2 - 5a$ –1, 6

2. $12x + 4 = -9x^2$ $-\frac{2}{3}$

3. $4y^2 = 25$ $\frac{5}{2}, -\frac{5}{2}$

4. $b^3 + 3b = -4b^2$ 0, –1, –3

Math ToolboX

This toolbox gives students another method for solving quadratic equations while reinforcing the skills they learned in Lessons 7-5 and 10-5. Begin by solving $x^2 = 6$ with the class. Emphasize that there are two solutions. Show students that $(x - 3)^2 = 6$ may be solved using the same method. Introduce solving by competing the square as a way to rewrite equations in this form. Then work through the example with the class. Briefly discuss the limitations of solving by completing the square by beginning a problem with an odd coefficient of x or a coefficient of x^2 that must be eliminated before proceeding. This will help students recognize problems that are suitable for solving by completing the square.

ERROR ALERT! Some students forget to add the square to both sides of the equation. **Remediation:** Remind students that an equation must remain balanced. Adding a number to both sides does not change the values of the solutions.

WRITING Exercise 18 A quadratic formula example is given on page 497.

ADDITIONAL PROBLEM Solve by completing the square.

1. $x^2 + 14x + 43 = 0$ $x = 7 \pm \sqrt{6}$
2. $x^2 - 2x - 6 = 0$ $x = 1 \pm \sqrt{7}$

Math ToolboX — Skills Review

Completing the Square

After Lesson 10-6

In Chapter 7, you solved quadratic equations by taking the square root of each side of an equation. In Lesson 10-5 you factored perfect square trinomials. Completing the square allows you to combine these skills to solve any quadratic equation that has real solutions.

What is the value of *c* needed to create a perfect square trinomial?

1. $y^2 + 4y + c$ 4
2. $a^2 - 10a + c$ 25
3. $n^2 + 14n + c$ 49
4. $x^2 - 20x + c$ 100
5. Explain the steps you used to find c in Questions 1–4.

Answers may vary. Sample: Divide the coeff. of the 1st. deg. term in half and square the result.

To solve an equation using completing the square, you create a perfect square trinomial on one side of the equation so that you can take the square root of both sides.

Example

Solve by completing the square: $x^2 + 6x + 4 = 0$.

$x^2 + 6x + 4 = 0$

$x^2 + 6x = -4$ ← Subtract 4 from each side.

$x^2 + 6x + 3^2 = -4 + 3^2$ ← Find half of 6. Square it, and add the result to both sides.

$x^2 + 6x + 9 = 5$ ← Simplify.

$(x + 3)^2 = 5$ ← Write the left side in factored form.

$\sqrt{(x + 3)^2} = \pm\sqrt{5}$ ← Take the square root of each side.

$x + 3 = \pm\sqrt{5}$ ← Simplify.

$x = -3 + \sqrt{5}$ or $x = -3 - \sqrt{5}$ ← Solve for x.

The solutions are $-3 \pm \sqrt{5}$.

Solve each equation by completing the square.

6. $x^2 - 2x - 3 = 0$ 3, −1
7. $x^2 + 8x + 12 = 0$ −6, −2
8. $x^2 + 2x = 8$ −4, 2
9. $x^2 + 6x = 16$ −8, 2
10. $x^2 + 10x = 16$ $-5 \pm \sqrt{41}$
11. $x^2 + 4x = 12$ −6, 2
12. $x^2 - 4x = 3$ $2 \pm \sqrt{7}$
13. $x^2 - 4x - 45 = 0$ 9, −5
14. $x^2 - 12x = 4$ $6 \pm 2\sqrt{10}$
15. $x^2 + 10x = 10$ $-5 \pm \sqrt{35}$
16. $x^2 - 6x = 10$ $3 \pm \sqrt{19}$
17. $x^2 + 8x + 3 = 0$ $-4 \pm \sqrt{13}$
18. *Writing* Solve one of the equations in Exercises 6–17 using the quadratic formula. Which method do you prefer, completing the square or the quadratic formula? Why? Answers may vary. Sample: The quadratic formula because can be used with any quadratic equation.

PROBLEM OF THE DAY

The PTA is buying cake for the entire senior class. Each cake costs \$23.35 and serves 18 people. How much money should the PTA budget for cake if there are 273 seniors? \$373.60

Problem of the Day is also available in Transparencies.

CONNECTING TO PRIOR KNOWLEDGE Have a volunteer write a quadratic equation on the board. Ask students to brainstorm a list of all the strategies they know for solving this equation. Have them include less efficient strategies, such as Guess and Check.

THINK AND DISCUSS

Have students look at the quadratic equation and the three solution methods shown. Point out that all the methods produce the same solution. Have students vote for the method they would prefer to use. Record the results. Have volunteers explain why they chose a particular method.

Question 1 If you have block scheduling or extended class periods, you may wish to have your students do the following exercise. After students have chosen their methods and written their justifications, put them into groups of four. Have students compare answers and discuss the reasons for their choices with each other. Have the members vote on which method they prefer as a group. This should be done for each of the equations. Then have each student solve the equations

Connections  City Parks . . . and more

10-7 Choosing an Appropriate Method for Solving

What You'll Learn

- Choosing the best way to solve a quadratic equation

...And Why

To choose efficient ways to solve construction problems

What You'll Need

- graphing calculator

THINK AND DISCUSS

In this chapter and in Chapter 7, you learned many methods for solving quadratic equations. You can always use the quadratic formula to solve a quadratic equation, but sometimes another method may be easier. Other methods include graphing and factoring.

Methods for Solving Quadratic Equations

Example: $2x^2 - 4x - 6 = 0$

Graphing

Graph the related function.

$y = 2x^2 - 4x - 6$

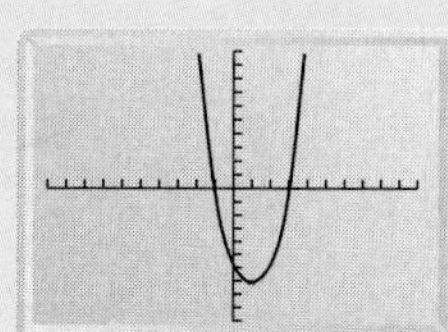

The x-intercepts are -1 and 3.

Quadratic Formula

If $ax^2 + bx + c = 0$ and $a \neq 0$,

$$x = \frac{-b \pm \sqrt{b^2 - 4ac}}{2a}.$$

$$x = \frac{-(-4) \pm \sqrt{(-4^2) - (4)(2)(-6)}}{2(2)}$$

$$= \frac{4 \pm \sqrt{64}}{4}$$

$$= \frac{12}{4} \text{ or } \frac{-4}{4}$$

$$= 3 \text{ or } -1$$

Factoring

Factor the equation. Use the zero-product property.

$$2x^2 - 4x - 6 = 0$$
$$(2x - 6)(x + 1) = 0$$
$$2x - 6 = 0 \text{ or } x + 1 = 0$$
$$2x = 6$$
$$x = 3 \text{ or } \quad x = -1$$

When you have an equation to solve, first write it in standard form. Then decide which method to use.

Method	When to Use
Graphing	Use if you have a graphing calculator handy.
Square Roots	Use if the equation has only an x^2 term and a constant term.
Factoring	Use if you can factor the equation easily.
Quadratic Formula	Use if you cannot factor the equation or if you are using a scientific calculator.

1. *Open-ended* Which method(s) would you choose to solve each equation? **Justify** your reasoning. See margin for samples.
   - a. $2x^2 - 6 = 0$
   - b. $9x^2 + 24x + 16 = 0$
   - c. $25x^2 - 36 = 0$
   - d. $6x^2 + 5x - 6 = 0$
   - e. $2x^2 + 7x - 15 = 0$
   - f. $16t^2 - 96t + 145 = 0$

Lesson Planning Options

Prerequisite Skills

- Factoring equations
- Using the quadratic formula

Assignment Options for Exercises On Your Own

Core 1–10, 13–28
Extension 11, 12, 29–33

Use Mixed Review to maintain skills.

Resources

 **Student Edition**

Skills Handbook, pp. 574, 582
Extra Practice, p. 565
Glossary/Study Guide

Teaching Resources

Chapter Support File, Ch. 10
- Practice 10-7 (two worksheets)
- Reteaching 10-7
- Checkpoint

Classroom Manager 10-7
Glossary, Spanish Resources
Two-Year Algebra Handbook 10-7

 **Transparencies**
11, 118

according to their group's preference. Have the groups compare their answers to stress that any method may be used; some methods are just easier to use for certain types of equations.

DIVERSITY Question 1 Ask volunteers who have mastered one of the methods to teach the method to others who may be less confident. Have four volunteers stand in four different areas of the classroom and name the method they will explain. Allow students to join the discussion they need the most.

Example 1 *Relating to the Real World* 

MAKING CONNECTIONS The water level in the world's tallest fountain, in Fountain Hills, Arizona, can reach 625 ft. The fountain cost $1.5 million.

Ask a volunteer to explain the procedure used to write the equation in standard form. **Divide each term by 2; subtract 750 from both sides.**

Ask: *Why is the answer approximate and not exact?* **The value of the square root was rounded to the nearest tenth.**

Ask: *How else could you solve this equation ?* **by graphing**

Question 4 Have students discuss the methods they chose to use and explain their choices .

Example 2

Stress to students that they may use either method to solve equations that contain only an x^2 term and a constant.

Additional Examples

FOR EXAMPLE 1

The City Council has decided to replace the fountain in the example with a circular fountain surrounded by a walkway 4 ft wide. If the total structure maintains the 1500 ft^2 area of the original fountain, what should the fountain's radius be? Note: The area of a circle is πr^2. **17.9 ft**

FOR EXAMPLE 2

Solve $-3x^2 + 27 = 0$. **3**

Discussion: *Which method did you use to solve this problem? Explain.*

FOR EXAMPLE 3

Solve $-4x - 3x^2 = 14$. **no real solutions**

Discussion: *Why does this equation have only one solution?*

FOR EXAMPLE 4

Solve $9x^2 + 6x - 8 = 0$. **$\frac{2}{3}, -\frac{4}{3}$**

PROBLEM SOLVING HINT

Draw a diagram.

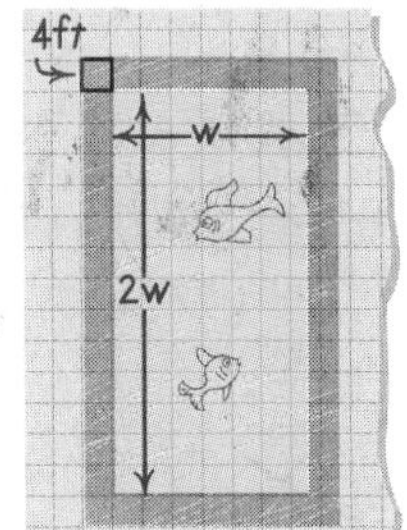

Example 1 *Relating to the Real World* 

City Parks A fountain has dimensions w and $2w$. The concrete walkway around it is 4 ft wide. Together, the fountain and walkway cover 1500 ft^2 of land. Find the dimensions of the fountain.

Define $w =$ width of fountain $\quad 2w =$ length of fountain
$w + 2(4) = w + 8 =$ width of fountain and walkway
$2w + 2(4) = 2w + 8 =$ length of fountain and walkway

Relate	width of fountain and walkway	times	length of fountain and walkway	=	area of fountain and walkway
Write	$(w + 8)$	$\cdot$	$(2w + 8)$	=	1500

$$(w + 8)(2w + 8) = 1500$$
$$2w^2 + 24w + 64 = 1500 \quad \leftarrow \text{Expand the product.}$$
$$2w^2 + 24w - 1436 = 0 \quad \leftarrow \text{Write in standard form.}$$
$$w = \frac{-24 \pm \sqrt{24^2 - 4(2)(-1436)}}{2(2)} \quad \leftarrow \text{Use the quadratic formula.}$$
$$w = \frac{-24 + 109.8}{4} \quad \leftarrow \text{Take the positive square root.}$$
$$w \approx 21.5 \text{ and } 2w \approx 43$$

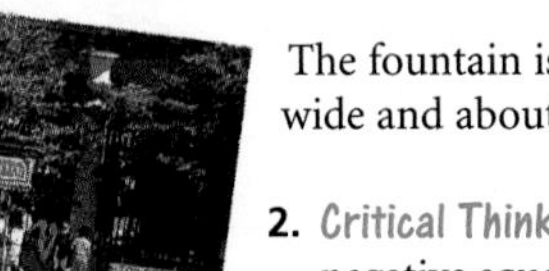

The fountain is about 21.5 ft wide and about 43 ft long.

2. **Critical Thinking** Why would using the negative square root have led to an unreasonable solution of Example 1? **Width cannot be negative.**
3. Which of the methods for solving quadratic equations would *not* be appropriate for solving Example 1? Explain. **Factoring; $w^2 - 12w - 718$ cannot be factored.**

Example 2

Solve $7x^2 - 175 = 0$.

Because this equation has only an x^2 term and a constant, try taking the square root of both sides or factoring a difference of two squares.

Method 1 Square roots
$$7x^2 - 175 = 0$$
$$7x^2 = 175$$
$$x^2 = 25$$
$$x = \pm 5$$
$$x = 5 \text{ or } x = -5$$

Method 2 Factoring
$$7x^2 - 175 = 0$$
$$7(x^2 - 25) = 0$$
$$7(x - 5)(x + 5) = 0$$
$$x - 5 = 0 \text{ or } x + 5 = 0$$
$$x = 5 \text{ or } x = -5$$

Both methods give the same solutions.

QUICK REVIEW

Zero-Product Property

For all real numbers a and b, if $ab = 0$, then $a = 0$ or $b = 0$.

Exercises MIXED REVIEW

PORTFOLIO Share with students the criteria you will use to assess their work in portfolios, as well as how you plan to use the results. Help students understand how you will use the rubrics to assess their work, how you will grade the portfolio, and how the scores they get in their portfolios will affect their overall evaluation.

Wrap Up

THE BIG IDEA Ask students to name four methods for solving quadratic equations and to tell which they find easiest to use and why.

RETEACHING ACTIVITY Students write equations in standard form, then choose an appropriate method to solve. (Reteaching worksheet 10-7)

A Point in Time

If you have block scheduling or extended class periods, you may wish to have students investigate the following events that occurred during the last decade of the 13th century and the beginning of the 14th century:

- *When did Edward I of England standardize the measures of* yard *and* acre*?* **1305**
- *In what year were spectacles invented?* **1290**
- *In what year did Marco Polo return to Italy and begin his memoirs from a Genoese jail?* **1298**

31. *Surveying* To find the distance across a marsh, a surveyor marked off a right triangle and measured two sides. Solve the equation $d^2 = 150^2 + 75^2$ to find this distance. Explain what solution method you used and why.
about 167.7 ft; take square roots.

32. *Geometry* Find all values of x such that rectangle $ACDG$ at the right has an area of 70 square units *and* rectangle $ABEF$ has an area of 72 square units. **5**

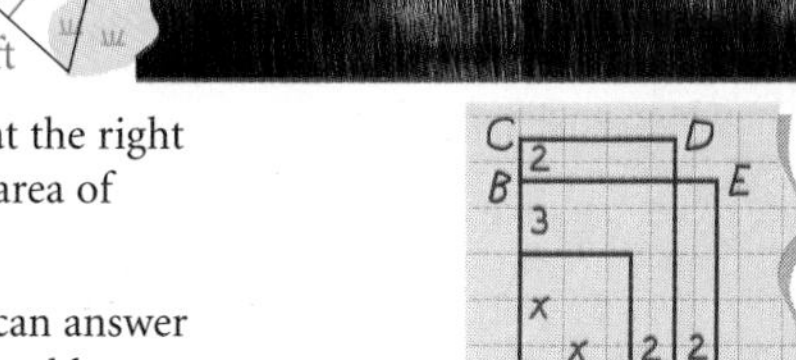

33. *Open-ended* Create an area situation and a question you can answer by solving a quadratic equation. Illustrate and solve your problem.
Answers may vary. Sample: A rectangle is 1 ft longer than it is wide. Its area is 20 ft². Find its dimensions. width = 4; length = 5

Exercises MIXED REVIEW

Find the minimum and maximum of each equation.

34. $x \geq 0, y \geq 0$
$x \leq 5$
$y \leq 4$
$C = x + 2y$
0; 13

35. $x + y \leq 10$
$x \geq 2$
$y \geq 3$
$C = 2x + 3y$
13; 28

36. $2x + y \leq 7$
$x \geq 0$
$y \geq 0$
$C = 4x + y$
0; 14

37. Describe the graph of $y = x^2 - 4x + 3$. Where is the axis of symmetry? What is the y-intercept? What are the roots? **the parabola $y = x^2$ shifted 2 units right, 1 unit down; $x = 2$; 3; 1 and 3.**

38. Find the product of $3x$ and $4x^2 - 2x + 5$.
$12x^3 - 6x^2 + 15x$

PORTFOLIO

For your portfolio, select one or two items from your work for this chapter:
- corrected work
- diagrams, graphs, or charts
- a journal entry

Explain why you have included each selection that you make.

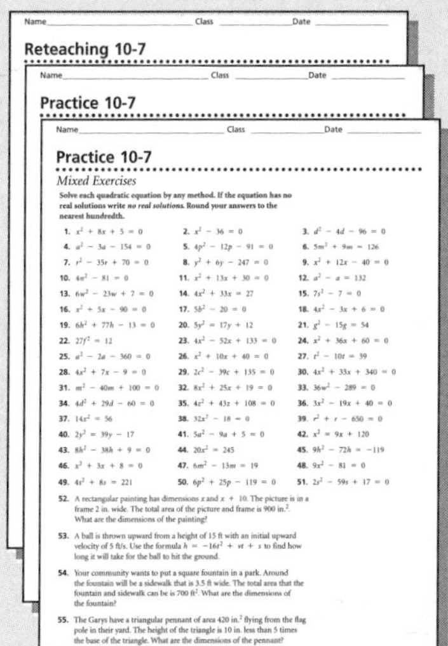

Reteaching 10-7

Practice 10-7

Practice 10-7

Mixed Exercises

A Point in Time

Chu Shih-Chieh

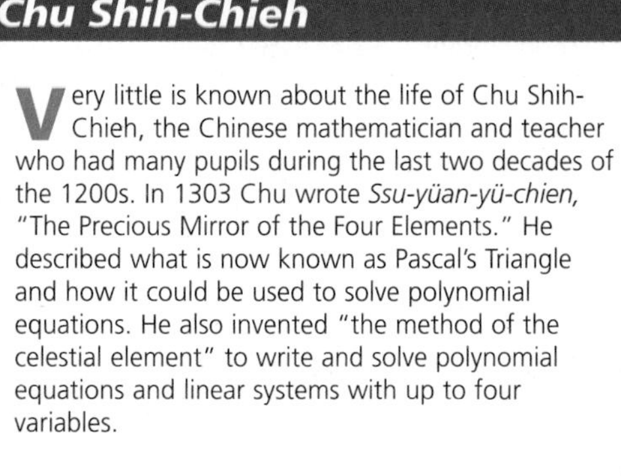

Very little is known about the life of Chu Shih-Chieh, the Chinese mathematician and teacher who had many pupils during the last two decades of the 1200s. In 1303 Chu wrote *Ssu-yüan-yü-chien,* "The Precious Mirror of the Four Elements." He described what is now known as Pascal's Triangle and how it could be used to solve polynomial equations. He also invented "the method of the celestial element" to write and solve polynomial equations and linear systems with up to four variables.

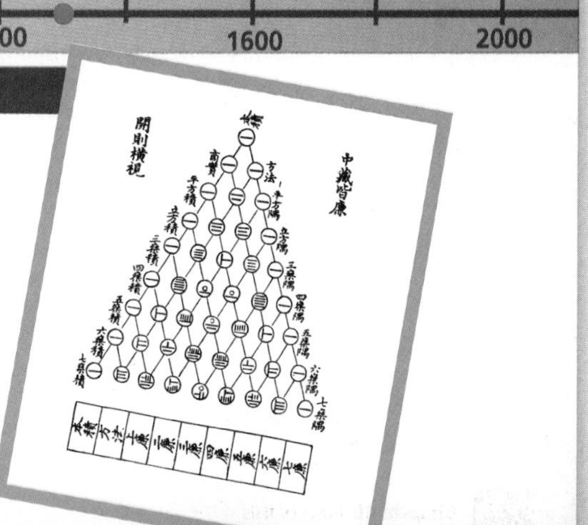

Lesson Quiz

Lesson Quiz is also available in Transparencies.

Select a method to solve each quadratic equation. Explain why you chose each method. If the equation has no real solutions, write *no real solutions.*

1. $12x^2 - 2x = 0$ $\frac{1}{6}$
2. $3j^2 + 4j - 7 = 0$ **−2.3, 1.0**
3. $6x^2 + 11x - 3 = 0$ $-\frac{1}{3}, -\frac{3}{2}$
4. $5x^2 + 4x - 3 = 3x^2 - 1$ $-3, \frac{1}{2}$

Finishing the Chapter Project

PROJECT DAY You may wish to plan a project day on which students share their completed projects. Encourage groups to explain their process as well as their product.

PROJECT NOTEBOOK Have students review their project work and bring their notebooks up to date.

- Have students review the research, equations, graphs, and explanations needed for the project.
- Ask groups to share any insights they found for completing the project, such as any shortcuts they found in doing research or solving equations and making graphs.

SCORING RUBRIC

3 Student demonstrates thorough research techniques by accessing many different sources of information. Calculations are correct. Graphs are neat, accurate, and clearly show differences between the formulas. Graphs have appropriate scales. Explanations are complete and well thought out.

2 Student demonstrates good research techniques to find information. Calculations are mostly correct but have minor errors. Graphs are neat and somewhat accurate with minor errors in scale.

1 Student locates needed information with some help. Calculations contain major errors. Graphs could be more accurate.

0 Major elements of the project are incomplete or missing.

CHAPTER PROJECT

Finishing the Chapter Project

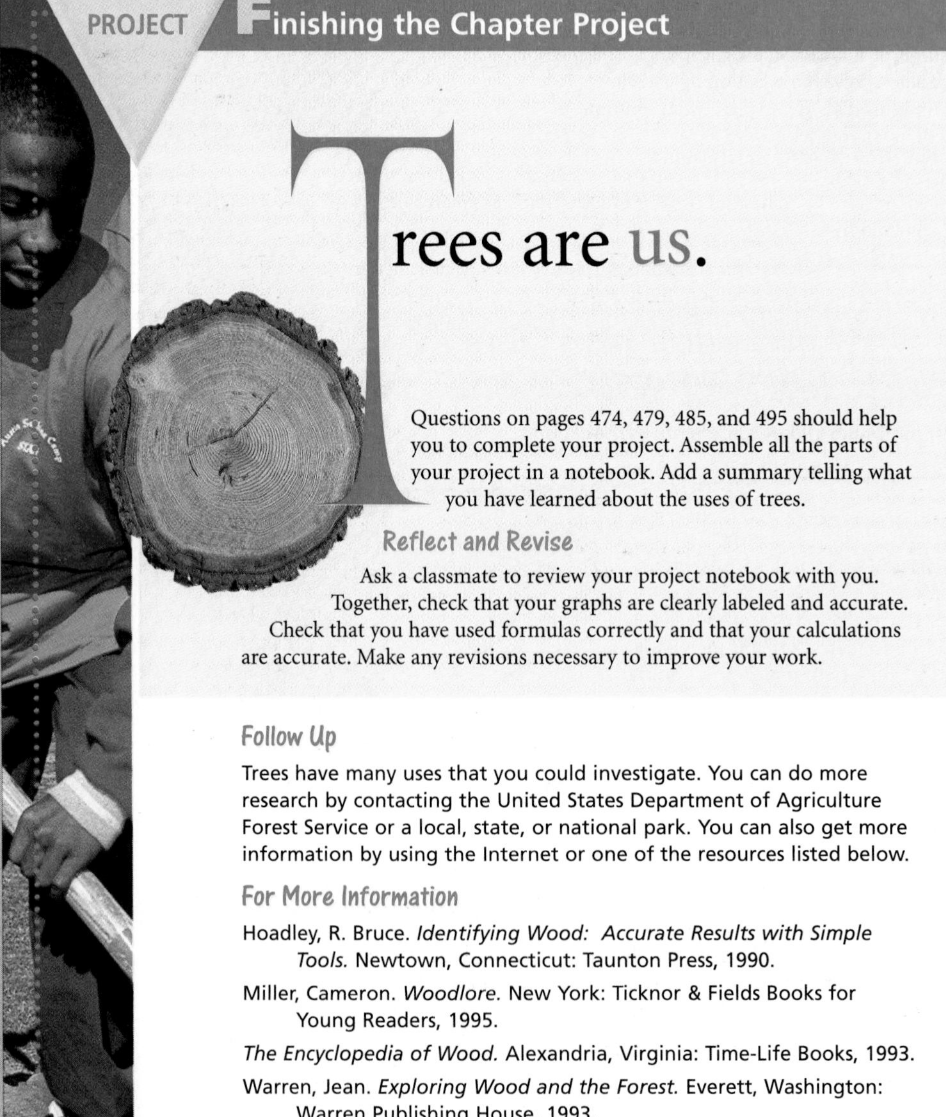

Trees are us.

Questions on pages 474, 479, 485, and 495 should help you to complete your project. Assemble all the parts of your project in a notebook. Add a summary telling what you have learned about the uses of trees.

Reflect and Revise

Ask a classmate to review your project notebook with you. Together, check that your graphs are clearly labeled and accurate. Check that you have used formulas correctly and that your calculations are accurate. Make any revisions necessary to improve your work.

Follow Up

Trees have many uses that you could investigate. You can do more research by contacting the United States Department of Agriculture Forest Service or a local, state, or national park. You can also get more information by using the Internet or one of the resources listed below.

For More Information

Hoadley, R. Bruce. *Identifying Wood: Accurate Results with Simple Tools.* Newtown, Connecticut: Taunton Press, 1990.

Miller, Cameron. *Woodlore.* New York: Ticknor & Fields Books for Young Readers, 1995.

The Encyclopedia of Wood. Alexandria, Virginia: Time-Life Books, 1993.

Warren, Jean. *Exploring Wood and the Forest.* Everett, Washington: Warren Publishing House, 1993.

Wrap Up

HOW AM I DOING? Have students work in small groups. Ask each group to prepare a short presentation explaining one of the methods used to factor quadratic equations. Each presentation must include a visual aid and an example problem.

KEY TERMS The numbers in parentheses direct students to the pages where the terms (or symbols) are used or defined. Students should be able to (1) write a simple explanation of each term, (2) illustrate the term with a diagram, or (3) show an example that uses the term.

Exercises 1–4 Remind students that when they subtract polynomials, the sign of *every* term in the second polynomial must be changed.

Exercises 12–14 Remind students that the GCF can include both variables and numbers.

Exercises 15–22 Point out to students that when they multiply two polynomials, the number of terms they have after multiplying—but before simplifying—is equal to the number of terms in the first polynomial multiplied by the number of terms in the second polynomial.

Exercise 32 Remind students that they can predict the sign of the final term of a polynomial by looking at the two signs of the original trinomial.

10 Wrap Up

Key Terms

binomial (p. 465)
binomial factors (p. 481)
degree of a polynomial (p. 465)
degree of a term (p. 465)
difference of two squares (p. 487)
monomial (p. 465)
perfect square trinomial (p. 487)
polynomial (p. 465)
standard form (p. 465)
trinomial (p. 465)
zero-product property (p. 491)

How am I doing?

- State three ideas from this chapter that you think are important. Explain your choices.
- Explain the different methods you can use to factor $2x^2 + 7x - 15$.

Adding and Subtracting Polynomials — 10-1

The **degree of a term** with one variable is the exponent of the variable. A **polynomial** is one term or the sum or difference of two or more terms. The **degree of a polynomial** is the same as the degree of the term with the highest degree. A polynomial can be named by its degree or by the number of its terms. You can simplify polynomials by adding the coefficients of like terms.

Find each sum or difference.

1. $(3x^3 + 8x^2 + 2x + 9) - (-4x^3 + 5x - 3)$ $7x^3 + 8x^2 - 3x + 12$
2. $(3g^4 + 5g^2 + 5) + (5g^4 - 10g^2 + 11g)$ $8g^4 - 5g^2 + 11g + 5$
3. $(-4b^5 + 3b^3 - b + 10) + (3b^5 - b^3 + b - 4)$ $-b^5 + 2b^3 + 6$
4. $(2t^3 - 4t^2 + 9t - 7) - (t^3 + t^2 - 3t + 1)$ $t^3 - 5t^2 + 12t - 8$
5. Open-ended Write a polynomial using the variable z. What is the degree of your polynomial? Sample: $3z^4 - 5z^2 + 1$; 4

Multiplying and Factoring — 10-2

You can use the distributive property or tiles to multiply polynomials. You can factor a polynomial by finding the greatest common factor (GCF) of the terms of the polynomial, or by using tiles.

Find each product. Write it in standard form.

6. $8x(2 - 5x)$ $-40x^2 + 16x$
7. $5g(3g + 7g^2 - 9)$ $35g^3 + 15g^2 - 45g$
8. $8t^2(3t - 4 - 5t^2)$ $-40t^4 + 24t^3 - 32t^2$
9. $5m(3m + m^2)$ $5m^3 + 15m^2$
10. $-2w^2(4w - 10 + 3w^2)$ $-6w^4 - 8w^3 + 20w^2$
11. $b(10 + 5b - 3b^2)$ $-3b^3 + 5b^2 + 10b$

Find the greatest common factor (GCF) of the terms of each polynomial. Then factor the polynomial.

12. $9x^4 + 12x^3 + 6x$ $3x$; $3x(3x^3 + 4x^2 + 2x)$
13. $4t^5 - 12t^3 + 8t^2$ $4t^2$; $4t^2(t^3 - 3t + 2)$
14. $40n^5 + 70n^4 - 30n^3$ $10n^3$; $10n^3(4n^2 + 7n - 3)$

Resources

Student Edition

Extra Practice, p. 565
Glossary/Study Guide

Teaching Resources

Student Study Survival Handbook
Glossary, Spanish Resources

pages 503–505 Wrap Up

49. Use factoring. The standard form shows the difference of two squares.
50. Use the quadratic formula because the trinomial does not factor easily.
51. Use the quadratic formula because the standard form trinomial does not factor easily.

Exercises 42–47 Remind students to divide a polynomial expression by its greatest common factor before attempting to find its remaining factors.

Remind students that the new mathematical terms in this chapter are defined in the Glossary/Study Guide in the back of the book.

Multiplying Polynomials — 10-3

You can use tiles or the distributive property to multiply polynomials. You can use the FOIL method (First, Outer, Inner, Last) to multiply two binomials.

Find each product.

15. $(x + 3)(x + 5)$ $x^2 + 8x + 15$
16. $(5x + 2)(3x - 7)$ $15x^2 - 29x - 14$
17. $(2x + 5)(3x - 2)$ $6x^2 + 11x - 10$
18. $(x - 1)(-x + 4)$ $-x^2 + 5x - 4$
19. $(x + 2)(x^2 + x + 1)$ $x^3 + 3x^2 + 3x + 2$
20. $(4x - 1)(x - 5)$ $4x^2 - 21x + 5$
21. $(x - 4)(x^2 - 5x - 2)$ $x^3 - 9x^2 + 18x + 8$
22. $(3x + 4)(x + 2)$ $3x^2 + 10x + 8$
23. Geometry A rectangle has dimensions $2x + 1$ and $x + 4$. Write an expression for the area of the rectangle as a product and as a polynomial in standard form. $(2x + 1)(x + 4)$; $2x^2 + 9x + 4$

Factoring Trinomials — 10-4

Some quadratic trinomials are the product of two **binomial factors.** You can factor trinomials using tiles or by using FOIL with the strategy *Guess and Test.*

Factor each quadratic trinomial.

24. $x^2 + 3x + 2$ $(x + 2)(x + 1)$
25. $y^2 - 9y + 14$ $(y - 7)(y - 2)$
26. $x^2 - 2x - 15$ $(x - 5)(x + 3)$
27. $2w^2 - w - 3$ $(2w - 3)(w + 1)$
28. $-b^2 + 7b - 12$ $(-b + 4)(b - 3)$
29. $2t^2 + 3t - 2$ $(2t - 1)(t + 2)$
30. $x^2 + 5x - 6$ $(x - 1)(x + 6)$
31. $6x^2 + 10x + 4$ $(3x + 2)(2x + 2)$
32. Standardized Test Prep What is $21x^2 - 22x - 8$ in factored form? A
    A. $(7x + 2)(3x - 4)$
    B. $(21x + 8)(x - 1)$
    C. $(3x + 2)(7x - 4)$
    D. $(3x - 2)(7x + 4)$
    E. $(x - 8)(21x + 1)$

Factoring Special Cases — 10-5

When you factor a **difference of two squares,** the two binomial factors are the sum and difference of two terms.

$$a^2 - b^2 = (a + b)(a - b)$$

When you factor a **perfect square trinomial,** the two binomial factors are the same.

$$a^2 + 2ab + b^2 = (a + b)^2 \qquad a^2 - 2ab + b^2 = (a - b)^2$$

Factor each polynomial.

33. $q^2 + 2q + 1$ $(q + 1)^2$
34. $b^2 - 16$ $(b - 4)(b + 4)$
35. $x^2 - 4x + 4$ $(x - 2)^2$
36. $4t^2 - 121$ $(2t + 11)(2t - 11)$
37. $4d^2 - 20d + 25$ $(2d - 5)^2$
38. $9c^2 + 6c + 1$ $(3c + 1)^2$
39. $9k^2 - 25$ $(3k + 5)(3k - 5)$
40. $x^2 + 6x + 9$ $(x + 3)^2$
41. Critical Thinking Suppose you are using tiles to factor a quadratic trinomial. What do you know about the factors of the trinomial if the tiles form a square? The factors are equal.

ALTERNATIVE ASSESSMENT To assess students' skill in factoring polynomial equations, write the solutions to Exercises 6–11 and 15–22 on the board in random order. Have students factor these expressions and match each solution to its original problem.

Getting Ready for Chapter 11

Students may work these exercises independently or in small groups. The skills previewed will help prepare students for working with rational expressions.

Solving Equations by Factoring 10-6

You can solve a quadratic equation by factoring and using the **zero-product property.** For all real numbers a and b, if $ab = 0$, then $a = 0$ or $b = 0$.

Simplify each equation if necessary. Then solve by factoring.

42. $x^2 + 7x + 12 = 0$ −3, −4 **43.** $5x^2 - 10x = 0$ 0, 2 **44.** $2x^2 - 9x = x^2 - 20$ 4, 5

45. $2x^2 + 5x = 3$ $-3, \frac{1}{2}$ **46.** $3x^2 - 5x = -3x^2 + 6$ $-\frac{2}{3}, 1\frac{1}{2}$ **47.** $x^2 - 5x + 4 = 0$ 1, 4

48. Gardening Alice is planting a garden. Its length is 3 feet less than twice its width. Its area is 170 ft^2. Find the dimensions of the garden. 10 ft by 17 ft

Choosing an Appropriate Method for Solving 10-7

You can solve a quadratic equation four different ways. You can graph the related function and find the x-intercepts. For $x^2 = c$, you can take the square root of each side of the equation. You can factor the equation and use the zero-product property. You can use the quadratic formula.

If the discriminant is a perfect square, there are two rational solutions, and the equation can be factored.

Writing **Solve each quadratic equation. Explain why you chose the method you used.** 49–51. See margin for method and reasoning.

49. $5x^2 + 10 = x^2 - 90$ −4.47, 4.47 **50.** $9x^2 + 30x - 29 = 0$ −4.12, 0.78 **51.** $8x^2 - 6x = 4x^2 + 6x - 2$ 0.18, 2.82

52. A square pool has length p. The border of the pool is 1 ft wide. The combined area of the border and the pool is 400 ft^2. Find the area of the pool. 324 ft^2

Getting Ready for CHAPTER 11

Rewrite each decimal as an improper fraction.

53. 5.7 $\frac{57}{10}$ **54.** −8.25 $-\frac{33}{4}$ **55.** 3.14 $\frac{314}{100}$ **56.** 10.4 $\frac{52}{5}$ **57.** −1.849 $-\frac{1849}{1000}$ **58.** 7.67 $\frac{767}{100}$

Evaluate each expression.

59. $\frac{3}{2x + 1}$ for $x = 4$ $\frac{1}{3}$ **60.** $\frac{3x^2}{5x + 2}$ for $x = -2$ $-1\frac{1}{2}$ **61.** $\frac{-3}{y^2 + 3}$ for $y = 2$ $-\frac{3}{7}$ **62.** $\frac{2x}{3x - 2}$ for $x = 5$ $\frac{10}{13}$

Solve each proportion.

63. $\frac{x}{3} = \frac{10}{4}$ $\frac{15}{2}$ **64.** $\frac{1}{y} = \frac{3}{7}$ $\frac{7}{3}$ **65.** $\frac{11}{3} = \frac{2r}{5}$ $\frac{55}{6}$ **66.** $\frac{6}{13} = \frac{12}{d}$ 26

67. Probability Make a tree diagram to show all the possible outcomes of rolling two number cubes. See margin.

page 505 Getting Ready for Chapter 11

67.

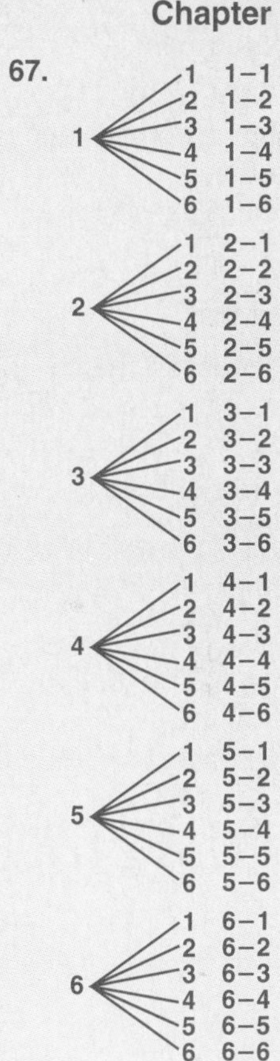

Assessment

ENHANCED MULTIPLE CHOICE QUESTIONS are more complex than traditional multiple choice questions, which assess only one skill. Enhanced multiple choice questions assess the processes that students use, as well as the end results. The questions are written so that students must use more than one strategy to solve the problem. Using multiple strategies is encouraged by the National Council of Teachers of Mathematics (NCTM). **Exercise 29** is an enhanced multiple choice question.

FREE RESPONSE QUESTIONS do not give answer choices. Some exercises have more than one possible answer. Students need to give only one correct response. **Exercises 1–4, 6–15, 17–28, and 30–38** are free response questions.

WRITING EXERCISES allow students to describe how they think about and understand the concepts they have learned. **Exercises 16 and 39** are writing exercises.

OPEN-ENDED PROBLEMS allow for more than one solution. Students must construct their own responses instead of choosing from possible answers. The student responses will help you determine the depth of their understanding and any possible areas of difficulty. **Exercise 5** is an open-ended problem. Make sure that the students' responses contain values for x and y and a written situation.

Resources

Teaching Resources

Chapter Support File, Ch. 10

- Chapter Assessment, Forms A and B
- Alternative Assessment

Chapter Assessment, Spanish Resources

Teacher's Edition

See also p. 462E for assessment options.

Software

Computer Item Generator

page 506 Assessment

16. Multiply each term of the first polynomial by each term of the second polynomial. Then combine like terms. Sample: $(x + 3)(x^2 - x + 1) = x^3 - x^2 + x + 3x^2 - 3x + 3 = x^3 + 2x^2 - 2x + 3$

39. See back of book.

10 Assessment

Simplify. Write each answer in standard form.

1. $(4x^2 + 2x + 5) + (7x^2 - 5x + 2)$ $11x^2 - 3x + 7$
2. $(9a^2 - 4 - 5a) - (12a - 6a^2 + 3)$ $15a^2 - 17a - 7$
3. $(-4m^2 + m - 10) + (3m + 12 - 7m^2)$ $-11m^2 + 4m + 2$
4. $(3c - 4c^2 + c^3) - (5c^2 + 8c^3 - 6c)$ $-7c^3 - 9c^2 + 9c$
5. Open-ended Write a trinomial with degree 6. Sample: $p^6 + p^2 + 1$

Write each product in standard form.

6. $8b(3b + 7 - b^2)$ $-8b^3 + 24b^2 + 56b$
7. $-t(5t^2 + t)$ $-5t^3 - t^2$
8. $3q(4 - q + 3q^3)$ $9q^4 - 3q^2 + 12q$
9. $2c(c^5 + 4c^3)$ $2c^6 + 8c^4$
10. $(x + 6)(x + 1)$ $x^2 + 7x + 6$
11. $(x + 4)(x - 3)$ $x^2 + x - 12$
12. $(2x - 1)(x - 4)$ $2x^2 - 9x + 4$
13. $(2x + 5)(3x - 7)$ $6x^2 + x - 35$
14. $(x + 2)(2x^2 - 5x + 4)$ $2x^3 - x^2 - 6x + 8$
15. $(x - 4)(6x^2 + 10x - 3)$ $6x^3 - 14x^2 - 43x + 12$
16. Writing Explain how to use the distributive property to multiply polynomials. Include an example. See margin.

Find the greatest common factor of the terms of each polynomial.

17. $21x^4 + 18x^2 + 36x^3$ $3x^2$
18. $3t^2 - 5t - 2t^4$ t
19. $-3a^{10} + 9a^5 - 6a^{15}$ $3a^5$
20. $9m^3 - 7m^4 + 8m^2$ m^2

Write an expression for each situation as a product and in standard form.

21. A plot of land has width x meters. The length of the plot of land is 5 m more than 3 times its width. What is the area of the land? $(3x + 5)x$; $3x^2 + 5x$
22. The height of a box is 2 in. less than its width w. The length of a box is 3 in. more than 4 times its width. What is the volume of the box in terms of w? $w(w - 2)(4w + 3)$; $4w^3 - 5w^2 - 6w$

Factor each expression.

23. $x^2 - 5x - 14$ $(x - 7)(x + 2)$
24. $x^2 + 10x + 25$ $(x + 5)^2$
25. $9x^2 + 24x + 16$ $(3x + 4)^2$
26. $x^2 - 100$ $(x - 10)(x + 10)$
27. $x^2 - 4x + 4$ $(x - 2)^2$
28. $4x^2 - 49$ $(2x - 7)(2x + 7)$
29. Standardized Test Prep Which of the following are perfect square trinomials? C
    - I. $x^2 + 14x + 49$
    - II. $16x^2 + 25$
    - III. $9x^2 - 30x + 25$
    - IV. $4x^2 - 81$

    A. I only B. II only
    C. I and III D. I, III, and IV
    E. I, II, III, and IV

Write each equation in standard form. Then solve the equation.

30. $4x^2 - 5x = -2x^2 + 2x + 3$ $6x^2 - 7x - 3 = 0; -\frac{1}{3}, 1\frac{1}{2}$
31. $2x^2 + 3x = x^2 + 28$ $x^2 + 3x - 28 = 0; -7, 4$
32. $3x^2 - 4 = x^2 - 5x + 12$ $2x^2 + 5x - 16 = 0; -4.34, 1.84$
33. $x^2 - 5 = -x^2 + 9x$ $2x^2 - 9x - 5 = 0; -\frac{1}{2}, 5$
34. Geometry The base of a triangle is 8 ft more than twice its height. The area of the triangle is 45 ft^2. Find the dimensions of the triangle. 18 ft base, 5 ft height

Use the quadratic formula to solve each equation to the nearest hundredth. If there are no real solutions, write *no real solutions*.

35. $4x^2 + 4x + 9 = 0$ no real solutions
36. $x^2 + 10x + 11 = 0$ −8.74, −1.26
37. $-2x^2 - x + 8 = 0$ −2.27, 1.77
38. $x^2 - 7x + 10 = 0$ 2, 5
39. Writing Explain when you would use the different methods of solving quadratic equations. Give examples. See margin.

Cumulative Review

Item	Review Topic	Chapter
1, 7	linear equations	5
2	multiplying polynomials	10
3	functions	2
4, 13	modeling equations	4
5, 9	using probability	3
6	Pythagorean theorem	9
8, 11	inequalities	4
10	naming polynomials	10
12	order of operations	1
14	quadratic functions	7

10 Cumulative Review

For Exercises 1–12, choose the correct letter.

1. A parachutist opens her parachute at 800 ft. Her rate of change in altitude is −30 ft/s. Which equation represents her altitude in feet *t* seconds after she opens her parachute? D
   A. $30t$ B. $-30t$ C. 800
   D. $800 - 30t$ E. $800 + 30t$

2. What is the standard form of the product $(3x - 1)(5x + 3)$? E
   A. $15x^2 + 2x - 3$ B. $15x^2 + 2x + 3$
   C. $15x^2 + 4x + 3$ D. $15x^2 - 4x - 3$
   E. $15x^2 + 4x - 3$

3. Which relations are functions? D

I.

x	1	−1	2	1
y	3	4	5	7

II.

x	1	2	3	4
y	1	1	3	5

III.

x	0	1	2	3
y	0	1	3	2

   A. I B. II C. III
   D. II and III E. I, II, and III

4. A rectangle has a perimeter of 72 in. The length is 3 in. more than twice the width. What is the length of the rectangle? C
   A. 11 B. 22 C. 25 D. 36 E. 50

5. *A* and *B* are independent events. If $P(A) = \frac{5}{6}$ and $P(A \text{ and } B) = \frac{1}{8}$, what is $P(B)$? B
   A. $\frac{1}{10}$ B. $\frac{3}{20}$ C. $\frac{1}{5}$ D. $\frac{1}{4}$ E. $\frac{3}{10}$

6. One leg of a right isosceles triangle is 8 cm long. What is the length of the hypotenuse? E
   A. 16 B. $16\sqrt{3}$ C. $8\sqrt{3}$ D. $4\sqrt{3}$ E. $8\sqrt{2}$

7. A truck traveling 45 mi/h and a car traveling 55 mi/h cover the same distance. The truck travels 4 h longer than the car. How far did they travel? D
   A. 880 B. 940 C. 960 D. 990 E. 900

8. Which compound inequality could the graph represent? B

−3 −2 −1 0 1 2 3

   A. $-1 \leq x \leq 2$ B. $x \leq -2$ or $x \geq 1$
   C. $-2 < x$ and $x \geq 1$ D. $-2 \leq x \leq 1$
   E. $x \leq -2$ and $x \geq 1$

9. Suppose you toss three coins. What is the probability of getting 2 heads and 1 tail? C
   A. $\frac{1}{8}$ B. $\frac{1}{4}$ C. $\frac{3}{8}$
   D. $\frac{1}{2}$ E. $\frac{5}{8}$

10. Which of the following is a monomial? A
   I. $5x^2$ II. $17x^{-2}$ III. $\frac{1}{x}$
   A. I B. II C. III
   D. I and III E. II and III

Compare the boxed quantity in Column A with the boxed quantity in Column B. Choose the best answer.
   A. The quantity in Column A is greater.
   B. The quantity in Column B is greater.
   C. The two quantities are equal.
   D. The relationship cannot be determined on the basis of the information supplied.

	Column A	Column B	
	$0 < x < 1$		
11.	x^2	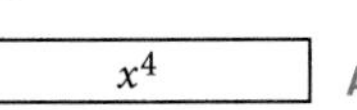 x^4	A
12.	$a^2 - b^2$	$a^2 + b^2$	B

Find each answer.

13. The product of two positive integers is 45. The first is 4 less than the second. Find the integers.
   5 and 9

14. *Open-ended* Write a quadratic function for a graph that opens downward with vertex (−2, 3).
   Sample: $y = -x^2 - 4x - 1$

15. The length of a rectangular garden is 3 m less than twice its width. The area of the garden is 35 m². Find the dimensions of the garden.
   5 m by 7 m

Resources

Teaching Resources

Chapter Support File, Ch. 10
- Cumulative Review
- Standardized Test Practice

Teacher's Edition

See also pp. 507E for assessment options.

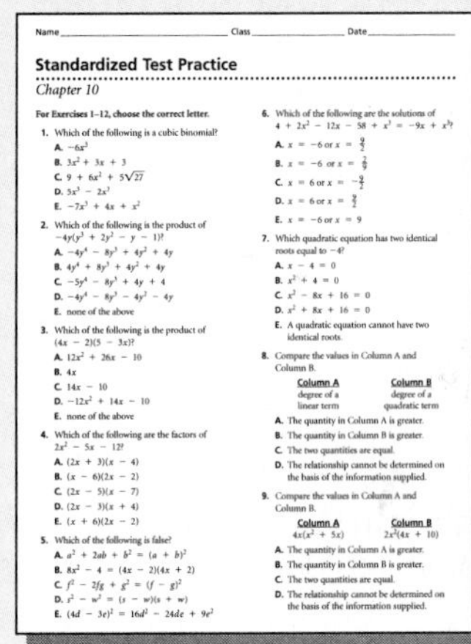

Name______ Class______ Date______

Standardized Test Practice

Chapter 10

For Exercises 1–12, choose the correct letter.

1. Which of the following is a cubic binomial?
   A. $-6x^3$
   B. $3x^2 + 3x + 3$
   C. $9 + 6x^2 + 5\sqrt{27}$
   D. $5x^3 - 2x^7$
   E. $-7x^3 + 4x + x^2$
2. Which of the following is the product of $-4y(y^3 + 2y^2 - y - 1)$?
   A. $-4y^4 - 8y^3 + 4y^2 + 4y$
   B. $4y^4 + 8y^3 + 4y^2 + 4y$
   C. $-5y^4 - 8y^3 + 4y + 4$
   D. $-4y^4 - 8y^3 - 4y^2 - 4y$
   E. none of the above
3. Which of the following is the product of $(4x - 2)(5 - 3x)$?
   A. $12x^2 + 26x - 10$
   B. $4x$
   C. $14x - 10$
   D. $-12x^2 + 14x - 10$
   E. none of the above
4. Which of the following are the factors of $2x^2 - 5x - 12$?
   A. $(2x + 3)(x - 4)$
   B. $(x - 6)(2x - 2)$
   C. $(2x - 5)(x - 7)$
   D. $(2x - 3)(x + 4)$
   E. $(x + 6)(2x - 2)$
5. Which of the following is false?
   A. $a^2 + 2ab + b^2 = (a + b)^2$
   B. $8x^2 - 4 = (4x - 2)(4x + 2)$
   C. $f^2 - 2fg + g^2 = (f - g)^2$
   D. $s^2 - w^2 = (s - w)(s + w)$
   E. $(4d - 3e)^2 = 16d^2 - 24de + 9e^2$
6. Which of the following are the solutions of $4 + 2x^2 - 12x - 58 + x^3 = -9x + x^3$?
   A. $x = -6$ or $x = \frac{9}{2}$
   B. $x = -6$ or $x = \frac{2}{9}$
   C. $x = 6$ or $x = -\frac{9}{2}$
   D. $x = 6$ or $x = \frac{9}{2}$
   E. $x = -6$ or $x = 9$
7. Which quadratic equation has two identical roots equal to −4?
   A. $x - 4 = 0$
   B. $x^2 + 4 = 0$
   C. $x^2 - 8x + 16 = 0$
   D. $x^2 + 8x + 16 = 0$
   E. A quadratic equation cannot have two identical roots.
8. Compare the values in Column A and Column B.

Column A	Column B
degree of a linear term	degree of a quadratic term

   A. The quantity in Column A is greater.
   B. The quantity in Column B is greater.
   C. The two quantities are equal.
   D. The relationship cannot be determined on the basis of the information supplied.
9. Compare the values in Column A and Column B.

Column A	Column B
$4x(x^2 + 5x)$	$2x^2(4x + 10)$

   A. The quantity in Column A is greater.
   B. The quantity in Column B is greater.
   C. The two quantities are equal.
   D. The relationship cannot be determined on the basis of the information supplied.

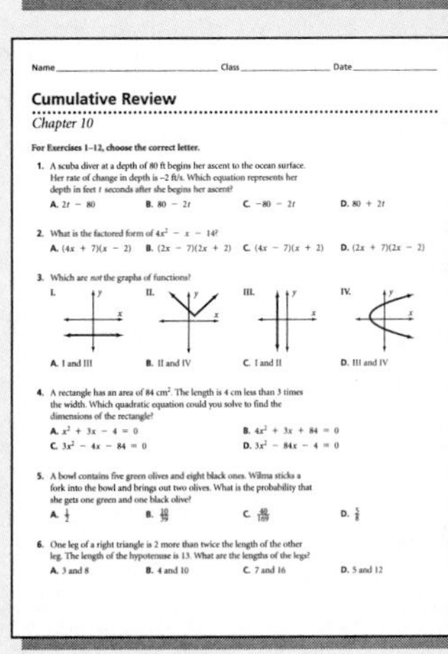

Name______ Class______ Date______

Cumulative Review

Chapter 10

For Exercises 1–12, choose the correct letter.

1. A scuba diver at a depth of 80 ft begins her ascent to the ocean surface. Her rate of change in depth is −2 ft/s. Which equation represents her depth in feet *t* seconds after she begins her ascent?
   A. $2t - 80$ B. $80 - 2t$ C. $-80 - 2t$ D. $80 + 2t$
2. What is the factored form of $4x^2 - x - 14$?
   A. $(4x + 7)(x - 2)$ B. $(2x - 7)(2x + 2)$ C. $(4x - 7)(x + 2)$ D. $(2x + 7)(2x - 2)$
3. Which are *not* the graphs of functions?
   I. II. III. IV.
   A. I and III B. II and IV C. I and II D. III and IV
4. A rectangle has an area of 84 cm². The length is 4 cm less than 3 times the width. Which quadratic equation could you solve to find the dimensions of the rectangle?
   A. $x^2 + 3x - 4 = 0$ B. $4x^2 + 3x + 84 = 0$
   C. $3x^2 - 4x - 84 = 0$ D. $3x^2 - 84x - 4 = 0$
5. A bowl contains five green olives and eight black ones. Wilma sticks a fork into the bowl and brings out two olives. What is the probability that she gets one green and one black olive?
   A. $\frac{1}{2}$ B. $\frac{10}{39}$ C. $\frac{40}{169}$ D. $\frac{5}{8}$
6. One leg of a right triangle is 2 more than twice the length of the other leg. The length of the hypotenuse is 13. What are the lengths of the legs?
   A. 3 and 8 B. 4 and 10 C. 7 and 16 D. 5 and 12

CHAPTER 11 Rational Expressions

To accommodate flexible scheduling, some lessons are divided into parts. Assignment Options are given in the Lesson Planning Options for each lesson.

PACING OPTIONS

This chart suggests pacing only for the core lessons and their parts, and it is provided merely as a possible guide. It will help you determine how much time you have in your schedule to cover other features, such as the Chapter Project, Math Toolboxes, Wrap Up, and Assessment.

	1 Class Period	1 Class Period	1 Class Period	1 Class Period	1 Class Period	1 Class Period	1 Class Period	1 Class Period	1 Class Period
Traditional (40–45 min class periods)	11-1 Part 1	11-1 Part 2	11-2 Part 1	11-2 Part 2	11-3 Part 1	11-3 Part 2	11-4	11-5	11-6 Part 1
Algebra Over 2 Years (40–45 min class periods)	11-1 Part 1	11-1 Part 1	11-1 Part 2	11-1 Part 2	11-2 Part 1	11-2 Part 1	11-2 Part 2	11-2 Part 2	11-3 Part 1
Block Scheduling (90 min class periods)	11-1 Part 1, 11-1 Part 2	11-2 Part 1, 11-2 Part 2	11-3 Part 1, 11-3 Part 2	11-4	11-5	11-6 Part 1, 11-6 Part 2	11-7		

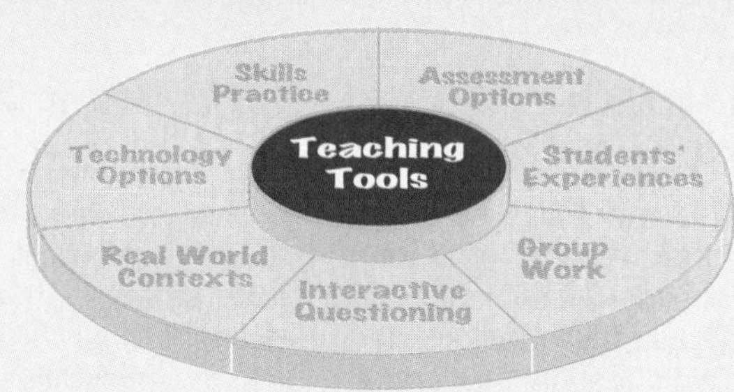

What Students Will Learn and Why

In this chapter, students will build on their knowledge of factoring, learned in Chapter 10, by learning to solve rational equations. They will learn to find inverse variation, and they will use rational expressions to solve real-world problems. They will learn to count outcomes and permutations and to determine combinations. Finally, students will learn how rational expressions allow the application of formulas to statistical situations.

Discussing the Chapter/Building on Experience

The concept map below relates chapter topics to real-world applications. You and your class may wish to add to the map or develop maps of your own. The center oval describes the topic of the chapter. The next level displays topics within the lessons. The outer ovals reflect applications of the content. As you and your class build a concept map, invite students to discuss applications with which they are familiar.

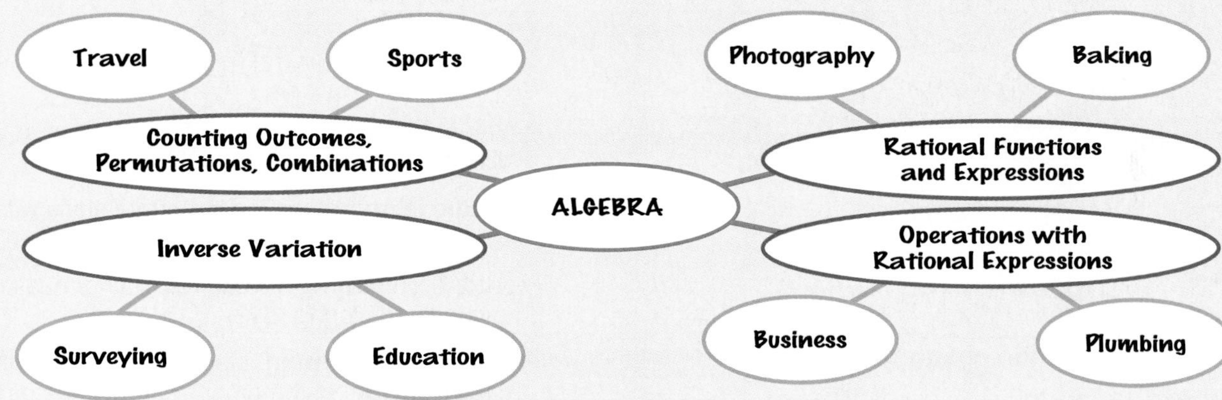

1 Class Period	1 Class Period	1 Class Period	1 Class Period	1 Class Period	1 Class Period	1 Class Period	1 Class Period	1 Class Period	1 Class Period	1 Class Period
11-6 ▽2	11-7									
11-3 ▽1	11-3 ▽2	11-3 ▽2	11-4	11-4	11-5	11-5	11-6 ▽1	11-6 ▽2	11-7	11-7

Interactive Questioning Tips

A question is interactive when there is “give and take” between the questioner (teacher or student) and the respondent. When a critical thinking question is asked, it is important for the questioner to phrase the question clearly and specifically. When given a direct question, students can use time efficiently to formulate responses rather than taking time to interpret the question.

Skills Practice

Every lesson provides skill practice with Exercises On Your Own and Exercises Mixed Review. The Student Edition includes Checkpoints (pp. 521, 536) and Cumulative Review (pp. 554–555). In the Teacher’s Edition, the Lesson Planning Options section for each lesson lists Prerequisite Skills students should know for that lesson. At the back of the Student Edition is the Skills Handbook—mini-lessons on math the students may need to review, The Chapter Support File for Chapter 11 in the Teaching Resources box includes two Practice worksheets per lesson, a worksheet for two Checkpoints, and worksheets for Cumulative Review and Standardized Test Preparation.

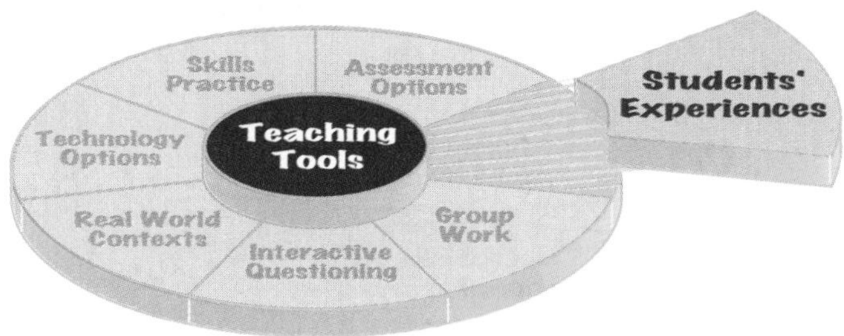

Diverse Learning and Teaching Styles

In your Teacher’s Edition, you will find suggestions as to how you can help students complete mathematical tasks in Chapter 11 by reinforcing various learning styles. Here are some examples.

- **Visual learning** drawing a tree diagram to visualize different permutations (p. 528)
- **Tactile learning** exploring permutations with three tiles or blocks of different colors (p. 539)
- **Auditory learning** asking each other questions about inverse variation (p. 512)
- **Kinesthetic learning** standing at a wall with one shoe directly in front of the other to form a line parallel to the wall (p. 517)

Alternative Activity for Lesson 11-1

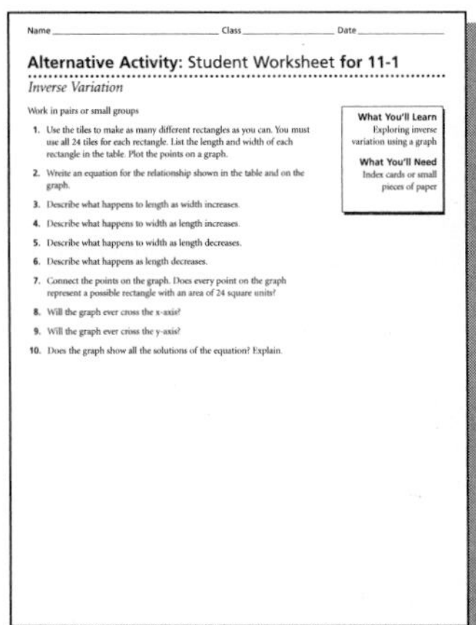

Name ______ Class ______ Date ______

Alternative Activity: Student Worksheet **for 11-1**

Inverse Variation

Work in pairs or small groups

What You’ll Learn Exploring inverse variation using a graph

What You’ll Need Index cards or small pieces of paper

1. Use the tiles to make as many different rectangles as you can. You must use all 24 tiles for each rectangle. List the length and width of each rectangle in the table. Plot the points on a graph.
2. Write an equation for the relationship shown in the table and on the graph.
3. Describe what happens to length as width increases.
4. Describe what happens to width as length increases.
5. Describe what happens to width as length decreases.
6. Describe what happens as length decreases.
7. Connect the points on the graph. Does every point on the graph represent a possible rectangle with an area of 24 square units?
8. Will the graph ever cross the x-axis?
9. Will the graph ever cross the y-axis?
10. Does the graph show all the solutions of the equation? Explain.

for use with Work Together, addresses tactile and visual learning by exploring inverse variation using tiles

Alternative Activity for Lesson 11-5

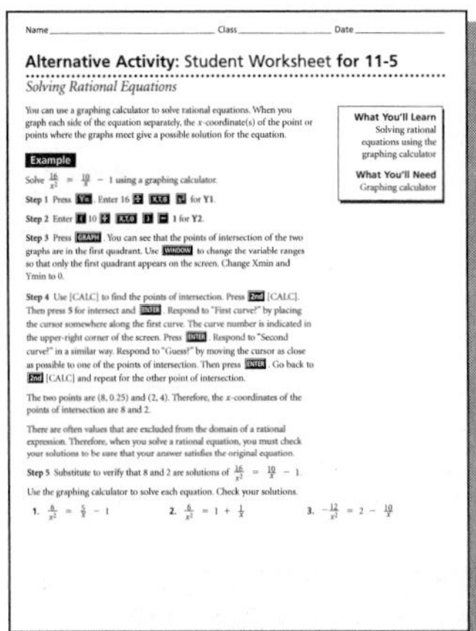

Name ______ Class ______ Date ______

Alternative Activity: Student Worksheet **for 11-5**

Solving Rational Equations

You can use a graphing calculator to solve rational equations. When you graph each side of the equation separately, the x-coordinate(s) of the point or points where the graphs meet give a possible solution for the equation.

What You’ll Learn Solving rational equations using the graphing calculator

What You’ll Need Graphing calculator

Example

Solve $\frac{16}{x^2} = \frac{10}{x} - 1$ using a graphing calculator.

Step 1 Press [Y=]. Enter 16 [÷] [X,T,θ] [x²] for **Y1**.

Step 2 Enter [(] 10 [÷] [X,T,θ] [)] [−] 1 for **Y2**.

Step 3 Press [GRAPH]. You can see that the points of intersection of the two graphs are in the first quadrant. Use [WINDOW] to change the variable ranges so that only the first quadrant appears on the screen. Change Xmin and Ymin to 0.

Step 4 Use [CALC] to find the points of intersection. Press [2nd] [CALC]. Then press 5 for intersect and [ENTER]. Respond to “First curve?” by placing the cursor somewhere along the first curve. The curve number is indicated in the upper-right corner of the screen. Press [ENTER]. Respond to “Second curve?” in a similar way. Respond to “Guess?” by moving the cursor as close as possible to one of the points of intersection. Then press [ENTER]. Go back to [2nd] [CALC] and repeat for the other point of intersection.

The two points are (8, 0.25) and (2, 4). Therefore, the x-coordinates of the points of intersection are 8 and 2.

There are often values that are excluded from the domain of a rational expression. Therefore, when you solve a rational equation, you must check your solutions to be sure that your answer satisfies the original equation.

Step 5 Substitute to verify that 8 and 2 are solutions of $\frac{16}{x^2} = \frac{10}{x} - 1$.

Use the graphing calculator to solve each equation. Check your solutions.

1. $\frac{6}{x^2} = \frac{5}{x} - 1$
2. $\frac{6}{x^2} = 1 + \frac{1}{x}$
3. $-\frac{12}{x^2} = 2 - \frac{10}{x}$

for use with Example 1, addresses visual learning by graphing rational equations with a graphing calculator.

Alternative Activity for Lesson 11-7

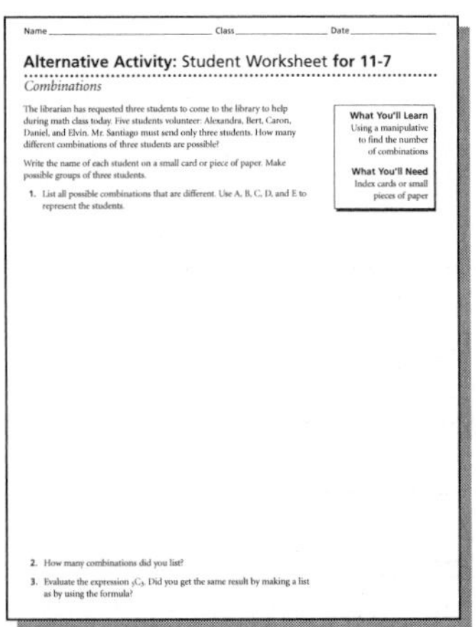

Name ______ Class ______ Date ______

Alternative Activity: Student Worksheet **for 11-7**

Combinations

The librarian has requested three students to come to the library to help during math class today. Five students volunteer: Alexandra, Bert, Caron, Daniel, and Elvin. Mr. Santiago must send only three students. How many different combinations of three students are possible?

Write the name of each student on a small card or piece of paper. Make possible groups of three students.

What You’ll Learn Using a manipulative to find the number of combinations

What You’ll Need Index cards or small pieces of paper

1. List all possible combinations that are different. Use A, B, C, D, and E to represent the students.
2. How many combinations did you list?
3. Evaluate the expression ${}_5C_3$. Did you get the same result by making a list as by using the formula?

for use with Example 1, addresses tacti[le] learning by counting combinations usin[g] index cards or small pieces of paper

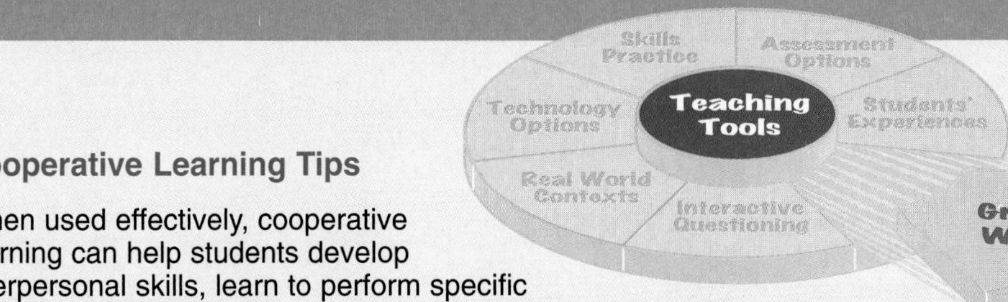

Cooperative Learning Tips

When used effectively, cooperative learning can help students develop interpersonal skills, learn to perform specific roles in a group, and learn to carry out specific responsibilities. The components of Chapter 11 provide a range of cooperative learning opportunities.

- In the Student Edition, the ***Work Together*** parts of lessons are specifically designed for cooperative learning activities.
- In the Teacher's Edition, you will find helpful hints for addressing diverse learning styles (see page C for Chapter 11). For every lesson, you will find a ***Reteaching Activity***, which may involve cooperative learning.

Materials and Manipulatives

The authors expect all students to use scientific calculators. Calculator use is integrated throughout the course.

- calculator (11-6, 11-7)
- graph paper (11-2)
- graphing calculator (11-2)
- index cards (11-2)

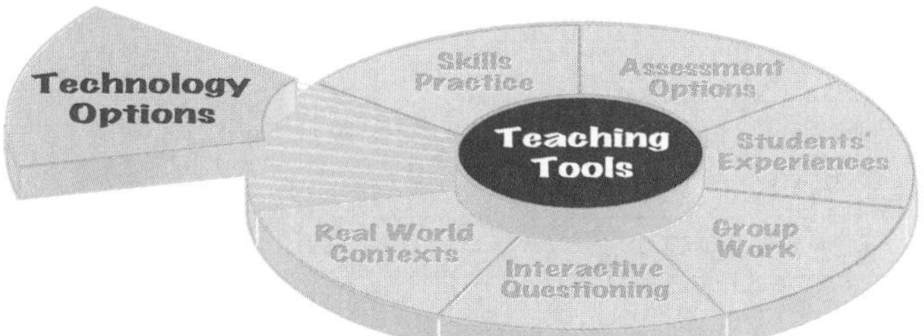

TECHNOLOGY OPTIONS

Technology Tools		Chapter Project	11-1	11-2	11-3	11-4	11-5	11-6	11-7
Calculator		Assumed that students have scientific calculators at any time.							
Graphing Calculator	Handbook			✔				✔	✔
	Student Edition			✔				✔	✔
Software	Secondary Math Lab Toolkit™		✔	✔	✔	✔	✔	✔	✔
	Integrated Math Lab							✔	✔
	Computer Item Generator		✔	✔	✔	✔	✔	✔	✔
	Student Edition								
Video	Video Field Trip	✔							
CD-ROM	Multimedia Algebra Lab		✔	✔		✔	✔		
Internet		See the Prentice Hall site. (http://www.phschool.com)							

The Prentice Hall Algebra program offers you a rich variety of technology options. Be assured that all these options are provided as a means of enriching the program and are not essential for the successful completion of the course.

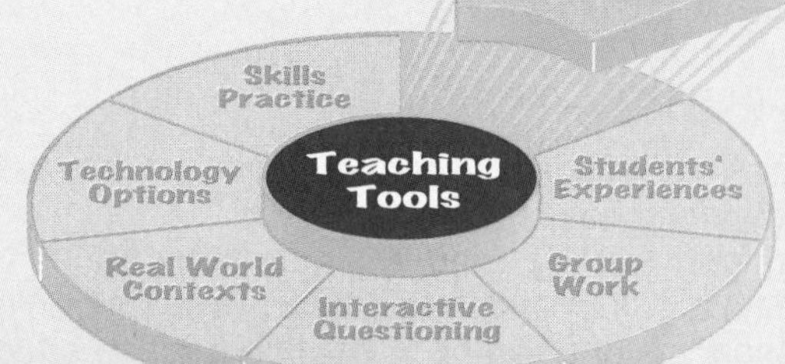

Assessment Options

The Prentice Hall Algebra Program provides you with many options. From these options, you may choose instructional materials and techniques appropriate for your students, or those necessary to meet your district's curriculum requirements. As the chart indicates, the program also supports your teaching efforts by offering you many choices for assessment.

ASSESSMENT OPTIONS

Assessment Support Materials	Chapter Project	11-1	11-2	11-3	11-4	11-5	11-6	11-7	Chapter End
Chapter Project	▲■●	▲■		▲■	▲■	▲■	▲■	▲■	▲■
Checkpoints			▲●			▲●			
Self-Assessment		▲■					▲■	▲■	▲■
Writing Assignment		▲	▲■	▲■	▲●	▲	▲	▲■	▲●
Chapter Assessment									▲●
Alternative Assessment		■	■	■	■	■	■	■	■●
Cumulative Review									▲●
Standardized Test Prep		▲■		▲■		▲■	▲■	▲■	●
Chapters 1–11 Assessment									●
Computer Item Generator	Can be used to create custom-made practice or assessment at any time.								

▲ = Student Edition ■ = Teacher's Edition ● = Teaching Resources

Checkpoints

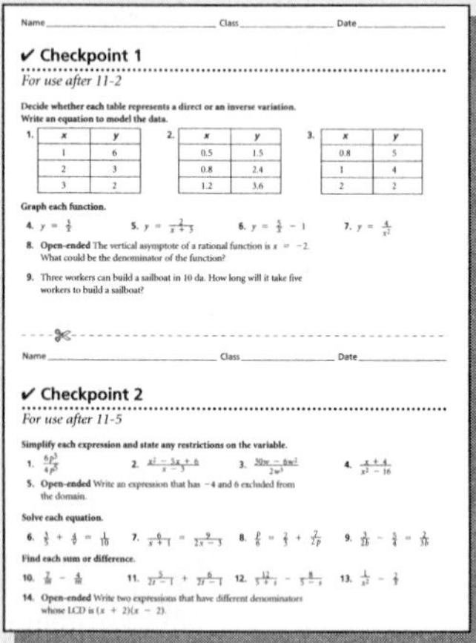

Name_____ Class_____ Date_____

✔ Checkpoint 1

For use after 11-2

Decide whether each table represents a direct or an inverse variation. Write an equation to model the data.

1.

x	y
1	6
2	3
3	2

2.

x	y
0.5	1.5
0.8	2.4
1.2	3.6

3.

x	y
0.8	5
1	4
2	2

Graph each function.

4. $y = \frac{1}{x}$ 5. $y = \frac{2}{x+3}$ 6. $y = \frac{5}{x} - 1$ 7. $y = \frac{4}{x^2}$

8. **Open-ended** The vertical asymptote of a rational function is $x = -2$. What could be the denominator of the function?

9. Three workers can build a sailboat in 10 da. How long will it take five workers to build a sailboat?

Name_____ Class_____ Date_____

✔ Checkpoint 2

For use after 11-5

Simplify each expression and state any restrictions on the variable.

1. $\frac{6p^3}{4p^5}$ 2. $\frac{x^2 - 5x + 6}{x - 3}$ 3. $\frac{30w - 6w^2}{2w^3}$ 4. $\frac{x + 4}{x^2 - 16}$

5. **Open-ended** Write an expression that has -4 and 6 excluded from the domain.

Solve each equation.

6. $\frac{3}{5} + \frac{4}{9} = \frac{1}{10}$ 7. $\frac{6}{x+1} = \frac{9}{2x-3}$ 8. $\frac{p}{6} = \frac{2}{3} + \frac{7}{2p}$ 9. $\frac{3}{2b} - \frac{5}{4} = \frac{2}{3b}$

Find each sum or difference.

10. $\frac{7}{m} - \frac{4}{m}$ 11. $\frac{5}{2t-1} + \frac{6}{2t-1}$ 12. $\frac{12}{5+x} - \frac{8}{5-x}$ 13. $\frac{1}{x^2} - \frac{2}{9}$

14. **Open-ended** Write two expressions that have different denominators whose LCD is $(x + 2)(x - 2)$.

Alternative Assessment

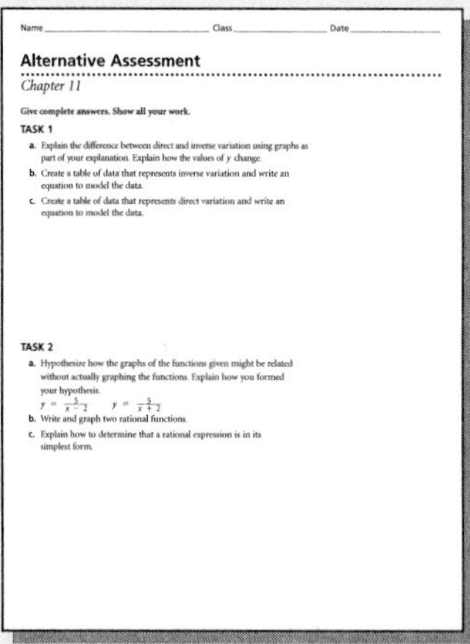

Name_____ Class_____ Date_____

Alternative Assessment

Chapter 11

Give complete answers. Show all your work.

TASK 1

a. Explain the difference between direct and inverse variation using graphs as part of your explanation. Explain how the values of y change.

b. Create a table of data that represents inverse variation and write an equation to model the data.

c. Create a table of data that represents direct variation and write an equation to model the data.

TASK 2

a. Hypothesize how the graphs of the functions given might be related without actually graphing the functions. Explain how you formed your hypothesis.

$y = \frac{5}{x-2}$ $y = \frac{5}{x+2}$

b. Write and graph two rational functions.

c. Explain how to determine that a rational expression is in its simplest form.

Chapter Assessment

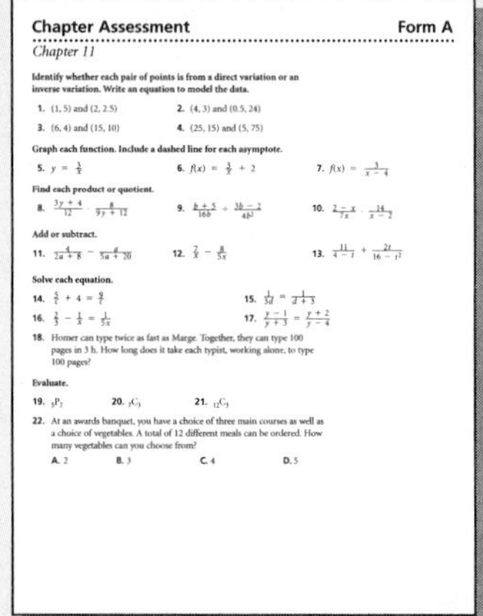

Chapter Assessment Form A

Chapter 11

Identify whether each pair of points is from a direct variation or an inverse variation. Write an equation to model the data.

1. (1, 5) and (2, 2.5) 2. (4, 3) and (0.5, 24)

3. (6, 4) and (15, 10) 4. (25, 15) and (5, 75)

Graph each function. Include a dashed line for each asymptote.

5. $y = \frac{3}{x}$ 6. $f(x) = \frac{3}{x} + 2$ 7. $f(x) = \frac{3}{x-4}$

Find each product or quotient.

8. $\frac{3y + 4}{12} \cdot \frac{8}{9y + 12}$ 9. $\frac{b + 5}{16b} \div \frac{3b - 2}{4b^2}$ 10. $\frac{2 - x}{7x} \cdot \frac{14}{x - 2}$

Add or subtract.

11. $\frac{4}{2a + 8} - \frac{a}{5a + 20}$ 12. $\frac{7}{x} - \frac{8}{5x}$ 13. $\frac{11}{4 - t} + \frac{2t}{16 - t^2}$

Solve each equation.

14. $\frac{5}{t} + 4 = \frac{9}{t}$ 15. $\frac{1}{3d} = \frac{1}{d + 3}$

16. $\frac{2}{3} - \frac{1}{x} = \frac{1}{5x}$ 17. $\frac{y - 1}{y + 3} = \frac{y + 2}{y - 4}$

18. Homer can type twice as fast as Marge. Together, they can type 100 pages in 3 h. How long does it take each typist, working alone, to type 100 pages?

Evaluate.

19. ${}_5P_2$ 20. ${}_7C_3$ 21. ${}_{12}C_3$

22. At an awards banquet, you have a choice of three main courses as well as a choice of vegetables. A total of 12 different meals can be ordered. How many vegetables can you choose from?

A. 2 B. 3 C. 4 D. 5

Available in both Form A and Form B

Making the Right Connections

Mathematics is imbedded in nearly every walk of life. The National Council of Teachers of Mathematics (NCTM) encourages educators to recognize these connections and to emphasize them for the purpose of better educating students for success in life and in a global economy. The ***Connections*** chart below highlights these connections for Chapter 11.

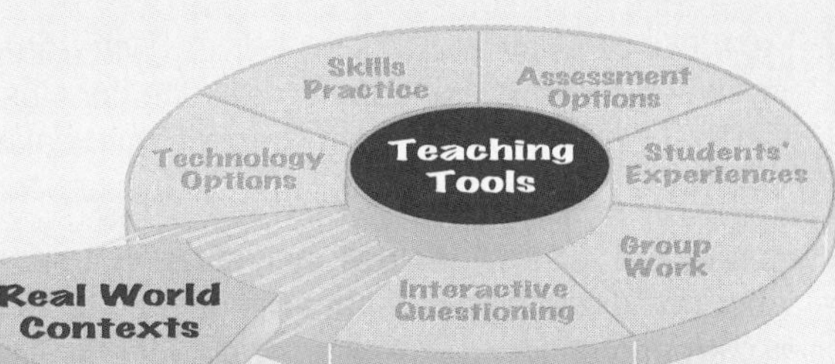

CONNECTIONS

Lesson	Interdisciplinary Connections	Career Prep	Other Real World Connections	Math Integration	NCTM Standards
Chapter Project	Social Studies		Music		Problem Solving
11-1	Physics Writing	Surveying Building Teaching	Construction Travel Education	Geometry	Algebra Communication Problem Solving
11-2	Physics Writing	Photography Marketing	Delivery Services		Algebra Communication Problem Solving Functions
11-3	Writing Home Economics	Chef Distribution	Baking	Geometry	Algebra Communication Problem Solving
11-4	Writing Health History	Electrician Transportation	Air Travel Population Surveys Exercise	Statistics	Algebra Communication Problem Solving Statistics
11-5	Writing Accounting	Sales Computers	Business Plumbing	Geometry	Algebra Communication Problem Solving
11-6	Writing Music Physical Education	Jobs School Paper	Entertainment Travel Sports Computers Telephones Camping	Probability	Algebra Communication Problem Solving Statistics
11-7	Economics Government	Manufacturing Sales Industrry	Juries Recreation Spelling Bees Sports Weather	Geometry	Algebra Communication Problem Solving Statistics

CONNECTING TO PRIOR LEARNING Elicit from students a definition for *ratio*. Help them draw the conclusion that a rational expression is a term or mathematical model with which they are already familiar.

CULTURAL CONNECTIONS Ask students to share information about different rhythmic patterns. Invite students from various cultures to share their native music with the class. Ask students what the different rhythmic patterns suggest to them. Have students explain the meaning of the patterns in their native music.

INTERDISCIPLINARY CONNECTIONS As sound waves vibrate the air, it is possible to determine how far away an object is. Musical instruments make sounds based on vibrations. Have students experiment with sounds and distances to see if they can identify what is making a sound and how far away the source is.

ABOUT THE PROJECT The Chapter Project gives students an opportunity to learn more about the connection between music and mathematics. In the Find Out questions, students do research, use formulas and graphs to explore different mathematical models related to sound and music.

Technology Options

Prentice Hall Technology

Video • Video Field Trip 11, "No Strings Attached—Yet," a look at the making of a guitar.

CHAPTER 11 Rational Expressions and Functions

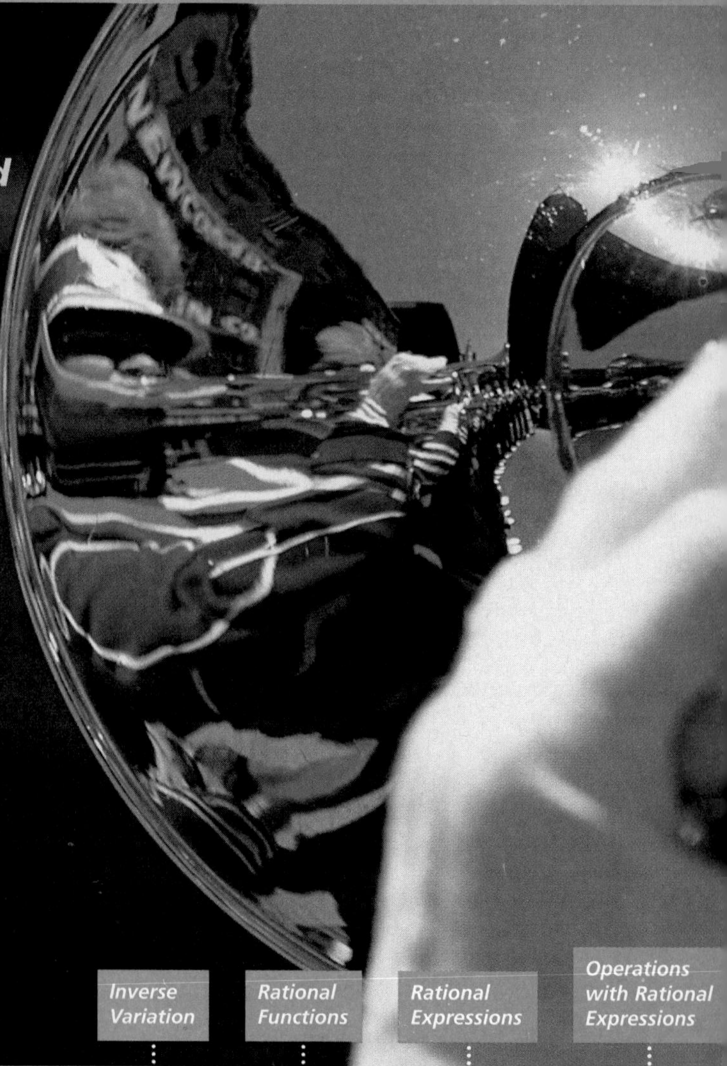

Relating to the Real World

Sixty dollars will buy you six $10 pizzas or four $15 pizzas. This simple relationship leads to fractions with variables in the denominators, called rational expressions. Rational expressions give social scientists and others who use statistics flexibility in applying formulas to their work.

Launching the Project

PROJECT NOTEBOOK Encourage students to keep all project-related materials in a separate folder or notebook. See Chapter Project and Scoring Rubric in Chapter Support File.

- Set the stage for the project by asking students how many of them play a musical instrument.
- Using students' musical knowledge and background, have them explain how a string's vibration results in a sound.
- Have students explain why the length of the string is important.
- Have students share their explanations with the rest of the class.
- Have the students look at the Find Out by Calculating section on page 514 of their textbooks. Explain that, when they begin the Chapter Project, they will use the table to find the string lengths for a C-Major scale.

TRACKING THE PROJECT You may wish to have students read Finishing the Chapter Project on page 548 to help them get an overview of the project. Set benchmark deadlines for students to show you their work in progress.

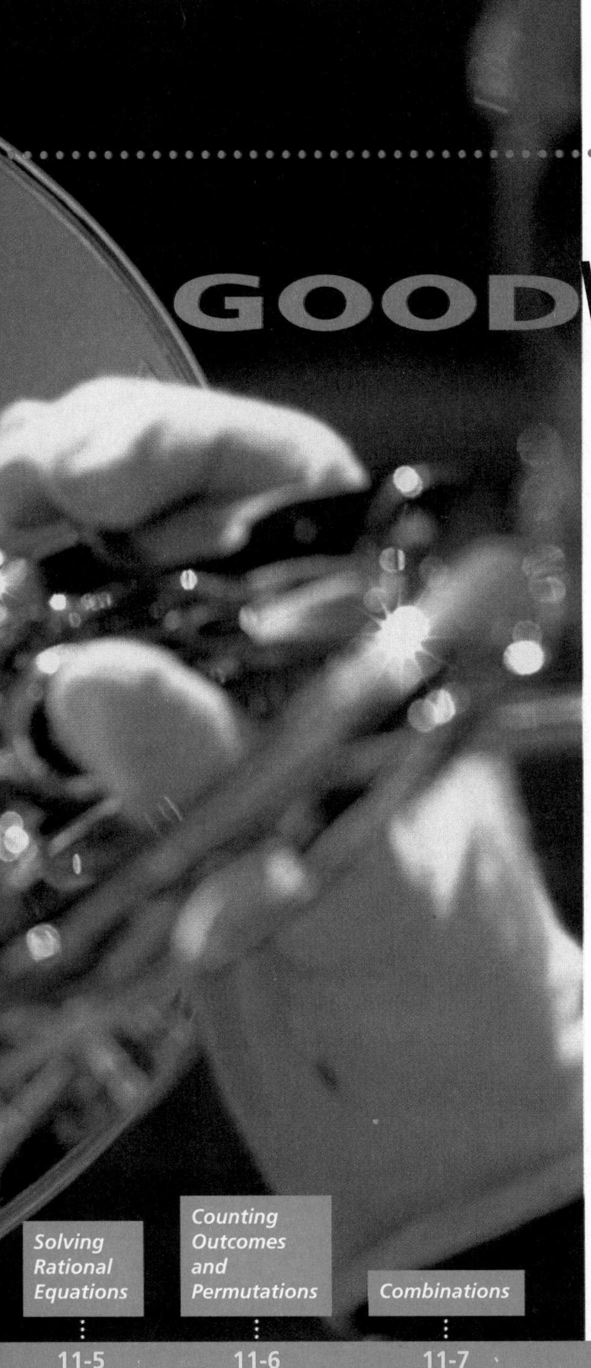

CHAPTER PROJECT

GOOD VIBRATIONS

Sounds are caused by vibrations—for example, a string vibrating on a violin. When the string is shortened it vibrates faster, and a higher pitch results. Pitch is also affected by tension (an example is your vocal cords).

As you work through the chapter, you will investigate a variety of musical pitches. You will use inverse variation to find pitch. You will create simple musical instruments to compare ratios of lengths to different pitches. Finally, you will choose a musical instrument and explain how it produces sounds at different pitches.

To help you complete the project:

- p. 514 *Find Out by Calculating*
- p. 526 *Find Out by Analyzing*
- p. 531 *Find Out by Creating*
- p. 536 *Find Out by Interviewing*
- p. 547 *Find Out by Communicating*
- p. 548 *Finishing the Project*

Project Resources

Teaching Resources

Chapter Support File, Ch. 11
- Chapter Project Manager and Scoring Rubric

Transparencies
124

Using the Rubric

Sharing the scoring rubric for the project with your students will alert them to your expectations before they begin work on the project.

As students complete each Find Out question in the chapter, you may wish to have them evaluate their own work or a partner's work, based on the scoring rubric. Students should have the opportunity to revise their work after it has been reviewed.

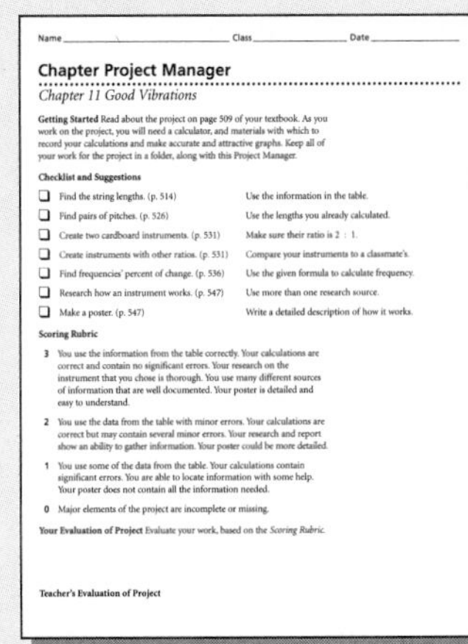

Name________ Class________ Date________

Chapter Project Manager

Chapter 11 Good Vibrations

Getting Started Read about the project on page 509 of your textbook. As you work on the project, you will need a calculator, and materials with which to record your calculations and make accurate and attractive graphs. Keep all of your work for the project in a folder, along with this Project Manager.

Checklist and Suggestions

❑ Find the string lengths. (p. 514)	Use the information in the table.
❑ Find pairs of pitches. (p. 526)	Use the lengths you already calculated.
❑ Create two cardboard instruments. (p. 531)	Make sure their ratio is 2 : 1.
❑ Create instruments with other ratios. (p. 531)	Compare your instruments to a classmate's.
❑ Find frequencies' percent of change. (p. 536)	Use the given formula to calculate frequency.
❑ Research how an instrument works. (p. 547)	Use more than one research source.
❑ Make a poster. (p. 547)	Write a detailed description of how it works.

Scoring Rubric

3 You use the information from the table correctly. Your calculations are correct and contain no significant errors. Your research on the instrument that you chose is thorough. You use many different sources of information that are well documented. Your poster is detailed and easy to understand.

2 You use the data from the table with minor errors. Your calculations are correct but may contain several minor errors. Your research and report show an ability to gather information. Your poster could be more detailed.

1 You use some of the data from the table. Your calculations contain significant errors. You are able to locate information with some help. Your poster does not contain all the information needed.

0 Major elements of the project are incomplete or missing.

Your Evaluation of Project Evaluate your work, based on the *Scoring Rubric*.

Teacher's Evaluation of Project

11-1 Teaching Notes

PROBLEM OF THE DAY

A flower bed is in the shape of an equilateral triangle. If the perimeter is 32.7 m, how much would it cost to fence one side at the cost of $2.60/m? $28.34

Problem of the Day is also available in Transparencies.

CONNECTING TO PRIOR KNOWLEDGE Have students describe events that involve two related quantities in the following relationships: as one increases, the other increases; as one increases, the other decreases.

WORK TOGETHER

Question 4 Ask students to compare the time 2 people would take to build a house versus the time 160 people would take. Ask students if they think either situation is realistic. Help students understand the necessity of considering real-world circumstances when trying to model a situation using math.

THINK AND DISCUSS

Emphasize that k is called a constant because it represents a number that will constantly remain the same value.

For students who are not familiar with the word *inverse*, have them look up the word in a dictionary or thesaurus. Two synonyms for *inverse* are *opposite* and *reverse*. Lead students

Lesson Planning Options

Prerequisite Skills

- Graphing ordered pairs
- Analyzing data from tables

Assignment Options for Exercises On Your Own

To provide flexible scheduling, this lesson can be subdivided into parts.

 **Core** 3–14, 16–22
Extension 1, 2, 15, 23–25

 **Core** 26–34
Extension 35, 36

Use Mixed Review to maintain skills.

Resources

 **Student Edition**

Skills Handbook, pp. 575–579
Extra Practice, p. 566
Glossary/Study Guide

 **Teaching Resources**

Chapter Support File, Ch. 11
- Practice 11-1 (two worksheets)
- Reteaching 11-1
- Alternative Activity 11-1

Classroom Manager 11-1
Glossary, Spanish Resources
Two-Year Algebra Handbook 11-1

 **Transparencies**
125, 129

Connections  Construction . . . and more

What You'll Learn

- Solving inverse variations
- Comparing direct and inverse variation

...And Why

To investigate real-world situations, such as those relating time and rate of work

11-1 Inverse Variation

WORK TOGETHER

Construction Suppose you are part of a volunteer crew constructing low-cost housing. Building a house requires a total of 160 workdays. For example, a crew of 20 people can complete a house in 8 days.

1. How long should it take a crew of 40 people? 4 days
2. Copy and complete the table.

Crew Size (x)	Construction Days (y)	Total Workdays
2	80	160
5	■ 32	160
8	■ 20	■ 160
■ 10	16	■ 160
20	8	160
40	■ 4	■ 160

3. Graph the (x, y) data from the table. See margin.
4. Describe what happens to construction time as the crew size increases. Construction time decreases.

Who? Using volunteers, Habitat for Humanity has helped build thousands of homes for low-income families around the world.

Source: Habitat for Humanity

THINK AND DISCUSS

Part 1 *Solving Inverse Variations*

When the product of two quantities remains constant, they form an **inverse variation.** As one quantity increases, the other decreases. The product of the quantities is called the **constant of variation** k. An inverse variation can be written $xy = k$, or $y = \frac{k}{x}$.

5. a. In the Work Together, what two quantities vary? See below.
   b. What is the constant of variation? 160
   c. Write an equation that models this variation. $xy = 160$

   5a. number of people, construction time

6. a. Complete the table for the inverse variation $xy = 100$.

x	1	2	4	5	10	20	50	100
y	100	■ 50	■ 25	■ 20	■ 10	■ 5	■ 2	■ 1

   b. Describe how the values of y change as the values of x increase. Values of y decrease.

to understand that when x changes, y changes in the opposite way.

OPEN-ENDED Question 7b Suggest that students find points by choosing values for x and then substituting these values into the equations to find the corresponding values for y.

VISUAL LEARNING Draw the graph of $xy = 8$ on the board. Label any point on the graph with its coordinates. Show students that the coordinates of the point you labeled fit the equation $xy = 8$. Label a second point and substitute the coordinates into the equation. Repeat for at least four points on the graph.

Example 1 ***Relating to the Real World*** 

Make sure students understand the meaning of *fulcrum*.

TACTILE LEARNING Encourage students to model the lever by placing a ruler across a pencil and then balancing tiles on each end of the ruler.

CONNECTING TO THE STUDENTS' WORLD If there is a park in your city that has a seesaw, encourage students to experiment by sitting closer to and farther from the fulcrum. Students may also be able to build a lever using yardsticks and then experiment using different weights.

EXTENSION Question 8 Encourage students to observe what happens when they move the fulcrum off-center of a lever they have created.

Question 8a For inverse variations, $x_1 \cdot y_1 = x_2 \cdot y_2$, point out that x_1, y_1, and y_2 are given. To find x_2, divide both sides of the equation by y_2.

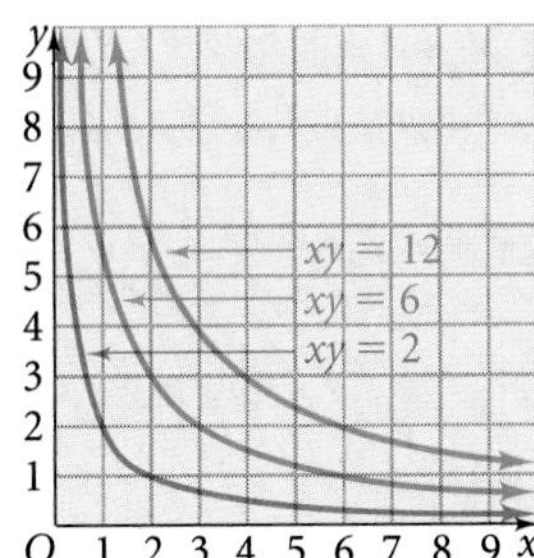

Inverse variations have graphs with the same general shape.

7a. For $xy = 12$, $k = 12$. For $xy = 6$, $k = 6$. For $xy = 2$, $k = 2$.

7. a. Name the constant of variation k for each graph shown above.
   b. *Open-ended* Name three points that lie on the graph of each inverse variation. Sample: For $xy = 12$, (1, 12), (2, 6), (3, 4); for $xy = 6$, (1, 6), (2, 3), (3, 2); for $xy = 2$, (1, 2), (2, 1), $(4, \frac{1}{2})$.

Suppose (x_1, y_1) and (x_2, y_2) are two ordered pairs in an inverse variation. Since each ordered pair has the same product, you can write the *product* equation $x_1 \cdot y_1 = x_2 \cdot y_2$. You can use this equation to solve problems involving inverse variation.

Example 1 ***Relating to the Real World*** 

Physics The weight needed to balance a lever varies inversely with the distance from the fulcrum to the weight. Where should Julio sit to balance the lever?

Relate A weight of 120 lb is 6 ft from the fulcrum. A weight of 150 lb is x ft from the fulcrum. Weight and distance vary inversely.

Define
$$\text{weight}_1 = 120 \text{ lb}$$
$$\text{weight}_2 = 150 \text{ lb}$$
$$\text{distance}_1 = 6 \text{ ft}$$
$$\text{distance}_2 = x \text{ ft}$$

Write
$$\text{weight}_1 \cdot \text{distance}_1 = \text{weight}_2 \cdot \text{distance}_2 \quad \leftarrow \text{Use a product equation.}$$
$$120 \cdot 6 = 150 \cdot x \quad \leftarrow \text{Substitute.}$$
$$720 = 150x$$
$$x = \frac{720}{150}$$
$$x = 4.8$$

Julio should sit 4.8 feet from the fulcrum to balance the lever.

8. **Try This** Solve each inverse variation.
   a. When $x = 75$, $y = 0.2$. Find x when $y = 3$. 5
   b. What weight placed on a lever 6 ft from the fulcrum will balance 80 lb placed 9 ft from the fulcrum? 120 lb
   c. A trip takes 3 h at 50 mi/h. Find the time when the rate is 60 mi/h. 2.5 h

Additional Examples

FOR EXAMPLE 1

Jeff and Tracy balance on a lever similar to that used in Example 1. Jeff weighs 130 lb and is 5 ft from the lever's fulcrum. If Tracy is 7 ft from the fulcrum, how much does she weigh? 93 lb

Discussion: *What effect does the weight of the lever have on this problem?*

FOR EXAMPLE 2

Decide if each data set represents a direct variation or an inverse variation. Then write an equation to model the data.

a.

x	3	5	10
y	10	6	3

inverse variation; $xy = 30$

b.

x	2	4	8
y	3	6	12

direct variation; $y = 1.5x$

Question 8b Encourage students to draw a sketch to illustrate their answer.

ESTIMATION Question 8c Have students make an estimate before they calculate the answer. Ask them whether the second trip will take more or less time. Then ask them to compare their answers to the estimate.

Example 2

ERROR ALERT! Students may not understand why they need to check each of the values in the table. **Remediation:** Draw the following table on the board:

x	2	4	6
y	4	16	36

Ask students whether this is an inverse or a direct variation. neither Use this example to help students understand why they must test each of the values.

AUDITORY LEARNING Question 9 Encourage students to read each statement to a partner. Then have them ask their partner:

- *What could variables represent here?*
- *How would those variables be related to each other?*

Exercises ON YOUR OWN

CRITICAL THINKING Exercise 2 Suggest that students write the equations for a direct variation and an inverse variation. Then have them use the point (4, 2) to find the equations for the graphs.

Technology Options

For Exercises 26–30, students may use spreadsheet software or a graphing calculator to complete the table and model the data.

Prentice Hall Technology 

Software
- Secondary Math Lab Toolkit™
- Computer Item Generator 11-1

CD-ROM
- Multimedia Algebra Lab 11

Internet
- See the Prentice Hall site. (http://www.phschool.com)

Part 2

Comparing Direct and Inverse Variation

This summary will help you recognize and use direct and inverse variations.

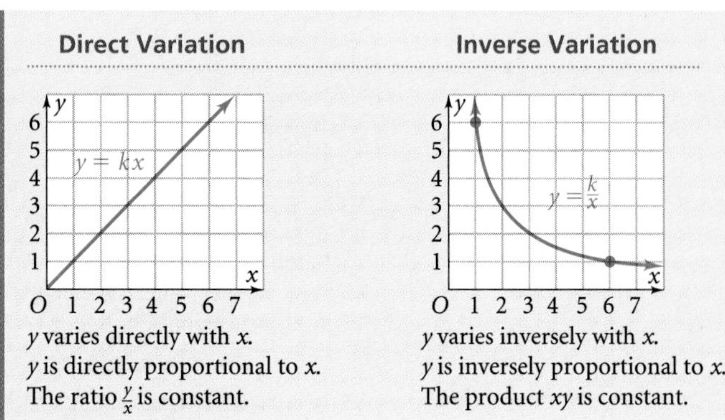

Direct Variation	Inverse Variation
y varies directly with x.	y varies inversely with x.
y is directly proportional to x.	y is inversely proportional to x.
The ratio $\frac{y}{x}$ is constant.	The product xy is constant.

Example 2

Do the data in each table represent a *direct variation* or an *inverse variation?* For each table, write an equation to model the data.

a.

x	2	4	10
y	5	10	25

The values of y seem to vary directly with the values of x. Check the ratio $\frac{y}{x}$.

$\frac{y}{x} = \frac{5}{2} = 2.5$

$\frac{10}{4} = 2.5$

$\frac{25}{10} = 2.5$

The ratio $\frac{y}{x}$ is the same for each pair of data. So, this is a direct variation and $k = 2.5$. The equation is $y = 2.5x$.

b.

x	5	10	25
y	20	10	4

The values of y seem to vary inversely with the values of x. Check the product xy.

$xy = 5(20) = 100$

$10(10) = 100$

$25(4) = 100$

The product xy is the same for each pair of data. So, this is an inverse variation and $k = 100$. The equation is $xy = 100$.

9. Match each situation with the equation that models it. Is the relationship between the data *direct* or *inverse*?
   - **a.** The cost of \$20 worth of gasoline is split among several people. III, inverse
   - **b.** You buy several markers for 20¢ each.
   - **c.** You walk 5 miles each day. Your pace (speed) and time vary from day to day.
   - **d.** Several people buy souvenirs for \$5 apiece.

   I. $y = 5x$
   II. $xy = 5$
   III. $y = \frac{20}{x}$
   IV. $y = 20x$

   b. IV, direct c. II, inverse d. I, direct

ALTERNATIVE ASSESSMENT **Exercises 3–6** To assess understanding of inverse variations, ask students to find the constant of variation for each pair of data points.

STANDARDIZED TEST TIP **Exercise 15** Since there can be only one correct answer for this question, immediately eliminate duplicate expressions. Cross multiplying will show that answers A and B represent the same expression, as do choices C and E. Therefore, answer choice D must be the correct answer.

CARTOON **Exercise 22** Point out that parts (a), (b), and (c) are three separate scenarios.

Encourage students to make up their own humorous laws involving direct variation and inverse variation and illustrate them with cartoons.

Chapter Project FIND OUT BY CALCULATING Students use the table to find string lengths for a C-Major scale. This task is essential to their work on the Chapter Project, introduced in the Chapter Opener. You may wish to check students' understanding of the project by having each student write a sentence or two describing the objectives of this project.

Exercises ON YOUR OWN

1. a. Suppose you want to earn \$80. How long will it take you if you are paid \$5/h; \$8/h; \$10/h; \$20/h?
   b. What are the two variable quantities in part (a)?
   c. Write an equation to represent this situation. $xy = 80$

   1a–b. See margin.

2. Critical Thinking The graphs p and q represent a direct variation and an inverse variation. Write the equation for each graph.
   p: $y = \frac{1}{2}x$; q: $xy = 8$

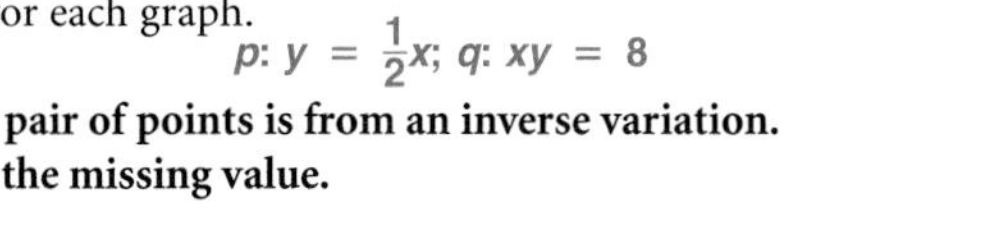

Each pair of points is from an inverse variation. Find the missing value.

3. (6, 12) and (9, y) 8
4. (3, 5) and (1, n) 15
5. (x, 55) and (5, 77) 7
6. (9.4, b) and (6, 4.7) 3
7. (24, 1.6) and (c, 0.4) 96
8. ($\frac{1}{2}$, 24) and (6, y) 2
9. (x, $\frac{1}{2}$) and ($\frac{1}{3}$, $\frac{1}{4}$) $\frac{1}{6}$
10. (500, 25) and (4, n)
11. ($\frac{1}{2}$, 5) and (b, $\frac{1}{8}$) 20
12. (x, 11) and (1, 66) 6
13. (50, 13) and (t, 5) 130
14. (4, 3.6) and (1.2, g)

10. 3125 14. 12

15. Standardized Test Prep Which proportion represents an inverse variation? D

    A. $\frac{x_2}{y_2} = \frac{y_1}{x_1}$ B. $\frac{x_1}{y_2} = \frac{y_1}{x_2}$ C. $\frac{x_2}{y_2} = \frac{x_1}{y_1}$ D. $\frac{x_1}{x_2} = \frac{y_2}{y_1}$ E. $\frac{x_1}{y_1} = \frac{x_2}{y_2}$

Find the constant of variation *k* for each inverse variation.

16. $y = 8$ when $x = 4$ 32
17. $r = 3.3$ when $t = \frac{1}{3}$ 1.1
18. $a = 25$ when $b = 0.04$ 1
19. $x = \frac{1}{2}$ when $y = 5$ 2.5
20. $p = 10.4$ when $q = 1.5$ 15.6
21. $x = 5$ when $y = 75$ 375

22. According to the First Law of Air Travel, will the distance to your gate be *greater* or *less* for this trip than for your last trip?
    a. You have more luggage. greater
    b. You have less time to make your flight. greater
    c. You have less luggage. less

CLOSE TO HOME by John McPherson

The First Law of Air Travel: The distance to your connecting gate is directly proportional to the amount of luggage you are carrying and inversely proportional to the amount of time you have.

23. Travel The time to travel a certain distance is inversely proportional to your speed. Suppose it takes you $2\frac{1}{2}$ h to drive from your house to the lake at a rate of 48 mi/h.
    a. What is the constant of variation? What does it represent?
    b. How long will your return trip take at 40 mi/h? 3 h

    a. 120; distance from your house to the lake

Solve each inverse variation.

24. Surveying Two rectangular building lots are each one-quarter acre in size. One plot measures 99 ft by 110 ft. Find the length of the other plot if its width is 90 ft. 121 ft

25. Construction If 4 people can paint a house working 3 days each, how long will it take a crew of 5 people? 2.4 days

page 510 Work Together

3.

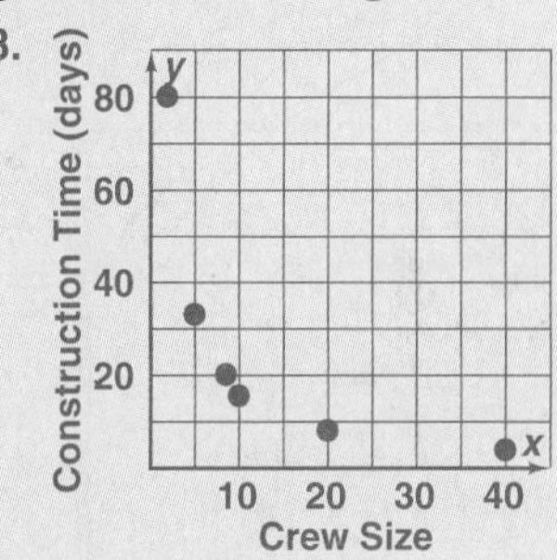

pages 513–514 On Your Own

1a. 16 h; 10 h; 8 h; 4 h

b. hourly wage and time worked

36. See back of book.

Exercises MIXED REVIEW

JOURNAL Ask students to include a sample situation that can be modeled by a direct variation or an inverse variation.

Exercise 38 If students are having difficulty finding common factors, remind them to think about squares.

GETTING READY FOR LESSON 11-2 These exercises prepare students for evaluating rational functions.

Exercises 41–46 Make sure students correctly recall the definition of a reciprocal.

Wrap Up

THE BIG IDEA Ask students to describe how direct variations and inverse variations are different.

RETEACHING ACTIVITY Students solve inverse variations by making a table. Then they write an equation and solve. (Reteaching worksheet 11-1)

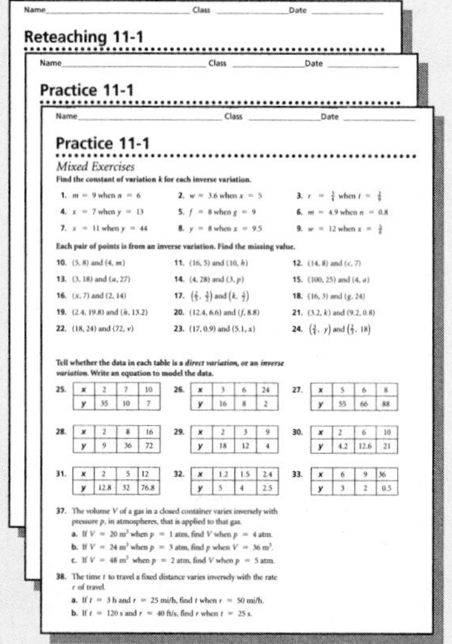

Reteaching 11-1

Practice 11-1

Practice 11-1

Mixed Exercises

Lesson Quiz

Lesson Quiz is also available in Transparencies.

1. The points (5, 1) and (10, y) are from an inverse variation. Find the value of y. 0.5
2. Find the constant of variation k for the inverse variation $a = 2.5$ when $b = 7$. 17.5
3. Write an equation to model the following data and complete the table.

x	y
1	$\frac{1}{3}$
2	$\frac{1}{6}$
■	$\frac{1}{9}$
6	■

$xy = \frac{1}{3}$; $x = 3$; $y = \frac{1}{18}$

Do the data in each table represent a *direct* or an *inverse* variation? Write an equation to model the data. Then complete the table.

26.

x	y
5	6
2	15
10	■ 3

$xy = 30$

27. inverse

x	y
0.4	28
1.2	84
0.9 ■	63

$y = 70x$

28. direct

x	y
10	4
20	■ 8
8	3.2

$y = 0.4x$

29. direct

x	y
1.6	30
4.8	10
0.5 ■	96

$xy = 48$

30. inverse

x	y
3	1
1	3
9	■ $\frac{1}{3}$

$xy = 3$

Does each formula represent a *direct* or an *inverse* variation? Explain.

30. inverse

31. the perimeter of an equilateral triangle: $P = 3s$ Direct; as s increases, P increases.
32. a rectangle with area 24 square units: $lw = 24$ Inverse; as ℓ increases, w decreases.
33. the time t to travel 150 mi at r mi/h: $t = \frac{150}{r}$ Inverse; as r increases, t decreases.
34. the circumference of a circle with radius r: $C = 2\pi r$ Direct; as r increases, C increases.
35. *Writing* Explain how the variable y changes in each situation.
    a. y varies directly with x. The value of x is doubled. The value of y is doubled.
    b. y varies inversely with x. The value of x is doubled. The value of y is halved.
36. *Open-ended* Write and graph a direct variation and an inverse variation that use the same constant of variation. Sample: $y = 3x$; $xy = 3$; see margin for graphs.

Find Out by Calculating

Under equal tension, the frequency of a vibrating string varies inversely with the string length. Violins and guitars use this principle to produce the different pitches of a musical scale. Find the string lengths for a C-Major scale.

C-Major Scale

Pitch	C	D	E	F	G	A	B	C
Frequency (cycles/s)	523	587	659	698	784	880	988	1046
String length (mm)	420	■	■	■	■	■	■	■

Exercises MIXED REVIEW

Factor each polynomial.

37. $x^2 + 10x + 25$ $(x + 5)^2$
38. $4t^2 - 9m^2$ $(2t - 3m)(2t + 3m)$
39. $x^2 - 6x + 9$ $(x - 3)^2$
40. *Education* The number of high school students taking advanced placement exams increased from 177,406 in 1984 to 459,000 in 1994. What percent increase is this? about 159%

FOR YOUR JOURNAL

Describe differences and similarities between direct variation and inverse variation. Include equations and graphs.

Getting Ready for Lesson 11-2

Find the reciprocal of each number.

41. 5 $\frac{1}{5}$
42. -4 $-\frac{1}{4}$
43. $\frac{8}{3}$ $\frac{3}{8}$
44. $3\frac{1}{7}$ $\frac{7}{22}$
45. -1 -1
46. $\frac{3}{4}$ $\frac{4}{3}$

Math ToolboX

In Lesson 11-2, students will explore rational functions. This toolbox uses the graphing calculator to preview rational functions and correct the problem of false connections in the calculator graphs. Exercise 13 introduces the concept of asymptotes by asking students to describe the trend of the y-values as they approach an asymptote.

First, work through the example with the class. Then have students work in pairs or small groups to complete the exercises.

ERROR ALERT! Some students forget to use parentheses when entering binomials. **Remediation:** Point out that the calculator is programmed to perform the correct order of operations. In order to enter $\frac{1}{x+2}$, parentheses must be used so that the calculator does not calculate $\frac{1}{2} + 2$.

WRITING Exercise 14 Suggest that students use TRACE to watch the change in y-values, before answering the questions.

ADDITIONAL PROBLEM Use a graphing calculator to graph the function $y = \frac{x+4}{x-3}$. Then sketch the graph.

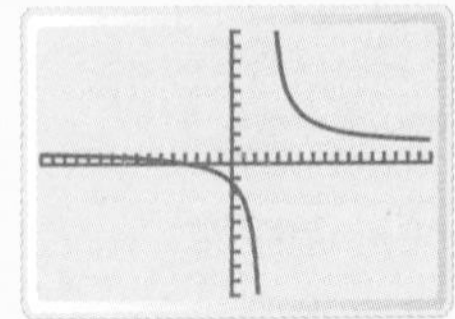

Math ToolboX Technology

Graphing Rational Functions

Before Lesson 11-2

When you use a graphing calculator to graph a rational function, sometimes false connections appear on the screen. When this happens, you need to make adjustments to see the true shape of the graph.

False connection

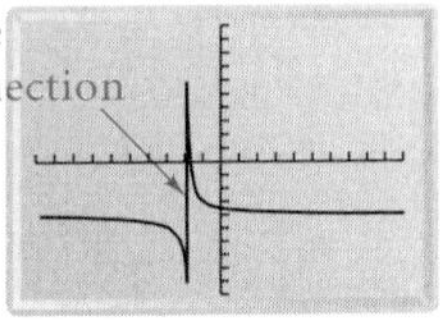

For example, on your graphing calculator the graph of the function $y = \frac{1}{x+2} - 4$ may look like the graph at the right. The highest point and lowest point on the graph are not supposed to connect. If you trace the graph, no point on the graph lies on this connecting line. So, this is a false connection.

Here's how you can graph a rational function and avoid false connections.

First press the MODE key. Then scroll down and right to highlight the word "Dot." Then press ENTER.

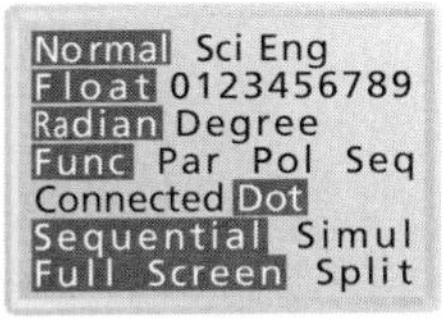

Normal Sci Eng
Float 0123456789
Radian Degree
Func Par Pol Seq
Connected Dot
Sequential Simul
Full Screen Split

Graph again. Now the false connection is gone!

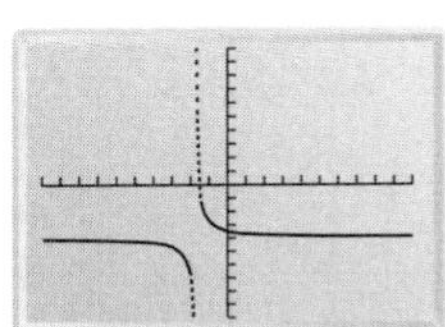

Use the TRACE key or TABLE key to find points on the graph. Sketch the graph.

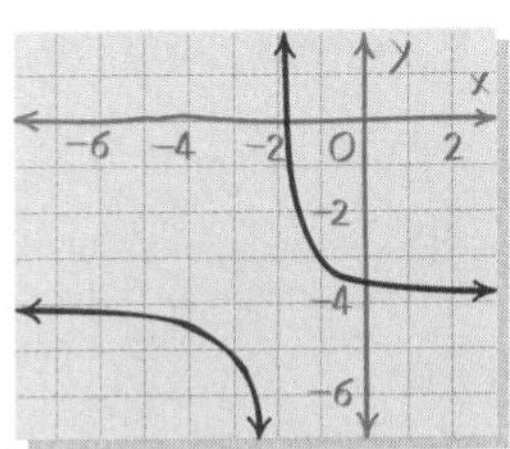

Graphing Calculator **Use a graphing calculator to graph each function. Then sketch the graph. (*Hint*: Use parentheses to enter binomials.)** 1-12. See margin.

1. $y = \frac{1}{x-3}$
2. $y = \frac{1}{x} + 3$
3. $y = \frac{5}{3x+2}$
4. $y = \frac{4}{3x+6}$
5. $y = \frac{x+3}{x-2}$
6. $y = \frac{1}{x-2} + 1$
7. $y = \frac{6}{x^2 - x - 6}$
8. $y = \frac{1}{x-4} + 2$
9. $y = \frac{8}{x}$
10. $y = \frac{4}{x+1} - 3$
11. $y = \frac{3x}{x+3}$
12. $y = \frac{2x+1}{x-4}$

13. a. Graphing Calculator Graph $y = \frac{1}{x}$, $y = \frac{1}{x-3}$, and $y = \frac{1}{x+4}$. What do you notice? See margin for graphs. The shapes of the graphs are the same, but they are in different positions along the x-axis.
    b. Graphing Calculator Graph $y = \frac{1}{x}$, $y = \frac{1}{x} - 3$, and $y = \frac{1}{x} + 4$. What do you notice? The shapes of the graphs are the same, but they are in different positions along the y-axis.
14. Writing Graph $y = \frac{1}{x}$. See margin.
    a. Examine both negative and positive values of x. Describe what happens to the y-values when x is near zero. See right.
    b. Describe what happens to the y-values as $|x|$ increases. y-values get closer to 0.

14a. As negative values of x get near 0, y decreases. As positive values of x get near 0, y increases.

Materials and Manipulatives

- Graphing calculator

Resources

Transparencies
1, 11

page 515 Math Toolbox

1.
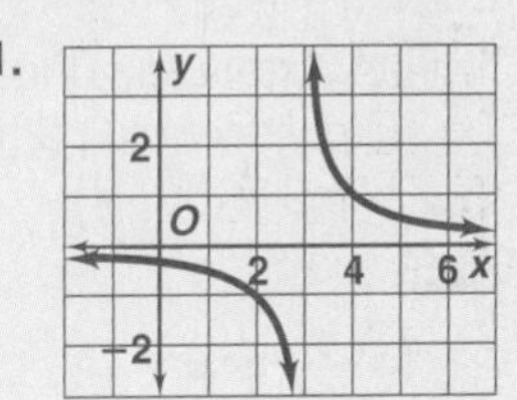

2.
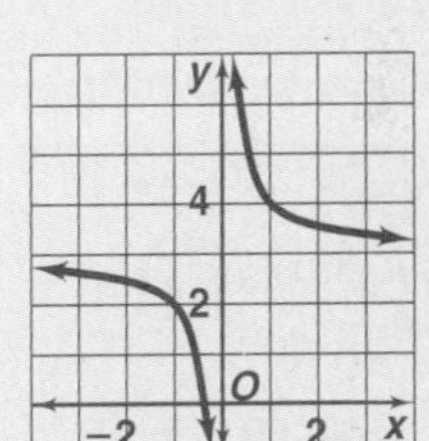

3.
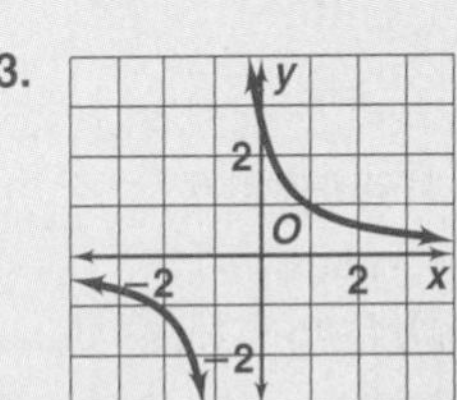

4–13. See back of book.

515

11-2 Teaching Notes

PROBLEM OF THE DAY

Four families plan a camping trip together and decide to evenly split the cost of food and drinks. The Seay family purchases the beverages for $41. The Zente family spends $107 on the food. The Smeltzer and Flores families bring the camping gear. How should the food and beverage costs be settled? **The Smeltzer and Flores families each owe $37. The Zentes should receive $4. The Seays should receive $70.**

Problem of the Day is also available in Transparencies.

CONNECTING TO PRIOR KNOWLEDGE Have students graph a direct variation and an indirect variation. Ask: *How are these graphs similar? How are they different?*

THINK AND DISCUSS

DIVERSITY Some students may not be familiar with automatic cameras. Ask a student who knows about these cameras to describe to the class how they work.

Question 1 Have a student define *polynomials*. Point out that 1 is a polynomial; therefore, $\frac{1}{x-4}$ meets the definition of a rational function.

Have some students take turns going to the board and writing any rational function they wish. After students have written

Lesson Planning Options

Prerequisite Skills

- Graphing functions
- Understanding functions

Assignment Options for Exercises On Your Own

To provide flexible scheduling, this lesson can be subdivided into parts.

1 **Core** 1–8
Extension 9–12, 26

2 **Core** 13–24, 29–44
Extension 25, 27, 28

Use Mixed Review to maintain skills.

Resources

 **Student Edition**

Skills Handbook, pp. 575–579
Extra Practice, p. 566
Glossary/Study Guide

 **Teaching Resources**

Chapter Support File, Ch. 11
- Practice 11-2 (two worksheets)
- Reteaching 11-2

Classroom Manager 11-2
Glossary, Spanish Resources
Two-Year Algebra Handbook 11-2

 **Transparencies**

1, 2, 11, 125, 130, 131

516

11-2 Rational Functions

What You'll Learn

- Evaluating rational functions
- Graphing rational functions

...And Why

To solve problems using rational functions, such as those involving light intensity

What You'll Need

- graph paper
- graphing calculator
- index cards

QUICK REVIEW

Two numbers are *reciprocals* of each other if their product is 1.

THINK AND DISCUSS

Part 1

Exploring Rational Functions

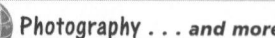

Photography Automatic cameras calculate shutter speed based on the amount of available light. The relationship between shutter speed and the amount of light can be modeled with a rational function. A **rational function** is a function that can be written in the form $f(x) = \frac{\text{polynomial}}{\text{polynomial}}$.

1. Evaluate each rational function for $x = 3$.
   a. $f(x) = \frac{1}{x-4}$ **−1** b. $y = \frac{2}{x^2}$ $\frac{2}{9}$ c. $g(x) = \frac{x^2 - 3x + 2}{x + 2}$ $\frac{2}{5}$

The function $y = \frac{1}{x}$ is an example of a rational function. You can use the graph of $y = \frac{1}{x}$ to show the relationship between reciprocals.

Table

Number (x)	-1	$-\frac{1}{3}$	1	1.5	3
Reciprocal (y)	-1	-3	1	$\frac{2}{3}$	$\frac{1}{3}$

Graph

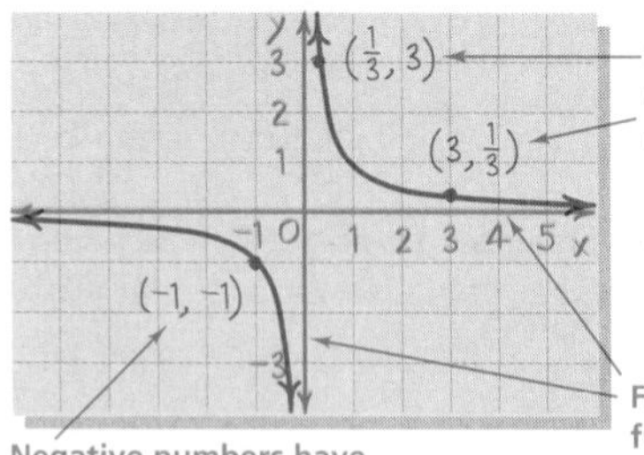

y
x
$(\frac{1}{3}, 3)$
$(3, \frac{1}{3})$
$(-1, -1)$
Since 3 and $\frac{1}{3}$ are reciprocals of each other, both $(3, \frac{1}{3})$ and $(\frac{1}{3}, 3)$ are on the graph.
For $x = 0$ and $y = 0$ the function is undefined. So the graph never intersects either axis.
Negative numbers have negative reciprocals.

2. Does each point lie on the graph of $y = \frac{1}{x}$? Why or why not?
   a. $(-100, -0.01)$ b. $(-5, 0.2)$ c. $(1{,}000{,}000, 0)$ d. $(0.04, 25)$
   a, d. Yes; the two numbers are reciprocals.
   b, c. No; the two numbers are not reciprocals.

A line is an **asymptote** of a graph if the graph of the function gets closer and closer to the line, but does not cross it.

3. The y-axis is a vertical asymptote of the function $y = \frac{1}{x}$. Is there a horizontal asymptote of the graph? Explain.
   The x-axis; the graph approaches the x-axis but does not intersect it.

their functions, make sure the functions are all in the form $\frac{\text{polynomial}}{\text{polynomial}}$.

AUDITORY LEARNING Ask students to describe reciprocals in their own words. Accept all reasonable answers, such as *a number turned upside-down.*

Write $y = \frac{1}{x}$. Ask students:

- *When $x = 0$, what is the value of y?* Answers may vary. Sample: y has no value because dividing by zero is undefined.
- *Can y ever have a value of zero?* no
- *What is the smallest positive value that x can have?* Answers may vary. Sample: 0.000001

Use answers to lead students to the concept of asymptotes.

KINESTHETIC LEARNING Have students stand at the wall with one shoe directly in front of the other to form a line parallel to the wall. Challenge students to stand with their shoes actually touching the wall. They will not be able to stand in this way without support. Point out that the wall models an asymptote. Students can stand away from or close to the wall. However, they cannot stand actually touching the wall without falling over.

Example 1

ERROR ALERT! Students may think that the asymptote moves four places in some direction because of the 4 in the numerator. **Remediation:** Point out that they need only look at the denominator when finding the asymptotes. Have them compare the graphs of $y = \frac{1}{x + 3}$ and $y = \frac{4}{x + 3}$ to see that the 4 has no effect on the asymptotes.

Part 2

Graphing Rational Functions

When you evaluate a rational function, some values of x may lead to division by zero. For the function $y = \frac{1}{x - 3}$, the denominator is zero for $x = 3$. So, the function is undefined when $x = 3$, and the vertical line $x = 3$ is an asymptote of the graph of $y = \frac{1}{x - 3}$.

Example 1

Graph $y = \frac{4}{x + 3}$.

Step 1 Find the vertical asymptote.

$$x + 3 = 0$$
$$x = -3 \leftarrow \text{vertical asymptote}$$

Step 2 Make a table using values of x near -3.

Step 3 Draw the graph.

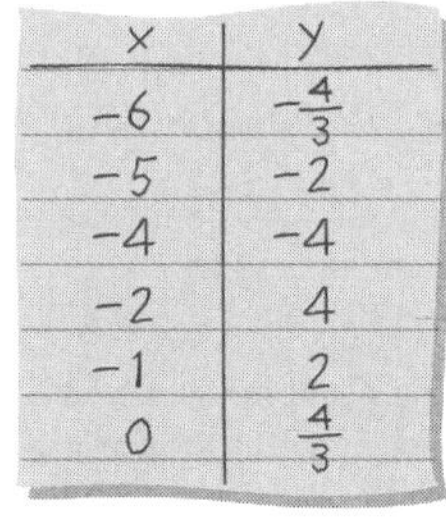

x	y
−6	$-\frac{4}{3}$
−5	−2
−4	−4
−2	4
−1	2
0	$\frac{4}{3}$

Use a dashed line for the asymptote $x = -3$.

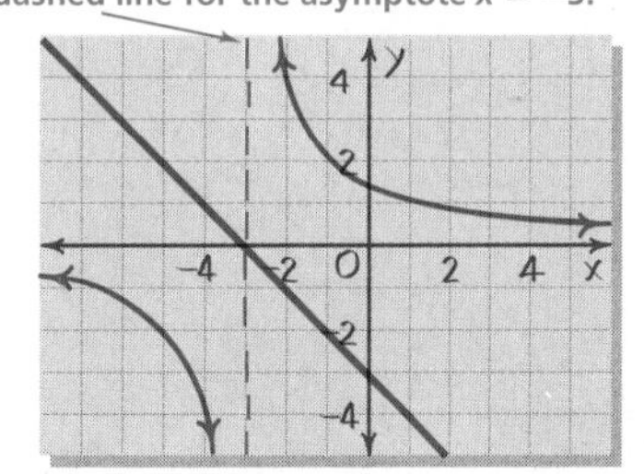

4a. See graph.

4. a. Copy the graph above and add the graph of the line $y = -x - 3$.
   b. Critical Thinking Fold the graph along the line you drew. What is true about the two parts of the graph of $y = \frac{4}{x + 3}$? See margin.

5. a. Evaluate $y = \frac{4}{x + 3}$ for $x = 1000$ and $x = -1000$. $\frac{4}{1003}$; $-\frac{4}{997}$
   b. Is there a horizontal asymptote for the graph? Explain. Yes; the graph approaches the x-axis for large abs. values of x.

PROBLEM SOLVING

Look Back How does the graph of a rational function compare with the graph of an exponential function? See margin.

The graphs of many rational functions are related to each other.

$y = \frac{1}{x}$

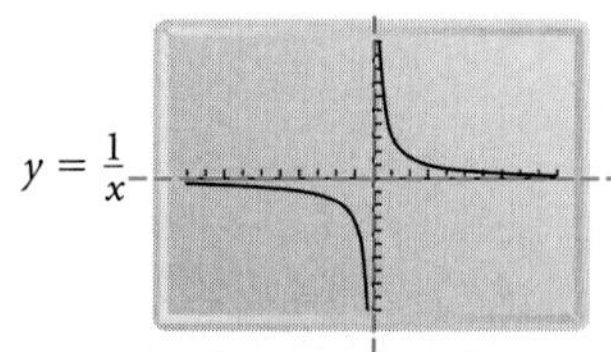

vertical asymptote when $x = 0$
horizontal asymptote along the x-axis

$y = \frac{1}{x - 2}$

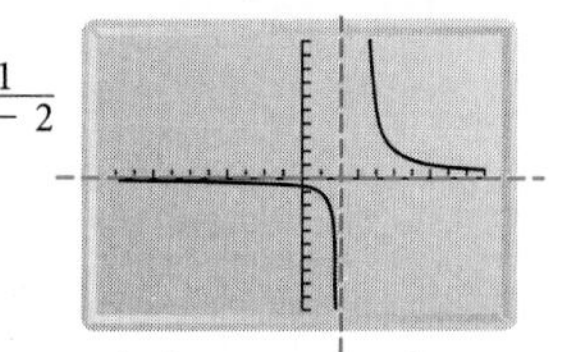

vertical asymptote when $x = 2$
horizontal asymptote along the x-axis

The graphs are identical in shape, but the second graph is shifted two units to the right.

Additional Examples

FOR EXAMPLE 1

Graph $y = \frac{4}{x - 3}$.

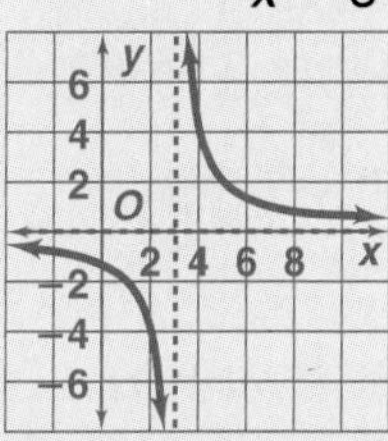

Discussion: *How is the graph of $y = \frac{4}{x - 3}$ similar to the graph of $y = \frac{4}{x + 3}$? Can you account for these similarities? Explain.*

FOR EXAMPLE 2

Graph the function $y = \frac{50{,}000}{x^2}$ and use your graph to find the light intensity 5 ft from a 50,000 lumen light source.

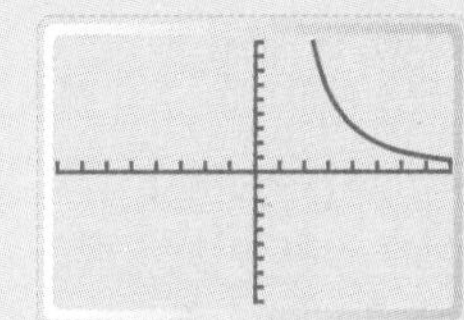

2,000

CRITICAL THINKING Question 4b Students may interpret this as a question about symmetry. They may assume that the graph is symmetrical along the line $y = x + 3$, and thus draw the line without reading the equation in Question 4a. In this case, either line of symmetry is correct.

Ask students how they can use symmetry to help them draw graphs.

CRITICAL THINKING Question 8 Ask students to make generalizations about $y = \frac{3}{x} + k$. Ask students what changes will occur in the graph if the values of k are negative.

Example 2 **Relating to the Real World** 

ESL Explain that the *shutter* is the part of the camera that exposes the film by opening and covering a small hole.

EXTENSION Light meters (also called actinometers) are instruments used primarily in photography to measure the intensity of light. Challenge interested students to find out about different kinds of light meters and how the meters detect light. Invite students to share their findings with the class.

MAKING CONNECTIONS Have students contact a professional photographer. Suggest that they ask the photographer questions such as the following: *How would you light a skater to eliminate glare in the photograph? What is an f-stop? What determines the f-stop setting?* Using this information, lead students to recognize that they can model f-stops with inverse variations.

Technology Options

For Exercises 13–24 and 33–44, students may find it helpful to graph the functions with a graphing calculator or with graphing software.

Prentice Hall Technology 

Calculator
- Graphing Calculator Handbook: Procedure 9

Software
- Secondary Math Lab Toolkit™
- Computer Item Generator 11-2

CD-ROM
- Multimedia Algebra Lab 11

Internet
- See the Prentice Hall site. (http://www.phschool.com)

6. Where is the vertical asymptote of the graph of each function?
   a. $y = \frac{6}{x}$ $x = 0$
   b. $y = \frac{6}{x - 3}$ $x = 3$
   c. $y = \frac{6}{x + 1}$ $x = -1$
7. Try This Graph each function in Question 6. See margin.
8. Critical Thinking Graph the functions $y = \frac{3}{x}$ and $y = \frac{3}{x} + 2$. Are the graphs identical in shape? What shift occurs? See margin for graphs. The second graph is identical to the first but shifted up 2 units.

You can use rational functions to describe relationships in the real world.

Example 2 **Relating to the Real World** 

Photography The output from the photographer's lighting system is 72,000 lumens. To get a good photo, the intensity of light at the circus performers must be at least 600 lumens. The light intensity y is related to their distance x in feet from the light by the function $y = \frac{72,000}{x^2}$. How far can the circus performers be from the light?

Relate light intensity at performers is at least 600 lumens

Write $\frac{72,000}{x^2} \geq 600$

Use a graphing calculator. Enter $y = \frac{72,000}{x^2}$ and $y = 600$. Find the point of intersection.

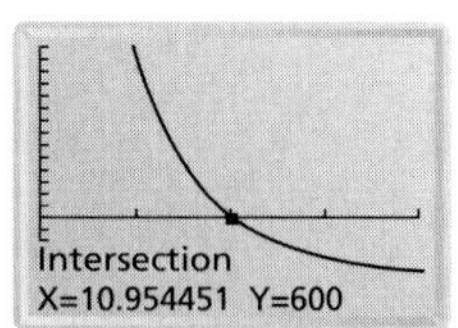

Xmin=0 Ymin=0
Xmax=20 Ymax=2000
Xscl=5 Yscl=100

The curved graph shows that the light intensity y decreases as the distance x increases.

The light intensity is about 600 lumens when $x \approx 11$. The circus performers should be within about 11 ft of the light.

GRAPHING CALCULATOR HINT
Use a viewing window for Quadrant I, since distance and lumens are positive.

9. a. Use the function in Example 2 to find the light intensity at each distance in the table. See right.
   b. Describe how the light intensity changes when the distance doubles. Light intensity becomes $\frac{1}{4}$ as great.

Distance (x)	Intensity (y)
3 ft	■
6 ft	■
12 ft	■

8000 lumens
2000 lumens
500 lumens

CRITICAL THINKING Question 10b Remind students to consider the negative values of *x* and *y*.

WORK TOGETHER

TACTILE LEARNING If you have block scheduling or extended class periods, consider grouping the class into six teams. Assign each team a function family and have them make an informational poster for the class.

EXTENSION Question 11 As well as the information listed in this question, have the appropriate team include the following information on their posters.

- Linear function: how the values of *m* and *b* change the graph of the function
- Absolute value function: how the values of *b* change the graph of the function
- Quadratic function: how the values of *a, b,* and *c* change the graph of the function
- Exponential function: how the values of *a* and *b* change the graph of the function
- Radical function: how the values of *b* and *c* change the graph of the function
- Rational function: how the values of *k, b,* and *c* change the graph of the function

AUDITORY LEARNING After teams have completed their posters, ask each team to describe their function family to the class.

10. *Graphing Calculator* Graph $y = \frac{1}{x}$ and $y = \frac{1}{x^2}$.

a. What is the vertical asymptote of the graph of each function? $x = 0$

b. *Critical Thinking* What is the range of $y = \frac{1}{x}$? of $y = \frac{1}{x^2}$?
all real $y \neq 0$; pos. real nos.

WORK TOGETHER

You have studied six families of functions this year. Their properties and graphs are shown in this summary.

Families of Functions

Linear function
$y = mx + b$

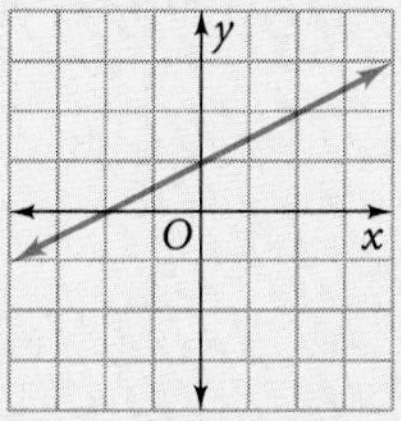

slope = m
y-intercept = b

Absolute value function
$y = |x - b|$

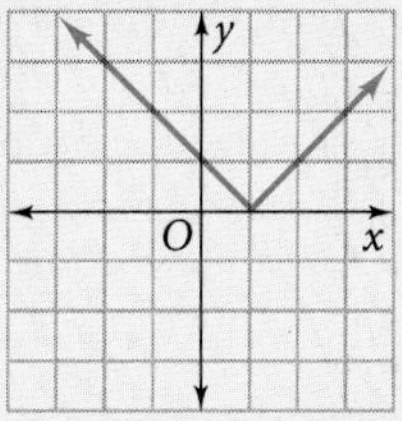

V-shape with vertex at $(b, 0)$

Quadratic function
$y = ax^2 + bx + c$

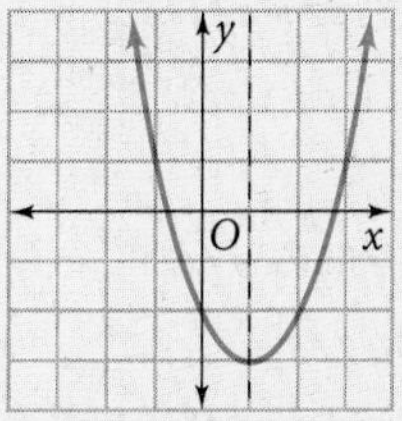

parabola with axis of symmetry at $x = -\frac{b}{2a}$

Exponential function
$y = ab^x$

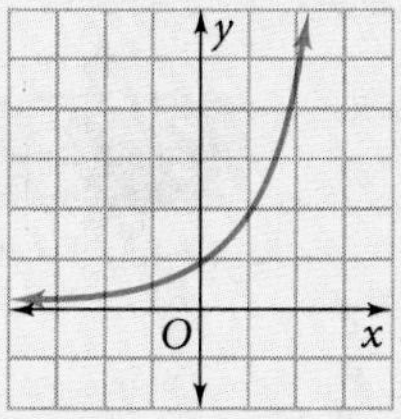

growth for $b > 1$
decay for $0 < b < 1$

Radical function
$y = \sqrt{x - b} + c$

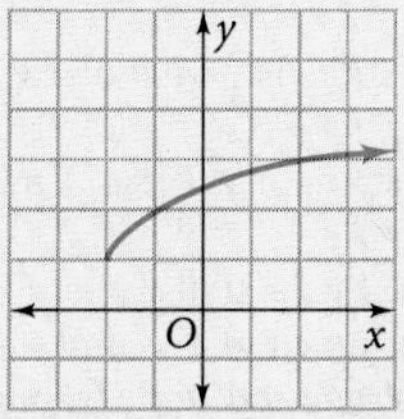

shift horizontally b units
shift vertically c units

Rational function
$y = \frac{k}{x - b} + c$

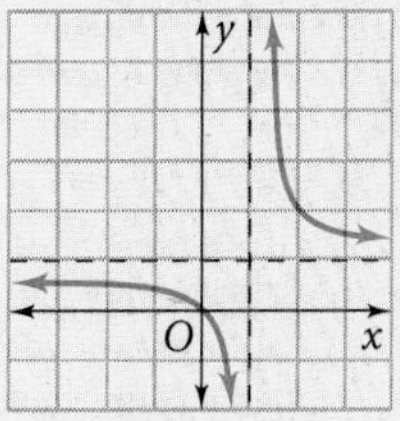

vertical asymptote at $x = b$

11. Work with a partner to prepare a note card for each family of functions. Include this information: Check students' work.

- one or more examples for that family
- a graph for each example
- notes about each function you have graphed (for instance, how to find the slope, the axis of symmetry, or an asymptote)

12. Make duplicate cards so that you and your partner each have a full set.
Check students' work.

pages 516–519 Think and Discuss

Problem Solving The graph of a rational function has at least two pieces. Each end of each piece either approaches a horizontal asymptote or becomes infinite. The graph of an exponential function always has only one piece. One end always approaches 0. The other end always becomes infinite.

4a.

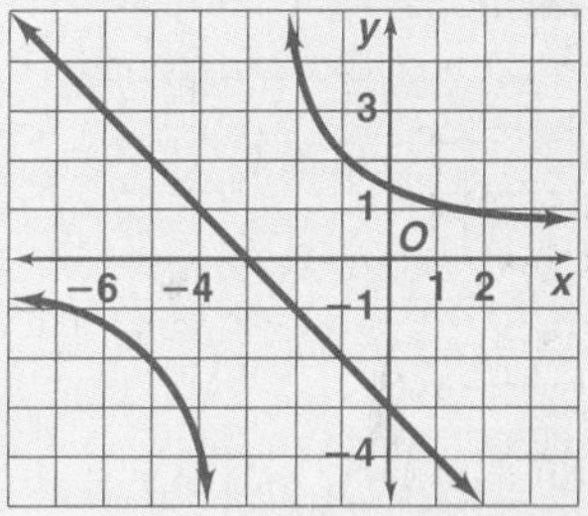

4b. They match exactly.

6a.

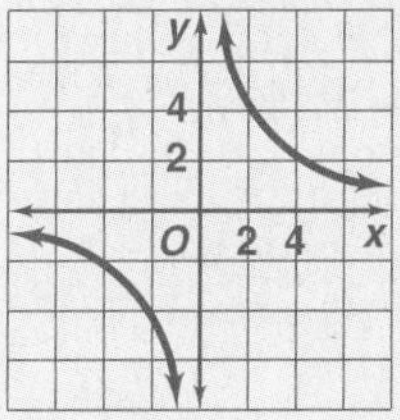

Exercises ON YOUR OWN

ALTERNATIVE ASSESSMENT **Exercises 9–12** Ask students to write rational functions which represent each of these graphs. This will help you assess students' understanding of rational functions.

Exercises 9–12 Ask students to describe techniques they use for identifying asymptotes.

OPEN-ENDED **Exercise 26** Ask students to think carefully about whether the asymptotes of the second graph also shift three units or whether the asymptotes for both graphs are the same.

WRITING **Exercise 27** Suggest that students first describe how to graph $y = \frac{1}{x}$. Then have them explain how the 5, 1, and 2 affect the graph.

Exercise 28 Suggest that students turn on a small table lamp in a dark room and notice how the light intensity diminishes as the distance from the lamp increases.

Exercise 29–32 For students using computer graphing software, point out that some software programs do not display asymptotes. Students will need to study the shape of graphs to learn where the asymptotes are.

6b. 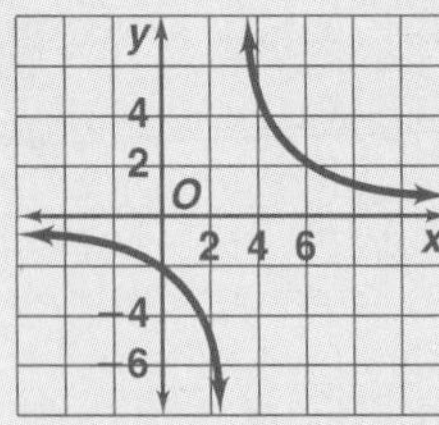

6c. 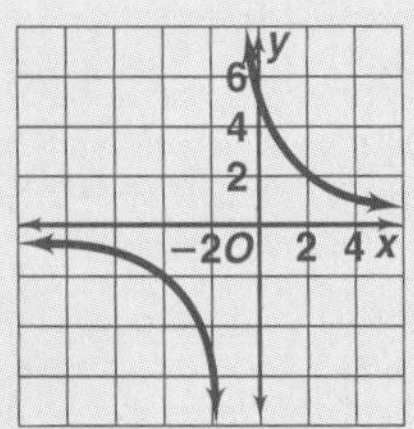

8. See back of book.

pages 520–521 On Your Own

13–24, 25a, 27, 29–44. See back of book.

page 521 Checkpoint

4–7. See back of book.

Exercises ON YOUR OWN

Evaluate each function for $x = -1$, $x = 2$, and $x = 4$.

1. $y = \frac{x-2}{x}$ 3; 0; $\frac{1}{2}$
2. $f(x) = \frac{3}{x-1}$ $-1\frac{1}{2}$; 3; 1
3. $y = \frac{2x}{x-3}$ $\frac{1}{2}$; −4; 8
4. $g(x) = \frac{12}{x^2}$ 12; 3; $\frac{3}{4}$

What value of x makes the denominator of each function equal zero?

5. $f(x) = \frac{3}{x}$ 0
6. $y = \frac{1}{x-2}$ 2
7. $y = \frac{x}{x+2}$ −2
8. $h(x) = \frac{3}{2x-4}$ 2

Describe the asymptotes in each graph.

9. 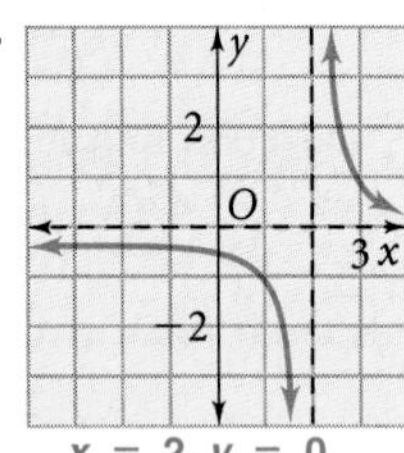

$x = 2$, $y = 0$

10. 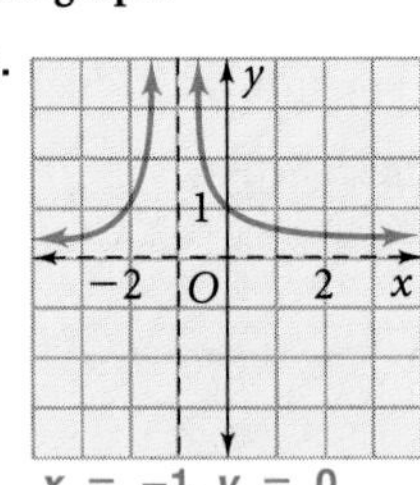

$x = -1$, $y = 0$

11. 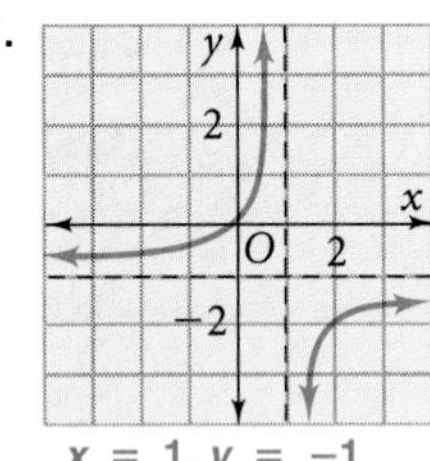

$x = 1$, $y = -1$

12. 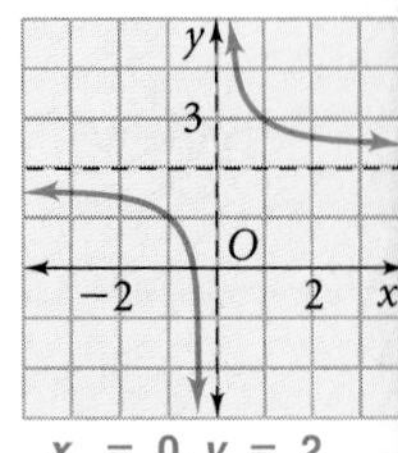

$x = 0$, $y = 2$

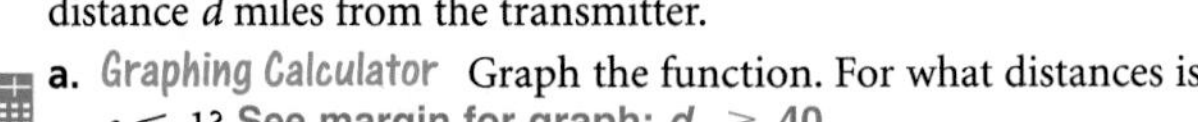

Graph each function. Include a dashed line for each asymptote. 13–24. See margin.

13. $y = \frac{4}{x}$
14. $f(x) = \frac{4}{x-5}$
15. $y = \frac{4}{x+4}$
16. $g(x) = \frac{4}{x} + 2$
17. $g(x) = \frac{12}{x}$
18. $y = \frac{12}{x+2}$
19. $f(x) = \frac{12}{x} - 3$
20. $h(x) = \frac{12}{x+2} - 3$
21. $f(x) = \frac{-1}{x}$
22. $y = \frac{-4}{x}$
23. $g(x) = \frac{4}{x-2} + 1$
24. $y = \frac{x+2}{x-2}$

25. *Physics* As radio signals move away from a transmitter, they become weaker. The function $s = \frac{1600}{d^2}$ relates the strength s of a signal at a distance d miles from the transmitter.
    a. *Graphing Calculator* Graph the function. For what distances is $s \leq 1$? See margin for graph; $d \geq 40$.
    b. Use the function to find the signal strength at 10 mi, 1 mi, and 0.1 mi. 16; 1600; 160,000
    c. *Critical Thinking* Suppose you drive by the transmitter for one radio station while your car radio is tuned to a second station. The signal from the first station can interfere and come through your radio. Use your results from part (b) to explain why. Answers may vary. Sample: The signal from the first station is very strong close to its transmitter.

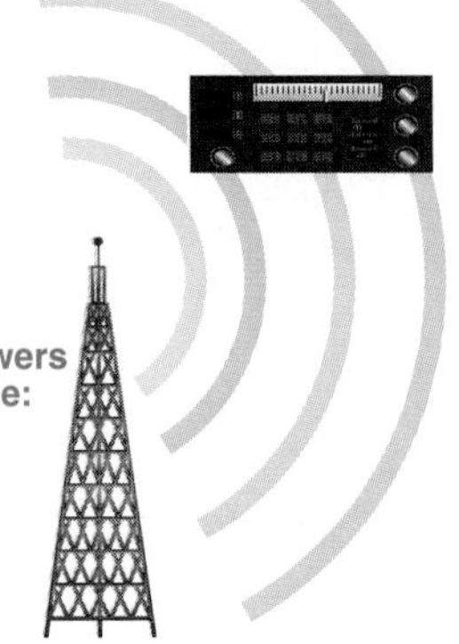

26. *Open-ended* Write two rational functions whose graphs are identical except that one has been shifted vertically 3 units from the other. See below.
27. *Writing* Suppose a friend missed class. How would you explain how to graph $y = \frac{5}{x-1} + 2$? See margin.
28. In the formula $I = \frac{445}{x^2}$, I is the intensity of light at a distance x feet from the light bulb. What the intensity of light 5 ft from the light bulb? 15 ft from the light bulb? 17.8; about 1.98

26. Sample: $y = \frac{6}{x}$, $y = \frac{6}{x} - 3$

Exercises MIXED REVIEW

Exercise 49 For students having difficulty with this problem, suggest that they review problems involving percentages larger than 100%.

GETTING READY FOR LESSON 11-3 These exercises prepare students for simplifying rational expressions.

Wrap Up

THE BIG IDEA Ask students: *What is an asymptote? What does an asymptote tell you about a function?*

RETEACHING ACTIVITY Students find x- and y-intercepts and vertical asymptotes to graph rational functions. (Reteaching worksheet 11-2)

Exercises CHECKPOINT

Students will assess their own progress on Lessons 11-1 to 11-2.

ESTIMATION Exercise 9 Suggest that students first draw a sketch of the lever and make an estimate before calculating the answer.

Graphing Calculator **Graph each function on a graphing calculator. Then sketch the graph. Include a dashed line for each asymptote.** 29–32. See margin.

29. $y = \frac{x+4}{x}$ 30. $y = \frac{x+1}{x+3}$ 31. $y = \frac{4}{x^2 - 1}$ 32. $y = \frac{6}{x^2 - x - 2}$

Graph each function. 33–44. See margin.

33. $y = x^2 + 3$ 34. $y = \sqrt{x+3}$ 35. $y = x + 3$ 36. $y = |x - 3|$

37. $y = 3x$ 38. $y = 3^x$ 39. $y = \frac{3}{x}$ 40. $y = \frac{1}{x+3} + 3$

41. $y = x^2 + 3x + 2$ 42. $y = \left(\frac{1}{3}\right)^x$ 43. $y = \frac{3}{x-3}$ 44. $y = \frac{3}{x} + 3$

Exercises MIXED REVIEW

Factor.

45. $g^2 - 12g + 35$ $(g - 7)(g - 5)$ 46. $9h^2 + 24h + 16$ $(3h + 4)^2$ 47. $a^2 + 6a - 7$ $(a + 7)(a - 1)$ 48. $25x^2 - 4$ $(5x + 2)(5x - 2)$

49. **Delivery Services** During the blizzard of 1996, a pizza delivery driver in Alexandria, Virginia, got tips that were 1000% of the usual \$1 for each delivery. How much did the delivery driver receive in tips for a delivery during the blizzard? \$10

Getting Ready for Lesson 11-3

Find the value(s) of x that makes each expression equal zero.

50. $10 - x$ 10 51. $x^2 - 3x$ 0, 3 52. $x + 4$ -4 53. $x^2 - 16$ 4, -4

Exercises CHECKPOINT

Each pair of points is from an inverse variation. Find the missing value.

1. $(9, 2)$ and $(x, 6)$ 3 2. $(8.2, 3)$ and $(12.3, y)$ 2 3. $(0.5, 7.2)$ and $(0.9, y)$ 4

Graph each function. 4–7. See margin.

4. $y = \frac{5}{x}$ 5. $y = \frac{8}{x-4}$ 6. $y = \frac{6}{x^2}$ 7. $y = \frac{1}{x} + 4$

8. **Writing** Describe the similarities and differences between the graphs of $y = \frac{3}{x}$ and $y = \frac{-3}{x}$. Answers may vary. Sample: The graphs have the same shape and the same asymptotes, but they are graphed in different quadrants.

9. **Physics** What weight placed on a lever 9 ft from the fulcrum will balance 126 lb placed 6.5 ft from the fulcrum? 91 lb

10. The graphs of a direct variation and an inverse variation both pass through the point (3, 5). Write an equation for each graph. $y = \frac{5}{3}x$; $y = 15x$

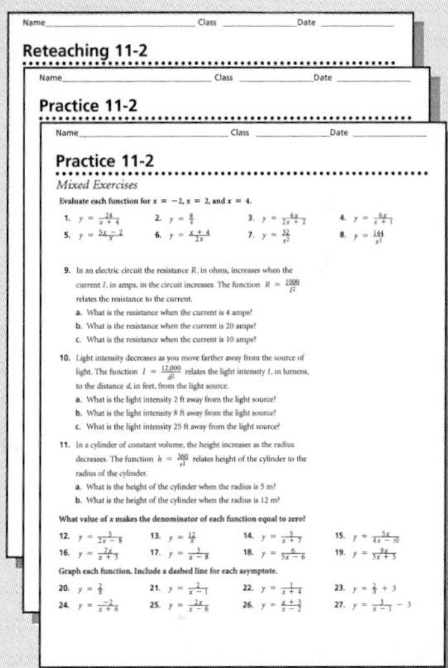
Reteaching 11-2

Practice 11-2

Practice 11-2

Mixed Exercises

Lesson Quiz

Lesson Quiz is also available in Transparencies.

Graph each function. Include a dashed line for each asymptote.

1. $y = \frac{3}{x-3}$

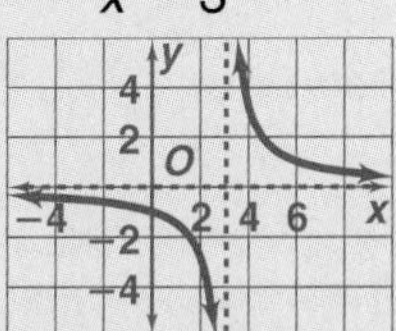

2. $y = \frac{5}{x} - 1$

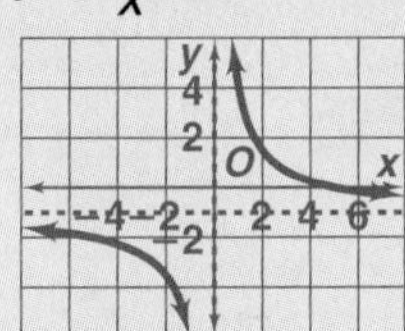

11-3 Teaching Notes

PROBLEM OF THE DAY

A florist buys carnations for \$.29 each and sells them in bouquets of eight for \$5.00. The cost of the wrapping is \$.76. How much does the florist profit per flower? \$.24

Problem of the Day is also available in Transparencies.

CONNECTING TO PRIOR KNOWLEDGE Write $\frac{1}{2} + \frac{1}{3}$ on the board. Ask students to recall what they know about adding, subtracting, multiplying, and dividing fractions. Then ask them to explain how the methods they use might change if they replaced the numbers in the denominator with algebraic expressions such as $\frac{1}{2 + x} + \frac{1}{3 + x}$.

WORK TOGETHER

AUDITORY LEARNING Group students in pairs. Have one person explain the steps while the partner records the steps. Have partners take turns explaining and writing.

THINK AND DISCUSS

Explain that for an expression to be a rational expression, it must have a variable in the denominator. Rational expressions may or may not have variables in the numerator.

VISUAL LEARNING To provide students with practice in recognizing rational expressions, write several expressions on the board in the form $\frac{\text{polynomial}}{\text{polynomial}}$. Then write several expressions that are not rational. Have students identify the rational expressions.

Lesson Planning Options

Prerequisite Skills

- Simplifying expressions
- Multiplying and dividing integers

Assignment Options for Exercises On Your Own

To provide flexible scheduling, this lesson can be subdivided into parts.

1 **Core** 1–16
Extension 17–19

2 **Core** 20–33, 34, 35, 36
Extension 37–39

Use Mixed Review to maintain skills.

Resources

Student Edition

Skills Handbook, pp. 575–579
Extra Practice, p. 566
Glossary/Study Guide

Teaching Resources

Chapter Support File, Ch. 11
- Practice 11-3 (two worksheets)
- Reteaching 11-3

Classroom Manager 11-3
Glossary, Spanish Resources
Two-Year Algebra Handbook 11-3

Transparencies
25, 126

11-3 Rational Expressions

What You'll Learn

- Simplifying rational expressions
- Multiplying and dividing rational expressions

...And Why

To solve real-world problems that involve ratios of unknown quantities

WORK TOGETHER

Work in groups to review how to simplify, multiply, and divide rational numbers.

1. a. Simplify each expression.
$\frac{8}{2}$ 4 $\qquad -\frac{15}{24}$ $-\frac{5}{8}$ $\qquad \frac{25}{35}$ $\frac{5}{7}$
b. Consider the steps you took to simplify each expression in part (a). Write the steps you use to simplify rational numbers.
Answers may vary. See left for sample.
2. a. Express each product in simplest form.
$-\frac{8}{21} \cdot \frac{7}{4}$ $-\frac{2}{3}$ $\qquad \frac{3}{5} \cdot \frac{2}{7}$ $\frac{6}{35}$ $\qquad -\frac{3}{4} \cdot (-2)$ $\frac{3}{2}$
b. Consider the steps you took to find each product in part (a). Write the steps you use to multiply rational numbers.
Answers may vary. See left for sample.
3. a. Express each quotient in simplest form.
$6 \div \frac{3}{8}$ 16 $\qquad \frac{2}{3} \div \left(-\frac{4}{5}\right)$ $-\frac{5}{6}$ $\qquad \frac{3}{4} \div 2$ $\frac{3}{8}$
b. Consider the steps you took to find each quotient in part (a). Write the steps you use to divide rational numbers.
Answers may vary. See left for sample.

QUICK REVIEW

Rational numbers are numbers that can be represented as the ratio of two integers.

1b. Sample: Find common factors of the numerator and denominator. Remove common factors.

2b. Sample: Multiply numerators. Multiply denominators. Simplify the fraction.

3b. Sample: Multiply by the reciprocal of the divisor. Simplify the fraction.

THINK AND DISCUSS

Part 1

Simplifying Rational Expressions

A **rational expression** is an expression that can be written in the form $\frac{\text{polynomial}}{\text{polynomial}}$, where a variable is in the denominator.

4. Evaluate each rational expression for $x = 2$. What do you notice?
a. $\frac{1}{x - 2}$ undefined $\qquad$ b. $\frac{3}{x^2 - 4}$ undefined $\qquad$ c. $\frac{2}{x^2 + x - 6}$ undefined

The **domain** of a rational expression is all real numbers excluding the values for which the denominator is zero. The values that are excluded are restricted from the domain. For the expression $\frac{1}{x - 2}$, 2 is restricted from the domain.

5. What other values are restricted from the domain in parts (b) and (c) of Question 4? (*Hint:* Solve an equation to find the values for which the denominator equals zero.) −2; −3

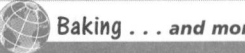

Example 1

ERROR ALERT! Students may think that they can find the domain of 6 instead of the domain of $\frac{6x+12}{x+2}$. **Remediation:** Make sure students understand that when they are asked to find the domain of an expression, they should find the domain before simplifying the expression.

EXTENSION Challenge students to think about how the graph of $y = \frac{6x+12}{x+2}$ differs from the graph of $y = 6$. Ask them to explain how they would show that x is excluded from the domain. Invite interested students to investigate to find out how mathematicians exclude points from a graph.

For students using computer graphing software, point out that the software may not exclude the values that are restricted from the domain. Some graphing programs may display the graph of $y = 6$ when they are asked to display the graph of $y = \frac{6x+12}{x+2}$. Remind students that they must be aware of the limitations of such graphing software.

Question 6c Remind students that there are two roots of x^2. Therefore, students can expect two values to be excluded from the domain of $\frac{x+3}{x^2-9}$.

Example 2 Relating to the Real World 

CONNECTING TO THE STUDENTS' WORLD Challenge students to think of a rule they might use when cooking. Have them

A rational expression is in *simplest form* if the numerator and denominator have no common factors except 1.

Example 1

Simplify $\frac{6x+12}{x+2}$ and state any values restricted from the domain.

$x + 2 = 0, x = -2$ ← Find the values restricted from the domain.

$\frac{6x+12}{x+2} = \frac{6(x+2)}{x+2}$ ← Factor the numerator. The denominator cannot be factored.

$= \frac{6}{1} \cdot \frac{x+2}{x+2}$ ← Rewrite to show a fraction equal to 1.

$= 6$ ← Simplify.

The solution is 6. The domain does not include $x = -2$.

6. Try This Simplify each expression and state any restrictions on the domain of the variable.
   a. $\frac{15b}{25b^2}$ $\frac{3}{5b}$; $b \neq 0$
   b. $\frac{12c^2}{3c+6}$ $\frac{4c^2}{c+2}$; $c \neq -2$
   c. $\frac{x+3}{x^2-9}$ $\frac{1}{x-3}$; $x \neq 3, -3$

Example 2 Relating to the Real World

Baking The baking time for bread depends, in part, on its size and shape. A good approximation for the baking time, in minutes, of a cylindrical loaf is $\frac{60 \cdot \text{volume}}{\text{surface area}}$, where the radius r and height h of the baked loaf are in inches. Rewrite this expression in terms of r and h.

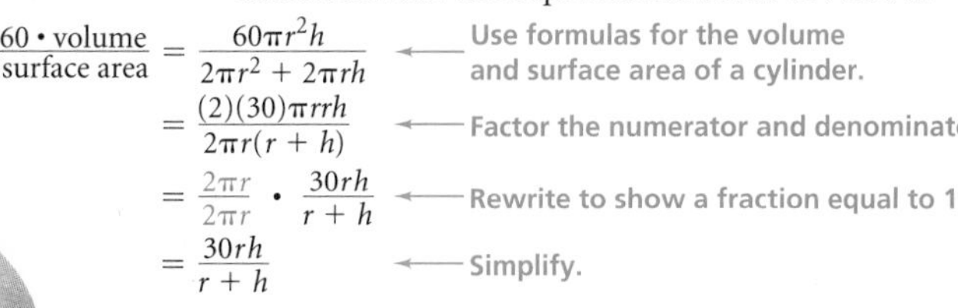

$\frac{60 \cdot \text{volume}}{\text{surface area}} = \frac{60\pi r^2 h}{2\pi r^2 + 2\pi rh}$ ← Use formulas for the volume and surface area of a cylinder.

$= \frac{(2)(30)\pi rrh}{2\pi r(r+h)}$ ← Factor the numerator and denominator.

$= \frac{2\pi r}{2\pi r} \cdot \frac{30rh}{r+h}$ ← Rewrite to show a fraction equal to 1.

$= \frac{30rh}{r+h}$ ← Simplify.

You can approximate the baking time using the expression $\frac{30rh}{r+h}$.

7. Check the answer to Example 2 by substituting values for r and h in both the original expression and the simplified expression. See margin.

8. *Critical Thinking* What values of r and h make sense in this situation? Would any values you would reasonably choose for r and h be restricted from the domain? Explain. See margin.

Additional Examples

FOR EXAMPLE 1

Simplify $\frac{6-2x}{x-3}$. State any values restricted from the domain. -2; domain does not include $x = 3$

Discussion: *If you graph the equation $y = -2$, the line passes through the point $(3, -2)$. Why is this not a legitimate solution to the equation $y = \frac{3-2x}{x-3}$?*

FOR EXAMPLE 3

Multiply $\frac{3x-2}{2x^2} \cdot \frac{4x}{9x^2-4}$. $\frac{2}{3x^2+2x}$

FOR EXAMPLE 4

Divide $\frac{2x^2}{3} \div \frac{4x^3}{6}$. $\frac{1}{x}$

express their rule as a mathematical equation. Invite students to share their equations with the class.

MAKING CONNECTIONS In March 1988, in Johannesburg, South Africa, a team baked the world's largest pan loaf. The loaf weighed 3,163 lb 10 oz and had dimensions of 9 ft 10 in. by 4 ft 1 in. by 3 ft 7 in.

Draw a sketch of a loaf and label it with the given dimensions. Tell students to assume that the loaf is rectangular. Challenge students to calculate the surface area and volume of this solid to find an approximate baking time. Remind students to convert all measurement to inches before they begin.
volume = 248,626 in.^{3};
surface area = 25,926 in.^{2};
baking time = 575.4 min or 9.6 h

CRITICAL THINKING Question 8 For students having difficulty with this question, ask:

- *When is the denominator zero?* **when $r + h = 0$**
- *If $r + h = 0$, what can you say about the relationship between r and h?* **$r = -h$**
- *Can r, the radius, or h, the height, ever be negative?* **no**
- *Will the denominator ever be zero?* **no**

Example 3

ERROR ALERT! Students may multiply 3 with $4x^2 - 1$ to get $12x^2 - 3$. **Remediation:** Point out that they are looking for factors common to the numerator and the denominator that can be divided out. Suggest that students write all expressions as factors instead of multiplying them.

Technology Options

For Exercises 1–16, students may be helped by using algebra software to simplify the expressions.

Prentice Hall Technology

Software
- Secondary Math Lab Toolkit™
- Computer Item Generator 11-3

Internet
- See the Prentice Hall site. (http://www.phschool.com)

10. Answers may vary. Sample: For $x = 4$,
$\frac{2x+1}{3} \cdot \frac{6x}{4x^2-1} = \frac{8}{7}$;
For $x = 4$, $\frac{2x}{2x-1} = \frac{8}{7}$.

Part 2

Multiplying and Dividing Rational Expressions

To multiply the rational expressions $\frac{a}{b}$ and $\frac{c}{d}$, where $b \neq 0$ and $d \neq 0$, you multiply the numerators and multiply the denominators. Then write the product in simplest form.

$$\frac{a}{b} \cdot \frac{c}{d} = \frac{ac}{bd}$$

Example 3

Multiply $\frac{2x+1}{3} \cdot \frac{6x}{4x^2-1}$.

$= \frac{2x+1}{3} \cdot \frac{6x}{(2x+1)(2x-1)}$ ← Factor the denominator.

$= \frac{\overset{1}{\cancel{2x+1}}}{\cancel{3}_1} \cdot \frac{\overset{2}{\cancel{6}}x}{(\cancel{2x+1})_1(2x-1)}$ ← Divide out the common factors 3 and (2x + 1).

$= \frac{2x}{2x-1}$ ← Simplify.

9. What values are restricted from the domain in Example 3? **$\frac{1}{2}, -\frac{1}{2}$**

10. Check the solution to Example 3 by substituting a value for x in the original expression and the simplified expression. **See left.**

11. Robin's first step in finding the product $\frac{2}{w} \cdot w^5$ was to rewrite the expression as $\frac{2}{w} \cdot \frac{w^5}{1}$. Why do you think Robin did this? **Robin wanted to make the terms in the denominator clearer.**

To divide the rational expression $\frac{a}{b}$ by $\frac{c}{d}$, where $b \neq 0$, $c \neq 0$, and $d \neq 0$, you multiply by the reciprocal of $\frac{c}{d}$.

$$\frac{a}{b} \div \frac{c}{d} = \frac{a}{b} \cdot \frac{d}{c}$$

12. Find the reciprocal of each expression.
   a. $\frac{-6d^2}{5}$ **$-\frac{5}{6d^2}$**
   b. $x^2 - 1$ **$\frac{1}{x^2-1}$**
   c. $\frac{1}{s+4}$ **$s + 4$**

Example 4

Divide $\frac{3x^3}{2}$ by $(-15x^5)$.

$\frac{3x^3}{2} \div (-15x^5) = \frac{3x^3}{2} \cdot \frac{1}{-15x^5}$ ← Multiply by the reciprocal of $-15x^5$.

$= \frac{\overset{1}{\cancel{3x^3}}}{2} \cdot \frac{1}{\underset{-5x^2}{\cancel{-15x^5}}}$ ← Divide out the common factor $3x^3$.

$= -\frac{1}{10x^2}$ ← Simplify.

13. **Try This** Find each product or quotient.
   a. $\frac{8y^3}{3} \cdot \frac{9}{y^4}$ **$\frac{24}{y}$**
   b. $\frac{m-2}{3m+9} \cdot (2m+6)$ **$\frac{2(m-2)}{3}$**
   c. $\frac{y+3}{y+2} \div (y+2)$ **$\frac{y+3}{(y+2)^2}$**

ERROR ALERT! Question 12b Students may think that the reciprocal of $x^2 - 1$ is $\frac{1}{x^2} - 1$. **Remediation:** Suggest that students begin by writing $x^2 - 1$ as $\frac{x^2 - 1}{1}$.

Example 4

Students may need to review how to multiply and divide expressions containing exponents.

Exercises ON YOUR OWN

ALTERNATIVE ASSESSMENT Exercises 1–16 Number four sets of index cards each from 1 through 6 (24 cards total). Randomly distribute the cards to the class. Ask students to write their own rational expressions that are restricted for the domain of the index card. Students should use Exercises 1–16 as a guide. Then have students exchange and solve these expressions. This will help you assess students' understanding of rational expressions.

WRITING Exercise 19 If students have difficulty getting started, suggest that they rewrite the numerator as $-1(x - 7)$.

Exercises 20–33 Encourage students to work carefully and neatly. Have them write the quotients using large, clear handwriting so that simplified expressions are easy to recognize.

GEOMETRY Exercise 35 Students explore the relationship between the formulas for the volume of a sphere and the surface area of a sphere.

Exercises ON YOUR OWN

Simplify each expression and state any values restricted from the domain.

1. $\frac{5a^2}{20a}$ $\frac{a}{4}$, $a \neq 0$
2. $\frac{3c}{12c^3}$ $\frac{1}{4c^2}$, $c \neq 0$
3. $\frac{4x - 8}{4x + 8}$ $\frac{x - 2}{x + 2}$, $x \neq -2$
4. $\frac{24y + 18}{36}$ $\frac{4y + 3}{6}$
5. $\frac{6a + 9}{12}$ $\frac{2a + 3}{4}$
6. $\frac{5c - 15}{c - 3}$ 5, $c \neq 3$
7. $\frac{4x^3}{28x^4}$ $\frac{1}{7x}$, $x \neq 0$
8. $\frac{5 - 2m}{15 - 6m}$ $\frac{1}{3}$, $m \neq 2\frac{1}{2}$
9. $\frac{24 - 2p}{48 - 4p}$ $\frac{1}{2}$, $p \neq 12$
10. $\frac{b - 4}{b^2 - 16}$ $\frac{1}{b + 4}$, $b \neq 4, -4$
11. $\frac{2x^2 + 2x}{3x^2 + 3x}$ $\frac{2}{3}$, $x \neq 0, -1$
12. $\frac{2s^2 + s}{s^3}$ $\frac{2s + 1}{s^2}$, $s \neq 0$
13. $\frac{3x^2 - 9x}{x - 3}$ $3x$, $x \neq 3$
14. $\frac{3x + 6}{3x^2}$ $\frac{x + 2}{x^2}$, $x \neq 0$
15. $\frac{w^2 + 7w}{w^2 - 49}$ $\frac{w}{w - 7}$, $w \neq 7, -7$
16. $\frac{a^2 + 2a + 1}{a + 1}$ $a + 1$, $a \neq -1$

17. **Critical Thinking** Explain why $\frac{x^2 - 9}{x + 3}$ is not the same as $x - 3$. $\frac{x^2 - 9}{x + 3}$ does not have -3 in its domain.

18. **Baking** Use $\frac{30rh}{r + h}$, where r is the radius and h is the height, to estimate the baking times for each type of bread shown. (Although bread is not exactly cylindrical, it is cylindrical enough to estimate the baking time.)

a.

about 13 min

biscuit: r = 1 in., h = 0.75 in.

b.

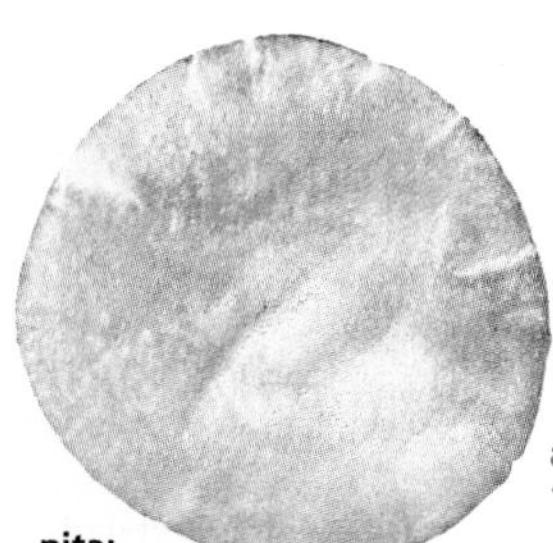

about 13 min

pita: r = 3.5 in., h = 0.5 in.

c.

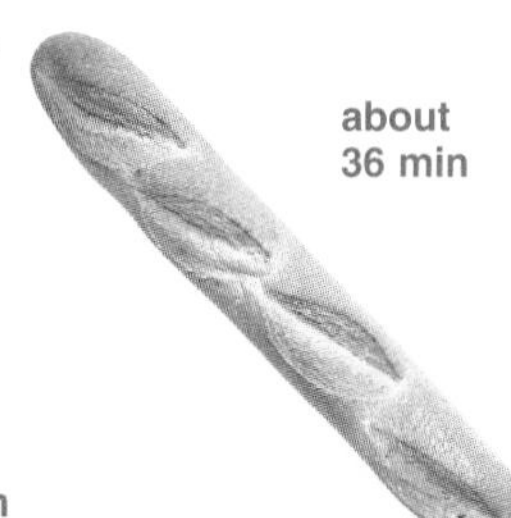

about 36 min

baguette: r = 1.25 in., h = 26 in.

19. **Writing** Explain why the simplified form of $\frac{7 - x}{x - 7}$ is -1 when $x \neq 7$.
Answers may vary. Sample: The numerator and denominator are opposites.

Find each product or quotient.

20. $\frac{7}{3} \cdot \frac{6}{21}$ $\frac{2}{3}$
21. $\frac{25}{4} \div \left(-\frac{4}{5}\right)$ $-\frac{125}{16}$
22. $\frac{7b^2}{10} \div \frac{14b^3}{15}$ $\frac{3}{4b}$
23. $\frac{6x^2}{5} \cdot \frac{10}{x^3}$ $\frac{12}{x}$
24. $15x^2 \div \frac{5x^4}{6}$ $\frac{18}{x^2}$
25. $\frac{-x^3}{8} \div \frac{-x^2}{16}$ $2x$
26. $\frac{3}{a^2} \div 6a^4$ $\frac{1}{2a^6}$
27. $\frac{5x^3}{x^2} \cdot \frac{3x^4}{10x}$ $\frac{3x^4}{2}$
28. $\frac{x}{x + 4} \div \frac{x + 3}{x + 4}$ $\frac{x}{x + 3}$
29. $\frac{3t + 12}{5t} \div \frac{t + 4}{10t}$ 6
30. $\frac{3x + 9}{x} \div (x + 3)$ $\frac{3}{x}$
31. $\frac{y - 4}{10} \div \frac{4 - y}{5}$ $-\frac{1}{2}$
32. $\frac{4x^2 + x}{5x} \cdot \frac{15}{2x - 2}$ $\frac{3(4x + 1)}{2(x - 1)}$
33. $\frac{11k + 121}{7k - 15} \div (k + 11)$ $\frac{11}{7k - 15}$

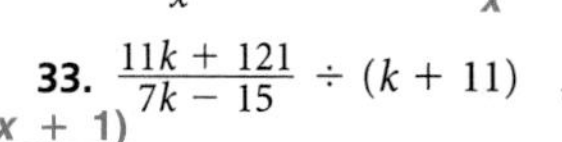

34. **Open-ended** Write an expression that has 2 and -3 restricted from the domain. Answers may vary. Sample: $\frac{1}{(x - 2)(x + 3)}$

35. **Geometry** Write and simplify the ratio for the $\frac{\text{volume of sphere}}{\text{surface area of sphere}}$. The formula for the volume of a sphere is $\frac{4}{3}\pi r^3$, and the formula for the surface area of a sphere is $4\pi r^2$, where r is the radius of the sphere. $\frac{\frac{4}{3}\pi r^3}{4\pi r^2} = \frac{r}{3}$

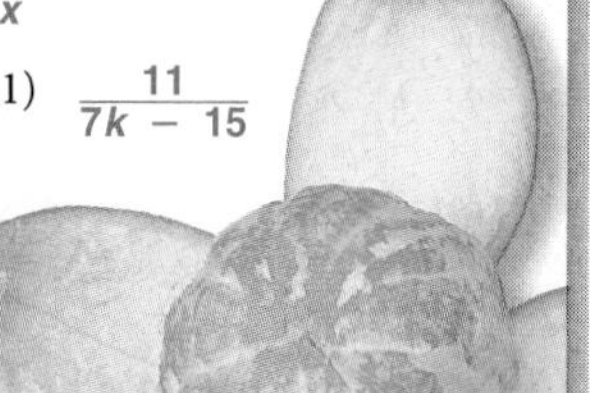

STANDARDIZED TEST TIP Exercise 36 Encourage students to simplify each expression as much as possible before comparing them.

Exercises 37–39 If students have difficulty getting started, suggest that they ask themselves these questions: *What do I know? What can I find out by using what I know? What am I trying to find?*

Chapter Project **FIND OUT BY ANALYZING** Students will use the lengths they calculated from the first Find Out question to find pairs of pitches that match each ratio. This information is essential to their work on the Chapter Project. Have students add this task to work they have already completed for the project. Check students' progress by having each student describe what they have done for the project thus far.

Exercises MIXED REVIEW

WRITING Exercise 44 Suggest that students review the definition of a function before writing an answer.

GETTING READY FOR LESSON 11-4 These exercises prepare students to add and subtract rational expressions.

Wrap Up

THE BIG IDEA Ask students: *How do you multiply and divide rational expressions?*

RETEACHING ACTIVITY Students simplify rational expressions. Then find values restricted from the domain. (Reteaching worksheet 11-3)

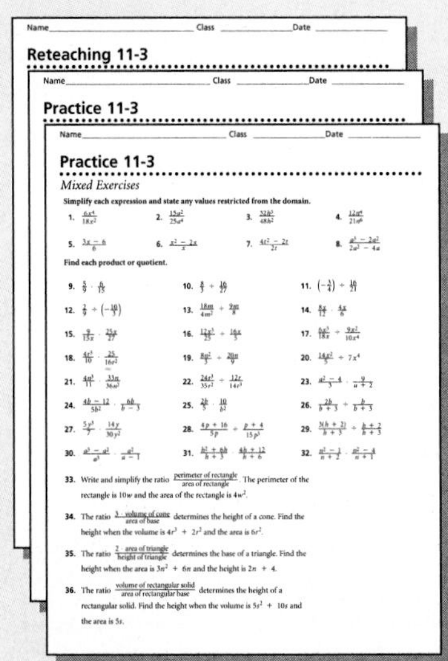
Name ______ Class ______ Date ______

Reteaching 11-3

Practice 11-3

Practice 11-3

Mixed Exercises

Simplify each expression and state any values restricted from the domain.

Find each product or quotient.

Lesson Quiz

Lesson Quiz is also available in Transparencies.

Simplify each expression. State any values restricted from the domain.

1. $\dfrac{12x+8}{18x^2+8}$ $\dfrac{2}{3x-2}$

2. $\dfrac{3a}{5} \div \dfrac{6}{15a^2}$ $\dfrac{3a^2}{2}$

3. $\dfrac{2x^2}{5x(x+3)} \cdot \dfrac{3x^3}{10x}$ $\dfrac{3x^3}{25(x+3)}$ $x \neq -3$

4. $\dfrac{4t+8}{3t} \cdot \dfrac{9t^2}{t+2}$ $12t$; $t \neq 0$ or -2

36. *Standardized Test Prep* Compare the quantities in Column A and Column B. Assume $x \neq -1, 0$. C

Column A	Column B
$\dfrac{-(5x+5)}{x+1}$	$-10x \cdot \dfrac{2x}{4x^2}$

A. The quantity in Column A is greater.
B. The quantity in Column B is greater.
C. The quantities are equal.
D. The relationship cannot be determined from the information given.

Geometry Write an expression in simplest form for $\dfrac{\text{area of shaded figure}}{\text{area of larger figure}}$.

37. $\dfrac{5w}{5w+6}$ 38. $\dfrac{1}{4}$ 39.

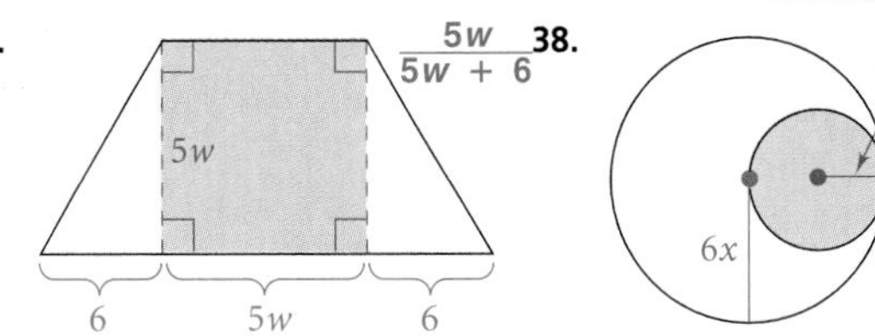

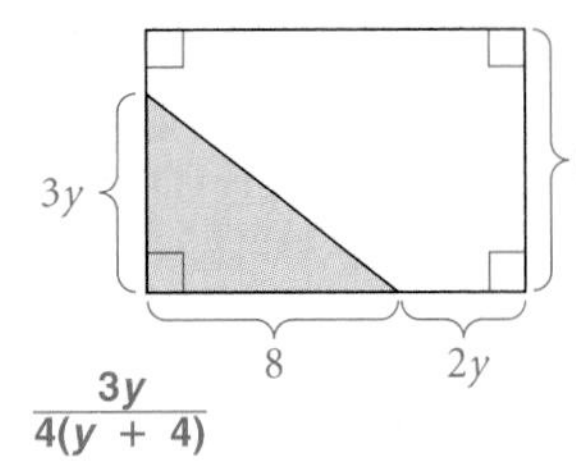

$\dfrac{3y}{4(y+4)}$

Chapter Project ***Find Out by Analyzing***

Pythagoras (540 B.C.) discovered that simple ratios of length produce pleasing combinations of musical pitches.

- Use the lengths you calculated in the Find Out question on page 514 to find pairs of pitches near each ratio.

  2 : 1 3 : 2 4 : 3 5 : 4 6 : 5

- A C-major chord consists of the pitches C, E, G, and C. What ratios are between the pitches of this chord?

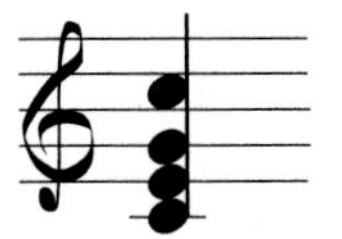

Exercises MIXED REVIEW

Find each product.

40. $y^2(5-y)$ $5y^2 - y^3$
41. $3^x \cdot 3^5$ 3^{x+5}
42. $4x^7z^2 \cdot 9xz^{-6}$ $\dfrac{36x^8}{z^4}$
43. $-6m(2m^3+3)$ $-12m^4 - 18m$

44. *Writing* Explain why a vertical line cannot be the graph of a function. Answers may vary. Sample: On a vertical line, the value of *x* is paired with many values of *y*.

Getting Ready for Lesson 11-4

Express each sum or difference in simplest form.

45. $\frac{1}{3} + \frac{1}{3}$ $\frac{2}{3}$
46. $\frac{1}{2} - \frac{2}{3}$ $-\frac{1}{6}$
47. $\frac{4}{5} - \frac{1}{5}$ $\frac{3}{5}$
48. $\frac{4}{9} + \frac{1}{6}$ $\frac{11}{18}$

11-4 Teaching Notes

PROBLEM OF THE DAY

A clothing store marks up the price of its shirts by 40%. What would be the selling price of a shirt that cost the store $23.00? $32.20

Problem of the Day is also available in Transparencies.

CONNECTING TO PRIOR KNOWLEDGE Ask students:

- *How do you add two fractions with different denominators?*
- *How do you find the least common denominator?*
- *How would you find the least common denominator of two fractions that had expressions instead of numbers in the denominators?*

WORK TOGETHER

Point out that it does not matter if the denominators contain variables so long as the denominators are the same.

THINK AND DISCUSS

Make sure students recall the meaning of a monomial denominator.

Example 1

KINESTHETIC LEARNING Have each student write a rational expression with a denominator of $x + 2$, $x + 3$, or $x + 4$ on a piece of paper. Have them exchange the papers randomly.

Connections  Air Travel . . . and more

What You'll Learn

- Adding and subtracting rational expressions
- Finding the LCD of two rational expressions

...And Why

To investigate real-world situations, such as groundspeed for air travel

11-4 Operations with Rational Expressions

WORK TOGETHER

Work with a partner to complete the following.

1. Add.
   a. $\frac{4}{9} + \frac{2}{9}$ $\frac{2}{3}$ b. $\frac{4x}{9} + \frac{2x}{9}$ $\frac{2x}{3}$ c. $\frac{4}{9x} + \frac{2}{9x}$ $\frac{2}{3x}$
2. Subtract.
   a. $\frac{7}{12} - \frac{1}{12}$ $\frac{1}{2}$ b. $\frac{7x}{12} - \frac{x}{12}$ $\frac{x}{2}$ c. $\frac{7}{12x} - \frac{1}{12x}$ $\frac{1}{2x}$
3. If $\frac{a}{b}$ and $\frac{c}{b}$ are rational expressions, where $b \neq 0$, write rules for adding and subtracting the two expressions. $\frac{a}{b} + \frac{c}{b} = \frac{a+c}{b}; \frac{a}{b} - \frac{c}{b} = \frac{a-c}{b}$

THINK AND DISCUSS

In the Work Together activity, you added and subtracted rational expressions with the same monomial denominator. You use the same method when you have a denominator that is a polynomial.

Example 1

Simplify $\frac{2}{x+3} + \frac{5}{x+3}$.

$\frac{2}{x+3} + \frac{5}{x+3} = \frac{2+5}{x+3}$ ← Add the numerators.

$= \frac{7}{x+3}$ ← Simplify the numerator.

4. **Try This** Simplify each expression.
   a. $\frac{3}{x+2} - \frac{2}{x+2}$ $\frac{1}{x+2}$ b. $\frac{y}{y-5} + \frac{3y}{y-5}$ $\frac{4y}{y-5}$ c. $\frac{5}{n+1} - \frac{2}{n+1}$ $\frac{3}{n+1}$

To add or subtract rational expressions with different denominators, you must write the expressions with a common denominator. Your work will be simpler if you find the least common denominator (LCD), which is the least common multiple of the denominators.

5. Look at the table below. How is finding the LCD of $\frac{3}{4}$ and $\frac{1}{6}$ like finding the LCD of $\frac{3}{4x}$ and $\frac{1}{6x^2}$? Each LCD includes all the factors of both denominators.

LCD of Numbers	*LCD of Variable Expressions*
$4 = 2 \cdot 2$	$4x = 2 \cdot 2 \cdot x$
$6 = 2 \cdot 3$	$6x^2 = 2 \cdot 3 \cdot x \cdot x$
$\text{LCD} = 2 \cdot 2 \cdot 3 = 12$	$\text{LCD} = 2 \cdot 2 \cdot 3 \cdot x \cdot x = 12x^2$

Lesson Planning Options

Prerequisite Skills

- Combining like terms
- Simplifying expressions

Assignment Options for Exercises On Your Own

Core 1–36
Extension 37–51

Use Mixed Review to maintain skills.

Resources

 **Student Edition**

Skills Handbook, pp. 575–579
Extra Practice, p. 566
Glossary/Study Guide

Teaching Resources

Chapter Support File, Ch. 11
- Practice 11-4 (two worksheets)
- Reteaching 11-4

Classroom Manager 11-4
Glossary, Spanish Resources
Two-Year Algebra Handbook 11-4

Transparencies
24, 126

Then direct students to arrange themselves into groups that have the same denominators. Have students check with each other to make sure that they are in the correct group. Have each group calculate the sum of all their expressions.

Review how to find LCDs if students need practice.

ESL Some students might confuse *least common* with *most unusual* and strive to find the most unusual denominator. Have a few volunteers demonstrate how to find the LCD for different sets of numbers.

Example 2

Point out that the expression is easier to simplify using the LCD, but other common denominators will also work. Suggest that students redo the example using a common denominator of $12x$.

In Step 2, remind students that if they multiply the numerator and denominator of a fraction by the same amount, the resulting fraction is equivalent to the one with which they started.

Example 3

ERROR ALERT! Students may think that $(y + 1)$ is the denominator because it is the factor that is common to both expressions. **Remediation:** Have students review and practice finding the LCDs of simpler fractions and monomials.

VISUAL and TACTILE LEARNING Encourage students to make themselves a card listing the three steps used to simplify Examples 2 and 3. For each step, have them use an example to illustrate the method.

Additional Examples

FOR EXAMPLE 1

Simplify $\frac{3}{x-2} + \frac{2}{x-2}$. $\frac{5}{x-2}$

FOR EXAMPLE 2

Simplify $\frac{5}{6} + \frac{2}{3x}$. $\frac{5x+4}{6x}$

FOR EXAMPLE 3

Simplify $\frac{2}{6x+9} - \frac{1}{4x^2-9}$. $\frac{4x-9}{3(4x^2-9)}$

PROBLEM SOLVING

Look Back Why is multiplying both the numerator and the denominator of a fraction by $2x$ equivalent to multiplying the fraction by 1?

Multiplying both numerator and denominator by $2x$ is equivalent to multiplying the fraction by $\frac{2x}{2x}$, and $\frac{2x}{2x} = 1$.

Example 2

Simplify $\frac{2}{3x} + \frac{1}{6}$.

Step 1: Find the LCD of $\frac{2}{3x}$ and $\frac{1}{6}$.

$3x = 3 \cdot x$ and $6 = 2 \cdot 3$ ← Factor each denominator.

The LCD is $3 \cdot x \cdot 2$ or $6x$.

Step 2: Rewrite the original expression and add.

$\frac{2}{3x} = \frac{2 \cdot 2}{3x \cdot 2} = \frac{4}{6x}$ ← Multiply numerator and denominator by 2.

$\frac{1}{6} = \frac{1 \cdot x}{6 \cdot x} = \frac{x}{6x}$ ← Multiply numerator and denominator by x.

Step 3: Rewrite the original expression and add.

$\frac{2}{3x} + \frac{1}{6} = \frac{4}{6x} + \frac{x}{6x}$ ← Replace each expression with its equivalent.

$= \frac{4+x}{6x}$ ← Add the numerators.

6. **Try This** Simplify.

a. $\frac{5}{12b} + \frac{1}{36b^2}$ $\frac{15b+1}{36b^2}$

b. $\frac{3}{7y^4} + \frac{2}{3y^2}$ $\frac{14y^2+9}{21y^4}$

c. $\frac{4}{25x} - \frac{49}{100}$ $\frac{-49x+16}{100x}$

You can factor to find the LCD when the denominators are polynomials.

Example 3

Simplify $\frac{1}{y^2+5y+4} - \frac{3}{5y+5}$.

Step 1: Find the LCD.

$y^2 + 5y + 4 = (y+1)(y+4)$ and $5y + 5 = 5(y+1)$ ← Factor each denominator.

The LCD is $5(y+1)(y+4)$.

Step 2: Write equivalent expressions with denominator $5(y+1)(y+4)$.

$\frac{1}{(y+1)(y+4)} = \frac{5}{5(y+1)(y+4)}$ ← Multiply numerator and denominator by 5.

$\frac{3}{5(y+1)} = \frac{3(y+4)}{5(y+1)(y+4)}$ ← Multiply numerator and denominator by $(y+4)$.

Step 3: Subtract.

$$\frac{5}{5(y+1)(y+4)} - \frac{3(y+4)}{5(y+1)(y+4)} = \frac{5-3(y+4)}{5(y+1)(y+4)}$$

$$= \frac{5-3y-12}{5(y+1)(y+4)}$$

$$= \frac{-7-3y}{5(y+1)(y+4)}$$

7. Explain why the LCD in Example 3 is not $5(y+1)(y+1)(y+4)$.
The greatest power of $(y + 1)$ in either denominator is $(y + 1)$, not $(y + 1)^2$.

Example 4 **Relating to the Real World** 

Make sure students understand what a *round trip* is. Point out that not enough information is provided for them to calculate the exact time for the round trip. Tell students to expect their answer to be an expression containing a variable.

Tell students to begin solving problems like this example by choosing a variable and writing down what it represents. In this example, r is chosen to represent the speed of the jet.

Make sure students are clear on how time, distance, and speed are related. If students are unsure, have them practice writing formulas using simple values.

MAKING CONNECTIONS The record time for an airplane flight from Los Angeles to Washington, D.C., is 68 min 17 s. The plane was an SR-71 Blackbird spy plane being flown to the Smithsonian Institution in 1990 to be retired. The plane averaged 2145 mi/h during the cross-country flight.

You can combine rational expressions to investigate real-world situations.

Example 4 **Relating to the Real World** 

QUICK REVIEW

15% more than a number is 115% of the number. 115% = 1.15.

Air Travel The groundspeed for jet traffic from Los Angeles to New York City is about 15% faster than the groundspeed from New York City to Los Angeles. This difference is due to a strong westerly wind at high altitudes. If r is a jet's groundspeed from New York to Los Angeles, write an expression for the round-trip air time. The two cities are about 2500 mi apart.

NYC to LA time: $\frac{2500}{r}$ ← time = $\frac{\text{distance}}{\text{rate}}$

LA to NYC time: $\frac{2500}{1.15r}$ ← time = $\frac{\text{distance}}{\text{rate}}$

Write an expression for the total time.

$\frac{2500}{r} + \frac{2500}{1.15r} = \frac{2875}{1.15r} + \frac{2500}{1.15r}$ ← Rewrite using the LCD.

$= \frac{5375}{1.15r}$ ← Add.

$\approx \frac{4674}{r}$ ← Simplify.

The expression $\frac{4674}{r}$ approximates the total time for the trip, where r is the speed of the jet.

8. Suppose a jet flies from Los Angeles to New York City at 420 mi/h. How long will the round-trip take? **about 11.13 h, or 11 h, 8 min**

Technology Options

For Exercises 1–12 and 25–36, students may use algebra software to simplify the expressions.

Prentice Hall Technology 

Software
- Secondary Math Lab Toolkit™
- Computer Item Generator 11-4

CD-ROM
- Multimedia Algebra Lab 11

Internet
- See the Prentice Hall site. (http://www.phschool.com)

Exercises ON YOUR OWN

ALTERNATIVE ASSESSMENT Exercises 25–36 To assess understanding of operations with rational expressions have students find the least common denominator of each of these expressions.

Exercise 37 Ask students why Juan's return rate is slower. Students will likely suggest that Juan walks slower because his grandfather walks slowly.

DIVERSITY Exercise 37 Use this example for a discussion of stereotypes of older people. Ask students to describe the stereotype and then name people they know who do not fit the stereotype. Older people are often stereotyped as weak, inactive, and slow. In contrast, the *Guinness Book of Records* describes a 98-year-old man and an 82-year-old woman who each completed marathons.

OPEN-ENDED Exercise 51 Ask students: *Are there any two denominators for which there is no Least Common Denominator?*

Chapter Project **FIND OUT BY CREATING** Students construct instruments out of cardboard tubes and evaluate different sounds based on ratios. This task is essential to their work on the Chapter Project. Have students add this part of the project to work they have already completed.

ESL To avoid confusion with a *pitch* in baseball, ask how students think the *pitch* of a flute would differ from that of a pipe organ.

530

Exercises ON YOUR OWN

Simplify the following expressions.

1. $\frac{4}{5}+\frac{3}{5}$ $\frac{7}{5}$
2. $\frac{6}{x}+\frac{1}{x}$ $\frac{7}{x}$
3. $-\frac{2}{3b}-\frac{4}{3b}$ $-\frac{2}{b}$
4. $\frac{5}{h}-\frac{3}{h}$ $\frac{2}{h}$
5. $\frac{7}{11g}-\frac{3}{11g}$ $\frac{4}{11g}$
6. $\frac{3x}{7}+\frac{6x}{7}$ $\frac{9x}{7}$
7. $\frac{5}{6d}+\frac{7}{6d}$ $\frac{2}{d}$
8. $\frac{5x}{9}-\frac{x}{9}$ $\frac{4x}{9}$
9. $\frac{3n}{7}+\frac{2n}{7}$ $\frac{5n}{7}$
10. $\frac{6}{17p}-\frac{9}{17p}$ $-\frac{3}{17p}$
11. $\frac{12r}{5}+\frac{14r}{5}$ $\frac{26r}{5}$
12. $-\frac{8}{7k}-\frac{9}{7k}$ $-\frac{17}{7k}$

Find the LCD.

13. $\frac{1}{2x}; \frac{1}{4x^2}$ $4x^2$
14. $\frac{b}{6}; \frac{2b}{9}$ 18
15. $\frac{6}{2m^2}; \frac{1}{m}$ $2m^2$
16. $\frac{1}{z}; \frac{3}{7z}$ $7z$
17. $\frac{-5}{6t^5}; \frac{3}{2t^2}$ $6t^5$
18. $\frac{3}{7s^5}; \frac{-4}{5s^2}$ $35s^5$
19. $\frac{24}{23d^3}; \frac{25}{2d^4}$ $46d^4$
20. $-\frac{3y^2}{15}; \frac{11y^5}{6}$ 30
21. $\frac{6a}{5}; \frac{-5}{a}$ $5a$
22. $\frac{7}{2k^4}; \frac{7}{9k^{11}}$ $18k^{11}$
23. $\frac{8}{5b}; \frac{12}{7b^3}$ $35b^3$
24. $\frac{6}{h^7}; \frac{1}{k^3}$ h^7k^3

Simplify.

25. $\frac{7}{3a}+\frac{2}{5a^4}$ $\frac{35a^3+6}{15a^4}$
26. $\frac{4}{x}-\frac{2}{3x^5}$ $\frac{12x^4-2}{3x^5}$
27. $\frac{6}{5x^8}+\frac{4}{3x^6}$ $\frac{18+20x^2}{15x^8}$
28. $\frac{12}{k}-\frac{5}{k^2}$ $\frac{12k-5}{k^2}$
29. $\frac{3}{6b}-\frac{4}{2b^4}$ $\frac{b^3-4}{2b^4}$
30. $\frac{2}{y}+\frac{3}{5y}$ $\frac{13}{5y}$
31. $\frac{3}{8m^3}+\frac{1}{12m^2}$ $\frac{9+2m}{24m^3}$
32. $\frac{1}{5x^3}+\frac{3}{20x^2}$ $\frac{4+3x}{20x^3}$
33. $-\frac{5}{4k}-\frac{8}{9k}$ $-\frac{77}{36k}$
34. $\frac{27}{n^3}-\frac{9}{7n^2}$ $\frac{189-9n}{7n^3}$
35. $\frac{9}{4x^2}+\frac{9}{5}$ $\frac{45+36x^2}{20x^2}$
36. $\frac{5}{12m^3}+\frac{7}{6m^8}$ 36. $\frac{5m^5+14}{12m^8}$

37. a. *Exercise* Suppose Jane walks one mile from her house to her grandparents' house. Then she returns home walking with her grandfather. Her return rate is 70% of her normal walking rate. Let r represent her normal walking rate. Write an expression for the amount of time Jane spends walking. $\frac{17}{7r}$
    b. Suppose Jane's normal walking rate is 3 mi/h. How much time does she spend walking? about 0.81 h, or 49 min

Simplify.

38. $\frac{10}{x-1}-\frac{5}{x-1}$ $\frac{5}{x-1}$
39. $\frac{7}{m+1}+\frac{3}{m+1}$ $\frac{10}{m+1}$
40. $\frac{4}{6m-1}+\frac{3}{6m-1}$ $\frac{7}{6m-1}$
41. $\frac{m}{m+3}+\frac{2}{m+3}$ $\frac{m+2}{m+3}$
42. $\frac{3n}{n+4}-\frac{n-8}{n+4}$ 2
43. $\frac{5}{t^2+1}-\frac{3}{t^2+1}$ $\frac{2}{t^2+1}$
44. $\frac{y}{y-1}-\frac{1}{y-1}$ 1
45. $\frac{s}{4s^2+2}+\frac{2}{4s^2+2}$ $\frac{s+2}{4s^2+2}$
46. $\frac{1}{2-b}-\frac{4}{2-b}$ $\frac{3}{b-2}$
47. $\frac{1}{m+2}+\frac{1}{m^2+3m+2}$ $\frac{1}{m+1}$
48. $\frac{4}{x-5}-\frac{3}{x+5}$ $\frac{x+35}{(x-5)(x+5)}$
49. $\frac{5}{y+2}+\frac{4}{y^2-y-6}$ $\frac{5y-11}{(y+2)(y-3)}$

50. *Writing* When adding or subtracting rational expressions, will the answer be in simplest form if you use the LCD? Explain. No; the combined numerators may contain a factor of the LCD.

51. *Open-ended* Write two rational expressions with different denominators. Find the LCD and add the two expressions.
    Answers may vary. Sample: $\frac{1}{x^2-9}, \frac{1}{x^2-6x+9}$; $(x+3)(x-3)^2$; $\frac{2x}{(x+3)(x-3)^2}$

Exercises MIXED REVIEW

GETTING READY FOR LESSON 11-5 These exercises prepare students for solving equations that contain rational expressions.

MENTAL MATH Exercise 58 Point out that the variable x represents the same value in both denominators.

Wrap Up

THE BIG IDEA Ask students: *How do you add and subtract rational expressions?*

RETEACHING ACTIVITY Students add and subtract rational expressions by finding the LCD. (Reteaching worksheet 11-4)

Algebra at Work

For further information about the skills and training necessary for becoming an electrician, contact the following resources:

- Community colleges that address the electrician's certification test. (TSTC, Electrical Systems Technology, 3801 Campus Drive, Waco, Texas 76705) 1-817-867-4837
- A local union representative of the Brotherhood of Electrical Workers

Students can learn from discussions with experts. Here are some possible sources.

- Invite an electrician to speak to the class.
- Plan a field trip to a construction site to investigate the wiring aspects of the job.

Chapter Project — *Find Out by Creating*

How does a flute or a pipe organ create different pitches? Get two cardboard tubes used to hold wrapping paper or to mail posters. Cut one tube into two lengths *A* and *B* whose ratio is 2:1. Hold your hand tightly over the longer piece (*A*) and blow over the open end until you get a pitch. Now try the shorter piece. What do you hear?

Cut two more pieces from the other tube by measuring them against piece *A*. Make one $\frac{2}{3}$ of *A* and the other $\frac{4}{5}$ of *A*. (You should have a small piece left over.) Get some friends together and play the first phrase of "The Star Spangled Banner."

Exercises MIXED REVIEW

Find the hypotenuse of a right triangle with the given legs.

52. $a = 9, b = 12$ 15

53. $a = 10, b = 7$ about 12.2

54. $a = 12, b = 5$ 13

55. $a = 6, b = 9$ about 10.8

56. Population Surveys The 1990 census may have missed 4 million people out of about 250 million people living in the United States. What percent of the people may have been missed? 1.6%

Getting Ready for Lesson 11-5

Mental Math **Find the value of each variable.**

57. $\frac{1}{2} + \frac{1}{m} = \frac{3}{4}$ 4

58. $\frac{3}{x} + \frac{1}{x} = 1$ 4

59. $\frac{2}{n} + \frac{1}{n} = \frac{1}{2}$ 6

60. $\frac{4}{3} - \frac{2}{x} = \frac{2}{3}$ 3

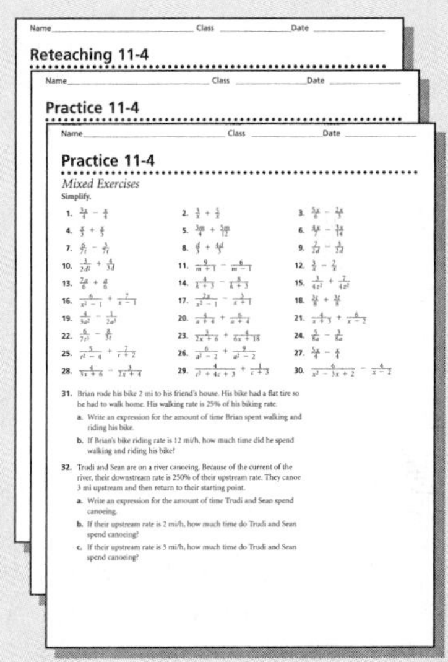

Reteaching 11-4

Practice 11-4

Practice 11-4

Mixed Exercises

Simplify.

Lesson Quiz

Lesson Quiz is also available in Transparencies.

Simplify.

1. $\frac{3}{4x^2} + \frac{2}{x}$ $\frac{8x + 3}{4x^2}$

2. $\frac{2x}{3x + 1} + \frac{x}{3x + 1}$ $\frac{3x}{3x + 1}$

3. $\frac{1}{3x + 3} - \frac{1}{x^2 + 2x + 1}$ $\frac{x - 2}{3(x + 1)^2}$

4. $\frac{2}{z - 3} + \frac{7}{z^2 - 6z + 9}$ $\frac{2z + 1}{z^2 - 6z + 9}$

Algebra at Work

Electrician

More than half a million men and women make their livings as electricians. All are highly skilled technicians licensed by the states in which they work. Electricians use formulas containing rational expressions. For example, when a circuit connected "in parallel" contains two resistors with resistances r_1 and r_2 ohms, the total resistance R (in ohms) of the circuit can be found using the formula $\frac{1}{R} = \frac{1}{r_1} + \frac{1}{r_2}$.

Mini Project: Research the difference between circuits connected *in parallel* and *in series*. Write a brief report of your findings. Include diagrams with your report.

11-5 Teaching Notes

PROBLEM OF THE DAY

A square area with sides of 12 ft is roped off in a parking lot. What would be the width of a rectangular area made using the same rope if the length is 16 ft? 8 ft

Problem of the Day is also available in Transparencies.

CONNECTING TO PRIOR KNOWLEDGE Ask students to describe the method they use for solving equations that contain fractions. Write notes on the board as they describe the steps. Leave the steps on the board so that students can refer to them as they work through the lesson.

THINK AND DISCUSS

Make sure students are clear on the difference between $\frac{1}{2}x$ and $\frac{1}{2x}$. Suggest that they substitute values for x to see that $\frac{1}{2}x$ and $\frac{1}{2x}$ are not equivalent.

Remind students to be neat and careful when canceling terms. They can easily cross out a wrong term by being too hasty.

When students have found a solution, remind them to substitute the value back into the original equation to check their work. Remind them to check for values that should be excluded from the domain.

Lesson Planning Options

Prerequisite Skills

- Solving equations
- Simplifying expressions

Assignment Options for Exercises On Your Own

Core 2–15, 20–29
Extension 1, 16–19, 30–31

Use Mixed Review to maintain skills.

Resources

 **Student Edition**

Skills Handbook, pp. 575–579
Extra Practice, p. 566
Glossary/Study Guide

 **Teaching Resources**

Chapter Support File, Ch. 11
- Practice 11-5 (two worksheets)
- Reteaching 11-5
- Checkpoint
- Alternative Activity 11-5

Classroom Manager 11-5
Glossary, Spanish Resources
Two-Year Algebra Handbook 11-5

 **Transparencies**

127

What You'll Learn

- Solving equations involving rational expressions

...And Why

To solve real-world situations, such as those involving mail processors

Connections  Business . . . and more

11-5 Solving Rational Equations

THINK AND DISCUSS

A **rational equation** contains rational expressions. A method for solving rational equations is similar to the method you learned in Chapter 3 for solving equations with rational numbers.

Solving Equations with Rational Numbers

$\frac{1}{2}x + \frac{3}{4} = \frac{1}{5}$ ← The denominators are 2, 4, and 5. The LCD is 20.

$20\left(\frac{1}{2}x + \frac{3}{4}\right) = 20\left(\frac{1}{5}\right)$ ← Multiply each side by 20.

$\overset{10}{\cancel{20}}\left(\frac{1}{\cancel{2}_1}x\right) + \overset{5}{\cancel{20}}\left(\frac{3}{\cancel{4}_1}\right) = \overset{4}{\cancel{20}}\left(\frac{1}{\cancel{5}_1}\right)$ ← Use the distributive property.

$10x + 15 = 4$ ← No fractions! This equation is much easier to solve.

Solving Equations with Rational Expressions

$\frac{1}{2x} + \frac{3}{4} = \frac{1}{5x}$ ← The denominators are $2x$, 4, and $5x$. The LCD is $20x$.

$20x\left(\frac{1}{2x} + \frac{3}{4}\right) = 20x\left(\frac{1}{5x}\right)$ ← Multiply each side by $20x$.

$\overset{10}{\cancel{20x}}\left(\frac{1}{\cancel{2x}_1}\right) + \overset{5}{\cancel{20}}x\left(\frac{3}{\cancel{4}_1}\right) = \overset{4}{\cancel{20x}}\left(\frac{1}{\cancel{5x}_1}\right)$ ← Use the distributive property.

$10 + 15x = 4$ ← No rational expressions! Now you can solve.

1. Find the LCD of each equation.
   a. $\frac{3}{4n} - \frac{1}{2} = \frac{2}{n}$ $4n$ b. $\frac{1}{3x} + \frac{2}{5} = \frac{4}{3x}$ $15x$ c. $\frac{1}{8y} = \frac{5}{6} - \frac{1}{y}$ $24y$

There are often values that are excluded from the domain of a rational expression. When you solve rational equations, check your solutions to be sure that your answer satisfies the original equation.

Example 1

Solve $\frac{5}{x^2} = \left(\frac{6}{x} - 1\right)$.

$x^2\left(\frac{5}{x^2}\right) = x^2\left(\frac{6}{x} - 1\right)$ ← Multiply each side by the LCD, x^2.

$\overset{1}{\cancel{x^2}}\left(\frac{5}{\cancel{x^2}_1}\right) = \overset{x}{\cancel{x^2}}\left(\frac{6}{\cancel{x}_1}\right) - x^2(1)$ ← Use the distributive property.

$5 = 6x - x^2$ ← Simplify.

$x^2 - 6x + 5 = 0$ ← Collect terms on one side.

$(x - 5)(x - 1) = 0$ ← Factor the quadratic expression.

Example 1

Before students begin studying the example, point out that x cannot have a value of zero because division by zero is undefined. Suggest that students note the excluded values as they work each problem.

Example 2 Relating to the Real World 

Some students may feel that they are unable to solve the problem because they do not know what an electronic mail processor is. Point out that the purpose of the machine does not relate to the math and that the problem can be solved without knowing what the machine does. Alternatively, have students speculate on what a mail processor might do.

ERROR ALERT! Students may define the time of the new machine to be $3n$ because it is three times faster.
Remediation: Point out that n is defined to be the time taken by the new machine to process 1000 pieces. Ask students:

- *Is the old machine faster or slower?* **slower**
- *Will the old machine take more time or less time to do the same work?* **more time**
- *If n is the time taken by the new machine, and the old machine takes three times as long, how do you express the time taken by the old machine?* **3n**

Point out that students could have defined t to be the time taken by the old machine. Then the time taken by the new machine would be $\frac{1}{3}t$. Defining the process rates and then solving the problem using these definitions would result in the

$(x - 5) = 0$ or $(x - 1) = 0$ ← Use the zero-product property.

$x = 5$ $\quad$ $x = 1$ ← Solve.

Check $\frac{5}{5^2} \stackrel{?}{=} \frac{6}{5} - 1$ $\qquad$ $\frac{5}{1^2} \stackrel{?}{=} \frac{6}{1} - 1$

$\frac{1}{5} = \frac{1}{5}$ ✔ $\qquad$ $5 = 5$ ✔

Since 5 and 1 check in the original equation, they are the solutions. ■

Example 2 Relating to the Real World 

Business The Fresia Company owns two electronic mail processors. The newer machine works three times as fast as the older one. Together the two machines process 1000 pieces of mail in 25 min. How long does it take each machine, working alone, to process 1000 pieces of mail?

Define

	Time to process 1000 pieces (min)	Processing rate (pieces/min)
New machine	n	$\frac{1000}{n}$
Old machine	$3n$	$\frac{1000}{3n}$
Both machines	25	$\frac{1000}{25} = 40$

Relate Processing rate of new machine + Processing rate of old machine = Processing rate of both machines

Write $\frac{1000}{n} + \frac{1000}{3n} = 40$

$3n\left(\frac{1000}{n} + \frac{1000}{3n}\right) = 3n(40)$ ← Multiply each side by the LCD, $3n$.

$\overset{3}{3n}\left(\frac{1000}{n_1}\right) + \overset{1}{3n}\left(\frac{1000}{3n_1}\right) = 3n(40)$ ← Use the distributive property.

$3000 + 1000 = 120n$ ← Simplify.

$4000 = 120n$

$33.3 \approx n$ ← Divide each side by 120.

The new machine can process 1000 pieces of mail in about 33 min. The old machine takes $3 \cdot 33.3$, or 100 min, to process 1000 pieces of mail. ■

Additional Examples

FOR EXAMPLE 1

Solve $\frac{6}{x^2} = -\frac{5}{x} - 1$. **−3; −2**

FOR EXAMPLE 2

Jeff and Bianca are proofing articles for the school newspaper. Bianca works one and a half times as fast as Jeff. Together they edit 2500 words in 3 h. How long does it take each editor, working alone, to edit a 2500-word newspaper? **Jeff 7.5 h; Bianca 5 h**

Discussion: *Is it always possible to reduce the time required for a complex task such as editing by adding additional help?*

FOR EXAMPLE 3

Solve $\frac{1}{(x - 1)} = \frac{3}{3x - 3} + \frac{1}{3}$.

undefined

same answer. Challenge students to rework the example using these definitions.

Example 3

VISUAL LEARNING Some students may think there is an error if an equation produces no solution. On the board, write $x + 1 = x + 2$. Ask students to study the equation and suggest values for x that solve the equation. Once students recognize that no such value exists, use this to help them remember that equations with no solutions are reasonable.

Exercises ON YOUR OWN

CRITICAL THINKING Exercise 1c Ask students which of the equations in Exercises 2–13 can be solved using Ingrid's method.

Exercise 14 Suggest that students begin by defining a variable to represent the rate at which the new machine scans documents.

ESTIMATION Exercise 14 Point out that if the new machine scans at the same rate as the old machine, together they would halve the time taken to scan 10,000 documents. Half of 25 min is 12.5 min. The two machines take less than 12.5 min to scan the documents, so the new machine must be slightly faster than the old machine.

ALTERNATIVE ASSESSMENT Exercise 14 Group students into pairs and have each pair write a real world problem similar to this exercise. Then have each pair of students exchange their problem with another pair to solve. This will help you assess students' skills in solving rational expressions.

Technology Options

For Exercises 2–13 and 20–28, students may find it helpful to check their answers by graphing the equations with a graphing calculator or graphing software.

Prentice Hall Technology

Software

- Secondary Math Lab Toolkit™
- Computer Item Generator 11-5

CD-ROM

- Multimedia Algebra Lab 11

Internet

- See the Prentice Hall site. (http://www.phschool.com)

534

2. Try This Solve each equation.

a. $\frac{1}{3} + \frac{1}{3x} = \frac{1}{7}$ $-\frac{7}{4}$

b. $\frac{2}{3x} = \frac{1}{5} - \frac{3}{x}$ $\frac{55}{3}$

The process of multiplying by the LCD can sometimes lead to a solution that does not check in the original equation.

Example 3

Solve $\frac{-1}{x-2} = \frac{x-4}{2(x-2)} + \frac{1}{3}$.

The LCD is $6(x - 2)$. Multiply each side by the LCD, then solve.

$$6(x-2)\left(\frac{-1}{x-2}\right) = 6(x-2)\left(\frac{x-4}{2(x-2)} + \frac{1}{3}\right)$$

$$6(x-2)\left(\frac{-1}{x-2}\right) = 6(x-2)\left(\frac{x-4}{2(x-2)}\right) + 6(x-2)\left(\frac{1}{3}\right)$$

$-6 = 3(x - 4) + 2(x - 2)$ ← Simplify.

$-6 = 3x - 12 + 2x - 4$ ← distributive property

$-6 = 5x - 16$ ← Combine like terms.

$10 = 5x$ ← Add 16 to each side.

$2 = x$ ← Divide each side by 5.

Check $\frac{-1}{2-2} \stackrel{?}{=} \frac{2-4}{2(2-2)} + \frac{1}{3}$

$\frac{-1}{0} = \frac{-2}{0} + \frac{1}{3}$ ← Undefined!

The equation has *no* solution because 2 makes a denominator equal 0.

3. Try This Solve each equation. Be sure to check your answers!

a. $\frac{2}{x} - \frac{8}{x^2} = -1$ $-4, 2$

b. $\frac{1}{w+1} = 1 - \frac{1}{w(w+1)}$ 1

Exercises ON YOUR OWN

1. Carlos studied the problem at the right and said, "I'll start by finding the LCD." Ingrid studied the problem and said, "I'll start by cross-multiplying."

   Solve $\frac{40}{x} = \frac{15}{x-20}$.

   a. Solve the equation using Carlos's method, and then using Ingrid's method. **32**
   b. *Writing* Which method do you prefer? Why? **Check students' work.**
   c. *Critical Thinking* Will Ingrid's method work for all rational equations? Explain. **No. An equation like $\frac{2}{x} + 3 = \frac{4}{x+1}$ cannot be cross multiplied because 3 is not part of the fraction $\frac{2}{x}$.**

Solve each equation. Be sure to check your answers!

2. $\frac{1}{2} + \frac{2}{x} = \frac{1}{x}$ -2
3. $\frac{1}{20} + \frac{1}{30} = \frac{1}{x}$ 12
4. $5 + \frac{2}{p} = \frac{17}{p}$ 3
5. $\frac{3}{a} - \frac{5}{a} = 2$ -1
6. $\frac{5}{n} - \frac{1}{2} = 2$ 2
7. $\frac{1}{x+2} = \frac{1}{2x}$ 2

GEOMETRY Exercises 17–19 Students use rational equations to calculate the areas of geometric figures.

Exercise 17 Lead students step by step through this exercise.

- Write the expression that represents the area of the rectangle.
- Write the expression that represents the area of the unshaded triangle. Remind students of the equation for the area of a triangle.
- Subtract the unshaded area from the area of the rectangle to get an expression representing the area of the shaded region.

The area of the shaded region is given as 64 square units. By setting the expression equal to 64, students can find the value of x. Have them use similar logic to solve Exercises 18 and 19.

STANDARDIZED TEST TIP Exercise 29 Remind students that a quadratic equation can have zero, one, or two solutions.

Chapter Project **FIND OUT BY INTERVIEWING** Students interview someone who plays a stringed instrument to learn how the instrument is tuned. This task is essential to their work on the Chapter Project. You may wish to check on students' progress by examining their project notebooks.

8. $y - \frac{6}{y} = 5$ −1, 6
9. $\frac{30}{x+3} = \frac{30}{x-3}$ no sol.
10. $\frac{4}{a+1} = \frac{8}{2-a}$ 0
11. $\frac{5}{y} + \frac{3}{2} = \frac{1}{3y}$ $-\frac{28}{9}$
12. $\frac{2}{3b} - \frac{3}{4} = \frac{1}{6b}$ $\frac{2}{3}$
13. $\frac{5}{2s} + \frac{3}{4} = \frac{9}{4s}$ $-\frac{1}{3}$

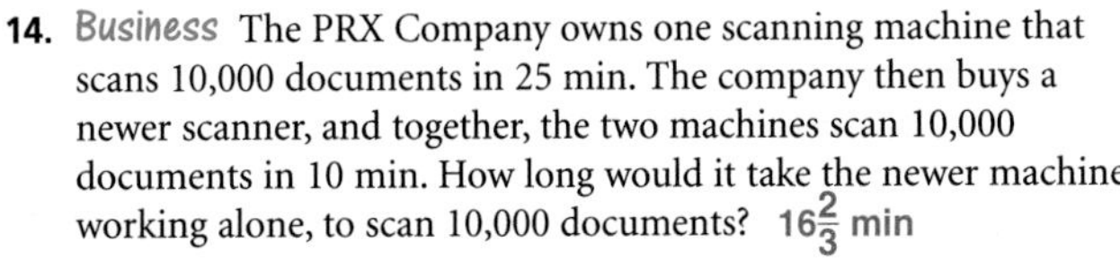

14. *Business* The PRX Company owns one scanning machine that scans 10,000 documents in 25 min. The company then buys a newer scanner, and together, the two machines scan 10,000 documents in 10 min. How long would it take the newer machine, working alone, to scan 10,000 documents? $16\frac{2}{3}$ min

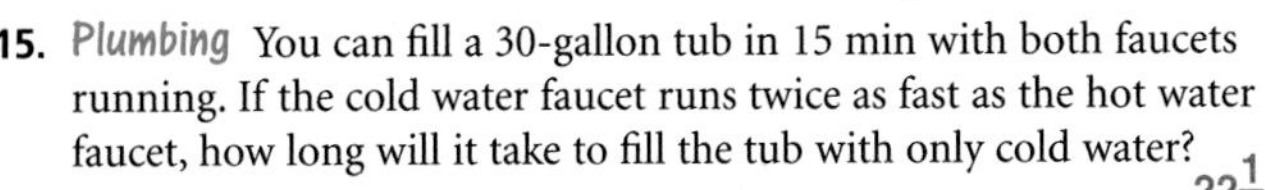

15. *Plumbing* You can fill a 30-gallon tub in 15 min with both faucets running. If the cold water faucet runs twice as fast as the hot water faucet, how long will it take to fill the tub with only cold water? $22\frac{1}{2}$ min

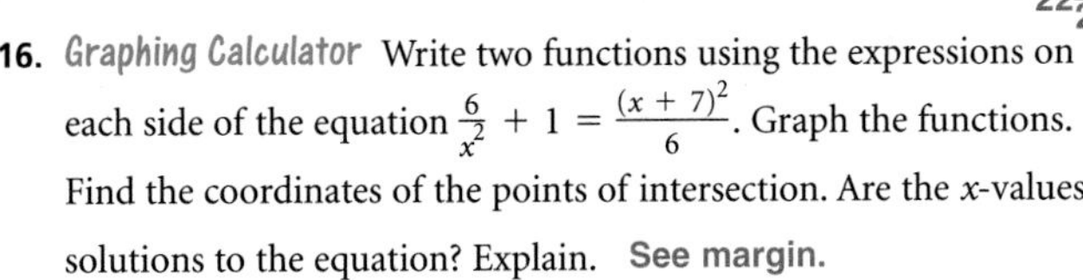

16. *Graphing Calculator* Write two functions using the expressions on each side of the equation $\frac{6}{x^2} + 1 = \frac{(x+7)^2}{6}$. Graph the functions. Find the coordinates of the points of intersection. Are the x-values solutions to the equation? Explain. See margin.

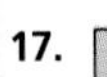

Geometry **Find each value of x if the area of each shaded region is 64 square units. Assume that each quadrilateral shown is a rectangle.**

17. 2

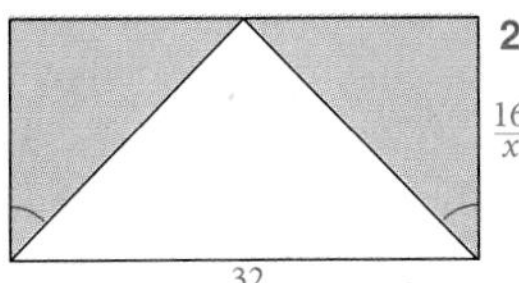

18. $1\frac{1}{8}$

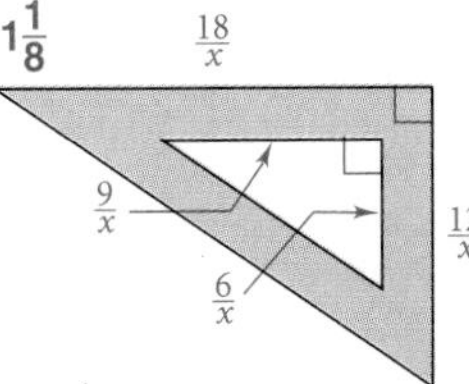

19. $\frac{1}{2}$

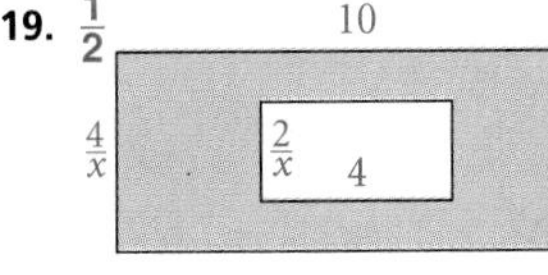

Solve each equation. Be sure to check your answers!

20. $\frac{1}{2} = -\frac{1}{3(x-3)}$ $\frac{7}{3}$
21. $\frac{1}{t-2} = \frac{t}{8}$ −2, 4
22. $\frac{x-11}{3x} = \frac{x-19}{5x}$ 1
23. $\frac{4}{3(c+4)} + 1 = \frac{2c}{c+4}$ $\frac{16}{3}$
24. $\frac{8}{x+3} = \frac{1}{x} + 1$ 1, 3
25. $\frac{x+2}{x+4} = \frac{x-2}{x-1}$ 6
26. $\frac{2}{c-2} = 2 - \frac{4}{c}$ 1, 4
27. $\frac{5}{3p} + \frac{2}{3} = \frac{5+p}{2p}$ 5
28. $\frac{3}{s-1} + 1 = \frac{12}{s^2-1}$ −5, 2
29. *Standardized Test Prep* Which inequality contains both of the solutions of the equation $5x = \frac{7}{2} + \frac{6}{x}$? A

    A. $-1 < x < 3$
    B. $0 < x < \frac{3}{2}$
    C. $-2 \leq x < 0$
    D. $-3 \leq x \leq -1$
    E. none of the above
30. A plane flies 450 mi/h. It can travel 980 mi with a wind in the same amount of time as it travels 820 mi against the wind. Solve the equation $\frac{980}{450+s} = \frac{820}{450-s}$ to find the speed s of the wind. 40 mi/h

pages 534–535 On Your Own

16. $y = \frac{6}{x^2} + 1$, $y = \frac{(x+7)^2}{6}$

    The graphs intersect at approximately (−4.16, 1.35), (−1.12, 5.76); and (0.81, 10.2); The x-values are approximate solutions because they make both sides of the equation approximately equal.

page 536 Mixed Review

36. Sample: the amount of time to solve each problem vs. the number of problems on a 30-min test; $y = \frac{30}{x}$

Exercises MIXED REVIEW

OPEN-ENDED Exercise 36 Encourage students to write humorous inverse equations. Each quantity, however, must be represented by a variable and be carefully defined.

GETTING READY FOR LESSON 11-6 These exercises prepare students for counting outcomes and permutations.

Wrap Up

THE BIG IDEA Ask students to describe how to solve an equation containing a rational expression.

RETEACHING ACTIVITY Students find the LCD of rational expressions. Then use it to simplify and solve equations. (Reteaching worksheet 11-5)

Exercises CHECKPOINT

Students will assess their own progress on Lessons 11-3 to 11-5.

OPEN-ENDED Exercise 5 If students have difficulty getting started, ask: *If the variable does equal 3, what is true about the denominator of the expression?*

STANDARDIZED TEST TIP **Exercise 14** Factoring the denominator of each expression makes the LCD easier to find.

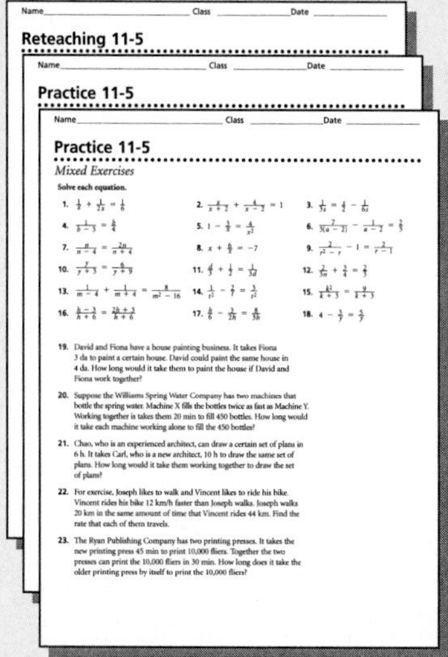
Reteaching 11-5

Practice 11-5

Practice 11-5

Mixed Exercises

Solve each equation.

Lesson Quiz

Lesson Quiz is also available in Transparencies.

Solve each equation. Be sure to check your work.

1. $\frac{2}{5} + \frac{3}{10} = \frac{1}{x}$ $\frac{10}{7}$

2. $\frac{2}{x+1} = \frac{1}{x}$ 1

3. $\frac{1}{3} = -\frac{2}{3(2+x)}$ −4

4. $\frac{1}{2x} + \frac{1}{2} = \frac{4}{x+3}$ 3, 1

31. Find the value of each variable. $\begin{bmatrix} \frac{5a}{3} & \frac{7}{3b} \\ \frac{2c-15}{35c} & \frac{5}{2d} + \frac{3}{4} \end{bmatrix} = \begin{bmatrix} 2 + \frac{7a}{6} & 9 \\ \frac{1}{5c} & \frac{9}{4d} \end{bmatrix}$
$a = 4$; $b = \frac{7}{27}$; $c = 11$; $d = -\frac{1}{3}$

Chapter Project *Find Out by Interviewing*

The pitch produced by a vibrating string is affected by how tightly it is stretched, or its tension. Find out how a violin, guitar, or other stringed instrument is tuned by talking with someone who plays it.

Exercises MIXED REVIEW

Solve each equation. Round to the nearest hundredth, where necessary.

32. $x^2 = 3x + 8$ 4.70, −1.70
33. $x^2 - 4 = 0$ 2, −2
34. $2x^2 + 6 = 4x$ no solution
35. $7x = 5x - 11$ −5.5

36. Open-ended Give an example of two quantities that vary inversely. Write an equation to go with your example. See margin.

Getting Ready for Lesson 11-6

List the possible outcomes of each action.

37. rolling a number cube once 1, 2, 3, 4, 5, 6
38. tossing a coin twice HH, HT, TH, TT
39. tossing a coin and rolling a number cube H1, H2, H3, H4, H5, H6, T1, T2, T3, T4, T5, T6

Exercises CHECKPOINT

Simplify each expresssion and state any restrictions on the variable.

1. $\frac{8m^2}{2m^3}$ $\frac{4}{m}$; $m \neq 0$
2. $\frac{6x^2 - 24}{x + 2}$ $6(x - 2)$; $x \neq -2$
3. $\frac{3c + 9}{3c - 9}$ $\frac{c + 3}{c - 3}$; $c \neq 3$
4. $\frac{3z^2 + 12z}{z^4}$ $\frac{3z + 12}{z^3}$; $z \neq 0$

5. Open-ended Write an expression where the variable cannot equal 3.
Answers may vary. Sample: $\frac{5}{x - 3}$

Solve each equation.

6. $\frac{9}{t} + \frac{3}{2} = 12$ $\frac{6}{7}$
7. $\frac{10}{z + 4} = \frac{30}{2z + 3}$ −9
8. $\frac{1}{m - 2} = \frac{5}{m}$ $\frac{5}{2}$
9. $c - \frac{8}{c} = 10$ about 10.74, −0.74

Find each sum or difference.

10. $\frac{5}{c} + \frac{4}{c}$ $\frac{9}{c}$
11. $\frac{9}{x - 3} - \frac{4}{x - 3}$ $\frac{5}{x - 3}$
12. $\frac{8}{m + 2} - \frac{6}{3 - m}$ $\frac{14m - 12}{(m + 2)(m - 3)}$
13. $\frac{6}{t} + \frac{3}{t^2}$ $\frac{6t + 3}{t^2}$

14. Standardized Test Prep What is the LCD of $\frac{8}{n}$, $\frac{5}{3 - n}$, and $\frac{1}{n^2}$? B

A. $n^2 + 3n$ B. $3n^2 - n^3$ C. $n^2 + 3$ D. $3n^3 - n^4$ E. $n^3 - 3n^2$

Math ToolboX

This toolbox introduces students to algebraic proofs. The logic in proofs is a skill used frequently in geometry, upper level math courses, and many other fields of study such as psychology or law.

Begin by introducing the three properties of equality. Make the properties memorable by comparing the mathematical examples to definitions of words with the same prefix: reflection and reflexive, or transfer and transitive. Then work through the example proofs with the class. Have students work in pairs or small groups to discuss and complete the exercises.

ERROR ALERT! Some students have difficulty choosing the correct property. **Remediation:** Have students work with a partner and verbalize the changes made from one step to the next.

WRITING Exercise 6 For students who have trouble starting, suggest that they substitute a phrase or expression for *a*, *b*, and *c* in the transitive property.

ADDITIONAL PROBLEM Supply the missing reasons.

$3(x + 2) - 7 = 20$	Given
$3(x + 2) = 27$	? Addition prop. of equality
$3x + 6 = 27$	? Distributive prop. of equality
$3x = 21$	? Addition prop. of equality
$x = 7$	? Division prop. of equality

Math ToolboX — Problem Solving

Algebraic Reasoning

After Lesson 11-5

You can use the properties you have studied and the three below to prove algebraic relationships and to justify steps in the solution of an equation.

Properties of Equality

For all real numbers *a*, *b*, and *c*:

Reflexive Property: $a = a$ — Example: $5x = 5x$

Symmetric Property: If $a = b$, then $b = a$. — Example: If $15 = 3t$, then $3t = 15$.

Transitive Property: If $a = b$ and $b = c$, then $a = c$. — Example: If $d = 3y$ and $3y = 6$, then $d = 6$.

Example 1

Prove that if $a = b$, then $ac = bc$.

Statements	Reasons
$a = b$	← Given
$ac = ac$	← Reflexive prop.
$ac = bc$	← Substitute *b* for *a*.

Example 2

Solve $-32 = 4(y - 3)$. Justify each step.

Steps	Reasons
$-32 = 4(y - 3)$	← Given
$-32 = 4y - 12$	← Distributive prop.
$-20 = 4y$	← Addition prop. of equality
$-5 = y$	← Division prop. of equality
$y = -5$	← Symmetric prop. of equality

Name the property that each exercise illustrates.

1. If $3.8 = z$, then $z = 3.8$. symmetric
2. If $x = \frac{1}{2}y$, and $\frac{1}{2}y = -2$, then $x = -2$. transitive
3. $-4r = -4r$ reflexive

Supply the missing reasons.

4. Solve $-5 = 1 + 3d$. Justify each step.

Steps	Reasons
$-5 = 1 + 3d$	← Given
$-6 = 3d$	← **a.** ? Subt. prop. of eq.
$-2 = d$	← **b.** ? Div. prop. of eq.
$d = -2$	← **c.** ? Symmetric prop.

5. Prove that if $ax^2 + bx = 0$, then $x = 0$ or $-\frac{b}{a}$.

Statements	Reasons
$ax^2 + bx = 0$	← Given
$x^2 + \frac{b}{a}x = 0$	← **a.** ? Div. prop. of eq.
$x(x + \frac{b}{a}) = 0$	← **b.** ? Dist. prop.
$x = 0$ or $x + \frac{b}{a} = 0$	← Zero-product prop.
$x = 0$ or $x = -\frac{b}{a}$	← **c.** ? Subt. prop. of eq.

6. Writing If Cal is Mia's cousin, then Mia is Cal's cousin. This relationship is symmetric. Describe a relationship that is transitive. Answers may vary. Sample: If John is the brother of Lou, and Lou is the brother of Gretchen, then John is the brother of Gretchen.

11-6 Teaching Notes

PROBLEM OF THE DAY

Three consecutive odd integers have a sum of 57. Find the integers. 17, 19, 21

Problem of the Day is also available in Transparencies.

CONNECTING TO PRIOR KNOWLEDGE Ask students to describe typical license plates for the state you live in. Then have students make conjectures about ways to determine how many possible license plates of each design can be made without repeating a combination.

WORK TOGETHER

Some CD players allow users to choose the order in which the CD tracks are played. Point out to students that this is more detail than is being asked in the text. Students can assume that the CDs will be played from beginning to end.

Question 1 Make sure students understand that they need to count only the number of ways in which the CDs can be ordered.

THINK AND DISCUSS

Point out that the order of choosing pants and a shirt does not matter in this example. The tree diagram could be drawn with pants in the left column and still be correct.

Lesson Planning Options

Prerequisite Skills

- Using proportions
- Multiplying integers

Assignment Options for Exercises On Your Own

To provide flexible scheduling, this lesson can be subdivided into parts.

1 **Core** 1–5, 26, 27
Extension 18, 28

2 **Core** 6–17, 22–25
Extension 19–21

Use Mixed Review to maintain skills.

Resources

Student Edition

Skills Handbook, pp. 569, 573
Extra Practice, p. 566
Glossary/Study Guide

Teaching Resources

Chapter Support File, Ch. 11
- Practice 11-6 (two worksheets)
- Reteaching 11-6

Classroom Manager 11-6
Glossary, Spanish Resources
Two-Year Algebra Handbook 11-6

Transparencies
127

Connections Computer Passwords . . . and more

11-6 Counting Outcomes and Permutations

What You'll Learn

- Using the multiplication counting principle to count outcomes
- Using permutations to count outcomes
- Finding probability

...And Why

To investigate the number of arrangements in situations such as the arrangement of players on a team

What You'll Need

- calculator

WORK TOGETHER

Entertainment Suppose you play the following CDs:

- *cracked rear view* by Hootie and the Blowfish
- *Design of a Decade* by Janet Jackson
- *Destiny* by Gloria Estefan

1. In how many different orders can you play the CDs? 6 ways
2. Describe how you determined the number of different ways to order the CDs. See left.
3. What is the probability that the CDs will be played in alphabetical order? $\frac{1}{6}$

2. Answers may vary. Sample: There are 3 choices for the first CD. For each such choice, there are 2 choices for the second and third CDs; so there are $3 \times 2 = 6$ ways to choose all three CDs.

THINK AND DISCUSS

Part 1

Using the Multiplication Counting Principle

One way you could find outcomes is to make an organized list or a tree diagram. Both help you to see if you have thought of all possibilities.

Suppose you have three shirts and two pair of pants that coordinate well together. You can use a tree diagram to find the number of possible outfits you have.

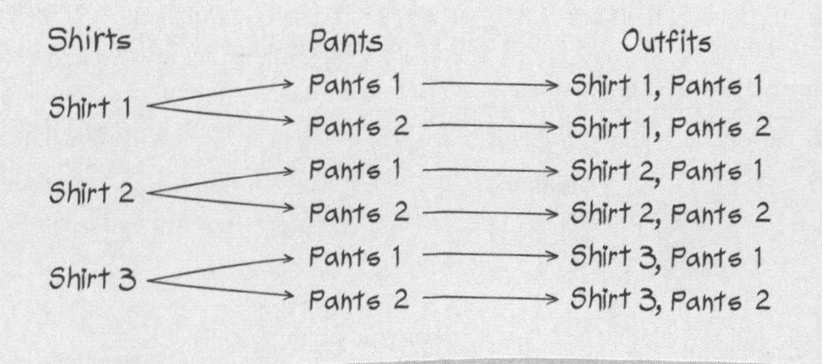

There are six possible outfits.

4. Would you want to use a tree diagram to find the number of outfits for five shirts and eight pairs of pants? Explain.
No; it would take too long to draw because of the large number of branches.

When two events are independent, the order in which they occur has no effect on the result. To find the number of possible outfits, students can calculate $3 \cdot 2 = 6$ or $2 \cdot 3 = 6$.

ESL Ensure students are familiar with *outfit* in this context. It means "a combination of shirt and pants."

Example 1 Relating to the Real World 

VISUAL LEARNING Encourage students to draw a tree diagram to help them visualize the scenario.

Point out that the word *outcome* is used when the problem is about making combinations of choices. The word *permutation* is used when the problem is about the number of ways to order a group of things. The multiplication counting principle applies to both situations.

TACTILE LEARNING To demonstrate permutations, give students three differently-colored tiles or blocks and have them find the number of different ways the three blocks can be arranged. Have them record their findings.

Example 2 Relating to the Real World 

Ask a student who is familiar with baseball to explain what *batting order* means.

VISUAL LEARNING To model the baseball problem, first write the letters A through I on the board. Then choose nine students. Tell the first student to choose a letter from the board. Ask how many choices there are. nine After the first student has chosen a letter, erase it from the board. Tell the second student to choose a letter. Ask how many choices there are. eight Continue through the other seven students.

QUICK REVIEW

When one event does not affect the result of a second event, they are *independent events.*

When events are independent, you can find the number of outcomes by using the multiplication counting principle.

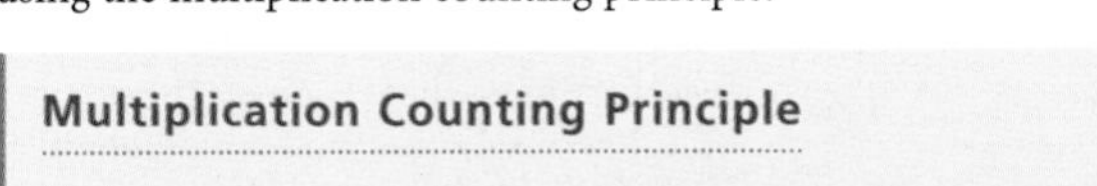

Multiplication Counting Principle

If there are m ways to make a first selection and n ways to make a second selection, there are $m \times n$ ways to make the two selections.

Example: For 3 shirts and 2 pairs of pants, the number of possible outfits is $3 \cdot 2 = 6$.

PIZZAZZ
TONIGHT'S TOPPINGS
vegetable: mushrooms, broccoli, onions, peppers, eggplant
meat: chicken, ground beef, pepperoni

Example 1 Relating to the Real World 

Travel Suppose there are two routes you can choose to get from Austin, Texas, to Dallas, Texas, and four routes from Dallas to Tulsa, Oklahoma. How many routes are there from Austin to Tulsa through Dallas?

$2 \cdot 4 = 8$ ← routes from Austin to Tulsa through Dallas

routes from Austin to Dallas (2); routes from Dallas to Tulsa (4)

There are eight possible routes from Austin to Tulsa.

5. **Try This** At the neighborhood pizza shop, there are five vegetable toppings and three meat toppings for a pizza. How many possible pizzas can you order with one meat and one vegetable topping? 15x possible pizzas

Finding Permutations

Part 

ABC	ACB
BAC	BCA
CAB	CBA

A common kind of counting problem is to find the number of arrangements of a set of objects. The list at the left shows the possible arrangements for the letters A, B, and C without repeating any letters in an arrangement.

Each of the arrangements is called a permutation. A **permutation** is an arrangement of some or all of a set of objects in a specific order.

Example 2 Relating to the Real World 

Sports In how many ways can nine baseball players be listed for batting order?

There are 9 choices for the first batter, 8 for the second, 7 for the third, and so on.

$9 \cdot 8 \cdot 7 \cdot 6 \cdot 5 \cdot 4 \cdot 3 \cdot 2 \cdot 1 = 362880$ ← Use a calculator.

There are 362,880 possible arrangements of the batters.

CALCULATOR HINT

The notation 9!, read as *nine factorial*, means the product of the integers from 9 to 1. Use this sequence to find 9!:

9 MATH PRB! ENTER

6. **Critical Thinking** Why are there only 8 choices for the second batter and 7 choices for the third batter?

6. After the first batter is chosen, 8 players remain from which to choose a second batter. Once the second batter is chosen, there are only 7 batters remaining.

Additional Examples

FOR EXAMPLE 1

The school cafeteria offers four kinds of bread, five kinds of meat, and three kinds of cheese on their sandwiches. How many unique sandwiches are offered by the cafeteria? 60

Discussion: *How many additional toppings would the school have to offer to bring the total possible sandwich combination to 100? 1000? What do your answers suggest about the multiplication counting principal?*

FOR EXAMPLE 2

In how many ways can 25 students enter a classroom? 1.55×10^{25}

FOR EXAMPLE 3

A bag contains 26 alphabet blocks, one for each letter of the alphabet. If you pick out 4 blocks at random, how many possible letter combinations are possible? 358,800

Discussion: *How would your answer change if you replaced each block in the bag before drawing the next one?*

CRITICAL THINKING Question 6 If students have difficulty with this question, help them make the connection between the visual learning activity and the baseball problem.

Ask a student to draw a baseball diamond on the board. Have them use the drawing to explain where right, center, and left fielders play.

AUDITORY LEARNING Have students read aloud $_8P_3$ as "the permutations of eight things chosen three at a time." If they read this notation in full each time, they will soon learn its meaning.

CONNECTING TO THE STUDENTS' WORLD Challenge students to think of a permutation problem they encounter in their daily lives, such as choosing which clothes to pack for a weekend trip. Have students describe the problem and then express it using the correct $_nP_r$ notation.

Example 3 *Relating to the Real World* 

ESL ERROR ALERT! Students who speak other languages may make mistakes when counting the number of outcomes from letters in the alphabet because not all alphabets have 26 letters. **Remediation:** If students have difficulty with this concept, make sure they are counting the letters in the English alphabet. Give them an example of the alphabet if they do not remember it.

Question 9b Point out that the letters can be arranged in any order.

EXTENSION Challenge interested students to find out more about creating passwords that are difficult to break. Invite students to share their findings with the class. Also suggest that they tell the class what they find out about common errors

Technology Options

For Exercises 1–5, students may find it helpful to use a graphing calculator to evaluate the permulations.

Prentice Hall Technology 

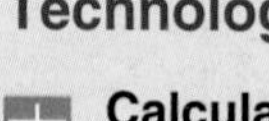

Calculator
- Graphing Calculator Handbook: Procedure 16

Software
- Secondary Math Lab Toolkit™
- Integrated Math Lab 21
- Computer Item Generator 11-6

Internet
- See the Prentice Hall site. (http://www.phschool.com)

CALCULATOR HINT

You can use the following calculator steps to find $_8P_3$:

8 MATH PRB $_NP_R$ 3 ENTER

In how many ways can you select a right, center, and left fielder from eight people? This means finding the number of permutations of 8 objects (players) arranged 3 at a time. You can use $_nP_r$ to express the number of permutations where n equals the number of objects and r equals the number of selections to make. Eight players arranged three at a time is $_8P_3$.

You can use the multiplication counting principle to evaluate permutations.

$_8P_3 = 8 \cdot 7 \cdot 6$, or 336 arrangements of 8 players arranged 3 at a time.

$_nP_r = n(n-1)(n-2)\ldots$

first factor — number of factors

7. a. Use $_nP_r$ to express the number of permutations of twelve players for five positions on a baseball team. $_{12}P_5$

b. *Calculator* Evaluate the number of permutations for part (a). **95,040 permutations**

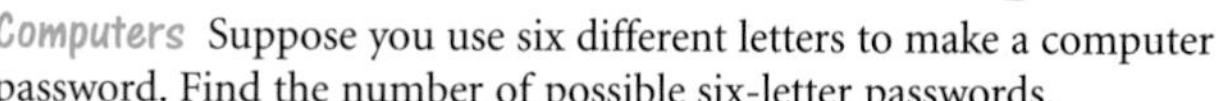

Example 3 *Relating to the Real World*

Computers Suppose you use six different letters to make a computer password. Find the number of possible six-letter passwords.

There are 26 letters in the alphabet. You are finding the number of permutations of 26 letters arranged 6 at a time.

$_{26}P_6 = 26 \cdot 25 \cdot 24 \cdot 23 \cdot 22 \cdot 21 = 165{,}765{,}600$ ← Use a calculator.

There are 165,765,600 six-letter passwords in which letters do not repeat. ■

8. What would $_{26}P_8$ mean if you were making a password? **the number of 8-letter passwords in which letters do not repeat**

9. a. *Probability* Lena wants a password that uses the four letters of her name. How many permutations are possible using each letter in her name only once? **24 permutations**

b. Suppose a four-letter password with no repeated letters is assigned randomly. What is the probability that it uses the letters L, E, N, and A? $\frac{1}{14{,}950}$

c. *Critical Thinking* Is creating a password based on your name a good idea? Explain. **See margin.**

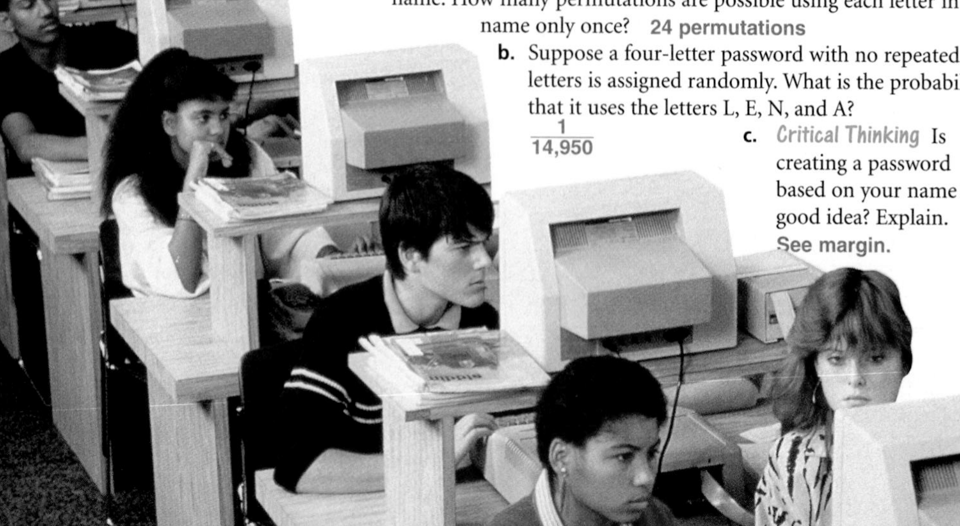

people make when creating passwords (example: choosing names of pets or family members, leaving a copy in an easily-accessible place like a desk drawer).

Exercises ON YOUR OWN

CRITICAL THINKING Exercise 5b To help students get started, suggest that the three-digit number will be even only if there is a 2 in the one's place.

Exercise 5d Remind students that probability is calculated using the formula $\frac{\text{number of favorable outcomes}}{\text{number of possible outcomes}}$.

ALTERNATIVE ASSESSMENT **Exercises 12–15** To assess understanding of permutations, ask students to write a word problem to correspond to each of these exercises. Students may wish to look at Exercises 1–4 for guidelines.

OPEN-ENDED Exercise 20b Make sure students understand why everyone's answer is the same for this exercise.

STANDARDIZED TEST TIP **Exercise 25** You can make a rough estimate of the number of objects in a permutation by taking the *n*th root of the number of permutations. For example, $\sqrt[3]{210} \approx 5.9$ which is close to the *middle* term of the correct answer $7 \times 6 \times 5$. Thus the correct answer choice must be C.

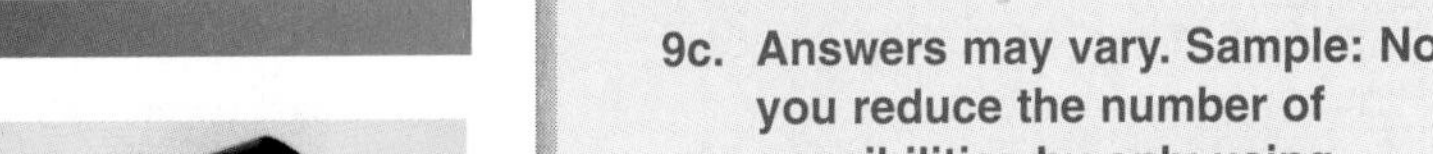

Exercises ON YOUR OWN

1. *Jobs* James must wear a shirt and tie for a job interview. He has two dress shirts and five ties. How many shirt-tie choices does he have? 10 choices
2. *Telephones* A seven-digit telephone number can begin with any digit except 0 and 1.
   a. How many possible choices are there for the first digit? the second digit? the third digit? the seventh digit? 8; 10; 10; 10
   b. How many different seven-digit telephone numbers are possible? 8,000,000 possible telephone numbers
3. On a bookshelf there are five novels, two volumes of short stories, and three biographies. In how many ways can you select one of each? 30 ways
4. A student council has 24 members. A 3-person committee is to arrange a car wash. Each person on the committee will have a task: One person will find a location, another will organize publicity, and the third will schedule workers. In how many different ways can three students be selected and assigned a job? 12,144 ways
5. a. How many different 3-digit numbers are possible using 2, 3, 5, and 7 if you do *not* repeat any digits? 24 different numbers
   b. *Critical Thinking* How many of the 3-digit numbers are even? 6 even numbers
   c. *Writing* Explain how you found how many numbers are even in part (b). There are 6 ways to arrange 3, 5, and 7 to the left of 2.
   d. What is the probability that a number selected at random is even? $\frac{1}{4}$

Calculator **Use a calculator to evaluate.**

6. ${}_8P_3$ 336	7. ${}_7P_3$ 210	8. ${}_6P_3$ 120	9. ${}_5P_3$ 60	10. ${}_4P_3$ 24	11. ${}_3P_3$ 6
12. ${}_7P_7$ 5040	13. ${}_7P_6$ 5040	14. ${}_7P_5$ 2520	15. ${}_7P_4$ 840	16. ${}_7P_2$ 42	17. ${}_7P_1$ 7

18. *School Paper* For an article in the school newspaper, Cora took a poll in which she asked students to rank the top four basketball players on the high school team. There are fifteen members on the team. How many possible outcomes are there? 32,760
19. *Writing* Write a problem for which ${}_{24}P_3$ would be the solution. See margin.
20. a. *Open-ended* Use the letters of your last name. If any letters are repeated, use only one of them. For example, for the last name Bell, use the letters BEL. In how many ways can the resulting letters be arranged?
    b. In how many ways can two different letters be selected and arranged from your last name? a–b. Answers may vary. Sample: J, O, H, N, S from Johnson can be arranged in 120 ways; two of its letters in 20 ways.
21. *Sports* In ice skating competitions, the order in which competitors skate is determined by a draw. If there are eight skaters in the finals, how many different orders are possible for the final program? 40,320 orders

pages 538–540 Think and Discuss

9c. Answers may vary. Sample: No; you reduce the number of possibilities by only using letters from your name. Someone could guess the password easily.

pages 541–542 On Your Own

19. Sample: There are 24 students in homeroom. One will read announcements, one will take attendance, and one will pass out notices. In how many ways can three student helpers be chosen to do these tasks?

page 542 Mixed Review

29\.

30\.

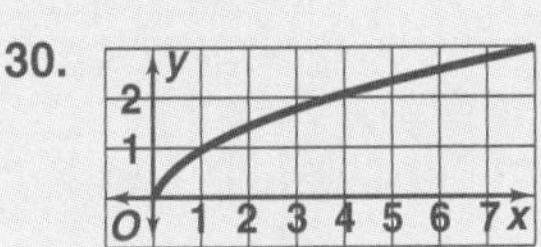

31–32. See back of book.

541

Exercises MIXED REVIEW

SELF ASSESSMENT **JOURNAL** Ask students if they can think of a situation where it might be difficult for someone to decide which meaning is intended in a problem.

Exercise 33 Suggest that students begin by having *n* represent the cost of a night and *p* represent the cost of a one-day pass. Then have students write equations that model the two camping plans.

GETTING READY FOR LESSON 11-7 These exercises prepare students for solving probability problems that include combinations.

Wrap Up

THE BIG IDEA Ask students: *How can you use the counting principle to find the number of possible outcomes? How do you use a tree diagram?*

RETEACHING ACTIVITY Students make a table and use a calculator to find outcomes using permutations. (Reteaching worksheet 11-6)

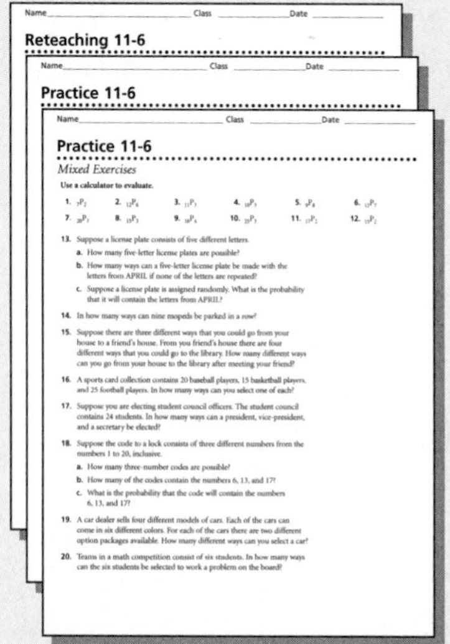

Lesson Quiz

Lesson Quiz is also available in Transparencies.

1. Jeff has five shirts, three pairs of pants, and two pairs of socks. How many possible clothes combinations can he make? 30
2. Social Security numbers consist of ten digits, each of which can have any value between 0 and 9. How many unique Social Security numbers are possible? 1×10^{10}
3. Calculate ${}_6P_3$. 120

Calculator Which is greater?

22. ${}_8P_6$ or ${}_6P_2$ ${}_8P_6$ **23.** ${}_9P_7$ or ${}_9P_2$ ${}_9P_7$ **24.** ${}_{10}P_3$ or ${}_8P_4$ ${}_8P_4$

25. *Standardized Test Prep* For $r = 3$ and ${}_nP_r = 210$, find n. C

A. 3 **B.** 6 **C.** 7 **D.** 8 **E.** 70

Use the tiles at the right for Exercises 26 and 27.

26. a. What is the number of possible arrangements in which you can select four letters? 840

b. Find the probability that you select C, A, R, and then E. $\frac{1}{840}$

27. a. What is the number of possible arrangements in which you can select five letters? 2520

b. Find the probability that you select B, R, A, I, and then N. $\frac{1}{2520}$

A E I B N C R

28. *Safety* A lock, such as the ones used on many bikes and school lockers, uses permutations. This is because the order of the numbers *is* important. Suppose you have a lock with the digits 0–9. A three-digit sequence opens the lock, and no numbers are repeated.

a. How many different sequences are possible? 720

b. How many sequences use a 7 as the first digit? 72

c. What is the probability that the sequence of numbers that opens your lock uses a 7 as the first digit? $\frac{1}{10}$

d. *Critical Thinking* What is the probability that the sequence of numbers that opens your lock includes a 7? $\frac{3}{10}$

Exercises MIXED REVIEW

Graph each function. 29–32. See margin.

29. $y = 6x$ **30.** $y = \sqrt{x}$

31. $y = \frac{5}{x}$ **32.** $y = |x| + 3$

33. *Camping* Your family is planning a camping trip. The state park has two camping plans available. The first costs \$95 for four nights at a campsite with family trail passes for three days. The second costs \$70 for three nights at a campsite with family trail passes for two days. There is one daily charge for the campsite and one daily charge for the trail pass. Find each daily charge. campsite \$20; family trail pass \$5

SELF ASSESSMENT **FOR YOUR JOURNAL**

The letter P is used to indicate permutations and probability. Explain how you can tell what P means in a given problem.

Getting Ready for Lesson 11-7

A and B are independent events. Find P(A and B) for the given probabilities.

34. $P(A) = \frac{1}{3}, P(B) = \frac{3}{4}$ $\frac{1}{4}$ **35.** $P(A) = \frac{1}{8}, P(B) = \frac{5}{9}$ $\frac{5}{72}$ **36.** $P(A) = \frac{9}{10}, P(B) = \frac{5}{6}$ $\frac{3}{4}$

11-7 Teaching Notes

PROBLEM OF THE DAY

At a take-out restaurant, six chicken wings cost $2.10, a dozen wings cost $3.96, and two dozen wings cost $6.96. If this pattern continues, how much would you expect to pay for five dozen wings? (Hint: Find the unit prices.) **$10.20**

Problem of the Day is also available in Transparencies.

CONNECTING TO PRIOR KNOWLEDGE Ask students to describe a permutation question they can recall from Lesson 11-6. Ask students how they think the number of possibilities would be different if the order in which items were chosen no longer mattered.

THINK AND DISCUSS

DIVERSITY Students who follow Kosher dietary rules may point out that turkey and cheese combined in a sandwich is not an acceptable combination for them. Invite students of different cultural backgrounds to compare and contrast different eating practices. Ensure that students show respect for each other's differences.

For students calculating the number of combinations with paper and pencil, point out that most of the factors cancel out, and that greatly simplifies the mathematics.

KINESTHETIC LEARNING Question 2 Select five students to come to the front of the class and represent five presidents. Have them form as many pairs as they can. Have a volunteer keep track of the pairs by writing students' names on the

Connections Restaurants . . . and more

11-7 Combinations

What You'll Learn

- Finding combinations
- Solving probability problems involving combinations

...And Why

To investigate situations like selecting a jury

What You'll Need

- calculator

THINK AND DISCUSS

Food Suppose you are making a sandwich with three of these ingredients: turkey, cheese, tomato, and lettuce. Below are the permutations of the four ingredients taken three at a time.

•turkey, cheese, tomato
•turkey, tomato, cheese
•cheese, tomato, turkey
•cheese, turkey, tomato
•tomato, turkey, cheese
•tomato, cheese, turkey

•turkey, cheese, lettuce
•turkey, lettuce, cheese
•cheese, lettuce, turkey
•cheese, turkey, lettuce
•lettuce, turkey, cheese
•lettuce, cheese, turkey

•turkey, tomato, lettuce
•turkey, lettuce, tomato
•tomato, lettuce, turkey
•tomato, turkey, lettuce
•lettuce, turkey, tomato
•lettuce, tomato, turkey

•cheese, tomato, lettuce
•cheese, lettuce, tomato
•tomato, lettuce, cheese
•tomato, cheese, lettuce
•lettuce, cheese, tomato
•lettuce, tomato, cheese

For sandwiches the *order* of the ingredients does not matter. So there are only four types of sandwiches. Each sandwich type is a **combination,** a collection of objects without regard to order.

The number of combinations of sandwiches equals the number of permutations divided by the number of times each type of sandwich is repeated.

$$\text{number of combinations} = \frac{\text{total number of permutations}}{\text{number of times the objects in each group are repeated}}$$

You can use the notation $_nC_r$ to write the number of combinations of n objects chosen r at time. The number of times each group of objects is repeated depends on r.

$$_4C_3 = \frac{_4P_3}{_3P_3} = \frac{4 \cdot 3 \cdot 2}{3 \cdot 2 \cdot 1} \quad \begin{matrix} \leftarrow \text{4 factors chosen 3 at a time} \\ \leftarrow \text{3 factors chosen 3 at a time} \end{matrix}$$

$$_nC_r = \frac{_nP_r}{_rP_r} = \frac{n(n-1)(n-2)\ldots}{r(r-1)(r-2)\ldots} \quad \begin{matrix} \leftarrow r \text{ factors starting with } n \\ \leftarrow r \text{ factors starting with } r \end{matrix}$$

CALCULATOR HINT

You can use the following calculator steps to find $_4C_3$:

4 MATH PRB $_NC_R$ 3 ENTER

1. *Calculator* Evaluate each expression.
   a. $_4C_2$ 6 b. $_8C_5$ 56 c. $_7C_3$ 35 d. $_{10}C_4$ 210
2. a. *Try This* For your history report, you can choose to write about two of five Presidents of the United States. Use $_nC_r$ notation to write the number of combinations possible for your report. $_5C_2$
   b. Calculate the number of combinations of presidents on whom you could report. 10

Lesson Planning Options

Prerequisite Skills

- Solving equations by combining
- Solving equations with multiplication and division

Assignment Options for Exercises On Your Own

Core 1–15, 20–28, 30–33
Extension 16–19, 29, 34

Use Mixed Review to maintain skills.

Resources

Student Edition

Skills Handbook, pp. 567, 569, 571
Extra Practice, p. 566
Glossary/Study Guide

Teaching Resources

Chapter Support File, Ch. 11
- Practice 11-7 (two worksheets)
- Reteaching 11-7
- Alternative Activity 11-7

Classroom Manager 11-7
Glossary, Spanish Resources
Two-Year Algebra Handbook 11-7

Transparencies
128

board. When students have completed this part of the task, have students explain how this number is different from the number of permutations they could have formed.

Example 1 Relating to the Real World 

If students have difficulty accepting the answer to Example 1, refer students to Question 2 or have them perform the kinesthetic activity for that question. Once students understand how there are ten combinations for just two presidents, they will be better able to grasp the magnitude of the answer to Example 1.

Example 2 Relating to the Real World 

ALTERNATIVE METHOD Before students work this example, have them estimate the probability that the group orders only mild items. Once students have found the answer, have them think about why their estimates were wrong. Use this as a starting point for a discussion of how permutations and combinations cause problems for many people because the answers are so different from what one would estimate.

Students may need to review the definitions of favorable outcomes and possible outcomes.

Question 4 Students might assume that if they like spicy food, the only possible outcomes are spicy items. Show students that the number of possible outcomes is equal to the number of ways to choose five items from 12 items. The number of favorable outcomes is equal to the number of ways to choose five spicy items from five spicy items.

Additional Examples

FOR EXAMPLE 1

A science classroom contains 24 students. Lab groups of 3 students are formed. How many different ways can the first group be chosen? **2024; 1932 groups**

Discussion: *After the first group is chosen, are the total number of possible combinations reduced? Explain.*

FOR EXAMPLE 2

Your upcoming history exam will consist of five questions randomly taken from a chapter that contains ten topics. Each topic can appear only once on the test. What is the probability that the test will consist of questions drawn from only the first five topics? **0.397%**

Discussion: *Explain why there is only one possible combination that can result in a test containing questions drawn from the first five topics, but 120 permutations for the same five topics.*

Example 1 Relating to the Real World

Juries Twenty people report for jury duty. How many different twelve-person juries can be chosen?

The order in which jury members are listed does not distinguish one jury from another. You need the number of combinations of 20 objects chosen 12 at a time.

$$_{20}C_{12} = \frac{20 \cdot 19 \cdot 18 \cdot 17 \cdot 16 \cdot 15 \cdot 14 \cdot 13 \cdot 12 \cdot 11 \cdot 10 \cdot 9}{12 \cdot 11 \cdot 10 \cdot 9 \cdot 8 \cdot 7 \cdot 6 \cdot 5 \cdot 4 \cdot 3 \cdot 2 \cdot 1}$$

$= 125970$  ← Use a calculator.

There are 125,970 different twelve-person juries possible.

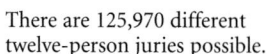

3. For some civil cases, at least nine of twelve jurors must agree on a verdict. How many combinations of nine jurors are possible on a twelve-person jury? **220 possible combinations**

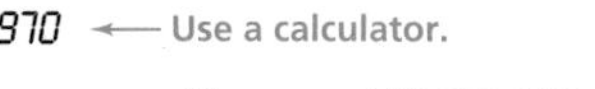

You can use combinations to solve probability problems. When finding the number of favorable outcomes, use the total number of objects that may give you a favorable outcome as the n value in $_nC_r$.

 **Example 2** Relating to the Real World 

Restaurants You and four friends visit a Thai restaurant. There are twelve different items on the menu. Seven are mild and the rest are spicy. You each order a different item at random. What is the probability that your group orders only mild items?

number of favorable outcomes $= {}_7C_5$ ← number of ways to choose 5 mild items from 7 mild items

number of possible outcomes $= {}_{12}C_5$ ← number of ways to choose 5 items from 12 items

$$P(\text{5 mild menu items}) = \frac{\text{number of favorable outcomes}}{\text{total number of outcomes}} = \frac{_7C_5}{_{12}C_5} = \frac{7}{264}$$

The probability your group orders only mild items is $\frac{7}{264}$, or about 3%.

4. Suppose you and your friends like spicy food. What is the probability that you choose five different spicy items at random? $\frac{1}{792}$

Exercises ON YOUR OWN

ALTERNATIVE ASSESSMENT **Exercise 1** To assess students' understanding of combinations, ask students to find the number of three person combinations possible for their algebra class. Then ask them to calculate the number of possible combinations in two algebra classes.

WRITING **Exercise 3** Ask students to include examples of permutations and combinations.

ERROR ALERT! **Exercises 4–15** If students make errors in these exercises, they will not be able to describe any patterns in Exercise 16. **Remediation:** Have students work together in pairs. Have one student use a calculator while the other student either calculates the problem using pencil and paper or helps the partner enter the numbers correctly into the calculator.

GEOMETRY **Exercise 17** Point out that all the points should be connected to each other by lines, including diagonal lines.

TACTILE LEARNING **Exercise 18** If students have difficulty visualizing the problem, direct them to draw Figure 2 enlarged on a piece of paper. Make sure their drawings have ten dots. Then, have them connect the pairs of dots and count how many line segments they draw. Suggest that students draw five segments in one color ink and then change colors so that the number of segments is easier to count. Alternatively, students can write the number of the segment next to each line segment.

Exercise 19 Point out that choosing a fruit and a cup of fruit juice does not count as repeating a fruit.

Exercises ON YOUR OWN

1. a. *Spelling Bee* Every spring the National Spelling Bee Championships are held in Washington, D.C. Prizes are awarded to the top three spellers. In 1995, 247 students competed. How many different arrangements of winners were possible? **14,886,690 different arrangements**
   b. Many students in the competition hoped to make it to the final round of 10 students. How many combinations of 10 students were possible? **about 1.94×10^{17} combinations**

2. *Sports* A basketball team has 11 players. Five players are on the court at a time. Your little brother doesn't know much about basketball and randomly names 5 players on the team. What is the probability that your brother's line-up lists the same 5 players as the coach's line-up for the next game? $\frac{1}{462}$

3. *Writing* Explain the difference between a permutation and a combination. **A permutation is a set of objects in a specific order. A combination is a set of objects without regard for order.**

Calculator **Use a calculator to evaluate.**

4. $_6C_6$ 1	5. $_6C_5$ 6	6. $_6C_4$ 15	7. $_6C_3$ 20	8. $_6C_2$ 15	9. $_6C_1$ 6
10. $_{15}C_{11}$ 1365	11. $_{15}C_4$ 1365	12. $_8C_5$ 56	13. $_8C_3$ 56	14. $_7C_4$ 35	15. $_7C_3$ 35

16. a. Describe any patterns you see in the answers to Exercises 4–15. $_nC_r = {}_nC_{n-r}$
    b. *Critical Thinking* Explain why the pattern you described is true. **See margin.**
    c. *Open-ended* Write two more combinations that have the pattern you described in part (a). **Sample: $_{11}C_9 = {}_{11}C_2$; $_{10}C_6 = {}_{10}C_4$**

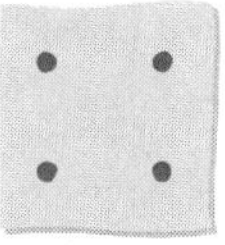

17. a. *Geometry* Draw four points on your paper like those in Figure 1. Draw line segments so that every point is joined with every other point. **See margin.**
    b. How many segments did you draw? **6 segments**
    c. Now find the number of segments you need for the drawing using combinations. You are joining four points, taking two at a time.
    d. How many segments would you need to join each point to all the others in Figure 2? **45 segments** **c. 6 segments**

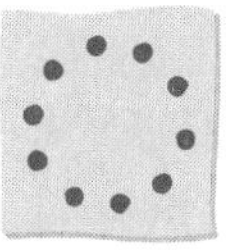

Figure 1

Figure 2

18. A famous problem, known as the Handshake Problem asks, "If ten people in a room shake hands with everyone else in the room, how many different handshakes occur?"
    a. Karen thinks that the Handshake Problem is a permutation problem while Tashia thinks it is a combination problem. What do you think? **Justify** your answer. **Combination; order is not important.**
    b. Solve the problem. **45 different handshakes**
    c. *Critical Thinking* Is the Handshake Problem similar to the problem in Exercise 17 (d)? Explain. **Yes; each problem involves the combinations of 10 objects taken two at a time.**

Technology Options

For Exercises 25–28, students may use a graphing calculator or algebra software to find the number of combinations.

Prentice Hall Technology

Calculator
- Graphing Calculator Handbook: Procedure 16

Software
- Secondary Math Lab Toolkit™
- Integrated Math Lab 22
- Computer Item Generator 11-7

Internet
- See the Prentice Hall site. (http://www.phschool.com)

STANDARDIZED TEST TIP Exercise 24 You can often compare combination expressions without multiplying them. First write out each expression: $\frac{10 \cdot 9 \cdot 8 \cdot 7 \cdot 6 \cdot 5 \cdot 4}{7 \cdot 6 \cdot 5 \cdot 4 \cdot 3 \cdot 2 \cdot 1}$, $\frac{7 \cdot 6 \cdot 5 \cdot 4}{4 \cdot 3 \cdot 2 \cdot 1}$ and then simplify each expression. After simplification, the expressions become $10 \cdot 4 \cdot 3$ and $7 \cdot 5$, which can be compared easily.

Exercise 31 Invite students to make conjectures about why the manager would choose to test five screens when only three are defective.

CRITICAL THINKING Exercise 32d Once students have calculated how unlikely it is that a committee of all boys could be chosen by chance, use the answer to lead a discussion on discrimination. Ask students whether they think it would be reasonable to accuse the people selecting the committee of discriminating against girls if they choose a committee of only boys.

RESEARCH Exercise 34 Tell students that they can find the number of members of their state legislature in the library in books about their state government or on their state government's home page on the World Wide Web.

Chapter Project **FIND OUT BY COMMUNICATING** Students choose a musical instrument and find out how it works. You may wish to check students' progress on the project by having each student write a short progress report telling what has been done so far and their plans for completion.

pages 545–547 On Your Own

16b. Answers may vary. Sample: Each combination of r objects corresponds to a combination of the other $n - r$ objects.

17a. 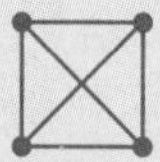

29. Answers may vary.

Sample permutation problem: The 8 finalists in the dog show are to be paraded single file in front of the judges. In how many ways can this be done?

Sample combination problem: Seven greyhounds are available for adoption. In how many ways can two be chosen?

14

HEALTH NOTES

The U.S. Department of Health and Human Services and the Department of Agriculture guidelines recommend that daily food intake should include at least 3–5 servings of vegetables, at least 2–4 servings of fruit, 6–11 servings of grains, 2–3 servings of milk or cheese, and 2–3 servings of meat or poultry.

19. *Math in the Media* Suppose you want to have the highest number of recommended daily servings of fruit without repeating a fruit. Find how many different combinations of these products you can have. 126 different combinations

One Serving of Fruit

apple	1
apple juice	1 cup
banana	1
cherries	10
grapefruit	$\frac{1}{2}$
orange	1
orange juice	1 cup
raisins	1 cup
strawberries	1 cup

Classify each of the following as a permutation or a combination problem. Explain your choice.

20. A locker contains eight books. You select three books at random. How many different sets of books could you select? Combination; the order of the books is not important.
21. You rent three videos to watch during spring vacation. In how many different orders can you view the three videos? Permutation; the order of viewing is important.
22. The security guard at an auto plant must visit 10 different sections of the building each night. He varies his route each evening. How many different routes are possible? Permutation; the order of sections visited is important.
23. A committee of four people needs to be formed from your homeroom of 25 students. How many four-person committees are possible? Combination; the order of committee members is not important.
24. *Standardized Test Prep* Compare the quantities in Columns A and B. B

Column A	Column B
${}_{10}C_7$	${}_7P_4$

A. The quantity in Column A is greater.
B. The quantity in Column B is greater.
C. The quantities are equal.
D. The relationship cannot be determined from the given information.

Find the number of combinations of letters taken three at a time from each set of letters.

25. A B C D E 10
26. P Q R 1
27. E F G H I J K 35
28. M N O P 4

29. *Open-ended* Write and solve two problems: one that can be solved using permutations and one that can be solved using combinations. See margin.

30. For your birthday you received a gift certificate from a local music store for three CDs. There are eight that you would like to have. In how many different ways can you select your CDs? 56 different ways

Exercises MIXED REVIEW

PORTFOLIO Share with students the criteria you will use to assess their work in portfolios, as well as how you plan to use the results. Students should understand how the rubrics are used to assess their work, how each piece in the portfolio counts, and how the scores they get in their portfolios will affect their overall evaluation.

Exercise 47a Ask students whether they need the dimensions of the driveway to answer part (a) of the problem.

Exercise 47b Tell students to assume that the snow falls evenly on the driveway.

Wrap Up

THE BIG IDEA Ask students: *How can you find the number of combinations? How do you solve problems that require distinguishing between permutations and combinations?*

RETEACHING ACTIVITY Students use calculators to solve problems involving combinations. (Reteaching worksheet 11-7)

31. **Manufacturing** Twelve computer screens are stored in a warehouse. The warehouse manager knows that three of the screens are defective, but the report telling him which ones are defective is missing. He selects five screens to begin testing.
    a. How many different choices of five screens does the manager have? **792 different choices**
    b. In how many ways could he select five screens that include the defective ones? **36 ways**
    c. What is the probability that he finds the three defective screens when he tests the first five screens? $\frac{1}{22}$
32. a. A committee of three people is to be selected from three boys and five girls. How many different committees are possible? **56 different committees**
    b. How many committees could have all boys? **1 committee**
    c. What is the probability that a committee will have all boys? $\frac{1}{56}$
    d. **Critical Thinking** What is the probability that the committee will have no boys? $\frac{5}{28}$
33. The letters A, B, C, D, E, F, G, H, I, and J are written on slips of paper and placed in a hat. Two letters are drawn from the hat.
    a. What is the number of possible combinations of letters? **45**
    b. How many combinations consist only of the vowels A, E, or I?
    c. What is the probability the letters chosen consist only of vowels?
    d. What is the probability the letters chosen consist only of B, C, D or F? **b. 3 combinations** **c.** $\frac{1}{15}$ **d.** $\frac{7}{15}$
34. **Research** Find the number of members in your state legislature. Find how many ways a committee of ten people may be formed from the members of the legislature. **Check students' work.**

Find Out by Communicating

Every musical instrument has a part that vibrates to make different pitches. Choose an instrument and find out how it works. Make a poster that describes what vibrates and how different pitches are produced.

Exercises MIXED REVIEW

Evaluate.

35. $_4P_2$ **12**
36. $_8P_3$ **336**
37. $_7P_6$ **5040**
38. $_8P_6$ **20,160**
39. $_9P_1$ **9**
40. $_7P_4$ **840**

Express each sum or difference in simplest form.

41. $\frac{4x^2}{3x} + \frac{5}{3x}$ $\frac{4x^2+5}{3x}$
42. $\frac{5c}{8m} - \frac{7m}{c}$ $\frac{5c^2-56m^2}{8mc}$
43. $\frac{9x}{x^2-4} + \frac{4}{x+2}$
44. $\frac{9}{z} + \frac{10}{3z}$ $\frac{37}{3z}$
45. $\frac{12}{5n} - \frac{15}{4n}$ $-\frac{27}{20n}$
46. $\frac{8w^2}{3-w} - \frac{5w}{18-2w^2}$

47. a. **Weather** After a heavy snowfall, a homeowner had to clear about four tons of snow off a 20 ft by 20 ft driveway. A full shovel of snow weighs about 7 lb. How many times would the homeowner have to fill his shovel to clear the driveway? **about 1143 times**
    b. How many tons of snow would be on a 20 ft by 30 ft driveway? **6 t**

PORTFOLIO

Select one or two items from your work for this chapter. Consider:
- corrected work
- diagrams, graphs, or charts
- a journal entry

Explain why you have included each selection you make.

43. $\frac{13x-8}{(x+2)(x-2)}$

46. $\frac{w(16w^2+48w-5)}{(3-w)(3+w)}$

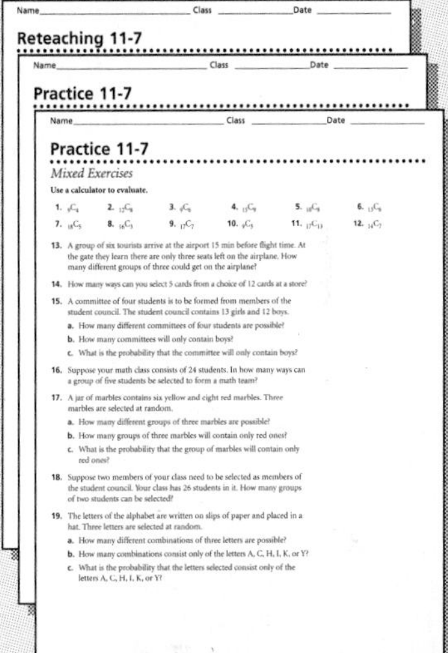
Name____ Class ____ Date ____

Reteaching 11-7

Name____ Class ____ Date ____

Practice 11-7

Name____ Class ____ Date ____

Practice 11-7

Mixed Exercises

Use a calculator to evaluate.

1. $_9C_4$ 2. $_{12}C_8$ 3. $_9C_6$ 4. $_{15}C_9$ 5. $_{10}C_9$ 6. $_{13}C_9$
7. $_{18}C_5$ 8. $_{16}C_3$ 9. $_{17}C_7$ 10. $_9C_5$ 11. $_{17}C_{13}$ 12. $_{14}C_7$

13. A group of six tourists arrive at the airport 15 min before flight time. At the gate they learn there are only three seats left on the airplane. How many different groups of three could get on the airplane?
14. How many ways can you select 5 cards from a choice of 12 cards at a store?
15. A committee of four students is to be formed from members of the student council. The student council contains 13 girls and 12 boys.
    a. How many different committees of four students are possible?
    b. How many committees will only contain boys?
    c. What is the probability that the committee will only contain boys?
16. Suppose your math class consists of 24 students. In how many ways can a group of five students be selected to form a math team?
17. A jar of marbles contains six yellow and eight red marbles. Three marbles are selected at random.
    a. How many different groups of three marbles are possible?
    b. How many groups of three marbles will contain only red ones?
    c. What is the probability that the group of marbles will contain only red ones?
18. Suppose two members of your class need to be selected as members of the student council. Your class has 26 students in it. How many groups of two students can be selected?
19. The letters of the alphabet are written on slips of paper and placed in a hat. Three letters are selected at random.
    a. How many different combinations of three letters are possible?
    b. How many combinations consist only of the letters A, C, H, I, K, or Y?
    c. What is the probability that the letters selected consist only of the letters A, C, H, I, K, or Y?

Lesson Quiz

Lesson Quiz is also available in Transparencies.

1. When ordering a meal in a restaurant, you are given a choice of three side items. If there are seven possible choices, and you choose each item no more than once, how many different side combinations are possible? **35**
2. What is the probability that both you and a friend order the same three side items from the choices listed in Question 1? **2.9%**
3. Evaluate $_4C_2$. **6**
4. Evaluate $_3C_1$. **3**

Finishing the Chapter Project

PROJECT DAY You may wish to plan a project day on which students share their completed projects. Encourage groups to explain their process as well as their product.

PROJECT NOTEBOOK Have students review their project work and bring their notebooks up to date.

- Have students review their research, equations, graphs, and explanations that they needed for the project.
- Ask groups to share any insights they found for completing the project, such as any shortcuts they found in doing research, solving equations, or what they learned while making cardboard instruments.

SCORING RUBRIC

3 Student participated in each activity. Calculations are accurate and contain only minor errors. Research on the chosen instrument is thorough and well documented.

2 Student participated in most of the activities. Calculations are accurate but may contain several minor errors. Research and report show effort but could be more thorough.

1 Student participated in some of the activities. Calculations could be more accurate. Research could be more thorough.

0 Major elements of the project are incomplete or missing.

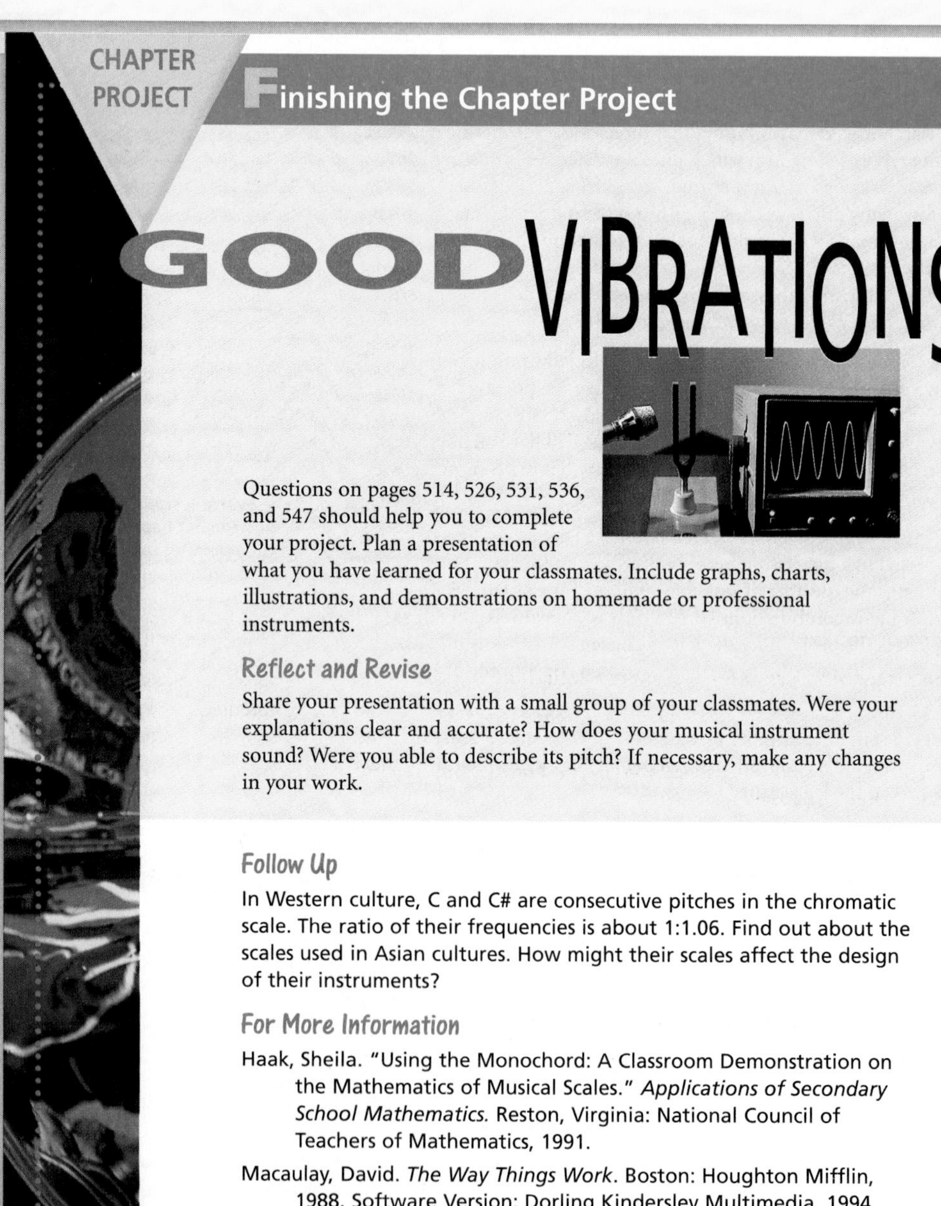

CHAPTER PROJECT

Finishing the Chapter Project

GOOD VIBRATIONS

Questions on pages 514, 526, 531, 536, and 547 should help you to complete your project. Plan a presentation of what you have learned for your classmates. Include graphs, charts, illustrations, and demonstrations on homemade or professional instruments.

Reflect and Revise

Share your presentation with a small group of your classmates. Were your explanations clear and accurate? How does your musical instrument sound? Were you able to describe its pitch? If necessary, make any changes in your work.

Follow Up

In Western culture, C and C# are consecutive pitches in the chromatic scale. The ratio of their frequencies is about 1:1.06. Find out about the scales used in Asian cultures. How might their scales affect the design of their instruments?

For More Information

Haak, Sheila. "Using the Monochord: A Classroom Demonstration on the Mathematics of Musical Scales." *Applications of Secondary School Mathematics.* Reston, Virginia: National Council of Teachers of Mathematics, 1991.

Macaulay, David. *The Way Things Work*. Boston: Houghton Mifflin, 1988. Software Version: Dorling Kindersley Multimedia, 1994.

Wilkinson, Scott R. *Tuning In: Microtonality in Electronic Music: A Basic Guide to Alternate Scales, Temperaments, and Microtuning Using Synthesizers*. Milwaukee, Wisconsin: H. Leonard Books, 1988.

Wrap Up

HOW AM I DOING? Separate students into seven groups, one for each lesson in the chapter. Ask each group to prepare for presentation to the class a *non-verbal* explanation of the main idea from their section. Encourage students to use illustrations, modeling, and movement in their presentations.

KEY TERMS The numbers in parentheses direct students to the pages where the terms (or symbols) are used or defined. Students should be able to (1) write a simple explanation of each term, (2) illustrate the term with a diagram, or (3) show an example that uses the term.

Exercises 6, 8 and 9 Remind students that the domain of a quadratic function cannot have a value of zero for the function's denominator.

Exercise 12–15 Emphasize that factoring the numerator and denominator of a rational function often provides important clues to simplifying it.

Exercises 19 Remind students that in order to divide two fractions, they should multiply the first fraction by the reciprocal of the second fraction.

Exercises 22 and 23 Remind students that factoring the denominator will often make the LCD easier to find.

Exercises 29–34 Point out that the permutation formula assumes that after choosing a given object, that specific object cannot be chosen again.

11 Wrap Up

Key Terms

asymptote (p. 516)
combination (p. 543)
constant of variation (p. 510)
domain (p. 522)
inverse variation (p. 510)
multiplication counting principle (p. 539)
${}_nC_r$ (p. 543)
${}_nP_r$ (p. 540)
permutation (p. 539)
rational equation (p. 532)
rational expression (p. 522)
rational function (p. 516)

How am I doing?

- State three ideas from this chapter that you think are important. Explain your choices.
- Explain the importance of restrictions on the variable in rational expressions.

Inverse Variation — 11-1

When two quantities are related so that their product is constant, they form an **inverse variation.** An inverse variation can be written $xy = k$, where k is the **constant of variation.**

Find the constant of variation for each inverse variation.

1. $m = 6$ when $n = 1$ 6
2. $g = 90$ when $h = 0.1$ 9
3. $s = 88$ when $t = 0.05$ 4.4

For each table, tell whether it represents an inverse variation. If so, write an equation to model the data.

4.

x	1	1.4	3	7
y	4	5.6	12	28

no

5.

x	4	1	2	2.5
y	2.65	10.6	5.3	4.24

yes; $xy = 10.6$

Rational Functions — 11-2

A **rational function** can be written in the form $f(x) = \frac{\text{polynomial}}{\text{polynomial}}$. The graph of a rational function may have asymptotes. An **asymptote** is a line that a graph approaches, but does not cross.

Graph each function. Use a dashed line for each asymptote. 6–9. See margin.

6. $y = \frac{8}{x}$
7. $y = \frac{20}{x}$
8. $y = \frac{6}{x - 5}$
9. $y = \frac{3}{x} + 2$

10. *Open-ended* Write the equation of a rational function with a graph that is in three quadrants only. Answers may vary. Sample: $y = \frac{1}{x + 1}$

11. *Writing* Explain why the function $f(x) = \frac{5}{x + 3}$ has asymptotes.
The graph of $f(x) = \frac{5}{x + 3}$ gets closer and closer to $x = -3$ and to $y = 0$.

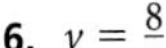

Resources

Student Edition

Extra Practice, p. 566
Glossary/Study Guide

Teaching Resources

Study Skills Handbook
Glossary, Spanish Resources

pages 549–551 Wrap Up

6.

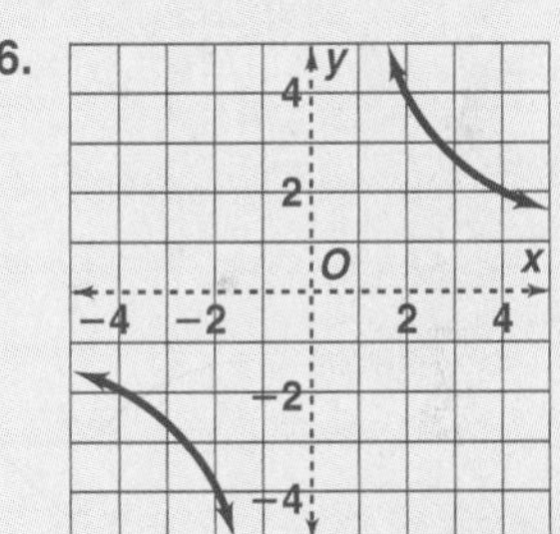

7.

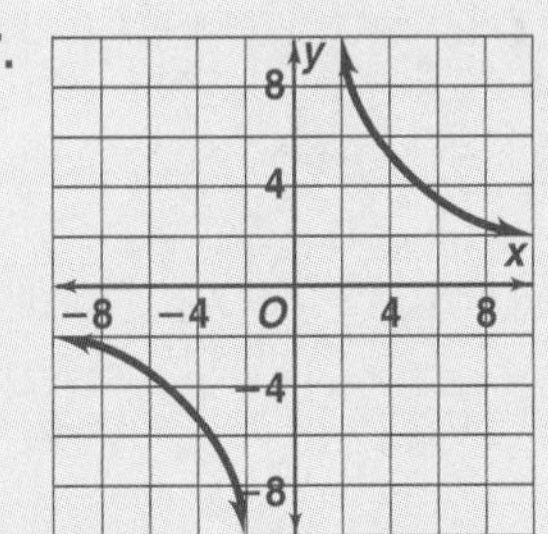

Remind students that the new mathematical terms in this chapter are defined in the Glossary/Study Guide in the back of the book.

8. 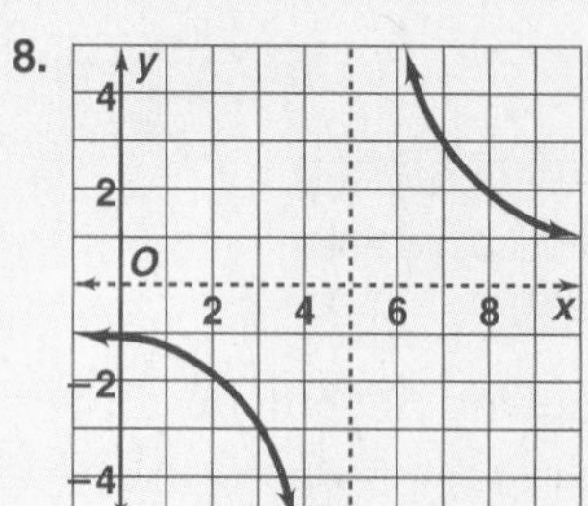

9. 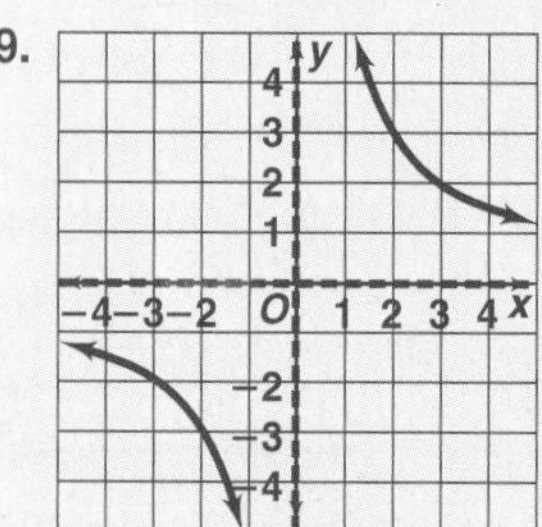

Rational Expressions 11-3

A **rational expression** can be written in the form $\frac{\text{polynomial}}{\text{polynomial}}$ where a variable is in the denominator. The **domain** of a variable is all real numbers excluding the values for which the denominator is zero. A rational expression is in simplest form when the numerator and denominator have no common factors other than 1.

You can multiply and divide rational expressions.

$\frac{a}{b} \cdot \frac{c}{d} = \frac{ac}{bd}$, where b and d are nonzero

$\frac{a}{b} \div \frac{c}{d} = \frac{a}{b} \cdot \frac{d}{c}$, where b, c, and d are nonzero

Simplify each expression and state any restrictions on the variable.

12. $\frac{x^2 - 4}{x + 2}$ $x - 2; x \neq -2$

13. $\frac{5x}{20x + 15}$ $\frac{x}{4x + 3}; x \neq -\frac{3}{4}$

14. $\frac{-3t}{t^3 - t^2}$ $\frac{-3}{t(t - 1)}; t \neq 0, 1$

15. $\frac{z + 2}{2z^2 + z - 6}$ $\frac{1}{2z - 3}; z \neq -2, \frac{3}{2}$

Find each product or quotient.

16. $\frac{8}{m - 3} \cdot \frac{3m}{m + 1}$ $\frac{24m}{(m - 3)(m + 1)}$

17. $\frac{2e + 1}{8e - 4} \div \frac{4e^2 + 4e + 1}{4e - 2}$ $\frac{1}{2(2e + 1)}$

18. $\frac{5c + 3}{c^2 - 1} \cdot \frac{c + 1}{2c}$ $\frac{5c + 3}{2c(c - 1)}$

19. $\frac{4n + 8}{3n} \div \frac{4}{9n}$ $3(n + 2)$

Operations with Rational Expressions 11-4

You can add and subtract rational expressions. Restate each expression with the LCD as the denominator, then add or subtract the numerators.

Add or subtract.

20. $\frac{5}{k} + \frac{3}{k}$ $\frac{8}{k}$

21. $\frac{8x}{x - 7} - \frac{4}{x - 7}$ $\frac{4(2x - 1)}{x - 7}$

22. $\frac{9}{3x - 1} + \frac{5x}{2x + 3}$ $\frac{15x^2 + 13x + 27}{(3x - 1)(2x + 3)}$

23. $\frac{7m}{(m + 1)(m - 1)} - \frac{10}{m + 1}$ $\frac{-3m + 10}{(m + 1)(m - 1)}$

24. Standardized Test Prep What is the LCD of $\frac{1}{4}$, $\frac{2}{x}$, $\frac{5x}{3x - 2}$, and $\frac{3}{8x}$? C

A. $96x^3 - 64x^2$ B. $12x + 2$ C. $24x^2 - 16x$ D. $27x^2 - 6x - 8$ E. $6x + 2$

Solving Rational Equations 11-5

You can use the least common denominator (LCD) to solve **rational equations.** Check possible solutions in the original equation to make sure that they do not make the denominator equal to zero.

Solve each equation.

25. $\frac{1}{2} + \frac{3}{t} = \frac{5}{8}$ 24

26. $\frac{3}{m - 4} + \frac{1}{3(m - 4)} = \frac{6}{m}$ 9

27. $\frac{2c}{c - 4} - 2 = \frac{4}{c + 5}$ −14

28. Business A new photocopier can make 72 copies in 2 min. When an older photocopier is working, the two photocopiers can make 72 copies in 1.5 min. How long does it take the older photocopier working alone to make 72 copies? 6 min

ALTERNATIVE ASSESSMENT After students have completed these exercises, ask each to write a series of problems to illustrate the major ideas of the chapter. Each student should also prepare a corresponding answer key. You may also wish to have students take each other's practice "tests" to reinforce the chapter's main concepts.

Counting Outcomes and Permutations 11-6

You can find a number of different outcomes using the **multiplication counting principle.** If there are m ways to make a first selection and n ways to make a second selection, there are $m \times n$ ways to make the two selections.

A **permutation** is an arrangement of objects in a definite order. To calculate $_nP_r$, the number of permutations of n objects taken r at a time, use the formula

$_nP_r = n(n - 1)(n - 2)\ldots$ ← r factors starting with n

Evaluate.

29. $_5P_3$ 60 **30.** $_8P_4$ 1680 **31.** $_6P_4$ 360 **32.** $_5P_2$ 20 **33.** $_9P_4$ 3024 **34.** $_7P_2$ 42

35. a. *Telephones* Before 1995, three-digit area codes could begin with any number except 0 or 1. The middle number was either 0 or 1, and the last number could be any digit. How many possible area codes were there? **160 possible area codes**

b. Beginning in 1995, area codes were not limited to having 0 or 1 as the middle number. How many new area codes became available? **640 new area codes**

Combinations 11-7

A **combination** is an arrangement of objects without regard to order. The

$$\text{number of combinations} = \frac{\text{number of permutations}}{\text{number of times the objects in each group are repeated}}.$$

To find $_nC_r$, the combinations of n objects taken r at a time, use the formula

$$_nC_r = \frac{_nP_r}{_rP_r} = \frac{n(n-1)(n-2)\ldots}{r(r-1)(r-2)\ldots} \quad \begin{matrix}\leftarrow r \text{ factors starting with } n\\ \leftarrow r \text{ factors starting with } r\end{matrix}$$

Evaluate.

36. $_6C_5$ 6 **37.** $_{10}C_4$ 210 **38.** $_9C_2$ 36 **39.** $_{11}C_3$ 165 **40.** $_5C_4$ 5 **41.** $_6C_4$ 15

42. *Nutrition* You want to have three servings of dairy products without repeating a food. Milk, yogurt, cottage cheese, and cheddar cheese are in the refrigerator. How many different combinations can you have? **4 different combinations**

43. A group of 10 friends go to an amusement park. There are 8 available seats on the roller coaster car. In how many different ways can the friends sit in the seats? **1,814,400 ways**

44. You subscribe to eight monthly magazines. Two are news magazines, three are on sports, two are on health and fitness, and one is about gardening. What is the probability that the next two magazines you receive in the mail are sports magazines? $\frac{3}{28}$

Assessment

ENHANCED MULTIPLE CHOICE QUESTIONS are more complex than traditional multiple choice questions, which assess only one skill. Enhanced multiple choice questions assess the processes that students use, as well as the end results. The questions are written so that students must use more than one strategy to solve the problem. Using multiple strategies is encouraged by the National Council of Teachers of Mathematics (NCTM). **Exercise 5** is an enhanced multiple choice question.

FREE RESPONSE QUESTIONS do not give answer choices. Some exercises have more than one possible answer. Students need to give only one correct response. **Exercises 1–4, 6–10, 12–15, and 17–36** are free response questions.

WRITING EXERCISES allow students to describe how they think about and understand the concepts they have learned. **Exercise 11** is a writing exercise.

OPEN-ENDED PROBLEMS allow for more than one solution. Students must construct their own responses instead of choosing from possible answers. The student responses will help you determine the depth of their understanding and any possible areas of difficulty. **Exercise 16** is an open-ended problem. Make sure that the students' responses contain values for x and y and a written situation.

Resources

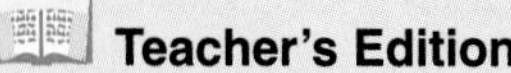

Teaching Resources

Chapter Support File, Ch. 11
- Chapter Assessment, Forms A and B
- Alternative Assessment

Chapter Assessment, Spanish Resources

Teacher's Edition

See also p. 508E for assessment options.

Software

Computer Item Generator

page 552 Assessment

7.

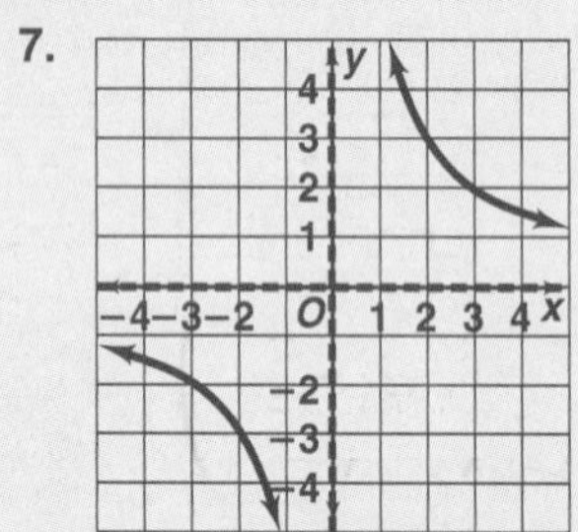

8.

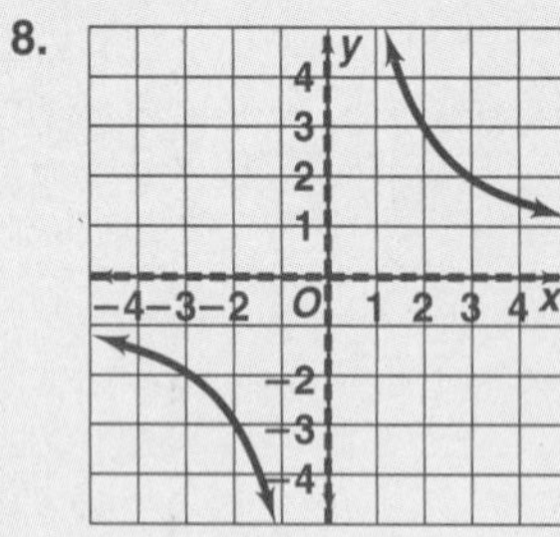

9–11. See back of book.

11 Assessment

Find the constant of variation for each inverse variation.

1. $y = 5$ when $x = 6$ 30
2. $y = 2.4$ when $x = 10$ −24
3. $y = 78$ when $x = 0.1$ 7.8
4. $y = 5.3$ when $x = 9.1$ 48.23

5. *Standardized Test Prep* Which point is *not* on the same graph of an inverse variation as the others? D
   A. (3, 12) B. (9, 4) C. (6, 6)
   D. (16, 2) E. (2, 18)

6. It took you 1.5 h to drive to a concert at 40 mi/h. How long will it take you to drive back driving at 50 mi/h? 1.2 h

Graph each function. Include a dashed line for each asymptote. 7–10. See margin.

7. $y = \frac{6}{x}$
8. $y = \frac{15}{x}$
9. $y = \frac{1}{x} + 3$
10. $y = \frac{4}{x} + 3$

11. *Writing* Explain how direct variations and inverse variations are similar and different. Include an example of each. See margin.

Find each product or quotient.

12. $\frac{3}{x-2} \cdot \frac{x^2-4}{12}$ $\frac{x+2}{4}$
13. $\frac{5x}{x^2+2x} \div \frac{30x^2}{x+2}$ $\frac{1}{6x^2}$
14. $\frac{4w}{3w-5} \cdot \frac{7}{2w}$ $\frac{14}{3w-5}$
15. $\frac{6c-2}{c+5} \div \frac{3c-9}{c}$ See below.

16. *Open-ended* Write a rational expression for which 6 and 3 are restricted from the domain. Answers may vary. Sample: $\frac{x+2}{(x-6)(x-3)}$

Solve each equation.

17. $\frac{v}{3} + \frac{v}{v+5} = \frac{4}{v+5}$ about 1.29, −9.29
18. $\frac{16}{x+10} = \frac{8}{2x-1}$ 4
19. $\frac{2}{3} + \frac{t+6}{t-3} = \frac{18}{2(t-3)}$ no solution

15. $\frac{2c(3c-1)}{3(c+5)(c-3)}$

20. If three people can clean an apartment working two hours each, how long will it take a crew of four people? $1\frac{1}{2}$ h

Simplify.

21. $\frac{5}{t} + \frac{t}{t+1}$ $\frac{t^2+5t+5}{t(t+1)}$
22. $\frac{9}{n} - \frac{8}{n+1}$ $\frac{n+9}{n(n+1)}$
23. $\frac{2y}{y^2-9} - \frac{1}{y-3}$ $\frac{1}{y+3}$
24. $\frac{4b-2}{3b} + \frac{b}{b+2}$ See below.

Classify each of the following as a combination or permutation problem. Explain your choice. Then solve the problem.

25. The 30-member debate club needs a president and treasurer. How many different pairs of officers are possible? perm.; order is important; 870 diff. pairs
26. How many different ways can you choose two books from the six books on your shelf? comb.; order is not important; 15 different ways
27. You have enough money for two extra pizza toppings. If there are six possible toppings, how many different pairs of toppings can you choose? comb.; order is not important; 15 different pairs of toppings

Find the number of combinations of letters taken four at a time from each set.

28. A E I O U Y 15
29. E Q U A T I O N 70
30. L E A R N 5

Evaluate.

31. ${}_4C_3$ 4
32. ${}_8P_6$ 20,160
33. ${}_{10}P_7$ 604,800
34. ${}_5C_2$ 10

35. You have 5 kinds of wrapping paper and 4 different colored bows. How many different combinations of paper and bows can you have? 20 different combinations
36. There are 15 books on your summer reading list. Three of them are plays, one is poetry, and the rest are novels. What is the probability that you will choose three novels if you choose the books at random? $\frac{33}{91}$

24. $\frac{7b^2+6b-4}{3b(b+2)}$

Preparing for Standardized Tests

Standardized tests, such as those administered for state assessment, the SAT, or the ACT, include regular math questions, quantitative comparison questions, open-ended problems, and free response questions (which the SAT calls *grid-ins*).

MULTIPLE CHOICE QUESTIONS are followed by five answer choices, one of which is correct. **Exercises 1–9** are multiple choice questions.

QUANTITATIVE COMPARISON QUESTIONS ask students to compare two quantities. **Exercises 10 and 11** are quantitative comparison questions.

FREE-RESPONSE QUESTIONS do not give answer choices. Students must provide the one correct answer on their own. **Exercises 12–16** are free-response questions.

OPEN-ENDED PROBLEMS allow for more than one solution. Students must construct their own responses instead of choosing a single answer. The responses students give will help you determine the depth of their understanding and what difficulties, if any, they are experiencing. This standardized test does not have an open-ended problem.

STANDARDIZED TEST TIPS Exercise 4 Point out that the negative sign in front of the 10 suggests that the final form of the equation will be (x + *a*)(x − *a*). Thus, answer choices C and D may be eliminated.

11 Preparing for Standardized Tests

For Exercises 1–11, choose the correct letter.

1. Subtract $(3x^2 - 7x - 2) - (8x - 3)$. D
   A. $3x^2 + x - 5$ B. $-5x^3 + 1$
   C. $3x^2 - 5x - 5$ D. $3x^2 - 15x + 1$
   E. none of the above

2. The value of x varies inversely with y, and $x = 2$ when $y = 5$. What is x when $y = 10$? E
   A. 0 B. 4 C. 5 D. 7 E. 1

3. Multiply $(2x - 3)(5x + 4)$. C
   A. $10x + 1$
   B. $10x^2 - 12$
   C. $10x^2 - 7x - 12$
   D. $10x^2 + 23x - 12$
   E. $10x^2 - 7x + 12$

4. Factor $x^2 + 3x - 10$. A
   A. $(x - 2)(x + 5)$
   B. $(x + 2)(x - 5)$
   C. $(x - 2)(x - 5)$
   D. $-(x + 2)(x + 5)$
   E. $(x - 10)(x + 1)$

5. Which expression is equal to $8x^3 - 12x^2 + 4x$? D
   A. x^6
   B. $2x(4x^2 + 6x + 2)$
   C. $x(8x^3 - 12x + 4)$
   D. $4x(2x^2 - 3x + 1)$
   E. $4x(4x^2 - 3x + 1)$

6. Solve $4x^2 - 4x - 3 \leq 0$. C
   A. $x \leq \frac{1}{2}$ or $x \geq \frac{3}{2}$
   B. $x \leq -\frac{1}{2}$ or $x \geq \frac{3}{2}$
   C. $-\frac{1}{2} \leq x \leq \frac{3}{2}$
   D. $-\frac{1}{2} \geq x \geq \frac{3}{2}$
   E. $x = -\frac{1}{2}$ and $x = \frac{3}{2}$

7. Simplify $\frac{3x - 9}{x^2 - 6x + 9}$. B
   A. $\frac{1}{3}x - \frac{1}{3}$ B. $\frac{3}{x - 3}$
   C. $\frac{1}{x + 3}$ D. $x^2 + \frac{1}{3}x - \frac{1}{3}$
   E. cannot be simplified

8. Find the restriction on the variable in $\frac{x - 8}{x - 10}$. D
   A. $x \neq 7$ B. $x \neq 8$ C. $x \neq 9$
   D. $x \neq 10$ E. $x \neq 11$

9. Divide $\frac{x - 4}{x^2 + 2x} \div \frac{x^2 - 16}{x^2 - x}$. A
   A. $\frac{x - 1}{x^2 + 6x + 8}$ B. $\frac{1}{x^2 + x - 4}$
   C. 1 D. $\frac{x^2 - 6^x + 4}{x^2 + x - 4}$
   E. $\frac{x^2 + 6x + 8}{x^2 - 1}$

Compare the boxed quantity in Column A with the boxed quantity in Column B. Choose the best answer.

A. The quantity in Column A is greater.
B. The quantity in Column B is greater.
C. The two quantities are equal.
D. The relationship cannot be determined on the basis of the information supplied.

	Column A	Column B	
10.	the degree of $4x^3 - 8x^2 - 7x$	the GCF of $6x^3, 9x^2, 12$	C
11.	the solution of $x^2 - 8x + 16 = 0$	the solution of $\frac{x - 4}{2} = \frac{x - 2}{4}$	B

Find each answer.

12. Simplify $\frac{4x + 10}{2x^2 + 7x + 5}$. $\frac{2}{x + 1}$

13. Factor $4x^2 - 12x + 9$. $(2x - 3)^2$

14. One baker can shape 24 bagels in 15 min. When the baker works with an apprentice, they can shape 24 bagels in 10 min. How long does it take the apprentice to shape 24 bagels working alone? 30 min

15. Find the asymptotes, the x-intercept, and the y-intercept of the graph of $y = \frac{5}{x - 2} + 1$. $x = 2, y = 1; -3; -1\frac{1}{2}$

16. Find the values of ${}_{10}P_3$ and ${}_8C_4$. 720; 70

Resources

Teaching Resources

Chapter Support File, Ch. 11
- Standardized Test Practice
- Cumulative Review

Teacher's Edition

See also p. 508E for assessment options.

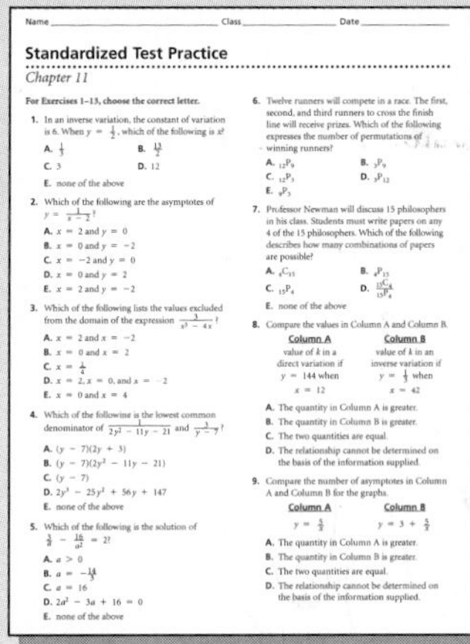

Name Class Date

Standardized Test Practice

Chapter 11

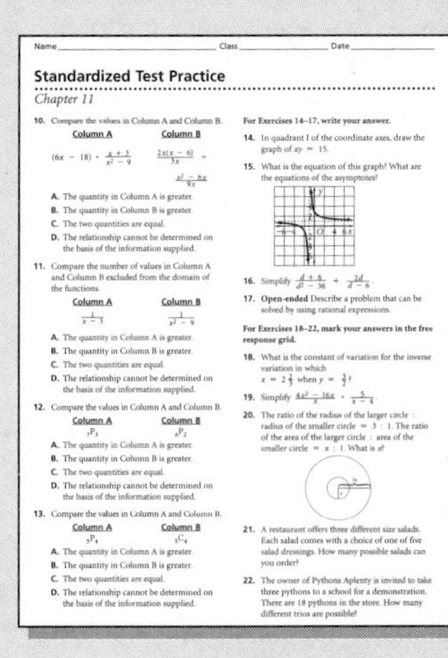

Name Class Date

Standardized Test Practice

Chapter 11

Cumulative Review

Item	Review Topic	Chapter
1	linear equations	1
2, 3, 30	modeling and solving equations	3
4	compound inequalities	4
5, 22, 23	factoring polynomials	10
6, 32	FOIL	10
7	square roots	7
8	solving inequalities	4
9, 20, 27	slope-intercept form	5
10, 23	graph of a function	2
11	linear inequalities	6
12	exponents	1
13, 14	exponential functions	8
15	scientific notation	8
16	multiplying exponents	8
17, 28	Pythagorean theorem	9
18	trigonometric ratios	9
21	multiplying polynomials	10
24	solving systems of equations	6
25	graphing systems of inequalities	6
26	probability	2
26	combinations and permutations	11
29	solving rational equations	11
19	evaluating functions	2

Resources

Teaching Resources

Chapter Support File, Ch. 11
- Cumulative Review
- Standardized Test Practice

Teacher's Edition

See also p. 508E for assessment options.

Name______ Class______ Date______

Cumulative Review

Chapter 11

For Exercises 1–12, choose the correct letter.

1. Find the difference. $\begin{bmatrix} 5 & 2 \\ -11 & 9 \end{bmatrix} - \begin{bmatrix} -3 & 10 \\ -1 & -9 \end{bmatrix}$
   A. $\begin{bmatrix} 4 & 2 \\ -1 & 0 \end{bmatrix}$ B. $\begin{bmatrix} 8 & -8 \\ -10 & 18 \end{bmatrix}$ C. $\begin{bmatrix} 2 & 12 \\ -12 & 0 \end{bmatrix}$ D. $\begin{bmatrix} -9 & 10 \\ 5 & -11 \end{bmatrix}$
2. Which of the following is *not* an example of continuous data?
   A. your height from age 8 to age 12 B. distance a train travels in one day
   C. altitude of a helium balloon on a windy day D. your shoe size from age 6 to age 10
3. What percent of 60 is 5.1?
   A. 11.8% B. 10.2% C. 8.5% D. 5.1%
4. Suppose your scores on math quizzes are 7.5, 8.5, 9, 8, and 9. What is the lowest score you can get on your next quiz and still have an average of at least 8.5?
   A. 7.5 B. 8 C. 9 D. 10
5. Which system of inequalities is represented by the graph?
   A. $y < -3x + 4$, $y \geq \frac{1}{2}x - 4$ B. $y < \frac{1}{2}x - 4$, $y \geq -3x + 4$ C. $y \leq -3x - 4$, $y > \frac{1}{2}x + 4$ D. $y > -\frac{1}{2}x - 4$, $y < 3x - 4$
6. Which of the following sets of numbers does *not* determine a right triangle?
   A. 3, 4, 5 B. 5, 12, 13 C. 9, 40, 41 D. 7, 9, 11
7. What is the LCD of $\frac{1}{45ab^2}$, $\frac{5a}{18b}$?
   A. $5ab - 3$ B. $90ab^2$ C. $9ab^2$ D. $45ab^2$
8. What is $24x^2 + 5x - 36$ in factored form?
   A. $12(x - 1)(2x + 3)$ B. $(8x + 9)(3x - 4)$
   C. $(8x - 9)(3x + 4)$ D. $6(4x - 1)(x + 6)$

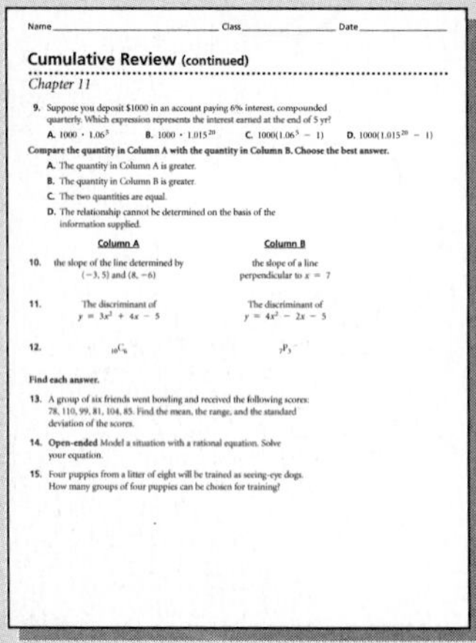

Name______ Class______ Date______

Cumulative Review (continued)

Chapter 11

9. Suppose you deposit \$1000 in an account paying 6% interest, compounded quarterly. Which expression represents the interest earned at the end of 5 yr?
   A. $1000 \cdot 1.06^5$ B. $1000 \cdot 1.015^{20}$ C. $1000(1.06^5 - 1)$ D. $1000(1.015^{20} - 1)$

Compare the quantity in Column A with the quantity in Column B. Choose the best answer.

A. The quantity in Column A is greater.
B. The quantity in Column B is greater.
C. The two quantities are equal.
D. The relationship cannot be determined on the basis of the information supplied.

	Column A	Column B
10.	the slope of the line determined by (−3, 5) and (8, −6)	the slope of a line perpendicular to $x = 7$
11.	The discriminant of $y = 3x^2 + 4x - 5$	The discriminant of $y = 4x^2 - 2x - 5$
12.	${}_{10}C_6$	${}_{7}P_3$

Find each answer.

13. A group of six friends went bowling and received the following scores: 78, 110, 99, 81, 104, 85. Find the mean, the range, and the standard deviation of the scores.
14. **Open-ended** Model a situation with a rational equation. Solve your equation.
15. Four puppies from a litter of eight will be trained as seeing-eye dogs. How many groups of four puppies can be chosen for training?

11 Cumulative Review

For Exercises 1–21, choose the correct letter.

1. Evaluate $2y + 7$ for $y = 4$. E
   A. 31 B. 12 C. 11
   D. 22 E. 15
2. Suppose you buy 2 items costing d dollars each and you give the cashier \$10. Which expression models the change you should receive? C
   A. $10 + 2d$ B. $2d - 10$ C. $10 - 2d$
   D. $2d$ E. $\frac{10}{2d}$
3. A manufacturing company spends \$1200 each day on plant costs plus \$7 per item for labor and materials. The items sell for \$23 each. How many items must the company sell in one day to equal its daily costs? E
   A. 160 B. 52 C. 200
   D. 150 E. 75
4. Which compound inequality represents the set of numbers shown on the number line? B

   −4 −3 −2 −1 0 1 2 3 4

   A. $3 \leq x < 2$ B. $-2 < x \leq 2$
   C. $3 \geq x$ and $x < 2$ D. $2 \geq x$ or $x < -2$
   E. $-2 \leq x \leq 2$
5. Find the greatest common factor (GCF) of the terms of the polynomial $12x^5 + 4x^3 - 16x^2$. A
   A. $4x^2$ B. $-2x$ C. x^3 D. $4x^3$ E. $16x^2$
6. Find the product of $(w - 5)(w + 7)$. E
   A. $w^2 + 12w - 35$ B. $w^2 + 2w - 2$
   C. $w^2 - 2w - 2$ D. $w^2 - 35$
   E. $w^2 + 2w - 35$
7. If $x^2 = 36$ and $y^2 = 16$, choose the least possible value for $y - x$. D
   A. 2 B. −2 C. 10 D. −10 E. 0
8. Which of the following points is *not* a solution of $y \leq 2x^2 - 5x + 3$? D
   A. (3, 6) B. (0, −2) C. (−1, 4)
   D. (2, 18) E. (−2, −5)
9. Which of the following is an equation of a line passing through (5, 1) and (3, −3)? D
   A. $y = \frac{1}{2}x - \frac{3}{2}$ B. $y = \frac{3}{2}x - 2$
   C. $y = 2x - \frac{3}{2}$ D. $y = 2x - 9$
   E. $y = 2x + 9$
10. The function $y = |2x + 6|$ belongs to which family of functions? E
    A. polynomial function
    B. quadratic function
    C. linear function
    D. rational function
    E. absolute value function
11. Which of the following describes the shaded region of the graph? A

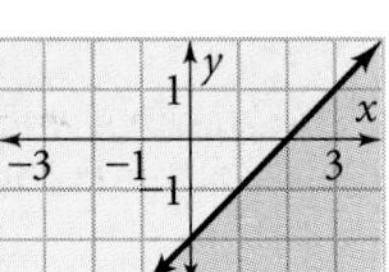

    A. $y \leq x - 2$ B. $y \geq x - 2$
    C. $y = x - 2$ D. $y < x - 2$
    E. $y > x - 2$
12. Evaluate $a^2b^3c^{-1}$ for $a = 2$, $b = -1$, and $c = -2$. A
    A. 2 B. −2 C. 8 D. −10 E. 4
13. Suppose you deposit \$1000 in an account paying 5.5% interest, compounded annually. Which expression represents the value of the investment after 10 years? B
    A. $1000 \cdot 1.55^{10}$ B. $1000 \cdot 1.055^{10}$
    C. $1000 \cdot 0.055^{10}$ D. $1000 \cdot 10^{1.055}$
    E. $1000^{10} \cdot 0.055$
14. Potassium-42 has a half-life of 12.5 h. How many half-lives are in 75 h? B
    A. 150 B. 6 C. 25 D. 4 E. 8
15. Simplify $(2.5 \times 10^4)(3.0 \times 10^{-15})$. D
    A. 7.5×10^{19} B. 7.5×10^{-19}
    C. 7.5×10^{11} D. 7.5×10^{-11}
    E. 7.5×10

31	percent	3
33	mean, range, standard deviation	9

16. Simplify $-3a^8 \cdot cb^{-3} \cdot b^{12} \cdot 9c^5$. C
   A. $6a^9b^5c^6$
   B. $-27a^8b^9c^5$
   C. $-27a^8b^9c^6$
   D. $-3abc$
   E. $-27a^8b^{15}c^6$

17. A support wire from the top of a tower is 100 ft long. It is anchored at a spot 60 ft from the base of the tower. Find the height of the tower. A
   A. 80 ft
   B. 40 ft
   C. 160 ft
   D. 20 ft
   E. $4\sqrt{10}$ ft

18. Triangle *DEF* is a right triangle with a right angle at *F*. Which of the following is *false*? C
   A. $\sin D = \frac{EF}{DE}$
   B. $\tan D = \frac{EF}{DF}$
   C. $\cos D = \frac{DE}{DF}$
   D. $\cos E = \frac{EF}{DE}$
   E. $\sin E = \frac{DF}{DE}$

Compare the boxed quantity in Column A with the boxed quantity in Column B. Choose the best answer.

A. The quantity in Column A is greater.
B. The quantity in Column B is greater.
C. The two quantities are equal.
D. The relationship cannot be determined on the basis of the information supplied.

	Column A	Column B	
19.	$f(4)$ when $f(x) = 2x^2 - 23$	$f(3)$ when $f(x) = x^2 + 5$	B

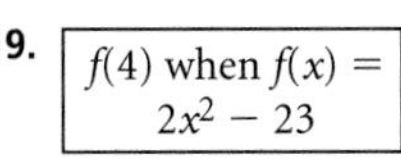

Use the equation $y = -12x + 2$.

	Column A	Column B	
20.	the slope	the *y*-intercept	B

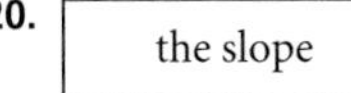

Use $2x - y = 3$ and $0.5(4x - 2y) = 3$.

	Column A	Column B	
21.	*x*	*y*	D

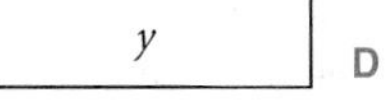

Find each answer.

22. **Open-ended** Write two quadratic trinomials that cannot be factored. Answers may vary. Samples: $x^2 + x + 3$; $4x^2 - 3x - 2$

23. a. Find the *x*-intercepts of the function $y = 4x^2 - 9x + 2$. $2, \frac{1}{4}$
   b. Make a sketch of the function. See margin.

24. Solve the system using elimination: (25, 9)
   $x + y = 34$
   $x - y = 16$

25. Graph the system: $y > x^2 - 3x + 5$
   $y \leq 3x + 2$ See margin.

26. A bag contains 5 green cubes and 7 yellow cubes. You pick two cubes without replacing the first one.
   a. What is the probability of choosing a yellow cube and then choosing a green cube? $\frac{35}{132}$
   b. What is the probability of choosing two yellow cubes? $\frac{7}{22}$

27. **Writing** Explain how you can use slope to determine the equation of a line perpendicular to a given line. The lines are perp. if the slopes are negative reciprocals.

28. What is the perimeter of a right triangle with legs measuring 8 ft and 15 ft? 40 ft

29. Solve $\frac{-2}{x-3} = \frac{x+4}{2(x-3)} + \frac{1}{4}$. $-\frac{13}{3}$

30. **Open-ended** Describe a situation that you can model with a two-step equation. Define what the variable represents, then solve the equation. See margin.

31. Russell wanted to purchase a computer that costs \$1575. With a different disk drive, the cost of the computer rose 8%. How much will the upgraded computer cost? \$1701

32. **Writing** Explain how to use the distributive property to multiply binomials. See margin.

33. Workers at a local charity worked the following numbers of hours: 3, 5, 9, 5, 8, 4, 2, 10, 7, 3, 4, 6.
   a. Find the mean. 5.5
   b. Find the range of the data. 8
   c. Find the standard deviation. about 2.43

23b.

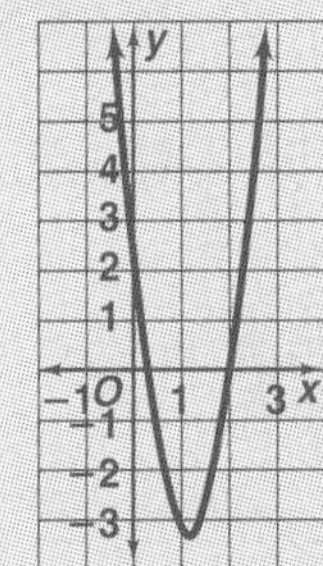

25.

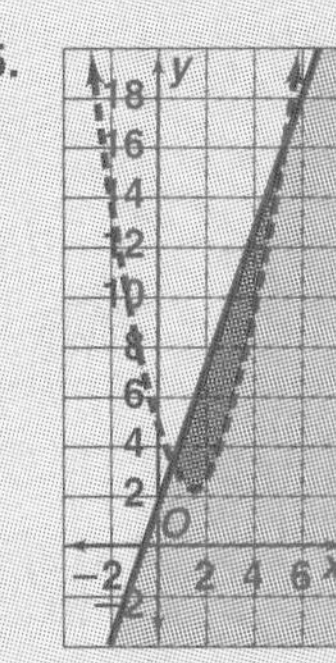

30, 32. See back of book.

Extra Practice

Find the mean, median, and mode for each set of data. ■ Lesson 1-1

1. 36, 42, 35, 40, 35, 51, 41, 35 **39.375; 38; 35**
2. 1.2, 0.9, 0.7, 1.1, 0.8, 1.3, 0.6 **about 0.94; 0.9; no mode**
3. 5, 8, 6, 8, 3, 5, 8, 6, 5, 9 **6.3; 6; 5 and 8**
4. A student surveyed the members of the drama club. She included a question about age. Her results are at the right.
   a. Make a line plot for the data. **See margin.**
   b. What is the median age of the drama club members? **15**

Ages of Drama Club Members								
14	18	16	15	17	14	15	18	15
13	14	15	18	14	17	16	14	16

■ Lesson 1-2

5. Write an equation to model each situation.
   a. The length of three cars is 24 feet. **c = length of car; $3c = 24$**
   b. The weight of two books equals 39 pens. **b = weight of a book, p = weight of a pen; $2b = 39p$**
6. a. Write a sentence and an equation to describe the relationship between the number of notebooks and the price. **See margin.**
   b. If you have \$5, can you buy a notebook for each of your classes? Explain. **Check students' work. \$5 is enough money for 5 notebooks.**

Notebooks	Price
1	\$.99
2	\$1.98
3	\$2.97

Simplify each expression. ■ Lessons 1-3 to 1-6

7. $4 + 3 \cdot 8$ **28**
8. $2 \cdot 3^2 - 7$ **11**
9. $6 \cdot (5 - 2) - 9$ **9**
10. $2 - 12 \div 3$ **−2**
11. $\left(\frac{1}{3}\right)^2 + 8 \div 2$ **$4\frac{1}{9}$**
12. $\frac{1}{2} \div \frac{4}{3}$ **$\frac{3}{8}$**
13. $-6 \cdot 4.2 - 5 \div 2$ **−27.7**
14. $9 - (3 + 1)^2$ **−7**
15. $\frac{1}{3} - \frac{5}{6}$ **$-\frac{1}{2}$**
16. $-4 + 9 \div 3 - 2$ **−3**
17. $(8 - 1.5) \cdot -4$ **−26**
18. $\frac{3}{4} \cdot 6 - \frac{1}{2}$ **4**

Evaluate each expression for $a = 8$, $b = -3$, and $c = \frac{1}{2}$.

19. $a - b - c$ **$10\frac{1}{2}$**
20. $c - a^2$ **$-63\frac{1}{2}$**
21. $8c + ab$ **−20**
22. $c(a - 2c)$ **$3\frac{1}{2}$**
23. $9b - 4a$ **−59**
24. $\frac{1}{2}b - ac$ **$-5\frac{1}{2}$**
25. $(2b)^2 - 2b^2$ **18**
26. $\frac{1}{a} + \frac{1}{c}$ **$2\frac{1}{8}$**

The results of rolling a number cube 54 times are at the right. Use the results to find each experimental probability. ■ Lesson 1-7

27. $P(3)$ **$\frac{7}{27}$**
28. $P(4)$ **$\frac{5}{54}$**
29. $P(\text{not } 5)$ **$\frac{22}{27}$**
30. $P(7)$ **0**
31. $P(\text{even number})$ **$\frac{23}{54}$**
32. $P(\text{not } 1)$ **$\frac{47}{54}$**

6 3 4 5 1 1 5 5 3 6 3 2 1 3 3 3 2 1
2 3 6 3 3 4 5 1 2 2 6 3 3 6 5 4 5 3
2 5 1 4 5 2 6 2 5 2 1 2 5 3 2 4 6 3

Find the sum or difference. **33–35. See margin.** ■ Lesson 1-8

33. $\begin{bmatrix} -3 & 0 \\ 11 & -5 \end{bmatrix} + \begin{bmatrix} -4 & 6 \\ -8 & 13 \end{bmatrix}$
34. $\begin{bmatrix} 6 & 12 \\ -9 & 7 \end{bmatrix} - \begin{bmatrix} 8 & -6 \\ 15 & 0 \end{bmatrix}$
35. $\begin{bmatrix} 4.2 & 0.6 \\ 1.7 & 9.5 \end{bmatrix} - \begin{bmatrix} 5.8 & -3.5 \\ 0.2 & 4.9 \end{bmatrix}$

■ Lesson 1-9

36. You can find the surface area of a cube using the formula $A = 6s^2$, where s is the length of a side.
   a. Write a spreadsheet formula for cell B2. **6 * B1^2**
   b. Find the values for cells B2, C2, and D2. **See below.**
   c. Write a spreadsheet formula to find the sum of the surface areas of the three cubes.

   b. 54; 162.24; 486 **c. B2 + C2 + D2**

	A	B	C	D
1	Side Length	3	5.2	9
2	Surface Area	■	■	■

Extra Practice

Use the table for Exercise 1. ■ Lesson 2-1

Cover Price and Number of Pages of Some Magazines

Cover Price	\$2.25	\$2.50	\$3.75	\$3.00	\$4.95	\$1.95	\$2.95	\$2.50
Number of Pages	208	68	122	124	234	72	90	90

1. a. Draw a scatter plot of the data. a–b. See margin.
b. Draw a trend line on the scatter plot. What is the relationship between the cover price and the number of pages in a magazine?
c. How many pages would you expect a \$2.00 magazine to have? Explain. About 75 pages; you can find the y-coordinate of the point on the line with x-coordinate 2.00.

Sketch a graph to describe each situation. Explain the activity in each section of the graph. 2–5. See margin. ■ Lesson 2-2

2. the amount of milk in your bowl as you eat cereal
3. the energy you use in a 24-h period
4. your distance from home plate after your home run
5. the number of apples on a tree over one year

Graph the data. Then write a function rule for each table of values. ■ Lessons 2-3, 2-5, 2-6
6–9. See margin for graphs.

6.

x	$f(x)$
−3	−1
−1	1
1	3
3	5

$f(x) = x + 2$

7.

x	$f(x)$
0	0
3	6
6	12
9	18

$f(x) = 2x$

8.

x	$f(x)$
21	14
25	18
29	22
33	26

$f(x) = x - 7$

9.

x	$f(x)$
−8	−4
−6	−3
−4	−2
−2	−1

$f(x) = \frac{1}{2}x$

Find the range of each function when the domain is {−4, −1, 0, 3, 8}. Each equation belongs to what family of functions? ■ Lessons 2-4, 2-7

10. $y = 6x - 5$ {−29, −11, −5, 13, 43}; linear
11. $y = |x| - 2$ {2, −1, −2, 1, 6}; abs. val.
12. $y = x^2 + 3x + 1$ {5, −1, 1, 19, 89}; quad.
13. $y = \frac{1}{2}x + 8$ {6, $7\frac{1}{2}$, 8, $9\frac{1}{2}$, 12}; lin.
14. $y = -x^2 - x$ {0, −12, −72}; quad.
15. $y = \frac{2}{3}x$ See below.
16. $y = |x - 2|$ {6, 3, 2, 1}; abs. val.
17. $y = 2x^2 - 5x$ {52, 7, 0, 3, 88}; quad.
18. $y = |4 - x|$ {8, 5, 4, 1}; abs. val.
19. $y = x + 5$ {1, 4, 5, 8, 13}; linear
20. $y = \frac{4}{9}x^2$ $\{\frac{64}{9}, \frac{4}{9}, 0, 4, \frac{256}{9}\}$; quad.
21. $y = |\frac{3}{5}x|$ See below.

Find each theoretical probability for one roll of a number cube. ■ Lesson 2-8

22. P(an odd number) $\frac{1}{2}$
23. P(a negative number) 0
24. P(an integer) 1
25. P(a factor of 6) $\frac{2}{3}$

At the Sock Hop, socks are sold in three sizes and six colors. The sizes are small, medium, and large. The color selection consists of white, gray, blue, red, black, and purple. Find the probability of choosing each kind of sock randomly.

15. $\{-\frac{8}{3}, -\frac{2}{3}, 0, 2, \frac{16}{3}\}$; linear
21. $\{\frac{12}{5}, \frac{3}{5}, 0, \frac{9}{5}, \frac{24}{5}\}$; abs. val.

26. P(large and gray) $\frac{1}{18}$
27. P(blue or red) $\frac{1}{3}$
28. P(small and purple) $\frac{1}{18}$
29. P(medium) $\frac{1}{3}$
30. Make a tree diagram to show all the kinds of socks. See margin.

557

Extra Practice

Solve each equation. ■ Lesson 3-1 to 3-5

1. $h - 4 = 10$ 14
2. $8p - 3 = 13$ 2
3. $8j - 5 + j = 67$ 8
4. $6t = -42$ −7
5. $-n + 8.5 = 14.2$ −5.7
6. $6(t + 5) = -36$ −11
7. $m + 9 = 11$ 2
8. $\frac{1}{2}(s + 5) = 7.5$ 10
9. $\frac{s}{3} = 8$ 24
10. $7h + 2h - 3 = 15$ 2
11. $\frac{7}{12}x = \frac{3}{14}$ $\frac{18}{49}$
12. $3r - 8 = -32$ −8
13. $8g - 10g = 4$ −2
14. $-3(5 - t) = 18$ 11
15. $3(c - 4) = -9$ 1
16. $\frac{3}{8}z = 9$ 24
17. $0.1(h + 20) = 3$ 10
18. $\frac{3m}{5} = 6$ 10
19. $4 - y = 10$ −6
20. $8q + 2q = -7.4$ −0.74

Write an equation to solve each problem.

21. School Your test scores for the semester are 87, 84, and 85. Can you raise your test average to 90 with your next test? n = next test score; $\frac{87 + 84 + 85 + n}{4} = 90$; no
22. You spend $\frac{1}{2}$ of your allowance each week on school lunches. Each lunch costs $1.25. How much is your weekly allowance? a = weekly allowance; $\frac{1}{2}a = 5 \times 1.25$; $12.50

You pick two balls from a jar. The jar has five blue, three yellow, six green, and two purple balls. Find each probability. ■ Lesson 3-6

23. P(purple and blue), with replacement $\frac{5}{128}$
24. P(yellow and purple), without replacement $\frac{1}{40}$
25. P(green and yellow), with replacement $\frac{9}{128}$
26. P(green and blue), without replacement $\frac{1}{8}$
27. P(purple and green), without replacement $\frac{1}{20}$
28. P(green and blue), with replacement $\frac{15}{128}$
29. P(blue and yellow), without replacement $\frac{1}{16}$
30. P(blue and purple), with replacement $\frac{5}{128}$

Write and solve an equation to answer each question. ■ Lesson 3-7

31. What is 10% of 94? $n = 0.1 \times 94$; 9.4
32. What percent of 10 is 4? $n \times 10 = 4$; 40%
33. 147 is 14% of what? $147 = 0.14 \times n$; 1050
34. What percent of 1.2 is 6? $n \times 1.2 = 6$; 500%
35. 13.2 is 55% of what? $13.2 = 0.55 \times n$; 24
36. What is 0.4% of 800? $n = 0.004 \times 800$; 3.2
37. What is 75% of 68? $n = 0.75 \times 68$; 51
38. 5 is 200% of what? $5 = 2 \times n$; 2.5
39. What percent of 54 is 28? $n \times 54 = 28$; 52%
40. 114 is 95% of what? $114 = 0.95 \times n$; 120
41. What percent of 20 is 31? $n \times 20 = 31$; 155%
42. What is 35% of 15? $n = 0.35 \times 15$; 5.25

Find each percent of change. Describe each as a percent of increase or decrease. Round to the nearest percent. ■ Lesson 3-8

43. $4.50 to $5.00 11% inc.
44. 56 in. to 57 in. 2% inc.
45. 18 oz to 12 oz 33% dec.
46. 1 s to 3 s 200% inc.
47. 8 lb to 5 lb 38% dec.
48. 6 km to 6.5 km 8% inc.
49. 39 h to 40 h 3% inc.
50. 7 ft to 2 ft 71% dec.
51. 0.2 mL to 0.45 mL 125% inc.
52. $\frac{1}{2}$ tsp to $\frac{1}{8}$ tsp 75% dec.
53. 18 kg to 20 kg 11% inc.
54. 55 min to 50 min 9% dec.
55. In 1988, the average resident of the United States ate about 2.4 lb of bagels per year. In 1993, bagel consumption had increased to about 3.5 lb annually. What percent increase is this? about 46%

Extra Practice

Solve each proportion. ■ Lesson 4-1

1. $\frac{3}{4} = \frac{-6}{m}$ −8
2. $\frac{t}{7} = \frac{3}{21}$ 1
3. $\frac{9}{j} = \frac{3}{16}$ 48
4. $\frac{2}{5} = \frac{w}{65}$ 26
5. $\frac{s}{15} = \frac{4}{45}$ $\frac{4}{3}$
6. $\frac{9}{4} = \frac{x}{10}$ $\frac{45}{2}$
7. $\frac{10}{q} = \frac{8}{62}$ $\frac{155}{2}$
8. $\frac{3}{2} = \frac{18}{y}$ 12
9. $\frac{5}{9} = \frac{t}{3}$ $\frac{5}{3}$
10. $\frac{6}{m} = \frac{3}{5}$ 10
11. $\frac{c}{8} = \frac{13.5}{36}$ 3
12. $\frac{7}{9} = \frac{35}{x}$ 45
13. Architecture A blueprint scale is 1 in. : 4 ft. On the plan, the garage is 2 in. by 3 in. What are the actual dimensions of the garage? 8 ft × 12 ft

Solve and check. If the equation is an identity or if it has no solution, write *identity* or *no solution*. ■ Lessons 4-2, 4-3

14. $|t| = 6$ 6, −6
15. $5m + 3 = 9m - 1$ 1
16. $8d = 4d - 18$ −4.5
17. $4h + 5 = 9h$ 1
18. $|k| - 4 = -7$ no sol.
19. $7t = 80 + 9t$ −40
20. $|w - 9| = 4$ 13, 5
21. $|m + 3| = 12$ 9, −15
22. $-b + 4b = 8b - b$ 0
23. $8 - |p| = 3$ 5, −5
24. $|h + 17| = -8$ no sol.
25. $6p + 1 = 3p$ $-\frac{1}{3}$
26. $10z - 5 + 3z = 8 - z$ $\frac{13}{14}$
27. $3(g - 1) + 7 = 3g + 4$ id.
28. $17 - 20q = -13 - 5q$ 2
29. Transportation A bus traveling 40 mi/h and a car traveling 50 mi/h cover the same distance. The bus travels 1 h more than the car. How many hours did each travel? The bus traveled 5 h; the car traveled 4 h.

Solve each equation for the given variable. ■ Lesson 4-4

30. $A = lw; w$ $w = \frac{A}{\ell}$
31. $c = \frac{w + t}{v}, t$ $t = cv - w$
32. $h = \frac{r}{t}(p - m); r$ $r = \frac{ht}{p - m}$
33. $P = 2l + 2w; l$ $\ell = \frac{P - 2w}{2}$
34. $v = \pi r^2 h; h$ $h = \frac{v}{\pi r^2}$
35. $m = \frac{t}{b - a}; t$ $t = m(b - a)$
36. $y = bt - c; b$ $b = \frac{y + c}{t}$
37. $g = 1.9\frac{m}{r^2}; m$ $m = \frac{gr^2}{1.9}$

Solve each inequality and graph the solutions on a number line. ■ Lessons 4-5 to 4-8

38–56. See margin for graphs.

38. $-8w < 24$ $w > -3$
39. $9 + p \le 17$ $p \le 8$
40. $\frac{r}{4} > -1$ $r > -4$
41. $7y + 2 \le -8$ $y \le -1\frac{3}{7}$
42. $t - 5 \ge -13$ $t \ge -8$
43. $9h > -108$ $h > -12$
44. $|8w + 7| > 5$ $w < -1.5$ or $w > -0.25$
45. $\frac{s}{6} \le 3$ $s \le 18$
46. $\frac{6c}{5} \ge -12$ $c \ge -10$
47. $-8l + 3.7 \le 31.7$ $\ell \ge -3.5$
48. $9 - t \le 4$ $t \ge 5$
49. $|m + 4| \ge 8$ $m \le -12$ or $m \ge 4$
50. $y + 3 < 16$ $y < 13$
51. $|n - 6| \le 8.5$ $-2.5 \le n \le 14.5$
52. $12b - 5 > -29$ $b > -2$
53. $4 - a > 15$ $a < -11$
54. $6m - 15 \le 9$ or $10m > 84$ $m \le 4$ or $m > 8.4$
55. $9j - 5j \ge 20$ and $8j > -36$ $j \ge 5$
56. $37 < 3c + 7 < 43$ $10 < c < 12$
57. The booster club raised $102 in their car wash. They want to buy $18 soccer balls for the soccer team. How many can they buy? no more than 5 balls

Solve and graph each inequality. The replacement set is the positive integers. 58–65. See margin for graphs. ■ Lesson 4-9

58. $t - 5 \le -3$ {1, 2}
59. $-6m + 2 > -19$ {1, 2, 3}
60. $|3c + 1| \ge 7$ {integers ≥ 2}
61. $8 - w < 8$ {integers > 0}
62. $2b + 3 < 7$ {1}
63. $-c - 5 \le 6$ {integers ≥ −11}
64. $|n| + 4 \le 5$ {1}
65. $\frac{3}{5}t > 6$ {integers ≥ 11}
66. You are solving a problem involving weight. Find the replacement set. nonnegative real nos.

Extra Practice

Find the slope of the line passing through each pair of points. Then write the equation of the line. ■ Lessons 5-1, 5-5

1. (2, 5) and (4, 8) $\frac{3}{2}$; $y = \frac{3}{2}x + 2$
2. (1, 6) and (7, 3) See below.
3. (−2, 4) and (3, 9) 1; $y = x + 6$
4. (1, 6) and (9, −4) $-\frac{5}{4}$; $y = -\frac{5}{4}x + 7\frac{1}{4}$
5. (−5, −7) and (−1, 3) $\frac{5}{2}$; $y = \frac{5}{2}x + 5\frac{1}{2}$
6. (7, 0) and (3, −4) 1; $y = x - 7$
7. (0, 0) and (−7, 1) $-\frac{1}{7}$; $y = -\frac{1}{7}x$
8. (10, −5) and (−2, 7) −1; $y = -x + 5$

2. $-\frac{1}{2}$; $y = -\frac{1}{2}x + 6\frac{1}{2}$

Write an equation for a line through the given point with the given slope. Then graph the line. 9–16. See margin for graphs.

9. (4, 6); $m = -5$ $y = -5x + 26$
10. (3, −1); $m = 1$ $y = x - 4$
11. (8, 5); $m = \frac{1}{2}$ $y = \frac{1}{2}x + 1$
12. (0, −6); $m = \frac{4}{3}$ 12. $y = \frac{4}{3}x - 6$
13. (−2, 7); $m = 2$ $y = 2x + 11$
14. (−5, −9); $m = -3.5$ $y = -3.5x - 26.5$
15. (4, 0); $m = 7$ $y = 7x - 28$
16. (6, −4); $m = -\frac{1}{5}$ $y = -\frac{1}{5}x - 2\frac{4}{5}$

Find the rate of change for each situation. ■ Lesson 5-2

17. growing from 1.4 m to 1.6 m in one year 0.2 m/yr
18. bicycling 3 mi in 15 min and 7 mi in 55 min 0.1 mi/min
19. walking 3 blocks in 10 min and 12 blocks in 55 min 0.2 blocks/min
20. reading 8 pages in 9 min and 22 pages in 30 min $\frac{2}{3}$ page/min

Draw the graph of a direct variation that includes the given point. Write the equation of the line. 21–32. See margin for graphs. ■ Lesson 5-3

21. (5, 4) $y = \frac{4}{5}x$
22. (7, 7) $y = x$
23. (−3, −10) $y = \frac{10}{3}x$
24. (4, −8) $y = -2x$
25. (−2, 9) $y = -\frac{9}{2}x$
26. (11, 1) $y = \frac{1}{11}x$
27. (8, −2) $y = -\frac{1}{4}x$
28. (−5, 9) $y = -\frac{9}{5}x$
29. (6, 8) $y = \frac{4}{3}x$
30. (1, −4) $y = -4x$
31. (−3, 3) $y = -x$
32. (1, 12) $y = 12x$

Find the slope and *y*-intercept. Then graph each equation. ■ Lessons 5-4, 5-7

33–40. See margin for graphs.

33. $y = 6x + 8$ 6; 8
34. $3x + 4y = -24$ $-\frac{3}{4}$; −6
35. $-2y = 5x - 12$ $-\frac{5}{2}$; 6
36. $6x + y = 12$ −6; 12
37. $y = \frac{-3}{4}x - 8$ $-\frac{3}{4}$; −8
38. $2y = 8$ 0; 4
39. $y = \frac{1}{2}x + 3$ $\frac{1}{2}$; 3
40. $y = -7x$ −7; 0

■ Lesson 5-6

41. a. Graph the ages and grade levels of some students in a school below.
    b. Draw a trend line. a–b. See margin.
    c. Find the equation of the line of best fit. $y = 0.7203196347x - 1.117579909$

(10, 6), (16, 10), (15, 10), (18,12), (17, 11), (17, 12), (19, 12), (16, 11), (11, 7), (15, 9), (13, 8)

Write an equation that satisfies the given conditions. ■ Lesson 5-8

42. parallel to $y = 4x + 1$, through (−3, 5) $y = 4x + 17$
43. perpendicular to $y = -x - 3$, through (0, 0) $y = x$
44. perpendicular to $3x + 4y = 12$, through (7, 1)
45. parallel to $2x - y = 6$, through (−6, −9) $y = 2x + 3$
46. perpendicular to $y = -2x + 5$, through (4, −10)
47. parallel to $2y = 5x + 12$, through (2, −1) $y = \frac{5}{2}x - 6$

44. $y = \frac{4}{3}x - 8\frac{1}{3}$ 46. $y = \frac{1}{2}x - 12$

Solve each equation by graphing. ■ Lesson 5-9

48. $x + 5 = 3x - 7$ 6
49. $-4x + 1 = 2x - 5$ 1
50. $9x - 2 = 7x + 4$ 3
51. $6x + 1 = 3x - 5$ −2
52. $8x - 2 = 7x + 9$ 11
53. $12x + 4 = x - 29$ −3
54. $-x - 4 = x + 18$ −11
55. $5x = 6x - 19$ 19

Extra Practice

Choose your own method to solve each system. ■ Lessons 6-1 to 6-3

1. $x - y = 7$; $3x + 2y = 6$ (4, −3)
2. $x + 4y = 1$; $3x - 2y = -25$ (−7, 2)
3. $4x - 5y = 9$; $-2x - y = -29$ (11, 7)
4. $x - y = 13$; $y - x = -13$ inf. many solutions
5. $3x - y = 4$; $x + 5y = -4$ (1, −1)
6. $x + y = 4$; $y = 7x + 4$ (0, 4)
7. $x + y = 19$; $x - y = -7$ (6, 13)
8. $-3x + 4y = 29$; $3x + 2y = -17$ (−7, 2)
9. $4x - 9y = 61$; $10x + 3y = 25$ (4, −5)
10. $6x + y = 13$; $y - x = -8$ (3, −5)
11. $3x + y = 3$; $-3x + 2y = -30$ (4, −9)
12. $4x - y = 105$; $x + 7y = -10$ (25, −5)

Write a system of equations to solve each problem. ■ Lesson 6-4

13. Suppose you have 12 coins that total \$.32. Some of the coins are nickels and the rest are pennies. How many of each coin do you have? $n + p = 12$, $5n + p = 32$; 5 nickels, 7 pennies
14. Your school drama club will put on a play you wrote. Royalties are \$50 plus \$.25 for each ticket sold. The cost for props and costumes is \$85. The tickets for the play will be \$2 each.
    a. Write an equation for the expenses. $y = 135 + 0.25x$
    b. Write an equation for the income. What is the break-even point? $y = 2x$; $77\frac{1}{7}$
    c. How much will the club earn if 200 tickets are sold? \$215

Graph each linear inequality. 15–22. See margin. ■ Lesson 6-5

15. $y \geq 4x - 3$
16. $y < x - 4$
17. $y > -6x + 5$
18. $y \leq 14 - x$
19. $x < -8$
20. $2x + 3y \leq 6$
21. $y \leq 12$
22. $y > -3x + 1$

Graph each system. 23–34. See margin. ■ Lessons 6-6, 6-8

23. $y \leq 5x + 1$; $y > x - 3$
24. $y > 4x + 3$; $y \geq -2x - 1$
25. $y = |2x| - 3$; $y = x + 4$
26. $y < -2x + 1$; $y > -2x - 3$
27. $y = |6x| - 2$; $y = x^2 - 2$
28. $y = 4x^2 - 5$; $y = x$
29. $y \leq 8$; $y \geq |-x| + 1$
30. $y \leq 5x - 2$; $y > 3$
31. $y = -2x - 7$; $y = |x + 3|$
32. $y > -3x - 9$; $y < 5x + 7$
33. $y = x^2 + 2$; $y = |\frac{1}{2}x| + 4$
34. $y \geq x$; $y \leq x + 1$

Graph each system of restrictions. Find the coordinates of each vertex. Evaluate the equation to find the maximum and minimum values of *B*. ■ Lesson 6-7

35–38. See margin for graphs.

35. $x \geq 0$; $y \geq 0$; $x \leq 5$; $y \leq 4$; $B = 2x + 5y$ 30; 0
36. $x \geq 2$; $x \leq 6$; $y \geq 1$; $y \leq 4$; $B = x + 3y$ 18; 5
37. $x \geq 2$; $y \geq 1$; $x + y \leq 10$; $B = 3x + 2y$ 29; 8
38. $x \geq 0$; $y \geq 0$; $y \leq -2x + 10$; $y \leq 4x + 2$; $B = x + y$ $8\frac{2}{3}$; 0

Extra Practice

Without graphing, describe how each graph differs from the graph of $y = x^2$. ■ Lessons 7-1 to 7-3

1. $y = 3x^2$ narrower
2. $y = -4x^2$ opens down, narrower
3. $y = -0.5x^2$ opens down, wider
4. $y = 0.2x^2$ wider
5. $y = x^2 - 4$ down 4
6. $y = x^2 + 1$ up 1
7. $y = 2x^2 + 5$ narrower, up 5
8. $y = -0.3x^2 - 7$ wider, opens down, down 7

Graph each quadratic function. Label the axis of symmetry, the vertex, and the y-intercept. 9–20. See margin.

9. $y = 3x^2$
10. $y = -2x^2 + 1$
11. $y = 0.5x^2 - 3$
12. $y = -x^2 + 2x + 1$
13. $y = 3x^2 + 5x$
14. $y = \frac{3}{4}x^2$
15. $y = 2x^2 - 9$
16. $y = -5x^2 + x + 4$
17. $y = x^2 - 7x$
18. $y = x^2 - 3x + 8$
19. $y = -x^2 + x + 12$
20. $y = -\frac{1}{2}x^2 + x - 3$

Graph each quadratic inequality. 21–26. See margin.

21. $y > x^2 - 4$
22. $y \geq -x^2 + 3x + 10$
23. $y < 2x^2 + x$
24. $y \leq 2x^2 + 5$
25. $y > -\frac{1}{2}x^2 - 3x$
26. $y \leq x^2 + x - 2$

Find the square roots of each number. ■ Lesson 7-4

27. 25 5, −5
28. $\frac{4}{9}$ $\frac{2}{3}, -\frac{2}{3}$
29. 64 8, −8
30. $\frac{25}{36}$ $\frac{5}{6}, -\frac{5}{6}$
31. 0.81 0.9, −0.9
32. 900 30, −30
33. 2.25 1.5, −1.5
34. 16 4, −4
35. $\frac{1}{25}$ $\frac{1}{5}, -\frac{1}{5}$
36. 169 13, −13
37. $\frac{4}{36}$ $\frac{1}{3}, -\frac{1}{3}$
38. 289 17, −17

Solve each equation. Round solutions to the nearest hundredth when necessary. If the equation has no real solution, write *no real solution.* ■ Lessons 7-5, 7-6

39. $x^2 = 36$ 6, −6
40. $x^2 + x - 2 = 0$ 1, −2
41. $c^2 - 100 = 0$ 10, −10
42. $9d^2 = 25$ $\frac{5}{3}, -\frac{5}{3}$
43. $(x - 4)^2 = 100$ 14, −6
44. $3x^2 = 27$ 3, −3
45. $2x^2 - 54 = 284$ 13, −13
46. $7n^2 = 63$ 3, −3
47. $h^2 + 4 = 0$ no real sol.
48. $x^2 + 6x - 2 = 0$ 0.32, −6.32
49. $x^2 - 5x = 7$ 6.14, −1.14
50. $x^2 - 10x + 3 = 0$ 9.69, 0.31
51. $2x^2 - 4x + 1 = 0$ 1.71, 0.29
52. $3x^2 + x + 5 = 0$ no real sol.
53. $\frac{1}{2}x^2 - 8 = 3x$ 8, −2
54. $x^2 + 8x - 5 = -9$ −0.54, −7.46
55. $2x^2 - 5x = x^2 - 3x + 6$ 3.65, −1.65
56. $-3x^2 + x - 2 = 5$ no real sol.

Evaluate the discriminant. Determine the number of real solutions of each equation. ■ Lesson 7-7

57. $x^2 - x + 5 = 0$ −19; 0
58. $3x^2 + 4x = -3 - 2x$ 0; 1
59. $-2x^2 - x + 7 = 0$ 57; 2
60. $3x^2 + 8x = 9$ 172; 2
61. $3x^2 + 5 = 6x$ −24; 0
62. $6x^2 + 11x - 4 = 0$ 217; 2
63. $-x^2 + x - 4 = 3$ −27; 0
64. $6x - x^2 = 4$ 20; 2
65. $x^2 = 5x - 1$ 21; 2

Extra Practice

Graph each function. Label each graph as *exponential growth* or *exponential decay*. 1–12. See margin for graphs.

■ Lessons 8-1 to 8-3

1. $y = 3^x$ growth
2. $y = \left(\frac{3}{4}\right)^x$ decay
3. $y = 1.5^x$ growth
4. $y = \frac{1}{2} \cdot 3^x$ growth
5. $y = 3 \cdot 7^x$ growth
6. $y = 4^x$ growth
7. $y = 3 \cdot \left(\frac{1}{5}\right)^x$ decay
8. $y = 2^x$ growth
9. $y = 2 \cdot 3^x$ growth
10. $y = (0.8)^x$ decay
11. $y = 2.5^x$ growth
12. $y = 4 \cdot (0.2)^x$ decay

Identify the growth factor or decay factor for each exponential function.

13. $y = 8^x$ 8
14. $y = \frac{3}{4} \cdot 2^x$ 2
15. $y = 9 \cdot \left(\frac{1}{2}\right)^x$ $\frac{1}{2}$
16. $y = 4 \cdot 9^x$ 9
17. $y = 0.65^x$ 0.65
18. $y = 3 \cdot 1.5^x$ 1.5
19. $y = \frac{2}{5} \cdot \left(\frac{1}{4}\right)^x$ $\frac{1}{4}$
20. $y = 0.1 \cdot 0.9^x$ 0.9

Write an exponential function to model each situation. Find each amount after the specified time.

21. \$200 principal
    4% compounded annually
    5 years
    $y = 200 \cdot 1.04^x$; \$243.33
22. \$1000 principal
    3.6% compounded monthly
    10 years
    $y = 1000 \cdot 1.003^{12x}$; \$1432.56
23. \$3000 investment
    8% loss each year
    3 years
    $y = 3000 \cdot 0.92^x$; \$2336.06

Simplify each expression. Use only positive exponents.

■ Lessons 8-4, 8-6 to 8-8

24. $(2t)^{-6}$ $\frac{1}{64t^6}$
25. $5m^5m^{-8}$ $\frac{5}{m^3}$
26. $(4.5)^4(4.5)^{-2}$ 20.25
27. $(m^7t^{-5})^2$ $\frac{m^{14}}{t^{10}}$
28. $(x^2n^4)(n^{-8})$ $\frac{x^2}{n^4}$
29. $(w^{-2}j^{-4})^{-3}(j^7j^3)$ w^6j^{22}
30. $(t^6)^3(m^2)$ $t^{18}m^2$
31. $(3n^4)^2$ $9n^8$
32. $\frac{r^5}{g^3}$ $\frac{r^5}{g^3}$
33. $\frac{1}{a^{-4}}$ a^4
34. $\frac{w^7}{w^{-6}}$ w^{13}
35. $\frac{6}{t^{-4}}$ $6t^4$
36. $\frac{a^2b^{-7}c^4}{a^5b^3c^{-2}}$ $\frac{c^6}{a^3b^{10}}$
37. $\frac{(2r^5)^3}{4t^8t^{-1}}$ $\frac{2r^{15}}{t^7}$
38. $\left(\frac{a^6}{a^7}\right)^{-3}$ a^3
39. $\left(\frac{c^5c^{-3}}{c^{-4}}\right)^{-2}$ $\frac{1}{c^{12}}$

Evaluate each expression for $m = 2, t = -3, w = 4$, and $z = 0$.

40. t^m 9
41. t^{-m} $\frac{1}{9}$
42. $(w \cdot t)^m$ 144
43. $w^m \cdot t^m$ 144
44. $(w^z)^m$ 1
45. w^mw^z 16
46. $z^t(m^t)^z$ 0
47. $w^{-t} \cdot t^t$ $-\frac{64}{27}$

Write each number in scientific notation.

■ Lesson 8-5

48. 34,000,000 3.4×10^7
49. 0.000 63 6.3×10^{-4}
50. 1500 1.5×10^3
51. 0.0002 2×10^{-4}
52. 360,000 3.6×10^5
53. 6,200,000,000 6.2×10^9
54. 0.05 5×10^{-2}
55. 0.000 000 000 891 8.91×10^{-10}
56. 910,000,000,000 9.1×10^{11}
57. 0.38 3.8×10^{-1}
58. 0.000 000 07 7×10^{-8}
59. 5,070,000,000,000 5.07×10^{12}

Write each number in standard notation.

60. 8.05×10^6 8,050,000
61. 3.2×10^{-7} 0.000 000 32
62. 9.0×10^8 900,000,000
63. 4.25×10^{-4} 0.000 425

Extra Practice

Could each set of three numbers represent the lengths of the sides of a right triangle? Explain your answers. ■ Lesson 9-1

1. 4, 5, 7 no; $4^2 + 5^2 \neq 7^2$
2. 6, 8, 10 yes; $6^2 + 8^2 = 10^2$
3. 6, 9, 13 no; $6^2 + 9^2 \neq 13^2$
4. 10, 13, 17 no; $10^2 + 13^2 \neq 17^2$
5. 15, 36, 39 yes; $15^2 + 36^2 = 39^2$
6. 3, 7, 10 no; $3^2 + 7^2 \neq 10^2$
7. 8, 15, 17 yes; $8^2 + 15^2 = 17^2$
8. 9, 12, 15 yes; $9^2 + 12^2 = 15^2$

Find the length of the diagonal of a rectangle with sides of the given lengths *a* and *b*. Round to the nearest tenth.

9. $a = 6, b = 8$ 10
10. $a = 5, b = 9$ 10.3
11. $a = 4, b = 10$ 10.8
12. $a = 9, b = 1$ 9.1

Find the distance between the endpoints of each segment. Round your answers to the nearest tenth. Then find the midpoint of each segment. ■ Lesson 9-2

13. $A(1, 3), B(2, 8)$ 5.1; (1.5, 5.5)
14. $R(6, -2), S(-7, -10)$ 15.3; (−0.5, −6)
15. $G(4, 0), H(5, -1)$ 1.4; (4.5, −0.5)
16. $A(-4, 1), B(3, 5)$ 8.1; (−0.5, 3)
17. $G(11, 7), H(-7, -11)$ 25.5; (2, −2)
18. $R(1, -6), S(4, -2)$ 5; (2.5, −4)
19. $R(-8, -4), S(5, 7)$ 17.0; (−1.5, 1.5)
20. $A(0, 6), B(-2, 9)$ 3.6; (−1, 7.5)
21. $G(5, 10), H(0, 0)$ 11.2; (2.5, 5)

Use $\triangle ABC$ to evaluate each expression. ■ Lesson 9-3

22. $\sin A$ $\frac{4}{5}$
23. $\cos A$ $\frac{3}{5}$
24. $\tan A$ $\frac{4}{3}$
25. $\sin B$ $\frac{3}{5}$
26. $\cos B$ $\frac{4}{5}$
27. $\tan B$ $\frac{3}{4}$

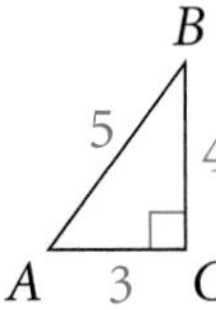

Simplify each radical expression. ■ Lessons 9-4, 9-5

28. $\frac{\sqrt{27}}{\sqrt{81}}$ $\frac{\sqrt{3}}{3}$
29. $\sqrt{\frac{25}{4}}$ $\frac{5}{2}$
30. $\sqrt{\frac{50}{9}}$ $\frac{5\sqrt{2}}{3}$
31. $\frac{\sqrt{72}}{\sqrt{50}}$ $\frac{6}{5}$
32. $\sqrt{75} - 4\sqrt{75}$ $-15\sqrt{3}$
33. $\sqrt{5}(\sqrt{20} - \sqrt{80})$ -10
34. $\sqrt{25} \cdot \sqrt{4}$ 10
35. $\sqrt{6}(\sqrt{6} - 3)$ $6 - 3\sqrt{6}$
36. $3\sqrt{300} + 2\sqrt{27}$ $36\sqrt{3}$
37. $5\sqrt{2} \cdot 3\sqrt{50}$ 150
38. $\sqrt{8} - 4\sqrt{2}$ $-2\sqrt{2}$
39. $\sqrt{27} \cdot \sqrt{3}$ 9

Solve each radical equation. Check your solutions. ■ Lesson 9-6

40. $\sqrt{3x + 4} = 1$ -1
41. $6 = \sqrt{8x - 4}$ 5
42. $2x = \sqrt{14x - 6}$ $\frac{1}{2}, 3$
43. $\sqrt{2x + 5} = \sqrt{3x + 1}$ 4
44. $2x = \sqrt{6x + 4}$ 2
45. $\sqrt{5x + 11} = \sqrt{7x - 1}$ 6

Find the domain of each function. Then graph the function. 46–51. See margin for graphs. ■ Lesson 9-7

46. $y = \sqrt{x + 5}$ $x \geq -5$
47. $y = \sqrt{x} - 2$ $x \geq 0$
48. $y = \sqrt{x + 1}$ $x \geq -1$
49. $y = \sqrt{x} - 4$ $x \geq 0$
50. $y = \sqrt{x - 3}$ $x \geq 3$
51. $y = \sqrt{x} + 6$ $x \geq 0$

Find the standard deviation of each set of data to the nearest tenth. ■ Lesson 9-8

52. 11, 14, 10, 13, 15 1.9
53. 2, 4, 3, 5, 3, 7 1.6
54. 21, 20, 26, 18, 30 4.4
55. 15, 13, 10, 20, 17 3.4
56. 32, 33, 30, 37 2.5
57. 7, 10, 4, 8, 2, 11 3.2

Extra Practice

Find each sum or difference. Write in standard form. ■ Lesson 10-1

1. $(5x^3 + 3x^2 - 7x + 10) - (3x^3 - x^2 + 4x - 1)$ $2x^3 + 4x^2 - 11x + 11$
2. $(x^2 + 3x - 2) + (4x^2 - 5x + 2)$ $5x^2 - 2x$
3. $(4m^3 + 7m - 4) + (2m^3 - 6m + 8)$ $6m^3 + m + 4$
4. $(8t^2 + t + 10) - (9t^2 - 9t - 1)$ $-t^2 + 10t + 11$
5. $(-7c^3 + c^2 - 8c - 11) - (3c^3 + 2c^2 + c - 4)$ $-10c^3 - c^2 - 9c - 7$
6. $(6v + 3v^2 - 9v^3) + (7v - 4v^2 - 10v^3)$ $-19v^3 - v^2 + 13v$
7. $(s^4 - s^3 - 5s^2 + 3s) - (5s^4 + s^3 - 7s^2 - s)$ $-4s^4 - 2s^3 + 2s^2 + 4s$
8. $(9w - 4w^2 + 10) + (8w^2 + 7 + 5w)$ $4w^2 + 14w + 17$

Find each product. ■ Lessons 10-2, 10-3

9. $4b(b^2 + 3)$ $4b^3 + 12b$
10. $(5c + 3)(-c + 2)$ $-5c^2 + 7c + 6$
11. $(3t - 1)(2t + 1)$ $6t^2 + t - 1$
12. $9c(c^2 - 3c + 5)$ $9c^3 - 27c^2 + 45c$
13. $8m(4m - 5)$ $32m^2 - 40m$
14. $(w - 1)(w^2 + w + 1)$ $w^3 - 1$
15. $5k(k^2 + 8k)$ $5k^3 + 40k^2$
16. $(3t + 5)(t + 1)$ $3t^2 + 8t + 5$
17. $(2n - 3)(2n + 4)$ $4n^2 + 2n - 12$
18. $5r^2(r^2 + 4r - 2)$ $5r^4 + 20r^3 - 10r^2$
19. $(b + 3)(b + 7)$ $b^2 + 10b + 21$
20. $2m^2(m^3 + m - 2)$ $2m^5 + 2m^3 - 4m^2$
21. **Geometry** A rectangle has dimensions $3x - 1$ and $2x + 5$. Write an expression for its area as a product and in standard form. $(3x - 1)(2x + 5)$, $6x^2 + 13x - 5$

Find the greatest common factor of each expression.

22. $t^6 + t^4 - t^5 + t^2$ t^2
23. $3m^2 - 6 + 9m$ 3
24. $16c^2 - 4c^3 + 12c^5$ $4c^2$
25. $8v^6 + 2v^5 - 10v^9$ $2v^5$
26. $6n^2 - 3n^3 + 2n^4$ n^2
27. $5r + 20r^3 + 15r^2$ $5r$
28. $9x^6 + 5x^5 + 4x^7$ x^5
29. $4d^8 - 2d^{10} + 7d^4$ d^4
30. $5t^2 + 3t - 8t^4$ t
31. $4m^2 + 16m - 20$ 4
32. $7n + 14n^2 + 21n^3$ $7n$
33. $5w - 8w^2 + 2w^3$ w

Factor each polynomial. ■ Lessons 10-4, 10-5

34. $x^2 + 6x + 9$ $(x + 3)^2$
35. $x^2 - 25$ $(x + 5)(x - 5)$
36. $4t^2 + t - 3$ $(4t - 3)(t + 1)$
37. $9c^2 - 169$ $(3c + 13)(3c - 13)$
38. $2c^2 - 5c - 3$ $(2c + 1)(c - 3)$
39. $t^2 - 6t + 9$ $(t - 3)^2$
40. $x^2 - 8x + 16$ $(x - 4)^2$
41. $4d^2 - 12d + 9$ $(2d - 3)^2$
42. $4m^2 - 121$ $(2m + 11)(2m - 11)$
43. $3v^2 + 10v - 8$ $(3v - 2)(v + 4)$
44. $4g^2 + 4g + 1$ $(2g + 1)^2$
45. $w^2 + 3w - 4$ $(w - 1)(w + 4)$
46. $9t^2 + 12t + 4$ $(3t + 2)^2$
47. $12m^2 - 5m - 2$ $(4m + 1)(3m - 2)$
48. $36s^2 - 1$ $(6s + 1)(6s - 1)$
49. $c^2 - 10c + 25$ $(c - 5)^2$

Solve each quadratic equation. Round answers to the nearest hundredth, if necessary. ■ Lessons 10-6, 10-7

50. $x^2 + 5x + 6 = 0$ $-2, -3$
51. $d^2 - 144 = 0$ $12, -12$
52. $c^2 + 6 = 2 - 4c$ -2
53. $x^2 + 4x = 2x^2 - x + 6$ $2, 3$
54. $3x^2 + 2x - 12 = x^2$ $2, -3$
55. $r^2 + 4r + 1 = r$ $-0.38, -2.62$
56. $d^2 + 2d + 10 = 2d + 100$ $9.49, -9.49$
57. $3c^2 + c - 10 = c^2 - 5$ $1.35, -1.85$
58. $t^2 - 3t - 10 = 0$ $-2, 5$
59. $4x^2 - 5x - 5 = 2x^2 + 4x$ $5, -0.5$
60. $4m^2 + 6m + 1 = 6m + 82$ $4.5, -4.5$
61. $d^2 - 5d = 3d^2 + 1$ $-0.22, -2.28$
62. **Agriculture** You are planting a rectangular vegetable garden. It is 5 feet longer than 3 times its width. The area of the garden is 250 ft^2. Find the dimensions of the garden. $8\frac{1}{3}$ ft × 30 ft

CHAPTER 11

Extra Practice

Find the constant of variation for each inverse variation. ■ Lesson 11-1

1. $y = 10$ when $x = 7$ 70
2. $y = 8$ when $x = 12$ 96
3. $y = 0.2$ when $x = 4$ 0.8
4. $y = 4$ when $x = 5$ 20
5. $y = 0.1$ when $x = 6$ 0.6
6. $y = 3$ when $x = 7$ 21

Each pair of points is from an inverse variation. Find the missing value.

7. $(5.4, 3)$ and $(2, y)$ 8.1
8. $(x, 4)$ and $(5, 6)$ 7.5
9. $(3, 6)$ and $(9, y)$ 2
10. $(100, 2)$ and $(x, 25)$ 8
11. $(6, 1)$ and $(x, 2)$ 3
12. $(8, y)$ and $(2, 4)$ 1
13. $(7, 35)$, and $(49, y)$ 5
14. $(x, 32)$ and $(16, 1)$ 0.5

Graph each function. Include a dashed line for each asymptote. 15–26. See margin. ■ Lesson 11-2

15. $y = \frac{6}{x}$
16. $y = \frac{8}{x + 2}$
17. $y = \frac{4}{x} - 3$
18. $y = \frac{5}{x + 1} + 3$
19. $y = \frac{-1}{x}$
20. $y = \frac{5}{x} - 1$
21. $y = \frac{-2}{x - 3}$
22. $y = \frac{2}{x - 1}$
23. $y = \frac{3}{x} + 4$
24. $y = \frac{5}{x}$
25. $y = \frac{2}{x + 1} - 1$
26. $y = \frac{x - 3}{x + 3}$

Simplify each expression and state any values restricted from the domain. ■ Lessons 11-3, 11-4

27. $\frac{4t^2}{16t}$ $\frac{t}{4}$; 0
28. $\frac{c - 5}{c^2 - 25}$ $\frac{1}{c + 5}$; 5, −5
29. $\frac{4m - 12}{m - 3}$ 4; 3
30. $\frac{a^2 + 2a - 3}{a + 3}$ $a - 1$; −3
31. $\frac{4}{x} - \frac{3}{x}$ $\frac{1}{x}$; 0
32. $\frac{6t}{5} + \frac{4t}{5}$ 2t, none
33. $\frac{6}{c} + \frac{4}{c^2}$ $\frac{6c + 4}{c^2}$; 0
34. $\frac{6}{3d} - \frac{4}{3d}$ $\frac{2}{3d}$; 0
35. $\frac{5s^4}{10s^3}$ $\frac{s}{2}$; 0
36. $\frac{4n^2}{7} \cdot \frac{14}{2n^3}$ $\frac{4}{n}$; 0
37. $\frac{8b^2 - 4b}{3b} \div \frac{2b - 1}{9b^2}$ $12b^2$; 0, $\frac{1}{2}$
38. $\frac{v^5}{v^3} \cdot \frac{4v^{-1}}{v^2}$ $\frac{4}{v}$; 0
39. $\frac{5}{t + 4} + \frac{3}{t - 4}$
40. $\frac{8}{m^2 + 6m + 5} + \frac{4}{m + 1}$
41. $\frac{3y}{4y - 8} \div \frac{9y}{2y^2 - 4y}$
42. $\frac{4}{d^2} - \frac{3}{d^3}$

39. $\frac{8t - 8}{t^2 - 16}$; 4, −4 40. $\frac{4m + 28}{m^2 + 6m + 5}$; −1, −5
41. $\frac{y}{6}$; 0, 2 42. $\frac{4d - 3}{d^3}$; 0

Solve each equation. Check your answers. ■ Lesson 11-5

43. $\frac{1}{4} + \frac{1}{x} = \frac{3}{8}$ 8
44. $\frac{4}{m} - 3 = \frac{2}{m}$ $\frac{2}{3}$
45. $\frac{1}{b - 3} = \frac{1}{4b}$ −1
46. $\frac{4}{x - 1} = \frac{3}{x}$ −3
47. $\frac{4}{n} + \frac{5}{9} = 1$ 9
48. $\frac{x}{x + 2} = \frac{x - 3}{x + 1}$ −3
49. $t - \frac{8}{t} = \frac{17}{t}$ 5, −5
50. $\frac{x + 2}{x + 5} = \frac{x - 4}{x + 4}$ −5.6
51. $\frac{4}{c + 1} - \frac{2}{c - 1} = \frac{3c + 6}{c^2 - 1}$ −12
52. $\frac{4}{m + 3} = \frac{6}{m - 3}$ −15
53. $\frac{4}{t + 5} + 1 = \frac{15}{t^2 - 25}$ 6, −10

Evaluate. ■ Lessons 11-6, 11-7

54. ${}_6C_4$ 15
55. ${}_7P_2$ 42
56. ${}_{10}C_5$ 252
57. ${}_8C_7$ 8
58. ${}_{12}P_6$ 665,280
59. ${}_9C_7$ 36
60. ${}_6P_4$ 360
61. ${}_9C_3$ 84
62. ${}_5P_2$ 20
63. ${}_{12}P_3$ 1320
64. ${}_8C_3$ 56
65. ${}_7P_6$ 5040

66. You are choosing a personal identification number using the digits 1, 3, 5, and 6 exactly once each. How many different numbers can you choose from? 24 numbers
67. The 18 members of the debate team need to form a committee of five people. How many different five-person committees are possible? 8568 committees

Problem Solving Strategies

You may find one or more of these strategies helpful in solving a word problem.

STRATEGY	WHEN TO USE IT
Draw a Diagram	The problem describes a picture or diagram.
Guess and Test	The needed information seems to be missing.
Look for a Pattern	The problem describes a relationship.
Make a Table	The problem has data that need to be organized.
Solve a Simpler Problem	The problem is complex or has numbers that are too cumbersome to use at first.
Use Logical Reasoning	You need to reach a conclusion using given information.
Work Backward	You need to find the number that led to the result in the problem.

Problem Solving: Draw a Diagram

Example **Two cars started from the same point. One traveled east at 45 mi/h, the other west at 50 mi/h. How far apart were the cars after 5 hours?**

Draw a diagram: West ← 50 mi/h • 5 — Start — 45 mi/h • 5 → East

The first car traveled: $45 \cdot 5$ or 225 mi. The second car traveled: $50 \cdot 5$ or 250 mi.

The diagram shows that the two distances should be added:
$225 + 250 = 475$ mi

After 5 hours, the cars were 475 mi apart.

EXERCISES

1. Jason, Lee, Melba, and Bonnie want to play each of the others in tennis. How many games will be played? **6 games**
2. A playground, a zoo, a picnic area, and a flower garden will be in four corners of a new park. Straight paths will connect each of these areas to all the other areas. How many pathways will be built? **6 paths**
3. Pedro wants to tack 4 posters on a bulletin board. He will tack the four corners of each poster, overlapping the sides of each poster a little. What is the least number of tacks that Pedro can use? **9 tacks**

Problem Solving: Guess and Test

When you are not sure how to start, guess an answer and then test it. In the process of testing a guess, you may see a way to revise your guess to get closer to the answer or to get the answer.

Example **Maria bought books and CDs as gifts. Altogether she bought 12 gifts and spent $84. The books cost $6 each and the CDs cost $9 each. How many of each gift did she buy?**

Guess: 6 books, 6 CDs

Test: $6 \cdot \$6 = \$36$
$6 \cdot \$9 = +\$54$
$\$90$

Revise your guess. You need fewer CDs to bring the total cost down.

Guess: 7 books, 5 CDs

Test: $7 \cdot \$6 = \$42$
$5 \cdot \$9 = +\$45$
$\$87$

The cost is still too high.

Guess: 8 books, 4 CDs

Test: $8 \cdot \$6 = \$48$
$4 \cdot \$9 = +\$36$
$\$84$

Maria bought 8 books and 4 CDs.

EXERCISES

1. Find two consecutive odd integers whose product is 323. **17 and 19, or –19 and –17**
2. Mika bought 9 rolls of film to take 180 pictures on a field trip. Some rolls had 36 exposures and the rest had 12 exposures. How many of each type did Mika buy? **3 36-exp. rolls and 6 12-exp. rolls**
3. Tanya is 18 years old. Her brother Shawn is 16 years younger. How old will Tanya be when she is 3 times as old as Shawn is then? **24 years old**
4. Steven has 100 ft of fencing and wants to build a fence in the shape of a rectangle to enclose the largest possible area. What should be the dimensions of the rectangle? **25 ft wide by 25 ft long**
5. The combined ages of a mother, her son, and her daughter are 61 years. The mother is 22 years older than her son and 31 years older than her daughter. How old is each person? **mother: 38 yr; son: 16 yr; daughter: 7 yr**
6. Kenji traveled 40 mi in a two-day bicycle race. He biked 8 mi farther on the first day than he did on the second day. How many miles did Kenji travel each day? **Kenji traveled 24 mi on the first day and 16 mi on the second day.**

Problem Solving: Look for a Pattern and Make a Table

Some problems describe relationships that involve regular sequences of numbers or other things. To solve the problem you need to be able to recognize and describe the *pattern* that gives the relationship for the numbers or things. One way to organize the information given is to *make a table.*

Example **A tree farm is planted as shown at the right. The dots represent trees. The lot will be enlarged by larger squares. How many trees will be in the fifth square?**

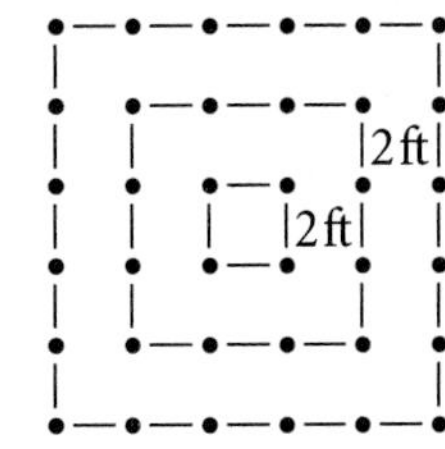

Make a table to help find a pattern.

Square position	1st	2nd	3rd	4th	5th
Number of trees	4	12	20	■	■

Pattern: 8 more trees are planted in each larger square.

The fourth square will have 28 trees. The fifth square will have 36 trees.

EXERCISES

1. Kareem made a display of books at a book fair. One book was in the first row and the other rows each had two more books than the row before it. How many books does Kareem have if he has nine rows? 81 books
2. Chris is using green and white tiles to cover her floor. If she uses tiles in the pattern, G, W, G, G, W, G, G, W, G, G, what will be the color of the twentieth tile? white
3. Jay read one story the first week of summer vacation, 3 the second week, 6 the third week, and 10 the fourth week. He kept to this pattern for eight weeks. How many stories did he read the eighth week? 36 stories
4. Jan has 6 coins, none of which is a half dollar. The coins have a value of $.85. What coins does she have? 2 quarters, 3 dimes, 1 nickel
5. Sam is covering a wall with rows of red, white, and blue siding. The red siding is cut in 1.8-m strips, the white in 2.4-m strips, and the blue in 1.2-m strips. What is the shortest length that Sam can cover by uncut strips to form equal length rows of each color? 7.2 m
6. A train leaves a station at 8:00 A.M. and averages 40 mi/h. Another train leaves the same station one hour later and averages 50 mi/h traveling in the same direction. At what time will the second train catch up with the first train? How many miles would each train have traveled by that time? 1 P.M.; 200 mi
7. The soccer team held a car wash and earned $200. They charged $7 per truck and $5 per car. In how many different ways could the team have earned the $200? 6 ways

Skills Handbook

Problem Solving: Solve a Simpler Problem

By solving one or more simpler problems you can often find a pattern that will help solve a more complicated problem.

Example **How many different rectangles are in a strip with 10 squares?**

Begin with one square, then add one square at a time. Find out whether there is a pattern.

Squares in the strip:	1	2	3	4	5
Number of rectangles:	1	3	6	10	15
Pattern:		$1 + 2 = 3$	$3 + 3 = 6$	$6 + 4 = 10$	$10 + 5 = 15$

Now continue the pattern.

Squares in the strip:	6	7	8	9	10
Number of rectangles:	21	28	36	45	55
Pattern:	$15 + 6 = 21$	$21 + 7 = 28$	$28 + 8 = 36$	$36 + 9 = 45$	$45 + 10 = 55$

There are 55 rectangles in a strip with 10 squares.

EXERCISES

1. Lockers in the east wing of Hastings High School are numbered 1–120. How many contain the digit 8? **21 lockers**
2. What is the sum of all of the numbers from 1 to 100? (*Hint:* What is $1 + 100$? What is $2 + 99$?) **5050**
3. Suppose your heart beats 70 times per minute. At this rate, how many times had it beaten by the time you were 10? **about 370,000,000 times**
4. For a community project you have to create the numbers 1 through 148 using large glittered digits, which you have to make by hand. How many glittered digits do you have to make? **336 digits**
5. There are 63 teams competing in the state soccer championship. If a team loses a game it is eliminated. How many games have to be played in order to get a single champion team? **62 games**
6. You work in a supermarket. Your boss asks you to arrange oranges in a pyramid for a display. The pyramid's base should be a square with 25 oranges. How many layers of oranges will be in your pyramid? How many oranges will you need? **5 layers; 55 oranges**
7. Kesi has her math book open. The product of the two page numbers on the facing pages is 1056. What are the two page numbers? **32 and 33**

Problem Solving: Use Logical Reasoning

Some problems can be solved without the use of numbers. They can be solved by the use of logical reasoning, given some information.

Example **Joe, Melissa, Liz, and Greg each play a different sport. Their sports are running, basketball, baseball, and tennis. Liz's sport does not use a ball. Joe hit a home run in his sport. Melissa is the sister of the tennis player. Which sport does each person play?**

Make a table to organize what you know.

	Running	Basketball	Baseball	Tennis
Joe	✗	✗	✓	✗
Melissa				✗
Liz	✓	✗	✗	✗
Greg				

← A home run means Joe plays baseball.

← Melissa cannot be the tennis player.

← Liz must run, since running does not involve a ball.

Use logical reasoning to complete the table.

	Running	Basketball	Baseball	Tennis
Joe	✗	✗	✓	✗
Melissa	✗	✓	✗	✗
Liz	✓	✗	✗	✗
Greg	✗	✗	✗	✓

← The only option for Greg is tennis.

Greg plays tennis, Melissa plays basketball, Liz runs, and Joe plays baseball.

EXERCISES

1. Juan has a dog, a horse, a bird, and a cat. Their names are Bo, Cricket, K.C., and Tuffy. Tuffy and K.C. cannot fly or be ridden. The bird talks to Bo. Tuffy runs from the dog. What is each pet's name? **dog K.C., horse Bo, bird Cricket, cat Tuffy**
2. A math class has 25 students. There are 13 students who are only in the band, 4 who are only on the swimming team, and 5 who do both activities. How many students are not in either activity? **3 students**
3. Annette is taller than Heather but shorter than Garo. Tanya's height is between Garo's and Annette's. Karin would be the shortest if it weren't for Alexa. List the names in order from shortest to tallest. **Alexa, Karin, Heather, Annette, Tanya, Garo**
4. The Colts, Cubs, and Bears played each other two games of basketball. The Colts won 3 of their games. The Bears won 2 of their games. How many games did each team win and lose? **Colts: 3W, 1L; Bears: 2W, 2L; Cubs: 1W, 3L**
5. The girls' basketball league has a calling chain when it needs to cancel its games. The leader takes 1 min to call 2 players. These 2 players take 1 min to call 2 more players, and so on. How many players will be called in 6 min? **126 players**

Skills Handbook

571

Problem Solving: Work Backward

To solve some problems you need to start with the end result and work backward to the beginning.

Example **On Monday, Rita withdrew \$150 from her savings account. On Wednesday, she deposited \$400 into her account. She now has \$1000. How much was in her account on Monday before she withdrew the money?**

Money in account now: \$1000

Undo the deposit: − \$400
\$600

Undo the withdrawal: + \$150
\$750

Rita had \$750 in her account on Monday before the withdrawal.

EXERCISES

1. Ned gave the following puzzle to Connie: I am thinking of a number. I doubled it, then tripled the result. The final result was 36. What is my number? **6**

2. Fernando gave the following puzzle to Maria: I am thinking of a number. I divided it by 3. Then I divided the result by 5. The final result was 8. What is my number? **120**

3. This week Sandy withdrew \$350 from her savings account. She made a deposit of \$125, wrote a check for \$275, and made a deposit of \$150. She now has \$225 in her account. How much did she have in her account at the beginning of the week? **\$575**

4. Jeff paid \$12.50 for a taxi fare from his home to the airport including a \$1.60 tip. City Cab charges \$1.90 for the first mile plus \$.15 for each additional $\frac{1}{6}$ mile. How many miles is Jeff's home from the airport? **11 miles**

5. Ben sold $\frac{1}{4}$ as many tickets to the fund-raiser as Charles. Charles sold 3 times as many as Susan. Susan sold 4 fewer than Tom. If Tom sold 12 tickets, how many did Ben sell? **6 tickets**

6. Two cars start traveling towards each other. One car is averaging 30 mi/h and the other 40 mi/h. After 4 h the cars are 10 mi apart. How far apart were the cars when they started? **290 miles**

7. Nina has a dentist appointment at 8:45 A.M. She wants to arrive 10 min early. Nina needs to allow 25 min to travel to the appointment and 45 min to dress and have breakfast. What is the latest time Nina could get up? **7:25 A.M.**

Prime Numbers and Composite Numbers

A *prime number* is a whole number greater than 1 that has exactly two factors, the number 1 and itself.

Prime number:	2	5	17	29
Factors:	1, 2	1, 5	1, 17	1, 29

A *composite number* is a number that has more than two factors. The number 1 is neither prime nor composite.

Composite number:	6	15	26	48
Factors:	1, 2, 3, 6	1, 3, 5, 15	1, 2, 13, 26	1, 2, 3, 4, 6, 8, 12, 16, 24, 48

Example 1 **Is 51 prime or composite?**

$51 = 3 \cdot 17$ ← Try to find factors other than 1 and 51.

51 is a composite number.

Every composite number can be factored into prime factors using a factor tree. When all the factors are prime numbers, it is called the *prime factorization* of the number.

Example 2 **Use a factor tree to write the prime factorization of 28.**

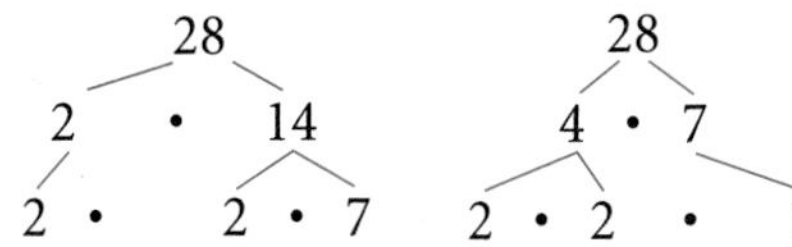

The order of listing the factors may be different, but the end result is the same.

The prime factorization of 28 is $2 \cdot 2 \cdot 7$.

EXERCISES

Is each number prime or composite?

1. 9 composite
2. 16 composite
3. 34 composite
4. 61 prime
5. 7 prime
6. 13 prime
7. 12 composite
8. 40 composite
9. 57 composite
10. 64 composite
11. 120 composite
12. 700 composite
13. 39 composite
14. 23 prime
15. 63 composite
16. 19 prime
17. 522 composite
18. 101 ~~composite~~ prime

List all the factors of each number.

19. 46 1, 2, 23, 46
20. 32 1, 2, 4, 8, 16, 32
21. 11 1, 11
22. 65 1, 5, 13, 65
23. 27 1, 3, 9, 27
24. 29 1, 29

Use a factor tree to write the prime factorization of each number.

25. 18 $2 \cdot 3 \cdot 3$
26. 20 $2 \cdot 2 \cdot 5$
27. 27 $3 \cdot 3 \cdot 3$
28. 54 $2 \cdot 3 \cdot 3 \cdot 3$
29. 64 $2 \cdot 2 \cdot 2 \cdot 2 \cdot 2 \cdot 2$
30. 96 $2 \cdot 2 \cdot 2 \cdot 2 \cdot 2 \cdot 3$
31. 100 $2 \cdot 2 \cdot 5 \cdot 5$
32. 125 $5 \cdot 5 \cdot 5$
33. 84 $2 \cdot 2 \cdot 3 \cdot 7$
34. 150 $2 \cdot 3 \cdot 5 \cdot 5$
35. 121 $11 \cdot 11$
36. 226 $2 \cdot 113$

Factors and Multiples

A common factor is a number that is a factor of two or more numbers. The *greatest common factor* (GCF) is the greatest number that is a common factor of two or more numbers.

Example 1 **Find the GCF of 14 and 42.**

Factors of 24: 1, 2, 3, 4, 6, 8, 12, 24 — Find the common factors: 1, 2, 4, 8
Factors of 64: 1, 2, 4, 8, 16, 32, 64 ← The greatest common factor is 8.
GCF (24, 64) = 8

Another way to find the GCF is to first find the prime factors of the numbers.

$24 = 2 \cdot 2 \cdot 2 \cdot 3$ ← Multiply the common prime factors.
$64 = 2 \cdot 2 \cdot 2 \cdot 2 \cdot 2 \cdot 2$
$\text{GCF} = 2 \cdot 2 \cdot 2 = 8$ ← Use the factor the number of times it appears as a common factor.

A common multiple is a number that is a multiple of two or more numbers. The *least common multiple* (LCM) is the least number that is a common multiple of two or more numbers.

Example 2 **Find the LCM of 12 and 18.**

Multiples of 12: 12, 24, 36,. . . — List a number of multiples until you find the first common multiple.
Multiples of 18: 18, 36,. . .
LCM = (12, 18) = 36

Another way to find the LCM is to first find the prime factors of the numbers.
$12 = 2 \cdot 2 \cdot 3$
$18 = 2 \cdot 3 \cdot 3$
$\text{LCM} = 2 \cdot 2 \cdot 3 \cdot 3 = 36$ ← Use each prime factor the greatest number of times it appears in either number.

EXERCISES

Find the GCF of each set of numbers.

1. 12 and 22 **2**
2. 7 and 21 **7**
3. 24 and 48 **24**
4. 17 and 51 **17**
5. 9 and 12 **3**
6. 10 and 25 **5**
7. 21 and 49 **7**
8. 27 and 36 **9**
9. 14 and 42 **14**
10. 20 and 30 **10**
11. 27 and 15 **3**
12. 12 and 28 **4**
13. 10, 30, and 25 **5**
14. 56, 84, and 140 **28**
15. 42, 63, and 105 **21**
16. 20, 28, and 40 **4**

Find the LCM of each set of numbers.

17. 16 and 20 **80**
18. 14 and 21 **42**
19. 11 and 33 **33**
20. 8 and 9 **72**
21. 5 and 12 **60**
22. 54 and 84 **756**
23. 48 and 80 **240**
24. 25 and 36 **900**
25. 54 and 80 **2160**
26. 75 and 175 **525**
27. 10 and 25 **50**
28. 24 and 28 **168**
29. 10, 15, and 25 **150**
30. 6, 7, and 12 **84**
31. 5, 8, and 20 **40**
32. 18, 21, and 36 **252**

Simplifying Fractions

A *fraction* can name a part of a group or region. This region is divided into 10 equal parts and 6 of the equal parts are shaded.

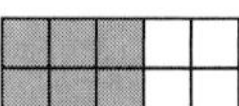

$\frac{6}{10}$ ← Numerator, ← Denominator Read: *six tenths*

A fraction can have many names. Different names for the same fraction are called *equivalent fractions.* You can find an equivalent fraction for any given fraction by multiplying the numerator and denominator of the given fraction by the same number.

Example 1 Write five equivalent fractions for $\frac{3}{5}$.

$\frac{3}{5} = \frac{3 \cdot 2}{5 \cdot 2} = \frac{6}{10}$ $\frac{3}{5} = \frac{3 \cdot 3}{5 \cdot 3} = \frac{9}{15}$ $\frac{3}{5} = \frac{3 \cdot 4}{5 \cdot 4} = \frac{12}{16}$ $\frac{3}{5} = \frac{3 \cdot 5}{5 \cdot 5} = \frac{15}{25}$ $\frac{3}{5} = \frac{3 \cdot 6}{5 \cdot 6} = \frac{18}{30}$

The fraction $\frac{3}{5}$ is in *simplest form* because its numerator and denominator are *relatively prime*, that is, their only common factor is the number 1. To write a fraction in simplest form, divide its numerator and denominator by their greatest common factor (GCF).

Example 2 Write $\frac{6}{24}$ in simplest form.

First find the GCF of 6 and 24.

$6 = 2 \cdot 3$ ← Multiply the common prime factors.

$24 = 2 \cdot 2 \cdot 2 \cdot 3$ ← GCF $= 2 \cdot 3 = 6$

Then divide the numerator and denominator of $\frac{6}{24}$ by the GCF, 6.

$\frac{6}{24} = \frac{6 \div 6}{24 \div 6} = \frac{1}{6}$ ← simplest form

EXERCISES

Complete.

1. $\frac{3}{7} = \frac{■}{21}$ 9
2. $\frac{5}{8} = \frac{20}{■}$ 32
3. $\frac{11}{12} = \frac{44}{■}$ 48
4. $\frac{12}{16} = \frac{■}{4}$ 3
5. $\frac{50}{100} = \frac{1}{■}$ 2
6. $\frac{5}{9} = \frac{■}{27}$ 15
7. $\frac{3}{8} = \frac{■}{24}$ 9
8. $\frac{5}{6} = \frac{20}{■}$ 24
9. $\frac{12}{20} = \frac{■}{5}$ 3
10. $\frac{75}{150} = \frac{1}{■}$ 2

Which fractions are in simplest form?

11. $\frac{4}{12}$ no
12. $\frac{3}{16}$ yes
13. $\frac{5}{30}$ no
14. $\frac{9}{72}$ no
15. $\frac{11}{22}$ no
16. $\frac{24}{25}$ yes

Write in simplest form.

17. $\frac{8}{16}$ $\frac{1}{2}$
18. $\frac{7}{14}$ $\frac{1}{2}$
19. $\frac{6}{9}$ $\frac{2}{3}$
20. $\frac{20}{30}$ $\frac{2}{3}$
21. $\frac{8}{20}$ $\frac{2}{5}$
22. $\frac{12}{40}$ $\frac{3}{10}$
23. $\frac{15}{45}$ $\frac{1}{3}$
24. $\frac{14}{56}$ $\frac{1}{4}$
25. $\frac{10}{25}$ $\frac{2}{5}$
26. $\frac{9}{72}$ $\frac{1}{8}$
27. $\frac{45}{60}$ $\frac{3}{4}$
28. $\frac{20}{35}$ $\frac{4}{7}$
29. $\frac{27}{33}$ $\frac{9}{11}$
30. $\frac{18}{72}$ $\frac{1}{4}$
31. $\frac{45}{85}$ $\frac{9}{17}$
32. $\frac{63}{126}$ $\frac{1}{2}$
33. $\frac{125}{150}$ $\frac{5}{6}$
34. $\frac{256}{320}$ $\frac{4}{5}$

Fractions, Decimals, and Mixed Numbers

Fractions can be written as decimals.

Example 1 **Write $\frac{3}{5}$ as a decimal.**

$$5\overline{)3.0} = 0.6, \quad -3.0$$

Divide the denominator into the numerator.

The decimal for $\frac{3}{5}$ is 0.6.

Decimals can be written as fractions.

Example 2 **Write 0.38 as a fraction.**

$0.38 = 38 \text{ hundredths} = \frac{38}{100} = \frac{19}{50}$

A fraction for 0.38 is $\frac{38}{100}$ or $\frac{19}{50}$.

An *improper fraction* is a fraction in which the numerator is greater than or equal to the denominator. An improper fraction can be written as a *mixed number*, that is as a whole number and a fraction.

Example 3 **Write $\frac{13}{5}$ as a mixed number.**

Divide the denominator into the numerator and write the remainder over the denominator.

$\frac{13}{5} = 5\overline{)13}$ (2r3), -10, 3 $= 2\frac{3}{5}$ ← remainder (3), ← denominator (5)

A mixed number can be written as an improper fraction.

Example 4 **Write $5\frac{3}{4}$ as an improper fraction.**

$5\frac{3}{4} = \frac{23}{4}$

First multiply the denominator by the whole number: $4 \cdot 5 = 20$.
Next add the numerator of the fraction to the answer: $3 + 20 = 23$.
Then write the result over the denominator: $\frac{23}{4}$.

EXERCISES

Write as a decimal.

1. $\frac{3}{10}$ 0.3
2. $\frac{1}{5}$ 0.2
3. $\frac{4}{20}$ 0.2
4. $\frac{25}{75}$ $0.\overline{3}$
5. $\frac{2}{3}$ $0.\overline{6}$
6. $\frac{1}{6}$ $0.1\overline{6}$

Write as a fraction.

7. 0.3 $\frac{3}{10}$
8. 0.25 $\frac{1}{4}$
9. 0.37 $\frac{37}{100}$
10. 0.13 $\frac{13}{100}$
11. 0.07 $\frac{7}{100}$
12. 0.875 $\frac{7}{8}$

Write as a mixed number.

13. $\frac{12}{7}$ $1\frac{5}{7}$
14. $\frac{23}{9}$ $2\frac{5}{9}$
15. $\frac{21}{10}$ $2\frac{1}{10}$
16. $\frac{30}{21}$ $1\frac{3}{7}$
17. $\frac{22}{5}$ $4\frac{2}{5}$
18. $\frac{27}{13}$ $2\frac{1}{13}$

Write as an improper fraction.

19. $2\frac{1}{2}$ $\frac{5}{2}$
20. $3\frac{1}{4}$ $\frac{13}{4}$
21. $5\frac{1}{6}$ $\frac{31}{6}$
22. $3\frac{4}{5}$ $\frac{19}{5}$
23. $4\frac{1}{7}$ $\frac{29}{7}$
24. $6\frac{3}{8}$ $\frac{51}{8}$

25. Celia answered $\frac{7}{8}$ of the decimal test correctly. Write this fraction as a decimal. 0.875

Skills Handbook

Adding and Subtracting Fractions

You can add and subtract fractions when they have the same denominator. Fractions with the same denominator are called *like fractions.*

Example 1 Add $\frac{4}{5} + \frac{3}{5}$.

$\frac{4}{5} + \frac{3}{5} = \frac{4+3}{5} = \frac{7}{5} = 1\frac{2}{5}$ ← Add the numerators and keep the same denominator.

Example 2 Subtract $\frac{5}{9} - \frac{2}{9}$.

$\frac{5}{9} - \frac{2}{9} = \frac{5-2}{9} = \frac{3}{9} = \frac{1}{3}$ ← Subtract the numerators and keep the same denominator.

Fractions with unlike denominators are called *unlike fractions.* To add or subtract fractions with unlike denominators, find the least common denominator (LCD) and write equivalent fractions with the same denominator. Then add or subtract the like fractions.

Example 3 Add $\frac{3}{4} + \frac{5}{6}$.

$\frac{3}{4} + \frac{5}{6} =$ ← Find the LCD. The LCD is the same as the least common multiple (LCM). The LCD(4, 6) is 12.

$\frac{9}{12} + \frac{10}{12} = \frac{9+10}{12} = \frac{19}{12}$ or $1\frac{7}{12}$ ← Write equivalent fractions with the same denominator.

Example 4 Subtract $\frac{5}{12} - \frac{2}{9}$.

$\frac{5}{12} - \frac{2}{9} =$ ← Find the LCD. The LCD (12, 9) is 36.

$\frac{15}{36} - \frac{8}{36} = \frac{15-8}{36} = \frac{7}{36}$ ← Write equivalent fractions with the same denominator.

To add or subtract mixed numbers, add or subtract the fractions. Then add or subtract the whole numbers. Sometimes when subtracting mixed numbers you may have to regroup.

Example 5 Subtract $5\frac{1}{4} - 3\frac{2}{3}$.

$5\frac{1}{4} - 3\frac{2}{3}$ ← Write equivalent fractions with the same denominator.

$5\frac{3}{12} - 3\frac{8}{12} =$ ← Write $5\frac{3}{12}$ as $4\frac{15}{12}$ so you can subtract the fractions.

$4\frac{15}{12} - 3\frac{8}{12} = 1\frac{7}{12}$ ← Subtract the fractions. Then subtract the whole numbers.

EXERCISES

Add. Write each answer in simplest terms.

1. $\frac{2}{7} + \frac{3}{7}$ $\frac{5}{7}$
2. $\frac{3}{8} + \frac{7}{8}$ $1\frac{1}{4}$
3. $\frac{6}{5} + \frac{9}{5}$ 3
4. $\frac{4}{9} + \frac{8}{9}$ $1\frac{1}{3}$
5. $6\frac{2}{3} + 3\frac{4}{5}$ $10\frac{7}{15}$
6. $1\frac{4}{7} + 2\frac{3}{14}$ $3\frac{11}{14}$
7. $4\frac{5}{6} + 1\frac{7}{18}$ $6\frac{2}{9}$
8. $2\frac{4}{5} + 3\frac{6}{7}$ $6\frac{23}{35}$
9. $4\frac{2}{3} + 1\frac{6}{11}$ $6\frac{7}{33}$
10. $3\frac{7}{9} + 5\frac{5}{27}$ $8\frac{26}{27}$

Subtract. Write each answer in simplest terms.

11. $\frac{7}{8} - \frac{3}{8}$ $\frac{1}{2}$
12. $\frac{9}{10} - \frac{3}{10}$ $\frac{3}{5}$
13. $\frac{17}{5} - \frac{2}{5}$ 3
14. $\frac{11}{7} - \frac{2}{7}$ $1\frac{2}{7}$
15. $\frac{5}{11} - \frac{4}{11}$ $\frac{1}{11}$
16. $8\frac{5}{8} - 6\frac{1}{4}$ $2\frac{3}{8}$
17. $3\frac{2}{3} - 1\frac{8}{9}$ $1\frac{7}{9}$
18. $8\frac{5}{6} - 5\frac{1}{2}$ $3\frac{1}{3}$
19. $12\frac{3}{4} - 4\frac{5}{6}$ $7\frac{11}{12}$
20. $17\frac{2}{7} - 8\frac{2}{9}$ $9\frac{4}{63}$

Multiplying and Dividing Fractions

To multiply two or more fractions, multiply the numerators, multiply the denominators, and simplify the product, if necessary.

Example 1 **Multiply $\frac{3}{7} \cdot \frac{5}{6}$.**

$$\frac{3}{7} \cdot \frac{5}{6} = \frac{3 \cdot 5}{7 \cdot 6} = \frac{15}{42} = \frac{15 \div 3}{42 \div 3} = \frac{5}{14}$$

Sometimes you can simplify before multiplying.

$$\frac{\cancel{3}^{1}}{7} \cdot \frac{5}{\cancel{6}_{2}} = \frac{5}{14}$$ ← Divide a numerator and a denominator by a common factor.

To multiply mixed numbers, change the mixed numbers to improper fractions and multiply the fractions. Write the product as a mixed number.

Example 2 **Multiply $2\frac{4}{5} \cdot 1\frac{2}{3}$.**

$$2\frac{4}{5} \cdot 1\frac{2}{3} = \frac{14}{\cancel{5}_{1}} \cdot \frac{\cancel{5}^{1}}{3} = \frac{14}{3} = 4\frac{2}{3}$$

To divide fractions, change the division problem to a multiplication problem. Remember that $8 \div \frac{1}{4}$ is the same as $8 \cdot 4$.

Example 3 **Divide $\frac{4}{5} \div \frac{3}{7}$.**

$$\frac{4}{5} \div \frac{3}{7} = \frac{4}{5} \cdot \frac{7}{3}$$ ← Multiply by the reciprocal of the divisor. Simplify the answer.

$$= \frac{4}{5} \cdot \frac{7}{3} = \frac{28}{15} = 1\frac{13}{15}$$

To divide mixed numbers, change the mixed numbers to improper fractions and divide the fractions.

Example 4 **Divide $4\frac{2}{3} \div 7\frac{3}{5}$.**

$$4\frac{2}{3} \div 7\frac{3}{5} = \frac{14}{3} \div \frac{38}{5} = \frac{14}{3} \cdot \frac{5}{38} = \frac{70}{114} = \frac{35}{57}$$

EXERCISES

Multiply. Write your answers in simplest form.

1. $\frac{2}{5} \cdot \frac{3}{4}$ $\frac{3}{10}$
2. $\frac{3}{7} \cdot \frac{4}{3}$ $\frac{4}{7}$
3. $\frac{5}{4} \cdot \frac{3}{8}$ $\frac{15}{32}$
4. $\frac{6}{7} \cdot \frac{9}{2}$ $3\frac{6}{7}$
5. $\frac{7}{3} \cdot \frac{4}{7}$ $1\frac{1}{3}$

6. $2\frac{3}{4} \cdot \frac{5}{8}$ $1\frac{23}{32}$
7. $1\frac{1}{2} \cdot 5\frac{3}{4}$ $8\frac{5}{8}$
8. $3\frac{4}{5} \cdot 10$ 38
9. $12 \cdot 1\frac{2}{3}$ 20
10. $5\frac{1}{4} \cdot \frac{2}{3}$ $3\frac{1}{2}$

Divide. Write your answers in simplest form.

11. $\frac{3}{5} \div \frac{1}{2}$ $1\frac{1}{5}$
12. $\frac{4}{5} \div \frac{9}{10}$ $\frac{8}{9}$
13. $\frac{4}{7} \div \frac{2}{3}$ $\frac{6}{7}$
14. $\frac{5}{8} \div \frac{7}{3}$ $\frac{15}{56}$
15. $\frac{4}{7} \div \frac{4}{3}$ $\frac{3}{7}$

16. $1\frac{4}{5} \div 2\frac{1}{2}$ $\frac{18}{25}$
17. $2\frac{1}{2} \div 3\frac{1}{2}$ $\frac{5}{7}$
18. $3\frac{1}{6} \div 1\frac{3}{4}$ $1\frac{17}{21}$
19. $9\frac{1}{2} \div 4\frac{1}{4}$ $2\frac{4}{17}$
20. $6\frac{3}{5} \div 2\frac{3}{5}$ $2\frac{7}{13}$

Fractions, Decimals, and Percents

Percent means *per hundred.* 50% means 50 per hundred. $50\% = \frac{50}{100} = 0.50$

You can use a shortcut to write a decimal as a percent and a percent as a decimal.

Example 1 **Write each number as a percent.**

a. 0.47 b. 0.8 c. 2.475

Move the decimal point two places to the right and write a percent sign.

a. 0.47 = 47% b. 0.80 = 80% c. 2.475 = 247.5%

Example 2 **Write each number as a decimal.**

a. 25% b. 3% c. 360%

Move the decimal point two places to the left and drop the percent sign.

a. 25% = 0.25 b. 03% = 0.03 c. 360% = 3.6

You can write fractions as percents by writing the fraction as a decimal first. Then move the decimal point two places to the right and write a percent sign.

Example 3 **Write each number as a percent.** a. $\frac{3}{5}$ b. $\frac{7}{20}$ c. $\frac{2}{3}$

a. $\frac{3}{5} = 0.6 = 60\%$ b. $\frac{7}{20} = 0.35 = 35\%$ c. $\frac{2}{3} = 0.66\overline{6} = 66.\overline{6}\%$ or 66.7% (rounded to the nearest tenth of a percent)

You can write a percent as a fraction by writing the percent as a fraction with a denominator of 100 and simplifying if possible.

Example 4 **Write each number as a fraction or mixed number.**

a. $43\% = \frac{43}{100}$ b. $\frac{1}{2}\% = \frac{\frac{1}{2}}{100} = \frac{1}{2} \div 100 = \frac{1}{2} \cdot \frac{1}{100} = \frac{1}{200}$ c. $180\% = \frac{180}{100} = \frac{9}{5} = 1\frac{4}{5}$

EXERCISES

Write each number as a percent.

1. 0.56 **56%** 2. 0.09 **9%** 3. 6.02 **602%** 4. 5.245 **524.5%** 5. 8.2 **820%** 6. 0.14 **14%**

7. $\frac{1}{5}$ **20%** 8. $\frac{9}{20}$ **45%** 9. $\frac{1}{9}$ **11.1%** 10. $\frac{5}{6}$ **83.3%** 11. $\frac{3}{4}$ **75%** 12. $\frac{7}{8}$ **87.5%**

Write each number as a decimal.

13. 7% **0.07** 14. 8.5% **0.085** 15. 0.9% **0.009** 16. 250% **2.5** 17. 83% **0.83** 18. 110% **1.1**

Write each number as a fraction or mixed number in simplest form.

19. 19% $\frac{19}{100}$ 20. $\frac{3}{4}\%$ $\frac{3}{400}$ 21. 450% $4\frac{1}{2}$ 22. $\frac{4}{5}\%$ $\frac{1}{125}$ 23. 64% $\frac{16}{25}$ 24. $\frac{2}{3}\%$ $\frac{1}{150}$

Exponents

You can express $2 \cdot 2 \cdot 2 \cdot 2 \cdot 2$ as 2^5. The raised number, 5, shows the number of times 2 is used as a factor. The number 2 is the *base*. The number 5 is the *exponent*.

2^5 ← exponent
↑ base

Factored Form		Exponential Form		Standard Form
$2 \cdot 2 \cdot 2 \cdot 2 \cdot 2$	=	2^5	=	32

A number with an exponent of 1 is the number itself: $8^1 = 8$.
Any number, except 0, with an exponent of 0 is 1: $5^0 = 1$.

Example 1 **Write using exponents.**

a. $8 \cdot 8 \cdot 8 \cdot 8 \cdot 8$ **b.** $2 \cdot 9 \cdot 9 \cdot 9 \cdot 9 \cdot 9 \cdot 9$ **c.** $6 \cdot 6 \cdot 10 \cdot 10 \cdot 10 \cdot 6 \cdot 6$

Count the number of times the number is used as a factor.

a. 8^5 **b.** $2 \cdot 9^6$ **c.** $6^4 \cdot 10^3$

Example 2 **Write each product.**

a. 2^3 **b.** $8^2 \cdot 3^4$ **c.** $10^3 \cdot 5^2$

Write in factored form and multiply.

a. $2 \cdot 2 \cdot 2 = 8$ **b.** $8 \cdot 8 \cdot 3 \cdot 3 \cdot 3 \cdot 3 = 5184$ **c.** $10 \cdot 10 \cdot 10 \cdot 5 \cdot 5 = 25{,}000$

In powers of 10, the exponent tells how many zeros are in the standard form.

$10^1 = 10$
$10^2 = 10 \cdot 10 = 100$
$10^3 = 10 \cdot 10 \cdot 10 = 1000$
$10^4 = 10 \cdot 10 \cdot 10 \cdot 10 = 10{,}000$
$10^5 = 10 \cdot 10 \cdot 10 \cdot 10 \cdot 10 = 100{,}000$

You can use exponents to write numbers in *expanded form.*

Example 3 **Write 739 in expanded form using exponents.**

$739 = 700 + 30 + 9 = (7 \cdot 100) + (3 \cdot 10) + (9 \cdot 1) = (7 \cdot 10^2) + (3 \cdot 10^1) + (9 \cdot 10^0)$

EXERCISES

Write using exponents.

1. $6 \cdot 6 \cdot 6 \cdot 6$ 6^4 **2.** $7 \cdot 7 \cdot 7 \cdot 7 \cdot 7$ 7^5 **3.** $5 \cdot 2 \cdot 2 \cdot 2 \cdot 2$ $5 \cdot 2^4$ **4.** $3 \cdot 3 \cdot 3 \cdot 3 \cdot 3 \cdot 14 \cdot 14$ $3^5 \cdot 14^2$

Write in standard form.

5. 4^3 64 **6.** 9^4 6561 **7.** 12^2 144 **8.** $6^2 \cdot 7^1$ 252 **9.** $11^2 \cdot 3^3$ 3267

Write in expanded form using exponents.

10. 658 $(6 \cdot 10^2) + (5 \cdot 10^1) + (8 \cdot 10^0)$ **11.** 1254 See above. **12.** 7125 $(7 \cdot 10^3) + (1 \cdot 10^2) + (2 \cdot 10^1) + (5 \cdot 10^0)$ **13.** 83,401 $(8 \cdot 10^4) + (3 \cdot 10^3) + (4 \cdot 10^2) + (1 \cdot 10^0)$ **14.** 294,863 See above.

11. $(1 \cdot 10^3) + (2 \cdot 10^2) + (5 \cdot 10^1) + (4 \cdot 10^0)$
14. $(2 \cdot 10^5) + (9 \cdot 10^4) + (4 \cdot 10^3) + (8 \cdot 10^2) + (6 \cdot 10^1) + (3 \cdot 10^0)$

Angles

An *angle* is a geometric figure formed by two rays with a common endpoint. The rays are *sides* of the angle and the endpoint is the *vertex* of the angle. An angle is measured in degrees. The symbol for an angle is ∠.

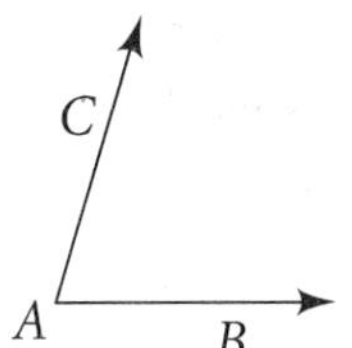

The angle pictured at the right can be named in three different ways: $\angle A$, $\angle BAC$, or $\angle CAB$.

Angles can be classified by their measures.

Acute angle	**Right angle**	**Obtuse angle**	**Straight angle**
less than 90°	**90°**	**greater than 90° but less than 180°**	**180°**

Example **Measure the angle and tell if it is *acute, right, obtuse,* or *straight.***

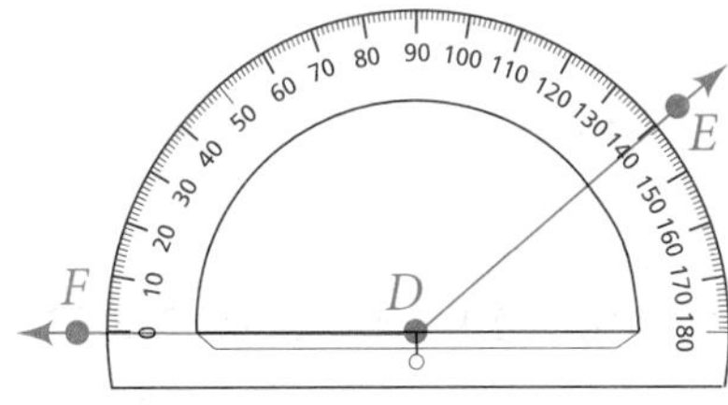

Line up side *DF* through 0° with the vertex at the center of the protractor. Read the scale number through which side *DE* passes.

The measure of the angle is 140°. The angle is obtuse.

EXERCISES

Measure each angle and tell whether it is *acute, right, obtuse,* or *straight.*

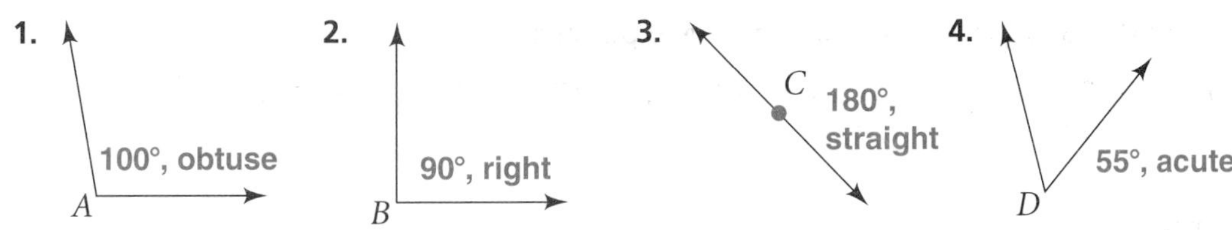

Draw an angle with the given measure. 5–8. Check students' drawings.

5. 45° **6.** 95° **7.** 120° **8.** 170°

9. *Open-ended* Draw a triangle. Use a protractor to find the measure of each angle of your triangle. **Check students' work.**

Perimeter, Area, and Volume

The *perimeter* of a figure is the distance around the figure. The *area* of a figure is the number of square units contained in the figure. The *volume* of a space figure is the number of cubic units contained in the space figure.

Example 1 **Find the perimeter of each figure.**

a.

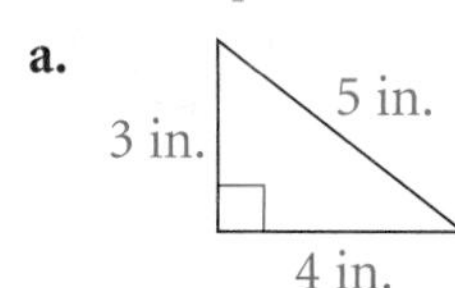

Add the measures of the sides.
$3 + 4 + 5 = 12$
The perimeter is 12 in.

b.

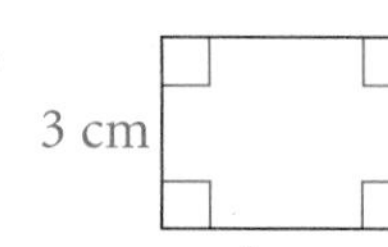

Use the formula $p = 2\ell + 2w$.
$p = 2(3) + 2(4)$
$= 6 + 8 = 14$
The perimeter is 14 cm.

Example 2 **Find the area of each figure.**

a.

Use the formula $A = lw$.

$A = 3 \cdot 5 = 15 \text{ cm}^2$

b.

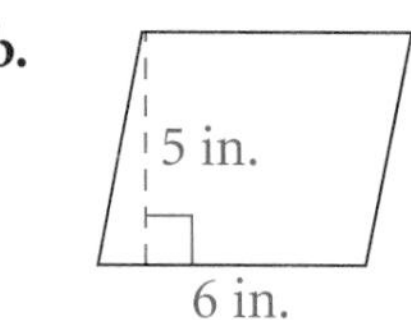

Use the formula $A = bh$.

$A = 6 \cdot 5 = 30 \text{ in.}^2$

c.

6 in.
7 in.

Use the formula $A = \frac{1}{2}(bh)$.

$A = \frac{1}{2}(7 \cdot 6) = 21 \text{ in.}^2$

Example 3 **Find the volume of each figure.**

a.

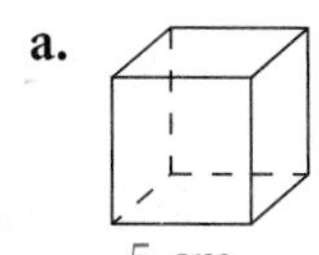

$V = e^3$
$V = 5^3 = 125 \text{ cm}^3$

b.

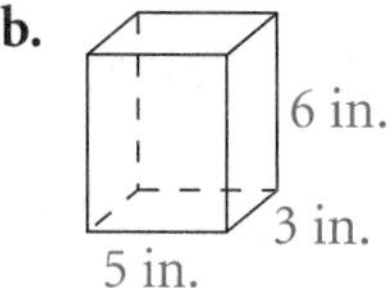

$V = Bh$ (B = area of the base)
$V = 3 \cdot 5 \cdot 6 = 90 \text{ in.}^3$

c.

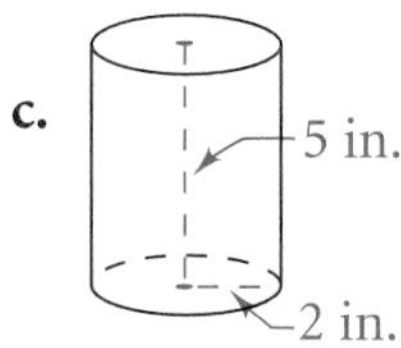

$V = \pi r^2 h$
$V = 3.14 \cdot 2^2 \cdot 5$
$= 3.14 \cdot 4 \cdot 5 = 62.8 \text{ in.}^3$

EXERCISES

Find the perimeter of each figure.

1. **22 cm**

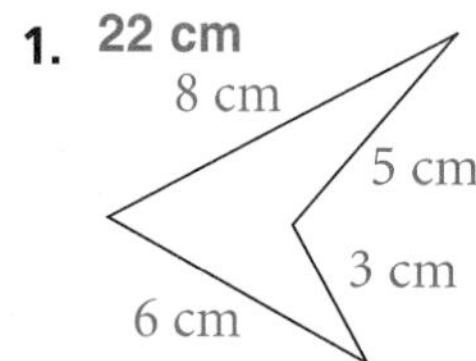

2. **22 in.**

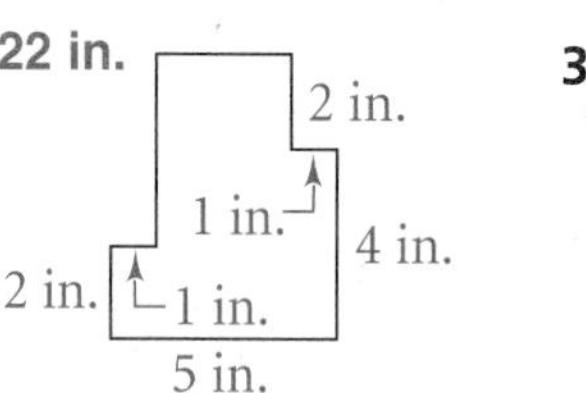

Find the area of each figure.

3. **24 cm²**

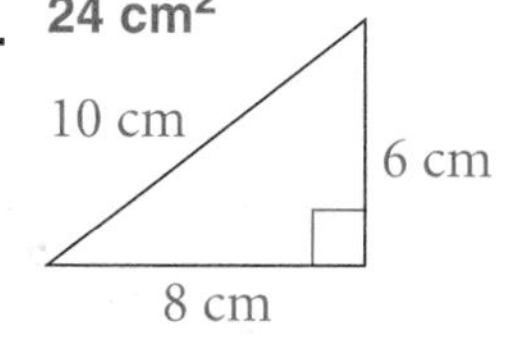

4. 8 in. 7 in. **56 in.²**

Find the volume of each figure.

5. **216 cm³**

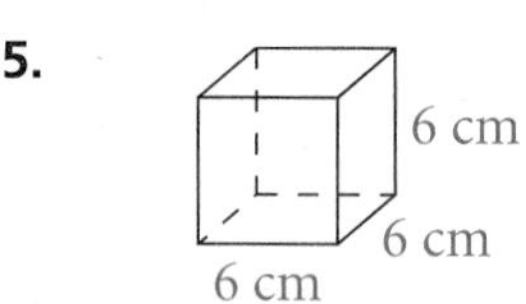

6. **48 in.³** 2 in 4 in. 6 in.

7. **351.68 cm³** 7 cm 4 cm

Tables

Measures

United States Customary	Metric
Length	
12 inches (in.) = 1 foot (ft)	10 millimeters (mm) = 1 centimeter (cm)
36 in. = 1 yard (yd)	100 cm = 1 meter (m)
3 ft = 1 yard	1000 mm = 1 meter
5280 ft = 1 mile (mi)	1000 m = 1 kilometer (km)
1760 yd = 1 mile	
Area	
144 square inches (in.^{2}) = 1 square foot (ft^2)	100 square millimeters (mm^2) = 1 square centimeter (cm^2)
9 ft^2 = 1 square yard (yd^2)	10,000 cm^2 = 1 square meter (m^2)
43,560 ft^2 = 1 acre (a)	10,000 m^2 = 1 hectare (ha)
4840 yd^2 = 1 acre	
Volume	
1728 cubic inches (in.^{3}) = 1 cubic foot (ft^3)	1000 cubic millimeters (mm^3) = 1 cubic centimeter (cm^3)
27 ft^3 = 1 cubic yard (yd^3)	1,000,000 cm^3 = 1 cubic meter (m^3)
Liquid Capacity	
8 fluid ounces (fl oz) = 1 cup (c)	1000 milliliters (mL) = 1 liter (L)
2 c = 1 pint (pt)	1000 L = 1 kiloliter (kL)
2 pt = 1 quart (qt)	
4 qt = 1 gallon (gal)	
Mass	
16 ounces (oz) = 1 pound (lb)	1000 milligrams (mg) = 1 gram (g)
2000 pounds = 1 ton (t)	1000 g = 1 kilogram (kg)
	1000 kg = 1 metric ton (t)
Temperature	
32°F = freezing point of water	0°C = freezing point of water
98.6°F = normal body temperature	37°C = normal body temperature
212°F = boiling point of water	100°C = boiling point of water
Time	
60 seconds (s) = 1 minute (min)	365 days = 1 year (yr)
60 minutes = 1 hour (h)	52 weeks (approx.) = 1 year
24 hours = 1 day (da)	12 months = 1 year
7 days = 1 week (wk)	10 years = 1 decade
4 weeks (approx.) = 1 month (mo)	100 years = 1 century

Symbols

Symbol	Meaning	Page
=	equals	p. 5
≈	is approximately equal to	p. 5
·	multiplication sign, times (×)	p. 5
( )	parentheses for grouping	p. 15
a^n	*n*th power of *a*	p. 15
[ ]	brackets for grouping	p. 17
. . .	and so on	p. 19
$\|a\|$	absolute value of *a*	p. 20
$-a$	opposite of *a*	p. 22
°	degree(s)	p. 23
<	is less than	p. 28
>	is greater than	p. 28
π	pi, an irrational number, approximately equal to 3.14	p. 30
$\frac{1}{a}$	reciprocal of *a*	p. 32
P(event)	probability of the event	p. 36
$\begin{bmatrix} 1 & 2 \\ 3 & 4 \end{bmatrix}$	matrix	p. 40
^	raised to a power (in a spreadsheet formula)	p. 46
*	multiply (in a spreadsheet formula)	p. 46
/	divide (in a spreadsheet formula)	p. 46
(x, y)	ordered pair	p. 58
$f(x)$	*f* of *x*; the function value at *x*	p. 79
x_1, x_2, etc.	specific values of the variable *x*	p. 89
y_1, y_2, etc.	specific values of the variable *y*	p. 89
$\stackrel{?}{=}$	is the statement true?	p. 109
$\angle A$	angle *A*	p. 111
%	percent	p. 139
$a : b$	ratio of *a* to *b*	p. 158
≠	is not equal to	p. 159
$\triangle ABC$	triangle *ABC*	p. 159
AB	length of $\overline{AB}$; distance between points *A* and *B*	p. 159
≤	is less than or equal to	p. 179
≥	is greater than or equal to	p. 179
{ }	set braces	p. 202
$\overleftrightarrow{AB}$	line through points *A* and *B*	p. 218
m	slope of a linear function	p. 231
b	*y*-intercept of a linear function	p. 231
$\sqrt{x}$	nonnegative square root of *x*	p. 332
±	plus or minus	p. 332
a^{-n}	$\frac{1}{a^n}, a \neq 0$	p. 380
$\overline{AB}$	segment with endpoints *A* and *B*	p. 420
$\sin A$	sine of $\angle A$	p. 425
$\cos A$	cosine of $\angle A$	p. 425
$\tan A$	tangent of $\angle A$	p. 425
$m\angle A$	measure of angle *A*	p. 429
$\bar{x}$	mean of data values of *x*	p. 452
$n!$	*n* factorial	p. 539
$_nP_r$	permutations of *n* things taken *r* at a time	p. 540
$_nC_r$	combinations of *n* things taken *r* at a time	p. 543

Properties and Formulas of Algebra

CHAPTER 1

Order of Operations

1. Perform any operation(s) inside grouping symbols.
2. Simplify any terms with exponents.
3. Multiply and divide in order from left to right.
4. Add and subtract in order from left to right.

The Identity Property

For every real number *a*:

$a + 0 = a$ and $0 + a = a$

$a \cdot 1 = a$ and $1 \cdot a = a$

The Commutative Property

For all real numbers *a* and *b*:

$a + b = b + a \qquad a \cdot b = b \cdot a$

The Associative Property

For all real numbers *a*, *b*, and *c*:

$(a + b) + c = a + (b + c)$

$(a \cdot b) \cdot c = a \cdot (b \cdot c)$

Property of Opposites

The sum of a number and its **opposite,** or **additive inverse,** is zero. $a + (-a) = 0$

Property of Reciprocals

The **reciprocal,** or **multiplicative inverse,** of a rational number $\frac{a}{b}$ is $\frac{b}{a}$. $a \cdot \frac{1}{a} = 1$

CHAPTER 2

The Probability Formula

$$P(\text{event}) = \frac{\text{number of favorable outcomes}}{\text{number of possible outcomes}}$$

CHAPTER 3

Properties of Equality

For all real numbers *a*, *b*, and *c*:

Addition: If $a = b$, then $a + c = b + c$.

Subtraction: If $a = b$, then $a - c = b - c$.

Multiplication: If $a = b$, then $a \cdot c = b \cdot c$.

Division: If $a = b$, and $c \neq 0$, then $\frac{a}{c} = \frac{b}{c}$.

Probability of Two Events

If *A* and *B* are independent events, then

$P(A \text{ and } B) = P(A) \cdot P(B)$.

If *A* and *B* are dependent events, then

$P(A \text{ and } B) = P(A) \cdot P(B \text{ after } A)$.

The Distributive Property

For all real numbers *a*, *b*, and *c*,

$a(b + c) = ab + ac \qquad (b + c)a = ba + ca$

$a(b - c) = ab - ac \qquad (b - c)a = ba - ca$

CHAPTER 4

Properties of Inequality

For all real numbers *a*, *b*, and *c*:

Addition: If $a > b$, then $a + c > b + c$.
If $a < b$, then $a + c < b + c$.

Subtraction: If $a > b$, then $a - c > b - c$.
If $a < b$, then $a - c < b - c$.

Multiplication: $c > 0$: If $a > b$, then $ac > bc$.
If $a < b$, then $ac < bc$.

$c < 0$: If $a > b$, then $ac < bc$.
If $a < b$, then $ac > bc$.

Division: $c > 0$: If $a < b$, then $\frac{a}{c} < \frac{b}{c}$.
If $a > b$, then $\frac{a}{c} > \frac{b}{c}$.

$c < 0$: If $a < b$, then $\frac{a}{c} > \frac{b}{c}$.
If $a > b$, then $\frac{a}{c} < \frac{b}{c}$.

CHAPTER 5

Slope-Intercept Form of a Linear Equation

The slope-intercept form of a linear equation is $y = mx + b$, where *m* is the slope and *b* is the *y*-intercept.

CHAPTER 7

Quadratic Formula

If $ax^2 + bx + c = 0$ and $a \neq 0$, then

$$x = \frac{-b \pm \sqrt{b^2 - 4ac}}{2a}.$$

Property of the Discriminant

For the quadratic equation $ax^2 + bx + c = 0$, where $a \neq 0$, the value of the discriminant $b^2 - 4ac$ tells you the number of solutions.

$b^2 - 4ac > 0$	two solutions
$b^2 - 4ac = 0$	one solution
$b^2 - 4ac < 0$	no solution

CHAPTER 8

Zero as an Exponent

For any nonzero number a, $a^0 = 1$.

Negative Exponents

For any nonzero number a and any integer n,

$$a^{-n} = \frac{1}{a^n}.$$

Multiplying or Dividing Powers

For any nonzero number a and any integers m and n:

$$a^m \cdot a^n = a^{m+n}$$

$$\frac{a^m}{a^n} = a^{m-n}$$

Raising a Power to a Power

For any number $a > 0$ and any integers m and n,

$$(a^m)^n = a^{mn}.$$

Raising a Product or a Quotient to a Power

For any nonzero numbers a and b and any integer n:

$$(ab)^n = a^n b^n$$

$$\left(\frac{a}{b}\right)^n = \frac{a^n}{b^n}$$

CHAPTER 9

Properties of Square Roots

For any numbers $a \geq 0$ and $b \geq 0$,

$$\sqrt{ab} = \sqrt{a} \cdot \sqrt{b}.$$

For any numbers $a \geq 0$ and $b > 0$, $\sqrt{\frac{a}{b}} = \frac{\sqrt{a}}{\sqrt{b}}$.

The Pythagorean Theorem

In a right triangle, the sum of the squares of the lengths of the legs is equal to the square of the length of the hypotenuse.

$$a^2 + b^2 = c^2$$

The Converse of the Pythagorean Theorem

If a triangle has sides of lengths a, b, and c, and $a^2 + b^2 = c^2$, then the triangle is a right triangle with hypotenuse of length c.

The Distance Formula

The distance d between any two points (x_1, y_1) and (x_2, y_2) is

$$d = \sqrt{(x_2 - x_1)^2 + (y_2 - y_1)^2}.$$

The Midpoint Formula

The midpoint M of a line segment with endpoints $A(x_1, y_1)$ and $B(x_2, y_2)$ is

$$\left(\frac{x_1 + x_2}{2}, \frac{y_1 + y_2}{2}\right).$$

Trigonometric Ratios

sine of $\angle A = \frac{\text{length of side opposite } \angle A}{\text{length of hypotenuse}}$

cosine of $\angle A = \frac{\text{length of side adjacent to } \angle A}{\text{length of hypotenuse}}$

tangent of $\angle A = \frac{\text{length of side opposite } \angle A}{\text{length of side adjacent to } \angle A}$

CHAPTER 10

Factoring Special Cases

For all real numbers a and b:

$a^2 - b^2 = (a + b)(a - b)$

$a^2 + 2ab + b^2 = (a + b)(a + b) = (a + b)^2$

$a^2 - 2ab + b^2 = (a - b)(a - b) = (a - b)^2$

Zero-Product Property

For all real numbers a and b, if $ab = 0$, then $a = 0$ *or* $b = 0$.

CHAPTER 11

Multiplication Counting Principle

If there are m ways to make a first selection and n ways to make a second selection, there are $m \cdot n$ ways to make the two selections.

Properties of Equality

For all real numbers a, b, and c:

Reflexive Property: $a = a$

Symmetric Property: If $a = b$, then $b = a$.

Transitive Property: If $a = b$ and $b = c$, then $a = c$.

Formulas from Geometry

You will use a number of geometric formulas as you work through your algebra book. Here are some perimeter, area, and volume formulas.

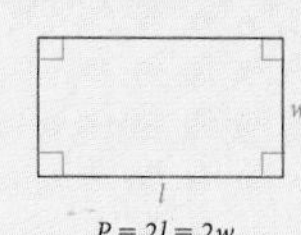

$P = 2l + 2w$
$A = lw$

Rectangle

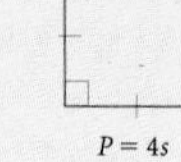
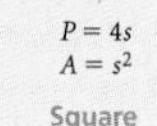

$P = 4s$
$A = s^2$

Square

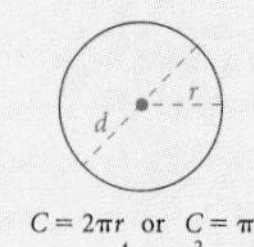

$C = 2\pi r$ or $C = \pi d$
$A = \pi r^2$

Circle

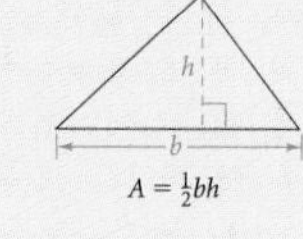

$A = \frac{1}{2}bh$

Triangle

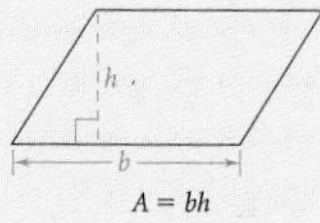

$A = bh$

Parallelogram

$A = \frac{1}{2}(b_1 + b_2)h$

Trapezoid

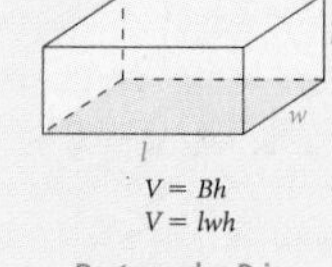

$V = Bh$
$V = lwh$

Rectangular Prism

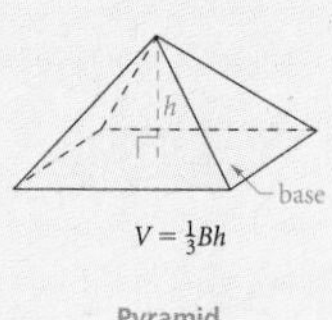

$V = \frac{1}{3}Bh$

Pyramid

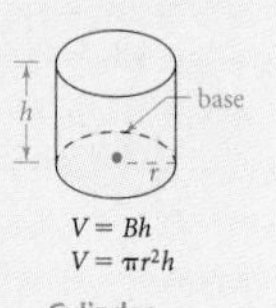

$V = Bh$
$V = \pi r^2 h$

Cylinder

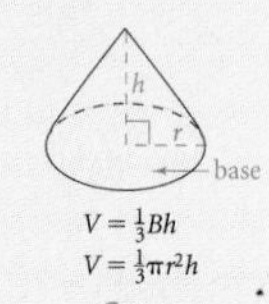

$V = \frac{1}{3}Bh$
$V = \frac{1}{3}\pi r^2 h$

Cone

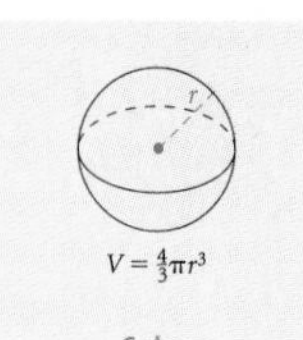

$V = \frac{4}{3}\pi r^3$

Sphere

Tables

Squares and Square Roots

Number n	Square n^2	Positive Square Root $\sqrt{n}$	Number n	Square n^2	Positive Square Root $\sqrt{n}$	Number n	Square n^2	Positive Square Root $\sqrt{n}$
1	1	1.000	51	2601	7.141	101	10,201	10.050
2	4	1.414	52	2704	7.211	102	10,404	10.100
3	9	1.732	53	2809	7.280	103	10,609	10.149
4	16	2.000	54	2916	7.348	104	10,816	10.198
5	25	2.236	55	3025	7.416	105	11,025	10.247
6	36	2.449	56	3136	7.483	106	11,236	10.296
7	49	2.646	57	3249	7.550	107	11,449	10.344
8	64	2.828	58	3364	7.616	108	11,664	10.392
9	81	3.000	59	3481	7.681	109	11,881	10.440
10	100	3.162	60	3600	7.746	110	12,100	10.488
11	121	3.317	61	3721	7.810	111	12,321	10.536
12	144	3.464	62	3844	7.874	112	12,544	10.583
13	169	3.606	63	3969	7.937	113	12,769	10.630
14	196	3.742	64	4096	8.000	114	12,996	10.677
15	225	3.873	65	4225	8.062	115	13,225	10.724
16	256	4.000	66	4356	8.124	116	13,456	10.770
17	289	4.123	67	4489	8.185	117	13,689	10.817
18	324	4.243	68	4624	8.246	118	13,924	10.863
19	361	4.359	69	4761	8.307	119	14,161	10.909
20	400	4.472	70	4900	8.367	120	14,400	10.954
21	441	4.583	71	5041	8.426	121	14,641	11.000
22	484	4.690	72	5184	8.485	122	14,884	11.045
23	529	4.796	73	5329	8.544	123	15,129	11.091
24	576	4.899	74	5476	8.602	124	15,376	11.136
25	625	5.000	75	5625	8.660	125	15,625	11.180
26	676	5.099	76	5776	8.718	126	15,876	11.225
27	729	5.196	77	5929	8.775	127	16,129	11.269
28	784	5.292	78	6084	8.832	128	16,384	11.314
29	841	5.385	79	6241	8.888	129	16,641	11.358
30	900	5.477	80	6400	8.944	130	16,900	11.402
31	961	5.568	81	6561	9.000	131	17,161	11.446
32	1024	5.657	82	6724	9.055	132	17,424	11.489
33	1089	5.745	83	6889	9.110	133	17,689	11.533
34	1156	5.831	84	7056	9.165	134	17,956	11.576
35	1225	5.916	85	7225	9.220	135	18,225	11.619
36	1296	6.000	86	7396	9.274	136	18,496	11.662
37	1369	6.083	87	7569	9.327	137	18,769	11.705
38	1444	6.164	88	7744	9.381	138	19,044	11.747
39	1521	6.245	89	7921	9.434	139	19,321	11.790
40	1600	6.325	90	8100	9.487	140	19,600	11.832
41	1681	6.403	91	8281	9.539	141	19,881	11.874
42	1764	6.481	92	8464	9.592	142	20,164	11.916
43	1849	6.557	93	8649	9.644	143	20,449	11.958
44	1936	6.633	94	8836	9.695	144	20,736	12.000
45	2025	6.708	95	9025	9.747	145	21,025	12.042
46	2116	6.782	96	9216	9.798	146	21,316	12.083
47	2209	6.856	97	9409	9.849	147	21,609	12.124
48	2304	6.928	98	9604	9.899	148	21,904	12.166
49	2401	7.000	99	9801	9.950	149	22,201	12.207
50	2500	7.071	100	10,000	10.000	150	22,500	12.247

Trigonometric Ratios

Angle	Sine	Cosine	Tangent	Angle	Sine	Cosine	Tangent
1°	0.0175	0.9998	0.0175	46°	0.7193	0.6947	1.0355
2°	0.0349	0.9994	0.0349	47°	0.7314	0.6820	1.0724
3°	0.0523	0.9986	0.0524	48°	0.7431	0.6691	1.1106
4°	0.0698	0.9976	0.0699	49°	0.7547	0.6561	1.1504
5°	0.0872	0.9962	0.0875	50°	0.7660	0.6428	1.1918
6°	0.1045	0.9945	0.1051	51°	0.7771	0.6293	1.2349
7°	0.1219	0.9925	0.1228	52°	0.7880	0.6157	1.2799
8°	0.1392	0.9903	0.1405	53°	0.7986	0.6018	1.3270
9°	0.1564	0.9877	0.1584	54°	0.8090	0.5878	1.3764
10°	0.1736	0.9848	0.1763	55°	0.8192	0.5736	1.4281
11°	0.1908	0.9816	0.1944	56°	0.8290	0.5592	1.4826
12°	0.2079	0.9781	0.2126	57°	0.8387	0.5446	1.5399
13°	0.2250	0.9744	0.2309	58°	0.8480	0.5299	1.6003
14°	0.2419	0.9703	0.2493	59°	0.8572	0.5150	1.6643
15°	0.2588	0.9659	0.2679	60°	0.8660	0.5000	1.7321
16°	0.2756	0.9613	0.2867	61°	0.8746	0.4848	1.8040
17°	0.2924	0.9563	0.3057	62°	0.8829	0.4695	1.8807
18°	0.3090	0.9511	0.3249	63°	0.8910	0.4540	1.9626
19°	0.3256	0.9455	0.3443	64°	0.8988	0.4384	2.0503
20°	0.3420	0.9397	0.3640	65°	0.9063	0.4226	2.1445
21°	0.3584	0.9336	0.3839	66°	0.9135	0.4067	2.2460
22°	0.3746	0.9272	0.4040	67°	0.9205	0.3907	2.3559
23°	0.3907	0.9205	0.4245	68°	0.9272	0.3746	2.4751
24°	0.4067	0.9135	0.4452	69°	0.9336	0.3584	2.6051
25°	0.4226	0.9063	0.4663	70°	0.9397	0.3420	2.7475
26°	0.4384	0.8988	0.4877	71°	0.9455	0.3256	2.9042
27°	0.4540	0.8910	0.5095	72°	0.9511	0.3090	3.0777
28°	0.4695	0.8829	0.5317	73°	0.9563	0.2924	3.2709
29°	0.4848	0.8746	0.5543	74°	0.9613	0.2756	3.4874
30°	0.5000	0.8660	0.5774	75°	0.9659	0.2588	3.7321
31°	0.5150	0.8572	0.6009	76°	0.9703	0.2419	4.0108
32°	0.5299	0.8480	0.6249	77°	0.9744	0.2250	4.3315
33°	0.5446	0.8387	0.6494	78°	0.9781	0.2079	4.7046
34°	0.5592	0.8290	0.6745	79°	0.9816	0.1908	5.1446
35°	0.5736	0.8192	0.7002	80°	0.9848	0.1736	5.6713
36°	0.5878	0.8090	0.7265	81°	0.9877	0.1564	6.3138
37°	0.6018	0.7986	0.7536	82°	0.9903	0.1392	7.1154
38°	0.6157	0.7880	0.7813	83°	0.9925	0.1219	8.1443
39°	0.6293	0.7771	0.8098	84°	0.9945	0.1045	9.5144
40°	0.6428	0.7660	0.8391	85°	0.9962	0.0872	11.4301
41°	0.6561	0.7547	0.8693	86°	0.9976	0.0698	14.3007
42°	0.6691	0.7431	0.9004	87°	0.9986	0.0523	19.0811
43°	0.6820	0.7314	0.9325	88°	0.9994	0.0349	28.6363
44°	0.6947	0.7193	0.9657	89°	0.9998	0.0175	57.2900
45°	0.7071	0.7071	1.0000	90°	1.0000	0.0000	

Tables

Glossary/Study Guide

A

Examples

Absolute value (p. 20) The distance that a number is from zero on a number line.

−7 is 7 units from 0, so $|-7| = 7$.

Absolute value function (p. 92) Function whose graph forms a "V" that opens up or down.

$y = |x - 3|$

Addition Property of Inequality (p. 180) For all real numbers a, b, and c: if $a > b$, then $a + c > b + c$ and if $a < b$, then $a + c < b + c$.

$4 > -2$, so $4 + 3 > -2 + 3$.
$5 < 9$, so $5 + 2 < 9 + 2$.

Additive inverses (p. 20) A number and its opposite. Additive inverses sum to 0.

5 and −5 are additive inverses because $5 + -5 = 0$.

Angle of elevation (p. 426) Used to measure heights indirectly. An angle from the horizontal up to a line of sight.

Associative Properties of Addition and Multiplication (p. 35) Changing the grouping of the addends or factors does not change the sum or product.
$(a + b) + c = a + (b + c)$ and $(a \cdot b) \cdot c = a \cdot (b \cdot c)$

$(9 + 2) + 3 = 9 + (2 + 3)$.
$(5 \cdot 8) \cdot 2 = 5 \cdot (8 \cdot 2)$.

Asymptote (p. 516) A line the graph of a function gets closer and closer to, but does not cross.

Example: The y-axis is a vertical asymptote for $y = \frac{1}{x}$. The x-axis is a horizontal asymptote for $y = \frac{1}{x}$.

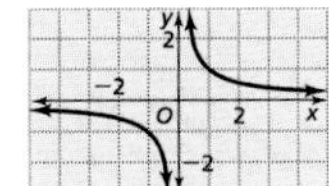

Axis of symmetry (p. 318) The line about which you can reflect a parabola onto itself.

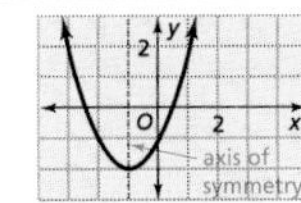

B

Examples

Bar graph (p. 6) A bar graph is used to compare amounts.

Example: The bar graph compares the number of students for grades 9, 10, and 11 over a three-year period.

Base (p. 15) The number that is multiplied repeatedly.

$4^5 = 4 \cdot 4 \cdot 4 \cdot 4 \cdot 4$. The base 4 is used as a factor 5 times.

Binomial factors (p. 481) Some quadratic trinomials are the products of two binomial factors.

$(x + 2)(x + 1) = x^2 + 3x + 2$
binomial factors quadratic trinomial

C

Cell (p. 45) A cell is a box where a row and a column meet in a spreadsheet.

A2 is the cell where row 2 and column A meet. The entry 1.50 is in cell A2.

	A	B	C	D	E
1	0.50	0.70	0.60	0.50	2.30
2	1.50	0.50	2.75	2.50	7.25

Certain event (p. 96) A certain event always happens, and has a probability of 1.

Rolling an integer less than 7 on a number cube is a certain event.

Coefficient (p. 119) The numerical factor when a term has a variable.

In the expression $2x + 3y + 16$, 2 and 3 are coefficients.

Combination (p. 543) An arrangement of some or all of a set of objects without regard to order. The number of combinations

$$= \frac{\text{total number of permutations}}{\text{number of times the objects in each group are repeated}}$$

You can use the notation $_nC_r$ to write the number of combinations of n objects chosen r at a time.

The number of combinations of 10 things taken 4 at a time is:

$_{10}C_4 = \frac{_{10}P_4}{_4P_4} = \frac{10 \cdot 9 \cdot 8 \cdot 7}{4 \cdot 3 \cdot 2 \cdot 1} = 210$

Common factors (p. 471) Numbers, variables, and any products formed from the prime factors that appear in all the terms of an expression.

x, 2, and $2x$ are common factors of $2x^2 + 4x$.

Examples

Commutative Properties of Addition and Multiplication (p. 35) Changing the order of the addends or factors does not change the sum or product. For all real numbers a and b:
$a + b = b + a$ and $a \cdot b = b \cdot a$.

$5 + 7 = 7 + 5$ and $9 \cdot 3 = 3 \cdot 9$.

Complement of an event (p. 96) All possible outcomes that are not in the event.
$P(\text{complement of event}) = 1 - P(\text{event})$

The complement of rolling a 1 or a 2 on a number cube is rolling a 3, 4, 5, or 6.

Compound inequalities (p. 195) Two inequalities that are joined by *and* or *or*.

$5 < x$ and $x < 10$
$x < -14$ or $x \geq 3$

Constant of variation (p. 225) The constant k in a direct variation.

For the function $y = 24x$, 24 is the constant of variation.

Constant term (p. 119) A term that has no variable factor.

In the expression $4x + 13y + 17$, 17 is a constant term.

Continuous data (p. 65) Have measurements that change between data points, such as temperature, length, and weight.

The height of a tree changes between annual measurings.

Conversion factors (p. 219) Used to change from one unit of measure to another.

Convert 2 h to minutes.
Since 60 min = 1 h, $\frac{60\text{ min}}{1\text{ h}} = 1$.
Use $\frac{60\text{ min}}{1\text{ h}}$ to convert 2 h to minutes.
$\frac{2\text{ h}}{1} \cdot \frac{60\text{ min}}{1\text{ h}} = 120$ min

Coordinate plane (p. 58) Formed when two number lines intersect at right angles. The x-axis is the horizontal axis and the y-axis is the vertical axis. The two axes meet at the origin, $O(0, 0)$.

Converse of the Pythagorean Theorem (p. 416) If a triangle has sides of lengths a, b, and c, and $a^2 + b^2 = c^2$, then the triangle is a right triangle with hypotenuse of length c.

Correlation (p. 60) A trend between two sets of data. A trend shows positive, negative, or no correlation.

Positive Negative No Correlation

Examples

Correlation coefficient (p. 242) Tells how well the equation of best fit models the data. The value r of the correlation coefficient is in the range $-1 \leq r \leq 1$.

Example: The correlation coefficient for the data points (68, 0.5), (85, 0.89), (100, 0.9), (108, 1.1) is approximately 0.94.

LinReg
y=ax+b
a=.0134039132
b=-.3622031627
r=.9414498267

Cosine (p. 425) In $\triangle ABC$ with right $\angle C$,
cosine of $\angle A = \frac{\text{length of side adjacent to } \angle A}{\text{length of hypotenuse}}$, or $\cos A = \frac{b}{c}$.

$\cos A = \frac{4}{5}$

Cross products (p. 159) In a proportion, the product of the numerator of the first ratio and the denominator of the second ratio; also the product of the denominator of the first ratio and the numerator of the second ratio. These products are equal.

$\frac{3}{4} = \frac{6}{8}$
The cross products are $3 \cdot 8$ and $4 \cdot 6$.
$3 \cdot 8 = 24$ and $4 \cdot 6 = 24$

Cubic equation (p. 494) An equation in the form $ax^3 + bx^2 + cx + d = 0$, where a, b, c, and d are real numbers and $a \neq 0$.

$14x^3 + 2x^2 - 8x - 2 = 0$

D

Degree of a polynomial (p. 465) The highest degree of any of its terms.

The degree of $3x^2 + x - 9$ is 2.

Degree of a term (p. 465) For a term that has only one variable, the degree is the exponent of the variable. The degree of a constant is zero.

The degree of $3x^2$ is 2.

Dependent events (p. 135) When the outcome of one event affects the outcome of a second event, the events are dependent events.

If you pick a marble from a bag and pick another without replacing the first, the events are dependent events.

Dependent variable (p. 69) A variable is dependent if it relies on another variable.

In the equation $y = 3x$, the value of y depends upon the value of x.

Difference of two squares (p. 487) A quadratic binomial of the form $a^2 - b^2$.

$x^2 - 16$

Dimensional analysis (p. 219) The process of analyzing units to decide which conversion factors to use to solve a problem.

$0.5\text{ mi} = \frac{0.5\text{ mi}}{1} \cdot \frac{5280\text{ ft}}{1\text{ mi}} = 2640$ ft

	Examples
Direct variation (p. 225) A linear function that can be expressed in the form $y = kx$, where $k \neq 0$.	$y = 18x$ is a direct variation.
Discrete data (p. 65) Involves a count of data items, such as number of people or objects.	the number of books on a shelf in the library
Discriminant (p. 349) The quantity $b^2 - 4ac$ for a quadratic equation of the form $ax^2 + bx + c = 0$.	The discriminant of $2x^2 + 9x - 2 = 0$ is $9^2 - 4(2)(-2) = 97$.
Distance Formula (p. 420) The distance d between any two points (x_1, y_1) and (x_2, y_2) is $d = \sqrt{(x_2 - x_1)^2 + (y_2 - y_1)^2}$.	The distance between $(-2, 4)$ and $(4, 5)$ is $d = \sqrt{(4 - (-2))^2 + (5 - 4)^2}$ $= \sqrt{(6)^2 + (1)^2}$ $= \sqrt{37}$
Distributive Property (p. 124) For all real numbers a, b, and c: $a(b + c) = ab + ac$ $(b + c)a = ba + ca$ $a(b - c) = ab - ac$ $(b - c)a = ba - ca$	$3(19 + 4) = 3(19) + 3(4)$ $(19 + 4)3 = 19(3) + 4(3)$ $7(11 - 2) = 7(11) - 7(2)$ $(11 - 2)7 = 11(7) - 2(7)$
Division Property of Inequality (p. 187) For all real numbers a and b, and for $c > 0$: If $a < b$, then $\frac{a}{c} < \frac{b}{c}$. If $a > b$, then $\frac{a}{c} > \frac{b}{c}$. $c < 0$: If $a < b$, then $\frac{a}{c} > \frac{b}{c}$. If $a > b$, then $\frac{a}{c} < \frac{b}{c}$.	$2 < 8$, so $\frac{2}{4} < \frac{8}{4}$. $5 > 1$, so $\frac{5}{2} > \frac{1}{2}$. $-2 < 6$, so $\frac{-2}{-3} > \frac{6}{-3}$. $-2 > -4$, so $\frac{-2}{-5} < \frac{-4}{-5}$.
Division Property of Square Roots (p. 431) For any numbers $a \geq 0$ and $b > 0$, $\sqrt{\frac{a}{b}} = \frac{\sqrt{a}}{\sqrt{b}}$.	$\sqrt{\frac{4}{9}} = \frac{\sqrt{4}}{\sqrt{9}} = \frac{2}{3}$
Domain (p. 74) The set of all possible input values.	In the function $f(x) = x + 22$, the domain is all real numbers.

E

Elimination (p. 280) A method for solving a system of linear equations. You add or subtract the equations to eliminate a variable.	$3x - y = 19$ $2x - y = 1$ $x + 0 = 18$ ← Subtract the second equation from the first. $x = 18$ ← Solve for x. $2(18) - y = 1$ ← Substitute 18 for x in the second equation. $36 - y = 1$ $y = 35$ ← Solve for y. The solution is (18, 35).

	Examples
Entry (p. 40) An item in a matrix.	2 is an entry in the matrix $\begin{bmatrix} 9 & 2 \\ 5.3 & 1 \end{bmatrix}$.
Equation (p. 11) An equation shows that two expressions are equal.	$x + 5 = 3x - 7$
Equivalent equations (p. 109) Equations that have the same solution.	$a = 3$ and $3a = 9$ are equivalent equations.
Equivalent inequalities (p. 180) Equivalent inequalities have the same set of solutions.	$n < 2$ and $n + 3 < 5$ are equivalent inequalities.
Event (p. 36) In probability, any group of outcomes.	When rolling a number cube, there are six possible outcomes. Rolling an even number is an event with three possible outcomes: 2, 4, and 6.
Experimental probability (p. 36) The ratio of the number of times an event actually happens to the number of times the experiment is done. $P(\text{event}) = \frac{\text{number of times an event happens}}{\text{number of times the experiment is done}}$	A baseball player's batting average shows how likely it is that a player will get a hit based on previous times at bat.
Exponent (p. 15) Shows repeated multiplication.	$3^4 = 3 \cdot 3 \cdot 3 \cdot 3$ The exponent 4 indicates that 3 is used as a factor four times.
Exponential decay (p. 373) For $a > 0$ and $0 < b < 1$, the function $y = ab^x$ models exponential decay. b is the decay factor.	$y = 5(0.1)^x$
Exponential function (p. 363) A function that repeatedly multiplies an initial amount by the same positive number. You can model all exponential functions using $y = ab^x$ where $b > 0$ and $b \neq 1$.	$y = 4.8(1.1)^x$
Exponential growth (p. 368) For $a > 0$ and $b > 1$, the function $y = ab^x$ models exponential growth. b is the growth factor.	$y = 100(2)^x$
Extraneous solution (p. 443) An apparent solution of an equation that does not satisfy the original equation.	$\frac{b}{b+4} = 3 - \frac{4}{b+4}$ Solving the equation by multiplying by $(b + 4)$ gives b as -4. Replacing b with -4 in the original equation makes the denominator 0, so -4 is an extraneous solution. The equation has no solution.

Glossary/Study Guide

F

	Examples
Families of functions (p. 92) Similar functions can be grouped into families of functions. Some families of functions are linear functions, quadratic functions, and absolute value functions.	$y = 3x^2$, $y = -9x^2$, and $y = \frac{3}{4}x^2$ belong to the quadratic family of functions.
Favorable outcomes (p. 95) In a probability experiment, favorable outcomes are the possible results that you want to happen.	In a board game you advance more spaces if you roll an even number on a number cube. The favorable outcomes are 2, 4, and 6.
Function (p. 73) A relation that assigns exactly one value of the dependent variable to each value of the independent variable.	Earned income is a function of the number of hours worked. If you earn \$4.50/h, then your income is expressed by the function $f(n) = 4.5n$.
Function notation (p. 79) To write a rule in function notation, you use the symbol $f(x)$ in place of y.	$f(x) = 3x - 8$ is in function notation.
Function rule (p. 74) An equation that describes a function.	$y = 4x + 1$ is a function rule.

H

Histogram (p. 4) A bar graph that shows the frequency of data. The height of the bars shows the number of items in each interval.	The histogram shows the birth months of 27 students in one math class.

Hypotenuse (p. 414) In a right triangle, the side opposite the right angle. It is the longest side in the triangle.	c is the hypotenuse.

I

Identity (p. 166) An equation that is true for every value.	$5 - 14x = 5(1 - \frac{14}{5}x)$ is an identity because it is true for any value of x.

	Examples
Identity Property of Addition (p. 35) The sum of any number and 0 is that number. For every real number a: $a + 0 = a$, $0 + a = a$.	$9 + 0 = 9$, $0 + 9 = 9$
Identity Property of Multiplication (p. 35) The product of any number and 1 is that number. For every real number a: $a \cdot 1 = a$, $1 \cdot a = a$.	$7 \cdot 1 = 7$, $1 \cdot 7 = 7$
Impossible event (p. 96) An impossible event never happens, and has a probability of 0.	Getting a decimal when rolling a number cube is an impossible event.
Independent events (p. 134) Events are independent when the outcome of one does not affect the other.	Picking a colored marble from a bag, replacing it, and picking another are two independent events.
Independent variable (p. 69) A variable is independent if it does not depend on another variable.	In the function $y = -2x$, the value of x does not depend on the value of y.
Integers (p. 19) The whole numbers and their opposites.	$\ldots -3, -2, -1, 0, 1, 2, 3, \ldots$
Inverse operations (p. 108) Operations that undo one another are called inverse operations.	Addition and subtraction are inverse operations. Multiplication and division are inverse operations.
Inverse variation (p. 512) A function that can be written in the form $xy = k$ or $y = \frac{k}{x}$. The product of the quantities remains constant, so as one quantity increases, the other decreases.	The length x and the width y of a rectangle with a fixed area vary inversely. If the area is 40, $xy = 40$.
Irrational number (p. 30) A number that cannot be written as a ratio of two integers. Irrational numbers in decimal form are nonterminating and nonrepeating.	$\sqrt{11}$ and 3.141592653. . . are irrational numbers.

L

Legs of a right triangle (p. 414) The sides that form the right angle.	a and b are legs.

Like terms (p. 119) Terms with exactly the same variable factors in a variable expression.	$4y$ and $16y$ are like terms.

Glossary/Study Guide

Examples

Line graph (p. 7) A graph that shows how a set of data changes over time.

The line graph at the left shows the change in the number of listeners to station KXXX during the day.

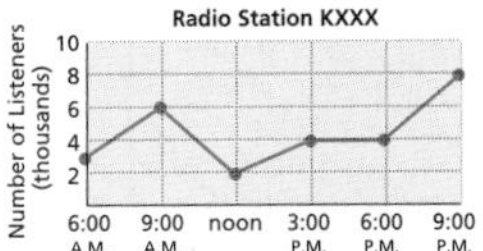

Line of best fit (p. 242) A trend line on a scatter plot that comes as close as possible to the data points.

Example: A line of best fit for the data points (1, 3), (−2, −5), (3, 4), (−3, −2), and (4, 7).

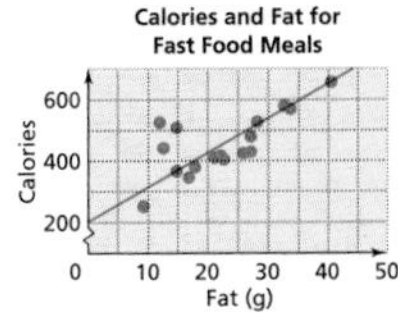

Line plot (p. 4) A graph that shows the number of times a data item appears.

The line plot at the left shows the data collected on birth months for 20 people.

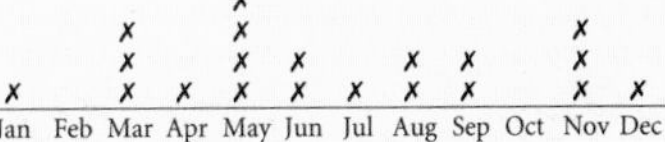

Linear function (pp. 92, 221) A function whose graph forms a straight line. Its rule is an equation that has 1 as the greatest power of x.

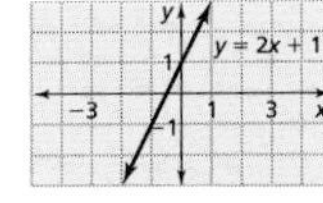

Linear inequality (p. 290) Describes a region of the coordinate plane that has a boundary line. Each point in the region is a solution of the inequality. A sign of ≤ or ≥ indicates a solid boundary line. A sign of < or > indicates a dashed boundary line.

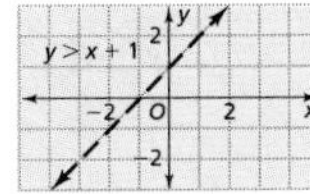

Examples

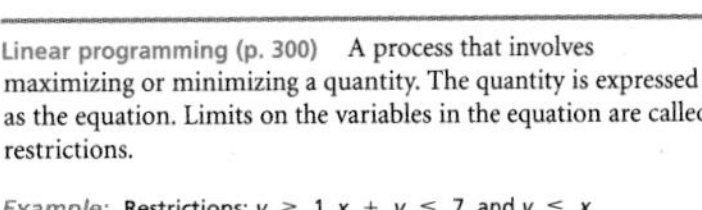

Linear programming (p. 300) A process that involves maximizing or minimizing a quantity. The quantity is expressed as the equation. Limits on the variables in the equation are called restrictions.

Example: Restrictions: $y \geq 1$, $x + y \leq 7$, and $y \leq x$
Equation: $B = 2x + 4y$
Graph the restrictions and find the coordinates of each vertex.
Evaluate $B = 2x + 4y$ at each vertex.
$B = 2(1) + 4(1) = 6$
$B = 2(3.5) + 4(3.5) = 21$
$B = 2(7) + 4(0) = 14$
The minimum value of B occurs at (1, 1).
The maximum value of B occurs at (3.5, 3.5).

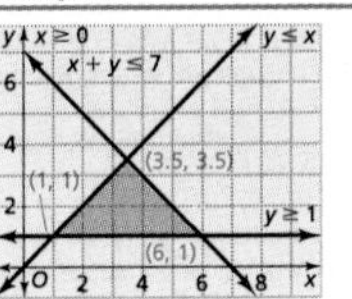

Literal equation (p. 176) An equation involving two or more variables.

$ax + b = c$ is a literal equation.

M

Matrix (p. 40) A rectangular arrangement of numbers. The number of rows and columns of a matrix determines its size. Each item in a matrix is an entry.

$\begin{bmatrix} 2 & 5 & 6.3 \\ -8 & 0 & -1 \end{bmatrix}$ is a 2×3 matrix.

Maximum value (p. 320) If a parabola opens downward, the y-coordinate of the vertex is the function's maximum value.

Example: Since the parabola opens downward, the y-coordinate of the vertex (0, −2) is the function's maximum value. The maximum value is −2.

Mean (p. 5) To find the mean of a set of numbers, find the sum of the numbers and divide the sum by the number of items.

mean is $\frac{\text{sum of the data items}}{\text{number of data items}}$

In the set 12, 11, 12, 10, 13, 12, and 7, the mean is $\frac{12 + 11 + 12 + 10 + 13 + 12 + 7}{7} = 11$.

Median (p. 5) The middle value in an ordered set of numbers.

In the set 7, 10, 11, 12, 15, 19, and 27, the median is 12.

Midpoint (p. 422) The point that divides a segment into two congruent segments.

M is the midpoint of $\overline{XY}$.

Examples

Midpoint Formula (p. 422) The midpoint M of a line segment with endpoints $A(x_1, y_1)$ and $B(x_2, y_2)$ is $\left(\frac{x_1 + x_2}{2}, \frac{y_1 + y_2}{2}\right)$.

The midpoint of a segment with endpoints $A(3, 5)$ and $B(7, 1)$ is (5, 3).

Minimum value (p. 320) If a parabola opens upward, the y-coordinate of the vertex is the function's minimum value.

Example: Since the parabola opens upward, the y-coordinate of the vertex (0, 1) is the function's minimum value. The minimum value is 1.

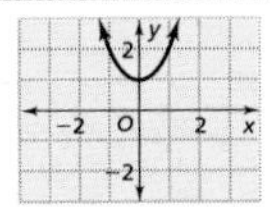

Mode (p. 4) The data item that occurs in a data set the greatest number of times. A data set may have no mode, one mode, or more than one mode.

In the data set 7, 7, 9, 10, 11, and 13, the mode is 7. The data set 5, 3, 2, 1.5, 9, 3, 6, 3, 2, 1, 4, 2 has two modes: 2 and 3.

Multiplicative inverse (p. 32) The multiplicative inverse, or reciprocal, of a rational number $\frac{a}{b}$ is $\frac{b}{a}$. The product of a nonzero number and its multiplicative inverse is 1.

$\frac{3}{4}$ is the multiplicative inverse of $\frac{4}{3}$ because $\frac{3}{4} \times \frac{4}{3} = 1$.

Multiplication Property of Inequality (p. 185) For all real numbers a and b, and for $c > 0$:
If $a > b$, then $ac > bc$.
If $a < b$, then $ac < bc$.
For all real numbers a and b, and for $c < 0$:
If $a > b$, then $ac < bc$.
If $a < b$, then $ac > bc$.

$5 > 1$, so $5(2) > 1(2)$.
$2 < 4$, so $2(3) < 4(3)$.
$-2 > -4$, so $-2(-5) < -4(-5)$.
$-2 < 6$, so $-2(-3) > 6(-3)$.

Multiplication Property of Square Roots (p. 430) For any numbers $a \geq 0$ and $b \geq 0$, $\sqrt{ab} = \sqrt{a} \cdot \sqrt{b}$.

$\sqrt{54} = \sqrt{9 \cdot 6} = \sqrt{9} \cdot \sqrt{6} = 3 \cdot \sqrt{6} = 3\sqrt{6}$

N

Negative square root (p. 332) $-\sqrt{b}$ is the negative square root of b.

$-\sqrt{49} = -7$ is the negative square root of 49.

O

Outcomes (p. 95) The possible results of a probability experiment.

The outcomes of rolling a number cube are 1, 2, 3, 4, 5, and 6.

Opposites (p. 19) Two numbers are opposites, or additive inverses, if they are the same distance from zero on the number line. The sum of two opposites is 0.

−3 and 3 are opposites.

Examples

Order of operations (p. 15)
1. Perform any operation(s) inside grouping symbols.
2. Simplify any terms with exponents.
3. Multiply and divide in order from left to right.
4. Add and subtract in order from left to right.

$6 - (4^2 - [2 \cdot 5]) \div 3$
$= 6 - (4^2 - 10) \div 3$
$= 6 - (16 - 10) \div 3$
$= 6 - 6 \div 3$
$= 6 - 2$
$= 4$

Ordered pair (p. 58) An ordered pair of numbers identifies the location of a point.

The ordered pair (4, −1) identifies the point 4 units to the right and 1 unit down from the origin.

Origin (p. 58) The axes of the coordinate plane intersect at the origin.

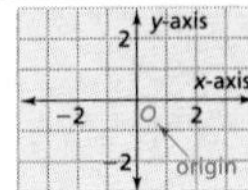

P

Parabola (p. 318) The graph of a quadratic function is a parabola.

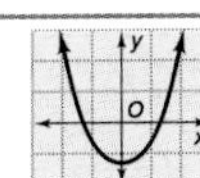

Parallel lines (p. 250) Two lines that are always the same distance apart. They do not intersect and the lines have the same slope.

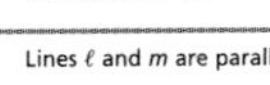

Lines ℓ and m are parallel.

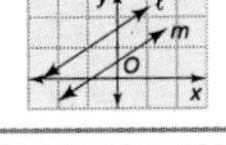

Percent of change (p. 146) The percent an amount changes from its original amount.

percent of change $= \frac{\text{amount of change}}{\text{original amount}}$

The price of a meal at a restaurant was \$7.95 last week; this week it is \$9.95.
percent of change $= \frac{9.95 - 7.95}{7.95} = \frac{2.00}{7.95}$
≈ 0.25 or 25%.

Percent of increase (p. 146) When a value increases from its original amount, you call the percent of change the percent of increase.

The average class size will increase from 30 to 33.
percent of increase $= \frac{3}{30} = 0.1$ or 10%

Percent of decrease (p. 147) When a value decreases from its original amount, you call the percent of change the percent of decrease.

The number of oranges in the bag decreased from 12 to 9.
percent of decrease $= \frac{3}{12} = 0.25$ or 25%

Glossary/Study Guide

Examples

Perfect square trinomial (p. 487) A quadratic expression whose factorization contains two factors that are the same. — $x^2 + 6x + 9 = (x + 3)(x + 3) = (x + 3)^2$

Perfect squares (p. 333) Numbers whose square roots are integers. — The numbers 1, 4, 9, 16, 25, 36, . . . are perfect squares because they are the squares of integers.

Permutation (p. 539) An arrangement of some or all of a set of objects in a definite order. You can use the notation ${}_nP_r$ to express the number of permutations, where n equals the number of objects available and r equals the number of selections to make. — How many ways can 5 children be arranged three at a time? ${}_5P_3 = 5 \cdot 4 \cdot 3 = 60$ arrangements

Perpendicular lines (p. 251) Lines that form right angles. Two lines are perpendicular if the product of their slopes is -1. — Lines ℓ and m are perpendicular.

Polynomial (p. 465) A sum of monomials. A quotient with a variable in the denominator is not a polynomial. — $2x^2$, $3x + 7$, 28, and $-7x^3 - 2x^2 + 9$ are all polynomials.

Power (p. 391) Any expression in the form a^n. — 5^4

Principal square root (p. 332) The expression $\sqrt{b}$ is called the principal square root of b. — $\sqrt{25} = 5$ is the principal square root of 25.

Probability of two dependent events (p. 135) If A and B are dependent events, then $P(A \text{ and } B) = P(A) \cdot P(B \text{ after } A)$. — You have 4 red marbles and 3 white marbles. The probability that you select one red marble, and then, without replacing it, randomly select another red marble is $P(\text{red and red}) = \frac{4}{7} \cdot \frac{3}{6} = \frac{2}{7}$.

Probability of two independent events (p. 134) If A and B are independent events, you multiply the probabilities of the events to find the probability of both events occurring. $P(A \text{ and } B) = P(A) \cdot P(B)$ — The probability of rolling a 1 on a number cube is $\frac{1}{6}$. The probability of rolling an even number on a number cube is $\frac{1}{2}$. The probability of rolling a 1 and then an even number is $P(1 \text{ and even number}) = \frac{1}{6} \cdot \frac{1}{2} = \frac{1}{12}$.

Properties of Equality (pp. 109, 110) For all real numbers a, b, and c:
Addition: If $a = b$, then $a + c = b + c$.
Subtraction: If $a = b$, then $a - c = b - c$.
Multiplication: If $a = b$, then $a \cdot c = b \cdot c$.
Division: If $a = b$, and $c \neq 0$, then $\frac{a}{c} = \frac{b}{c}$.

Since $\frac{2}{4} = \frac{1}{2}$, then $\frac{2}{4} + 5 = \frac{1}{2} + 5$.
Since $\frac{9}{3} = 3$, then $\frac{9}{3} - 6 = 3 - 6$.
Since $\frac{10}{5} = 2$, then $\frac{10}{5} \cdot 15 = 2 \cdot 15$.
Since $6 + 2 = 8$, then $\frac{6+2}{4} = \frac{8}{4}$.

Examples

Proportion (p. 158) A statement that two ratios are equal. — $\frac{2}{3} = \frac{10}{15}$ is a proportion.

Pythagorean Theorem (p. 414) In a right triangle, the sum of the squares of the lengths of the legs is equal to the square of the length of the hypotenuse. $a^2 + b^2 = c^2$. — $3^2 + 4^2 = 5^2$

Q

Quadrants (p. 58) The coordinate plane is divided by its axes into four quadrants.

Quadratic equation (p. 337) An equation you can write in the form $ax^2 + bx + c = 0$, where a, b, and c are real numbers and $a \neq 0$. — $4x^2 + 9x - 5 = 0$

Quadratic Formula (p. 343) If $ax^2 + bx + c = 0$ and $a \neq 0$, then $x = \frac{-b \pm \sqrt{b^2 - 4ac}}{2a}$.

$$2x^2 + 10x + 12 = 0$$
$$x = \frac{-b \pm \sqrt{b^2 - 4ac}}{2a}$$
$$x = \frac{-10 \pm \sqrt{10^2 - 4(2)(12)}}{2(2)}$$
$$x = \frac{-10 \pm \sqrt{4}}{4}$$
$$x = \frac{-10 + 2}{4} \text{ or } \frac{-10 - 2}{4}$$
$$x = -2 \text{ or } x = -3$$

Quadratic function (pp. 92, 319) A function with an equation of the form $y = ax^2 + bx + c$, where $a \neq 0$. The graph of a quadratic function is a parabola, which is a U-shaped curve that opens up or down. — $y = 5x^2 - 2x + 1$

Quadratic trinomial (p. 481) A quadratic trinomial is an expression of the form $ax^2 + bx + c$, where a, b, and c are nonzero real numbers. — $4x^2 + 2x + 9$

R

Radical equation (p. 440) An equation that has a variable under a radical. — $\sqrt{x} - 2 = 12$

Examples

Range (p. 74) The set of all possible output values of a function. — In the function $f(x) = |x|$, the range is the set of all positive numbers and 0.

Rate of change (p. 220) Allows you to see the relationship between two quantities that are changing. The rate of change is also called slope.

$$\text{Rate of change} = \frac{\text{change in dependent variable}}{\text{change in independent variable}}$$

Video rental for 1 day is \$1.99. Video rental for 2 days is \$2.99.
$$\text{rate of change} = \frac{2.99 - 1.99}{2 - 1} = \frac{1.00}{1} = 1$$

Ratio (p. 158) A comparison of two numbers by division. — $\frac{5}{7}$ and 7 : 3 are ratios.

Rational expression (p. 522) An expression that can be written in the form $\frac{\text{polynomial}}{\text{polynomial}}$. The value of the variable cannot make the denominator equal to 0. — $\frac{3}{x^2 - 4}$, $x \neq 2, -2$.

Rational function (p. 520) A function that can be written in the form $f(x) = \frac{\text{polynomial}}{\text{polynomial}}$. The value of the variable cannot make the denominator equal to 0. — $y = \frac{x}{x - 5}$, $x \neq 5$

Rational number (p. 30) A real number that can be written as a ratio of two integers. Rational numbers in decimal form are terminating or repeating. — $\frac{2}{3}$, 1.548, and 2.292929. . . are all rational numbers.

Rationalize the denominator (p. 432) Make the denominator of a fraction a rational number without changing the value of the expression. — $\frac{2}{\sqrt{5}} = \frac{2}{\sqrt{5}} \cdot \frac{\sqrt{5}}{\sqrt{5}} = \frac{2\sqrt{5}}{\sqrt{25}} = \frac{2\sqrt{5}}{5}$

Real number (p. 30) A number that is either rational or irrational. — 5, −3, 9.2, −0.666. . . , $5\frac{4}{11}$, 0, and $\frac{15}{2}$ are all real numbers.

Reciprocal (p. 32) The reciprocal, or multiplicative inverse, of a rational number $\frac{a}{b}$ is $\frac{b}{a}$. The product of a nonzero number and its reciprocal is 1. — $\frac{2}{5}$ and $\frac{5}{2}$ are reciprocals because $\frac{2}{5} \times \frac{5}{2} = 1$.

Relation (p. 73) Any set of ordered pairs. — {(0, 0), (2, 3), (2, −7)} is a relation.

Replacement set (p. 202) The set of possible values for the variable in an inequality. — When the replacement set is all integers, the solution of the inequality $x > -2.5$ is {−2, −1, 0, 2, . . . }.

Restrictions (p. 522) Values that make the denominator of a rational expression equal to 0 are restrictions on the variable. — For $\frac{4}{x - 3}$, the restriction is $x \neq 3$.

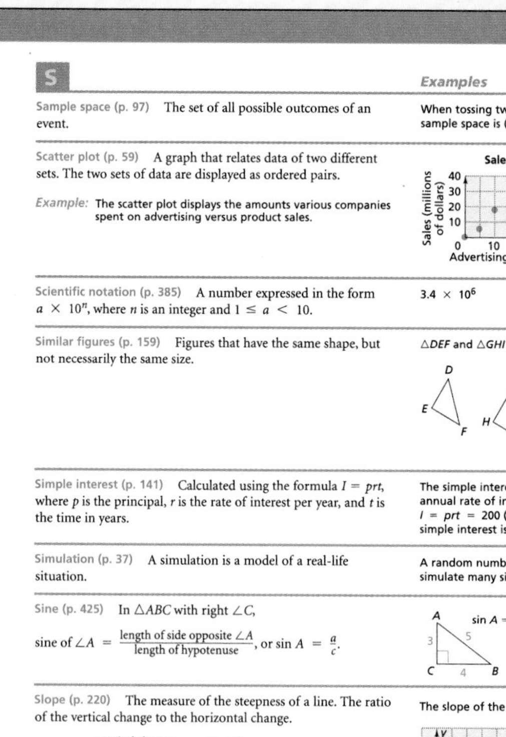

S

Examples

Sample space (p. 97) The set of all possible outcomes of an event. — When tossing two coins one at a time, the sample space is (H,H), (T,T), (H,T), (T,H).

Scatter plot (p. 59) A graph that relates data of two different sets. The two sets of data are displayed as ordered pairs.

Example: The scatter plot displays the amounts various companies spent on advertising versus product sales.

Scientific notation (p. 385) A number expressed in the form $a \times 10^n$, where n is an integer and $1 \leq a < 10$. — 3.4×10^6

Similar figures (p. 159) Figures that have the same shape, but not necessarily the same size. — $\triangle DEF$ and $\triangle GHI$ are similar.

Simple interest (p. 141) Calculated using the formula $I = prt$, where p is the principal, r is the rate of interest per year, and t is the time in years. — The simple interest on \$200 with an annual rate of interest of 4% for 3 years is $I = prt = 200\,(0.04)\,(3) = 24$. The simple interest is \$24.

Simulation (p. 37) A simulation is a model of a real-life situation. — A random number table can be used to simulate many situations.

Sine (p. 425) In $\triangle ABC$ with right $\angle C$,

$$\text{sine of } \angle A = \frac{\text{length of side opposite } \angle A}{\text{length of hypotenuse}}, \text{ or } \sin A = \frac{a}{c}.$$

$\sin A = \frac{4}{5}$

Slope (p. 220) The measure of the steepness of a line. The ratio of the vertical change to the horizontal change.

$$\text{Slope} = \frac{\text{vertical change}}{\text{horizontal change}} = \frac{y_2 - y_1}{x_2 - x_1}, \text{ where } x_2 - x_1 \neq 0$$

The slope of the line is $\frac{2 - 0}{4 - 0} = \frac{1}{2}$.

Examples

Slope-intercept form (p. 231) The slope-intercept form of a linear equation is $y = mx + b$ where m is the slope and b is the y-intercept. | $y = 8x + 2$

Solution (p. 108) Any value or values that make an equation true. | For the equation $y + 22 = 11$, the solution is -11.

Solution, no (pp. 165, 192, 271) (1) An equation has no solution if no value makes the statement true.
(2) An inequality has no solution if it is false for all values of the variable.
(3) A linear system has no solution if the graphs of the equations in the system are parallel. | $2a + 3 = 2a + 5$; $4n + 1 > 4n + 7$; $y = 3x + 9$ and $y = 3x + 28$

Solution of the inequality (p. 179) Any value or values of a variable in the inequality that makes an inequality true. | The solution of the inequality $x < 9$ is all numbers less than 9.

Solutions, infinitely many (p. 271) A linear system has infinitely many solutions when the equations are equivalent. | $2y = x + 7$ and $6y = 3x + 21$

Spreadsheet (p. 45) Organizes data in rows and columns. A cell is a box where a row and column meet. | In the spreadsheet, column C and row 2 meet at the shaded box, cell C2. The value in cell C2 is 2.75.

	A	B	C	D	E
1	0.50	0.70	0.60	0.50	2.30
2	1.50	0.50	2.75	2.50	7.25

Square root (p. 332) If $a^2 = b$, then a is a square root of b. $\sqrt{b}$ is the principal square root. $-\sqrt{b}$ is the negative square root. | -3 and 3 are square roots of 9. $\sqrt{9} = 3$; $-\sqrt{9} = -3$.

Standard deviation (p. 452) Shows how spread out a set of data is from the mean. | The standard deviation for the data set {2, 12, 6, 10, 9} is about 3.49.

Standard form of a polynomial (p. 465) When the degree of the terms in a polynomial decrease from left to right, it is in standard form, or descending order. | $15x^3 + x^2 + 3x - 9$

Standard form of a quadratic equation (p. 319) When a quadratic equation is in the form $ax^2 + bx + c = 0$. | $-x^2 + 2x + 9 = 0$

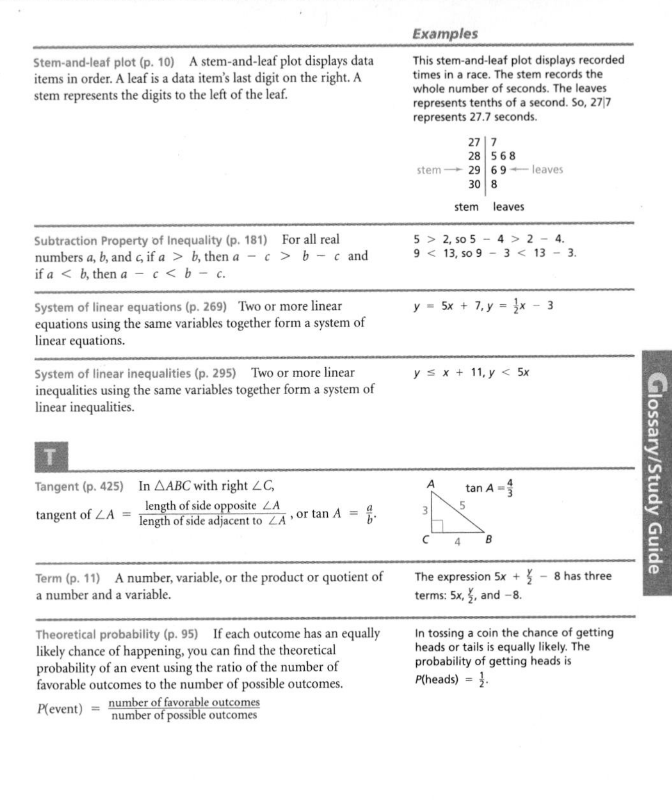

Examples

Stem-and-leaf plot (p. 10) A stem-and-leaf plot displays data items in order. A leaf is a data item's last digit on the right. A stem represents the digits to the left of the leaf. | This stem-and-leaf plot displays recorded times in a race. The stem records the whole number of seconds. The leaves represents tenths of a second. So, 27|7 represents 27.7 seconds.

Subtraction Property of Inequality (p. 181) For all real numbers a, b, and c, if $a > b$, then $a - c > b - c$ and if $a < b$, then $a - c < b - c$. | $5 > 2$, so $5 - 4 > 2 - 4$. $9 < 13$, so $9 - 3 < 13 - 3$.

System of linear equations (p. 269) Two or more linear equations using the same variables together form a system of linear equations. | $y = 5x + 7, y = \frac{1}{2}x - 3$

System of linear inequalities (p. 295) Two or more linear inequalities using the same variables together form a system of linear inequalities. | $y \le x + 11, y < 5x$

T

Tangent (p. 425) In $\triangle ABC$ with right $\angle C$, tangent of $\angle A = \frac{\text{length of side opposite } \angle A}{\text{length of side adjacent to } \angle A}$, or $\tan A = \frac{a}{b}$. | $\tan A = \frac{4}{3}$

Term (p. 11) A number, variable, or the product or quotient of a number and a variable. | The expression $5x + \frac{y}{2} - 8$ has three terms: $5x$, $\frac{y}{2}$, and -8.

Theoretical probability (p. 95) If each outcome has an equally likely chance of happening, you can find the theoretical probability of an event using the ratio of the number of favorable outcomes to the number of possible outcomes.

$P(\text{event}) = \frac{\text{number of favorable outcomes}}{\text{number of possible outcomes}}$ | In tossing a coin the chance of getting heads or tails is equally likely. The probability of getting heads is $P(\text{heads}) = \frac{1}{2}$.

Glossary/Study Guide 607

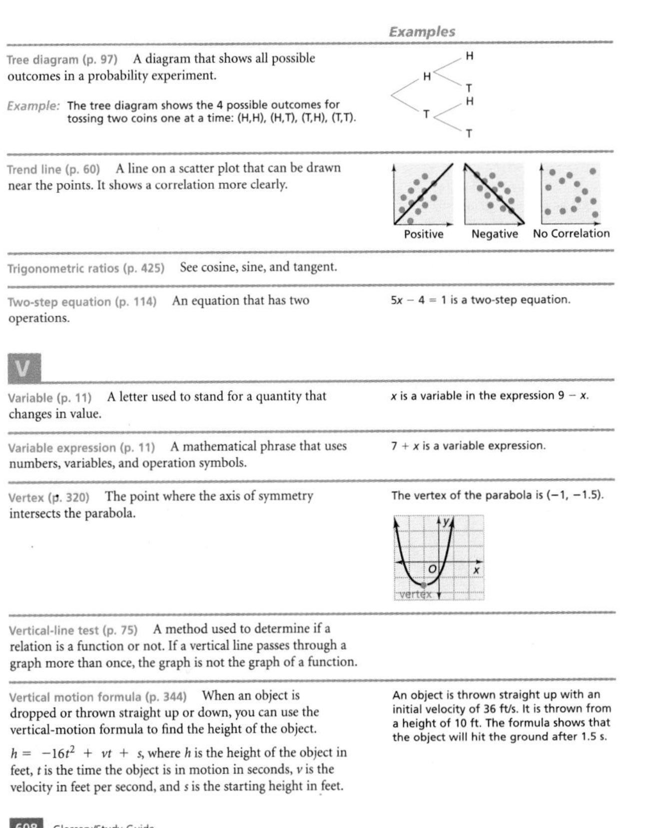

Examples

Tree diagram (p. 97) A diagram that shows all possible outcomes in a probability experiment.

Example: The tree diagram shows the 4 possible outcomes for tossing two coins one at a time: (H,H), (H,T), (T,H), (T,T).

Trend line (p. 60) A line on a scatter plot that can be drawn near the points. It shows a correlation more clearly.

Trigonometric ratios (p. 425) See cosine, sine, and tangent.

Two-step equation (p. 114) An equation that has two operations. | $5x - 4 = 1$ is a two-step equation.

V

Variable (p. 11) A letter used to stand for a quantity that changes in value. | x is a variable in the expression $9 - x$.

Variable expression (p. 11) A mathematical phrase that uses numbers, variables, and operation symbols. | $7 + x$ is a variable expression.

Vertex (p. 320) The point where the axis of symmetry intersects the parabola. | The vertex of the parabola is $(-1, -1.5)$.

Vertical-line test (p. 75) A method used to determine if a relation is a function or not. If a vertical line passes through a graph more than once, the graph is not the graph of a function.

Vertical motion formula (p. 344) When an object is dropped or thrown straight up or down, you can use the vertical-motion formula to find the height of the object.

$h = -16t^2 + vt + s$, where h is the height of the object in feet, t is the time the object is in motion in seconds, v is the velocity in feet per second, and s is the starting height in feet. | An object is thrown straight up with an initial velocity of 36 ft/s. It is thrown from a height of 10 ft. The formula shows that the object will hit the ground after 1.5 s.

608 Glossary/Study Guide

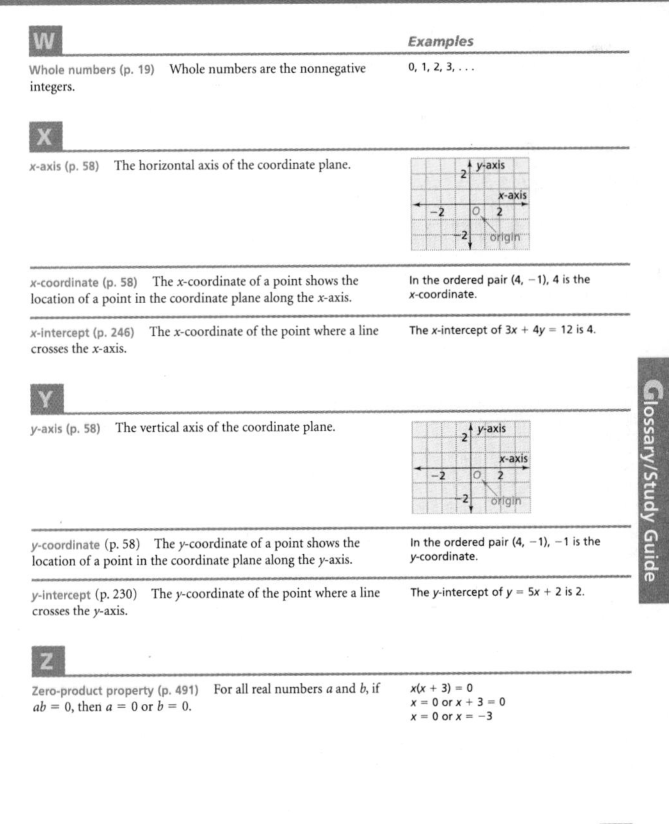

W **Examples**

Whole numbers (p. 19) Whole numbers are the nonnegative integers. | 0, 1, 2, 3, . . .

X

x-axis (p. 58) The horizontal axis of the coordinate plane.

x-coordinate (p. 58) The x-coordinate of a point shows the location of a point in the coordinate plane along the x-axis. | In the ordered pair $(4, -1)$, 4 is the x-coordinate.

x-intercept (p. 246) The x-coordinate of the point where a line crosses the x-axis. | The x-intercept of $3x + 4y = 12$ is 4.

Y

y-axis (p. 58) The vertical axis of the coordinate plane.

y-coordinate (p. 58) The y-coordinate of a point shows the location of a point in the coordinate plane along the y-axis. | In the ordered pair $(4, -1)$, -1 is the y-coordinate.

y-intercept (p. 230) The y-coordinate of the point where a line crosses the y-axis. | The y-intercept of $y = 5x + 2$ is 2.

Z

Zero-product property (p. 491) For all real numbers a and b, if $ab = 0$, then $a = 0$ or $b = 0$. | $x(x + 3) = 0$; $x = 0$ or $x + 3 = 0$; $x = 0$ or $x = -3$

Glossary/Study Guide 609

Glossary/Study Guide

CHAPTER 1

Lesson 1-1 pages 4–9

ON YOUR OWN **1a.**

```
      X
    X X
    X X X
  X X X X
  X X X X
  X X X X X
  X X X X X   X
  0 1 2 3 4 5 6
```

1b. 1.96; 2; 2 **3.** D **5.** 10; 10; 10 **7.** 25; 25; 25 **9.** 38.6; 37.9; 41.0 **11a.** 10 **13b.** English: 4.68; 4, 3; Spanish: 4.93, 5, 2

MIXED REVIEW **19.** 0.4; $\frac{2}{5}$ **21.** $\frac{1}{4}$; 0.25 **23.** 1.8; 180% **25.** 7, 9, 11 **27.** 108, 324, 972

Toolbox page 10

1. 27 students **3.** 8 students **5a.** 1.9, 2.0, 2.1, 2.2, 2.3, 2.4, 2.5 **5b.** Sample: 2.2 | 5 means $2.25

5c.

Stem	Leaves
1.9	5 7 9
2.0	0 5 9
2.1	6
2.2	1 5 9
2.3	6
2.4	0 9
2.5	0 0 7

5d. $2.23; $2.50

Lesson 1-2 pages 11–14

ON YOUR OWN **1.** 3 **3.** 2 **7.** c = cost, n = number of cans; $c = 0.7n$ **9.** r = amount of rope, t = number of tents; $r = 60t$ **11.** $d = 50h$ **13.** $g = 0.165d$ **15.** $r = 13w$ **17.** $e = 0.4s$ **21a.** Sample: Statement (a) is better because it indicates that each lawn takes 2 h to mow. **21b.** 1 lawn: 2 hours; 2 lawns: 4 hours; 5 lawns should take 10 hours. **23.** s = height of second bounce; $s = \frac{1}{4}d$ **25a.** n = number of quarters, m = money in bag (in dollars); $m = 0.25n$ **25b.** $3.25

MIXED REVIEW **27.** Complete list: 23, 29, 31, 37, 41, 43, 47 **29.** 8 **31.** 7 **33.** 11

Lesson 1-3 pages 15–18

ON YOUR OWN **1.** 23 **3.** 1 **5.** 64 **7.** 50.43 **9.** 75 **11.** 14 **15.** 1 **17.** 12.8 **19.** 104.58 **21.** 26 **23.** 2.4 **25.** $14 - (2 + 5) - 3 = 4$ **29.** F; simplify the power before multiplying. **31.** F; simplify within parentheses first.

MIXED REVIEW **33.** $\frac{2}{3}$ **35.** $4\frac{1}{2}$ **37.** $\frac{4}{5}$ **39.** 2 units **41.** 3 units

Lesson 1-4 pages 19–24

ON YOUR OWN **1.** $-227 + 319$ is greater because the sum is positive and $227 + (-319)$ is negative. **3.** positive **5.** negative **7.** positive **9.** 15 **11.** 5 **13.** −35 **15.** 11 **17.** −28 **19.** 4 **21.** −43.9 **23.** 1.5 **25.** −26.9 **27.** 45.7 **29.** 0 **31.** Q; it is the farthest point from the midpoint between R and T. **33.** Yes; $3 - 4 = -1, 4 - 3 = 1; |-1| = |1|$. So $a - b$ and $b - a$ are opposites. Absolute values of opposites are equal. **37.** −11,331 ft **39.** 8 **41.** −9 **43.** −2 **45.** 1 **47.** −5 **49.** 14 **51.** 7 **53.** −1 **55.** B

MIXED REVIEW **57b.** A line graph shows change over time. **59.** 4 **61.** $\frac{1}{4}$ **63.** $\frac{1}{3}$ **65.** −3, −6, −9 **67.** 0, −2, −4

Lesson 1-5 pages 25–29

ON YOUR OWN **1.** −120 **3.** −81 **5.** −6 **7.** −4 **9.** −1 **11.** 30 **13.** 44 **15.** $5\frac{1}{2}$ **17a.** The product is positive when a and b have the same sign. **17b.** The product is negative when a and b have different signs. **19.** −12 **21.** −15 **23.** −8 **25.** −64 **27.** $-1\frac{1}{3}$ **29.** −27 **33.** positive; even number of factors **35.** negative; odd number of negative factors **37.** negative; $(-3)^2$ is positive. Its opposite is negative. **39.** negative; 5^{10} is positive. Its opposite is negative. **47a.** odd **47b.** even

MIXED REVIEW **49.** 5 people **51a.** 2.8; 2.5; 1 **51b.** Sample: The median is the most useful because it is not affected by extreme values. **53.** $\frac{1}{4}$ **55.** $\frac{1}{3}$

CHECKPOINT **1.** 20 **2.** 4 **3.** 6 **4.** 7 **5.** 7 **6.** 49 **7.** −3 **8.** 11 **9.** 11 **11.** −0.4; −3; −4 **12.** $t = 175b$

Lesson 1-6 pages 30–34

ON YOUR OWN **1.** > **3.** > **5.** > **7.** > **9.** $\frac{5}{3}$ **11.** $\frac{8}{7}$ **13.** −2 **15.** $\frac{2}{3}$ **17.** 1 **19.** $-6\frac{3}{4}$ **21.** $2\frac{1}{6}$ **23.** $-2\frac{1}{12}$ **25.** $-\frac{2}{3}, -\frac{1}{2}, \frac{1}{4}$ **27.** $-9\frac{3}{4}, -9.7, -9\frac{7}{12}$ **29a.** −44°F **29b.** −4°F **31.** $-\frac{9}{16}$ **33.** $-\frac{5}{48}$ **37.** $2\frac{3}{5}$ **39.** −4.5 **41.** $-\frac{1}{10}$ **43.** false; $-\frac{3}{4}$ **45.** true; since integers are real numbers, some real numbers are integers.

MIXED REVIEW **47.** c = change, p = purchase; $c = 10 - p$ **49.** $1,290,000

Toolbox page 35

1. identity prop. of mult. **3.** associative prop. of mult. **5.** commutative prop. of mult. **7.** 830 **9.** 7400 **11.** $9m + 15$ **13.** 13 **15.** $72pqr$

Lesson 1-7 pages 36–39

ON YOUR OWN **3.** $\frac{8}{15}$ **5.** $\frac{7}{15}$ **7.** 8.3% **9.** 62.5% **11.** P(test) = 0.04%; P(control) = 0.08%; control group is twice as likely to develop polio. **13.** $\frac{1}{9}$ **15.** $\frac{7}{18}$ **17.** $\frac{7}{9}$ **19.** 0.04, 4%

MIXED REVIEW **23.** −19 **25.** −49 **27.** 53.6

Lesson 1-8 pages 40–43

ON YOUR OWN **1.** 2×3 **3.** 3×2 **5.** 3×4 **7.** $\begin{bmatrix} 0 & 0 \\ 0 & 0 \end{bmatrix}$ **9.** –6; 5; 0

11. $\begin{bmatrix} 81 & -32 \\ 27 & 100 \end{bmatrix}; \begin{bmatrix} 3 & -40 \\ -47 & 8 \end{bmatrix}$ **13.** $\begin{bmatrix} 8.3 & -5.8 \\ -2.6 & 2.9 \\ 13 & -7.8 \end{bmatrix}; \begin{bmatrix} -1.5 & 10 \\ -10.2 & 8.5 \\ 4.6 & -10.8 \end{bmatrix}$ **15.** $\begin{bmatrix} 1\frac{1}{6} & 1\frac{1}{5} \\ \frac{7}{8} & 1\frac{1}{6} \end{bmatrix}; \begin{bmatrix} \frac{1}{6} & \frac{2}{5} \\ -\frac{5}{8} & -\frac{5}{6} \end{bmatrix}$

17. $\begin{bmatrix} 2.5 & 0.3 \\ -8.1 & 2.7 \end{bmatrix}; \begin{bmatrix} 4.5 & -0.7 \\ 8.1 & -16.7 \end{bmatrix}$ **21.** $\begin{bmatrix} 1\frac{1}{4} & -6\frac{1}{2} \\ 2\frac{1}{2} & 0 \end{bmatrix}$

23a. $\begin{bmatrix} 13 & 5 & 6 & 2 \\ 18 & 4 & 2 & 2 \\ 6 & 2 & 0 & 2 \end{bmatrix}$ **23b.** 4 employees

25. $\begin{bmatrix} -350 & -150 & 50 \\ -400 & -100 & 250 \end{bmatrix}$

MIXED REVIEW **27.** −7.9 **29.** 17.4 **31.** 24 in.^{2} **33.** 68.08 m^2

CHECKPOINT **1.** $\frac{5}{8}$ **2.** 12.1 **3.** $\frac{25}{32}$ **4.** $\begin{bmatrix} 3\frac{1}{2} & -1 \\ 4 & -9 \end{bmatrix}$ **5.** $\begin{bmatrix} 5.6 & -6.8 & 11.5 \\ -5.8 & 3.3 & -15.7 \end{bmatrix}$

Toolbox page 44

1. $\begin{bmatrix} 4.9 & 0.7 \\ 2.1 & 0.7 \\ -3.1 & 2.7 \end{bmatrix}; \begin{bmatrix} 2.7 & 3.5 \\ 0.9 & -2.3 \\ 0.3 & 1.1 \end{bmatrix}$

3. $\begin{bmatrix} 0 & 4 & 20 & -25 \\ 8 & 2 & -3 & 15 \\ -9 & -1 & -14 & 9 \end{bmatrix}; \begin{bmatrix} 10 & -12 & -6 & -1 \\ 0 & -20 & -17 & -5 \\ 21 & -13 & 2 & 15 \end{bmatrix}$

5a. $\begin{bmatrix} 4.9 & 0.7 \\ 2.1 & 0.7 \\ -3.1 & 2.7 \end{bmatrix}; \begin{bmatrix} -2.7 & -3.5 \\ -0.9 & 2.3 \\ -0.3 & -1.1 \end{bmatrix}$

Lesson 1-9 pages 45–49

ON YOUR OWN **3.** 4 * A2 ^ 2; 2 * A2 − 5; 3 * A2 + 7 **5.** 16, −9, 1 **7a.** −3 **7b.** −1.9 **9a.** A2 * B2 * C2 **9b.** D2: 1920; E2: 725.2; F2: 5334 **11a.** C2/B2 **11b.** D2: $18,569.23; D3: $24,259.50; D4: $27,702; D5: $30,861.03 **13a.** 1.766 or 1.76625 **13b.** 1562.665 **15a.** 52 **15b.** 118.72 **17a.** Jan.: $1136.65; Feb.: $1557.30; March: $1396.20 **17b.** Jan.: $279.40; Feb.: $382.80; March: $343.20 **19.** C

MIXED REVIEW **21.** −3 **23.** 2 **25.** 26 **27.** 3 **29.** −71

Wrap Up pages 51–53

1. 30.6; 27; 24 **2.** 84.3; 83; 80 and 87 **3.** 2.3; 2.3; 2.3 **4.** 4 recorders **5.** March **6.** 4.7 recorders; 8 recorders **7.** 8 **8.** 64 **9.** 4 **10.** 9 **12a.** $r = 140d$ **12b.** 2940 records; 4200 records **13.** −17 **14.** −12 **15.** 9.9 **16.** 24.9 **17.** −5 **18.** −12 **19.** 15 **20.** 13 **21.** 2 **22.** −1 **23.** 0

24. −18 **25.** −40 **26.** −8 **27.** 2 **28.** B **29.** $2\frac{5}{6}$ **30.** $5\frac{1}{6}$ **31.** $-3\frac{1}{3}$ **32.** $-1\frac{1}{4}$ **33.** $1\frac{5}{18}$ **34a.** $\frac{3}{14}$ **34b.** $\frac{5}{14}$ **34c.** $\frac{3}{7}$ **34d.** $\frac{4}{7}$

35. $\begin{bmatrix} 7 & 7 \\ -2 & -10 \end{bmatrix}; \begin{bmatrix} -3 & -13 \\ 20 & 8 \end{bmatrix}$

36. $\begin{bmatrix} 4.3 & 0.6 \\ 1.0 & -0.3 \\ 1.7 & 1.2 \end{bmatrix}; \begin{bmatrix} -1.9 & -2.8 \\ 4.4 & 6.1 \\ -4.7 & -3.4 \end{bmatrix}$ **37.** $\begin{bmatrix} \frac{1}{8} \\ -\frac{7}{20} \end{bmatrix}; \begin{bmatrix} -\frac{9}{8} \\ 1\frac{23}{20} \end{bmatrix}$

39. 5.6 **40.** −11 **41.** 80 **42.** 0; −4; 4; −8 **43.** 2; 0; 0; 0 **44.** 6; 8; 16; 64

Preparing for Standardized Tests page 55

1. D **3.** A **5.** D **7.** B **9.** B **11.** A **13.** 3

CHAPTER 2

Toolbox page 58

1. H **3.** E **5.** (4, 5) **7.** (−5, 0) **13.** II **15.** IV

17a.

17b. It looks like a four-pointed star.

Lesson 2-1 pages 59–63

ON YOUR OWN **1.** Negative; with each class taken, you need to study more, so the amount of free time decreases. **3.** Positive; you expect to sell more shovels as the amount of snow increases. **5.** Negative; as the temperature rises, people use less oil for heating. **7.** No correlation; shoes are not priced by size. **9a.** Negative correlation; everything else being equal, voters are more likely to stay home in bad weather. **9b.** Sample: Yes; a low voter turnout may favor the candidates whose supporters feel strongly about the issues in the campaign and the turnout may be low because of bad weather.

11a. How Big Is a Bird Egg?

11b. There is a positive correlation between the length and the weight of a bird egg, because the larger eggs are heavier. **11c.** Sample: The egg is about 5.8 cm long. From the graph, its weight should be about 55 g. **13.** positive correlation **15.** positive correlation

17a. Calories and Fat in Common Foods

17b. positive correlation **19.** Each data point represents the total toll charged (y-coordinate) for a vehicle that traveled a specified number of miles (x-coordinate). **21a.** The corresponding points have the same y-coordinate.

MIXED REVIEW **23.** 0 **25.** −9 **27.** $\frac{3}{10}$ **29.** 33 **31.** Mode; the mean and the median need to be computed. You can read the mode off the plot. **33.** 1981–1989 and 1991–1995

Lesson 2-2 pages 64–68

ON YOUR OWN **1.** Sample: Height While Jumping Rope

3. Sample:

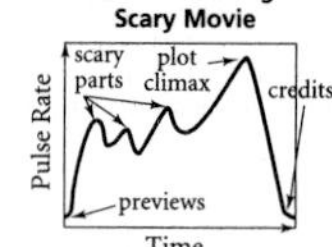

5. The graph is continuous but data are discrete. To fix the graph, you must remove all lines connecting data points. **7.** Sample: Continuous; between any two weights you measure, any value is possible.

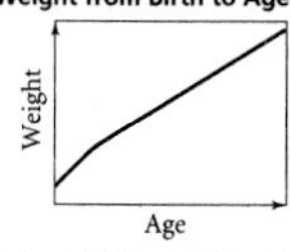

9. Sample: Discrete; there is only one time for each morning.

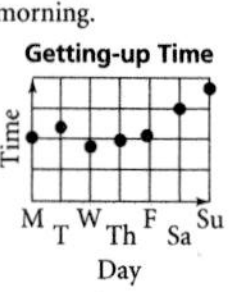

11. Sample: Discrete; you cannot have half a book. **13.** Sample: Continuous; walking speed can be measured at all times.

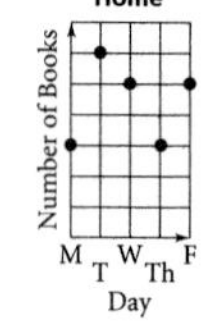

Walking Speed from Class to Locker

15a. blue; red **15b.** Growing at a steady rate, a puppy becomes heavier than a baby but stops growing soon after that. A baby's weight gain increases with time, so it outgrows the puppy. **17.** Graph II; graph I; at constant speed, distance increases with time.

MIXED REVIEW **21.** $\begin{bmatrix} -14 & 11 \\ -2 & -51 \end{bmatrix}$ **23a.** f = amount of fruit used per day; $f + 200$ **23b.** 15,400 lb

CHECKPOINT **1a–b.**

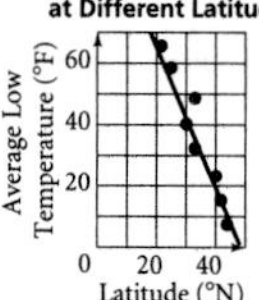

There is a negative correlation. **1c.** Sample: about 70°F **3.** C; this is the only graph where the distance from home drops back to 0, indicating that you came home.

Lesson 2-3 pages 69–72

ON YOUR OWN **1a.** Sample: Horizontal axis from 0 to 200 with a line of the graph every 25 lb; vertical axis 0 to 30 with a line of the graph every 5 lb.

1b.

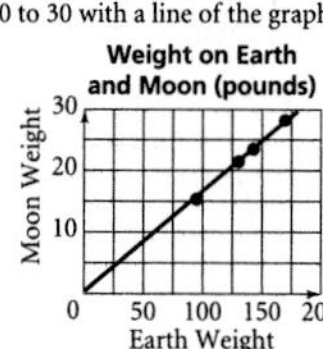

1c. Find the point on the graph that corresponds to your weight on Earth. Use this point to identify the coordinate for the moon. This is your weight on the moon. **3.** dependent; independent **5.** independent; dependent **7.** B **9.** C

11a.

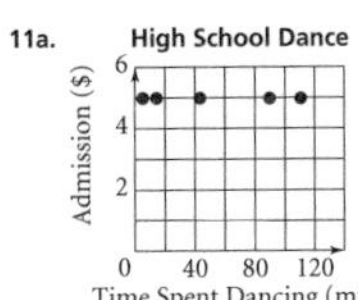

11b. Sample: The horizontal range is 0 to 120 with the scale labeled every 20 min on the axis. The vertical range is 0 to 6 with the scale labeled at every dollar on the axis.

MIXED REVIEW **15.** Sample:

17a.

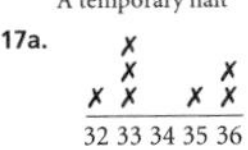

17b. 33 **19.** −1 **21.** 108 **23.** 3 **25.** 9

Lesson 2-4 pages 73–78

ON YOUR OWN **1.** 4 **3.** 9 **5.** 18 **7.** $-\frac{3}{4}$ **9.** No; graph fails the vertical-line test. **11.** Yes; graph passes the vertical-line test. **13.** function; amount of money, number of tickets **15.** function; monthly payment, number of years **17.** function; speed, time **19.** no **21.** yes **25.** {4, 5, 11} **27.** $\{-225\frac{1}{2}, -220, -187\}$ **29.** {−11, −7, 17} **31.** {6, 0, 216} **33.** {7, 6, −210} **35.** {2, 0, 30} **37a.** dependent variable **37b.** Input value is to output value as domain is to range. **39.** 2.13 **41.** −4.3 **43.** 0.04 **45.** 0.84 **47.** {0.25, 121} **49.** {25, 1, 0, 4, 100} **51.** The same y-value may correspond to any number of x-values. **53.** D

MIXED REVIEW **55.** 6 **57.** 80 **59.** $-\frac{1}{10}$ **61.** $s = \frac{1}{3}e$

Lesson 2-5 pages 79–83

ON YOUR OWN **1.** 16 **3.** 28 **5.** 2 **7.** 21 **9a.** amount of water; number of loads **9b.** $w(n) = 42n$ **9c.** 294 gal **9d.** 13 loads; you divide 546 gal by 42 gal/load. **13.** A **15a.** $T(a) = 10a + 1$ **15b.** \$31 **19.** {18, 15, 13.8, 10} **21.** {−9, 0, −1.44, −25} **23.** {10, 1, 2.44, 26} **25.** {−22, −4, 3.2, 26} **27.** $y = 4x$ **29.** $f(x) = x^3$ **31a.** $R(K) = 8 - K$ **31b.** {0, 1, 2, 3, 4, 5, 6, 7, 8}; {8, 7, 6, 5, 4, 3, 2, 1, 0} **33.** $e(h) = 4.25h$

MIXED REVIEW **35.** 23 **37.** 10 **39.** −2.5

41. **43.**

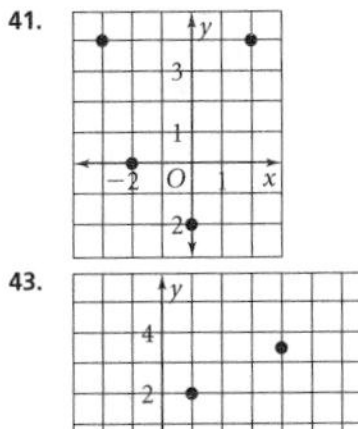

CHECKPOINT **1a.** number of minutes; number of words

1b.

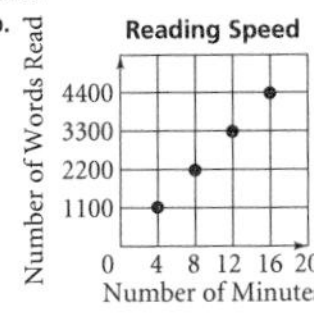

The x-value range is from 0 to 20. It is easiest to label every 4 minutes, to correspond to the given data. The y-value range is from 0 to 5500 with labels every 1100 words.

3. You multiply 3 by 4, then subtract 12 from the result. **4.** $p(r) = 0.3r$ **5.** $c(d) = 1.15d$ **6.** {−120, −36, −24, 0} **7.** {−40, 9, 8, 0} **8.** {−25.5, 2.5, 6.5, 14.5} **9.** {−3, 4, 5, 7}

Lesson 2-6 pages 84–88

ON YOUR OWN

1. Sample table:

x	y
−2	6
−1	3
0	0
1	−3
2	−6

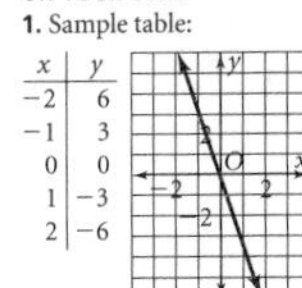

3. Sample table:

x	$f(x)$
0	−7
1	−5
2	−3
3	−1
4	1

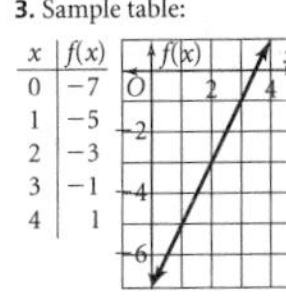

5. Sample table:

x	$f(x)$
−3	−1
−1	$-\frac{1}{3}$
0	0
1	$\frac{1}{3}$
3	1

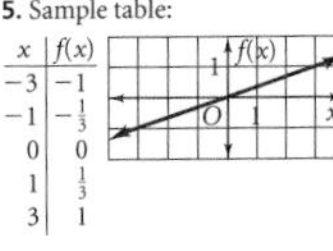

7. Sample table:

x	$f(x)$
−2	2
−1	−1
0	−2
1	−1
2	2

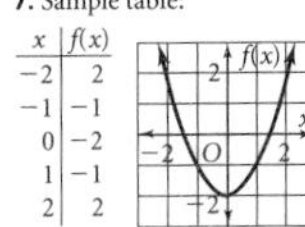

9a. C **9b.** 9 **9c.**

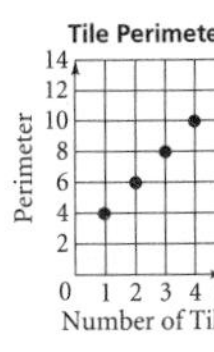

11a. $s(t) = 6t$ **11b.** $w(t) = 3t$ **11c.** 73.2 gal; 36.6 gal

11d.

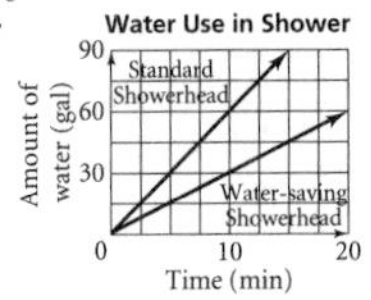

13. **15.**

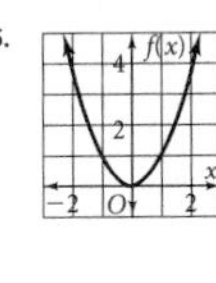

17.

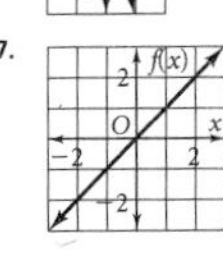

19.

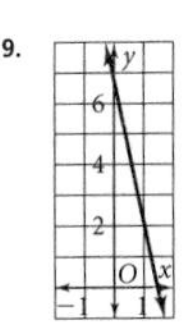

21.

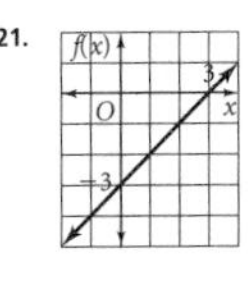

23.

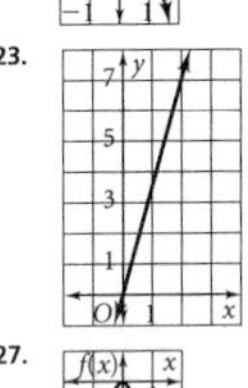

25. **27.**

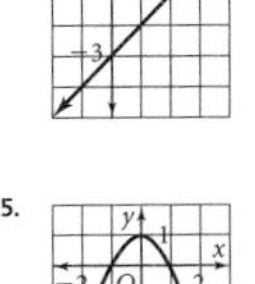

29. Sample table:

x	y
−2	2
0	2
2	2
4	2

31a. \$.71

31b. Sample table:

a	$C(a)$
0	\$.27
1	\$.38
2	\$.49
3	\$.60
4	\$.71
5	\$.82

Cost of Call (Boston to Worcester)

31c. at most 12 min **33.** B

MIXED REVIEW **39.** {3, 7, 9, 15} **41.** {4, 2, 1, −2} **43.** −20 **45.** 4 **47.** −1

Toolbox page 89

1. **3.** **5.** **7.**

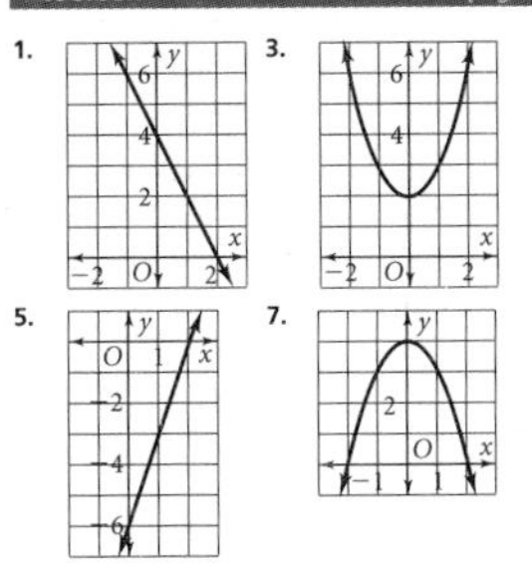

Lesson 2-7 pages 90–94

ON YOUR OWN **1.** Linear function family; the highest power of the variable is 1. **3.** Absolute value family; there is a variable expression inside the absolute value symbol. **5.** Quadratic family; the highest power of the variable is 2. **7.** Quadratic family; the highest power of the variable is 2. **9.** Absolute value family; there is a variable expression inside the absolute value symbol. **11.** Absolute value; the graph forms a "V" that opens up or down. **13.** Quadratic; the highest power of the variable is 2. **15.** Sample: A family of functions is a collection of functions whose graphs and equations are alike. **17.** No; it fails the vertical line test so it is not a function. **19.** U-shaped curve **21.** V-shape **23.** linear

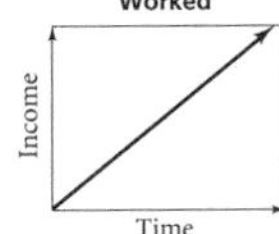

25. line; slants down from left to right

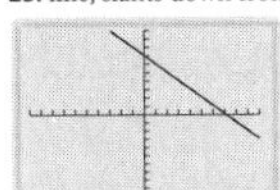

27. V-shaped; opens up **29.** V-shaped; opens down

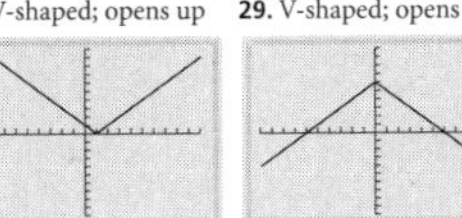

31. U-shaped; opens up **33.** V-shaped; opens up

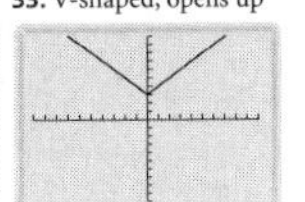

35. Quadratic family; the curve is U-shaped, opening down. **37.** Absolute value family; the path of the ball is V-shaped, opening up. **39a.** V-shaped; opens up

39b.

MIXED REVIEW

41. **43.** **45.** **47.**

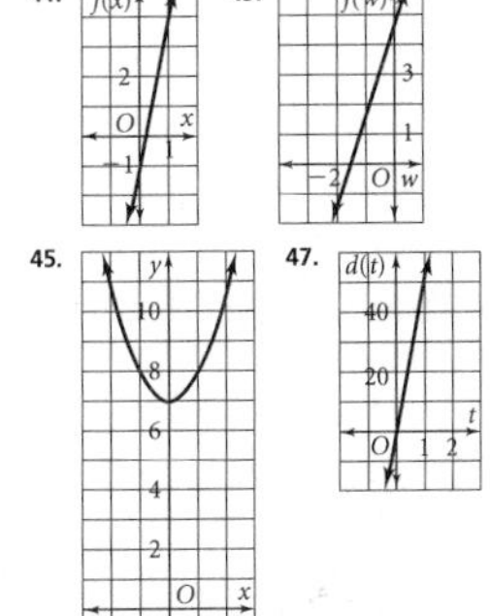

49a. −116 **49b.** Square −8; multiply by −2; add 12 to the result. **51.** $\frac{5}{16}$ **53.** $\frac{15}{16}$ **55.** $\frac{11}{16}$

Lesson 2-8 pages 95–99

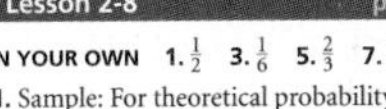

ON YOUR OWN **1.** $\frac{1}{2}$ **3.** $\frac{1}{6}$ **5.** $\frac{2}{3}$ **7.** $\frac{1}{2}$ **9.** 0 **11.** Sample: For theoretical probability, you know all the possible outcomes and you want to find out how likely a specific event is to happen. For experimental probability, you know the results of an experiment and you want to find out how frequently a specific event has happened. **13b.** $\frac{1}{12}$ **13c.** $\frac{1}{2}$ **13d.** $\frac{1}{4}$ **15.** $\frac{1}{2}$ **17.** 0 **19.** $\frac{4}{9}$ **21b.** $\frac{1}{12}$ **21c.** $\frac{1}{6}$ **23.** $\frac{1}{6}$ **25.** $\frac{17}{18}$ **27.** $\frac{1}{3}$ **29.** 0 **31.** $\frac{3}{8}$ **33.** $\frac{11}{16}$ **35.** $\frac{1}{4}$ **37.** 0 **39.** B **41.** $\frac{2}{6}$

MIXED REVIEW **45.** 5 **47.** 0.6 **49.** $-\frac{1}{9}$

Wrap-Up pages 101–103

1. Positive correlation; people drink more cold drinks in warm weather. **2.** Negative correlation; a better-trained swimmer takes less time to swim 100 m. **4.** Sample: A computer rental costs \$2.50 per hour. If you start with a fixed amount of money, the longer you plan to work on the computer, the less money you will have left. **5.** Sample: A residential thermostat senses when the temperature in the room fall below the set level, turns the heater on until the temperature reaches 3°F above the set level, then turns the heater off. The graph shows the air temperature rising while the heater is working and then falling after the heater is turned off. **6.** Sample: An elevator is on the second floor. Someone gets in, goes to the eleventh floor, and gets off.

7. Sample:

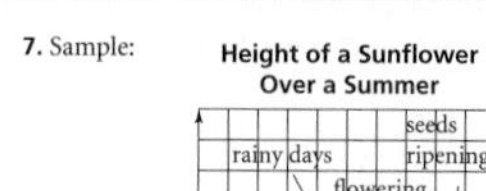
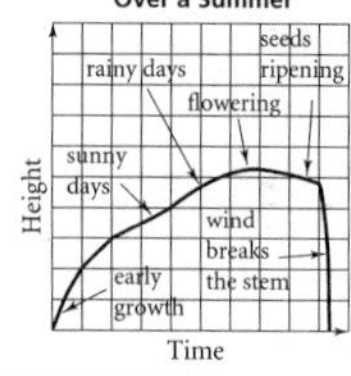

8.

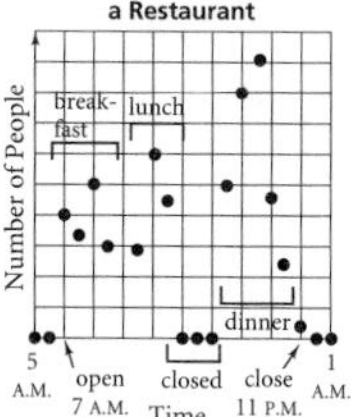

9a. year; price

9b.

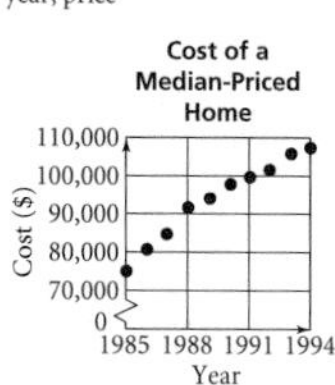

9c. Sample: Years from 1985 to 1994 on the x-axis with labels every 3 years; amounts from \$0 to \$110,000 with a break from \$0 to \$70,000 on the y-axis with labels every \$10,000; this graph accurately represents all the data without taking too much space. **10.** $\{-23, -7, -3, 13\}$ **11.** $\{1, 3, 3.5, 5.5\}$ **12.** $\{17, 1, 2, 26\}$ **13.** not a function **15.** $f(x) = x + 1$ **16.** $f(x) = -x$ **17.** $f(x) = x + 3.5$

18. Sample table:

x	$f(x)$
−1	1
0	−3
1	1

19. Sample table:

x	$f(x)$
−6	1
−5	0
−2	3
0	5

20. Sample table:

x	$f(x)$
−2	−3
−1	−1
0	1
1	3

21. Sample table:

x	$f(x)$
−2	−5
−1	−6
0	−7
1	−6
2	−5

22. quadratic, highest power of x is 2; absolute value; linear, highest power of x is 1; absolute value **23.** $\frac{1}{6}$ **24.** $\frac{5}{6}$ **25.** $\frac{1}{3}$ **26.** 0 **27a.** $\frac{1}{4}$ **27b.** P(not 3 heads) means the chances of getting 1, 2, or 4 heads. **28.** E **29.** 8 **30.** −3 **31.** 6 **32.** $\frac{1}{2}$ **33.** 5 **34.** $-\frac{1}{2}$ **35.** −1 **36.** 7 **37.** 4.5 **38.** $5\frac{1}{4}$

Preparing for Standardized Tests page 105

1. B **3.** D **5.** C **7.** B **9.** A **11.** $\{13, 7, 5\}$

CHAPTER 3

Lesson 3-1 pages 108–113

ON YOUR OWN **1.** Kendra's method is better; it needs only one operation and Ted's needs two. **3.** 6 **5.** 8 **7.** 14 **9.** 500 **11.** C **13.** $B + 109 = 180$; 71° **15.** $38\frac{3}{4} + g = 41\frac{1}{2}$; $2\frac{3}{4}$ in. **17.** −1 **19.** −1.57 **21.** −196 **23.** −4.5 **25.** −20 **27.** $-5\frac{5}{6}$ **29.** $4\frac{1}{4}$ **31.** −245 **33.** subtraction property **39.** 31.5 **41.** −99 **43.** 13 **45.** $16\frac{1}{2}$ **47.** $-1\frac{1}{4}$ **49.** 11.5 **51.** $330 = 280 + x$; \$50 **53a.** $3L = 166.5$; \$55.50 **53b.** $1.5r = 55.5$; \$37 **55.** −18 **57.** −8 **59.** $-3\frac{3}{4}$ **61.** −262 **63.** −4.5 **65.** $-\frac{3}{8}$ **67.** 5.12 **69.** $-\frac{1}{24}$ **71.** $5s = 1.6$; 0.32 km

MIXED REVIEW **73.** 8 **75.** about 500,000 **77.** 42.3 **79.** 31 **81.** 28

Lesson 3-2 pages 114–118

ON YOUR OWN **1.** $3x + 2 = 5$ **3.** $3x + 1 = -2$ **5.** −1 **7.** −2 **9.** 1 **11.** −2 **13.** 21 **15.** 7 **17.** −6 **19.** 27 **21.** 0.382 **23.** −60 **25.** $18 + 2t = 60$; 21 min **27.** −6 **29.** 4.5 **31.** 16 **33.** 30 **35.** −6 **37.** 75 **39.** 0 **41.** −19 **43.** 2.2 **45.** 3.864 **47a.** Yes; with this estimate, the total bill is \$55, which is close to \$60. **47b.** 125 mi **49.** 14 bulbs **51.** Multiplying by 100 changes all the numbers in the equation to integers and it might be easier to work with integers. **53.** $r = (2.8 - 1.34) \cdot 2$; 2.92 **55.** $n = (8 + 0.5) \div 0.05$; 170 **57.** $n = (7 - 5.3) \div (-0.8)$; −2.125 **59.** $t = (-7.06 - 3)(-2.5)$; 25.15

MIXED REVIEW **63.** 12 **65.** −9 **67.** 9 **69.** −20 **71.** 3

Lesson 3-3 pages 119–123

ON YOUR OWN

1.

$2y + 1$; **3.** $-n$ **5.** $8b$ **7.** $-4x + 8n$ **9.** $-6x + m + 8$ **11.** $2x - x = 4$; 4 **13.** $-x - 2 + 3x + 1 = 3$; −1 **15.** 8 **17.** 3 **19.** 7 **21.** 2 **23.** \$6.50 **27.** 6 **29.** −4 **31.** $5\frac{1}{2}$ **33.** $5\frac{4}{7}$ **35.** −0.48 **37.** 2.4 **39.** 78 **41.** $4\frac{1}{4}$ in. **43.** 1986, 1987, 1988 **45.** 60 **47.** 50 **49.** 50

MIXED REVIEW **51.** $y = -2x$

53. Sample:

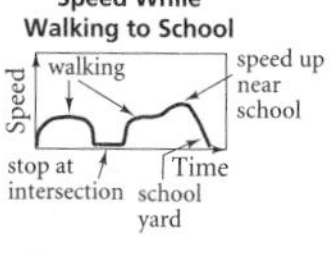

55. −9 **57.** −48

Lesson 3-4 pages 124–128

ON YOUR OWN **1.** $2(t - 2) = -6$; −1 **3.** $4(z + 1) = 4$; 0 **5.** No; $2a \cdot 2b = 4ab$. You cannot use the distributive property because ab is a product, not a sum. **7.** $-2n + 12$ **9.** $2b - \frac{8}{5}$ **11.** $4y + 6$ **13.** $18n - 42$ **15.** $-4.5b + 13.5$ **17.** $-36 + 16n$ **19.** 900° **21.** 3 **23.** 9 **25.** 1 **27.** −1 **29.** 8 **31.** −2 **33.** −1 **35.** 1 **37.** 380 mi; 280 mi **39.** 9 aluminum bats and 6 wooden bats **43.** \$3.96 **45.** \$29.55 **47.** \$98.97

MIXED REVIEW **49.** $-\frac{2}{3}$ **51.** −7 **53a.** B2/250 **53b.** 2.9; 0.9; 0.5 **53c.** \$504 million **55.** −17

CHECKPOINT **1.** 1 **2.** 33 **3.** 20 **4.** $2\frac{1}{2}$ **5.** 120 **6.** −2 **7.** −2 **8.** 10 **9.** $18\frac{1}{3}$ **10.** −2.7 **11.** −16 **12.** 2 **13.** 63 in. by 83 in. **14.** B

Lesson 3-5 pages 129–132

ON YOUR OWN **1.** 12 **3.** −10 **5.** −1 **7.** −48 **9.** In using the distributive property, the student did not multiply 1 by 8. **11.** A **13.** \$4/yd **15.** −45 **17.** −3 **19.** 15 **21.** −7 **23.** $-14\frac{1}{7}$ **25.** −10 **27.** 32 **29.** $\frac{6}{13}$ **31.** −60 **33.** 54 **35.** about 61.5 mi/h **37.** 33.7 **39.** −21 **41.** 5 **43.** −33 **45.** −1 **47.** $\frac{3}{5}$ **49.** −9 **51.** $1\frac{5}{6}$ **53.** $1\frac{1}{14}$ **55.** \$140 million **57.** $2\frac{1}{2}$ h

MIXED REVIEW **59.** −7 **61.** $\frac{1}{2}$ **63.** 12 **65.** 1 **67.** $\frac{1}{5}$ **69.** $\frac{2}{5}$

Toolbox page 133

1. Spain—gold; Poland—silver; Ghana—bronze **3.** Coretta lives in New York; David lives in Houston; Nando lives in Dallas; Helen lives in Chicago.

Lesson 3-6 pages 134–138

ON YOUR OWN **1.** independent **3.** dependent **5.** $\frac{1}{6}$ **7.** $\frac{1}{12}$ **9.** $\frac{5}{12}$ **11.** $\frac{3}{50}$ **13.** $\frac{3}{20}$ **15.** $\frac{1}{10}$ **17.** $\frac{1}{6}$ **19.** $\frac{1}{9}$ **21.** 0 **23.** D **27.** about 1.2% **29.** 0.0036 **31.** $\frac{1}{5}$ **33.** $\frac{3}{5}$ **35.** $\frac{4}{5}$ **37.** $\frac{3}{4}$ **39.** A

MIXED REVIEW **41.** −60 **43.** 10 **45.** 45% **47.** 32.8% **49.** 100%

Lesson 3-7 pages 139–144

ON YOUR OWN **1.** $n = 0.04 \times 150$ **3.** $24 = 1.5 \times n$ **5.** $0.1 \times n = 8$; 80 **7.** $0.06 \times n = 36$; 600 **9.** $n \times 45 = 18$; 40% **11.** \$62.40 **15.** \$20 **17.** 50% **19.** 39.2 **21.** 4.5% **23.** 330 **25.** 60% **27.** \$84 **29.** \$320 **31.** about 12.6% **33.** 40% **35.** The number of people 65 and over is projected to be 80 million, which is 20% of the total population, so the projected population is 400 million people. **37.** \$1250 **39.** 2 yr

MIXED REVIEW **41.** Sample table:

x	$f(x)$
0	−2
2	1
3	2.5
4	5

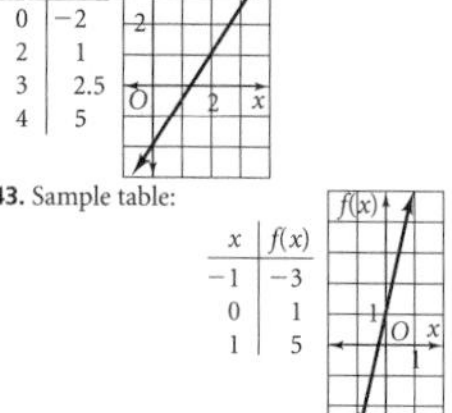

43. Sample table:

x	$f(x)$
−1	−3
0	1
1	5

45. $\{-1, -2, -3\}$ **47.** 0.75 **49.** 0.88

CHECKPOINT **1.** $-\frac{14}{5}$ **2.** 9.6 **3.** $3\frac{11}{15}$ **4.** $-11\frac{1}{4}$ **5.** 3 **6.** $\frac{1}{9}$ **7.** $\frac{1}{24}$ **8.** $\frac{1}{27}$ **9.** $\frac{1}{36}$ **10.** The probability increases, because the first draw is the same, but for the second the number of marbles decreases and the number of favorable outcomes stays the same. **11.** 218 members

Toolbox page 145

1.

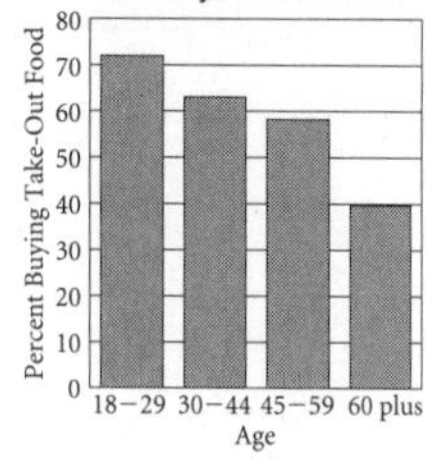

3.

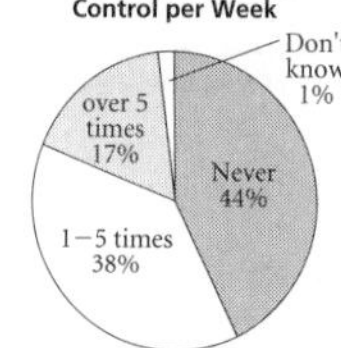

Lesson 3-8 pages 146–149

ON YOUR OWN **1.** 25% decrease **3.** 5% increase **5.** 3% decrease **7.** 29% decrease **9.** about 39% **11.** 100% **13.** 60% **15.** 33% **17.** 200% **21.** Less; the 20% you subtracted is 20% of an amount greater than the original amount, so it is greater than the 20% you added. **23.** 6% increase **25.** 22% decrease **27.** 4% increase **29.** 120% increase **31.** 22% increase **33.** 28% decrease **35a.** 100% **35b.** 100 **35c.** 50% **35d.** 50% **35e.** 100 **35f.** 100 **37.** C **39.** 60% **41.** 14% **43.** about 50% **45a.** 25% **45b.** 20%

MIXED REVIEW **47.** 1.5 **49.** 13 **51.** 60%

Wrap-Up pages 151–153

1. 16 **2.** −36 **3.** 31 **4.** 24 **5.** 44 **6.** 4.25 **7.** 12 **8.** 2 **9.** $2x + 5 = 9$; 2 **10.** $3x - 2 = 7$; 3 **11.** $-4x + 2 = -6$; 2 **12.** $\frac{1}{2}$ **13.** 3 **14.** 3 **15.** 11 **16.** 18 **17.** 4 **18.** −2 **19.** −9 **21.** $4m + 3$ **22.** $b + 6$ **23.** $-5w + 20$ **24.** $36 - 27j$ **25.** −9 **26.** 4 **27.** −0.5 **28.** $1\frac{1}{2}$ **29.** −2 **30.** 1.5 **31.** 2 **32.** 1 **33.** $2[l + (l - 6)] = 72$; 15 cm × 21 cm **34.** C **35.** −45 **36.** 24 **37.** −11 **38.** −4 **39.** \$40 **40.** dependent; $\frac{2}{45}$ **41.** independent; $\frac{1}{20}$ **43.** $n = 0.15 \times 86$; 12.9 **44.** $n \times 5 = 40$; 800% **45.** $1.8 = 0.72 \times n$; 2.5 **46.** 3.75% **47.** \$220 **48.** about 13% increase **49.** 25% decrease **50.** about 33% decrease **51.** 27.2% **53.** −2 **54.** 1.4 **55.** 19 **56.** 7 **57.** 4.5 **58.** 5 **59.** $\frac{1}{4}$ **60.** $1\frac{1}{7}$ **61.** 3 **62.** $\frac{4}{9}$ **63.** −4 **64.** 5 **65.** 52 **66.** −3

Preparing for Standardized Tests page 155

1. E **3.** D **5.** A **7.** C **9.** $g(x) = x - 3$ **11.** 0 **13.** $5a - 5b + 4$

CHAPTER 4

Lesson 4-1 pages 158–162

ON YOUR OWN **1.** yes; $6 \cdot 20 = 8 \cdot 15$ **3.** yes; $-0.12 \cdot 0.5 = -0.4 \cdot 0.15$ **5.** no; $-3 \cdot 25 \neq -1 \cdot 100$ **7.** about 433 times **9.** \$54 **11.** 20% **13.** 50 **15.** 27.6% **17.** 307.1 **19.** 15.6 **21.** 10 ft by 12 ft **23.** E **25.** 5 **27.** 12.5 **29.** 4 **31.** 5 **33.** 12.5 **35.** −0.864 **37a.** No; For an item that originally costs \$12, $\frac{1}{3}$ off results in the price \$8; 50% off results in the price \$6. **39.** 9 **41.** 5.5 in. **43.** 7.1%

MIXED REVIEW **45.** $\begin{bmatrix} -1.2 & 14.6 \\ -1.4 & -5.7 \end{bmatrix}$ **47.** −11.5 **49.** 16 paperbacks **51.** −14.5

Lesson 4-2 pages 163–168

ON YOUR OWN **1.** $2x + 2 = x - 8$; −10 **3.** $2x + 3 = 3x - 7$; 10 **5.** 7 **7.** 3 **9.** no solution **11.** no solution **13.** identity **15.** You should add y, not subtract, on the third line; 5.3. **17.** 0 **19.** no solution **21.** 2 **23.** no solution **25.** −41 **27.** $-\frac{1}{2}$ **29.** identity **31.** $a = -\frac{1}{4}$; $w = -4$; $x = -1$; $y = 0$ **35.** 20 h **43.** $DF = 7$, $DE = 10$, $EF = 6$ **45.** 7 **47.** −3 **49.** −0.5 **51.** 10 **53.** C

MIXED REVIEW **55.** −28 **57.** 7.7 **59.** 15 **61.** 34 **63.** 11 **65.** −19 **67.** −2

Toolbox page 169

1a. $y = 2x - 1$; $y = x + 1$ **1b.** 2 **1c.** 2
3. 3 **5.** 30

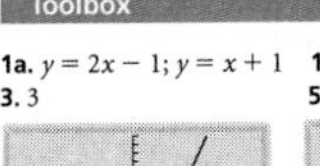

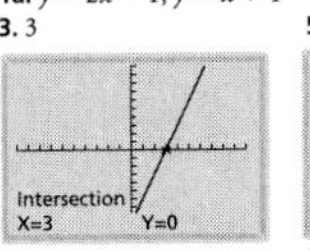

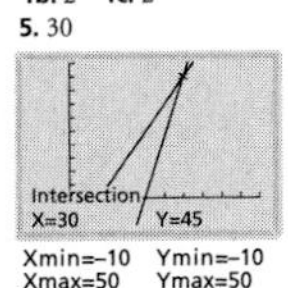

Xmin=–10 Ymin=–10
Xmax=50 Ymax=50
Xscl=5 Yscl=5

7. no; $-\frac{3}{4}$ **9.** no; $-\frac{2}{3}$

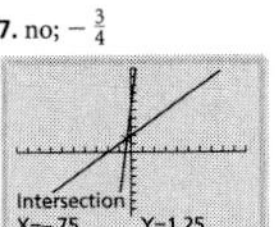

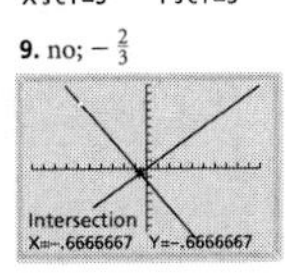

Lesson 4-3 pages 170–174

ON YOUR OWN **1.** 3 **3.** 2 **5.** 6 **7.** −4 **11.** −6, 6 **13.** −4, 3 **15.** 3, 13 **17.** −16, 2 **19.** −9, 5 **21.** −0.72, 0.72 **23.** $|x| = 3$ **25.** $|x| = 0$ **29.** True; absolute value cannot be negative. **31.** False; $z = 0$ or $z = -6$. **33.** −12, 12 **35.** −48 **37.** −10, 10 **39.** −9, 9 **41.** 39°F, 11°F **43a.** 7.265 oz; 7.235 oz **43b.** No; one coin may weigh much more than 0.18 oz, while another may weigh much less.

MIXED REVIEW **45.** −6 **47.** −0.5 **49.** −2.5 **51.** Ted earns $5.50 per hour; Elio earns $7.50 per hour. **53.** 24 ft

Lesson 4-4 pages 175–178

ON YOUR OWN **1.** $r = \frac{s-4}{3}$ **3.** $p = \frac{mq}{n}$ **5.** $r = \frac{C}{2\pi}$ **7.** $t = \frac{3m}{2} - 4$ **9a.** 5.85 **9b.** $b = 2m - a$ **9c.** 5.9 **11.** $n = 4m - 6$ **13.** $z = a + y$ **15.** $b = \frac{7a}{2}$ **17.** $b = \frac{2A}{h}$ **19.** $h = \frac{V}{lw}$ **21.** $p = \frac{b-r}{a}$ or $p = \frac{b+r}{a}$ **23a.** $A = \frac{s^2}{R}$ **23b.** 27 ft^2 **27a.** $y = -\frac{1}{2}x + 2$

27b. **27c.** $\frac{1}{2}$; 0; $-\frac{1}{2}$

MIXED REVIEW **29.** 560 **31a.** $69 **31b.** Sample: p = total pay; r = hourly rate; h = number of hours; $p = rh$ **33.** > **35.** >

CHECKPOINT **1.** $n = s + 90$ **2.** $b = y - mx$ **3.** $a = \frac{2b^2}{L}$ **4.** $f = \frac{W}{d}$ **5.** 16.8 **6.** no solution; $-4 \neq 4$ **7.** no solution; absolute value cannot be negative **8.** 0 **9.** −8, 8 **10.** $9\frac{2}{3}$ **11.** 5 **12.** 1.5, 7.5 **13.** $17\frac{1}{3}$ km **14.** C

Lesson 4-5 pages 179–184

ON YOUR OWN **1.** n = number of students; $n \leq 48$ **3.** a = age of applicant; $a \geq 16$ **5.** w = light bulb wattage; $w \leq 60$ **7a.** $6 < 9$ **7b.** $10 < 13$ **7c.** $-1 < 2$
7d.

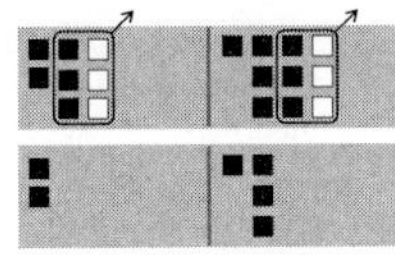

9. Sample: −20, −1, 0, 0.2 **11.** Sample: 1, 2, 3, 26 **13.** Add 4 to each side. **15.** Add $\frac{1}{2}$ to each side. **17.** b is greater than 0. **19.** m is at most −1. **21.** z is at least −4. **23.** 40 points

25. $x > 11$ (9 10 11 12 13 14)
27. $h \geq -\frac{1}{4}$ (−2 −1 0 1 2 3)
29. $x \geq -1.7$ (−4 −3 −2 −1 0 1)
31. $m > 5.5$ (3 4 5 6 7 8)
33. $y < 3.1$ (0 1 2 3 4 5)
35. $t \geq 12.9$ (10 11 12 13 14 15)

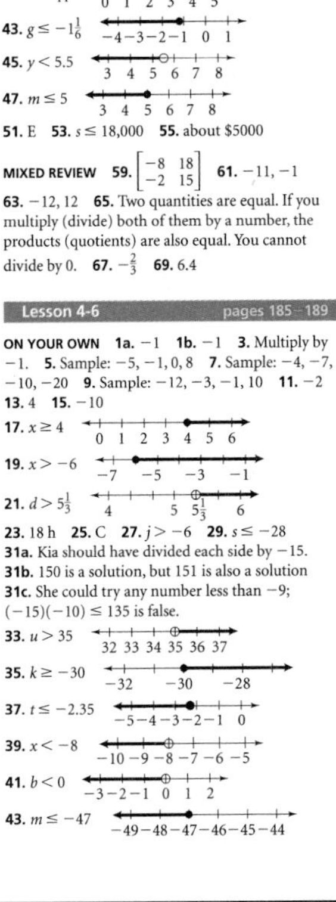

37. $n < -5\frac{4}{5}$ (−8 −7 −6 −5 −4 −3)
39. $x \neq -5.7$ (−8 −7 −6 −5 −4 −3)
41. $c \neq 2\frac{11}{14}$ (0 1 2 3 4 5)
43. $g \leq -1\frac{1}{6}$ (−4 −3 −2 −1 0 1)
45. $y < 5.5$ (3 4 5 6 7 8)
47. $m \leq 5$ (3 4 5 6 7 8)
51. E **53.** $s \leq 18{,}000$ **55.** about $5000

MIXED REVIEW **59.** $\begin{bmatrix} -8 & 18 \\ -2 & 15 \end{bmatrix}$ **61.** −11, −1 **63.** −12, 12 **65.** Two quantities are equal. If you multiply (divide) both of them by a number, the products (quotients) are also equal. You cannot divide by 0. **67.** $-\frac{2}{3}$ **69.** 6.4

Lesson 4-6 pages 185–189

ON YOUR OWN **1a.** −1 **1b.** −1 **3.** Multiply by −1. **5.** Sample: −5, −1, 0, 8 **7.** Sample: −4, −7, −10, −20 **9.** Sample: −12, −3, −1, 10 **11.** −2 **13.** 4 **15.** −10

17. $x \geq 4$ (0 1 2 3 4 5 6)
19. $x > -6$ (−7 −5 −3 −1)
21. $d > 5\frac{1}{3}$ (4 5 $5\frac{1}{3}$ 6)

23. 18 h **25.** C **27.** $j > -6$ **29.** $s \leq -28$ **31a.** Kia should have divided each side by −15. **31b.** 150 is a solution, but 151 is also a solution **31c.** She could try any number less than −9; $(-15)(-10) \leq 135$ is false.

33. $u > 35$ (32 33 34 35 36 37)
35. $k \geq -30$ (−32 −30 −28)
37. $t \leq -2.35$ (−5 −4 −3 −2 −1 0)
39. $x < -8$ (−10 −9 −8 −7 −6 −5)
41. $b < 0$ (−3 −2 −1 0 1 2)
43. $m \leq -47$ (−49 −48 −47 −46 −45 −44)

45a. 480 tiles **45b.** Sample: they may need to cut some tiles.

MIXED REVIEW **47.** $y = \frac{6-2x}{3}$ **49.** $y = \frac{15-4x}{-5}$ **51a.** 1750 tons **51b.** about 0.4 oz **53.** −4.4 **55.** −2.4 **57.** $-18\frac{6}{7}$

Lesson 4-7 pages 190–194

ON YOUR OWN **1.** Subtract 5 from each side. Then divide each side by 4. **3.** Subtract 8 from each side. **5.** Subtract y from each side. Then add 5 to each side. **7.** Add s to each side. Then subtract 6 from each side. **9.** Multiply or divide each side by −1. Reverse the order of the inequality. **11.** E **13.** A **15.** C

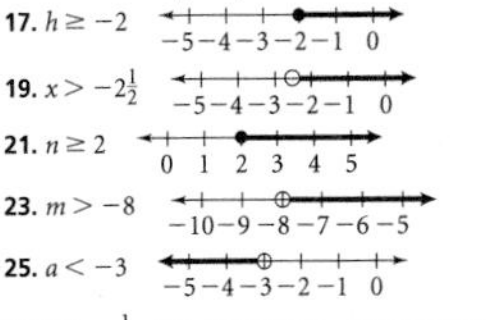

17. $h \geq -2$ (−5 −4 −3 −2 −1 0)
19. $x > -2\frac{1}{2}$ (−5 −4 −3 −2 −1 0)
21. $n \geq 2$ (0 1 2 3 4 5)
23. $m > -8$ (−10 −9 −8 −7 −6 −5)
25. $a < -3$ (−5 −4 −3 −2 −1 0)

29. $k > -\frac{1}{4}$ **31.** all numbers **33.** no solutions **35.** $k \leq -33$ **37.** $s \geq -\frac{22}{37}$ **41.** at least $7000 **43.** $15n - (490 + 45 + 65) \geq 1200$

MIXED REVIEW **45.** $m \geq 1850$ **47.** Absolute value; variable expression is inside the absolute value symbol. **49.** Quadratic; highest power of x is 2. **51a.** about 1.125 tons **51b.** about 10.7 million tons

53. (−6 −5 −4 −3 −2 −1)
55. (−2 −1 0 1 2 3)

Lesson 4-8 pages 195–200

ON YOUR OWN **1.** e = elevation anywhere in North America; $-282 \leq e \leq 20{,}320$ **3.** $-2 < x < 3$ **5.** $-4 \leq x \leq 3$ **7.** Sample: $x < 1$ or $x > 0$ **9.** $|x| < 3$ **11.** $|x - 6| > 2$ **13.** $-5 < j < 5$
(−6 −4 −2 0 2 4 6)

15. $k < -5$ or $k > -1$
(−6 −5 −4 −3 −2 −1 0)
17. $1 \leq r \leq 2$
(−1 0 1 2 3 4)
19. $-2 < r \leq -1.5$
(−4 −3 −2 −1 0 1)
21. $-1.5 < w < 3.5$
(−2 −1 0 1 2 3 4)
23. $4 \leq t \leq 14$
(3 4 5 6 7 8 9 10 11 12 13 14 15)
27. 7585 cm $\leq l \leq$ 7615 cm
29. $f < -2.5$ or $f > 2.5$
(−4 −3 −2 −1 0 1 2 3 4)
31. $n \leq -13$ or $n \geq -3$
(−14 −13 −12 −11 −10 −9 −8 −7 −6 −5 −4 −3 −2)
33. $w < -2$ or $w > 2$
(−3 −2 −1 0 1 2 3)
35. $-6 < x < 6$
(−7 −6 −5 −4 −3 −2 −1 0 1 2 3 4 5 6 7)
37. $y \leq 1$ or $y \geq 3$
(0 1 2 3 4 5)
39. $t \leq -2.4$ or $t \geq 4$
(−4 −3 −2 −1 0 1 2 3 4 5)
41. $t < -3$ or $t > 2\frac{1}{3}$
(−4 −3 −2 −1 0 1 2 3 4)
43. all numbers
(−3 −2 −1 0 1 2)

45. $2.5 < x < 7.5$ **47.** $7 < x < 49$ **49.** $66 \leq C \leq 88$ **51.** Charlotte: $29 \leq C \leq 90$; Detroit $15 \leq D \leq 83$

MIXED REVIEW **53.** $-\frac{1}{8}$ **55.** −1 **57.** −0.9, 0.9 **59.** $-\frac{1}{4}$, $1\frac{1}{4}$ **61.** −3, −2, −1, 0, 1 **63.** −3, −2, −1, 0, 1, 2

CHECKPOINT
1. $c > 6$ (4 5 6 7 8 9)
2. $x \leq -8$ (−10 −8 −6)
3. $m \leq -4$ (−7 −6 −5 −4 −3 −2)
4. $c \geq -5$ (−8 −7 −6 −5 −4 −3)
5. $b > \frac{1}{3}$ (−2 −1 0 1 2 3)
6. $n \leq 3$ (0 1 2 3 4 5)
7. $r \leq 7$ (4 5 6 7 8 9)
8. $w \leq -2$ (−5 −4 −3 −2 −1 0)
9. $h \geq -4$ (−7 −6 −5 −4 −3 −2)
10. $0 \geq t \geq -0.5$
(−3 −2 −1 0 1 2)
11. $x < -27$ or $x > 4$
(−30 −27 −24 −21 −18 −15 −12 −9 −6 −3 0 3 6)
12. $g < 1$ or $g > 5$
(−2 −1 0 1 2 3 4 5 6)
13. $d \leq -3$ or $d \geq 3$
(−4 −3 −2 −1 0 1 2 3 4)
14. $-10 < x < 4$
(−10 −5 0 5)
15. $-1 < x < 4$
(−2 −1 0 1 2 3 4 5)
16. $y \leq -4$ or $y \geq 5$
(−5 −4 −3 −2 −1 0 1 2 3 4 5 6)
17. $2\frac{2}{5} \leq x \leq 5\frac{3}{5}$
(2 3 4 5 6 7)
18. $36 < t < 37.2$

Toolbox page 201

1. 6

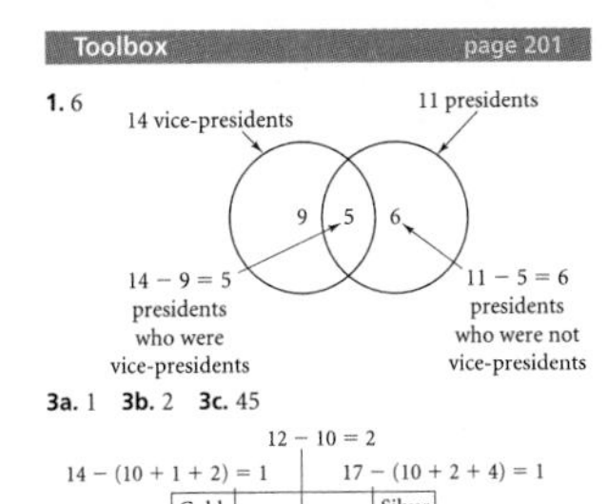

3a. 1 **3b.** 2 **3c.** 45

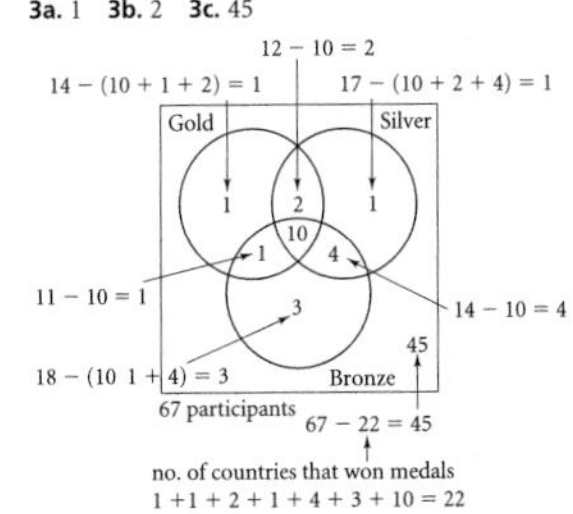

Lesson 4-9 pages 202–205

ON YOUR OWN

1. $p \leq 6$ (0 1 2 3 4 5 6 7)
3. no solutions **5.** no solutions
7. $-5 < k < 6$
(−1 0 1 2 3 4 5 6)
9. $y > 4.2$ (3 4 5 6 7 8 …)
11. $u < 2$ or $u \geq 3$
(0 1 2 3 4 5 …)
13. $-4 \leq q < 2.25$
(−5 −4 −3 −2 −1 0 1 2 3)
15. $-3 \leq m < -2$; no solutions

17. $12 \leq a < 65$
(10 20 30 40 50 60 70)
19. $23 < c < 23.5$
(21 22 23 24 25 26)

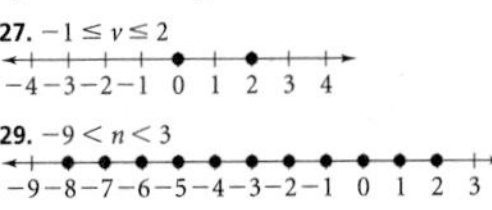

21.

23a. 6 h **23b.** Divide 175 by the hourly rate and subtract the hours worked so far. Round the result up to the next integer.

27. $-1 \leq v \leq 2$
(−4 −3 −2 −1 0 1 2 3 4)
29. $-9 < n < 3$
(−9 −8 −7 −6 −5 −4 −3 −2 −1 0 1 2 3)

MIXED REVIEW **31.** 5 **33.** 50% **35.** 150%

Wrap Up pages 207–209

1. 2 **2.** 2.3 **3.** −6 **4.** 10 **5.** −0.4 **6.** 1.5 **7.** 36 ft **8.** $2x - 6 = -3x + 4$; 2 **9.** $-x + 8 = -2x - 9$; −17 **10.** −11 **11.** identity **12.** 0 **13.** −2 **14.** −5, 5 **15.** −12.5, 6.5 **16.** No solution; absolute value cannot be negative. **17.** 8 **19.** $c = 3m - a - b$ **20.** $d = \frac{C}{\pi}$ **21.** $x = \frac{y-b}{m}$ **22.** $h = \frac{2A}{b}$ **23a.** $I = \frac{E}{R}$ **23b.** 40 amperes

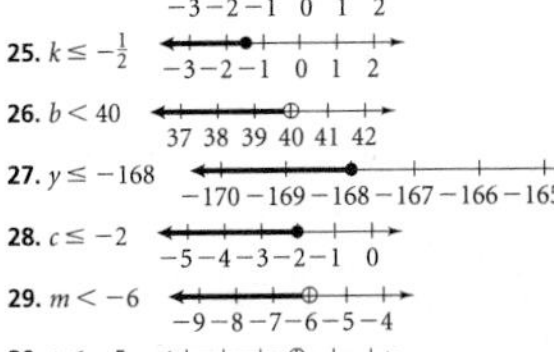

24. $h > -1$ (−3 −2 −1 0 1 2)
25. $k \leq -\frac{1}{2}$ (−3 −2 −1 0 1 2)
26. $b < 40$ (37 38 39 40 41 42)
27. $y \leq -168$ (−170 −169 −168 −167 −166 −165)
28. $c \leq -2$ (−5 −4 −3 −2 −1 0)
29. $m < -6$ (−9 −8 −7 −6 −5 −4)
30. $t < -5$ (−8 −7 −6 −5 −4 −3)
31. $x \geq 9$ (7 8 9 10 11 12)
32. There is no solution because $\frac{3}{4} \leq \frac{1}{4}$ is false.

33. $n \leq -6$ or $n \geq 2$

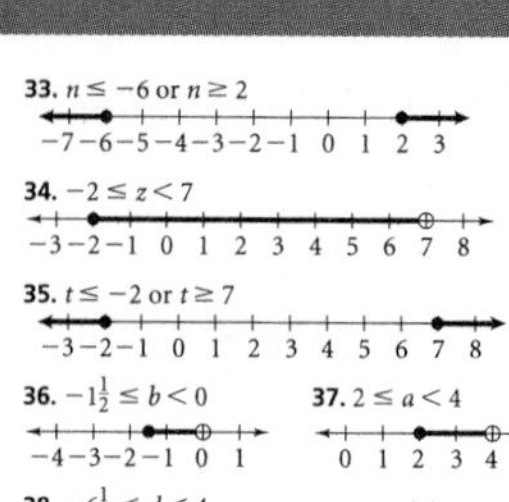

34. $-2 \leq z < 7$

35. $t \leq -2$ or $t \geq 7$

36. $-1\frac{1}{2} \leq b < 0$ **37.** $2 \leq a < 4$

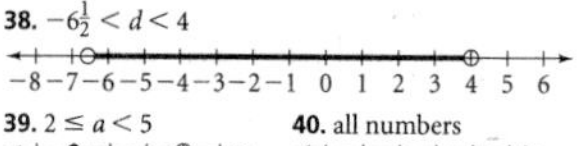

38. $-6\frac{1}{2} < d < 4$

39. $2 \leq a < 5$ **40.** all numbers

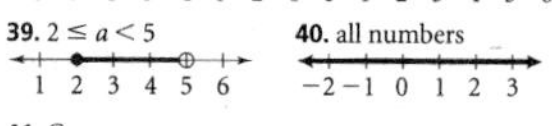

41. C

42. $h \leq -1.5$ or $h \geq 2$

43. $t \leq 3\frac{1}{3}$

44. no solution

45. $-3 \leq m < 0$

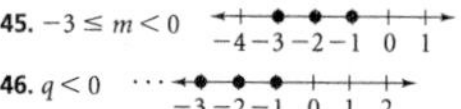

46. $q < 0$

47. $\frac{1}{4} \leq c \leq 4$

48. $l > 1000$

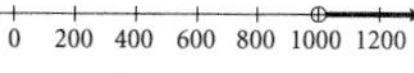

50.

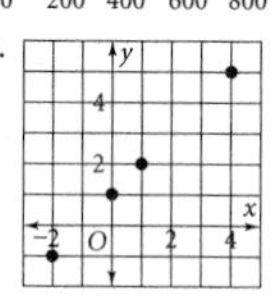

51.

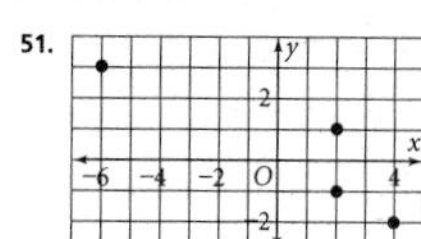

52.

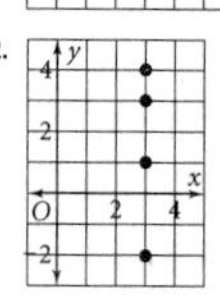

53. $f(x) = x - 5$ **54.** $f(x) = x + 9$ **55.** $f(x) = \frac{x}{2}$ **56.** $f(x) = 3x$

Cumulative Review page 211

1. D **3.** C **5.** E **7.** B **9.** B **11.** $3x + 17 = 32$; $x = 5$ **13.** 20%

CHAPTER 5

Lesson 5-1 pages 214–218

ON YOUR OWN **1.** 1; -1 **3.** $\frac{5}{9}$ **5.** $-\frac{6}{5}$ **7.** undefined

9. **11.**

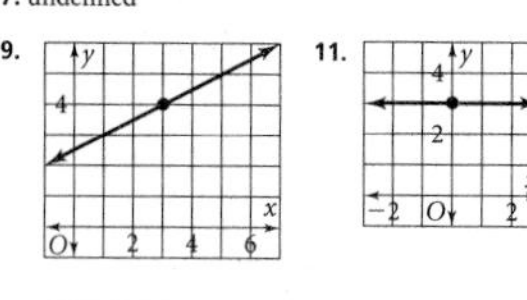

13.

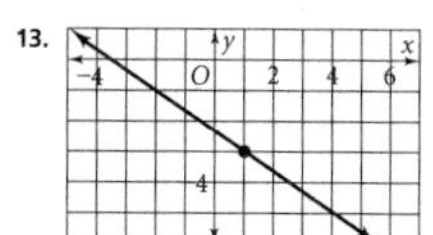

15.

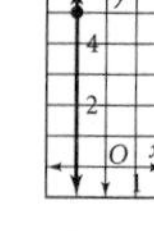

17. T; all horizontal lines have zero slope. **19.** T; two lines can have the same slope but will pass through different points. **21.** increase **23.** increase **25.** AB: 0; BC: $\frac{5}{6}$; AC: $\frac{5}{3}$ **27.** JK: $-\frac{1}{2}$; KL: 2; LM: $-\frac{1}{2}$; JM: 2 **29.** no **31a.** About 110; the rent is \$110 per hour. **31b.** 55 customers per hour **33.** E

MIXED REVIEW **35.** $\frac{1}{2}$ **37.** 1 **39.** 75% **41.** independent: temperature; dependent: amount of fuel used

Toolbox page 219

1. A **3.** B **5.** $3\frac{1}{3}$ **7.** 270 **9.** about 15.83 mi/h **11.** 20,000 mi/h

Lesson 5-2 pages 220–224

ON YOUR OWN **1.** 0.9 in./mo **3.** 30 mi/h **5.** $\frac{1}{4}$; \$1 buys 4 oz of oregano. **7a.** Yes; for every 6°F decline in temperature, the number of calories burned increases by 100. **7b.** Subtract 20°F. Multiply the difference by the rate of change. Then add the result to 3330 calories. The number of calories burned at 20°F is 3830. **9.** true **11.** F; the boiling temperature of water decreases as altitude increases. **13.** no **15.** no **17.** yes **19a.** A: fastest; C: slowest **19b.** C; you can record more if you use less tape per unit of time.

21a.

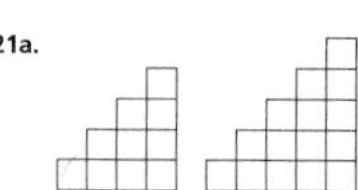

21b.

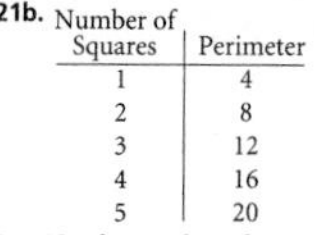

Number of Squares	Perimeter
1	4
2	8
3	12
4	16
5	20

21c. No; the number of squares increase by 1, 2, 3, etc.; but the perimeter increases 4 units with each figure.

MIXED REVIEW **25.** 4 **27.** 3 **29.** about 21%

31. Sample:

x	y
−3	−2.4
−1	−0.8
0	0
1	0.8

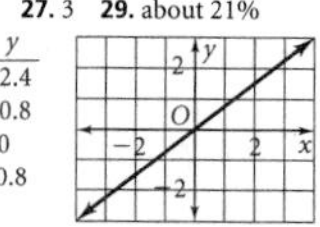

33. Sample:

x	y
−1	−7.2
−0.5	−3.6
0	0
0.5	3.6
1	7.2

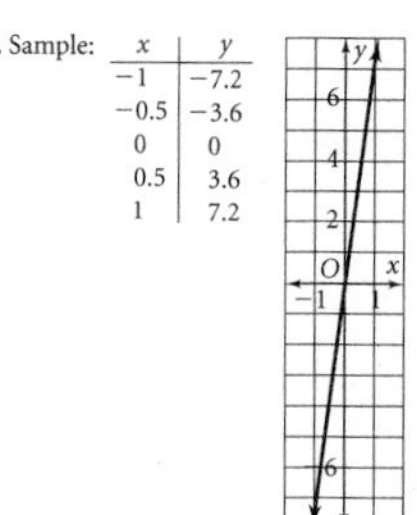

Lesson 5-3 pages 225–229

ON YOUR OWN **1.** A; B; the graph of a direct variation must pass through the origin. **3.** no **5.** yes; $\frac{1}{4}$ **7.** no **9.** yes; $y = 1.8x$ **11.** yes; $y = -1.5x$ **13.** $y = \frac{1}{3}x$ **15.** $y = -\frac{5}{2}x$ **17.** 52 lb **19.** False; the graph must pass through (0, 0). **21.** False; a direct variation is a function. Vertical lines have more than one y-value for an x-value. **23a.** $\frac{1}{32}$; $b = \frac{1}{32}w$ **25.** 12 **27.** 8 **29.** 5 **31.** -2 **33a.** $d = 7p$ **33b.** 84 dog years

MIXED REVIEW **35.** $t > -1\frac{1}{3}$ **37.** B

39.

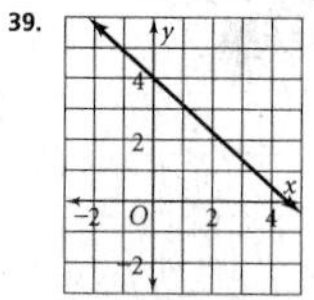

Lesson 5-4 pages 230–234

ON YOUR OWN **1.** $-\frac{3}{4}$; -5 **3.** 3; -9 **5.** 0; 3 **7.** III **9.** II **11a.** 12 in. **11b.** $h = \frac{2}{15}t + 12$ **11c.** 90 min

13.

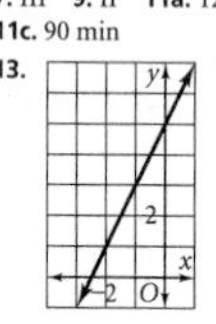

15.

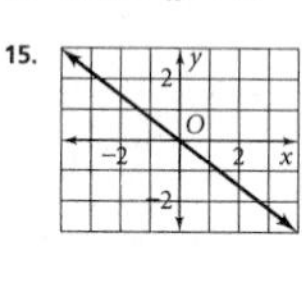

17. **19.**

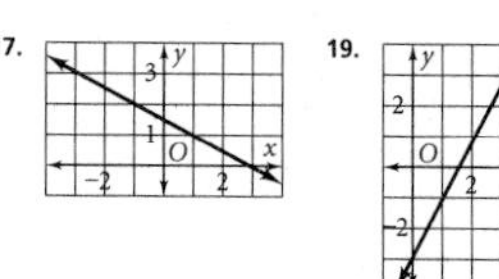

21. $-\frac{2}{3}$; 2 **23.** 0; 1 **27.** $y = \frac{2}{3}x + 3$ **29.** $y = -\frac{5}{4}x$

MIXED REVIEW **31.** $-\frac{9}{2}$ **33.** $-\frac{1}{5}$ **35.** 2 **37.** 1

CHECKPOINT **1.** yes; $\frac{y}{x} = m$ **2.** no; $\frac{y}{x} \neq m$ **3.** No; a vertical line cannot be a direct variation. **4.** yes; $\frac{y}{x} = m$ **5.** \$121.75 billion/yr

Toolbox page 235

1a. Divide Xmax and Ymax by the number of ticks between 0 and the edge; 5, 2 **1b.** $y = 2x + 5$

3.

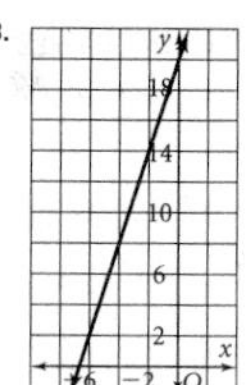

5.

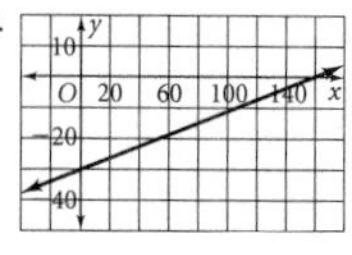

7.

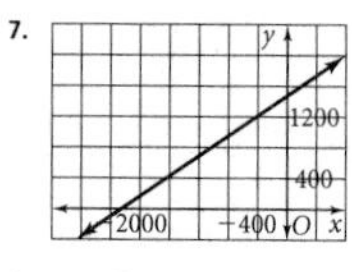

9.

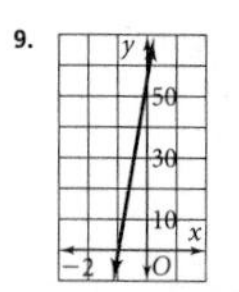

Lesson 5-5 pages 236–239

ON YOUR OWN **1.** $y = 2x - 9$ **3.** $y = \frac{4}{3}x + 3\frac{1}{3}$ **5.** $y = x + 3$ **7.** $y = 2$ **9.** $y = -\frac{1}{4}x + 3\frac{3}{4}$ **11.** $y = -\frac{3}{4}x - 3\frac{1}{2}$ **13a.** $L = 0.025M + 7.25$

13b. The y-intercept is the length of the spring when no mass is attached. **13c.** 9 cm **15.** $y = \frac{1}{5}x + 1\frac{3}{5}$ **17.** $y = \frac{1}{9}x + 2\frac{8}{9}$ **19.** $y = 2x + 50$ **21.** $y = \frac{2}{5}x + 2\frac{4}{5}$ **23a.** For 86 corresponding to 1986, $r = 0.5t - 39.1$ **23b.** \$15.9 billion **25.** yes; $y = -2x + 1$ **27.** yes; $y = 3x + 25$ **29a.** $c = 2.5l + 2$ **29b.** slope: cost/mi; y-intercept: initial cost of ride

MIXED REVIEW **31.** $c < 8$ **33.** $m \leq -1\frac{3}{4}$ **35.** $-1\frac{4}{7}$ **37.** 928.6% increase **39.** positive **41.** positive

Toolbox page 240

1. positive

Lesson 5-6 pages 241–245

ON YOUR OWN **1.** Sample: yes; $y = 11.25x + 200$ **3.** Sample: yes; $y = 0.6x + 330$ **5.** no **9.** A **11b.** Sample: $y = 6.404761905x - 392.7261905$ **11c.** about 280 million TV sets **13a.** Sample: $y = 0.2278305085x - 11.71566102$ **13b.** about 12.2%

19a.

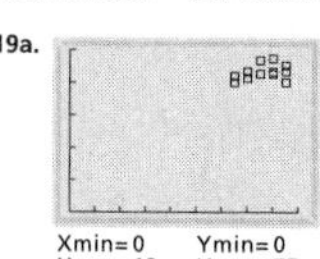

19b. $y = 1.206586826x + 45.71556886$ **19c.** No; $r \approx 0.448$, which is not close to 1.

MIXED REVIEW **21.** 23 **23.** -6 **25.** -3 **27.** $-\frac{8}{3}$ **29.** 20

Lesson 5-7 pages 246–249

ON YOUR OWN **1.** A **3.** C

5.

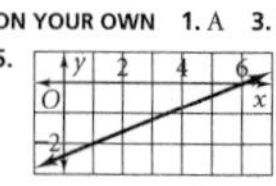

7. **9.** **11.**

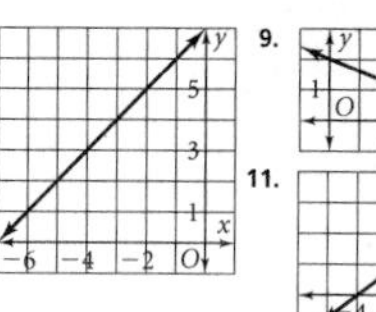

15.

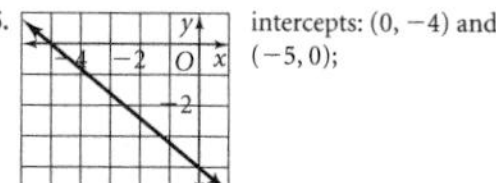

intercepts: (0, −4) and (−5, 0);

17.

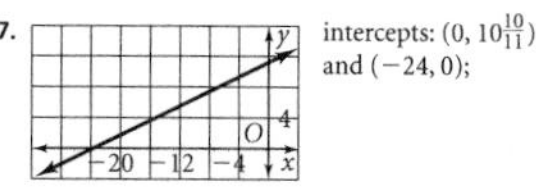

intercepts: $(0, 10\frac{10}{11})$ and (−24, 0);

19.

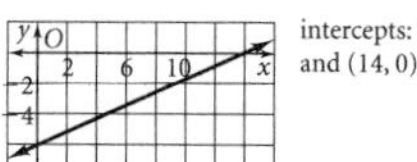

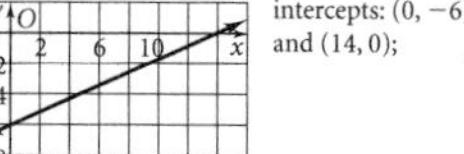

intercepts: (0, −6) and (14, 0);

21. intercepts: (0, 30) and $(23\frac{1}{3}, 0)$;

23. $y = \frac{A}{B}x + \frac{C}{B}$ **25.** $x + 3y = 26$ **27.** $3x + 2y = -20$ **29.** $3x + y = 7$ **31.** $x + 5y = -39$

MIXED REVIEW **33.** $\frac{1}{36}$

35.

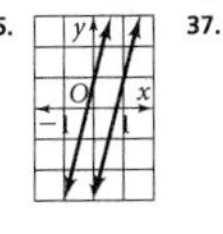

37.

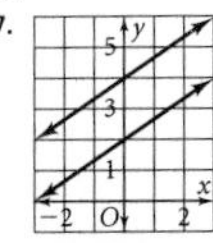

Lesson 5-8 pages 250–255

ON YOUR OWN **1.** $\frac{1}{2}$ **3.** $-\frac{3}{4}$ **5.** 0 **7.** 2 **9.** -3 **11.** $\frac{5}{4}$ **13.** $-\frac{1}{2}$ **15.** $-\frac{5}{7}$ **17.** $\frac{3}{2}$ **19.** undefined **21.** $-\frac{1}{2}$ **23.** $-\frac{2}{9}$ **25.** perpendicular **27.** perpendicular **29.** perpendicular **31.** The vertical and horizontal scales use units of different length. **33a.** Yes; they have the same slope, $-\frac{3}{5}$ **33b.** Since the coefficients of x and y are the same, the ratio for the slope will be the same. **35.** No; the slopes are $\frac{5}{6}$, -2, and 8. No pair of slopes has product -1. **37.** The slope of AD and BC is undefined so they are parallel. The slope of AB and CD is $\frac{1}{2}$ so they are parallel. The quadrilateral is a parallelogram. **39.** The slope of PQ and RS is $-\frac{1}{2}$. The slope of QR and SP is $-\frac{3}{2}$. The quadrilateral is a parallelogram. **41.** $y = 6x$ **43.** $y = -x + 10$ **45.** $y = -\frac{2}{3}x + \frac{1}{3}$ **47.** True; the definition of perpendicular lines says that horizontal and vertical lines are perpendicular. **49.** False; two positive numbers need not be equal. **51.** The slope of AB and CD is $\frac{2}{5}$. The slope of BC and AD is $-\frac{5}{2}$. The product is -1, so the quadrilateral is a rectangle. **53.** The slope of PQ and RS is $\frac{1}{2}$. The slope of BC and AD is -2. The product is -1, so the quadrilateral is a rectangle.

MIXED REVIEW **55.** $m < -2$ **57.** $y > 1$ **59.** $-\frac{8}{3}$ **61.** 20

CHECKPOINT **1.** $-2x + 3y = 6$ **2.** $-4x + y = -7$ **3.** $-\frac{1}{3}x + y = 5$ **4.** $-9x + 15y = -12$ **5.** $y = -\frac{1}{4}x + 4\frac{3}{4}$ **6.** $y = 18x - 3$ **7.** $y = \frac{5}{7}x$ **8.** $x = 5$ **9.** $y = 5.4x + 1.3$ **10.** $y = -6.1x + 62.7$ **15.** D

Lesson 5-9 pages 256–259

ON YOUR OWN **1.** $y = 3x - 8$ **3.** $y = -x - 4$ **5.** A **7.** $-\frac{3}{2}$ **9.** 3 **11.** $5\frac{2}{3}$ **13.** 25 **15.** 36 **17.** $7\frac{1}{4}$ **19a.** $15x = 100$ **19b.** $6\frac{2}{3}$ **19c.** \$50 **23.** -17.5 **25.** 3.75 **27.** False; a vertical line has no y-intercept. **29.** True; the x-intercept is the x-value for y-value 0.

MIXED REVIEW **31.** -11.6 **33.** 4.84

Wrap Up pages 261–263

1. Sample: (10, 0.5), (10, 1); $\frac{1}{40}$ **2.** Sample: (0, 150), (40, 150); 0 **3.** (1, 7.5) (3, 5); -1.25 **4.** $\frac{1}{2}$ page/min **5.** $\frac{1}{27}$ mi/min **6.** 3 goals/game **7.** 12 pages/min **8.** 6 **9.** 4 **10.** -4 **11.** $y = \frac{3}{4}x + 8$ **12.** $y = -7x + \frac{1}{2}$ **13.** $y = \frac{2}{5}x$ **14.** $y = -3$ **15.** $y = \frac{1}{3}x + 1\frac{2}{3}$ **16.** $y = \frac{6}{5}$ **17.** $y = \frac{1}{2}x + \frac{1}{2}$ **18.** $y = \frac{3}{10}x + 6\frac{2}{5}$ **19.** $y = -\frac{1}{2}x - \frac{1}{2}$ **20.** $y = -3x$ **21.** $y = \frac{1}{4}x - 3$ **22a.** Sample: $y = 1.3x - 59$ **22b.** Using equation in (a): 77.5 lb/person **23.** 2; 5 **24.** 8; -13 **25.** 15; -5 **26.** 13; -104 **27.** $y = 5x - 11$ **28.** $y = \frac{1}{5}x + 4$ **29.** $y = 9x - 5$ **30.** $y = \frac{1}{8}x + 10\frac{1}{2}$ **32a.** $30x = 175$ **32b.** $5\frac{5}{6}$ lawns, or 6 lawns to make a profit **33.** The break-even point is the amount you need to produce to make \$0 profit. **34.** C

35.

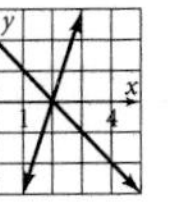

36.

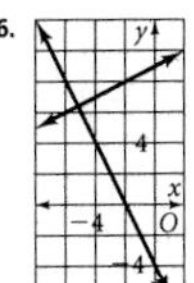

37.

38.

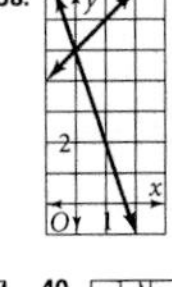

39.

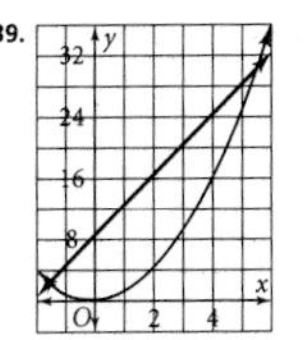

40.

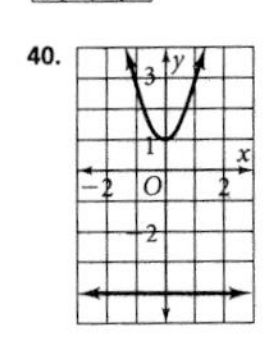

Preparing for Standarized Tests page 265

1. B **3.** B **5.** C **7.** A **9.** D **11.** B **13.** $\frac{3}{5}$ **15.** 4, 5

CHAPTER 6

Toolbox page 268

1. 4 and 8 **3a.** about 2:13 P.M. **3b.** about 13.5 mi

Lesson 6-1 pages 269–274

ON YOUR OWN

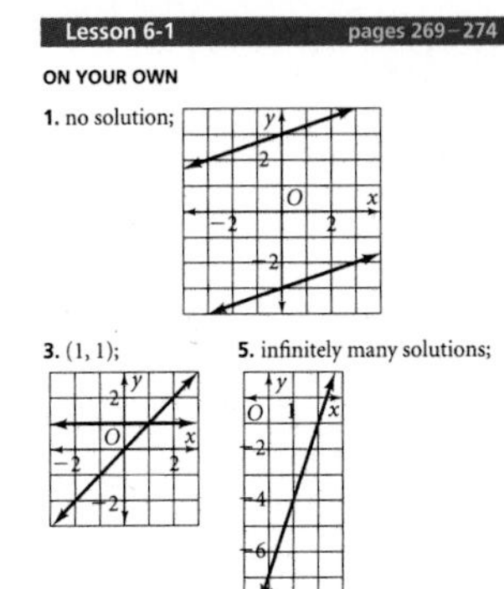

1. no solution; **3.** (1, 1); **5.** infinitely many solutions;

7. yes **9.** yes **11a.** No; the graphs are not straight lines. **11b.** The solution of the system is the point in time when the fear of commitment is equal to the fear of baldness. **11c.** The solution on Lily's chart shows the point in time when the level of hope of meeting Brad Pitt is equal to the level of fear of cellulite. **11d.** feelings; time

13. (1, 3);

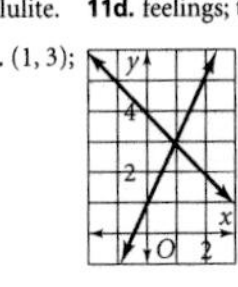

15. (0, 0);

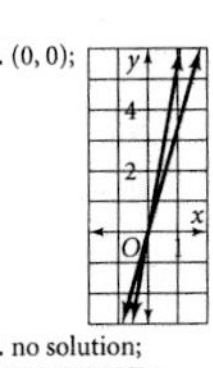

17. $(3, -4)$;

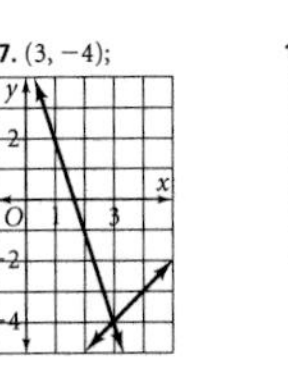

19. no solution;

21. $(0, -1)$;

23. no solution;

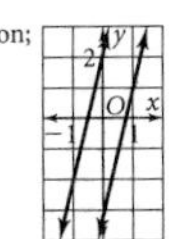

25. True; the graphs of lines can intersect once, not intersect, or be the same line. **27.** False; the system may have no solutions. **29.** parallel; no solutions **31.** intersect; one solution **33.** same line; infinitely many solutions **35.** intersect; one solution **41a.** (2, 200) **41b.** Both studios charge \$200 rent for 2 h. **43.** (4.5, 6.5) **45.** $(4, -6)$

MIXED REVIEW **47.** $\frac{1}{36}$ **49.** 1 **51.** $b \le 6$ **53a.** about 138.5 mi^2 **53b.** about 72 million **55.** $x = 8y$ **57.** $x = \frac{3}{2}y + \frac{5}{2}$

Lesson 6-2 pages 275–279

ON YOUR OWN **1.** one **3.** no solution **5.** infinitely many solutions **7.** one **9.** 12 cm by 3 cm **11.** (2, 4) **13.** $(-\frac{1}{2}, -\frac{1}{2})$ **15.** $(2, \frac{1}{2})$ **17.** $(-\frac{1}{2}, 0)$ **19.** (3, 5) **21.** infinitely many solutions **23.** $(\frac{1}{2}, 1)$;

25. $(-1, 1)$;

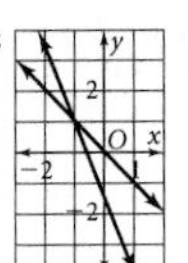

27. no solution;

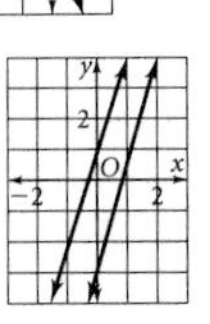

29. D **31.** C **33.** A **35.** infinitely many solutions **37.** no solution **39.** 160 acres of soybeans; 80 acres of corn **41.** $(3, -3)$ **43.** $(\frac{1}{4}, 0)$ **45.** $(12, -8)$ **47.** $(-2, -5)$

MIXED REVIEW **49.** 7; -4 **51.** 9; 0 **53.** $y = \frac{3}{2}x$ **55.** $-7x$ **57.** $15x$

Lesson 6-3 pages 280–284

ON YOUR OWN **1.** Add equations; $(5, -6)$ **3.** Multiply the first equation by 4; (5, 4) **5.** Add equations; $(2, -3)$ **7.** Subtract first equation from second; (4, 2) **9.** Brass parts cost \$6; steel parts cost \$3. **11.** (5, 2) **13.** (1, 3) **15.** $(-4.5, 8)$ **17.** (14, 14) **19.** $(-\frac{2}{3}, 2)$ **21.** (2, 0) **23.** infinitely many solutions **25.** $(2\frac{1}{2}, 3\frac{1}{2})$ **27.** 3 V, 1.5 V **29.** Agree; you do not need to solve an equation for y before substituting the values. **31.** infinitely many solutions **33.** (18, 52) **35.** (3, 4) **37.** $(1\frac{1}{2}, -\frac{1}{2})$ **39a.** (81.25, 8.125) **39b.** Room for one night costs \$81.25 per person; the average cost of one meal is about \$8.13.

MIXED REVIEW **41.** $x = 0$ **43.** $x = 10$ **45.** $2s + d = 6.50$ **47.** $5p + 2n = 32$

CHECKPOINT **1.** $(-1\frac{1}{9}, 3\frac{5}{9})$ **2.** $(-\frac{15}{17}, -1\frac{8}{17})$ **3.** (9, 10) **4.** $(1\frac{9}{13}, -1\frac{3}{13})$ **5.** Sample: $2x - y = 3$; $\frac{1}{2}x + \frac{1}{2}y = 6$ **6a.** (33, 20)

Lesson 6-4 pages 285–288

ON YOUR OWN **1.** 72.5° and 107.5° **3.** 7 dimes and 3 nickels **5.** $(\frac{1}{3}, 2\frac{1}{3})$ **7.** (5, 1) **9.** no solution **11.** (0, 0) **13a.** $m = 254 + 400 + 1.2n$ **13b.** $m = 4n$ **13c.** 234 tickets **15.** 577 games **17.** C **19.** Sample: Substitution; at least one equation is solved for one of the variables. **21.** Sample: Elimination; coefficients match well for subtraction. **23.** Sample: Substitution; coefficients -1 and 1 make it easy to solve either equation for one of the variables. **25a.** $2s + t = 12$ **25b.** $s = 3t$ **25c.** $5\frac{1}{7}, 5\frac{1}{7}, 1\frac{5}{7}$

MIXED REVIEW **27.** -2 **29.** 25 or -17 **31.** \$130 billion

33.

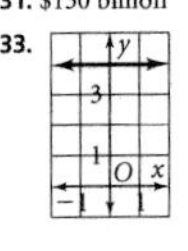

35.

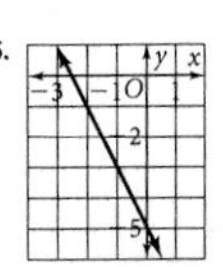

Lesson 6-5 pages 289–293

ON YOUR OWN **1.** A **3.** B **5.** 1 and 3

9. **11.** **13.** **15.** **17.** **19.**

21.

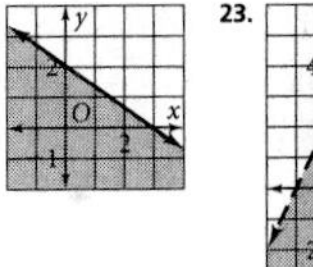

23.

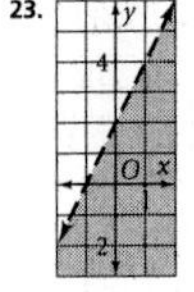

25. $y < x + 2$ **27.** $y > 2x + 1$ **29.** $y \le \frac{1}{3}x - 2$

31. $x > 0$;

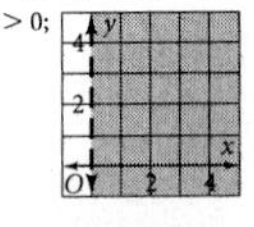

33. $y \ge 0$;

35a. $2x + 2y \le 50$

35b.

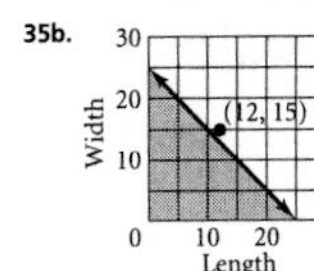

35d. No; $2(12) + 2(15) \le 50$ is false. **37.** yes **39.** yes

MIXED REVIEW **41.** 5 **43.** $\frac{8}{3}$ **45.** 1.2 h **47.** (3, 5) **49.** $(4\frac{2}{3}, 37\frac{1}{3})$

Toolbox page 294

1.

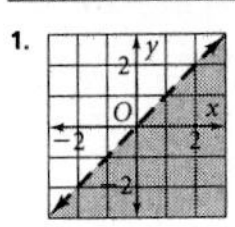

3. 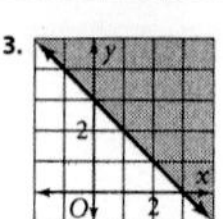

5. 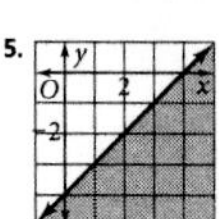

7. 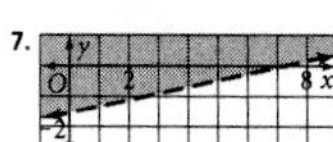

Lesson 6-6 pages 295–299

ON YOUR OWN **1.** C **3.** The point is on the boundary line of $2x + y > 2$, so it is not a solution of $2x + y > 2$.

5.  **7.** no solution; 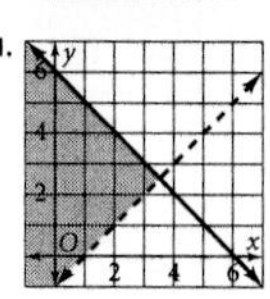

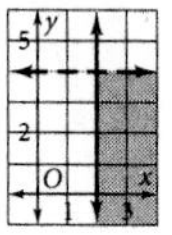

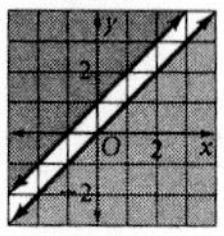

9. **11.**

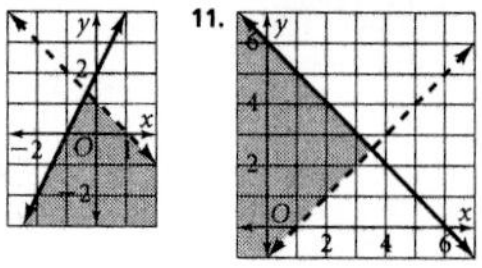

13. 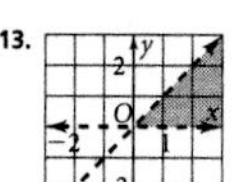 **15.**

17a. $0.6f + 0.55c \le 33; f \ge 9; c \ge 12$

17b. 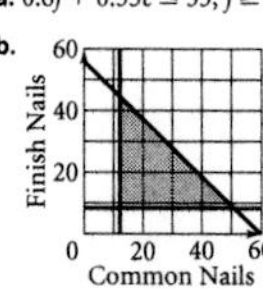

17c. Sample: (20, 20) **17d.** Sample: (40, 25)

23. 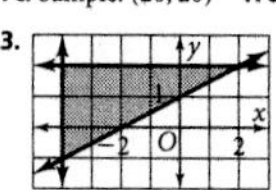

23a. triangle **23b.** (−4, −1), (−4, 2), (2, 2)
23c. 9 sq. units

25. 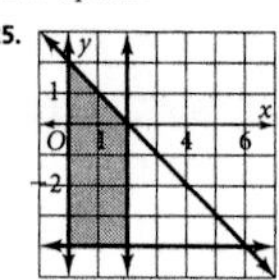

25a. trapezoid **25b.** (0, −4) (0, 2), (2, 0), (2, −4)
25c. 10 sq. units **27a.** $5.99x + 9.99y \le 50; x \ge 0; y \ge 0$

27b. 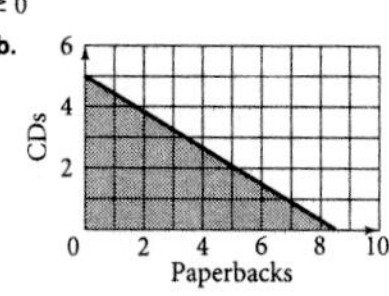

27c. 2 books and 6 CDs; (2, 6) is not a solution of the system because 2 books and 6 CDs cost $71.92.
27d. 5 CDs, no books **27e.** 8 books

MIXED REVIEW **29.** 1 **31.** 4 **33.** 4 **35.** 22 **37.** 62 **39.** 1350

CHECKPOINT

1. **2.** 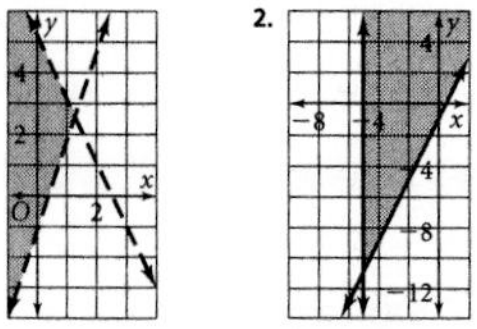 **3.** **4.**

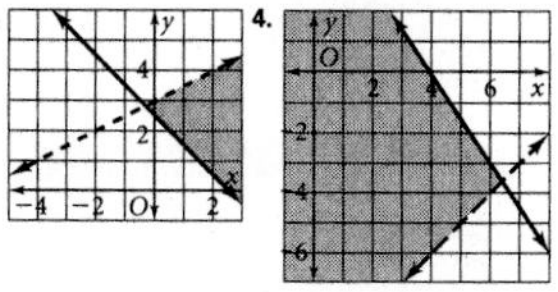

5. D **6a.** $p = 7 + 0.75t$; $p = 8 + 0.5t$ **6b.** (4, 10); a pizza with 4 toppings costs $10 in each restaurant.

Lesson 6-7 pages 300–304

ON YOUR OWN **1.** 24, 32, 34, 30; (3, 5), (8, 0) **3.** 0, 100, 400, 800; (20, 10), (0, 0) **5.** (6, 0) **7.** (4, 0) **9.** 0; 36 **11.** 1; 8 **13.** $c = 8d + 20t$
15b. $N = 30x + 40y$
15c. 400 ft² at Location A and 100 ft² at Location B

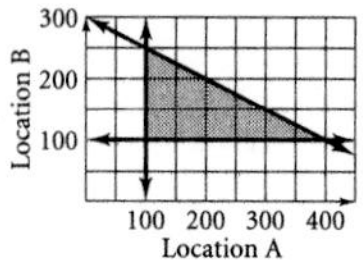

MIXED REVIEW **17.** $y = -\frac{1}{2}x + 2\frac{1}{2}$ **19.** $y = \frac{8}{3}x$
21. quadratic **23.** linear

Lesson 6-8 pages 305–309

ON YOUR OWN **1.** (0, 0), (2, 4) **3.** (0, 0)
5. (−2, −2), (2, −2) **7.** (−2, 1), (2, 1)

9. (−1.94, 0.15), (1.94, 0.15) 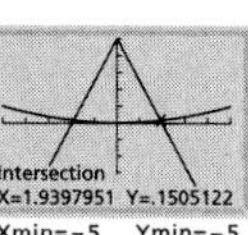

11. A; (−1, −2), (1, −2) **13.** C; no solution **15.** D
21. no solution; 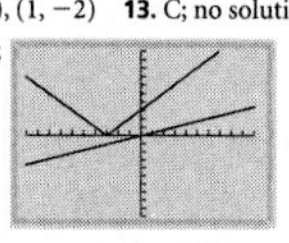

23. (−2, 3), (1, 0); 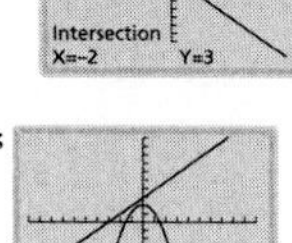

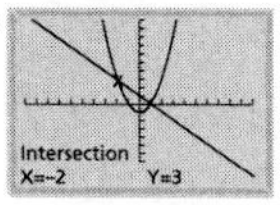

25. no solution; 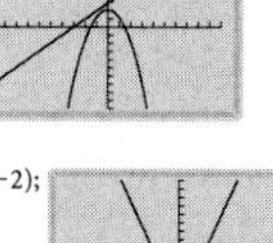

27. (−1, −2), (1, −2); 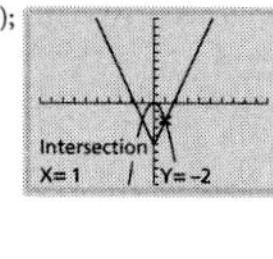

29. no solution; 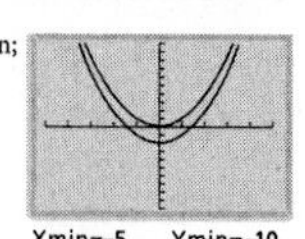

31. (−2, 1), (4, 4); 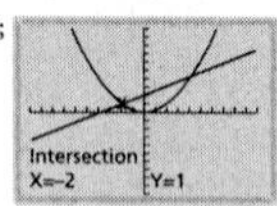

MIXED REVIEW **35.** 106.25% **37.** 6.5%

39. 91,800 **41.** $\begin{bmatrix} 7-x & x-5 \\ -3x & -y \\ 8z & 0 \end{bmatrix}$

43a. 1000000*B2/C2 **43b.** D2: 1576; D3: 2423

Wrap-Up pages 311–313

1. $y = x, y = 1$; (1, 1) **2.** $y = -2x - 2, y = 2x + 2$; (−1, 0) **3.** $y = -x - 1, y = -x + 3$; no solution
4. $x = -1, y = -x + 1$; (−1, 2)

5. (1, 2); 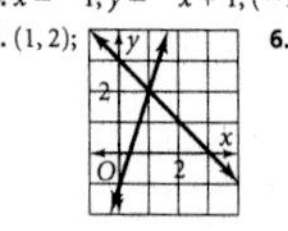 **6.** (−1, 2);

7. no solution; 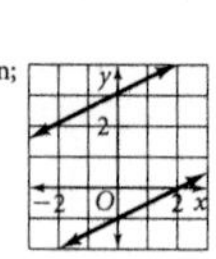

8. (1, −1);

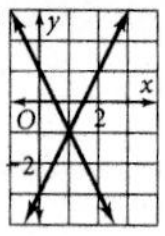

9. $(-1\frac{3}{5}, 4\frac{1}{5})$ **10.** $(-4\frac{1}{2}, -6)$ **11.** (2, 1)
12. $(-1\frac{1}{9}, \frac{5}{9})$ **14.** (−6, 23) **15.** (1, −1) **16.** (6, 4)
17. (12, 6) **18.** $10\frac{2}{3}$ fl oz **19.** 63° and 27°

20. 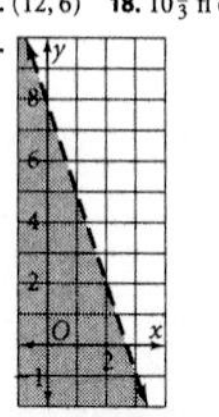 **21.** 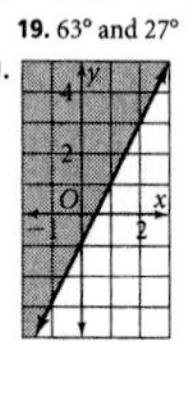

22. 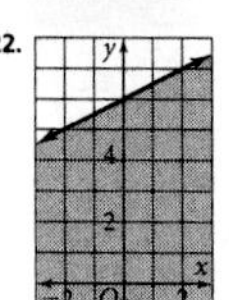 **23.** 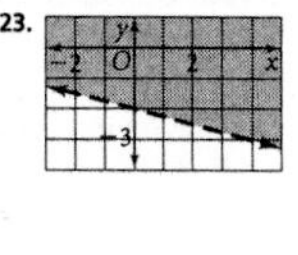

24. **25.** 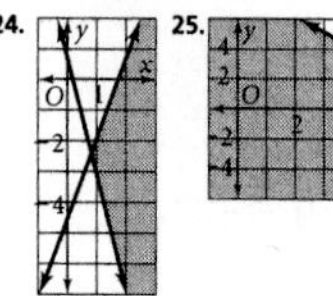

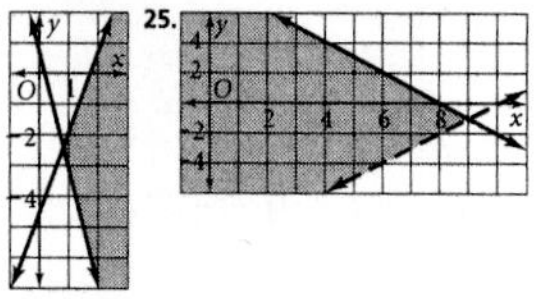

26. **27.**

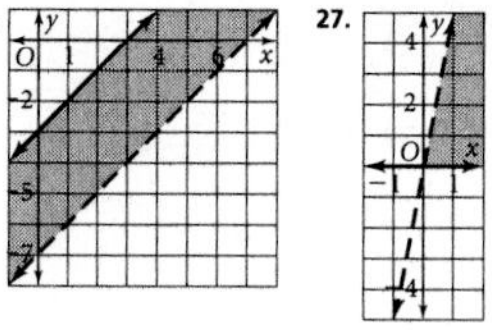

29. 0; 21; 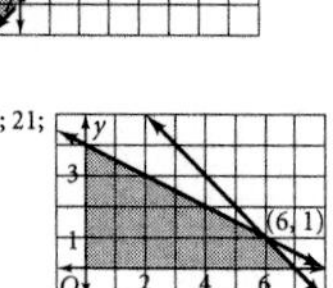

30. 3; 7.5; 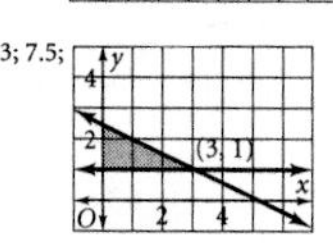

31. 0; 18; 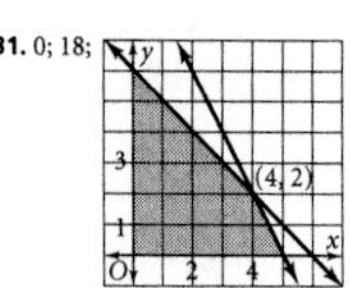

32. 7; 16; 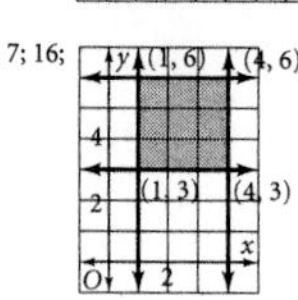

33. E

34. (2, 0), (−3, 5);

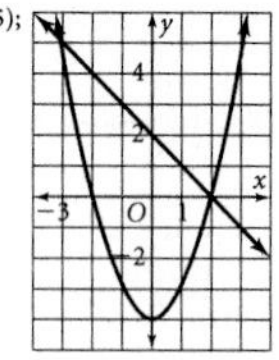

35. (2, 1), $(-\frac{2}{3}, -1\frac{2}{3})$;

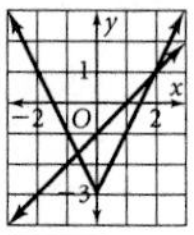

36. (0, 3), (2, 1);

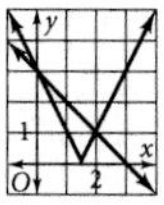

37. (−1, 4), (3, 12) 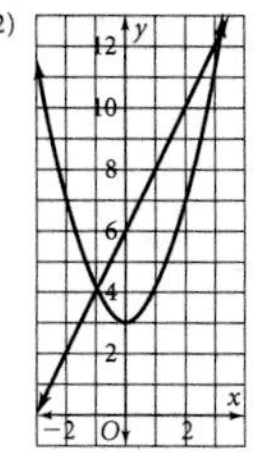

38. 0, 1, 2, 3, or 4; the number of solutions of a system with an absolute value equation and a quadratic equation correspond to the number of times the graphs of the equations intersect. It is possible for the graphs to intersect 0, 1, 2, 3, or 4 times. The graphs cannot intersect more than 4 times. **39.** 9 **40.** 49 **41.** 20.25 **42.** $\frac{1}{4}$ **43.** 121 **44.** 67.24

45. 46.

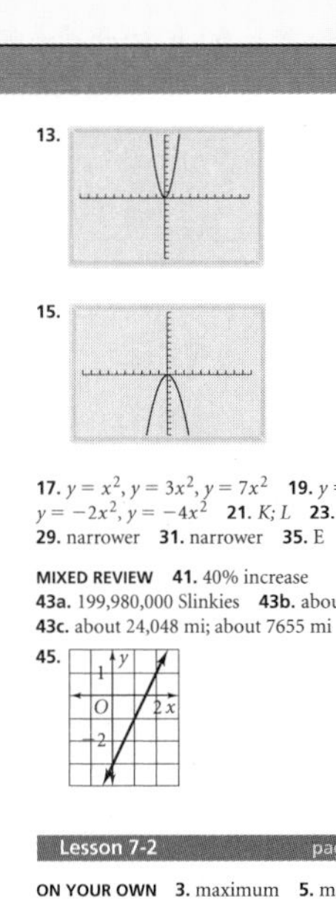

47.

48. 1 **49.** 109 **50.** 64 **51.** 17 **52.** 11 **53.** 170.24

Cumulative Review — page 315

1. E **3.** E **5.** A **7.** D **9.** E **11.** B
13. $y = -\frac{5}{2}x + 2$ **15.** about 3.4%

CHAPTER 7

Lesson 7-1 — pages 318–322

ON YOUR OWN **1.** 1; 2; 4 **3.** −1, −3, −9
5. upward; min. **7.** downward; max.

9.

11.

13.

15.

17. $y = x^2, y = 3x^2, y = 7x^2$ **19.** $y = -\frac{2}{3}x^2$, $y = -2x^2, y = -4x^2$ **21.** *K*; *L* **23.** *K*
29. narrower **31.** narrower **35.** E **37.** F **39.** C

MIXED REVIEW **41.** 40% increase
43a. 199,980,000 Slinkies **43b.** about 33 lb
43c. about 24,048 mi; about 7655 mi

45.

Lesson 7-2 — pages 323–326

ON YOUR OWN **3.** maximum **5.** maximum

7.

9.

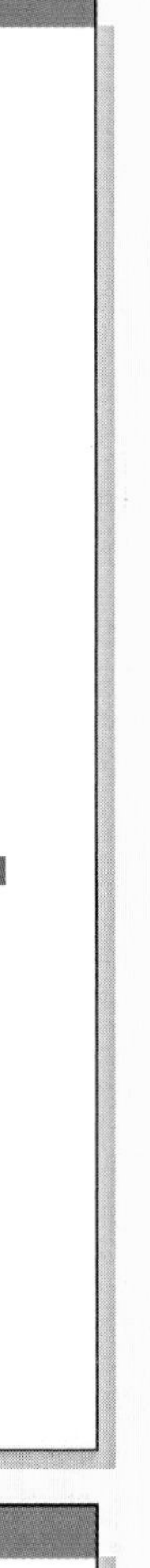

11.

13.

15.

17.

21. *E* **23.** *F* **25.** *C* **27.** *E* **29.** *E*, *F* **31.** *G*

33a.

33b. $0 < x < 12$; the length of the side of the garden must be positive and shorter than 12 ft, the shorter side of the patio. **33c.** $96 < y < 240$; the larger the garden the smaller the area of the patio. As the length of the side of the garden changes from 0 to 12 ft, the value of the function changes from 240 to 96.

MIXED REVIEW **35.** $y = -\frac{1}{5}x + \frac{7}{5}$
37. $y = -2x - 18$ **39.** $y = -\frac{1}{4}x + 3$
41. about $129,067 **43.** $\frac{1}{16}$

Lesson 7-3 — pages 326–331

ON YOUR OWN **1.** $x = 0$; (0, 4) **3.** $x = -1$; (−1, −7) **5.** $x = 0$; (0, −3) **7.** $x = -2$; (−2, −1)
9. $x = 0$; (0, 12) **11.** E **13.** F **15.** D

17. **19.** **21.** **23.** **25.**

27.

29. 20 ft; 400 ft² **31.**

33. **35.**

37. **39.**

MIXED REVIEW **43.** $\frac{1}{4}$ **45.** $\frac{1}{4}$ **47a.** 10.1 mi
47b. 569 strokes **49.** −1 **51.** −36

Lesson 7-4 — pages 332–336

ON YOUR OWN **1.** ±13 **3.** $\pm\frac{1}{3}$ **5.** ±0.5 **7.** ±1.1
9. irrat. **11.** irrat. **13.** 5 and 6 **15.** −16 and −15

17. 3.46 **19.** 107.47 **21a.** about 0.76 **21b.** about 4.6 ft **23.** 0.6 **25.** $\frac{5}{4}$ **27.** 20 **29.** 25 **31.** $\frac{1}{9}$
33. 27 **35a.** about 8500 km **35b.** about 7684 km
37. 4 **39.** $-\frac{2}{5}$ **41.** $\frac{1}{6}$ **43.** −12.53 **45.** −3.61
47. 33 **49.** 6.40 **53.** irrat. **55.** rat. **57.** irrat.
59. undefined **61.** true **63.** true **65.** false; $-17 < -\sqrt{280} < -16$ **67.** false; $-37 < -\sqrt{1300} < -36$ **69.** true

MIXED REVIEW

71. $s < -2$;

73. $b < -6$;

75. $c \le \frac{1}{2}$;

77. $t > 1\frac{2}{3}$;

79. $3\frac{2}{3}$ **81.** 0

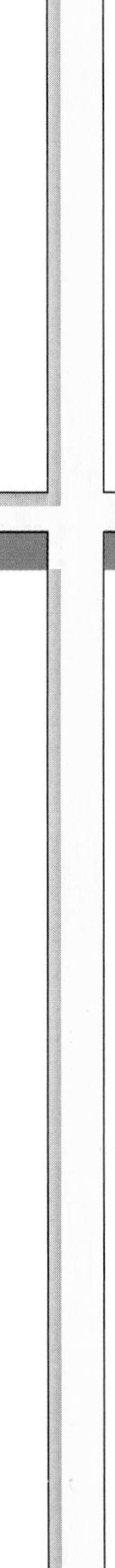

CHECKPOINT

1. (0, 0); **2.** (0, 7);

3. (−1, 11);

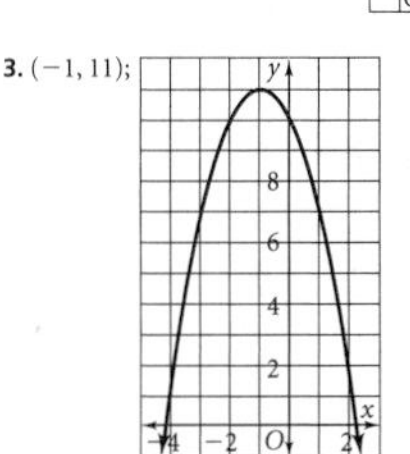

4. 2.65 **5.** −10 **6.** 4.80 **7.** 12 **8.** −12.25 **9.** $-\frac{1}{3}$

Lesson 7-5 — pages 337–341

ON YOUR OWN **1.** −2, 2 **3.** no solution **5.** no solution **7.** $-\frac{3}{7}, \frac{3}{7}$ **9.** $-\frac{5}{2}, \frac{5}{2}$ **11.** 0 **13.** Sample: Michael used −5 and 5 as square roots of −25, which does not have a real number square root.
15. −2.8, 2.8 **17.** −1.4, 1.4 **19.** −4.6, 4.6
21. 3.6 in. **23.** False; there are no solutions.
25. False; there are two solutions **27.** true; −8, 8
29. true **31.** 11.3 ft **33.** 2.5 s **37.** −2.5, 2.5
39. −2.4, 2.4 **41.** −2, 2

MIXED REVIEW **43.** (2, 3) **45.** (15, −15) **47.** 3.16
49. −9

Toolbox — page 342

1. about 10.78, 13.22 **3.** $-2, 1\frac{1}{2}$ **5.** about 0.28, 17.72 **7.** −12, 6

Lesson 7-6 — pages 343–347

ON YOUR OWN **1.** $3x^2 + 13x - 10 = 0$
3. $x^2 - 5x - 7 = 0$ **5.** $12x^2 - 25x + 84 = 0$
7. −1.67, 0.5 **9.** −4.65, 4.65 **11.** −2.2, 3
13. 0, 0.56 **15.** −1.78, 0.28 **17.** −1.5, −1
19b. about 356.9 million **19c.** 2007 **21.** −6, 6
23. −1.41, 1.41 **27.** −2, 3 **29.** −0.39, 1.72
31. −1.43, 2.23 **35a.** 630 ft **35b.** 0 ft
35c. about 6.3 s

MIXED REVIEW **37.** $\frac{1}{4}$ **39.** 0 **41.** 41 **43.** 24
45. 49

CHECKPOINT **1.** −8.06, 8.06 **2.** −2, 2 **3.** −11, 11
4. 1, 3 **5.** −1, 0.4 **6.** 0.23, 3.27 **7.** D

Toolbox — page 348

1. $y = x^2 + 6x + 8$ **3.** $y = 2x^2 + 6x - 8$

Lesson 7-7 — pages 349–353

ON YOUR OWN **1.** C **3.** A **5.** 1 **7.** 2 **9.** 2
11. 0 **13.** 1 **15.** 2 **17a.** no value of x **17b.** Yes; $266 = -x^2 + 3x + 270$ transforms to $-x^2 + 3x + 4 = 0$. The discriminant is positive. **19.** A
21. Rational; the square root of the discriminant is a positive integer. **23.** 2 **25.** 2 **27.** 0 **29.** 0 **31.** B

33. no **35.** yes; $(\frac{2}{3}, 0)$, (−1, 0) **37.** yes; (−1, 0), $(2\frac{1}{2}, 0)$ **39.** no **41a.** A2^2 − 4; A2^2 − 8

41b. for integer values ≤ −2 and ≥ 2 **41c.** integer values between −2 and 2

MIXED REVIEW **43.** **45.** 44,372 ft

Wrap-Up — pages 355–357

5. **6.**

7. **8.**

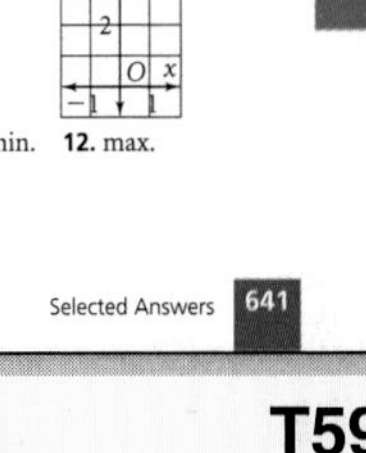

9. min. **10.** max. **11.** min. **12.** max.

13.

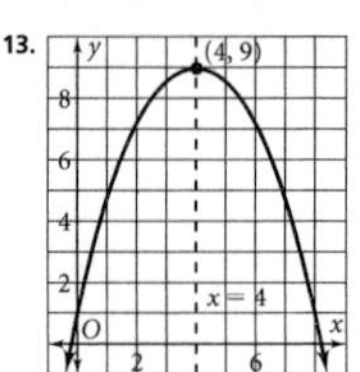

14.

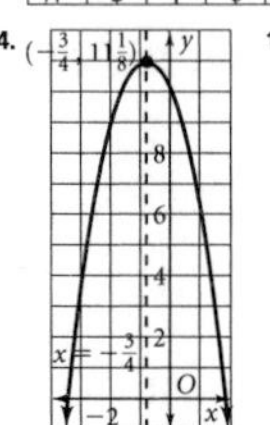

15.

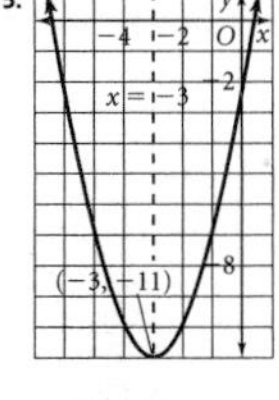

16. 17.

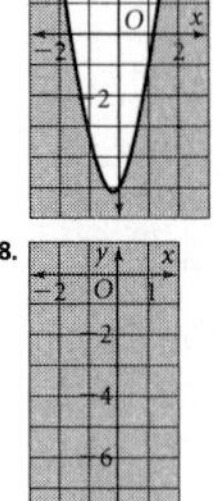

18.

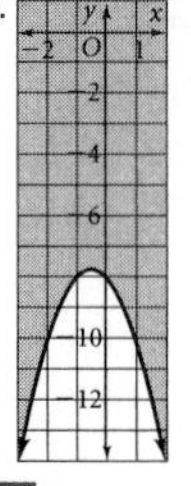

19. irrat. **20.** rat. **21.** irrat. **22.** rat. **23.** irrat. **24.** rat. **25.** 3 **26.** −6.86 **27.** 0.6 **28.** 11.83 **29.** −1 **30.** 14 **31.** B **32.** False; there are two solutions. **33.** true; −3, 3 **34.** −2, 2 **35.** −5, 5 **36.** 0 **37.** no solution **38.** 2.3 in. **39.** −1.84, 1.09 **40.** 0.5, 3 **41.** 0.13, 7.87 **42.** −5.48, 5.48 **43.** 1.5 s **44.** 49; 2 **45.** 112; 2 **46.** −39; 0 **48.** $3^2 5^3$ **49.** $8^3 x^4$ **50.** $h^5 w^2$ **51.** 5200 **52.** 0 **53.** 80,000 **54.** 9 **55.** −720 **56.** 0.7 **57.** 36 **58.** 12 **59.** $\frac{1}{3}$ **60.** 0 **61.** 4 **62.** 2560

Preparing for Standardized Tests page 359

1. E **3.** C **5.** D **7.** A **9.** B **11.** B **13.** $\left(\frac{3}{8}, -\frac{9}{16}\right)$; $x = \frac{3}{8}$

15. −8 −4 0 4 8 12

CHAPTER 8

Lesson 8-1 pages 362–366

ON YOUR OWN

1.

Time	Time Periods	Pattern	Number of Bacteria Cells
Initial	0	75	75
20 min	1	$75 \cdot 2$	$75 \cdot 2^1 = 150$
40 min	2	$75 \cdot 2 \cdot 2$	$75 \cdot 2^2 = 300$
60 min	3	$75 \cdot 2 \cdot 2 \cdot 2$	$75 \cdot 2^3 = 450$
80 min	4	$75 \cdot 2 \cdot 2 \cdot 2 \cdot 2$	$75 \cdot 2^4 = 600$

There will be more than 30,000 bacteria cells after 3 h. **3.** equal **5.** $100x^2$ **7.** 4, 16, 64, 256, 1024; increasing **9.** 1, 1, 1, 1, 1; neither **11.** 0.5, 0.25, 0.125, 0.0625, 0.03125; decreasing **13.** 40, 400, 4000, 40,000, 400,000; increasing **15.** B **19.** 1.5625 **21.** 0.2 **23.** I: A; II: C; III. B

25a.

x	y
1	−2
2	4
3	−8
4	16
5	−32
6	64

25b. Sample: The positive and negative values alternate. The absolute values of the output are the same as the absolute values of the output of 2^x. **25c.** No; the shape of the graph cannot be similar to the shape of the graph of an exponential function.

27.

29. **31.**

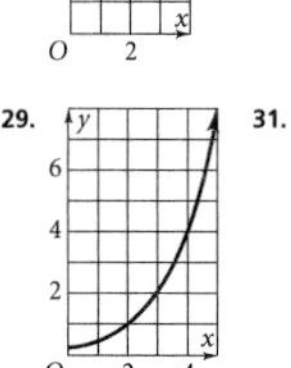

MIXED REVIEW **33.** 1, −4 **35.** 2, 3 **37.** {2, 4, 8, 16, 32} **39.** {1, 2, 4, 8, 16} **41.** {6, 12, 24, 48, 96}

Lesson 8-2 pages 367–372

ON YOUR OWN **1.** 20; 2 **3.** 10,000; 1.01 **5.** 50% **7.** 4% **9.** 1.04 **11.** 1.037 **13.** 1.005 **15.** x = number of years; y = population; $y = 130{,}000 \cdot 1.01^x$ **17.** x = number of months; y = deposit with interest; $y = 3000 \cdot \left(1 + \frac{0.05}{12}\right)^x$ **19a.** \$355 **19b.** 8% **19c.** about \$766 **21.** E

23. linear **25.** exponential

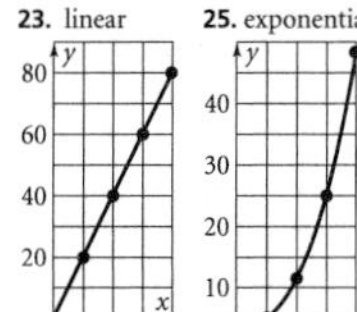

27. Linear; graph is a straight line. **29.** Exponential; graph curves upward. **31.** $y = 20{,}000 \cdot \left(1 + \frac{0.035}{4}\right)^{4x}$; \$28,338.18 **33.** $y = 2400 \cdot (1 + 0.07)^x$; \$4721.16

MIXED REVIEW **35.** $\left(-\frac{1}{3}, -8\frac{1}{3}\right)$; minimum **37.** (0, −11); minimum **39.** $\frac{1}{4}$ **41.** $\frac{3}{4}$ **43.** 0.9604

Lesson 8-3 pages 373–377

ON YOUR OWN **1.** exponential growth **3.** exponential growth **5.** 0.5 **7.** $\frac{2}{3}$ **9.** 10% **11.** 50% **13.** 4 da **15.** 2.5 yr

17. exponential decay

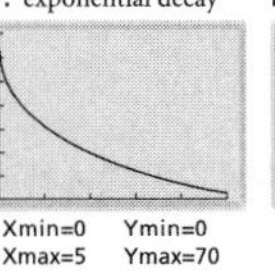

Xmin=0 Ymin=0
Xmax=5 Ymax=70
Xscl=1 Yscl=10

19. exponential decay

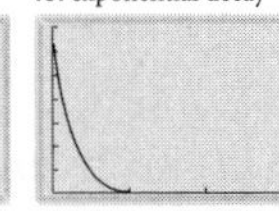

Xmin=0 Ymin=0
Xmax=3 Ymax=3.5
Xscl=1 Yscl=0.5

21. exponential decay

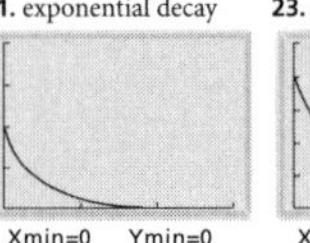

Xmin=0 Ymin=0
Xmax=3 Ymax=0.2
Xscl=1 Yscl=0.05

23. exponential decay

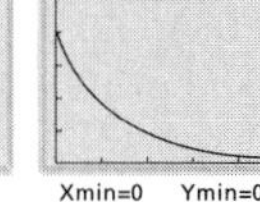

Xmin=0 Ymin=0
Xmax=5 Ymax=0.5
Xscl=1 Yscl=0.1

27. $y = 900 \cdot 0.8^x$; \$235.93 **29a.** $y = 161{,}000 \cdot 0.99^x$ **29b.** 145,606 **29c.** 119,092 **31.** 0.97 **33.** 0.974 **35.** 0.993 **37a.** about 4 h **37b.** about $\frac{1}{4}$

37c. $\frac{15 \cdot 0.84^8}{15} = 0.247875891 \approx \frac{1}{4}$ **39.** 4 half-lives

MIXED REVIEW **41.** $x < -3$ **43.** −3, 3 **45.** 4 **47.** 4; 16; $\frac{1}{4}$

CHECKPOINT **1.** $y = 65{,}000 \cdot 1.032^x$; \$104,257.86 **2.** $y = 200{,}000 \cdot 0.95^x$; 71,697 **3.** $y = 300 \cdot 2^x$; x = ten-year period; \$9600 **4.** B

5. exponential growth

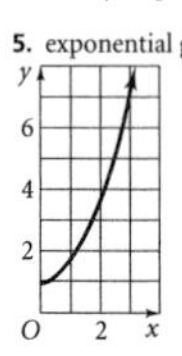

6. exponential decay

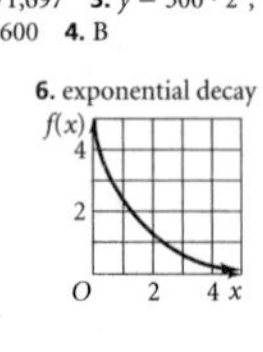

7. exponential growth

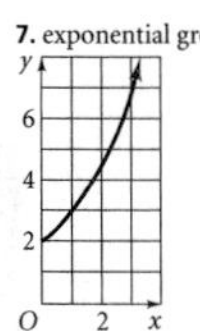

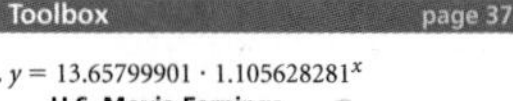

Toolbox page 378

1. $y = 13.65799901 \cdot 1.105628281^x$

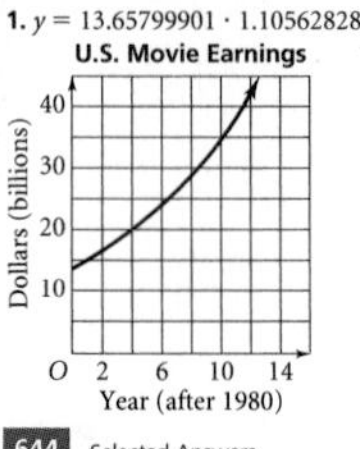

3a. Use $x = 0$ for 1970; $y = 1.040248033 \cdot 1.01999578^x$; 2.297 million

Population of New Mexico

Population (millions)

(population in 2010)

Year (after 1970)

Lesson 8-4 pages 379–384

ON YOUR OWN **1.** −1 **3.** $-\frac{1}{5}$ **5.** 8 **7.** 9 **9.** $\frac{1}{64}$ **11.** 45 **13.** 6 **15.** $\frac{1}{20}$ **17.** $\frac{1}{4}$ **19a.** 5^{-2}; 5^{-1}; 5^0 or 1; 5^1; 5^2 **19b.** 5^4 **19c.** a^n **21.** x^7 **23.** $\frac{1}{625x^4}$ **25.** $\frac{1}{a^4}$ **27.** $\frac{12x}{y^3}$ **29.** $\frac{5a}{c^5}$ **31.** $\frac{8c^3}{a^5d^3}$ **33.** $\frac{7t^3}{5s^5}$ **35.** $\frac{a^3}{b^3}$ **37.** $\frac{mq^3}{n^4}$ **39.** $\frac{2}{9}$ **41.** $\frac{1}{16}$ **43.** −64

45. **47.**

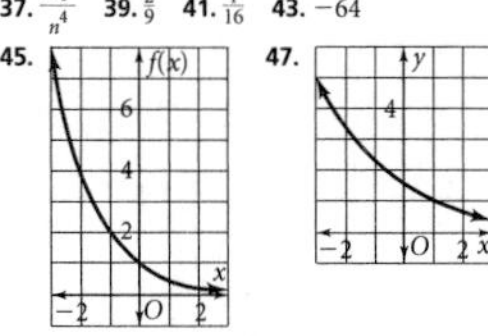

49. I

51.

a	4	0.2	$\frac{1}{3}$	6	$\frac{7}{8}$	2
a^{-1}	$\frac{1}{4}$	5	3	$\frac{1}{6}$	$\frac{8}{7}$	0.5

53. positive **55.** negative **57.** positive **59.** −3 **61.** 0; −3 **63.** −3; −2 **65.** −2; −3 **67.** A, B, D **69a.** 10^{-15} m **69b.** 10^{-6} m **69c.** 10^{-10} m **71.** about 2 students; about 16 students; about 29 students

73.

75a. $\frac{1}{2}, \frac{1}{4}, \frac{1}{8}, \frac{1}{16}$ **75b.** $2^{-1}, 2^{-2}, 2^{-3}, 2^{-4}$ **75c.** $r = 2^{-n}$ **75d.** 2^{-10} or $\frac{1}{2^{10}}$

MIXED REVIEW

77.

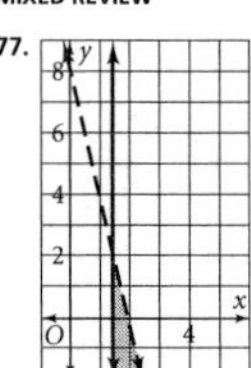

79a. $b = 31.17p$ **79b.** \$124.68 **81.** 0.07 **83.** 0.003 **85.** 3067.85

Lesson 8-5 pages 385–389

ON YOUR OWN **1.** $10^{-3}, 10^{-1}, 10^0, 10^1, 10^5$ **3.** $4.1 \times 10^4, 4 \times 10^5, 4.02 \times 10^5, 4.1 \times 10^5$ **5.** 2.3×10^8 **7.** 2.7×10^{-7} **9.** 2.579×10^{-4} **11.** 9.04×10^9 **13.** 9.2×10^{-4} **15.** 3.7×10^{11} **17.** 4.18×10^{10} **19.** 6.29×10^{16} **21.** 3.3×10^{-8} **23.** 8.35×10^{-5} **25.** 4.18×10^5 **27.** 1.28×10^{-2} **29.** 4×10^{-6} **33.** 0.000 004 69 **35.** 50,000,000,000 **37.** 5.6×10^{-2} **39.** 6×10^2 **41.** 2×10^{-8} **43.** about 1.5×10^{-4} **45.** about 5×10^2 s

47. 7×10^{-3} **49.** 9×10^1 **51.** 5×10^7 **53.** about \$3426 per person

MIXED REVIEW **55.** $\begin{bmatrix} -3.2 & 2.4 \\ 4.1 & -1.2 \end{bmatrix}$ **57.** t^7 **59.** 5^3s^3

Toolbox page 390

1. 6 **3.** 3 **5.** 1.02×10^{-4} **7.** 2.53×10^5 **9.** 7.25×10^5

Lesson 8-6 pages 391–395

ON YOUR OWN **1.** $x^4 \cdot x^3 = x^7$ **3.** $x^{-2} \cdot x^{-5} = x^{-7}$ **5.** −4 **7.** 5 **9.** −4 **11.** 2; −3 **13.** 8×10^5 **15.** 6×10^9 **17.** 4×10^3 **21.** 1 **23.** $3r^5$ **25.** a^8b^3 **27.** $3x^4$ **29.** -0.99^3 **31.** b^3 **33.** −7 **35.** $-45a^4$ **37.** $\frac{b^3}{c^6}$ **39.** $12a^6c^8$ **41.** a^8b **43.** a^5 **45.** $6a^3 + 10a^2$ **47.** $-8x^5 + 36x^4$ **49a.** Jerome's work; Jeremy added exponents having different bases. **49b.** a^3b^6 **51.** $6x^3 + 2x^2$ **53.** $4y^5 + 8y^2$ **55.** $3(-2)x^{2+4} = -6x^6$ **57.** $x^{6+1+3} = x^{10}$ **59a.** about 10^{-7} m **59b.** longer

MIXED REVIEW **61.** irrational **63.** rational; $\frac{1}{2}$ **65.** rational; $\frac{3}{4}$ **67.** 3^6 **69.** 5^{28}

CHECKPOINT **1.** 1 **2.** $\frac{1}{9}$ **3.** $-\frac{1}{8}$ **4.** 125 **5.** −16 **6.** $-\frac{1}{8}$ **7.** s^3 **8.** $21ab^2$ **9.** −1 **10.** $\frac{g^3}{h^4}$ **11.** $\frac{y^6}{x^5}$ **12.** 1.5×10^7 **13.** 8×10^{-8} **14.** 3×10^{11} **15.** 2.4×10^4 **16a.** 3×10^6; 9×10^4 **16b.** about 462

18. **19.**

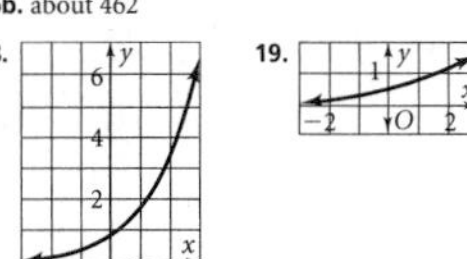

20.

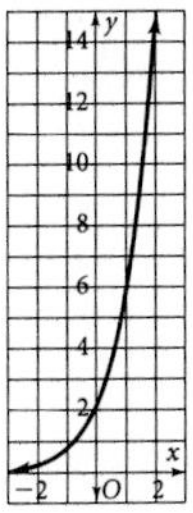

Lesson 8-7 pages 396–400

ON YOUR OWN **1.** x^9y^9 **3.** a^6b^{12} **5.** $\frac{1}{g^{40}}$ **7.** $x^{12}y^4$ **9.** 16 **11.** g^{10} **13.** s^{11} **15.** $\frac{m}{g^4}$ **17.** 1 **19.** $64a^9b^6$ **21.** 1 **23.** 10,000 **25.** b^{3n+2} **27.** 1 **29a.** about 5.15×10^{14} m^2 **29b.** about 3.6×10^{14} m^2 **29c.** about 1.37×10^{18} m^3 **31.** 8×10^{-9} **33.** 3.6×10^{25} **35.** 6.25×10^{-18} **41.** 3 **43.** 4 **45.** 0 **47.** 6 **49.** 3 **51.** 1.1×10^{18} insects **53.** $(ab)^5$ **55.** $(\frac{2}{xy})^2$

MIXED REVIEW **57.** -1 **59.** $\frac{1}{x}$ **61.** $\frac{1}{t^5}$

Lesson 8-8 pages 401–405

ON YOUR OWN **1.** $\frac{1}{4}$ **3.** $\frac{9}{16}$ **5.** $\frac{1}{9}$ **7.** y **9.** a^6 **11.** $-\frac{27}{8}$ **13.** $5^3 = 125$ **15.** $(2c)^4 = 16c^4$ **17.** $x^0 = 1$ **19.** 5×10^7 **21.** 9×10^2 **23.** about 1.52×10^{-6} **25a.** about 1642 h **25b.** about 4.5 h/da **27.** 4 **29.** 729 **31.** $2c^6$ **33.** $\frac{a^5}{b^3}$ **35.** $\frac{d^8}{c^5}$ **37.** $\frac{1}{25}$ **39.** $\frac{1}{c^{12}}$ **41.** $\frac{9}{25}$ **43.** $\frac{125n^3}{m^6}$ **45.** $\left(\frac{3}{5}\right)^5$ **47.** d^3 **49.** $\left(\frac{3x}{2y}\right)^3$ **51a.** about \$4013 **51b.** about \$17,846 **51c.** about 345% **53.** division and negative exponent properties; $\frac{1}{2^3}$ **55.** multiplication property; $\frac{1}{2^3}$

57a.

Mercury	≈ 1.52
Venus	≈ 1.01
Earth	≈ 1.03
Mars	≈ 1.21
Jupiter	≈ 1.10
Saturn	≈ 1.12
Uranus	≈ 1.10
Neptune	≈ 1.02
Pluto	≈ 1.67

57b. Pluto; its circularity is farthest from 1; Venus; its circularity is closest to 1.

MIXED REVIEW **59.** 0.6, -0.6 **61.** -6

63.
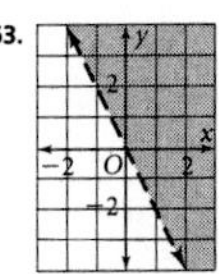

65. 3.26×10^{11} gal

Wrap-Up pages 407–409

1. 6, 12, 24, 48 **2.** 7.5, 5.625, 4.21875 **3a.** 2430 bacteria **3b.** about 178 min **4.** growth; 3 **5.** growth; $\frac{3}{2}$ **6.** decay; 0.32 **7.** decay; $\frac{1}{4}$

8.
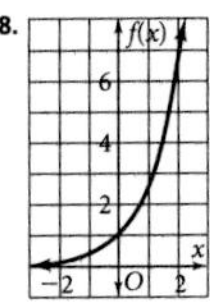

9.

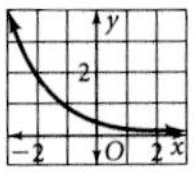

10.

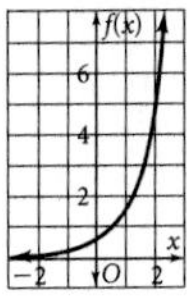

11.
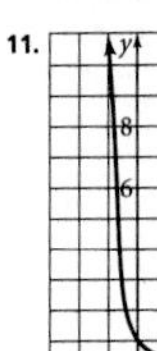

12. 2.5% **13.** 25% **14.** 100% **15.** 10% **16a.** $y = 100{,}000 \cdot 1.07^x$ **16a.** $y = 100{,}000 \cdot 1.07^x$ **16b.** 542,743 **17a.** \$1250.18; \$1267.37 **17b.** 6%; it pays more interest. **18.** 2.5 mCi **19.** $\frac{d^6}{b^4}$ **20.** $\frac{y^8}{x^2}$ **21.** $\frac{7h^3}{k^8}$ **22.** $\frac{q^4}{p^2}$ **23.** $\frac{5^4}{2^4}$ or $\frac{625}{16}$ **24.** no; $(-3b)^4 = 81b^4$ **25.** C **26.** 9.5×10^7 **27.** 7.235×10^9 **28.** yes **29.** 8.4×10^{-4} **30.** 2.6×10^{-4} **31.** 2.793×10^6 **32.** 1.89×10^8 **33.** 1.606×10^8 **34.** 4.6×10^{-9} **35.** 1.4168×10^7 **36.** 1×10^{-12} **37.** $2d^5$ **38.** $q^{12}r^4$ **39.** $-20c^4m^2$ **40.** 1.7956 **41.** $\frac{243}{64}x^2y^{14}$ **42.** Sample: about 7.8×10^3 pores **44.** $\frac{1}{w^3}$ **45.** $\frac{1}{64}$ **46.** yes **47.** $\frac{n^{35}}{v^{21}}$ **48.** $\frac{c^3}{e^{11}}$ **49.** $s^{15}x^{45}$ **50.** 2×10^{-3} **51.** 2.5×10^1 **52.** 5×10^{-5} **53.** 3×10^3 **55.** 2 **56.** 11 **57.** 6 **58.** $77r + 28rx$ **59.** $24m - 16mt$ **60.** $8b + 2b^2$ **61.** $5p + 6d$ **62.** n **63.** 61 **64.** 65 **65.** $13t^2$ **66.** $4y^2$ **67.** \$19; about \$48.86 **68.** 5 kg; 4.5 kg **69.** 0.4 min; 1.24 min

Cumulative Review page 411

1. D **3.** D **5.** D **7.** A **9.** D **11.** A **13.** A

15.
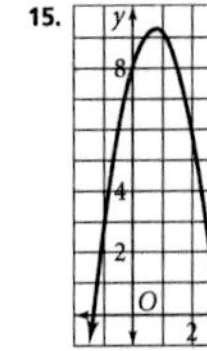

17. Yes; sides are parallel and meet in right angles.

CHAPTER 9

Lesson 9-1 pages 414–418

ON YOUR OWN **1.** 10 **3.** 2.2 **5.** 7.1 **7.** 5.7 **9.** 20 **11.** 66.1 **13a.** $6^2 + 8^2 = 100$ and $10^2 = 100$

13b.

a	b	c
3	4	5
5	12	13
7	24	25
9	40	41

13c. Sample: 8, 15, 17 **15.** 9.7 **17.** 559.9 **19b.** 6 in.^{2}, $\sqrt{10^2 - 8^2}$; about 12.8 in.^{2}, $\sqrt{10^2 + 8^2}$ **21.** yes; $9^2 + 12^2 = 15^2$ **23.** no; $2^2 + 4^2 \neq 5^2$ **25.** no; $4^2 + 4^2 \neq 8^2$ **27.** yes; $1.25^2 + 3^2 = 3.25^2$ **29.** yes; $14^2 + 48^2 = 50^2$ **31.** no; $1^2 + 1.5^2 \neq 2^2$ **33.** no; $4^2 + 5^2 \neq 6^2$ **35.** yes; $18^2 + 80^2 = 82^2$ **37a.** about 1340 ft **37b.** about 67 ft/min; about 0.76 mi/h **39.** 4.3 cm

MIXED REVIEW **41.** $\frac{1}{2}, \frac{1}{2}$ **43.** 5 **45.** 8.5

Toolbox page 419

1. yes **3.** yes **5.** yes **7.** yes **9.** no **11.** yes **13.** yes **15.** yes

Lesson 9-2 pages 420–424

ON YOUR OWN **1.** 10.6 **3.** 8.2 **5.** 11.3 **7.** 2.2 **9.** 11.2 **11.** 8.2 **13.** 14 **15.** 1.4

17a.

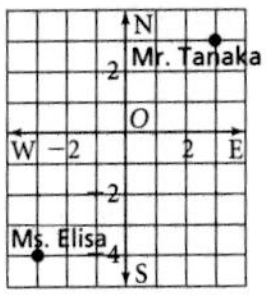

17b. $(0, -0.5)$ **17c.** The meeting point is 0.5 mi south of the substation. **19.** $(1\frac{1}{2}, 7\frac{1}{2})$ **21.** $(-2\frac{1}{2}, 3)$ **23.** $(-4, 4)$ **25.** $(8\frac{1}{2}, -9)$ **27.** $(7, 7)$ **29.** $(0, 0)$ **31.** 16 **33.** 22.2 **35.** x-coordinates are opposites; y-coordinates are opposites.

MIXED REVIEW **37.** $2 - \frac{7}{2}y$ **39.** $\frac{V}{\pi r^2}$ **41.** \$720 **43.** $\frac{5}{13}, \frac{12}{13}, \frac{12}{5}$ **45.** 6.8 **47.** 7.28

Lesson 9-3 pages 425–429

ON YOUR OWN **1.** Sample: Kentucky and Virginia are adjacent states. **3.** $\frac{4}{5}$ **5.** $\frac{4}{5}$ **7.** $\frac{4}{3}$ **9.** 0.7660 **11.** 0.9962 **13.** $\frac{21}{29}, \frac{20}{29}, \frac{21}{20}$ **15.** 0.9231, 0.3846, 2.4 **17.** 10.4 **19.** about 250 ft **21.** about 10,000 ft **23.** 78.4 **25.** 514.3

MIXED REVIEW

29.
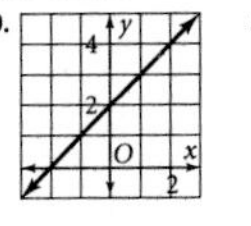

31.
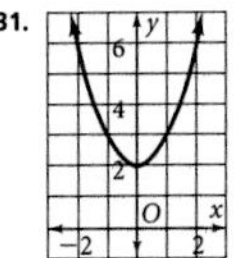

33. 250 mi **35.** 13 **37.** -6

CHECKPOINT **1.** $c \approx 4.5$ **2.** $c \approx 5.8$ **3.** $c \approx 7.1$ **4.** $c = 5$ **5.** $b = 24$ **6.** $a \approx 8.5$ **7.** $b \approx 8.9$ **8.** $a = 15$ **10.** about 270 ft **11.** 3.2 **12.** 6.7 **13.** 9.8 **14.** 3.2 **16.** about 2.7 **17.** about 4.5

Lesson 9-4 pages 430–434

ON YOUR OWN **1.** no **3.** no **5.** no **7.** 16 **9.** 20 **11.** $10\sqrt{2}$ **13.** $875\sqrt{7}$ **15.** 72 **17.** $3\sqrt{17}$ **19.** $4\sqrt{6}$ **21.** 10 **23.** $5\sqrt{10}$ **25.** $\frac{3}{2}$ **27.** $\frac{5}{2}$ **29.** $\frac{2\sqrt{30}}{11}$ **31.** $\frac{3\sqrt{2}}{4}$ **33.** $\frac{13}{12}$ **35.** 1.2 m **37a.** $\sqrt{18} \cdot \sqrt{10} = \sqrt{180} = \sqrt{36 \cdot 5} = 6\sqrt{5}$ **39.** D **41.** $\frac{2\sqrt{a}}{a^2}$ **43.** $ab^2c\sqrt{abc}$ **45.** $2y^2$ **47.** $\frac{\sqrt{2x}}{3}$ **49.** $2\sqrt{3}$ **51.** $\frac{9\sqrt{2}}{4}$ **53.** $5\sqrt{5}$ **55.** $\frac{8\sqrt{6}}{3}$

MIXED REVIEW **57.** 3; 2 **59.** -5; 3 **61.** For c = crab and t = turkey, $5c + 3t \leq 30$. **63.** -1 **65.** 10

Lesson 9-5 pages 435–439

ON YOUR OWN **1.** 42 **3.** $3\sqrt{2} + \sqrt{3}$ **5.** $6 - 2\sqrt{3}$ **7.** $3\sqrt{3} - 3\sqrt{2}$ **9.** $\sqrt{7}$ **11.** $3\sqrt{5} + 2\sqrt{3}$ **13.** $2\sqrt{3} + 2\sqrt{6} - 6$ **15.** $4 - 4\sqrt{2}$ **17.** 3.1 **19.** 15.8 **21.** 6.0 **23.** 8.5 **25.** -1.0 **27.** $2\sqrt{3} + 4\sqrt{2}$ **29.** -24 **31.** $6\sqrt{2} + 6\sqrt{3}$ **33.** $2\sqrt{17} + 17$ **37a.** $2\sqrt{6} + 4\sqrt{3}$ **37b.** Bill simplified $\sqrt{48}$ as $2\sqrt{24}$ instead of $2\sqrt{12}$ or $4\sqrt{3}$. **39.** $8\sqrt{2}$ **41.** $6\sqrt{10}$ **43.** E

MIXED REVIEW **45.** -3; 5; rational **47.** 1 **49.** 0

Lesson 9-6 pages 440–444

ON YOUR OWN **1.** 3 **3.** 27 **5.** 625 **7a.** 25 **7b.** $11\frac{1}{4}$ **9.** no solution **11.** $-\frac{1}{4}$ **13.** 0, 12 **15.** 7 **17.** 3 **19.** about 98.2 in.^{3} **21.** 2 **23.** 3, 6 **25.** no solution **27.** 4 **29.** 1, 6 **31a.** $V = 10x^2$ **31b.** $x = \frac{\sqrt{10V}}{10}$ **33.** -4 **35.** $-\frac{1}{2}$ **37.** none

MIXED REVIEW **39.** $\frac{1}{a^5b^5}$ **41.** $\frac{2}{xy^2}$ **43.** 0 **45.** 2

Lesson 9-7 pages 445–449

ON YOUR OWN **1.** $x \geq 2$ **3.** $x \geq 0$ **5.** $x \geq -7$ **7.** $x \geq -3$ **9.** $x \geq 5$ **11.** B **13.** A

15.
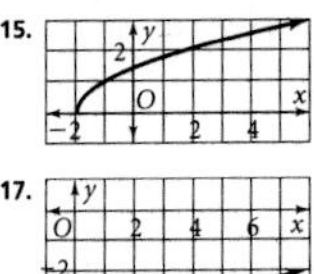

17.

19.
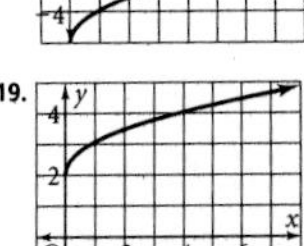

21.
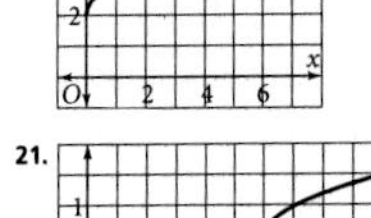

23.
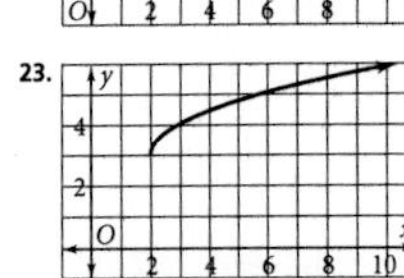

25.
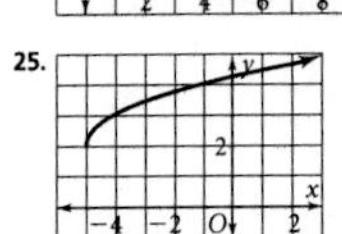

27. shift left 8 **29.** shift up 12

MIXED REVIEW

33.
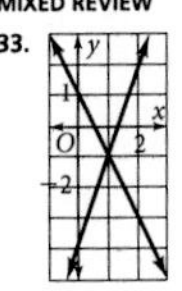

(1, −1) **35.** 7.8 **37.** -1.8

CHECKPOINT **1.** 9 **2.** 14 **3.** $8b^2\sqrt{b}$ **4.** $\sqrt{2}$ **5.** $3 + 4\sqrt{3}$ **6.** $2\sqrt{2}$ **7.** $\frac{x\sqrt{x}}{2}$ **8.** $2\sqrt{3}$ **9.** 68 **10.** 6 **11.** 8 **12.** 18

13.
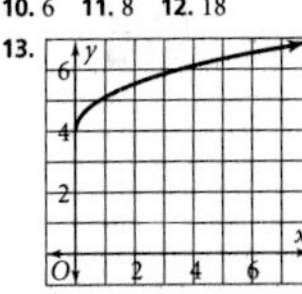

14.
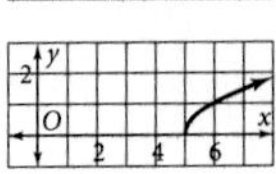

15.
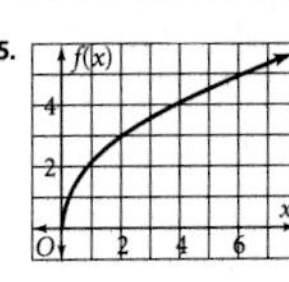

16.
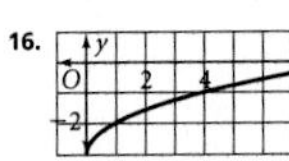

18. D

Toolbox page 450

1.
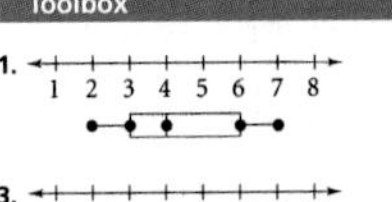

3.

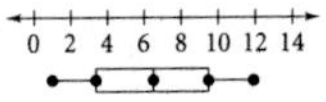

5.
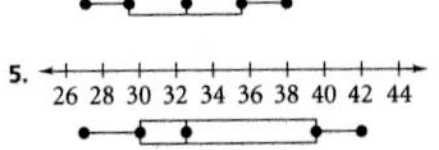

7. 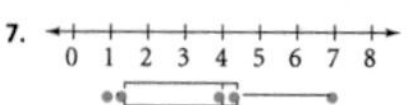

9b. 2; 33; 33 **9c.** 21 h; 23 h; both are much less than the typical 29-h week.

Lesson 9-8 pages 451–455

ON YOUR OWN

1.

x	$\bar{x}$	$x-\bar{x}$	$(x-\bar{x})^2$
5	5	0	0
3	5	−2	4
2	5	−3	9
5	5	0	0
10	5	5	25
		Sum:	38

standard deviation: 2.8

3.

x	$\bar{x}$	$x-\bar{x}$	$(x-\bar{x})^2$
3.5	4	−0.5	0.25
4.5	4	0.5	0.25
6.0	4	2	4
4.0	4	0	0
2.5	4	−1.5	2.25
2.5	4	−1.5	2.25
5.0	4	1	1
	Sum:		10

standard deviation: 1.2

5a. The data are spread far apart **5b.** Most data are near 25. **9a.** Sample: Lyman should have asked about the range and the standard deviation or the median salary and the mode. The range and the standard deviation would have told him that the salaries are widely distributed. The mode and the median, both \$265, would have told him that the mean does not represent the data well. **9b.** \$1165 **9c.** Large; all the data are far from the mean. **9d.** 321.36 **11.** 0.98 **13.** 25.47 **15a.** \$20 **15b.** 5.83 **15c.** 6; 39

MIXED REVIEW **17.** 3.47×10^{-3} **19.** 8.25×10^{-4} **21.** (1, 2) **23.** (5, 2) **25.** $x \geq 5$ **27.** \$14,961.04

Wrap-Up pages 457–459

1. 5.83 **2.** 17.80 **3.** 14.76 **4.** 9.85 **6.** (0.5, 5.5) **7.** (5.5, 6) **8.** (8.5, 3.5) **9.** (2.5, 13) **11.** $BC \approx 6.71$; $AC \approx 9.95$ **12.** $AC \approx 27.7$; $AB \approx 29.12$ **13.** $BC \approx 6.25$; $AB \approx 10.15$ **14.** C **15.** $48\sqrt{2}$ **16.** $\frac{2\sqrt{21}}{11}$ **17.** $20\sqrt{6}$ **18.** $\frac{10}{13}$ **19.** $10\sqrt{2}$ cm by $70\sqrt{2}$ cm **20.** $2\sqrt{7}$ **21.** $30\sqrt{5}$ **22.** $\sqrt{6}$ **23.** $2\sqrt{5}$ **24.** $10 - 10\sqrt{2}$ **25.** $25\sqrt{10}$ **26.** $18\sqrt{2}$ **27.** $17\sqrt{7}$ **28.** $\sqrt{29}$ in. **29.** 2 **30.** 16 **31.** 8 **32.** 81 **33.** 2.93 in.

34. $x \geq 0$; 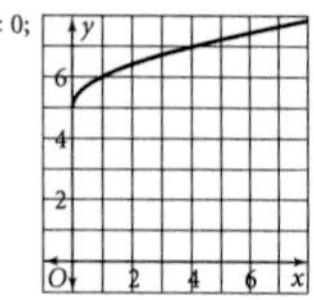

35. $x \geq 2$; 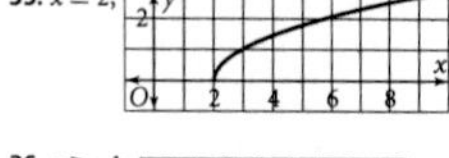

36. $x \geq -1$;

37. $x \geq 0$; 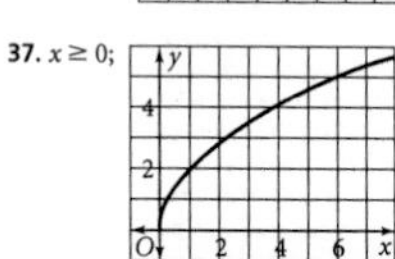

38. 5.44 **39.** 4.69 **40.** 6.15 **41.** 7.44 **42.** 0.64, −3.14 **43.** 1.90, −1.23 **44.** 4.70, −1.70 **45.** no solution **46.** 2, −1.2 **47.** 1.47, −1.75

Preparing for Standardized Tests page 461

1. B **3.** B **5.** A **7.** C **9.** A **11.** B **13.** D **15.** 85; 5.40 **17.** 31 or 32

CHAPTER 10

Lesson 10-1 pages 464–469

ON YOUR OWN **1.** C **3.** E **5.** A **7.** $4x + 9$; linear binomial **9.** $-11z^3 + 9z^2 + 5z - 5$; cubic with four terms **11.** $c^2 + 4c - 2$; quadratic trinomial **13.** $4q^4 + 3q^2 - 8q - 10$; fourth degree polynomial with four terms **15.** $x^2 - 1$ **17.** $-6x^3 + 3x^2 - x - 4$ **19.** $7a^3 + 11a^2 - 4a - 2$ **21.** $6x^2 - 12$ **25.** $8y$ **27.** $28c - 16$ **31.** $5x + 18$ **33.** $6y + 2$ **35.** $-2x^2 - 7x - 11$ **37.** $2r^2 - r - 7$ **39.** $9b^4 - 3b^2 + 7b - 6$

MIXED REVIEW **41.** 4^4 **43.** $\left(\frac{2}{3}\right)^5$ **45.** 5.0 **47.** 1.9 **49.** $2x - 6$ **51.** $15a - 21$

Lesson 10-2 pages 470–474

ON YOUR OWN **1.** $3x + 12$ **3.** $4x^2 + 2x$ **5.** $8x^2 + 12x$

7. 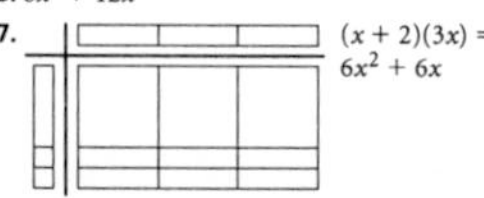 $(x + 2)(3x) = 6x^2 + 6x$

9. $8x + 28$ **11.** $-54x^4 + 36x^2 - 48x$ **13.** $-12a^3 + 15a^2 - 27a$ **15.** $6p^5 - 15p^3$ **17.** $-4x^2 + 13x$ **19.** $x^2 + x$ **21.** 3 **23.** 12 **25.** 5 **27.** $3x$ **29.** $5s$ **31.** $8x^2$ **35a.** $V = 64s^3$ **35b.** $V = 48\pi s^2$ **35c.** $V = 64s^3 - 48\pi s^2$ **35d.** $V = 16s^2(4s - 3\pi)$ **35e.** about 182,071 in.^3 **37.** $s(s^3 + 4s^2 - 2)$ **39.** $2x^2(1 - 2x^2)$ **41.** $7k(k^2 - 5k + 10)$ **43.** $3x(3 + 4x)$ **45.** $6m^2(m^4 - 4m^2 + 1)$ **47.** $m^2(5m - 7)$ **49.** $(x^2 - 7)(2x + 1)$ **51a.** 7; 4 **51b.** $n - 3$ **51c.** $\frac{1}{2}n^2 - \frac{3}{2}n$

MIXED REVIEW **53.** $\frac{4}{5}$ **55.** $\frac{3}{5}$ **57.** $\frac{3}{4}$ **59.** $28x^3 - 48x^2 + 128x$

Lesson 10-3 pages 475–479

ON YOUR OWN **1.** $x - 4$; $-x + 2$; $-x^2 + 6x - 8$ **3.** $x - 3$; $-x + 3$; $-x^2 + 6x - 9$ **5.** $x^2 - x - 20$ **7.** 30 **9.** 7 **11.** $x^2 + x - 42$ **13.** $2y^2 - y - 15$ **15.** $5y^2 + 12y - 32$ **17.** $a^3 - 6a^2 + 9a - 4$ **19.** $4t^3 - 22t^2 + 36t - 15$ **23.** $x^2 + 8x + 15$ **25.** $a^2 - 8a + 7$ **27.** $5a^2 + 13a - 6$ **29.** $-6y^2 + 20y - 14$ **33.** 1; rational **35.** $26 + 10\sqrt{5}$; irrational

MIXED REVIEW **37.** 2, −2 **39.** 2, −1 **41.** $(2x + 4)(3x + 5)$ **43.** $(-3x + 5)(4x + 1)$

CHECKPOINT **1.** $3x^2 + 5x - 4$ **2.** $2a^3 + a^2 - 7$ **3.** $-2m^2 - 5m + 8$ **4.** $9x^2 - x - 8$ **5.** $-w^2 + 9$ **6.** $-6t^3 - 4t^2 + 12t$ **7.** $m^2 + m - 12$ **8.** $b^2 - 9b + 18$ **9.** $-3c(c^2 - 5c + 1)$ **10.** $5a(2a^2 + a + 1)$ **11.** $4p(2p^2 - 5p - 6)$ **12.** $x(x^2 + 4x + 7)$

Toolbox page 480

1. **3.** 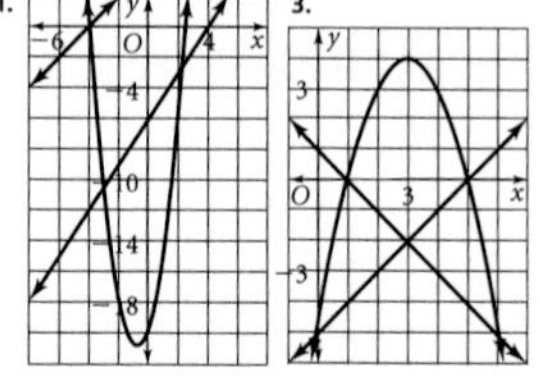

5. 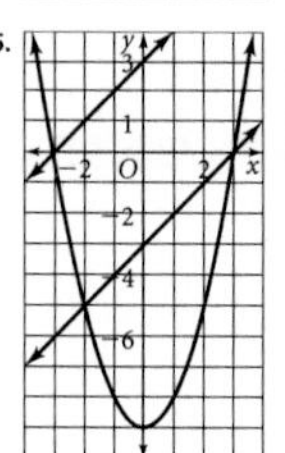

7. The x-intercepts are the same. **9.** $-x + 2$; $x - 2$; $-x^2 + 4x - 4$ **11.** $-2x - 6$; $\frac{1}{2}x - 1$; $-x^2 - x + 6$

Lesson 10-4 pages 481–485

ON YOUR OWN **1.** $x + 3$; $x + 2$; $x^2 + 5x + 6$ **3.** $2 + x$; $x + 2$; $x^2 + 4x + 4$ **5.** No; no two numbers with product 8 can have sum 5. **7.** Yes; $(x + 8)(x + 1) = x^2 + 9x + 8$

9. **11.** 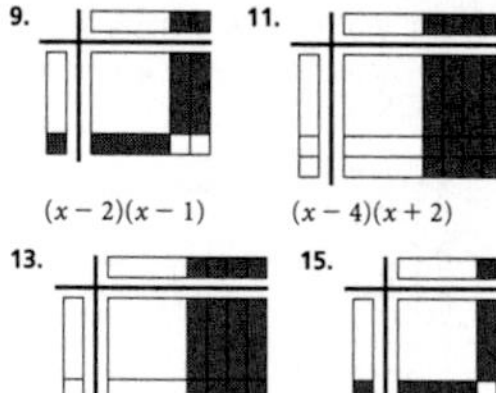

$(x - 2)(x - 1)$ $(x - 4)(x + 2)$

13. **15.**

$(x - 4)(x + 1)$ $(x - 1)(x - 1)$

17. 2 **19.** −1 **21.** −3 **23a.** $x^2 + 15x + 56$ **23b.** $(x + 7)(x + 8)$ **25.** $(a - 2)(a - 3)$ **27.** $(k + 1)(k + 8)$ **29.** $(r - 11)(r + 1)$ **31.** $(x + 5)(x - 3)$ **33.** $(x + 5)(x + 7)$ **35.** $(a - 7)(a - 2)$ **37.** $(y + 16)(y - 3)$ **39.** $(w - 6)(w + 4)$ **43.** $(2x - 1)(x - 7)$ **45.** $(2x - 3)(x + 1)$ **47.** $(2x + 3)(x - 7)$ **49.** $(2x - 1)(x - 3)$ **51.** $(7x + 1)(x - 3)$ **53.** $(3x + 5)(x + 4)$

MIXED REVIEW **55.** 5 **57.** about 8.1

59. 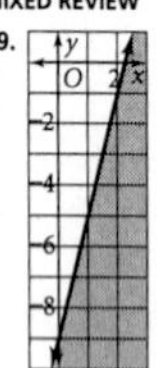

61. 84 **63.** $4x^2 + 12x + 9$ **65.** $x^2 - 64$ **67.** $4x^2 - 49$ **69.** $4x^2 - 4x + 1$

Lesson 10-5 pages 486–490

ON YOUR OWN **1.** $x^2 + 10x + 25$; $x + 5$; $x + 5$ **3.** $9x^2 + 6x + 1$; $3x + 1$; $3x + 1$ **5.** $(x + 1)^2$ **7.** $(x - 9)^2$ **9.** $3(x - 1)^2$ **11.** $6(x + 5)(x - 5)$ **13.** $(x + 7)(x - 7)$ **15.** $(2x - 1)^2$ **17.** $(3x + 1)^2$ **19.** $(x + 20)(x - 20)$ **23a.** $3.14n^2 - 3.14m^2$; $3.14(n + m)(n - m)$ **23b.** 285.74 in.^2 **25.** 891 **27.** 1591 **29.** 384 **31.** 396 **33.** $\left(\frac{1}{2}m + \frac{1}{3}\right)\left(\frac{1}{2}m - \frac{1}{3}\right)$ **35.** $\left(\frac{1}{3}n + \frac{1}{5}\right)\left(\frac{1}{3}n - \frac{1}{5}\right)$ **37a.** $t - 3$; 4 **37b.** $(t + 1)(t - 7)$

MIXED REVIEW **39.** $5 + 6\sqrt{2}$ **41.** $\sqrt{41}$ **43.** 30% **45.** $-\frac{4}{3}$ **47.** $\frac{5}{3}$

Lesson 10-6 pages 491–495

ON YOUR OWN **1.** 3, 7 **3.** $-\frac{2}{7}, -\frac{4}{5}$ **5.** −2, 7 **7.** 5, 11 **9.** −4, 3 **11.** 0, 10 **13.** $-4, \frac{2}{5}$ **15.** −1, 10 **17.** 5 cm **19.** $x^2 - 10x + 24 = 0$; 4, 6 **21.** $2q^3 + 22q + 60 = 0$; −6, −5 **23.** $6n^2 - 5n - 4 = 0$; $-\frac{1}{2} \cdot \frac{4}{3}$ **25.** $20p^2 - 80 = 0$; −2, 2 **27.** $4n^2 - 100 = 0$; 5, −5 **29.** $9c^2 - 36 = 0$; 2, −2 **31.** 5 in. **33a.** 10 ft base; 22 ft height **33b.** about 24.2 ft **35.** 0, 4, 6 **37.** 0, 3 **39.** 0, 1, 9 **41.** 3, 4, 5 **43.** 0, −1 **45.** 8 in. × 10 in.

MIXED REVIEW **47.** $\frac{1}{3y^3}$ **49.** 1 **51.** $\frac{25}{3}$ **53.** 60 students **55.** $\frac{1}{5}, -\frac{3}{2}$

CHECKPOINT **1.** $(x - 3)(x - 2)$ **2.** $(n - 3)^2$ **3.** $(g - 10)(g + 2)$ **4.** $(3x + 7)(3x - 7)$ **5.** $-3, \frac{2}{3}$ **6.** $6, -\frac{3}{2}$ **7.** −7, −2 **8.** 2, −2 **9.** $-\frac{4}{3}$ **10.** $\frac{3}{2}, -\frac{2}{5}$ **11.** A

Toolbox page 496

1. 4 **3.** 49 **5.** Sample: Divide the coefficient of the first degree term in half and square the result. **7.** −6, −2 **9.** −8, 2 **11.** −6, 2 **13.** 9, −5 **15.** $-5 \pm \sqrt{35}$ **17.** $-4 \pm \sqrt{13}$

Lesson 10-7 pages 497–501

ON YOUR OWN **1.** −0.08, −13 **3.** 1.25, −1.33 **5.** no real solutions **7.** 3, −3 **9.** 2, −2 **11a.** $-\frac{1}{2}, 3$ **11b.** $-\frac{1}{2}, 3$ **11c.** $(2x + 1)(x - 3) = 0$, so $x = -\frac{1}{2}$ or $x = 3$. **13.** −4, 2 **15.** no real solutions **17.** no real solutions **19.** 4 **21.** $0, \frac{4}{5}$ **23.** −1.85, 4.85 **25.** 1.75 **27.** $\frac{13}{7}, -\frac{13}{7}$ **29.** B **31.** about 167.7 ft

MIXED REVIEW **35.** 13; 28 **37.** the parabola $y = x^2$ shifted 2 units right, 1 unit down; $x = 2$; 3; 1 and 3.

Wrap-Up pages 503–505

1. $7x^3 + 8x^2 - 3x + 12$ **2.** $8g^4 - 5g^2 + 11g + 5$ **3.** $-b^5 + 2b^3 + 6$ **4.** $t^3 - 5t^2 + 12t - 8$ **6.** $-40x^2 + 16x$ **7.** $35g^3 + 15g^2 - 45g$ **8.** $-40t^4 + 24t^3 - 32t^2$ **9.** $5m^3 + 15m^2$ **10.** $-6w^4 - 8w^3 + 20w^2$ **11.** $-3b^3 + 5b^2 + 10b$ **12.** $3x$; $3x(3x^3 + 4x^2 + 2x)$ **13.** $4t^2$; $4t^2(t^3 - 3t + 2)$ **14.** $10n^3$; $10n^3(4n^2 + 7n - 3)$ **15.** $x^2 + 8x + 15$ **16.** $15x^2 - 29x - 14$ **17.** $6x^2 + 11x - 10$ **18.** $-x^2 + 5x - 4$ **19.** $x^3 + 3x^2 + 3x + 2$ **20.** $4x^2 - 21x + 5$ **21.** $x^3 - 9x^2 + 18x + 8$ **22.** $3x^2 + 10x + 8$ **23.** $(2x + 1)(x + 4)$; $2x^2 + 9x + 4$ **24.** $(x + 2)(x + 1)$ **25.** $(y - 7)(y - 2)$ **26.** $(x - 5)(x + 3)$ **27.** $(2w - 3)(w + 1)$ **28.** $(-b + 4)(b - 3)$ **29.** $(2t - 1)(t + 2)$ **30.** $(x - 1)(x + 6)$ **31.** $2(3x + 2)(x + 1)$ **32.** A **33.** $(q + 1)^2$ **34.** $(b - 4)(b + 4)$ **35.** $(x - 2)^2$ **36.** $(2t + 11)(2t - 11)$ **37.** $(2d - 5)^2$ **38.** $(3c + 1)^2$ **39.** $(3k + 5)(3k - 5)$ **40.** $(x + 3)^2$ **41.** The factors are equal. **42.** −3, −4 **43.** 0, 2 **44.** 4, 5 **45.** $-3, \frac{1}{2}$ **46.** $-\frac{2}{3}, 1\frac{1}{2}$ **47.** 1, 4 **48.** 10 ft by 17 ft **49.** −10, 10 **50.** −4.12, 0.78 **51.** 0.18, 2.82 **52.** 324 ft^2 **53.** $\frac{57}{10}$ **54.** $-\frac{33}{4}$ **55.** $\frac{314}{100}$ **56.** $\frac{52}{5}$ **57.** $-\frac{1849}{1000}$ **58.** $\frac{767}{100}$ **59.** $\frac{1}{3}$ **60.** $-1\frac{1}{2}$ **61.** $-\frac{3}{7}$ **62.** $\frac{10}{13}$ **63.** $\frac{15}{2}$ **64.** $\frac{7}{3}$ **65.** $\frac{55}{6}$ **66.** 26

Cumulative Review page 507

1. D **3.** D **5.** B **7.** D **9.** C **11.** A **13.** 5 and 9 **15.** 5 m by 7 m

CHAPTER 11

Lesson 11-1 pages 510–514

ON YOUR OWN **1a.** 16 h; 10 h; 8 h; 4 h **1b.** hourly wage and time **1c.** $xy = 80$ **3.** 8 **5.** 7 **7.** 96 **9.** $\frac{1}{6}$ **11.** 20 **13.** 130 **15.** D **17.** 1.1 **19.** 2.5 **21.** 375 **23a.** 120; distance from your house to the lake **23b.** 3 h **25.** 2.4 da **27.** $y = 70x$ **29.** $xy = 48$ **31.** Direct; as s increases, P increases. **33.** Inverse; as r increases, t decreases.

MIXED REVIEW **37.** $(x + 5)^2$ **39.** $(x - 3)^2$ **41.** $\frac{1}{5}$ **43.** $\frac{3}{8}$ **45.** −1

Toolbox page 515

1. 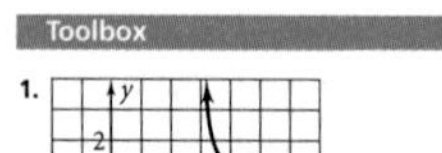

3. 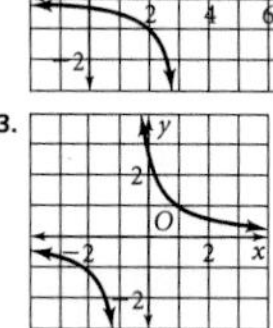

5. 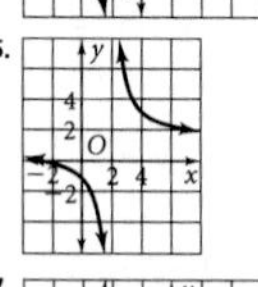

7. 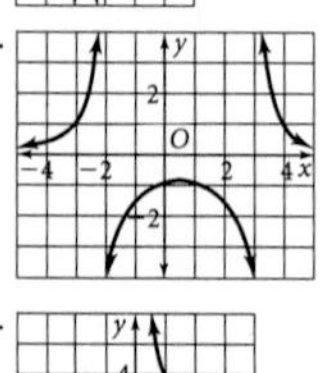

9. 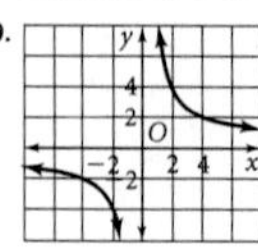

11.

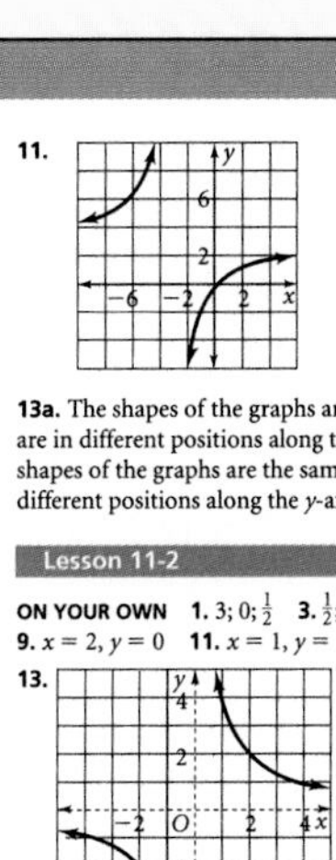

13a. The shapes of the graphs are the same, but they are in different positions along the x-axis. 13b. The shapes of the graphs are the same, but they are in different positions along the y-axis.

Lesson 11-2 pages 516–521

ON YOUR OWN **1.** 3; 0; $\frac{1}{2}$ **3.** $\frac{1}{2}$; −4; 8 **5.** 0 **7.** −2 **9.** $x = 2, y = 0$ **11.** $x = 1, y = -1$

13.

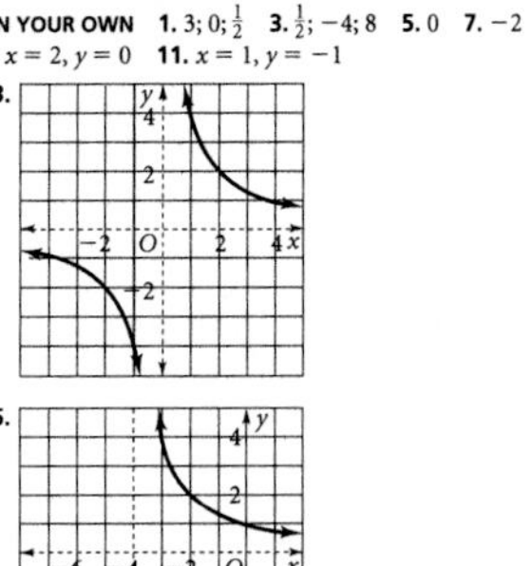

15.

17.

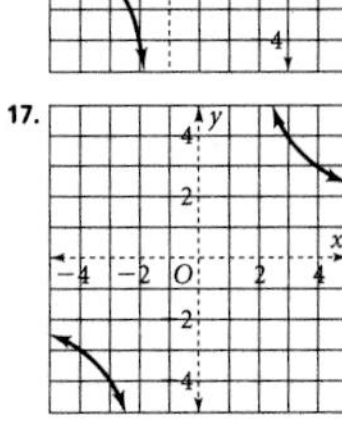

19.

21.

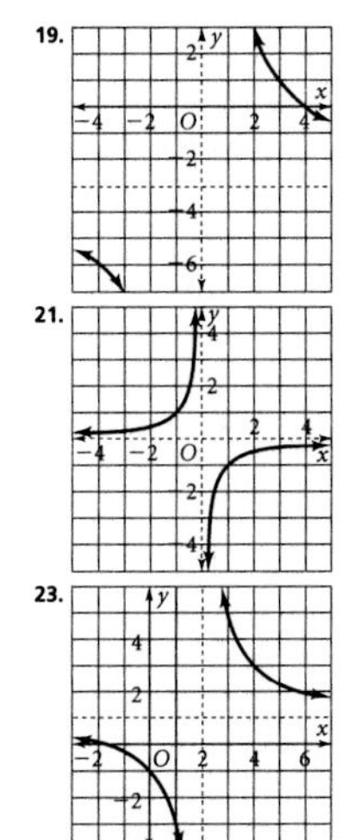

23.

25a.

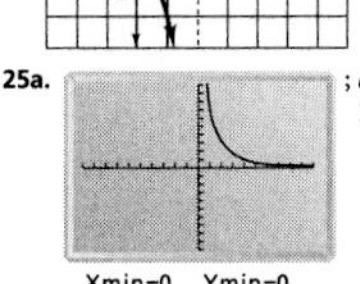

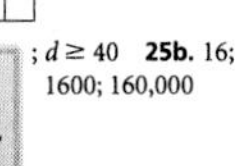

Xmin=0 Ymin=0
Xmax=6 Ymax=2000
Xscl=1 Yscl=200

; $d \geq 40$ **25b.** 16; 1600; 160,000

29.

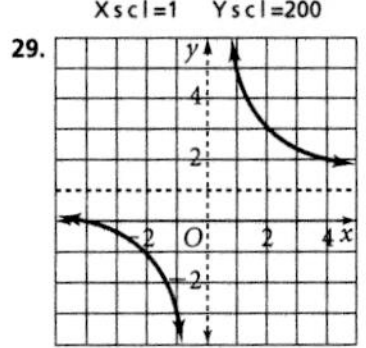

31.

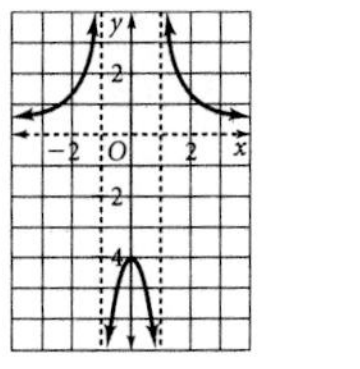

33.

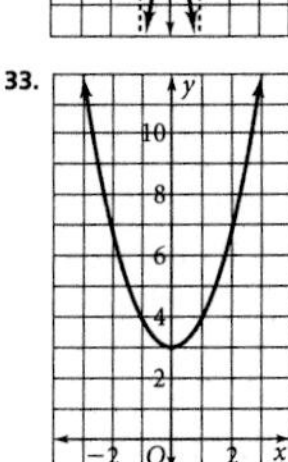

35.

37.

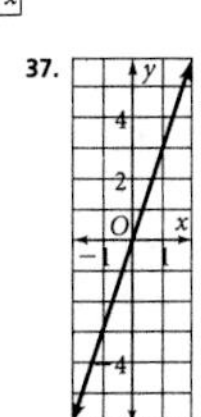

39.

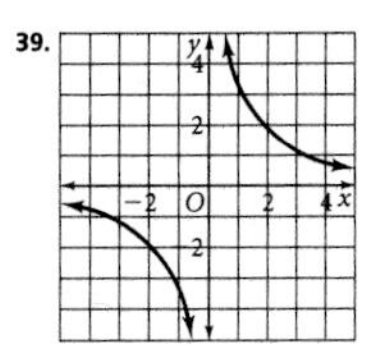

41.

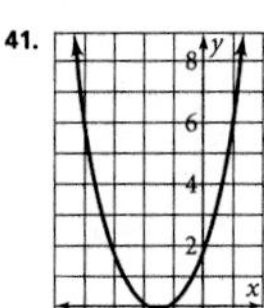

43.

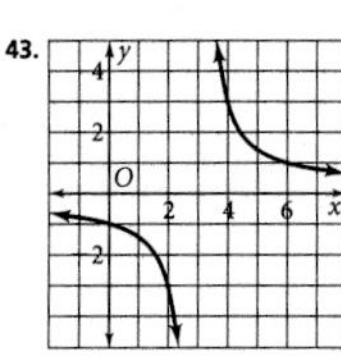

MIXED REVIEW **45.** $(g - 7)(g - 5)$ **47.** $(a + 7)(a - 1)$ **49.** \$10 **51.** 0, 3 **53.** −4, 4

CHECKPOINT **1.** 3 **2.** 2 **3.** 4

4.

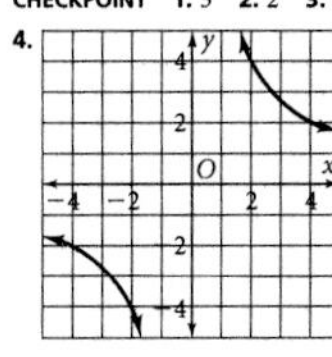

5.

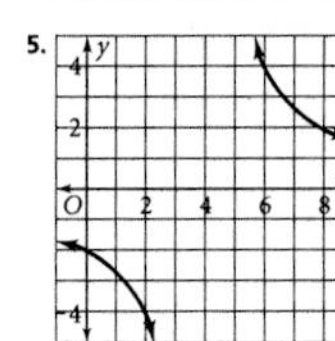

6.

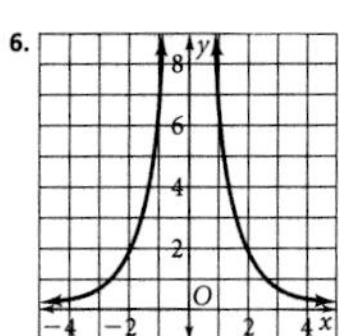

7.

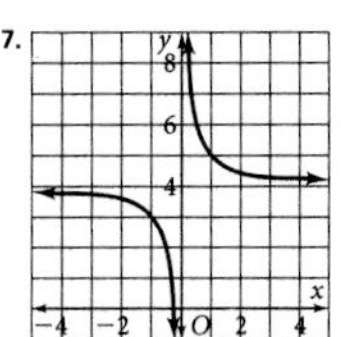

9. 91 lb

Lesson 11-3 pages 522–526

ON YOUR OWN **1.** $\frac{a}{4}$; $a \neq 0$ **3.** $\frac{x-2}{x+2}$; $x \neq -2$ **5.** $\frac{2a+3}{4}$ **7.** $\frac{1}{7x}$; $x \neq 0$ **9.** $\frac{1}{2}$; $p \neq 12$ **11.** $\frac{2}{3}$; $x \neq 0$, −1 **13.** $3x$; $x \neq 3$ **15.** $\frac{w}{w-7}$; $w \neq -7, 7$ **17.** $\frac{x^3-9}{x+3}$ does not have −3 in its domain. **21.** $-\frac{125}{16}$ **23.** $\frac{12}{x}$ **25.** $2x$ **27.** $\frac{3x^4}{2}$ **29.** 6 **31.** $-\frac{1}{2}$ **33.** $\frac{11}{7k-15}$ **35.** $\frac{\frac{4}{3}\pi r^3}{\pi r^2} = \frac{r}{3}$ **37.** $\frac{5w}{5w+6}$ **39.** $\frac{3y}{4(y+4)}$

MIXED REVIEW **41.** 3^{x+5} **43.** $-12m^4 - 18m$ **45.** $\frac{2}{3}$ **47.** $\frac{3}{5}$

Lesson 11-4 pages 527–531

ON YOUR OWN **1.** $\frac{7}{5}$ **3.** $-\frac{2}{b}$ **5.** $\frac{4}{11g}$ **7.** $\frac{2}{d}$ **9.** $\frac{5n}{7}$ **11.** $\frac{26r}{5}$ **13.** $4x^2$ **15.** $2m^2$ **17.** $6t^5$ **19.** $46d^4$ **21.** $5a$ **23.** $35b^3$ **25.** $\frac{35a^3+6}{15a^4}$ **27.** $\frac{10+20x^2}{15x^8}$ **29.** $\frac{b^2-4}{2b^4}$ **31.** $\frac{9+2m}{24m^3}$ **33.** $-\frac{77}{36k}$ **35.** $\frac{45+36x^2}{20x^2}$ **37a.** $\frac{17}{7r}$ **37b.** about 0.81 h, or 49 min **39.** $\frac{10}{m+1}$

41. $\frac{m+2}{m+3}$ **43.** $\frac{2}{t^2+1}$ **45.** $\frac{s+2}{4s^2+2}$ **47.** $\frac{1}{m+1}$ **49.** $\frac{5y-11}{(y+2)(y-3)}$

MIXED REVIEW **53.** about 12.2 **55.** about 10.8 **57.** 4 **59.** 6

Lesson 11-5 pages 532–536

ON YOUR OWN **1a.** 32 **1c.** no **3.** 12 **5.** −1 **7.** 2 **9.** no solution **11.** $-3\frac{1}{9}$ **13.** $-\frac{1}{3}$ **15.** $22\frac{1}{2}$ min **17.** 2 **19.** $\frac{1}{2}$ **21.** −2, 4 **23.** $5\frac{1}{3}$ **25.** 6 **27.** 5 **29.** A **31.** $a = 4$; $b = \frac{7}{27}$; $c = 11$; $d = -\frac{1}{3}$

MIXED REVIEW **33.** −2, 2 **35.** −5.5 **37.** 1, 2, 3, 4, 5, 6 **39.** H1, H2, H3, H4, H5, H6, T1, T2, T3, T4, T5, T6

CHECKPOINT **1.** $\frac{4}{m}$; $m \neq 0$ **2.** $6(x - 2)$; $x \neq -2$ **3.** $\frac{c+3}{c-3}$; $c \neq 3$ **4.** $\frac{3z+12}{z^3}$ **6.** $\frac{6}{7}$ **7.** −9 **8.** $\frac{5}{2}$ **9.** −0.74, 10.74 **10.** $\frac{9}{c}$ **11.** $\frac{5}{x-3}$ **12.** $\frac{2(7m-6)}{(m+2)(m-3)}$ **13.** $\frac{6t+3}{t^2}$ **14.** B

Toolbox page 537

1. symmetric **3.** reflexive **5a.** division property of equality **5b.** distributive property **5c.** subtraction property of equality

Lesson 11-6 pages 538–542

ON YOUR OWN **1.** 10 choices **3.** 30 ways **5a.** 24 different numbers **5b.** 6 even numbers **5d.** $\frac{1}{4}$ **7.** 210 **9.** 60 **11.** 6 **13.** 5040 **15.** 840 **17.** 7 **21.** 40,320 orders **23.** ${}_9P_2$ **25.** C **27a.** 2520 **27b.** $\frac{1}{2520}$

MIXED REVIEW

29.

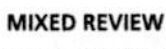

31.

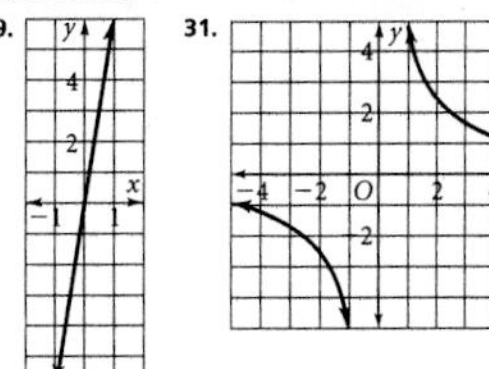

33. campsite \$20; family trail pass \$5 **35.** $\frac{5}{72}$

Lesson 11-7 pages 543–547

ON YOUR OWN **1a.** 14,886,690 different arrangements **1b.** about 1.94×10^{17} combinations **5.** 6 **7.** 20 **9.** 6 **11.** 1365 **13.** 56 **15.** 35 **17a.**

17b–c. 6 segments

17d. 45 segments **19.** 126 different combinations **21.** Permutations; the order of videos is important. **23.** Combinations; the order of committee members is not important. **25.** 10 **27.** 35 **31a.** 792 different choices **31b.** 36 ways **31c.** $\frac{1}{22}$ **33a.** 45 possible combinations **33b.** 3 combinations **33c.** $\frac{1}{15}$ **33d.** $\frac{2}{15}$

MIXED REVIEW **35.** 12 **37.** 5040 **39.** 9 **41.** $\frac{4x^2+5}{3x}$ **43.** $\frac{13x-8}{(x+2)(x-2)}$ **45.** $-\frac{27}{20n}$ **47a.** about 1143 times **47b.** 6 t

Wrap-Up pages 549–551

1. 6 **2.** 9 **3.** 4.4 **4.** no **5.** yes; $xy = -10.6$

6.

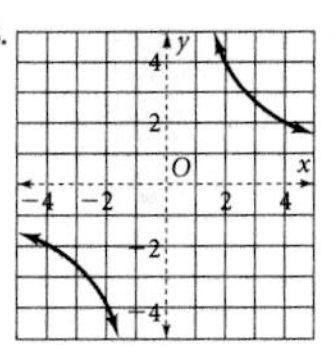

7.

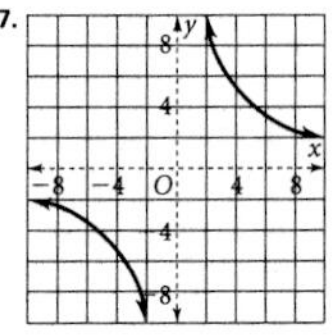

8.

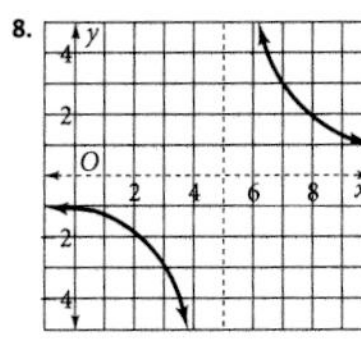

9.

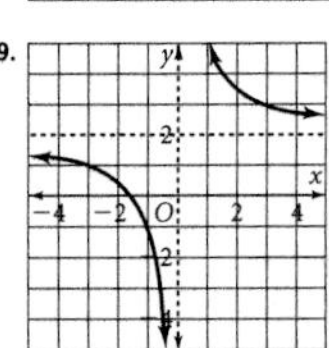

12. $x - 2$; $x \neq -2$ **13.** $\frac{x}{4x+3}$; $x \neq -\frac{3}{4}$ **14.** $\frac{-3}{t(t-1)}$; $t \neq 0, 1$ **15.** $\frac{1}{2z+3}$; $z \neq -2, 1\frac{1}{2}$ **16.** $\frac{24m}{(m-3)(m+1)}$ **17.** $\frac{1}{2(2e+1)}$ **18.** $\frac{5c+3}{2c(c-1)}$ **19.** $3n + 6$ **20.** $\frac{8}{k}$

21. $\frac{8x-4}{x-7}$ **22.** $\frac{15x^2+13x+27}{(3x-1)(2x+3)}$ **23.** $\frac{-3m+10}{(m-1)(m+1)}$

24. C **25.** 24 **26.** 9 **27.** −14 **28.** 6 min **29.** 60 **30.** 1680 **31.** 360 **32.** 20 **33.** 3024 **34.** 42 **35a.** 160 possible area codes **35b.** 640 new area codes **36.** 6 **37.** 210 **38.** 36 **39.** 165 **40.** 5 **41.** 15 **42.** 4 different combinations **43.** 1,814,400 ways **44.** $\frac{3}{28}$

Preparing for Standardized Tests page 553

1. D **3.** C **5.** D **7.** B **9.** A **11.** B **13.** $(2x-3)^2$ **15.** $x = 2, y = 1; -3$

Cumulative Review pages 554–555

1. E **3.** E **5.** A **7.** D **9.** D **11.** A **13.** B **15.** D **17.** A **19.** B **21.** D **23a.** $2\frac{1}{4}$ **29.** $-\frac{13}{3}$ **31.** $1701 **33a.** 5.5 **35b.** 8 **35c.** about 2.4

EXTRA PRACTICE

Chapter 1 page 556

1. 39.375; 38; 35 **3.** 6.3; 6; 5 and 8 **5a.** c = length of a car; $3c = 24$ **5b.** b = weight of a book, p = weight of a pen; $2b = 39p$ **7.** 28 **9.** 9 **11.** $4\frac{1}{9}$ **13.** −27.7 **15.** $-\frac{1}{2}$ **17.** −26 **19.** $10\frac{1}{2}$ **21.** −20 **23.** −59 **25.** 18 **27.** $\frac{7}{27}$ **29.** $\frac{22}{27}$ **31.** $\frac{23}{54}$ **33.** $\begin{bmatrix}-7 & 6\\ 3 & 8\end{bmatrix}$ **35.** $\begin{bmatrix}-1.6 & 4.1\\ 1.5 & 4.6\end{bmatrix}$

Chapter 2 page 557

1a.

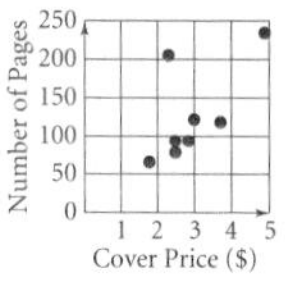

1b. The number of pages increases as the price increases. **1c.** About 75 pages; you can find the y-coordinate of the point on the line with x-coordinate 2.00.

7. $f(x) = 2x$

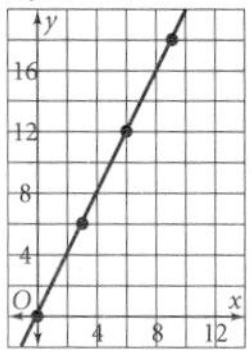

9. $f(x) = \frac{1}{2}x$

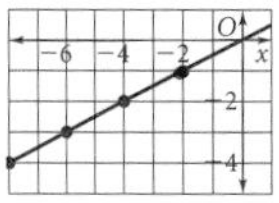

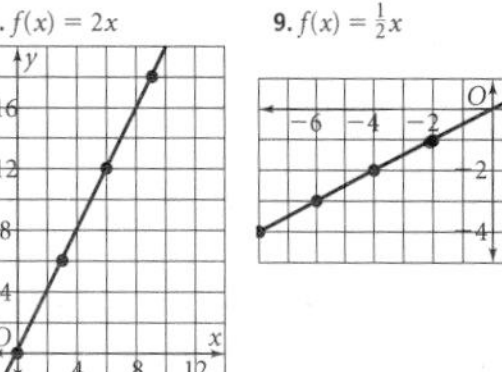

11. {−2, −1, 1, 2, 6}; absolute value **13.** {6, $7\frac{1}{2}$, 8, $9\frac{1}{2}$, 12}; linear **15.** {$-\frac{8}{3}$, $-\frac{2}{3}$, 0, 2, $\frac{16}{3}$}; linear **17.** {0, 3, 7, 52, 88}; quadratic **19.** {1, 4, 5, 8, 13}; linear **21.** {0, $\frac{3}{5}$, $\frac{9}{5}$, $\frac{12}{5}$, $\frac{24}{5}$}; absolute value **23.** 0 **25.** $\frac{2}{3}$ **27.** $\frac{1}{3}$ **29.** $\frac{1}{3}$

Chapter 3 page 558

1. 14 **3.** 8 **5.** −5.7 **7.** 2 **9.** 24 **11.** $\frac{18}{49}$ **13.** −2 **15.** 1 **17.** 10 **19.** −6 **21.** n = next test score; $\frac{87+84+85+n}{4} = 90$; no **23.** $\frac{5}{128}$ **25.** $\frac{9}{128}$ **27.** $\frac{1}{20}$ **29.** $\frac{1}{16}$ **31.** $n = 0.1 \times 94$; 9.4 **33.** $147 = 0.14 \times n$; 1050 **35.** $13.2 = 0.55 \times n$; 24 **37.** $n = 0.75 \times 68$; 51 **39.** $n \times 54 = 28$; 52% **41.** $n \times 20 = 31$; 155% **43.** about 11% increase **45.** about 33% decrease **47.** 38% decrease **49.** about 3% increase **51.** 125% increase **53.** about 11% increase **55.** about 46%

Chapter 4 page 559

1. −8 **3.** 48 **5.** $\frac{4}{3}$ **7.** $\frac{155}{2}$ **9.** $\frac{5}{3}$ **11.** 3 **13.** 8 ft × 12 ft **15.** 1 **17.** 1 **19.** −40 **21.** 9, −15 **23.** 5, −5 **25.** $-\frac{1}{3}$ **27.** identity **29.** The bus traveled 5 h; the car traveled 4 h. **31.** $cv - w$ **33.** $\frac{P-2w}{2}$ **35.** $m(b-a)$ **37.** $\frac{gr^2}{1.9}$ **39.** $p \le 8$; −8 −4 0 4 8 12

41. $y \le -1\frac{3}{7}$; −4 −2 0 2 4

43. $h > -12$; −12 −8 −4 0 4 8

45. $s \le 18$; −12 −6 0 6 12 18

47. $l \ge -3.5$; −4 −3 −2 −1 0 1

49. $m \le -12$ or $m \ge 4$; −12 −8 −4 0 4 8

51. $-2.5 \le n \le 14.5$; −4 −2 0 2 4 6 8 10 12 14 16

53. $a < -11$ −12 −8 −4 0 4 8

55. $j \ge 5$; 0 2 4 6 8 10

57. no more than five balls

59. {1, 2, 3}; −2 −1 0 1 2 3

61. integers > 0; −2 −1 0 1 2 3 4

63. integers ≥ −11; −12 −11 −10 −9

65. integers ≥ 11; 0 4 8 12

Chapter 5 page 560

1. $\frac{3}{2}$; $y = \frac{3}{2}x + 2$ **3.** 1; $y = x + 6$ **5.** $\frac{5}{2}$; $y = \frac{5}{2}x + 5\frac{1}{2}$ **7.** $-\frac{1}{7}$; $y = -\frac{1}{7}x$ **17.** 0.2 m/yr **19.** 0.2 blocks/min

9. $y = -5x + 26$

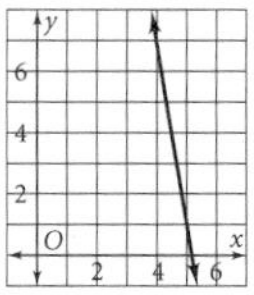

11. $y = \frac{1}{2}x + 1$

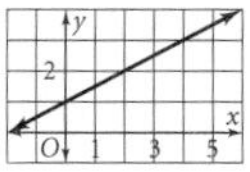

13. $y = 2x + 11$

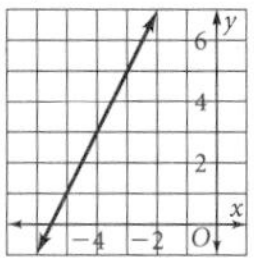

15. $y = 7x - 28$

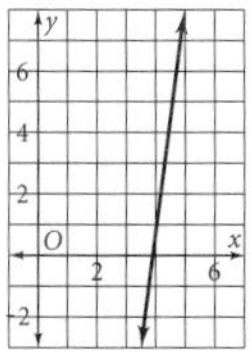

21. $y = \frac{4}{5}x$

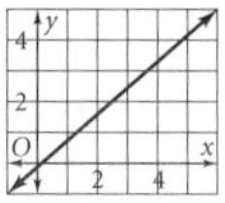

23. $y = \frac{10}{3}x$

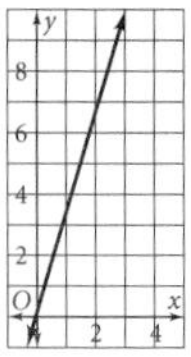

25. $y = -\frac{9}{2}x$

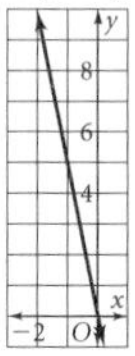

27. $y = -\frac{1}{4}x$

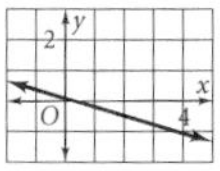

29. $y = \frac{4}{3}x$

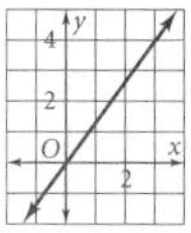

31. $y = -x$

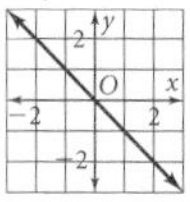

33. 6; 8;

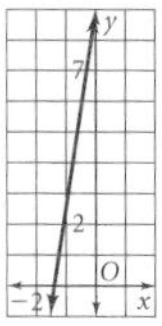

35. $-\frac{5}{2}$; 6;

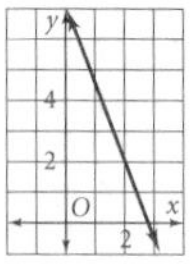

37. $-\frac{3}{4}$; −8;

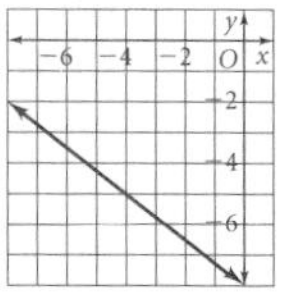

39. $\frac{1}{2}$; 3;

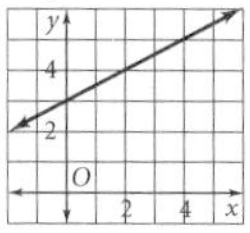

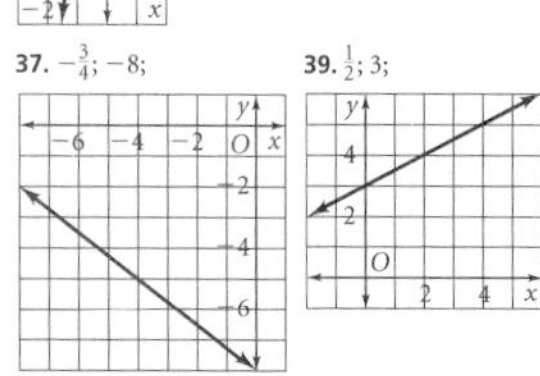

41a–b.

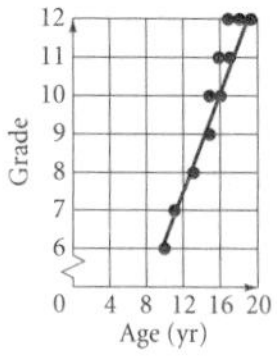

41c. $y = 0.7203196347x - 1.117579909$ **43.** $y = x$ **45.** $y = 2x + 3$ **47.** $y = \frac{5}{2}x - 6$ **49.** 1 **51.** −2 **53.** −3 **55.** 19

Chapter 6 page 561

1. (4, −3) **3.** (11, 7) **5.** (1, −1) **7.** (6, 13) **9.** (4, −5) **11.** (4, −9) **13.** $n + p = 12$, $5n + p = 32$

15.

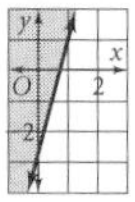

17.

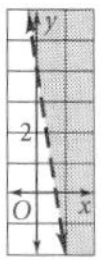

19.

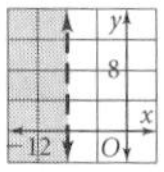

21.

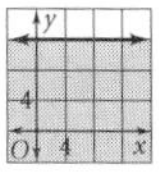

23.

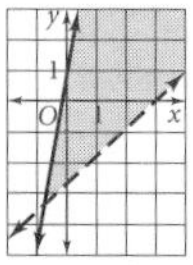

25.

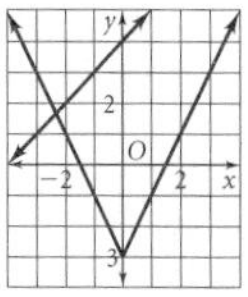

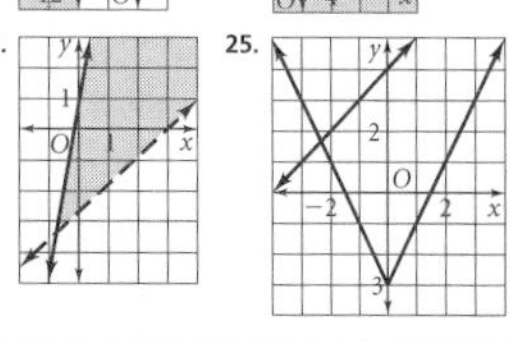

27.

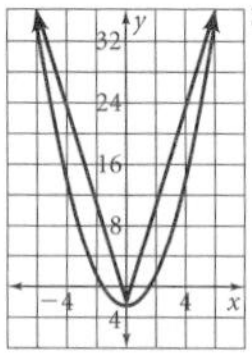

29.

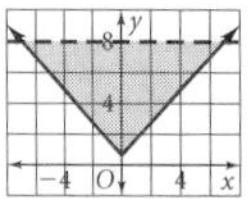

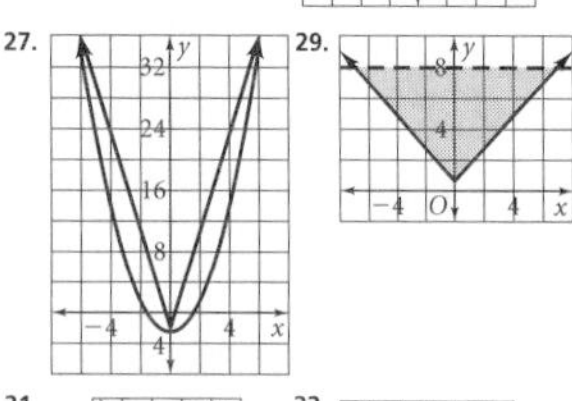

31.

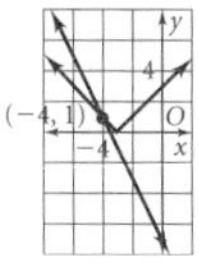

33.

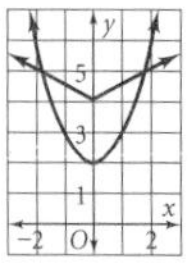

35. 30; 0

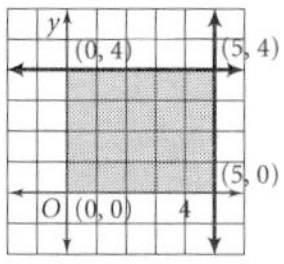

37. 29; 8

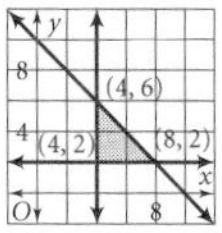

Chapter 7 page 562

1. narrower **3.** opens down, wider **5.** down 4 **7.** narrower, up 5

9.

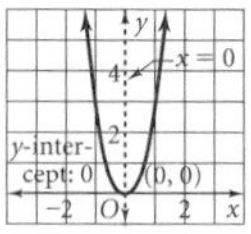

11.

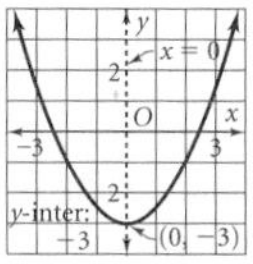

13.

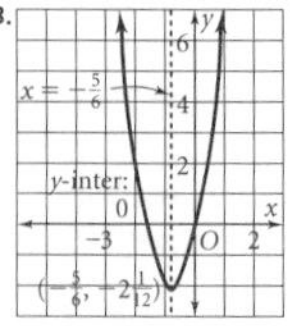

15.

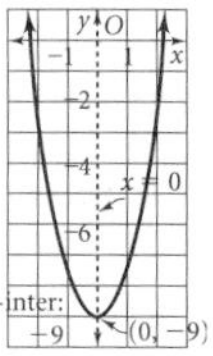

17.

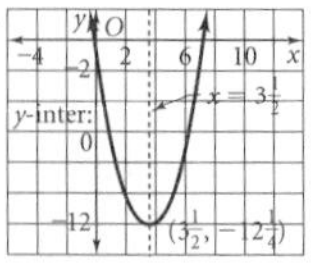

21.

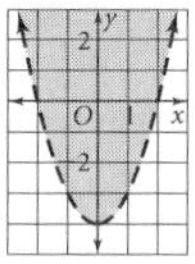

23. **25.**

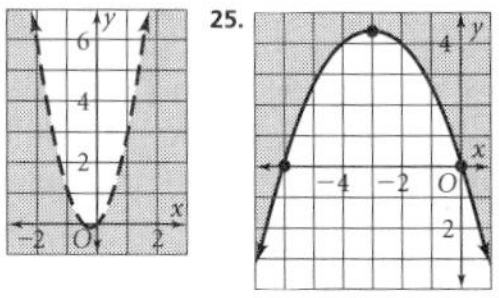

27. 5, −5 **29.** 8, −8 **31.** 0.9, −0.9 **33.** 1.5, −1.5 **35.** $\frac{1}{5}$, $-\frac{1}{5}$ **37.** $\frac{1}{3}$, $-\frac{1}{3}$ **39.** 6, −6 **41.** 10, −10 **43.** 14, −6 **45.** 13, −13 **47.** no real solution **49.** 6.14, −1.14 **51.** 1.71, 0.29 **53.** 8, −2 **55.** 3.65, −1.65 **57.** −19; 0 **59.** 57; 2 **61.** −24; 0 **63.** −27; 0 **65.** 21; 2

Chapter 8 page 563

1. exponential growth **3.** exponential growth

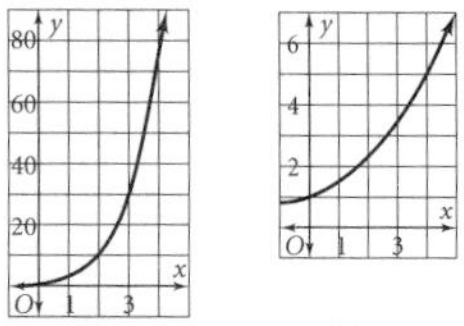

5. exponential growth **7.** exponential decay

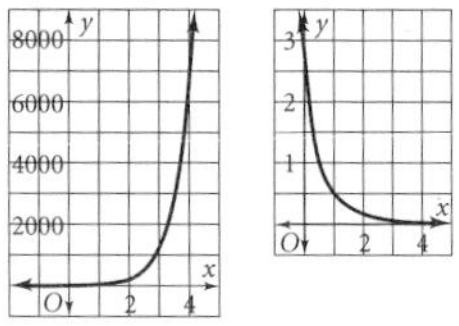

9. exponential growth

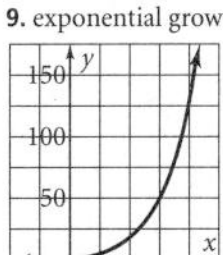

11. exponential growth

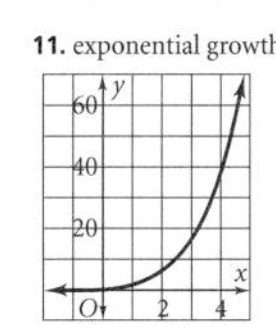

13. 8 **15.** $\frac{1}{2}$ **17.** 0.65 **19.** $\frac{1}{4}$ **21.** $y = 200 \cdot 1.04^x$; \$243.33 **23.** $y = 3000 \cdot 0.92^x$; \$2336.06 **25.** $\frac{5}{m^3}$
27. $\frac{m^{14}}{j^{10}}$ **29.** $w^6 j^{22}$ **31.** $9n^8$ **33.** a^4 **35.** $6t^4$
37. $\frac{2r^{15}}{t^7}$ **39.** $\frac{1}{c^{12}}$ **41.** $\frac{1}{9}$ **43.** 144 **45.** 16 **47.** $-\frac{64}{27}$
49. 6.3×10^{-4} **51.** 2×10^{-4} **53.** 6.2×10^9 **55.** 8.91×10^{-10} **57.** 3.8×10^{-1} **59.** 5.07×10^{12} **61.** 0.000 000 32 **63.** 0.000 425

Chapter 9 page 564

1. no; $4^2 + 5^2 \neq 7^2$ **3.** no; $6^2 + 9^2 \neq 13^2$ **5.** yes; $15^2 + 36^2 = 39^2$ **7.** yes; $8^2 + 15^2 = 17^2$ **9.** 10 **11.** 10.8 **13.** 5.1; (1.5, 5.5) **15.** 1.4; (4.5, −0.5) **17.** 25.5; (2, −2) **19.** 17.0; (−1.5, 1.5) **21.** 11.2; (2.5, 5) **23.** $\frac{3}{5}$ **25.** $\frac{3}{5}$ **27.** $\frac{3}{4}$ **29.** $\frac{5}{2}$ **31.** $\frac{6}{5}$
33. −10 **35.** $6 - 3\sqrt{6}$ **37.** 150 **39.** 9 **41.** 5 **43.** 4 **45.** 6

47. $x \geq 0$; **49.** $x \geq 0$;

51. $x \geq 0$;

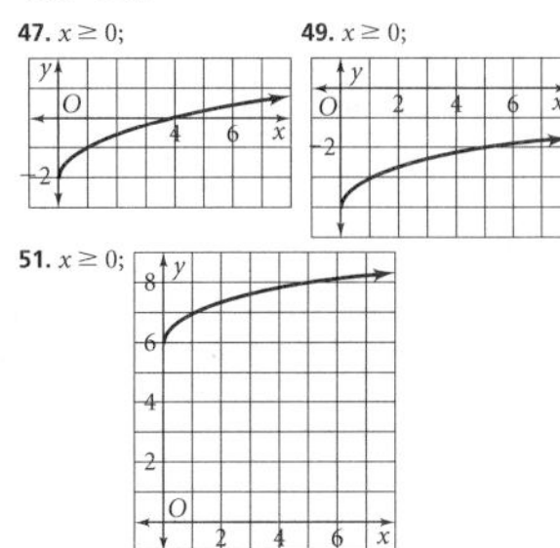

53. 1.6 **55.** 3.4 **57.** 3.2

Chapter 10 page 565

1. $2x^3 + 4x^2 - 11x + 11$ **3.** $6m^3 + m + 4$ **5.** $-10c^3 - c^2 - 9c - 7$ **7.** $-4s^4 - 2s^3 + 2s^2 + 4s$ **9.** $4b^3 + 12b$ **11.** $6t^2 + t - 1$ **13.** $32m^2 - 40m$ **15.** $5k^3 + 40k^2$ **17.** $4n^2 + 2n - 12$ **19.** $b^2 + 10b + 21$ **21.** $(3x - 1)(2x + 5), 6x^2 + 13x - 5$ **23.** 3 **25.** $2v^5$ **27.** $5r$ **29.** d^4 **31.** 4 **33.** w
35. $(x + 5)(x - 5)$ **37.** $(3c + 13)(3c - 13)$ **39.** $(t - 3)^2$ **41.** $(2d - 3)^2$ **43.** $(3v - 2)(v + 4)$ **45.** $(-w + 1)(w - 4)$ or $(-w + 4)(w - 1)$ **47.** $(4m + 1)(3m - 2)$ **49.** $(c - 5)^2$ **51.** 12, −12 **53.** 2, 3 **55.** −0.38, −2.62 **57.** 1.35, −1.85 **59.** 5, $-\frac{1}{2}$ **61.** −0.22, −2.28

Chapter 11 page 566

1. 70 **3.** 0.8 **5.** 0.6 **7.** 8.1 **9.** 2 **11.** 3 **13.** 5

15. **17.**

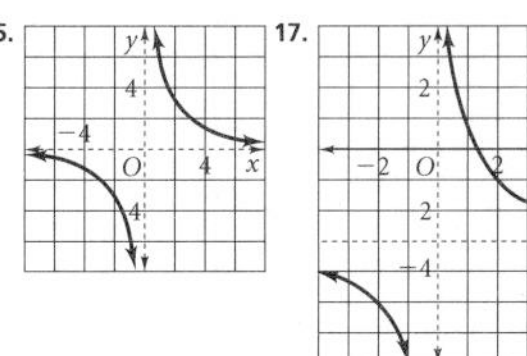

19. **21.**

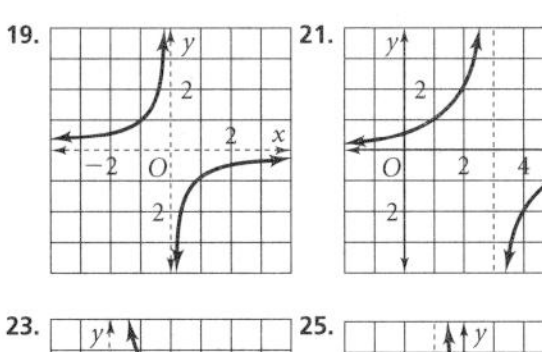

23. **25.**

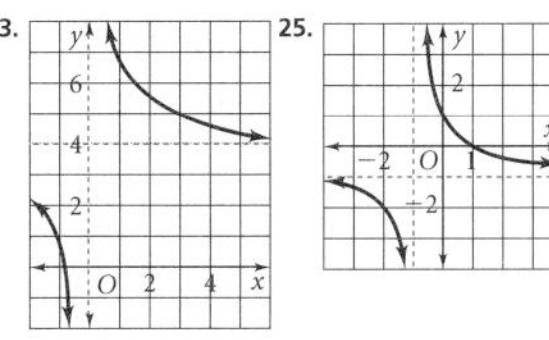

27. $\frac{t}{4}$; 0 **29.** 4; 3 **31.** $\frac{1}{x}$; 0 **33.** $\frac{6c + 4}{2}$; 0 **35.** $\frac{s}{2}$; 0
37. $12b^2$; 0, $\frac{1}{2}$ **39.** $\frac{8(t - 1)}{(t - 4)(t + 4)}$; 4, −4 **41.** $\frac{y}{6}$; 0, 2
43. 8 **45.** −1 **47.** 9 **49.** 5, −5 **51.** −12 **53.** 6, −10 **55.** 42 **57.** 8 **59.** 36 **61.** 84 **63.** 1320 **65.** 5040 **67.** 8568 committees

SKILLS HANDBOOK

page 567

EXERCISES **1.** 6 games **3.** 9 tacks

page 568

EXERCISES **1.** 17, 19 or −19, −17 **3.** 24 **5.** mother 38, son 16, daughter 7

page 569

EXERCISES **1.** 81 books **3.** 36 stories **5.** 7.2 m **7.** 6 ways

page 570

EXERCISES **1.** 21 lockers **3.** about 370,000,000 times **5.** 62 games **7.** 32 and 33

page 571

EXERCISES **1.** dog K.C., horse Bo, bird Cricket, cat Tuffy **3.** Alexa, Karin, Heather, Annette, Tanya, Garo **5.** 126 players

page 572

EXERCISES **1.** 6 **3.** \$575 **5.** 6 tickets **7.** 7:25 A.M.

page 573

EXERCISES **1.** composite **3.** composite **5.** prime **7.** composite **9.** composite **11.** composite **13.** composite **15.** composite **17.** composite **19.** 1, 2, 23, 46 **21.** 1, 11 **23.** $2 \cdot 3 \cdot 3$ **25.** $2 \cdot 3 \cdot 3$ **27.** $3 \cdot 3 \cdot 3$ **29.** $2 \cdot 2 \cdot 2 \cdot 2 \cdot 2 \cdot 2$ **31.** $2 \cdot 2 \cdot 5 \cdot 5$ **33.** $2 \cdot 2 \cdot 3 \cdot 7$ **35.** $11 \cdot 11$

page 574

EXERCISES **1.** 2 **3.** 24 **5.** 3 **7.** 7 **9.** 14 **11.** 3 **13.** 5 **15.** 21 **17.** 80 **19.** 33 **21.** 60 **23.** 240 **25.** 2160 **27.** 50 **29.** 150 **31.** 40

page 575

EXERCISES **1.** 9 **3.** 48 **5.** 2 **7.** 9 **9.** 3 **11.** no **13.** no **15.** no **17.** $\frac{1}{2}$ **19.** $\frac{2}{3}$ **21.** $\frac{2}{5}$ **23.** $\frac{1}{3}$ **25.** $\frac{2}{5}$ **27.** $\frac{3}{4}$ **29.** $\frac{9}{11}$ **31.** $\frac{9}{17}$ **33.** $\frac{5}{6}$

page 576

EXERCISES **1.** 0.3 **3.** 0.2 **5.** $0.\overline{6}$ **7.** $\frac{1}{4}$ **9.** $\frac{13}{100}$ **11.** $\frac{7}{100}$ **13.** $1\frac{5}{7}$ **15.** $2\frac{1}{10}$ **17.** $4\frac{2}{5}$ **19.** $\frac{5}{2}$ **21.** $\frac{31}{6}$ **23.** $\frac{29}{7}$ **25.** 0.875

page 577

EXERCISES **1.** $\frac{5}{7}$ **3.** 3 **5.** $10\frac{7}{15}$ **7.** $6\frac{2}{9}$ **9.** $6\frac{7}{33}$ **11.** $\frac{1}{2}$ **13.** 3 **15.** $\frac{1}{11}$ **17.** $1\frac{7}{9}$ **19.** $7\frac{11}{12}$

page 578

EXERCISES **1.** $\frac{3}{10}$ **3.** $\frac{15}{32}$ **5.** $1\frac{1}{3}$ **7.** $8\frac{5}{8}$ **9.** 20 **11.** $1\frac{1}{5}$ **13.** $\frac{6}{7}$ **15.** $\frac{3}{7}$ **17.** $\frac{5}{7}$ **19.** $2\frac{4}{17}$

page 579

EXERCISES **1.** 56% **3.** 602% **5.** 820% **7.** 20% **9.** 11.1% **11.** 75% **13.** 0.07 **15.** 0.009 **17.** 0.83 **19.** $\frac{19}{100}$ **21.** $4\frac{1}{2}$ **23.** $\frac{16}{25}$

page 580

EXERCISES **1.** 6^4 **3.** $5 \cdot 2^4$ **5.** 64 **7.** 144 **9.** 3267
11. $(1 \cdot 10^3) + (2 \cdot 10^2) + (5 \cdot 10^1) + (4 \cdot 10^0)$
13. $(8 \cdot 10^4) + (3 \cdot 10^3) + (4 \cdot 10^2) + (1 \cdot 10^0)$

page 581

EXERCISES **1.** 100°; obtuse **3.** 180°; straight

page 582

EXERCISES **1.** 22 cm **3.** 24 cm^2 **5.** 216 cm^3 **7.** 351.68 cm^3

Additional Answers

CHAPTER 1

LESSON 1-1

pages 7–9 On Your Own

13a. English Word Length Frequency

```
     X
   X X
   X X
   X X
   X X
   X X   X
   X X   X
   X X   X   X
   X X   X   X
   X X   X   X
   X X   X X X
   X X   X X X
   X X X X X X
   X X X X X X X   X
   X X X X X X X   X X
   X X X X X X X   X X
 1 2 3 4 5 6 7 8 9 10 11 12
```

Spanish Word Length Frequency

```
   X
   X
   X
   X
   X
   X
   X
   X             X
   X             X
   X X           X
 X X X         X X X
 X X X         X X X
 X X X         X X X
 X X X   X     X X X
 X X X   X X X X X X
 X X X   X X X X X X
 X X X X X X X X X X     X
 X X X X X X X X X X X   X X
 1 2 3 4 5 6 7 8 9 10 11 12 13
```

page 9 Mixed Review

26a. TV Viewing

Number of Programs Watched	0	1	2	3	4	5	6
Number of Students	3	5	3	4	3	5	2

LESSON 1-3

pages 17–18 On Your Own

28b. Cable TV Subscribers 1970-1990

LESSON 1-8

pages 41–43 On Your Own

10. $\begin{bmatrix} 25 & 21 & -6 \\ 7 & -2 & -5 \end{bmatrix}$; $\begin{bmatrix} 7 & 1 & -34 \\ 37 & 18 & -15 \end{bmatrix}$

11. $\begin{bmatrix} 81 & -32 \\ 27 & 100 \end{bmatrix}$; $\begin{bmatrix} 3 & -40 \\ -47 & 8 \end{bmatrix}$,

12. $\begin{bmatrix} 4 & 0 & 4 \\ 0 & 7 & -6 \\ 7 & -10 & 14 \end{bmatrix}$; $\begin{bmatrix} 2 & 4 & -2 \\ 8 & -3 & 6 \\ -7 & 6 & -4 \end{bmatrix}$

13. $\begin{bmatrix} 8.3 & -5.8 \\ -2.6 & 2.9 \\ 13 & -7.8 \end{bmatrix}$; $\begin{bmatrix} -1.5 & 10 \\ -10.2 & 8.5 \\ 4.6 & -10.8 \end{bmatrix}$

14. $\begin{bmatrix} 10.7 & 5.1 \\ -2.8 & -5 \\ 7.7 & 10.5 \end{bmatrix}$; $\begin{bmatrix} -0.7 & -8.5 \\ 5.2 & -2.6 \\ -4.7 & -5.7 \end{bmatrix}$

15. $\begin{bmatrix} 1\frac{1}{6} & 1\frac{1}{5} \\ \frac{7}{8} & 1\frac{1}{6} \end{bmatrix}$; $\begin{bmatrix} \frac{1}{6} & \frac{2}{5} \\ -\frac{5}{8} & -\frac{5}{6} \end{bmatrix}$

16. $\begin{bmatrix} 357 & -184 & 167 \\ -273 & 318 & -145 \end{bmatrix}$; $\begin{bmatrix} 355 & -196 & 175 \\ -239 & 324 & -155 \end{bmatrix}$

17. $\begin{bmatrix} 2.5 & 0.3 \\ -8.1 & 2.7 \end{bmatrix}$; $\begin{bmatrix} 4.5 & -0.7 \\ 8.1 & -16.7 \end{bmatrix}$

18b. $\begin{bmatrix} 1.8 & -0.1 & -0.2 \\ 0.4 & -0.2 & 0 \\ 0.6 & -0.3 & 0.4 \\ -0.3 & -1.4 & 0.4 \end{bmatrix}$

e. Answers may vary. Sample: Invest in volleyball because it has the sharpest increase in participation from the 7–11 category to the 12–17 category. Also, the numbers decline only a small amount in the 18–24 category.

20a. Sample: [A] 5

$$\begin{bmatrix} 1 & 2 & 3 & 0 & -1 \\ -5 & 0 & 0 & 3 & -1 \\ 1 & 1 & 4 & -3 & -1 \end{bmatrix}$$

2); [B] =

$$\begin{bmatrix} 3 & -2 & 1 & -2 & 1 \\ 2 & 1 & -3 & -1 & 4 \\ 4 & -2 & 0 & 0 & 0 \end{bmatrix}$$

b. for sample in (a): [A] + [B] =

$$\begin{bmatrix} 4 & 0 & 4 & -2 & 0 \\ -3 & 1 & -3 & 2 & 3 \\ 5 & -1 & 4 & -3 & -2 \end{bmatrix}$$

c. for sample in (a): [A] − [B] =

$$\begin{bmatrix} -2 & 4 & 2 & 2 & -2 \\ -7 & -1 & 3 & 4 & -5 \\ -3 & 3 & 4 & -3 & -2 \end{bmatrix}$$

23a. $\begin{bmatrix} 13 & 5 & 6 & 2 \\ 18 & 4 & 2 & 2 \\ 6 & 2 & 0 & 2 \end{bmatrix}$

24. $\begin{bmatrix} -27 & 6 \\ 18 & -36 \end{bmatrix}$

25. $\begin{bmatrix} 36 & -72 \\ 27 & 45 \\ -99 & 18 \end{bmatrix}$

26. $\begin{bmatrix} -350 & -150 & 50 \\ -400 & -100 & 250 \end{bmatrix}$

Checkpoint page 43

6. Sample: Suppose a basketball player scores on 80% of his free throws. What is the probability that he will score on the next two free throws?

   Use a random number table. Let 0–7 represent hits and 3–9 represent misses. Count the number of pairs with two hits and divide by the number of pairs in the sample.

Preparing for Standardized Tests
page 55

15. For 2(5 + 4), add the values inside the parantheses first. For 2(5) + 4, multiply first.

CHAPTER 2

LESSON 2-1

pages 61–63 On Your Own

11a. How Big Is a Bird Egg?

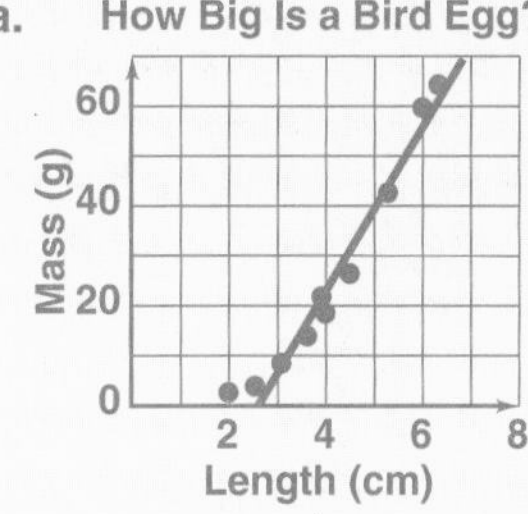

b. There is a positive correlation between the length and the weight of a bird egg, because the larger eggs are heavier.

c. Answers may vary. Sample: The egg is about 5.8 cm long. From the graph, its weight should be about 55 g.

16. Samples: Negative correlation—as the price of a specific model of bicycle increases, the number of these bicycles sold decreases.

Positive correlation—popularity of a candidate in pre-election polls corresponds to the percentage of votes this candidate receives.

No correlation—amount of precipitation on each day is not related to the number of homework problems assigned on that day.

17a. Calories and Fat in Common Foods

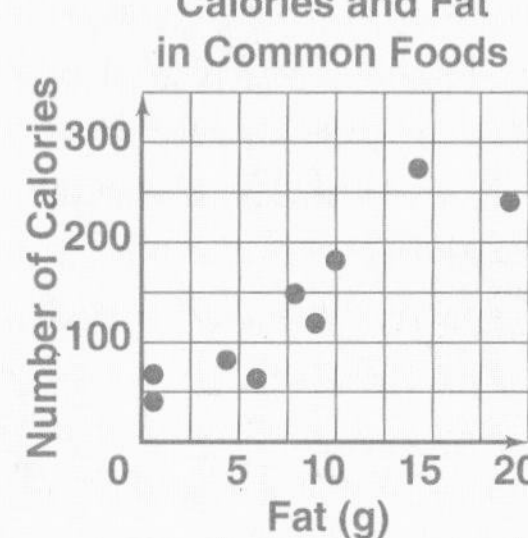

page 63 Mixed Review

19. Each data point represents the total toll charged (y-coordinate) for a vehicle that traveled a specified number of miles (x-coordinate).

31. Mode; the mean and the median need to be computed. You can read the mode off the plot.

TE page 63 Lesson Quiz

1.

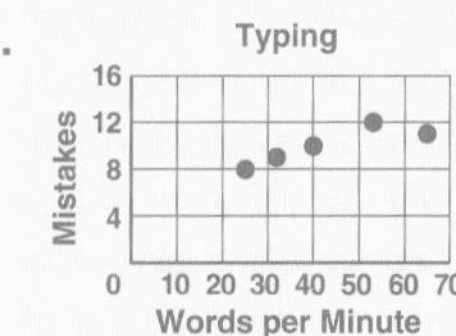

LESSON 2-2

pages 66–68 On Your Own

2. Graphs may vary. Sample:

Energy Level during Gym Class

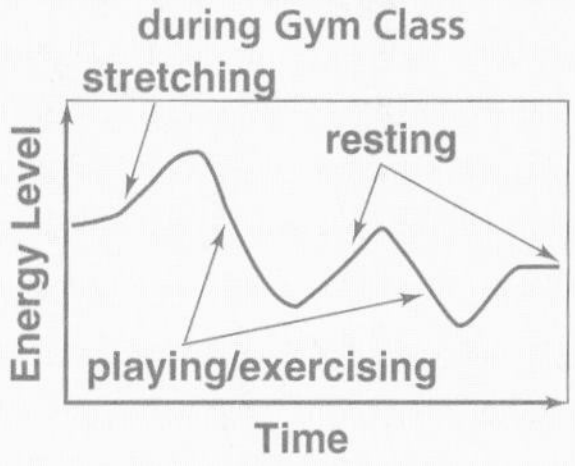

stretching
resting
Energy Level
playing/exercising
Time

3. Graphs may vary. Sample:

Height While Jumping Rope

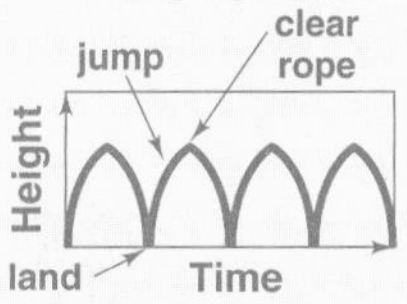

jump
clear rope
Height
land
Time

4. Graphs may vary. Sample:

Speed Skateboarding Downhill

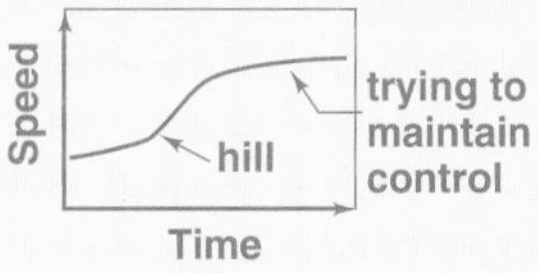

Speed
hill
trying to maintain control
Time

6. From 7 A.M. to 7 P.M., as the storm moved in, the pressure dropped. From 7 P.M. to 7 A.M. the next day, the pressure increased as the blizzard ended.

7. Answers may vary. Sample: Continuous; between two weights you measure, any value is possible.

Weight from Birth to Age 14

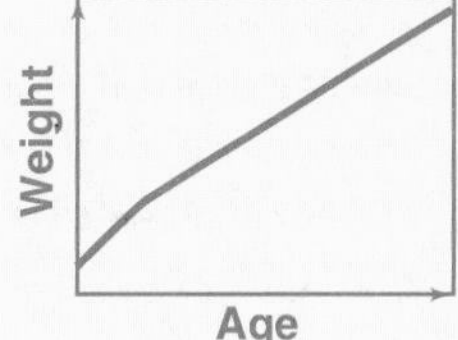

8. Answers may vary. Sample: Discrete; there are no test grades between tests.

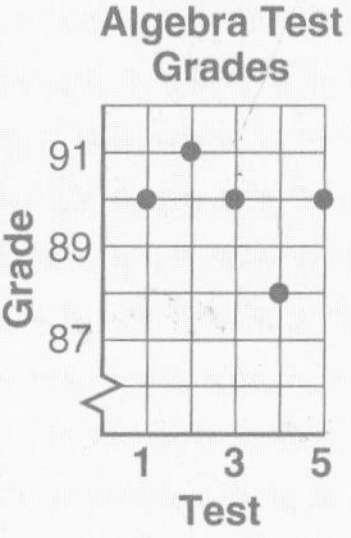

9. Answers may vary. Sample: Discrete; there is only one time for each morning.

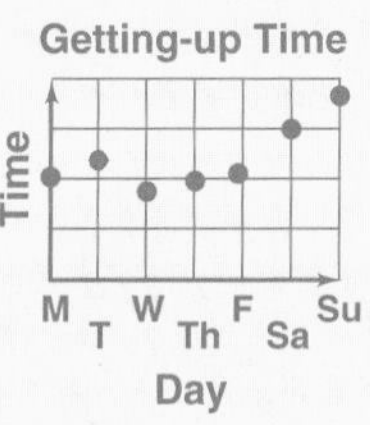

10. Answers may vary. Sample: Continuous; between any two temperatures you measure any value is possible.

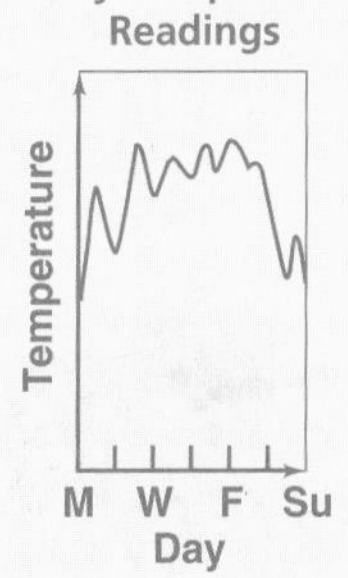

11. Answers may vary. Sample: Discrete; you cannot have half a book.

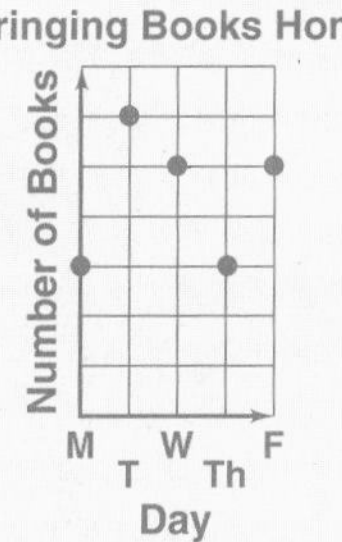

12. Answers may vary. Sample: Continuous; your hair is always growing.

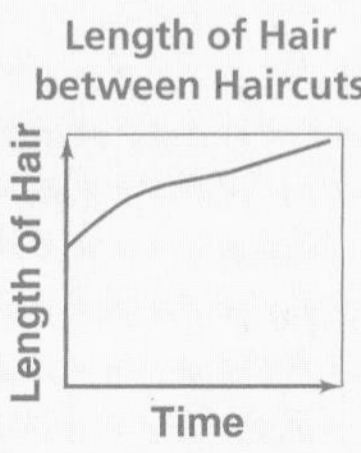

13. Answers may vary. Sample: Continuous; walking speed can be measured at all times.

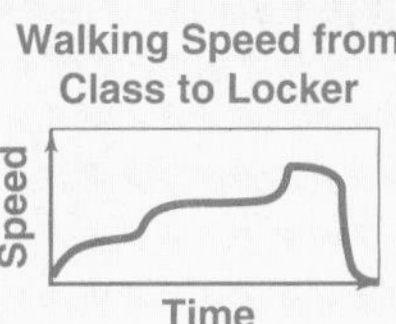

14. Answers may vary. Sample: Discrete; there is only one total for each day.

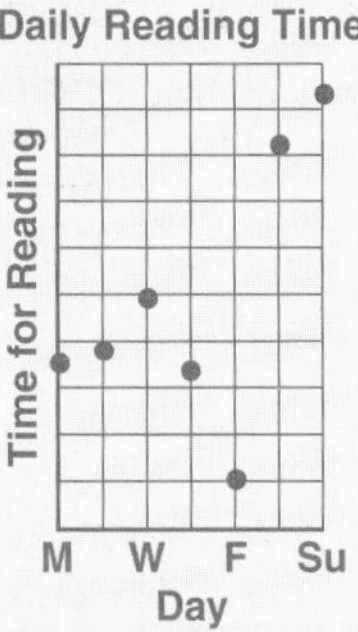

15b. Growing at a steady rate, a puppy becomes heavier than a baby but stops growing soon after that. A baby's weight gain increases with time, so it outgrows the puppy.

16. Answers may vary. Sample: The number of buses stopping at the intersection for each hour varies with time. More buses are scheduled to arrive during peak commute hours, fewer during the day and very few or none at night.

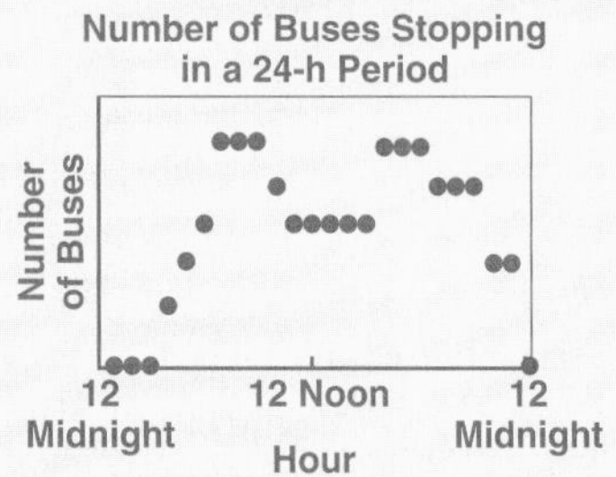

18a. Answers may vary. Sample: monthly profit and months

b. The line in the graph is flat because the profit does not change.

c. Answers may vary. Sample: The axes are not viewed in a normal manner. It is customary to have a horizontal and a vertical axis. The profits are actually declining.

d. Answers may vary. Sample: He is trying to disguise the fact that the company is not doing well.

page 68 Mixed Review

20. $\begin{bmatrix} -3 & 6 \\ 7 & 2 \end{bmatrix}$

21. $\begin{bmatrix} -14 & 11 \\ -2 & -51 \end{bmatrix}$

22. $\begin{bmatrix} -4.5 & -1.7 \\ -0.4 & 7.2 \end{bmatrix}$

23a. f = amount of fruit used per day; $f + 200$.

24–27.

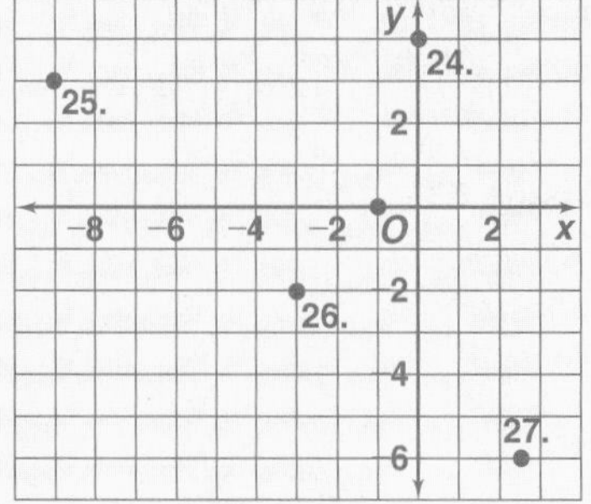

Checkpoint page 68

1a–b. There is a negative correlation.

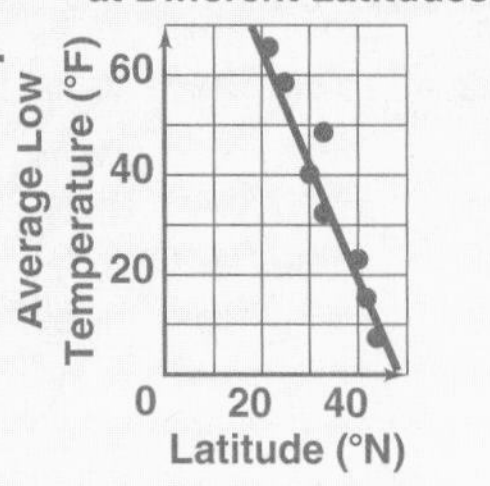

3. C; this is the only graph where the distance from home drops back to 0, indicating that you came home.

TE page 65 Additional Example 2

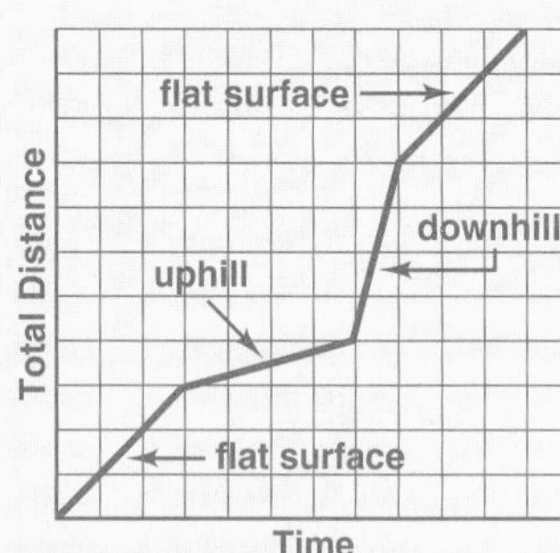

LESSON 2-3

page 69 Work Together

2a.

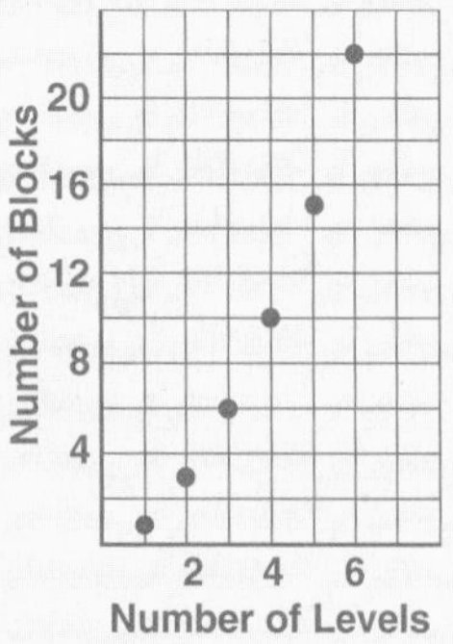

pages 69–70 Think and Discuss

10a. Sample: How much does the temperature change going up from 800 to 3100 ft?

b. Answers may vary. Answer to sample in part (a): The temperature drops by about 5.8°C.

TE page 70 Additonal Example

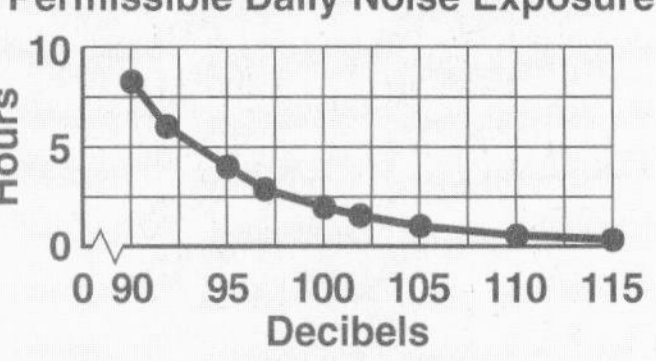

pages 71–72 On Your Own

1a. Answers may vary. Sample: Horizontal axis from 0 to 200 labeled every 25 lb; vertical axis 0 to 30 labeled every 5 lb.

b.

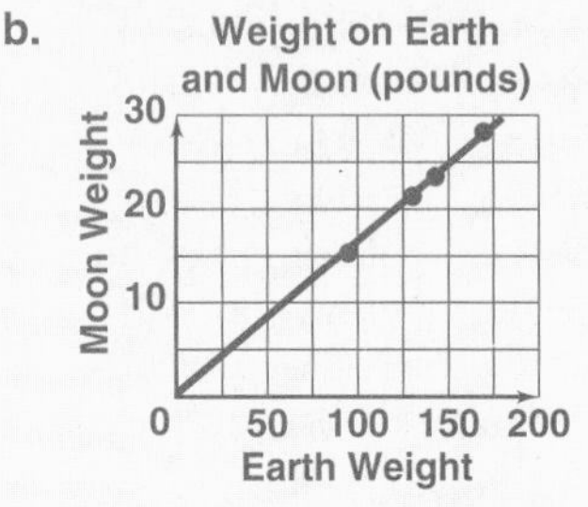

c. Find the point on the graph that corresponds to your weight on Earth. Use this point to identify the coordinate for the moon. This is your weight on the moon.

6. The amount paid depends on the number of hours worked; the area of a square depends on the length of its side; the weight of a book depends on the number of pages in it; the cost of apples depends on their weight.

10a. As the length of the loan increases, monthly payments decrease.

b. length of loan; monthly payments

c.

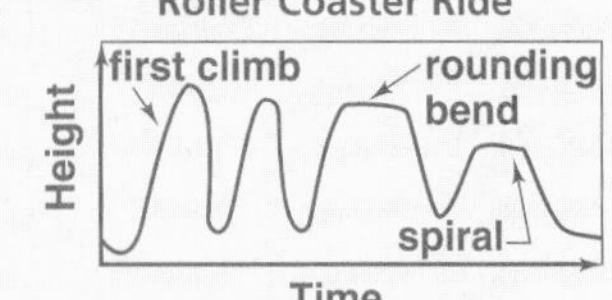

d. Answers may vary. Sample: Graph; you can see more easily on the graph how the monthly payments vary with the length of loan.

11a. High School Dance

b. Answers may vary. Sample: The horizontal range is 0 to 120 with the scale labeled every 20 min on the axis. The vertical range is 0 to 6 with the scale labeled at every dollar on the axis.

12c. As temperature increases, the volume of gas also increases.

13. Answers may vary. Sample: A dependent variable is one whose value is based on or determined from the values of other variables. An independent variable is a variable whose values are not influenced or determined by other given variables. For example, monthly life insurance premiums depend on age—age is the independent variable and the monthly premium is the dependent variable.

14a.

page 72 Mixed Review

15. Answers may vary. Sample:

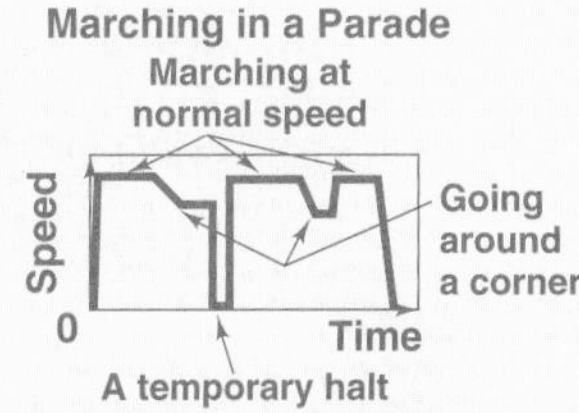

16. Answers may vary. Sample:

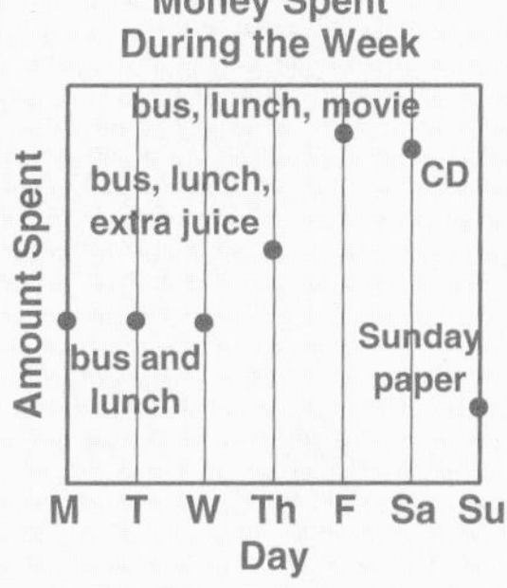

17a.

LESSON 2-4

pages 73–75 Think and Discuss

5d.

LESSON 2-5

page 83 Mixed Review

43.

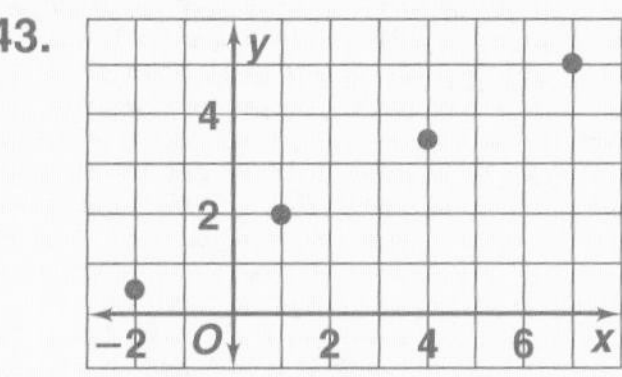

Checkpoint page 83

1b.

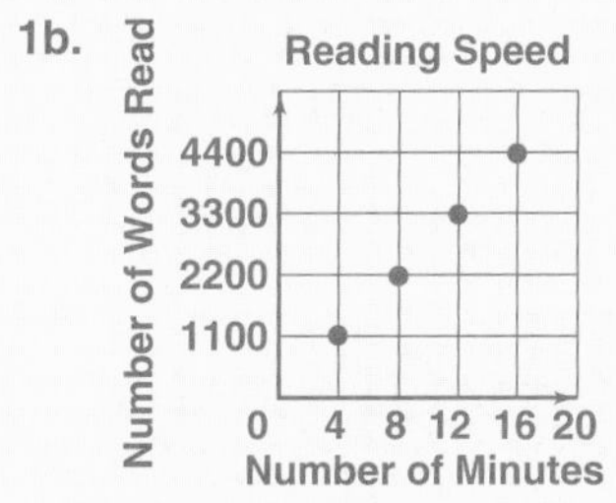

The *x*-value range is from 0 to 20. It is easiest to label every 4 minutes, to correspond to the given data. The *y*-value range is from 0 to 5500 with labels every 1100 words.

2. A *relation* is a set of ordered pairs. A *function* is a relation in which each input value corresponds to exactly one output value.

LESSON 2-6

pages 84–86 Think and Discuss

6a. b.

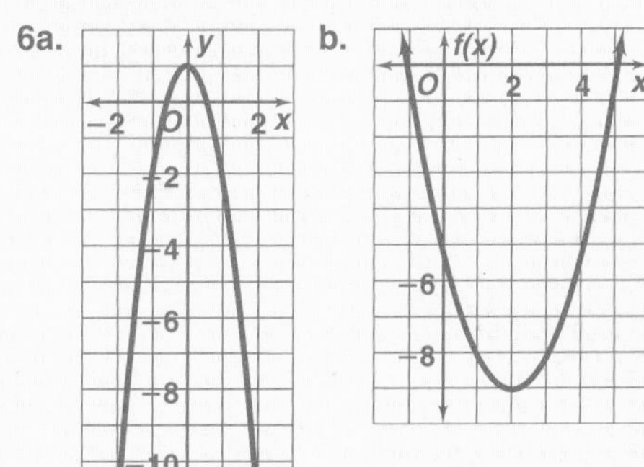

pages 86–88 On Your Own

1. Tables may vary. Sample:

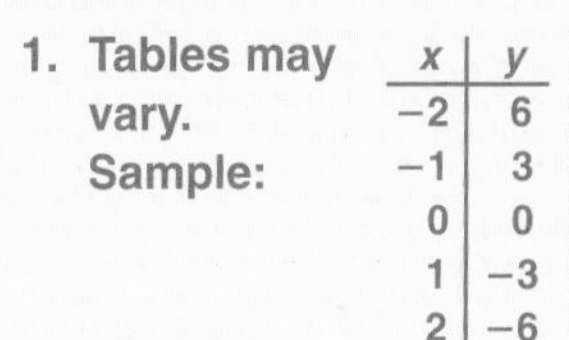

x	y
−2	6
−1	3
0	0
1	−3
2	−6

2. Tables may vary. Sample:

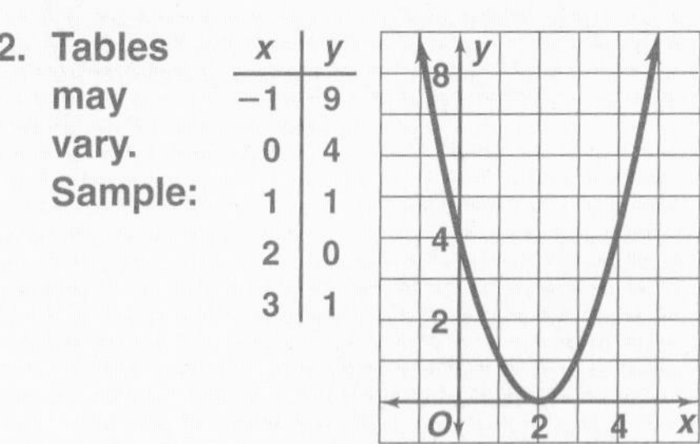

x	y
−1	9
0	4
1	1
2	0
3	1

3. Tables may vary. Sample:

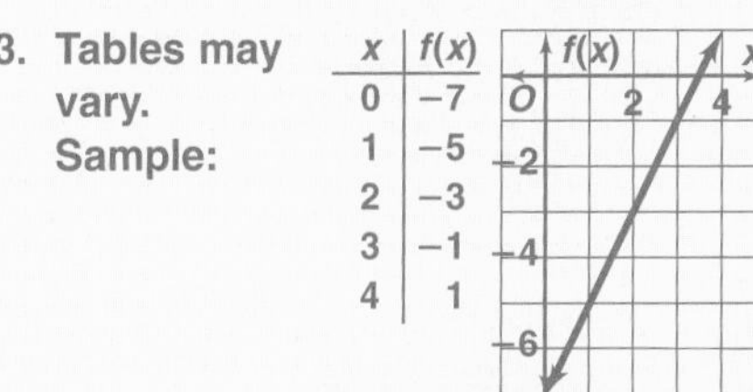

x	f(x)
0	−7
1	−5
2	−3
3	−1
4	1

4. Tables may vary. Sample:

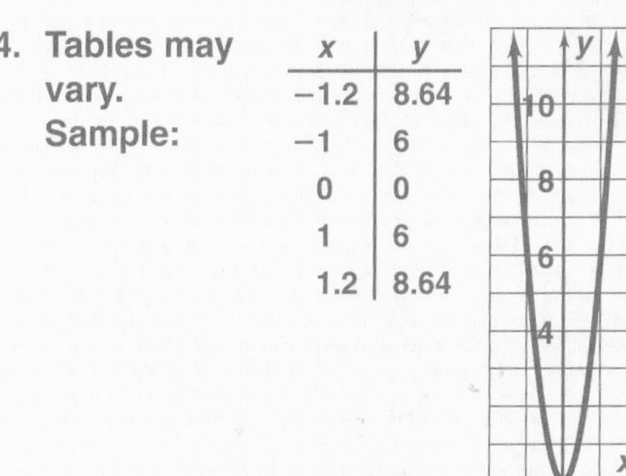

x	y
−1.2	8.64
−1	6
0	0
1	6
1.2	8.64

5. Tables may vary. Sample:

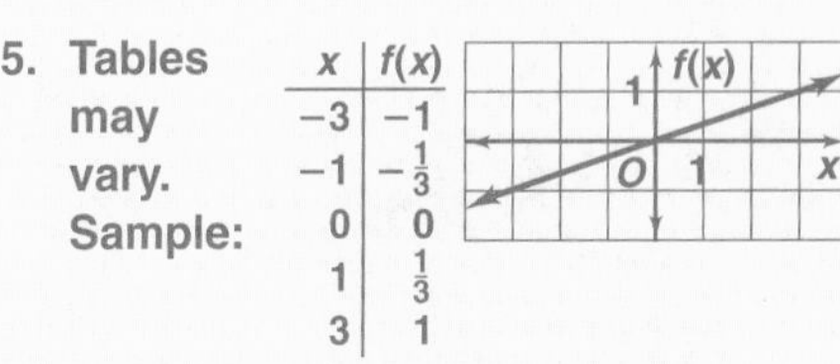

x	f(x)
−3	−1
−1	$-\frac{1}{3}$
0	0
1	$\frac{1}{3}$
3	1

6. Tables may vary. Sample:

x	f(x)
−1	9
0	8
2	6
4	4
8	0

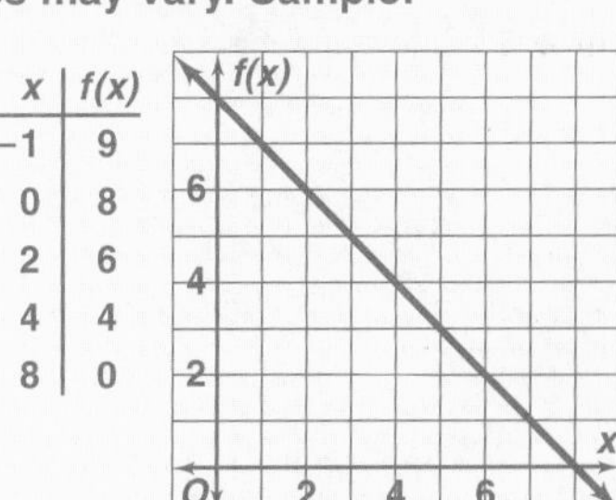

Additional Answers

7. Tables may vary. Sample:

x	$f(x)$
−2	2
−1	−1
0	−2
1	−1
2	2

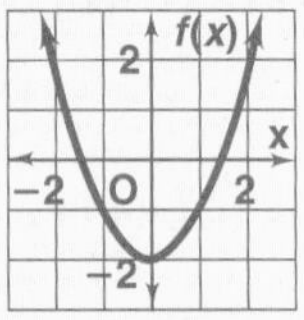

8. Tables may vary. Sample:

x	y
−2	−3
−1	1
0	5
1	9
2	13

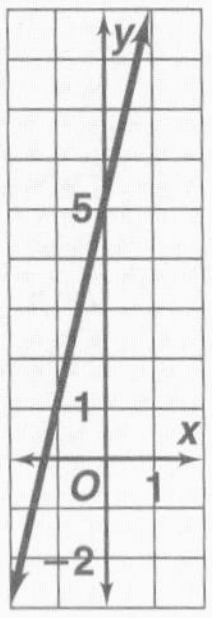

9c.

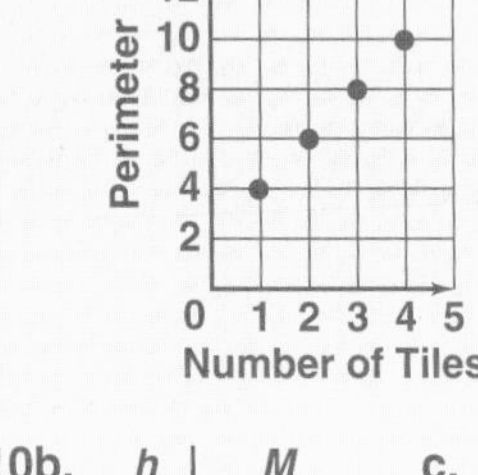

10b.

h	M
0	\$0.00
$\frac{1}{2}$	\$1.75
1	\$3.50
2	\$7.00
3	\$10.50

c.

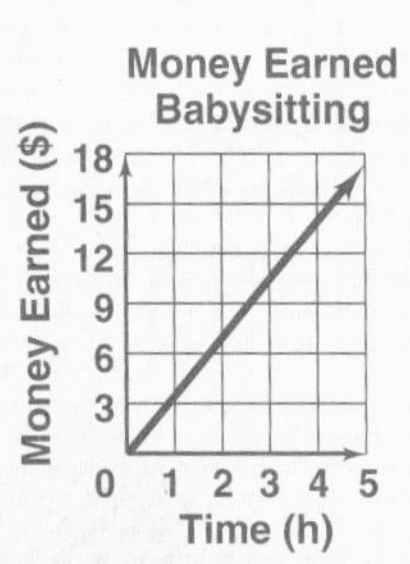

e. Answers may vary. Sample: Juan may charge for portions of 1 hour but he cannot charge fractions of 1 cent.

11d.

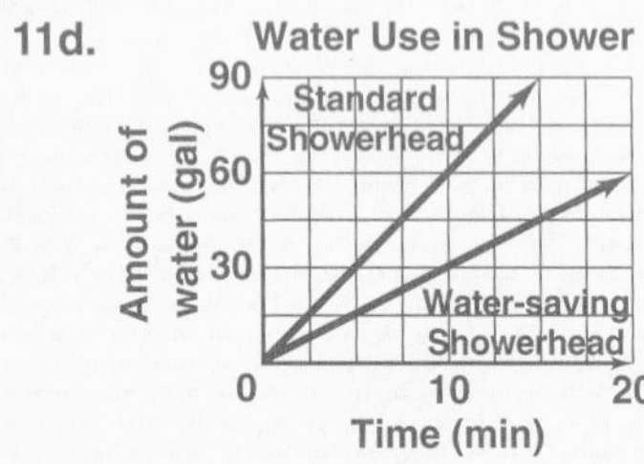

12.

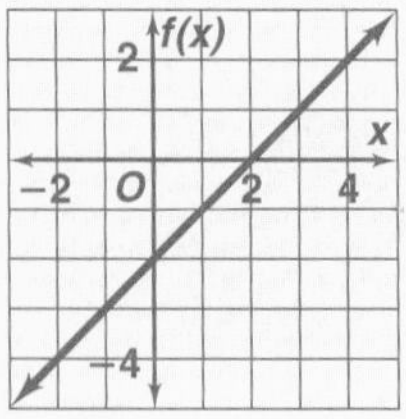

13.

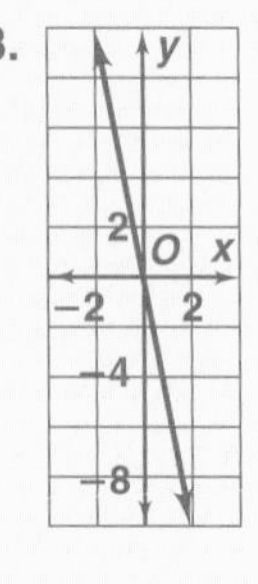

14.

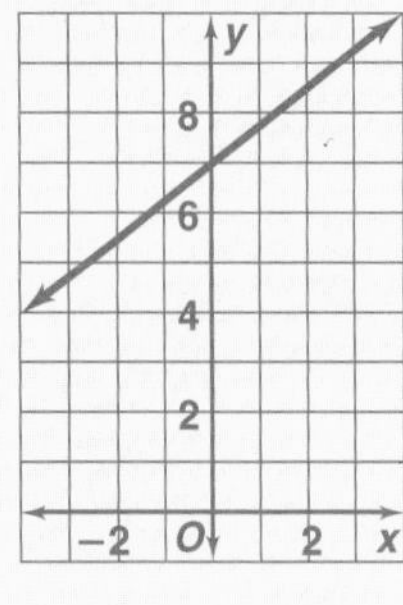

15.

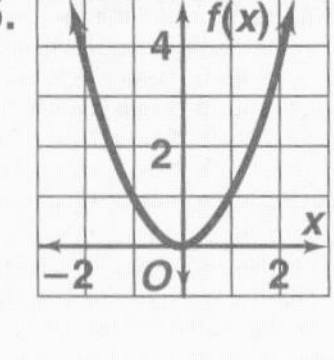

16.

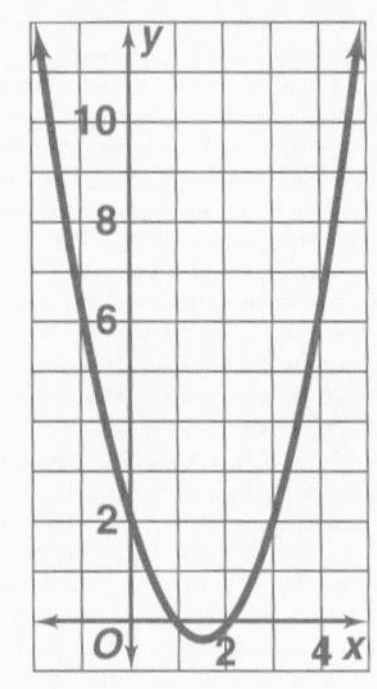

17.

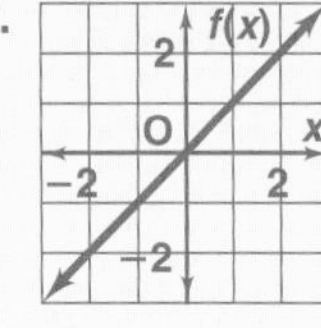

18.

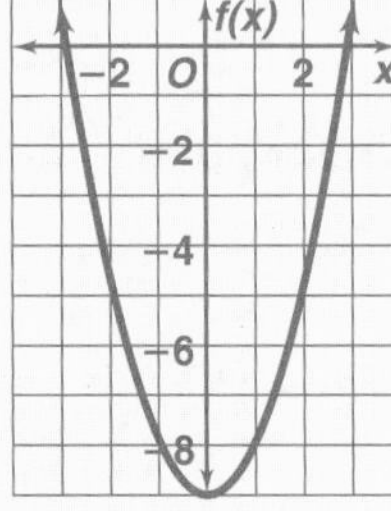

19.

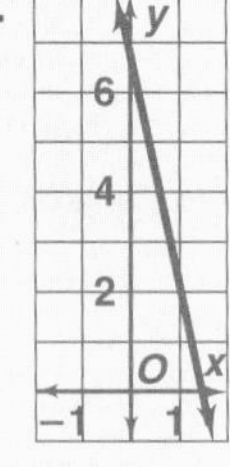

20.

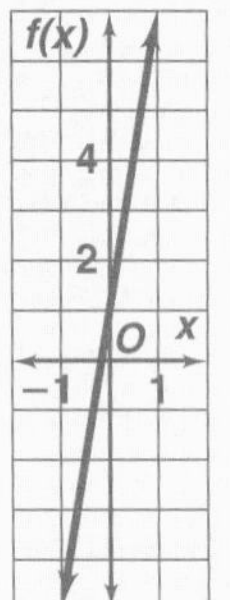

21.

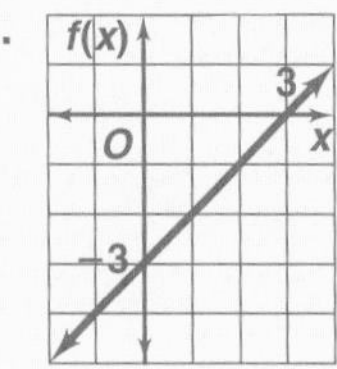

22.

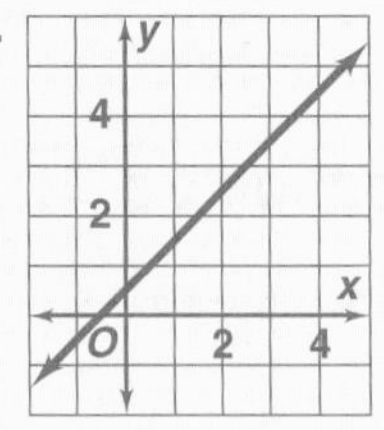

23.

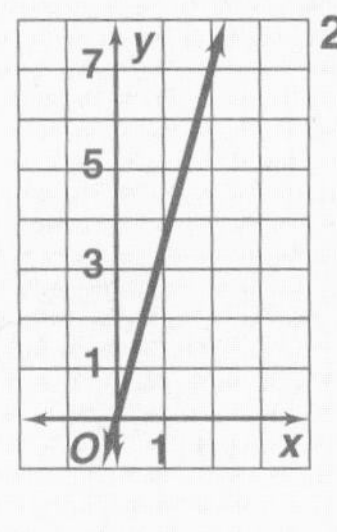

24.

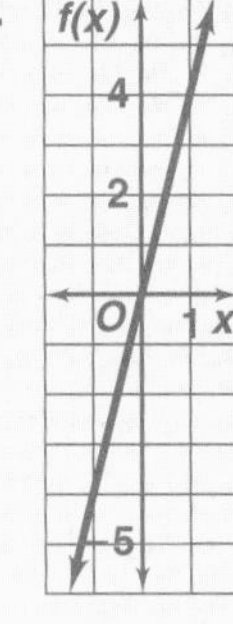

25.

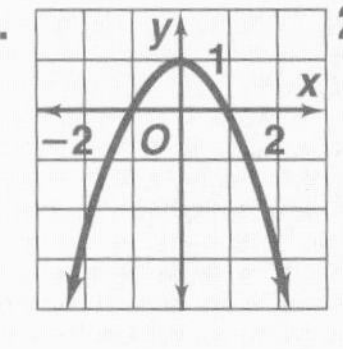

26.

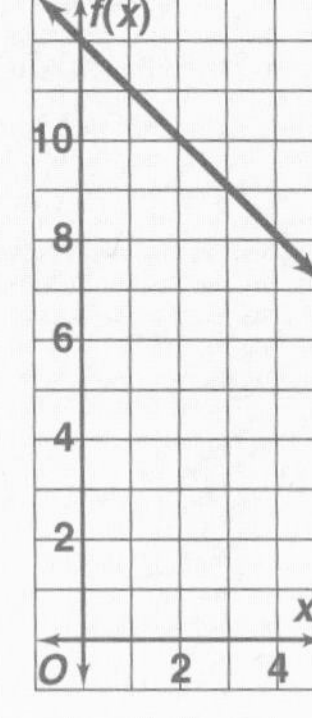

27.

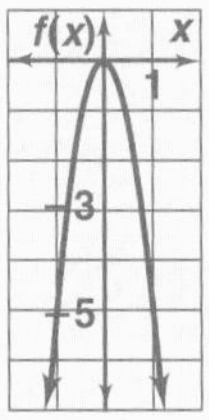

28. Answers may vary. Sample:

x	$f(x)$
−2	−2
0	−1
2	0
4	1

29. Answers may vary. Sample:

a	$C(a)$
0	\$.27
1	\$.38
2	\$.49
3	\$.60
4	\$.71
5	\$.82

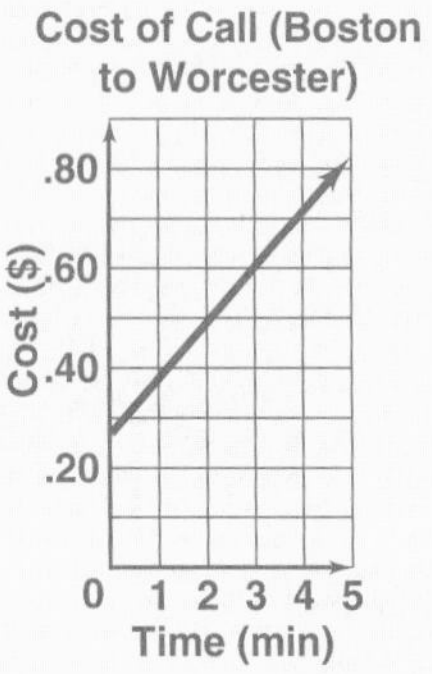

Cost of Call (Boston to Worcester)

Cost (\$)

Time (min)

30. Answers may vary. Sample:

x	$f(x)$
−3	4.5
−2	2
−1	0.5
0	0
1	0.5
2	2
3	4.5

31.

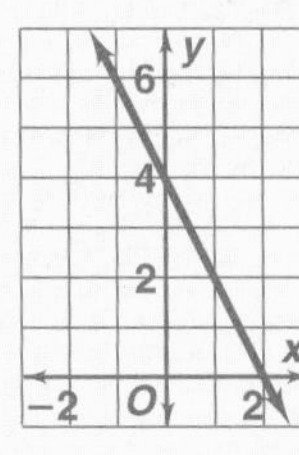

34a.

l	$A(l)$
1	0.5
2	2
3	4.5
4	8

b.

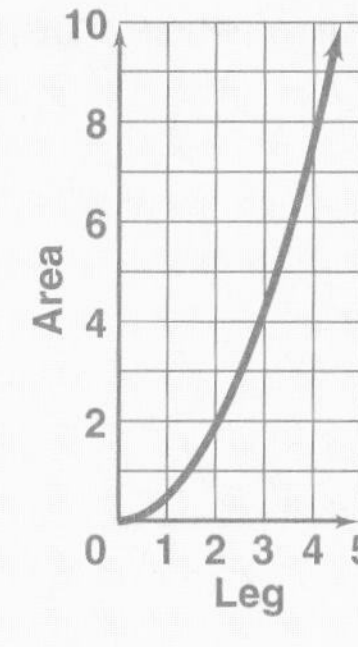

page 88 Mixed Review

37a. Sample: Students' scores for the first two tests in the algebra class have a positive correlation.

b. For sample in part (a): scores on second test vs. scores on first test

Math Toolbox page 89

4.

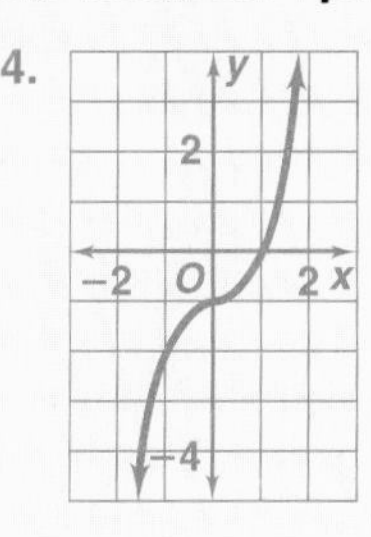

5.

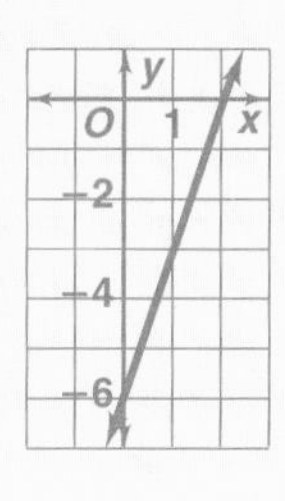

6.

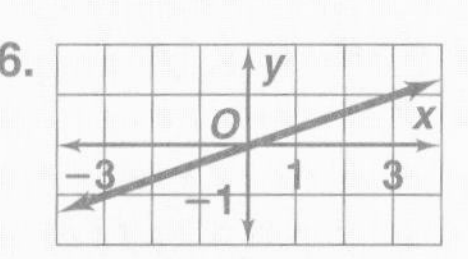

7.

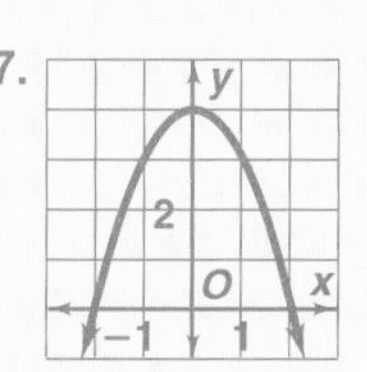

8.

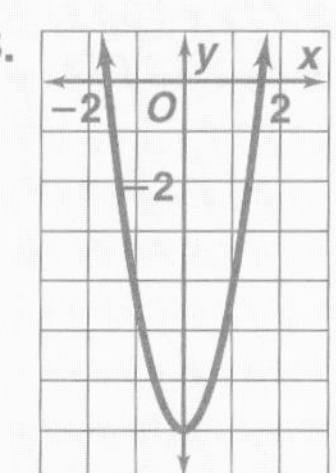

LESSON 2-7

page 90 Work Together

1.

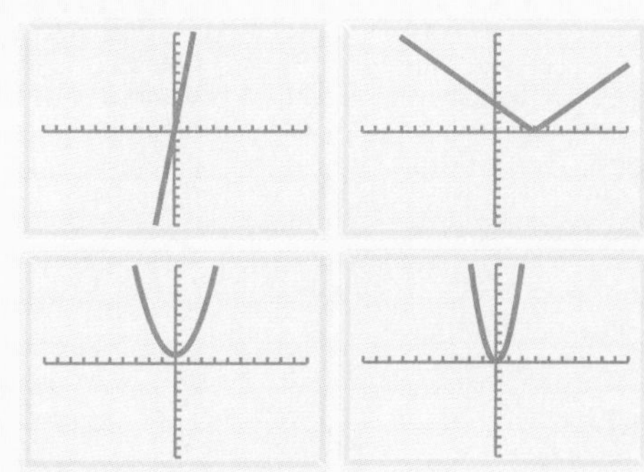

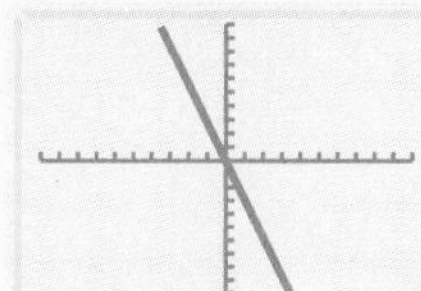

pages 90–92 Think and Discuss

8b.

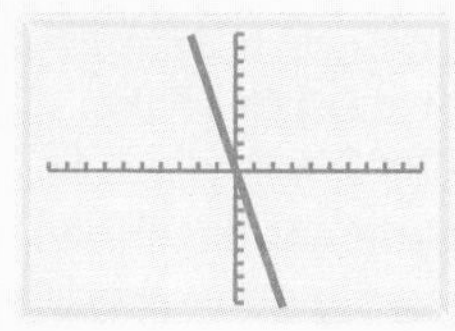

9. Samples:

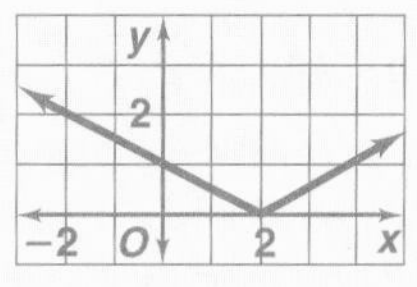

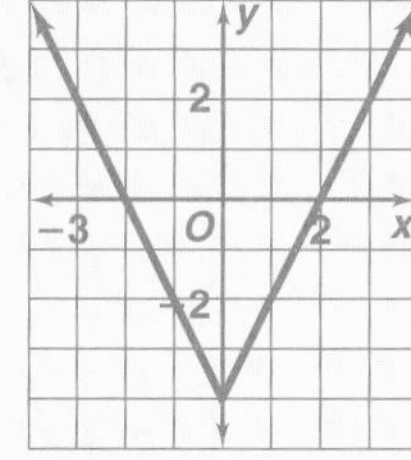

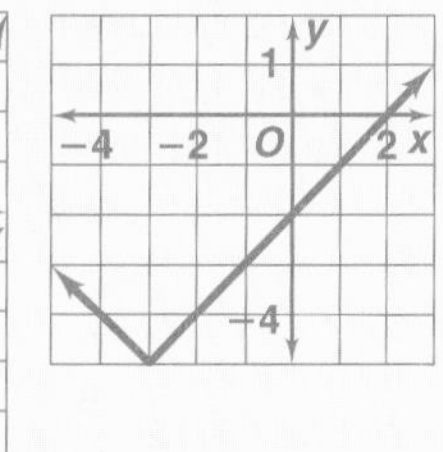

pages 92–94 On Your Own

1; 6: Linear function family; the highest power of the variable is 1.

2; 4; 5; 7: Quadratic family; the highest power of the variable is 2.

3; 8; 9: Absolute value family; there is a variable expression inside the absolute value symbol.

14. Graphs of linear and abs. value functions are composed of pieces of straight lines; graph of a linear function is a single line while graph of an abs. value function is two rays going in different non-opposite directions.

23. Total Income as a Function of HoursWorked

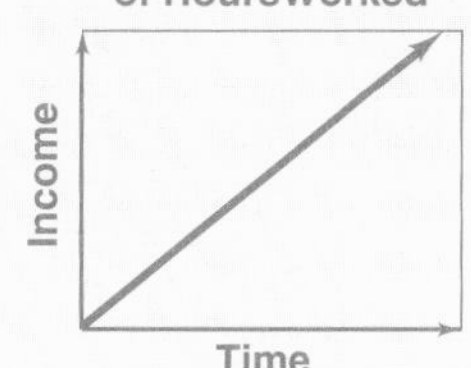

24. Height of a Baseball as a Function of Time

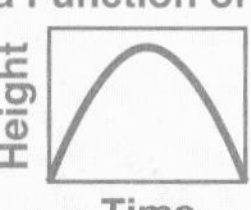

25.

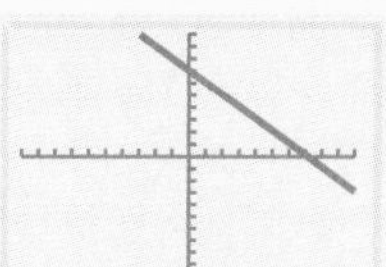

26.

27.

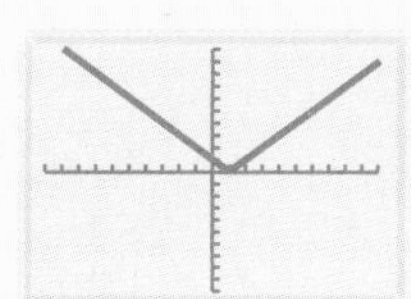

28.

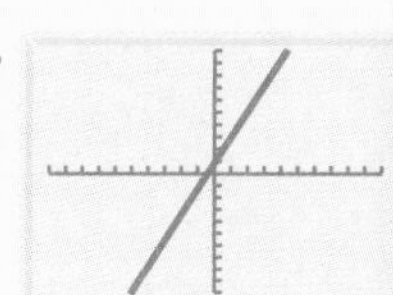

29.

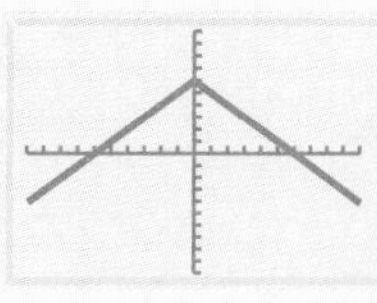

30.

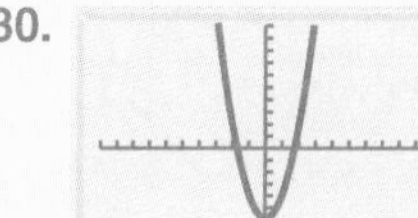

31.

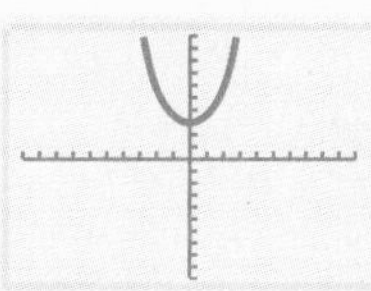

32.

33.

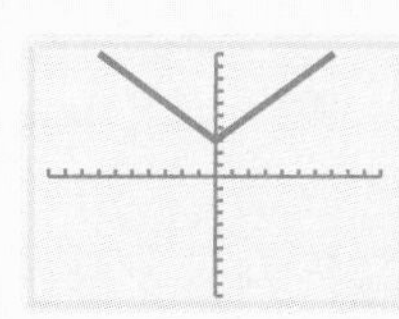

34. Absolute value family; the picture shows a V-shape opening down.

35. Quadratic family; the curve is U-shaped, opening down.

36. Linear family; the path is a straight line.

37. Absolute value family; the path of the ball is V-shaped, opening up.

38a. Answers may vary. Sample:

x	y
−2	2
−1	1
0	0
1	1
2	2

x	y
−2	−4
−1	−1
0	0
1	−1
2	−4

b. not negative; not positive

c. I and II; III and IV

d.

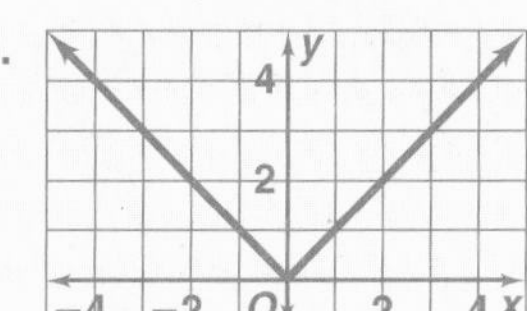

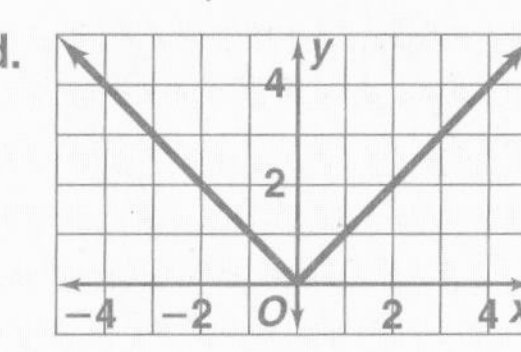

Additional Answers

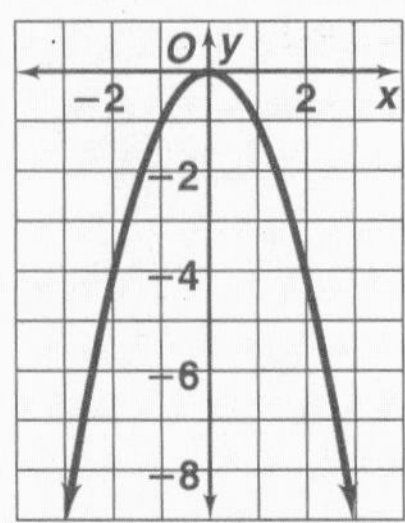

e.

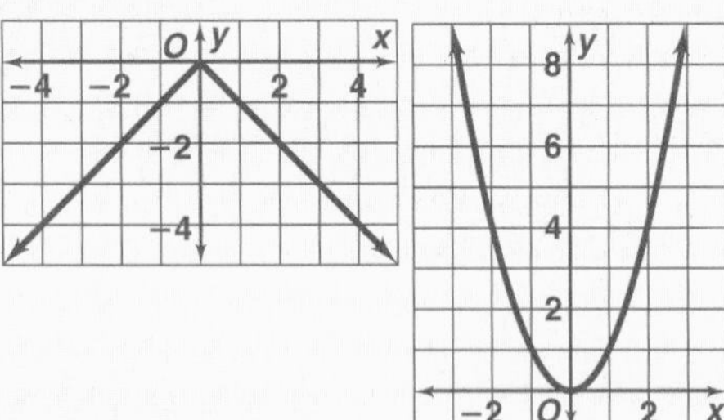

39b.

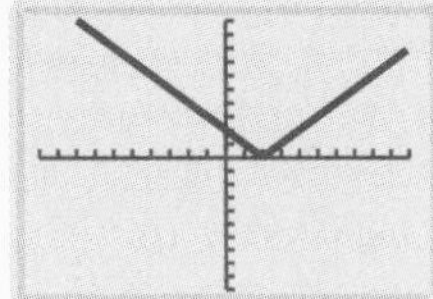

page 94 Mixed Review

41.

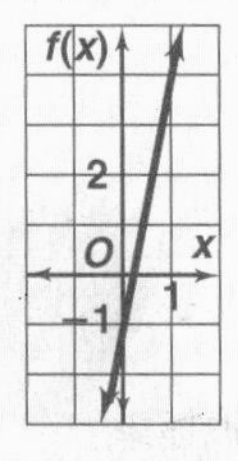

42.

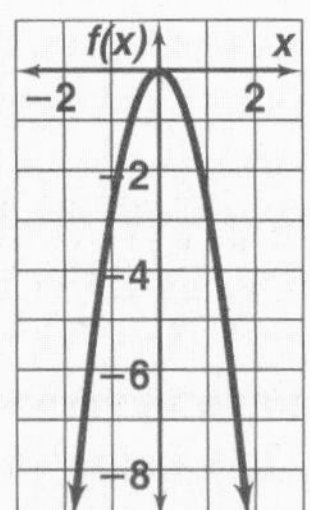

43.

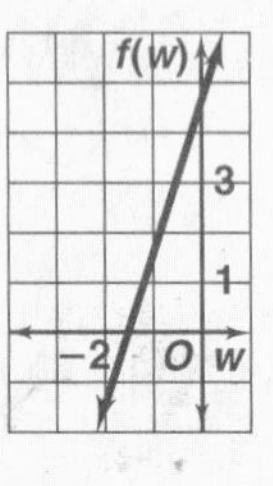

44.

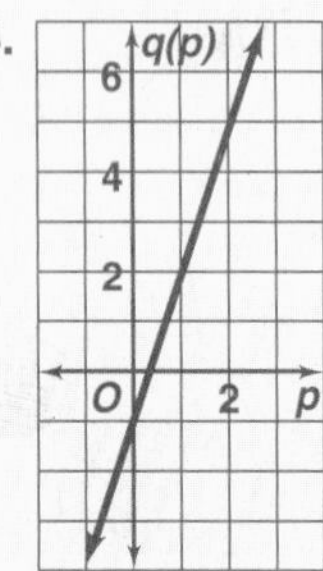

45.

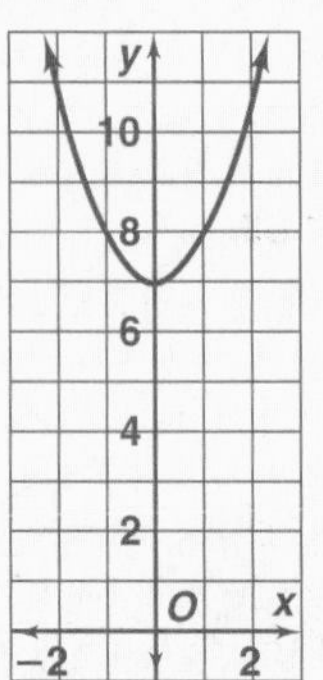

46.

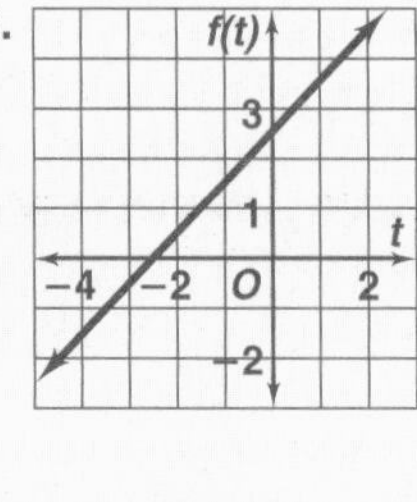

47.

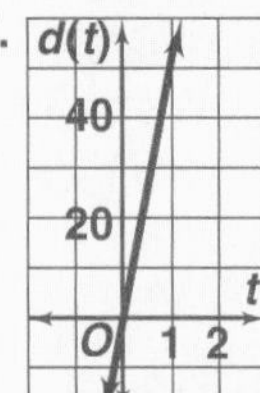

48.

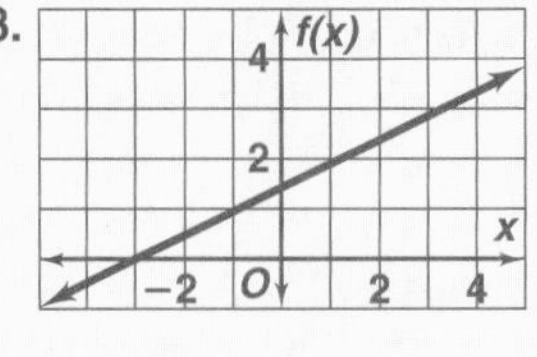

50a. Answers may vary. Sample: odd numbers for white socks; even numbers for blue

b. In the experiment described in part (a), the experimental probability is about 50%.

pages 97–99 On Your Own

31–38.

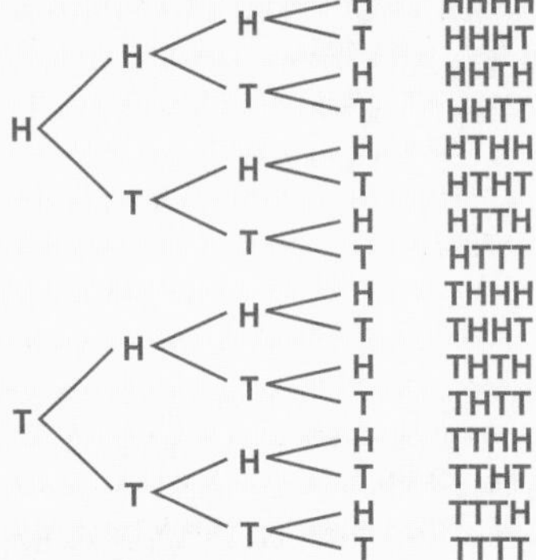

Wrap Up pages 101–103

19. Tables may vary. Sample:

x	$f(x)$
−6	1
−5	0
−2	3
0	5

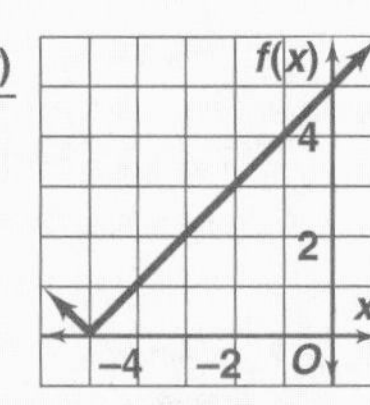

20. Tables may vary. Sample:

x	$f(x)$
−2	−3
−1	−1
0	1
1	3

21. Tables may vary. Sample:

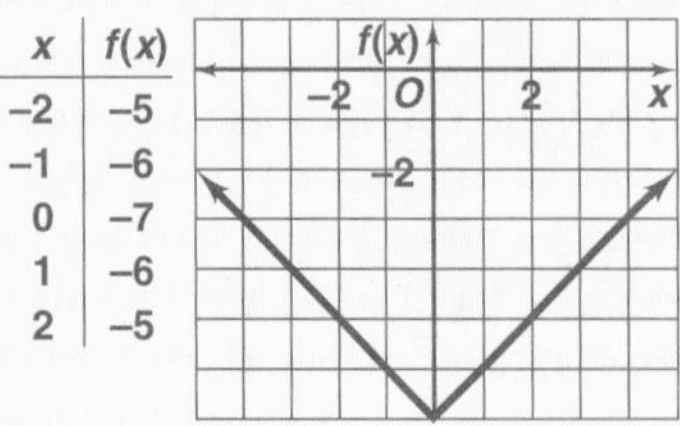

x	$f(x)$
−2	−5
−1	−6
0	−7
1	−6
2	−5

Chapter Assessment page 104

9. Answers may vary. Sample:

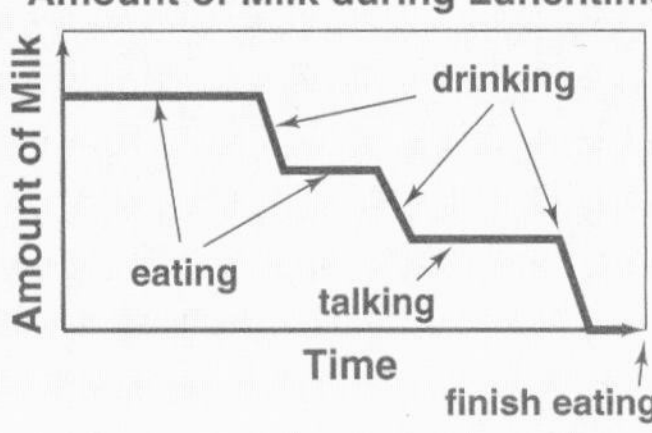

18. Answers may vary. Sample:

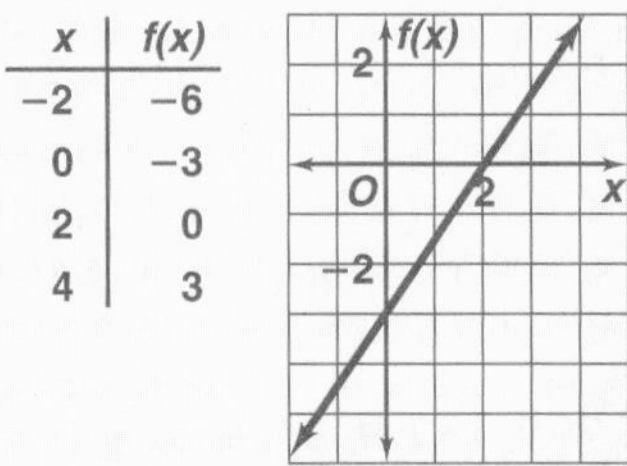

x	$f(x)$
−2	−6
0	−3
2	0
4	3

19. Answers may vary. Sample:

x	$f(x)$
−2	0
−1	3
0	4
1	3
2	0
3	−5

20. Sample: money earned vs. number of cars washed at $5 each

Color	Size	Sample Space
green	S	green-S
	M	green-M
	L	green-L
blue	S	blue-S
	M	blue-M
	L	blue-L
red	S	red-S
	M	red-M
	L	red-L
purple	S	purple-S
	M	purple-M
	L	purple-L
black	S	black-S
	M	black-M
	L	black-L

Cumulative Review page 105

10. No, data points do not lie on a straight line.

CHAPTER 3

LESSON 3-2

pages 116–118 On Your Own

47a. Yes; with this estimate, the total bill is \$55, which is close to \$60.

50. Sample: A company in Ohio charges \$100 for preparation of the recording plus \$7.50 per copy to print a CD. If you have \$2000, how many CDs can you afford to print?
n = number of CDs
$100 + 7.5n = 2000$; 253 copies

60.

Fahrenheit	Celsius
212°F	100°C
99°F	37°C
68°F	20°C
45°F	7°C
32°F	0°C
19°F	−7°C

LESSON 3-3

page 119 Work Together

d. 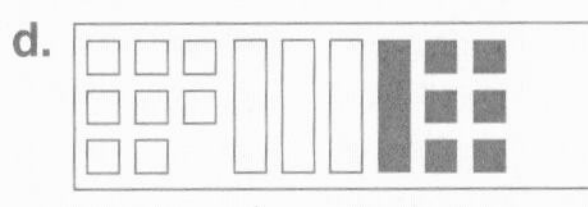

e. 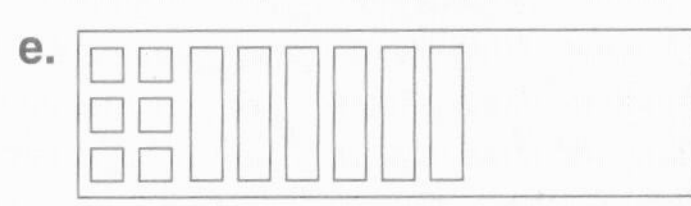

f. 

g. 

h. 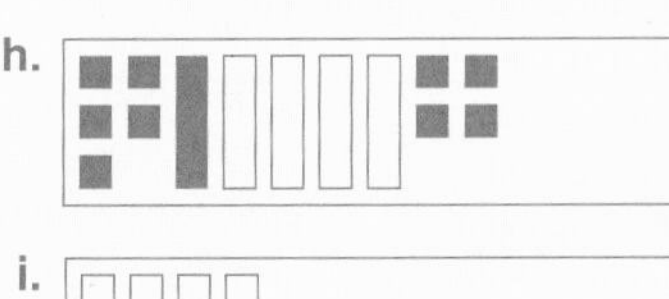

i.

pages 119–121 Think and Discuss

3. 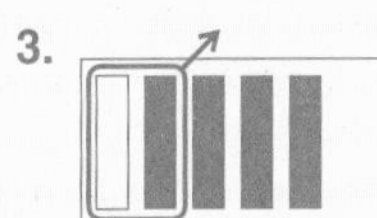

pages 121–123 On Your Own

1. 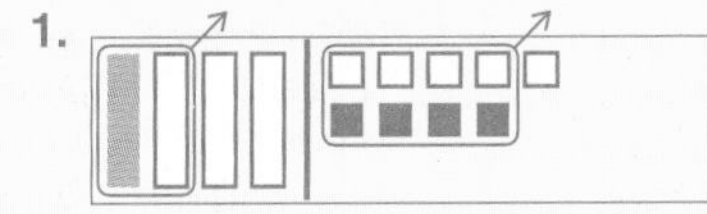

page 123 Mixed Review

53. Sample:

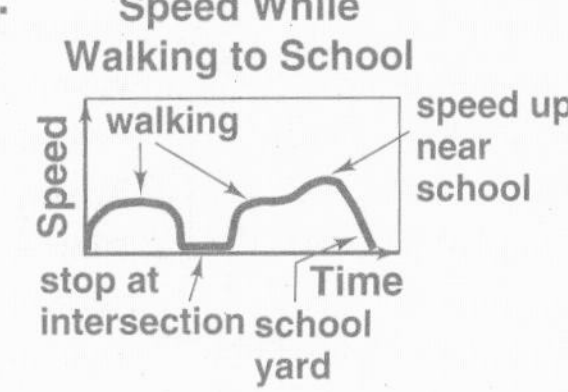

LESSON 3-5

pages 131–132 On Your Own

38. Answers may vary. Sample: You can simplify the equation to an equivalent equation with only integers by multiplying every term by a common denominator. Then use the usual methods for solving equations.

$$\frac{3}{2}x - \frac{1}{2} = \frac{1}{3}x$$
$$30 \cdot \left(\frac{3}{2}x - \frac{1}{2}\right) = 30 \cdot \frac{1}{3}x$$
$$45x - 15 = 10x$$
$$x = \frac{3}{7}$$

Math Toolbox page 145

38.

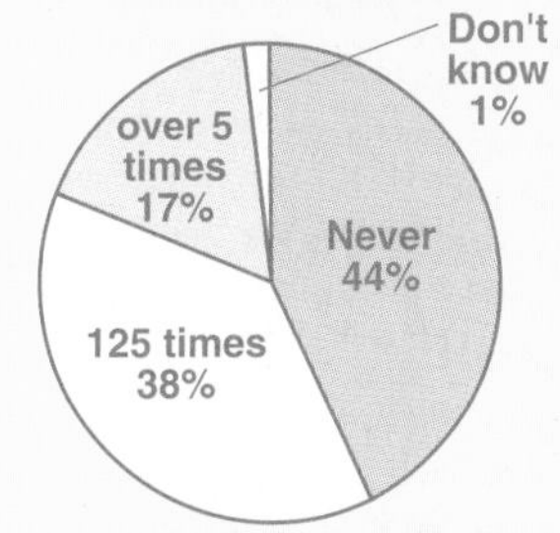

4. Answers may vary. Samples:
(1) A bar graph allows you to compare amounts in each category.
(2) A line graph shows changes over time.
(3) A circle graph displays parts of a whole.

Preparing for Standardized Tests page 155

12. Sample: 68 students and chaperones went on a field trip. An equal number of people went in each of 3 buses and the remaining 5 people went in the teacher's car. How many people were on each bus?

CHAPTER 4

LESSON 4-2

pages 166–168 On Your Own

7. 

33. An equation with the variable only on one side cannot be an identity and cannot be an equation with no solutions. In these two kinds of equations the variable must appear on both sides of the equation with the same coefficient:
$2x - 3 = 2(x - 1) - 1$ identity
$3x + 1 = 9 + 3x$ no solution
$4x + 3 = 15$ one solution.

42b.

	A	B	C
1	x	5(x − 3)	4 − 3(x + 1)
2	−5	−40	16
3	−4	−35	13
4	−3	−30	10
5	−2	−25	7
6	−1	−20	4
7	0	−15	1
8	1	−10	−2
9	2	−5	−5
10	3	0	−8
11	4	5	−11
12	5	10	−14

e.

	A	B	C
1	n	5.2n − 9	11.2n + 3
2	−5	−35	−53
3	−4	−29.8	−41.8
4	−3	−24.6	−30.6
5	−2	−19.4	−19.4
6	−1	−14.2	−8.2
7	0	−9	3

Math Toolbox page 169

6.

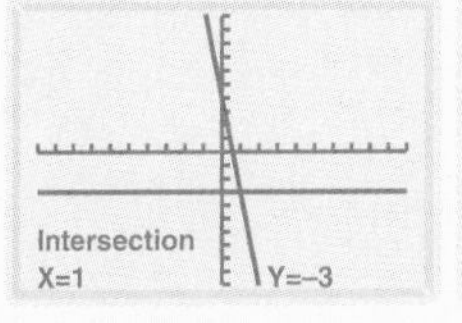

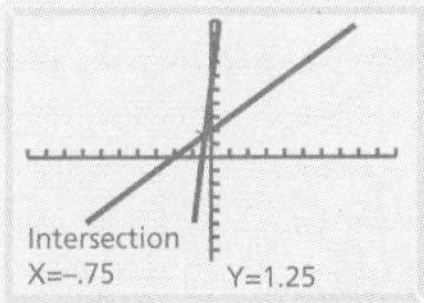

8. 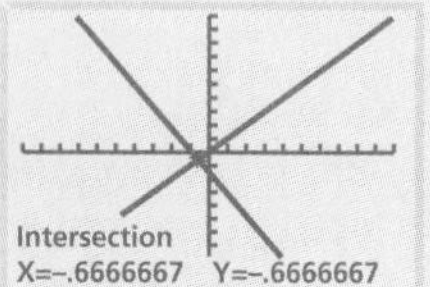

9. 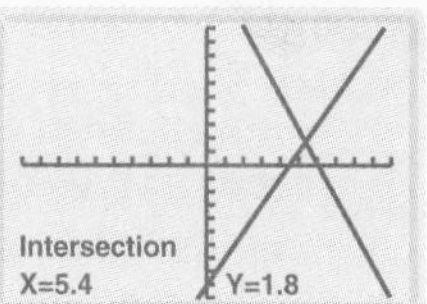

LESSON 4-4

pages 175–176 Think and Discuss

Problem Solving

$C = \frac{5}{9}(F - 32)$

$C = \frac{5}{9}F - \frac{5}{9} \cdot 32$

$C = \frac{5}{9}F - \frac{160}{9}$

$C + \frac{160}{9} = \frac{5}{9}F - \frac{160}{9} + \frac{160}{9}$

$\frac{9}{5}\left(C + \frac{160}{9}\right) = \frac{9}{5}\left(\frac{5}{9}F\right)$

$\frac{9}{5}C + 32 = F$

5a. You divide each side by 7. Next, you subtract 3 from each side. Then, you multiply each side by –1.

b. Multiply each side by 6. Add 5 to each side.

c. Multiply each side by $\frac{3}{2}$. Next, subtract 4 from each side. Then, divide each side by 2.

pages 177–178 On Your Own

27b. 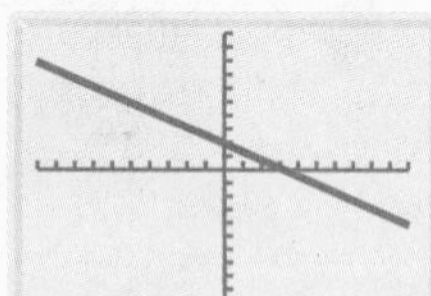

LESSON 4-5

pages 182–184 On Your Own

42. –2 –1 0 1 2 3

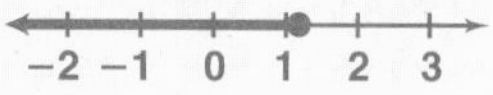

43. –4 –3 –2 –1 0 1

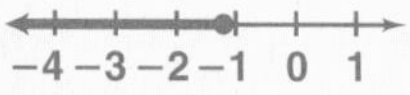

44. 0 1 2 3 4 5

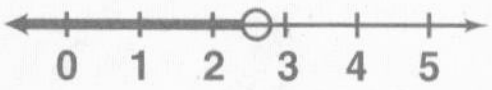

45. 3 4 5 6 7 8

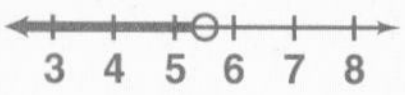

46. 8 9 10 11 12 13

47. 3 4 5 6 7 8

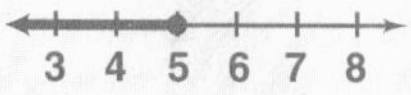

48. –2 –1 0 1 2 3

49b. Answers may vary. For samples in part (a):

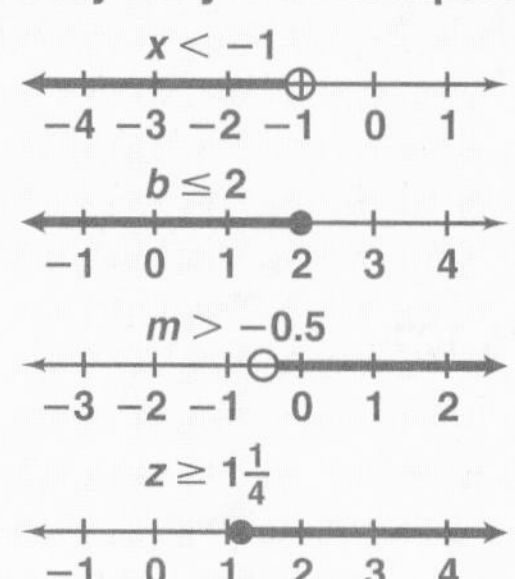

LESSON 4-6

pages 188–189 On Your Own

38. –7 –6 –5 –4 –3 –2

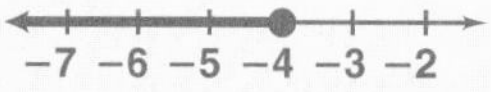

39. –10 –9 –8 –7 –6 –5

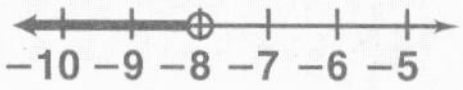

40. 0 1 2 3 4 5

41. –3 –2 –1 0 1 2

42. –5 –4 –3 –2 –1 0

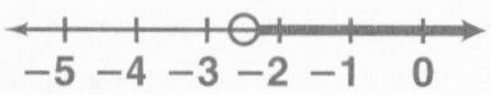

43. –49 –48 –47 –46 –45 –44

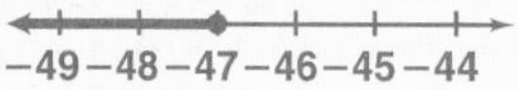

44. If you multiply or divide each side of an inequal. by a pos. no., the order of the inequal. does not change. If you multiply or divide each side of an inequal. by a neg. no., the order of the inequal. is reversed. If you multiply each side of an inequal. by 0, you no longer have an inequality.

LESSON 4-7

pages 192–194 On Your Own

17. –5 –4 –3 –2 –1 0

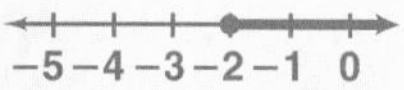

18. –1 0 1 2 3 4

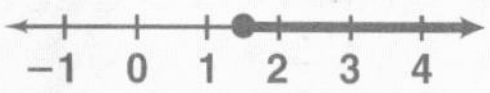

19. –5 –4 –3 –2 –1 0

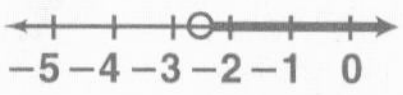

20. 5 6 7 8 9 10

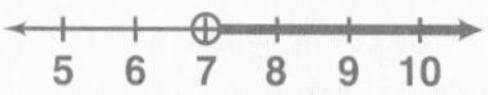

21. 0 1 2 3 4 5

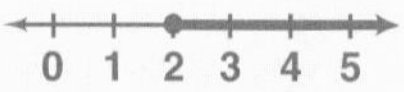

22. –2 –1 0 1 2 3

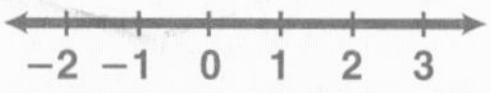

23. –10 –9 –8 –7 –6 –5

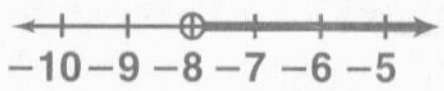

24. –2 –1 0 1 2 3

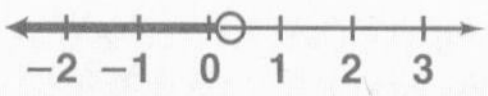

25. –5 –4 –3 –2 –1 0

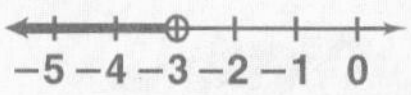

27. You use the distributive property on each side and collect like terms; $2.5p - 10 > 3p + 6$. Next, use the Add. or Subt. Prop. of Inequal. to subtract 6 from each side; $2.5p - 16 > 3p$. Then, subtract $2.5p$ from each side; $-16 > 0.5p$. Use the Mult. Prop. to multiply each side by 2; $-32 > p$.

page 194 Mixed Review

52.

53. –6 –5 –4 –3 –2 –1

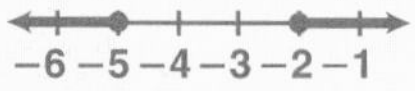

54. 6 7 8 9 10 11 12 13

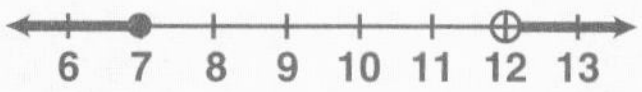

55. –2 –1 0 1 2 3

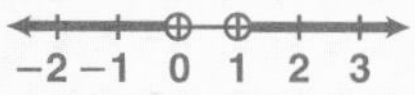

LESSON 4-8

pages 198–200 On Your Own

28. To solve an abs. val. inequality, first, change it to a compound inequality. To solve $|2x - 1| < 3$, change it to $-3 < 2x - 1 < 3$. If the abs. val. is less than or less then or equal to a positive number, use an *and* inequality.
$|2x - 1| \leq -1$ has no solution because abs. val. cannot be negative. If the abs. val. is less than a negative number or 0 or less than or equal to a negative number, the inequality has no solution.
To solve $|2x - 1| \geq 3$, change it to $2x - 1 \leq -1$ or $2x - 1 \geq 1$. If the abs. val. is greater than or greater then or equal to a number, use an *or* inequality.
Then solve the compound inequality.

29. $f < -2.5$ or $f > 2.5$

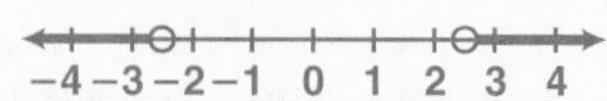

30. $-8 < x < 2$

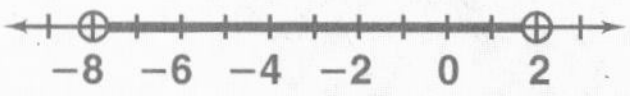

–8 –6 –4 –2 0 2

31. $n \leq -13$ or $n \geq -3$

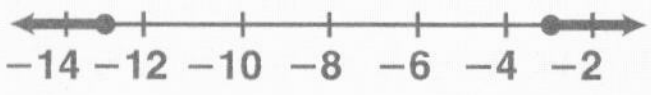

–14 –12 –10 –8 –6 –4 –2

32. $y \leq -2$ or $y \geq 5$

33. $w < -2$ or $w > 2$

34. $-5 \leq n \leq 5$

35. $-6 < x < 6$

36. $d \leq -2$ or $d \geq 2$

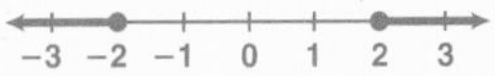

37. $y \leq 1$ or $y \geq 3$

38. $-3 < c < 6$

39. $t \leq -2.4$ or $t \geq 4$

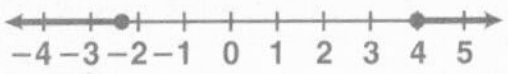

40. $-2 < p < 8$

41. $t < -3$ or $t > 2\frac{1}{3}$

42. no solution

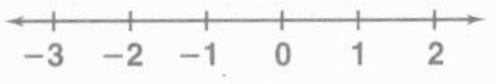

43. all numbers

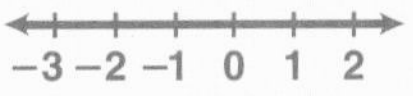

44. all numbers

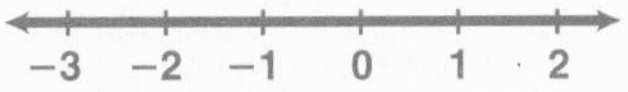

Checkpoint page 200

1. $c > 6$

2. $x \leq -8$

3. $m \leq -4$

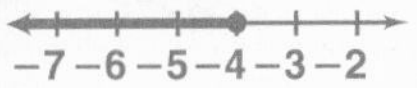

4. $c \geq -5$

5. $b > \frac{1}{3}$

6. $n \leq 3$

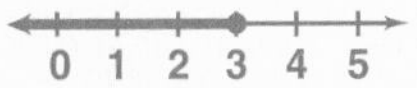

7. $r \leq 7$

8. $w \leq -2$

9. $h \geq -4$

10. $0 \geq t \geq -0.5$

11. $x < -27$ or $x > 4$

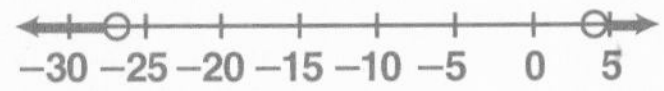

12. $g < 1$ or $g > 5$

13. $d \leq -3$ or $d \geq 3$

14. $-10 < x < 4$

15. $-1 < x < 4$

16. $y \leq -4$ or $y \geq 5$

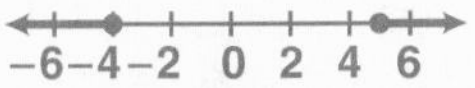

17. $2\frac{2}{5} \leq x \leq 5\frac{3}{5}$

20. A solution of an "and"-inequal. is a no. that makes both inequals. true. On a number line the solution is the part that is common to both graphs. A solution of an "or"-inequal. is a no. that makes either inequal. true. On a number line, the graph of the compound inequal. includes the graphs of each inequal.

Math Toolbox page 201

2.

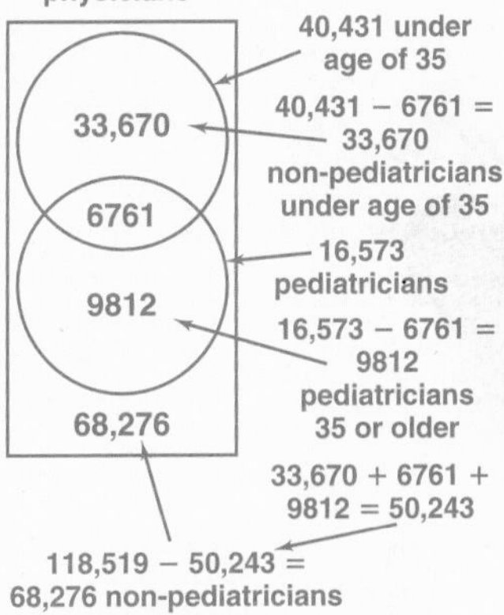

3.

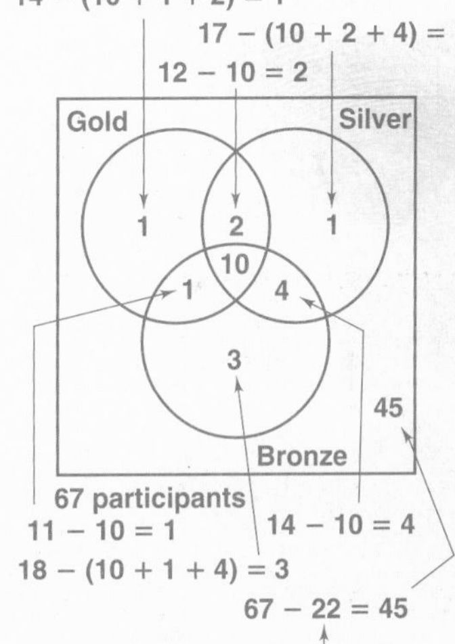

4. Sample: Lincoln High School has 173 students in its first-year class. 69 of them speak Spanish and 52 speak French. 18 first-year students speak both French and Spanish. How many first-year students do not speak either French or Spanish?
70 first-year students do not speak French or Spanish.

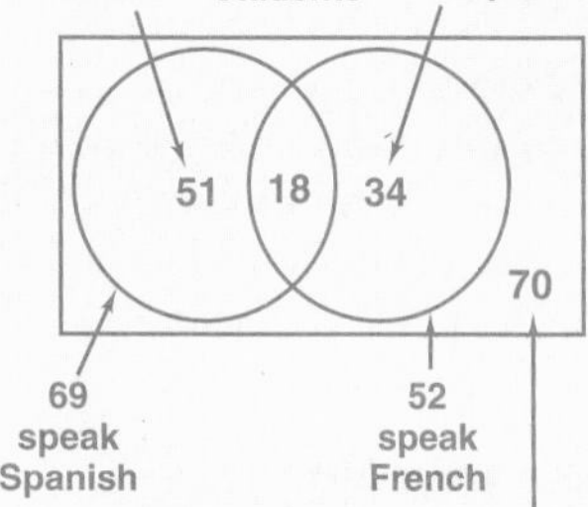

69
speak
Spanish

52
speak
French

51 + 18+ 34 = 103 ⟶ 173 − 103 = 70
speak French or
Spanish

do not speak
French or
Spanish

LESSON 4-9

pages 204–205 On Your Own

1.
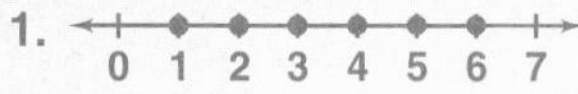

2.  0 1 2 3 4 5 6 7 8 . . .

4.  0 1 2 3 4 5 6 7 8

6.  0 1 2 3 4 5 6 7 . . .

7. −1 0 1 2 3 4 5 6

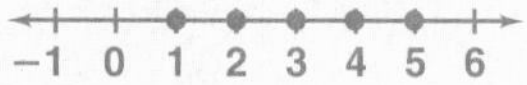

9.  3 4 5 6 7 8 . . .

10. 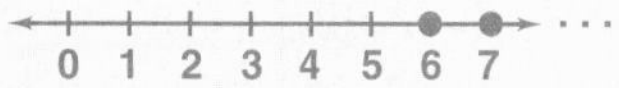 0 1 2 3 4 5 6 7 . . .

11.  0 1 2 3 4 5 . . .

12.  −1 0 1 2 3 4 . . .

13. 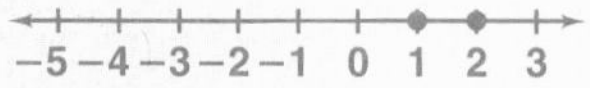 −5 −4 −3 −2 −1 0 1 2 3

14. 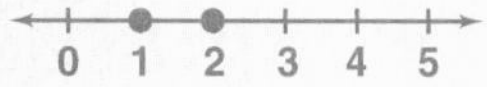 0 1 2 3 4 5

16. −5 −4 −3 −2 −1 0 1 2 3 4

17. 10 20 30 40 50 60 70

18. 17 18 19 20 21 22 23 24 25 26

19. 21 22 23 24 25 26

20. 14 15 16 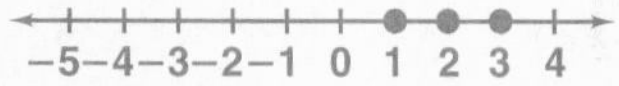 17  18  19  20  21

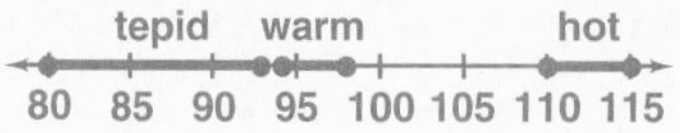

21. tepid warm hot
80 85 90 95 100 105 110 115

22. Sample: A TV news meteorologist guarantees tomorrow's high temp. to be within 5°F of 68°F. What are the possible high temperatures that she predicts for tomorrow?

24. 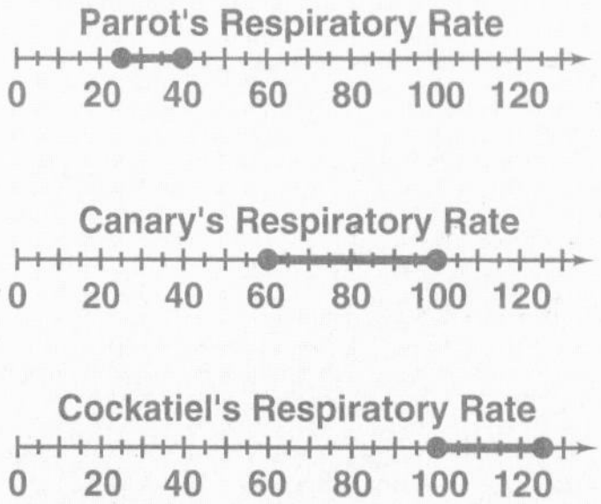 Parrot's Respiratory Rate
0 20 40 60 80 100 120

Canary's Respiratory Rate
0 20 40 60 80 100 120

Cockatiel's Respiratory Rate
0 20 40 60 80 100 120

25. The replacement set is the set of possible values for a variable that are consistent with the situation in the problem. The solutions to an inequality usually are real numbers. In a problem asking about a number of people a solution makes sense only if it is an integer, so the replacement set would be integers.

26.

27.  −4 −3 −2 −1 0 1 2 3 4

28. 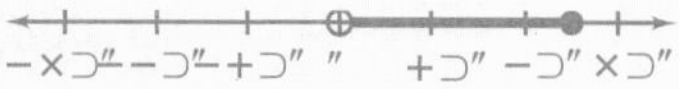 −×⊃″ −⊃″ +⊃″ ″ +⊃″ −⊃″ ×⊃″

29. 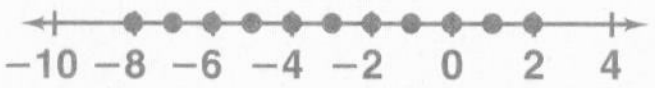 −10 −8 −6 −4 −2 0 2 4

Chapter Assessment **page 210**

31. $m > -3\frac{5}{7}$
−6 −5 −4 −3 −2 −1

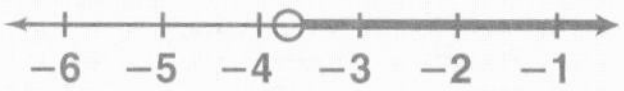

32. $x \leq -5$ or $x \geq 15$

33. $-3 < h < 2$
−4 −3 −2 −1 0 1 2 3

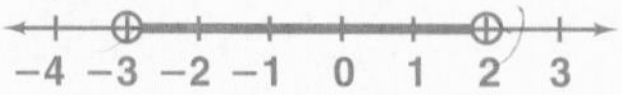

34. $-6 < b \leq -3$

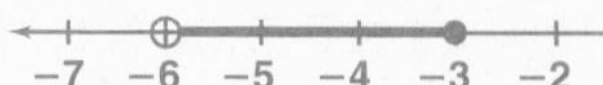

35. $-2\frac{1}{2} < q < 3$

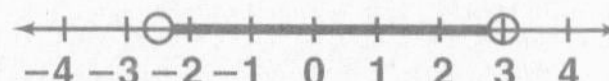

36. $n < -5$ or $n \geq -1$
−6 −5 −4 −3 −2 −1 0 1

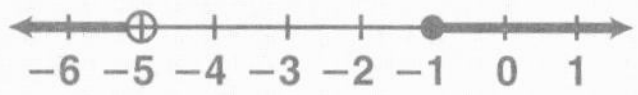

37. $k \leq 4$

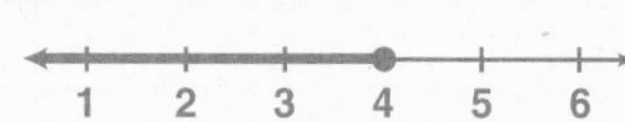

39a.  −5 −4 −3 −2 −1 0 1 2 3 4

b.

52.
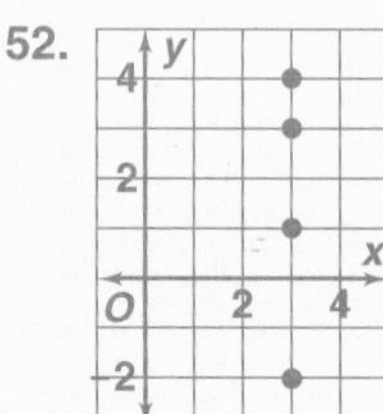

CHAPTER 5

LESSON 5-1

pages 217–218 On Your Own

11.
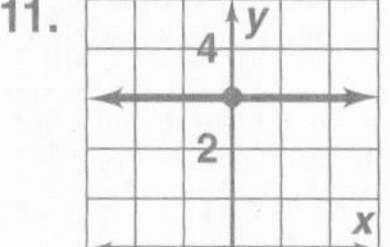

12.
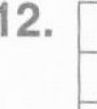
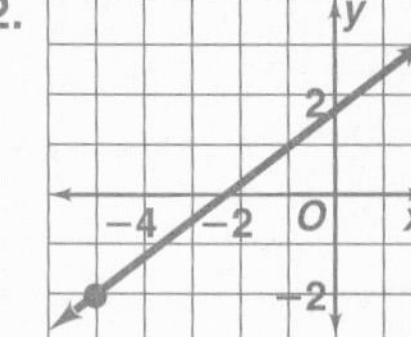

13.
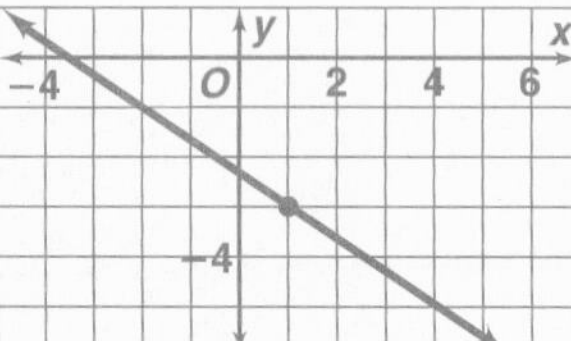

14.
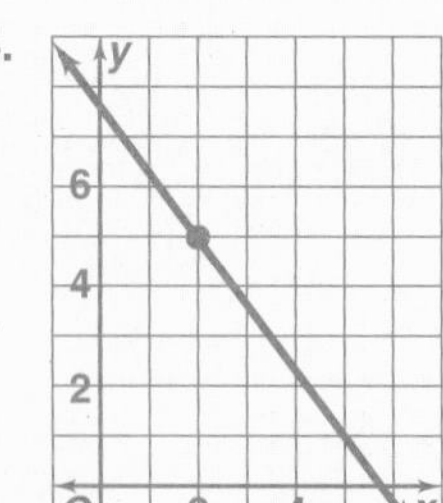

15.
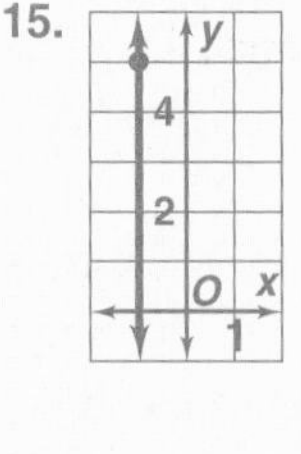

16.
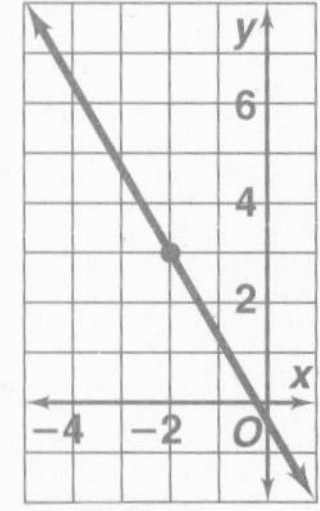

TE page 218 Lesson Quiz

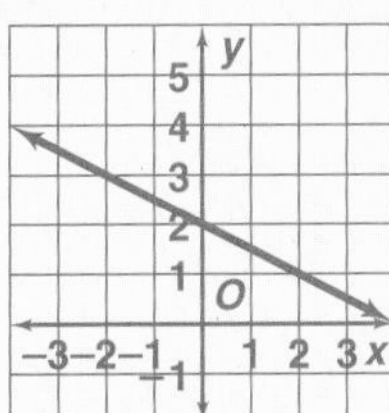

LESSON 5-2

page 224 Mixed Review

31. Tables may vary. Sample:

x	y
−3	−2.4
−1	−0.8
0	0
1	0.8

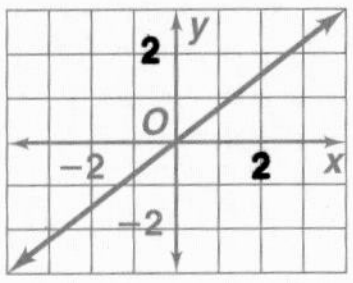

32. Tables may vary. Sample:

x	y
−2	−1.5
−1	−0.75
0	0
1	0.75
2	1.5
4	3

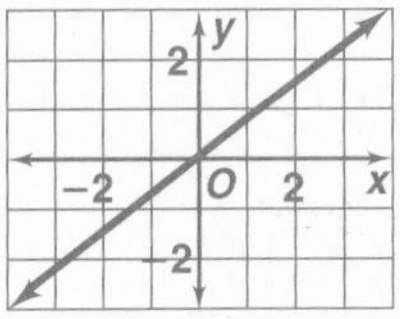

33. Tables may vary. Sample:

x	y
−1	−7.2
−0.5	−3.6
0	0
0.5	3.6
1	7.2

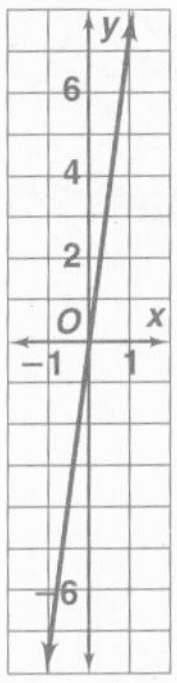

LESSON 5-3

pages 228–229 On Your Own

16.

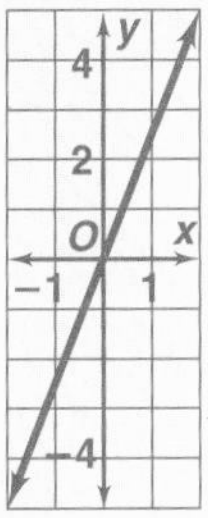

23b.

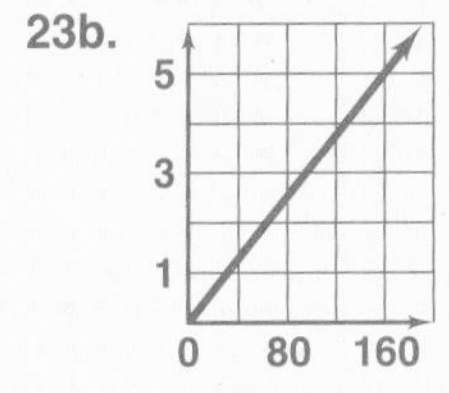

page 229 Mixed Review

34.

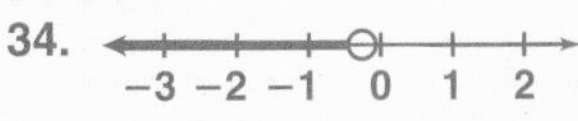

35.

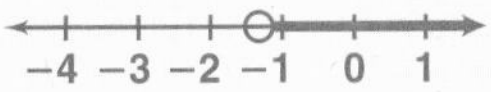

36.

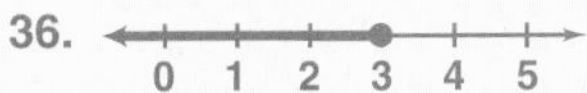

38.

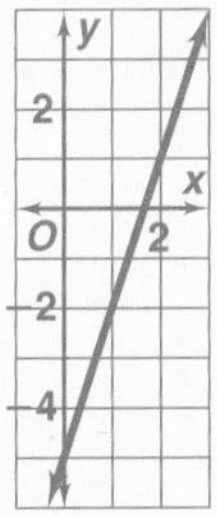

39.

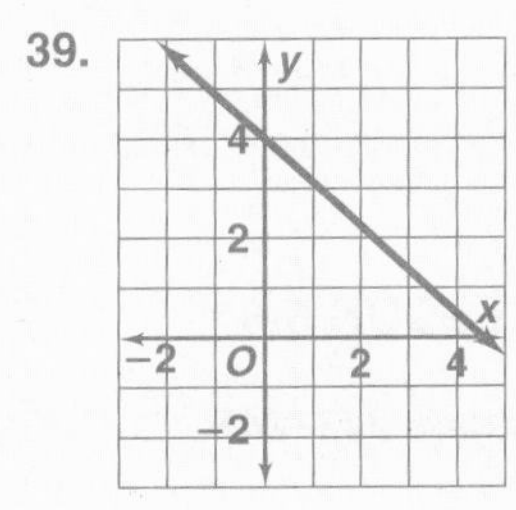

LESSON 5-4

pages 232–234 On Your Own

10a.

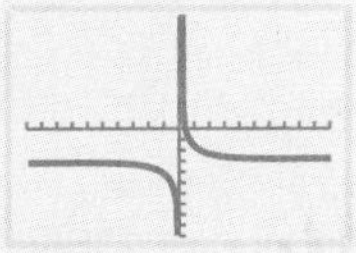

b.

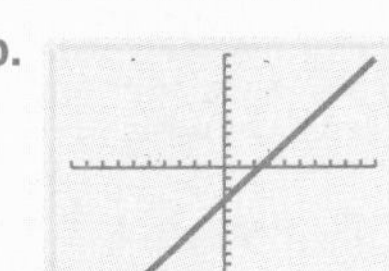

12.

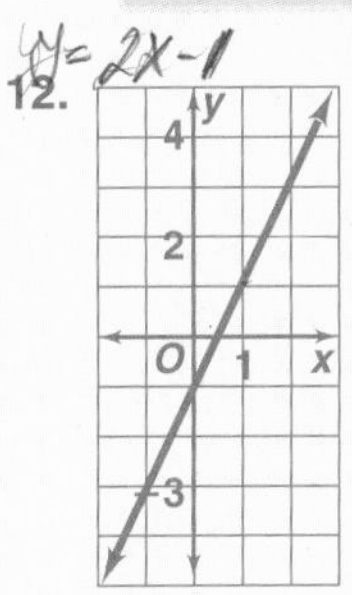

13.

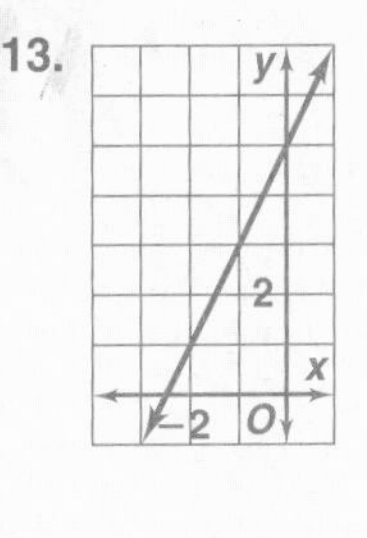

14.

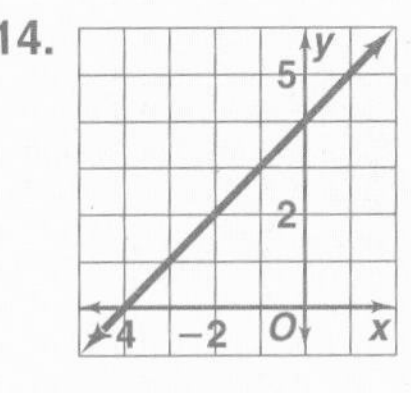

15.

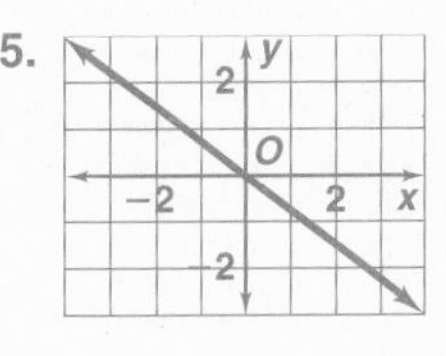

16.

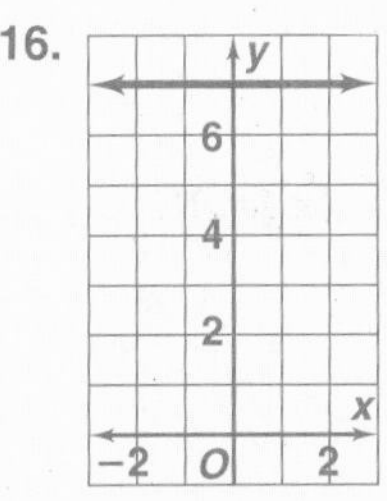

17.

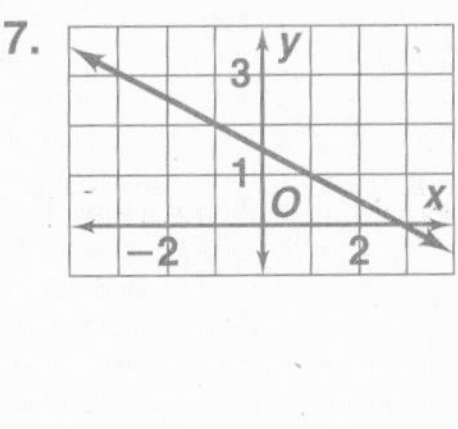

18.

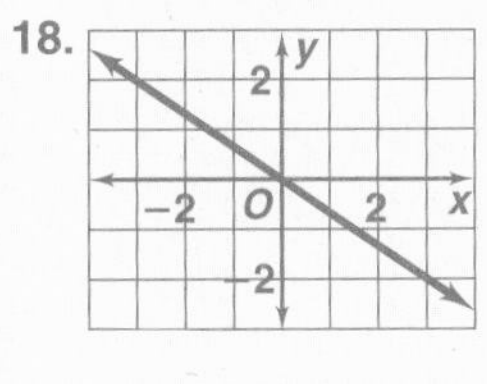

19.

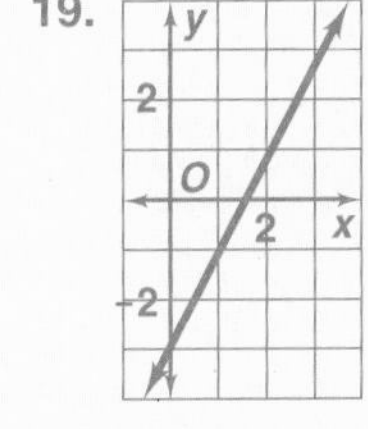

20.

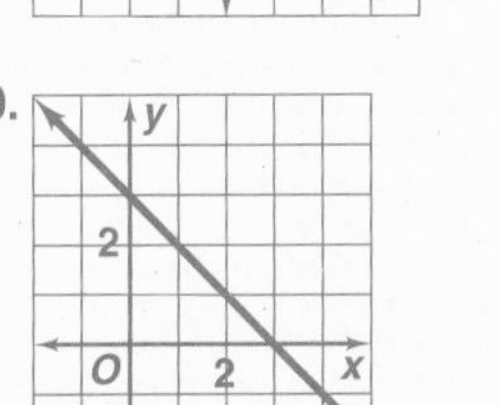

24b.

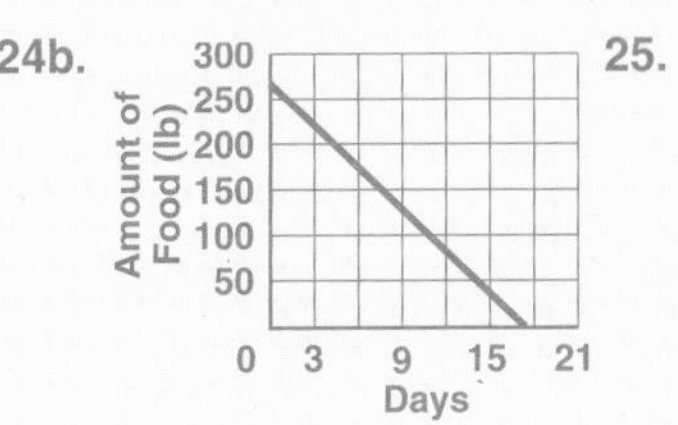

25.

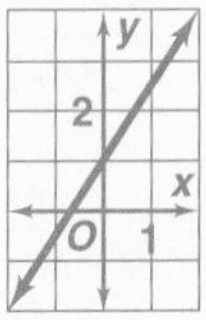

26. Answers may vary. Sample: Find (0, 5). Then find the point 4 units to the right and 3 units up. Draw the line through these points.

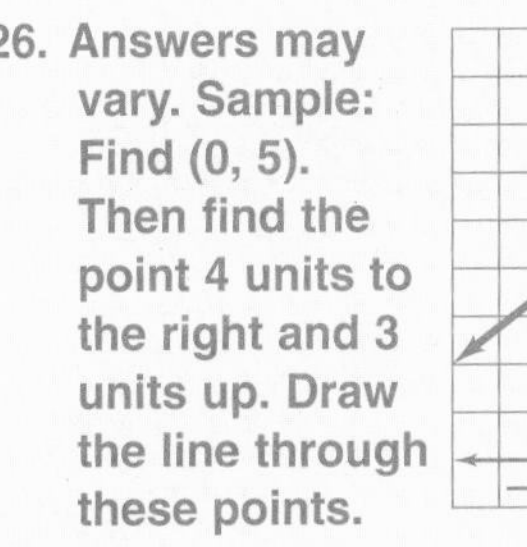

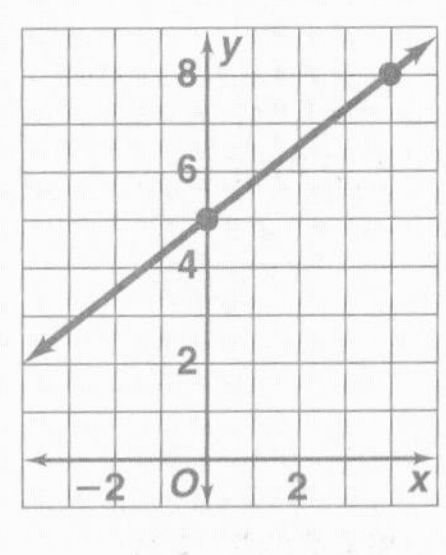

page 234 Mixed Review

35.

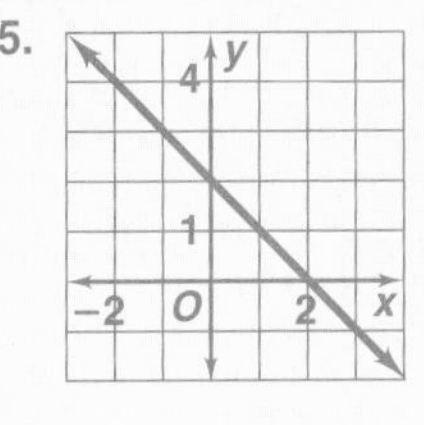

36.

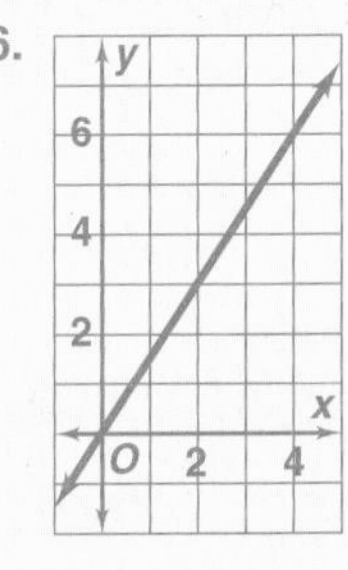

37.

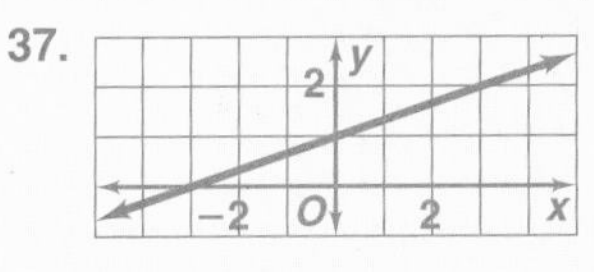

38.

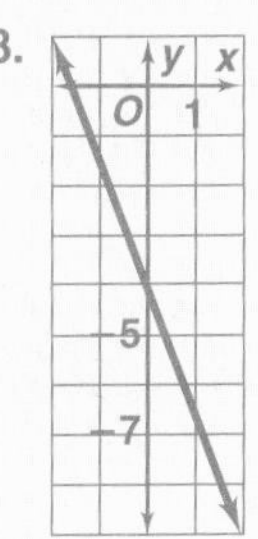

TE page 234 Lesson Quiz

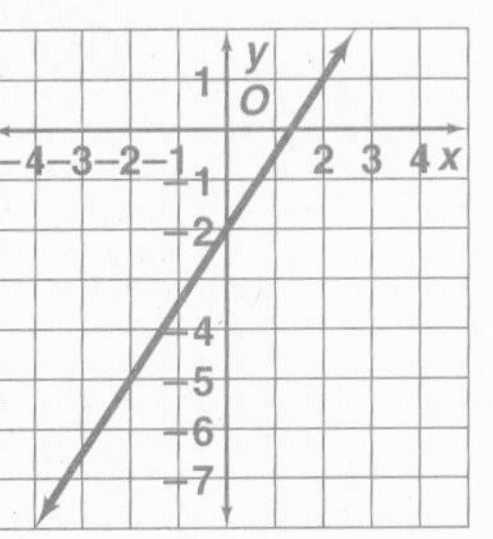

Math Toolbox **page 235**

2b.

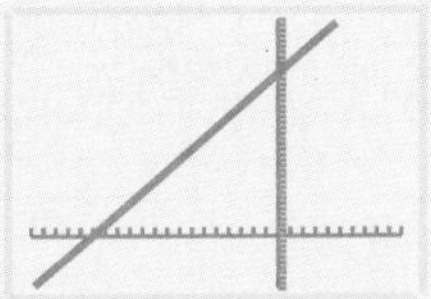

Xmin=−20 Ymin=−10
Xmax=10 Ymax=40
Xscl=1 Yscl=1

3. Xmin = −10
Xmax = 2

Ymin = −2
Ymax = 24

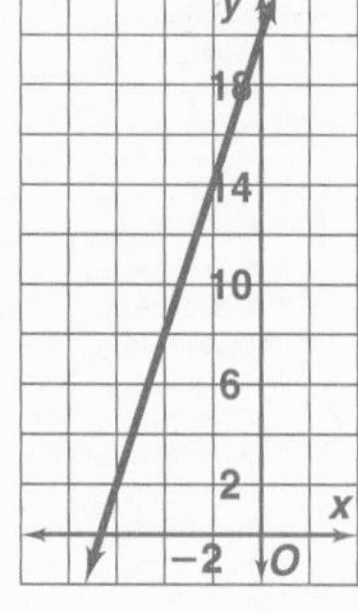

4. Xmin = −20
Xmax = 120

Ymin = −20
Ymax = 120

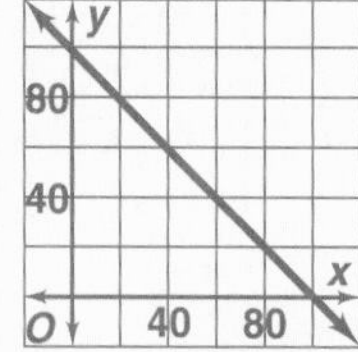

5. Xmin = −20
Xmax = 160

Ymin = −40
Ymax = 20

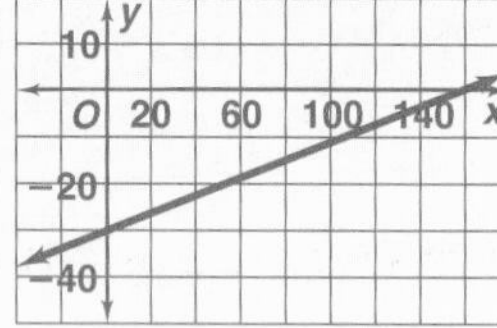

6. Xmin = −2
Xmax = 6

Ymin = −0.2
Ymax = 0.2

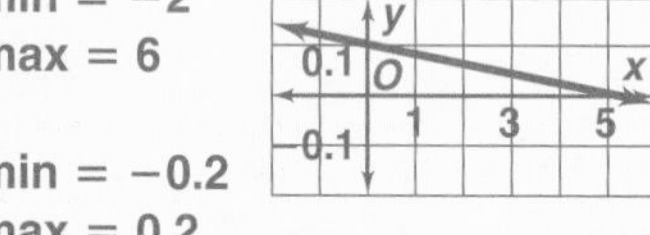

7. Xmin = −2800 Ymin = −400
Xmax = 800 Ymax = 1600

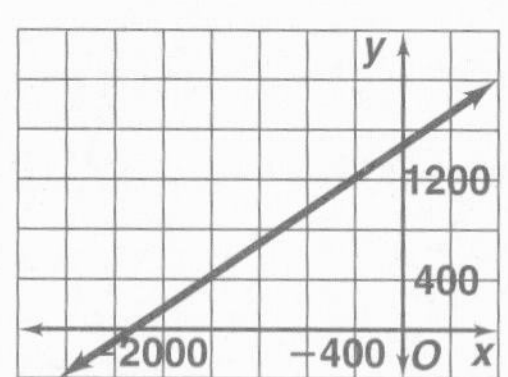

8. Xmin = −240
Xmax = 40

Ymin = −70
Ymax = 10

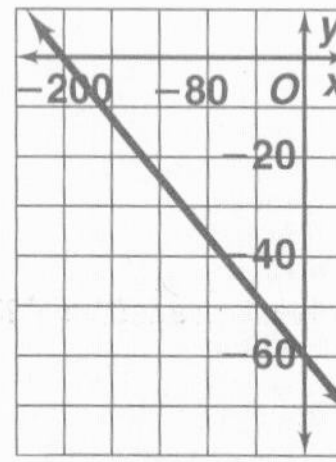

9. Xmin = −2
Xmax = 2

Ymin = −10
Ymax = 60

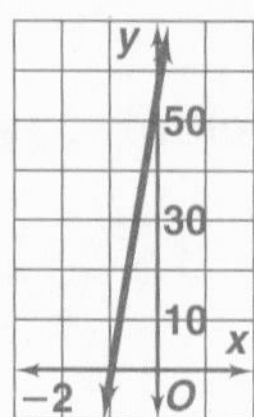

10. Xmin = −0.01
Xmax = 0.01

Ymin = −0.05
Ymax = 0.02

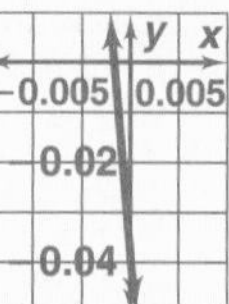

LESSON 5-6

pages 243–245 On Your Own

13a.

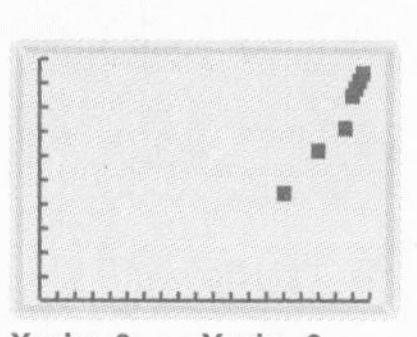

Xmin=0 Ymin=0
Xmax=95 Ymax=10
Xscl=5 Yscl=1

b. Answers may vary. Sample:
$y = 0.2278305085x - 11.71566102$

c. about 12.2%

14a.

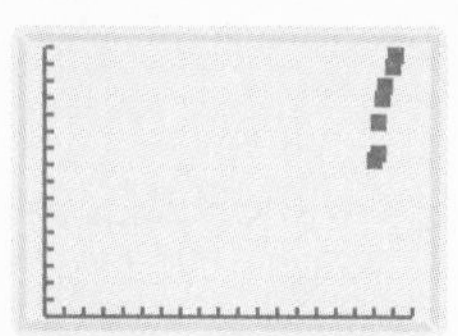

Xmin=0 Ymin=0
Xmax=95 Ymax=80
Xscl=5 Yscl=5

b. Answers may vary. Sample:
$y = 5.432142857x - 415.4714286$

c. about $154.9 billion

15a. $y = 2x - 1$

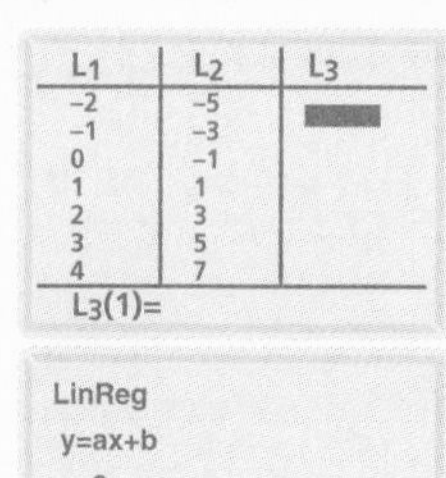

L1	L2	L3
−2	−5	
−1	−3	
0	−1	
1	1	
2	3	
3	5	
4	7	

L3(1)=

LinReg
y=ax+b
a=2
b=−1
r=1

16a.

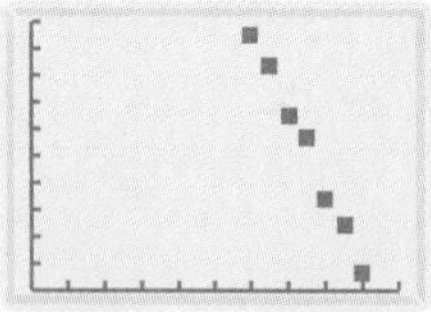

Xmin=0 Ymin=0
Xmax=100 Ymax=100
Xscl=10 Yscl=10

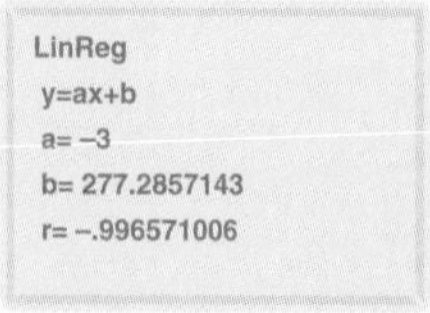

LinReg
y=ax+b
a=−3
b=277.2857143
r=−.996571006

$y = -3x + 277.2857143$

19a.

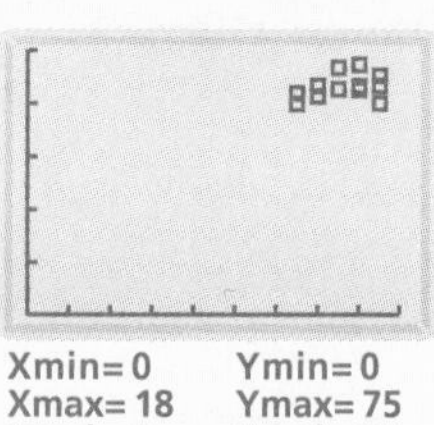

Xmin=0 Ymin=0
Xmax=18 Ymax=75
Xscl=2 Yscl=15

pages 246–248 Think and Discuss

3. x-intercept = 13.5;
y-intercept = 4.5

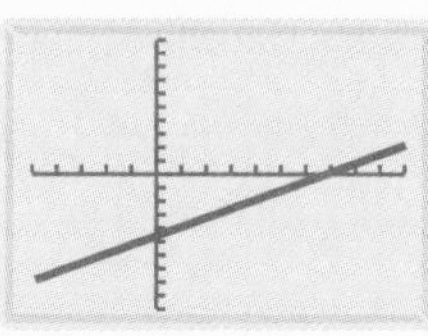

Xmin=−10 Ymin=−10
Xmax=20 Ymax=10
Xscl=2 Yscl=1

4a.

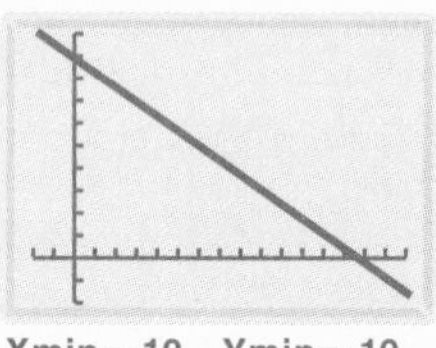

Xmin=−10 Ymin=−10
Xmax=80 Ymax=50
Xscl=5 Yscl=5

pages 248–249 On Your Own

4.

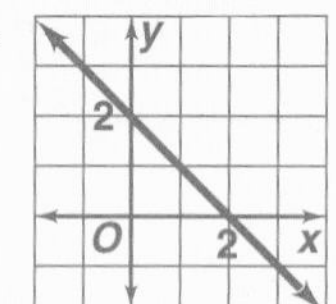

5.

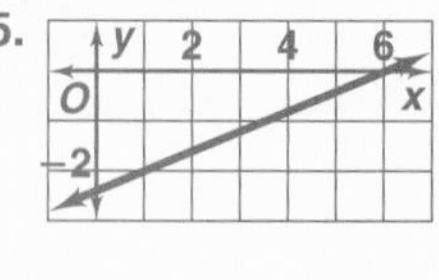

6.

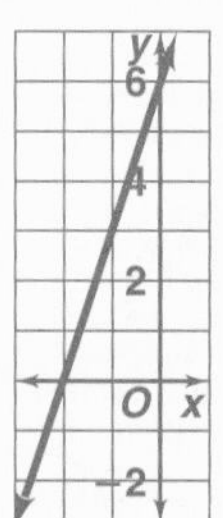

7.

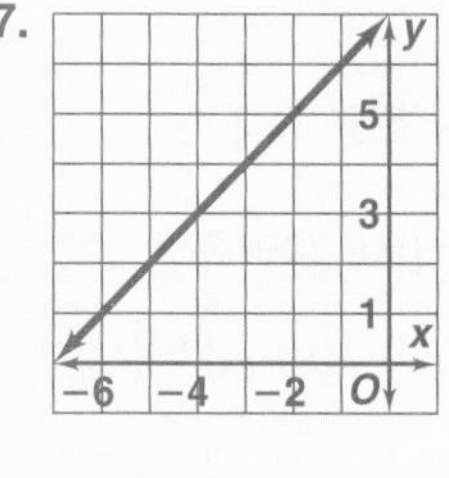

8.

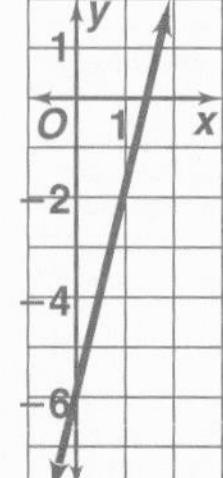

9.

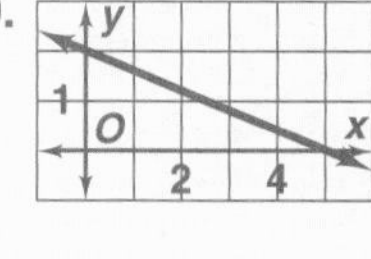

10. 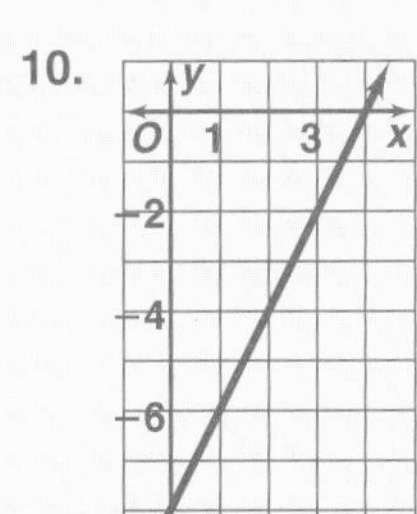

11. 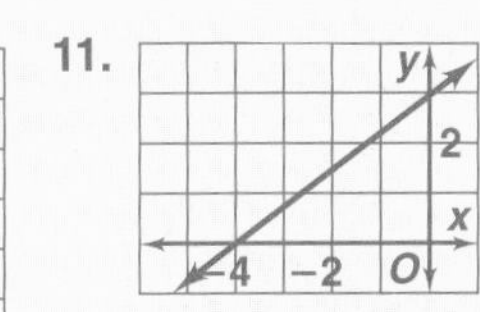

12b. Answers may vary. Sample: $.75 and $5, $1.50 and $4, $3 and $2; $1.50 and $4 is the best choice because you want to charge children less than adults, but not too little.

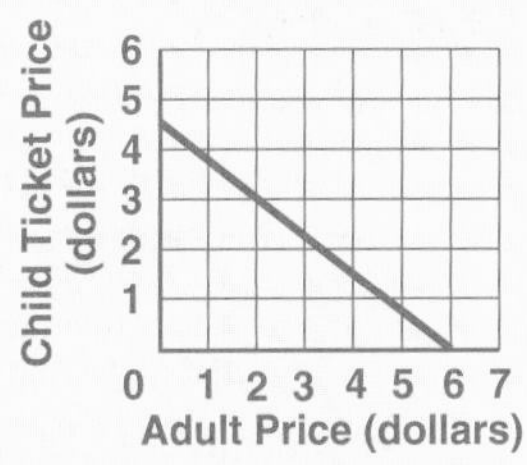

13. Answers may vary. Sample: The slope-intercept form provides more information about the rate of change of the corresponding linear function. You can use the $Ax + By = C$ form when modeling a real-world situation, like the one in Exercise 12.

15.

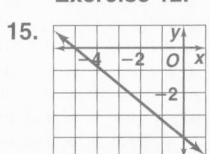

16. 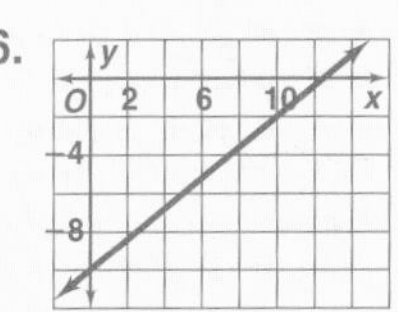

17.

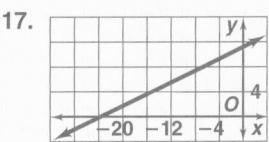

18. 

19. 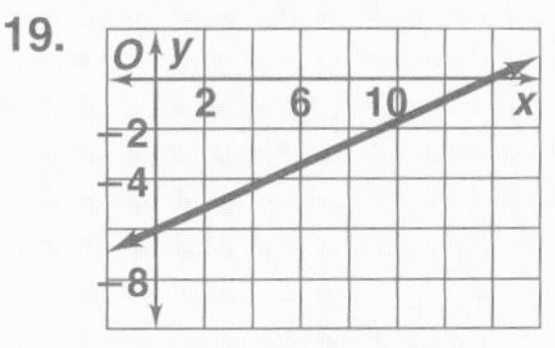

20.

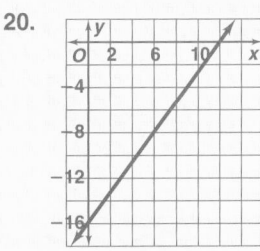

21. 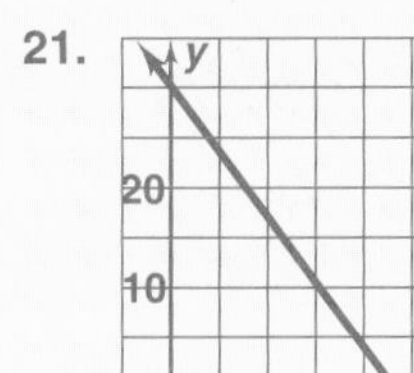

22. 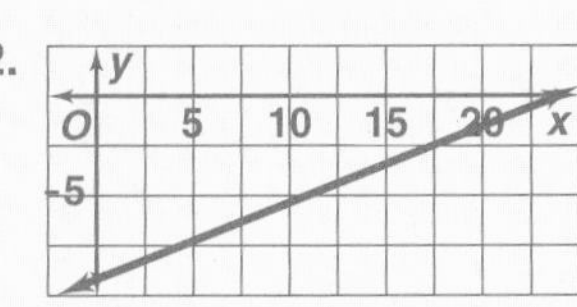

32b. 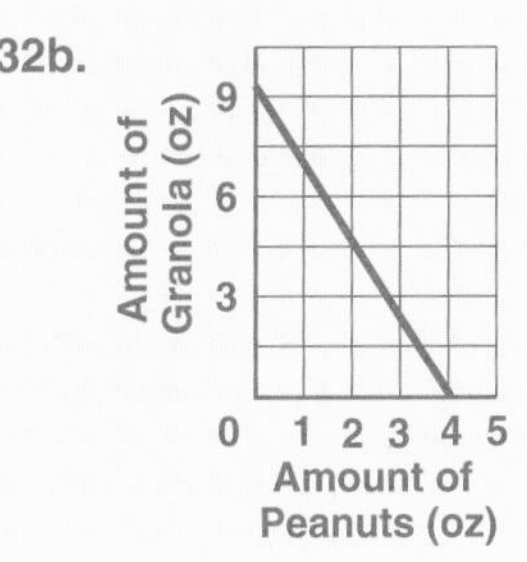

page 249 Mixed Review

35. 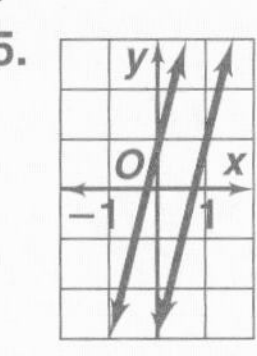

36. 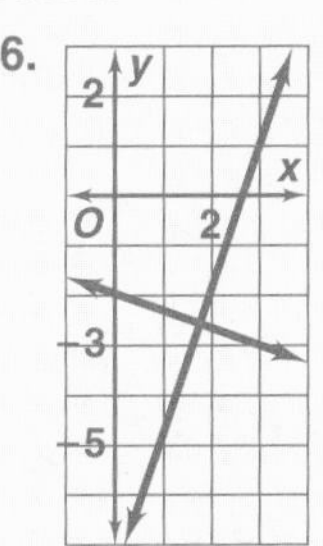

37. 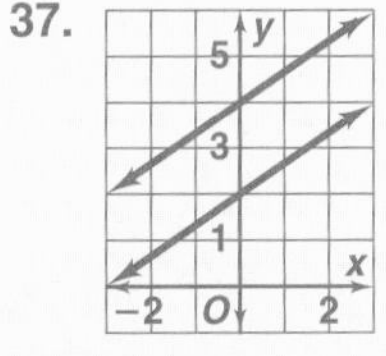

38. 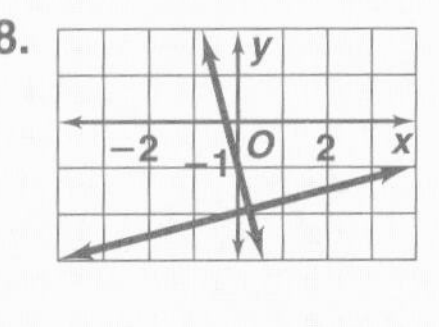

LESSON 5-9

page 256 Work Together

2a. 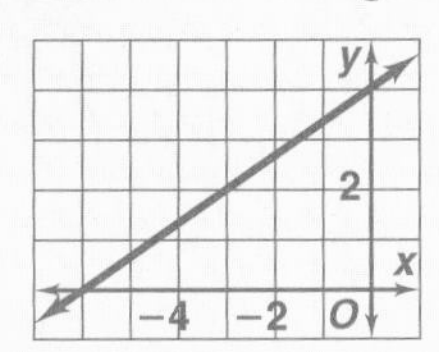

pages 258–259 On Your Own

19b. 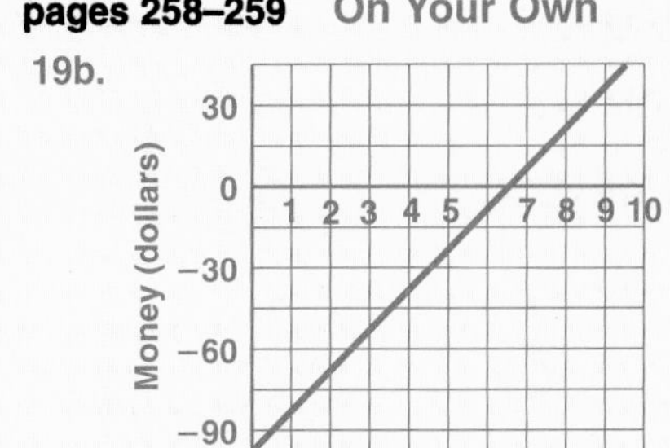

21. 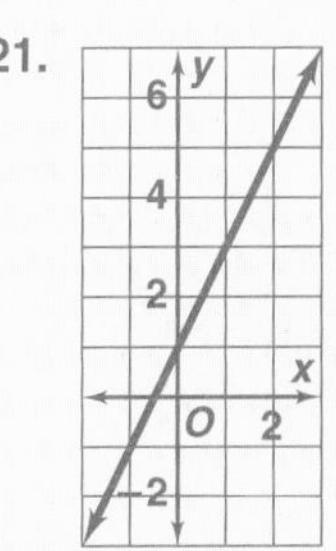

Wrap Up pages 261–263

36. 37. 38.

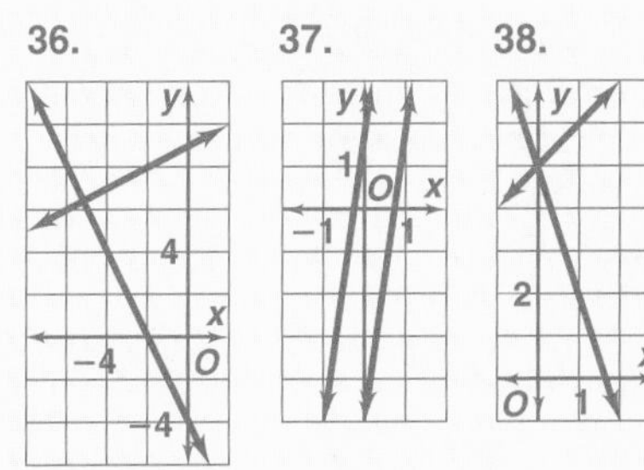

39. 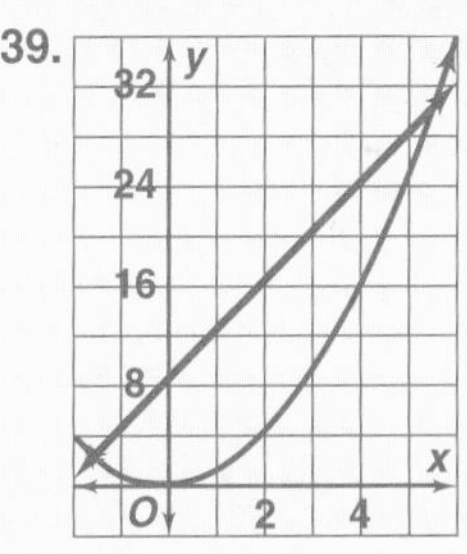

40. 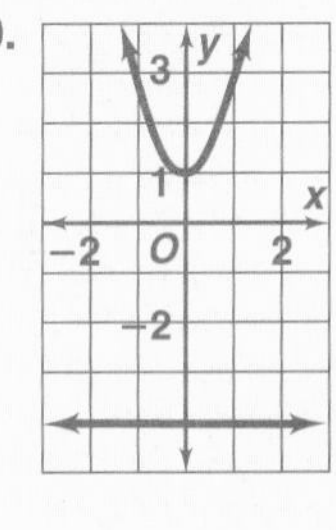

Chapter Assessment page 264

8. 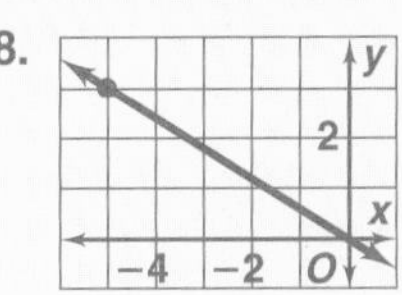

17. For a direct variation the y-value is a multiple of the x-value, so the y-value is 0 when the x-value is 0.

32b. 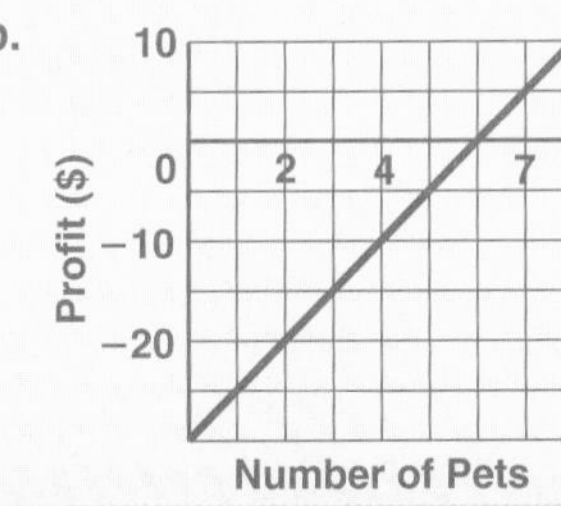

CHAPTER 6

LESSON 6-1

pages 272–274 On Your Own

3.

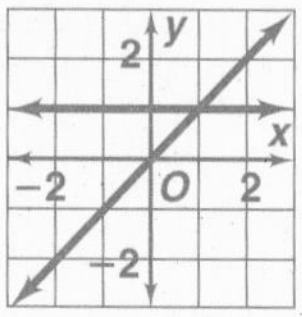

4. 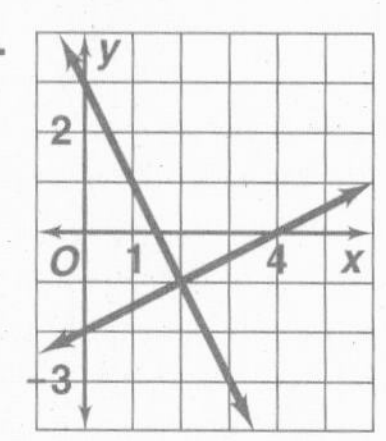

5. 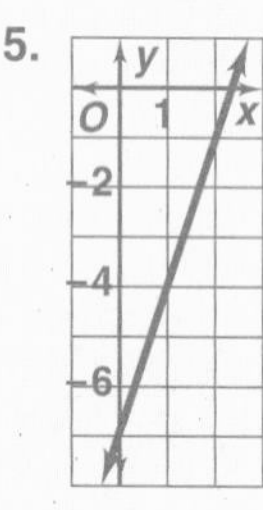

6b. 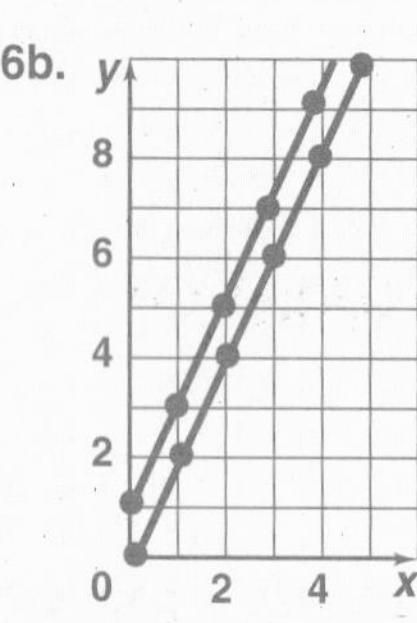

11b. The solution of the system is the point in time when the fear of commitment is equal to the fear of baldness.

c. The solution on Lily's chart shows the point in time when the level of hope of meeting Brad Pitt is equal to the level of fear of cellulite.

13.

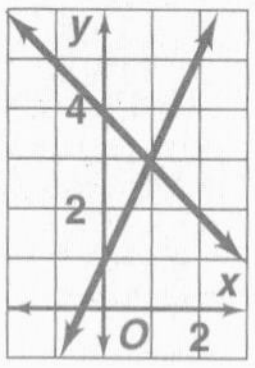

14. 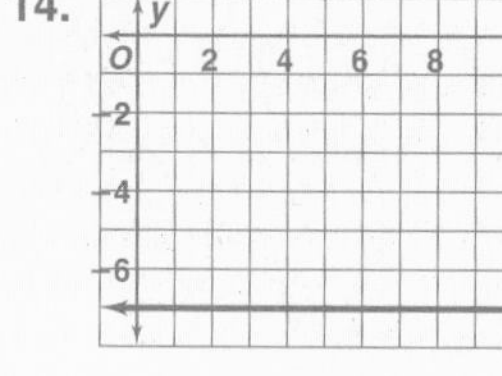

15. 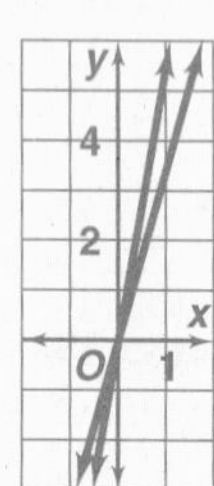

16. 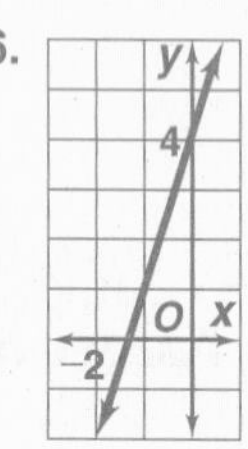

17. 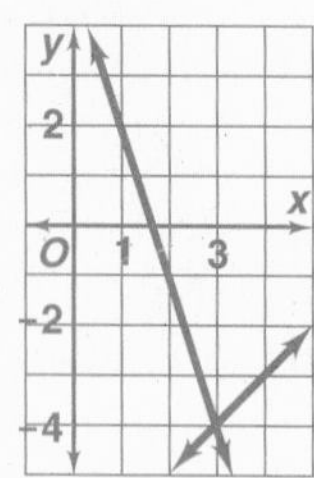

18. 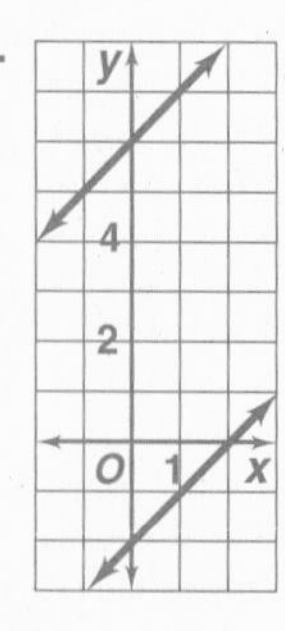

19. 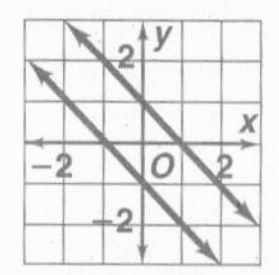

20. 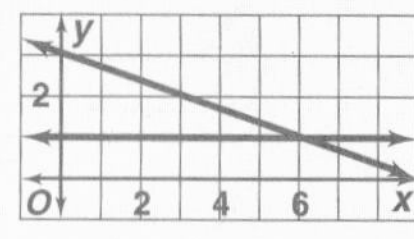

21. 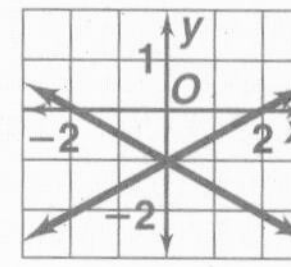

22. 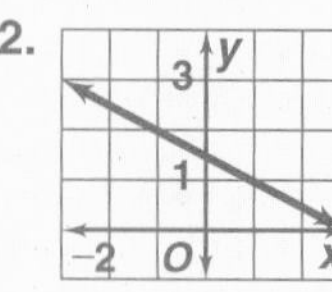

23. 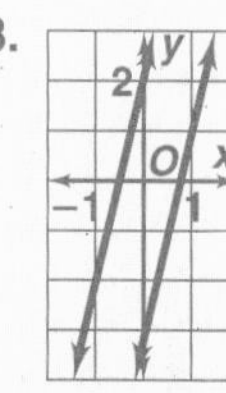

24. 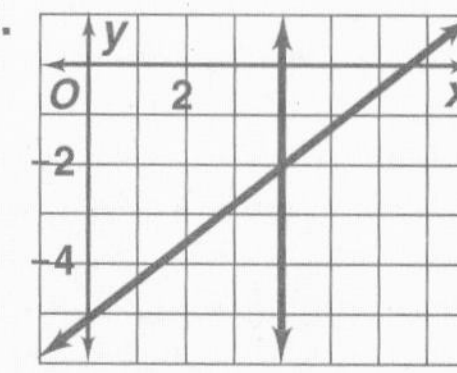

25. Answers may vary. Sample: The two lines can cross at 1, 0 or infinitely many points.

37. $y = -x + 2$
$y = x + 1$

38. $y = 2x + 1$
$y = 2x + 5$

39. $y = -x - 4$
$y = 2x - 4$

40. $y = 4x$
$2y - 8x = 0$

LESSON 6-2

pages 278–279 On Your Own

23. 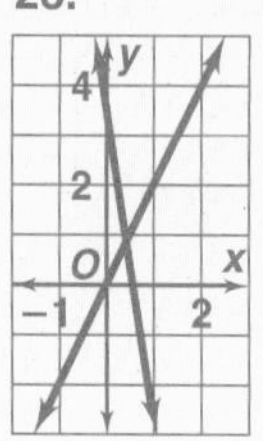

24. 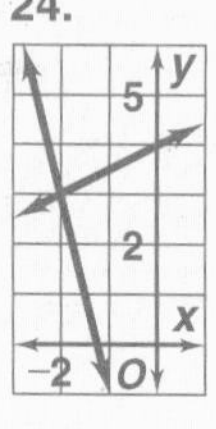

25. 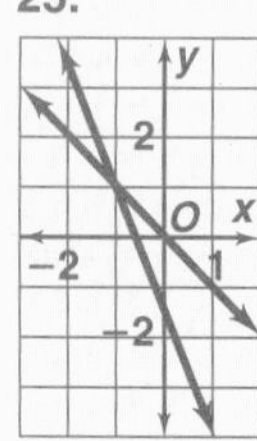

26. 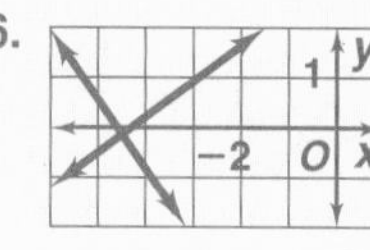

27. 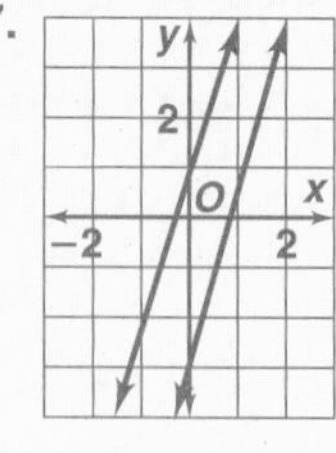

28. Answers may vary. Sample: You can use substitution to solve any system. If there is one solution, you can find the exact values for x and y. To solve
$y = x + 1$
$11y = 17x$
use substitution to find the solution $(\frac{11}{6}, \frac{17}{6})$. Graphing will not give the exact solution.

LESSON 6-4

pages 287–288 On Your Own

19. Substitution; at least one equation is solved for one of the variables.
20. Substitution; at least one equation is solved for one of the variables.
21. Elimination; coefficients match well for subtraction.
22. Substitution; at least one equation is solved for one of the variables.
23. Substitution; coefficients -1 and 1 make it easy to solve either equation for one of the variables.
24. Substitution; at least one equation is solved for one of the variables.

page 288 Mixed Review

32. 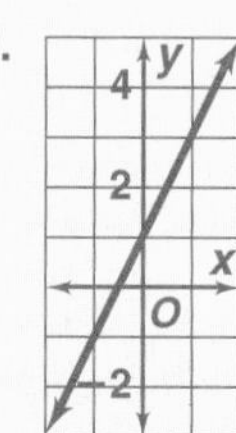

33. 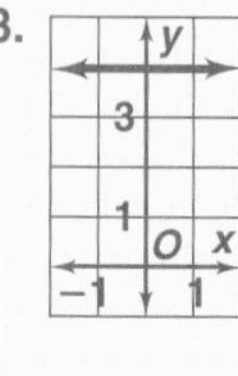

34. 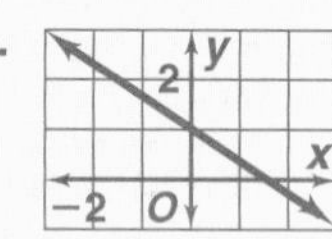

35. 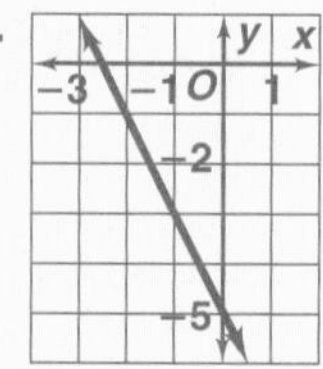

LESSON 6-5

pages 292–293 On Your Own

9. 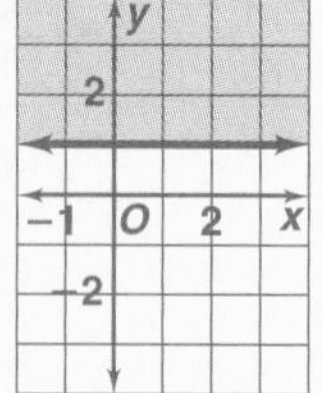

10. 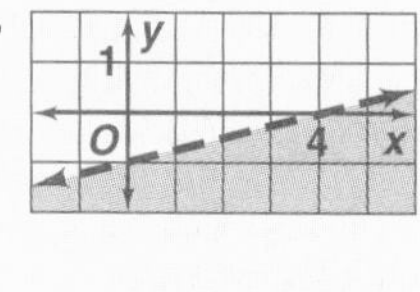

11. 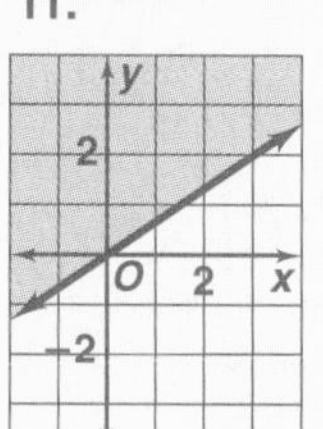

12. 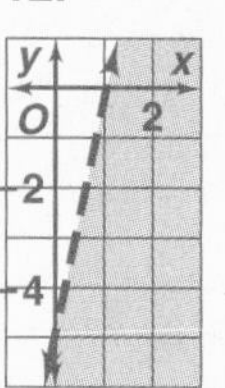

13. 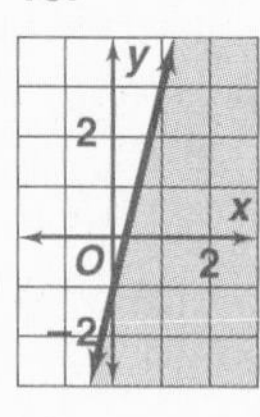

14.

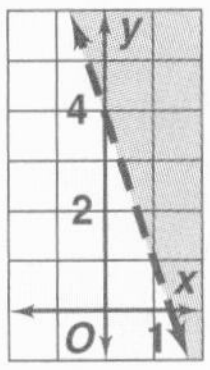

15.

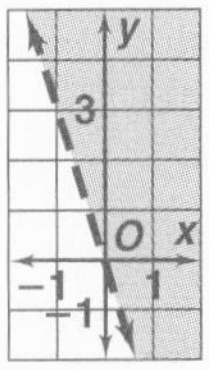

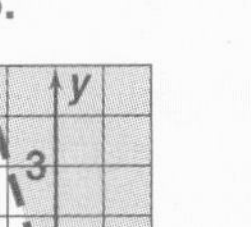

16.

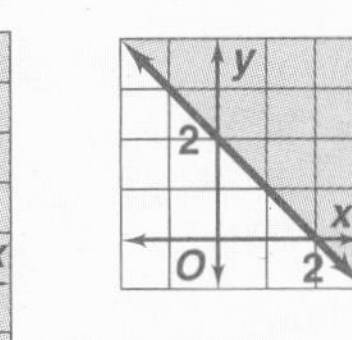

17.

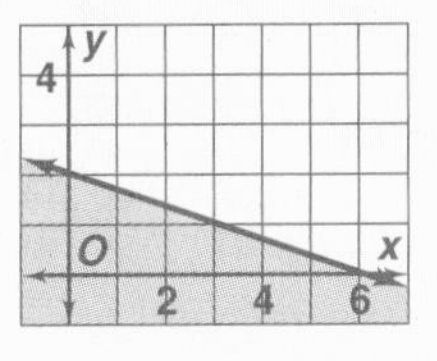

18.

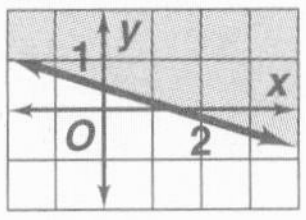

19.

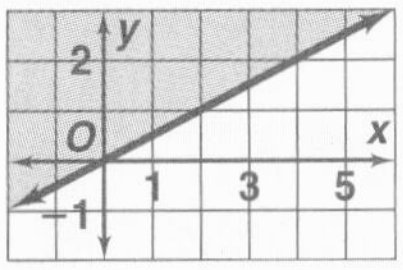

20.

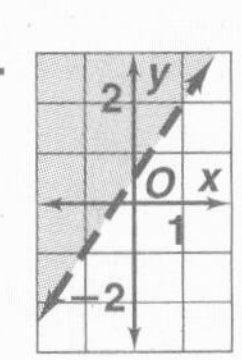

21.

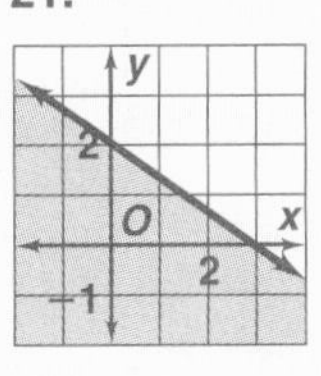

22.

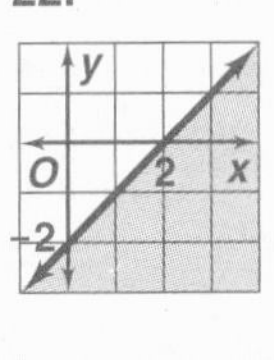

23.

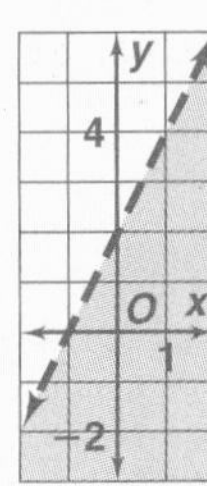

31.

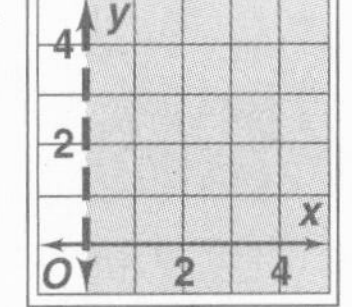

32.

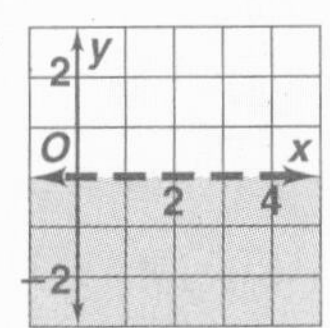

33.

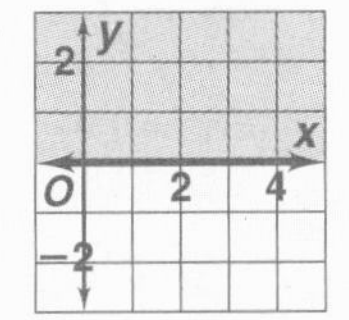

34.

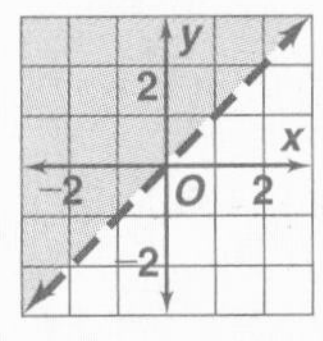

35b.

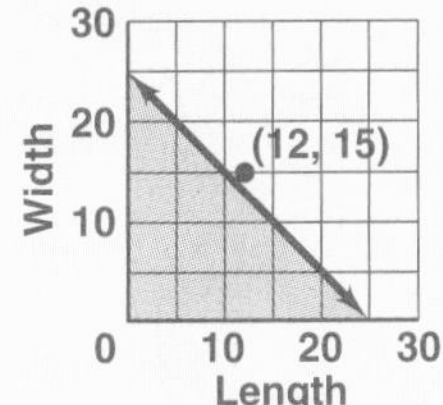

Math Toolbox page 294

4.

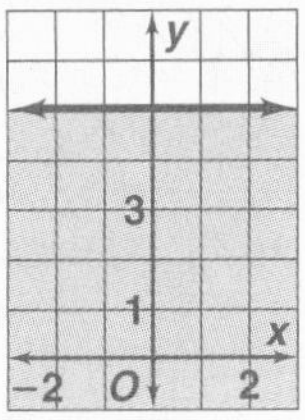

5.

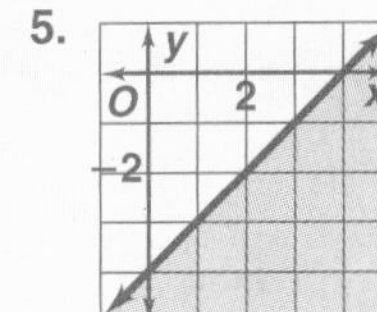

6.

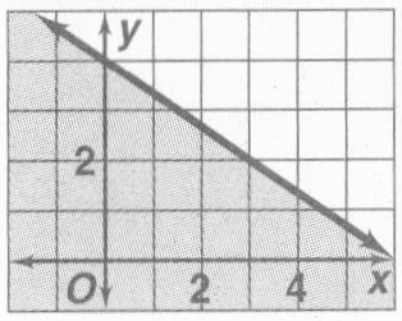

7.

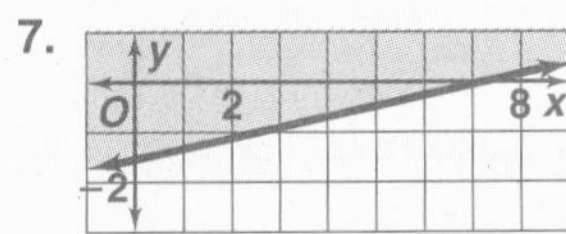

8.

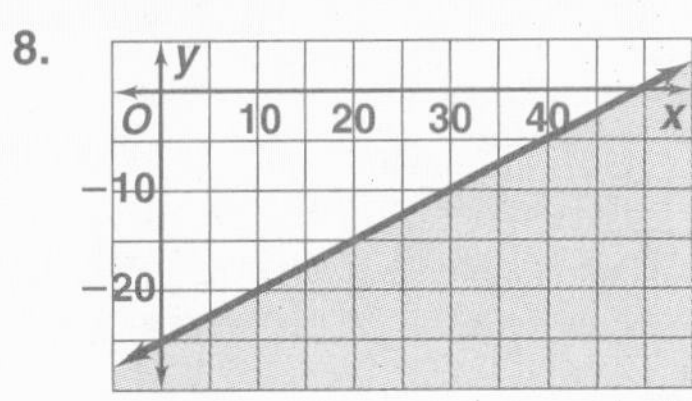

LESSON 6-6

page 295 Work Together

2c.

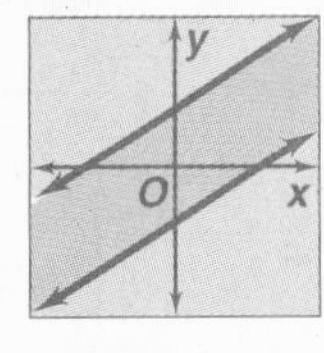

d.

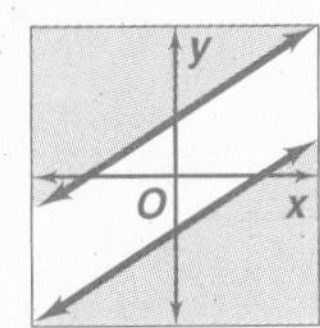

3. Answers may vary. Sample: If the boundary lines intersect at a point, the graphs of the inequalities intersect at a region. If the boundary lines do not intersect, the graphs of the inequalities can intersect at a region or not at all. If the boundary lines are the same, the graphs can intersect at a line, a region or not at all.
4. A point, a line, no intersection; sample graphs:

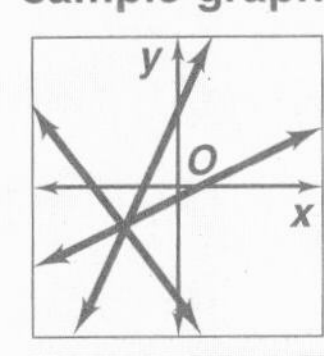

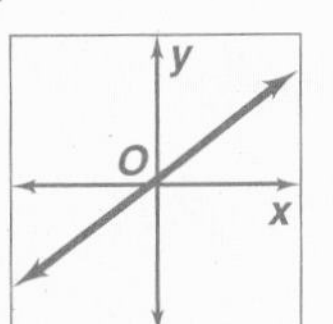

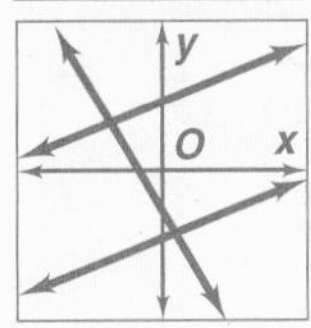

5. A line, a ray with or without the endpoint, a region, and no intersection are possible; a point is not possible.

pages 295–297 Think and Discuss

9.

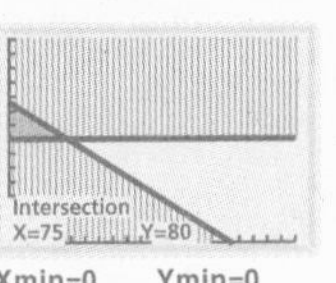

Xmin=0 Ymin=0
Xmax=200 Ymax=200
Xscl=10 Yscl=10

10a.

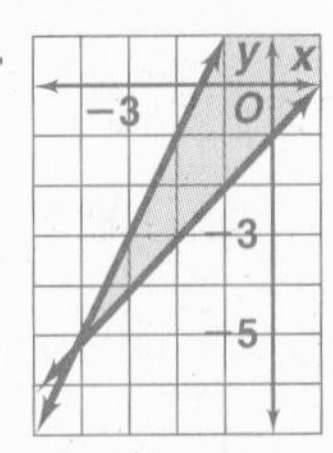

b.

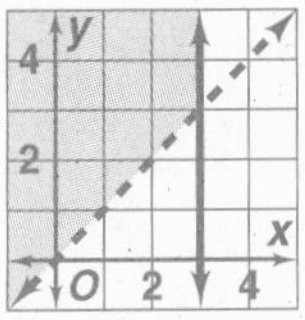

c.

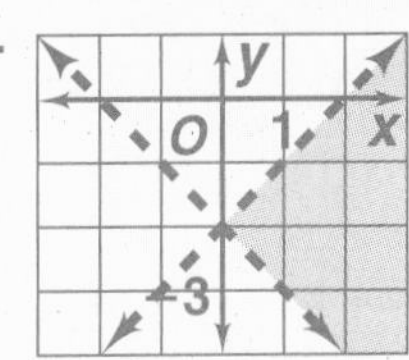

pages 297–299 On Your Own

5.

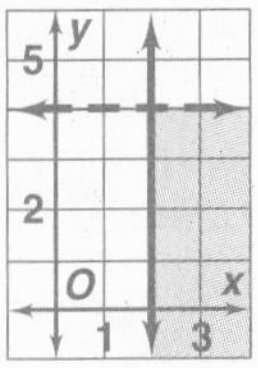

6.

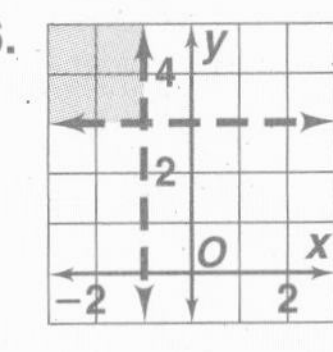

7.

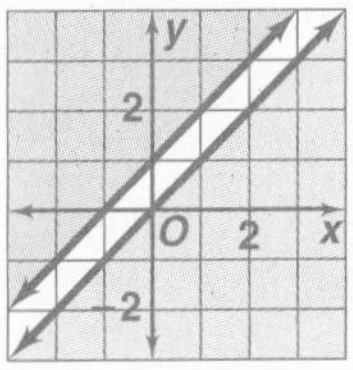

8.

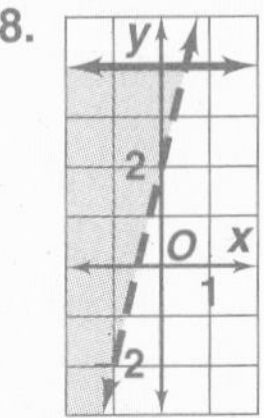

9.

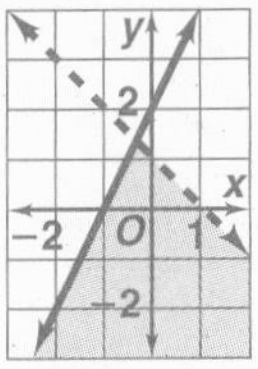

10.

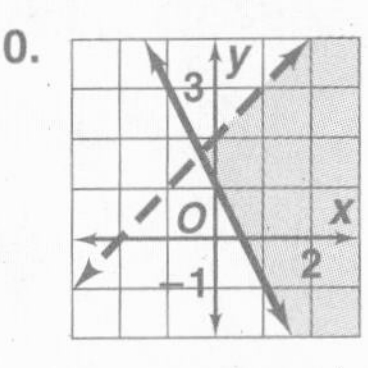

11.

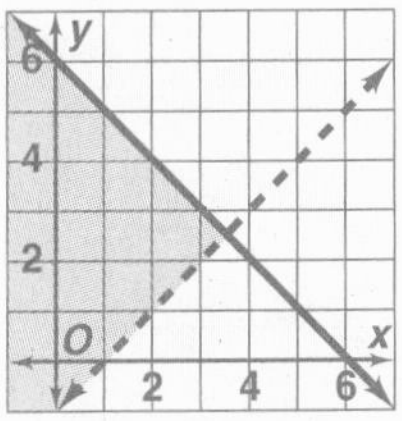

12.

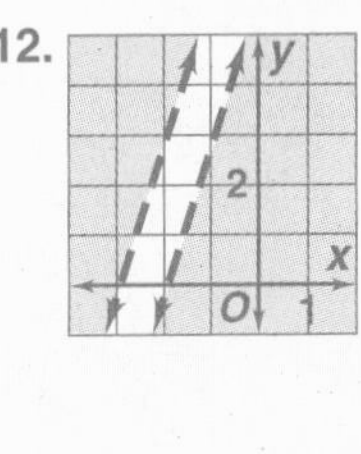

13.

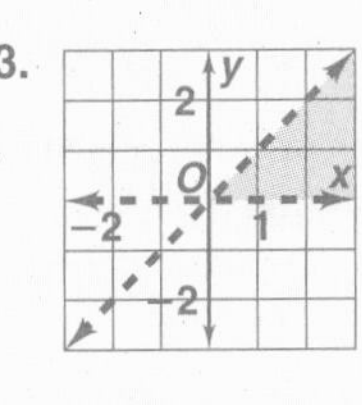

14.

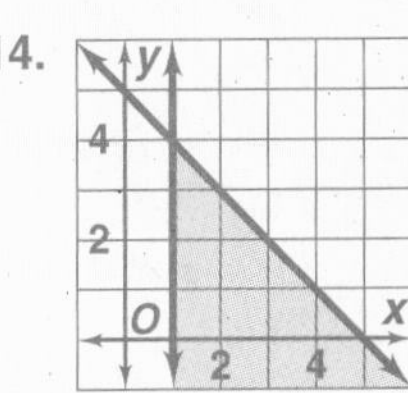

15. 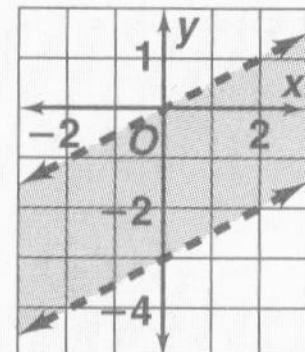

16. 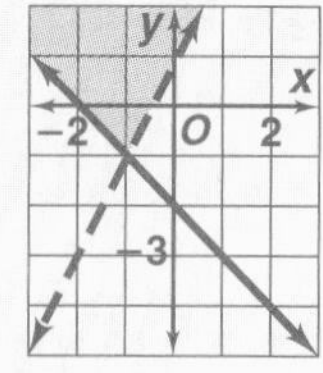

17a. $0.6f + 0.55c \leq 33$
$f \geq 9$
$c \geq 12$

b. 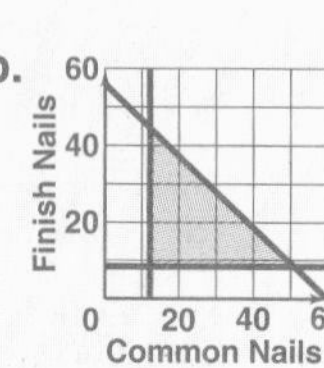

22a. 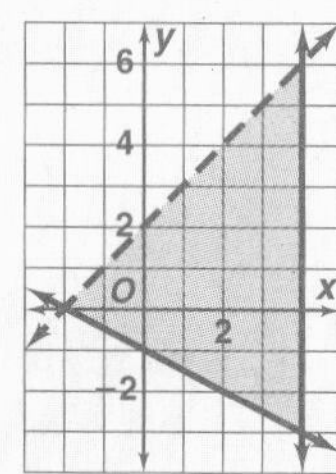

23. 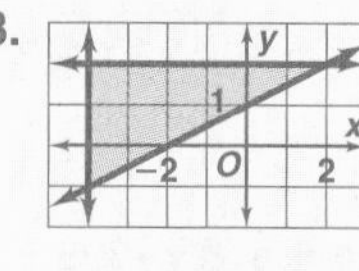

24. 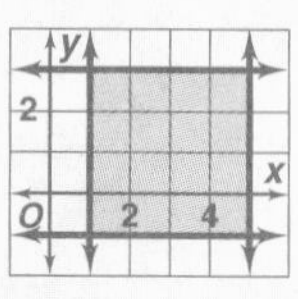

25. 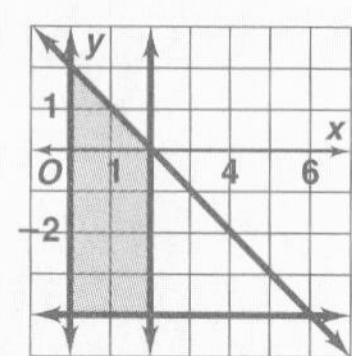

26. 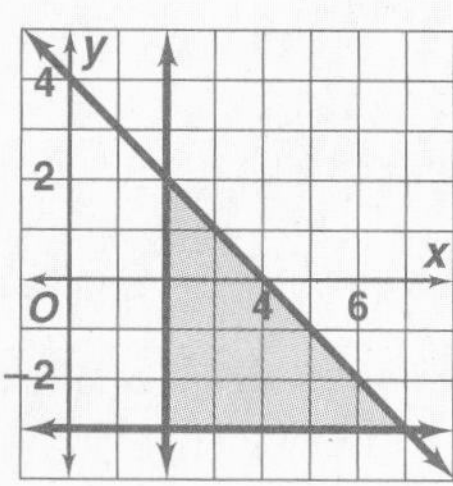

27a. $5.99x + 9.99y \leq 50$
$x \geq 0$
$y \geq 1$

b. 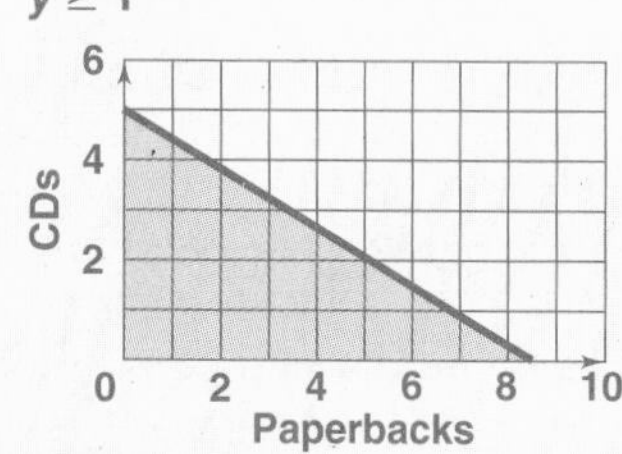

c. 2 books and 6 CDs; (2, 6) is not a solution of the system because 2 books and 6 CDs cost $71.92.

page 299 Mixed Review

36b. 

c. Answers may vary. Sample: Use $0 to $8 on the x-axis with labels every $2 because the largest amount is $7. Use $0 to $.40 on the y-axis with labels every $.10 because the largest amount is $.35.

Checkpoint page 299

1. 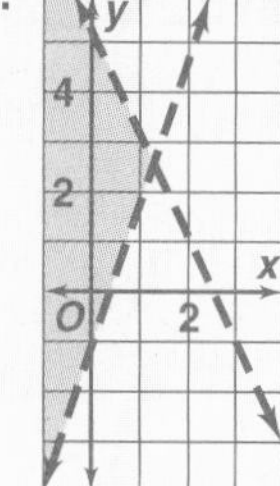

2. 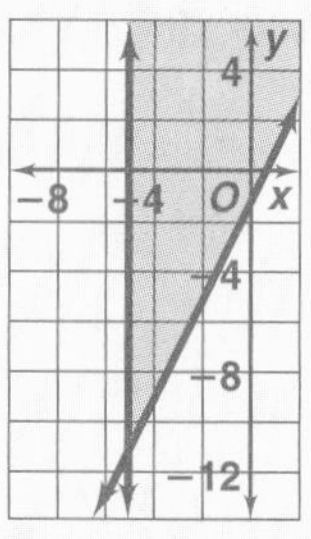

3. 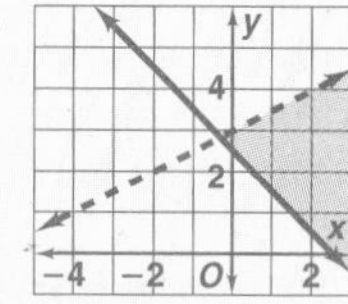

4. 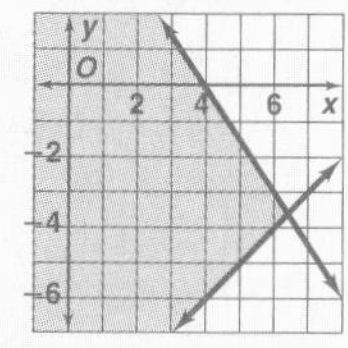

6b. (4, 10); a pizza with 4 toppings costs $10 at each restaurant.

LESSON 6-8

pages 307–309 On Your Own

9. 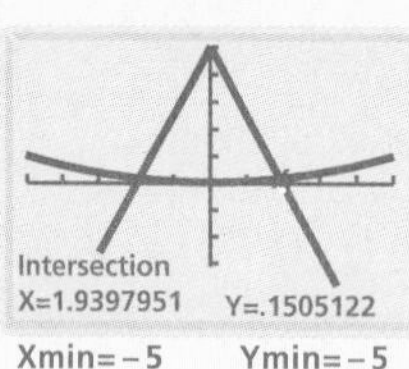

Xmin=–5 Ymin=–5
Xmax=5 Ymax=5
Xscl=1 Yscl=1

16. 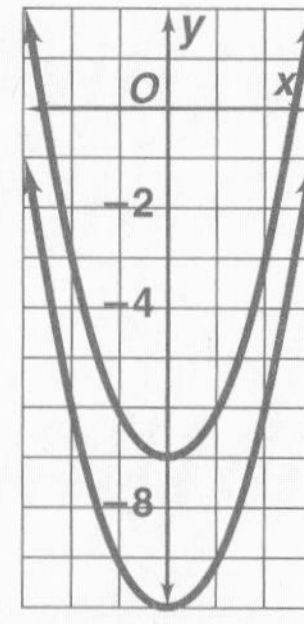

17. 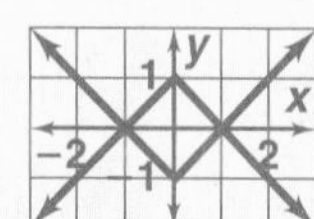

18. 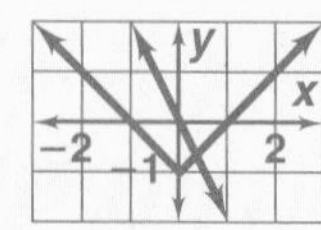

19. 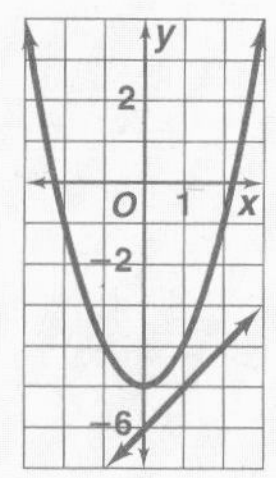

20. 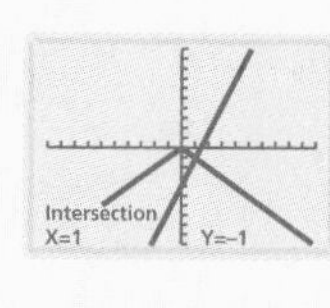

21. 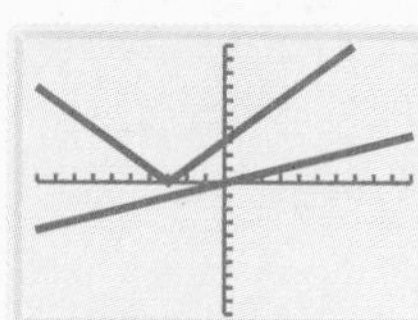
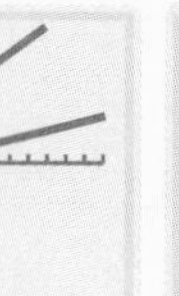
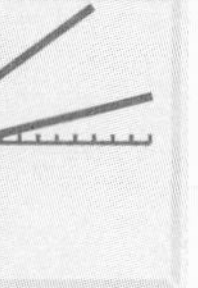

22. 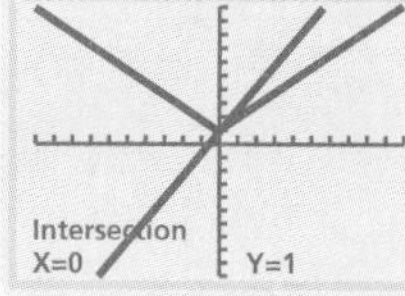

23. 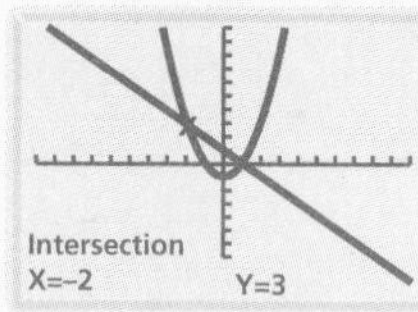

24. 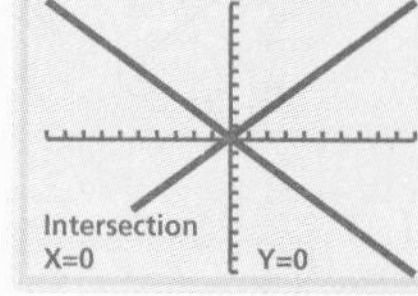

25. 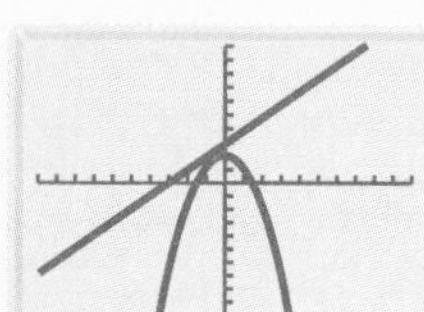

26. 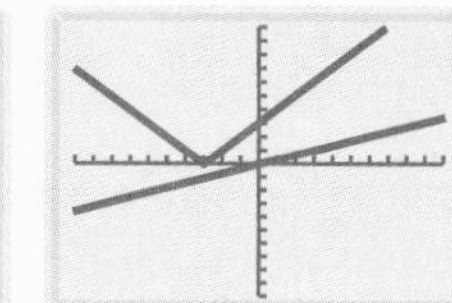

27. 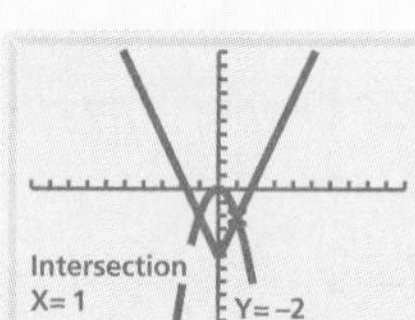

28. 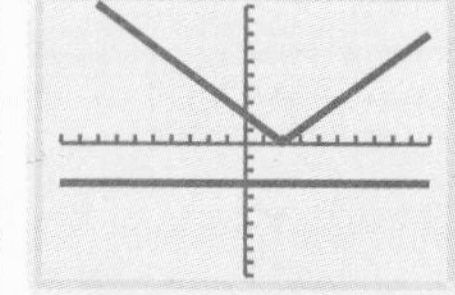

29. 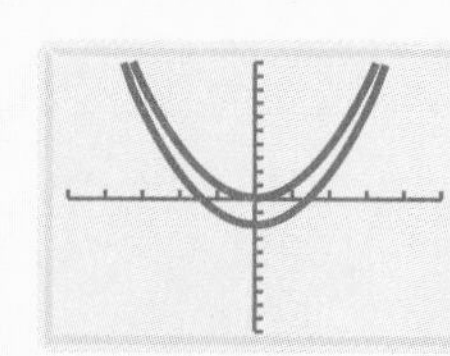
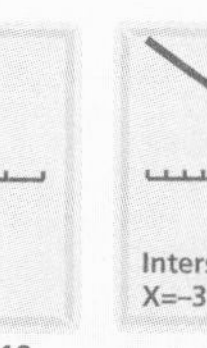

Xmin=–5 Ymin=–10
Xmax=5 Ymax=10
Xscl=1 Yscl=1

30. 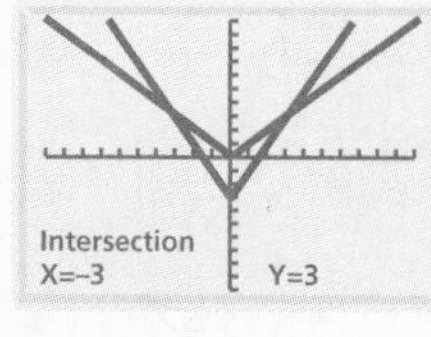

31. 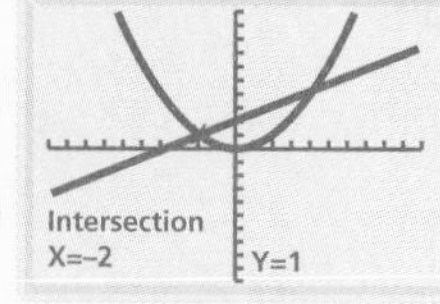

32a. 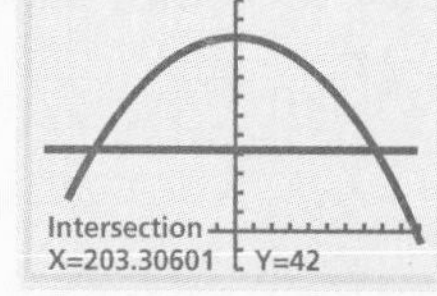

Xmin=–250 Ymin=–20
Xmax=250 Ymax=120
Xscl=25 Yscl=10

page 309 Mixed Review

42. The solution of a single equation is a number or an ordered pair that makes the equation true. A solution of a system of two equations is an ordered pair that makes both equations true at the same time.

Wrap Up pages 311–313

24\.

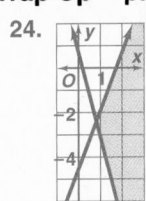

25\.

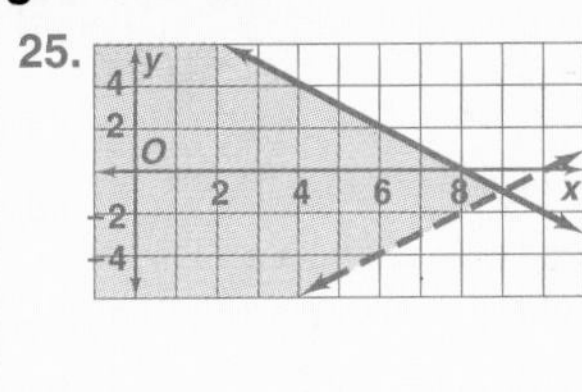

26\.

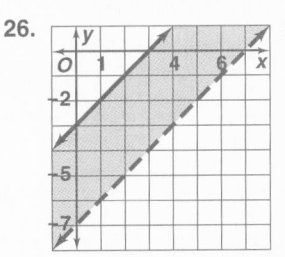

27\.

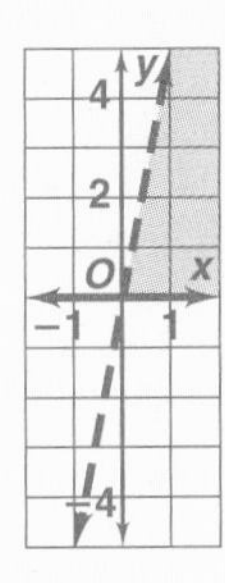

29\.

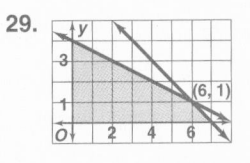

30\.

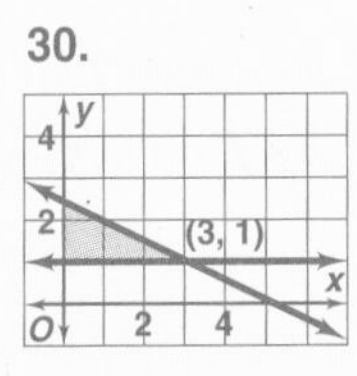

31\.

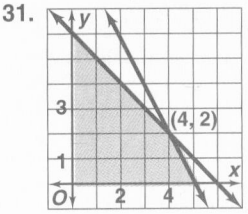

32\.

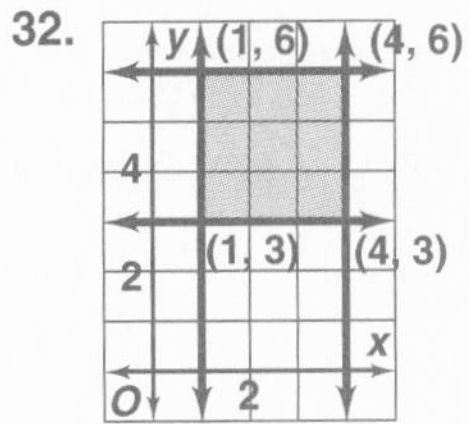

34\.

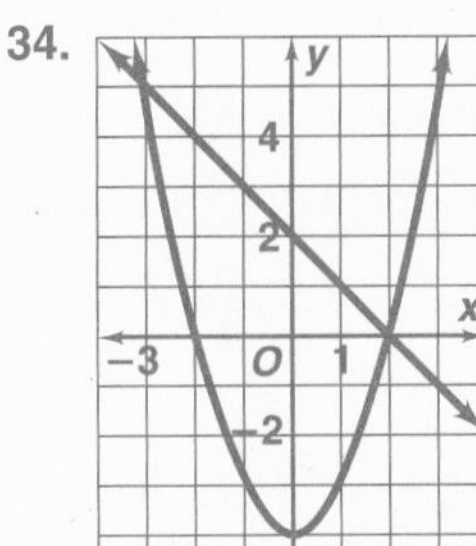

35\.

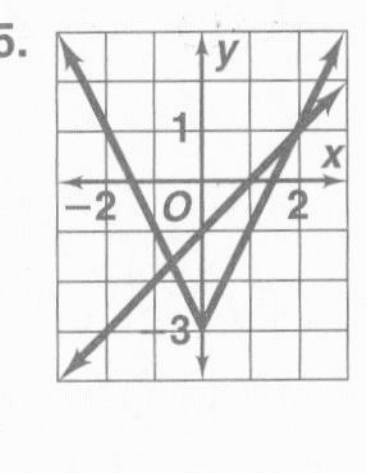

36\.

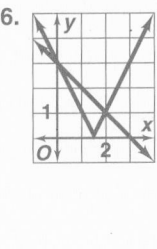

37\.

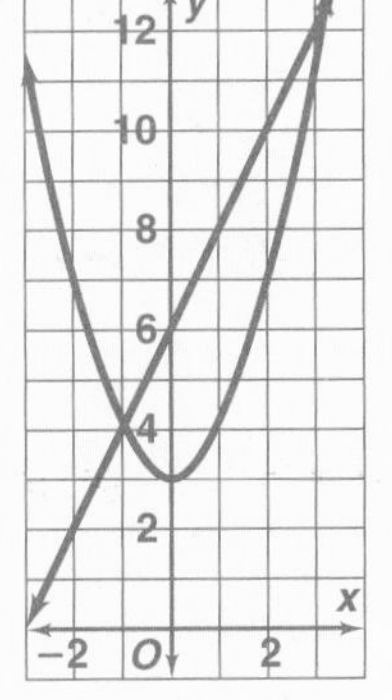

38\.

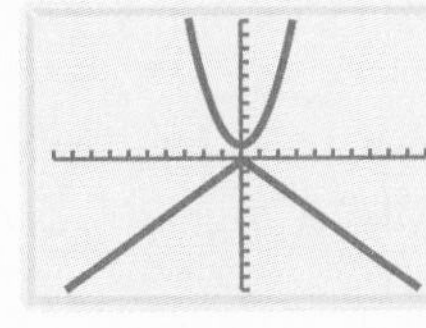

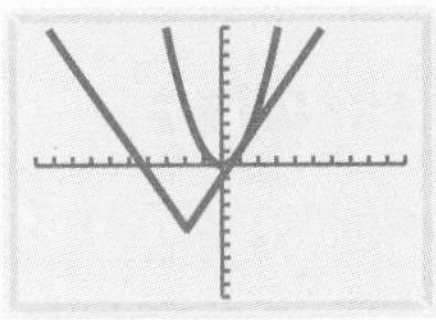

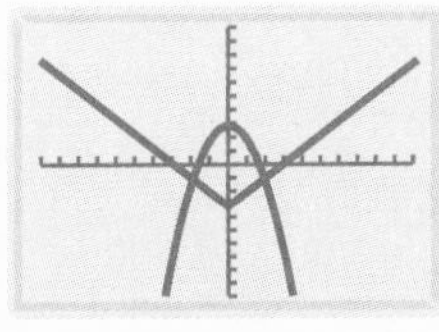

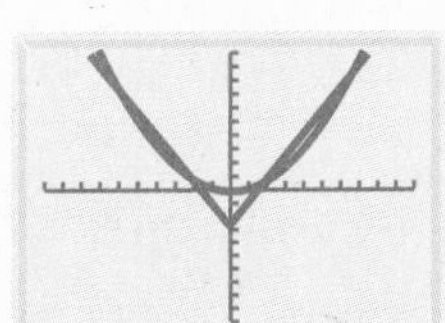

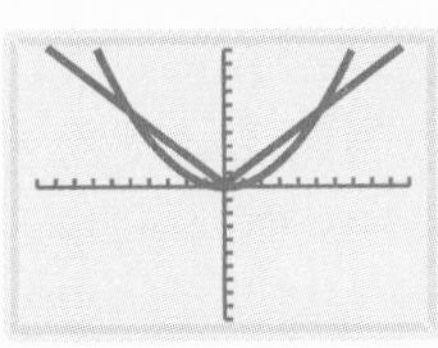

The graphs cannot intersect more than 4 times.

Getting Ready for Chapter 7 page 313

45\.

x	y
−1	8
0	3
1	0
2	−1
3	0

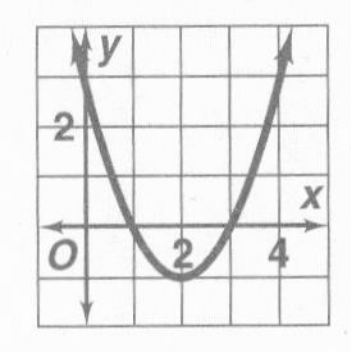

46\.

x	y
−3	4
−2	0
−1	−2
0	−2
1	0
2	4

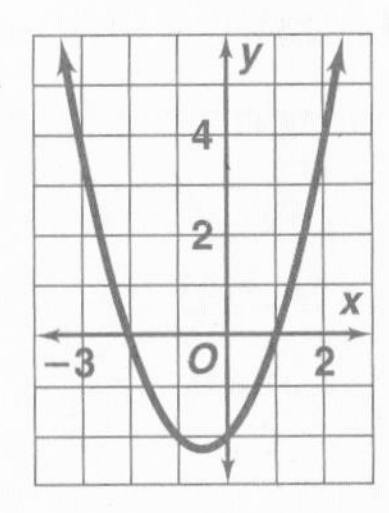

47\.

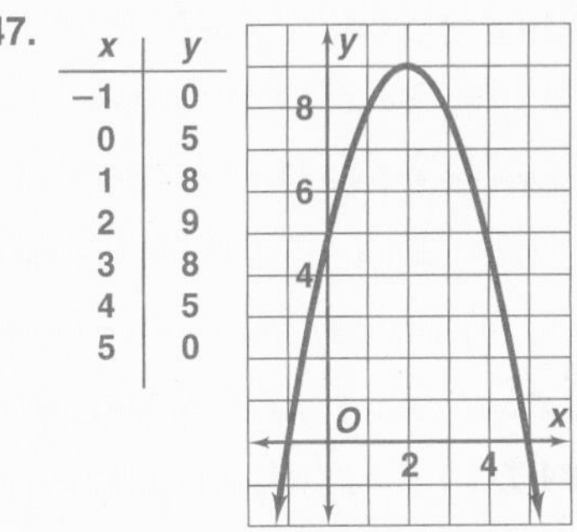

x	y
−1	0
0	5
1	8
2	9
3	8
4	5
5	0

Chapter Assessment page 314

18. Answers may vary. Sample: You solve a linear inequality by solving a related linear equation. A linear equation usually has a single number as a solution. The solution of a linear inequality usually is a part of a number line.

20\.

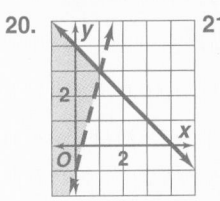

21\.

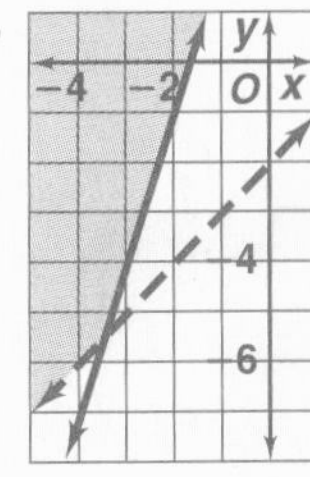

22\.

23\.

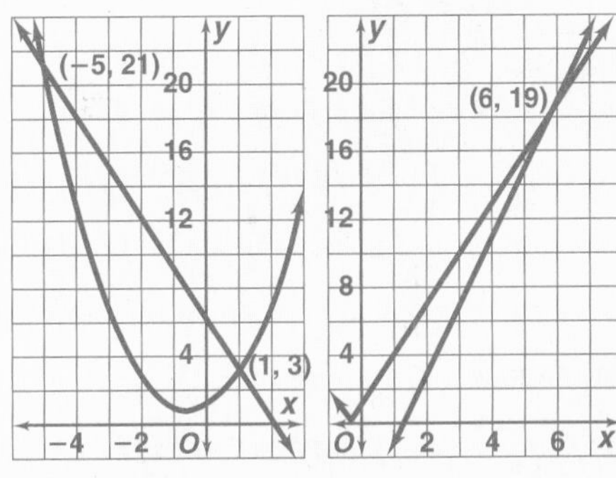

27a. $2x + 3y \leq 12$: The combined amount of flour is no more than 12 cups.
$2x + y \leq 8$: The combined number of eggs is no more than 8 eggs.
$x \geq 0$; $y \geq 0$: The number of cups of flour and the number of eggs cannot be negative.

b. $P = x + 2y$

c. 4 loaves of banana bread and no oatmeal bread

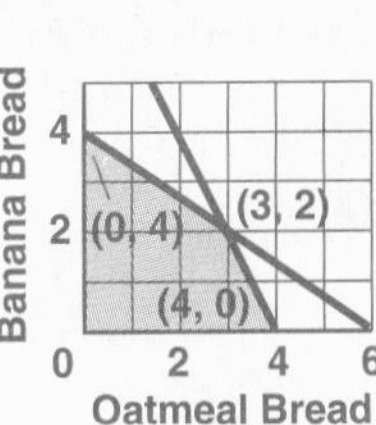

Cumulative Review **page 315**

14. $-2 \le x < 3$

CHAPTER 7

LESSON 7-1

pages 321–322 On Your Own

10.

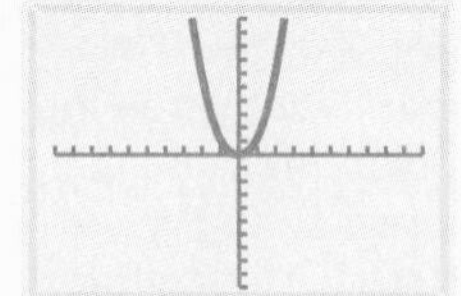

11.

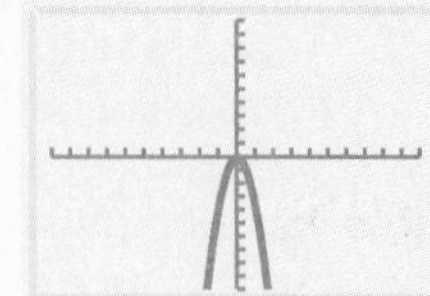

12.

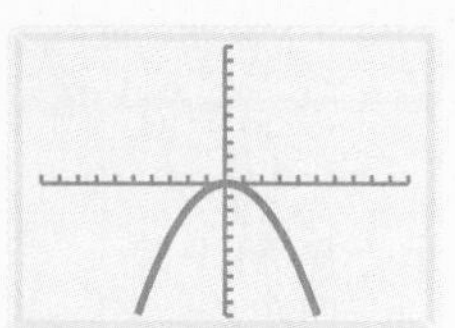

13.

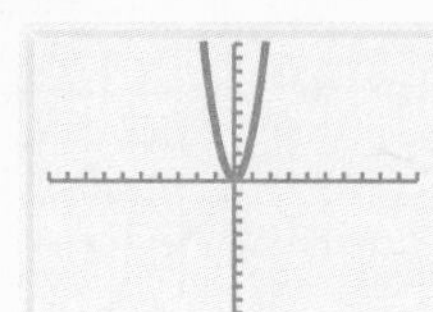

14.

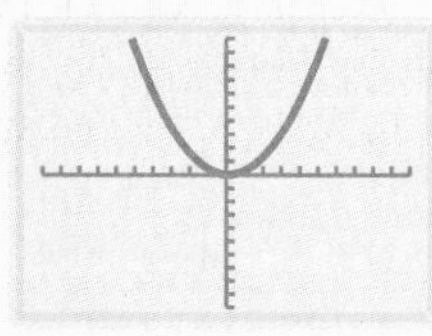

15.

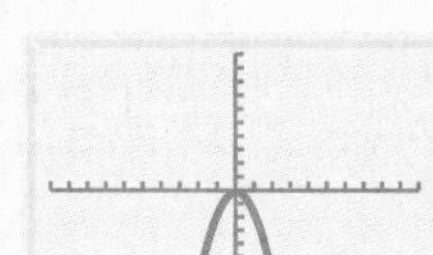

16.

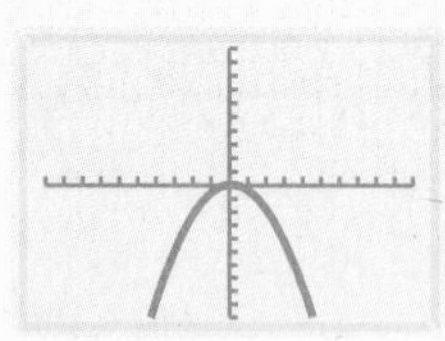

17.

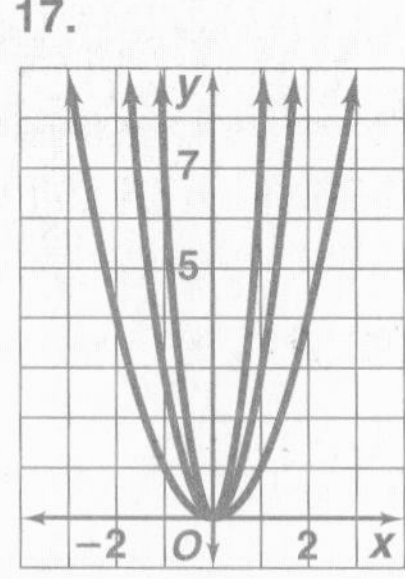

18.

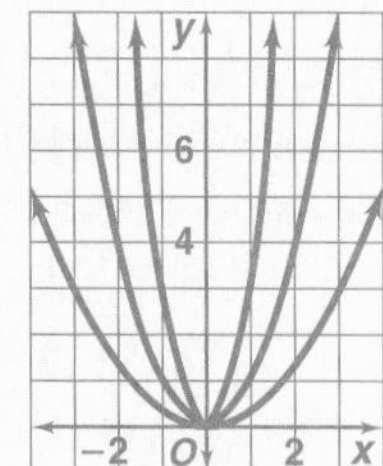

19.

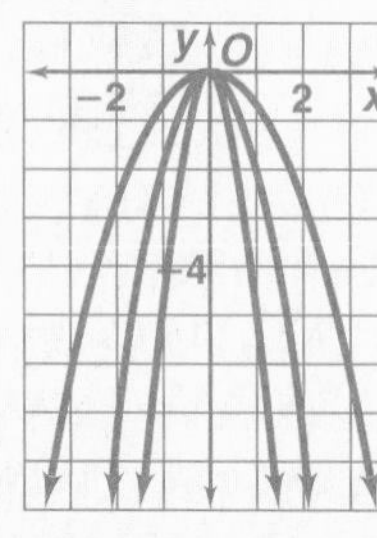

20.

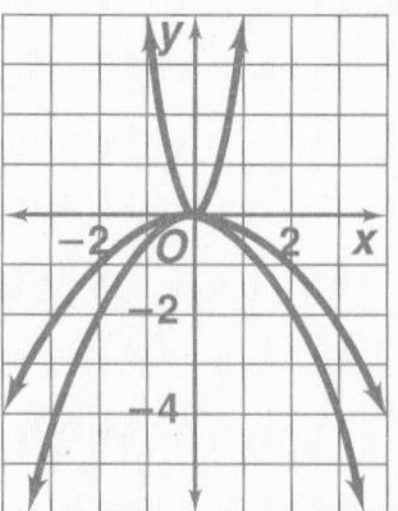

page 322 Getting Ready for Lesson 7-2

44.

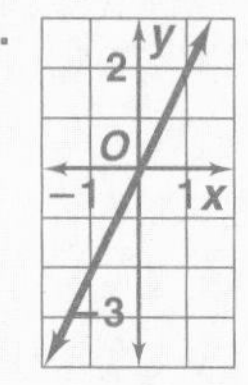

45.

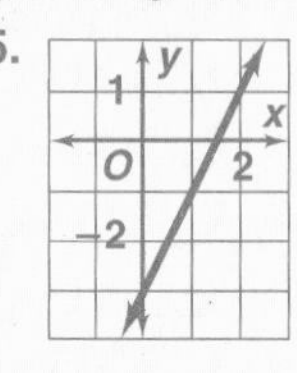

46.

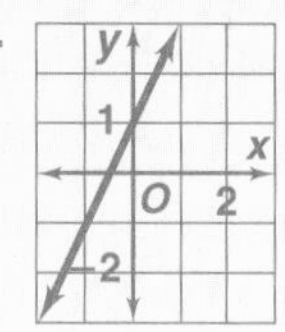

LESSON 7-2

page 323 Work Together

1a.

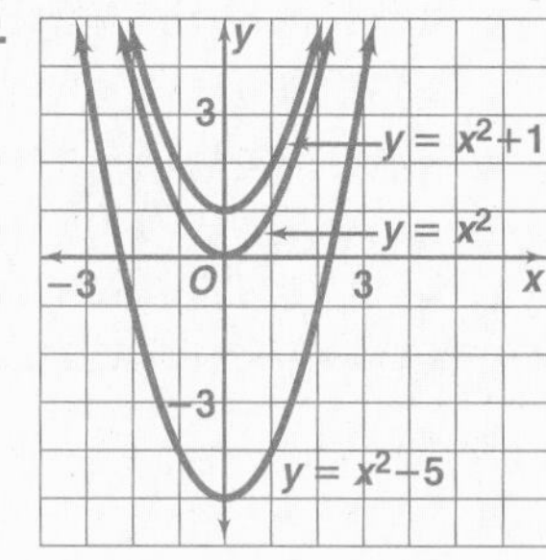

2a.

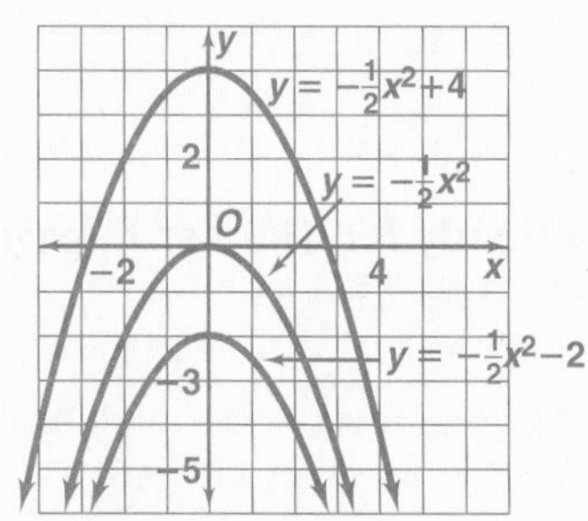

pages 323–324 Think and Discuss

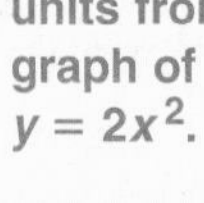

3. The graph of $y = 2x^2 - 4$ is shifted down 4 units from the graph of $y = 2x^2$.

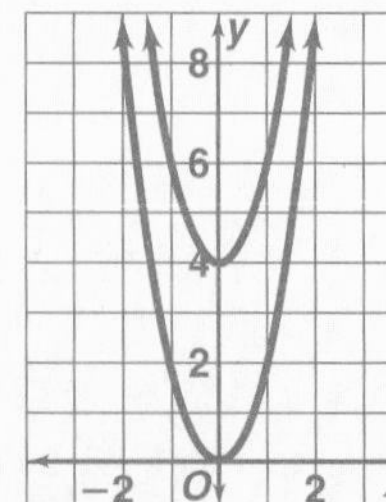

pages 325–326 On Your Own

6.

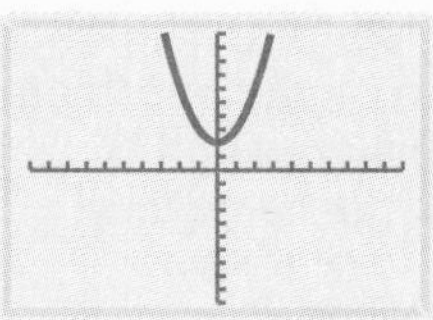

7.

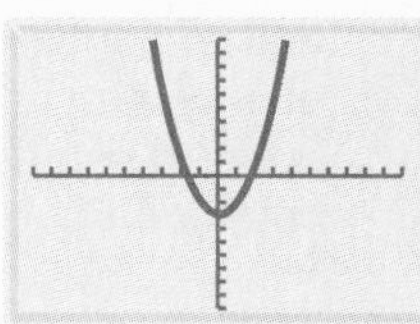

8.

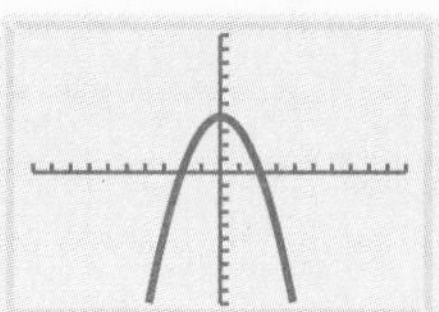

9.

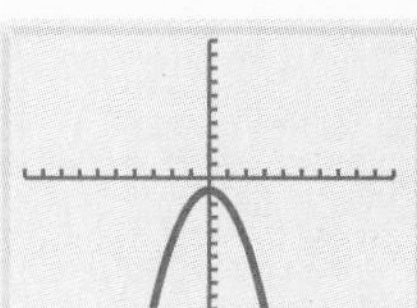

10.

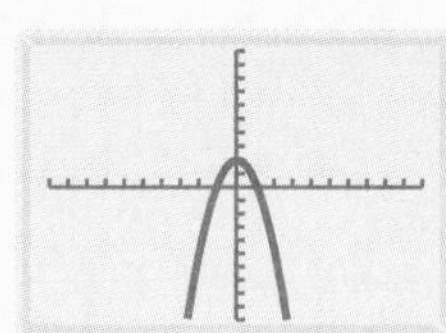

11.

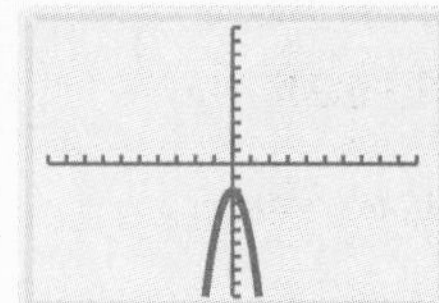

12.

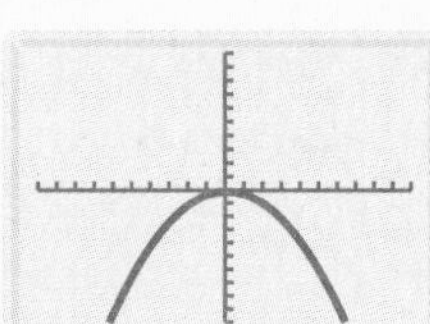

13.

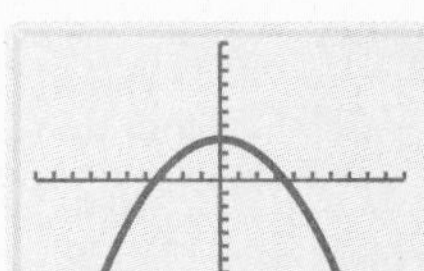

14.

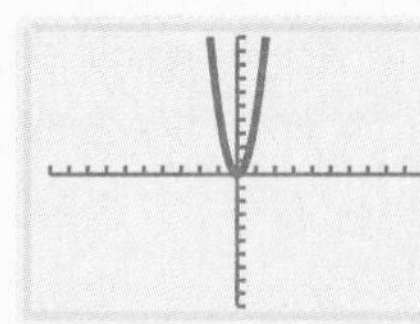

15.

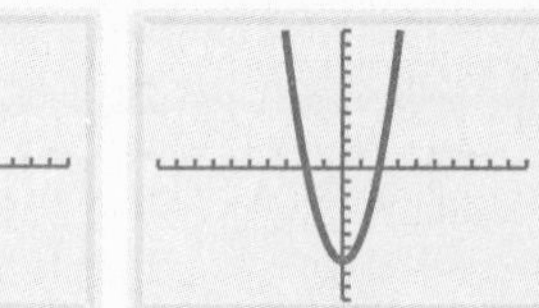

16.

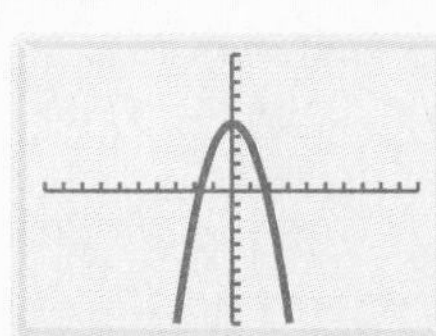

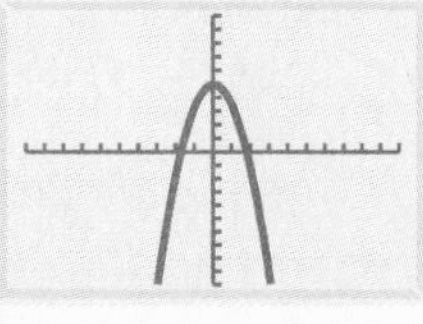

17.

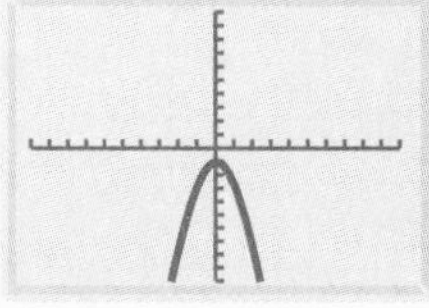

20c.

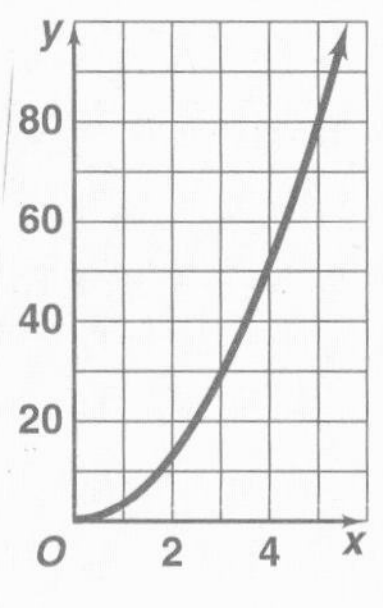

33a.

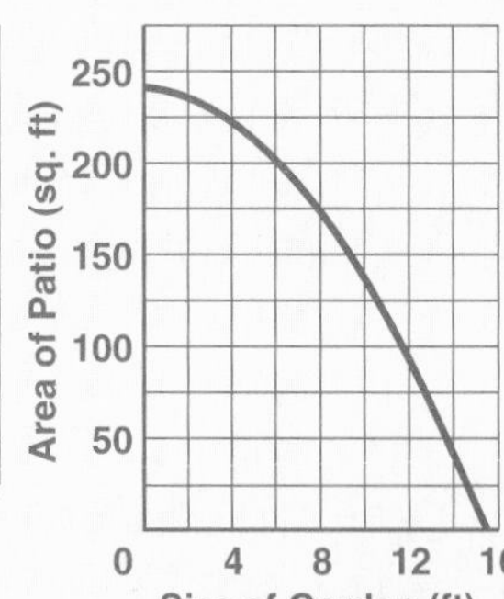

33b. The length of the side of the garden must be positive and shorter than 12 ft, the shorter side of the patio.

c. The larger the garden the smaller the area of the patio. As the length of the side of the garden changes from 0 to 12 ft, the value of the function changes from 240 to 96.

34a. 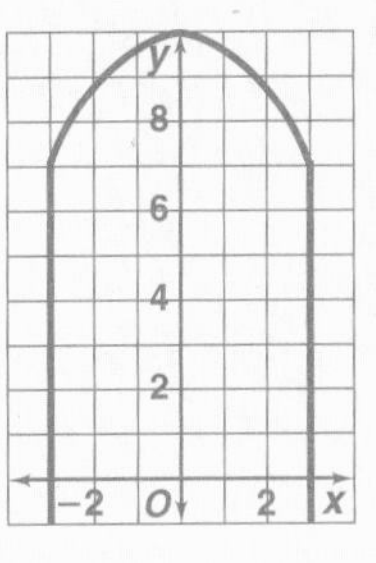

b. 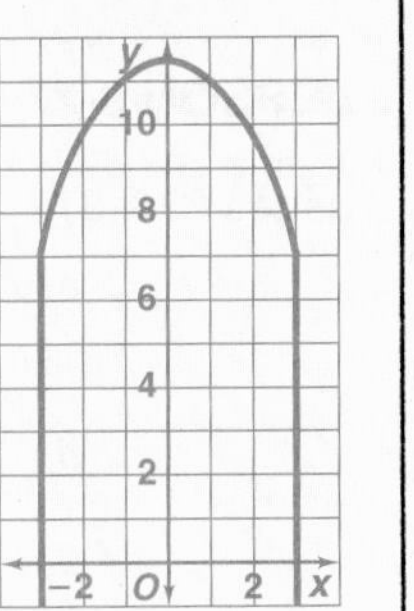

LESSON 7-3

pages 330–331 On Your Own

17. 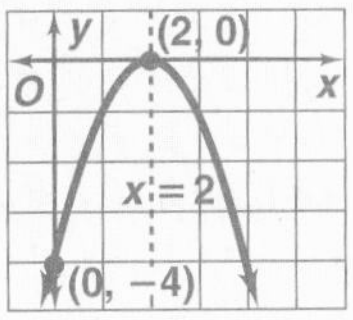 (2, 0); $x = 2$; (0, −4)

18. 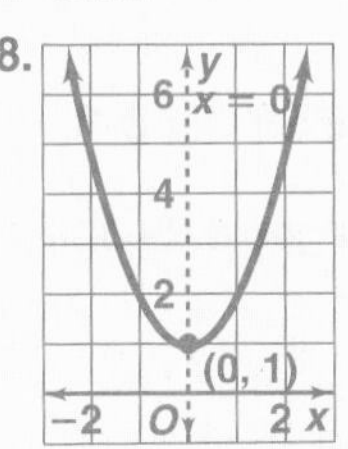

19. 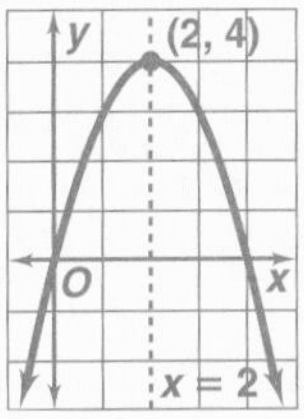 (2, 4); $x = 2$

20. 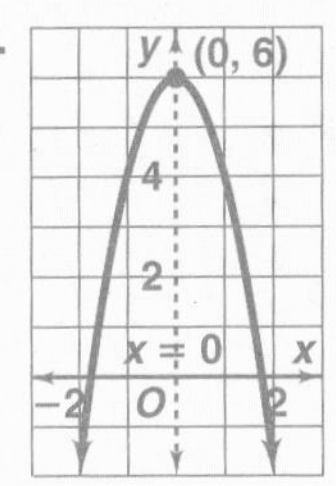

21. 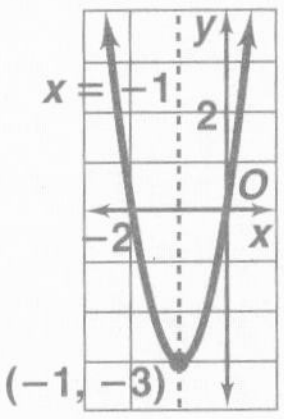 $x = -1$; (−1, −3)

22. 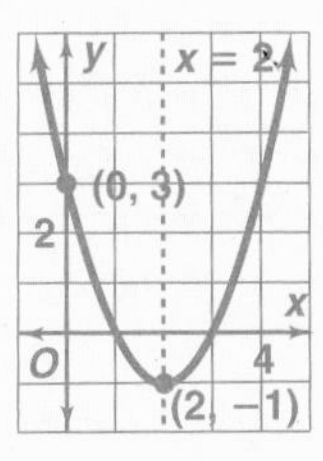

23. 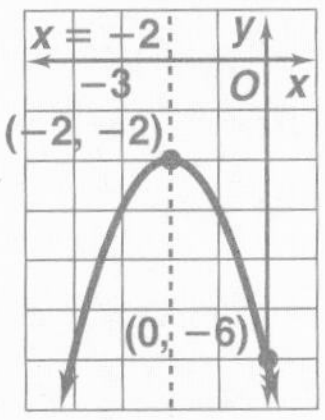 $x = -2$; (−2, −2); (0, −6)

24. 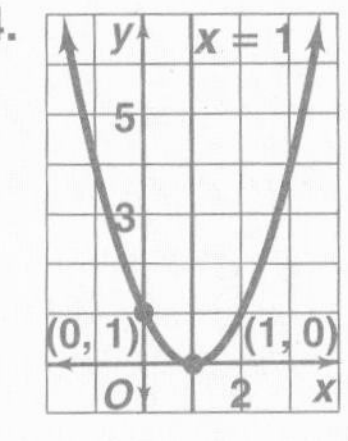

25. 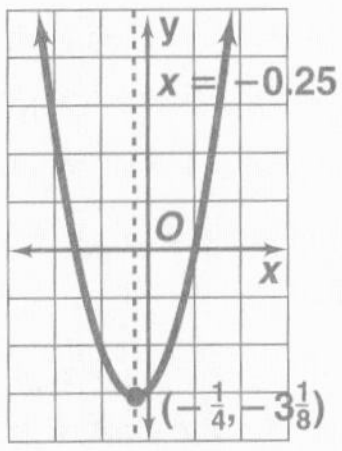 $x = -0.25$; $(-\frac{1}{4}, -3\frac{1}{8})$

26. 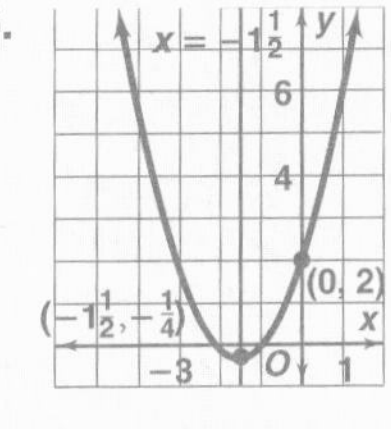

27. 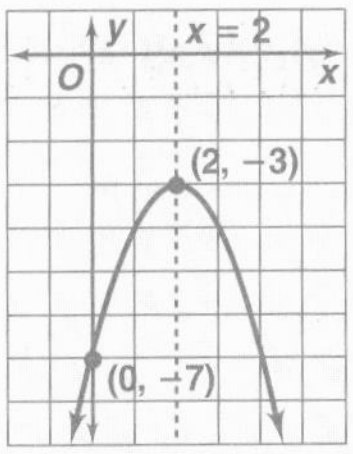 $x = 2$; (2, −3); (0, −7)

30. The sign of a determines whether the graph open upward or downward. The larger the abs. val. of a the narrower the graph. Use $x = \frac{-b}{2a}$ to find the axis of symmetry and the x-coordinate of the vertex of the graph. Use c to find the y-intercept.

31. 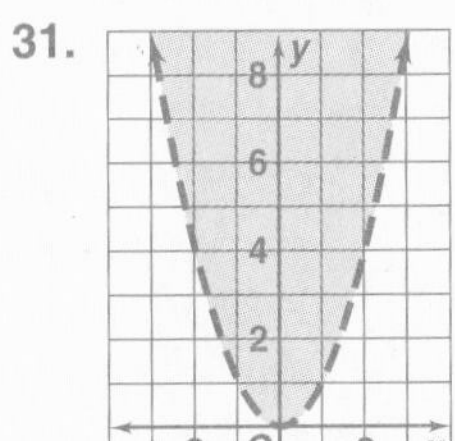

32. 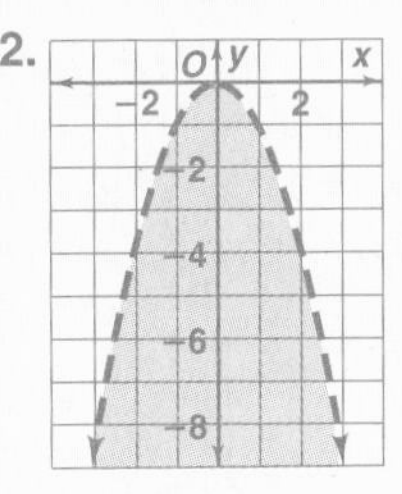

33. 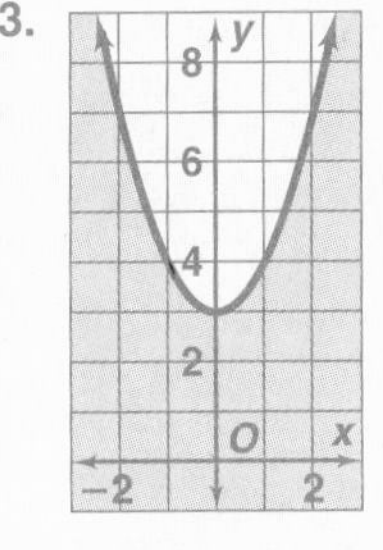

34. 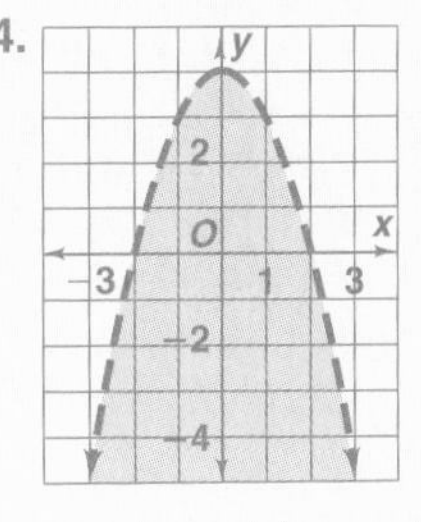

35. 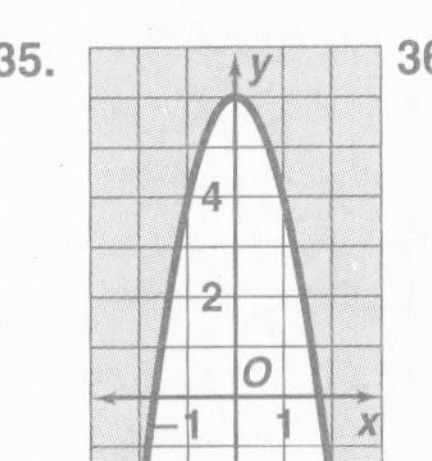

36. 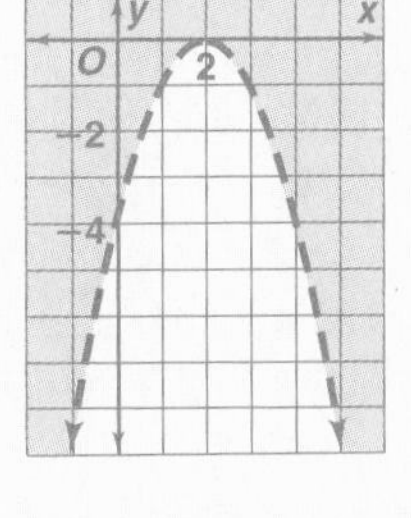

37. 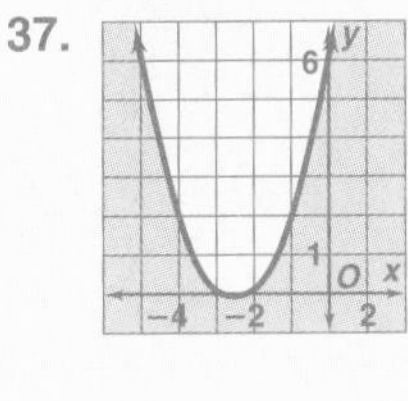

38. 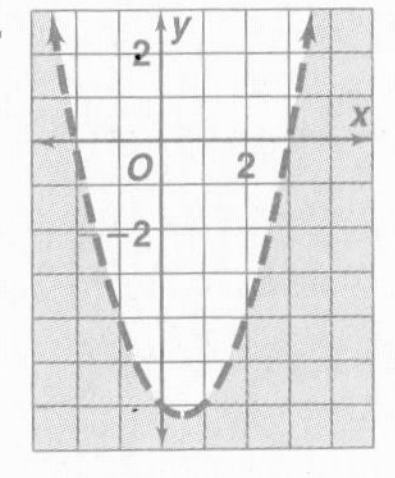

39. 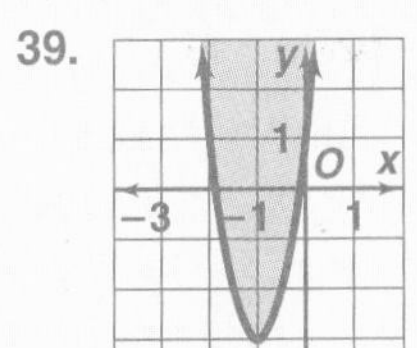

TE page 331 Lesson Quiz

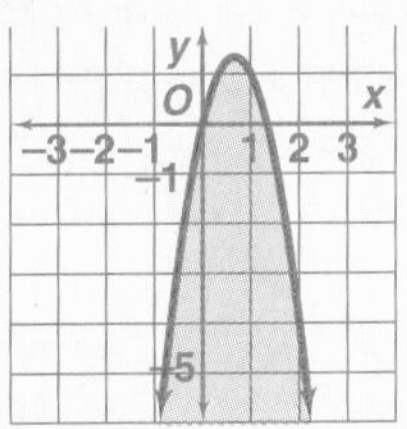

LESSON 7-4

Checkpoint **page 336**

1. 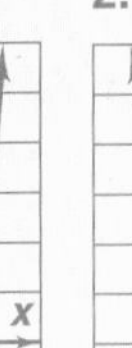

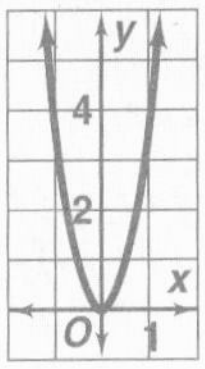

2. 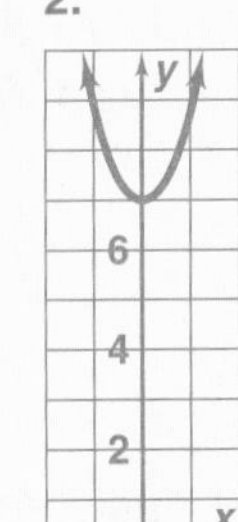

3. 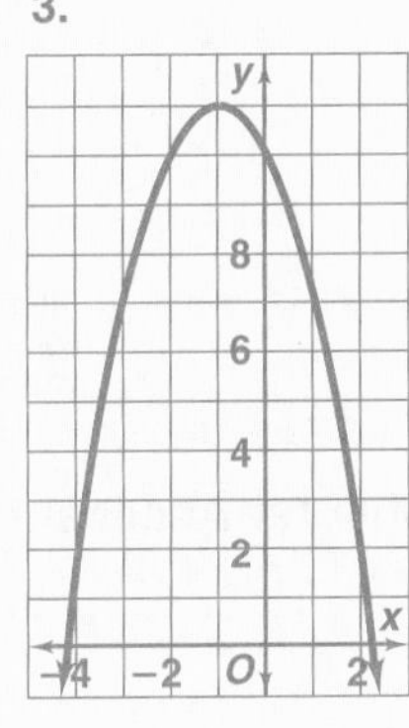

10. For $y = ax^2 + bx + c$, $-\frac{b}{2a}$ is the x-coordinate of the vertex, and $x = -\frac{b}{2a}$ is the axis of symmetry. The y-intercept is c.

LESSON 7-6

pages 343–345 Think and Discuss

6. The rocket reaches the altitude of 128 ft on the way up and on the way down.

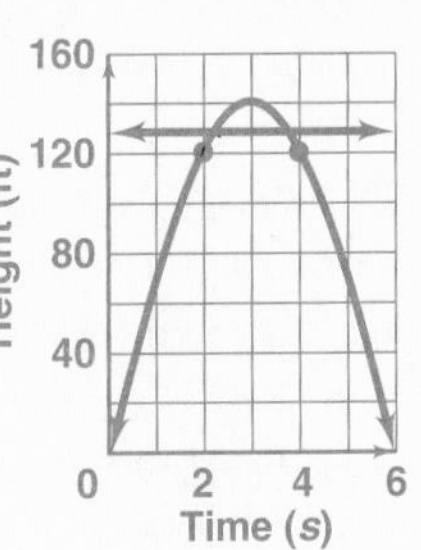

pages 345–347 On Your Own

33. Answers may vary. Sample: You use addition, subtraction, multiplication and division to solve a linear equation. You use addition, subtraction, multiplication, division, squaring and square roots to solve a quadratic equation.

Wrap Up pages 355–357

15. 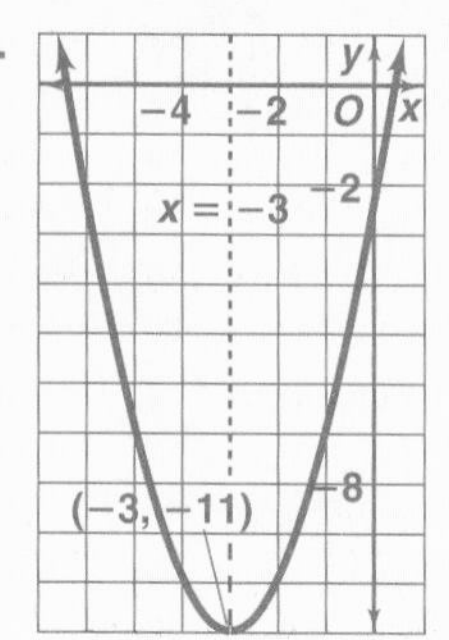
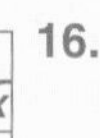

16. 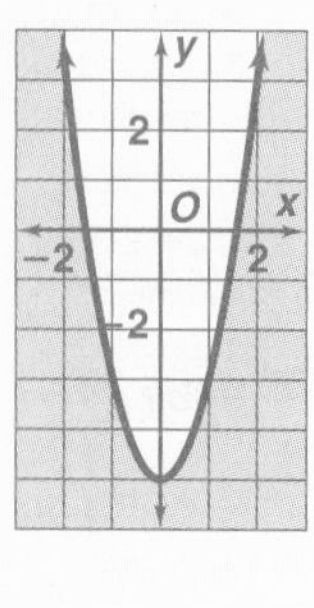

17. 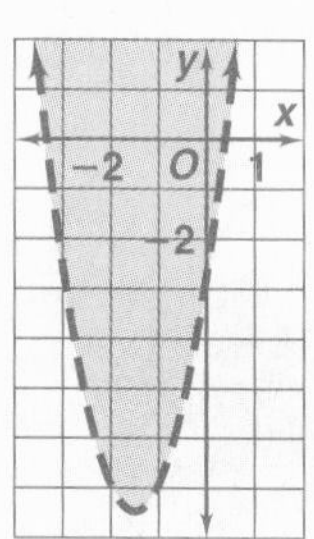

18. 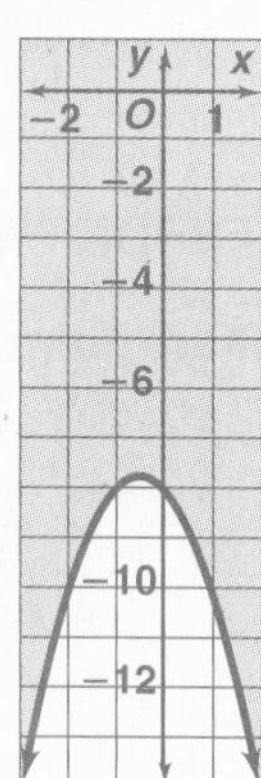

Chapter Assessment page 358

11.

x	y
−1	5
−0.5	1.25
0	0
0.5	1.25
1	5

12.

x	y
−4	6
−3	2.5
−2	0
−1	−1.5
0	−2
1	−1.5

18. 19.

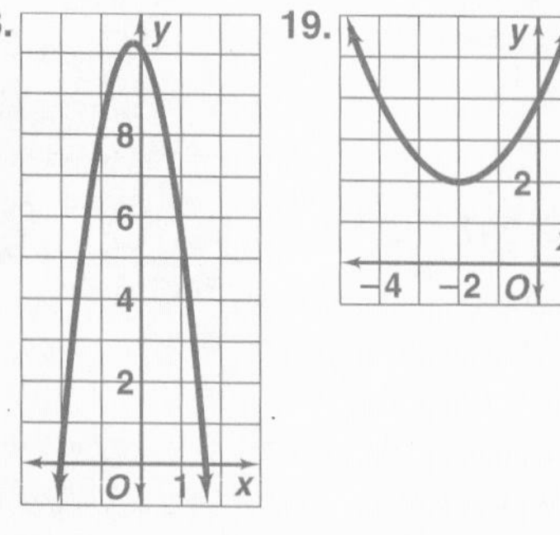

20. 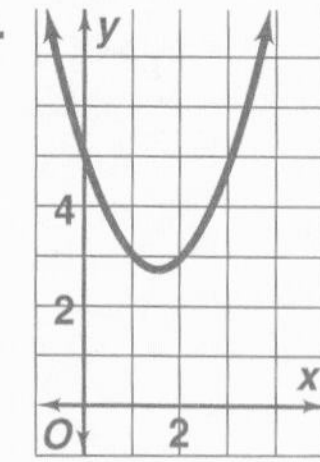

31. 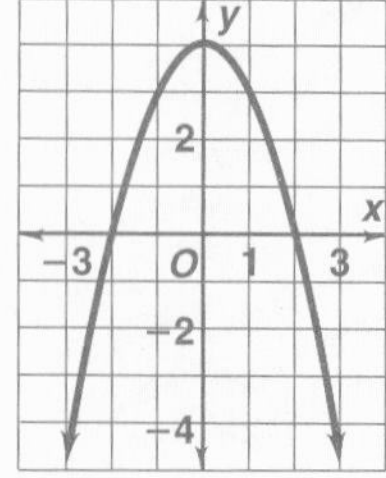

Preparing for Standardized Tests page 359

15. 

CHAPTER 8

LESSON 8-1

pages 365–366 On Your Own

24. Two 150% enlargements result in a 225% enlargement because $1.5 \cdot 1.5 = 2.25$.

25a.

x	y
1	− 2
2	4
3	− 8
4	16
5	− 32
6	64

b. Answers may vary. Sample: The positive and negative values alternate. The absolute values of the output are the same as the values of the output of 2^x.

26. 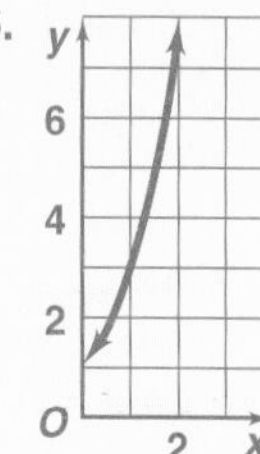

27. 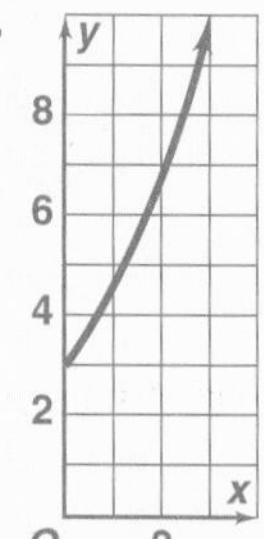

28. 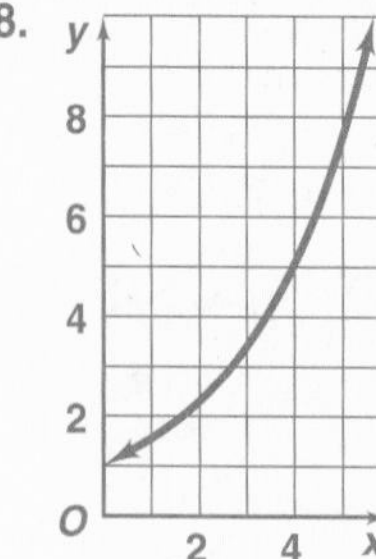

29. 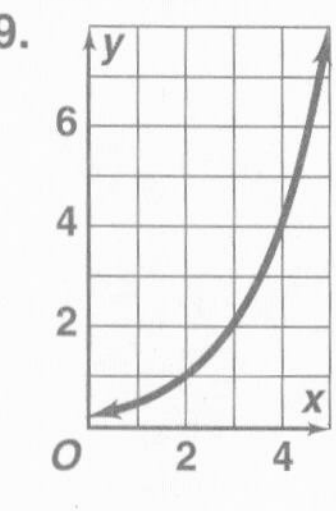

30. 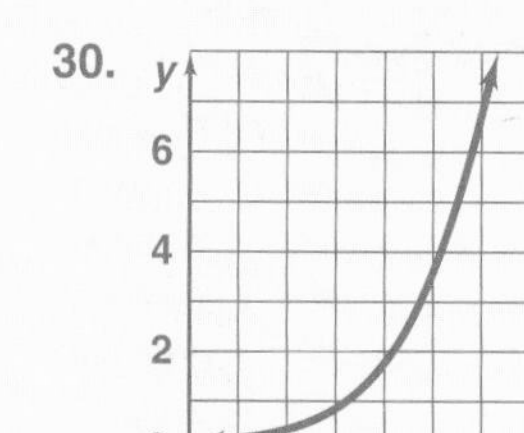

31.

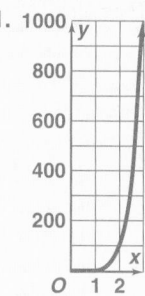

LESSON 8-2

pages 370–372 On Your Own

26. 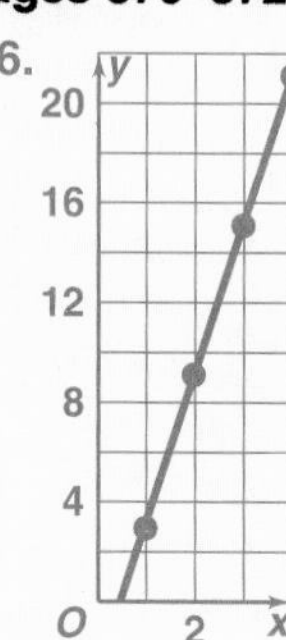

30a. 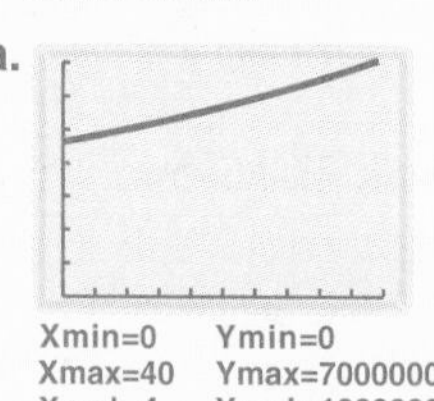

Xmin=0 Ymin=0
Xmax=40 Ymax=7000000
Xscl=4 Yscl=1000000

LESSON 8-3

pages 375–377 On Your Own

19. 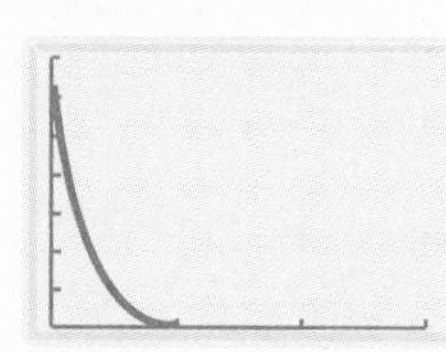

Xmin=0 Ymin=0
Xmax=3 Ymax=3.5
Xscl=1 Yscl=0.5

20. 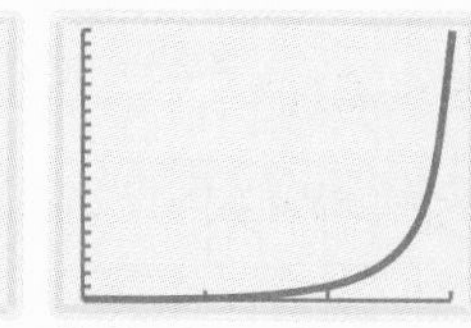

Xmin=0 Ymin=0
Xmax=3 Ymax=2000
Xscl=1 Yscl=100

21. 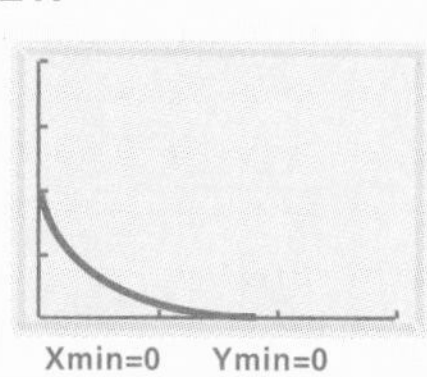

Xmin=0 Ymin=0
Xmax=3 Ymax=0.2
Xscl=1 Yscl=0.05

22. 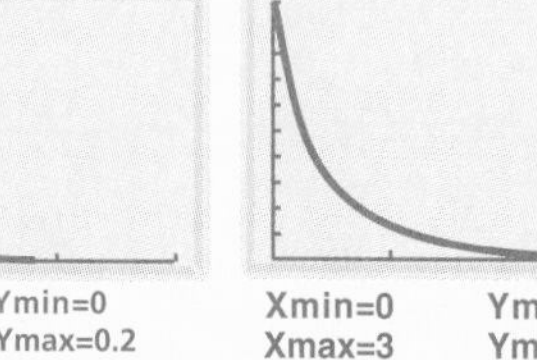

Xmin=0 Ymin=0
Xmax=3 Ymax=10
Xscl=1 Yscl=1

23. 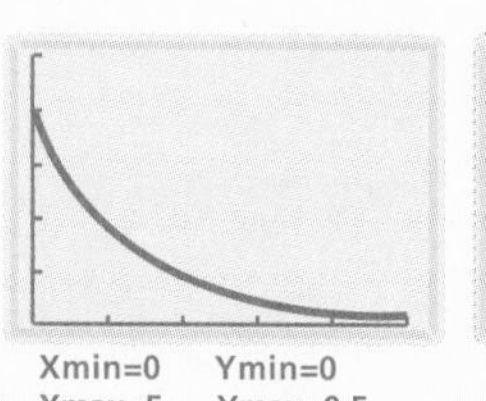

Xmin=0 Ymin=0
Xmax=5 Ymax=0.5
Xscl=1 Yscl=0.1

24. 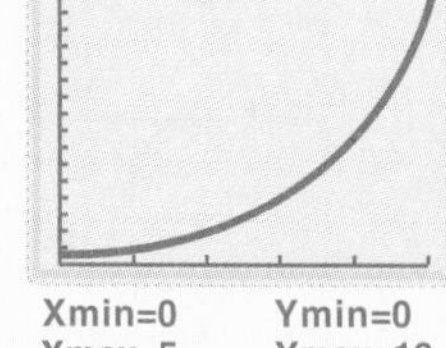

Xmin=0 Ymin=0
Xmax=5 Ymax=16
Xscl=1 Yscl=1

30c. 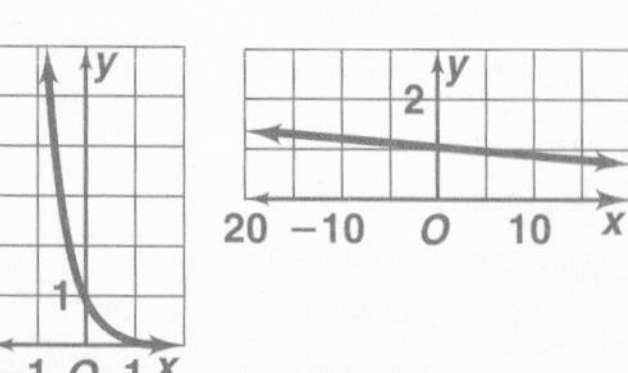

Checkpoint page 377

5. 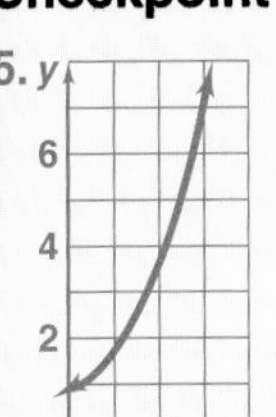
6. 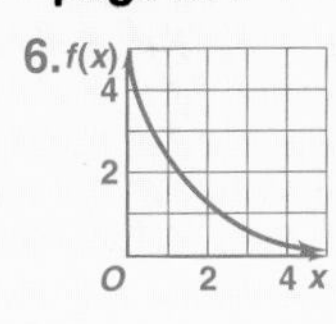
7. 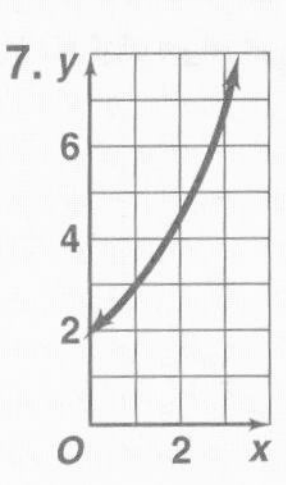

8a. Answers may vary. Sample: 6.7% interest; even compounded monthly the second account has an annual interest of only 5.1%.

Math Toolbox page 378

3a. Use $x = 0$ for 1970; $y = 1.04 \cdot 1.02^x$; 2.297 million

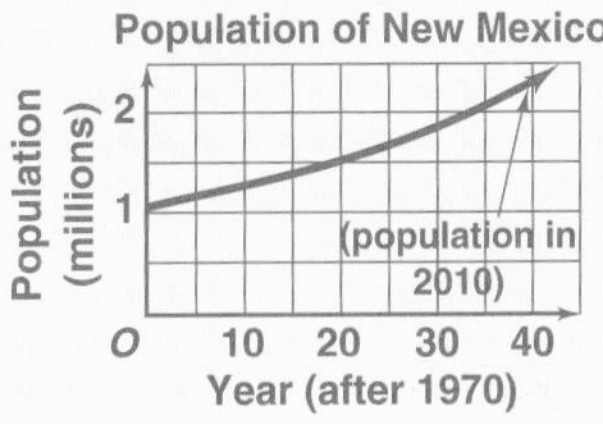

b. Answers may vary. Sample: The closer r is to 1 the more accurate the prediction is likely to be.

LESSON 8-4

pages 382–384 On Your Own

46. 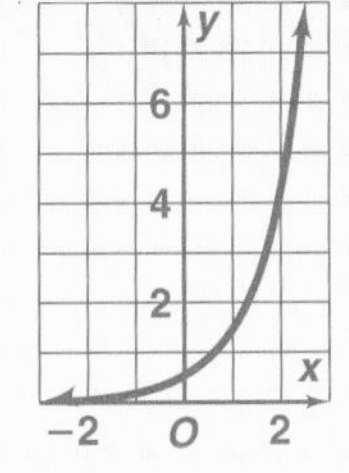
47. 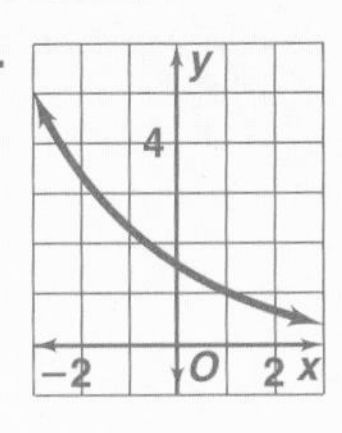
73. 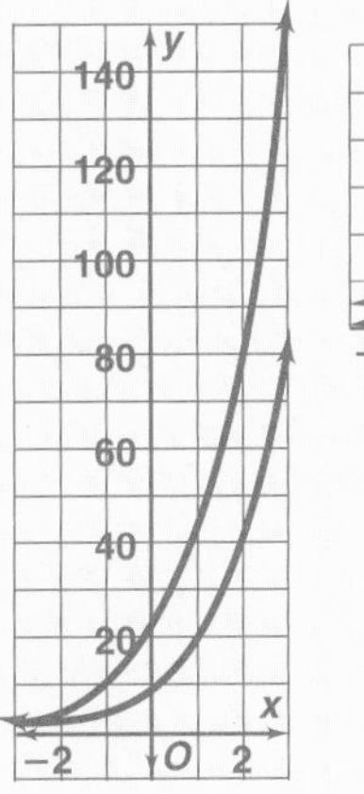
74. 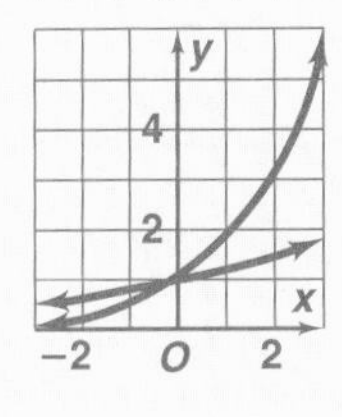

page 384 Mixed Review

77. 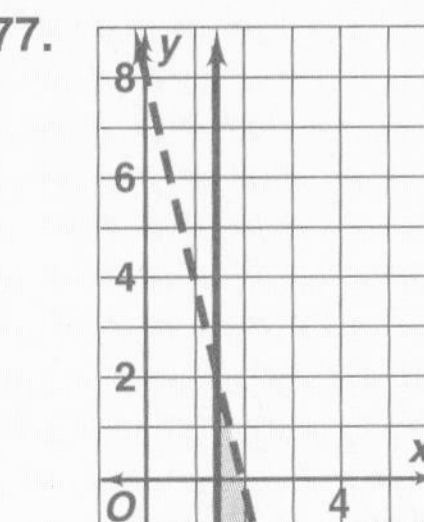

LESSON 8-6

Checkpoint page 395

20. 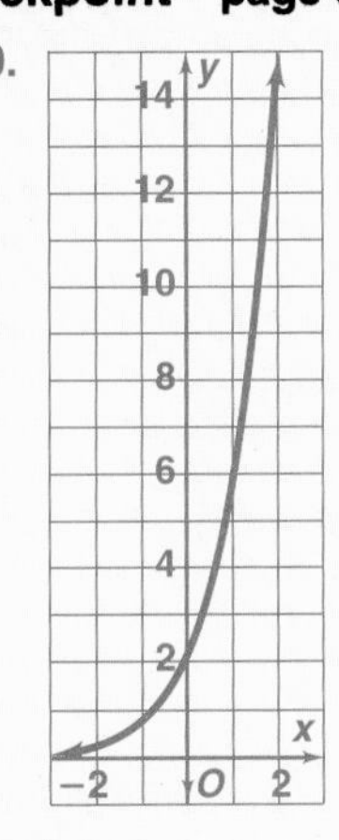

LESSON 8-8

page 405 Mixed Review

63.

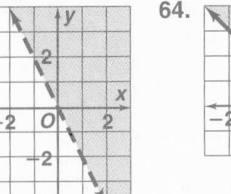

64. 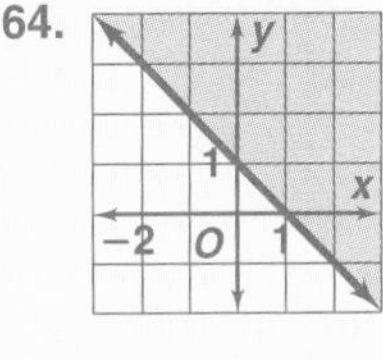

Chapter Assessment page 410

6c. Check students' work. Samples:
2008—23.85 billion kW•h;
2009—25.52 billion kW•h;
2010—27.30 billion kW•h;
2011—29.21 billion kW•h

d. Check students' work. Samples:
1988—6.16 billion kW•h;
1989—6.59 billion kW•h;
1990—7.06 billion kW•h;
1991—7.55 billion kW•h

7. Sample: A computer loses 20% of its value each year. How much will a $3500 computer be worth in 3 years? $1792

10. 

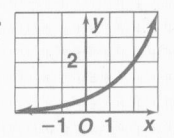

11.

12. 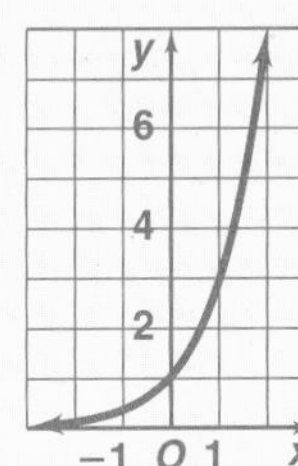

Cumulative Review page 411

16. Sample: Kalafia had $17. He bought several melons at $3 each. He has $5 left.

CHAPTER 9

LESSON 9-2

pages 422–424 On Your Own

36. To find the distance between two points, first, find the differences between the corresponding coordinates. For the points $(-1,4)$ and $(3,-2)$, $-1 - 3 = -4$ and $4 - (-2) = 6$. Then square each difference and add the results, $(-4)^2 + 6^2 = 52$. The distance between the points is the square root of the sum, $\sqrt{52}$.

For a midpoint, use the average of the x-coordinates to find the x-coordinate of the midpoint. Use the average of the y-coordinates to find the y-coordinate of the midpoint. Sample: The midpoint of the segment between $(-1,4)$ and $(3,-2)$ is $\left(\frac{-1+3}{2}, \frac{4+(-2)}{2}\right)$, or $(1, 1)$.

LESSON 9-3

page 429 Mixed Review

31. 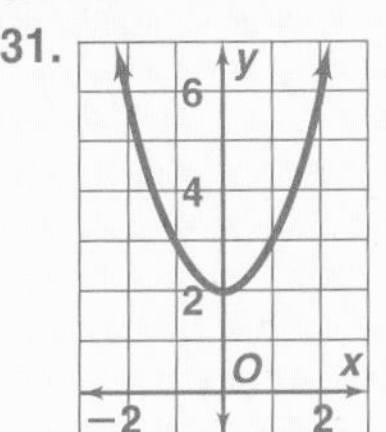
32. 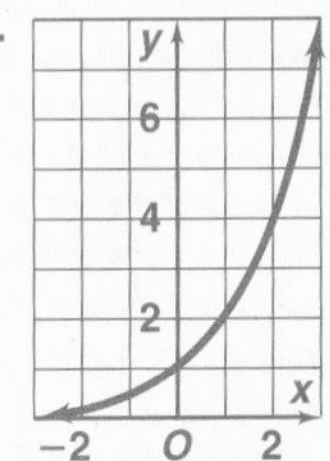

page 429 Checkpoint

9. Answers may vary. Sample: Find the difference between the x-coordinates, $x_2 - x_1$, and the

difference between the corresponding y-coordinates, $y_2 - y_1$. For $(-2,3)$ and $(10, 8)$, $10 - (-2) = 12$ and $8 - 3 = 5$. Square each difference, $12^2 = 144$ and $5^2 = 25$. Add the squares, $144 + 25 = 169$. Then take the square root to find the distance, $\sqrt{169} = 13$.

LESSON 9-7

pages 447–449 On Your Own

19.

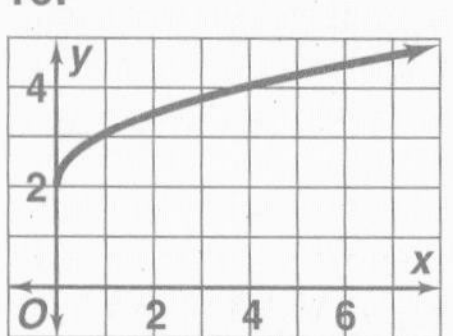

20.

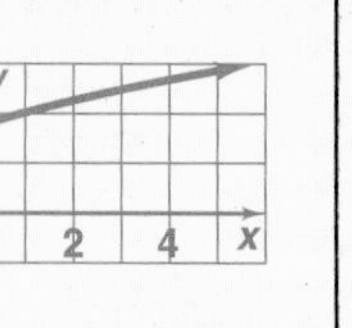

21.

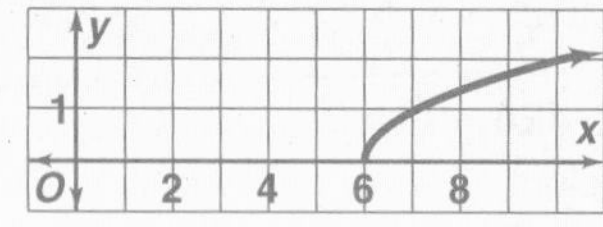

22.

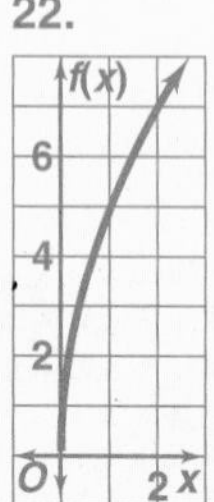

23.

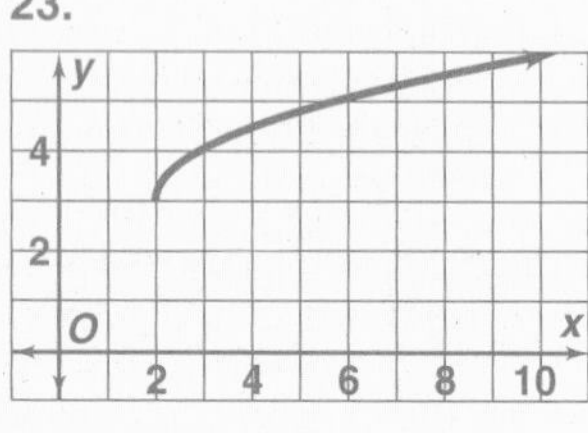

24.

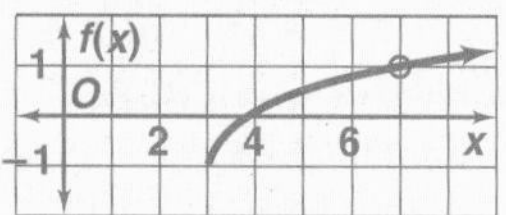

25.

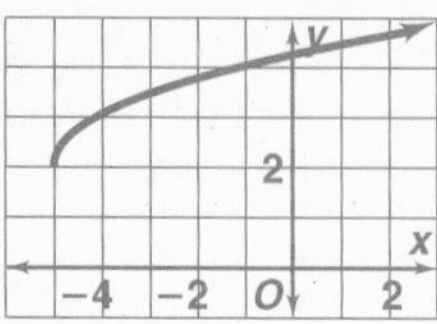

26.

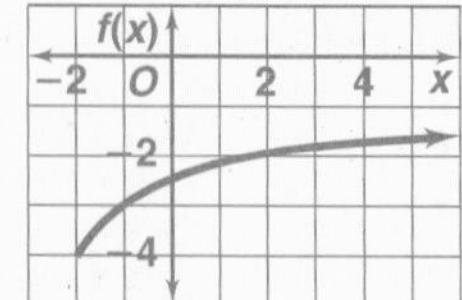

31b. For sample in part (a)

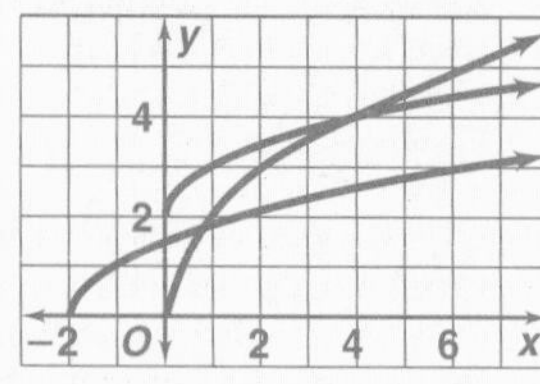

page 449 Mixed Review

32.

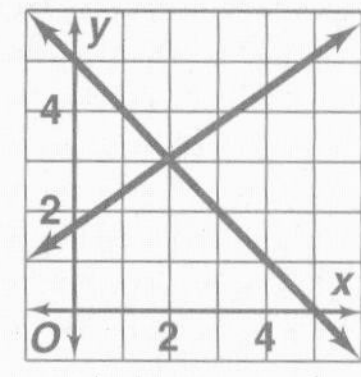

33.

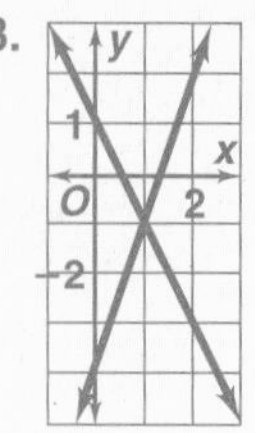

Checkpoint page 449

13.

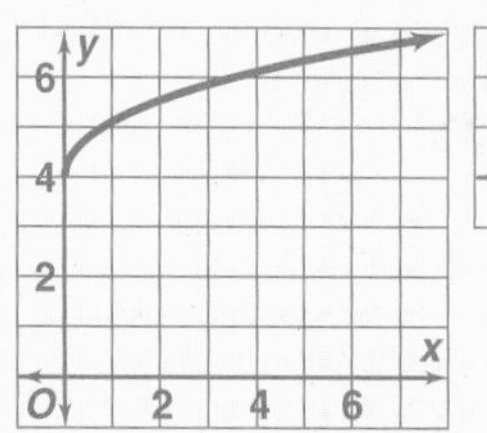

14.

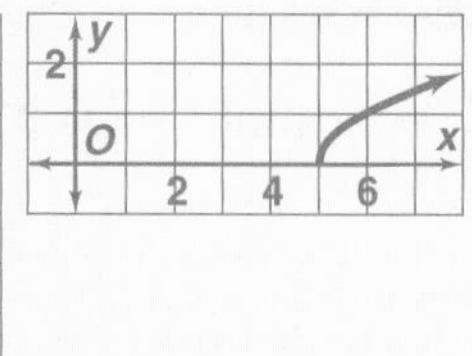

15. 16.

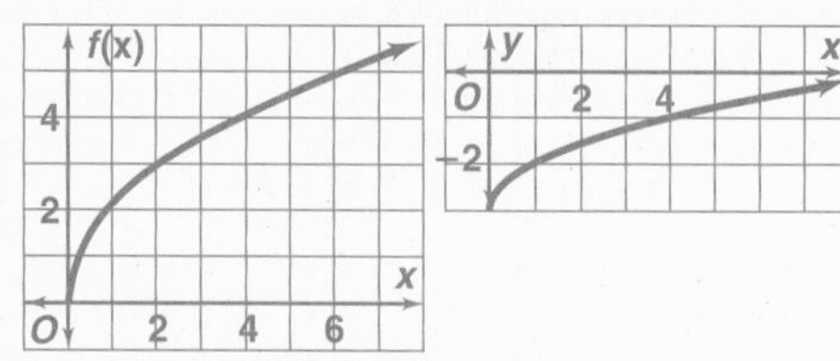

17. For both, the distributive property applies and coefficients are combined.

TE page 449 Lesson Quiz

Math Toolbox page 450

7.

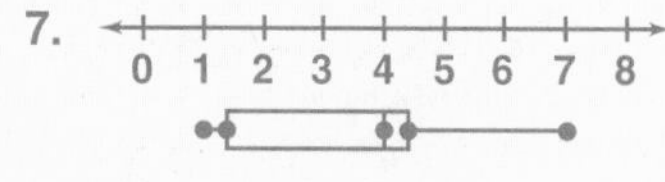

8.

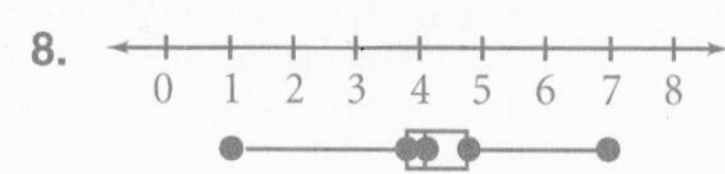

9a.

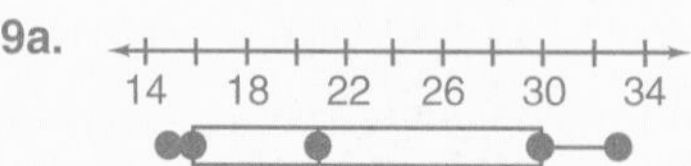

LESSON 9-8

pages 454–455 On Your Own

1.

x	x	$x - x$	$(x - x)^2$
5	5	0	0
3	5	−2	4
2	5	−3	9
5	5	0	0
10	5	5	25
		Sum:	38

2.

x	$\bar{x}$	$x - \bar{x}$	$(x - \bar{x})^2$
10	11	−1	1
9	11	−2	4
10	11	−1	1
12	11	1	1
11	11	0	0
14	11	3	9
		Sum:	16

3.

x	$\bar{x}$	$x - \bar{x}$	$(x - \bar{x})^2$
3.5	4	−0.5	0.25
4.5	4	0.5	0.25
6.0	4	2	4
4.0	4	0	0
2.5	4	−1.5	2.25
2.5	4	−1.5	2.25
5.0	4	1	1
		Sum:	10

4.

x	$\bar{x}$	$x - \bar{x}$	$(x - \bar{x})^2$
11	13.5	−2.5	6.25
11	13.5	−2.5	6.25
17	13.5	3.5	12.25
17	13.5	3.5	12.25
10	13.5	−3.5	12.25
11	13.5	−2.5	6.25
12	13.5	−1.5	2.25
19	13.5	5.5	30.25
		Sum:	88

7. A very large value increases the mean. A very small value decreases the mean. Each increases the standard deviation by a large amount.

9a. Answers may vary. Sample: Lyman should have asked about the range and the standard deviation or the median salary and the mode.

The range and the standard deviation would have told him that the salaries are widely distributed. The mode and the median, both \$265, would have told him that the mean does not represent the data well.

Chapter Assessment **page 460**

35. 36.

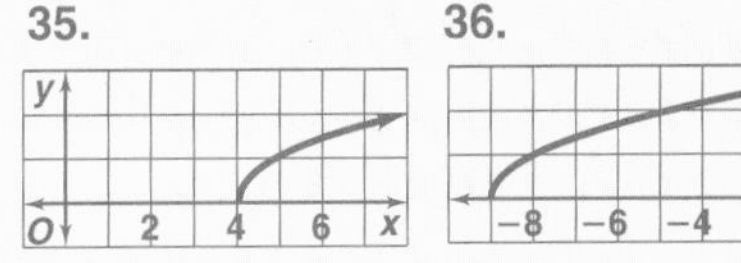

42. Answers may vary. Sample: Set B has a few values far from the mean, but most data points are very close to the mean. Set A has no values far from the mean, but the data points are not close to the mean.

CHAPTER 10

LESSON 10-2

pages 473–474 On Your Own

6. 7.

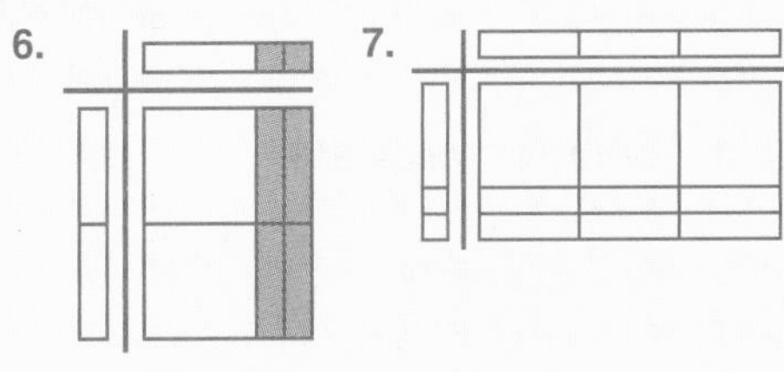

8.

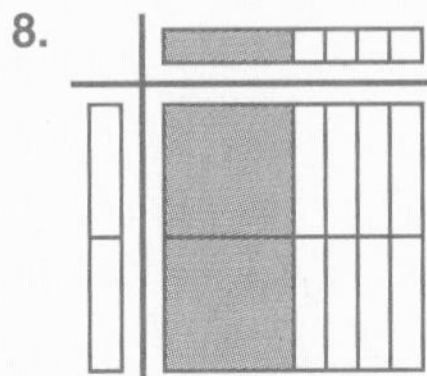

33a. Samples:

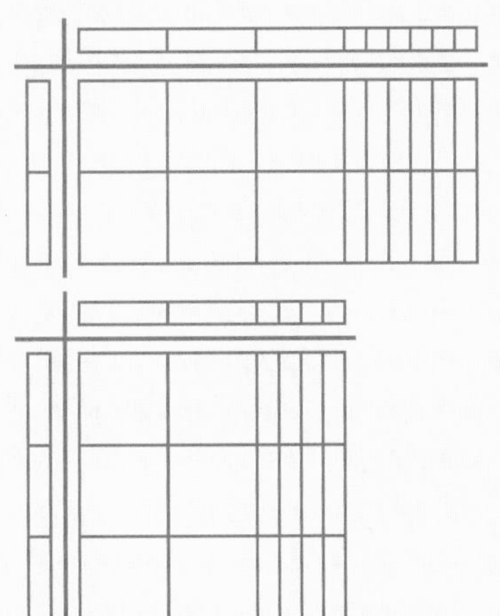

b. For samples in part (a): $(2x)(3x + 6)$; $(3x)(2x + 4)$

34b. $n^2 - n = n(n - 1)$ is a product of two consecutive integers. One of them must be even, so the product is always even.

LESSON 10-3

pages 475–477 Think and Discuss

7. $y = 6x^2 + 23x + 20$ $y = 6x^2 + 7x - 20$

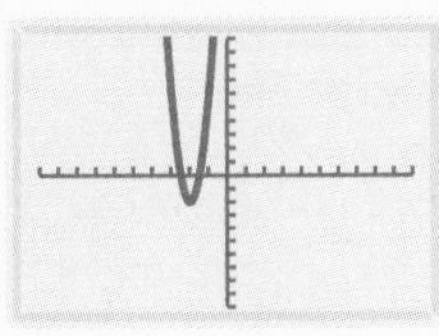

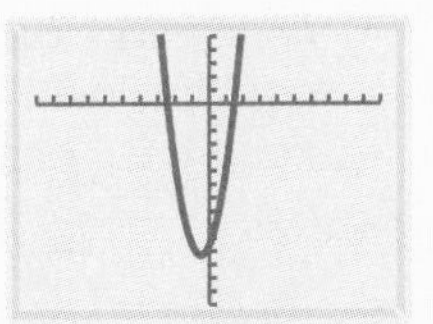

Xmin = −10 Ymin = −30
Xmax = 10 Ymax = 10
Xscl = 1 Yscl = 2

$y = 6x^2 - 7x - 20$ $y = 6x^2 - 23x + 20$

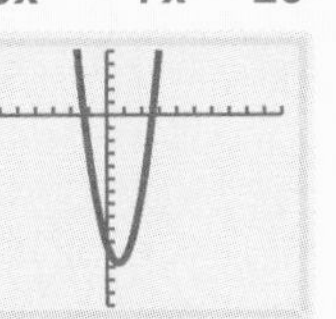

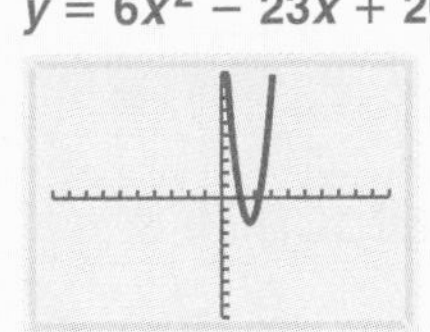

Xmin = −10 Ymin = −30
Xmax = 10 Ymax = 10
Xscl = 1 Yscl = 2

pages 477–479 On Your Own

4. 5.

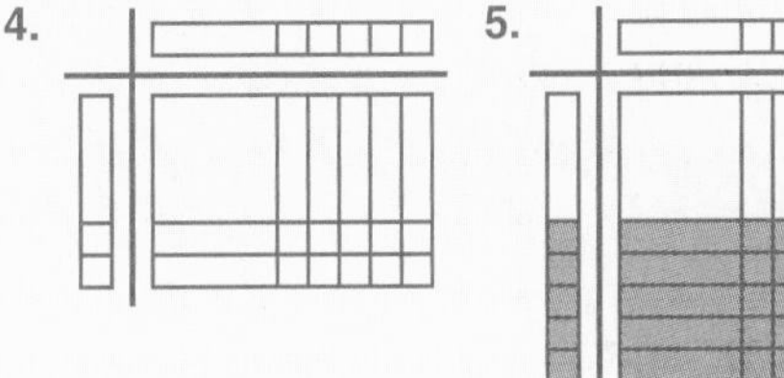

6.

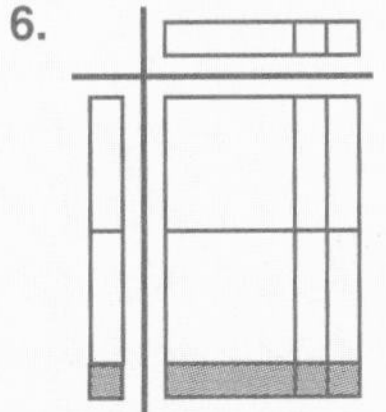

Math Toolbox **page 480**

3.

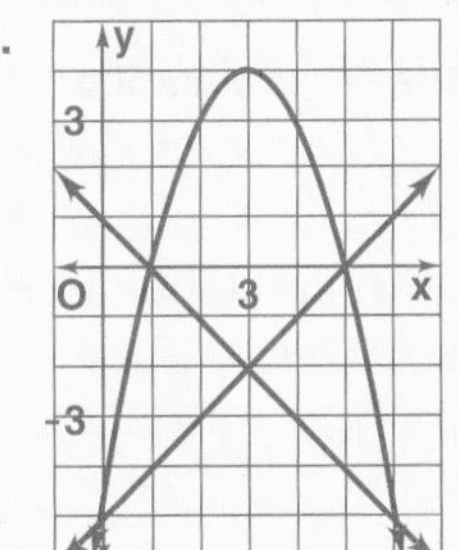

4.

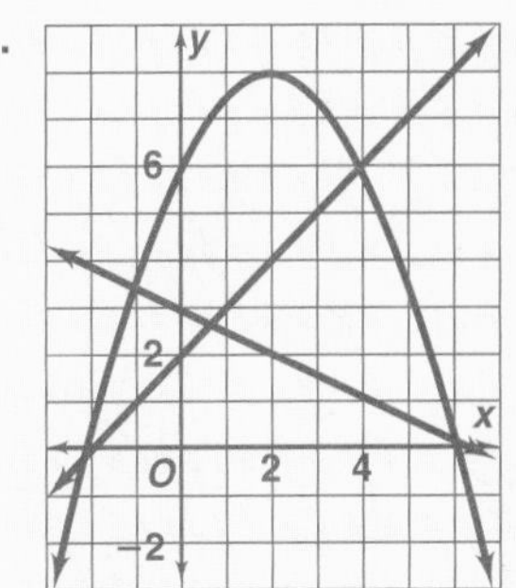

5.

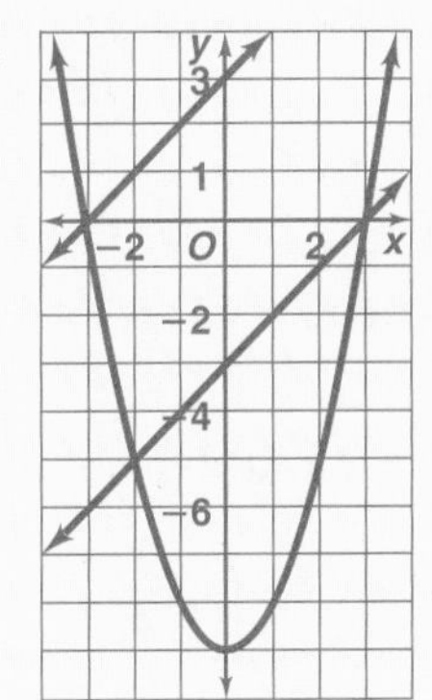

6. 8d.

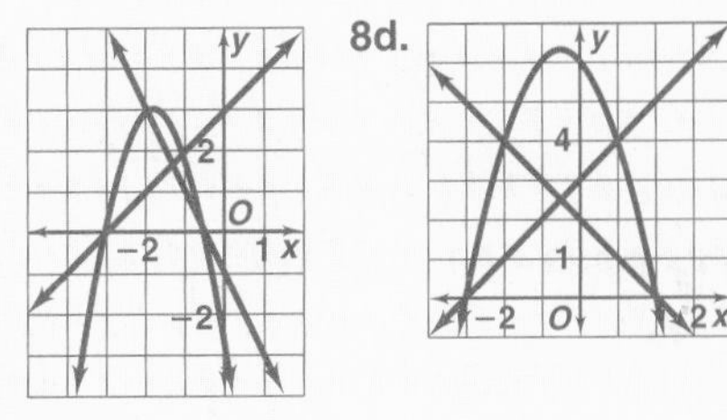

LESSON 10-4

pages 484–485 On Your Own

10. 11. 12. 13. 14. 15.

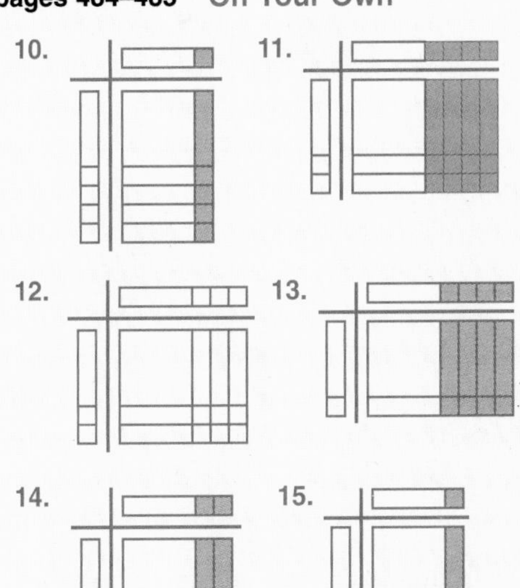

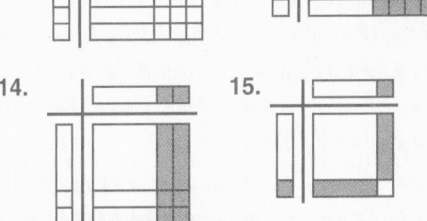

22a. Factors have the same operation.

b. Factors have opposite operations.

40. Samples: 4, 10, 18; $(x + 1)(x - 4)$, $(x + 2)(x - 5)$, $(x + 3)(x - 6)$

41. Samples: 2, 6, 12; $(x + 2)(x - 1)$, $(x + 3)(x - 2)$, $(x + 4)(x - 3)$

42. Samples: 13, 7, −7; $(x + 1)(x + 12)$, $(x + 3)(x + 4)$, $(x - 3)(x - 4)$

page 485 Mixed Review

59.

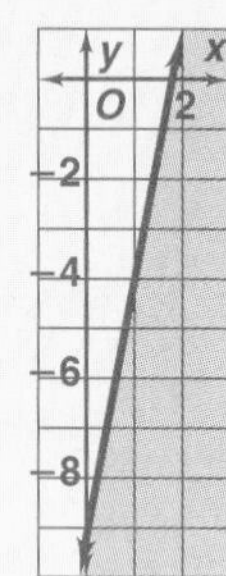

60.

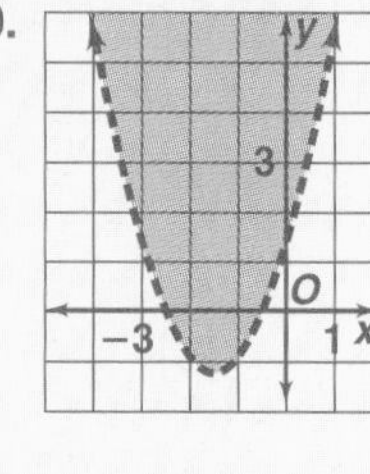

LESSON 10-5

pages 486–488 Think and Discuss

8a.

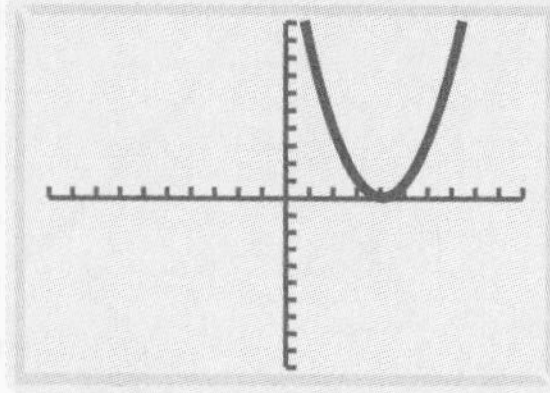

pages 489–490 On Your Own

4a.

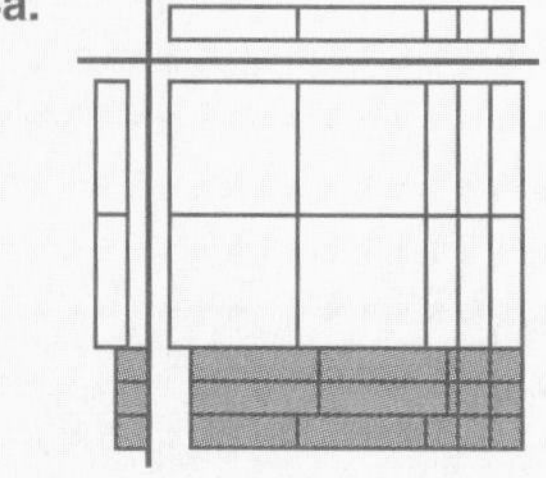

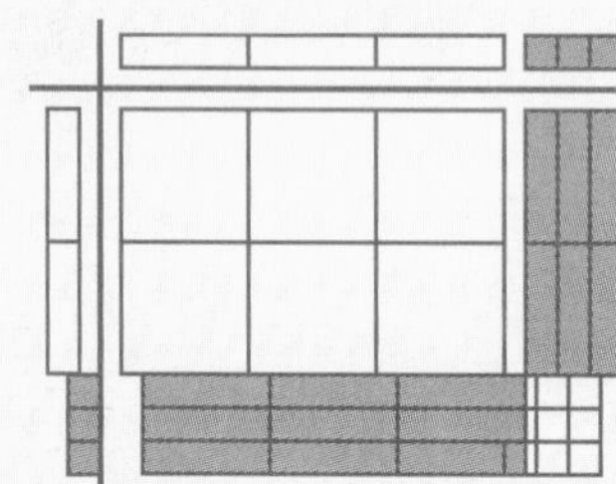

b. $(2x - 3)(2x + 3) = 4x^2 - 9$; $(3x - 3)^2 = 9x^2 - 18x + 9$

21. Answers may vary. Sample: Compute the square roots of the quadratic term and the constant term. If the linear term has a positive coefficient then the trinomial is the square of the sum of these square roots, $4x^2 + 12x + 9 = (2x + 3)^2$. If the linear term has a negative coefficient then the trinomial is the square of the difference of the square roots, $4x^2 - 4x + 1 = (2x - 1)^2$.

22a. View $(\pi d)^2 - (\sqrt{15w})^2$ in $(\pi d)^2 - (\sqrt{15w})^2 \geq 0$ as the difference of two squares and replace it in the inequality with its factorization $(\pi d - \sqrt{15w})(\pi d + \sqrt{15w}) \geq 0$.

b. $(\pi d - \sqrt{15w})(\pi d + \sqrt{15w}) \geq 0$ means that the factors have the same sign. Since d and w are both positive, neither factor can be negative. Then $\pi d - \sqrt{15w} \geq 0$ and $\pi d \geq \sqrt{15w}$.
So $d \geq \frac{\sqrt{15w}}{\pi}$.

38a.

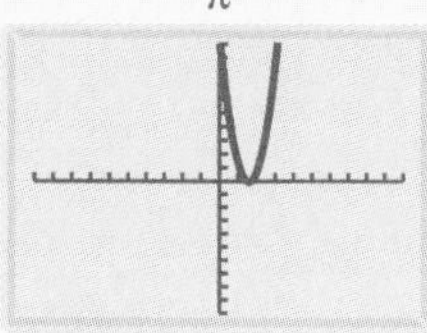

LESSON 10-6

pages 493–495 On Your Own

32b.

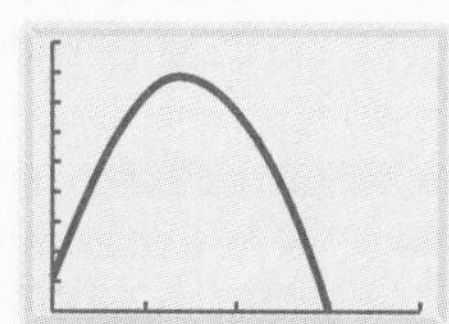

Xmin = 0 Ymin = 0
Xmax = 4 Ymax = 45
Xscl = 1 Yscl = 5

LESSON 10-7

pages 500–501 On Your Own

26. Use the quadratic formula because the trinomial does not factor easily.

27. Use factoring. Standard form shows the difference of two squares.

28. Use the quadratic formula because the standard form trinomial does not factor easily.

Chapter Assessment **page 506**

39. Use factoring when the quadratic factors easily, such as a perfect square trinomial or a difference of two squares. Samples:
$x^2 - 2x - 8 = 0$
$x^2 - 2x - 8 = (x - 4)(x + 2)$
$4x^2 + 40x + 25 = 0$
$4x^2 + 40x + 25 = (2x + 5)^2$
$16x^2 - 81 = 0$
$16x^2 - 81 = (4x - 9)(4x + 9)$
Use square roots if the function has no linear term and does not factor easily. Sample:
$x^2 - 200 = 0$
$x = \pm\sqrt{200}$
Use the quadratic formula if the trinomial in standard form does not factor easily. Sample:
$2x^2 + 5x + 1 = 0$
$x = \frac{-5 \pm \sqrt{5^2 - 4 \cdot 2}}{2 \cdot 2}$
Use graphing if a graphing calculator is available.

CHAPTER 11

LESSON 11-1

pages 513–514 On Your Own

36.

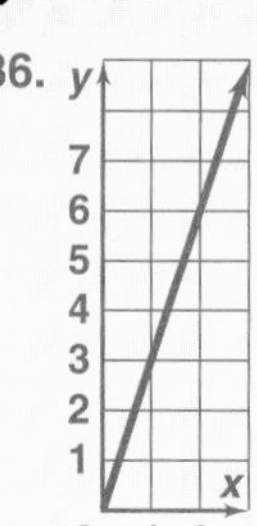

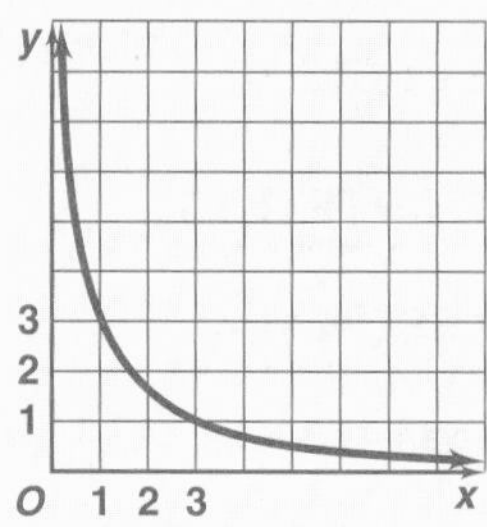

Math Toolbox **page 515**

4.

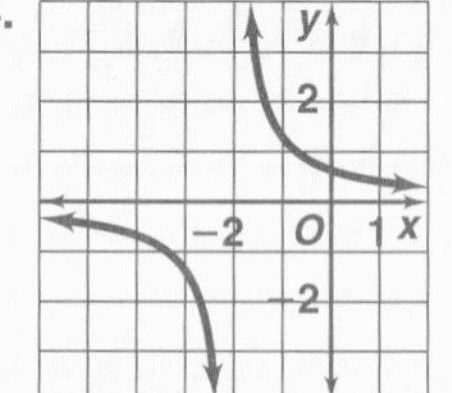

5.

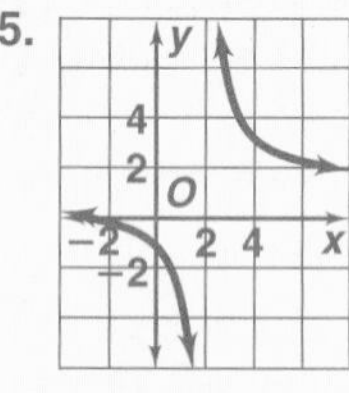

6. 7.

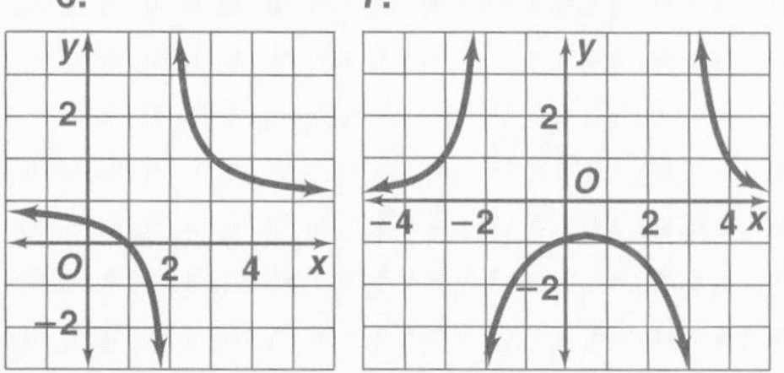

8. 9.

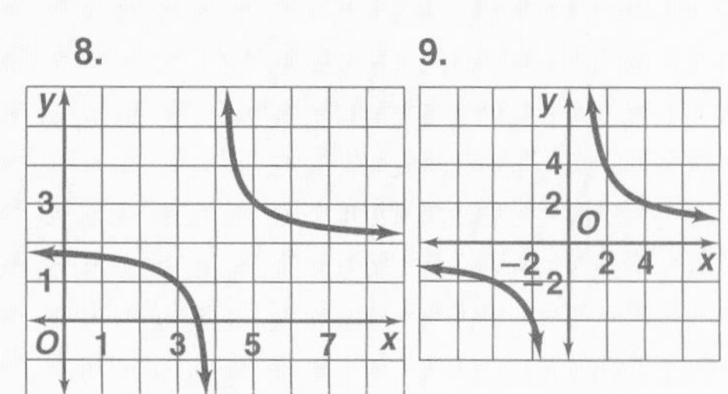

10. 11.

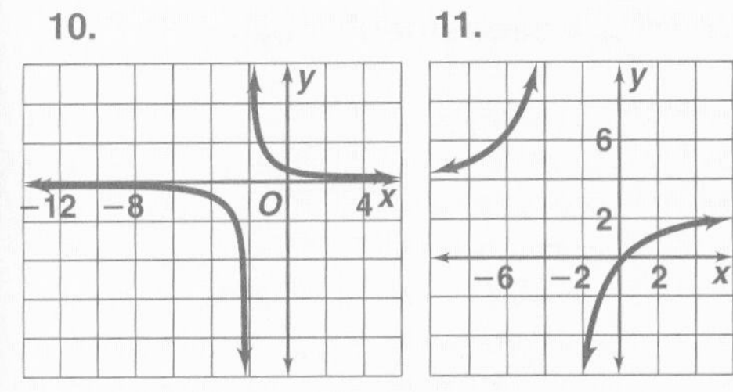

12. 13.

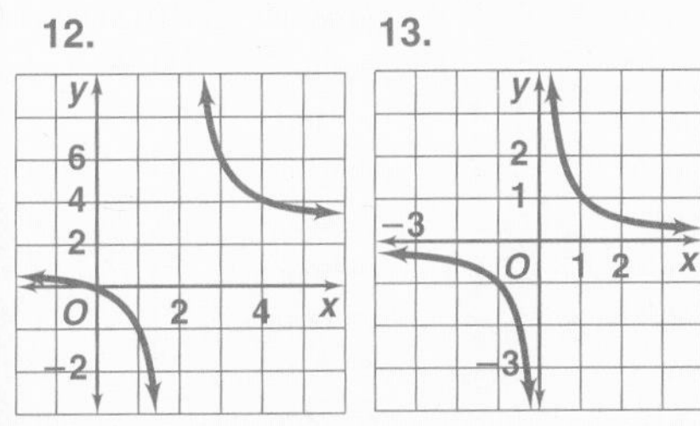

LESSON 11-2

pages 516–519 Think and Discuss

8.

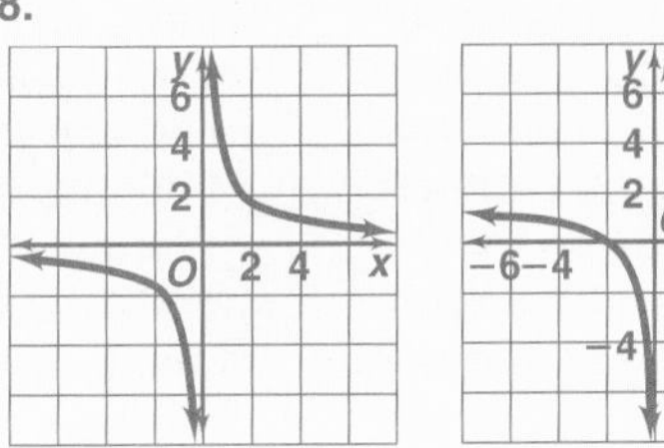

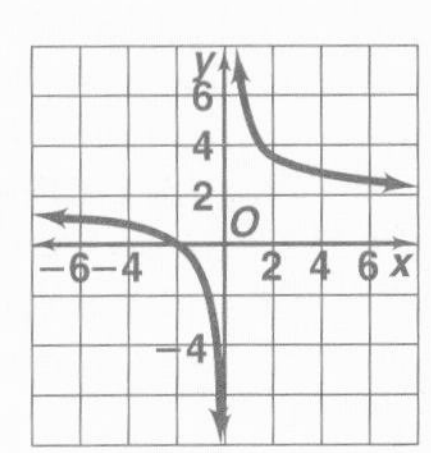

pages 520–521 On Your Own

13.

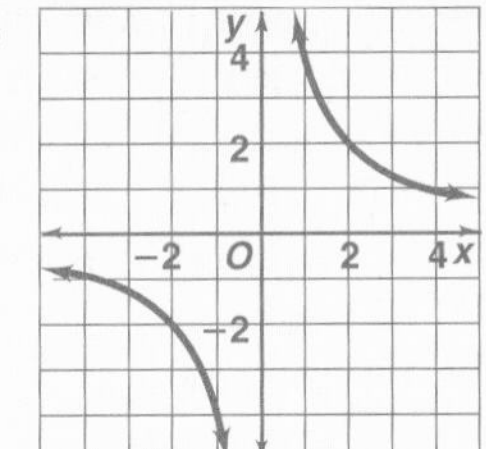

14.

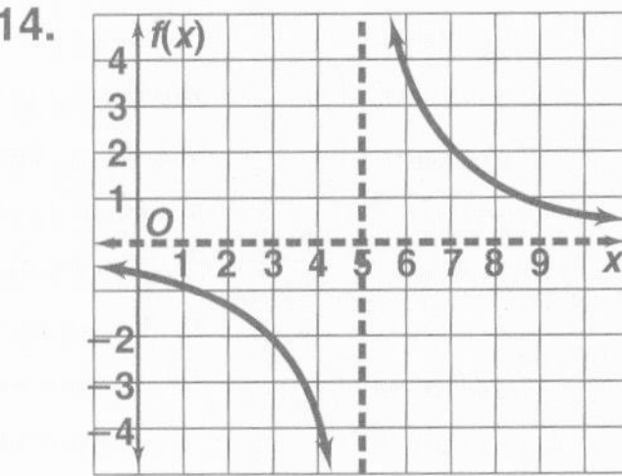

15. 16.

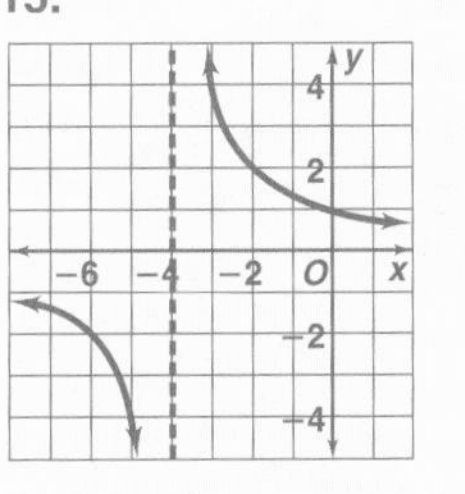

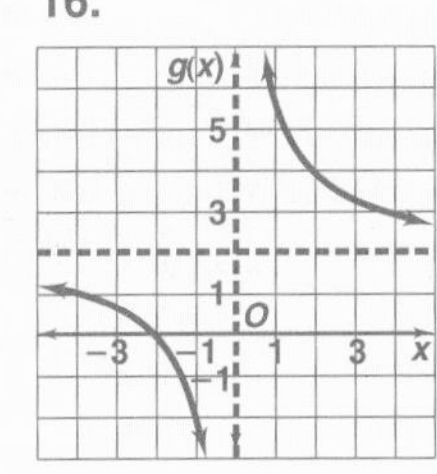

17. 18.

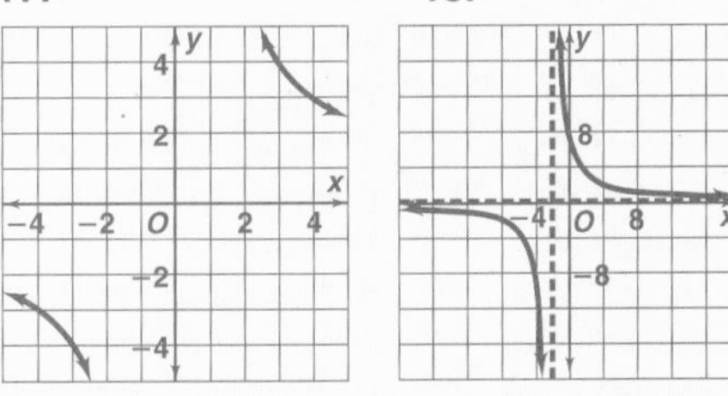

19. 20.

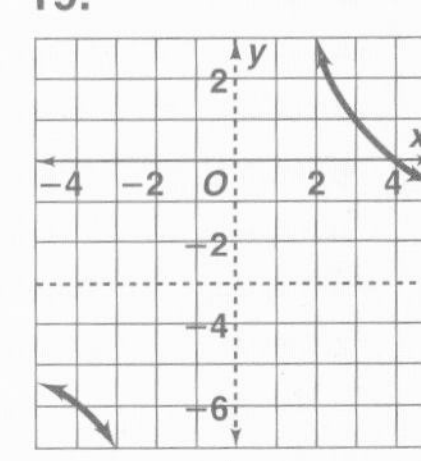

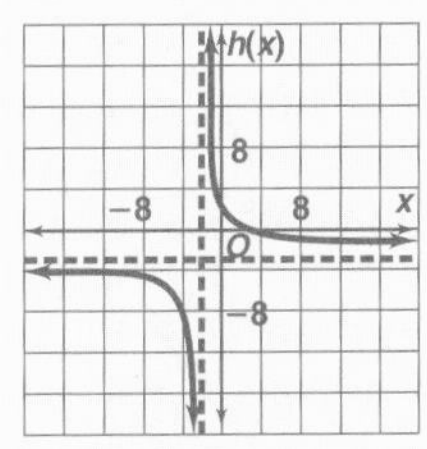

21. 22.

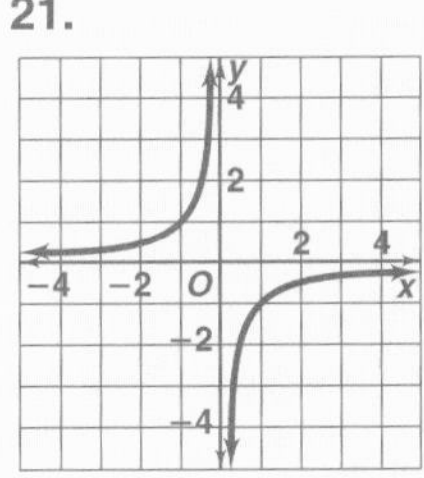

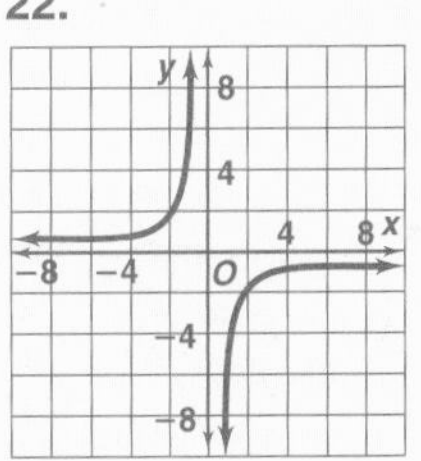

23. 24.

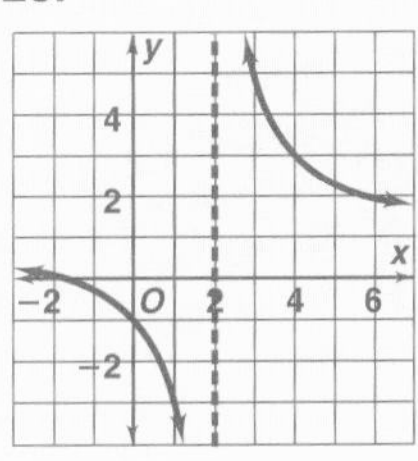

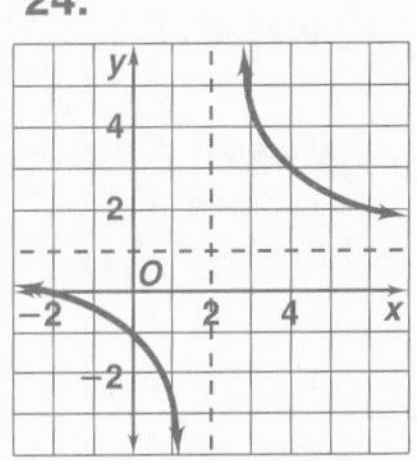

25a.

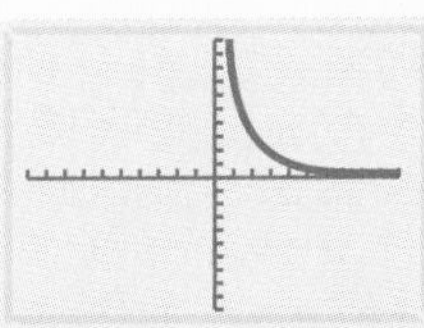

Xmin=0 Ymin=0
Xmax=6 Ymax=2000
Xscl=1 Yscl=200

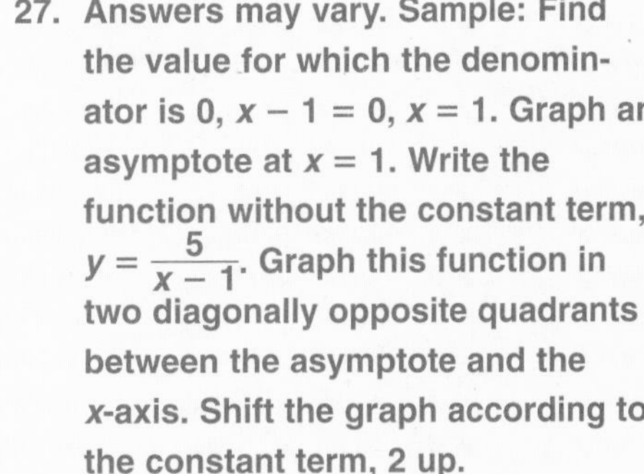

27. Answers may vary. Sample: Find the value for which the denominator is 0, $x - 1 = 0$, $x = 1$. Graph an asymptote at $x = 1$. Write the function without the constant term, $y = \frac{5}{x - 1}$. Graph this function in two diagonally opposite quadrants between the asymptote and the x-axis. Shift the graph according to the constant term, 2 up.

29.

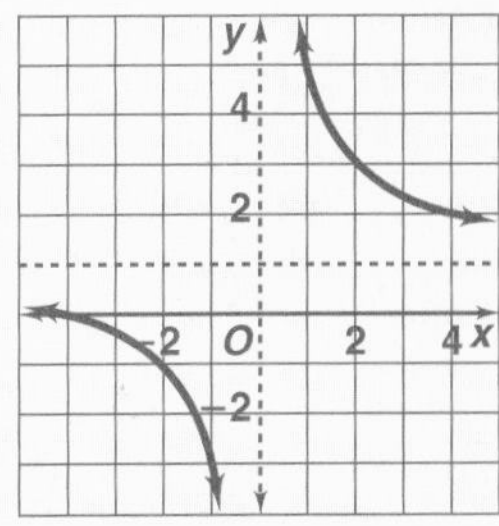

30. 31.

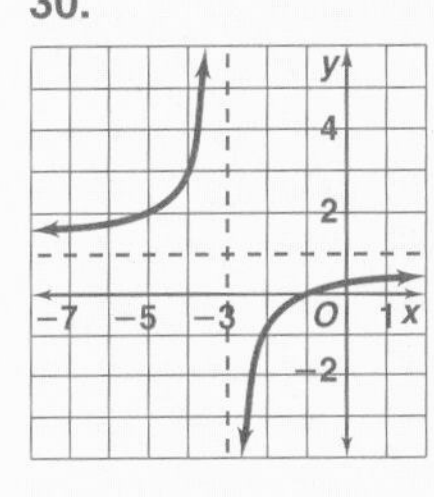

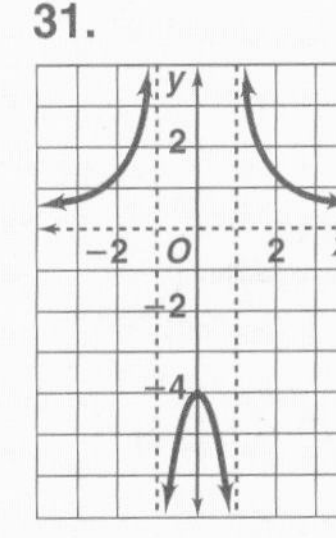

32. 33.

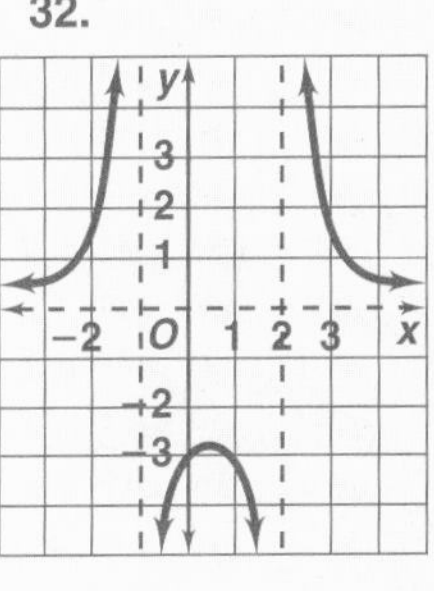

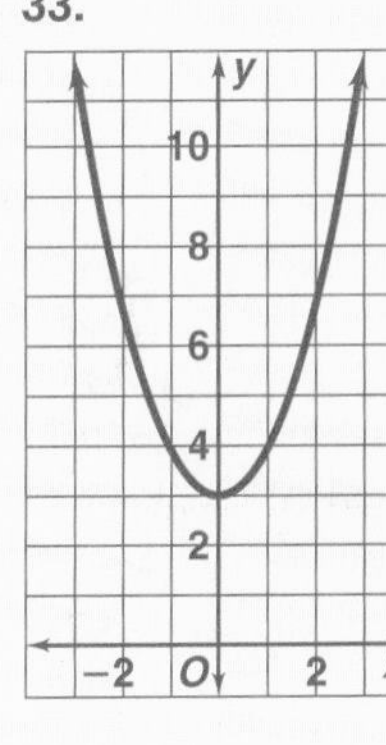

34. 35.

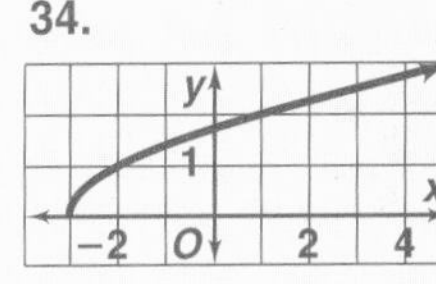

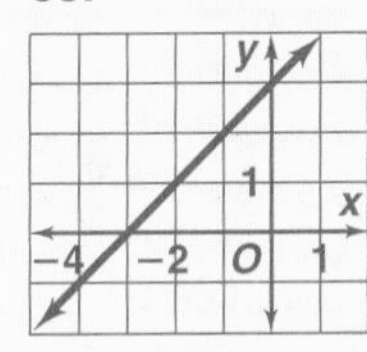

36. 37.

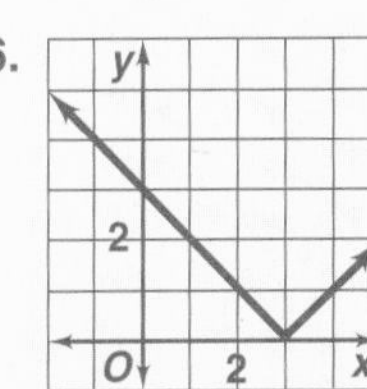

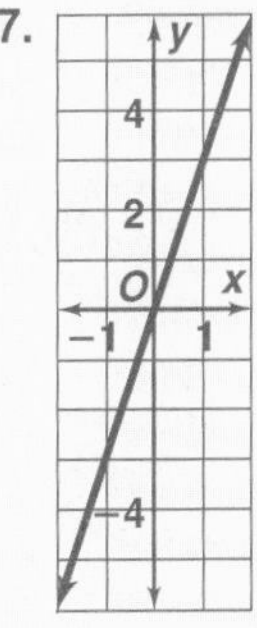

38.

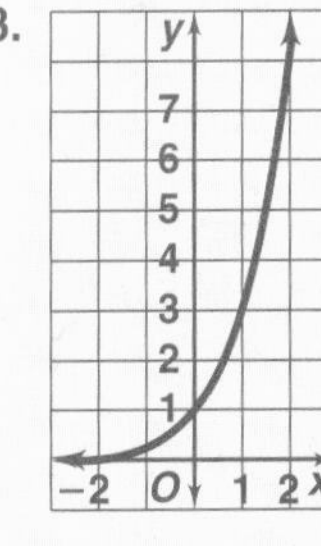

39.

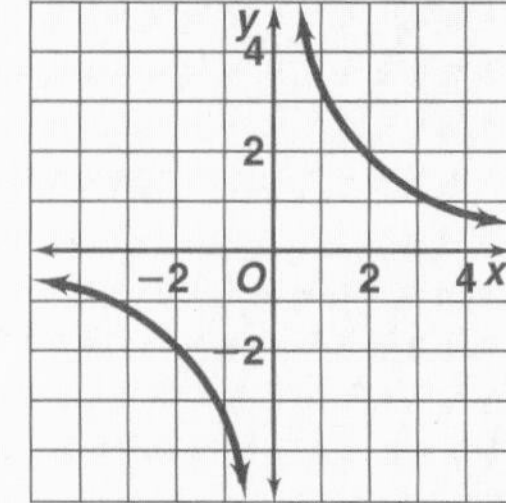

40.

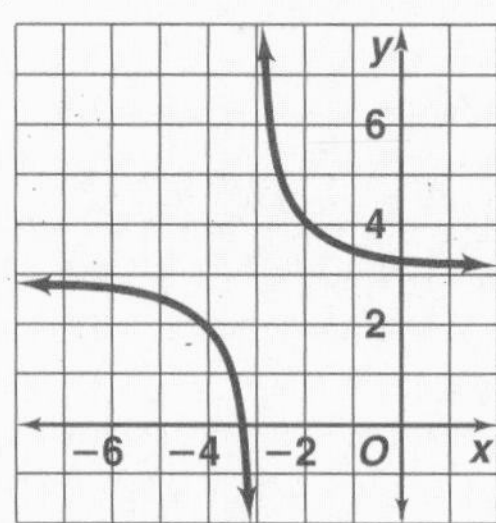

41.

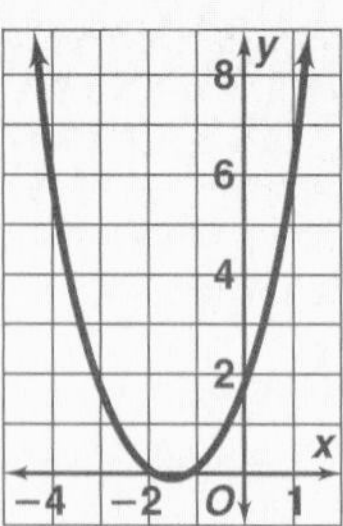

42.

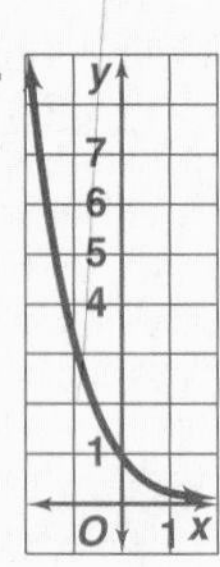

43.

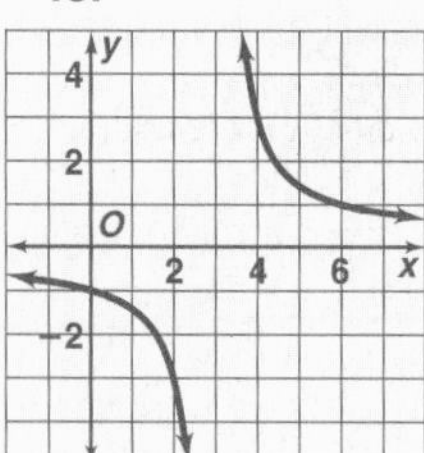

44.

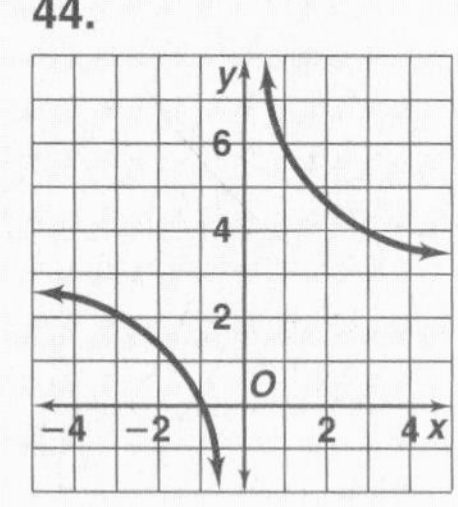

Checkpoint **page 521**

4.

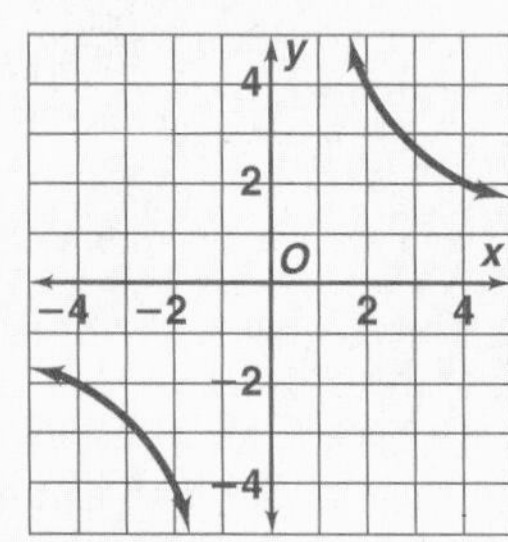

5.

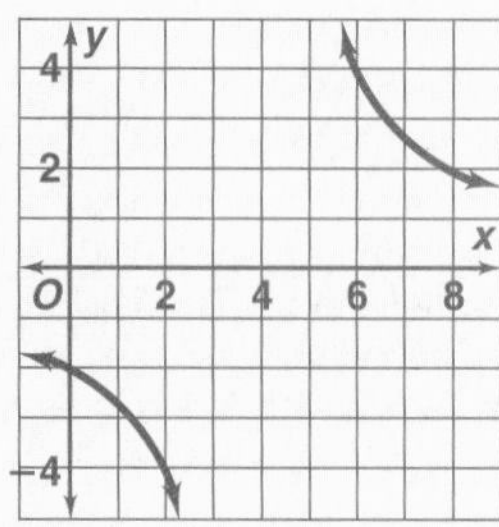

6.

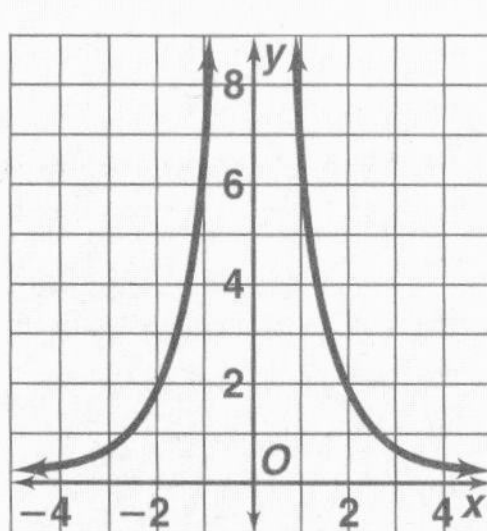

7.

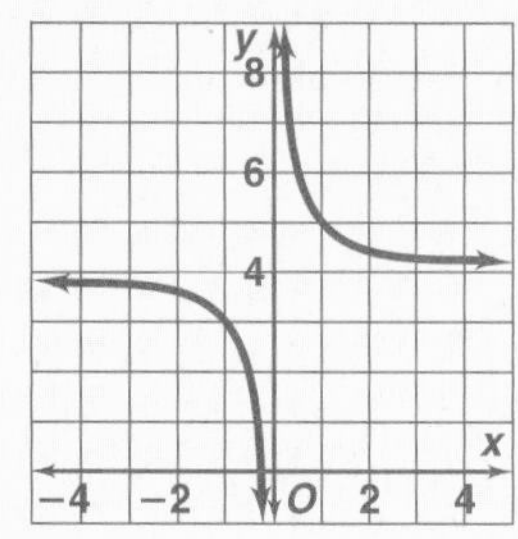

LESSON 11-6

page 542 Mixed Review

31.

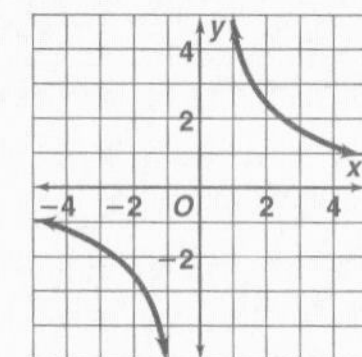

32.

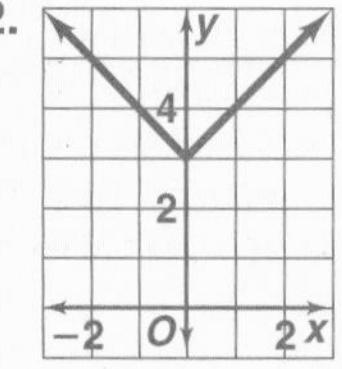

Chapter Assessment page 552

9.

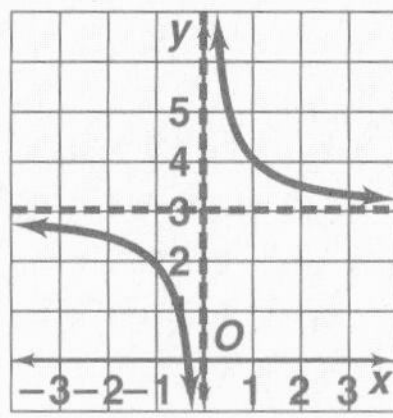

10.

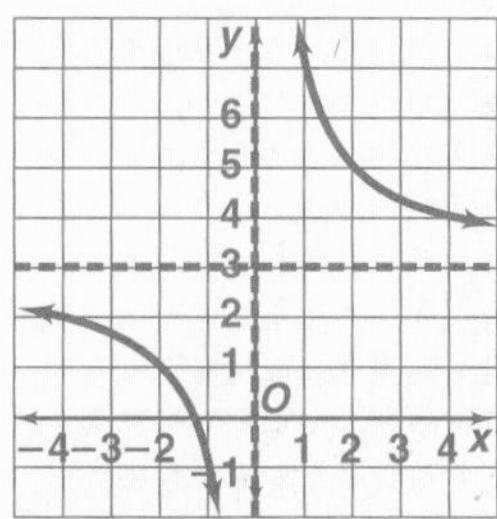

11. Answers may vary. Sample: In direct variation and in inverse variation the variables are related to each other by a constant. But in direct variation that number is the ratio of any corresponding pair of input and output values. In inverse variation that number is the product of any corresponding pair of input and output values.

Cumulative Review pages 554–555

30. Sample: Juan's salary is \$300 per week. On January 1, he gets a raise that gives him an annual (52-week) salary of \$18,200. How much more money will Juan receive per week?
x = weekly increase
$18{,}200 = 52(300 + x)$
\$50 a week

32. Multiply each term of the first binomial by each term of the second binomial. Then combine like terms.

Index

Index

Cover Design Bruce Bond

Cover Photos Globe and background, Bill Westheimer; jet plane, Joe Towers/Check Six

Book Design Linda Johnson, Eve Melnechuk, Stuart Wallace, Alan Lee Associates

Technical Illustration ANCO/Outlook

Illustration

Leo Abbett: 67

ANCO/Outlook: 16, 18, 26, 82, 87, 95, 193, 227, 230, 249, 253, 270, 274, 279, 289, 326, 399, 439

David Frazier: 116, 117, 143, 167, 170

Kathleen Dempsey: 273, 298

Dave Garbot: 161

Barbara Goodchild: 12

Keith Kasnot: 392

Ellen Korey-Lie: 3, 57, 61, 107, 157, 213, 267, 317, 361, 413, 463, 509

Seymour Levy: 107, 150, 291, 346, 371, 455, 464, 489, 490, 513, 539

Andrea G. Maginnis: 147, 221, 241, 286, 287, 292, 298, 299, 302, 369, 429, 464, 468, 529, 546

Gary Phillips: 129

Matthew Pippin: 122, 124

Pond Productions: 9

Gary Torrisi: 276-277, 306

Photography

Photo Research Toni Michaels

Abbreviations: KO = Ken O'Donoghue; MT = Mark Thayer; JC = Jon Chomitz; TSW = Tony Stone Images; SM = Stock Market; FPG = Freelance Photographer's Guild; SB = Stock Boston; PH = Prentice Hall File Photo

Front matter: Page vii, ©Markus Amon/TSW; **viii,** KO; **ix,** MT; **x,** ©Mark Burnett/SB; **xi,** ©Paramount Studios/The Kobal Collection; **xii,** ©David Young Wolff/TSW; **xiii,** Johnny Johnson/Animals Animals; **xiv,** ©Nigel Cattlin/Photo Researchers; **xv,** ©John Gilmoure/The Stock Market; **xvi,** MT; **xvii,** ©Charles West/SM; **xix,** FPG; **xxi,** JC; **xxii,** ©John Madere/SM; **xxv,** ©Andy Sacks/TSW.

Chapter 1: Pages 2-3, O. Louis Mazzatenta/National Geographic Image Collection; **3 inset,** JC; 4, ©David Noble/FPG; **5 t,** ©Tony Freeman/PhotoEdit; **5 b,** ©Marcel Ehrhard; **6, 8 t, 8 b,** ©Andy Sacks, Lonnie Duka, Darrell Gulin all TSW; **9,** Superstock, **12 l,** ©Markus Amon/TSW; **12 r,** KO; **14,** PH; **15,** NASA; **16,** ©Toni Michaels; **17,** KO; **19 l,** Karl Kreutz; **19 r,** ©Bruce Coleman; **20,** ©Uniphoto/Alan Lardman; **21,** ©Fotoconcept/Bertsch/Bruce Coleman; **23,** FRANK & ERNEST reprinted by permission of Newspaper Enterprise Association, Inc.; **23 t,** ©Janice Rubin/Black Star; **25,** NASA; **28,** Courtesy, Cedar Point/Photo by Dan Feicht; **30 all,** KO; **33,** ©Nicholas Devore III/Bruce Coleman; **34,** ©Michael Holford; **37,** ©Bob Daemmrich/The Image Works; **38,** ©Toni Michaels; **41,** ©Paul Souders/TSW; **42,** PH; **46,** KO; **48,** ©Gary Geer/TSW.

Chapter 2: Pages 56-7, ©Larry Lawfer; **57 inset, 59,** JC; **61,** PH; **62,** ©Leonard Lees Rue III/Bruce Coleman; **64,** ©Richard Wood/The Picture Cube; **66,** Courtesy, Professor Bernie Phinney, University of California; **69,** Martha Cooper/Peter Arnold; **71,** NASA; **72,** PH; **74-5,** Steve Greenberg; **77,** ©JC; **80,** The Computer Museum, Boston; **82,** JC; **84,** KO; **88 t,** PH; **88 b,** UPI/Bettmann; **90-1,** JC; **93 tl, br,** ©David McGlynn, Ralph Cowan both FPG; **95,** ©Winter/The Image Works; **96,** Courtesy, Wheel of Fortune; **98,** JC.

Chapter 3: Pages 106-7, JC; **108,** MT; **112,** ©Baloo/Rothco; **114, 115 all,** KO; **117,** PH; **120,** MT; **124,** ©Kevin Morris/Allstock; **126,** Schomburg Center for Research in Black Culture, New York Public Library; **130,** ©Michael Newman/Photo Edit; **135, 137,** MT; **140,** ©M. Siluk/The Image Works; **143,** ©David Young-Wolff/TSW; **144,** ©Jose L. Pelawz/SM; **146,** Courtesy, Dr. Graciela S. Alarcon; **147,** ©Arthur C. Smith III/Grant Heilman Photography; **148,** ©Charles Krebs/SM.

Chapter 4: Pages 156-57, ©Telegraph/FPG; **157 inset,** PH; **158 t,** ©Superstock; **158 b,** ©Bonnie Kamin/PhotoEdit; **160 l,** ©Fred Whitehead/Animals Animals; **160 r,** ©Mark Burnett/SB; **163, 164 all,** KO; **167,** ©Michelle Bridwell/PhotoEdit; **170 t,** ©Bob Daemmrich; **170 b,** ©Anthony Edgeworth/SM; **171,** MT; **173,** JC; **174,** ©Bob Daemmrich; **175,** ©Bill Luster/NCAA Photos; **176 all,** The Granger Collection; **177,** ©Guido Alberto Rossi/Photo Researchers; **179 br,** JC; **179 TR,** ©David Frazier/Photo Researchers; **179 l,** JC; **182,** PH; **184,** ©George Disarid/SM; **187,** JC; **189,** ©Stephen Dalton/Animals Animals; **190,** JC; **193,** TSW; **195,** ©Melinda Berge/Bruce Coleman; **196,** ©Bob Daemmrich/SB; **199,** ©Lawrence Migdale/Photo Researchers; **203,** ©John Madere/SM; **204,** Custom Medical Stock Photo.

Chapter 5: Pages 212-13, Larry Lawfer; **213 inset,** PH; **214 l,** ©Dean Siracusa/FPG; **214 r,** ©Chris Cheadle/TSW;

215 t, ©Valder Tormey/Picture Perfect; **215 inset,** ©John McGrail/FPG; **219,** ©E&P Bauer/Bruce Coleman; **220,** ©Gary Bigham/International Stock; **222,** ©Paul Silverman/Fundamental Photographs; **223,** ©Bruce Hands/TSW; **225,** JC; **226,** ©Telegraph/FPG; **229 tr,** ©David Madison/Bruce Coleman; **229 b,** Reprinted by permission: Tribune Media Services; **230 both,** ©Jean-Pierre/Sygma; **232,** MT; **234,** ©Peter Vanderworker/SB; **236,** ©LOL/FPG; **237,** PH; **239,** ©C. C. Lockwood/Bruce Coleman; **241,** ©Paramount Studios/The Kobal Collection; **244,** ©Chris Jones/SM; **247,** ©Nathan Bilow/Allsport; **248 l,** ©Lynn Karlin/FPG; **248 br,** Jeff Spielman/Stockphotos/The Image Bank; **250,** ©Tony Freeman/PhotoEdit; **257,** MT; **258,** ©The Granger Collection; **259,** NASA.

Chapter 6: Pages 266-67, ©David Austen/SB; **269,** ©Mary Kate Denny/TSW; **270,** ©Jose L. Pelaez/SM; **272,** ©ROBOTMAN reprinted by permission of Newspaper Enterprise Association, Inc.; **274,** ©Robert Caputo/SB; **278,** ©David Young Wolff/TSW; **279,** Courtesy, Rubber Stamps of America; **281,** JC; **283,** PH; **284,** ©Will & Deni McIntyre/TSW; **285,** ©Bob Daemmrich/The Image Works; **286 tl,** JC; 286 br, ©Mark Burnett/SB; **286 tr,** ©David Young Wolff/PhotoEdit; **288,** ©Don Johnson/SM; **291 r,** Spencer Grant/The Picture Cube; **292,** ©D. Well/The Image Works; **296,** ©Stephen Kline/Bruce Coleman; **298 both,** PH; **300-301,** JC; **300 l,** ©Lawrence Migdale/Photo Researchers; **301 tr,** ©Bob Daemmrich; **301 b, 302,** ©John Eastcott, Frank Pedrick/The Image Works; **304,** ©Brian Smith/SB; **308,** From American Indian Design & Decoration by LeRoy H. Appleton, Dover Books; **309,** ©Charles Mercer Photography.

Chapter 7: Pages 316-17, Larry Lawfer; **317 inset,** Courtesy, Indiana Department of Motor Vehicles; **318,** ©Richard Megna/Fundamental Photographs; **321,** PH; **324,** ©Johnny Johnson/Animals Animals; **326,** ©Granitsas/The Image Works; **327 all,** JC; **328,** ©Mark Burnett/SB; **329,** ©Joe Szkodzinski/The Image Bank; **330,** ©Richard Pasley/SB; **333,** ©David Austen/TSW; **334,** NASA; **335,** Reprinted by permission: Tribune Media Services; **338,** ©Gene Peach/The Picture Cube; **340,** ©Nebraska State Historical Society; **341,** ©John Lund/TSW; **344 l, r** ©Chuck Savage, Jon Feingersh both SM; **346,** ©Hank del Espinasse/The Image Bank; **350 b,** ©Yellow Dog Productions/The Image Bank; **350 t,** ©Tony Page/TSW; **353 inset,** The Granger Collection; **353,** UPI/Corbis/Bettmann.

Chapter 8: Pages 360-61, ©Bob Daemmrich/SB; **361 inset,** ©Sidney Moulds/Science Photo Library/Photo Researchers; **363,** ©Andrew Henley/Auscape; **364,** PH; **365,** ©Chris Rogers/SM; **367,** ©Mark Burnett/SB; **371 l,** ©Alan McFee/FPG; **371 r,** ©Chip Henderson/TSW; **372,** ©Miro Vintoniv/SB; **373,** ©Corbis/Bettmann; **374** ©Grant Heilman Photography; **380 both,** ©Nigel Cattlin/Photo Researchers; **383,** ©Jack Stein Grove/Tom Stack and Assoc.; **385,** ©Bob Daemmrich/TSW; **386,** ©Telegraph/FPG; **387,** Martucci Studio; **388 tl,** Ray Ng; **388 br,** ©Paul S. Howell/Gamma Liaison; **389,** FOXTROT ©1991 Bill Amend. Reprinted with permission of UNIVERSAL PRESS SYNDICATE. All rights reserved; **390,** ©Dan McCoy/Rainbow; **392,** ©Nathan Benn/SB; **394,** ©Dr. Jeremy Burgess/Science Photo Library/Photo Researchers; **398,** The Granger Collection; **400,** Courtesy Dr. Jewel Plummer Cobb; **402 all,** MT.

Chapter 9: Pages 412-13, ©Paul Chesley/TSW; **415,** ©Spencer Jones/FPG; **417,** ©Jose Fuste Raga/SM; **418,** ©Mark Antman/The Image Works; **421,** ©Bob Daemmrich; **423,** ©Torleif Sënsson/SM; **424,** Cindy Loo/The Picture Cube; **426,** ©Joe McDonald/Bruce Coleman; **427,** T. Kevin Smyth/SM; **428,** D&J Heaton/SB; **431,** ©Dave Watters/TSW; **433,** ©Bob Daemmrich/The Image Works; **434,** PH; **436,** ©ABC/Mondrian Estate/Holtzman Trust/Haags Gemeentemuseum; **438,** Reprinted by permission: Tribune Media Services; **439 t,** ©Porterfield-Chickering/Photo Researchers; **439 b,** ©Ron Sherman/Stock Boston; **441,** ©Jim Tuten/FPG; **443,** ©Chris Jones/SM; **445,** ©John Gilmore/SM; **448,** KO; **451 tr,** ©James Shaffer/PhotoEdit; **451 bl,** ©Bob Daemmrich/SB; **453,** ©Anton Want/Allsport; **453 inset,** PH; **454,** ©Kindra Clineff/The Picture Cube; **455,** ©Jonathan Rawle/Stock Boston.

Chapter 10: Pages 462-63, ©Bob Daemmrich/TSW; **463 inset,** ©Don Mason/SM; **464 all,** MT; **472,** ©Patrick Ingrand/TSW; **476,** ©Bob Daemmrich; **479,** Courtesy, Southern Forest Products Association; **481,** ©Stacy Pick/SB; **484,** ©Kevin R. Morris/TSW; **485,** ©Tom & Pat Leeson/Photo Researchers; **486 both,** MT; **489,** ©Frank Gordon/FPG; **490,** ©Michael Newman/PhotoEdit; **492,** ©Robert Essel/SM; **493,** ©Henryk T. Kaiser/The Picture Cube; **494,** ©Michael Keller/FPG; **498,** Joachim Messerschmidt/FPG; **500,** CLOSE TO HOME copyright John McPherson. Reprinted with permission of UNIVERSAL PRESS SYNDICATE. All rights reserved; **501 t,** ©Mike Surowiak/TSW; **501 b,** ©Needham Research Institute.

Chapter 11: Pages 508-09, ©Jean-Claude Lejeune/SB; **509 inset,** ©Leonard Lessin/Peter Arnold; **510,** ©Ron Sherman/TSW; **513,** CLOSE TO HOME copyright John McPherson. Reprinted with permission of UNIVERSAL PRESS SYNDICATE. All rights reserved.; **514 r,** ©Lawrence Migdale/SB; **514 l,** ©Richard Laird/FPG; **516,** ©Michael Keller/SM; **518,** ©Michael Grecco/SB; **523,** ©Charles West/SM; **525 all,** KO; **530,** ©Bill Bachmann/PhotoEdit; **531 b,** ©Gerald Gscheidle/Peter Arnold; **531 t,** ©Greg Mancuso/Stock Boston; **533 c, l,** ©Spencer Jones/FPG; **533 r,** Lee Snider/The Image Works; **535,** Courtesy, Yield House; **540,** ©Steve Dunwell/The Image Bank; **541,** ©Christian Michaels/FPG; **544;** Ron Chapple/FPG.

Teacher's Edition

Editorial Services Publishers Resource Group, Inc.

Design Coordination Susan Gerould/Perspectives